BIOLOGY TODAY

An Issues Approach

SECOND EDITION

BIOLOGY TODAY

An Issues Approach

SECOND EDITION

Eli C. Minkoff
Pamela J. Baker

Garland Publishing
New York and London

Vice President: Denise Schanck
Commissioning Editor: Jane Mackarell
Text Editor: Elmarie Hutchinson
Managing Editor: Sarah Gibbs
Editorial Assistant: Mark Ditzel
Production Editor: Emma Hunt
Production Assistant: Angela Bennett
Copy Editor: Bruce Goatly
Illustrator: Nigel Orme
Designer: Joan Greenfield
Indexer: Jill Halliday

THE COVER
Leaf by Lester Lefkowitz/The Stock Market Photo Agency Inc.; electric man courtesy of Gary Kaemmer/Image Bank; nautilus by Lester Lefkowitz/The Stock Market Photo Agency Inc.; sheep by Sanford/Agliolo/The Stock Market Photo Agency Inc.; acid rain effect by Richard Gross/Biological Photography.

Visit the *Biology Today* web site:
http://www.garlandscience.com/biologytoday

Library of Congress Cataloging-in-Publication Data
Minkoff, Eli C.
Biology today: an issues approach/Eli C. Minoff, Pamela J. Baker.—2nd ed.
p. cm.
Includes bibliographical references (p.).
ISBN 0-8153-2760-9
1. Biology. I. Baker, Pamela J., 1947–II. Title.

QH315.M63 2000
570—dc21 00-060061

Published by Garland Publishing, a member of the Taylor & Francis Group
29 West 35th Street, New York, NY 10001-2299

Printed in the United States of America
15 14 13 12 11 10 9 8 7 6 5 4 3 2 1

Preface

Our book takes an issues-oriented approach to the teaching of biology, one that emphasizes coherent understanding of selected issues rather than an attempt to cover everything. The issues we have chosen to include are current topics that students are likely to see in the news or are of general interest. It is our belief that students (especially those not majoring in biology) are more likely to remember this material if it is meaningfully related to issues of concern to them. Our approach accordingly helps students to experience the connections among the fields of biology, the interdisciplinary nature of today's biology, and the intimate connections between biological and social issues. We also hope to instill in students the feeling that biology is both interesting and relevant to their lives, and that a further understanding of biology can be a delight rather than a burden.

One of the aims of our approach is to educate good citizens, biologists and nonbiologists alike, with an understanding that will enable them to evaluate scientific arguments and make appropriate decisions affecting their own lives and the well-being of society. We are committed to teaching science as a human activity that impinges upon other aspects of society and gives rise to social issues that require discussion. Citizens are increasingly called upon to deal with science-based issues throughout their lives, in the foods they choose to eat, the medicines they take, and the very air they breathe. Legislators, juries, and corporate managers need to make important decisions, affecting many lives, based in part on the findings of science. We think that all good citizens need to be aware of science and the way that scientists work, and they also need to know how science can be used and how it can be misused.

We are committed as teachers to fostering understanding of biology, what some now call 'biological literacy.' Thus, we have chosen to teach 'facts' in a context that emphasizes how they are produced, organized, and used to solve problems. Subsequently, the issues we have selected are ones that are not only of current importance, but ones that lend themselves as vehicles for teaching the major concepts of biology.

Biology as a discipline has become fragmented to the extent that different perspectives on the same problem, for example, molecular perspectives and environmental perspectives, are often taught in separate courses with no reference to each other. We aim for a more comprehensive view of each issue. The current understanding of each issue is covered from different perspectives, which often include cellular and molecular perspectives, organismal or individual perspectives, and global or population perspectives, combined as appropriate. Coverage of each issue also includes its social context, both historical and contemporary. Phrasing ideas as 'our current understanding' will help students to realize the ongoing nature of discovery and to identify the processes that are necessary for new ideas to be accepted.

Those who have been teaching introductory biology for the last two or three decades have been burdened with the supposition that *absolutely everything* needs to be covered in a single course. Textbooks written in this tradition are weighty, encyclopedic works with a thousand or more pages and hefty pricetags. Students are exposed to the *results* of biology without gaining understanding of biology as a *process of discovery*. It is an understanding of this *process* that we hope to instill in students. To help students appreciate this process, we have presented multiple interpretations or points of view as much as possible. Societal and ethical issues are mentioned wherever relevant, and part of the initial chapter is devoted to an examination of ethical principles. We encourage teachers to set aside time for class discussions to further stimulate student thought, or for students to set up such discussions among themselves informally. With *Biology Today* we aim to stimulate critical thinking and questioning rather than memorization.

We would like to take this opportunity to thank the many people who reviewed portions of this text and provided us with helpful suggestions. In alphabetical order; they were: Lee Abrahamsen, Bates College; Gregory Anderson, Bates College; Andrew N. Ash, Pembroke State University; David Baker; Virginia Bliss, Framingham State College; Bruce Bourque, Bates College; Joe W. Camp, Jr., Purdue University North Central; William J. Campbell, Louisiana Tech University; Phillip D. Clem, University of Charleston; Mary Colavito-Shepanski, Santa Monica College; Diane Cowan, The Lobster Conservancy; Martha Crunkleton, Pitzer College; David Cummiskey, Bates College; Christine P. Curran, University of Cincinnati; Mark Dixon, formerly of Bates College; Elizabeth Eames, Bates College; Lynn A. Ebersole, Northern Kentucky University; Edward Goldin, Columbia University School of Dentistry; Helen Greenwood, University of Southern Maine; David Handley, University of Maine Agricultural Research Station; Pat Hauslein, St. Cloud State University; Susan Hutchins, Itasca Community College; Alan R. P. Journet, Southeast Missouri State University; Donald E. Keith, Tarleton State University; John Kelsey, Bates College; Sharon Kinsman, Bates College; Virginia G. Latta, Jefferson State Community College; Laura Malloy, Hartwick College; Richard J. Meyer, Humboldt State University; Glendon R. Miller; Wichita State University; Nancy Minkoff; Neil Minkoff, Deaconess-Waltham Hospital; Sandra L. Mitchell, Western Wyoming College; Jane Noble-Harvey, University of Delaware; Mark Okrent, Bates College; Lois Ongley, Bates College; Joseph G. Pelliccia, Bates College; Karen Rasmussen, Maine Cancer Research and Education Foundation; Larry G. Sellers, Louisiana Tech University; Gary Shields, Kirkwood Community College; Thomas P. Sluss, Fort Lewis College; Barbara Stewart, Swarthmore College; Gregory J. Stewart, West Georgia College; Robert Thomas, Bates College; Robert M. Thornton, University of California-Davis; Robin W. Tyser, University of Wisconsin; James E. Urban, Kansas State University; Aaron Wallack, Cognex Corporation; Linda Wallack; Thomas Wenzel, Bates College; Anne Williams, Bates College; Thomas M. Wolf, Washburn University of Topeka; and H. Elton Woodward, Daytona Beach Community College.

Special thanks are due to Denise Schanck and Jane Mackarell, who saw us through the work on two editions. Elmarie Hutchinson provided numerous helpful suggestions, both large and small, and Nigel Orme drew most of the illustrations. We also thank Angela Bennett, Mark Ditzel, Sarah Gibbs, Emma Hunt, Angela Kao, Richard Woof and the rest of the staff at Garland Publishing and the Taylor & Francis Group who helped us throughout the process of bringing the second edition to completion.

Eli C. Minkoff
Pamela J. Baker

ABOUT THE BOOK

This second edition of *Biology Today* has undergone considerable revision and reorganization. The basic biology content has been expanded on several topics while current issues and recent science have been updated.

Here is an overview of the study aids and features included in this second edition.

Chapter outlines: Chapters begin with an outline detailing major section headings. The outline enables students to assess the full range of biological and issue-related content covered within the chapter.

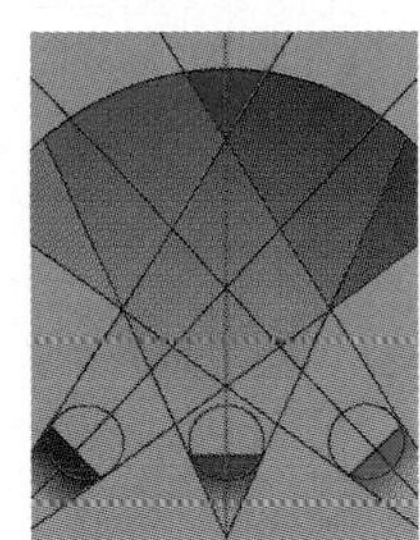

Issues: Each chapter in *Biology Today* encourages the student to perceive biology as a process, where the biologist is constantly questioning the facts they are presented with. Every chapter raises a number of critical issues, and asks the student to consider a number of important, pivotal questions. These underlying questions within the chapter are listed in the Issues section. The perspectives icon links these underlying questions with other critical thinking sections within the chapter, such as the Thought Questions and the concluding paragraph for each chapter.

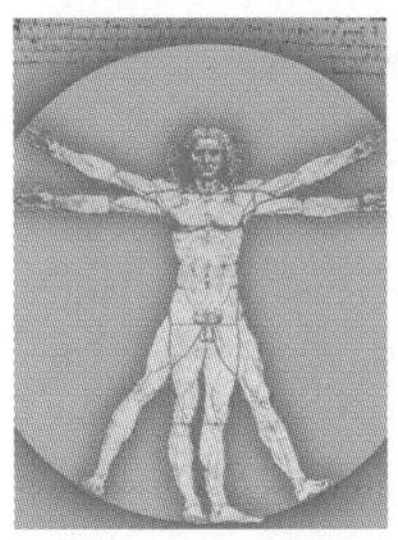

Biological Concepts: When teaching an issues-oriented introductory biology course, it is still vital to cover key biological concepts. These sections summarize the biology that is covered within the chapter. The Biological Concepts sections are directly related to key biological concepts developed by the Biological Sciences Curriculum Study. This table of concepts is fully cross referenced to the material in the book and can be found on the *Biology Today* web sites.

Opening Narrative: The opening narrative scenario immediately introduces the science covered in the chapter by placing it within an interesting and relevant societal context.

Thought Questions: Sets of Thought Questions appear at the end of each chapter section. They are linked to the chapter opening Issues section and the concluding paragraph of each chapter by the perspectives icon. Although a few thought questions have factual 'right' answers, most do not. Some questions require students to do further reading and many encourage students to think about the limitations of available data or the applications and implications of science. We encourage students with differing viewpoints to discuss these questions among themselves and to ask, "What further information would help us resolve our differences or reach decisions?". These questions can form the basis for discussion either in class or in informal study groups.

Illustrations: Since many of the concepts of biology can be understood and remembered visually, the book is well illustrated with photographs and drawings. The captions to these illustrations often provide another important avenue to understanding.

Tables: Some of the tables included summarize important sections of text while others provide specific examples.

Boxes: Boxes provide additional material that supplements the text.

Box 2.1 Science and Fair Play: The Case of Rosalind Franklin
Box 3.1 Ethical Issues in Medical Decision-making Regarding Genetic Testing
Box 3.2 Who Decides What is Considered a 'Defect'?
Box 4.1 Procaryotic and Eucaryotic Cells Compared
Box 4.2 The Six Kingdoms of Organisms
Box 5.1 Is Intelligence Heritable?
Box 5.2 The Hardy–Weinberg Equilibrium
Box 7.1 The Sociobiology Paradigm
Box 8.1 How Does Sugar Contribute to Tooth Decay?
Box 9.1 The Ames Test
Box 13.1 ELISA and Western Blot Tests for Detection of Antibodies to HIV
Box 13.2 Can Mosquitoes Transmit AIDS?

Key Terms to Know: Important vocabulary is highlighted in bold type in the text. All of the Key Terms from a chapter are listed at the end of that chapter with page references. Key Terms are mainly concepts rather than specific examples (that is, neurotransmitter rather than the name of a specific neurotransmitter). Additional terms are highlighted in blue type in the text. These terms are defined in an expanded glossary that can be found on the *Biology Today* web sites (see below).

Connections to Other Chapters: *Biology Today* aims for an integrated comprehensive view of biology. The Connections to Other Chapters section at the end of each chapter helps reinforce this integrated view of biology as a science. Icons placed throughout the book indicated cross-references to other material in the text.

Practice Questions: Learning biological concepts still requires some memorization of the facts. End-of-chapter Practice Questions are straightforward review questions for students to answer.

Glossary: Each of the Key Terms is defined in an alphabetically arranged glossary at the end of the book. An expanded glossary which contains definitions of the Key Terms and the blue-highlighted terms can be found on the *Biology Today* web sites.

Bibliography: Suggestions for further readings are presented on the web site, organized by chapter.

BIOLOGY TODAY WEB SITE

Biology Today offers two complementary websites that serve as a complete teaching and learning resource—an essential supplement to an issues-oriented biology course.

Where the web site icon is featured in the book, students are encouraged to go to the *Biology Today* web site at http://www.garlandscience.com/biologytoday for additional textual content as well as a bibliography and expanded glossary.

The Garland Science Classwire web site, located at http://www.classwire.com/garlandscience offers extensive instructional resources. In addition to containing all materials on the first site, it provides a testbank, additional critical thinking questions, web links and bibliography, as well as curriculum advice and assistance for those teaching an issues-oriented biology course for the first time. It also contains all the images from the textbook available in a downloadable, web-ready, as well as Power Point-ready, format. Instructors can choose whether they wish to make these resources available to students.

Garland Science Classwire also does much more than offer supplementary teaching resources. It is a flexible and easy to use course management tool that allows instructors to build web sites for their classes. It offers features such as a syllabus builder, a course calendar, a message center, a course planner, virtual office hours and a resource manager. No programming or technical skills are needed. Garland Science Classwire is offered free of charge to all instructors who adopt *Biology Today* for their course.

Contents

Chapter 1	Biology: Science and Ethics	2
Chapter 2	Genes, Chromosomes, and DNA	36
Chapter 3	Human Genetics	70
Chapter 4	Evolution and Classification	122
Chapter 5	Human Variation	186
Chapter 6	The Population Explosion	230
Chapter 7	Sociobiology	272
Chapter 8	Nutrition and Health	314
Chapter 9	Cancer	358
Chapter 10	The Nervous System	408
Chapter 11	Drugs and Addiction	450
Chapter 12	Mind and Body	492
Chapter 13	HIV and AIDS	534
Chapter 14	Plants and Crops	580
Chapter 15	Biodiversity and Threatened Habitats	634
Glossary		687
Credits		695
Index		699

List of Headings

1 Biology: Science and Ethics **2**
Science Develops Theories by Testing Falsifiable Hypotheses **5**
Hypotheses 5
Theories 8
A theory to describe living systems 9
Hypothesis testing in experimental science 10
Hypothesis testing in naturalistic science 13
Revolutionary Science Differs from 'Normal Science' **14**
Scientific revolutions 15
Molecular genetics as a paradigm in biology 15
The scientific community 17
Scientists Often Consider Ethical Issues **20**
Ethics 20
Resolving moral conflicts 21
Deontological ethics 22
Utilitarian ethics 23
How societies make ethical decisions 25
Ethical Questions Arise in Decisions about Experimental Subjects **27**
Uses of animals 27
The animal rights movement 28
Humans as experimental subjects 29

2 Genes, Chromosomes, and DNA **36**
Mendel Observed Phenotypes and Formed Hypotheses **39**
Traits of pea plants 39
Genotype and phenotype 41
The Chromosomal Basis of Inheritance Explains Mendel's Hypotheses **45**
Mitosis 46
Meiosis and sexual life cycles 48
Gene linkage 50
Confirmation of the chromosomal theory 51
The Molecular Basis of Inheritance Further Explains Mendel's Hypotheses **52**
DNA and genetic transformation 53
The chemical composition of DNA 56
The three-dimensional structure of DNA 57
DNA replication 59
Transcription and translation of genes 61
Mutations 63

3 Human Genetics **70**
Genes Carried on Sex Chromosomes Determine Sex and Sex-linked Traits **73**
Sex determination 74
Sex-linked traits 75
Chromosomal variation 76
Some Diseases and Disease Predispositions Are Inherited **79**
Identifying genetic causes for traits 79
Some hereditary diseases that are associated with known genes 85
Molecular Techniques Have Led to New Uses for Genetic Information **89**
The Human Genome Project 89
Using DNA markers to identify individuals 92
Genetic engineering 94
Other spin-off technologies 97
Genetic Information Can Be Used or Misused in Various Ways **99**
Genetic testing and counseling 100
Altering individual genotypes 106
Altering the gene pool of populations 111
Changing the balance between genetic and environmental factors 114

4 Evolution and Classification **122**
The Darwinian Paradigm Reorganized Biological Thought **125**
Pre-Darwinian thought 126
The development of Darwin's ideas 126
Natural selection 128
Descent with modification 133
Fossils and the fossil record 137
Creationists Challenge Evolutionary Thought **142**
Creationism and Darwin's response 142
Early twentieth-century creationism 143
Creationism today 144
Species are Central to the Modern Evolutionary Paradigm **146**

Populations and species 146
How new species originate 148
Higher taxa 149
Life Originated on Earth by Natural Processes 151
Evidence of early life on Earth 152
Procaryotic and eucaryotic cells 154
Kingdoms of organisms 158
Procaryotic organisms 158
Eucaryotic Diversity Dominates Life Today 162
Kingdom Protista 162
Kingdom Plantae 163
Kingdom Mycota 164
Kingdom Animalia 164
Humans are Products of Evolution 176
The genus *Australopithecus* 178
The genus *Homo* 179
Evolution as an ongoing process 180

5 Human Variation 186
There is Biological Variation Both Within and Between Human Populations 189
Continuous and discontinuous variation 189
Variation between populations 191
Concepts of race 192
The study of human variation 200
Population Genetics Can Help us to Understand Human Variation 202
Human blood groups and geography 202
Isolated populations and genetic drift 206
Reconstructing the history of human populations 209
Malaria and Other Diseases Are Agents of Natural Selection 211
Malaria 211
Sickle-cell anemia and resistance to malaria 212
Other genetic traits that protect against malaria 217
Population genetics of malaria resistance 219
Other diseases as selective factors 220
Natural Selection by Physical Factors Causes More Population Variation 221
Human variation in physiology and physique 221
Natural selection, skin color, and disease resistance 223

6 The Population Explosion 230
Demography Helps to Predict Future Population Size 233
Growth rate 234
Exponential (geometric) growth 236
Malthus' views on population 237
Logistic growth 239
Demographic transition 241
Human Reproductive Biology Helps us to Understand Fertility and Infertility 247
Reproductive anatomy and physiology 247
Reproductive technologies 251
Can We Diminish Population Growth and its Impact? 254
Contraceptive measures 254
Post-fertilization birth control methods 259
Cultural and ethical opposition to birth control 260
The abortion debate 261
Population control movements 263
The education of women 264
Controlling population impact 266

7 Sociobiology 272
Sociobiology Deals With Social Behavior 275
Learned and inherited behavior 275
The paradigm of sociobiology 277
Research methods in sociobiology 279
Instincts 280
Social Organization Is Adaptive 283
The biological advantages of social groups 283
Simple forms of social organization 284
Altruism: an evolutionary puzzle 285
The evolution of eusociality 288
Reproductive Strategies Can Alter Fitness 292
Asexual versus sexual reproduction 292
r-selection and *K*-selection 293
Differences between the sexes 294
Mating systems 296
Primate Sociobiology Presents Added Complexities 298
Primate social behavior and its development 298
Social organization among primates 301
Reproductive strategies among primates 303
Some examples of human behaviors 305

8 Nutrition and Health 314
Digestion Processes Food into Chemical Substances that the Body Can Absorb 317
Chemical and mechanical processes in digestion 317
The digestive system 317
Absorbed Nutrients Circulate Throughout the Body 325
Circulatory system 325
The heart 326

All Humans Have Dietary Requirements for Good Health **328**
Carbohydrates 329
Lipids 331
Proteins 336
Conversion of macronutrients into cellular energy 339
Fiber 340
Vitamins 342
Minerals 346
Malnutrition Contributes to Poor Health **349**
Eating disorders 349
Protein deficiencies 350
Ecological factors contributing to poor diets 351
Effects of poverty and war on health 352
Micronutrient malnutrition 353

9 Cancer **358**
Multicellular Organisms Are Organized Groups of Cells and Tissues **361**
Compartmentalization 361
Specialization 362
Cooperation and homeostasis 363
Cell Division is Closely Regulated in Normal Cells **364**
The cell cycle 364
Regulation of cell division 365
Regulation of gene expression 367
Cellular differentiation and tissue formation 369
Limits to cell division 375
Cancer Results When Cell Division is Uncontrolled **377**
Properties of cancer cells 377
Oncogenes and proto-oncogenes 379
Tumor suppressor genes 381
Accumulation of many mutations 381
Progression to cancer 382
Cancers Have Complex Causes and Multiple Risk Factors **384**
Inherited predispositions for cancers 386
Increasing age 387
Viruses 388
Physical and chemical carcinogens 389
Dietary factors 393
Tumor initiators and tumor promoters 394
Internal resistance to cancer 396
Social and economic factors 397
We Can Treat Many Cancers and Lower our Risks for Many More **399**
Surgery, radiation, and chemotherapy 399
New cancer treatments 400
Cancer detection and predisposition 401
Cancer management 403
Cancer prevention 404

10 The Nervous System **408**
The Nervous System Carries Messages Throughout the Body **411**
The nervous system and neurons 411
Nerve impulses: how messages travel along neurons 413
Neurotransmitters: how messages travel between neurons 416
Dopamine pathways in the brain: Parkinsonism and Huntington's disease 419
Messages are Routed To and From the Brain **422**
Message input: sense organs 422
Message processing in the brain 428
Message output: muscle contraction 432
The Brain Stores and Rehearses Messages **435**
Learning: storing brain activity 435
Memory formation and consolidation 437
Alzheimer's disease: a lack of acetylcholine 439
Biological rhythms: time-of-day messages 440
Dreams: practice in sending messages 444
Mental illness and neurotransmitters in the brain 445

11 Drugs and Addiction **450**
Drugs are Chemicals that Alter Biological Processes **453**
Drugs and their activity 453
Routes of drug entry into the body 454
Distribution of drugs throughout the body 457
Elimination of drugs from the body 457
Drug receptors and drug action on cells 462
Side effects and drug interactions 463
Psychoactive Drugs Affect the Mind **465**
Opiates and opiate receptors 466
Marijuana and THC receptors 467
Nicotine and nicotinic receptors 467
Amphetamines: agonists of norepinephrine 468
LSD: an agonist of serotonin 468
Caffeine: a general cellular stimulant 469
Alcohol: a CNS depressant 469
Most Psychoactive Drugs are Addictive **472**
Dependence and withdrawal 472
Brain reward centers and drug-seeking behaviors 473
Drug tolerance 477
Drug Abuse Impairs Health **478**
Drug effects on the health of drug users 478

Drug effects on embryonic and fetal development 483
Drug abuse as a public health problem 485

12 Mind and Body 492
The Mind and the Body Interact 495
The Immune System Maintains Health 497
The immune system and the lymphatic circulation 497
Development of immune cells 499
Acquiring specific immunity 501
Mechanisms for removal of antigens 503
Turning off an immune response 505
Passive immunity and innate immunity 505
Inflammation and healing 507
Harmful immune responses 508
Plasticity of the immune responses 511
The Neuroendocrine System Consists of Neurons and Endocrine Glands 513
The autonomic nervous system 613
The stress response 516
The relaxation response 518
The placebo effect 519
The Neuroendocrine System Interacts with the Immune System 521
Shared cytokines 522
Nerve endings in immune organs 522
Studies of cytokine functions 523
Stress and the immune system 524
Individual variation in the stress response 526
Conditioned learning in the immune system 528
Voluntary control of the immune system 528

13 HIV and AIDS 534
AIDS is an Immune System Deficiency 537
AIDS is caused by a virus called HIV 538
Discovery of the connection between HIV and AIDS 539
Establishing cause and effect 542
Viruses and HIV 545
Evolution of virulence 549
HIV Infection Progresses in Certain Patterns, Often Leading to AIDS 551
Events in infected helper T cells 551
Progression from HIV infection to AIDS 552
Tests for HIV infection 555
A vaccine against AIDS? 558
Drug therapy for people with AIDS 560
Knowledge of HIV Transmission Can Help You to Avoid AIDS Risks 562
Risk behaviors 563
Communicability 564
Susceptibility versus high risk 567
Public health and public policy 569
Worldwide patterns of infection 573

14 Plants and Crops 580
Plants Capture the Sun's Energy and Make Many Useful Products 583
Plant products of use to humans 583
Photosynthesis 584
Nitrogen for plant products 590
Plants Use Specialized Tissues and Transport Mechanisms 596
Tissue specialization in plants 596
Water transport in plants 599
Crop Yields Can Be Increased by Overcoming Various Limiting Factors 604
Fertilizers 604
Soil improvement and conservation 607
Irrigation 608
Hydroponics 609
Chemical pest control 610
Integrated pest management 614
Altering plants through artificial selection 617
Altering strains through genetic engineering 619

15 Biodiversity and Threatened Habitats 634
Biodiversity Results from Ecological and Evolutionary Processes 637
Factors influencing the distribution of biodiversity 638
Interdependence of humans and biodiversity 639
Extinction Reduces Biodiversity 642
Types of extinction 642
Analyzing patterns of extinction 643
Species threatened with extinction today 650
Some Entire Habitats Are Threatened 652
Tropical rainforest destruction 653
Desertification 664
Valuing habitat 666
Pollution Threatens Much of Life on Earth 672
Detecting, measuring, and preventing pollution 672
Air pollution 674
Acid rain 674
Polluted Habitats Can be Restored 677
Bioremediation of oil spills 677
Bioremediation of wastewater 679
Treatment of drinking water 681
Costs and benefits 682

1 Biology: Science and Ethics

CHAPTER OUTLINE

Science Develops Theories by Testing Falsifiable Hypotheses

- Hypotheses
- Theories
- A theory to describe living systems
- Hypothesis testing in experimental science
- Hypothesis testing in naturalistic science

Revolutionary Science Differs from 'Normal Science'

- Scientific revolutions
- Molecular genetics as a paradigm in biology
- The scientific community

Scientists Often Consider Ethical Issues

- Ethics
- Resolving moral conflicts
- Deontological ethics
- Utilitarian ethics
- How societies make ethical decisions

Ethical Questions Arise in Decisions about Experimental Subjects

- Uses of animals
- The animal rights movement
- Humans as experimental subjects

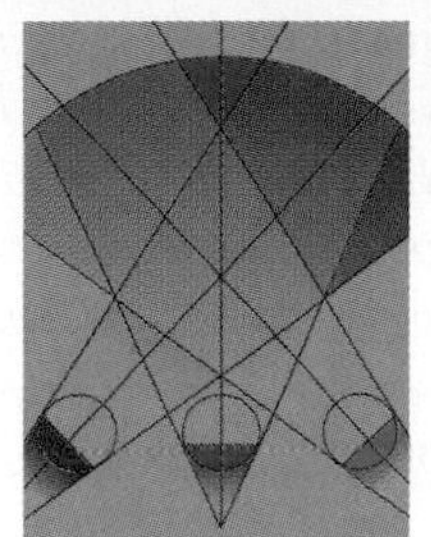

ISSUES

How do we know what we know?

How do scientists make discoveries and advance our knowledge?

What constitutes a 'discovery' in science?

How is science creative?

Does science contain absolute truths?

How do ethics and morals fit into science?

How do scientists make ethical decisions in a social context?

How are decisions made on social issues, and to what extent can science help in these decisions?

What rights do animals have? How do we safeguard those rights?

How do we safeguard the rights of experimental subjects?

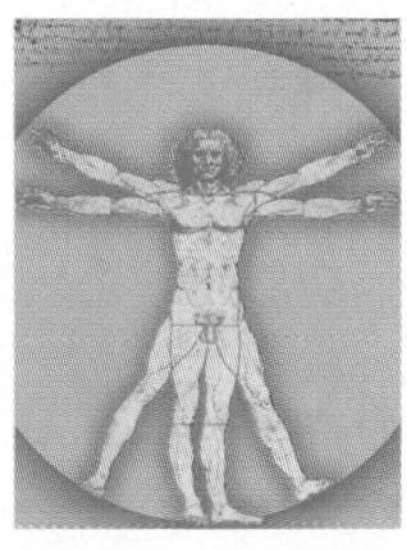

BIOLOGICAL CONCEPTS

Properties of living organisms (metabolism, selective response, homeostasis, genetic material, reproduction, population)

Hypotheses and theories

Experimental science versus naturalistic science

Normal science and paradigm shifts

Science and society

Biological ethics

eth-ic \'e-thik\ n

. *ethikos*] (14c) **1**

with what is good

a set of moral pri

values (the present

: the principles of

sional~*s*) **d** : a gui

eth-i-cal \'e-thi-k

Chapter 1: Biology: Science and Ethics

Biology *is the scientific study of living systems. Our gardens, our pets, our trees, and our fellow humans are all examples of living systems. We can look at them, admire them, write poems about them, and enjoy their company. The Nuer, a pastoral people of Africa, care for their cattle and attach great emotional value to each of them. They write poetry about—and occasionally to—their cattle, they name themselves after their favorite cows or bulls, and they move from place to place according to the needs of their cattle for new pastures. They come to know individual cattle very well, almost as members of the family. The Nuer have also acquired a vast store of useful knowledge about the many animal and plant species in their region. Many other people who live close to the land have a similar familiarity with their environment and the many species living in it. Scientific understanding of the world around us grew out of this kind of familiarity with nature, supplemented by a tradition of systematic testing. In this chapter we examine the methods of science in general and the application of those methods to the study of living systems.*

Because living systems are complex and continually changing, an understanding of these systems often requires special methods of investigation or ways of formulating thoughts. This chapter describes the special methods that have come to be called ***science****. Many people think that science is defined by its subject matter, but this is not correct.* Science is defined by its methods.

The scientific method does not answer questions about values and, therefore, cannot by itself answer questions such as whether certain types of research should be done, or to what uses scientific results should be put. Such decisions often involve a branch of philosophy called ethics. Many issues confronting societies today have a science and technology dimension. Policy decisions on such issues involve both science and ethics.

SCIENCE DEVELOPS THEORIES BY TESTING FALSIFIABLE HYPOTHESES

The essence of science is the formulation and testing of certain kinds of statements called hypotheses. At the moment of its inception, a hypothesis is a tentative explanation of events or of how something works. What makes science distinctive is that hypotheses are subjected to rigorous testing. Many hypotheses are falsified (rejected as false) by such testing. Eliminating one hypothesis often helps us to frame the next hypothesis. If a hypothesis is repeatedly tested and not falsified, it may be put together with related hypotheses that have also withstood repeated testing. Such a group of related hypotheses may become recognized as a theory.

Hypotheses

Hypotheses must be statements about the observable universe, formulated in such a way that they can be tested. To be a hypothesis, a statement must be either **verifiable** (confirmable) or **falsifiable** (capable of being falsified). Observations gathered for testing any hypothesis are generally called **data**. Certain types of statements cannot be used as scientific hypotheses. Moral judgments and religious concepts differ from scientific statements because they are not falsifiable. For example, the statement, "there is a God," cannot be disproven or falsified by any possible demonstration of empirical fact or observation. Similarly, judgements about what ideas or things are valuable, beautiful, or likable are not subject to falsification by hypothesis testing.

Specific versus general hypotheses. Hypotheses that are easy to verify generally tell us very little. For example, the hypothesis "the sun will rise in the east tomorrow morning" can be tested by awakening early, facing east, and observing what happens. If the sun does rise, then our hypothesis is verified or confirmed; if the sun does not rise, then our hypothesis is falsified or disconfirmed. However, the confirmation of this hypothesis about sunrise on a certain specified day is far from an important scientific discovery. It is relatively unimportant because it is too specific, which is exactly what makes it verifiable.

Suppose, now, that we examine the much bolder hypothesis "the sun will rise in the east *every* morning." We can test this second hypothesis in the same way that we tested the first hypothesis, by rising early and facing east, and we could also declare that the hypothesis would be falsified if the sun failed to rise. But what if the sun does rise? Does this verify that the sun will rise *every* morning? Suppose we decide to watch the sunrise 5 days in a row, or 5000? A single failure of the sun to rise will absolutely falsify the hypothesis, but no finite number of sunrises would be sufficient to verify the hypothesis for all time. This is the kind of hypothesis that science usually examines: hypotheses that are absolutely falsifiable, but not absolutely verifiable.

Falsified hypotheses are rejected, and new hypotheses (which may in some cases be modifications of the original hypotheses) are suggested in their place. If testing a hypothesis does not reject it, we may want to generalize the hypothesis. For example, if a hypothesis tested using rats has not been falsified, we might want to apply the hypothesis to people as well, or to all animals. However, we can never know how far we can extrapolate (generalize) results unless we continue to try to falsify our premise under different conditions. In this way, the testing of hypotheses allows us to draw conclusions about the observable world, but only to the extent that we have tested many possible circumstances and conditions.

Ways of devising hypotheses. Deduction is reasoning from the general to the specific. Deductive logic of the 'if...then' form is frequently used to set up testable hypotheses: "*If* organisms of type X require oxygen to live, *then* this individual of type X will die if I put it in an atmosphere without oxygen." Often contrasted with deduction is another type of reasoning called **induction** (or, more properly, 'inductive generalization'), reasoning from the specific to the general. This type of reasoning is commonly used in everyday life: "I like the pizza in restaurants A, B, C, and D; therefore I will like pizza in any other restaurant." Induction never guarantees the truth of any conclusions drawn—the next restaurant may serve pizza that I do not like. As we have seen above, science also uses inductive methods to generalize from specific hypotheses.

Deduction and induction are only two of the many ways in which scientists go about the business of formulating hypotheses. Other ways include (1) intuition or imagination, (2) esthetic preferences, (3) religious and philosophical ideas, (4) comparison and analogy with other processes, and (5) serendipity, or the discovery of one thing while looking for something else. Moreover, these ways may be mixed or combined. For example, Albert Einstein declared that he arrived at his hypotheses about the physics of the universe by considering esthetic qualities such as beauty or simplicity and by asking, "if I were God, how would I have made the world?" Einstein also said that "imagination is more important than knowledge," a remark that is particularly true for the formulation of hypotheses (Figure 1.1). Nobel prize-winning physicist Niels Bohr said that his hypothesis of atomic structure (the heavy nucleus in the center, with the electrons circling rapidly around it, "like a miniature solar system") first occurred to him by analogy with our solar system. Alexander Fleming found the first antibiotic as the result of a laboratory accident: on dishes of bacteria that should have been thrown away earlier, he observed clear areas where fungi had overgrown the bacteria. His hypothesis, that a product of the fungi had killed the bacteria, was validated by tests, and that fungal product is what we now know as penicillin. As these several examples show, *hypotheses are formed by all kinds of logical and extralogical processes*, which is one more reason why they must be subjected to rigorous testing afterward.

Biology: hypothesis testing in living systems. Animals, plants, and bacteria are complex and variable. So are other living systems, large and small, from ecosystems to individual cells. No individual animal or plant is exactly like any other animal or plant. At any moment, living systems that are otherwise similar may differ in external conditions, internal conditions, or in the way in which these conditions interact. Further, the same individual is not exactly the same from one day to the next. Because living systems vary, tests must be repeated. If the hypothesis is tested in one animal, or one cell, and a particular response occurs, the result is far less reliable as a means of prediction than if 10 animals, or 100 cells, all responded in the same way. What often happens, however, is that 9 out of 10 animals, or 94 of 100 cells, respond in one way and the remainder in another way. The differing responses may come from a source of variation that has not yet been identified. Scientists who study the anomalous cases sometimes discover new, previously overlooked phenomena.

Interpretation of the results from tests on variable systems usually requires statistical treatment to ascertain whether the observed differences are 'real' or can be explained by random variation. For example, scientists who suspected that dietary fats were contributing to the risks for heart disease obtained important evidence on this hypothesis by comparing heart attack rates in populations with low fat consumption with heart attack rates in populations with high fat consumption. Once the rates of heart attack in the populations under study had been determined,

FIGURE 1.1

Imaginative hypotheses may originate from various logical or extralogical processes, especially from young scientists. Does the idea shown here qualify as a scientific hypothesis? Why or why not? Is it falsifiable?

the results were analyzed statistically to determine whether the preliminary findings—that high fat consumption was associated with increased risk of heart attack—were meaningful or could have arisen by chance or from sampling error. (Sampling error arises when people in a sample are not representative of the general population, either because the sample is too small or because it was not chosen randomly.) Studies of the relationship between high fat diets and heart disease have indeed found significantly higher heart attack rates (and also significantly higher rates of some cancers) in populations with a higher consumption of dietary fat.

A definition of science. Science may now be defined as a method of investigation based on the testing of falsifiable hypotheses that are generalizations that can be falsified but never absolutely verified. Notice that this makes scientific statements tentative, or provisional, and subject to possible falsification by the next test. Repeated exposure of our hypotheses to possible falsification increases our confidence in these hypotheses when they are not falsified, but no amount of testing can guarantee absolute truth.

Any hypothesis that is tested again and again without ever being falsified is considered well supported and comes to be generally accepted. It may be used as the basis for formulating further hypotheses, so there is soon a cluster of related hypotheses, supported by the results of many tests, which is then called a theory.

Theories

A **theory** is a cluster of related hypotheses that share a common language and a common subject matter. Theories develop after the results of many tests have accumulated. One of the most important features of a good theory is that it may suggest new and different hypotheses. A theory of this kind is a stimulus to further research and is sometimes called a productive theory. A theory may be productive for a while and then no longer stimulate new research. The theories that last are the ones that remain productive the longest, while the less productive theories are often abandoned. Sometimes they are abandoned without ever being fully disproved. In other cases, it is the falsification of one of its hypotheses (or the failure of a crucial test) that causes a theory to be rejected. Remember that the hypotheses that make up a theory are always subject to possible falsification. Even a long-cherished theory may be abandoned (or greatly modified) if it no longer holds predictive or explanatory power.

Theoretical models. Many theories can be communicated by using a simplified mathematical or visual form, called a **model**. Such a model, while not a formal part of the theory, can nevertheless be an important teaching tool in helping communicate the theory to other people. For example, Bohr's conceptualization of the atom in terms of electrons circling around the nucleus "like a miniature solar system" was the model of atomic structure for generations of students. However, models are analogies. Like other analogies, models are comparable to the phenomena they describe only so far, and no further. Attempts to determine *how far* an analogy holds often suggest new hypotheses to test or new ways to test old hypotheses. The planetary model of atomic structure is a case in point. With the development of quantum physics, it became clear that the solar system model was inadequate to explain the behavior of subatomic particles. Scientific theories are tentative. Even the best-cherished theoretical models can be supplanted by other models—either because an important hypothesis is falsified or because a more satisfactory explanation or model is proposed.

A theory to describe living systems

Animals, plants, and the cells of which they are composed are examples of living systems that share many properties distinguishing them from nonliving systems. Each of these properties is really a hypothesis about living systems; together, they constitute a theory of how living and nonliving systems differ.

- *Metabolism*. All living systems take energy-rich materials from their environment and release other materials that, on average, have a lower energy content. Some of the energy fuels life processes, but some accumulates and is released only upon death.
- *Motion*. Most (but not all) living systems convert some of the energy that they use into motion of some sort, including internal motion within cells.
- *Selective response*. All living systems can respond selectively to certain external stimuli and not to others. Many organisms respond to offensive stimuli by withdrawing. All organisms can distinguish needed nutrients from other chemicals.
- *Homeostasis*. All living systems have at least some capacity to change potentially harmful or threatening conditions into conditions more favorable to their continuing existence, e.g. by metabolizing certain toxic chemicals into less harmful ones.
- *Growth and biosynthesis*. All living systems go through phases during which they make more of their own material at the expense of some of the materials around them.
- *Genetic material*. All living systems contain hereditary information derived from previously living systems. This genetic material is a nucleic acid (either DNA or RNA) in all known cases.
- *Reproduction*. All living systems can produce new living systems similar to themselves by transmitting at least some of their genetic material.
- *Population structure*. Living organisms form populations. Populations can be defined retrospectively as groups of individual organisms related by common descent. Among organisms capable of sexual processes, a population is all those organisms that can interbreed with one another.

Anything is considered to be living if it exhibits growth, metabolism, homeostasis, and selective response at some time during its existence. Living systems include both single-celled and multicellular organisms. Organisms belong to populations of similar organisms, of which at least some are capable of reproducing.

Each of these hypotheses has been tested repeatedly. As new organisms were discovered, the hypotheses have been modified but never falsified. For example, the invention of the microscope in about 1700 led to the discovery of bacteria, which caused us to expand our concepts of motion and selective response. The discovery of viruses early in the twentieth century have also strained this theory: all of the eight hypotheses apply, yet viruses cannot reproduce on their own. They must use the cellular machinery of other organisms to reproduce themselves. Together, these hypotheses form a working theory about the characteristics of living things that continues to be productive and to suggest new hypotheses to test.

FIGURE 1.2

Scientists at work.

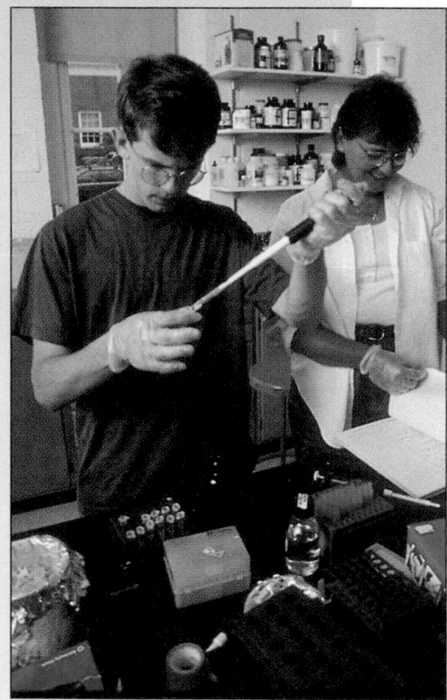

Transferring and examining solutions in a biochemistry laboratory.

Preparing cultures in a bacteriology laboratory.

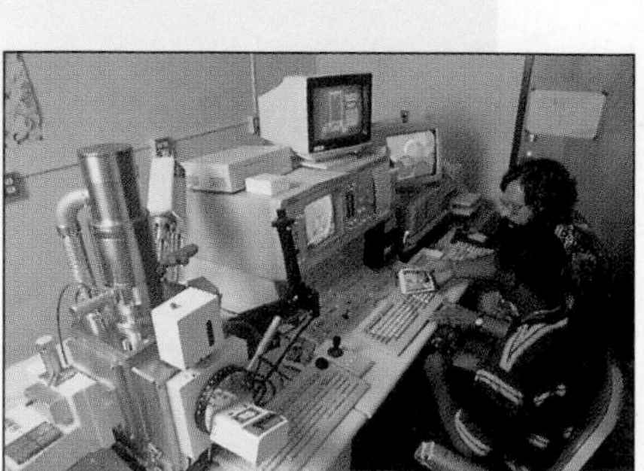

Examining cells with an electron microscope.

Surveying diversity in each square meter along a transect line perpendicular to a rocky shore.

Scientists test hypotheses by comparing them with the real world through empirical observations. Scientists differ from one another, however, in the ways in which hypotheses are tested. Some scientists do all their work in laboratories with specially designed equipment; other scientists gather data and specimens in the field for analysis and interpretation (Figure 1.2).

Hypothesis testing in experimental science

Some scientists test hypotheses by conducting **experiments**—artificially contrived situations set up for the express purpose of testing some hypothesis. Most **experimental sciences** aim to answer questions of the form "How does X work?" The scientist designs an experiment such that, if the hypothesis is true, a certain outcome is expected (or not expected). Then, the results of the experiment are determined **objectively**, which means, in this context, *without* *bias either for or against the hypothesis being tested, and without any limitation that would impair the falsifiability of the hypothesis.*

In many experiments an experimental situation or group is compared with a control situation or **control group**; ideally, the control group exactly matches the experimental group in all variables except the one being tested. For example, animals given a new drug are compared with a similar group of animals—the control group—that are not given the drug. The control group is given a substance similar to whatever is given to the experimental group, but lacking the one ingredient being tested. The two groups are selected and handled so as to be equivalent in every way other than receiving the drug: similar animals, similar cages, similar temperatures, similar diets, and so on.

As an example of the experimental approach, consider the following experiment in bacterial genetics that was conducted by Joshua and Esther Lederberg. (This experiment was part of the basis for Joshua's subsequent Nobel Prize.) Most bacteria are killed by streptomycin, but the Lederbergs exposed the common intestinal bacterium *Escherichia coli* to this drug and were able to isolate a number of streptomycin-resistant bacteria. They allowed these bacteria to reproduce and were able to show that resistance to streptomycin was inherited by their offspring. In other words, the change to streptomycin resistance was a permanent genetic change; such changes are called mutations (see Chapter 2, pp. 63–66). The discovery of resistance gave the Lederbergs two hypotheses to test. The first hypothesis was that the mutation had been induced, or caused, by exposure to the streptomycin. The second was that the bacteria had mutated before being exposed to the streptomycin, in which case the mutation would be independent of the exposure. To distinguish between these hypotheses, the Lederbergs devised the experiment shown in Figure 1.3. In this experiment, a copy, or replica, of the original plate of bacteria was made. Only the replica, not the original, was exposed to streptomycin, and the position of each bacterial colony was noted. The induced-mutation hypothesis predicted that bacteria exposed to streptomycin would mutate, but that unexposed bacteria would not. In fact, most of the bacteria died, but a few survived and were thus identified as being streptomycin resistant. The prior-mutation hypothesis predicted that the mutation conferring streptomycin resistance had occurred before the exposure to streptomycin. To test this second hypothesis, the Lederbergs went back and tested the colonies from the original plate. They discovered that the same colonies that were streptomycin resistant on the replica plate were also streptomycin resistant on the original plate. This finding was consistent with the prior-mutation hypothesis for this particular sample of bacteria.

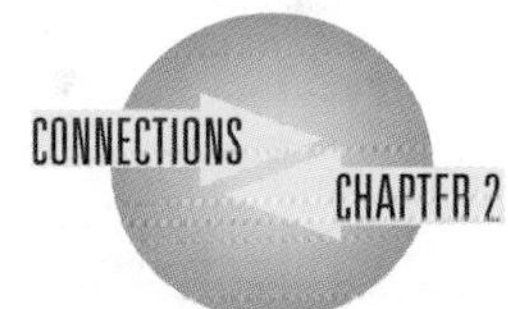

The prior-mutation hypothesis for drug resistance had been tested and not falsified in the case of one mutation in one species of bacteria. How far could the finding be generalized? From this one experiment alone, one cannot tell. However, other investigators repeated the experiment for other mutations and other species of microorganisms. So far, the hypothesis of prior mutation has not been falsified. It is difficult to test the hypothesis in large or long-lived organisms, but most scientists are willing to assume the truth of the hypothesis for *all* organisms. There are many species (and thousands of mutations for each species) that have never been tested in this way, which leaves opportunities for the hypothesis to be falsified in the future.

FIGURE 1.3

The replica-plating experiment of Lederberg and Lederberg.

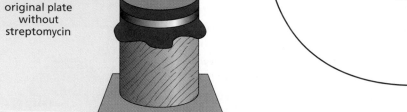

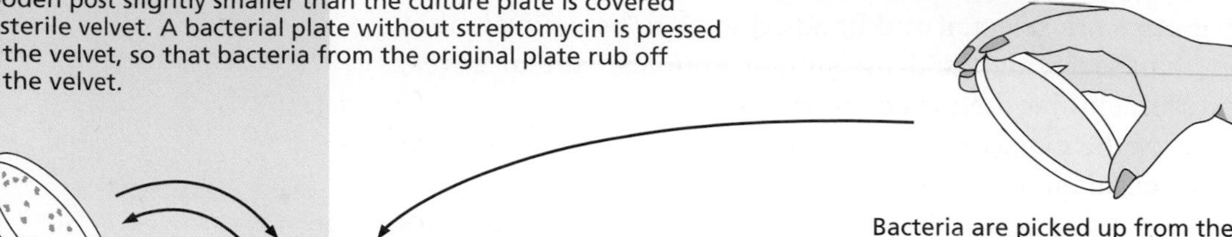

STEP 1
A wooden post slightly smaller than the culture plate is covered with sterile velvet. A bacterial plate without streptomycin is pressed onto the velvet, so that bacteria from the original plate rub off onto the velvet.
STEP 2
A new plate containing streptomycin is pressed onto the velvet.
original plate without streptomycin

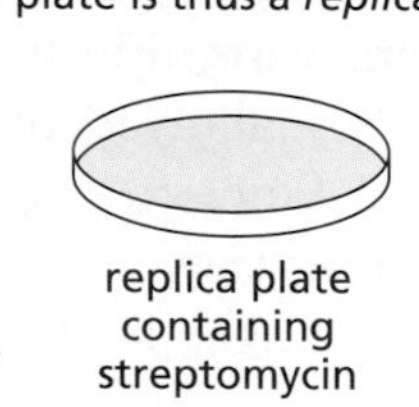

Bacteria are picked up from the velvet, and their locations on the new plate match the colony locations on the original plate. The new plate is thus a *replica* of the original.
replica plate containing streptomycin
incubation

STEP 3
The replica plate is covered and incubated under conditions that stimulate bacterial growth. Only an occasional colony grows. Because the plate contained streptomycin, any colony that grows must be composed of streptomycin-resistant bacteria.

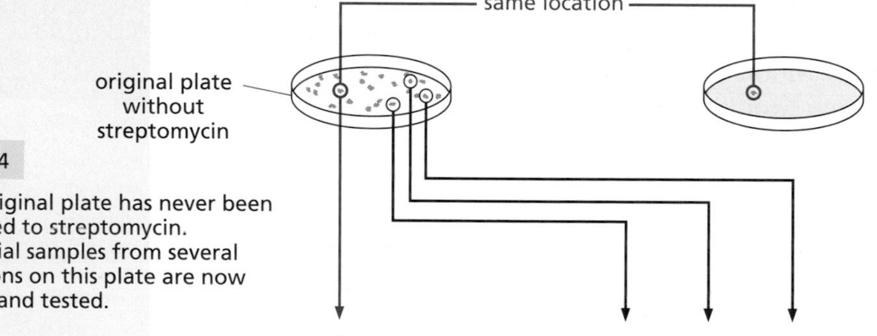

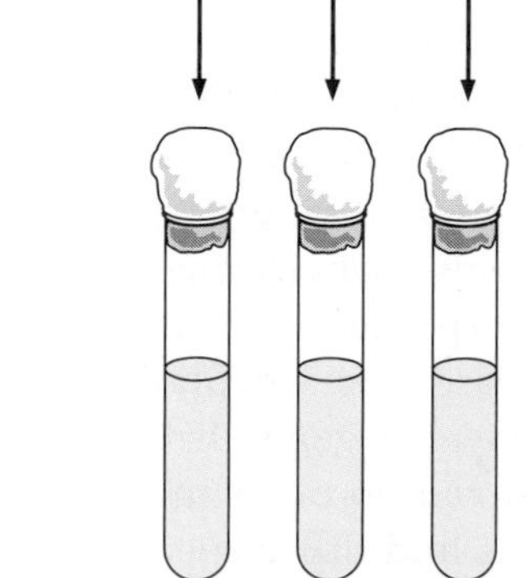

The sample from the location where a streptomycin-resistant colony grew on the replica plate grows in a test tube with streptomycin, showing that some bacteria in this location on the original plate were streptomycin resistant before they were exposed to the streptomycin in the experiment.
Samples from other locations do not grow in test tubes containing streptomycin, showing that other colonies on the original plate are not streptomycin resistant.
CONCLUSION
The result falsifies the hypothesis of induced mutations, but is consistent with the hypothesis that the bacteria mutated before they were exposed to streptomycin.

Hypothesis testing in naturalistic science

Another type of hypothesis testing is one in which direct experimental manipulation is either impossible or undesirable. For example, if an animal behaviorist wishes to study mating behavior *under natural conditions*, then any experimental manipulation that alters these natural conditions must be avoided. Thus we see ornithologists hiding in blinds to study birds, while other naturalists photograph their subjects using telephoto lenses. The stars, which are the subject matter of astronomy, are too large and too distant to be experimentally manipulated. The extinct species studied by paleontologists cannot be recreated in the laboratory to permit an experiment. We can refer to these sciences as **naturalistic sciences** because their method is based primarily on naturalistic observation rather than experiment. Naturalists attempt to falsify hypotheses, but they do so by patient observation and record keeping. The major difference between the experimental and the naturalistic sciences is that experimentalists set up and control the experiments, while naturalists can only observe 'experiments' that occur in nature.

The naturalistic sciences, moreover, are **historical sciences**. Any scientist seeking to understand why mammals differ from reptiles, or why the U.S. economy differs from the Japanese economy, will soon realize that the histories of animals, or of economies, form an important part of the explanation.

There are many types of questions in the naturalistic or historical sciences. The most characteristic type of question in these sciences is, "How did X get to be that way?" For example, a scientific team led by Rebecca Cann examined the DNA inside the mitochondria (the major energy-producing cell parts) of a large number of human populations. Mitochondrial DNA is always inherited from the mother, never from the father. Cann and her co-workers found that the chemical structure of the mitochondrial DNA in certain populations was very similar to its structure in other populations, allowing groups of related populations to be recognized. These scientists hypothesized that populations with similar mitochondrial DNA sequences share a close common descent through female lineages. This hypothesis explains the patterns of similarity among mitochondrial DNA sequences by a series of progressively 'smaller' hypotheses about the past histories of a given set of populations: that the populations of the Americas all share a common descent, that the populations of New Guinea all share a common descent, and so on. Some of these smaller hypotheses are falsified by the data, and must be replaced by modified hypotheses: New Guinea, for example, forms two clusters, and we can set up the hypothesis that it was colonized twice, with each line of descent forming a separate cluster. As modified in this manner, these smaller hypotheses of geographic dispersal are now interpreted as part of a common pattern of descent (see Figure 5.7, p. 209), with an area of origin in Africa. The data are consistent with a hypothesis that all human populations are descended from an ancestral African population, or from a single ancestral female, nicknamed 'Eve' in the popular press. Like most other explanations in the natural sciences, the 'Eve hypothesis' explains present conditions on the basis of their past history, an evolutionary or historical mode of explanation.

THOUGHT QUESTIONS

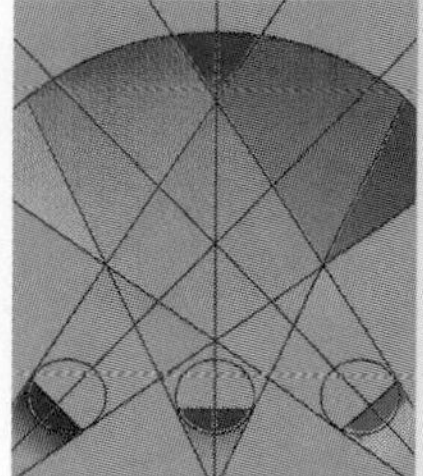

1. In a group, discuss the hypothesis shown in Figure 1.1. Is it falsifiable? If you believe so, then explain what sorts of observations might falsify it. How could we go about testing such an idea?

2. Viruses strain any definition of living systems: they contain genetic material, yet they do not exist as cells, and they replicate only inside and with the help of some other organism. Should we think of viruses as alive and write a definition of life that includes them, or should we think of them as lifeless and write a definition that excludes them? Would one of these definitions be right and the other wrong, or are we free to choose either option?

3. Which of the following are experimental tests, and which are naturalistic observations?
a. Measurements made on the bones of an extinct species are compared with similar measurements made on the bones of a related living species.
b. The activity of white blood cells in a blood sample taken from stressed rats is compared with the activity of white blood cells taken from unstressed rats.
c. A group of animals is fed a certain chemical to see whether they will get cancer as a result.
d. A list of the species found in a particular square meter near the coast is compared with another list of species found 20 meters farther inland.

REVOLUTIONARY SCIENCE DIFFERS FROM 'NORMAL SCIENCE'

In a book that was itself considered revolutionary when it was first published in 1968, Thomas Kuhn proposed a new method of looking at the ways in which science accommodates to new discoveries. Kuhn's observations were based on his studies of historical revolutions in science.

According to Kuhn, everything that we have described thus far is part of **normal science**, science that proceeds by the piecemeal discovery and gradual accumulation of new but small findings. Normal science in Kuhn's theory is always channeled by what he calls a **paradigm** (pronounced 'para-dime'). A paradigm is much more than a theory; it includes a strong belief in the truth of one or more theories and shared opinions as to what problems are important, what problems are unimportant or uninteresting, what techniques and research methods are useful, and so on. The research methodology and sometimes the instrumentation are important parts of the paradigm. Normal science proceeds cumulatively, in small steps, within the context of an existing paradigm. Paradigms, according to Kuhn, are best represented by science textbooks, which are written for the purpose of training new scientists within the paradigm. Students trained by these textbooks are taught not just facts, they are taught attitudes, approaches, values, and a vocabulary that teach them to think in certain ways.

Scientific revolutions

Once in a great while, says Kuhn, science proceeds in a very different way. A **scientific revolution** occurs: an entire paradigm is discarded and a new one replaces it. Few scientists educated in the old paradigm support the new paradigm at first. Most support for the new paradigm comes from new scientists just beginning their careers, and the founder of the revolution is usually either young or a new entrant into that particular scientific field. Once a scientific revolution occurs, its new paradigm opens up a new field of investigation or rejuvenates an old one. Such an infrequent event is called a **paradigm shift**. A paradigm shift requires that the new paradigm explain everything that the older paradigm explained, and more besides. Usually this means that the vocabulary of the old paradigm must either be adopted by the new paradigm or translated into newer terms. A paradigm shift is not just a triumph of logic or of experimental evidence. It is decided, at least in some measure, by a political-style process in which allegiances and influences shift. New paradigms succeed when scientists find them to be fruitful or productive of new approaches to research. The triumph of the Darwinian paradigm over its competitors, described in Chapter 4, is a very good example of a paradigm shift.

Paradigms are sometimes so powerful as to allow anomalies—observations that 'do not fit'—to be ignored. To scientists working within a paradigm, anomalies are small problems that they agree to ignore, believing that the integrity and success of the paradigm are more important than trying to accommodate the unexplained anomaly. Scientists who become interested in the anomaly must work outside the paradigm, which means that they may become founders of their own new paradigms. Paradigms become successful in large measure by the students that they attract. Paradigms that no longer attract students die out.

Molecular genetics as a paradigm in biology

As an example of a scientific paradigm in biology, we describe in this section the field of molecular genetics as it has existed since about 1950. Other examples of scientific paradigms are described in subsequent chapters, including Darwinian evolution in Chapter 4, sociobiology in Chapter 7, and the connection between the mind and the body in Chapter 12.

The paradigm of molecular genetics (or molecular biology) emerged in the decades following the determination of the structure of DNA by James Watson and Francis Crick in 1953. The structure of DNA was itself simply a hypothesis whose increasing ability to make falsifiable predictions made it into a theory. The 'central dogma' of molecular biology (so named by the molecular biologists themselves) was that DNA was used to make RNA and RNA was used to make protein. (Further details of this process are described in Chapter 2.) Both DNA and RNA were said to contain information, and the making of one molecule from another was said to be a form of information transfer. As in other paradigms, the central dogma was more than just a theory because it also suggested a new vocabulary and drove a new research program. The language used within a paradigm often reveals much about how the paradigm is understood by the scientists working within it. How did DNA make copies of itself? This was called *replication* long before any of its details became known. How was information from DNA transferred to RNA? This was called *transcription*. How was information from RNA transferred to protein? This was called *translation*. How replication, transcription, and translation occurred were among the major problems to be solved. The terminology in molecular biology, like that in many fields, was part of an elaborate analogy that drew its inspiration from a comparison with linguistics and included such new vocabulary words as *code* (the language itself), *codons* (items in the code), and *reading frame* (the pieces in which the code was presented). Also, words such as *transcription* (rewriting within the same language) and *translation* (changing from one language to another) were deliberately chosen for literal meanings that matched the biological theory. Textbook descriptions were replete with verbs like *read*, *copy*, and *translate*. There were also a number of laboratory methods, inherited from the field of biochemistry, plus a few extra technical advances, such as the use of high-speed centrifuges. Together, this all formed an orderly paradigm that outlined not only what was known, but also what remained to be discovered, what was thought to be important, and how the details were to be investigated and described. DNA was championed as the most important 'master molecule,' RNA was almost as important, and protein was important only until its synthesis was completed. Protein that was completely synthesized was no longer deemed interesting, except for a few enzymes that helped in the working of DNA or RNA. The paradigm thus defined the boundaries of the field.

The paradigm of molecular genetics guided research on DNA, RNA, and protein synthesis throughout the 1950s and 1960s; much of the work begun in those decades continues today. For its workers, the paradigm defined a set of shared beliefs (including the central dogma), a vocabulary, a set of research techniques, and, most of all, a set of problems to be solved. These problems included the mechanisms by which replication, transcription, and translation took place, as well as how to crack the genetic code. Once this last problem had been solved, the 'coding dictionary' (i.e. the list of correspondences between RNA sequences and protein sequences) was given a prominent place in every genetics book and most general biology texts.

As the molecular genetics paradigm matured, some of its early tenets were modified. Information flow, once thought to be unidirectional, is now thought to be bidirectional. Also, the idea of 'master' molecules that 'make' or 'control' other molecules is slowly being replaced by a vocabulary that speaks in terms of cells 'communicating' with other cells (sending and receiving signals) or 'influencing' other cells (in both directions). Likewise, attention has shifted to new questions, such as how the environment of a cell influences that cell to transcribe certain portions of its DNA at certain times. The molecular biology paradigm, like other paradigms before it, has gradually changed over time, although its core beliefs remain unshaken.

The scientific community

Is science something that only scientists can do? On the contrary, many people use scientific methods in their everyday lives. For example, if my car fails to start, I might formulate one hypothesis after another as to the possible cause. To test the hypothesis that the car is out of gas, I would examine the gas gauge. Additionally, I could add some gasoline to the tank and then try to start the car. If the car starts, I conclude that it was out of gas. The Swiss child psychologist Jean Piaget has written that children often behave as little scientists, formulating possibilities (hypotheses) in their minds and then testing them. 'I can take the toy away from my little brother' can be tested by trying to take it away; the hypothesis would be falsified if brother successfully resisted or if an adult intervened.

The testing of hypotheses is an ancient discipline. Examples of hypothesis testing are found in the writings of the Greek historian Herodotus (fifth century B.C.). Before that, agricultural practices were developed by trial and error. Practices that gave poor results were abandoned; those that seemed to give good results were adopted more widely. Non-Western civilizations have frequently developed sophisticated systems of medicine, metallurgy, and other technologies on the basis of a solid scientific foundation. In those cultures where written communication did not exist, other mechanisms ensured the transmission of the information across the generations: traditions of apprenticeship, formulations that were chanted or sung (and were thus more easily memorized), and so on.

It is unusual for a single person to formulate a hypothesis, test it, and then critically evaluate the results. For this reason, it is important for scientists to communicate with one another so that all these steps can be performed. Historians often date the beginning of modern Western science from seventeenth-century England, specifically from the founding of the Royal Society in 1660. Individual scientists certainly existed in many countries before this time: Copernicus, Galileo, Descartes, and Newton were all earlier. Also, much older traditions of scientific investigation existed in India, China, and elsewhere. However, the formation of the Royal Society marked the first time in history that a permanent, *organized community* of scientists had communicated with one another and shared their results in a scientific journal (*Philosophical Transactions of the Royal Society*). For the first time, there was a written and permanent record of experiments performed and conclusions reached—a shared record that encouraged scientists to check one another's work in a systematic way. Because of this written record, scientists of the past continue to be part of the scientific community when their ideas are tested, even generations later. The scientists shown in Figure 1.4, whose accomplishments are each described elsewhere in this book, are part of this scientific community even though they published over a timespan of about 150 years.

Many of the ways in which today's scientists behave toward one another may be viewed as efforts to maintain their ability to do the kind of systematic checking described above, including the ability to falsify

FIGURE 1.4

A few notable scientists of the nineteenth and twentieth centuries.

Charles Darwin (1809–1882), evolutionary biologist. Darwin's theories are described in Chapter 4.

Gregor Mendel (1822–1884), botanist and geneticist. Mendel's experiments in genetics are described in Chapter 2.

Ernest Everett Just (1883–1941), marine biologist and embryologist.

Barbara McClintock (1902–1992), agricultural geneticist and Nobel Prize winner. Some of her contributions to genetics are described in Chapter 2.

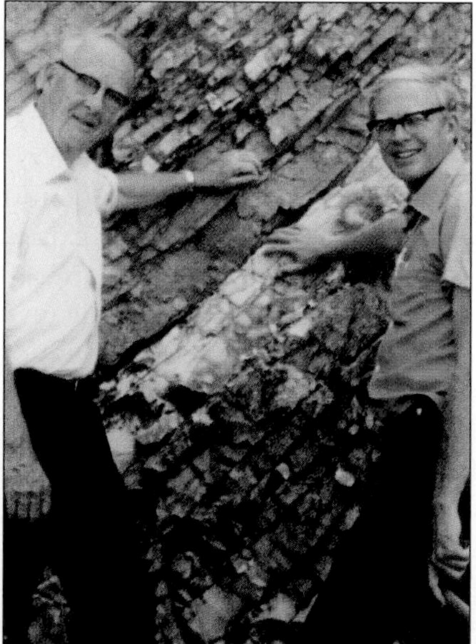

Luis W. Alvarez (1911–), Nobel Prize-winning physicist, and Walter Alvarez (1940–), geologist. Walter's right hand rests on a layer of clay 65 million years old, at the boundary between the Age of Reptiles and the Age of Mammals. The Alvarez's hypothesis to account for the extinction of dinosaurs and many other species across this boundary is described in Chapter 15.

Sarah B. Hrdy (1946–), sociobiologist, helping care for a friend's baby, a form of behavior whose importance she emphasizes among primates in general and humans in particular. Some of Hrdy's work in sociobiology is described in Chapter 7.

hypotheses. Every test must be conducted in such a way as to make it possible for the hypothesis to be falsified, if indeed it is false, and the testing of hypotheses should be described as publicly as possible so as to permit the test to be repeated by other scientists. As David Hull points out, "Scientists rarely refute their own pet hypotheses, especially after they have appeared in print, but that is all right. Their fellow scientists will be happy to expose these hypotheses to severe testing." (David Hull. *Science as a Process*. Chicago: University of Chicago Press, 1988, p. 4.)

The process of science is conducted in the public forum as well as in the laboratory; the publishing and dissemination of results and the repetition of observations and experiments by others are thus valued among scientists. Scientists are expected not to work in isolation, but to discuss their results with other interested scientists, allowing them to build upon the results of previous scientists. They can repeat experiments and confirm the results, but they do not need to start from scratch and repeat *all* earlier work in their field. Skeptics who doubt a particular result unless and until they have seen it themselves can best be won over by a tradition that allows them to hear about repetitions of the test or to repeat the test themselves and to make their own observations. For example, Galileo, the astronomer and early scientist, invited critics who doubted his observations to look for themselves through his telescope. Science is a cumulative process in which it pays for individual scientists to begin with some of the groundwork laid by others, rather than to start always from scratch. As Isaac Newton once said, "If I have seen further than others who have gone before me, it is because I have stood on the shoulders of giants."

THOUGHT QUESTIONS

1. Why is it so difficult for a scientist to work outside the prevailing paradigm? Give at least three reasons.
2. In what ways is the paradigm of molecular genetics more than just a scientific theory?
3. In what ways has Western science progressed more rapidly and more efficiently since the seventeenth century than it did before that time?
4. Many companies conduct what they call research and development, yet many of these companies zealously guard their results and do not publish them. Are they doing science?

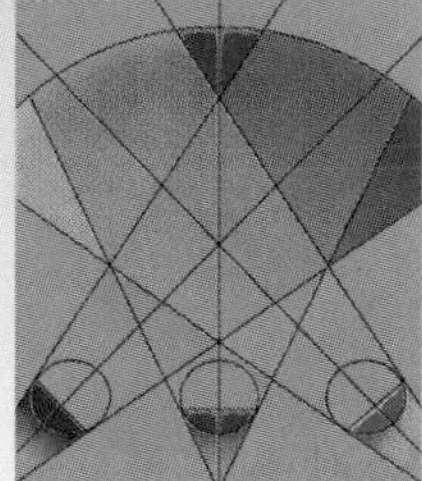

SCIENTISTS OFTEN CONSIDER ETHICAL ISSUES

Science itself sometimes cannot tell us whether certain research should be done or how the results should be used by society; for those answers we turn to the branch of philosophy called ethics. Many topics in this book have an ethical dimension. Are some applications of specific biological research morally right and other applications morally wrong? Should society place legal restrictions on scientific research? Should biologists concern themselves with the ethics, applications, and implications of their work? This section describes some of the ways in which individuals and societies make ethical decisions.

We each use beliefs concerning what is right or wrong, proper or improper, to guide our own behavior. It is right to come to class at the scheduled time and in general to keep appointments that one has agreed to. It is wrong to steal, to lie, to murder, or to park in the NO PARKING zone. It is proper to wait for the traffic light to turn green and to wait for one's turn in line. All these moral rules, or **morals**, are products of societies. Anthropologists who have compared societies from around the world tell us that moral rules differ from one society to the next; they also change over time as society changes.

Any personal decision about whether to follow a moral rule may be called a moral decision: for example, should I park in the NO PARKING zone? Moral decisions are often made with the knowledge that society will attempt to enforce the rules with penalties or sanctions. Formal sanctions include fines and jail time; informal sanctions, which operate more often, include being criticized or avoided by others and ending up with fewer friends.

Ethics

The analysis of moral rules is part of a discipline known as **ethics**. In particular, ethics is concerned with the basis for moral judgement, or the ways in which moral judgements are made and justified. Philosophers distinguish **normative ethics**, the study of how ethical judgements *should* be made, from **descriptive ethics**, the study of how ethical judgements *are* in fact made. Descriptive ethics is studied by observing human behavior by using the scientific methods just discussed. Normative ethics is a theoretical discipline rooted in logical analysis, an analysis for which observational data are insufficient. Normative ethics cannot be reduced to a set of data, and no quantity of data can either confirm or refute a moral law such as "Thou shalt not kill."

In its simplest form, normative ethics is an attempt to reduce moral codes to a minimum set of basic rules (maxims). For example, I should come to class on time because my signing up for a course is like making an appointment. Appointments should be kept because they are promises or contracts. An ethic of keeping appointments is part of a larger ethic of keeping promises.

Some rules of conduct are simply inventions of a society for the convenience of its members, such as waiting for the green light, driving (in North America) on the right side of the street, and the observing of NO PARKING zones (Figure 1.5). We cannot all drive through the intersection at the same time, and traffic lights are a convenient (if arbitrary) contractual way of arranging whose turn is next. The contractual nature of such agreements is obvious because there are usually publicly controlled processes (like city council meetings) to decide where to put NO PARKING zones. We promise to observe traffic laws when we apply for a driver's license, so following these laws may be viewed as another form of promise keeping.

FIGURE 1.5

Would you park here? Give reasons to explain your decision.

Waiting one's turn in a line or waiting for a green traffic light are both ways of introducing order and fairness into a situation that would otherwise be chaotic and conducive to unnecessary disputes. A major difference, however, is that waiting one's turn in line is not enforced by law or traffic code. It is enforced informally by the tacit agreement of those who are present.

We have thus developed a simple moral code: keep your promises, do not interfere with the rights of others, and observe the common social conventions. This could easily be expanded into a more general code of benevolence, cooperation, and mutual aid.

Resolving moral conflicts

There are occasions when conflicts arise within sets of moral rules. I know I should obey the traffic laws, but what if I am taking an injured person to the hospital and the person's life is in danger? Does the duty to save a life justify driving above the speed limit, driving through a red light, or parking in a loading zone? Can I justify disobeying traffic laws to keep an appointment? Does it matter how important I think the appointment is? Resolving conflicts of this kind is one of the major goals of ethics.

In most cases, the resolution of such moral conflicts is made by determining that one rule or goal is more important than another: saving a life is more important than obeying traffic rules, for example. Thus, there are *exceptions* to most moral rules: obey traffic rules and other useful conventions *except* when obeying them causes greater harm or violates a more important rule. Notice that this ranks certain rules as more important than other rules, allowing us to justify an exception to one rule by invoking a 'higher' rule.

Although ethics is a branch of philosophy, ethical arguments arise in everyday life and also in science. More and more scientific endeavors are raising ethical issues that are of practical interest to people in all walks of life. The United States government has sponsored research, meetings, and publications in the field of biological ethics. Many government programs, notably the Human Genome Project (see Chapter 3), have set aside portions of their budgets for the examination of the ethical implications of science. Our grounding in ethics in this chapter will support our

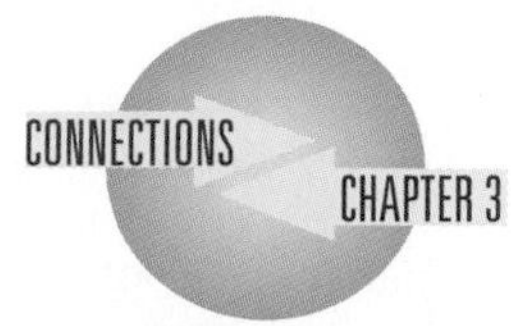

examination of many issues with far-reaching ethical implications in the chapters that follow.

Deontological ethics

An **ethical system** is a set of rules for resolving ethical questions or for judging moral rules. We will describe two major types of ethical systems.

A **deontological** system of ethics is one in which the rightness or wrongness of an act depends on the act itself and not on its consequences. To a deontologist, the wrongness of murder is in the act itself, not stemming from its results or its effects on society. Similarly, a deontologist who believes in keeping promises does so apart from any consequences.

Traditionally, most deontologists have developed moral codes based on religious traditions. The Bible, the Koran, and the sacred texts of other religions have been the source of many moral codes. Deontology grounded in religion has the advantage of commanding agreement among those who share the same religious beliefs, although differences in the interpretation of sacred texts often arise. However, a deontological system based on a particular religion may have less influence on people not belonging to that religion.

The German philosopher Immanuel Kant (1724–1804) devised a deontological system without a religious basis. According to Kant, all ethical statements are based on a single precept: act only according to rules that you could want everyone to adopt as general legislation. This rule is called the **categorical imperative**, and it is formulated (as are most ethical rules) as a universal law. Kant's system, based on this rule, is consistent with many religious teachings. Under Kant's system, the test of the morality of an act is whether the act can be universalized, that is, applied to all people at all times and under all conditions. Thus, killing is (always) a wrongful act because I could not possibly want people always to kill one another—I would be willing my own death and the death of my loved ones. Keeping promises can be universalized, and promise-keeping is therefore (always) moral. Kant said that lying could not be moral because you could never want everyone to lie.

Concepts of 'rights.' Most deontological systems include a formulation of certain **rights**, based on respect for the dignity and autonomy of all persons. Respect for all human beings is an important part of Kant's system. If you respect the dignity of all human beings, then you cannot ever will the death of any person, nor can you deny them their fundamental rights, nor can you use them as objects for your own personal gratification in any way. If you respect their selfhood, then you cannot morally abridge their freedom. Of course, it remains to be determined exactly what rights we do and do not have.

There has been considerable disagreement among philosophers, and even more variations in historical practice, over the types of beings to which various rights apply. At various times in the past, certain groups of persons (including women, children, slaves, the lower classes of stratified societies, impoverished people, foreigners, members of various races, mental patients, and persons unable to speak for themselves) were denied the rights that were afforded to other members of society (Figure 1.6). Many people now invoke dignity and autonomy criteria in discussions of whether certain rights should also now be extended to unborn fetuses or to animals.

Criticisms of deontology. A criticism of rights-based deontology is that there are many circumstances in which one right conflicts with another, resulting in a moral dilemma. Unless there is a clear way of deciding between conflicting rights, moral dilemmas are inevitable. An obvious way out is to declare one particular right (like the right to life or the right to freedom of action) supreme over all others. Aside from the problem that different deontologists would choose different rights to take precedence over the others, there is the more serious objection that insistence on a single right leads to the dangers of absolutism. Historically, many atrocities have been perpetrated by the followers of systems that put absolute adherence to a single principle above all others.

FIGURE 1.6

Do you have a deontological reason for agreeing or disagreeing with the premise that all persons share the same basic rights? This photograph shows a woman protesting the fact that U.S. President Woodrow Wilson supported the rights of poor Germans in World War I, while women in the United States did not have the right to vote.

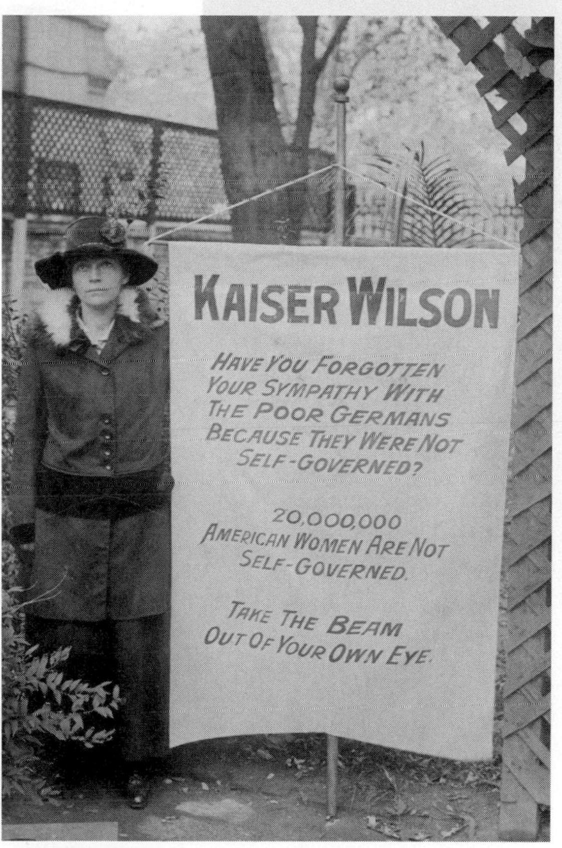

Utilitarian ethics

In a **utilitarian** system of ethics, acts are judged right or wrong according to their consequences: rightful acts are those whose consequences are beneficial, whereas wrongful acts are those with harmful consequences. To a utilitarian, murder is wrong because the death of the victim is an undesirable outcome under most circumstances. Also, on a larger scale, murder is additionally wrong because it produces a society in which people live in fear.

Cost–benefit analyses. A challenge to all utilitarian systems is to find a way of measuring the goodness or badness of consequences. Over the years, utilitarian philosophers have come up with different criteria by which to judge consequences: the greatest happiness for the greatest number of individuals, the greatest excess of pleasure over pain, and so on. All utilitarian systems require that value judgements be made between outcomes that are difficult to measure and quantify.

Utilitarianism takes two forms. In **act utilitarianism**, individual acts are judged as right or wrong on a case-by-case basis. In **rule utilitarianism**, general rules are judged on utilitarian principles, and individual acts are judged by whether they follow an ethical rule. Act utilitarianism allows for an occasional exception. The murder of a traitor (or a gangster or a drug dealer) might be judged as a rightful act if it could be shown that greater harm would result from the victim's continued activities than from his or her death. Under rule utilitarianism with a rule against murder, an exception could be justified only by appeal to a 'higher' rule. Under either form of utilitarianism, ethical decisions are made by comparing, in what is essentially a cost–benefit analysis, all the good and bad consequences of an act with the consequences of other alternative courses of action, including the failure to act (Figure 1.7).

FIGURE 1.7

How would a utilitarian argue in favor of this nuclear power plant? How might another utilitarian argue against it?

Utilitarianism can be summarized by the rule 'always act so as to maximize the amount of good in the universe.' The first major utilitarian philosopher was Jeremy Bentham (1748–1832), who said that we should always strive to bring about "the greatest good for the greatest number." To decide which actions produce the greatest good, Bentham suggested a type of cost–benefit analysis that he called "a calculus of pleasures and pains." Other notable utilitarian philosophers include John Stuart Mill (1806–1873) and G.E. Moore (1873–1958).

Criticisms of utilitarianism. One of the major criticisms of utilitarianism is that its cost–benefit approach reduces the status and dignity of human beings, and in some cases violates their rights. Deontologists argue that certain individual rights must be protected regardless of whether society as a whole benefits. To do any less, they argue, deprives human beings of their fundamental dignity as individuals and makes them nothing more than a cluster of costs and benefits. Even if a larger benefit can be demonstrated in a given instance, say these critics, it is still unethical to violate an individual's fundamental right, because "the ends do not justify the means."

A criticism of utilitarianism frequently raised by philosophers deals with the probability that, under most formulations of utilitarian ethics, the killing of one person would be justified if it resulted in saving the lives of other people. The discussions often center on certain hypothetical situations. In one situation, several people exploring a cave at the edge of the ocean are trapped inside because their leader is stuck in the mouth of the cave; they could escape by blasting him free with a stick of dynamite they happen to have with them, but, if they do nothing, the rising tide will drown them all. Do the calculations of costs and benefits demand that they kill their own leader to avoid the deaths of the entire team? The essential point is that *in some situations* utilitarian ethics might require us to violate even our most strongly held ethical norms. Even in this case, however, a utilitarian could argue that greater harm to society at large would come *in the long run* from the violation of a strongly held norm against the killing of innocent people, and that the world would be a better place if the norms were upheld. History and folklore would, for example, honor the cave explorers who perished rather than blowing up their leader, and the example would benefit future generations brought up to follow the norms.

Although deontological and utilitarian ethical systems have generally attracted the most followers, other ethical systems also exist. Visit this book's web site to learn about several of these systems.

How societies make ethical decisions

Ethical systems differ from scientific theories in that scientific hypotheses are made to be falsifiable, while moral judgements and religious concepts are not. This means that a devout person's belief in God cannot be shaken by any demonstration of an empirical fact or observation. To a devout believer, no such demonstration is even possible. The same is true of strongly held beliefs about the goodness of human equality or the wrongfulness of inflicting death.

In culturally homogeneous societies in which people share common values and religious beliefs, it may be possible to reach consensus about which acts are wrong and which are right. However, most societies today are *pluralistic* in the sense that their populations include people with differing cultural and religious backgrounds. Reaching collective decisions on ethical issues is more difficult in pluralistic societies. Most pluralistic societies are democracies or are becoming more democratic. We therefore examine how people in modern pluralistic democracies reach agreement on issues with moral dimensions.

In a pluralistic society, some people come to the public forum with utilitarian assumptions, others come with deontological assumptions, and others may have different positions. Even within these ethical traditions, people differ in their outlook. Utilitarians have differing evaluations of costs and benefits, and deontologists have differing rankings of rights. One way of reconciling these different views is to have a public debate (so that all views are heard) and then vote. The voting procedure should be structured in a way that all parties recognize as fair. **Fairness** is the principle that states that all people should be treated impartially and equally. Fairness is a way of ensuring that rights are not violated. In most cases, a system that works in this way results in the least displeasure with the decision. Of course, total agreement on a decision is hard to achieve in any large pluralistic society.

Social policy includes all those laws, rules, and customs that people follow in making individual decisions in a society. The making or changing of a new social policy is called a **policy decision**. Most social policies and policy decisions involve ethical considerations. An increasing number of policy decisions today also involve some aspect of science or technology.

It is often convenient to divide policy decisions involving science into three phases: scientific issues, science policy issues, and policy issues.

- *Scientific issues*. What possible explanations (hypotheses) exist that might explain the available data? Can these hypotheses be tested? Do the tests support or falsify each hypothesis? Are alternative explanations available? What additional data are needed to evaluate these hypotheses? These are often characterized as 'purely' scientific issues on which scientists of different political or ethical persuasions may be expected to agree if sufficient data are available.

- *Science policy issues*. What would be the consequences of this or that particular legislation or policy change? What would be lost or gained from each proposed plan of action? Can the probability of uncertain consequences be estimated? The probability of any outcome, especially one not desired, is called a **risk**. Can we calculate or estimate the risks? In a cost–benefit analysis, what are the costs and what are the benefits? How certain are we of the estimated values? These are still scientific issues in the sense that they are evaluated from data, but disagreements on the data or their significance are expected between experts with different political or ethical viewpoints. If the experts disagree, how shall we evaluate their respective positions?
- *Policy issues*. Once we have evaluated the possible consequences of various possible policies, which one should we choose? These are ethical decisions in which values have a prominent role: is it worth spending billions of dollars and risking the lives of crew members to explore a distant continent or an even more distant planet? Is it worth the pain and suffering of a certain number of experimental animals to develop a drug that could save a certain number of human lives each year? In general, is some predicted but uncertain benefit worth the calculated risks? If a proposed change can be made only by incurring a certain cost to society, is this social cost worth the intended benefit? In all these cases, the cost–benefit analysis is used not to make the final decision, but rather to provide the necessary data on which a policy decision can be intelligently based.

Who makes the decisions? In most societies, scientific issues are frequently decided by scientists with little input from interested citizens. Science policy issues are often decided in the court of public opinion by an interplay of scientists and other 'experts' under the scrutiny of policy advocates for one side or the other. Although evidence is used, it is more like courtroom evidence, obtained and evaluated by cross-examining witnesses, than like the evidence of the laboratory, obtained by the formulation and testing of alternative hypotheses.

Misinformation is an important hazard in public policy discussions. Efforts are therefore made to expose and eliminate any faulty information. For example, open hearings are often held, and evidence is made available to all sides of any dispute; at the very least, this affords an opportunity for data or interpretations to be challenged by the opposing sides.

As for the final policy decisions, who makes them: the scientists, the public, the media, or the government? In most democracies, the decisions are made either by the public or by government agencies acting (in theory at least) in the public interest. Decision makers are often influenced by scientists, the media, particular interest groups, and public pressures of various forms—market-place decisions (decisions by individuals about where to spend their money), letters and e-mailings, enthusiasm at public gatherings, public opinion surveys, and direct votes on referendum questions. All of these influences certainly have a role in what is essentially a political process.

THOUGHT QUESTIONS

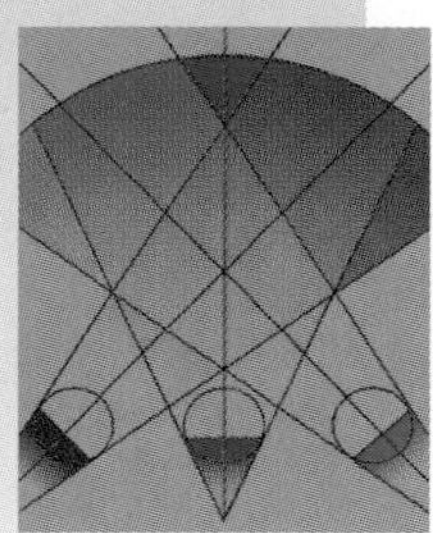

1. Together with other students, make a list of five to ten laws or rules that are generally followed on your campus or in your community. For each, try to discover: (a) why such a rule is considered important (or why it *was* considered important when the rule was adopted), (b) whether there is a more general moral concept of which this rule is just a special case, and (c) whether society is better off with rules of this general kind than without them, and, if so, why?
2. Try to justify the wrongness of the following acts under the two ethical systems discussed in this section:
   murder
   rape
   bank robbery
   failure to repay a debt
   racial segregation
   driving over the speed limit
   parking in the handicapped space shown in Figure 1.5 if you are not handicapped
   Which ethical system makes it easy to explain the wrongness of these acts? Under which ethical system are these explanations difficult?
3. Is it ethical to infect a few people with a deadly disease to study its effects in the hopes of saving many more lives in the future? How do you justify your answer?
4. What scientific issues are raised by the nuclear power plant pictured in Figure 1.7? What science policy issues? What data would you seek on which to base a policy decision one way or the other?

ETHICAL QUESTIONS ARISE IN DECISIONS ABOUT EXPERIMENTAL SUBJECTS

So far we have discussed ethical issues in very general terms. We now turn, for illustrative purposes, to a specific and timely issue, that of the use of experimental subjects in biological research. First, we consider animal rights and the uses of animals in scientific experiments. Second, we contrast animal experimentation with experiments on humans. Of the many ethical issues surrounding biology today, few are as divisive as those touched upon here.

Uses of animals

Human societies have kept animals at least since the origin of agriculture. There are few societies in which animals are not kept as pets, as workmates, or as food. Most societies that practice agriculture use animals for all three purposes. Love of animals and use of animals can go hand in hand.

By far the largest number of animals used by most societies are raised for consumption as food for humans. Animal products are used for clothing. Animals are also the targets of recreational hunting, fishing, and trapping, even in many industrial societies. Many people keep pets or 'companion animals.' Work animals pull and carry loads, help in police work, and help handicapped people. Finally, animals are often used in research, although the number is only a tiny fraction of the numbers consumed as food and used for other purposes.

Nearly all new drugs, cosmetics, food additives, and new forms of therapy and surgery are tested on animals before they are tested on humans. Many people regard animal research as critical to continued progress in human health. Over 40 Nobel Prizes in medicine and physiology have been awarded for research that used experimental animals. Organ transplants, open-heart surgery, and various other surgical techniques were performed and perfected on animals before they were performed on humans. All vaccines were tested on animals before they were used on human patients. In most cases, animals are used in research as stand-ins for humans. If we did not use animals for these purposes, humans would be the experimental subjects tested. Most animal testing is limited to the initial development of a drug or surgical procedure, but the human benefit continues for many generations or longer.

Few people object to a use of animals that saves human lives. Most people also agree that animals should not suffer unnecessarily, whether they are pets, work animals, or research subjects. Scientists need to use healthy, well-treated animals in research, and the U.S. Guide for the Care and Use of Laboratory Animals reflects this concern. There are standards such as cage sizes that must be followed by scientists using animals. All research using live animals must, by law, be scrutinized and approved by supervisory committees, and the committees require the investigators to minimize both the number of animals used and the amount of pain that those animals experience, and to substitute other types of tests where possible. According to statistics from the U.S. Department of Agriculture, 62% of animals used in research experienced no pain, and another 32% were given anesthesia, pain killers, or both, to alleviate pain. Only 6% of animals suffered pain without benefit of anesthesia. Federal law in the United States requires the use of anesthesia or pain killers in animal research wherever possible. Exceptions are allowed only when the use of anaesthesia would compromise the experimental design and when no alternative method is available for conducting the test.

The animal rights movement

Those concerned with animal rights vary from traditional humane societies such as the Society for the Prevention of Cruelty to Animals (S.P.C.A.) and various national, state and local humane societies, through groups such as People for the Ethical Treatment of Animals (PETA, founded in 1980), to groups such as the Animal Liberation Front (ALF, founded in 1972). Some animal rights advocates use a utilitarian ethic to advocate their position; others are deontologists. Bernard Rollin, an American philosopher who supports animal rights, has observed:

> The main problem...is polarization and irrationality on the animal ethics issues by both sides... The American Medical Association's recent paper on animal rights labels all animal advocates as 'terrorists,' and scientific and medical researchers continue to equate animal rights supporters with lab trashers, Luddites, misanthropes, and opponents of science and civilization. Animal rights activists continue to label all scientists as sadists and psychopaths. Thus an unhealthy *pas de deux* is created that

blocks rather than accelerates the discovery of rational solutions to animal ethics issues. (B.E. Rollin. *Animal Rights and Human Morality,* Revised ed. Buffalo, NY: Prometheus Books, 1992, p. 10)

The organization most often labeled as 'terrorist' is the ALF. Since its founding, the ALF has concentrated on such tactics as breaking into animal research laboratories, destroying facilities and equipment, and 'liberating' the animals. Because these activities are illegal, leading members of the ALF are sought by law enforcement authorities, and most of them are now in hiding.

Do animals have rights? Bernard Rollin and a few other philosophers argue that there is no good reason for drawing an ethical distinction between mentally competent adult humans, other human beings (including children, comatose patients, mentally ill or brain-damaged persons), and animals. Animals are therefore, in his view, worthy of any moral consideration that would normally be given to babies, comatose patients, and other people unable to speak for themselves or to articulate their own viewpoints.

Historically, animals have been treated legally as property. Animal owners have property rights such as the right to sue for damages if their animals are killed or injured, but the animals themselves have no legal rights.

The question of animal rights is actually part of a broader question: how far do we extend the scope of any rights that we recognize? Many societies have historically denied even the most basic of rights to certain classes of persons on the basis of economics, gender, race, ethnicity, or religious beliefs. The extension of certain basic rights to all humans, including children and convicted criminals, is now considered so fundamental to the ethical sensibilities of most people that we refer to these as *human* rights. International agreements, such as the Geneva Convention on the treatment of prisoners of war and the International Convention on Human Rights (the Helsinki Accord), attest to the importance given to these human rights in world affairs. But should we stop there? Animal rights advocates say that we should extend these same rights to all beings capable of sensing pleasure and pain. A few people go even further, asserting that even trees have such rights as the right to go on living or not to have their air and soil poisoned. To go still further, a few people assert that habitats themselves, including mountains and forests, have rights not to be despoiled.

This book's web site contains further discussion of the philosophical arguments about animal rights and current debates regarding the use of experimental animals.

Humans as experimental subjects

Many animal experiments are undertaken to determine the effects of some new drug or other therapy. The real question is usually about what the effects would be in humans, and the animals are merely used as stand-ins. Is it safe to extrapolate to humans results obtained from nonhuman species? In most cases in which data are adequate to answer this question, human physiological reactions have turned out to be comparable to those of experimental animals. Even when differences between humans and other species are known, they are often known in sufficient detail that the different responses to testing can help us understand the human system better, which still makes the animal tests valuable.

A few instances are known in which humans and certain commonly used experimental animal species respond differently. Saccharin, for example, causes cancer in rats, but has never been shown to cause harm to humans.

Direct experimentation on humans avoids the question of comparability between species, meaning that the results can be used more directly than results obtained from other species. Also, results obtained from psychologists, epidemiologists, and others who study humans with naturalistic methods can be applied even more directly. For example, one could not ethically force-feed cholesterol to an experimental group of human subjects, but one could observe the diets that different people choose on their own and study how people with high-cholesterol diets differ from people with low-cholesterol diets. The diets in such a study are more directly comparable to the diets of other humans than are the diets of experimental animals fed with different amounts of cholesterol in their food. As explained in many later chapters, more than one type of method must be applied to answer many scientific questions. We should not view animal studies versus naturalistic studies of humans as either/or choices; in most cases, both approaches are needed.

Among possible experimental subjects, humans have a special status. On the one hand, inflicting pain on human subjects, or exposing them to experimental risks, raises more ethical objections than does the similar treatment of nonhuman animals. On the other hand, human subjects can tell us how they feel, or when and where they experience discomfort or pain. Certain types of drug side effects, like headaches or impairment of problem-solving ability, are difficult to assess without using human subjects.

Voluntary informed consent. A purely ethical consideration is that human subjects can voluntarily consent to serve as experimental subjects, which is something that nonhuman animals cannot do. Humans are considered to be autonomous beings who have the right to consent to putting themselves at risk, whether in a space capsule, a bungee jump, or an experiment. An important consideration, however, is that the person serving as a subject must give consent voluntarily. This is a legal as well as a moral issue, because persons who did not consent voluntarily can sue for damages if any harm comes to them. Consent is usually obtained in writing on a form that informs the person of the possible benefits and risks of the experimental procedure and is therefore called **informed consent**. If an experiment may bring direct benefit to a subject (as when a disease or its symptoms are being treated), potential subjects may be more willing to undergo certain risks than they would otherwise.

Special questions arise in the case of persons who may not have the full capacity to understand all the possible risks and benefits, including mentally deficient persons, unconscious persons, or children. Most people would now consider it unethical to use such a person as an experimental subject unless there was obvious great promise of direct benefit and only minimal risk of harm. In most jurisdictions, parents are considered to have the legal right to make such decisions on behalf of their children. In past decades, prison populations were often used as sources for experimental subjects, but this practice is now frowned upon because the consent of a prisoner may not be truly voluntary if she or he thinks—rightly or wrongly—that cooperating might result in a sentence reduction.

Guidelines for human experimentation. As a safeguard against possible abuses, research on human subjects is now usually reviewed by institutional committees set up for that purpose. As is true in reviews of animal experimentation, the review process ensures that someone other than the researchers evaluates the ethics of the proposed experiment. Federally sponsored research in the United States and in many other countries requires that such committees authorize all experiments in which humans are used as subjects. In addition to ensuring that proper voluntary consent has been obtained, such committees also have the obligation to suggest ways in which risks can be reduced or benefits increased without impairing the validity of the experiment. Many scientists work within the ethical tradition in which exposing humans to experimental risks is more objectionable than exposing animals of other species to those same risks. Guidelines have been developed that specify testing to be carried out on animals first, then on small numbers of carefully chosen and carefully monitored human subjects, and only last on large and diverse human populations. In the United States, federally sponsored research and research on new drugs seeking federal approval are required to follow this procedure.

Avoiding gender bias. Before the 1990s, many animal and clinical trials were done only on male subjects; one reason frequently given for this practice was to avoid the variable of hormonal fluctuations of the female reproductive cycles. Many test results based on male-only studies were extrapolated to women, and drug doses and other treatments were prescribed for women on this basis. In several cardiovascular conditions, including heart attacks and strokes, it now appears that men and women respond differently to certain drugs, and that drug doses calculated for men may be inappropriate for many women. When she was head of the National Institutes of Health, Dr. Bernadette Healy criticized a number of studies that had been done on men only that she thought would have been more appropriately done on women alone or on both sexes. One of the studies, for example, was based on the observation that pregnant women almost never have heart attacks. To test whether estrogen was the cause of this protective effect, the effects of estrogen therapy on heart attack rates was measured in *men*! Largely as the results of Dr. Healy's efforts, National Institutes of Health guidelines now require studies to be done on both sexes when appropriate. Although this is an issue of good experimental procedure, the issue was first raised as an ethical question of unfairness to women. Dr. Healy and others claimed that women were getting potentially substandard medical care if they were treated with drugs that had been tested on men only and with doses calibrated for male patients.

THOUGHT QUESTIONS

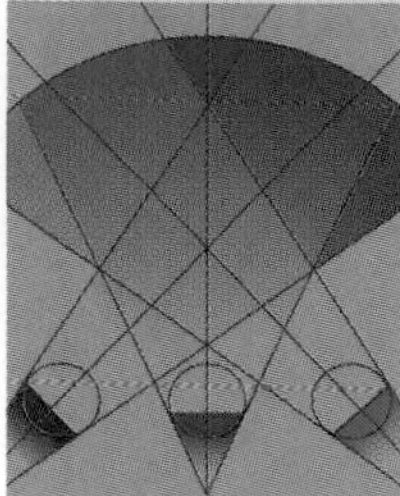

1. For each of the following acts, try to give:
a. a deontological argument against the act
b. a deontological argument justifying the act
c. a utilitarian argument against the act
d. a utilitarian argument justifying the act
• beating your horse
• taking a canary into a coal mine so that, if it dies from toxic gases, miners would be warned to evacuate
• raising broiler chickens or beef cattle for human consumption
• testing a drug on rats (or cats) before giving it to humans
• testing a drug on human prison convicts

2. How widely can experimental results be extrapolated? If a drug is tested on inbred male rats, is it certain that the results are applicable to humans? Is it likely? Is the drug likely to have similar effects on both sexes? What issues of methodology or of ethics are raised by experiments that used only inbred male rats?

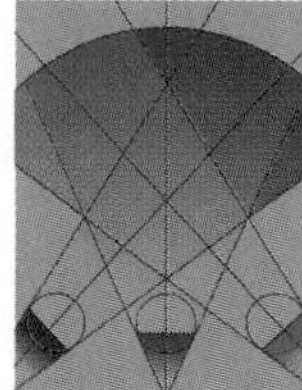

IN SCIENCE, WE KNOW WHAT WE KNOW THROUGH A PROCESS often called the scientific method. Scientists formulate tentative ideas (hypotheses) about living systems and about the world in general, and submit these ideas to extensive testing by comparing their ideas with observations made in the material world around them. Scientific knowledge is forever tentative and is never 'proved' because it is always subject to change if new observations do not fit our existing theories. The language of science is often metaphorical. Scientists often use words with specialized meanings. Scientific paradigms give science its vocabulary, imagery, attitudes, and value judgments. Science is conducted in a social context that includes a community of scientists sharing their ideas and testing each other's hypotheses.

Ethical decisions can be made either by judging actions themselves (deontological ethics) or by judging actions on the basis of their consequences (utilitarian ethics). Many scientific issues have social implications. All citizens, not just scientists, should help make these decisions collectively. However, scientists should bear some responsibility for educating others about the science issues and science policy issues that can inform these decisions. Scientists should educate themselves as to the ethical dimensions of their work, including both the treatment of experimental subjects and the possible uses and misuses of scientific findings.

CHAPTER SUMMARY

- **Science** is based on the testing of **falsifiable hypotheses**.
- Hypotheses are derived by both **inductive** and **deductive** reasoning.
- A group of well-tested hypotheses forms a **theory**. A theory may be communicated by a descriptive analogy called a **model**.
- **Biology** is a science because biologists use hypothesis testing to study living systems.
- Biologists test hypotheses either by studying natural conditions as they occur (**naturalistic science**) or by **experiments** under conditions that the

scientists help to create (**experimental science**). Interpretation of results usually involves comparison with **control groups**, often with the help of statistical methods.

- **Normal science** proceeds by testing hypotheses one at a time, but **scientific revolutions** occur when many widely held ideas are discarded as a new **paradigm** emerges to take their place.
- Scientific methods are used in everyday life, but scientists use these methods more often and more systematically.
- Science has existed for centuries in various countries. In seventeenth-century England, scientists first began to organize as a community into scientific societies and to publish their findings in scientific journals.
- Certain values are held by members of the scientific community that ensure them the continuing ability to test, falsify and change each other's hypotheses.
- **Morals** are rules that guide our conduct. **Ethics** is the discipline that examines moral rules and attempts to explain or justify them.
- Two major types of ethical systems are **deontological** and **utilitarian**. Deontologists judge the rightness or wrongness of an act by characteristics of the act itself, apart from its consequences. Utilitarians judge the rightness or wrongness of an act on the basis of its consequences. Utilitarian analysis often includes a comparison of the undesirable effects (costs) of an act with its desirable effects (benefits).
- In facing ethical issues involving scientific questions, it is often useful to distinguish between scientific issues, science policy issues, and policy issues.
- Animals are used in our society for food, for labor, for companionship, and for laboratory experimentation. Biological experiments often use living organisms as subjects. In many cases, laboratory animals are used as stand-ins for humans, and their use is often justified on a utilitarian basis (the cost–benefit ratio is lower if animals are tested before humans) or on a deontological basis (humans have rights and animals do not, or human rights supersede animal rights).
- Before any experimentation on animals or humans can take place, the proposed experiments must pass an ethics review. If humans are used, their voluntary **informed consent** must be obtained.

KEY TERMS TO KNOW

biology (p. 4)
control group (p. 11)
data (p. 5)
deduction (p. 6)
deontological (p. 22)
ethics (p. 20)
experiments (p. 10)
experimental sciences (p. 10)
fairness (p. 25)
falsifiable (p. 5)
hypotheses (p. 5)
induction (p. 6)
informed consent (p. 30)
model (p. 8)
morals (p. 20)
naturalistic sciences (p. 13)
normal science (p. 14)
paradigm (p. 14)
rights (p. 22)
science (p. 4)
scientific revolution (p. 15)
theory (p. 8)
utilitarian (p. 23)
verifiable (p. 5)

CONNECTIONS TO OTHER CHAPTERS

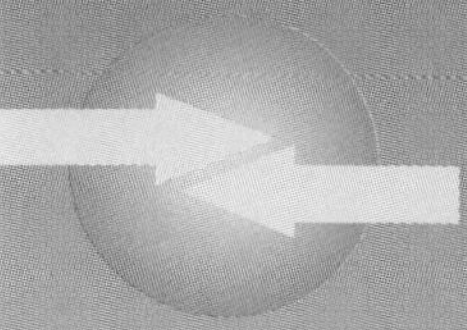

This chapter connects to the remainder of the book because the methods of discovery outlined in this chapter were used to explore all of the topics described in subsequent chapters. The characteristics of life listed at the beginning of this chapter are referred to throughout the book. Also, many applications of science have ethical dimensions, including the following:

Chapter 3: The Human Genome Project and human genetic testing raise ethical questions.

Chapter 5: Ethical objections have been raised against the ways in which biology has supported racism.

Chapter 6: The need for population control conflicts with the ethic of allowing reproductive freedom.

Chapter 7: Evolution has resulted in both moral and immoral behaviors.

Chapter 8: Patterns of food consumption and distribution raise ethical issues.

Chapter 9: Cancer research often involves the use of animal experimentation.

Chapter 10: Brain research involves animal experimentation and also the use of fetal tissue.

Chapter 11: Drugs are usually tested on animals before giving them to people.

Chapter 13: Many ethical issues surround the transmission of AIDS, testing for AIDS, and the prevention of AIDS.

Chapter 14: Some people object to genetically engineered foods on ethical grounds.

Chapter 15: Many ethical issues are raised by global patterns of pollution, habitat destruction, and species extinction.

PRACTICE QUESTIONS

1. Which property of life is exhibited by each of the following?

   a. The frog jumps around when I touch it.

   b. The bread rises because the yeast have given off carbon dioxide bubbles.

   c. Blood samples from healthy humans always have about the same pH and salt concentration.

   d. Wherever I find one mosquito, I usually find many.

   e. Only a few rabbits were brought to Australia, but now there are millions.

   f. Puppies usually resemble their parents.

   g. Baby animals get bigger and become adults.

   h. A bright light at night always attracts moths.

2. Which of the following are falsifiable statements? For each statement that you think is falsifiable, explain what sort of observation might falsify it.

   a. The Backstreet Boys are a better musical group than the Rolling Stones.

   b. In a maze that they have never seen before, rats will turn right just about as often as they will turn left.

   c. If these two plants are crossed, approximately half of the offspring will resemble one parent and half will resemble the other.

   d. It is wrong to inflict pain on a cat.

   e. Restaurant A is better than restaurant B.

   f. The average science major at this school gets better grades than the average humanities major.

3. Which of the following are inductive? Which are deductive?

   a. If all adult female birds lay eggs, then this female chick will lay eggs if raised to maturity.

   b. If all known species of birds are egg-laying, then

the next species to be discovered will be egg-laying too.

c. If all enzymes are made of protein, and I discover a new enzyme, then it, too, will be made of protein.

d. If the amounts of protein X are increased under stress, then I should be able to increase the amount of protein X in these frogs by subjecting them to stressful conditions.

e. If this species mates in April, then I should be able to observe more mating on April 10 than on June 10.

4. Which of the following reflect the community nature of science?

a. A scientist presenting a talk at a scientific meeting.

b. Another scientist asking a question at that same meeting.

c. A field naturalist tracking a rare species.

d. The same field naturalist publishing her findings.

e. A scientist feeding a new chemical to mice to study its effects.

f. A bacteriologist using techniques developed by Louis Pasteur and Robert Koch for growing bacteria in laboratory cultures.

g. A scientist displaying his experimental results over the Internet.

5. For each of the following, identify (a) whether the argument is based on utilitarian or deontological ethics, and (b) whether any assumptions are made about whether animals have no rights, some rights, or rights equal to those of human beings.

a. Hunting is wrong because the victim is part of nature and it is wrong to interfere with nature.

b. Hunting is justified because the death of one animal makes such a small difference to most hunted species.

c. Hunting is wrong because it makes the hunter more prone to future violence.

d. Hunting is justified if the animal is used as food but not as a trophy.

e. Raising beef cattle for human consumption is justified because people need to eat.

f. Raising beef cattle for human consumption is wrong because cows are sacred.

g. Raising beef cattle for human consumption is wrong because people would be healthier if they ate more plant foods instead.

h. Raising beef cattle for human consumption is wrong because it causes pain and suffering to the animals.

i. Be kind to your pet because you will be rewarded with loving companionship.

2 Genes, Chromosomes, and DNA

CHAPTER OUTLINE

Mendel Observed Phenotypes and Formed Hypotheses

Traits of pea plants

Genotype and phenotype

The Chromosomal Basis of Inheritance Explains Mendel's Hypotheses

Mitosis

Meiosis and sexual life cycles

Gene linkage

Confirmation of the chromosomal theory

The Molecular Basis of Inheritance Further Explains Mendel's Hypotheses

DNA and genetic transformation

The chemical composition of DNA

The three-dimensional structure of DNA

DNA replication

Transcription and translation of genes

Mutations

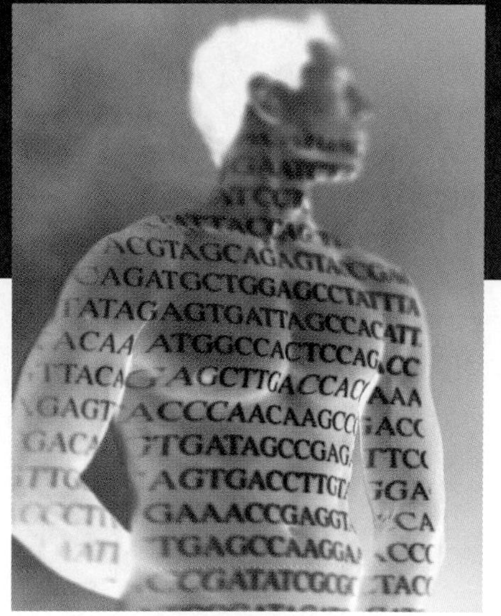

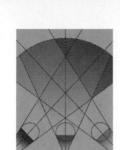

ISSUES

How have our concepts of genes developed?

How did molecular genetics grow out of Mendel's hypothesis? How do they help explain each other?

What are the limitations of Mendelian genetics?

Do Mendelian and molecular genetics explain inheritance in all species?

What do we *not* understand about genes, chromosomes and DNA?

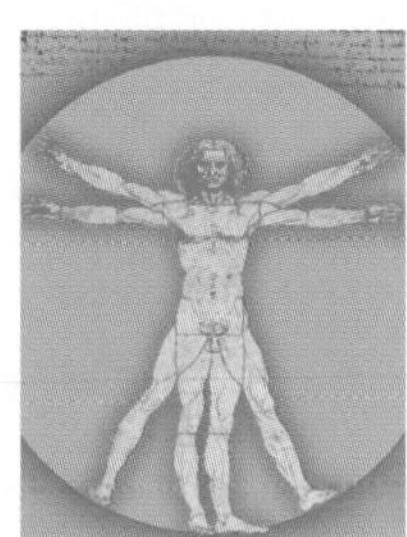

BIOLOGICAL CONCEPTS

The gene

Patterns of inheritance

Genotype and phenotype

DNA (the genetic material, DNA structure, DNA replication, mutation)

Mitosis and meiosis

Molecular genetics (gene action, RNA, transcription, translation)

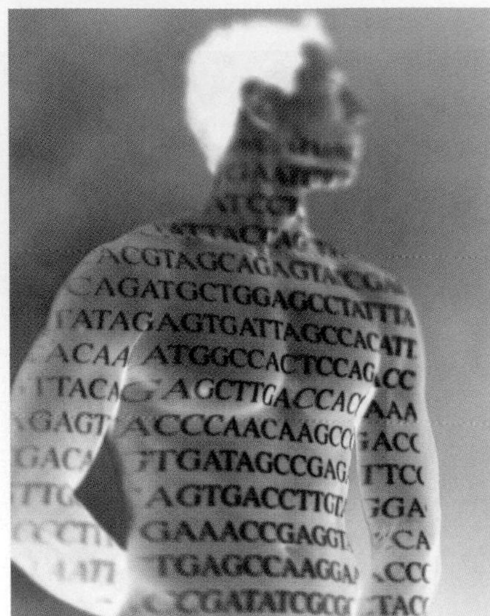

Chapter 2: Genes, Chromosomes, and DNA

How do offspring come to resemble their parents physically? This is the major question posed by the field of biology called ***genetics****, the study of inherited traits. Genetics begins with the unifying assumption that biological inheritance is carried by structures called* ***genes****. The discovery of what genes are and how they work has been the subject of many years of research. Among the earliest findings was the fact that the same basic patterns of inheritance apply to most organisms.*

MENDEL OBSERVED PHENOTYPES AND FORMED HYPOTHESES

No two individual organisms are exactly alike. Folk wisdom going back to ancient times taught people that a child or an animal resembles both its mother and its father by showing a mixture of traits derived from the two sides of the family. This suggested a concept that came to be called blending inheritance, in which heredity was compared to a mixing of fluids, often identified as 'blood.' The research that we are about to describe caused this theory to be abandoned.

Traits of pea plants

During the nineteenth century, Gregor Mendel, a Czech scientist living under Austrian rule, worked out the principles of inheritance for simple traits that he described in 'either/or' terms. Mendel was a priest who grew pea plants (*Pisum sativum*, kingdom Plantae, phylum Anthophyta also called Angiospermae) in the garden of his monastery. Mendel was curious about differences that he observed among different varieties of pea plants, so he decided to breed them and keep careful records.

Why were the peas a good species for Mendel's experiments? Pea plants have many distinctive traits (Figure 2.1), and other practical advantages: Mendel could easily grow them in the monastery garden, and many varieties were locally available, including some with yellow peas and others with green peas, and some with round and others with

FIGURE 2.1

The seven traits studied by Mendel in peas.

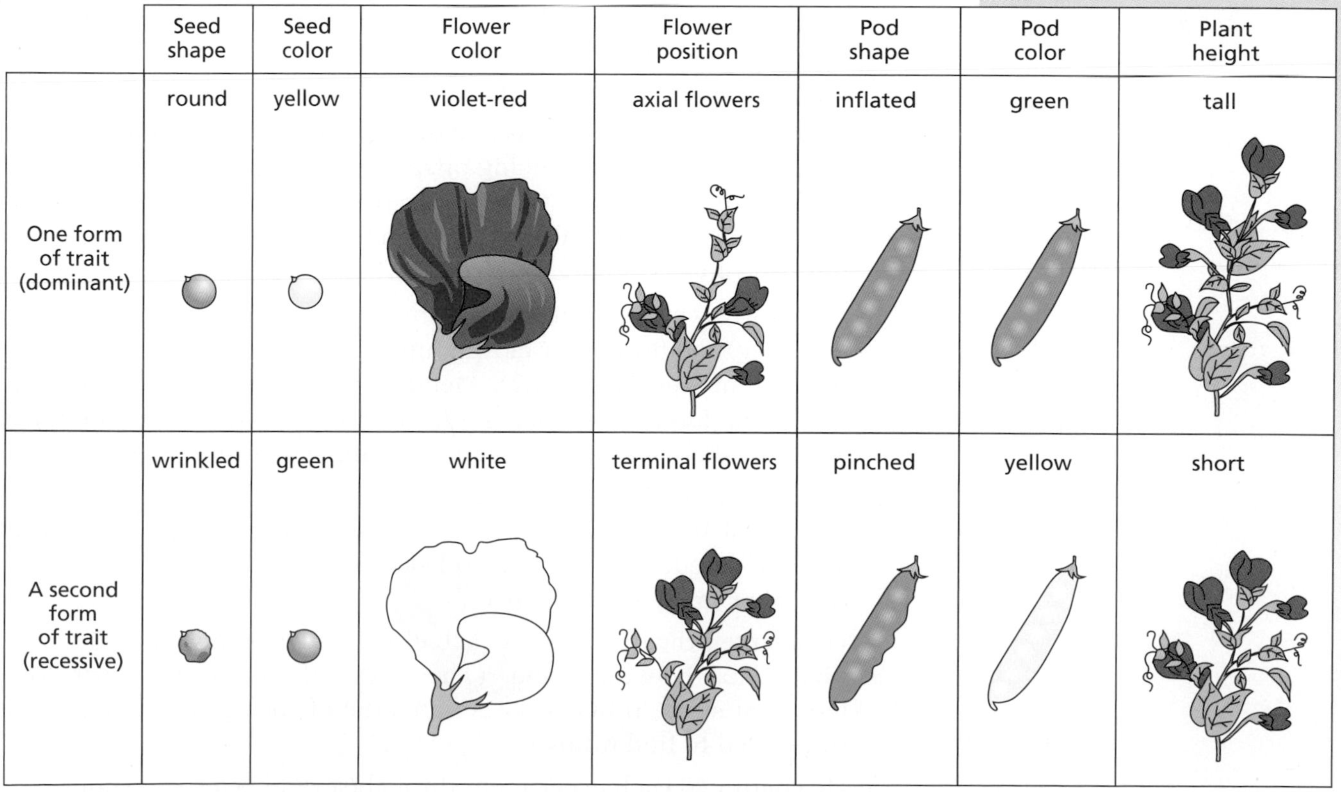

	Seed shape	Seed color	Flower color	Flower position	Pod shape	Pod color	Plant height
One form of trait (dominant)	round	yellow	violet-red	axial flowers	inflated	green	tall
A second form of trait (recessive)	wrinkled	green	white	terminal flowers	pinched	yellow	short

wrinkled peas. Mendel knew that peas reproduce sexually—that is, a new individual forms when an egg from a female unites with a sperm from a male, an event called **fertilization**. Unlike most animals, an individual plant may have both female and male reproductive organs. This is true of many plant species, including peas. Within the pea flower are male structures called anthers that produce pollen grains, each of which contains a sperm. The pea flower also has a female structure called a stigma that receives pollen grains (a step called **pollination**) and permits the sperm to travel to the ovary to reach the eggs (Figure 2.2). Self-pollination occurs when pollen from a plant is deposited on a stigma of the same plant. In some of his experiments, Mendel sewed together the margins of one large petal to enclose the flower and ensure self-pollination. At other times, he cross-pollinated his peas by dusting pollen from a flower of one plant onto the stigma of a flower of another plant, after first removing the anthers from the recipient plant.

FIGURE 2.2

The structure of a pea flower. A more complete view of the sexual reproductive structures of flowering plants is shown in Figure 14.10.

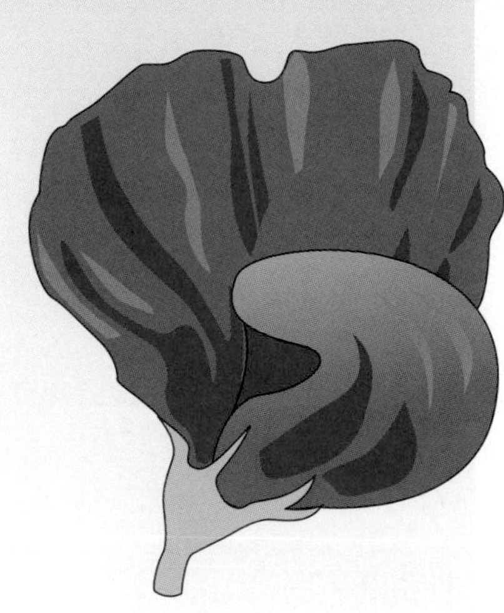

pea flower with petals enclosing sexual parts inside

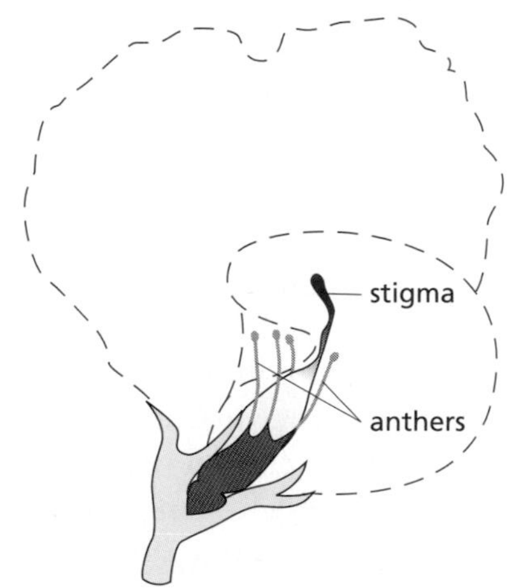

pea flower with petals removed, to show male and female flower parts

Mendel organized his work so as to answer specific questions, a procedure that we recognize today as good experimental design. Unlike most of his predecessors who failed to discover how plant offspring inherit their parents' traits, Mendel followed certain careful procedures:

1. First, for each of the traits he studied, Mendel used peas belonging to pure lines. A **pure line** is one that breeds true from generation to generation, always producing offspring that express the same form of the trait as the parents. For example, tall parents from one pure line always produce tall offspring and short parents from another pure line always produce short offspring.
2. He chose which plants would mate by either cross-pollinating (crossing) the flowers or closing up the flower parts to ensure self-pollination (see Figure 2.2). Mendel became skillful at sewing up one of the large petals, sealing in the stigma and anthers, to make sure that there was no unwanted crossing with other plants.
3. He first studied only one trait at a time, until he understood its pattern of inheritance. Later, he studied two and three traits at a time. His predecessors, on the other hand, often began by examining several or many traits at once.
4. He counted the offspring of each cross, and was thus able to recognize ratios between them. (Those few of his predecessors who looked at single traits never counted the offspring of each type and thus failed to find ratios.)
5. He continued each experiment through several pea generations.

Mendel studied one trait at a time. For example, he crossed plants having white flowers with plants having violet-red flowers, and found that all the first generation offspring (symbolized as F_1) had violet-red flowers. When he crossed tall and short plants, all the offspring were tall. Mendel introduced the term **dominant** for the form of the trait that appeared in the first-generation offspring of his initial cross; the trait that did not show up he called **recessive**. Thus, violet-red flowers are dominant to white, and tall is dominant to short. Mendel found for each of the seven pairs of either/or traits that one form of the trait was dominant and one was recessive. The forms of the traits shown in the upper row of Figure 2.1 are dominant; those in the lower row are recessive. The traits did not blend. Violet-red-flowered plants crossed with white-flowered plants produced violet-red-flowered offspring, not pink-flowered offspring. Tall plants crossed with short plants produced offspring of the height of the tall parents, not halfway between.

Genotype and phenotype

The tall F_1 plants in Mendel's experiments were just as tall as their tall parents, which is expressed in genetics by saying that the **phenotype** for height of both parents and offspring is tall. But Mendel realized that the hereditary makeup of the F_1 plants was different from that of their parents: the tall parents had come from a pure line, so all their hereditary make-up had been tall, but each F_1 plant had both tall and short parents. Could this difference in hereditary make-up, or **genotype**, be made visible? Would it show up in future generations? Mendel found out by mating the plants of the F_1 generation with themselves (self-pollinating them) and raising a second generation, symbolized as F_2. He obtained both plants with violet-red flowers and plants with white flowers in the F_2, but no plants of intermediate color. When he counted them, he found that the plants with violet-red flowers were approximately three times as numerous as the white-flowered ones. Mendel conducted similar experiments with other traits, such as plant height, and in each experiment he produced an F_2 generation. In each case he discovered the same thing: in the F_2 generation, the dominant phenotype outnumbered the recessive phenotype in the approximate ratio of 3:1, so that, three-quarters of the F_2 plants had the dominant phenotype and one-quarter had the recessive one.

Mendel's explanation for inheritance of single traits. Mendel proposed a multipart hypothesis to explain his results. Using modern terminology and some modern understanding, we can list the points covered by his hypothesis.

1. The inheritance of traits is controlled by hereditary factors; today these factors are called genes.
2. Each individual has two copies of the gene for each trait. If each gene is represented by a letter, then the genetic makeup (genotype) of an individual for a trait can be represented as two letters.
3. Each gene exists in different forms; these variant forms of the same

gene are called **alleles**. For example, the gene for flower color in peas has an allele that produces white flowers and another allele that produces violet-red flowers. Alleles that are dominant produce only the dominant phenotype even when the recessive allele is present. Mendel designated the dominant and recessive alleles controlling the same trait by a single letter of the alphabet, using a capital letter for the dominant allele and the lowercase of the same letter for the recessive allele. For example, the allele *T* for tall plants is dominant to the allele *t* for short plants. (Today, many genes are designated by two- and three-letter combinations, and different alleles of the same gene by superscripts.)

4. An individual whose genotype contains two identical alleles, such as *TT* or *tt*, is said to be **homozygous** for the trait. An individual whose genotype combines dissimilar alleles, such as *Tt*, is said to be **heterozygous**.
5. Dominant alleles always show up in the phenotype, but recessive alleles are masked by their dominant counterparts. When a dominant and a recessive allele for the same trait are present in a heterozygous individual, the dominant allele produces the phenotype. Recessive alleles produce the phenotype only when they are homozygous—only, in other words, when the corresponding dominant allele is absent.
6. The genes behave as particles that remain separate instead of blending. Recessive genes are masked during the F_1 generation but pass unchanged through the heterozygous F_1 individuals, which show the dominant phenotype.
7. When the F_1 individuals produce eggs and sperm, the dominant and recessive alleles separate from one another, or 'segregate.' The separation of alleles is called the **law of segregation**, or Mendel's first law. The eggs and sperm are called **gametes**, and each gamete receives only one allele of each gene.
8. During sexual reproduction, the gametes combine so that each F_2 individual has two alleles, one contributed by the egg and one by the sperm. This explains both why the recessive phenotype disappears in the F_1 and why it reappears in about one-quarter of the F_2 individuals.

Figure 2.3 shows Mendel's multipart hypothesis applied to one of the traits he studied. A pure-line plant with violet-red flowers and the genotype *VV*, which produces *V* gametes, is crossed with a pure-line, white-flowered *vv* plant, which produces *v* gametes. All F_1 plants, having received *V* from one parent and *v* from the other, have violet-red flowers and the heterozygous genotype *Vv*. Half of the eggs produced by the F_1 carry the dominant allele *V* and half carry the recessive allele *v*, and the same is true for the sperm, as indicated in the margins of the square in Figure 2.3. When the gametes from the self-pollinated F_1 combine at random to form the F_2 generation, one out of four new individuals is homozygous for the dominant allele *VV* and has violet-red flowers, two out of four are heterozygous *Vv* with violet-red flowers, and one out of four is homozygous for the recessive allele *vv* and has white flowers. The ratio of violet-red to white phenotypes is thus 3:1. A square like this, showing how gametes combine to produce genotypes and phenotypes, is called a Punnett square.

Independent assortment. After Mendel had investigated the inheritance of single traits, he proceeded to study the inheritance of two traits together. In one of his experiments, the parents differed in both seed shape and seed color. The parents in one group had yellow, round seeds, were homozygous for both traits (*YYRR*), and thus produced all *YR* gametes. The other parents had green, wrinkled seeds, were homozygous for both traits (*yyrr*), and thus produced all *yr* gametes. The first generation offspring (F_1) were all *YyRr*, heterozygous for both traits. All these F_1 plants produced seeds that were yellow (because yellow is dominant to green) and round (because round is dominant to wrinkled). No new principles were involved so far.

Mendel expected half of the gametes produced by the F_1 to contain the dominant allele *Y* and the remainder the recessive allele *y*. He also expected half of the gametes to contain allele *R* and the other half the allele *r*. But would a gamete's receiving *Y* rather than *y* influence whether it received *R* or *r*? To find out, Mendel raised an F_2 generation by self-pollinating some F_1 plants. He obtained the 9:3:3:1 ratio of phenotypes shown in Figure 2.4, with 9/16 of the F_2 individuals showing both dominant traits (they were both yellow and round); 3/16 were round but green, not yellow; 3/16 were yellow but wrinkled, not round; and only 1/16 showed both recessive traits (green and wrinkled). Mendel then reasoned that this ratio could be explained if he assumed that all four possible types of gametes (*YR*, *Yr*, *yR*, and *yr*) were produced in equal proportions, as shown in the margins of the square in Figure 2.4. This means that the inheritance

FIGURE 2.3

One of Mendel's experiments with peas differing in a single trait.

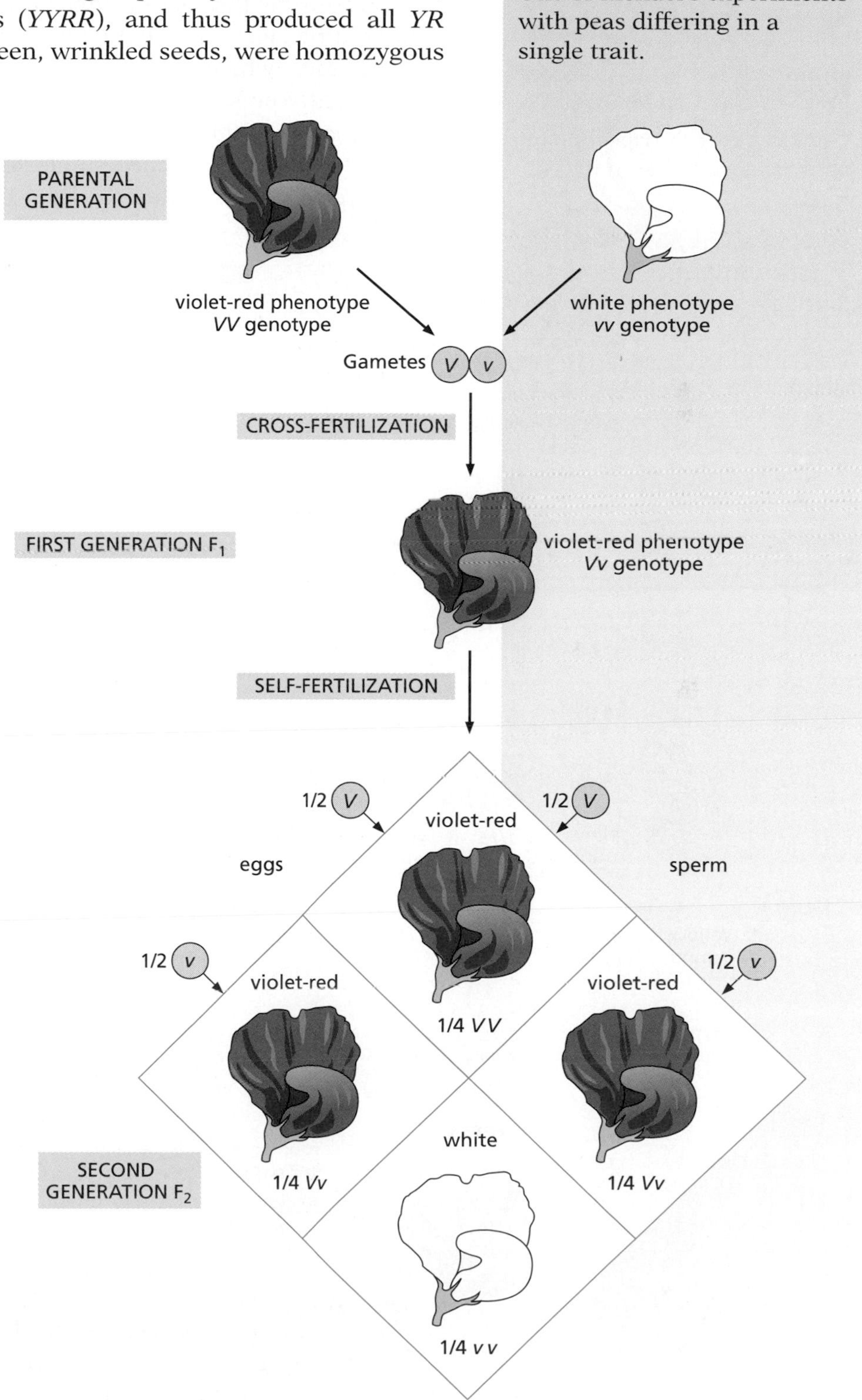

of the trait of seed color has no influence on the inheritance of the trait of seed shape. This principle is called the **law of independent assortment** or Mendel's second law.

All of the traits that Mendel worked with assorted independently in this way, but, as we will soon see, there are exceptions to Mendel's second law.

Mendel's results, published in 1865, were ignored by most scientists. The reasons for the lack of impact of his theories are many, but a main contributing factor was that he presented his theories in the language of mathematics, which was not a language in which his fellow scientists were fluent. The same problem blocks interdisciplinary efforts in many fields today. In 1900 each of three other scientists conducted experiments similar to Mendel's, reached similar conclusions, and then subsequently discovered Mendel's earlier work.

FIGURE 2.4

One of Mendel's experiments with peas differing in two traits.

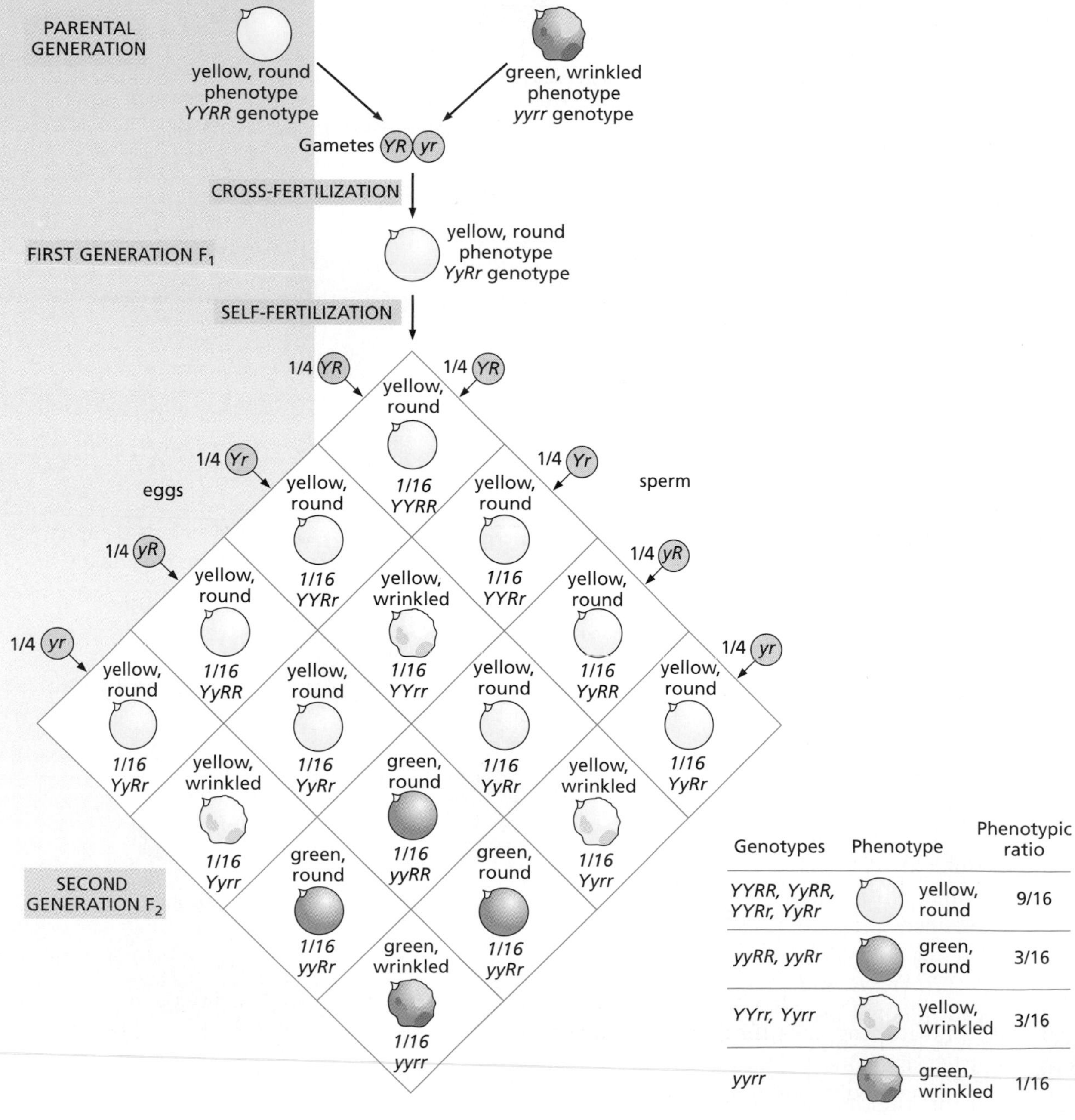

PARENTAL GENERATION
yellow, round phenotype *YYRR* genotype
green, wrinkled phenotype *yyrr* genotype
Gametes *YR* *yr*
CROSS-FERTILIZATION
FIRST GENERATION F_1
yellow, round phenotype *YyRr* genotype
SELF-FERTILIZATION
eggs
sperm
1/4 *YR* 1/4 *Yr* 1/4 *yR* 1/4 *yr*
yellow, round 1/16 *YYRR*
yellow, round 1/16 *YYRr*
yellow, round 1/16 *YYRr*
yellow, round 1/16 *YyRR*
yellow, wrinkled 1/16 *YYrr*
yellow, round 1/16 *YyRR*
yellow, round 1/16 *YyRr*
yellow, round 1/16 *YyRr*
yellow, round 1/16 *YyRr*
yellow, round 1/16 *YyRr*
yellow, wrinkled 1/16 *Yyrr*
green, round 1/16 *yyRR*
yellow, wrinkled 1/16 *Yyrr*
green, round 1/16 *yyRr*
green, round 1/16 *yyRr*
green, wrinkled 1/16 *yyrr*
SECOND GENERATION F_2

Genotypes	Phenotype	Phenotypic ratio
YYRR, YyRR, YYRr, YyRr	yellow, round	9/16
yyRR, yyRr	green, round	3/16
YYrr, Yyrr	yellow, wrinkled	3/16
yyrr	green, wrinkled	1/16

THOUGHT QUESTIONS

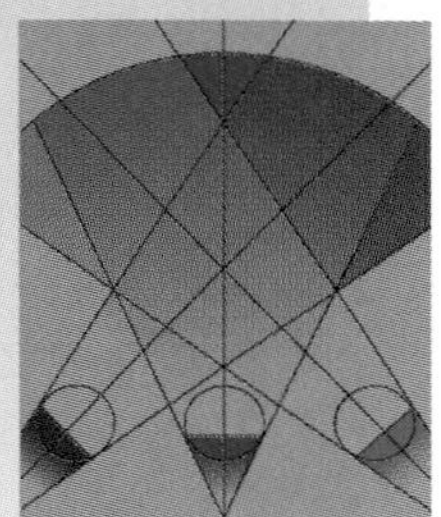

1. Mendel's experiments distinguished between two alternative hypotheses: either traits blend, so that the offspring have traits intermediate between those of their parents, or traits are inherited as discrete particles that do not blend. Which hypothesis is favored by Mendel's results? Did he disprove one hypothesis? Did he prove one hypothesis?

2. An experimenter cross-pollinates flowers of tall pea plants with pollen from flowers from short pea plants. She harvests the seeds and plants them, and all of the plants that grow in the F_1 generation are as tall as the tall parent plants. The F_1 plants produce flowers, each containing many sperm and many eggs. Each sperm and each egg carries only one allele of each gene. What are the possible genotypes for plant height among the sperm from the F_1 plants? What are the genotypes among the eggs? What fractions of the gametes possess each genotype? Would the distribution of types of gametes in the F_1 plants be the same or different if they were produced by cross-pollinating flowers of short pea plants with pollen from tall pea plants?

3. To study two traits at a time, Mendel first crossed one line of plants that bred true for two traits with another line that bred true for different alleles of the same two genes. F_1 flowers were then self-pollinated and the F_2 peas produced as before. If you counted 16 of the F_2 peas, how many would you expect to be the doubly dominant phenotype, round and yellow? If you counted 1600 F_2 peas, how many round and yellow ones would you expect?

4. Why are certain traits studied in some species and not in others?

THE CHROMOSOMAL BASIS OF INHERITANCE EXPLAINS MENDEL'S HYPOTHESES

Notice that some of Mendel's assumptions raise questions that Mendel himself did not answer:

1. Where are the genes located?
2. Why do the genes exist in pairs?
3. Why do different traits assort independently?
4. What are the genes made of?

Answers to the first three of these questions were suggested by a young American scientist, Walter Sutton, who read about the rediscovery of Mendel's work in 1900. By this time, it was already well known that the cells that make up the bodies of plants and animals contain a central portion called the **nucleus** and a surrounding portion called the **cytoplasm** (Figure 2.5; for more on cell structure see Chapters 4, 8, and 9). Plants and animals are able to develop and grow because their cells divide. One cell divides to become two cells. Scientists who looked at dividing cells through their microscopes saw that the division of the cytoplasm is a very simple affair but that the dividing nucleus undergoes a

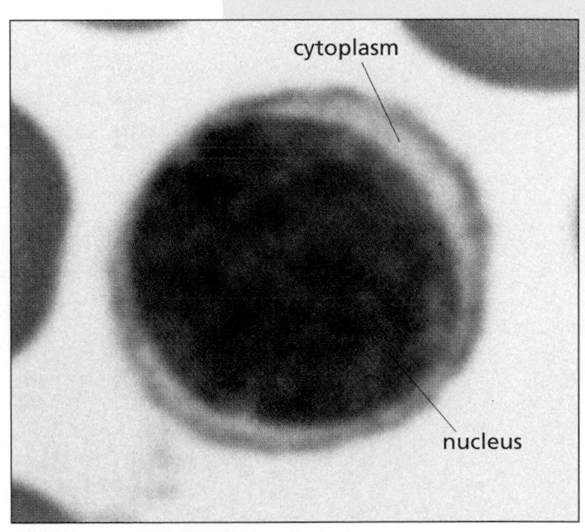

FIGURE 2.5

Structure of a nucleated cell. This particular cell is a type of white blood cell called a lymphocyte (see Chapter 12) that has been stained with a dye to make its nucleus show distinctly.

complex rearrangement of the rod-shaped bodies called **chromosomes**. Chromosomes cannot usually be seen, except when cells divide. The chromosomes of dividing cells are usually paired. The only exceptions are gametes (eggs and sperm). Gametes are **haploid**, with one chromosome from each pair. All other body cells, called somatic cells, have a **diploid** chromosome number with pairs of chromosomes (Figure 2.6). Each set of two chromosomes that look alike is called a **homologous pair**. Sutton noticed that eggs in most species are many times larger than sperm because they have a greater amount of cytoplasm (see Figure 6.8, p. 248). The nuclei of egg and sperm are approximately equal in size, and these nuclei fuse during fertilization. From these facts, Sutton reasoned as follows:

FIGURE 2.6

Chromosomes in a gamete and a somatic cell of a species having two pairs of chromosomes in somatic cells.

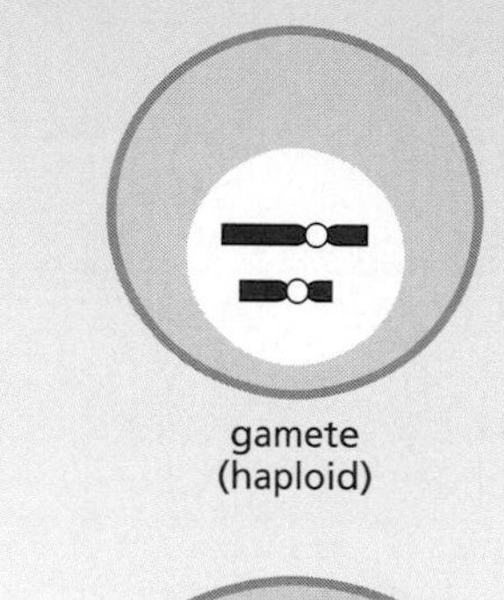

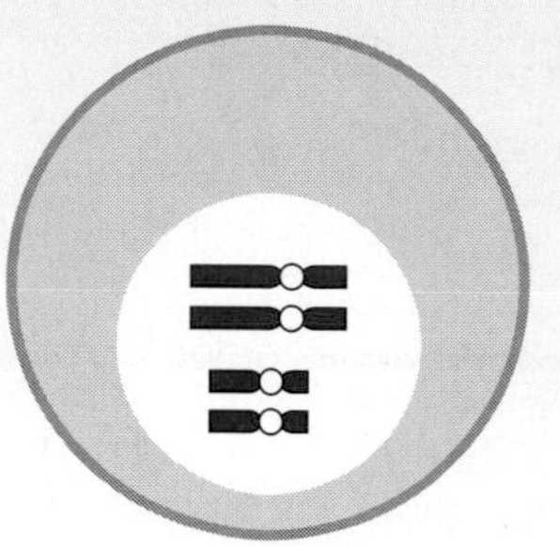

1. The genes are probably in the nucleus, not the cytoplasm, because the nucleus divides carefully and exactly, whereas the cytoplasm divides inexactly. Also, if genes were in the cytoplasm, the larger amount of cytoplasm in the egg would lead one to expect the egg's contribution always to be much greater than the sperm's, contrary to the observation that parental contributions to heredity are usually equal.
2. No other structures in cells, apart from chromosomes, are known to exist in pairs in somatic cells and singly in gametes. *The known behavior of chromosomes exactly parallels the behavior that Mendel postulated for genes*. Genes must be located on chromosomes.
3. Mendel's genes assort independently because they are located on different chromosomes. However, there are only a limited number of chromosomes (4 pairs in fruit flies, 23 pairs in humans), but hundreds or thousands of genes. Sutton predicted that Mendel's law of independent assortment would apply only to genes located on different chromosomes. Genes located on the same chromosome would be inherited together as a unit, a phenomenon now known as **linkage**.

Sutton's idea that genes are located on chromosomes came to be called the **chromosomal theory of inheritance**. To understand the chromosomal theory of inheritance and Sutton's prediction that some pairs of traits would not follow Mendel's law of independent assortment, we must understand how chromosomes behave in dividing cells. There are two types of division: **mitosis** is the process by which somatic cells divide and **meiosis** is the process of division that produces haploid gametes. In this section we consider each type of cell division in turn.

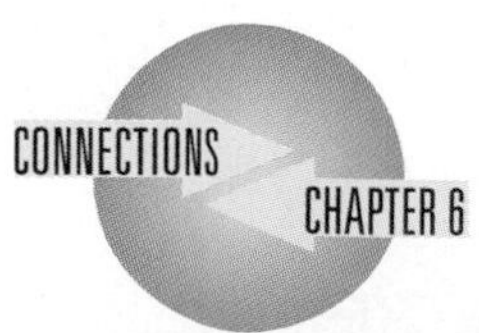

Mitosis

In mitosis, a somatic cell divides in two in a way that leaves each new cell with a diploid set of chromosomes, no more and no less. Each offspring cell that results from mitosis is thus genetically identical to the beginning cell.

Mitosis proceeds in five stages: interphase, prophase, metaphase, anaphase, and telophase, which are shown in Figure 2.7. During

INTERPHASE

The cell increases in size and the chromosomes are duplicated to twice the diploid number, although they are not visible in the nucleus at this stage.

nuclear membrane

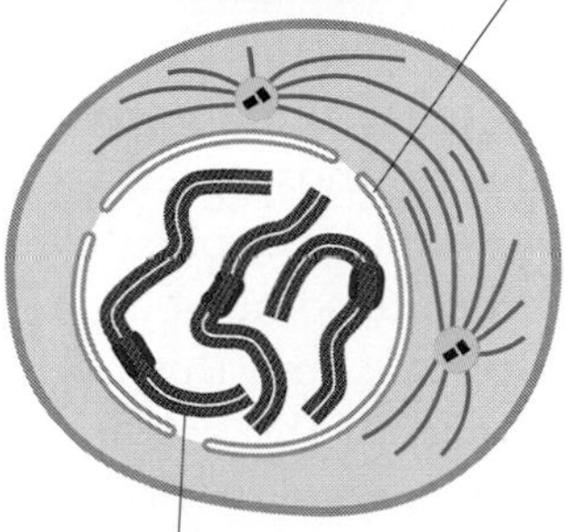

PROPHASE

The chromosomes condense and shorten, becoming visible under a microscope as attached pairs.

nuclear membrane fragments

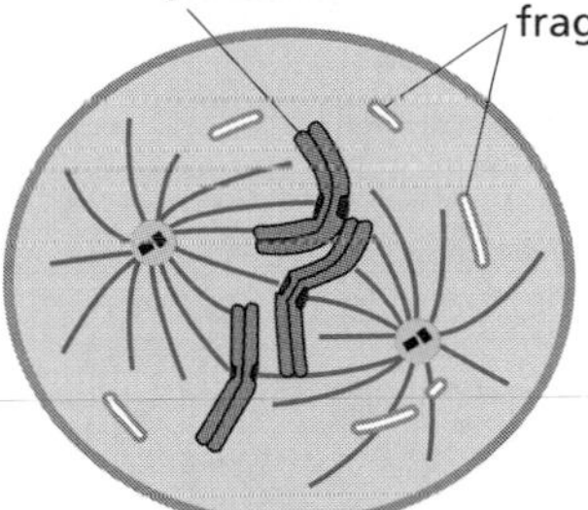

METAPHASE

The membrane surrounding the nucleus breaks down; each chromosome and its attached duplicate line up along the midline of the cell.

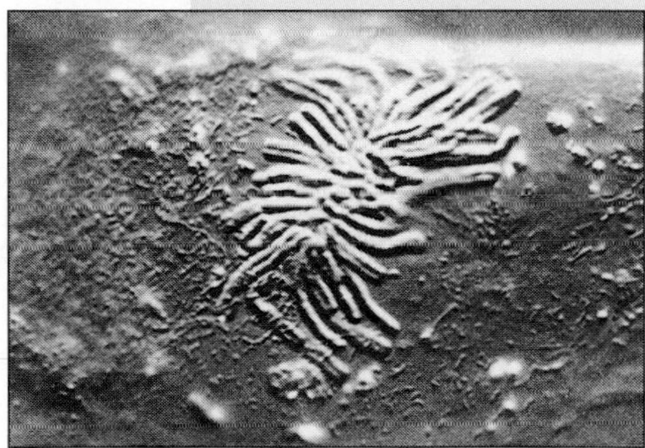

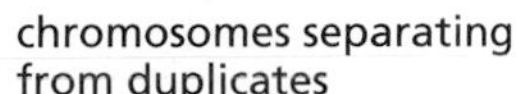

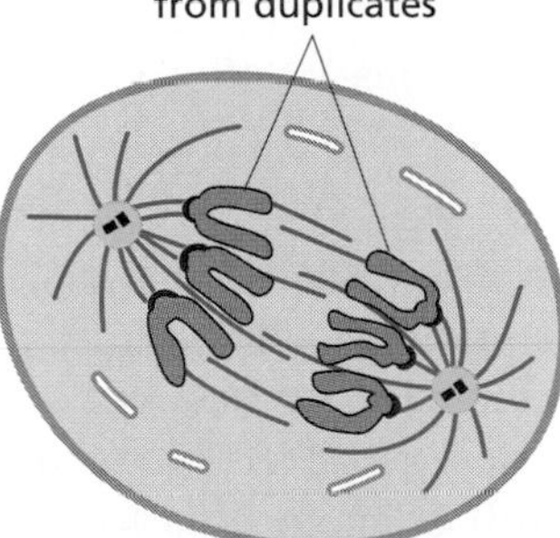

ANAPHASE

Each chromosome separates from its duplicate; the chromosome and its duplicate move toward opposite ends of the cell.

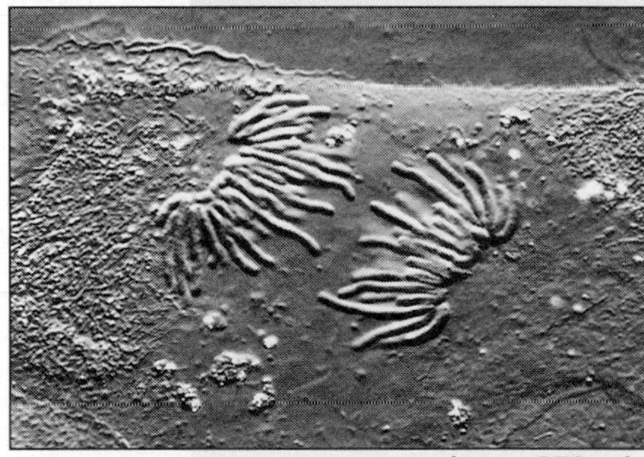

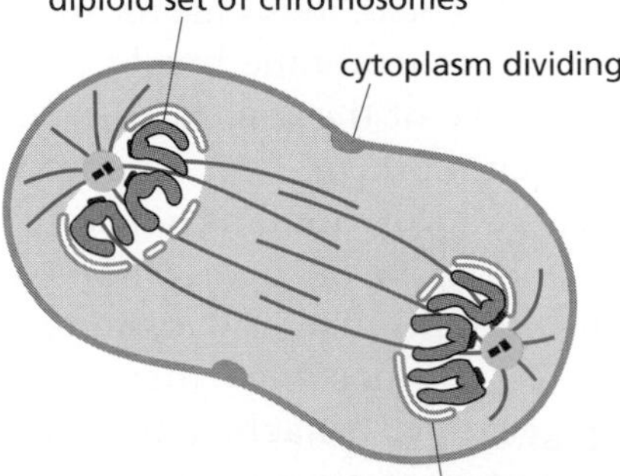

TELOPHASE

A complete diploid set of chromosomes arrives at each end of the cell, a nuclear membrane reassembles around each set, forming two nuclei, and the cell begins to separate in half.

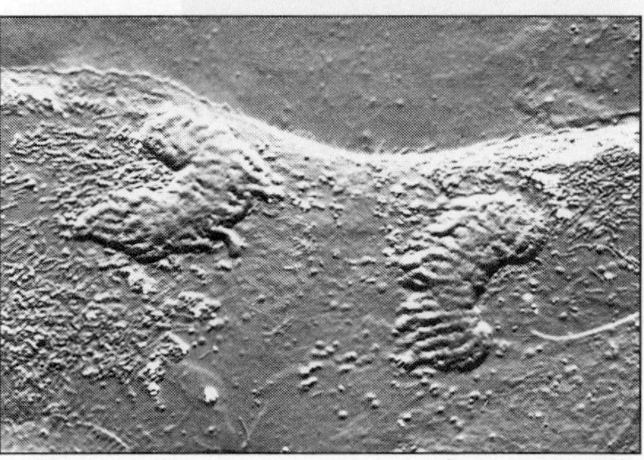

FIGURE 2.7

Mitosis. The cell divides in such a way that each of the two offspring cells contains the same (diploid) number of paired chromosomes as the parent cell did.

interphase, the cell enlarges in size, increasing its cytoplasm. It also makes a copy of each of its chromosomes, so that the cell has twice its diploid number of chromosomes, or four chromosomes of each type. However, none of the chromosomes are visible in the nucleus during interphase. With the start of **prophase** the chromosomes condense and become structures visible under a microscope. Each chromosome can now be seen to be attached to its new duplicate, which was made during interphase. By the end of prophase, the membrane that surrounds the nucleus breaks down. During **metaphase**, each chromosome, attached to its new duplicate, lines up along the center of the cell. The other chromosome of each type is also attached to its newly synthesized duplicate, and also goes to the midline of the cell. The two homologous sets of attached duplicates are not in contact with each other, but are randomly spread along the midline. In **anaphase**, each chromosome separates from its attached duplicate, and the chromosome and its duplicate begin to move to opposite ends of the cell. The other chromosome of each type does the same, separating from its attached duplicate, and this homologous chromosome also moves to the opposite end of the cell from its duplicate. In **telophase** the chromosomes complete their move to the two ends of the elongated cell. Because each chromosome and its newly made duplicate have moved to opposite ends of the cell, each end of the cell now has the diploid number of chromosomes composed of two chromosomes of each homologous pair. At the end of telophase, a nuclear membrane reappears around each diploid set of chromosomes, forming two nuclei, one at each end of the cell, and completing mitosis. After mitosis is finished, the cytoplasm divides between the two nuclei, finalizing the division of one diploid cell into two diploid cells.

Meiosis and sexual life cycles

Recall that pea plants reproduce sexually. In all species that reproduce sexually, two haploid gametes join to form a diploid cell called a **zygote** that can become a new individual organism. The gametes are produced in specialized cells in a process called meiosis.

Meiosis. In meiosis, as in mitosis, the chromosomes are first duplicated to twice the diploid number and the cell divides, but then in meiosis, a second division follows the first. The end result of meiosis is four cells, each with a haploid number of chromosomes.

As in mitosis, each chromosome is duplicated, the chromosomes condense and become visible, then the nuclear membrane breaks down and the chromosomes line up along the midline of the cell. At this step, however, there is a major difference from the mitotic process. In mitosis, the homologous pairs of chromosomes come to the midline separately. As shown in Figure 2.8, in the first division of meiosis, the two homologous chromosomes, each with its attached duplicate, come together as quadruplicates. The homologous pairs of chromosomes then go to opposite ends of the cell, each remaining with its attached duplicate. Nuclear membranes form and the cell separates into two offspring cells, completing the first meiotic division. Each cell has a diploid number of chromosomes, the same number as there are at the completion of mitosis, but they have been divided up differently. At this stage of meiosis, each new cell contains one chromosome of each homologous pair, and each chromosome remains attached to its new duplicate, synthesized at the beginning of meiosis.

The first meiotic division is followed by a second division in each of the two new cells. The nuclear membranes break down once again and each chromosome now separates from its attached duplicate. Nuclear membranes form and the cells divide. The final result is four cells. Because there was no further replication of chromosomes between the first and second meiotic divisions, each of the four new cells, or gametes, ends up with the haploid number of chromosomes.

During the first meiotic division, when the homologous pairs of chromosomes are together as quadruplicates, **crossing over** can occur. One chromosome from each attached pair can exchange part of itself with the corresponding part on the homologous attached pair, as indicated by the exchange of red and blue segments in Figure 2.8. This phenomenon allowed the discovery of gene linkage, which is discussed below.

Sexual life cycles. After the haploid gametes are produced, they can be brought together by sexual reproduction, forming a diploid zygote. Figure 2.9 shows a sexual life cycle. After the gametes fuse to form a zygote, the zygote of a multicellular organism undergoes repeated mitosis to become an adult. Each somatic cell contains the full diploid set of chromosomes and thus all somatic cells are genetically identical. As the organism develops, different somatic

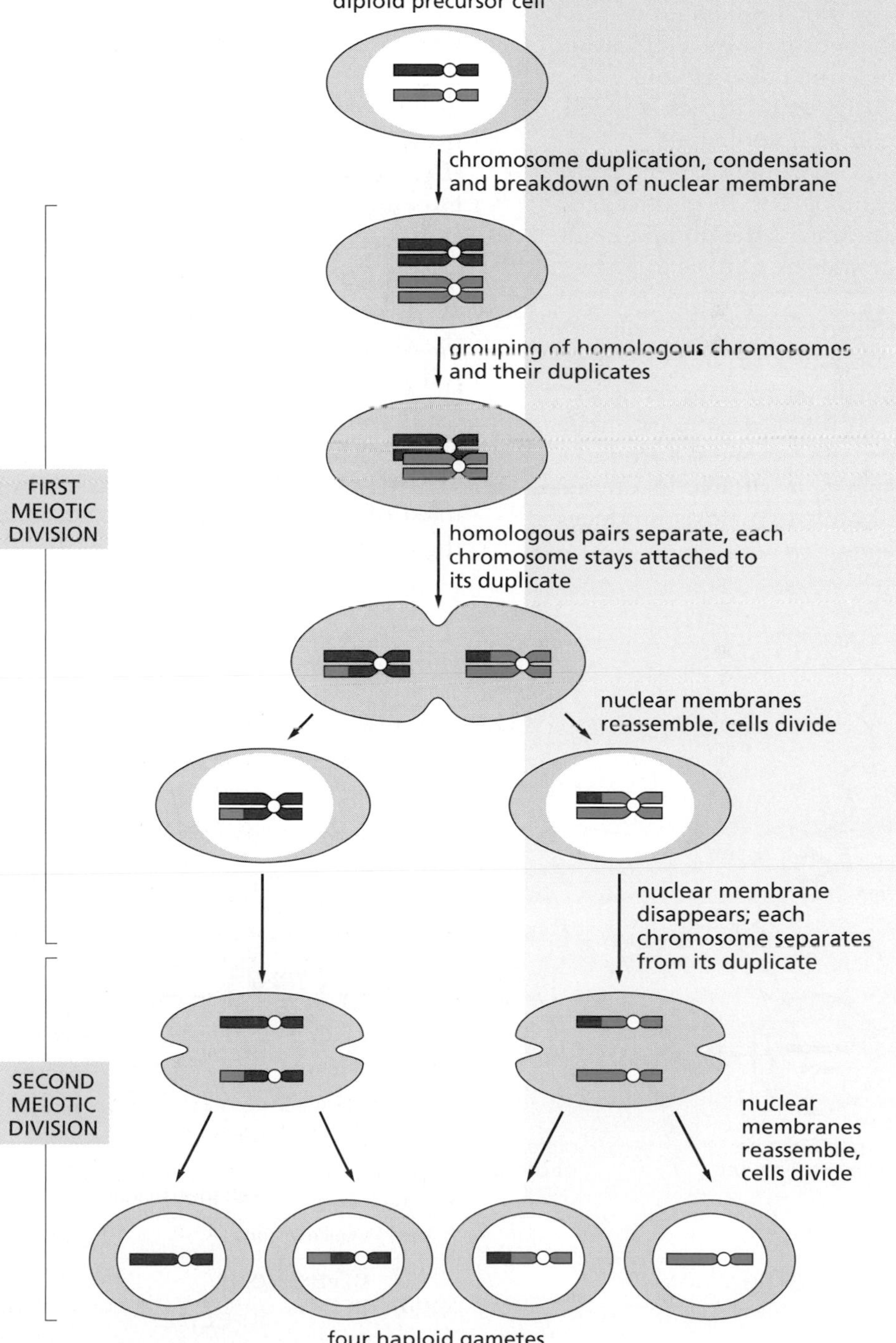
diploid precursor cell
chromosome duplication, condensation and breakdown of nuclear membrane
grouping of homologous chromosomes and their duplicates
FIRST MEIOTIC DIVISION
homologous pairs separate, each chromosome stays attached to its duplicate
nuclear membranes reassemble, cells divide
nuclear membrane disappears; each chromosome separates from its duplicate
SECOND MEIOTIC DIVISION
nuclear membranes reassemble, cells divide
four haploid gametes

FIGURE 2.8

Meiosis. The cell divides twice to produce four haploid gametes. For simplicity, only one pair of homologous chromosomes is shown.

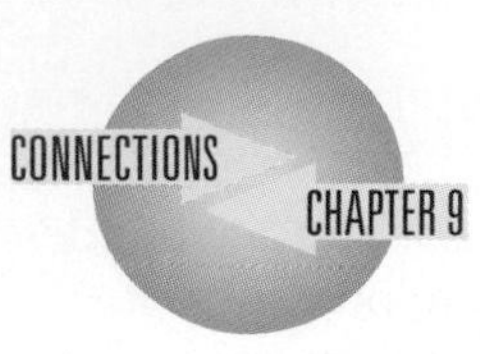

cells specialize for different functions (see Chapter 9, p. 369), although they all retain the full diploid set of chromosomes. Some of these cells specialize to undergo meiosis, producing new haploid gametes, and completing the sexual life cycle. Many, but certainly not all, kinds of organisms have a sexual life cycle, alternating between haploid gametes and diploid somatic cells.

Gene linkage

Sutton predicted that independent assortment would not apply to all pairs of genes and that some genes would be found to be linked. Other investigators quickly confirmed his prediction. A British geneticist, William Bateson, described linked genes in a cross of varieties of garden peas. Other investigators soon discovered similar examples in fruit flies (*Drosophila*), corn (*Zea mays*), and other species, showing that Mendel's laws and Sutton's theory were not unique to peas or to plants. Figure 2.10 shows a cross revealing that two genes are linked in corn. Plants of genotype *CCSS* (colored, full seeds) crossed with *ccss* plants (colorless, shrunken seeds) produced all seeds of the colored, full phenotype among the F_1. In this experiment, the F_1 plants (*CcSs* heterozygotes) were not self-fertilized as in Mendel's experiments such as in Figures 2.3 and 2.4. Rather, the F_1 heterozygotes were fertilized by plants of the doubly recessive *ccss* genotype. This type of cross, in which an F_1 is crossed with one of the parental types, is called a backcross. In the backcross offspring, most of the seeds that were colored were also full, and most of the seeds that were colorless were also the shrunken phenotype. The genes for

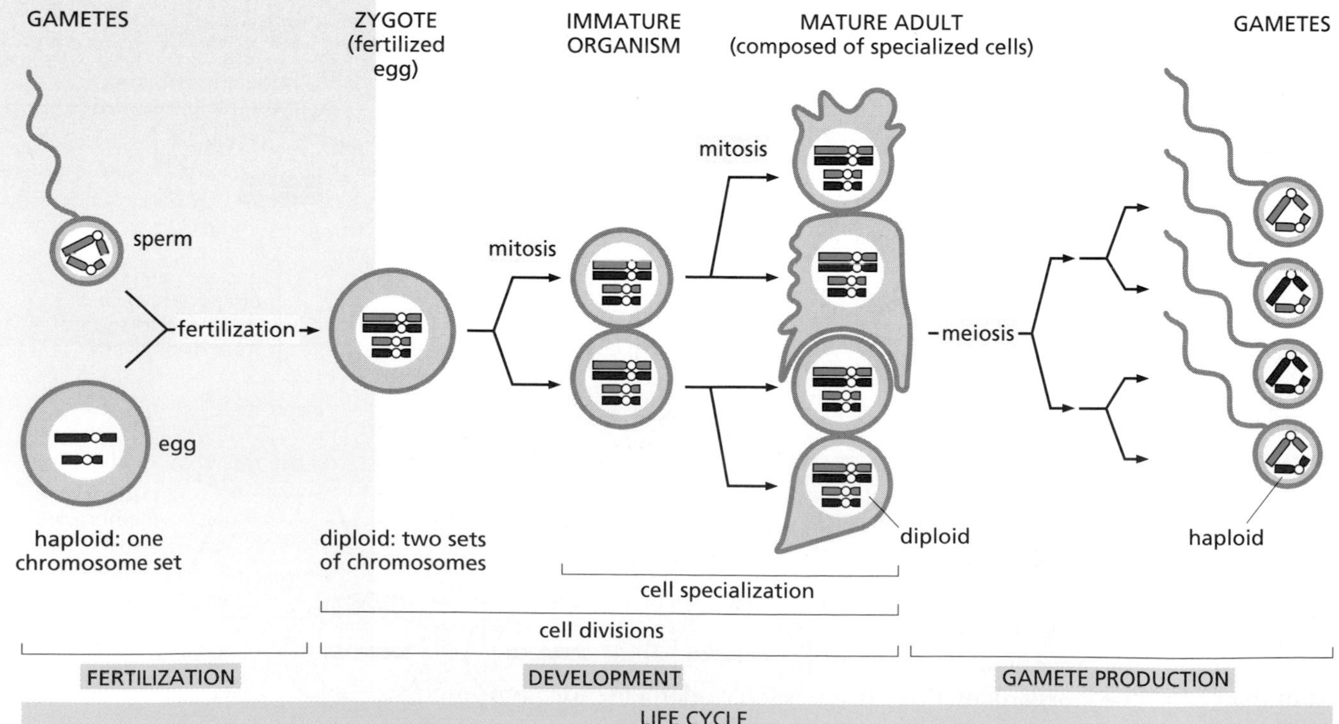

FIGURE 2.9 Sexual life cycles. In sexual reproduction, haploid gametes join by fertilization to form a new diploid individual with one of each pair of homologous chromosomes coming from each parent. In multicelled organisms the diploid zygote divides by mitosis to form the adult organism. Each of the somatic (body) cells contains a set of chromosomes the same as that in the zygote. In a male organism, meiosis produces sperm, as shown. In a female organism, meiosis produces eggs.

these two traits were said to be linked. That they were linked, rather than independently assorting, could be explained by assuming that the genes for the two traits were on the same chromosome.

Most of the progeny in Figure 2.10 show the parental linked phenotypes. However, notice also in Figure 2.10 that small numbers of the backcross progeny were not of the colored and full or colorless and shrunken seed phenotypes but instead had colored and shrunken or colorless and full seeds. These atypical plants had new (nonparental) combinations of the phenotypes for the traits; the underlying recombinant genotypes were inferred to have arisen from the process of crossing over in which chromosomes break and recombine by exchanging pieces. Some microscopists thought they had observed X-shaped arrangements of the chromosomes during meiosis (see Figure 2.8) that looked like crossing over, but many scientists were unsure.

FIGURE 2.10

A cross between pure-line corn plants having different alleles for two linked genes.

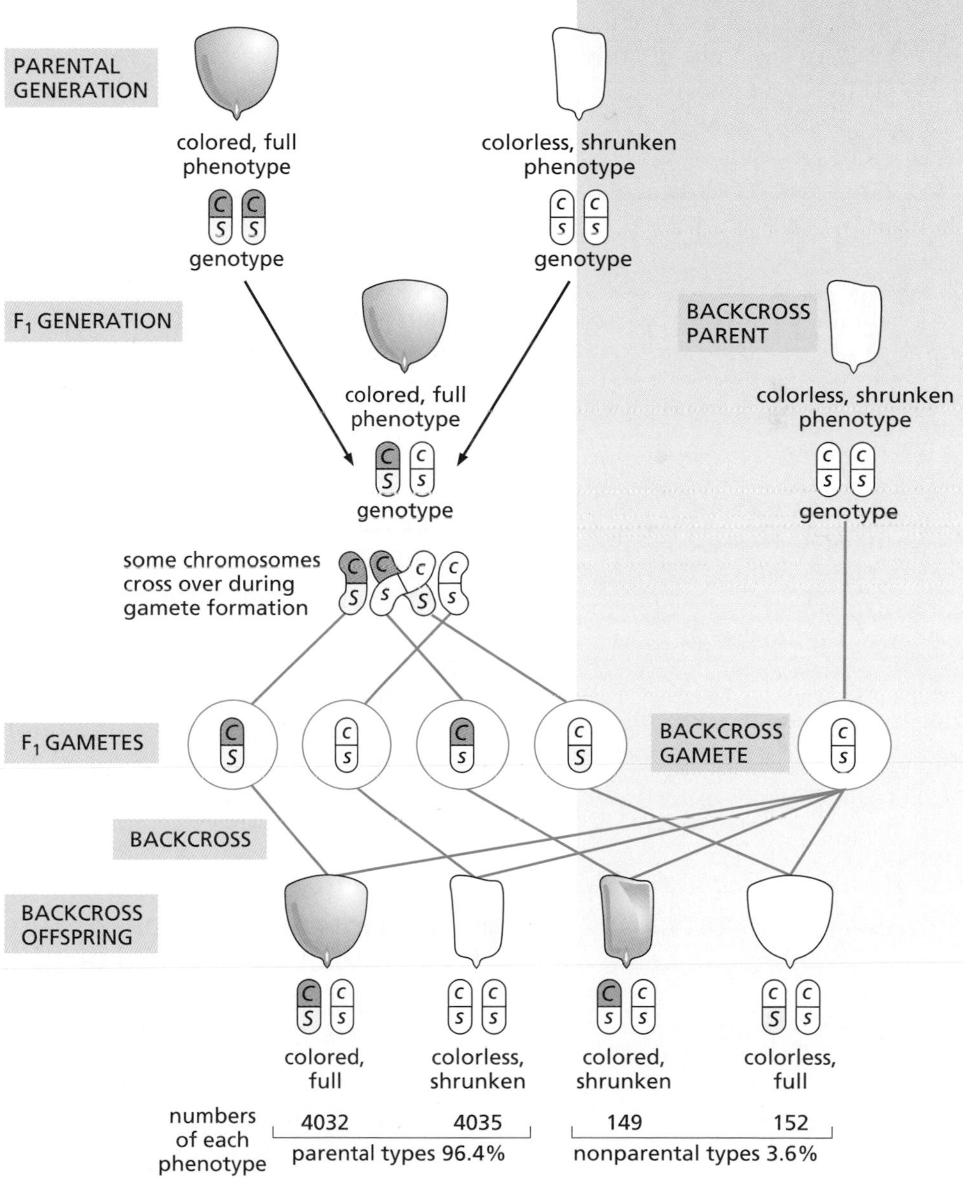

Confirmation of the chromosomal theory

Sutton's theory had to wait three decades for confirmation by other researchers in genetics. In 1931, Harriet Creighton and Barbara McClintock confirmed the chromosomal theory of inheritance in corn; later that year Curt Stern observed the same thing in fruit flies. Creighton and McClintock used plants whose chromosomes had structural abnormalities on either end, enabling them to recognize the chromosomes under the microscope. What they were able to demonstrate was that *genetic recombination* (the rearranging of genes) *was always accompanied by crossing-over* (the rearranging of chromosomes). Barbara McClintock went on to discover genes that move from place to place, the so-called transposable elements or jumping genes, a discovery for which she later received the Nobel Prize.

THOUGHT QUESTIONS

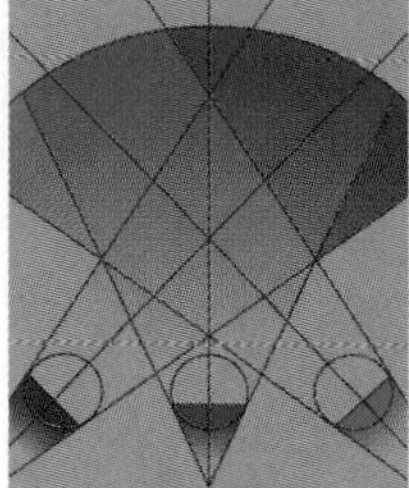

1. In a cross between pea plants of genotype *YYRR* (yellow, round seeds) and *yyrr* (green, wrinkled seeds), the F_1 plants are all *YyRr*. Make a series of large drawings showing the movements during mitosis of the chromosomes carrying these genes in cells of an F_1 heterozygous plant.

2. For the same F_1 heterozygous plants (*YyRr*) described in question 1, make a series of drawings showing the way in which the chromosomes separate in the first division of meiosis and in the second division of meiosis. Label each diagram with the symbols *Y*, *y*, *R*, and *r* to show how all four types of gametes originate.

The frequency of recombination between linked genes is very roughly a measure of the distance between them along the chromosome. Recombination between closely linked genes is a rare event, while recombination between genes farther apart is more frequent. By making crosses between individuals having different alleles for pairs of linked genes, geneticists were able to determine the linear arrangement of many genes and the approximate distances between genes on the chromosomes of many species.

An interesting footnote to Mendel's work was provided in 1936 by the British geneticist R.A. Fisher, who noticed that garden peas have seven pairs of chromosomes. Mendel had picked seven traits that assorted independently in all possible pairwise combinations! Because the probability of this outcome's occurring by chance is extremely remote, Fisher concluded that Mendel may have studied many more traits and only reported the results for the seven independently assorting traits whose inheritance he could understand (see Figure 2.1).

THE MOLECULAR BASIS OF INHERITANCE FURTHER EXPLAINS MENDEL'S HYPOTHESES

We now address the last of the four questions posed earlier. What are genes made of? In other words, what chemical substance transfers genotypes from parents to offspring?

The story begins with a curious experiment carried out in 1928 by Frederick Griffith, a U.S. Army medical officer attempting to develop a vaccine against pneumonia. Griffith worked with two strains of bacteria that differed in their outer coats. Strain S had an outer coat that gave a smooth appearance when masses of bacteria (called colonies) were grown on agar in dishes. The smooth-colony bacteria were **virulent** (see Chapter 13, p. 549), which means that the bacteria cause a disease (pneumonia in this case). Strain R of the same bacterial species lacked the outer coat, which gave the colonies of this strain a rough appearance, and they were nonvirulent. (S stands for 'smooth,' R for 'rough.') When strain S was injected into mice, all the mice died of pneumonia. Strain R, when injected, did not kill mice, nor did bacteria of strain S that had been killed by heat. The surprising result, shown in Figure 2.11, was that a mixture of live bacteria of strain R and heat-killed bacteria of strain S did kill mice. Furthermore, living S bacteria were isolated from all mice that died this way. Griffith interpreted this experiment as showing that something in the dead S bacteria had somehow transformed the living R bacteria into virulent S bacteria. The bacteria resulting from this **transformation** had been altered genetically, not just phenotypically. A change that was only phenotypic would not be passed on to future generations, but Griffith demonstrated that descendents of the transformed bacteria were also of strain S and continued to kill mice.

What Griffith had done in his experiment was transfer a genetic trait from one bacterial strain to another. What was the chemical substance that had been transferred? Griffith's use of bacteria as an experimental species and the unequivocal evidence that genetic material had been transferred in his experiment is significant because it was the background for work two decades later that demonstrated what the chemical substance is.

This section describes the research that revealed the chemical substance of the gene, the composition of this substance, and how it replicates. We also learn how the chemical substance directs the synthesis of proteins and how changes in the substance can produce changes in proteins.

DNA and genetic transformation

From the beginning of the twentieth century into the 1940s, most researchers thought that proteins were the most likely candidates to be the chemical substance of genes. They thought this because proteins were known to be complex and varied, while most other molecules were thought not to be. In an attempt to discover whether protein was indeed the chemical substance, three bacteriologists, Oswald Avery, Colin MacLeod, and Maclyn McCarty, in 1944 conducted a chemical study of the bacteria Griffith had injected into his experimental mice. First, they were able to show that a chemical extract of heat-killed S bacteria transformed strain R into strain S. They then separated the strain-S extract into different fractions, each containing different types of chemical molecules. They found that the fraction containing the **nucleic acids** transformed strain R into strain S, but that the protein fraction did not.

Finally, they distinguished between the two major types of nucleic acids (DNA and RNA) by using enzymes. **Enzymes** are biological molecules (nearly always proteins) capable of speeding up chemical reactions without themselves getting used up in those reactions. Enzymes control many biological processes, and the action of most enzymes is very restricted in that specific enzymes act on specific types of molecules. The enzyme deoxyribonuclease (DNase) specifically breaks down DNA. DNase treatment destroyed the ability of the strain-S extract to transform the strain-R bacteria, demonstrating that the chemical that

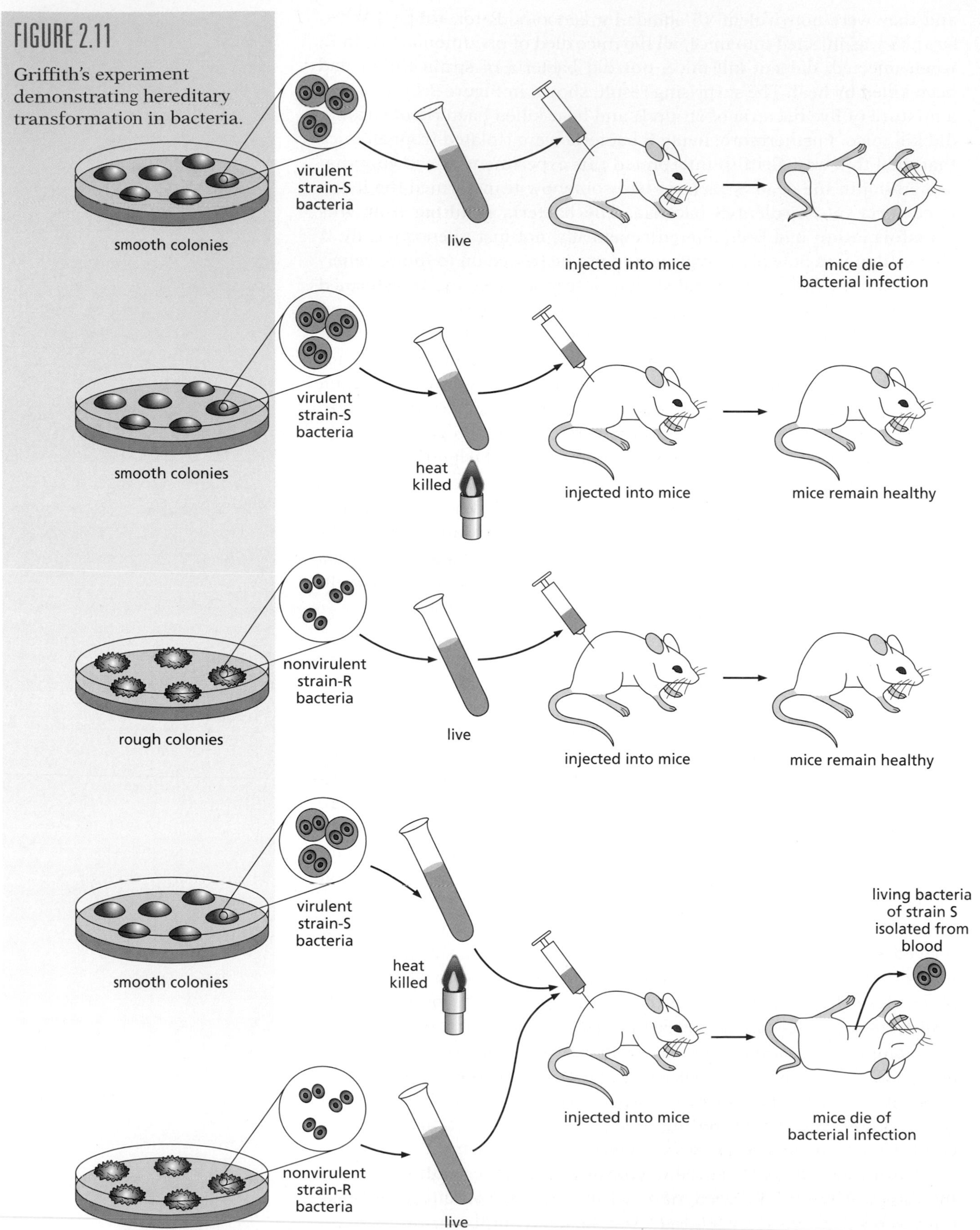

FIGURE 2.11

Griffith's experiment demonstrating hereditary transformation in bacteria.

carries the genetic material is, in fact, **deoxyribonucleic acid** (**DNA**). On the other hand, the enzyme ribonuclease, which breaks down RNA, had no effect on the ability to transform; **ribonucleic acid** (**RNA**) is therefore not the genetic material.

The discovery of Avery and his co-workers did not get the attention it deserved. Doubters still remained. It took about a decade for many scientists to accept that DNA is the genetic material. Experimental work on the problem had meanwhile shifted from using bacteria to using viruses.

In 1952, two American virologists, Alfred Hershey and Martha Chase published the results of a landmark experiment that confirmed the finding that DNA, not protein, is the genetic material. For their experiment, Hershey and Chase used a virus that infects bacteria and reproduces within them. They infected bacteria with the viruses and studied the viral offspring. It was known that this virus consisted only of protein and DNA. Was it the protein or the DNA that carries the genetic material?

In preparation for their experiment, Hershey and Chase grew some viruses in a medium containing radioactive phosphorus (^{32}P) and others in a medium containing radioactive sulfur (^{35}S), atoms that the virus needs to make new viruses. Because DNA contains phosphorus but protein does not, the new viruses grown with ^{32}P had radioactive phosphorus in their DNA but no radioactivity in their protein. In contrast, the proteins, but not the DNA, of viruses grown with ^{35}S were radioactive because proteins contain sulfur but DNA does not. Because radioactivity is easily detected, material prepared in this way is said to be radioactively labeled. Hershey and Chase applied the radioactive labels so they would be able to 'see' what happened to the viral DNA and protein when the viruses infected bacteria.

Hershey and Chase exposed *Escherichia coli* bacteria to the radioactively labeled viruses for long enough to permit the viruses to infect the bacteria. Part of the virus enters the cells that they infect and directs the replication of more viruses, producing thousands of new virus particles and eventually killing the bacteria and breaking them open to release the viruses (Figure 2.12A). Was the injected material that was carrying the viral genotype DNA or protein? Hershey and Chase devised a way to interrupt the viral cycle after the infection period by using a kitchen blender to knock the attached virus capsules off the bacterial surfaces. These detached capsules could easily be separated from the bacteria by spinning the mixture in a centrifuge. Viral material that had been injected into the bacteria continued the process of viral reproduction, eventually lysing the bacteria. When ^{35}S-labeled viruses were used, the radioactive proteins remained outside the bacteria (Figure 2.12B); the viruses eventually released were not radioactive so they had not used radioactive protein to replicate themselves. However, when the ^{32}P-labeled viruses were used, the radioactive DNA entered the bacteria, making the bacteria radioactive. The viral offspring released when these bacteria broke open were also radioactive (Figure 2.12C), showing that they had used some of the radioactive DNA in their reproduction.

This experiment upheld the hypothesis that the genetic material of the virus was DNA. It falsified the hypothesis that the viral genetic material was made of protein. DNA had been identified as the genetic material, but its composition and structure were still a mystery.

The chemical composition of DNA

Chemical breakdown of DNA into its parts showed that it was made of phosphate groups, deoxyribose (a sugar), and four nitrogen-containing bases called adenine (A), guanine (G), cytosine (C), and thymine (T) (Figure 2.13A). Biochemists soon realized that the deoxyribose could form a middle link between the phosphate groups and the nitrogenous bases, creating units called nucleotides (Figure 2.13B). Beyond this, the structure of DNA was unclear, and it was not at all obvious how the structure could give DNA the ability to carry genetic information.

Some researchers suggested that the order of bases in DNA repeated regularly: AGTCAGTCAGTC.... If this were true, then the amounts of the

FIGURE 2.12

The Hershey–Chase experiment. This experiment confirmed DNA as the genetic material.

(A) Pattern of viral infection of *E. coli* bacteria.

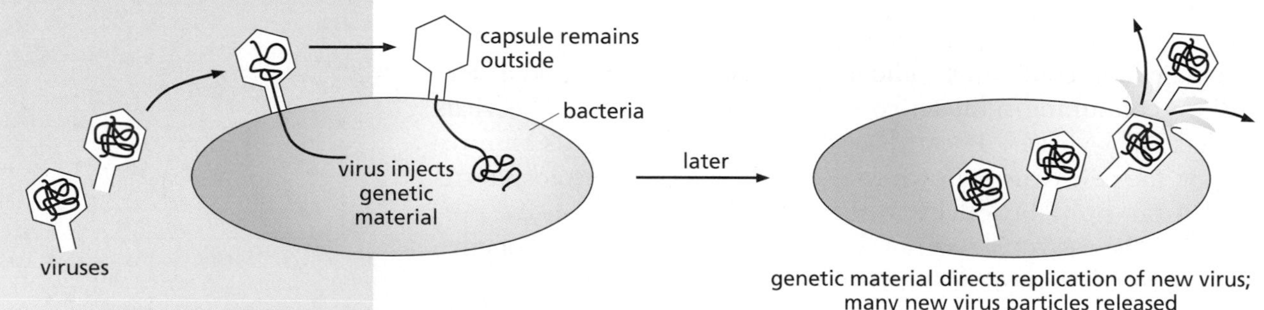

(B) Viruses grown with radioactive sulfur.

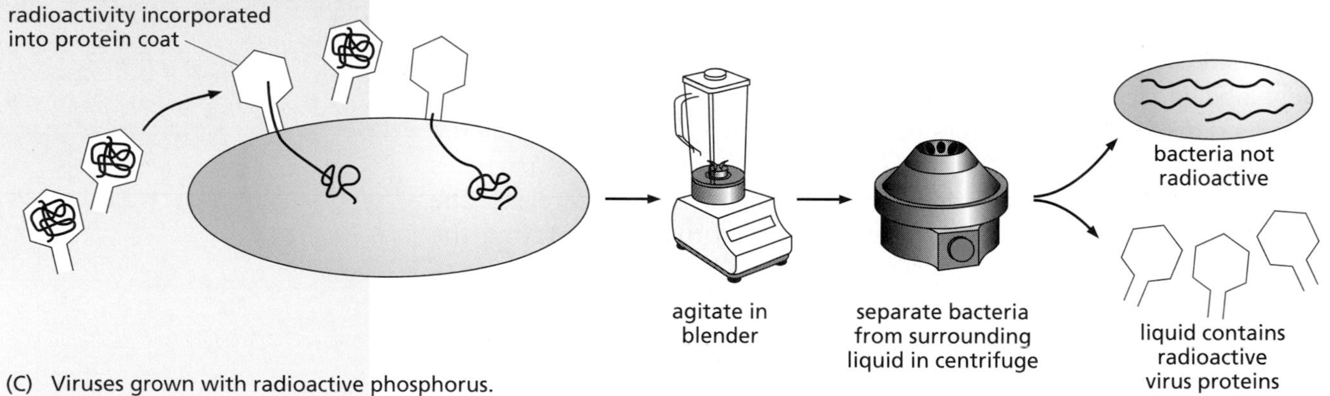

(C) Viruses grown with radioactive phosphorus.

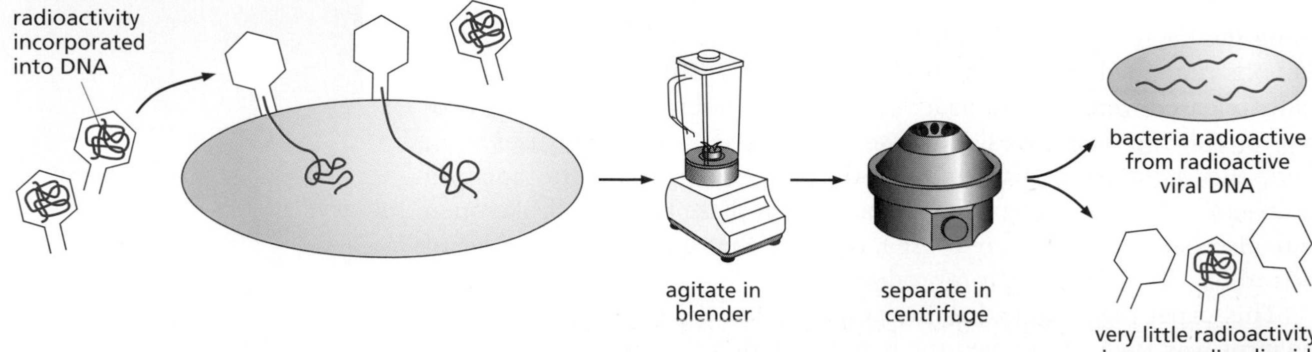

four nitrogenous bases should be equal: there should be 25% of each of the bases in a DNA molecule. To test this hypothesis, biochemist Erwin Chargaff of Columbia University took DNA from various sources, broke it down using enzymes, and measured the relative amounts of the four nitrogenous bases. His findings were as follows.

1. The proportions of the four nitrogenous bases are constant for all cell types within a species. For example, all human cells contain about 31% adenine (A), 19% guanine (G), 19% cytosine (C), and 31% thymine (T), regardless of whether the DNA is from brain cells, liver cells, kidney cells, or skin cells.
2. Although the proportions are constant within a species, they differ from one species to another. All humans, for example, have the same proportions of the four bases. Proportions are different in rats and in bread molds, but all rats have the same proportions as one another and so do all bread molds.
3. The most unexpected finding, and the hardest to explain, was that the proportion of adenine was always the same as the proportion of thymine (within the limits of experimental error), and the levels of cytosine and guanine were also equal. These findings (symbolized as A=T and G=C) became known as Chargaff's rules.

FIGURE 2.13

The nucleotides of DNA.

(A) Two of DNA's nitrogenous bases, adenine (A) and guanine (G), are larger than the other two, thymine (T) and cytosine (C).

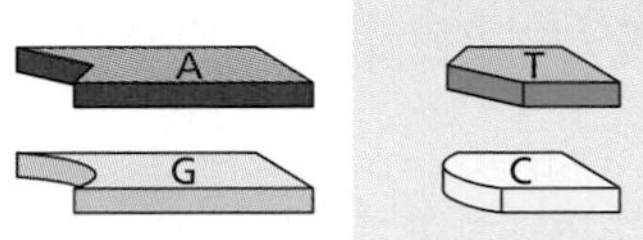

(B) With the addition of deoxyribose (a five-sided sugar molecule) and a phosphate group, each of these nitrogenous bases forms a nucleotide.

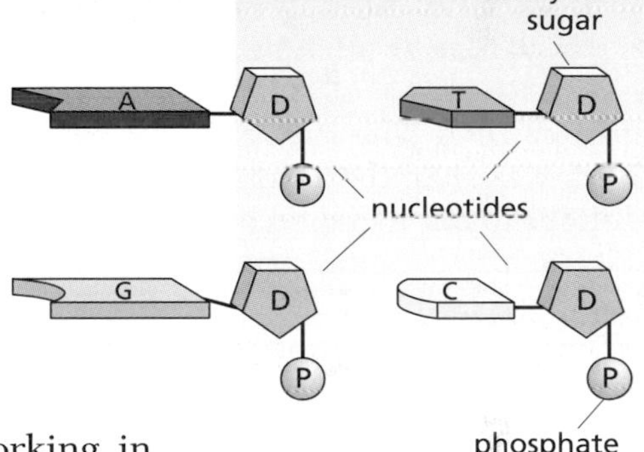

The three-dimensional structure of DNA

In 1953 James Watson and Francis Crick, two geneticists working in Cambridge, England, proposed a structure for DNA that explains Chargaff's rules and also explains how DNA carries genetic information. They did this with the help of some X-ray diffraction information obtained from the Cambridge laboratory of biochemist Maurice H.F. Wilkins, with whom they later shared a Nobel Prize. The data from Wilkins' laboratory were gathered and interpreted by Rosalind Franklin, a biochemist, whose contribution never received the recognition it deserved (Box 2.1).

The X-ray diffraction information suggested certain dimensions and distances for the repetition of structures within the DNA molecule. From this information, Watson and Crick constructed the model of DNA structure summarized in the following list:

1. Each phosphate in a DNA molecule is attached a deoxyribose sugar, which in turn is attached to a nitrogenous base. The three parts together constitute a nucleotide (see Figure 2.13B).
2. The phosphate group of one nucleotide is also connected to the deoxyribose sugar of the next nucleotide. The alternation of phosphates and sugar units thus forms a backbone that holds the entire strand together, while the nitrogenous bases point inward (Figure 2.14A).
3. Each strand is a linear sequence of bases (it does not branch) that,

BOX 2.1 Science and Fair Play: The Case of Rosalind Franklin

A large part of the effort to reveal the molecular structure of DNA centered on a technique known as X-ray diffraction. In this technique, the molecules being examined must first be crystallized; for studies of DNA, the sodium salt of deoxyribonucleic acid was used instead of the acid itself because it was more easily crystallized. The crystals are then exposed to X-rays and the reflections produced by the atoms in the crystal are examined on photographic film. The images produced by X-ray diffraction are very difficult to interpret and require a sharp mind and a strong mathematical background. Rosalind Franklin (1920–1958), who worked in the laboratory of Maurice H.F. Wilkins, was an expert in reading and interpreting such X-ray diffraction photographs, a process that required many hours of careful measurement and calculation. (Today, there are powerful computers that process X-ray diffraction patterns, but in the 1950s this work was all done by hand.) Watson and Crick used Rosalind Franklin's data to figure out the structure of DNA. The 1962 Nobel Prize for this discovery was shared by Watson, Crick, and Wilkins. Rosalind Franklin had died by then, so she did not share in the prize (Nobel prizes are not given posthumously). Nonetheless, she did not receive the credit that her contribution to Watson and Crick's discovery deserved.

In his 1968 book, Watson admits that he got hold of Rosalind Franklin's results (including a critical X-ray photo) without her knowledge or permission. He writes:

"[Maurice Wilkins] revealed that with the help of his assistant Wilson he had quietly been duplicating some of Rosy's and Gosling's X-ray work.... When I asked what the pattern was like, Maurice went into the adjacent room to pick up a print of the new form...." (*The Double Helix*, 1968, p. 98).

"By then it had been checked out with Rosy's precise measurements. Rosy, of course, did not directly give us her data. For that matter, no one at King's realized they were in our hands." (*The Double Helix*, 1968, p. 104–105).

Watson claims that his action was justified because the race to find the structure of DNA was highly competitive and because Franklin was, in his estimation, proceeding too slowly and in the wrong direction. Undoubtedly, there were other issues involved. Watson never liked Rosalind Franklin, and he makes this abundantly clear in his book. Franklin's biographer, Ann Sayre, talks of Watson's "rationalization which implies that Rosalind, as an impediment standing squarely in the path of scientific progress, deserved to be pushed aside." (*Rosalind Franklin and DNA*, 1975, p. 143)

In a review of Watson's book, Andre Lwoff wrote: "His portrait of Rosalind Franklin is cruel. His remarks concerning the way she dresses and her lack of charm are quite unacceptable. At the very least the fact that all the work of Watson and Crick starts with Rosalind Franklin's X-ray pictures and that Jim has exploited Rosalind's results should have inclined him to indulgence." (quoted in Sayre, *Rosalind Franklin and DNA*, p. 194–195).

Ann Sayre has said, "Rosalind has been robbed, little by little; it is a robbery against which I protest" (p. 190). "Her work was appropriated and used without proper credit" (p. 194).

Among the questions that you might want to ponder are the following.

1. How would you describe Watson's use of Franklin's data? Was it 'robbery', or just 'looking' at her data? Is there a difference? In either case, do you think that competition in the race to discover the structure of DNA justified his actions?

2. Does high-pressure competition do more good for science, or more harm? Try to list both good and bad consequences before you decide.

3. Although everyone agreed that Franklin was a brilliant researcher, she was often criticized for not wearing makeup and for her lack of interest in her clothing. What kind of treatment would a scientist like Rosalind Franklin be likely to receive today? Franklin differed in social background, politics, and religion from her male colleagues. Do a scientist's looks, religion, politics, or personality affect how her or his data are regarded?

4. Can the science that scientists produce be separated from the other aspects of their lives? Can we monitor discriminatory behavior without restricting academic freedom?

because of the angles of the chemical bonds, is twisted in the shape of a corkscrew (a helix).

4. DNA has two strands wound around each other, forming a **double helix**, with the bases arranged in the interior, like steps in a spiral staircase (Figure 2.14B).
5. The strands run in opposite directions and are so arranged that an adenine on one strand is always paired with a thymine on the other strand, and vice versa. Also, cytosine on one strand is always paired with guanine on the other strand, and vice versa (Figure 2.14C). These pairings of **complementary** (matching) bases explain Chargaff's rules.
6. Because the base pairings are mandated by the sizes and shapes of the four nucleotides (see Figure 2.13), each strand contains all the information necessary to determine the structure of the complementary DNA strand.

DNA replication

Watson and Crick's model for the structure of DNA led quickly to an understanding of the mechanism of **replication**, the process by which DNA molecules make copies of themselves. Before DNA replicates, the two strands of the double helix unwind and separate from each other, as

FIGURE 2.14

The three dimensional structure of DNA.

(A) Thousands of nucleotides are strung together by a phosphate–sugar backbone.

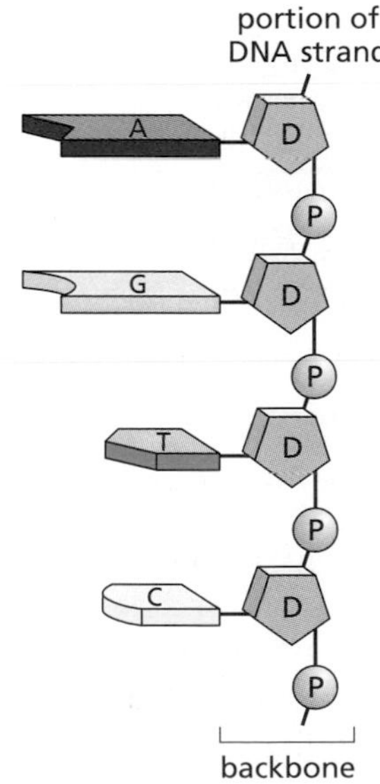

(B) Two strands of DNA twist around one another to form a double helix. A straightened portion of this double helix resembles a ladder with the paired (complementary) bases forming the rungs.

(C) Two nucleotide sequences running in opposite directions pair with one another, with each adenine (A) pairing with a thymine (T), and each guanine (G) pairing with a cytosine (C).

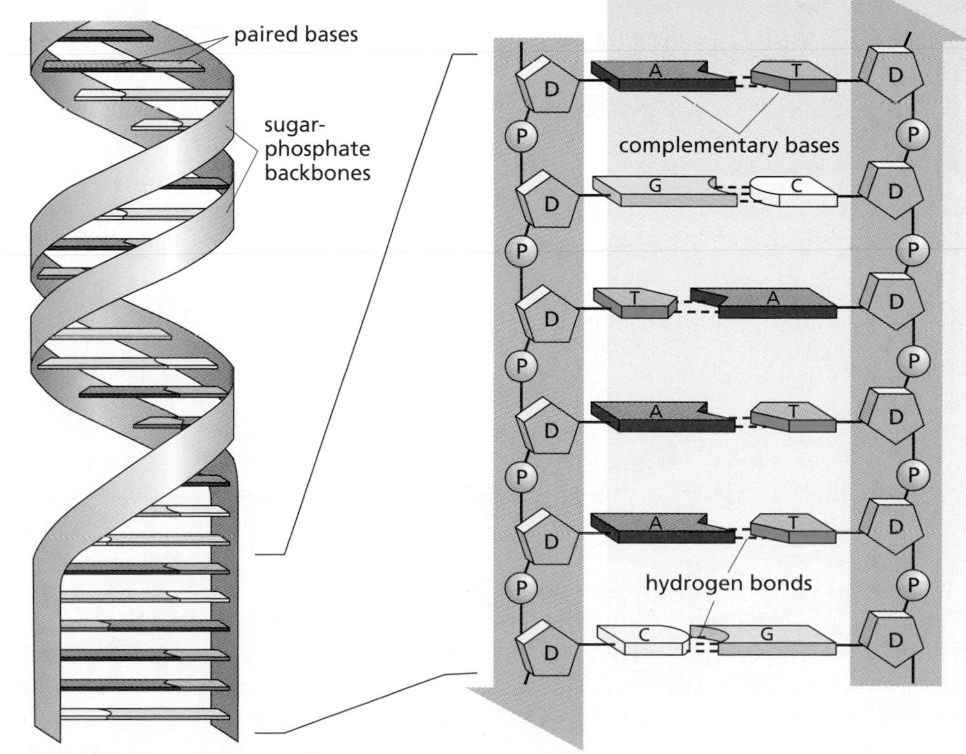

shown in Figure 2.15A. Notice the orientation of the 5-sided deoxyribose molecules, which show that the two strands run in opposite directions. After separating, the strands are bound by an enzyme called **DNA polymerase**, which actually begins the replication process. New strands of DNA are synthesized one nucleotide at a time, with one of the existing strands serving as a template (pattern to be copied). If the next unmatched base on the template is adenine (A), then a thymine (T) nucleotide (adenine's complementary base) is added to the growing new strand opposite the adenine. In like manner, G is added to match C, C to match G, and A to match T (Figure 2.15B). The other parental (preexisting) strand is simultaneously acting as a second template and other complementary bases pair with it.

Each nucleotide is actually added in the form of a triphosphate such as ATP (adenosine triphosphate) or GTP (guanosine triphosphate). Two of the three phosphate groups are split off from each triphosphate to provide energy for the replication process, while the remaining phosphate becomes part of the growing DNA molecule by joining to the deoxyribose sugar of the adjacent nucleotide (Figure 2.15C).

Recall that the two parental DNA strands run in opposite directions. Consequently, synthesis on the two template strands proceeds in opposite directions. Yet, the parental stands are separating at only one location, the

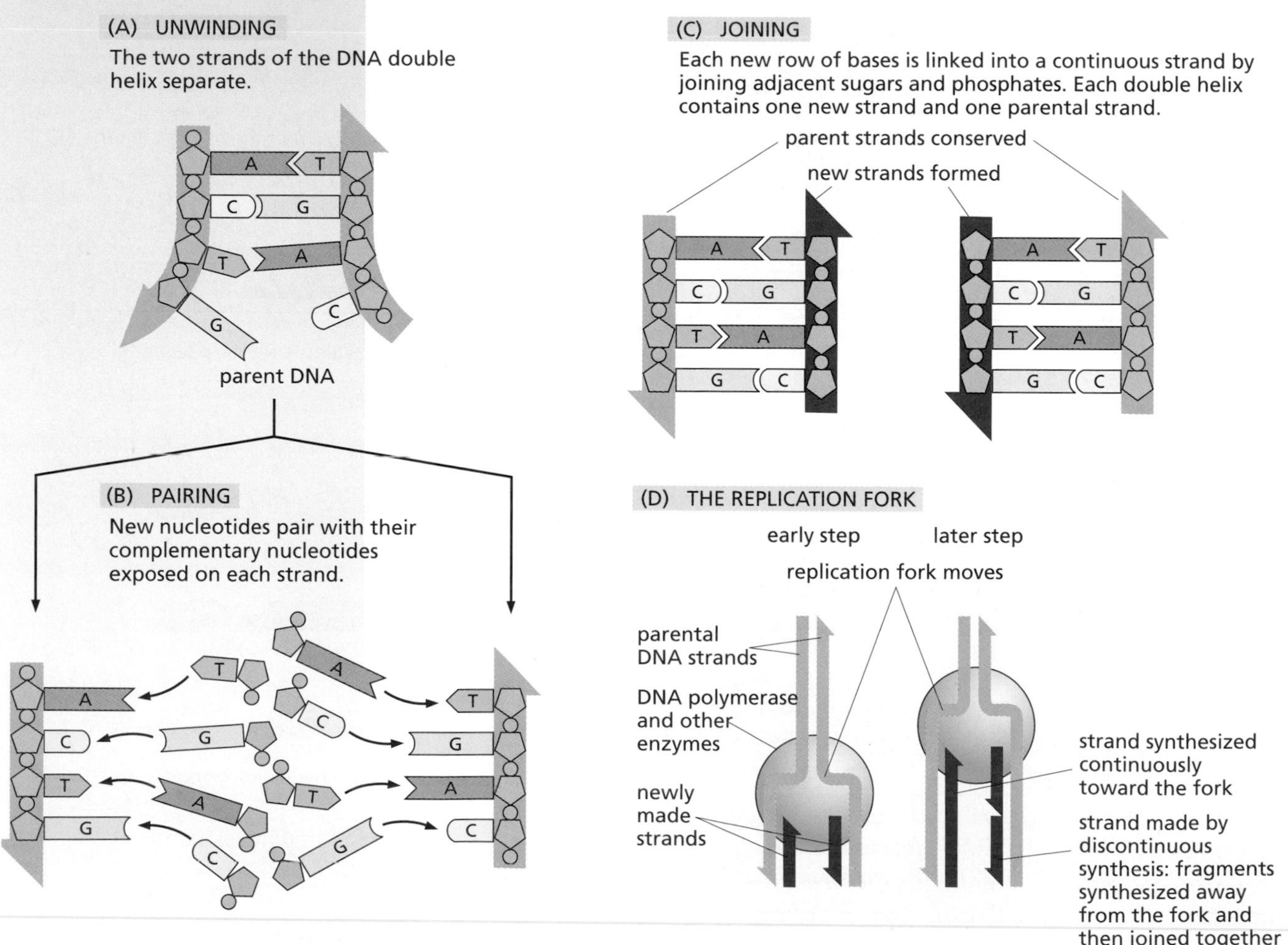
(A) UNWINDING
The two strands of the DNA double helix separate.
A
T
C
G
T
A
G
C
parent DNA
(B) PAIRING
New nucleotides pair with their complementary nucleotides exposed on each strand.
(C) JOINING
Each new row of bases is linked into a continuous strand by joining adjacent sugars and phosphates. Each double helix contains one new strand and one parental strand.
parent strands conserved
new strands formed
(D) THE REPLICATION FORK
early step
later step
replication fork moves
parental DNA strands
DNA polymerase and other enzymes
newly made strands
strand synthesized continuously toward the fork
strand made by discontinuous synthesis: fragments synthesized away from the fork and then joined together

FIGURE 2.15

DNA replication. The two strands run in opposite directions, as denoted by the arrowheads on the backbones. Note that either strand contains all the information needed to synthesize the other (complementary) strand.

so-called replication fork. One new strand of DNA is thus synthesized continuously, with the direction of synthesis running towards the replication fork. The other strand, however, is synthesized in the opposite direction, away from the replication fork. This strand must be synthesized in short fragments that are later joined together (Figure 2.15D).

Transcription and translation of genes

The DNA containing the hereditary information is packaged in chromosomes. Each chromosome contains two very long strands of DNA (in humans the length of DNA in *each cell* is about 1 m or 3 feet). Chromosomal DNA is surrounded (in most organisms) by some protein. A gene is a segment of the DNA strand, that is, a subset of bases within the linear sequence of the whole DNA. Each person or pea plant (or any diploid individual) has two chromosomes of each type, and thus has two genes for each hereditary trait, one on each chromosome of a homologous pair. The location of the gene on the chromosome is called its **locus**. Within a species, there may be any number of possible alleles for the gene at a given locus, but each individual can only have two, one on each chromosome of a homologous pair.

The nucleic acids DNA and RNA can be compared to blueprints that contain instructions for building proteins. Genes (made of DNA) are expressed as phenotypes by providing information for the synthesis of RNA, which in turn provides information for the synthesis of protein. Often several proteins (and therefore several genes) interact to produce a phenotype.

RNA (Figure 2.16) differs from DNA in that (1) the backbone contains the sugar ribose rather than deoxyribose, hence the name ribonucleic acid, (2) it is mostly single-stranded, and (3) the nitrogen-containing bases include uracil (U) instead of thymine (the other three nitrogen bases, adenine, guanine, and cytosine, are the same). In RNA, C pairs with G and A pairs with U. Genes undergo **transcription** into RNA, meaning that information is transferred from DNA to RNA, still within the language of nucleic acids with their linear sequence of bases. *The linear sequence of nucleotides in DNA determines the linear sequence of nucleotides in RNA*. Some gene products stop here, as special types of RNA that have functions described below. The product of transcription for most genes is **messenger RNA (mRNA)**, which then goes through a second information transfer (**translation**), in which the information is changed into another language, the language of amino acids. *The linear sequence of nucleotides in messenger RNA determines the linear sequence of amino acids in the protein*. Transcription and translation together are called **gene expression**.

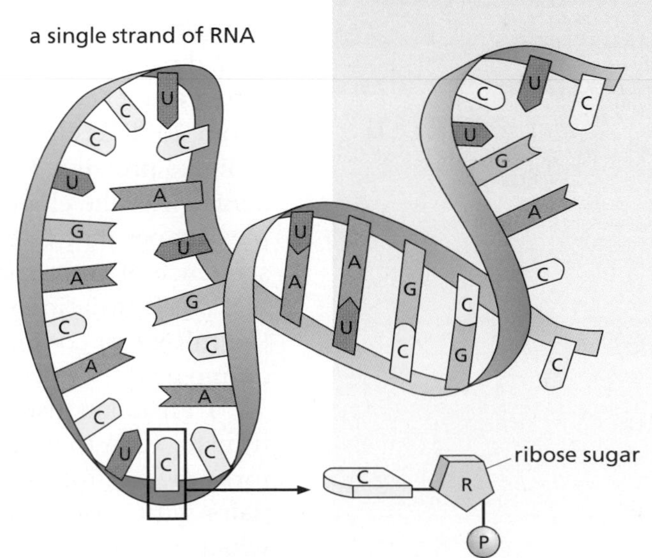

FIGURE 2.16
The structure of RNA. The molecule as a whole is usually single-stranded, but short portions of some RNA molecules can base-pair with other portions of the same molecule.

Information flows from DNA to RNA to protein. This summary statement has been called the central dogma of molecular biology (Figure 2.17A). As we see in Chapters 9 and 13, this central dogma has been considerably modified by the finding that information flow is not only in one direction. Proteins can affect the transcription of DNA and translation of RNA, and in some viruses RNA can be a template for DNA synthesis. Moreover, one protein is not always the product of a single gene. Our current concepts of information flow are more accurately represented as a network (Figure 2.17B).

FIGURE 2.17

Changing concepts of the flow of genetic information.

(A) The central dogma of molecular biology, as it was understood in the 1960s: information flows from DNA to RNA and then to protein.

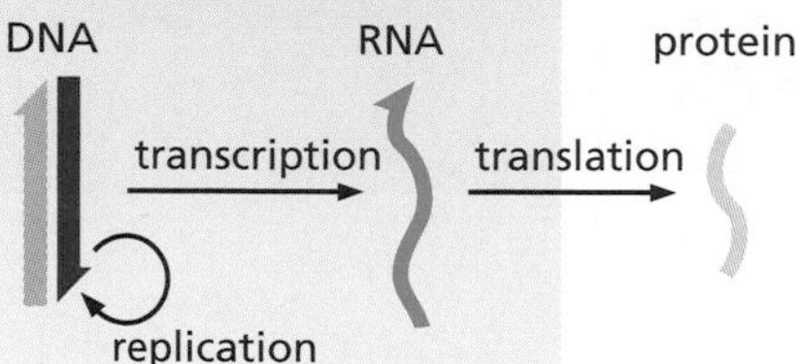

(B) Many exceptions are now known. For example, some viruses, including the one that causes AIDS, have a 'reverse transcription' process in which RNA is used to make DNA. Also, many proteins influence the timing and amount of gene transcription and RNA translation.

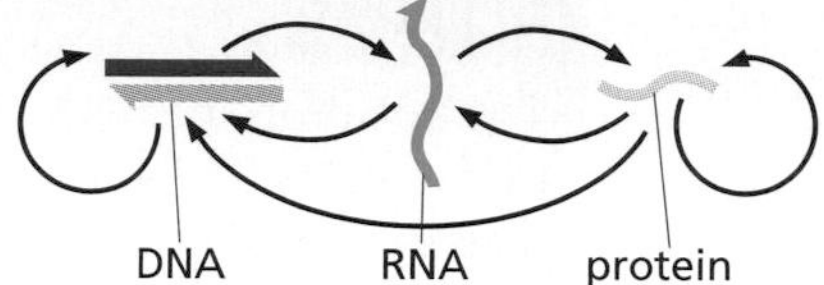

Transcription to RNA. During transcription, a portion of DNA is used as a template to make a single-stranded mRNA. There are several differences between transcription and DNA replication (see Figure 2.15). In DNA replication, the whole length of both DNA strands is copied to make two new strands of DNA. In transcription, a small, discrete part of a single DNA strand is the template for the synthesis of RNA. The portions of a DNA strand that contain the necessary information to make different proteins are the genes. A particular gene is transcribed from only one of the DNA strands but, at a different place on the same chromosome, a different gene may be transcribed from the other DNA strand. Transcription begins when a molecule of **RNA polymerase** combines with a portion of the DNA molecule known as a **promoter** (Figure 2.18). The product of transcription is usually mRNA, which carries the information for the second step of gene expression. Two other forms of RNA, transfer RNA and ribosomal RNA, are also transcribed from DNA, but are never translated into proteins. Several proteins are known that can either inhibit or enhance transcription, providing a means by which the expression of the gene can be controlled (see Figure 9.7, p. 368).

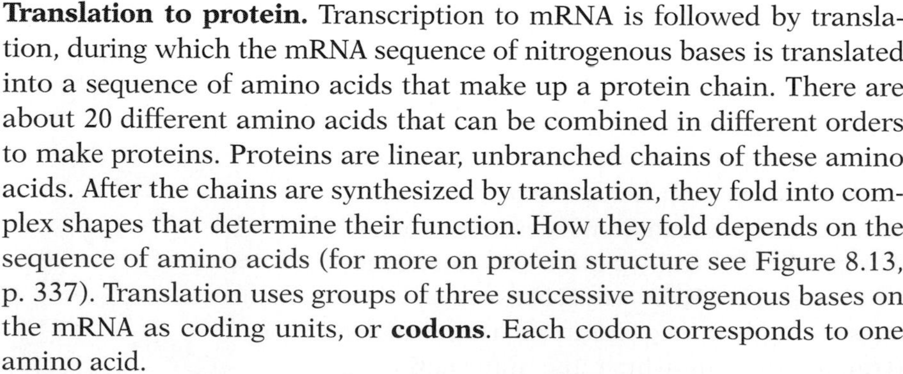

Translation to protein. Transcription to mRNA is followed by translation, during which the mRNA sequence of nitrogenous bases is translated into a sequence of amino acids that make up a protein chain. There are about 20 different amino acids that can be combined in different orders to make proteins. Proteins are linear, unbranched chains of these amino acids. After the chains are synthesized by translation, they fold into complex shapes that determine their function. How they fold depends on the sequence of amino acids (for more on protein structure see Figure 8.13, p. 337). Translation uses groups of three successive nitrogenous bases on the mRNA as coding units, or **codons**. Each codon corresponds to one amino acid.

Translation uses all three forms of RNA: mRNA as the template, transfer RNA to match a codon to an amino acid, and ribosomal RNA to form the 'scaffold' on which the process takes place. Each mRNA codon pairs with a complementary three-base sequence called an **anticodon**, which is part of a clover leaf form of RNA called transfer RNA (tRNA)

(Figure 2.19). Each tRNA molecule carries a specific amino acid molecule that can be added to the growing protein chain.

The mRNA and tRNAs are held in the proper relation by a particle called a **ribosome**, containing both protein and ribosomal RNA (rRNA). As each mRNA codon is read and translated by the binding of its complementary tRNA anticodon, one amino acid is added at a time to the growing protein until the end of the mRNA is reached and the protein is complete (Figure 2.20).

Mutations

DNA sequences occasionally undergo sudden but permanent heritable changes known as **mutations**. Some of these can result from mistakes during DNA replication, which may be caused in turn by chemicals or by

FIGURE 2.18

The transcription of DNA into RNA.

STEP 1

RNA polymerase binds to a promoter, part of a DNA strand at the beginning of a gene.

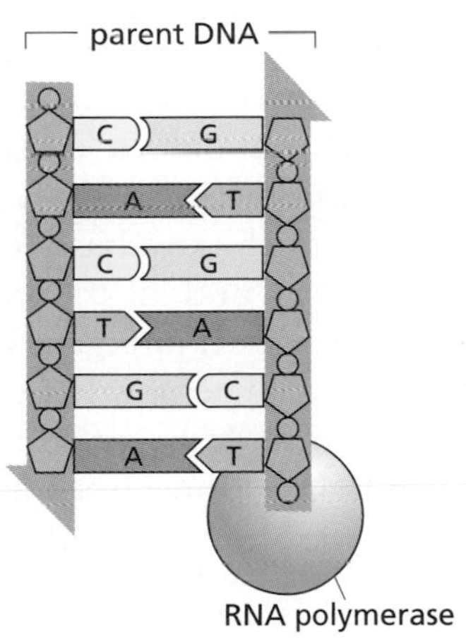

STEP 2

Polymerase unwinds a portion of the double helix, separating the strands locally. RNA nucleotides pair one at a time with the complementary nucleotides on *one* DNA strand.

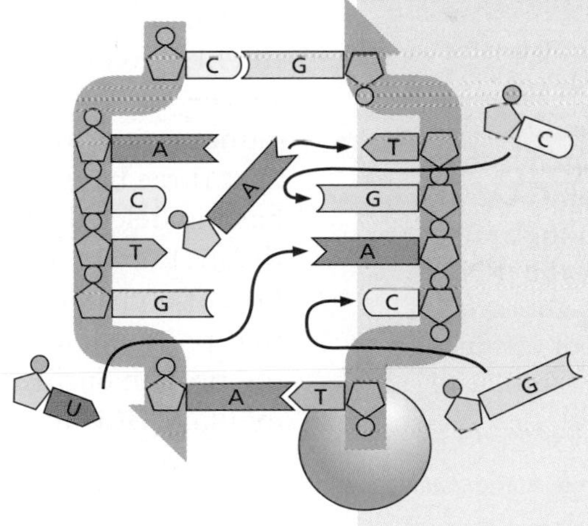

STEP 3

The RNA nucleotides join to form a continuous RNA strand, complementary to one of the DNA strands.

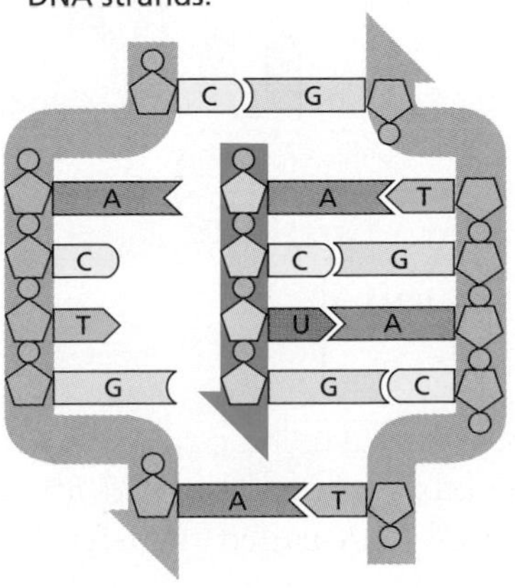

STEP 4

The RNA molecule separates, polymerase comes off, and the DNA strands rejoin.

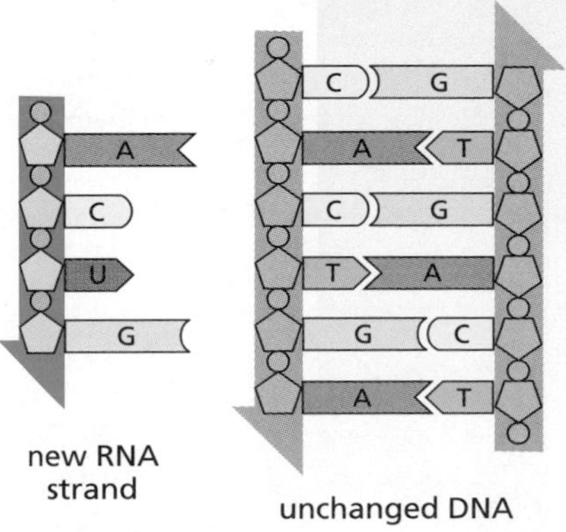

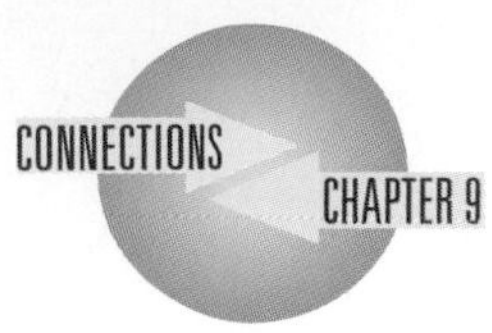

radiation (see Chapter 9). Most mistakes and damage are immediately fixed by various self-correction mechanisms and so do not persist. Mistakes that persist in somatic (body) cells can cause problems in the individual carrying the affected genetic material (see Chapter 9), but the change will not be passed on to the next generation. But mutations in cells that give rise to gametes may be passed on to future generations.

Point mutations. Ultimately, the different alleles of a gene originate as mutations. The simplest kind of mutation is a single-base mutation, or **point mutation**, such as the substitution of one nitrogenous base (A, G, C, T) for another. Because substitutions of this kind change the mRNA codon, they may result in the wrong amino acid's being inserted into a protein sequence (Figure 2.21A). As mentioned earlier, protein function depends on the shape of the folded protein molecule. The substituted amino acid may alter the protein shape and therefore change or impair the protein's function. Phenotypic consequences of inserting the wrong amino acid run the gamut from those that are undetectable to those that are fatal.

Frameshift mutations. **Frameshift mutations** occur when one or two extra nitrogenous bases are inserted into a DNA sequence, or when one or two bases are deleted from the sequence. The three bases that form the first codon determine the **reading frame**, or starting point, for each succeeding codon. The DNA or RNA code contains no 'commas' or other 'punctuation' to signify where a new codon starts. Each new codon is the next three bases after the previous codon. If an extra base is inserted or a base is deleted, the reading frame is shifted, and all codons that follow the mutation are changed. In the example shown in Figure 2.21B, a deletion of a guanine nucleotide (G) causes a change in mRNA codons from that point on, leading to the wrong amino acids' being added to a growing protein during translation. CGT in DNA is transcribed to the complementary GCU in mRNA, the codon for the amino acid alanine (Ala). By

FIGURE 2.19

The structure of transfer RNA and its role in translation. The mRNA codon CUU matches the tRNA anticodon GAA, which corresponds to the amino acid carried by the tRNA, phenylalanine, abbreviated Phe. The symbol ψ stands for pseudouridine (similar to uracil).

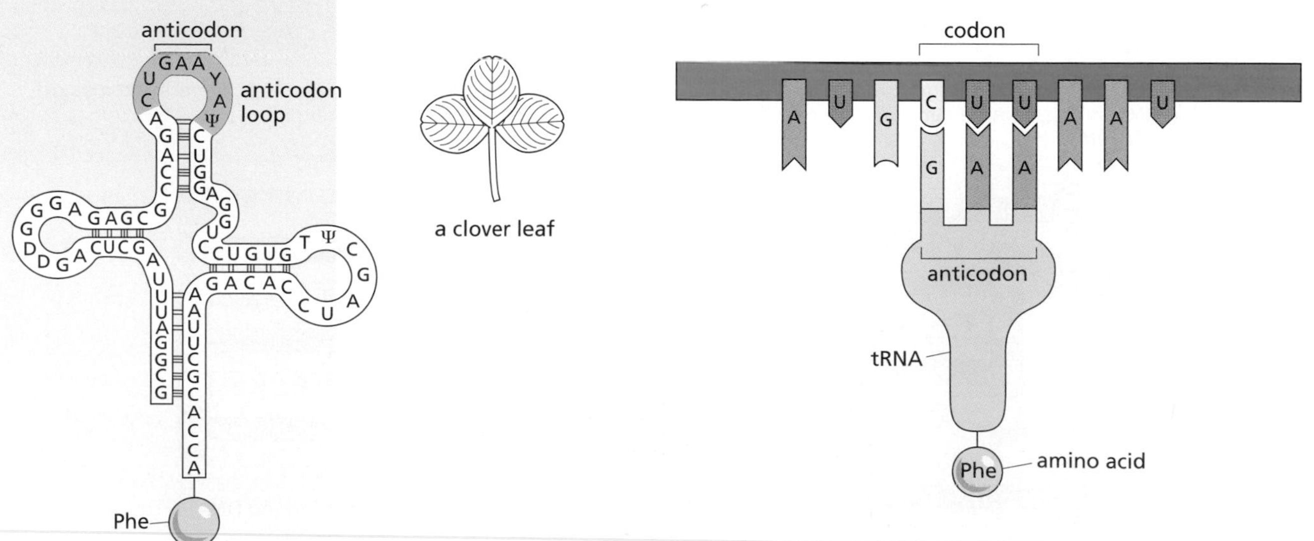

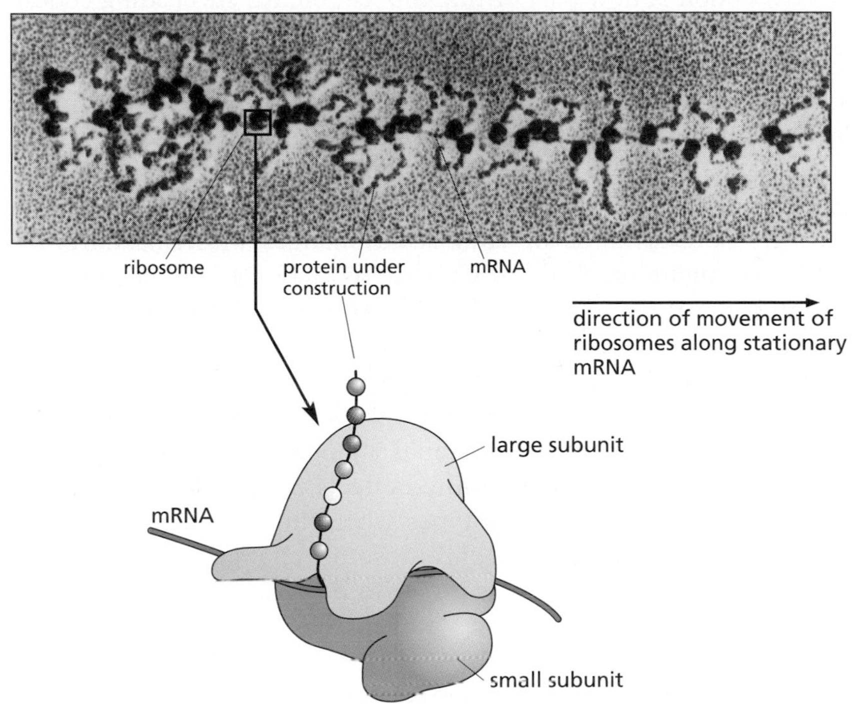

FIGURE 2.20

Translation of a nucleic acid sequence into a protein. Each group of three bases in messenger RNA (mRNA) serves as a codon to determine what amino acid is to be inserted next into the protein sequence. The ribosome holds the mRNA while tRNAs bring successive amino acids to the growing protein chain.

STEP 1 tRNA-4 binds to its mRNA codon,bringing amino acid 4 to the growing protein.

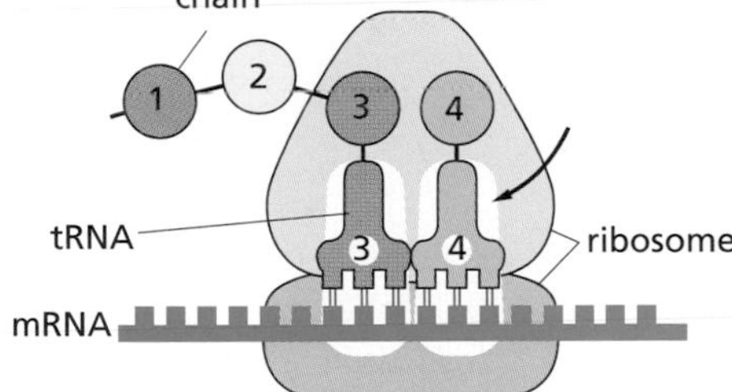

STEP 2 Amino acid 4 gets attached to the chain.

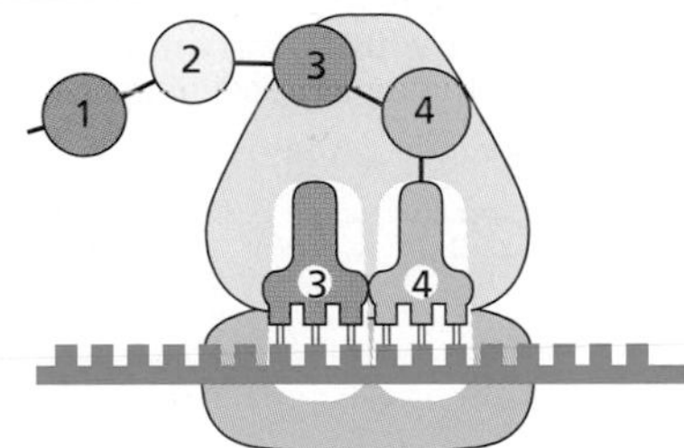

STEP 3 The previous tRNA (tRNA-3) is released and the ribosome moves down one codon on the mRNA.

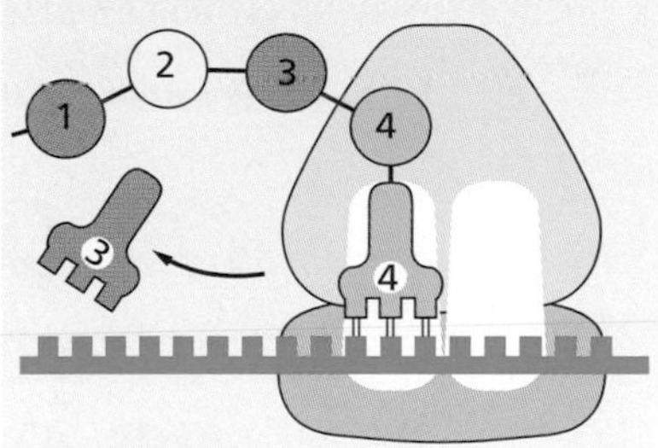

STEP 1 The process repeats with the next tRNA (tRNA-5) binding to its mRNA codon.

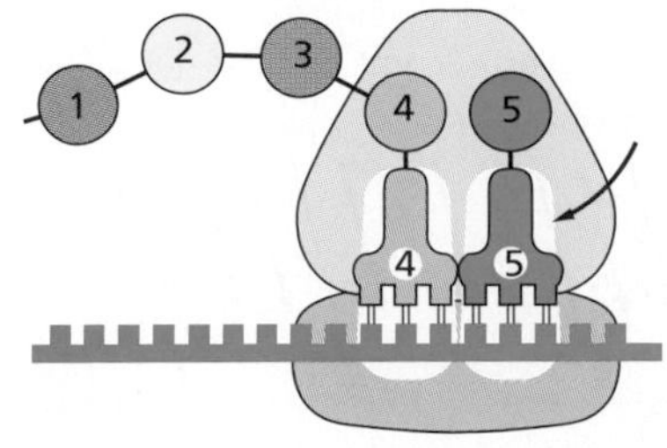

STEP 2 The protein chain elongates.

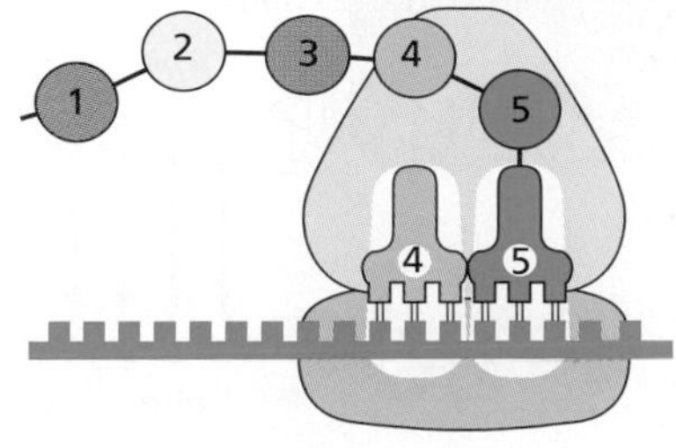

removing G, the DNA sequence becomes CTC, which is transcribed to the complementary mRNA codon GAG. GAG is the codon for the amino acid glutamine (Glu), not Ala. Note that the 'C' that would have been part of the next codon is now part of the 'Glu' codon. All succeeding codons are likewise shifted, leading to a protein with a different amino acid sequence from that coded from the unmutated DNA strand. When the codons are altered, the amino acids added to the growing protein are different and most frameshift mutations result in nonfunctional proteins.

Chromosomal aberrations. In addition to these small-scale mutations, there are several kinds of large-scale changes involving chromosome fragments, including both DNA strands and the associated proteins. Chromosome fragments may become duplicated (repeated); they may become attached to a new location, possibly on a different chromosome; or they may be lost entirely. A chromosome fragment may also be turned end-to-end and reinserted at its former location, forming an **inversion**. Of these four types of chromosomal changes, inversions are the most frequent, and have the most limited effects, while the other three types may result in nonviable phenotypes when the rearranged fragments of DNA are long. Although these chromosomal changes are often classified as mutations, many geneticists prefer to limit the use of the term 'mutation' to changes in single genes and to refer to the larger changes as **chromosomal aberrations**.

Another kind of chromosomal aberration is a change in chromosome number. Examples of such change are explained in the next chapter.

FIGURE 2.21

Examples of two types of mutations.

(A) Point mutation. When, for example, a C is substituted for a G in the DNA strand, the mRNA codon matches with the tRNA carrying the amino acid glycine rather than the tRNA carrying the amino acid alanine.

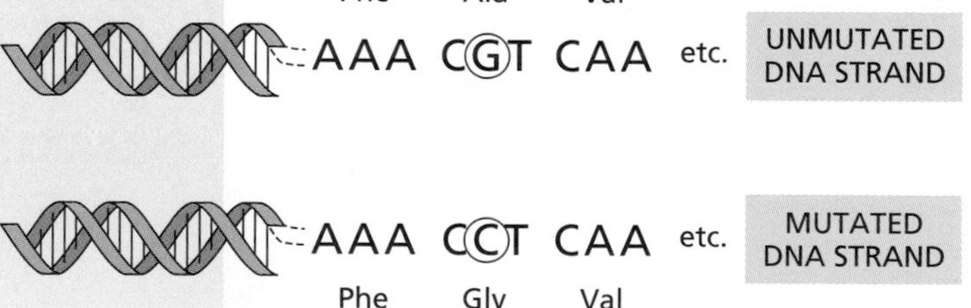

(B) Frameshift mutation. When, for example, a G is deleted from the DNA strand, the codon that had the G and all subsequent codons are misread and different amino acids are placed in the chain.

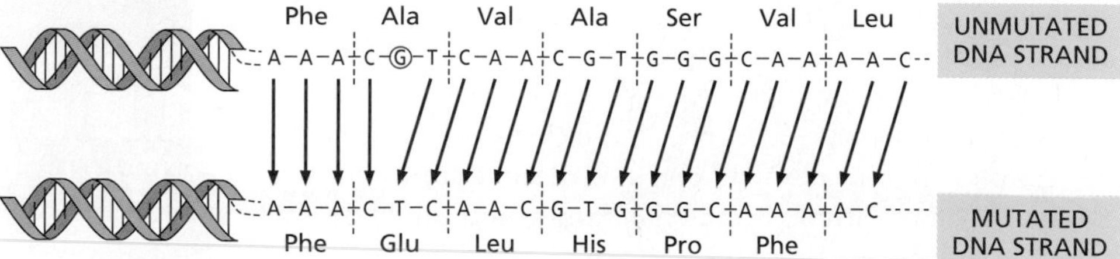

THOUGHT QUESTIONS

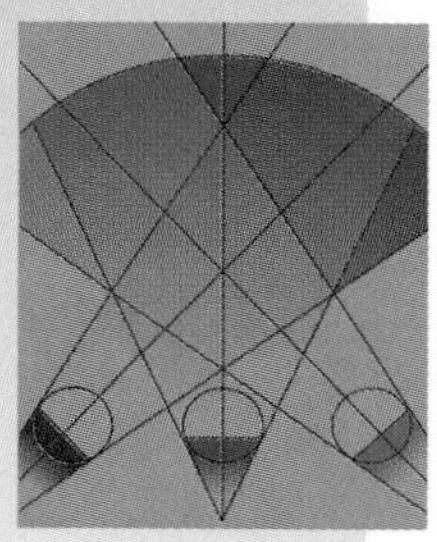

1. DNA has been called the master molecule because it controls (or determines) RNA sequences, which control protein sequences, which (as enzymes) control all the cell's other activities. Is the term 'master molecule' an accurate description? Does the language of control (e.g., DNA controls the type of RNA produced) say more about the molecules or about the scientists? Does the use of a word such as 'master' suggest a hierarchical approach in which information flows in one direction only?
2. Before either mitosis or meiosis, the DNA in each chromosome is replicated, forming two identical chromosomes that are attached as a pair at the beginning of mitosis or meiosis. Does one of these contain the old DNA and the other the newly replicated DNA? Or does each contain some of the old DNA and some of the newly replicated DNA?
3. Is the DNA sequence in the two chromosomes of an attached pair identical? Is the DNA sequence in the two chromosomes of a homologous pair identical?
4. DNA is a double helix, two complementary strands that wind around each other. A particular gene is on one of the two strands. A different gene at a different locus may be on the same strand or on the other strand. In transcription, the strand with the gene acts as a template for mRNA synthesis. What happens on the other strand during transcription? What happens on the other strand during DNA replication?
5. How would our concepts of genetics be different if Mendel had never published his study of pea plants?

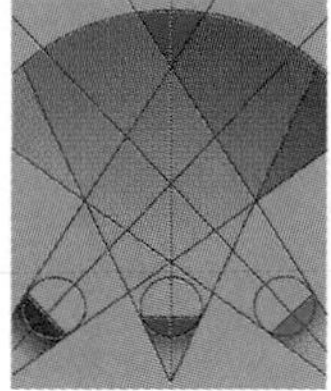

IN THIS CHAPTER WE HAVE SEEN HOW SCIENTISTS, beginning with Mendel, used observation and experimentation to understand the patterns of inheritance of simple, either/or traits. The same rules that work for pea plants work for other species, including humans. The units that assort and segregate in inheritance have come to be known as genes. Genes are located on chromosomes, a hypothesis that was first suggested because the numbers, locations, and movements of chromosomes could explain the observed patterns of inheritance. Genes were later found to be composed of DNA, a molecule that consists of long chains of nucleotides. The double-stranded structure of DNA accounts for its ability to be replicated accurately. Genes go through transcription to RNA and translation to protein, a sequence once thought to lead directly to phenotypes. Scientists now realize that no gene works independently of its cellular environment and that phenotypes for most traits are modifiable. Many traits in many species follow much more complex patterns of inheritance than the simple Mendelian either/or traits that we have seen in this chapter. In the next chapter we discuss some human traits that can be described as simple Mendelian traits, and many others whose inheritance is much more complex.

CHAPTER SUMMARY

- Hereditary information is carried in the form of DNA segments known as **genes**, which are parts of chromosomes. The location of a gene on the chromosome is called its **locus**.
- Most cells have the **diploid** number of chromosomes, and the chromosomes exist in pairs. The gametes are an exception because they have the **haploid** number, including only one chromosome of each pair.
- An **allele** is a variant of a gene. **Dominant** alleles show up in the phenotype when either **homozygous** or **heterozygous**. **Recessive** alleles are only expressed when they are homozygous. A **genotype** is the sum of an organism's alleles.
- **Chromosomes** are replicated and separated in **mitosis** during cell division, maintaining the diploid number and the full genotype.
- During **gamete** formation, **meiosis** halves the chromosome number to the haploid value and **segregates** the alleles into different gametes.
- When two genes are located on different chromosomes, their alleles segregate independently during meiosis, undergoing **independent assortment**. When two genes are on the same chromosome, their alleles show **linkage**, staying together in meiosis unless crossing-over occurs.
- **DNA**, a **nucleic acid**, is the template for the synthesis of another nucleic acid, **RNA**, in a process called **transcription**.
- **mRNA** is the template for protein synthesis in **translation**. Three mRNA bases form a **codon**, directing the addition of one amino acid to a protein.
- Changes in the DNA sequence (**mutations**) are reflected as changes in an mRNA sequence that are usually expressed as changes in the amino acid sequence of proteins. Such changes may affect the folded shape and therefore the function of the proteins. Gene mutations create new alleles.
- **Phenotypes** result from the activity of proteins, often of several proteins acting together.

KEY TERMS TO KNOW

allele (p. 42)
chromosomes (p. 46)
codon (p. 62)
cytoplasm (p. 45)
deoxyribonucleic acid (DNA) (p. 55)
diploid (p. 46)
dominant (p. 41)
enzyme (p. 53)
gamete (p. 42)
gene (p. 38)
genetics (p. 38)
genotype (p. 41)
haploid (p. 46)
heterozygous (p. 42)
homozygous (p. 42)
independent assortment (p. 44)
linkage (p. 46)
locus (p. 61)
meiosis (p. 46)
messenger RNA (mRNA) (p. 61)
mitosis (p. 46)
mutation (p. 63)
nucleic acids (p. 53)
nucleus (p. 45)
phenotype (p. 41)
recessive (p. 41)
replication (p. 59)
ribonucleic acid (RNA) (p. 55)
segregation, law of (p. 42)
transcription (p. 61)
translation (p. 61)
zygote (p. 48)

CONNECTIONS TO OTHER CHAPTERS

Chapter 1: Mendel gave genetics its first paradigm; the structure of DNA is the basis for the current paradigm.

Chapter 3: The same laws of inheritance that apply to pea plants apply to humans. Many diseases have a genetic component.

Chapter 4: Mutations and other genetic changes supply the raw material for evolutionary change. Evolution takes place whenever the frequencies of different alleles change.

Chapter 5: Human populations differ in the frequencies of many alleles.

Chapter 7: Mating patterns and sexual strategies can alter the frequencies of different alleles in populations.

Chapter 15: Conserving genetic diversity is an important aspect of protecting biodiversity.

PRACTICE QUESTIONS

1. A pea plant that is homozygous tall (*TT*) produces male gametes of what genotype? What is the genotype of the female gametes it produces?
2. If pollen from a homozygous tall pea plant fertilizes eggs from a homozygous short pea plant, what are the genotype(s) of the F_1 offspring? What are their phenotype(s)? What are the F_1 genotype(s) and phenotype(s) if the pollen is from a homozygous short plant and the eggs are from a homozygous tall plant?
3. What are the F_1 genotypes if eggs from a homozygous tall pea plant are fertilized by pollen from a heterozygous tall pea plant? What are the F_1 phenotypes?
4. What are the offspring genotype(s) and phenotype(s) if pollen from the F_1 in question 3 fertilizes eggs from a homozygous short pea plant?
5. Is a homozygous tall pea plant taller if it is grown in nutrient-rich soil than if it is grown in nutrient-poor soil? What about a heterozygous tall pea plant?
6. What is the height of the F_1 plants when plants that are homozygous for both tall height and yellow peas are crossed with plants that are homozygous for both short height and green peas ? What is their pea color?
7. If the F_1 plants from question 6 are self-fertilized, what are the genotypes, phenotypes, and phenotypic ratios of the F_2 plants?
8. If plants that are heterozygous for tall height and yellow peas are fertilized with pollen from plants homozygous for short height and green peas, what are the genotypes, phenotypes, and phenotypic ratios of the F_2 plants?
9. If *N* is the number of structurally different chromosomes in a mammalian species, how many chromosomes does a liver or skin cell from this species have before it enters mitosis? How many chromosomes does one of the offspring cells have when mitosis is finished?
10. If *N* is the number of structurally different chromosomes in a reptilian species, how many chromosomes does a cell of this species have before it enters meiosis? How many chromosomes does one of the offspring cells have when meiosis is finished?
11. How many of the two attached chromosomes go to each end of the cell during anaphase of mitosis? How many of the two attached chromosomes go to each end of the cell during the first cell division of meiosis?
12. Do homologous pairs of chromosomes line up together on the midline of the cell during metaphase of mitosis? Do homologous pairs of chromosomes line up together on the midline of the cell during the first cell division of meiosis?
13. How many genes undergo DNA replication before mitosis? How many genes undergo DNA replication before meiosis?
14. A point mutation that substitutes a single base for another changes that codon but not the succeeding ones. If two bases that are next to each other are replaced by two different bases, how many codons will be altered? On what might your answer depend?

3 Human Genetics

CHAPTER OUTLINE

Genes Carried on Sex Chromosomes Determine Sex and Sex-linked Traits

- Sex determination
- Sex-linked traits
- Chromosomal variation

Some Diseases and Disease Predispositions Are Inherited

- Identifying genetic causes for traits
- Some hereditary diseases that are associated with known genes

Molecular Techniques Have Led to New Uses for Genetic Information

- The Human Genome Project
- Using DNA markers to identify individuals
- Genetic engineering
- Other spin-off technologies

Genetic Information Can Be Used or Misused in Various Ways

- Genetic testing and counseling
- Altering individual genotypes
- Altering the gene pool of populations
- Changing the balance between genetic and environmental factors

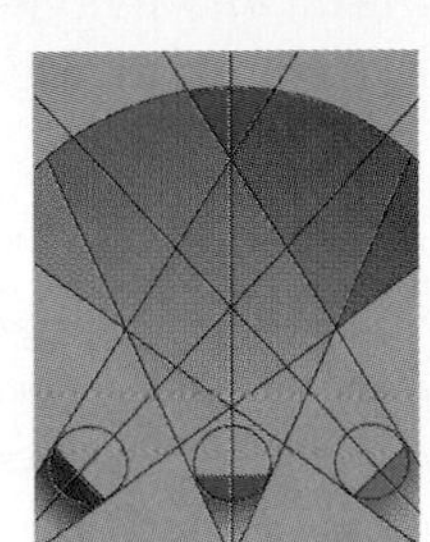

ISSUES

How can we use the study of genetics to fight disease?

Does gene therapy work?

Are there any inherent dangers of genetic research that we should be aware of?

Is every genetic variant a defect?

Will the Human Genome Project benefit mankind?

Why does the U.S. government fund the Human Genome Project?

Why are we racing to complete the sequencing of the human genome and whom are we racing against?

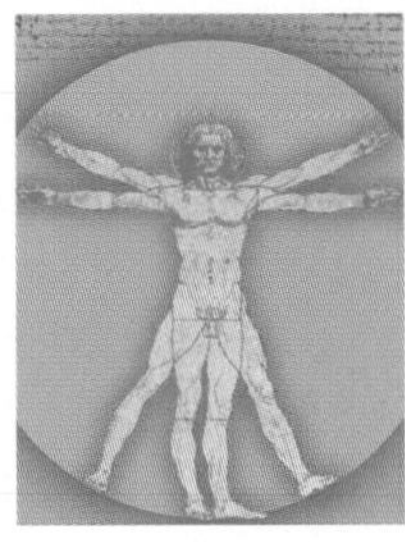

BIOLOGICAL CONCEPTS

Patterns of inheritance (trait, phenotype; sex-linked traits; sex determination)

Molecular genetics (genetic markers and genes)

Human health and disease (disease predisposition; genetic testing)

Biotechnology (The Human Genome Project; genetic engineering)

A nervous couple sits in the waiting room, anxiously anticipating the results of a test. Their last child had lived her short life in almost constant pain, and had died, blind, at age three, a victim of Tay–Sachs disease. The couple wants another child, but their previous experience was a heart-wrenching nightmare that they don't want to go through again. They are awaiting the results of an amniocentesis, a technique that you will read about later in this chapter. A doctor enters the room with good news: the enzyme that her technicians were testing for is present in the amniotic fluid. The mother-to-be is carrying a child who will not get Tay–Sachs disease. The couple can look forward to raising a healthy child in a happy home.

Scenes like the one just described are happening more often with each passing year. An increasing number of couples are undergoing medical procedures that did not exist when they themselves were born, seeking assurances that would have been unthinkable a mere 25 years ago. Tay–Sachs disease is one of a growing number of conditions that can now be diagnosed before birth. Along with physical characteristics, these conditions (traits) are passed on from one generation to the next.

Some traits follow the simple Mendelian patterns of inheritance that we studied in the previous chapter. Many more traits, however, follow much more complex patterns because they are governed by multiple genes, each contributing a small amount to the trait. How do scientists find which genes are associated with which trait? Molecular biology has greatly changed how scientists go about the search. It has also changed what we mean by the concepts of 'trait' or 'phenotype' and 'mutation,' and has led to many new uses for genetic information.

GENES CARRIED ON SEX CHROMOSOMES DETERMINE SEX AND SEX-LINKED TRAITS

Much has been learned about human genetics from the study of chromosomes. During mitosis and meiosis, the double-helical DNA becomes further coiled around supporting proteins and then coiled around itself until the structure becomes thick enough to be visible under a light microscope. This structure is generally what we think of as a chromosome.

During mitosis, when the chromosomes are visible, cells can be squashed onto a glass slide. Photographs of the chromosomes can be made through a microscope and the photographs can then be cut up. Using chromosome lengths and banding patterns, geneticists can line up the photos, putting the homologous chromosome pairs together. Such an arrangement is called a **karyotype**. The karyotypes of a female human and a male human are shown in Figure 3.1. The chromosomes of nearly every person can be arranged in a karyotype similar to one of those shown in Figure 3.1, with 46 chromosomes arranged in 23 pairs. Of the 23 pairs of chromosomes, 22 pairs, called the **autosomal** chromosomes, are the same in both sexes. The two chromosomes in each of these autosomal pairs are **homologous**, meaning that they carry the same set of genes (although possibly two different alleles of any gene). The 23rd pair of chromosomes (labeled X and Y in Figure 3.1) are called the **sex chromosomes** because they differ in males and females and have a role in determining a person's sex. The X and Y chromosomes are only partly homologous: they do pair during meiosis, but most of the genes on the X

FIGURE 3.1

Human female and male karyotypes.

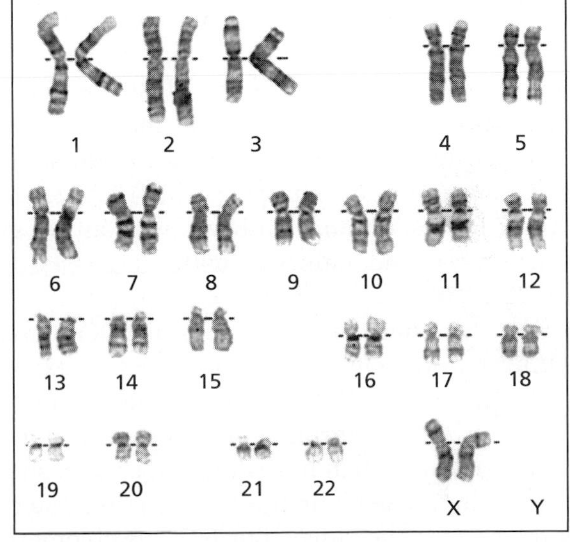

the XX karyotype (female)

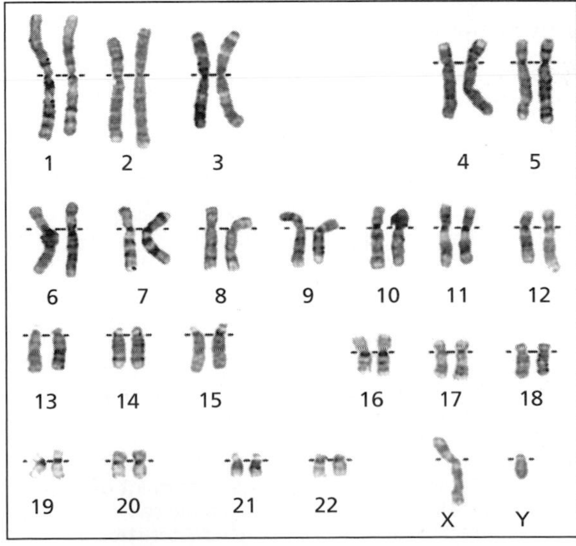

the XY karyotype (male)

chromosome are not present on the Y chromosome, and some of the genes on the Y chromosome are not present on the X chromosome.

Sex determination

Human females typically have two similar sex chromosomes, symbolized as XX. Human males typically have one X chromosome and one different sex chromosome, the Y chromosome, and thus are symbolized as XY (see Figure 3.1).

Not all human females are XX and not all human males are XY. There are unusual situations in which a cross-over during meiosis between an X and a Y chromosome is followed by an exchange of DNA pieces known as a **translocation**. Approximately 1 in 20,000 normal males is chromosomally XX, but one of his X chromosomes contains a small translocated piece of the Y chromosome. About the same frequency of normal females are chromosomally XY but are *missing* the same small piece of Y chromosome. One such XY female had 99.8% of the Y chromosomal DNA, indicating that a male-determining factor, or testis-determining factor (TDF), was located in the 0.2% portion of the Y chromosome that she did not have. Examination of this 0.2% portion of the Y chromosomal DNA led to the identification of a gene now called *sry*, hypothesized to induce development as a male. Other Y chromosomal genes had previously been hypothesized to code for the TDF until male individuals lacking these genes were found. This story is typical of the way in which science proceeds, with today's 'facts' being supplanted by additional evidence. The current hypothesis of sex determination by *sry* should be considered just that: a hypothesis that explains all of the current evidence but may be supplanted when new information becomes available.

Research suggests that the *sry* gene helps determine sex by producing a protein, the SRY protein, that binds to another DNA region, changing the way in which the DNA is folded. DNA folding regulates whether genes can be transcribed. The SRY protein folds the DNA in a way that allows the transcription of the gene for an enzyme, the testis-determining factor (TDF), that converts progesterone into testosterone (Figure 3.2). Both progesterone and testosterone are hormones, small molecules by means of which cells communicate with one another. Testosterone acts on cells to induce the development of male organs in embryos. If progesterone is not converted to testosterone, it is instead converted to estrogen, which triggers the development of female organs. Embryos with an *sry* gene thus become males, and embryos without an *sry* gene become females.

FIGURE 3.2

The hypothesized relationship between the *sry* gene and sexual development in humans.

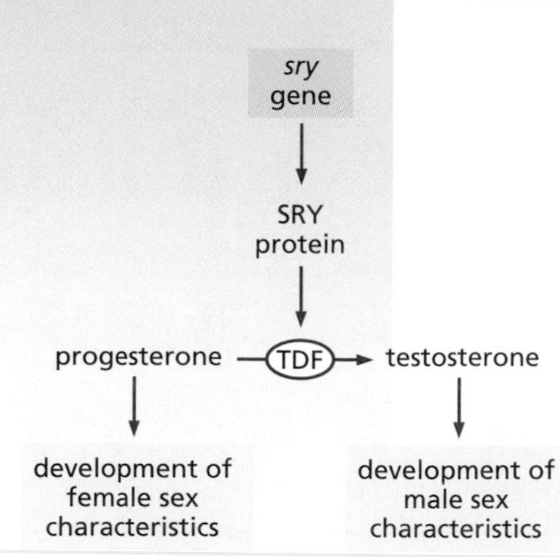

Realization that there are XX males and XY females forced the International Olympic Committee to reexamine the stipulation that only XX individuals could compete in female sporting events. If chromosome appearance is not sufficient to determine which individuals are male, should the presence of the *sry* gene or the hormone testosterone be used as the test? Either turns out to be problematical.

Females also have testosterone, although generally in smaller amounts than males. Also, there are rare XY individuals who have the functional *sry* gene and produce male concentrations of testosterone but are nevertheless phenotypically female because they lack a functional allele of a different gene, the gene for the receptor for testosterone. Cells cannot respond to a hormone during development or during adult life unless they possess receptor molecules to bind that hormone. People with variations in the relation between their karyotype and their phenotype are not defective individuals; they just do not fit a previously held view of what determines the sex of an individual. The International Olympic Committee now uses the presence or absence of the functional *sry* gene to decide the sex of Olympic athletes, but clearly sex is not determined by one single gene. Other genes such as the testosterone receptor gene and other genes on autosomal chromosomes also have a role in sexual development. Thus, even though we refer to X and Y as the sex chromosomes, many genes on many other chromosomes are also involved.

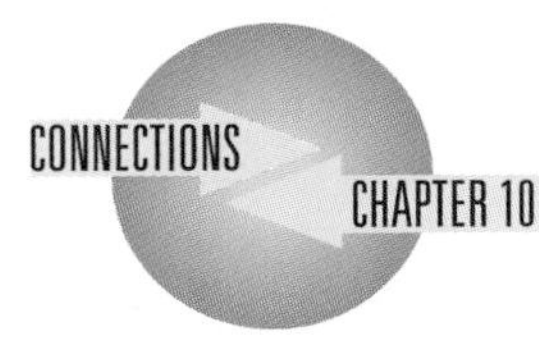

Sex-linked traits

Very few genes are, like *sry*, located on the Y chromosome. Many more genes are located on the X chromosome. Genes that are on the X chromosome and not on the Y chromosome, such as the gene for **red–green colorblindness**, are said to be **sex-linked**. Females can be either homozygous or heterozygous for sex-linked traits because they have two X chromosomes and therefore two alleles of every sex-linked gene. If the allele for a trait is recessive, as with red–green colorblindness, then a heterozygous woman has the dominant phenotype but is said to be a carrier of the trait. Because a male has only a single X chromosome, he has only a single allele for each sex-linked trait, and this allele determines his phenotype for that trait. The inheritance of a recessive sex-linked trait is shown in Figure 3.3. Note that a single sex-linked allele is phenotypically displayed in a male regardless of whether it is dominant or recessive.

Females generally possess two X chromosomes, but in a given cell only one of them has active genes that make a product or express a phenotype. Females thus express one phenotype (from their mother's X chromosome) in some cells and another phenotype (from their father's X chromosome) in other cells. This expression of two phenotypes at the cellular level is called **mosaicism**. All females are mosaics for X-linked genes. For example, in a female whose father carries an X-linked allele for colorblindness, patches of cells within the retina of the eye (see Chapter 10, pp. 424–425) cannot respond to color. Other patches of cells, which express the normal allele on the X-chromosome from the mother, respond normally.

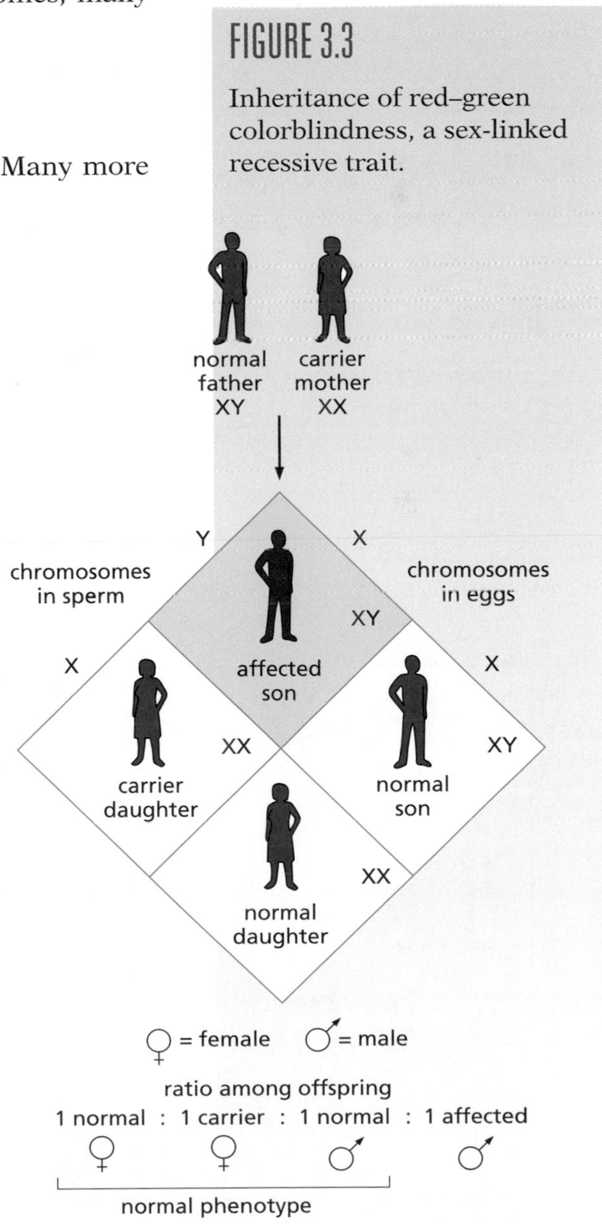

FIGURE 3.3
Inheritance of red–green colorblindness, a sex-linked recessive trait.

Chromosomal variation

Most humans have 46 chromosomes, consisting of 22 pairs of autosomal chromosomes and a pair of sex chromosomes. Other chromosomal constitutions have multiple consequences, called syndromes, usually named after the physicians who identified them. For example, several variations of the sex chromosome number are known. The XXY chromosomal type (Figure 3.4A) results in **Klinefelter's syndrome**; persons with this condition have male phenotypes but are sterile. Some of the symptoms of Klinefelter's syndrome can be successfully treated with hormones. In contrast, **Turner's syndrome** results from the XO chromosomal type, in which only one X chromosome is present, the O representing its missing partner (Figure 3.4B). Persons with Turner's syndrome develop as females; however, their ovaries do not produce female hormones. Puberty does not take place and gametes do not develop, resulting in infertility. The infertility cannot be overcome at present, but the other symptoms of Turner's syndrome can now be treated hormonally with much success. Clearly male and female are not either/or categories. There are persons who fall between, either because of chromosomal variation or variation in alleles at particular genes such as *sry* or the testosterone-receptor gene.

Turner's and Klinefelter's syndromes are believed to result from the same cause, a **nondisjunction** (abnormal cell division) of the sex chromosomes. In the nondisjunction, the two sex chromosomes fail to separate during meiosis, resulting in some egg cells' having two of the mother's X chromosomes and some having none. Nondisjunction during gamete production has in fact been observed, partly confirming the hypothesized

FIGURE 3.4

Two variations in human X and Y karyotypes.

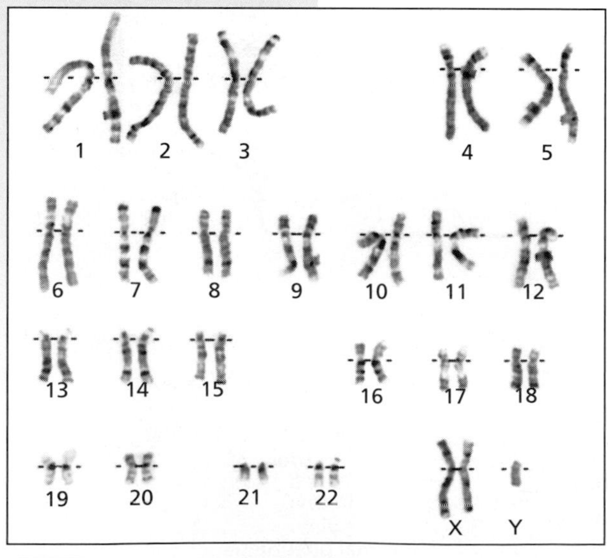

(A) Klinefelter's syndrome (XXY)

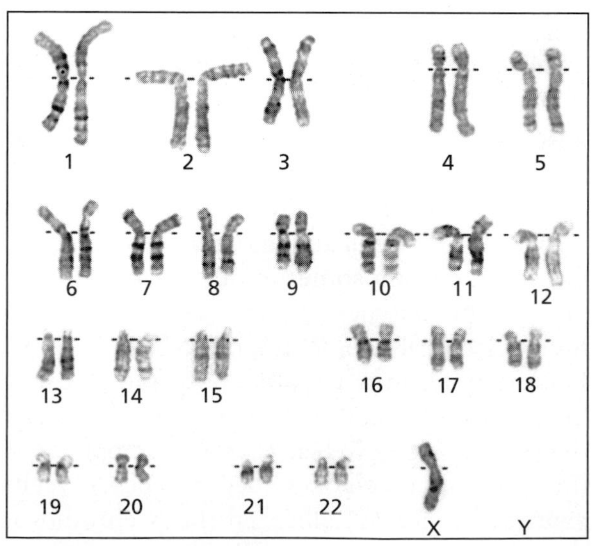

(B) Turner's syndrome (XO)

series of events shown in Figure 3.5. Also supporting the hypothesis is the very rare XXX chromosomal abnormality; most XXX females are mentally retarded and sterile. The Y-only type of embryo, also predicted by the hypothesis, has never been observed, presumably because it dies at a very early stage of development. Nondisjunction can also take place

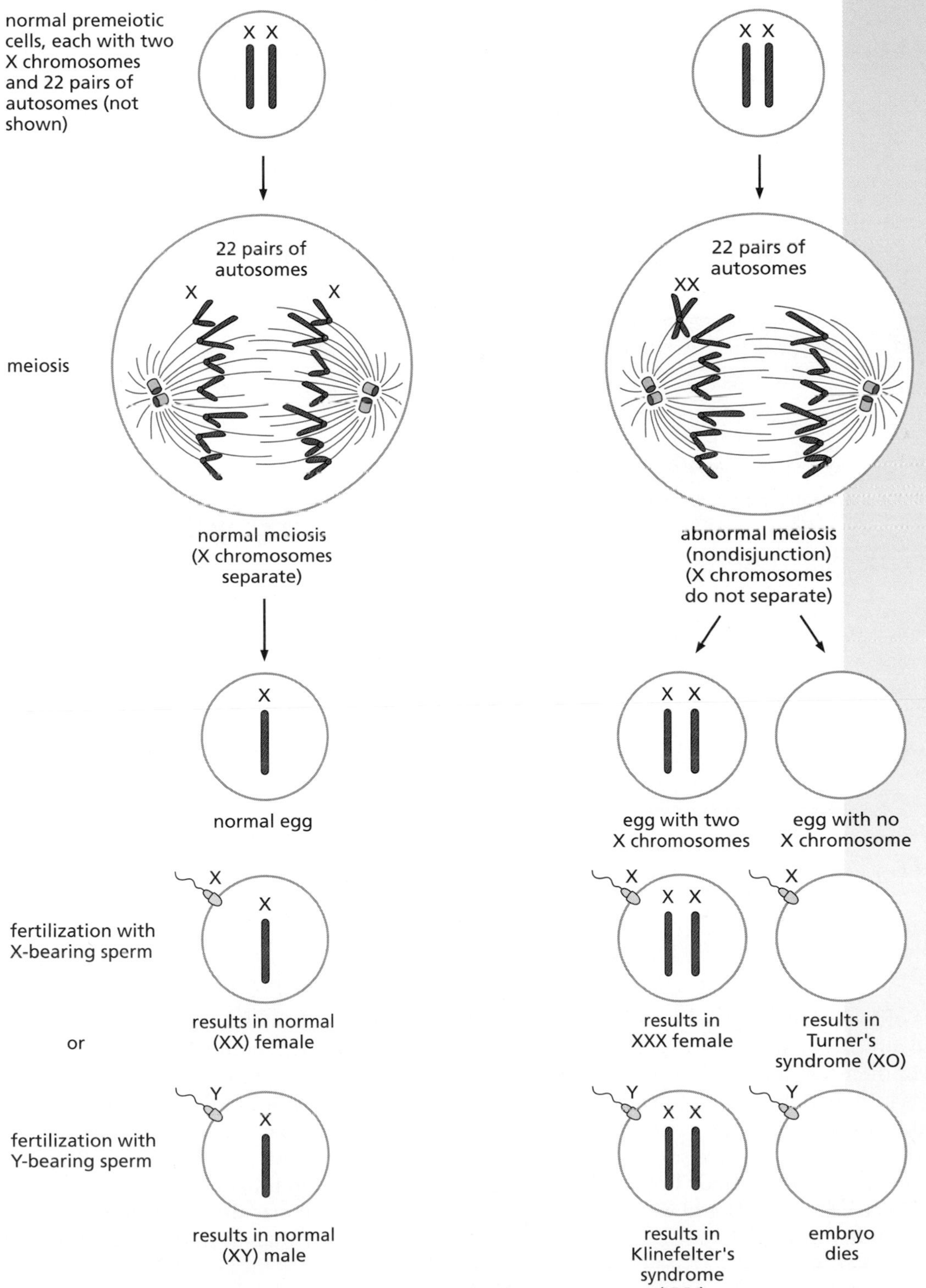

FIGURE 3.5

Nondisjunction during egg production, showing how certain chromosomal variations may arise.

during the formation of sperm cells, resulting in XY sperm and O sperm. When these fertilize a normal X egg, again either Klinefelter's syndrome (XXY) or Turner's syndrome (XO) can result.

There are also variations in number for chromosomes other than the sex chromosomes. The most common of these is associated with **Down's syndrome**, marked by facial characteristics (including an epicanthic fold over the eyes), heart abnormalities, and a variable amount of mental retardation. Down's syndrome usually results from an extra chromosome 21, a condition known as **trisomy** 21, in which three of these chromosomes are present instead of two (Figure 3.6). Other chromosome abnormalities are less common. For example, Patau's syndrome (trisomy 13) results in severe mental retardation, a small head, extra fingers and toes, and usually death by one year of age. In addition to extra chromosomes, part or all of a chromosome can be missing. The *cri du chat* syndrome is caused by deletion of the short arm of chromosome 5 and results in a small head, a catlike cry, and mental retardation.

FIGURE 3.6

Down's syndrome.

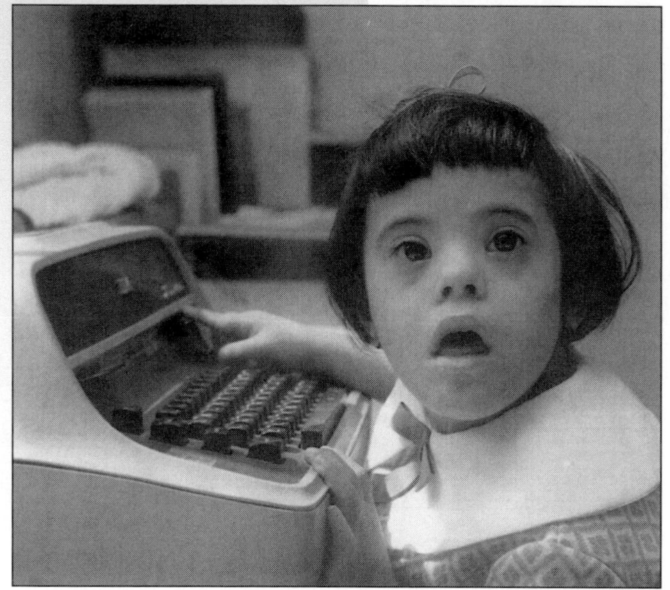

A child with Down's syndrome

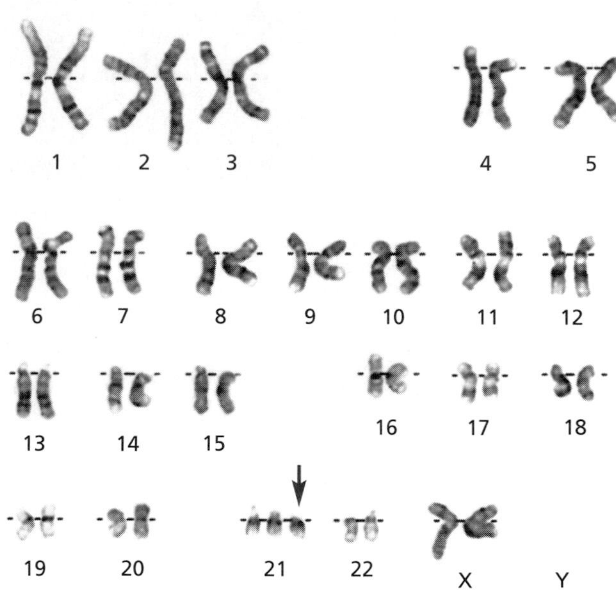

The karyotype with an extra chromosome 21 associated with most cases of Down's syndrome

THOUGHT QUESTIONS

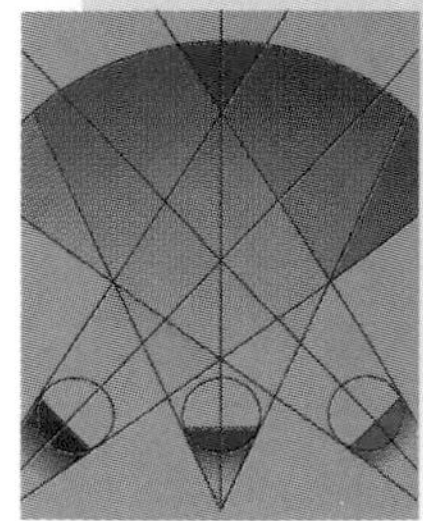

1. Sports officials have repeatedly asked female athletes to submit to testing to confirm their femaleness, but comparable proof is seldom demanded of males. Why do you think this disparity exists? Can you think of other ways besides sex in which athletes might be classified? Could we use age as a criterion? Could we use fat-to-muscle ratios?

2. Individuals with Klinefelter's syndrome or Turner's syndrome are fully functional aside from their infertility, yet many are 'treated' with hormones to give them a less indeterminate appearance of secondary sexual characteristics such as breasts. Why do humans tend to think of such differences as 'abnormalities' needing 'treatment' rather than as normal variation within a trait?

SOME DISEASES AND DISEASE PREDISPOSITIONS ARE INHERITED

Early in the twentieth century, pioneering geneticists discovered that Mendel's rules, formulated on the basis of experiments with pea plants, could also explain the inheritance of many human traits. For example, **albinism** is a total lack of melanin pigment in the skin, eyes, hair, and the body's internal organs, and the inheritance of albinism follows Mendel's rules. The skin of albinos is white and their hair is white as well. One of the normal functions of melanin is to block ultraviolet light, and albinos sunburn easily and are very sensitive to bright lights. All geographical races of humans have albino individuals, as do many other species.

The inheritance of albinism is shown in Figure 3.7. A recessive allele is responsible for this condition and thus it can be transmitted without detection through many successive generations of normally pigmented individuals (Figure 3.7A and C). However, matings between heterozygous individuals may produce albino children (Figure 3.7B). The likelihood that both parents are heterozygotes is increased if mates are chosen from among related persons such as cousins (see Figure 3.7C). This is because the same rare recessive alleles are more likely to be present in other family members than they are in the general public. Diagrams giving the pattern of mating and descent, as in Figure 3.7C, are called **pedigrees**.

Shortly after the rediscovery of Mendel's laws, an English physician, Archibald Garrod, made an important discovery: Mendel's laws applied not only to visible characteristics such as eye colors and albinism, but also to certain medical conditions. Garrod's identification of the genetic basis of a condition called alkaptonuria is described later in this section.

Identifying genetic causes for traits

Before searching for genes that cause a particular disease or trait, we first need to know whether there is any basis for thinking that the disease is inherited. As we will see, some diseases are the direct result of particular gene mutations. In other diseases, genetic effects are indirect and contribute to disease **susceptibility**, the likelihood that a person will get a disease. Susceptibility to many diseases seems to have a hereditary component. However, certain disease susceptibilities and many nondisease traits are the result of the interaction of multiple genes, not the result of mutations in single genes.

Several kinds of studies are used in answering the question of whether a disease or other trait is inherited.

Pedigrees. Geneticists have several ways of studying human hereditary traits. One of the most basic methods is to present the available data in pedigrees, as in Figure 3.7C. Pedigrees are most useful when they span many generations and hundreds of people, or when separate pedigrees are available for hundreds of different families. The study of pedigrees can help to identify whether a condition is inherited, and permit us to determine which traits are dominant, which are recessive, and which have a more complex genetic basis. If the genetic basis of a trait is complex, or not fully known, then pedigrees can also help in an empirical determination of risks, including medical risks. For example, a child has a greatly increased risk of having insulin-dependent diabetes mellitus (IDDM) if one or both of the child's parents has the disease. The term **risk** has a precise statistical meaning: it is the probability that a particular condition will occur or that a particular condition will be inherited.

Studies of twins and of adopted children. Studies of twins are sometimes useful in suggesting the extent to which the presence of a trait can be explained genetically, rather than environmentally. In such studies, numerous twin pairs are located in which at least one twin has the condition being investigated. For pairs of this kind, the frequency with which

FIGURE 3.7

An example of simple Mendelian inheritance in humans. Albinism, a recessive trait, arises in most cases from matings between heterozygotes, although it could also arise from matings between a heterozygous person and someone who is homozygous recessive.

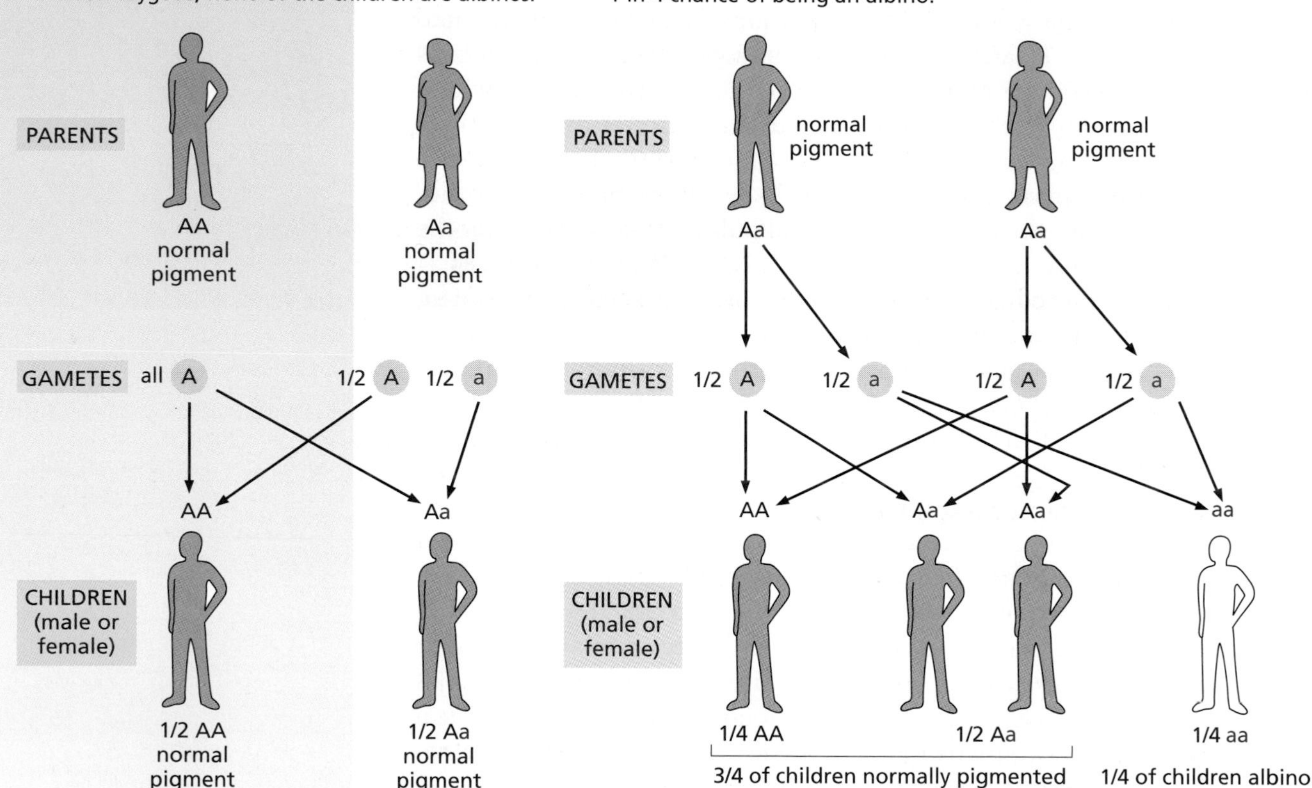

(C) Human pedigree for a family with albinism; the short horizontal lines between a male and female represent matings that produced the children in the row below.

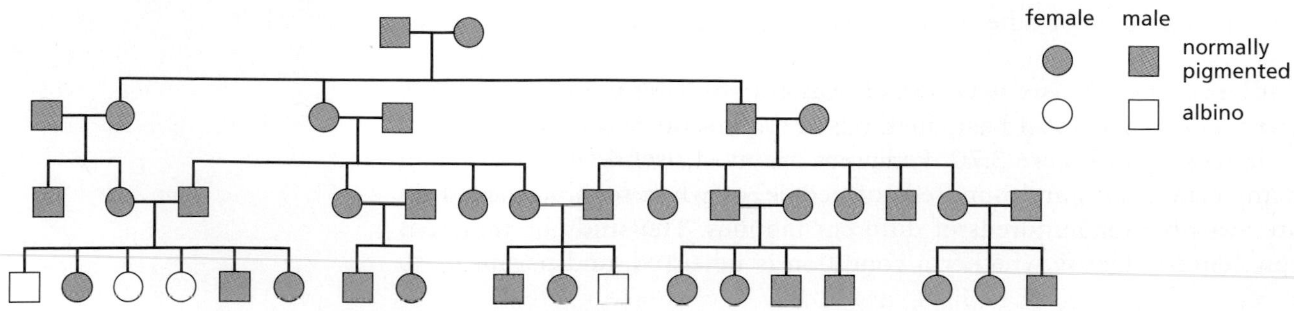

the other twin has the condition is called the **rate of concordance**. For the vast majority of traits, studies on twins find that a mixture of both heredity and environment is involved. Traits under strong genetic control usually have higher concordance rates for **monozygous** twins (identical twins, derived from a single fertilized egg) than for **dizygous** twins (fraternal twins, derived from two separate eggs). In contrast, traits with mainly environmental causes have similar concordance rates for both types of twins.

Adoption studies can also provide important clues about whether a particular trait is heritable: if adopted children show a higher rate of concordance with their birth parents than with their adoptive parents, then the hypothesis of a genetic cause is made stronger. Some researchers have, however, criticized this type of study because adoption agencies do not place children at random but purposely try to match children with adopting parents whose backgrounds are similar to the backgrounds of the children's birth parents. This practice introduces a bias that raises the concordance rates between adopted children and the parents who adopted them. Another complication is that many children are adopted by relatives, and such adoptions make it very difficult to sort out which similarities are environmental and which are genetic.

Linkage studies. Once evidence has been found that there is a genetic basis for a trait, linkage studies can help locate the relevant gene or genes. To do this, we first need a set of **markers**, pieces of chromosomes that are visibly different under the microscope or short sequences of DNA that can be revealed by the molecular techniques discussed later in this chapter. If we can locate a marker whose pattern of inheritance is the same as the pattern of inheritance of the trait, then we can conclude that a gene associated with the trait is located near the marker. One problem is that we may at first not know where to look, so we need many markers, scattered across all the chromosomes. Also, large pedigrees are needed to carry out this type of analysis. After a linkage between markers and a trait is found, other molecular techniques are used to close in on the actual gene. Finally, after the gene has been located, its full sequence can be determined.

In the past, genes that did not assort independently were said to be linked, and the frequency of crossing-over was used as a measure of the distance between genes (see Chapter 2, pp. 50–52). If linkage to a visible chromosomal abnormality could be established, then the group of genes could be assigned to a particular chromosome. These classical genetic techniques were developed in other species, using hundreds or thousands of offspring in each generation to assess cross-over frequencies. These techniques could not be used on humans because humans have such small family sizes and such a long generation time, and because humans cannot be bred for experimental purposes. Before 1980, very few human genes had been mapped to their chromosomal locations.

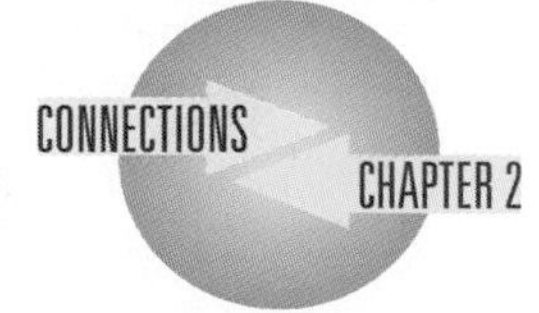

Several marker systems have now been discovered for studying human DNA and human genetics. The first of these marker systems that was found, restriction-fragment length polymorphisms, is discussed next. More recently other markers, with names such as expressed sequence tags, microsatellites, and single-nucleotide polymorphisms, have been discovered. Geneticists use all of these in developing linkage maps.

Restriction-fragment length polymorphisms. In 1980 a new mapping technique was devised that could readily be used in human studies, as well as in studies on other species. DNA contains, in addition to genes, non-coding regions that vary in length from one individual to another. Short sequences of nucleotides, 3–30 bases long, are repeated over and over anywhere from 20 to 100 times. These are called short tandem repeats; several thousand different such repeats are now known in humans, each with a unique sequence not found elsewhere in the genome. When DNA containing variable numbers of repeats is cut with a special enzyme called a restriction enzyme (discussed in detail on p. 95), fragments of DNA of various lengths are produced (Figure 3.8A). Variations (also called polymorphisms) in the lengths of the fragments produced with restriction enzymes are known as **restriction-fragment length polymorphisms**, or **RFLPs** (pronounced 'riflips'). The fragments of different lengths are separated by a technique called electrophoresis (Figure 3.8B). Because DNA carries an electric charge it moves in an

FIGURE 3.8

Restriction-fragment length polymorphisms (RFLPs).

(A) CUTTING DNA WITH RESTRICTION ENZYMES

The pieces differ in length depending on the number of repeats that exist within a piece. In this example, the piece from the father is shorter because it has fewer repeats than the piece from the mother, which is longer because it has more repeats.

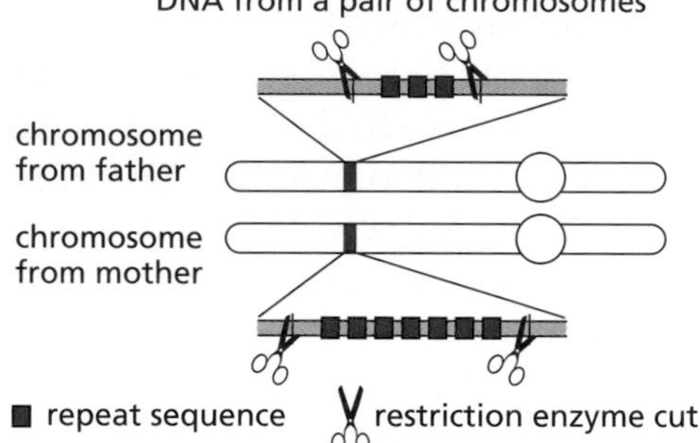

(B) SEPARATION BY ELECTROPHORESIS

The mixture of pieces is placed on a gel and exposed to an electric field. Because DNA has a charge, the pieces move toward the electrode of opposite charge. In the time that the current is on, smaller pieces travel further through the gel than the larger ones do. None of these pieces is visible yet.

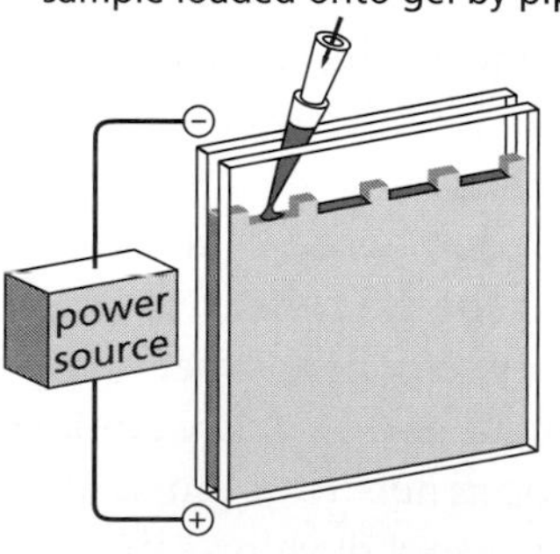

(C) DETECTION WITH A PROBE

None of the pieces can be seen; however, they can be detected with a variable-repeat probe (bands shown in color). The probe is a small piece of DNA with a sequence complementary to the sequence of that variable repeat, so the probe will bind to those pieces of DNA containing that variable repeat. The probe thus does two things: it identifies pieces with that specific repeat and it indicates whether the sequence is repeated a few times (to give a short DNA piece) or many times (to give a long piece). Other probes will find other sequences that are repeated in other chromosomal locations.

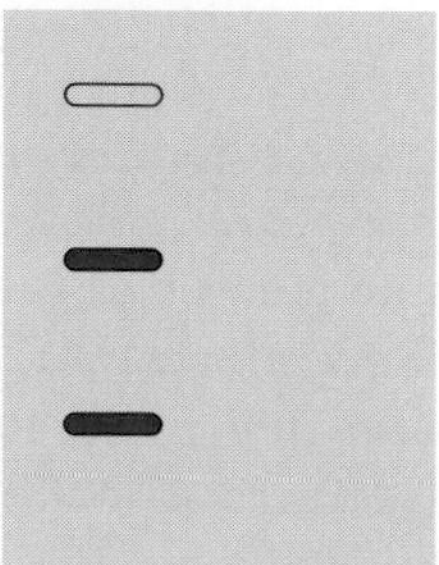

electric field. When a DNA sample that has been cut into fragments is loaded onto a gel and electric current is applied, the fragments move. The gel material retards the movement of the fragments somewhat, and the larger the fragment, the more its movement is retarded by the gel. In the time that the electric current is on, smaller fragments will therefore move further than large fragments. Because the nucleotide sequence of each short tandem repeat is unique, each can be detected by a specific **probe**, a piece of DNA with a sequence complementary to the repeat sequence (Figure 3.8C). Probes are specific and cause only those fragments to show up that have sequences complementary to the probe sequence. If a radioactive DNA probe is added to some DNA that has been digested into fragments with a restriction enzyme, only those fragments that can pair with the probe DNA pick up the radioactivity.

Because RFLPs are numerous and are scattered throughout the genome, they can be used as genetic markers. Once the chromosomal locations and genetic map positions of particular RFLPs have been established, geneticists can determine the positions of presumed genes located near the RFLPs by finding a pattern of linkage between the trait and the RFLP. DNA samples are collected from members of a family with a pedigree in which the trait appears. Molecular geneticists then search in the DNA samples for a RFLP, among the thousands of RFLP markers known, that is inherited within the family in the same pattern as the trait of interest. If all the individuals within a large pedigree who have a particular trait have a RFLP of a certain length, and all those without the trait have a RFLP of another length, the gene for the trait is presumed to be near that RFLP (Figure 3.9A). For the other RFLP markers that are not in the vicinity of a gene linked to the trait, the mother's and father's RFLPs can be found in the children, but the pattern of bands does not follow the pattern of inheritance of the trait (Figure 3.9B). In this way, an increasing number of presumed gene locations are being discovered at an accelerating rate.

Identifying a specific gene as the cause of a trait. It is important to understand that RFLP markers are not the genes themselves. In other words the linkage indicates the approximate *location* of a gene of interest. It does not tell us what the gene is, or what its function is, or what alleles are associated with disease or nondisease.

Often when such a location has been identified, news reports are published claiming that 'the gene for X' has been discovered, but no gene has even been investigated, only a DNA region that maps with the trait. When a linkage area has been identified, sequence databases can be consulted to see what genes are known in this area of the chromosome. If genes are known, their protein products can sometimes be deduced from the nucleotide sequence and those proteins can be further investigated for their connection with the trait or disease. Often, however, the linkage area has not yet been sequenced, and then sequencing must proceed before a gene can be identified.

The first gene to be located by using DNA molecular markers was the gene for Duchenne's muscular dystrophy, which is located on the X chromosome. Other diseases for which linkage areas were located and identified during the 1980s and 1990s include Huntington's disease (chromosome 4), cystic fibrosis (chromosome 7), Alzheimer's disease (chromosome 21), one form of colon cancer (chromosome 2), and two forms of manic depression (chromosome 11 and the X chromosome). Of these linkage areas, genes have so far been found for muscular dystrophy, Huntington's disease, and cystic fibrosis. Among these diseases, only cystic fibrosis seems to be inherited on the basis of single-gene Mendelian genetics. Even in cystic fibrosis, how the gene product produces the disease is not fully understood.

Quantitative-trait locus analysis. It is becoming increasingly apparent that most phenotypes, including disease phenotypes, are not caused by single genes. Each of several genes may contribute to a phenotype under investigation. Using animal models, researchers quantify the contribution of each gene relative to the others. In a new method of analysis called quantitative-trait locus analysis, several traits associated with a disease can be compared statistically against a large number of DNA markers simultaneously. Where there are high correlations between phenotypes and markers, contributing genes are assumed to be located in the vicinity of the markers. Traits such as bone density or obesity in

FIGURE 3.9

Using RFLPs to establish a linkage between DNA markers and inherited phenotypes.

(A) Probe 1 detects a RFLP linked to the trait

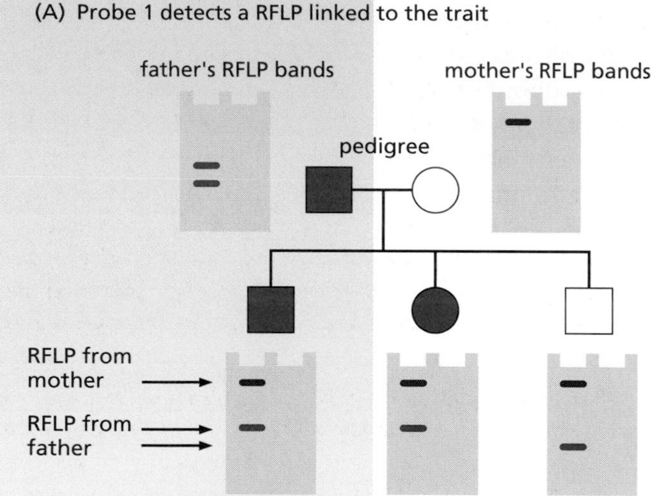

unaffected female
unaffected male
affected individual

The father has the trait and is heterozygous at the RFLP detected by probe 1. The mother does not have the trait and is homozygous at the RFLP detected by probe 1; the mother's RFLP has a greater number of repeats than either of the fragments from the father's DNA.

The children who have the trait have one of their father's length of RFLP. The child without the trait does not. The RFLP lengths detected with probe 1 thus follow the pattern of inheritance of the trait, indicating that an associated gene may be nearby.

(B) Probe 2 detects a different RFLP, and shows that it is not linked to the trait

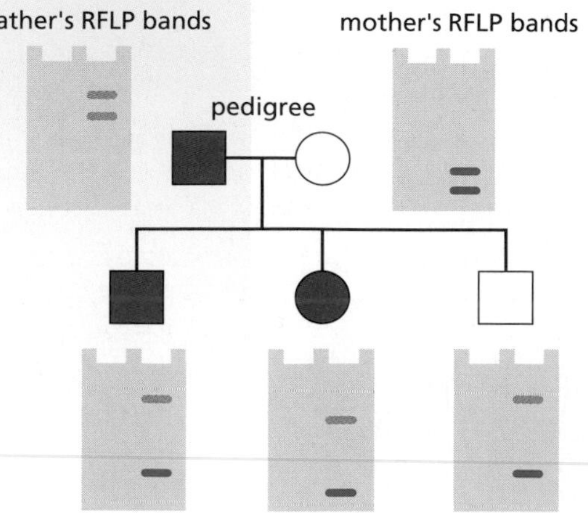

A child with the trait and a child without the trait have the same band pattern for the RFLP detected by probe 2. The RFLPs detected by probe 2 are therefore not near any gene associated with the trait.

mice, and behavioral traits such as ethanol consumption by rats, have been found to be associated with DNA markers by this approach.

While the genes themselves have not yet been identified, news stories have overplayed such statistical associations between traits and markers by, for example, referring to the 'gene for obesity,' but such terms are premature and surely oversimplified. All of these traits are influenced by multiple genes, so we cannot think of them as we think of Mendelian either/or traits. These traits vary along a continuum and are not just present or absent like the traits studied by Mendel. Thus the concept of a single gene determining a single trait does not apply. A single gene codes for a single protein, but the amount of that protein produced depends on what other genes and proteins are present. It is the interaction of them all that results in the trait.

Some hereditary diseases that are associated with known genes

Some human diseases that follow simple Mendelian genetics were identified early in the twentieth century. Genes for some other diseases have been found recently by using DNA marker systems such as RFLPs. In this section we consider some hereditary diseases whose underlying genes are known. In some cases the mechanism by which the gene mutation produces the disease is known, but for others, although the full DNA sequence of the gene and its mutations may be known, the mechanism by which these result in disease is not.

Alkaptonuria. **Alkaptonuria** is a rare condition in which a patient's face and ears may be discolored and in which their urine turns black upon exposure to air. Archibald Garrod, mentioned earlier in this chapter, tested the urine of these patients and discovered that the color is caused by an acid. We now know that this substance, homogentisic acid, is formed in the course of breaking down the amino acid tyrosine. In most individuals, the homogentisic acid can be broken down harmlessly with the help of an enzyme. However, in patients with alkaptonuria, the necessary enzyme is missing or defective. Garrod realized that an error in an important biochemical (metabolic) process was responsible, and he called this type of condition an **inborn error of metabolism**. Garrod studied the families of individuals with alkaptonuria and two other such conditions, including albinism, and found a common pattern: each of these inborn errors of metabolism was inherited as a simple Mendelian trait, and in each case the lack of a functional enzyme was recessive. Many other inborn errors of metabolism have since been discovered and their biochemical defects identified. Each of these inborn errors is caused by a recessive allele, the product of a DNA mutation that, when transcribed and translated to its protein product, changed a functional enzyme into a nonfunctional one. Alkaptonuria, albinism, and the condition called phenylketonuria, which is described next, all arise from errors in a series of closely related metabolic pathways (Figure 3.10).

Phenylketonuria. **Phenylketonuria (PKU)** is a genetically controlled defect in amino acid metabolism. The amino acid phenylalanine, which is present in most proteins, is normally converted by an enzyme into another amino acid, tyrosine; the tyrosine is then broken down by the pathway shown in Figure 3.10. A defect in the enzyme that usually converts phenylalanine to tyrosine causes the phenylalanine to be processed by an alternative pathway. A product of this alternative pathway accumulates in the blood and in all cells, acting as a poison that causes most of the debilitating symptoms of the disease: insufficient development of the insulating layer (myelin) around nerve cells, uncoordinated and hyperactive muscle movements, mental retardation, defective tooth enamel, retarded bone growth, and a life expectancy of 30 years or less. Thus a change in one gene (and one enzyme) can have many phenotypic consequences throughout the body.

Fortunately for people carrying this genetic defect, a simple test for its presence exists. If phenylketonuria is detected at birth or earlier, it is possible to avoid the symptoms of the disease by greatly limiting those foods that contain phenylalanine (nearly all proteins, including breast

FIGURE 3.10

Biochemical pathways for three inborn errors of metabolism: phenylketonuria, alkaptonuria, and albinism.

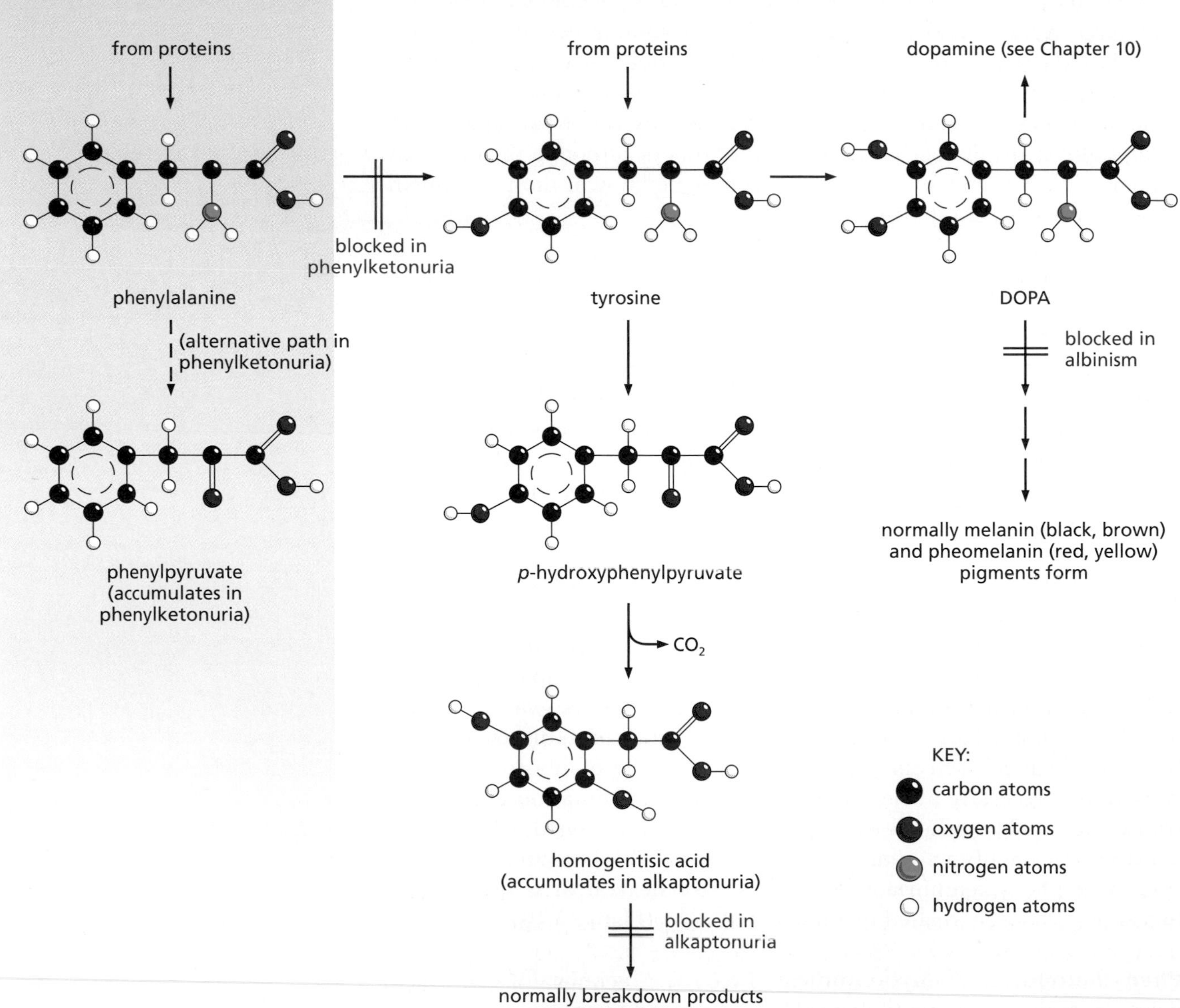

milk), and diet foods and soda containing the artificial sweetener aspartame, which is metabolized to phenylalanine. Small amounts of phenylalanine are essential in protein synthesis (see Chapter 8), but the diet must be carefully monitored to guard against the larger amounts of phenylalanine whose breakdown products would be in toxic amounts that cause the disease symptoms.

Alkaptonuria and phenylketonuria are inherited via single recessive alleles that follow Mendel's laws. Yet the trait is a medical condition, rather than a visible trait like those studied by Mendel. New findings thus broadened the concept of a 'trait.' The concept has been broadened further by the discovery of a genetic basis for diseases that are not just present or absent but that show a range of severity in different people.

Duchenne's muscular dystrophy. **Duchenne's muscular dystrophy** is a sex-linked genetic disorder that causes muscles to become weak and nonfunctional. In most cases, the inability of the muscles of the diaphragm to keep the patient breathing leads to death during the teenage years or in the early twenties. After the gene responsible for this disease was located by RFLP linkage studies, its protein product, dystrophin, was found. So far, this discovery has not led to new therapies for the disease, but the research on dystrophin is greatly adding to our understanding of normal muscle contraction, and it is hoped that this knowledge will lead to effective treatments.

Cystic fibrosis. **Cystic fibrosis** is the most common genetic disease among the white population in the United States and much of western Europe. Cystic fibrosis is characterized by thickened fluids, especially the fluids that line the lungs. Normally, as these fluids are cleared from the lungs, they remove respiratory bacteria, preventing infections. The thicker-than-normal fluids of cystic fibrosis are not cleared from the lungs as they should be, so people with cystic fibrosis have breathing difficulties and frequent lung infections.

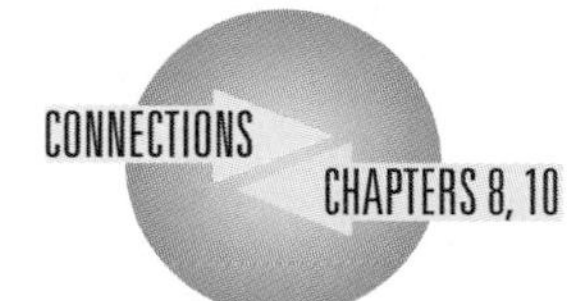

Cystic fibrosis was mapped by linkage studies to a region on chromosome 7 and later a gene was identified. The product of that gene is a membrane transporter, a protein that carries ions, chloride in this case, into or out of a cell. A very large number of different mutations have now been found in the gene for this transporter protein. Some mutations lead to few or no symptoms; some are associated with more severe disease. However, no particular mutation is completely predictive of disease onset or severity. Although identification of the gene and its product has increased our understanding of the disease mechanism, it is still not known how the mutations lead to the many symptoms of the disease.

Huntington's disease. **Huntington's disease** (or Huntington's chorea) is a neurological disorder that begins between the ages of 40 and 50 with uncontrollable spasms or twitches of the hands or feet (see Chapter 10, p. 420). As the disease progresses, the spasms become more pronounced, and the patient gradually loses conscious control of all motor functions and of mental processes. The disease progresses slowly, but is invariably fatal. American song writer and balladeer Woody Guthrie died of Huntington's disease. Although it is always lethal, Huntington's disease does not appear until after its victims have lived through their prime reproductive years, during which they may have passed the mutation to their children. Studies of family trees show that Huntington's disease is inherited as a dominant trait. The gene responsible for Huntington's disease was located on chromosome 4 in 1983. The gene was fully isolated and identified in 1993, but its normal function remains unknown.

Subsequent studies have shown that Huntington's disease is associated with a different type of DNA mutation, called a dynamic mutation. Most mutations are rare, stable, and inherited unchanged from one parent or the other. In dynamic mutations, a three-nucleotide segment is repeated many times, increasing in number during mitosis in different tissues. The number of copies can also change during meiosis, so a child may inherit more copies of the repeat than is present on either parent's chromosome.

The allele that causes Huntington's disease differs from the other alleles of the gene in having many extra repetitions of the three-nucleotide sequence AGC. Persons with fewer than 30 repeats are unlikely to get the disease. Persons with more than 38 repeats are almost certain to get the disease, with the age of onset being younger when there are more than 50 repeats. A test for the allele responsible for Huntington's disease has since been devised, but there is a problem in interpreting it since there is an overlap in the number of repeats associated with a particular outcome. The 'normal' range in number of repeats is 9–37, but some people with as few as 30 repeats have become ill. Because the number of repeats can increase during meiosis, people with 30–37 repeats may pass the disease to their offspring.

Huntington's is one of several diseases, called trinucleotide repeat diseases, associated with dynamic mutations. Another is fragile X syndrome, the most common type of hereditary mental retardation, in which the repeat is CCG in a gene on the X chromosome. The gene has been identified but its normal function remains unknown.

Genes increasing susceptibility to disease. The transmission of human traits controlled by single genes follows the rules that Mendel developed for peas. The situation is more complex when we study phenotypic traits such as height or skin color that are controlled by many genes at once and that are also influenced by environmental variables such as nutrition.

Genes have now been found that do not lead directly to a trait (such as a disease) but increase the probability that some trait or disease will develop. Having a particular allele of these genes does not mean that a person is certain to get a particular disease, only that the likelihood of their doing so is higher than that in people with other alleles. These genes, like all genes, code for proteins. The proteins associated with susceptibility are often regulatory proteins that change the body's response to some environmental factor. The genes have sometimes been called susceptibility genes, although 'genes associated with increased susceptibility' is a more accurate description. The genes associated with predisposition to cancer, discussed in Chapter 9, are examples. As we mentioned earlier for other multigene traits, we must be cautious not to think of these 'susceptibility genes' as simple, either/or, Mendelian alleles.

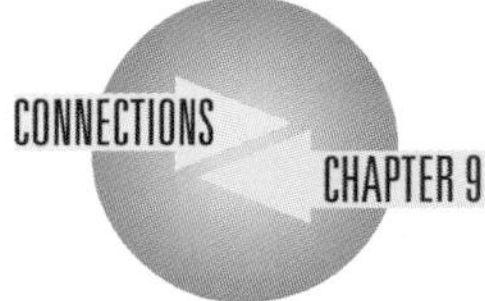

THOUGHT QUESTIONS

1. What is the difference between a genetic marker and a gene?
2. Is research on genetic diseases important for society even if it does not lead to new methods of treatment?
3. With our thinking influenced by genetic diseases like PKU or Huntington's disease, we have been accustomed to thinking of 'mutations' as 'defects.' Not every mutation is a defect, however. With genetics research now turning to the identification of disease susceptibility, it is likely that we will all turn out to have a hereditary predisposition to something. Do new research directions necessitate reconceptualizing the term mutation as 'variation,' rather than 'defect'?

MOLECULAR TECHNIQUES HAVE LED TO NEW USES FOR GENETIC INFORMATION

Molecular biology is an interdisciplinary field that focuses on DNA. Although there are many other kinds of molecules, molecular biologists are mostly concerned with DNA. Molecular biology techniques are telling us a lot about human genetics, as we saw in the previous section. These techniques have also turned out to have wider applications. Many of the techniques and their applications have developed as a result of a research effort called The Human Genome Project.

The Human Genome Project

The complete hereditary material of an entire organism is known as its **genome**. There are an estimated 50,000 to 100,000 human genes in a genome of 3 billion nucleotide base pairs. Some types of heritable human variation have been known for centuries, before their genetic basis was known. Many more human genes have been identified throughout the twentieth century, but the chromosomal locations of these genes were not always known. The Human Genome Project was first proposed in 1986 as a program to make a genetic map, or catalogue, of a prototypical human, including the chromosomal location of all human genes and the complete DNA sequence of the genome.

It should be noted, however, that the DNA sequence of each person is unique. There is no one DNA sequence that is representative of every human, just as no one person could be said to represent all humans in any other method of describing people. It is estimated that one person differs from another in about 0.1% of the 3 billion base pairs. People share the same genes but the nucleotide sequences of those genes vary in different alleles. There is even more sequence variation in the regions of the genome that are not genes. When people are grouped in different ways (by race, ethnicity, or sex, for example), there is always more genetic variation within a group than there are differences between groups. (The topic of the genetics of populations is covered in greater detail in Chapter 5.)

The Human Genome Project was funded by the U.S. Congress to begin work in the fall of 1989, and James Watson, co-discoverer of the double helix structure of DNA, was appointed as the first director. Watson stated his belief that the Human Genome Project will tell us what it means to be human.

Sequencing the human genome. One of the goals of the Human Genome Project is to determine the human DNA sequence. When we read in the newspaper or hear on TV about the genome being sequenced, what does this mean? The 'sequence' of DNA is the order in which the four nucleotide bases (see Chapter 2, pp. 56–59) appear from one end of the DNA molecule to the other. Because DNA is an unbranched molecule, the sequence of bases can be 'read' from one end to the other.

Because the amount of DNA in even one chromosome is enormous, it is not practical to work with the whole length of a chromosome in determining sequences. The maximum size of pieces that can be sequenced is currently 500–700 bases long. The chromosomes are therefore cut into pieces with restriction enzymes (as described below), and bacteria are used to make large quantities of one piece at a time. The DNA to be sequenced is used as a template for synthesis of new DNA strands, as outlined in Figure 3.11.

Each of the pieces is then separated by electrophoresis as shown in Figure 3.11. The pieces are made visible by dyeing them with a fluorescent dye; unlike the specific DNA probes used with RFLPs, fluorescent dyes make all of the pieces visible. The sequence of bases in the DNA fragment can thus be read from the gel: the base found at the end of the shortest piece is first, followed by the base found at the end of the next longer piece, and so forth.

Mistakes can occur in either copying or sequencing, and repeating the process does not always give the same answer, so the technique must be repeated several times by different laboratories until a consensus sequence is established. After the sequence of each piece has been determined, the pieces must be arranged in their original order to get the overall sequence. This latter task is done by huge computers.

Genes are those DNA sequences that are transcribed and then translated into proteins. Many scientists and physicians think that many medical and other benefits could flow from knowing the location and sequence of all the genes. These genes, however, constitute only about 3% of the entire genome in most species; most of the chromosomal DNA is in the regions that do not code for genes, and the Human Genome Project includes the sequencing of these noncoding regions. The nongene DNA consists of 'spacer' sequences that are never transcribed or other kinds of sequences that are transcribed but never translated. The function of most of these nongene sequences is currently unknown, and the wisdom of spending an estimated $15 billion on their sequencing is a question on which opinion, even among scientists, differs widely. These noncoding regions, however, have turned out to be the locations of the various DNA markers, such as the RFLPs discussed earlier, that have allowed us to find where specific genes are located. Other scientists suggest that these noncoding regions will also turn out to be important for other reasons. For example, the noncoding regions are the binding sites for proteins, such as the SRY protein, that regulate DNA folding, and thus regulate when a gene is transcribed.

FIGURE 3.11

Sequencing DNA.

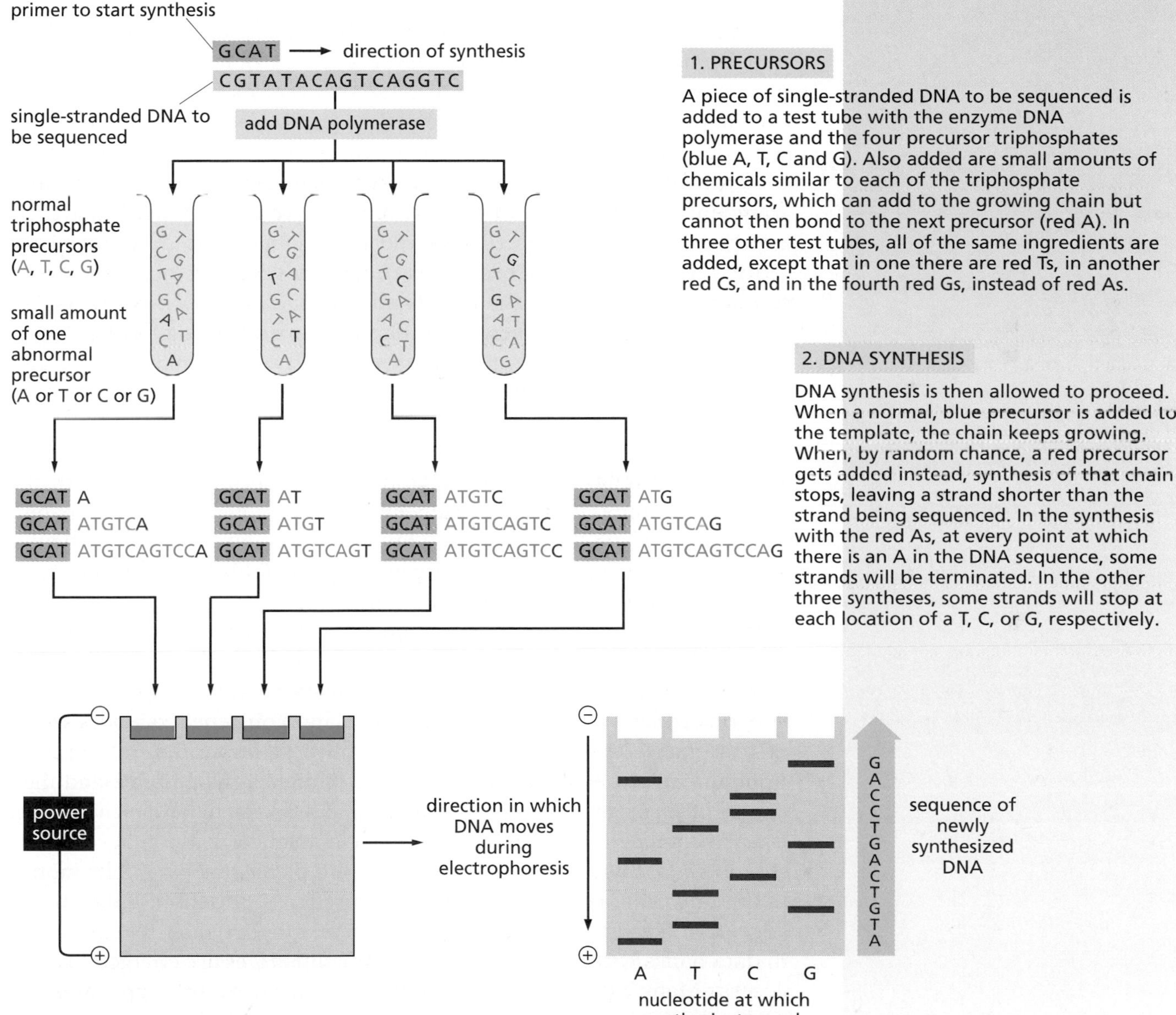
primer to start synthesis
GCAT
direction of synthesis
CGTATACAGTCAGGTC
single-stranded DNA to be sequenced
add DNA polymerase
normal triphosphate precursors (A, T, C, G)
small amount of one abnormal precursor (A or T or C or G)
1. PRECURSORS
A piece of single-stranded DNA to be sequenced is added to a test tube with the enzyme DNA polymerase and the four precursor triphosphates (blue A, T, C and G). Also added are small amounts of chemicals similar to each of the triphosphate precursors, which can add to the growing chain but cannot then bond to the next precursor (red A). In three other test tubes, all of the same ingredients are added, except that in one there are red Ts, in another red Cs, and in the fourth red Gs, instead of red As.
2. DNA SYNTHESIS
DNA synthesis is then allowed to proceed. When a normal, blue precursor is added to the template, the chain keeps growing. When, by random chance, a red precursor gets added instead, synthesis of that chain stops, leaving a strand shorter than the strand being sequenced. In the synthesis with the red As, at every point at which there is an A in the DNA sequence, some strands will be terminated. In the other three syntheses, some strands will stop at each location of a T, C, or G, respectively.
GCAT A
GCAT ATGTCA
GCAT ATGTCAGTCCA
GCAT AT
GCAT ATGT
GCAT ATGTCAGT
GCAT ATGTC
GCAT ATGTCAGTC
GCAT ATGTCAGTCC
GCAT ATG
GCAT ATGTCAG
GCAT ATGTCAGTCCAG
power source
direction in which DNA moves during electrophoresis
GACCTGACTGTA
sequence of newly synthesized DNA
A T C G
nucleotide at which synthesis stopped

3. ELECTROPHORESIS

The pieces can then be separated by size using electrophoresis. In the time that the current is on, the fragment that consists of the primer plus a single nucleotide (A in this illustration) will travel the furthest. The fragment that is the primer plus two nucleotides (A + T) will travel not quite as far and so forth.

4. READING THE SEQUENCE

None of the pieces can be seen, but they can be made to show up with a fluorescent dye. Any band in lane A is DNA with a sequence that ends in A. Any band in lane T is DNA with a sequence that ends in T. By reading the bands from the bottom of the gel (the smallest piece) up to the top of the gel, the sequence can be deduced. The sequence shown in the green arrow is complementary to that of the original DNA strand.

Mapping the human genome. Another goal of the Human Genome Project is to map the genome. Mapping a species' genome means identifying the chromosomal location of each gene and the order of the genes relative to one another. Just determining the sequence of a piece of DNA does not tell you its location in the genome. In an earlier section, we mentioned the use of linkage maps in finding the location of genes associated with diseases. The molecular techniques developed as part of the Human Genome Project have accelerated the mapping and identification of genes.

A few chromosomal features are visible under the microscope. These had been used for decades as markers for chromosomal locations of genes (see Chapter 2). Because there are not very many of these markers, the maps devised were never very detailed. Molecular biologists have now developed several new types of markers that are enabling much more finely detailed mapping. The first of these were the restriction-fragment length polymorphisms (RFLPs). More recently devised are the expressed sequence tags (ESTs) and single-nucleotide polymorphisms (SNPs). Maps have now been constructed showing the locations on the human genome of all of the RFLP and EST and other DNA markers. We have already seen how these new map markers are being used to locate genes associated with diseases and other traits. These markers are also used in diagnostic testing (p. 100), and, as we will see in the next section, in the identification of individuals for various purposes.

Ethical and legal issues. Legal and ethical issues associated with the Human Genome Project include questions of ownership and patent rights. Who owns the human genome or the sequence of any particular gene? If a researcher localizes a gene to a particular chromosome, can that researcher patent the information? Can a gene sequence be copyrighted in the manner of a book? Can the genes themselves be patented? Certain biotechnology companies stand to profit greatly from the marketing of gene sequences, tests for gene sequences, or cures for various genetic diseases, but the sharing of information on gene sequences seems at first glance to threaten their competitive position. Several corporations intend to determine as many gene sequences as possible and then copyright them and sell the information at a profit. Other scientists feel that the human genome should be public information, and that scientists should share this information cooperatively, particularly if public money in the form of research grants has been used in production of the knowledge. A middle ground is developing, wherein most sequences are posted in data banks with public access, but sometimes fees are charged for that access. Many other ethical questions are raised by the application of molecular biology techniques to human health and disease. We discuss those questions in the final section of this chapter.

Using DNA markers to identify individuals

Each person has a unique DNA sequence. If it were practical to sequence a whole genome, a person could be definitively identified by his or her DNA. The human genome is far too long for it to be useful for such identification, but the DNA marker techniques that have been so useful in mapping and sequencing gene regions have also proved useful in distinguishing, with a high probability, any person from another except for identical twins. Two frequent uses of this technique are in the identification of suspects in police investigations and in disputes over paternity; an unusual use of this technique helped shed new light on a historical controversy involving Thomas Jefferson, as described below.

Comparison of band patterns. Using the same marker techniques that we saw earlier (see Figure 3.8, RFLPs), geneticists can compare DNA samples from different persons. The samples are cut with restriction enzymes. Pieces are separated by size by electrophoresis and then transferred to a paper material. Radiolabeled probes complementary to known DNA sequences are then used to detect the fragments containing particular variable repeats. These fragments appear as bands, with their location indicating the fragment length. Thus bands at the same position indicate fragments of the same length in samples being compared. If the band patterns are not the same, then it can be stated with certainty that two samples did not come from the same person. In the example from a criminal investigation shown in Figure 3.12, person 1 can be eliminated as a suspect because the band pattern from the evidence is not the same as that from sample 1. The reverse is not true, however; band patterns that are the same are not an absolute guarantee that the samples came from the same individual. What is being visualized are chunks of DNA of variable lengths, not the DNA sequences of the chunks. A score is calculated that indicates how likely it is that a randomly chosen person, other than the one tested, could have the same band pattern.

The likelihood that another, randomly selected person could have the same banding pattern is made very small in two ways. First, the probes selected are those that pick up specific RFLPs or other DNA markers that are rare in a given population. Also, several probes are used, one after another, to produce a composite banding pattern. The probability that the bands produced with just one probe are the same for two people is equal to the frequency of that probe in the population. If more than one probe is used, the probability of both band patterns' matching is equal to the population frequency of the first multiplied by the population frequency of the second.

There are many ways in which the banding pattern can yield flawed or ambiguous results if samples are

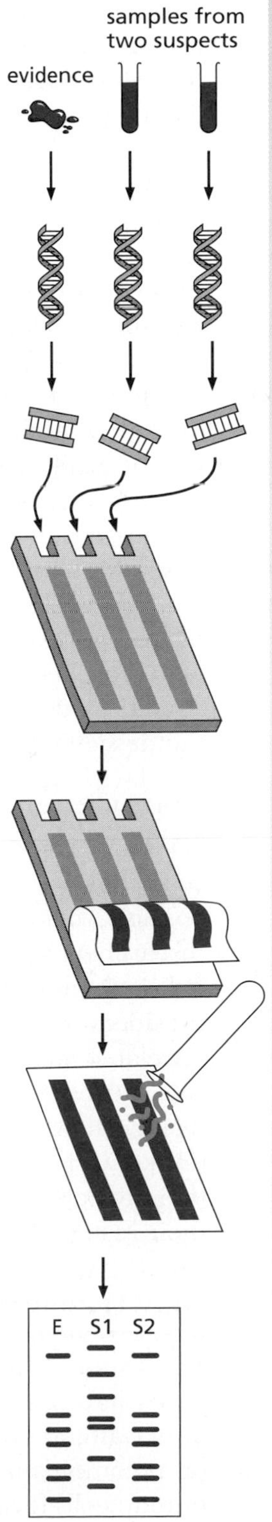

FIGURE 3.12

Forensic DNA technology. In this example, the evidence sample shows the same pattern of bands as DNA from suspect 2. There is therefore a high probability that the DNA in the evidence is from that suspect. The person from whom sample 1 was taken can be eliminated as a suspect.

not properly processed. In samples from crime scenes, there is often DNA from mixed sources, including DNA from several people and from bacteria or fungi. Protein material in the sample may slow the movement of a restriction fragment in the electrophoresis, making the DNA fragment appear as if it were larger than it is. Other chemicals in the samples, such as the dyes in cloth, can interfere with the restriction enzymes cutting the DNA. However, when the tests are done properly and with the proper controls, they can be very reliable. In addition to linking suspects to material taken from crime scenes, the methods can be used to identify dead bodies, or to settle questions of disputed parentage.

Using DNA testing in historical controversies. RFLP markers were used in a recent study that investigated whether Thomas Jefferson could have been the father of children borne by one of his slaves, Sally Hemings. Two oral traditions exist: descendants of Hemings's sons, Eston Hemings Jefferson and Thomas Woodson, believe that Jefferson was their ancestor, while descendants of Jefferson's sister believe that one of her children, Jefferson's nephew, fathered Sally Hemings's later children. Researchers compared Y chromosomal DNA from descendants of two of Sally Hemings's sons with DNA from descendants of one of Jefferson's uncles. No Y chromosomal DNA was available from Thomas Jefferson's direct descendants because he had no sons who survived to have children.

The DNA data show that a set of 19 markers (collectively called the haplotype) is shared by all five of the descendants of Jefferson's uncle who were tested and by Eston Hemings Jefferson. The haplotype is not shared by descendants of Hemings's other son, Thomas Woodson, or by the descendants of Jefferson's nephew, nor was it found in almost 1900 unrelated men. Thus, Jefferson may definitively be ruled out as the father of Thomas Woodson. In the case of the positive match, however, the evidence supports, but does not prove, the idea that Thomas Jefferson could have been Eston Hemings Jefferson's father. As we explained earlier, positive matches indicate probabilities, not definite identity. The researchers state that because "the frequency of the Jefferson haplotype is less than 0.1%," their results are "at least 100 times more likely if the president was the father of Eston Hemings Jefferson than if someone unrelated was the father." They also state that they "cannot completely rule out other explanations of our findings," but that "in the absence of historical evidence to support such possibilities, we consider them to be unlikely." Interestingly, after the authors are very precise in the text of their article, the title, "Jefferson fathered slave's last child," overstates their results (E.A. Foster et al. *Nature* 396: 27, 1998).

Genetic engineering

The same technologies that have allowed mapping and sequencing of genes have also allowed us to alter genomes in various ways. Any such direct alteration of individual genotypes is called **genetic engineering**, or recombinant DNA technology. Human genes can be inserted into human cells for therapeutic purposes (pp. 106–109). In addition, because all species carry their genetic information in DNA, genes can be moved from one species to another. The uses of genetic engineering in plants are discussed in Chapter 14. Here we see some of the applications of genetic engineering for human medicine.

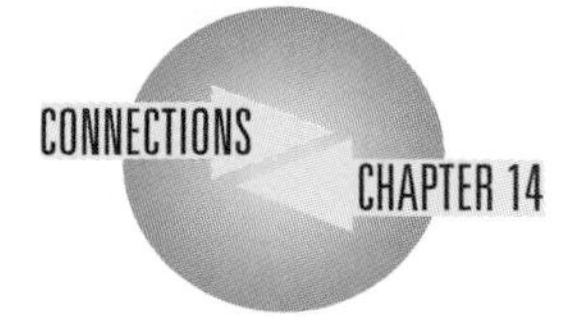

Restriction enzymes. We have previously mentioned **restriction enzymes**, special enzymes used to cut DNA at specific sites. As we have seen, these are used in sequencing and mapping genomes, and in the identification of individuals. There are over 100 restriction enzymes currently known and each cuts DNA at a different nucleotide sequence; these target sites are generally about four to eight nucleotides long (Figure 3.13). Each of these restriction enzymes is a normal product of a particular bacterial species, and most are named for the bacteria from which they are derived. Thus, in Figure 3.13, Hae III is an enzyme from the bacteria *Haemophilus aegypticus* and Eco R1 is from *Escherichia coli*. They are called restriction enzymes because their normal function within the bacteria is to restrict the uptake of DNA from another bacterial species. Each species' restriction enzyme cuts the DNA from other species, but not its own, because its own DNA does not contain the nucleotide sequence that is the target site for its own enzyme.

Several enzymes are known that can break apart a DNA molecule, but an enzyme that acts indiscriminately is of little use in genetic engineering. Restriction enzymes act specifically. Each restriction enzyme generally cuts a sample of DNA in several places, wherever the sequence contains the target site, forming a series of pieces called **restriction fragments**. A given restriction enzyme mixed with the same sequence of DNA always produces the same number of fragments. The length of the pieces may vary if there are variable repeat sequences, for example, but the number of pieces and the places cut are always the same. Before the discovery of restriction enzymes, breaking chromosomal DNA into pieces was done mechanically, producing different numbers of pieces every time the procedure was done, making the results of sequencing and other DNA techniques impossible to reproduce from one experiment to the next. Because restriction enzymes always cut at the same sites, they can be used in genetic engineering.

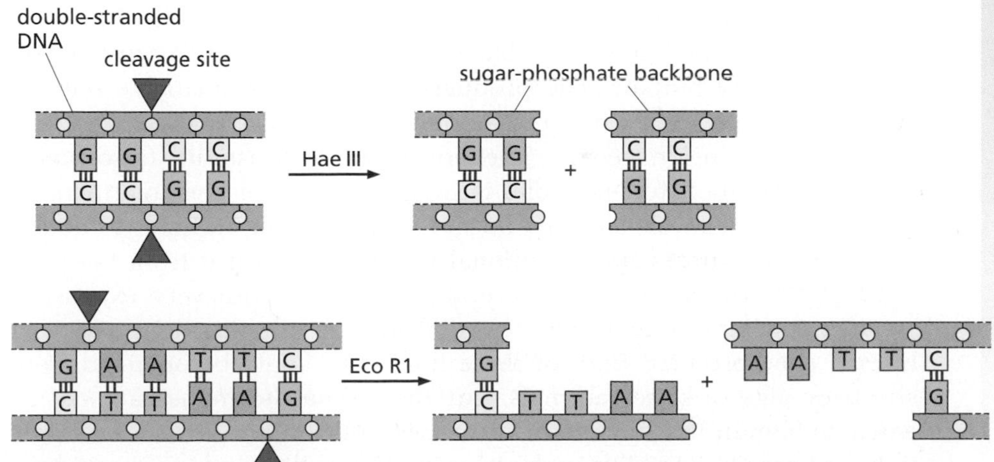

FIGURE 3.13

Restriction enzymes. The nucleotide sequences recognized and cut by the restriction enzymes Hae III and Eco R1 are shown.

The Hae III target site is 4 bases long and the enzyme cuts the DNA strands at sites directly across from each other, leaving double-stranded ('blunt') ends.

The Eco R1 target is 6 bases long and it cuts the DNA between G and AATTC. The sites on the two strands are not directly across from each other, leaving short single-stranded ('sticky') ends.

Restriction enzymes in genetic engineering. The first step in inserting a gene into a genome through recombinant DNA technology is to use a restriction enzyme to snip out a desired segment of DNA. Each restriction enzyme cuts the DNA at specific places, defined by their DNA sequences. The most useful restriction enzymes are those that cut the two DNA strands at locations that are not directly across from each other, producing short sequences of single-stranded DNA known as **sticky ends** (see Figure 3.13). For example, the commonly used restriction enzyme Eco R1 always targets the sequence GAATTC, cutting it between G and AATTC. The complementary strand (opposite GAATTC) also reads GAATTC, so this short sequence is a **palindrome**, a sequence that reads the same on either strand. The Eco R1 enzyme recognizes and cuts the GAATTC sequences on each strand, breaking the two-stranded sequence into fragments that have sticky ends. The ends are called 'sticky' because they stick together spontaneously to form complementary double strands. In fragments cut with Eco R1, the single-stranded AATT sequences can pair with one another, stick together, and then join permanently. (An enzyme such as Hae III that cuts at sites directly across from each other forms 'blunt', rather than sticky ends, as shown in Figure 3.13.)

With a particular restriction enzyme that produces sticky ends, a fragment cut can be replaced with another fragment cut with the same enzyme. This makes it possible to use restriction enzymes to cut a DNA sequence and insert a functional gene with matching sticky ends. (Restriction enzymes that produce blunt ends are useful in DNA sequencing and for RFLPs, but are not useful for genetic engineering because the fragments cannot be put back together.)

Cutting an entire chromosome with a restriction enzyme produces many fragments, only one of which contains the gene to be isolated. A DNA probe specific for that gene isolates the fragments containing the gene of interest. As we have seen before, such a probe is a complementary DNA strand that is radioactive. The probe allows geneticists to isolate the radioactive sequences, then separate the desired genes from the DNA probes that paired with them.

A functional gene isolated in this way can then be inserted into another genome that has been cut with the same restriction enzyme. So far, most genetic engineering of human genes has involved the introduction of these human genes into bacteria. The reasons for this are largely practical: many human gene products are useful in medicine but are more readily produced in large amounts inside genetically engineered bacteria than inside people. The hormone somatostatin, (also called growth hormone), for example, is highly valued for the treatment of certain types of dwarfism. The hormone is, however, difficult to obtain from human sources (the traditional way is to extract it from the pituitary glands of dozens of cadavers) and is therefore very expensive. **Insulin**, the hormone needed by diabetics, is another example of a human gene product. Both of these hormones could be obtained from sheep or pigs or other animals, but the animal hormones are not as active in humans as the human hormones, and some patients are allergic to hormones obtained from other species. Genetic engineering provides a cost-effective way of manufacturing large amounts of these human hormones in bacteria.

Genetically engineered insulin. Human insulin was the first commercially produced genetically engineered product. The first step in making genetically engineered insulin is to grow human cells in tissue culture. DNA extracted from the cell nuclei is then exposed to a restriction enzyme, and the same restriction enzyme is used on nonchromosomal DNA molecules, called **plasmids**, from the recipient bacteria.

Bacteria have a single chromosome and many also have a number of plasmids, short circular DNA pieces that detach from the bacterial chromosome, exist on their own for a long while, and then become incorporated into the bacterial chromosome. Plasmids are used in genetic engineering because, being short, they have fewer sites at which a given restriction enzyme can cut. Cutting a DNA sequence in the plasmid with the same restriction enzyme that was used on the human DNA creates sticky ends that match the DNA fragment taken from the human cell. This allows incorporation of the human gene for insulin identified by a DNA probe into a bacterial plasmid. After the plasmid has been taken up by the bacteria, it inserts itself into the bacterial chromosome. In most cases, the plasmid either contains or is given another DNA sequence that can be used to select those bacteria that have incorporated the plasmid. For example, the plasmid might contain the gene for an enzyme that gives the bacteria resistance to a common antibiotic; the antibiotic can then be used to select the bacteria that have incorporated this gene while killing the majority that are still susceptible. The procedures sound easy and straightforward, but each step of the process is technically difficult and only a small proportion of the attempts succeed.

The bacteria can now be **cloned**, that is, allowed to multiply asexually, which produces vast numbers of genetically identical copies of itself. The resultant bacteria then transcribe and translate the human gene to produce human insulin (Figure 3.14). Because DNA has the same chemical structure in all species, a gene from one species can be used in mRNA and protein synthesis by the cells of another species. The human insulin extracted from these bacteria, called recombinant human insulin, can be given to diabetic patients.

Other spin-off technologies

In addition to the new applications just mentioned, money that has gone into funding for the Human Genome Project is driving the development of new computer technologies, and with them, new fields of employment. DNA sequencing and mapping would not have been practical before the advent of large computers. Although the techniques for determining sequences of short pieces of DNA are rather simple (see Figure 3.11), finding the overlaps that indicate how the small sequenced pieces were originally arranged requires massive computer power. Then, when the longer sequences have been determined, storing the data has necessitated the development of larger and larger computer databases and new methods for searching them. Comparing sequences of genes from one species to another, or comparing new sequences with those already known, is providing the impetus for the development of new types of computer software. The need for people who are trained in both molecular biology and computer science who can work with these data has led to a fast-growing new field of employment called bioinformatics.

Just as the NASA space program led to many unexpected 'spin-off' technologies in the 1960s and 1970s, the Human Genome Project is doing so as well, with new computer technologies and genetic engineering having wide applications outside genetics.

FIGURE 3.14

Production of genetically engineered insulin.

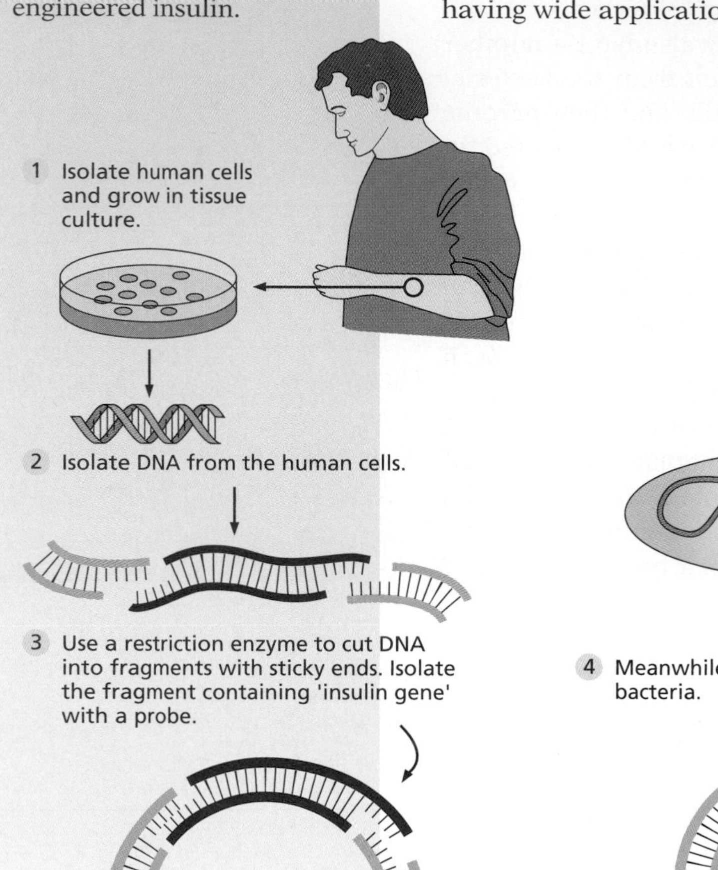

1 Isolate human cells and grow in tissue culture.
2 Isolate DNA from the human cells.
3 Use a restriction enzyme to cut DNA into fragments with sticky ends. Isolate the fragment containing 'insulin gene' with a probe.

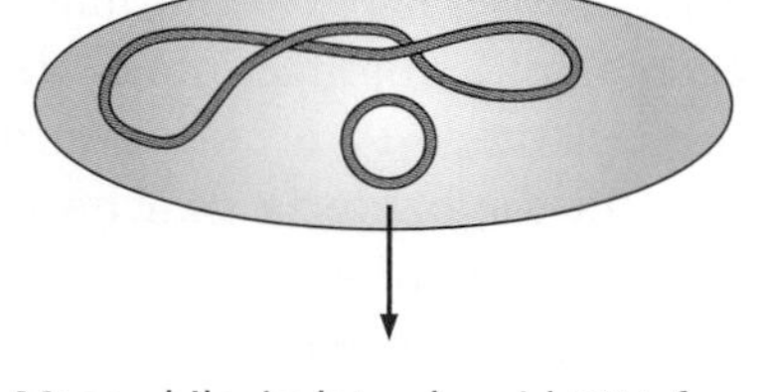

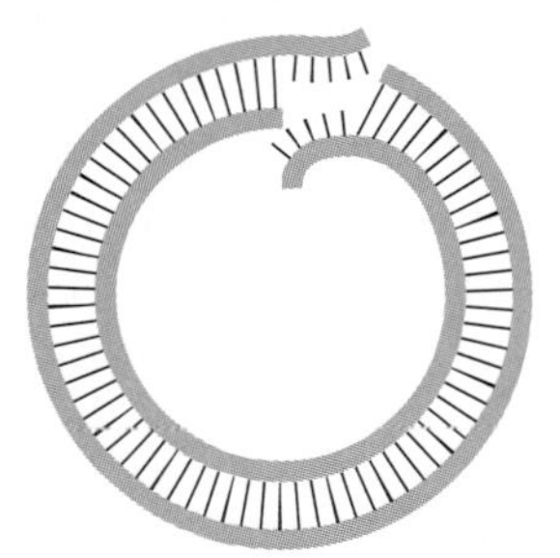

4 Meanwhile, isolate plasmid DNA from bacteria.

5 Use the same restriction enzyme to cut the plasmid DNA, creating matching sticky ends.
6 Combine plasmid and human DNA; some of the plasmids will recombine with the human DNA fragment containing the insulin gene

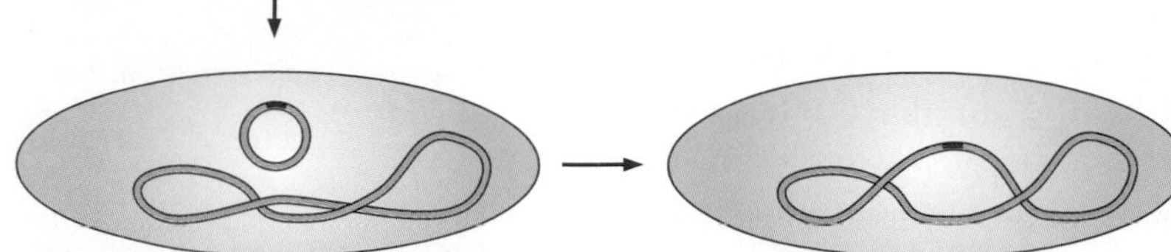

7 Allow new bacteria to incorporate the recombinant plasmid into the chromosomal DNA, then screen bacteria to find the ones that have incorporated the human gene for insulin.

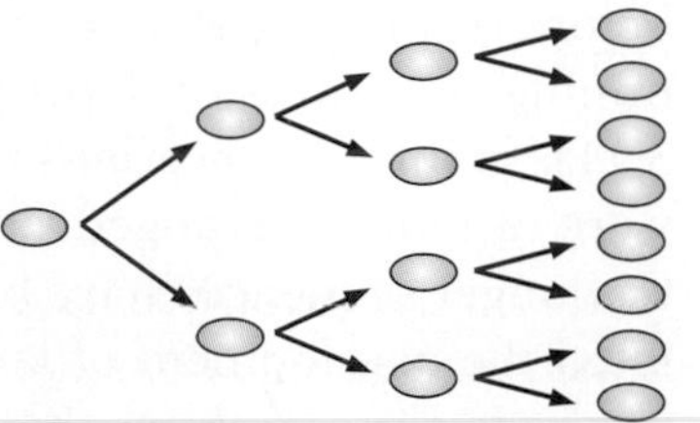

8 Grow trillions of new insulin-producing bacteria.

THOUGHT QUESTIONS

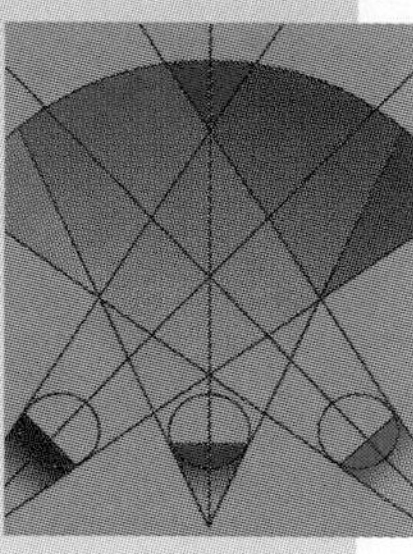

1. In what ways are humans poor subjects for genetic research? In what ways are humans good subjects? Which of your reasons are purely biological, and which have ethical components?
2. Why are certain traits studied in some species and not others?
3. If only stretches of DNA 500–700 bases long can be sequenced at a time, how many of these small sections of DNA must be sequenced to determine the sequence of the entire human genome? (Think also about the overlaps required to piece the sequences together; assume an average of 10% overlap.)
4. The use of growth hormone for the treatment of shortness (not dwarfism) in otherwise healthy children is controversial, but its testing for this purpose was approved in 1993 by the Food and Drug Administration. When does a phenotypic condition unwanted by its bearer become a disease to be treated? Who decides? Is gene therapy to increase someone's height simply another form of cosmetic surgery, similar to breast implants or face-lifts?
5. If a person dissatisfied with his or her phenotype suffers from lack of self-esteem on that account, does the lack of self-esteem justify a procedure to correct the phenotype? (This same argument is raised to justify traditional forms of cosmetic surgery.) Do parents have the right to anticipate for a child what the future effects on self-esteem will be with and without corrective procedures? For a phenotype such as height that develops over a period of years, at what age is it appropriate (if ever) to evaluate the phenotype and decide upon corrective measures?
6. To what extent do you agree with Watson's statement that sequencing the human genome will tell us what it means to be human? Suppose you knew the exact gene sequence of part or all of your genome, what would you really know about yourself?

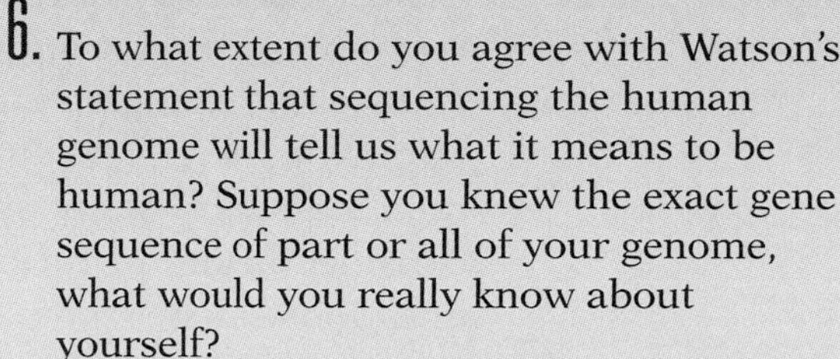

7. Will the DNA sequence of the human genome tell us what traits are controlled by each part of the sequence? Will it tell us which sequences represent genes and which sequences represent spacers?

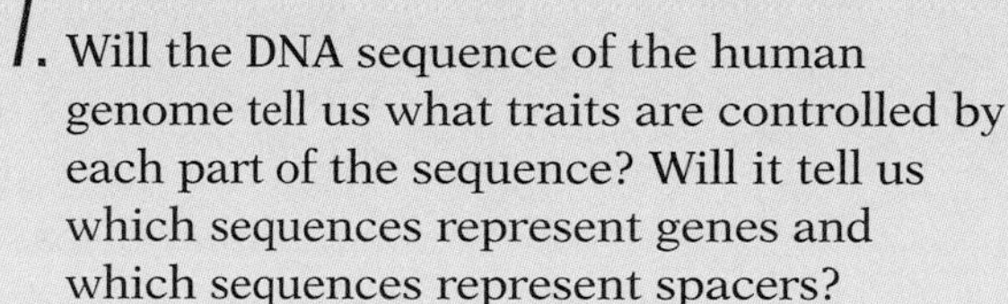

8. If you have a certain rare genetic condition, and scientists use cell samples from your body to determine the gene's DNA sequence, what rights (if any) does this give you to the information? Do the scientists have the right to publish your gene sequence, or any part of it? Is it an invasion of your privacy? Can the scientists sell the information? If they do, are you entitled to a share of the profits?
9. Thomas Jefferson had daughters who survived to have children. Why was the DNA of their descendants not used in the study to determine the paternity of Eston Hemings Jefferson and Thomas Woodson?
10. The authors of the Jefferson study state that they "cannot completely rule out other explanations of our findings." What other explanations are biologically possible?

GENETIC INFORMATION CAN BE USED OR MISUSED IN VARIOUS WAYS

After a genetic basis has been identified for a particular trait, what happens next depends in a large part on the values that individuals and society place on that trait. We have considered in this chapter many traits that at least some people consider undesirable, although not all of them impair health or longevity. In many cases, but not others, we can identify a specific gene that fails to make a functional protein of some sort. These are often called genetic defects, a category that includes all inborn errors of metabolism such as those shown on p. 86. The term genetic defect thus means that a specific allele and its product are defective; it does not mean that the person bearing the gene is defective. For this reason, many people prefer terms such as genetic disease or genetic condition, rather than genetic defect.

Humans can deal with hereditary conditions and hereditary risks in many ways. We can conveniently describe four broad categories of response: gathering and sharing information through genetic testing and counseling, changing individual genotypes, changing the gene pool at the population level, and changing the balance between genetic and environmental factors. Many of the methods in these four categories, which we consider below in turn, raise important ethical questions. The ethical questions can be summarized as follows.

- Who decides who should be tested?
- Who has access to the results of the test?
- What are the responsibilities of a person who carries a gene for a hereditary disease?
- Do we have a responsibility to maintain genetic diversity?
- Who determines what traits (if any) are called 'defects'?

Think about these ethical issues as you read the rest of the chapter.

Genetic testing and counseling

Advances in medical genetics have led to better ways of detecting genetic diseases and to ways of detecting them earlier. Identification of chromosomal variations, mutated alleles of genes, or products of mutated alleles may allow the detection of a disease at the earliest possible stage.

Prenatal detection of genetic conditions. Some conditions can be detected before birth, *in utero* (literally, 'in the womb'). From conception until about the eighth week of pregnancy, a pregnant woman is carrying an embryo; from the eighth week until birth, the term fetus is used rather than embryo. In the technique of **amniocentesis** (Figure 3.15A), a small amount of fluid (amniotic fluid) is withdrawn from the sac in which the fetus is developing in the mother's uterus (see Chapter 6, p. 249). The

FIGURE 3.15

Techniques for prenatal detection of genetic conditions.

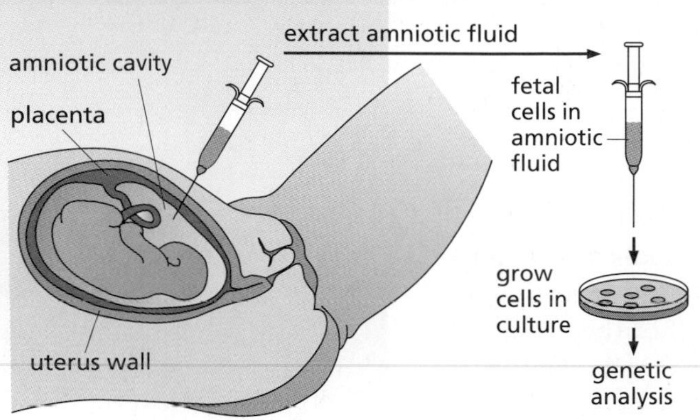

fluid itself is analyzed for the presence or absence of certain enzymes that might indicate a genetic defect in the fetus. Also, amniotic fluid usually contains cells that have been shed from the surface of the fetus. Growing the cells in the laboratory can reveal additional information.

Instead of amniocentesis, **chorionic villus sampling** is sometimes used for prenatal detection. This technique is a type of biopsy (removal of living tissue for examination) of the placenta (Figure 3.15B). The placenta is the structure by which the fetus attaches to the wall of the uterus (see Chapter 6, p. 251 and Chapter 11, p. 458). Because the part of the placenta biopsied is tissue derived from the fetus, not from the mother, the cells sampled by this technique are fetal cells. Chorionic villus sampling can be performed earlier in pregnancy than amniocentesis can. Certain low but nonzero risks are associated with amniocentesis or chorionic villus sampling, including a risk of mechanical injury to the growing fetus and a risk that the pregnancy will be prematurely terminated. Because of these risks and other reasons, these tests are not performed routinely on every expectant mother.

After fetal cells have been obtained by either amniocentesis or chorionic villus sampling, chromosomes from these cells are analyzed for evidence of Down's, Turner's, or Klinefelter's syndromes. Also, if there is a reason to study a particular gene, its sequence can be determined by comparing it with a known sequence for that gene. This cannot be done on the minute amounts of DNA that exist in the cells unless these amounts are first increased, or amplified. Amplification of DNA is accomplished by the **polymerase chain reaction** (**PCR**) (Figure 3.16). The polymerase chain reaction is often used to detect genetic conditions by using DNA from eight-cell embryos prior to implantation. These embryos are derived from *in vitro* fertilization (literally 'in glass'), meaning that the fertilization of the egg by the sperm took place in laboratory glassware rather than inside the body (*in vivo*). One of the eight cells can be removed for genetic testing and the other seven can be implanted into a woman's uterus to grow to term.

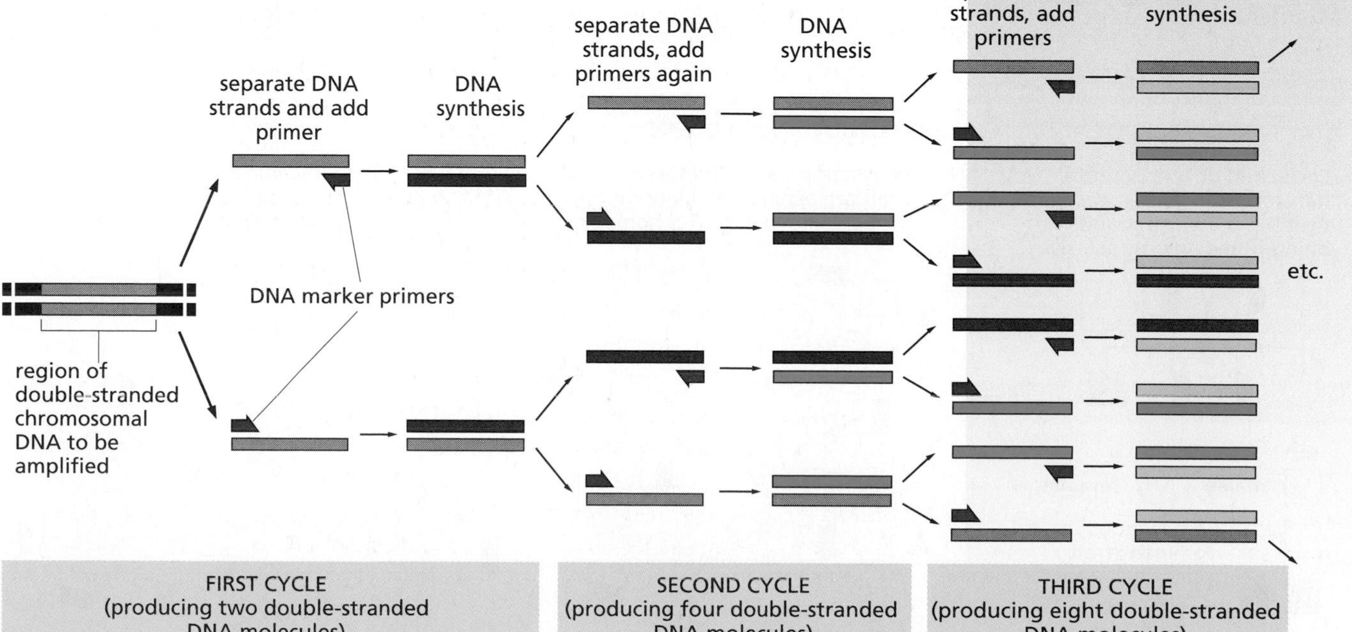

FIGURE 3.16

The polymerase chain reaction (PCR).

Several dozen genetic diseases are now detectable through prenatal tests, including Tay–Sachs disease, cystic fibrosis, and phenylketonuria.

Testing newborns or adults. Other tests are done on newborns or on adults. Tests that are simple and inexpensive can be used for mass screening. For example, many hospitals routinely screen all infants at birth for phenylketonuria (by a blood test to detect the amino acid phenylalanine, not by a DNA test). Such screening is considered ethical because it is done on all infants and it is a clear benefit to the infant for the information to be known. Tests that detect heterozygosity for defective alleles (e.g., testing for carriers of the allele causing sickle-cell anemia; Chapter 5) can be performed on adults before they become parents. Persons undergoing this type of screening must first give their **informed consent** and sign a form stating that they understand the nature of the test, the possible outcomes (including the conditions that the test can detect and the likelihood that the genotype will result in a disease phenotype), the possible risks of the procedure, and the possible benefits.

CONNECTIONS CHAPTER 5

People in these situations often consult a genetic counselor to help them understand the test and the risks before giving their consent. For example, they can be advised that if they are identified as being heterozygous and they have children with a person also heterozygous for the same allele, each of their offspring has a 25% chance of being homozygous for the recessive condition. (To understand why, refer back to Figure 3.7B.) The extent to which homozygosity predicts disease severity differs with the disease.

Who should be tested? Genetic testing is expensive and it would not be reasonable to test everyone for all genetic disorders. So an effort is made to identify persons at higher risk of having certain defective alleles. Figure 3.17 shows two ways in which risk is estimated. First, family history identifies persons at higher risk and may prompt testing because having a family history of a genetic condition increases the probability that a family member carries the defective allele (see Figure 3.17A). However,

FIGURE 3.17

Two ways of assessing risk for a recessive trait with single-gene, simple Mendelian inheritance.

(A) FAMILY PEDIGREE

Each child of two parents heterozygous for a recessive trait has a 25% probability of being homozygous recessive

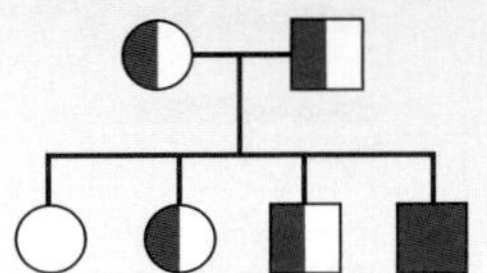

KEY:

males females

homozygous for dominant trait

homozygous for recessive trait

heterozygous

(B) POPULATION FREQUENCY

If 1% of a population expresses a trait known to be recessive (meaning that those who express the trait are assumed to be homozygous), 18% can be assumed to be heterozygous. Out of 100 individuals shown here, 1 is homozygous for the recessive trait and 18 are heterozygous.

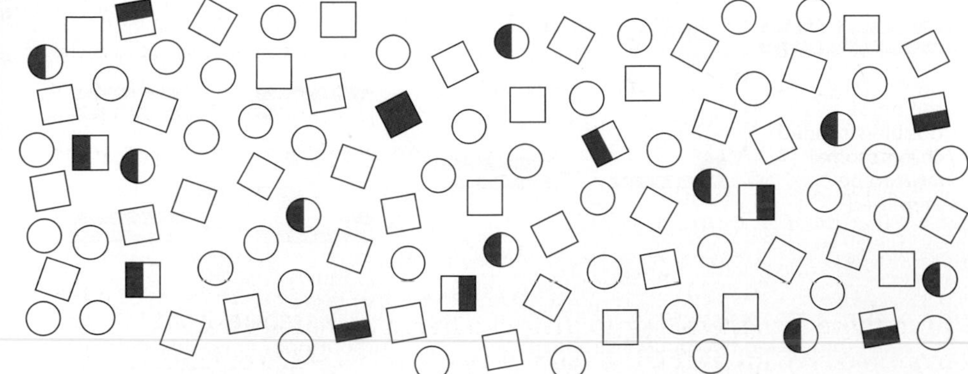

because recessive traits show up phenotypically only when the genotype is homozygous or sex-linked, a recessive trait may not show up in a pedigree and a person may not know the family history. A second way of estimating risk is by the population frequency of a trait (that is, how many people have the trait in a given population). The likelihood of carrying a recessive allele for a particular condition is higher for a person from a population in which the allele is more frequent (Figure 3.17B). Diagnostic testing for a particular genetic trait is sometimes recommended specifically to persons from those populations or ethnic groups known to have a greater prevalence of the trait. When the frequency of an allele is higher in a particular group, the probability of having an offspring homozygous for the trait is higher if both parents are from the same group, than if one marries outside the group.

Examples of genetic testing for recessive alleles because of within-group risk are:

- Many African-Americans now seek testing to see whether they carry a sickle-cell allele because the frequency of sickle-cell anemia is higher among African-Americans (see Chapter 5, pp. 214–216).
- People of Mediterranean or southeast Asian descent may seek testing to see whether they carry an allele for thalassemia (see Chapter 5, pp. 217–218) because the frequency of this disorder is higher in these groups.
- Ashkenazi Jews (those of eastern European descent) commonly seek testing for the recessive allele that causes Tay–Sachs disease, a fatal disorder of brain chemistry, because the frequency of the disease is higher in their group.
- People of western European (especially Irish) descent may seek testing for mutations in the membrane transporter gene responsible for cystic fibrosis because the frequency of the disease is higher in this group.

Each of these four diseases is a single-gene trait with recessive inheritance. There may be more than one defective allele for a disease and a range of disease severity depending on the exact mutation in the allele; cystic fibrosis is an example. Therefore, determining by genetic testing that a person is homozygous recessive most often does not tell you the severity of future disease, or even whether disease will actually develop.

People outside any higher-risk group can also inherit each of these diseases but their probability of doing so is lower simply because the frequency of the recessive allele is lower in their groups. These traits are rare even in the higher-risk groups where they are 'more frequent,' far more rare than the 1% shown in Figure 3.17B. Therefore most people, even those in a higher-risk group, are not heterozygous carriers of these rare recessive traits. (The frequency of heterozygous carriers can be calculated from the frequency of the recessive trait by a method we discuss in Chapter 5, pp. 206–207).

Genetic testing of this sort should only be done on a voluntary, informed-consent basis and, in general, only when it is of potential benefit to those being tested or to their children. Community leaders of various ethnic groups, including many clergy, have helped to organize genetic testing programs and have encouraged people to participate.

Using information from genetic tests. When a genotype for a disease is detected, the decision about what to do is left up to the person, or to his or her parents if the person is a child. A patient's decision should be based on a clear knowledge of the possible choices, their consequences, and the extent to which an outcome can or cannot be predicted by the test. Genetic counselors help people to understand these choices but the code of ethical conduct of genetic counselors prohibits them from making a decision on behalf of their clients: clients could rightfully resent any counselor who has pressured them into a decision.

Some decisions that must be made after genetic testing are difficult for the people making them. Couples who know they are at risk of bearing children with a genetic disease may decide to adopt children instead. In other cases, knowledge of a genetic condition permits medical intervention at the earliest possible stages, when chances of successful treatment may be better. For conditions that cannot be treated, some couples may choose to abort the fetus bearing the genetic defect. However, people committed to a pro-life position believe that the potential benefits to those being tested or to any future children can never justify what they view as the murder of a fetus.

Genetic testing has already led to some highly inventive mixtures of tradition and modern technology. The Hasidic Jews of Brooklyn, New York, who are mostly descended from the Ashkenazi Jews of eastern Europe, have a relatively high population frequency of the gene for Tay–Sachs disease. Marriages are traditionally arranged within the Hasidic community, and marriages outside the community are rare; this pattern generally increases the rate at which recessive alleles come together and produce recessive phenotypes. Because they are ethically opposed to all abortions, the Hasidim do not permit genetic testing *in utero*. The availability of a test that detects Tay–Sachs heterozygotes has, however, allowed the Hasidic community to set up a computerized registry under their strict control. Testing of all persons within the community is encouraged, and the results are entered into the registry under a code number that guarantees confidentiality. The registry permits the traditional matchmakers to check potential couples before proposing a match; if both partners are carriers for Tay–Sachs disease, the matchmaker is warned of this fact and the match is never made. Before this registry was set up in 1984, the Kingsbrook Jewish Medical Center in Brooklyn, which serves the Hasidic community, had an average of 13 Tay–Sachs children under treatment at any one time; after just 5 years, the number of Tay–Sachs children under treatment in the hospital dropped to two or three.

The ethics of genetic testing. Genetic testing is sometimes a mixed blessing. If a genetic defect can either be cured or phenotypically suppressed, or if heterozygote detection permits at-risk couples to decide against having children, then a genetic test can be justified on the grounds that it relieves future suffering. However, most genetic defects cannot be cured. What is the point of testing a person for a condition such as Huntington's disease that can neither be controlled or cured? One reason is that it permits people who carry a genetic condition to decide whether or not to have children. Will a person who tests positive for such a genetic disease be denied insurance or employment on the basis of the test results? Will a woman choose to abort a fetus if a genetic disease is detected *in utero*? Box 3.1 examines some of the ethical questions that arise in connection with various forms of prenatal and at-birth testing.

Another ethical issue concerns the use of prenatal screening not for the purpose of detecting a disease-associated allele but to find out whether the fetus is a boy or a girl, something easily determined from examining the chromosomes. Will couples use this technology to select the sex of their offspring? This already happens in India, where abortion is legal and determination of the sex of the fetus by ultrasound is widely available to those who can afford (or can borrow) a fee of a few hundred dollars. A 1988 study of 8000 abortions in India's clinics showed that

BOX 3.1 Ethical Issues in Medical Decision-making Regarding Genetic Testing

Should society influence the private decisions of individuals? To what extent do (or should) financial considerations limit the choices available?

Suppose a child is born with a birth defect or other congenital condition. Is it ever ethical to withhold treatment? (Similar ethical issues are raised by conditions resulting from injuries, infectious diseases, poor maternal nutrition, or other causes.) What if the same disease is diagnosed in a fetus *in utero*—is it ethical to abort the fetus? The decision to abort a fetus or to withhold treatment from a child with a genetic disease raises important ethical questions. Here are some questions to consider:

1. Tay–Sachs disease is a genetically controlled disease whose victims are in constant pain and never survive beyond about 4 years of age. Does it make sense to spend thousands of dollars on the medical care of a child who has no chance of living beyond age 4, or even of enjoying those few years free from pain? Would it make a difference if a few people with the disease were capable of surviving? What if we were dealing with a disease that people could survive, but only with some disability?

2. The involvement of third-party insurance policies raises more issues. Should insurance policies pay for genetic testing? Should insurance policies pay medical expenses for genetic diseases that could have been avoided after screening? Some insurance policies will pay for medical treatment, but not for the testing that might have avoided the need for the treatment. Do you think insurance policies should cover genetic screening?

3. Should genetic screening be covered for certain ethnic groups but not others, just because the risks differ? For example, thalassemia is more prevalent among Italians, Greeks, and certain southeast Asians; should insurance cover testing for this condition in a person of Italian descent, but not in a person of English or Danish descent? Or in the instances common in the United States, in which descent is either mixed or unknown?

4. If a genetic disease is detected during pregnancy, should insurance policies cover termination of the pregnancy if desired by the parents? If parents elect not to terminate a pregnancy, and a child is born with a genetic disease, should insurance policies cover any specialized medical care that might be necessary? Should insurance companies be allowed to deny coverage or raise premiums if a genetic disease is discovered?

5. What role might science have in answering questions of medical ethics that are related to genetics? For example, can science help in assessing the benefits and risks of genetic screening? To what extent does the detection of an allele predict a harmful phenotype? What constitutes sufficient evidence that a condition is genetically determined? What are the limits of the ability of science to contribute answers to these questions of medical ethics?

7997 were female and only 3 were male. In the United States, clinics offering prenatal genetic testing have found that over one-fourth of the couples who come to them are motivated by the possibility of choosing their baby's sex.

As genetic testing becomes more common, it is inevitable that test results will occasionally be misused. In one case, school officials were told that a child needed to be kept on a special diet because he had phenylketonuria (PKU). Although he was functioning normally, the child was placed in a class for the learning disabled. The school officials apparently knew that PKU could cause mental retardation, but were unaware that this outcome could be averted by the special diet. As this case shows, the misuse of genetic information may result from ignorance.

Discrimination by employers or insurers represents another possible misuse of genetic information. Employers may refuse to hire or promote, or insurance companies may refuse to insure, persons who have or who are suspected of having a genetic condition. In some cases, benefits have been denied to heterozygous carriers of recessive conditions or to persons at risk for other reasons whose phenotype was unaffected. The practice is still uncommon, but it is growing and is likely to continue to grow as more and more genetic information becomes available through medical testing. One of the best safeguards against this kind of discrimination is a strict adherence to rules governing the confidentiality of medical records and other personal information held by health care providers and medical testing laboratories. Recent rulings state that the Americans with Disabilities Act protects people from discrimination on the basis of their genetic profile.

The language used by geneticists and genetic counselors may be misleading when only those persons who are homozygous for the nondisease allele are reported as 'normal.' Heterozygotes are called 'carriers' although phenotypically they show no disease. In Figure 3.17A, for example, only the child on the far left of the diagram would be reported as normal, even though 75% of the children and both of the parents are phenotypically normal with no sign of disease.

Altering individual genotypes

Some rare genetic traits impair health. In the future, it may become possible to correct certain genetic defects by direct alteration of the individual genotype. A more realistic possibility is a form of gene splicing in which the functional allele is added to the DNA of persons with defective alleles.

Instead of growing human insulin in bacteria, (see Figure 3.14), gene splicing could theoretically be used to introduce the insulin gene into human cells that do not possess a functional copy. (That would still not cure diabetes unless these cells were also capable of appropriately increasing or decreasing their output of insulin according to conditions.) This type of genetic engineering is called **gene therapy**, the introduction of genetically engineered human cells into a human body for the purpose of curing a disease or a genetic defect.

Treatment for hereditary immune deficiency. Human gene therapy has been used successfully to treat **severe combined immune deficiency syndrome (SCIDS)**, a severe and usually fatal disease in which a child is born without a functional immune system. Unable to fight infections, these children will die from the slightest minor childhood disease unless they are raised in total isolation: the 'boy [or girl] in a bubble' treatment. The enzyme that controls one form of SCIDS has been identified; it is called adenosine deaminase (ADA) and is located on chromosome 20. A rare homozygous recessive condition results in a deficiency of this enzyme, which in turn causes the disease. Gene therapy for this condition consists of the following procedural steps, shown in Figure 3.18.

1. Normal human cells are isolated. The cells most often used are T-lymphocytes, a type of blood cell that is easy to obtain from blood and easy to grow in tissue culture.
2. The isolated cells are grown in tissue culture.
3. The DNA from these cells is isolated.
4. A restriction enzyme is used to cut out a DNA fragment containing the functional gene for ADA and two sticky ends. A probe with a complementary DNA sequence is then used to isolate and identify fragments bearing the gene.
5. The same restriction enzyme is used to create matching sticky ends in viral DNA isolated from a virus known as LASN. This virus was chosen because it can be used as a vector to transfer the gene into the desired host cells. (Other vector viruses have also been used; each virus type varies in the size of DNA fragment that can be inserted and the type of cell that it can enter.)
6. The viral DNA is then mixed with the human DNA fragments and allowed to combine with them.
7. The virus is allowed to reassemble itself; it is then ready for further use.
8. Blood is drawn from the patient to be treated and T lymphocytes are isolated from this blood. These lymphocytes, like all of the other cells from this person, are ADA-deficient because they do not possess a functional ADA allele.
9. The virus is now used as a **vector** to transfer the functional gene. The virus must get the gene not only into the lymphocyte but also into its nucleus. The gene must incorporate into the cell's DNA in a location where it will be transcribable and where it does not break up some other necessary gene sequence.
10. The lymphocytes are tested to see which ones are able to produce a functional ADA enzyme, showing that they have successfully incorporated the functional ADA allele.
11. The genetically engineered lymphocytes are injected into the patient, where they are expected to outgrow the genetically defective lymphocytes because the ADA-deficient cells do not divide as fast as cells with the ADA enzyme.

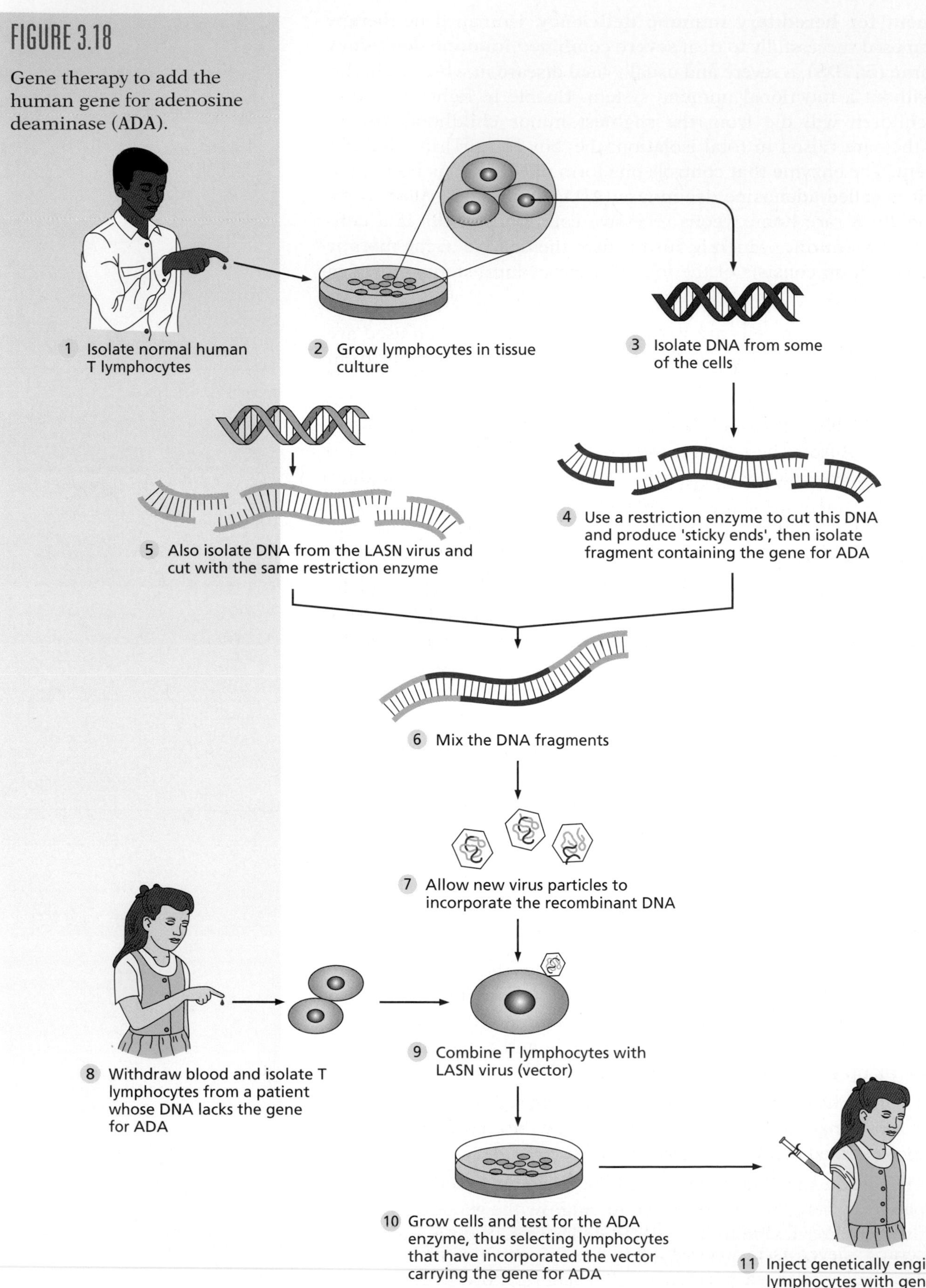
1 Isolate normal human T lymphocytes
2 Grow lymphocytes in tissue culture
3 Isolate DNA from some of the cells
4 Use a restriction enzyme to cut this DNA and produce 'sticky ends', then isolate fragment containing the gene for ADA
5 Also isolate DNA from the LASN virus and cut with the same restriction enzyme
6 Mix the DNA fragments
7 Allow new virus particles to incorporate the recombinant DNA
8 Withdraw blood and isolate T lymphocytes from a patient whose DNA lacks the gene for ADA
9 Combine T lymphocytes with LASN virus (vector)
10 Grow cells and test for the ADA enzyme, thus selecting lymphocytes that have incorporated the vector carrying the gene for ADA
11 Inject genetically engineered lymphocytes with gene for ADA enzyme into the patient

FIGURE 3.18

Gene therapy to add the human gene for adenosine deaminase (ADA).

The gene therapy described above provides a functional gene that is transcribed and translated by the body cells, producing the missing enzyme in lymphocytes. Because lymphocytes are not the only cells that need the ADA enzyme, the patient must also receive injections of the ADA enzyme coupled to a molecule that permits it to enter cells. (This last step might not be necessary for the treatment of other enzyme defects.) The enzyme controls the symptoms of the disease, but it is not a cure because the underlying disease is still present. Gene therapy for ADA was first successfully used on a 4-year-old girl in 1990. A second patient, a 9-year-old girl, began receiving treatments in 1991. Both patients are being closely monitored, and their immune systems are now working properly. However, because the genetically engineered cells are mature lymphocytes, which have only a limited lifetime, repeated injections of genetically engineered cells are needed.

To get around this problem, in the hope of bringing about a more lasting cure, some Italian researchers have tried using both genetically engineered lymphocytes (as described above) and genetically engineered bone marrow stem cells. Stem cells divide to form all the developed types of blood cells (see Chapter 12) and they maintain this ability throughout life. Therefore, after repaired lymphocytes die off, stem cells with repaired DNA could divide to provide new, ADA-functional lymphocytes, possibly for the lifetime of the individual. This type of therapy was begun on a 5-year-old boy in 1992, and since then several other children have received this treatment.

Correcting DNA defects in cells. In the strategy just described, a new gene is inserted into a cell that also continues to carry the mutated allele. Technical difficulties with this approach are numerous. Getting large pieces of DNA into cells (most genes are large) is difficult. Getting them to insert in a location in the genome where they would be normally expressed and regulated is far more difficult.

In contrast, many of the mutations in alleles are very small: changes in single nucleotides or very short sequences. Some scientists are now working on ways to repair the mutations. Once a specific mutation and its location are known, short pieces of DNA with the correct sequence can be attached to some flanking sequences that are complementary to the bases around the mutation. After a cell takes up such a piece, these flanking sequences locate the piece in the genome and align the corrective piece in its proper site. DNA-correcting enzymes cut out the old, mutated piece and splice in the new, correct piece. At present this work remains experimental.

Changing genotypes in gametes. We have been discussing examples of altering genotypes expressed by somatic cells, the diploid body cells that do not give rise to gametes. Whatever happens to these somatic cells, they will eventually die, but the germ cells that are passed on to succeeding generations will still contain the defective genes that produced the troublesome genotypes in the first place. Still in the future is the possibility of gene therapy for the germ cells, those that give rise to gametes. If gene therapy is ever used successfully to repair a genetic defect in germ cells, the result will be a permanent change in genotype that will be transmitted to future generations.

Questions of safety and ethics. There are legitimate safety concerns with human gene therapy. For example, any virus used as a vector must be capable of entering human cells. Might such a virus cause a disease of its own? To preclude this possibility, the viruses used so far in most human gene therapy have been from viral strains with genetic defects that render them incapable of reproducing and spreading to other cells. Might random insertion into the host DNA destroy some other gene? Methods are being developed for directing the insertion location, but it is still largely a random event. In 1999, gene therapy clinical trials were halted in the United States after an 18-year-old died after receiving a viral vector for gene therapy for a metabolic disease. The reasons for his death were not immediately apparent, so clinical studies were halted until issues of safety can be addressed. The boy's father has testified at a U.S. Senate hearing that the boy and his family were not fully informed of the dangers of the experiment. Others have raised ethical objections to the use of the term 'gene therapy' in clinical trials when most of the experiments that have been done so far have not been designed to cure any condition, only to alleviate symptoms (or to test the safety of the procedure itself).

Gene therapy also raises other ethical concerns. The price of genetically engineered proteins, initially high when they were first introduced, has fallen as they have become more widely available. New recombinant DNA procedures continue to be very expensive to develop. This raises ethical issues of fairness: Will the benefits of genetic engineering be available only to those who can afford them? Should government programs provide them through Medicare and Medicaid? Should insurance cover their use? How can society's health care resources best be distributed? If medical resources are limited, should an expensive procedure used on one person take up needed resources that could cover inexpensive treatments of other diseases for many people? These particular questions are not unique to recombinant DNA; they apply to any expensive form of medical treatment.

Recombinant DNA therapy may someday become commonplace in human cells as well as bacteria. In theory, recombination therapy could be practiced either on somatic cells or on gametes. If it were performed on somatic cells, the effects of the recombination therapy would last as much as a lifetime, but no longer. For example, insertion of the functional allele for insulin into the pancreatic cells of patients with diabetes might cure them of the disease, but they would still pass on the defective alleles to their children. A general consensus has been reached that using gene therapy on somatic cells has an ethical value if it is used for the purpose of treating a serious disease.

If successful recombination therapy is performed on germ cells, then the genetic defect will be cured in the future generations derived from those germ cells. In addition to all the ethical questions raised earlier, gene therapy on germ cells raises many additional ethical questions. Most medical ethicists today advise caution and waiting in the case of germ-cell gene therapy on humans until we have more experience with gene therapy on somatic cells or in other species.

Altering the gene pool of populations

Some people have proposed that, instead of treating people one at a time, we should alter the genetic makeup of populations (the entire gene pool) by changing the frequencies of certain genotypes. One difference between this approach and the approaches already described has to do with who is perceived to reap the benefits. Genetic testing, counseling, and the altering of individual genotypes are justified in terms of the pain and suffering that may be spared to individuals. In contrast, all attempts to alter the gene pool carry with them notions of harm or benefit to society rather than to the individual.

Positive eugenics. The altering of the gene pool through selection is called **eugenics**, from the Greek words meaning 'good birth.' This idea is not at all new: Plato's *Republic* (book 5) suggests that the best and healthiest individuals of both sexes be selected to be the parents of the next generation, much as we breed our horses and cattle. Plato's type of eugenics is called **positive eugenics**, meaning an attempt to alter the gene pool by selectively *increasing* the genetic contributions of certain chosen individuals or genotypes. Positive eugenics was also proposed in the twentieth century by the Nobel prize-winning geneticist H.J. Muller, who advocated setting up sperm banks to which selected male donors would contribute. Muller thought that women would eagerly seek artificial insemination with these sperm in the hopes of producing genetically superior children. Several entrepreneurs have established sperm banks (and a smaller number of egg banks) offering to infertile couples (and others) the gametes of people thought to carry desirable traits. The system is not regulated, however, by any public agency, and many sperm banks and egg banks seem to be motivated more by profit than by any desire to change the gene pool. In 1999, a photographer in California began advertising the eggs of several fashion models, offering them at auction to the highest bidder over the Internet.

Ethical and other questions raised by positive eugenics usually center on the lack of an agreed-upon standard for human excellence. The traits most often discussed by those who favor eugenics are intelligence and athletic ability. However, these traits are genetically complex and are highly influenced by education, training, and other environmental variables. Studies attempting to demonstrate a genetic influence on these and other traits were in many cases poorly done, leading many scientists to doubt the existence of any reliable evidence concerning the genetic control of human intelligence and other complex traits.

The complexity of the human genotype raises other issues: What if Einstein had been heterozygous for some genetic disease? If a society wanted to use his germ cells to breed people of superior intelligence, they would also unwittingly be selecting whatever other traits he happened to possess, possibly including a genetic defect in the process. Suppose a person inherited the manic-depressive disorder of Robert Schumann or Vincent Van Gogh, instead of their creative talents? What liability or what responsibility would a sperm bank face if a descendant were born with a genetic defect? What constitutes 'superiority' in an individual, and who should have the power to make such choices?

Cloning. Another possible technology for achieving eugenic aims is the creation of clones. A **clone** is defined as a cell or individual and all its asexually produced offspring (i.e., those produced without sexual reproduction). All members of a clone are genetically identical, except when a mutation occurs. No mammal or other complex animal reproduces naturally by cloning, though several insects and many plants do in some circumstances and a few do so regularly. In 1997, scientists in Scotland succeeded in cloning a sheep; other mammalian species (mice and cows) have since been cloned. In this procedure the nucleus from a fertilized egg (zygote) is removed and the nucleus from a cell of a fully developed individual is inserted in its place. The altered zygote is then implanted in a suitable womb and brought to term. The new individual formed in this way is a genetically identical clone of the individual whose nucleus was used. In theory, cloning could make multiple copies of a desired genotype.

Another type of cloning is the division of a single egg or early embryo into two or more separate embryos, the same process that normally creates identical twins. Offspring from this type of cloning are genetically identical to each other but carry chromosomes from each of the two parents. In 1993, two scientists from George Washington University succeeded for the first time in cloning a human embryo under laboratory conditions (*in vitro*) by stripping away a jellylike protective coating that separated the individual cells of the embryo, and then furnishing a new artificial coating for each cell. The resulting embryos were never implanted in a uterus or brought to term, but coming this close to cloning a human has set off wild speculations and a series of ethical debates. If cloning ever became possible on a large scale, it would result in many of the same ethical problems as positive eugenics. Many people, frightened by these prospects, staged a demonstration near the laboratory where the human cloning experiments were carried out. Various ethicists have pointed out an array of possible misuses or dubious uses of the technology, such as raising a child to serve as a transplant donor for an identical twin born years earlier. Public opinion polls also show strong opposition to unrestricted cloning: 75% of people questioned in one poll disapproved of human cloning, and 58% thought that cloning was morally wrong (*Time*, 8 November 1993, pp. 65–70).

The second type of cloning has already been used to produce genetically identical cattle and other farm animals; the cells of early embryos were separated as described above and implanted. The technique is expensive and has a low success rate. It has been available for over a decade for domestic cattle, and has become an accepted part of breeding practice for these animals, though it has never proved popular; one company founded to take advantage of the practice has gone out of business. Cloning techniques are most likely to be used for valuable animals of known pedigree, such as racehorses.

Negative eugenics. Most discussions of eugenics have centered on **negative eugenics**, the prevention of reproduction among people thought to be genetically defective or inferior. Founded by Francis Galton (1822–1911), a cousin of Charles Darwin, the modern eugenics movement has generally tended to emphasize negative measures. Galton and his supporters were very much interested in measuring intelligence, and they developed some of the early versions of what we now call IQ tests. Through the use of these and other tests, supporters of eugenics have long sought scientific respectability for their attempts to label certain people as genetically defective or inferior.

The Nazis instituted a program of negative eugenics in Germany, beginning with the forced sterilization of mental 'defectives,' deaf people, homosexuals, and others. The eugenics program soon grew into a program for the mass killing of all those millions who did not belong to Hitler's 'master race.' By 1945, the Nazis had killed millions in the name of racial purity and Aryan superiority. The Nazis also practiced positive eugenics by encouraging German women with certain traits to have more children.

In the United States, the eugenics movement started as a series of attempts to identify, segregate, and sterilize mental 'defectives.' The movement soon found allies among racists and especially among those who sought to curb the new waves of immigration during the period from about 1890 to 1920. During the 1890s, one Kansas doctor sterilized 44 boys and 14 girls at the Kansas State Home for the Feeble-Minded, while Connecticut passed a law prohibiting marriage or sexual relations between any two people 'either of whom is epileptic, or imbecile, or feeble-minded.' A 1907 Indiana law required the sterilization of 'confessed criminals, idiots, imbeciles, and rapists in state institutions when recommended by a board of experts.' Fifteen other states passed similar laws, as often for punitive as for eugenic reasons. From 1909 to 1929, 6255 people were sterilized under such laws in California alone.

The writings of the American eugenicists became increasingly racist and antiimmigrationist in tone during this period. One eugenicist, for example, wrote in 1910 that "the same arguments which induce us to segregate criminals and feebleminded and thus prevent breeding apply to excluding from our borders individuals whose multiplying here is likely to lower the average [intelligence] of our people". In the 1960s, H.J. Muller wrote several articles warning against the practice of protecting and extending the lives of the 'genetically unfit,' those whom natural selection would tend to eliminate from the population. According to Muller, our medical intervention would only perpetuate genetic defects in our gene pool. Muller spoke pessimistically of a population divided into two groups, one so enfeebled from genetic defects that their very lives had to be sustained by extraordinary means, and the other group consisting of phenotypically normal people who had to devote their entire lives to the care and sustenance of the first group. Muller's views have not been substantiated by any evidence.

Biological objections to eugenics. Biological arguments against negative eugenics are based on the realization that eugenic measures could be expected to produce only small changes at great cost. Most known genetic defects are both rare and recessive, and selection against rare, recessive traits can only proceed very slowly no matter what the circumstances. As the trait gets increasingly rare, selection against it becomes increasingly ineffective. For example, the gene for albinism has a frequency of about 1 in 2000 in many human populations. If a eugenic dictator ordered all albinos to be killed or sterilized, theoretical calculations show that it would require about 2000 generations (about 50,000 years) of constant vigilance just to reduce the frequency of this trait to half of its present value. The reason why the process works so slowly is that most individuals carrying the gene for a rare, recessive trait are heterozygous and their phenotype does not reveal the presence of the gene. In contrast, modern techniques that allow the detection of the gene in heterozygous form would greatly increase the effectiveness (and hence the dangers) of negative eugenic measures.

For characteristics such as height or IQ, which are controlled by many genes and influenced strongly by environmental factors, estimates are that eugenic selection would be so slow as to be barely perceptible. One geneticist calculated that it would take about 400 years of constant, unrelenting and totally efficient selection to raise IQs by about 4 points; the same improvement could be achieved through education in as little as four years, and with far less cost. This topic is addressed in more detail in Chapter 5.

CONNECTIONS CHAPTER 5

Finally, eugenic measures can at best address only a small percentage of undesirable conditions, because most physical disabilities and medical conditions result from accidents, from infectious illnesses, or from exposure to toxic substances in the environment, not from inherited genetic makeup, and eugenic measures are powerless to alter these non-genetic causes. Genetic conditions may also result from new mutations, rather than from the inheritance of defective genes, and eugenic measures have no capacity to eliminate newly mutated genes or to depress the frequency of any gene below the mutation rate.

There is no biological basis for the claims of any eugenics movement that their methods could in any way improve humankind other than at great cost. The risks of negative eugenics are especially great, and include the possibility of genocide—the attempted extermination of a race or ethnic group. In addition there are no biological benefits. We are coming to know that population health and stability depends on genetic diversity; thus, narrowing a gene pool eugenically makes the population more vulnerable to infectious disease. Infectious disease is a far more widespread and common cause of sickness and death than is genetic disease.

Changing the balance between genetic and environmental factors

Although many traits are inherited, most are also influenced by the environment. Phenotypes result not just from genes but from the interactions between genes and their environment. In addition, far more disability is caused entirely environmentally through accidents and illnesses than is caused genetically. Even for most conditions that are caused genetically, elimination of an allele from the population is hardly the only option. Most genetic conditions can be modified or accommodated in several different ways.

Euphenics. **Euphenics** (literally, 'good appearance') includes all those techniques that either modify genetic expression or alter the phenotype to produce a modified phenotype known as a **phenocopy**. Plastic surgery, such as to repair body parts, is a form of medical intervention that alters the phenotype. Other examples include the installation of pacemakers in defective hearts, the giving of insulin to diabetics, and the dietary control of phenylketonuria. Although the genes remain unchanged, euphenics modifies or compensates for their phenotypic expression is in such a way that they no longer cause harm.

Many leading geneticists have argued, as an alternative to eugenics, that there is nothing wrong with altering the phenotype or the environment so that formerly disabling genotypes are no longer so harmful or debilitating. A leading advocate of this viewpoint was Theodosius Dobzhansky (1893–1975), who favored measures to permit people with hereditary 'defects' to overcome their handicaps and become phenotypic copies (phenocopies) of normal, healthy human beings. Once phenotypes could be controlled culturally, said Dobzhansky, the presence of formerly defective genotypes would cease to be the subject of any great concern. Euphenic intervention is already common practice for a number of genetic conditions. As our ability to modify phenotypes increases (e.g., with advances in corrective surgery), this type of practice is likely to become more common.

Euthenics. Another type of intervention is called **euthenics**. In this form of intervention, both genotype and phenotype remain unchanged, but the environment is modified or manipulated so that the phenotype is no longer as disabling as before. (In euphenics, by contrast, the phenotype is modified.) Examples of euthenic measures include canes, crutches, wheelchairs, and wheelchair ramps for those who cannot walk unaided, guide dogs and Braille for the sight-impaired, eyeglasses for the near-sighted, and so on. Conditions that are improved or assisted by euthenics may be either genetic or not.

Most people with disabilities support research that would prevent the recurrence of their condition in other people, especially if pain or paralysis are involved. However, medical research is expensive, its results are uncertain, and its benefits may take many years to become widely available. Euthenic measures are often less expensive and more quickly made available once they are developed. Many of the people who use euthenic devices feel that they would be better served by simple improvements in the devices (e.g., better wheelchairs) than they would be if medical research were our only emphasis.

Eupsychics. Many people with uncommon traits or conditions (whether genetic or not) feel that they are best served by being accepted as they are and do not necessarily want to be 'cured' (Box 3.2). Social and behavioral measures, or **eupsychics**, may lessen the impact of or compensate for disabling conditions. Included are the special education of handicapped individuals, mainstreaming (education of the handicapped in a regular public school setting), and the education and social conditioning of non-handicapped members of society so that they will better understand and accommodate the needs of all citizens.

Box 3.2 Who Decides What is Considered a 'Defect'?

Very often, groups of people that society wants to 'help' are given little or no voice in how society will treat them or 'help' them. The deaf are a case in point. Some forms of deafness are inherited, yet many deaf people consider it abhorrent if genetic counseling is used to avoid having deaf children.

The following excerpt is from a speech by I. King Jordan delivered in 1990 to an international symposium on the Genetics of Hearing Impairment. Research funded by the Human Genome Project is aimed at identifying genetic causes of deafness and there is an underlying assumption, on the part of geneticists who are not deaf, that people would then want to avoid this trait. A deaf man, Jordan is president of Gallaudet University in Washington D.C., established in the 1800s to educate deaf people. He explains that he is deaf both medically and culturally. He speaks for most deaf people when he tries to explain that deafness is not a trait that he would want to eliminate.

For about 18 years, I have taught a course on the psychology of deafness. One of the first things we discuss in the class is the difference between viewing deafness as a pathology that should be cured or prevented and viewing it as a human condition to be understood. I call these two perspectives the medical and cultural points of view. Individuals from these two groups agree on audiological definitions, but disagree on the emphasis that should be given to social and rehabilitative services. I adhere to the social or cultural point of view.

What I mean by this is that I, personally, and many of the people I know well, have accepted the fact that deafness is one aspect of my individuality. I do not spend any time or energy thinking about curing my deafness or restoring my hearing, but I do spend substantial time and energy trying to improve the quality of life for all people who are deaf.

For some reason, people who hear have a very difficult time understanding this concept. If you will permit me to digress for a moment, I will give you an example. I was interviewed by Ms. Meredith Vieira for the television show '60 Minutes.' During the interview, she asked me this question: "If there was a pill that you could take and you would wake up with normal hearing, would you take it?" I told her that her question upset me. I told her that it was something I spent virtually no time at all thinking about, and I asked her if she would ask me the same question about a 'white' pill if I were a black man. Then I asked if, as a woman, she would take a 'man' pill. Our conversation continued long after the videotaping was done, and we have had several subsequent conversations. But she never understood. She still does not. She still thinks only from her own frame of reference and imagines that not hearing would be a terrible thing. Deafness is not simply the opposite of hearing. It is much more than that, and those of us who live and work and play and lead full lives as deaf people try very hard to communicate this fact.

As you can see, this is an emotional issue for me. It is a much more emotional issue for many other deaf people. Is that relevant here? Yes, I believe it is, because the genetic study of deafness and genetic counseling have a great deal of significance for the deaf community generally. Many deaf people, particularly those who consider themselves members of the deaf community, do not consider themselves to be defective, rather, they consider themselves to be different, normal but different. In particular, this difference has a cultural or sociological basis and is expressed most saliently in the use of sign language. If deaf people are not defective or dysfunctional then, at least in their own eyes, it follows that they would be suspicious of attempts to eradicate deafness....

Genetic counseling and screening with respect to potential deafness must differ, therefore, in a fundamental way from screening for 'birth defects.'

(I. King Jordan, 'Ethical Issues in the Genetic Study of Deafness,' *Annals N. Y. Acad. Sci.* 630: 236–237.)

Genetic research seeks to understand the molecular mechanisms underlying normal physiology and health, but genetic testing and counseling emphasize diseases. The assumptions underlying genetic testing have at times included viewing variations as defects and desiring to apply to humans some arbitrary standard of perfection. Society may be better served by an emphasis on the abilities, rather than the disabilities, of each individual. All individuals should be encouraged to develop their talents and abilities to the fullest. Whenever a person is discouraged from trying to develop a certain skill, ability, or talent, both the individual and the society are the losers in the long run.

THOUGHT QUESTIONS

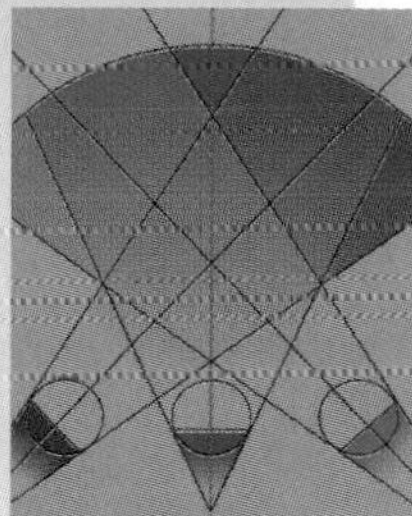

1. Are all 'birth defects' genetic defects? Do the same ethical questions regarding diagnosis and counseling apply to both genetic and nongenetic traits?
2. In some hospitals, screening for phenylketonuria is often performed on all infants. Does this violate the principle of informed consent? Is this practice ethical or not? How is it commonly justified? Do you feel that the justification is adequate?
3. If cloning were ever perfected, to what degree would the cloned individuals be identical to the original? How strong a role would upbringing play?
4. Some groups opposed to abortions have also begun to object to certain kinds of genetic testing. What good, they ask, can come from knowing that a fetus suffers from a particular genetic or chromosomal defect if the parents are opposed to abortion of the fetus on religious or similar grounds? For such situations, discuss the costs, benefits, and ethical status of genetic testing. Does it matter what kind of testing is performed? Does it matter what genetic or chromosomal defect is being tested for?
5. A procedure such as gene therapy is expensive. Who should pay for it? Is gene therapy a limited resource? Does giving gene therapy to one patient thereby deprive another of medical care?
6. Do unborn children have a 'right' to inherit an unmanipulated set of genes? Do they have a right to inherit 'corrected' genes if such a possibility exists? What kind of informed consent can we expect on behalf of unborn generations? Can a person make decisions that affect the genotype of all of his or her progeny? Do we need safeguards to protect future generations against the selfish interests of the present generation?
7. In what sense is 'positive' eugenics positive? In what sense is it negative? Should human populations be bred as we breed domesticated animals?
8. If public funds will be spent for the care and treatment of a person with a genetic condition, does that alter the ethical balance between the rights of the individual and the rights of society? If a euphenic measure is available, should public funds be used to 'correct' the condition? Should the person have the right to refuse such treatment? Should public funds be withheld from a person who refuses such euphenic measures?
9. How much access to genetic information about a subscriber or employee *should* an insurance company or an employer have? How much access *do* they have now?

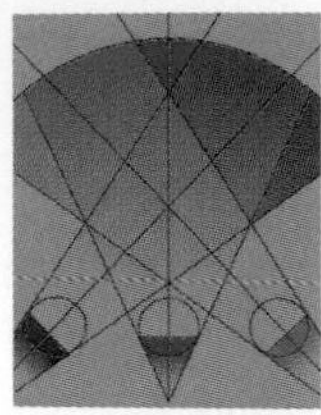

AS WE HAVE SEEN IN MANY CONTEXTS IN THIS CHAPTER, most of human genetics is more complex than the either/or traits that Mendel studied in pea plants. It is not that Mendel's laws do not apply to humans; they do. There are some human conditions that do follow simple Mendelian inheritance with a mutation in a single gene leading to the trait. Rather, since Mendel's time, we have learned that the inheritance of most traits in all organisms, not just in humans, is more complex than Mendel and scientists early in the twentieth century envisioned. Many traits, including sex determination and disease susceptibility, are influenced by multiple genes. The expression of the genotype as a phenotype is also influenced by the environment. Most biologists today would agree that phenotypes are far more malleable than was assumed in the past, contradicting earlier ideas of **biological determinism** that assumed that genotypes solely determined phenotypes. In addition, phenotypes can be modified by technologies and other aspects of cultures.

We are entering an era in which people will be able to find out a lot about their own genotype and the genotypes of their children. Identifying a genetic predisposition to a chronic disease may allow a person to make healthier choices about his or her diet and lifestyle. As we saw with the couple in the introduction to this chapter, people will have the power to find out whether a fetus carries a genotype for a fatal illness. This new knowledge will give people choices that they have not had in the past, but none of those are likely to be easy choices.

CHAPTER SUMMARY

- One pair of human chromosomes differs between males and females; the chromosomes of this pair are called **sex chromosomes**.
- Sexual development is coded for by many genes, including a gene on the Y chromosome.
- Genes that are on the X chromosome and not on the Y chromosome are called **sex-linked** genes.
- Human genes are now being identified at an increasingly rapid pace through **pedigrees** and linkage of molecular markers, and a project has been started to sequence and map the entire human **genome**.
- Many genetic diseases result from alleles that code for nonfunctional proteins. Some of these are inherited as simple Mendelian alleles; others show more complex patterns of inheritance.
- **Risk** is the probability of a condition's occurring; some genotypes are associated with increased risk.
- **Restriction enzymes** cut DNA into fragments; variations in the lengths of these fragments are called **restriction-fragment length polymorphisms** or **RFLPs**. RFLPs have helped in finding the location of many genes, as well as in the identification of individuals and in genetic engineering.
- **Genetic engineering** consists of inserting functional genes into cells, thereby altering the cell's genotype. The recipient cells may be bacterial cells that may then acquire the ability to make certain human proteins, or they may be human cells that acquire a functional allele and are injected into a patient as **gene therapy**.

- Genetic diseases can be diagnosed prenatally by amniocentesis, by chorionic villus sampling, and by other techniques, including those that use the **polymerase chain reaction** (**PCR**) to amplify DNA sequences into many copies. Testing for genetic diseases can also be done on children or adults. In any case, testing requires prior **informed consent**.
- Altering the gene pool at the population level is called **eugenics**. **Cloning** is a type of eugenics.
- For some genetic diseases, an enzyme product can be supplied artificially, or some substitute or mechanical compensation can be engineered.

KEY TERMS TO KNOW

biological determinism (p. 118)
clone (p. 112)
eugenics (p. 111)
gene therapy (p. 106)
genetic engineering (p. 94)
genome (p. 89)
informed consent (p. 102)
karyotype (p. 73)
pedigree (p. 79)
polymerase chain reaction (**PCR**) (p. 101)
restriction enzyme (p. 95)
restriction-fragment length polymorphism (RFLP) (p. 82)
risk (p. 80)
sex chromosomes (p. 73)
sex-linked (p. 75)

CONNECTIONS TO OTHER CHAPTERS

Chapter 1: Molecular genetics is forcing a change in our concepts of phenotypes and traits.

Chapter 1: Manipulating human heredity has raised several ethical concerns.

Chapter 2: The same basic laws of genetics apply to humans as to other animals and to plants.

Chapter 4: Evolution takes place whenever allelic frequencies change.

Chapter 5: Human populations differ in the frequencies of many alleles.

Chapter 6: Population control seeks to manage the size of populations; eugenics, in contrast, seeks to alter the gene pool.

Chapter 7: Mating patterns and sexual strategies can alter allelic frequencies in populations.

Chapter 9: Predispositions for some cancers are hereditary.

Chapter 10: Huntington's disease is one of several brain disorders for which a genetic basis has been identified.

Chapter 14: Genetic engineering can be used to improve the traits of commercially important plant species.

Chapter 15: Conserving genetic diversity is an important aspect of protecting biodiversity.

PRACTICE QUESTIONS

1. If one individual human differs from another in 0.1% of the genome, how many bases are different?
2. The sample of repeated DNA in Figure 3.8 is from a heterozygote who has different numbers of repeats on the two chromosomes of the homologous pair that carries the sequence. How many bands would show up if the person were homozygous? What size would the fragment or fragments of a homozygote be?
3. What is the risk of a child's having a recessive genetic trait when both parents are from a population in which the frequency of the recessive allele is 1 in 1000 (0.1%)? What is the risk when one parent is from a population with an allelic frequency of 0.1% and the other parent is from one in which the frequency is 1 in 10,000 (0.001%)?
4. Think about the study done on DNA from descendants of Jefferson's family and Sally Hemings's sons. Why is the title of the study, "Jefferson fathered slave's last child," an overstatement of the results?
5. In the following stretch of DNA, how many fragments will result from digestion with the Hae III restriction enzyme shown in Figure 3.13? How many will result from digestion with Eco R1?

strand 1

```
A T C C G T A G G C C T A A C C A T C C T A G T G C
T A G G C A T C C G G A T T G G T A G G A T C A C G
```

strand 2

6. How many copies of a DNA fragment are synthesized in 20 rounds of amplification by PCR (assuming that each step works correctly)? How many will be synthesized in 30 rounds?
7. Why are restriction enzymes that produce fragments with 'sticky ends' more useful in genetic engineering than restriction enzymes that produce fragments with 'blunt ends'?
8. Because a G on one strand of DNA always binds to a C on the complementary strand (and A binds to T), what characteristic of double-stranded DNA makes palindromic sequences possible?
9. Can RFLP band patterns be used to identify maternity, as well as paternity?
10. Can RFLP testing be used to identify individual organisms in other species besides humans?
11. Could the following sequence be used as an insert into genomic DNA? Why or why not?

strand 1

```
A A G C T T A A C G G A T T A G C A A G C
      C G A A T T G C C T A A T C G T T C G A A
```

strand 2

12. Could the following sequence be used as an insert into genomic DNA? Why or why not?

strand 1

```
A A G C U U A A C G G A U U A G C A A G C
      C G A A U U G C C U A A U C G U U C G A A
```

strand 2

13. When a plasmid is being cut with a restriction enzyme in preparation for inserting a DNA fragment, the plasmid needs to be cut with the same restriction enzyme as was used to make the DNA fragment. Why?
14. In the study on Jefferson's descendants, why did the researchers test DNA at 19 marker sites, rather than just at one or two sites?

4 Evolution and Classification

CHAPTER OUTLINE

The Darwinian Paradigm Reorganized Biological Thought
- Pre-Darwinian thought
- The development of Darwin's ideas
- Natural selection
- Descent with modification
- Fossils and the fossil record

Creationists Challenge Evolutionary Thought
- Creationism and Darwin's response
- Early twentieth-century creationism
- Creationism today

Species are Central to the Modern Evolutionary Paradigm
- Populations and species
- How new species originate
- Higher taxa

Life Originated on Earth by Natural Processes
- Evidence of early life on Earth
- Procaryotic and eucaryotic cells
- Kingdoms of organisms
- Procaryotic organisms

Eucaryotic Diversity Dominates Life Today
- Kingdom Protista
- Kingdom Plantae
- Kingdom Mycota
- Kingdom Animalia

Humans are Products of Evolution
- The genus *Australopithecus*
- The genus *Homo*
- Evolution as an ongoing process

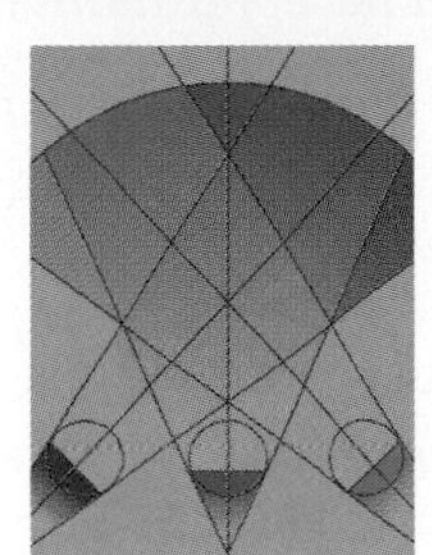

ISSUES

Did life evolve?

Is science compatible with religion?

How does evolution relate to genetics?

How do new species originate?

How do our classifications reflect evolution?

Did humans evolve?

Is life still evolving?

Will life continue to evolve?

Is the science of evolution static or changing?

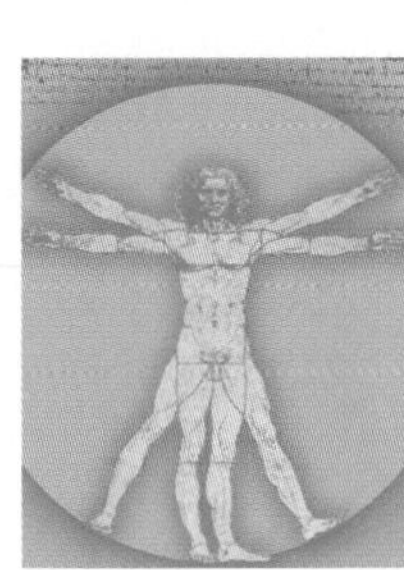

BIOLOGICAL CONCEPTS

Evolution

Fossils and geologic time

Natural selection

Environmental influences on species

Adaptation

Form and function

Descent with modification

Species and speciation

Products of evolutionary change (phylogenetic classification, biodiversity, biogeography)

Origin of life

Types of cells (procaryotic, eucaryotic)

Hierarchy of organization

Classification of organisms (six kingdoms)

Chapter 4: Evolution and Classification

Ask any biologist to name the most important unifying concepts in biology, and the theory of evolution is likely to be high on the list. As geneticist Theodosius Dobzhansky explained, "nothing in biology makes sense, except in the light of evolution." However, many people in the United States are unaware of the importance of evolution as a unifying concept: public opinion surveys reveal that 25–40% of Americans either do not believe in evolution or think that evidence for it is lacking. (The percentage varies depending on how the question is worded.) In this chapter we examine both the theory of evolution and the opposition to it.

As explained in Chapter 1, scientists use the word **theory** *for a coherent cluster of hypotheses that has withstood many years of testing. In this sense, evolution is a thoroughly tested theory that has withstood nearly a century and a half of rigorous testing. Scientific evidence for evolution is as abundant as, and considerably more varied than, the evidence for nearly any other scientific idea. To refer to evolution as 'just a theory' is thus a grave misunderstanding of both scientific theories in general and evolutionary theory in particular. When physicists speak of the atomic theory or the theory of relativity, or when medical professionals speak of the germ theory of disease, they are speaking of great unifying principles. These principles are now well established, but they have withstood repeated testing for somewhat fewer years than the theory of evolution has. Educated people no longer doubt the existence of atoms or of germs, and nobody refers to any of these concepts as 'just a theory.' In the way that the atomic theory is a unifying principle for much of physics and chemistry, the theory of evolution is a unifying principle for all of the biological sciences.*

THE DARWINIAN PARADIGM REORGANIZED BIOLOGICAL THOUGHT

Arguably the most influential biology book of all time was published in 1859. *On the Origin of Species by Means of Natural Selection*, written by the English naturalist Charles Darwin (1809–1882), contains at least two major hypotheses and numerous smaller ones, along with an array of evidence that Darwin had already used to test these hypotheses. Both hypotheses deal with **evolution**, the process of lasting change among biological populations. The first major hypothesis, **branching descent**, is that species alive today came from species that lived in earlier times and that the lines of descent form a branched pattern resembling a tree (Figure 4.1). Using this hypothesis, Darwin explained similarities among groups of related species as resulting from common inheritance. The second major hypothesis is that parents having genotypes that favor survival and reproduction leave more offspring, on average, than parents having less favorable genotypes for the same traits. Darwin called this process **natural selection**, and he hypothesized that major changes within lines of descent had been brought about by this process. Both of these two hypotheses are falsifiable, and they have been tested hundreds if not thousands of times, without being falsified, since Darwin first proposed them in 1859. Darwin's two hypotheses made sense of several previously noticed but unexplained regularities in anatomy, classification, and geographic distribution. As both a unifying theory and a stimulus to further research, Darwin's *Origin of Species* fits the concept of a scientific paradigm expounded by Thomas Kuhn and explained in Chapter 1 (p. 14) of this book. Modern evolutionary thought is still largely based on Darwin's paradigm of branching descent and natural selection, expanded to include the findings of genetics.

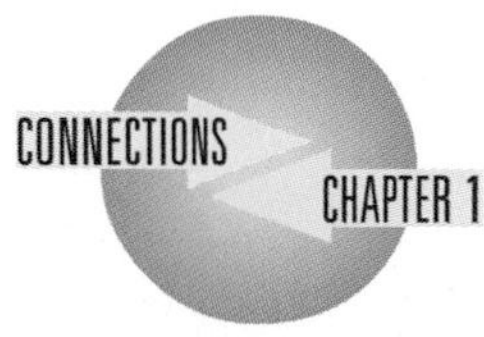

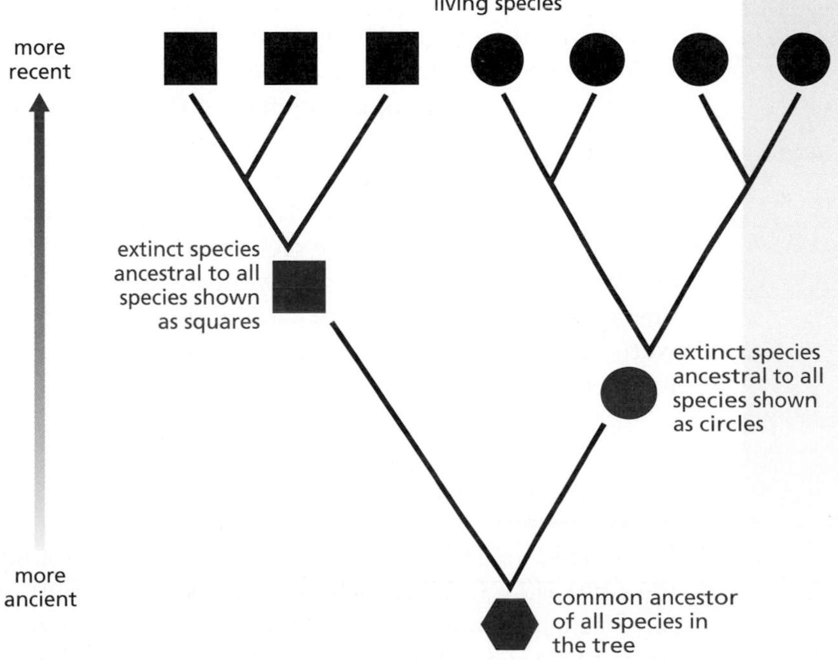

FIGURE 4.1

The pattern of branching descent. Living species in the top row are descended from the ancestors below them. The red circle represents the common ancestor to all other circles, and the red square is likewise ancestral to all the squares. The red hexagonal shape at the bottom is ancestral to all species shown in this phylogenetic tree. In a classification, all the squares would be placed in one group and all the circles in another.

Pre-Darwinian thought

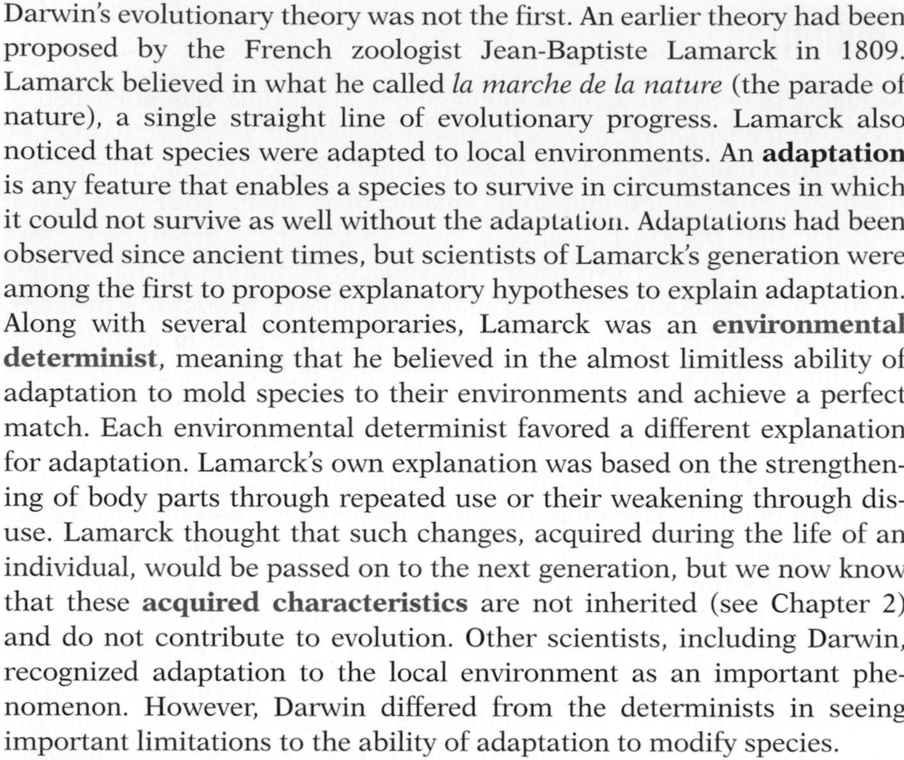

Darwin's evolutionary theory was not the first. An earlier theory had been proposed by the French zoologist Jean-Baptiste Lamarck in 1809. Lamarck believed in what he called *la marche de la nature* (the parade of nature), a single straight line of evolutionary progress. Lamarck also noticed that species were adapted to local environments. An **adaptation** is any feature that enables a species to survive in circumstances in which it could not survive as well without the adaptation. Adaptations had been observed since ancient times, but scientists of Lamarck's generation were among the first to propose explanatory hypotheses to explain adaptation. Along with several contemporaries, Lamarck was an **environmental determinist**, meaning that he believed in the almost limitless ability of adaptation to mold species to their environments and achieve a perfect match. Each environmental determinist favored a different explanation for adaptation. Lamarck's own explanation was based on the strengthening of body parts through repeated use or their weakening through disuse. Lamarck thought that such changes, acquired during the life of an individual, would be passed on to the next generation, but we now know that these **acquired characteristics** are not inherited (see Chapter 2) and do not contribute to evolution. Other scientists, including Darwin, recognized adaptation to the local environment as an important phenomenon. However, Darwin differed from the determinists in seeing important limitations to the ability of adaptation to modify species.

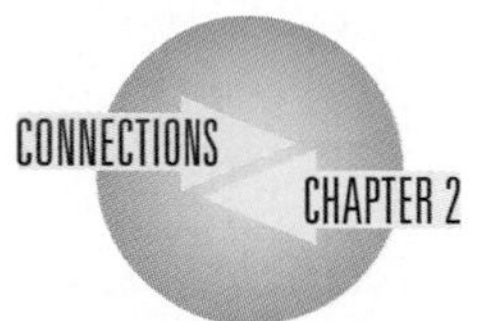

British naturalists had quite different explanations for adaptation. The Natural Theology movement, led by the Reverend William Paley, sought to prove the existence of God by examining the natural world for evidence of perfection. By careful examination and description, British scientists found case after case of organisms with anatomical structures so well constructed, so harmoniously combined with one another, and so well suited in every detail to the functions that they served that one could only marvel at the degree of perfection achieved. Such harmony, design, and detail, they argued, could only have come from God. Paley offered well-planned adaptation as proof of God's existence: "The marks of *design* are too strong to be gotten over. Design must have a designer. That designer must have been a person. That person is God." (Paley. *Natural Theology*, end of Chapter 23; page numbers vary among many editions.) In a nation in which many clergymen were also amateur scientists, it became quite the thing to dissect organisms down to the smallest detail, all the better to marvel at the wondrously detailed perfection of God's design. A large series of intricate and sometimes amazing adaptations were thus described, which Darwin would later use as examples to argue for an evolutionary explanation based on natural selection.

The development of Darwin's ideas

From 1831 to 1836, Charles Darwin traveled around the world aboard the ship H.M.S. *Beagle*. His observations in South America convinced him that the animals and plants of that continent are vastly different from those inhabiting comparable environments in Africa or Australia.

For example, all South American rodents are relatives of the guinea-pig and chinchilla, a group found on no other continent. South America also had llamas, anteaters, monkeys, parrots, and numerous other groups of animals, each with many species inhabiting different environments throughout the continent, but different from comparable species elsewhere (Figure 4.2). This was definitely not what Darwin had expected! Environmental determinist theories such as Lamarck's had led Darwin to expect that regions in South America and Africa that were similar in climate would have many of the same species. Instead, he found that most of the species inhabiting South America had close relatives living elsewhere on the continent under strikingly different climatic conditions. They had no relationship, however, to species living in parts of Africa or Australia with similar climates. The animals inhabiting islands near South America were related to species living on the South American continent. Fossilized remains showed that extinct South American animals were related to living South American species. "We see in these facts some deep organic bond, prevailing throughout space and time, over the same

FIGURE 4.2

An assortment of South American mammals. These species are very different from the mammals found on other continents, even where climates are similar.

areas of land and water, and independent of their physical conditions. The naturalist must feel little curiosity, who is not led to inquire what this bond is. This bond, on my theory, is simply inheritance, that cause which alone, as far as we positively know, produces organisms quite like, or...nearly like each other." (Darwin. *Origin of Species*, 1859, p. 350.)

The Galapagos Islands. The Galapagos Islands are a series of small volcanic islands in the Pacific Ocean west of Ecuador. Darwin's visit to these islands proved especially enlightening to him. In this archipelago, a very limited assortment of animals greeted him: no native mammals or amphibians were present; instead there were several species of large tortoises and a species of crab-eating lizard. Most striking were the land birds, now often called 'Darwin's finches' (phylum Chordata, class Aves, order Passeres): a cluster of more than a dozen closely related species, each living on only one or a few islands (Figure 4.3). The tortoises also differed from island to island, despite the clear similarities of climate throughout the archipelago. Darwin hypothesized that each species cluster had arisen through a series of modifications from a single species that had originally colonized the islands. The islands, Darwin noted, were similar to the equally volcanic and equally tropical Cape Verde Islands in the Atlantic Ocean west of Senegal, which Darwin had also visited, but the inhabitants were altogether different. Darwin concluded that the Galapagos had received its animal colonists (including the finches) from South America, while the Cape Verde Islands had received theirs from Africa, and that each group of colonists had given rise to a cluster of related species. Darwin was the first evolutionary theorist to emphasize that clusters of related species indicated a branching pattern of descent, a pattern that Darwin called "descent with modification."

Patterns of distribution. Darwin continued to find large groups of related animal species inhabiting each continent. These groups were unrelated to the very different groups inhabiting similar climates on other land masses. Several large land areas had flightless birds, but they differed strikingly from one continent or island to the next: rheas in South America, kiwis and extinct moas in New Zealand, emus and cassowaries in Australia, extinct elephant birds on Madagascar, and ostriches in Africa. Each land mass had its own distinct type of flightless bird, although they all lived in regions of similar climate. Theories of environmental determinism (such as Lamarck's) could not explain these differences, nor could theories of divine creation explain why God had seen fit to create half a dozen distinct types of flightless birds where one might have sufficed.

Natural selection

When Darwin returned to England, he began reading about the ways in which species could be modified. How, he wanted to know, could a single colonizing species produce a whole cluster of related species on a group of islands? During the preceding hundred years, Darwin observed,

British animal breeders had produced many new varieties of dogs, sheep, and pigeons, and had greatly improved wool yields in sheep, and milk yields in cattle, by careful breeding practices. By methodically selecting the individuals in each generation with the most desired traits and breeding these individuals with each other, British animal breeders had modified a number of domestic species through a process that Darwin called **artificial selection**. This process simply took advantage of the natural variation that was present in each species, yet it produced breeds that were strikingly different from their ancestors. Darwin remarked that

FIGURE 4.3

Some of Darwin's finches from the Galapagos Islands.

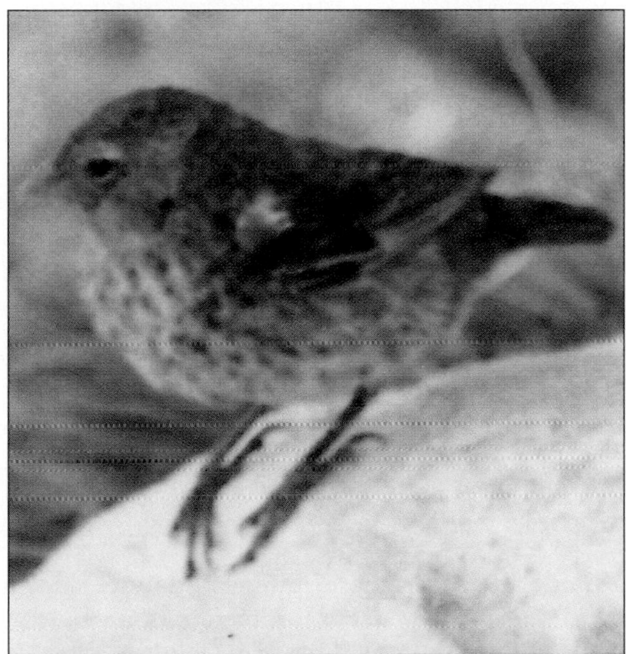

Small ground finch (*Geospiza fulginosa*)

Medium ground finch (*Geospiza fortis*)

Warbler finch (*Certhidea olivacea*)

Large cactus finch (*Geospiza conirostris*)

some of the domestic varieties of pigeons or dogs differed from one another as much as did natural species, despite the fact that the domestic varieties had been produced within a short time from a known group of common ancestors. Could a similar process be at work in nature?

At about this same time, Darwin read Thomas Robert Malthus' *Essay on Population* (see Chapter 6, p. 238). Malthus emphasized that, in the natural world, each species produces more young than are necessary to maintain its numbers. This overproduction is followed in each generation by the premature death of many individuals and the survival of only a few. When Darwin compared this process with the actions of the animal breeders, he concluded that nature was slowly bringing about change in each species. Individuals varied in every species, and those that died young in each generation differed from those that survived to maturity and mated to produce the next generation. In this 'struggle for existence,' Darwin hypothesized that

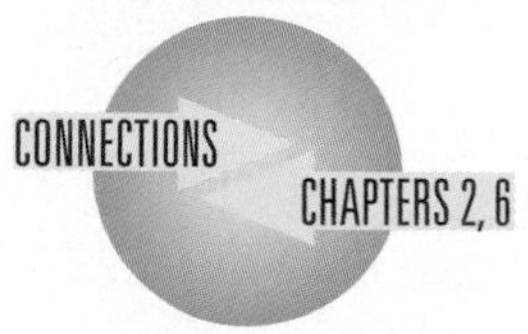

> ...individuals having any advantage, however slight, over others, would have the best chance of surviving and of procreating their kind.... On the other hand, we may feel sure that any variation in the least degree injurious would be rigidly destroyed. This preservation of favourable variations and the rejection of injurious variations, I call Natural Selection.... Natural selection...is a process incessantly ready for action, and is as immeasurably superior to man's feeble efforts, as the works of Nature are to those of Art. (Darwin. *Origin of Species*, 1859, pp. 61, 81.)

Natural selection defined. All modern descriptions of natural selection are stated in terms of the concepts of genetics outlined in Chapter 2. New genotypes originate by mutation and recombination, both of which act prior to any selection. Darwin, of course, knew nothing of mutations or of modern genetics, but he did realize that heritable variation had to come first and that "any variation which is not heritable is unimportant to us."

Natural selection may be defined as consistent differences in what Darwin called "success in leaving progeny," meaning the contributions of different genotypes to future generations. The relative number of viable individuals that each genotype contributes to the next generation is called its **fitness**, and *natural selection favors any trait that increases fitness*. Darwin's theory of natural selection is the basis for all modern explanations of adaptation.

Mimicry. In the years since Darwin first proposed the hypothesis of natural selection as an evolutionary mechanism, many tests of the hypothesis have been conducted. One of the earliest tests involved the phenomenon of **mimicry**, in which one species of organisms deceptively resembles another. In one type of mimicry, a distasteful or dangerous prey species, called the model, gives a very unpleasant and memorable experience to any predator that attempts to eat it. Predators always avoid the model following such an unpleasant experience. A palatable prey species, the mimic, secures an advantage if it resembles the model enough to fool predators into avoiding it as well.

Selection by predators explains mimicry rather easily. Any slight resemblance that might cause a predator to avoid the mimic as well as its model is favored by selection and passed on to future generations of the mimic species, while individuals not protected in this way would be eaten in greater numbers. Predator species differ greatly in their abilities to distinguish among prey species, so a resemblance that fools one predator might not fool another. Any advantage that increases the number of predators fooled is favored by selection, causing closer and closer resemblance to evolve with the passage of time.

Sometimes several species that resemble each other are all distasteful to predators. Predators learn to avoid distasteful species, but a certain number of prey individuals are killed for each predator individual that learns its lesson. Without mimicry, each prey species must sustain this loss separately. Mimicry allows predators to learn the lesson with fewer individuals of each prey species dying in the process. The mimicry therefore benefits each prey species and is thus favored by natural selection.

Mimicry often varies geographically. Some wide-ranging tropical species mimic different model species in different geographic areas. The deceptive resemblance is always to a species living in the same area, never to a far-away species. Environmentalist theories such as Lamarck's had no way to account for the evolution of mimicry, and the patterns of geographic variation could not be explained by either environmentalist theories or by Paley's natural theology. Natural selection, however, explains the variation as resulting from selection by predators.

In a well-known case of mimicry, the model is the monarch butterfly, a distasteful species that feeds on milkweed plants. An unrelated species, the viceroy, is similar in superficial appearance, and is thus avoided by many predators (Figure 4.4); some of these predators may find the viceroy distasteful as well.

Industrial melanism. The power of natural selection is also shown in **industrial melanism**, in which species that are usually light in color evolve darker colors in areas polluted by industrial soot. In the British Isles, a species known as the peppered moth (*Biston betularia*) (kingdom Animalia, phylum Arthropoda, class Insecta, order Lepidoptera) had long been recognized by an overall light gray coloration with a salt-and-pepper pattern of irregular spots. A black variety of this species was discovered in the 1890s. The black moths increased until they came to outnumber the original forms in some localities (Figure 4.5). The British naturalists E.B. Ford and H.B.D. Kettlewell studied this case of industrial melanism. Downwind from the major industrial areas, the woods had become

FIGURE 4.4

Warning coloration and mimicry. (A) *Limenitis arthemis*, a nonmimic relative of (B) *Limenitis archippus*, the viceroy. The viceroy resembles the unrelated monarch butterfly (C, *Danaus plexippus*), the model. The monarch is avoided by predators after just a single unpleasant experience (D, E). The warning color pattern of the monarch helps predators learn to avoid it; the viceroy is protected because its color pattern mimics that of the monarch.

(A) Butterfly closely related to one from which the viceroy evolved

(B) Viceroy

(C) Monarch

(D) Blue jay eating monarch

(E) Jay vomiting after eating monarch

polluted with black soot that killed the lichens growing on the tree trunks. Most of the moths living on the darkened tree trunks in these regions were black. However, where the woods were untouched by the industrial soot, the tree trunks were still covered with lichens and the moths had kept the light-colored pattern. Ford and Kettlewell hypothesized that the moths resembling their backgrounds would be camouflaged and thus harder for predators to see. To test this hypothesis, they pinned both light and dark moths on dark tree trunks in polluted woods, and they also pinned both types on lichen-covered tree trunks in unpolluted woods. They observed that birds ate more of the dark moths in the unpolluted woods (favoring the survival of the light-colored pattern), but birds in the polluted woods ate more of the light-colored moths, not the dark ones. These observations and the geographical patterns of variation (see Figure 4.5) were easily explained in terms of natural selection by predators. In addition, since the experiments were first conducted, laws to control smokestack emissions and other forms of pollution were passed and enforced, and many of the woods affected by pollution have returned to their former state. In these woods, the lichens have returned to the tree trunks, and most of the moths in these places now have the original light-colored pattern.

Agents of selection. In the preceding examples of natural selection, the selecting agents are predators. There are many other kinds of selecting agents operating in nature. Any cause of death contributes to natural selection if it reduces the opportunity for reproduction and if some genotypes are more likely to die. Some genotypes may be more susceptible to particular diseases or parasites, and die in greater numbers from these

FIGURE 4.5

Industrial melanism in peppered moths in the British Isles.

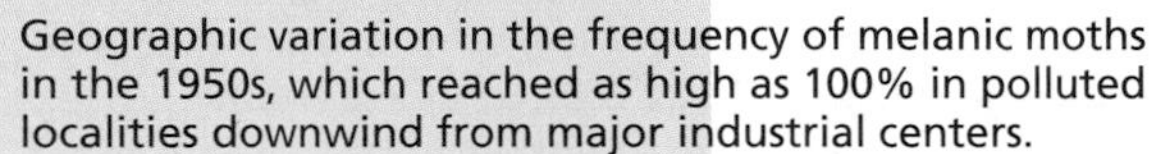
Geographic variation in the frequency of melanic moths in the 1950s, which reached as high as 100% in polluted localities downwind from major industrial centers.

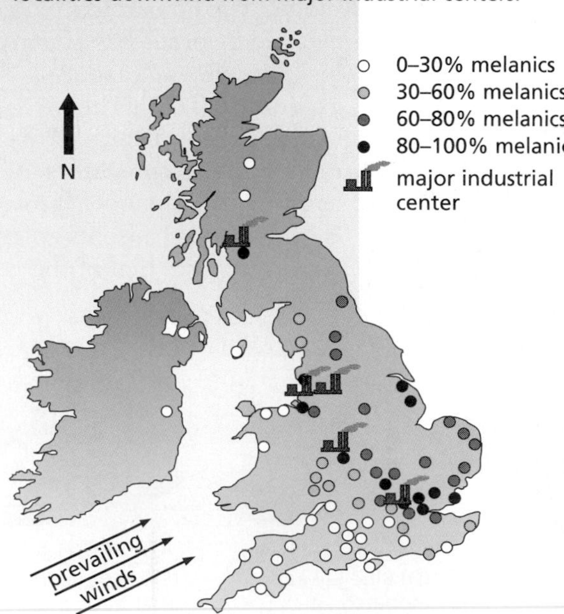

The melanic (black) variety and the original 'peppered' variety (below the right wing-tip of the melanic moth) on a light, lichen-covered tree trunk.

The same two varieties on a dark, soot-covered tree trunk.

causes, while other genotypes might be more resistant and thus survive more readily. Starvation and weather-related extremes of cold, dryness, or precipitation may also be agents of selection if some genotypes can survive these conditions better than others. These and other causes of mortality are all agents of selection if there are differences in the death rates for different genotypes.

Not every agent of selection causes death, however. Natural selection also favors those genotypes that reproduce more and leave more offspring. A special type of selection, called **sexual selection**, operates on the basis of success (or lack of success) in attracting a mate and reproducing. For example, animals of many species attract their mates with mating calls (such as bird songs), visual displays (as in peacocks; see Figure 7.2, p. 282), or special odors (as in silkworm moths or many other invertebrate animals). Individuals that do not perform well enough to attract a mate may live long lives but leave no offspring.

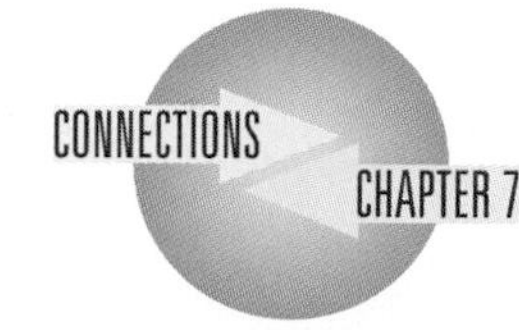

Descent with modification

Darwin's concept of descent with modification has been used to make sense out of a variety of observations not easily explained by other means. For example, Darwin recognized that the species clusters found on islands such as the Galapagos had their closest relatives on the nearest continent, not on geologically similar or climatically similar but distant islands. Geographic proximity, in other words, was often more important than climate or other environmental variables in determining which species occurred in a particular place.

At least a century before Darwin, biological classifications had already taken their modern hierarchical form (described later in this chapter). Darwin explained this hierarchy as the natural result of descent with modification, a process that results in a series of treelike branchings in which species correspond to the finest twigs, groups of species to the branches from which these twigs arise, larger groups to larger branches, and so forth (see Figure 4.1). Darwin predicted that classifications would increasingly become genealogies (that is, maps of descent) as more and more details about the evolution of each group of organisms became known. In this section we consider similarities and differences that biologists have used in classifying organisms.

Homologies. The construction of family trees is based in large measure on the study of shared structures or gene sequences. Under Darwin's paradigm, shared similarities are evidence that the organisms in question share a common ancestry. In a sense, a shared similarity is a falsifiable hypothesis that the several species sharing it are related to one another by descent. By itself, one such similarity reveals very little, but a large number of similarities that fit together into a consistent pattern strongly suggest shared ancestry. When the evidence for shared ancestry is compelling, the similarity is called **homology**.

Darwin noted the similarities among the forelimbs of mammals: "what can be more curious than that the hand of a man, formed for grasping, that of a mole for digging, the leg of the horse, the paddle of the porpoise, and the wing of the bat, should all be constructed on the same pattern, and should include the same bones, in the same relative positions?" (Figure 4.6). Darwin wondered why similar leg bone structures appeared in the wings and legs of a bat, used as they are for such totally different purposes. Why should a crustacean that has more mouthparts have correspondingly fewer legs, or why should those with more legs have fewer mouthparts? Darwin's answer is that all these structures arose as homologies by modification of the same type of repeated part. Crustaceans, for example, have evolved their mouthparts, legs, and certain other structures from a set of common leglike appendages because natural selection favored different structures for different uses. Thus, animals that evolved more mouthparts, because more appendages were used for mouth-related activities, had fewer appendages to be used as legs. An omnipotent God, however, would be subject to no such limitation, leaving Darwin to declare, "How inexplicable are these facts on the ordinary view of creation!" (Darwin. *Origin of Species*, 1859, p. 437.)

FIGURE 4.6

Homologies among mammalian forelimbs adapted to different functions.

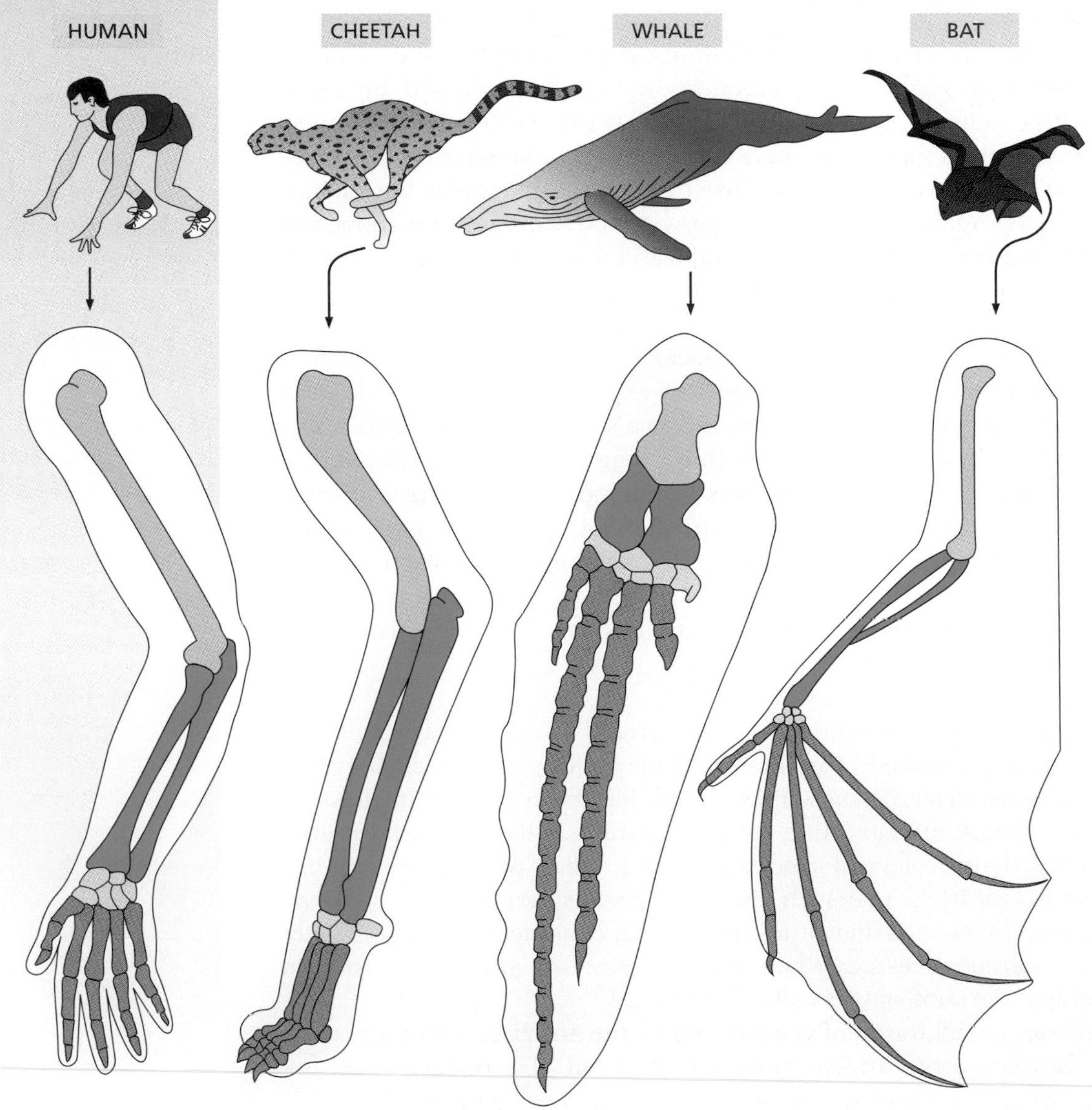

Vestigial structures. Structures whose function has been lost in the course of evolution tend to diminish in size. Often, they persist as small, functionless remnants, called **vestigial structures**. A good human example is the coccyx, a set of two or three vestigial tail bones at the base of the spinal column, homologous to the tails of other mammals. The Darwinian paradigm of natural selection and branching descent explains these vestigial structures as the remnants of structures that had once been functional. Neither Lamarck nor the creationists had any explanation for the presence of vestigial structures, and certainly not for the homologies between many such structures and the functional structures of related species.

Convergence. Similarities that result from common ancestry (that is, true homologies) should also be similar at a smaller level of detail, and they should be similar in embryological derivation as well. A hypothesis of homology can thus be falsified if two similar structures turn out to be dissimilar in detailed construction or in embryological derivation. There are also cases in which several hypotheses of homology are in conflict because they require different patterns of relationship for different characters. In such cases, evolutionists examine all the similarities more closely and repeatedly to see whether a reinterpretation is possible for one set of similarities.

Convergence is an evolutionary phenomenon in which similar adaptations evolve independently in lineages not closely related. Similarities for which the hypothesis of homology is falsified by more careful scrutiny are often reinterpreted as convergent adaptations, meaning structures that evolved independently in unrelated lineages. Resemblance resulting from convergent adaptation is called **analogy**. Distinguishing homology from analogy is an ongoing aim of evolutionary classification. For example, the wings of bats and insects are analogous, rather than homologous, structures. They are constructed in different ways and from different materials, and their common shapes (which they also share with airplane wings) reflect adaptation to the aerodynamic requirements of flying. Although bat wings are not homologous to insect wings, they are homologous to human arms, whale flippers, and the front legs of horses and elephants. These all have similar bones, muscles, and other parts in similar positions despite their very different shapes and uses, while insect wings have no bones and their muscles are very differently located.

Evolution of complex structures. Paley and his supporters had paid much attention to complex organs such as the human eye. The eye, they pointed out, was composed of many parts, each exquisitely fashioned to match the characteristics of the other parts. What use would the lens be without the retina, or the retina without a transparent cornea? An eye, they argued, would be of no use until all its parts were present, thus it could never have evolved in a series of small steps, but must have been created, all at once, by God. To counter this sort of argument, Darwin pointed out that the eyes of various invertebrates can be arranged into a series of gradations, ranging in complexity from "an optic nerve merely coated with pigment" to the elaborate visual structures of squids, approaching those of vertebrates in form and complexity. A large range of variation in the complexity of visual structures is found within a single group, the Arthropoda, the group that includes barnacles, shrimp, crabs, spiders, insects, and their many relatives. All the visual structures, regardless of their degree of complexity, are fully functional adaptations, advantageous to their possessors. It would therefore be quite reasonable, argued Darwin, to imagine each more complex structure to have evolved from one of the simpler structures found in related animals. Eyes, in other words, could have evolved through a series of small gradations.

The explanatory power of branching descent. Tests of the hypothesis of branching descent have also taken several other forms. One type of test identifies a group of organisms that share some particular character, such as an anatomical peculiarity. The general hypothesis of branching descent then gives rise to a more specific hypothesis, that this particular group of organisms share a common descent from a common ancestor.

An example of this type of reasoning can be illustrated by the Cephalopoda, a group of mollusks that includes the squids and their relatives. All cephalopods can be recognized by the presence of a well-developed head and a mantle cavity beneath (Figure 4.7). The mantle cavity contains the gills, the anus, and certain other anatomical structures.

FIGURE 4.7

Three living types of cephalopod mollusks (kingdom Animalia, phylum Mollusca, class Cephalopoda): the cuttlefish, the squid, and the chambered nautilus.

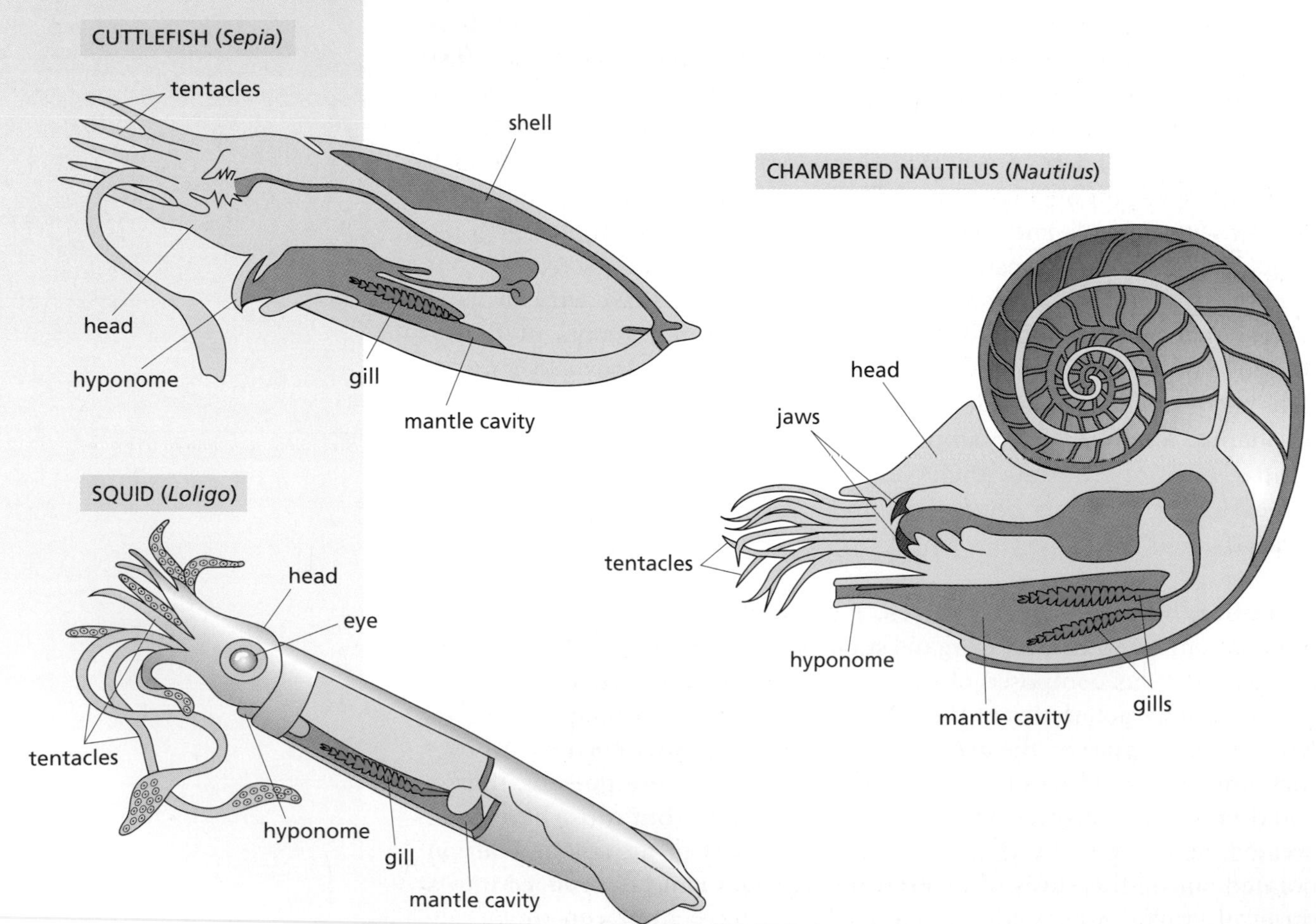

Other mollusks have mantle cavities, but only in the Cephalopoda is the mantle cavity located beneath the head and prolonged into a nozzlelike opening known as the hyponome. Knowing this, we can formulate the specific hypothesis that all cephalopods share a common descent.

If our hypothesis is true, we should be able to find some additional similarities among cephalopods not shared with other mollusks. Such similarities do exist: all cephalopods have beaklike jaws at the front of the mouth, and a muscular part, called the foot in other mollusks, subdivided into a series of tentacles (see Figure 4.7). Moreover, all cephalopods have an ink gland that secretes a very dark, inky fluid. When a squid or octopus feels threatened by a predator, it releases this fluid into its mantle cavity and quickly squirts the contents of the mantle cavity forward through its nozzlelike hyponome. This action propels the animal backwards, in a direction not expected by its predator, while the predator's attention is held by the puff of black, inky fluid. By the time the inky fluid dissipates, the squid or octopus has vanished. All members of the Cephalopoda have this elaborate and unusual escape mechanism, including squids, cuttlefishes, octopuses, and the chambered nautilus. The hypothesis of a common descent for all the Cephalopoda is thus consistent with the known data, meaning that the hypothesis has been tested and not falsified.

In the same way that a hypothesis of common descent has been tested for the cephalopod mollusks, hundreds of similar hypotheses have been tested for other groups of animals and plants. This increases our confidence in the larger hypothesis that all species of organisms have evolved from earlier species in patterns of branching descent. The many facts of comparative anatomy, comparative physiology, embryology, biogeography, and animal classification all make much more sense if we hypothesize that modern species have evolved from ancestors that lived in the remote past.

Since Charles Darwin published his evolutionary ideas in 1859, thousands of tests have been made of his twin hypotheses of branching descent and natural selection. Because these thousands of tests have failed to falsify either hypothesis, both now qualify as scientific theories that enjoy widespread support. The Darwinian paradigm continues to this day as a major guide to scientific research.

Fossils and the fossil record

The history of life on Earth is measured on a time scale encompassing billions of years. This geological time scale (Figure 4.8) was first established by studying **fossils**, the remains and other evidence of past life forms. Scientists had recognized since 1555 that most fossils were the remnants of species no longer living, thus providing clear-cut evidence for extinction.

Stratigraphy. The geological time scale was established first through **stratigraphy**, the study of layered rocks. One of the first observations in stratigraphy was that when rock layers have not been drastically disturbed, the oldest layers are on the bottom and successively newer layers are on top of them. Using this principle, geologists can identify the rock formations in a particular place as part of a local sequence, arranged chronologically from bottom to top.

Local sequences from different places can be matched with one another in several ways, but the most reliable of these proved to be the study of their fossil contents. **Correlation by fossils** is a technique for judging that two rock formations are contemporaneous (from the same time period) because they contain many of the same fossil species. The rocks do not need to be similar in composition or rock type—one can be

FIGURE 4.8

The geological time scale.

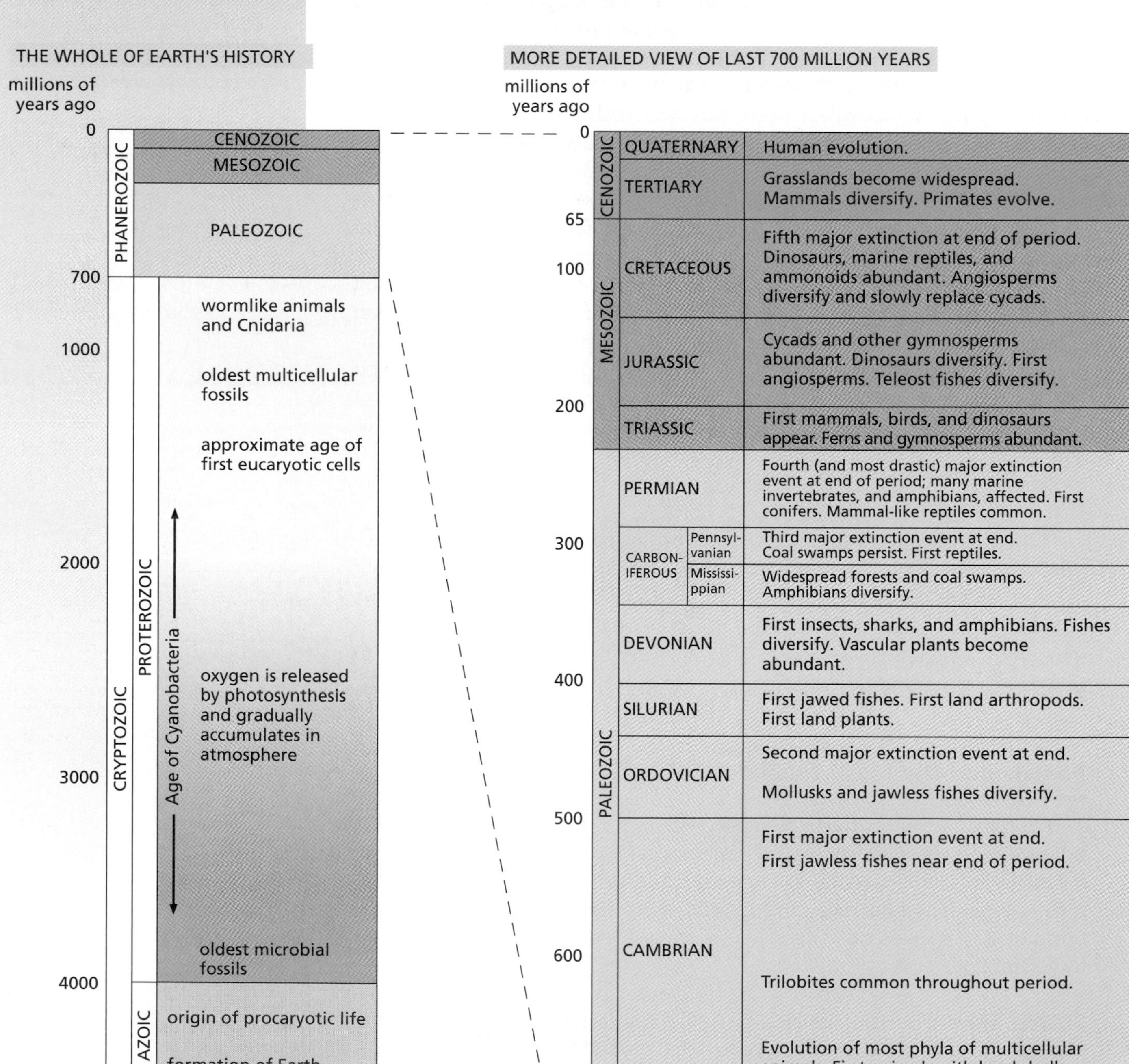

a limestone and the other a shale—but if their fossil assemblages are similar, they are judged to be contemporaneous. A single species of fossil is never sufficient; an assemblage of several fossil species is needed. Using this technique of correlation in the manner shown in Figure 4.9, paleontologists (scientists who study fossils) were able to assemble the world's various local sequences into a 'standard' worldwide sequence, which is the basis for the complete sequence of time periods shown in Figure 4.8. The dates assigned to these periods are determined by measuring the rate of radioactive decay in certain rocks.

Phylogeny. The age of a fossil, by itself, tells us very little about its place in any family tree. The relative ages of fossils only begin to have meaning when we study a group of organisms represented by many fossils. The family tree or genealogy of any group, called its **phylogeny**, fits into a pattern like that shown in Figure 4.1. Any such family tree is a hypothesis that biologists use to explain the anatomical and other characteristics of each species, and the classification of the group as a whole. In any family tree, the known fossils must fit into a consistent framework.

For example, there are many fossils of cephalopod mollusks, permitting further tests of the hypothesis of a common descent for all the Cephalopoda. Living and extinct cephalopods can be arranged into a family tree consistent with our knowledge of the characteristics of each species and the relations among them (Figure 4.10). Differences among

FIGURE 4.9

How correlation by fossils is used to establish a single worldwide stratigraphic sequence. The letters represent different types of fossils found in each rock formation. Correlation by fossils allows the lines to be drawn connecting those formations judged to be contemporaneous. The result is the composite rock sequence shown on the right.

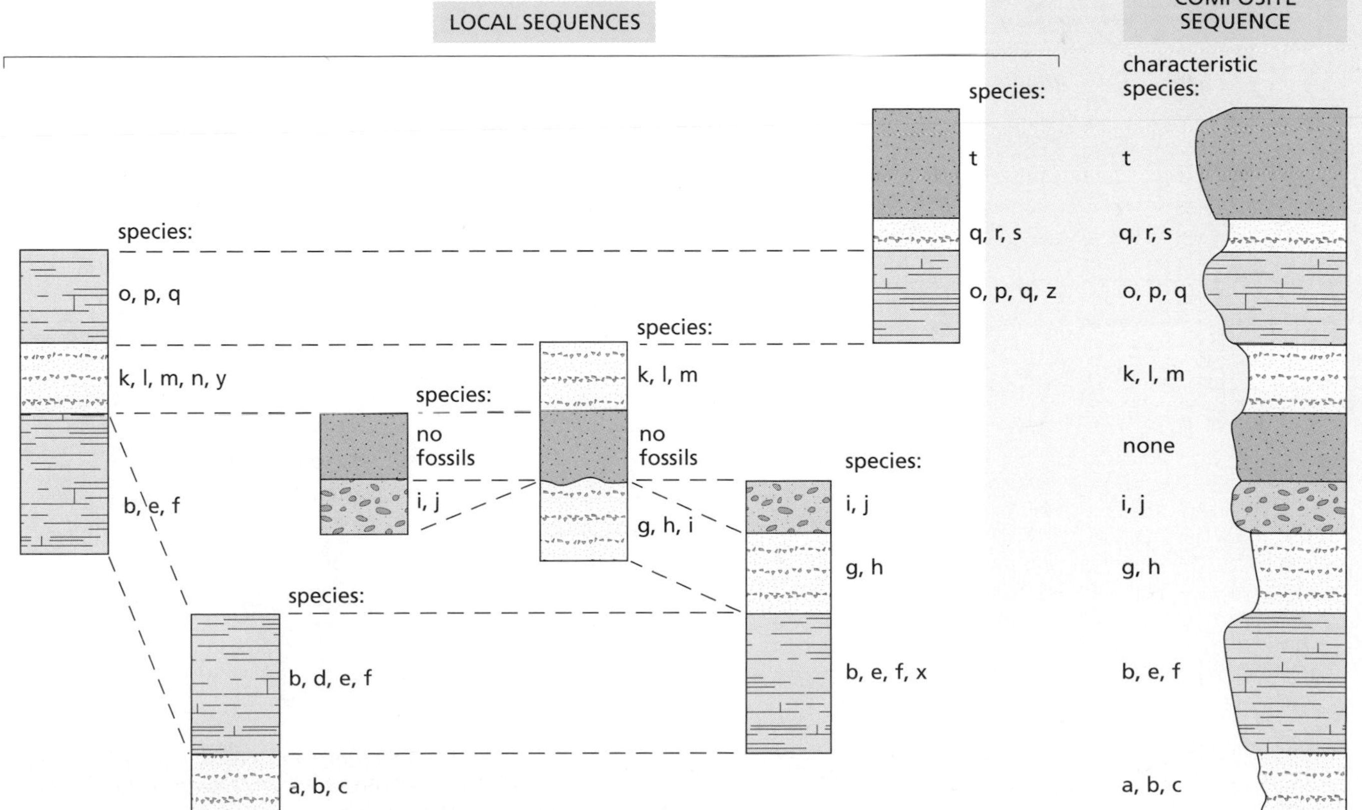

the living cephalopods can be explained with reference to this phylogeny. The chambered nautilus is very different from other living cephalopods because it is fully housed within a coiled shell and has four gills, while the squids and octopuses have only two gills and a very small, reduced shell or else none at all. One would therefore imagine a family tree in which octopuses and squids have a common ancestor that the chambered nautilus does not share. The fossil Cephalopoda conform to these expectations. The group of cephalopods with the oldest fossil record are the nautiloids, of which the chambered nautilus is the only living remnant. A second group of cephalopods, called the ammonoids (see Chapter 15, p. 646), flourished in Mesozoic times, during the age of dinosaurs. A small, third group had an internal shell that became reduced in size. When the ammonoids became extinct, this third group the Dibranchiata, persisted and is represented today by the squids and octopuses. Thus, the fossil record of the cephalopod mollusks, including both the anatomy and age relationships of fossil forms, confirms the relationships hypothesized on the basis of the anatomy of the living forms.

The fossil record has repeatedly confirmed hypotheses of descent for particular living species. For example, Thomas Henry Huxley, one of Darwin's early supporters, studied the anatomy of birds and declared them to be "glorified reptiles." The interpretation of birds as descendents

FIGURE 4.10

Family tree of the class Cephalopoda (phylum Mollusca), showing branching descent over time. Horizontal width represents number of species in each group; vertical distance represents time.

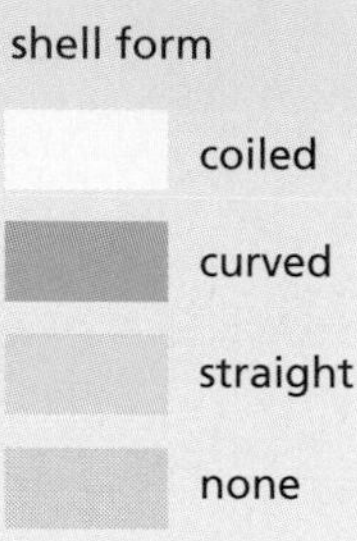

of the reptiles was strengthened by the discovery of *Archaeopteryx*, a fossil with many birdlike and also many reptilian features. Among the reptilian features of *Archaeopteryx* were a long tail, simple ribs, a simple breastbone, and a skull with a small brain and tooth-bearing jaws (Figure 4.11). Despite these reptilian features, *Archaeopteryx* had well-developed feathers and was probably capable of sustained flight. The discovery of transitional forms like *Archaeopteryx* strengthens our confidence in the hypothesis that birds evolved from reptiles. Other transitional forms are known, such as those between older and more modern bony fishes, between fishes and amphibians, and between reptiles and mammals. Instead of being exactly intermediate in each trait, transitional forms like *Archaeopteryx* usually exhibit a mix of some advanced characteristics and some primitive characteristics.

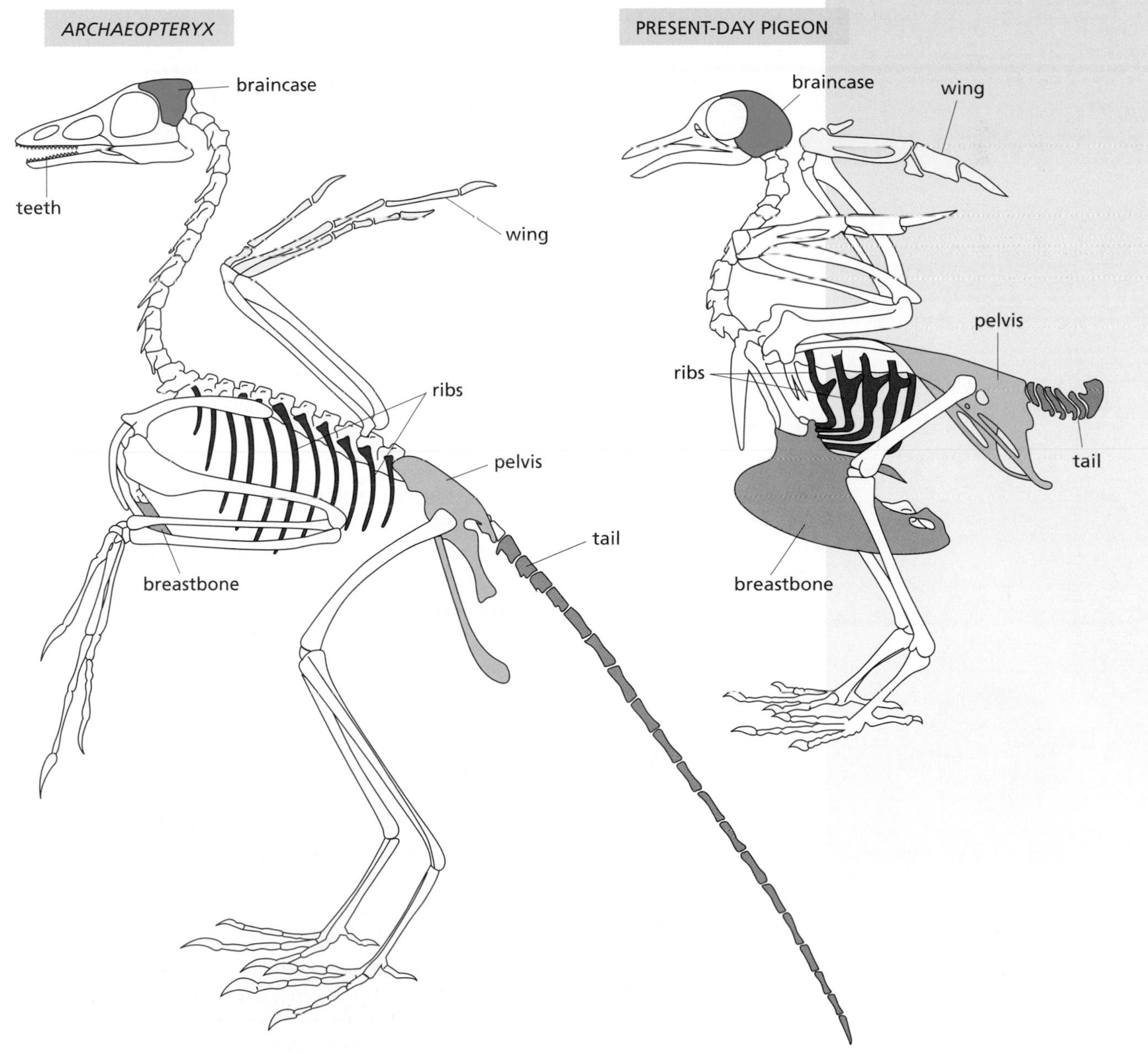

FIGURE 4.11

The early bird *Archaeopteryx*, compared with a modern pigeon. Modern birds lack teeth, and evolution has enlarged the braincase and strengthened other parts (wing, ribs, breastbone, pelvis, tail) highlighted here.

THOUGHT QUESTIONS

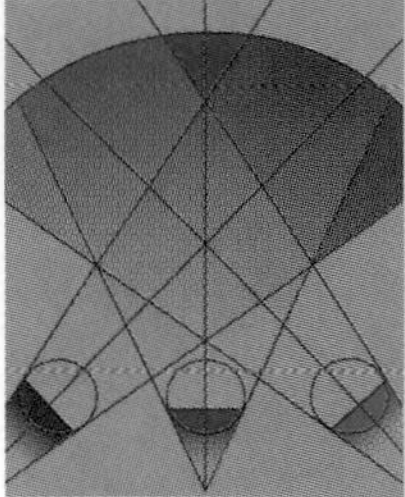

1. For a family tree such as that shown in Figure 4.10, what kinds of fossil evidence (be specific) would falsify the descent pattern shown? What kinds of evidence would cause paleontologists to modify the family tree but continue to believe in a process of descent with modification? What kinds of evidence would falsify the hypothesis of descent with modification?
2. One of the Galapagos finches studied by Darwin has woodpeckerlike habits and certain woodpeckerlike features: it braces itself on vertical tree trunks with stiff tail feathers in the manner of true woodpeckers and drills holes for insects with a chisel-like bill. However, it lacks the long, barbed tongue with which true woodpeckers spear insects; it uses cactus thorns for this purpose instead. How would Lamarck have accounted for this set of adaptations? How would Paley? How would Darwin? Which of these explanations accounts for the absence of the barbed tongue in the woodpecker finch? How would each hypothesis account for the absence of true woodpeckers on the Galapagos Islands?

CREATIONISTS CHALLENGE EVOLUTIONARY THOUGHT

Opposition to the idea that life evolves has come from various quarters. Many opponents of evolution have been **creationists**, people who believe in the fully formed creation of species by God. Most creationists base their beliefs on the biblical book of Genesis. In this section we discuss creationist ideas of the nineteenth century, of the early twentieth century, and of our own time.

Creationism and Darwin's response

In the eighteenth and nineteenth centuries, many creationists were also scientists who proposed and tested hypotheses. For example, Reverend William Paley and his supporters proposed that biological adaptations were the work of a benevolent God. Paley pointed to the structure of the heart in human fetuses as containing features that adaptation to the local environment could not account for. In adult mammals, including humans, the blood on the left side of the heart is kept separate from the blood on the right side of the heart (see Chapter 8, p. 326). In fetal mammals, the blood runs across the heart from the right side to the left, bypassing the lungs, which are collapsed and nonfunctional before birth. As the blood enters the left side of the heart, it passes beneath a flap that is sticky on one side. When the baby is born, its lungs fill, and blood flows through them. The blood returning to the heart from the lungs now builds up sufficient pressure on the left side that the flap closes. Because it is sticky on one side, it seals shut. No amount of adaptation to the environment, said Paley, could endow a fetus with a valve that was sticky on one side so that it would seal shut at birth. Only a power with foresight could have realized that the fetus would need a heart whose pattern of blood flow would change at birth, and thus designed the sticky valve. Paley attributed the foresight to God, and he insisted that no other hypothesis could explain such an adaptation to future conditions.

What is most interesting to a modern reader is that Paley and his many supporters understood the nature of science and used the methods of science to argue their case. Paley in particular sought scientific proof of God's existence and benevolence by arguing that no other hypothesis could explain the evidence as well. This example shows that good science is certainly compatible with a belief in God or a rejection of evolution. In fact, the best scientists of the period from 1700 to 1859 were, with few exceptions, devout men who rejected the pre-Darwinian ideas of evolution on scientific grounds.

Darwin was quite familiar with Reverend Paley's arguments, and he offered evolutionary explanations for many of the intricate and marvelous adaptations that Paley's supporters had described. In each case, Darwin argued that the hypothesis of natural selection could account for the adaptation as well as the hypothesis of God's design could. There were also some adaptations that were *less* than perfect, or that seemed to be 'making do' with the materials at hand. The gills in barnacles are modified from a brooding pouch that once held the eggs. The milk glands of mammals are modified sweat glands. The giant panda, evolved from an ancestor that had lost the true thumb, developed a new thumblike structure made from a little-used wrist bone. (This last example was not known in Darwin's time, but fits well into Darwin's argument.) These many adaptations seem more easily explained by natural selection than by God's design because God could presumably have 'done better.' Natural selection is limited to the use of the materials at hand, and then only if there is variation; an omnipotent God could have made barnacle gills from entirely new material without taking away the brood pouches, and could have given pandas a true thumb instead of modifying a wrist bone. Darwin and his supporters used examples like these to show that the evolutionary explanation fitted the available evidence better than Paley's explanation of divine planning. For example, natural selection perpetuates only those hearts whose flaps seal properly at birth.

Early twentieth-century creationism

In the early twentieth century, most opposition to evolution came from certain Protestants in the United States (but not in Europe), who declared that evolution conflicted with the account of creation given in the Bible. These people founded a number of societies, including the Society for Christian Fundamentals (the origin of the term *fundamentalist*). They were creationists who insisted on the biblical account of creation. The fundamentalists persuaded several state legislatures to pass laws restricting or forbidding the teaching of evolution in schools. Some of these state laws remained on the books until the 1960s.

In 1925, a famous court case was brought in Tennessee by the fledgling American Civil Liberties Union. A teacher, John H. Scopes, was arrested for reading a passage about evolution to his high school class. The trial attracted worldwide attention. Scopes lost and was assessed a $100 fine. Upon appeal, the case was thrown out because of the way in which the fine had been assessed; the merits of the case were never really debated. The Scopes trial did, however, have a chilling effect on the textbook publishing industry: books that mentioned evolution were revised to take the subject out, and most high school biology texts published in the United States between 1925 and 1960 made only the barest reference, if any, to Charles Darwin or any of his theories.

Creationism today

The Soviet launch of the Earth-orbiting satellite Sputnik in 1958 set off a wave of self-examination in American education. Groups of college and university scientists began examining high school curricula with renewed vigor, and several new high school science texts were written. Most of the new biology texts emphasized evolution, or at least gave it prominent mention.

Alarmed in part by the new textbooks, a new generation of creationists began a series of attacks on the teaching of evolution. These new creationists tried to portray themselves as scientists, calling their new approach 'creation science' even though they never conducted experiments or tested hypotheses. Instead of making their studies falsifiable, the new creationists claimed that they held the absolute truth:

> Biblical revelation is absolutely authoritative.... There is not the slightest possibility that the *facts* of science can contradict the Bible and, therefore, there is no need to fear that a truly scientific comparison...can ever yield a verdict in favor of evolution.... The processes of creation...are no longer in operation today, and are therefore not accessible for scientific measurement and study. (H.M. Morris (Ed.). *Scientific Creationism*. San Diego: Creation-Life Publishers, 1974, pp. 15–16 and 104.)

> We do not know how the Creator created, what processes He used, for He used processes which are not now operating anywhere in the natural universe.... We cannot discover by scientific investigations anything about the creation process used by the Creator. (D.T. Gish. *Evolution: The Fossils Say No!*. San Diego: Creation-Life Publishers, 1978, p. 40.)

In contrast, Charles Darwin knew that his theories were—rightly—subject to the principle of falsifiability:

> If it could be demonstrated that any complex organ existed, which could not possibly have been formed by numerous successive, slight modifications, my theory would absolutely break down. (C. Darwin. *Origin of Species*. London: John Murray, 1859, p. 189.)

Some creationist writings also contain faulty explanations of many scientific concepts, including the second law of thermodynamics. According to this law, a closed system (one in which energy neither leaves nor enters) can only change in the direction of less order and greater randomness. Thus, a building may crumble into a pile of stones, but a pile of stones cannot be made into a building without the expenditure of energy. Creationists have claimed that this law precludes the possibility of anything complex ever evolving from something simpler. The second law of thermodynamics does not, however, rule out the building up of complexity; rather, it states that making something complex out of something simple requires an input of energy. The second law of thermodynamics *does* apply to all biological processes. If the Earth were a thermodynamically closed system, life itself would soon cease. However, the Earth is not a thermodynamically closed system because energy is constantly being received from the sun, and this energy allows life to persist and evolve.

In the 1960s, because many of the laws in the United States forbidding the teaching of evolution had been declared unconstitutional, the new creationists, led by Henry Morris, Duane Gish, and John Slusher, decided on a new approach. Evolution could be taught in the schools, they argued, but only if 'creation science' was taught along with it and given equal time. (The concept of 'equal time,' was originally a measure to ensure fairness in political campaigns.) A few state legislatures passed laws inspired by this new group of creationists. An Arkansas law known as the Balanced Treatment Act (Public Law 590) was finally declared unconstitutional in 1981, and a similar Louisiana law was declared unconstitutional a few years later. Interestingly, in the challenges to these laws, the scientific issues *were* raised in court, and prominent scientists were called upon to testify. Specifically, in the Arkansas and Louisiana cases, the U.S. Court of Appeals was asked to rule on what is scientific and what is not. The court finally ruled that evolution is a scientific theory and may be taught, whereas 'creation science' is not science at all because it involves no testing of hypotheses and because its truths are considered to be absolute rather than provisional. Instead, 'creation science' was found to be a religion, or to include so many religious concepts (creation by God, Noah's flood, original sin, redemption, and so forth) that it could not be taught in a public school without violating the U.S. Constitution's historic separation of church and state.

Creationists continue to exert influence today. Despite state laws that have been declared unconstitutional, creationists have vowed to continue pressuring each local school board and each state legislature or Department of Education to support their approach. These efforts have sometimes been successful. In 1999, the state school board in Kansas approved a statewide science curriculum that included no mention of evolution. They also approved a statewide program for testing scientific knowledge and understanding, but decided that an understanding of evolution should not be part of this testing program, not even with such qualifiers as "according to the theory of evolution." These decisions were all made at the urging of creationists.

Opposition to the teaching of evolution is largely an American phenomenon. Biologists in most countries other than the United States have not faced similar opposition.

THOUGHT QUESTIONS

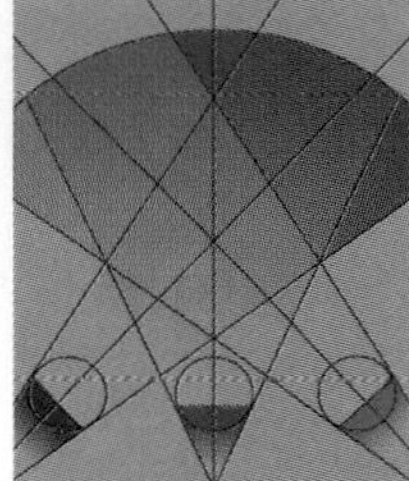

1. In what ways did William Paley use scientific evidence? Did he use falsifiable hypotheses? Do today's creationists use falsifiable hypotheses to support their claims?
2. How much time should be devoted in science classes to alternative explanations or theories that have been falsified? Should time be given to explanations that are not falsifiable hypotheses? Should all explanations be given equal time? How much (if any) of a science curriculum would you devote to divine creation as an alternative to evolution? To astrology as an alternative to astronomy? To the theory that disease is caused by demons or evil spirits?
3. Astronomer and mathematician Pierre Simon LaPlace is one of the authors of the idea that galaxies and solar systems form from swirling masses as the result of natural gravitational forces. When he published his book on this 'nebular hypothesis,' he presented a copy to the emperor Napoleon, who asked him why he had not mentioned God in his book. LaPlace replied, "I have no need of that hypothesis." Does this suggest a way in which scientists can reconcile their faith with the practice of science?
4. Do you agree with the Kansas education officials in their decision not to include evolution in their state science curriculum? What effect do you think their decisions will have on the future citizens of Kansas? Do you think there will be an influence beyond the study of evolution, on citizens' general level of scientific understanding? If you were in charge of a high-technology company, would a decision like the one in Kansas influence your choice of where to locate your company?
5. Does the teaching of unpopular or rejected theories encourage students to think critically? Does it encourage attitudes of fairness? Does it increase students' understanding of what science is and how science works?

SPECIES ARE CENTRAL TO THE MODERN EVOLUTIONARY PARADIGM

Although Darwin's theories about natural selection and branching descent continue to guide biological research to this day, the early 1940s saw an expansion of the evolutionary paradigm called the **modern synthesis**. Most of the Darwinian paradigm was retained in this expanded paradigm, but, to explain the source of heritable variation, the findings of genetics were also incorporated. The cornerstone of the modern synthesis paradigm is a theory of **speciation**, the process by which one species branches into two species.

Populations and species

A biological **population** consists of those individuals within a species that can mate with one another in nature. If we look backward in time, we realize that any two individuals in a population share at least some of their alleles because of common descent. If we look into the future, we see that any two opposite-sex individuals in a population are potential mates. Membership in a population is determined by descent and by the capacity to interbreed.

Biological populations within a species may exchange hereditary information (alleles) with one another. The combining of genetic information from different individuals or the exchange of genetic information between populations is called **interbreeding**. The existence of biological barriers to such exchange is called **reproductive isolation**. Interbreeding between populations of the same species takes place when members of different populations mate and produce offspring; reproductive isolation inhibits such matings to varying degrees.

Species are defined as *reproductively isolated groups of interbreeding natural populations*. There are several points to note in this definition. Physical characteristics (morphology) are not part of the definition of species; species are defined by breeding patterns instead. Populations belonging to the same species will interbreed whenever conditions allow them to. Populations belonging to different species are reproductively isolated from one another and will thus not interbreed. Any biological mechanism that hinders the interbreeding of these populations is called a **reproductive isolating mechanism**, as explained below. Species are composed of natural populations, not of isolated individuals. Thus, the mating behavior of individuals in captivity can only serve as *indirect evidence* of whether *natural populations* would interbreed under natural conditions.

The many reproductive isolating mechanisms fall into two broad categories, **premating mechanisms** and **postmating mechanisms**. Premating mechanisms that prevent the exchange of gametes are more efficient and are therefore favored by natural selection over postmating mechanisms that act later. Premating mechanisms include **ecological isolation**, in which potential mates never encounter each other, either because they live in different habitats, or because they are active at different times of day or in different seasons, or because they are not physiologically capable of reproduction at the same time. Figure 4.12 shows that wood frogs are fully isolated ecologically from tree frogs and bullfrogs by breeding at different seasons; they are partly isolated from pickerel frogs because the breeding seasons overlap only slightly. Premating reproductive isolating mechanisms also include **behavioral isolation**, in which potential mates live together in the same place but do not mate because their courtship rituals differ. For example, different species of fireflies (phylum Arthropoda, class Insecta, order Coleoptera, family Lampyridae) recognize their mates on the basis of different flashing

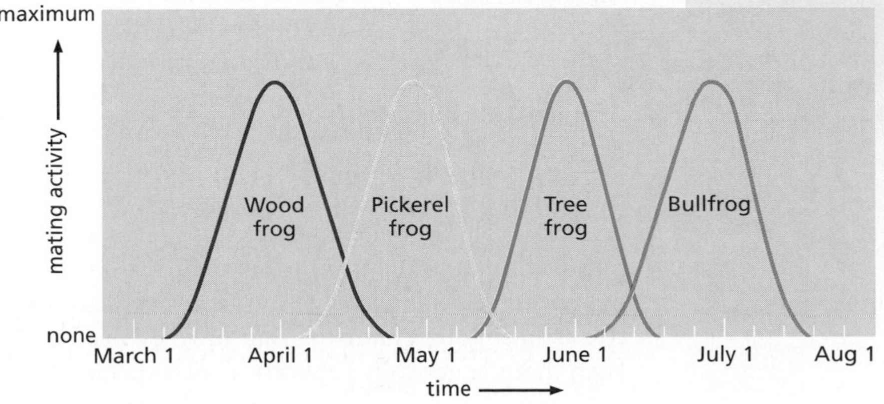

FIGURE 4.12

Reproductive isolation of several frog species by season of mating, an example of premating ecological isolation.

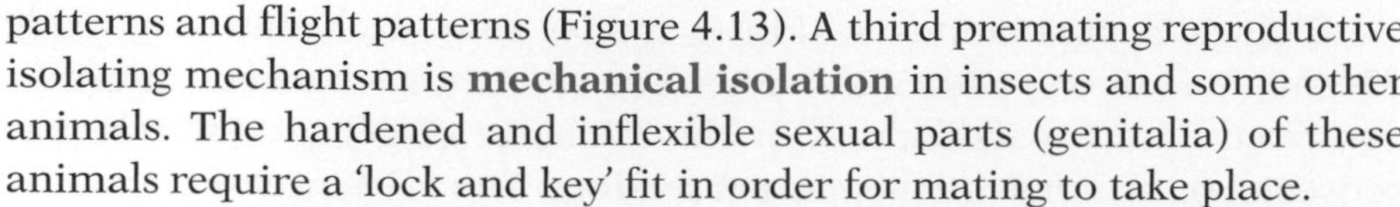

patterns and flight patterns (Figure 4.13). A third premating reproductive isolating mechanism is **mechanical isolation** in insects and some other animals. The hardened and inflexible sexual parts (genitalia) of these animals require a 'lock and key' fit in order for mating to take place.

Various postmating reproductive isolating mechanisms also exist. In animals, sperm from a male of another species may die before fertilization takes place. In plants, the pollen may fail to germinate on the flowers of another species. If a mating takes place between species, the fertilized egg may die after fertilization. Incompatible chromosomes may disrupt cell divisions and developmental rearrangements, leaving the embryo or larva to die. Alternatively, hybrid individuals may live for a while but not reach reproductive age, or they may be sterile. The mule, a sterile hybrid between a horse and a donkey, is an example.

FIGURE 4.13

Flashing patterns used as mating signals by different species of fireflies. The species 1 to 9 are reproductively isolated from one another by the behavioral differences shown in these patterns. Details in this form of behavioral isolation include the duration of each flash, the number of repetitions, and the location of the insect when it flashes. A firefly will respond only to the flashing pattern of its own species.

How new species originate

To explain how a new biological species has come into existence, we need to explain how it has become reproductively isolated from closely related species. The origin of a species is thus the origin of one or more reproductive isolating mechanisms.

In the vast majority of cases, new species have come into existence through a process of speciation that includes a period of **geographic isolation** in which populations are separated by some sort of barrier such as a mountain range or simply an uninhabited area that the organisms do not cross. The essence of the theory is that reproductive isolating mechanisms originate during times when such barriers separate populations geographically. Geographic isolation is not by itself considered to be a reproductive isolating mechanism; rather, it sets up the conditions under which the separated populations may evolve along different lines, resulting in reproductive isolation.

What happens depends in part on the length of time for which the populations are geographically isolated—more time allows more chances for reproductive isolating mechanisms to evolve. Another factor is that natural selection must favor different traits on the two sides of the geographic barrier. That is, conditions must be different enough for one set of traits to increase fitness in one locale and for a different set of traits to increase fitness in another locale. If the populations on opposite sides of the barrier are selected in different directions for a long enough period, then one or more reproductive isolating mechanisms may evolve between the two groups of populations and separate them into different species (Figure 4.14). If the populations later come into geographical contact again, the reproductive isolating mechanisms that have evolved during their separation will keep them genetically separate as two species. For example, frog or cricket populations isolated on opposite sides of a mountain chain or a large body of water may develop different mating calls. Because the animals respond only to the mating calls of their own population, the two populations will be reproductively isolated.

The geographic theory of speciation predicts that examples of incomplete speciation may be discovered. If two populations are separated for a very long time (or if selective forces on opposite sides of a barrier differ greatly), then the populations are likely to split into two species. If the separation is brief, then speciation is unlikely. These two situations lie at opposite ends of a continuum. Somewhere along this continuum lies the situation in which populations have been separated by a geographic barrier long enough for reproductive isolation to begin evolving, but not yet long enough for the reproductive isolation to be perfected. Partial or imperfect reproductive isolation between two populations would lessen the chances of interbreeding between them, but not prohibit it entirely. Such situations have indeed been found, for example, among the South American fruit flies known as *Drosophila paulistorum*. Crosses among divergent populations of *Drosophila paulistorum* produce fertile hybrid females but sterile hybrid males. The geneticists studying these flies referred to them as "a cluster of species *in statu nascendi*" ("in the process of being born").

Higher taxa

People intuitively group all insects together and all birds together. We all have names for collective groups of similar species: birds, snakes, insects, pines, orchids, and so forth. Some collective groups, such as beetles, are contained within larger groups, such as insects. Biologists organize these collective groups into a **classification**, an arrangement of larger groups that are subdivided into smaller groups each reflecting their degree of evolutionary relatedness.

Any one of these collective groups, such as insects or orchids, is

FIGURE 4.14

Geographic speciation: the evolution of reproductive isolation during geographic isolation. Genetically variable populations that spread geographically can develop locally different populations that are capable of interbreeding with one another initially. If the populations are separated for a long enough time by a barrier such as a mountain range or a deep canyon, they may develop differences that prevent interbreeding even after contact is resumed.

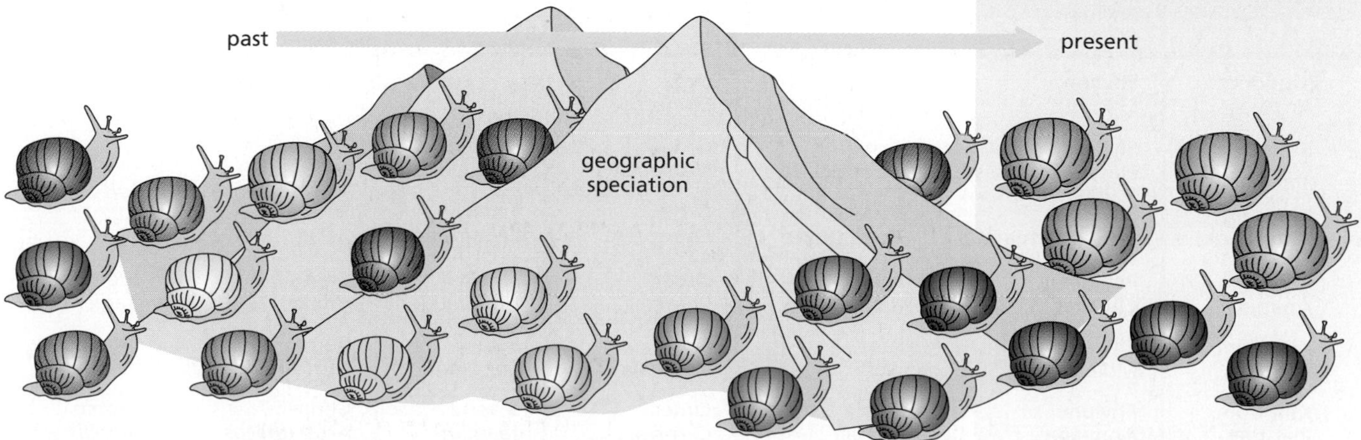
past
present
geographic speciation
Initial population has lots of genetic variation
Mountain range arises, separating population into two groups
Environment becomes different on the two sides
Two populations diverge as mutation and selection fit organisms to environment
When populations come into contact again, reproductive isolating mechanisms keep species genetically separate

called a **taxon**, and **taxonomy** is the study of how these taxa (plural of *taxon*) are recognized and how classifications are made. In a biological classification, species are grouped into successively more inclusive groups, or higher taxa: related species are grouped into **genera** (singular, **genus**), related genera into **families**, related families into **orders**, related orders into **classes**, related classes into **phyla**, and related phyla into **kingdoms** such as the animal or plant kingdoms. All these are arranged as groups within groups, with the less inclusive (smaller) groups sharing more characters and the more inclusive (larger) groups sharing fewer characters. (Thus, species within a genus share more characters in common than do families within an order.)

For example, human beings constitute the species *Homo sapiens*. Figure 4.15 shows, beginning on the right, that *Homo sapiens* is grouped together with *Homo erectus* and certain other fossil species into the genus *Homo*. (A genus always has a one-word name that is capitalized; a species has a two-word name in which the first word is the name of the genus; after the two-word name has been introduced, the genus may subsequently be abbreviated, for example, *H. sapiens*.) The genus *Homo* is grouped together with the extinct genera *Australopithecus* and *Ardipithecus* into the family Hominidae. This family is included in the order Primates, which also includes apes (Pondigae), monkeys (Cebidae and Cercopithecidae), and lemurs (Lemuridae). The primates are grouped together with rodents (Rodentia), carnivores (Carnivora), bats (Chiroptera), whales (Cetacea), and over two dozen other orders into the class Mammalia, including all warm-blooded animals with hair or fur that feed milk to their young. Mammals are one of several classes in the phylum Chordata, a group that includes all vertebrates (animals with backbones) and a few aquatic relatives such as the sea squirts and amphioxus. The Chordata and several dozen other phyla are together placed in the animal kingdom (Animalia). Animals are one of the several kingdoms currently recognized; these kindgoms are the topic of a later section in this chapter.

FIGURE 4.15

The place of the species *Homo sapiens* in the classification of organisms. Reading from right to left shows the increasingly more inclusive taxa to which our species belongs. Reading from left to right focuses in on taxa of increasingly narrow scope.

KINGDOMS	PHYLA	CLASSES	ORDERS	FAMILIES	GENERA	SPECIES
	Phylum Porifera					
Kingdom Archaea	Phylum Cnidaria	Class Agnatha	Order Monotremata			
Kingdom Eubacteria	Phylum Platyhelminthes	Class Chondrichthyes	Order Marsupialia	Family Cebidae		
Kingdom Protista	Phylum Echinodermata	Class Osteichthyes	Order Chiroptera	Family Lemuridae		
Kingdom Animalia	Phylum Chordata	Class Mammalia	Order Primates	Family Hominidae	Genus *Homo*	*Homo sapiens*
Kingdom Mycota	Phylum Mollusca	Class Aves	Order Rodentia	Family Pongidae	Genus *Australopithecus*	*Homo erectus*
Kingdom Plantae	Phylum Annelida	Class Reptilia	Order Carnivora	several other families	Genus *Ardipithecus*	*Homo habilis*
	Phylum Arthropoda	Class Amphibia	Order Artiodactyla			
	many other phyla	several other classes	many other orders			

THOUGHT QUESTIONS

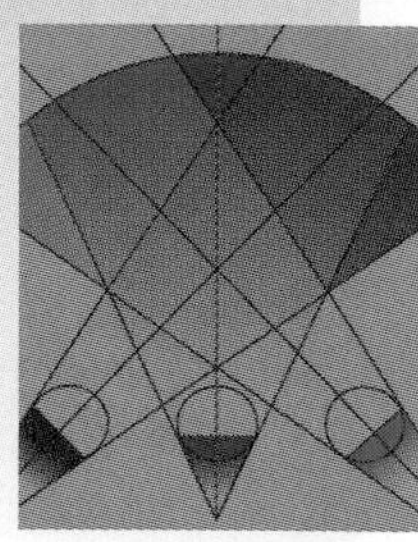

1. What kinds of reproductive isolating mechanisms might prevent related species of antelopes from interbreeding? Answer the same question for related species of birds, related species of trees, and related species of butterflies.

2. Several kinds of organisms reproduce asexually, producing offspring without combining gametes from two parents. Can asexually reproducing organisms belong to species? Can the definition of species be modified to apply to asexually as well as sexually reproducing organisms?

LIFE ORIGINATED ON EARTH BY NATURAL PROCESSES

In addition to explaining how species change and how new species arise, modern evolutionary theory also accounts for the origins of life on Earth.

Modern thoughts on the origin of life on Earth began with the biological theories of the early twentieth-century Russian biochemist Aleksandr Oparin. Most scientists at the time accepted the French microbiologist Louis Pasteur's conclusions that living organisms always came from preexisting organisms. But where had the first living organisms come from? Oparin, a Marxist, rejected the possibility of a divine or other miraculous creation. Had life always existed? The evidence from astronomy was that the Earth originated under conditions that could not have supported life. Oparin also knew that the conditions in interplanetary space, such as extreme cold (around –270 °C), utter dryness, and constant bombardment by high levels of ultraviolet radiation, were incompatible with all forms of life. These facts convinced Oparin that life could not have come through space from anywhere else. Life must have originated on planet Earth.

Pasteur had said, "there is now no circumstance known in which it can be affirmed that microscopic beings came into the world without...parents similar to themselves." From this, Oparin reasoned that, if present conditions do not permit organisms to originate from nonliving matter, then life must have originated under conditions very different from those that prevail today. Oparin searched for clues about what the conditions on the primitive Earth might have been. From studying what was known about the chemical composition of the solar system, Oparin postulated an early atmosphere consisting of the gases hydrogen (H_2), ammonia (NH_3), methane (CH_4), and water vapor (H_2O), with hydrogen in greatest abundance. Oparin worked out some of the chemical reactions that he thought might have produced simple biological molecules such as sugars and organic acids. He theorized that these molecules would have built up in the primitive oceans once the Earth's temperature permitted water to become liquid. By the slow accumulation of these molecules, the world's oceans would have become like a "hot, dilute soup." Oparin published his detailed hypothesis in book form in 1935, and the book was soon translated into English under the title *The Origin of Life*.

In 1952, an American biochemist named Stanley Miller decided to test Oparin's model experimentally. He built several types of apparatus, such as the one shown in Figure 4.16. The apparatus was filled with hydrogen, ammonia, methane, and water vapor, the four gases that had been postulated by Oparin. The gases reacted in a chamber in which electric sparks simulated atmospheric lightning. Reaction products were cooled so that water became liquid, simulating rain. Compounds that formed in the reaction chamber dissolved in the water, which collected in the lower part of the apparatus, simulating the ponds and oceans of the primitive Earth. A heat supply vaporized some of the liquid and returned it to the reaction chamber. Miller circulated his reaction mixture for several days and withdrew samples to analyze the results. Among the compounds that had been formed he found amino acids (building blocks for proteins), simple sugars, and most of the building blocks for DNA and RNA. The building blocks for all major biological molecules had thus been formed, without the aid of any organisms, under conditions simulating those of the primitive Earth.

Miller then proceeded to repeat his experiment under somewhat altered conditions, and several other scientists have also done so. They found that differences in the starting materials did not greatly affect the experimental results, as long as sources of hydrogen, oxygen, nitrogen, and carbon were all present. For example, carbon dioxide or another carbon compound could substitute for methane, and nitrogen gas or nitrogen oxides could substitute for ammonia. An energy source was also needed, but the particular kind of energy was not important. An ultraviolet light source could successfully substitute for the electric sparks; so could natural sunlight, even with no other heat source.

To show that the reaction products were produced by the experimental reactions, and not by any biological contaminants, Miller also did a control experiment: he repeated all experimental conditions except for the electric sparks. Without the energy source provided by these electric sparks, no reaction products were detected.

Experiments simulating primitive Earth conditions have succeeded in producing so many biologically important products that the following conclusion is now inescapable: all of the molecules important to life *could have been produced* in a lifeless environment on the primitive Earth, starting with nothing more than a few basic gases mixed together as a simulated atmosphere. However, while these experiments show us what *could* have happened, they cannot show us what actually *did* happen.

FIGURE 4.16

Stanley Miller's experiment, in which amino acids and other molecules used by living organisms were produced. Heating the flask at the lower left boils the water and keeps the mixture circulating in the direction shown by the arrows. Reactions take place in the spark chamber and reaction products are condensed and recirculated. Valve A is used to sterilize the apparatus and to introduce the starting materials; valve B is used to withdraw samples of the reaction products.

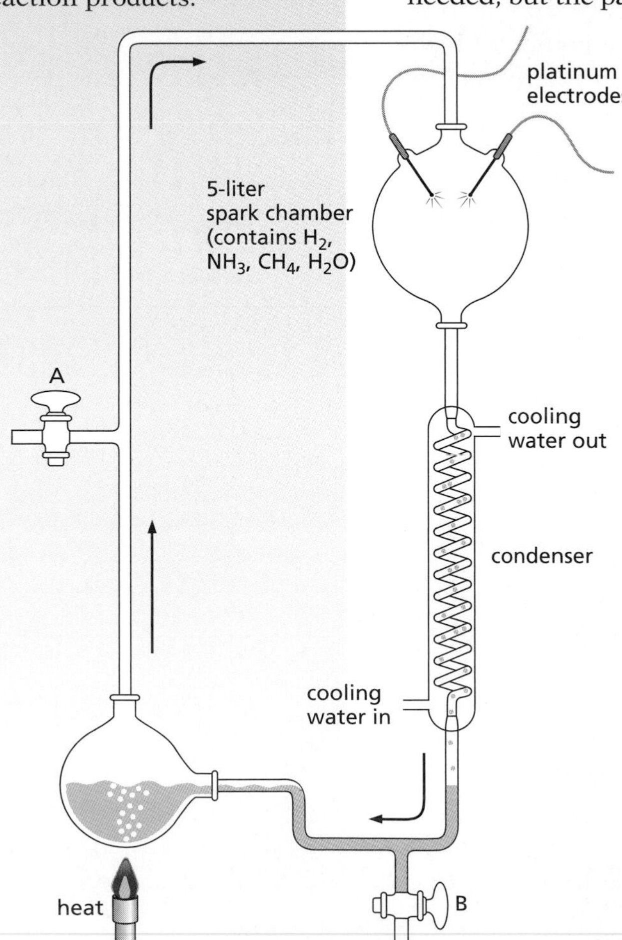

Evidence of early life on Earth

Is there any evidence of the events that actually took place on the primitive Earth? There is some evidence, but it is incomplete and indirect. Ancient rocks on Earth over a billion (10^9) years old contain various kinds of 'chemical fossils,' compounds that give us clues to the conditions that prevailed at the time when these compounds were formed. Our present atmosphere differs greatly from the primordial atmosphere postulated by Oparin because free oxygen (O_2) is now abundant. The oldest rocks on Earth contain compounds that would not have persisted under oxygen-rich conditions, and thus seem to have been deposited at a time when the atmosphere contained little or no oxygen.

Evolution of our atmosphere. How did the present oxygen-rich atmosphere of the Earth originate? Most scientists who have investigated this question have concluded that life itself was primarily responsible. Most important is the finding that chemicals formed from the breakdown of chlorophyll molecules are not present in the oldest rocks. Chlorophyll is a key molecule in photosynthesis, a reaction by which plants and blue-green bacteria (Cyanobacteria) produce oxygen (see Chapter 14, pp. 586–589). The breakdown products of chlorophyll first appear in the geological record at about the time that oxygen in the atmosphere began to increase slowly, over a period of about a billion years.

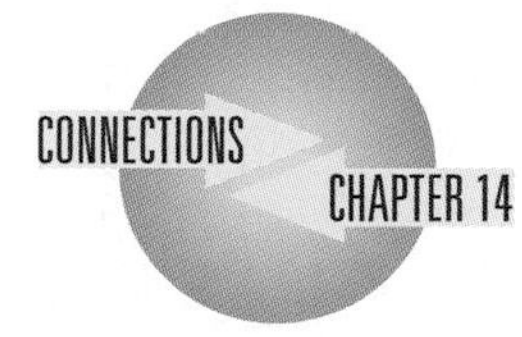

The first forms of life had to live under conditions in which no oxygen was present, conditions called **anaerobic**. A variety of bacteria are capable of living under anaerobic conditions. Scientists now hypothesize that the first bacteria would have had to derive all their energy from the high-energy molecules that they found in their surroundings. Nowadays, such molecules are in most cases produced by other organisms, but the first organisms would have had to rely on molecules that had formed without life (abiotically), under conditions like those simulated in experiments like Miller's. For many thousands or maybe millions of years, the supply of these molecules may have been adequate for life to expand and perhaps to flourish. (We cannot tell for sure, because such organisms leave very few fossil traces of their existence.) Eventually, however, the populations of organisms expanded to the point that the limited amount of energy-rich chemicals in the environment was just not enough. This was possibly the first global environmental crisis in the history of life on Earth.

We can imagine several possible outcomes of this crisis. Some organisms may have discovered a way to attack and devour other organisms, getting their nutrients from their prey. Organisms of this type continued to eat one another, but the total quantity of living organisms (the total biomass) that the planet could support remained limited. The problem may even have become worse because the waste products of organisms include gases such as carbon dioxide (CO_2), which would simply have escaped into the atmosphere, taking the carbon that organisms need out of reach.

Evolution of photosynthesis. A more permanent solution to the limited supply of energy-rich chemicals came into place much later, when some bacteria produced the first chlorophyll-like molecules, about 4 billion years ago. Such molecules enable life forms to use the sun's energy rather than energy derived from high-energy molecules or from other organisms. All forms of photosynthesis require some chemical that supplies hydrogen atoms and removes electrons. Bacterial photosynthesis uses hydrogen sulfide (H_2S) for this purpose, or iron compounds, or organic molecules derived from nucleic acids, but none of these compounds was abundant 4 billion years ago.

The greatest change to the Earth and its atmosphere resulted from the evolution of a new and more efficient kind of photosynthesis, using a new source of hydrogen: water, the most abundant hydrogen source on Earth, a source readily available in most environments. The splitting of water in photosynthesis generated a new atmospheric gas: oxygen (O_2) (see Chapter 14, pp. 586–589). The first organisms to evolve this kind of photosynthesis were blue-green bacteria (kingdom Eubacteria, phylum Cyanobacteria). During the next 2 billion years or so, these blue-green bacteria became the dominant form of life on Earth, reducing the abundance of atmospheric CO_2 to a small fraction of its former level, and slowly generating more and more O_2. Calculations of the photosynthetic capabilities of these blue-green bacteria show that they were quite capable of generating all the oxygen gas (O_2) in the Earth's atmosphere within half a billion years or so. Not only did all of the life forms on Earth arise by evolution, but our very atmosphere is a product of biological evolution.

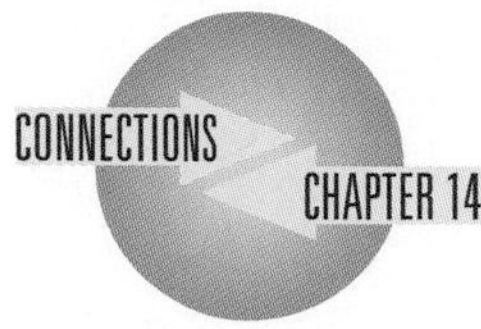

Procaryotic and eucaryotic cells

A great gulf separates the major types of organisms on the basis of the structure of their cells. The evolutionarily ancient Archaebacteria as well as the true bacteria and the related blue-green photosynthetic organisms (Cyanobacteria) are single-celled and procaryotic. Plants, animals, fungi, algae, and certain one-celled organisms such as amoebas have the more complex eucaryotic cells. All procaryotic organisms are single celled, while some eucaryotic organisms are single celled and others are multicellular. Thus the distinction between the two groups is based not on unicellularity versus multicellularity but rather on the structures of the cells themselves.

Procaryotic cells. The first organisms were simple cells with no internal compartments. A membrane called the plasma membrane formed an outer boundary of the cell and kept its contents inside. Simple cells of this type are called **procaryotic** cells (Box 4.1). The term *procaryotic* means 'first nucleus,' reflecting the theory that organisms of this type existed before the evolution of the true (eucaryotic) nucleus. All bacteria and cyanobacteria are procaryotic. Procaryotic cells lack most of the complex internal structures possessed by more advanced (eucaryotic) cells. Procaryotic cells have only a single chromosome consisting of nucleic acid only and no protein. The DNA double helix of this chromosome is joined end to end in a circular form resembling a closed necklace. The region of the cell containing this procaryotic chromosome is not surrounded by a membrane or set apart from the rest of the cell in any other way. Many procaryotes have fragments of DNA, called **plasmids**, that can detach from the chromosome, lead an independent existence for a long while, and then reincorporate into the chromosome (see Chapter 3, p. 97 and Chapter 14, p. 621).

Eucaryotic cells. Plants, animals, fungi, and certain other organisms called *protists* are composed of more complex **eucaryotic** cells (see Box 4.1). Eucaryotic cells have various internal parts, called **organelles**, that are bounded by intracellular membranes separating the various functions of the cell into different compartments. A defining characteristic of eucaryotic cells is the presence of a nucleus surrounded by a double membrane called the nuclear envelope (the name *eucaryotic* means 'true nucleus'). The nucleus contains rod-shaped chromosomes composed of DNA and proteins. Many types of cell organelles, such as mitochondria, Golgi apparatus, plastids, and endoplasmic reticulum, are shown and explained in Box 4.1. Eucaryotic cells also have an internal network of protein filaments called a **cytoskeleton**. These filaments determine the shape of the cell and keep many organelles in their positions. Contraction of these filaments may help to move the whole cell.

Keep in mind that there are a vast number of variations on these cell types, both among species and among the various specialized cells within multicelled organisms; there is probably no actual cell that exactly matches the accompanying diagrams.

Endosymbiosis. How did eucaryotic cells evolve? According to a theory first championed by the American cell biologist Lynn Margulis in the 1970s, eucaryotic cells arose from procaryotic cells by a process called **endosymbiosis** (literally, 'living together inside'). According to this theory, large procaryotic cells incapable of performing certain energy-producing chemical reactions (those of the Krebs cycle, described in Chapter 8, pp. 339–340) engulfed smaller procaryotic cells able to carry out these reactions. The larger (host) cells could obtain energy by digesting the smaller cells, but they could obtain even more energy if they allowed the smaller cells to go on living inside them and then used the products of the energy-producing reactions. In this situation, host cells that allowed the smaller cells to persist were favored by natural selection over host cells that digested the smaller cells. Over time, the smaller cells became energy-producing cellular organelles now known as **mitochondria**.

Those eucaryotic cells capable of photosynthesis have additional organelles called **plastids**, including **chloroplasts** and several other kinds. The pigments that carry out photosynthesis are contained within these plastids. Plastids are believed to have evolved by a process similar to that described above for mitochondria. In this case, however, the smaller cells were cyanobacteria capable of photosynthesis. The larger cells achieved greater growth potential by harboring these smaller cells rather than digesting them. The large cells containing these plastids did better and reproduced in greater numbers than similar cells that had no plastids, and cells with plastids ultimately persisted while those without plastids died out.

In support of the theory of endosymbiosis is the fact that both plastids and mitochondria have their own types of membranes and their own DNA, separate and different from the DNA of the eucaryotic host cells

BOX 4.1 Procaryotic and Eucaryotic Cells Compared

The great differences between procaryotic (bacterial) cells and eucaryotic cells (including both animal and plant cells) is shown in the accompanying drawings and also in chart form.

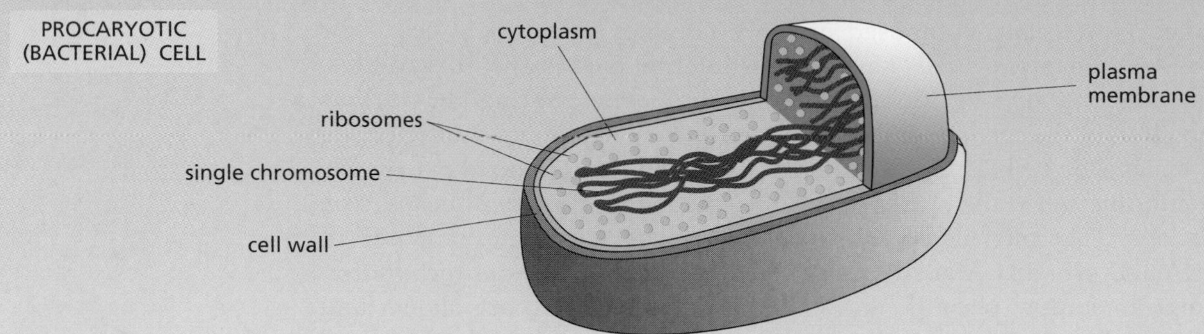

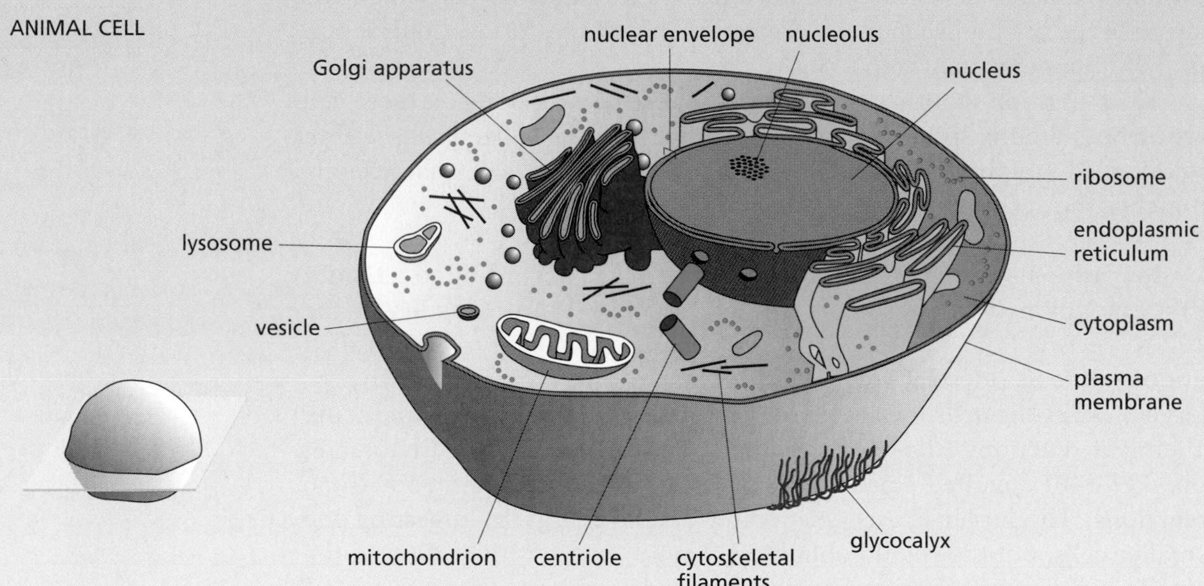

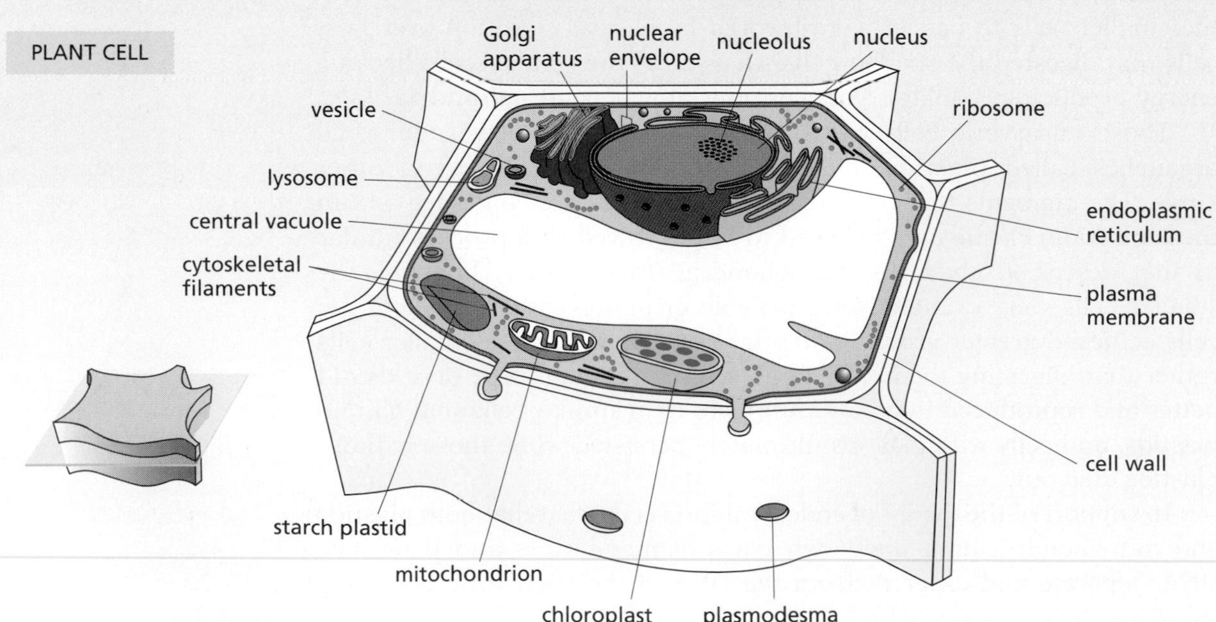

structure	function	present in procaryotic cells	present in eucaryotic cells	
			plant cells	animal cells
plasma membrane	protection; communication; regulates passage of materials	✓	✓	✓
DNA	contains genetic information	✓	✓	✓
nuclear envelope	surrounds genetic material		✓	✓
several linear chromosomes	contain genes that govern cell structure and activity		✓	✓
cytoplasm	gel-like interior of cell	✓	✓	✓
cytoskeleton	aids in cell and organelle movement and in maintaining cell shape		✓	✓
endoplasmic reticulum	transport and processing of many proteins		✓	✓
Golgi apparatus	adds sugar group to proteins and packages them into vesicles		✓	✓
ribosomes	protein synthesis (translation) along mRNA	✓	✓	✓
lysosomes	contain enzymes; aid in cell digestion; have a role in programmed cell death		✓	✓
mitochondria	provide cellular energy		✓	✓
chloroplasts and other plastids	capture sunlight; produce energy for cell		✓	
central vacuole	maintains cell shape; stores materials and water		✓	
flagella (whiplike appendages)	cell movement	simple propeller type	complex undulating type	complex undulating type
cilia (hairlike appendages)	cell movement; present only in certain types of cells			✓
cell wall	protects cell; maintains cell shape	✓	✓	
glycocalyx	surrounds and protects cell			✓
pili (hairlike appendages)	mating; adherence	✓		
plasmodesmata	cell-to-cell communication		✓	

that contain them, but similar to the membranes and DNA of procaryotic organisms. The presence of plastids is used in this book (and many others) as the defining attribute that determines the boundaries of the plant kingdom.

Kingdoms of organisms

Early classifications of organisms recognized only two kingdoms: plants and animals. Plants were distinguished as being nonmotile organisms with rigid cell walls, capable of using sunlight as a source of energy. Animals, in contrast, were recognized for their ability to move, their lack of cell walls, and their inability to derive energy directly from sunlight. This two-kingdom classification continued in use despite the discovery of animals that do not move and many bacteria and other organisms that do not fit well into either the plant or the animal kingdom.

Advances in our knowledge made possible by electron microscopy, particularly the discovery of the profound structural differences between procaryotic and eucaryotic cells, led to major classification changes. A five-kingdom classification system that became widely followed was first proposed in 1963.

Organisms did not change, but our arrangement of them changed because classifications are **socially constructed**, that is, devised by humans and agreed upon as a matter of social convention. To say that a classification scheme is socially constructed does not mean that the process of establishing a classification scheme is arbitrary or that all schemes are equally valid. Classifications are now usually understood as hypotheses about how organisms are related by patterns of descent. As more and more knowledge is accumulated about organisms, that knowledge is used to test the hypotheses and to replace rejected hypotheses with new ones. The division of living things into five kingdoms must be viewed as a widely accepted theory, not an unchanging set of facts. Indeed, most biologists now add a sixth kingdom, the Archaebacteria, more recently discovered but evolutionarily ancient. We use the six-kingdom classification in this book.

The kingdoms of organisms now recognized by biologists are listed in Box 4.2, along with a family tree showing how these kingdoms are thought to be related.

Procaryotic organisms

The Archaebacteria are one of the two procaryotic kingdoms. The Archaebacteria live only in very special environments. Some of them, the methane producers, live only under strictly anaerobic conditions (with absolutely no free oxygen present), such as inside cows' guts, where, in the course of consuming energy, they perform chemical reactions that produce methane gas (CH_4) as a byproduct. Other Archaebacteria, discovered only recently, live in extreme environments once thought incompatible with life. Some of these environments have very high salt concentrations. Others are at the edges of thermal vents in the ocean floor, where both heat and pressure are extreme. Still others are in hot springs. Scientists reasoned that adaptation for such extreme environments must require very different enzymes from those previously known, which proved to be so. One enzyme that can operate on DNA at very high temperatures is now used in the polymerase chain reaction, central to much work in biotechnology (see Chapter 3, p. 101).

The second procaryotic kingdom, now called Eubacteria, includes the more commonly known bacteria and the blue-green cyanobacteria, many of which have coevolved with multicellular host species. Many bacteria are free-living (in soil, for example), but many other species live only in a narrow range of host species. For example, most bacteria that use dogs as hosts cannot live in humans. We often think of bacteria as the 'germs' that cause disease, as many of them do. However, a far greater number of bacterial species are beneficial to humans, to other organisms, or to entire ecosystems. For example, bacteria decompose dead material into chemical forms that other organisms can then use to sustain life. Without such decomposition by bacteria, all other forms of life would soon cease. Certain bacteria (and a few cyanobacteria) are also important in the reactions of the nitrogen cycle (see Chapter 14, p. 591), which also sustains nearly all other species on Earth. Most of biotechnology depends on the use of bacterial plasmids and bacterial enzymes, and many industrial processes, including the making of cheese, yogurt, and sauerkraut, depend on chemical reactions performed by bacteria.

All eubacteria are single-celled organisms, but they often grow in colonies or filaments of many individual cells attached to a substrate or to one another. Some of these colonies have characteristics that differ from characteristics of single cells of the same species; they thus exhibit what may have been the first step in the evolution of multicellularity.

THOUGHT QUESTIONS

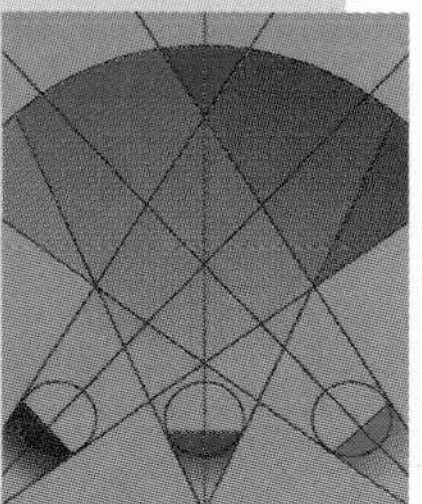

1. Which step in the origin of life do you think was the most important? Which step marks the boundary between the nonliving and living worlds? How much do we know about each step in the process of life's origin? What means of investigation do you think will bring us additional knowledge?
2. Are there limitations on the types of organisms that could evolve on Earth? How would we investigate such limitations? What limitations, if any, would apply to organisms on other planets or in other solar systems? How similar to organisms on Earth would such extraterrestrial organisms necessarily be?
3. Find four or more books on botany or general biology. List the phyla or divisions of the plant kingdom that each book recognizes. What similarities do you find? What differences do you find? How do you account for the differences?
4. How would a biologist decide which of several possible classifications is best? What would she or he look for? What kind of evidence is relevant? (These are among the most basic questions of taxonomy.)

BOX 4.2 The Six Kingdoms of Organisms

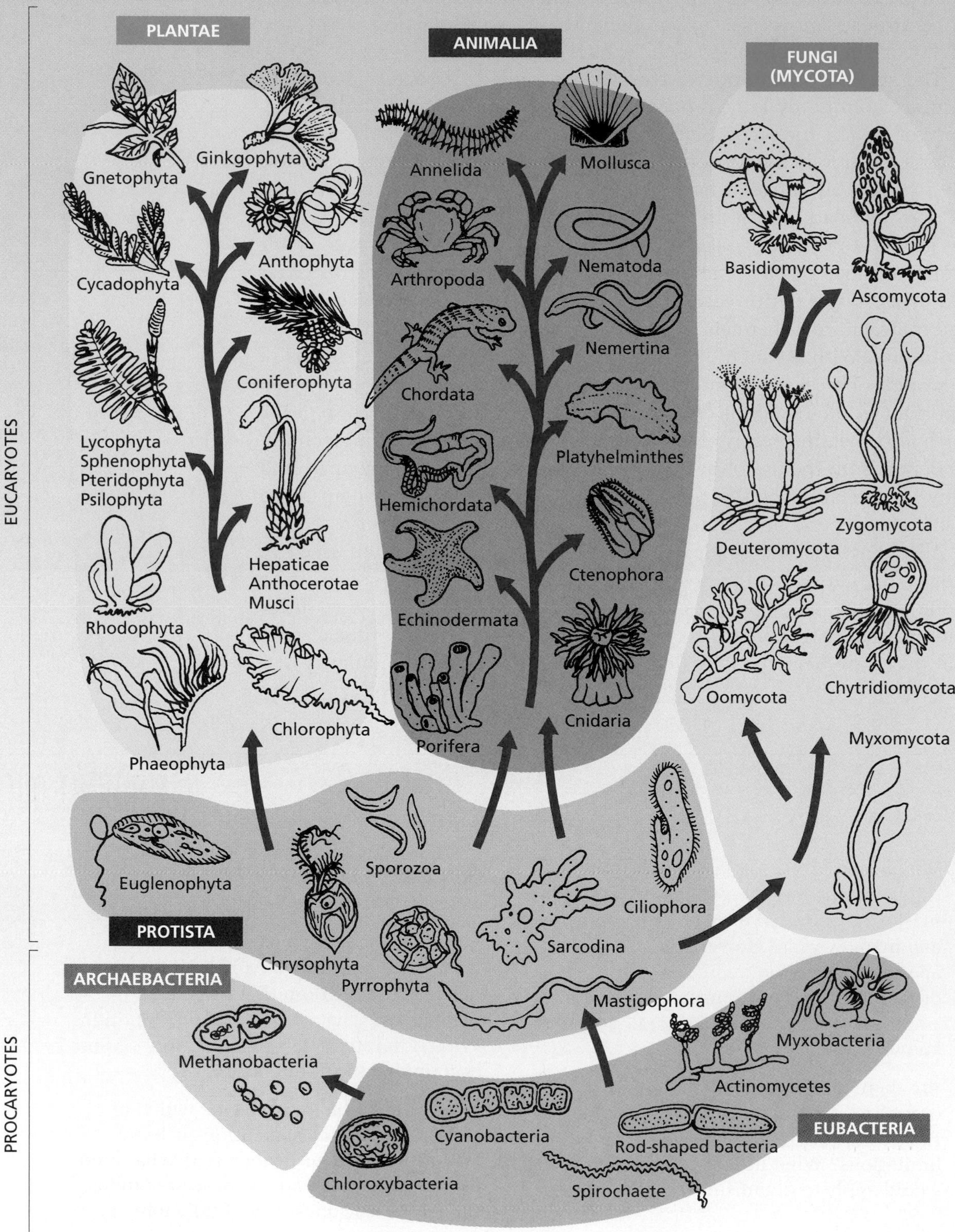

The placement of certain organisms differs among experts. In particular, the algae are sometimes included with plants, sometimes with protists, and sometimes divided between these two kingdoms. For further details, see the classification on this book's web site.

ARCHAEBACTERIA

A small group of organisms, some of them adapted to extremely hot environments and many producing methane as a product of metabolism. Nucleic acid sequences of these organisms show them to be only distantly related to Eubacteria, with which they share the procaryotic type of cell structure shown in Box 4.1.

EUBACTERIA

The vast majority of procaryotic organisms, including the typical bacteria and Cyanobacteria (blue-green bacteria). No well-defined nucleus or nuclear envelope, nor any type of organelle (such as mitochondria or endoplasmic reticulum) that requires internal membranes.

PROTISTA

Eucaryotic unicells without plastids or cell walls. Various adaptations for locomotion may be present (cilia in one group, whiplike flagella in another group, protoplasmic extensions called 'pseudopods' in the largest group), but one group lacks motility and resembles the fungi in reproducing by spores. Some authorities list the algae here rather than among the plants. This and the remaining three kingdoms all have eucaryotic cells, as explained in Box 4.1.

PLANTAE

Eucaryotic organisms with plastids, including various algae, mosses, liverworts, ferns and fern allies, conifers and a vast array of flowering plants from buttercups to orchids and from grasses to trees. Most plants have nonmotile life stages and cells surrounded by cell walls, whose presence strengthens plant tissues. Nearly all possess chlorophyll *a* and are capable of carrying out photosynthesis using sunlight.

FUNGI OR MYCOTA

Nonphotosynthetic eucaryotic organisms with cell walls and absorptive nutrition, reproducing by means of spores. Includes slime molds, yeasts, mushrooms, and various other forms.

ANIMALIA

Eucaryotic organisms without plastids, usually possessing a life stage with at least some locomotor capabilities, and developing by means of an embryonic stage consisting of a hollow ball of cells (blastula). No chlorophyll or photosynthesis; no cell walls.

EUCARYOTIC DIVERSITY DOMINATES LIFE TODAY

As we have seen, evolutionary change occurs when natural selection acts on genetic diversity. Species evolve when new reproductive isolating mechanisms arise. Eucaryotic organisms such as animals and plants have speciated much more often than procaryotic organisms, so that most of the species alive today are eucaryotic. In this section we examine the evolutionary advances that have taken place among eucaryotic organisms.

Key to the success of eucaryotic organisms are a number of important evolutionary advances. Of these, sexual reproduction and multicellularity evolved early, and perhaps repeatedly, among organisms whose bodies were still small and simple. One group of organisms evolved chloroplasts, and these organisms became plants. Protective structures for the egg cells evolved among later plants, as did vascular tissue, seeds, and flowers (see Chapter 14). Another large group of eucaryotes developed motile cells, and some of these became multicellular animals. Animals subsequently evolved bilateral symmetry, body cavities, segmented body plans, and, in our own phylum, backbones.

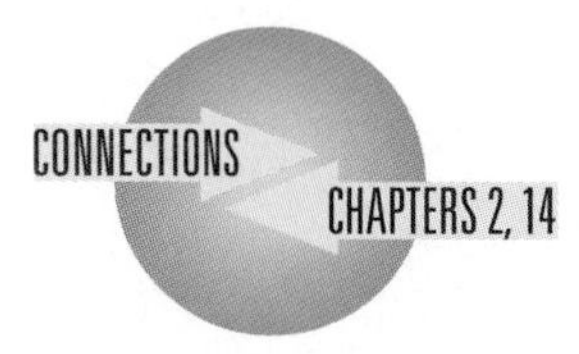

Eucaryotic organisms evolved via natural selection: their interactions with their environments brought about differential survival and differential reproductive success (fitness) of some genotypes over others. The genotype provided the inheritable variation, and the environment provided the selection.

Kingdom Protista

The earliest eucaryotes were simple, single-celled organisms (unicells). These species and their immediate descendents, those lacking the characteristics of the plant, animal, or fungal kingdoms, are placed in the kingdom Protista (Figure 4.17). Among protists, mechanisms evolved to ensure that, when cells divided, all their chromosomes would be present in the offspring cells; mitosis and meiosis (see Chapter 2) first evolved among the Protista, as did sexual recombination. By having haploid gametes that joined during fertilization to produce diploid **zygotes** (fertilized eggs), the eucaryotes became able to generate new genetic

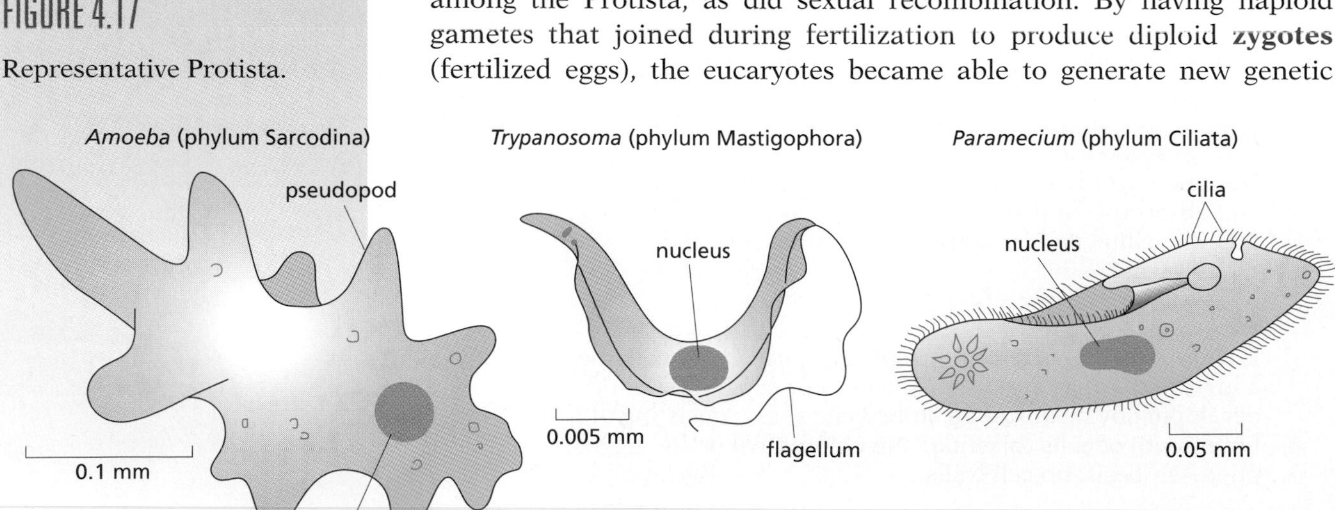

FIGURE 4.17
Representative Protista.

variation in every generation. Because variation is the raw material on which natural selection works, sexual recombination accelerated the rate of evolution among eucaryotes, bringing about a much greater evolutionary diversity than had ever been possible among procaryotes.

Eucaryotic organisms initially evolved in aquatic habitats; only much later did several eucaryotic groups independently colonize the land. Among the early eucaryotes, there must have been a great selective advantage in being able to move from place to place to find food or to escape from unfavorable conditions. Several phyla of protists evolved mechanisms for motility (movement), and the major kinds of protists are distinguished by these adaptations. The earliest protists had contractile protein filaments that allowed the cells to change shape and creep through their surroundings. One large group of protists, the phylum Sarcodina, includes those protists, such as *Amoeba* and its relatives, whose body shape continually changes as they creep along. These protists move by sending out extensions called **pseudopods** in the direction of movement; as cytoplasm flows into a pseudopod, the cell moves forward and its other end retracts. Another large group of protists, the phylum Mastigophora (or Flagellata), achieve motion through a whiplike structure called a **flagellum**. A third group of protists move by means of numerous hairlike structures called **cilia**. A fourth group of protists are the nonmotile Sporozoa, a group that includes the parasites that cause malaria (see Chapter 5, pp. 211–213). Representative Protista are shown in Figure 4.17.

Kingdom Plantae

One of the great achievements of the early eucaryotic organisms was the acquisition of chloroplasts, probably by endosymbiosis (see p. 155). Organisms possessing chloroplasts but lacking specialized tissues are called **algae**. Some experts place the single-celled algae among the Protista, and other experts place all the algae there, but many more experts include all algae in the plant kingdom, as we do in this book. The Plantae can then be defined as eucaryotic organisms possessing chloroplasts and other plastids, chlorophyll, and usually a cell wall containing cellulose. Different groups of algae are distinguished by the types of photosynthetic pigments that they contain. Representative algae of different groups are shown in Figure 4.18.

Within several groups of algae, independently of one another, single-celled organisms evolved into multicellular aggregations. At first, these aggregations were just colonies of similar cells, similar to those formed by the green alga *Volvox* (see Figure 4.18). Then colonies began to function as multicellular single organisms. Cells located in different places within evolving organisms began to develop differently. Evolved differences between surface cells and those in the interior, or between cells near the top and the bottom of a former colony, allowed the organisms to take advantage of the differences in environment between the various locations. The organisms whose cells differed in different locations within the organism were favored by natural selection over organisms whose

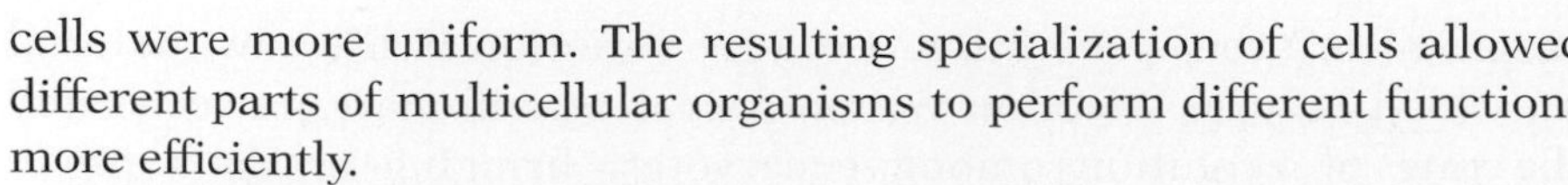

cells were more uniform. The resulting specialization of cells allowed different parts of multicellular organisms to perform different functions more efficiently.

One group of multicellular plants, the Bryophyta (mosses and liverworts), evolved a layer of sterile, nonreproductive cells that surrounded and protected their egg cells. This adaptation permitted plants to emerge from aquatic environments and colonize the land. Most botanists believe that these land plants evolved from green algae. At first, the land plants were small and restricted to moist habitats, but some plants evolved vascular tissues (see Chapter 14, p. 596), which allowed them to grow much taller. The vascular plants (Tracheophyta) are now the largest and most conspicuous organisms in most terrestrial habitats. Their further evolution, including their development of seeds and flowers, is described in Chapter 14. An assortment of plants is shown in Figure 4.19.

FIGURE 4.18

Representative algae.

Dulse (*Palmaria*), a red alga (phylum Rhodophyta).

Ascophyllum, a brown alga (phylum Phaeophyta)

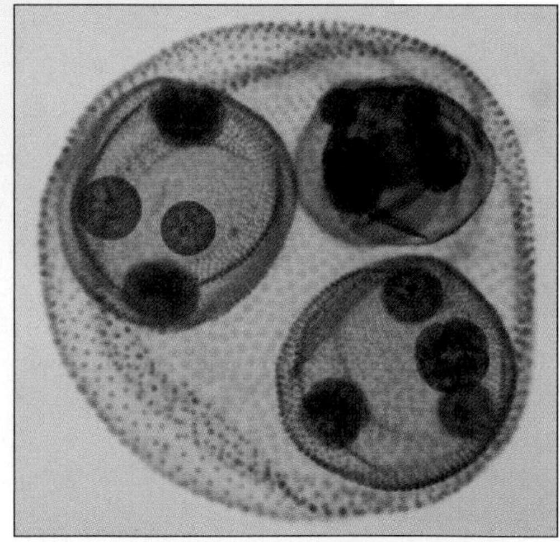

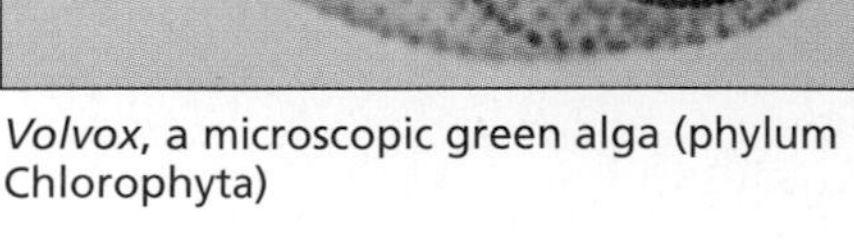

Volvox, a microscopic green alga (phylum Chlorophyta)

Kingdom Mycota

Organisms of the kingdom Mycota are commonly known as fungi. These organisms live on dead or decaying organic matter that they absorb through threadlike extensions called **hyphae**.

Typical fungi (subkingdom Eumycota) have cell walls and are nonmotile, but a few primitive fungi (subkingdom Myxomycota) have motile stages in their life cycles. For example, the slime molds such as *Dictyostelium* (see Chapter 9, p. 362) have multicellular reproductive stages that look like fungi and carry out absorptive nutrition, but at other times they live as motile, amoebalike individual cells. Yeasts and bread molds belong to the phylum Ascomycota, while mushrooms belong to the phylum Basidiomycota. Two of the many types of fungi are shown in Figure 4.20.

Kingdom Animalia

Animals are multicellular organisms with eucaryotic cells and an embryonic life stage consisting of a hollow ball of cells called a **blastula**. Most animals have at least some motility during at least some stage of their life cycle.

As we have seen, there are many highly successful forms of life that are not animals. Even within the animal kingdom, most animals are very different from the vertebrate animals of the phylum Chordata, the group to which humans belong. There is a great diversity of life strategies, and each is biologically successful within the habitat that it occupies. In this section we look at the major evolutionary divisions among animals.

The animal kingdom is divided into about 30 phyla; experts differ on the exact number because they are not in agreement about how to classify some animals. Some small phyla are not included in the following survey, but they are listed in the classification on this book's web site.

Minimal organization and the phylum Porifera (sponges). Multicellular organization in animals takes several forms. The simplest animals are sponges (phylum Porifera). Various cell types are present in these aquatic animals, including wandering amoebalike cells, barrel-shaped

FIGURE 4.19

A bryophyte and several vascular plants.

Polytrichum, a moss (phylum Bryophyta)

Nephrolepis, a fern (phylum Pterophyta), showing reproductive structures (sori) on the underside of the leaf

The lady's slipper orchid *Cypripedium*, a flowering plant (phylum Anthophyta)

Equisetum, a horsetail (phylum Arthrophyta)

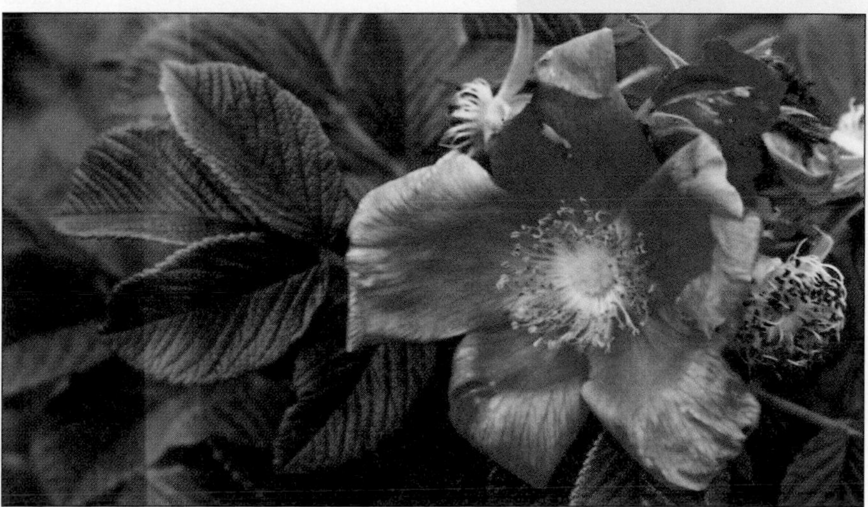

Rosa, another flowering plant or angiosperm (phylum Anthophyta)

cells with hollow interiors, and cells with a whiplike flagellum surrounded by a 'collar.' These cells, however, are not organized into different tissue layers, as they are in all other animal phyla. Sponges have evolved sharp, needlelike structures (spicules) and poisonous chemicals that deter predators who would otherwise feed upon them.

FIGURE 4.20

Two of the many types of fungi (kingdom Mycota).

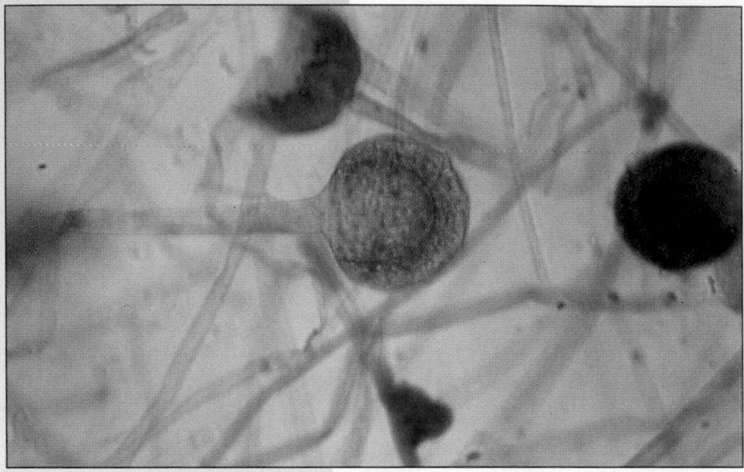

The black bread mold, *Rhizopus* (phylum Zygomycota)

A mushroom, poison *Amanita* (phylum Basidiomycota)

Tissue layers and the phylum Cnidaria. The simplest animals having cells organized into tissues are in the phylum Cnidaria. Like the sponges, these are aquatic animals. In the Cnidaria and in the embryos of all other animals except sponges, one portion of the hollow blastula puckers inward and turns inside-out. The resultant cup-shaped structure, called a **gastrula**, contains two distinct layers of cells: those that have turned inward are called **endoderm**, and those that remained on the outside are called **ectoderm** (Figure 4.21).

The ectoderm and endoderm form two distinct tissue types, the beginnings of the differentiation of multicellular animals into a variety of such tissues. **Tissues** are groups of similar cells that form sheets or other integrated structures, each specialized to perform a different function. We discuss tissues further in Chapter 9. In addition to the ectoderm and endoderm, the gastrula contains an endoderm-lined central cavity, open to the outside. The fact that all animals (except for sponges) go through such a gastrula stage in their development is strong evidence that they all share a common ancestry. Some of the great variety of invertebrate animals (those without backbones) is shown in Figure 4.22.

The two tissue layers are arranged in two basic body plans among the Cnidaria. One plan, called a **polyp**, has the central cavity opening upward. Polyps of most species grow attached to the ocean bottom or to other animals. Many polyps live in large colonies that we recognize as corals. The other body plan, called a **medusa**, has the central cavity opening downward. Medusas float freely in the water, and most can contract portions of their body to control their movement. Because of the large amount of jellylike material that lies between the outer and inner layer of cells, most medusas are popularly known as 'jellyfish.' The major subgroups of Cnidaria are distinguished on the basis of whether their life cycle includes a polyp stage, a medusa stage, or both. Both cnidarian body plans have a series of tentacles surrounding the opening of their central cavity. These tentacles contain specialized stinging cells, which are used to defend the animal against predators.

Bilateral symmetry and the phylum Platyhelminthes (flatworms). Most sponges, and a few other types of animals that live attached to the ocean bottom, have irregular body shapes that show no symmetry. Other sponges, and all members of the Cnidaria, have a **radially symmetrical** body plan: if you turn their bodies around a central axis, you will see the same anatomical details again and again. A cnidarian, whether a polyp

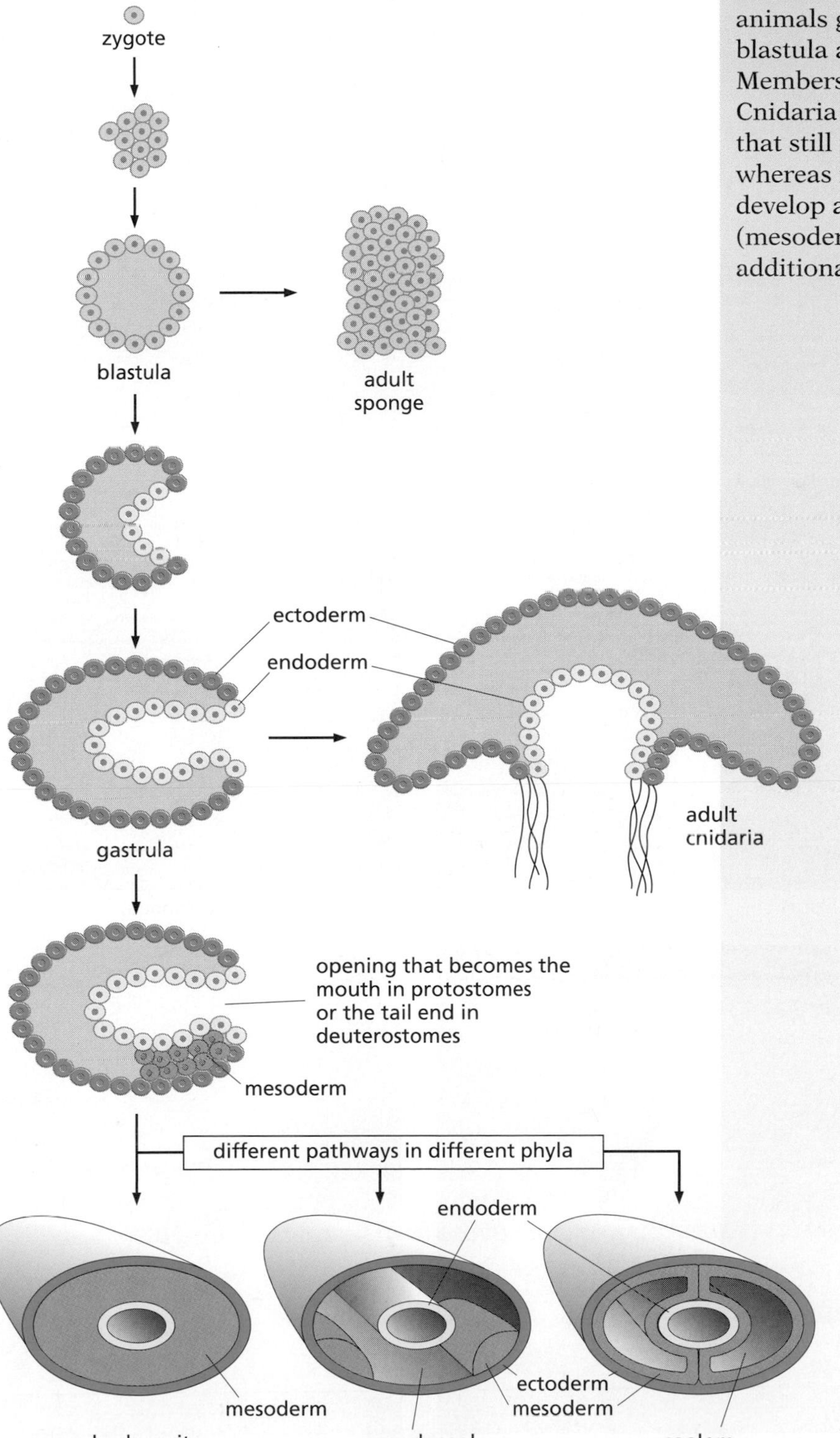

FIGURE 4.21

Early stages in the embryology of animals. Adult sponges develop directly from a modified blastula stage. All other animals go through both blastula and gastrula stages. Members of the phylum Cnidaria form adult stages that still resemble gastrulas, whereas most other animals develop a middle layer (mesoderm) that gives rise to additional internal organs.

FIGURE 4.22

An assortment of invertebrate animals belonging to various phyla.

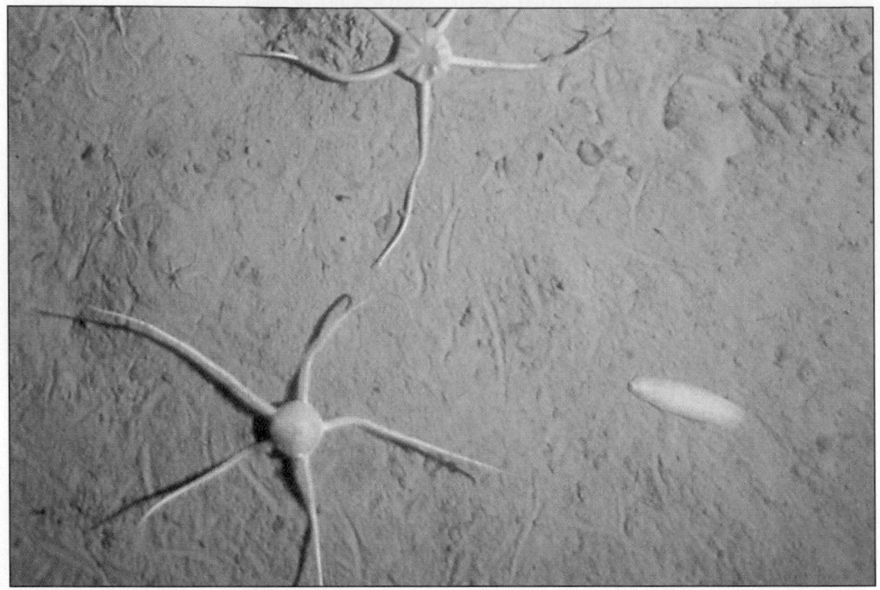

Brittle stars (phylum Echinodermata)

Tree snail (phylum Mollusca)

Shrimp (phylum Arthropoda)

Copepod (phylum Arthropoda)

Crinoid or sea lily (phylum Echinodermata)

Sponge (phylum Porifera)

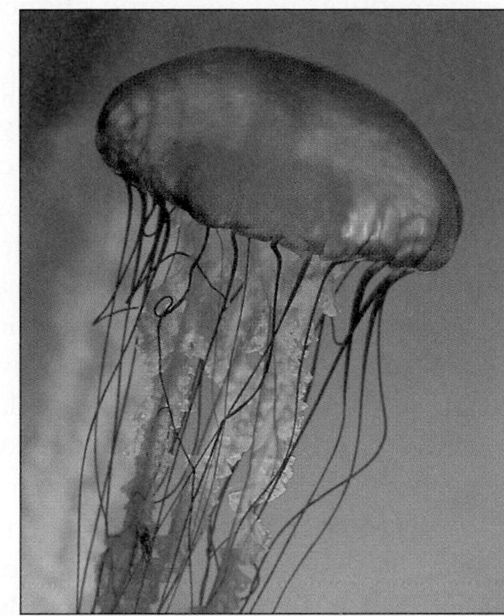
Jellyfish or medusa (phylum Cnidaria)

Hydra, a polyp (phylum Cnidaria)

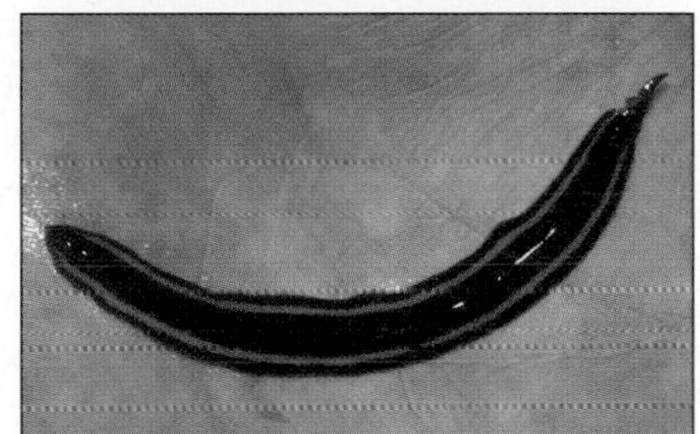
Giant flatworm (phylum Platyhelminthes)

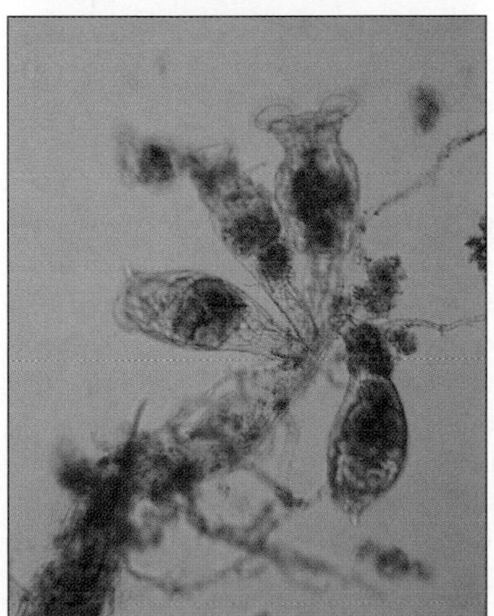
Rotifers (phylum Rotifera)

Tropical earthworm (phylum Annelida)

or a medusa, has the same chance of finding something nutritious, or something dangerous, in any direction.

The vast majority of animals, belonging to over two dozen phyla, are **bilaterally symmetrical**, meaning that their body can be divided by a central plane such that structures on the left side of this plane are mirror images of corresponding structures on the right side. Bilateral symmetry is believed to have evolved in animals as an adaptation that came along with forward movement. Imagine an animal that creeps along the ocean bottom in such a way that one end of its body is in a forward position. Movement would be made easier by a streamlined or elongated body. New discoveries, whether of food or of danger, would be more likely to be made with the front end. Waste products would be discharged more efficiently if they were released from the hind end and left behind as the animal moved forward. Under these conditions, natural selection would favor the development of sense organs (eyes, feelers, taste organs, sound and motion detectors) at the front end rather than at the rear, and the development of feeding organs or aggressive weapons also at the front.

An animal that creeps along the ocean bottom may also be expected to react differently to the water above than to the sediment below, and so natural selection would tend to favor organisms having structures on the top (dorsal) surface that differ from those on the underside (ventral). However, any structure or ability that is adaptive on the right side is equally adaptive on the left, and organisms would have no selective advantage if their right side differed from their left. The result of selection under this regime is a body plan that is bilaterally symmetrical. Such body plans are common in the animal kingdom and present in many different phyla.

The simplest animals with bilateral symmetry are the flatworms of the phylum Platyhelminthes (see Figure 4.22). They have somewhat elongated bodies, with sense organs concentrated at one end, which is recognizable as a **head**. The body is flattened, with broad dorsal and ventral surfaces that in many species differ from one another in coloration and in other ways. The body plan is bilaterally symmetrical, with right and left halves of the body being mirror images of one another. Flatworms have a middle layer of tissue, called **mesoderm**, in addition to the ectoderm and endoderm (see Figure 4.21). All the animals yet to be described in this chapter have tissues derived from these three basic layers.

Assembly-line digestion and the phylum Nematoda (roundworms). One way in which flatworms are similar to cnidarians is in their digestive system, which is just a sac with a single opening that serves as both entrance and exit. With this arrangement, much of what is discarded as waste is immediately taken in again as food, making the system very inefficient. A further inefficiency is that every region of the digestive tract, and every group of cells in the digestive lining, must be capable of performing the entire digestive process from beginning to end. With this arrangement, cells cannot specialize to carry out early or late steps of digestion. Neither can selection favor organisms with cells that specialize in protein digestion only, because much of the food contains other nutrients such as carbohydrates.

A more efficient arrangement, which first evolved in roundworms (phylum Nematoda) and several related phyla, is an assembly-line digestive tract with an entrance, the **mouth**, at the front end and an exit, the **anus**, at the hind end. With this arrangement, selection can favor organisms in which the cells near the front end can perform the early stages of digestion more efficiently and those near the hind end are more efficient in completing the later stages. Also, certain parts of the digestive tract can now specialize in the processing of proteins or other individual nutrients, or of hard substances requiring mechanical break-up (see Chapter 8).

The evolution of body cavities. In the course of creeping forward along the ocean floor, some animals occasionally found reason to burrow into the bottom sediment. Burrowing is usually accomplished by a mechanical process of wedging the front of the body farther forward, then forcefully inflating part of the body to make it wider, then repeating the process. Forceful widening of the body thus alternates with forceful elongation, both in time and space. At any moment when the narrow portions of the body are pushing forward, the other parts of the body are widening to give the parts in front of them something against which to push.

This alternation of widening and elongating can be done much more efficiently if the body contains one or more fluid-filled cavities. Any fluid-filled cavity can be forcefully widened by contracting muscles running front to back, while the same cavity can be forcefully elongated by contracting muscles that encircle its girth. The exact construction and embryologic origin of the cavity does not matter, as long as muscles can act upon it in the manner just described.

In the course of evolution, a variety of animal phyla evolved fluid-filled cavities of different constructions and different embryological derivations. If such a cavity is entirely lined with mesoderm, it is called a **coelom**; otherwise, it is called a **pseudocoel** (see Figure 4.21). Roundworms, horsehair worms, rotifers (see Figure 4.22), and a handful of other animal phyla are characterized by pseudocoels. The remaining phyla described below all have coeloms.

Protostome phyla and the evolution of segmentation. The remaining animal phyla are separated into two large groups, the protostomes and deuterostomes, which evolved in different directions. The **protostome** phyla include many familiar phyla, such as the Mollusca, Annelida, and Arthropoda, as well as many smaller phyla. All of them share certain embryological similarities, such as the way in which their coelom forms or the way in which the mouth develops.

At some point in protostome evolution, the mesoderm and its coelomic cavity became subdivided into a series of individual blocks or pouches called **somites**. The somites were arranged from front to rear, setting the stage for the evolution of segmentation of the body. Some protostome animals, such as the annelid worms, are thoroughly segmented in both larval and adult stages, but other protostomes, including many mollusks, have lost most of their segmentation as adults.

Animals of the phylum Mollusca are in most cases protected by a hard outer shell, secreted by a special layer called the mantle. Part of the mantle is retracted at the rear of the animal to form a **mantle cavity** that contains aquatic respiratory organs (**gills**) and an anus. Anyone who has admired seashells has some idea of the tremendous variety of species of mollusks. In addition to the familiar snails, clams, and oysters, mollusks also include cephalopods such as the squid and octopus, in which the shell is hidden inside or has been lost entirely (see Figure 4.7).

Animals of the phylum Annelida, of which the earthworm is a familiar example, are anatomically arranged as a series of repeated segments; all the body segments are similar to one another in size and in anatomical structure. Segmentation permits parts of the body to work as self-contained units, allowing rhythmic swimming or crawling motions. The worms crawl through soil or sediment by using rhythmic waves of muscle contraction. A few body segments elongate while the next few widen. Waves of elongation alternate with waves of widening of the body segments, and these alternating waves pass down the length of the body from the front end to the rear. The annelid worms have no legs, but they do have tiny bristles, called setae, that stick out of their sides and anchor the widened, nonmoving segments of the body to the surroundings, giving the elongated segments something to push against as they move forward.

The phylum Arthropoda is the largest and most diverse phylum of the entire animal kingdom. The Arthropoda include lobsters, crabs, shrimp (see Figure 4.22), barnacles, spiders, scorpions, centipedes, millipedes, and insects. Insects alone account for over two-thirds of the animal kingdom and over half of all living species on Earth. Arthropods evolved from annelid ancestors. The body segments of arthropods are fewer than in annelids, and these segments are more specialized, differing from one another in both size and anatomy. Most notably, the arthropods have a series of leglike structures that differ in most cases from segment to segment. Arthropods have a strong, protective outer coating called an **exoskeleton**, with rigid segments separated by flexible hinge regions. The legs are also arranged as a series of rigid segments separated by flexible, hinged joints, giving the phylum its name (*arthro* means 'hinged' or 'jointed'; pod means 'leg' or 'foot').

Evolution of the deuterostome phyla. The remaining phyla of the animal kingdom also have coeloms but have evolved separately from the protostomes described above; they are called **deuterostomes**. Deuterostomes differ from protostomes in the manner in which the coelom usually develops. Deuterostomes also differ from protostomes in the embryologic formation of the mouth. In protostomes, the gastrula opens to the outside by an opening that becomes the future mouth. That same opening in deuterostomes ends up near the hind end of the animal, just above the anus, while the mouth develops as a secondary structure at the other end. (*Protostome* means 'first mouth,' while *deuterostome* means 'secondary mouth'.) Thus, in a very real sense, your head corresponds to an insect's hind end (they are homologous), and an insect's head corresponds to your rear.

The deuterostomes and protostomes evolved separately, but some convergent adaptations have appeared. For example, a form of segmentation of the muscles and certain other body systems evolved in the deuterostomes independently of its evolution in protostomes.

The deuterostomes include four phyla; two of these, the echinoderms and chordates, are large groups. All deuterostomes evolved from bilaterally symmetrical ancestors, and all go through bilaterally symmetrical early (larval) stages of development, but a few have evolved irregular (asymmetric) or radially symmetrical adults.

The phylum Echinodermata includes sea stars (starfish), brittle stars (see Figure 4.22), sea urchins, sand dollars, crinoids (sea lilies), and sea cucumbers. The living species of echinoderms show a fivefold (pentameral) symmetry as adults, but their larvae are bilaterally symmetrical. Their other characteristics include a bumpy or spiny skin protected by calcium carbonate deposits and a water-vascular system through which sea water circulates in a series of tubes.

The ancestral deuterostomes probably lived attached to the ocean bottom, either directly or by means of a stalk. Many extinct groups of echinoderms grew this way (crinoids still do; see Figure 4.22), but most living deuterostomes have evolved a free-living, unattached way of life. The transition from attached to free-living is best shown by the tunicates (phylum Chordata, class Urochordata), also called sea-squirts (Figure 4.23). These small animals generally spend their adult lives attached to a rocky bottom. Here they sit and pump water through a large basketlike structure called the **pharynx**, whose numerous slits strain the water through while suspended food particles collect on a sticky, ciliated surface coated with mucus. The attached, filter-feeding existence of adult tunicates is an ancestral way of life for deuterostomes. Larval tunicates, by contrast, are actively free-swimming animals that resemble tadpoles. They swim by means of a long tail that sweeps back and forth like the tail of a fish, and the muscle blocks and nerves of this tail are segmentally organized. Free-swimming chordates, including all fishes, have more in common with the tunicate 'tadpole' than with the filter-feeding adult.

Besides tunicates, the phylum Chordata (see Figure 4.23) includes amphioxus (subphylum Cephalochordata) and all vertebrates (subphylum Vertebrata, a group that includes humans). Although humans are obviously different from tunicates in many ways, they share major characteristics not found in any other group of organisms. Characteristics shared by all chordates include (1) a body axis containing a stiff, flexible rod called a **notochord**, (2) a dorsal, hollow nerve cord, and (3) a series of openings called gill slits, located behind the mouth region. These three characteristics originate early in life and are not retained in the adults of all species. For example, fish keep their gill slits throughout life and use them to breathe, but humans lose their gill slits long before birth, keeping only a few remnants here and there, such as the eustachian tube connecting the throat to the middle ear (see Figure 10.10, p. 426).

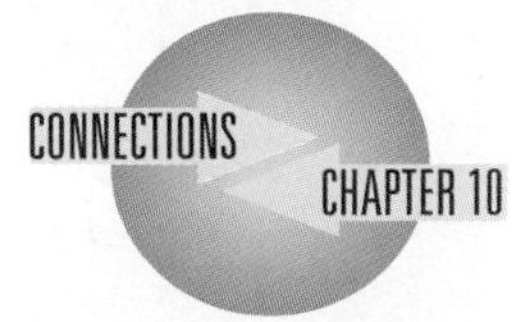

Animals with backbones (Vertebrata). Among the members of the phylum Chordata, the majority are backboned animals that make up the subphylum Vertebrata. The notochord is functionally replaced in adults of this group by a backbone made of a series of individual bones or cartilages. Included in the vertebrates are four classes of fishes: the jawless

FIGURE 4.23

Representatives of the phylum Chordata.

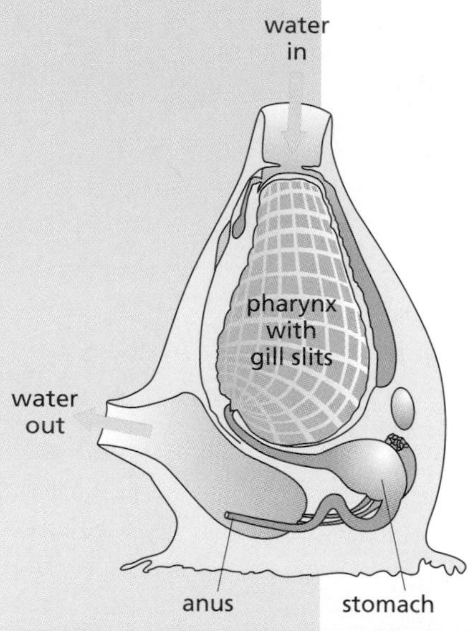

Drawing of a tunicate (class Urochordata)

Tree frog (class Amphibia)

Queen angle fish (class Osteichthyes)

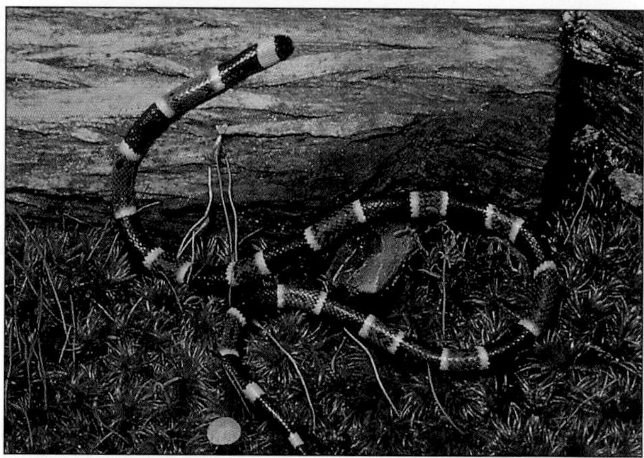

Coral snake (class Reptilia)

Penguins (class Aves)

Kodiak brown bear (class Mammalia)

fishes (Agnatha), the extinct, armored Placodermi, the cartilage fishes (Chondrichthyes, including the sharks and rays), and the bony fishes (Osteichthyes, the group to which most fishes belong). There are more living species of bony fishes than of all the other vertebrate classes combined. All fishes are aquatic vertebrates that use gills to breathe and that swim by waving their hind end from side to side.

The four remaining vertebrate classes evolved, directly or indirectly, from the bony fishes. First of these are the amphibians (class Amphibia), which include the frogs, toads, and salamanders. These animals lay eggs in water, and the eggs develop into aquatic, gill-breathing larvae, commonly called tadpoles. After a while, the tadpoles undergo a rapid developmental change (metamorphosis) into adults that have legs and in most cases lungs. Fossil amphibians are known from the Devonian period to the present day.

Derived from the amphibians are the reptiles (class Reptilia), which include turtles, snakes, lizards, crocodiles, and many extinct species including dinosaurs. Unlike the amphibians, the reptiles have dry, scaly skin, and they lay their eggs on dry land (except for a few species that retain the egg inside the mother and give birth to live young). Fossil reptiles are known from the Pennsylvanian period to the present, but the Mesozoic era (Triassic, Jurassic, and Cretaceous periods) was populated by so many reptiles that it is often called the Age of Reptiles.

One group of reptiles, the Archosauria, included the dinosaurs and other dominant reptiles of the Mesozoic era. Derived from this group of reptiles are the birds (class Aves), distinguished by their possession of feathers. Most bird adaptations have to do with flying, including adaptations (such as hollow bones and loss of one ovary) that lighten the body, and the high metabolism (and thus the high internal body temperature) that flying requires. Feathers do double duty as a flight surface and as insulation.

Another group of ancient reptiles had mammal-like features, and the class Mammalia, to which we belong, evolved from them. Mammals maintain a high and fairly constant body temperature, made possible by an insulating layer of hair or fur, supplemented in some cases by fat or blubber. A four-chambered mammalian heart prevents oxygen-rich blood from the lungs from mixing with oxygen-poor blood returning from other parts of the body. Also characteristic of mammals is the fact that they supply their young with milk, a secretion of the female's mammary glands. The word 'mammal' comes from *mamma*, the Latin word for breast. Mammals include kangaroos, shrews, monkeys, humans, bats, rats, squirrels, rabbits, whales, dogs, cats, bears, seals, elephants, horses, pigs, sheep, cattle, and many other species.

Humans are mammals because we share such mammalian characteristics as hair, a four-chambered heart, and the feeding of milk to our young. We are also chordates because we share in the embryonic gill slits and other characteristics that unite us with tunicates, amphibians, and other Chordata. We also share with all chordates and echinoderms the characteristics that define us as deuterostomes. We share with many more phyla the possession of a coelom, and with all animals the presence of motile cells and development from a blastula. We share the eucaryotic type of cell with four of the six kingdoms of organisms. The evolution of species over billions of years accounts for these patterns of shared characteristics.

THOUGHT QUESTIONS

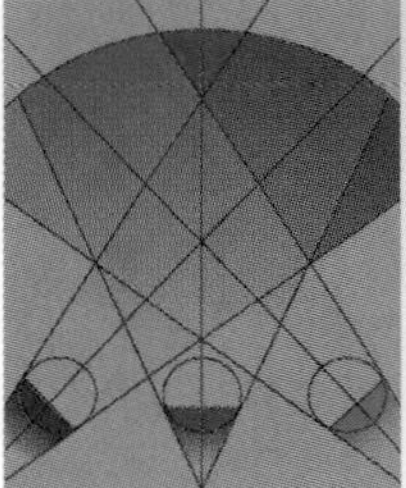

1. Why are algae sometimes considered protists? Why are they sometimes considered plants? How would you decide which is the better approach? If plastids evolved only once, how would this affect your answer? What if plastids evolved many times, independently?

2. Many bilaterally symmetrical animals have a long, thin 'wormlike' body shape. What advantages do you think such a body shape can confer? What problems do you think can arise from such a body shape?

HUMANS ARE PRODUCTS OF EVOLUTION

As we saw in the last section, humans are one species among many. At what point in evolutionary history did our ancestors evolve into something we could call 'human?' Answers to this question are reconstructed from fossils. The fossils all help us to reconstruct our family tree, although there are frequent disagreements among scientists as to where a particular new fossil fits in.

Along with monkeys, apes, and lemurs, humans belong to the mammalian order Primates (Figure 4.24). We share many characteristics with other primates, but we did not evolve from any present-day species. Most adaptations shared by primates are related to the requirements of life in the trees. Most primates live in trees today, and those that do not had ancestors that did. Nonprimate mammals whose ancestors never lived in trees do not share these adaptations. Primate characteristics directly related to the requirements of arboreal locomotion (meaning locomotion in trees) are the independent and individual mobility of the fingers, the ability of the thumb to oppose the action of the other fingers, and the presence of friction ridges on the palm of the hand and the sole of the foot. Primates have also retained some primitive characteristics that many other mammals have lost in the course of their evolution, including the five-fingered hand, the collar-bone (clavicle), and the ability to rotate the two bones of the forearm. Unlike those mammals that rely heavily on the sense of smell, primates rely heavily on vision. Primates have binocular vision that merges images from both eyes to give three-dimensional information. This binocular vision is possible because the eyes rotated forward in the skull during early primate evolution, so that the optic fields overlap. The portions of the brain related to vision are expanded in primates (compared with other mammals), and the outer surface of the brain (the cerebral cortex) is more complex. The increased complexity of the primate brain is associated with an increased complexity of behavior, most of which is learned (see Chapter 7). The reliance on learned behavior

would be impossible without a lengthy period of very intensive parental care. Primates typically give birth to one young at a time. Primate nipples are restricted to a single pair in the chest region (other mammals have many pairs). The uteri, paired in other mammals, are fused into a single uterus. Primates include lemurs, lorises, galagos, tarsiers, monkeys,

FIGURE 4.24

Representative primates (phylum Chordata, class Mammalia, order Primates).

Ring-tailed lemur *(Lemur catta)*

Slow loris (*Nycticebus coucang*)

Squirrel monkey (*Saimiri sciureus*)

Mandrill (*Mandrillus sphinx*)

Chimpanzee (*Pan troglodytes*)

apes, and humans. Among these, humans are most closely related to apes, but differ primarily in walking upright.

The genus *Australopithecus*

In 1925, a fossilized child's skull was discovered in a cave near Taung, South Africa, and was named *Australopithecus africanus*. Although the skull had both apelike and human features, most experts treated it as just another ape. Additional fossils of *A. africanus* were discovered subsequently (Figure 4.25). These fossils enabled the anatomist W.E. LeGros Clark to show that these primates walked upright and were therefore more like humans than like apes. Primates that walk upright are placed in the family Hominidae and referred to as **hominids**.

Scientists have since unearthed the remains of several other species of *Australopithecus*. The oldest species is *A. anamensis*, discovered in 1995, which lived about 4 million years ago in Kenya. *A. anamensis* is thought to be the ancestor of all later species of *Australopithecus* and a close relative of the small hominid *Ardipithecus ramidus*, which lived about 3.7 million years ago. Another species, *Australopithecus afarensis*, about 3.5 million years old, is represented by the well-known skeleton known as Lucy, a female about 1.3 meters in height, slightly over 4 feet. Enough of the dimensions of Lucy's brain are known to permit us to say that these dimensions are consistent with the hypotheses that *A. afarensis* was the common ancestor of the genus *Homo* and of several later species of *Australopithecus*. Two of these later species, *A. robustus* and *A. boisei*, which lived from 2.3 to 1.7 million years ago, were considerably larger than the better-known *Australopithecus africanus*, which lived from approximately 3.0 to 2.0 million years ago. Of the species we have described, *Australopithecus anamensis* and *Australopithecus afarensis* were probably along the line leading to *Homo*; the other species of *Australopithecus* and *Ardipithecus* were probably side branches of the family tree that died out without leaving any surviving descendents.

FIGURE 4.25

Fossils of the genus *Australopithecus*.

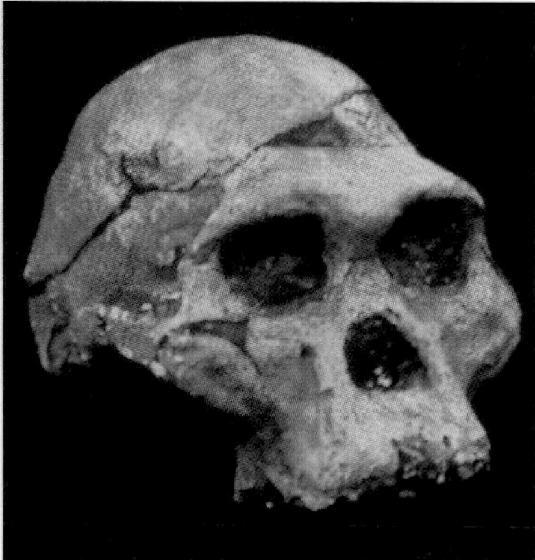

Adult *Australopithecus africanus* (Sterkfontein, South Africa)

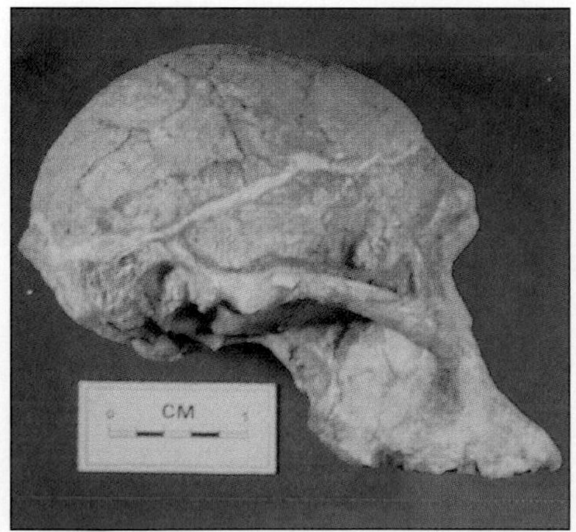

Side view of *Australopithecus africanus* (Sterkfontein, South Africa)

It is now certain that *Australopithecus* walked upright, that their brains were about as big as those of modern chimpanzees, and that at least some of them used tools. Evidence for upright walking comes from the anatomical structure of the foot, the pelvis, and the lower part of the spinal column, and also from the discovery of a set of footprints at Laetoli, Kenya, approximately 4.5 million years old. The earliest *Australopithecus* thus came well before the earliest known *Homo* (about 3.5 million years ago, or a full million years after the Laetolil footprints), but later species of *Australopithecus* persisted side by side with *Homo*, at least in East Africa.

The genus *Homo*

Modern humans (*Homo sapiens*) and at least two extinct species are placed in the genus *Homo* (Figure 4.26). The oldest species of *Homo* was *H. habilis*, which lived in East Africa from about 3.5 to 1.7 million years ago, coexisting with *Australopithecus boisei* and perhaps with other *Australopithecus* species as well. *H. habilis* had a brain that was small in absolute terms (about 400 cm^3, compared with 1200–1500 cm^3 for most modern humans), but the proportions of the brain to body size were more comparable to those of *Homo* than to those of *Australopithecus*. *H. habilis* has been found contemporaneously with certain types of tools, including simple stone tools. It is generally presumed that *H. habilis* was the maker of these tools.

A later species, *H. erectus* (see Figure 4.26), is now known from fossils from about 1.5 million to 300,000 years old in China, Java, Europe, and

FIGURE 4.26

Representatives of the genus *Homo*.

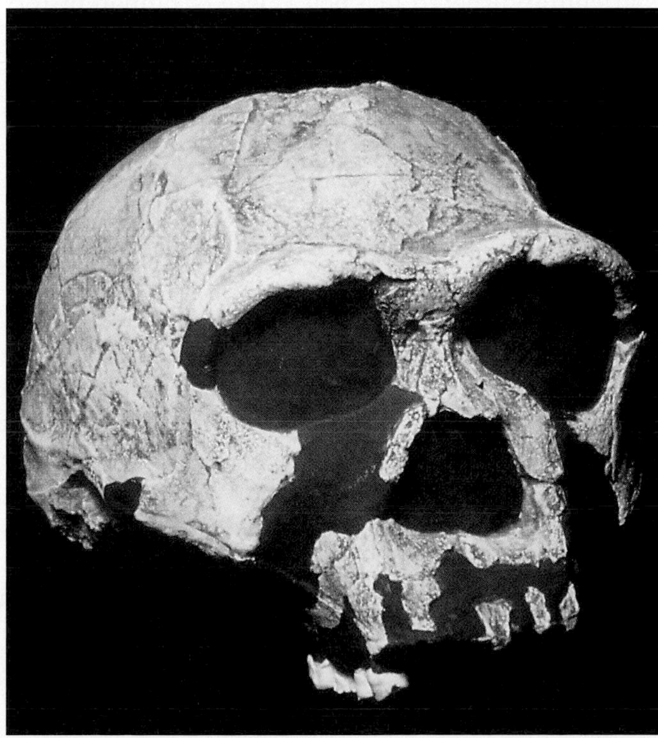

Homo erectus (Koobi Fora, Kenya)

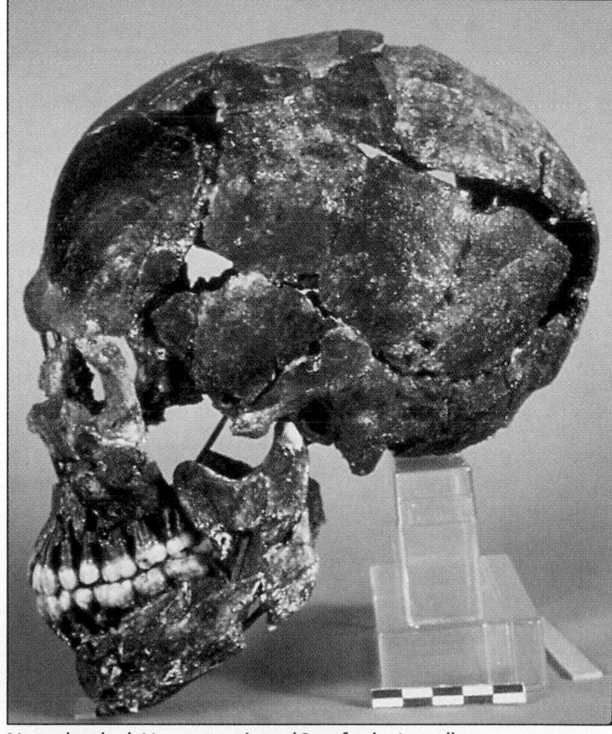

Neanderthal *Homo sapiens* (Quafzeh, Israel)

several parts of Africa. A cave site at Choukoudian, China (near Beijing), has heat-fractured stones indicative of the use of fire. There is also evidence of round or oval tents supported by poles and held down along the margins by a circle of stones.

H. erectus was the ancestor of *H. sapiens*, the species to which living humans belong. With the advent of *H. sapiens*, tools became more sophisticated, and were in many cases mounted on wooden shafts. The *H. sapiens* that lived in Europe from about 150,000 to 50,000 years ago are called Neanderthals (see Figure 4.26); some scientists place them in a separate species. Neanderthals hunted deer, horses, and other species as large as rhinoceroses. Healed surgical wounds show that these skilled hunters took care of sick companions, set broken bones, and even performed simple brain surgery. They buried their dead and decorated the graves with flowers of preferred colors, mostly white or cream-colored. The decoration of graves is thought by several anthropologists to indicate a belief in an afterlife.

The more modern *H. sapiens* that replaced the Neanderthals were Upper Paleolithic people (including the Cro-Magnons) who lived from about 50,000 to 15,000 years ago. They had an even greater variety of tools, including fishhooks and harpoons. They hunted wooly mammoths and large herd animals. They also left records of their activities in the form of cave paintings, showing their interest in hunting and their understanding of both animal anatomy and physiology. By prominently drawing the heart and singling it out as a target, these hunters showed that they understood how vital this organ was. Their drawings of pregnant deer and of mating rituals show that they knew enough reproductive biology to understand the relationships between mating, birth, and subsequent herd sizes.

The discovery of agriculture ushered in a new phase of human history called the Neolithic. With the planting and harvesting of crops, humans began to settle down into villages, which later grew into towns and cities. Civilization has greatly changed the ways in which we live our lives. Human evolution has not stopped, however, as we discuss in the next section.

Evolution as an ongoing process

Evolution is a process that takes place within species as well as between species. The process continues in the present as it has in the past. Within the twentieth century, the peppered moths of some locations in England changed from predominantly light-colored to almost all dark and back again. In one species of Galapagos ground finches, *Geospiza fortis*, the average bill size changes back and forth. Small-beaked birds that eat soft seeds survive and produce the most offspring in years when rainfall is adequate, but birds with larger beaks are at an advantage in drought years because they can open large, tough old seeds. The average bill size of birds within the population thus increases in drought years and decreases in wet years. In fruit flies, different chromosomal variations are favored in different seasons. We see that evolution responds adaptively to fluctuating environmental conditions. Different alleles are selected by different environmental conditions at different times because their phenotypes are more adaptive in those conditions.

The rapid pace and power of cultural change leaves many people wondering whether biological evolution of *H. sapiens* has become a thing of the past. If we need to travel faster, the argument goes, our species tames horses or builds automobiles instead of evolving longer legs for faster running. Evolution by natural selection is much slower than cultural innovation. In this view, the future development of our species resides more in our technology than in our bodies.

No one questions that cultural changes in human beings have far outstripped biological ones as the most rapid and far-reaching changes taking place today. Cultural innovation spreads rapidly, in part because there are no species barriers to prevent transmission from one human group to another. (Language barriers and geographic barriers can always be crossed, especially in the age of television and jet travel.) The ease of travel and the global spread of people and their culture has brought us to an era in which there is no significant geographic isolation of human populations. Without geographic barriers, no reproductive isolating mechanisms will evolve, and all humans will remain one species. Cultural change is also more rapid than biological evolution because new inventions and other culturally acquired characteristics *are* inherited, although not genetically. Each generation inherits the stored knowledge of past generations (in libraries and museums, for example), along with tools (from tractors to telephones to satellites) and the technology needed to design and build new and better tools in the future.

Although natural selection continues to take place, the environment, and therefore the traits favored by selection, have changed because of our own culture. Many of the selection forces that shaped human evolution in the past, including famines, epidemics, and predators, have been greatly diminished in modern times (see Chapter 6). Many traits that were once disadvantageous have become much less so. For example, poor eyesight is no longer an important barrier to survival and reproduction in societies that supply eyeglasses.

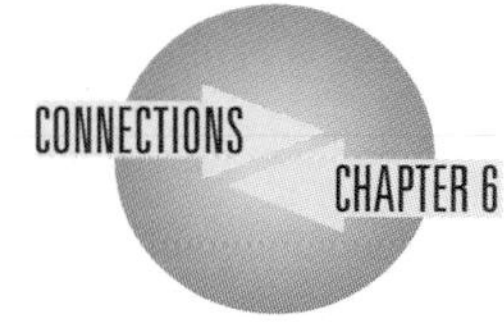

However, there continue to be situations in which the chance of survival or reproduction differs among people as a consequence of their genotypes. In our own species, genetic conditions such as cystic fibrosis, Tay–Sachs disease, muscular dystrophy, and others continue to cause numerous deaths before reproductive age, despite the best that medical technology has to offer. Other diseases that are generally survivable may reduce reproductive capacity, which decreases fitness. For example, chondrodystrophy is a rare disease, controlled by a dominant gene, in which the cartilage tissue turns bony at an early age, resulting in a form of dwarfism. Most chondrodystrophic dwarfs enjoy fairly normal health as adults, but have only about one-fifth as many children as their nondwarf siblings. Lowered reproductive rates are also found in diabetics. As these examples show, natural selection continues to affect the human species. Biological evolution thus continues to operate and to interact with cultural evolution in all human populations.

THOUGHT QUESTIONS

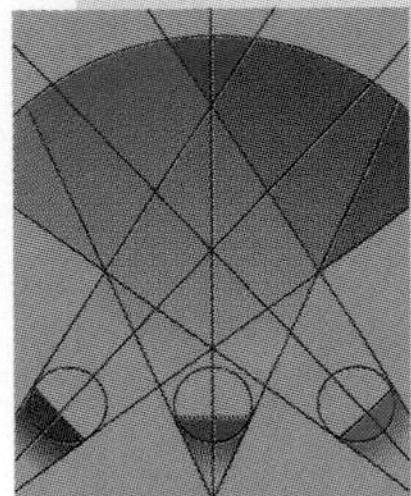

1. The Neanderthals were similar to us in many respects, though their skulls had a somewhat more 'rugged' appearance, with brow ridges and cheek bones protruding. How should we decide whether Neanderthals should be placed in their own species, separate from *Homo sapiens*?
2. In Europe, Upper Paleolithic culture replaced the culture of the earlier Neanderthal populations rather suddenly. Do you think that the replacement of one set of tools and traditions by another took place mostly by conquest, by intermarriage, or by some combination of the two? What evidence would you look for to test one hypothesis against the others?
3. Which would have a greater effect in terms of natural selection: 100 deaths of people in their seventies, or 100 deaths of people in their twenties? Why? List five or more causes of death of humans aged 30 or younger. What traits or abilities might raise or lower someone's odds of dying from each of these causes? How do you think natural selection might work on the traits you have listed?
4. Why have disease organisms evolved? Why do human diseases continue to exist?
5. Is the study of evolution static or changing? Find some recent news articles dealing with new fossil discoveries or other new findings that deal with evolution.

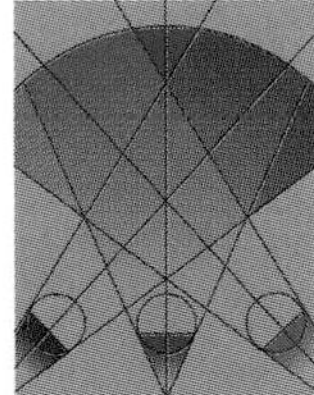

CONSIDERABLE EVIDENCE NOW SHOWS THAT EVOLUTION has taken place in the past and that organisms continue to evolve today, though often slowly. The ways in which species resemble one another and are related to one another reflect branching patterns of descent. Evolutionary change is brought about by natural selection, a process that operates whenever some genotypes leave more offspring than others. Species are reproductively isolated from one another, and the splitting of a species therefore requires the evolution of a new reproductive isolating mechanism. All species, including humans, arose by speciation and are products of evolution.

Although it now interacts with cultural evolution and technological revolution, biological evolution continues to act today in humans as it does in other species. Natural selection continues to operate by both differential mortality and differential reproduction, and continued selection will result in biological changes within our species. In Chapter 5, we examine further how these processes and others continue to bring about biological change within human populations.

CHAPTER SUMMARY

- **Evolution** is the central, unifying concept of biology.
- **Evolution** produces branching patterns of descent.
- Only inherited traits contribute to evolution and bring about **adaptation**; acquired characteristics do not.
- **Evolution** operates through **natural selection**: there is heritable variation in all **species**, and different genotypes differ in **fitness** by leaving different numbers of surviving offspring.
- Forces of **natural selection** include predators, disease, and **sexual selection**.
- **Speciation** occurs through the build-up of genetic differences between **populations** arising primarily during times of geographic isolation, resulting over time in **reproductive isolation**.
- Branching descent with modification accounts for **homology** between species.
- **Taxonomy** groups species together on the basis of their evolutionary history.
- Life originated on Earth under conditions in which no free oxygen was present. Our present oxygen-rich atmosphere is a product of the evolution of photosynthesis.
- Early cells were **procaryotic** and contained no internal membrane limited compartments.
- **Eucaryotic** cells, with a variety of internal organelles, originated by **endosymbiosis**.
- Highlights of eucaryotic evolution include the origin of chloroplasts and other plastids, the evolution of multicellularity, the origin of bilateral symmetry in animals, the evolution of body cavities, the development of segmented body plans, and the evolution of the backbone.
- **Fossils** showing the evolutionary history of humans are placed in an earlier genus, *Australopithecus*, and a later genus, *Homo*, which includes all living people.
- **Evolution** continues today in many species.
- In humans, biological evolution interacts with cultural evolution.

KEY TERMS TO KNOW

adaptation (p. 126)
artificial selection (p. 129)
classification (p. 149)
endosymbiosis (p. 155)
eucaryotic (p. 155)
evolution (p. 125)
fitness (p. 130)
fossils (p. 137)
homology (p. 133)
interbreeding (p. 147)
mimicry (p. 130)
natural selection (p. 125)
population (p. 146)
procaryotic (p. 154)
reproductive isolating mechanism (p. 147)
reproductive isolation (p. 147)
sexual selection (p. 133)
speciation (p. 146)
species (p. 147)
taxon (p. 149)
taxonomy (p. 149)
vestigial structures (p. 135)

CONNECTIONS TO OTHER CHAPTERS

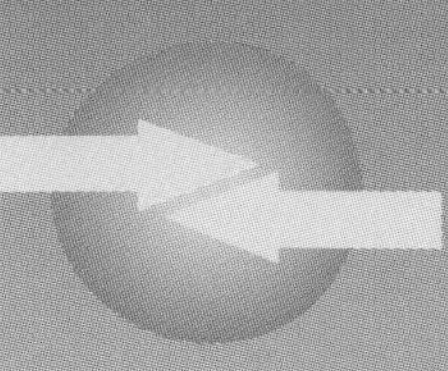

Chapter 1: Darwinian evolution and modern evolutionary theory are both good examples of successful paradigms.

Chapter 1: Presenting creationist ideas in school classrooms raises several social policy issues.

Chapter 2: Gene mutations provide the raw material for evolution.

Chapter 5: Differences have evolved and continue to evolve both within and among human populations.

Chapter 6: Successful species may increase so rapidly in numbers that they outstrip the available resources.

Chapter 7: Social behavior and reproductive strategies are, in part, products of evolution.

Chapter 10: Differences in brain anatomy in different species provide good evidence of evolution.

Chapter 13: Viruses and other microorganisms may evolve disease-causing strains, as well as strains resistant to certain medicines.

Chapter 14: Plant characteristics resulting from evolution include the presence of chloroplasts and vascular tissues.

Chapter 15: Speciation increases biodiversity, whereas extinction diminishes biodiversity.

PRACTICE QUESTIONS

1. Match the ideas in the left column with the people on the right. One name needs to be used twice.

a. Evolution is a branching process.

b. Adaptations should be studied carefully as a way of understanding God's creation.

c. Evolution should never be taught.

d. New species originate by a process that includes geographic isolation.

e. Adaptation occurs by use and disuse.

f. Organisms with successful adaptations will be perpetuated, whereas those with unfavorable characters will die out.

g. Evolution and creation science should be given equal time in science classes.

i. Creationist supporters of the 'balanced treatment act'

ii. Creationists of past decades

iii. Charles Darwin

iv. J.B. Lamarck

v. William Paley

vi. Modern evolutionary biologists

2. The Bahamas are a group of islands in the Atlantic, made mostly of coral fragments. The closest mainland is North America, but political ties are to Great Britain. According to Darwin's reasoning, the birds and other species living on these islands should have their closest relatives in:

   a. other islands of similar composition in the Pacific

   b. islands such as the Canary Islands, in the Atlantic at a similar latitude

   c. North America

   d. England

3. Which theory had no way of explaining the sticky flap in the fetal heart?

   a. Darwin's

   b. Paley's

   c. Lamarck's

4. In mimicry, the mimics and their models always:

   a. live in similar climates, although they may be far away

   b. live close together

   c. taste the same to predators

   d. are camouflaged to resemble their backgrounds

5. Which of these is NOT considered a reproductive isolating mechanism?

   a. two geographically separated species

   b. two species breeding in different seasons

   c. two species that produce infertile hybrids when they mate

   d. two species with different mating calls

   e. two species whose external genitalia cannot fit together

6. Name the four gases used in Miller's initial experiment. Also name the four elements needed to produce biological molecules. Did the gases provide all of these elements?

7. Name the six kingdoms of organisms currently recognized.

8. Name five organelles that eucaryotic cells possess but procaryotic cells do not.

9. Give a clear definition of the term *species*.

10. What is the major anatomical difference between humans and apes? What is the major anatomical difference between the genus *Australopithecus* and the later members of the genus *Homo*?

5 Human Variation

CHAPTER OUTLINE

There is Biological Variation Both Within and Between Human Populations
- Continuous and discontinuous variation
- Variation between populations
- Concepts of race
- The study of human variation

Population Genetics Can Help us to Understand Human Variation
- Human blood groups and geography
- Isolated populations and genetic drift
- Reconstructing the history of human populations

Malaria and Other Diseases Are Agents of Natural Selection
- Malaria
- Sickle-cell anemia and resistance to malaria
- Other genetic traits that protect against malaria
- Population genetics of malaria resistance
- Other diseases as selective factors

Natural Selection by Physical Factors Causes More Population Variation
- Human variation in physiology and physique
- Natural selection, skin color, and disease resistance

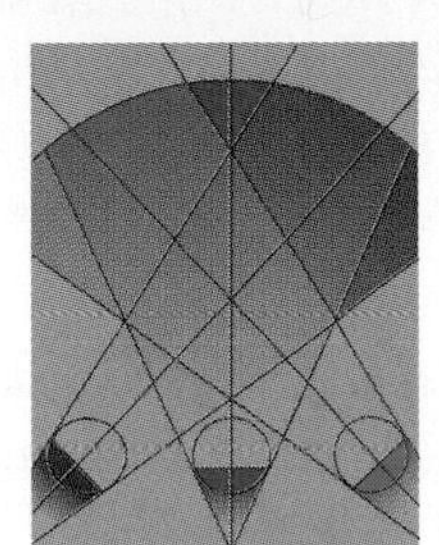

ISSUES

How can we describe and compare variation within and variation between populations?

How is the study of population genetics related to human variation?

Why do human populations differ biologically?

Do human races exist?

Is there a biological basis for the idea of race? Is biology the most accurate descriptor of race?

Will changing biological concepts of race diminish racism? Why or why not?

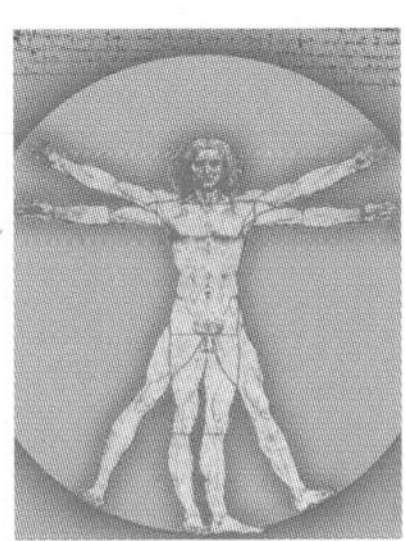

BIOLOGICAL CONCEPTS

Populations and population ecology

Population genetics (genetic variation, Hardy–Weinberg equilibrium, blood groups, genetic drift)

Patterns of evolution (adaptation, physiology)

Forces of evolutionary change (natural selection, environmental factors, communicable diseases, parasitism, human health)

Gene action (molecular structure, genetic polymorphism)

Scaling (body size and shape)

Chapter 5: Human Variation

The human species is highly variable in every biological trait. Humans vary in their physiology, body proportions, skin color, and body chemicals. Many of these features influence susceptibility to disease and other forces of natural selection. Continued selection over time has produced adaptations of local populations to the environments in which they live. Much of human biological variation is geographic; that is, there are differences between population groups from different geographical areas. For example, northern European peoples differ in certain ways from those from eastern Africa, and those from Japan differ in some ways from those from the mountains of Peru. Between these populations, however, lie many other populations that fill in all degrees of variation between the populations we have named, and there is also a lot of variation within *each of these groups.*

Central to the study of human variation is the concept of a biological population, as defined in Chapter 4, p. 146 and as explained again below. Both physical features and genotypes vary from one person to another within populations, but there is also a good deal of variation between human populations from different geographic areas as the result of evolutionary processes. How do populations come to differ from one another? How do alleles spread through populations? How do environmental factors such as infectious diseases influence the spread? Why are certain features more common in Arctic populations and other features more common in tropical populations? Why do we think of some of these variations as 'races'? These are some of the questions that are explored in this chapter.

THERE IS BIOLOGICAL VARIATION BOTH WITHIN AND BETWEEN HUMAN POPULATIONS

All genetic traits in humans and other species vary considerably from one individual to another. Some of this variation consists of different alleles at each gene locus; other variation results from the interaction of genotypes with the environment. The simplest type of variation governs traits such as those discussed in Chapter 3, p. 85 in which an enzyme may either be functional or nonfunctional. The inheritance of these traits follows the patterns described in Chapter 2, which you may want to review at this time. In particular, be sure that you understand the meaning of *dominant* and *recessive* alleles and of *homozygous* and *heterozygous* genotypes. Many other traits, as we saw in Chapter 3, have a more complex genetic basis. In this section we examine how biological variation is described.

Continuous and discontinuous variation

Many human traits vary over a range of values, with all intermediate values being possible. Some of this **continuous variation** is for traits, such as height, that can be measured in an individual and expressed as a numerical value. Other continuous traits, such as hair curliness or skin color, are seldom expressed numerically, although theoretically they could be. A description of continuous variation in a population requires the use of statistical concepts such as average (mean) values.

Continuous variation can result from the cumulative effects of multiple genes, each of which by itself contributes a small effect. Dozens of known genes, perhaps even hundreds, influence height in one direction or another. If we make the simplifying assumption that these effects are independent of one another and that they add up, we can predict that a population of individuals will show a range of variation in height similar to the bell-shaped curve of Figure 5.1. When we measure heights in any large population, we do in fact get a curve that closely matches this predicted curve. Many other continuous traits vary in much the same way as height. For most of these traits, a strong environmental component also exists. Height, for example, is strongly influenced by childhood nutrition as well as by genes. Environmental components of traits can consist of thousands of small influences (such as a few extra glasses of milk or a few meals missed each week). In a

FIGURE 5.1

Continuous variation in a single population: all intermediate values are possible.

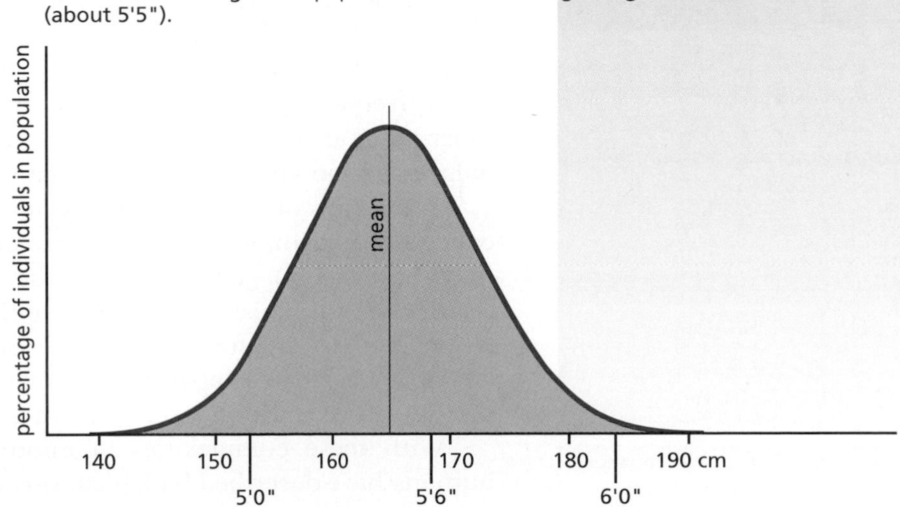

population, such small environmental and genetic differences from person to person contribute to the formation of the bell-shaped curve.

For a particular group of people, we can calculate an *average* height, weight, or head breadth, but these averages are just statistical abstractions—there are perfectly normal individuals that differ from the average, perhaps even greatly. As can be seen in Figure 5.1, average values conceal a large amount of variation for each trait. Any statement about the height, hair, or skin color of any group of people is at best an average value surrounded by lots of variation. Additional statistics can describe the amount of variation around the average, and these statistics reveal more information than the average alone can convey.

For most continuous traits such as height, the group average tells us little about any individual. Your individual traits result from the influences of alleles from your mother, alleles from your father, and the environmental factors to which you are exposed. What you inherit from your parents is a predisposition for a range of possible future variations in phenotype. For example, when you are born, your exact height as an adult cannot be predicted, but if your mother and father are both significantly taller than average, you will, if you receive adequate nutrition, probably also be taller than average. The average values of continuous variables are characteristic of the population as a whole, not of any individual member within the group. Also, whereas height can actually be measured (and average height computed), concepts such as 'tall' are relative: a height that is average in England may be considered tall in India or the Philippines.

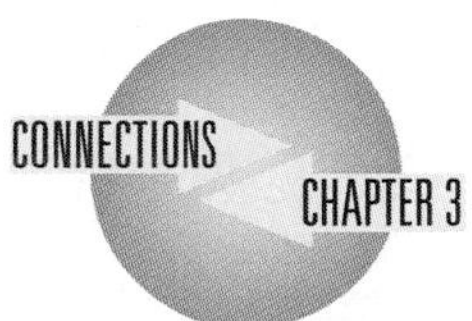

Discontinuous variation is represented by traits that are either present or absent, with no intermediate values possible. Most of these traits have a simple genetic basis, so that a person's phenotype may allow us to infer his or her genotype. Discontinuous traits include blood groups and the presence or absence of conditions such as albinism or Tay–Sachs disease (see Chapter 3). A particular phenotype of such a discontinuous trait is either present or not in a particular individual. To characterize a group value for a discontinuous trait, we divide the number of people who have a particular phenotype by the total size of the population; the resulting fraction is the frequency of that phenotype. From these phenotypic frequencies, scientists use formulas to calculate the frequencies of the alleles responsible. These **allele frequencies** (also called gene frequencies) are characteristic of the entire population, not of any one individual. These frequencies are most easily studied for traits whose patterns of inheritance are known and simple. It is important to realize that all individuals have genotypes, but only populations can have allele frequencies.

The study of variation in the allele frequencies of populations is called **population genetics**. The measuring of allele frequencies requires that the different genotypes, and the alleles responsible for them, can readily be distinguished from one another. It is for this reason that population geneticists often concentrate on those genes whose phenotypic effects are easy to tell apart from one another. Most of those genes control discontinuously variable traits that are either present or not.

With these complexities in mind, we now begin to look at how humans have described biological variations between populations.

Variation between populations

One of the central tenets of modern biology is that evolution can occur only if populations are genetically varied. However, biologists did not always think in terms of evolving and variable populations. For over 2000 years, biologists believed that species were constant, unvarying entities. Plato and Aristotle had declared that each species was designed according to an ideal form that they called an *eidos*, often translated as 'type' or 'archetype.' Biologists following this view developed the **morphological species concept**. Each species was described as having certain fixed and invariant physical characteristics (morphology). The whole 'type' of that species was believed to be a cluster of 'essential' characteristics inherited as a single unit.

Biologists now recognize that species are constantly evolving, largely as the result of natural selection working on the genetic variation that is present within populations (Chapter 4). For us to describe variation within and among human populations, we must first have some clear way of recognizing population groups. We could group people by some physical trait, such as distinguishing between people who are tall, short, or average in height. If we chose some other physical trait, such as eye color or hair curliness, we would find that *each physical characteristic results in a different grouping of the same people*. In addition, we find that groupings based exclusively and strictly on any single trait always group together people who are quite dissimilar in many other respects (especially on a worldwide basis). For these reasons primarily, biologists prefer not to base the definition of population groups on physical characteristics.

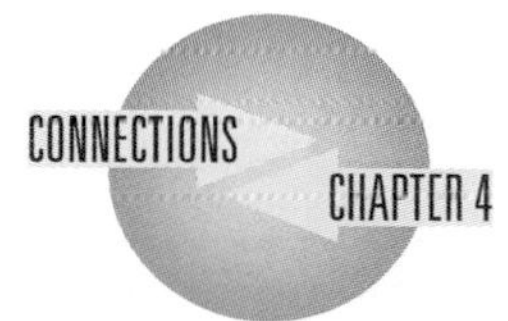

Instead of using physical characteristics to define groups of organisms, biologists recognize groups whose members mate with one another under natural conditions. Naturally occurring groups of organisms that are able to mate with one another in nature are assigned to the same species. A biological **population** is part of a species; it consists of those members of a species from which mates are actually chosen. Membership in a population is determined by mating behavior, not by physical characteristics. Population membership therefore depends very strongly upon geographical location.

Populations that interbreed with one another under natural conditions belong to the same species (see Chapter 4, p. 147). All humans have the capacity to mate with one another and produce fertile offspring; for this reason, all humans are placed in a single species, *Homo sapiens*. Not all humans belong to the same population, however. People in different geographical locations form different populations. Genetic variation within any population is usually less than in the species as a whole. In past centuries, geographic isolation kept many human populations more distinct than they are now with worldwide transportation and migration. Population boundaries are not the same as national boundaries. Several different populations often live in the same geographic area; sometimes, these populations are distinguishable by cultural factors or by their derivation from geographically separate earlier populations.

Human populations in different places differ from one another in many physical traits. The average Canadian is taller than the average Southeast Asian, and the average African has darker skin than the average European. For natural selection, however, the characteristics that matter the most are those with the greatest impact on health and disease (or life and death). For example, cystic fibrosis and skin cancer are more frequent among people of European descent, but people of African descent have a higher risk of sickle-cell anemia and are more susceptible to frostbite if exposed to very cold temperatures. Most single-gene traits that are examined closely show some small average difference in allele frequency among human populations. For continuous traits, the *difference between the averages of two populations is much less than the variation within either population* (Figure 5.2). For example, the average height in the United States is taller than in China, but many Americans are shorter than the Chinese average and many Chinese are taller than the American average.

Although it is easy to find human populations that differ from one another in both physical features (morphology) and genetic traits, it is usually very difficult to find sharp boundary lines dividing these populations from one another. If you were to walk from Asia to Europe and then to Africa, you would see populations differing only slightly, in most cases imperceptibly, from their neighbors, and you would meet representatives of the three largest population groups on Earth *without finding any abrupt boundaries between them*. Another way to say this is that *variation between human populations is always continuous*. This is so even when the trait in one individual is discontinuous. The population frequency of the allele responsible for the trait can vary continuously between zero (no one has the phenotype) and 100% (everyone has the phenotype). For discontinuous traits such as blood type, the allele frequencies of adjacent populations are generally close, just as is true for the average values of continuous traits, as seen in Figure 5.2.

FIGURE 5.2

Continuous variation in two populations with different mean values.

Distribution of height in two populations whose average values are 165 and 180 cm respectively. The variation within each population is greater than the difference between the average values of the two populations. Note that one of these populations is identical to the one shown in Figure 5.1.

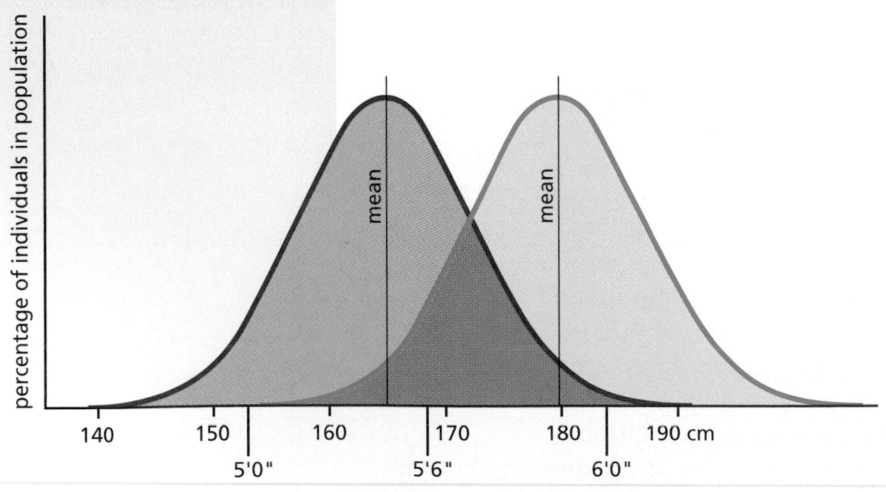

Concepts of race

Humans have developed various ways of describing both themselves and the other human populations with which they have had contact. Biologists (who study all forms of life) and anthropologists (social scientists who study human populations and human cultures) have assisted in these descriptions by studying and measuring certain physical traits and allele frequencies. There are many ways in which human variation can be described, and there are many uses to which these descriptions have been put. One of the most problematic has been the attempt to separate people into different **races**. As we will soon see, there are various different meanings to this term, all of them different from the term 'population.' The term 'population' always describes smaller and more cohesive units than the term 'race.' No physical features are used in defining populations, but some race concepts have been based on physical features. In this section we describe four different concepts of race in the order in which they originated. The older concepts have not entirely died out; they have in many cases persisted side by side with the concepts that came later.

Races based on cultural characteristics. In the Bantu languages of Africa, the word for 'people' is *Bantu*. Likewise, the Inuit word for 'people' is *Inuit*. Every group of people has a name for itself and its members, and the name often means *people* or *human*. Names that people apply to other groups of people may simply be descriptive, but value judgements are often implied as well. In some instances, the value judgment implicit in the choice of name has been used to justify widespread abuses against the negatively labeled population. Such was the case when land and labor shortages resulted from large-scale cereal agriculture, a problem that arose independently in many places. A commonly developed solution to these shortages was to conquer neighboring people (the 'other') and confiscate their land. Slavery and several other systems of coercion were developed to secure the labor of conquered peoples. Slavery, oppression, and conquest all call upon the victorious people to practice certain atrocities on others that they would never tolerate within their own group. To justify these atrocities to themselves, and to protect their own members from practicing similar atrocities on one another, just about every conquering group has found it expedient to distinguish themselves from the 'other,' and furthermore to depict the conquered people as somehow inferior, subhuman, or deserving of their fate. Many of the groups that were culturally defined as races in the past are really language groups, cultural groups, or national groups that are hardly distinguishable on any biological basis from the group that traditionally oppressed them.

The imposition of social inequalities between 'Us' and 'Them' is now recognized as racism. **Racism** has many connotations, but all of them are based on the belief that some groups of people are better than others, and that it is somehow justified or proper for the more powerful group to subdue and oppress the less powerful. In most cases, the motivation to conquer and oppress others came first; the racist ideology came later.

The 'races' identified by the conquering group are socially constructed to serve the interests of the oppressors only. The distinctions and values of the oppressors are forcibly imposed on the oppressed, who are often taught to believe in their own inferiority. Most people now regard racism as unethical because it denies basic rights to many people and because it results in frequent crime, violence, and social conflict.

Separation based on race serves better the political and economic causes that have engendered it if the distinctions recognized are declared to be biologically based and therefore 'natural' and irremediable, as opposed to characteristics that can easily be changed by education or religious conversion. Scientists belonging to racist societies have therefore sometimes attempted to 'prove' the **hereditarian** assertion that the traits characteristic of an 'inferior' race have an inherited basis that cannot easily be changed. Behind these assertions is the view that a group identity (an 'essence' or Platonic *eidos*) can be inherited, a view for which there is no basis in genetics. Anthropologist Eugenia Shanklin documents several instances in which scientists conducted 'scientific' studies to help 'prove' the values and prejudices of their own social group. In their genocidal campaigns of the 1940s, the Nazis exterminated many millions of Jews, gypsies, Slavs, and other groups, but not until they had declared each of them to be an inferior 'race.'

Racism and hereditarianism are not synonymous, but they often go together as attitudes shared by many of the same people. The supporters of eugenics (see Chapter 3, p. 111) had many followers, including Nazis in Germany and anti-immigrationists in the United States. These followers sought ways to prove the inferiority, and especially the biologically unchangeable inferiority, of other people.

The other race concepts that we discuss below differ from this earliest concept, and resemble one another, in their avoidance of language, customs, and other cultural traits in the delineation of races. However, racism is not confined to those societies that embrace the cultural concept of race. Many biologists and anthropologists have pointed out that racism is also built into the the next concept, race delineated by body features.

The morphological or typological race concept. Biologists who study plant and animal species often describe the geographical variation within a species by subdividing the larger species into smaller and more compact subgroups, each of which is less variable than the species as a whole. These subgroups are generally called **subspecies**, but within our own species they are called races. To bring the study of human variation more in conformity with that of other species, scientists began to restrict their attention to characters that could be studied biologically and to exclude personality traits, languages, religions, and customs more influenced by culture than by biology.

Before the days of ocean-going vessels, most of the world's people had only a limited awareness of human variation on a worldwide scale. Each population, of course, knew about other populations nearby, but in most cases adjacent populations differ only slightly from one another. When trade extended over great distances, it usually did so in stages, so that none of the traders ever had to go more than a few hundred miles from home. The trade routes were also in most cases traditional, meaning that traders and migrants had generally come and gone over the same routes for centuries. This contributed to a gene flow or mixing of alleles that lessened the degree of difference between populations that would be noticed along the trade routes.

When explorers began to sail directly to other continents, they found people in other lands who differed more sharply from themselves in physical features. Many scientists subsequently became curious about the origin of these physical differences. Discussions of racial origins from about 1750 to 1940 tended to dwell on the origin of physical differences. A morphological definition of each race, based on physical features (morphology), was an outgrowth of the same thinking that had earlier resulted in a morphological species concept. At least initially, the major founders of this tradition were scientists who had no interest in oppressing the newly discovered peoples, so finding an excuse for racial oppression was less of a motive than was scientific curiosity. The emphasis was no longer on distinguishing only 'us' from 'them,' but on distinguishing among many different racial groups.

By the 1700s, biologists were actively describing and categorizing the variation in all living species. The eighteenth century naturalist Linnaeus (Carl von Linné) divided the biological world into kingdoms, classes, orders, genera, and species (see Chapter 4). He also divided humans into four subspecies: white Europeans, yellow Asians, black Africans, and red (native) Americans. The use of physical features such as skin color and hair texture to define subspecies was common among biologists using a morphological race concept. Other scientists in this same tradition recognized more races or fewer, but each race was always described on the basis of morphological characteristics such as skin color, hair color, curly or straight hair, and the occurrence of epicanthic folds of skin over the eyes.

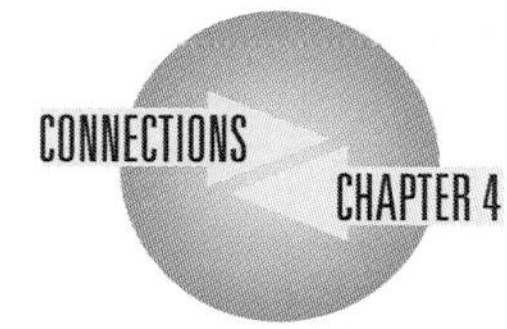

Under the morphological concept of race, each race was defined by listing its common physical features *as though* they were invariant. For example, when describing a feature such as color, only one color was given, as if this color were invariant throughout the group and throughout time. This approach, which classified races on the basis of 'typical' or 'ideal' characteristics, ignoring variation, is called typology. *Morphological definitions of race were always typological*. Africans, for example, were declared to have black skins and curly hair, overlooking the fact that both skin color and hair form vary considerably from place to place within Africa and even within many African populations. All of the morphological characteristics were assumed to be inherited as a whole; a person was assumed to inherit a Platonic *eidos* (a 'type' or 'essence') for whiteness or redness, not just a white or red skin. Supporters of the typological concept of races were also supporters of a typological concept of species.

Years after morphological races had been defined, closer scrutiny revealed both variation within the morphological races and intergradation between them across their common boundaries. A few Europeans tried to save the morphological definitions by proposing that each race had originally been 'pure' and invariant, and that present-day variation within any population was the result of mixture with other races. One zoologist, Johann Blumenbach (1752–1840), divided up humans into American, Ethiopian, Caucasian, Mongolian, and Malayan races. He thought that each of these races was originally homogeneous (that is, 'pure'), and he named each after the place that he identified as its ancestral homeland. For example, white-skinned people are called 'Caucasian' because Blumenbach thought that this race originated in the Caucasus Mountains, east of the Black Sea.

There is no scientific support nowadays for the concept of originally pure races or for the concept of ancestral centers of origin of different races; human populations have never been homogeneous and have always been quite variable. In some cases, however, Europeans and others who feared for the 'purity' of their own group sought to pass laws limiting contacts, especially sexual contacts, between the races that they recognized. Most of these laws were brutal but still ineffective in stopping what were viewed as interracial matings. There is no scientific basis for the belief that such matings are in any way harmful. On the contrary, variation within any species confers a long-term evolutionary advantage because it provides the raw material that natural selection can use to adjust to changing environmental conditions.

But hereditarian assumptions were even more strongly embedded in the morphological race concepts than they are in the culturally based race concepts. Lest one think that science has long since banished such attitudes in educated people, it is only necessary to point to the great storm of controversy that flourished over the subject of race and IQ in the 1970s. Arthur Jensen attempted to convince his readers that the mental abilities of African Americans were below those of other races and that these differences were fixed by heredity and unchangeable by educational means. A number of scientists, including Leon Kamin, Richard C. Lewontin, and Stephen Jay Gould, showed that his claims were unsupportable and based on fallacies and fabricated evidence. As recently as 1994, a book by Richard Herrnstein and Charles Murray once again brought up many of the same hereditarian arguments that had earlier been debunked (Box 5.1).

One of the strange ironies of a racist past is that many attempts at remediation, such as affirmative action, continue to require, at least for a time, the identification and naming of the same groups that were used previously for racially divisive purposes. Attempts to ensure fair and nondiscriminatory treatment for members of different socially recognized racial groups (in housing, employment, schooling, and so forth) require that we first identify and study the groups that we wish to compare. In this way, societies trying to overcome a history of racism find themselves using the very racial classifications of their racist past in order to redress the injustices of past generations.

The population genetics concept. Biologists who study geographic variation are well aware that all boundaries between groups of populations are arbitrary and that all transitions are gradual. A continuous increase or decrease in the average value, allele frequency or phenotypic frequency of any one trait is called a **cline**, after a Greek word meaning 'slope' (as in words like 'incline' or 'recline'). Clines are an accurate (but lengthy) way of describing the variation in each trait, one trait at a time. For a dozen characteristics, a dozen different maps would be needed, because the patterns of variation would in general not coincide.

The maps in Figure 5.3 show the clinal variation in the allele frequencies of three blood group alleles. Before such maps of allele frequencies can be drawn, local populations must first be identified and sampled. For example, blood groups must be studied in many local populations before

BOX 5.1 Is Intelligence Heritable?

To address a question such as this, we must first define intelligence. Intelligence is not easily defined, but it includes the ability to reason and the ability to learn new ideas and new forms of behavior, the measurement of which is far from simple. The biological bases for these abilities are likely to be multifaceted (Chapter 10), and genetic factors are likely to be the result of the interaction of many, many genes. Most discussions on the inheritance of human intelligence deal only with a single measure of this very complex trait, the IQ score, obtained from a test. IQ is not the same thing as intelligence and is at best an imperfect measure of mental abilities.

Also, to address this question, we must define the word 'heritable.' Heritability is defined in statistical terms as the proportion of the population's *variation* in some trait associated with genetic as opposed to environmental variation. Statistical association, or correlation, does not imply causation, and it certainly cannot be used to justify the claim that 'there is a gene for' the trait in question. One way to determine heritability of a trait in a domesticated species is to compare the variability of that trait in the population at large with the variability of the trait among highly inbred, genetically uniform individuals. Another way to determine heritability is to compare the variability that a trait exhibits at large with the variability of that trait among individuals raised in a standardized, experimentally controlled environment. Neither of these methods can be applied to humans, and the measures that are used to study humans are all indirect, complicated, and subject to criticism on technical grounds. For these reasons, *there is no agreement* on the heritability of any important human ability, including 'intelligence.' Moreover, because variation is a characteristic of populations and not of individuals, a term such as '60% heritable' would simply be a ratio of variation in one population to variation in another and would not tell us anything about the genotype or phenotype of any individual.

Numerous studies on IQ scores have shown the following:

It is difficult to devise IQ tests that are free from cultural bias and from bias based on the language of the test, the gender and race of the test subjects, and the circumstances in which the test is administered.

IQ scores seem to have both genetic and non-genetic components. Children's IQ scores correlate strongly with those of their parents. The IQ scores of adopted children usually agree more closely with their adoptive parents than with their birth parents, although studies on adopted children have been criticized for a variety of reasons (see Chapter 3, p. 81).

IQ scores can be greatly improved by environmental enrichment. They can also be adversely affected by poor nutrition, poor prenatal conditions, and a number of other environmental circumstances.

Populations historically subject to discrimination, such as African Americans in the United States, Maoris in New Zealand, and Buraku-Min in Japan, have average IQ scores about 15 points below those of the surrounding majority populations. However, these lower average scores do not always persist in people who migrate elsewhere: descendants of Buraku-Min living in the United States have, on average, IQ scores on a par with those of other people of Japanese descent.

In the United States, IQ scores of whites and also of blacks (African Americans) vary from state to state, in some cases more than the average 15-point difference between blacks and whites. Among African Americans born in the South but now living in the North, IQ scores vary in proportion to the number of years spent in northern school systems.

Transracial adoption studies show that African American children adopted at birth and raised by white families had IQ scores close to (in fact, slightly higher than) the white average.

Careful studies of matched samples in schools in Philadelphia failed to show significant average differences in IQ scores between black and white schoolchildren if differences in background were controlled. 'Matched samples' mean that children in the study were compared only with other children of comparable age, gender, family income level, parents' occupation, and similar variables.

Taken together, these data indicate that there is, at most, a small degree of heritability for IQ. They provide little support for the hereditarian claim that IQ is fixed and immutable, or that observed differences in scores cannot be diminished. They provide no support whatever for predicting any individual's IQ score on the basis of their inclusion in any group.

FIGURE 5.3

The clinal distribution of alleles for the ABO blood groups in indigenous populations of the world. Indigenous populations are those that have lived for hundreds or thousands of years in approximately the same region, to which they have had time to evolve adaptations. This includes Native Americans in the Western Hemisphere, Bantu and Xhoisan peoples in southern Africa, and Aborigines in Australia, but not the European colonists who came to these places after A.D. 1500.

(A) The distribution of allele *A*

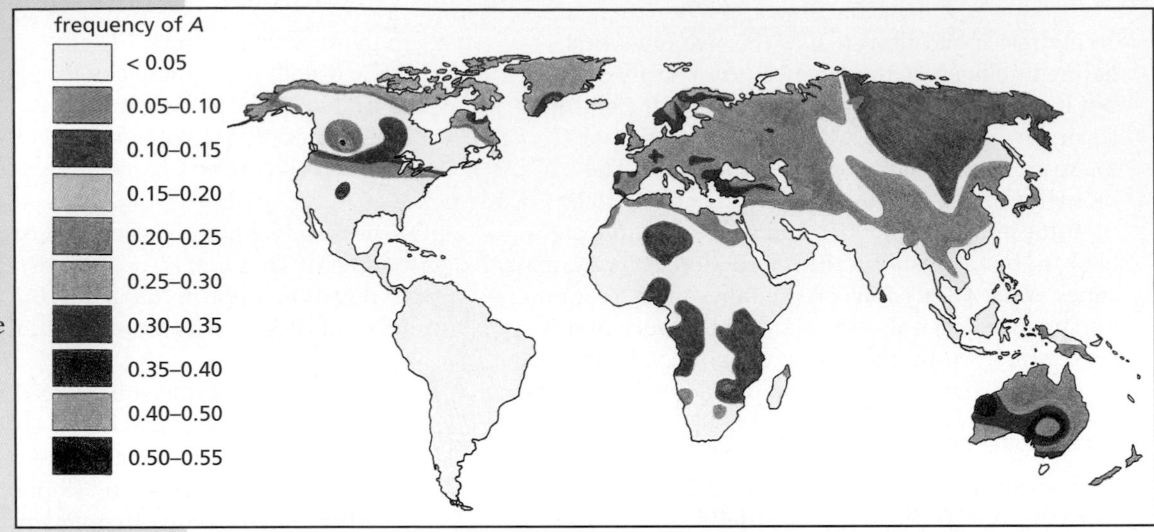

(B) The distribution of allele *B*

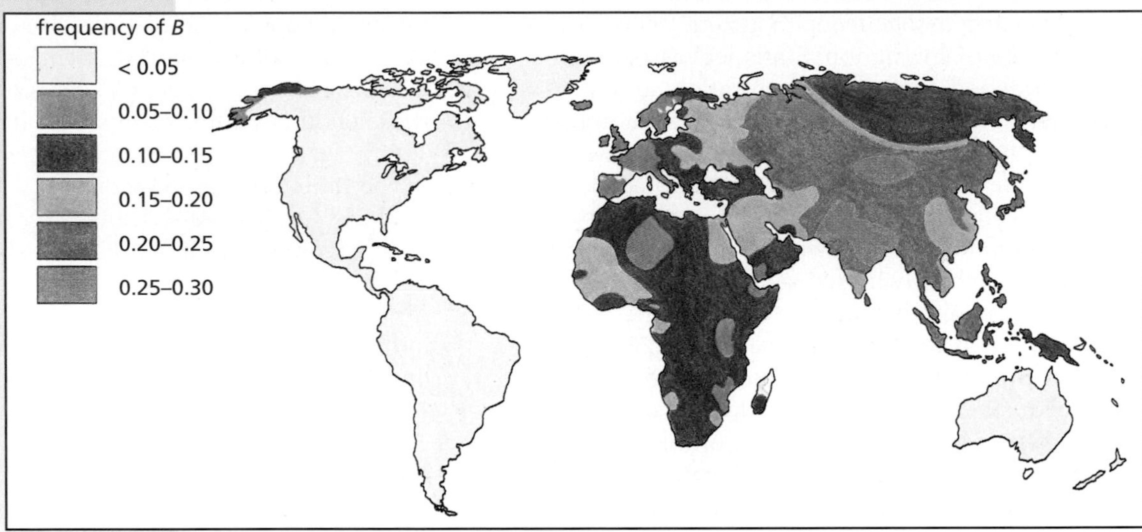

(C) The distribution of allele *o*

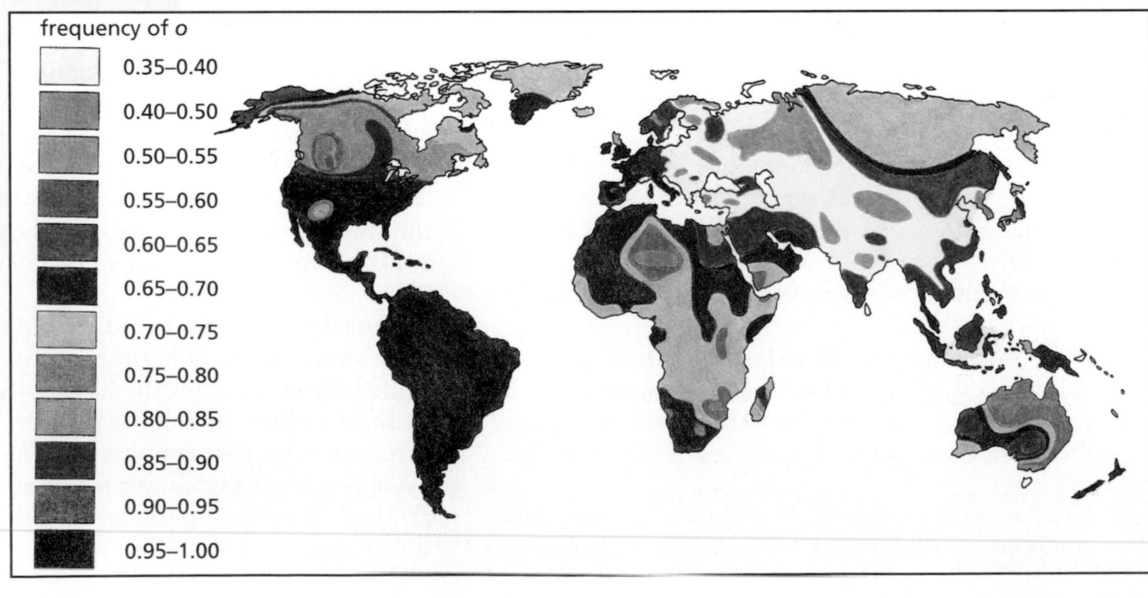

drawing maps such as those in Figure 5.3. From maps such as these, we learn that large continental areas usually show gradual clines. Thus, Figure 5.3A shows a gradual south-to-north increase in the frequency of allele *A* across North America, and Figure 5.3B shows a gradual west-to-east increase in the frequency of allele *B* across most of Eurasia. Abrupt changes are uncommon, and when they do occur they generally coincide with geographic barriers to migration, which are also barriers to gene flow. Examples of such barriers include the Sahara Desert, the Himalaya Mountains, and the Timor Sea north of Australia.

As can be seen in Figure 5.3, the frequencies of the blood group alleles *A*, *B*, and *o* vary greatly from one human population to another. The variations, however, do not necessarily coincide with other traits or with the groups recognized on the basis of morphology. Allele *B*, for example, reaches its highest frequency on mainland Asia, but is nearly absent from Native American populations or among Australian Aborigines. The frequency of allele *A* decreases from west to east across Asia and Europe. In Native American populations, allele *A* occurs mostly in Canada, and is mostly absent from indigenous Central or South American populations. The allele for blood group O has a frequency of 50% or more in most human populations, but its frequency approaches 100% in Native American populations south of the United States. African populations generally have all three of the alleles for ABO blood groups at levels close to worldwide averages.

Since the clinal variation concept was introduced in 1939, it has become customary to describe human variation by drawing maps of one cline after another. In addition to cline maps of phenotypes and allele frequencies, the techniques of molecular genetics (such as the RFLP technique described in Chapter 3, p. 82) are now being used to study clines at the molecular level. Clinal maps can also be drawn for continuous traits, in which case average values for the trait are calculated in each population. To describe the geographic variation in *Homo sapiens* or any other species, we could draw one map showing clinal variation in average body height, another showing variation in skin color or hair form, and so on. The population genetics approach encourages the scientific study and description of populations, including studies on the origins and former migrations of populations.

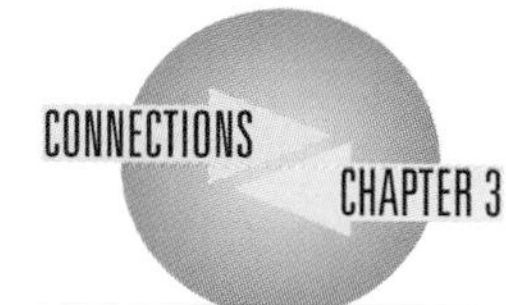

After the Holocaust (1933–1945), the fledgling United Nations felt the need to refute many Nazi claims about race. The result was the 1948 *Statement on Race and Racism*, written by a committee that included several prominent anthropologists and geneticists. The statement, which has been revised several times since 1948, correctly pointed out that nations, language groups, and religions have nothing to do with race, and that no group of people can claim any sort of superiority over another. The statement went further, however, to proclaim a new definition of race that replaced older, morphological definitions based on the inheritance of Platonic 'ideal types' with a new definition based on population genetics.

Under the population genetics definition, a race is a geographic subdivision of a species distinguished from others by the allele frequencies of a number of genes. A race could also be defined as a coherent group of populations possessing less genetic variation than the species as a whole.

Either definition means that blood group frequencies are now considered more important than skin color in describing race, and that races are groups of similar populations whose boundaries are poorly defined. It also means that one cannot assign an individual to a race without first knowing what interbreeding population that individual belongs to. 'Race' is no longer a characteristic feature of any individual, because allele frequencies, like average phenotype values, characterize populations only, not individuals. Allele frequencies are consequences of population membership; they cannot be used to assign someone to a particular population or group. For this reason, racially discriminatory laws cannot and do not use population genetics; such laws rely invariably on the older morphological definitions or the still older social definitions of race.

Some writers maintain that racism is still contained in the population genetics race concept. They contend that studies that describe allele frequencies in geographic populations are merely reinscribing the racism of earlier concepts. Although far fewer people see racism in population genetics than in the earlier race concepts, some wish to go even further than the U.N. statement goes.

The 'no races' concept. Some scientists went even further in rejecting the heritage of the racist past: in the 1960s, led by the British anthropologist M.F. Ashley Montagu (who had earlier contributed to the U.N. definition), they declared that they would not recognize races at all. Among their arguments, one of the most compelling is that race concepts have always been misused by racists of the past and that the only way to rid the world of racism was to reject the entire concept of race. History is replete with examples of slavery, apartheid, discrimination, genocide, and warfare between racial groups. It is therefore easy to argue that the naming of races has in past generations done far more harm than good.

One stimulus to the 'no races' approach arises from the great increase in international travel and migration that has occurred especially since World War II. To a certain extent, human populations have always mated with one another whenever there has been geographic contact between them. This is one reason why human population groups do not differ more than they do and why neighboring populations are so often similar. Since the advent of the jet age, frequent migrations have allowed more extensive contact and more opportunities for mating between people of different genetic backgrounds than ever existed before. Such matings have always occurred and always will; they even occur in societies that have tried to outlaw them. This type of mating will slowly but inevitably diminish the differences in the mix of alleles (the so-called **gene pools**) of populations, making it progressively more difficult to identify *any* significant differences between populations.

The study of human variation

All studies of human variation run the risk that they can be misused or misinterpreted by racists. Nevertheless, there are many good reasons for studying human variation, and this study serves as the basis for the entire field of 'human factors engineering.' To take a simple example, the design of a passenger compartment (for automobiles, aircraft, etc.) must accommodate a certain range in the size, sitting height, arm length, and other dimensions of its possible occupants. These and other accommodations must take into account the total range of human variation, including all races and both sexes. In airline cockpits and similar enclosures, controls should be both visible and reachable by persons of different sizes. Moreover, these features are often matters of safety as well as comfort. Vehicle seat belts and airbags, sports equipment, surgical equipment, wheelchairs and similar aids, boots, helmets, kitchen counters, telephone receivers, gas masks, toilets, and doorways all need to accommodate the range of dimensions of the human body. Variation in other human characteristics (breathing rates, sweating) must also be considered in the design of space suits, diving equipment, respiratory equipment for fire fighting, or protective clothing for other situations. Most of the variation relevant for human factors engineering is found *within* each population group, including variation by age and sex; variation between human populations is generally minor by comparison.

A further reason for studying genetic variation among human populations is that it can help us to understand evolution. Population genetics has helped us to recognize geographic patterns of disease resulting from natural selection acting on human populations. Studies of this kind can also help us to reconstruct the past history of particular human populations, or of the human species as a whole. In succeeding sections of this chapter, we examine some of these studies.

THOUGHT QUESTIONS

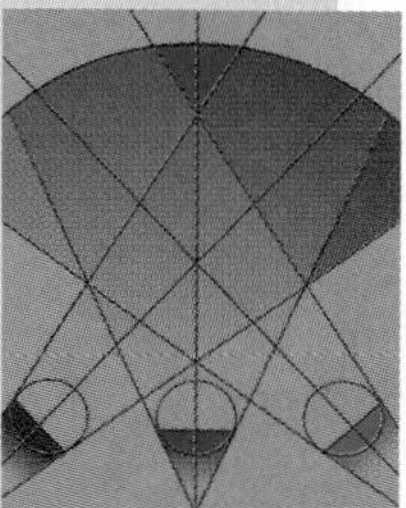

1. Twentieth-century approaches to the description of human variation have in large measure been revolts against the earlier approaches. Against which of the earlier approaches was the 'no races' approach primarily directed? Against which earlier approach was the population genetics approach directed?

2. African Americans more often have high blood pressure and more often die from their first heart attacks than do white Americans. How would you decide whether this is the result of a difference in genes, in diets, in the availability of medical care, or in the lasting effects of discrimination in U.S. society? If people in rural Africa seldom have heart attacks or high blood pressure, what possible hypotheses are falsified?

3. To produce research results of the kind referred to in Thought Question 2, one must have a way of assigning an individual to a population group. How does one determine a person's membership in a biological population? Is it sufficient to know that they live in a particular place? Will asking people to name the racial or ethnic group in which they claim membership (self-identification) produce biologically meaningful results?

POPULATION GENETICS CAN HELP US TO UNDERSTAND HUMAN VARIATION

The geographic variation shown in Figure 5.3 deals with human blood groups. We know a lot about the genetic basis of blood groups, and a person's blood group is easily determined, making blood groups good candidates for study by population geneticists. We now look in more detail at human blood groups and what their study has taught us about our own species.

Human blood groups and geography

In the days before reliable blood banks, blood transfusions were much riskier than they are today. Soldiers wounded in battle were generally treated in the field. If a transfusion was needed, it was done directly from the blood donor to a patient lying on an adjacent stretcher. Some transfusions were successful, but others resulted in death of the patient. Studies on the reasons for these different outcomes led to our knowledge of the existence of blood groups.

ABO blood groups. During the Crimean war (1854–1856), a British army surgeon kept careful records of which transfusions succeeded and which did not. From his notes he was able to identify several types of soldiers, including two types that he called A and B. Transfusions from type A to type A were nearly always successful, as were transfusions from type B to type B, but transfusions from A to B or B to A were always fatal. Also discovered at this time was a third blood type, O, which was initially called 'universal donor' because people with this blood type could give transfusions to anyone. These results were put to immediate practical use in treating battlefield injuries.

A German doctor, Karl Landsteiner, discovered the reason for these distinctions. Persons with blood type A make a carbohydrate of type A, which appears on the surfaces of their blood cells. Persons with blood type B make a carbohydrate of type B; persons with type AB make both type A and B carbohydrates; and persons with blood type O make neither of these carbohydrates. The A and B carbohydrates are also called antigens because they are capable of being recognized by the immune system (see Chapter 12). The immune system of each individual also makes antibodies against the blood group antigens that their own body does not make. In a person receiving a transfusion with incorrectly matched blood, these antibodies bind to the type A or B antigens, causing the blood cells to clump together within the blood vessels (Figure 5.4), often with fatal results. For explaining these immune reactions, Landsteiner received the Nobel Prize in 1930.

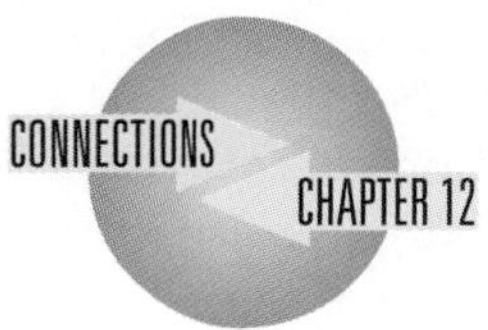

The A and B antigens allow all people to be classified into the four blood groups A, B, AB, and O. These blood groups are controlled by a gene that has three alleles: allele *A* is dominant and it contains information for producing antigen A (its phenotype); allele *B* is dominant and it contains information for producing antigen B (its phenotype); allele *o* is recessive and it functions as a 'place-holder' on the DNA but produces neither functional antigen. The *AA* and *Ao* genotypes both produce antigen A and are therefore assigned to blood group A. Likewise, both *BB* and *Bo* genotypes produce antigen B and result in the B blood type. Genotype *oo* produces neither A nor B antigens, which results in the O blood type (universal donor). Finally, genotype *AB* allows both alleles *A* and *B* to produce their respective antigens, resulting in the AB blood type. When they occur together, the *A* and *B* alleles are said to be **codominant** because the heterozygote shows the effects of both phenotypes.

For the purpose of matching blood donors and recipients, any person who shares your blood type is a good donor. It is therefore possible to collect blood in advance from many donors, sort the blood by blood type, and store it under refrigeration for use in an emergency. It is ironic that the doctor who developed this concept, an African American named Dr. Charles Drew (1904–1950), was denied its full benefits because many hospitals at the time kept separate blood banks for whites and nonwhite patients, a practice that has no biological foundation. Because the chemical composition of the allele products does not vary, type A antigen from an African American is identical to type A antigen from a Native American or from anyone else. A person with blood type A is therefore a good donor for almost any other person with blood type A.

Other human blood groups. Another totally independent system of blood groups, called the Rh system, actually has three genes located very close together on the same chromosome: the first gene has alleles *C* and *c*, the second has either *D* and *d*, and the third has *E* and *e*. Unlike the ABO system, in which alleles are codominant, *c*, *d*, and *e* are recessive to

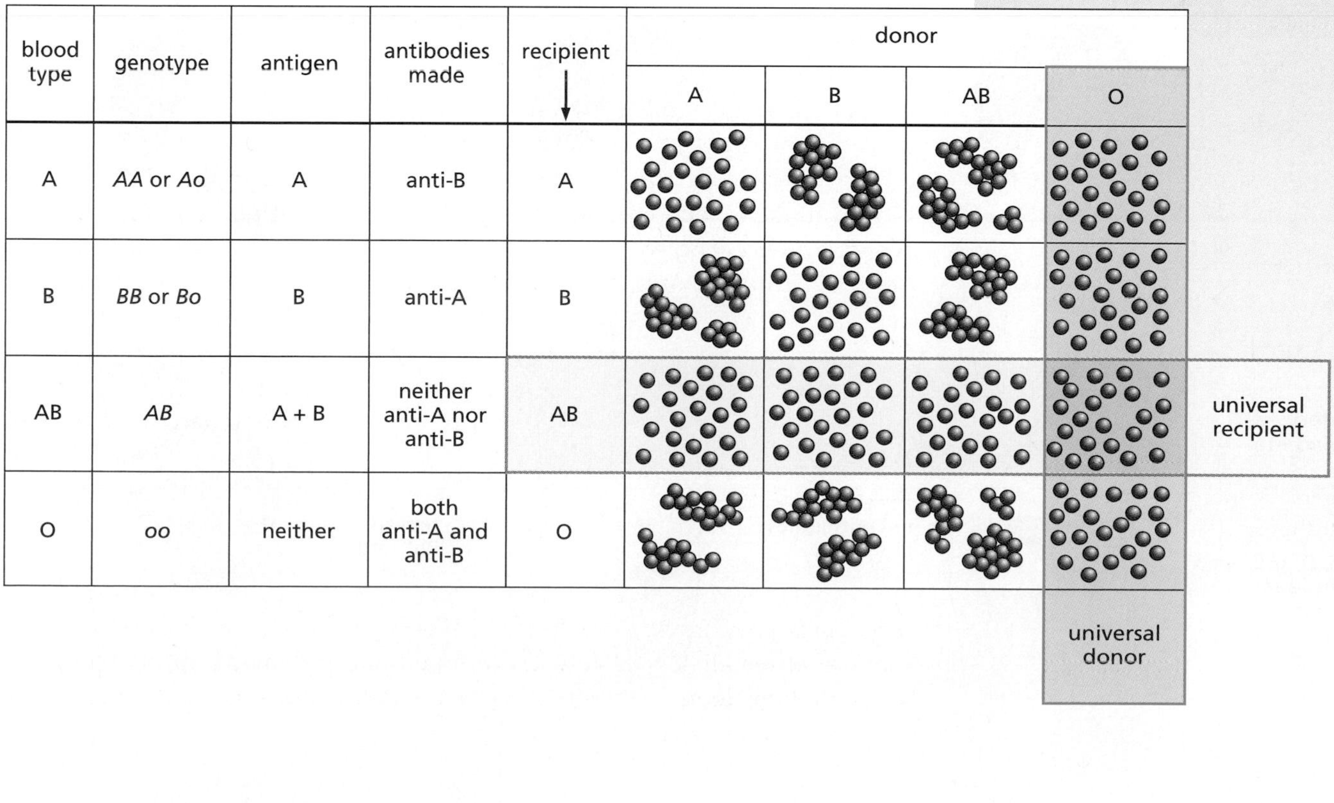

blood type	genotype	antigen	antibodies made	recipient ↓	donor A	donor B	donor AB	donor O	
A	*AA* or *Ao*	A	anti-B	A					
B	*BB* or *Bo*	B	anti-A	B					
AB	*AB*	A + B	neither anti-A nor anti-B	AB					universal recipient
O	*oo*	neither	both anti-A and anti-B	O					
								universal donor	

FIGURE 5.4

The human ABO blood groups. If a person of blood type A, who makes antibodies against blood type B, receives a transfusion of type B or AB blood, those antibodies cause the donated blood cells to clump together. Transfusion with matched blood or with blood type O (no A or B antigens) does not cause clumping

C, *D*, and *E*. In all, there are eight phenotypic possibilities, of which phenotype cde (genotype *ccddee*, homozygous recessive for all three genes) is sometimes called Rh-negative and the others Rh-positive. The Rh-positive phenotype CDe is the most frequent phenotype in most populations, except in Africa south of the Sahara, where cDe predominates. The Rh-negative phenotype cde is the second most common Rh phenotype in Europe and Africa, but is rare elsewhere.

Problems arise when a mother with the cde Rh-negative phenotype is pregnant with a baby who has a dominant *C* or *D* or *E* allele and is therefore Rh-positive. In this case, the mother makes antibodies against the C, D, or E antigens on the baby's blood cells, especially in response to the tearing of blood vessels during the process of birth. Because these antibodies are made at the end of pregnancy, they usually don't affect the first Rh-positive fetus that the mother carries. However, once these antibodies are made, the mother's immune system attacks any subsequent pregnancy with an Rh-positive fetus, destroying many of the fetus' immature red blood cells, which can cause the death of the fetus (Figure 5.5). This problem can now be prevented by giving the Rh-negative mother gamma globulin (e.g. RhoGAM) at the time of the birth of any Rh-positive child; the globulin inhibits the formation of antibodies against Rh antigens, thereby protecting future pregnancies.

FIGURE 5.5

Rh incompatibility arising in an Rh-negative mother pregnant with an Rh-positive child.

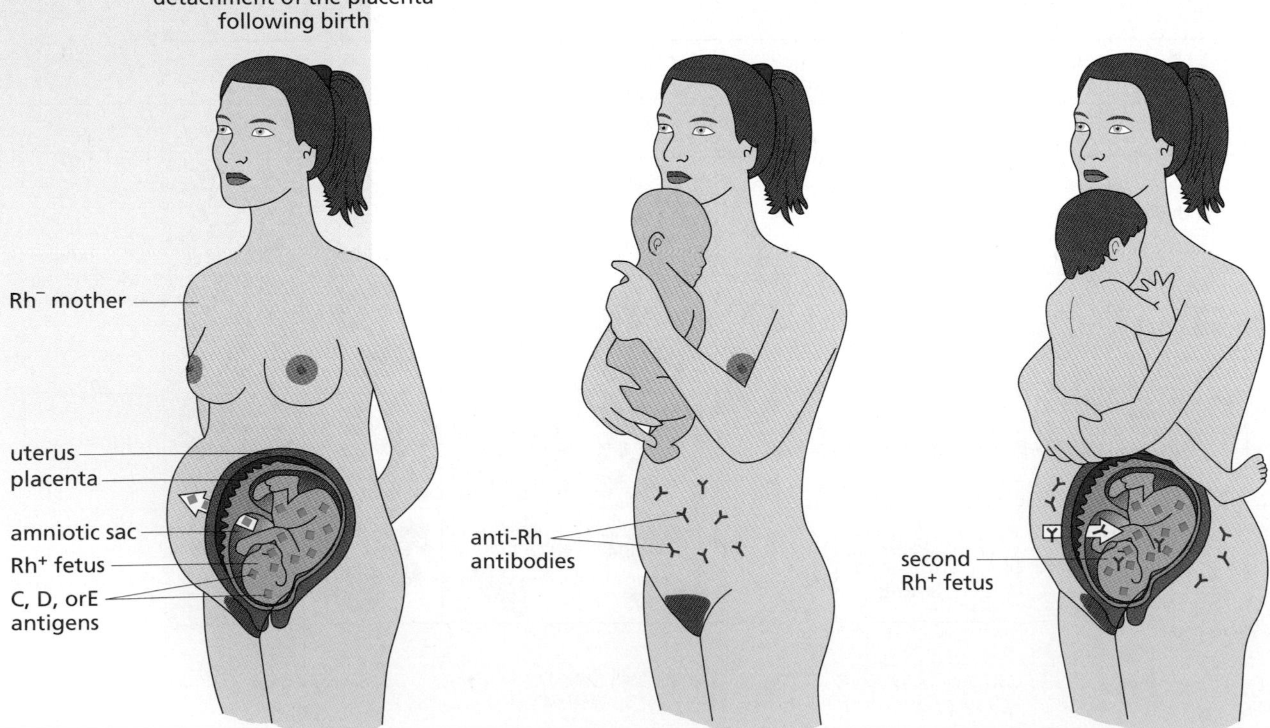
FIRST RH$^+$ PREGNANCY
When an Rh-negative mother has her first Rh-positive pregnancy, C, D, or E antigens from the baby enter the mother's circulation during the detachment of the placenta following birth
SOON AFTER BIRTH
The mother's immune system soon makes antibodies against Rh antigens C, D, or E
SECOND RH$^+$ PREGNANCY
Antibodies made after the first pregnancy can endanger any subsequent Rh-positive fetus unless protective measures are taken
Rh$^-$ mother
uterus
placenta
amniotic sac
Rh$^+$ fetus
C, D, orE antigens
anti-Rh antibodies
second Rh$^+$ fetus

Separate from the ABO and Rh blood group systems are an MN system (with *M* most frequent among Native Americans and *N* among Australian Aborigines), a Duffy blood group system (with alleles *Fy*, Fy^a, and Fy^b), and many others.

Geographic variation in blood group frequencies. We saw earlier that the alleles for the ABO blood groups vary in frequency in different geographic locations (see Figure 5.3). Table 5.1 shows how the major geographic subgroups of *Homo sapiens* differ in the frequencies of various blood groups and other genetic traits. It is important to remember that allele frequencies characterize populations only, not individuals. No blood group is unique to any population, so a person's blood type cannot identify them as a member of any population.

Frequencies of blood group alleles also vary on a smaller geographic scale. This is especially true among rural people who remain in their native villages or districts all their lives. The geneticist Luigi Cavalli-Sforza has documented variation in the ABO, MN, and Rh blood group frequencies from one locality to another across rural Italy. Similar results have been observed in rural populations in the valleys of Wales, in African Americans from city to city across the United States, and among the castes and tribes of a single province in India. These studies emphasize the hazards of assigning all people in a single country to a single population, especially when cultural barriers discourage random mating. However, populations that have become more mobile experience less of this microgeographic variation. As stated earlier, these are variations in allele frequencies and therefore cannot be used to establish clear-cut boundaries between populations.

TABLE 5.1

Allele frequencies in the major geographic population groups.

AFRICAN POPULATIONS

Frequencies of blood group alleles *A*, *B*, and *o* in the ABO system and *M* and *N* in the MN system close to world averages; Rh blood group system with allelic combination *cDe* most frequent and *cde* second; allele *Fy* most common in the Duffy blood group system; allele P_1 more common than P_2 in the P blood group system; hemoglobin alleles Hb^S and Hb^C more frequent than in most other populations.

EUROPEAN OR CAUCASIAN POPULATIONS

Frequencies of *M* and *N* in the MN system close to world averages; allele *A* in the ABO blood group system somewhat more frequent than in African or Asian populations; Rh blood group system with allelic combination *CDe* most frequent and *cde* second; Duffy blood group system with allele Fy^a most frequent and Fy^b second; alleles P_1 and P_2 both common in P blood group system; alleles for G6PD deficiency and thalassemia more frequent than in most other population groups.

ASIAN POPULATIONS

High frequencies of allele *B* (and correspondingly less of allele *A*) in the ABO blood group system; *M* and *N* alleles at frequencies close to world averages; Rh blood group system with allelic combination *CDe* most frequent and *cde* rare or absent; Fy^a especially common in the Duffy blood group system; allele P_2 more common than P_1 in the P blood group system; some populations with high frequencies of alleles for thalassemia or ovalocytosis.

NATIVE AMERICAN (AMERINDIAN) POPULATIONS

Very high frequencies of allele *o* and virtually no *B* in the ABO blood group system; very high frequencies of allele *M* in the MN system; high frequencies of Di^a in the Diego blood group system; Rh blood groups with allelic combination *CDe* most frequent and *cde* rare or absent.

AUSTRALIAN AND PACIFIC ISLAND POPULATIONS

High frequencies of allele *N* in the MN blood group system; Rh blood groups with allelic combination *CDe* most frequent and *cde* rare or absent.

NOTE: For the Rh blood group system, the three genes *C*, *D*, and *E* lie very closely linked on the same chromosome, so that the unit of inheritance is usually a combination of one allele from each of the three genes, such as *cde* or *CDe*, inherited as a unit.

Isolated populations and genetic drift

In large, randomly mating populations in which selection and migration are not operating, the frequencies of the genotypes in the population tend to remain the same. This principle, which operates in all sexually reproducing species, is called the **Hardy–Weinberg principle** (or law), and the predicted equilibrium is called the **Hardy–Weinberg equilibrium** (Box 5.2).

One of the criteria for a Hardy–Weinberg equilibrium is that the population be large. In small populations, allele frequencies tend to vary erratically, in unpredictable directions, from the expectations of the Hardy–Weinberg equilibrium. This phenomenon is usually called **genetic drift**, defined as changes in allele frequencies in small to medium-sized populations due to chance alone. (We can now define a 'large' population

BOX 5.2 The Hardy–Weinberg Equilibrium

The Hardy–Weinberg principle can be stated as follows:

In a large, random-mating population characterized by no immigration, no emigration, no unbalanced mutation, and no differential survival or reproduction (that is, no selection), *the frequencies of the alleles (genotypes) tend to remain the same.*

Allele frequencies are fractions of the total number of alleles present. If a population of 500 individuals (or 1000 alleles at a single genetic locus) contains 400 alleles of type *A* and 600 alleles of type *a*, then we say that the frequencies of the two alleles are 0.40 and 0.60 respectively, or 40% and 60% of the total number of alleles in the gene pool. At a given locus, the allele frequencies always add up to 1, or 100% of the population's gene pool.

Under the conditions specified in the Hardy–Weinberg principle, as stated above, there is a simple equilibrium of unchanging allele frequencies. Let us consider the case of a gene locus that contains two alleles, *A* and *a*. If the frequency of allele *A* is called p and the frequency of allele *a* is called q (where $p + q = 1$), then the equilibrium frequencies of all three diploid genotypes is given by the Hardy–Weinberg formula:

Genotypes	*AA*		*Aa*		*aa*
Frequencies	p^2	+	$2pq$	+	$q^2 = 1$

This formula predicts that the frequency of the homozygous dominant genotype *AA* will be p^2, the frequency of the heterozygous genotype *Aa* will be $2pq$, and the frequency of the homozygous recessive genotype *aa* will be q^2.

To show that these equilibrium frequencies remain stable over successive generations and do not tend to change in either direction, consider the production of gametes in a population already at equilibrium. All of the gametes produced by the dominant homozygotes *AA* carry allele *A*, so the frequency of *A* gametes from *AA* homozygotes is p^2. Half of the gametes produced by the heterozygotes *Aa* also carry allele *A*, so the frequency of *A* gametes from heterozygotes is half of $2pq$, which equals pq. The total proportion of *A* gametes is thus $p^2 + pq$. We can now use simple algebra, separating out the common factor and then applying the equation $p + q = 1$ to calculate the frequency of *A* gametes:

Frequency of *A* gametes:

$$\begin{aligned} p^2 + pq &= p(p + q) \\ &= p(1) \\ &= p \end{aligned}$$

In similar fashion, the proportion of gametes carrying allele *a* is equal to pq (the other half of $2pq$) from the heterozygotes plus q^2 from the recessive homozygotes *aa*.

Frequency of *a* gametes:

$$\begin{aligned} pq + q^2 &= (p + q)q \\ &= (1)q \\ &= q \end{aligned}$$

So the frequency of *A* and *a* gametes corresponds to the frequency of *A* and *a* alleles.

as one of sufficient size for genetic drift not to occur, and hence one in which each generation receives a representative sample of alleles from the previous generation.)

The original model of genetic drift dealt with populations that remained small all the time, but other types of genetic drift were found to apply in particular situations. For example, a large population that became temporarily small and then large again would experience a **bottleneck effect**: the random changes that occurred when the population was small—the bottleneck—would be reflected in the allele frequencies of subsequent generations (Figure 5.6). Another type of genetic drift is known as the **founder effect**. If a small number of individuals become the founders of a new population, then the allele frequencies in the new population—whatever its subsequent size—will reflect the allele composition of this small group of founders.

Combining the gametes in all possible combinations (to simulate random mating) produces the following results:

Female gametes

Male gametes	A p	a q	← Gametes / ← Frequencies
A p	AA p^2	Aa pq	← Genotypes / ← Frequencies
a q	Aa pq	aa q^2	

Taking the resulting genotypes from the chart above (and adding the two heterozygous combinations together), we obtain:

$$\begin{matrix} AA & & Aa & & aa \\ p^2 & + & 2pq & + & q^2 = 1 \end{matrix}$$

This is the same equation that we started with, which shows that the frequencies have not changed. It can also be shown that a population that does not start out at equilibrium will establish an equilibrium in a single generation of random mating.

Notice all the assumptions of the model: the population must be closed to both emigration and immigration, and there must be no unbalanced mutation and no selection. The population must be large enough to permit accurate statistical predictions, and the population members must mate at random. In reality, most natural populations are subject to mutation, selection, and nonrandom mating (including inbreeding), and most usually experience emigration and immigration as well. The Hardy–Weinberg model, in other words, describes an idealized situation that is seldom realized in practice. The Hardy–Weinberg equilibrium is important to population genetics as an ideal situation with which real situations can be compared; if a population is *not* in Hardy–Weinberg equilibrium, one can ask why and then seek to measure the extent of the deviation from equilibrium. The same procedure is followed in other sciences as well. For example, 'freely falling bodies without air resistance' are an ideal situation in physics, and air resistance can be measured as a deviation from this ideal.

The Hardy–Weinberg equation is useful in estimating allele frequencies for traits controlled by a single gene. For example, if a population of 1000 has 960 individuals showing the dominant phenotype (such as normal pigmentation) and 40 displaying the recessive phenotype (such as albinism), then q^2, the proportion of homozygous recessive individuals, is equal to 40/1000, or 0.04. From this, we can calculate $q = \sqrt{0.04} = 0.2$.

From the fact that $p + q = 1$, we can calculate $p = 1 - q$. Substituting the value of 0.2 that we found for q gives us $p = 1 - 0.2$ or $p = 0.8$. Then the proportion of homozygous dominants in the population is $p^2 = (0.8)^2 = 0.64$ and the proportion of heterozygous individuals is $2pq = 2(0.8)(0.2) = 0.32$.

Several cases of genetic drift have been studied in isolated human populations. One well-studied example concerns the German Baptist Brethren, or Dunkers, a religious sect that originated in Germany during the Protestant Reformation. Forced to flee their native Germany, a few dozen Dunkers came to Pennsylvania in the 1600s and started a colony that now numbers several thousands, mostly in rural Pennsylvania and neighboring Ohio. Because their strict religious code forbids marriage outside the group, they have remained a genetically distinct population.

Allele frequencies among the Dunkers have been influenced by genetic drift, particularly by the founder effect. If the Dunkers were a representative sample of seventeenth-century German populations, we would expect similar allele frequencies to those of present-day German populations derived from the same source. If, however, natural selection had changed the Dunker populations as the result of adaptations to their new location, then we would expect their allele frequencies to come closer to those of neighboring populations of rural Pennsylvania. Neither of these predictions is correct. Allele frequencies among the Dunkers differ from populations of *both* western Germany and rural Pennsylvania in a number of traits that have been studied. Blood group B, for example, hardly occurs at all among the Dunkers, although the frequency of the *B* allele is around 6–8% in most European-derived populations, including those of both Germany and Pennsylvania. Other genetically determined traits show similar patterns, including the nearly total absence of the *Fy*ᵃ allele (from the Duffy blood group system) among Dunkers. The explanation that best agrees with the data is that the original founder population, known to have been made up of only a few dozen individuals, happened not to include anyone carrying *Fy*ᵃ or the allele for blood group B. Additional alleles may have been lost by genetic drift while the population remained small. The result was a population that derived its allele frequencies from the assortment of alleles that happened to be present in the founders. We can test this assumption by looking for the rare Dunkers who do possess an allele such as *Fy*ᵃ. In every case that has been investigated, the occurrence of such an allele among the Dunkers can be traced to a person who joined the group as a religious convert within the last few generations.

Because they are genetically isolated, except for occasional religious conversions, the Dunkers have kept a unique combination of unusual allele frequencies. In the absence of blood group B, they resemble Native American populations; in the absence of *Fy*ᵃ, they resemble African populations. In

FIGURE 5.6

The bottleneck effect, a form of genetic drift that operates when populations are temporarily small.

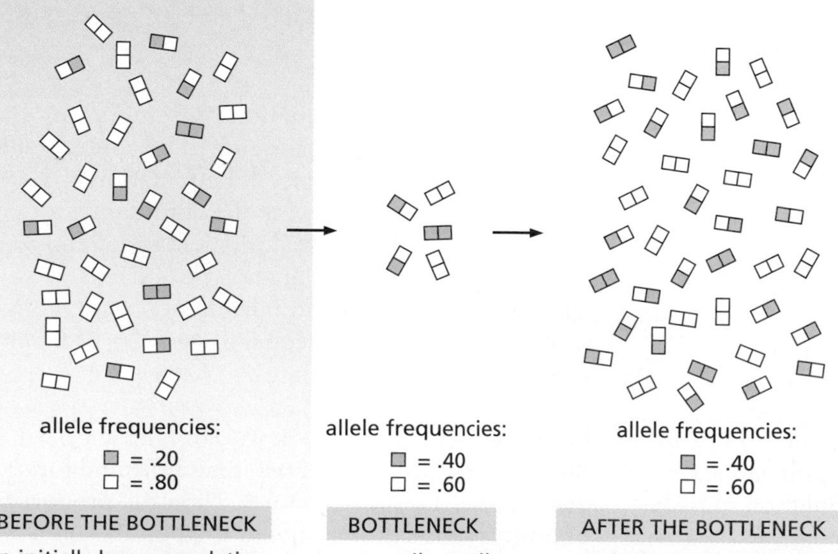
allele frequencies:
■ = .20
□ = .80
BEFORE THE BOTTLENECK
an initially large population (actually much larger than shown here)
allele frequencies:
■ = .40
□ = .60
BOTTLENECK
temporarily smaller population: allele frequencies can drift in random directions
allele frequencies:
■ = .40
□ = .60
AFTER THE BOTTLENECK
population once again large, but allele frequencies now reflect genetic drift that happened when the population was small

most traits, however, their derivation from a European source population is evident. These findings show that population resemblances based on a single blood group or gene system may often be misleading, and that distinctions among human populations, if used at all, should be based on a multiplicity of genetic traits.

The bottleneck effect has been used as a hypothesis to explain the near-total absence of blood group B among Native Americans and of cde (in the Rh blood groups) among Pacific Islanders. When the ancestors of these people first migrated from Asia, the random changes in allele frequency that occurred when the groups were small gave rise to distinct, isolated populations whose allele frequencies differed from those of the ancestral populations. Genetic drift of this kind would apply primarily to groups of people, like the Polynesians or Native Americans, whose founder populations were initially small. The effects of genetic drift are minimal in the larger and more widespread population groups of Africa, Europe, and mainland Asia.

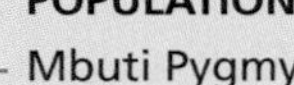

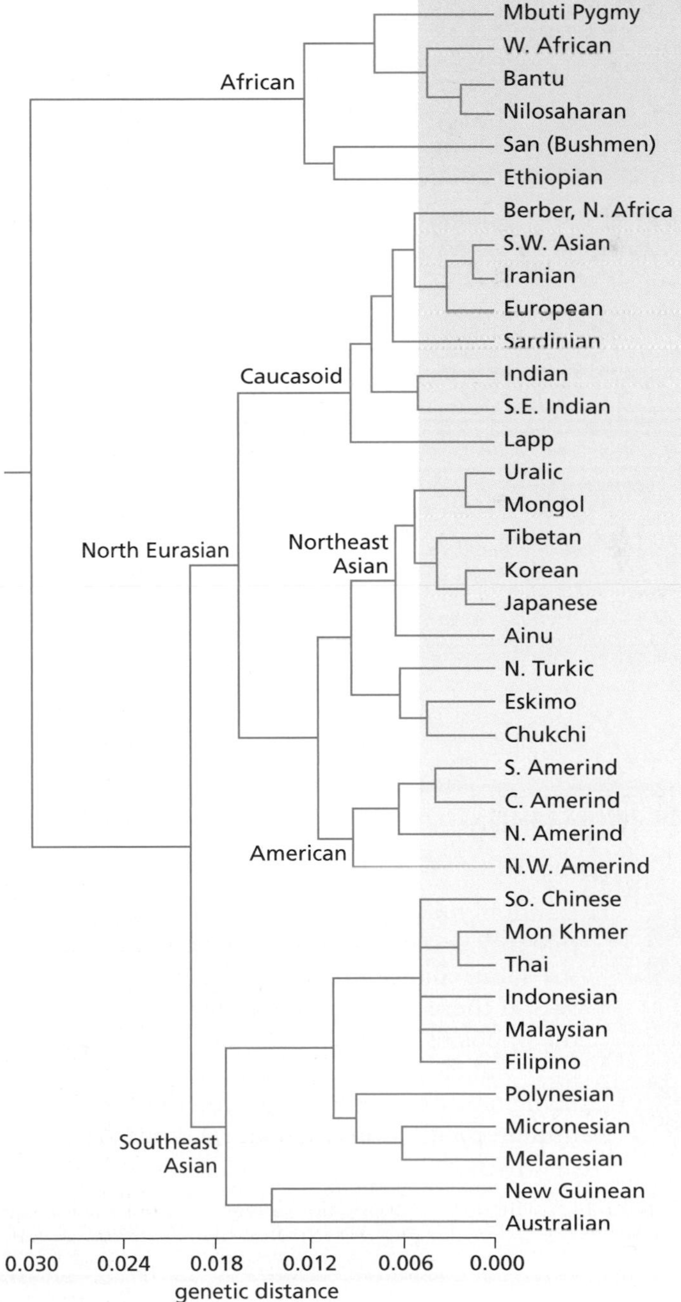

FIGURE 5.7

A family tree of human populations constructed on the basis of mitochondrial DNA sequences. 'Genetic distance' refers to the fraction of mitochondrial DNA sequence not shared by two populations, so that a fork at a genetic distance of 0.006 means that the populations share 99.4% of their mitochondrial DNA sequences.

Reconstructing the history of human populations

Allele frequencies and DNA sequences in modern populations can be used as clues to their origins. For example, American molecular biologist Rebecca Cann and her co-workers studied mitochondrial DNA sequences in samples from over 100 human populations. Mitochondria are organelles in the cytoplasm of eucaryotic cells (Chapter 4, p. 156) that produce much of the cell's energy and that also contain small strands of DNA independent of the DNA in the nucleus. Mitochondrial DNA is transmitted only maternally, from mother to both male and female offspring. Sperm from the father contain almost no cytoplasm and do not transmit mitochondrial DNA. Because mitochondrial DNA is smaller than chromosomal DNA in the nucleus, it is ideal for tracing evolutionary patterns. On the basis of these DNA sequences, Cann and her colleagues proposed a family tree of human populations using a maximum-parsimony computer model: of all possible family trees, the one shown in Figure 5.7 requires there to have been fewer mutational changes than any other tree. Another research team, headed by Luigi

Cavalli-Sforza, used alleles at 120 loci to study the genetic similarities among 42 populations representing all the world's major population groups and many small ones as well. The findings of these two studies (and others) support the hypothesis of a divergence in the distant past between African and non-African populations, with the non-African populations later splitting into North Eurasian and Southeast Asian subgroups (see Figure 5.7). Australian Aborigines and Pacific Islanders are descended from the Southeast Asian subgroup, whereas Caucasians (Europeans, West Asians) and Native Americans (Amerind) are both descended from the North Eurasian group, which also includes Arctic peoples. The groups suggested by this study are geographically coherent and confirm certain well-documented patterns of migration. Existing linguistic evidence also matches these groupings, except for a few cases of cultural borrowing, which can be documented historically. Cavalli-Sforza's group estimates, largely on the basis of archaeological evidence, that the split between African and non-African populations took place 92,000 or more years ago. Other estimates have placed this split much earlier, back to the time of *Homo erectus*.

The study of allele frequencies has also been used to determine the origins of particular groups of people. One such study, for example, showed that Koreans are derived from a group that includes the Mongolians and Japanese but not the Chinese. Also, several studies have provided evidence for a Middle Eastern contribution (perhaps via Phoenecian sailors) to the populations of both Sicily and Sardinia.

How did these various population differences in allele frequencies come about? The next two sections attempt to provide some of the answers to this question.

THOUGHT QUESTIONS

1. Random mating in a sexual species means that any two opposite-sex individuals have the same chance of mating as any other two. If there are a million individuals of the opposite sex, then each should have an identical chance (one in a million) of being chosen as a mate. Do you think human populations mate at random? Why or why not?

2. Is there ever a real population (of any species) in which the conditions specified by the Hardy–Weinberg equilibrium exist? How close do particular populations come?

3. If language has nothing to do with race, why do you suppose that researchers attempting to reconstruct the past history of human populations use linguistic evidence?

MALARIA AND OTHER DISEASES ARE AGENTS OF NATURAL SELECTION

Diseases are among the selective forces that can result in differences among populations. In this section we consider some genetic traits that confer partial resistance to malaria. In malaria-ridden areas, natural selection acts to increase the frequency of alleles that confer partial resistance to malaria while decreasing the frequency of alleles that leave people susceptible to malaria. Many other selective forces have also operated over the course of human history, but resistance to malaria provides a series of well-studied examples.

New traits are produced by mutation (see Chapter 2, pp. 63–66) and are then subjected to natural selection, a process in which many traits die out in populations. The traits that survive natural selection are adaptive traits, or **adaptations** (see Chapter 4), that is, traits that increase a population's ability to persist successfully in a particular environment. A good deal of human variation consists of adaptations that have resulted from natural selection operating over time, disease being a significant agent of that selective process.

Malaria

On a worldwide basis, malaria causes over 110 million cases of illness each year and causes close to 2 million deaths, more than most other diseases. (Only malnutrition and tuberculosis cause more deaths each year, and measles causes about the same number.) Malaria also has a greater impact than most other diseases on the average human life expectancy because most of its victims are young, so that many more years of life are lost for each death that occurs. Malaria is more prevalent in tropical and subtropical regions than in temperate climates. The threat of malaria has largely been eliminated in the industrially developed countries through mosquito eradication programs and the draining of swamps, but as late as the first half of the twentieth century, malaria claimed many thousands of victims in Florida, Louisiana, Mississippi, and Virginia.

Historical and anthropological evidence confirms that malaria was rare (and therefore not a significant selective force) before the invention of agriculture. Even today, the disease is rare in undisturbed forests or in hunting-and-gathering societies. The clearing of forests for agricultural use opens up more swampy areas, and the building of irrigation canals or drainage ditches creates additional pools of stagnant water. The mosquitoes that carry malaria breed best in stagnant water open to direct sunlight. Agriculture therefore did much to change, in unintended directions, the agents of death (and thus the selective pressures) that act on human populations.

Life cycle of *Plasmodium*. Malaria is caused by one-celled protozoan parasites belonging to the genus *Plasmodium* (kingdom Protista, phylum Sporozoa), which live in human blood and liver cells. Of the four species of *Plasmodium* that cause malaria, *Plasmodium falciparum* is the most virulent. All species of *Plasmodium* have a complex life cycle, spending different parts of their life cycle in two different host species, mosquitoes and humans. The *Plasmodium* sexual stages (male and female gametocytes) are intracellular parasites that inhabit human red blood cells. When a female mosquito of the genus *Anopheles* is ready to lay her eggs, she first takes a blood meal from a person during which she ingests large numbers of red blood cells. (Mosquitoes rarely bite otherwise.) If the red blood cells contain *Plasmodium*, the male and female gametocytes combine in the mosquito's gut to form zygotes (fertilized eggs). The zygotes develop asexually through several stages within the mosquito, culminating in the infective forms (sporozoites), which migrate into the mosquito's salivary glands (Figure 5.8).

The mosquito's thin mouthparts function like a tiny soda straw or hypodermic needle. Shortly before consuming a blood meal, the female mosquito injects her saliva into her victim. The saliva contains anticoagulants that prevent the human blood from clotting inside the mosquito's mouthparts. When the mosquito injects saliva into a new human host, any sporozoites present in her salivary glands are injected along with it. These sporozoites enter the human bloodstream and are taken up by the liver. Each parasite then develops into thousands more, which may remain in the liver for years. Some parasites periodically escape from the liver into the bloodstream and invade the red blood cells. The parasites reproduce asexually within the red blood cells, producing the disease symptoms. The parasites digest the cell's oxygen-carrying hemoglobin molecules, and one stage also ruptures the red blood cells. Any impairment of the ability of the blood to carry oxygen to the body's tissues is called an **anemia**; all anemias leave their victims run-down and weakened. In malaria, the anemia is caused by destruction of both the hemoglobin and the red blood cells. Cell rupture also brings on fevers, headache, muscular pains, and liver and kidney damage. Within a given host, the asexual cycle of *Plasmodium* continues again and again until the patient either recovers or dies. In the red cells, the parasites can also develop into the sexually reproducing gametocyctes, which may be picked up by another mosquito in its next blood meal, spreading the disease.

Sickle-cell anemia and resistance to malaria

One of the symptoms of malaria is anemia. There are many other types of anemia. A very serious type was first discovered in 1910 by a Chicago physician named Charles Herrick. This strange and usually fatal disease also produced abnormally shaped red blood cells that sometimes resembled sickles. For this reason, Herrick called the disease **sickle-cell anemia**.

A simple blood test was soon devised to test for the condition: a glass slide containing a bowl-shaped depression is used, and a drop of the patient's blood is placed inside the depression. A ring of vaseline is placed around the margins of the depression and a cover glass is then applied, forming an airtight seal with the vaseline. As the red blood cells use up the available oxygen in the depression, the oxygen level decreases. Under these conditions, the red blood cells of a person with sickle-cell anemia assume their characteristic sickle-like shape, while normal red blood cells retain a circular biconcave shape (Figure 5.9). This blood test also

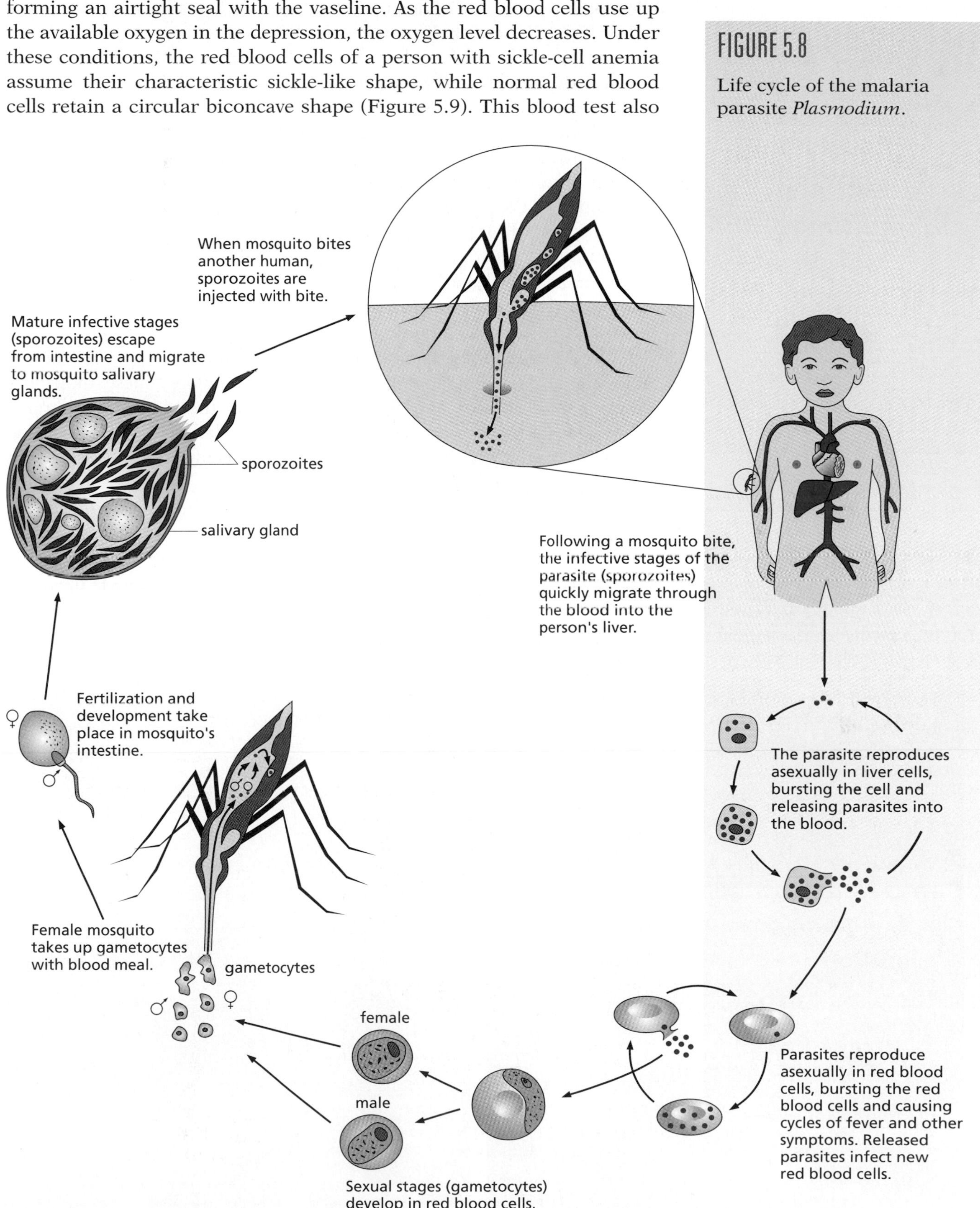
When mosquito bites another human, sporozoites are injected with bite.
Mature infective stages (sporozoites) escape from intestine and migrate to mosquito salivary glands.
sporozoites
salivary gland
Following a mosquito bite, the infective stages of the parasite (sporozoites) quickly migrate through the blood into the person's liver.
Fertilization and development take place in mosquito's intestine.
♀
♂
The parasite reproduces asexually in liver cells, bursting the cell and releasing parasites into the blood.
Female mosquito takes up gametocytes with blood meal.
gametocytes
♂
♀
female
male
Parasites reproduce asexually in red blood cells, bursting the red blood cells and causing cycles of fever and other symptoms. Released parasites infect new red blood cells.
Sexual stages (gametocytes) develop in red blood cells.

FIGURE 5.8

Life cycle of the malaria parasite *Plasmodium*.

allows the recognition of heterozygous carriers, half of whose blood cells sickle while the other half remain round.

Normal and abnormal hemoglobins. Sickle-cell anemia is caused by an abnormality in the **hemoglobin** molecules that carry oxygen within the red blood cells. The hemoglobin molecule consists of four protein chains (two each of two different proteins) surrounding a ringlike 'heme' portion. The heme structure is responsible for hemoglobin's red color. Suspended in the middle of this ring is an iron atom that can bind one oxygen molecule (O_2), giving the hemoglobin its ability to transport oxygen.

A change in a single amino acid, number 6 in one of the protein chains, is responsible for sickle-cell anemia. Normal adult hemoglobin (hemoglobin A) has glutamic acid in this position in the chain, while sickle-cell hemoglobin (hemoglobin S) has valine instead. This minute change makes the hemoglobin S molecules stickier; these molecules adhere to one another and also to the inside of the red cell membrane, deforming the cells into the characteristic sickled shapes. The sickled shape strains the ringlike heme part of the molecule so that hemoglobin S does not carry oxygen as well as hemoglobin A. The difference in the proteins is hereditary and is caused by an altered codon in the hemoglobin gene on the DNA.

The genetics of sickle-cell hemoglobin. Sickle-cell anemia is inherited as a simple Mendelian trait. People who die from sickle-cell anemia are always homozygous and their parents are almost always heterozygous, as are a certain number of siblings and other relatives. The gene for

FIGURE 5.9

Normal red blood cells and red blood cells from a patient with sickle-cell anemia.

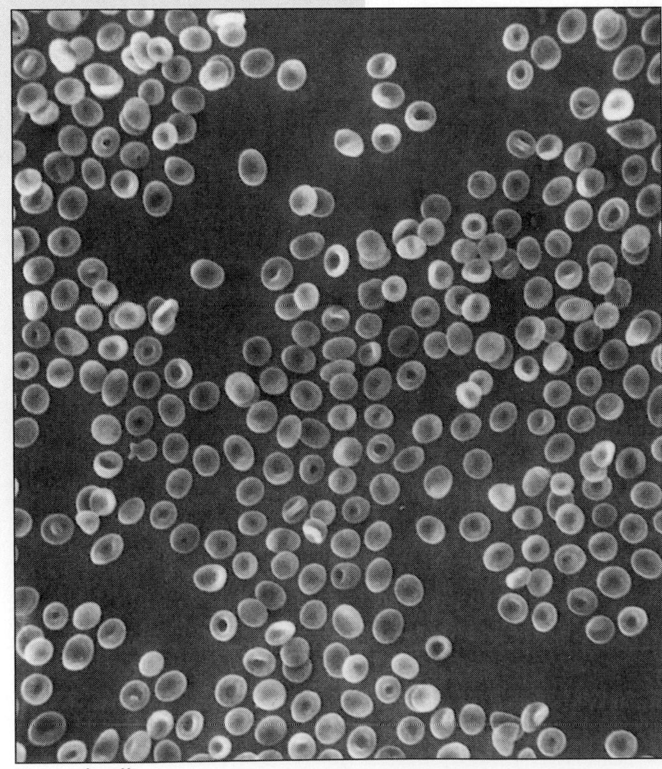

Normal cells

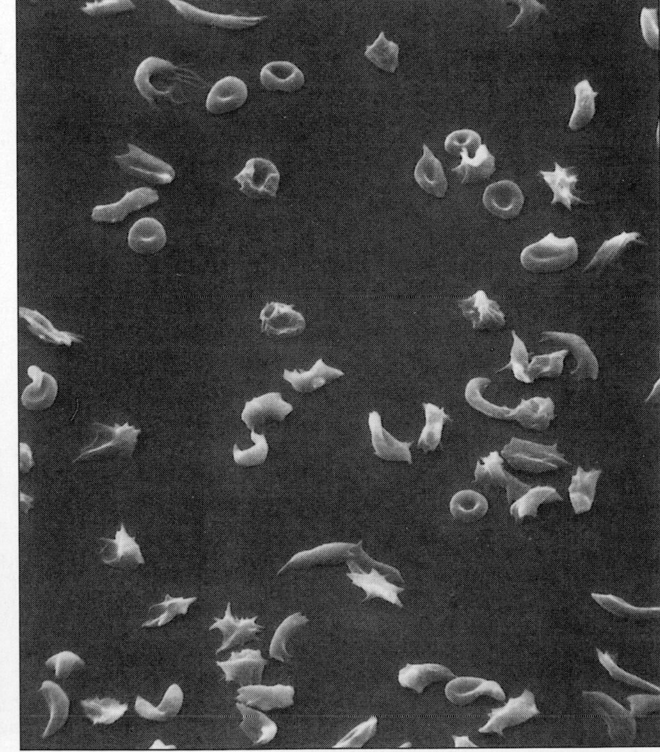

Sickle cells

hemoglobin is designated *Hb* and the different alleles are designated by superscripts: Hb^A is the allele for normal hemoglobin and Hb^S is the allele for sickle-cell hemoglobin.

In U.S. and Caribbean populations, the vast majority of people carrying the Hb^S allele for sickle-cell hemoglobin are blacks of African ancestry. Tests of African populations also show high frequencies of the sickling allele, up to 25% in certain populations. In homozygous individuals (Hb^SHb^S), all the red blood cells are sickled at low oxygen concentrations, as commonly occurs during heavy exertion. Heterozygous individuals (Hb^AHb^S) have both types of hemoglobin and about half of their red blood cells are sickled and the rest are normal in shape. Because both alleles produce a phenotypic result in heterozygotes, they are codominant, as we described earlier in connection with the AB blood type.

Symptoms of sickle-cell anemia. Most of the debilitating symptoms of the disease are consequences of the deformed, sickle-shaped cells. The smallest blood vessels, capillaries, have a diameter only slightly larger than the diameter of blood cells. Because of their sickle shape and changed diameter, sickled cells cause flow resistance in the capillaries and thus impair microcirculation. In most of the body's organs, impaired microcirculation brings about hypoxia (reduced oxygen levels), which results immediately in a severely painful condition known as sickle-cell crisis. These crises begin in infancy. Damaged cells collect in the capillaries of the joints and result in painful swelling. The sickled cells are also more easily disrupted and destroyed than the normal-shaped round ones, resulting in a decreased oxygen carrying capacity (anemia). The anemia and impaired circulation results in tissue damage to many organs, eventually resulting in death (Figure 5.10). In African populations, the death of homozygous Hb^SHb^S individuals often occurs before adulthood, but in the United States and the Caribbean, survival to reproductive age is now increasingly common. The reduction in red blood cell number and the sickle-cell crises also occur among heterozygotes, but not as severely.

Population genetics of sickle-cell anemia. When geneticists realized that sickle-cell anemia in the United States and Jamaica was largely confined to people of African descent, they began to investigate other populations. Using the blood test described earlier in this chapter, researchers investigated the frequency of the allele for hemoglobin S in many African and Eurasian populations. Over large parts of tropical Africa, researchers found remarkably high frequencies of the Hb^S allele, up to 25% or more. At first this appeared puzzling, because sickle-cell anemia was nearly always fatal before reproductive age. An allele whose effects are fatal in homozygous form should long ago have been eliminated by natural selection because people having sickle-cell children would have fewer children surviving to reproductive age.

Maps were made of the frequency of the sickle-cell allele. From these maps and from other evidence, it was noticed that the areas where the sickle-cell allele was frequent were also areas with a high incidence of malaria, particularly the variety caused by *Plasmodium falciparum* (Figure 5.11A and B).

Subsequent research confirmed the basic fact that the Hb^S allele, even in heterozygous form, confers important resistance to the most virulent form of malaria. Tests in which volunteers were exposed to *Anopheles* mosquitoes showed that the mosquitoes are far less likely to bite heterozygous Hb^A/Hb^S individuals than homozygous Hb^A/Hb^A individuals. Tests with the *Plasmodium falciparum* parasites showed that they thrive on the red blood cells of Hb^A/Hb^A individuals, who nearly always come down with a serious case of malaria after infection. However, when Hb^A/Hb^S heterozygotes or Hb^S/Hb^S individuals with sickle-cell anemia are infected with *Plasmodium falciparum*, their malaria symptoms are mild and they recover quickly because the parasite cannot complete its asexual cycle in their sickled blood cells. The protection that the Hb^S allele affords against malaria is sufficient to explain its persistence in those populations in which the incidence of malaria is high.

Hemoglobin S thus decreases the fitness of homozygotes by causing sickle-cell disease, but it increases the fitness of heterozygotes in areas where malaria occurs. In this way, malaria acts as an instrument of natural selection and has a dramatic influence on the allele frequencies of populations.

In addition to hemoglobin A and hemoglobin S, several other genetic variants of hemoglobin have been discovered. Some of these, such as Hb^C, also occur principally in areas where malaria is present and are thought to confer some resistance to malaria.

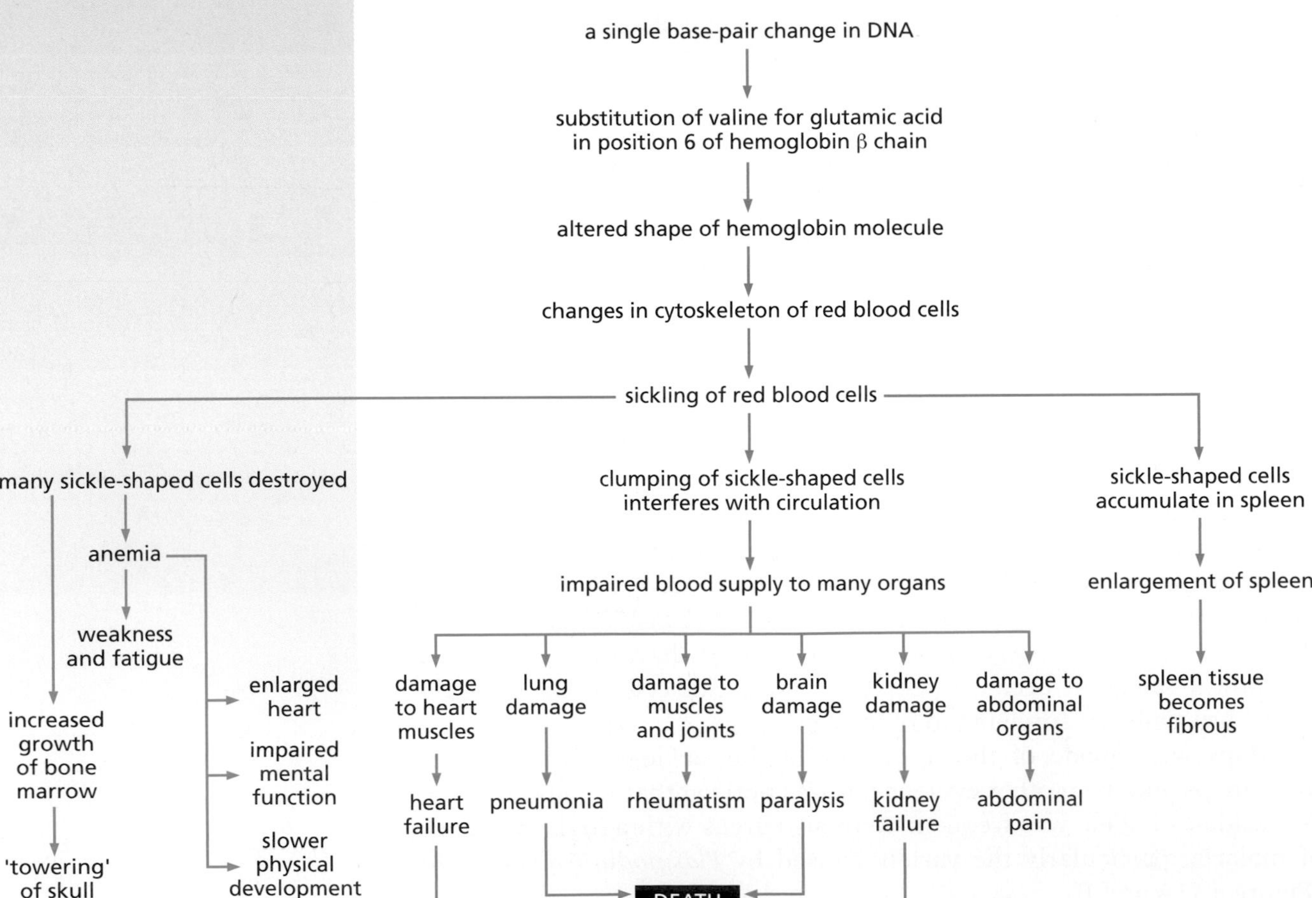

FIGURE 5.10

Development of the consequences of the Hb^S mutation in the hemoglobin gene. A small change in a gene can have many phenotypic consequences.

Other genetic traits that protect against malaria

Sickle-cell anemia is not the only heterozygous condition that protects against malaria. Two others are thalassemia and G6PD deficiency.

Thalassemia. In many countries bordering the Mediterranean Sea (including Spain, Italy, Greece, North Africa, Turkey, Lebanon, Israel, and Cyprus), many people have suffered from a different debilitating

(A) Occurrence of *P. falciparum* malaria

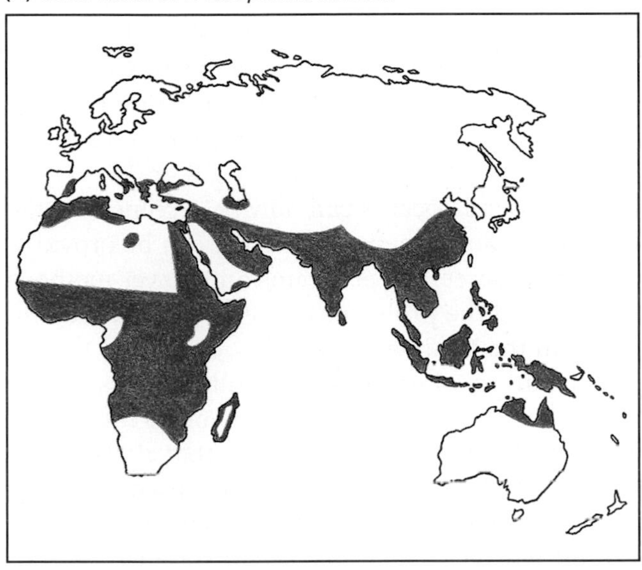

(B) Frequency of Hb^S allele

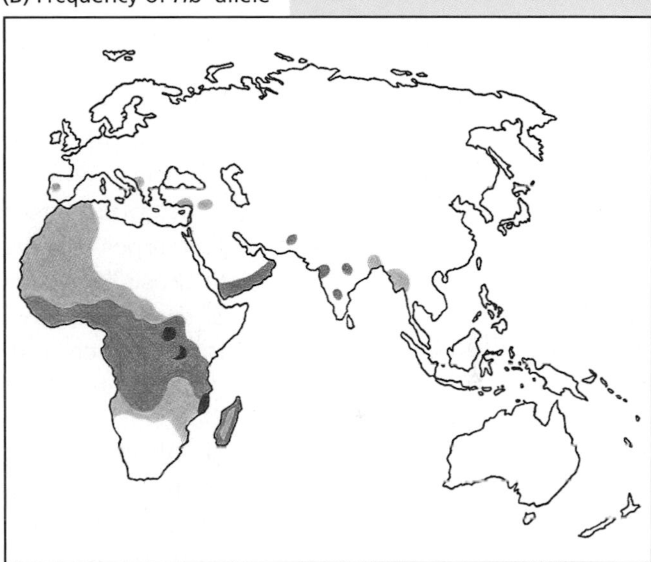

(C) Frequency of alleles for thalassemia (actually the sum for alleles that lead to several different forms of thalassemia)

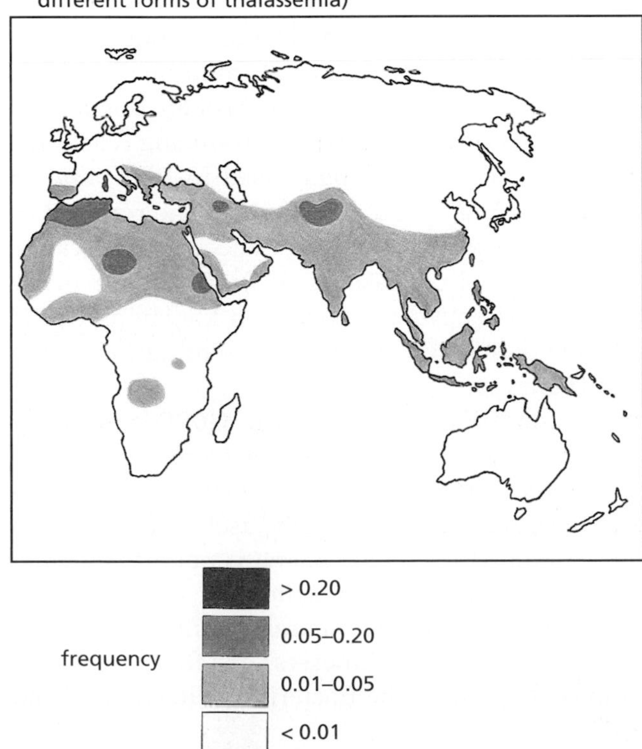

(D) Frequency of allele for G6PD deficiency

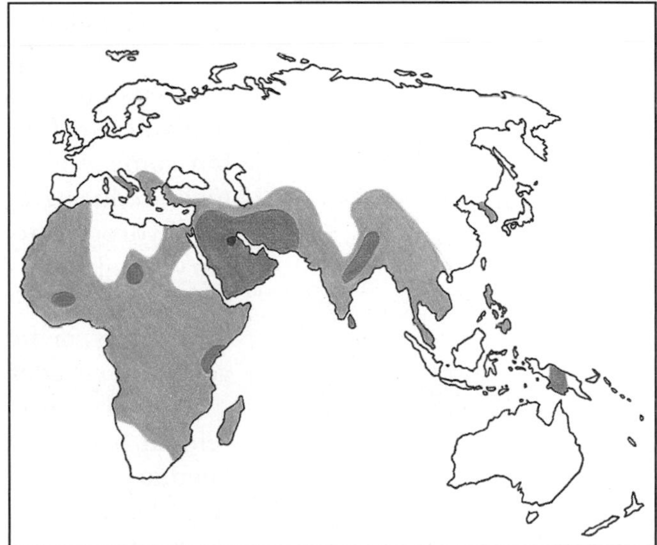

FIGURE 5.11

Distributions in the Eastern Hemisphere of *Plasmodium falciparum* malaria and several genetic conditions that protect against it.

type of anemia known as **thalassemia** (literally, 'sea blood' in Greek). The disease also occurs further east, especially in Southeast Asian countries such as Laos and Thailand (Figure 5.11C).

Thalassemia is marked by a reduced amount of one or more of the protein chains in the hemoglobin molecule. The disease exists in a more serious, often fatal, homozygous form called thalassemia major and a less severe heterozygous form called thalassemia minor. Red blood cells containing nonfunctional hemoglobin are destroyed in the spleen, producing anemia.

The symptoms of thalassemia vary, but all forms result in some decrease in oxygen transport in the blood; blood cell volume and hemoglobin levels are also usually reduced. The bone marrow compensates by overproducing red blood cells, and this overproduction robs the body of much-needed protein and results in stunted growth and smaller stature.

Populations in which thalassemia occurs can now be screened for the genotypes that cause the disease, and genetic counseling can be provided to those found to carry the trait. Screening programs and newer methods of treatment have greatly reduced the problems caused by this disease in Italy, Greece, and elsewhere in the Mediterranean.

The geographical distribution of thalassemia follows closely the distribution of malaria in countries where sickle-cell anemia is infrequent or absent. For this reason, it has long been suspected that thalassemia confers a protective resistance to malaria, similar to that caused by sickle-cell anemia. The evidence is indirect: if heterozygous individuals (those with thalassemia minor) did not have *some* selective advantage such as malaria resistance, then the deaths caused by thalassemia major would have caused the genes for this trait to die out long ago.

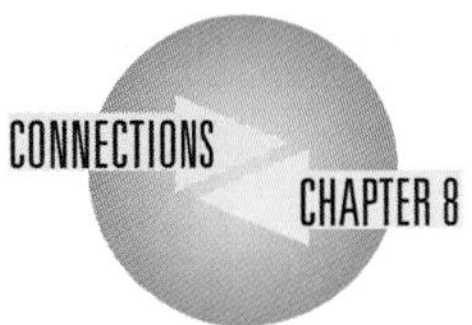

G6PD deficiency. Blood sugar (glucose) is normally broken down within each cell in a series of reactions that begin with the formation of glucose 6-phosphate. Most of the glucose 6-phosphate is broken down into pyruvate (see Chapter 8, p. 339) in a series of energy-producing reactions, but some is also used to make ribose (the sugar used in RNA) and to make reducing agents such as NADPH and glutathione. The removal of two hydrogen atoms from the glucose 6-phosphate molecule requires the enzyme glucose 6-phosphate dehydrogenase (G6PD). There are many people who have too little of this enzyme, a condition known as **G6PD deficiency**, or **favism**. G6PD deficiency results from a mutation in the gene that encodes the G6PD enzyme.

Under many or most conditions, people with G6PD deficiency remain perfectly healthy, but they occasionally suffer from a hemolytic anemia in which the red blood cells rupture, spilling their hemoglobin into the blood plasma (where it is physiologically useless but easy to detect by simple lab tests). Hemolytic anemia, which is potentially fatal, can occur in G6PD-deficient people as a reaction to certain drugs (aspirin, quinine, quinidine, chloroquine, chloramphenicol, sulfanilamide, and others), in response to certain illnesses, or after eating fava beans (*Vicia faba*), a common legume of the Eastern Mediterranean and

Middle East. The anemia may also exist chronically in a nonfatal form in people with G6PD deficiency.

G6PD deficiency has been shown to offer protection against *P. falciparum* malaria. It affects some 10 million people, and is thus the most common disorder offering protection against malaria. Most importantly, heterozygous carriers of the deficiency are also malaria-resistant, but the exact mechanism of the resistance has yet to be worked out.

G6PD deficiency occurs mostly in Mediterranean populations from Greece to Turkey and from Tunisia to the Middle East, and among Sephardic Jews. It also occurs south of this area into Africa and eastward across Iran and Pakistan to Southeast Asia and southern China (Figure 5.11D). The Greek mathematician Pythagoras may have suffered from this disorder, for his aversion to beans (one of the triggers of hemolytic anemia in G6PD-deficient people) has become legendary. Pythagoras founded a religious cult in which the avoidance of beans was an important belief. Opponents of his cult once captured Pythagoras by chasing him toward a bean field, which they knew he would not cross.

Population genetics of malaria resistance

Genetic variation within a population is known as **polymorphism**, a condition in which two or more alleles are present (at the same genetic locus) at frequencies higher than new mutations could possibly explain. Some alleles of polymorphic genes have harmful effects when homozygous, but they persist in populations because the same alleles also confer some important benefit (such as malaria resistance) when heterozygous. If the polymorphism persists for many generations, it is likely to be a **balanced polymorphism**. The conditions that lead to balanced polymorphism are that all homozygous genotypes suffer from some selective disadvantage or reduction in fitness, while the heterozygotes have the maximum fitness. For example, in a country in which malaria is present, $Hb^{A}Hb^{A}$ homozygotes have lower fitness because they are susceptible to malaria, and most $Hb^{S}Hb^{S}$ homozygotes die young from sickle-cell anemia. The $Hb^{A}Hb^{S}$ heterozygotes have maximal fitness because they are malaria-resistant and because they have enough normal red blood cells for them not to suffer from fatal sickle-cell anemia. Under conditions like these, natural selection brings about and perpetuates a situation in which both alleles persist.

The selection by malaria for genetic traits that offer resistance to it is at least as old as the open, swampy conditions (ideal for the breeding of mosquitoes) brought about by agriculture in warm climates. Evidence for this exists in the form of human bones found at a Neolithic archaeological site along the coast of Israel. Cultural remains found at this site show that it was an early farming community, one of the first in the area. Pollen analysis shows the presence of many plants characteristic of swampy areas. Some of the bones show characteristic increases in porosity (due to the increased production of red blood cells in the bone marrow) indicative of thalassemia.

Other diseases as selective factors

CONNECTIONS CHAPTER 3

Hereditary diseases that confer some advantage in the heterozygous state are not confined to those that protect against malaria. In European populations of past centuries, tuberculosis, an infection caused by a bacterium called *Mycobacterium tuberculosis*, was an important force of selection, especially in crowded cities from the Middle Ages to the early twentieth century. One scientist has proposed that people heterozygous for the alleles that cause cystic fibrosis (an inherited lung disorder discussed in Chapter 3, p. 87) were protected against tuberculosis; they therefore survived tuberculosis epidemics in greater numbers than did people without cystic fibrosis alleles. As the heterozygotes increased in number, some of them married one another, and, on average, one out of four of their children became afflicted with cystic fibrosis.

What about the geographic variation in blood groups and other genetic traits? There is evidence that at least some of this variation may also result from the natural selection brought about by various medical conditions. In a smallpox epidemic in Bihar province, India, researchers found that those who died were more often of blood group A, while survivors were more often of blood group B. In similar fashion, cholera selects against blood group O and favors blood group B. (Note that these studies demonstrated a difference in fitness, but did not explain the mechanism.) Other studies have shown statistical correlation of various blood types with other diseases: blood group O is correlated with an increased risk of duodenal ulcers and ovarian cancers, and blood group A with a slightly increased risk of stomach cancer. Associations of particular blood groups with cancers of the duodenum and the colon have also been postulated. Such statistical associations do not necessarily indicate a cause-and-effect relationship between the associated factors.

Fatal diseases are among the most striking agents of natural selection, but there are many other selective forces. We examine some of these other forces of natural selection in the next section.

THOUGHT QUESTIONS

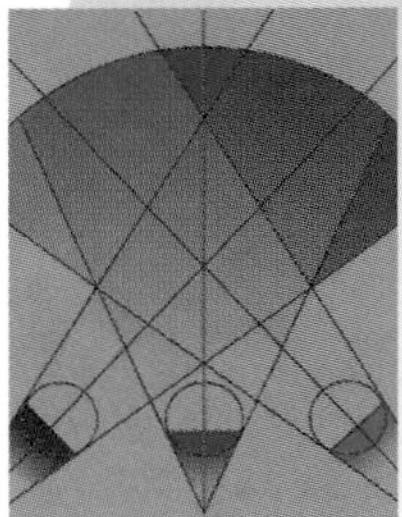

1. How is an average life expectancy measured? Why is the average life expectancy of a population more affected by the deaths of children (e.g., from malaria) than by the deaths of elderly people?

2. All heterozygous carriers of the allele for G6PD deficiency are female. What does this tell you about the location of the G6PD gene? (You may need to review Chapter 3 to answer this question.)

NATURAL SELECTION BY PHYSICAL FACTORS CAUSES MORE POPULATION VARIATION

There are other agents of natural selection in addition to diseases. Among them are climatic factors such as temperature or sunlight, as well as climatic variation that makes food more scarce at some times of year or from one year to another. Like the genetically based traits that confer protection against disease, other genetic variation between populations has arisen in response to these other selective factors. In this section we look at some of these other factors and how they have selected in different geographic regions for differences in the genetically regulated aspects of physiology and of body shape and size.

Human variation in physiology and physique

During part of the Korean War (1950–1953), American soldiers were exposed to the fierce, frigid conditions of the Manchurian winter. Many soldiers were treated for frostbite. Most of the Euro-American (Caucasian) soldiers responded well to the medical treatment that was given, but a disproportionate number of African American soldiers did not and many of them lost fingers and toes as a result. Disturbed by these findings, the U.S. Army ordered tests on resistance to environmental extremes in soldiers of different racial backgrounds.

In one series of tests, army recruits were required to perform strenuous tasks (such as chopping wood) under a variety of climatic conditions. In a hot, humid climate, the African American soldiers were able to continue working the longest and performed the best as a group; Asian-American and Native American soldiers performed nearly as well as the African Americans, and Euro-American recruits lost excessive fluids through sweating and became easily fatigued and dehydrated. Under dry, desert conditions, the Asian-American and Native American soldiers did best, the African Americans were second best, and again the Euro-American soldiers became dehydrated. Under extremes of cold, it was the Euro-American soldiers who did best, followed closely by the Native American and Asian-American soldiers; the African Americans shivered the most and some became too cold to continue. These tests demonstrated definite differences between groups in bodily resistance to physiological stress under a variety of environmental extremes. The significance of these differences was enhanced by the fact that, in other respects, the recruits represented a fairly homogeneous population: 18- to 25-year-old males who had all been screened by the army as being physically fit and free from disease and who had passed the same army physical and mental exams.

Other physiologists outside the Army conducted tests in which adult male volunteers immersed their arms in ice water almost to the shoulders. African Americans in general shivered the most and suffered the most rapid loss of body heat, as measured by a decline in body temperature. Euro-Americans and Asian-Americans lasted longer without shivering, but they, too, eventually suffered loss of body heat. Only the Inuit (Eskimo) volunteers were able to keep their arms immersed indefinitely without any discomfort and without shivering. Subsequent studies that replicated these results made the additional finding that diet is also a factor: Inuit volunteers who ate high-protein, high-fat diets (traditional for the Inuit) did far better than other Inuit who had become acculturated to American dietary habits. It would be a mistake, however, to extrapolate findings from studies such as these beyond the groups used for the tests (adult males in good health) without further investigation. Many traits vary with age or sex or both.

Bergmann's rule. Genetically based differences in physiology that correlate with climate are the basis for a number of ecogeographic rules. Adaptations can also work indirectly, through variables such as body physique. Biologists have long noticed certain general patterns of geographic variation among mammals and birds. In one such pattern, called **Bergmann's rule**, body sizes tend to be larger in cold parts of the range and smaller in warm parts. This can be explained by the relationship of body size to mechanisms of heat generation and heat loss. For example, an animal twice as long in all directions as another animal has eight times the volume of muscle tissue generating heat ($2 \times 2 \times 2 = 8$) as the smaller animal but only four times the surface area over which heat is lost ($2 \times 2 = 4$). Thus, the larger animal is twice as efficient as the smaller ($8/4 = 2$) in conserving heat under cold conditions. A survey of human variation confirms that the largest average body masses are found among people living in cold places (like Siberia), while most tropical peoples within all racial groups are of small body mass, even when their limbs are long. These relationships of body size to climate result from natural selection acting on genetic variation within populations over long periods. It does not mean that a person of a certain genotype will grow larger if they move to a cold climate.

FIGURE 5.12

Bergmann's and Allen's rules illustrated by comparisons between arctic and tropical body forms.

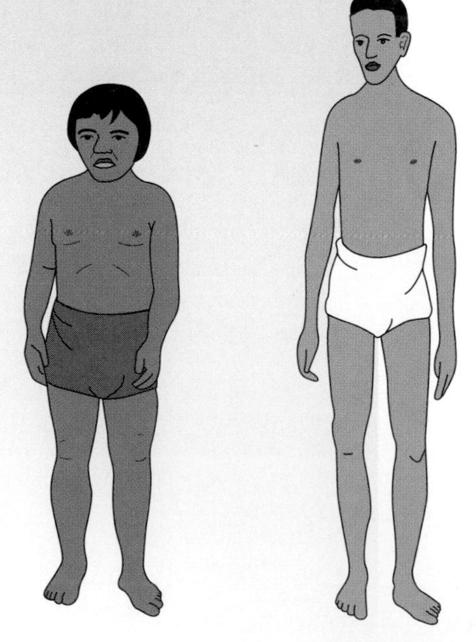

Allen's rule. Another broad, general phenotypic pattern in most geographically variable species of mammals and birds is **Allen's rule**: protruding parts like arms, legs, ears, and tails are longer and thinner in the warm parts of the range and shorter and thicker in cold regions. This rule is usually explained as an adaptation that conserves heat in cold places by reducing surface area and dissipates heat more effectively in warm places by increasing surface area. Human populations generally follow this rule: Inuit people have shorter, thicker limbs, while most tropical Africans have longer, thinner limbs (Figure 5.12). There are exceptions, however: a number of forest-dwelling populations along the Equator are much smaller than Allen's rule would predict, although they are usually thin-legged. Also, the tallest (Tutsi) and shortest (Mbuti) people on Earth live near one another in the Democratic Republic of Congo (formerly called Zaire), showing that climate is not the only instrument of natural selection influencing limb length or overall height within populations.

Diabetes and thrifty genes. Diabetes, a potentially life-threatening illness in many populations, may be an indirect result of one or more of the so-called 'thrifty genes' that protected certain people from starvation in past centuries. Ancestral Polynesians, for example, endured uncertain journeys over vast stretches of Pacific Ocean waters. Uncertain food supplies during such voyages selected for people who could withstand longer and longer periods of starvation and still remain active. The postulated 'thrifty genes' may have caused excess food, when it was available, to be converted into body fat that could be used for energy in times of famine. The result was a population that was stocky in build and resistant to starvation in periods when food supplies were low but that was also more susceptible to diabetes under modern conditions, when physical exhaustion is rare and food is always available. Diabetics fed 'ordinary' diets have excess sugar in their blood, much of which is converted to fat and stored. Although diabetes is itself an unhealthy condition, the storage of fat may have been, under conditions like those described for the early Polynesians, an adaptive trait. Perhaps diabetes is an unfortunate modern consequence of having one or more alleles originally selected for their ability to convert sugar to body fat.

A similar history of selection for 'thrifty genes' (not necessarily the same ones) might also explain the late twentieth-century upsurge of diabetes in certain Native American populations, notably the Navajo and Pima of the southwestern United States. The risks that selected for 'thrifty genes' in the past were more significant in barren environments than in places in which the food supply was more assured. However, the commercial introduction of sugar-rich foods and a change from an active to a sedentary lifestyle have both raised the risks of diabetes, which are higher for sedentary people eating carbohydrate-rich diets. Because of these environmental changes, the genes that were once advantageous have in some cases turned into a liability, putting people of these genotypes at greater risk of diabetes. The Navajo and Pima have discovered that a return to frequent long-distance foot racing (a traditional activity they had nearly abandoned) has kept their populations healthier and has significantly lowered the incidence of diabetes in the runners. Not enough time has yet elapsed for the allele frequencies of the 'thrifty genes' to have again changed in this population, but the partial return to an earlier lifestyle has changed the environmental stresses and decreased the incidence of diabetes.

Natural selection, skin color, and disease resistance

The skin is the largest organ of the body and a major surface across which the body makes contact with the forces of natural selection in its environment. Human populations vary widely in skin color. Could these differences in skin color be adaptive?

Geographic variation in skin color. Skin color is one of the most visible human characteristics, and the one to which Americans have always paid the most attention when identifying race. Long-standing patterns of geographic variation are easier to understand if we ignore the population movements of the years since A.D. 1500 and consider only those populations still living where they did before that time.

Europe has always been inhabited by light-skinned peoples, Africa and tropical southern Asia by dark-skinned peoples, and the drier, desert regions of Asia and the Americas by people with reddish or yellowish complexions. What is even more remarkable is that we find geographic variation along the same pattern *within* most continents, and in fact greater variation within the larger population groups than between such groups. For example, among the group of populations spread continuously from Europe across Western Asia to India, we find the lightest skin colors (also eye and hair colors) in Scandinavia and Scotland, progressively darker average colors (and darker hair) closer to the Mediterranean Sea, further darkening as we move through the Middle East and across Iran to Pakistan and India, and the darkest at the southern tip of India and on the island of Sri Lanka. A similar gradient (a cline) for skin color can be found among Asians, from northern Japan south through China into the Philippines and Indonesia.

Why would it be adaptive for people to be light-skinned in Europe but dark in Africa, Sri Lanka, and New Guinea? Notice that there are some very dark-skinned people outside Africa, and they generally have few other physical or genetic characteristics in common with Africans other than their dark skin colors. The natives of Sri Lanka, for example, have very straight hair and blood group frequencies totally different from those of Africa. One clue to this puzzle is that all very dark-skinned peoples have lived for millennia in tropical latitudes.

Sunlight as an agent of selection. Tropical regions receive on a year-round basis more direct sunlight than do temperate regions. In fact, the amount of sunlight received at ground level decreases with increases in latitude and corresponds more closely to belts of latitude than to variations in temperature. This is especially true for light in the ultraviolet region of the sun's spectrum.

If we exclude places where few people live, Europe receives the least sunlight of all the inhabited regions of the world. This is first and foremost a function of latitude: Europe includes populated regions of higher latitudes than on any other continent: London and fourteen other European capitals are located north of latitude 50°, while North America and Asia above this latitude contain few large cities and a great deal of sparsely inhabited land. Europe also has a frequent cloud cover that screens out even more of the Sun's rays. As a result of both high latitude and cloud cover, people in Europe receive much less exposure to ultraviolet light than most other people.

That sunlight levels select for body coloration is described by a third ecogeographic rule, **Gloger's rule**. While Bergmann's and Allen's rules, described earlier, take only temperature into account, Gloger's rule takes into account sunlight and humidity as well. Under Gloger's rule, most geographically variable species of birds and mammals have pale-colored or white populations in cold, moist regions, dark-colored or black populations in warm, moist regions, and reddish and yellowish colors in arid regions. We do not know all the reasons for this variation. Camouflage has been suggested as a cause, but vitamin D synthesis also plays an important role.

Vitamin D is needed for the proper formation of bone (see Table 8.2, p. 343). Children who do not receive adequate vitamin D during growth suffer from a condition called **rickets**, a disease of bone formation that may result in weakness and curvature of the bones (especially those of the legs) and in crippling bone deformities if left untreated. Sunlight is necessary for vitamin D synthesis. Many foods are rich in vitamin D, such as egg yolks and whole milk, but most vitamin D found in foods is in a biologically inactive form. The final step of vitamin D biosynthesis takes place just beneath the skin, with the aid of the ultraviolet rays of natural sunlight. This is why vitamin D is sometimes called the 'sunshine vitamin.' To get adequate amounts of vitamin D, a population must have both adequate intake of the vitamin in the diet and adequate exposure to sunlight. European populations have the lightest skin colors (and they get lighter the farther north you go) as an adaptation that allows maximum sunlight penetration into the skin. Europeans also have many cultural adaptations related to vitamin D intake, such as the eating of cheeses and other fat-rich milk products containing vitamin D. Northern Europeans place great value on outdoor activity at all times of the year, including such occasional extremes as nude dashes into the snow after the traditional sauna.

In northern Europe, people with dark skins could be at a very high risk of vitamin D deficiency because melanin pigment blocks out a large proportion of the Sun's ultraviolet rays. Very few dark-skinned people lived in northern Europe even as immigrants. This has changed since World War II when synthetic vitamin D became widely available. Because this prepared vitamin D is already in its active form, sunlight is no longer needed for its activation. Dark-skinned people can now live and remain healthy in northern latitudes without developing deficiency diseases.

At latitudes closer to the Equator another problem exists: the same wavelengths of ultraviolet that are needed in the final step of vitamin D synthesis are also cancer-causing. Skin cancer (malignant melanoma; see Chapter 9) is generally a disease of those white-skinned people who are overexposed to the Sun's direct rays. Populations of all racial groups living closer to the Equator have been selected over the millennia to have darker skins. Those individuals who had lighter skins in the past more often got skin cancer and died, in many cases before their reproductive years had ended. In the tropics, the sunshine is more than enough for all people to get adequate ultraviolet rays for vitamin D synthesis, even though melanin pigment absorbs much of the ultraviolet light.

Nutritional source of vitamin D in the far north. For the reasons given in the preceding sections, populations living in the high latitudes are generally light-skinned and populations that are adapted to living in tropical latitudes are generally dark-skinned. There is one very interesting exception: the Inuit populations of Arctic regions, sometimes known as Eskimos. (These people have always called themselves Inuit; the name

'Eskimo' was a pejorative name used by their enemies.) The Inuit are not very light-skinned, nor do they expose themselves much to sunlight. Most Inuit people live in places so cold that the exposure of bare skin poses a greater danger than any benefit of ultraviolet rays could overcome, and most Inuit are fully protected by clothing that offers hardly any exposure to the Sun. So how do they get enough vitamin D? The Inuit have discovered their own way of staying healthy. One of the world's richest sources of vitamin D is fish livers, especially those of cold-water fishes. (Cod liver oil is a very rich source of both A and D vitamins.) Moreover, the vitamin D in fish oils is fully synthesized and needs no sunlight to activate it. So, instead of having pale skins and traditions of exposing their skins to sunlight, the Inuit have traditions of catching cold-water fish (Figure 5.13) and eating them whole, liver and all. These traditions have allowed them to stay healthy in a climate that is too cold and too sunless for most other populations.

In all of the above examples, a population that has lived in a particular geographic area for long periods has become adapted to the temperature, humidity, sunlight, and other conditions of their environment. The evidence presented in this chapter and in Chapter 4 suggests that natural selection is largely responsible for these adaptations.

FIGURE 5.13

An Inuit woman fishing. The Inuit get most of their vitamin D from eating whole fish, including the liver.

THOUGHT QUESTIONS

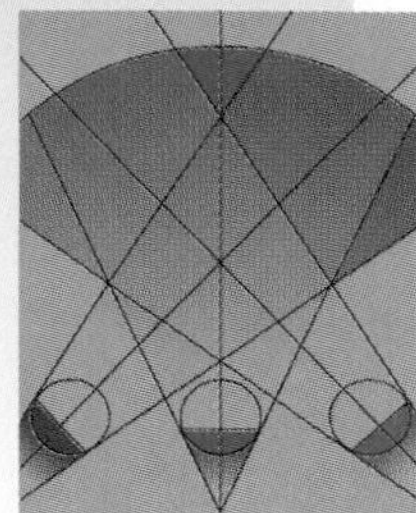

1. If people differ in their resistance to extreme cold or heat, does this mean that the difference is genetic? What would you need to know to answer this question? How could an experiment be arranged to test this?
2. Blood type O is statistically associated with duodenal ulcers, one of many such correlations between a blood type and a disease. Does a correlation demonstrate a cause? Does a correlation imply a mechanism of some kind? Does a correlation suggest new hypotheses? How can scientists learn more about whether there is a causal connection between the blood type and the disease?

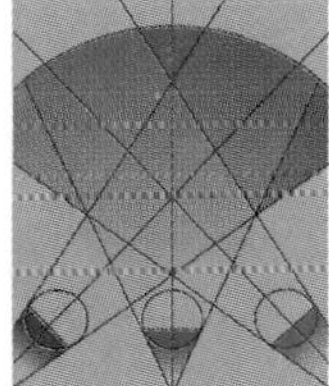

THROUGHOUT THE HISTORY OF BIOLOGY, SCIENTISTS HAVE developed various ways of describing groups of people. Some of these groupings have been known as races. Some concepts of race have attempted to find biological explanations for the racial groupings already established by various societies. Morphological concepts of race divided humans on the basis of their physical appearance. Biologists and anthropologists of the past gathered descriptive data about the physical characteristics of different populations and assumed that each group was distinct and unchanging. More recently, biologists have abandoned these concepts, in part because of the racism that has flowed from them, but also because these ideas no longer fit the data that we now have. The population genetics theory of human variation views human populations as varying continuously, with no group being uniquely different from any other. Differences among human populations are products of evolution. Like any other species, humans can evolve only when genetic variation is present in a population. When a population encounters some agent of natural selection, such as disease or climate, people with certain genotypes survive in greater numbers and leave more offspring than those with other genotypes. Over long periods, this process results in the adaptation of a population to its environment, with allele frequencies differing from one population to the next. This evolution continues today, although the increased mobility of people and technological alterations of the environment are slowly making populations less distinct than in past centuries. Populations vary in the frequencies of traits; they do not carry any unique traits. There is no biological phenotype, genotype, or DNA sequence that can assign an individual to a race or to a population. Although our biological concepts about race and other human variation have changed over time, racism will continue to exist if one group of people is held to be more valuable than another.

CHAPTER SUMMARY

- Human **populations** vary geographically. Phenotypic and genotypic variation within populations usually exceeds variation between them.
- Differences among populations have historically been described in terms of culturally defined or morphological **races**.
- **Population genetics** allows us to describe groups of populations that differ from one another by certain characteristic **allele frequencies**.
- Most allele frequencies vary gradually and continuously among populations. Continuous variation of this sort is best described in terms of geographic gradients, also called **clines**. Clines can be plotted on maps for average values of continuous characters such as height or for population frequencies of particular blood groups or alleles or DNA sequences.
- When more than one allele of a gene persists in a population this is called a **polymorphism**.
- The **Hardy–Weinberg equilibrium** describes the conditions under which allele frequencies remain constant in a population.
- Populations that were at one time small may have allele frequencies that have been shaped in part by **genetic drift**. Populations that grew from a small number of individuals reflect the alleles of those individuals, an effect known as the **founder effect**.
- Aside from genetic drift, most geographic variation among human populations has resulted from natural selection producing **adaptation** to the environment.
- Disease is an important force of natural selection. Malaria, a widespread parasitic infection, has selected in different regions for high frequencies of alleles associated with sickle-cell anemia, thalassemia, and G6PD deficiency, all of which protect heterozygous individuals against malaria. Malaria and other diseases result in **balanced polymorphism** whenever the heterozygous genotype enjoys maximum fitness.
- Temperature selects for geographic variation in the alleles influencing body size (**Bergmann's rule**) and body shape (**Allen's rule**).
- Ultraviolet light at different latitudes selects for geographic variation in the alleles influencing skin color (**Gloger's rule**). Pale skin is favored in high latitudes as an adaptation to absorb more ultraviolet light and prevent vitamin D deficiency. Dark skin is favored near the Equator as a protection against skin cancer from too much ultraviolet exposure.

KEY TERMS TO KNOW

adaptation (p. 211)
allele frequency (p. 190)
Allen's rule (p. 222)
anemia (p. 212)
balanced polymorphism (p. 219)
Bergmann's rule (p. 222)
cline (p. 196)
founder effect (p. 207)
genetic drift (p. 206)
Gloger's rule (p. 224)
Hardy–Weinberg equilibrium (p. 206)
hereditarian (p. 194)
polymorphism (p. 219)
population (p. 191)
population genetics (p. 190)
race (p. 193)

CONNECTIONS TO OTHER CHAPTERS

Chapter 1: Every study of human variation is conducted in a cultural context.

Chapter 1: Studies of human variation have ethical implications, including those arising from inappropriate use of the results.

Chapter 3: Many human variations have a genetic basis; such alleles arise ultimately from mutations.

Chapter 4: Human population variations reflect evolutionary processes, including mutation, natural selection, and genetic drift, all of which continue to work in modern populations.

Chapter 6: Nearly all human populations are growing, and some are growing much faster than others. Population growth and migrations change various allele frequencies.

Chapter 8: Different populations sometimes have different ways of meeting their nutritional requirements.

Chapter 9: Some types of cancer are more frequent in some human populations and less frequent in others.

Chapter 15: Human variation is an example of biodiversity at the population level.

PRACTICE QUESTIONS

1. How many different genotypes can code for the blood group B phenotype? What are they? Are they heterozygous or homozygous?
2. How many different genotypes can code for the blood group AB phenotype? What are they? Are they heterozygous or homozygous?
3. How many different genotypes can code for the blood group O phenotype? What are they? Are they heterozygous or homozygous?
4. How many different genotypes can code for the Rh^- phenotype? Are they heterozygous or homozygous?
5. How many different genotypes can code for the Rh^+ phenotype? Are they heterozygous or homozygous?
6. If the allele frequency of Hb^S in a population is 0.1, how many people in that population will be heterozygous Hb^AHb^S? (Review the Hardy–Weinberg equation.)
7. How many different host species does the *Plasmodium* parasite need to complete its life cycle?
8. Why do people who are heterozygous for sickle-cell anemia have less severe anemia than people who are homozygous Hb^SHb^S ?
9. How does the bottleneck effect alter the allele frequencies of a population?
10. What is a balanced polymorphism? Give an example.

6

The Population Explosion

CHAPTER OUTLINE

Demography Helps to Predict Future Population Size

Growth rate

Exponential (geometric) growth

Malthus' views on population

Logistic growth

Demographic transition

Human Reproductive Biology Helps us to Understand Fertility and Infertility

Reproductive anatomy and physiology

Reproductive technologies

Can We Diminish Population Growth and its Impact?

Contraceptive measures

Post-fertilization birth control methods

Cultural and ethical opposition to birth control

The abortion debate

Population control movements

The education of women

Controlling population impact

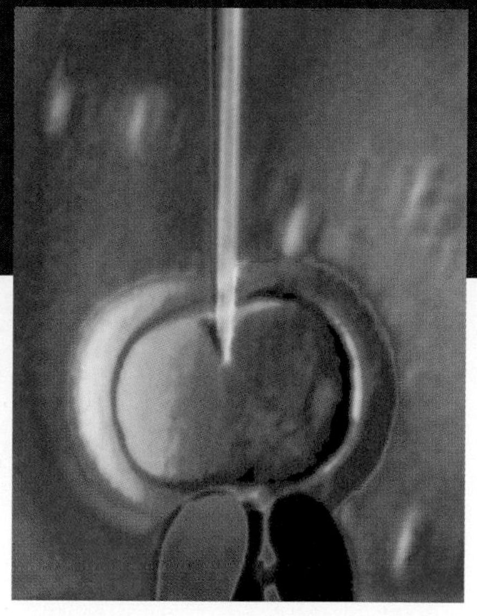

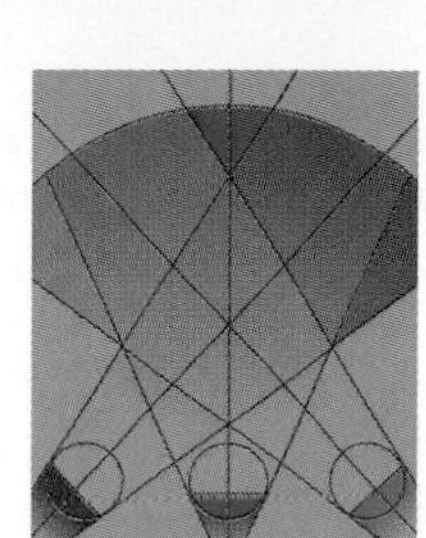

ISSUES

Is the Earth overpopulated?

How fast are human populations growing?

Why might a population explosion be detrimental to the social good?

What methods are available to people who want to control their reproduction?

Can we restrict reproduction without violating people's rights?

Can we diminish population growth and its impact?

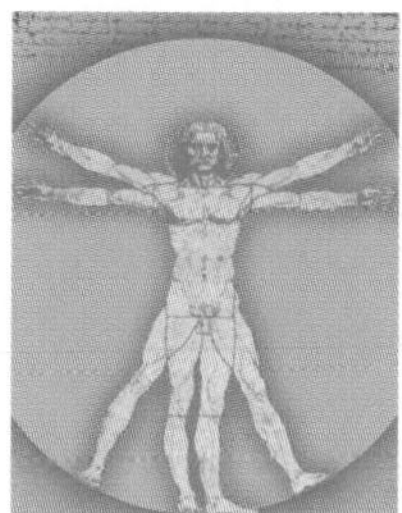

BIOLOGICAL CONCEPTS

Population ecology (populations, population density, growth rates, carrying capacity, population regulation)

Reproductive biology (reproductive anatomy, reproductive physiology, reproductive cycles, hormonal controls)

Biosphere (human influences, overpopulation and its effects, resource uses, habitat alteration)

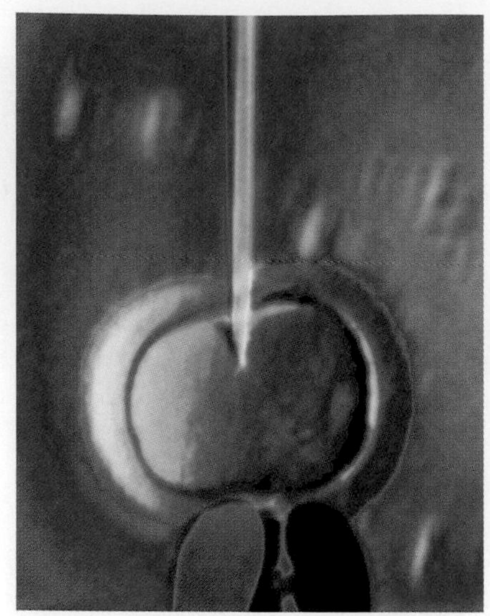

Imagine a world where people must share a room with 4 to 12 others. A room of one's own is a rare luxury. In fact, people who have any housing at all consider themselves fortunate, because so many people have none. Drinking water is in short supply each summer, and overworked sewer systems are breaking down all the time; many millions have no sewer system at all. Jobs are scarce, and well-paying jobs are almost unheard of. Beggars crowd every street, and each garbage can is searched through time after time by starving people looking for something to sustain them.

Some parts of the world already experience these conditions. Some experts predict that a future like this may be in store for all of us unless something is done soon, and on a massive scale, to control population growth.

The Earth is currently experiencing the most rapid population increase in all of human history. From 2.5 billion people in 1950, the total world population more than doubled to 6 billion in 1999 (Figure 6.1). At current rates of increase, the global population will double again in about 38 years. Each year, the world's population increases by some 94 million people.

In this chapter we consider the factors that control the size and the rate of growth of populations, including the biological controls on populations that operate independently of any conscious planning. The ecological principles that we discuss include models of population growth, limits to growth, and some of the consequences of excessive growth. Although we focus on human populations, all these ecological principles apply equally well to populations of other organisms.

As with many of the other issues in this book, population growth cannot be looked at as a purely biological issue. There are political, religious and ethical dimensions to population growth and its control, and so we consider some of these factors also. As we have seen earlier (see Chapter 1), the boundaries between the individual good and the social good are one of the subjects of ethics. Biology can inform ethical debate by assessing, for different scenarios, the biological risks to the individual and to society.

DEMOGRAPHY HELPS TO PREDICT FUTURE POPULATION SIZE

The study of the biological factors that affect the sizes of populations is called **population ecology**; the study of human populations in numerical terms is known as **demography**. Recall that a **population** is defined as a set of potentially interbreeding individuals in a certain geographical location at a certain time (see also Chapter 4, p. 146). Our ability to understand population growth depends on our ability to make predictions, based on both population ecology and demography. Because populations are large aggregates, we need mathematical models to make these predictions and to study the factors that might influence and possibly curb population growth. These mathematical models were initially developed from the study of bacterial populations, but they pertain to the growth of all other species, including humans.

Mathematical models of population growth begin with the gathering

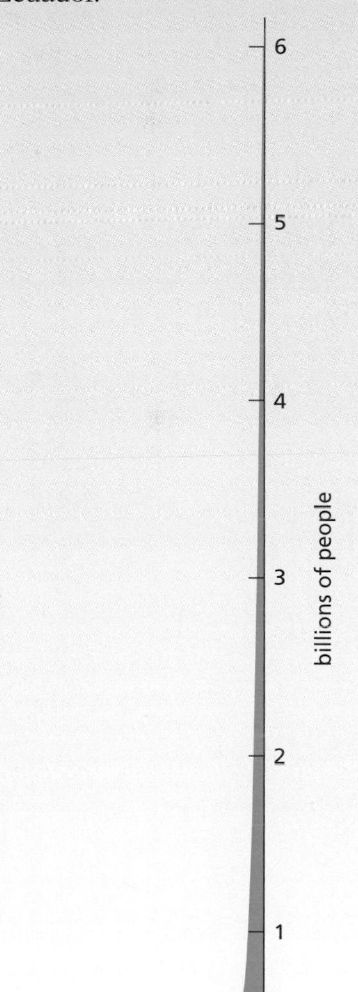

FIGURE 6.1

Graph showing the growth of the world's human population. The inset shows a street scene in Quito, Ecuador.

of census data. A **census** is, at minimum, a head count of all the individuals living in a specified area, usually within recognized political boundaries. Early censuses of human populations were often inaccurate, and the opposition of the censused populations (who did not want to be taxed) only compounded the inaccuracy. Moreover, these early censuses were hardly ever repeated over the same stable boundaries at a later time, the only conditions under which population growth could be accurately assessed.

The United States implemented the first nationwide census of any modern nation in 1790, for the purpose of achieving proportional representation of the different states in Congress. (This remains one of the major functions of the U.S. census today.) In the first few decades of the nineteenth century, most European nations began to census their populations and have continued to do so at regular intervals. In addition to the registration and counting of births, deaths, marriages, and divorces as they occur, modern censuses also record information about age, sex, marital status, age at marriage, and often such additional details as income and employment status as well as cause of death.

Growth rate

The rate of change of the size of a population depends on its birth rate, its death rate and the relation between the two. The rate of change of population size is called the **growth rate**.

Birth rate. The **birth rate**, B, of a population for a given period is found by dividing the number of births during that period by the number of people already in the population, N:

$$B \text{ (each year)} = \frac{\text{number of births per year}}{N}$$

The birth rate, like any other rate, is always a fraction; the word 'rate' always means that one number is divided by another. To illustrate with some actual numbers, if there are 10,000 people in a population (N = 10,000) and 1000 babies are born that year, then the birth rate is

$$B = \frac{1000}{10{,}000} = \frac{1}{10} = 0.1 \text{ per year}$$

From the calculation we obtain a decimal fraction, but the rate itself is then generally expressed as a percentage, rather than as a fraction. To convert a fraction to a percentage, multiply by 100; the above fraction thus becomes $0.1 \times 100 = 10\%$.

Death rate. The **death rate** is found in a similar manner to the birth rate. Given that a certain fraction, D, of the population dies within the time interval in question, we have

$$D = \frac{\text{number of deaths per year}}{N}$$

The term D is thus the death rate (also called the mortality rate). If 100 people die that year, what is the death rate? Notice that N, the population size, appears in the equations for both the birth rate and the death rate because both are proportional to the size of the population, that is, both are fractions of N.

Change in population size. At the end of the year, N will have increased by the number of births and decreased by the number of deaths. To express this mathematically, the above equations can be rearranged and then combined.

By rearranging the equation for the birth rate, we get

$$\text{number of births per year} = B\,N$$

and by rearranging the equation for the death rate, we get

$$\text{number of deaths per year} = D\,N$$

Thus, to find the change in N per year, we add the births and subtract the deaths, to get

$$\text{change in } N \text{ per year} = B\,N - D\,N$$

We can rewrite this equation using the notation standard in mathematics for rates of change: if the change in N, called dN, is divided by the change in time, dT, we have

$$dN/dT = B\,N - D\,N$$

or

$$dN/dT = (B - D)\,N$$

Growth rate. Rearranging the above rate equation gave us a new quantity $(B - D)$, which is the difference between the birth rate and the death rate. This quantity $(B - D)$ is called the **growth rate**, and is symbolized as r. The equation becomes

$$dN/dT = r\,N$$

A population increases if its birth rate, B, exceeds its death rate, D, and in this case r is positive. If $B = D$, then $r = 0$, and the population is stable, neither increasing nor decreasing. A population whose death rate exceeds its birth rate decreases and r is negative.

The growth rates of 130 different nations are shown in Figure 6.2A. Growth rate does not correlate with population density (numbers of people per square mile, Figure 6.2B). Some nations with high growth rates have low population densities, although density can be assumed to be increasing wherever the growth rate is positive. Current data on population density and population growth rates are shown on our web site.

Immigration and emigration. Population ecology tells us that populations do not change only by births and deaths, but also by individuals migrating into or leaving a population. The preceding discussion assumed a 'closed' population unaffected by migration. This is a safe assumption for the world as a whole, but for a single country or region we must also add terms for both immigration (people entering) and emigration (people leaving). The United States, for example, currently has a birth rate not much higher than the death rate, but the population continues to grow because more people move to the United States each year from other countries than move away from the United States. To include migration rates in the calculation of r, we must write

$$r = B - D + i - m$$

where i is the rate of immigration (the number of immigrants in a year divided by the population size) and m is the similarly defined rate of emigration. In most nations today, population growth results mostly from the excess of births over deaths rather than from the excess of immigration over emigration. However, one-third or more of the yearly population growth in the United States comes from immigration.

Exponential (geometric) growth

Although the definition of the term r may include immigration and emigration, the overall equation for the growth rate has not changed, and remains

$$dN/dT = r\,N$$

The type of growth described by this equation is an example of geometric growth. A geometric series is one in which each number is multiplied by a constant to produce the next number in the series. For example, 1, 2, 4, 8, 16, ..., is a geometric series in which each number is multiplied by 2 to give the next number. The growth rate equation is a geometric series because each new value of N results from multiplying the previous

FIGURE 6.2

Population growth rates (r) and population densities around the world.

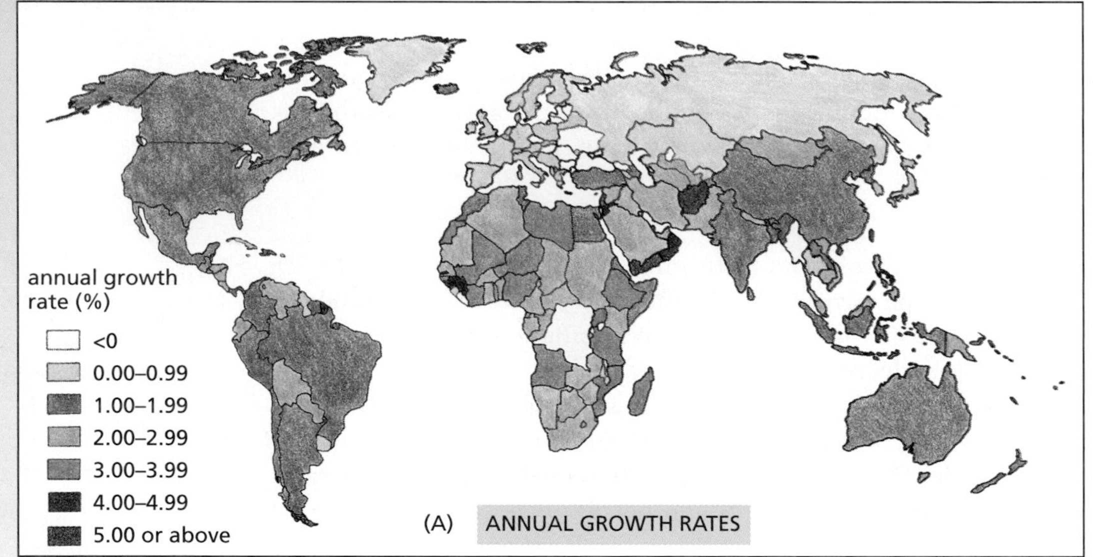

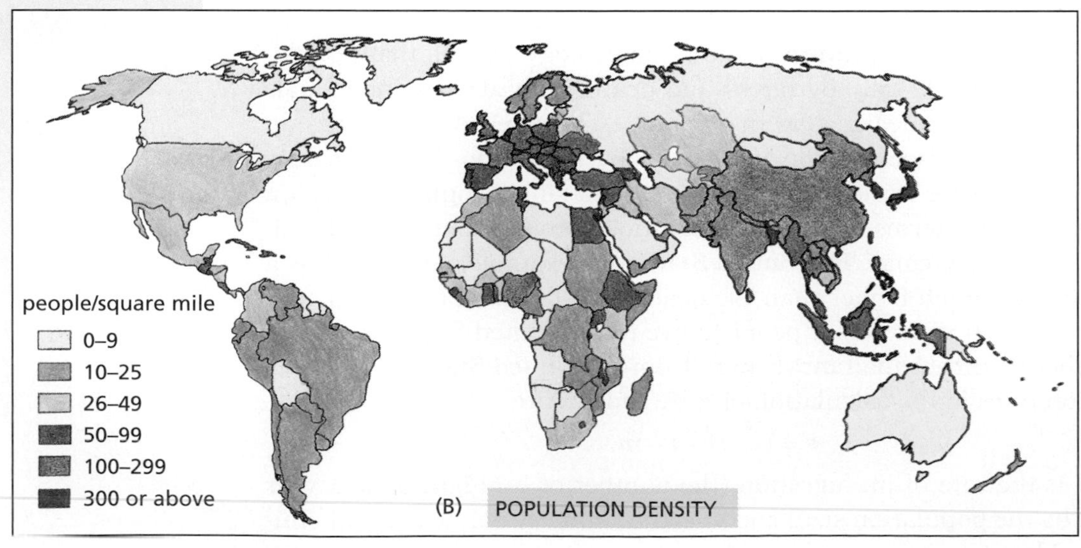

value by 1 + r. (Another familiar example of geometric increase is the growth of money by compound interest.)

Geometric growth is also called **exponential growth**. When demographers work with the growth rate equation, they most often use a branch of mathematics called calculus to express the equation as

$$N = N_0 e^{rT}$$

where N is the population size at time T, N_0 is the initial population size (at time T = 0), e is the base of natural logarithms (approximately 2.71828), and r is the growth rate.

As you can see, when the equation is put in calculus form, the constant by which each number is multiplied (r) appears as an exponent, hence the term exponential growth. The same equation can be used to describe populations of bacteria, fish, or humans, although different time units may be used in each case. Graphing such an equation gives a growth curve such as the blue line in Figure 6.3. (Logistic growth, also shown on Figure 6.3, is explained later.)

For exponentially growing populations, we can calculate the **doubling time**, the length of time it will take for the population to double. Mathematically, doubling time is expressed as

$$\text{doubling time} = 0.69315/r$$

where the number 0.69315 is the natural logarithm of 2.

In 1993, the United Nations calculated that the growth rate for the world's population was 1.8% per year, which will cause a doubling every

$$0.69315/0.018 = 38.5 \text{ years}$$

As we saw earlier, the value of r varies from place to place (see Figure 6.2). The fastest-growing nations have growth rates near 4%, which will cause them to double their population every 17.3 years; many more nations are growing at about 3% per year and will thus double their population every 23.1 years.

Malthus' views on population

The world's human population has long been on the increase, but the *rate* of increase was slow before modern times (see Figure 6.1). During the seventeenth and eighteenth centuries, several European countries began to experience a great upswing in their populations, adding to the many motivations for sending forth expeditions to find and settle new lands. The need to clothe growing populations encouraged innovations that brought greater efficiency to textile production, marking the onset of the industrial revolution.

During the early part of the industrial revolution, philosophers and economists began to pay attention to the phenomenon of population growth. David Hume and Benjamin Franklin each wrote about populations, generally with the attitude that a population increase was a blessing for civilization. The first person to emphasize the negative consequences of population

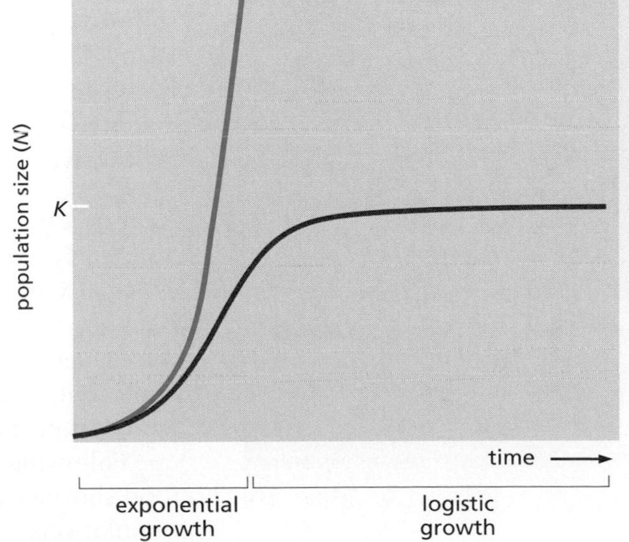

FIGURE 6.3

Exponential growth compared with logistic growth. The steeper (more vertical) the slope, the more rapid the rate of growth.

growth was Thomas Robert Malthus. In his *Essay on the Principle of Population* (1798), Malthus explained the following dilemma:

1. A population tends to increase *geometrically* if its growth is unchecked. (As we saw above, a geometric series is one in which each number is *multiplied* by a constant to obtain the next number.)
2. The available food supply, in Malthus' view, increased only *arithmetically*. (An arithmetic series, like 3, 4, 5, 6, 7, ..., is one in which a constant, in this case 1, is *added* to each number to obtain the next number.)
3. Because the population increases faster than the food supply, the increasing population compounds human misery and poverty, especially among the lower classes.

Positive checks and preventive checks. Malthus divided the factors controlling the population increase into two broad categories that he called preventive checks and positive checks. **Preventive checks** were those that could prevent births from occurring. These were usually voluntary measures, operating on an individual level. They included delayed marriage (also called 'moral restraint'), reduced family size, and several forms of 'vice' (practices that Malthus condemned, including birth control and homosexuality). The **positive checks** on population were those that would operate automatically after births had taken place, whenever the preventive checks were not sufficient. Malthus identified as positive checks overcrowding, poverty, epidemic diseases, rising crime rates, warfare, starvation, and famine. Malthus opposed the social welfare legislation of his time because he thought that any measure to improve the condition of the poor would only encourage them to reproduce faster, further outstripping their food supply and compounding their own misery. One outgrowth of this type of thought was the nineteenth-century theory attributing warfare to the economic needs confronting the population of each nation, the so-called economic theory of war.

Malthus noticed that, even in European countries with stable populations, the *rate* of population growth temporarily rose in the years following a famine or plague until the population was restored to its pre-disaster level. This phenomenon also takes place after many wars and economic hard times (witness the 'baby boom' in the years following World War II). Such events show that a population's potential for increase is much greater than is usually realized. The actual rate of population growth is kept lower than the potential rate of growth by positive and preventive checks.

Recent changes in population pressures. Throughout the nineteenth and early twentieth centuries, technological progress in agriculture, especially in mechanized farming and in the use of chemical fertilizers, increased crop yields among the wealthy nations. European population crises were in many cases also relieved by large-scale emigration to other continents. Malthus and his gloomy predictions of starvation and misery were largely forgotten.

Following World War II, attempts to deal with poverty, disease, and food shortages around the world ran into the harsh reality that burgeoning populations were exacerbating all these problems. Public health improvements (in public sanitation, in mosquito control, in vaccination against infectious diseases, and in the delivery of medical care generally) were diminishing the death rates, especially among the young. The result was in many cases a staggering population growth. Many nations of the developing world were trying to diversify their economies, modernize their industries, and improve their living standards. These nations suddenly found their financial and other resources stretched thin as their populations continued to increase. Economic development, in other words, was being slowed by population growth, and many leaders of the developing nations became increasingly concerned. The positive checks that Malthus had foreseen were operating. The developing world was coming face to face with a population crisis (see Figure 6.1).

The population crisis in developed countries was taking a different form: wealth was diverted into providing additional housing, roads, sewers, and needed services. However, this development may not be sustainable in the long run, because it depends on the use of nonrenewable resources such as fossil fuels and soil (see Chapter 14, p. 608), and on resources and products imported from less developed countries, leaving less for them. The diversion of nonrenewable resources cannot indefinitely support stable or growing populations. By diverting resources, wealthy nations can postpone, but not avoid, the effects of global population growth.

Logistic growth

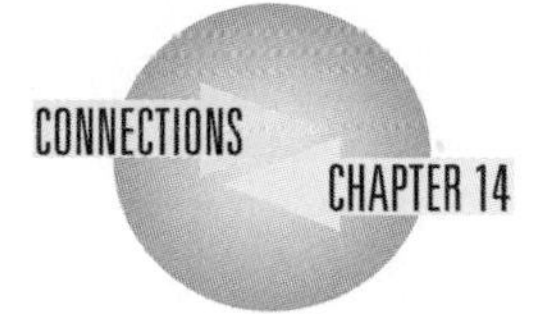

Malthus's assumption of an arithmetical increase in the food supply has been questioned. Although the limited data available to Malthus were consistent with an arithmetical increase, there is no biological or other theory that explains why this should be so. However, most biologists do agree with Malthus' point that the growth in food supplies is slower than the rise in population. One reason this is so is because there is a loss of energy at each stage in which one type of plant or animal becomes food for another, a topic that we develop more fully in Chapter 14. Because food and other resources increase more slowly than population, any population growing exponentially will outstrip its food supply and other resources, including the available space in its habitat. Clearly, no population can continue growing exponentially. After growing exponentially for a while, a population usually follows a pattern called **logistic growth**. Logistic growth has been demonstrated in all types of experimental populations, and we know of no biological reason why this would not also apply to human populations.

Logistic growth can be modeled mathematically by the equation

$$dN/dT = r\,N\,(K - N)/K$$

In this equation, K is a new quantity called the **carrying capacity** of the environment. Like N, K is a number, not a rate. K refers to the maximum size of the population that can be sustained indefinitely by the environment. As population size approaches this carrying capacity, the population growth (symbolized by dN/dT) slows down, and when $N = K$ the population growth is zero (Figures 6.3 and 6.4). When the population reaches whatever carrying capacity is imposed by the environment (see

Figure 6.4C), the birth and death rates are balanced. If the population size overshoots *K*, then a population crash follows in which deaths outnumber births and the population size declines to the carrying capacity.

K is related to the amount of space in an environment and the other resources available, including the amount of energy that is in a form that can be used by the organisms living there. For animal or plant species living in an unchanging environment, *K* is generally constant. For our own species, *K* varies with changes in technology, especially technology that gets more usable energy from the same environment, for example by increases in the efficiency of energy use or of food production. A certain amount of land may support a particular population size of hunters and gatherers at a low carrying capacity (a low value of *K*). The development of agriculture generally results in an increased carrying capacity, and agricultural mechanization (including the use of tractors and chemical fertilizers) in most cases increases it still further.

Humans can avoid an environmentally imposed population crash by limiting the birth rate before the population reaches its carrying capacity; in addition, we need to consider the effects of human populations on the populations of other species. When two species interact, either one may modify the effective carrying capacity of the environment for the

FIGURE 6.4

A variety of logistic growth curves.

(A) Shape of a logistic growth curve

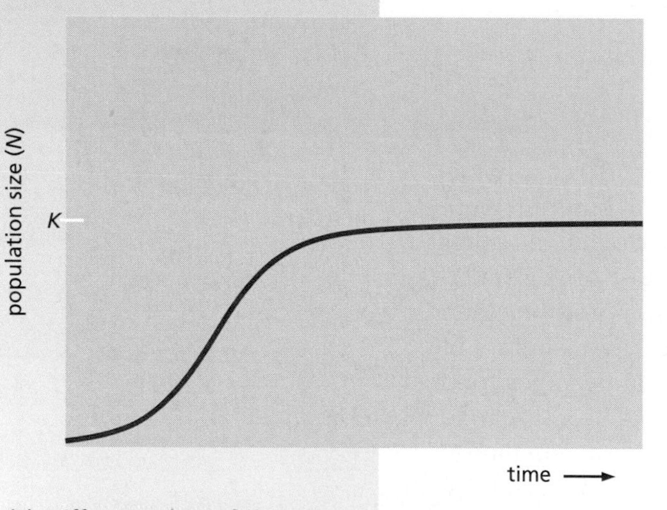

(B) Different initial population sizes (N_0)

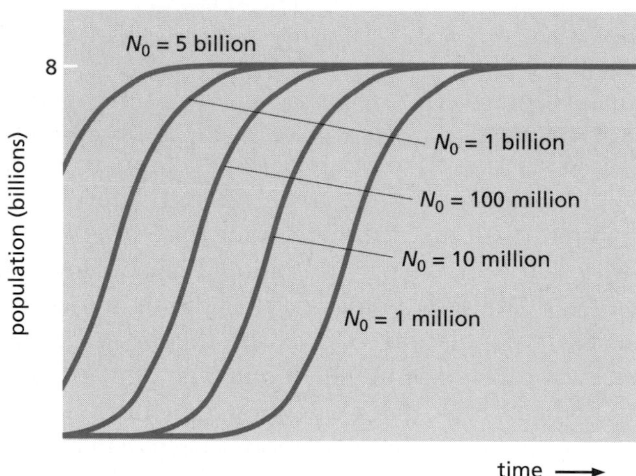

(C) Different values of *K*

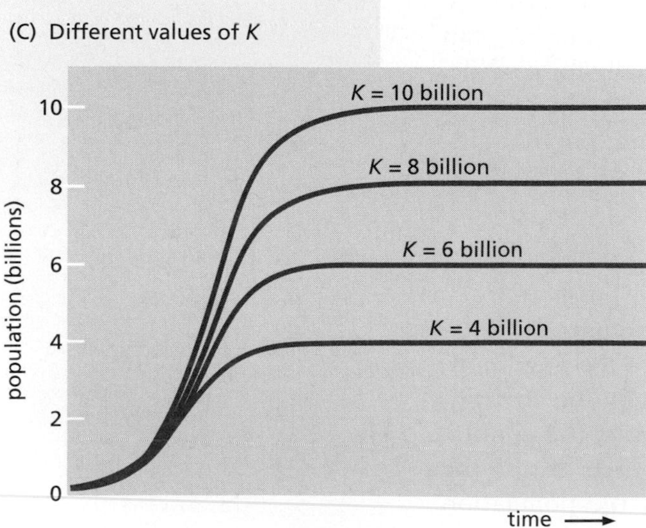

(D) Different rates of increase (*r*)

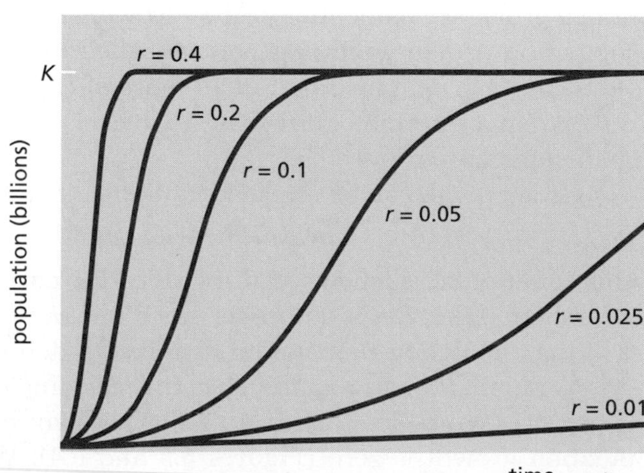

population size of the other. Population ecologists define competition as a type of interaction in which a species diminishes the population size of another. In such cases, the logistic growth equation carries an additional term in the numerator, subtracted from the carrying capacity. An increasing human population diminishes the sustainable population size of all species with which we compete, and is driving many of these other species toward extinction (see Chapter 15).

Demographic transition

Population increases in the past have coincided with major advances in technology that have allowed the carrying capacity to reach new levels. The development of agriculture made it possible for human populations to increase well above the size permitted by hunting and gathering. Demographers estimate that the world's population was about 50 million in 7000 B.C. and increased to about 250 million by the time of Christ. The earliest date for which there are reliable population estimates is 1650, at which time the world's population stood at 500 million. By 1850, when actual census figures were available for most of the industrialized nations, the world's population stood at an estimated one billion and has increased rapidly since (see Figure 6.1).

Most of our understanding of the changes that accompany a population increase come from studying the increases brought about by the industrial revolution or by the spread of industrially based technology to the nonindustrial world. Our current model of this process is that it occurs in an orderly succession of stages known as a **demographic transition**.

Stages of a demographic transition. As shown in Figure 6.5, the first stage of a demographic transition is a stable population in which a high death rate is balanced by a high birth rate. Over the centuries, traditional societies with high death rates (especially from infant mortality and childhood infections) developed customs that encouraged high birth rates. In other words, high mortality rates encourage high birth rates.

The second stage of the process is a period of exponential growth in which the population size climbs to a new high. This is brought about by technological changes that result in falling mortality rates, which is termed **death control**. War, starvation, violence, and disease all cause death. Even when they do not cause death directly, they add to the risk of death from other causes. The same may also be said of poverty, crime, overcrowding, crop failures, and unsanitary living conditions. In Malthusian terms, they are positive checks to population size, meaning factors that increase the death rate, D.

FIGURE 6.5

Idealized stages of a demographic transition. Some authorities recognize only three stages by combining the middle two.

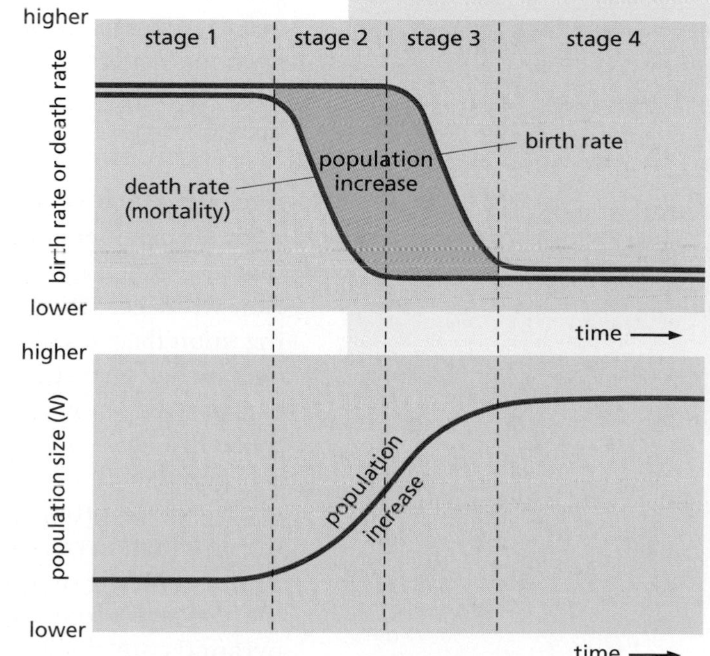

To the extent that any of these factors are controlled, the consequent rate of death is also controlled. In modern times, the emphasis is on alleviation of these positive checks and thus a lowering of D. However, it always takes at least a few generations for the cultural values and customs to change so as to permit a matching decline in birth rates. In the meantime, the memory of high death rates in the recent past continues to encourage high birth rates. In some cases, the birth rate may even rise as the result of better nutrition and the improved physiological condition and reproductive health of prospective mothers. The combination of high birth rates and lower death rates causes the population growth that characterizes the middle of the demographic transition.

The third stage of the transition is marked by a decline in birth rates as the population adjusts to new conditions and as the incentives for high birth rates are removed. Population growth follows the logistic growth equation during this stage. As the birth rate declines to match the new lower mortality rate, the population once again stabilizes, but at a larger population size (larger K) than before. The demographic transition is complete.

Demographers estimate that overall mortality rates held steady or declined slowly in pre-industrial Europe, with occasional but temporary upsurges during wars and epidemics such as the great bubonic plague (the black death) that decimated European populations in the 1300s. England's demographic transition began in the early 1700s and took some 250 to 300 years to complete. Other industrialized countries in North America, Europe, and Japan took closer to 200 years to complete their demographic transitions. The remaining countries of the world began their demographic transitions only during the twentieth century, and did so much more suddenly, often going from high traditional mortality rates to low modern rates in a single generation.

It now seems that the United States has completed its demographic transition because birth and death rates are now approximately equal, a condition known as **zero population growth**. The U.S. population continues to increase, however, because of the excess of immigration over emigration. In addition, there can be short-term increases, for example when baby boomers have additional children many years after they first became parents, or when people remarry and start second families.

Age structure of populations. The models we have examined so far treat all members of a population as being the same. However, we know that the probability of death and the probability of reproduction both vary with age. Thus, the true rate of population growth may vary depending upon the ages of individuals within the population. Grouping individuals by age gives the **age structure** of the population. The age structure can best be shown by a population pyramid, or **age pyramid**, such as those in Figure 6.6. Each horizontal layer on such a diagram represents the percentage of the population in a particular age group, with the youngest age groups on the bottom. Altogether, the age pyramid shows the distribution of individuals among the various age groups. To get a feeling for the scale of the age pyramids in Figure 6.6, notice that in Tanzania the 0–4 age group constitutes about 20% of the population. Most age pyramids are divided by a vertical midline, with male age distribution shown on the left and female age distribution on the right. Notice in Figure 6.6 that the 0–4 age group has approximately equal numbers of males and females in all countries, but the 80+ age group in Denmark has about five females for every male. Among human populations, a pyramid with sloping sides and a wide base (many children) characterizes an expanding population. A shape maintaining more or less the same width throughout (except for the oldest few age classes) indicates a stable population. A stable age distribution is reached when the pyramid keeps the same shape as each age group grows older so that the numbers in each age group are replaced by an equal number advancing from the next younger group.

Sometimes there are age bulges, as with the post World War II baby boom, when the birth rate increased temporarily. The United States and some other countries experienced a second baby boom (or a baby boom 'echo') in the late twentieth century as members of the earlier baby boom reached their prime childbearing years. In these countries, schools must now cope with the largest generation of school-age children that has ever lived.

Predictions of future values of r can sometimes be made on the basis of age structure. Clearly, a population of 10,000 individuals with 4000 females of reproductive age has much more potential for increase than

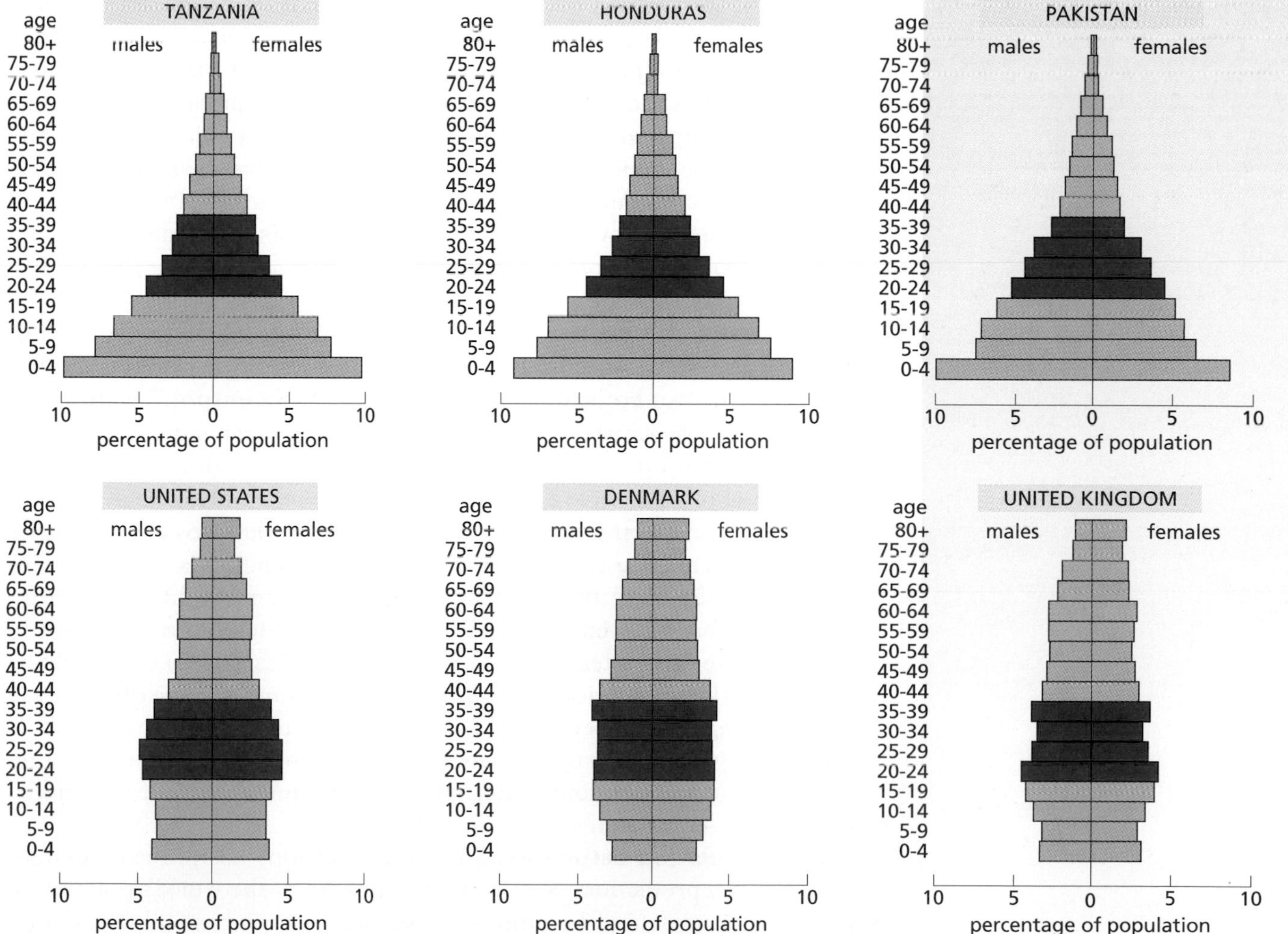

FIGURE 6.6

Age pyramids for three rapidly growing populations (top row), a slowly growing population (United States), and two very stable populations (Denmark and United Kingdom), from United Nations (1993) data. Age groups from 20 to 39 are darker to emphasize that most reproduction occurs in these age groups.

one with only 400 females of reproductive age. Calculations of the potential for future increase are often carried out by multiplying the number of females in each age group by the number of children that each of those females is likely to bear, then adding up these products for all age groups.

Life expectancy. In most of the industrialized nations, the control of many infectious diseases since the late 1800s and improvements in sanitation have resulted in decreased infant mortality. A few twentieth-century changes, such as increases in smoking, auto accidents, and hand guns, have increased death rates, but these have generally been offset by much greater declines in the death rates from famines and infectious diseases. One consequence of declining death rates, especially among the young, is a greater **life expectancy**, meaning the average maximum age that people attain in life, or the age to which a person born in a particular place and at a particular time can expect to live (Figure 6.7). Life expectancy is calculated on a statistical basis, taking into account the probability at any age of a person's proceeding on to the next age group. Therefore, both decreased mortality and increased longevity (the maximum age achieved by some individuals in the population) contribute to an increased life expectancy, with the largest increases resulting from reductions in childhood mortality. Life expectancy in the United States has risen from somewhere near 50 years in colonial times to over 75 years today, and the age group over 80 is the most rapidly growing segment of the U.S. population. This 'graying' of America and of many other countries results in a decrease in the proportion of the population in the most fertile age group and an increase in the proportion of older people, many of whom are dependent on the younger generation for their care. Numerous sociological changes follow the shift toward more people of advanced age and fewer children—more emphasis on medical care and less emphasis on schools, for example.

Demographic momentum. Even after the birth rate falls to the level of the mortality rate, the population may continue to increase for another generation or two because of a **demographic momentum**. The momentum is caused by an age structure opposite to the one just described for a graying population. A population in which a large fraction of individuals is not yet of reproductive age (see Figure 6.6), while only a small fraction is past the age of reproduction, has a future reproductive capacity larger than one in which a higher percentage of the population is beyond reproductive age. Because the younger age groups are more likely to reproduce in the next 20 years and less likely to die, a temporary increase in the birth rate (and a temporary decrease in the death rate) can easily be predicted. The population will continue to grow until the age distribution readjusts itself more evenly, that is, until a stable age distribution is reached. With a stable age distribution, the birth and death rates will no longer change, unless some external factor disrupts the stable situation.

World population estimates. The United Nations has published tables with detailed predictions for the further growth of the world's population under different sets of assumptions. According to the model that demographic experts consider most likely, the world's population will grow to about 9.4 billion in 2050 and will stabilize at around 11 billion by the year 2200. Other models, with different sets of assumptions, predict population values in the year 2150 as low as 3.6 billion or as high as 27.0 billion. Each of these assumes logistic growth, except for the model in which population growth continues exponentially at its present rate—that model predicts a population size of 296 billion, nearly 50 times the

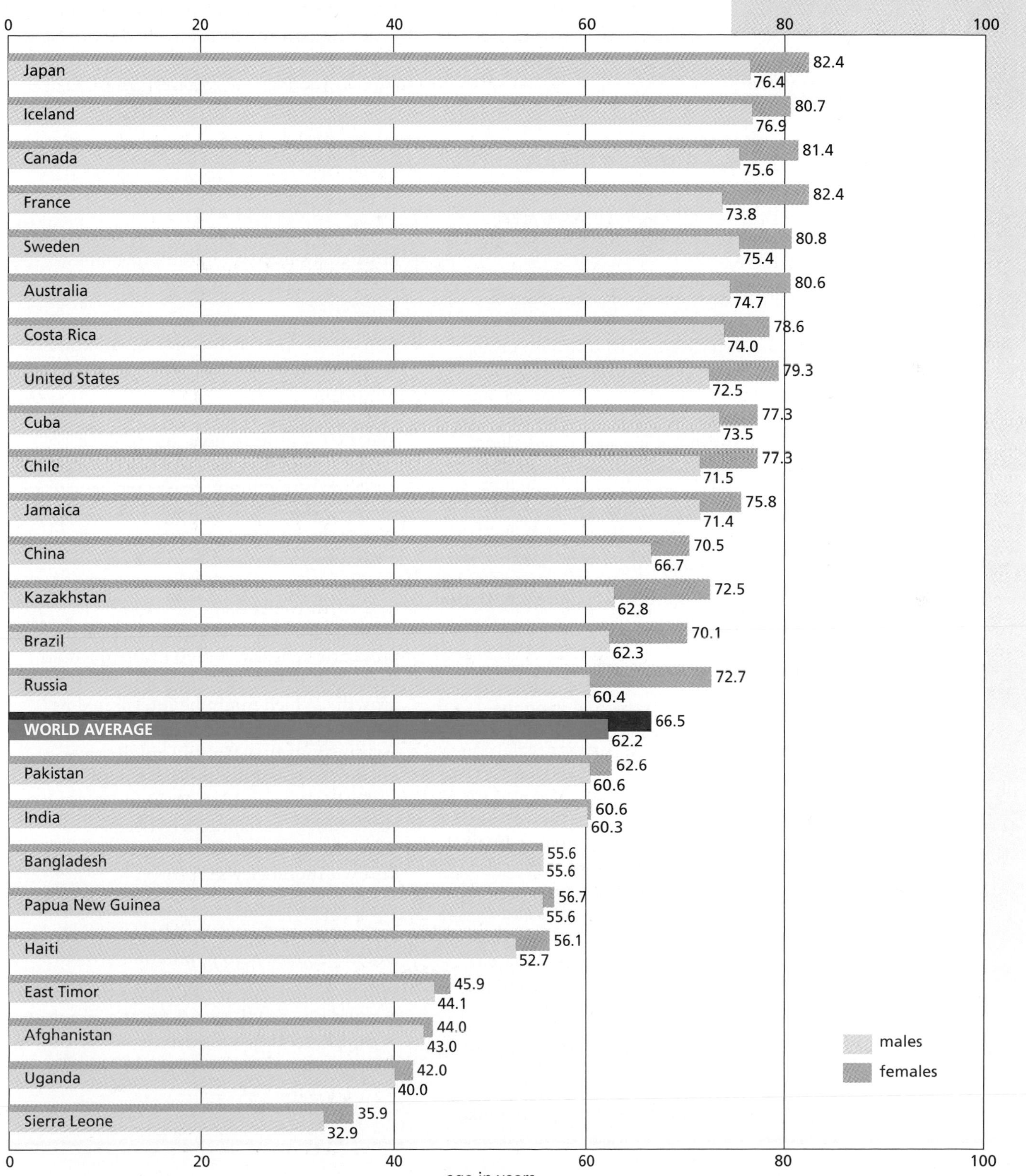

FIGURE 6.7

Human life expectancy in various countries, from United Nations (1999) data.

present value! Few, if any, biologists think that the latter is in any way sustainable; population growth will have to level off from its present rates.

Because people do not generally find raising the death rate to be an acceptable method of lowering population growth, decreasing the birth rate is the only other option. In the next section we study reproductive biology and the factors that contribute to fertility. In the final section of the chapter we see how fertility can be controlled to lower the birth rate.

THOUGHT QUESTIONS

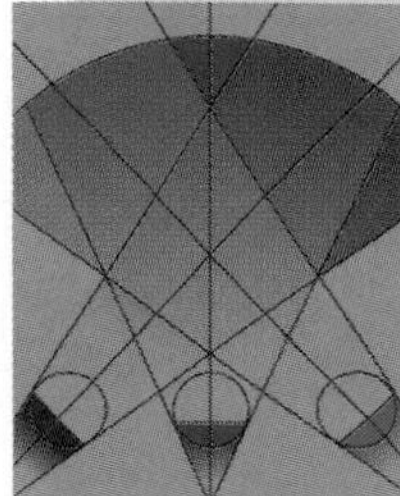

1. What changes (biological, social or economic) could increase the carrying capacity of the entire world or of one nation? Is it more important to modify *r* or to modify *K*?
2. Suppose that you count the number of people in a given town every year for five years and you discover that *N* has stayed about the same during that time period; say, at 10,000. You also know that no one has moved into the town or away from the town in that time. What can you deduce about *B* and *D*? Confirm this for yourself by solving the equations given earlier in the chapter. Can you tell from this information how many people were born in the town in any of the five years?
3. Study the age pyramids in Figure 6.6. Does the age distribution of the females or that of the males have a greater impact on future population size? Why?
4. What values of *r* are typical during the several stages of a demographic transition? See Figure 6.5, and review the way in which *r* is defined.
5. Which is likely to produce a greater increase in the number of people, a small population with a high *r* or a large population with a small *r*? Try some calculations with any values of *N* and *r* that you would like to examine. Use the simplest model that seems appropriate, then ask what changes the more complex models would bring.
6. From the data presented in this chapter, can we estimate the value of *K* (the carrying capacity) for the human population on planet Earth? Can we make a minimum or maximum estimate? Have we already reached the carrying capacity of the planet?
7. Which would you think would contribute more to an increase in life expectancy, decreased infant mortality or increased longevity?
8. Among the 'preventive checks,' Malthus listed several forms of 'vice,' including the following:
   a. heterosexual intercourse outside marriage ("promiscuous intercourse"; "violations of the marriage bed");
   b. sexual attraction to one's own sex (homosexuality), to animals (zoophilia), or to inanimate objects (fetishism);
   c. "Improper arts to conceal the consequences of irregular connections" (i.e., birth control).
   Malthus listed all these practices as preventive checks. Which of them would actually function to limit population growth? Which might be interpreted as preventive checks under certain assumptions that Malthus might have made? Do you think these assumptions are realistic? Do you think Malthus condemned these practices because of their effects on population or because of his own moral views?
9. What factors would need to be included in an equation to estimate what birth rate would produce zero population growth?
10. Which countries currently have the highest population growth rates? Are they wealthy or poor? Are they influential? Where do they stand in the current world order?

HUMAN REPRODUCTIVE BIOLOGY HELPS US TO UNDERSTAND FERTILITY AND INFERTILITY

In sexual reproduction, a male haploid gamete unites with a female haploid gamete. The two types of gametes are produced by individuals of different sexes. We begin this section by comparing the anatomy and physiology of the two sexes in humans.

Reproductive anatomy and physiology

Reproductive anatomy includes those structures that allow for sexual intercourse. It also includes those structures that secrete hormones involved in reproductive physiology, including the production of gametes. Although the same hormones occur in both males and females, they are present in different amounts and act in different ways to induce males and females to mature differently, both in appearance and in the type of gamete that they produce.

Maturation. The reproductive organs are formed during embryonic development, but remain immature until **puberty**, at which stage sex hormones bring about many changes in the body. Hormones are small molecules that are used for chemical communication throughout the body (see Chapter 12, p. 513), and at puberty a group of hormones produce bodily changes. At an age that varies greatly around an average of about 12–13 years, increasing levels of the **follicle-stimulating hormone (FSH)** stimulate the final maturation of the egg-producing **ovaries** in females and the sperm-producing **testes** in males. These, in turn, step up their secretion of other hormones (estrogen and testosterone), stimulating (among other changes) the development of **secondary sexual characteristics**. Such characteristics include the widening of the hips, growth of breasts, and redistribution of body fat in females, the growth of facial hair (including a beard) and the deepening of the voice in males, and the growth of long bones and pubic hair in both sexes. The term secondary sexual characteristics means that these features, while characteristic of mature men and women, do not have a direct role in the production of gametes. After puberty, FSH, estrogen, and testosterone continue to have other roles throughout the lifetime of the individual.

Male reproductive organs and sperm production. The sperm are male gametes, formed by the process of meiosis (see Chapter 2, p. 48) in which each gamete receives half the adult number of chromosomes—one chromosome from each pair. Sperm have very little cytoplasm surrounding their nucleus, but they have a long tail that moves rapidly back and forth to propel the sperm along the reproductive tract of the female after intercourse (Figure 6.8).

Sperm are produced in the testes of the male (Figure 6.9). The hormone **testosterone** is secreted by the interstitial cells that are crowded into the spaces between the sperm-producing tubules in the testes.

(Testosterone is also made in smaller amounts by the brain in both sexes and by the ovaries in females.) The sperm accumulate in a series of wrinkled ducts that form the epididymis. The epididymis also secretes a fluid (the seminal fluid) that carries the sperm through a long duct called the vas deferens through which the sperm leave the testes. The left vas deferens and right vas deferens merge within the prostate gland, where they join the urethra carrying urine from the urinary bladder. The secretions of the prostate gland add certain nutrients (including the sugar fructose) that help the sperm to swim more vigorously. Near the base of the penis, the bulbourethral gland helps squirt the seminal fluid and sperm through the length of the penis during ejaculation. Between ejaculations, sperm and seminal fluid are held in extensions of the prostate gland called seminal vesicles.

FIGURE 6.8

An egg surrounded by sperm.

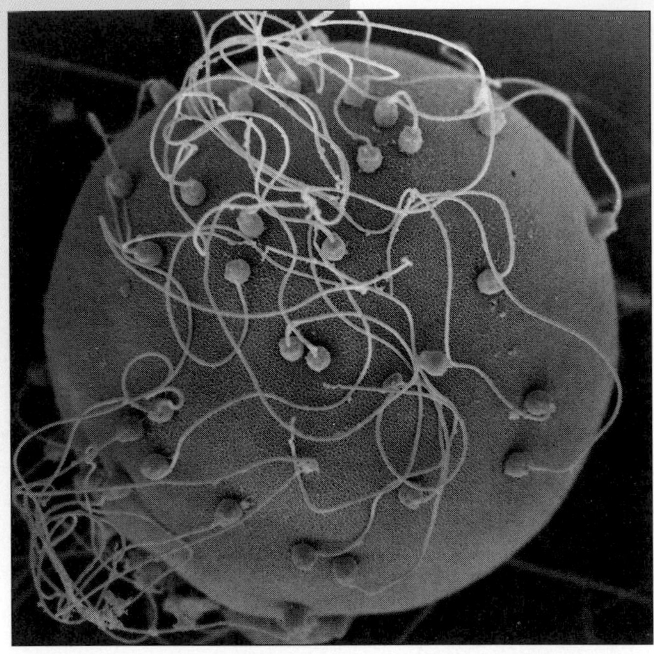

Female reproductive organs and ovulation. The female reproductive organs (Figure 6.10) include the uterus and a pair of ovaries. The female gamete, or egg, is formed within the ovary by meiosis. The first meiotic division occurs before

FIGURE 6.9

The human male reproductive system.

STRUCTURES IN THE MALE PELVIC AREA

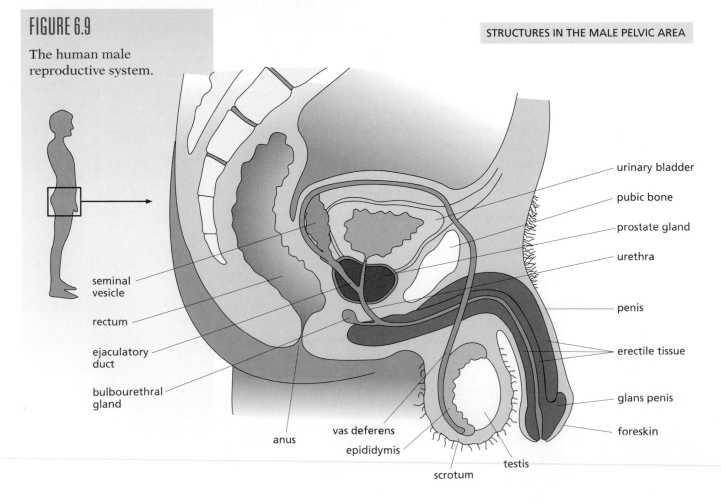

urinary bladder
pubic bone
prostate gland
urethra
penis
erectile tissue
glans penis
foreskin
seminal vesicle
rectum
ejaculatory duct
bulbourethral gland
anus
vas deferens
epididymis
scrotum
testis

the egg is released from the ovary, but the second meiotic division, in which the chromosome number becomes haploid, does not occur until after the egg is fertilized. During the two cell divisions during meiosis in egg formation, the division of the cytoplasm between the offspring cells is extremely unequal. Only one of the four resultant cells becomes a mature egg with a large amount of cytoplasm to nourish the zygote after fertilization. The other three become polar bodies with very little cytoplasm, and these are usually lost. The nonreproductive cells surrounding the egg enlarge to form an ovarian follicle. The rupture of this follicle and the release of its egg are called **ovulation**.

The menstrual cycle. Egg production and ovulation in mammals is a hormonally regulated cycle. In humans, this cycle lasts about 28 days and is called the **menstrual cycle**.

The cycle of egg production and changes in the uterus is controlled by two ovarian hormones, estrogen and progesterone, and by three hormones secreted by the pituitary gland at the base of the brain (Figure 6.11). At the start of each cycle, which begins with menstruation, the hypothalamus stimulates the pituitary to secrete small amounts of follicle-stimulating hormone (FSH). The FSH stimulates two processes within the ovary: growth of an ovarian follicle, and production of the hormone **estrogen**, which reaches a peak concentration during the second week of the cycle. The estrogen stimulates the thickening of the lining of the uterus and the release of a second pituitary hormone, luteinizing hormone (LH), that induces the release of the egg (ovulation), after which the tissue that surrounded the egg is left behind to form a scar tissue called the **corpus luteum**. A third pituitary hormone stimulates the corpus luteum to secrete the hormone **progesterone**, which maintains the uterine lining in a thickened and receptive condition, ready for the

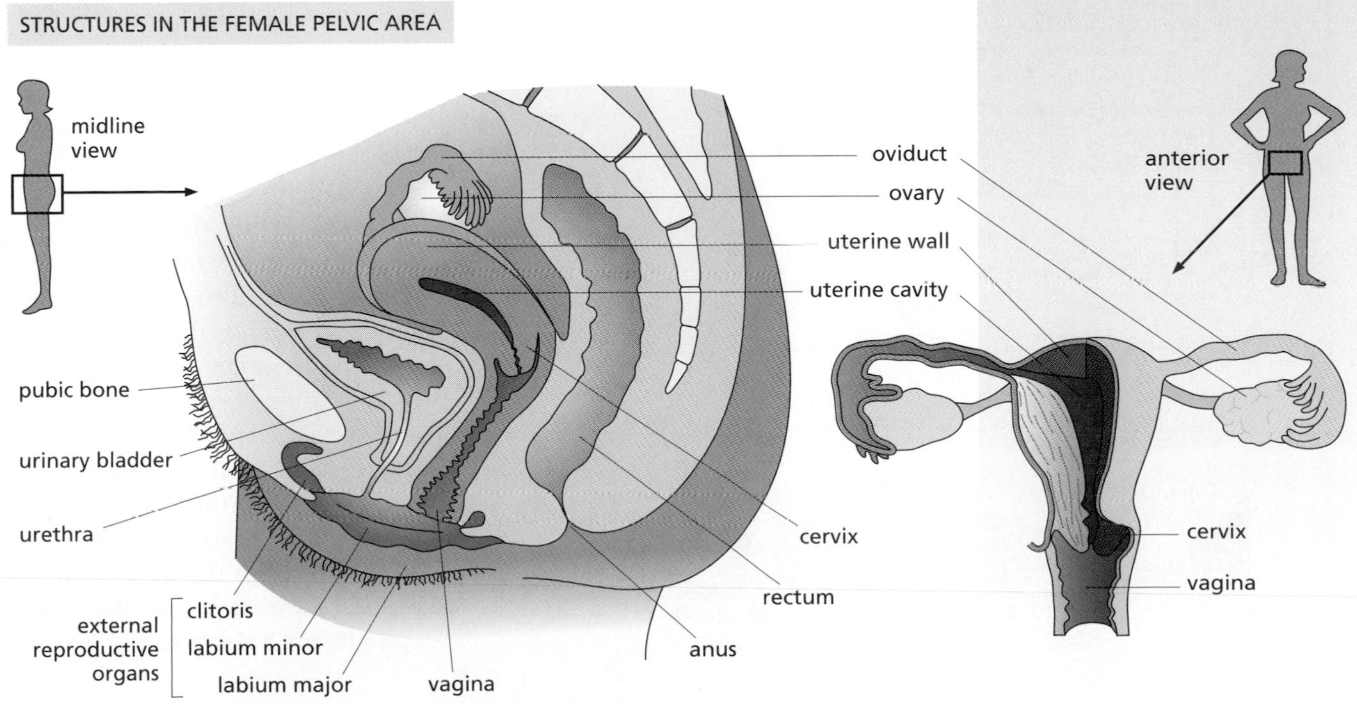

FIGURE 6.10

The human female reproductive system.

implantation of an embryo should the egg be fertilized. If fertilization does not occur and no implantation takes place, the corpus luteum degenerates and the supply of progesterone drops sharply, causing the uterine lining to break down. The egg and the uterine lining are then sloughed off in the form of menstrual bleeding. The absence of progesterone also releases the pituitary to begin secreting follicle-stimulating hormone once again, initiating a new cycle.

FIGURE 6.11

Reproductive cycles in the human female.

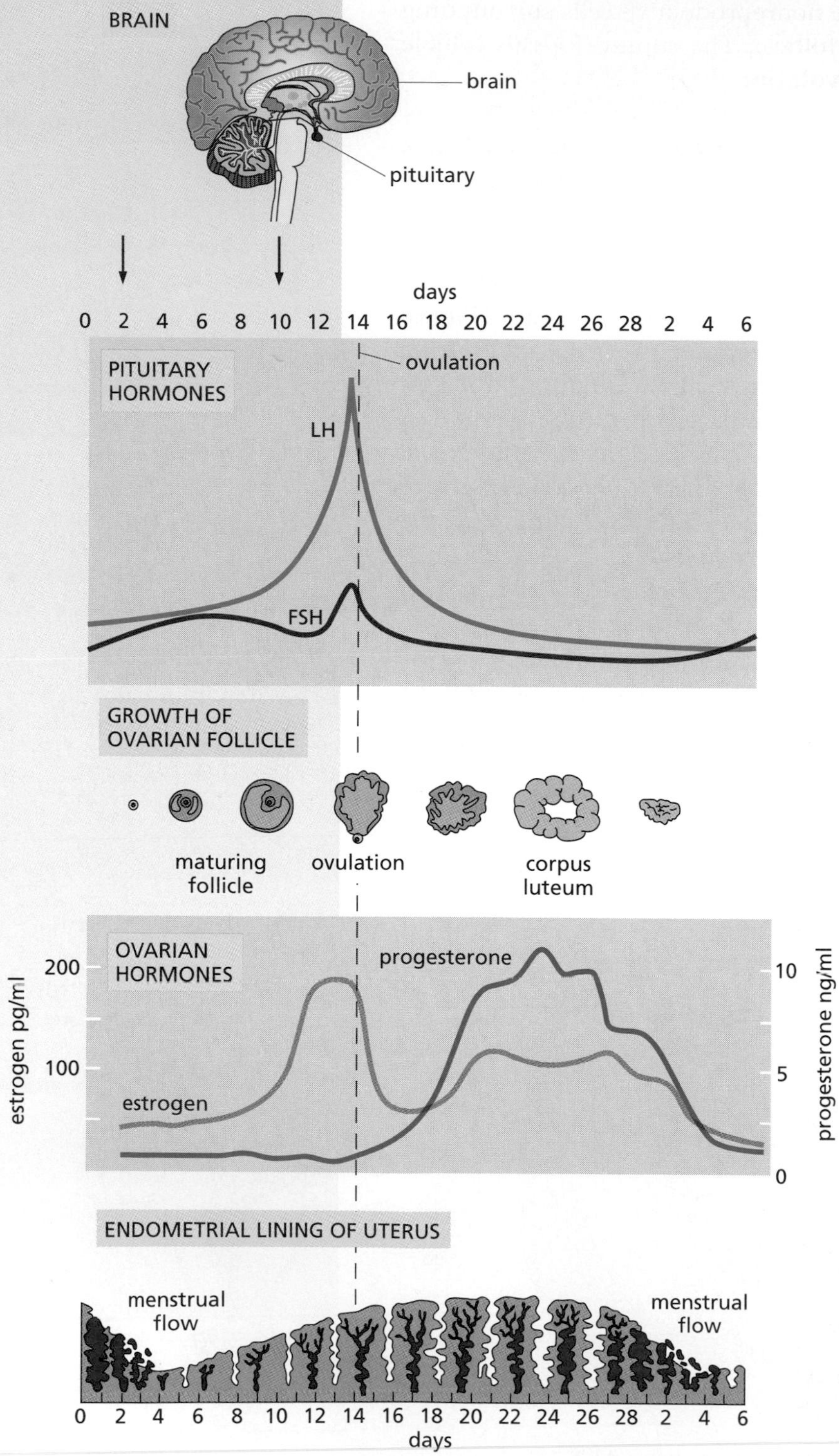

During the menstrual cycle, changes in the concentration of each hormone stimulate the production of the next hormone in the sequence. Hormones from the ovaries and hormones from the pituitary regulate each other. In several cases, the presence of a later hormone has an inhibiting effect on the secretion of the previous hormone. This regulation of a previous step of a cycle by a later step of the cycle is called a **feedback mechanism**. In this instance, feedback prevents the overproduction of any hormone and stops the production of a hormone once it has done its job.

Fertilization and implantation. Neither the sperm nor the egg live very long. After its release from the follicle, the egg travels along the uterine (Fallopian) tubes or oviducts, and it is here that the egg is fertilized if sperm are present.

When the male ejaculates during sexual intercourse, approximately 300 million sperm are inserted into the vagina of the female. Of these, only about 3000 will successfully swim through the cervix and the uterus and into the oviducts, following hormonal signals released by the egg. Several sperm need to contact the egg to dissolve the coating that surrounds it. After the coating has dissolved, allowing a sperm to reach the plasma membrane of the egg, the sperm and egg plasma membranes fuse and the egg draws the sperm nucleus inside. The coating closes again, preventing the entry of any more sperm nuclei. The egg completes its second meiotic division and its haploid chromosomes join with the haploid chromosomes from the sperm, a process called **fertilization**.

If an egg is fertilized by a sperm, the resultant zygote continues travelling along the oviduct for 4–5 days, during which it undergoes several cell divisions and becomes an embryo. When it reaches the uterus, at a stage called the blastocyst, the embryo adheres to the inner (endometrial) lining of the uterine wall, a process called **implantation** (Figure 6.12). If the embryo does not implant, it does not develop further and is shed from the uterus. Implantation triggers the growth of the placenta, consisting of intertwined tissues of the embryo and the endometrium of the mother, through which the developing embryo is nourished (see also Figure 11.2, p. 458). Most of the organs of the growing embryo form during the first month, after which the embryo is known as a fetus.

In this section we have described the normal events of human reproduction, in which egg and sperm from a fertile female and a fertile male combine to make an embryo. Not all individuals are fertile, however. Infertility can result from lack of production of sperm or eggs, from low motility of sperm, shortened viability of egg or sperm, or anatomical abnormalities preventing the sperm from contacting the egg. In the next section we describe some of the methods that have been developed for overcoming infertility.

Reproductive technologies

At the population level, there are many motivations for decreasing the birth rate, but on an individual level the urge to have children is very strong. Often couples who cannot conceive a child seek treatment for infertility. In parallel with research on birth control technologies, much

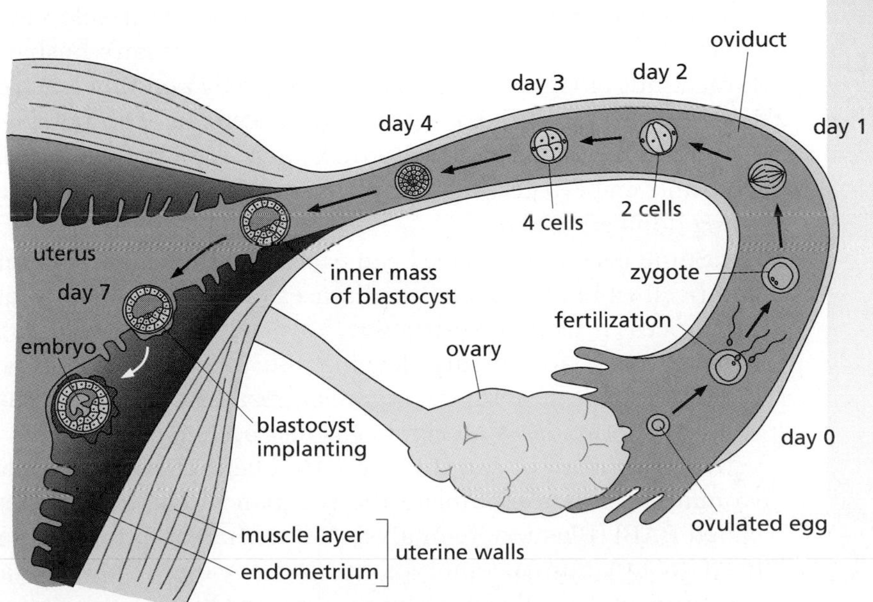

FIGURE 6.12

Fertilization in the oviduct and implantation in the uterus.

research has gone into the development of several new reproductive technologies. Much of the research on infertility has been done in countries such as India and China, which also have active population control programs, a reminder that individuals and societies may have conflicting goals.

Artificial insemination. Artificial insemination means introducing sperm into a female's reproductive tract other than through sexual intercourse. The sperm could be derived from a woman's husband or from another man. The procedure is fairly simple (and is routinely performed on cattle and certain other domesticated species as are *in vitro* fertilization and surrogate pregnancy). However, the legal rights and responsibilities of a sperm donor other than the recipient's husband are unclear in a number of jurisdictions.

Hormonal methods. Various hormonal treatments are available to increase fertility in either men or women. Often these methods are used in conjunction with another reproductive technology, particularly in treating women to increase the probability of an egg's being available for fertilization. Usually only one or a small number of eggs is released in a monthly cycle. Fraternal twins result when two eggs are fertilized in the same cycle. (Identical twins come from a single fertilized egg that later divides into two embryos.) Hormonal treatments for infertility may induce the release of many more eggs than normal. The recent increase in the frequency of multiple births has been largely due to these treatments.

***In vitro* fertilization.** An *in vitro* process is one that takes place in laboratory glassware rather than inside the body (*in vivo*). In the case of *in vitro* fertilization, eggs are harvested from a woman and fertilized in a glass or plastic dish, using sperm contributed either by her husband or by another man. Fertilized eggs are then allowed to develop to approximately the 64-cell stage, after which one or more of these embryos are implanted in the woman's uterus and allowed to develop to term. Using sperm donated by someone other than the woman's husband raises the same kinds of legal issues (including custody and financial responsibility issues) as does artificial insemination. Persons who possess alleles that they do not wish to pass on to their offspring may seek *in vitro* fertilization and embryo testing before implantation or artificial insemination using donated sperm.

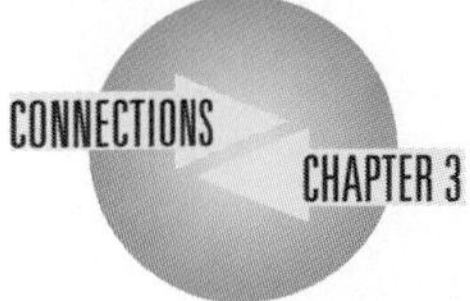

Some researchers have been experimenting with techniques that test eight-cell embryos for genetic diseases (see Chapter 3). Because the testing process is usually destructive, only one of the eight cells is separated from the rest for testing, leaving the other seven cells available for implantation if desired; the organism resulting from a seven-cell embryo is just as normal as if all eight cells had been used. Embryos obtained by *in vitro* fertilization can thus be tested before they are implanted into a woman's uterus to complete the pregnancy. This technique, sometimes called BABI (Blastomere Analysis Before Implantation), has already been used to test human embryos *in vitro* for cystic fibrosis, allowing the selection of only those embryos that are free of the disease. Selected embryos can then be implanted, and the couple can be free of the fear that their child will be born with cystic fibrosis, a disease that is usually fatal. Once this technique becomes readily available for a wider variety of human conditions, it will become possible to avoid certain genetic diseases without performing any abortions, or to choose certain other characteristics, such as the child's sex.

Surrogate pregnancy. Surrogate pregnancy is the use of another woman's womb to carry a baby to term on behalf of a woman who cannot undergo the pregnancy herself, usually for medical reasons. In most cases, the baby is conceived by *in vitro* fertilization, using egg and sperm cells donated by a couple unable to conceive themselves. The resulting embryo, which is the genetic offspring of the donor couple, is then implanted into another woman who agrees to act as a surrogate mother, usually for a fee. In addition, medical expenses are generally paid by the donor couple. The legal status and rights of the surrogate mother are subject to many ethical and legal questions. Surrogacy contracts have been outlawed or held invalid in a number of jurisdictions that view the birth mother (i.e. the surrogate) as the legal parent who is therefore 'selling' her baby if she receives any payment. Among the ethical issues raised are the exploitation of poor women by wealthy couples. Financial need is often a factor (one of many) in a woman's decision to become a surrogate. Other issues include the amount of compensation that can ethically or legally be given to the surrogate and the ways in which this situation is distinguished from 'baby-selling.' A final issue concerns the available alternative of adoption, which is generally less expensive and raises fewer legal and ethical objections.

The drive to reproduce is a strong force among all species, including humans. Not all humans want to have children, but for many it is something they desire greatly. When some people seek out reproductive technologies to overcome infertility, it shows their motivation to have children that are genetically their own. In contrast, many people wish to limit the number of children that they have. In the next section we discuss the ways in which the timing and number of births can be controlled.

THOUGHT QUESTIONS

1. Why might it be an evolutionary advantage for males to produce many millions of sperm, but for females to produce very small numbers of eggs?

2. Why do we sometimes call testosterone the male hormone and estrogen the female hormone even though both are found in both sexes?

3. Is the brain a reproductive organ?

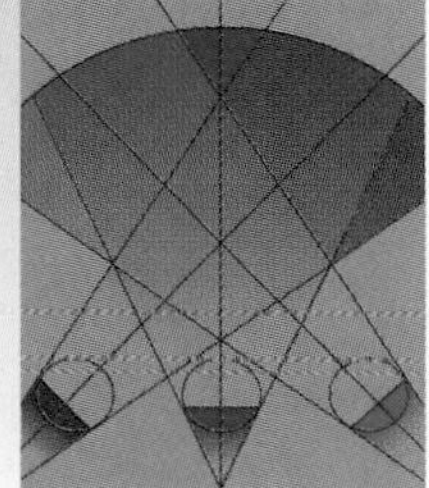

CAN WE DIMINISH POPULATION GROWTH AND ITS IMPACT?

As Malthus realized, many factors influence population growth. Controlling the death rate, as we have seen, increases population. Improving nutrition also generally increases population by increasing fertility and decreasing infant mortality. Deaths from accidents are being decreased by safety measures. To arrive at our carrying capacity without involuntarily increasing the death rate, there are few options available other than voluntarily decreasing the birth rate. As we will see, birth control has always been practiced among populations, but many more birth control options are available today.

We should make note of a distinction at this point: **population control** is usually understood to operate on the level of populations, while **birth control** mechanisms generally operate by preventing births one at a time. A birth control method is not a successful population control method unless it is widely adopted. As we have seen, individual motivation for wanting children is very strong.

Birth control methods allow the spacing and timing of the birth of children and so are also called family planning methods. The majority of these methods seek to control births by preventing pregnancy. Many prevent pregnancy by interfering with the reproductive anatomy or physiology of either the female or her male partner. For each method, a series of questions can be asked:

1. How does it work?
2. How effective is it in birth control?
3. What costs or risks are involved?
4. What kinds of objections have been raised against it?
5. Does it have any benefits apart from birth control, e.g., in the prevention of sexually transmitted disease?

Contraceptive measures

The various birth control methods form a spectrum of possibilities that are listed in Figure 6.13. Of these methods, those that act to prevent pregnancy before conception, that is, before the joining of an egg and a sperm, can be called **contraceptive** measures.

Sterilization methods. Sterilization, the elimination of reproductive capacity, usually involves surgery and is usually permanent, although some methods are potentially reversible. One method of permanent male sterilization is the removal of the testes (castration). The testes (see Figure 6.9) produce the male gametes or sperm, so the removal of the testes permanently prevents reproduction by that individual. The testes also produce the hormone testosterone, so that removal of the testes also has many other consequences, depending on the age of the individual. One way to achieve male sterilization but allow hormone secretion to continue is by vasectomy, the surgical cutting and tying off of the sperm duct (the vas deferens, Figure 6.9). Males with vasectomies continue to produce both testicular hormones and sperm for some time, but the sperm cannot reach the penis for release.

Among female sterilization methods, tubal ligation (tying off of the oviduct) is the only one done primarily as a birth control measure. Tubal ligation is analogous to male vasectomy; eggs and hormones are still produced, but the eggs are blocked from traveling to the uterus. Surgical removal of the uterus (hysterectomy) is performed for medical reasons other than birth control, but the removal of the uterus results in permanent sterility because the uterus is where the developing embryo grows. Surgical removal of the ovaries, the organs that produce the female gametes, also results in sterility, but this is usually avoided for the same reason as male castration is avoided: the ovaries produce many hormones and their removal has widespread effects on the individual.

All these sterilization methods involve surgery, which makes them expensive to implement on a very large scale, especially in poor or medically underserved areas. As with all forms of surgery, there are risks, such

FIGURE 6.13

Methods of birth control.

I. CONTRACEPTIVE METHODS	effectiveness*	relative cost
A. STERILIZATION METHODS		
1. irreversible sterilization: (castration, hysterectomy, or ovariectomy)	100%	high
2. semipermanent sterilization:		
tubal ligation	99.6%	high
vasectomy	99.6%	medium
B. ABSTINENCE METHODS		
3. long-term sexual abstinence: (celibacy, delayed marriage)	100%	none
4. timed abstinence ('rhythm' methods)	general use 76%; experienced users 98%	none
5. withdrawal during intercourse (coitus interruptus)	general use 77%; experienced users 84%	none
C. BARRIER METHODS		
6. condom	general use 90%; experienced users 98%	low
7. vaginal diaphragm with spermicide	general use 81%; experienced users 98%	low
8. cervical cap with spermicide	general use 87%; experienced users 98%	medium
9. sponge with spermicide	general use 80%; experienced users 98%	low
D. SPERMICIDAL METHODS		
10. spermicidal creams, foams, and jellies alone	general use 82%; experienced users 97%	low
E. HORMONAL METHODS		
11. estrogen alone	general use 98%; experienced users 99.5%	medium
12. estrogen plus progesterone	general use 98%; experienced users 99.5%	medium
13. progestin only (injectible or 'minipill')	general use 97.5%; experienced users 99%	medium
II. POST-FERTILIZATION METHODS		
14. post-coital pills	not known	medium
15. intrauterine devices (IUDs)	general use 95%; experienced users 98.5%	medium
16. abortion	100%	high

* A 99% rate of effectiveness means that 1% of couples using that method will become pregnant in a year of use.

as those of infection or from the use of anesthesia. However, both costs and risks are experienced on a one-time basis only and do not recur. All sterilization methods are completely effective as birth control methods without any further action on the part of the individual.

One of the greatest objections to all these methods are that they are permanent. Many people, even people who want birth control, do not want to become permanently sterile. In the United States, about 60% of men who undergo vasectomy later regret that they did so. Vasectomy and tubal ligation can in some cases be reversed, but success depends on microsurgical techniques, and reported success rates vary greatly. A new method for male sterilization has been developed in China, and is reversible under local anesthesia. In this method, a polyurethane elastomer is injected into the sperm duct, where it solidifies to form a plug that effectively blocks the passage of sperm.

Abstinence methods. Abstinence methods have the distinct advantage of being available to all people free of charge, but their success depends upon the determination of the people using them. Total voluntary abstinence (celibacy) has long been practiced as part of a regimen of religious devotion, but only by small numbers of people. Delayed marriage (with no other sexual activity) greatly reduces the birth rate, especially inasmuch as the years before age 30 are the most fertile period for a majority of women.

The so-called **rhythm method** is a form of partial abstinence, based upon the fact that a woman is fertile only for several days after ovulation, while the egg is in the oviduct or uterus; sexual intercourse performed at other times will generally not result in conception. Several versions of the rhythm method are practiced. The simplest version is a calendar method, in which a couple abstains from day 10 to day 17 (or, to be more certain, from day 7 to day 17) of a 28-day cycle in which the onset of menstrual bleeding is counted as day 1 (see Figure 6.11). Another version, with a higher success rate, is called the Billings method. In this method, the woman feels the mucus just inside her vagina with her fingers. Estrogen causes this mucus to become more slippery and elastic just before ovulation, when estrogen reaches a peak; after that, the secretion becomes scant and dry. A couple using this method abstains from intercourse from the onset of slippery mucus until 4 days after the peak day, but may have intercourse on the remaining dry days.

If practiced correctly, the rhythm method is highly effective, but its effectiveness depends on several conditions, including the regularity of a woman's menstrual cycles (this varies individually), the ability of the couple to keep a calendar and count the days without making a mistake, and the willingness of the couple to refrain from sex (or else to practice another method of birth control) during the woman's fertile period. The effectiveness of the rhythm method can be increased by monitoring the woman's vaginal temperature, since a rise in temperature indicates the time of ovulation more precisely.

Sexual intercourse is also called coitus. **Coitus interruptus** is the withdrawal of the penis before ejaculation occurs. Some couples have used this method effectively, but the majority find it unsatisfying or difficult to follow. Since some sperm can be released before ejaculation, coitus interruptus is not reliable. On a population-wide scale, it is generally not as effective as other methods.

Barrier methods. **Barrier methods** are those that impose a barrier to the passage of sperm. Most **condoms** are designed for males and cover the penis, but a condom that is worn by women has also been developed. The male condom is the oldest of the barrier methods. First developed in England, traditional condoms were constructed of animal membranes (usually sheep intestine) and were therefore considered a luxury item. The development of rubber and then latex made condoms more widely available and also more reliable. Condoms have the added advantage of protecting against AIDS and other sexually transmitted diseases.

Other barrier methods include such vaginal inserts as cervical caps, vaginal diaphragms, and sponges, all worn by women. Vaginal diaphragms must initially be individually fitted by a physician or other trained medical worker, and must be inserted correctly into the vagina before intercourse and left in place for several hours thereafter. When properly placed, vaginal diaphragms block the movement of sperm from the vagina to the uterus, thereby preventing their joining with the egg. (One study in Brazil reported a higher-than-usual success rate if the diaphragm was left in place nearly all the time, but this result remains to be confirmed in other populations.) Cervical caps are made to fit over the narrow portion (cervix) of the uterus, where they also block sperm.

Spermicidal agents, chemicals that can kill sperm, in creams, foams, jellies, or suppositories, are often used together with a barrier method; the combination of barrier plus spermicide is much more effective than either method used alone. One of the newest methods is a sponge impregnated with spermicidal fluid. (Spermicides should not, however, be used with many types of condoms; the spermicide partly dissolves the condom, making it ineffective as a barrier to sperm or to sexually transmitted diseases.)

Barrier methods used with spermicides have extremely low failure rates when used by people familiar with their proper use; most pregnancies occurring with barrier methods are the result of improper use. Barrier methods are widely used in many countries.

Hormonal administration methods. Several birth control methods depend upon alterations of the female reproductive cycle. Hormonal birth control methods take advantage of the feedback mechanisms shown in Figure 6.11. For example, estrogen and progesterone both inhibit the secretion of FSH, so that supplying these hormones (or a similar compound) prevents ovarian follicles from reaching maturity and releasing their eggs. The hormones can be given as birth control pills, as injections, as implants (such as Norplant) just under the skin, or as patches on the skin. Regardless of the method by which the drug is delivered, all hormonal methods work by preventing the egg from maturing and being released. Because hormones have many effects throughout the body, hormones used in birth control have many side effects, which include the possibility of medical problems such as blood clots. For this reason, medical supervision is recommended and hormonal methods require a prescription in most countries.

Early birth control pills contained estrogen alone, but progesterone was later added (producing the combination birth control pills) to reduce the levels of estrogen and also its side effects. The continuous levels of these hormones prevent the usual hormonal cycling from taking place. The expense and the requirement of obtaining a prescription limit the use of birth control pills in many populations. Some developing countries have made birth control pills available *without* prescription, to encourage their more widespread use, but cost is still a problem. Birth control pills have become a commonly used contraceptive method among the middle and upper classes in many countries. Newer types of birth control pills were developed in the 1980s. These include the 'minipill', which uses progestin (a progesterone-like compound) only.

One of the newest methods of birth control is a male contraceptive pill that uses a drug called gossypol, derived from cottonseed. This drug was first developed around 1970 in China and is said to be about 99% effective. Although it does not affect the hormone testosterone, gossypol does somehow interfere with sperm production. Tests of gossypol have reported some toxic side effects, so an effort is now being made to develop a synthetic substitute.

Delayed weaning. An older hormonal method of birth control is the practice of delayed weaning, which is common in many traditional societies in African countries. Children in Africa are almost always breast-fed, and many are not weaned until they are 4–6 years old. While a woman is breast-feeding, she is producing hormones that stimulate milk production. These same hormones also inhibit the rise and fall of the hormones produced by the ovary, thus interrupting the menstrual cycle and preventing egg maturation. African women who use this method of birth control do not wean their youngest child from the breast until they feel they are ready to have another child. This method of birth spacing is a widespread and seemingly effective practice in many parts of Africa, although studies have shown that it is unreliable among women living in the industrialized world at high caloric intake levels. For example, prolonged breast-feeding has a contraceptive effect among the !Kung San (Bushmen) of South Africa and Namibia. These women have a low caloric intake and walk 4–6 miles a day, conditions that seldom occur among North American women. Among Hutterite women (belonging to a Protestant sect of mostly farming communities in the north central United States and Canada), breast-feeding has been shown to have a delaying effect on the interval between pregnancies, but the effect is less than among African women. Because the ability of breast-feeding to delay the return of the menstrual cycle is greatest among women who are physically active but who have a low caloric intake, improvements in maternal nutrition may actually decrease the effectiveness of delayed weaning as a method of birth control. Birth spacing by pronged breast-feeding is also most effective when babies suck vigorously and often. Any contribution to infant nutrition other than breast milk (e.g., by bottled milk or cereal) reduces the effect. In most third-world countries, the use of bottled milk reduces the effectiveness of birth spacing by delayed weaning. The closer spacing of births leads to an increase in birth rates at the population level, possibly offsetting the effects of birth control programs.

Post-fertilization birth control methods

Another category of birth control methods work after fertilization of the egg.

Hormonal methods. The morning-after pill is an example; it hormonally prevents implantation if taken within 72 hr after intercourse. One of the newest is a drug called mifepristone (or RU-486), developed in France and now available also in Great Britain, Sweden, and China. In most countries, it requires a prescription, but school nurses in France can now dispense this drug to teenage girls upon request. In the United States, clinical testing of mifepristone, completed in 1995, found the drug safe and effective in preventing pregnancy. The drug was recommended for approval by the Food and Drug Administration (FDA), but is still not widely available in the United States. Mifepristone blocks the action of the hormone progesterone, which is necessary for maintenance of a pregnancy. RU-486 has its greatest potential use as a morning-after pill to prevent implantation from occurring during the first 5 days after intercourse.

Prevention of implantation. An **intrauterine device (IUD)** is a small piece of plastic or wire, in one of several shapes (e.g., a loop or a coil), that is inserted by a physician into a woman's uterus, where it remains until removed by the physician. IUDs prevent pregnancy by preventing implantation, although the exact mechanism by which this occurs is not known. (Desert Bedouins have long practiced a similar method of birth control on their camels by inserting stones into the uteri of female camels to prevent pregnancy and removing the stone when breeding was again wanted.) A major advantage to this method is that, once inserted, the IUD works on its own with no need of further action on the part of the woman. The use of IUDs trails far behind the use of birth control pills and even sterilizations, and the method should not be used by women who have never been pregnant. However, women who use IUDs have a higher rate of satisfaction than with any other form of birth control, including pills. There are some possible side effects with IUDs, such as the possibility of uterine bleeding, but the method is effective and the failure rate is low. Earlier IUDs were not as safe; one early model, the Dalkon Shield, caused infections and severe bleeding problems in many women, resulting in hysterectomies, large lawsuits, and the corporate bankruptcy of its manufacturer. Many people have shown renewed interest in IUDs in the last decade because of the development of newer, safer types. About 50 million IUDs are used worldwide. The steroid-releasing Progestasert and a version of the copper-T are the most-used IUDs in the United States.

Abortion. The termination of a pregnancy, including the cleaning out of the uterine lining and the expulsion of the embryo or fetus, is called **abortion**. The traditional method of dilation and curettage (enlarging the cervix, then scraping out the uterine interior with a spoonlike instrument) has now been supplemented by newer techniques such as vacuum aspiration (using a machine that uses suction to clean out the uterine contents). There are medical risks to the woman, including excessive bleeding, the chances of infection, and uterine injury, which can result in sterility. The sum total of risks to the life and health of the woman is less than the sum total of risks associated with completing the pregnancy and giving birth. The medical risks are especially low if the abortion is done early, during the first trimester, meaning the first three-month portion of a nine-month pregnancy. Second-trimester abortions using saline injections can also be done safely in most cases. At whatever time an abortion is performed, the risks are much higher if it is done by an untrained person. The drug mifepristone, previously mentioned as a morning-after pill, can also be used to induce abortion after the embryo has become implanted in the uterus. Successful abortion has also been achieved in 96% of a group of 178 women given two drugs (methotrexate, followed days later by misoprostol) that are already legally available (by prescription) in the United States.

Of course, birth control methods can be combined, and a later-acting method can be used if an earlier-acting method fails in a particular case. Most population planners have advocated that abortion be made a widely available option as a backup when the other, less costly methods have failed.

Infanticide. Though not technically a means of birth control, **infanticide** has long been practiced as a means of population control in many parts of the world, especially in times of famine. There are records of infanticide from Medieval Europe, and the practice was still widespread in China, India, and other parts of Asia well into the twentieth century. In most cases, the infant is not directly killed, but is instead allowed to die through lack of care. In societies in which infanticide is practiced, female infanticide is more common than male infanticide. In China, infanticide is now officially outlawed. However, the government's strict population control policy allows only one child per couple (Figure 6.14), and social scientists suspect that female infanticide is still widely practiced by couples who want a boy but instead have a girl.

Cultural and ethical opposition to birth control

No single method of population control is best for all societies. Abortions and sterilizations, for example, require medically trained personnel. They are more expensive and more labor-intensive than other methods, and are therefore unlikely to become the methods most widely used even among populations that have no objections to them. All methods need to be adapted to the customs of the people using them, and education in the use of certain methods will meet with resistance of various kinds. For instance, women in many Muslim societies are generally forbidden to discuss reproductive matters with anybody outside their families, including health care workers.

The attitudes of the Catholic church toward birth control have varied over the centuries; official opposition to most forms of birth control is historically recent. A considerable debate about birth control took place within the Catholic hierarchy in the 1960s, resulting in two Papal encyclicals, *Populorum progressio* (1967) and *Humanae vitae* (1968). The first of these acknowledges the population problem and the need for family planning in underdeveloped areas; the second denounces abortion, sterilization, and all forms of birth control except for the rhythm method. Surveys in many countries show that a large majority of Catholics use various forms of birth control despite the Church's official position. Attempts to spread birth control information have often been opposed by the Church, especially in Latin America, but Church teachings have not stopped Italy, a country over 98% Catholic, from achieving a stable (nongrowing) population, with one of the lowest birth rates in Europe.

Other religious groups have generally been more tolerant of contraceptive methods. Abortion, however, is opposed by Catholics, Protestant fundamentalists, Orthodox Jews, and Muslims. In its simplest terms, the principal argument voiced by these groups is that a fetus is a living human being and that killing it is an act of murder. People who wish to keep abortion legally available have argued several major points, including a woman's right to choose, a child's right to be wanted by his or her parents, and the need to control the world's population. This is a highly charged issue.

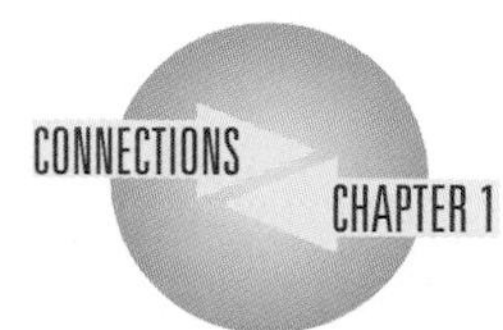

The abortion debate

Laws on abortion vary greatly from place to place and sometimes from one time period to another. In the United States, many states outlawed abortions until the Supreme Court ruling in *Roe v. Wade*. As a result of this court opinion, women in the United States have a legally recognized right to an abortion under certain conditions. During the first trimester of pregnancy, this right can be exercised by the woman in consultation with her physician, without any interference from state laws. During the second trimester of pregnancy, state governments can impose waiting periods or certain other conditions, and during the third trimester they can limit abortions more strictly or outlaw them entirely. As a result, abortion practices vary from state to state. Practices also vary elsewhere: Ireland and most Moslem countries outlaw abortions, whereas most other European and Asian countries permit them.

FIGURE 6.14

A poster promoting birth control in China. The poster reads, "birth control benefits the nation and benefits the people."

Many of the questions usually raised in the course of the abortion debate revolve around matters of definition: When does 'life' begin? Is a fetus a 'person'? Is it a 'human being'?

Biological definitions of life. The usual definitions of 'life' (see Chapter 1, p. 9) mention properties such as motility, metabolism (internal chemical reactions), homeostasis (the ability to maintain certain conditions), irritability (the ability to respond to stimuli), and the presence of genetic material that is inherited. By these criteria, a fetus has the characteristics of life as long as it remains in the womb, but it will quickly lose many of these characteristics if removed. Although the fetus is 'alive,' it is alive in the same way that the fertilized egg is alive, or the unfertilized egg, or the sperm, or, for that matter, the appendix or the gall bladder. Opponents of abortion may argue that the fetus is 'alive,' and that abortion is therefore a form of murder. A pro-choice counter-argument might be that an organ such as the appendix is also 'alive,' and that an abortion is therefore the moral equivalent of an appendectomy.

Legal definitions of personhood. Is the fetus a person in a way that the appendix is not? 'Personhood' is a social or legal concept, not a biological one, and it is differently defined in each culture or legal system. In the legal system followed in the English-speaking world, a 'person' is defined as a legal entity having certain legally recognized rights and duties. In this tradition, corporations and estates have the legal rights of 'persons.' The 'personhood' of a fetus is therefore a matter of legal definition and not of biology, and, because legislators can define personhood in various ways, the legal rights of a fetus can vary from one jurisdiction to another. Other cultures have their own ways of defining the rights or personhood of a fetus or newborn:

1. in Japanese tradition, a baby is not considered a person until it utters its first cry, so killing it is not considered homicide;
2. in parts of West Africa, a child is not considered human until it is a week old;
3. the Ayatal aborigines of Formosa had no punishment for killing a child before it was given a name at age two or three years;
4. natives of the Pacific island of Truk considered deformed infants to be ghosts and either burned or drowned them.

The point is that different cultures and different legal systems can reach remarkably different conclusions. Notice again that these are really not scientific questions, capable of falsification.

Biological definitions of humanness. To think about whether a fetus is human, we must first decide what we mean by 'human.' If we mean anything that possesses human genetic material in the form of DNA, then a fetus is definitely human, but so are white blood cells, haploid sperm and egg cells, and the many organs such as the appendix that are removed and discarded each year as 'medical waste.' If we distinguish the fetus from these others as being 'potentially' capable of forming a human life capable of independent existence, then we need to examine what we mean by 'potential' and when an 'independent existence' begins.

Can biology help us define 'humanness'? One book, *The Facts of Life*, by Harold Morowitz and James Trefil (New York, Oxford University Press, 1992) examines several possible definitions and argues that all of them point to an 'acquisition of humanness' at around 25 weeks of age, the time at which most of the connections between nerve cells in the cerebral cortex are made. Electrical waves in the brain (EEGs) (see Chapter 10, pp. 441–442) begin at about this time, providing evidence that the fetus can respond to certain stimuli and can be described as having 'experiences.' Only after the 25th week are the lungs sufficiently well endowed with a vascular blood supply to allow the baby's tissues to receive enough oxygen. Before the 25th week, the lungs tend to collapse

when empty, and their internal mucous linings tend to stick together, preventing the lung from refilling. Morowitz and Trefil also place the upper limit of abortion safety at around the 25th week, which they say is also close to the current technological limit on the minimum age of 'viability.' The survival of premature babies is 80% or higher beginning with the 26th week of gestation, but 50% or less in the 25th week or earlier because of the underdeveloped state of the lungs and brain. A medical study published in 1993 by M.C. Allen and colleagues confirms low rates of survival before 24 weeks of age. According to this study, 69% of infants born at 25 weeks of age, but only 21% of infants born at 24 weeks of age, survive without severe abnormalities. According to the U.S. Centers for Disease Control and Prevention, only 1% of abortions are performed past week 21, and only a small fraction of these beyond the 24th week.

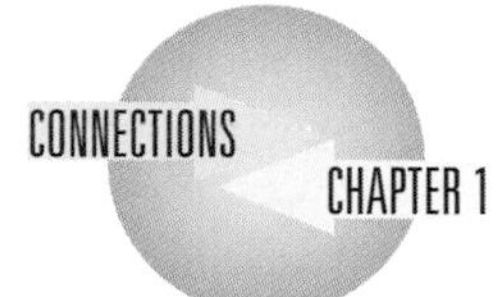

Ethical considerations. Using the ethical principles discussed in Chapter 1, a deontologist who opposed abortions would simply argue that it is a wrongful act regardless of any type of medical or other evidence. Such a person would regard the matter as totally outside the bounds of science, because no possible observation or experimental evidence could change the wrongness of what they regard as a wrongful act. A utilitarian would weigh the possible consequences of an abortion (including the monetary costs and the medical risks) against the consequences of the birth if carried to term. Among the latter consequences are medical risks to the mother's health from the delivery, medical expenses, and the costs to the mother of raising the child (or the costs to society, if for some reason the mother does not raise the child properly). From a utilitarian viewpoint, the mother's wishes, abilities, and financial circumstances are all important in the evaluation; to the deontologist, they are all irrelevant. Much of the frustration of the whole abortion debate is that deontologists and utilitarians talk in such different terms and use such different arguments that neither has much hope of convincing the other of anything.

In the meantime, a small number of anti-abortion extremists have resorted to bombing clinics and terrorizing or shooting doctors who perform abortions. Many hospitals have discontinued performing abortions to avoid being the targets of such intimidation. One result is the increasing concentration of abortions performed at clinics that do little else, making them easier targets for these extremists. Other results are that fewer doctors are being trained to perform abortions, while many other doctors who might otherwise perform abortions are refusing to do so for reasons of personal safety. These developments further reduce the pool of doctors and hospitals willing to help someone seeking an abortion. In some states it has become increasingly difficult for a woman seeking an abortion to find a doctor willing to perform one.

Population control movements

Organizations to promote birth control and control population growth were first formed in the nineteenth century. In England, several of these organizations called themselves 'Malthusian,' even though Malthus, a curate in the Anglican Church, was opposed to nearly all of the birth control methods available in his day. Knowledge about reproduction and birth control was not widely available before the mid-twentieth century. An 1832 book on the subject by an American physician, Charles Knowlton, was banned as immoral in the United States and elsewhere. Two British reformers, Charles Bradlaugh and Annie Besant, and an American, Margaret Sanger, worked through the late 1800s and early 1900s to make information and contraceptives more widely available. Besant strongly influenced Mahatma Gandhi and later migrated to India. Sanger always viewed birth control from the perspective of giving individual women more control over their own lives. The availability of birth control in the United States owes much to her tireless campaigns. She also traveled widely, spreading the message of birth control to India, China, and Japan.

The influence of Gandhi, Besant, and Sanger led India to become the first modern nation to institute a government-funded campaign to control its population. Beginning in 1951, India implemented population control measures that featured easy access to both contraception and abortion and an information campaign to encourage their widespread voluntary use. During the 1960s, nine other nations, including China, Egypt, and Pakistan, also implemented population control programs.

The most ambitious population control program in history was adopted in China in 1962, at a time when their population was around 700 million and was growing at just above a 2% annual rate. The campaign for "only one child for one family" (see Figure 6.14) was waged with special vigor, and parents who bore more children than this were fined and sometimes imprisoned. The goal of this campaign was not just to limit population growth, but to reduce the population to its 1962 level of 700 million as quickly as possible. Because of a large demographic momentum (that is, an age structure with many children), China's officials realized that they would have to cut the birth rate to somewhat *below* the mortality rate for a time in order to achieve a stable population. China's 1990 population was around 1.1 billion, but the annual growth rate has been cut to 1.45%.

The education of women

One of the most effective methods of reducing the number of children borne by each woman is to educate women. Studies in many parts of the world have shown that the population birth rate falls with each rise in the education level of women, even in the absence of any program aimed specifically at birth control (Figure 6.15). Around the world, the countries with the lowest rates of female literacy also have the highest population growth rates, whereas those with higher female literacy have lower growth rates. Countries with female literacy rates below 15% include Mali, with a population growth rate of 3.17%, and Yemen and Afghanistan, both with population growth rates exceeding 5.0%. Third-world countries with female literacy rates above 80% include Colombia, with a population growth rate of 1.88%, and Jamaica, Sri Lanka, and Thailand, all with population growth rates below 1.0%.

Third-world women with a seventh-grade education or higher tend to marry later than other women (four years later, on average, and these are among the most fertile years). They also use voluntary means of birth control more often, have fewer (and healthier) children, and suffer far less often from either maternal or infant mortality in childbirth. The empowering of women (by giving them more education and more control over their reproductive lives) also raises the educational level of their children and results in more rapid economic development (at lower cost) than many other programs aimed more specifically at development.

In the United States, educational efforts are also an important part of most efforts to reduce pregnancy rates in teenagers. The rate of teenage pregnancy is lowest among women with the most years of schooling and highest among those with only a grade-school education or less.

Although most population control programs are carried out by national governments on their own populations, several programs are international in scope, including those run by the World Health Organization (a branch of the United Nations) and by the U.S. Agency for International Development (USAID). The United Nations has sponsored many conferences on population. Some past conferences emphasized the environmental impacts of population growth, but the 1994 conference in Cairo, Egypt, shifted the attention to the education of women. Women's rights advocates from a variety of countries stressed the need to improve both the education and legal status of women. Many people cautioned that overzealous government-sponsored programs aimed at population control could restrict the reproductive freedom of individual women as much as the earlier lack of birth control information. Although these people generally see the need to reduce the rate of population growth, they are deeply suspicious of programs that coerce individual women or restrict their freedom, and they are especially suspicious of programs urged upon third-world nations by male-dominated institutions in the industrial world. Instead, they favor programs to educate women and improve their legal status and reproductive choices. They point out that

FIGURE 6.15

The education of women reduces the average number of children per family. Data for the graph are from the World Bank.

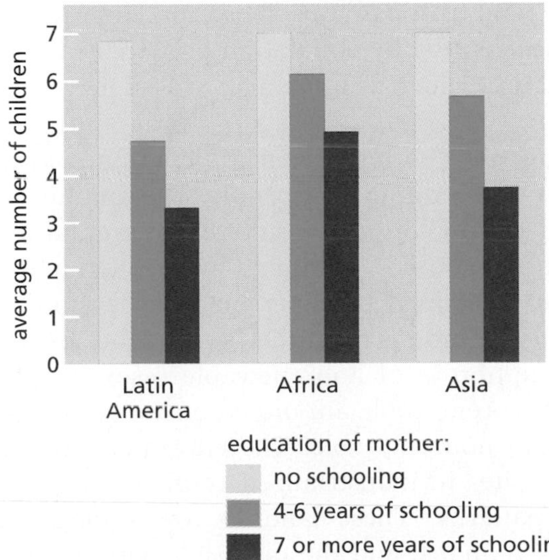

(B) Teenage women in a Papua New Guinea classroom

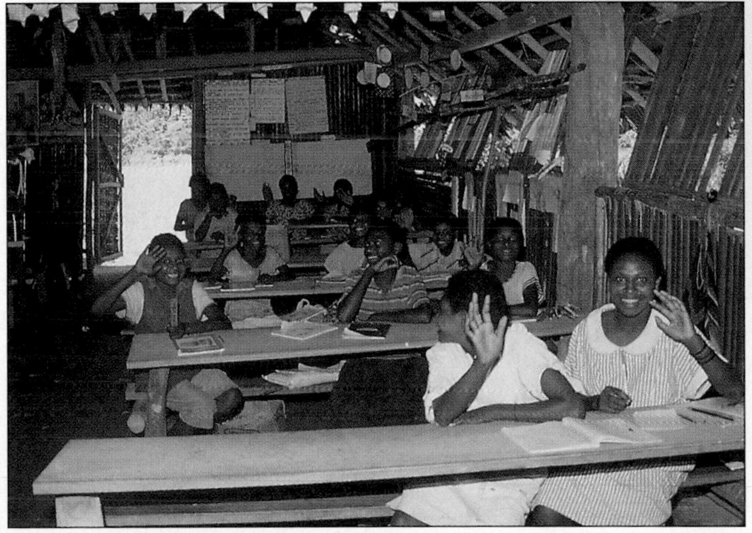

birth rates have diminished whenever the education, legal rights, and reproductive choices of women have improved.

Controlling population impact

Demographic transition brings about a marked increase in human population. In most parts of the world, the excess population tends to migrate to the cities, producing an overcrowding that strains the resources of those urban areas. In Europe and Japan, this process occurred gradually over a period of several hundred years (roughly, 1600 to 1900), giving cities a chance to adjust to their changing conditions. Many cities have accommodated high population densities (meaning large numbers of people per square mile) without widespread misery. Since World War II, urbanization in the third world has taken place much more rapidly than it did in Europe. Rapid, unplanned growth has strained most urban services to the point at which many of the newly arrived migrants have inadequate housing. Crowded slum areas or shanty towns often lack safe drinking water and may also suffer from chronic water shortage; sanitation and waste disposal are also frequent problems. It is often not crowding in itself that results in these problems, but crowding without sufficient facilities to support the population. Crime often increases and may become difficult to control, although many other factors beside population contribute to crime rates. Unemployment and economic hardship, when they occur, may compound these problems. The hardships of urban crowding usually fall disproportionately on the poor.

Pollution (see Chapter 15) tends to increase approximately in proportion to population, most obviously because of corresponding increases in household garbage and waste water. Densely populated areas are dependent on food, water and fuel coming in from a much wider radius; consequently their environmental impact is felt far beyond their political borders. As population increases, more forests are cleared for agricultural use and more trees are cut down to build houses. The destruction of habitat for other organisms, particularly of forests, is one result (see Chapter 15). The loss of arable land (through topsoil erosion, desertification, and other processes) is accelerated by population growth, as is the depletion of nonrenewable resources such as minerals or fossil fuels.

Effects of consumption patterns. The impact on the environment is not, however, solely a function of the number of people. The amount of the world's resources that each person consumes is not equal around the globe. On average, the amount of resources consumed by a person in the United States is 54 times that consumed by a person in a developing country. The impact of this consumption is magnified still further by the fact that much of this consumption is of nonrenewable resources. In addition, resources that might be renewable are often consumed or discarded in ways that make them nonrenewable. The enormous size of municipal solid waste disposal sites in industrialized countries is testament to these consumption patterns. These landfills are among the largest structures ever built by humans, and the materials within them are unavailable for reuse or biodegradation.

If the rate of energy use does not exceed the rate at which energy is captured from the sun in photosynthesis (see Chapter 14), then the energy use is considered **sustainable**. In many industrialized countries, however, present patterns of energy consumption are already unsustainable because they remove far more energy from global ecosystems than they produce.

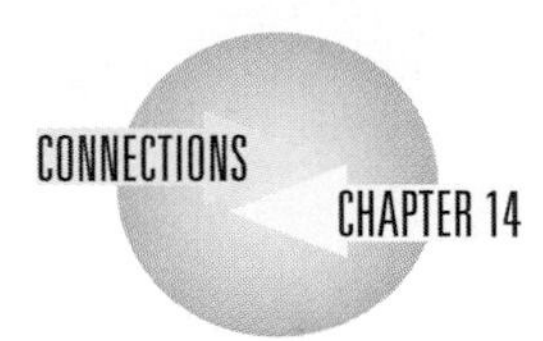

Discussions of the world's population crisis frequently become linked to discussions of the environmental crisis. Many people, especially in the third world, believe that the population crisis is only a small part of a greater environmental crisis. This environmental crisis, they say, is made worse by the industrial world's overconsumption more than by the third world's population increase. Frances Moore Lappé is one of several American writers holding such views. Some analysts even question whether the industrial world's concerns over population are misdirected (and possibly racist). Third-world countries, they say, could well support far larger populations than they do now if it were not for the export of so many of their resources to support the patterns of overconsumption that have become so typical of the industrial world. If the industrial countries, they say, were to give up their lavish patterns of consumption, then the third world could well support a larger human population (at a larger carrying capacity) than it does in current circumstances.

Others take what may be called a neo-Malthusian position. Paul Ehrlich, for example, views most other problems as consequences of overpopulation. If the population were smaller, he argues, most environmental problems would diminish or even disappear. Because some countries (mostly in Europe) have already limited their population growth, the greatest efforts should be directed at those nations (mostly in the third world; see Figure 6.2) that have the highest population growth rates.

Overpopulation and overconsumption need not be viewed as opposing viewpoints. Population growth and profligate consumption are both widely recognized as problems, and each makes the other worse. Some people see one of these as the bigger problem; some people see the other. Efforts directed at addressing either problem can only help to ameliorate both.

Limits on carrying capacity. Many scientists tell us that we will soon reach or even exceed the carrying capacity of the planet. In fact, this is one point on which people concerned with overpopulation and those concerned with overconsumption agree, although they postulate different causes for this condition. One of the few dissenters, economist Julian Simon, observes that the technological revolutions of past centuries have repeatedly brought about demographic shifts, each of which has increased the carrying capacity. He predicts that future technological revolutions will continue to enlarge the planet's carrying capacity indefinitely. Nearly all other scientists and writers who have contemplated the subject of population believe instead that the planet's carrying capacity has a limit.

Can the global carrying capacity be increased further? The answer is not known with certainty, but it depends in part on whether we assume the Earth's natural resources to be renewable and unlimited (Julian

Simon's view) or limited and nonrenewable (the majority viewpoint). Those who accept the limits imposed by nonrenewable resources will be driven to the conclusion that carrying capacity cannot be increased very much. In fact, if we maintain our present patterns of consumption, we may not even be able to sustain the present population levels forever.

Human populations, like all other populations, are biological entities, requiring energy flow to survive. Populations are therefore subject to the laws of physics (energy is neither created nor destroyed), and populations cannot exceed the limits imposed by the availability of energy. When they approach K, populations are controlled by biological factors, such as starvation and disease. So the question "Should populations be controlled?" is academic because populations will be controlled by the forces of biology and physics, regardless of our answer. The relevant questions are, "Should we exercise preventive efforts at population control?" and, if we should, "How should we do so?"

THOUGHT QUESTIONS

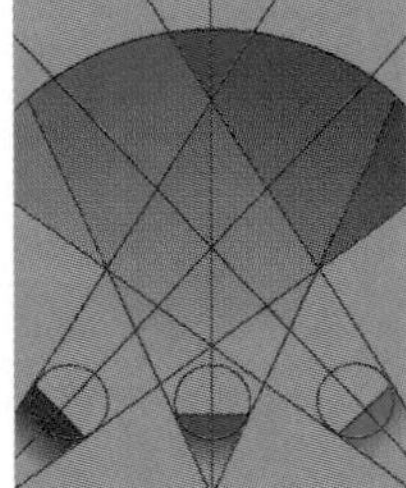

1. Is an increase in population the only factor that puts resources in limited supply? Does population growth affect the availability of housing or medical care in the same way that it affects the availability of drinking water, sanitation controls, and food?
2. Find out whether your college makes birth control information and birth control itself available to the student population. Are certain methods favored over others? Why?
3. Why are most forms of birth control aimed at women rather than men? Why are most population control campaigns aimed at women? In societies in which women have little or no control over their lives, are they likely to be able to carry out family planning? Will family planning give them greater control?
4. Do you think it is proper to view birth control information as a freedom-of-speech issue? Do you think birth control methods should be taught in the public schools? Why or why not?
5. What social benefits are likely if population growth is controlled? What social and ethical problems need to be considered? What individual rights are at risk? In your opinion, what is the best way for a government to control its country's population growth without restricting the reproductive freedom of its women?
6. In what countries is abortion legal? Where is abortion illegal? Can you suggest reasons for these differences?
7. Can biology have any useful role in the abortion debate? What role? Would any biological data be persuasive to a person who opposed abortion on the basis of deontological principles? Would data be persuasive to someone using utilitarian ethics?
8. Most stem cells (useful in cancer research or cancer therapies) come from discarded embryos and aborted fetuses. Fertilized eggs are the ultimate stem cells. However, U.S. law prohibits the National Institutes of Health (NIH) from funding any experiments that deliberately create or destroy a human embryo. How does this restriction on research relate to the abortion debate in the United States? Do you agree with the restriction? If you do not, how would you change the law?
9. In Practice Question 15 on p. 271, what considerations have been ignored? Do you think they may safely be ignored?

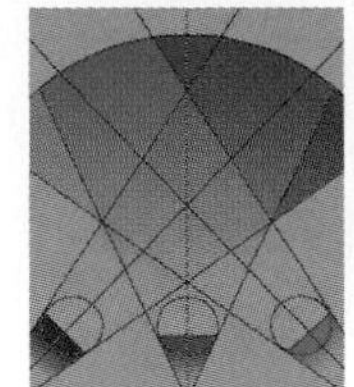

INDIVIDUAL DECISION MAKING IN FAMILY PLANNING IS sometimes at odds with government decisions aimed at population control and also sometimes with various religious teachings. There are also many other reasons why people might resent strangers urging them to modify their most personal behaviors in one way or another. All of these factors, moreover, vary from place to place, and the lessons learned in one country or population cannot necessarily be applied uncritically to other populations elsewhere. However, any attempts to implement change based on biological data will necessarily take place in a context of many, often competing, social values. Many scientists think that moving to sustainable consumption levels may partly alleviate the problems of population growth, but only temporarily. In addition, motivating people to decrease their consumption may be just as difficult as motivating them to have fewer children. Even well-planned efforts to address social, economic, or environmental problems may prove to be inadequate as resources are stretched to the breaking point in the face of increasing population pressure. "Whatever your cause," says one slogan, "it's a lost cause unless we can control population."

CHAPTER SUMMARY

- The principles of **population ecology** are applicable to all species.
- A new **population** shows rapid **exponential growth** at first, but its **growth rate** levels off when it approaches the **carrying capacity** of its environment, a phenomenon called **logistic growth**.
- Human populations have grown markedly after each major advance in technology. Each major population increase has taken the form of a **demographic transition**, beginning with declining mortality and ending when the **birth rate** declines to match the **death rate**.
- Understanding reproductive biology can help us to treat infertility and also to control the birth rate.
- Many methods of **birth control** are available. They differ in their biological mechanisms, their costs, their medical risks, and their acceptance by different groups of people.
- Many studies have found that improvements in the education and legal status of women lowers the birth rate in a cost-effective manner and brings other benefits besides.
- The effects of population growth beyond the carrying capacity are numerous: consumption of resources is increased, pollution is increased and any inefficiency in the utilization of resources results in starvation and death.

KEY TERMS TO KNOW

abortion (p. 259)
age pyramid (p. 242)
birth control (p. 254)
birth rate (p. 234)
carrying capacity (*K*) (p. 239)
contraceptive (p. 254)
death control (p. 241)
death rate (p. 234)
demographic momentum (p. 244)
demographic transition (p. 241)
demography (p. 233)
doubling time (p. 237)
exponential growth (p. 237)
growth rate (*r*) (p. 234)
logistic growth (p. 239)
ovulation (p. 249)
population (p. 233)
population ecology (p. 233)

CONNECTIONS TO OTHER CHAPTERS

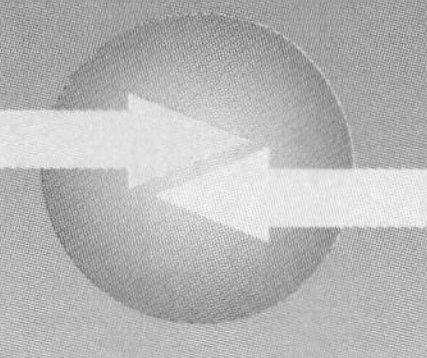

Chapter 1: Abortion and other forms of birth control raise important ethical issues.

Chapter 4: Population size responds to evolutionary forces, such as competition from other species.

Chapter 7: Different species have different reproductive strategies that are either *r*-selected or *K*-selected. Reproductive strategies control birth rates.

Chapter 8: Undernutrition and malnutrition increase the death rate.

Chapter 12: Overcrowded conditions cause stress in many species.

Chapter 13: AIDS has greatly increased the death rate in many countries.

Chapter 14: Plants are the main biological energy producers on which the life of humans and other consumer organisms depends. Populations cannot be sustained above levels that can be supported by producer organisms.

Chapter 15: Increasing population size usually makes pollution problems worse. Needs for water and sewage treatment increase with population size.

PRACTICE QUESTIONS

1. What is the birth rate B in a nation with 4 million men and 3.9 million women, in which 200,000 children are born in one year? What is the birth rate B in a nation with 3 million men and 4.9 million women, in which 200,000 children are born in one year?
2. What is the death rate D in a nation of 100 million people in which 500,000 people died in a year and 200,000 of those who died were children below the age of 2 years? What is the death rate D in a nation of 100 million people in which 500,000 people died in a year and 50,000 of those who died were children below the age of 2 years?
3. What is the growth rate r in a nation of 500 million people if the birth rate is 2% and the death rate is 1% (assuming no immigration or emigration)? What is the growth rate r in a nation of 50 million people if the birth rate is 2% and the death rate is 1% (assuming no immigration or emigration)?
4. How many people will be added in one year to a population of 500 million people with a growth rate of 2%? How many will be added if the growth rate is 4%? How many will be added at growth rates of 2% or 4% if the initial population was 5 billion?
5. What is the doubling time of a population of 500 million people if the growth rate is 2%? What is the doubling time of a population of 5 billion people if the growth rate is 2%?
6. If a population is at its carrying capacity K of 500 million and its birth rate is 3%, what is its death rate (assuming no immigration or emigration)? If a population is at its carrying capacity K of 5 billion and its birth rate is 3%, what is its death rate (again assuming no immigration or emigration)?
7. A population of 4.5 million has a birth rate of 0.067 (or 6.7%) and a death rate of 0.024 (2.4%). Find the growth rate (r) and the number of years that it will take for the population to double.
8. For the population in the previous question, find:

   a. the population increase this year;

   b. the size of the population after a year of increase;

   c. the population increase in the second year;

   d. the population size after two years of increase.
9. Which will increase more rapidly this year: a population of 3.3 million with a growth rate of $r = 2.0\%$, or a population of 5 million with a growth rate of 1.1%?
10. Where in the body is testosterone produced? Where in the body is estrogen produced?
11. During the menstrual cycle, what hormones are secreted by the ovaries? Other hormones are involved, in addition to the ovarian hormones. Where in the body are these other hormones produced?
12. How do hormonal contraceptive methods induce infertility?
13. How do barrier contraceptives prevent fertilization?
14. A particular birth control method is 98% effective. In a nation of 8 million people, if 400,000 women use this method, how many of them will become pregnant this year?
15. Suppose you are working for a family-planning program in a nation whose population is growing rapidly. Among several available methods of birth control are the following.

    Method A costs $1600 per woman, lasts an average of 20 years, and is 99% effective.

    Method B costs $20 per month for each woman and is 78% effective.

    Method C costs $200 per year for each woman and is 88% effective.

    Which method is most cost-effective (reduces pregnancies by the largest amount per $1000 spent)?
    (See also Thought Question 10 on p. 268.)

7 Sociobiology

CHAPTER OUTLINE

Sociobiology Deals With Social Behavior
- Learned and inherited behavior
- The paradigm of sociobiology
- Research methods in sociobiology
- Instincts

Social Organization Is Adaptive
- The biological advantages of social groups
- Simple forms of social organization
- Altruism: an evolutionary puzzle
- The evolution of eusociality

Reproductive Strategies Can Alter Fitness
- Asexual versus sexual reproduction
- *r*-selection and *K*-selection
- Differences between the sexes
- Mating systems

Primate Sociobiology Presents Added Complexities
- Primate social behavior and its development
- Social organization among primates
- Reproductive strategies among primates
- Some examples of human behaviors

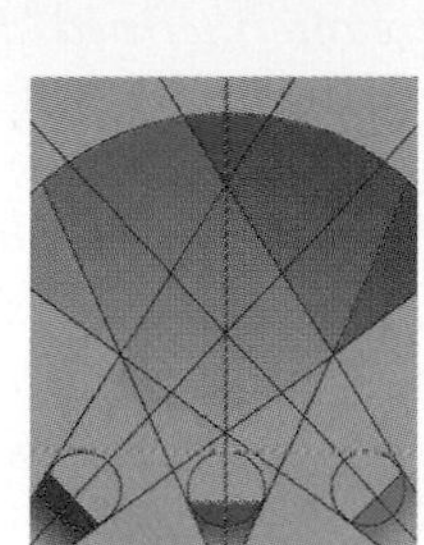

ISSUES

What is sociobiology? How is it different from sociology?

Why does sociobiology have so many critics? Who is objecting, and what are they objecting to?

Is most behavior learned or inherited?

What are the differences between instincts and other innate behaviors?

How do learned behaviors relate to evolutionary change?

To what extent can social behavior be modified?

Why do the sexes behave differently in so many species?

How different are humans from other species in social behavior? To what extent can findings in other species be extrapolated to humans?

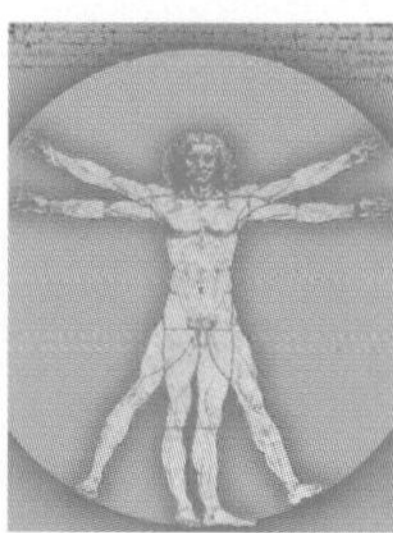

BIOLOGICAL CONCEPTS

Evolutionary change (variation, natural selection, nonrandom mating, specialization and adaptation, human evolution)

Population ecology (populations, regulation of population size)

Learning and instinct (interaction of genotype and environment)

Reproduction (asexual reproduction, sexual reproduction, mating systems, sexual dimorphism, reproductive strategies)

Behavior (social behavior, communication, courtship and mating)

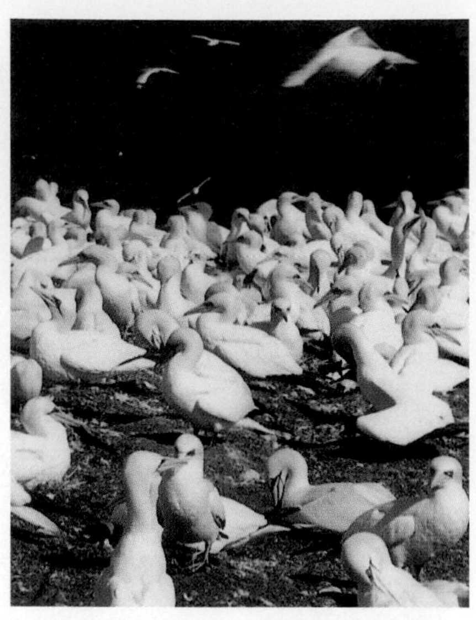

Behavior that influences the behavior of other individuals of the same species is called ***social behavior****. Examples of social behavior in animals include cooperative feeding, cooperative defense, aggression within the species, courtship, mating, and various forms of parental care. People also practice many forms of social behavior: nurturing their young, helping their neighbors, defending their possessions, and providing both material help and emotional support to their loved ones and to others. The population crisis discussed in Chapter 6 is a direct result of reproductive behavior. Some types of social behavior are often termed 'antisocial' behavior and result in problems for society; examples include violence, crime, racist acts, sexist acts, and child abuse and neglect. All of the above are social behaviors because they affect the behavior of other individuals.* ***Sociobiology*** *is the comparative study of social behaviors and social groupings among different species. The study of social behaviors in complex human societies is a separate discipline called sociology.*

Can behaviors that cause problems be changed easily? Can beneficial behaviors (however defined or recognized) be substituted for destructive behaviors? Is behavior in general, and human behavior in particular, rigid and unchangeable, or plastic and easily molded? Are we governed more strongly by our genetic background (nature) or by our upbringing (nurture)? The debate is at least as old as Shakespeare's Tempest *(4:1), in which Caliban is characterized as "a born devil on whose nature/Nurture can never stick." If human behavior were strongly determined by genes, then cultural influences, including education and training, would have only limited power to bring about changes in human behavior. Social reformers of all kinds usually support the opposite viewpoint, that human behavior can be modified almost at will, subject to few if any restrictions. Debates about alcoholism or homosexuality are often unproductive because some people assume that these are behaviors that could easily and voluntarily be changed, while others assume that these are permanent and deeply rooted in biological differences that may or may not be genetic. Differences in behavior between the sexes are likewise seen by some researchers as genetically constrained and by others as culturally controlled and easily changeable.*

We have spoken of 'human' behaviors, which provoke the most heated discussions. However, sociobiology is a broad field of study and humans are but a single species. Most people doing research in sociobiology focus on nonhuman animals in their research. Altruism, for example, poses a major research question in the sociobiology of all species. Among other broad-spectrum issues within sociobiology are the advantages of sociality itself, the kind of social organization found in each species, and the manner in which it evolved. Another issue is that of social relations between the sexes of each species, including the concept of reproductive strategies; in this chapter we show that parental care, infanticide, adultery, and altruism can all be viewed as components of reproductive strategies. The evolution of these strategies is an important field of investigation for sociobiology. In this chapter we examine the sociobiology paradigm and some of the major issues within the paradigm.

SOCIOBIOLOGY DEALS WITH SOCIAL BEHAVIOR

Sociobiology means different things to different people. To scientists working in sociobiology, it is a field of study that deals with social behavior and its evolution. Sociobiologists usually explain behaviors in evolutionary terms. Although sociobiologists are more interested in the inherited components of behavior, they all acknowledge that much of behavior can also be modified by learning. They also acknowledge that natural selection can act only on those components of behavior that are inherited. One of the important research goals of sociobiology has therefore become the investigation of the relative importance of learned and inherited influences on particular behaviors. Sociobiology also has a number of critics who challenge the emphasis on inherited behavior patterns. These critics prefer to emphasize learning, including cultural learning in humans, as a strong influence on behavior. We will examine both viewpoints.

Learned and inherited behavior

Many behavioral patterns may be strongly influenced by experience in dealing with the environment, i.e., by **learning**. Nearly every behavior that has been carefully investigated also has some genetic component. Learned behavior may increase fitness, but only the genetic components (or predispositions) underlying the behavior can be influenced by natural selection (see Chapter 4). Natural selection can operate on the capacity for learning particular kinds of things, such as how to find one's way through the maze of one's surroundings. The character favored by selection in such cases is not the behavior itself, but rather the capacity to learn the behavior.

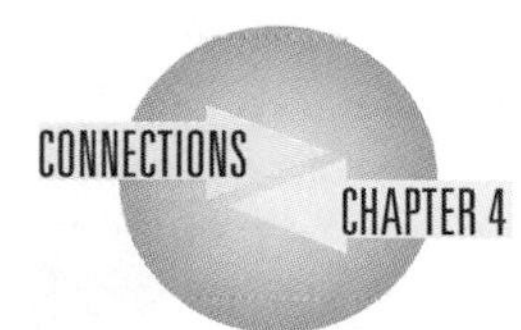

This is true of the ability to run through mazes, one of the most often studied types of learned behavior. Rats were tested for their ability to learn certain mazes, and the number of training sessions before the rats learned the maze was recorded. Their littermates, who were never tested themselves, were then selectively bred for several generations. Breeding the littermates of fast maze-learners resulted in a strain in which the average number of training sessions needed was low, while a strain of slow learners was bred from the littermates of individuals who needed more repetitions. The use of littermates in this experiment eliminated learning experience or other influences as determinants of the differences between the two selected strains. Notice that the behavior was not fully determined by inheritance; it still had to be learned. This behavioral trait is determined by many genes and environmental influences acting together. The *difference* between the two strains resulted from the buildup of gene combinations.

Furthermore, to say that variation between groups is inherited *does not mean that the behavior is inherited as a fixed and unchangeable trait*. There are extremely few behaviors in any species (and none at all in humans) that are not subject to modification through learning. For example, nobody learns to play basketball like Michael Jordan without years of practice, or to play the cello like Yo Yo Ma, or any other skilled activity, so these are clearly not completely inherited behaviors. (Nobody can become even a mediocre basketball or cello player without lots of practice.) However, there is surely some talent and ability at work, or else any of us would be able to become a great basketball star or a world-renowned cellist simply by practicing enough.

Thus, it is important to emphasize that the oft-posed question of learned versus inherited behavior is a false dichotomy. Every learned behavior is based in part on some inherited capacity to learn, which may include the capacity to learn certain kinds of behaviors and not others, to respond to some stimuli and not others, to learn up to a certain level of complexity, and so on. Similarly, most behavior patterns with an inherited component can be modified to some extent by learning. These observations give rise to the falsifiable hypothesis that *nearly every behavior pattern is at least partly learned and at least partly inherited*. Behaviors that precede any learning are called **innate**, and innate behaviors are assumed to have an inherited component. No behavior is 100% learned, and few are 100% inherited in any species (Figure 7.1). The methods

FIGURE 7.1

Learned versus instinctive behavior patterns.

Nursing and suckling behaviors have strong instinctive components in most mammals (but this doesn't prevent bottle-feeding of many human infants).

Novel behaviors can be learned by many species.

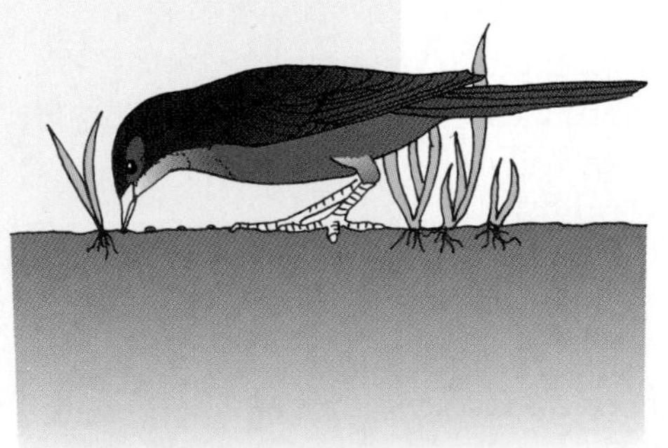

Most forms of behavior show both learned and inherited components. Robins and many other birds instinctively peck at certain stimuli and thereby gain learning experiences about how to hunt more effectively and how to distinguish food objects from other objects.

used to distinguish between learned and innate components of behavior are described later in this chapter.

The paradigm of sociobiology

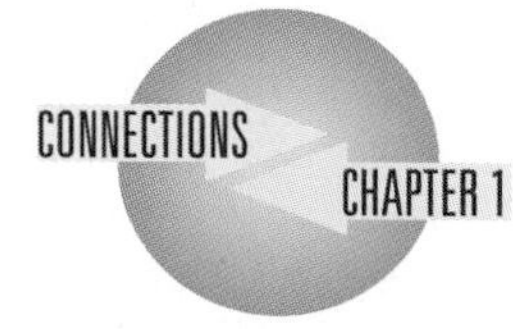

Sociobiology, the study of social behavior among different species, uses a scientific paradigm of the kind described in Chapter 1: one or more theories, plus a set of value-laden assumptions, a vocabulary, and a methodological approach (Box 7.1). The formulation of sociobiology as a paradigm dates from the publication of the book, *Sociobiology: the New Synthesis*, by Edward O. Wilson (1975). Nothing in this paradigm was without antecedents—every idea had been expressed before, including the use of the term 'sociobiology.' Many of the ideas can be traced to Charles Darwin's writings. What was new in 1975 was the way in which these ideas were put together to form the paradigm.

If people outside the paradigm had viewed sociobiology as no more than the study of social behavior, few objections would have been raised to it. However, sociobiology was frequently criticized by sociologists, philosophers, feminists, and many others for its emphasis on inherited behavior. These critics pointed out what no one had ever doubted, that a good deal of behavior is learned, and that learning is especially important for those species most similar to ourselves. The many critics of sociobiology have raised the following points. Much of behavior is learned, and nearly all behavior can be modified by learning. This is particularly so in mammals. By presupposing that innate components of behavior *must* exist and must be sought for, the critics charge that sociobiologists have already decided the issue in favor of innate control in every case. By making innate behavior the focus of their research, sociobiologists have devalued the importance of learning. The sociobiology paradigm makes no distinctions between the study of social behavior of insects, fish, or birds and that of people. This offends many critics because humans have both language and culture as distinctive characteristics, and human behavior is strongly influenced by both. Many scientists who are otherwise sympathetic to sociobiology consider any extrapolation to humans of sociobiological findings based on other animals to be strongly suspect.

The objections just listed arise within the scientific community. Other criticisms originate outside the sciences from those who assume that anything 'innate' is unmodifiable, a view that sociobiologists do not generally share. Thus, some critics of sociobiology equate sociobiology with genetic determinism, the assertion that our individual characteristics are determined before birth and cannot be changed. Genetic determinism is feared for at least two reasons. First, throughout history, people in power have often sought to control other people (other social classes, other races, and women) by teaching that existing inequalities were 'natural' because they were based on innate and unchangeable differences. The fear of such oppression motivates many of the critics of sociobiology. Second, many people fear that the mere claim that a behavior is innate will discourage people from trying to change that behavior through

BOX 7.1 The Sociobiology Paradigm

Research activity in science is often organized around paradigms (see Chapter 1). Here, in brief outline form, are some of the major points of the sociobiology paradigm:

1. Behavior is interesting to observe and to study. (This is a value judgement; people who do not share it will never be attracted to the paradigm.)

2. Much of the interesting behavior influences the behavior of other individuals, and is called 'social.' (This is a definition with an implied value judgement that people within the paradigm are expected to share.)

3. Social behavior has evolved and continues to evolve. (This is a central theory whose rejection would bring down the entire paradigm.)

4. The evolution of social behavior takes place by natural selection, along the lines outlined by Darwin: variations occur, and the variations that increase fitness persist more often than those that do not. (This is again a theory; it includes theoretical concepts like 'fitness' and 'variation.')

5. Behavior is often modified by individual experience ('learning'). However, this learning takes place within limits set by the biology of the organism: the eyes limit what can be seen (likewise with other sense organs); the muscles and skeleton limit the possible responses; the structure of the brain limits the learning capacity, and so on. There are also many preexisting predispositions to respond to certain types of stimuli, to react in certain ways, and so on. These predispositions may have been learned at an earlier time, but *at least some of them* precede any learning and may be called 'innate.' (This is a central tenet of the paradigm, forming the basis for its further research.)

6. In the evolution of behavior, learned modifications are not directly inherited. Learned behaviors can contribute to fitness, but cannot be inherited. Only the innate predispositions and their biological underpinnings can be inherited, and only these inherited components can evolve. Natural selection can only work on the inherited aspects of behavior. (These ideas follow in part from the ways in which 'learned' and 'innate' are defined, and in part from the findings of evolutionary theory.)

7. It is therefore important to distinguish the learned and innate components of behavior, and to focus attention on the latter. This is a value judgement about the aims of research within the paradigm. It does not mean that learned behaviors are unimportant; it just means that sociobiologists would rather identify what is learned so that they can ignore it and spend the rest of their time studying the innate components. *It is this preference for studying the innate components of behavior that makes the sociobiology paradigm so controversial; most critics of sociobiology have the opposite preference.*

8. We can use (modifications of) Darwinian methods of investigation to study those components of behavior that evolve. One method is to measure variations in fitness by observing many individuals and studying the number of viable offspring successfully reared by each. Another method is to study the results of past evolution by comparing social behaviors among different populations or different species. (These are the basic research methods.)

9. Before comparisons can be made, however, there must first be an often lengthy period of observation and description. However, we realize that the presence of observers might modify the behavior that we wish to study. Because we are interested in behavior under 'natural conditions,' it follows that we should conduct most observations at a distance and interfere as little as possible. (These are more research methods.)

education or similar means. The claim that behavior is innate can be particularly threatening to social reformers who pin their hopes for the future on the ability of people to modify their behavior.

Among biologists, those who believe in genetic determinism are decidedly in the minority. Most biologists, especially those who study animal behavior, are impressed with the plasticity of behavior, meaning its ability to change in response to environmental circumstances, including the behavior of other individuals. To be sure, there are genetic constraints on what can and cannot be learned, but, within these limits, behavior is remarkably plastic in most animal species, especially those that are similar to ourselves. It is misleading to say that a particular behavior is 'determined' either by genetics or by environment—almost every behavior is influenced by both of these factors throughout the lifetime of the individual.

We turn now to an examination of research methods used by sociobiologists and the results that they have produced. Think about whether these results support the sociobiology paradigm (see Box 7.1). Think also about whether these results negate the points raised by the critics of sociobiology.

Research methods in sociobiology

No behavior can be analyzed by any method until it has been adequately described. Sociobiology therefore includes a great many observational field studies of nonhuman animals. How does one distinguish between the learned and innate components of a particular behavior? Sociobiologists use the following methods to investigate these components.

Rearing animals in isolation. A classic type of experiment is to raise an animal in isolation, in a soundproof room with bare walls and minimal opportunities for learning, including no opportunity to learn the behavior from others. Behavior that the animal exhibits under these conditions is assumed to be largely innate. Experiments of this sort cannot ethically be performed on humans.

Rearing animals under different conditions. If the behavior is performed in the same way by animals or humans reared under strikingly different circumstances, then the behavior is largely innate. If, in contrast, the behavior varies according to the circumstances of rearing, then the variation can be attributed to environmental influences, although this does not rule out inherited influences, which may also be present. Cross-cultural studies are used to compare the behaviors of people raised in different societies or under different customs; innate behaviors are expected to be constant across various cultures, while learned behavior patterns are expected to vary.

Studying behavior in different genetic strains. If different strains or breeds differ behaviorally in a consistent and characteristic way, then a strong inherited component exists. (This does not rule out learned components, which might also be present.)

Conducting adoption studies. If two populations differ in a particular behavior, it may be useful to study individuals from one group who are adopted early in life and raised by the other group. Under these conditions, behavior consistently resembling the population of birth demonstrates an inherited influence, while behavior resembling the population of rearing demonstrates a learned influence. Mixed or inconsistent results may indicate that both influences are present.

Conducting twin studies. If a trait is under strong genetic control, then identical twins should usually both exhibit the trait whenever either one does, while fraternal twins more often exhibit differences. Twin studies in humans are frequently criticized because the effects of learning cannot easily be separated from those of inheritance unless the twins are reared separately in families randomly chosen, conditions that are rarely even approximated.

Instincts

A subset of innate behaviors are called **instincts**. Instincts differ from other innate behaviors in being complex behavior patterns that are under strong genetic control. The classical test for whether a particular behavior is an instinct is whether the behavior appears at the appropriate time of life in an animal reared in isolation since birth or hatching. For example, if a songbird reared in a soundproof room sings the song of its species and sex upon reaching maturity, then the song is considered to be instinctive. By this test, many behaviors that have been studied in fishes, birds, and many invertebrates (including insects) have been shown to be largely instinctive. In general, behaviors related to courtship and mating have strong instinctive components in most species. Other behavior patterns that are frequently instinctive include automatic 'escape' behavior, nest-building behavior, orb-weaving in spiders, and various gestures used to threaten other individuals of the same species. When instinctive behavior leaves a lasting product, such as a nest or a spider's web, these products are often so distinctive that they can be used to identify the species that created them.

Mammals generally rely more on learned behavior than on instinct. This is especially true of primates. Many behaviors, such as mating behaviors and aggressive threats, that are instinctive in other species, have strong learned components among monkeys and apes and may vary greatly among human societies.

Advantages of instincts. Short-lived animals rely heavily on instincts. Mayflies, for example, have an adult life span of less than 24 hours. During this brief period, they do not feed, but have just enough time to find a mate, copulate, lay their eggs, and die. There is no time for learning to take place, nor is there any time for mistakes. The mayflies that accomplish their mission successfully are those that can perform their behavior correctly on the first try; they will probably never get a second chance. Selection over millions of years has therefore produced a series of adult behaviors that are instinctive and automatic, allowing no room for diversity or innovation. This is typical of instincts generally: behavior is instinctive in contexts in which uniformity and automatic response are adaptive and where innovation and diversity might be maladaptive.

(Adaptation in populations is explained in Chapter 4.) A greater complexity of behavior is possible with a simpler brain if the behavior is instinctive; learned behavior of equal complexity requires a more elaborate nervous system and also a long learning period during which many mistakes are made.

Mating behavior. Mating behavior includes both courtship (attracting a mate and becoming accepted as a mate) and the actual release or transfer of gametes. Mating behavior has a strong instinctive component in nearly all species, except in higher primates. Scientists can demonstrate the instinctive component of most forms of mating behavior by raising individuals in isolation until they are sexually mature, then testing them to see whether they can perform the behavior typical of their species.

As just mentioned, behavior tends to be instinctive in situations where natural selection favors uniformity rather than diversity. Such unvarying behavior is called **stereotyped** behavior and is used for mate location and recognition in many species. The behavior that evolved in each species matches the type of signal that each is able to sense, so that visual mating signals are used by species with good vision, chemical signals by species with good chemical reception, and sounds by species with good sound discrimination. Many species of birds, frogs, and insects use sounds as mating signals, and the noncalling sex (usually female) responds only to mating calls of the proper pitch, duration, and pattern of repetition. Both the mating signals and the behavioral response to them are instinctive. Members of each sex know exactly what to listen for in the other sex and usually avoid nonconformers who deviate from the instinctive pattern. Sexual selection thus penalizes the nonconformers, who generally fail to mate and therefore leave no offspring. The flashing patterns of fireflies, though visual, are sexually selected in the same way. Because of sexual selection, mating calls or visual displays are precisely controlled within a narrow range for each species. Closely related species often differ in their mating calls and courtship patterns. Most often, it is the males that emit audible calls or put on conspicuous displays, while females respond to these calls or displays and are very sensitive to any variations. Differences in mating calls and other courtship displays can therefore serve as reproductive isolating mechanisms that prevent interbreeding between species (see Chapter 4).

Male birds of many species display conspicuously colored parts during courtship. Mating rituals that include beautiful, ornate displays evolve as a consequence of sexual selection in those species where the discriminating sex (i.e., the one doing the choosing) consistently prefers the most conspicuous displays. Peacocks, lyre-birds, and birds-of-paradise are renowned for their beautiful and ornate male plumage (Figure 7.2). Male birds of species with less conspicuous plumage may concentrate instead on building an elaborate nest. The South Pacific bowerbirds build their nests within a large framework (a bower) that also serves as a place of mating. A few species even build an 'avenue' lined with colorful stones leading to the entrance of the bower. Generally, bowerbird species with fancy and ornate plumage do not build elaborate bowers, and the species that build impressive bowers do not have elaborate plumage.

Territorial behavior. In many species, one or both sexes may show **territorial behavior** by defending a territory, either throughout the year or only during the mating season. The defense of a territory against intruders of the same species is common in many animal species. In some species, only males are territorial. Territorial behavior spaces individuals apart and encourages the losers to strike out in search of new territory, thus extending the range of the species wherever possible. Each territory must have sufficient food resources for a mating couple and their offspring, places for hiding and refuge, and at least one suitable nest site. Males without any territory are usually unable to attract mates and thus leave no offspring in that particular season.

Territorial species may use gestures to threaten territorial rivals. Mammals who establish territories may mark their territory with their own scent. The intimidation of rivals by gestures or by the presence of odors serves to space individuals apart without causing injury or loss of life. Such ritualized forms of territorial defense are much more common than any form of fighting in which injuries are likely.

FIGURE 7.2

An example of a conspicuous mating display in a peacock. Females of this species prefer the males with the most conspicuous displays.

Nesting behavior. The choice of a nesting site may be an important part of territorial behavior. In some bird species, the male builds the nest and then offers it to the female as part of the mating ritual. In other species, male and female may cooperate in building the nest together as part of the mating ritual. Females may incubate the eggs alone, but males may provide other forms of assistance by bringing food or by defending the area against predators. In other species, the males and females take turns in guarding the nest and sitting on the eggs. Feeding the hatchlings may similarly be either a solitary or a shared task.

The behaviors just described are performed by individuals. Behaviors can also be performed by groups of organisms, a subject that we take up in the next section.

THOUGHT QUESTIONS

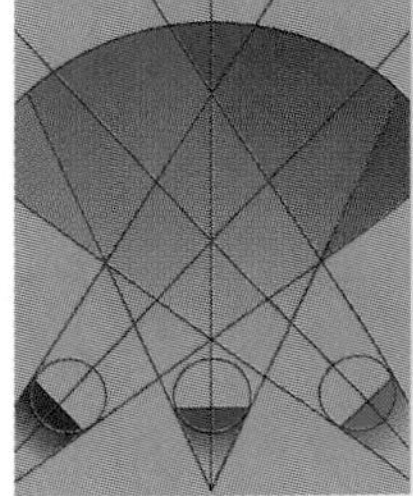

1. Does 'antisocial' behavior (such as assaulting others and causing them injury) fit the definition of social behavior? Do you think the definition should be modified? In what way?
2. Is sociobiology a subject area with room for many viewpoints, or is it a single viewpoint that enshrines both genetic determinism and sexism? Can sociobiology be studied without the assumptions of genetic determinism?
3. Can the methods used for gathering or analyzing data in sociobiology be the same for different species? To what extent do size (small versus large animals) or habitat (above ground, underground, underwater, in trees, etc.) require differences in field methods? What special problems in methodology arise when humans are being studied? Can the methods used for other species be applied to *Homo sapiens*?

SOCIAL ORGANIZATION IS ADAPTIVE

Very few animal species consist of solitary individuals that spend all their time alone. Even in species whose members are solitary much of the time, individuals must come together for sexual reproduction. Most species, however, are far more social than this, and species that form social groups greatly outnumber those that consist primarily of solitary individuals. Social groups vary greatly in both size and cohesiveness. Simple pairs and family groups have only a few individuals. Larger social groupings include antelope herds, baboon troops, and fish schools, all of which may include up to a few hundred members. Still larger are the colonies of social insects, which may include many thousands or in some cases millions of individuals. Some social groups are loosely organized, with individuals staying together but seldom interacting, while others are organized into social hierarchies within which interactions are complex, as they are among humans and social insects.

The biological advantages of social groups

Some advantages of social grouping are related to the obtaining of food. A large group of individuals searching for food together has a higher probability of finding food than a single individual. If food tends to be discovered in quantities much greater than a single individual needs, selection favors the formation of social groups.

Other advantages of social grouping relate to defense against predation. Social groups can often defend themselves more effectively than individuals can. Musk oxen, for example, respond to threats by standing close together with individuals facing outward in different directions (Figure 7.3). Even in species that do not practice group defense, members of a group may warn one another by giving alarm signals, or simply by fleeing as soon as a predator is spotted. Thus, belonging to a group gives all group members the advantage of greater (and earlier) alertness against predator attacks. For this reason, large but loosely organized flocks, schools, or herds are common among birds, fishes, and ungulates (hoofed mammals such as wildebeest and zebra) (Figure 7.4). Other advantages to group membership arise from the sharing of risks: a predator attacking the entire herd may capture one of its members at most, while the rest escape, so that each individual in a herd of 500 is exposed to only 1/500 of the risk of capture faced by a solitary individual. Actually, the risk may be even smaller because predators can more easily capture solitary individuals. Most herd animals taken by predators are individuals that have strayed from the herd.

FIGURE 7.3

Musk oxen in a defensive formation. When musk oxen stand close together and face in different directions, no predator can surprise them.

Simple forms of social organization

Social organization refers to the ways in which social groupings are structured. The fact that social organization sometimes varies among closely related species suggests that social organization evolves. Studies on the inheritance of social status (dominance) within organized social groups point to a complex interplay of learned and inherited behavioral components in the establishment of social organization.

Groups without dominant individuals. Perhaps the simplest form of social organization is shown by brittle stars (see Figure 4.22, p. 168), marine organisms distantly related to sea stars. On encountering one another, brittle stars tend to stay together in clumps, even though there is no evidence of any more complex interaction.

The schooling behavior of fish is another very simple form of social organization. There are hydrodynamic advantages to schooling—swimming is made slightly easier by certain changes in water pressure caused by the swimming of the other fish—but these effects are small. The major advantage to schooling behavior may be that the fish hide behind one another in such a way that most escape predators. Most fish school closer

FIGURE 7.4

Social groups in various species.

Pelicans

Minnows schooling in the presence of predators

Wildebeest on the plains of Africa

Gannets on the coast of Quebec

together when a predator is nearby (see Figure 7.4). When attacked, schools of fish or flocks of birds tend to scatter in every direction, a reaction that confuses many predators and that gives the individuals a chance to escape.

The size of social groups can vary greatly, often in response to ecological factors. For instance, the weaver birds live in many parts of Africa and Asia. In humid, forested regions, most weaver birds nest in pairs and feed on insects, while those species inhabiting grasslands and other drier habitats build large communal nests and eat a diet rich in seeds.

Groups with dominant individuals. One form of social organization is called a **linear dominance hierarchy** or 'pecking order.' Such hierarchies are found among domestic fowl and certain other captive animals (Figure 7.5). The top-ranked individual, usually a strong male, can successfully bully or threaten all the other individuals in the group, literally pecking at them in the case of birds. The second-ranked individual can intimidate all others except for the top-ranked individual. The third-ranked individual can intimidate all except for the first two, and so on. Occasionally, two closely ranked individuals may be tied for status, so that neither one can dominate the other. For the most part, however, this type of organization results in the biggest bully getting whatever he wants, the second biggest getting whatever he wants as long as he steers clear of the top-ranked individual, and so on. Some feminist critics of sociobiology suggest that such male-dominated forms of social organization exist more in the minds of male sociobiologists than in the animals that they study. In at least some studies, linear hierarchies may reflect the artificial conditions of captivity and confinement.

Altruism: an evolutionary puzzle

Efforts to solve human social problems such as pollution often call for individuals or corporations to sacrifice their own interests for the common good, a practice called **altruism**. Altruistic behavior exists in other species as well. As an example, consider the 'broken wing' display of certain female birds such as nighthawks. When guarding her nest against a predator, a female nighthawk may lead it away from the nest location, distracting its attention by limping or pretending to have a broken wing. Once she has drawn the predator sufficiently far from the nest, she flies away, leaving the predator confused. While protecting her young, however, she has increased her own danger. Thus, in evolutionary biology, altruism is more precisely defined as behavior that *decreases* the fitness of the

FIGURE 7.5

Domestic fowl showing a linear dominance hierarchy or 'pecking order.'

performer while it increases the fitness of another individual. In this example, the female bird has decreased her own fitness by putting her life at risk for the sake of her offspring. Remember that fitness is defined as the relative number of fertile offspring produced by an individual (see Chapter 4). Only changes that increase fitness are favored by natural selection.

Altruistic behavior poses a problem in evolutionary theory because natural selection might be expected to work against it. How could altruism evolve if it decreases fitness? Various hypotheses have been developed to explain this. In this section we examine several hypotheses that act at different levels of selection.

Selection at the species level. One early hypothesis for the evolution of altruism is that it benefits the species as a whole. However, careful examination of this hypothesis shows it to be unsatisfactory. If a species had both altruists and selfish individuals ('cheaters'), and if some part of this behavioral difference was controlled genetically, then selection would work against the altruists and in favor of the cheaters. Altruism may benefit all recipients of another individual's altruistic behavior, but the advantage is greater to selfish individuals than to other altruists. Under these conditions, natural selection should favor selfishness and eliminate altruism from the population.

Group selection. Another possible explanation for the evolution of altruism was proposed by the British ecologist V.C. Wynne-Edwards. If a species is subdivided into populations or social groups, then selection among these groups, called **group selection**, may favor one group over another. In particular, a group containing altruists is favored *as a group* over other groups composed of selfish individuals only.

As we explained earlier, the defense of territory prevents excessive population density by spacing individuals apart and limiting population size. Wynne-Edwards describes the losers of territorial disputes as altruists who forgo mating for the benefit of the group as a whole. He argues that the mating of individuals without territories would lead to overpopulation, increased mortality, and a smaller resulting population size. Selection between groups would thus favor altruism. Other biologists who have examined this claim with mathematical models have shown that a loser who cheats (mates anyway) would greatly increase its fitness over one who does not mate, and that cheating behavior would thus be favored over altruism in every territorial species. Similar arguments have been advanced to show that other behaviors that achieve spacing or population control would also not be favored by group selection because cheaters would tend to leave more offspring than altruists.

Kin selection. Many biologists dissented from the group selection hypothesis because they thought a simpler and more satisfactory explanation would be based on individual selection instead of group selection. An explanation of altruism acceptable to many of these biologists is based on the concept of **inclusive fitness**, defined as the total fitness of all copies of a particular genotype, including those that exist in relatives.

Relatives are listed according to their degree of relationship, symbolized by R. For sexually reproducing organisms with the common types of mating systems, an individual shares half of its genotype ($R = 1/2$) with its parents, its children, and, on average, with its brothers or sisters (who share two parents). Also, an individual shares one-fourth of its genotype ($R = 1/4$) with grandchildren, half-siblings (who share one parent only), uncles, aunts, nieces, and nephews. The inclusive fitness of your genotype is the sum total of your individual fitness *plus* one-half the fitness of your parents, children, and full siblings who share half your genotype, *plus* one-fourth the fitness of those relatives who share one-fourth of your genotype, *plus* one-eighth of the fitness of your cousins who share one-eighth of your genotype, and so on. This concept allows us to define **kin selection** as the increased frequency of a genotype in the next generation on the basis of its inclusive fitness.

The conditions under which kin selection favors the evolution of altruism were specified by the British sociobiologist William D. Hamilton. Assume that altruistic behavior results in a certain reduction in fitness or 'cost' to the altruist, and a corresponding gain in fitness or 'benefit' to another individual who shares a fraction of the altruist's genotype. Hamilton reasoned that natural selection would favor altruism whenever the gain in inclusive fitness to the altruist's genotype exceeds the cost. If I perform an altruistic act that diminishes my individual fitness by a certain cost but raises my child's fitness or my sister's fitness (with whom I share half my genotype) by more than twice that cost, then the net effect on my inclusive fitness is positive. The probability that my genotype will be represented in future generations is increased because the benefit to my relatives (or to the fraction of my genotype that they share) exceeds the cost, so the net result is an increase in my inclusive fitness.

The above explanation, however, gives rise to an interesting prediction: kin selection only favors altruism if close relatives are more likely to benefit from altruistic acts than more distant relatives or nonrelatives. Studies of many species have confirmed this prediction: the beneficiaries of altruism are often close relatives of the altruist, and the frequency of altruistic acts varies in almost direct proportion to the degree of the relationship. In the Florida scrub jay, the offspring of the previous year are not mature enough to mate. Instead, they help the nestlings who are their own brothers and sisters (Figure 7.6). In so doing, they contribute to the survival (and thus the fitness) of these near relatives, who share a portion of their own genotype. Ground squirrels emit an alarm call when a predator is spotted. The alarm call decreases the fitness of the caller, but increases inclusive fitness by warning the caller's kin (see Figure 7.6).

For kin selection to operate, it is not necessary that the altruist be able to distinguish relatives from nonrelatives; it is only necessary that close relatives are more likely to benefit from altruistic acts. Although kin selection does not *require* kin recognition, can animals assess the degree to which other organisms are related to themselves? In some social animals, individual recognition (based on growing up together) can be used. Other species, including mice, use odor cues. The odor of each animal is genetically influenced, and the diversity of genotypes results in a diversity of odors. Mice can detect by odor which individuals are the most closely related to themselves. Mice seem to use odor-based kin-recognition when they establish communal nests. Several females share a nest and nurse each others' offspring. A mother's inclusive fitness is maximized if she nurses only offspring that are closely related to her. Females who share a communal nest are usually related genetically.

Another explanation of altruism, based on **game theory**, is described on the web site.

The evolution of eusociality

The highest degree of social cooperation is developed among the truly social, or **eusocial** insects. Eusocial species are recognized by the possession of three characteristics: strictly defined subgroups called **castes**, cooperative care of the eggs and young larvae (cooperative brood care), and an overlap between generations. Eusociality occurs in the insect order Isoptera (termites) and particularly often in the order Hymenoptera (bees, wasps, and ants). A few bird species and one mammal (the naked mole rat, a burrowing type of rodent) approach eusociality in having 'helper' individuals who assist in caring for their siblings, but these helpers do not form a distinct caste.

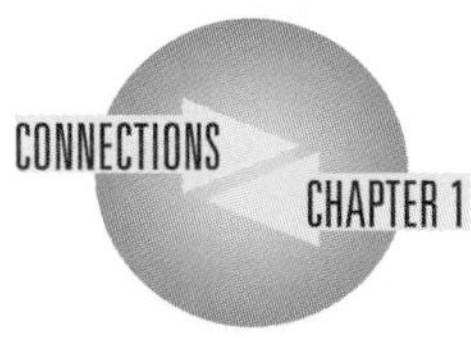

Humans show some of the characteristics of eusocial behavior, but not to the extent shown by the eusocial insects. Humans have overlapping generations but often do not cooperatively care for their young and do not usually form castes. According to Sarah Hrdy, **alloparental behavior** (assisting in the care of another person's children) is an important characteristic of our species (see Figure 1.4, p. 18), but the eusocial insects far surpass us in this behavior.

Eusociality in termites. Termites (Isoptera) are a group of insects related to the cockroaches. Termite colonies are founded by a single reproductive pair called the 'king' and 'queen.' The queen grows many times larger

FIGURE 7.6

Two examples of altruism favored by kin selection.

A year-old Florida scrub jay (right) assists in the care and feeding of its younger siblings.

A female ground squirrel (*Spermophilus beldingi*) stands guard against predators. If a coyote or hawk is spotted, the guard female emits an alarm call that attracts the predator and thus endangers the caller, but the alarm also warns the caller's next of kin and thus raises her inclusive fitness.

than the other colony members, her offspring, who continually feed her and raise her additional offspring.

An important termite characteristic central to the understanding of their evolution is their chewing and digesting of wood. Termites can digest wood only with the help of symbiotic microorganisms (mostly flagellated unicellular organisms of the kingdom Protista) that live in their guts. The termites transmit these protists through regurgitated food passed to other members of the colony. This habit not only spreads the wood-digesting microorganisms throughout the colony, it also feeds those members of the colony, such as the queen, who do not feed themselves. In the evolution of eusociality among termites, the chewing of wood led to selection favoring the retention and transfer of the symbiotic microorganisms.

Along with food and microorganisms, termites also pass chemical secretions that communicate social information to other colony members. Such chemicals are called **pheromones**. Some pheromones are similar to hormones, except that the pheromone secreted by one individual produces its effects in other individuals. One termite pheromone, secreted by the queen, inhibits most other individuals in the colony from becoming reproductively mature. Thus, the passing of food and symbiotic microorganisms throughout the colony was a precondition that probably led to the evolution of termite eusociality by providing each queen with the means to disseminate her pheromones and control the reproduction of other individuals. These other individuals form several types of sterile castes, depending on the species. Many of the nonreproductive individuals are workers that feed the queen, tend her larvae, and enlarge the colony's living space. Other individuals serve as soldiers, fending off potential enemies that pose threats to the colony.

At seasonally timed intervals, winged reproductive individuals of both sexes are produced; these winged individuals emerge from the colony all at once and embark on nuptial flights during which mating takes place. Newly mated pairs become the founders of new colonies. Meanwhile, the original colony persists for the lifetime of the queen, a period of some 10–12 years.

Eusociality in the Hymenoptera. The insect order Hymenoptera (bees, wasps, and ants) has a much larger number of social species, of which an estimated 12,000 are species of ants. The American evolutionary biologist E.O. Wilson, a specialist on ants, has estimated that eusociality has evolved among the Hymenoptera as many as a dozen times and perhaps more. Why has eusociality evolved so many times in this one insect order, and so seldom in other animals? The clue seems to be found in a unique form of sex determination and in its effects on inclusive fitness. The social Hymenoptera have a unique form of sex determination called **haplodiploidy**, in which fertilized eggs produce diploid females and unfertilized eggs produce haploid males. Each reproductive female mates only once, for life, with a single male, who contributes the same haploid set of chromosomes to all his offspring. The daughters therefore share 100% of their father's genes, rather than the 50% that daughters share in more common sexual reproduction. The sisters are therefore very closely related to one another, for they share 100% of their father's genes, plus an average of 50% of their mother's genes. Because the average of 50% and 100% is 75%, a female shares an average of 3/4 of her genotype with her sisters, much more than the half of her genotype that she shares with her daughter or her mother (Figure 7.7). By neglecting her own daughters (to whom she is related only by 1/2) and raising her sisters instead (to whom she is related by 3/4), she is increasing her genetic fitness. For this reason, haplodiploidy favors the evolution of eusociality in the Hymenoptera because most females can gain greater inclusive fitness by becoming sterile workers and by helping their mother (the queen) to raise her offspring (their sisters) than by raising offspring of their own. Ancestral Hymenoptera were solitary (and many solitary species still exist), but sociality has evolved repeatedly and independently in this group of insects (Figure 7.8).

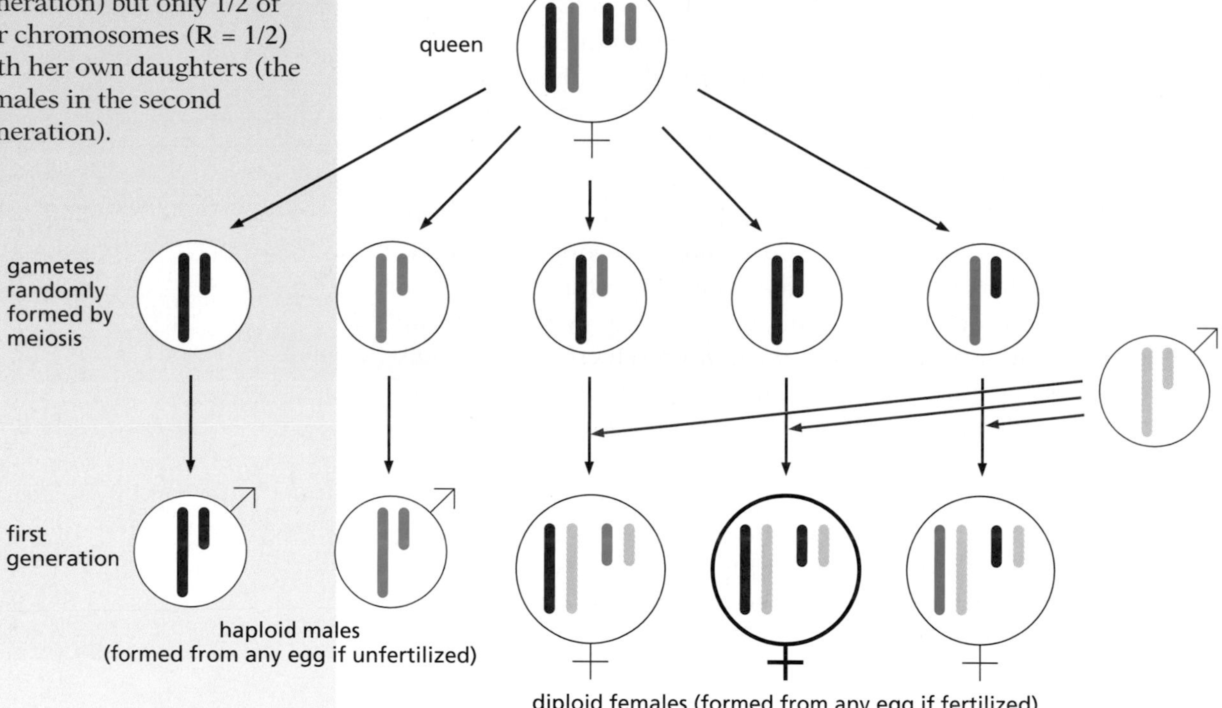

FIGURE 7.7

Haplodiploidy in a species with a haploid chromosome number of 2. Notice that the female represented by the cell with the heavy black border shares an average of 3/4 of her chromosomes (R = 3/4) with her sisters, (the other females in the first generation) but only 1/2 of her chromosomes (R = 1/2) with her own daughters (the females in the second generation).

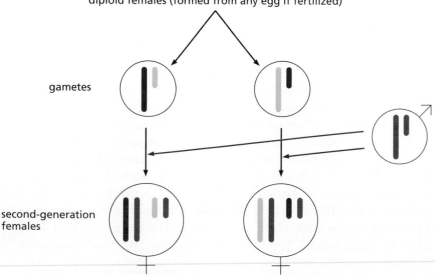

The queen bee or wasp usually secretes pheromones that inhibit the sexual development of other females in the colony. Other mechanisms determine which larvae develop into queens and which into sterile workers. For example, future queens are fed a nutritious 'royal jelly' that also contains pheromones that stimulate their reproductive development. Also, whenever new queens emerge, one of them (usually the one emerging first) stings the others to death and thus emerges as the undisputed queen.

Most of the social behavior of eusocial insects is under instinctive control; in fact, the eusocial insects represent the height of complexity that instinctive behavior can reach. Antisocial behavior (meaning behavior that decreases the fitness of others) does not exist in these societies because antisocial individuals are quickly eliminated.

In the next section we look at reproductive behaviors in species that are not eusocial.

FIGURE 7.8

Eusocial insects.

Honeybees swarming

A colony of ants

THOUGHT QUESTIONS

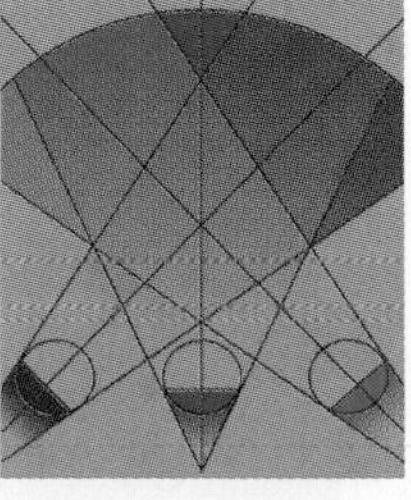

1. Are the behaviors of individuals within a species more alike than the behaviors of individuals from different species?
2. As noted in this section, individuals share, on average, one half of their genotype with their siblings. Refer to the discussion of meiosis in Chapter 2 (pp. 48–49), and explain this statement.
3. Why should humans be interested in the social behavior of insects?

REPRODUCTIVE STRATEGIES CAN ALTER FITNESS

Natural selection favors those genotypes that are able to leave more copies of themselves in subsequent generations than other genotypes do. The manner in which these copies are produced can be called a **reproductive strategy**. Reproductive strategies include such behaviors as the laying of many eggs or few, the presence or absence of parental care, the presence or absence of sexual recombination, and, if there is a mating system, whether it is predominantly monogamous, polygamous, or promiscuous. Sexual behavior is an important part of reproductive strategy in many social species.

Asexual versus sexual reproduction

Reproduction of organisms can be either sexual or asexual. Many species (including all mammals and birds) are exclusively sexual, while bacteria are predominantly asexual, and certain other species (yeasts, aphids, and a variety of plants) can reproduce either way depending on the circumstances.

Asexual reproduction may be defined as reproduction without any genetic recombination. This type of reproduction has certain advantages over sexual reproduction. Within a group of organisms that includes both species reproducing sexually and species reproducing asexually, those reproducing asexually can generally do so faster and with lower energy costs. Asexual reproduction allows reproduction at an earlier age and a smaller body size, and it also avoids the costs associated with sexual reproduction. For an individual that discovers a large but finite or perishable supply of food or some other resource, asexual reproduction is an advantage because more offspring, and more generations of offspring, can be produced in a minimum of time, without any need of finding or courting a mate. Moreover, the numerous offspring are genetically identical to the original parent or founder, ensuring that favorable combinations of genes are perpetuated exactly. The genetically identical asexual offspring of a single individual are referred to as a **clone**.

In contrast, **sexual reproduction**, reproduction with genetic recombination, is more costly than asexual reproduction because of the time and energy expended in searching, finding, and courting a mate, and in transferring sperm. Energy is also used in synthesizing structures that attract mates, and in the mating act itself. Mate attraction also makes a sexually reproducing individual more visible to predators, exposing that individual to increased risks. A major genetic cost is giving up half of one's genes (during meiosis) in favor of someone else's. In view of these costs, it is amazing that sexual reproduction would be so widespread in both the animal and plant kingdoms. Sexual reproduction must have some great advantage.

The great advantage of sexual reproduction is genetic variety among the offspring. In the most common type of sexual reproduction, males produce sperm cells that contain the haploid number of chromosomes, and females produce eggs that are also haploid. Each sex cell (gamete) produced by an individual carries only half of that individual's genetic material, formed during meiosis by a random choice of one chromosome from each pair. Because each gamete-forming cell undergoes meiosis independently, the chromosomal choices are different each time, and the gametes thus vary among themselves. The combination of gametes with the gametes of the opposite sex is also random. The result is that *sexually produced offspring vary greatly in all genetically controlled traits*. This may be a disadvantage if tomorrow's (or next year's) conditions are identical to today's, and unchanging conditions do in fact favor asexual reproduction. However, if tomorrow's (or next year's) environmental conditions are uncertain, then the best hedge against this type of uncertainty is to produce many *different kinds* of offspring, and sexual reproduction achieves this very efficiently. What we have just said pertains not only to the common forms of sexual reproduction, but also to other forms (such as the haplodiploidy described earlier): however much they differ in detail, all forms of sexual reproduction are characterized by greater variation among offspring than any form of asexual reproduction.

The hypothesis that sexual reproduction derives its adaptive advantage from the greater variation among the resultant offspring receives support from the study of certain insect species (such as aphids, also called plant lice) that are capable of producing either sexual or asexual generations. During the summer, when maturing crops offer dependable food supplies for several months in a row, these insects produce several asexual generations in quick succession. At the end of the season, however, these insects reproduce sexually, and the sexually produced eggs overwinter. When they emerge the following spring, diverse genotypes of offspring find their way to the new stands of plants under new weather conditions, neither of which could have been predicted during the previous fall. Many genotypes perish, but a few survive and prosper by reproducing asexually during the new season. The important point is that the genotype that proves most fit in the spring is not necessarily the same one that produced successful offspring asexually during the preceding year. *Sexual reproduction is favored whenever future conditions are uncertain*, and experiments confirm that individuals laying eggs in the fall have more surviving offspring the next year if they reproduce sexually than if they reproduce asexually.

r-selection and *K*-selection

In their reproductive strategies, species can also differ in population size, life cycle, and resource allocation. Two basic types of strategies exist: *r*-selection and *K*-selection. Each is favored by natural selection, but in different sets of circumstances.

Species that have high reproductive rates are said to have evolved through ***r*-selection**. This term refers to *r*, the intrinsic rate of population increase or growth rate, which was discussed in Chapter 6 (p. 235). High values of *r* lead to very rapid population growth under favorable conditions. However, most *r*-selected species live in unstable environments in which very favorable and very unfavorable conditions alternate with one

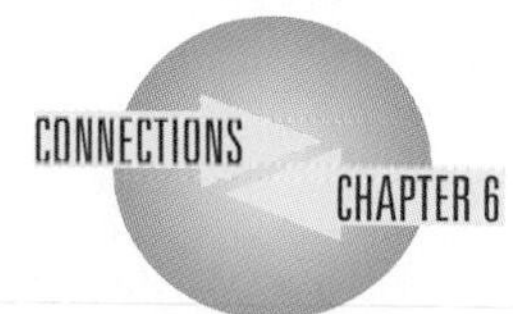

another erratically and unpredictably. Their populations suffer frequent and unpredictable, devastating losses. Rapid increase under favorable conditions must compensate for rapid decrease under unfavorable conditions, and population size fluctuates greatly over time. Most *r*-selected species have a high capacity for dispersal (spread of the population) to new habitats and locations. Most new locations are unsuitable, but the occasional favorable one permits a rapid explosion in numbers. Reproduction in *r*-selected species is prodigious and may be either sexual or asexual. Because the emphasis is on rapid reproduction, selection favors reproduction at a small body size and a young age. Eggs, seeds, or other reproductive stages are produced in great numbers, are released at a small size, and are widely scattered, with no provisions made for their care. Most die or are eaten, but *r*-selected species tend to compensate by producing offspring in even greater numbers. Carrion-feeding beetles, tapeworms, and weeds are examples of *r*-selected species.

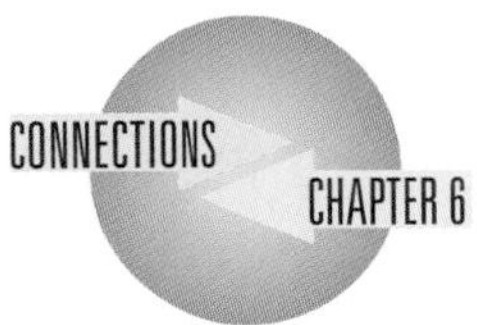

The opposite of *r*-selection is ***K*-selection**. *K*-selected species live in populations of stable size at or near the carrying capacity, which is symbolized by *K* (see Chapter 6, p. 239). Selection in these species favors reproduction (always sexual) at a large body size, producing a small number of offspring for which parental care or some other form of 'parental investment' is provided. Offspring may themselves be large, and each of them represents a greater proportion of its parent's reproductive output. *K*-selection favors the efficient use of resources, especially energy. The advantage generally goes to whomever can most efficiently convert food into new adults of the next generation. Humans and cattle are *K*-selected species.

Differences between the sexes

In sexually reproducing species, the two sexes are not necessarily different. Some species, such as the green alga *Chlamydomonas* (kingdom Plantae, phylum Chlorophyta), have male and female haploid gametes that look identical, a condition called **isogamy** ('equal gametes'). But a pair of gametes may be at an advantage if at least one of them is capable of finding the other over greater distances, thus allowing more mating or mating from a wider choice of potential mates. In some cases there may also be an advantage for the resultant fertilized egg (zygote) if it contains stored food or protective layers, each of which increases bulk. The advantages of motility and of large size can best be balanced if one of the gametes is large and the other is small and motile, a condition called **anisogamy** ('unequal gametes'). The larger, nonmotile gamete is called an **egg**, and the smaller, motile gamete is known as a **sperm**.

Males and females. Although it is possible that different-sized gametes could be produced by identical organisms, this does not usually happen. Instead, reproductive anatomy and behavior differ between the sexes in most species of animals and plants.

Most of the familiar differences between males and females are explained within evolutionary theory as the consequences of anisogamy.

Selection among sperm-producers, or **males**, favors the release of numerous gametes, each of which is of minimal size and maximum motility. The minimal size means that each individual sperm represents a trivial investment (in energy and materials) for the male that produces it, for he can easily produce thousands (or millions) more, and can compensate for a poor choice of mates by mating more often. Competition among males usually favors whichever one can produce the most gametes that combine successfully with the most eggs.

Selection among egg-producers, or **females**, generally favors a larger investment of parental resources, such as stored food, in each egg. Among numerous eggs, those with the most stored food or the strongest protective layers generally have the best chance of survival. This necessarily limits the number of eggs that a female can produce, and places a premium on egg *quality* rather than number. A further consequence is that females, having fewer eggs, can produce more surviving offspring if they invest more care and protection in each one. This is especially true in mammalian females, which devote much time and energy to gestation, intrauterine feeding, and lactation.

Parental investment. There are further consequences of **parental investment**. If male parental investment is low in both energy and material costs, the price that a male pays for mating with a given female is very small. If their offspring are low in fitness (i.e. have a small chance of survival), the male can simply mate again with other females. Low parental investment produces non-discriminating males.

If, on the other hand, a female's parental investment is high, each of her offspring is more costly to her. The price that she pays for mating with a low-fitness male greatly reduces her fitness. She cannot simply make up for a poor choice by mating again because her capacity for repeated mating is generally limited by the large investment she must make in each of her offspring. Females thus have more at stake in each mating, and stand to gain more by choosing a mate who will father offspring who are more fit, or to lose more by choosing a mate who will father offspring who are less fit. In social species in which males vary in social status, a female can generally maximize her fitness by mating with a high-status male who can provide her and her offspring with a greater degree of protection. Females, in other words, tend to become more discriminating in their mate choices, both as to social status and genetic fitness. Females of many species have a remarkable capacity to discern variation in male fitness (and social status, if it also varies), and to mate preferentially with males of high fitness and high status.

Evolution has produced some interesting exceptions to these generalizations about differences between the sexes. Male Mormon crickets, for example, mate only once, and they do so by offering a large clump of sperm to a female of their choice. Because a male Mormon cricket's parental investment is high, he has reason to be careful and discriminating in his choice of mates. In fact, male Mormon crickets lift the females during the mating ritual, estimating the number of eggs that they contain by weighing them! Heavier, more egg-laden females are more likely to be chosen as mates and to receive the prized sperm.

Mating systems

There are many types of mating systems known within sexual reproduction. In species in which care of the young requires the cooperation of both parents, parental investment tends to be high for both sexes. These conditions favor **monogamy**, or mating between one male and one female (Figure 7.9). If the rearing of their common offspring takes a long time, the formation of a permanent pair-bond (i.e., mating for life) is favored.

Another common situation is one in which only females care for the young, but males provide protection to both female and offspring. This situation generally favors the development of one form of **polygyny**, a mating system in which one male mates with several females. Many mammals form polygynous mating units; for example, male fur seals come ashore during the breeding season and establish territories, which they defend against other males (see Figure 7.9). The strongest male defends the best territory, an area where females can rear their pups within easy reach of the sea. Females are attracted to the territory (rather than to the male himself) and mate with males that hold territories.

FIGURE 7.9

Examples of different mating systems.

Promiscuity: baboons

Monogamy: a family of Canada geese

Polygyny: a large male fur seal surrounded by females

Males who lose territorial contests go off in search of other suitable territories. If they find none, they will not mate during that season.

Red deer, bighorn sheep, and certain other species of hoofed mammals (ungulates) form polygynous mating units in a different way. Adult males establish a dominance hierarchy, either through ritualized threat displays or through actual fighting. The females are most attracted to the dominant male. The dominant male gathers together as many females as he can, forming a **harem**. Male social status in harem-forming species often correlates with fighting ability and with the size of horns, antlers, or other conspicuous features, so females can see at a glance which male is dominant. Females can ensure better protection against predators for themselves and their offspring by following and mating with the dominant male. Any genetic component of the characteristics correlated with social dominance is passed on to their offspring, who will thus inherit such characteristics as fighting ability, size, and the size of horns or other weapons. Nondominant (subordinate) males have fewer mating opportunities than dominant males. Many subordinates are simply young adults who will get their turn to become dominant the following year.

In addition to monogamy and polygyny, other types of mating systems include **polyandry**, an uncommon type in which one female mates with multiple males. The term 'polygamy' is sometimes used to include both polygyny and polyandry. Another mating system is **promiscuity**, in which members of both sexes mate with multiple partners and generally avoid forming permanent partnerships (see Figure 7.9).

We have seen that reproductive behaviors, as well as other social behaviors, vary greatly between species. We have primarily looked at examples from the animal kingdom, but even bacteria show some social behavior; bacteria can, for example, influence one another in the timing of their cell cycles and metabolic events. Plants can be either *r*-selected or *K*-selected. They also show some ability for kin recognition in that some plants can assess the 'match' between proteins or other molecules derived from the male pollen and the female stigma. In the next section we look more closely at social behaviors and reproductive strategies in primates, including humans.

THOUGHT QUESTIONS

1. What are the biological definitions of 'male' and 'female?' How do these compare with cultural definitions of the same words? Do 'male' and 'female' (or 'masculine' and 'feminine') mean different things in different cultures, or at different times in history?
2. In humans and other species, males tend to have greater muscle mass than females. Under what conditions would you expect anatomical differences (in muscle mass, antlers, or size) to evolve? Is there a reason why such differences would be favored by natural selection?
3. Does the difference in gamete size in humans and other mammals tell us anything about our sexual behavior? Are human males 'destined' to be promiscuous?

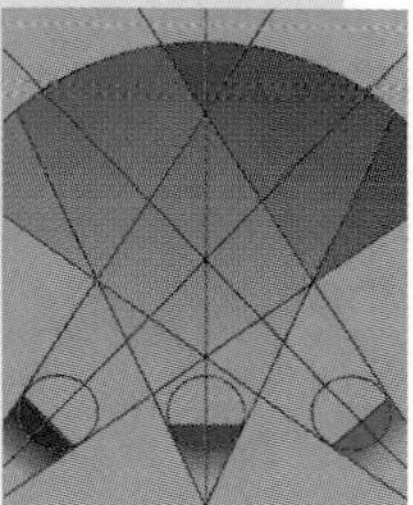

PRIMATE SOCIOBIOLOGY PRESENTS ADDED COMPLEXITIES

Primates are an order of mammals that includes monkeys, apes, lemurs, tarsiers, and humans. Primates are all extremely social animals. Primates are so interested in interacting with other members of their species that they sometimes go to great lengths to maintain an interaction or merely to look. We can experimentally set up a window, a partition that completely obstructs the view through the window, and a lever that raises the partition for a predetermined length of time, affording a temporary view through the window. Most primates will then spend hours repeatedly pressing the lever and looking through the window, then going back to press the lever again for another view almost as soon as the partition falls. The rate of lever-pressing is higher if the window affords a view of moving objects (such as electric trains) rather than nonmoving objects such as furniture. The rate is higher still if an actively moving animal is visible through the window, and it is highest of all if the view includes other primates of the same species. Is it any wonder that people also spend hours looking through windows at the world around them, especially when other people's movements and interactions are visible? Think of a television screen as a window through which we can watch people interact.

Primate social behavior and its development

Social skills in both human and nonhuman primates depend strongly on learning that takes place early in life. All parents and future parents should be aware of the paramount importance of early childhood experiences in all later aspects of human life.

Early development of behavior. As we stated earlier, the standard test for an instinct requires that an animal be raised in isolation. Raising a primate in isolation, however, results in abnormal behavior resembling that of abused children. Siegmund Freud claimed that a baby's attachment to its mother is based initially on its need for nutrition. To test this hypothesis, Harry Harlow of the Wisconsin Primate Research Center raised infant rhesus monkeys (*Macaca mulatta*) with various forms of care but with no live mothers. Instead, dummy 'mothers' with colorful wooden 'heads' held baby bottles mounted in wire frames. Although the infant monkeys drank the milk, their behavior grew progressively more abnormal with time. The infants frequently cowered in the corner and were easily frightened. They formed no emotional attachments and seemed to ignore their 'mothers' except when they were hungry. Freud's hypothesis was falsified because the young monkeys failed to treat the wire model as a mother. Something more than the milk supply was needed for infants to form a bond with their mothers.

Harlow noticed that young monkeys liked the feel of terrycloth towels. He tried wrapping the wire mothers in a few layers of terrycloth to make them soft and clingy. The terrycloth retained the infant's own body heat during periods of clinging. The infant monkeys enjoyed clinging to these cloth-covered dummies. Harlow raised infant monkeys with two dummy 'mothers,' with and without terrycloth, one of them holding a baby bottle. Young monkeys spent countless hours clinging to the terrycloth 'mother,' regardless of which dummy held the bottle (Figure 7.10). When exposed to a novel or frightening stimulus, the infant monkeys would run to their terrycloth 'mothers' to cling for reassurance. After clinging for a while, the young monkeys were sufficiently reassured that they became brave enough to inspect the previously frightening stimulus in many cases, their curiosity finally overcoming their fear. Wire dummies, in contrast, never provided the behavior-changing reassurance.

FIGURE 7.10

An infant rhesus monkey raised by two dummy 'mothers,' one made of bare wire and the other covered with soft terrycloth. Note that the infant maintains contact with the terrycloth 'mother,' even while nursing from the wire dummy.

Development of adult behavior. Rhesus monkeys raised with terrycloth 'mothers' seem to function normally until they become sexually mature, but behavioral deficits eventually do appear. A normally raised male rhesus monkey 'mounts' a female during her reproductive cycle if she 'presents' to him (Figure 7.11), but the motherless males never mounted any females and seemed not to know how to behave in this situation. Motherless females did come into their reproductive cycles (their genitals swelled up and became bright pink), but they never 'presented' to any test males, and they consistently rejected all sexual advances. A few such females were artificially inseminated under anaesthesia and became pregnant. When their babies were born, they showed no signs of maternal behavior, such as picking up their infants and holding them to the breast. Instead, they either ignored or rejected their infants, in some cases so forcibly that the infants had to be removed for their own safety. Sexual behavior and maternal behavior had never been learned in these monkeys, even though their behavior had seemed normal up to the time of sexual maturity. Adult social behavior has very strong learned components in rhesus monkeys and in other higher primate species as well.

Harlow continued his experiments, seeking to pinpoint what the motherless monkeys were failing to learn from the terrycloth dummies. Could the young monkeys receive a proper upbringing without a live mother? What conditions were minimally necessary? Remembering that wild juvenile primates associate with one another in play groups, Harlow let some of the young, motherless monkeys play together. He found that motherless monkeys who had opportunities both to cling to a terrycloth dummy and to play with one another developed normal adult social behaviors. By varying the length of the play period, Harlow was able to show that as little as half an hour of play per day was adequate to ensure that young monkeys would acquire normal adult behaviors. Harlow concluded that instincts were not sufficient to produce the proper sexual behavior or maternal behavior in these monkeys, but that a youthful period of social learning was also required.

Rough and tumble play. Most play in primates is **rough and tumble play**, in which there is frequent and repeated body contact, including pushing, pulling, and climbing—just watch young children in a schoolyard to see examples. Primate play also includes a good deal of chasing and dodging, usually followed by more rough and tumble play. Although rough and tumble play is neither sexual nor maternal, it seems to teach many lessons, such as how to handle and perhaps restrain other individuals without hurting them. Hurting another individual, whether accidentally or not, brings squeals of pain, generally causing adults to intervene and break up the activity. Play also teaches taking turns at different roles: pursuer and pursued, restrainer and restrained, climber and support, etc. In the context of play, the players learn how strong or weak other individuals are, and how much rough play each will tolerate. These lessons are later refined into dominance and submission relationships with other individuals and into sexual behaviors such as those in which male monkeys mount females. Mounting behavior arises during rough and tumble play, without regard to the sex of either individual; only after sexual maturity does it take on an explicitly sexual meaning. The defense and protection of smaller individuals, including picking them up and delicately cradling them, is also learned in play. In large, mixed-age social groups, there is usually an opportunity for subadult animals to practice the behaviors related to child care.

There are parallels in human behavior. Children learn many lessons in play, including cooperation, turn-taking, role-playing, counting and score-keeping, setting and following rules, and settling arguments and disputes. They also learn a good deal about each other's personalities:

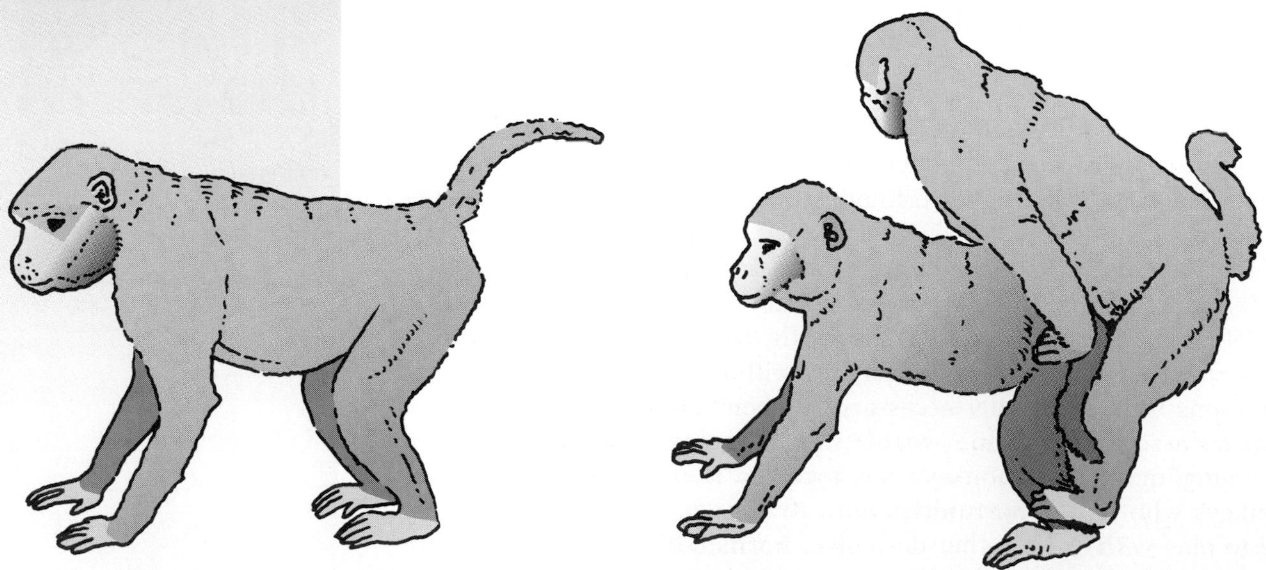

FIGURE 7.11

Normal mating behavior in rhesus monkeys.

who plays fair, who cheats, who is a bully, who cries if they do not get their way, and so forth. Children often imitate adult roles in play, practicing many of the skills that they see adults using and that they may themselves use later in life: hunting, digging, child care, food preparation, useful and artistic crafts, and so on. Abused children and those deprived of the opportunities of exploratory and rough and tumble play with other children often fail to develop the proper adult social behaviors, including both marital behavior (which is much more than just sexual) and parental care.

Social organization among primates

Most primates are extremely social, but the size and complexity of social groupings vary greatly. In this section we examine baboons, monkeys of the genus *Papio*, closely related to the species that Harry Harlow studied. *Papio hamadryas* is a harem-forming, polygynous species that lives in the rocky highlands of Ethiopia. The other baboons, *Papio cynocephalus* and related species, live on the open, grassy savannas of Africa.

The savanna baboons all share a complex form of social organization. In the wild, they hardly ever fight. They express dominance largely through gestures such as staring at an opponent, showing their teeth, or slapping the ground. We can study dominance by observing pairwise encounters (between two individuals at a time) and noting which baboon more often gets what it wants. Dominance status generally follows size and fighting strength, although it is rarely contested and outright fighting is rare. The situation becomes more complex in encounters of more than two individuals. One group of males, called the central males, support each other, in effect 'ganging up' on any threat to one of their number or to an infant or juvenile member of the group. Because they support one another, these males form the stable core of the group. Individual strength does not ensure membership in this central group, for the individually strongest male (the one who ranks highest in pairwise encounters) is usually not a member of the central group unless the troop is small. This individually strongest male (the 'scout') generally travels in the very front of the group, the most vulnerable position in the face of danger. It is the central males, however, who determine the group's direction of movement. The central males also keep order in the group; their mere arrival breaks up fights. Their superior fighting abilities protect the entire group from external threats such as predators.

The females help hold the group together in other ways. Baboons, like other monkeys, are forever grooming one another—picking burrs and parasites from each other's fur (Figure 7.12). Any baboon may groom any other, but females generally do the most. As a gesture of friendliness, it is generally reciprocated, with groomer and recipient taking turns. Infants and juveniles are often groomed by their mothers. Females who are not yet mothers themselves often practice at grooming behavior and infant care. This 'mother-in-training' behavior, called 'allomothering' or 'aunt behavior,' is very important in many primate species. Human examples include holding and feeding other people's children, playing with children,

FIGURE 7.12

Social behavior in monkeys.

Threat display of a male Hamadryas baboon, showing his large canine teeth

Grooming behavior in rhesus monkeys

Baboons grooming one another

and, or course, baby-sitting. Through such experiences, young primates of both sexes learn the behavior patterns essential to parenting. Alloparenting benefits young primates by providing them with social experiences, learning experiences, and even substitute parents in the event of the parent's death or temporary removal.

Females primates go through reproductive cycles when they are not pregnant or nursing. These reproductive cycles are marked, as in most female mammals other than humans, by a conspicuous **estrous period** that coincides with the time of ovulation. The female's sexual status is advertised to males by swelling and reddening of her genital area, as well as by 'presenting,' a behavior in which a female displays her genital area to interested males (see Figure 7.11).

The characteristics of the central males are perpetuated by a form of selection in which they gain access to estrous females at the time when sexual swellings are maximal and ovulation is most likely. Other males 'take what they can get,' meaning that their access to estrous females is at times when ovulation is less likely. As a consequence, high-ranking males are likely to leave more offspring than low-ranking ones, and their genes are thus favored by selection.

Sometimes a male and female form a 'consort pair' for up to several days. Female savanna baboons also copulate with males frequently and promiscuously without necessarily forming consort pairs; after mating, they often assert their independence by immediately running away from their partner. In this, they differ strikingly from the females of the cliff-dwelling Hamadryas baboons, *Papio hamadryas*, a species in which females are herded into harems.

Observations over time have led to the discovery of turnover and replacement in the social system. When one of the central males dies, the remainder of the central group generally carries on without him. However, when the central group falls below a certain minimum size, the entire group dissolves and a new central group takes over. The new group usually includes the 'scout' male (who is replaced by a new scout), and the other males in the new central group are generally his age-mates. The social cohesiveness of this central group was formed years earlier in a juvenile play group in which dominance relationships and future alliances are formed. There are dominance interactions among female baboons, and high-ranking females generally have high-ranking offspring. Juvenile baboons at play frequently look back to their mothers for back-up, and a higher-ranking mother generally provides more reassurance.

Reproductive strategies among primates

Primatologists of the 1960s and earlier decades usually emphasized social dominance relationships among males. Beginning with the early work of Jeanne Altmann, Phyllis Jay, and Jane Goodall, relationships among female primates began to receive equal or greater attention. The primatologists of the subsequent generation conducted many important new studies that focused on the social behavior of female primates.

One primatologist who has changed our views of primate sexual biology is Sarah Hrdy, whose sociobiology is influenced by a feminist outlook. Female primates, according to Hrdy, are much more sophisticated than previous researchers had ever imagined. Whereas the adult males use rather obvious means to maximize their inclusive fitness, Hrdy discovered that the means used by females were considerably more subtle and usually involved influencing the behavior of the males.

In her work on langur monkeys in India, Hrdy discovered the important ways in which female monkeys, although subordinate in power and strength to males, nevertheless managed to influence male reproductive choices and male social behavior to the female's own advantage. Male primates differ from one another in the number of offspring that they leave, but female primates frequently modify what males must do to achieve reproductive success. Female primates can often maximize their own reproductive success by the ways in which they influence male social behavior. Hrdy identified at least five ways in which female primates can maximize their reproductive fitness:

1. by choosing their mates,
2. by influencing males to support and protect them,
3. by competing with other females for resources,
4. by cooperating with other females (usually close relatives),
5. by increased efficiency in daily activities such as locomotion and obtaining food.

Studies of animal behavior and parental investment before about 1970 were in most cases written by male scientists and tended to emphasize male behavior and dominance relations among males. Males were often described as making choices, while females were often depicted as either passive or 'coy.' We now know that females make important choices of their own and solicit male attention for a variety of reasons not always related to the production of offspring, showing that it often pays for them to be flirtatious rather than coy. For example, females of many species can mate with males who are not their usual partners, and they have often been observed to mate at times when they were already pregnant or otherwise unable to produce new offspring. Males can generally increase their reproductive fitness by mating with as many females as they can, indiscriminately. The optimal behavior for a female, however, depends on her own fitness and social status, as well as that of her possible mates. If a female is of high status herself, and is mated to a high-status male, then she has nothing to gain from mating with a lower-ranking male. In contrast, a female of low status, or one mated to a low-status male, could potentially increase her fitness by mating additionally with a high-status male. If he sires one of her offspring, then she has produced a higher-status offspring and raised her own fitness as a result. That is because the offspring of higher-status males have more opportunities to mate; therefore, females can maximize their fitness (leave more grandchildren) if they raise the offspring of high-ranking males. Moreover, even if their mating produces no offspring at all, the high-status male who has mated with her will maximize *his* fitness by protecting any female that he has mated with, as well as her subsequent offspring, because he would be operating under the assumption that these offspring might be his.

Thus, females can gain important advantages from liaisons with high-ranking males, even at times when ovulation has not occurred and when subsequent pregnancy and childbirth are not possible.

Hrdy also discovered that male primates are sometimes infanticidal, and that female willingness to mate with powerful males was sometimes a strategy to discourage their infanticidal tendencies. Infanticide may occur among certain primate species whenever a new dominant male takes over a group. The new male can increase his fitness if he kills infants that are not his, especially if their mothers are lactating. Lactation inhibits the female reproductive cycle in most mammalian species; infanticide by the male causes lactation to end. Females enter estrus and the male gains access sooner to reproductive females. Once he has mated and produced offspring, however, the male will maximize his fitness if he defends all his mates and their offspring.

One of the many consequences of primate reproductive strategies is a difference between the sexes in how they pay attention to social rankings. Males in socially ranked species must pay attention to their own rank and status—they must remember who has ever threatened them or been intimidated by them. Females, however, must know much more, because each female must not only know her own status, but also that of every male in the group. In order to know whether one potential mate ranks higher than another, she must pay attention to *all* the social interactions among the males. In social species, females therefore generally take more interest than males in knowing about the social interactions of all other members of the group and in learning the status of all the males. Those who are better at paying attention to male–male interactions and correctly judging each male's social status and genetic fitness are at a selective advantage because they are better able to maximize their fitness by their behavior toward these males.

Both Hrdy and Jane Goodall have observed several instances in which competition between females produced outright hostility, even infanticide. Arguing from a sociobiological perspective, Hrdy explained that competition among unrelated females should be expected when their genetic self-interests are in conflict. A universal sisterhood, in which all females cooperate as a unit, would therefore never evolve. In evolutionary terms, such a universal sisterhood would not be a stable strategy, because an individual female would always be able to 'cheat' by refusing to cooperate, and by doing so she would raise her fitness and be favored by natural selection. Because evolution would never be expected to produce cooperative sisterhoods among unrelated females, Hrdy suggests that women who share her desire for such cooperative sisterhoods should strive to create them socially. Humans are not prisoners of biological destiny and are able to create social groupings and social behaviors that have not evolved.

Some examples of human behaviors

Much interest has focused on human behaviors of which large segments of society disapprove and the extent to which these behaviors are learned or inherited. People who wish to change behavior that has a strong learned component generally seek to find *how* it is learned and how an alternative form of behavior can be learned instead. If the behavior has a strong inherited component, it will be more difficult to bring about change through education. Other forms of intervention that might be more appropriate in such cases include trying to identify any genes involved in the behavior. However, most of the behaviors of interest are complex and are probably influenced by many genes, making it harder to identify any of them or to modify them in any meaningful way.

Aggression. Konrad Lorenz was a German scientist who studied animal behavior. He first won recognition (including a Nobel Prize) for his studies on imprinting, a form of learning that occurs early in life. In his later years, Lorenz wrote several controversial books in which he claimed that many human behaviors are instinctive. For example, in his book *On Aggression*, Lorenz claimed that aggression is largely instinctive in humans as well as other animals. As evidence, Lorenz argued first that aggression is widespread in many animal species and in various human societies. Second, he argued that the facial expressions and other gestures that accompany aggression and aggressive threats are similar in humans and animals and are also similar across many human societies.

Other scientists, however, have marshalled considerable evidence that aggression in humans has strong learned components.

1. Aggression takes many different forms in different societies, which use different weapons and different fighting traditions. If aggression were entirely instinctive, one would not expect it to be so variable.
2. Aggression is more prevalent in those societies that encourage it, and it generally takes the form that the society encourages. In the many societies that encourage aggression only in males or only in certain age groups, it occurs primarily in the groups in which it is encouraged. In societies that discourage aggression, it is much less common.
3. Within any society, some individuals are more aggressive than others. Individuals trained to be aggressive become aggressive, while most people raised to be less aggressive become less aggressive. We would not expect such large individual differences if aggression were inherited.
4. When aggressive behavior is desired, as in the military or in sports such as boxing and judo, it must be taught and practiced frequently.

Alcoholism. Alcoholism is a complex form of behavior that seems to have both learned and inherited components. To complicate matters, there are various degrees of alcoholism, and many individuals are classified as alcoholics by some criteria and not others. However, the greatest complication arises from the heterogeneity of the disorder itself: alcoholism manifests itself differently in men and in women, and it may also have different characteristics in different social classes.

Recent studies show that alcoholism may in fact exist in two or more separate forms. Type I alcoholism, also called late-onset or milieu-limited, typically arises after age 25 and is common in both sexes. It is characterized by psychological or emotional dependence (or loss of control), by guilt, and by fear of further dependence. This type of alcoholism frequently responds well to treatment. By contrast, type II alcoholism, also called male-limited, early-onset, or antisocial alcoholism, typically arises during the teenage years and is common in males only. It manifests itself in alcohol-seeking behavior, in novelty-seeking or risk-taking behavior generally, and in frequent impulsive and antisocial behavior including alcohol-related fighting and arrests. This type of alcoholism responds poorly to conventional forms of treatment.

Adoption studies in Denmark, Sweden, and the United States suggest that a predisposition for type II alcoholism may be inherited. The largest study, of 1775 adoptees in Sweden, found that the rate of alcoholism among the biological sons of type II alcoholic fathers raised in families without alcoholics was nine times the rate among other adoptees, including those adopted into type II alcoholic households. Type I alcoholism, however, shows a much smaller hereditary influence and may instead be subject to strong environmental influences. Some experts suggest that type I alcoholism is still heterogeneous and should be subdivided further.

Twin studies on alcoholism show a higher rate of concordance in identical twins than in fraternal twins, meaning that among identical twins, if one twin is an alcoholic that there is greater probability that the other is also an alcoholic. (As described in Chapter 3, pp. 80–81, a rate of concordance is the fraction of people who match in a certain trait.) The concordance is greater for type II alcoholism than for type I.

Sexual orientation. Some people regard variations in sexual orientation, including homosexuality, as innate and unchangeable, while others view them as learned behavior patterns that are subject to change. The available evidence, which is not very extensive, was summarized and reviewed in two books by the neurobiologist Simon LeVay. Some small differences were observed between the brain structures of homosexual men and heterosexual men, but many of the homosexual men in the study had died from AIDS, so it is uncertain whether these differences resulted from AIDS or pre-dated the onset of that disease. If a difference in brain structure could be demonstrated between homosexual and heterosexual men, other questions would remain to be answered: did the structural difference precede the sexual orientation, or might the structural change have resulted from some aspect of a lifestyle difference? Scientists are only just beginning to examine such questions in homosexual men; studies examining lesbian women are even rarer.

Twin studies have been conducted on homosexual males who have twin brothers. The rate of concordance is higher for identical twins than for fraternal twins, meaning that, if one twin is homosexual, there is a much higher probability that the other twin is also homosexual if he is an identical (monozygous) twin than if he is a fraternal (dizygous) twin. Such a result is suggestive of at least some genetic influence, but the very real methodological problems of such twin studies makes it very difficult to rule out other possible influences. The biggest shortcoming of twin studies is that the environments in which the twins are raised are never chosen at random and are usually very similar, even in cases of adoption.

Rape. The subject of forcible sexual intercourse, or rape, has received much attention recently. Discussion has centered on ideas put forth by two American sociobiologists, Randy Thornhill and Craig T. Palmer. Feminist writers such as Susan Brownmiller had earlier portrayed rape as primarily a crime of violence rather than of sex, an attempt by the rapist to dominate and control his victim. Against this idea, Thornhill and Palmer argue that rape is very much about sex. They use statistical records from rape crisis centers to show that victims are most often in the prime reproductive age range and that married rape victims feel more heavily traumatized than unmarried ones. They claim that a predisposition to rape persists because rape does occasionally produce children who perpetuate the genes of the rapist. Therefore, these authors argue, rape is natural, though they hasten to add that it is still reprehensible behavior. Their hypothesis does not, however, explain why the overwhelming majority of men are *not* rapists.

Many studies show that most women prefer as mates men who are good-looking, healthy, strong, skillful, kind, respected by others, and high in social status and wealth. However, Thornhill and Palmer say that "men might resort to rape when they are socially disenfranchised, and thus unable to gain access to women through looks, wealth, or status."According to this hypothesis, the men who become rapists can leave more offspring if they rape than if they do not, because they are usually the men that women are unlikely to choose as mates.

Barbara Ehrenreich, a critic of the Thornhill–Palmer hypothesis, emphasizes that rapists make inferior husbands and fathers, and that the children of rape are thus far less fit than other children. The mothers of these children have been traumatized, and their fathers are in most cases gone, and when present they are neither good fathers nor good role models. Compared with the men that women would choose as mates, rapists are inferior in social standing, inferior in fitness, and inferior in their ability to raise fit children. This may explain why most men are not rapists: they can produce more children, and contribute as fathers to their children's fitness, by cultivating the behaviors that women value. The children of these men and the women who choose to marry them usually attain higher social status and are more socially and psychologically equipped to enter into normal and stable relationships themselves. They tend to leave more future children and are thus far more fit than the children born of rape.

Child abuse. There is now a growing body of evidence that child abuse in humans follows the same patterns as infanticide in other primate species. In particular, stepfathers (who are genetically unrelated to the children who live with them) are up to 100 times more likely to abuse or kill the children in their care than are genetic fathers. This and related findings are the subject of numerous studies on criminal records in various countries around the world. Much of the evidence is reviewed in a recent book by Canadian sociobiologists Martin Daly and Margo Wilson.

THOUGHT QUESTIONS

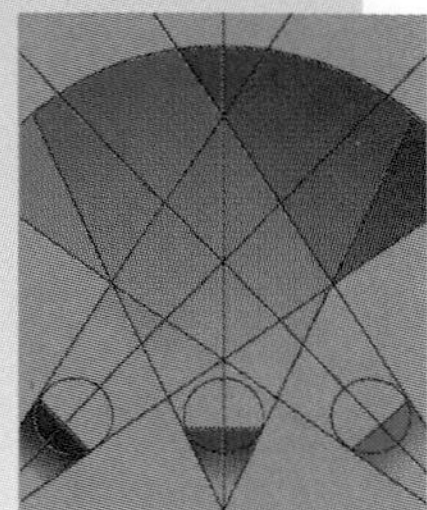

1. To what extent can sociobiological findings on animals be extrapolated to humans? Are animal studies relevant at all to the study of human behaviors such as alcoholism or homosexuality?

2. How important are fathers in early childhood development? What important *social* skills do children learn from interacting with their mothers? With their fathers? What do children learn from watching their parents interact with one another? What happens in families in which no father is present? What happens when no mother is present?

3. Think of the many ways in which humans learn (and subsequently practice) the social skill of evaluating the social status and motives of others. How much do we learn (or what skills do we exercise and practice) from play, from small-group discussions, from gossip, from novels, or from television? Do males and females participate in these activities in the same way? Why or why not?

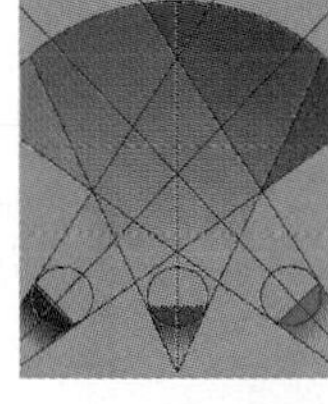

SOCIOBIOLOGY, THE COMPARATIVE STUDY OF SOCIAL behavior and social groups among organisms, is a subfield within evolutionary biology. Much of social behavior is learned, but only those aspects of social behavior that are inherited are subject to natural selection and therefore to evolutionary change. Sociobiology therefore focuses on inherited behaviors or capacities, although all sociobiologists agree that learning can modify those behaviors in many species. Often it is a predisposition for a behavior, or a capacity to learn a behavior, that is inherited, not the behavior itself. Sociobiology can predict when natural selection will favor altruism, social groups of differing sizes, behavior that is stereotyped as opposed to plastic, or behavior that differs between the sexes. Many such hypotheses have already withstood repeated testing. Because of differences in parental investment, natural selection usually favors different behaviors in males and females.

In humans, even though some components of behavior are inherited, every behavior can also be modified by learning. Twin studies, adoption studies, cross-cultural studies, and studies of other species can all provide important clues to the understanding of human behavior patterns. Many human behaviors vary across cultures; others are strongly influenced by early childhood experiences. One of the most effective and cost-efficient ways in which we can improve human society is to provide each and every child with a safe childhood full of experiences from which to learn.

CHAPTER SUMMARY

- **Sociobiology** is the biological study of **social behavior** and **social organization** among all types of organisms.
- Organisms live in social groups because it affords such advantages as group defense, help in finding and exploiting food resources, and greater reproductive opportunities.
- **Altruistic** behavior is favored if it contributes to **inclusive fitness**.
- Among **reproductive strategies**, **asexual reproduction** is favored by natural selection in situations in which a quickly produced series of uniform offspring are advantageous, but **sexual reproduction** is favored whenever future conditions are uncertain and a greater variety of offspring is a greater advantage.
- Species living in unstable environments are often ***r*-selected**, meaning that they reproduce rapidly and prolifically but provide little or no parental care. By contrast, ***K*-selected** species living in stable environments provide more parental care and are selected for greater efficiency in exploiting environmental resources.
- There are many different mating systems, including **monogamy**, **polyandry**, **polygyny** and **promiscuity**.
- In many species, different levels of **parental investment** favor different reproductive strategies in **females** (**egg** producers) and in **males** (**sperm** producers).
- All behavioral characteristics that have been closely studied are influenced by both genetic and environmental influences to various degrees. In animal species, behaviors related to mating and courtship are more often instinctive, while locomotor behaviors are more often **learned**.
- Learned behavior is highly important among primates, especially among humans.

KEY TERMS TO KNOW

altruism (p. 285)
anisogamy (p. 294)
asexual reproduction (p. 292)
egg (p. 294)
eusocial (p. 288)
females (p. 295)
group selection (p. 286)
haplodiploidy (p. 289)
inclusive fitness (p. 286)
instinct (p. 280)
isogamy (p. 294)
***K*-selection** (p. 294)
kin selection (p. 287)
learning (p. 275)
linear dominance hierarchy (p. 285)
males (p. 295)
monogamy (p. 296)
parental investment (p. 295)
pheromones (p. 289)
polyandry (p. 297)
polygyny (p. 296)
promiscuity (p. 297)
***r*-selection** (p. 293)
reproductive strategy (p. 292)
rough and tumble play (p. 300)
sexual reproduction (p. 292)
social organization (p. 284)
social behavior (p. 274)
sociobiology (p. 274)
sperm (p. 294)

CONNECTIONS TO OTHER CHAPTERS

Chapter 1: Sociobiology is a good example of a paradigm.

Chapter 2: Social behavior can differ among different genotypes.

Chapter 4: Social behavior can affect the fitness of each genotype. Social behavior also evolves, and the evolution of behavior is of prime interest to sociobiologists.

Chapter 5: Social behavior can greatly affect fitness and thus alter allele frequencies.

Chapter 6: Mating is one of the most important kinds of social behavior. Population growth is a result of mating behavior on a large scale.

Chapter 8: Access to good nutrition is one important motivating force in social behavior.

Chapter 10: Social behavior happens as the result of brain activity.

Chapter 11: Drugs can alter social behavior.

Chapter 12: Social support can promote healing; stress can interfere with healing.

Chapter 13: AIDS is spread by certain social behaviors.

PRACTICE QUESTIONS

1. For each of the following human behaviors, state at least one piece of evidence pointing to an important learned component for the behavior:
   a. eating with utensils
   b. speaking English
   c. hunting
2. For each of the following behaviors, present an argument for an important instinctive component of the behavior:
   a. tail wagging in dogs
   b. mooing in cows
   c. smiling in humans
3. Present at least one argument supporting each of the following assertions:
   a. that piano playing ability has important learned components
   b. that piano playing ability has important innate components
4. State at least six research methods used in sociobiology.
5. Which of the following behaviors is most likely to have strong instinctive components in a wide variety of species?
   a. attracting a mate
   b. climbing a tree
   c. finding and capturing food
   d. moving about one's habitat
   e. none of the above
6. Natural selection favors the instinctive control of behavior in all of the following situations *except*:
   a. in species with short life spans
   b. in outsmarting prey
   c. in courtship displays or mating calls
   d. in escaping from sudden danger
   e. in building a nest or weaving a web
7. Which of the following does NOT fit the definition of an altruistic act?
   a. a millionaire leaves money to charity in her will
   b. a taxi driver runs through red lights to get a pregnant woman to the hospital in time to deliver her baby
   c. a man runs in front of oncoming traffic to save a small child
   d. a firefighter runs into a burning building to rescue people who may be trapped inside

8. Which of the following belongs in a different group from the rest?

   a. ants b. bees c. termites d. wasps

9. Which of the following situations favors sexual reproduction?

   a. microorganisms reproducing inside a human host

   b. fungi growing in a fallen tree as it decays

   c. insects or worms exploiting a large and dependable food supply

   d. insects colonizing new food supplies by laying eggs in them

10. What are the three conditions that define eusociality?

11. How is kin selection defined?

12. The central hierarchy in a baboon troop usually includes:

    a. males of mixed ages

    b. females as well as males

    c. age-mates who played together as juveniles

    d. one or more consort pairs

    e. the 'scout'

8 Nutrition and Health

CHAPTER OUTLINE

Digestion Processes Food into Chemical Substances that the Body Can Absorb
- Chemical and mechanical processes in digestion
- The digestive system

Absorbed Nutrients Circulate Throughout the Body
- Circulatory system
- The heart

All Humans Have Dietary Requirements for Good Health
- Carbohydrates
- Lipids
- Proteins
- Conversion of macronutrients into cellular energy
- Fiber
- Vitamins
- Minerals

Malnutrition Contributes to Poor Health
- Eating disorders
- Protein deficiencies
- Ecological factors contributing to poor diets
- Effects of poverty and war on health
- Micronutrient malnutrition

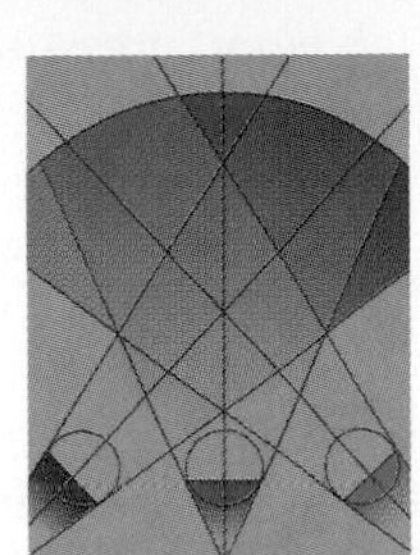

ISSUES

How do organisms get the energy that they need to sustain life? How similar are humans to other organisms in this regard?

How does the body process food?

Do all humans have the same dietary requirements?

How are human diets related to good health? How are they related to chronic diseases?

What is malnutrition? What are its causes and consequences?

Why do some people deliberately starve themselves?

BIOLOGICAL CONCEPTS

Organ systems (digestive system, circulatory system)

Cell membranes (diffusion, active transport)

Acidity

Molecular structure (chemical and physical basis of biology, water, carbohydrates, lipids, proteins, enzymes, polar and nonpolar molecules)

Energy and metabolism (energy conversion and storage, chemical bond energy, ATP, calories, Krebs cycle, oxidation–reduction reactions)

Evolution (lactose intolerance)

Homeostasis

Health and disease (macronutrient malnutrition, micronutrient malnutrition, vitamins and minerals, eating disorders, ecological factors)

Species interactions (mutualism)

When she weighed 140 pounds, Melanie thought of herself as fat and ugly. Her menstrual periods stopped when her weight dropped to 100 pounds. Now that she weighs 90 pounds, all her friends tell her she is too skinny, but she is sure they are wrong because she still thinks of herself as chubby. She wants to lose even more weight. Melanie has an eating disorder known as anorexia nervosa. Her body is not getting the nutrition it needs. She could die if the situation remains untreated.

Melanie's father went in last week for a routine checkup. Although he was feeling fine, the doctor told him he had high blood pressure and needed to control his fat intake. Unless he lowers the fat content of his diet, he faces an increased risk of having a heart attack. He must now learn to eat a low-fat, high-fiber diet, which will lower his chances for getting heart disease, the number one cause of death in most industrialized countries.

All of us need food, but our dietary requirements vary according to our body size, age, sex, level of activity, and previous state of health. In addition, there are variations caused by hereditary differences in body constitution, metabolic rates, and other factors. The world's populations have found many different ways of meeting these nutritional needs. Different diets arose in different parts of the world because different kinds of plants grew best in each climate and in each type of soil, and each culture has its own preferences and prohibitions that limit their uses of the available foods in their environment—no culture makes use of all foodstuffs available to them.

In this chapter we examine the body's use of food, human dietary requirements, correlations of diet with the incidence of chronic diseases, and the effects of malnutrition that can result from eating too little or from eating the wrong foods. Malnutrition is one of the major health problems of the world, particularly among the poor and in areas of turmoil. Malnutrition can also result from eating disorders among people with access to sufficient foods.

DIGESTION PROCESSES FOOD INTO CHEMICAL SUBSTANCES THAT THE BODY CAN ABSORB

All organisms need energy to carry out life processes. Plants get this energy from sunlight through photosynthesis (see Chapter 14, pp. 584–590). Most other organisms, including humans, get their energy from the foods that they eat. In addition, food provides the chemical building blocks needed to make and repair body materials.

Most of what we call food can be classified chemically into three types of major constituents and several minor constituents. The major constituents, which are called **macronutrients**, include carbohydrates, proteins, and lipids (fats); the minor constituents, which are called **micronutrients**, include vitamins and minerals. Digestion turns the food we eat into chemical forms that can be utilized by our bodies. To be useful to our bodies, food must first be converted into substances that the body can absorb. Digestion is the process that breaks down food into absorbable products.

CONNECTIONS
CHAPTERS 2, 14

Chemical and mechanical processes in digestion

Digestion has two aspects: chemical digestion and mechanical digestion. **Chemical digestion** breaks foods down chemically using **enzymes**, which are substances that promote or speed up a chemical reaction without themselves being used up in the reaction (Figure 8.1). This speeding up of reactions is known in chemistry as **catalysis**, and enzymes are therefore biological catalysts. Nearly all enzymes are proteins. In Chapter 2 (p. 60) we saw that the enzyme DNA polymerase helps to synthesize molecules. The enzymes of digestion, in contrast, help to break down molecules. Chemical digestion works on the surfaces of food fragments. **Mechanical digestion** exposes new surface areas to chemical digestion by breaking fragments into smaller fragments and by removing partly digested surface material.

The digestive system

The **digestive system** is one of the organ systems in the body. An organ is a group of tissues that are integrated structurally and functionally. An organ system is a group of organs that perform different parts of the

FIGURE 8.1

Enzymes: biological catalysts.

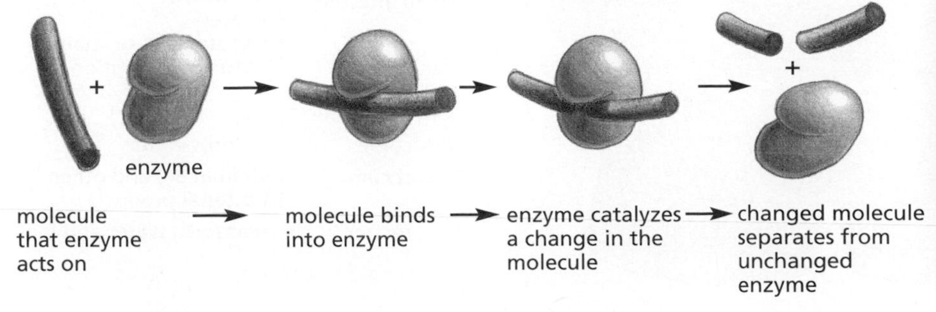
+
enzyme
+
molecule that enzyme acts on
molecule binds into enzyme
enzyme catalyzes a change in the molecule
changed molecule separates from unchanged enzyme

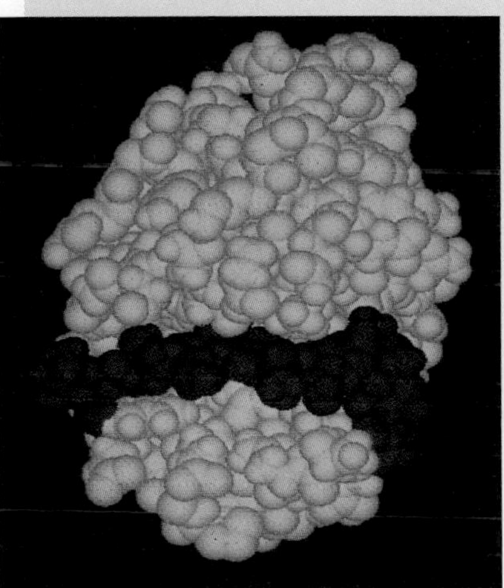

space-filling model of the enzyme lysozyme (blue) with a large sugar molecule (red) bound to it same process. Thus, the digestive system is a group of organs that together digest food. The body plan of the human digestive system is a common one: animals from the phylum Nematoda (roundworms) to humans have digestive systems that are a continuous hollow tube, called a **gut**, with an entrance at one end and an exit at the other. As we go through the next sections, locate the human digestive organs on Figure 8.2.

The mouth. Food is taken in through the mouth, and mechanical digestion begins in the mouth when the food is chewed. Chemical digestion of starches (carbohydrates) begins in the mouth with the enzyme **salivary amylase**. This enzyme, present in saliva, breaks down starches into smaller units (sugars). Starches are usually not in the mouth for long enough to be completely digested, however, and their digestion is completed later. Another salivary enzyme called lysozyme catalyzes the breakdown of large sugar molecules (polysaccharides) into smaller units (see Figure 8.1).

FIGURE 8.2

The human digestive system.

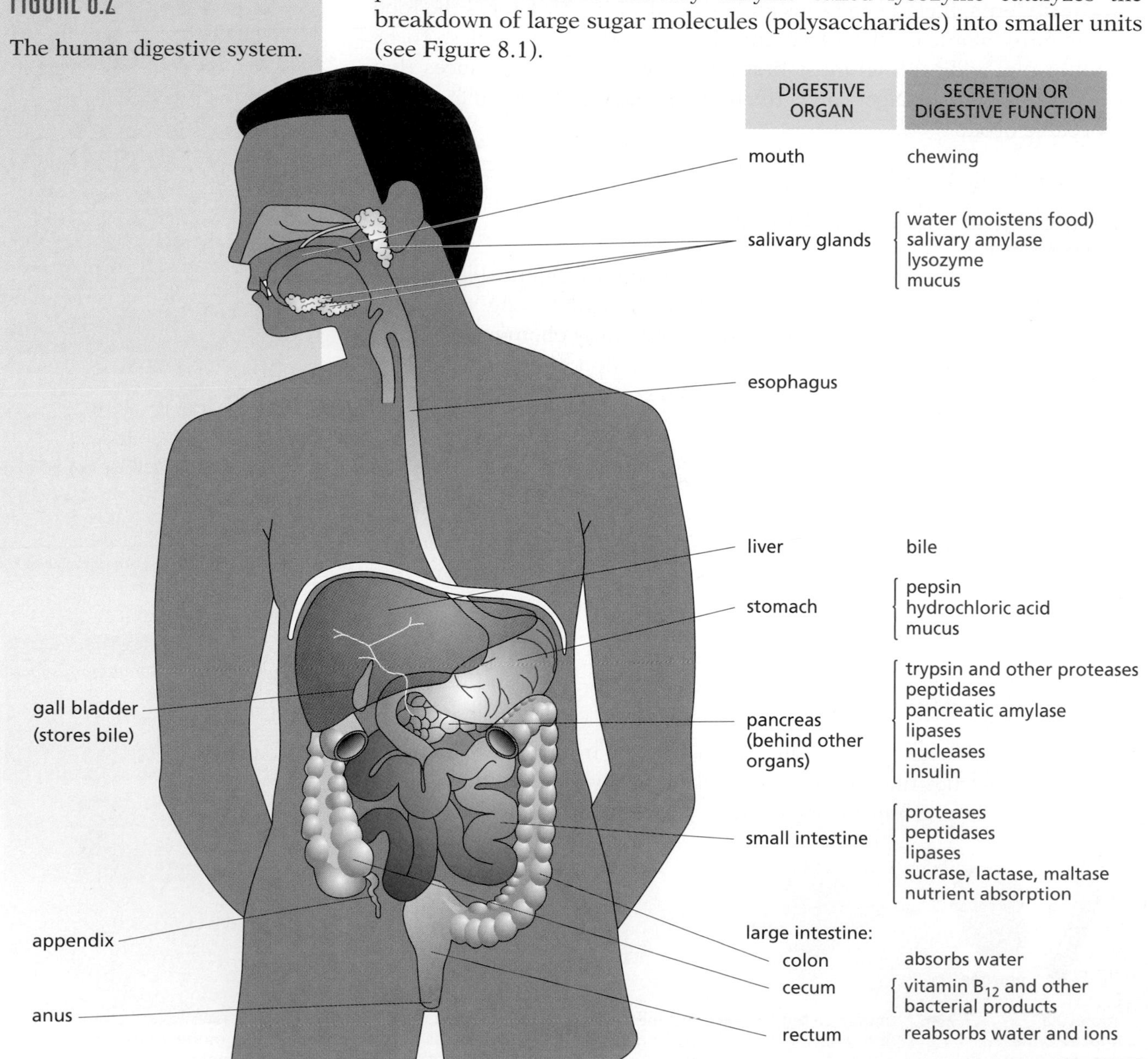

The stomach. Once food is swallowed, it passes quickly through the **esophagus** and into the **stomach**. The stomach performs mechanical digestion through rhythmic contractions that knead the food back and forth, mixing it thoroughly, rubbing food particles against one another, and exposing new surface areas. The main activity in the stomach is the digestion of protein, accomplished with the aid of the enzyme pepsin, which breaks large protein molecules up into smaller fragments called **peptides**. Like many other protein-digesting enzymes, pepsin is secreted in an inactive form, which protects the glands that secrete the enzyme from digesting themselves. The inactive form is converted into active pepsin by other digestive enzymes. Pepsin works best in an acidic solution, which the stomach provides by secreting hydrochloric acid (HCl). Acidity is measured by a scale called the pH scale; the lower the pH, the more acidic the solution (Figure 8.3). Fluids in the stomach are among the most acidic in biological systems, with a typical pH of 2. The stomach also secretes a mucus that protects the stomach lining (which is partly protein) from the pepsin and acid.

The small intestine: processing of fat. The lower end of the stomach empties into the **small intestine**, where the pH is no longer acidic. The term 'small' refers to the diameter, which is about 3 cm; the small intestine is actually very long (20 feet, or 6 m). Here, the food receives the secretions of the **liver**, called **bile**. In the watery environment of the intestine, fats tend to come out of solution and form large globules that coalesce to form even larger globules whenever they collide. Bile breaks up fat globules into smaller droplets and keeps these small droplets

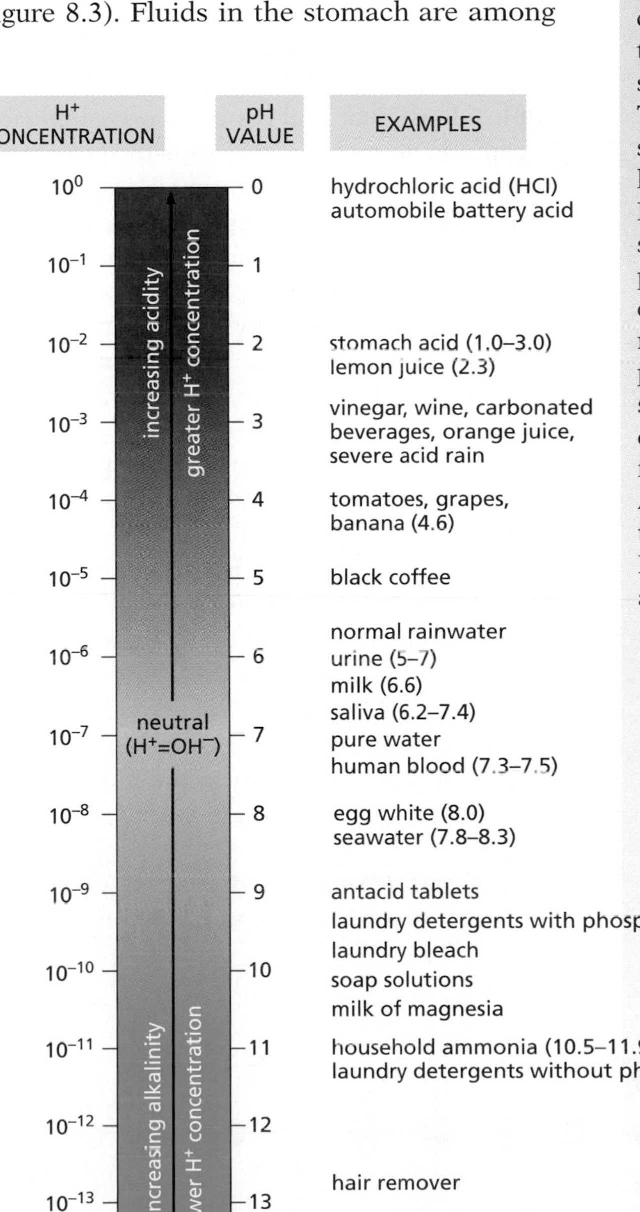

FIGURE 8.3

The pH scale. The pH of a solution tells us the concentration of hydrogen ions (H^+) in the solution. The H^+ concentration can be expressed as moles of H^+ per liter of solution, as shown to the left of the bar. It is more common, however, to express the concentration as pH, shown to the right of the bar. The pH scale is a reciprocal scale: the lower the pH, the higher the concentration of H^+ (and the more acidic the solution) and the higher the pH, the lower the concentration of H^+ (and the more basic the solution). The pH scale is also a logarithmic scale; that is, each number differs by a factor of ten from the next number. A solution with a pH of 2 thus has 10 times more H^+ ions than a solution with a pH of 3.

separate. Fats are insoluble in water because of an important difference in chemical structure of fats and of water (Figure 8.4). Chemical bonds holding atoms together to form a molecule consist of moving electrons (carriers of negative charge) from the individual atoms. The electrons can be shared equally by the atoms forming the bond, in which case the negative charges are evenly distributed and the bond is **nonpolar**. In a **polar** bond, the electrons are shared unequally, spending more time around one atom in the bond than around the other atom, making one atom slightly negative in comparison to the other. When certain atoms (oxygen, nitrogen, phosphorus) make bonds, they tend to attract the electrons toward themselves, making the bond polar. Water (H_2O) is one of the most polar liquids, with electrons unequally shared between hydrogen and oxygen atoms. Fats have few or no polar bonds, making them nonpolar (see Figure 8.4). Polar molecules tend to be soluble in water, and nonpolar molecules tend to be insoluble.

Globule formation is prevented by bile. One portion of the bile molecule is polar and consequently stable in contact with water; another portion is nonpolar and is thus unstable in contact with water but stable in contact with fat. The nonpolar portions of the bile molecules dissolve in the fat droplets, leaving the polar portions of these molecules exposed on the surface, in contact with the watery intestinal fluids. The polar coating helps the fat droplets to mix with the water and also prevents the small droplets formed by mechanical action from coming back together to form large globules. This overall process is called **emulsification** (Figure 8.5). Emulsification maintains the surface area of fat droplets and this increases the efficiency of digestion because only the surface is accessible for digestion and absorption.

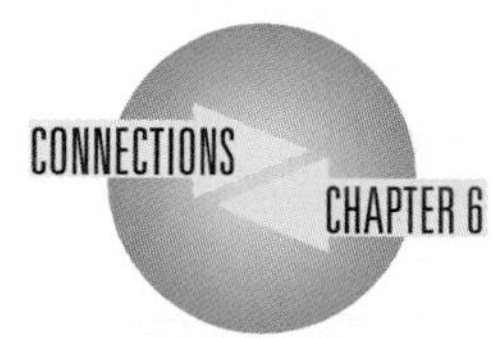

Bile is secreted by the liver in a steady dribble, but is used in large amounts when fats or oils are present in the intestine. Bile from the liver accumulates in the **gall bladder** until it is needed, and is then released all at once under the stimulus of a digestive hormone. A **hormone** is a chemical messenger that causes a specific physiological change in one or more target organs. There are many kinds of hormones with very different functions; in Chapter 6 we saw the functions of some of the reproductive hormones. Here, a digestive hormone is secreted by the intestinal lining whenever fats are present in the intestine; it acts on the gall bladder—its target organ—stimulating the release of bile into the intestine.

The small intestine: digestive enzymes. Farther down the small intestine, chemical digestion is completed by enzymes secreted by the

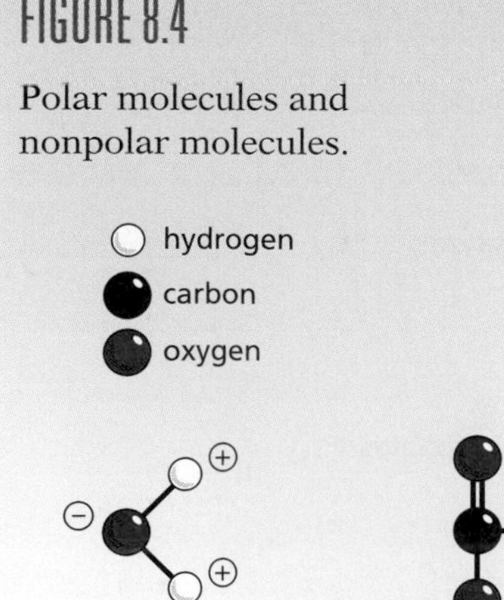

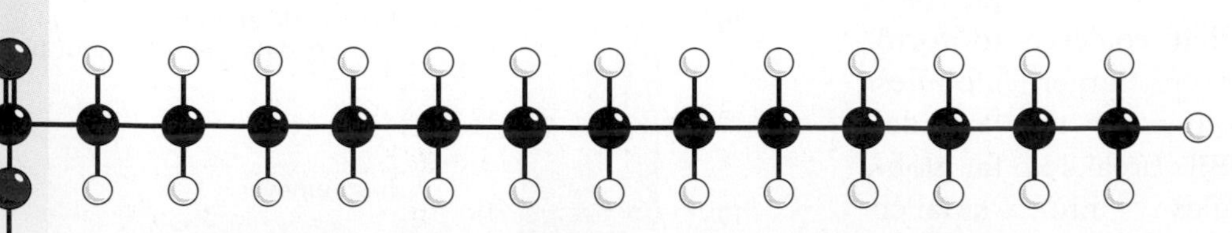

a fatty acid: a nonpolar molecule held together by nonpolar bonds

FIGURE 8.4 Polar molecules and nonpolar molecules.

pancreas and by the intestine's own lining. Enzymes are often named by combining the suffix '-ase' with the name of the molecule on which the enzyme works: proteases break down proteins, lipases break down lipids, and so forth. Among the intestinal enzymes are the following.

- **Proteases**, protein-digesting enzymes such as trypsin and chymotrypsin, secreted by the pancreas. Other proteases are secreted by the small intestine. Like the pepsin in the stomach, these enzymes break the chemical bonds between certain amino acids, thus breaking the proteins into smaller chains of amino acids called peptides. Each protease is specific and breaks only the bonds between certain specific amino acids.
- **Peptidases**, enzymes that complete the final stages of protein digestion by breaking peptides down into individual amino acids. Both the pancreas and intestinal lining secrete peptidases.
- Pancreatic amylase, an enzyme secreted by the pancreas. This enzyme continues the job, begun in the mouth, of breaking starches down into sugars.
- **Lipases**, fat-digesting enzymes, secreted by both the pancreas and the intestinal lining. These enzymes break down fats and oils into glycerol and fatty acids, molecules small enough to be absorbed.
- Sugar-digesting enzymes such as sucrase and lactase, secreted by the small intestine, which break down larger sugars (sucrose or lactose) into simple sugars such as glucose and fructose.

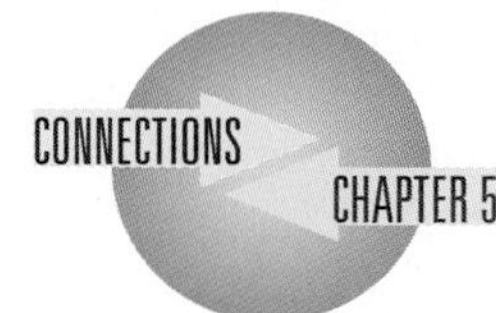
CONNECTIONS
CHAPTER 5

The presence of digestive enzymes may vary in human populations. We saw in Chapter 5 that northern Europe receives less ultraviolet radiation than other regions of the world, and populations living in northern Europe therefore have less sunlight to help them synthesize vitamin D. To supply this vitamin, most Europeans consume dairy products rich in vitamin D, and these dairy products also contain significant amounts of lactose, the sugar in milk. Thus, natural selection acted on European populations to favor those individuals who possessed the enzyme lactase, needed to digest lactose. Outside of Europe, ultraviolet radiation is usually sufficient for the synthesis of large quantities of vitamin D, so dairy products are not needed in the adult diet. Because they do not need to digest lactose, people in these populations often do not have the enzyme lactase. When a person without lactase consumes most dairy products, the unused

FIGURE 8.5

The action of bile salts in the emulsification of fats.

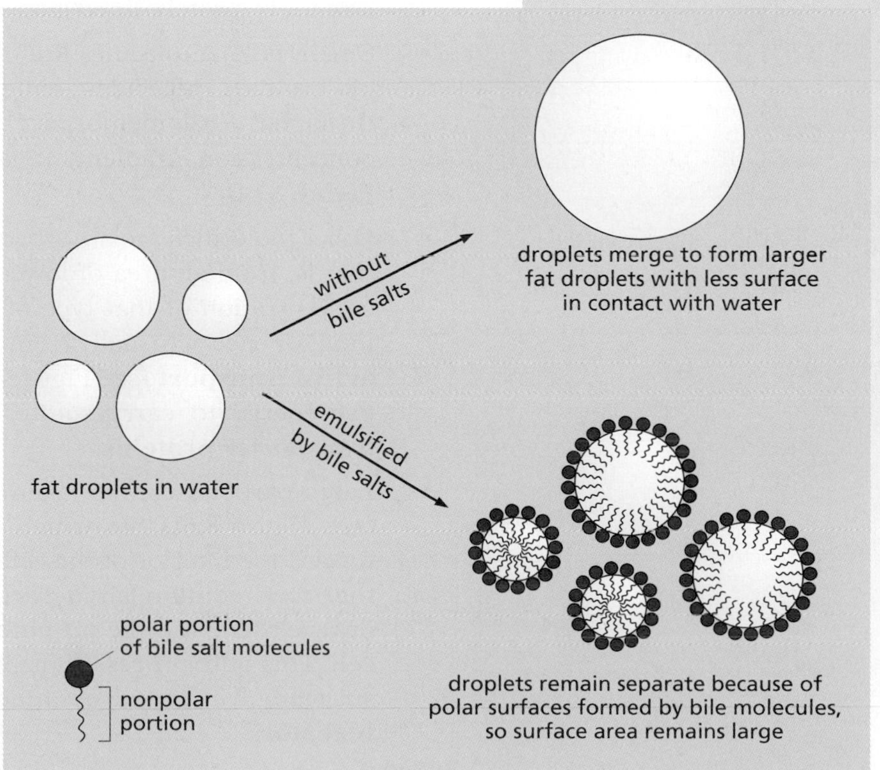
droplets merge to form larger fat droplets with less surface in contact with water
without bile salts
fat droplets in water
emulsified by bile salts
polar portion of bile salt molecules
nonpolar portion
droplets remain separate because of polar surfaces formed by bile molecules, so surface area remains large

lactose is fermented by gut bacteria, producing large amounts of carbon dioxide gas that results in painful cramps, diarrhea, and sometimes vomiting. This condition is called **lactose intolerance**.

The small intestine: nutrient absorption. The absorptive part of the intestine, the **ileum**, is lined on the inside with thousands of tiny fingerlike tufts called **villi**, which greatly increase the surface area through which the products of digestion are absorbed. Absorbable products include simple sugars, glycerol, fatty acids, and amino acids. Also absorbed by the villi are water and mineral salts, including dissolved ions (charged atomic particles) of sodium, calcium, and chloride, which do not require digestion to make them absorbable.

Absorption takes place through the cell membrane of the cells lining the intestine. The polar chemical structure of most products of digestion means that they cannot enter the cell directly, because the interior of a cell membrane is nonpolar. The membrane thus acts as a controller for what enters (or leaves) the cell. Chemicals are absorbed by one of four mechanisms (Figure 8.6); these mechanisms also bring chemicals into cells elsewhere throughout the body.

- Some small molecules enter the cells by **diffusion**, a process that requires no added energy and is therefore sometimes called **passive diffusion** (see Figure 8.6A). Diffusion works only if a **concentration gradient** exists; each substance diffuses from a place where it is more concentrated to a place where it is less concentrated. Small, uncharged molecules such as water (H_2O) or oxygen (O_2) diffuse directly through the cell membrane. Charged molecules cannot cross the membrane but may diffuse through **channels**, protein-lined 'holes' in the membrane.
- Small, polar molecules are often moved into cells with the help of protein molecules called **carrier proteins** that extend through the cell membrane. When this transport proceeds with a concentration gradient, it is called **facilitated diffusion** (see Figure 8.6B).
- Other molecules are absorbed *against* the concentration gradient, that is, from an area of lower concentration to an area of higher concentration of that type of molecule. This process requires an input of energy (usually from the breakdown of ATP) and is called **active transport** (see Figure 8.6C). The membrane proteins that use energy to carry solutes across the membrane are called **transporter proteins**.
- Large particles can be taken in by a process called **endocytosis** (see Figure 8.6D), in which the plasma membrane is pulled in toward the interior of the cell. The plasma membrane forms a pit that may contain large particles, then the margins of the pit draw closed, and the pit pinches off to form a vesicle inside the cell. This bulk process transports many molecules at once, either suspended in liquid or attached to membrane proteins called **receptors**.

(A) passive diffusion

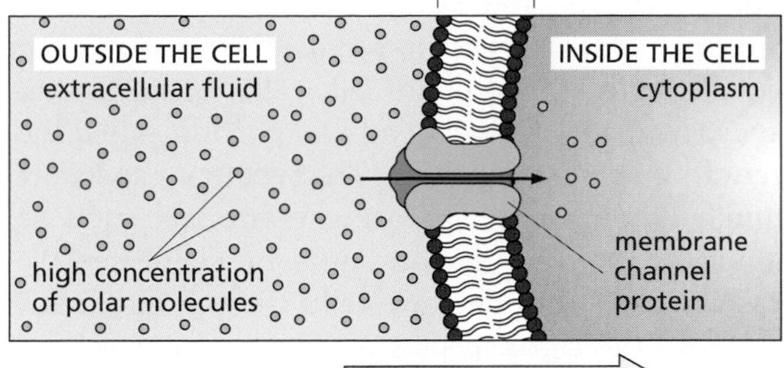

(B) facilitated diffusion

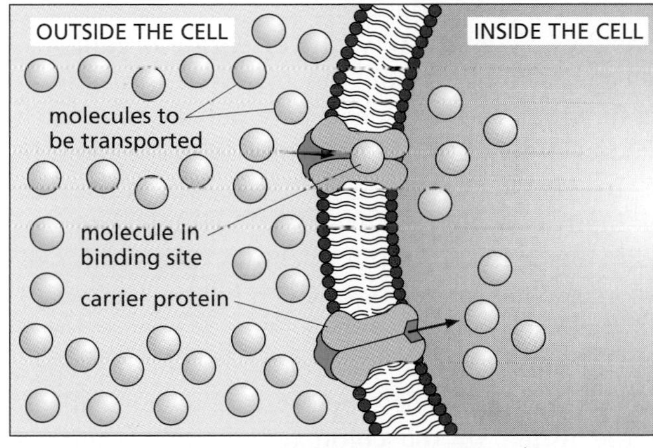

(C) active transport

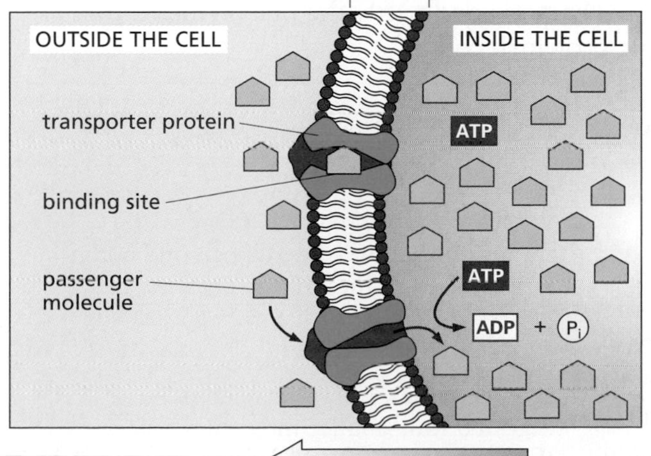

FIGURE 8.6

Membrane transport mechanisms.

(D) two forms of endocytosis

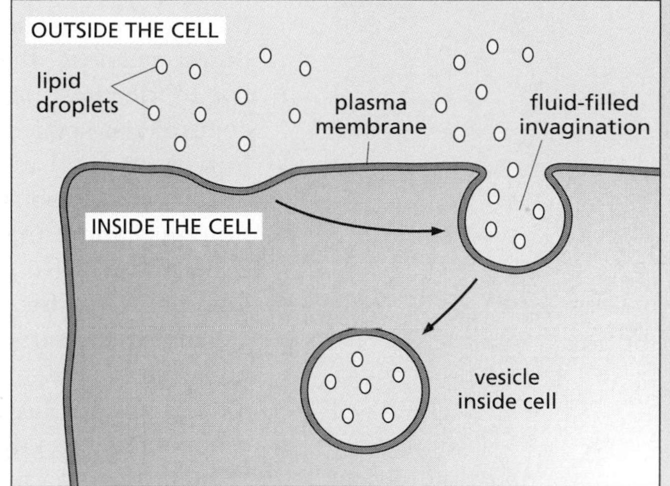

uptake without receptors

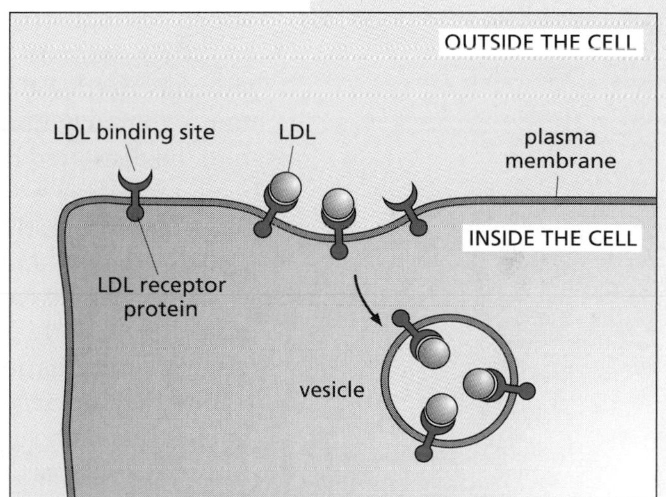

receptor-mediated endocytosis

ATP adenosine triphosphate
ADP adenosine diphosphate
P_i inorganic phosphate
LDL low-density lipoprotein

The large intestine: our mutualistic relationship with intestinal bacteria. The material that has not been absorbed by the ileum passes into the **large intestine**, most of which is also called the **colon**. In comparison with the small intestine, the large intestine has a larger diameter (2.5 inches or 6–7 cm), but is much shorter (4 feet, or 1.2 m). This part of the intestine is inhabited by bacteria, and certain nutrients produced by the bacteria are absorbed here. Mammals cannot make the enzymes that digest cellulose, the major constituent of plant cell walls. Bacteria that live in the intestine, and especially in a small dead-end portion called the **cecum** (also spelled caecum), must do it for them. The cecum is especially important (and also much larger) in plant-eating mammals such as horses and rabbits, which consume large amounts of cellulose. We humans cannot make the enzymes that degrade cellulose (no mammals can), and we also do not have the right species of intestinal bacteria to digest cellulose for us, so we cannot digest cellulose at all. We do, however, get some necessary nutrients from our intestinal bacteria. Humans cannot make vitamin K and biotin, needed for the synthesis of blood-clotting factors and for fatty acid synthesis, but our gut bacteria can synthesize them and we can then absorb these micronutrients.

Gut bacteria are considered **symbiotic** with vertebrate organisms. Symbiosis means simply that two organisms live together; **mutualism** is the form of symbiosis in which the two species are beneficial to each other. Mutualistic gut bacteria derive nutrients from the food taken in by their human host. In exchange, they synthesize vitamins that humans need and break many complex molecules into simpler components that are more easily absorbed from the intestine. The symbiosis may be disrupted by factors such as antibiotics, which kill the bacteria.

The large intestine: water absorption. The remainder of the large intestine consists of a straight portion called the **rectum**, which leads to a final opening called the **anus**. In the colon and rectum water is absorbed, mostly by diffusion, from the material passing through the gut, making this material a firmer consistency. Much of this material, which is called feces, is undigested food, but more than half is intestinal bacteria, which are rapidly replaced by bacterial cell division in the intestine. It is partly these bacteria and partly the bile pigments that give feces their characteristic brownish color.

THOUGHT QUESTIONS

1. Why is it important for food to spend the proper length of time in the stomach? What would happen if food left the stomach too soon?
2. The contents of the digestive tract are pushed along rather slowly by rhythmic muscular action (peristalsis). How does this relate to the body's need for a long intestinal tract?
3. What would be the consequences of a mutation that prevented a person's body from making a membrane protein necessary for the active transport of sugar from the intestine?
4. Why do food substances need to be digested into smaller molecules before they can be absorbed from the intestine?

ABSORBED NUTRIENTS CIRCULATE THROUGHOUT THE BODY

After nutrient molecules have been absorbed in the small intestine, they circulate throughout the body. Other materials, including dissolved ions, oxygen, and cells of the immune system, also circulate throughout the body.

Circulatory system

The circulation of these materials is carried out by the **circulatory system**. The circulatory system consists in all vertebrate animals of **blood**, contained within a series of **blood vessels**, and a muscular pump, the **heart**, that keeps the blood circulating.

The blood consists of a fluid material, the **plasma**, containing a series of cells and platelets. The cells include the red blood cells or **erythrocytes**, which contain the oxygen-carrying molecule **hemoglobin** (see Chapter 5, p. 214), and several types of white blood cells or **leucocytes**, which form the immune system (see Chapter 12, p. 499). The **platelets** are cell fragments that release materials important in blood clotting. Also important in clotting is a soluble protein, fibrinogen, one of several soluble proteins that circulate within the plasma. Fibrinogen can turn into an insoluble form, fibrin. A clot consists of tangled fibers of fibrin in which cells become trapped.

The blood circulates through a series of larger and smaller blood vessels. The vessels leading away from the heart are called **arteries**. The vessels leading back toward the heart are called **veins**. The arteries branch into finer and finer vessels. The veins are arranged as a series of tributaries that flow into larger vessels and eventually back to the heart. The thinnest vessels, called **capillaries**, carry blood from the smallest arteries to the smallest veins (Figure 8.7). The capillary walls are a single layer of cells. Materials diffuse from capillaries into tissues and from tissues into capillaries across the cells of the capillary walls. The very large capillary surface area permits diffusion on a large scale. Throughout the body, no cell of any tissue is very far from a capillary.

Nutrients absorbed in the gut enter the capillaries of the gut lining and flow to the liver via the hepatic portal vein. If the blood contains more glucose than the body needs immediately, the excess is converted into the storage molecule **glycogen**. Glycogen storage takes place in most body cells, but the largest amount is stored in the liver. As the body uses up blood glucose, the liver cells convert glycogen back into glucose and release it into the bloodstream as needed, a mechanism that ensures a dependable but moderately low concentration of glucose in the blood. The storage of glycogen and the efficient use of glucose both require the hormone **insulin**, secreted by special clumps of cells, the islets of Langerhans, within the pancreas. Persons in whom the islets have degenerated cannot produce enough insulin; their condition is known as insulin-dependent diabetes mellitus (IDDM, or type I diabetes). The symptoms of diabetes can be controlled by supplying insulin or by controlling weight and diet.

Blood from the liver and the body's other organs flows through veins to the heart. Most veins have thin, flexible walls. The blood within them is propelled by the massaging action of nearby muscles and other organs. Valves within the veins keep the blood flowing in one direction and prevent it from flowing backwards.

The heart

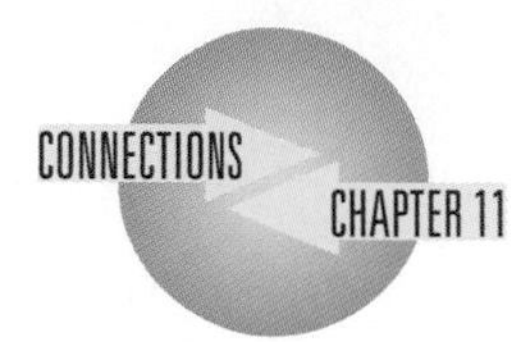

In addition to distributing nutrients, the circulatory system also distributes oxygen and removes carbon dioxide (a waste product of metabolism) from the body. The way in which blood circulates through the heart keeps the oxygen and carbon dioxide from mixing.

The heart is a muscular organ whose rhythmic contractions keep the blood circulating throughout the body. In all mammals, the heart contains four chambers (Figure 8.8). Oxygen-poor blood from the body's various organs enters the **right atrium** and is pumped into the **right ventricle**. Contraction of the right ventricle propels the blood out through the pulmonary arteries and into the lungs. Here, the oxygen in the lung's tiny pockets, or alveoli (see Chapter 11, p. 455), diffuses into the blood. Oxygen-rich blood from the lungs returns to the left side of the heart, where it enters the **left atrium**. Contraction of the left atrium pushes the blood into the **left ventricle**, the heart's largest chamber. Contraction of the left ventricle propels the blood throughout the arteries and into the body's various organs. Oxygen diffuses from the blood into the body's many cells across the thin capillary walls, and cellular wastes, including carbon dioxide, diffuse from the cells into the blood. The veins collect this oxygen-poor blood

FIGURE 8.7

The human circulatory system.

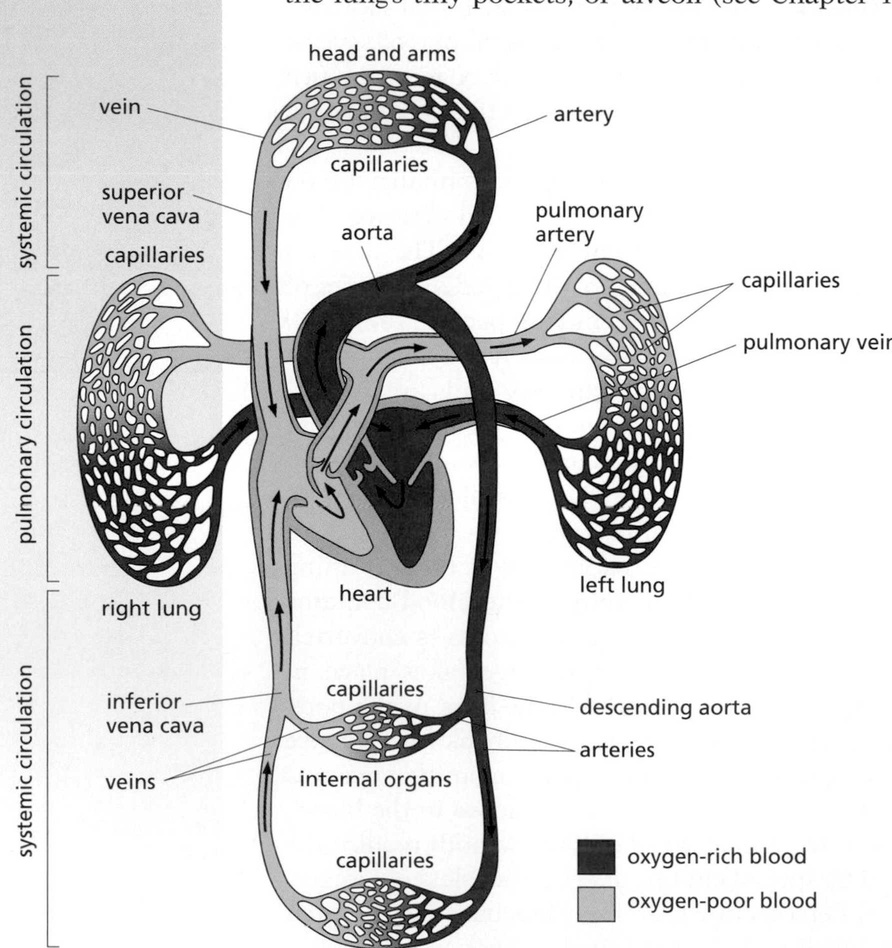

from the body's various organs and carry it back to the heart, where the pattern of circulation repeats (see Figure 8.7).

The heart maintains its own rhythmic pattern of contractions. A heart cut from a living animal and placed in a salt solution will continue to beat for many hours. The rhythm is maintained even in the absence of any nerve input, showing that the heart's rhythm originates in the heart itself.

Cardiovascular disease includes both heart disease and diseases of the blood vessels. In the United States each year 500,000 people die of heart disease, making it the number one cause of death. Another 1.5 million have nonfatal heart attacks. Each year almost another 500,000 die of strokes, a blood vessel disease. Men have more cardiovascular disease earlier than women (2:1 male:female ratio overall), but cardiovascular disease is nevertheless the major killer of both men and women. As we see in the next section, dietary fat can promote cardiovascular disease in several ways. In fact, proper diet, along with exercise and stress reduction, are the major ways of reducing the risks of cardiovascular disease.

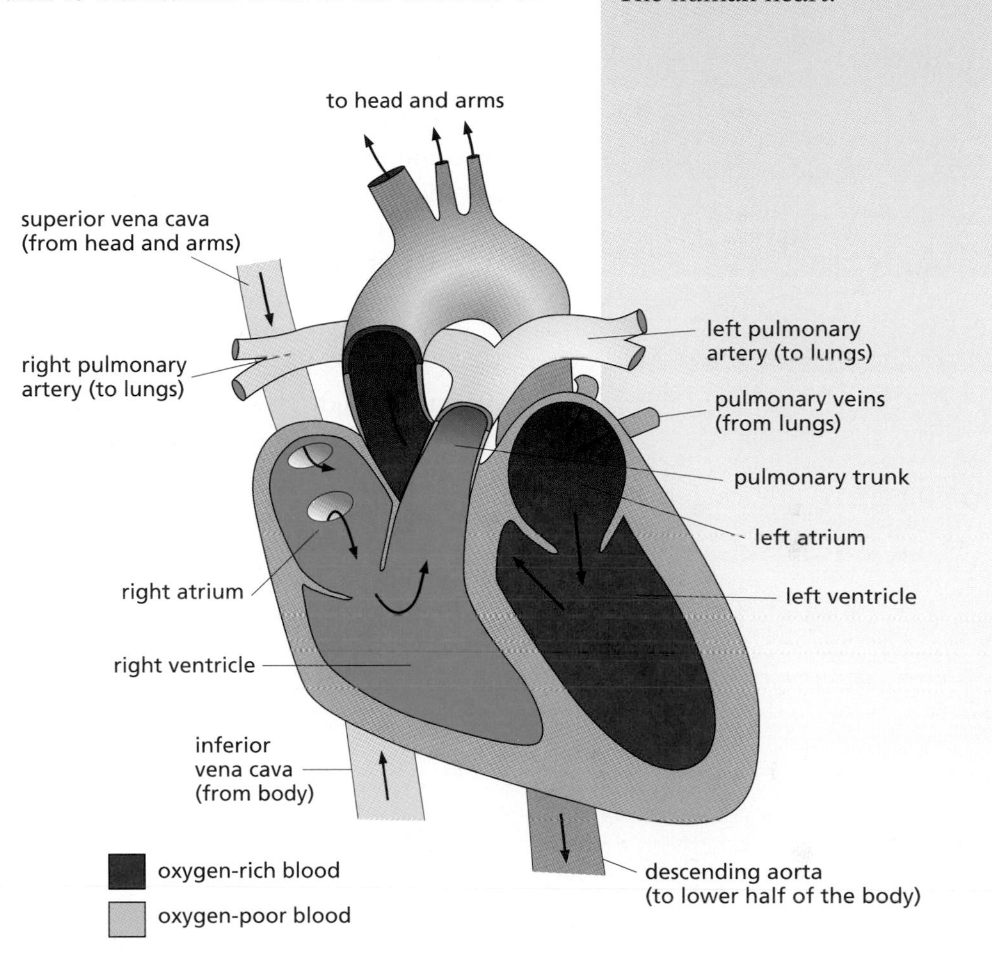

FIGURE 8.8

The human heart.

THOUGHT QUESTIONS

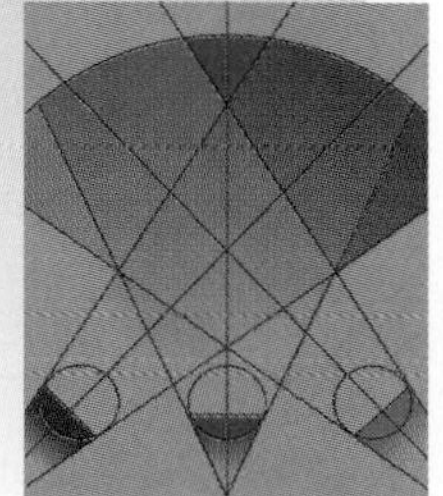

1. Cancer is not as frequent a cause of death as is heart disease, yet cancer research seems to get more publicity than research on heart disease. Examine some newspapers and magazines to see whether this is a correct impression. If it is, what factors might contribute to this difference?

2. The human circulatory system is a continuous, closed system; in other words, none of the blood vessels are open-ended. Why must this be so?

ALL HUMANS HAVE DIETARY REQUIREMENTS FOR GOOD HEALTH

Foremost among our dietary needs is a need for energy, as measured in kilocalories (kcal). A **kilocalorie** is the amount of energy required to raise the temperature of a kilogram of water one degree celsius. The 'calories' that dieters count are actually kilocalories. Your body's need for caloric energy depends on many factors. A certain number of calories are required even by a person who is completely inactive. This basal amount depends on the **basal metabolic rate**, the rate at which a person utilizes energy. Other factors include body weight, level of activity, and whether the person is male or female (Table 8.1).

Caloric intake is the most important measure of dietary sufficiency. In most industrialized countries, most people are adequately nourished or overnourished. This is certainly true in the United States, where one-third of the adult population is estimated to be obese. **Obesity** is defined as a body weight 20% or more above the ideal for the particular subject's sex and height.

Worldwide, inadequate caloric intake is the most widespread nutritional problem. Starvation kills millions of people each year, most of them children. Starvation and malnutrition are most noticeable in the nonindustrialized, or third world, countries, but are also present in impoverished areas, both rural and urban, within many industrial

TABLE 8.1

Calculating your body's caloric needs.

A. BASAL METABOLIC RATE

First, find the number of kilocalories required to maintain basal metabolism, that is, to keep you alive when you are inactive and lying down:

Average adult woman	21.6 kcal/kg body weight per day
Average adult man	24.0 kcal/kg body weight per day

This works out to about 1225 kcal daily for a 125-pound woman or 1850 kcal for a 170-pound man. (1 pound = 0.454 kg)

B. LEVEL OF ACTIVITY

Multiply the figure obtained above by a factor depending on your normal level of activity:

1.35 for sedentary activity (e.g., telephone sales, TV viewing)
1.45 for light activity (e.g., college studies, office work with occasional errands, light housekeeping)
1.55 for moderate activity (e.g., nursing, vigorous housekeeping, waiting on tables, light carpentry)
1.65 for heavy activity (e.g., pick-and-shovel work, bricklaying, full-time competitive athletics)

C. OTHER FACTORS

Figures obtained above need to be increased by as much as 10% for any of the following conditions:

Growth (children 15 years old and younger)
Pregnancy
Recovery from a major illness or injury

D. INDIVIDUAL DIFFERENCES

Figures calculated above are only guidelines or averages. Your individual need may either be greater or smaller. If you maintain a steady caloric intake on a day-to-day basis and you gain weight, your caloric intake is greater than your caloric needs. Conversely, if you lose weight, your intake is less than your caloric needs.

nations. Many other nutritional problems, such as vitamin deficiencies, exist in undernourished people; most of these other problems are hard to treat if the caloric intake remains inadequate.

The components of food can be classified into three major chemical groupings and several smaller ones. The major groupings are carbohydrates, lipids, and proteins; these are the macronutrients, or major sources for calories. In addition, people require fiber and many micronutrients such as vitamins and minerals, which are not energy sources, but have other vital functions. We look first at the macronutrients and how they are converted into cellular energy.

Carbohydrates

Most people around the world derive the majority of their calories from **carbohydrates**, which include starches and sugars. Plants store energy as carbohydrates, so they are a good dietary source for carbohydrates. Cereal grains such as wheat, rice, oats, and corn are the most nutritious source of carbohydrates because they also contain important vitamins, protein, and fiber. Breads, pastas, and other foods made from cereal grains retain all their nutritional value as long as the whole grain is used. Fruits and fruit products (including juices) generally contain sugars such as fructose or sucrose, together with important vitamins, minerals, and fibers. However, refined sucrose (table sugar) lacks these other nutrients and can also contribute to tooth decay (Box 8.1). Most vegetables contain carbohydrates but are even more important as sources of vitamins and minerals.

Carbohydrates are molecules formed principally of three types of atoms: carbon, hydrogen and oxygen (Figure 8.9). Because of the large number of oxygen atoms, a large proportion of the bonds in carbohydrates are polar; carbohydrates are therefore polar molecules, and most are soluble in water. A single carbohydrate unit is called a **monosaccharide** or simple sugar. Simple sugars differ from each other by their number of carbon atoms and the placement of their chemical bonds. More complex

BOX 8.1 How Does Sugar Contribute to Tooth Decay?

The sugar that we add to coffee or cereal is known chemically as sucrose. There are many other sugars: fructose (fruit sugar), lactose (milk sugar), and dextrose (a synonym for glucose). Many bacteria live in our mouths and use these dietary sugars for their metabolic energy. One type of mouth bacteria make a gluelike substance that attaches them to the tooth surface, and to make this substance they require sucrose. Once the bacteria are glued to the tooth they can use other sugars (including sorbitol, the sugar in 'sugarless gum') as energy sources. When bacteria extract energy from sugars, acids are produced, and these acids dissolve tooth enamel, resulting in cavities. Without sucrose, the bacteria cannot make the glue and the acids are not trapped so closely against the enamel surface.

FIGURE 8.9

Chemical structure of selected carbohydrates. The abundance of oxygen atoms (especially in -OH groups) makes most of these substances polar and therefore stable in water.

carbohydrates are built by hooking these monosaccharides together in groups of two **(disaccharides)** or of many **(polysaccharides)**.

In the gut, all carbohydrates are digested into simple sugars such as glucose and fructose. Because these sugars are polar, they are soluble in the watery digestive fluids but are unable to cross the nonpolar interior part of cell membranes and thus cannot enter cells without the help of transporter proteins (see Figure 8.6C).

After the body has absorbed simple sugars, their further processing is

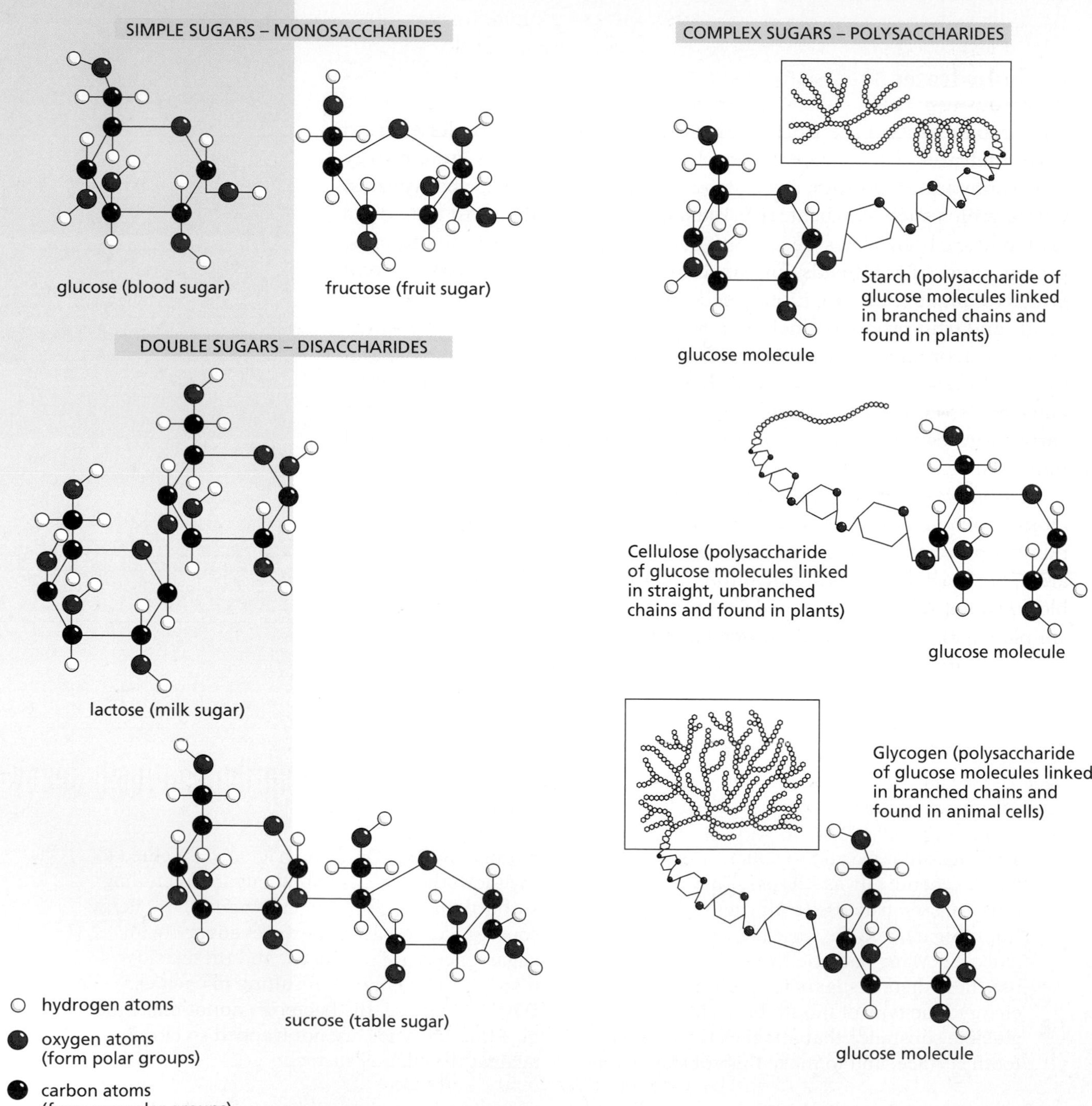

essentially the same regardless of which carbohydrate was initially present or which simple sugar was absorbed. The human body's daily need for carbohydrates is measured in terms of total caloric intake, as indicated in Table 8.1. In terms of caloric content, all carbohydrates, both sugars and starches, are the same, providing 4 kilocalories per gram (kcal/g). From an energy standpoint, it makes little difference if the carbohydrates are eaten in the form of sugar or starch, or whether the sugar is fructose or sucrose. There is, however, a difference in the rate of absorption: starches generally take a few hours to be digested into absorbable sugars, while dietary sugars are capable of being absorbed within minutes. A meal containing both sugars and starches will therefore maintain the body's energy level (or blood glucose) more evenly over a longer period.

In most populations, an increase in the consumption of carbohydrate-rich foods (especially whole grains) is desirable. Many third world diets supply inadequate calories; carbohydrates provide the most efficient and most economical means of improving these diets. Fewer kilocalories of labor, or fewer dollars, are needed to produce a kilocalorie of carbohydrate food than a kilocalorie of most fat-rich or protein-rich foods. In the United States, the replacement of dietary fats by carbohydrates would have certain indirect health benefits, including a reduction in risks for heart attacks and certain forms of cancer.

Lipids

Lipids are organic compounds that do not dissolve in water because they are made mostly of hydrogen and carbon atoms organized into nonpolar hydrocarbon chains. Dietary lipids are mostly **triglycerides**, molecules in which glycerol (a three-carbon molecule) is linked to three long chains of carbons and hydrogens called **fatty acids** (Figure 8.10). Triglycerides that are solid at room temperature are commonly called **fats**; those that are liquid at room temperature are commonly called **oils**. As sources of caloric energy, fats and oils contain almost 9 kcal/g, which is over twice as much as carbohydrates. A small amount of lipid is a dietary necessity, in part because the fat-soluble vitamins (especially A and D) cannot be absorbed without it. Lipids are also a source of fatty acids, which are the nonpolar portion of the phospholipid molecules that form the cell membrane. Two particular fatty acids (linoleic and arachidonic acids) are required from dietary sources because they cannot be made by the body, but they are required only in very small amounts (about 3 g or one tablespoonful per person per day). Most nonstarving people have an adequate intake of lipids. In the United States, many people consume too much lipid. The body tends to store excess lipid (and some excess carbohydrate) as fatty deposits within numerous adipose (fat-storing) cells.

Dietary lipids and atherosclerosis. Excess dietary fat can result in fat deposits that build up in the arteries, causing **atherosclerosis** (Figure 8.11). The fat deposits obstruct the blood vessels, making the passages narrower; eventually these deposits may calcify and make the vessels more rigid. Atherosclerosis contributes to high blood pressure (hypertension), although a person can have hypertension without having atherosclerosis.

Several researchers have found that different types of fats in the diet have different effects on health. **Saturated fats** are fats whose fatty acids have only single bonds (see Figure 8.10). Most saturated fats are derived from animal sources (or from a few tropical plants such as palm and coconut), and most are solid at room temperature. There is some evidence that saturated fats contribute more readily to atherosclerosis than unsaturated fats. **Unsaturated fats**, often derived from plant sources, have double bonds as well as single bonds in their fatty acid chains, causing the molecule to bend (see Figure 8.10). Those containing only one double bond are sometimes called monounsaturated while those with multiple double bonds are **polyunsaturated**. Both types are usually liquid at room temperature because the bends made by the double bonds prevent the molecules from packing too tightly together and solidifying.

FIGURE 8.10

Chemical structure of three types of lipids. Notice that the general lack of -OH groups (or other polar groups in which electric charges can partly separate) makes most parts of these molecules nonpolar and tending to separate away from water.

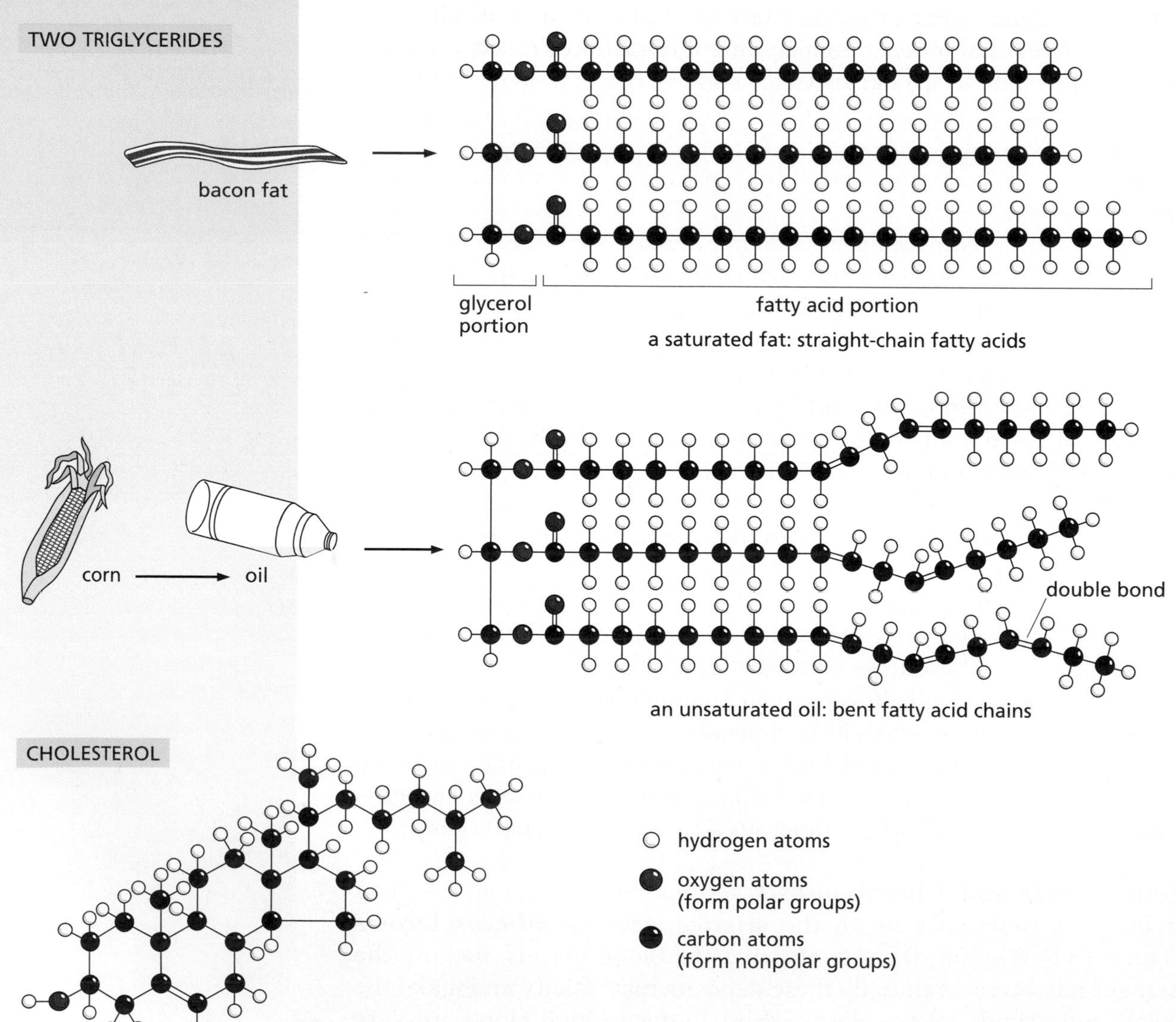

The study of disease factors in large populations is called **epidemiology**. Epidemiological studies point to a connection between certain types of dietary fats and cardiovascular disease (heart disease and strokes). The United States, Australia, and New Zealand—all meat-producing countries—have a high consumption rate of meat products per person and also high incidences of cardiovascular diseases. Most meats are high in saturated fats. People in Mediterranean countries consume much of their lipid in the form of olive oil, an unsaturated fat, and their cardiovascular disease rates are lower. Heart attacks are very rare among the Inuit (Eskimos), whose diet contains large amounts of cold-water fish, a good source of a type of fatty acid called omega-3 fatty acid that has been shown to guard against the production of chemicals that damage cell membranes. The Japanese also tend to have low consumption rates of saturated fats and low rates of cardiovascular disease.

Epidemiology also provides clues about whether the association between dietary fats and cardiovascular disease is more closely related to diet or to genetically inherited traits: Japanese people in Japan have much lower rates of heart disease or stroke than do Japanese living in Hawaii or California, whose rates are similar to those of their non-Japanese neighbors. These findings (and similar ones on other immigrant groups) all point to diet, not heredity, as the major difference responsible for the different disease rates between populations.

Because saturated fats have been linked to a greater risk of atherosclerosis, many experts recommend that saturated fats be replaced with unsaturated fats in most diets. Advertising has convinced many people that unsaturated fats—especially the polyunsaturated kind—are desirable, but this is true only if those fats replace saturated fats. Most experts recommend that the quantities of all dietary fats be reduced to lower the risk of heart disease and atherosclerosis.

Fatty acids, cholesterol, and cell membranes. Fats are important in cell membranes. Fatty acids from dietary fats become incorporated into cell membranes as part of molecules called phospholipids. The fatty acid

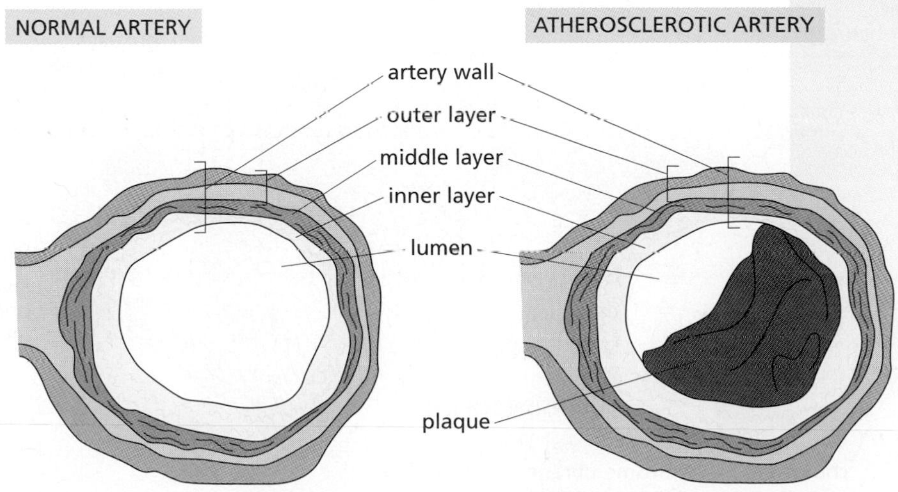

FIGURE 8.11

Atherosclerotic plaque reducing the effective diameter of an artery. Although the outer diameter of the vessel has not changed, the inner diameter (lumen) through which the blood flows has become smaller by the formation of lipid deposits on the inside of the vessel, making the blood pressure higher. These deposits may also become calcified, further increasing blood pressure.

portion is nonpolar, as we have seen, but the other end of a phospholipid is polar. In water, phospholipids orient spontaneously to form bilayer membranes (Figure 8.12). Because of the way in which the phospholipids orient, membranes have polar surfaces facing the watery fluids outside the cell as well as the watery cytoplasm inside the cell. The interior of the bilayer is composed of the nonpolar fatty acids, making the interior of the bilayer nonpolar. Membrane proteins are embedded in this phospholipid bilayer. When unsaturated fatty acids are incorporated into the phospholipid cell membrane, the bends prevent their tight packing in the membrane, keeping the membrane more fluid. The phospholipid molecules need to be fluid to allow the embedded proteins to function.

FIGURE 8.12

The structure of phospholipids and cell membranes. The cell membrane shown is an animal cell membrane, and therefore contains cholesterol.

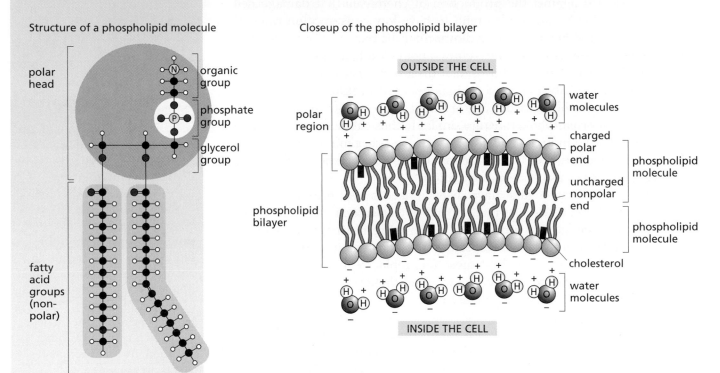

Structure of a phospholipid molecule

polar head
N
organic group
P
phosphate group
glycerol group
fatty acid groups (nonpolar)

Closeup of the phospholipid bilayer

OUTSIDE THE CELL
polar region
water molecules
charged polar end
phospholipid molecule
uncharged nonpolar end
phospholipid bilayer
phospholipid molecule
cholesterol
water molecules
INSIDE THE CELL

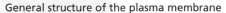

General structure of the plasma membrane

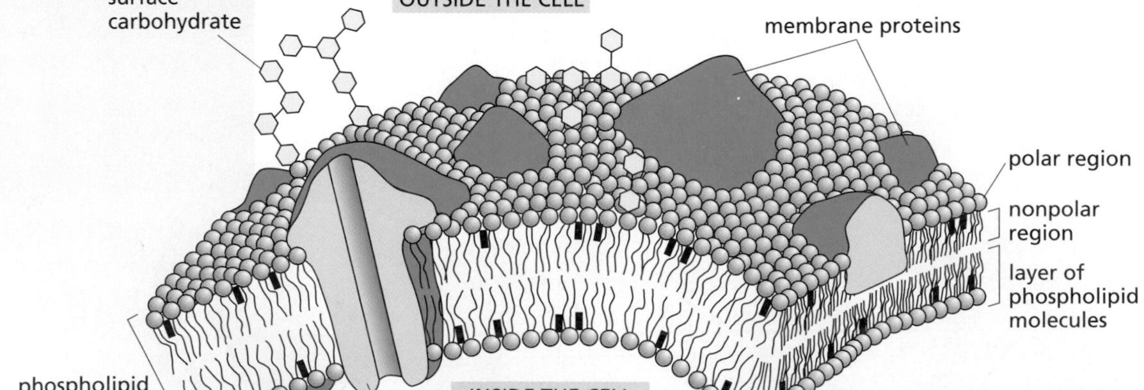

Cells remove the fatty acid chains from dietary triglycerides and incorporate the chains into membrane phospholipids. Thus, if the diet is high in saturated fats, the membrane is less fluid, which reduces molecular absorption because it impairs the functioning of membrane proteins such as carrier and transporter proteins (see Figure 8.6B and C). It is hypothesized that when lipids cannot be properly transported across cell membranes (see Figure 8.6D) they may tend to build up on blood vessel walls to form atherosclerotic plaques (see Figure 8.11).

Another dietary lipid is **cholesterol** (see Figure 8.10), a fat-soluble molecule that is an important constituent of animal cell membranes. Along with unsaturated fatty acids, cholesterol helps to keep the membranes fluid, thereby keeping the cell and the organism functioning properly. Cholesterol is also the precursor of several important hormones.

We need cholesterol in small quantities, but our bodies can usually synthesize this amount, so little or none is needed from food. Plant cell membranes do not contain cholesterol, so plant products are always cholesterol-free, although some (like coconut oil) contain saturated fatty acids that are easily converted into cholesterol by the body. All dietary fatty acids are broken down into several molecules of acetoacetate, one of the major starting materials of cholesterol synthesis; cholesterol synthesis is thus increased by nearly all fatty foods, even if they are advertised as 'cholesterol free.'

Because the body makes about 75–80% of its own cholesterol, and makes it from dietary fats, most of the cholesterol circulating in the bloodstream comes from dietary fats (especially saturated fats), not from dietary cholesterol. Excess cholesterol, like excess amounts of other lipids, can build up on blood vessel walls and increase your risk of atherosclerosis. Most foods that contain cholesterol are also high in saturated fats, so avoiding either also helps you to avoid the other. Eggs are exceptional in having a lot of cholesterol with few other fats.

LDLs and HDLs: lipid transport particles. In contrast with carbohydrate molecules, lipid molecules contain few oxygen and nitrogen atoms and have mostly nonpolar bonds in which electrons are shared equally around carbon and hydrogen atoms (see Figure 8.10). Because the bonds in lipids are nonpolar, lipids are not water soluble. Blood plasma is mainly water, so lipids must be transported through the blood from one part of the body to another by transport particles such as **low-density lipoproteins (LDLs)** and **high-density lipoproteins (HDLs)**. These transport particles are proteins that bind lipids in such a way that they can move through body fluids.

People eating identical diets may not have the same cholesterol level. This difference seems to have a genetic component and relates in part to each person's ability to make LDLs and HDLs. No one has very much free (unbound) cholesterol because cholesterol is very nonpolar, so what is called serum cholesterol is actually the total of all the cholesterol bound to HDLs and LDLs. HDLs transport cholesterol, phospholipids, and triglycerides out of tissues to the liver, where they are used in the synthesis of bile acids, while LDLs transport cholesterol and lipids into tissues and cells. The HDL/LDL ratio is the ratio of outbound to inbound lipids. A high HDL/LDL ratio thus indicates that the proportion of 'good cholesterol' (lipids on their way out) is higher than the proportion of 'bad cholesterol' (lipids on their way into cells). A very low ratio, meaning a preponderance of inbound lipids, correlates strongly with an increased risk of atherosclerosis in the arteries of the heart (coronary arteries), a condition that can precipitate a heart attack.

Some people are genetically prone to high cholesterol levels because their cells lack LDL receptors on their surfaces. When cells need cholesterol, they get the cholesterol from the LDLs in the blood stream by binding these particles to LDL receptors and internalizing the receptors and LDLs by endocytosis (see Figure 8.6D). If the LDL receptors are missing or nonfunctional, the cells manufacture their own cholesterol even when LDL levels are already high, because they cannot take up LDL from the blood. Thus, diet is not the only factor leading to atherosclerosis; problems with lipid transport and lipid uptake are also factors.

Proteins

The body uses **proteins** for tissue growth and repair, including the healing of wounds, replacement of skin and mucous membranes, and manufacture of antibodies (see Chapter 12). Proteins are important components of all cell membranes and function as transporters, receptor molecules, and channels (see Figures 8.6 and 8.12). Many proteins of the cellular interior provide structure, motility, and contractility to muscles and other cells. Other proteins outside the cell, such as collagen and elastin, give connective tissues their strength and thus help to support the entire body, while keratin is essential for healthy skin and is the main constituent of hair and fingernails. A much larger assortment of proteins function as the enzymes described at the beginning of this chapter. Some enzymes (such as those used in digestion) function outside cells (extracellularly); many others function inside cells.

Some body proteins are needed only in small quantities, but our muscles, blood, skin, and connective tissues need proteins in large amounts. Tendons and certain other body parts are made of proteins that are relatively stable once they have been formed, but blood, skin, bone tissue, bone marrow, and many internal membrane surfaces all undergo constant reworking, repair, and replacement, requiring new protein supplies throughout life. Protein requirements are even higher, per unit of body weight, in growing infants and children, and in persons recovering from a major illness or injury.

Dietary amino acids. Proteins are built from chains of smaller chemicals called amino acids (see Chapter 2, pp. 62–64). The digestive system breaks down the proteins in food into individual amino acids. After they have been absorbed by the body, these amino acids can then be used to build the body's own proteins. How a protein functions depends to a large extent on its three-dimensional shape after the linear sequence of amino acids has folded. The way in which a protein folds, and whether it is stable in the watery cytoplasm of the cell or in the nonpolar cell membrane, is determined by the arrangement of the polar and nonpolar side groups of its amino acids (Figure 8.13).

Because proteins are synthesized by adding one amino acid at a time to the end of a growing chain (see Chapter 2, pp. 62–64), if one type of amino acid is missing from the cell, the synthesis of any protein needing that amino acid stops. An amino acid that is present in small quantities and is used up before other amino acids is called a **limiting amino acid**.

Proteins are necessary in the diet. The daily requirement is 0.8 g per kilogram of body weight, for example, about 45 g for a 125-pound (57 kg) woman, or 64 g for a 175-pound (80 kg) man. Each species has its own capacities for making certain of the amino acids and therefore has its own dietary requirements for those it cannot make. Eight of the twenty standard amino acids cannot be synthesized by the human body and are therefore considered essential in the human diet; a ninth amino acid is essential in human infants. The human body can make the remaining amino acids from these **essential amino acids**.

Complete and incomplete proteins. Most animal proteins are **complete proteins** in that they contain all the amino acids essential in the human diet. Soy protein is also complete, but most plant proteins lack at least

FIGURE 8.13

Chemical structure of part of a protein.

A tripeptide of three amino acids held together by peptide bonds (red lines)

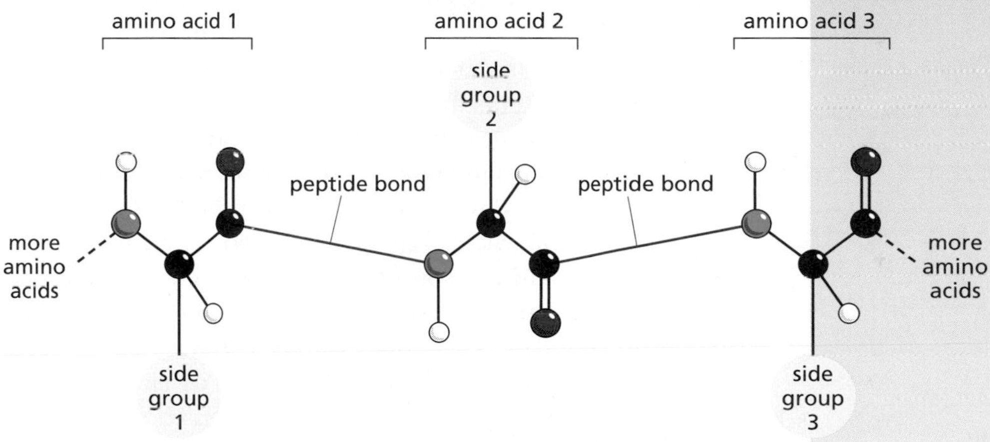

the breaking of the peptide bonds releases the individual amino acids

H H O
N—C—C
H O—H
H

H H O
N—C—C
H O—H
CH_2

H H O
N—C—C
H O—H
CH_2
SH

Glycine has the simplest side group, a hydrogen atom.

Seven of the 20 amino acids in living things have side groups that are insoluble in water. Phenylalanine is such an amino acid.

Nine amino acids have side groups that are soluble in water. Cysteine is such an amino acid.

one essential amino acid needed by humans. When an incomplete protein is eaten, the body uses all the amino acids until one of them, the limiting amino acid, becomes depleted. After the limiting amino acid is used up, the body uses the remaining amino acids to produce energy instead of making proteins, because dietary protein cannot be stored for later use in the way that carbohydrates and lipids can.

To get around the problem of incomplete proteins in the human diet, we can eat plant proteins in combinations in which one protein supplies an essential amino acid missing in another one. The Iroquois and many other Native Americans commonly obtained complete protein by combining beans, squash, and corn in their diets. Most bean proteins, for example, are deficient in the amino acids valine, cysteine, and methionine, while corn proteins are deficient in the amino acids lysine and tryptophan. Alone, neither one of these proteins is nutritionally complete for humans, but in combination (as in corn tortillas with a bean filling, or succotash, a mixture of beans and corn cooked together) the two plant sources provide a nutritionally complete assortment of amino acids because each has the essential amino acids that the other lacks.

Vegetarian diets. The amino acid inadequacy of plant proteins poses special problems for **vegetarian** (meat-avoiding) diets. Vegetarian diets are generally rich in carbohydrates, fiber, vitamins, and minerals, but they may be deficient in certain amino acids unless care is taken to combine several plant proteins at once, such as those just described above. Some vegetarians avoid meat but consume fish or milk or eggs; these 'ovolacto vegetarians' can usually meet their protein needs without much difficulty, especially if they combine proteins from both plant and animal sources in the same meal. Strict vegetarians, also called **vegans**, who do not eat food from any animal source (including milk or eggs) need to carefully combine plant proteins sources so as to supply their bodies with nutritionally complete protein. For example, legumes (beans, peas, and peanuts) can be combined with whole grains (such as rice, corn, or wheat). Nuts and seeds contain protein and can be used to supplement amino acids missing from plant proteins from other sources. Some vegetables also contain individual amino acids that can serve the same function. Our web site contains vegetarian recipes and additional references.

Because animal cells store energy principally as fat, proteins obtained from animals are accompanied by fat. Plant cells, in contrast, store energy in the form of complex carbohydrates such as starch, and plant proteins are thus accompanied by very little fat.

Plant-rich diets have other advantages. A given amount of arable land can support a larger human population if that land is used for raising crops for human consumption, including sources of plant proteins, than if the same land is used to raise food for animals that humans can eat. It takes 5–16 pounds of grain protein to produce one pound of meat protein. In well-fed countries with plenty of land, such as Australia or the United States, large tracts can be used for grazing or for the raising of crops primarily for animal consumption. However, poor countries of high population density can ill afford to feed crops to animals. Most of the world's poor eat little meat and get most of their protein from vegetable sources, or in some cases from fish. All whole-grain cereals contain some protein. If this protein is eaten with beans or other legumes, a high-quality protein source is created that is much less expensive than meat and contains far less saturated fat.

Conversion of macronutrients into cellular energy

Carbohydrates, fats, and proteins are all macronutrients, the principal sources of calories for the body. After large macromolecules are broken down into simple subunits in the digestive system, the subunits are absorbed into the cells. They are absorbed first by the cells lining the digestive tract, and are then transported via the blood to all of the other cells of the body.

After absorption, these simple subunits are then broken down into even smaller molecules, as can be followed on Figure 8.14.

FIGURE 8.14

How the major products of digestion are broken down in a series of energy-yielding reactions.

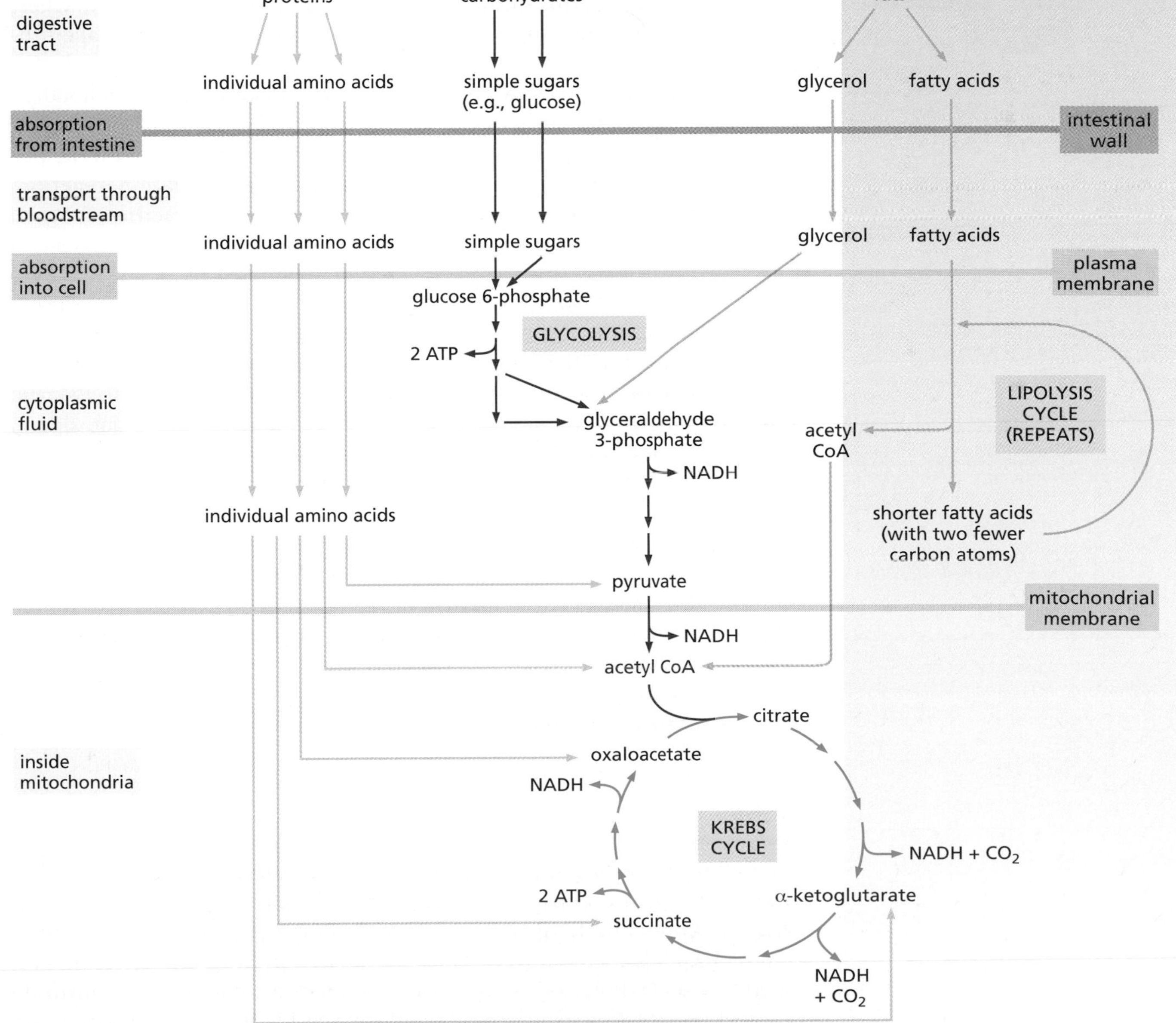

- The various amino acids are degraded into either pyruvate or acetyl CoA or one of the molecules in the Krebs cycle (see below).
- Sugars such as glucose are converted by a process called **glycolysis** into a three carbon molecule called pyruvate. Pyruvate is then converted into acetyl CoA. During glycolysis, limited amounts of the energy-rich molecule **ATP** (adenosine triphosphate) are synthesized. ATP is one of the principal molecules in which chemical energy is stored for later use by the cell.
- The long carbon chains of fatty acids are broken down in stages. In each stage, two carbons at a time are broken off from the long carbon chains to make an acetyl group. These acetyl groups are each put onto a carrier molecule called coenzyme A, so that the complex is called acetyl coenzyme A (acetyl CoA).

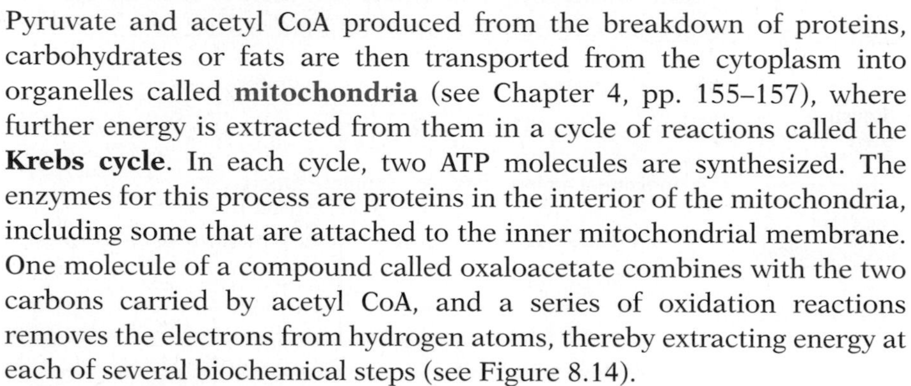

Pyruvate and acetyl CoA produced from the breakdown of proteins, carbohydrates or fats are then transported from the cytoplasm into organelles called **mitochondria** (see Chapter 4, pp. 155–157), where further energy is extracted from them in a cycle of reactions called the **Krebs cycle**. In each cycle, two ATP molecules are synthesized. The enzymes for this process are proteins in the interior of the mitochondria, including some that are attached to the inner mitochondrial membrane. One molecule of a compound called oxaloacetate combines with the two carbons carried by acetyl CoA, and a series of oxidation reactions removes the electrons from hydrogen atoms, thereby extracting energy at each of several biochemical steps (see Figure 8.14).

The hydrogens from these oxidation reactions are taken up by a molecule called NAD^+ (forming NADH) and then donated to a series of mitochondrial membrane protein complexes that make up the **electron transport chain**. These protein complexes are oriented in the membrane so that each time an electron passes from one electron carrier to the next electron carrier in the chain, a proton (H^+) is pumped from one side of the mitochondrial membrane to the other. An unequal distribution of protons on opposite sides of the membrane, called a **proton gradient**, is thus formed. This proton gradient is both a chemical gradient and a gradient of electrical charge, two important forms in which energy can be stored for later use by the cell. Some of this stored energy is used by an enzyme called ATP synthetase in a process that produces additional ATP. After their energy has been extracted in steps, the electrons finally end up combining with oxygen, which is why we need to breathe in oxygen from the atmosphere. The energy-extracting processes in the mitochondria are shown in Figure 8.15.

Fiber

Not all nutrients are required as sources of calories. One example is fiber, material that the body cannot digest and absorb. Human diets should include both soluble fiber (pectin, gums, mucilages) and insoluble fiber (mostly cellulose), and both types are present in most fruits, vegetables, legumes, and whole grains. Many of these fibers are complex carbohydrate molecules (for the structure of cellulose, see Figure 8.9). An increase in dietary fiber reduces the incidence of several cancers, especially those of the colon and rectum, but scientists are not sure about the exact mechanism of this effect. One intriguing possibility is that protection against these cancers depends on the rate of movement of food through the colon, and that fiber maintains the optimal rate of food movement. Higher rates cleanse the colon of potentially toxic chemicals, while lower rates allow these chemicals to remain in one place long enough to undergo fermentation by bacteria into cancer-causing substances (carcinogens; see Chapter 9). Another possibility is that harmful carcinogens are frequently present inside the intestine for whatever reason, but a mucus protects the intestinal lining from them; the insoluble fiber rubbing against the intestinal lining stimulates the lining to secrete more of this protective mucus. Soluble fiber such as oat bran may reduce the level of serum cholesterol and the risk of heart disease. The mechanism for this effect is not known with certainty, but one hypothesis is that certain soluble fibers bind strongly to the bile secreted into the intestinal tract. Without the soluble fiber, the bile would be reabsorbed by the intestinal lining and reused, but the soluble fiber prevents this reabsorbtion and ensures that the bile is eliminated with the stools. Without recycled bile, new bile must be synthesized, and it is synthesized from cholesterol, which the body withdraws from the blood, lowering blood cholesterol levels. Diets that are high in fiber are statistically associated with lower rates of coronary heart disease and stroke.

CONNECTIONS
CHAPTER 9

FIGURE 8.15

Processes that take place inside the mitochondria after glycolysis. Electrons are shown as e⁻, and protons (hydrogen ions) as H⁺.

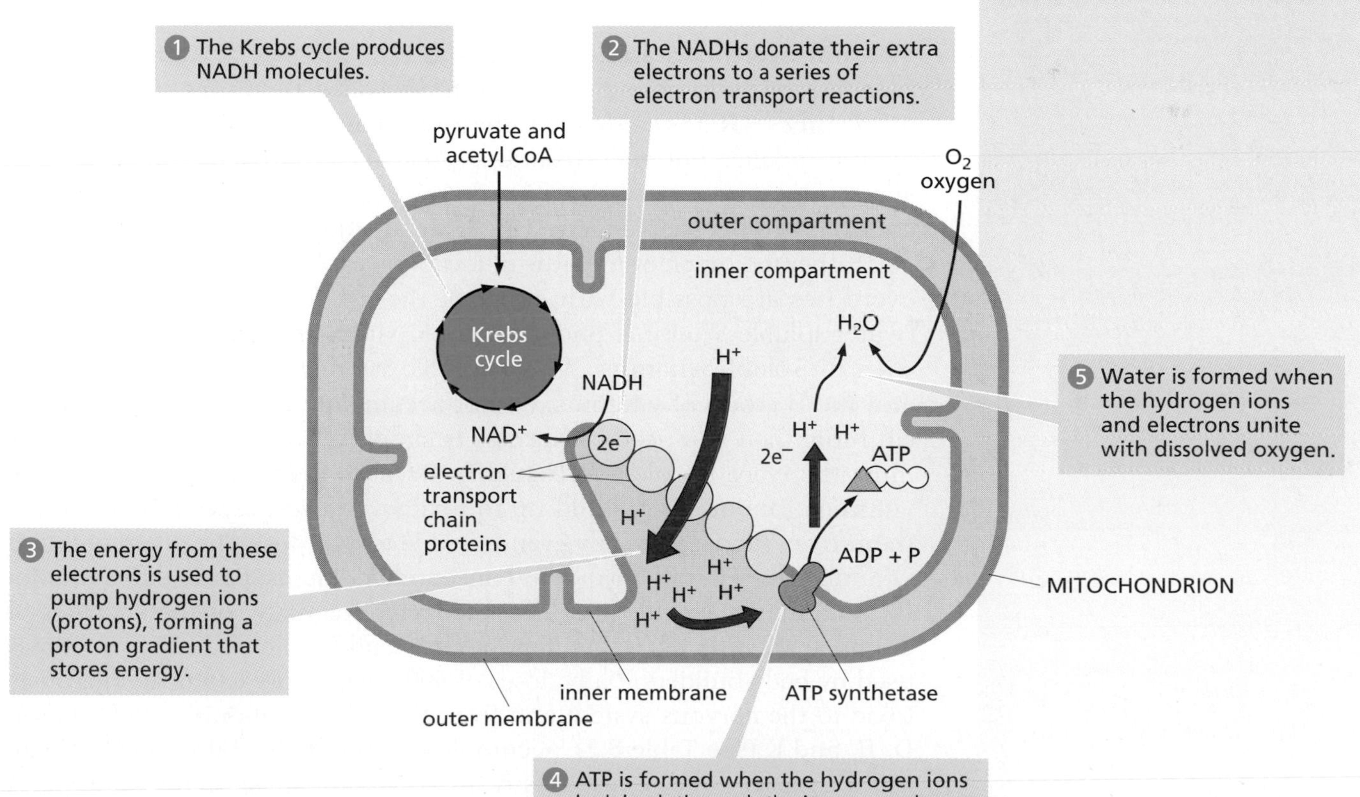
1 The Krebs cycle produces NADH molecules.
2 The NADHs donate their extra electrons to a series of electron transport reactions.
pyruvate and acetyl CoA
O2 oxygen
outer compartment
inner compartment
Krebs cycle
H2O
H+
NADH
NAD+
2e−
H+ H+
2e−
ATP
electron transport chain proteins
H+
ADP + P
H+
H+ H+ H+
H+
3 The energy from these electrons is used to pump hydrogen ions (protons), forming a proton gradient that stores energy.
5 Water is formed when the hydrogen ions and electrons unite with dissolved oxygen.
MITOCHONDRION
inner membrane
ATP synthetase
outer membrane
4 ATP is formed when the hydrogen ions leak back through the inner membrane via the ATP synthetase enzyme.

Vitamins

Some other nutrients that the body needs but cannot make are collectively called micronutrients and include vitamins and minerals. Plants are good sources of micronutrients, because plants need these substances for their own metabolism.

Vitamins are complex nutrients needed only in very small quantities. Most vitamins are **coenzymes**, the nonprotein portions of enzymes needed for the enzymes to function as catalysts. Enzymes (and their coenzymes) are needed only in very small quantities because they are used and reused in the chemical reactions that they regulate. Vitamins, listed in Table 8.2, may be obtained either from pills or from food. The reasons for preferring vitamins in food are as follows.

- They are much less expensive this way.
- Foods rich in vitamins are also rich in other important substances, including minerals, fiber, and protein. We do not know the complete nutritional requirements of any organism more complex than bacteria, and undoubtedly our food contains many unknown but needed nutrients. These other nutrients, known and unknown, are not obtained from vitamin pills.
- Some vitamins are more easily absorbed by the body in the combinations with other ingredients that exist in food than they are in the combinations that exist in vitamin pills.
- Purified vitamins can be toxic if taken in excessive amounts, an unlikely danger with vitamins contained in foods.

Vitamin overdoses and deficiencies. The amounts of vitamins recommended for maintaining good health are called **recommended dietary allowances (RDAs)**. Most of these amounts are the same for most healthy adults, but menstruation, pregnancy, and lactation can alter some values in women. Nutritional requirements also differ for growing children and for people recovering from a major illness or injury.

Either too much or too little of a vitamin can result in disease. Vitamin overdoses are possible with either of the two large classes of vitamins (water-soluble vitamins and fat-soluble vitamins), but are more likely with fat-soluble vitamins. Water-soluble vitamins, including vitamin C and the B group of vitamins, do not accumulate in the body. When you eat more than you need, the excess is simply excreted in the urine. They must therefore be taken in regularly. Because they are not stored, these vitamins cannot easily build up to toxic overdoses, especially if you get them from foods. It is, however, possible to overdose on water-soluble B vitamins taken in pill form or as concentrated liquids, particularly vitamin B_6. Vitamin B_6 (pyridoxine) is a coenzyme for many of the enzymes of amino acid synthesis; it therefore helps to build proteins and is sometimes used by body-builders. Daily doses of 500 mg or more can be dangerously toxic to the nervous system and liver. Fat-soluble vitamins, including A, D, E, and K (see Table 8.2), accumulate in the body's fat tissues and can build up over time. Overdoses of these vitamins, especially vitamins A and D, can be toxic.

People can have vitamin deficiencies. Disorders of fat absorption (LDL defects, for example) often cause deficiencies in fat-soluble vitamins because these vitamins are transported and absorbed along with dietary fats. People with such disorders may have plenty of the vitamins in their blood, but their cells are unable to absorb them. Diseases result from

TABLE 8.2

Vitamins and minerals in human health.

	IMPORTANCE FOR GOOD HEALTH	GOOD FOOD SOURCE
WATER-SOLUBLE VITAMINS		
Vitamin B_1 (thiamine)	Helps to break down pyruvate; maintains healthy nerves, muscles, and blood vessels; prevents beri-beri	Meat, whole grains, legumes
Vitamin B_2 (riboflavin)	Important in wound healing and in metabolism of carbohydrates; prevents dryness of skin, nose, mouth, and tongue	Yeast, liver, kidney
Vitamin B_3 (niacin)	Maintains healthy nerves and skin; prevents pellagra	Legumes, fish, whole grains
Vitamin B_6 (pyridoxine)	Coenzyme used in amino acid metabolism; prevents microcytic anemia	Whole grains (except rice), yeast, liver, mackerel, avocado, banana, meat, vegetables, eggs
Vitamin B_{12} (cyanocobalamin)	Required for DNA synthesis and cell division; prevents pernicious anemia (incomplete red blood cell development)	Meat, liver, eggs, dairy products, whole grains
Folic acid	Used in synthesis of hemoglobin, DNA, and RNA; prevents megaloblastic anemia and spina bifida	Asparagus, liver, kidney, fresh greens, vegetables, yeast
Pantothenic acid	Needed to make coenzyme A for carbohydrate and lipid metabolism	Liver, eggs, legumes, dairy products, whole grains
Biotin	Used in fatty acid synthesis and other reactions using CO_2	Eggs, liver, tomatoes, yeast
Vitamin C (ascorbic acid)	Antioxidant; used in synthesis of collagen (in connective tissues) and epinephrine (in nerve cells); promotes wound healing; protects mucous membranes; prevents scurvy	Fresh fruit (especially citrus and strawberries), fresh vegetables, liver, raw meat
FAT-SOLUBLE VITAMINS		
Vitamin A (retinol)	Antioxidant; precursor of visual pigments; prevents night blindness and xerophthalmia	Yellow and dark green vegetables, some fruits, fish oils, creamy dairy products
Vitamin D (calciferol)	Promotes calcium absorption and bone formation; prevents rickets and osteomalacia	Eggs, liver, fish, cheese, butter
Vitamin E (tocopherol)	Antioxidant; protects cell membranes against organic peroxides; maintains health of reproductive system	Whole grains, nuts, legumes, vegetable oils
Vitamin K	Essential for blood clotting; prevents hemorrhage	Green leafy vegetables
MINERALS		
Electrolytes (Na^+, K^+, Cl^-)	Maintain balance of fluids in body; maintain cell membrane potentials	Raisins, prunes; K^+ also in dates
Calcium	Part of crystal structure of bones and teeth; maintains muscle and nerve membranes	Dairy products, peas, canned fish with bones (sardines, salmon), vegetables
Phosphorus	Part of crystal structure of bones and teeth	Dairy products, corn, broccoli, peas, potatoes, prunes
Magnesium	Maintains muscle and nerve membranes	Meat, milk, fish, green vegetables
Iron	Part of hemoglobin; used in energy-producing reactions	Meat, egg yolks, whole grains, beans, vegetables
Iodine	Maintains thyroid gland; prevents goiter	Fish and other seafood products
Fluorine	Strengthens crystal structure of tooth enamel	Drinking water, tea
Zinc	Promotes bone growth and wound healing	Seafood, meat, dairy products, whole grains, eggs
Copper	Cofactor for enzymes used to build proteins, including collagen, elastin, and hair	Nuts, raisins, shellfish, liver
Selenium	Statistically associated with lower death rates from heart disease, stroke, and cancer	Vegetables, meat, grains, seafood

deficiencies of each of the vitamins; in fact, research on the cause of these diseases led to the discovery of vitamins.

Vitamin B_1. Vitamin B_1 (thiamine) was the first vitamin to be chemically characterized. While stationed on the island of Java in the 1890s, the Dutch physician Christiaan Eijkman noticed that polyneuritis, a neurological disease in chickens, had many symptoms similar to those of a human disease called beri-beri. Both diseases caused muscle weakness and leg paralysis, resulting in an inability to stand up; both diseases were fatal if they persisted. Eijkman noticed that the chickens got polyneuritis only when they were fed on polished white rice, but the disease cleared up when rice polishings were added to their feed or when whole brown rice was used. From the rice polishings, thiamine was later isolated and was found effective in both treating and preventing beri-beri in humans. Because thiamine is a vital (necessary) substance and is also an amine (a chemical containing an $-NH_2$ group), it was called a 'vital amine.' This term was later shortened to 'vitamine' and then 'vitamin', and the name was applied to the entire class of substances needed only in small quantities and capable of curing deficiency diseases such as beri-beri. Beri-beri occurs primarily in people whose dietary carbohydrates come from a single, highly refined source such as white rice or white (unenriched) flour. Many countries now have laws requiring the addition of thiamine (and other B vitamins) to refined flour. For this reason, beri-beri is now rare in the industrialized world, although it does occur in severe alcoholics whose dietary intake is inadequate.

As an example of how a vitamin functions as a coenzyme, consider the role of thiamine in the breakdown of the molecule pyruvate. Remember that pyruvate is a product of carbohydrate metabolism and a major input into the energy-producing Krebs cycle (see Figure 8.14). An enzyme that contains thiamine as one of its constituent parts combines with the pyruvate molecule, releases CO_2, then emerges from a later reaction in its original form. Because the thiamine is not used up, it can participate in the reaction again and again. For this reason, only minute amounts of thiamine are needed to facilitate the breakdown of large quantities of pyruvate formed in carbohydrate metabolism.

Other B vitamins. Other water-soluble vitamins are described in Table 8.2. Many are coenzymes or form portions of other molecules. Vitamins B_2 and B_3 form parts of the larger molecules that carry electrons from the Krebs cycle to the electron transport chain (see Figure 8.15). Many vitamins owe their discovery to research on **vitamin deficiency diseases** such as beri-beri. Vitamin B_6 deficiency (microcytic anemia) is seen frequently in people whose diets consist mostly of rice. A deficiency of niacin (Vitamin B_3) causes pellagra, a disease of the skin and nervous system. Because corn is particularly deficient in niacin, populations in which corn is the only protein source are often subject to pellagra.

Vitamin C. A deficiency of vitamin C causes scurvy, a disease once common in sailors at sea and among prisoners. A British naval surgeon discovered in the 1600s that limes and other fresh fruits would both prevent and cure scurvy. British ships then began carrying limes and became so well-known for this practice that British sailors came to be called 'limeys.' An inflammation of the mucous membranes, as in a cold, increases the body's need for vitamin C. Vitamin C may therefore decrease the severity of the symptoms of such an infection, but it cannot on this account be considered a 'cure' or a prevention for the common cold, as has sometimes been claimed.

People who take large doses of vitamin C can suffer the symptoms of scurvy when they stop taking the vitamin. In addition, megadoses can produce hemolytic anemia (red blood cell deficiency caused by the rupture of red blood cells) in people with the G6PD metabolic deficiency (see Chapter 5, p. 218) found in African American, Asian, and Sephardic Jewish populations. In addition, individuals who are genetically predisposed to gout find that high doses of vitamin C can sometimes bring on the condition by raising blood levels of uric acid. Vitamin C megadoses can produce deficiencies of another vitamin, B_{12}, in people who are iron deficient. Even in healthy people, megadoses of vitamin C can irritate the bowel sufficiently to result in diarrhea.

Antioxidant vitamins. Vitamin A (retinol) is essential in the synthesis of the light-sensitive chemicals (retinal) used in vision. Vitamin A is also an **antioxidant**, meaning that it protects body tissues from chemicals that would rob those tissues of electrons. The removal of electrons from other molecules is a process that chemists call **oxidation**. Chemicals that bring about oxidation by taking up electrons are called oxidizing agents; among the most highly reactive oxidizing agents are a group of chemicals, called **free radicals**, that have one or more unpaired electrons and a strong tendency to remove electrons from other molecules. Free radicals can thus damage many cellular molecules and are hypothesized to play a role in initiating some cancers (see Chapter 9). Vitamin A and other antioxidants protect the body by destroying free radicals.

Vitamin A can be obtained from animal sources such as dairy products or fish. Many vegetables also contain a vitamin A precursor, the orange–yellowish pigment **beta carotene**, which is split after ingestion to produce two molecules of vitamin A. High consumption rates of foods rich in beta carotene are statistically associated with lower rates of lung cancer, but the causal link between the two is unclear. Laboratory studies on beta carotene, mostly in rodents, have shown that it suppresses or retards the growth of chemically induced cancers of the skin, breast, bladder, esophagus, pancreas, and colon.

Vitamin E (tocopherol) is another antioxidant vitamin that is especially important in breaking down a group of strong oxidizing agents called peroxides. Vitamin E also helps to prevent spontaneous abortions and stillbirths in pregnant rats, and for this reason it has acquired a reputation as an antisterility vitamin. However, health claims related to the effects of this vitamin on sexual function remain unproved. Vitamin E overdose results in low blood sugar and headache, fatigue, blurred vision, muscle weakness, and intestinal upset. Vitamin E occurs in several forms, of which alpha-tocopherol is the most potent. It is destroyed by freezing and also by cooking food.

Other fat-soluble vitamins. Vitamin D (calciferol) is discussed at greater length in Chapter 5 (p. 225). It is essential to the body's use of calcium in bone formation. Vitamin K is essential to blood clotting because it serves as a cofactor in reactions that produce blood-clotting factors from their inactive precursors. Most people get adequate amounts of vitamin K from the bacteria that live in their intestines. However, newborn infants, whose intestines have not yet been colonized by bacteria, and persons whose intestinal bacteria have been killed off by antibiotics, need more dietary vitamin K until gut bacteria have become established or reestablished.

Minerals

Minerals are inorganic (noncarbon containing) ions and atoms necessary for proper physiological functioning. The ions of sodium (Na^+), potassium (K^+), and chloride (Cl^-) are the principal **electrolytes** (charged particles) of the body. Differences in the concentration of ions on opposite sides of a cell membrane are both a type of concentration gradient and a type of electrical gradient, which together are called a **membrane potential**. Like chemical bonds, membrane potentials are a means by which cells store energy in a usable form. One example of a membrane potential is the proton gradient in the mitochondria (see Figure 8.15); another example is the electrical potential of nerve cell membranes (see Chapter 10, p. 413) created by the distribution of sodium and potassium ions. Because the body's electrically excitable cells (nerve and muscle cells) respond to changes in these membrane potentials (see Chapter 10, p. 416), maintenance of these electrolytes within a very narrow concentration range is very important. If the number of ions, particularly of sodium ions, gets too high, the body compensates by retaining water that would otherwise be excreted in the urine. Because blood pressure is related to the volume of fluid in the circulatory system, too high a concentration of sodium in the body tissues results in **high blood pressure (hypertension)**. This is an otherwise symptomless condition that increases the risks for vascular (blood vessel) diseases such as stroke and coronary artery disease. Overuse of salt (sodium chloride) makes this condition worse, but is seldom the original cause. Many people in the United States consume too much sodium and not enough potassium. Potassium must be present in the proper amounts; either too much or too little can lead to heart failure and death. Raisins, prunes, dates, and bananas are good sources of potassium.

Other minerals important for human health are iron, calcium, fluoride, and a group of trace minerals.

Iron. Soluble iron is needed for the formation of blood hemoglobin and as a cofactor for many other enzymes. A deficiency of iron in cells causes an anemia that is more common in older people with poor dietary habits and in menstruating women. Iron deficiency anemia is in fact the single most common nutritional deficiency in most industrialized countries, including the United States. Menstruating women need almost twice as much iron as men, and pregnant women need even more for the proper synthesis of hemoglobin in the fetus's blood. Vitamin C increases the cellular absorption of iron; therefore if vitamin C supplies are inadequate, cellular uptake of iron is inadequate too. Some people may have iron deficiencies resulting from low levels of the proteins that transport iron in and out of cells.

Calcium. Calcium (Ca^{2+}) is needed as an intracellular messenger for many processes, including muscle contraction (see Chapter 10, pp. 432–433). In addition, the crystal structure of bones and teeth is composed of calcium combined with phosphate and other minerals. Vitamin D promotes calcium absorption, so most people suffering from vitamin D deficiency have symptoms of calcium deficiency as well. High dietary levels of protein can sometimes contribute to an increase in the rate of excretion of calcium by the kidneys.

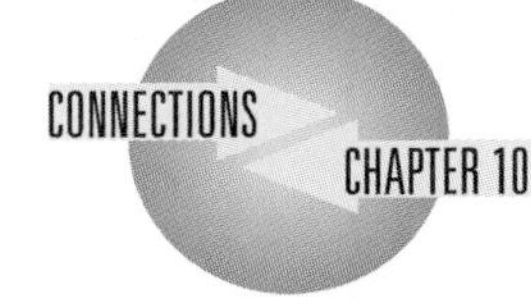

Many older women suffer from low bone density and bone brittleness (**osteoporosis**). Although bone and teeth seem to be very unchanging because of their solidity, they are actually living tissues that are constantly exchanging molecules with the surrounding fluids. There is a balance between the calcium in bone and the calcium in blood; in osteoporosis the balance shifts and calcium dissolves out of bone. Although low blood calcium levels are involved, the problem is not so simple that it can be solved by increasing the dietary intake of calcium later in life. Estrogenic hormones are important, and so is vitamin D, which promotes calcium absorption, but the exact processes are poorly understood. Supplementary doses of both calcium and vitamin D are recommended for postmenopausal women, although most bone loss within the first five years after menopause is caused by estrogen withdrawal, not by any nutritional deficiency. Higher levels of exercise in women aged 18 to 25 can increase their bone density and forestall the development of osteoporosis later in life.

Fluoride. Fluoride (the ion F^-) is important in the growth of strong teeth during the childhood years. Insufficient fluoride results in a greater incidence of tooth decay. Drinking water is the most important dietary source of fluoride. In some areas the natural sources of drinking water contain high levels of fluoride. The observation that people in these areas had lower incidences of cavities led to a search for possible factors. Epidemiological research showed that nearby areas had similar diets and climate but higher rates of cavities. Analysis of the water showed higher fluoride levels in the low-cavity areas than in areas with a higher incidence of cavities. Many municipalities now add fluoride (in carefully measured amounts) to the drinking water supply as a preventive measure against tooth decay. Fluoride is also available as drops for breast-fed infants and others who do not have access to fluoridated water. Fluoride is toxic in very high doses. If high doses are accidentally ingested, milk can neutralize the fluoride.

Trace minerals. Most of the remaining mineral nutrients are sometimes called **trace minerals** because they are needed by the body only in very small quantities. Deficiencies of these trace minerals were more common in the past, when vegetables were grown locally in soils deficient in one or another trace minerals, and when domestic animals grazing on plants growing in the same soil were the main supply of animal food. In the industrial world, such nutritional deficiencies are much less likely because our food supply comes from numerous sources grown in a variety of different soils and climates.

In addition to the trace minerals listed in Table 8.2, chromium and manganese are needed in carbohydrate metabolism; cobalt is an important part of the vitamin B_{12} molecule; molybdenum and nickel are required in the metabolism of nucleic acids; and silicon, tin, and vanadium are needed in trace amounts for proper growth including the development of bone and connective tissue. Diets adequate in other nutrients usually supply sufficient amounts of these trace minerals.

Because of regional variations in the mineral content of soils, the mineral content of foods that grow in those soils also varies. Therefore, mineral deficiencies often vary geographically. Zinc deficiency, for example, is common in the Middle East. Iodine deficiency is found primarily in certain inland locations, such as the high Andes, the Himalayas, and parts of central Africa.

Most vitamin and mineral needs can be met economically, even in poor countries, from grain and vegetable sources. Grains contain most B vitamins and also vitamin E and several important minerals including zinc. Fresh vegetables contain additional vitamins (including A and C) and several important minerals including calcium and iron. Fresh fruits provide additional vitamin C. Legumes provide calcium and iron in addition to proteins. In general, all mineral nutrients can be supplied from plant sources, except vitamin B_{12}, the one important vitamin that cannot be supplied from plant sources alone.

THOUGHT QUESTIONS

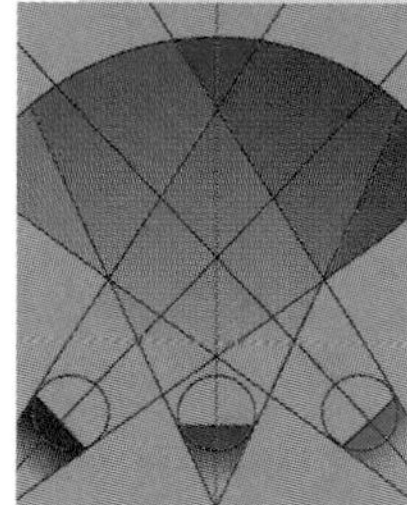

1. If two people drink soft drinks every day, but one chews sugarless gum and the other chews regular gum, would you expect they would have a difference in the number of cavities at their next dental check-up? Would it matter whether the soft drinks were sugar-free or not? (see Box 8.1 for help with this question.)
2. Why are sugars absorbed from the intestine faster than starches?
3. Think about the definition of obesity: weight 20% or more above the ideal weight for a person's sex and height. What is meant by 'ideal?' Who sets these 'ideals?' Are there cultural differences in what is considered 'ideal?' How do cultural definitions of 'ideal' relate to biological definitions?
4. How is it possible for different species (such as rats and people) to have different vitamin requirements? What does this mean on a biochemical level?
5. How is it possible for some studies to show that calcium supplements forestall osteoporosis whereas other studies do not?
6. In Mexico before European contact, the diet consisted mostly of corn, beans, chili peppers, and squash; these same foods are still the major elements of most Mexican diets, especially in rural areas. Can you think of any biological reasons why this diet has proved so stable?

MALNUTRITION CONTRIBUTES TO POOR HEALTH

The term **malnutrition** literally means 'bad nutrition.' Malnutrition can result from eating too little or from eating the wrong foods, and is one of the major health problems of the world, particularly among the poor and in areas of turmoil. Malnutrition also exists in people with access to sufficient foods, sometimes as a result of eating disorders.

Eating disorders

In the middle and upper classes of the industrialized nations, some people suffer from a condition called **anorexia nervosa**. The ratio of anorexic women to anorexic men is about 9:1. Anorexic individuals suffer from a mistaken perception of their body size. They imagine themselves to be heavier than they really are, and desire to be thinner as a result, a feature that clearly distinguishes anorexia from all other forms of undernourishment (Figure 8.16). Anorexics also respond poorly to body cues of hunger and satiety. The misperception of hunger, satiety, and body size are early symptoms that precede the most noticeable feature of the disease, what Dr. Hilde Bruch has called a "relentless pursuit of thinness," a self-imposed undernourishment that borders on starvation. At the same time, there is usually an absorbing or obsessive interest in food, which may include talking or reading about food, preparing food, collecting recipes, or serving food to others, while all the time avoiding eating.

FIGURE 8.16

One of the earliest signs of anorexia is misperception of one's own body.

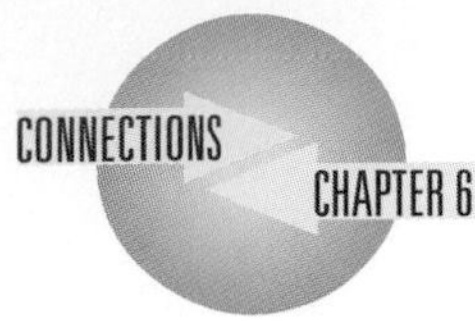

One of the surest signs of the disease in anorexic women is that they usually stop menstruating because of a lack of the cholesterol needed for synthesis of the hormones that regulate the menstrual cycle (see Chapter 6, pp. 249–250). Other symptoms include changes in brain activity. The brain must be constantly supplied with glucose to provide cellular energy for nerve cell function; when it is not, many mental functions may be impaired. These impairments may manifest themselves in anorexic persons as deception (hiding things, keeping secrets) and a distrust of others, but only late in the process, after starvation has already set in. Untreated anorexia is usually fatal.

Anorexia is most common among Caucasian women between the ages of 15 and 30 years with an average or above-average level of education. The disease also occurs in Japan and in parts of Southeast Asia, but only in the uppermost social strata. Anorexia is virtually unknown among people living in poverty or in undernourished populations anywhere. It never seems to occur where food is scarce, or in times of famine. Even in countries where it occurs, it seems to vanish during economic hard times, such as the Depression of the 1930s.

Anorexia was once extremely rare. Its marked increase in the United States since World War II has been attributed by several experts to a general standard of beauty that has increasingly glorified thinness, as measured by such criteria as waist and hip measurements of Miss America contestants, *Playboy* centerfolds, and models and ballet dancers more generally. In fact, women in professions such as modeling and ballet dancing are particularly likely to develop anorexia. Also, female athletes in sports such as rowing (where competition is organized by weight classes) are at high risk for developing a **female athlete triad**, consisting of eating disorders such as anorexia, combined with loss of menstruation and a loss of bone mass that may cause osteoporosis later in life.

Many anorexics also suffer from a related condition called **bulimia**, although bulimia can also occur independently of anorexia. Bulimia is characterized by occasional binge eating of everything in sight, usually including large quantities of high-calorie 'forbidden' foods, in total disregard of any concept of a balanced diet. Immediately after a binge, bulimics typically force themselves to vomit or else purge themselves with an overdose of a laxative. Both the binge and the purge are usually done in secret; bulimics (like anorexics) become extremely skillful at hiding their condition from others. Persistent bulimia can lead to ulcers and other problems of the digestive tract, and also to chemical erosion of the teeth from the frequent contact of the teeth with the acidic secretions of the stomach. Bulimia occurs in both sexes, but more often in women. Bulimia is especially common among educated women who have easy access to unrestricted amounts of food, a situation common on many college campuses.

Protein deficiencies

Inadequate caloric intake can result in protein deficiencies that take several forms, including both low total protein and low levels of particular amino acids. If protein intake is inadequate for either reason, a protein deficiency develops called **kwashiorkor**. Kwashiorkor occurs when carbohydrate intake is adequate but protein intake is not.

In all organisms, cells and proteins are constantly being broken down, and in a healthy, adequately nourished body, they are constantly being replenished. When protein intake is insufficient for this replenishment to occur, there is considerable loss of muscle tissue, and the death of many cells releases numerous dissolved ions into the surrounding tissue. These ions retain water and contribute to tissue swelling (edema) that makes the loss of muscle tissue harder to see. Children suffering from kwashiorkor have swollen abdomens, a fact often noticed in photographs from protein-deficient areas. Their large bellies conceal the fact that these children are actually starving to death.

If protein intake is inadequate and carbohydrate intake is also inadequate, the combined deficiency produces a condition called **marasmus** in which the body slowly digests its own tissues and wastes away. When carbohydrate is not available as an energy source, amino acids are used to produce energy. Because amino acids are not stored except in the form of the body's structural and functional proteins, the use of amino acids for metabolic energy degrades these proteins. Once the body's muscle mass falls below a certain minimum, marasmus is always fatal.

Ecological factors contributing to poor diets

Malnutrition is recognized as a worldwide disease with regional differences in its cause but with similar outcomes everywhere. Here we take Africa as an example, although many other examples exist. Many populations in Africa experience either chronic or periodic protein deficiency; Africa has therefore been described as a protein-starved continent. Fresh vegetables are widely available, so diets are high in fiber and most vitamins. Vegetables (including legumes) can fill nearly all of a population's nutritional needs as long as supplies are adequate and as long as some form of nutritionally complete protein is obtainable from animal sources (including fish) or from a combination of plant proteins. There are lakes and rivers where fishing provides adequate protein resources, and there are cattle-herding groups, such as the Masai and Fulani, who can meet their protein needs from their animals. Across most of the continent, however, animal protein sources are not readily available, and the supplies of grains that could offer protein by amino acid balancing are not always adequate.

Grain supplies are inadequate owing to a combination of ecological and social factors. Tsetse flies, which spread blood-borne diseases, have made much of the land uninhabitable to most domestic animals. This not only limits the availability of meat proteins, but also limits the supply of draft animals that can pull plows and till the soil. Tractors and fuel are too expensive for most farmers in Africa, and many places are either too dry or too wet to support agriculture. The world's largest desert, the Sahara, occupies most of Africa's northern half; other deserts exist in Somalia and Namibia. Most of these deserts, including the vast Sahara, are growing larger each year as animals such as sheep and goats overgraze and destroy the plants on the desert fringes, a process known as

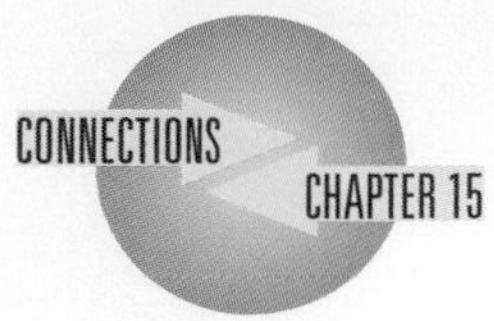

desertification (see Chapter 15, pp. 664–666). In other parts, high rainfall leaches important minerals from the soil, leaving soils deficient in the minerals necessary for plant growth (see Chapter 15, pp. 662–663). Rice, wheat, and most other grains grow rather poorly in many African soils. Some success has been achieved with millet and corn (maize), but raising a sufficient quantity and variety of grains to provide complete protein is difficult. In most places in Africa, populations can maintain adequate nutrition in years when there are ample harvests and efficient distribution. Unfortunately, these two conditions are not always met. Protein deficiencies are made worse by political and military upheavals that drive people from their farms or that prevent the planting, harvesting and distribution of crops. Protein starvation (marasmus and kwashiorkor) is all too common, particularly among children.

Effects of poverty and war on health

Climatic, economic, and political factors all contribute to the unequal production and unequal distribution of food across the planet. Poverty exists in all nations of the world. Many populations are marginally nourished and are therefore more vulnerable to a year of drought or a bad harvest, but malnutrition exists in every part of the globe, especially among the poor. Even in regions in which nourishment is usually adequate, people can become undernourished because of the disruptions of life associated with war.

Wartime starvation. The effects of starvation on humans have been studied retrospectively in people who were, at an earlier time, subjected to starvation by war. Such a study on the short-term and long-term effects of starvation was conducted by using the birth records of infants born in the Netherlands in the winter of 1944–1945, when many people in certain cities experienced starvation. The food shortages varied from city to city, but the other effects of war were equal across the country. It was found that when food rations dropped below a threshold level, fertility in the population decreased and the decrease was greatest among people in the lower classes. Fetuses carried by women who experienced starvation in the first trimester had a higher rate of abnormal development of the central nervous system. Maternal undernourishment in that period also carried forward to premature births, very low birth weight, and an increase in infant death rate immediately after birth. Maternal undernourishment in the third trimester produced the greatest increase in the infant death rate in the three months after birth. Brain cells were depleted in infants who died. Among survivors, however, undernourishment in infancy was not correlated with long-term effects on mental development when males were tested at age 18. Other retrospective studies have found similar results.

Long-term effects of childhood undernutrition. There are several indications that undernutrition during childhood decreases brain development and capacity. Infants who died of kwashiorkor or marasmus had less DNA, protein, and lipid in their brain cells compared with infants who died at similar ages of causes not related to nutrition. Evidence suggests that both the severity and the length of the period of malnutrition affect intellectual development. Iron deficiency, particularly in the early years of life, can also impair mental functions.

Malnutrition can also result from children's failure to eat properly because of a depressed mental state. Malnutrition of this type is called 'failure to thrive.' Infants must learn that their needs will be met, and this learned capacity is termed 'basic trust.' When circumstances are such that a child is not regularly fed and nurtured, it fails to develop trust. Loss of a care-giver or a diminution or loss of nurturing care can create depression in infants, who then lose trust and become malnourished. Medical intervention cannot rescue the child unless a trusting relationship is established. Adults recover their physical and mental capacities if they are rescued from starvation. For children, however, nutritional replacement will not lead to full recovery unless emotional and psychological support is also provided. The recovery of children from famine requires much more than food.

Micronutrient malnutrition

Malnutrition can still exist when total caloric intake is sufficient or even when it is excessive. Because the roles of micronutrients include the proper functioning of enzymes, micronutrient deficiencies prevent the proper utilization of other foods. As we saw earlier, mineral deficiencies can result when diets are restricted to those foods grown in mineral deficient soils. Diets that include only a few types of foods, particularly processed 'junk foods,' may be lacking in necessary micronutrients. Limited incomes, limited mobility, and limited availability of fresh foods all make nutritional problems worse. Micronutrient malnutrition may be overlooked among people who lose interest in eating, including elderly people, people with chronic diseases including cancer, and people who have mental illnesses such as depression. Some elderly people may also forget what they have or have not eaten.

THOUGHT QUESTIONS

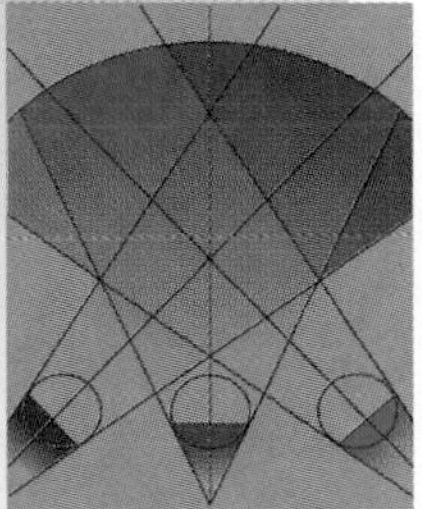

1. Why do you suppose anorexia occurs only among well-fed populations, and never among the poor or in times of famine? What does this imply about the possible causes of the condition?
2. Will efforts aimed at improving nutrition in people with various forms of malnourishment be readily accepted by the people they are meant to help? What factors determine acceptance?
3. If sufficient food can be produced, will every person have good nutrition? Why or why not?
4. Why might starvation have greater long-term effects in children than in adults?

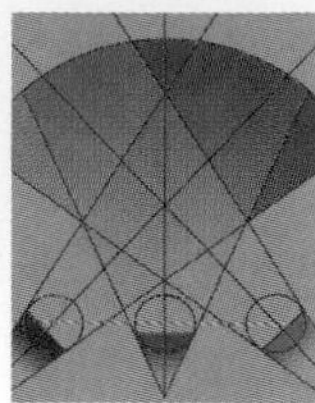

DURING THE FIRST HALF OF THE TWENTIETH CENTURY, vitamin deficiencies were major public health concerns. Vitamin and mineral deficiencies can lead to various deficiency diseases, birth defects, or neurological damage. Nutritional deficiencies and deficiency diseases have declined since then in the industrial world, the results of better eating habits, more varied diets, and vitamin-fortified foods. Greater interest now centers on whether certain foods can promote good health and reduce the risks of chronic diseases. Since the 1960s, public health officials and nutritionists have increasingly turned their attention to heart disease, stroke, cancer, and chronic health problems such as obesity and high blood pressure that can influence the risks for these fatal diseases. Heart disease is the number one cause of death in many industrialized countries and is significantly associated with a diet that is too high in fat, particularly saturated fats.

Malnutrition still exists, however, and it can take many forms. Inadequate caloric intake can result from crop failures, from poverty, or from eating disorders. Poverty and war usually worsen nutrition and contribute to stress among civilians. Adequate nutrition depends on getting all the necessary nutrients without taking in too much of any one of them. Much of the world's population gets inadequate food and inadequate protein. In many industrialized countries, problems are caused instead by high-fat diets with inadequate fiber. The best way to avoid both types of problems is to eat a varied diet while keeping fat intake low. Sensible eating habits such as these are important for good health. If we want to promote health for ourselves, our families, and the rest of humankind, then a very important step is ensuring that each person has an adequate and balanced diet.

CHAPTER SUMMARY

- Digestion is a process in which materials consumed as food are broken down into substances that the body can absorb. Digestion consists of **chemical digestion** with the aid of **enzymes**, and **mechanical digestion**.
- The macronutrients are the major sources for energy, measured in **kilocalories**, and including **carbohydrates**, **lipids** (fats and oils), and **proteins**. Lipids are also needed for the formation of cell membranes and proteins are needed as enzymes, receptors and membrane channels and transporters.
- Digested food enters cells by passive **diffusion**, facilitated diffusion, **active transport** or **endocytosis**.
- Cellular energy is derived in the mitochondria by **oxidation** reactions in the **Krebs cycle** and electron transport to form **ATP**.
- Material that the body cannot absorb constitutes fiber.
- Micronutrients (**vitamins** and **minerals**, including electrolytes) are needed for good health.
- Good nutrition reduces such chronic conditions as high blood pressure and lowers the risks for cardiovascular disease and certain cancers.

KEY TERMS TO KNOW

active transport (p. 322)
antioxidant (p. 345)
atherosclerosis (p. 331)
ATP (p. 340)
basal metabolic rate (p. 328)
carbohydrates (p. 329)
chemical digestion (p. 317)
cholesterol (p. 335)
complete protein (p. 337)
concentration gradient (p. 322)
diffusion (p. 322)
endocytosis (p. 322)
enzymes (p. 317)
epidemiology (p. 333)
glycogen (p. 325)
hormone (p. 320)
kilocalories (p. 328)
Krebs cycle (p. 340)
limiting amino acid (p. 337)
lipids (p. 331)
mechanical digestion (p. 317)
minerals (p. 346)
nonpolar (p. 320)
obesity (p. 328)
oxidation (p. 345)
polar (p. 320)
proteins (p. 336)
saturated fats (p. 332)
unsaturated fats (p. 332)
vitamins (p. 342)

CONNECTIONS TO OTHER CHAPTERS

Chapter 3: Some genetic differences exist in the body's ability to digest certain substances (as in phenylketonuria) or to use certain nutrients after their absorption (as in diabetes).

Chapter 5: Nutritional requirements may differ among human populations for various inherited and environmental reasons.

Chapter 6: Unchecked population growth puts more people at risk for malnutrition.

Chapter 9: Certain cancers have frequencies that vary according to diet: high-fat diets promote certain cancers, whereas high-fiber diets lower many cancer risks.

Chapter 12: Poor nutritional status impairs the immune system.

Chapter 13: People with HIV infection stay less sick if they have good nutrition, but appetite is often suppressed (and nutrition suffers) as AIDS progresses.

Chapter 14: Crop improvements may help to alleviate starvation in many populations.

Chapter 15: Biodiversity is threatened by the need to clear more land for farming to feed the world's population. Developing and conserving better and more varied crop plants will feed more people and support greater biodiversity at the same time.

PRACTICE QUESTIONS

1. Explain the differences between chemical digestion and mechanical digestion.
2. How many times greater is the hydrogen ion concentration in stomach acid than in blood?
3. What makes a molecule polar? What makes a molecule nonpolar? Can some molecules be both polar and nonpolar?
4. What kinds of food molecules can supply precursors to the Krebs cycle?
5. What type of transport is used to bring glucose into a cell?
6. What hormone regulates the uptake of glucose by cells?
7. What type of transport brings lipids into cells? How are nonpolar lipids able to travel through the blood to get to cells?
8. How does atherosclerosis result in an increase in blood pressure?
9. What is the HDL/LDL ratio and why is it important?
10. What are the functions of cholesterol in the human body?
11. Which food molecules produce the most ATP?
12. Where in the cell is ATP produced?
13. What are some of the functions of the water-soluble vitamins in the body?
14. How is an electrolyte different from other mineral micronutrients? What are some of the functions of each in the body?
15. Which can lead to health problems: vitamin deficiencies or vitamin megadoses?

Smokin'

20 PREMIUM CIGARETTES

Smokin'

SMOKING MAY CAUSE CANCER

9 Cancer

CHAPTER OUTLINE

Multicellular Organisms Are Organized Groups of Cells and Tissues

- Compartmentalization
- Specialization
- Cooperation and homeostasis

Cell Division is Closely Regulated in Normal Cells

- The cell cycle
- Regulation of cell division
- Regulation of gene expression
- Cellular differentiation and tissue formation
- Limits to cell division

Cancer Results When Cell Division is Uncontrolled

- Properties of cancer cells
- Oncogenes and proto-oncogenes
- Tumor suppressor genes
- Accumulation of many mutations
- Progression to cancer

Cancers Have Complex Causes and Multiple Risk Factors

- Inherited predispositions for cancers
- Increasing age
- Viruses
- Physical and chemical carcinogens
- Dietary factors
- Tumor initiators and tumor promoters
- Internal resistance to cancer
- Social and economic factors

We Can Treat Many Cancers and Lower our Risks for Many More

- Surgery, radiation, and chemotherapy
- New cancer treatments
- Cancer detection and predisposition
- Cancer management
- Cancer prevention

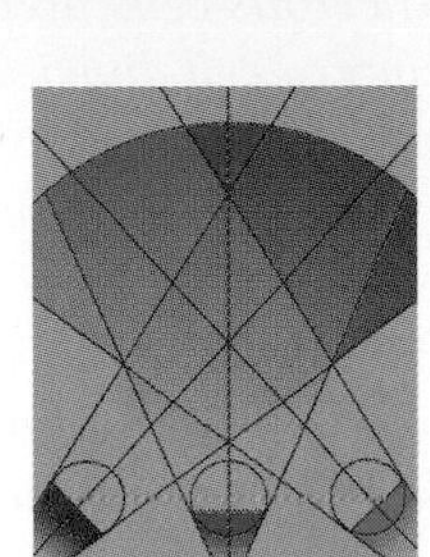

ISSUES

Why is cancer so important?

What are stem cells and why are they so important?

What is cloning? Why has it already been banned in some countries?

What are the causes of cancer? Are they more genetic or more environmental?

What are the treatments for cancer?

Can we test people for predisposition to disease? Should we?

Can we prevent cancer?

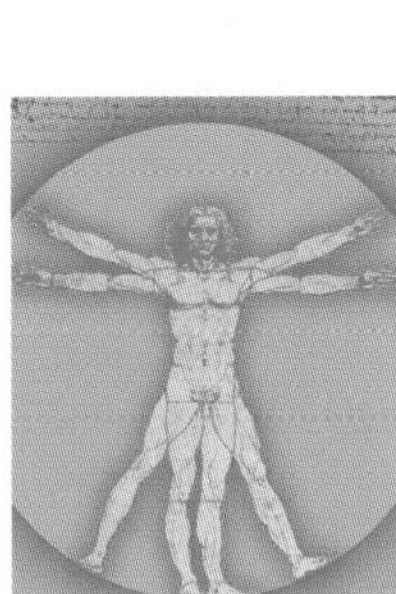

BIOLOGICAL CONCEPTS

The cell cycle and its regulation (receptors, growth factors)

Levels of organization (cells, tissues)

Gene expression and regulation (cell differentiation in embryos and adult animals, potentiality)

Interaction of genotype and environment (carcinogens, mutagens, diet)

Scaling (ratio of surface area to volume)

Health and disease (homeostasis, risk)

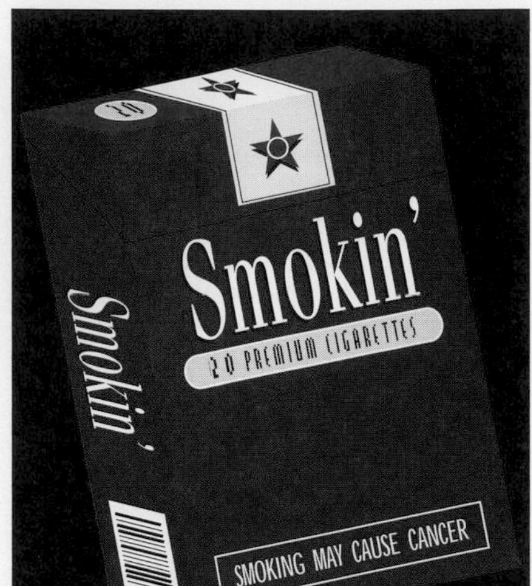

Chapter 9: Cancer

Cancer is now the second leading cause of death in most industrialized countries, second only to heart disease. In the Netherlands and some other countries, cancer ranks first. Cancer is also one of the most dreaded illnesses, and people who learn that they have cancer often suffer additionally from fear of the disease.

Because there are many forms of cancer, it is more accurate to speak of cancers in the plural. **Cancers** *of all types result from the same problem: cell division that is out of control. The cancerous cells no longer respond to the signals that normally limit the frequency of cell division. These signals keep normal cells functioning in an* integrated *manner, a necessity in all multicellular organisms whose cells are organized into specialized tissues. The activity of normal cells in tissues is sometimes compared to the behavior of animals in social groups (see Chapter 7): like the behavior of one animal, the behavior of one cell influences that of others. In contrast with normal cells, cancer cells do not behave in an integrated, social way. Instead, they grow chaotically, under their own direction, gradually pushing normal cells aside or growing right over them. Research on cancer has taught us a lot about the biology of normal cells. The basic rules of normal cellular behavior are common to all multicellular species. In this chapter we consider first how cell growth and behavior are regulated in normal cells. Then we see what goes wrong in cancer and what brings it about. We will also consider how we can most effectively reduce our risks for various cancers.*

MULTICELLULAR ORGANISMS ARE ORGANIZED GROUPS OF CELLS AND TISSUES

Most organisms larger than a certain microscopic size are subdivided into compartments called **cells**. The first organisms consisted of only single cells performing all life functions in that single compartment, with little or no spatial separation among their functions. Bacteria continue to live very successfully as single-celled organisms. So why should multicellular life forms have evolved at all? Multicellularity evolved because it offers living things several advantages over unicellularity.

Compartmentalization

Compartmentalization into cells permits organisms to become much larger than they could be as single cells. Physical restrictions are imposed on living things by the ratio of their surface area to their volume. The requirements for energy and the production of wastes both increase in proportion to the volume of an organism, while the organism's ability to absorb nutrients and to release wastes varies with its surface area. However, as an organism enlarges, its volume grows faster than its surface area. Look at the cubes shown in Figure 9.1. What is the volume of cube A and of cube B? What is the surface area of cube A and of cube B? When you calculated the volume of cube A you multiplied three numbers: height times width times depth. (Another way to say this is that volume is proportional to length cubed.) When you calculated surface area you multiplied only two numbers—height times width—and then added up the number of surfaces. (Another way to say this is that area is proportional to length squared.) As you found out, the volume of cube B is 27 times the volume of cube A, yet the surface area of cube B is only 9 times the surface area of cube A. By subdividing a cube, we make more surface area. Subdivided cube C has the same volume as cube B, but three times the surface area.

Just as with the cubes in Figure 9.1, compartmentalization of the interior volume of an organism into cells keeps the volume the same, but increases the effective surface area. This maintains an efficient ratio of surface area to volume so that nutrient intake stays balanced with metabolism. Again in Figure 9.1, the ratio of surface area to volume of cube A is 6, but for the larger cube B the ratio is only 2. Subdividing cube C returns the ratio of surface area to volume to 6, the same as it was in the small cube A. Subdividing an organism into cells achieves the same effect.

During the course of evolution, several kinds of unicellular organisms began to grow as collections or communities. Such aggregates have greater fitness than a single, large, undivided organism, for the reasons just discussed. Some present-day species can change back and forth between unicellular and aggregate forms. The cellular slime molds, such as *Dictyostelium discoideum* (kingdom Fungi, subkingdom Myxomycota), are independent mobile unicells as long as conditions are favorable for feeding. When conditions are no longer favorable, the individual cells

FIGURE 9.1

Subdividing a volume into cells increases the effective surface area. Cube A is one unit on each side; cube B is three units on each side. Cube C is also three units on each side but is subdivided into smaller cubes, each 1 unit per side. What is the total volume of each cube? What is the total surface area? What is the ratio of surface area to volume?

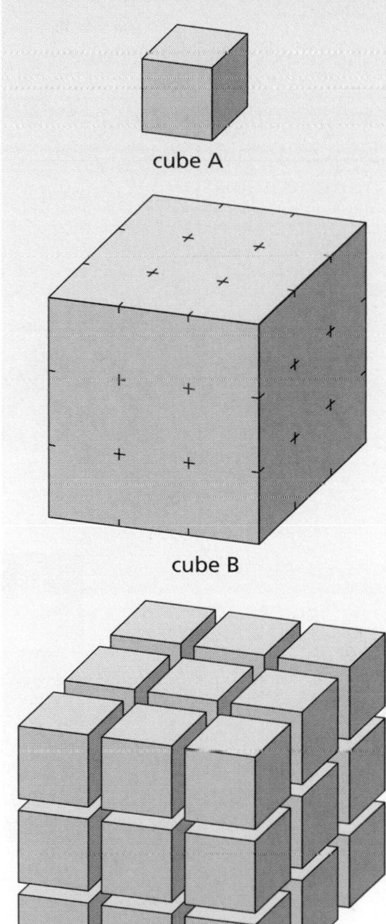

creep toward one another to form a multicellular aggregate in which some of the cells become specialized for different functions (Figure 9.2).

Specialization

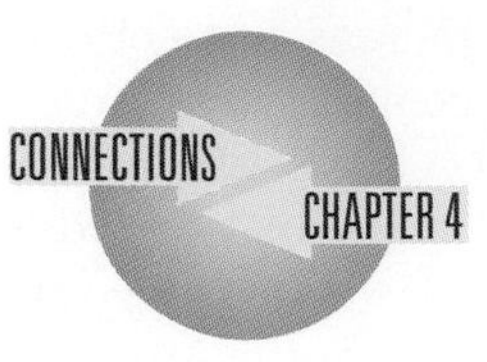

An advantage that multicellular aggregates and organisms have is that not every cell needs to perform every function. This allows specialization. In sponges (kingdom Animalia, phylum Porifera), cells are specialized but are not organized into tissues (see Figure 4.22, p. 169). The cells of all other animals form **tissues**. A tissue consists of similar cells and their products located together (structurally integrated) and functioning together (functionally integrated). The inner and outer cell layers of animals in the phylum Cnidaria, also called coelenterates (see Figure 4.22, p. 169), are separate tissues with different functions. Although the cells in the two layers are different, each cell is still changeable, so that the Cnidaria are able to regenerate an entire organism from a small piece. The cellular flexibility of Cnidaria (and other organisms capable of regeneration from parts) contrasts with the situation in more complex multicellular organisms. In these complex organisms, the fate of a cell becomes more and more restricted as it divides, through a process called **differentiation**. Initially, a cell is capable of performing a variety of functions, but as the cell differentiates (literally, 'becomes different'), it progressively loses some of its abilities and becomes specialized at doing only a few things very well. Some cells, such as human muscle or nerve cells, lose so many important abilities during differentiation that they may become incapable of further cell division. Other cells, such as those

FIGURE 9.2

The life cycle of the slime mold *Dictyostelium discoideum*, alternating between unicellularity and multicellularity.

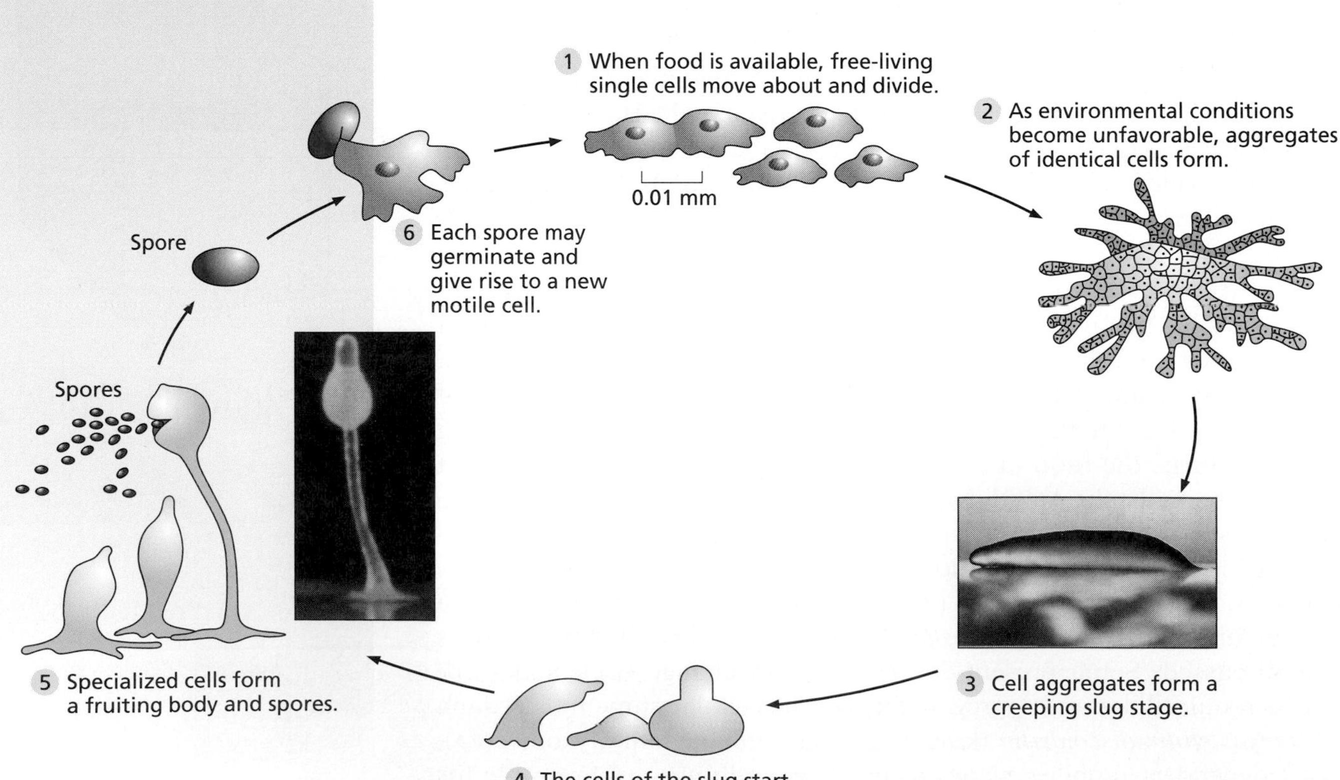

of human bone marrow, retain throughout life a good deal of the flexibility characteristic of cells in developing organisms. How cells 'know' what type of tissue to become has long been one of the major questions in biology. Much of what we currently know about normal cell division and differentiation has been aided by the comparison of normal cells and cancer cells.

Cooperation and homeostasis

The organization of cells into tissues allows multicellular organisms to specialize their functions while maintaining an efficient ratio of surface area to volume. For specialization to be beneficial, the behavior of one type of cell must be integrated with the behavior of other cells. Tissues are further integrated into organs and organ systems, in which two or more types of tissues coordinate to perform more complex functions, such as reproduction (see Chapter 6), digestion (see Chapter 8), external sensing (see Chapter 10), or circulation (see Chapter 8). A multicellular organism can thus be considered a complex ecosystem: a human organism, for example, is an ecosystem of some ten trillion individual cells. In the course of evolution, specialization based on cooperation worked well and was favored by natural selection. All multicellular organisms—fungi, plants, and animals—have continued this basic plan.

The proper functioning of the whole organism depends on the continued integration and cooperation of all the cells. When this integration is functioning properly, that is, when the ecosystem of cells is stable, we may consider that organism to be in a state of health. According to French physiologist Claude Bernard (1813–1878), cells are responsible for maintaining a 'milieu intérieur' within each cell and within the body as a whole. Good health is defined as the maintenance of more or less constant conditions within this internal environment, a process that Bernard named **homeostasis**. This does not mean that during homeostasis there are no changes within the organism. Just the opposite is true: molecules and cells are constantly being made and being broken down, but these changes occur around a balance point. Homeostasis is the ability to return to that balance point. Disruption of homeostasis produces illness. As we will see, cancer is a disruption of the cellular homeostasis in which cell division is no longer in balance with cell death.

THOUGHT QUESTIONS

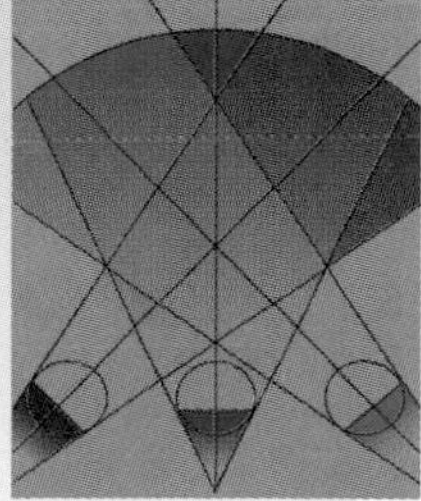

1. When the weight of some material in solution remains the same, how do changes in cell volume influence the concentration of the material?
2. What would an organism be like if all of its cells were the same?
3. Many animals, including insects, fishes, amphibians, and reptiles, do not maintain a constant internal temperature but allow their internal temperature to change with the external temperature. Are these animals in homeostasis? Why or why not? (Many of these animals can regulate their temperature behaviorally by moving to different locations.)
4. How could multicellularity in animals or plants have evolved through natural selection?

CELL DIVISION IS CLOSELY REGULATED IN NORMAL CELLS

The proper functioning of multicellular organisms depends on the regulation and integration of the processes of all their cells, particularly the process of cell division. We learned about mitosis in Chapter 2. Mitosis is one step of a larger process of cell division called the cell cycle. Scientists are finding that the molecular signals that regulate cell division and the cell cycle are remarkably similar in highly diverse organisms, from fungi to humans.

CONNECTIONS CHAPTER 2

The cell cycle

Normal cells grow only a small fraction of the time. They continually make new proteins and other cellular chemicals to replace ones that have been used or damaged, but most of the time they do not increase in size. When cells do grow, they soon reach the size at which their ratio of surface area to volume makes them inefficient. Instead of becoming increasingly inefficient, the cells divide. When we talk about how fast cells grow, we really mean how frequently they divide, not how fast they enlarge.

The complex process of eucaryotic cell division recurs whenever the cell divides; it is called the **cell cycle**. You can follow the steps of the cell cycle in Figure 9.3. The cell cycle begins with a phase called G_1 in which protein synthesis is increased. Then, if the cell receives signals, it enters the synthesis, or S, phase, marked by DNA synthesis and the replication of both DNA strands (see Figure 2.15, p. 60). Entry into the S phase is the decision point at which a cell is committed and must go on to complete the cycle. When DNA synthesis is complete, the cell enters the G_2 phase in which preparations are made for mitosis. Mitosis itself (see Chapter 2, pp. 46–48) constitutes the M phase, at the end of the cell cycle.

FIGURE 9.3

The cell cycle. Hours shown are the approximate length of time for each phase in a cell with a 24-hour cell cycle, typical of many eucaryotic cells.

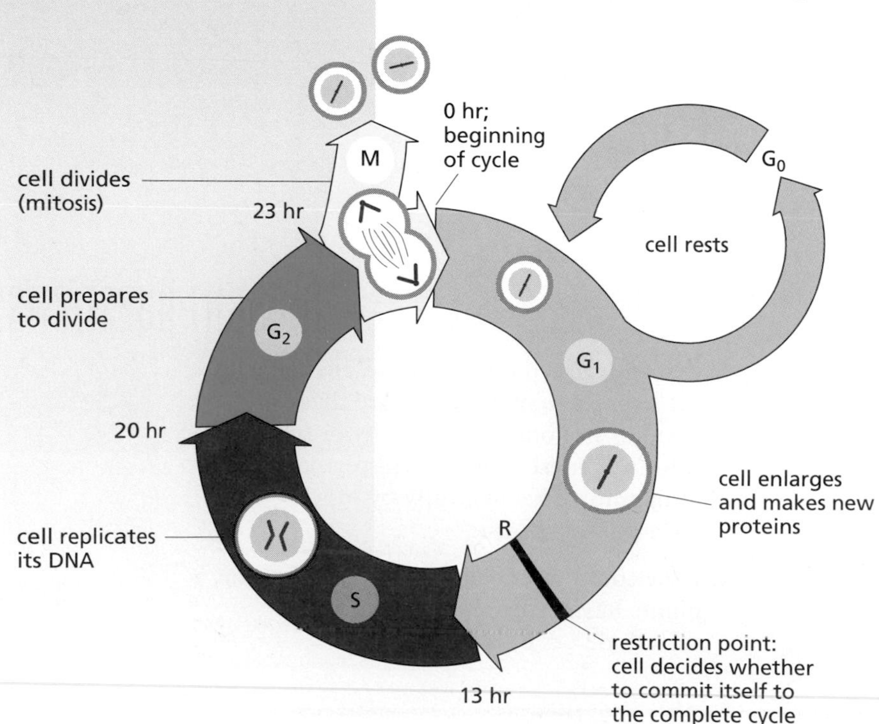

A cell spends most of its time in a resting stage, or G_0, between cell divisions. During the resting stage other cellular metabolic processes proceed but the cell does not re-enter the cell cycle to divide again unless it is signaled to do so. The duration of the cell cycle (G_1 through M) is fairly constant with a species, but G_0 varies greatly. For single-celled organisms, the length of time in the resting stage depends on the availability of nutrients. The length of G_0 in multicellular organisms varies with the developmental stage. When an individual animal or plant is developing, the rate of increase in the number of cells can be very rapid and cells spend little or no time in G_0. In most species, most types of cells spend more time in G_0 once adulthood is reached. The neighboring cells only signal a cell to enter the G_1 and then the S phase when some cell has died and needs to be replaced. Replacements are obviously needed if there has been an injury; but even in uninjured tissues some cells die and others divide to replace them. How often this occurs depends on the tissue.

Regulation of cell division

The cell cycle (and thus cell division) is a tightly controlled process in all types of organisms, both single-celled and multicellular. As an example of how well this process is controlled, consider the human liver. A normal liver grows until it reaches a certain size, and then it stops growing. If a piece of the liver is removed, the liver cells divide to replace what is missing; then they stop. Body tissues have some way of determining how big they should be, and they stop growing when they reach that size. We will soon see that the major difference between normal tissues and cancerous tissues is that cancers grow without any such limits.

In multicellular organisms, the size of most cells is restricted not only by surface area but also by the cell's being confined to a space of a certain size within a tissue. In adult organisms, cells do not divide unless a previous cell has died or been damaged, opening a space for a new cell. Many types of cells, including mature blood cells, are no longer capable of dividing at all. When old or damaged blood cells are removed, they are replaced by cells exiting from the bone marrow, the hollow interior spaces in bones. Once these cells have left, other cells in the bone marrow divide, producing new cells to fill the spaces.

Contact inhibition and anchorage dependence. Contact with neighboring cells suppresses cell division in normal cells. This is called **contact inhibition**. Contact inhibition can be demonstrated in artificial tissue cultures. The cells divide until they form a continuous monolayer (a layer one cell thick); they then stop dividing because they are in contact with other cells. At this point some cells can be removed and put into fresh containers with fresh nutrients. The transferred cells adhere to the new containers and, because they do not cover the surface, they begin to divide and continue until they form a complete monolayer. In addition, cells in the original location resume dividing if enough neighboring cells are removed, and continue until they again form a continuous monolayer (Figure 9.4). Thus, one requirement for cell division in most normal cells is that they not be completely surrounded by other cells.

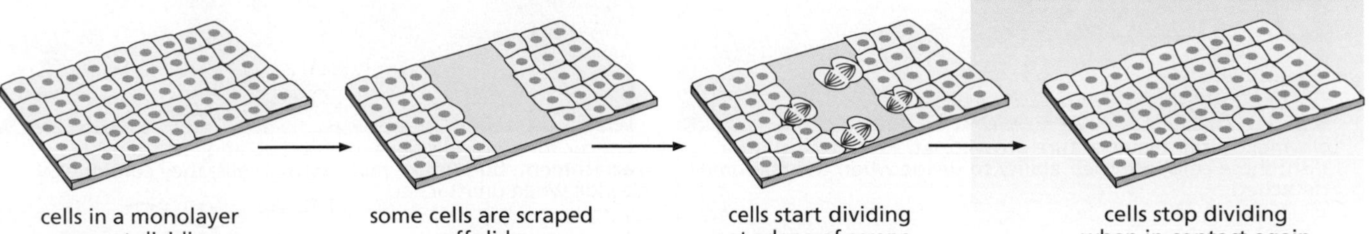
cells in a monolayer are not dividing
some cells are scraped off slide
cells start dividing at edges of scrape
cells stop dividing when in contact again

FIGURE 9.4 Contact inhibition stops cell division once cells are in contact with each other.

Normal cells from many types of tissues have an additional requirement for division called **anchorage dependence**: they divide only when they are attached to a surface. In tissue culture they attach to the plastic dish or to a coating on the dish (Figure 9.5). In multicellular organisms, cells attach to complex organic molecules outside the cells called the **extracellular matrix**. A normal cell may be prevented from dividing if it loses its ability to adhere to the matrix or if the matrix changes in a way that prevents adherence.

CONNECTIONS CHAPTERS 6, 12, 13

External receptors and internal second messengers. We have said that cell division is a tightly controlled process, but *how* is this process controlled? In general, cells receive signals of various kinds from their external environment and do not divide unless they receive signals that send them out of the G_0 resting phase and into the G_1 phase of the cell cycle (see Figure 9.3). These signals are usually small molecules, called **cytokines**, secreted by other cells. Signaling by means of cytokines is one important way in which cells communicate with one another. Cytokine is a general term for the molecules that regulate many important body functions, including reproduction and defense against diseases (see Chapters 6, 12, and 13). The cytokines that regulate entry into the cell cycle are called **growth factors**. Growth factors are secreted by cells into the spaces between cells, yet like most molecules they cannot cross the cell membrane. So how do they tell cells to divide?

The multistep process that tells cells to divide is shown in Figure 9.6. The first messengers, cytokines, bind to specific **receptors**, molecules that extend through the membrane of the cell. The term *specific* means that a given receptor can bind to only one particular type of molecule. Binding of a growth factor to its specific receptor on the exterior of the membrane changes the portion of the receptor molecule that is on the interior of the membrane. The changes are then passed along through the cytoplasm to the nucleus of the cell by a network of molecules

FIGURE 9.5

Anchorage dependence.

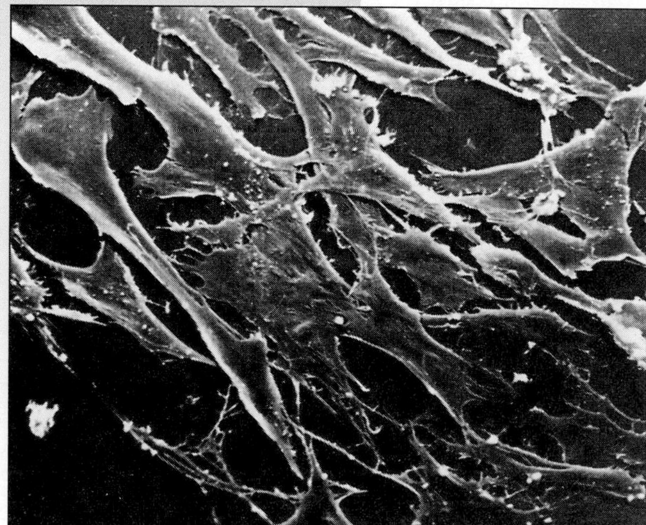

Normal cells in tissue culture growing attached to a culture dish; these cells lose their ability to divide when they become detached.

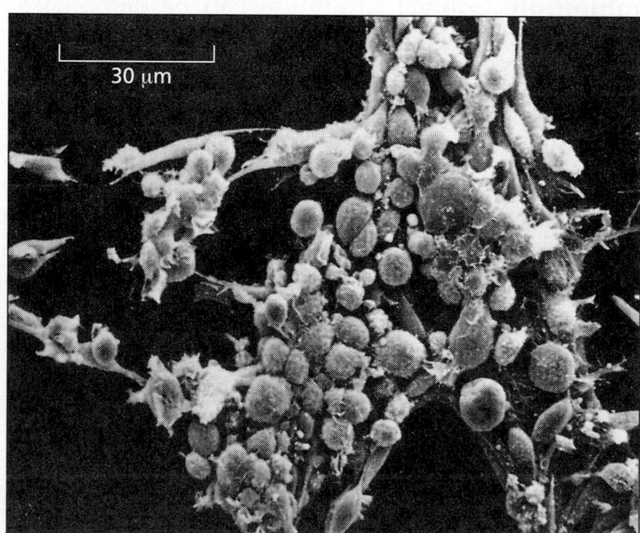

Cancer-forming cells have rounded up and lost their attachment, but, unlike most normal cells, they continue to divide when unattached.

called **second messengers**. In response to second messengers, the concentrations in the nucleus of proteins called **cyclins** change. When the concentration of cyclin D is high, the cell enters the S phase, committing it to division. The growth factor remains outside the cell, but *information* has been transmitted across the cellular membrane and to the cell nucleus, triggering cell division.

Even in the presence of growth factors, however, tissue cells do not divide unless they are attached to the matrix. The response of a cell to external signals thus depends on the presence and normal functioning of signal molecules, receptors for these signals, second messengers and cyclin nuclear proteins, and anchorage to a normal extracellular matrix.

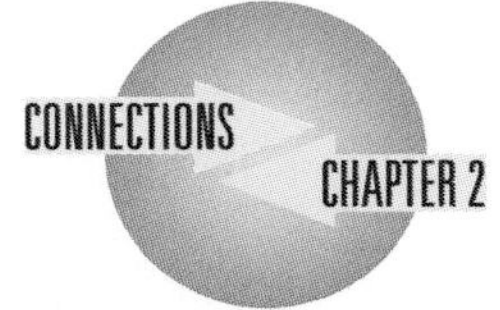

Regulation of gene expression

Cell division, like other cellular processes, depends on the presence of the right proteins at the right time and in the right amounts. The process of using the DNA of the genes to make RNA and then protein is called **gene expression**, a complex process with two major steps called transcription (using DNA to make RNA) and translation (using RNA to make protein), as described in Chapter 2. Gene expression is regulated to control whether a protein is produced and how much of the protein is made.

The regulation of transcription is summarized in Figure 9.7. Transcription begins when a molecule called RNA polymerase binds to a special DNA sequence known as a **promoter** sequence (see Figure 9.7A). Each gene has its own promoter. Promoter sequences can be called strong or weak, according to how strongly they bind RNA polymerase. Proteins needed only in small amounts are coded for by genes present in the genome as single copies and controlled by weak promoters. Many proteins needed in very large amounts are coded for by multiple copies of the same gene and controlled by a strong promoter. The stronger the promoter, the faster the rate of transcription, and thus the greater the amount of messenger RNA (mRNA) produced. The transcription of some cell division genes is turned on by the binding of RNA polymerase to their promoters after the receipt of second messengers resulting from cytokine signals.

On the DNA near the promoter there are regulatory gene sequences that can either enhance or repress transcription by changing how RNA polymerase binds to the promoter. If enhancers bind, RNA polymerase binds more strongly and more copies of mRNA are transcribed from that gene (see Figure 9.7B). If repressors bind to the regulatory sequences, RNA polymerase is blocked from the

FIGURE 9.6

How growth factor cytokines signal cell division in normal cells.

1 External signaling molecules (cytokines, or first messengers) bind to specific receptors present in the cell membrane.

growth factor cytokines

specific receptor

plasma membrane

2 Changes are produced on the cytoplasmic side of the receptor.

3 Second messenger molecules are activated.

cytoplasm

nucleus

Nuclear proteins (cyclins)

DNA

4 When the first messenger is a growth factor, second messengers stimulate the activity of proteins in the nucleus, triggering entry into the cell cycle.

promoter and transcription is halted (see Figure 9.7C). Transcription repressors include the tumor suppressor proteins that slow or prevent cell division. These proteins inhibit RNA polymerase from binding to the promoters of the genes for certain proteins that signal the cell to enter the cell cycle. If the genes for these signal proteins are not transcribed and translated, the cell does not divide.

Repressors themselves are also regulated. A repressor can be inhibited from binding either by an inhibitor that blocks its DNA binding site (see Figure 9.7D), or by mutations that change the shape of the repressor protein. In either case, transcription is once again allowed. If the gene product was a cell division signal, the cell divides, possibly dividing repeatedly until a tumor results.

Synthesis of the proteins needed for cell division (or any other proteins) can be regulated at several steps, beginning with whether a gene is transcribed into mRNA (Figure 9.8, step 1). We have just seen how transcription is regulated (see Figure 9.7). In eucaryotic cells the mRNA synthesized during transcription needs to leave the nucleus before it can be translated into proteins because the ribosomes are in the cytoplasm. Many mRNAs must be chemically modified before they can leave the nucleus (Figure 9.8, step 2); if they are not modified, the mRNA is not translated, and gene expression is halted. Once the mRNA is in the cytoplasm, the rate of translation can be changed. Rapid translation produces more copies of a protein (Figure 9.8, step 3). The amino acid sequence first produced in translation (the primary translation product) is often not the amino acid sequence of the final protein. Some amino acids may need to be removed, or other chemical groups added, before the protein can fold properly into its functional shape. Without this post-translational processing, functional proteins are not produced (Figure 9.8, step 4). Finally, protein activity can be regulated by the binding of other molecules, called effector molecules, which change the protein shape to either slow down or speed up the activity of the protein (Figure 9.8, step 5).

FIGURE 9.7

Regulation of transcription.

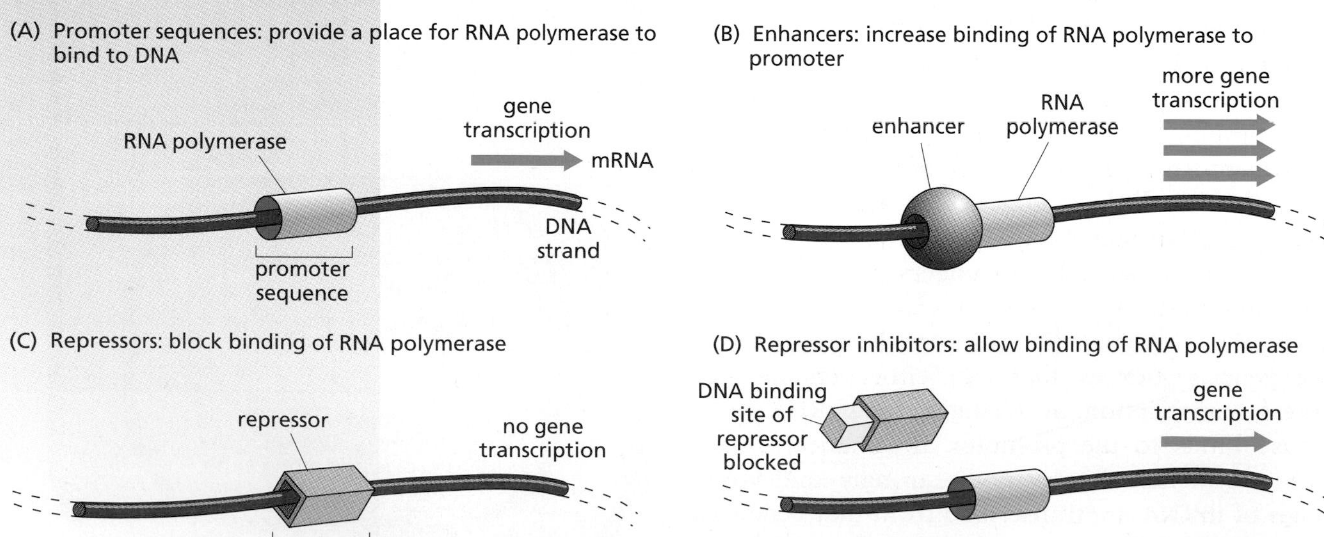

Cell division is thus regulated by all of the mechanisms shown in Figures 9.7 and 9.8. These mechanisms control the concentration and activity of regulatory proteins, including cytokine growth factors, nuclear cyclins, and some second messengers. Many of the genes that control and coordinate cell division were identified by research on cancer cells.

Cellular differentiation and tissue formation

A fertilized egg (zygote) is a single cell whose cellular descendants are capable of forming all the different cell types within the body. The long list of possibilities includes skin cells, muscle cells, glandular cells, bone cells, liver, and so forth. Such a list of possibilities constitutes the **potentiality** of a cell or group of cells. Within a developing multicellular organism, cells that are dividing also become different. Differentiation takes place in steps. At each successive cell division and differentiation, the range of possible future identities for that cell lineage is narrowed, until the potentiality narrows to a single cell type. Once a cell lineage has differentiated as muscle cells, for example, all progeny cells are committed to being muscle cells.

Like cell division, differentiation is tightly regulated by the control of gene expression. Much of what we know about cell differentiation has come from embryology, the study of the development of an organism from a zygote. Studies of normal differentiation have taught us much about the abnormal conditions that exist in cancer.

Potentiality and determination. The zygote has maximum potentiality because it gives rise to all cell types and may thus be called **totipotent**. The potentiality of cells has been investigated by transplanting cells from the embryos of experimental animals. Up until the eight-cell stage, each of the cells in a mammalian embryo is totipotent and could develop into a complete organism. As the cells continue to divide, they first form a hollow ball called a **blastula** (see Figure 4.21, p, 167). Cells in a blastula are no longer totipotent. Soon they begin to differentiate and form tissue layers (endoderm, mesoderm, and ectoderm), a process that begins at a landmark called the dorsal lip. Cells in each layer are restricted to becoming certain types of tissues. At the stage where the hollow-ball embryo has formed differentiated cell layers, it is called a **gastrula**. These steps in embryonic development are shown in Figure 9.9A.

A group of cells removed from the ectodermal layer of the embryo at the gastrula stage and

FIGURE 9.8

Five steps in the process of gene expression at which regulation can take place.

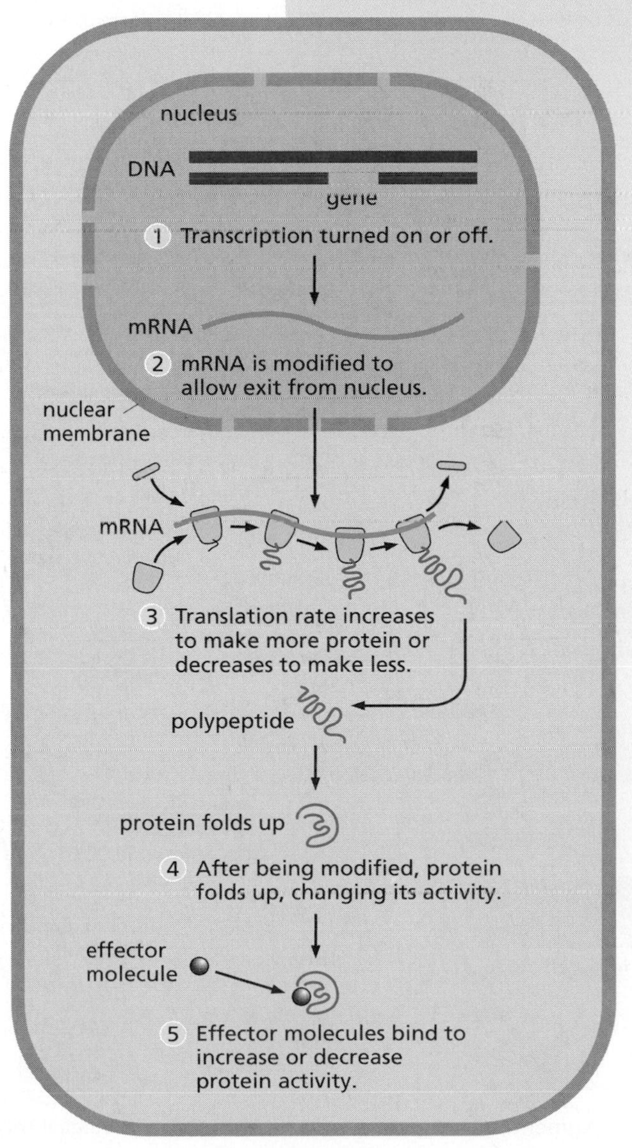

transplanted elsewhere on the same embryo can form various tissue types, but only types that are ectodermal. Such cells are said to be **pluripotent** because their potentiality is still quite broad, but not as broad as that of the zygote. As shown in Figure 9.9B, each of the gastrula cell layers is pluripotent, destined to become certain types of cells. Cells

FIGURE 9.9

Differentiation of various cell types.

(A) Cell layers form as a blastula develops into a gastrula.

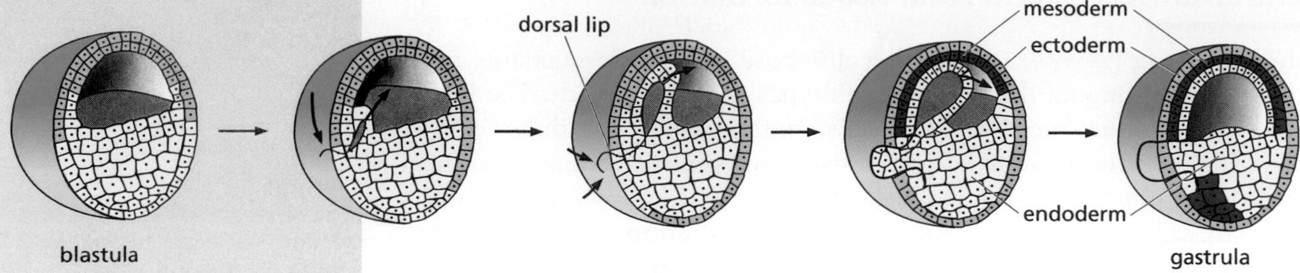

(B) As cells in the ectoderm, mesoderm, and endoderm divide, they differentiate, eventually becoming specialized cells.

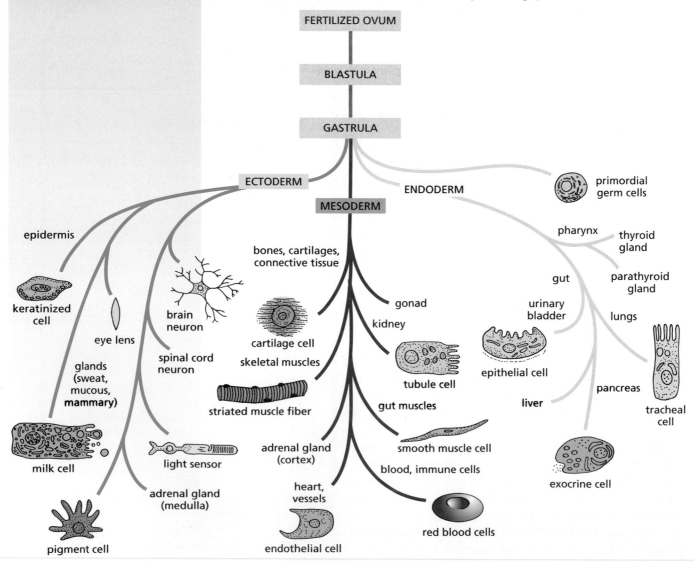

FERTILIZED OVUM
BLASTULA
GASTRULA
ECTODERM
MESODERM
ENDODERM
primordial germ cells
epidermis
keratinized cell
eye lens
brain neuron
spinal cord neuron
glands (sweat, mucous, **mammary)**
milk cell
light sensor
pigment cell
adrenal gland (medulla)
bones, cartilages, connective tissue
cartilage cell
skeletal muscles
striated muscle fiber
adrenal gland (cortex)
heart, vessels
endothelial cell
gonad
kidney
tubule cell
gut muscles
smooth muscle cell
blood, immune cells
red blood cells
pharynx
thyroid gland
parathyroid gland
gut
urinary bladder
epithelial cell
lungs
tracheal cell
liver
pancreas
exocrine cell

transplanted at a later time have a further narrowed potentiality. An ectodermal cell is restricted to one of two groups, epidermal cells or cells of the nervous system (see Figure 9.9B). Finally, at a still later stage, the fate of these cells is completely **determined**; so that eye lens cells, for example can form only eye lens tissue (see Figure 9.9B).

We have seen that as a cell becomes differentiated its potentiality becomes restricted and that the cell somehow seems to 'know' what these restrictions are. Do these restricted potentialities result from a loss of genes as a cell differentiates? To find out, a British cell biologist named J.B. Gurdon exposed some frog eggs (phylum Chordata, class Amphibia) to ultraviolet radiation. Because ultraviolet radiation is absorbed by DNA, a sufficient dose of ultraviolet can be used to destroy the egg nucleus without damaging its cytoplasm. Gurdon then carefully inserted into each of these eggs the nucleus of a differentiated cell type, such as a skin cell. The resulting cell thus had cytoplasm from a totipotent cell but a nucleus from a differentiated cell. Gurdon was able to show that this cell, like a zygote, produces an entire tadpole (Figure 9.10). The various types of cells of the body do not differ in the genes that they contain. Each nucleus usually keeps its full genome, and it is thus not the loss of genes that restricts potentiality.

Cloning. Gurdon's experiments were a type of **cloning**, the production of a new individual having the complete genome of another individual. Animals produced by nuclear transfer are genetically identical to the adult animal that donated the nucleus. It was only recently (1997) that technical difficulties of producing a mammal in this way were overcome and a sheep was cloned. 'Dolly' became famous (and infamous) overnight. Differentiated cells from a sheep's udder were incubated in conditions that relaxed the differentiation signals on their DNA. (What these signals are is not yet fully known.) The nucleus from one of these cells was then put into an egg cell whose own nucleus had been destroyed. The resulting egg cell was allowed to divide *in vitro* for a few cell generations before being implanted in a surrogate mother sheep. Still more recently (1999), calves were cloned from nuclei taken from mammary cells that had been shed into the cow's milk.

Tissue differentiation. If every cell contains all the genes, how do cells become specialized in tissues? Differentiation, the process of becoming different, is, like cell division, coordinated by the regulation of gene expression. To a great extent, differentiation is a process of controlling what proteins (including enzymes) are made by a particular cell at a particular time. Although the DNA of all cells continues to carry the instructions for building all of the different cell types that make up that organism, each cell is somehow restricted to expressing only certain of its genes. Gene expression varies during the lifetime of the individual; some proteins are needed only by the developing organism and are not made in later stages of development.

FIGURE 9.10

Gurdon's experiment demonstrating that a differentiated cell contains all the genes needed for the development of a complete organism. The nucleus of a frog egg was destroyed by ultraviolet irradiation and was replaced by the nucleus from the fully differentiated skin cell of another frog. The egg with its transplanted nucleus was allowed to grow and it developed into a normal tadpole.

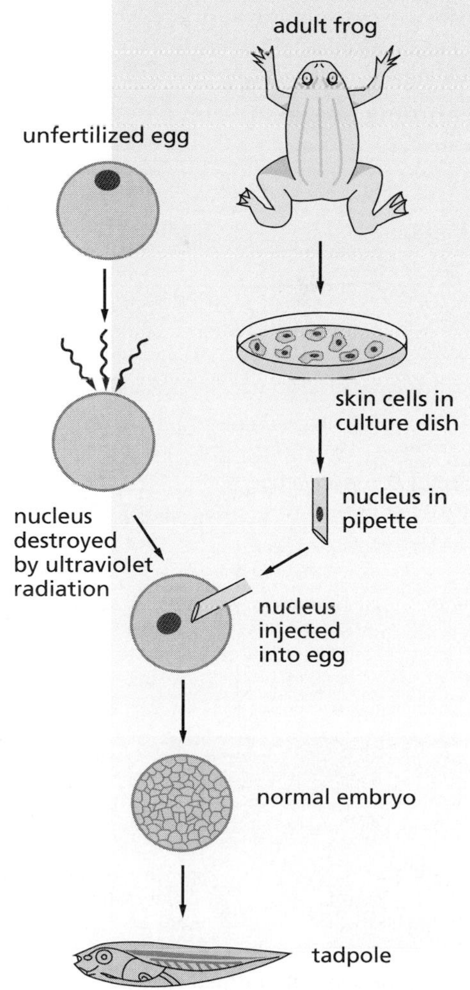

Proteins in the nucleus and cytoplasm regulate the expression of genetic information and thus determine the future identity of the cell. One example of the regulation of gene expression by factors in the cytoplasm is a group of regulatory proteins that are in the egg cytoplasm even before fertilization produces a zygote. Much of what we know about these proteins and the genes that code for them was studied first in fruit flies of the genus *Drosophila* (phylum Arthropoda, class Insecta). The earliest-acting of the developmental control genes are the **egg-polarity genes.** The action of these egg-polarity genes establishes the basic pattern of the body. As the egg develops in the female fly, meiosis produces a haploid number of chromosomes (see Chapter 2, pp. 48–49). These chromosomes from the female fly are called maternal chromosomes. A few genes are transcribed from the maternal chromosomes in the unfertilized egg. The protein products of these genes are present before fertilization in the cytoplasm of the egg, in unequal concentrations in different locations. By their locations they define the 'head' (anterior) and 'tail' (posterior) ends of the future embryo. The development of head and tail ends can be altered experimentally by adding cytoplasm from another egg. Cytoplasm from the tail end of another egg induces the growth of a second tail end in place of the normal head end after the egg has been fertilized (Figure 9.11). Although the DNA donated by the sperm also contains the egg-polarity gene, the sperm do not contain any of its protein product. Therefore it is the proteins already present in the egg as products of the maternal DNA that determine the tail end. Genes expressed only from maternal chromosomes are called **maternal effect genes**.

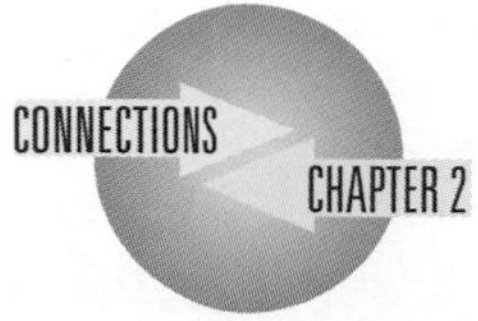
CONNECTIONS
CHAPTER 2

FIGURE 9.11

Determination of the tail end of a developing *Drosophila* embryo by the protein product of the egg-polarity gene. Replacement of some anterior cytoplasm with posterior cytoplasm from another egg induces the beginning of two tail ends.

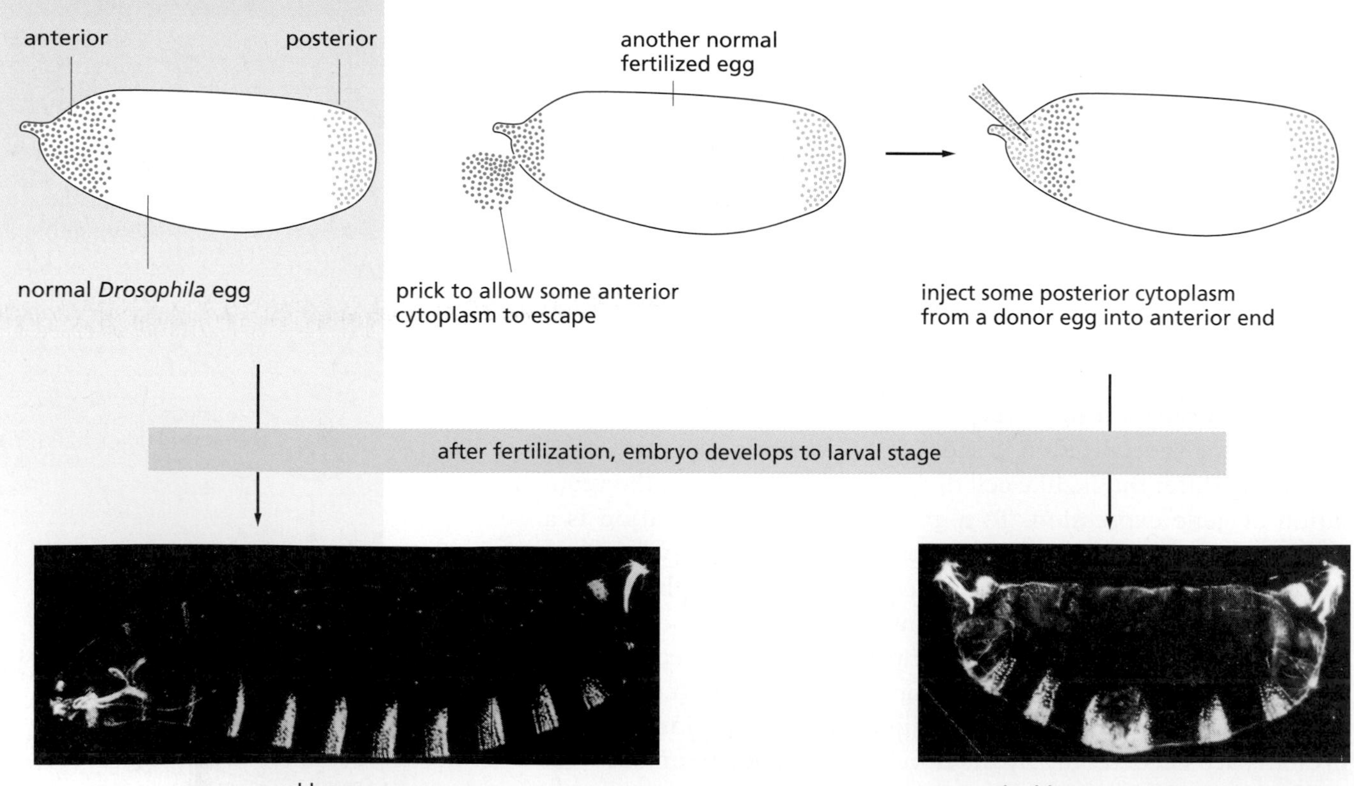
anterior
posterior
normal *Drosophila* egg
another normal fertilized egg
prick to allow some anterior cytoplasm to escape
inject some posterior cytoplasm from a donor egg into anterior end
after fertilization, embryo develops to larval stage
normal larva
double-posterior larva

Tissue induction by organizer cytokines. At each cell division, a developing cell receives cytokine signals that determine whether it will differentiate and what type of cell it will form. Any region of an embryo that produces cytokines that cause cells to differentiate is called an **organizer**. The effect is local, meaning that most of the influences on cell differentiation come from neighboring cells. The concept of an organizer was first developed by the German embryologists Hans Spemann and Hilde Mangold, work for which Spemann later won the Nobel Prize. In their early experiments, Spemann and Mangold transplanted various parts of frog gastrulas into different positions on other frog gastrulas and observed developmental changes. Through such experiments, they were able to show that a part of the embryo called the dorsal lip acts as an organizer that induces (causes) the overlying tissue to form a neural plate, the earliest part of the nervous system to be formed. As Figure 9.12 shows, a second neural plate forms above the transplanted dorsal lip cells in addition to the normal neural plate that forms above the dorsal lip of the host.

Organizers do not actually contribute cells to the tissues that they stimulate. Spemann demonstrated this by transplanting cells from a chick embryo into the embryo of a duck. Ducks and chickens are closely enough related for a transplant of embryo cells from one to grow in the other. Because these species have different chromosome numbers, the origin of cells in a structure developing after a transplant can be ascertained. In all vertebrate embryos, the neural tube forms from ectodermal cells above a structure called the notochord. Spemann transplanted notochord tissue from a chick to a duck and it induced the formation of a second neural tube in the duck embryo. He was able to show that the second neural tube was made of duck cells, even though the transplanted tissue consisted of chick cells (Figure 9.13). The transplanted chick cells, in other words, formed no part of the neural tube

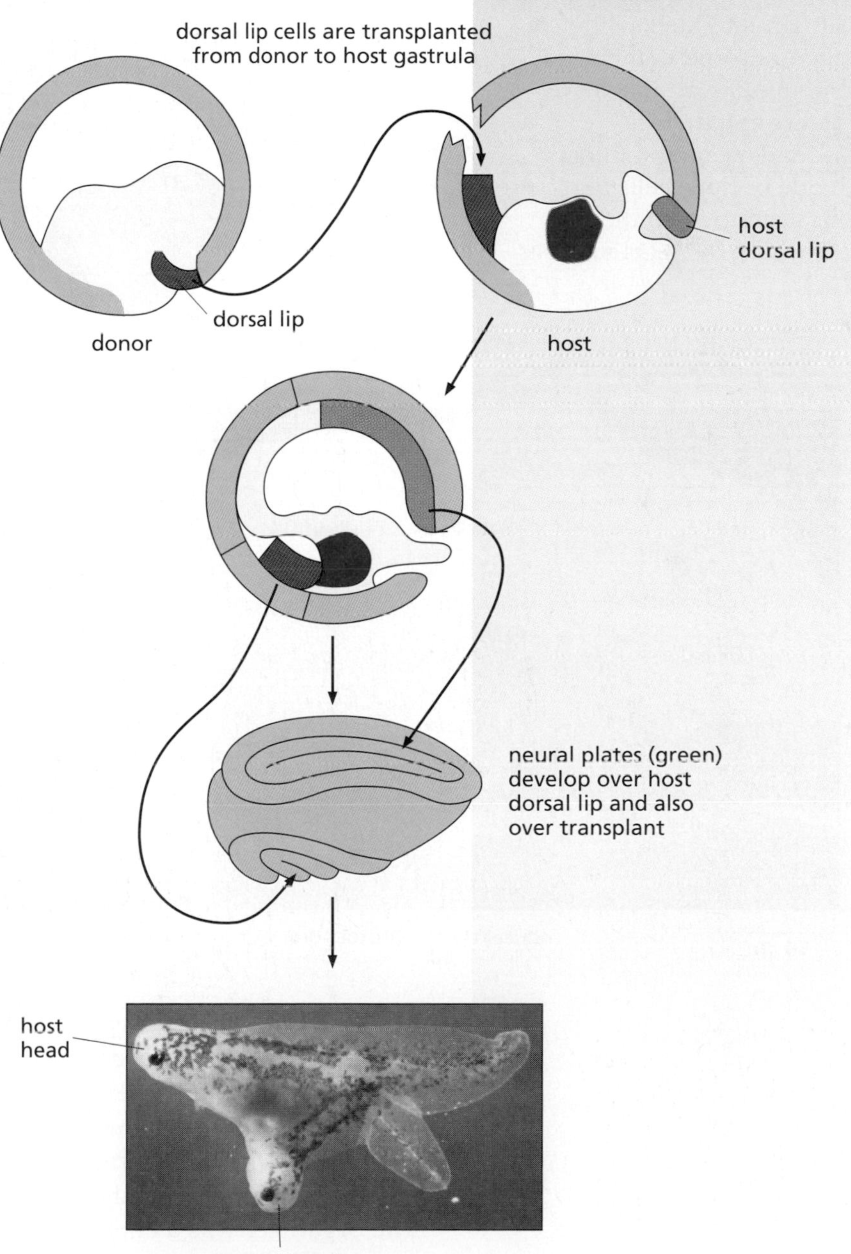
dorsal lip cells are transplanted from donor to host gastrula
host dorsal lip
dorsal lip
donor
host
neural plates (green) develop over host dorsal lip and also over transplant
host head
second head

FIGURE 9.12

Spemann and Mangold's experiments on tissue formation. By manipulating cells from the gastrula stage of frog embryos, these scientists showed that neural plates form at the locations of both host and transplanted dorsal lip cells. Both neural plates developed to form brains and other head structures.

FIGURE 9.13

Chick–duck transplants to distinguish between cell donation and cell induction. Transplanting chick notochord tissue induces the formation of a second neural tube in a duck embryo. Do the transplanted chick notochord cells become the secondary neural tube or do they induce neural tube formation by duck cells? Chicks and ducks differ in chromosome number, so Spemann was able to determine that the secondary neural tube is made of duck cells that were induced to form the structure by the chick cells.

DEVELOPING NEURAL TUBE IN A TWO-DAY-OLD CHICK EMBRYO

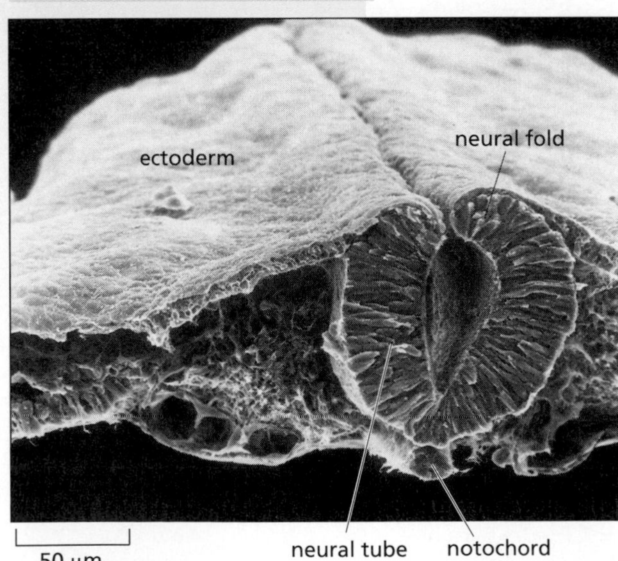

DEVELOPMENT FOLLOWING TRANSPLANTATION OF CHICK NOTOCHORD INTO DUCK EMBRYO

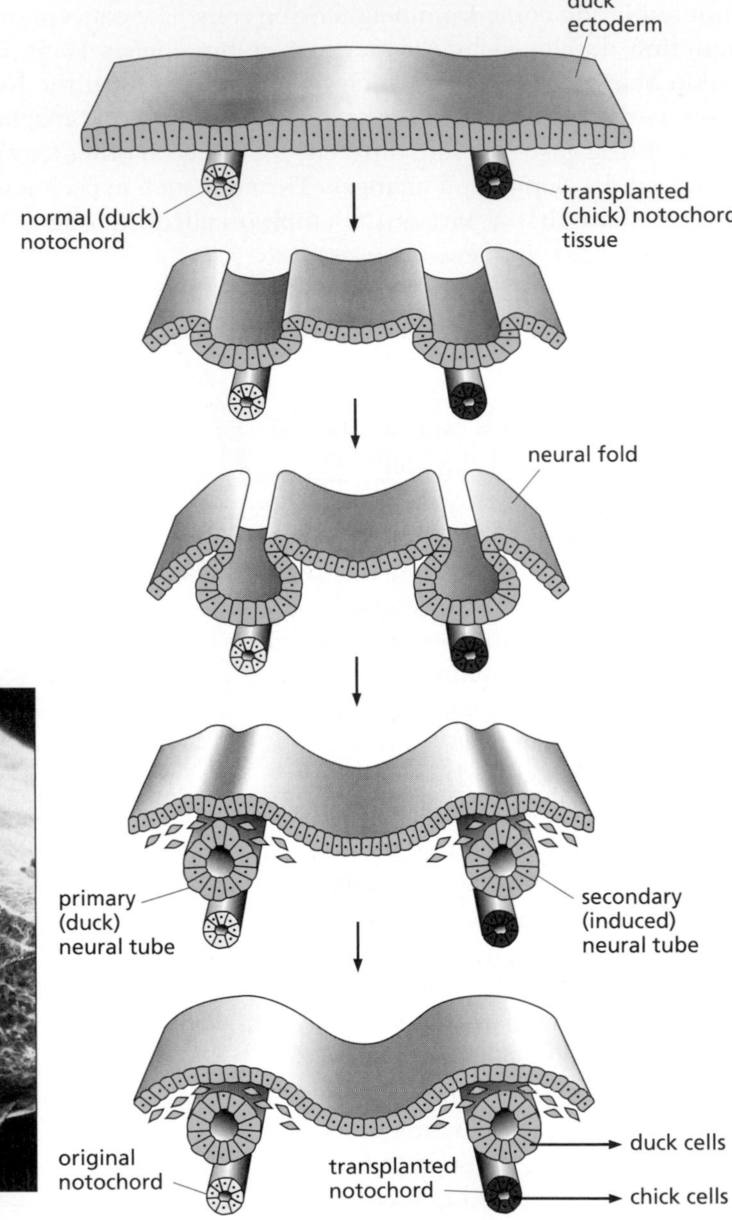

that they induced, thus falsifying the hypothesis that neural tube cells arise from notochord cells. Duck cells had divided to form the tube in response to a signal from the organizer region transplanted from the chick. In later experiments, a chemical extract of the notochord was combined with egg white; this extract was able to induce a neural tube in precisely the way that the notochord had, suggesting that the signal from the organizer was a chemical substance rather than a group of cells.

In the development of normal tissues, cells need not arise in their end location; tissue formation relies on many migrating cells that travel through the organism until they find their 'proper' location, where they adhere and join the tissue. They are generally partly differentiated at the time of their migration and become fully differentiated when exposed to growth factors in their new microenvironment. Such molecular 'addresses' can be in the form of membrane receptors that bind specifically to molecules expressed only on certain types of tissues. Abnormalities in cellular adhesion and cell migration are pertinent to the spread of cancer to other tissues.

Gene regulation is thus important for development. Is it important in adult animals? The answer is yes, in part because very few types of cells are permanent. Cells die; they must be replaced by cell division and differentiation throughout the lifetime of the organism (Table 9.1). These processes are normally tightly coordinated in adult organisms so that cells lost from specific tissues are replaced by the correct number and type of cells.

TABLE 9.1

Average life spans of human differentiated cell types.

CELL TYPE	LIFE SPAN (DAYS)
Intestinal lining	1.3
Stomach lining	2.9
Tongue surface	3.5
Cervix	5.7
Stomach mucus	6.4
Cornea	7
Epidermis: abdomen	7
Epidermis: cheek	10
Lung alveolus	21
Lung bronchus	167
Kidney	170
Bladder lining	333
Liver	450
Adrenal cortex	750
Brain nerve	27,375 + (75+ years)

Limits to cell division

Normal cells that have already differentiated into various cell types seem to have a limit to the number of additional times that they can divide. As we saw in Figure 9.4, cells in tissue culture whose growth has been stopped by contact inhibition can be induced to start dividing again by separating them from contact with their neighboring cells. However, this process cannot be repeated indefinitely. After a certain number of divisions the cells die rather than divide, even when optimal conditions exist.

The exact number of divisions possible in culture dishes depends on the species and the cell type. This limit was first discovered by American biologist Leonard Hayflick in 1965. In general, the maximum number of times that a cell can divide is called the **doubling number**, or the Hayflick limit. The doubling number is proportional to the life span—the maximum possible length of life—for that species. Life span is different from life expectancy, which is the length of time that an individual in a given environment is expected to live. In humans, for example, the life expectancy for white males in the United States is about 72 years, while the life span for all humans is around 110 years. Normal human fibroblast cells divide approximately 50 times and then divide no more, after which they age and die. In comparison, the maximum number of cell divisions for comparable mouse cells in culture is 20 and a mouse's life span is 3.5 years, while for tortoises the maximum is 90–125 divisions and the life span 150 years. If cells are frozen and stored for a prolonged period, then thawed and returned to tissue culture, the maximum doubling number is not shortened or lengthened. There seems to be a biological clock of some sort that keeps track of the number of cell divisions.

One candidate for this 'clock' is the end portion of each chromosome, called a **telomere**. Telomeres are thought to function in maintaining the integrity of chromosomes. Each time a cell divides, a few dozen base pairs are lost from the telomere. The chromosome becomes progressively shorter on a molecular scale, although it does not lose enough length for this to be visible microscopically. When the telomere has shortened to a certain length, the cell can no longer divide.

In bacteria and in the cells that produce gametes in eucaryotes, an enzyme called telomerase restores the bases lost from telomeres, thus maintaining the length of the chromosome. These cells show no Hayflick limit and can continue to divide indefinitely. Telomerase is inactive in differentiated cells, so they have a maximum number of doublings that they can undergo in tissue culture, and it is thought that similar limits exist in the living body. As we will see, telomerase is active in cancer cells, maintaining the telomeres no matter how many times the cell divides. Cancer cells thus have no limit to the number of times that they can divide and are therefore considered **immortal** (Figure 9.14).

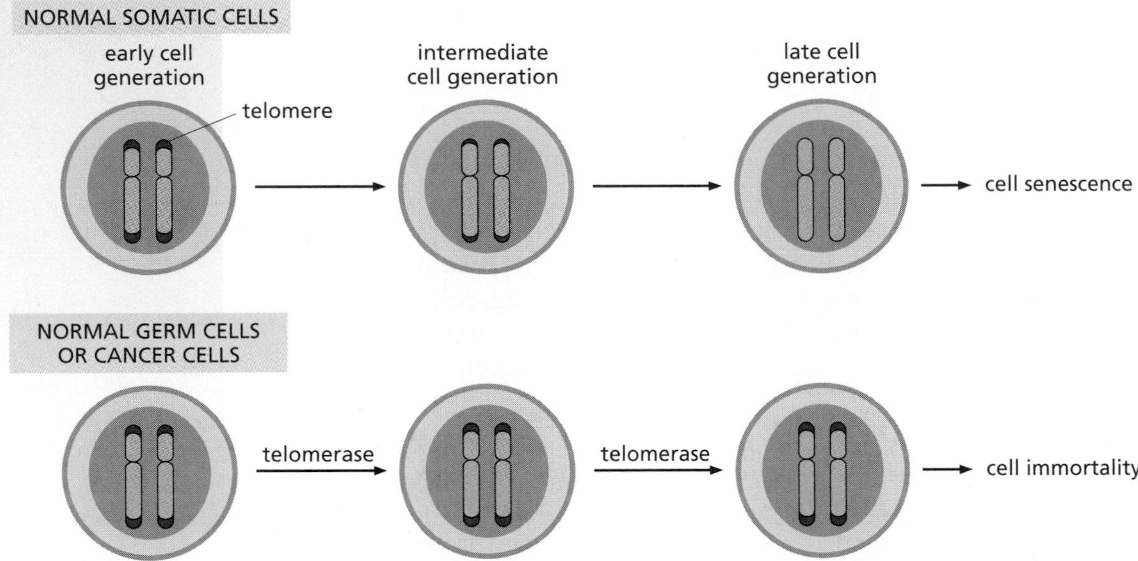

FIGURE 9.14

Loss of telomeres and preservation of telomeres by telomerase. In normal somatic cells, the loss of part of the telomeres from the ends of chromosomes at each cell division gradually leads to cell aging and loss of the ability to divide. In cancer cells, the enzyme telomerase is active and restores the telomeres, allowing the cells to continue dividing without limit.

THOUGHT QUESTIONS

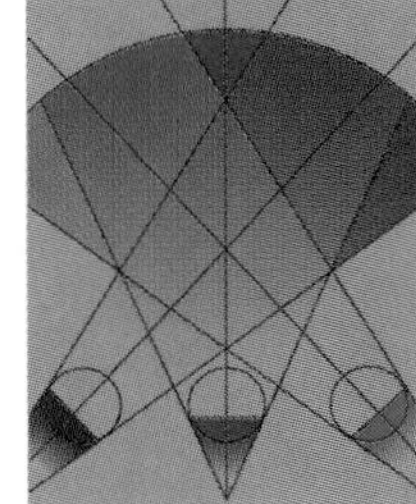

1. Why is it an adaptive advantage for an organism to have certain proteins such as insulin produced by just one type of cell rather than produced by all cells throughout the organism?
2. Does a cell's DNA determine what type of cell it becomes? What other factors, if any, are involved?
3. In what way might research on cancer also lead to a better understanding of the aging process? How might it further our understanding of birth defects?
4. Should cloning of humans be allowed? Why or why not?
5. Would a human clone and his or her DNA donor be identical just because they have identical DNA? Identical twins share the same genome. Are they identical people?
6. What effect does the cytoplasm have on gene expression? If two identical nuclei were transferred into denucleated eggs from different individuals, would the clones develop differently?

CANCER RESULTS WHEN CELL DIVISION IS UNCONTROLLED

Now that we have seen how cell division and differentiation are controlled in normal cells, we can examine these processes in cancer cells. Cancer is more than new growth of cells; it is the growth of cells that have escaped from the controls that operate in normal cells. Cancers can arise in any tissue whose cells are dividing. All multicellular organisms can develop cancer. In this section we discuss primarily human cancers, although much of what follows also applies to cancer in other species.

Properties of cancer cells

In cancer cells, control of cell division has been lost. The process that a cell undergoes in changing from a normal cell to an unregulated, less differentiated, immortal cell is called **transformation**. The transformed state is traceable to changes in the DNA, and is therefore passed on to all progeny cells. Cancer can result from the transformation of just a single cell.

After cells have been transformed, they exhibit many characteristics that differ from those of normal cells (Table 9.2). Cancer cells do not have a Hayflick limit and continue to divide indefinitely. Some cancer cells have been maintained in tissue culture for decades. In many cases cancer cells are less differentiated than the cells from which they arose. The membrane transport systems of transformed cells carry nutrient molecules into the cell at a higher rate. In the body this gives transformed cells a competitive advantage over normal cells. Transformed cells are not inhibited by contact with other cells. In tissue culture, their growth does not stop when they have formed one-cell-thick monolayers, but instead continues, forming piles of cells growing over and on top of each other. Cancer cells grow this way inside organisms, and the growing piles of cells are called **tumors** (Figure 9.15).

Transformed cells grow without the need to be attached (see Figure 9.15B–D); in fact, this is the characteristic that best predicts whether a cell growing in culture will form a tumor if put into an animal. Changes

TABLE 9.2

Characteristics of normal cells and transformed (cancer) cells.

CELLULAR BEHAVIOR	NORMAL	TRANSFORMED
Hayflick limit	Finite	Immortal
Differentiation	Present	Inhibited
Transport of nutrients across cell membrane	Slower	Faster
Nutrient requirement	Higher	Lower
Contact inhibition	Present	None
Anchorage dependence	Present	None
Adhesiveness	High	Low
Secretion of protein-degrading enzymes	Low	High
Genetic material	Stable	Unstable

back and forth from attached growth to unattached growth are believed to spread some tumors to new locations.

Cells become transformed when they are dividing; therefore, cells that are terminally differentiated and will never again divide cannot become transformed, for example nerve cells in the brain and muscle cells of the heart. In contrast, many types of cancers arise from the transformation of the **stem cells**. These are partly differentiated (pluripotent) cells whose normal function is to divide and replace cells that are lost through routine physiological processes. These stem cells are located in areas where cells are continually being lost: the skin, gut lining, uterine cervix, bone marrow, and many glands. Because stem cells divide often, they respond more easily to cell division signals than do highly differentiated cells. When stem cells divide normally, one daughter cell remains undifferentiated as a stem cell and the other differentiates and is therefore less likely to continue dividing. However, a transformed stem cell divides into two daughter cells that remain undifferentiated

FIGURE 9.15

The growth of a tumor in the lining of the cervix of the human uterus. Very few cells in the normal tissue show the mitotic spindles characteristic of dividing cells, but they are common among the transformed cells.

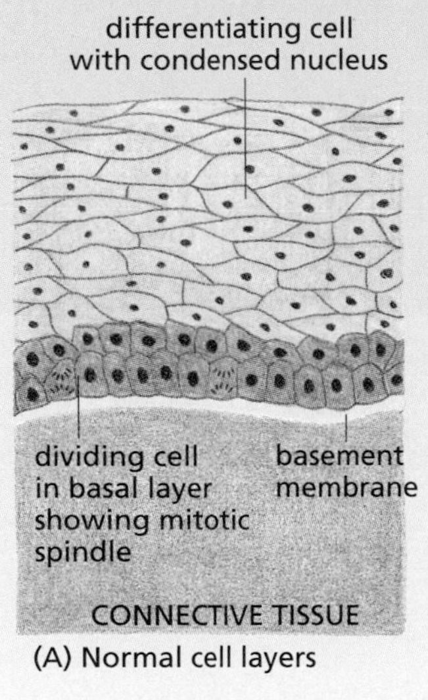

(A) Normal cell layers

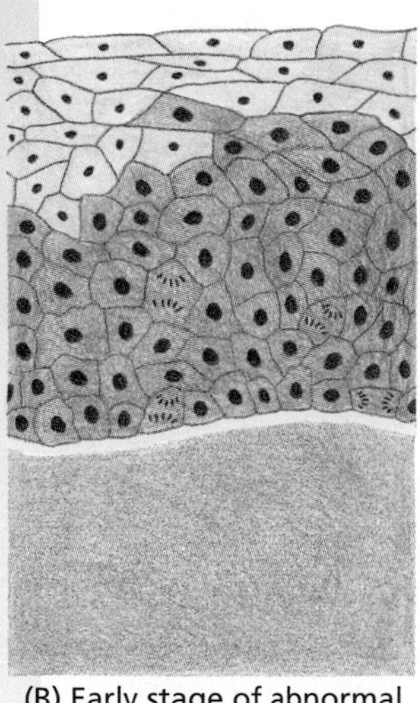
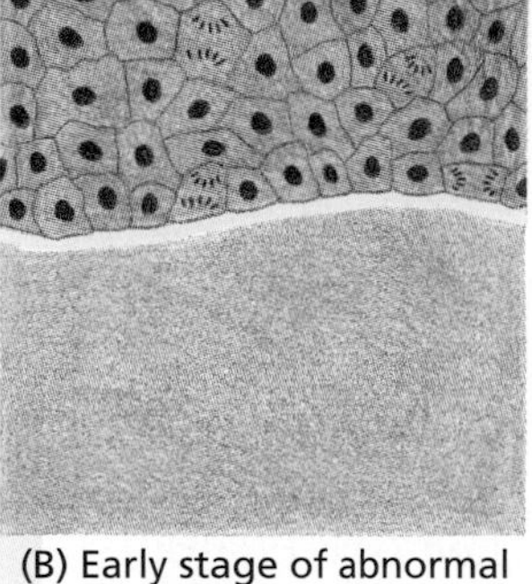

(B) Early stage of abnormal growth

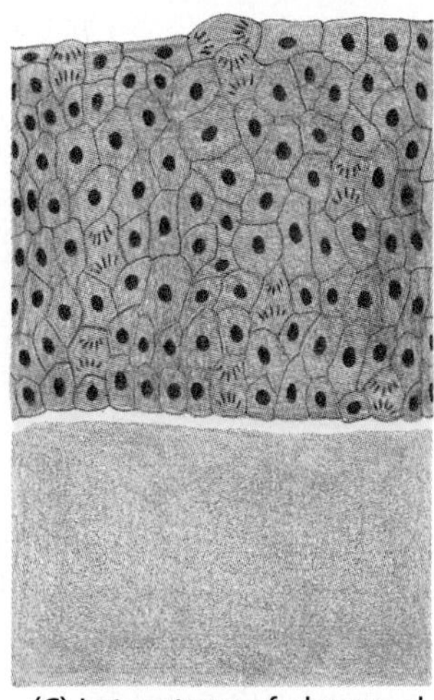

(C) Later stage of abnormal growth

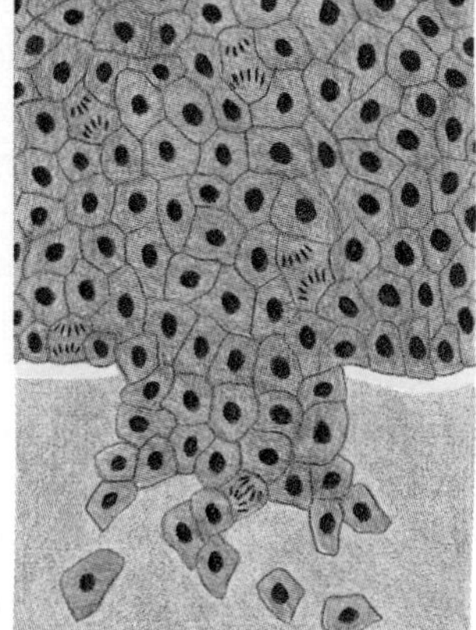

(D) Cancerous tumor spreading to new locations

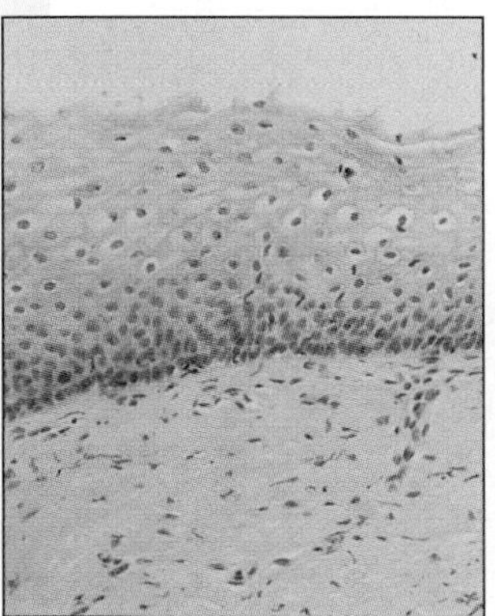

(E) Photograph of normal lining of the cervix

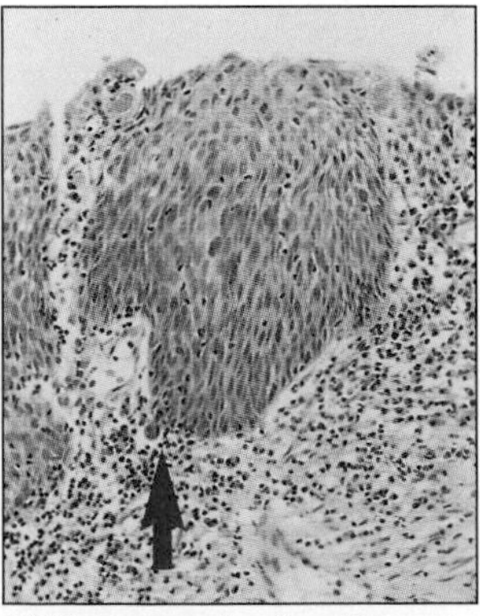

(F) Photograph of malignant cervical cells beginning to spread (at arrow) into a lower layer and each can continue to proliferate. The less differentiated (or more pluripotent) the stem cell that has been transformed, the more aggressive the cancer. One concept of experimental therapy for stem cell cancers is to give drugs that promote cellular differentiation and thus slow down the progression of the cancer.

A transformed cell is a less-differentiated 'immortal' cell that no longer responds to the signals that normally regulate cell division and cell differentiation. These signals have not been completely identified, but this much seems certain: cancerous growth signals are aberrant forms of the normal growth signals, and the aberration is located in the cell's DNA. The normal growth regulatory genes fall into two categories: genes encoding proteins that promote cell division (proto-oncogenes) and genes whose protein products normally inhibit cell division (tumor suppressor genes). Mutations in either type of regulatory gene increase the probability that cancer will arise.

Oncogenes and proto-oncogenes

Signaling cell division in normal cells was discussed earlier in this chapter and is summarized in Figure 9.6. The genes whose products are the growth factors, receptors, second messengers, and cyclins shown in Figure 9.6 are the **proto-oncogenes**. Proto-oncogenes are thus genes whose products signal and regulate normal cell division. American cell biologists J. Michael Bishop and Harold Varmus received the 1989 Nobel Prize for their discovery of proto-oncogenes, leading to a new era in our knowledge of cell division. The abnormal, mutated forms of these proto-oncogenes that lead to cell transformation and cancer are called **oncogenes**. Oncogenes, the cancer-causing mutants, were actually discovered before the normal proto-oncogenes. In this way, research on cancer has led to a better understanding of the ways in which normal cell division is controlled.

Oncogenes differ from proto-oncogenes in any of three basic ways: the timing and quantity of their expression, the structure of their protein products, and the degree to which their protein products are regulated by cellular signals. The expression of a proto-oncogene responds to cellular controls (see Figure 9.6), but the expression of an oncogene does not. The protein product of an oncogene may differ by as little as a single amino acid from the protein product of a corresponding proto-oncogene, but this small change in structure can be enough to remove the protein from control by the cell's regulatory mechanism.

Mutation of a proto-oncogene to an oncogene can alter the cell division signals at any of five steps and trigger uncontrolled cell division. One type of oncogene codes for a modified growth-factor receptor that, unlike its normal proto-oncogene counterpart, continuously activates second messengers without having bound its growth factor, thus triggering cell division without cytokine signals (Figure 9.16, type 1). Another type of oncogene causes a cell to secrete growth factors for which it has receptors, allowing the cell to stimulate itself to divide rather than needing signals from its neighbors (Figure 9.16, type 2). A third type of oncogene codes for altered cellular second messenger molecules that carry 'activate cell division' commands across the cytoplasm in the absence of any growth factor signal from outside the cell (Figure 9.16, type 3). Still other oncogenes alter the regulatory steps inside the nucleus, affecting the concentrations of cyclins (Figure 9.16, type 4) or of certain DNA-binding proteins (Figure 9.16, type 5). In other cases there may be an increase in the number of copies of the unmutated proto-oncogene in the DNA (gene amplification), causing it to stimulate uncontrolled cell division by producing abnormally large amounts of the normal protein.

Sometimes, continuous cell division can be signaled by normal proto-oncogenes if they are moved to a different chromosome location. Chromosome rearrangement can result in an unmutated proto-oncogene's being moved near a strong enhancer or being fused to a gene that is transcribed continuously or very frequently, as was found in a cancer called Burkitt's lymphoma. Burkitt's lymphoma affects B lymphocytes (B cells), a type of white blood cell involved in immunity (see Chapter 12). This cancer is associated with a gene whose product is a DNA-binding protein (see Figure 9.16, type 5) that can immortalize a cell by preventing it from leaving the cell cycle and entering the G_0 resting phase (see Figure 9.3). The proto-oncogene counterpart is normally located on chromosome 8 in humans, but, in the transformed cells of Burkitt's lymphoma, sections of this chromosome have traded places with a small section of DNA from another chromosome. Such an exchange of DNA material is called a **translocation**; translocations are visible microscopically because the banding pattern on each chromosome is distinctive (see Chapter 3, pp. 73–74). In its new location, the proto-oncogene is deregulated, that is, it has escaped from its normal control, so that it is expressed more often. The translocated proto-oncogene is now located near the genes that code for different parts of the antibody molecule, a variable protein that is synthesized when the body is challenged by an infectious agent (see Chapter 12). Because of its altered location near the antibody genes, any infectious challenge that stimulates the transcription and translation of the antibody genes now also causes high levels of the DNA-binding protein to be synthesized, sending the cell into cell division. Infection with Epstein–Barr virus can insert this proto-oncogene

FIGURE 9.16

Continuous signaling of cell division in transformed cells by the protein products of five types of oncogenes that lead to the cell's escape from the regulation of cell division.

1 Altered growth-factor receptor (coded for by *erbB* oncogene) sends signal without binding its cytokine.

2 Self-stimulation by altered growth factor (product of oncogene) produced by same cell.

3 Altered second messengers (products of oncogenes) activated without receptor signal.

4 Altered level of cyclins produces continuous cell division.

5 Altered amounts or shapes of DNA-binding proteins (products of oncogenes).

into this abnormal location in the genome. The hypothesis that antibody genes are associated with this proto-oncogene also offers an explanation of why this cancer arises more often in patients infected with Epstein–Barr virus who receive a second infective challenge such as malaria.

Tumor suppressor genes

The protein products of tumor suppressor genes normally repress cell division. If these genes are altered, their repressor activity may be removed. Inactivation of a cell-division repressor leads to cell division (see Figure 9.7C and D). A tumor suppressor gene, *p53*, is mutated in as many as 55% of noninherited cancers. When something goes wrong inside a cell, the normal p53 protein halts cell division and causes the abnormal cells to die. When the *p53* gene is mutated, the altered p53 protein does not halt cell division, and cells with damaged DNA continue to live and divide, passing on their accumulated mutations to their progeny cells.

Accumulation of many mutations

The transformation of cells may require a combination of changes in several proto-oncogenes and tumor suppressor genes rather than a change in just one. If single mutations were the primary cause of cancer, the rate of incidence for new cancers would be the same for individuals of every age. That is clearly not the case for most cancers; cancer rates increase with age, particularly in advanced age. Some mutations may bring about cell immortalization. Other mutations may cause the cells to become unattached, lose their anchorage dependence, change shape from flat to rounded, or begin to secrete growth factors. It is estimated that five or six such mutations must occur *in a single cell* before it becomes transformed to a cancer cell. Most cancers arise in somatic cells (body cells), rather than in gametes. Somatic mutations are passed along to the progeny cells in that individual but are not passed on to the individual's offspring. In the rarer inherited cancers, some mutation in gamete DNA is passed on to sons or daughters. Because these individuals have inherited one or more of the mutations needed for transformation, fewer somatic mutations need to accumulate for some cells to become fully transformed and to progress to cancer. Inheriting such mutations thus predisposes the individual to get cancer more readily than someone else living in the same environment.

In a human lifetime, there are on the order of 10^{16} cell divisions. Because of the limitations on the accuracy of DNA replication, at every cell division each gene has about a 1 in 10^6 chance of being copied wrongly. This gives, for the approximately 10^{16} cell divisions, a mutation rate of 10^{16} divided by 10^6, or about 10^{10} mutations per gene in a human lifetime just from mistakes in replication.

Several mechanisms prevent the overwhelming majority of these mutations from initiating a cancer. Most mutations are corrected as they occur by 'spell-checking' proteins (Figure 9.17). Cells with mutations that are not successfully corrected are usually induced to die by the p53 protein. To contribute to transformation, uncorrected mutations must be in a proto-oncogene or tumor suppressor gene. Because these genes are a tiny fraction of the whole genome, the probability that one of them will mutate at random and that the mutation will be uncorrected is low. Additionally, for cancer to arise, mutations in many growth regulatory genes must all accumulate in the same cell.

Progression to cancer

We see the characteristics of transformed cells when we study them at the cellular level. However, cancer occurs in whole, multicellular organisms, not in isolated cells. As mentioned earlier, an organism can be considered an ecological system of many billions of cells. The interacting growth and differentiation signals keep the cellular ecosystem stable. Cancer is a disease in which a single cell, as a result of DNA mutations, escapes from normal cell division controls, gaining a competitive advantage over its neighbors. After cells are transformed, progression to a tumor depends on many ecological factors. Mutated progeny cells may be killed or they may not outgrow the normal cells and thus never progress to a tumor. Alternatively, the transformed cell and its progeny may continue to divide, taking up space and nutrients required by their neighbors, passing on the mutation to each new progeny cell. Normal cells begin to die off, not because they are killed outright by cancer cells, but because they are deprived of space and nutrients. With the decline in the number of normal cells comes a reduction of their normal function, and the organism begins to show signs of illness. The particular symptoms depend on the type of cancer and the type of normal cells that are lost.

FIGURE 9.17

DNA mismatch mutations and their repair. (A) A mismatch leads to a permanent mutation on one DNA strand if not corrected. (B) A mismatch repair protein (spell-checking protein) acts on a mutated strand, cutting out that DNA and allowing the correct base to be added.

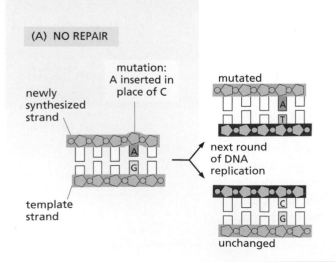

(A) NO REPAIR
mutation: A inserted in place of C
newly synthesized strand
template strand
A
G
next round of DNA replication
mutated
A
T
C
G
unchanged

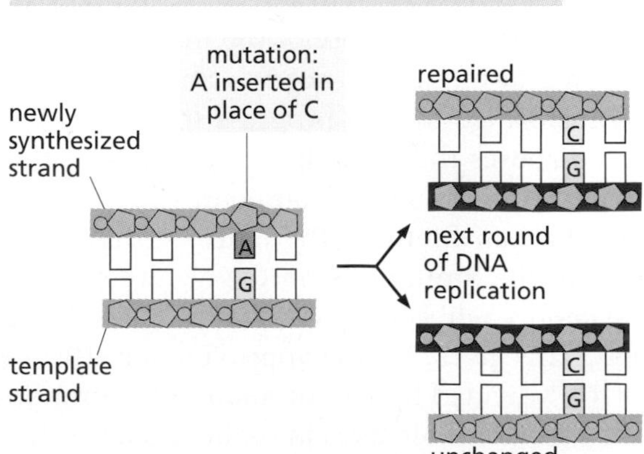

Transformed cells within organs may form solid tumors within those organs. For a tumor to be visible on X-ray, the original transformed cell must divide repeatedly until there are about 10^8 cells in the tumor. For a tumor to be large enough to be felt (about 1 cm in diameter), approximately 10^9 cells are needed. By the time that tumors are this size they have begun to influence their environment. The tumor cells may secrete **angiogenic growth factors** that induce nearby blood vessels to develop new branches that grow into the tumor. These are normal blood vessel cells whose growth is induced by the tumor cells. A tumor may thus contain normal cells as well as transformed cells.

A tumor is said to be **benign** if it is contained in one location and has not broken through the basement membrane to which normal cells are attached. Benign tumors, as their name suggests, often cause no health problems for the individual. Benign tumors can become large enough to interrupt the functioning of normal tissues, but their removal by surgery is generally successful because they have not intermingled with normal tissue. Tumor cells that invade normal tissues, rather than just pushing them out of the way, are said to be **malignant** (see Figure 9.15D). For cells to be invasive, they must produce protein-degrading enzymes such as collagenase, an enzyme that dissolves the collagen connective tissue that holds groups of cells together (see Table 9.2). The term 'cancer' is generally reserved for malignant tumors.

Because malignant tumors produce enzymes that allow them to invade other tissue, they often spread to new locations, a process known as **metastasis**. In this process, one or more of the transformed cells lose their attachment to the other cells of the tumor, break through the basement membrane, and spread via the circulation to other areas of the body (see Figure 9.15D). In the new location they regain attachment and continue to divide, forming new tumors. The new tumors are of the same type as the original tumor and thus when viewed with a microscope are seen to be different from the cells around them. Cancers that have begun to metastasize are far more serious and more resistant to treatment than those that have not, because no amount of surgery can eliminate all the cancerous cells that have spread.

THOUGHT QUESTIONS

1. How do stem cells differ from transformed cells? How do stem cells differ from muscle cells or blood cells?
2. In what ways are benign and malignant tumors the same and in what ways are they different?
3. What does the statement "Cancer is a disease of the genes, but it is not a genetic disease" mean?

CANCERS HAVE COMPLEX CAUSES AND MULTIPLE RISK FACTORS

The study of disease at the population level constitutes the science of **epidemiology**. The basic epidemiological data for various forms of cancer have been compiled for the United States since 1950. The incidence for various cancers in the United States is given in Table 9.3.

Epidemiology uses descriptive statistics to find patterns in the incidence of diseases. Those patterns indicate possible risk factors that can suggest hypotheses that can be further tested in other ways. In general, and with a number of exceptions, the causes of adult cancers seem to be mainly environmental, not genetic. Evidence to support this conclusion comes from epidemiological data for the United States and several European countries, showing in each case a marked increase in cancer rates throughout the twentieth century. Most of this increase in cancer pertains to a single type, cancer of the lung (Figure 9.18A). The increase coincided with advancing industrialization and other changes in the environment, but very little change in the gene pool. In Germany, cancer caused only 3.3% of all deaths in 1900, but 20.9% of all deaths in 1967, a more than six-fold increase. From 1950 to 1979, lung cancer death rates more than tripled for both U.S. males and U.S. females, with higher rates of increase among nonwhites than among whites. In each case, the death rate increases paralleled an increase in the consumption of cigarettes, with a lag of about 15–20 years (Figure 9.18B). Since 1960, cigarette consumption by males has decreased and in 1990 the lung cancer death rate among males began to decrease. Cancers of the pancreas and large intestine increased more slowly over the same period, while stomach and rectal cancers declined. The rapid change and irregular pattern of

TABLE 9.3

Cancer Cases, Deaths and Five Year Survival Rates in the United States (totals estimated by the National Cancer Institute from incidence rates found in the Surveillance, Epidemiology and End Results [SEER] Program).

TYPE OF CANCER	NEW CASES*	DEATHS*	FIVE YEAR SURVIVAL RATE (%)‡
Breast	175,000†	43,700†	83.2
Colon	94,700	47,900	61.0
Leukemia	30,200	22,100	68.6
Lung	171,600	158,900	13.4
Lymphoma	64,000	27,000	51.0
Oral cavity and pharynx	29,800	8100	–
Pancreas	28,600	28,600	3.6
Prostate	179,300	37,000	85.5
Rectum	34,700	8700	61.0
Skin	1,000,000	9200	–
melanoma	44,200	7300	86.6
Urinary bladder	54,200	12,100	80.7
Uterus cervix	12,800	4800	68.3
endometrium	37,400	6400	83.2

* 1999 American Cancer Society Facts and Figures.

† 1999 Includes 1300 cases and 400 deaths in men.

‡ Percentage of people treated for some cancers who are still alive after five years later, compared with perecentage in a comparable, cancer-free population; *Scientific American,* September 1996.

change both fit much better the hypothesis of environmental causes than the alternative hypothesis of a genetic cause or causes. We present the evidence for various hypotheses of cancer causation in the next section.

In this section we examine the evidence from epidemiological studies and animal studies that have suggested many possible causes of cancer. We will see how these seemingly disparate causes may be working by very similar pathways at the cellular and molecular levels.

Human cancers are named according to the type of cell from which the cancer is derived. A cancer that arises in epithelial tissue (sheetlike tissue or glandular tissue) is called a carcinoma; a cancer that arises in connective tissue is called a sarcoma. There are also subtypes of tumors: a mesothelioma, for example, is a sarcoma of the lining of the abdominal cavity. A cancer that arises among white blood cells (leucocytes) is called a leukemia if the cells are circulating throughout the body via the bloodstream, but it is called a lymphoma if it is a solid tumor in lymphoid (leucocyte-containing) tissue.

The causes of childhood cancers may differ from the causes of adult cancers. Some blood cell cancers (leukemias) are more common among children. Approximately 85% of childhood cancers are acute lymphocytic leukemias that arise from stem cells in the bone marrow. Although these are very aggressive, they have a good cure rate because children still have many normal cells to take over after therapy.

About 85% of adult cancers are carcinomas, including cancers of the lungs, breast, colon, rectum, pancreas, skin, prostate, and uterus (see Table 9.3). The incidence of these cancers (and many others) increases with age, so that cancers become more and more significant as causes of death with advancing age. Environmental or lifestyle factors are believed to affect most of these adult cancers. The following factors

FIGURE 9.18

Deaths from cancers in the United States.

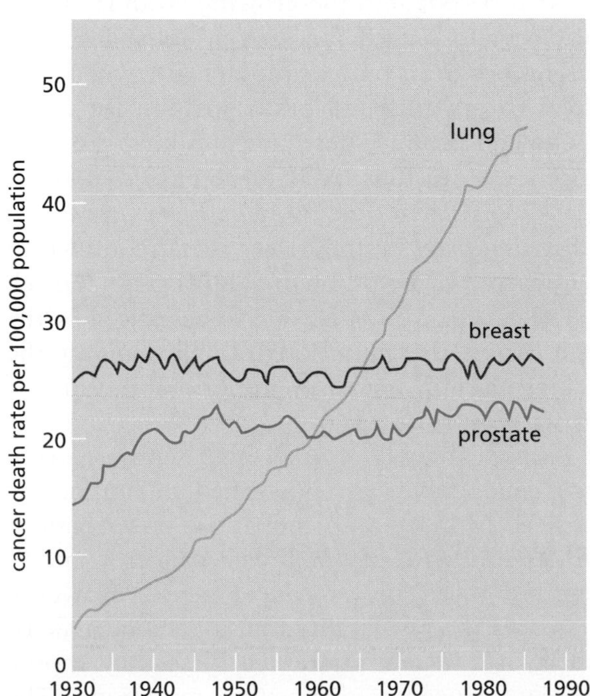

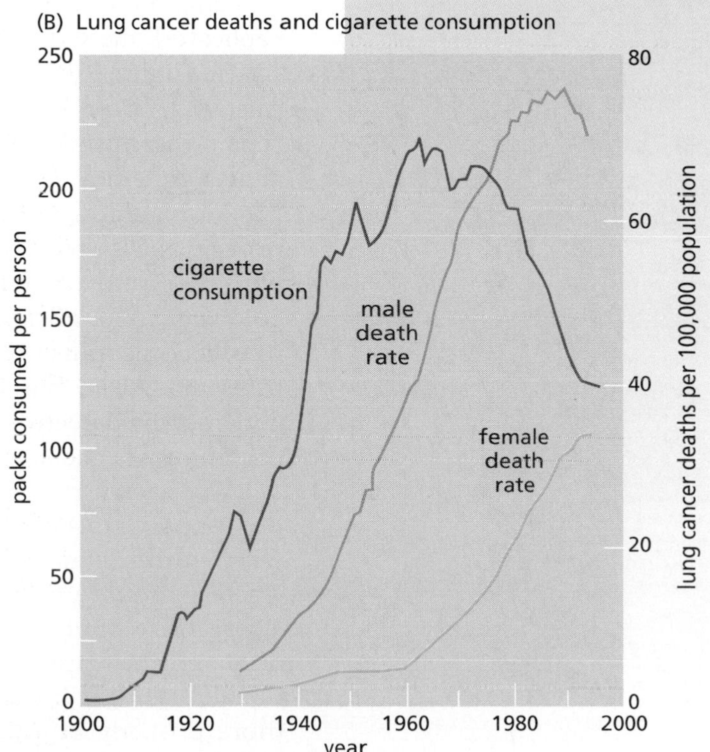

have been suspected, on the basis of epidemiological evidence, of causing at least one type of cancer or of increasing the rate at which at least some cancers occur: genes, increasing age, viruses, ionizing radiation, ultraviolet radiation, diet, stress, mental state, weak immune systems, sexual behavior, hormones, alcohol, tobacco, and some chemical substances.

Keep in mind that when we speak of 'causes' of cancer we often mean factors that are associated in epidemiological studies with increased incidence in populations. As such, these factors are more properly called 'risk factors,' not causes. A multitude of factors contributes to whether any particular person gets cancer. We generally cannot say that one thing 'caused' a particular cancer. As Clark Heath of the American Cancer Society has said, "Cancer cases are clinically nonspecific—you can't look at a leukemia case clinically and say, 'Ah, this a radiation-caused leukemia.' " (*Scientific American*, September 1996, p. 86.)

Inherited predispositions for cancers

Cancer is not inherited, but a predisposition for some cancers can be. For example, retinoblastoma is a rare cancer of the eye, caused by defective alleles of the gene for a protein called pRB, which helps control the cell cycle. Of the people who carry the defective alleles, 80–90% develop retinoblastoma. Another rare cancer is xeroderma pigmentosum, a skin cancer that results from a defect in the mechanism of DNA repair. Almost all persons with mutant alleles for this DNA repair protein develop cancer because of their extreme sensitivity to ultraviolet radiation. (Most cancer-related genes have much lower rates of associated cancers than do retinoblastoma or xeroderma pigmentosum.) Recently, a genetic mutation associated with a rare form of colon cancer was located and identified as being in a gene coding for a DNA-checking protein. When DNA is replicated, the wrong bases can be put into the growing strands. As we saw in Figure 9.17B, specific proteins called checking proteins check for mistakes, like a spell-checking program on a computer. Enzymes then repair the mistake. People with an inherited predisposition for colon cancer inherit a mutated gene and produce defective checking proteins. Mutations throughout the genome are thus likely to be passed along to progeny cells (see Figure 9.17A). Other cancers have been associated with this same defective checking protein; the defect is not confined to colon cells. If a mistake is uncorrected in a growth control gene, the cell may become transformed. The cancer cannot be said to be caused by the inherited allele, but by increasing the mutation rate throughout the genome the defective checking protein increases the probability that a mutation will occur in a growth control gene.

Five to ten percent of breast cancers are associated with a genetic predisposition. At least two genes have been identified in families in which multiple members have early onset (premenopausal) breast cancer. These genes are called *BRCA1* and *BRCA2*, for breast cancer 1 and 2. These have often been referred to as 'breast cancer genes.' However, although their presence increases the probability that a person (male or female) will get breast cancer, it does not guarantee it, so they should more properly be called 'breast cancer predisposition mutations.' The incidence of *BRCA1* mutations in the general U.S. population is estimated to be one person in 1000. These mutations were found before the normal genes were known. At present the sequence of the normal genes is known but the functions of their normal protein products are not, although they are assumed to be tumor suppressor proteins. Tumor suppressors normally inhibit cell division, so if the genes are mutated in a way that makes the suppressor proteins nonfunctional, cell division is not inhibited and cells divide when they normally would not (see Figure 9.7D).

Increasing age

Far more cancers seem to be environmentally caused than genetically caused, even after taking into account genetic predispositions for some cancers. The more common type of breast cancer, a late-onset disease of postmenopausal women, is not linked to inheritance. The strongest risk factor for these breast cancers is age. The incidence of all cancers increases with age, presumably because there has been more time for environmental exposures to produce accumulated mutations. In fact, one of the reasons for the present-day higher incidence of cancers (and chronic diseases such as heart disease) is that people are living much longer because mortality from infectious diseases is lower.

From data on the incidence of cancers, the probability of acquiring cancer at different ages can be calculated. Table 9.4 gives data for breast cancer. As can be seen, the probability increases with age. Because age is such a strong factor in all health studies, epidemiologists must 'control for age;' that is, they must either compare groups of the same ages, or use mathematical formulas to 'age-adjust' the data.

Cancer data are often shown as the probability of acquiring cancer by age 75. Examples are shown in Figure 9.19. Here data of the type shown in Table 9.4 have been added up to give the 'lifetime probability' of acquiring cancer. Note that these numbers do not mean, for example, that a white woman's chances of acquiring breast cancer are 10%. The

TABLE 9.4

Age-specific probabilities of developing breast cancer.

AT AGE	PROBABILITY OF DEVELOPING BREAST CANCER IN THE NEXT TEN YEARS	
	%	1 IN:
20	0.04	2500
30	0.40	250
40	1.49	67
50	2.54	39
60	3.43	29

Based on 1997 data from the American Cancer Society

chances for each age group are the numbers shown in Table 9.4, which are much lower.

Data of the type shown in Figure 9.19 are useful in comparing the probabilities for different types of cancers. They are also useful in comparing the probabilities for different segments of the population. In the United States, health statistics are summarized by sex and by race. These are data for populations and do not mean that any individual's probability is the number shown. Individual risk is increased or decreased by all of the factors mentioned in this section.

Viruses

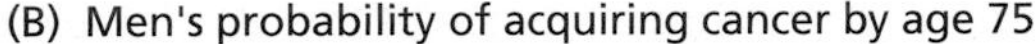

Several cancers are known to be associated with viruses and other infectious agents. In 1911, American pathologist Peyton Rous showed that a tumor of connective tissues (a sarcoma) in chickens was caused by a virus that was later named Rous sarcoma virus (Rous received a Nobel Prize for this work but not until 1966). Chickens infected with this virus develop sarcomas. Another cancer-causing virus, feline leukemia virus, causes a blood cancer in cats. By the late 1960s and 1970s, cancer-causing viruses had been identified in several species, but not in humans. Then, in 1980, Robert Gallo found a human T-cell leukemia virus, HTLV (see Chapter 13, p. 541), that is associated with one form of human leukemia. (Many other human leukemias do not show evidence of HTLV or other viral infection.) Viruses seem to be associated with cancer in at least two different ways. First, a virus infection may cause a decrease in the activity of the immune system. Decreased immunity increases the likelihood that a transformed cell will progress to cancer. An example of this type is Kaposi's sarcoma, which occurs in people with AIDS (see Chapter 13, p. 539).

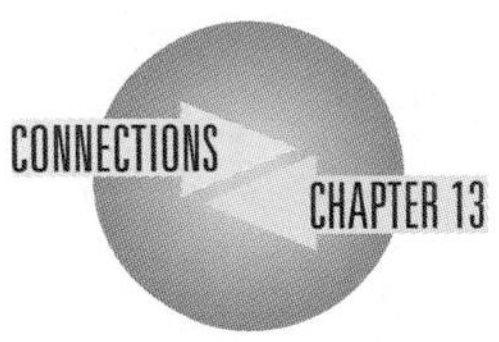

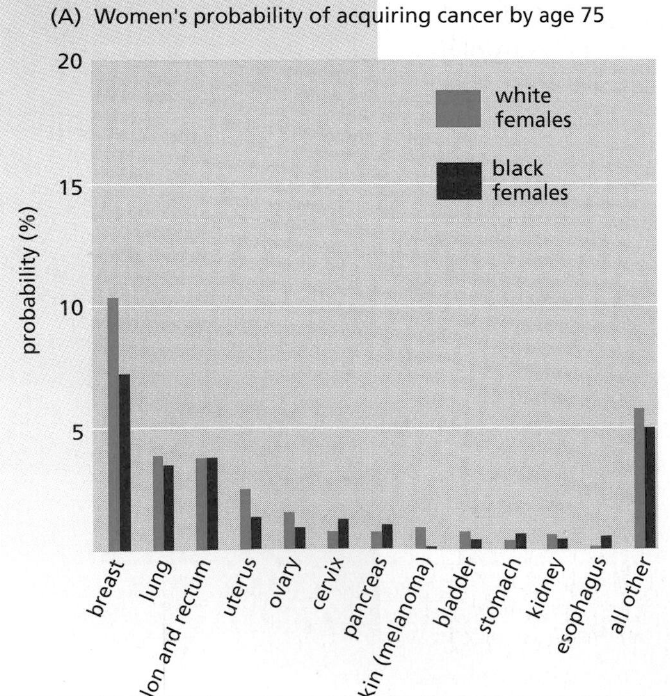

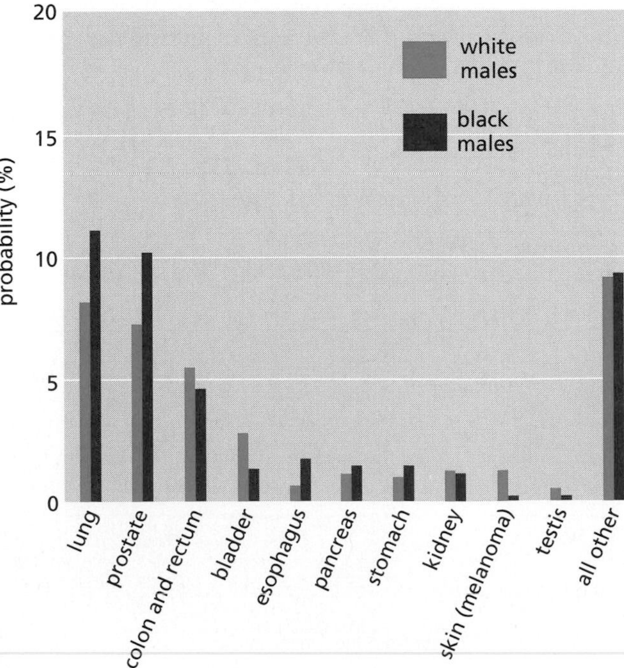

FIGURE 9.19

Lifetime probabilities of acquiring various types of cancers.

Second, some viruses carry genes that, when inserted into the host DNA, cause the host cell to become transformed into a cancerous cell. These genes are oncogenes, mutated forms of growth control genes. The viruses do not cause the mutation; rather, they carry an entire mutated gene into the cell they infect.

The prevalence of Burkitt's lymphoma in equatorial Africa is correlated with infection by Epstein–Barr virus (EBV) followed by infection with malaria. EBV is just as common in the United States as it is in Africa, but the incidence of malaria is very low in the United States. This pattern suggested the mechanism for Burkitt's lymphoma that we saw earlier. EBV carries an oncogene, but viral infection with EBV by itself does not produce cell transformation, as Rous sarcoma virus does. However, when, the immune system responds to another infection such as malaria, the translocated oncogene sends the cell into uncontrolled cell division.

The incidence of liver cancer is high in third-world countries. A very high proportion of the people who develop liver cancer have previously had hepatitis B, a viral infection of the liver. Some people, after recovering from the acute symptoms of hepatitis, remain infected carriers of the virus. Over 250 million people worldwide are carriers of hepatitis B virus; carriers have a 100-fold higher risk of developing liver cancer, although this may be as long as 40 years after having hepatitis. Hepatitis C virus is also associated with liver cancer, particularly in Japan. Worldwide, as many as 80% of liver cancers are caused by viral infections.

Cancers of the reproductive organs, especially cancer of the uterine cervix, are statistically related to both male and female sexual behavior. Incidence rates for cervical cancer are higher among women who were younger at the time of their first intercourse or who have had multiple sexual partners. The rates of cervical cancers are also high in those countries in which women tend to have few sexual partners and to marry as virgins but where a tradition of *machismo* often encourages men to seek multiple sexual partners. This epidemiological evidence argues that male promiscuity is an important risk factor for cervical cancer even though it is women who develop the disease. Sexually transmitted human papilloma viruses often cause genital warts, but some types also cause cancer of the cervix, which is thus a sexually transmitted form of cancer. Papilloma viruses account for more than 80% of cancers of the genitals and anus.

For the association of each of these viruses with cancer, the epidemiological evidence has been verified by animal and tissue-culture experimentation. Although the incidence of virally induced tumors is low in the United States, such tumors account for 20% of all cancers worldwide.

Physical and chemical carcinogens

A large and growing number of external agents are known to cause cancer; such agents are called **carcinogens**. Evidence that carcinogens cause cancer comes from studies in which animals are exposed to the chemicals. Evidence also comes from epidemiological studies of occupationally exposed persons, such as industrial workers who handle the agents. Still other carcinogens are discovered by recognizing epidemiological clusters of persons with unusually high incidence rates for particular cancers living in the area surrounding an industrial plant or waste disposal site. There are two types of carcinogens. The first are physical agents, energy sources with high enough power to damage DNA. The second are chemical agents; these too work by damaging DNA.

Physical carcinogens (radiation). Some carcinogens are physical agents, particularly certain types of energy sources. Ultraviolet (UV) radiation, for example from sunlight, can cause susceptible people to develop skin cancers, including malignant melanoma, a cancer of the pigment cells (melanocytes). Melanomas are dangerous because they readily spread around the body and thus are difficult to treat once they have metastasized. Melanomas kill more women in their twenties than does breast cancer. Light-skinned people are more susceptible to these cancers. Melanoma affects some 44,000 Americans annually and a total of 1 million people get some form of skin cancer in the United States each year. Most of these are squamous or basal cell carcinomas, which are less aggressive and more treatable than melanoma. Ninety percent of skin cancers are attributable to UV radiation, particularly UV B rays, which cause DNA damage. The adult incidence of skin cancer is correlated with the number of sunburns that a person received as a child. Epidemiological evidence shows that even exposures to natural levels of UV radiation increase the incidence rates for skin cancers, which are much higher in the southern half of the United States than in the northern half.

Ionizing radiation, such as that produced by radioactive substances, was clearly shown to be carcinogenic by studies on the Japanese survivors of the 1945 bombings of Hiroshima and Nagasaki. The French (Polish-born) chemist Marie Curie (1867–1934), a two-time winner of the Nobel Prize and pioneer in the study of radioactive elements, died of a leukemia induced by her frequent handling of these elements. X-ray machines also produce ionizing radiation, so the level of exposure is carefully controlled to minimize the exposure. Medical and dental diagnostic X-rays do not increase cancer incidence. The only medical uses of radiation associated with increased cancer risk are the very high radiation doses used in cancer therapy itself. Although these treatments are necessary to save a patient, they do induce DNA damage from which new cancers may arise a decade or more later. For example, about 5% of people treated with radiation for Hodgkin's lymphoma later develop leukemia. Radiation treatments, like all medical treatments, are based on estimates of risk–benefit ratios. When someone is very sick and would die without treatment, the probable benefit from the treatment may make a higher level of risk acceptable to the patient and her or his physician.

Ionizing radiation associated with cancer is also caused by radon. Radon is a radioactive element that occurs naturally in certain types of rocks. When such rock is uncovered, radon may be given off as a gas. A person with long-term exposure to radon gas may develop lung cancer. Radon accounts for less than 10% of lung cancers, however, whereas smoking accounts for more than 85%. Proper ventilation removes virtually all of the risk from radon.

Low-frequency electric and magnetic fields from power lines or household appliances and radio-frequency electromagnetic radiation from cell phones or microwaves are too low in energy to cause ionizing damage to DNA. Any time that charged particles move, whether they are electrons in a power line or ions across cell membranes, electric and magnetic energy fields are created. All living things, because of ions moving through them, are sources of electrical and magnetic energy. The ambient level of energy from a cell phone is less than one one-hundredth of the electromagnetic radiation given off by the person holding the phone.

Chemical carcinogens. Other carcinogens are chemicals. Exposure to the chemicals in tobacco smoke, including second-hand smoke, is the largest single risk factor for cancer in the industrialized world. People who begin to smoke when they are teenagers or in college are more than ten times more likely to develop lung cancer than people who have never smoked (Figure 9.20A). Although a person's risk of cancer decreases after he or she stops smoking, the risk never drops as low as that for people who have never smoked (Figure 9.20B). The danger of second-hand smoke is evident from the fact that nonsmoking women whose husbands smoke have higher cancer rates than nonsmoking women with nonsmoking husbands. Exposure to tobacco smoke is also a risk factor for heart disease and emphysema.

Tobacco smoke contains dozens of known carcinogens including nitrosamines, formaldehyde and other aldehydes, arsenic, nickel, cadmium, and benzo(α)pyrene. These are also present in smokeless tobacco (chewing tobacco and snuff). Both smoked and smokeless tobacco greatly increase the risk of cancers of the mouth and throat (oral cavity and pharynx).

A large number of industrial chemicals have been shown to be carcinogenic, including vinyl chloride (used in the making of many plastics), formaldehyde, asbestos, nickel, arsenic, benzene, chromium, cadmium, and polychlorinated biphenyls (PCBs) (Table 9.5). The aromatic amines used in dye manufacture and in the rubber industry can cause bladder cancer; benzene can cause bone marrow cancer. Vinyl chloride causes an

FIGURE 9.20

The effects of smoking on the incidence of cancer.

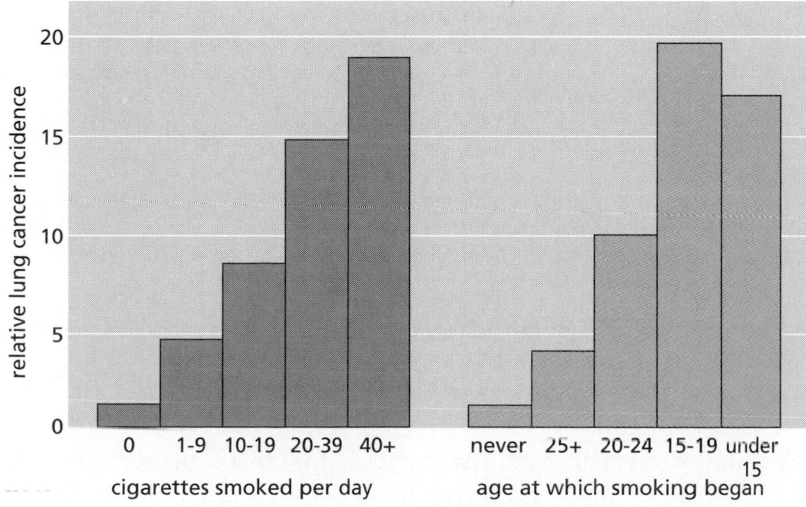

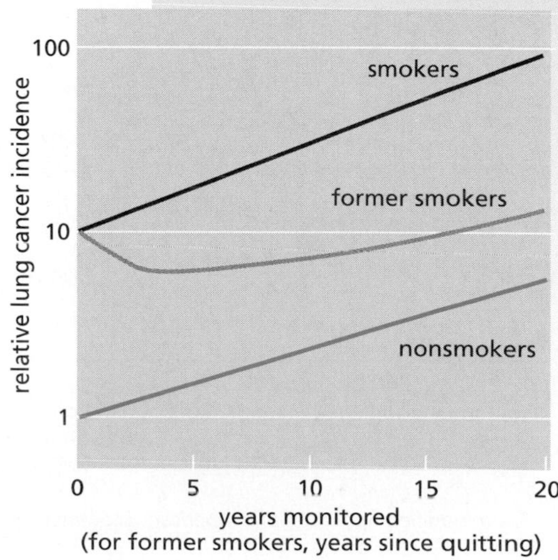

otherwise unusual liver cancer among plastics workers. Cadmium workers have an increased risk of prostate cancer, whereas asbestos workers have higher cancer risks at a number of sites including the lungs, peritoneal lining (mesothelioma), stomach, colon, and esophagus.

There is a clear correlation between the ability of an agent to act as a **mutagen** and induce the mutation of DNA (Box 9.1) and its ability to act as a carcinogen and induce cancer. The terms mutagen and carcinogen are not synonymous, however. Transformation is initiated by a mutation in the DNA; the mutation or mutations can be caused by a virus, a chemical, radiation, or by spontaneous mistakes in replication. The presence of mutagens greatly increases the mutation rate throughout the genome, increasing the likelihood that some mutation will be caricinogenic and turn a proto-oncogene into an oncogene

Some carcinogens exhibit a **synergistic effect**, meaning that the increased risk due to two causes is much more than the additive combination of their effects taken separately. For example, smokers have an eleven-fold greater risk of dying from lung cancer than nonsmokers, and nonsmoking asbestos workers have a five-fold greater risk for this than people who do not work with asbestos. Asbestos workers who smoke have a 53-fold greater mortality risk than nonsmokers who do not work with asbestos, an effect that is multiplicative rather than additive.

TABLE 9.5

Carcinogens in the workplace.

CARCINOGEN	CANCER TYPE	EXPOSURE OF GENERAL POPULATION	EXAMPLES OF WORKERS FREQUENTLY EXPOSED OR EXPOSURE SOURCES
CHEMICAL AGENT			
Arsenic	Lung, skin	Rare	Insecticide and herbicide sprayers; oil refinery workers
Asbestos	Lung, other sites	Uncommon	Brake-lining; shipyard; insulation and demolition workers
Benzene	Bone marrow	Common	Painters; distillers and petrochemical workers; dye users; furniture finishers; rubber workers
Diesel exhaust	Lung	Common	Railroad and bus-garage workers; truck operators; miners
Formaldehyde	Nose, pharynx	Rare	Hospital laboratory workers; manufacture of wood products, paper, textiles, garments and metal products
Heavy metals (cadmium, uranium, nickel)	Prostate	Rare	Metal workers
Man-made mineral fibers	Lung	Uncommon	Wall and pipe insulation; duct wrapping
Hair dyes	Bladder	Uncommon	Hairdressers and barbers (inadequate evidence for customers)
Mineral oils	Skin	Common	Metal machining
Nonarsenical pesticides	Lung	Common	Sprayers; agricultural workers
Painting materials	Lung	Uncommon	Professional painters
Polychlorinated biphenyls (PCBs)	Liver, skin	Uncommon	Heat transfer and hydraulic fluids and lubricants; inks; adhesives; insecticides
Soot	Skin	Uncommon	Chimney sweeps and cleaners; bricklayers; insulators; firefighters; heating-unit service workers
Vinyl chloride	Liver	Uncommon	Plastic workers
PHYSICAL AGENT			
Ionizing radiation	Bone marrow, several others	Common	Sunlight; nuclear materials; medicinal products and procedures
Radon	Lung	Uncommon	Mines; underground structures

Data modified from *Scientific American*, September 1996.

Dietary factors

Evidence that dietary factors contribute to the development of cancer is best established for cancers of the digestive tract, including the colon and rectum. The evidence comes from laboratory studies of animals exposed to experimentally controlled diets, from clinical studies on human patients, and from epidemiological studies of large populations.

Dietary fiber and fats. Diets high in fiber and low in fats are associated with a lower incidence of cancers of the intestinal tract (including the colon and rectum) and also those of the pancreas and breast. In countries where fiber consumption is high and fat consumption is very low, as in most of equatorial Africa, incidence rates of colon and rectal cancer are only a fraction of what they are in the industrialized world. Australia, New Zealand, and the United States, where diets are lower in fiber and higher in fats, have high rates of colon and rectal cancers. In fact, diet is more strongly correlated with cancer incidence than is industrial pollution. Studies comparing the cancer rates in Iceland and New Zealand, where diets are similar to those in the United States but where there is far less industrialization, have shown that the cancer incidence is the same in these countries as it is in the United States. In contrast, cancer rates overall, and for many specific types of cancer, are much lower in Japan, which, like the United States, is an industrialized nation, but one in which dietary fat intake is very low. The incidence of cancers among Seventh-Day Adventists who are vegetarian and do not smoke or drink is much lower than the incidence in their neighbors, despite both groups' living in the same conditions and being exposed to the same environmental pollutants. Several studies have also shown that eating fresh vegetables, particularly those rich in vitamins A, C, E, and beta-carotene (a vitamin A precursor), reduces the incidence of many cancers.

Salty and pickled foods. Cancer of the stomach follows a different epidemiological pattern correlated with a different set of dietary factors. This cancer is most frequent in Japan and in certain Latin American countries, where it seems to be correlated with the eating of very salty foods and pickled vegetables. The incidence of this cancer in Japanese immigrants to Hawaii and California decreases after a generation or two, while that for cancers of the colon, rectum, and breast increases. Among Japanese-Americans in Hawaii, the incidence of stomach cancers correlates closely with the retention of other aspects of Japanese culture: that segment of the Japanese-American population who maintain more of their traditional culture have higher rates of stomach cancer than those who adopt more Western cultural practices. This evidence suggests that diet has a larger role than genetics in the incidence of stomach cancer.

Alcohol. Ethyl alcohol has been identified as a risk factor for cancer by a number of researchers, but the increased cancer risk is largely confined to people who also smoke. The risk of developing a cancer of the mouth or throat, for example, is much higher in people who both smoke and drink (Figure 9.21, p. 396). This is another example of a synergistic effect; the increased risk is much greater than would be expected from

BOX 9.1 The Ames Test

Tens of thousands of known chemical substances have never been tested as possible carcinogens in animals. Animal testing is expensive and slow; it would take many, many decades (and many billions of research dollars) to test all these substances. Clearly, we need a quick screening method that tells us which substances are more likely to be carcinogenic; these substances can be tested first, while the testing of less likely carcinogens can wait.

The Ames test, devised by cell biologist Bruce Ames of Cornell University, is a screening method that detects mutagens capable of causing a particular type of mutation in a culture of *Salmonella* bacteria. The bacteria used are from a strain called *his*$^-$, which are unable to synthesize histidine, an amino acid required for the manufacture of bacterial proteins and hence for bacterial growth. Most bacteria are *his*$^+$, meaning that they can make their own histidine from other materials. In the Ames test, *his*$^-$ bacteria are grown in a medium containing just a small amount of histidine, which allows just enough growth for mutations to have a chance to occur. Soon, however, the histidine is used up, and the bacteria die unless they have mutated from *his*$^-$ to *his*$^+$ and thus have become able to make their own histidine. The rate of spontaneous mutation is very low. If a chemical is added to the culture medium and many more bacterial colonies grow than in a culture without this addition, the chemical can be assumed to have caused the mutations—i.e., to be a mutagen. Counts of the numbers of colonies also identify stronger and weaker mutagens.

Remember that the Ames test was designed as a *screening method* for carcinogens. The basis of the Ames test is the observation that many known carcinogens are also mutagenic. This is simply a statement of *correlation*; it does not necessarily indicate a causal relation between mutagenesis and carcinogenesis. Not every mutagen is a carcinogen, so the Ames test is only preliminary. It focuses our attention on chemicals that are mutagenic in bacteria, and thus more likely to be carcinogenic in animals. We can then proceed with the animal testing of these substances.

Bruce Ames, the originator of the Ames test, has also pointed out that nearly *any* substance is mutagenic in a sufficiently high dose. We are surrounded with thousands of naturally occurring carcinogens (mostly weak ones), and yet we do not all get cancer from them. Perhaps, he argues, we should study our mechanisms of defense against these natural carcinogens rather than concentrating on merely identifying one carcinogen after another.

---

just adding the risks from the two separate activities. Alcohol and tobacco are also synergistic in producing other forms of cancer.

Tumor initiators and tumor promoters

Tumor initiators are agents that begin the process of transformation by causing permanent damage in the DNA. Mutagens, including tobacco smoke, are tumor initiators. In a cell whose DNA is damaged in this way,

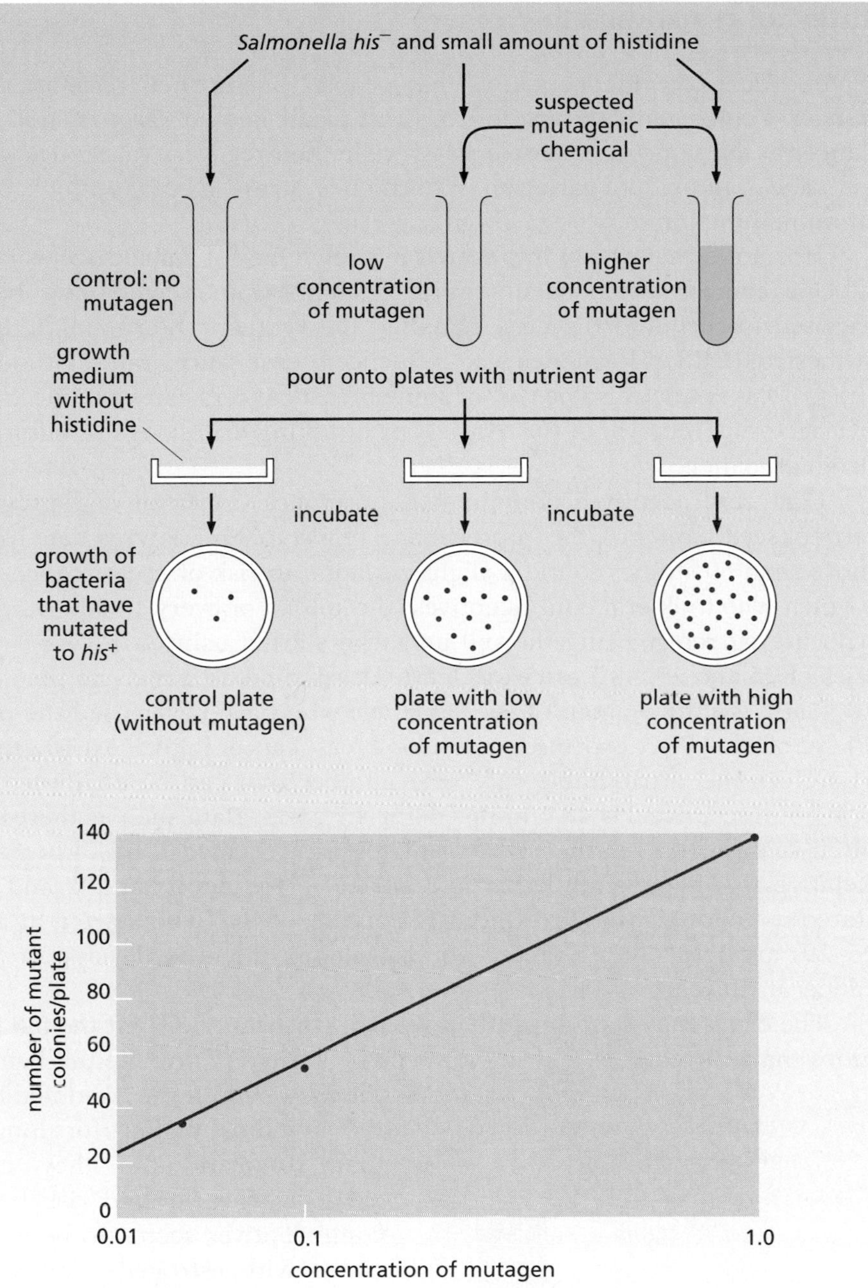

transformation can be completed by exposure at a later time to a **tumor promoter**. Tumor promoters by themselves do not cause mutation in the Ames test described in Box 9.1. They can induce cell division, and if a dividing cell contains a mutation from an earlier exposure to an initiator, its chance of acquiring additional mutations that lead to complete transformation are increased. Because the DNA damage from an initiator is permanent, a tumor promoter can have its effect years after exposure to the initiator. Tumor promoters include alcohol, phenobarbital, dioxin, saccharin, asbestos, and tobacco.

Internal resistance to cancer

A good deal of evidence shows that people vary in their resistance to cancer. People with the same exposure to all known risks do not get cancer at the same rate. People also vary in their recovery rates once they get cancer. Individual variation in hormones, stress, mental outlook, and immune function may be involved.

Hormones have been implicated in some types of cancers, especially uterine cancer, which occurs more often in women who have been exposed to certain estrogens, including the synthetic hormone diethylstilbestrol (DES). Hormones also influence breast cancer rates, although the process is unclear. The risk of some breast cancers can be reduced by ovariectomy (removal of the ovaries) or by taking the estrogen-inhibiting drug tamoxifen.

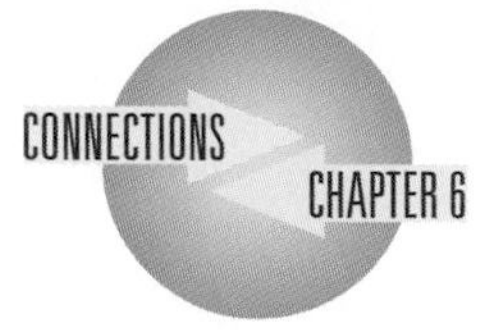

Oral contraceptives contain two hormones, estrogen and progesterone (see Chapter 6). A comprehensive report that reanalyzed data from more than 50 studies found a slight elevation in risk of breast cancer in women who took oral contraceptives (the 'pill'). For every 10,000 women who are currently using the pill and who started using it between the ages of 25 and 29, 48.7 are expected to develop breast cancer in the next 10 years. Among women of the same age who have never used the pill, 44 out of 10,000 are expected to develop breast cancer in the next 10 years. Therefore the 'attributable risk' (the number that can be attributed to oral contraceptives) is 48.7 minus 44 or 4.7 cases. Data such as these are often reported as 'relative risk' calculated as 48.7 divided by 44 or 1.16. Relative risk is generally what is reported to the public and would be stated as follows: women using oral contraceptives (who started at age 25–29) are 1.16 times as likely (or 16% more likely) to develop breast cancer in 10 years.

The elevated risk disappears in women who have been off the pill for more than 10 years, except in women who started taking the pill before the age of 20. In these women the risk remains higher for longer than 10 years after they stop using it. The elevated risk from contraceptives seems to be associated with estrogen's ability to promote the growth of breast cancer cells that have arisen from other causes. Estrogen itself does not seem to initiate cancerous transformation. Smoking is synergistic with birth control pills in increasing cancer, possibly owing to the presence of both tumor initiators and tumor promoters in tobacco.

In contrast, oral contraceptives are associated with *decreased* rates of ovarian cancer. A decrease by as many as 1700

FIGURE 9.21

Synergism between alcohol and cigarettes in producing cancers of the mouth and throat.

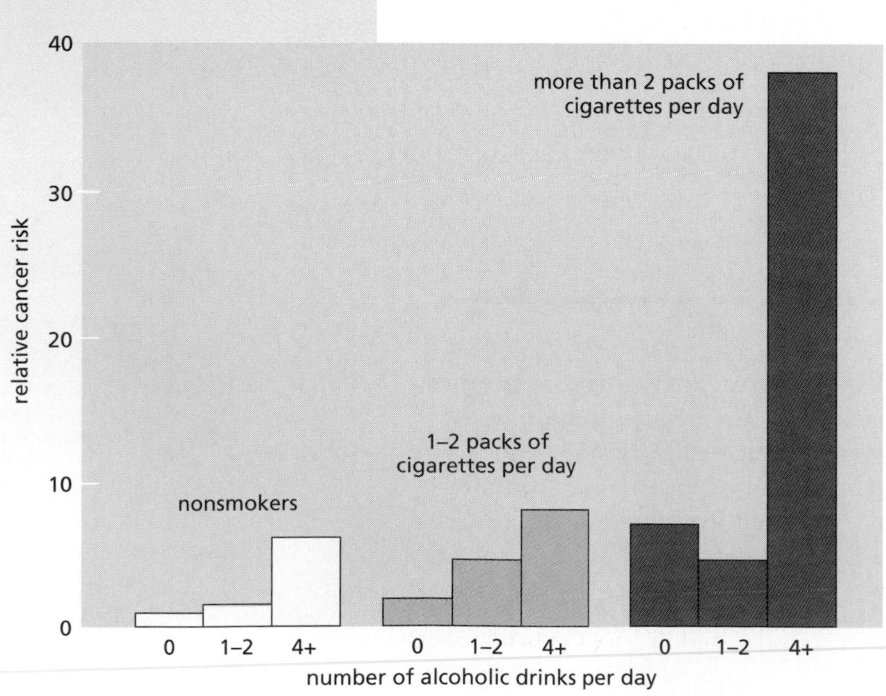

cases in the United States per year may be associated with oral contraceptive use. The decreased risk continues in women for as long as 15 years after they stop taking contraceptives, according to the National Cancer Institute.

Certain cancers are more common among people under chronic stress. Stress is a difficult variable to measure, and stress levels are usually reported simply as 'high' or 'normal.' Many studies of stress are flawed in that researchers failed to take into account other variables in addition to stress. For example, night workers and daytime workers in the same industry may differ in many other ways besides stress levels. Psychologists who have focused their attention on the means by which people deal with their stress have found lower cancer rates among people with better coping skills.

Evidence is increasing that people with weakened immune systems develop cancers more frequently. This factor is striking in conditions that severely damage the immune system, such as AIDS (see Chapter 13), but is also present in people whose immune systems are weakened less drastically from other causes, including chronic stress, sleep disorders, and so forth. Cells that are fully transformed may not survive to proliferate into tumors. Far more cells are mutated and transformed than ever develop into cancers. A healthy and active immune system eliminates most of these cells as they arise. Any weakening of the immune system increases the number of transformed cells that grow and proliferate, and a higher rate of cancer is one of the results. People's immune systems also weaken with age, which is consistent with the finding that the incidence of new cancers increases with age.

Social and economic factors

Social and economic factors have also been shown to be correlated with incidence rates and especially with survival rates for various cancers. In the United States, some studies have shown that whites and blacks (African Americans) have comparable incidence rates for certain cancers, but the mortality rates for blacks are higher because they receive far fewer routine medical exams that would detect the common cancers in their earliest and most treatable stages.

More recent studies have found higher incidence and mortality rates for most types of cancers for urban black populations than for urban white populations. Much of the excess cancer rate among blacks was in those with low income levels and low educational attainment. In other words, a large amount of the difference in cancer incidence rates could be explained by differences in income and factors related to income. When blacks and whites of comparable socioeconomic status were compared, many of these differences disappeared or were reversed. Colon cancer showed no difference in incidence by race, and rectal cancer was more common in whites than in blacks. Blacks had higher rates for cancers of the stomach, prostate, and uterine cervix. Female breast cancer showed a higher rate in white women than in black women, and among white women it showed a higher incidence in the higher income brackets than in the lower ones.

TABLE 9.6

Summary of various causes of cancer worldwide.

CAUSE	RELATIVE PERCENTAGE OF CANCER DEATHS*
Smoking	30
Diet	30
Alcohol	3
Food additives (salt)	1
Sedentary lifestyle	3
Radiation	2†
Pollutants (air, water)	2
Viruses	20
Chemical carcinogens	Variable‡
Genetic susceptibility	<10

* Derived from statistical analysis of epidemiological data. A figure of 30 means that 30% of all cancer deaths worldwide are attributable to that particular cause.

†Over 90% of skin cancers are caused by UV radiation.

‡ The percentage of deaths in the general population attributable to carcinogen exposure is low, but can be locally very high, for example in industries with high or prolonged exposure.

As we have seen, many factors may contribute to cancer incidence rates, including genetic predisposition, lifestyle, and exposure to environmental carcinogens. A summary of the relative contributions of various biological causes is given in Table 9.6. These are rough averages for global cancer incidence. The percentage of cancers attributable to viruses is 20% worldwide, but is much lower in the United States. The percentage attributable to diet also varies from one part of the world to another, as does the percentage attributable to chemical carcinogens in the environment. With the exception of the low percentage of cancers that may be attributable to genetic predisposition, most cancers are preventable by changes that can be made by individuals or by societies, a topic we explore further in the next section.

THOUGHT QUESTIONS

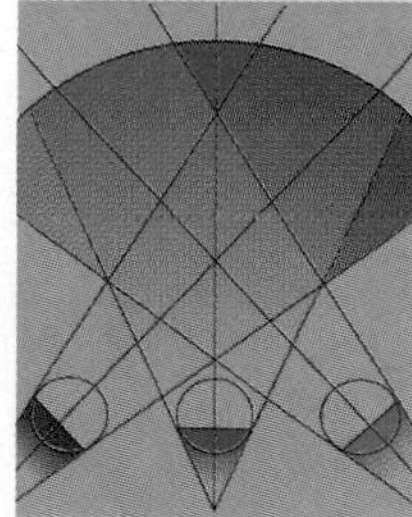

1. Not everyone who smokes gets cancer. Does this mean that smoking is not a risk factor for cancer?
2. What is the difference between a risk factor and a cause? Can we say what caused cancer in a given individual?
3. Why do different individuals respond differently to cancer risk factors?
4. Does an increase in the percentage of deaths due to cancer necessarily mean that cancer rates have increased? What else could explain such findings? How could you go about determining which of the possible explanations best fits the data?
5. Recently the genetic defect associated with an inherited form of colon cancer was identified as a defect in a DNA-checking protein. This finding was reported in the lay press as the discovery of 'the colon cancer gene.' Is this name misleading? To what extent can a defective repair mechanism be considered to be the same thing as a cause of a cancer?
6. Tobacco smoke contains chemicals that are tumor initiators and other chemicals that are tumor promoters. How does this combination contribute to the carcinogenicity of tobacco smoke?
7. Tobacco and alcohol act synergistically in increasing cancer risks. Can you explain this in terms of what is happening inside cells?

WE CAN TREAT MANY CANCERS AND LOWER OUR RISKS FOR MANY MORE

An understanding of the mechanisms that produce cancer has greatly increased our understanding of how to prevent it. Treatments have improved and continue to improve so that survival rates for most cancers have increased. In this section we examine medicine's current strategies for treatment and for prevention.

Surgery, radiation, and chemotherapy

Most of the present-day treatments of cancer use one or more of three types of treatments: surgery, radiation, and chemotherapy. Surgery is limited to those cancers that produce visible tumors. Because single cancer cells that metastasize can lead to later cancer recurrence, surgery is often combined with radiation or chemotherapy. In either radiation therapy or chemotherapy, the strategy is the same: cancer cells are dividing cells; therefore agents that interfere with cell division should stop cancer cells. Radiation causes breaks in the DNA of dividing cells that are so large that the cell cannot repair them and the cell cannot live with the damage. Chemotherapeutic drugs prevent DNA synthesis at several steps. Some of the drugs inhibit the synthesis of the nucleotides needed to build DNA; some substitute for certain nucleotides in newly synthesized DNA, preventing its further replication; and some inhibit an enzyme needed to unwind and rewind the double helix during its replication. Other chemotherapeutic drugs, some of which are natural plant products, prevent RNA synthesis or block mitosis. Still others act by damaging the DNA strands, thus preventing cell division and killing the cell.

An important drawback to radiation and chemotherapy treatments is that, because they damage DNA, they increase the risk for the development of secondary cancers from damaged cells that survive the treatments. Another drawback is that both radiation and chemotherapy are nonspecific: both kill any type of dividing cell. A high proportion of cells in hair follicles are dividing, so hair loss frequently accompanies these treatments. A large percentage of cells of the immune system are also dividing and so are killed. Not all hair follicle cells and immune cells are killed because not all were dividing at the time of treatment and so they repopulate. Hair grows back and people regain their immune cells. During the time when people's immune systems are compromised, they need to avoid exposure to infectious diseases. Both radiation and chemotherapy may destroy a particular type of immune cell, memory cells, which 'remember' which diseases the person has been exposed to or vaccinated against (see Chapter 12). If these memory cells are killed, even a person who has regained the ability to form new immune responses has lost previous immunities and therefore may need to be revaccinated.

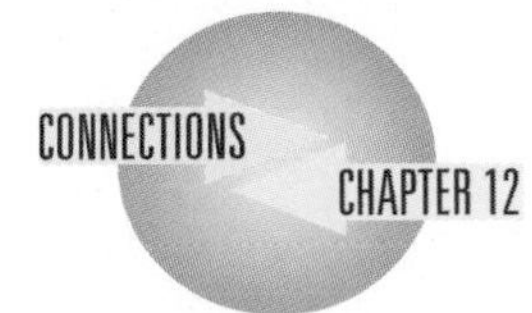

A further risk from chemotherapeutic drugs is that they put a selective pressure on the population of transformed cells. As a result, any cells that become resistant to the drug quickly overgrow the drug-susceptible cells.

Several drugs, each of which works by a different mechanism of action, are often used in combination to minimize the development of drug resistance.

Despite the drawbacks and risks, surgery, radiation, and chemotherapy have been very effective, increasing the survival rates of many types of cancers. Moreover, in most patients, leukemias, Hodgkin's lymphoma, and testicular cancer can now be cured by these treatments.

New cancer treatments

Research on cancer treatments continues. New chemotherapeutic agents, both natural and artificial, continue to be sought and developed. New strategies are being developed to make chemotherapeutic agents target cancer cells more specifically and to reduce damage to normal cells.

The SERMs (selective estrogen receptor modulators) are new chemotherapeutic drugs with more specific action. The hormone estrogen binds to receptors inside cells and causes some types of cells to divide, including some types of breast cancer cells. However, estrogen has many normal functions, including maintenance of bone density and cholesterol regulation. The SERMs block estrogen's ability to promote breast cancer cell division but do not block its desirable effects. Tamoxifen is one such drug and has proved effective in reducing mortality from breast cancer and in decreasing its onset in women at high risk due to genetic predisposition or age (over 60).

Another new type of chemotherapeutic drug starves tumors. As mentioned earlier, many tumors have the ability to induce the body to grow new blood vessels, which then bring needed blood to the tumors. Drugs called angiogenesis inhibitors block this process, thereby cutting off the nutritional supply of the tumors. Treating tumors with these drugs shows great promise and has received much publicity. At present several angiogenesis inhibitors are in clinical trials.

In addition to treatments with direct effects on tumors, new treatments are being developed to boost the immune system's ability to fight off tumors. If the immune system gets rid of many transformed cells, why can we not vaccinate people against cancer as we do against many infectious diseases? One reason is that the immune system can only act against cells it perceives to be nonself (see Chapter 12). The surface molecules of cancer cells are often the same as those on normal cells; they are just expressed in the wrong amounts or at the wrong times. The immune system cannot distinguish the last two possibilities as differing from the normal: it can detect only new or different cell surface molecules. Some cancers, especially cancers initiated by chemical carcinogens, do have new or altered molecules at their surface. Unfortunately these tumor-associated antigens are different in each person and even in two different chemically induced tumors in the same person. What new molecules will be present cannot be predicted, so vaccines cannot be developed against them. Some tumors induced by viruses do have common tumor-associated antigens, and vaccines can be developed. Feline leukemia is the most successful example so far of a cancer that can be prevented by vaccination.

Antibodies can block growth factor receptors. About 25–30% of breast cancers have cells with abnormally high levels of the growth factor receptor molecule called HER2/neu. More receptors mean that the cell can be triggered to divide by lower concentrations of growth factor. A new drug called Herceptin is being considered for approval by the FDA in the United States for use along with conventional chemotherapy for the treatment of metastatic breast cancer. This drug is an antibody that binds specifically to the receptor, blocking binding of the growth factor and thus inhibiting cell division. Initial studies have found that it shrinks tumors in some women and delays tumor progression in others.

Another entirely new type of treatment, called photodynamic therapy, has been approved for esophageal cancer in the United States, and some other countries have approved its use for other cancers as well. The patient is given an intravenous injection of a photosensitizing dye. Over the next 24–48 hr most of the dye accumulates in tumor cells (normal tissues remove the drug). Then the patient's tumor is illuminated by using fiber optics for 30–60 min. Light reacts with the dye to kill many tumor cells and sensitize others to being killed by the immune system.

A good deal of cancer research is aimed at assessing the efficacy of new therapies. After extensive laboratory research, new therapies are tested on patients who have given their informed consent (see Chapter 1) to being part of the studies. Only by gathering data in properly designed and controlled clinical trials can the risks and benefits of new treatments be demonstrated.

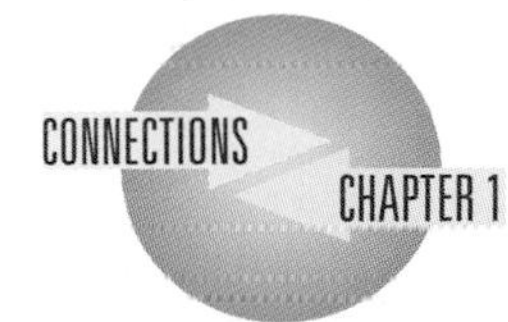

Because cancer is greatly feared and not always curable, some people put their hopes in unproven remedies. Various unconventional cancer therapies and treatments have been publicized in the last few decades. Some of these, such as laetrile, achieved a large and devoted following. The supporters of laetrile finally became so influential that the National Cancer Institute conducted careful clinical trials and announced in 1981 that laetrile had proved to be worthless as a cancer treatment.

Cancer detection and predisposition

Early detection greatly increases the probability of successful cancer treatment. Breast self-examination is very effective at finding tumors while they are treatable. Diagnostic breast X-rays (mammograms) detect smaller tumors than can be felt, but are less effective in younger women than in postmenopausal women whose breast tissue is less dense. Microscopic examination of tissue from the cervix taken during a medical examination is effective at early detection of cervical cancer. The test is called a Pap smear. Testicular cancer, although rare, is the most frequent cancer in men between the ages of 15 and 34; monthly self-examinations to detect lumps in the testes are an important method of early detection.

An emphasis on early detection has contributed to the increased survival rate from some cancers. For other cancers, it is not so clear what is meant by an increase in survival rate and how this relates to a 'cure.'

When we try to evaluate the meaning of statistical statements like 'survival rates,' we need to know a lot about how the numbers were gathered and what definitions are being used for certain terms. What is often reported as survival rate refers to the proportion of persons with cancer still living after 5 years compared with the proportion of surviving persons without cancer. For example, it is estimated that it takes 9 years for a breast cancer to develop, spread, and kill a person. If past methods of detection led to discovery of the cancer 7 years after its inception, very few people would have been alive 5 years after its discovery. Now, with better detection methods and better public education for breast self-examination, the 5-year survival rate looks much improved. Does this mean that therapies have improved, or does it simply mean that people are now finding the cancers at 2 years into their development rather than at 7 years? It is not always easy to distinguish advances made through better or earlier diagnosis from advances made in the treatment of cancers once they have reached comparable stages of development.

In addition to the tests mentioned above, two types of laboratory cancer tests currently exist. One type is for the early detection of existing cancers. The other type is genetic testing for cancer predisposition.

An example of the first type is the PSA test for prostate cancer. This test measures the level of PSA (prostate-specific antigen) in a person's blood. This protein is elevated when prostate cancer begins to develop. Thirty percent of people with elevated PSA are found to have cancer when the test is followed by a biopsy (tissue sample examined by microscope). The PSA test can detect cancer up to 5 years before there are other symptoms. The clinical question then becomes what to do about it. Cancers detected by PSA tests are generally still localized and can therefore be successfully removed surgically. However, prostate cancer is very slow-growing, and it has been said that most men die with prostate cancer, not because of it. One-third of men have some form of prostate cancer by the time they are over the age of 50, but only 3% die from it, so, for many individuals, the recommended treatment after a biopsy has confirmed the presence of cancer is to do nothing.

The second type of laboratory test does not detect cancer, but instead identifies DNA sequences that are statistically correlated with an increased probability of some day acquiring the disease. Tests of this type are available for *BRCA1* and *BRCA2* (breast cancer predisposition mutations), DNA mismatch repair genes (predisposing to some kinds of colon and uterine cancers), *p53* (predisposing to brain and other tumors) and a few other tumor suppressor genes and oncogenes.

These tests for genetic predispositions for cancer are controversial for many reasons. They are very expensive and do not give much more useful information than is gained from knowing your family medical history. A negative test does not mean that a person will not get cancer; cancers arise spontaneously in the same way as they do in people with no family history of the disease, so the recommended preventive measures discussed below should still be followed. A positive test is not a guarantee that a person will get cancer. The probability is increased but we cannot tell by how much. The increase in risk is different for different mutations, but none increases the probability to 100%. The genetic mutation cannot be repaired, so there is little that a person with increased risk can or should do beyond what is recommended for everyone (regular check-ups and the lifestyle choices summarized below). Still, because some of these cancers are difficult to detect early, being aware of a predisposition for them can ensure that physical examinations are done even more thoroughly than usual, and perhaps more often.

Some women with increased breast cancer risk due to mutations in *BRCA1* (or family history) have opted for 'prophylactic mastectomies,' that is, removal of their breasts before there is any evidence of disease. A recent study was widely publicized as showing that women who had their breasts removed reduced their risk of dying by 90%. While this statement is not untrue, it is only part of the story and is an example of reporting 'relative risk' instead of 'absolute risk.' In the study, 639 women had their breasts removed. On the basis of calculations made from the number of deaths among their sisters who faced the same increased susceptibility but who had not had their breasts removed, it was estimated that 20 of the 639 women would have died. Only 2 actually did die, so the relative risk was decreased by 90% $[(20 - 2)/20 \times 100 = 90\%]$. When absolute risk is considered, it can also be correctly stated that 97% of the women had their breasts removed unnecessarily $[(639 - 20)/639 \times 100 = 97\%]$. The difficulty for any woman faced with such a choice is that there is no way to predict whether she will be one of the 18 saved by the procedure or one of the 619 who did not need it.

Cancer management

Some people feel that more research dollars should be spent on cancer management rather than cancer treatment. Cancer management includes the development of drugs or strategies to minimize the side-effects of cancer treatments. Examples include cold-capping, a procedure in which a cold pack is applied to the scalp during chemotherapy so as to slow cell division in hair follicle cells, thus decreasing hair loss. Another possibility is the development of anti-nausea drugs. The possible medicinal use of marijuana to overcome nausea and restore appetite in chemotherapy patients is being studied (see Chapter 11).

Cancer management also includes support groups and grief therapy. The aim of these approaches is to improve the quality of life—to treat the person, not the disease. Women who were in support groups after recurrent breast cancer lived longer than those who were not. Survival rates have been found in some cases to be influenced by mental attitude (see Chapter 12). Patients who were optimistic, who were aggressive, or who were 'determined fighters' had statistically longer survival rates and higher cure rates than those who were pessimistic or who resigned themselves early to their fate. There is a large body of psychological literature on 'learned helplessness,' a phenomenon in which a person or an animal experiences repeated stresses from which there is no escape and for which no remedy is available. Such individuals 'learn' that there is nothing they can do to change anything, a lesson that they then apply to other areas of their lives. When such people get cancer, their learned helplessness results in a much lower survival rate and a shorter life span.

Cancer prevention

As we learn more about the causes of cancer, it seems that one of the more successful strategies may be cancer prevention. Preventing cancer may be far easier than curing it.

Smoking remains a major cause of cancer. The Centers for Disease Control and Prevention state that cigarette smoke causes 30 times more lung cancer deaths than all regulated air pollutants combined. Exposure to secondhand smoke, for example as a result of living in a home with someone who smokes, causes the deaths from lung cancer of 3000 non-smokers a year in the United States. Tobacco smoke contains both tumor initiators and tumor promoters, and it suppresses the immune system. In addition to its cancer risks, exposure to secondhand smoke is also responsible for nearly 300,000 infections per year in infants younger than 18 months, and has been implicated as an important contributing factor to Sudden Infant Death Syndrome (SIDS).

CONNECTIONS CHAPTER 8

Some dietary regimens have been associated with a decreased risk of cancer: lower total calories and low fat, for example. Food containing antioxidants may help; the antioxidants include beta-carotene and vitamins A, C, and E (see Chapter 8). The American Cancer Society has issued guidelines promoting fresh foods high in vitamins A and C, especially fresh vegetables of the plant family Cruciferae. Cruciferous vegetables include cabbage, radish, turnip, broccoli, cauliflower, kale, kohlrabi, mustard greens, and brussels sprouts. Clinical studies have shown that some vitamin supplements are not as effective in preventing cancer as the same vitamins obtained from foods, probably because other food ingredients are also at work.

High-fat diets may be an important risk factor in cancers of the colon and rectum, and in postmenopausal cancer of the breast. There is some evidence that high-fiber diets lower the risks of colon and rectal cancers.

Given the available evidence, the most important actions you can take to lower your cancer risks are the following.

1. *Don't smoke*! Also, avoid secondhand smoke from poorly ventilated rooms where others smoke. *These are the single greatest steps you can take to reduce your cancer risk, far outweighing all other possible measures*.
2. Follow a diet low in fats, high in fiber, and high in antioxidants such as beta-carotene and vitamins A and C.
3. Avoid occupational exposures to potential carcinogens; minimize exposure through the appropriate use of safety equipment.
4. Avoid exposure to radioactive substances and X-rays above necessary minimum levels; avoid needless exposure to ultraviolet radiation from the sun or tanning booths.
5. As you age, be sure to get checkups at regular intervals, including screening that detects the common cancers in their earliest and most easily treated stages. If you are a woman, learn to practice breast self-examination and, if you are a man, testicular self-examination.

THOUGHT QUESTIONS

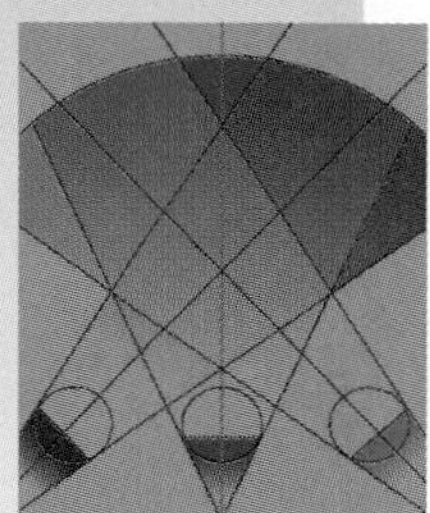

1. Explain how the theory of evolution accounts for the development of cancer cells that are resistant to chemotherapy.
2. What characteristics would you look for in an ideal chemotherapeutic drug for the treatment of cancer?
3. Secondhand smoke is a cancer risk. Does this biological reality change the ethical debate about smoking in restaurants or smoking in the workplace? What rights are in conflict on either side?
4. Do you agree with the following statements? When someone's chances for survival are predicted to be very low, any and all treatments are justified. In other words, any treatment is good as long as it is not harmful. Try to apply this thinking to such unproven remedies as laetrile.
5. Is it ever ethically permissible to give up on treatment? Are there things other than treatments that can be done for a dying person? Do you think people who have terminal cancer should be told of their condition? Try to justify your answer. What ethical assumptions underlie your argument?

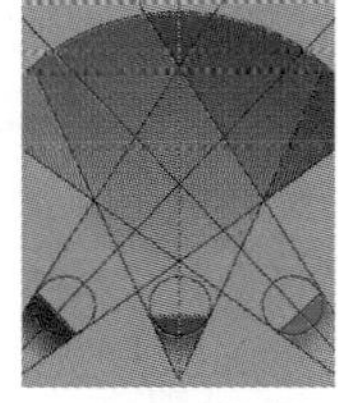

RESEARCH DIRECTED TOWARD UNDERSTANDING THE BASIC biology of cancer continues. Some people feel that treatment cannot be rationally designed unless the underlying biology is known, while others feel that such basic understanding is not important. The former group point to the development of new treatments such as tamoxifen and Herceptin. The latter group use arguments such as the following: we still do not know the basic biology underlying the disease polio, but development of a vaccine for its prevention has eliminated our need to know. The real dilemma is a problem in the allocation of resources. How much money should we spend on treating cancer patients, how much on improving methods of treatment, how much on laboratory research to discover more information on the causes of cancer, and how much on cancer prevention activities? How much funding should be aimed at particular types of cancers, such as breast cancer as compared with colon cancer? There are no clear answers here because we cannot accurately predict how well or how soon funds spent on certain activities (especially research) will translate into a reduction of cancer incidence rates or cancer deaths. Cancer prevention is clearly very cost-effective, but the cost-effectiveness of the other alternatives may be very difficult to assess.

On an individual level, we can reduce our exposure to cancer risk factors by many choices that we make in our lives. However, not all types of exposure among those listed above are matters of personal choice. Dumping of carcinogens on the land and water of poor people with little or no political power has become a global environmental issue. Air pollution affects people at great distances from the source. Therefore, as part of any effort to prevent cancer, people need to work together to prevent or remove environmental hazards from work places and communities.

CHAPTER SUMMARY

- All organisms are built of **cells** that maintain an efficient ratio of surface area to volume for the organism.
- Normal cells occasionally enter the **cell cycle** and divide. They always stop dividing when enough cells are present because of such phenomena as contact inhibition and other processes that maintain **homeostasis** of cell number.
- Cells also **differentiate** as they divide, becoming more and more restricted in their potential to form different kinds of cells that in most species are organized into **tissues**.
- **Stem cells** retain the ability to differentiate into many different kinds of cells.
- Both cell division and cell differentiation are the result of differential **gene expression**, and both are influenced by molecular signals (cytokines) secreted by other cells. Cytokines that trigger cell division are called growth factors.
- Cancer cells are cells that have undergone **transformation** and differ from normal cells in abnormal responses to growth control signals—they do not stop dividing (they are 'immortal'), and they remain less differentiated.
- A few **cancers** have genetic predispositions, but most are caused by environmental factors. These factors include exposures to certain viruses and infective agents, dietary and other behavioral factors, and exposure to a long list of **carcinogens** including ionizing radiation (radioactivity), ultraviolet radiation, tobacco smoke, and a variety of occupational carcinogens.
- Cancerous **tumors** can be removed surgically, but other forms of cancer therapy target any cells that are dividing, destroying many healthy cells along with the cancer. New, more specific chemotherapies are being developed, including therapies based on boosting the body's immune system.
- We can best reduce our risks for cancer by avoiding tobacco smoke and other carcinogens (including ultraviolet and ionizing radiations) and by eating a diet low in fats, high in fibers, and rich in beta-carotene and vitamins A and C.

KEY TERMS TO KNOW

cancer (p. 360)
carcinogen (p. 389)
cell (p. 361)
cell cycle (p. 364)
cytokine (p. 366)
determined (p. 371)
differentiation (p. 362)
gene expression (p. 367)
homeostasis (p. 363)
mutagen (p. 392)
potentiality (p. 369)
second messenger (p. 367)
stem cell (p. 378)
synergistic effect (p. 392)
tissue (p. 362)
transformation (p. 377)
tumor (p. 377)

CONNECTIONS TO OTHER CHAPTERS

Chapter 2: Cancers are caused by mutated growth-control genes.

Chapter 3: Cancers can result from changes that bring normal genes to abnormal locations in the genome.

Chapter 5: Several cancers have different incidence rates in different human populations.

Chapter 8: High-fat diets with low fiber content increase the risks for several cancers.

Chapter 12: Good mental outlook and immunological health can improve cancer survival rates.

Chapter 13: Certain otherwise rare cancers occur more frequently in AIDS patients.

Chapter 14: Many cancer-fighting drugs are plant products.

PRACTICE QUESTIONS

1. What are the four phases of the cell cycle and what happens in the cell during each phase? Is G_0 part of the cell cycle?
2. Which cells show contact inhibition, normal cells or cancer cells?
3. Compartmentalizing an organism by dividing it into cells increases which of the following: the volume or the surface area?
4. Which of the following properties do slime molds show in some part of their life cycle? Totipotent unicells; cell aggregations; cellular differentiation; motility.
5. Which of the following develop by differentiation from the zygote: tracheal cells in the lungs, muscle cells, or cells of the eye? Which develop from the gastrula? Which develop from the endoderm?
6. Which of the following two statements is true?

   a. Maternal effect genes are present in the genome of the sperm.

   b. Maternal effect genes are transcribed from the genome of the sperm.
7. How long does an average human tongue cell live? How long do human liver cells live on average? Human nerve cells?
8. When the telomere region of the chromosomes becomes too short, which of the following happens?

   a. The cell can no longer divide.

   b. The cell becomes cancerous.

   c. The cell can no longer differentiate.
9. Do the protein products of proto-oncogenes induce cells to divide or do they prevent cells from dividing? What about the protein products of tumor suppressor genes?
10. How many cells need to be transformed for cancer to develop? Does every transformed cell result in cancer? Why or why not?
11. Which of the following causes more cases of cancer: heredity, smoking, or viruses?
12. What processes are induced in a cell by binding of a growth factor to its receptor? Does a growth factor induce these processes in every cell?
13. Are the same genes transcribed and translated in every cell of the body?
14. Does cellular differentiation take place in adult animals or only in embryos?

10 The Nervous System

CHAPTER OUTLINE

The Nervous System Carries Messages Throughout the Body

The nervous system and neurons

Nerve impulses: how messages travel along neurons

Neurotransmitters: how messages travel between neurons

Dopamine pathways in the brain: Parkinsonism and Huntington's disease

Messages are Routed To and From the Brain

Message input: sense organs

Message processing in the brain

Message output: muscle contraction

The Brain Stores and Rehearses Messages

Learning: storing brain activity

Memory formation and consolidation

Alzheimer's disease: a lack of acetylcholine

Biological rhythms: time-of-day messages

Dreams: practice in sending messages

Mental illness and neurotransmitters in the brain

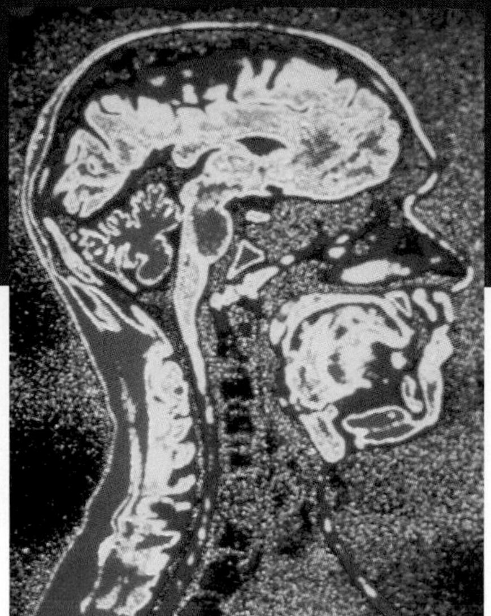

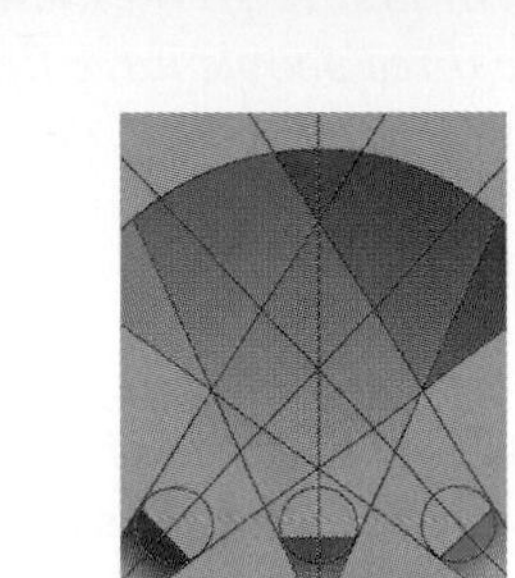

ISSUES

What happens when the nervous system malfunctions?

How can changes in brain chemistry interfere with message transmission?

How can many dissimilar malfunctions have similar causes?

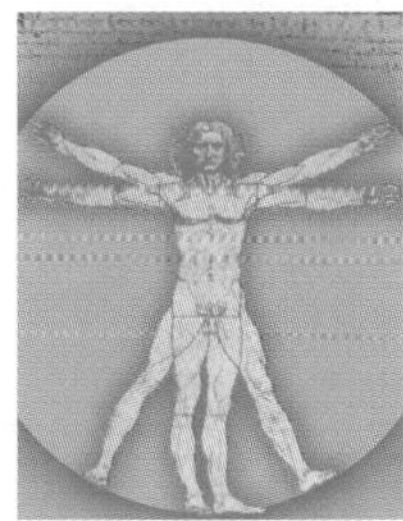

BIOLOGICAL CONCEPTS

Evolution (comparative anatomy, specialization and adaptation, form and function)

Hierarchy of organization (ions, neurons, brain, behavior)

Matter and energy (membranes, ion potentials, action potentials)

Detection of environmental stimuli (receptors, neurotransmitters, sense organs)

Movement (muscular system)

Homeostasis (feedback mechanisms, nervous system)

Health and disease

Behavior (sleep, learning)

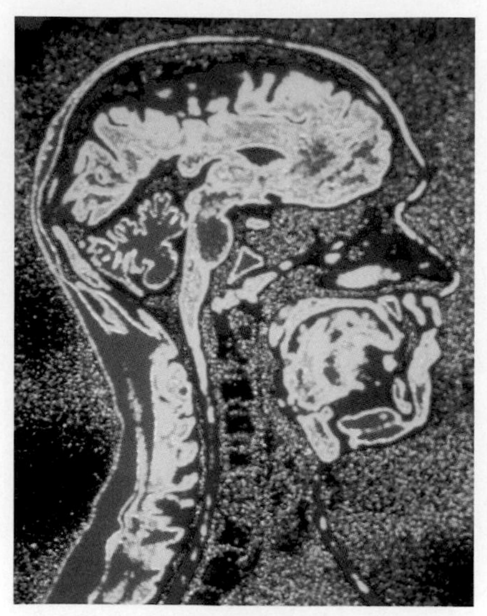

Chapter 10: The Nervous System

A man in his sixties finds it difficult to walk and shuffles his feet along the floor, just an inch or two at a time. Last week he tried to walk by lifting his feet, but he stumbled after only two quick but awkward steps. A woman in her seventies used to work crossword puzzles and solve math problems with ease, but she has forgotten how to do these tasks. Her friends, whom she no longer recognizes, say that she is not the same person that they knew a few years ago. A younger woman has lost all motivation to work, to keep herself clean, or just to go on living. "What for?" she asks, "none of it matters." These three people have three very different diseases: Parkinsonism, Alzheimer's disease, and depression. Could it be that all their assorted symptoms are caused by chemical imbalances in the brain?

The brain coordinates the activities of the rest of the body, including moving, sleeping, eating, and breathing. Changes in the brain can thus produce diseases such as Parkinsonism, whose symptoms include uncoordinated movements.

Does the brain also produce activities that we associate with the mind? Does biochemical activity in the brain produce perceptions, emotions, moods, and personality? Does it produce thoughts, dreams, and hopes? A branch of biology called neurobiology is the scientific study of the brain and nervous system. A central theory of neurobiology is that the mind and the brain are one and the same. As a framework that guides (and limits) research, the assumption that mind equals brain is a good example of a research paradigm (see Chapter 1). There is probably no theory in biology that is more controversial, both among biologists and between biologists and the public.

There is now considerable evidence in support of this theory. Electrical and biochemical activities have been measured in the brain during dreams and thought. Some diseases, such as Alzheimer's disease, in which there is brain degeneration, are accompanied by changes in personality. Mental illnesses such as depression are associated with changes in brain chemistry and can be treated with drugs.

In this chapter we examine the workings of the brain and the nervous system and consider the extent to which the mind is another name for the brain. The nervous system is adapted for sending and receiving messages. You will learn how nervous system messages travel electrically along cell membranes and how messages are carried from one cell to another by chemicals called neurotransmitters. You will also learn how the brain is organized, how sense organs handle incoming messages, how muscles react by contracting, how the brain processes and stores messages, and how mental processes in both health and disease derive from activity at the chemical and cellular levels.

THE NERVOUS SYSTEM CARRIES MESSAGES THROUGHOUT THE BODY

Central to the science of neurobiology is the theory that all functions and dysfunctions of the nervous system, including mental activities and mental illnesses, result from the actions and interactions of cells and chemicals of the nervous system. The diseases described in this chapter all involve malfunctions related to chemicals in the brain; several are also characterized by the degeneration of particular groups of cells.

In 1817, Dr. James Parkinson first described a disease marked by muscle tremors, including a distinctive 'pill-rolling' movement of the thumb and forefinger. The disease is also characterized by a walking gait in which the body above the waist leans forward while the feet shuffle slowly in small steps and are barely lifted from the ground. People with Parkinsonism have difficulty in initiating voluntary movements. This difficulty is not a true paralysis because, by stopping other activities and concentrating on their voluntary movements, Parkinson patients can temporarily improve their performance.

For voluntary movements of any kind to take place, the brain must initiate and send messages to the muscles. Parkinsonism presents some of the clearest evidence of a malfunction in this message-sending system. However, before we can understand this malfunction in greater detail, we must first describe the cells of the nervous system and how the sending of messages usually works.

The nervous system and neurons

Messages flow in and out of the **central nervous system**, which includes the **brain** and **spinal cord**. The rest of the nervous system is called the **peripheral nervous system**, which includes both the **peripheral nerves** and the special sense organs such as the eyes and ears (Figure 10.1).

The functions of the nervous system, including the brain, are carried out largely by nerve cells, also called **neurons**, cells that carry **nerve impulses** along their membrane surfaces. In this chapter we focus on humans, but nearly all animals (and only animals) have neurons arranged in some type of nervous system. Each neuron contains a cell body that includes a nucleus and surrounding cytoplasm. Extending out from this cell body are the branches called **dendrites**, which conduct nerve impulses toward the cell body. Each neuron also has another extension, called an **axon**, that conducts impulses away from the cell body. Nerve impulses pass from neuron to neuron (or from neuron to muscle) across a gap known as a **synapse**. **Sensory neurons** are oriented so that their axons conduct impulses from a sense organ inward toward the central nervous system, whereas the axons of **motor neurons** conduct impulses out from the central nervous system to an **effector organ**, which is usually either a muscle or a gland (Figure 10.2A).

FIGURE 10.1

Major divisions of the nervous system.

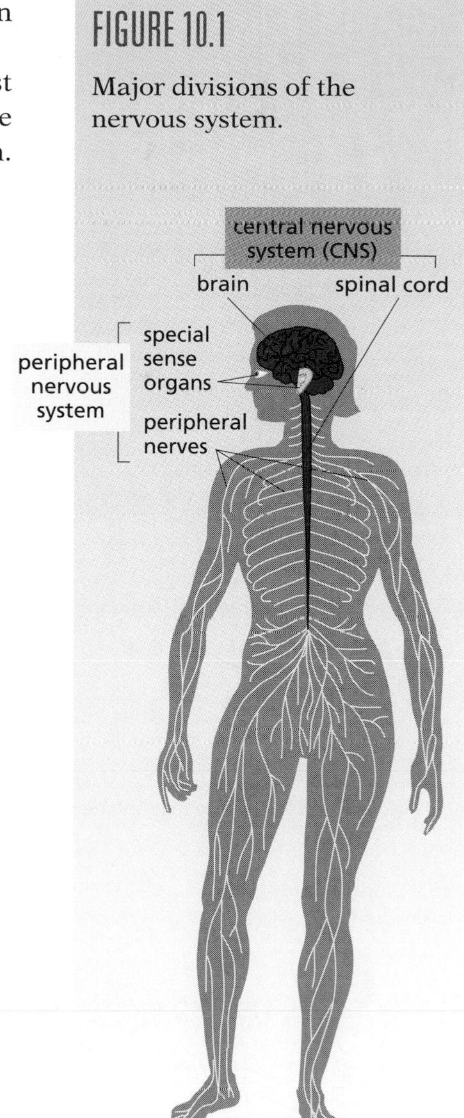

Many, but not all axons, are surrounded by a series of special cells whose rolled-up plasma membranes form a structure called the **myelin sheath** (Figure 10.2B and C). The spaces between the rolled-up myelin sheath cells are called the **nodes of Ranvier**, which make a myelinated axon look like a string of sausages. The myelin sheath acts as an insulator that prevents nerve impulses from spreading sideways from one axon to another; it keeps the impulses traveling along the length of the axon. The disease called multiple sclerosis (MS) results from destruction of the myelin sheath by cells of the body's own immune system. Myelin disruption produces disordered transmission of nerve impulses, resulting in

FIGURE 10.2

The neuron and its myelin sheath.

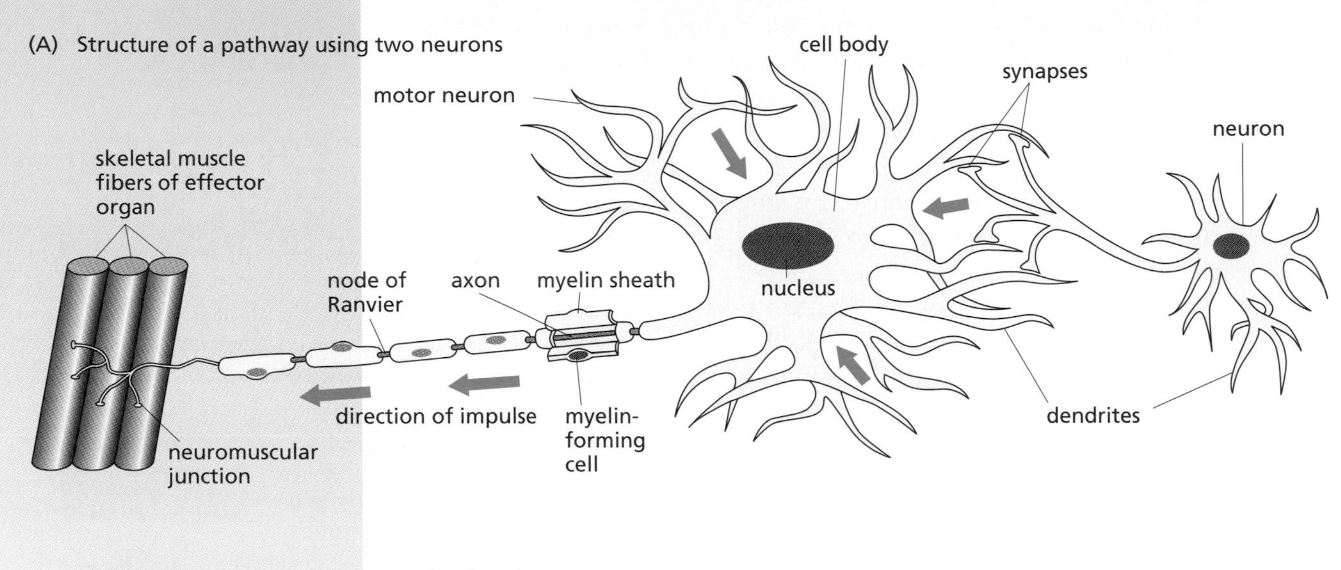

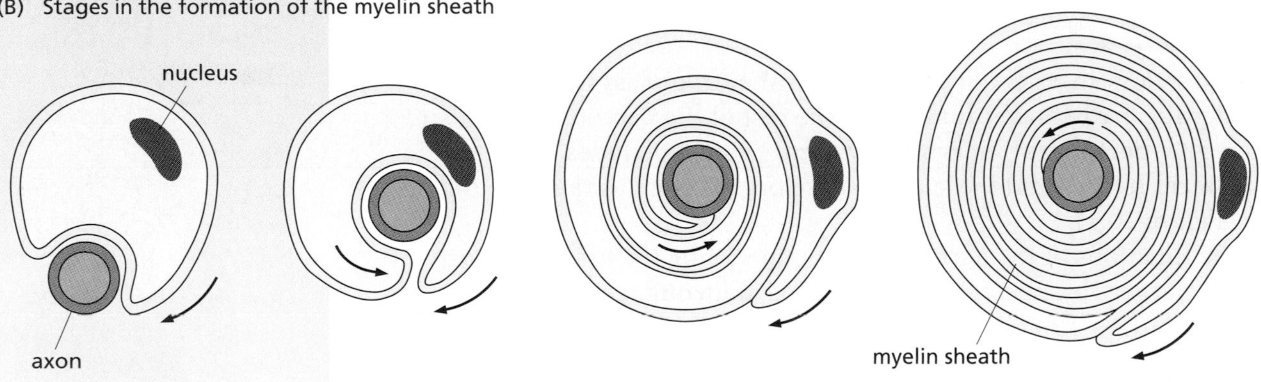

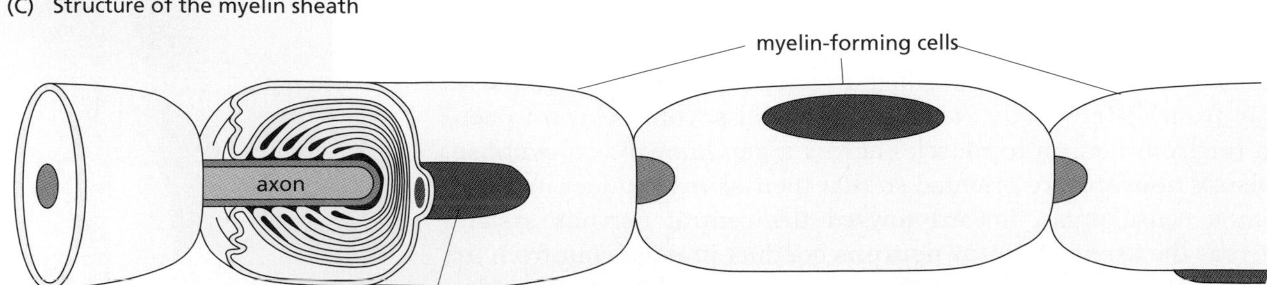

weakness or trembling of arms or legs and in hazy or double vision, among other symptoms.

Aggregations of neurons or their parts have distinctive names in the nervous system. Bundles of axons are called **nerves** throughout the peripheral nervous system and are called **tracts** within the brain and spinal cord. Clumps of cell bodies are called **ganglia** throughout the peripheral nervous system and **nuclei** within the central nervous system, except for one group of brain structures called the basal ganglia.

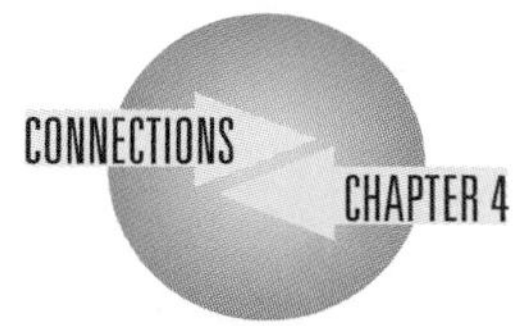

In addition to neurons, the nervous system contains other types of cells called **neuroglia**. Neuroglia in the brain have cellular extensions that wrap around the neurons in the brain and other cellular extensions that wrap around small blood vessels on the surface of the brain. Through these extensions, the neuroglia carry nourishment from the small blood vessels to the neurons. The neuroglia also provide structural support for the brain tissue.

Nerve impulses: how messages travel along neurons

Much of what we know about nerve impulses was learned from studying the giant axons of the squid (Figure 10.3A), a member of a group called the Cephalopoda (which includes the octopus and chambered nautilus) which is a class of the phylum Mollusca (see Figure 4.7, p. 136). The animals of this class need to react both quickly and forcefully to stimuli that could signal danger; they all have giant axons as part of their quick-response system. These axons are many times the diameter of typical axons in humans and so are ideal for studying nerve impulses, as we will see below.

FIGURE 10.3

Resting potentials in the giant axons of the squid.

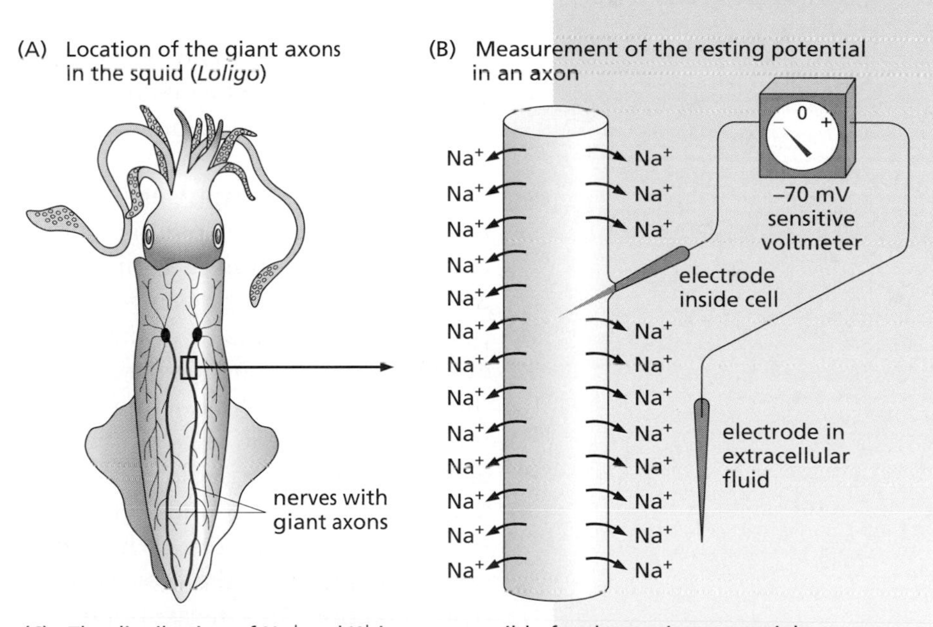

Electrical potentials. One way to study nerve impulses within a single neuron is by using a very sensitive voltmeter attached to needlelike probes (electrodes) that conduct electricity. Voltmeters read differences in the amount of electric charge in contact with its two electrodes. A giant axon has a large enough diameter that one of the electrodes can be placed inside the axon and the other electrode outside. When placed this way, voltmeter readings show a difference in charge between the inside and outside of the axon, a difference called an **electrical potential**, measured in units called millivolts (mV). Unlike electrical potentials in nonliving systems (such as electrical wiring), which are due to electrons, electrical potentials in living systems are due to differences in the concentration of charged particles called ions. Electrical potentials in the nervous system are produced by **sodium–potassium pumps**, a series of membrane proteins engaged in the active transport of sodium ions (Na^+) and potassium ions (K^+) in opposite directions across the cell membrane. (Active transport and ion gradients are described in Chapter 8.) For every two potassium ions transported into the cell, three sodium ions are transported across the membrane to the outside. Because each sodium ion and each potassium ion carry a single positive charge, this also moves three positive charges out of the cell for every two that are moved in. Active transport thus produces two chemical gradients (a buildup of sodium outside the cell and a buildup of potassium inside) and an electrical potential (more positive charges outside the cell than inside). The excess of positive charge outside the cell creates an electrical potential of about –70 mV across the cell membrane of the axon, meaning that the inside of the cell is negatively charged in comparison with the outside. (The magnitude of this potential is different in different types of neurons.) When a neuron is not conducting an impulse, the ions are unable to cross through the cell membrane, so the electrical potential across the cell membrane remains fairly constant and is called a **resting potential** (Figure 10.3B and C).

Measuring nerve impulses. We say that a membrane is **polarized** when there is an electrical potential across it. The nerve impulse is a wave of depolarization traveling along the cell membrane. The easiest way to demonstrate this is by using a voltmeter with electrodes applied to two points along the outside of the axon. The voltmeter compares the charge at the second site outside the axon with the charge at the first site. When the neuron is not conducting an impulse, the sodium concentration is the same at the two sites so the two electrodes detect the same charge, and the voltmeter needle reads zero. When the sodium concentration, and thus the charge, is less at the first electrode than at the second, the voltmeter reads negative. When the sodium concentration, and thus the charge, is higher at the first electrode than at the second, the voltmeter reads positive. As a nerve impulse passes, the needle deflects first one way and then the other, as shown in Figure 10.4.

Along the length of an axon, the nerve impulse travels at a relatively rapid rate that varies with the diameter of the axon and with the presence or absence of a myelin sheath. Unmyelinated neurons generally have small diameters, between 0.3 and 1.3 micrometers (1 μm = 10^{-6} m) and conduction velocities of 0.5 to 2.3 meters per second (abbreviated m/sec). Myelinated neurons vary from 3 to 20 μm in diameter; conduction velocities vary from 3 m/sec in the smallest fibers to 120 m/sec in the largest. The giant squid axons, so named because of their large diameters, have a very high conduction velocity.

FIGURE 10.4

Detection of a nerve impulse by using a sensitive voltmeter.

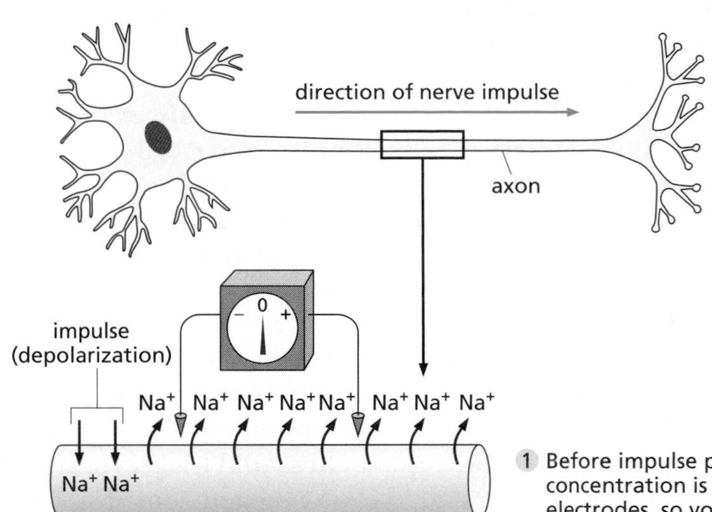

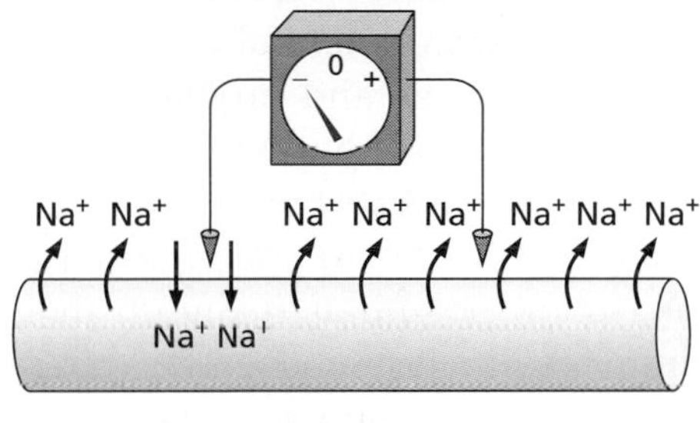

1 Before impulse passes, sodium ion concentration is equal at both electrodes, so voltmeter reads zero.

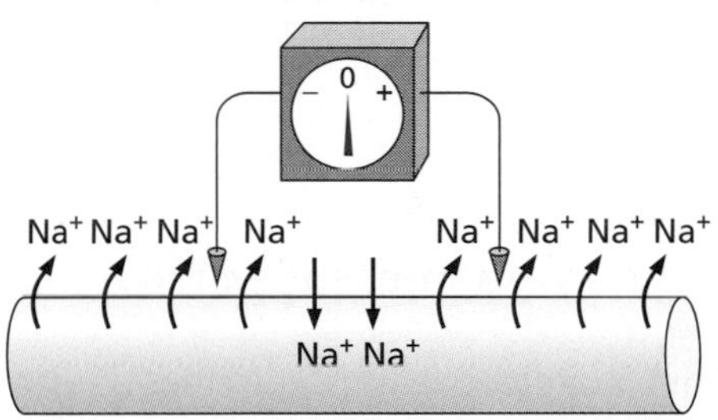

2 Impulse causes drop in sodium ion concentration at first electrode, so needle moves to minus.

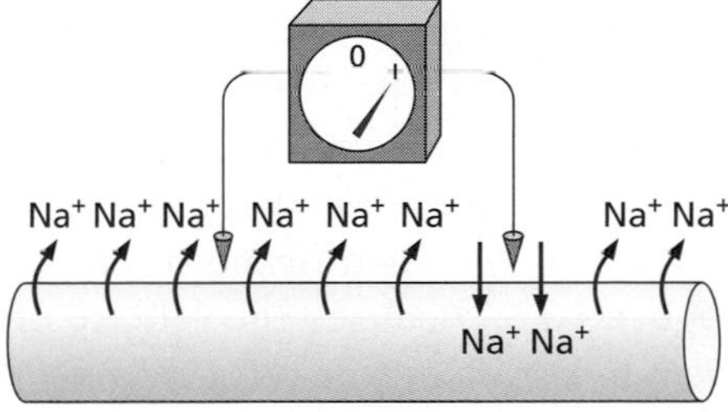

3 Sodium ion concentration recovers at first electrode, so voltmeter again reads zero.

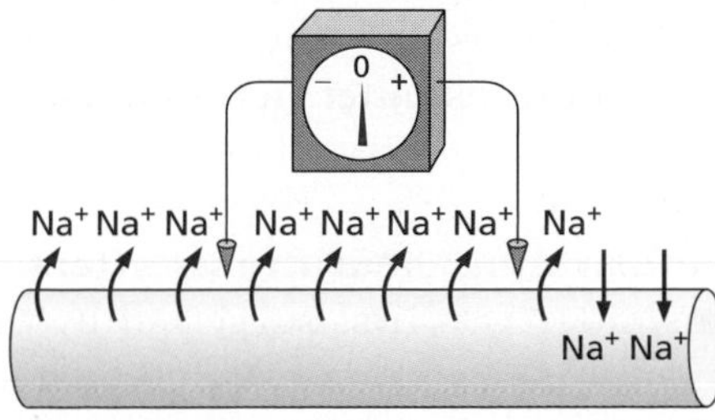

4 Impulse arrives at second electrode; drop in sodium ion concentration causes needle to read plus.

5 Impulse has passed; sodium ion concentration is again equal at both electrodes, so voltmeter again reads zero.

Neurons are stimulated to depolarize in a small portion of the axon; however, not every localized depolarization leads to a nerve impulse that travels the length of the axon. Depolarizations that are strong enough to exceed a threshold will trigger sustained impulses called action potentials.

Action potentials. Neurons are **electrically excitable**; that is, they can be stimulated electrically. When a nerve cell membrane is sufficiently stimulated, channels through the membrane open and let sodium ions flow back in, locally depolarizing the membrane for a short time. Changes in voltage over time can be measured with electrodes placed as in Figure 10.3 and recorded as graphs. A slight **depolarization** of the membrane (a voltage change toward zero) or a **hyperpolarization** (an increase in the voltage difference) is transmitted a very short distance to adjacent portions of the cell membrane; these effects decrease rapidly with distance and may thus disappear quickly. However, a depolarization that reaches or exceeds some **threshold** triggers nearby membrane channels to open. The rapid inflow of sodium ions (Na^+) results in a characteristic type of electrical discharge known as an **action potential**, or 'spike' (Figure 10.5). The action potential causes the adjacent portion of the cell membrane to depolarize and produce another action potential equal in strength to the first. In this way, the action potential rapidly spreads along the entire cell membrane with no reduction in its size or intensity. It is this propagation of an action potential that we recognize as a nerve impulse. Action potentials maintain their direction of travel because once sodium channels are opened, they are prevented from reopening for a short period, during which the outflow of potassium ions (K^+) repolarizes the membrane and the sodium–potassium pumps reestablish the resting potential.

Neurotransmitters: how messages travel between neurons

Even though neurons are electrically excitable, the transfer of information from cell to cell (across a synapse) is generally chemical, not electrical. Any chemical substance that can stimulate or inhibit an action potential is called a **neurotransmitter**. The vast majority of synapses between nerve cells are chemical synapses, in which a chemical neurotransmitter crosses the narrow space separating two adjacent cells. In a chemical synapse, depolarization of the membrane in the presynaptic or transmitting neuron (Figure 10.6A) causes the release of a neurotransmitter (Figure 10.6B). The neurotransmitter released into the synapse binds to receptors on the next cell (the postsynaptic cell), opening channels in its membrane through which ions flow, thus altering its electrical potential (Figure 10.6C). The neurotransmitter never enters the postsynaptic cell (Figure 10.6D).

Experiments demonstrating neurotransmitters. The concept of a chemical neurotransmitter was first suggested by the British physiologist Henry H. Dale but was first demonstrated by his German-American

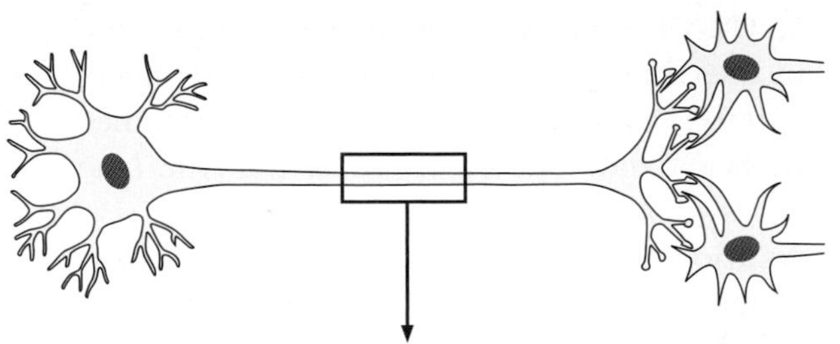

time

(A) Resting potential when no impulse is present

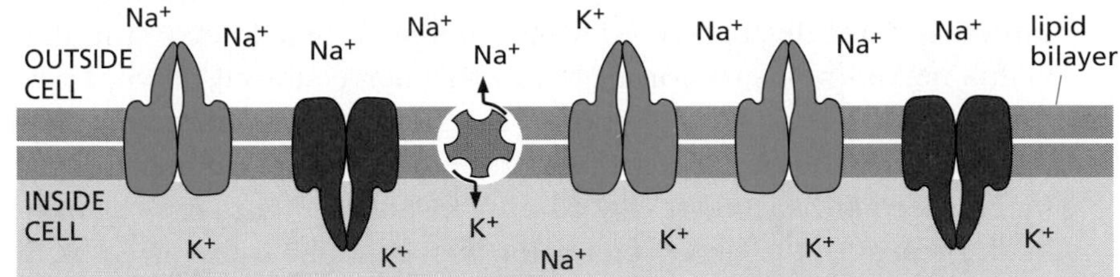

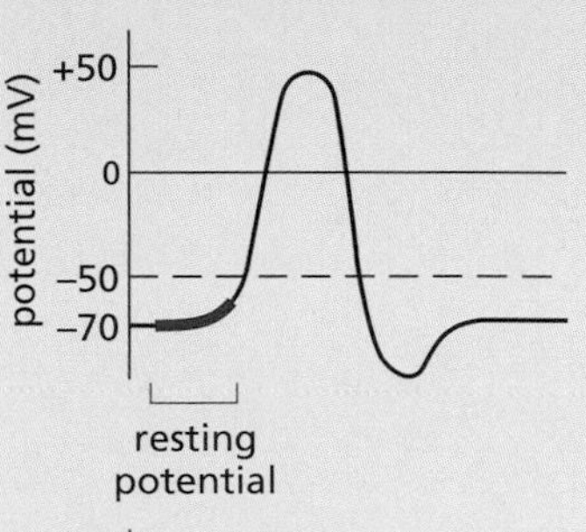

(B) Stimulus causes a few channels to open; sodium ions leak in, depolarizing the membrane

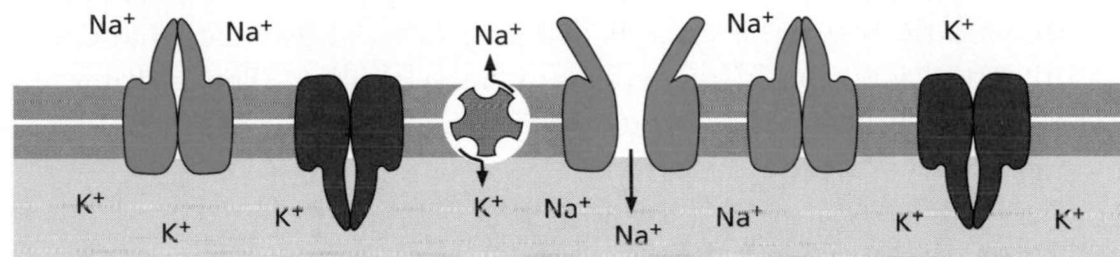

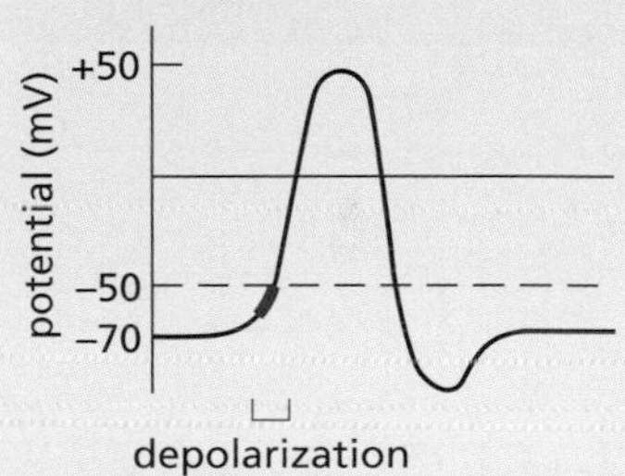

(C) At −50 mV, all sodium channels open; sodium rushes in, causing charge to reverse and form an action potential

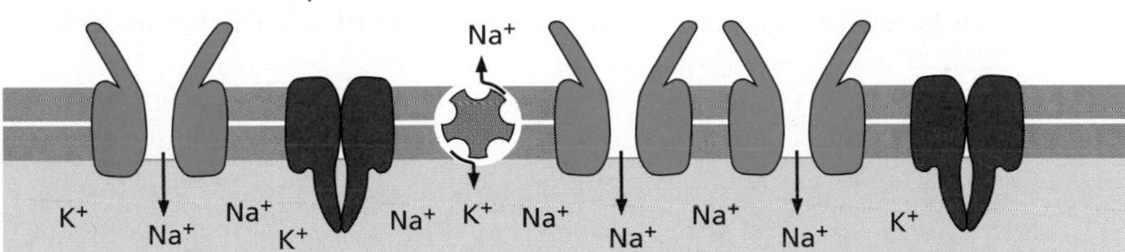

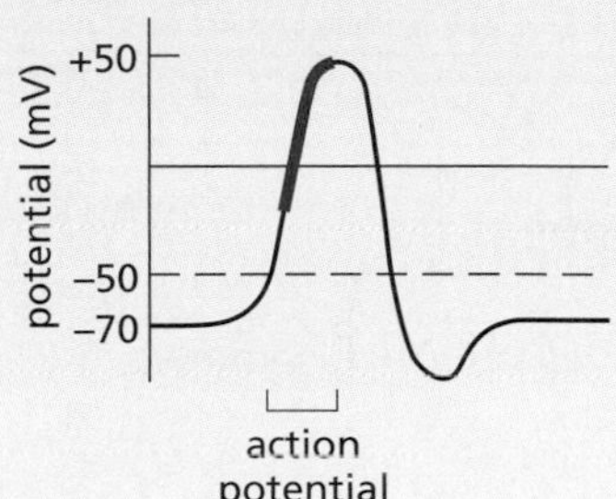

(D) Sodium channels close as potassium channels open; potassium ions rush out, repolarizing the membrane

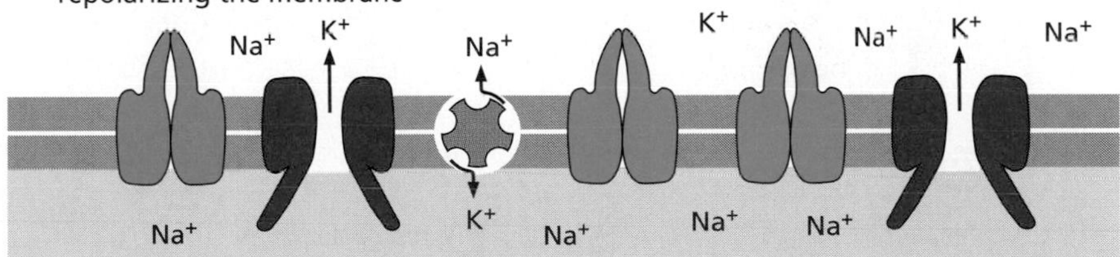

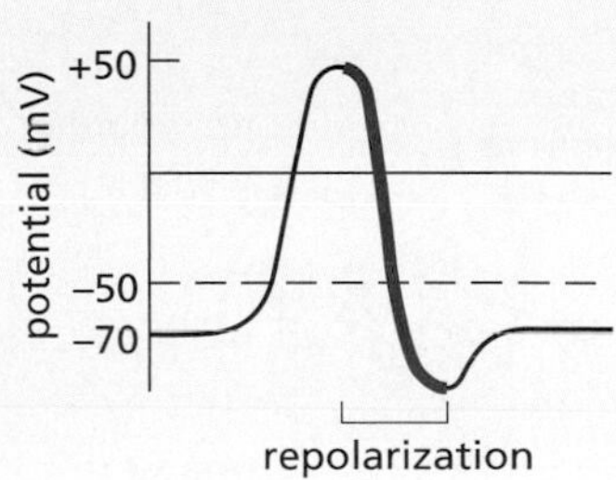

(E) Closing of potassium channels restores resting potential; sodium–potassium pump redistributes ions

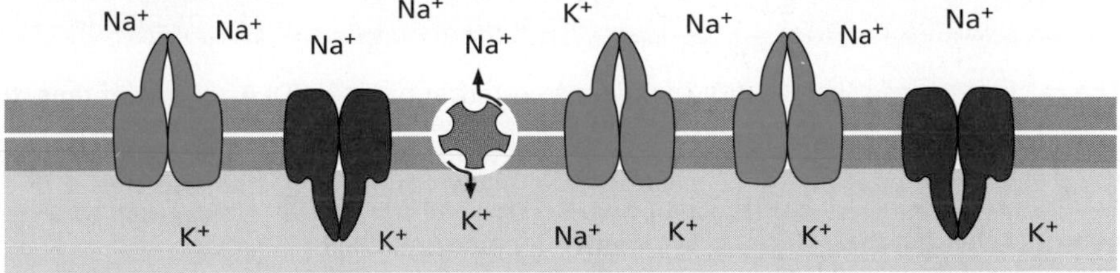

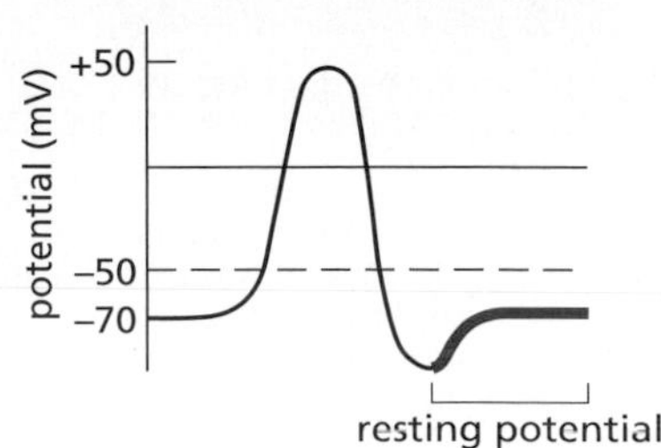

FIGURE 10.5

Generation of an action potential in a nerve cell.

colleague Otto Loewi, with whom Dale shared the Nobel Prize in 1936. Dale and Loewi already knew that the vagus nerve, running down from the brain into the abdominal cavity, sends branches to the heart that slow down the heartbeat. They also knew that the heart of a frog separated from the rest of the animal and maintained in a physiological salt solution (a solution containing various ions at concentrations close to those in the intact organism) would keep beating for hours. Loewi dissected out the beating hearts of two frogs and placed them in salt solutions. One of his preparations contained nothing but the heart, but the other contained a carefully preserved vagus nerve. When Loewi stimulated the vagus nerve, the heart connected to this nerve slowed down. Loewi then used a dropper to take some of the fluid surrounding this heart and transfer it to the other heart, which no longer had any nerves leading to it. The second heart also slowed down, showing that some chemical had been present in the first preparation that could be transferred by dropper to the second, for this was the only connection between the two. The chemical was then isolated and identified as **acetylcholine**, the first neurotransmitter to be studied experimentally. Loewi's preparation also allowed the testing of various drugs or other substances that could block or otherwise modify the effect of acetylcholine. These drugs, in turn, could often be used to study whether a particular synapse used acetylcholine as a neurotransmitter.

Types of neurotransmitters. A large number of chemicals are now known to be able to function as neurotransmitters; some of these are listed in Table 10.1, grouped by their chemical structure. Most neurons in the peripheral nervous system use the neurotransmitter acetylcholine; a

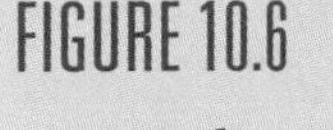

FIGURE 10.6

How a chemical synapse works.

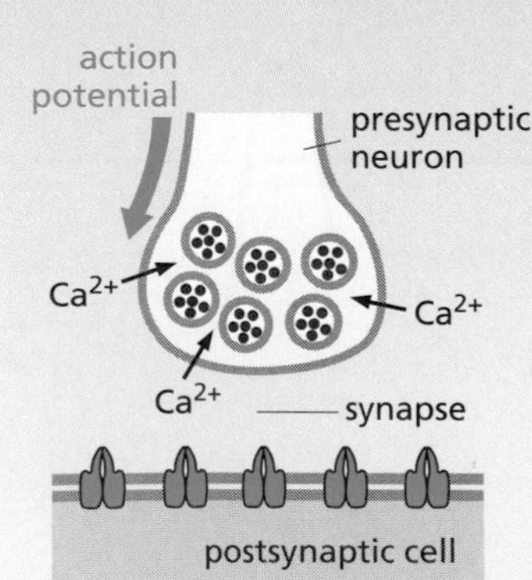

(A) Depolarization by an action potential triggers entry of calcium ions into the cytoplasm of the presynaptic neuron.

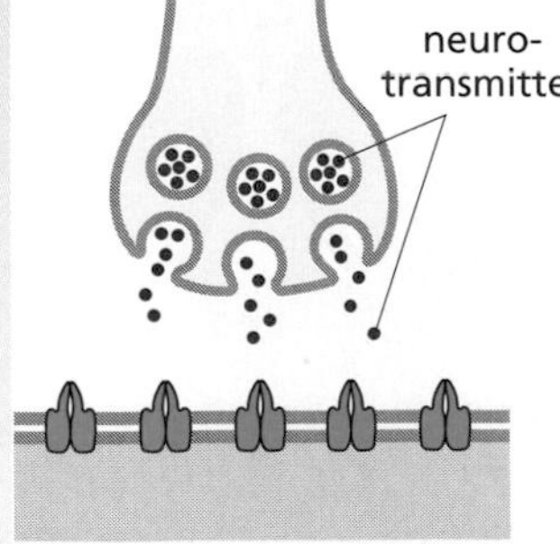

(B) The calcium triggers the cytoplasmic vesicles containing neurotransmitters to fuse with the plasma membrane of the presynaptic neuron, releasing the neurotransmitters into the synapse.

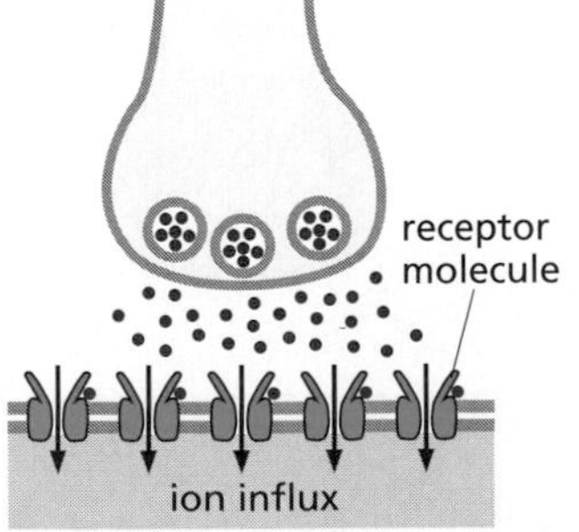

(C) The neurotransmitter binds to receptor molecules on the postsynaptic cell, opening channels that allow positive ions to enter that cell's cytoplasm, depolarizing its membrane.

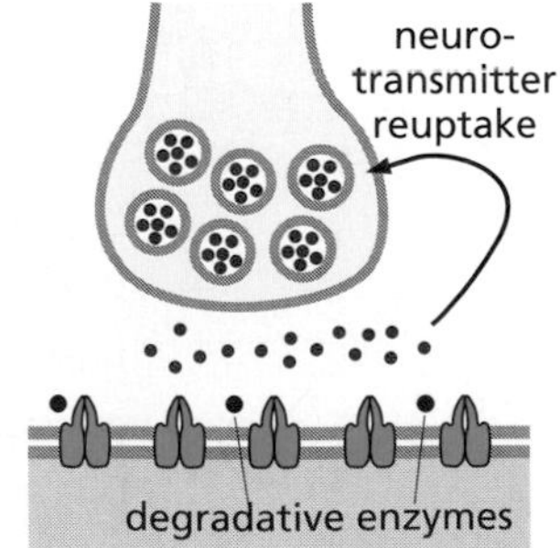

(D) After a brief time, the neurotransmitter is either taken back into the cell that secreted it, or else it is degraded by an enzyme.

few use norepinephrine. But neurons in the brain use many different neurotransmitters, including all the ones in Table 10.1. Each particular neuron secretes one type of neurotransmitter primarily or exclusively. Each cell that responds to a neurotransmitter has a specific receptor for that neurotransmitter.

Removal of neurotransmitters from the synapse. After a neurotransmitter evokes its response, there are mechanisms that stop further transmission by removing loose (unbound) neurotransmitter molecules from the synapse (see Figure 10.6D). Many neurotransmitters are reabsorbed by the cell that secreted them, permitting the same molecules to be recycled and reused. This process, called **reuptake**, is typical of many synapses in the brain. Also, several neurotransmitters can be chemically degraded by enzymes. The amine neurotransmitters can be degraded by the enzyme MAO (monoamine oxidase). Another enzyme, cholinesterase, breaks down acetylcholine molecules after they have stimulated postsynaptic cells outside the brain. Any process interfering with the chemical breakdown or reuptake of neurotransmitters lets the neurotransmitter stay in the synapse and thus excessively stimulate the postsynaptic cell; likewise, any enhancement of chemical breakdown or reuptake may decrease neurotransmission. As we shall see, changes in neurotransmission can result in disease.

TABLE 10.1

Neurotransmitters.

AMINE NEUROTRANSMITTERS
Epinephrine
Norepinephrine
Dopamine
Serotonin
Histamine
AMINO ACID NEUROTRANSMITTERS
Aspartic acid (aspartate)
Glutamic acid (glutamate)
Gamma-aminobutyric acid (GABA)
Glycine
PROTEIN OR PEPTIDE NEUROTRANSMITTERS (NEUROPEPTIDES)
Somatostatin
α-endorphin
β-endorphin
Leu-enkephalin
Met-enkephalin
Substance P
GAS NEUROTRANSMITTER
Nitric oxide (NO)
ESTER NEUROTRANSMITTER
Acetylcholine

Dopamine pathways in the brain: Parkinsonism and Huntington's disease

One of the amine neurotransmitters (see Table 10.1) is **dopamine**. Dopamine transmits messages across synapses between neurons in the brain. Some of the neurons that are stimulated by dopamine act, in turn, on peripheral neurons that stimulate voluntary muscle cells. Dopamine does not act directly on muscle cells; rather it acts within the brain to smooth and coordinate signals to the muscles. Two diseases have helped us to understand what can happen when dopamine secretion is out of balance. In Parkinsonism there is too little dopamine and in Huntington's disease there is too much.

Parkinsonism: too little dopamine. When autopsies are performed on the brains of Parkinson patients, a consistent finding is the degeneration of a bundle of darkly pigmented neurons called the **substantia nigra**. Experimental staining shows that these neurons secrete the neurotransmitter dopamine. Among the cells stimulated by this dopamine are those of the basal ganglia. The degeneration of the substantia nigra in Parkinson patients thus deprives the basal ganglia of dopamine. The basal ganglia stimulate and coordinate muscle movements by acting on the acetylcholine-secreting neurons that trigger muscle contraction. If sufficient dopamine is not present, the cells of the basal ganglia do not function normally, and the person suffers from muscle tremors and has difficulty walking.

Evidence for the hypothesis that dopamine underproduction has a large role in Parkinsonism comes from the effectiveness of the drug L-DOPA (levodopa) in temporarily alleviating many of the symptoms of Parkinsonism. Because DOPA is a precursor of dopamine, it is hypothesized that supplying L-DOPA (a synthetic form of DOPA) can increase dopamine production and thus relieve the symptoms of dopamine deficiency. Unfortunately, this form of treatment increases dopamine production everywhere, not just to the portion of the brain that needs it. The excess of dopamine in other places may cause serious side effects, such as schizophrenialike symptoms.

One promising form of therapy that has already been tested is the implantation of fetal tissue into the brains of Parkinson patients. The hypothesis is that the fetal tissue will grow and replace the missing or damaged cells. Adult tissue is unsuitable for this purpose because the brain loses much of its capacity to regenerate new neurons at an early age; even tissue from newborn babies or infants has much less regenerative capacity than fetal tissue. This is one reason why many people in the medical community welcomed the lifting in 1993 of an earlier ban on the use of fetal tissues in medical therapy and in medical research. However, the use of fetal tissue raises objections from opponents of abortion because the tissue is obtained in most cases from aborted fetuses.

Huntington's disease: too much dopamine. If Parkinsonism is a disease in which voluntary movements are made difficult by a lack of dopamine, then too much dopamine might cause excessive movements. This is precisely what happens in **Huntington's disease**, a degenerative neurological disease whose genetic basis was discussed in Chapter 3 (p. 87). Huntington's disease is marked by uncontrollable spasms or twitches of many muscles, beginning between ages 40 and 50. As the disease progresses, the spasms become more pronounced, and the patient gradually loses control of all motor functions and of mental processes. A slow death occurs within a few years of the disease's onset. Autopsies of Huntington patients reveal that some of the brain cells in the basal ganglia have been destroyed. Because these cells normally inhibit the production and release of dopamine, one major effect of their destruction (and one sign of the disease) is an overproduction of dopamine. Drugs known to inhibit the action of dopamine will temporarily reduce the symptoms of Huntington's disease.

Feedback systems. How does the destruction of some of the cells in the basal ganglia result in an overproduction of dopamine? One answer may lie in another neurotransmitter, gamma-aminobutyric acid (GABA). GABA functions in many neuronal pathways by inhibiting a neuron that has fired from firing again unless another stimulus, larger than the first, is received. Because the GABA-secreting neuron inhibits a neuron at an earlier point in the pathway, the information is 'feeding back' and is thus described as a **feedback system**. Feedback occurs in many biological systems whenever a later step regulates an earlier step in any process. Feedback systems function to keep certain variables such as body temperature within narrow limits, turning metabolic processes higher when the body is cold or taking measures (such as sweating) to dissipate heat when the body is warm. We saw in Chapter 6 (p. 250) that levels of certain hormones are regulated by feedback from other hormones. In the nervous system, feedback can often be seen literally as neurons that feed information back to the earlier neurons that stimulated them.

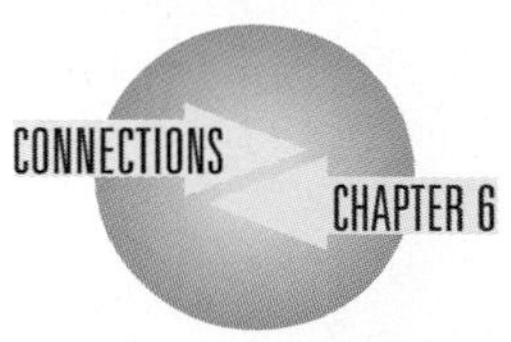

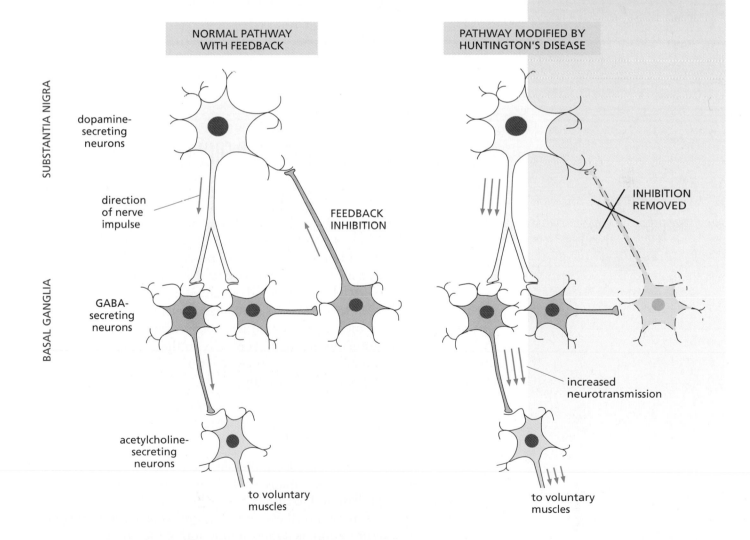

GABA normally has this type of feedback effect on the dopamine-secreting pathways of the substantia nigra. Because it acts to inhibit dopamine, it is an example of feedback inhibition. The destruction of the GABA-secreting neurons in Huntington's disease removes this inhibition. Because the dopamine-secreting neurons are then no longer inhibited, they begin to overproduce dopamine (Figure 10.7). Too much dopamine leads to overstimulation of other GABA-secreting neurons, ones that trigger the acetylcholine-secreting neurons, overproducing muscle contractions.

FIGURE 10.7

Feedback inhibition usually prevents dopamine overproduction, but impairment of the feedback pathway in Huntington's disease allows dopamine overproduction.

THOUGHT QUESTIONS

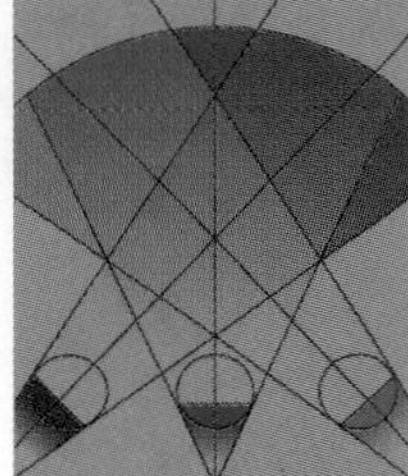

1. To test the clinical benefits of a technique such as the surgical implantation of fetal cells into the brains of Parkinson patients, a double-blind comparison needs to be made with a group of control subjects. The control subjects in such a study are generally given sham surgery, meaning that a hole is drilled in their skull and a probe inserted into their brains, but no fetal cells are introduced. Because of a strong placebo effect, patients receiving this treatment often improve at least temporarily. Do you think it is ethical to conduct tests in this way? Do you think it is ethical to adopt such a procedure without this kind of clinical testing? Construct both utilitarian and deontological arguments for your position.

2. Suppose a scientist proposes the hypothesis that, in a select group of hard-to-reach cells, a particular amino acid functions as a neurotransmitter. How would you test this hypothesis? What kinds of drugs or poisons would you look for to help you study the properties of the hypothesized neurotransmitter?

3. Explain how Loewi's experiment demonstrated that neurotransmitters are chemicals, not electric current.

4. What are the arguments for and against the use of fetal tissue for the treatment of Parkinsonism?

5. If a neuron is stimulated halfway along its axon, will an impulse travel in both directions on the axon?

MESSAGES ARE ROUTED TO AND FROM THE BRAIN

Some types of neuron messages are internal. As seen in the feedback system illustrated in Figure 10.7, neurons can carry information about the internal workings of the organism. They can also regulate internal processes, both within the brain (see p. 435) and in other parts of the body (see Chapter 12). Many other messages processed by the brain originate outside the organism. The nervous system is one of the major body systems through which the individual receives information about its environment. External stimuli cause specialized cells to trigger depolarizations (changes in membrane potentials) that travel to other neurons by means of neurotransmitters, and eventually to the brain, where huge numbers of incoming messages are processed. Some, but not all, of these messages become conscious perceptions.

Message input: sense organs

The brain receives sensory input from the outside world through a variety of sense organs, all containing many specialized sensory cells. Different types of sensory cells respond to different types of stimuli. Although sensory **reception** is a function of these sense organs, sensory **perception**, which involves interpretation, is largely a function of the brain.

The skin: reception of touch, temperature and pain. The main functions of skin, our body's largest organ, are protective: it protects us from temperature variations, harmful chemicals, water loss, and pathogens that might cause disease. Our skin also protects us by sending messages to the rest of the body from its various sensory nerve endings. Five types of specialized sensory endings are known, corresponding to five different skin senses: light touch, deep pressure, warmth, cold, and pain (Figure 10.8). These endings are microscopic in size, much too small to be stimulated individually in an experiment. So how do we know which type of ending serves which function? Our knowledge is largely based on the spatial distribution of the endings over the body's surface. For example, the places most sensitive to touch (such as the lips and the fingertips) have the highest densities of the endings thought to sense light touch, while the places most sensitive to cold (such as the cornea of the eye or the tip of the penis) have the highest densities of the nerve endings thought to sense cold.

Several of these nerve endings can report anomalous sensations when they are overstimulated. For example, temperatures a few degrees

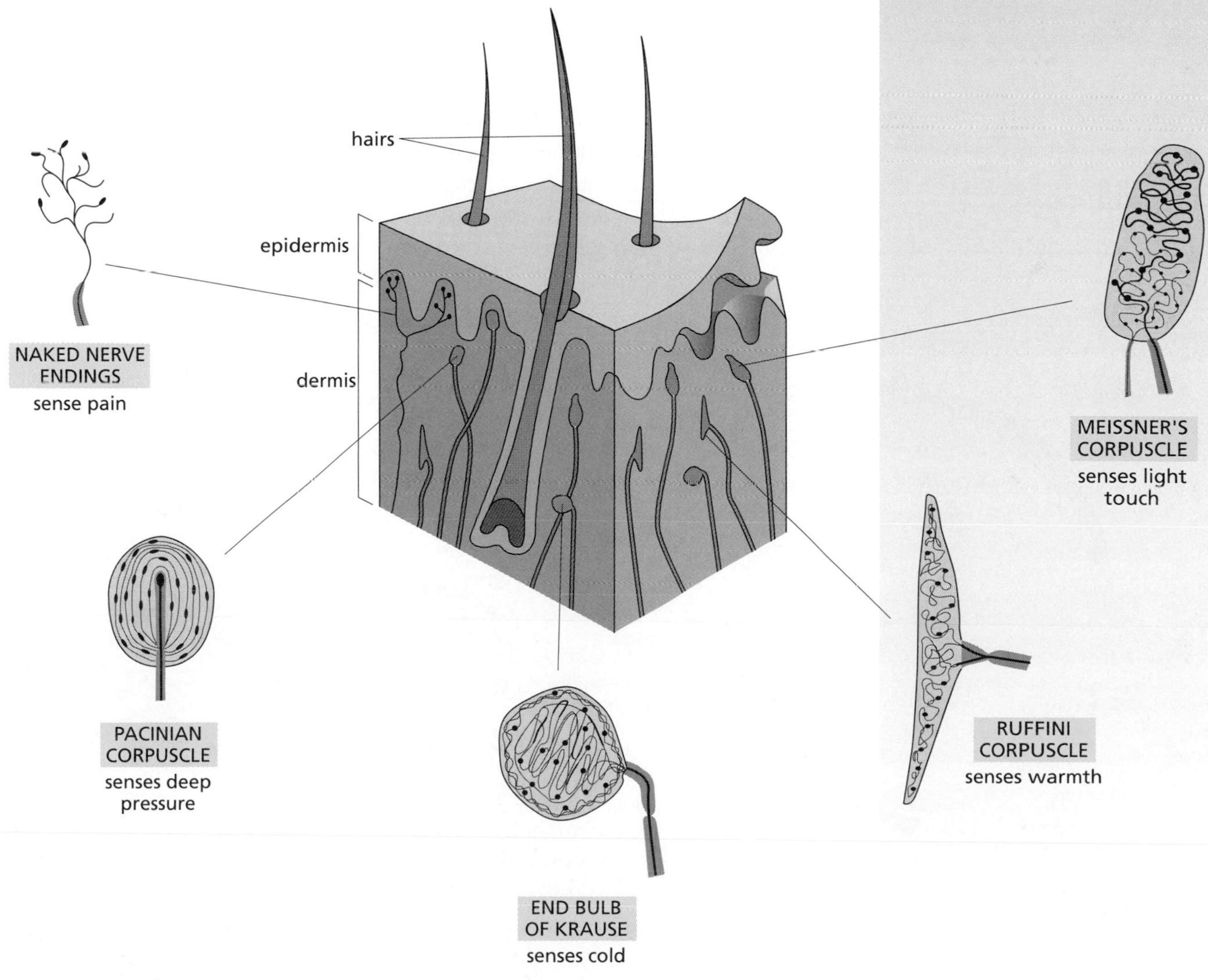

FIGURE 10.8

Nerve endings in the skin.

lower than the surroundings are perceived as cold, but much colder temperatures may be perceived as heat or burning, and temperatures still lower are simply perceived as pain. In fact, a sufficiently large stimulus of any sort, even a sound much too loud or a light much too bright, is perceived by the brain as pain. It is perhaps for this reason that the nerve endings that are specifically pain receptors are naked nerve endings; nearly any stimulus to these naked endings is likely to overstimulate them and be perceived as pain. At the other extreme are the onionlike layers that surround the pressure receptors, ensuring that the nerve ending within can only be stimulated to produce changes in membrane potential if there is enough pressure to deform the shape of these layered capsules deep inside the skin. The altered membrane potential produced in sensory cells is called a **generator potential**. Unlike an action potential, which operates on an all-or-none principle, the magnitude of a generator potential depends on the strength of the stimulus.

Our sense of touch requires direct pressure on the skin. Fish have an additional sense organ for touch called the **lateral line system** that is sensitive to very small changes in water pressure caused by nearby obstacles or by the swimming movements of other fish.

The eye: reception of light. The organ of vision in all vertebrates is the eye (Figure 10.9A), which is surrounded by two protective layers, the choroid and the sclera. Light reaches the interior of the eye through a transparent front layer, the cornea. The light is concentrated by a transparent lens, which changes shape to bring into focus objects at varying distances from the eye. A suspensory ligament holds the lens in place, while a series of smooth muscles change the shape of the lens. The iris diaphragm, which gives each of us our individual eye color, controls the

FIGURE 10.9

The human eye.

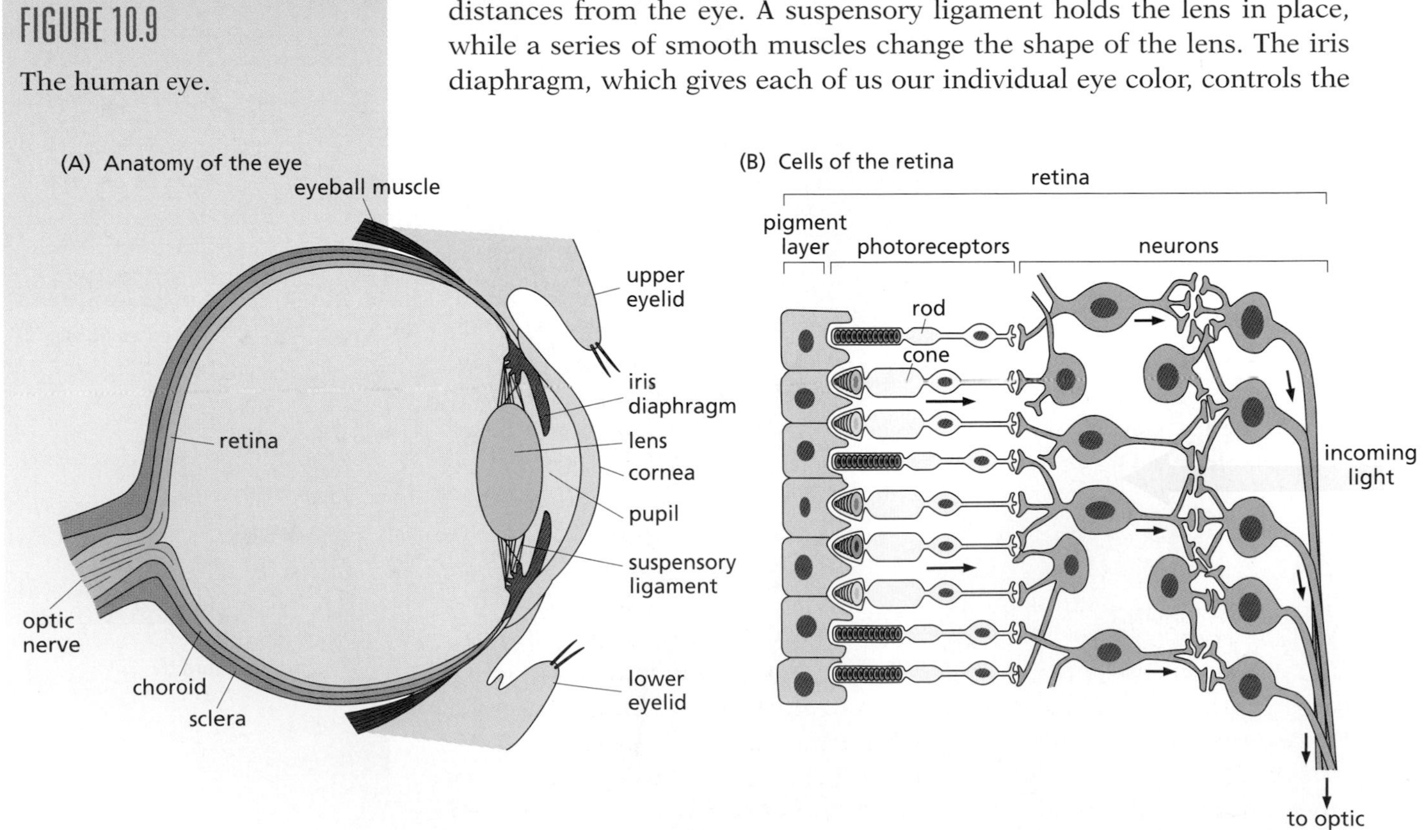

amount of light that enters the lens through the pupil, an opening that enlarges in response to dim light and becomes smaller in response to bright light.

The lens focuses light on the retina, the sensory sheet along the rear of the eye that responds to light. Because the retina is too far from the lens in a nearsighted person, the lens of a nearsighted person is not able to focus light on the retina. In a farsighted person, the opposite is true—the lens is too close to the retina. While the person is young the lens itself can change shape enough to accommodate for this problem, focusing the light onto the retina. As the person ages and the lens is less responsive, however, this accommodation is no longer possible. For both nearsighted and farsighted people, glasses or contact lenses bring the focal point of the light to the retina.

The retina contains photoreceptor (light-receiving) cells called rods and cones, and light that falls on the retina causes changes in the chemical structure of pigment molecules in these cells. These pigment changes alter the membrane potential in the rods and cones, inducing impulses. The impulses travel along the retina to the optic nerve, which carries the information to the areas of the brain responsible for vision (Figure 10.9B). Rod cells respond in dim light. Cone cells require brighter light to trigger impulses, and different ones respond to different wavelengths of light. The brain receives messages from cones sensitive to different wavelengths and processes these messages, creating our perception of color. Insect eyes are sensitive to a wider range of wavelengths, especially in the ultraviolet range.

The ear: reception of sound and balance. The ear is a highly elaborate structure. In humans and other mammals, it is arranged in outer, middle, and inner parts (Figure 10.10A). The outer ear, in which sound waves travel as vibrations in air, consists of an external flap, the pinna, and a tube called the ear canal. The pinna acts as a funnel to focus the sound waves into the tube leading to the tympanic membrane (eardrum), which marks the boundary between the outer ear and middle ear. The middle ear is connected to the back of the throat by the auditory (Eustachian) tube, which allows pressure to equalize on the two sides of the eardrum. If the Eustachian tube swells shut, as can happen in middle-ear infections or changes in altitude, pressure builds up inside the middle-ear preventing the vibrations of the ear bones. This can lead to partial hearing loss, which is usually a temporary loss that is alleviated when the tube opens.

Sound waves in the middle ear travel as vibrations through a series of small bones, the hammer, anvil and stirrup bones, to the oval window marking the beginning of the inner ear. The inner ear is divided into a cochlea, responsible for sensing sound, and the semicircular canals, responsible for sensing gravity and balance. Both portions of the inner ear are filled with fluid and respond to the movement of fluids. The fluid in the coiled cochlea picks up the vibrations from the bones of the middle ear. The cochlea contains a sensory membrane whose hairlike cells vibrate in response to these vibrations (Figure 10.10B and C). Very loud noises (as from industrial sources or rock concerts) can damage these

FIGURE 10.10
The human ear.

(A) Overall view

Outer
Middle
Inner
semicircular canals
oval window
sensory nerves
ear canal
cochlea
auditory (Eustachian) tube
hammer
stirrup
anvil
tympanic membrane (eardrum)
middle ear cavity
pinna

(B) Interior structure of cochlea

cochlea
sensory nerves

(C) Detail of cochlea

cochlear duct
organ of Corti

(D) Normal sensory hair cells from organ of Corti (left) and hair cells damaged by exposure to toxic chemicals or loud sound (right)

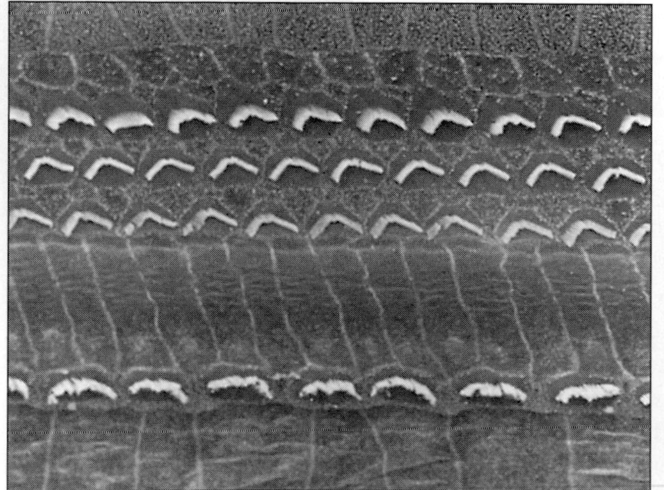

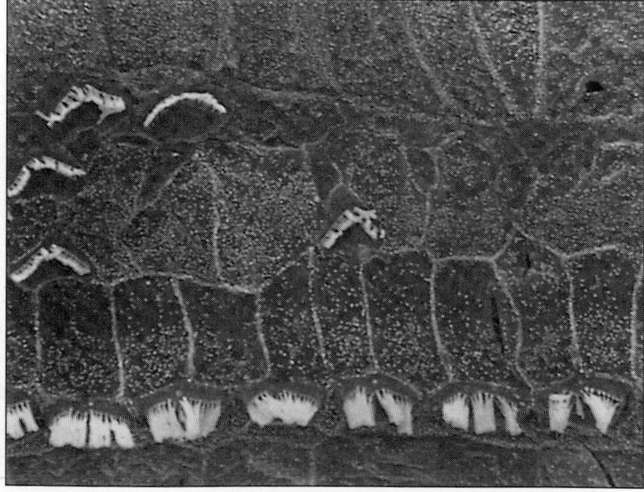

hair cells and result in a permanent loss of hearing (Figure 10.10D). Hair cells at different positions along the cochlea trigger action potentials in response to different pitches or frequencies of sound. The range of pitch detectable by different species varies greatly from that detectable by humans. Whales detect extremely low-pitched sound waves; dogs detect much higher-pitched sounds than we can, and bats higher still.

The other portion of the inner ear includes the three semicircular canals, oriented at right angles to one another. Accelerations due to gravity and to body movements result in the movement of the fluid within these canals, and this information is sensed by a patch of sensory cells near the end of each canal. People with damage to the nerve serving these patches have difficulty in standing up and maintaining balance.

The tongue: reception of chemical signals as taste. Humans sense taste through a series of taste buds (Figure 10.11A) located along the tongue (Figure 10.11B) and the roof of the mouth. There are four different basic tastes in humans, usually described as sweet, salty, sour, and bitter. The taste buds for these different tastes all look the same, but differ in their sensitivity to different chemicals: the sweetness receptors trigger action potentials in response to sugars, as do the salt receptors in response to sodium ions (Na^+), the sour receptors in response to the hydrogen ions (H^+) present in acids, and the bitter receptors in response to various other chemicals, including bitter-tasting plant compounds called alkaloids.

The nose: reception of chemical signals as smell. Although we are sensitive to thousands of different odors, the mechanism of smell is poorly understood. There is even disagreement as to the number of basic odors, with some experts naming as few as seven while others list 20 or 30. Nerve endings sensitive to smell are most abundant in the lining of the nose (Figure 10.11C), although most vertebrates also have a similar sense organ opening into the roof of the mouth. These nerve endings trigger action potentials in response to chemicals just as the taste buds do. We

FIGURE 10.11

Organs of taste and smell.

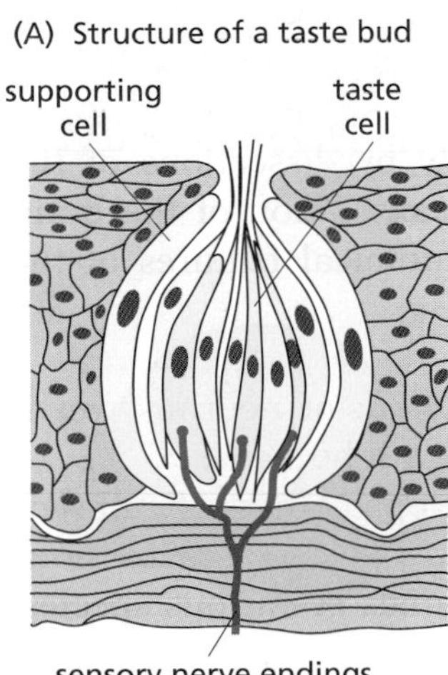

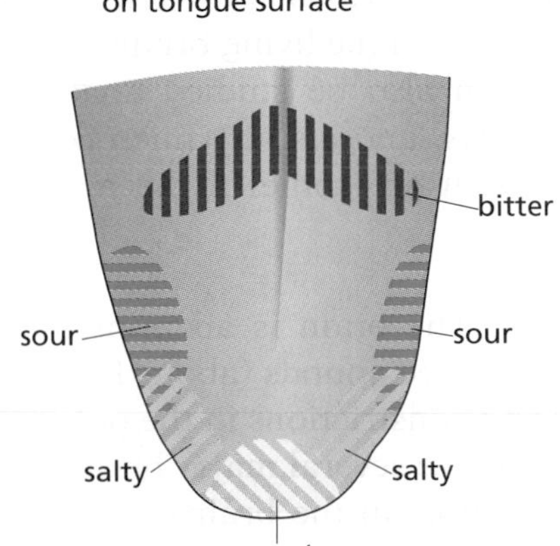

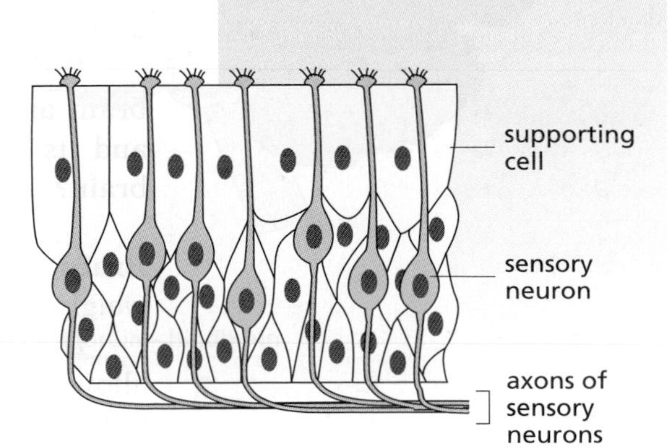

species that rely primarily on learned behavior. Other changes in proportions are related to the senses that each species uses: brain regions concerned with vision are larger in those species that rely upon vision, whereas species that rely on smell or hearing have larger brain regions devoted to those functions.

Next we look in more detail at the anatomy of the human brain (Figure 10.13).

Forebrain. The forebrain includes the cerebrum, olfactory bulbs, hippocampus, and diencephalon. In early vertebrates, the forebrain was concerned with the sense of smell, and much of it still is through the olfactory bulbs, which process nerve impulses from sensory cells in the nose. Smells can influence both hormonal secretions and emotional responses such as anger, fear, and sexual response, which are therefore also processed in the forebrain, specifically in the diencephalon. In many cases, these responses can take place on an unconscious level, meaning that we may not even be aware that the changes are taking place. The **hippocampus**, concerned with certain types of learning, is located off center in the forebrain.

The **cerebrum** is divided into two halves, the right and left **cerebral hemispheres**. Thoughts and actions originate in the cerebral hemispheres, as do most of our 'higher' functions of intellectual thought and reasoning. The cerebral hemispheres of mammals are much larger than those of other vertebrates. This is particularly true in humans and closely related primates, in which the cerebral hemispheres make up the largest part of the brain. Conscious activity and higher thought originate

FIGURE 10.13

Section through the human brain. Only structures located in the midline plane are visible in this view; off-center structures (such as the olfactory bulbs) cannot be seen.

Major regions of the brain

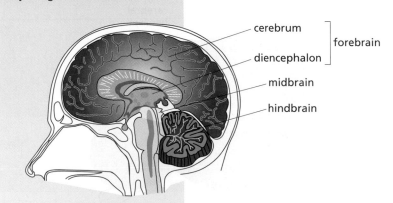

Forebrain and hindbrain structures

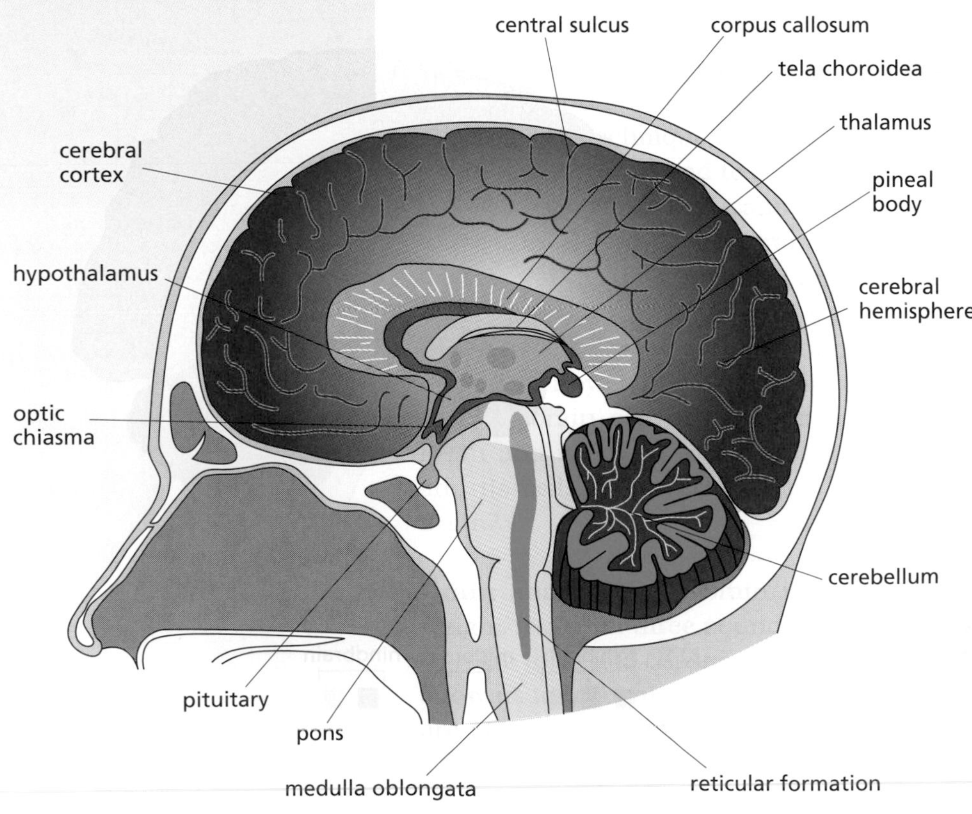

in the highly folded surface layer of the cerebral hemispheres known as the **cerebral cortex**. Below the cortex lie many series of neuron interconnections that constitute the cerebral white matter. Deeper still lie several clumps of neuron cell bodies known as basal ganglia. Several of these structures are involved in the disorders described elsewhere in this chapter.

Midbrain. The midbrain includes a **ventral tegmental area**, containing the brain's **positive reward centers** (see Chapter 11). Also located in the midbrain, but extending into part of the hindbrain, is the **reticular formation**, important in keeping us awake and alert.

Hindbrain. The hindbrain includes the **cerebellum** and the **medulla oblongata**. Functionally, the cerebellum is concerned primarily with balance, processing neuron impulses from the semicircular canals of the inner ear. The cerebellum also coordinates complex muscle movements. The medulla is concerned with such involuntary functions as breathing, functions that must continue even in sleep.

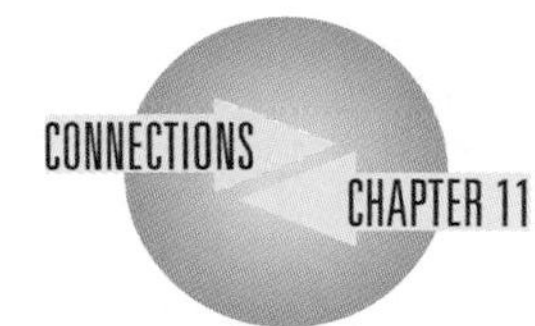

Blood–brain barrier. The brain has few internal blood vessels; most of the arteries and veins that supply the brain run along the brain surface only. The cells deep in the interior must therefore receive most of their nutrition through the **cerebrospinal fluid**, which fills the brain's interior cavities. There is no direct flow of fluid from the blood to the cerebrospinal fluid, hence the name blood–brain barrier. This cerebrospinal fluid communicates with the blood supply across a thin membrane, the **tela choroidea**. Nutrients and other small molecules cross this membrane or move through the neuroglia mentioned earlier, but many types of molecules, particularly larger molecules, cannot cross the blood–brain barrier. This barrier thus functions to prevent many molecules that have access to the rest of the body from entering the brain (see Chapter 11).

Epilepsy: abnormal message processing. Messages come to the brain from the peripheral nervous system and the sense organs. These signals trigger action potentials in a series of interconnected neurons in different parts of the brain. It is this message processing that produces our perceptions of the world.

In some disorders, neurons in the brain trigger action potentials spontaneously, without an outside signal. **Epilepsy** is a disorder marked by brain seizures, usually mild, characterized by uncontrolled electrical activity in the cerebral cortex. Both the cerebral cortex and the hippocampus contain many feedback pathways involving GABA-secreting neurons. As you probably recall, GABA inhibits a neuron that has fired from firing again (see Figure 10.7). An impairment of one or more of these GABA feedback pathways is hypothesized to allow a stimulated neuron to keep firing, perhaps causing an epileptic seizure. Some of the drugs that block GABA synthesis or GABA receptors can bring on epileptic seizures, thus lending support to the hypothesis. What is not fully explained by this hypothesis is why the seizures are temporary, why long periods intervene between them, and why certain events precipitate the onset of seizures.

In epilepsy the processing of messages by the brain is abnormal. Some people with epilepsy experience 'auras'. These take many forms depending on the part of the brain in which the spontaneous neuron signals occur. The person might have a visual hallucination or smell a bad smell when no actual source of such a sight or smell is present. A specific thought may be triggered, or a vague feeling of a place being familiar, even if it is not. Another of the symptoms of uncontrolled brain activity in epilepsy is muscle seizures in which muscles are continually stimulated to contract. To understand how changes in brain activity could affect muscle activity we next examine the pathways of normal muscle contraction.

Message output: muscle contraction

In addition to message processing by neurons forming synapses to other neurons within the brain and nervous system (a subject we will return to on p. 435), the brain also coordinates the functioning of other structures within the body. Neurons, as well as forming synapses with other neurons, can also form synapses with cells of other types, such as muscle cells and gland cells. Glandular secretions are regulated by the autonomic nervous system, a part of the peripheral nervous system that will be described in Chapter 12.

Voluntary movements are produced when the brain sends nerve impulses to the body's skeletal muscles. Because the muscles are attached to the skeleton, contraction of the muscles brings about movement of the skeleton and of the body as a whole. Most movements would be jerky and uncontrolled if only a single muscle were involved; controlled, steady movements usually require the simultaneous contraction of several muscles that pull in different directions to smooth and steady the movement. Coordination of messages to thousands of cells is a function of the brain. Earlier we saw that diseases (Parkinsonism and Huntington's disease) can result when the brain does not function properly. Now we will examine in more detail how neurons induce muscle contraction in health.

The neuromuscular junction. Neurons whose axons form synapses with muscle cells rather than with other neurons are called motor neurons. The synapse between the motor neuron and the muscle cell membrane is known as the neuromuscular junction (Figure 10.14A). Muscle cell membranes, like the cell membranes of neurons, are electrically excitable. They have sodium pumps that maintain an electrical potential by actively transporting positively charged sodium ions to the outside of the cell. The muscle cell electrical potential can be depolarized after the cell receives a neurotransmitter signal from the motor neuron. Motor neurons generally secrete the neurotransmitter acetylcholine, and muscle cell membranes have specific receptors for this neurotransmitter. When a motor neuron releases acetylcholine into the neuromuscular junction, acetylcholine binds to its receptors on the muscle cell membrane, causing it to depolarize.

Muscle contraction due to actin and myosin. This depolarization of the muscle cell membrane causes a release of calcium ions from vesicles within the cell. These calcium ions trigger reactions that allow cross bridges to form between the two major muscle proteins, **actin** and **myosin**. The cross bridges pull the actin filaments, increasing their overlap with the myosin filaments and producing the muscle contraction (Figure 10.14B and C). Many filaments of actin are pulled over many filaments of myosin to produce a forcible contraction of the muscle fiber as a whole.

Actin and myosin may be arranged in orderly bands (see Figure 10.14A), giving certain types of muscle fiber a striated (cross-banded) appearance common to both skeletal and heart muscle tissue. Another type, smooth muscle tissue, found in blood vessels and internal organs, lacks this cross-banding because the actin and myosin fibers are arranged at random intervals.

Muscle cell contraction is a neuron-stimulated process; muscle cell relaxation is not. When the muscle cell membrane is no longer receiving depolarization signals, calcium ions are removed from the cell cytoplasm. Removal of calcium breaks the cross-bridges between actin and myosin, allowing them to return automatically to their original positions relative to each other (see Figure 10.14B). They do not receive another neuron signal that pulls them back apart. Removal of calcium ions does require energy, however. This energy is supplied as ATP. When an organism dies, no more ATP is made, so calcium cannot be removed and the muscle cells stay contracted. This produces rigor mortis.

Within the whole organism, controlled muscle contraction may bend an arm. Relaxation of those same muscles does not return the arm to its original position; for that movement another set of muscles must contract. Skeletal muscles are therefore arranged in opposing sets, with flexor muscles bending limbs and extensor muscles straightening them (Figure 10.15).

FIGURE 10.14

Structure of a portion of skeletal muscle.

(A) Overall structure of skeletal muscle, showing cross-striations

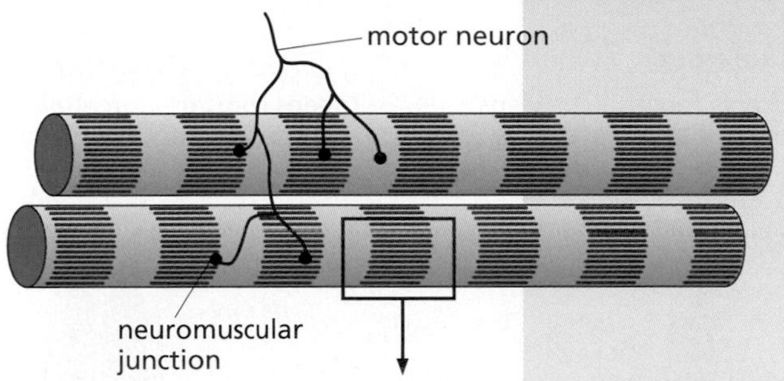

(B) Enlargement showing arrangement of actin and myosin filaments in a relaxed portion of skeletal muscle

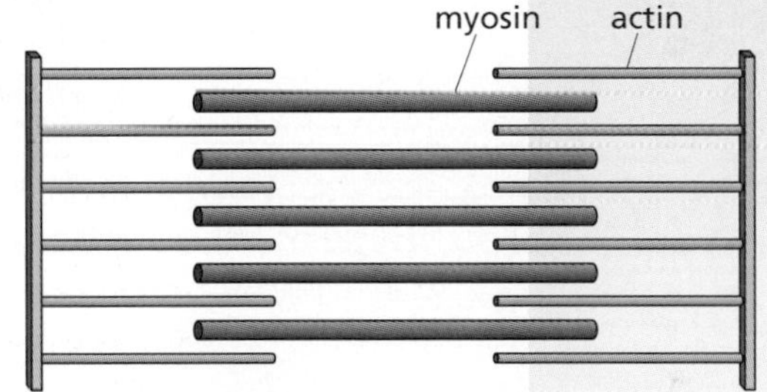

(C) The same portion of muscle, contracted

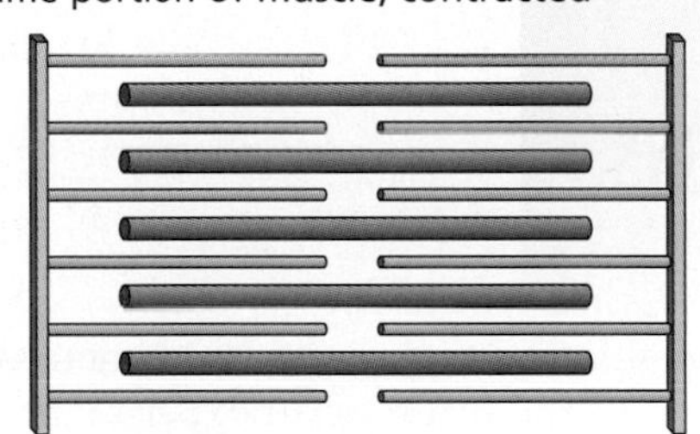

Blocking of muscle relaxation by some insecticides. As we saw earlier (see Figure 10.6), neurotransmitters released into a synapse are normally taken back up by the presynaptic neuron or degraded. Motor neurons release acetylcholine into the neuromuscular junction. Normally acetylcholine is broken down by the enzyme cholinesterase, so muscle contraction ceases. Some insecticides work by blocking the action of

FIGURE 10.15

The major muscles that flex and extend the arm. Most muscles act together with other muscles that steady their action by pulling in the opposite direction.

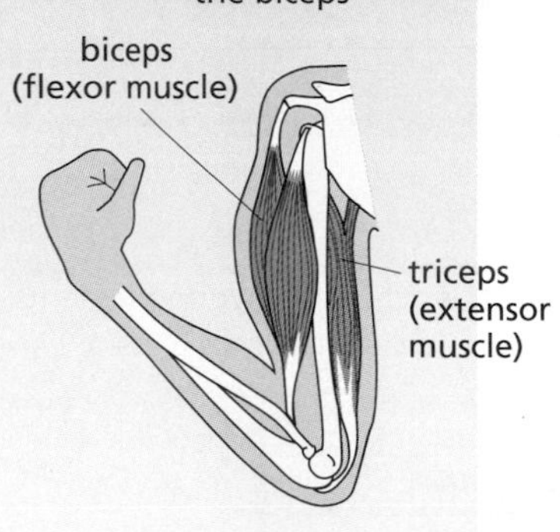
FLEXION
Biceps contracts forcefully; triceps contracts only slightly, to steady the biceps
biceps (flexor muscle)
triceps (extensor muscle)

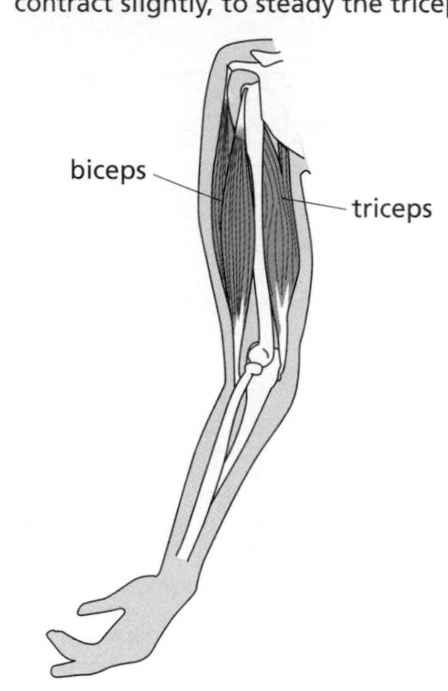
EXTENSION
Triceps contracts forcefully; biceps may contract slightly, to steady the triceps
biceps
triceps

cholinesterase at neuromuscular junctions and elsewhere. Then any synapse stimulated by acetylcholine remains continually stimulated (because the acetylcholine is never broken down), and the insect receiving the insecticide dies with most of its muscles in a state of rigid contraction. Because most animal nervous systems use acetylcholine as a neurotransmitter, such pesticides are toxic to all animals, great and small, including insects, pets, and humans. The toxic dose, however, depends on body size, so an amount that might kill most insects would only make your pet sick and might not noticeably affect you at all.

Many messages to and from the brain pass through neuron synapses connecting the peripheral and central nervous systems. We have seen that sense organs in the peripheral nervous system send messages to the brain. The brain also sends messages to the peripheral nervous system, for example through motor neurons that induce muscle contraction. Many other messages, and message storage, pass through synapses within the brain itself, a subject we take up in the next section.

THOUGHT QUESTIONS

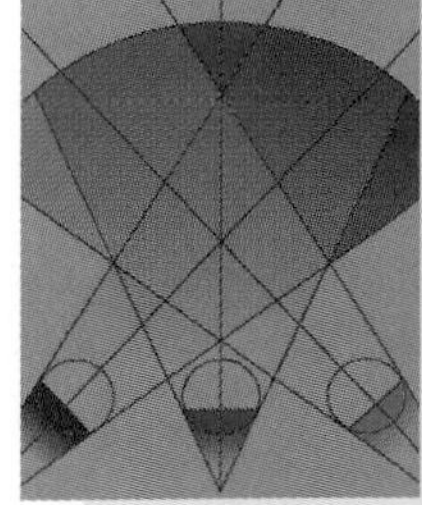

1. In the ablation technique, the function of a portion of the brain of an experimental animal is investigated by destroying it and studying the defect produced. What are this technique's limitations? Could you study how a piano or an automobile works by inserting a probe, destroying some local region, and studying the resulting defects? Could you study the function of a radio in this way, or a computer? Which of these is more comparable to the brain in complexity?

2. What does it mean to say that there are four basic tastes, not five or three? Compare this with the several different types of receptors in the skin.

3. Why would your being on a roller coaster or in a stunt airplane confuse your inner ear with misleading stimuli?

4. What causes muscle contraction to stop? What happens to an organism if muscle contraction does not stop?

5. Why does the body have so many different kinds of neurotransmitters?

THE BRAIN STORES AND REHEARSES MESSAGES

Neuron activity within the brain produces many of the functions that we associate with the mind. Some messages are stored as memory for future retrieval. Some of these modify our future behavior, a process that constitutes learning. At times, but especially when we sleep, the brain may rehearse sending and interpreting certain messages without remembering them. Brain activity produces personality, emotions, thoughts, and dreams; and, as we will see, mental illnesses can result from abnormal brain activity.

Learning: storing brain activity

When we change in response to changes in the world around us, this is known as **adaptation** (a separate meaning, unrelated to the adaptations that arise from the operation of natural selection). Adaptation can take place through many physiological, immunological, and neurological mechanisms and can be either conscious or unconscious. **Learning**, which consists of lasting changes in behavior or knowledge in response to experience, is an important form of adaptation brought about by the nervous system. Different types of learning are distinguished by the types of information that are learned and by ways in which the nervous system processes and stores the information. **Declarative learning** is mostly conscious remembrance of persons, places, things, and concepts, requiring the actions of neurons in the hippocampus and certain parts of the cerebral cortex. Memory of how to do things, **procedural learning**, does not require the hippocampus or the temporal lobe of the cerebral cortex, and is not necessarily conscious. People with hippocampal damage can still learn how to do new things, although they will not consciously recall that they can.

Procedural learning. Procedural learning can be very simple—in fact, simpler animals such as mollusks and insects are capable of procedural learning because it does not require the forebrain structures that evolved in vertebrates. In human development, procedural learning becomes possible earlier than declarative learning. Infants at first learn procedurally: how to eat, how to move, how to respond to gravity.

The three simplest kinds of procedural learning are habituation, sensitization, and classical conditioning. These simple kinds of learning are distinguished from declarative learning in that they can be involuntary: the learner changes his or her behavior without showing any awareness of the learning process. Most animals, no matter how minimal their nervous systems, can learn in these simple ways.

When a stimulus is presented repeatedly, an animal may learn not to respond to it. This is known as **habituation**. Habituation occurs both behaviorally and at the level of the neuron. Single neurons can stop making action potentials owing to changes in their receptors or to an increase in the threshold for generating action potentials. When a stimulus does not result in harm, an organism may learn to change its behavior and not expend energy in responding to the stimulus (Figure 10.16A). Humans habituate to all kinds of signals: if you move to a new location, you see and hear many things that people who have lived there for a while have learned not to notice any more. After a time, you no longer see or hear them either. If something unusual happens, the habituation can be overcome. Habituation is thus context-specific to some extent.

Sensitization (Figure 10.16B) is the opposite of habituation. An organism becomes sensitized when an intense and aversive stimulus, such as a loud gunshot, increases subsequent responses to other stimuli. On the cellular level, for a prolonged time, neurons become more capable of generating an action potential, a change known as long-term potentiation. Several hypotheses have been suggested to account for long-term potentiation. One recent hypothesis involves a type of glutamate receptor known as an NMDA receptor (*N*-methyl-D-aspartate receptor). The secretion of glutamate (a neurotransmitter) causes these receptors to respond by having the postsynaptic cell increase its production of nitric oxide. The nitric oxide then diffuses from the postsynaptic cell to the presynaptic cell, where it enhances the future release of glutamate. Glutamate secretion thus enhances future glutamate secretion in a positive feedback loop. Extreme overstimulation can lead to glutamate poisoning (characterized by convulsions), a condition made worse by the food additive MSG (monosodium glutamate).

A third type of simple procedural learning is called **classical conditioning** (Figure 10.16C), a change in which an organism learns to associate a stimulus with a particular response. Classical conditioning is also called Pavlovian conditioning because it was first demonstrated by the

FIGURE 10.16

Three kinds of procedural learning.

(A) A person becomes habituated to a ticking clock

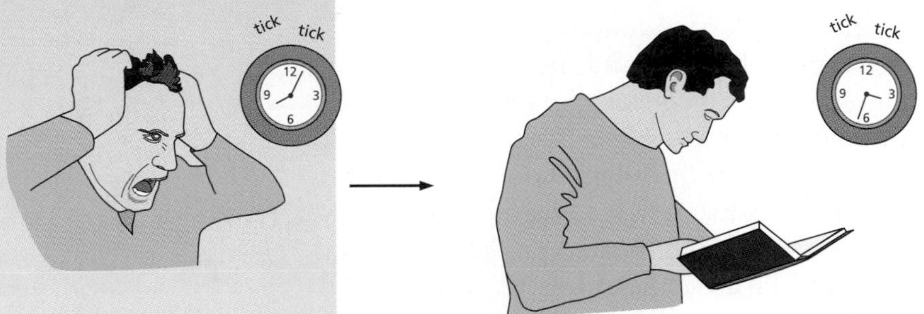
tick tick
tick tick

(B) A deer becomes sensitized by a loud noise

bang
crack
deer is startled by loud noise
deer is now more sensitive to other sounds

(C) A dog undergoes Pavlovian conditioning

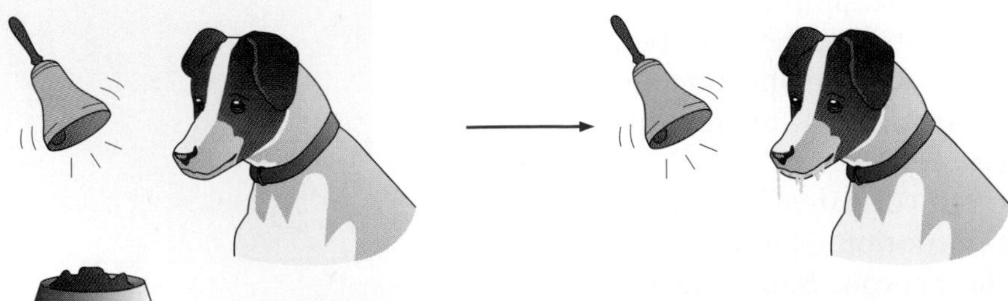
ringing of bell is followed or accompanied by food
dog now salivates in response to bell alone

Russian physiologist Ivan Pavlov around 1900. Dogs salivate when they see or smell food. If a bell is rung each time food is presented, a dog learns after a very few repetitions to salivate when it hears the bell ring, even if no food is present.

Humans can learn through classical conditioning, very often without being consciously aware of it. Fears sometimes become associated with various objects, colors, or smells when we are young because those stimuli were present when we were hurt in some way, or because they remind us of other unpleasant stimuli. One person we know grew up with a decades-long aversion to gelatin desserts. It seems that at the kindergarten he attended, such desserts were brought in on a tray, piled in cubes whose wiggling movements reminded him of certain caterpillars. Caterpillars are something that most children would avoid eating, so the shaking movements in this case conditioned an aversion to the desserts. Adults too can become conditioned. In one case, several people became nauseated every time they saw the carpeting in a hospital. It turned out that the carpet was the same color as the chemotherapeutic drugs that these people had taken for treatment of their cancers, drugs that had made them sick to their stomachs. The people had become conditioned, associating the color of the carpet with the cause of their nausea.

Declarative learning. In contrast to procedural learning, more complex types of learning require the activity of the hippocampus and cerebral cortex. There is evidence that complexity of experience actually contributes to the size of the cortex. Rats raised in 'enriched environments'—in large cages with other rats and with 'toys'—develop a thicker and more elaborated cortex. In humans, at about the age of two, the age at which the number of brain cells reaches its maximum and a critical level of complexity of connections is achieved, declarative learning begins. Although the exact age varies from child to child, the capacity for declarative learning always develops later than the capacity for procedural learning.

Memory formation and consolidation

To become a **memory**, a piece of information must be acquired, stored, and retrieved. Many acquired pieces of information can be retrieved for only a short period. For example, you may be able to recall having heard a particular sound if someone asks you about the sound within a few minutes of the time you heard it, but not after that. Information that is quickly forgotten is said to be part of **short-term memory**. Short-term memory is mostly chemical, having to do with temporary changes in neurotransmitters and their receptors. Long-term storage of information requires actual structural changes (the formation of new synapses) within many parts of the brain.

Long-term memory. One part of the brain that is essential to change a sensory input from a short-term into a **long-term memory** is the hippocampus (Figure 10.17). People who have suffered damage to the hippocampus are unable to form new long-term memories. They can recall things from before the time of the damage, indicating that the storage sites themselves are not in the hippocampus. Their recall of newly acquired information or experiences is limited to the short term; thus, long-term retention of a memory may be said to require a 'lack of forgetting.' This is especially true of **declarative memories**, meaning memories that can be consciously known.

FIGURE 10.17

Horizontal section through the brain, showing the hippocampus, an important structure in the formation of certain types of memory.

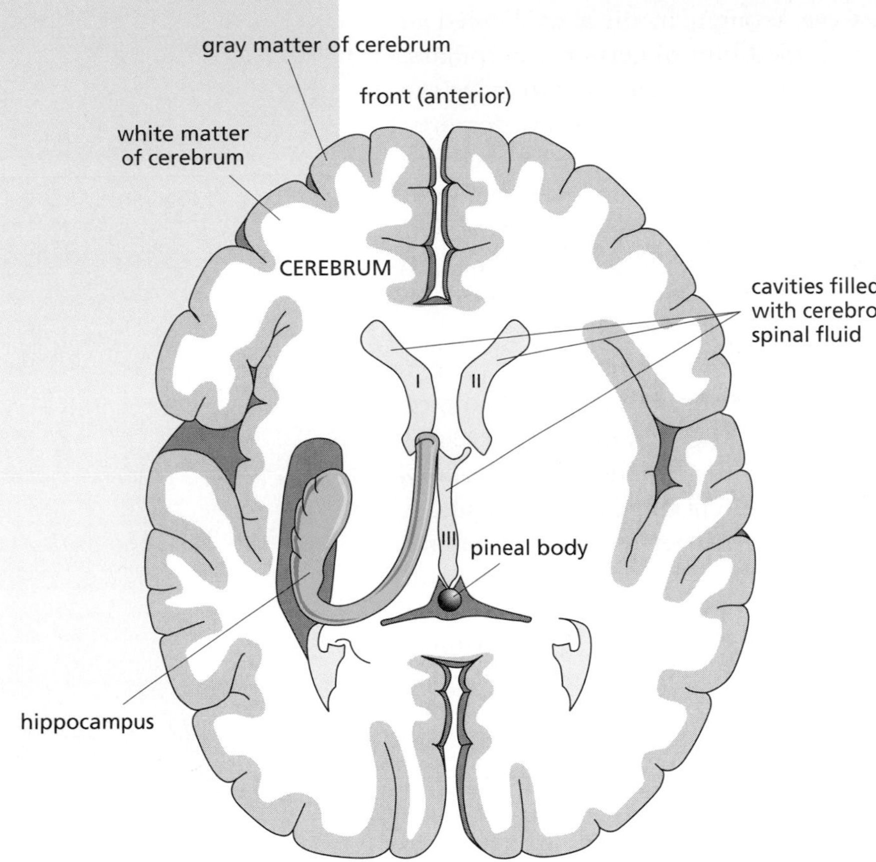

The hippocampus turns short-term memories into long-term ones by making new cellular connections with other parts of the brain. Neurons form new synapses connecting several neurons in loops with each cell synapsing on the next cell in the loop. The stimulation of any one of these neurons results in information transfer around the whole loop. Stimulation of these assemblies of cells may need to continue for years before a memory is permanently stored.

As time passes after the acquisition, **memory consolidation** occurs. While long-term memories are forming, and even after they have been stored, they are organized and restructured on the basis of even more recent experiences. Existing knowledge is constantly being reordered in the light of new knowledge. Memory consolidation relies on information processing, one innate aspect of which is the capacity to make generalizations from specific experiences. If we live in a city and have walked in a forest only once, we mentally picture all forests as being like the one we walked in. Further experience, either 'in person' or acquired through seeing pictures or reading stories, enables us to reformulate the initial generalization that we made, replacing it with another generalization. In addition, we can still recall some very specific aspects of the particular forest that we first walked in.

Emotional states can sometimes modify the process of memory consolidation: we are more apt to remember something that we would normally not consider worth remembering if we associate it with an event that had great emotional meaning for us (either positive or negative). People who were old enough at the time of the assassination of John F. Kennedy, the explosion of the space shuttle *Challenger*, or the tearing down of the Berlin Wall can remember vivid details of where they were and what they were doing. The same is generally true of events with great personal meaning, such as weddings, deaths, or natural disasters such as earthquakes or floods.

Abstraction and generalization. Memory consolidation also relies on the capacity to conceptualize, an extension of the ability to generalize. Researchers can demonstrate the capacity for generalization by using what are called **oddity problems**. Monkeys are shown three or four objects, all alike except one, and they are rewarded for picking the different one. If a set of objects consists of two toy trucks and a car, they must learn to pick the car. Presented with a totally different set, two oranges and an apple, they must learn to pick the apple. After many such sets, each set different, have been presented, the monkeys develop a concept of 'oddness' and pick the single object immediately. This type of declarative learning requires activity in the temporal lobe of the cerebral cortex. Much of human learning and memory is processed by the cerebral cortex, consolidating memories and forming concepts, although other parts of the brain are also involved.

Memory formation requires the hippocampus, but memory consolidation requires more parts of the brain, particularly the cerebral cortex. The cortex is also important for memory retrieval. In a series of experiments in the 1940s, a neurosurgeon named Wilder Penfield electrically stimulated the cerebral cortex of conscious patients who were undergoing brain surgery for various neurological diseases. Such stimuli cause a person to recall a memory so vividly that they feel they are reliving the experience.

Alzheimer's disease: a lack of acetylcholine

Alzheimer's disease is a form of progressive mental deterioration in which there is memory loss and a loss of control of body functions that ends in complete dependence and death.

Clinical evidence suggests that nerve transmission across certain synapses that use acetylcholine is impaired in Alzheimer patients. Acetylcholine, in addition to being secreted by motor neurons, is also secreted by some neurons in the brain in synapses to other neurons, especially in the hippocampus and cerebral cortex. Drugs that inhibit cholinesterase can temporarily improve memory and other brain functions in Alzheimer patients, although they cannot arrest the course of the disease. Inhibiting cholinesterase amplifies the effect of any existing acetylcholine. This fact, together with the gradual and progressive nature of the mental decline, has caused many workers to hypothesize that the disease symptoms are caused by a gradual loss of acetylcholine receptors in postsynaptic cells; others believe that the primary defect is in the synthesis of acetylcholine itself. Recently, receptors for another neurotransmitter, glutamate, have also been implicated. Drugs that enhance glutamate reception also improve long-term memory formation in elderly patients.

In addition to changes in neurotransmitters, Alzheimer patients have abnormal deposits of a protein called amyloid in their brains. In 1993, a gene called *apo-e4* was identified as being present in a majority of patients with the most common form of Alzheimer's disease. The product of this gene is a lipoprotein related to the formation of amyloid. Autopsies reveal that amyloid deposits occur throughout the brains of people who had Alzheimer's disease, but are especially prevalent in the hippocampus and the parts of the cortex that are involved in memory.

Biological rhythms: time-of-day messages

In addition to the conscious processes of declarative learning, the brain sends itself many unconscious messages. Our bodies respond all the time to messages that tell us what time of day it is. In response to these messages, we establish a biological rhythm that governs our pattern of sleep and wakefulness.

Circadian rhythms and their control. There are many kinds of biological rhythms, biological processes that repeat at somewhat predictable intervals. Those of approximately 24 hours' duration are called **circadian rhythms** (Latin *circa*, 'about,' and *die*, 'day'). Various biological functions can be monitored on a 24-hour basis, and most of them show some recurrent circadian rhythm.

Where do circadian rhythms originate? If the rhythms are **endogenous** (internal in origin), how do they keep tuned to the 24-hour cycle of the world around us, and how can they adjust to different time zones, seasons, and work shifts? If, in contrast, the rhythms are **exogenous** (external in origin), how do external cues regulate the body's cycles, and can these external cues be easily manipulated? The first serious attempts to answer these questions began with **isolation experiments**, such as those conducted in Mammoth Cave in Kentucky. In these experiments, volunteers who were kept from all sources of natural light were allowed to set their own daily routines. Nearly all individuals maintained fairly constant circadian rhythms in their sleep–wake cycles and also in body temperature, activity, and a variety of physiological measurements. Circadian rhythms were all maintained despite the constant environment and were rather uniform for each individual. Most significantly, the circadian rhythms maintained in these isolation experiments were generally slightly more than 24 hours, between 25 and 26 hours for most individuals. The maintenance of such a rhythm after its drift from synchrony with the day–night cycle of the outside world showed that endogenous rhythm-keeping mechanisms existed.

How, then, does the external world exert its influence, causing most of us to maintain a 24-hour daily rhythm? Our current understanding is that the external environment provides us with certain time-related clues. The most important of these clues is the natural rise and fall of light intensity throughout the day, which imposes a 24-hour rhythm on the secretion of melatonin; other important clues include our own activity and our bombardment with external stimuli (including social stimuli) during daylight hours.

Important to circadian rhythms are a group of neurons located in the **hypothalamus** (see Figure 10.13) and called the **suprachiasmatic nucleus**. Destruction of this nucleus abolishes all circadian rhythms. Under experimental conditions of continuous darkness or continuously dim lighting, the suprachiasmatic nucleus maintains an endogenous circadian rhythm with a cycle of slightly more than 24 hours in length. Under more natural conditions, the light received by the **pineal body** adjusts the rhythm of the suprachiasmatic nucleus to follow a 24-hour cycle, with a peak of activity in the late morning and with greatly reduced activity throughout the hours of darkness. About the size of a pencil eraser, the pineal body is located in the brain on the roof of the diencephalon (see Figures 10.13 and 10.17). During times of darkness, the pineal body secretes a hormone called **melatonin**, which is not secreted during times of illumination. By illuminating various parts of the body separately from the rest, biologists have experimentally determined that the pineal body is sensitive to the light that it receives right through the skull and brain! If a laser (a highly focused beam of light) is focused on the pineal body through the head, and is turned on and off in a 24-hour rhythm, all parts of the body follow the established circadian rhythms, even when the rest of the body is in darkness. If, instead, the head is kept in darkness while other parts of the body are illuminated, the effect is the same as if the body and head were both in total darkness.

Gradual or slight disturbances in a person's 24-hour rhythm can result from short-distance travel, seasonal changes, and the semiannual change of clocks at the beginning and end of daylight savings time. The effects of these changes are minimal in most cases. More drastic effects are felt in the phenomenon known as jet lag, the disturbance of our 24-hour rhythms as a result of travel through several successive time zones. The major symptom of jet lag is fatigue, plus a desire to sleep or remain awake at inappropriate times for a few days until the body readjusts to the new cycle. The rigors of travel add to the fatigue, but travel north to south within a time zone is much less fatiguing than travel east to west across time zones. People who have traveled long distances by air are also statistically more susceptible to infection. While some of this may be due to other conditions of air travel, the increase in susceptibility is greater among people who have flown east–west than among those who have flown south–north, suggesting that interrupted circadian rhythms are involved. People whose bodies are not able to adjust to any particular rhythm because of irregular work schedules also show an increase in fatigue-related events such as the number of accidents.

Sleep. One mental state that shows a strong circadian rhythm is sleep. At the end of each day, we usually have a strong urge to sleep; we can postpone this urge, but only to a limited extent. People deprived of sleep do not function well when awake (they make more mistakes, for example); and people who awake from a 'good night's sleep' feel refreshed and alert.

As we have seen, the electrical activities of single cells take place across their membranes. The cumulative effect of ions flowing across the membranes of many cells can be seen in an **electroencephalogram** (EEG), a graph of electric activity obtained from electrodes pasted onto the scalp (Figure 10.18A). By using an EEG, we can detect different patterns of ion flow in the brain. EEGs show several levels or stages of sleep. Stage 1 is characterized by regular respiration, slowed heart rate, drifting mental imagery (similar to daydreaming), and low-voltage EEG wave patterns of about 7 to 10 Hz. (Hz stands for **Hertz**, a measure of frequency equivalent to cycles, or waves, per second.) Subjects aroused from stage 1 sleep will often say, "I wasn't sleeping." Sleep stages 2, 3, and 4 also have characteristic EEG patterns (Figure 10.18B). A further sleep stage is characterized by **rapid eye movements (REMs)** noticeable as movements of the eyeball beneath the closed eyelids of sleeping subjects, including cats and dogs as well as humans. About 80% of human subjects awakened during REM sleep tell of some dream that they were having, often in vivid detail, while subjects awakened during other sleep phases do not remember any dreams.

During a typical night's sleep, an adult goes through about four or five sleep cycles. As shown by the steps in Figure 10.18C, each cycle goes through stages 1, 2, 3, and 4, in that order, then in reverse order through

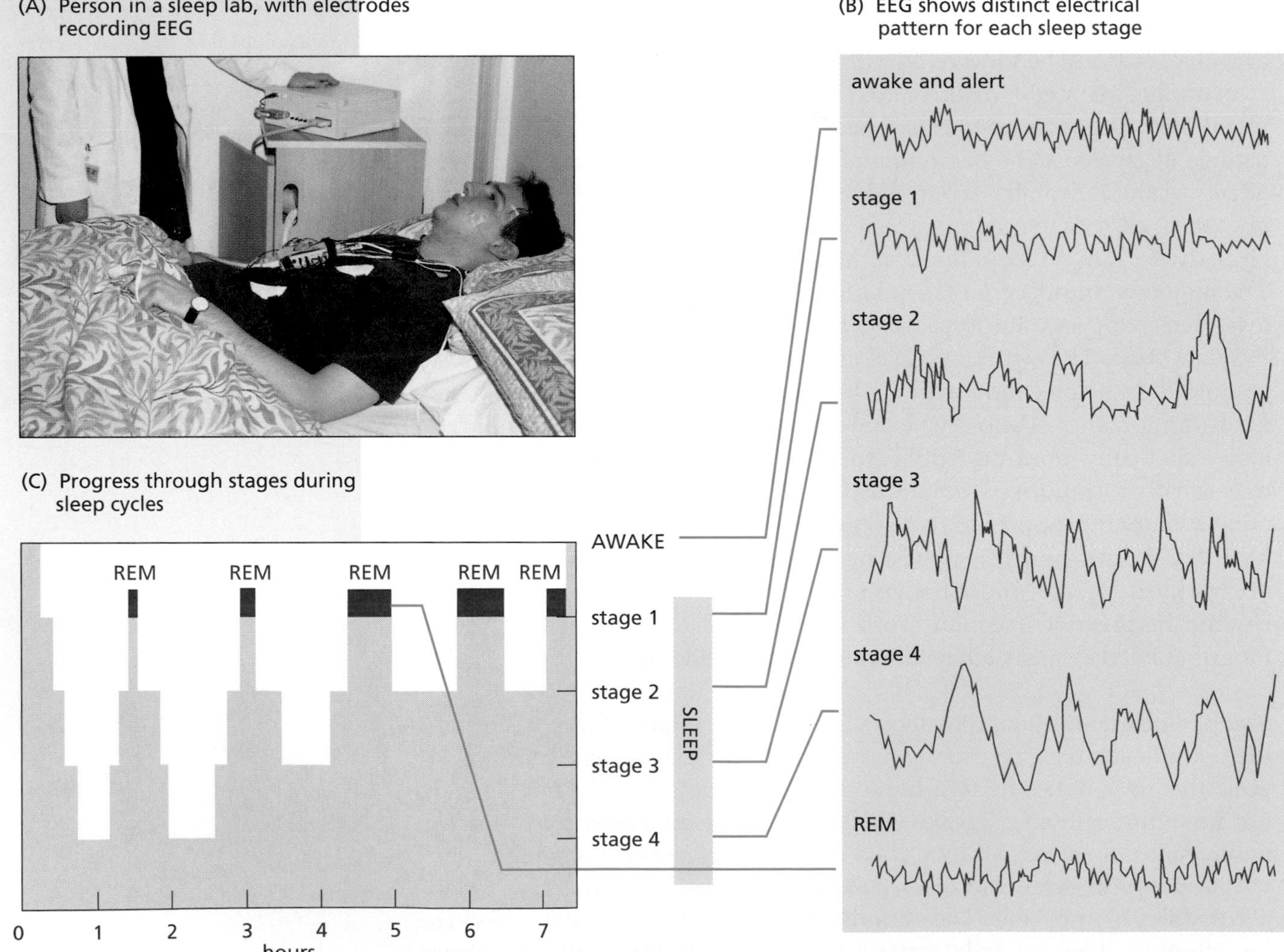

FIGURE 10.18

Electrical activity in the brain during different stages of sleep.

stages 3, 2 and 1, followed by a REM period, after which the next cycle begins. The proportion of REM sleep increases with each successive cycle, as shown by the increasing width of the REM bars in the later sleep cycles in Figure 10.18C. The later sleep cycles may also skip some of the stages, for example proceeding only to stage 3 before reversing back to 2, 1 and REM, and at a still later cycle, proceeding only to stage 2 before returning to 1 and REM. If subjects are allowed to awaken by themselves (with no alarm clock or other external stimulus), they usually awaken near the end of a cycle—that is, following a REM stage or during stage 1 (see Figure 10.18C). Although this pattern is fairly typical, sleep cycle patterns vary widely with the person and the circumstances.

Evidence continues to mount that REM sleep is extremely important. Volunteers who are deprived of REM sleep for a day or two begin to day-dream and have their thoughts wander. They also make more mistakes and have more accidents, even if their total amount of sleep and their amounts of all the other sleep stages are normal or above normal. If finally allowed to sleep as long as they wish, REM-deprived subjects sleep longer than usual, and a higher than usual proportion of their sleep is REM sleep. People deprived of any of the non-REM stages do not show any abnormal symptoms.

Some drugs can alter the natural occurrence of sleep. Caffeine, amphetamine, and other stimulants may interfere with the onset of sleep, although sensitivity to this effect varies with the person. Alcohol and barbiturate drugs bring on sleep more readily, especially in persons already tired, but this drug-induced sleep is less restful because it has longer stage 3 and stage 4 intervals and shorter REM episodes. Muscle relaxants have no effect on sleep except to overcome muscle tension, which may sometimes inhibit the onset of sleep. (Most drugs taken as aids to sleep are either barbiturates or muscle relaxants.) There is no totally safe sleeping pill, and all drugs that alter sleep patterns usually have other effects as well.

The role of neurotransmitters in the control of sleep and wakefulness is unclear. The rate at which neurotransmitters or certain other chemicals are produced or degraded during sleep can be studied by labeling these chemicals radioactively. For example, studies in which animals are fed radioactive tryptophan (a chemical precursor from which the neurotransmitter serotonin is synthesized) show that the rate at which tryptophan is converted into serotonin increases during sleep. Likewise, studies have found that serotonin breaks down more rapidly during sleep. Such experiments have enabled us to locate areas of neurotransmitter activity during sleep and wakefulness. In these studies, as in other biological studies with radioactively labeled substances, the radioactive dosage is kept low to minimize the risks to the experimental subjects and experimenter.

An important part of the brain governing sleep and wakefulness is a group of neurons called the **reticular activating system**, which radiates outward and upward from the reticular formation of the brain stem (Figure 10.19). These neurons send 'alertness' signals through the diencephalon to widely scattered parts of the cerebral hemispheres. These signals seem to accompany most types of sensory input but do not seem to vary depending on the type of stimulus. Their message seems to be simply "pay attention!" The reticular activating system is usually more active by day and quiescent at night. Low-level activity allows sleep, but higher activity maintains wakefulness. Fortunately, it is also possible for the reticular activating system to awaken us to an emergency in the middle of the night or to deviate in other ways from its usual 24-hour rhythm, as conditions demand.

Dreams: practice in sending messages

Dreams are an important phenomenon of sleep. Philosophers and poets, fortune-tellers and psychiatrists have each had their ways of interpreting dreams. Modern-day dream researchers have used several techniques to study dreams. One method is to have the subject keep a 'dream diary' in which they record as much of a dream as they can remember upon awakening. Among the findings using this technique are the following: most dreams are visual, including those of people who became blind after age five or six, but excluding those of people who were blind from birth or infancy. Other senses (hearing, touch, smell, taste) also are mentioned in dream reports, but less often than vision. (These other senses predominate, however, in the dreams of people blind since birth or infancy.) Familiar persons, places, and types of events appear in most dreams, but not always congruously: the dreamer sometimes experiences familiar activities in the wrong setting, and places that are supposedly far away or

FIGURE 10.19

The reticular activating system.

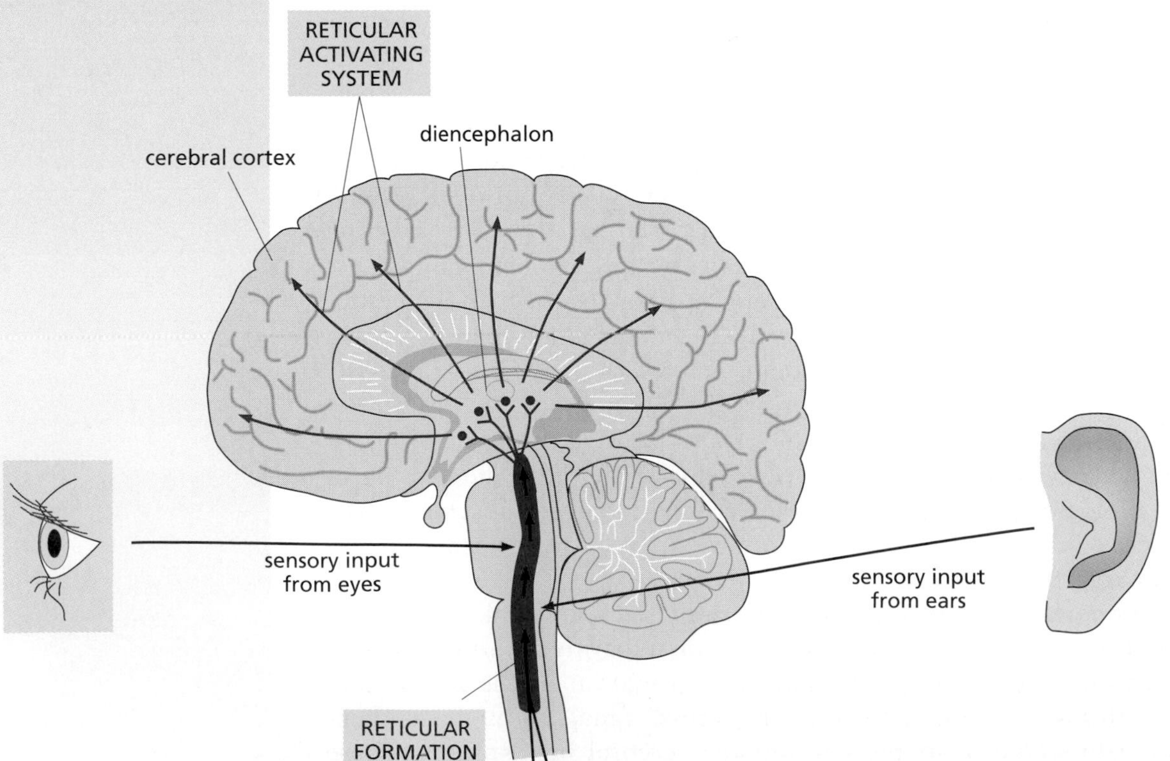

unfamiliar often look very familiar in the dream. There are frequent changes of scene, of mood, or of persons present. Stimuli in the dreamer's environment are often incorporated in the dream imagery itself. Actual sounds, smells, flashes of light, and other sensations from the sleeper's environment may be worked into the dream.

Perhaps one of the most important and most constant feature of our dreams is that we remember so little about them during our waking hours. There seems to be a good reason for this. The EEGs recorded during REM sleep show spontaneous electrical activity in the brain. This is our brain's way of rehearsing its message-sending functions, and it might be related to the strengthening (through practice) of existing synaptic connections or the establishing of new ones.

Our brains are usually programmed to interpret any such electrical activity as a coherent picture of the world around us. When we are awake, our brains usually record these coherent pictures so that we can remember them at some later time. In dreams, the integrating mechanisms are still at work, so a coherent or semicoherent picture of the world is drawn, but the mechanism whereby these pictures are remembered for later use is suppressed, at least most of the time. When people have disturbing nightmares, what is usually most disturbing about them is that they are remembered as if they had been real. Of all people, schizophrenics have the greatest difficulty in distinguishing reality from dream activity.

Mental illness and neurotransmitters in the brain

We have described brain functioning during normal mental states. Several mental illnesses are associated with chemical imbalances in the brain, such as abnormalities of neurotransmitters.

Depression and serotonin. Depression is a disorder marked by feelings of total helplessness, despair, and frequent thoughts of suicide. Many people suffering from depression attempt suicide, and some succeed. The smallest task, such as getting out of bed in the morning, can seem overwhelming. Everyone has unhappy or pessimistic feelings from time to time, but a person suffering from depression has these feelings nearly all the time and to a severe degree. Depression is about twice as prevalent among women as among men.

Patients suffering from depression have smaller amounts of several neurotransmitters than do other people. In particular, the brains of depressed patients who commit suicide are found at autopsy to contain low concentrations of serotonin.

A number of drugs are effective in treating the symptoms of depression. These drugs act in either of two ways: some drugs act by blocking the action of the enzyme MAO, which degrades many neurotransmitters; other drugs, including Prozac, Elavil, and Tofranil, inhibit the reuptake of serotonin and several other neurotransmitters by presynaptic cells. In either case, the effect is the same: the neurotransmitter remains active for a longer time after it is secreted and results in a greater (or a more lasting) stimulus being passed on to the next cell. Drugs that inhibit the body's production of serotonin can reverse the effects of antidepressant drugs, but drugs that inhibit the production of several other neurotransmitters do not have this effect.

How might decreased amounts of a neurotransmitter bring about the symptoms of depression? One possible explanation is that, when something good happens, most people receive a pleasurable stimulus (called a **reinforcement**) in the form of a stimulation to the brain's positive reward centers, and this stimulation makes them more likely to repeat whatever behavior led to the reinforcement (see Chapter 11, p. 473). However, the reinforcement mechanism is not working in patients who are depressed, so they are never rewarded, nor do they learn to repeat whatever behavior led them to the sensation.

Schizophrenia and dopamine. Schizophrenia is a disorder characterized by frequent delusions and auditory or visual hallucinations. Schizophrenia seems to result from an excess of the neurotransmitter dopamine; drugs that stimulate or mimic dopamine (such as amphetamines or L-DOPA) make schizophrenic symptoms worse, and drugs that block dopamine (such as chlorpromazine and haloperidol) generally lessen symptoms. Further evidence comes from schizophrenic side effects observed when drugs are used to treat Parkinsonism, and Parkinsonian side effects observed when drugs are used to treat schizophrenia. Recently, it has been found that serotonin may also be involved in schizophrenia.

THOUGHT QUESTIONS

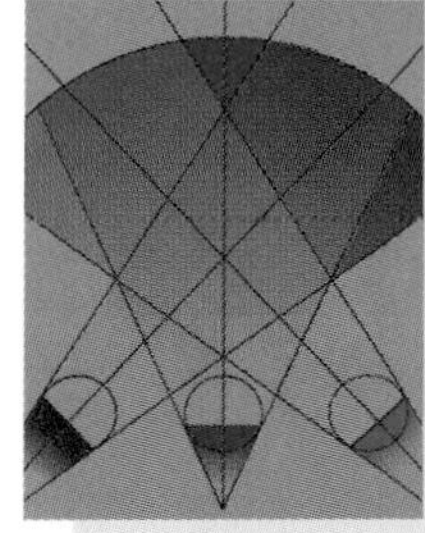

1. Suppose that you were investigating why depression occurs more often in women than in men. How might you test whether differences in upbringing and other cultural influences were at work? Would it be useful to study the prevalence of depression in different cultures? What methodological, ethical, or social problems would such a study face? Would you examine many people or few? What additional difficulties would you face if depression were defined differently in each culture? Could you investigate depression by studying an outcome such as suicide? What new ethical or social problems might arise from the results of such a study?

2. Is it ethical to give a drug that changes someone's personality? If a drug such as Prozac is given to a patient, and the patient kills someone, can the doctor be held responsible?

3. If a person's social situation may be contributing to a problem such as depression, is it proper for a doctor to treat the condition with drugs without also addressing the social situation?

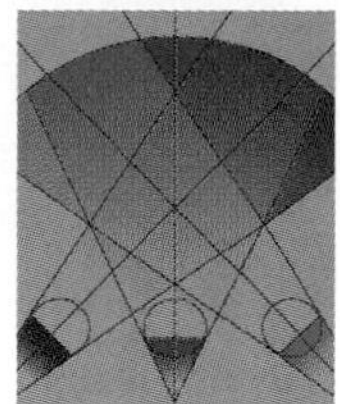

MANY OF THE FUNCTIONS THAT PEOPLE HAVE HISTORICALLY attributed to the mind have been shown by neurobiology to originate in activity in the brain. Neurons transmitting action potentials and stimulating other neurons across synapses can account for mental activities such as sensations, dreams, learning, and memories. Mental illness can also be seen to have a neurochemical basis, at least in part. However, whether these brain and nervous system activities actually *are* the mind, or whether they are the mechanisms by which the mind becomes material, is a question that philosophers will continue to debate. Certainly a person is not an isolated collection of biochemicals, but exists in a social relationship to other people and is necessarily linked to the rest of the world. So, although much progress has been made in drug treatments of mental illness by thinking of the mind and the brain as synonymous, it is a concept that could be carried too far if it leads us to ignore the importance of the whole person and her/his relation to others and to the environment.

CHAPTER SUMMARY

- The work of the brain is carried out by nerve cells (**neurons**).
- Using **sodium–potassium pumps**, these neurons maintain differences in ion concentration across their cell membrane and thus a difference in electrical charge known as the **resting potential**.
- If a **threshold** of depolarization is exceeded, neurons carry nerve impulses in the form of **action potentials** down the length of their **axons**.
- Neurons communicate with one another across spaces called **synapses** by means of such **neurotransmitters** as acetylcholine, norepinephrine, serotonin, dopamine, and GABA.
- Decreased dopamine neurotransmission may lead to Parkinsonism, while excessive dopamine neurotransmission may produce the uncontrolled movements that characterize Huntington's disease.
- Neurons control the activity of other neurons by **feedback**.
- Sense organs in the **peripheral nervous system** are responsible for receiving external stimuli such as visual images, sounds, tastes, smells, touch, pressure, heat, cold, and pain.
- Sensory messages are carried to the **central nervous system** and processed in the brain; such processing produces our perceptions of the stimuli.
- The brain also sends out messages such as those that stimulate muscle contraction.
- Learning requires activity in the brain but not necessarily consciousness of the activity.
- **Procedural learning** includes **habituation**, **sensitization**, and **classical conditioning**, none of which require conscious awareness or activity in the hippocampus or the temporal lobe of the cerebral cortex.
- **Declarative learning** is conscious learning, requiring the hippocampus and cerebral cortex for stimulus processing, **memory** consolidation, and the formation of generalizations and abstractions.
- Many biological functions follow a **circadian rhythm** that is set endogenously and fine-tuned by light acting on the pineal body.

- Sleep follows definite stages, each with a distinctive EEG pattern. Dreams coincide with periods of rapid eye movements (REMs) and result from the brain's own practice at sending messages.
- Abnormal neurotransmitter concentrations are associated with some mental illnesses.

KEY TERMS TO KNOW

action potential (spike) (p. 416)
adaptation (p. 435)
axon (p. 411)
central nervous system (p. 411)
circadian rhythm (p. 440)
classical conditioning (p. 436)
declarative learning (p. 435)
dendrites (p. 411)
electrical potential (p. 414)
feedback system (p. 420)
habituation (p. 435)
learning (p. 435)
memory (p. 437)
myelin sheath (p. 412)
nerve (p. 413)
neuron (p. 411)
neurotransmitter (p. 416)
peripheral nervous system (p. 411)
procedural learning (p. 435)
resting potential (p. 414)
sensitization (p. 436)
sodium–potassium pump (p. 414)
synapse (p. 411)
threshold (p. 416)

CONNECTIONS TO OTHER CHAPTERS

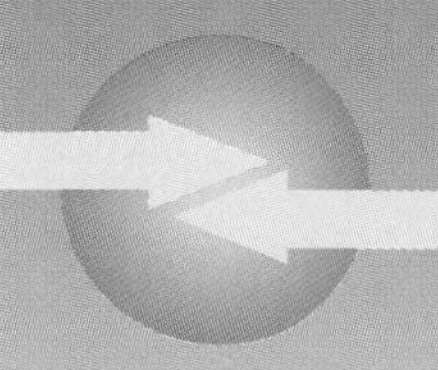

Chapter 1: Equating the mind and the brain is an example of a research paradigm.

Chapter 1: Ethical issues are raised by the use of fetal tissue for research and for therapy.

Chapter 3: Some brain disorders have a genetic basis.

Chapter 4: Similarities in the brains of different species have resulted from evolution.

Chapter 7: Social behavior is, in part, learned.

Chapter 8: Neurotransmitters are made from dietary amino acids.

Chapter 11: Psychoactive drugs alter the functions of the brain, in some cases by affecting neurotransmission.

Chapter 12: Brain activity can influence our general health. Sleep deprivation, for example, can impair the immune system.

Chapter 15: Brain disorders may arise from environmental pollution.

Chapter 15: Many drugs that influence brain activity are obtained from tropical rainforest plants.

PRACTICE QUESTIONS

1. In a neuron that is not conducting an impulse, there is an excess of _____ ions inside the cell and an excess of _____ ions just outside the cell membrane.
2. An action potential is caused by a large number of _____ ions moving across the cell membrane toward the _____.
3. Which neurotransmitter(s):

   a. is a gas?

   b. is broken down by cholinesterase?

   c. is overproduced in Huntington's disease?

   d. was the first to be experimentally investigated?

   e. do depressed patients often have in insufficient amounts?

   f. are amines?
4. _____ is a neurotransmitter but not an amine; the enzyme _____ breaks it down in the synapse.
5. Name three parts of the forebrain.
6. For each of the following, name the sense organ that contains it:

   a. anvil bone

   b. iris diaphragm

   c. cones

   d. naked nerve endings

   e. tympanic membrane

   f. cochlea

   g. an area sensitive to acidic compounds
7. Each of the following is an example of what form of learning? Please be as specific as you can.

   a. Learning to associate horses with the word 'horse'.

   b. Being more alert to other stimuli when the fire alarm rings.

   c. Learning not to pay attention to traffic noises in one's neighborhood.

   d. Learning to recognize a TV show by its theme music.
8. How does the removal of GABA produce more dopamine?
9. What is the effect of monoamine oxidase inhibitors on neurotransmission, and why?
10. Name the types of sensory input that induce generator potentials in neurons of each of the following sense organs: skin, eye, ear, tongue, nose.
11. What are the functions of the blood–brain barrier?
12. What are the major ways in which brain anatomy has changed during evolution?
13. What are the major functions of each of the three divisions of the brain: forebrain, midbrain and hindbrain?

11 Drugs and Addiction

CHAPTER OUTLINE

Drugs are Chemicals that Alter Biological Processes

Drugs and their activity
Routes of drug entry into the body
Distribution of drugs throughout the body
Elimination of drugs from the body
Drug receptors and drug action on cells
Side effects and drug interactions

Psychoactive Drugs Affect the Mind

Opiates and opiate receptors
Marijuana and THC receptors
Nicotine and nicotinic receptors
Amphetamines: agonists of norepinephrine
LSD: an agonist of serotonin
Caffeine: a general cellular stimulant
Alcohol: a CNS depressant

Most Psychoactive Drugs are Addictive

Dependence and withdrawal
Brain reward centers and drug-seeking behaviors
Drug tolerance

Drug Abuse Impairs Health

Drug effects on the health of drug users
Drug effects on embryonic and fetal development
Drug abuse as a public health problem

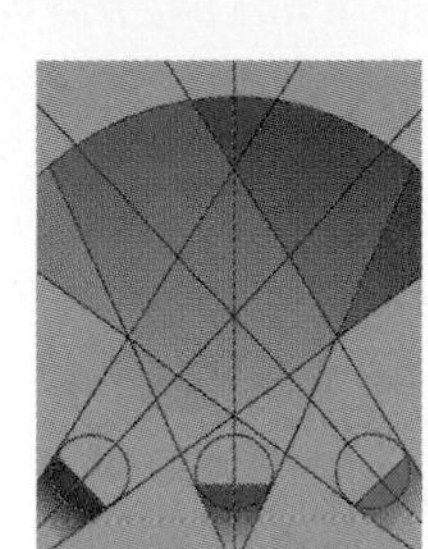

ISSUES

What is a drug?

How do drugs affect the body's functions? How do they affect the mind and senses?

How do drugs interact with one another?

How does the body change drug molecules?

What is addiction?
Why are some drugs addictive?
Is drug abuse the same as addiction?
Does drug abuse lead to addiction?

What are the social effects of drug abuse?

Are all forms of drug addiction treated as crimes?

Why do some argue that marijuana should be legalized, while others think that it should not?

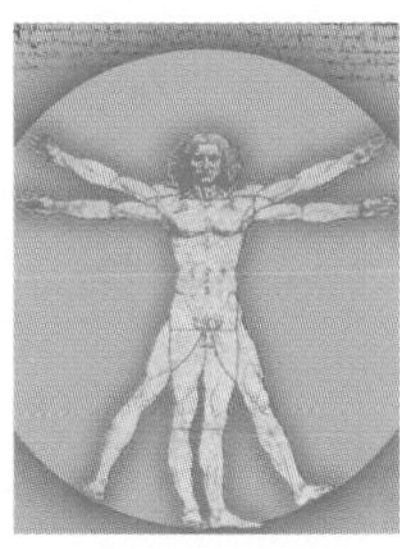

BIOLOGICAL CONCEPTS

Molecules (structure, diffusion)

Membranes and cell-surface receptors (agonists, antagonists)

Energy and metabolism

Organ systems (respiratory system, excretory system, placental circulation)

Homeostasis (drug tolerance, drug metabolism, excretion)

Central nervous system (brain)

Behavior (reinforcement, addiction, behavior modification)

Health and disease (public health issues)

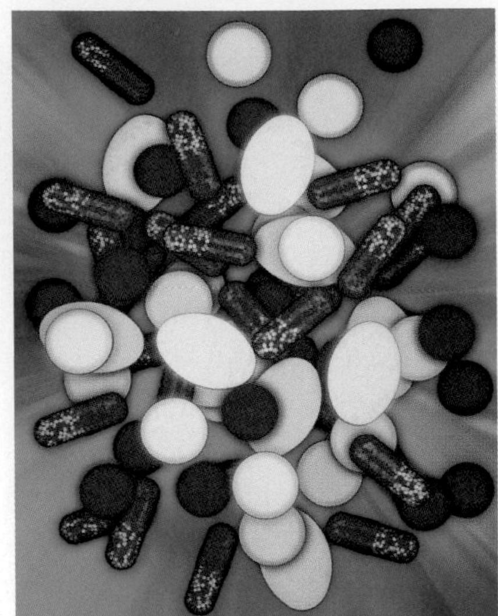

Chapter 11: Drugs and Addiction

Most of us are accustomed to the idea of taking drugs when we are sick. Many of these drugs are prescribed to fight bacterial infections, to fight cancer, or to regulate the body's physiological processes. Many other drugs are taken without medical supervision and for a wide variety of purposes. Many of these drugs are legal; some are not. The United States is the number one drug-producing and the number one drug-using country in the world. There is said to be a 'drug problem' that has led to a 'war on drugs;' yet there does not seem to be societal agreement on the answers to some very basic questions: What is a drug? What are the various legitimate uses of drugs? What is addiction and what makes a drug addictive? In this chapter we examine some recent research in biology and in related fields relevant to these questions. We also examine the basic biological principles that underlie our understanding of how drugs work. An understanding of the respiratory, circulatory, and excretory systems is needed to see how drugs enter and become distributed around the body, how long they stay in the body, and how they are eventually removed.

DRUGS ARE CHEMICALS THAT ALTER BIOLOGICAL PROCESSES

The term **drug** can have many meanings, depending on the context. To biologists, a drug is any chemical substance that alters the function of a living organism other than by supplying energy or needed nutrients. In a medical context, a drug may be thought of as any agent used to treat or prevent disease. Those who work in the field of drug addiction define a **psychoactive drug** as any chemical substance that alters consciousness, mood, or perception. As with all definitions in science, these are open-ended; no one definition can cover every possible situation. Each of these definitions expresses a slightly different concept, and each is correct within its contextual field.

This chapter emphasizes psychoactive drugs, that is, those that alter consciousness, mood, or perception. All psychoactive drugs alter biological functions, and are thus considered to be drugs on the basis of the first definition. Many, but not all, are also drugs by the medical definition.

Drugs and their activity

To understand how psychoactive drugs work, we need to know some general principles that apply all types of drugs. The study of drugs, their properties, and their effects is called **pharmacology**. Pharmacological principles explain how drugs can be both effective and dangerous, how both concentration and time affect the activity of drugs, and how interactions between drugs can change drug activity.

The activity of any drug varies with its **dose**, meaning the amount given at one time. There is an **effective dose**, the amount 'effective' in producing the desired change; in medicine, the effective dose is also called a therapeutic dose. For almost all drugs there is also a **toxic dose**, the amount at which the drug produces harmful effects, and a **lethal dose**, the amount that kills the organism. The more commonly used term, **overdose**, includes both toxic and lethal doses. Some drugs have a wide **margin of safety**; that is, the toxic dose is many hundreds or thousands of times as much as the therapeutic dose. For other drugs, the toxic dose or even the lethal dose may be very close to the effective dose.

The exact amount that is effective, toxic, or lethal differs with the chemical structure of the drug. It also differs from one individual to the next depending on body size and many other physiological variables. We discuss some of these variables later in the chapter.

The drug concentration achieved at any given location in the body depends on several factors: how the drug enters the body, how it is carried around the body, how it is transformed by the body's cells, and how it is removed from the body.

Routes of drug entry into the body

Most drugs enter the body through the digestive system or the respiratory system; a few are injected directly into the bloodstream. Drugs that are taken orally and swallowed must be able to withstand the acidic environment of the stomach and then be absorbable by the cells of the intestinal lining in the same ways in which food substances are taken up (see Chapter 8, p. 322). Drugs that enter through the respiratory system face different obstacles.

The respiratory system. The lungs are the body organ by which vertebrate land animals take up oxygen and give off carbon dioxide, a waste product of their metabolism (see Figure 8.14, p. 339). This process is controlled by the **respiratory system** and has two parts: the mechanics of breathing and the exchange of gases.

Breathing is accomplished, not by the lungs themselves, but by the diaphragm, a muscular layer below the lungs. When the muscles of the diaphragm contract, they pull the diaphragm downward, expanding the chest cavity. The laws of physics tell us that the pressure of a gas depends on the number of gas molecules in a given volume; therefore, when the volume of the chest cavity becomes larger, the pressure inside it decreases. The air pressure inside the lungs is then less than the air pressure outside the body. Physics also tells us that gases flow from areas of higher pressure to areas of lower pressure, so the enlargement of the chest cavity causes air to enter the lungs (inhalation). (Hiccups result when the downward movement of the diaphragm is more rapid than normal.) When the diaphragm relaxes, it returns to its higher position, decreasing the volume of the chest cavity and pushing air back out (exhalation) (Figure 11.1A).

Air passes into the lungs from the **trachea**, or windpipe, through branching airways called **bronchi** into thin-walled sacs called **alveoli**. Oxygen gas and carbon dioxide are exchanged between alveoli and the blood vessels of very small diameter (**capillaries**) that surround them (Figure 11.1B). The gases move by passive diffusion (see Chapter 8, pp. 322–323) across the cell membranes of the alveolar cells and the cells of the capillaries. Oxygen is in higher concentration in the inhaled air in the alveoli than it is in the capillaries, so oxygen diffuses *into* the blood. Carbon dioxide is in higher concentration in the blood than it is in the inhaled air, so it diffuses *out of* the blood into the alveoli, to be exhaled. The rate of diffusion depends on the difference in concentration and on the amount of surface area over which the diffusion can occur. The huge number of capillaries and the compartmentalized, saclike structure of the alveoli provide an enormous surface area (Figure 11.1C), making diffusion very rapid (the effects of compartmentalization on surface area are discussed in Chapter 9, p. 361).

Gaseous drugs, including many anesthetics, can enter the lungs, diffuse across cell membranes, and enter the blood by the same rapid mechanism as does oxygen. Toxic inhalants, including various solvents and propellants, can enter in the same way. Particulate matter, such as the chemicals in smoke, can adhere to the inner surfaces of the alveoli, causing damage to those surfaces while delivering drugs to the bloodstream (see below).

The route of entry and effective drug dose. The way in which a drug enters the body often affects its resulting concentration in body tissues. As an example, consider cocaine, a product of the coca plant, *Erythroxylon*

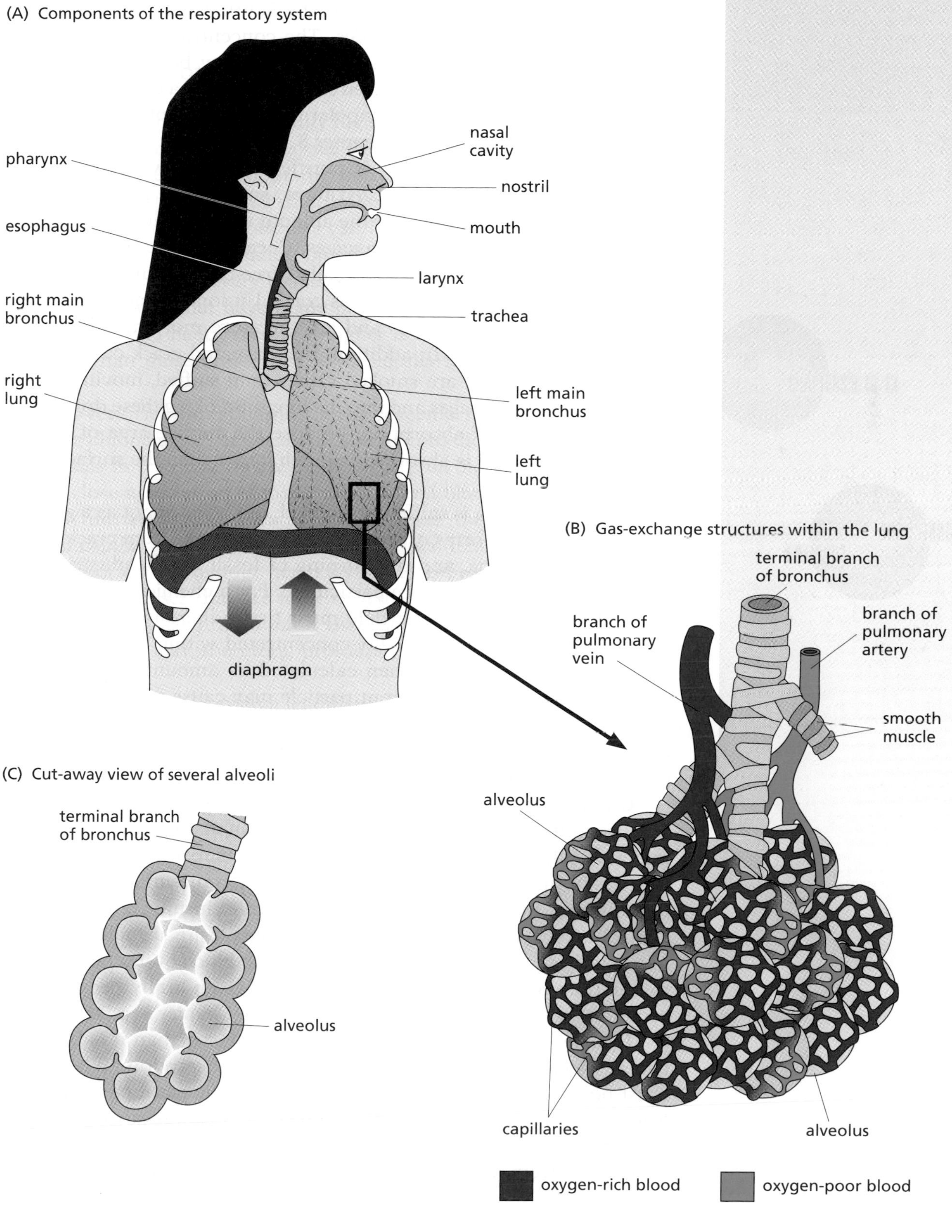

FIGURE 11.1

The human respiratory system.

by the **excretory system**. Another way is for the active form of a drug to be chemically altered, in a process called **drug metabolism**, so that it is no longer active. Drug metabolism can occur in a central location such as the liver, or in scattered locations, as in the inactivation of neurotransmitters within synapses between nerve cells (see Chapter 10, p. 419). In this section we look at these two mechanisms by which drugs can leave the body.

Metabolic elimination. Many drugs are chemically altered by the body into substances that no longer produce the drug's effects, although some of these breakdown products may have effects of their own. One example of a drug that is eliminated by drug metabolism is ethyl alcohol (ethanol), which is metabolized by the cells of the liver. The metabolic breakdown of ethyl alcohol involves several steps, but the overall rate is

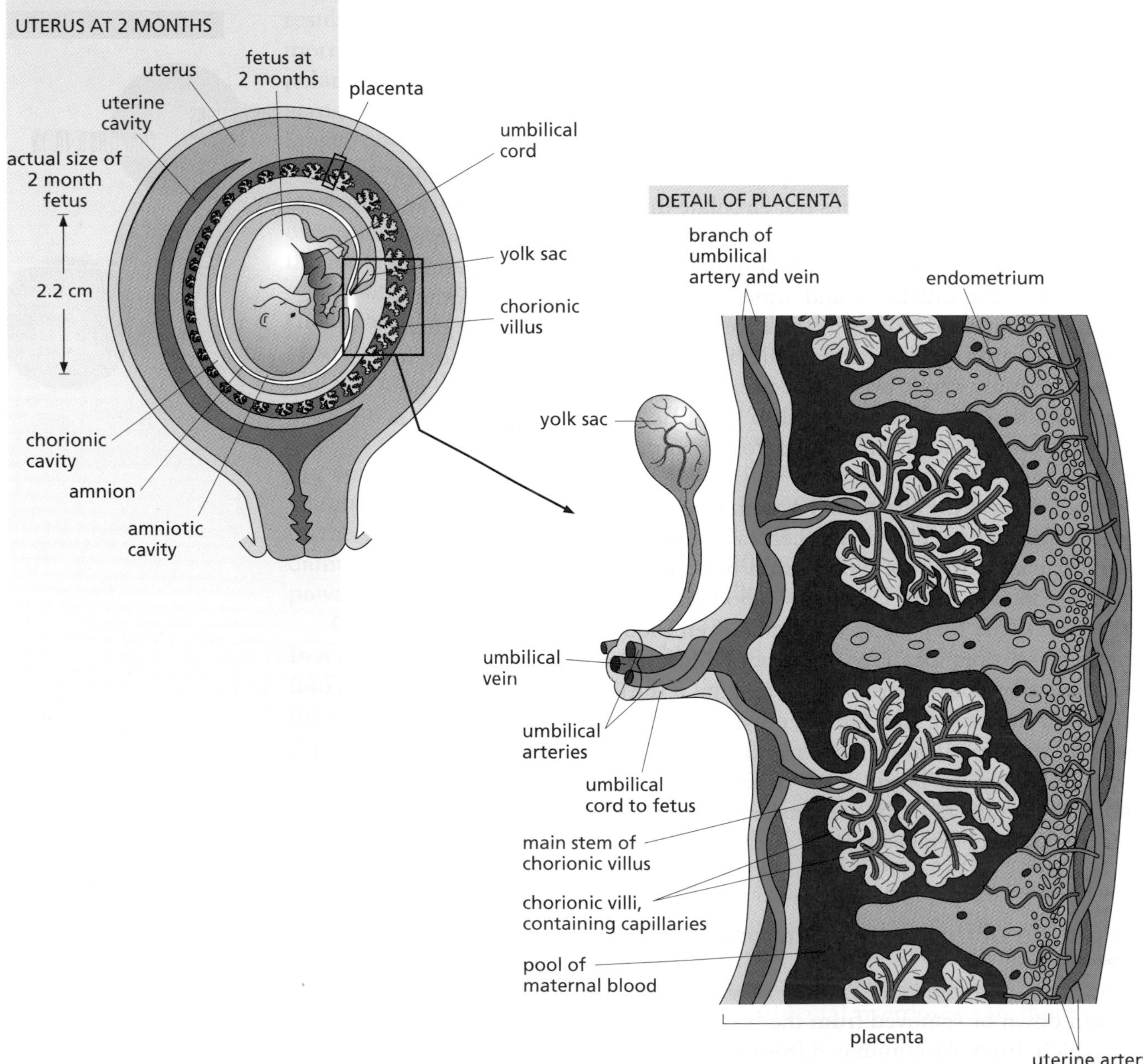

FIGURE 11.2

Maternal and fetal circulation in the placenta. Fetal blood flows through vessels in the umbilical cord to the chorionic villi, which are in close contact with maternal blood.

TABLE 11.1

Beverages containing equivalent amounts of ethyl alcohol (ethanol).

BEVERAGE	ALCOHOL CONTENT (%) (VARIES SOMEWHAT)	AMOUNT OF BEVERAGE EQUIVALENT TO 'ONE DRINK'	QUANTITY OF ETHANOL
Beer and ale	4	12 oz. can (350 ml)	1/2 oz. (15 ml)
Table wine	12–15	4 oz. glass (100 ml)	1/2 oz. (15 ml)
Dessert wine (sherry or fortified wine)	20	2/3 glass (2.5 oz. or 70 ml)	1/2 oz. (15 ml)
Distilled liquor (whiskey, vodka, etc.)	40	1.25 oz. (35 ml)	1/2 oz. (15 ml)
Liqueur	40	1.25 oz. (35 ml)	1/2 oz. (15 ml)

limited by the amount of the proton carrier NAD available to be reduced to NADH (see Chapter 8, p. 340) and by the levels of the enzyme **alcohol dehydrogenase**. The liver can metabolize 7–10 ml (about 0.25–0.33 fluid ounces) of alcohol per hour; if the rate of intake exceeds the rate of metabolism, intoxication results. As Table 11.1 shows, the intake of most forms of alcoholic beverages can easily exceed this limit.

One step in the metabolic breakdown of alcohol involves the production of acetaldehyde, which is immediately broken down by another enzyme, acetaldehyde dehydrogenase. Some people become sick from small amounts of alcohol because they lack this second enzyme, leading to a buildup of acetaldehyde, a toxic chemical. In some populations, the proportion of people who lack acetaldehyde dehydrogenase is quite high; for example, as many as 50% of all people in Japan and China lack this enzyme. Disulfiram (Antabuse), a therapeutic drug used in the treatment of alcoholism, works by inhibiting acetaldehyde dehydrogenase in people with normal levels of this enzyme. People who take this drug and then drink alcohol become very sick.

The excretory system. In addition to drug metabolism, drugs are also eliminated by the same system that rids the body of the normal waste products of metabolism, such as the nitrogen compounds derived from the breakdown of proteins. The process of removing these waste materials from the body is called **excretion**. The major route for the excretion of drugs and other substances is through the urine. The excretion of urine is vital to maintaining the blood and tissues at the proper concentrations of many types of ions. For example, calcium (Ca^{2+}) is an important second messenger in muscle contraction, and potassium (K^+), sodium (Na^+) and chloride (Cl^-) each function in maintaining charge gradients across membranes and in the propagation of nerve action potentials (see Chapter 10, pp. 413–417). Levels of these ions are monitored and controlled by feedback loops in the kidneys.

Urine is produced in three steps—filtration, reabsorption and secretion—all in the nephrons of the kidneys. The first step, filtration, takes place in the many thousand glomeruli within the nephrons. Each glomerulus consists of a mesh of capillaries surrounded by the Bowman's capsule (Figure 11.3). The circulatory system brings blood in, and low-molecular-weight substances, including ions, amino acids, glucose, urea, and water, are filtered out of the blood and into the Bowman's capsule, which empties into the proximal tubule. Larger molecules such as proteins are retained in the blood.

The second step, reabsorption, takes place in the proximal tubules and in the loops of Henle, both of which are surrounded by capillaries. Here the ions, amino acids, and glucose that the body can reuse are reabsorbed, meaning that they move back from the tubules into the blood. Reabsorption is by active transport (see Chapter 8, pp. 322–323) via transporter proteins in the membranes of the nephron cells. Much of the water is also reabsorbed.

FIGURE 11.3

The major organs of the human excretory system, responsible for the production of urine.

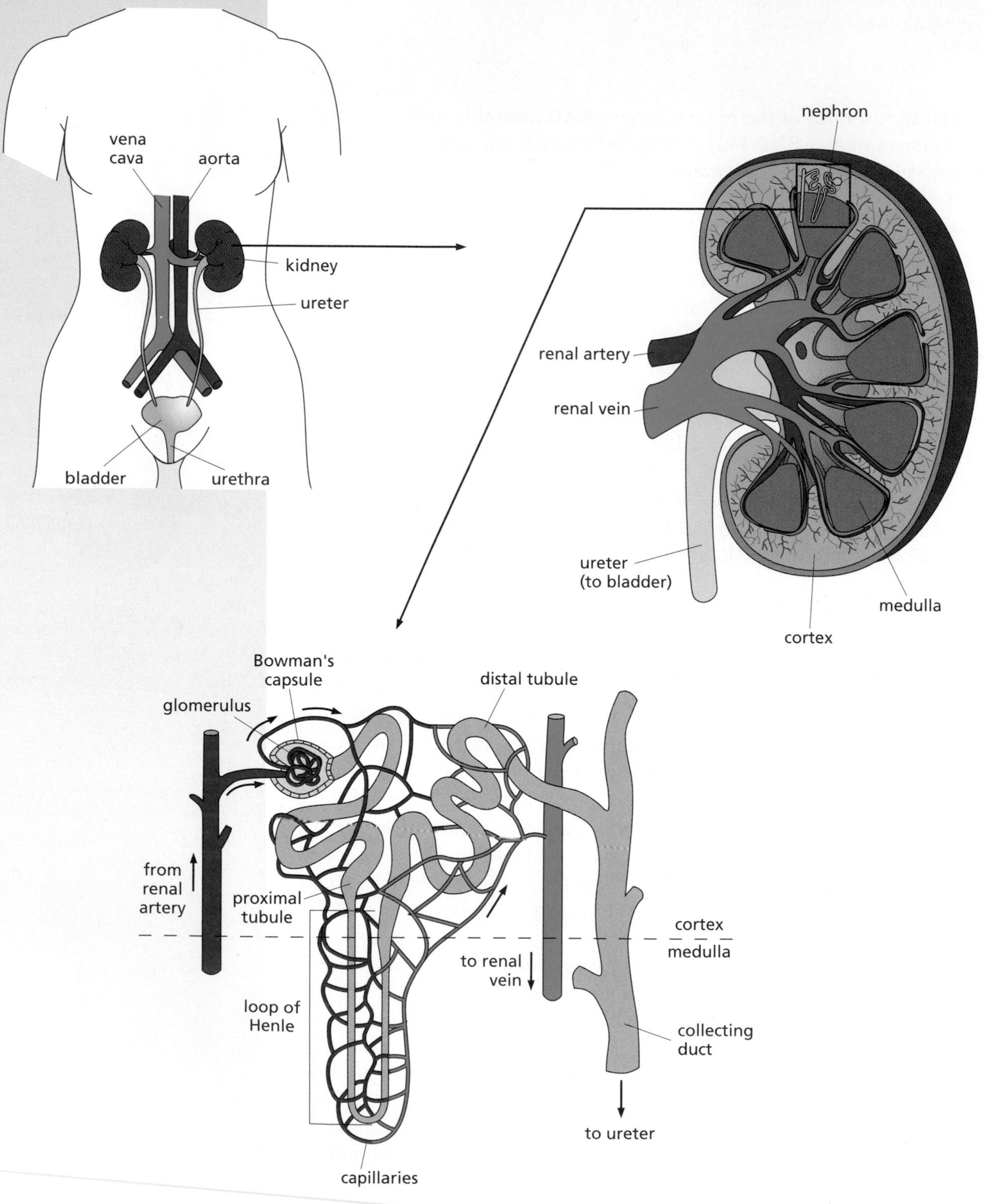

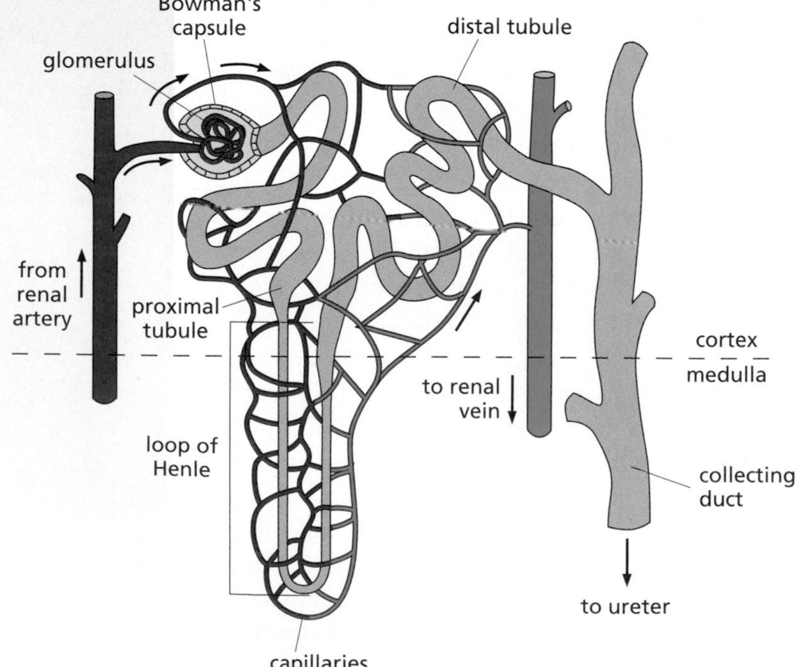

The third step, secretion, takes place farther along, in the distal tubules. Higher-molecular-weight substances including drugs and toxins are secreted from the blood into the tubules. After these three stages, the liquid in the tubules is called urine. Urine goes to the collecting ducts and then from the kidneys to the bladder, where it is stored or excreted (see Figure 11.3). Many drugs, or their breakdown products (metabolites), enter the nephrons at either the filtration or secretion steps and are excreted in the urine.

Drugs are also excreted in smaller amounts via saliva and sweat, and in air exhaled from the lungs. In nursing mothers, drugs may also be excreted into breast milk, with obvious consequences for the infant if the excreted drugs are still active, which they very often are.

Drug half-lives. The concentration of a drug in its active form can be measured, usually in the blood. The length of time required for the drug concentration to be decreased by half is called the half-life of the drug (Figure 11.4A). Cocaine, for example, has a **half-life** of 5–15 min, meaning that, regardless of the mode of intake of the drug or the initial starting dose, half of it will be gone in 5–15 min. A drug with a half-life of 10 min remains active for a shorter time than a drug with a half-life of 10 hr (Figure 11.4B), even if they start out at the same initial concentration in the blood. Drugs that are inactivated in a simple one-step process

FIGURE 11.4

Half-lives of various drugs. The half-life of a drug ($t_{1/2}$) is the time required for its active concentration to be reduced by half.

(A) Regardless of initial dose, a drug's concentration is reduced by half in one half life.

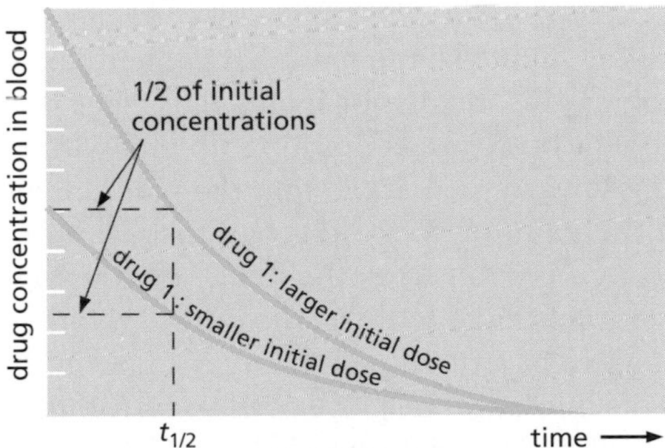

(B) Drugs with longer half-lives stay active for longer.

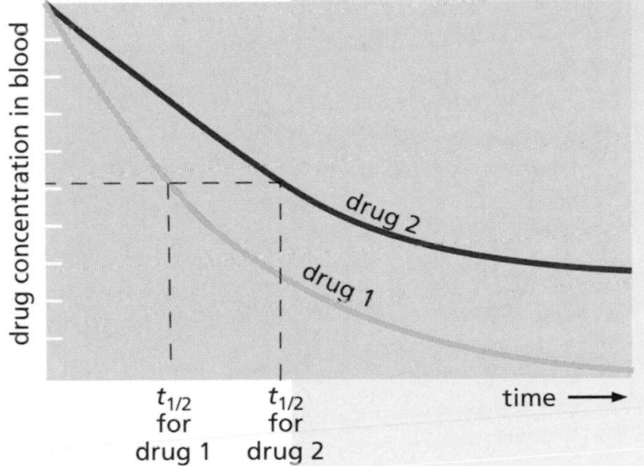

(C) Taking a second dose of drug before the first dose is completely eliminated raises the concentration, even if the two doses taken are equal.

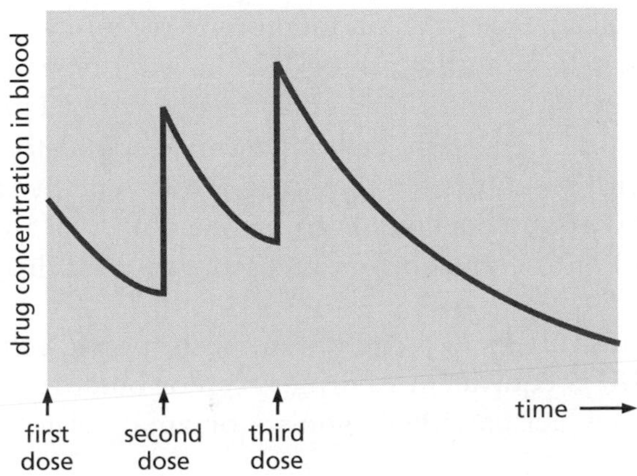

often have a shorter half-life than those whose inactivation pathways are more complex. For example, the half-life of alcohol is very short in most people, while marijuana remains in the blood for up to a week. Taking more of a drug before the original dose is completely eliminated increases the concentration of the drug to a level higher than the amount actually taken in the second dose (Figure 11.4C).

Drug receptors and drug action on cells

The activity of a drug depends on its dosage and its concentration in tissue. In many cases, drug activity also depends on the existence of specific receptors for the drug.

Specific receptors for drugs. After a drug has reached the site of its action, it acts on individual cells at that site. The ability of a drug to act on a particular cell most often depends on the presence of a cellular **receptor** for that drug. Picture a receptor as being like a glove. For a drug to bind to that receptor, the drug must have the shape of a hand. A small hand fits into a large glove, but, for any glove, there is some hand size that fits best. A large hand does not fit into a small glove. Those substances that have the best fit to a receptor have the greatest activity. Those that do not fit have no activity.

Many chemical molecules can have right-handed or left-handed shapes depending on the directions in which their bonds are arranged. That is, two molecules composed of identical atoms can differ in their shape. Right-handed molecules are said to be mirror images of left-handed molecules. As an aid in visualizing this, hold your hands flat. Place them in front of you, with the fingers of both hands pointing up and with the thumbs of both hands pointing toward the left (so that the palm of your left hand is towards you and the palm of your right hand is away from you). Look at your right hand in a mirror. Does the image of your right hand in the mirror look like your left hand without the mirror? Now put your right hand into the right glove of a fitted pair, then try to put it into the left glove. Your right hand fits into only one glove of the pair. Similarly, biological receptors are specific for right-handed or left-handed molecules, not both.

Tissue locations of receptors and locations of drug actions. Receptors can be located on the cell surface (plasma membrane receptors) or within the cytoplasm of the cell. Not all types of cells have all types of receptors, and the number of copies of a particular receptor molecule on a cell may change over time. The only cells that can respond to a particular drug are those with receptors for that drug. Thus, whether a drug is active in a particular tissue depends both on the ability of the drug to get to the tissue and on the presence of receptors for that drug on the cells of that tissue.

The binding of neurotransmitters to receptors on postsynaptic cells (see Chapter 10, pp. 416–418) is equivalent to drug–receptor binding. In fact, many psychoactive drugs act as either agonists or antagonists of neurotransmitters. An **agonist** is a substance that elicits a particular response or stimulates a receptor. For example, an agonist of a neurotransmitter is any other substance that produces the same response as that neurotransmitter (Figure 11.5A and B). An **antagonist** is a substance that inhibits a response (Figure 11.5C). The interactions of the activities induced in all the receptor-bearing cells produce functional effects on the body as a whole.

Side effects and drug interactions

Because drugs are not taken by isolated cells, many effects are likely to be produced, especially for drugs that circulate throughout the body. Although a drug is usually intended to produce a specific effect, drugs almost always have effects other than the intended ones; these other effects are called **side effects**. Side effects may be weak or strong and can vary from person to person. Side effects can also vary from beneficial or harmless to harmful or even lethal. Calling something a side effect simply means that it is not the main effect or the reason for which the drug was used.

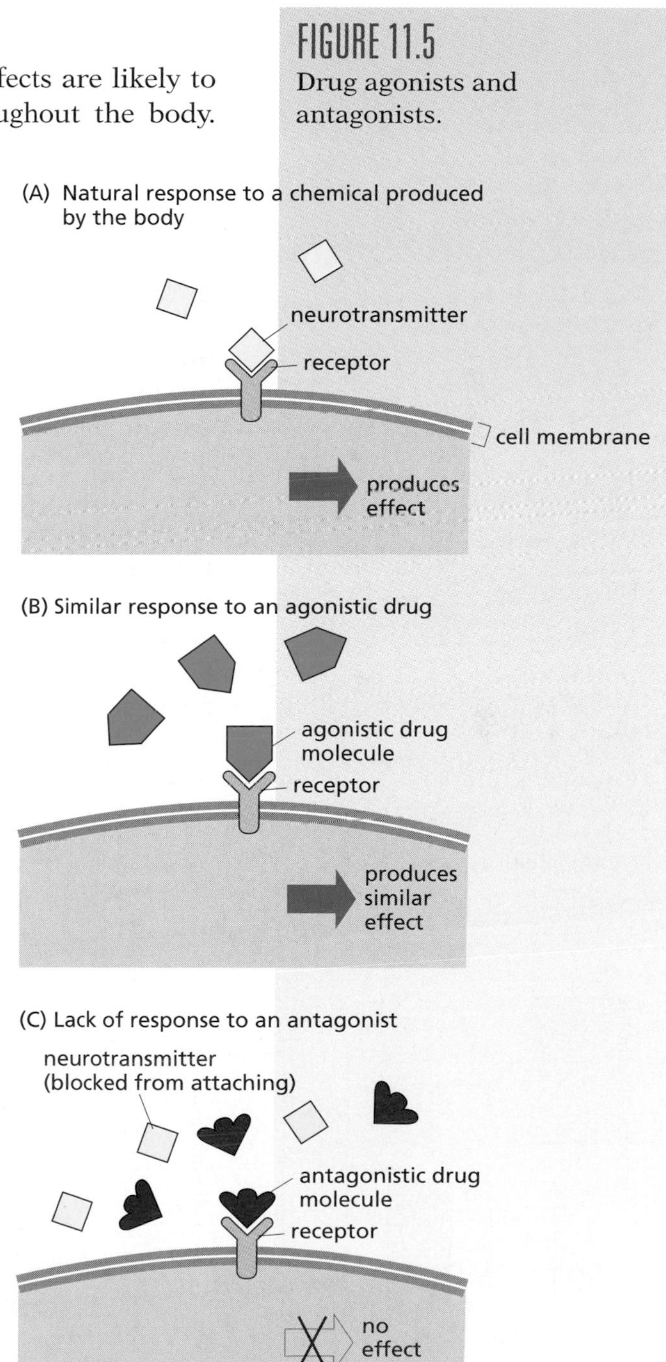

FIGURE 11.5 Drug agonists and antagonists.

Types of drug–drug interactions. When two or more drugs are taken, they may interact in various ways. The simplest interaction is called an **additive effect**. When two drugs have an additive effect, the response to taking them together equals the sum of the responses produced by each drug individually (Figure 11.6A). In a **synergistic interaction**, one drug **potentiates** the action of the other so that the total response is *greater* than the sum of the responses to each drug separately (Figure 11.6B). An **antagonistic interaction** is one in which one drug inhibits the action of another so that the total response is *less* than the response to the two drugs individually (Figure 11.6C).

Mechanisms of interactions. One mechanism for drug interactions is for one drug to change the threshold for response to the other drug. The **threshold** is a value (in this case, a drug dose) below which no effect is detectable. One drug may lower the threshold for response to the second drug, making the body responsive to a lower concentration of the second drug. This could happen, for example, if one drug increased the number of receptors or the affinity of the cell receptors for the other drug. In contrast, one drug may antagonize the action of a second drug by raising the threshold for response to

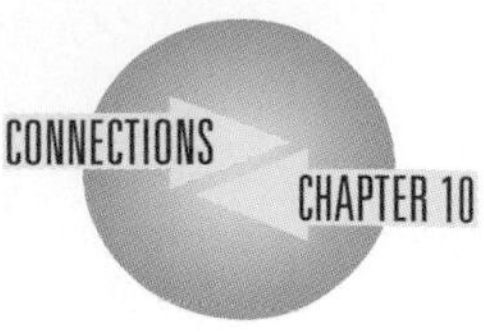

the second drug or decreasing the number or affinity of receptors for it. Again, this is similar to the actions of neurotransmitters (see Chapter 10, pp. 416–418).

Differences in drug half-lives have implications for drug dosage and interactions between drugs. Because drugs with long half-lives remain in the body for a long time, a second dose adds to the remaining fraction of the first dose to produce a concentration higher than would result from either dose separately (see Figure 11.4C). Drugs with long half-lives also have the potential to interact with other drugs taken a long time after the first drug. Thus, it is not necessary that two drugs be taken at the same time for them to interact.

Another way in which drugs can interact is by altering the body's ability to metabolize other drugs. Barbiturates are eliminated from the body by enzymes contained in liver cells, and the use of barbiturates causes more of these enzymes to be made. Because these same enzymes eliminate other drugs, barbiturate use lowers the body concentrations of other medications, including steroid hormones. Body concentrations of estradiol, a steroid in birth control pills, is decreased by barbiturates, so taking birth control pills along with using barbiturates may result in an unwanted pregnancy.

Frequency of drug–drug interactions. Drug–drug interactions are an increasingly important problem, because many people take medication on a lifetime basis for the control of chronic conditions such as high blood pressure. A recent study on an elderly population demonstrated that the probability of having some adverse interactions from medications was 75% if they were taking five different medicines and 100% if they were taking eight or more medications on a regular basis. However, potentially harmful drug interactions are not solely a problem for elderly people. There can also be harmful interactions between various street drugs, between street drugs and medications, or between any of these and alcohol.

FIGURE 11.6

How drugs interact.

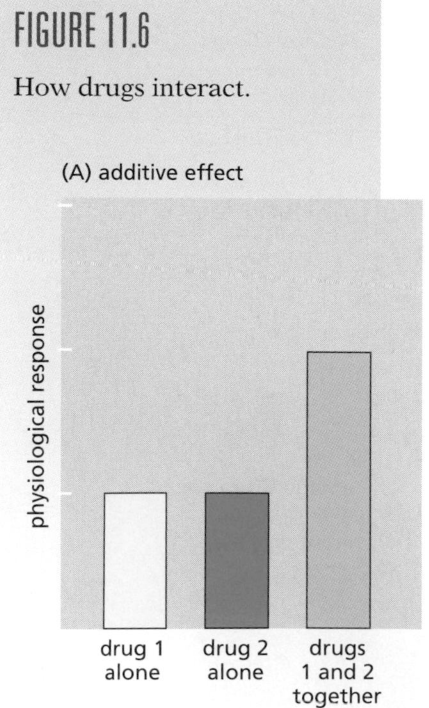

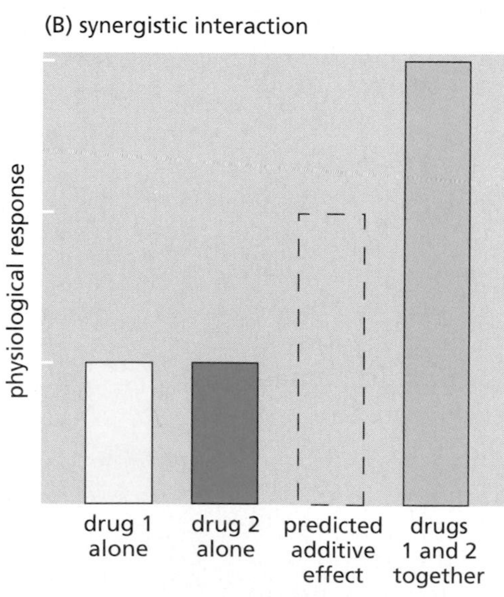

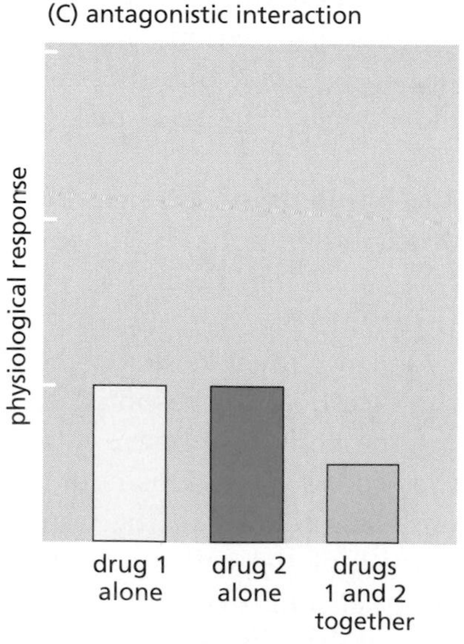

THOUGHT QUESTIONS

1. The physiological effects of a drug usually depend on its concentration in body tissues. If two people, one of whom weighs 60 kg (about 130 pounds) and the other 90 kg (about 200 pounds), take the same *amount* of a drug (one beer each, for example, or two aspirins), will the concentration reached in their body tissues be the same? If other factors were equal, which person do you think would be more strongly affected by the drug?
2. How would you set up an experiment to monitor the rate at which a drug is delivered to various body tissues? Consider both animal and human test subjects.
3. Why do left-handed and right-handed forms of the same molecule frequently differ in their ability to act as drugs? When a difference of this kind affects the activity of a drug molecule, what does it indicate?
4. How long will it take for the blood concentration of a drug whose half-life is 20 hr to be reduced to one-eighth of its peak concentration?
5. Devise a model of the mechanism of action of two drugs with their receptors that would account for additivity between the two drugs. How would you modify this model to account for antagonism? How would you modify it to account for synergism?

PSYCHOACTIVE DRUGS AFFECT THE MIND

The brain coordinates the activities of the rest of the body (moving, sleeping, eating, breathing), but does it also produce activities that we associate with the mind? Does biochemical activity in the brain produce perceptions, emotions, moods, and personality? Does it produce thoughts, dreams, and hopes? The central theory of neurobiology is that the mind and the brain are one and the same. As a framework that guides (and limits) research, the assumption that mind equals brain is a good example of a research paradigm (see Chapter 1, p. 14). There is probably no theory in biology that is today more controversial, both among biologists and between biologists and the public.

As we saw in Chapter 10, there is now considerable evidence in support of this theory. Electrical and biochemical activities have been measured in the brain during dreams and thought. Some diseases in which there is brain degeneration can be accompanied by changes in personality. Mental illnesses such as depression are associated with changes in brain neurotransmitters and can be treated with drugs (see Chapter 10). Additional evidence for the 'mind equals brain' theory comes from research on psychoactive drugs. Psychoactive drugs act directly on the nerve cells of the brain or central nervous system (CNS) to produce changes in consciousness, mood, or perception in highly drug-specific ways. Unlike **sensation**, which is the mere receipt of stimuli, **perception**, a higher-order brain function, includes the processing and interpretation of stimuli coming from the physical world. Both sensation and perception can be altered, sometimes permanently, by drugs. People may become paranoid due to permanent damage from drugs; that is, they perceive danger that is not present in their incoming sensations. They may also experience hallucinations, chemically induced changes in sensation.

Some psychoactive drugs (opiates, marijuana, and nicotine) work through specific receptors, and some (like amphetamines and hallucinogens) work because they are structurally similar to a neurotransmitter and bind to receptors for that neurotransmitter. Drugs that do not work via receptors (e.g., caffeine and alcohol) have a more generalized effect because they act on many types of cells.

Opiates and opiate receptors

Opiates are **narcotics**, that is, drugs that cause either a drowsy stupor or sleep. Most narcotics also cause some degree of **euphoria** (literally, 'good feeling'), and most are highly addictive. In low doses, most narcotics can be used as painkillers (analgesics). High doses of narcotics can produce coma and death. Most of our common narcotics, including heroin, morphine and codeine, are derived from the opium poppy (*Papaver somniferum*) and thus are known as **opiates**. The painkilling and euphoria-inducing properties of opiates have been known for thousands of years in Asia, as has their ability to cause addiction. Pain reduction results from changes in the release of the neurotransmitters acetylcholine, norepinephrine, dopamine, and a pain-related substance called substance P. In people who are not in pain, opiates produce a euphoria by action on neurons in certain locations in the brain.

Opiates and similar chemicals (opioids) act on cells via one or more specific opiate receptors to produce their various effects. Synthetic antagonist drugs have been made that block the binding of opiates to their receptors. These narcotic antagonists, of which naloxone (Narcan) and naltrexone are examples, are useful in the treatment of opiate overdoses. Some other synthetic drugs act as both agonists and antagonists: they bind to opiate receptors, thus blocking the effects of narcotics like morphine (antagonistic action), but their own binding produces some euphoric effect (agonistic action), sometimes leading to their abuse.

Opiate receptors are found on the membranes of many other types of cells in addition to neurons. Opiates consequently relax various muscles, including those of the colon. Consequently, morphine is an ingredient in some prescription anti-diarrheal medications (Paregoric), and severe constipation is an effect of long-term opiate use.

Several neurobiologists hypothesized that opiate receptors would never have evolved unless they served some adaptive function other than just allowing addictive behaviors to be learned. This type of thinking led to the hypothesis that opiate receptors must have some normal physiological molecules that could bind to them. The search for such molecules led to the discovery of **endorphins** and **enkephalins**, sometimes called **endogenous opiates** because they are produced within the body (endogenous), rather than being taken in from the outside (exogenous). These are peptides (short chains of amino acids) that have a molecular shape very similar to that of a portion of certain opiate molecules. The current hypothesis is that the endogenous opiates act as inhibitory neurotransmitters, decreasing the activity of the neurons that normally signal pain and stress. These same endogenous opiates have other effects throughout the body, including actions on the immune system, which are discussed in Chapter 12.

Marijuana and THC receptors

A number of psychoactive drugs are contained in marijuana smoke, the most active of which is Δ9–tetrahydrocannabinol (THC). There are receptors for THC in the parts of the brain that influence mood. Binding of THC to these receptors produces an altered sense of time, an enhanced feeling of closeness to other people, and an intensity of sensory stimuli. In higher doses, marijuana can also cause hallucinations. Unlike the opiate receptors, for which endogenous brain chemicals have been found, endogenous substances that bind to THC receptors have not yet been found. However, researchers have found that certain compounds found in chocolate may bind to THC receptors. People who refer to their love of chocolate as an addiction may thus be close to the truth.

Stimulation of the THC receptors causes a release of norepinephrine by the nerves in the median forebrain bundle, producing euphoric effects. There are also THC receptors on the cells of the hypothalamus (see Figure 10.13, p. 430), a secretory part of the brain that regulates the steroid sex hormones. Long-term marijuana use decreases testosterone levels and numbers of sperm in males and alters the menstrual cycle in females.

Marijuana has long been recognized as a potentially dangerous and addictive drug, and its use is illegal in most countries. Later in this chapter we examine the possible medical uses of marijuana and the efforts by some people to relax legal restrictions on such medical use.

Nicotine and nicotinic receptors

Cigarette smoke contains over 1000 drugs, a large number of which are carcinogens (see Chapter 9, p. 391). The primary psychoactive drug among them is nicotine. In the brain, nicotine acts to stimulate the cerebral cortex, possibly by a direct effect on the cortical neurons, which have a series of nicotinic receptors. Nicotine also acts by stimulating nicotinic receptors on the neurons in the sympathetic ganglia, releasing the neurotransmitters acetylcholine, epinephrine and norepinephrine. In the brain, norepinephrine produces increased awareness. In the rest of the body, these neurotransmitters produce a variety of physiological effects: increased heart rate and blood pressure, constriction of blood vessels, and changes in carbohydrate and fat metabolism.

Amphetamines: agonists of norepinephrine

Amphetamines are an example of a type of drug called CNS stimulants. All amphetamines are derivatives of ephedrine, a drug originally obtained from the mah huang plant (*Ephedra sinica*). CNS stimulants increase behavioral activity by increasing the activity of the reticular formation. The reticular formation is the portion of the brain that normally maintains a baseline level of neuronal activity in the brain as a whole, thus keeping the body at a baseline level of wakefulness and awareness (see Chapter 10, pp. 443–444). CNS stimulants have side effects on organs outside the brain. Because of their effects on judgment and their effects on other organ systems, such as the heart (see Chapter 8, pp. 326–327) and diaphragm (see Figure 11.1), they are dangerous drugs, accounting for 40% of all drug-related trips to the emergency room and 50% of all sudden deaths due to drugs.

Amphetamines mimic the effects of the neurotransmitter norepinephrine by binding to norepinephrine receptors (see Chapter 10, pp. 418–419). They can also indirectly increase norepinephrine activity by blocking its reuptake from the synapse and by inhibiting monoamine oxidase (MAO), the enzyme that normally breaks down norepinephrine. Either mechanism results in more norepinephrine remaining in the synapse to act on the post-synaptic cell (see Figure 10.6, p. 418). Prolonged use of high doses of amphetamines can induce a form of psychosis that includes aggressiveness, delusions, and hallucinations, possibly because of an oversupply of an enzyme involved in norepinephrine synthesis.

A controversial drug of this group is methylphenidate (Ritalin). In most human subjects, it has mild amphetaminelike effects similar to those described above. However, the drug has quite the opposite effect (called a 'paradoxical effect') in children who have attention deficit hyperactive disorder (ADHD, formerly called ADD)—it reduces their hyperactivity. The reason for this paradoxical effect seems to be related to the fact that hyperactive children actually have a lower than normal function in the reticular formation, the area of the brain that keeps the rest of the brain alert. Such children may need to be constantly moving around to arouse their reticular formation. Ritalin's chemical stimulation of this area obviates the children's need for movement.

One currently abused drug of this group is methylenedioxymethamphetamine (MDMA), also known as 'ecstasy.' MDMA triggers the release of all stores of the neurotransmitter serotonin from brain neurons (see Figure 10.6, p. 418). It also blocks the reuptake of serotonin, producing prolonged alterations of sensory perceptions. In addition, one of the normal targets of serotonin is the hypothalamus, the part of the brain that regulates body temperature. Prolonged triggering of the hypothalamus raises body temperature to damaging, and even fatal, levels.

LSD: an agonist of serotonin

Lysergic acid diethylamide (LSD) is derived from the fungus *Claviceps purpurea*, which grows on rye. MDMA triggers release of serotonin inducing sensory alterations. In contrast, LSD, and the related compound psilocybin from mushrooms found in Central and South America, are structurally similar to serotonin. These drugs therefore act as serotonin agonists and activate the nerve cells that normally respond to serotonin, leading to altered sensory perception and hallucinations. 'Altered perception' means a change in the awareness of stimuli that actually exist. Colors may become more bright, or sounds more clear. Perceptions of the sizes of objects and of speed or time may be altered. **Hallucination**, in contrast, is the perception of things for which no outside physical stimuli have been received. Hallucinations can be visual, auditory, olfactory, or cognitive. Heavy use of these serotonin agonists leads to permanent brain damage, with symptoms ranging from impairments of memory, attention span, and abstract thinking to severe, long-lasting psychotic reactions.

Caffeine: a general cellular stimulant

Not all psychoactive drugs produce their effects via action on CNS neurotransmitters and their receptors. Some produce their effects within cells. Caffeine, for example, works inside cells to increase their rate of metabolism by inhibiting the breakdown of the second messenger cyclic AMP. Caffeine thus has a general stimulatory effect on cells throughout the body, including neurons in the brain. Like other CNS stimulants, caffeine increases the general level of awareness via action on the neurons of the reticular formation of the brain. In higher doses, or in more susceptible individuals, it can also produce insomnia (inability to sleep), anxiety, and irritability. It increases the heart rate, the respiratory rate, and the rate of excretion of urine by the kidney. It dilates peripheral blood vessels, but constricts the blood vessels of the CNS, which produces headaches in some people at high concentrations.

Caffeine is derived from several types of plants, including the beans of the coffee plant (*Coffea arabica*), the leaves of the tea plant (*Thea sinensis* or *Camellia sinensis*), the seeds of the cocoa plant (*Theobroma cacao*, from which we get chocolate), and nuts from the kola tree (*Cola acuminata*, an African tree, the source of cola beverages). Table 11.2 shows the amounts of caffeine in different beverages and nonprescription medications.

Alcohol: a CNS depressant

Ethyl alcohol belongs to a category of drugs called **CNS depressants** because they depress the functioning of the CNS by inhibiting the transmission of signals in the reticular formation. Also included in this category are barbiturates and tranquilizers. Because these drugs lower the general level of awareness, they are also called sedatives or hypnotics. Because they all affect the reticular formation, when two or more CNS depressants are taken together, the effect is stronger (either additively or synergistically) than when either is used alone. The actions of barbiturates and tranquilizers are mediated through receptors on neurons in the brain,

TABLE 11.2

Caffeine content of beverages and nonprescription medications.

FORM OF INTAKE	APPROXIMATE CAFFEINE DOSE (mg)
HOT BEVERAGES, PER 6 OZ. CUP (170 ml):	
Coffee, brewed	100–180
Coffee, instant	100–120
Tea, brewed from bag or leaves	35–90
Cocoa	5–50
Coffee, decaffeinated	2–4
CARBONATED BEVERAGES, PER 12 OZ. CAN (350 ml):	
Cola drinks; also several others	35–60
NONPRESCRIPTION (OTC) DRUGS, PER TABLET:	
Stimulants	
Vivarin, Caffedrine	200
NoDoz	100
Analgesics (pain relievers)	
Excedrin extra strength	65
Midol maximum strength	60
Anacin, Bromo Seltzer, Cope, Emprin	32

whereas alcohol produces a more generalized effect because it acts by making all cell membranes more fluid.

Alcohol is soluble in both water and fat and is thus readily able to pass through the plasma membrane of the cells forming the blood–brain barrier. The portions of the brain affected depend on the dose: the higher the dose, the deeper into the brain the alcohol penetrates. Even low doses of alcohol can impair a person's response time, with devastating (often fatal) consequences if that person is driving a motor vehicle or boat. In the United States, alcohol is involved in over 60% of all motor vehicle fatalities and in over half of all drownings. Higher doses of alcohol suppress the reticular formation's stimulation of those portions of the brain involved in involuntary processes, such as the brain stem and the medulla oblongata. Depression of the medulla can result in the cessation of breathing (respiratory arrest) and death.

The effects of alcohol on behavior can be predicted by the blood alcohol level (Table 11.3). Blood alcohol level is measured as the number of grams of alcohol in each 100 ml of blood. Thus 1 g of alcohol per 100 ml equals a 1% blood alcohol content and 100 mg equals 0.1%. Since alcohol is evenly distributed throughout the body, the actual blood alcohol content that results from drinking a given amount of alcohol varies with the blood volume of the person, which is approximately proportional to the muscle weight of the person. A blood alcohol level of 0.1% is the level at which a person can be charged with Driving While Intoxicated (DWI) or Operating [a motor vehicle] Under the Influence (OUI) of alcohol in most states of the United States, and several states have amended their laws to make the limit even lower. At a blood alcohol content of 0.05%, the probability of being involved in an automobile accident is 2–3 times higher than for someone who has not been drinking.

TABLE 11.3

Brain and behavioral effects of alcohol consumption. The greater the amounts of alcohol consumed, the deeper within the brain the alcohol penetrates after it is absorbed across the blood–brain barrier. Blood alcohol content is given for a 150 pound (68 kg) person; in general, persons weighing less will experience higher blood alcohol levels after consuming the same amounts.

BLOOD ALCOHOL CONTENT (%)	NUMBER OF DRINKS*	EFFECTS
0.01–0.04	1–2	Slightly impaired judgment Lessening of inhibitions and restraints Alteration of mood
0.05–0.06	3–4	Disrupted judgment Impaired muscle coordination Lessening of mental function
0.07–0.10	5–6	Deeper areas of cortex affected Slower reaction time Exaggerated emotions Talkativeness or social withdrawal Mental impairment Visual impairment
0.11–0.16	7–8	Cerebellum affected Staggering Slurred speech Blurred vision Greater impairment of judgment, coordination, and mental function
0.17–0.20	9–10	Midbrain affected Inability to walk or do simple tasks Double vision Outbursts of emotion
0.21–0.39	11–15	Lower brain affected Stupor and confusion Increased potential for violence Noncomprehension of events
0.40–0.50	16–25	Activity of lower brain centers severely depressed Loss of consciousness; shock
0.51 or more	26 or more	Failure of brain to regulate heart and breathing Coma and death

* Based on equivalent amounts of alcohol: 12 oz. of beer, 4 oz. of wine, 1.25 oz. of liquor (see Table 11.1).

THOUGHT QUESTIONS

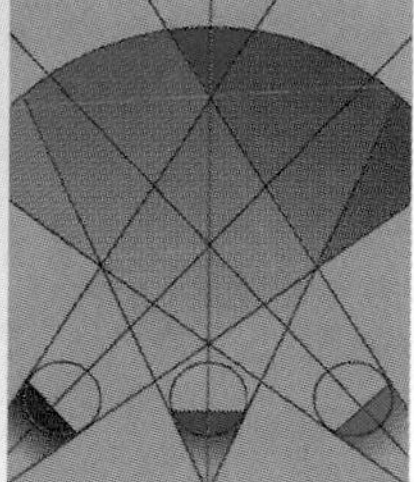

1. Does the finding that many drugs act directly on the cells of the brain to alter perception, mood, and consciousness necessarily mean that there is no 'mind' or 'spiritual essence' apart from the physical entity of the brain?

2. Can a person accept the scientific findings in neuroscience without accepting the central theory that the mind and the brain are one? Can the central theory of neurobiology ever be proved beyond doubt?

3. Alcohol concentrations in the body are measured as 'blood alcohol' levels. After alcohol is consumed, and before any of it is eliminated from the body, does all of it remain in the blood? Where else does it go?

4. In many countries the drinking age is the same as the driving age. In the United States, the driving age is lower than the drinking age. What kinds of impacts has this had in various societies?

MOST PSYCHOACTIVE DRUGS ARE ADDICTIVE

Some drugs that have uses as medicines, and many others that do not, are also used socially. All cultures have used at least some drugs, particularly psychoactive drugs, for nonmedical purposes that can be described as social. For example, in many cultures, wine is a common accompaniment to food, and wine is also used in many religious ceremonies.

Excessive or harmful social use of a drug is considered **drug abuse**, or **substance abuse**. Drug abuse is a major problem that affects many thousands of people, their families, and many other people with whom they interact. Most drugs that are abused socially are psychoactive drugs. These drugs have both direct effects on the user and indirect effects on other people owing to the user's altered behavior.

If psychoactive drugs have been used by all cultures, why not approve their use? In most cultures in which the social use of psychoactive drugs has been endorsed by tradition, the uses have been highly ritualized or ceremonial, and the decision of how much drug to use is not left up to the individual. Psychoactive drugs such as alcohol impair a number of higher-order mental functions (see Table 11.3). One of these is judgment, the very mental function needed for a person to be able to distinguish between use and abuse. In addition, most, but not all, drugs that are abused are addictive.

Addiction has been defined as a compulsive 'physiological and psychological' need for a substance, implying that there is both a biological basis and a mental basis for addiction. However, as more psychologists have accepted the neurobiology paradigm that 'mind equals brain' and that all brain functions are biochemically based, the distinction between physiological and psychological addition has become increasingly blurred. The above definition of addiction carries with it the assumption that all addictive drugs are psychoactive. The term 'psychoactive drug' is not a synonym for 'addictive drug,' however, because not all psychoactive drugs are physiologically addictive. Hallucinogens such as LSD, for example, are not known to be addictive. The numbers of people in the United States using addictive drugs are given in Figure 11.7.

FIGURE 11.7

Users of addictive drugs in the United States in 1991. People are listed as users if they use the drug at least once a week (or, for heroin, at least once in the past year).

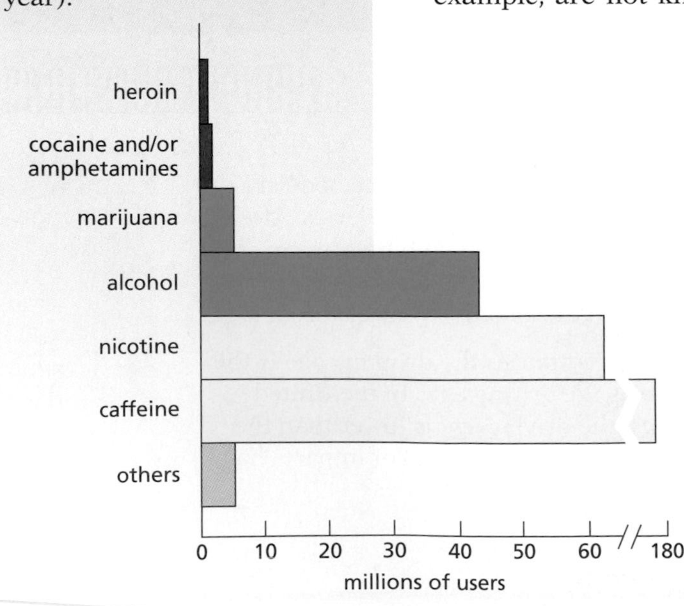

Dependence and withdrawal

Addictive drugs cause a physiological **dependence**, meaning that the person can no longer function normally without the drug. Once a person has become dependent on a drug, cessation of drug taking produces the biological symptoms of **withdrawal**. The length of time required for the development of dependence varies with the drug, as does the severity of withdrawal. Dependence on morphine and related drugs develops very quickly; withdrawal begins within 48 hr of the last dose and lasts for about 10 days. People are very ill during this time, but withdrawal is rarely fatal. Withdrawal from alcohol dependence is physically much more severe and can sometimes be fatal.

One form of addiction is psychological dependence. In this form of addiction, the physiology of the brain has become dependent on the drug in such a way that the person can no longer function normally without the drug. The sensations felt during withdrawal tend to be the opposite of the sensations produced by the original drug taking. For example, a person dependent on depressant drugs may become anxious and agitated during withdrawal. People who have become dependent on painkillers feel pain when the drug is withdrawn, long after the original, biological source of the pain has been eliminated, and these pain sensations often lead to resumption of the drug.

Cocaine, caffeine, nicotine, and marijuana all produce psychological dependence. Withdrawal from these drugs is typified by symptoms that affect the physiology of the brain rather than the physiology of the entire body, and because of this they were originally thought to be nonaddictive. In light of the neurobiology paradigm, it is now known that all four are addictive. In fact, if we define level of addiction as the degree of difficulty of getting through withdrawal without returning to the drug, nicotine must be considered as one of the most highly addictive drugs known.

One aspect of psychological drug dependence in humans involves the context—the places in which the drugs are taken and the people sharing the experience. **Conditioned withdrawal syndrome** refers to the fact that visual cues associated with drug taking (for example, seeing one of these people or places) can bring on the physiological symptoms of withdrawal in drug-dependent people even when the body is not actually in withdrawal, possibly leading them to seek the drug again. For this reason, many drug recovery programs recommend that recovering addicts stay away from specific people and locations associated with drug use.

Brain reward centers and drug-seeking behaviors

Neuroscientists have discovered both negative and positive reward, or reinforcement, systems in the brain. Certain things make us feel good (positive reward, or **positive reinforcement**). Stimulation of the nerves in the positive reinforcement system leads to a repetition of the behavior. Basic biological functions like eating and sexual activity are repeated because they activate these nerves. Throughout our lives we learn other experiences that stimulate these centers. Our ability to derive pleasure from certain experiences, such as the feeling of 'a job well done' or the feelings evoked by a beautiful painting or a sunset, is 'learned' to a large extent; our families, religions, cultures and other influences operate from birth to teach us to view certain experiences as positive and certain experiences as negative.

Two hypotheses of drug addiction. One hypothesis of drug addiction is that people use drugs to escape from some kind of pain. This hypothesis suggests that drug-taking behavior results from the attempt to inhibit or avoid the negative reward and to avoid withdrawal. The avoidance of a negative consequence, or removal of an unpleasant stimulus, is called **negative reinforcement**.

A newer hypothesis suggests that drug addiction results from stimulation of the positive reinforcement centers. Of the psychoactive drugs, the ones that cause addiction are those that stimulate the positive reward system in the brain. They directly stimulate these centers, neurochemically producing the positive sensation. This hypothesis is supported by the findings that people may become addicted very rapidly, well before physiological dependence has begun. The two hypotheses may be valid for different drugs or different individuals, meaning that there may be more than one mechanism for addiction.

Evidence from behavioral experiments with rats. Much of the research on drug-seeking behavior has been done with rats. Rats have been fitted with tubes so that they can give themselves drugs, either into their bodies or directly into specific parts of their brains. These rats quickly learn to self-administer addictive drugs, and they increase the frequency of self-administration if allowed. They do not, however, self-administer nonaddictive drugs. Although physiological dependence on the addictive drugs does develop, the experiments suggest that dependence is the result of addiction rather than its cause.

A part of the brainstem known as the **ventral tegmental area (VTA)** (Figure 11.8) is thought to be the positive reinforcement center, or 'pleasure center.' This has been demonstrated in several types of experiments. If electrodes are placed into the ventral tegmental area, rats will activate the electrodes, electrically stimulating that area of the brain. They quickly learn to repeat the behavior, giving themselves repeated electrical stimulation, preferring it even over food. The same type of experiment can be done on rhesus monkeys, with the same results: they refuse food if their choice is between eating (even after starvation) and stimulating their ventral tegmental areas electrically. Electrodes placed in other areas do not produce repeated self-stimulation, and destruction of the ventral tegmental area of the brain stops the behavior.

In another type of experiment on rats, tubes are placed into an area of the brain so that by pushing a lever the rat can administer drugs directly to the area. By their action on neurons or their receptors, psychoactive drugs stimulate nerve impulses in the brain, producing the same effects as elicited by electrical stimulation of those same neurons. The negative reinforcement centers are in the periventricular areas of the brain, while the positive reinforcement center, as previously stated, is in the ventral tegmental area. The nerves of the ventral tegmental area make synaptic connections to the nerves of another brain area, the nucleus accumbens (NA), which is involved in the processing or interpretation of the signal (see Figure 11.8). Self-administration of drug to the negative reward center is not reinforcing and is therefore not repeated, while self-administration to the ventral tegmental area or the nucleus accumbens is.

When rats are able to take drugs by a more normal route, rather than directly into their brains, the results are similar: they repeat the drug taking if the drug is addictive, and do not when the drug is nonaddictive.

Measurements of changes in electrical activity in different parts of the CNS have shown that activity increases in the ventral tegmental area (and generally nowhere else) after the administration of an addictive drug.

The ventral tegmental area of the brain includes the reticular formation

Brain pathways for effects of addictive drugs

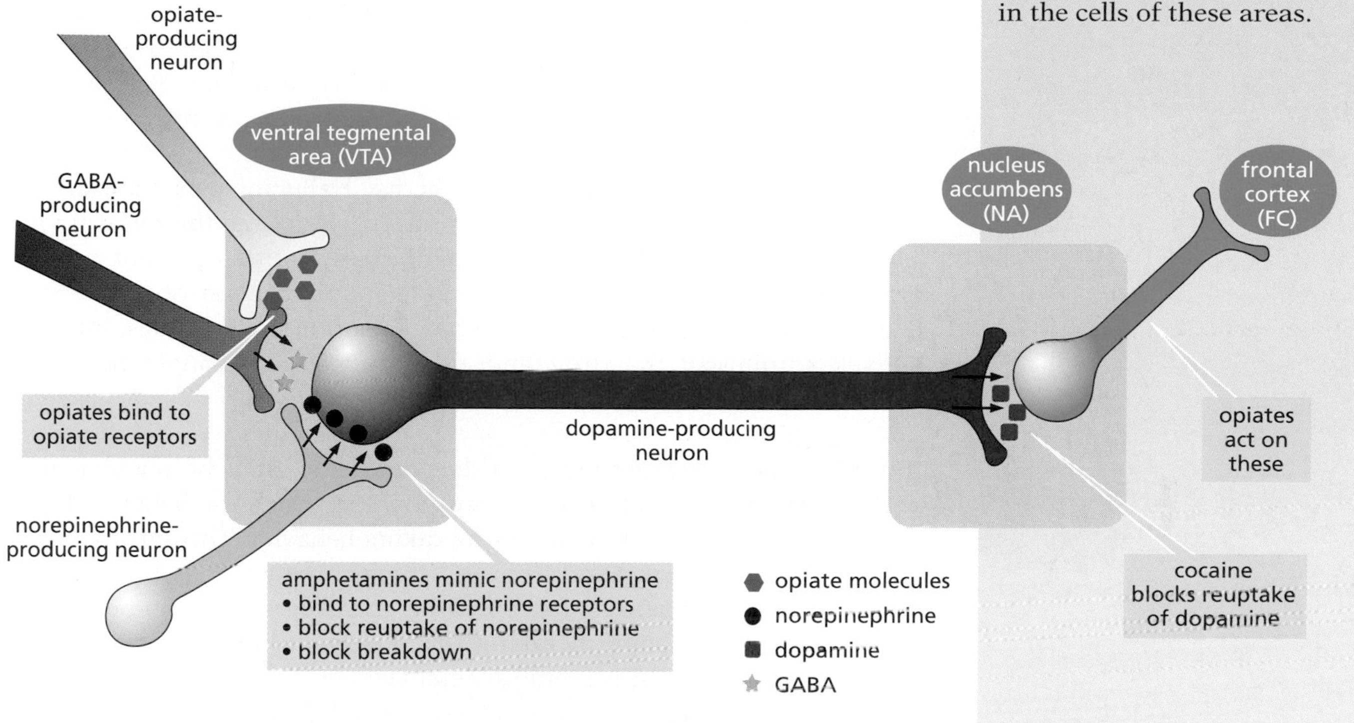

Locations within the brain of activity elicited by addictive drugs

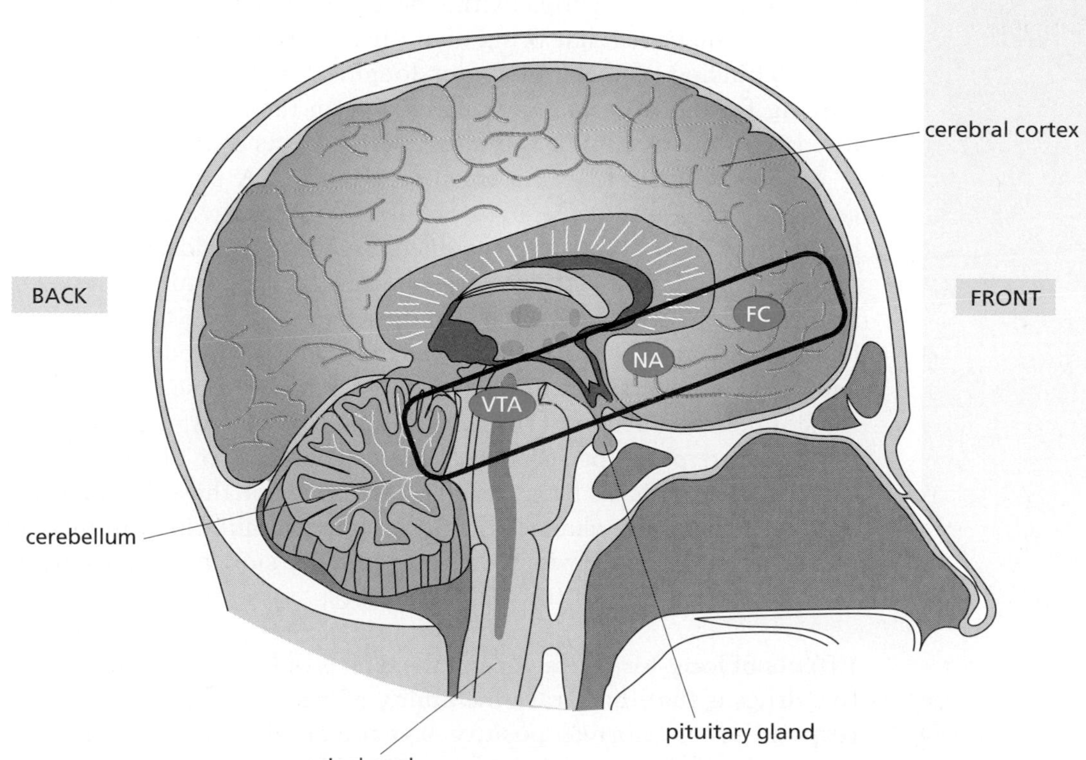

FIGURE 11.8

Positive reinforcement areas of the brain, and the locations of action of addictive drugs on synapses in the cells of these areas.

described above, so that many of the drugs that act on the reticular formation also act on the positive reinforcement area of the ventral tegmental area. Amphetamines indirectly stimulate the neurons of the ventral tegmental area, elevating mood. For this reason they have been successful in the treatment of depression. Cocaine acts on brain cells of the ventral tegmental area that secrete dopamine. Most researchers think that the euphoria produced by cocaine is due to its effects on the dopamine-secreting cells because the euphoria can be stopped with drugs that block dopamine receptors on postsynaptic cells (see Figure 11.8). Opiates, marijuana, caffeine, and alcohol all produce ventral tegmental self-reinforcing effects.

Not all psychoactive drugs are addictive. Hallucinogens, for example, do not produce repeated self-administration by rats and therefore are not considered addictive under this hypothesis. Nicotine does not initially produce self-administration in rats; in fact, initially it is strongly aversive. However, after a rat has been exposed to nicotine several times, it begins to self-administer the drug and this becomes a strongly persistent behavior, that is, a behavior that is hard to break.

Conditioned learning in drug addiction. Addiction is both a biological response and a learned behavioral response in which the behavior being learned is the drug-seeking and drug-taking behavior. This type of learning is called **operant conditioning**, in which behavior is learned as a result of its consequences. Drug seeking and drug taking can be learned by operant conditioning if taking the drug is usually associated with stimulation of the positive reinforcement centers in the brain. Contextual cues are important. As is true in classical conditioning, one stimulus can become associated with another (see Chapter 10, p. 436). Being in certain places or being with people with whom a person has taken drugs provide strong learned cues that bring on a physical sensation of craving for the drug. Thinking of those places or longing for those people brings on a craving for the drug and can also bring on the sensations produced by the drug itself. Seeing or thinking about aspects of the drug taking itself is often sufficient to bring on these feelings. A person dependent on cocaine reported that the sight of someone wearing a gold watch would bring on the sensations of a cocaine high because gold watches were one of the items he had often stolen to purchase his cocaine. The contribution of the 'drug culture' to the reinforcement of drug-taking behavior is enormous: many of the symbols and rituals associated with the culture become contextual cues. Attempts to block and reverse drug dependence must take into account these learned associations, which are called 'conditioned place preferences.' The rehabilitation of drug addicts is usually very difficult because of the strength of these learned associations. However, the chances of successful rehabilitation are increased if the addict can be helped to develop an aversion to the old behaviors while learning to substitute new behaviors.

Effects of long-term use. One of the effects of long-term use of psychoactive drugs is that they erase the ability of the ventral tegmental nerves to respond to the normal positive signals: appreciation of a good meal, enjoyment of the company of friends, and happiness from helping others may all disappear. We tend to interpret positive experiences as 'pleasure,' so the positive reinforcement centers have sometimes been called the 'pleasure centers.' Drugs compete with the normal neurotransmitters in these brain centers. Long-term use of addictive drugs decreases the number of receptors on the nerve cells (see below) so that these centers are only triggered in the drug-abusing person by taking drugs, and no longer by the experiences that used to be pleasurable.

Drug tolerance

One of the biological effects of addictive drugs is that they produce **tolerance**, which is also called 'homeostatic compensation' (i.e., the body adjusting to new conditions). This means that the same dose of drug exerts a decreased effect when administered repeatedly; greater amounts of the drug must be taken to produce the same effect. Some well-meaning people have suggested that addicted people be given 'all the drugs they want' to keep them off the street. The simple biological reason why such an approach would not work is that drug tolerance develops. A person needs more and more drug to produce the original psychoactive effect. Higher doses affect other body systems and begin to produce negative mental states such as hostility and paranoia. The increasing doses may often become toxic or lethal.

Drug tolerance can be shown graphically as a shift of the drug's dose–response curve to the right (Figure 11.9). There are two broad categories of mechanisms that produce tolerance: metabolic and cellular. Metabolic tolerance is the body's production of an increased amount of the enzymes for the breakdown of the drug. Tolerance to barbiturates develops at least in part because the drug stimulates the synthesis of the liver enzymes responsible for its elimination, so the rate of elimination increases with repeated use.

Cellular tolerance results from changes in the receptors for the drug, principally the receptors on the nerve cells in the brain. Either a drug-induced decrease in the number of receptors or an increase in their response threshold results in tolerance. Heroin produces these changes in the brain within a week or two of daily use. If use is more frequent than once a day, a higher level of tolerance develops; that is, even fewer receptors are present on the nerve cells so an even higher dose is required to produce an effect.

Some drugs can cause permanent damage to receptors, and therefore tolerance to these drugs becomes permanent. For most drugs, however, tolerance is not permanent but disappears gradually with time. The time period varies greatly from one drug to another.

So far we have emphasized the effects of psychoactive drugs on the brain. However, these drugs also have effects throughout the body, as we see in the next section.

FIGURE 11.9

Effect of drug tolerance on the dose–response curve. A higher dose (X_2) is required to produce the same response after drug tolerance develops.

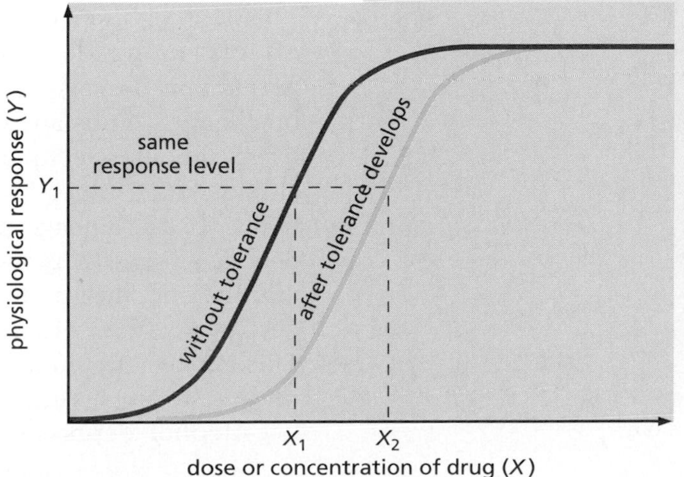

THOUGHT QUESTIONS

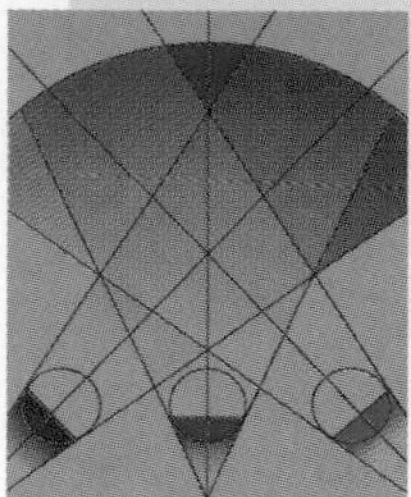

1. If a person has never learned to feel pleasure from daily activities, will they be more attracted to the 'artificial' pleasure offered by drugs? Can this suggest anything to us about drug prevention strategies? One slogan says that "hugs are better than drugs." Does this slogan correlate with neurobiological findings? Can this idea be applied in drug prevention programs?
2. Because psychoactive drugs work directly on the brain cells, are individuals exempted from responsibility for their own drug use? Apply this reasoning to alcohol and tobacco as well as illegal drugs.

DRUG ABUSE IMPAIRS HEALTH

In addition to addiction, the use of psychoactive drugs can impair the health of individual drug users in other ways. It can also affect their gametes, and thus the development of their children.

Drug effects on the health of drug users

Drug use that negatively affects the health of individuals or society is called drug abuse or substance abuse. In the United States, deaths from drug overdoses increased from 6500 in 1979 to 10,000 in 1988. Most of these deaths were from heroin and cocaine, with the remainder being mostly the result of the misuse of legal drugs such as alcohol. Many more deaths also resulted from chronic use of drugs.

Alcohol. Among the most harmful of drugs, particularly in view of the frequency of its use, is alcohol. While each of the other types of CNS depressants is designated as a controlled substance, subject to fines and/or imprisonment for possession or sale in the United States and in many other countries, alcohol is not. The reasons for the difference are certainly not biological, because ethyl alcohol is a powerful depressant comparable to the others in terms of both short-term and long-term risks to health.

Most of the biological effects of alcohol on the body are **acute effects**, reversible in a matter of hours or days. Over time, however, permanent damage results from the **chronic effects**. Alcohol-induced biochemical imbalances permanently damage tissues such as the brain, liver, and muscle tissue (including the heart), resulting in dementia, cirrhosis and other liver diseases, and cardiovascular disease, respectively. Tissue damage may in turn contribute to altered uptake of and decreased sensitivity to medically necessary drugs, including antibiotics, antidiabetic drugs, and other medications. Alcohol in the gut also destroys certain vitamins and interferes with the absorption of others. This is why vitamin deficiencies rarely seen in industrial countries occur in those countries among alcoholics.

Alcohol depresses the immune system, leaving alcoholics very susceptible to infectious disease, including tuberculosis. Because alcoholics frequently substitute alcohol for food, they are often malnourished either in total calories or in micronutrients, which can further depress the immune system (see Chapter 12, p. 511). Alcohol consumption is also associated with an increased risk for cancer (see Chapter 9, p. 393).

Caffeine. Caffeine, which is even more widely used than alcohol, can have significant negative effects on health. The consumption of more than ten cups of coffee a day increases chromosome damage (which can lead to birth defects), respiratory difficulties, and heart and circulatory problems.

Tobacco. Tobacco is carcinogenic in any form. Smoking tobacco is correlated with lung cancer, whereas chewing tobacco is correlated with cancer in the mouth, a form of cancer that often metastasizes to other parts of the body. Tobacco and alcohol together produce a synergistic increase in the number of cancer deaths, beyond what either would produce without the other (see Figure 9.21, p. 396). Nicotine suppresses the immune system; consequently smokers have a high incidence of other lung diseases, including emphysema and respiratory tract infections such as pneumonia and bronchitis. Passive smoke is related to an increase in sudden infant death syndrome as well as to the incidence of pneumonia and bronchitis in the first year of an infant's life. Because an infant's immune system is only partly developed, passive smoke can be much more damaging to infants than to adults.

In addition to direct effects on immune cells, tobacco smoke paralyzes the cilia on the lining of the respiratory tract. When we breathe, dust, bacteria, and other particles enter our respiratory tract along with the air. Many of these particles are trapped on a sticky fluid (mucus) that coats the respiratory lining. The cells lining the upper respiratory tract have many hairlike projections (cilia) on their surfaces. These cilia beat rhythmically in a coordinated way; the beating of the cilia moves the mucus and its trapped material upward, out of the respiratory tract. Tobacco smoke stops the beating of the cilia; inhaled bacteria and viruses consequently work their way down into the lungs instead of being eliminated. People who smoke therefore have a much higher incidence of respiratory tract infections and other infectious diseases of the lungs than do nonsmokers. The particulate matter from smoke, which can include cancer-causing chemicals in highly concentrated form, also works its way into the lungs and begins the process of transformation of normal cells into cancer cells (see Chapter 9, p. 377). Smokers therefore have higher rates of lung cancer than nonsmokers.

Marijuana. The present day controversy over the use and abuse of marijuana has led to a large amount of research on its effects. Drug tolerance does develop, as does physiological dependence in some heavy users. Although marijuana has a few possible medicinal uses, it is also known to have several biologically adverse effects. Possible medicinal uses include the reduction of eyeball pressure in glaucoma patients, the stimulation of appetite and suppression of nausea in cancer patients undergoing chemotherapy, and the stimulation of appetite in AIDS patients. In the United States, several states, including Maine and California, have legalized the medical use of marijuana under certain conditions, and similar measures have been proposed elsewhere. Nevertheless, federal laws still prohibit the sale or use of marijuana, and federal officials have vowed to enforce these laws.

The harmful effects of marijuana are much better understood than the benefits. When smoked in cigarette form, much of the particulate matter in marijuana smoke stays in the lungs and builds up to form tar. Marijuana smoke produces more tar per weight of plant material than does tobacco smoke, and the tar is equally carcinogenic. It also inhibits the immune cells that clear debris from the lungs and protect against air-borne infectious bacteria and viruses. All forms of marijuana alter the production of reproductive hormones, decreasing the production of sperm in men and ovulation in women. Men who use marijuana over long periods of time often develop fatty enlargement of the breasts (gynecomastia).

In 1999, the Institute of Medicine of the National Academy of Sciences released a report evaluating the health consequences of marijuana smoking and of the use of purified THC and other chemically related compounds (cannabinoids). Among their conclusions are the following:

- THC and other cannabinoids have some potential medical uses that deserve further study.
- "The accumulated data indicate a potential therapeutic value for cannabinoid drugs, particularly for symptoms such as pain relief, control of nausea and vomiting, and appetite stimulation."
- Cannabinoids can be addictive. The brain develops tolerance to them, dependence develops in some users, and withdrawal symptoms may occur when cannabinoid use is discontinued. These effects, however, are not as severe as with certain other drugs that are currently legal, such as diazepam (Valium) and other benzodiazepines, and should not preclude the development of useful medications.
- Smoked marijuana is "a crude THC delivery system that also delivers harmful substances" such as the tars and other compounds that impair the health of the respiratory and immune systems. Further and more carefully controlled studies are needed on the adverse health effects of smoked marijuana.
- Delivery of cannabinoid drugs through the smoking of marijuana is problematical at best. For any possible therapeutic uses of cannabinoids, a safer method of drug delivery needs to be developed, possibly by use of an inhaler.
- "Marijuana is not the most common, and is rarely the first, 'gateway' to illicit drug use;" tobacco and alcohol are both more common in this context. Despite some people's concern that medical users of marijuana-based drugs would become addicts or would proceed to abuse other drugs, "at this point there are no convincing data to support this concern."

Designer drugs. The term 'designer drugs' refers to those drugs that are slight structural alterations of existing drugs. Designer drugs are often made to get around laws that ban particular drugs by name. New designer drugs may be legal until laws are rewritten to cover them, but they can be dangerous nevertheless.

MPTP and MPPP are two designer derivatives of Demerol (meperidine), itself a derivative of opium. These two have a psychoactivity similar to other opiates, but are also potent neurotoxins (nerve cell poisons). They particularly destroy the nerve cells in the substantia nigra, the area of the brain that controls movement, causing movement defects similar to those found in Parkinson's disease, a condition that otherwise mostly affects people over age 50 (see Chapter 10, p. 419). In 1985, 400 cases of Parkinsonism in young people were found to be due to MPTP. After these cases, MPTP and MPPP were made illegal in the United States.

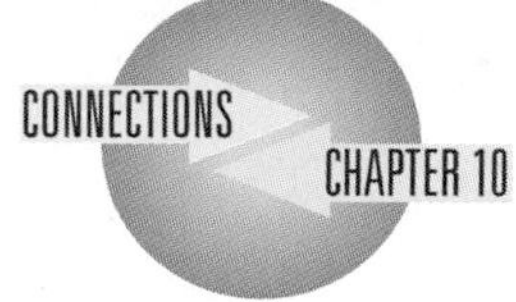

Over-the-counter drugs. Over-the-counter (OTC) drugs are those that can legally be sold without a prescription. Although people tend to view OTC drugs as 'safe' because a prescription is not required, there are actually thousands of deaths from OTC drugs each year and over a million cases of drug poisoning, generally from overdose. Aspirin is second only to barbiturates as the drug most frequently used in suicides and suicide attempts.

Several OTC preparations contain combinations of many drugs. Most of the OTC sleeping pills are combinations of aspirin and antihistamines. Cold remedies often have many ingredients and the liquid ones contain high concentrations (up to 25%) of alcohol. These combination drugs are marketed by urging people to 'cover the bases' by treating all possible symptoms. There is the unstated assumption in such advertising that the taking of unneeded drugs is harmless; there is no mention of possible negative effects of any of the ingredients.

While it is generally true that OTC drugs, when used as intended, are generally safe, use in the wrong dose or for purposes other than those designated on the package can be harmful. People in the United States often have cultural expectations that problems can be fixed rapidly and with little effort on the part of the patient. This cultural expectation, encouraged by advertising, is certainly a factor in the tremendous use of OTC drugs. Many of the OTC drugs, particularly those with psychoactivity, have a potential for abuse. There are an estimated 350,000 such products (brand names) in the United States alone, with annual sales of $5 billion. New drugs seeking entry into the U.S. market must now undergo an extensive review process by the Food and Drug Administration (FDA). The drug must be proved to be effective in its intended use and must be demonstrated to be free of adverse side effects. Many of the OTC drugs on the market have not undergone this process because they were 'grandfathered,' that is, exempted from testing because of their many years of prior use. Although official terminology lists these drugs as GRAS (Generally Regarded As Safe), many studies have shown that common OTC drugs, including aspirin, have side effects that cause harm in so many cases that they would not be able to meet the more stringent standards applied to new drugs.

Drug contaminants and additives. Although prescription drugs and OTC drugs are not always 'safe' unless taken as directed (and even then are not without their side effects), consumers can at least be assured that the product contains the drug it claims to contain and that the quantities of it are standardized from one batch to the next. Although no medication is sold as an unmixed, pure compound, the purchaser can be assured that harmful impurities are not present. None of these assurances exist, however, for the purchaser of street drugs. Many impurities are present from the chemical synthesis itself; others are added deliberately. Strychnine is sometimes added to LSD ('white acid'), supposedly to sensitize the nerves to the LSD. It is also sometimes added to marijuana without the knowledge of the purchaser. By acting on the nerve cells, strychnine causes abnormal muscle contractions (convulsions) and is therefore sometimes used as rat poison. There is no specific antidote for strychnine, so it cannot be counteracted once it is taken, and it can cause permanent nerve damage in the brain and elsewhere.

Herbal medicines. Herbal medicines have a long history of use in many cultures, and a majority of the medicines that we now use were originally derived from plant sources. Many people have been returning to the use of herbal medications, sometimes favoring them as an alternative to the medicines sold by drug companies. At least some of this trend stems from a belief that such 'natural' remedies are better or safer than drugs obtained in pill form. Although many of these herbs are very active as medications, it is not accurate to think of them as always being 'safe.' Unprocessed herbs have unpredictable variations in the quantities of both the desired ingredients and the potentially harmful ones. Moreover, many of the most potent psychoactive drugs are plant products. Valerian is a plant containing a chemical called chatinine, which is a tranquilizer. Scotch broom contains a strong sedative, cytisine. Other herbal products, flowers and some spices (nutmeg, sassafras, mace, saffron, crocus and parsley) are strong CNS stimulants.

Chemically, the active ingredient in an herbal medication is identical to the same active ingredient if it is in a medicine produced by a drug company. Herbal medications are present in an unpurified form; that is, the active ingredient is present along with many other chemicals whose effects are not known and many of which are unidentified. A potential problem with herbal medications is the inconsistency of dose, both in the amount present in a particular plant and in the amount present in a particular extract. For example, many herbal remedies are taken as teas, that is, extracted in hot water. The length of brewing time and the temperature of the water greatly affect the amount of drug extracted.

Unlike OTC drugs, which are regulated by law to contain predictable doses of labeled medications and to contain no harmful ingredients, herbal remedies are unregulated in most places. A customer buying an herbal remedy cannot be assured that harmful substances are absent, or that the product in question has been tested and shown to be effective, or even that the product contains any active ingredient at all. Also, knowing that a substance is 'natural' offers no assurance that it is harmless:

opium, strychnine, snake venom, and botulism toxin are all deadly poisons and all are perfectly natural.

In most industrial nations, prescription medications and OTC medications are generally regulated by a government agency such as the FDA. A regulatory loophole, however, permits substances to escape regulatory oversight in the United States if they are marketed as 'food supplements.' Unfortunately, the term 'food supplement' is so ill-defined that nearly any substance can be so designated by the company that sells it, even if it fits the biological definition of a drug. Many herbal formulations are marketed this way in the United States, including echinacea, *Ginkgo biloba*, saw palmetto, and St. John's wort. Without regulation, companies can market these drugs without testing them for dosage, efficacy, harmful side effects, or interactions with other drugs, and they can also make health claims that have never been substantiated by research (although they are prohibited from making therapeutic medicinal claims). For these reasons, people taking such substances should realize that they are dealing with untested risks.

Steroids. Although most commonly abused drugs are psychoactive drugs, not all drugs that are abused are psychoactive. Anabolic steroids are abused instead because of their hormonal effects on physical development. Certain athletes of both sexes (body builders, weight lifters, swimmers, runners, football players, and others) have used these drugs because they cause an increase in muscle mass, resulting in a bulkier and more powerful physique. These drugs are not addictive, but their many dangerous side effects include damage to the reproductive organs and the circulatory system, especially the heart, as well as increased hair in some places and premature baldness in others. Deaths have occurred among amateur and professional athletes from the abuse of steroids.

Drug effects on embryonic and fetal development

Aside from the biological effects on the person taking a drug, there are many effects on developing embryos. Many drugs can affect fetuses *in utero*. Additional harm can result from damaged gametes from either a mother or a father who has used drugs.

Caffeine. Some studies on rats show that caffeine intake comparable to 12–24 cups of coffee per day resulted in offspring with missing toes, while a dose comparable to as little as 2 cups per day delayed skeletal development.

One **retrospective study** on the effects of caffeine intake on pregnancy found that of 16 pregnant women whose estimated daily intake was 600 mg or more of caffeine (eight cups), 15 had miscarriages, stillbirths or premature births. In this study, the caffeine intake of men was also examined. In a subgroup of 13 fathers with a daily intake of 600 mg, although the mother's intake was less than 400 mg, only five of the births were normal. All of the births in which both the parents consumed less than 300–450 mg (four to six cups) were normal. Any paternal effects were presumably caused by chromosome damage before conception. Another retrospective study on pregnant women showed that an amount of caffeine equivalent to as little as half a cup a day increased the frequency of miscarriages; this study also demonstrated that some of the effects of caffeine might occur *before* pregnancy. However, other studies of caffeine have reached inconsistent and inconclusive results for a number of reasons. Among these reasons are the difficulty of measuring caffeine intake when people use different brewing methods and cup sizes, the lack of control for noncaffeine ingredients in caffeine-containing beverages, and a variety of other methodological differences among the studies such as inconsistencies in the number of people and the types of beverages studied. Another difficulty with retrospective studies is that it is often impossible for a person to reliably recall their drug intake, particularly so for the so-called 'soft' drugs such as caffeine.

Nicotine. Nicotine damages the placenta, increasing the likelihood of miscarriages, premature births, and damage to the fetus. Nicotine crosses the placenta very quickly and remains in the fetal circulation longer than it does in the mother's bloodstream (see Figure 11.2). Nicotine causes oxygen deprivation in the fetus, as do carbon monoxide and cyanide from cigarette smoke. Because oxygen is required as the terminal electron acceptor in the production of ATP (see Chapter 8, p. 340), cells deprived of oxygen have less ATP and are less able to perform the cell synthesis functions necessary to produce new cells in the growing fetus, especially in the brain. Oxygen deprivation is made worse by nicotine-induced damage to the blood vessels, including those of the placenta.

Alcohol. Alcohol that is consumed by a pregnant woman will be quickly distributed into the blood of the fetus at the same concentration as is present in the mother's blood, causing severe and permanent mental and physical birth defects called **fetal alcohol syndrome**. The prevalence in the United States in 1983 was 1–3 affected children per 1000 total births and 23–29 per 1000 births to alcohol-abusing mothers. The period during which the fetus is most sensitive to damage from alcohol is the first month, often before pregnancy is recognized. There is also evidence that alcohol abuse by women before conception correlates with decreased fetal growth, even when the mother abstains during pregnancy itself. There are few data on the fetal effects of heavy alcohol consumption on the part of the father. Alcohol is toxic to sperm and five or more drinks daily decrease the number of sperm produced.

Drug combinations. Alcohol, caffeine, and nicotine all increase the blood levels of the neurotransmitter acetylcholine, lowering placental blood flow. The effects increase with the dose of the drug (a positive dose–response correlation) and also with its duration in the body. Because the fetus lacks the enzymes for breaking down either alcohol or caffeine, the concentrations of these drugs stay higher longer in the fetus than in the maternal circulation. Because there is a higher incidence of smoking in people who abuse alcohol, the interactions of these drugs are also significant. Marijuana also crosses the placenta and is correlated with low birth weight and prematurity. Barbiturates readily cross the placenta, and the use of barbiturates by pregnant women can cause birth defects. The combination of marijuana or barbiturates with any of the other drugs mentioned increase the risks to the fetus.

Drug persistence after birth. Drugs passed on to the fetus *in utero* may remain in the child for a long time, particularly if the enzymes needed to metabolize these drugs are not present. Phencyclidine (PCP), known as 'angel dust,' can still be present in the blood of a five-year-old child of a PCP-using mother. Because children's brains and immune systems continue to develop after they are born, toxic drugs may continue to interfere with the development of the brain and the immune system in young children long after birth.

Drug abuse as a public health problem

Recall that drug abuse has been defined as drug use that negatively affects the health of individuals or society. We have examined the effects of drugs on cells and on individuals, but an ecological perspective on biology teaches us that actions in cells and organisms generally have consequences for populations.

There are different concepts of how to protect society from the consequences of drug use by some members of society. One concept views drug addiction as a crime and drug abuse as a law enforcement problem. The U.S. Comprehensive Drug Abuse Prevention and Control Act of 1970, commonly called the 'Controlled Substances Act,' established five categories of controlled substances according to their potential for abuse and whether they had a recognized medicinal use in 1970. Although the classification is not perfect, it represents an attempt to classify potentially dangerous drugs in a consistent manner. Table 11.4 is a partial listing of substances regulated by this law. The distinctions made by this classification are legal rather than medical. For example, drugs in Schedule I are illegal to possess or to prescribe. Drugs in Schedules II and III can be prescribed only by certain physicians registered with the U.S. Drug Enforcement Agency (DEA), and certain records must be kept and reported to the DEA periodically. Each schedule also specifies what kinds of researchers or medical professionals may possess or handle the drug, what kinds of records they must keep, and what penalties can be imposed on people who possess these substances illegally.

Another concept, referred to as **harm reduction**, is based on a view of addiction as a disease and drug abuse as a public health problem. Many European countries have followed harm reduction strategies, particularly in dealing with such drugs as heroin. The drug-abuse rates (and crime rates) of these countries are much lower than those of the United States, which has followed the crime concept for the most part. In England, for example, heroin can legally be prescribed to those who are addicted. The

TABLE 11.4

Controlled substances (as currently defined under U.S. law)

SCHEDULE I OR C-I

The law describes these as drugs with (1) a high potential for abuse, (2) no currently accepted medical use in the United States (as of 1970), and (3) no acceptable safe dosage standards even with medical supervision.

- LSD
- Marijuana (including hashish)
- Peyote
- PCP
- Heroin
- Cocaine

SCHEDULE II OR C-II

Drugs with a currently accepted medical use, but also a high potential for abuse that may lead to "severe psychologic or physical dependence."

- Amphetamines
- Morphine
- Codeine
- Barbiturates such as secobarbital and pentobarbital
- Methadone
- Meperidine
- Methaqualone
- Percodan

SCHEDULE III OR C-III

Drugs with a high potential for abuse that may lead to a moderate or low physical dependence or a high psychological dependence.

- Paregoric
- Doriden
- Noludar
- Tylenol or Emprin formulations containing codeine

SCHEDULE IV OR C-IV

Drugs with a lower potential for abuse than those listed above, and an abuse potential that can lead to 'limited' physical or psychological dependence.

- Phenobarbital
- Many common tranquilizers: chlordiazepoxide (Librium), chlorazepate (Tranxene), diazepam (Valium), flurazepam (Dalmane), oxazepam (Serax)
- Other sedative/hypnotic agents: chloral hydrate, glutethimide (Doriden), meprobamate (Equanil, Miltown), methyprylon (Noludar), paraldehyde

SCHEDULE V OR C-V

Drugs with a low potential for abuse (relative to those in Schedules I–IV) and an abuse potential that can lead to limited physical or psychological dependence compared with schedules I–IV.

- Terpin hydrate with codeine
- Cheracol with codeine
- Lomotil
- Robitussin A–C with codeine

approach seems to work in several ways: harm to the addicts from overdose or impure formulation is minimized. There is no incentive for the addict to commit crimes to pay for drugs or to recruit others into becoming addicts. (Many drug dealers are addicts who recruit others so that they will have a steady supply of customers and profits to support their own habit.) Under the harm-reduction approach, the rates of heroin use in England have dropped, while rates of heroin use in the United States have risen. The illegal heroin trade has withered in England because it is no longer profitable, and new cases of addiction are rare because the drug is available only to persons registered as already addicted. Most promising of all is the fact that about 25% of heroin addicts in England spontaneously give up the habit on their own.

The approach in several other European countries, including Germany, Switzerland, and the Netherlands, allows drug users freedom from arrest if they follow a few simple rules (staying in certain locations, for example). Instead of spending money mostly on enforcing drug laws (as in the United States), more government funds are spent in these countries on public health campaigns aimed at education, prevention, and rehabilitation. Although the unlicensed selling of drugs is illegal in these countries, the criminal justice system is used very little in attempts to minimize the harm done by addictive drugs, either to the addicts themselves or to society as a whole.

Education is at the heart of most measures aimed at preventing drug abuse, including most harm reduction strategies. Where prevention of addiction has been tried, it is cheaper and more successful than rehabilitation. Education about the risks of drug use does decrease drug use (Figure 11.10). European countries that emphasize education and prevention have much lower drug abuse rates than the rest of the industrialized world.

Occasional critics on both the political left and right have suggested legalizing various dangerous drugs, taxing them, and treating them as public health problems in the way that we treat alcohol and tobacco, with heavy reliance on education and prevention programs. Few people who work in the field of drug addiction favor this approach because these drugs are truly dangerous, and because addiction is easy to establish and very difficult to break. Legalization without the structures that are part of harm reduction policies would, they feel, result in increases in health risks.

The harm done by alcohol shows that a drug need not be illegal to cause considerable social harm. In addition to its negative effects on the user's health, alcohol abuse also causes harm to others. Alcohol use currently causes deaths in motor-vehicle accidents, boating accidents, drownings, and many other causes of accidental injury including industrial accidents. It also is responsible for much employee absenteeism, job loss, and school failure. Alcohol is frequently a factor in acquaintance rape (also called 'date rape'), child neglect, child abuse, spouse abuse, divorce, and suicide. Alcohol and other psychoactive drugs can also do great harm in safety-sensitive occupations such as commercial transportation (airline pilots, air traffic controllers, railroad engineers), power plant operations, the nuclear industry, and much of the military. We tend to hear about the increases in crime when alcohol use was illegal during the Prohibition on alcoholic beverages in the period from 1920 to 1933 in the United States, but it is also true that the incidence of alcohol-related accidents and disease was greatly *decreased* during Prohibition.

FIGURE 11.10

Influence of perception of risk on the use of marijuana. The availability of marijuana in the United States has remained uniform over the years covered by this graph, but use has decreased in inverse proportion to the perception of risk.

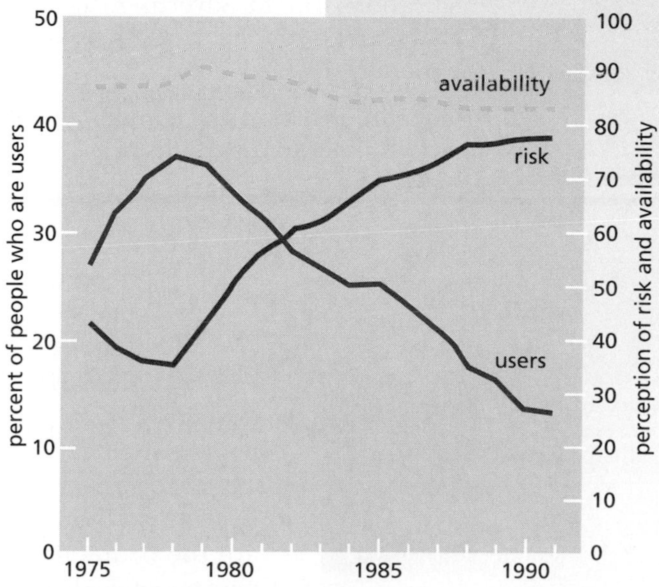

THOUGHT QUESTIONS

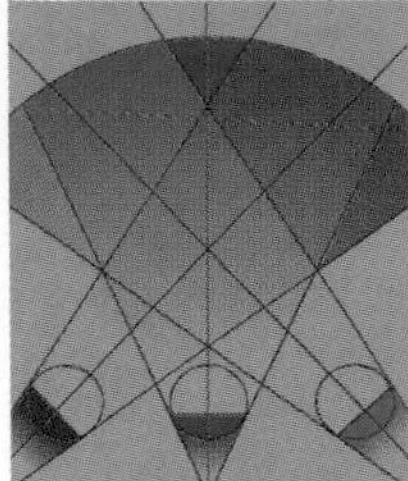

1. Divide a piece of paper into three columns. In one column list all the characteristics you can think of to describe addictive drugs; in the second, list the characteristics of psychoactive drugs; in the third column list the characteristics of drugs of abuse. Also list specific drugs in each column. Are all addictive drugs drugs of abuse? Are all drugs of abuse addictive? Are all psychoactive drugs addictive? Are all psychoactive drugs drugs of abuse?
2. Many factors contribute to making a drug dangerous. In what ways are street drugs more dangerous than a chemically similar drug obtained from a licensed manufacturer?
3. Does the American cultural expectation of a quick fix for life's pains contribute to our tremendous use of legal drugs? Does the widespread and somewhat casual use of legal drugs contribute to drug abuse?
4. Read the package inserts for some drugs that you have purchased over the counter. Do these inserts indicate that there are any potential negative effects? What kind of warnings do the package labels contain? Study the advertisements for these medications. Do the advertisements and the labels give you the same impression of the products?
5. Use the information given in this chapter to argue either for or against the legalization of marijuana for medical uses. What type of ethical reasoning are you using? What facts about marijuana are you using in your argument? What do we not yet know about marijuana that might cause you to change your position?
6. Select a substance sold as a 'food supplement' and find out what research (if any) has ever been conducted to test (a) whether it has any of the effects claimed for it, (b) whether it is safe (specify at what dose and under what conditions), (c) whether it has harmful side effects, and (d) whether it interacts with other drugs. If any research studies have been conducted, did they use adequately large samples? Was there a control group? Was the supplement compared with a placebo in a double-blind study?
7. Is information about the biological effects of drugs an effective prevention or a deterrent against drug use? See whether you can find any published information on the effectiveness of various educational programs.
8. What criteria could be used to distinguish between the use and abuse of caffeine? Can these same criteria be applied to other drugs?
9. What ethical considerations govern the use of animals in testing drugs for safety and effectiveness? What about the use of human volunteers?

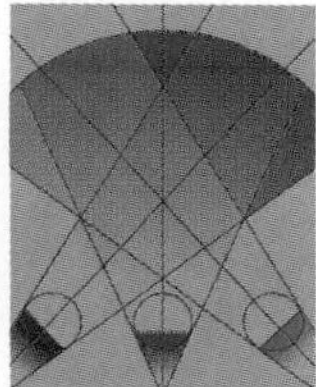

WHERE IS THE BOUNDARY BETWEEN THE RIGHTS OF THE individual and the rights of the group in relation to drug use? Many people are inclined to leave this matter up to the individual if there is no 'harm to society.' However, in many if not all forms of drug abuse, there is clearly a harm to society. Two obvious effects of drugs on populations are the bodily harm done to others by persons under the influence of a psychoactive drug and the commission of crimes to pay for the drugs. In some instances, once the societal effects of individual drug abuse have been documented, laws have been passed to limit these effects. The U.S.

public has accepted laws designed to protect the nondrug-using citizen from the harmful effects of others' use (or abuse) but has not accepted laws that are perceived as infringement on individual rights. For example, laws designed to protect others from exposure to secondhand smoke have been successful at limiting where or when people may smoke, while attempts to pass laws prohibiting individuals from smoking at all have not been successful. A similar approach is being tried on college campuses in educational efforts to raise awareness about the secondhand effects of binge drinking, which include higher risks for rape and for sexually transmitted diseases including AIDS.

In Chapter 1 we outlined steps for arriving at policy decisions on societal issues that are influenced by science. Have these methods been followed on the issues presented by drug use and abuse? Nicotine addiction causes far more deaths (from lung cancer) than all other drugs combined, yet it is legal! Alcohol abuse ruins more families and careers than do illegal drugs, and causes more fatal accidents, yet it is legal in most places. Marijuana users, on the other hand, cause far less harm to others, yet the substance is illegal in most places. Caffeine is not regarded by most people as a drug and is readily available even to children, yet it is certainly addictive. The inconsistencies go on and on. Clearly, decisions as to which drugs should be legal and which should be illegal are not always made on scientific criteria.

CHAPTER SUMMARY

- A **drug** is a chemical substance that produces one or more biological effects, and usually several.
- The effects of any drug depend on the molecular structure of the drug, on its **dose** or concentration, and in many cases on the bodily location of specific receptors for the drug.
- All drugs, whether used medically or nonmedically, must enter the body by some route, usually orally or via the respiratory system. The drug must then be distributed around the body, usually by the circulatory system, and eventually eliminated either by metabolic breakdown or by **excretion**. The length of time that a drug remains metabolically active in the body is measured by its **half-life**.
- Drugs have both **acute effects** and **chronic effects**.
- Drugs have their effects on cells either by direct action on the cell membranes or by stimulating **receptor** molecules to alter one or more cellular functions. Actions on cells in different tissues produce different physiological effects.
- There are always other effects in addition to that for which the drug was taken, and these **side effects** are every bit as real as the intended effect.
- Drugs can interact either **additively**, **synergistically**, or **antagonistically**.
- **Psychoactive drugs** cross the blood–brain barrier and act directly on the nerve cells of the central nervous system. Those that produce **addiction** do so by stimulating activity in the positive reinforcement center of the brain. Addictive drugs induce **dependence** and **tolerance** to the drug, as

well as **withdrawal** symptoms that appear when the drug is stopped. The possible side effects of psychoactive drugs include permanent damage to the brain cells and interference with normal physiological functioning of other organ systems, impairing the health of drug users.

- Many drugs can cross the placental barrier and cause damage to a fetus *in utero*; **fetal alcohol syndrome** is an example. Other drugs can be transmitted to infants through breast milk.
- **Drug abuse**, also called **substance abuse**, has a tremendous cost to society.

KEY TERMS TO KNOW

acute effects (p. 478)
addiction (p. 472)
additive effect (p. 463)
agonist (p. 463)
antagonist (p. 463)
antagonistic interaction (p. 463)
chronic effects (p. 478)
dependence (p. 472)
dose (p. 453)
drug (p. 453)
drug abuse (substance abuse) (p. 472)
excretion (p. 459)
fetal alcohol syndrome (p. 484)
half-life (p. 461)
negative reinforcement (p. 474)
positive reinforcement (p. 473)
psychoactive drug (p. 453)
receptor (p. 462)
side effect (p. 463)
synergistic interaction (p. 463)
threshold (p. 463)
tolerance (p. 477)
withdrawal (p. 472)

CONNECTIONS TO OTHER CHAPTERS

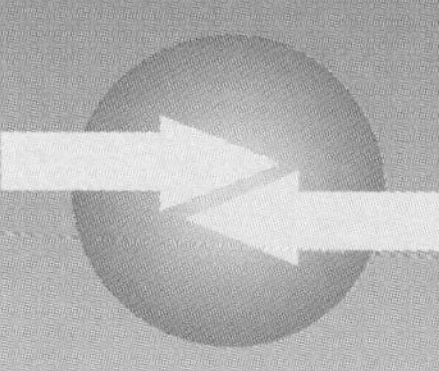

Chapter 1: 'The mind is the same thing as the brain' is an example of a research paradigm.

Chapter 1: Attempts to limit the effects of drug abuse raise numerous ethical issues.

Chapter 6: Drug use can affect the physiological regulation of sex hormones.

Chapter 8: Drugs interfere with nutrient pathways at many levels.

Chapter 9: Drugs such as tobacco and marijuana contain many cancer-causing agents. Alcohol and tobacco act synergistically as cancer causes.

Chapter 10: Drugs may interfere with the normal processes of the brain.

Chapter 12: The brain and the endogenous opiates interact with the immune system.

Chapter 13: Drug use by injection is a major risk factor in the transmission of HIV infection and AIDS.

Chapter 14: Most drugs are plant products or are derived from plant products.

PRACTICE QUESTIONS

1. If the toxic dose of caffeine is about 16 g for a person weighing 80 kg (about 180 pounds), how many cups of coffee, consumed at one time, would it take to realize this dose?
2. Which of the following drugs are addictive?

Cocaine	Heroin	Alcohol	Methadone
Morphine	Codeine	Tobacco	Meperidine

3. How many equivalent 'drinks' are there in a bottle of wine, which is typically 750 ml? How many 'drinks' are there in a quart of liquor? (1 ounce = 28.4 ml; 1 quart = 32 ounces).
4. Of the many drugs described in this chapter, which causes the largest number of deaths from people under its influence?
5. Name three drugs that can enter the body through the respiratory system.
6. If you take 100 mg of a drug, how much remains metabolically active in your body after one half-life?
7. If the half-life of the drug in question 6 is 1 hr, and you start with 100 mg of drug, how much drug is left after 2 hr?
8. If you take 100 mg of a drug that has a half-life of 1 hr, and after 1 hr you take another 100 mg of the drug, how much of the drug will be in your body after the second dose?
9. Of all addictive drugs, which is the most widely used in the United States?
10. If a drug is described as an opiate agonist, what does that mean?
11. If another drug is described as an estrogen antagonist, what does that mean?
12. If two drugs interact synergistically, what does that mean?
13. If the body develops a tolerance to a painkilling drug, what does that mean?

12 Mind and Body

CHAPTER OUTLINE

The Mind and the Body Interact

The Immune System Maintains Health

The immune system and the lymphatic circulation
Development of immune cells
Acquiring specific immunity
Mechanisms for removal of antigens
Turning off an immune response
Passive immunity and innate immunity
Inflammation and healing
Harmful immune responses
Plasticity of the immune responses

The Neuroendocrine System Consists of Neurons and Endocrine Glands

The autonomic nervous system
The stress response
The relaxation response
The placebo effect

The Neuroendocrine System Interacts with the Immune System

Shared cytokines
Nerve endings in immune organs
Studies of cytokine functions
Stress and the immune system
Individual variation in the stress response
Conditioned learning in the immune system
Voluntary control of the immune system

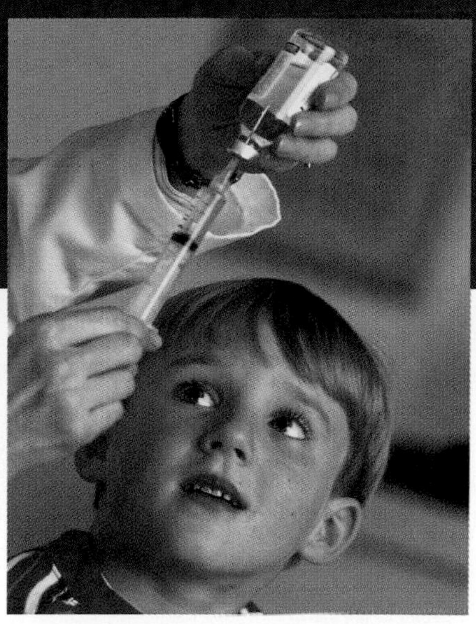

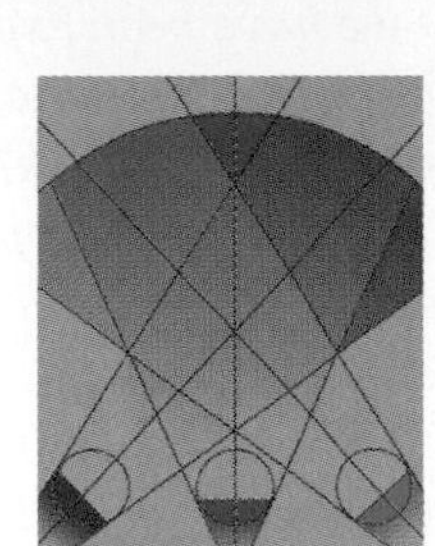

ISSUES

What are the causes of disease? Can we promote health, or is it simply the absence of disease?

Why does one person get sick and not another?

How do our mental and emotional states affect our physical health?

How do our body systems communicate with our emotions?

Can stress make you sick?

How has psychoneuroimmunology reworked our definitions of disease and its prevention?

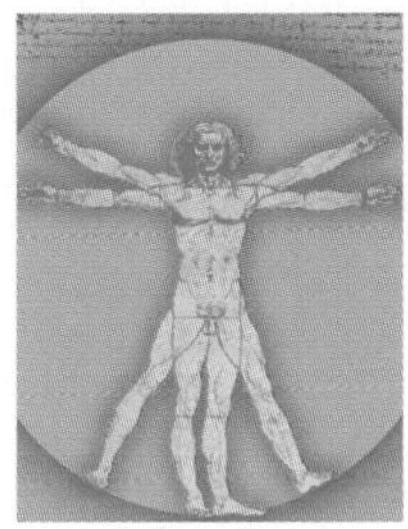

BIOLOGICAL CONCEPTS

Dynamic equilibrium (detection of environmental stimuli, homeostasis)

Organ systems (immune system, lymphatic circulation, autonomic nervous system, endocrine system)

Health (specific immunity, inflammation, tissue healing, stress response, relaxation response, placebo effect)

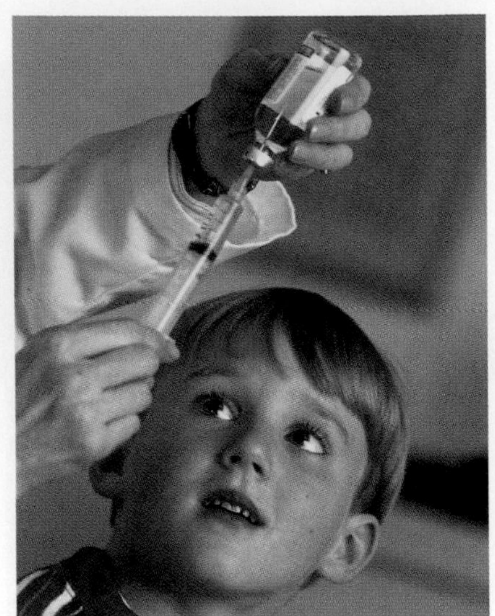

Chapter 12: Mind and Body

What do we mean by health, and how do we achieve it? What makes one person healthy and someone else frequently ill? Certainly there are many answers to these questions, and we have touched on some of them in previous chapters. A person's genetic heritage plays a role in some diseases (see Chapter 3), as does his or her diet (see Chapter 8). How and where a person lives are important because they influence the person's exposure to infectious microorganisms and to hazardous chemicals. Our immune systems help to remove damaged tissue and repair the body, preventing some diseases before we know we have been exposed and bringing us back to health after we have been sick. Genetics, nutrition, and exposure to chemicals and microorganisms all affect the functioning of the immune system. However, it often happens that some people in a particular area will get sick, while their relatives in the same area with about the same exposure, diet, and genetic background do not. In Chapter 10, we saw that many mental illnesses have mechanisms based in brain biochemistry. In this chapter we examine the theory that a person's mental and emotional states are factors in physical health or disease, and that the mind exerts its effect on the body because it interacts with the immune system.

THE MIND AND THE BODY INTERACT

In the preceding chapters we have discussed some biological fields in which theories have been debated, tested, and modified for 150 years—a long time in the science of biology. In this chapter we discuss a very new field of study, **psychoneuroimmunology**. This new subject area is built upon a *central organizing theory*: the premise that the mind, through the action of the nerves, affects the functioning of the immune system and therefore affects human health. Each part of the name contributes to the overall meaning: *psycho*, from the Greek word *psyche*, meaning 'the mind'; *neuro*, referring to the nerves and the brain; and *immunology*, the study of the immune system. Psychoneuroimmunology is the study of how the nervous and immune systems, which were previously assumed to be independent of one another, interact with and influence each other. Psychoneuroimmunology is thus a new paradigm because it embodies both a new theory and a new field of study (see Chapter 1).

Central to the paradigm are large and important issues with which people have struggled for millennia: What do we mean by **health**? Why do we get sick? The answers to these questions will help to define what areas of investigation are valid with regard to the cure and prevention of disease and the promotion of health.

No theories arise spontaneously, and neither did psychoneuroimmunology. Its roots lie in ancient observations that personality, emotional state, and attitude influence when and if people get sick and how sick they get. Doctors in China, India, and (later) Greece rejected supernatural forces (the gods, evil spirits, or magic) in favor of natural (biological) forces as the explanations for both health and disease. Each of these traditions maintained that there is a life force or life spirit, called *qi* (or *ch'i*) by the Chinese, *prana* by the Indians, and *pneuma* by the Greeks, that is present in humans and other organisms for the length of time that they are alive. A person is healthy when the life force is balanced and unhealthy when it is out of balance. A person's mental and emotional states alter the balance or imbalance of the life force.

The Greek physician and teacher Hippocrates advocated the 'rational' study of diseases. Because it was believed that nature followed a rational course, it followed that diseases had natural causes and that those causes should be discernible and knowable entirely by the mind. Hippocrates taught that the body contained four fluids, or 'humors' (from the same Greek word that gives us 'humid'). Each of the four humors corresponded to a personality type. Each personality type also predisposed people to a further excess of the corresponding humor, creating diseases, also classified into four types according to which humor was present in excess. A person with an excess of black bile, for example, would have a 'hot' personality and would be prone to 'hot' diseases accompanied by fevers. People who had an even balance of all four humors enjoyed good health and were thus said to be in 'good humor.'

As European science developed in the 1600s, the criteria for 'knowability' changed. In 1616 an English physician, William Harvey, described the circulation of the blood on the basis of dissection and observation, not purely on rational thought. At about the same time, Galileo invented the thermometer, and this was used by another Italian, Santorio, to demonstrate that people said to have an excess of black bile were no more hot than other people. Not only did these two discoveries undermine the notion of the four humors, but they ushered in an era in which hypothesis testing was added to the standards of 'knowing' about human health. Although the scientific method has vastly increased our knowledge of health and disease, a negative aspect has been that things that could not be seen or in some way quantified have come to be viewed as irrelevant to the explanation of health and disease.

Christianity viewed humans as unchanging reflections of God and therefore beyond the scope of study by the scientific method. The French philosopher René Descartes offered a way around this problem by positing that the mind was separate from the body: the mind was the seat of the spiritual essence, and hence belonged to the realm of the church, and the body was a purely physical essence, and hence suitable for scientific study. This split of mind from body, or **Cartesian dualism** as it is sometimes called, had a profound effect on the study of biology and medicine (and on some church doctrines). As medicine strove to become less of an art and more of a science, the split became wider. It is just this split, however, that psychoneuroimmunology postulates is artificial.

In seeking to rejoin the mind to the body, scientists working in this field test three major hypotheses: that the immune system maintains health, that the mind and emotions can affect the functioning of the immune system, and that mental states can therefore affect health. There is an underlying assumption here, as in Chapter 10, that all of the functions of the 'mind' (thoughts, emotions, hopes, and dreams) can be studied in terms of brain biochemistry. Some psychoneuroimmunologists take this assumption one step further, postulating that the mind is more than the brain: it is the integrated, inseparable network that includes the nervous system, the endocrine system, and the immune system. Other scientists in this field would be more likely to say that the 'psyche' (or mind or spirit) affects the brain and the body in ways that can be studied by biology. Think about these distinctions as you proceed through this chapter.

THOUGHT QUESTIONS

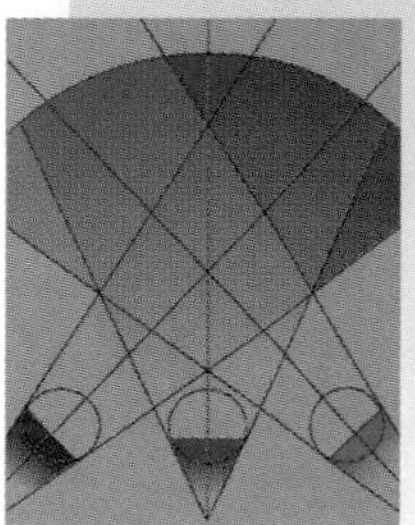

1. Think about the difference between the following two statements: (a) The mind *is* the neuroendocrine–immune system, and (b) The mind can be studied by studying the neuroendocrine–immune system. Are both scientific statements? Would both allow the formulation of falsifiable hypotheses?

2. Can a person accept the data produced by psychoneuroimmunologists without taking a stand on which of the assumptions stated in Thought Question 1 is correct?

THE IMMUNE SYSTEM MAINTAINS HEALTH

To understand how the central theory of psychoneuroimmunology can be tested, we need to learn about the biology of the immune system, the nervous system, and the endocrine system. These three systems in the body have communication as their primary function.

A healthy multicelled organism can be viewed as an ecosystem of cells in which the parts are in **dynamic equilibrium**, or **homeostasis** (see Chapter 1, p. 9). The immune system is the sense organ that detects whether this homeostasis exists and attempts to bring the organism back to this state if it does not exist. We can talk about homeostasis in many physiological contexts: temperature regulation in organisms that maintain a constant body temperature, for example. Immunological homeostasis is a more general state, suggesting that the cells have a way of asking, "Are we all together?" and "Are we in harmony?". The **immune system** can be viewed as a communication network that carries on a 'conversation' throughout the organism, checking to make sure that all the parts are contributing. In this view, the central function of the immune system is the maintenance of **self**, meaning the aggregate of cells forming a cooperative unit that we recognize as a multicellular organism. As we see in this chapter, the immune network is aided in this task by its ability to exchange chemical messages with the nervous and endocrine systems.

The immune system and the lymphatic circulation

The immune system is very diffuse, consisting of mobile cells that travel throughout the body and often are not confined to specific locations. This diffuse organization is in contrast to the nervous system, which is organized in distinct nerve pathways, and the endocrine system, in which cells are associated in distinct entities called **glands**. Immune cells spend much of their time checking tissues for nonself molecules. Cells and molecules that are not contributing to the homeostasis of the cellular ecosystem are called **nonself**. Thus, dead, damaged, or cancerous cells, as well as molecules from outside the organism, are nonself.

Immune cells develop in the bone marrow, spleen, and thymus and are then transported throughout the body by the circulatory system, particularly the areas of the body that contact the external environment: the skin, the nasal passages and lungs, and in areas of the intestinal lining called Peyer's patches. They leave the blood and 'crawl' through the spaces between the cells in tissues; later they are transported via a second circulatory system called the **lymphatic circulation** to the lymph nodes, from which they are returned to the blood. These structures of the immune system are shown in Figure 12.1.

The lymphatic circulation drains liquid from tissues. The spaces between cells in all tissues is filled with a water-based liquid called **interstitial fluid**. (Recall that all cells must be constantly in contact with water both inside and outside the cell, as it is the repulsion of lipid molecules by water that keeps cell membranes intact; see Figure 8.12,

p. 334.) Immune cells called **macrophages** move in the interstitial fluid, cleaning up any dead or damaged cells. Interstitial fluid diffuses from the intercellular spaces into lymphatic capillaries. Once inside the lymphatic capillaries, the fluid is called **lymph**. Lymph contains immune cells, but no red blood cells. From the lymphatic capillaries, lymph drains into larger collecting vessels, the lymphatic vessels. Lymph is returned to the blood via the thoracic duct, which empties into a large vein near the heart (see Figure 12.1). There is no pump to move fluids through the lymphatic circulation in the way that the heart moves fluids through the blood. Muscle contractions and movements of the individual provide what little push this system gets.

FIGURE 12.1

Macroscopically visible parts of the immune system. The parts of the lymphatic circulation are in blue.

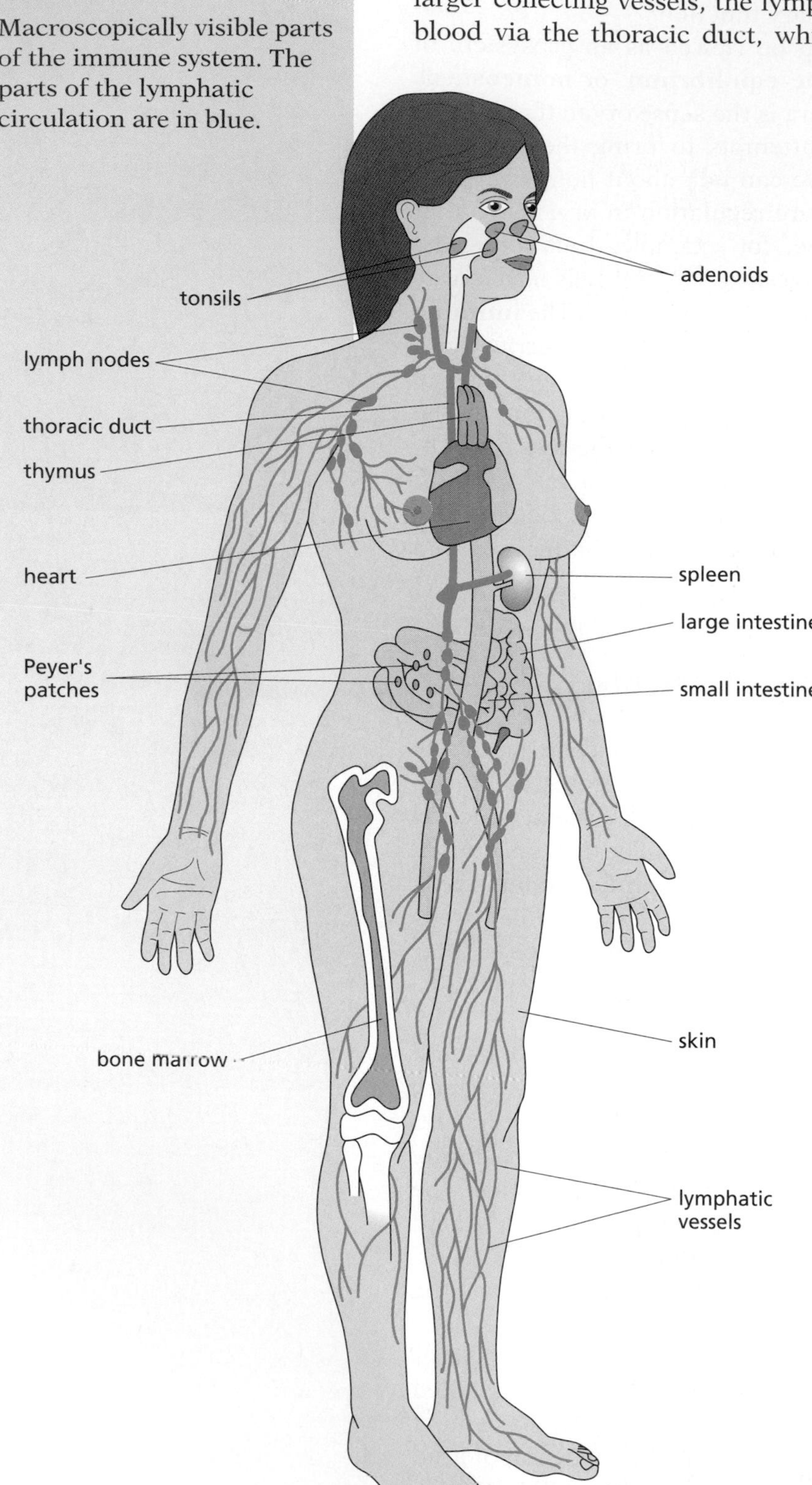

The lymphatic circulation not only helps to remove wastes and damaged cells from tissues, but it also solves the problem of how the immune system can monitor all of the cells and molecules in all of the tissues of the body. This is accomplished by macrophages bringing molecules to centralized locations (the lymph nodes, tonsils, and adenoids; see Figure 12.1) for checking by other immune cells called **lymphocytes**. In an active immune response, ten times the normal number of lymphocytes enter the nodes, causing the nodes to swell. Our lymph nodes are what we commonly refer to as 'swollen glands' when we are sick. The fact that they get larger during sickness indicates that an immune response is occurring; thus swollen glands are generally a sign that a return to health is under way (although lymph nodes that remain chronically swollen sometimes indicate other problems and should be checked by a physician).

In the mid-twentieth century, the function of tonsils and adenoids as tissues of the immune system was not known. Tonsils and adenoids were routinely removed from children who had repeated respiratory or middle ear infections because the swelling of these tissues during infections can make children's breathing difficult or block the eustachian tube, which connects the back of the throat to the middle ear (see Chapter 10, pp. 425–426). Fortunately, there are backup systems in the immune tissues so that removal rarely had serious consequences, but today tonsils and adenoids are left in place unless the blockage is extreme.

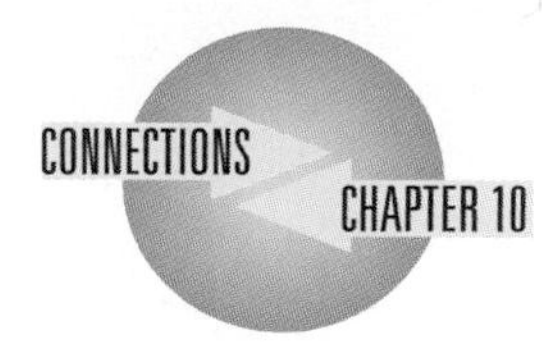

Development of immune cells

The immune system's capacity to distinguish self from nonself prevents the immune system from reacting against self while allowing it to eliminate nonself. In vertebrates, this distinction is made in an ingenious way. Individual lymphocytes have highly specific binding characteristics. In the jargon of immunology, we say that each lymphocyte can 'recognize' only one **antigen**, an antigen being any molecule that is detected by the immune system. The population of lymphocytes as a whole, however, can recognize a vast diversity of antigens. This combination of individual cell specificity, along with population diversity, is what allows the immune system to distinguish between self and nonself.

White blood cells. Cells of the immune system are primarily **white blood cells**. They are called that even though many spend as much time in the lymph as they do in the blood, because under the microscope they look clear or white by comparison to the oxygen-carrying red blood cells. Among the various types of white blood cells are the macrophages and the lymphocytes, mentioned above. There are two types of lymphocytes: the **B lymphocytes** (B cells), which make blood proteins called antibodies (explained below), and the **T lymphocytes** (T cells), some of which kill infected cells directly and some of which help other immune responses. Two other types of white blood cells with immune functions are the **neutrophils**, which are important in removing bacteria, and the **mast cells**, which help to initiate healing.

New blood cells are continuously produced throughout a person's life. In this ability, immune cells differ strikingly from nerve cells, which lose their capacity to divide when a person is still a child. Both white and red blood cells are produced in the bone marrow, the porous interior of the major bones, and in the spleen (see Figure 12.1). New blood cells are produced only to replace blood cells that have been lost through injury or that have reached the end of their lifespan. Both the types of cells and the numbers of each type that are produced are tightly regulated. These regulatory signals arrive in the form of **cytokines**, chemical messengers that deliver information between cells and control both cell division and cell differentiation, as well as many other cellular processes.

Differentiation of blood cells from stem cells. Blood cells differentiate in several steps, each step taking place during a cell division. Of the offspring cells, some become white blood cells that circulate throughout the body, while others remain in the bone marrow as undifferentiated or partly differentiated **stem cells**. Thus the bone marrow maintains a supply of cells with the ability to replenish the blood cells throughout a persons lifetime. Scientists are now studying how to transplant these less differentiated cells from one person to another. Such bone marrow transplants can reestablish blood cells and an immune system in individuals lacking them, as, for example, in person's whose immune cells have been killed by cancer therapy (see Chapter 9). The bone marrow also contains undifferentiated stem cells that can develop into types of cells other than blood cells; transplants of these stem cells are being investigated for their future potential to regenerate other types of tissues, such as muscle or nerves.

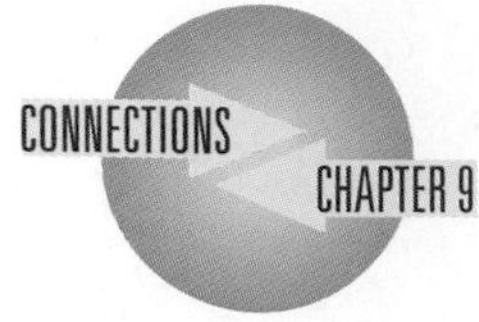

Many types of immune cells differentiate completely in the spleen and bone marrow. Other immune cells may differentiate in the Peyer's patches in the lining of the gut and in the lower layers of the skin. The skin and the gut lining are thus important organs of the immune system.

Lymphocyte population selection. Each B or T lymphocyte has receptors that can bind to only one specific antigen, yet the immune system as a whole is able to recognize over 10^{11} (100,000,000,000) different antigens. In each B and T lymphocyte during its differentiation, the DNA rearranges, and some portions are cut out of the DNA, making the rearrangements permanent in that cell and its offspring cells. Each mature T lymphocyte or B lymphocyte is thus able to synthesize a unique protein that functions as a receptor for only one specific antigen.

The DNA is rearranged randomly in each developing T or B lymphocyte so that the antigen receptor proteins made on one lymphocyte differ from those on another. The whole population of T and B lymphocytes thus contains cells that can bind to over 10^{11} different antigens. Not only are the rearrangements random, but they take place *independently* of any exposure to antigens. *Even before* being infected by a particular type of virus, for example, a person has a small number of lymphocytes that can bind to that virus.

Immature T lymphocytes develop in the bone marrow, then travel through the blood to an organ called the **thymus** (see Figure 12.1), where they receive the cytokine signals that stimulate their final differentiation into T lymphocytes. The developing T lymphocytes are tested and sorted on the basis of matching their antigen receptors with self antigens on the surfaces of the accessory cells of the thymus. Those T lymphocytes whose receptors match an accessory cell antigen would be capable of reacting against self. In a process called **population selection**, these cells are eliminated, an important process that usually protects you from reacting against the tissues of your own body. Cells with receptors that do not match self antigens go on to become fully mature T lymphocytes. Thus, although any one lymphocyte can match only one antigen, the total population of mature lymphocytes has the potential of reacting with any nonself antigen encountered later, but no lymphocytes react with self (Figure 12.2). The immune system's capability of discriminating self from nonself thus results from the selection of a population of responding cells from those that arose randomly. This capability is a characteristic of the *population* of cells, not a characteristic of any single cell.

Like T lymphocytes, the developing B lymphocytes undergo population selection to eliminate any B lymphocyte whose antigen receptors could bind to self molecules. The specific antigen receptors on B lymphocytes are cell surface-bound proteins called **antibodies**. The reaction of a B lymphocyte upon encountering its specific antigen is to secrete more of these antibodies, which then circulate in the blood or in secretions such as saliva and tears. Population selection thus also ensures that only those antibodies that are against nonself molecules will be secreted.

Acquiring specific immunity

A person is not born with immunity to specific antigens, but is born with the capacity to acquire it. This capacity is the result of the two processes just explained: production of the vast numbers of different antigen receptors by DNA rearrangements, followed by population selection of lymphocytes. Lymphocytes in the selected population are released into the bloodstream, where some circulate at all times, ready to go into action when needed. When some nonself antigen is detected, additional white blood cells are called in from the bone marrow. These white blood cells are carried by the bloodstream to the lymph nodes, tonsils, and adenoids (see Figure 12.1), where they encounter nonself antigens brought from the tissues by the lymphatic circulation. Most of the lymphocytes cannot bind to the antigen and so continue on their way. However, a small number of lymphocytes can bind, and they do so, which triggers them to begin to divide. This division of a small number of selected cells produces a subpopulation of identical cells, called a **clone**, able to recognize the antigen (Figure 12.3). This clone of cells then differentiates into two types as shown in Figure 12.3. As explained below, the antibodies produced by antibody-secreting B lymphocytes help to destroy the antigens, and the memory cells help to strengthen the immune response in subsequent attacks by the same antigen. If you are exposed to a particular virus (such as influenza virus), for example, a subpopulation of lymphocytes that recognize the virus develops and kills off the virus.

Immunological memory. Antigen recognition and cell division take some time, so the first time that a person encounters a particular virus, the virus has time to make that person sick before the immune system fights it off. In other words, on the first exposure, the immune system may not block the virus fast enough to prevent the disease, but then, as the specific lymphocyte clone grows, it is able to stop the virus, thereby ending the disease. The clone of specific lymphocytes produced in that first encounter remains in the body as **memory cells**, sometimes for the lifetime of the individual. The second

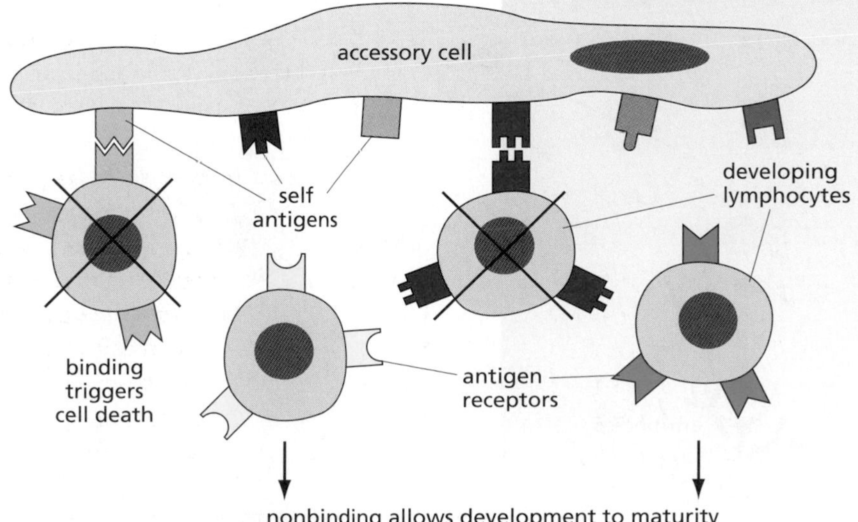

FIGURE 12.2

Lymphocyte population selection. Developing lymphocytes encounter characteristic self antigens on accessory cells. Any developing lymphocyte whose antigen receptor binds to an antigen on an accessory cell dies. Lymphocytes whose receptors do not bind develop to maturity.

time that the person is exposed to that same flu virus, this subpopulation of memory cells is ready to stop the virus more quickly than on the first encounter, in most cases quickly enough to prevent the illness. The response is also greater because the number of cells in the clone is greater than it was before the first exposure to that antigen. This accelerated response on the second or succeeding exposures to the same antigen is called **specific (acquired) immunity**.

Lymphocytes also encounter self molecules in the lymph nodes, but remember that in a fully functional immune system the lymphocytes that could have bound to self molecules have been eliminated from the lymphocyte population. If self molecules become altered, lymphocytes exist that can bind and thus remove cells bearing the altered molecules; self cells that have become transformed or damaged are removed in this way.

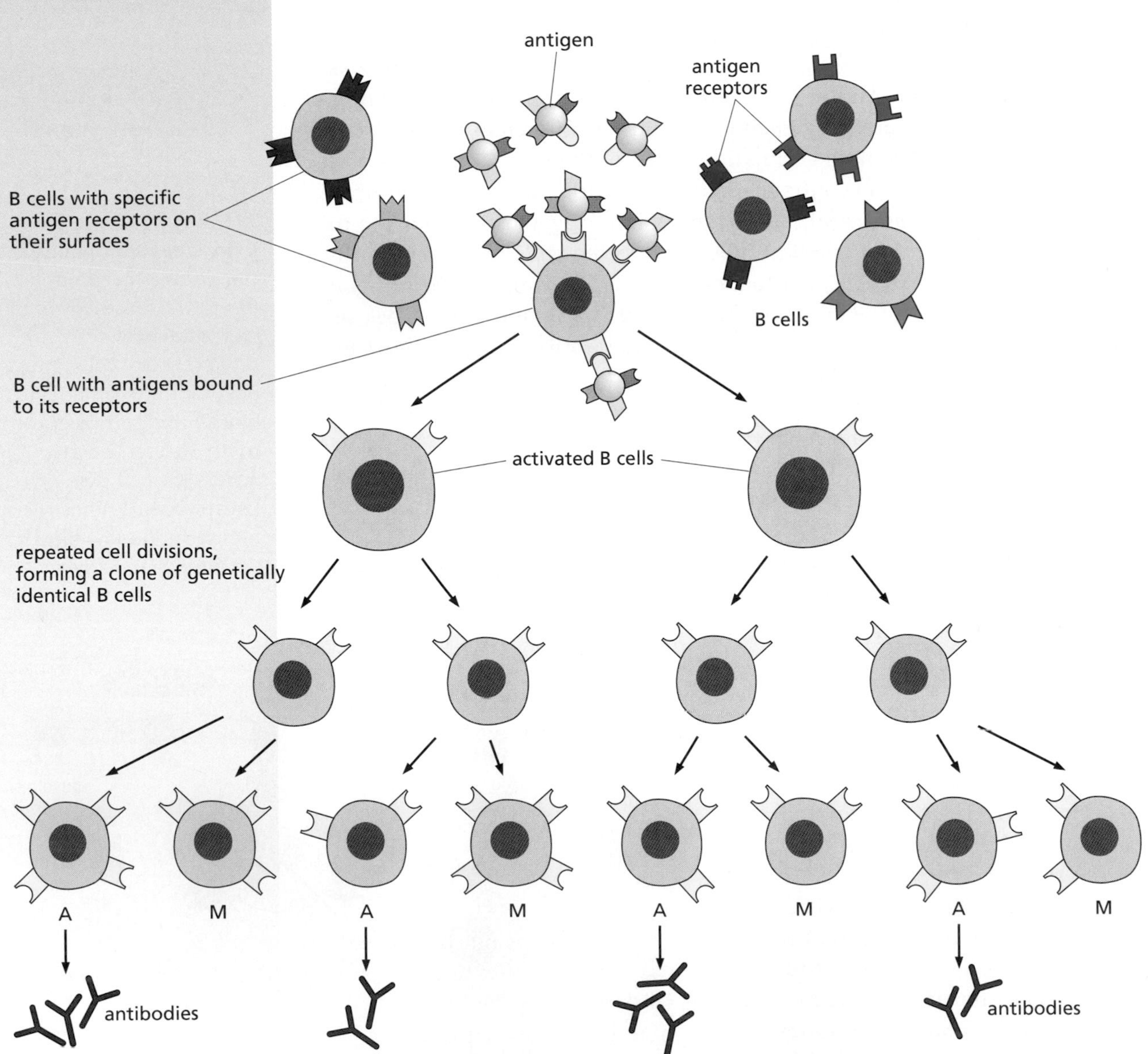

FIGURE 12.3

Cell division producing a clone of antigen-specific B lymphocytes.

Immunization. Remember that we are not born with specific immunity, but with the potential to develop it. We develop specific immunity only to those things to which we are exposed in our individual lifetimes, and so one person's 'immune repertoire' will not be the same as another's.

This is the basis for **immunization**, also called **vaccination**. A person is given molecules from various disease-causing bacteria or viruses, but in a form that will not cause disease. The person's immune system responds to this artificial challenge and establishes a clone of memory cells, which stand ready to protect the person from later exposure to the real bacteria or virus. Immunization, working as it does *with* the disease-preventative powers of the body, has become one of our most effective ways of preventing many infectious diseases.

Mechanisms for removal of antigens

Many antigens recognized by the immune system are parts of whole bacteria, viruses, or cancer cells. It is these whole cells or organisms that need to be removed from the body, and there are several mechanisms by which the activated immune system does so. These mechanisms are summarized in Figure 12.4.

T lymphocyte-mediated mechanisms. A type of T lymphocytes called cytotoxic T lymphocytes can kill target cells directly, removing cancer cells or cells infected with virus (Figure 12.4A). A particular T lymphocyte is able to kill only a target cell whose antigen matches the antigen receptors on that T lymphocyte.

Antibody-mediated mechanisms. B lymphocytes do not kill antigens directly. Instead they make and secrete antibodies, which then circulate in the body fluids. A clone of B lymphocytes secretes an antibody that is specific for the particular antigen that activated the formation of the clone (see Figure 12.3). Antibodies bind to their antigens on a bacterium or virus. This antigen–antibody complex can then combine with other blood proteins called **complement**. The antibody–complement combination can break apart bacterial cell membranes (Figure 12.4B) and can inactivate viruses that are not inside cells.

In addition, antibodies and complement can also coat bacteria, allowing the bacteria to be engulfed and killed by the white blood cells called neutrophils. Unlike lymphocytes, neutrophils do not have antigen receptors, so they cannot bind to most bacteria directly. Instead, they have receptors that bind to one end of antibody molecules. As a result, once a bacterium has been coated with antibodies, the other ends of the antibody molecules can be bound by a neutrophil, which will then take up the bacterium and digest it (Figure 12.4C).

Some antibodies work not by killing microorganisms but by preventing their adherence to the host. Most microorganisms cannot initiate disease without adhering to the host; this is especially true of respiratory viruses and oral bacteria. Each species of microorganism can adhere only to specific types of cells and only in a limited number of host species (see Chapter 13, p. 546). Antibodies bound to these organisms block their

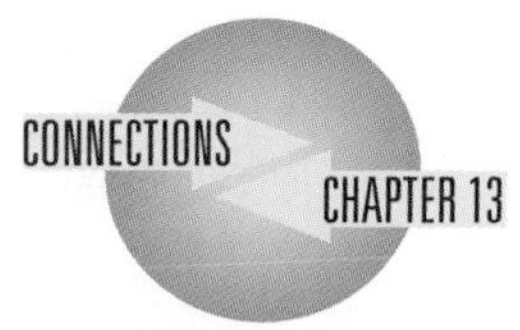

adherence, preventing the disease. Antibodies present in the mucous linings of the respiratory tract and the oral cavity are especially important in blocking the adherence of organisms trying to gain entry through those routes, causing them to pass harmlessly through the body and to be excreted as waste (Figure 12.4D).

FIGURE 12.4

Ways in which the immune system can eliminate nonself antigens.

(A) Antigen-specific cytotoxic T lymphocyte killing a cancer cell

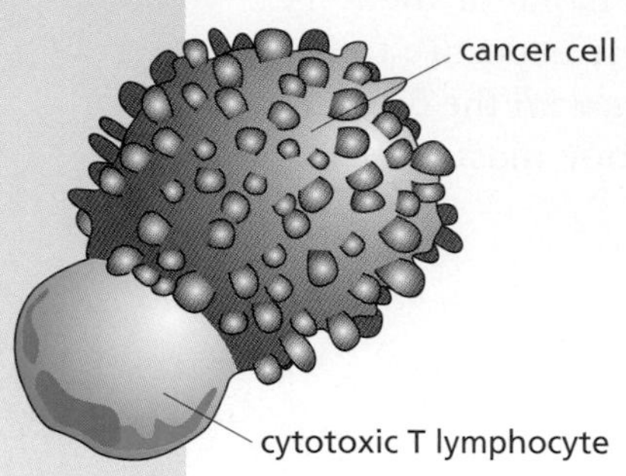

(B) Antibody and complement forming holes in a bacterial membrane

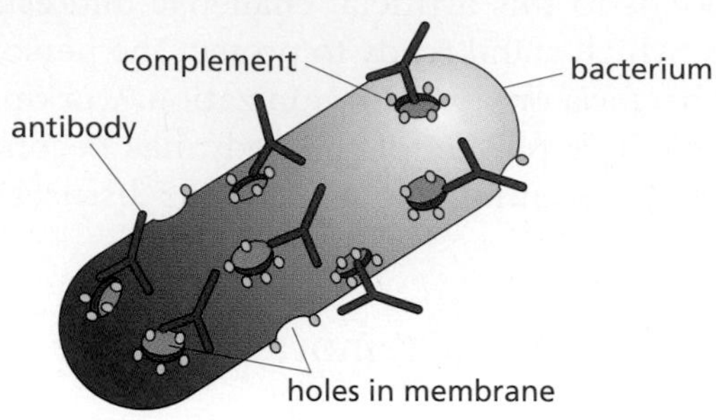

(C) Neutrophil white blood cell destroying antibody-coated bacterium

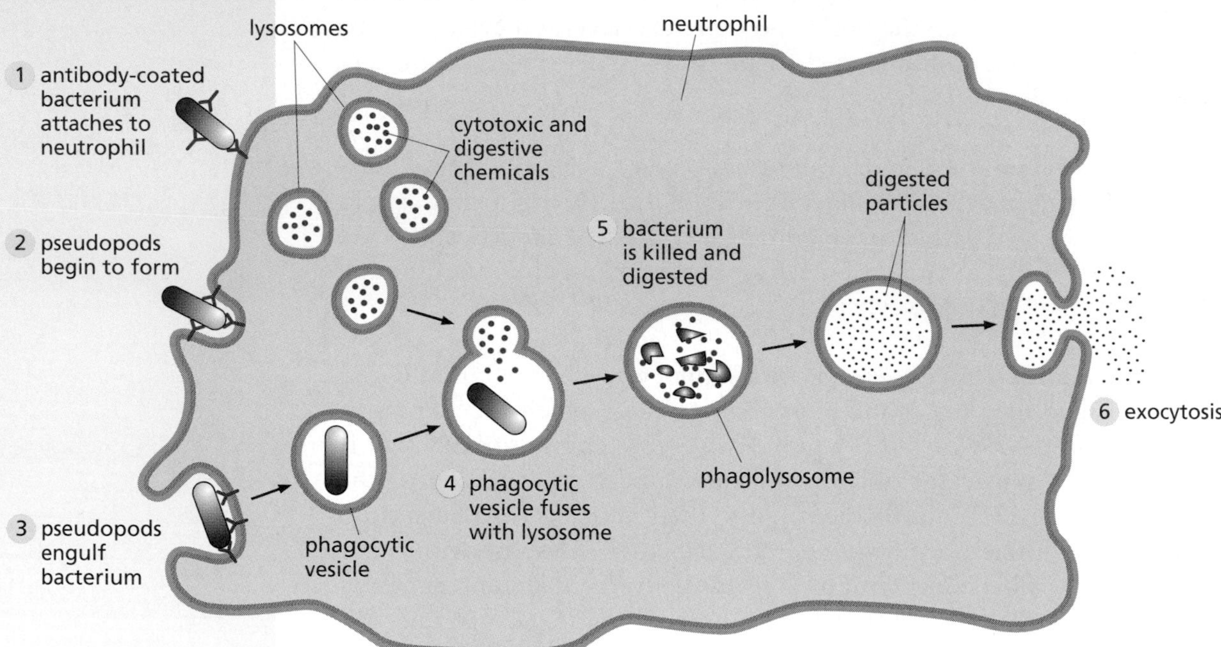

(D) Antibody blocking of bacterial adherence to host tissue, allowing the bacterium to be washed away

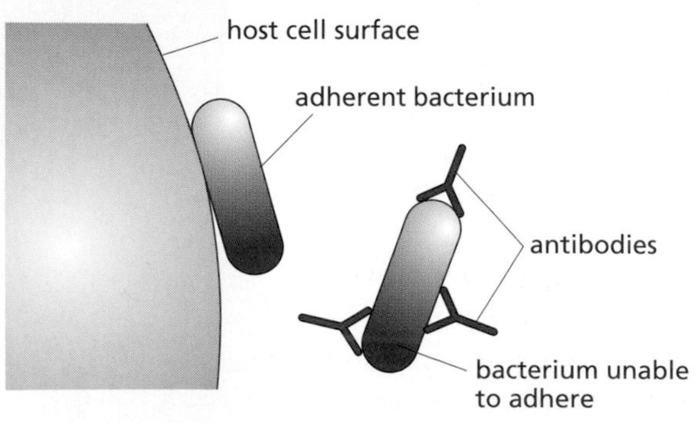

(E) Antibody inhibition of toxin activity

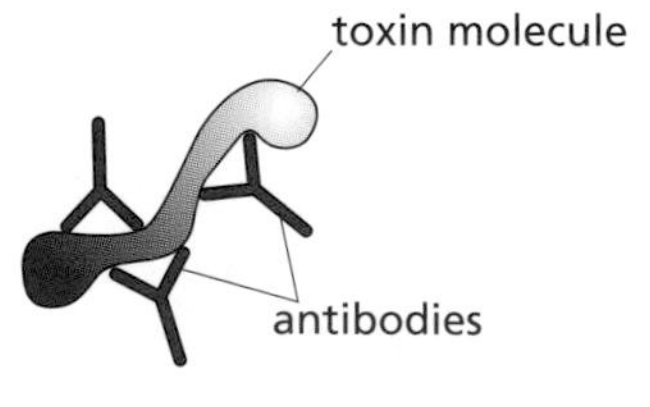

Other antigens, such as toxins, are soluble molecules secreted by bacteria, not parts of bacterial cells, and it is the toxin, rather than the whole bacteria, that induces the disease. Toxins can become inactivated by having specific antibody bind to them (Figure 12.4E). Many of our most successful vaccines actually stimulate the production of antitoxins, that is, antibodies against toxins. The lethal results of diphtheria, for example, result from the action of a bacterial toxin. Immunity conferred by diphtheria vaccine produces an antitoxin that prevents the disease.

Helper T lymphocytes. Cytotoxic T lymphocyte responses and responses involving antibodies are made stronger ('helped') by another class of antigen-specific T lymphocytes called helper T lymphocytes. Helper T lymphocytes do not get rid of antigen; instead, they secrete a cytokine called **interleukin-2** or **IL-2**, which boosts the strength of the responses of cytotoxic T lymphocytes and B lymphocytes to antigens (see Chapter 13, pp. 537–538). Without these cells, the immune response of the cytotoxic T lymphocytes and the B lymphocytes is often not strong enough to prevent disease. This is the situation when HIV infection results in AIDS (see Chapter 13).

Turning off an immune response

Once an immune response has been activated, it does its job of removing antigens, whether the antigens are soluble molecules such as toxins, or molecules that are part of bacteria, viruses, or cancer cells. As long as the nonself antigen is present, the immune response stays activated. Once the antigen is no longer present, it would be inefficient for the process to continue in high gear. As in many other physiological systems, **feedback inhibition** shuts off the response when it is no longer needed. Secreted antibody feeds back to suppress the production of more antibody by B lymphocytes. Cytotoxic T lymphocyte activity is also suppressed by feedback inhibition.

Activation and suppression are not on–off phenomena. Remember that we are dealing with subpopulations (clones) of lymphocytes. At any given instant, some of the lymphocytes able to react to a given antigen are being stimulated and others are being suppressed. Measured on a population basis, when the activity curve is going up, more cells are being activated than are being suppressed. This stage is called activation. An **equilibrium** is reached, with the same numbers of cells being activated as are being suppressed. The same individual cell is not being both activated and suppressed; rather, the number of cells in the antigen-specific clone that is being activated is equal to the number being suppressed. As the activity curve slopes back down, more cells are being suppressed than activated, a stage called suppression. When the curve achieves a new baseline level, another equilibrium state has been reached. These stages are illustrated in the left-hand graph in Figure 12.5.

It is important to realize that when the population activity has returned to the baseline, it does *not* mean that nothing is happening. A small number of cells are being activated and suppressed, keeping the immune system 'tuned up' and ready for its next response.

Activation of an immune response is antigen-specific; that is, not all of the B lymphocytes or T lymphocytes are activated, only those that match the antigen (see Figure 12.3). Feedback inhibition is also antigen-specific; the entire immune system is not shut off, only those cells of the clone that had been reacting to that antigen.

Because the activation and later suppression of each immune response are antigen-specific, several immune responses can be occurring at one time and be at different stages. As shown in the right-hand graph in Figure 12.5, a response to one antigen may be in the suppression phase, while a response to another antigen is just beginning.

The strength of the response to an antigen is shown by the maximum height of the activity curve. The strength of the response may be different to different antigens. Response strength depends on the number of cells in that lymphocyte clone; clones with more cells produce bigger responses. For example on Figure 12.5, clone 1 produced in response to antigen 1 contains more cells than clone 2 produced in response to antigen 2; consequently, the strength of the immune response is greater to antigen 1 than to antigen 2.

Passive immunity and innate immunity

If we have to *acquire* our own immunity, why are we not killed by our first exposure to bacteria when we are very young? There are many answers to this important question.

Passive immunity. First of all, newborns do temporarily have some specific immunity transferred from their mothers; some antibodies cross the placenta, enter the fetal circulation, and can protect an infant for six months or so after birth. Other antibodies can be passed from mother to child in breast milk, particularly in the first week or two of breast-feeding (Figure 12.6). Remember that these antibodies protect the infant against antigens for which the mother has acquired immunity, which may or

FIGURE 12.5

Antigen-specific activation and suppression of immune responses.

The activity measured in a subpopulation (clone) of specific lymphocytes is the sum of the actions of the individual lymphocytes.

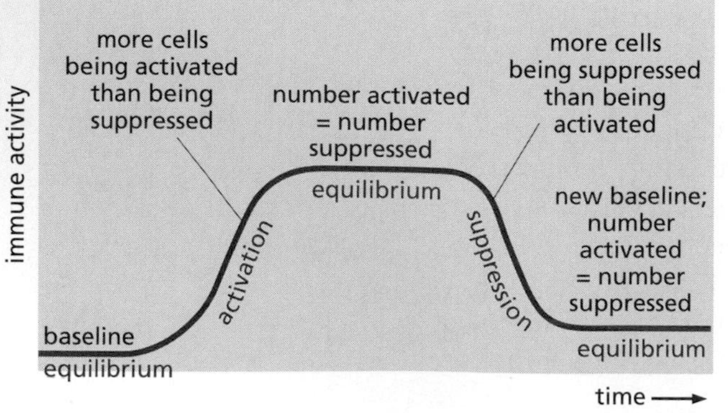

Activation and suppression to two antigens occurs independently but at the same time.

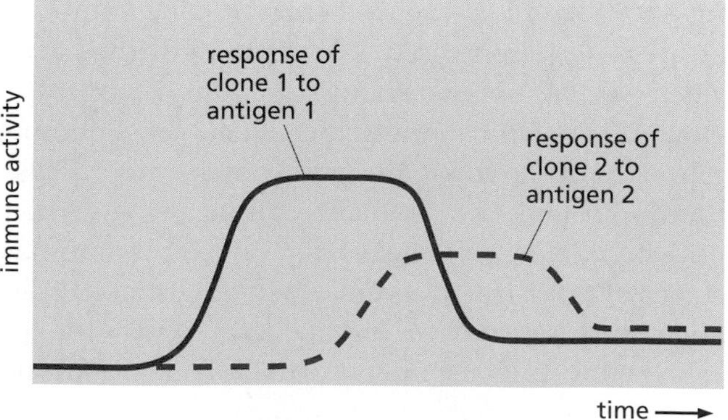

may not be the antigens to which the infant is exposed. This is a type of **passive immunity**, meaning antigen-specific immunity transferred from another individual. Passive immunity is only temporary. Antibody proteins, like all proteins in the body, are constantly being degraded, and the recipient of antibody has not acquired the antigen-specific B lymphocytes to produce more. Passive immunity can also be transferred from one adult to another by transfusing blood containing specific antibody or by giving an antibody-containing fluid called gamma globulin, derived from blood.

Antigen-nonspecific innate immunity. There are parts of the immune system that function without antigen specificity and that operate even on our first exposure to some, but not all, new antigens. These parts of the immune system are collectively called **innate immunity**. Innate immunity is inborn, not acquired, and is antigen-nonspecific. It does not involve cell division of antigen-specific clones of lymphocytes; therefore, innate immunity does not show memory and is not stronger on the second exposure to the same antigen.

The blood proteins called complement are able to bind to and kill some types of bacteria or to inactivate some types of viruses without the aid of antibodies. Viruses nonspecifically induce lymphocytes to secrete **interferon**, a cytokine molecule that then prevents the replication of other virus strains as well as the virus strain that induced its secretion. Cells capable of engulfing particles (macrophages and neutrophils) can engulf some types of bacteria and fungi without antibodies and thus seem to be homologous to the nonspecific immune system of invertebrates. In evolution, these cells have been retained, while antigen-specific immunity has been added. Other cells, the **natural killer cells**, kill tumor cells in a manner chemically similar to that of cytotoxic T lymphocytes; however, natural killer cells act without an antigen-specific immune response.

Inflammation and healing

The immune system does not just remove disease-causing microorganisms and cancer cells. Another important function of the immune system is to promote growth and repair after injury, whether the injury is due to microorganisms or physical damage to tissues.

The mobilization of immune cells and cytokines to get rid of microorganisms or damaged cells is called **inflammation**. In the first

FIGURE 12.6

Passive immunity transferred from mother to child.

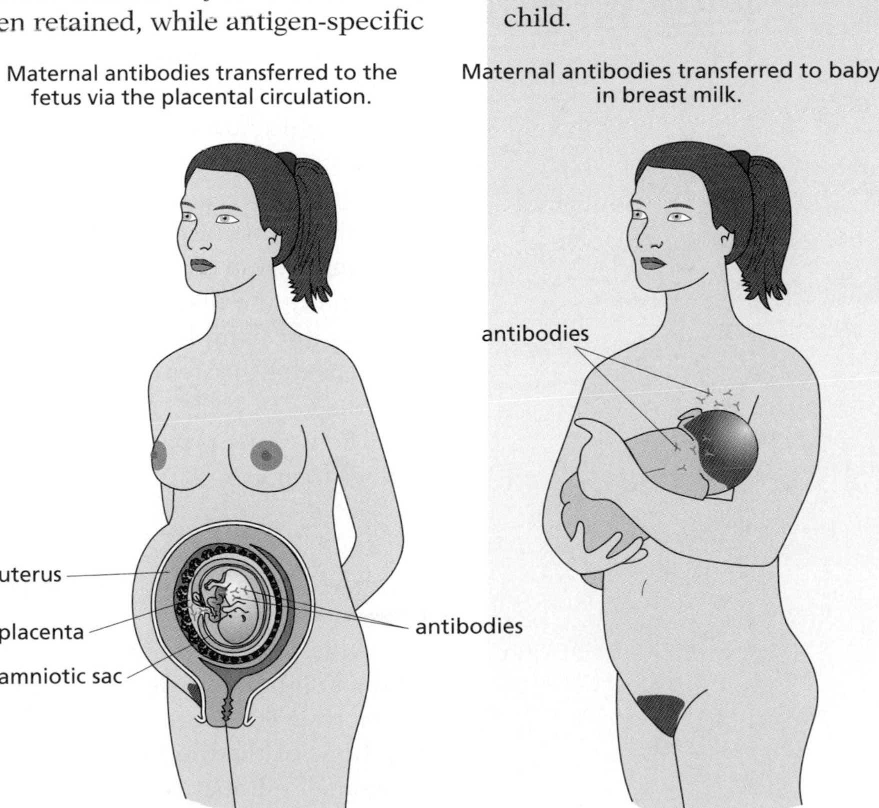
Maternal antibodies transferred to the fetus via the placental circulation.
uterus
placenta
amniotic sac
antibodies
Maternal antibodies transferred to baby in breast milk.
antibodies

century A.D., the Roman physician Celsus described the 'four cardinal signs of inflammation:' *rubor* (redness), *calor* (heat), *dolor* (pain), and *tumor* (swelling). If you have ever had a scraped knee or a splinter in your finger, you have no doubt experienced these cardinal signs. The injured area becomes red, hot, sore, and slightly swollen.

Inflammation is brought about by macrophage and neutrophil white blood cells. When tissue is wounded, the damaged cells change the acidity (pH; see Figure 8.3, p. 319) of the local fluids, activating various chemicals. Some of these factors attract macrophages and neutrophils to the area. These cells can discern concentration gradients of certain chemicals and move in response to those gradients, a behavioral response called **chemotaxis**. Other chemical factors constrict the blood vessels beyond the site of the wound, causing blood to build up in the capillaries close to the wound. These changes in blood flow result in the redness, heat, and swelling of inflammation. Still other chemicals (histamine) increase the permeability of the capillaries in the local area. Fluid escapes into the intercellular area, and the neutrophils and macrophages are able to crawl through the capillary walls and into the tissue. There the neutrophils remove the bacteria and the macrophages remove the damaged tissue. Macrophages also secrete the cytokines called **growth factors** that are the first step in wound healing. These growth factors stimulate cell division, providing offspring cells to replace the damaged cells, for example the skin cells lost because of the wound. Thus, when we talk about **healing**, we are referring to a process coordinated by immune cells.

In addition to local effects on blood vessels, inflammation produces effects throughout the body. One such effect, **fever**, is what we often recognize as a symptom of having an infection. During inflammation, macrophages secrete several cytokines (including **interleukin-1** or **IL-1**), which induce fever by acting on part of the brain called the hypothalamus (see Figure 10.13, p. 430). The increased body temperature inhibits the growth of bacteria and also enhances the immune response to the bacteria. This is one of many examples now known in which products of the immune system act on the cells of the nervous system.

Finally, the macrophages enter the lymphatic vessels and carry antigens to the lymph nodes. Here the macrophages show the bacterial antigens to the lymphocytes, triggering an antigen-specific immune response.

The various processes of inflammation that lead to healing are illustrated in Figure 12.7.

Harmful immune responses

The widespread metaphorical view of the immune system as our defender against disease may lead us to assume that the immune system is always protective. There are several situations in which it is not. The abnormal reactions of the immune system are generally against a specific antigen or a small number of antigens, while other antibodies and immune cells function normally.

FIGURE 12.7

The stages of inflammation.

(A) A wound, with or without bacteria, starts inflammation.

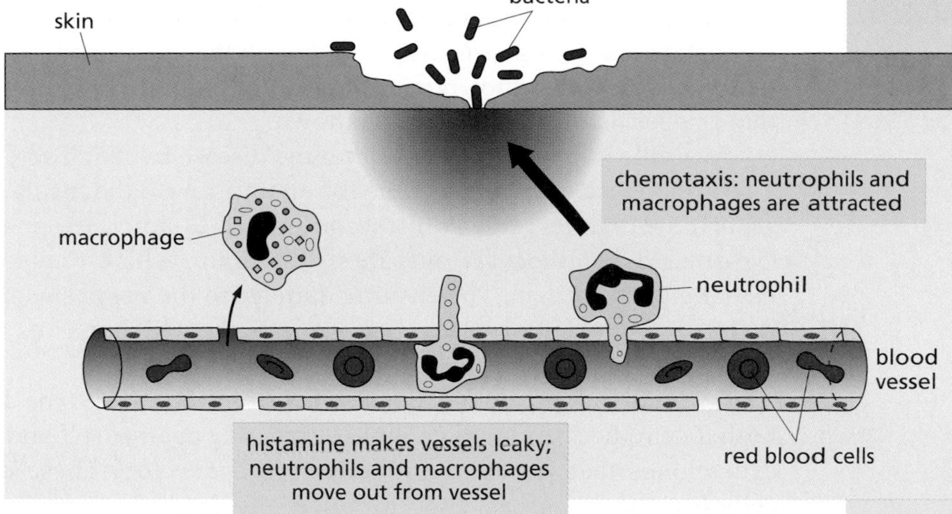
bacteria
skin
chemotaxis: neutrophils and macrophages are attracted
macrophage
neutrophil
blood vessel
histamine makes vessels leaky; neutrophils and macrophages move out from vessel
red blood cells

(B) Immune cells heal the wound.

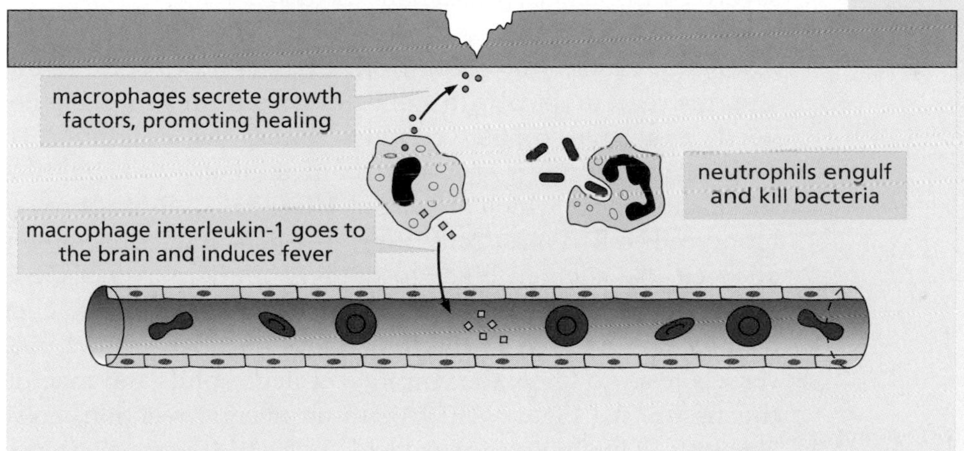
macrophages secrete growth factors, promoting healing
neutrophils engulf and kill bacteria
macrophage interleukin-1 goes to the brain and induces fever

(C) Macrophages initiate specific immunity.

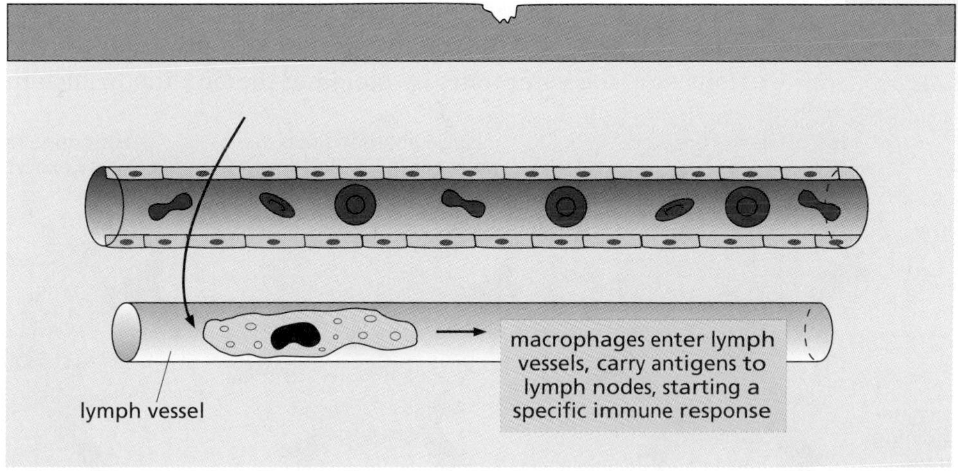
macrophages enter lymph vessels, carry antigens to lymph nodes, starting a specific immune response
lymph vessel

Autoimmune diseases. **Autoimmune diseases** result when the immune system begins to make an immune response to self, resulting in both antibodies and cytotoxic T lymphocytes that react with antigens in the body's own tissues. The immune cells or their antibodies then try to rid the body of these antigens as if they were nonself, resulting in damage to the body's own tissues. Although the mechanisms that produce tissue damage are known for some autoimmune diseases, the factors that trigger autoimmunity are unknown.

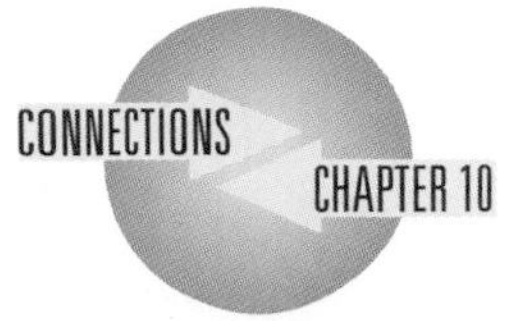

Multiple sclerosis is an autoimmune disease in which some cytotoxic T lymphocyte clones are specific for an antigen on the insulating myelin sheath around nerve cells in the brain (see Chapter 10, p. 412). These cytotoxic T lymphocytes migrate to the brain, where they kill the cells bearing their antigen. The ensuing damage to the nerve sheaths causes a variety of problems, depending on exactly which nerves have been affected.

In insulin-dependent diabetes mellitus (IDDM or type I diabetes), both T lymphocyte clones capable of reacting against self and B lymphocyte clones that produce antibody to self develop. These destroy the cells in the pancreas that produce the hormone insulin. Because insulin controls the cellular uptake of glucose, its absence produces severe consequences throughout the body.

Allergies. People who suffer from allergies do so because their immune systems react atypically to some antigens from which the host does not need protection (pollen or dust mites, for example). The atypical response produces a special type of antibody called IgE, specific for these substances which are called **allergens**. IgE binds to certain white blood cells called mast cells. When the person later encounters the same allergen, the allergen binds to the IgE on the mast cells, triggering the explosive release of histamine. Histamine is one of the chemicals that plays a positive role in the first stages of inflammation, making blood vessels leaky to allow the entrance of neutrophils and macrophages into the tissue (see Figure A12.7A). In an allergic reaction, however, larger amounts of histamine are suddenly released (Figure 12.8), producing the various symptoms of allergy. Whether an allergic response produces runny eyes, sneezing, or shortness of breath depends on the tissue in which the mast cells were triggered (the eyes, nasal lining, or the lungs). Because the symptoms are produced by histamine, antihistamine medications stop the symptoms by blocking the binding of histamine to cells

FIGURE 12.8

Allergic release of histamine.

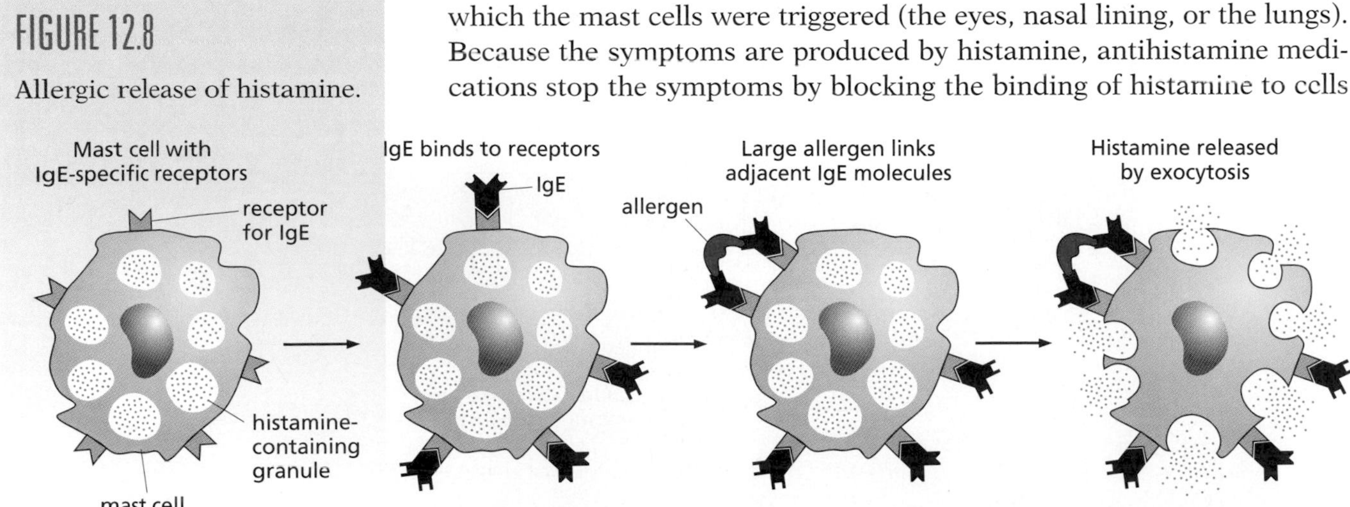

in the blood vessels. Antihistamines do not prevent the immune response or the release of histamine by the mast cells.

Because allergy is an antigen-specific immune response, it shows memory and a greater response on the next exposure, which is why people's allergies can worsen over time. Although there are thousands of different substances to which some people are allergic, each person is usually allergic to only a few. The severity of allergy to one substance does not predict the severity of allergy to some other substance, because each is a separate, antigen-specific response. We are not yet able to predict who will become allergic or what they will become allergic to, but there does seem to be some inherited component because allergies do run in families.

Plasticity of the immune system

Plasticity is a word used in biology to denote things that are changeable. Almost every biological system is plastic to some degree, but the immune system is probably among the most plastic. Some or all parts of the immune system can be either inhibited or strengthened.

Immunological tolerance. When the immune responses to a particular antigen are inhibited, it is called immunological **tolerance**. Such tolerance can be induced; for example, people who suffer from allergies can often be desensitized, that is, made nonresponsive to those particular antigens. The desensitization procedure consists of giving the person repeated small doses of the substance that he or she is allergic to. The procedure must be performed very carefully because giving the wrong dose, either too much or too little, will make the allergy worse, not better. Because both allergy and its desensitization are antigen-specific, the procedure must be performed for each separate allergen. Induced tolerance to one antigen does not change the ability of the individual to react to other antigens.

Immunosuppression. Other factors can inhibit the functioning of all or parts of the immune system to all antigens and are thus said to produce **immunosuppression**. Because immunosuppression inhibits the workings of all B lymphocytes and all T lymphocytes, rather than just antigen-specific clones, it is generally correlated with an increase in disease.

Many environmental pollutants and other chemicals suppress the whole immune system. Other immunosuppressive factors include alcohol, cocaine, and heroin. Chronic use of these leads to increased incidence of infectious disease and, with alcohol, an increased incidence of cancer. Particular foods have not been found to be immunosuppressive, but too much food (overnutrition) has been. People who are obese have a higher incidence of infection-related sickness and death. Chronic protein under-nourishment, even of a moderate nature, impairs the ability of the immune system to fight off infectious diseases. The high rates of mortality in undernourished infants is partly due to their high incidence of infections. Deficiencies of many of the micronutrients, including magnesium, selenium, zinc, copper, vitamin A, and vitamin C (see Chapter 8), impair

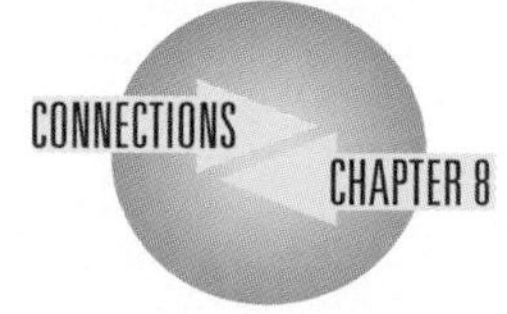

various aspects of the immune system. Many elderly people become deficient in one or more of these micronutrients and become immunosuppressed. Many of the infectious diseases of the elderly can be minimized in frequency and severity if nutrition is adequate.

In contrast, in autoimmunity, where the immune system has turned against itself, suppression of immune reactivity may be needed to prevent tissue damage. Because immunosuppression is nonspecific, however, it also increases susceptibility to infectious disease, and the risk–benefit ratio of performing immunosuppressive procedures must be carefully assessed in each case.

Immune potentiation. Strengthening antigen-nonspecific immune responses, usually called **immune potentiation**, is also possible. Tumor necrosis factor (TNF), a cytokine produced by macrophages in response to some bacterial infections, can cause the death of tumor cells. TNF was discovered in the 1970s by Lloyd Old as the result of observations made in the 1890s by a surgeon named William Coley. Coley had noted that some patients had a spontaneous remission from sarcoma, a type of cancer, if they had also had erysipelas, a severe infection with streptococcal bacteria. (Now such infections would be treated with antibiotics and would not progress to the point that would stimulate much TNF production.) Old found that when tumor-bearing mice were injected with Coley's streptococcal toxins, their tumors regressed, and the active factor, TNF, was eventually isolated. Clinical applications for TNF in cancer therapy are now being investigated.

Many factors can alter the strength of immune responses either positively or negatively. Psychological factors, such as stress, are among these. In the next sections we examine psychologically produced immunosuppression and immune potentiation in more detail, after we examine the workings of the neuroendocrine system.

THOUGHT QUESTIONS

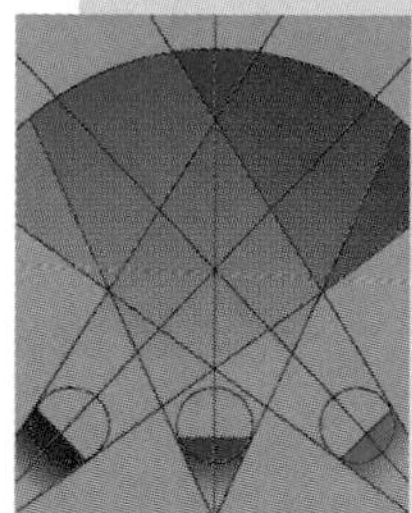

1. Since the beginnings of immunology at the turn of the twentieth century, its language has often been very military. The immune cells are said to 'protect us from invaders' or to 'kill off foreign antigens.' Locate an immunology textbook or an article on immunology from the popular press. Can you find examples of military language in these accounts?

2. We need words to convey what we imagine the immune system to be doing, based on experimentation and hypothesis testing. Do you think that the new imagery of the immune system as a communications system has taken hold at this time because we are in the 'information age,' or because new discoveries brought about a need for new terminology? To what extent are scientific terms metaphors for reality and to what extent are they models? Can the words we choose cloud our view or prevent us from being open-minded about new hypotheses?

3. Make a list of the thoughts that come to your mind when the word *self* is used in a nonimmunological context. Make a second list for your thoughts about what *self* means in immunology. Is there any overlap between your two lists?

4. When people get a bacterial or viral infection, they often get a fever. Why? Why do people sometimes get a fever after a vaccination?

THE NEUROENDOCRINE SYSTEM CONSISTS OF NEURONS AND ENDOCRINE GLANDS

The **endocrine glands** are a series of organs that secrete chemical products directly into the bloodstream. (In contrast, glands that release their secretions into the digestive tract or through the skin are called **exocrine glands**.) These secreted chemicals alter the function of target organs and thus can be said to carry messages from one organ to another. The chemicals secreted by endocrine glands are called hormones. Because hormones are distributed throughout the body by the bloodstream, the target or receiving cells can be far removed from the endocrine glands.

It was once common for the endocrine glands to be described as an endocrine system because the actions of these glands were thought to be separate from the other known communication system of the body, the nervous system. In the last few decades, biologists have discovered that several hormones originally thought to be secreted only by cells of the endocrine glands are also secreted by brain cells. Other endocrine secretions are chemically related to the neurotransmitter substances originally thought to be secreted only by the cells of the nervous system. Today, the endocrine and the nervous systems are considered to be so completely intertwined that many scientists now refer to them collectively as the **neuroendocrine system**.

The autonomic nervous system

All vertebrate nervous systems have a central nervous system (CNS), consisting of a brain and spinal cord, and a peripheral nervous system (see Chapter 10, p. 411). The peripheral nervous system connects the CNS to the more distant parts of the organism (the sensory receptors, muscles, glands, and organs) and is itself composed of the **somatic nervous system** and the **autonomic nervous system**. The somatic nervous system, largely under conscious, voluntary control, carries signals to and from the skeletal (voluntary) muscles, skin, and tendons. The autonomic nervous system has been considered, at least by Western scientists, to be largely involuntary, carrying signals to and from the gut, blood vessels, heart, and various glands, and thus regulating the internal environment of the body. 'Autonomic' literally means 'self-governing' or 'self-regulating,' reflecting the fact that the autonomic nervous system can work by itself without any input from the centers of conscious awareness in the brain. The autonomic nervous system can thus regulate body functions even while we are distracted or asleep.

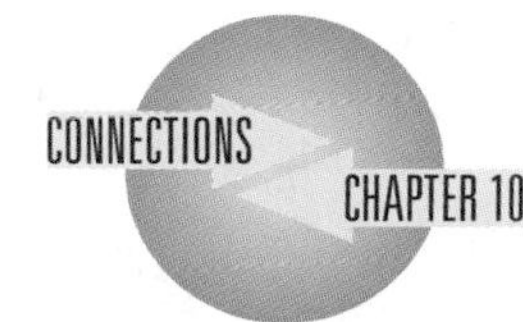

The sympathetic and parasympathetic nervous systems. The autonomic nervous system consists of two functionally separate divisions, called the sympathetic and the parasympathetic nervous systems. As is shown in Figure 12.9, the two divisions of the autonomic nervous system have opposite effects. In general, the neurons of the **sympathetic nervous system** ready the organism for heightened activity, while neurons of the **parasympathetic nervous system** do the opposite.

FIGURE 12.9

Functions of the sympathetic and parasympathetic nervous systems.

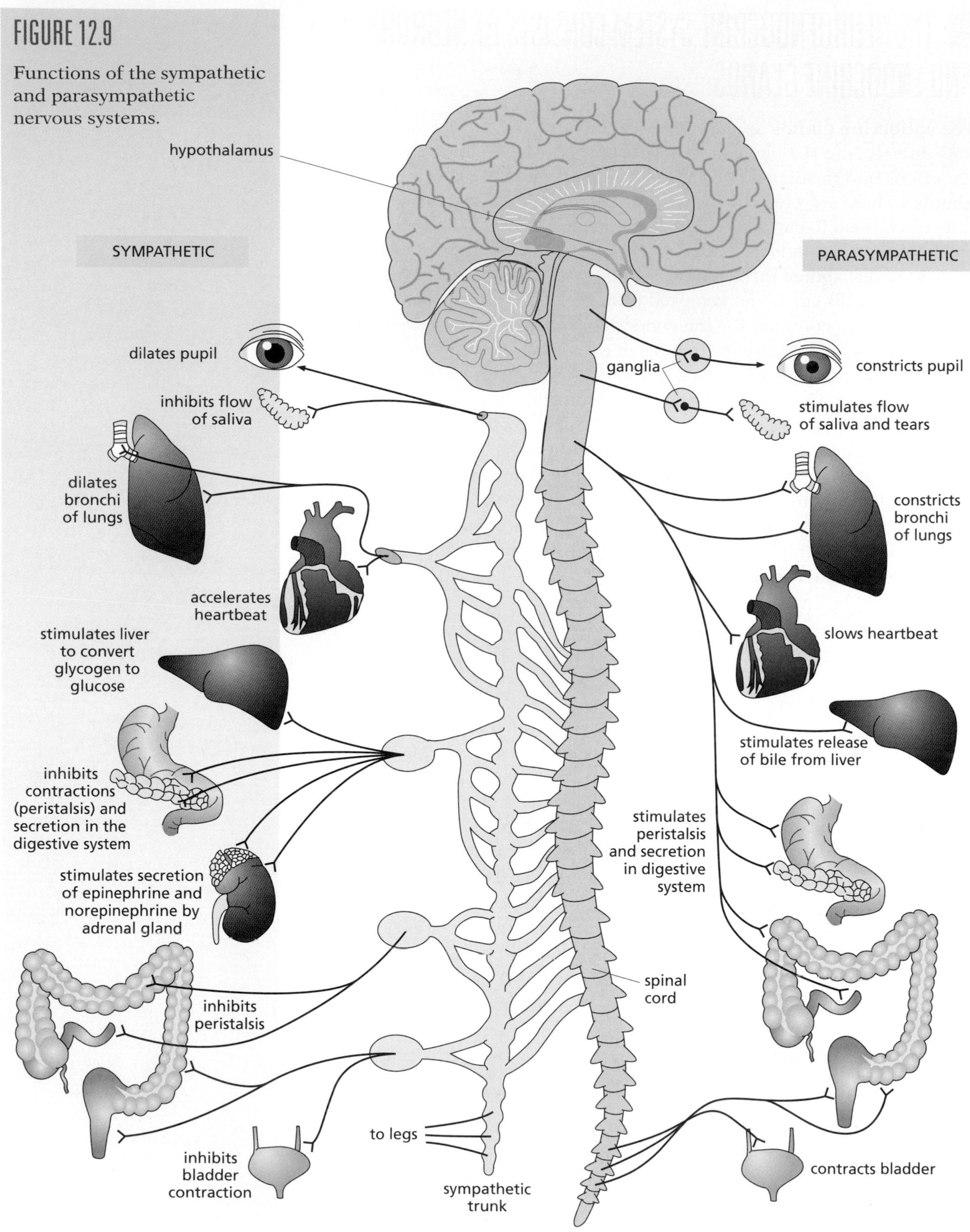

During moments of physical exertion or emergency, the hypothalamus at the base of the brain signals the sympathetic system to dominate, stimulating dilation of the pupils and bronchi, conversion of the storage molecule glycogen to glucose or conversion to energy, acceleration of the heartbeat, and slowing of the digestive processes. During rest, or when the organism is not receiving much sensory input, the parasympathetic system is predominant. Saliva, bile, and stomach enzyme secretion are stimulated, as is peristalsis (rhythmic muscle contraction) in the stomach and intestine, while the pupils of the eye constrict, the bronchi of the lungs constrict, and the heart rate slows. When the stimuli that produced the sympathetic response are no longer present, the parasympathetic system predominates once again. In extreme cases, this switch can be rapid, producing a **rebound effect**, as when a person feels woozy or faint after an emergency situation is over.

The neurons of the sympathetic and parasympathetic divisions differ from one another both anatomically and chemically. In a sympathetic nerve pathway, the first neuron runs a short distance from the spinal chord to a ganglion (collection of cell bodies) close to the spinal cord, where it makes a synapse onto a second neuron. Most of these ganglia also connect to one another, forming a sympathetic trunk. The second neuron is long, running from the ganglion to the target organ, and secretes norepinephrine (also called noradrenaline) as its principal neurotransmitter. In a parasympathetic pathway, the first neuron is very long, running from the spinal cord to a ganglion close to, or on, the target organ. The second neuron, after the synapse in the ganglion, is usually very short. The neurons of the parasympathetic division secrete the neurotransmitter acetylcholine. Tissues and organs throughout the body receive nerve endings and neurotransmitters from both of these divisions of the autonomic nervous system.

Fight or flight. Imagine that you are crossing a street. You hear a loud noise! You turn your head suddenly, and a large truck is heading right at you! Your heart begins to pound faster, your sweating increases, your voluntary muscles are stimulated (and their threshold for action is lowered), your breathing speeds up, and your digestive organs stop digesting your last meal. (You may even feel nauseous in extreme cases.) All these are the results of stimulation of different organs and tissues by the sympathetic nervous system and its neurotransmitter, norepinephrine. In general terms, the sympathetic nervous system prepares the body for reactions that require large amounts of energy and oxygen for voluntary muscle contraction, including the **fight-or-flight response**.

Rest and ruminate. Now imagine having finished a sumptuous candlelight dinner with your favorite food, elegant service, and quiet music playing softly in the background. You are relaxing in a comfortable chair or sofa with your favorite drink in your hand and wonderful company nearby. Just thinking about this scene (or reading this paragraph to yourself slowly and calmly) can relax you, cause your heartbeat and your breathing to slow down, your sweating to stop, your voluntary muscles to relax (and to raise their threshold for action), and your blood to be diverted to digestive organs, which are now digesting the sumptuous meal. These are all the effects of the parasympathetic nervous system and its neurotransmitter, acetylcholine. The parasympathetic division prepares the body to **rest and ruminate**, activities that use less oxygen while replenishing the body's store of energy supplies.

Triggering of the autonomic nervous system by emotional stimuli. As you may have been able to demonstrate to yourself as you imagined the scenes described above, the actual frightening or relaxing situation need not be present. The fight-or-flight response can be triggered by just thinking of tense situations or by watching a frightening movie. Similarly, a rest-and-ruminate response can be brought about just by relaxing comfortably and imagining a relaxing, pleasurable situation. The triggers for these responses can thus originate completely within the brain of the person imagining them. We say that the responses can be triggered by **emotional stimuli**.

The stress response

Research on the fight-or-flight response by Canadian physician Dr. Hans Selye showed that it is the first step of a larger series of physiological reactions termed the **general adaptation syndrome**. These reactions are produced by chemicals secreted by the nervous system and also by the endocrine and immune systems. The process begins when some stimulus or force, called a **stressor**, causes the body to deviate at least temporarily from its normal state of balance, homeostasis. The body's response to this deviation from homeostasis is called **stress** or the **stress response**. Stress consists of physiological and immunological changes that allow our bodies to fight off or remove ourselves from stressors and return to homeostasis, and thus stress can be a useful response. However, when stress persists too long, it can become harmful, causing disease or even death.

Alarm. Alarm, the first stage of the stress response, includes the fight-or-flight response. The hypothalamus stimulates the sympathetic neurons to secrete norepinephrine, stimulating an endocrine gland called the adrenal gland to secrete epinephrine (also called adrenaline). Epinephrine and norepinephrine together bring about the physiological changes known as fight-or-flight. In addition, sympathetic neurons release norepinephrine directly into the lymphoid organs (the spleen, thymus, and lymph nodes), stimulating these organs to release their store of lymphocytes into the bloodstream. As the lymphocytes are released, the organs in which they were stored decrease in size. The hypothalamus also secretes a hormone called adrenocorticotropic hormone (ACTH), which stimulates the adrenal gland to produce corticotropin-releasing hormone (CRH), which in turn stimulates **steroid hormones** (particularly cortisol) to be released from the cells of the outer layer (cortex) of the adrenal gland. The alarm phase of the stress response is shown in the upper part of Figure 12.10.

Resistance. If the stress continues, the body enters the second stage, **resistance**, in which resources are mobilized to overcome the stressor and to regain homeostasis. New stores of steroid hormones are synthesized, keeping blood levels of these hormones elevated. If the stressor is a disease or injury, inflammation begins and phagocytic cells are chemically attracted to the inflamed area. Epinephrine acts on the heart to increase the heart rate (the number of contractions per minute) and cardiac output (the strength of the contractions). Norepinephrine increases the flow of blood to the heart and muscles for possible increased activity, while constricting other vessels, diminishing the flow of blood to the gut, skin, and kidneys. Cortisol suppresses lymphocyte activity, thereby decreasing the level of the cytokine IL-2; it also suppresses inflammation and the inflammatory cytokine IL-1. The resistance phase of the stress response is shown in the middle part of Figure 12.10.

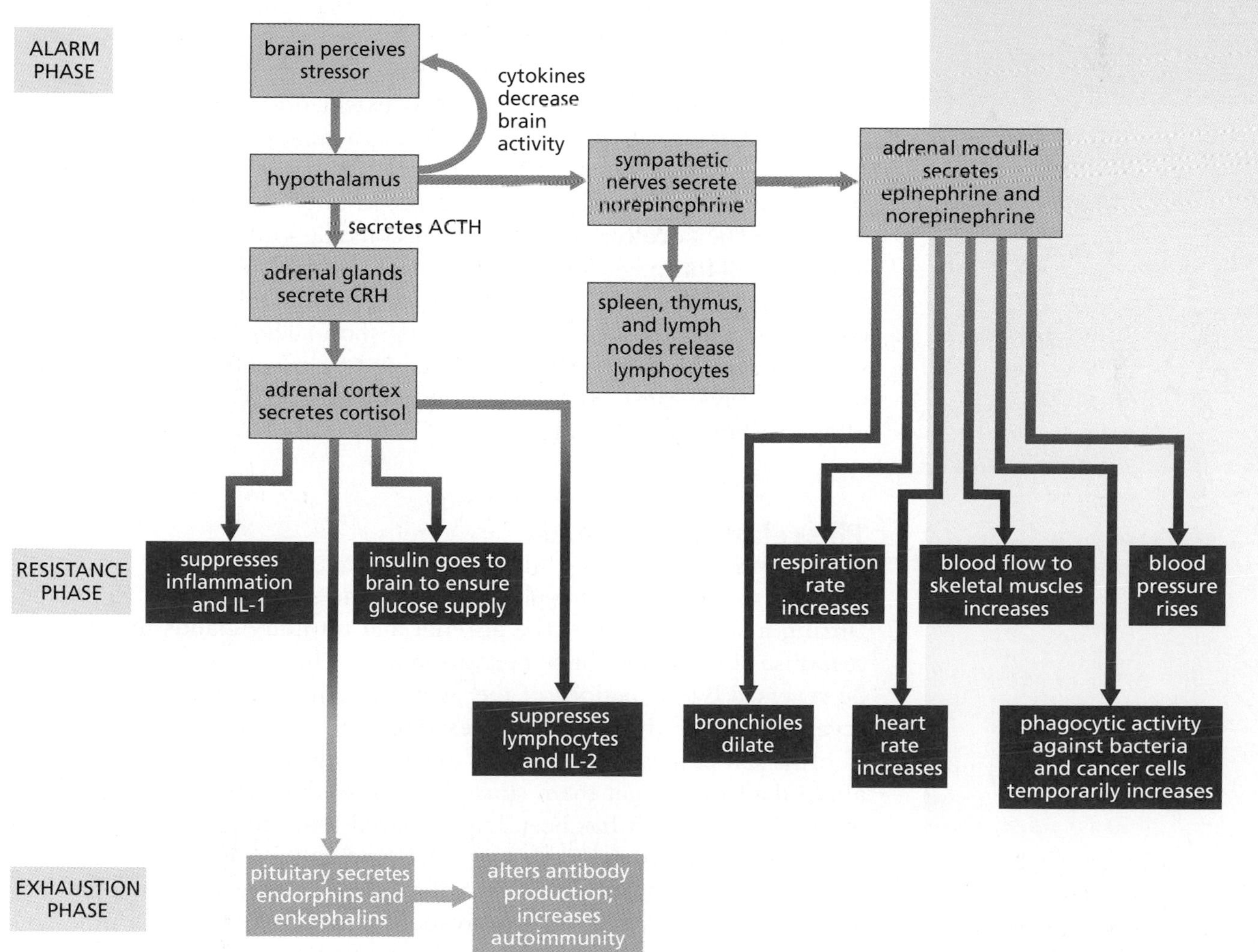

FIGURE 12.10

The phases of the stress response. Stressors induce a variety of physiological changes, including immunological changes, which are mediated by the sympathetic nervous system and adrenal hormones.

Exhaustion. If the stressor is not successfully overcome, the adaptation syndrome reaches its third phase, **exhaustion**. The steroids made in the resistance phase are used up and the animal is unable to make more. During the exhaustion phase, the action of the sympathetic nerves tapers off, while the endocrine organs take over. Adrenal hormones stimulate another endocrine gland, the pituitary gland, to secrete endorphins and enkephalins. These are chemically related to opioid drugs (see Chapter 11) and alter the activity of neurons and of immune cells, as shown in the lower part of Figure 12.10.

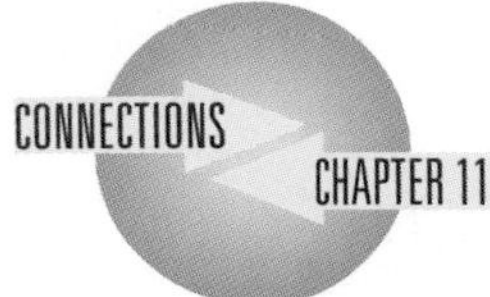

The pituitary was once called 'the master gland' because its hormone secretions controlled the activity of many other endocrine glands, but it is now known that the pituitary itself is under the control of the brain via the hypothalamus. Hypothalamic cytokines also act on other cells within the brain itself, changing the activity level of neurons, and this change affects behavior, heat production, and many other functions.

Some stress responses are due to actual physical danger; but stress responses can also result from the demands of work or school (e.g., deadlines or examinations), or the actions of the people in our lives. Whether these demands are real or perceived, they can have the same biological consequences: the physiological changes that characterize the stress response.

Effects of stress on health. To what extent does the stress response influence human health? The answer is: to a great extent. Stress (i.e., the stress response) is an important risk factor in heart disease, and there is considerable evidence that people exposed to chronically high levels of stress are statistically more likely to become ill with infectious diseases, to remain ill for longer periods, and to suffer more severe consequences, even death. Depression is much more common in people subject to chronic stress. Among cancer patients, those who have better coping skills for dealing with stress (as measured by psychological tests) have higher survival rates, compared with patients who have the same forms of cancer but poorer coping skills.

The relaxation response

When actual physical danger has passed, the stress response will abate. During a stress response, the adrenal and pituitary glands mediate the response of the sympathetic nervous system. This response will gradually be reversed by the actions of the parasympathetic nervous system. This reversal is called the **relaxation response**.

It is sometimes more difficult to get our mentally induced stresses to abate than it is to get them started. Because the stress response can be mentally induced, it has been hypothesized that the relaxation response should be mentally inducible as well. Some cultures and some religions have been more open to this idea than others. Many practices that are aimed at evoking this relaxation response, including Yoga and transcendental meditation, originated in Asian traditions. Hindu yogis have learned how to consciously control the actions of their autonomic nervous systems and to bring about levels of activity even below the normal levels for the resting state. Measurements made by Western scientists have shown that these yogis are able to lower their blood pressure, breathing, oxygen consumption, heart rate, and metabolic rates.

Other cultures, including many Western cultures, have been less open to the idea that the relaxation response can be controlled. The traditional definition of the autonomic nervous system emphasized that it governed involuntary functions over which we do not have conscious control and had the unintended effect of discouraging research on any possible interactions of the autonomic nervous system with our emotions and other conscious body states. Recently, however, some of the methods for conscious control of autonomic processes are being borrowed from other cultures. Some athletes have learned meditation, while others have learned a related technique called **imaging**. In imaging, one invokes specific mental imagery in order to put one's body as well as mind in a certain state of relaxed determination to succeed in sport. Western medicine has begun to use similar techniques to help cancer patients in fighting their cancers, as described further in a later section on mental imaging.

Less conscious strategies can also produce the relaxation response. Studies using measurements of blood pressure and other physiological indicators have shown that contact with pets can reduce stress and bring about relaxation. Older people who keep pets have also been shown to live longer than those who do not, even when comparison is made between people of comparable health status initially. Heart-attack victims are much less likely to have a second heart attack if they care for a pet. Studies such as these show a statistical correlation between two factors. From such data by themselves we cannot assign a causal relation between the two; that is, we cannot say that pets caused the increased longevity or improved health.

The placebo effect

In clinical trials testing new drugs, a common experimental design is for one group of people to receive the test drug and for another group, the **control group**, to receive a **placebo**, a preparation that is similarly colored and flavored but does not contain the test ingredient. To be in such a study, people must have given their informed consent (see Chapter 1); that is, each person, after being informed as to the nature of the test being conducted, must have signed a written consent form. Many studies are conducted **double blind**; that is neither the subject nor the experimenter knows who is receiving the placebo and who is receiving the actual drug. For the drug to be considered effective, there must be a statistically significant difference in outcome between the group receiving the experimental drug and the control group receiving the placebo.

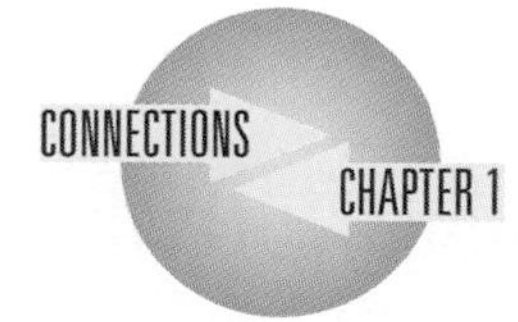

Studies like this were initially designed to demonstrate whether particular drugs were effective or not. They have also shown, over and over again, that the people who receive the placebo in such tests have a significant change from the baseline values of whatever parameters are being measured, an effect known as the **placebo effect** (Figure 12.11). They also experience many 'side effects,' although they have not received a drug. In experiments on pain perception, people who are given placebos instead of painkillers very often experience a reduction in pain. People who have purchased street drugs will often feel the reaction they seek even when the drugs are in such low concentration that no real effect could be produced.

Such placebo effects have often been treated as an annoyance in research; that is, they make it more difficult to demonstrate the 'real' effects of test compounds. The existence of such effects, however, is additional evidence that the mind can bring about physiological changes in the body, in addition to the effects of the stress response and the relaxation response.

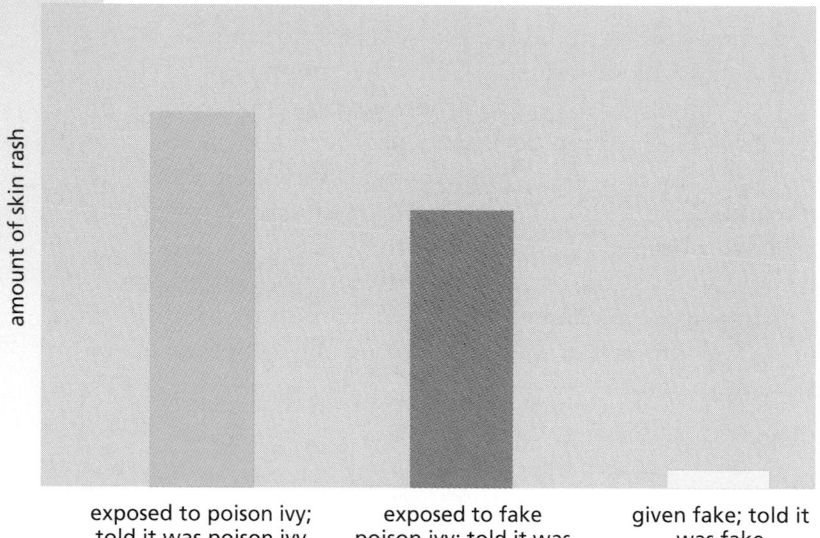

FIGURE 12.11
The placebo effect.

THOUGHT QUESTIONS

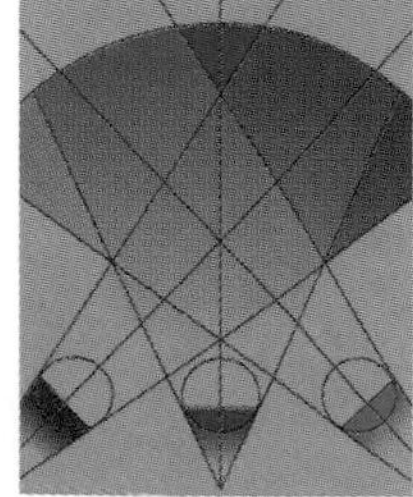

1. Is epinephrine (adrenaline) a hormone or a neurotransmitter?

2. How does thinking about an annoying or threatening event produce stress?

THE NEUROENDOCRINE SYSTEM INTERACTS WITH THE IMMUNE SYSTEM

We have discussed how the nervous system and the endocrine system communicate with each other. In recent decades, it has also become apparent that both of these systems also communicate with the immune system. Because, as we have seen, the immune system protects against disease, communication between the brain and the immune system suggests a mechanism by which the mind can affect health.

The emerging concept is that the nervous, endocrine, and immune systems do not exist as separate entities but are one interacting communications network. Extending the metaphor, we might say that the nervous system provides the hardwiring of the system, because nerve cells have distinct locations and make direct connection with one another via synaptic junctions. The immune cells, by analogy, are mobile synapses, responding and transmitting messages at the same time as they themselves are moving, the original cellular phones. Endocrine glands could be thought of as more like microwave transmission towers, having a fixed location themselves but sending out their messages chemically rather than over wires (nerves).

The above metaphor is intended to aid in conceptualizing how the immune system works and should not be taken too literally. Mental images of this kind can, however, provide a basis for discussion and can suggest hypotheses to test. When used this way and shared by a number of scientists, an image becomes a **model** (see Chapter 1). Psychoneuroimmunology proceeds by testing hypotheses suggested by the model of a system of interactive chemical communications linking the immune, nervous, and endocrine systems.

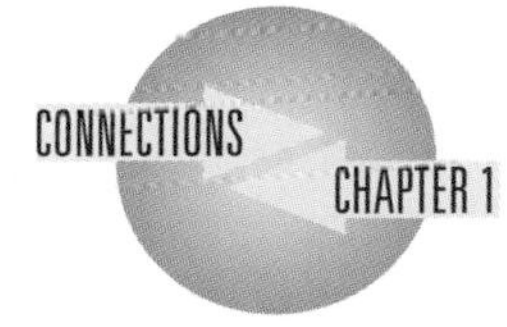

Unlike literary images and metaphors, which are unconstrained, those in science must suggest hypotheses that can be tested against reality. Models in science gradually become more limited, or more well defined, as data are gathered from observation and experimentation. Models are not just descriptions of reality but are evolving concepts, consistent with the body of data known at that time. The psychoneuroimmunology model as it exists today is already more limited than the communications network metaphor just described. The 'hardwired' nerves are not completely immobile because synaptic connections between cells are formed and lost throughout life. There are also known to be limits to the mobility of immune cells. Because the model is relatively new, its limitations are now in the process of being identified. Scientific explanations are always tentative, but nowhere is this more so than in new models that have not yet stood the test of time.

Psychoneuroimmunology has an appealing central model, suggesting that the mind and body are intertwined. However, models or theories that are appealing or that fit with our common sense do not necessarily stand up to scientific scrutiny. The reader is strongly encouraged to think critically about the information presented here. Where are the gaps? What still needs to be done? Is there any scientific evidence that supports the hypothesis that the nervous, endocrine, and immune systems are interconnected? Yes, there are. The first line of evidence comes from the discovery that these systems use the same cytokines for cellular communication. In addition, there is both structural and functional evidence, gathered from *in vitro* studies ('in glass,' that is, studies in test tubes), and *in vivo* studies ('in life,' that is, studies in animals and humans). In this section we examine each of these in turn.

Shared cytokines

The term 'cytokine' includes molecules that are secreted by cells and function in communication with other cells. The ability of any cell to respond to a particular cytokine is dependent on whether the cell has receptors for that cytokine. Therefore, for the cytokines secreted by the nervous, endocrine, or immune systems to have an effect on the other systems, one must ask whether there are receptors for them on cells of the other systems. The answer is yes, there are. American pharmacologist Candace Pert, working in the laboratory of Solomon Snyder in the early 1980s, discovered that there were receptors on immune cells for endorphins, cytokines produced by nerve cells. Since that time, receptors have been found for many other cytokines that interconnect the nervous, immune, and endocrine systems. B lymphocytes have receptors for many neurotransmitters, including norepinephrine and enkephalins. The presence on immune cells of receptors for these neurotransmitters suggests that immune cells could respond to these transmitters, as does the finding that the number of these neurotransmitter receptors per lymphocyte increases during immune activation. Lymphocytes possess receptors for several endocrine cytokines induced during stress (ACTH and steroid hormones) or in pain (beta-endorphin). Neurons in the hypothalamus have receptors for the immune cell cytokine IL-1. The neuroglial cells that feed the neurons in the brain (see Chapter 10, p. 413) have receptors for another immune cytokine, IL-2 (see Figure 13.1, p. 538), as do some endocrine cells.

Nerve endings in immune organs

Evidence of another sort came from studies done by another American scientist, David Felten, on the innervation of the organs of the immune system. Neurons of the sympathetic nervous system were found to terminate in the immune organs, such as the spleen, thymus, lymph nodes, bone marrow, and lymphoid tissue in the gut. The sympathetic nerve cells release norepinephrine, and the immune cells in these organs have receptors for norepinephrine.

These studies used a technique called immunohistochemistry. Antibodies ('immuno-') are used to stain very thin slices of tissue ('histology-') and the antibodies are then detected by chemically-induced color changes ('chemistry'). Such techniques borrow the exquisite antigen specificity of the immune system. Antibodies that recognize some molecule that you want to detect are produced artificially in cell cultures. The antibody used by Felten's group recognizes and binds to an enzyme used in the synthesis of norepinephrine but does not bind to other chemicals. Thus, the antibodies give a specificity to the technique, just as they do during an immune response in the body. A color-based detection process then highlights the location of the bound antibody. Some typical results are shown in Figure 12.12. While most nerve cells terminate in synapses with other nerve cells or on muscle cells (see Chapter 10, pp. 411–412 and pp. 432–433), Felten's immunohistochemical studies showed nerve cells terminating and synthesizing norepinephrine in proximity to the cells of the immune system.

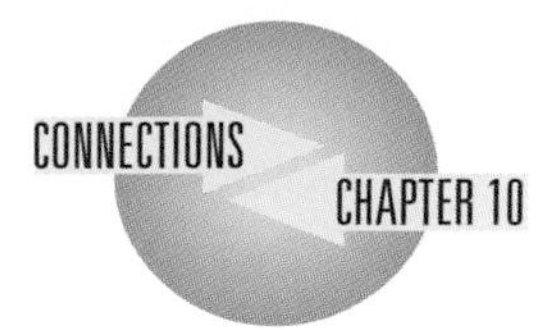

Studies of cytokine functions

Demonstrating that a receptor is present, or even that both the receptor and its cytokine are present together in a tissue, does not by itself show that any effect follows the binding of the cytokine to its receptor. For that, experiments known as functional assays (techniques used to measure a response) need to be done.

Immune cytokines and hormone secretion. If the secretion of one cytokine can be shown to happen after the administration of another cytokine, this is evidence of a functional connection between the two. Such studies can also give us clues as to the way in which an effect can be triggered, which we call the **mechanism** (or mechanisms, since more than one trigger is often possible).

Functional assays can also tell us the order in which events happen. For example, during acute bacterial infection, the secretion of adrenal and pituitary hormones increases. Functional studies showed that the effect is not a direct one. If pituitary cells are stimulated with bacterial molecules *in vitro* they do not secrete these hormones. When immune cells are exposed to these bacterial products, however, they secrete cytokines. If immune cell cytokines (including IL-1, IL-2, or TNF) are administered to animals, the level of pituitary hormone in the blood increases.

Neuroendocrine effects on immune cells. Among the cytokines secreted by the brain cells are the enkephalins and endorphins. When enkephalins are given to living rats, immune responses are altered, including antibody responses and T lymphocyte-mediated responses. Interestingly, low doses of enkephalins increase antibody production, whereas high doses suppress it. Low doses of enkephalins also increase some destructive aspects of the immune response, including both allergic and autoimmune responses, whereas high doses of enkephalins depress those responses (see Figure 12.10, exhaustion phase).

The glucocorticoid steroid hormones (cortisol, cortisone, corticosterone, and several related compounds secreted by the adrenal cortex during the stress response) can have inhibitory effects on the immune system.

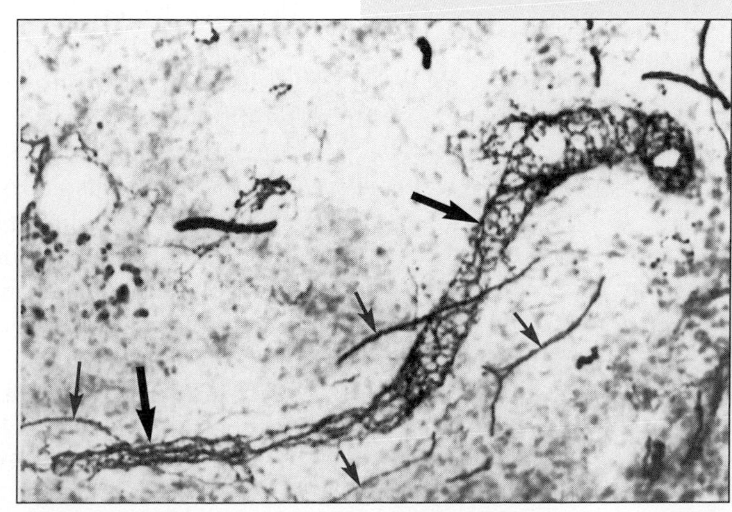

FIGURE 12.12

Immunohistochemical staining showing the presence of the neurotransmitter norepinephrine in the rat spleen. Large black arrows: blood vessels. Small red arrows: nerve fibers synthesizing norepinephrine.

Steroid hormones given to animals decrease the numbers of cells in lymphoid organs and suppress IL-1 and IL-2 secretion (see Figure 12.10, resistance phase). Steroids also increase the susceptibility of the animals to disease and activate latent infections (infectious organisms that have previously been present but have not brought about disease now do so).

A chemically similar compound, hydrocortisone, is used medicinally as an **anti-inflammatory** agent. Remember that normal inflammation is the healing phase of the immune response (see Figure 12.7). In insect bites, severe poison ivy, athletic injuries, and rheumatoid arthritis, the swelling and pain are the result of the inflammatory response. In these situations, the annoying symptoms are actually an indication that the immune system is at work. Thus, a person who chooses to take an anti-inflammatory drug is choosing to suppress the healing processes of the immune system with the aim of suppressing the negative symptoms of inflammation. The symptoms may be so severe that immunosuppression is needed, but long-term immunosuppression by corticosteroids are likely to have adverse consequences on other aspects of health.

Stress and the immune system

In the previous section, results demonstrating cytokine function must be extrapolated with caution because the intact living body is more complex than any experimental system. Another type of study, therefore, looks at changes in cytokine and immune functions after real-life stressful events. Rather than giving the cytokine directly to the experimental subjects, they are instead studied after exposure to stress.

Short-lasting stress may actually strengthen aspects of the immune system. In mice, short-term stress brought on by conflict increases the engulfment of bacteria by white blood cells (see Figure 12.4). The ability of the natural killer cells (described in an earlier section of this chapter) to kill tumor cells can be increased by restraining rats on a single day so as to increase their stress response. However, several days of restraint-induced stress causes a decrease in natural killer cell activity.

The production of IL-2 is decreased during stress (see Figure 12.10). This cytokine, secreted by helper T lymphocytes, is necessary for a full-strength antigen-specific immune response. So decreased levels of IL-2 decrease the response of antibody-producing B lymphocytes and cytotoxic T lymphocytes.

An increase in the hormone corticosterone causes the death of pre-T lymphocytes that are developing in the thymus. Daily fluctuations (circadian rhythms; see Chapter 10, p. 440) in the plasma levels of corticosterone are also correlated with the circadian rhythm of the numbers of B and T lymphocytes circulating in the blood. Stress can, however, suppress immune function even in rats whose adrenal glands (the source of corticosterone) have been removed, demonstrating that there must be other pathways as well.

Anti-inflammatory feedback inhibition. Why would organisms have a mechanism that could inhibit their immune systems? The anti-inflammatory activity of the steroid hormones may provide a natural feedback loop to keep the extent of inflammation within homeostatic limits. **Addison's disease** is a disease that results from insufficient secretion of glucocorticoid steroid hormones by the adrenal gland. People with Addison's disease have such low levels of these anti-inflammatory cytokines that they suffer severe inflammation when they get bacterial infections and therefore need to be given anti-inflammatory drugs. The opposite problem occurs in **Cushing's syndrome**, a disease in which persistently high levels of steroid hormones result in immunosuppression. People with Cushing's syndrome get frequent bacterial infections, and, because the normal inflammatory response is lacking, the infections can be fatal.

Effects of stress on disease incidence. The biochemical events of stress can be started by psychological factors, and prolonged stress can suppress the immune system. There have been many demonstrations of immune suppression in people undergoing various types of stress. It follows that long-term psychological stressors might produce conditions in which disease can develop. In one experimental design, blood samples are taken from young, basically healthy students during exams, and various immune parameters are compared to baseline levels measured in blood samples from the same students one month before exams. In this type of experiment, each person serves as his or her own control, minimizing differences due to factors other than the tension of exam situations (see below). Exam periods brought on an increase in adrenal cortex hormones and a decrease in natural killer cell activity. There were also changes in the numbers of T lymphocytes and a decrease in the ratio of helper T lymphocytes to suppressor T lymphocytes. Because helper T lymphocytes stimulate immune responses and suppressor T lymphocytes inhibit them, a decrease in the ratio of the two means less help and more suppression, for an overall decrease in T lymphocyte responsiveness. In these studies, such short-term stress was correlated with an increase in disease, primarily infections of the upper respiratory tract. Many college health centers report an increase in student admissions for infectious diseases during exam periods.

Another study examined the immune function and health status of men who had separated or divorced within the previous year. The experimental design was different from the one used with the students. One group of people, the divorced men, were compared with people in a control group of married men. Not only were the immune systems of the divorced men impaired, they experienced a greater number of illnesses than the controls. Comparisons were also made between those men who had not initiated the separation or divorce and those who had. Those who had not initiated the break were significantly more immunosuppressed and had more illnesses than those who did. Studies like this employ statistical methods for determining whether the differences between groups are greater than could have been predicted by chance only.

Other researchers found that elderly people who had been caring for a spouse with Alzheimer's disease demonstrated a decrease in three different measures of immune function. The elderly people undergoing prolonged stress got sick more often than those in the control group. The immune systems of these care givers stayed depressed after the death of the spouse.

Several studies have showed decreased immune function in people with clinical depression. A prospective study showed that people who were depressed had a higher incidence of cancer 17 years later. A **prospective experimental design** is one in which a group of people are examined first (using either physical or psychological exams) and then their outcomes are monitored at various later times. One strength of a prospective study is that no one knows ahead of time who will be sick and who will not; baseline data are taken before the outcomes are known. One weakness is that the percentage of people in any particular group who will get a particular disease may be very low, so that the number of people in the study must be very large. Many people will leave the study for unrelated reasons. Many other factors can influence the outcome; to some extent this can be corrected for by statistical methods, but only those factors that have been identified can be factored out by statistical methods.

What mechanisms could bring about cancer after clinical depression? Immune activity is compromised, and other functions are also impaired, including levels of an enzyme that repairs damaged DNA. Breaks and misreadings of DNA occur rather frequently, but normally several 'proofreading' mechanisms check the DNA and repair most of the mistakes; mistakes that remain uncorrected are capable of transforming cells. The suppressed immune system then fails to remove these transformed cells before they have become established as cancer (see Chapter 9).

Individual variation in the stress response

There are many types of stressors. The effect of stressors on health is highly variable, because the effects are modified by many additional factors. For example, of the people exposed to infectious mononucleosis, prevalent in college-age populations, not everyone becomes sick, and of those who do, some become sicker than others. It is important whether the stress occurs before or after the immune challenge started. The severity and duration of the stress are also important. Genetic factors play some role; in animal studies, different strains of mice (each inbred to minimize genetic variation within the strain) respond differently to stressors. Psychological factors are just as important.

Personality profiles and life events have some bearing on disease susceptibility and disease progression in humans as well. Testing methods have been developed for quantifying the psychosocial impacts of life events (Table 12.1). The use of such methods has indicated that certain life events can increase the probability that cancer will develop, although the results have been highly variable from study to study.

We cannot predict how much immunosuppression will be sufficient to result in disease in a given person. Both the degree and duration of suppression that result in disease are likely to be different for different people. Some studies suggest that coping styles can help to regulate the degree of impact that stressful events will have on individual health. When psychological tests are given to matched sets of cancer patients (with the same kind of cancer and in comparable stages of the disease), those with more optimistic or aggressive personalities show higher survival rates than those who are more easily resigned to what they perceive to be their fate. Other factors, such as environmental pollutants, drugs, alcohol, and malnutrition, may also weaken the immune system. If a person's immune system is already weakened by one or more of these factors, the additional immunosuppression effects of stress are more likely to result in disease.

TABLE 12.1

Stressful life events. The relative stressfulness of each event is indicated by the number on the left. The most stressful life event was assigned a value of 100, and other events were assigned lower values in proportion to their effects on stress. Divorce, for example, caused stress in 73% as many individuals that experience this event as was true for the death of a spouse.

STRESSFULNESS	LIFE EVENT
100	Death of spouse
73	Divorce
65	Marital separation
63	Jail term
63	Death of close family member (except spouse)
53	Major personal injury or illness
50	Marriage
47	Being fired from work
45	Marital reconciliation
45	Retirement
44	Change in health of family member (not self)
40	Pregnancy
39	Sex difficulties
39	Gain of new family member
39	Business readjustment
38	Change in financial state
37	Death of close friend
36	Change to different occupation
35	Change in number of arguments with spouse
31	Mortgage over $40,000
30	Foreclosure of mortgage or loan
29	Change in responsibilities at work
29	Son or daughter leaving home
29	Trouble with in-laws
28	Outstanding personal achievement
26	Spouse begins or stops work
26	Begin or end school
25	Change in living conditions
24	Change in personal habits (self or family)
23	Trouble with boss
20	Change in work hours or conditions
20	Change in residence
20	Change in schools
19	Change in recreation
19	Change in church activities
18	Change in social activities
17	Mortgage or loan less than $40,000
16	Change in sleeping habits
15	Change in number of family get-togethers
13	Change in eating habits
13	Vacation
12	Christmas
11	Minor violations of the law

Conditioned learning in the immune system

If brain cells can interact with immune cells and psychological factors can influence the onset and progression of disease, could a person learn how to control his or her own immune response? As odd as this idea may sound, evidence in support of it is accumulating.

CONNECTIONS CHAPTER 10

Classical or Pavlovian conditioning is a type of unconscious learning in which the organism learns to associate one stimulus with another. After such conditioning, presentation of the second stimulus will bring on the physical effects of the first stimulus (see Chapter 10, p. 436). Robert Ader, a psychiatrist, and Nicholas Cohen, an immunologist, worked together to try to explain why some of Ader's mice had been dying unexpectedly in his studies on a drug called cyclophosphamide. This drug suppresses the immune system and in fact is given to recipients of organ transplants so that their immune systems do not reject their transplants. Ader's mice had been receiving cyclophosphamide along with saccharin in their drinking water. Later the mice received only the saccharin, but their immune systems again became suppressed as they had when on the cyclosphosphamide, even though saccharin itself has no effect on the immune system. What Ader and Cohen demonstrated in several controlled studies is that the dying mice had been conditioned. After the mice learned to associate the immunosuppressant chemical, cyclophosphamide, with the saccharin, the immunosuppressant effect could be produced by giving them only the saccharin water without cyclophosphamide.

These experiments demonstrating conditioned immunosuppression have been repeated with several other paired stimuli. Such conditioning has been shown to improve the health of mice with autoimmune disease; preliminary results suggest that conditioned immunosuppression is also useful in the treatment of people with autoimmune disease.

If animals are challenged with an antigen paired with another stimulus, classical conditioning that boosts immunity can be demonstrated (Figure 12.13). In a normal immune response, a second exposure to the same antigen would produce an increase in specific immunity to that antigen. In a typical conditioning experiment, the animals are given their first exposure to antigen paired with a conditioned stimulus such as saccharin. Saccharin itself does not induce antibody for antigen 1 (see Figure 12.13A). When saccharin is paired with the first exposure to antigen 1, animals make antibody for antigen 1 (see Figure 12.13B). After conditioning, when the animals are exposed to saccharin, they react as though they were being exposed for a second time to antigen 1. Saccharin induces an increased secretion of antibody for antigen 1, without a second exposure to antigen 1 (see Figure 12.13C).

Voluntary control of the immune system

Other work has shown that people can learn to regulate voluntarily many of the physiological processes mediated by the autonomic nervous system. People can learn to regulate the temperature of their hands, their blood pressure, their heart rate, and their galvanic skin resistance (resistance to electrical conductivity, which is a measure of the amount of sweat on the skin). Because the immune system communicates with the autonomic system, these findings raised the hypothesis that parameters of the immune system may also be subject to voluntary control. Several studies have shown that this is possible by using different self-regulation procedures, including relaxation, mental imaging, biofeedback, and emotional support. Voluntary potentiation of several immune parameters has been shown, including white blood cell engulfment of bacteria, antibody production, lymphocyte reactivity, and natural killer cell activity. Women with metastatic breast cancer, all of whom were receiving medical treatment of their cancers, survived longer if they were part of support groups than if they were not.

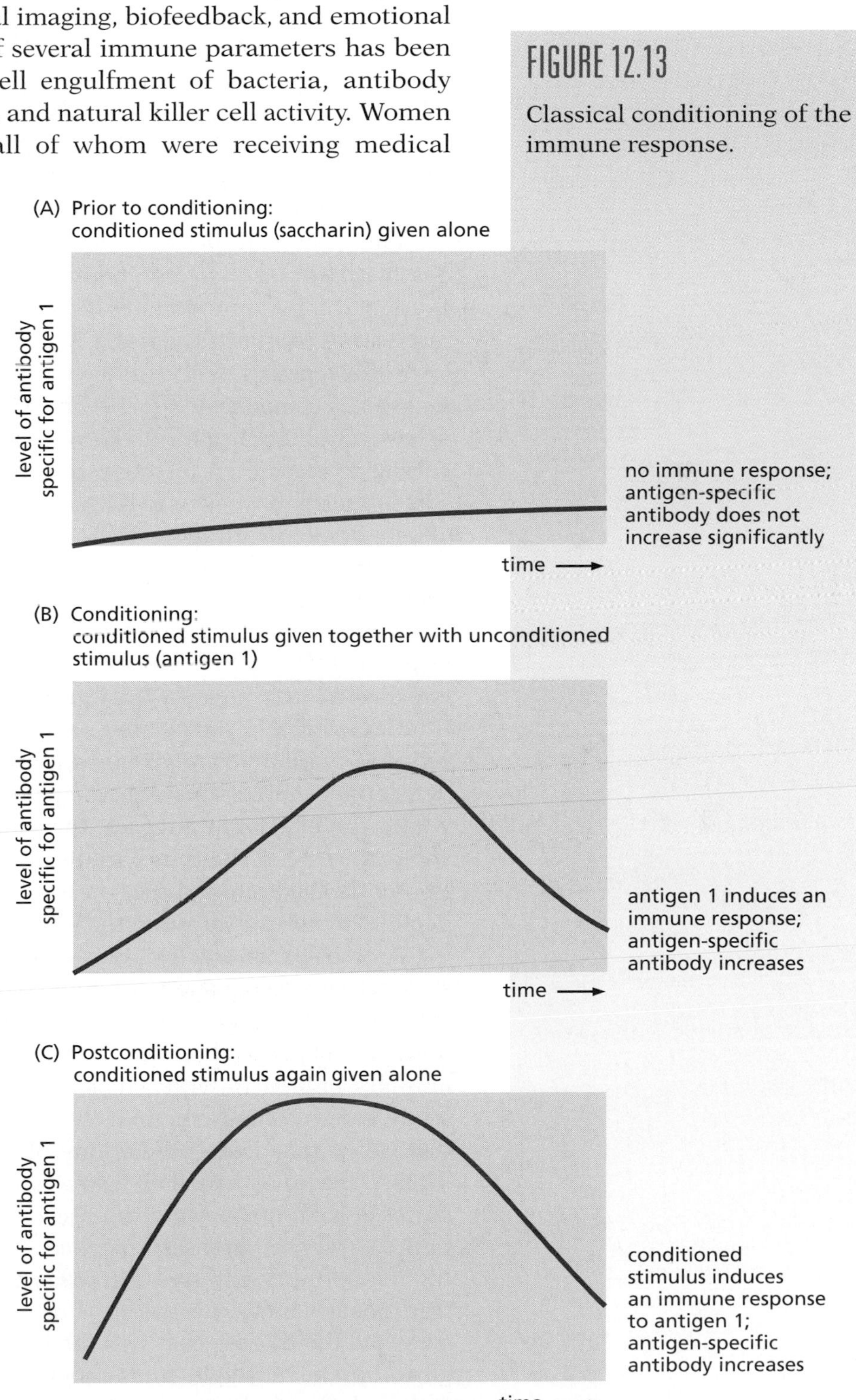
(A) Prior to conditioning:
conditioned stimulus (saccharin) given alone
level of antibody specific for antigen 1
no immune response; antigen-specific antibody does not increase significantly
time
(B) Conditioning:
conditioned stimulus given together with unconditioned stimulus (antigen 1)
level of antibody specific for antigen 1
antigen 1 induces an immune response; antigen-specific antibody increases
time
(C) Postconditioning:
conditioned stimulus again given alone
level of antibody specific for antigen 1
conditioned stimulus induces an immune response to antigen 1; antigen-specific antibody increases
time

FIGURE 12.13

Classical conditioning of the immune response.

Mental imaging. In one study, subjects were asked to make mental images of their neutrophils becoming more adherent, a cellular process that might make neutrophils more efficient at getting to a disease site. When the adherence of their neutrophils was measured after several sessions of practicing such imaging, it was significantly increased in comparison with the neutrophils from people who had simply relaxed without forming the mental image of their neutrophils. Adherence was also measured in neutrophils from a third group who made mental images but did not have the practice sessions. Neutrophils from this group did not show an increase in adherence, showing that effective imaging takes a period of training.

Biofeedback. In **biofeedback**, measurement devices are placed on people so that they can monitor the results of their self-regulation. Biofeedback has proved to be effective for some people in the management of chronic pain and migraine headaches. Although it is too soon for there to be conclusive data on whether voluntary regulation of immunity can translate into improved health, preliminary studies suggest that it does. People with HIV infection and AIDS have remained healthier when they have used these techniques.

Studies on populations. If the mind and the emotions can influence the immune system, and the immune system helps to fight off many diseases, how far can the disease-fighting process be controlled by the mind? In recent years, statistical evidence has been accumulating from studies in both the United States and China to show that dying patients can exercise control over their disease processes to the extent that they actually influence the time of their death.

Large-scale studies are often done on entire populations by studying death certificates and comparable records. When the date of death is examined for a large number of patients in a population at large, several interesting regularities appear. For instance, the overall mortality rate is lower for a period of several days to either side of each person's birthday, and this is compensated for by an increasing mortality rate about a week or two later. Such data make it seem that dying patients are eager to survive to reach their birthdays and that they can postpone the inevitable by as much as a week or two. Other studies have shown similar statistical effects demonstrating the ability of people to postpone the time of their death until after holidays or family events (e.g., weddings) of special importance to them. In China, this effect even has a name: the Harvest Moon phenomenon. The Harvest Moon Festival is a traditional family celebration in which the oldest and most respected woman in each family is expected to prepare a large feast to celebrate with her entire family. Studies of death certificates show that older Chinese women have a reduced mortality rate around the time of this festival. It is difficult to determine whether this effect is primarily the result of the activities in which the matriarch engages, the increased esteem or importance that she receives, her desire not to disappoint others, or simply the anticipation of the big event. Studies on Jewish populations have shown a similar decline in mortality around the time of Passover.

A complex interaction between expectation and mortality has also surfaced in a recent study in China. Traditional Chinese astrology divides the calendar into 12-year cycles. Each year is represented by a different animal, and people born in that year are said to be under control of that animal's influence, with which certain diseases are associated. In this study, elderly Chinese patients were surveyed to see whether or not their disease matched the predictions of Chinese astrology. Patients with a disease that matched their astrological year were then compared with patients having the same disease but a different astrological year. The patients whose diseases matched their astrological year experienced higher mortality, and this effect was proportional to the patient's belief in traditional Chinese astrology. Presumably, patients who believed that they had the disease that was fated for them in the stars more willingly gave up the struggle and resigned themselves to an earlier death. Although these phenomena are well documented, the mechanism(s) by which they are produced are not known.

THOUGHT QUESTIONS

1. Given that chronic or severe stress generally weakens the immune system, how might such an apparently harmful relationship have evolved?
2. Is stress always harmful?
3. List the types of experiments done in psychoneuroimmunology. Do the results of any of these falsify the hypothesis that the mind and the body interact? Do any of these results prove that the mind and the body interact?
4. Is the Cartesian concept of the mind the same thing as the brain? Is 'mind' simply the name that we give to the 'workings' (or functions) of the brain?

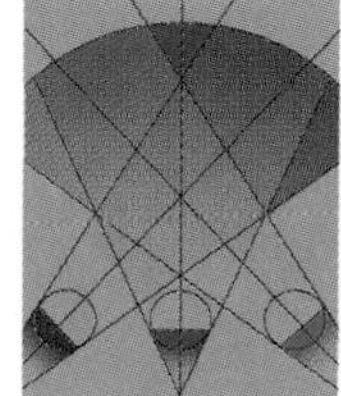

SINCE THE MID-NINETEENTH CENTURY, WESTERN MEDICINE has tended to view disease as having external causes. The ascendancy of this view can be traced back to the work of Louis Pasteur, who championed the theory of 'specific etiology' as part of the germ theory of disease. In this theory, each disease has one specific and identifiable cause. This theory led to much highly successful research that associated single species of microorganisms (bacteria, viruses, and parasites) with specific diseases. Such research ultimately produced vaccines and antibiotics, which have successfully controlled many infectious diseases. However, many of the diseases that today are still without effective cures are chronic diseases (cancer and heart disease, for example) that do not seem to have simple, single causes. Maybe a new concept of disease and of health is needed to find therapies and preventative measures for these diseases.

Psychoneuroimmunology is redefining our concepts of health and disease. Scientists in this field are using new technologies to reexamine some old concepts of disease causation. Working at the same time as Pasteur, another French scientist, Claude Bernard, questioned what it meant for a microorganism to 'cause' a disease. He observed that there were very great differences in individual response to microorganisms; some people got sick and even died, while other people who were also exposed did not get sick. Bernard, a physiologist, developed an alternative theory, that of the *milieu intérieur* or inner environment, as being an equal determinant in whether or not a person became sick. The past century of research in immunology and more recently in psychoneuroimmunology suggests that even the diseases for which an infectious agent is known are not caused by the microorganism alone, but rather by the outcome of a complex process in which the microorganism disturbs homeostasis while host mechanisms attempt to restore it. Bernard's theory is compatible with a view of the neural–endocrine–immune communication network as a sensory organ with which deviations from homeostasis are detected and corrected.

Psychoneuroimmunology also borrows from Chinese traditional medicine and ayurvedic medicine in India, as well as other Asian traditions that view health as the balance of life forces. Like these, the psychoneuroimmunology paradigm uses a *functional* model of the body, a model that regards the body as an entity in a constant state of change. Health is the state in which these forces are in balance, in homeostasis; disease is the state in which they are not. African traditions in which a person's health and well-being are seen to be influenced by the social environment in which the person lives coincide with the psychoneuroimmunology view that mental and emotional factors can affect health and disease. Within the psychoneuroimmunology paradigm, scientists are testing hypotheses suggested by these ancient traditions. Widening one's point of view and being open to new ideas are integral parts of science. New hypotheses are formed by the melding of ideas and must then be followed by the hard, and often slow, work of hypothesis testing.

CHAPTER SUMMARY

- The **immune system** is a system that works to detect and correct deviations from **homeostasis** within the organism. It is composed of white blood cells that travel throughout the body in the blood and in the **lymphatic circulation**.
- **Specific immunity** is acquired by the individual after exposure to specific **antigens** such as bacteria or viruses. By forming specific **clones** of **B lymphocytes** and **T lymphocytes**, the immune system retains a memory of the encounter so that it can react faster and more strongly to subsequent exposures to that antigen. This is the basis for **immunization**: an artificial first exposure gives protection against later natural exposure to the same disease.
- Products of the specific immune response (**antibodies** and cytotoxic T lymphocytes) rid the body of the specific antigen that induced their production.
- **Innate immunity** is present from birth and does not depend on exposure to develop. Innate immunity rids the body of some pathogens, but does not show antigen specificity. It also does not show memory; a second exposure induces the same response as the first exposure.
- The immune system interacts with the **autonomic nervous system** and **endocrine glands** in an integrated and multidirectional way. Communication among these three systems is mediated by chemicals called **cytokines**.
- Factors that interrupt this communication network can prevent the restoration of homeostasis within the organism, producing disease.
- Mental states can affect the functioning of the immune system and can either increase or decrease disease-fighting activity, **inflammation**, and healing.
- The **stress response**, through the action of the **sympathetic nervous system** and stress hormones, can produce **immunosuppression**, while the **relaxation response**, through the action of the **parasympathetic nervous system**, can reverse this process.

KEY TERMS TO KNOW

antibodies (p. 501)
antigen (p. 499)
autonomic nervous system (p. 513)
B lymphocytes (p. 499)
clone (p. 501)
cytokine (p. 499)
endocrine glands (p. 513)
health (p. 495)

homeostasis (p. 497)
immune system (p. 497)
immunization (vaccination) (p. 503)
immunosuppression (p. 511)
inflammation (p. 507)
innate immunity (p. 507)
lymphatic circulation (p. 497)
neuroendocrine system (p. 513)
parasympathetic nervous system (p. 513)
passive immunity (p. 507)
placebo (p. 519)
placebo effect (p. 520)
psychoneuroimmunology (p. 495)
relaxation response (p. 518)
specific (acquired) immunity (p. 502)
stress (stress response) (p. 516)
sympathetic nervous system (p. 513)
T lymphocytes (p. 499)
tolerance (p. 511)

CONNECTIONS TO OTHER CHAPTERS

Chapter 1: Psychoneuroimmunology is a good example of a paradigm.

Chapter 2: The great variety of antigen receptor proteins that can each recognize one of the huge variety of antigens in the world is based on the rearrangements of just a few genes.

Chapter 5: The ability to form an immune response to any particular antigen depends on cell surface proteins. The allele frequencies of the genes that code for these proteins vary among populations, making some populations more susceptible than others to a particular disease.

Chapter 7: The genes that allow kin selection in many species are also used by the immune system to recognize self.

Chapter 8: Poor nutrition suppresses the immune system.

Chapter 9: Suppressed immune function greatly increases the risks for cancer.

Chapter 10: The brain can affect many immune functions.

Chapter 11: Many drugs suppress the immune system. Many drugs mimic some of the activities of the autonomic nervous system.

Chapter 13: Immunosuppression is characteristic of AIDS.

PRACTICE QUESTIONS

1. What type of white blood cell secretes antibodies?
2. What type of white blood cell can kill cancer cells?
3. What type of white blood cell can engulf bacteria and kill them?
4. What type of white blood cell releases histamine?
5. What type of white blood cell releases IL-1?
6. What part of the body is acted on by IL-1 to induce fever?
7. Which part of the autonomic nervous system mediates the fight-or-flight response?
8. Which part of the autonomic nervous system mediates the relaxation response?
9. Which of the following are parts of the immune system: bone marrow, spleen, skin, intestines, bladder, tonsils, eyes?
10. What types of scientific evidence suggest that the brain and nervous system interact with the immune system?
11. How do the terms 'stress' and 'stressor' differ?
12. What are some diseases that result from either too much corticosteroid or too little?
13. Stress is part of what larger response? What are the stages of this response, and what characterizes each stage?

13 *HIV and AIDS*

CHAPTER OUTLINE

AIDS is an Immune System Deficiency

- AIDS is caused by a virus called HIV
- Discovery of the connection between HIV and AIDS
- Establishing cause and effect
- Viruses and HIV
- Evolution of virulence

HIV Infection Progresses in Certain Patterns, Often Leading to AIDS

- Events in infected helper T cells
- Progression from HIV infection to AIDS
- Tests for HIV infection
- A vaccine against AIDS?
- Drug therapy for people with AIDS

Knowledge of HIV Transmission Can Help You to Avoid AIDS Risks

- Risk behaviors
- Communicability
- Susceptibility versus high risk
- Public health and public policy
- Worldwide patterns of infection

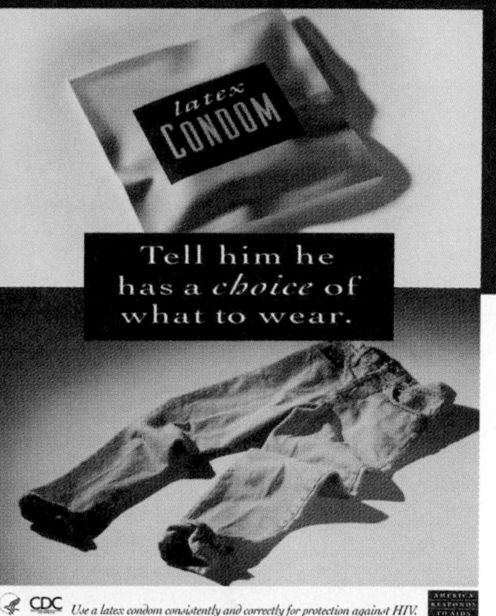

latex
CONDOM
Tell him he has a *choice* of what to wear.
CDC
Use a latex condom consistently and correctly for protection against HIV.

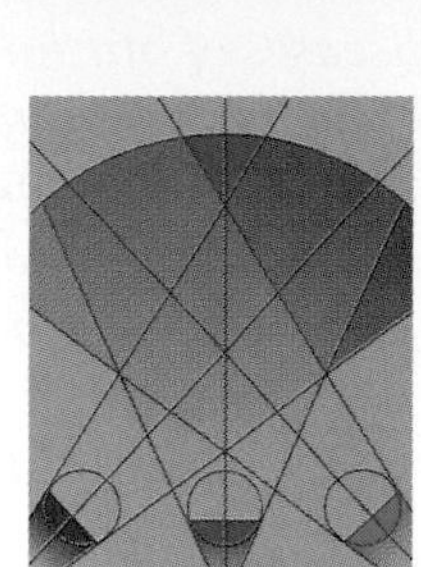

ISSUES

How did the discovery of AIDS and HIV develop? What other diseases helped or hindered scientists and doctors in the discovery?

What do HIV tests tell us?

Will there be a cure for AIDS? What about this disease makes a cure so difficult?

Will there be a vaccine to prevent AIDS? If a vaccine is possible, will it solve all the problems associated with AIDS?

Will studying HIV teach us all we need to know about the AIDS pandemic? What are the social factors in the global spread of AIDS?

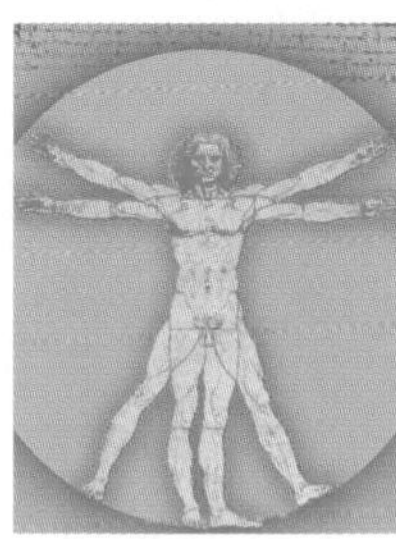

BIOLOGICAL CONCEPTS

Health and disease (immune system, pathogens and response, receptors, Koch's postulates, routes of transmission, behavior)

Evolution of virulence

Biodiversity (viruses)

Scales of size

Chapter 13: HIV and AIDS

AIDS is a disease caused by the virus HIV. HIV undermines the immune system, leaving the infected person vulnerable to other diseases. As we saw in Chapter 12, the immune system has several ways of protecting the body from disease. When people have AIDS, their immune systems no longer function properly, so they are at risk of becoming ill from infections that would barely affect a healthy person. Many people with AIDS, for which there is currently no cure, suffer long and painful deaths. AIDS first received public attention in 1981. It quickly became one of the most feared and widely discussed diseases of our time. HIV is spread from person to person in infected body fluids. Sexual contact is one of the main routes of transmission; thus AIDS is a sexually transmitted disease. Worldwide, nearly eleven million children have been orphaned because their parents have died of AIDS. The spread of AIDS has been accompanied by the spread of misconceptions concerning the disease. In this chapter we summarize what is known about this dreaded disease and also address certain misconceptions.

AIDS IS AN IMMUNE SYSTEM DEFICIENCY

The acronym **AIDS** stands for ***A**cquired **I**mmuno**D**eficiency **S**yndrome*. *Acquired* means that the illness is not genetically inherited as the result of a defective DNA message. *Immunodeficiency* means that some part of the immune system is not functional, and *syndrome* means that a wide range of symptoms are associated with the disease. People whose immune systems are deficient can become seriously ill with infectious diseases or cancers.

The body has many ways of protecting itself from diseases. The skin protects the body's surface against entry of bacteria, viruses, or other microorganisms. The mouth, vagina, and many other potential entry points are coated with mucous secretions that continually wash away adherent bacteria, inhibit bacterial growth, and promote healing.

A more specific type of protection is afforded by the immune system. An organism's immune system distinguishes between molecules that are part of the organism (self) and ones that are not (nonself; see Chapter 12). Nonself molecules include bacteria, viruses, and molecules made by cancer cells. Many nonself molecules trigger an immune response that inactivates or destroys the nonself molecules (see Figure 12.4, p. 504).

An **immunodeficiency** is an absence of one or more of the normal functions of the immune system. People who are immunodeficient get sick more often than people with healthy immune systems, and their illnesses last longer and are more severe.

How does someone become immunodeficient? Some of the many causes are inherited and some are environmental. One type of inherited immunodeficiency, the severe combined immune deficiency syndrome (SCIDS), caused by a lack of the enzyme adenosine deaminase (ADA), is discussed in Chapter 3 (p. 107). Inherited immunodeficiencies are rare; much more common are those that are acquired as a result of environmental exposures.

The functioning of the immune system can be depressed, for example, by alcohol and drugs such as cocaine and marijuana, psychological stress and depression, cigarette smoke and other pollutants, and malnutrition (either total calorie deficit or micronutrient malnutrition; see Chapter 12). Clearly, AIDS is not the only kind of immunodeficiency, but it is among the most severe. Many immunodeficiencies are temporary and reversible: if the causative factor is removed, the immune system recovers. AIDS is long-lasting; the immune system does not recover, and the disease is fatal.

AIDS specifically targets the lymphocytes called **helper (CD4) T cells**. These cells and the cytokine interleukin-2 that they secrete are necessary for both the B lymphocyte and **cytotoxic (CD8) T cell** responses of the immune system (Figure 13.1). As you probably recall from Chapter 12, B cells make antibodies, our main defense against bacteria and fungi. Cytotoxic T cells protect against viral infections and against cancer cells. When helper T cells are destroyed in AIDS, both the B cell and cytotoxic T cell arms of the immune system are lost, leaving the individual vulnerable to bacterial, fungal, and viral infections and to cancer. What would be a minor infection in a person with a healthy immune system can quickly become life-threatening in a person with AIDS.

AIDS is caused by a virus called HIV

We now know that AIDS is an infectious disease caused by a virus known as HIV (Human Immunodeficiency Virus). However, when the syndrome first began to appear in the United States at the very end of 1980, the cause, and even the fact that it was an immunodeficiency, was not known. In this section we trace the steps that led to the identification of this immunodeficiency and its causative agent. How were hypotheses suggested? How were these hypotheses tested? What types of evidence are necessary to call something the 'cause' of a disease? Does such evidence rule out other hypotheses? We then look at the virus itself and how it lives in human cells.

FIGURE 13.1

Interactions between lymphocytes and their cytokines.

Helper T cells, B cells, and cytotoxic T cells have receptors for nonself molecules. Within each cell population, individual cells have receptors for different nonself molecules.

B lymphocytes and cytotoxic T lymphocytes require two signals to become functional immune cells: (1) binding specific nonself molecule to their receptors, causing receptors for IL-2 to be brought to the cell surface, and (2) binding of IL-2 to IL-2 receptors.

Helper T cells that have bound their specific nonself molecule are the source of IL-2; thus removal of helper T cells shuts down both B-cell and cytotoxic T-cell activity.

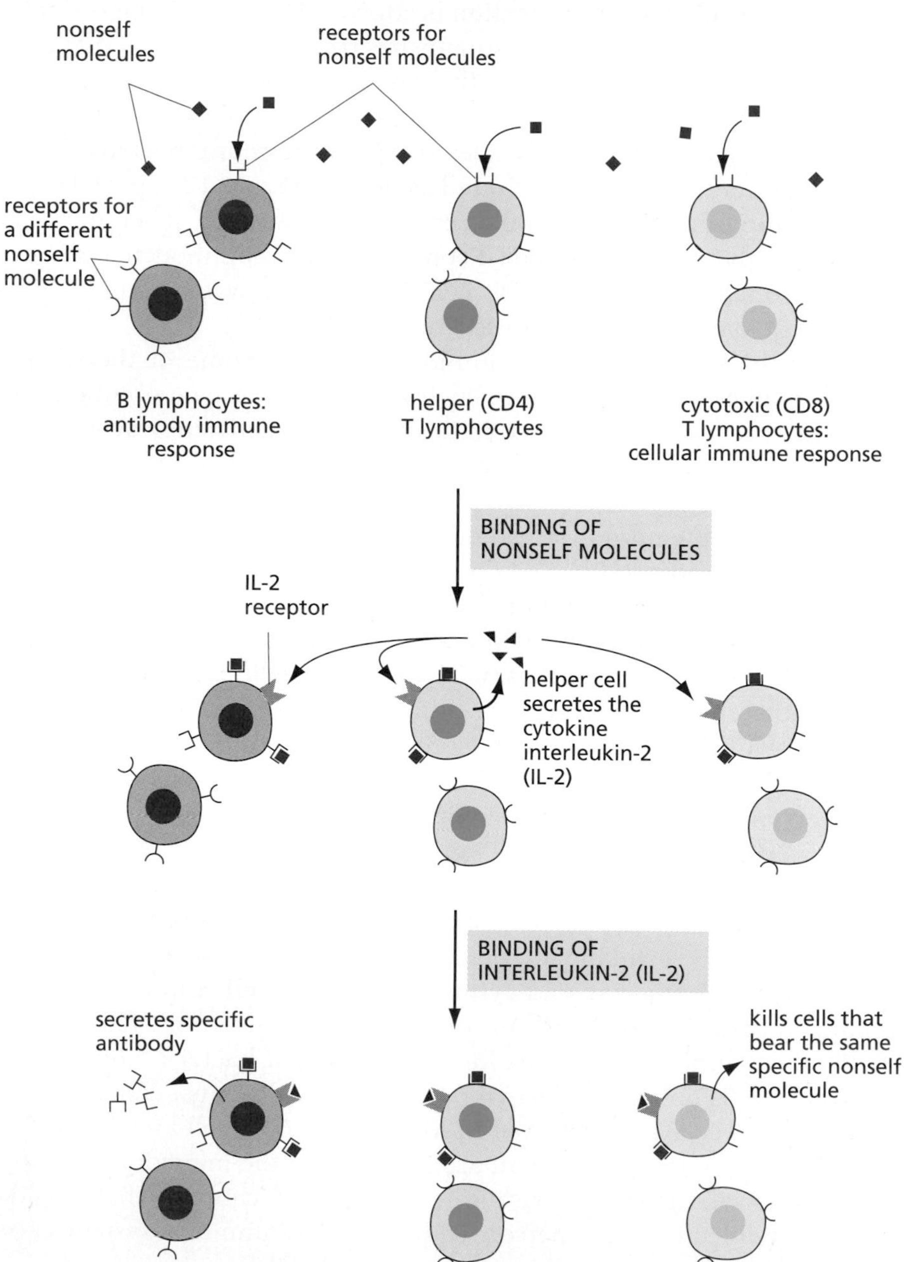

Discovery of the connection between HIV and AIDS

In June of 1981, five cases of pneumonia in San Francisco were reported to be associated with a microorganism called *Pneumocystis carinii*. The report appeared in *Morbidity and Mortality Weekly Review* (*MMWR*), a publication of the Centers for Disease Control and Prevention (CDC), which tallies all cases of sickness (morbidity) and death (mortality) in the United States due to certain kinds of diseases called reportable diseases. They are called 'reportable' diseases because a physician seeing a patient with one of these diseases is legally obliged to report it to the CDC. (The numbers of cases are reported; patients' names are not.) Reportable diseases listed in *MMWR* include most of the serious contagious diseases caused by microorganisms. Each issue of *MMWR* also contains articles written by alert clinicians who have observed patterns of disease that are unusual for a particular geographic area or season, or are too frequent, or are occurring in an age group that does not usually get the disease. Tallies of reportable illnesses are often the first indication of an unusual spread of a known disease or the appearance of a new disease. The statistical study of information about the occurrence and spread of diseases in whole populations is called **epidemiology**.

Pneumonia, a disease characterized by fluid in the lungs, can be caused by many different bacteria and viruses. The five cases of *Pneumocystis* pneumonia reported in *MMWR* were quite unusual because *Pneumocystis* is a parasite that is neither a bacterium nor a virus. All five cases were from a single geographic area and close together in time, and thus represented what epidemiologists call a **case cluster**.

Later in the summer of 1981, a dermatologist in New York City, Dr. Alvin Friedman-Kien, noticed an unusual cancer, called Kaposi's sarcoma, among many of his young homosexual male patients, a finding that he reported in *MMWR*. Kaposi's sarcoma, a cancer of the cells lining the walls of blood vessels, causes red or purple raised patches on the skin. Kaposi's sarcoma was rare in the United States and had previously been found only in elderly men of Italian or Eastern European Jewish descent. The Kaposi's sarcoma seen in the reported cluster was far more aggressive than that seen in elderly men, meaning that it spread much faster and was present in the internal organs, not just on the skin. Aggressive Kaposi's sarcoma is, however, seen in kidney transplant patients, who take medication to suppress their immune systems, which would otherwise reject the transplanted tissue. This fact suggested the hypothesis that the Kaposi's sarcoma becomes aggressive when the immune system is suppressed.

The immunodeficiency hypothesis. Were the Kaposi's cases in any way related to the unusual pneumonia cases? Did immunodeficiency underlie both the unusual pneumonias and the unusual cancers? Both the *Pneumocystis* pneumonia patients and the Kaposi's patients were found to have severely decreased numbers of helper T cells. As we have seen above, helper T cells are central in the immune system; a person lacking helper T cells has a suppressed immune system (see Figure 13.1). The evidence thus fit the hypothesis that the cancer and the pneumonia belong to a syndrome resulting from the same underlying mechanism, namely immunodeficiency. This syndrome was given the name AIDS. Reported cases accumulated quickly: 87 in the first six months of 1981, 365 in the first six months of 1982, and 1215 in the first six months of 1983. AIDS had been given a name, but its cause was still unknown. Funding for research was slow in coming, as was interest on the part of many scientists. Even after the cause was established, there was silence on the part of government agencies. Various projects, such as the AIDS quilt (started in 1987; Figure 13.2), were started by private individuals and nongovernmental organizations to increase public awareness and push for increased research.

Lifestyle hypotheses. Epidemiologists gathered information from AIDS patients, trying to establish any common links between the cases: had they all been exposed to the same chemical agent? Did they all live in the same geographical area or in the same household? Were the patients known to one another? For a while, because the first AIDS patients were homosexual men, the search was for common lifestyle factors, on the assumption that there is such a thing as a 'homosexual lifestyle,' one characterized by some drug or dietary factor that was shared by most or all homosexual men. Several researchers hypothesized that the immunodeficiency was due to an overload of the immune system by chronic exposure to nonself molecules via promiscuous sexual activity. Others doubted these hypotheses because the effects seemed to be specifically targeted on one type of cell, the helper T cell; they searched instead for infectious microorganisms that homed in on this type of cell and that might be transmitted by sexual contact.

The viral hypothesis. Was the infectious microorganism a bacterium, a fungus, a protozoan, or a virus? Support for the hypothesis of a viral agent came when some cases of AIDS were reported among hemophiliacs. People with hemophilia lack the genes that code for certain blood proteins necessary for forming blood clots after an injury. Hemophilia can be life-threatening because the person can bleed to death from a minor cut or scrape. As protection, hemophiliacs are given blood-clotting proteins from other people. This works well, but only temporarily; transferred clotting agents, like all proteins in the body, are eventually broken down by protein-degrading enzymes. The clotting factors must therefore be supplied repeatedly to hemophiliacs. These clotting factors are obtained from blood pooled from many donors and filtered to

FIGURE 13.2

The AIDS quilt, commemorating those who have died of AIDS. In 1999 the quilt had grown to be 42,960 panels commemorating 83,279 names.

remove bacteria and fungi. Viruses, however, can pass through the filters. Because hemophiliacs receiving a filtered blood product were contracting AIDS, it was reasoned that the infectious agent could be a virus.

One laboratory that was studying viruses at the time was the National Cancer Institute's Laboratory for Tumor Cell Biology, headed by Robert Gallo. The occurrence of AIDS in hemophiliacs convinced Gallo that the infectious agent must be a virus. Gallo's laboratory was studying **retroviruses**, a type of virus whose genetic information is RNA that is copied to make DNA. One retrovirus being studied was the human T-cell leukemia virus, HTLV-I. This retrovirus was known to cause a form of leukemia (a blood cell cancer) associated in some patients with a mild immunodeficiency. It could not be said, however, whether the immunodeficiency seen in the HTLV-I patients was caused by the virus or was the result of the cancer or some other factor. Nevertheless, because it was known that HTLV-I was transmitted from person to person by sexual contact and that it specifically attacked T cells, it fit the pattern seen for AIDS.

In 1983 Luc Montagnier and his co-workers at the Pasteur Institute in France found important new evidence in the tissues of a patient with chronically swollen lymph nodes, a condition common in the early stages of AIDS. (The lymph nodes are the structures that temporarily enlarge when the body is fighting an infection; people often refer to them as 'swollen glands.') The scientists found the enzyme reverse transcriptase in the lymph tissues. Reverse transcriptase is used by retroviruses to produce DNA from RNA. Montagnier's group had not yet found the virus, just one of its enzymes, but the reverse transcriptase was strong evidence of the presence of a retrovirus.

The presence of any retrovirus can be detected by finding reverse transcriptase (as Montagnier's group had done), but the identification of a specific retrovirus requires testing of large quantities of viruses, which are obtained by growing them in laboratory culture. Viruses cannot replicate outside a host cell, but these host cells may be grown in the laboratory rather than in an animal. Gallo's laboratory had developed a method for growing human T cells in the laboratory, and for growing HTLV-I in those cells. Antibodies were made to HTLV-I grown this way. These antibodies were used to show that the new retrovirus from AIDS patients was not HTLV-I.

When scientists tried to grow the new retrovirus from AIDS patients in human T cells in the laboratory, it killed the cells. Michael Popovic in Gallo's lab found a type of leukemia T cell in which the new retrovirus could be grown without killing the host cells. Once the method for growing quantities of the new virus had been developed, antibodies were made that were specific for it. These antibodies were then used by Gallo's group to test viruses isolated from three groups of people: a control group that consisted of healthy heterosexuals, a group of AIDS patients, and a group of patients with AIDS-related complex or ARC, a set of symptoms assumed to be an early stage of AIDS. The viruses isolated from AIDS patients and some ARC patients were identified as being the same as the new virus. None of the healthy subjects had this new virus.

Using the specific antibodies, scientists found the virus in 80–100% of AIDS patients, in varying percentages of people in certain defined risk groups, and only rarely in healthy individuals outside the risk groups. These results were strong evidence that this new retrovirus was associated with AIDS. The retrovirus found by Gallo and the retrovirus found earlier by Montagnier were determined to be two strains of the same virus. Each group had given their virus a different name and each group wanted the name they had chosen to become the standard. The International Committee on the Taxonomy of Viruses studied the naming problem and decided in 1986 that neither name should be used. They assigned a new name to the virus, **human immunodeficiency virus** or **HIV**.

Establishing cause and effect

Just because a microorganism is associated with a disease does not mean that it causes the disease. How do we know that HIV is the cause of AIDS? There is a set of rules that have traditionally been used to identify a microorganism as the cause of a particular disease. These rules were formulated in the late 1800s by Robert Koch, a German physician, and have come to be known as **Koch's postulates**:

- First, the microorganism suspected as the causative agent must be present in all (or nearly all) animals or people with the disease.
- Second, the microorganism must not be present in undiseased animals.
- Third, the microorganism must be isolated from a diseased animal and grown in pure culture (that is, a culture containing no other microorganisms).
- Fourth, the isolated microorganism must be injected into a healthy animal, the original disease must be reproduced in that animal, and the microorganism must be found growing in the infected animal and be reisolated from it in pure culture.

Koch used these rules to show that the bacterium *Bacillus anthracis* was the cause of anthrax, a fatal disease in sheep that was decimating European herds in the 1870s. Koch later used his postulates to identify a bacterium now known as *Mycobacterium tuberculosis* as the causative agent for human tuberculosis. Koch received a Nobel Prize for his demonstration of the causes of anthrax and tuberculosis.

Limitations to Koch's postulates. While these rules are straightforward to state, they are difficult to fulfill. Every animal is host to many bacteria, so finding one bacterial species that is present only in diseased animals is not easy. Koch and his contemporaries developed bacterial culture media and techniques that enabled them to grow pure cultures of certain bacteria, but the growing of many other bacterial species (such as those killed by exposure to air) required technology that did not exist in Koch's time. In studying anthrax, a disease in sheep, the fourth postulate—requiring the production of the disease in an experimentally infected animal—was straightforward. In Koch's later studies on tuberculosis, ethical considerations dictated that the fourth postulate could not be fulfilled by infecting a healthy human, so an animal was used instead as a model, or experimental patient.

Despite the difficulties associated with meeting the requirements of Koch's postulates, they have been very useful. About a dozen or so infectious diseases are controllable by vaccination, including rabies, poliomyelitis, whooping cough, tetanus and diphtheria. The infectious microorganisms responsible for these diseases and others were identified on the basis of Koch's postulates. A microorganism that has been shown to cause a disease is called a **pathogen**.

There are other diseases for which these postulates have not been demonstrated. Most bacteria can grow outside cells, so they can be grown in 'pure cultures' if the proper growth conditions can be found. As we have seen with human retroviruses, viruses grow only inside a host cell, so that difficulties in growing the infectious agent in culture became even more acute when viral, rather than bacterial, diseases were studied. Because many bacteria and viruses that infect one species do not infect other species, it is not always possible to find animal models for human diseases, as Koch did for tuberculosis. Further, even the best animal models can never reproduce all of the aspects of a human infection. Some diseases may be caused by the interactions of more than one species of bacteria or virus; in such cases, Koch's last postulate would not work because the organisms isolated in pure cultures would no longer have the same effect unless combined. In many diseases there are healthy carriers (people who are infected and can transmit the infection to someone else, but do not become ill themselves), so the second postulate is not met. Other diseases, such as some cancers, may be multifactorial, meaning that no single factor is causative by itself. Nevertheless, Koch's postulates are the standards by which scientists establish cause and effect for most *infectious* diseases, meaning those diseases that can spread by infection with a microorganism.

HIV and Koch's postulates. Koch's postulates are the accepted standard of proof for asserting cause and effect in infectious disease. To say that HIV *causes* AIDS thus carries with it the implication to many in the scientific community that Koch's postulates have been fulfilled. This implication has been contested by some scientists, most notably by the American virologist Peter Duesburg, who maintains that three out of the four postulates have not been met. The presence of the virus itself was difficult or impossible to demonstrate in some people. The basis for considering someone to be HIV-infected is the presence in their blood of antibodies to the virus, not the virus itself. Duesburg maintains that Koch's first postulate (that the suspected causative agent must be present in all people with the disease), has therefore not been fulfilled. Most scientists no longer find this particular point controversial because newer, more sensitive techniques (such as the polymerase chain reaction described in Chapter 3, p. 101) have been used to detect the virus directly. The virus has now been found in virtually all persons with AIDS. We also now realize that, during the asymptomatic phase of HIV infection when

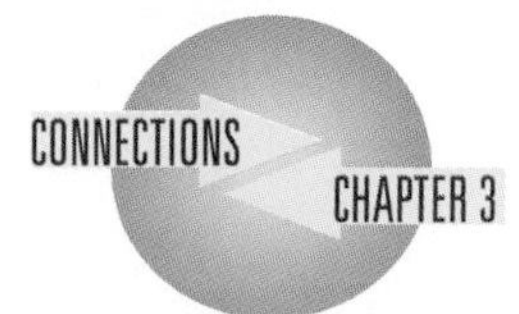

the number of viruses in the blood is low, the virus can be found in high numbers in the cells of the lymph nodes. The third postulate (that the suspected agent must be grown in pure cultures) has also been difficult to satisfy. Although methods were developed for growing HIV in human T cells in culture, Luc Montagnier demonstrated that HIV could only be made to replicate in those cells if there were other infectious agents present. The fourth postulate (that injecting the suspected agent into an animal must produce the disease) has not been fulfilled. Only two nonhuman animals have been found in which HIV will grow (chimpanzees and macaque monkeys), and when these animals are infected with HIV, they do not develop AIDS.

Although many scientists agree that Koch's postulates have not been entirely fulfilled (and perhaps cannot be), the failure to fulfill them does not prove that the suspected agent does *not* cause the disease. Gallo has pointed out that Koch's postulates cannot even be strictly applied to the diseases that Koch himself studied. Many people, for example, can be healthy carriers of the tuberculosis bacterium, and so the second rule is not always met.

Criteria other than Koch's postulates. Because of the limitation of Koch's postulates, other criteria have been suggested for establishing causality, particularly of viral diseases.

The criteria used by Gallo for stating that HIV is the sole cause of AIDS are as follows:

- HIV or antibody to HIV is found in the vast majority of persons with AIDS.
- HIV is found in a high percentage of people with ARC.
- HIV is a new virus and AIDS is a new disease.
- Wherever HIV is found, AIDS develops; where there is no HIV, there is no AIDS.
- People who received transfusions of blood contaminated with HIV developed AIDS.
- HIV infects CD4 helper T lymphocytes, a cell type depleted in AIDS.
- On autopsy, HIV is found in the brains of people who have died of AIDS, and dementia (loss of brain cell function) is a symptom of AIDS.

Necessary causes and sufficient causes. As the above discussion points out, it is not always easy to identify cause and effect. There are some scientists who do not agree that Gallo's criteria are sufficient to establish cause. There are others who say that Koch's rules are the correct criteria, but that, in the case of HIV and AIDS, the criteria have not been met, particularly rule 4. At present, the vast majority of scientists agree that HIV is a **necessary cause** of AIDS; that is, someone who is not infected with HIV will not get AIDS. However, not all scientists agree that HIV is the sole or **sufficient cause** of AIDS (that is, no other factors are required) because there is a great difference in the course of the infection among various persons with AIDS.

There are several lessons to be learned here. Hypotheses about cause and effect of diseases are debated both in written articles and at scientific meetings. Many additional hypotheses are suggested and tested, and the number of scientists involved is often (as in this case) very large. Eventually a consensus develops as to the explanation that best fits the data. Consensus is never 100% agreement, but public policy and public health decisions must be made nevertheless. The accumulation of data may settle a given controversy (e.g. in the case of HIV and AIDS), but at any time new data may arise that require a change in the consensus explanation. Scientific debate and research must always be open to new possibilities because the findings of science are always tentative or provisional.

Saying that a particular virus causes a disease, especially a disease with such a diffuse group of symptoms as AIDS, really tells us little or nothing by itself. All of the 'how' questions remain. How does the virus infect? How do cellular effects progress to clinical symptoms? How can the disease be prevented or stopped? How is the infection transmitted? How contagious is it? How likely is it that HIV infection will become AIDS? How do people cope with such a disease? It is interesting that the articles written by scientists, either for other scientists or for the public, are almost entirely devoted to answering the first three questions, while the literature written by nonscientists is much more concerned with the last four. We examine each of these questions later in this chapter.

Viruses and HIV

Many human diseases are caused by viruses. AIDS is one such disease; measles, mumps, polio, and herpes are also caused by viruses. In addition to this role in disease, viruses are interesting to biologists because they challenge our understanding of what it means for something to be alive (Chapter 1, p. 9). **Viruses** are bits of either DNA or RNA that cannot reproduce by themselves but can replicate inside a cell (called the host cell) by using the biochemical machinery of the host. Biologists define a living organism as one that can reproduce itself, which viruses cannot; yet once inside a host, viruses can cause the host to replicate the virus, something that is not a characteristic of any known nonliving thing. What is the structure of a virus and how do viruses accomplish this?

The viral life cycle. A virus consists of nucleic acid, an outer protein shell and, in some viruses including HIV, a phospholipid bilayer membrane called the viral envelope that also contains some viral proteins. A particular virus has either DNA or RNA, and the nucleic acid is either single-stranded or double-stranded, distinctions that are used in viral classification. Viral genomes vary in size: some have only enough nucleic acid to code for 3–10 proteins, while others code for 100–200 or more proteins. HIV, the virus that causes AIDS, is at the small end of this size range. Viruses are very small, even in comparison with bacteria. A single human cell is typically 10 μm (10 millionths of a meter) in diameter. Magnifying this human cell 100,000 times would make it a meter wide; at the same magnification a bacterium would be about the size of a football and a virus would be only the size of an M&M candy.

Some viruses can survive outside cells, but no virus can replicate unless it is inside a host cell. Human cells, animal cells, plant cells, and bacteria can all serve as hosts to viruses. For each virus there are only certain species that can serve as its host, and within an individual of the host species only certain types of cells can be host cells. This is because to enter a cell the virus must attach to some molecule on the host cell surface. Each type of virus is able to bind only to specific host cell molecules, which are usually membrane proteins. The species of virus that can adhere to dog or cat cells, for example, usually cannot adhere to human cells, which explains why we usually cannot catch viral diseases from our pets.

After a virus attaches to a host cell, the viral nucleic acid, sometimes with some viral proteins, enters the cell's cytoplasm, usually with the help of energy derived from the host cell. In viruses whose nucleic acid is DNA, many copies of the virus are made, using the host's molecular machinery for DNA replication. Viruses whose nucleic acid is RNA may replicate their RNA genome directly or convert their RNA into DNA by using the host's machinery to make more viral particles. The final stage of the viral life cycle consists of the release of viruses from the cell by the rupturing (also called lysis) of the cell or by the budding out of viruses through the host cell membrane. The new viruses can then infect other cells and repeat their life cycle.

FIGURE 13.3

The structure of HIV.

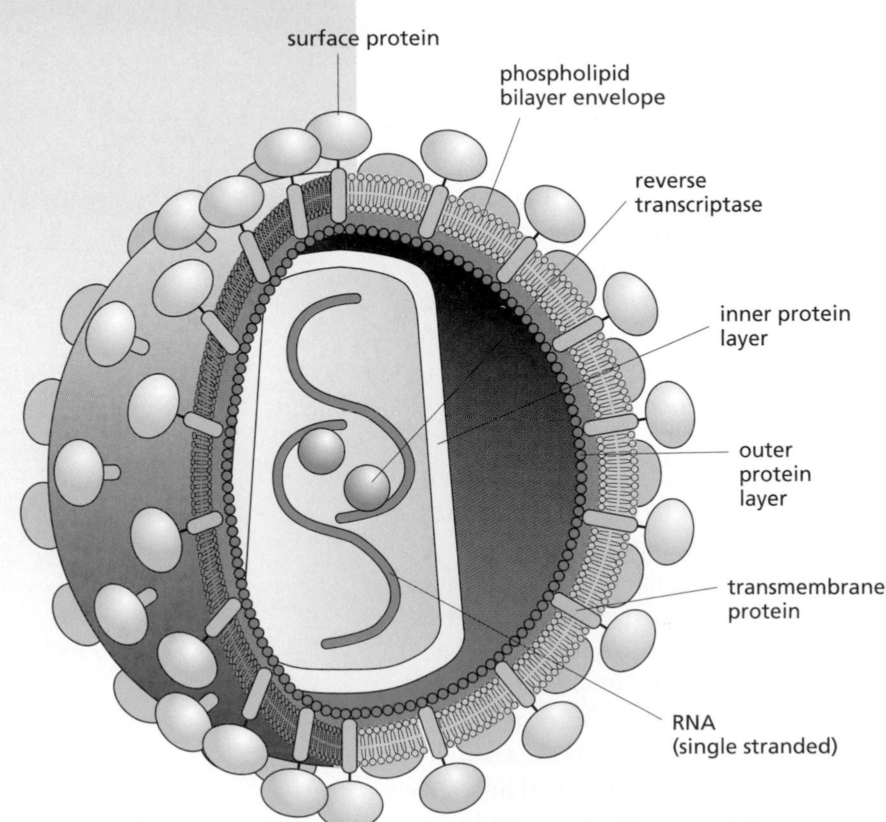

HIV structure and life cycle. HIV is an enveloped virus whose genome consists of two copies of a single strand of RNA. The structure of HIV is shown in Figure 13.3. The virus is surrounded by a phospholipid bilayer envelope. In the viral envelope are proteins that are important for attaching to and entry into specific host cells. Inside the viral envelope are two protein layers. Inside the inner protein layer is the viral genome consisting of two identical strands of RNA.

The HIV life cycle is shown in Figure 13.4 and is typical for many retroviruses. Proteins in the viral envelope bind to the CD4 protein found on only a few types of human cells: helper T cells and some macrophages, a type of white

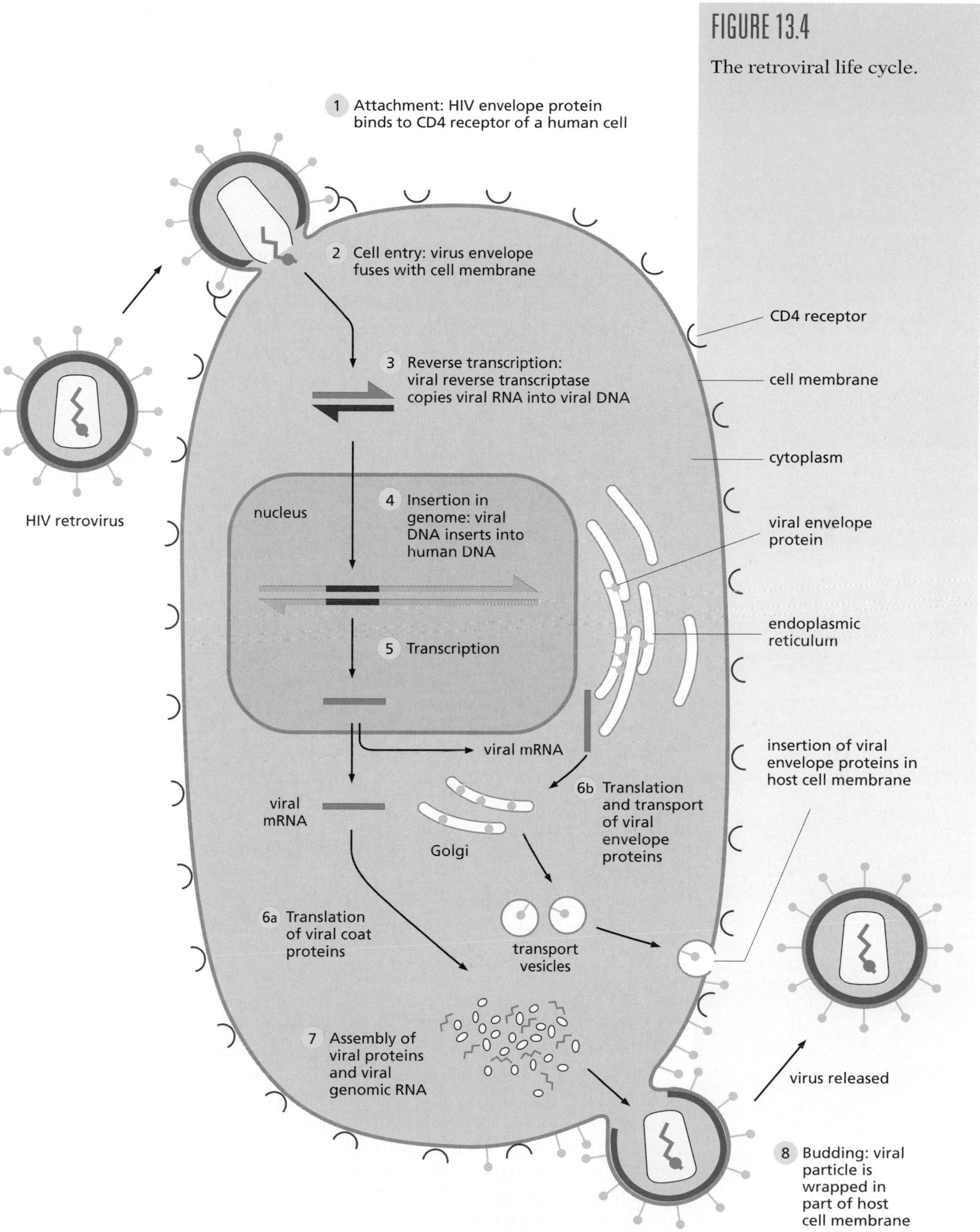

FIGURE 13.4

The retroviral life cycle.

blood cell that can engulf microorganisms. Helper T cells, macrophages, and cells related to macrophages are thus the only cells that become directly infected by the HIV virus. Other species of mammals also have CD4 and helper T cells but the structure of CD4 in each species is different so that HIV attaches only to human CD4. Once attached to the host cell, HIV enters by fusion of its viral envelope with the host's plasma membrane.

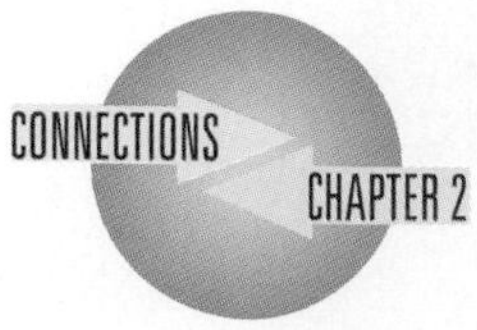

As we have already mentioned, HIV is a retrovirus, a type of virus whose genetic information is RNA. For reproduction of the virus, the RNA must first be used to make DNA; because this is the reverse of the usual DNA-to-RNA process of transcription (see Chapter 2, p. 61), the RNA-to-DNA process is called **reverse transcription**. After entry into the cell, the viral enzyme reverse transcriptase becomes activated. This enzyme uses the viral RNA as a template and synthesizes complementary DNA. The first DNA strand is, in turn, the template for synthesis of the second strand of DNA. The now double-stranded DNA is incorporated into the host's DNA. The viral RNA is meanwhile broken down or degraded.

Once in the host's DNA, the viral DNA can be transcribed by host enzymes into many copies of viral messenger RNA (mRNA). Some viral mRNA is then translated in the host cell cytoplasm to be the protein coats of new virus particles. Other viral mRNA is translated into viral envelope proteins. Like host cell membrane proteins, viral envelope proteins are made on the host cell's endoplasmic reticulum, transported through the Golgi, then carried via transport vesicles to the plasma membrane. Multiple copies of viral envelope proteins build up on the surface of the host cell (see Figure 13.4). Viral DNA is also transcribed as a whole to make copies of the viral RNA genome. The viral genomic RNA with its protein coat joins a portion of the host cell plasma membrane containing viral envelope proteins. The virus then buds out, carrying along a piece of the host cell membrane, which becomes the viral envelope. A photograph taken at very high magnification through an electron microscope shows HIV budding from a helper T cell (Figure 13.5A). On leaving the host cell, the virus is a mature, cell-free virus, ready to infect a new host cell. Many new viruses bud out of a single helper T cell (Figure 13.5B). HIV can also lyse (rupture) the T cell.

The HIV envelope and genome contain all of the molecular determinants that make the virus virulent (able to cause disease), infective (able to enter a cell), cell-specific (entering only certain types of cells), and cytopathic (able to kill or inactivate the host cell). The HIV genome also contains some regulatory genes, meaning genes that turn other genes 'off' or 'on.' All of the HIV regulatory genes seem to be essential for the life cycle; thus, they are possible targets for drugs or vaccines because blocking any essential step would inhibit the whole cycle.

Both types of HIVs currently known, HIV-1 and HIV-2, have the same basic structure, life cycle, and routes of transmission, but the rate of transmission of HIV-1 is 5–10 times higher than that for HIV-2. People infected with HIV-1 are 3–8 times more likely to have a decrease in helper T cells, lose immune function, and progress to AIDS than people infected with HIV-2.

Evolution of virulence

Virulence is the ability of a pathogen to overcome host defenses, thereby causing serious illness or death. From an evolutionary standpoint, virulence poses a severe problem for the pathogen: if it kills its host, it deprives itself of a suitable habitat and food supply. Clearly, a pathogen that causes minimal harm to its host is assured a longer time span for itself and its offspring to continue living in the same place than a pathogen that kills its host. HIV is related to several viruses that infect higher primates such as monkeys and apes. These viruses are nonvirulent—they spread from one host to another without causing serious illness or death. Thus, they have been around long enough—thousands of years at the very least—to have evolved symbiotic relationships with their hosts.

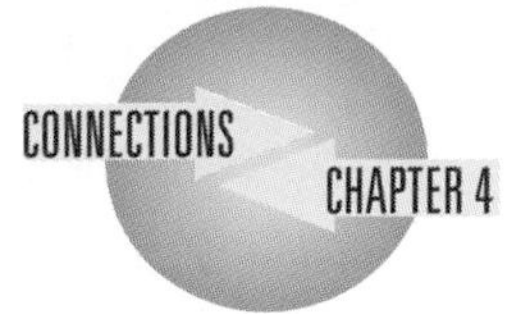

Evolutionary biologists who study bacteria and viruses believe that the evolution of virulence is related to the pathogen's **fitness**, meaning its capacity to leave offspring (see Chapter 4). When a new strain originates, it must compete with the older, nonvirulent strains. A virulent pathogen can proliferate rapidly in a host, but if it spreads from host to host at a slow rate, it might kill off its hosts before being able to colonize new ones; such a virulent strain will be less fit and will soon die out. A virulent strain that spreads rapidly from host to host will soon outcompete its

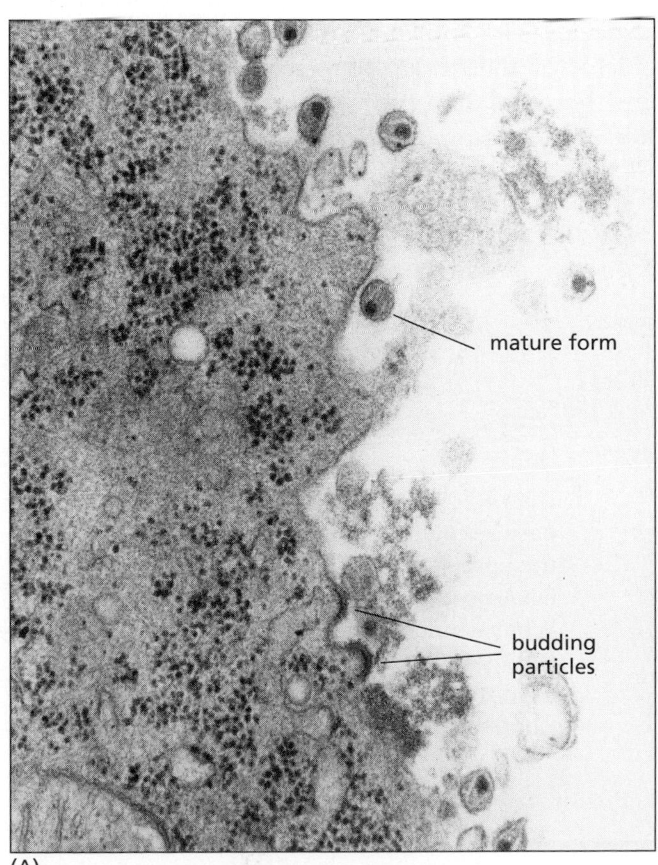

(A)

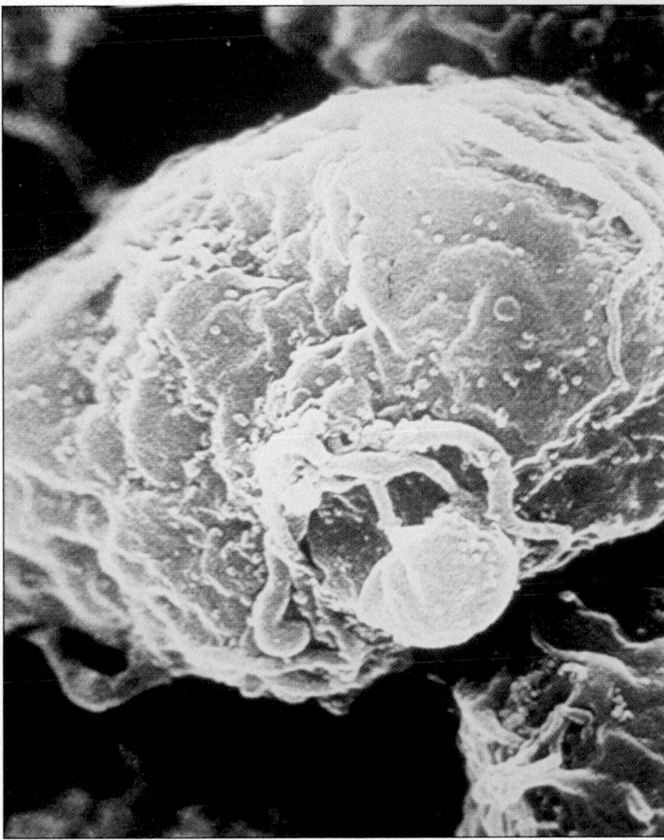

(B)

FIGURE 13.5

HIV budding from a helper T cell. (A) Budding and mature HIV. The mature viruses are free of the cell; the viral protein can be seen inside the viral envelope (a piece of the cell membrane that the viruses have taken with them as they budded out). (B) Many HIV particles bud from the same cell.

nonvirulent relatives. If the process is rapid enough, an epidemic occurs. Under this hypothesis, we suspect that between the late 1970s and 1981 increases in intravenous drug abuse and increases in the numbers of people having sex with multiple partners had given the HIV pathogen new opportunities to spread—rapidly enough to cause an epidemic.

In the long run, natural selection favors the evolution of host defenses against the pathogen, including both physiological and chemical defenses. These changes in the host reduce microbial virulence directly, and the adoption of host behaviors less conducive to the pathogen's spread also slows transmission. When transmission slows, less virulent strains once again become more fit than the virulent strains, and the cycle repeats.

You will recall that evolutionary change is dependent on genetic change. HIV changes genetically very rapidly. Reverse transcription is error-prone, with 1–5 mutations per round of reverse transcription. In part this is because the RNA is single stranded; there is no complementary strand on which to make corrections. In addition, there are no correcting and editing enzymes like those that keep the mutation rate low in DNA replication. Because there are two single strands of RNA per virus particle, these strands can recombine, further adding to genetic diversity. The virus thus evolves rapidly within a single host. Virulent strains will be selected in circumstances in which the rate of transmission is also rapid. Public health efforts to change human behaviors and slow transmission will, in contrast, select for less virulent virus mutations.

The amount of genetic change within the HIV in a single patient over a 10-year period of disease is estimated to equal millions of years of change in the human species. The enormous resulting genetic variation presents a major problem for the design of vaccines or drugs against HIV. Before we examine therapies, however, we need to look at how HIV infection actually produces the disease AIDS.

THOUGHT QUESTIONS

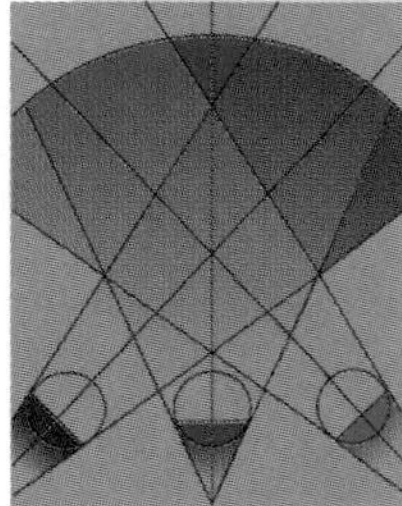

1. How might scientists decide whether or not two strains of virus are actually the same?
2. What are some of the reasons why evidence of the types called for by Koch's postulates may be impossible to obtain for some infectious diseases?
3. What are some differences between bacteria and viruses?
4. If an RNA strand from HIV contains the base sequence AAUGCA, what would be the base sequence on the first strand of DNA produced by reverse transcription? What would be the sequence of the second DNA strand transcribed from the first one? (You may need to review material from Chapter 2 to answer this question.)

HIV INFECTION PROGRESSES IN CERTAIN PATTERNS, OFTEN LEADING TO AIDS

We have already seen that HIV binds to cells that have a CD4 molecule on their surfaces. It then enters these cells, replicates, and goes on to infect more cells. HIV infection diminishes both the number and the activity of the CD4-bearing cells, thus reducing their ability to perform their disease-fighting functions. Because CD4-bearing helper T cells are central to both arms of the specific immunity (see Figure 13.1), their elimination results in immune deficiency, which in turn results in disease. This process can be described and studied at many levels. At the cellular level, just how does HIV eliminate helper T cells? At the organismal level (the person) how does infection progress to disease? At the population level, how is HIV transmitted? We look in this section at the cellular and organismal effects and in a later section at the population effects.

Events in infected helper T cells

How does HIV eliminate helper T cells? There are several different mechanisms.

1. Direct killing. As we have already seen, HIV can directly kill the cell it has entered by rupturing it. Repeated budding out of replicated viruses also eventually kills the cell (Figure 13.6).
2. Cell suicide (apoptosis). HIV may also change a helper T cell so that when it responds to another infection, it commits suicide instead of dividing. Healthy T cells, when activated by a pathogen, begin to synthesize DNA and divide. Under the same conditions, HIV-infected helper T cells undergo **apoptosis**, a process in which the DNA breaks up into small fragments and the cell dies.
3. Killing by cytotoxic T cells. Once viral proteins that are made in the helper T cell show on its membrane, they mark these helper T cells as targets for killing by cytotoxic T cells (see Chapter 12, p. 503), a process that eliminates many viral infections. With HIV, however, the target of cytotoxic T cells is the infected helper T cell, which adds to the decrease in helper T cells.

FIGURE 13.6

Many HIV particles emerging from a helper T cell. The dark circles are holes in the cell membrane left when the viruses bud out. This eventually kills the cell.

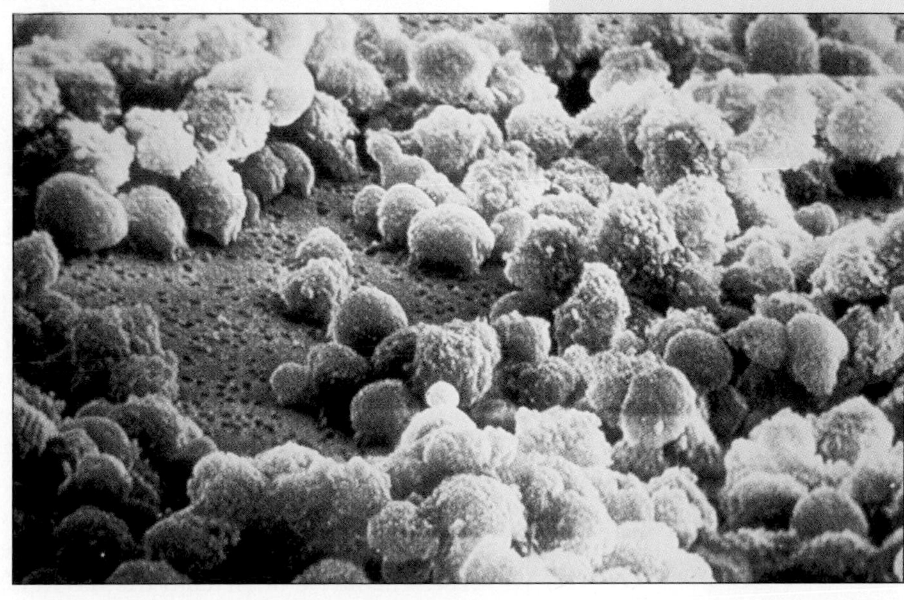

4. Cell fusion. HIV carries in its viral envelope a protein that helps it bind to the CD4 protein on the surface of the cell it infects. The infected host cell also expresses some of this viral envelope protein in its plasma membrane before new viruses bud out (see Figure 13.4.) The infected host cell can thus bind to the CD4 protein on the surface of other, uninfected helper T cells. The plasma membranes of the infected and uninfected host cells then fuse, bringing about cell fusion and spreading the virus to a new cell. This fusion can be repeated until a multinucleated 'giant cell' is formed. Although these giant cells are still alive, they can no longer perform the immunological activities of normal helper T cells.
5. Indirect inactivation. Certain strains of HIV cause the production of the wrong cytokines or inhibit the production of the cytokines needed for T-cell growth. Without these cytokines, the helper T cells cannot divide to perform their normal disease-fighting processes or to replace cells lost to HIV-induced lysis and apoptosis.

Although these mechanisms can be demonstrated in laboratory experiments, it is not certain that they all occur in infected human hosts. We do not yet know which of these mechanisms causes the greatest loss of T cells or of T-cell functions. The result, however, is the same: the loss of healthy, active helper T cells results in immune deficiency. Many of these events also occur in HIV-infected CD4-bearing macrophages, thus impairing the antigen non-specific, innate portion of the immune system as well.

Progression from HIV infection to AIDS

When a person becomes infected with HIV, the virus keeps spreading to more and more cells. When an HIV infection has progressed to AIDS, 1–10% of the helper T cells have become infected. How does infection of some cells progress to disease in a person?

Three stages of HIV infection. The progression from HIV infection to AIDS follows three stages: the initial infection, an asymptomatic phase, and a third phase called disease progression (Figure 13.7). An infected person can transmit HIV to another person at any of the three stages, but is most likely to do so in the first and third stages, when the numbers of cell-free viruses and infected cells in the body fluids is highest.

In the initial stage, virus levels in the blood are high. As more and more helper T cells are infected and killed, the helper T-cell count begins to drop. The initial infection may be accompanied by flulike symptoms—fever, swollen lymph nodes, and fatigue—which, because they are similar to the symptoms for many other diseases, are often not diagnosed as being an acute HIV infection. The initial infection also stimulates two types of immune response. Antibodies to HIV are produced by B cells and there is an increase in HIV-specific cytotoxic T cells that can kill cells containing HIV virus. These processes initially are able to contain the HIV. The levels of virus in the blood decrease. New helper T cells develop to replace those that were killed and helper T-cell counts return to normal.

With the decrease in virus in the blood and the return of helper T cells, the person enters the second, or asymptomatic, phase, which can last for a few months to many years, with 10 years being typical. The levels of virus in the blood decrease, but the viral population in the lymph nodes continues to grow by viral replication and infection of new helper T cells. By binding to the virus, the antibodies that were synthesized in the initial phase can neutralize the virus, that is, prevent it from infecting more cells. During the asymptomatic phase, the immune system is still able to keep the infection under control, so the person does not feel ill.

In the third phase, the levels of virus in the blood increase once more, while the numbers of helper T cells in the blood decrease. It is uncertain what triggers the onset of the third phase, although malnutrition, stress or other immunosuppressive factors seem to hasten the onset. Originally it was assumed that HIV infection inhibited the maturation of new helper T cells. It is now known that the maturation rate is actually normal or greater than normal, but the rate of cell death is so great that the overall helper T cell population decreases, particularly as an HIV-infected person progresses to AIDS. The virus mutates to forms that no longer match the antibodies produced in the acute phase, which thus cannot bind to the virus and neutralize it. As we have seen, the host's own cytotoxic T cells turn against the helper T cells. A person is defined as having progressed to AIDS when his or her CD4 helper T-cell count (also called the T4 count) falls from a normal value of 1000 cells per microliter of blood to less than 200. Death generally follows when the level of helper T cells declines still further in the third phase.

Other cells bearing CD4 may also be infected and inactivated by HIV. As mentioned previously, macrophages are white blood cells that engulf and remove pathogens and damaged cells or molecules, and secrete cytokines that strengthen immune responses. The elimination of

FIGURE 13.7

The course of HIV infection.

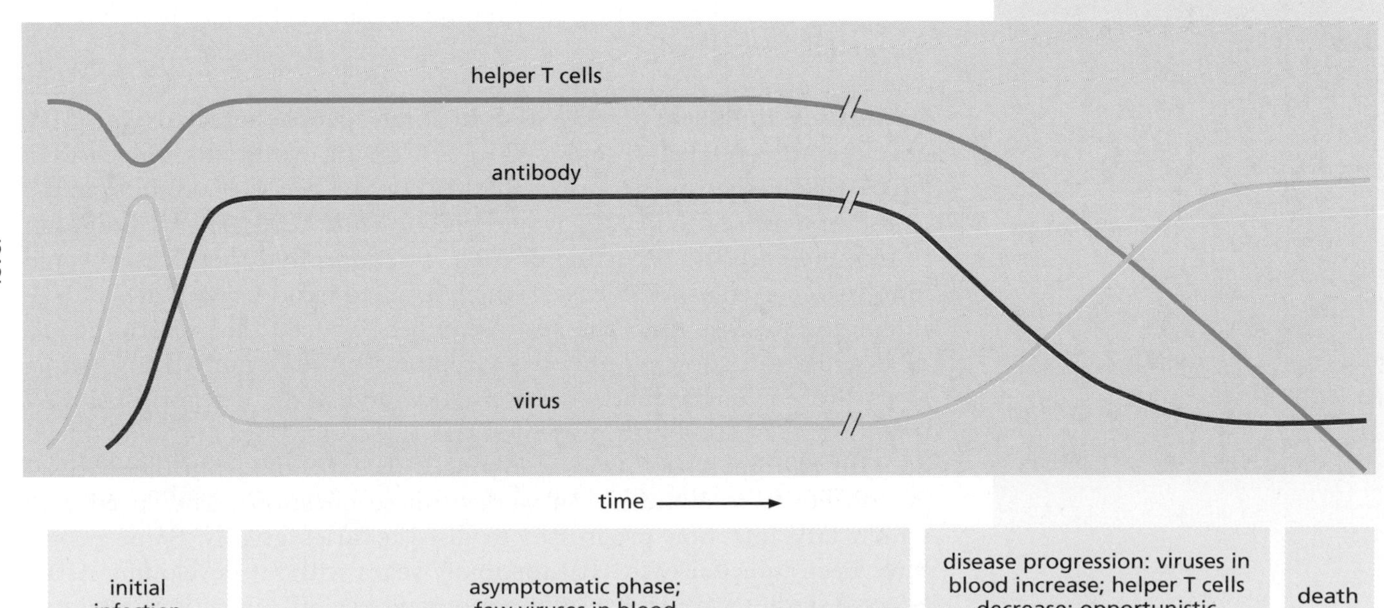

macrophage cells contributes to the risk for opportunistic infections that may, in turn, lead to the patient's death.

Opportunistic infections and other symptoms. There are microorganisms that are always present in a person or the environment but are kept in check by a healthy immune system, so that they seldom cause illness. Infections caused by these microorganisms in immunodeficient people are called **opportunistic infections** and are one of the primary symptoms of AIDS. These infections can be very severe and even fatal in the person with AIDS. People with intact immune systems need not fear catching these opportunistic infections from a person with AIDS. In the United States, typical opportunistic infections that accompany AIDS are *Pneumocystis* pneumonia, caused by *Pneumocystis carinii*, and fungal infections with *Toxoplasmodium* or *Histoplasmodium*. Recall that the appearance of a cluster of cases of this rare *Pneumocystis* pneumonia was the first hint of AIDS. Another fungus called *Candida* (a yeast) causes mild infections of the mouth, esophagus, or vagina in the absence of AIDS, but *Candida* infections in people with AIDS are much more severe. The same is true of viral diseases including shingles, cytomegalovirus eye infections, and herpes viruses. In Africa, the more common opportunistic infection accompanying AIDS is tuberculosis, a bacterial infection caused by *Mycobacterium tuberculosis*. Tuberculosis is a disease in which there are active periods and periods of remission; HIV infection increases the frequency of reactivation of tuberculosis and also the mortality rate. Worldwide, tuberculosis is the leading cause of death in HIV-infected people.

AIDS patients may also suffer from high fevers, night sweats, general weakness, mental deterioration (dementia), and severe weight loss, although these last two symptoms may not develop for a long time. Dementia may be related to the elimination of macrophagelike cells from the brain. In the gut, there is a type of CD4-bearing cell that has a role in the absorption of nutrients; elimination of these cells may be related to the weight loss.

Variations in disease progression. Many people infected with HIV develop AIDS-related complex (ARC), a set of symptoms milder than AIDS. Originally it was thought that ARC was a pre-AIDS condition and that everyone who had ARC would end up with AIDS. The CDC did not initially require the reporting of ARC, assuming that these cases would later be reported as AIDS cases, which resulted in underestimates of HIV infection rates. As time passed, researchers noticed that several people died while still showing only the symptoms of ARC, not AIDS. Distinctions are no longer made between ARC and other categories of HIV infection; they are all simply called HIV infection. Does everyone infected with HIV get AIDS? Does everyone with AIDS die from the disease? We do not have definitive answers to these questions. The speed with which HIV infection progresses to disease varies greatly. Some people have been infected with HIV for many years without developing AIDS. Several studies have shown that nonprogressive HIV infections are often characterized by a very small amount of the virus, but it is not clear whether this reflects a low infective dose initially or an immune system that has successfully kept the viral population low. A long-term study of HIV-infected homosexual men in San Francisco showed that, after 12 years, 65% had progressed to AIDS, but 35% had not. It may yet turn out that the progression from HIV infection to AIDS is not inevitable. Certainly, the avoidance of other immunosuppressive factors, including drugs, alcohol, and stress, can help to maintain health (see Chapters 11 and 12).

Researchers studied several dozen professional sex workers (prostitutes) in west Africa who were infected with the less virulent strain HIV-2. Significant findings of this study are that HIV-2 infection seemed to offer these women some protection against the more virulent strain HIV-l and that they were less sick than people infected with HIV-1. They did have high rates of infection for other sexually transmitted diseases, falsifying the hypothesis that the lower HIV-1 rates were simply the result of safer sex practices.

Molecules called chemokine receptors have been found to be co-receptors for cellular infection by HIV, meaning that the infection process is greatly enhanced when CD4 molecules and chemokine receptors are both present on host cells. People with heterozygous deletions of these co-receptors generally stay asymptomatic for longer, while those with homozygous deletions often remain HIV uninfected despite repeated exposures. However, it seems probable that deletions do not offer protection against all strains of HIV.

Tests for HIV infection

How can people tell whether they are infected with HIV? The B lymphocytes of a person infected with HIV respond to the virus. This response, which is the basis of most testing for HIV, takes a couple of weeks or months and results in the production of antibodies to HIV in the blood. The development of specific antibodies is called **seroconversion**, and once the antibodies have developed the person is said to be **HIV-positive** (HIV^+).

There are two common tests for HIV, the ELISA test and the Western blot test, as described in Box 13.1. Both of these tests detect antibodies to HIV. As mentioned earlier, there are now tests to detect the virus itself, but these tests are very expensive, so most HIV tests are still based on the presence of antibodies to HIV.

For every diagnostic test there exist the possibilities of **false positives**, test results that are positive when the person does not really have the condition, and **false negatives**, test results that are negative when the person really does have the condition. The frequency of false negatives determines the **sensitivity** of the test; the frequency of false positives determines the **specificity** of the test. The reliability of a given test depends on both its sensitivity and its specificity. The more sensitive a test, the less often it will miss a truly positive case; the more specific a test, the fewer will be the cases that are truly negative but that are reported as positive. Every diagnostic test must be thoroughly tried on samples

BOX 13.1 ELISA and Western Blot Tests for Detection of Antibodies to HIV

The ELISA and the Western blot test use immunological techniques to detect HIV-specific antibodies in a person's blood and thus are called immunodiagnostic techniques. In the ELISA (Enzyme-Linked ImmunoSorbent Assay), laboratory-grown HIV are immobilized onto a surface, generally small wells made of a plastic designed to bind protein molecules tightly (1). The rest of the plastic is coated with other proteins to block any non-specific binding of proteins used later in the assay. The immobilized virus particles then act as binding sites for specific antibodies: they are exposed to blood serum from the person being tested, and if that person's serum contains antibodies whose specific binding sites match molecules on the virus, the antibodies bind to the immobilized virus (2). There are many other antibody molecules in the person's blood that do not match any HIV molecule; these do not bind. The wells are then rinsed, removing any unattached antibody molecules. The viral molecules and any specific anti-HIV antibodies bound to them are so tightly attached that they do not wash away. Anti-HIV antibodies are then detected by a second antibody to which an enzyme has been attached (3). The binding sites on the enzyme-linked antibody match amino acid sequences on human antibody molecules; they bind, not to the plastic or to the HIV, but to any human antibodies present. Again any unbound antibody is washed away, then enzyme substrate is added (4). Enzymes are proteins that catalyze biochemical reactions; the enzymes used in these tests cause a color change in the medium. A well in which the medium has changed color (from clear to yellow in this photo, part 5) is thus a well in which the blood plasma used in the test contained antibody specific for HIV, an initial bit of evidence that the person is HIV-positive.

Rows A to G are sera from different people. Many dilutions of each serum are tested, with wells to the right in each row being the most dilute. The yellow color thus decreases from left to right across a row; the more wells that are yellow in a row, the more antibody was present in the sample. The bottom row is a negative control sample with no anti-HIV antibody.

If an ELISA suggests the presence of anti-HIV antibodies, the Western blot test is done. The viral proteins are separated by using a technique called electrophoresis (1) (see Figure 3.11, p. 91). The separated proteins are then transferred out of the gel onto special paper that has a high affinity for proteins (2). The paper is then exposed to blood serum from the person being tested. Specific antibodies bind, but in this case they bind not to the whole virus but to some individual viral protein. The serum may contain specific antibodies that bind to some viral proteins but may lack antibodies to other viral proteins (3). Unbound antibody is rinsed away and bound antibody is detected, as in the ELISA, with enzyme-linked antibody that binds to all human antibodies (4). If the enzyme-linked antibody finds human antibody to bind to, the enzyme makes dark bands on the paper where the blood contained antibody specific for that viral protein; where there is no specific antibody, no band appears (5). If antibody to specific proteins is present, the person has tested HIV-positive (also referred to as seropositive, because the test is done on serum, the liquid part of the blood after blood has been allowed to clot).

---

from thousands of individuals whose actual status is known before the test can be sold. These trials must be conducted blind; that is, the person doing the testing cannot know during the trials whether the test samples came from persons infected with HIV or not. Afterwards, the true infection status (known beforehand but concealed from the researchers) is

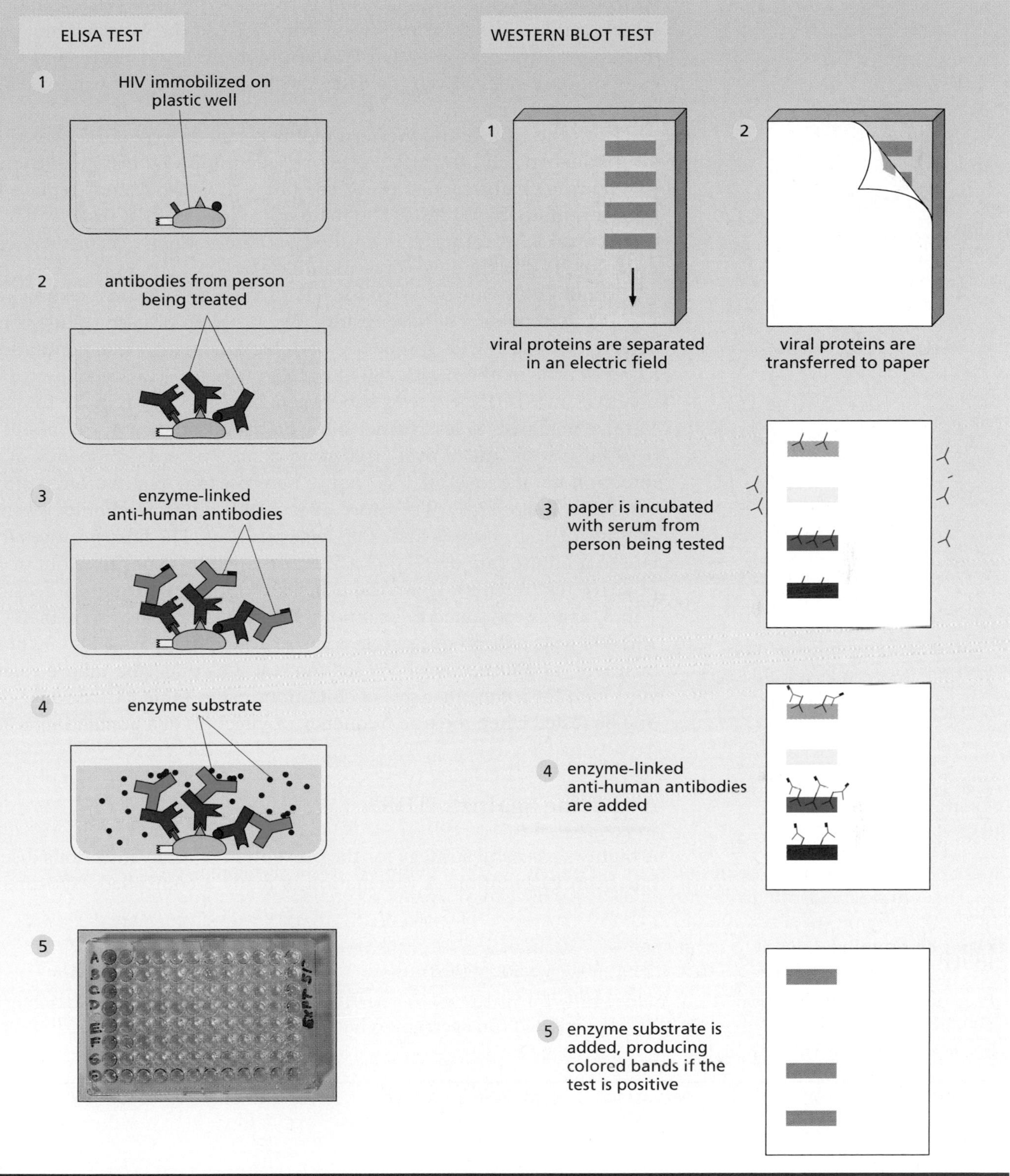
ELISA TEST
1
HIV immobilized on plastic well
2
antibodies from person being treated
3
enzyme-linked anti-human antibodies
4
enzyme substrate
5
WESTERN BLOT TEST
1
viral proteins are separated in an electric field
2
viral proteins are transferred to paper
3
paper is incubated with serum from person being tested
4
enzyme-linked anti-human antibodies are added
5
enzyme substrate is added, producing colored bands if the test is positive

compared with the status revealed by the test. In this way the frequency of false results can be quantified.

For HIV testing, the ELISA test is done first. The sensitivity of the ELISA test is high—less than 1% false negatives—but it is not very specific: there can be as many as 2 to 3% false positives. For this reason, when an

ELISA result is positive, the result is rechecked with a Western blot test, which rarely gives false positives. Why not use the Western blot as the initial test? The reason is that Western blots are more costly and technically more difficult. Even the ELISA test is too costly for widespread use in many countries.

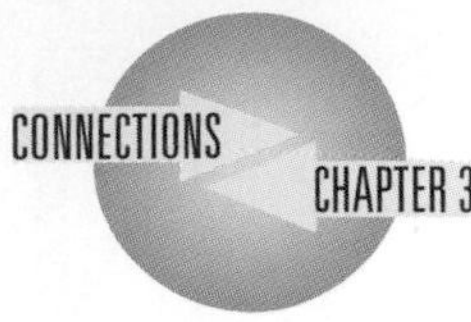

Both the ELISA and Western blot tests are based on the detection of antibodies specific for HIV. A second generation of tests based on the polymerase chain reaction (PCR; see Chapter 3, p. 101) may replace both these antibody-based tests. The main advantage of PCR tests is that they detect viral RNA rather than antibody, so they can give results soon after infection, rather than weeks or months later.

Should everyone be tested for HIV? One reason against testing everyone is the frequency of false results. The problem of false results is more severe when the true frequency of infection in the test population is lower. A little mathematics will illustrate the point. The frequency of false positives in HIV ELISA tests is between 2% and 3%, while the frequency of false negatives is less than 1% for an overall inaccuracy of about 3%, or 3 false tests out of every 100 tests done. The true frequency of HIV infection in the overall U.S. adult heterosexual population is 15 per 100,000 (Table 13.1). Therefore every 100,000 tests should reveal an average of 15 true cases and 3000 false positives. The false positives translate to a failure rate of 99.5% for the test (3000 false positives out of 3015 positive test results). If, on the other hand, the true rate of infection is 1 in 3, as it is estimated to be among U.S. injection drug users, then every 100,000 tests will produce an average of 33,000 true cases and 3000 false positives, a failure rate of 8% for the test. Calculate the failure rates on your own for some other sets of conditions using Table 13.1. Should everyone be tested when the true frequency of infection in a population is low?

A vaccine against AIDS?

A highly successful strategy for the prevention of many infectious diseases has been vaccination. A vaccination is really a controlled exposure of a

TABLE 13.1

Prevalence of HIV infection. A prevalence rate of 1 in 109, or 917 per 100,000 means that, in a population of 100,000, 917 people (or 1 in every 109) are expected (on average) to have the condition being discussed. Data based on 1998 statistics.

GROUP (ADULTS 15–49)	PREVALENCE	NUMBER PER 100,000	PERCENTAGE
U.S. general heterosexual	1 in 6666	15	0.015
U.S. college students	1 in 500	200	0.2
U.S. prison population	1 in 495	202	0.2
U.S. male homosexuals*	1 in 5	20,000	20
U.S. bisexuals, infrequent homosexuals	1 in 20	5000	5
U.S. injection drug users	1 in 3	33,300	33
Worldwide	1 in 109	917	0.9

*Rate among men who have sex with men and who were attending clinics for sexually transmitted diseases.

person to molecules similar or identical to those carried by the pathogen. The material to which the person is exposed is called the **vaccine**, so named because the first successful vaccine (which was against smallpox) used the vaccinia virus from the sores of infected cows (Latin *vacca*). Exposure to a vaccine stimulates the immune system to make an immune response to the molecules, and vaccination is therefore also called immunization. The pathogen itself is not used so that the person is not given the disease. The vaccine may be another microorganism, closely related to the pathogen but nonvirulent to humans, as when vaccinia from cows was used to protect against smallpox. (Smallpox vaccination succeeded in eliminating this disease from the globe; therefore smallpox vaccinations are no longer routinely given.) A vaccine may be the pathogen itself but treated so as to make it nonvirulent or kill it. Older vaccines used whole microorganisms, but today molecules vital to the pathogen's life cycle or to its ability to cause disease are more frequently used instead. Several laboratories are attempting to develop vaccines that would prevent HIV infection (preexposure immunization) or would prevent the progression of HIV infection to AIDS (postexposure immunization).

There are many biological barriers to developing vaccines against AIDS. These roadblocks include genetic variation of the virus, a lack of knowledge about which immune responses are protective against HIV, and a lack of animal models in which to test trial vaccines. The viral RNA and DNA sequences are changing rapidly. Is it possible to develop one vaccine that could stimulate a protective immune response in every person vaccinated and that would continue to protect infected people as the viral nucleic acid sequences changed? The answer right now is "maybe": maybe there are some sequences that do not change very much or for which changes have no effect on recognition by the immune system. The latter is possible because the immune system actually recognizes protein *shapes*, not sequences of nucleic acids or amino acids. A change in nucleic acid sequence may cause one amino acid to be substituted for another in the protein during its synthesis, but some substitutions do not alter the shape of the completed and folded protein. If the shape did not change, the immune system would still recognize the altered protein.

Not all immune responses against HIV are protective, as can be seen by the fact that HIV-infected people develop antibody and CD8 cytotoxic T-cell immune responses to HIV but still eventually get AIDS. Proteins that function in the viral life cycle are being targeted for vaccine development, but it is not known whether these will stimulate protective responses. Stimulating an immune response by vaccination may actually trigger progression to AIDS in someone already infected with HIV, as happened in one documented case in which an HIV-positive person rapidly progressed to AIDS after a smallpox vaccination.

The lack of animal models is a significant problem. The effects on each step in an immune response can be studied *in vitro*, but protection from disease can be evaluated only in an animal that gets the disease. Ethical considerations call for extreme caution in the testing of vaccines on human volunteers in a disease known to have a high percentage of fatalities and for which there is no known cure (see Chapter 1).

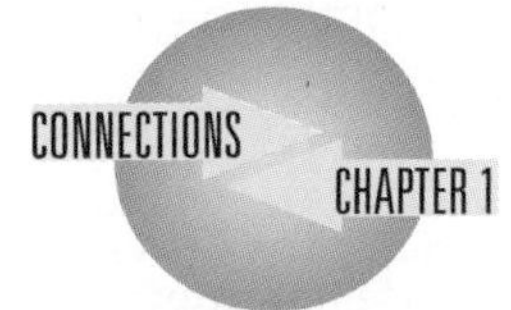

In 1994, several vaccines were being tested in small-scale trials on humans in Europe and North America. These vaccines did not prove to be entirely protective: a few individuals contracted HIV after vaccination. They did not get HIV from the vaccine; rather, the vaccine failed to protect them from transmission by the routes described in the next section of this chapter. The National Institutes of Health did not allow larger-scale tests to proceed in the United States. The World Health Organization took a different stand and has allowed vaccine tests to be conducted, with Uganda and Thailand chosen as the locations. These tests are continuing (as of the year 2000), so the results are not yet known. The governments in these countries have welcomed these tests because, if successful, the vaccines would confer protection against the HIV strains prevalent in Africa and Asia. So far, 90% of research has focused on the subtype common in Europe and North America; although present elsewhere, this subtype is not the one common in other parts of the world. Thailand's public health officials are additionally interested because drug therapies, which we look at in the next section, are too expensive. Prevention by education and vaccination remains the only affordable option.

Drug therapy for people with AIDS

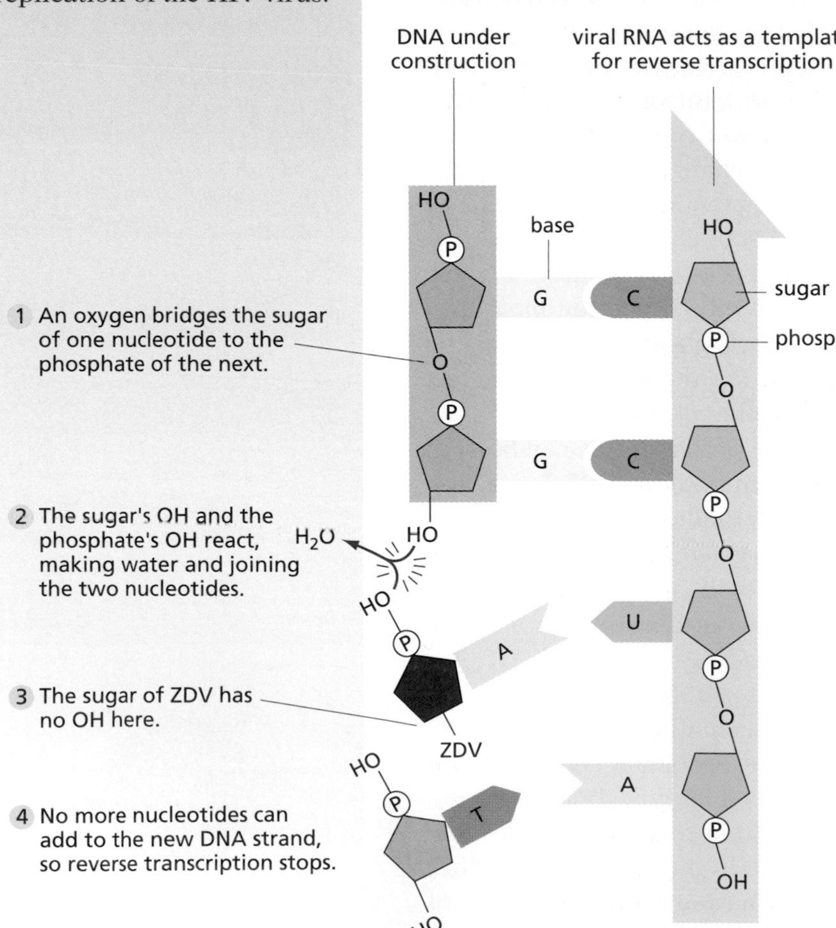
DNA under construction
viral RNA acts as a template for reverse transcription
HO
base
HO
G
C
sugar
P
phosphate
O
G
C
1 An oxygen bridges the sugar of one nucleotide to the phosphate of the next.
2 The sugar's OH and the phosphate's OH react, making water and joining the two nucleotides.
H_2O
HO
U
A
3 The sugar of ZDV has no OH here.
ZDV
A
T
4 No more nucleotides can add to the new DNA strand, so reverse transcription stops.
OH

FIGURE 13.8

How the drug zidovudine (ZDV) interrupts the replication of the HIV virus.

Very few drugs are helpful against viral diseases. Antibiotics, which are highly effective against bacterial diseases, do not work against viruses. As detailed knowledge becomes available about the few enzymes of its own that HIV has, new drugs may be developed to target these enzymes specifically.

Antiviral drugs. One type of drug consists of **reverse transcriptase inhibitors** that are similar in chemical structure to parts of DNA. These drugs block the reverse transcription of viral RNA to DNA (Figure 13.8), thus preventing the virus from making the DNA it needs to complete its infective cycle (see Figure 13.4). One such drug is **zidovudine** or **ZDV** (trade name Retrovir, formerly known as azidothimidine, or AZT). Zidovudine greatly reduces the rate of HIV transmission from pregnant women to their babies before and during birth. Didanosine (Videx, formerly called dideoxyinosine or DDI) is another drug that inhibits reverse transcriptase.

The **protease inhibitors** are a newer type of antiviral drug. Later in the viral life cycle than the reverse transcription step, an enzyme called a protease is required to trim newly translated proteins into their functional form (see Figure 13.4, step 6). Blocking this step stops viral replication and infectivity. HIV protease is very different from human protease enzymes, reducing the effects of the drug on the human host.

Combination therapy. With its rapidly changing genome, HIV has the potential to evolve resistance to any particular drug quickly. In **combination therapy** the ideal is to combine two drugs that work by different mechanisms. If resistance to one mechanism of action evolves, the other drug will still be effective. Combination therapy against HIV at first used two or more drugs with the same mechanism of action, that is, two or more reverse-transcriptase inhibitors, because they were the only drugs available. Now, more than two drugs are included in what is called the 'drug cocktail,' and the combination includes both reverse transcription inhibitors and protease inhibitors. These combinations have been very successful, reducing viral loads below detectable levels in a high percentage of HIV-infected people.

Drug combinations reduce the probability of HIV's becoming resistant and rebounding, but these drugs must be taken on schedule. A typical regimen may involve taking 16 pills at 6 different times during the day. Many regimens are even more complex. This complexity, combined with unpleasant side effects of various kinds, and the enormous costs of the drugs ($70,000 to $150,000 per year) makes therapy difficult for many patients. Yet many HIV-infected people—even some who have progressed to AIDS—have been restored to functional lives. It is not yet known how long a person would need to stay on therapy. Reducing virus to 'below detectable limits' does not necessarily mean that the virus is gone; some people's viral loads have returned when they stopped taking the drugs. So, although these drugs have certainly been a source of optimism, we cannot yet call them a 'cure' for HIV or AIDS.

Prevention and treatment of opportunistic infections in persons with AIDS. Nearly all AIDS deaths are caused by opportunistic infections. Attempting to prevent opportunistic infections in people with AIDS is thus highly important. During the phase of CD4 helper T-cell depletion, people are very susceptible to infectious diseases carried by people who are not infected by HIV. A cold or the flu can have grave consequences in an immunodeficient person. Bacteria picked up from food can be equally hazardous. In the developed world, therapeutic drugs are available for the treatment of many of the opportunistic infections, such as a combination of the drugs trimethoprim and sulfamethoxazole for *Pneumocystis* pneumonia. However, in many parts of the world, such drugs are unavailable because of their cost. The only other way to stop AIDS deaths in populations is by preventing its transmission, which is the topic of the next section.

THOUGHT QUESTIONS

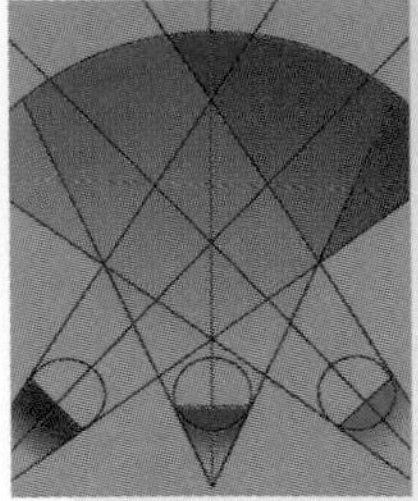

1. What is the difference between HIV infection and AIDS?
2. What lifestyle choices can a person make to decrease their chances of becoming immunodeficient? Would those choices also be important for an HIV-positive person?
3. What steps in the ELISA test determine its specificity for HIV? What steps determine its sensitivity?
4. If a vaccine against AIDS were developed, how would you go about testing it? Remember that AIDS develops only in people, so animals cannot reliably be used as subjects. Would your test have a control group? How would you ensure that the conduct of the test was ethical?

KNOWLEDGE OF HIV TRANSMISSION CAN HELP YOU TO AVOID AIDS RISKS

The general term for the transfer of a pathogen from one individual to another is **transmission**. How is HIV transmitted from one person to another? HIV does not have any other animal hosts and does not remain infective in water or in air. It can pass from one person to another only in certain body fluids—blood, semen, and vaginal fluids. To enter another person, these fluids containing HIV or HIV-infected cells must rapidly come in contact with the rectal mucosa or with the bloodstream via breaks in the mucous membranes or skin of the other person. HIV must make rapid contact with cells bearing the CD4 molecule. Cell-free HIV does not remain infectious very long if the fluids are outside a person, for example in a blood spill. Washing with ordinary soap and water kills cell-free HIV when it is outside a person, because soap dissolves the viral lipid envelope. A mother can transmit HIV to her unborn fetus or to her baby during delivery, and a few cases arc known in which HIV has been transmitted in breast milk. HIV is not found in feces or urine. There are small numbers of HIV particles in the saliva or tears from 1–2% of HIV-infected people, but saliva contains antiviral activity, and HIV has never been known to be transmitted through saliva, including human bites, or via tears. How do the fluids containing HIV get passed? The percentage of AIDS cases in the United States transmitted by various routes is shown in Figure 13.9. This figure represents total numbers of persons with AIDS in 1999.

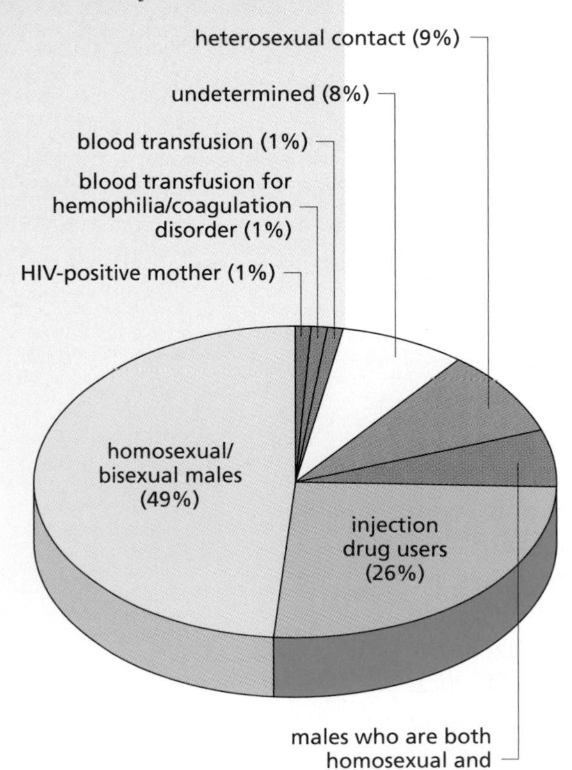

FIGURE 13.9 Routes of AIDS transmission in the United States, based on 1999 statistics from the Centers for Disease Control and Prevention. The percentages indicate the proportion of cases transmitted by each route.

Risk behaviors

When scientists use the term **risk** of HIV infection, they mean the mathematical probability of transmission of the infection. Various behaviors or activities have been grouped into 'risk categories,' based on what is known about transmission routes. High-risk behaviors are those that give a high probability of transmission from an infected person to an uninfected person. Notice that risk is now categorized by specific **risk behaviors** and not by population groups. The safest way to make choices about these behaviors is to assume that every person whose HIV status is unknown to us may be HIV infected and a potential source of HIV transmission.

Category I: high-risk behaviors. Transmission while engaging in a **high-risk behavior** is very likely. These behaviors, which account for over 97% of all AIDS cases, are:

1. Behaviors in which the passage of blood, vaginal fluids or semen is very likely, such as anal or vaginal intercourse with an infected person without the protection of a condom (unsafe sex).
2. Injection drug use in which needles or syringes are shared with an infected person.
3. An infected mother's going through pregnancy and giving birth without receiving zidovudine treatment.

Anal intercourse is more risky than vaginal intercourse because semen contains an enzyme (collagenase) that breaks down the lining of the rectum and exposes blood vessels, a form of injury to which the vagina is much more resistant. In addition, vaginal intercourse is not as likely to result in HIV infection as anal intercourse because the cells of the rectal mucosa have the CD4 molecule on their surfaces and so can be infected with HIV, whereas the intact epithelial lining of the vagina is a significant barrier. The risk of male-to-female transmission of HIV infection during vaginal intercourse is greater for than that for female-to-male transmission.

Category II: likely-risk behaviors. HIV transmission has been documented for routes in this category, but with lower frequency:

1. Anal or vaginal intercourse using a condom (safer sex, not safe sex). Condoms do not make intercourse completely safe. Condoms fail as birth control for about 10% of couples who use them; this means they can also fail to prevent HIV transmission. Although the rate of failure seems to be low (less than 5%) for condoms *used properly*, many users do not exercise proper care in putting condoms on or taking them off, so estimates of failure rates *in general use* can be as high as 20%. Although reliable data of this kind are difficult to obtain, one study found that 17% of women whose husbands were HIV-positive became infected despite proper and consistent condom use.
2. Breast-feeding (transmission to a baby from an infected mother).

3. Receiving a blood tranfusion or organ transplant. This risk was high prior to 1987, but is now very low in the United States because of careful screening of blood products and donated organs. The risk remains higher elsewhere, where blood donors and blood products are not screened as stringently (see below).
4. Artificial insemination. As with blood transfusion, the risks are now low when donated semen has been tested for HIV.
5. Infection of health care professionals by needle-stick injuries.
6. Dental care by an infected worker. There is a single cluster of six cases involving only one dentist; no other cases are known in which an infected health care worker has transmitted HIV to a patient.
7. Deep kissing. A single case has been documented and, because both people had severe periodontal disease (bleeding gums), HIV was probably transmitted through blood, not saliva.

Category III: low-risk behaviors. This category includes routes that are biologically plausible, but no cases have been confirmed.

1. Sharing toothbrushes or razors or other implements that may be contaminated with infected blood.
2. Being tattooed or body pierced to produce ornamental scars or for jewelry.
3. Receiving tears or saliva.
4. Oral sex. Although there is anecdotal evidence of transmission via oral sex, there are no cases in which transmission by this route is documented.

Category IV: no-risk behaviors. Transmission of HIV from engaging in the following behaviors with an infected person is considered not biologically possible:

1. Shaking hands
2. Sharing a toilet
3. Sharing eating utensils
4. Being sneezed on
5. Working in same room
6. Handling the same pets
7. Close-mouthed kissing (kissing with no exchange of saliva or blood).

Also, exposure to mosquitoes or other biting insects is a no-risk behavior, as explained in Box 13.2.

Communicability

Another question people ask about HIV is, "How contagious is it?" meaning, "If I am exposed, how likely is it that I will become infected?" The term used by the medical community to mean the likelihood of transmission

BOX 13.2 Can Mosquitoes Transmit AIDS?

Two frequently asked questions are, "Why don't mosquitoes transmit HIV?" and "How do we know that they don't?" Epidemiological evidence shows that the frequency of AIDS and of the number of unexplained cases is no higher in mosquito-infested areas of the United States than in other areas. In Africa, the people with AIDS are mostly babies and sexually active young adults; mosquitoes do not bite people in these groups more frequently than they bite other people. On all continents, children who are not yet sexually active often get mosquito bites, but they do not get AIDS unless they are infected from their mothers at birth. Laboratory experiments have shown that although HIV and HIV-infected cells may be taken up by mosquitoes who bite infected people, HIV is not transmitted to other people through mosquito bites. Several factors help to explain this: HIV cannot replicate in mosquitoes or survive long in their bodies (because mosquitoes are not a host for HIV), and the amount of blood ingested (3–4 μl, millionths of a liter) is too small to contain enough HIV or HIV-infected cells to infect a person. Saliva (including mosquito saliva) may also have substances that inhibit the virus. Furthermore, because of the many biological factors involved in the transmission of a disease by an insect, it is highly unlikely that a single mutation in either the mosquito or the virus would significantly alter this situation.

A study done at the Institute for Tropical Medicine in North Miami, Florida, proposed in 1986–1987 that the high rate of AIDS in the town of Belle Glade, Florida, could be attributed to the squalor and crowding of its people and to the mosquitoes breeding in nearby swampy lands, "where 100 insect bites a day are not unusual." The U.S. Centers for Disease Control and Prevention (CDC) studied this situation and concluded that the high incidence of AIDS in Belle Glade was attributable to sexual contact and shared needles, not insects.

Transmission of HIV by other blood-sucking animals such as bedbugs (which are insects) and ticks (which are more closely related to spiders) has also been ruled out. A tick that is endemic in the same parts of Africa where AIDS is common carries enough blood and live virus to make transmission theoretically possible. However, the possibility does not fit with the epidemiology: children below the age of sexual activity do get bitten in significant numbers but do not get AIDS.

---

after exposure to HIV is **communicability**. (**Contagious** simply means 'capable of being transmitted;' it does not refer to the probability of transmission.) The concept of communicability, or likelihood of transmission, is directly related to the concept of risk. High-risk behaviors (Category I above) increase the probability of transmission.

There are at least two ways to answer the question of communicability; one takes an epidemiological approach and the other a microbiological approach. The epidemiological approach compares the number of encounters with the number of infections throughout the population or within certain population subgroups. It is difficult to designate the probabilities for HIV transmission because there is a period of weeks or months before antibody develops, and there is often a period of years between infection and disease symptoms, during which people may not know they are HIV-infected. The number of encounters with HIV is often not known for an individual, and is even less known for all the people constituting a population.

Probabilities are, however, more accurately known for some routes of transmission than for others. For example, there is a 30–35% chance that an infected mother will transmit the virus to the fetus if she is not treated with the drug zidovudine. The efficiency of transmission of HIV by various routes is shown in Table 13.2. An efficiency of over 90% means that for every 100 exposures more than 90 will result in an HIV infection.

The microbiological approach to determining communicability is to quantify (measure numerically) what is called the **infective dose**, the number of pathogenic particles that must be transferred to result in an infection of an individual. This value is not known for HIV, although one estimate is that the transfer of 10,000–15,000 HIV particles can establish an infection. Very early in an HIV infection, in the weeks or months before antibody develops, and also very late, when the CD4 helper T-cell count is low and the antibody concentration has dropped, the number of HIV particles in blood and genital fluids is much higher than at other stages (see Figure 13.7). The number of HIV particles in general is higher in semen than in vaginal fluids, but each varies at different stages of infection. One study found 4.2 million HIV particles per milliliter of blood on average in people with AIDS, but this can vary tremendously.

From accidents in which health care workers have been exposed to infected blood, it is known that there is a higher probability of infection when a person has been splashed with large quantities of blood onto open sores in the skin and a lower probability when they have been pricked by a needle. For every 250 reported needlesticks, there has been one transmission (0.4% efficiency). The effect of blood volume can also be seen in Table 13.2, where the transfer of greater quantities of blood by intravenous injection or transfusion increases the efficiency of transmission.

Viral load is certainly a factor in determining the efficiency of HIV transmission, but only one of many factors. The precise infective dose is not known and probably varies from one person to another. For example, persons with open genital sores due to other sexually transmitted diseases such as syphilis or herpes are 10–20 times more likely to become infected than other people. Gonorrhea or chlamydia infections increase the probability of HIV transmission threefold or fourfold.

In general, it may be said that HIV is much less communicable than a virus such as hepatitis (another virus spread by contact with contaminated blood), but the communicability varies with the risk behavior. In comparison with the 0.4% efficiency of HIV transmission from a needlestick, the efficiency of transmission of hepatitis B virus by this route is 6–30%. Transmission of opportunistic microorganisms from a person not infected with HIV to an HIV-infected person is much more likely than

TABLE 13.2

Estimated efficiency of transmission of HIV by various routes.

ROUTE OF TRANSMISSION	EFFICIENCY (%)
Vaginal sexual intercourse (unprotected)	0.1
Receptive anal intercourse (unprotected)	0.5–3
Mother to child during pregnancy or childbirth (without use of zidovudine)	30–40
Transfusion of infected blood	>90
Transfusion of screened blood	0.00015–0.0002
Intravenous injection with infected needle	1–2
Needle stick with infected needle	0.4

transmission of HIV from an HIV-infected person to an HIV-uninfected person. Because people with AIDS have severely impaired immune systems, their risk of catching diseases from other people is very high.

Susceptibility versus high risk

What is the difference between the terms susceptibility and high risk? To examine this question, let us go back and look further at how knowledge of AIDS and HIV developed. The fact that early cases were reported in hemophiliacs suggested an infectious cause for AIDS. As more cases were reported, the affected individuals seemed to fall into five groups, which became known among epidemiologists as 'the five H's:' homosexual males, hemophiliacs, heroin addicts, Haitians, and hookers (prostitutes). From an epidemiological perspective, these categories served a useful purpose to describe groups within which cases were showing up. But what was useful scientifically turned out to have negative social consequences. The terms quickly became imprinted in the minds of scientists and the public, allowing complacency on the part of people who were not in these groups and prejudice against those in the groups. To epidemiologists, the five H's were merely a convenient way to designate clusters of reported cases of a mysterious, new syndrome. Such identification did not in itself imply anything about cause and effect or about transmission or about all of the people in the groups, but it was useful in suggesting hypotheses that could be tested.

To epidemiologists, the term **high-risk group**, as applied in connection with a particular disease, simply means that there is a higher **frequency** of the disease among members of that group (frequency equals the number of people with the infection divided by the number of people in the group). Use of the term implies nothing about the possible reasons for the increased frequency, which may stem from increased exposure or increased susceptibility or both. *Exposure* to a disease means coming in contact with the disease agent. Increased exposure can sometimes be due to shared behaviors, but there are many other possible explanations. A disease may have a higher frequency in a certain group of people if all the people in that group came from the same geographic location so that they were exposed to the same toxic chemical, or if they all ate food from the same source and so were exposed to the same food-borne pathogen. It does not mean that these people are more susceptible; anyone else exposed to the same factors would also have become sick. **Susceptibility** to a disease means the ability to contract that disease *if exposed*. Humans are susceptible to HIV and most other animals are not. Susceptibility can vary from one person to another, and it can be genetically or environmentally influenced (malnutrition, for example, may make a person more susceptible to many infectious diseases).

Several misperceptions about AIDS resulted from the early identification of specific high-risk groups. First, some people not in the identified high-risk group assumed they were not susceptible. Some people assumed that every person within a high-risk group was equally likely to be infected (and that people outside these groups were unlikely to carry the disease). Haitians, in particular, suffered adverse consequences by being classified as 'high risk.' In efforts to screen blood donors before the cause of the disease was known and before appropriate tests were available for screening blood, all Haitians were barred from donating blood in the United States. In the resultant hysteria, some Haitians were evicted from their homes and lost their jobs, Haiti's tourist trade collapsed, and Haitian dictator Jean-Claude Duvalier's state police rounded up and incarcerated homosexuals in Haiti. Haiti's ambassador to Washington wrote a letter published in the *New England Journal of Medicine* deploring the damage done by North American semantic carelessness. As he pointed out, being from a certain country does not contribute to disease in the way that socially acquired behaviors do (for example, having multiple sex partners or using intravenous drugs).

The fifth H, hookers, always seemed problematic because the other risk categories were predominantly or exclusively male. Why did so few women contract the disease at first? If women could contract the disease, why only prostitutes? There was a period of time when women were thought of as 'carriers' even though they were dying of AIDS themselves. Scientists now think that women are just as susceptible as men to HIV infection, but that the epidemic in the United States began among homosexual men and spread only slowly to women.

Once mechanisms of transmission are known, it becomes more appropriate to focus on high-risk behaviors than on high-risk groups. However, the frequency of infection within a discernible group of people can sometimes have a role in an individual's risk. People within a high-risk group are at risk to the extent that they engage in high-risk behaviors. Their risk may be increased to the extent that their partners in high-risk behaviors are also members of a group in which the frequency of infection is high. A higher population frequency of a disease increases risk by increasing the chance of encountering an infected person, not by altering any individual's susceptibility. (Remember that membership in a group, either a group with a high frequency of infected individuals or one with a low frequency, does not tell you whether a particular individual is or is not infected.)

As we have seen, in the United States infection rates were higher in homosexual men. The lower frequency of HIV infection in females in the United States led many people to assume that women were less susceptible. Therefore, when more women began to fall ill, there was a further misconception that the virus must have mutated to become more virulent, and if it could mutate once, it could mutate again, and heterosexuals would be susceptible. Women and heterosexual men have always been susceptible to HIV infection, as amply demonstrated by the pattern of the infection in Africa, where the numbers of men and women infected have been about equal (see Figure 13.12). The pattern of infection in the United States has changed over time, but it is possible to explain all of the changes on the basis of frequency of HIV in various subpopulations, not on the basis of changes in virulence of the virus.

Transmission of HIV by vaginal intercourse is not as likely as it is by anal intercourse, but this does not mean that vaginal sex is safe, only that the number of infections per number of encounters is lower. It also does not mean that women are less susceptible than men. It seems that, if there are breaks in the vaginal epithelium (for example, as a result of other sexually transmitted diseases), women are just as likely to be infected as men. So, again, the risk is related to particular practices, not to differences in susceptibility, and these practices carry comparable risks for all groups of people. For example, data collected in both the United States and Africa seem to show that anal sex is just as risky for females as for males.

Epidemiology shows that there is a positive correlation in both sexes between the rate of HIV infection and the number of sexual partners; that is, persons with greater numbers of partners have higher rates of HIV infection. There is also a positive correlation between the infection rate and the frequency of previous infections with other kinds of sexually transmitted diseases.

The use of illegal drugs other than injectable drugs also increases the rate of infection, particularly in women. In New York City, 32% of female crack cocaine users were HIV-positive, compared with 6% of other women. It has not been shown that crack is a cofactor (a factor that increases susceptibility), but the subculture in which crack is used is often one of a high incidence of sexual activity and of sexually transmitted diseases. On college campuses, where the 'drug of choice' is frequently alcohol, the impaired judgment that accompanies alcohol (or other drug use) is a factor working against sexual abstinence or the practice of 'safer sex.'

Another aspect to 'risk' is a person's ability or inability to say "no" to high-risk behaviors. The ability of a person to say no is termed his or her **refusal skills**. Economic and cultural factors can put severe limitations on a person's refusal skills. In some cultures, for example, women may not be able to insist that their male sexual partners use a condom. Education about the risks of HIV and AIDS must do much more than provide people with information about transmission routes, as we discuss below.

Public health and public policy

Whereas medicine deals with individual cases of disease, **public health** deals with populations and seeks to minimize the levels of particular diseases in those populations. Many public health efforts require legislation and most require funding, so they are most often conducted by governments or large organizations. Many nongovernmental organizations (NGOs) have been crucial in educating the public about AIDS and in caring for persons with AIDS and their families. Early in the epidemic, they were also crucial in pressuring governments for more funding for research on this disease.

History of public health responses to disease. In each nation in which the AIDS epidemic has spread, the governmental response was molded by the unique history and social customs of that nation. Some nations sought to restrict the immigration of HIV-infected people; others did not. Some jurisdictions segregated certain types of AIDS patients, while others did not. Hospital care and medical insurance for AIDS patients varied greatly from one country to another. Some nations instituted needle-exchange programs for drug addicts; others did not.

The response of the U.S. government to AIDS has been influenced partly by the nature of transmission of the disease (that is, it is not transmitted by casual contact), partly by the political organization of the 'AIDS community,' and partly by changes in civil rights laws during the period since 1950. Before 1950, U.S. law gave only weak support to individual rights. The Supreme Court decision in *Brown v. Board of Education*, the resultant Civil Rights Act of 1964, and the Voting Rights Act of 1965 have given much greater strength to the rights of individuals in the face of discrimination by the many. More recently, the Rehabilitation Act of 1973 mandated that employers receiving federal funds cannot discriminate against someone with a handicapping condition who is otherwise qualified, and a disease that does not endanger others is considered a handicapping condition. This was upheld in *School Board of Nassau County v. Arline* (1987), in which a person with tuberculosis won the right to continue work. These principles have been further extended in the United States by the Americans with Disabilities Act (1990). Under this act, persons with disabilities can not be denied "full and equal enjoyment" of goods, services and public accommodations. AIDS is considered a disability under this act. The U.S. Supreme Court has recently ruled that asymptomatic HIV infection can also be a disability. A disability is defined under this law as "a physical or mental impairment that substantially limits one or more of the major life activities of the individual." The court case involved an HIV-positive woman who was denied treatment by a dentist in his regular dental office. The court ruled that the woman was disabled because HIV infection limited her ability to have children (because unprotected intercourse would put her husband and a child they conceived at risk for infection). The dentist had wanted to offer the woman dental treatment in a hospital but the court also ruled (5 to 4) that the Universal Precautions for handling body fluids (see below) should make a dental office a safe environment for treatment.

A bioethical principle known as the harm principle provides a moral limit on the exercise of freedom of individuals when others may be injured. By this principle, a person with AIDS could morally be prevented from deliberately spreading the disease but could not be prevented from working or attending school or living in a particular place.

HIV testing and notification. AIDS is a reportable disease in all 50 of the United States. Twenty-seven states require the notification of public health officials of HIV infection, including the person's name, so that cases can be followed. The issue of mandatory testing for HIV antibodies remains very controversial. Issues of confidentiality are involved, because many people fear (correctly or incorrectly) they would be discriminated against if they were identified as being HIV-positive. At the same time, medical professionals would like to know the HIV status of their patients so that they can provide both better care for the infected person and better protection for health care workers against unintentional infection. Forty-four states permit notification of healthcare workers of patients' HIV status; two states (Arkansas and Missouri) require it.

Current laws generally prohibit testing a person without the person's consent. However, the U.S. government does test everyone who applies for immigration, the Peace Corps, the Job Corps, the military, or the Foreign Service, and also tests the spouses of Foreign Service applicants. The 1992 International Conference on AIDS, which was to have been held in the United States, was moved to Amsterdam in protest against the U.S. policy of denying entry to any HIV-positive person; that policy has since been relaxed. Fourteen states screen all prisoners and six segregate those who are HIV-positive.

Another issue relates to the notification of sexual partners of HIV-infected persons. For other sexually transmitted diseases, public health officials notify and test the sexual partners of all infected persons, then the partners of those sexual partners, and so on. Some doctors have argued against this practice for HIV-infected women, who are often diagnosed during prenatal testing when they are pregnant. Many HIV-positive women (and their unborn fetuses) become victims of domestic violence when their HIV status is reported to their partner. Partner notification is permitted ('duty to warn'), but not required, in most states. Notification obviously depends on infected persons' accurate disclosure of the names of all their partners.

Guidelines for handling blood. People likely to come in contact with blood—for example, dentists and surgeons as well as their auxiliary workers, sports coaches and trainers, and security personnel—are now required to follow a series of guidelines known in the United States as the *Universal Precautions for Blood-Borne Pathogens*. The term 'universal' refers to the fact that all blood must be handled as though it were infected. The guidelines, which are intended to limit the transmission of any blood-borne pathogen, include wearing gloves when handling blood (Figure 13.10), and further covering when handling large quantities of blood. The guidelines also specify procedures for cleaning blood spills, reporting accidents and injuries in which workers have come into contact with blood, and for educating workers about the risks in handling blood. The U.S. guidelines were developed by the CDC, and the Occupational Safety and Health Administration (OSHA) has been charged by the federal government with monitoring the compliance of employers with these guidelines. Every college and university, for example, must have an infection control plan.

The risk of transmission of HIV through blood transfusion depends a good deal on the methods of blood donor selection. In the United States, blood donors are volunteers. Blood is screened for antibody to HIV, but blood donors are not. The behaviors that transmit HIV are explained to blood donors, and they can anonymously tag

FIGURE 13.10

Health care workers, researchers, and others who handle human blood must follow the *Universal Precautions for Blood-Borne Pathogens*.

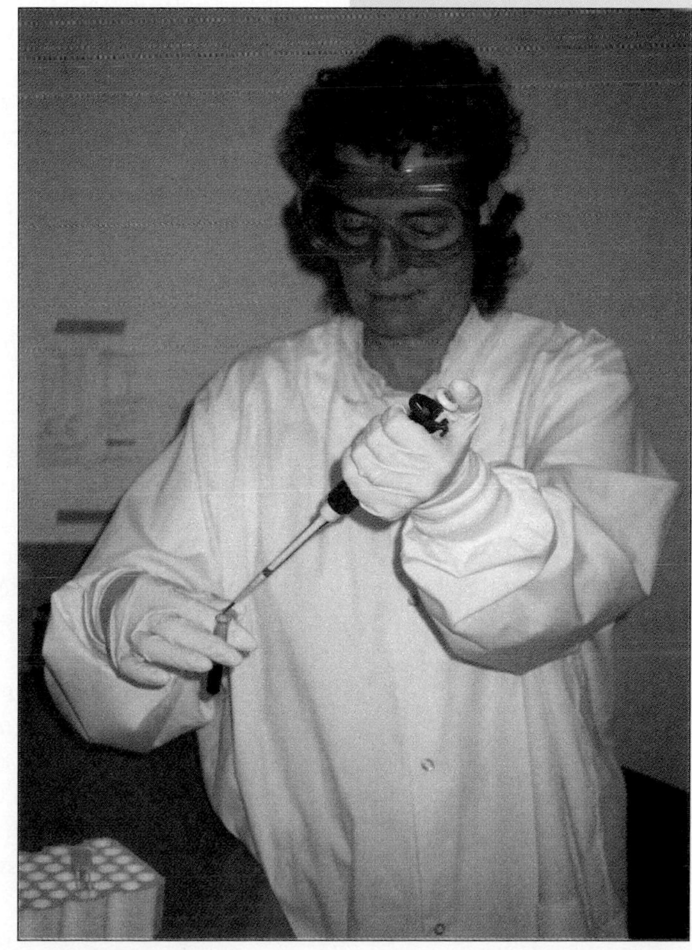

their donated blood as having come from a person involved in high-risk behaviors. In the United States, the risk of transmission from a blood transfusion is currently estimated to be 1 in 450,000 to 650,000 (see Table 13.2).

In countries in which donors are paid, the safety of the blood supply is much less assured than in countries in which donations are voluntary. In India, 30–50% of blood donors are professionals who sell their blood an average of 3.5 times per week. In one city in India, 200 professional donors were screened, and 86% were found to be HIV-positive.

Access to health care. In the United States, the CDC develops the criteria that define AIDS. The criteria have changed as new information has become known. The wording of the definition is important because people with AIDS are eligible for some types of care from the government that other people are not. AIDS was originally defined as a set of symptoms including particular opportunistic infections. The CDC definition now includes all those persons who are HIV-positive and have a CD4 helper T-cell count below 200.

HIV-positive women have a poorer prognosis (predicted outcome) than HIV-positive men, both in the United States and elsewhere in the world. This is probably a result of women's generally having poorer access to medical care for the infections that accompany AIDS. The care of people with AIDS has put a strain on public health monies and personnel. In some countries of Africa, more than half of all public health expenditures are for AIDS.

Educational campaigns. In educational campaigns aimed at increasing AIDS awareness, some organizations distribute free condoms and promote their use, while others emphasize abstinence. The former Surgeon General of the United States, C. Everett Koop, stated that the only safe sex is a faithfully monogamous relationship with a faithfully monogamous uninfected partner, and the next best thing is the use of a condom.

Education about HIV and its transmission changed the behavior of homosexual men so that the rates of infection within this group began to decline. This subgroup is generally well educated and has provided many model programs which have been copied in educational efforts to reach other groups. Because the factors guiding people's private behavior differ from one group to another (on the basis of language, income, geography, religion, and cultural background), educational campaigns need to be designed for each different locale and target group.

Information is not the same thing as education. Giving people information about how HIV is spread may not help unless the reasons underlying their high-risk behaviors are addressed. The motivations of people having consensual sex differ from the motivations of commercial sex workers (prostitutes) and street children having sex for survival. Many teenagers and young adults engage in sexual activity (often including high-risk activity), and those who do not are frequently subjected to very strong peer pressure to conform. Education often includes strategies for raising self-esteem and providing support for avoiding high-risk behaviors.

Worldwide patterns of infection

HIV is now a **pandemic**, a worldwide epidemic. As of 1999, HIV-infected adults totaled 55 million worldwide (up from 10 million in 1992). Five million children are infected, and in many countries AIDS has reversed the hard-won decreases in the infant mortality rate. In addition, 11 million children have been orphaned by the death of parents from AIDS. In the United States a cumulative total of 700,000 people have been diagnosed with AIDS. As many as 1 million may be HIV infected and 300,000 are currently living with AIDS in the United States.

HIV compared with other infections. How do the numbers of AIDS deaths compare with the numbers of deaths from other diseases? The answer varies from one country to another. In the United States there are more than ten times as many deaths per year from heart disease and five times as many deaths from cancers as there are deaths from AIDS. Most persons with AIDS are young (AIDS is the second leading cause of death among 25–44-year-olds), while most people dying of heart disease and cancers are older. Deaths from other infectious diseases remain very high in many countries of the world. Worldwide, 2 million children per year die of measles, for example. There are 8 million new cases and 3 million deaths from tuberculosis per year, compared with 5.8 million new HIV infections and 2.3 million deaths (460,000 deaths among children in 1997).

How does the incidence of AIDS compare with that of other diseases that are transmitted by sexual contact? In the United States and worldwide, the incidence of AIDS is lower than that of many other sexually transmitted diseases (Figure 13.11). Each year in the United States, 5–10 million people are infected with *Chlamydia*, and there are 120,000 new cases of syphilis, with incidence of the latter rising among teenagers and young adults. Also, about 31 million people carry type II herpes simplex (the most common genital herpes virus), and 500,000 new cases are reported to the CDC annually. Except for AIDS and untreated syphilis, most sexually transmitted diseases are not fatal, but they have other serious consequences including (for different diseases) sterility, paralysis, arthritis, and chronic pain in adults as well as severe disease in newborns when transmitted from the mother. Because chlamydia, gonorrhea and syphilis are bacterial diseases, most cases can be successfully treated with antibiotics if treatment is started early enough. In contrast, there is no drug to eradicate herpes, which

FIGURE 13.11

Worldwide incidence (new cases per year) of sexually transmitted diseases (STDs).

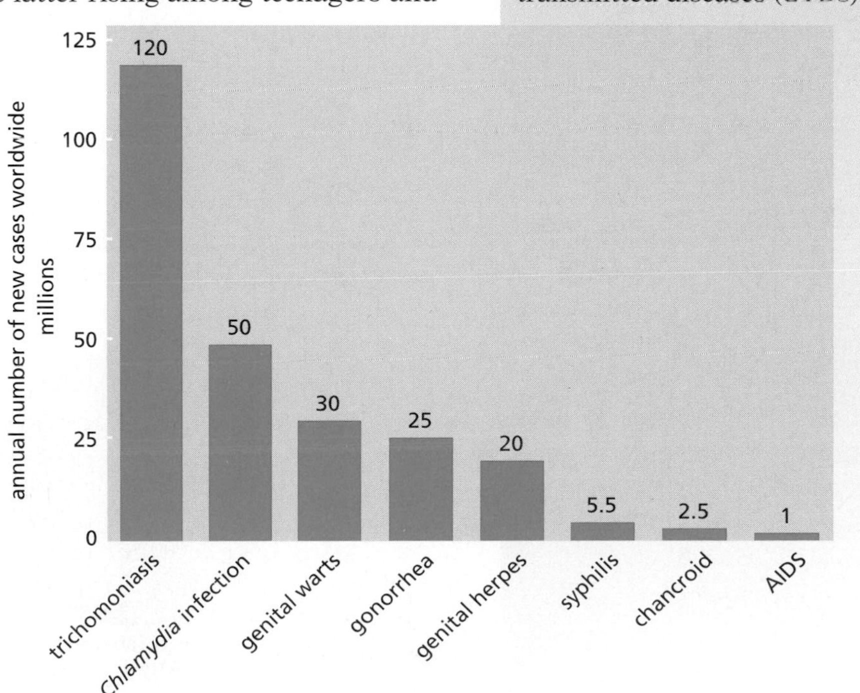

is a virus. Antibiotics put a strong selective pressure on bacterial populations, selecting for mutated variants that are resistant to the antibiotic. Many strains of bacteria, including those that cause syphilis and gonorrhea, are now resistant to many antibiotics, severely limiting the options for treatment of a disease caused by these strains.

People who have contracted any other sexually transmitted disease have already engaged in behavior that puts them at risk for HIV infection. Because properly used condoms can, in general, greatly reduce transmission rates for other sexually transmitted diseases as well as for HIV, someone who has contracted a sexually transmitted disease has probably not used a condom during intercourse. Moreover, they have further increased their risk for HIV infection because the presence of open genital sores greatly increases the probability that contact with HIV will result in HIV infection.

Worldwide HIV incidence. Obviously, statistics on AIDS vary according to the way in which AIDS is defined. The World Health Organization (WHO) criteria for diagnosing someone with AIDS are very different from the CDC criteria. The WHO criteria, based on symptoms, not on HIV status or T cell counts, are used in countries where monetary or technical considerations make testing for HIV infection impossible. Worldwide surveillance of numbers of AIDS cases is therefore not the same as surveillance of HIV infection, which must be estimated from the numbers of AIDS cases (Figure 13.12). Moreover, many people with

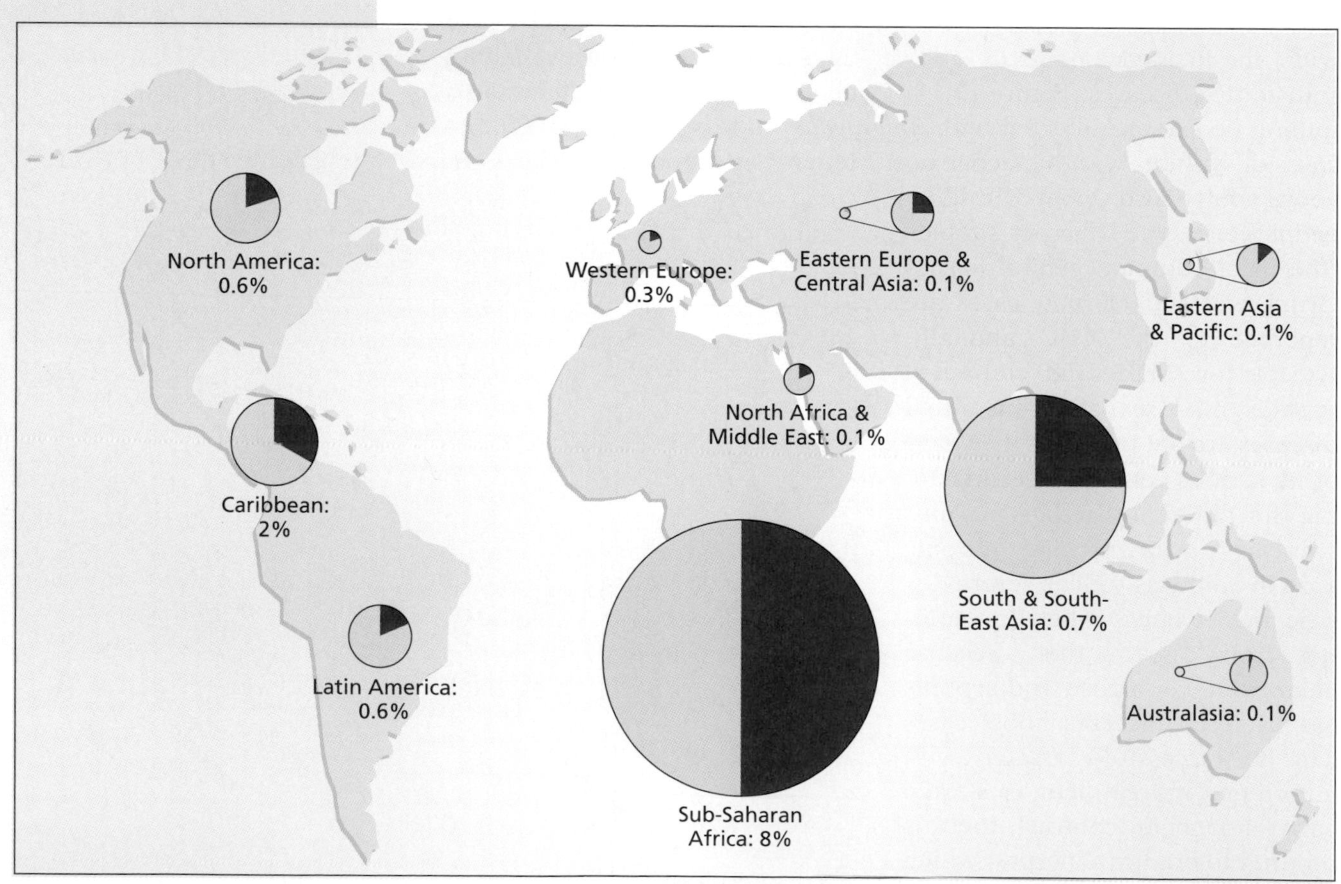

FIGURE 13.12

HIV/AIDS in different parts of the world. The percentages shown on the map indicate the proportions of the adult population (age 15 and older) currently living with HIV/AIDS. The area of each circle is proportional to the total number of people with HIV/AIDS (projected to the end of 1999; the global total is 65.3 million people). In each circle, the red segment represents women with HIV or AIDS, and the blue segment represents men.

AIDS are difficult to distinguish from people who are immunodeficient from other causes, such as undernourishment, and so may or may not be counted as AIDS cases. The numbers may thus underestimate the true AIDS incidence.

Africa has the highest prevalence of AIDS (the number of infected people divided by the total population), and in most areas the rate is still increasing. In the areas with the highest prevalence, such as Uganda, the percentage of the population that is infected has stabilized at around 20%. Infection rates stabilize when the rates of new infections are balanced by the death rate from the disease. HIV infection is not uniformly prevalent throughout Africa. Several countries in West Africa have lower rates owing to early and sustained implementation of prevention programs. Prevalence is also low in North Africa.

The area in which HIV infections are increasing most rapidly is Asia, with Thailand being the most severe. In Africa the rate of infection is about equal in men and women, while in North America, South America, Europe and Australia it is higher in men. In the United States, a majority of infected women are between the ages of 15 and 25, and 80% of newly infected women are either intravenous drug users or sexual partners of intravenous drug users.

Worldwide, half of all new infections are in people aged 15–24. In the United States, this age group has 15% of new HIV infections and 4% of new AIDS cases (1997–1998). However, there is some reason for optimism. From 1996 to 1997, new AIDS cases in the United States declined for the first time since the epidemic began (although the numbers of cases among women increased). The number of AIDS deaths also decreased, by 44%, largely owing to the new protease inhibitor drugs. AIDS was the number one killer of people aged 25–44 in the United States, but it has now dropped to number two. Drug development, and prevention and education strategies, have contributed. Whether the pandemic has turned a corner or merely temporarily slowed will become clear in the coming years.

THOUGHT QUESTIONS

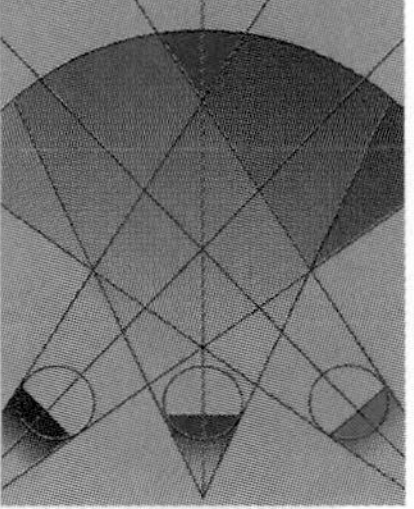

1. What advice would you give to college students about the best ways to avoid getting AIDS? How might you modify the advice for different groups of people?
2. What are the misconceptions surrounding HIV and AIDS? Have they changed over the years?
3. How are the Center for Disease Control and Prevention's criteria for AIDS different from those of the World Health Organization? Why are they different?
4. Is medical research disease-specific? In other words, will the knowledge gained from studying one disease be applicable only to that disease?
5. HIV infection is associated with 'risk behaviors.' To what extent is 'risk' the result of an individual's choice? To what extent do societal factors remove choice from individuals?

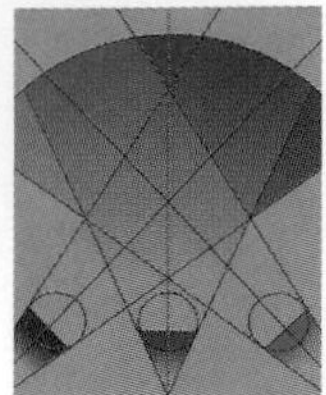

WHAT DOES THE FUTURE HOLD? PROMISING DRUG THERAPIES now exist and more treatments may be on the way. Prevention, however, remains the best hope for the control of HIV infection. Effective prevention programs will require that research scientists, medical professionals, and educators work together, rather than in isolation. Common language must be found in which these groups can communicate with each other and with the people that they serve. Education cannot be unidirectional: professionals educating the people on the street must also learn from those people.

As Jonathan Mann said in his opening remarks to the VIII International Conference on AIDS in 1992,

> We have seen important success in basic and applied research, yet that research in isolation from concern about access to its achievements has severely limited its impact on lives of people with HIV.... If we believe that the entire problem of AIDS is really only about a virus, then we really only need a virucide or a vaccine. Yet if AIDS is deeply, fundamentally about people and society and if societal inequity and discrimination fuel the spread of the pandemic then, to be effective against AIDS, we would have to address these issues.

(Jonathan Mann headed the WHO AIDS office from 1987 to 1990. He was killed in an airplane crash in 1998.)

CHAPTER SUMMARY

- The antibodies and **cytotoxic T cells** of the immune system protect the body from disease, but they are produced only in the presence of the cytokine interleukin-2 secreted by the **CD4 helper T cells**.
- When some part of the immune system fails to work, an **immunodeficiency** results. Immunodeficiency leads to an increased probability and severity of sickness.
- **Human immunodeficiency virus** (**HIV**) is the virus that causes **acquired immunodeficiency syndrome** (**AIDS**) by destroying CD4 helper T cells and other immune cells displaying CD4 molecules.
- **Epidemiology**, **Koch's postulates** and other types of evidence helped to establish HIV as the cause of AIDS.
- HIV is a retrovirus that uses **reverse transcription** to convert its RNA genome to DNA and then uses the host's cellular machinery to replicate itself. There are two strains of HIV (HIV-I and HIV-2), which seem to differ in their **virulence**.
- Microorganisms that are normally kept in check by healthy immune systems can cause serious and possibly lethal opportunistic infections when the immune system is compromised by AIDS.
- Most HIV testing detects antibodies to HIV; some tests detect virus itself. All tests have rates of **false negative** results (related to the **sensitivity** of the test) and **false positive** results (related to the **specificity** of the test) that must be considered when interpreting a test result.

- HIV **transmission** is via behaviors in which blood, semen, or vaginal fluids are passed from an infected person to an uninfected person. Essentially everyone is **susceptible** to HIV infection if they receive body fluids from an infected person. **High-risk behaviors** (those with the highest likelihood of transmission) include unprotected sexual intercourse and sharing injection drug needles.
- Since 1981, AIDS has spread to become a worldwide pandemic, burdening individuals and their loved ones and also public health resources.
- There are currently no cures, although new drugs seem promising as therapies. Vaccines are currently being tested.
- Prevention is key, but depends on many groups of people listening to each other, learning from each other and working together.

KEY TERMS TO KNOW

AIDS (acquired immunodeficiency syndrome) (p. 537)
cytotoxic (CD8) T cells (p. 537)
epidemiology (p. 539)
false negative (p. 555)
false positive (p. 555)
helper (CD4) T cells (p. 537)
high-risk behaviors (p. 563)
high-risk groups (p. 567)
human immunodeficiency virus (HIV) (p. 542)
immunodeficiency (p. 537)
Koch's postulates (p. 542)
reverse transcription (p. 548)
risk (p. 563)
risk behaviors (p. 563)
sensitivity (p. 555)
specificity (p. 555)
susceptibility (p. 567)
transmission (p. 562)
virulence (p. 549)
virus (p. 545)

CONNECTIONS TO OTHER CHAPTERS

Chapter 2: HIV carries out reverse transcription, which produces DNA from an RNA template, the opposite of normal transcription.

Chapter 4: Virulence follows a long-term pattern in its evolution.

Chapter 6: Condoms are useful for both contraception and preventing the transmission of sexually transmitted diseases, including AIDS.

Chapter 8: A decline in appetite (and therefore a decline in nutritional status) often occurs in the late stages of AIDS.

Chapter 9: A healthy immune system kills many cancer cells. Kaposi's sarcoma is an otherwise rare cancer that occurs more frequently in AIDS patients.

Chapter 10: Dementia from loss of brain cells is one of the late-developing symptoms of AIDS.

Chapter 11: The abuse of injectable drugs is a major risk factor for AIDS.

Chapter 12: The immune system that normally protects us from infection is greatly impaired by HIV infection.

PRACTICE QUESTIONS

1. What nucleic acid is in the genome of HIV?
2. Where does the phospholipid bilayer envelope of HIV originate?
3. What does the enzyme reverse transcriptase do? Is reverse transcriptase an enzyme used by the virus, the host cell or both?
4. Why does HIV infect only cells that carry human CD4 molecules on their surface?
5. What cellular machinery of the host cell does HIV use to replicate itself?
6. What criteria are used for saying that a person is HIV-positive?
7. What criteria are used to say that a person has gone from being HIV-infected to having AIDS?
8. Define risk.
9. What behaviors produce the greatest risk of HIV transmission? Why?
10. How do reverse transcriptase inhibitor drugs stop HIV replication?
11. How do protease inhibitor drugs stop HIV replication?
12. What species is/are susceptible to HIV infection? What species is/are susceptible to AIDS?
13. Among humans, who is susceptible to HIV infection?
14. When body fluids such as blood are spilled, can cell-free HIV be killed by soap and water? Why or why not?
15. What are the advantages to an ELISA test for HIV infection and what are the disadvantages? What are the advantages and disadvantages of the Western blot test? The PCR test?

Californian
TOMATO
PUREE
MADE WITH GENETICALLY
MODIFIED TOMATOES
DOUBLE CONCENTRAT

14 Plants and Crops

CHAPTER OUTLINE

Plants Capture the Sun's Energy and Make Many Useful Products

- Plant products of use to humans
- Photosynthesis
- Nitrogen for plant products

Plants Use Specialized Tissues and Transport Mechanisms

- Tissue specialization in plants
- Water transport in plants

Crop Yields Can Be Increased by Overcoming Various Limiting Factors

- Fertilizers
- Soil improvement and conservation
- Irrigation
- Hydroponics
- Chemical pest control
- Integrated pest management
- Altering plants through artificial selection
- Altering strains through genetic engineering

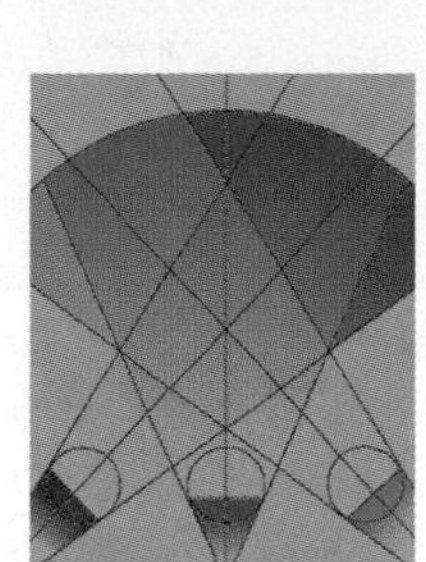

ISSUES

What is plant science?

How has plant science changed the world?

Can plant science feed the world?

Why is the 'law of unintended consequences' so important?

Are genetically engineered plants different from other plants?

BIOLOGICAL CONCEPTS

Photosynthesis

Osmosis

Nitrogen cycle

Plant structure and function (tissues, water transport, gas exchange, seeds and seed dispersal)

Ecosystems (trophic levels: producer, consumer, decomposer; soil; sustainable agriculture; pest control; integrated pest management)

Limiting factors (fertilizers, irrigation)

Plant genetic engineering

Chapter 14: Plants and Crops

Plants are essential to life for humans and most other organisms on Earth. The oxygen that we breathe is produced by plants. The food that we eat comes from plants, either directly or indirectly. Without plants, most other forms of life would soon die out.

Plants are the world's richest energy source. We all use plant energy as food. Wood is the most commonly used household fuel for much of the world's population. Fossil fuels such as coal and petroleum were formed from the remains of plants of past geological ages. On a worldwide basis, the amount of energy produced by plants is about 6×10^{17} kilocalories per year (abbreviated kcal/year), or the equivalent of a sugar cube 5 km (or 3 miles) on each side. The rate of energy use by humans in New York City is estimated at about 10^8 kcal/year per person. If all humans used energy at this rate, then plant life on Earth could support a maximum of 6 billion people, with no energy left over for other organisms. As the world population is 6 billion people now and increasing (see Chapter 6), we are rapidly outgrowing the world's chief energy resource. Population control (see Chapter 6) and energy conservation (see Chapter 15) are among the solutions to this problem. Another solution consists of growing more of the plants used by humans and growing them more efficiently. This requires a knowledge of how plants capture light energy and how we can help them to do so even more efficiently. Plants, in other words, are key to solving the world's energy crisis as well as the world's food shortages. This chapter discusses how plants make energy available to other organisms and how humans, through application of this knowledge, might increase crop yields.

PLANTS CAPTURE THE SUN'S ENERGY AND MAKE MANY USEFUL PRODUCTS

As we saw in Chapter 4, plants are a kingdom of living organisms that have eucaryotic cells containing specialized structures for capturing energy from the sun's light. Plants do not simply capture the energy of light and consume it. They are essential to life on Earth because they capture more energy than they use and convert it into a form that can readily be used by organisms unable to use light energy. And plants do even more: in the course of building and maintaining their own bodies, plants make products that serve many needs of other organisms.

Plant products of use to humans

Humans and other animals find many ways of benefiting from the use of plants. Most importantly, plant products are eaten as foods, that is, as sources of energy and nutrients (see Chapter 8). Many of these foods are the carbohydrates (and some are the oils) that plants have stored for their own use as energy sources for a later time. **Agriculture**, meaning the cultivation of plants as food for humans, is often regarded as the one critical achievement that led also to the establishment of human civilization. The cultivated plants most important in the development of civilization were the cereal grains (wheat, rice, corn, oats, and others). Today, wheat, rice, corn, and potatoes provide more of the world's food than all other crops combined. Of the almost 250,000 species of plants known, some 80,000 are edible by humans. Of these, however, only about 30 form the major crop plants.

People also use plants as sources of beverages, flavorings, fragrances, dyes, poisons, decorations, building materials, and medicines. Beverages made from plants include coffee, tea, cola, beer, wine, spirits, and many juices. Many plant parts are used as spices, fragrances, and flavorings, from barks such as cinnamon, seeds such as black pepper, and roots such as ginger to flowers and flower parts such as cloves, saffron, and vanilla (Table 14.1). Often, an essential oil or other ingredient is squeezed or extracted from the appropriate plant part and used in concentrated form. Some of these extracts are used as fragrances or food ingredients; others are used as animal poisons. For example, roots containing rotenone are used by native South Americans to help in capturing fish by temporarily paralyzing them. Rotenone is also used as a pesticide because it paralyzes and kills insects. Manioc, a tropical root, is the source of both an arrow poison and tapioca, a food that also serves as a thickening agent.

Wood is a commercially important plant product used the world over as a fuel and as a building material in the form of lumber. The history of human civilization would have been very different without spears, axes, hoes, boats, and many other objects made mostly of wood. Also, paper and paper products are made largely from wood.

Our modern arsenal of prescription medicines is derived largely from

TABLE 14.1

A few of the many vascular flowering plants used as spices and fragrances.

PLANT TAXON	PLANT NAME	PLANT USE	PLANT PART USED
CLASS DICOTYLEDONAE			
Mint family	*Mentha* spp.	Mint	Leaves or essential oil
	Ocimum basilicum	Basil	Leaves
	Origanum vulgare	Oregano	Leaves
Pepper family	*Piper nigrum*	Pepper	Seeds
Nightshade family	*Capsicum frutescens*	Chili peppers, paprika	Fruit
Laurel family	*Cinnamomum* spp.	Cinnamon	Inner bark
Myrtle family	*Eugenia aromatica*	Cloves	Whole, unopened flower buds
Ginger family	*Zingiber officianalis*	Ginger	Roots
Mustard family	*Brassica nigra*	Mustard	Seeds
Rose family	*Rosa* spp.	Rose	Essential oil
CLASS MONOCOTYLEDONAE			
Lily family	*Allium sativum*	Garlic	Bulbs
Iris family	*Crocus sativus*	Saffron	Petals
Orchid family	*Vanilla* spp.	Vanilla	Immature seed capsule

NOTES: In many cases, the plant part used is dried and ground into a powder. The abbreviation spp. means that various species in the same genus are used.

plant products, many of them tropical (Table 14.2). Other drugs such as aspirin were originally derived from plants (the bark of willow trees, *Salix* spp.), but now are manufactured synthetically. Willow bark, or tea-like infusions made from willow bark, were used to treat aches and pains for centuries by the Greeks and by Native Americans, among others. Many herbal medicines continue to be used in many parts of the world.

We could compile more lists of human uses of plants. But how exactly do plants trap and convert light energy, the ultimate source of the plant products that we depend on? We proceed to this topic next.

Photosynthesis

Plants use the energy that they capture from the sun to make energy-rich carbohydrates by a process called **photosynthesis**. They store the carbohydrates for their own use as energy sources at a later time. When humans and other animals eat plants, they harvest some of this energy.

TABLE 14.2

A few of the many medicinal plants.

PLANT		DRUG	
LATIN NAME	COMMON NAME	NAME	USE
Digitalis purpurea	purple foxglove	digitalis	strengthens heart contractions
Rauwolfia serpentia	India snakeroot	reserpine	lowers blood pressure
Atropa belladonna	deadly nightshade	atropine	blocks neurotransmitters, antispasmodic
		belladonna	blocks neurotransmitters, antispasmodic
Datura spp.	Jimson weed (thorn apple)	scopalamine	sedative, controls nausea
Papaver somniferum	opium poppies	codeine	cough suppressant, pain killer
		morphine	pain killer
Cinchon ledgeriana	cinchona tree bark	quinine	malaria prevention
Catharanthus roseus	Madagascar rose periwinkle	vinblastine	cancer chemotherapy
		vincristine	cancer chemotherapy
Taxus brevifolia	Pacific yew	taxol	cancer chemotherapy

Energy producers and energy harvesters. Organisms, such as plants, that can use light or other inorganic energy sources to make all of their own organic (carbon-containing) molecules from simpler molecules are called **autotrophs**. Nearly all plants use light energy from the sun as their energy source, making them **phototrophs**. Because they are both autotrophs and phototrophs, we can call them photoautotrophs. Most of the species in all the other kingdoms are **heterotrophs**, organisms that cannot make their own organic compounds from inorganic materials and are therefore absolutely dependent on the organic compounds made by plants and other autotrophs. The energy source used by most heterotrophs is chemical bond energy, making them **chemotrophs**; because they are both heterotrophs and chemotrophs, we can call them chemoheterotrophs.

Most autotrophs are also called **producers** because they produce compounds usable by other organisms, including ourselves. Most heterotrophs are **consumers** that must obtain their energy by eating other organisms. Plants are the ultimate source of food energy not only for the primary consumers that eat plants directly, but also for the secondary and higher-order consumers that eat other consumers. When organisms die, their complex organic molecules are broken down by other heterotrophs called **decomposer** organisms; the breakdown products can then be recycled and used by other living things.

Energy enters the biological world as sunlight and flows through producer, consumer, and decomposer organisms in turn (Figure 14.1). At each step, some energy is lost by being converted into forms that are not usable by organisms. Most of this unusable energy escapes as heat. Because of the loss of heat, the process would run down and stop altogether unless new energy were continually supplied. The new energy for the living world is sunlight. Plants are essential to the global energy flow because they are the principal means by which light energy is captured and changed into forms that other organisms can use.

Energy and pigments. By the process of photosynthesis, plants gather energy from sunlight and use this energy to make carbohydrates (sugars and starches; see Chapter 8, pp. 329–330) from atmospheric carbon dioxide and water. The overall process can be summarized by the following equation:

$$\underset{\text{carbon dioxide}}{6CO_2} + \underset{\text{water}}{12H_2O} \xrightarrow{\text{light energy}} \underset{\text{glucose (a sugar)}}{C_6H_{12}O_6} + \underset{\text{oxygen}}{6O_2} + \underset{\text{water}}{6H_2O}$$

The capture of light energy for photosynthesis takes place in certain light-sensitive molecules called pigments, of which **chlorophyll** is the most important. These molecules absorb some wavelengths of light and not others. Chlorophyll absorbs blue and red light but not green light. The colors that we see are those that are reflected rather than absorbed, which is why so much of the living world looks green (Figure 14.2). In addition to chlorophyll, plants and other photosynthetic organisms possess various other pigments, such as carotenes and xanthophylls, which absorb light of other colors and pass the energy on to chlorophyll. These pigments are useful to the organism because they enable it to use light energy of different wavelengths. Because many of these **accessory pigments** are found only in certain groups of photosynthetic organisms, they are used to identify these groups and reconstruct their evolution. For example, similarity between the pigments of green algae,

FIGURE 14.1

Energy flow through a biological system. Energy enters as sunlight. Producers convert the sunlight to chemical bond energy usable by other organisms, the consumers. Energy locked in the chemical bonds of dead producers and consumers is released by decomposer organisms. Energy is also lost as heat at each step, so more energy must continuously enter the system for the process to continue.

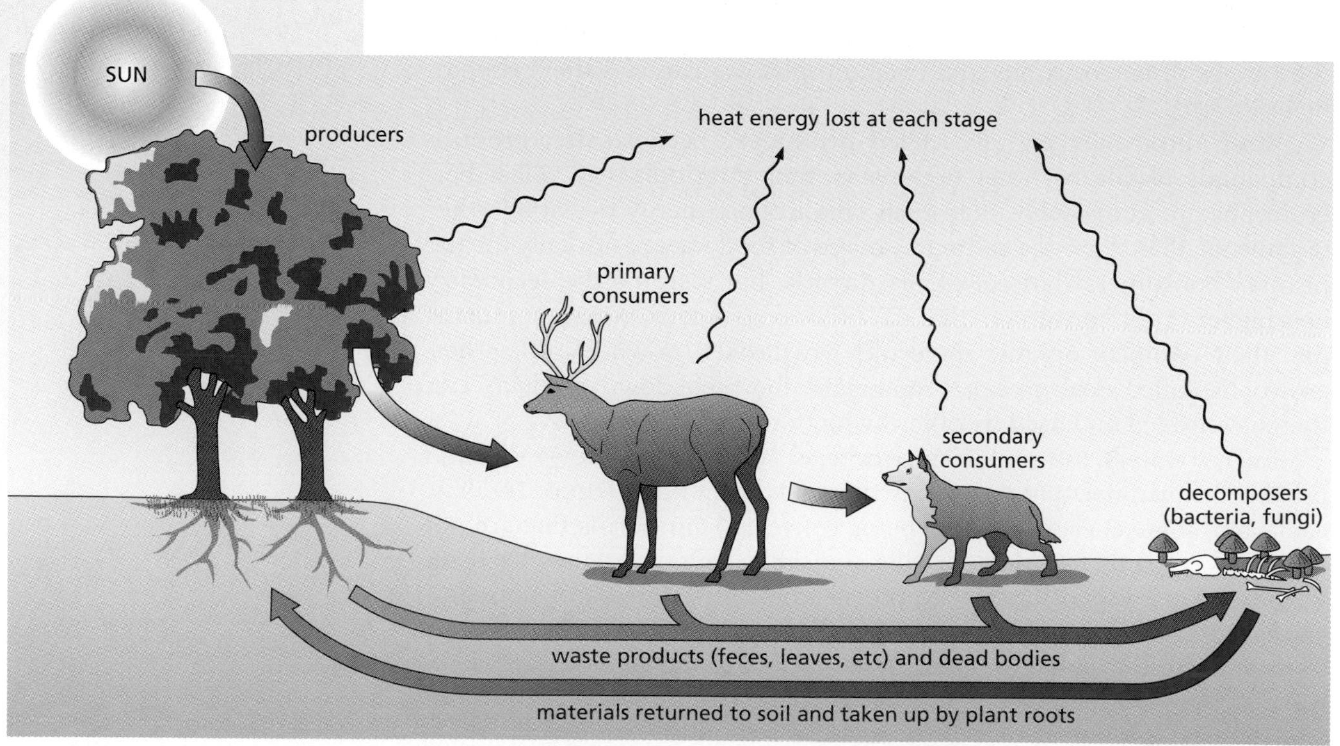

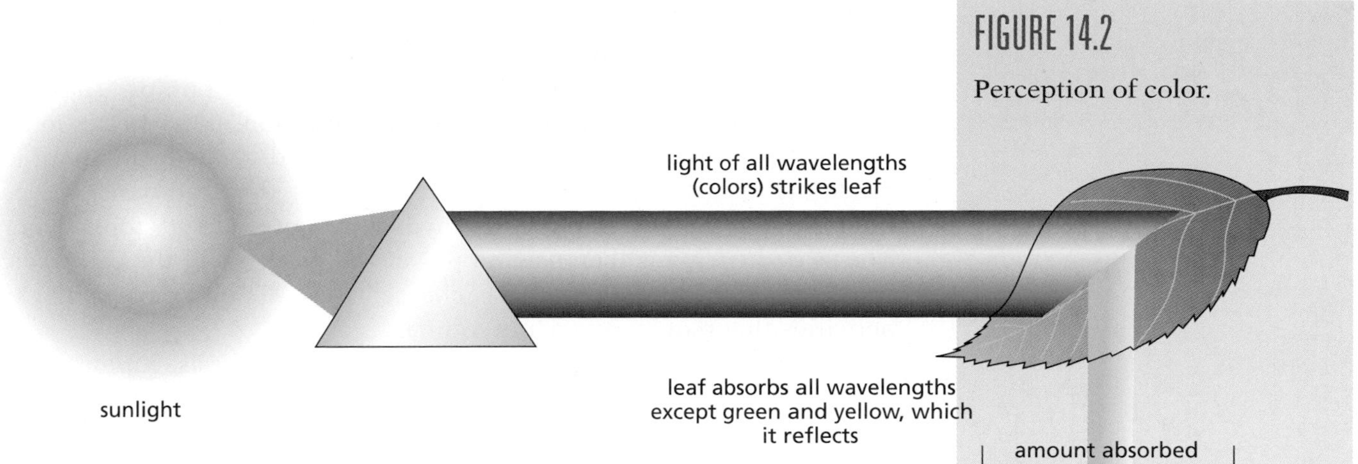

FIGURE 14.2

Perception of color.

bryophytes, and vascular plants has been used to argue that bryophytes and vascular plants probably evolved from green algae (see Chapter 4).

Many of the broad-leaved trees of temperate regions stop making chlorophyll in the autumn. In the absence of chlorophyll, the other pigments in the leaves become apparent. These pigments absorb green and blue and reflect many other colors of light, resulting in the fantastic rainbow of leaf colors in the fall foliage (see Figure 1.1, p. 7).

In all photosynthesizing organisms that have a eucaryotic cell structure (plants, including algae; see Chapter 4), the photosynthetic pigments are contained in organelles known as **chloroplasts**, which are located within the cells of the green parts, especially leaves. The fluid interior of these chloroplasts, called the **stroma**, contains stacks of flattened membrane vesicles called **thylakoids** (Figure 14.3). Photosynthesis takes place along the membranes of these thylakoids.

Although most photosynthesizing organisms are plants, there are also a few photosynthesizing species of true bacteria and many photosynthesizing species of cyanobacteria. These simpler photosynthesizing organisms have a procaryotic cell structure (see Box 4.1, p. 156) with no internal membranes or chloroplasts, so photosynthesis in these organisms takes place along the cell plasma membrane and in the cytoplasm.

Photosynthesis: light reactions. Photosynthesis takes place in two stages. The first stage consists of the reactions that take place in the membranes of the thylakoids; these require light energy and so are called **light reactions**. Light energy, represented by the gold rays in Figure 14.4, is captured by the photosynthetic pigments. The captured energy is used to split water molecules into hydrogen and oxygen. The oxygen is released to the atmosphere, where it is useful to oxygen-dependent organisms, including humans. Each hydrogen atom is further split into a hydrogen ion (also called a proton or H^+) and an electron (e^-). The electrons move down a chain of electron transport proteins in the thylakoid membrane (as shown by the turquoise arrows in Figure 14.4) and are ultimately delivered to a molecule called $NADP^+$, forming NADPH.

NADP+ is an energy-carrying molecule that is made in part of niacin, a vitamin described in Chapter 8, (p. 344).

Some of the electron transport proteins are ion pumps. These use some of the electrons' energy to pump hydrogen ions (gray circles in Figure 14.4) into the interior space of the thylakoids, thus forming an ion gradient. As with other ion gradients, such as those across nerve cell membranes (see Chapter 10), the hydrogen ion gradient stores energy. The stored energy is then used to power the synthesis of the energy-rich molecule ATP (see Figure 14.4). The cell uses ATP for most biological activities that require an energy source.

In summary, in the light reactions of photosynthesis, captured light energy is converted into high-energy chemical bonds: the bonds in NADPH and the phosphate bonds of ATP. This stored, light-derived energy is later used in the dark reactions that form the remainder of photosynthesis. Some light energy is also unavoidably transformed into heat.

Photosynthesis: dark reactions. The ATP and NADPH created in the thylakoids are not released from the chloroplasts; they move from the thylakoids into the stroma of the chloroplasts. In the stroma, the ATP and NADPH participate in the second of the two stages of photosynthesis. The ATP supplies energy and the NADPH provides hydrogen for the synthesis of glucose ($C_6H_{12}O_6$) from carbon dioxide (CO_2), the source of the carbon and oxygen atoms. Because the reactions that use ATP and NADPH do not directly require light, they are called the **dark reactions**. The net outcome of the dark reactions is that atmospheric carbon dioxide is 'fixed,' or incorporated, into plant organic material, first as the sugar glucose. Figure 14.5 summarizes the overall process of photosynthesis.

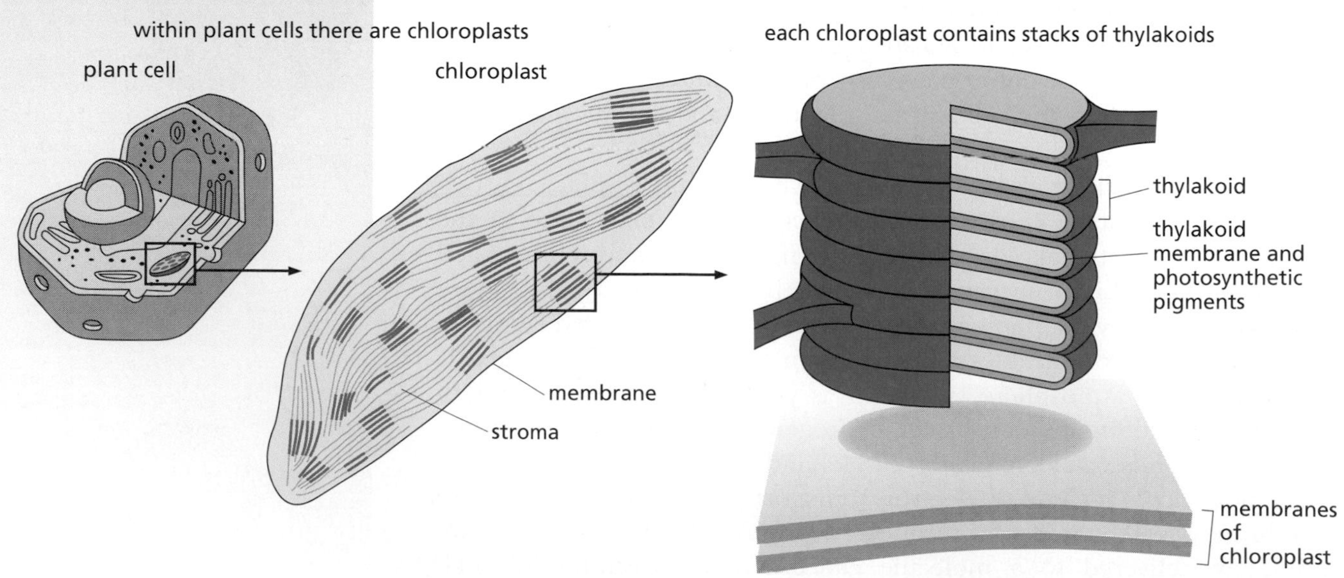

FIGURE 14.3

Location of photosynthesis in plant cells.

Under most conditions, glucose is immediately used as an energy source in metabolism or converted into sucrose, fructose, starch, or other carbohydrates (see Chapter 8, pp. 329–330) for long-term energy storage. If the energy is later needed by the plant at a time when photosynthesis is not possible or in a nonphotosynthetic part of the plant, storage compounds in plants can be converted back into glucose and broken down to supply energy. Table sugar is sucrose, the storage product found in sugar cane and sugar beets. Starch is the storage product in potatoes and cereal grains. In addition to these carbohydrate storage molecules, many plants (including corn, palm, and most nuts) use oils (see Chapter 8, pp. 331–332) as storage products in their seeds; the energy stored in seeds is used when the seed germinates. Growth of the new plant depends on energy from the seeds until enough new leaves have been produced to carry on photosynthesis for themselves.

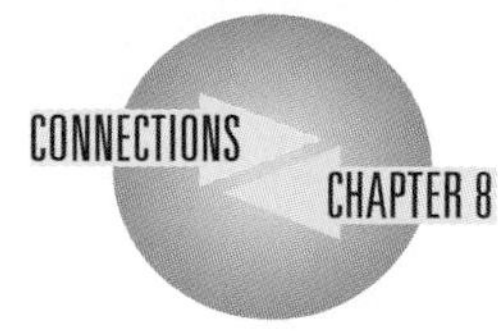

Plant growth requires other molecules besides carbohydrates and oils. Like all other organisms, plants must synthesize nucleotides for DNA and RNA and amino acids for structural and enzymatic proteins. Synthesis of these molecules requires nitrogen. Of the several chemical elements that plants need to make their essential biological molecules, carbon can be obtained by photosynthetic plants from atmospheric carbon

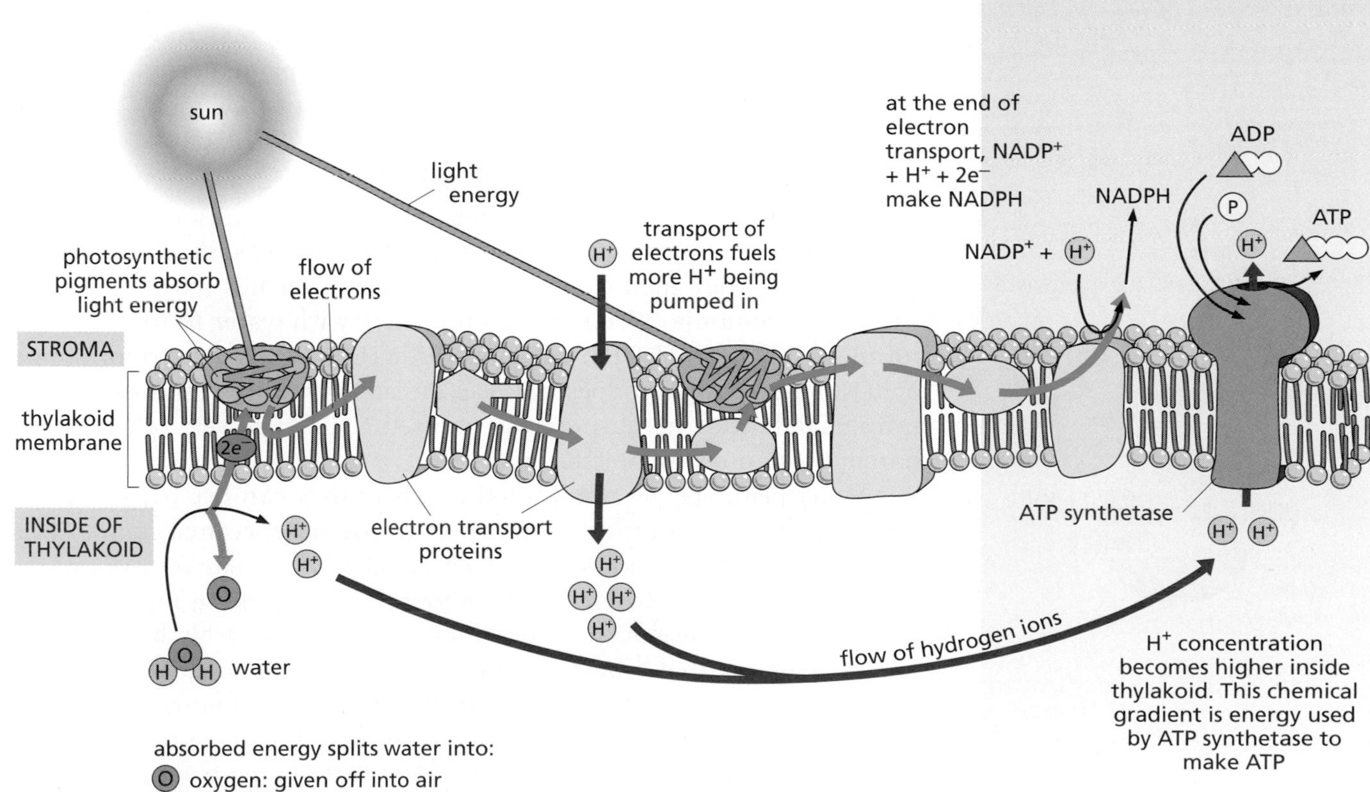

FIGURE 14.4

The light reactions of photosynthesis. The flows of energy, electrons, and hydrogen ions are shown, as are the syntheses of ATP and NADPH, both of which contain high-energy chemical bonds.

dioxide, and water can serve as the source for hydrogen and oxygen as well as dissolved ions. The source of nitrogen, however, is somewhat more complex. In the next section we see how plants obtain the nitrogen they need.

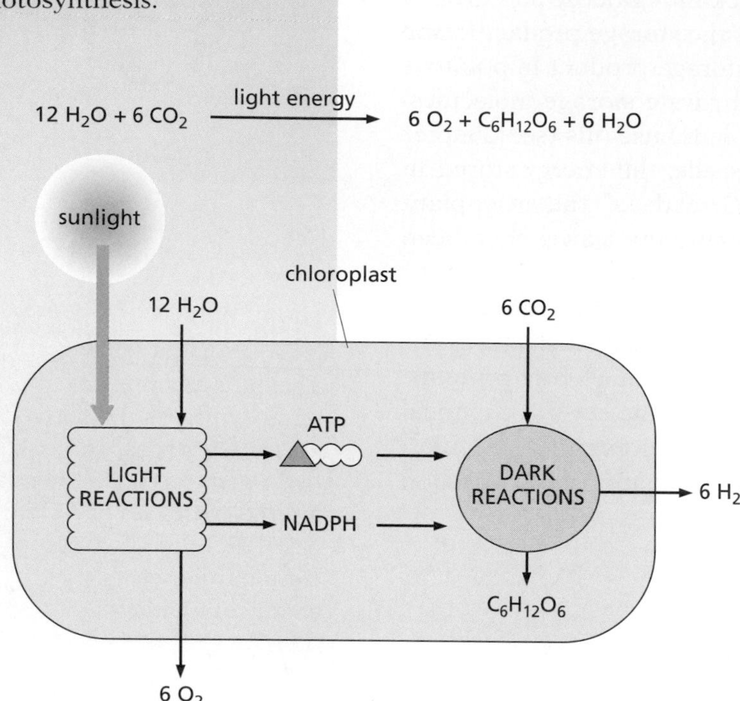

FIGURE 14.5

Summary of the reactions of photosynthesis.

Nitrogen for plant products

The carbohydrates produced by photosynthesis contain carbon, hydrogen, and oxygen only. Many other plant products contain other elements, notably nitrogen. Because proteins are made from amino acids containing nitrogen, the synthesis of proteins requires significant quantities of nitrogen. The simplest amino acids can be made from compounds in the Krebs cycle (see Chapter 8, pp. 339–340) by the addition of an amino group ($-NH_2$). The amino group can be supplied from soluble ammonium compounds (containing the NH_4^+ ion). Amino groups can also be transferred from one amino acid to another.

Once nitrogen-containing amino acids have been synthesized, they can be used as the starting materials for all the other nitrogen-containing compounds that the plant needs, including nucleic acids, vitamins, plant hormones, and pigments.

Nitrate: a limiting nutrient. Although plants need a source of amino groups to help make proteins, most plants cannot absorb NH_4^+ ions directly. Instead they get their nitrogen from the soil as dissolved nitrates (NO_3^- ions). They then convert NO_3^- first into nitrites (NO_2^- ions) and then into ammonia (NH_3). Ammonia reacts with water to form ammonium hydroxide (NH_4OH), the source of NH_4^+, as it is needed for amino acid synthesis. A plant does not accumulate any more ammonia than it immediately uses because excess, unused NH_4OH is very alkaline and damaging to most organic tissues.

Many plant species are limited in where they can live by the availability of dissolved nitrates. A nutrient whose absence makes a species unable to grow in a particular place or on a particular food source is called a **limiting nutrient**. If the amount of the limiting nutrient were increased (by the use of fertilizers, for example), more growth would take place, assuming that other nutrients were present in adequate amounts. For many plants in many places, nitrates are a limiting nutrient.

Are nitrates scarce in many places because Earth's supply of nitrogen is low? Where does the nitrogen in nitrates come from and how is it incorporated? To find answers to these questions we need to take a global view.

The nitrogen cycle. Nitrogen moves through the world's living and non-living systems, changing from one molecular form to another, forming a loop called the **nitrogen cycle**. Plants and other living organisms are an important part of this cycle: chemical end products released by one type of organism are used in biochemical reactions by other organisms. Some of these organisms can fix (incorporate) atmospheric nitrogen into molecular forms that other organisms can use, while other organisms release nitrogen as a gas into the atmosphere. In each ecosystem, living organisms are thus united with each other and with their physical surroundings (including the atmosphere) by the nitrogen cycle. The nitrogen cycle is a good example of a **nutrient cycle**; all such cycles are loops in which materials such as nitrogen are exchanged among producers, consumers, decomposers, and their surroundings, including the atmosphere.

The key stages of the nitrogen cycle are shown in Figure 14.6. Follow first around the outer circle, starting with free nitrogen (N_2) in the atmosphere.

Earth has an abundant source of nitrogen in atmospheric nitrogen gas (N_2). Plants need nitrogen, but, like other eucaryotic organisms, they are unable to make the enzymes necessary to use nitrogen from the

FIGURE 14.6

The nitrogen cycle.

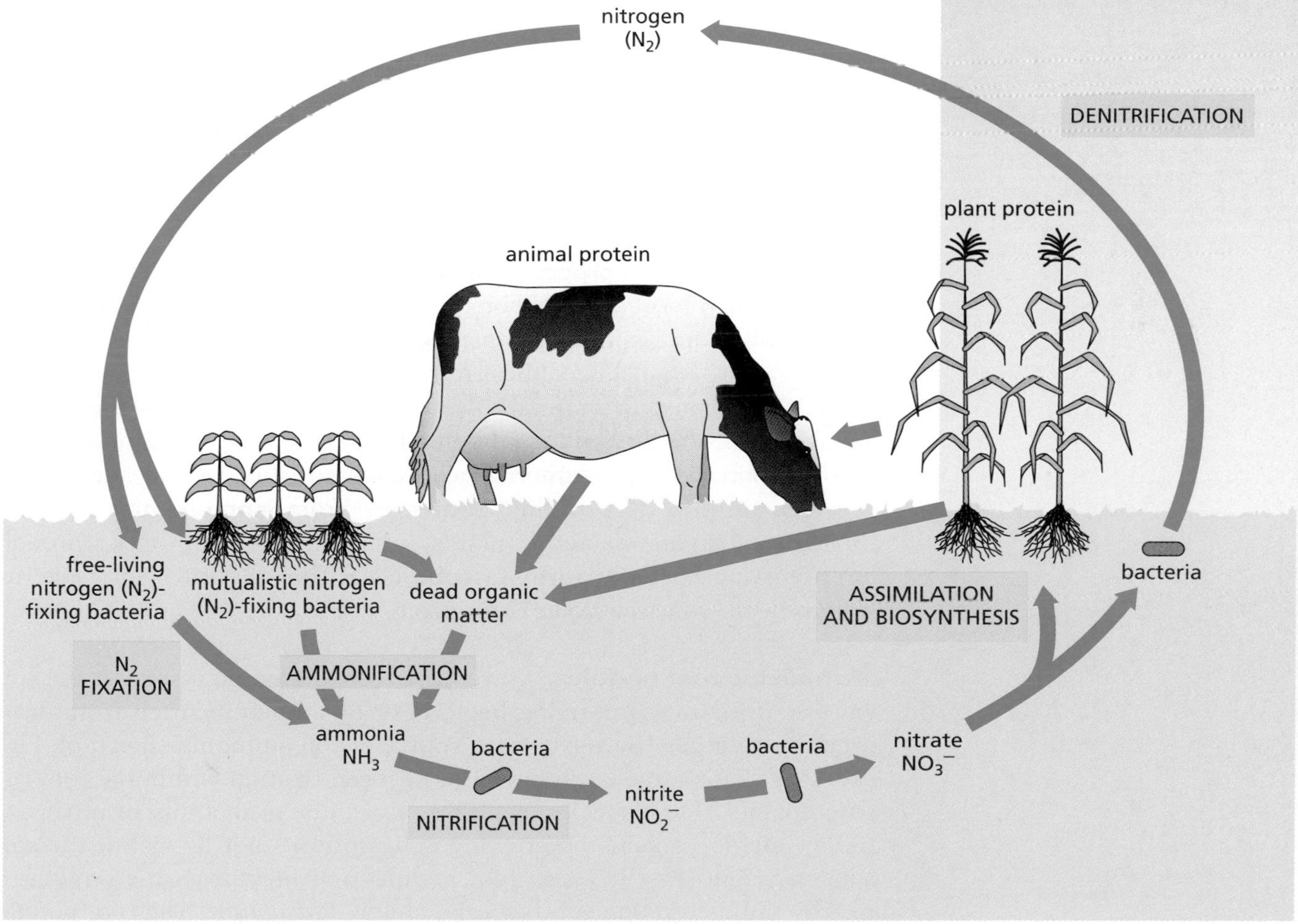

atmosphere. They are therefore dependent upon certain procaryotic microorganisms that have the ability to convert atmospheric nitrogen into ammonia (NH_3). Incorporation of atmospheric nitrogen into ammonia is called **nitrogen fixation**. There are several kinds of nitrogen-fixing organisms in the soil, including many cyanobacteria (see Chapter 4, p. 159) such as *Nostoc* and many true bacteria such as *Rhizobium*, *Azotobacter*, *Klebsiella*, and *Clostridium*.

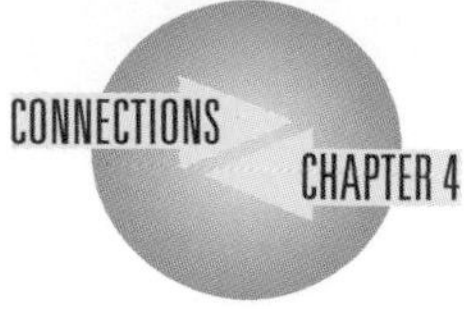

Most plants, however, cannot take up ammonia; they can only take up nitrogen as nitrite (NO_2^-) ions or as nitrate (NO_3^-) ions. Ammonium compounds produced by nitrogen-fixing bacteria in the soil can be converted into nitrites and then into nitrates by still other bacteria, in a process called **nitrification**. Bacteria such as *Nitrosomonas* convert ammonia to NO_2^- and obtain their energy from this conversion process. *Nitrobacter*, another bacterium, then converts the nitrites into nitrates. Plants can absorb both nitrites and nitrates. After the nitrites and nitrates have been absorbed (a process called **assimilation**), plants convert NO_2^- or NO_3^- back into NH_3 to be used in the **biosynthesis** (biological synthesis) of proteins and nucleic acids.

The nitrogen cycle is completed by still other soil bacteria in a process called **denitrification**. Excess NO_3^- that is not absorbed by plants is converted by denitrifying bacteria into nitrogen (N_2), which is then released into the atmosphere. In a balanced nitrogen cycle, the amount of nitrification equals the amount of denitrification. In waterlogged situations, where conditions may be anaerobic (lacking in oxygen), the balance may be lost. Denitrification is still possible because those bacteria do not require oxygen to live, but the nitrifying bacteria, which do require oxygen, die off. Denitrification then outstrips nitrification, depleting the soil of nitrates.

Thus, in the nitrogen cycle as a whole, N_2 is taken from the atmosphere, cycled through many organisms, and returned to the atmosphere. Most of these processes are carried out by bacteria. Nitrogen also cycles through more complex organisms, as when plants take up nitrites and nitrates and use the nitrogen to make amino acids, proteins, and nucleic acids. Animals get their nitrogen by eating plants or other animals. When plants and animals die or give off waste products, their nitrogen is returned to the soil by ammonifying bacteria. **Ammonification** is a second source of the ammonia needed for nitrification, in addition to nitrogen fixation. The cycling of nitrogen from ammonifying bacteria, through nitrifying bacteria, plants, and animals and back through ammonifying bacteria, forms a subcycle within the nitrogen cycle, as shown by the inner circle in Figure 14.6.

Mutualistic root nodules. As we have seen, ammonia can enter the soil via free-living nitrogen-fixing bacteria or by ammonification from dead organic material. There is a third source of soil ammonia shown on Figure 14.6; that source is via bacterial nitrogen fixation within the roots of some plants. Some species of plants ensure the availability of nitrogen-fixing microorganisms by growing **root nodules** not far below the soil surface (Figure 14.7). These root nodules actively attract the growth of nearby nitrogen-fixing soil bacteria, chiefly *Rhizobium*. The root nodule serves as a culture chamber for these microorganisms that then carry out nitrogen fixation inside the plant roots.

The relationship between the two organisms is called **mutualism**, an interaction from which both species benefit. Bacteria produce NH_3, which, because it is already inside the plant, is taken up directly without the need for conversion to NO_2^- or NO_3^-. The plants thus benefit from the mutualism because they have a built-in supply of nitrogen for biosynthesis. Bacteria have to spend a lot of energy to fix nitrogen. The reaction that fixes the nitrogen uses a great deal of energy, as does the synthesis of the nitrogenase enzyme needed to carry out the reaction. The bacteria benefit from the mutualism because they get the energy for the synthesis of nitrogenase and for nitrogen fixation from the plant in the form of organic compounds.

FIGURE 14.7

Underground nitrogen fixation in root nodules.

Plants of the family Leguminosae, including beans, peanuts, peas, locusts, and alfalfa, are all capable of adding nitrogen compounds to the soil because of their mutualistic association with nitrogen-fixing *Rhizobium* bacteria. This is the basis for the use of **green manure**, the planting and subsequent plowing under of a legume crop in a field that has been depleted of nitrates by the earlier growing of plants that absorb large quantities of nitrates from the soil. When the legume is plowed under, the nitrogen compounds that it has produced with the help of its mutualistic nitrogen-fixing bacteria are returned to the soil in a form that nitrifying bacteria can convert to nitrates. Green manure thus adds nitrogen compounds to the soil, reducing or eliminating the need for the addition of fertilizers containing nitrate or ammonia.

Most plants do not have mutualistic nitrogen-fixing bacteria living in their roots. They therefore need to absorb their nitrogen compounds from the soil, principally as nitrates. A few plant species, however, have evolved other means of obtaining nitrogen.

Plants living in nitrogen-poor soils. Plants that have evolved to live in nitrogen-poor soils have a number of different ways of coping with the scarcity. Some plants can absorb the ammonium ions (NH_4^+) that form in wet, decaying leaf litter, but they do so by exchanging the NH_4^+ ions for hydrogen ions (H^+), making the soil more acidic. Microbial decay of the leaves of the plants releases even more acid, eventually killing all those plant and microbial species that are not acid-tolerant. The bacteria that convert ammonium ions to nitrites and then to nitrates are not acid-tolerant and thus are inhibited in this environment. The habitat that results is an acid bog with high organic content but few nitrates.

One rather unusual solution to the problem of obtaining nitrogen in an acid bog is to digest animal proteins. Few plants, however, can be so fortunate as to have an animal die and leave its carcass in the soil just within reach of their root system. How, then, are they to obtain animal protein? Carnivorous plants have evolved adaptations to trap and kill

FIGURE 14.8

Plants that obtain nitrogen from animals.

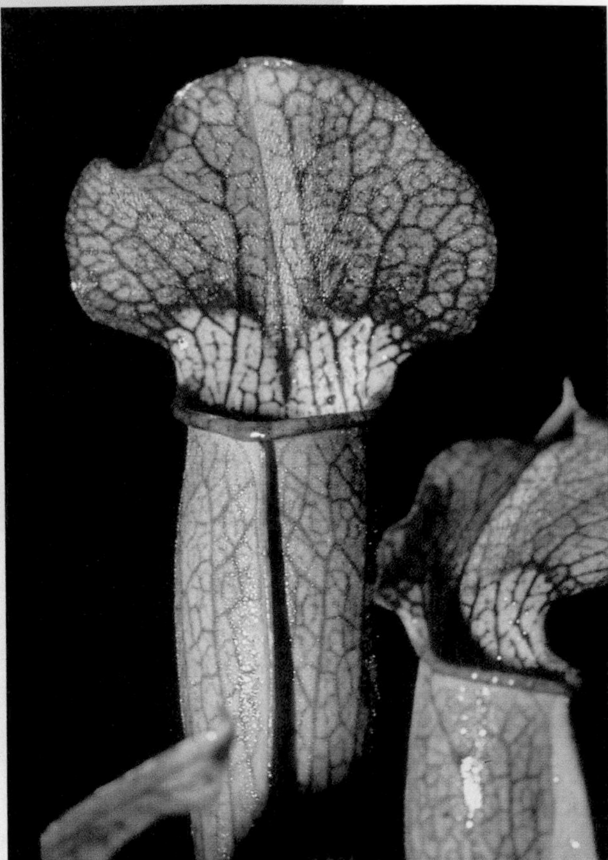

Pitcher plant (*Sarracenia oreophila*), showing a passive trap (pitfall type), with slippery inside surfaces and fluid pool at bottom containing digestive enzymes.

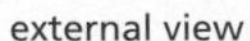

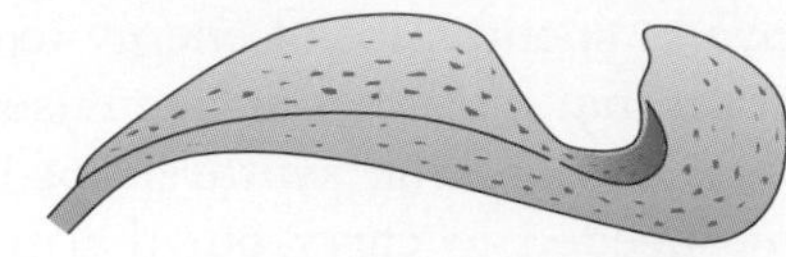

cutaway view

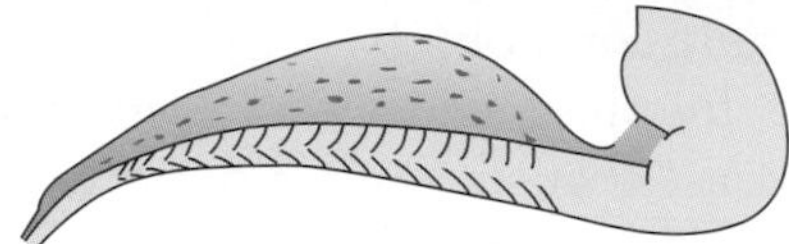

Sarracenia psittacina, a pitcher plant with two passive traps. The entrance works like a lobster trap because insects that have entered the chamber have difficulty finding the opening by which they came in. When they crawl into the long tube, the hairs inside form a second trap that allows them to crawl deeper in only one direction.

The sundew plant, *Drosera*, whose sticky hairs close over an insect, forming an active trap (flypaper type).

small animals (mostly insects) and derive nitrogen from the digested proteins. Most carnivorous plants live in such nitrogen-poor habitats as acid bogs, where moisture and insects are both usually abundant. A variety of mechanisms have evolved by which these plants trap their prey, digest their proteins, and absorb the resulting amino acids. The traps can either be passive or active (Figure 14.8). Passive mechanisms can include pitfalls (as in pitcher plants), sticky materials that form a passive flypaper, and traps shaped like lobster pots. Active mechanisms can include trapdoors, active flypapers (such as the sundew, whose hairs bend to enclose and further hold their victim), or pads that swing shut when an object touches them (such as the Venus flytrap; see also Figure 14.15).

It is clear that some plants have evolved some intriguing structures to obtain nitrogen, while others have developed complex mutualisms and niches within the nitrogen cycle. Most nitrogen, as well as water, comes in through the roots of the plants; most carbohydrates are made by photosynthesis in the leaves. Yet all of these compounds are needed throughout the plant. In the next section, we see how plants have specialized to carry out different functions in different parts of the plant, much as animals have, and how molecules are transported around the plant.

THOUGHT QUESTIONS

1. Consider the value judgements that may be hidden in the term *useful* (as in 'useful products'). Are plants that are useful to humans more valuable than plants that are not? Is utility the only criterion for making something valuable? Is economic value the only way of measuring value? What might some of the other ways be? How would you measure them?
2. What are some ways in which large plants (e.g., trees) are used by other plants? What are some ways in which plants are used by animals? List as many possible uses as you can.
3. Do animals have any molecules that can absorb light energy? Where might you expect to find them?
4. In addition to hydrogen ions pumped during electron transport in photosynthesis, what other ions do you know of that are pumped across membranes? What functions do these other ion gradients serve? In what ways are all ion gradients the same?
5. Some producer organisms are autotrophs, but are all producers autotrophs? Can a consumer organism also be considered to be a producer organism if it is eaten by another consumer organism?
6. Why do you think plants can take up NO_3^- or NO_2^- ions but not NH_4^+ ions?
7. Some plants have mutualistic associations with bacteria in root nodules. Do humans also have mutualistic associations with bacteria?
8. What is the definition of acidity? How does the uptake of NH_4^+ ions by some plants make the soil around them more acidic?

PLANTS USE SPECIALIZED TISSUES AND TRANSPORT MECHANISMS

As we have seen, in most plants photosynthesis is carried out principally in the leaves. Other plant parts are also specialized for particular functions. Roots are specialized for water absorption and fruits for reproduction and dispersal. The division of labor among different parts of the plant would not be possible without the specialization of plant tissues, nor would it be possible without efficient mechanisms for the transport of materials from one part of the plant to another. These adaptations are described in this section.

FIGURE 14.9

Different functions carried out by specialized parts of vascular plants. Detail at the top shows a cross-section through a leaf.

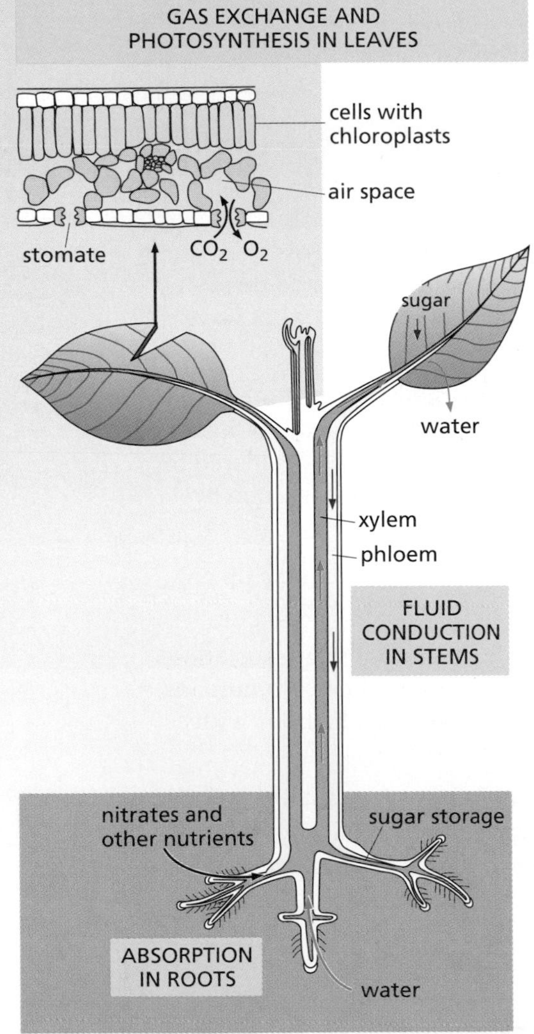

Tissue specialization in plants

In many algae and other simple plants, each cell carries out its own photosynthesis, absorbs its own nutrients, and gets rid of its own waste products. As was true for the evolution of animals, the evolution of plants on land has been characterized by an increasing specialization of parts, each for a different function. Most plants, like animals, have groups of similar cells organized into **tissues**, and groups of tissues organized into **organs** such as leaves and roots. For example, each leaf is an organ, while each cell layer within a leaf is a tissue. The simplest plants containing separate types of tissues are the mosses, liverworts, and hornworts, often grouped as Bryophyta and also known as nonvascular plants.

Tissues and organs of vascular and nonvascular plants. The most familiar and ecologically dominant group of plants are the **vascular plants**, including all plants that have vascular (conducting) tissues that transport fluids, generally through tubular cells surrounded by rigid cell walls. Vascular plants contain two types of conducting tissue: **xylem**, which usually conducts water and minerals upward, and **phloem**, which conducts a water solution of photosynthetic products (mostly sugars) in both directions but more often downward. The existence of these vascular tissues allows the parts of the plant to specialize (Figure 14.9). The **roots** grow underground, anchor the plant in the soil, and absorb water and dissolved nutrients. The vascular tissues conduct the water and nutrients from the roots through the xylem of the **stem** to the above-ground parts, where the water is needed for both photosynthesis and support of the upper parts of the plant. The roots can receive photosynthetic products through the phloem from chlorophyll-containing tissues above. Roots thus need not contain chlorophyll or carry out photosynthesis themselves, so they are not green and do not require light.

Vascular tissues have rigid cell walls, and, as we see in the next section, water pressure makes plant tissues even stronger. Thus, in addition to transport, these adaptations in stems allow many vascular plants to grow tall without having the skeletons that support vertebrate animals.

In most vascular plants, photosynthesis takes place only in specialized organs called **leaves**. The middle tissue layer of each leaf features many chloroplast-rich cells, where photosynthesis takes place, and air spaces that permit the diffusion of CO_2 into the photosynthesizing cells and O_2 out of them, a process called gas exchange (see Figure 14.9).

In contrast to vascular plants, nonvascular plants such as mosses and liverworts do not have deep underground parts because all parts of the plant carry out photosynthesis and therefore need to be in the light. They cannot grow very tall because they lack vascular tissue that would conduct fluids and because they lack the roots and stems that would provide anchorage and support.

Specializations of flowering plants. The most highly evolved vascular plants reproduce with the aid of **seeds**, which contain small diploid embryos capable of being dispersed to new locations away from the parent plant. These plants have branching roots and leaves with multiple veins. The largest and most diverse, as well as ecologically dominant, group of seed-producing plants are the **angiosperms**, or flowering plants (Angiospermae or Anthophyta). The seeds of angiosperms develop within elaborate reproductive structures called **flowers** (Figure 14.10). Eggs, each with a haploid set of chromosomes, are produced by the female part of the flower within the ovary. The male part of the flower produces haploid gametes (sperm) within pollen grains in the anthers. Pollination is the introduction of the pollen onto the stigma, the female receptive surface of the flower (see Figure 14.10). A pollen tube grows from the pollen grain to the ovary, where the sperm fertilizes the egg, resulting in a diploid zygote. Many flowers are pollinated by wind, but a much larger

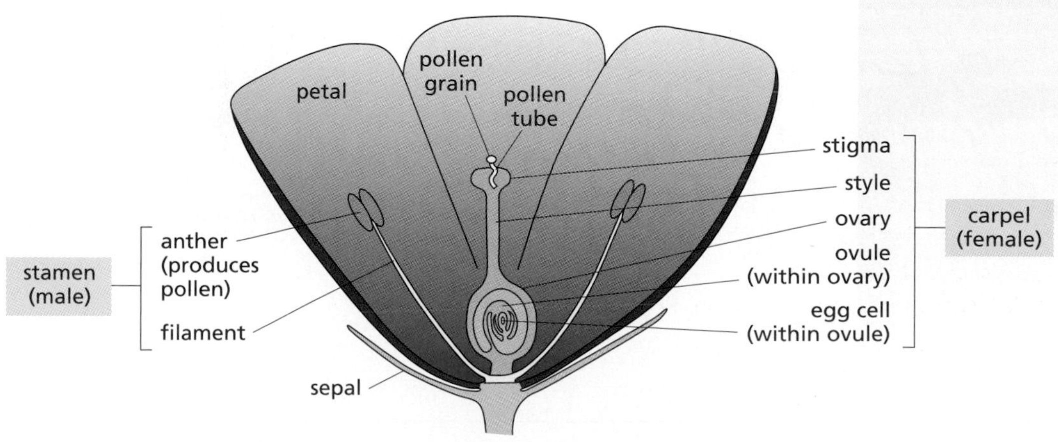

FIGURE 14.10

Diagram of a complete flower, containing both male and female parts together. After fertilization and ripening, the ovary becomes a fruit. Variations in the number and structure of the parts shown here are among the most useful characters in plant classification. Some plants have incomplete flowers in which there are separate male flowers with undeveloped female parts and female flowers with undeveloped male parts.

FIGURE 14.11

The dispersal of seeds in different kinds of fruits.

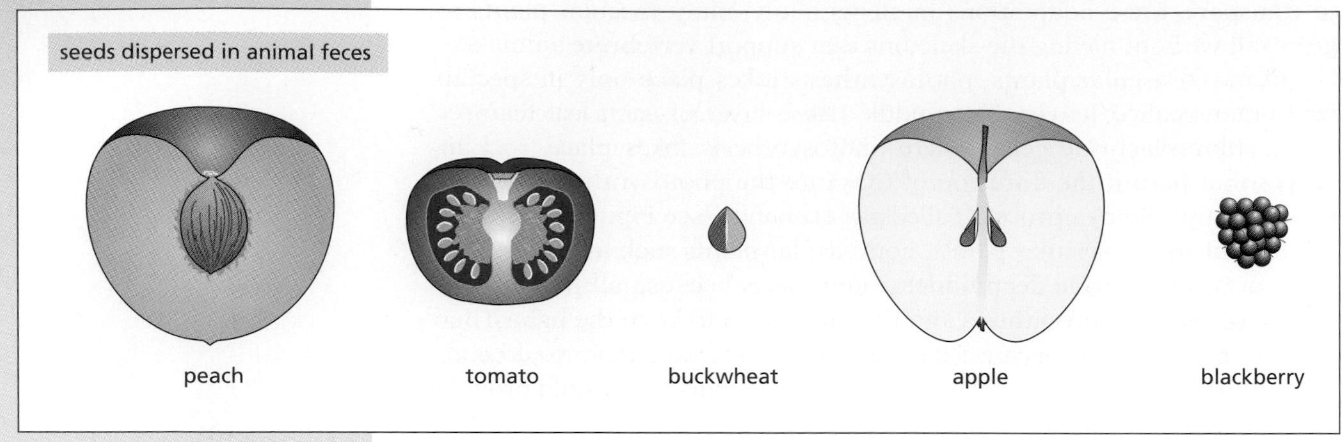

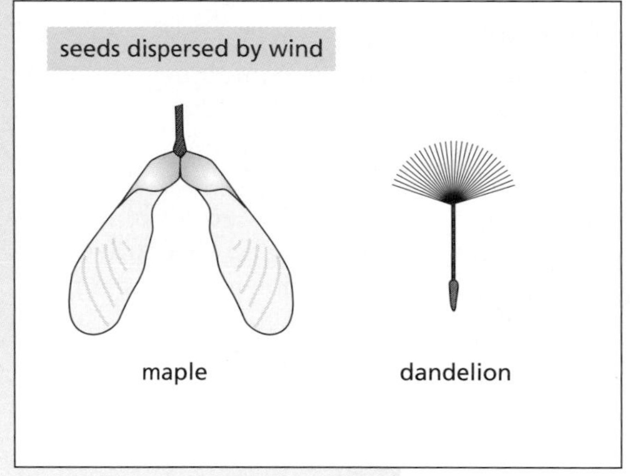

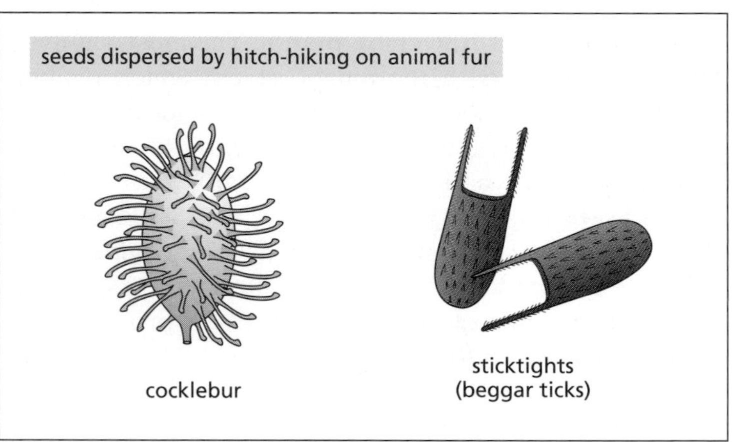

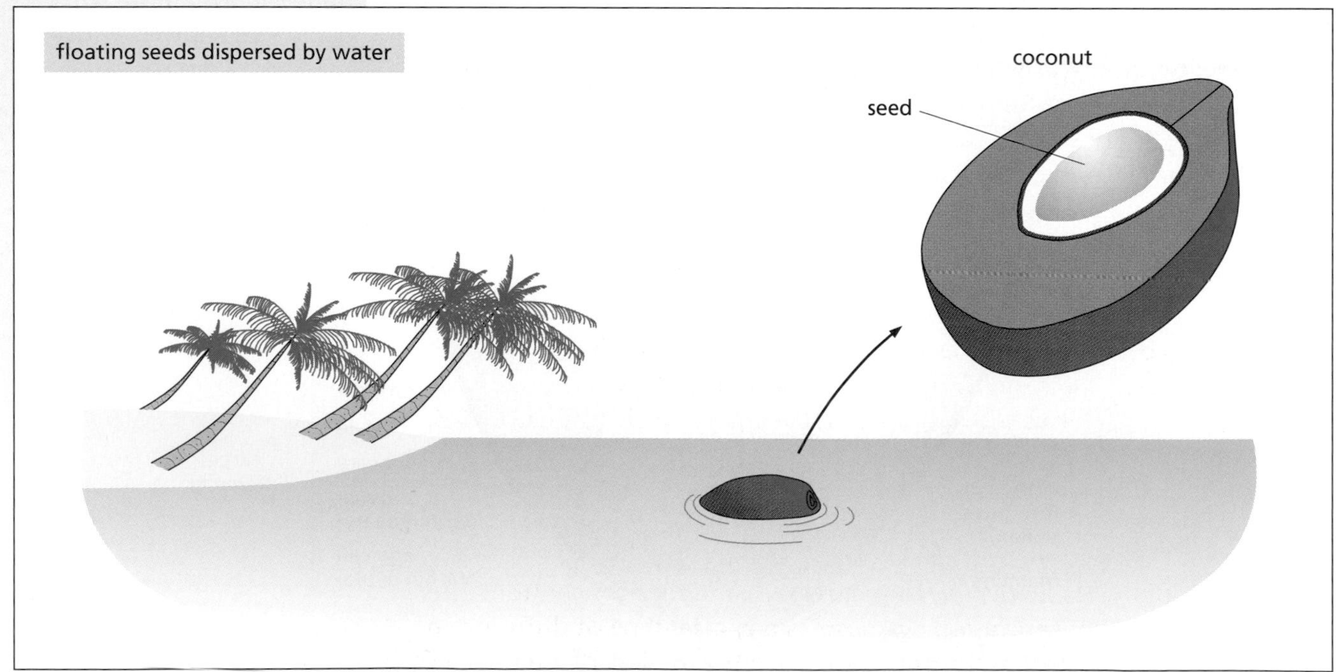

number are pollinated by insects. The relations between flowering plants and the insects that pollinate them are often quite elaborate and are the key to much of the diversity and also the evolutionary success of both flowering plants and insects (see Chapter 15, p. 637).

After fertilization, the zygote undergoes cell division and becomes an embryo, while the structures surrounding the zygote mature into a seed. The seed-bearing structures in a flower ripen into **fruits**, defined as ripened ovaries that contain seeds. In addition to the seeds themselves, fruits often contain tissues that attract various animals by means of conspicuous colors, special odors, carbohydrate nutrients, or a combination of these. Animals that eat the fruits may disperse the seeds in their feces, often far from the parent plant. Seeds are also dispersed in other ways (Figure 14.11). Among the fruits that humans eat are many that we commonly recognize as fruits (e.g., apples, peaches, melons) and others that we do not usually regard as fruits (e.g., nuts, grains, cucumbers, tomatoes, peppers, eggplant).

Recall that the source of energy used by plants to make fruits and do everything else that plants do comes from photosynthesis, and that carrying out photosynthesis requires water. Water pressure also helps plants to stand upright. We need to think about water in plants.

Water transport in plants

We have just seen that water moves over great distances within plant bodies. But plants would not be able to absorb nutrients, conduct photosynthesis, synthesize chemical compounds, or get rid of waste products if water were not able to move in and out of plant cells as well as plant bodies. We next discuss how water moves in and out of plant cells, then consider in more detail how water moves through the plant vascular tissue, and finally how water can make plant parts move.

Osmosis. Like all cell membranes, plant cell membranes stay intact only because they are surrounded by water. Lipid bilayer membranes are impermeable to ions, as we have seen in other chapters, yet are permeable to water. This allows water to move by a process called **osmosis** whenever opposite sides of the membrane contain solutions at different concentrations. We usually think of the concentration of a solution in terms of the number of dissolved ions or other particles in a given volume of water. We have seen that a membrane can store chemical potential energy in the form of differences in ionic concentration on the membrane's two sides (e.g., as the result of the light reactions of photosynthesis earlier in this chapter, or as a resting potential in nerve cells in Figure 10.3, p. 413). But we can also think of the solution in terms of the concentration of water, that is, the numbers of molecules of water per volume of solution: the higher the concentration of ions and other dissolved materials, the lower the concentration of water. All substances tend, according to the laws of physics, to diffuse from a region of higher concentration to a region of lower concentration. Osmosis is a special type of diffusion in which water passes through a membrane from a region of high water concentration (and low ion concentration) to a region of low water concentration (and high ion concentration). (The ions are prevented by their charge from moving through the membrane.) There is no attraction or repulsion in the process of osmosis, just a flow of water molecules from an area where their number is higher to an area where their number is lower (Figure 14.12).

FIGURE 14.12

Osmosis in plant cells. Water moves across cell membranes toward the side with the lower water concentration and higher ion concentration.

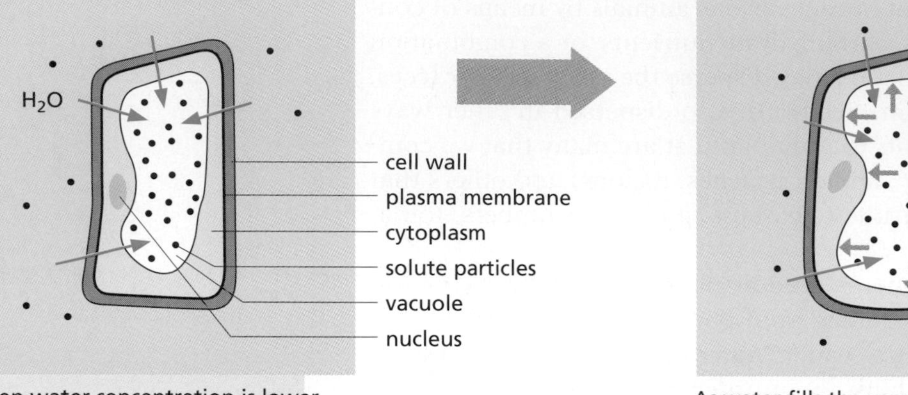

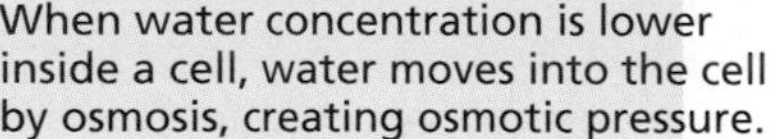

When water concentration is lower inside a cell, water moves into the cell by osmosis, creating osmotic pressure.

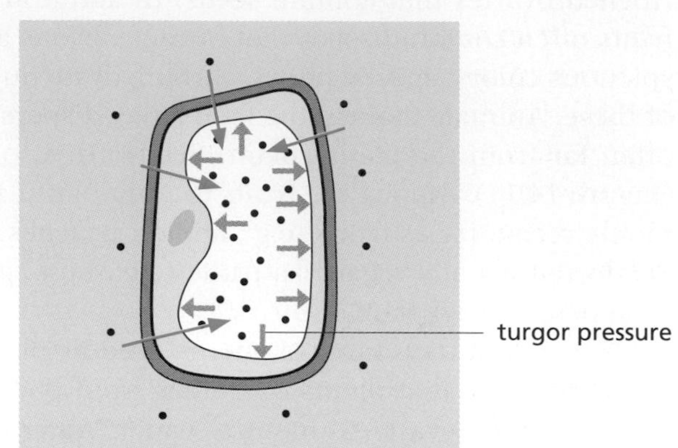

As water fills the vacuole, turgor pressure inside the cell becomes high; the cell becomes turgid.

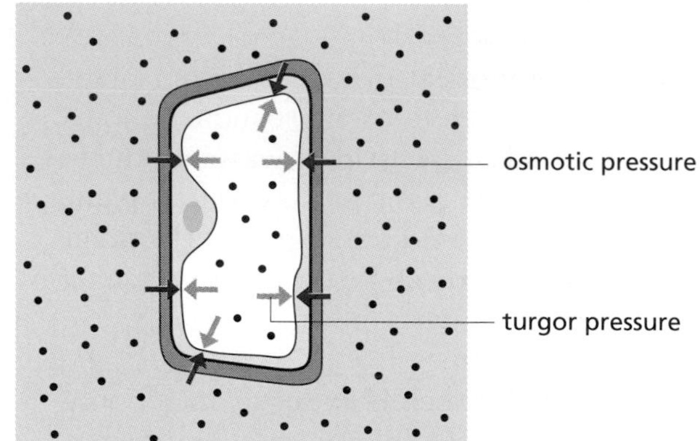

When concentration is the same inside and outside, there is no osmotic flow; osmotic pressure equals turgor pressure.

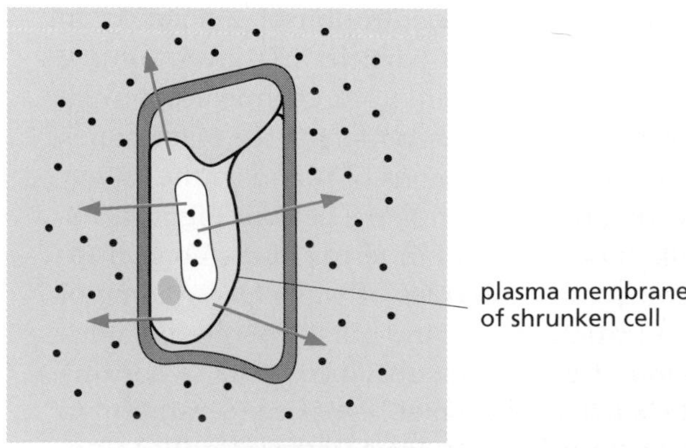

When water concentration is lower outside, a cell shrinks as water flows out.

Plant cells are surrounded by a rigid cell wall made of cellulose, which lies outside the plant cell membrane. When water flows into plant cells by osmosis, the cell membrane eventually pushes against this cell wall, producing **turgor** (see Figure 14.12), a form of fluid pressure that makes plant tissues stiff. It is thus water pressure, rather than a skeleton, that keeps nonwoody plants upright. In woody plants, a rigid biomaterial called **lignin** provides additional support.

Water transport. A plant needs to move water from its roots to all of its cells. There are several hypotheses for how this transport is accomplished, but the one that best fits existing data is the **transpiration-pull theory**, the idea that water is pulled up through a plant rather than being pushed up from its roots. Although water pressure does build up in the roots of plants, it is not of sufficient magnitude to push water against gravity to the heights of many plants. When xylem is cut, water continues to be pulled upwards above the cut, but none is pushed into the cut from below. If Queen Anne's lace or other similar white-flowered plants grow in your area, you can demonstrate this for yourself. Cut some flowers, accompanied by leaves and stem, and put them into water with food coloring. The flowers will turn the color of the food coloring as water is drawn up the xylem even though the flowers are separated from the roots. The rate of water movement is greatest when the plant is in sunlight, a time when **transpiration**, the water loss by evaporation from the leaves, is greatest.

Transpiration is controlled by openings called **stomates**, or **stomata**, in the undersides of leaves. The opening and closing of the stomates is accomplished by **guard cells**, which change shape in response to increasing and decreasing turgor (Figure 14.13). When turgor is low, which is also when water supplies are low, the guard cells are closed. When turgor increases in adjoining cells, ion channels in the guard cell membranes open, allowing K^+ and Cl^- ions to enter the guard cells. Water molecules flow by osmosis, increasing turgor in the guard cells and resulting in a change in their shape. In the new shape, a space between the two guard cells opens and allows the exchange of gases, including CO_2, O_2, and water vapor. When the stomates are open, gases flow in and out: carbon dioxide (CO_2), needed for photosynthesis, is taken up, and oxygen (O_2), a product of photosynthesis, is given off (see Figure 14.9). Water is also given off because the water concentration of the air is lower than that of the xylem.

FIGURE 14.13

Opening of a stomate.

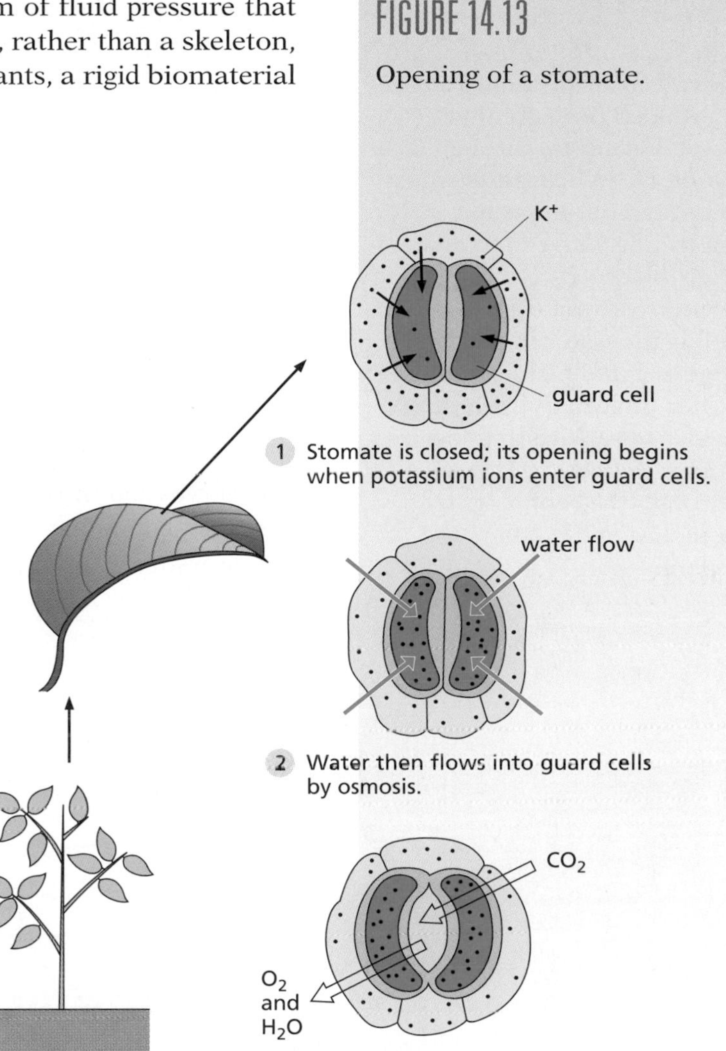
K^+
guard cell
1 Stomate is closed; its opening begins when potassium ions enter guard cells.
water flow
2 Water then flows into guard cells by osmosis.
CO_2
O_2 and H_2O
3 Guard cells swell, stomate opens.

The oxygen atom in a water molecule is partially negative and the hydrogen atoms are partially positive. Because of this, the oxygen of one water molecule forms a weak bond, called a **hydrogen bond**, with one of the hydrogen atoms of an adjacent water molecule. Water has many such hydrogen bonds, and thus a high degree of cohesion. As water molecules are lost through the stomates, cohesion pulls the next water molecules up to take their place. A tension is thus put on the thin column of water in the xylem, pulling the whole column upward (Figure 14.14).

In addition to moving water up through the plant, transpiration also serves to cool the plant. Some of the light energy absorbed by plants is transformed into heat and cannot be used in photosynthesis. Plants avoid overheating by evaporating water from their leaves through transpiration. Most of the water absorbed by any plant is eventually lost through transpiration.

Rapid movement in plants. While animals can use nerve cells to stimulate movement through muscle contraction, plants have neither muscles nor nerves and thus have only limited powers of movement. Some plants, however, can move their parts quite rapidly; their ability to regulate turgor pressure is the key to how plants such as the Venus fly trap (*Dionaea*

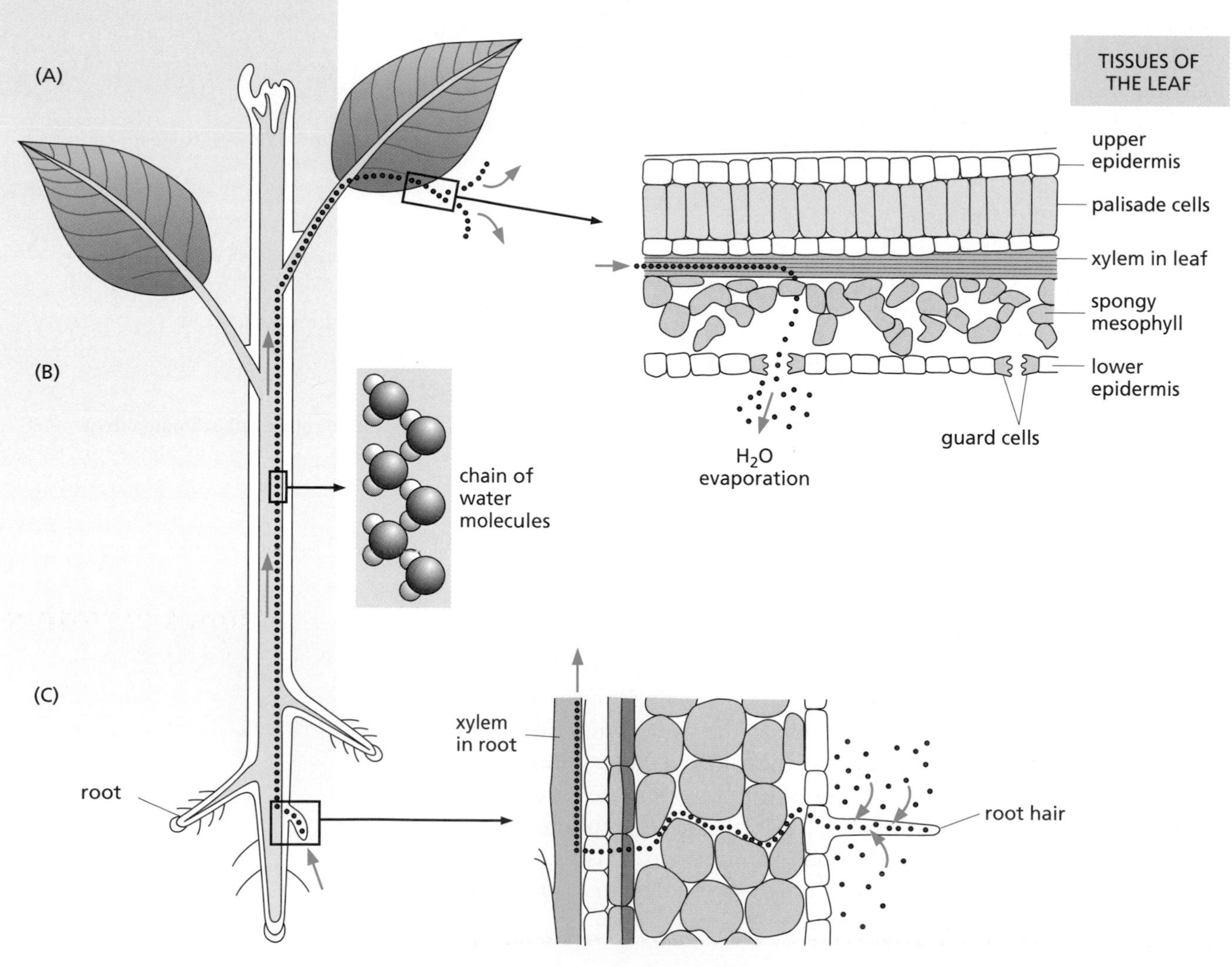

FIGURE 14.14

The transpiration-pull theory. Transpiration creates a reduced pressure that acts to pull water up through a plant. (A) When water is lost by transpiration, water molecules move in to take their place. (B) Cohesion from hydrogen bonding holds the water molecules together, pulling a chain of water molecules upward. (C) Water to replace that lost by transpiration is taken up through the root hairs by osmosis, and it enters the xylem.

FIGURE 14.15

Rapid and selective movement in plants.

When this sensitive plant (*Mimosa pudica*) is touched, the leaflets fold together within a fraction of a second.

When an insect touches the sensitive hairs of this Venus fly trap, the leaf halves snap together in less than half a second, trapping the insect.

muscipula) and the sensitive plant *Mimosa pudica* are capable of rapid movements (Figure 14.15). Touching one of these plants causes ion channels in the tissue touched to open, and ions to flow out from the cells. Water flows out in response, and turgor drops quickly. The cytoplasm of a plant cell is connected to the cytoplasm of adjoining cells at small openings called **plasmodesmata**. The changes of ion concentration are thus passed along to the next cell, where ion channels are stimulated, passing along the change in turgor. When turgor drops in the cells where a *Mimosa* leaflet joins the stem, the leaflets collapse towards each other (see Figure 14.15). When changes in turgor have propagated to the base of the leaf stem, which may take more than one touch, the whole leaf droops. The plant will return to its former position when water flow reverses, reestablishing turgor pressure. Thus, in plants, movement can be accomplished without contractile muscle fibers.

Now that we have learned some things about the structures of plants and how they obtain their nutrients and energy, we can return to one of the questions posed by this chapter: can plant science feed the world?

THOUGHT QUESTIONS

1. Do plants have sense organs?

2. How can plants move without nerves and muscles?

3. How are plants able to stand upright?

4. Why can plants grow without soil but not without water?

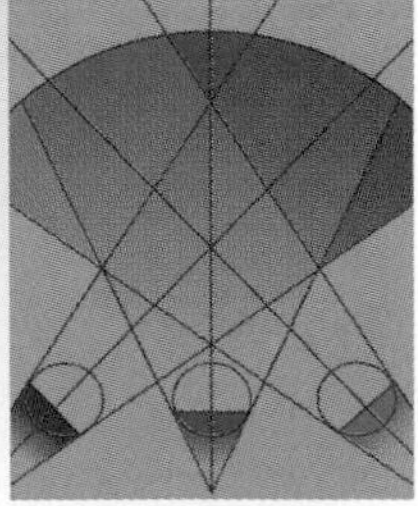

CROP YIELDS CAN BE INCREASED BY OVERCOMING VARIOUS LIMITING FACTORS

Hunger, starvation, and malnutrition are endemic in many parts of the world (see Chapter 8), and rapid increases in the world's population (see Chapter 6) have intensified these problems. One approach to addressing these problems is to increase crop production, but there are very few paths available for doing this. One is to increase the amount of land under cultivation; this path has its geographic limits as well as very high biological costs in terms of the destruction of natural ecosystems (see Chapter 15). Another possibility is to increase the yields or the nutrient content of crop species through such techniques as soil improvement, using fertilizers and advanced irrigation techniques, pest control, and the use of different plants or new genetic strains of plants. Each of these methods seeks to increase yield in a different way: by supplying a limiting nutrient, by supplying water (which is often limiting in the same way), by controlling natural enemies of the crop species, or by developing new crop plants such as drought-resistant strains, pest-resistant strains, or strains with diminished nutrient requirements, capable of growth on marginal or poor soils.

Fertilizers

In those locations where a nutrient is a limiting factor for the growth of plants, crop yields can often be dramatically increased by adding the appropriate fertilizer to the soil. A **fertilizer** is a substance that supplies organic or inorganic nutrients needed by plants. In the largest number of locations, nitrogen is the limiting nutrient for many plant species, and nitrates are among the most important fertilizers used in agriculture. In many places, crop yields can be dramatically increased by adding nitrates to the soil. Phosphorus (P), in the form of phosphates (PO_4^{3-}), can also be an important limiting nutrient, especially for ornamental flowers or crops whose edible portion is a seed, a flower, or a fruit. Phosphorus often signals the plant to begin setting flowers, seeds, and fruit. Potassium (K^+) is often important as well. It is used by plants to control the opening and closing of stomates (see Figure 14.13), and it may be scarce (and therefore limiting) in certain soils. The hazards as well as the benefits of applying fertilizers need to be considered.

Organic and inorganic fertilizers. Fertilizers can be either organic, including various manures and composts, or inorganic. The nutrients in organic fertilizers are complex molecules that break down slowly. The slow breakdown gives these fertilizers the advantage of slow release, a process that provides a steady supply of soluble nutrients that is more likely to be absorbed by growing plants than to be washed away as are many inorganic fertilizers. One disadvantage of organic fertilizers is their high bulk and cost of transportation, features that often limit their use to the immediate locality of their production. In China, human and animal wastes have been used as fertilizers for centuries. This practice provides needed nutrients to crops, but it also has the undesirable effect of spreading parasites (especially flukes of the animal phylum Platyhelminthes) and other infectious diseases.

Green manure. As described earlier in this chapter, crop rotation provides organic fertilizers without need of transportation whenever plants that accumulate nitrogen are planted and plowed under. In some cases plants are put in only to be plowed under; in other cases, some part of the plant is harvested and the remainder is plowed under. Lima beans are an example of the latter; the beans are mechanically harvested and the roots and stems are then plowed under. The alternation of cereal grains with soybeans or alfalfa (both legumes) is a form of crop rotation widely practiced in many large regions of North America and Asia. Organic farmers use green manuring, sometimes in combination with the use of various animal and plant wastes as fertilizer. The benefit of crop rotation for maintaining the fertility of the soil was recognized in ancient Rome and several other ancient civilizations.

Sources of phosphorus. Nutrients are also slowly released from fish meals and bone meals, which are dried, powdered bone. These fertilizers supply phosphorus from the bone mineral hydroxyapatite, an inorganic crystal that contains both calcium and phosphates. The phosphates provided by these fertilizers are insoluble; they are converted to soluble form only slowly, often by the very plants that use them. (The slow solubility is a disadvantage at first, especially if the practice is begun in soil that is nutritionally deficient; chemical fertilizers may be necessary during the initial phase-in period.) Fish or fish meals also contain additional phosphorus in the form of nucleic acids from the nuclei and organic phospholipids in the membranes of the cells of the dead animal. Native Americans throughout the eastern woodlands traditionally grew their corn in mounds, under which they customarily buried a fish that provided both phosphate and nitrogen as fertilizer for the corn.

Chemical fertilizers. Inorganic chemical fertilizers are usually nitrate or phosphate salts sold as powders or granules to be applied to fields, usually by mechanical equipment. When they were first introduced, chemical fertilizers were cheap and readily available. The benefits to crop production were immediately evident, and, during the twentieth century, the use of chemical fertilizers increased greatly in countries that could afford them, principally in North America, Europe, and Australia.

The problems associated with the prolonged use of inorganic chemical fertilizers did not become evident until they had been in general use for several decades, in accordance with what some call the **law of unintended consequences**. This has also been stated as the first law of ecology: You can never change just one thing.

Excessive use of inorganic fertilizers can kill off the soil organisms required for maintaining soil fertility, thus requiring the use of even more fertilizer for subsequent crops. Compared with organic fertilizers, inorganic minerals are relatively expensive to produce or extract, and most of them must be transported over great distances. Cheap supplies that formerly existed near places where fertilizers were in demand have in most cases been depleted, and many are now mined far from the places where they are used. For example, large quantities of nitrate minerals are mined in Chile for use as fertilizers in countries in the Northern Hemisphere, thousands of miles from Chile. Fossil fuels are used in the mining and transportation of fertilizers as well as by the tractors that are typically necessary to spread these fertilizers on fields. Most chemical fertilizers are costly to transport to where they are needed, and in most of the developing nations of the world they are prohibitively expensive. (Organic fertilizers can also be expensive to transport on a per-mile basis, but they are often available more locally.)

Runoff and eutrophication. Additional costs associated with the use of fertilizers include the problems of runoff and stream pollution. Fertilizers are concentrated nutrient sources and are not solely nutrients for the crop plants for which they are intended. When fertilizers run off from agricultural fields into bodies of water, they supply nutrients to algae in the water. The growth of the algae may have previously been held in check by the lack of some limiting nutrient. Once that nutrient is present, the growth of these algae can be so rapid that they use up all the oxygen in the water, causing the death of many fish. The decomposition of dead fish releases more nutrients that further accelerate the growth of the algae. This process of nutrient enrichment, algal growth, and oxygen depletion is called **eutrophication**. Eutrophication led to the deaths of most fish, and the commercial fishing industry, in Lake Erie in the 1960s. International regulations that cut down on runoff pollution have allowed the lake to recover to a great extent. Eutrophication remains a problem in more than half of the lakes in the United States (Figure 14.16).

FIGURE 14.16

Eutrophication: algal blooms accelerated by fertilizer runoff. The two portions of this lake in Ontario were separated by a plastic curtain, and phosphates were added to one side of the curtain only. An algal bloom occurred only on the far side, to which the phosphates had been added, making the water look more cloudy.

The problems of runoff and eutrophication can originate from animal manures (e.g., from dairy or hog farms or from human sewage) or from the use of inorganic chemical fertilizers, especially those applied to the soil in soluble form, as most nitrates are. Laundry detergents containing phosphates were also a major source of nutrients; however, public awareness pressured manufacturers into changing their formulations, so that many detergents are now made phosphate-free. The amount of rainfall and its seasonal distribution can greatly influence the severity of the problems. Runoff can sometimes be minimized by applying fertilizers at the proper time of year, just before the nutrients are most needed by the plants and are most likely to be absorbed.

Soil improvement and conservation

Soil is the loose material derived from weathered rocks and supplemented with organic material from decaying organisms. This organic material supports the growth of plants as well as that of many bacterial, protist, and animal species that, in turn, add to the soil-building process. The type of soil formed depends on the parent rock type, the climate, the removal of soluble materials by percolating groundwater, and the mechanical and chemical activities of living organisms such as bacteria and protists, earthworms, and plant roots. Soil is thus the product of both biological and geological processes (Figure 14.17).

FIGURE 14.17

Soil formation from the weathering of rock (in this case, limestone) and from biological processes.

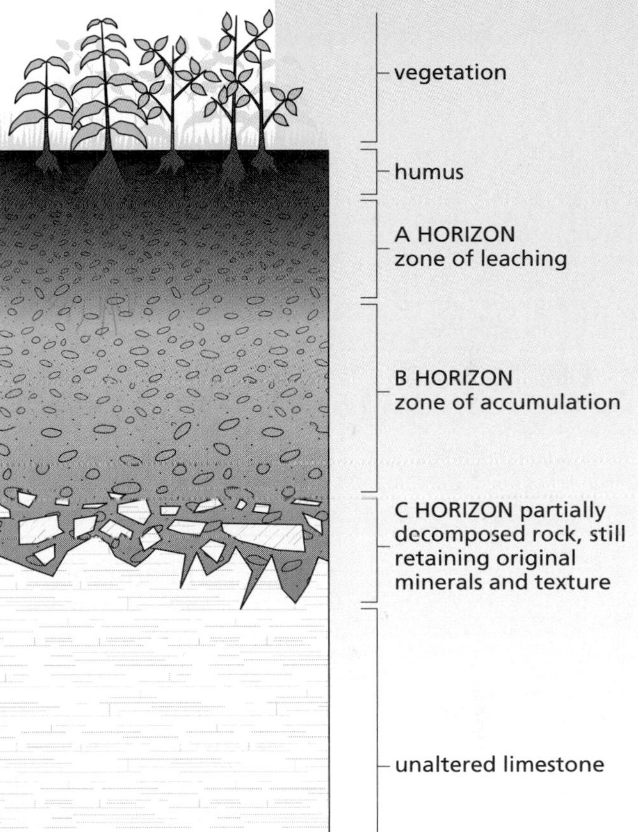

Humus. Organic material in the soil is broken down by bacteria and fungi, which function as decomposers. The partially decomposed organic matter is called **humus** and it serves several important functions in the top layer of the soil. It binds the particles of weathered rock together to form small aggregates, which give the soil its structure. The small holes between particles help to hold water and oxygen and provide space for the growth of minute root hairs, extensions of single cells on the roots of plants. Although the larger roots provide the anchorage for the plant and in some plants can reach deeply into the ground in search of water, it is the root hairs that provide the enormous surface area for absorption of nutrients, just as the microvilli of the intestine provide the surface area for absorption of nutrients from mammalian guts (see Chapter 8, p. 322). Because root hairs are so delicate, they cannot push into soil, but must grow into already existing spaces in the soil structure.

Humus has an overall negative charge, which helps to hold positively charged nutrient ions such as potassium (K^+), calcium (Ca^{2+}), and ammonium (NH_4^+) in the topsoil, where they are available to plants' roots. Inorganic fertilizers can provide chemical nutrients, but they do not contribute to humus and soil texture as organic fertilizers can. The plowing under of organic matter helps to build humus. Rather than being left bare, fields can be planted with grasses during the season when cash crops do not grow, or with legumes that will be plowed under when it is time to sow the crops. This both prevents erosion and adds to humus.

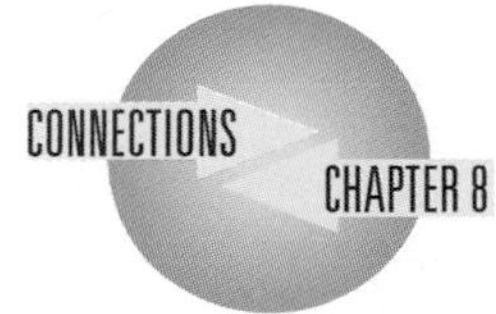

Humus can be produced by **composting**. Organic material (leaves, grass clippings, food waste) is layered with manure or soil. In a matter of months the decomposer bacteria from the soil turn the organic matter into humus. Many United States municipalities encourage composting, either by teaching people how to do it in their own backyards or by having centralized composting facilities to which people can bring their yard wastes. Even so, less than 1% of organic wastes are composted in the United States, while much more is done in Europe. It is estimated that solid waste in United States landfills could be reduced by 20% if organic material were composted.

Soil as a nonrenewable resource. Topsoil is eroding faster than it is being formed on approximately one-third of the world's croplands. Soil that is lost can be regenerated by biological processes, but only very slowly—one inch of topsoil takes between 200 and 1000 years to renew, depending on the climate and other factors. Groundwater pollution, excess fertilizer, and runoff from highway salt can kill soil microorganisms and stop the regeneration of topsoil. Because soil cannot be regenerated within a time period that is relevant within a human lifetime, it must be treated in the same way as a nonrenewable resource. Treating soil as a nonrenewable resource, farming and other land uses are best done in ways that are **sustainable** (see Chapter 15, pp. 669–670). Sustainable practices are those that lead to no net loss of a resource. Important practices in sustainable agriculture include crop rotation, use of organic, humus-building fertilizers, and the prevention of erosion.

Irrigation

The **irrigation** of crops is the process of adding water, a vital substance for the growth of plants. As we have seen, plants need water to supply the hydrogen ions used in photosynthesis. They also need water for transport of their nutrients and the products of their synthesis reactions. About 90% of the water absorbed by a plant is evaporated through its leaves during transpiration, dissipating excess heat. Also, the rigidity and strength of plants is based partly on the properties of water, as explained earlier.

Irrigation is expensive in regions where water is scarce and where irrigation is therefore most needed. Traditional methods of spray irrigation lose much of the water to evaporation. Newer methods include **drip irrigation**, in which irrigation tubes with tiny holes are laid at or below ground level. The tiny holes, spaced every few inches, are designed to leak or drip water slowly into the soil, providing moisture but minimizing evaporation. All irrigation systems, whether drip or traditional, require a large initial investment in pipes or ditches, pumping stations, and the like, plus a supply of fresh water or desalinized water (seawater from which the salts have been removed). Freshwater sources can become the subject of political disputes between neighboring governments, such as those between Israel and Jordan or between California and Arizona.

Hydroponics

Although plants cannot live without water, they can live without soil. Some plants grow naturally without soil. The growing of crop plants without soil is called **hydroponics** (Figure 14.18). In places where there is very little soil, or where the soil is unsuitable, hydroponics offers a possible alternative agricultural method. In a hydroponic system, the plants are grown with their roots immersed in tanks through which water carrying dissolved nutrients is allowed to flow. The water is recirculated (and therefore conserved), and its dissolved materials are frequently monitored and adjusted.

Advantages of hydroponics. A well-managed hydroponic system can produce greater yields than traditional soil-based systems. Because hydroponic systems are generally inside greenhouses, where they are protected from insects and other plant pests, the produce that results is free from the blemishes often caused by these pests. Much of the labor traditionally concentrated on soil care, including tilling, planting, fumigation, and irrigation, is eliminated. Water is provided to the plants directly and more efficiently, and is recycled so as to minimize its use. This is extremely important in arid climates, where sunshine may abound but where water is scarce. Instead of traditional application of fertilizers that can diffuse beyond the reach of plant roots, mineral nutrients are added directly to the water, where the excess not taken up by the plants remains available and can be recycled. In this way, the use of nutrient supplements is minimized, and runoff problems are also minimized. Disease and pest control can be handled easily by adding the necessary chemicals to the recycled hydroponic water, with far less danger that these substances will spread to local water supplies or to domestic animals and humans.

FIGURE 14.18

Hydroponics: growing plants in water tanks, without soil.

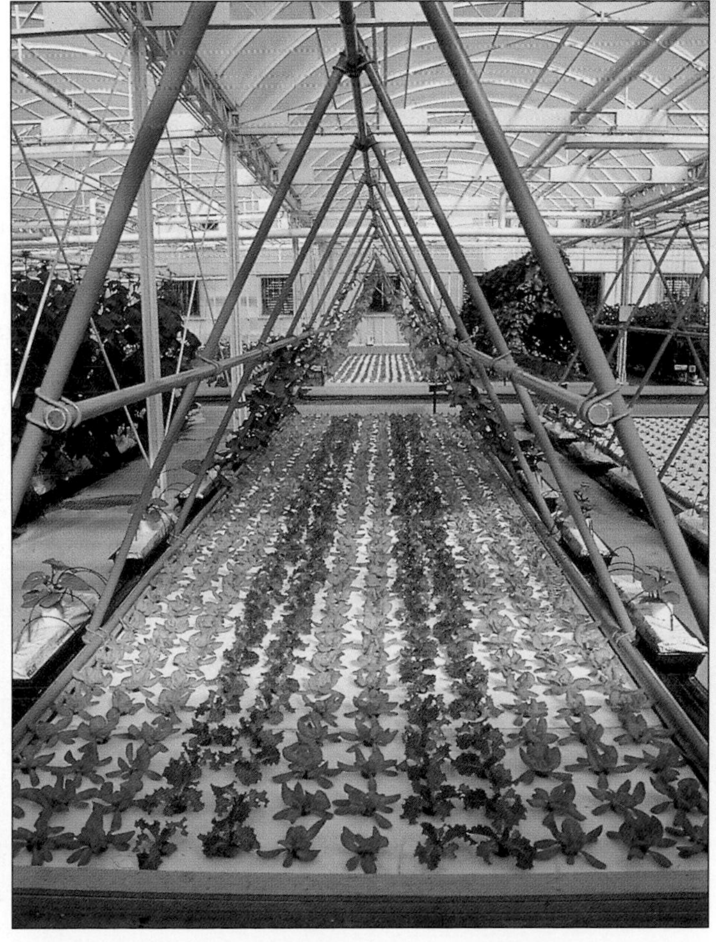

Disadvantages of hydroponics. Many of the disadvantages of hydroponic systems have to do with cost: the initial construction costs; the costs of maintaining equipment and greenhouse facilities; the costs of nutrients; and the salaries of trained personnel for constant monitoring for nutritional problems (such as nutrient depletion) and waterborne diseases. Hydroponic systems may be in locations distant from the Equator, where the provision of artificial light and heat can add to the cost.

Plants are subject to disease, as are all living organisms; waterborne diseases are those in which the pathogen (disease-causing organism) can live in water. Even if a pathogen does not grow well in water, its ability to survive in water might enable it to be transmitted from plant to plant in a hydroponic system. Most hydroponic farms concentrate on a single plant species, a practice known as **monoculture** that brings a risk of the rapid spread of disease, whether the monoculture is hydroponic or land-based. The risk relates to the fact that many pathogens are able to infect only one or a few host species. When a pathogen encounters a monoculture, it can spread very quickly because of the proximity of other individuals of the same host species.

For these reasons, hydroponic systems are easier to justify economically for plants that are economically profitable in small quantities (e.g., pharmaceutical plants or certain vegetables) rather than for staple crops such as cereal grains. Hydroponics is potentially useful where either growing space or soil are limited, as in proposed space stations. On Earth, hydroponics works best in warm climates such as Israel or southern Italy, where light energy is abundant and artificial heating is unnecessary, or in countries such as Japan, where people are willing to pay extra for produce that is 'picture perfect.'

Chemical pest control

Each year, 30% or more of many crops are destroyed by insects and other crop pests. In addition to insects and their larvae, which damage the plants themselves, other crop pests, including various rodents, damage many crops in the field and also consume stored grains. Fungi also destroy up to 25% of stored crops. Nutrients that are consumed by pests are not available to the plant or to humans for food.

Monoculture and pests. Some of the farming practices that have increased crop yields over the last few centuries have also increased the susceptibility of crops to damage by pests. For example, large, mechanized farms (farms that are heavily dependent on the use of machinery) may plant monocultures that are thousands of hectares in size. (The **hectare**, or metric unit of land area, is an area 100 m by 100 m, equivalent

FIGURE 14.19

Monoculture: the planting of a single species only.

A wheat field in Nebraska.

A lettuce field in Ontario.

to about 2.47 acres.) Monocultures are especially suitable for mechanized agriculture (Figure 14.19) and large-scale operations tend to be more economical than small-scale ones; as a result, monoculture is extremely profitable for a while. However, as we have seen, monocultures make it easier for a pest species to spread rapidly, especially if neighboring farms are also planted with the same crop. If neighboring fields were planted with different crops, the spread of pest species would be interrupted or made more difficult. Also, planting the same crop year after year enables pest species to survive from year to year in the form of eggs or larval stages. In contrast, planting different crops in rotation interrupts the life cycle of a pest that depends entirely on one particular crop species for its propagation.

Pesticides. Chemical pesticides were used in ancient times. For example, the Romans dusted sulfur, which we now know acts as a fungicide, on their grapes. The use of chemical pesticides increased enormously during the late nineteenth and twentieth centuries. Arsenic and copper compounds were widely used in the nineteenth century, but they were replaced in the twentieth century with pesticides derived from petroleum. During the first several decades of their use, chemical pesticides greatly reduced the amount of crop damage due to pests, and they continue to be used in many countries because they increase crop yields.

For much of the twentieth century, economic pressure encouraged the use of pesticides on crops and post-harvest treatment with fungicides. The post-harvest treatments have given many farm products longer shelf lives, allowing transportation across longer distances. As a result, people in the industrialized world have come to expect perfect, blemish-free produce in every store and at almost any time of the year, even for crops that do not grow at all in their local area.

A classic example of a chemical pesticide is **DDT**. First developed in the late 1930s, DDT was found to kill large numbers of different insect species. During World War II, DDT was sprayed on soldiers to control body lice and similar pests. Its effectiveness in insect control was followed by extensive spraying campaigns on crops after the war, first in the United States and Europe, and then in other countries. Because it killed so many species of insects, DDT reduced crop damage and greatly increased the food supply in many countries. In addition, DDT and similar chemicals sprayed onto the surface of bodies of water also controlled mosquitoes and other disease-carrying insects that have aquatic stages in their life cycles, thereby reducing human disease. Disease reduction and increased food supplies both contributed to population increases in many countries (see Chapter 6). The use of DDT marked the time of greatest optimism in the use of chemical pesticides. This era was brought to a close by the discovery of DDT's toxic effects on nontarget species, findings that were convincingly made public by Rachel Carson's book, *Silent Spring* (1962). The term **nontarget species** is an apt one for a phenomenon that is another example of the operation of the law of unintended consequences: the target species is the one you intended to kill, and the nontarget species are those that are killed unintentionally. Often, unintentional effects are not detected immediately, not because of ill-will,

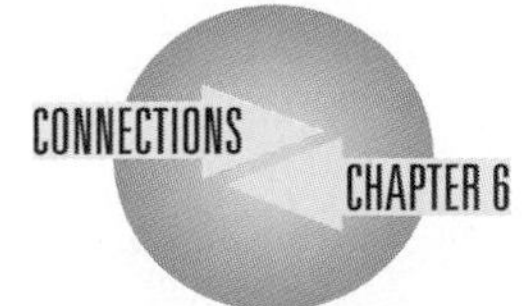

but because it can be difficult to predict where, when, and how the unintended effects will show up.

Negative consequences of pesticide use. There are many problems associated with the use of chemical pesticides such as DDT. The pesticides themselves are generally expensive; most of them are petroleum derivatives, and a great deal of energy is used in their extraction and further synthesis. Attempts to control pests with chemical pesticides have in several cases brought about increased levels of pest-related devastation several decades later. Pesticides like DDT are toxic to a wide variety of harmful and beneficial species alike. They may kill so many of the target species' natural enemies that the population size of the target species subsequently increases (after a time delay) above its earlier levels. Another problem with frequent pesticide use is that the target species develop pesticide-resistant mutations and are no longer killed by the spraying. Over 400 insect species, for example, are now DDT-resistant. Widespread use of the same pesticide year after year favors the evolution of pest populations with mutations that make them pesticide-resistant. Once they originate, these resistant strains of pest species spread rapidly because of selection by the pesticide itself.

As with fertilizers, there is also a runoff problem. A pesticide can find its way into groundwater supplies and from there into streams and lakes. Pesticides may contaminate drinking water supplies in this way, and they may also poison the fish in our lakes and streams.

Biomagnification. DDT and many other long-lasting insecticides that are applied to crops become concentrated in the bodies of the pests that eat those crops, and then further concentrated (and thus more toxic) in the bodies of animals that eat the pests, a principle known as **biomagnification**. Biomagnification does not apply just to pesticides. A wide range of pollutants including heavy metals such as mercury become concentrated in this way. To understand biomagnification, recall that producer organisms such as plants supply food energy to the primary consumers that eat them. The primary consumers, in their turn, supply food energy to secondary consumers, and so on. In each conversion, much of the food energy is lost in the form of heat, so that 10,000 kilocalories (kcal) of sunlight provided to the grass produces only 1000 kcal of grass energy to the cows that eat the grass, who in turn provide only 100 kcal of energy to the people that eat beef steak. These relations can be represented in a **trophic pyramid**, also called a food pyramid, as shown in Figure 14.20. Similar food pyramids could be drawn in proportion to biomass (quantity of biological tissues) or numbers of individual organisms rather than energy; most such diagrams would have basically the same pyramidal shape. The energy relations in a food pyramid are one reason why it is inefficient and wasteful to have intermediate steps between the producer plants and the top level consumers, and thus why a certain supply of crops will support more people if the people eat the crops directly instead of eating animals that eat the crops (see Chapter 8, p. 338).

When pesticides are used, an additional problem arises. Many pesticides are not broken down by decomposer organisms and thus persist in the environment. Also, many pesticides accumulate in biological tissues without being excreted. If a long-lasting pesticide is taken up by plants at the base of a food pyramid, the same amount of the pesticide is incorporated into a smaller and smaller amount of biological tissue with each successive conversion. The pesticide concentration at each step is the total quantity of the pesticide divided by the total biomass. With a reduction in biomass at each successive step, but with little reduction in the quantity of the pesticide, the concentration of the pesticide increases with each successive step in the food pyramid as the result of biomagnification.

CONNECTIONS
CHAPTER 10

As a case in point, Figure 14.21 shows the biomagnification of the long-lasting pesticide DDT. In Long Island Sound, New York, the DDT concentration was measured as 0.000003 parts per million (ppm) in the water, but was 25 ppm in the tissues of fish-eating birds such as ospreys and eagles, an increase in concentration of more than 8 million times.

DDT and other chemical pesticides that concentrate in fish and fish-eating birds interfere with calcium metabolism, causing a thinning of the shells of the eggs laid by the birds. The thin shells break before the chicks can develop, and the bird populations, unable to reproduce, decline. For this and other reasons, DDT is now banned in most of the industrialized countries, though it is still used in some parts of the world.

Pesticides and neurotransmitters. Many pesticides kill insect pests by blocking the removal of their neurotransmitters inducing continuous muscle contraction (see Chapter 10, p. 419). Because some neurotransmitters are the same in vertebrates (including humans) and in insects, these pesticides can also block neurotransmitter degradation in humans. Although the recommended concentrations properly used for pest control would not kill a person, they can cause permanent injury. The concentrated forms in which the pesticides are manufactured, transported

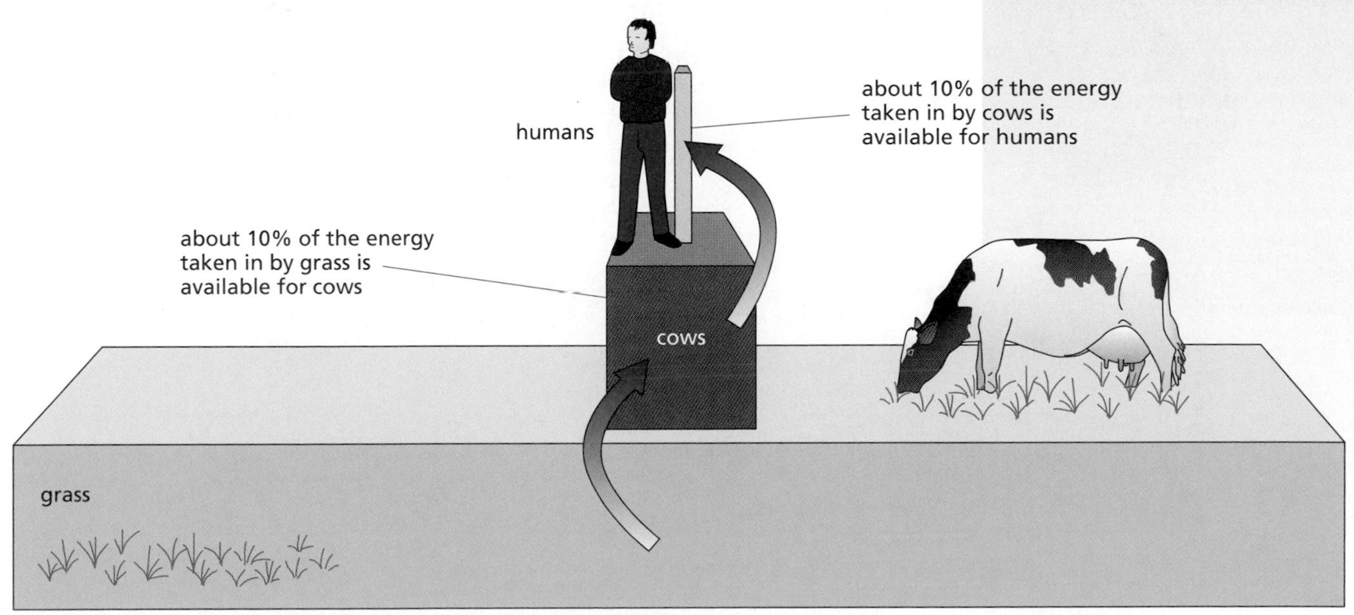
humans
about 10% of the energy taken in by cows is available for humans
about 10% of the energy taken in by grass is available for cows
cows
grass

FIGURE 14.20

A simple food pyramid. The width of each column is proportional to the amount of energy available as food for the next higher step up the pyramid. At each energy conversion, 90% or more of the energy is lost, leaving only 10% available as food, as expressed by the widths of each step.

and stored are more dangerous and may be lethal. Every year in the United States 1000 cases of pesticide poisoning are reported, primarily among agricultural workers, and the actual incidence is probably much higher. Containers with warning labels are sometimes not in the language of the people using the pesticide. Safety standards, where they exist, are often not enforced. High concentrations can also result from biomagnification. Unfortunately, the banning of DDT use in many countries has resulted in the development and use of other chemicals that are even more toxic to nontarget species, including humans.

Integrated pest management

Integrated pest management is a newer approach to crop pest management, one that uses a combination of techniques (Figure 14.22). The term 'management' is meant to convey the intent to keep pest populations under control, so that they stay below the levels at which they cause economic harm. Total pest eradication is in most cases viewed as a goal that can only be achieved at an unacceptably high cost (including the cost to the environment or to society as a whole) or that cannot be achieved at any cost. The term 'integrated' means that all available tools are used in a mix of strategies that includes chemical controls (such as

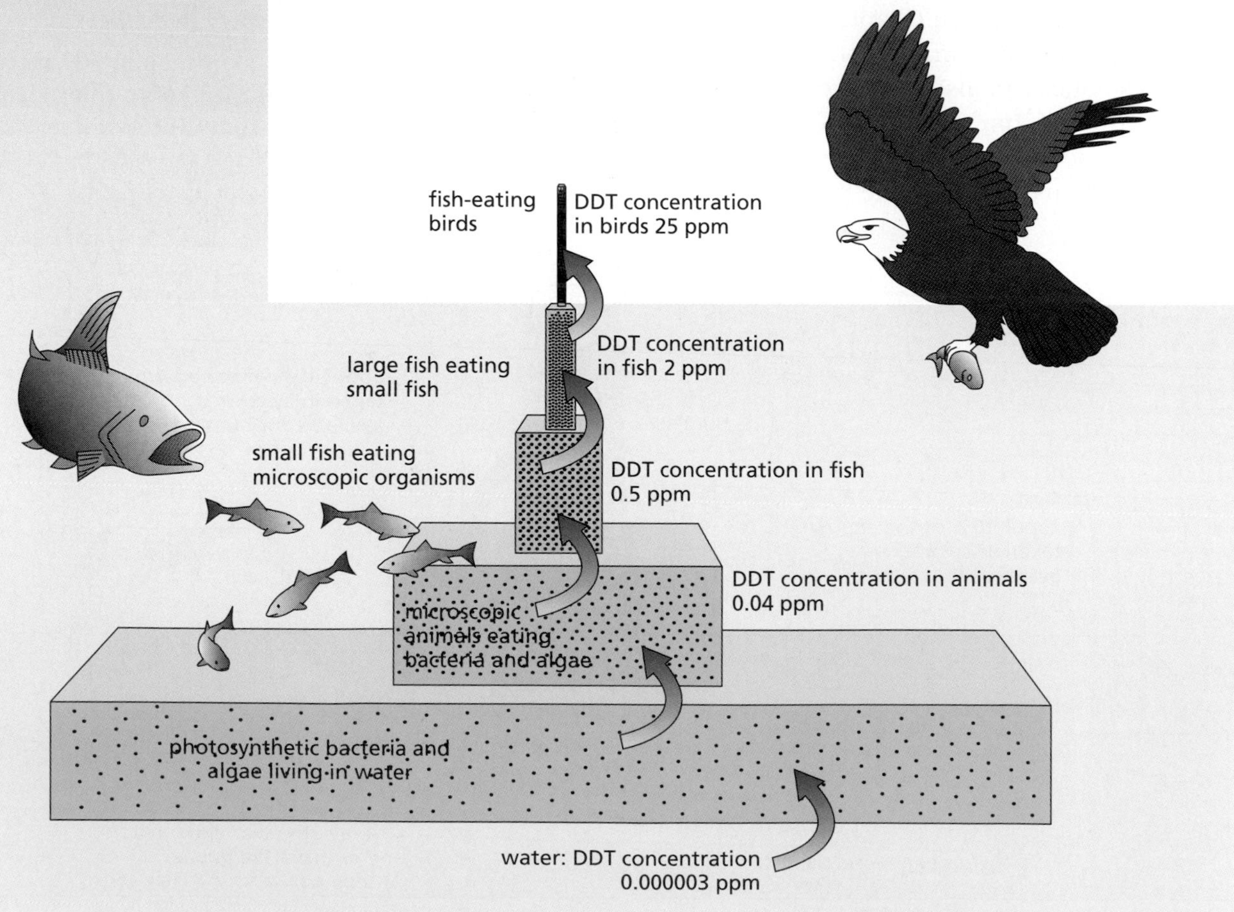

FIGURE 14.21

Biomagnification of DDT in Long Island Sound, New York, around 1970. The width of each column is proportional to the biomass.

pesticides), biological controls (such as maintaining a population of the pest's natural enemies), cultural control (such as public education), and regulatory control (such as public policy legislation). Integrated pest management avoids or reduces most of the risks of chemical pesticides.

Integrated pest management requires the monitoring of pest populations to assess the possible damage that they may do (see Figure 14.22A–C). This allows the use of just enough pesticide to reduce pest populations to acceptable levels, saving expense and reducing runoff and possible harm to nontarget species. Because integrated pest management relies more on biological controls than previous techniques, it requires a good working knowledge of the ecology of the pest species, including knowledge of its natural enemies (see Figure 14.22D).

FIGURE 14.22

Various techniques of integrated pest management.

(A) Several small traps set to monitor the levels of pest populations.

(B) Closeup view of a trap that uses pheromones to attract target species for monitoring.

(C) Corn earworm, an insect whose presence in corn ears can be monitored visually.

(D) A plant pest, the saddleback caterpillar, parasitized by a wasp. Larval wasps eat the caterpillar and kill it; new wasps emerge from the pupae (cocoons) seen here and lay their eggs in another saddleback caterpillar.

Economic impact level. An important feature of any integrated pest management program is the concept of an economically acceptable level of the pest population, which is generally termed an **economic impact level**. The economic impact level is the threshold level above which corrective action must be taken. Pest populations are constantly monitored, and as long as the populations stay below the economic impact level, they are left alone and the cost of countermeasures is saved. The cost of corrective measures includes the cost of expendable materials such as pesticides, the cost of using and maintaining necessary equipment, the cost of labor, and the costs to the environment (including cleanup). For integrated pest management to become widely adopted, it must result in a net saving most of the time. As compared with 'calendar' spraying (spraying at a particular time of year, without any regard to the level of the pest population or the need to spray), integrated pest management saves costs in chemicals and equipment, but there are costs in monitoring pest populations and in using biological controls.

Introduction of predator species. Planting crops in smaller, separated patches instead of larger, single-species blocks is one way in which the spread of pests can be controlled without the use of chemical pesticides. Planting seasons can sometimes be modified so as to interrupt the life cycles of the pests. The most important techniques in integrated pest management, however, are those that take advantage of the natural enemies that keep the pest species in check. If a predator that preys upon the pest can be identified, then measures that encourage the growth, development, and proliferation of the predator may be able to keep the pests in check. For example, the bacterium *Bacillus thuringiensis* (Bt) can attack the larvae of insect pests and prevent them from destructively feeding on crops; therefore, many plant growers use *B. thuringiensis* on their crops. There are many subtypes of *B. thuringiensis*, each producing a toxin specific for only certain types of insects, so the *B. thuringiensis* type needs to be matched with the pest. Integrated pest management can often cost less than the application of chemical pesticides. For example, a predator species need not be applied repeatedly because it will reproduce naturally on its own, particularly if conditions that support all life stages of the predator species are maintained. As another advantage, because only predators with a very specific and limited range of prey species should be selected as biological controls, there should be less damage to nontarget species.

For example, cotton has long been a crop of commercial importance in the southern United States, India, Egypt, and elsewhere. Cotton pests include the boll weevil and the pink bollworm. Spraying with chemical pesticides initially reduced the levels of these pests, but, by the 1960s, pesticide resistance had developed in both pest species. Despite increased spraying, pest populations continued to increase. Worse yet, the chemical sprays destroyed many of the pests' natural enemies, such as the spined soldier bug, and the destruction of the natural predators allowed other pest species, such as the tobacco budworm (previously unimportant as a pest of cotton), to become significant pests—in some cases more devastating than the original ones.

In both Texas and Peru, integrated pest management techniques have been used successfully to control cotton pests. Soldier bugs and other natural predators are collected, reared, and released on the cotton fields, while chemical spraying has been greatly reduced and, although not eliminated entirely, is used only selectively. The planting season is timed so as to disrupt the life cycle of the pink bollworm moth; when the moths emerge, they can find no cotton plants on which to lay their eggs. Stalks and other unused parts of the plants are shredded and plowed under soon after each harvest, denying the pest insects places to hide and overwinter until the next growing season. In some places, corn and wheat are interplanted with the cotton to help the growth of populations of natural predators and to reduce the ability of the cotton pests to spread from one field to the next.

Alfalfa is another important crop in which integrated pest management techniques have been used successfully. As a nitrogen-rich legume, alfalfa is useful in crop rotation, and its high-quality protein is valued as an animal feed for many domestic animal species. In the United States, alfalfa ranks fourth (behind corn, cotton, and soybeans, and ahead of wheat) in area under cultivation. The principal pests of alfalfa are two related species of alfalfa weevils (genus *Hypera*). At least nine natural enemies of these weevils have been identified, most of them wasps that parasitize either the weevil larvae or other stages of the weevil life cycle. The weevil's life cycle can also be disrupted by harvesting alfalfa early. Alfalfa pests were formerly controlled with chemical pesticides, but this practice was sharply curtailed when one such pesticide, heptachlor, showed up in the milk produced by cows that had eaten the treated alfalfa.

Use of pheromones. Also part of integrated pest management is the spraying of **pheromones**, hormones that function in animal communication (see Chapter 7, p. 289). The pheromone glossyplure is used by female pink bollworm moths to attract their mates. Spraying this pheromone on cotton fields confuses the male moths and interferes with their ability to locate the females, resulting in a natural birth control that is very specific to the pink bollworm and that has no effect on other species. It is, moreover, a chemical to which the bollworm can never develop a natural resistance without impairing its own ability to mate.

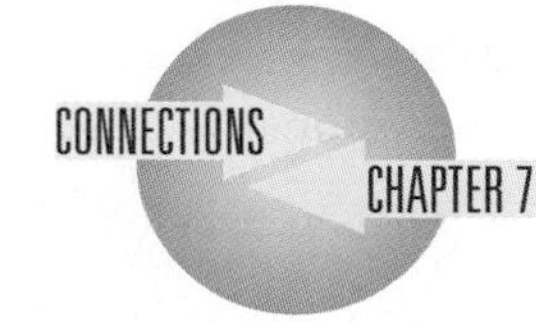

Altering plants through artificial selection

Many plant characteristics, including the size, texture, and sweetness of the edible portion, are at least partially genetically determined. Also under genetic influence are many factors that determine the hardiness of crop plants, their drought resistance, their rate of growth under different soil conditions, their dependence on artificial fertilizers, and their resistance to various pests and plant diseases. Therefore, the yield, both in terms of the amount of crop per acre and the amount of nutrition per unit of crop, can be increased by selective plant breeding.

Artificial selection. Selective breeding is also called **artificial selection**. As carried out by both animal and plant breeders, the practice was already well known in Charles Darwin's time, and served as a model for his theory of natural selection (see Chapter 4, pp. 128–130). Darwin realized that great changes in agriculturally important plants and animals had been made within his own lifetime by British animal and plant breeders. These breeders chose the individuals of the species that best exemplified the trait they desired. They allowed these individuals to mate, while preventing mating between individuals that did not have the desired trait.

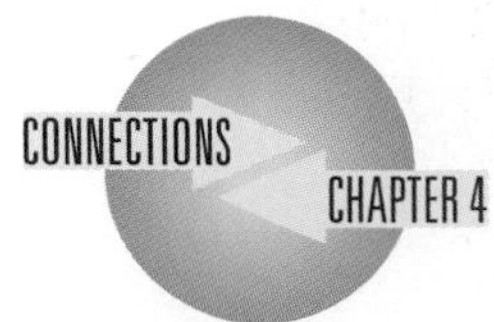

Artificial selection can be used to change almost any trait of a crop species in one direction or the other. A closely related wild species may offer a desired trait, such as a nutritionally more complete protein, in which case the wild species may be crossed with the crop plant as a first step toward the production of a nutritionally superior strain. If this makes you wonder how the concept of crossing members of different species can be reconciled with the biological species definition given in Chapter 4 (p. 147), remember that the definition refers to populations (not individuals) that do not *naturally* interbreed. It is often possible to get individuals under domesticated conditions to do what is not natural for entire populations, for example, by dusting pollen artificially from a cultivated plant species onto a wild relative.

Figure 14.23 shows the results of 50 years of selection to produce corn plants with high or low oil content, or high or low protein content. However, attempts to change only one trait at a time can often result in the production of an inferior strain. For example, it does no good to select for corn plants with larger kernels or larger ears unless the stalks and root systems are capable of supporting them and unless the plants are sufficiently drought resistant and disease resistant to survive under field conditions. Modern breeding practices include the selection for several traits at once, resulting in harmonious combinations of traits that are well adapted to function together as a whole.

By selectively planting strains with desirable traits and by avoiding the use of genetic strains with less desirable traits, agricultural scientists in many nations have dramatically increased crop yields in the last few hundred years. The seeds of high-yielding strains command a high price. Around the world, nations that have achieved the most efficient agricultural production (high cash value yields of major crops per worker-day or per cash unit invested) have generally become wealthy, while those countries with the least efficient agricultural production are generally among the poorest. Thus there is a high correlation between the affluence of a

FIGURE 14.23

The results of 50 years of selection on the oil content and protein content of corn (*Zea mays*). By selectively breeding only those plants that had the highest or lowest protein or oil content in each generation, plant scientists have changed the inherited characteristics of each strain.

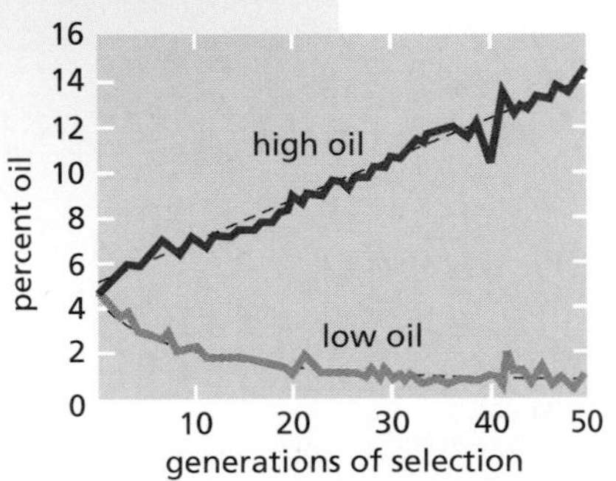

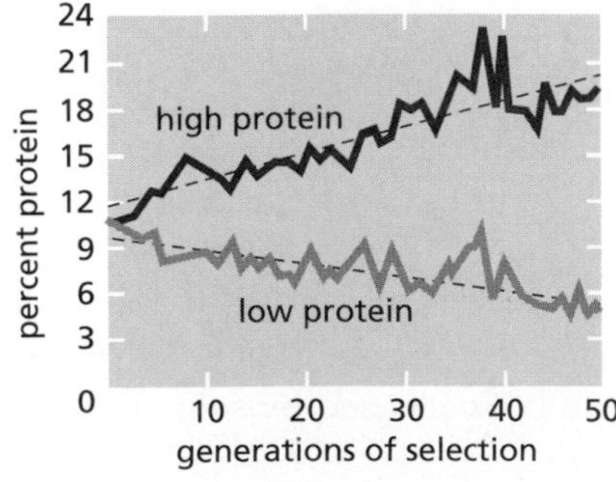

nation as a whole and the efficiency of its agricultural production. The development of new strains of crops, each suited to a particular climate and soil type, is among the most important components of agricultural efficiency, rivaling even the mechanization of agricultural work. Since about 1920 these strains have contributed to the increase in crop yields in industrialized countries. In the United States, for example, crop production doubled between 1940 and 1990 even though some land was taken out of cultivation.

The green revolution. In the 1960s and 1970s, an effort was made to export many new and improved genetic strains of plants from North America and Europe to other parts of the world. This effort, loosely termed the 'green revolution,' was aimed at improving both the agricultural yields and the nutritional content of crops in the recipient countries. For example, some agricultural scientists developed a more nutritious variety of corn, high in lysine, an amino acid in which corn is usually deficient. High-lysine corn provides more complete protein for human nutrition (see Chapter 8, pp. 337–338). Yield was increased by the development of wheat and rice strains with short stems. They produce more grain on less stem and mature earlier so that more than one crop can be planted in a year. Many of these strains were developed in third-world countries under the auspices of international plant breeding institutes established there.

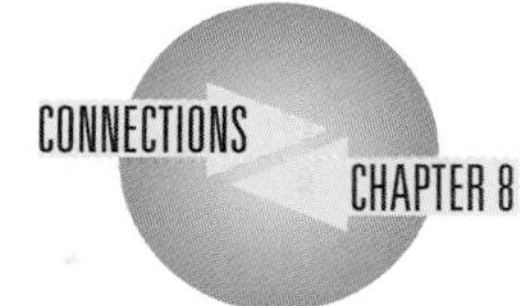

Greatly improved crop strains are, however, as subject to the law of unintended consequences as are other biological interventions. For example, many of these new strains grew well with mechanized agriculture, but getting comparably increased yields in the third world meant the adoption not just of the new plant strains but of irrigation and fertilizer use. Although production on farms was increased 50–100%, it never increased to the extent that it had on research stations. The results of the promised 'green revolution' have been mixed, as is summarized in this quote:

> Forty years after the first adoption cycles of the Green Revolution began in Mexico, and 15 years after they came to completion in Asia, we see that the world is not much better off. A similar percentage (10–15%) of the world population that was undernourished in the 1950s and 1960s is undernourished in the 1990s. The increases in agricultural production, while impressive, have kept just ahead of population growth, and not led to a more even distribution of food to all people. In addition, many problems plague sustainability in the high-input system, in the Third World just as well as in the developed countries (M.J. Chrispeels and D. Sadava, *Plants, Genes and Agriculture.* 1994).

Altering strains through genetic engineering

The kinds of changes to a species that can be accomplished by artificial selection are limited by the genetic variation that exists within the species or its close relatives. **Genetic engineering** offers a newer method for customizing food crops by giving them genetic traits that they normally lack. These may include the ability to live in nitrogen-poor soils and other marginal habitats, the ability to fix atmospheric nitrogen and make their own nitrates, and pest or herbicide resistance.

Plant genetic engineering follows the general concepts of genetic engineering that were described in Chapter 3 (pp. 94–97), but the specific methods of, for example, gene transfer are different. The genetic traits to be changed must first be identified. As in animals, many plant traits are not controlled by single genes and thus are not easily altered by genetic engineering. Most of the traits related to hardiness (drought resistance or cold resistance) are multigene traits. Other traits, for example, pest or herbicide resistance, have been successfully introduced by transferring single genes into plants, often a gene from another species. In this section we examine some techniques of genetic engineering in plants and the uses to which plant genetic engineering is put.

Genetic engineering is not simple. Whether or not a genetically altered plant actually makes a desired protein depends on whether the desired gene has been inserted into a portion of the genome that is transcribed into mRNA. This means that the gene must be located 'downstream' from the signals that control gene expression (see Chapter 9, p. 367). We also need to know how the altered plant will function at the ecological level, knowledge that we most often do not have until after the genetic engineering has been performed.

Insertion of new genes by plant viruses. Scientists can use viruses to insert a new or altered gene into a plant genome. First they use restriction enzymes to make 'sticky ends' on the DNA fragment and on the viral DNA (see Chapter 3, pp. 94–95). The sticky ends bind the gene to the viral DNA, and the scientists then make use of the ability of viruses to incorporate into the host genome (see Chapters 3, p. 107 and 13, p. 547). The virus enters the plant cells and adds its DNA and the new gene to the plant's DNA.

One virus used in genetic engineering experiments in plants is the tobacco mosaic virus. Like most viruses, the tobacco mosaic virus is restricted in its choice of hosts. Because the virus reproduces only in tobacco plants, it very unlikely that it could accidentally spread new traits to other species. Tobacco mosaic virus has the advantage that both its biology and that of its host, the tobacco plant (*Nicotiana tabaccum*), have been intensively studied for decades. In cases in which it does not matter what plant is used in producing a particular compound, the tobacco plant is a logical choice, because methods for its cultivation and for the growth and insertion of tobacco mosaic virus are well known.

One type of genetic engineering that uses tobacco plants is called molecular farming. The goal here is not to make a better tobacco plant, but to use the tobacco plant as a biological factory to produce, say, a medicine even small quantities of which would be valuable. The genetically engineered medicine would need to be purified to remove the nicotine and other tobacco plant molecules, but this might not be any more expensive than purifying the medicine from its original plant source. Further, the tobacco plant may grow in places where the original plant source will not, or it may grow faster. Success in such endeavors might encourage tobacco farmers to grow more plants for pharmaceutical uses and fewer for cigarettes. Especially if the pharmaceutical plants became more profitable, tobacco farmers would have reason to switch away from growing the crop that is currently the leading cause of lung cancer and other diseases such as emphysema.

A recent example of such molecular farming is the insertion into tobacco plants of a gene from cows that codes for lysozyme. Lysozyme is found naturally in the saliva of many animal species, where it has an important antibacterial function because of its ability to digest the cell walls of many bacterial species. Cow lysozyme produced by tobacco plants makes the tobacco leaves resistant to those bacteria. The lysozyme so produced can be used to treat seeds from many plant species. Disease-transmitting bacteria on seeds are a major agricultural problem. Treating seeds with dilute lysozyme produced by genetically engineered tobacco plants clears harmful bacteria from the seeds.

Insertion of new genes by bacterial plasmids. Because tobacco mosaic virus does not enter cells of species other than the tobacco plant, this virus is unsuitable for changing most crop species. For these species, genetic engineers have experimented principally with *Agrobacterium tumefaciens*, a bacterial species that causes tumors in many plant species. *Agrobacterium*, like many bacteria, can carry DNA fragments known as **plasmids**. Plasmids can be used to carry new genes into host species, as was described in Chapter 3 (p. 97). In the normal life cycle of *Agrobacterium*, plasmids can introduce bacterial genes into the cells of the host plants, and these genes cause tumors to form. The plasmids used in genetic engineering are modified so that they are still capable of introducing bacterial genes into the host plants, but they lack the gene that causes the tumor formation, so these plasmids no longer induce tumors. Many different plant species can serve as hosts to *Agrobacterium* and can incorporate the plasmid; several other plasmids are useful in only one or two host species.

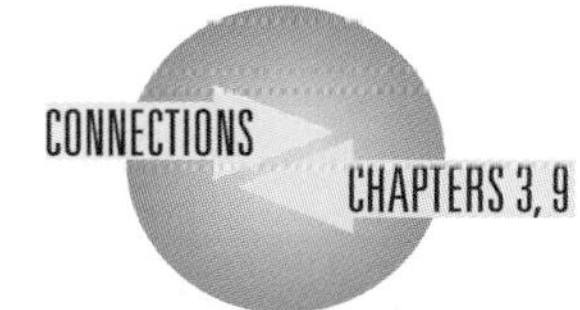

The plasmid can be engineered to carry the gene of interest (the gene associated with the trait being modified), as well as regulatory DNA sequences (promoters and enhancers) that will enable the gene of interest to be expressed (see Figure 9.7, p. 368). The regulatory DNA may also control the tissues of the plant in which the gene will be expressed. The engineered plasmid is then induced to enter the *Agrobacterium* cells. The bacteria are incubated with pieces of the plant to be transformed. The bacteria then insert the plasmid carrying the desired genes and regulatory DNA sequences into the genome of the recipient plant cells (Figure 14.24).

Other methods of gene insertion. In any kind of genetic engineering, getting the inserted DNA past the cell membrane and nuclear membrane of the recipient cell is one of the stumbling blocks. In addition to host-specific virus or plasmid transfer methods, less specific methods of gene transfer have been developed. In the particle-gun method, plant pieces are literally shot with tiny pellets whose surfaces carry the DNA for the

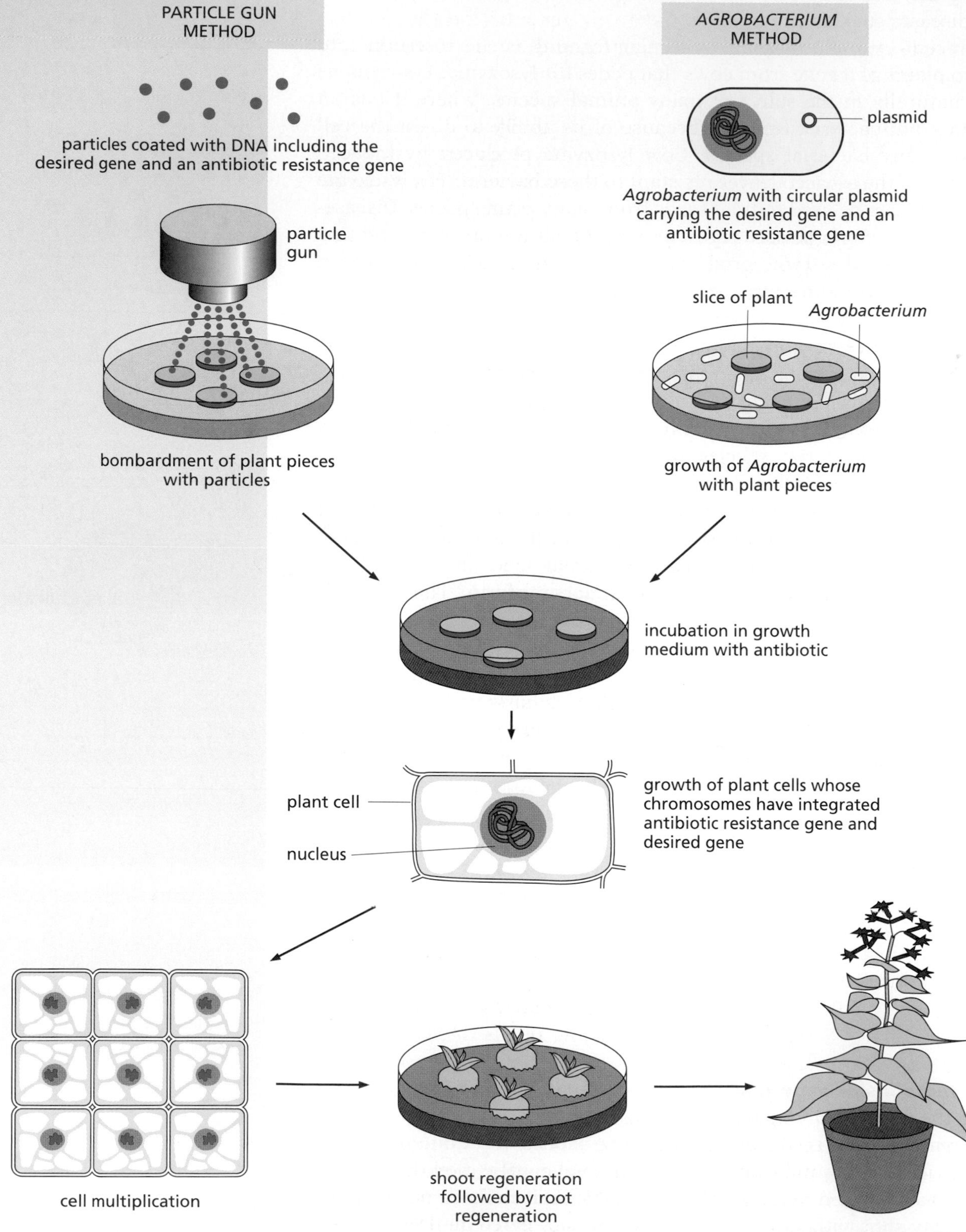

PARTICLE GUN METHOD
particles coated with DNA including the desired gene and an antibiotic resistance gene
particle gun
bombardment of plant pieces with particles
AGROBACTERIUM METHOD
plasmid
Agrobacterium with circular plasmid carrying the desired gene and an antibiotic resistance gene
slice of plant
Agrobacterium
growth of Agrobacterium with plant pieces
incubation in growth medium with antibiotic
plant cell
nucleus
growth of plant cells whose chromosomes have integrated antibiotic resistance gene and desired gene
cell multiplication
shoot regeneration followed by root regeneration
plant with new trait

FIGURE 14.24

Methods for inserting new genes into plant cells.

gene of interest and its regulatory DNA sequences (see Figure 14.24). The force of the bombardment rams some of the pellets through the plant cell walls and membranes. The 'naked DNA' carried by the pellets will sometimes incorporate into the DNA of the plant cells. In another nonspecific method called electroporation, temporary holes are made in plant cell membranes by disrupting them with electrical current. While these holes are open, DNA can enter the cytoplasm and then the nucleus, before the hole closes again.

Screening with antibiotic resistance genes. Regardless of the method of entry, the uptake and incorporation of new DNA do not happen in most plant cells that are treated. Scientists therefore need a method of screening to find out which cells actually have the new gene. This is done in most cases by incorporating another gene along with the gene for the trait being modified. This additional gene is usually one that confers resistance to some antibiotic, as shown in Figure 14.24. After treatment of the plant pieces by some insertion method, the pieces are incubated in growth medium containing the antibiotic to which the screening gene confers resistance. Those cells that have not taken up the transferred genes are killed by the antibiotic, while those cells in which the new genes have been successfully incorporated will survive and grow.

Cloning plants. Once it has been determined which cells have incorporated the gene of interest, those cells can be used to grow complete plants. Many plants can be grown from single cells containing the integrated DNA (see Figure 14.24). Growing many genetically identical plants from single cells is another form of **cloning**, which we have previously discussed in other contexts. Such methods involve growing the cells in small dishes in growth medium containing plant growth hormones, from which complete plants will develop without sexual reproduction or seed formation. Plants are particularly amenable to cloning, and methods have been developed for the cloning of many species of plants. These techniques are not restricted just to growing genetically engineered plants; they can also be used to clone plants with desirable traits that have been developed by traditional plant breeding (artificial selection). In fact, cloning of traditionally developed plant strains is in far more general use and has had a far greater impact on agriculture around the world than any form of genetic engineering.

Transgenic plants with altered nutritional content. Genetic engineering using *Agrobacterium* was first achieved in 1983. Since then, over a dozen plant species have received genetically engineered genes by using the techniques just described. In all, **transgenic** plants (meaning those with genes derived from another species) have been produced in over 20 species, including tomatoes, potatoes, carrots, alfalfa, peas, cotton, and sugar beets. Few of these have become commercially available.

Although genetic engineering holds the promise of producing plants strains with higher or more complete nutritional content, very few of the strains that have been developed have actually been changed in this way. One that has is the potato, in which the starch content has been increased by insertion of a bacterial gene. Potatoes are already 21–22%

starch, but an even higher starch content makes the potatoes better for processing into potato chips and frozen french fries, the major ways in which potatoes are consumed in the United States.

The canola plant (*Brassica napus*), whose seeds are the source of canola oil, has been engineered to produce lauric acid, a saturated fatty acid used in the food industry (for food additives such as coffee creamers and cake and candy coatings) and in the detergent industry. Lauric acid is not ordinarily made by *B. napus*. It is naturally synthesized by tropical plants such as palm and coconut and laurels, but by inserting a gene from the California bay laurel tree, scientists have modified a canola strain to produce lauric acid.

Pest-resistant transgenic plants. Genetic engineering does not always alter the nutritional content of crops. Sometimes genetic engineering is instead used to modify some other trait that makes growing agricultural crops more efficient. Currently, the most widespread transgenic plants are those in which the gene for a toxin from the bacterium *B. thuringiensis* (Bt) has been inserted. *B. thuringiensis* toxin is a natural insecticide and the bacteria are widely used for spraying onto plants as part of integrated pest management. Transgenic plants containing the toxin gene make the toxin themselves, doing away with the need for repeated spraying of the bacteria onto crops. The toxin protects the crop plants against damaging pests but does no harm to the natural enemies of those pests. Cotton, potatoes, and corn with Bt toxin genes are all commercially available.

Resistance to nematodes has also been engineered into potatoes, tomatoes, and sugar beets. Nematodes are roundworms (kingdom Animalia, phylum Nematoda), which do a tremendous amount of damage to the roots of plants, in many parts of the world, but particularly in tropical climates. NemaGene^{TM} is a patented gene for a protease inhibitor enzyme that kills nematodes as they feed on the plant roots. This gene has been engineered with regulatory DNA sequences that ensure that it is transcribed and translated only in the roots of the plant, not in the parts of the plant that humans eat.

Herbicide-resistant transgenic plants. Yet another use for genetic engineering in plants is the introduction of genes for herbicide tolerance. If a crop plant is given a gene that allows it to resist (or tolerate) a particular herbicide, then the herbicide can be used as a weedkiller to control weed species that would otherwise compete with the crop for water and other limited resources. A bacterial gene conferring resistance to Roundup^{TM}, a popular wide-spectrum herbicide, has been introduced into the most commonly planted strains of corn and soybeans, making them resistant to the herbicide. Fields can now be sprayed with the herbicide, killing the weeds but sparing the crops that carry the resistance gene. Over half of all soybeans grown in the United States now have this genetically engineered gene.

Products refined from transgenic crops include oil and corn syrup from corn, and protein, oil, lecithin, and several vitamins from soybeans, all of which are now a common part of the food supply in the United

States. The transgenes have to do with pest or herbicide resistance, not with the character of the food produced by the plant. This poses a problem for European countries that have sought to ban genetically engineered crops, because once a refined product like corn syrup has been extracted, there is no way of determining the genetic background of the plant source.

Crop plants modified by gene silencing. In 1994, the Flavr-Savr™ tomato was introduced into markets in the United States as the world's first commercially available genetically engineered fresh produce. The gene introduced into this tomato variety is an **antisense** gene. In double-stranded DNA, the strand that carries the gene is called the sense strand; the strand complementary to it is the antisense strand. Normally the antisense strand is not transcribed because it is not preceded by binding sites for RNA polymerase, but antisense strands can be engineered to carry regulatory elements enabling them to be transcribed. When the plant cells transcribe the antisense DNA into mRNA, this antisense mRNA is complementary to the normal mRNA. The single-stranded antisense mRNA binds to the complementary single-stranded normal mRNA decreasing its translation into protein. This strategy is known as **gene silencing** because the expression of the gene into its protein product is inhibited.

A gene had been identified that codes for a tomato enzyme that softens the tomatoes as they ripen. In production of the Flavr-Savr™, knowledge of the DNA sequence of this gene was used to synthesize its complementary, antisense DNA, and the antisense DNA along with regulatory DNA was inserted into a strain of tomato plants. The engineered plants secrete less of a softening enzyme, which allows the tomatoes to be vine ripened, thus becoming more flavorful, and still last through transportation and storage. The Flavr-Savr™ tomato also has less water in proportion to the amount of pulp, making it advantageous for the manufacture of ketchup, sauces, and soups by reducing the amount of energy needed to boil off the water. It is in far greater use in these cooked commercial products than it is as fresh produce (indeed, the Campbell's Soup Company commissioned its development by the Calgene Corporation).

***Ice*-minus bacteria.** Not all agriculturally important genetic engineering is done on the plants themselves, and not all genetic engineering involves gene insertions. Bacteria have been developed that can delay ice formation on plants, protecting them from frost damage, which costs farmers in the United States $40 billion a year. As the temperature drops, frost crystallizes on plants at spots where bacterial cell walls provide tiny regions of molecular regularity. A gene was found in the bacterium *Pseudomonas syringae* that contributes to cell wall regularity, and the gene was removed from one strain. These engineered bacteria have irregular cell walls and when the killed bacteria are sprayed onto plants, the temperature must be colder before frost forms. Because scientists had named the normal gene *ice*, the bacteria from which it was deleted are called *ice*-minus. These bacteria are used to protect frost-sensitive crops such as strawberries and tomatoes.

Genetically engineered plant products of the future. The above-mentioned genetically modified species are all now commercially available. Many more strains are in development but are not yet marketed. Plant strains have been engineered to make both soybeans and canola oil more nutritionally complete by the inclusion of methionine, an amino acid not normally produced in these plants (see Chapter 8, pp. 337–338). Plants resistant to viral and fungal infections are being developed, as well as plants with increased abilities to fix nitrogen or to be tolerant of high-salinity soils or drought conditions. Many edible vaccines are being developed; for example, bananas have been modified to make a viral molecule that, when eaten by people, immunizes them against the viral disease hepatitis B. Plants have also been engineered to produce nonfood products, such as polyhydroxybutyrate, which is used in the manufacture of plastics that can be broken down by decomposer organisms. The Monsanto Corporation has a research project that it calls the Blue Gene project, in which a transgene for a blue pigment inserted into cotton plants makes them produce blue cotton, decreasing the need for chemical dyeing to produce denim for blue jeans.

Opposition to genetically engineered agricultural crops. Whether these genetically modified plants will become commercialized, and whether those commercially available will gain widespread acceptance, will depend in part on overcoming opposition to genetic engineering. Part of the opposition is biological and part of it is ethical.

Ethical questions include both deontological and utilitarian concerns. A few critics, notably Jeremy Rifkin, have consistently opposed all biotechnology, especially transgenic research. We are, these critics say, attempting to alter nature by going considerably beyond the bounds that nature intended. Rifkin goes so far as to question whether *any* transgenic research can be ethical. In the terminology explained in Chapter 1, Rifkin might be described as a deontologist who believes that any transplantation of a gene from one species to another is inherently unethical, a stance that prevents the experimental measurement of certain risks (among other things). One possible answer to such criticisms—one grounded in natural law ethics (see Chapter 1)—is to point out that genomes are being rearranged all the time in nature. Gene transfer (introgression) between related plant species happens fairly often in plants, and cross-breeding has been transferring genes between domestic plant strains of the same species for centuries.

Other deontological objections include those from some vegetarians who are opposed to products that might contain even a gene from an animal. Religious groups have put different interpretations on the meaning of a transferred gene. Muslims consider the transgene to retain some fundamental aspect of its species of origin; therefore, foods with genes transferred from animals such as pigs may not be eaten. Jewish groups, in contrast, view the transgene as becoming part of the species into which it is transferred and therefore do not see an objection to the use of foods from plants that might have received a gene from a pig. Muslims also draw a stronger distinction between plants modified by traditional breeding methods and those that are 'engineered.' Christian and Jewish groups generally hold that different methods of modification are ethically equivalent, in keeping with the Judeo-Christian tradition of humans' having dominance over nature. The equivalence of modification methods is the view generally taken by scientists also, although they are more likely to consider this to be a scientific, rather than an ethical, issue. For example, the Scientific Committee on Problems of the Environment (SCOPE), a committee of the International Council of Scientific Unions, has stated that "there are no convincing scientific grounds for distinguishing engineered organisms from natural ones" (H.A. Mooney and G. Bernardi, *Introduction of Genetically Modified Organisms Into the Environment.* 1990). Also, "because organisms of either type could pose unforeseen hazards, some safety testing is desirable before large-scale propagation."

Utilitarian ethical arguments center more on the premise that the risks of genetic engineering are poorly understood and quantitatively uncertain. Risks probably vary from one plant species to another, making it important for the questions to be raised (and the research conducted to answer them) again and again for each new application. For example, the use of the *Agrobacterium* plasmid has been criticized because the original form of this plasmid stimulates plant tumor formation. Suppose that a genetically altered strain of a plant containing an *Agrobacterium* plasmid managed to reacquire the gene for plant tumor formation, either by mutation or (more likely) by genetic recombination with wild strains of *Agrobacterium*; plants carrying the plasmid might then grow tumors. The probability of such an event must be carefully estimated if reliable risk–benefit ratios (see Chapter 1) are to be obtained and if defensive measures against such mutant strains are to be planned in advance. Potential economic benefits from genetically engineered crops are generally easy to estimate; risks often are much more uncertain. Many utilitarians are not opposed to genetic engineering in principle, but argue that the risks are great and should be evaluated before we proceed.

The risks (or potential dangers) of biotechnology do exist, and some of them have been alluded to earlier. Many biologists, particularly ecologists, argue that there are biological questions that need to be answered. Most of these center on a perspective that says we cannot think only at the molecular level. When molecules operate in plants and when those plants may be introduced into the environment, we need to consider the functioning of the system as a whole. The Union of Concerned Scientists has recently published a summary of their concerns, which are in two areas, the possible escape of genetically altered strains as superweeds, and the spread of plant viruses. They argue that because agricultural crops cannot be isolated from their surrounding ecosystems, transgenic plants pose risks that other genetically engineered species, such as bacteria grown in factory vats for the production of medicines, do not. They caution that transgenic plants could escape from cultivation and become weeds. They give as examples various plants, such as kudzu and purple loosestrife in the United States, that have been introduced into new locales and have overrun the environment, altering the habitats of other plants and of animals. (Scientists in SCOPE do not see this as an equivalent example; genetically engineered crops, they say, would most probably be introduced into areas where the unengineered form of the crop had already been grown, so natural ecological balances should still apply.) A larger risk, though, is whether engineered plants would cross-pollinate or otherwise transfer their new genes to weeds. Plant species frequently do cross-pollinate with weeds of related species. This might be particularly worrisome if herbicide-resistant plants passed their transgenes to weeds, making the weeds herbicide resistant. The Union of Concerned Scientists caution that small field trials are not necessarily good predictors of full-scale agricultural conditions, particularly if the field trials have not been designed to examine ecological effects. They propose expanded testing protocols that should be done before crop species are approved.

The issue of the spread of plant viruses, or of the creation of new viruses from recombinations of the virus used to insert the transgenes with normal viruses already present in the plant, is also of concern. Although viruses are known to recombine, the risk is considered to be less, simply because viruses are not now frequently used for gene insertion into plants.

Another issue raised by biologists is a food safety issue, but again it has to do with thinking in a more integrated way about how systems function. The issue here has to do with the antibiotic resistance genes that are engineered into plants along with the gene of interest (see Figure 14.24). The unanswered question is whether these genes will contribute to humans and animals becoming resistant to the antibiotics. Many disease-causing bacteria are already becoming resistant to antibiotics owing to their overuse in animal feeds and their improper use medically (for example, in the treatment of colds, which are viral diseases against which antibiotics have no effect). If transgenes in food add to this growing problem, it could help undo the control of infectious diseases that has been achieved in the past fifty years.

Much of the opposition of member nations within the European Union (EU) to genetically engineered foods centers on this last point. The EU banned Ciba-Geigy Corporation's Bt transgenic corn in 1996 because it also contains an inserted gene for ampicillin resistance. While this would not pose any direct threat to human health, the possibility exists that the resistance gene could be transferred to gut bacteria. Most, but not all, of the DNA ingested in food is destroyed by strong acids present during digestion. The possibility of resistance transfer is further diminished in any food that is cooked, since the DNA and its proteins would be destroyed in the cooking. Because ampicillin is used to treat many human diseases, and because resistance to it often confers resistance to many other penicillin-type antibiotics, the development of resistant bacteria in the gut, which could possibly transfer resistance to pathogenic bacteria, is a risk that many scientists feel should be more thoroughly investigated. The EU reversed its outright ban, but now requires labeling on any food that contains live genetically modified organisms, or has modified ingredients that are not equivalent to, or materials that are not present in, the original, or has substances that might be objected to on ethical grounds (such as animal genes that might be opposed by vegetarians).

Another line of opposition to genetically engineered crops is that they do not contribute to efforts to develop sustainable agricultural practices. The Union of Concerned Scientists states that we should be developing sustainable practices that prevent environmental problems in the first place, rather than focussing on solving problems after they are created. The proponents of genetic engineering point to crops modified to be pesticide resistant as an example of ways in which plant engineering could cut down on pesticide use. On the other hand, herbicide-resistant crops have no benefit if they are not used in conjunction with the matching herbicide. (In fact farmers who use Monsanto Corporation's RoundupReady™ soybean seeds are required to sign an agreement with the company to use only Monsanto's Roundup™ herbicide, and are faced with heavy fines imposed by the company if they do not.)

Some see genetically engineered plants, in part because they are developed and marketed by large multinational corporations, as inherently contributing to monoculture practices, thereby also contributing to the loss of biological diversity, a topic that we examine in greater detail in Chapter 15.

The other half of a risk–benefit equation is the benefit side. Many see the benefits of engineered plants as potentially immense. Others see the profits as potentially immense, but the benefits to society as very small. The countries in which farmers will be able to afford these seeds are countries that are already awash in excess food. Examples are raised such as the engineering of lauric acid into canola, which will benefit North American farmers that grow canola, but this will be at the expense of tropical farmers who grow palm and coconut, the natural sources of lauric acid. Because engineering is seldom done on the primary food crops used in the third world it is unlikely to be of any direct benefit in these countries. Moreover, the use of patented or trademarked plant strains, or the increasing use of plant strains that require mechanized agricultural methods will probably make third-world farmers more dependent on imported seed supplies and on foreign debt. Many groups of scientists also caution that genetic engineering will not solve the world's food problems; they see these problems as being largely due to the unequal distribution of food, not to a lack of food production. They point to the example of the green revolution, which has tremendously increased production but has not done away with hunger.

The possible benefits of the genetic engineering of crops are very large. The monetary costs may also be large, but they are very difficult to estimate because of the great uncertainties involved. The biological dangers are not fully known. Given the law of unintended consequences, difficulties should not be underestimated. The risks may prove to be minimal, or they may prove to be significant, and the dangers may prove to be easily controlled even in worst-case scenarios—we will never know unless we undertake the relevant investigations for each species. Only if we have investigated the possible dangers will we be able to assess the possible risks. That is one reason why it may be desirable to proceed with testing, why all applications should be closely monitored, and why many people await the evaluation of risks as well as benefits.

THOUGHT QUESTIONS

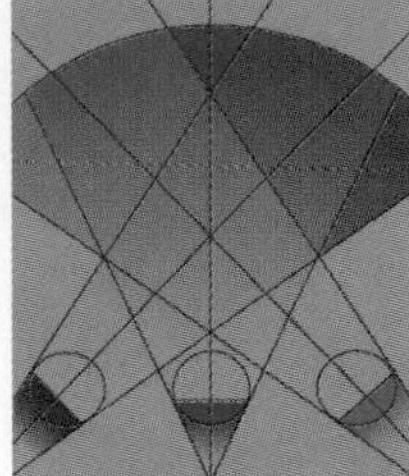

1. The runoff of fertilizers from agricultural fields often produces algal blooms. Can you explain why this would be so? (What limits algal growth under normal conditions?) Would the problem be greater with organic fertilizers or with inorganic ones? Why do you think so?
2. Compare the volume occupied by the same weight of commercially available potting soil and sand. Compare the amount of water that can be held by equal weights of soil and sand. Now compare those results with the amount of water that can be held by soil from your area. Does the soil in your area contain a lot of humus? What ways can you think of to improve the quality of your soil?
3. Can you see any evidence of erosion in the area where you live? What natural processes or human activities might contribute to soil erosion?
4. Once the toxic effects of DDT on nontarget species received widespread publicity, agricultural use of the chemical was banned in the United States and in many European countries. The United Nations considered imposing a worldwide ban, but this effort was stopped by the insistence of many third-world nations that they needed the pesticide to help control both crops pests and mosquitoes. Do you think DDT should have been banned in countries like the United States? Do you think the ban should have been extended worldwide?
5. What do you think would happen if DDT were allowed to be used in limited amounts? How might the limits be enforced? What would be done if a farmer found that he or she could kill more pests (and thus increase crop yields) by using more DDT and causing potential future harm to the environment? How intrusive would enforcement agencies need to be? Is a total ban more practical than a limited ban? Would rationing work? Would an 'agricultural prescription' system (similar to medical prescriptions for drugs) work?
6. Once pest resistance to a pesticide arises, it can spread quickly through any pest populations treated with the pesticide. Explain this fact using your knowledge of genetic mutations and natural selection.
7. Artificial selection and genetic engineering are both ways of modifying plants. In what ways are the two methods similar? In what ways are they different?
8. Experimenters attempting to alter plant strains by selecting for one trait sometimes end up changing not only that trait but some other trait along with it. Use your knowledge of genetics to explain why this is so.
9. Do genetically engineered foods increase nutritional quality for consumers or are they of more benefit to mechanized farming and the food processing industries? Develop arguments on both sides.
10. Jeremy Rifkin and other critics of genetic engineering have argued that the escape of a genetically engineered strain from cultivation would be a chaotic event whose consequences are inherently unknowable. Do you agree or disagree? Is there any way of planning for such events? If a genetically altered strain of plants (say, tomatoes that stay on the vine longer to develop better flavor or color) were found growing outside cultivated areas, what should our response be?

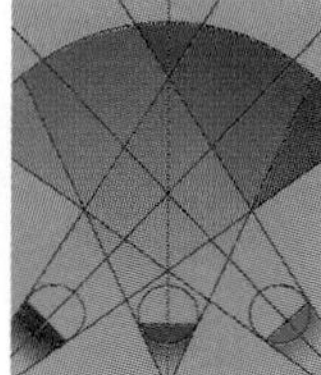

AS THE WORLD'S POPULATION CONTINUES TO INCREASE methods are being developed to make agricultural production more efficient. Overcoming various limiting factors, such as nutrients or water, increases food production. Sustainable agricultural practices aim at ensuring that production will remain high far into the future. Plants need light energy for photosynthesis, and light energy is more abundant in countries close to the Equator, including many poor countries. Wherever light energy is naturally abundant, the need for fertilizers can be diminished by planting crops that harbor symbiotic nitrogen-fixing organisms in their roots. Alternatively, the genes for nitrogen fixation or for other desirable traits can be genetically engineered into plants that do not naturally possess them. Genetic engineering can also be used to make plants more nutritious and more resistant to drought and to pests.

Although there is considerable opposition to genetic engineering, most groups of scientists conclude that the risk–benefit ratio still argues in favor of continuing research on genetic engineering of food crops. Genetically engineered crop plants may increase food yields and nutritional value. The possible benefits of the genetic engineering of crops are thus immense, although the dollar costs may also be immense. Many scientists caution that increased food production by itself will not solve the world's food problems. They see these problems as being also due to the social, economic and political forces that result in unequal distribution of food. Plant science will be an important part of any solution, but world food problems will not be solved by science alone.

CHAPTER SUMMARY

- Plants make carbohydrates by the process of **photosynthesis**, using water and atmospheric carbon dioxide as raw materials and sunlight as an energy source.
- Because plants can make all of their own organic compounds they are **autotrophs**; organisms that cannot, including all animals, are **heterotrophs**, dependent on autotrophs for their food.
- Proteins, nucleic acids, vitamins, and other plant products require nitrogen for their synthesis. Most plants get their nitrogen from dissolved nitrates, which limits the distribution of many plant species to soils that contain adequate nitrogen. Some plants form **mutualisms** with microorganisms that can fix atmospheric nitrogen and convert it into a form that the plants can use. The cycling of nitrogen through the biosphere and atmosphere is called the **nitrogen cycle**.
- The overwhelming majority of plants have vascular **tissues** that allow them to grow tall and allow other tissues to specialize further.
- **Vascular plants** absorb water through their roots and evaporate water through the stomates in their leaves. **Osmosis** generates water pressure, and this pressure contributes strength to most plant tissues; a lack of water pressure causes wilting.
- Crop yields can be increased by supplying **limiting nutrients** through **fertilizers** or soil improvement, by supplying water, by controlling pests that compete with the plant or with humans for the energy produced by

plants, and by altering the traits of the plants either by **artificial selection** or by **genetic engineering**.

- Pesticides and chemicals can become concentrated in biological tissues by **biomagnification**.
- **Monocultures** allow the rapid expansion of pest species.
- **Integrated pest management** and **transgenic** plants can reduce our dependence on chemically produced pesticides and herbicides.

KEY TERMS TO KNOW

artificial selection (p. 618)
autotroph (p. 585)
biomagnification (p. 612)
chlorophyll (p. 586)
chloroplast (p. 587)
economic impact level (p. 616)
fertilizer (p. 604)
genetic engineering (p. 619)
heterotroph (p. 585)
integrated pest management (p. 614)
limiting nutrient (p. 590)
monoculture (p. 610)
mutualism (p. 593)
nitrogen cycle (p. 591)
organ (p. 596)
osmosis (p. 599)
photosynthesis (p. 584)
tissue (p. 596)
transgenic (p. 623)
vascular plants (p. 596)

CONNECTIONS TO OTHER CHAPTERS

Chapter 1: Genetic engineering of crops raises ethical issues. Pest control and fertilizer uses have both costs and benefits.

Chapter 2: Artificial selection can be compared to natural selection.

Chapter 3: Genetic engineering techniques can be used on crop species as well as other species.

Chapter 4: Plants have adapted to their environments in the course of evolution.

Chapter 5: Some human populations have evolved that cannot digest certain plant crops that are useful nutrition for most humans.

Chapter 6: Feeding the world's growing population will be aided by increased crop yields and more nutritious crops.

Chapter 8: Undernutrition and malnutrition affect human health. Plants are the source of most nutrients that we need.

Chapter 9: Several anti-cancer drugs are plant products. The plasmid used in plant genetic engineering originally induced tumors in host plants.

Chapter 10: Plants do not possess nervous systems or contractile muscle fibers, in contrast with animals.

Chapter 11: Most drugs are plant products, and several are psychoactive in humans.

Chapter 15: Clearing more land for agriculture threatens biodiversity and destroys ecosystems.

PRACTICE QUESTIONS

1. What is the difference between an autotroph and a heterotroph?
2. Name at least:
   a. one microscopic heterotroph
   b. two heterotrophs larger than your thumb
   c. two autotrophs
3. The major chemical process in the light reactions of photosynthesis involves the splitting of _____ and the release of _____.
4. In the dark reactions of photosynthesis, _____ from the atmosphere is incorporated into organic molecules such as _____.
5. What form of energy enters photosynthesis? In what form is that energy stored at the end of photosynthesis? In what other forms is energy stored during photosynthesis?
6. What carbohydrates can be obtained by eating plants? What function(s) do these carbohydrates serve in the plant?
7. Name three types of compounds containing nitrogen that plants need to make. Do humans need nitrogen for these same compounds?
8. How do plants obtain nitrogen? How do humans obtain nitrogen? Can either plants or humans obtain nitrogen from the air?
9. What is a limiting nutrient? What are some examples of nutrients that can be limiting for the growth of plants? How can farmers supply limiting nutrients to plants?
10. Which part of a vascular plant is responsible for each of the following?
    a. uptake of water
    b. the bulk of photosynthesis
    c. the major portion of fluid transport
    d. holding the plant up
    e. pollination in angiosperms
    f. anchoring the plant in place
    g. seed dispersal in angiosperms
11. Are all vascular plants flowering plants? Are all flowering plants vascular plants?
12. What molecules move through membranes during osmosis?
13. What are the functions of humus?
14. Under integrated pest management, name:
    a. two items that cost more time or money than in traditional forms of pest management
    b. two items that cost less time or money than in traditional forms of pest management
15. Name four ways of introducing a gene into a strain of plants in which it is not already present.

15 Biodiversity and Threatened Habitats

CHAPTER OUTLINE

Biodiversity Results from Ecological and Evolutionary Processes
- Factors influencing the distribution of biodiversity
- Interdependence of humans and biodiversity

Extinction Reduces Biodiversity
- Types of extinction
- Analyzing patterns of extinction
- Species threatened with extinction today

Some Entire Habitats Are Threatened
- Tropical rainforest destruction
- Desertification
- Valuing habitat

Pollution Threatens Much of Life on Earth
- Detecting, measuring, and preventing pollution
- Air pollution
- Acid rain

Polluted Habitats Can be Restored
- Bioremediation of oil spills
- Bioremediation of wastewater
- Treatment of drinking water
- Costs and benefits

ISSUES

What is biodiversity? How is it measured?

What do we lose if we lose biodiversity? Why should humans be concerned?

How do humans contribute to loss of biodiversity or to habitat destruction?

How do economic disparities among people influence habitat destruction? How do economic disparities among nations influence habitat destruction?

Is it possible to live in the industrial world without polluting?

How can societies limit the threats to habitats and to biodiversity?

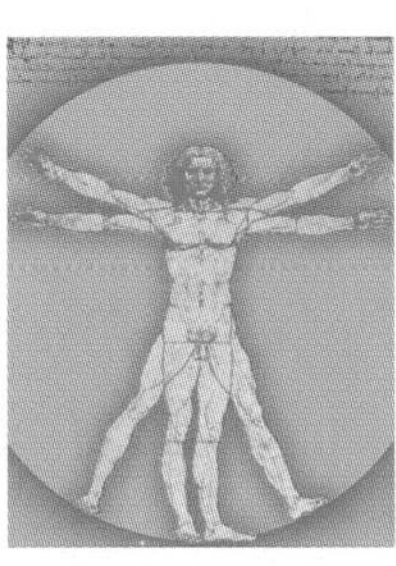

BIOLOGICAL CONCEPTS

Biodiversity (species diversity, species richness)

Scales of time (geological time, human experience)

Extinction

Specialization and adaptation

Community structure (mutualisms and other interactions between species)

Biogeography (biomes)

Ecosystems (habitats, habitat alteration, human influences, biosphere)

Conservation biology (renewable and non-renewable resources, sustainable and nonsustainable uses)

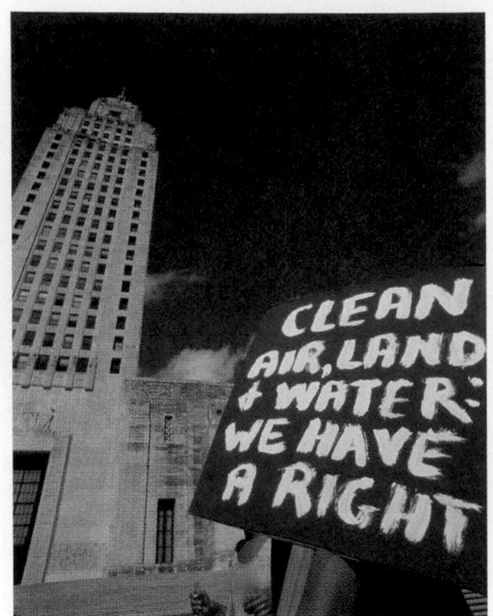

Chapter 15: Biodiversity and Threatened Habitats

In the shadow of trees over 60 m tall (more than 200 feet, or as high as a 17-story building), workers use bulldozers and other heavy equipment to clear a 100-m-wide (312-foot-wide) path through a tropical forest. They are building a new road that will bring commerce and communications to the people of the region and will enable them to send their agricultural products, crafts, and minerals to markets in faraway countries. For each kilometer (0.62 mile) of roadway, they are destroying 10 hectares (about 24.7 acres) of tropical rainforest. The building of the road brings many high-paying jobs to the workers who build it, and the road will also open up new land for agriculture and human settlement, a process that will destroy even more forest. The trees are important in themselves, and also because they provide ***habitat*** *(a set of environmental conditions that make up a place to live) to thousands of species.*

The number and variety of species in a place are referred to as biological diversity or, simply, ***biodiversity****. Biodiversity is measured most easily by the number of distinct species present. More broadly, biodiversity also includes genetic diversity within species and also ecological diversity within habitats or ecosystems. In this chapter we consider the importance of biodiversity, the conditions that support biodiversity, the many threats to biodiversity, and some ways in which humans can reduce these threats.*

Recall from Chapter 4 that species are reproductively isolated groups of interbreeding natural populations. Most new species originate by a process of geographical speciation, in which reproductive isolation evolves during a period of geographic separation. The processes that give rise to new species increase biodiversity, while the processes that result in the extinction of species decrease biodiversity.

BIODIVERSITY RESULTS FROM ECOLOGICAL AND EVOLUTIONARY PROCESSES

There are nearly 1,500,000 species of organisms currently known to science. More than half of these (53.1%) are insects, and another 17.6% (approximately 250,000 species) are vascular plants, so that over 70% of all known species are either vascular plants or insects (Figure 15.1). Animals other than insects make up 281,000 species, or about 19.9% of the total, and the remaining 9.4% are fungi, algae, protozoans, and various procaryotes, most of them microscopic.

Our knowledge is far from complete. Various estimates put the total number of species—known and not yet identified—between 5 and 30 million. Such estimates are extrapolations from the few studies in which

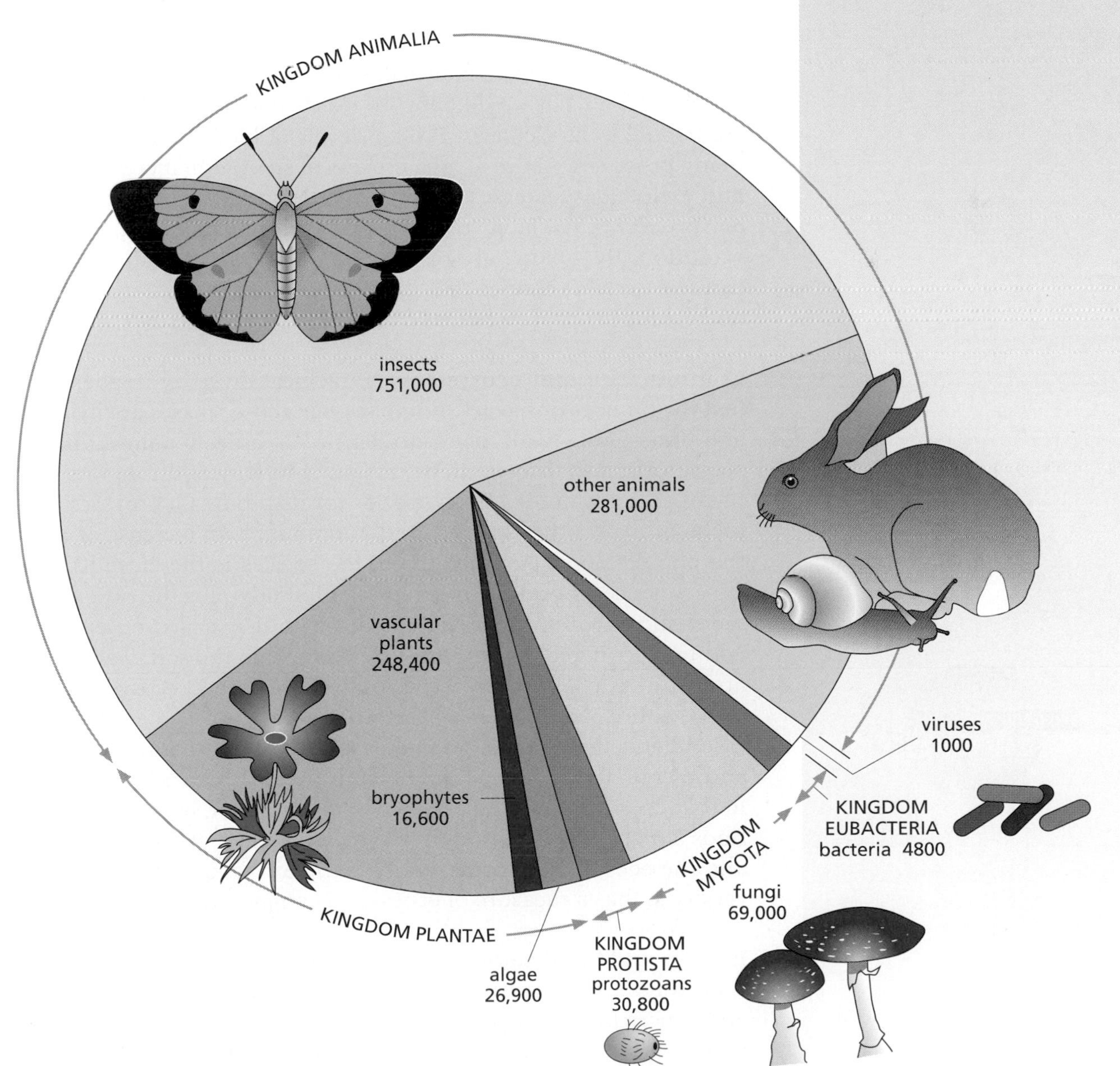

FIGURE 15.1

Numbers of species currently known in the major groups of organisms.

an effort has been made to identify every species in a given area. In one such study, between 4000 and 5000 species were found in a single gram of sand. In another study, 5000 species were found in a gram of forest soil, and these 5000 were almost completely different from the species found in the gram of sand. The more thoroughly we look, the more species we find.

Earth's biodiversity is studied by evolutionists and ecologists. To understand what they are learning, we need to become familiar with the ecological concepts of communities, ecosystems, climate, and energy. These concepts will help us to explain patterns in the distribution of biodiversity across the Earth's surface and also how our own species interacts with the myriad other species that inhabit this planet. Human activities affect biodiversity, and biodiversity affects our lives in return.

Factors influencing the distribution of biodiversity

Despite our incomplete knowledge of the extent of biodiversity, we have begun to use what we do know to test hypotheses about the factors that contribute to its richness. Present levels of biodiversity are the result of many processes. Above all, the process of speciation increases biodiversity and extinction decreases biodiversity. Because these are both evolutionary processes, the study of biodiversity depends on an understanding of evolution. By looking at present-day biodiversity in different places, we can test certain theories that give a better understanding of evolutionary and ecological processes that contribute to biodiversity.

Communities and ecosystems. Biological diversity implies more than just the number of species. It implies that those species are living together and interacting with one another in organized **communities**. Each species also has its own **niche**, meaning its way of life and its role in the life of the community. Each species (or its products) provides part of the niche of many other species in its community. An increase or decrease in the population size of any one of the species is therefore likely to have consequences for all the others. A community plus the physical environment surrounding it and interacting with it is called an **ecosystem**. In addition to living species, ecosystems include the soil, water, rocks, and atmosphere in which those species live. The largest ecosystem of all, that of the entire planet, is called the **biosphere**.

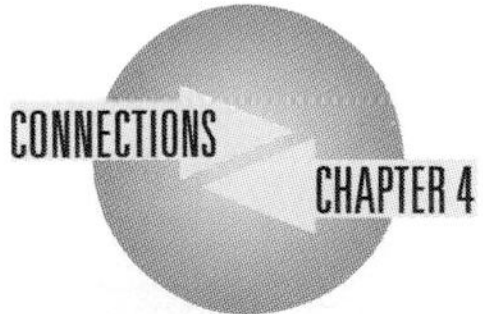
CONNECTIONS
CHAPTER 4

When a new species originates (see Chapter 4), it must find its niche and fit into the ecosystem in which it lives. Otherwise it fails to flourish in that ecosystem and quickly becomes extinct. If, however, a new species finds a niche that integrates it into the ecosystem, its population increases and the ecosystem becomes more complex owing to its presence. Biodiversity is thus a measure of ecosystem complexity.

Energy and biodiversity. One of the major determinants of biodiversity is latitude. Tropical ecosystems contain a much richer diversity of species and genera than temperate-zone ecosystems. The richest diversity on land is in tropical rainforests, while the richest marine ecosystems are those of warm-water coral reefs (Figure 15.2). Some large taxonomic groups, including many entire families and several orders, are confined to tropical ecosystems, and nearly every major group reaches its maximum diversity in the tropics. In marked contrast are the Arctic and Antarctic ecosystems, which are relatively sparse in biodiversity and in ecological complexity.

Why should this be true? One of the great theoretical problems in evolutionary biology is why species diversity is greater in the tropics. One possible explanation is provided by the **energy–stability–area theory** of biodiversity, which begins with the observation that the species-rich tropics receive the greatest amounts of solar energy and also a more continuous level of solar energy. Each biological population requires a certain minimum amount of energy to maintain a population size capable of reproducing itself. Photosynthesizing plants capture the energy that they need directly from sunlight; most other species obtain their energy from food (see Chapter 14). For an equal amount of nutrients (an important proviso), greater quantities of **biomass** (mass of living things) grow in areas that are the hottest (receive the most solar energy) and the most humid (have a constant supply of water for photosynthesis). Also in tropical regions, the climate varies little throughout the year and is also more stable over the centuries or across geological time. Within each unit of area, there are many more niches than would be present in an equal-sized portion of the temperate zone, each niche differing slightly from the next in the amount and type of energy that is available. Tropical species can specialize to fill these different stable niches in many ways, living at different heights in a vertically stratified forest, occupying different kinds of microhabitats, exploiting different food resources, or attracting different species of pollinators.

FIGURE 15.2

Coral reef diversity in the Red Sea, Egypt.

Interdependence of humans and biodiversity

The study of biological diversity is as old as our need for food, clothing, shelter, and medicines, because all of these things, as well as tools and weapons, are made from the millions of other species that inhabit our planet. At the same time, humans have altered many ecosystems, some of them profoundly, and have thus influenced biodiversity.

The biological value of preserving species. The preservation of biological diversity is important for many reasons, of which three broad types can be distinguished. First, our ignorance as to which species might be beneficial is a reason for simply preserving all species. We know that many plants have yielded important drugs; other plants have yielded important foods, dyestuffs, paper, and rubber. Among the species now living but poorly known, some probably possess a wealth of new possibilities for such uses; therefore we must preserve them all for the sake of those that may someday prove useful to humans.

A second reason for preserving biodiversity pertains to the wild relatives of our domesticated species. The store of genetic variation, and therefore the possible number of genetic traits from which to choose, is greatly reduced in each of our domestic species, and is much greater in their wild relatives. It is therefore in our long-range best interests to preserve the wild relatives of all domestic species and varieties, so that newly discovered desirable properties (or properties that become desirable) can be bred into domestic stocks from their wild relatives.

For example, corn (*Zea mays*, also called maize) is one of the world's most valuable domesticated species of plants, but the domesticated variety is an annual plant that must be replanted each year at considerable labor and expense. In the 1970s, however, a wild relative named *Zea diploperennis* was discovered growing in the Mexican state of Jalisco, confined to a small mountain tract. The discovery was made just days before the land was scheduled to be cleared, which would have wiped the species out. *Z. diploperennis* was found to be resistant to a number of diseases that afflict domestic varieties. Best of all, unlike all other species and varieties of corn in the world, the newly discovered species grows as a perennial, meaning that an individual plant produces corn year after year without replanting. If some of these genetic traits could be introduced into domestic corn, either by breeding or by genetic engineering, the new strains could represent billions of dollars' worth of savings for the farmers of all corn-producing regions. Had the Jalisco corn not been discovered in time, an important genetic reserve for this important domestic species would have been lost forever. This is just one instance; similar arguments can be given for the preservation of the genetic resources of other species in zoos, botanical reserves, and gene banks, but the most cost-effective way to preserve these genetic resources is to promote the survival of the wild species or varieties in their natural habitats.

Preserving ecosystem stability. A third reason for preserving biological diversity is that species affect one another. *There are no ecosystems that are made up of only one or a few species*. Recall that a community is a group of species whose needs are interdependent. Stable communities are stable in part because materials are recycled: many producer, consumer, and decomposer species (see Chapter 14, p. 585) are present. A small group of species is much less likely to form a complete and stable community than a larger one. Multiple species of each kind make the stability less likely to be disrupted. For example, multiple prey species provide a more stable food supply for predators because the predators can survive a scarcity of one prey species by switching to other species for their food.

Many communities are unstable in the sense that the removal of just one 'keystone' species can cause the balance among dozens of other species to collapse, so that the disappearance of one species causes other extinctions and leads to other drastic changes. Some of these changes may even affect the physical environment, as when the removal of beavers causes dams not to be built and allows water to flow more freely. Other communities, such as tropical rainforests, are thought to be more stable than this, a consequence of the large number and variety of species. In a typical rainforest there are hundreds of species of trees, with no single species constituting more than 5% or so of the total. There are also hundreds of bird species, thousands of insect species, and a large diversity of other animals and plants, some of them illustrated later in this chapter.

The health of animals, including humans, is promoted by the variety of plants available for them to eat or to climb or to nest in. Likewise, the health and well-being of many of the plants depends on the variety of animals that can pollinate them, disperse their seeds, or fertilize the ground near their roots with their feces and other remains. While it is obvious that the survival of any of these species depends on the survival of the ecosystem as a whole, it is equally true that the stability of the community (and often of the entire ecosystem) depends on the survival of certain key species or groups of species.

On a global scale, the stability of the Earth's very atmosphere depends on its major ecosystems. For example, photosynthesis by plants, especially rainforest plants, helps to limit the buildup of carbon dioxide that contributes to global warming. The same plants are also the principal source of the oxygen in our atmosphere. The health of our atmosphere thus depends on the continued health of major tropical ecosystems.

Many evolutionary and ecological processes increase biodiversity. Life as we know it depends on the rich biodiversity of the world's major ecosystems, yet we are currently in an era when biodiversity is rapidly declining. What forces are producing this decline? Will the current decline in biodiversity match those of the great mass extinctions of the past? These are some of the issues addressed in the next section.

THOUGHT QUESTIONS

1. Why would the perennial growth habit in the corn from Jalisco be considered a valuable trait? Under what agricultural conditions (and in what nations) would this trait be especially valuable?
2. How easily could the genes for perennial growth be identified? How easily could these genes be introduced from the Jalisco corn into the domestic varieties? (You will probably need to review parts of Chapters 2, 3, and 14 to answer this question.)
3. What would be the effects of clearing 100 square kilometers (100 km^2) from a small rainforest 5000 km^2 in size (about 2000 square miles, approximately the size of Delaware)? What would be the effect of clearing 1000 km^2 of this same forest?
4. In what way is the atmosphere part of the biosphere?

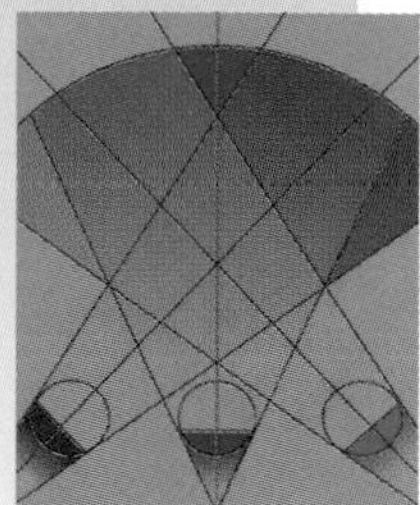

EXTINCTION REDUCES BIODIVERSITY

Evolutionary change often produces new species and thus increases biodiversity. But evolution can also lead to the disappearance of species, a phenomenon called **extinction**.

Species that have no living members are said to be **extinct**. Extinct species are known through the fossil record. As an example, consider the Age of Reptiles, or Mesozoic era, a time interval from approximately 200 to 65 million years ago (see Figure 4.8, p. 138). Of all the species that lived during that time, none are still alive today—they are all considered extinct. These species did not die out all at once but rather a few at a time, although many perished in a mass extinction at the end of the Cretaceous period. The reasons behind these extinctions differed from case to case and are imperfectly known for most species. In some cases, the fossil record shows that a competing group of species appeared on the scene shortly before the extinction occurred. In other cases no specific cause can be identified.

Types of extinction

If all Mesozoic species are now extinct, how did life manage to persist? To answer this question, we must distinguish between two major types of extinction. First we must recognize the concept of a **lineage**, which is an unbroken series of species arranged in ancestor-to-descendant sequence, with each later species having evolved from the one that immediately preceded it. If we had a complete record of the history of life on this planet, every lineage would extend back in time to the common origin of all earthly life. Working forward from earlier times, each lineage extends either to a species alive today or to one that has become extinct.

When an entire lineage has died out without issue, no living descendant species exist. We call this **true extinction**. Many groups of organisms have, according to current theories, undergone true extinction, meaning that no living species are descended from them—their lineages have ended. Among these groups are the trilobites and conodonts shown in Figure 15.3.

When a species no longer exists, its lineage may continue in the form of descendant species. This type of change, called either **pseudoextinction** or **phyletic transformation**, occurs when a species evolves into something recognizable as a different species. The ancestral horse, *Hyracotherium* (formerly called *Eohippus*), is extinct in this sense: there are none alive today, but they have living descendants, the modern horses. Many of the traits of *Hyracotherium*, and the genes that contributed to these traits, persist among modern horses. In some cases we do not know whether an extinct group has undergone true extinction or only pseudoextinction—the evidence does not permit us to make a clear choice. For example, the dinosaurs are no longer alive, but if living birds are their descendants, as many scientists believe, then the dinosaurs are only pseudoextinct.

Analyzing patterns of extinction

Has extinction occurred randomly? Does the probability of extinction remain constant over time and from place to place? Several biologists in the late twentieth century have hypothesized that extinction occurs at random over vast time periods. This hypothesis has been tested in a mathematical model that compares actual extinctions with theoretical predictions based on a model of random extinctions. Comparison of actual data with a theoretical model is a good research strategy (see Chapter 1) because it allows us to identify both circumstances for which the model holds and circumstances for which it does not (and for which additional explanations are therefore needed).

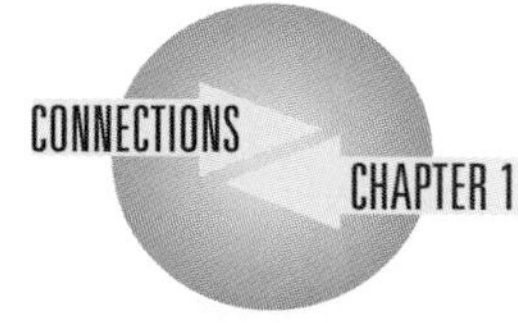

Several studies have compared extinction in the fossil record of particular animal groups with the random model. Most of these comparisons

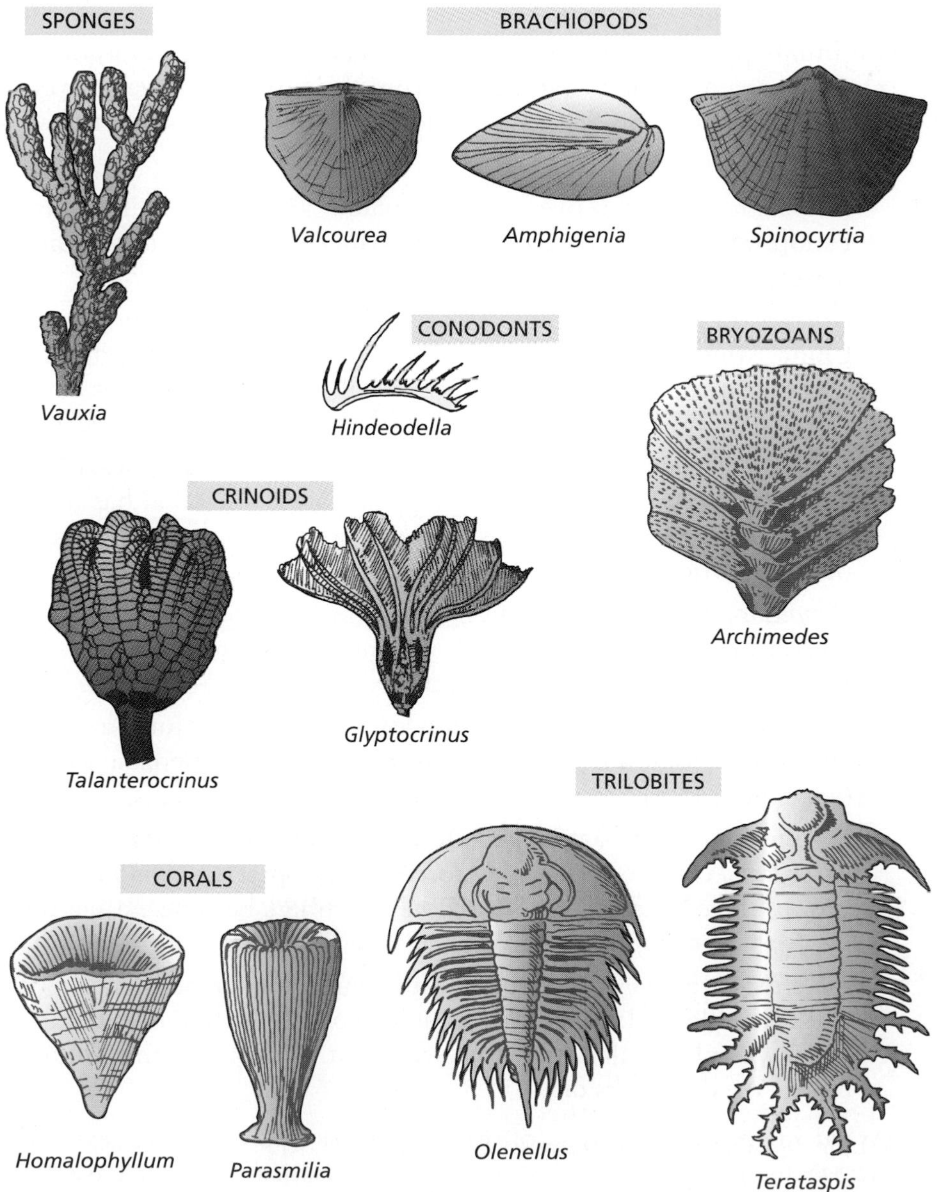

FIGURE 15.3 Extinct species of the Paleozoic Era. The fossils shown here represent groups that were abundant during Paleozoic times. All these groups suffered considerable extinction at the end of the Permian period. The trilobites (belonging to the phylum Arthropoda) and conodonts (probably belonging to the phylum Chordata, part of jaw structures are shown here) are truly extinct groups with no living descendants. The other groups have some living species, but the species alive today are not the same as the Paleozoic species.

counted all extinctions together, without distinguishing true extinction from pseudoextinction. The fossil record of mammals from the Tertiary period is one of several such groups that conforms to the random extinction model; that is, the actual extinctions match the rates predicted by the model. Many instances of nonrandom extinction have also been discovered. For example, many early invertebrate groups suffered most of their extinction early in their history rather than at a constant rate through time.

Comparison with mathematical models has revealed two major types of departure from randomness: situations in which fewer species become extinct than random models predict, and situations in which more species become extinct.

Living fossils. We first examine situations in which the frequency of extinctions is reduced. **Living fossils** are species or genera that have survived for many millions of years without true extinction and with only minimal pseudoextinction, meaning that very little morphological change separates the living species from their fossil relatives. Several of these living fossils are described below (the time periods mentioned are shown in Figure 4.8, p. 138).

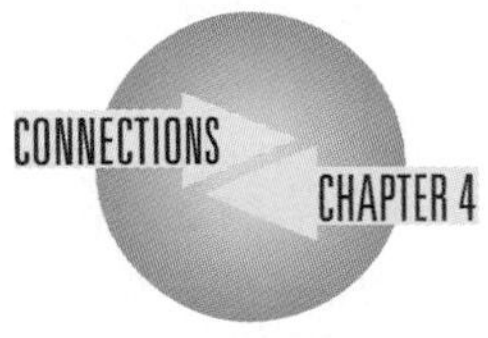

Psilophyton, a primitive vascular plant (kingdom Plantae, phylum Psilophyta) closely resembling the earliest land plants of the Silurian period.

Ginkgo, a tree (kingdom Plantae, phylum Ginkgophyta) native to China, which closely resembles its Mesozoic ancestors and which is planted in many urban areas around the world because it tolerates urban pollution (Figure 15.4).

Lingula, a type of brachiopod (kingdom Animalia, phylum Brachiopoda) with a wormlike body enclosed in a two-valved shell that has no hinge and with a feeding structure (called a lophophore) that strains suspended particles from the water (see Figure 15.4).

Neopilina, a deep-water mollusk (kingdom Animalia, phylum Mollusca) with a low-domed conical shell resembling that of the extinct genus *Pilina*, one of the most primitive mollusks.

Limulus, the horseshoe crab (kingdom Animalia, phylum Arthropoda, class Merostomata), which closely resembles its Paleozoic ancestors (see Figure 15.4).

Latimeria, a large, rare Indian Ocean fish (kingdom Animalia, phylum Chordata, class Osteichthyes, order Crossopterygii) belonging to a group (the coelacanths) whose other members became extinct during the Mesozoic era (see Figure 15.4).

Sphenodon, a lizard-like reptile (kingdom Animalia, phylum Chordata, class Reptilia, order Rhynchocephalia) confined to the northern island of New Zealand, the only living remnant of an order that flourished before the dinosaurs did.

What might make a taxon of organisms less likely to suffer extinction? These living fossils share several characteristics: they all have locally large populations (with a sufficient gene pool to maintain a large amount of genetic variation); they are all adapted to dependably persistent habitats (such as deep ocean waters); those that are animals do not depend on a narrow range of food species (some of them will eat anything within a certain size range); and they all have reproductive stages (pollen, spores, or larvae) that are dispersed mechanically by wind or ocean currents rather than by other species. If there is any secret to long-term survival, these species have surely stumbled upon it.

Mass extinctions. Departing in the other direction from the random extinction model are examples in which many more species became extinct within a geologically short interval of time than would have been predicted by the model. We call these mass extinctions. There was one such event at the end of the Cretaceous period, which was also the end of the Mesozoic era. There was another, even larger, mass extinction at the end of the Permian period.

Mass extinctions devastate biodiversity. Over half of the families and 85% of the genera became extinct in the last 5 million years of the Permian period, and much higher percentages were lost in some classes and phyla. Although many more species and genera perished in this mass extinction than in any other, the Permian event has attracted much less attention than other mass extinctions because nearly all the species were unfamiliar types of organisms (for example, crinoids and brachiopods; see Figure 15.3) that lived in underwater habitats such as the shallow inland seas that were abundant at that time.

The mass extinction at the end of the Cretaceous period has attracted the most attention because many well-known animals became extinct at this time: dinosaurs (the reptilian orders Saurischia and Ornithischia), flying reptiles (order Pterosauria), several types of marine reptiles (orders Sauropterygia, Ichthyopterygia, and others), ammonoids (phylum Mollusca, class Cephalopoda, mostly with large, coiled shells), and several groups of fish, plants, and other organisms (Figure 15.5).

Possible causes of mass extinctions. The fossil record shows at least five mass extinctions in which many families of marine organisms died out (Figure 15.6). The rates of extinction happening today are as great as the rates during these mass extinctions. Many

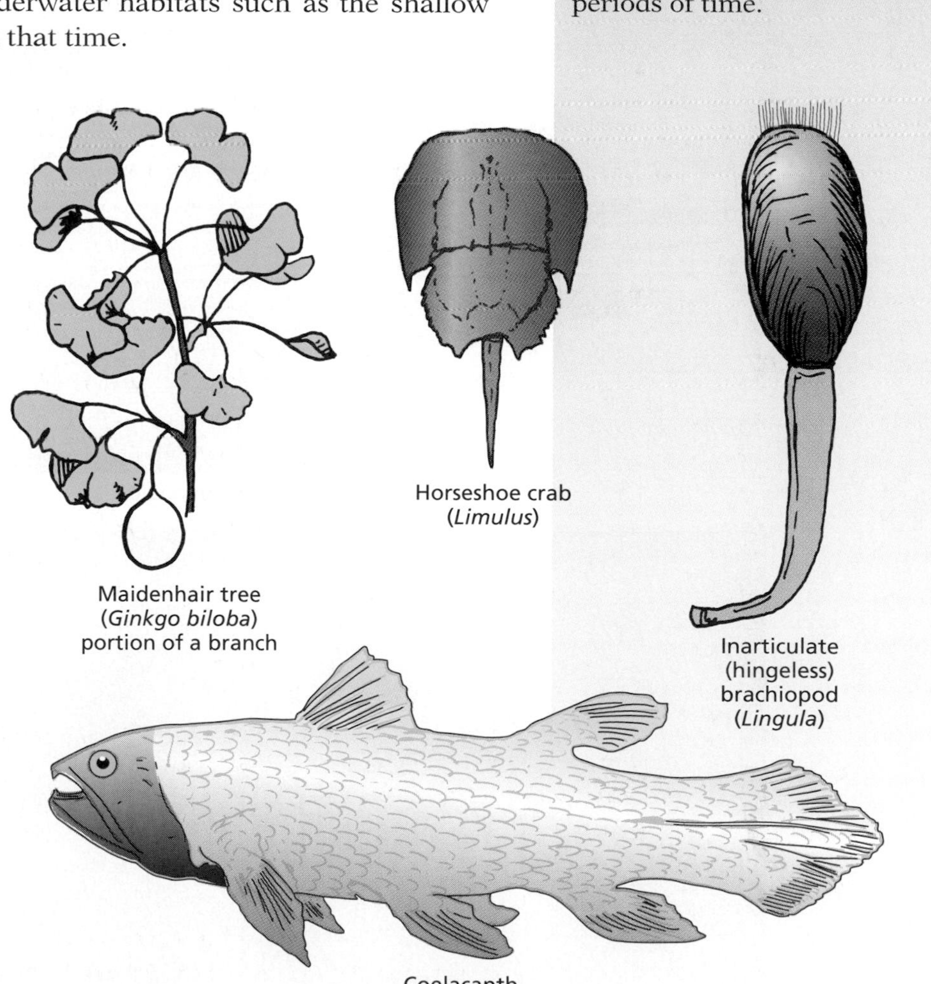

FIGURE 15.4
Living fossils that have avoided extinction over long periods of time.

scientists have therefore concluded that a sixth great mass extinction is currently in progress.

What could cause such high rates of extinction? There are several hypotheses including: warming or cooling of the Earth, changes in seasonal fluctuations or ocean currents, and changing positions of the continents (plate tectonics). Biological hypotheses include ecological changes brought about by evolution of cooperation between insects and flowering plants or of bottom-feeding predators in the oceans. Some of the proposed mechanisms require a very brief period during which all extinctions suddenly took place; other mechanisms would be more likely to have taken place more gradually, over an extended period, or at different times on different continents. Some hypotheses fail to account for simultaneous extinctions on land and in the seas. Each mass extinction may have had a different cause. Evidence points to human hunting and habitat destruction as the likely causes for the current mass extinction.

American paleontologists David Raup and John Sepkoski, who have studied extinction rates in a number of fossil groups, suggest that episodes of increased extinction have recurred periodically, approximately every 26 million years since the mid-Cretaceous period. The late Cretaceous extinction of the dinosaurs and ammonoids was just one of the more drastic in a whole series of such recurrent extinction episodes. The

FIGURE 15.5

Extinct species belonging to groups that died out completely at the end of the Cretaceous period.

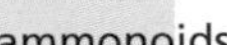

ammonoids

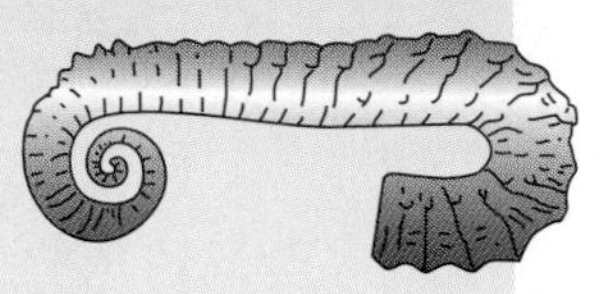

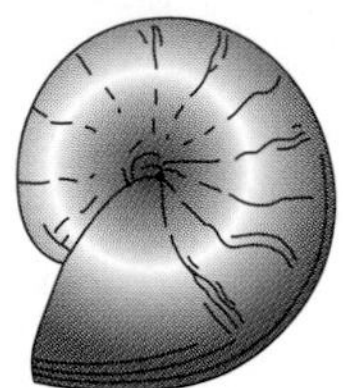

giant ammonoid

artist's view of various Mesozoic reptiles

possibility that mass extinctions may recur periodically has given rise to such hypotheses as that of a companion star with a long-period orbit deflecting other bodies from their normal orbits, causing some of them to fall to Earth as meteors, wreaking widespread devastation upon impact.

The asteroid impact hypothesis. Of the various hypotheses attempting to account for the late Cretaceous extinctions, the one that has attracted the most attention in recent years is the asteroid impact hypothesis first suggested by Luis and Walter Alvarez. According to this hypothesis, the Earth collided with an asteroid with an estimated diameter of 10 km, or with several asteroids, the combined mass of which was comparable. The force of collision spewed large amounts of debris into the atmosphere, darkening the skies for several years before the finer particles settled. The reduced level of photosynthesis led to a massive decline in plant life of all kinds, and this led to massive starvation first of herbivores and subsequently of carnivores. The mass extinction would have occurred very suddenly under this hypothesis.

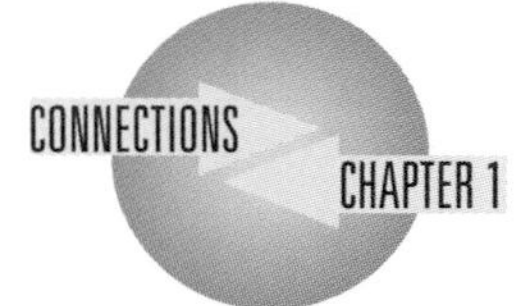

One interesting test of the Alvarez hypothesis is based on the presence of the rare-earth element iridium (Ir). The Earth's crust contains very little of this element, but most asteroids contain a lot more. Debris thrown into the atmosphere by an asteroid collision would presumably contain large amounts of iridium, and this material would be carried by atmospheric currents all over the globe. A search of sedimentary deposits that span the boundary between the Cretaceous and Tertiary periods shows that several do in fact show a dramatic increase in the abundance of iridium briefly and precisely at this boundary (visible in Figure 1.4, bottow row, middle panel, p. 18). This iridium anomaly offers strong support for the Alvarez hypothesis. However, no asteroid itself has ever been recovered.

An asteroid of this size would be expected to leave an immense

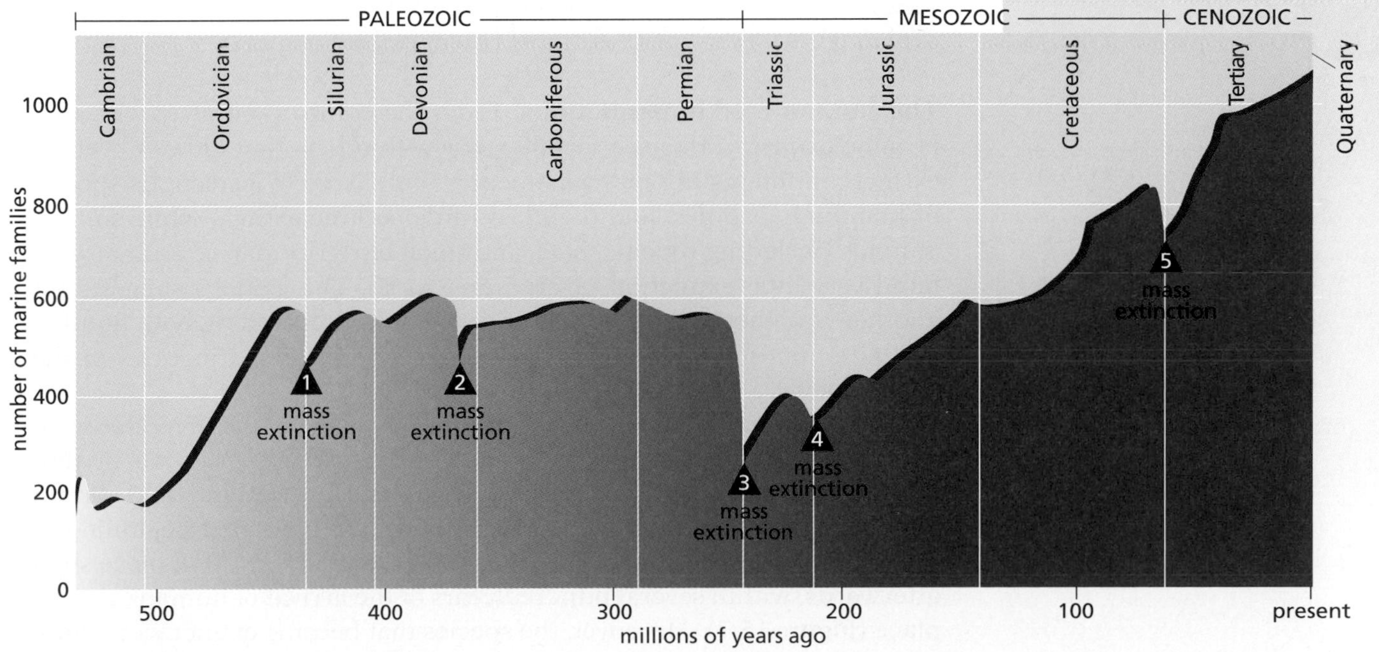

FIGURE 15.6

Changes through time in the number of families of marine organisms. Mass extinctions are indicated by numbered triangles.

crater, even if the asteroid itself was disintegrated by the impact. The intense heat of the impact would produce heat-shocked quartz in many types of rocks. Also, large blocks thrown aside by the impact would form secondary craters surrounding the main crater. To date, several such secondary craters have been found along Mexico's Yucatan peninsula, and heat-shocked quartz has been found both in Mexico and in Haiti. Several candidate locations for the primary impact site have been suggested, including one site called Chicxulub, near the coast of Yucatan.

Quaternary extinctions. There were many extinctions during the past two million years, an interval that includes the Pleistocene epoch and its several glacial episodes, plus the Recent epoch (beginning about 15,000 years ago). A number of species of large mammals, reptiles, and flightless birds became extinct during this time, and several hypotheses have been advanced as explanations. The most obvious hypothesis attributes the extinctions to changes in climate and the advance and retreat of Pleistocene glaciers. But paleontologists who have carefully examined the fossil record point out that this hypothesis fails to explain the timing of the extinctions, most of which did not coincide with the extremes of temperature.

A second hypothesis is that newly introduced species brought about extinction through increased competition. Widespread glaciation caused a decline in sea levels, which resulted in the emergence of land bridges, including those across Panama, the Bering Strait, and the English Channel. Many species were thus introduced from one landmass to another, and the newly introduced species competed with other species already present and established new predator–prey interactions. Many species could not adjust to the new conditions and became extinct as a result. This hypothesis has been used to explain many of the animal extinctions that followed the emergence of the land bridge across Panama, connecting South America (previously an island continent) with North America. Most of this group of extinctions, however, happened early in the Pleistocene, leaving another large group of later extinctions still to be explained. We examine these most recent extinctions next.

The human role in extinctions. From the comparison of species that became extinct in the past 50,000 years with others that did not become extinct, an interesting pattern emerges: only large, conspicuous species of mammals, reptiles, and flightless birds became extinct, while smaller animals (including rodents, bats, and small birds) or marine animals suffered very little extinction or else none at all. This pattern suggests yet another hypothesis: that the activities of humans, including both hunting and alterations of habitat, played a large role in the extinctions of the past 50,000 years.

A great deal of circumstantial evidence favors the hypothesis of extinction by human agency: in those places where the time of first human arrival can be dated (e.g., Madagascar, New Zealand, and certain Pacific islands), dense piles of animal bones accumulated beginning at the times of human arrival, and most of the extinctions took place soon afterwards, within several hundred years of the arrival of humans at each place (Figure 15.7). Moreover, the species that became extinct were those that humans would be apt to hunt, mostly large herbivores, whereas most species that would have been difficult to hunt, or too small to be worth hunting, survived. On the island continent of Australia, for example, the arrival of humans some 50,000 to 60,000 years ago was soon followed by the extinction of 20 species of giant kangaroos, along with a marsupial lion, a marsupial wolf or tiger (*Thylacinus*, surviving into the 1800s on Tasmania), and the giant, cow-sized herbivore *Diprotodon*. On New Zealand and the Hawaiian Islands, the extinction of flightless birds began with human arrival and was nearly complete by the time of European discovery.

Several vertebrate species have become extinct within historic time: the dodo (a large, flightless bird) and the passenger pigeon are two famous examples. Many other species of birds and also many plants and insects also became extinct. Hawaii was home to some 50 species of land birds at the time of its European discovery in 1778. Humans and the animals that they have introduced since that time have caused the extinction of one-third of these species, and archaeologists have shown that an additional 35 to 55 species had been hunted to extinction by the indigenous Hawaiians before the arrival of Captain Cook. On New Zealand, the giant moa (an ostrich-like flightless bird) was hunted to extinction by the Maoris, the indigenous people of that island nation. Much the same thing happened on Madagascar, where an even larger flightless bird, the elephant bird, had become extinct long before European colonists arrived.

Most recently, and on a worldwide scale, it is now estimated that more than 100,000 plant and animal species became extinct during the decade of the 1980s. Nearly all of these species were on land and were scattered among many families, so they do not show on the graph of Figure 15.6, which counts only marine families.

FIGURE 15.7

The extinction of many large mammals and flightless birds followed soon after human arrival in many parts of the world. In Africa, however, extinctions took place more gradually because humans had been present for a much longer period.

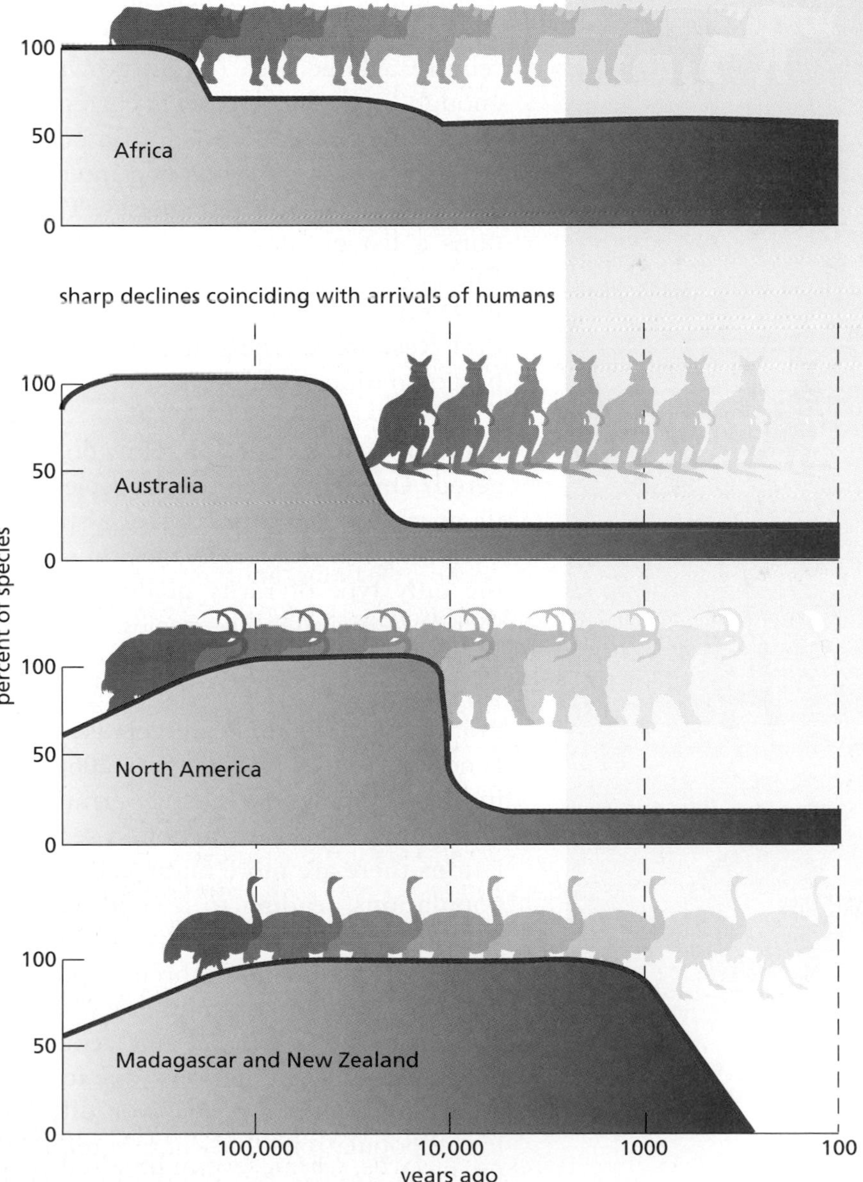

In the United States, many species of migratory songbirds have been greatly reduced in numbers and several have become extinct. Bachman's warbler, a bird species that was once common throughout the southeastern United States, was last seen in the 1950s. The Ohio River and Lake Erie once had dense populations of 78 species of freshwater mollusks, of which 19 are now extinct and another 29 are rare. Biologists now believe that a single dam, Wilson Dam, on the Tennessee River, caused the extinction of 44 species of freshwater mollusks when it was built. In Africa's Lake Victoria, over 100 species of fishes became extinct after the introduction of the Nile perch, a large, predatory sport fish introduced to the lake in 1959.

Species threatened with extinction today

A species threatened with extinction is called an **endangered species**. An example is the northern spotted owl (*Strix occidentalis*) of California, Oregon, and Washington. These owls nest only in old pine trees and require a large area of 'old growth' forest for each individual to find enough food. The species is currently endangered because of logging of the Pacific coastal forests for timber.

Various governments and international organizations maintain official lists of endangered species. The United States, for example, maintains a list of endangered species in the United States as part of the Endangered Species Act. International lists are maintained by such organizations as the International Union for Conservation of Nature and Natural Resources. These lists differ from one another because they are based on different criteria.

Predictors of extinction. How do we know whether a species is endangered? One indication that a species may be endangered is a reduction in its numbers. The extinction of a species is nearly always preceded by its becoming rare, and rarity may thus be the prelude to extinction. However, the only type of rarity qualifying as the harbinger of extinction, and therefore qualifying a species as endangered, is the kind in which the entire species is represented by only one or a few populations, all of which are small. Once the population size falls below a certain minimum, several factors can increase the risk of extinction. One of these is genetic drift (see Chapter 5, p. 206), a pattern in which gene frequencies in small populations change erratically and not necessarily adaptively, which may hasten extinction. A second factor is that in very small populations there are more matings between related individuals than in large populations, leading to a condition called **inbreeding depression**, in which homozygous recessive traits are more frequent in the offspring than they are in non-inbred populations. Because many homozygous recessive traits are harmful (see Chapter 3), they reduce the fitness of the population as a whole and reduce survival. A third factor is that environmental fluctuations (due to changing seasons, weather phenomena, and so forth) may favor different genotypes at different times. Large populations that contain many different genotypes are thus more likely to survive, but populations with less genetic diversity are more susceptible to extinction.

Extinction of a niche. Other indications that a species is threatened with extinction are the disappearance (or impending disappearance) of its habitat, or the disappearance of another species on which its niche depends. A remarkable example of the latter is the tambalacoque tree (*Sideroxylon* or *Calvaria grandiflorum*) of the island of Mauritius in the Indian Ocean. None of the seeds of these long-lived trees were ever observed to germinate, even when planted. A botanist who studied these seeds noticed that they have a very hard outer husk that mechanically prevents the seed within from breaking through. The seeds could be made to germinate by abrading the outer husk before planting the seeds. The same effect was obtained by feeding these seeds to turkeys, a bird about the size of the dodo. The dodo was a type of bird that lived on the island of Mauritius before Europeans and their domestic animals caused its extinction around 1681. We now believe that the dodos fed on the seeds of the tambalacoque tree, and the hard outer husk was an adaptation that permitted the seeds to survive the digestive action of the dodo's stomach. (Seed-eating birds often intentionally swallow and retain pebbles; these pebbles work like pulverizing machines in their muscular stomachs to abrade such things as tough seeds.) Once the dodos became extinct, the tambalacoque trees were unable to reproduce. Young tambalacoque trees are now growing with the aid of humans, saving the species from the threat of certain extinction.

Species currently in danger. The list of endangered species is long and growing. Among mammals, the list includes giant pandas, gorillas, orangutans, elephants, manatees, caribou, timber wolves, and dozens of less familiar species. There are also a large number of endangered fishes, birds, amphibians, reptiles, insects, and plants. The International Council for Bird Preservation estimates that close to 2000 species of birds have already become extinct in the past 2000 years, mostly driven by human agency, and that 11% of the living species (1029 of 9040) are endangered.

A few species that were once listed as endangered have, for the moment, been saved from the brink of extinction, but only when a concerted effort has been made to conserve the species and its habitat. For example, bald eagles (*Haliaeetus leucocephalus*) have been protected and in some cases released into the wild from captivity. Because their numbers are once again sufficient to maintain stable populations, they are no longer considered an endangered species.

Some species, of course, are endangered because of indiscriminate hunting, fishing, or poaching. These are mostly large and conspicuous organisms. However, a much larger threat to biodiversity lies in the destruction of natural habitats, a process that threatens thousands of species at once, including those that are inconspicuous and poorly known. Among these small and inconspicuous organisms, the number of endangered species is certainly many more than those on any official list, and many of these inconspicuous organisms will undoubtedly become extinct before they have even been discovered and named.

The whole strategy of saving individual species is being rethought. In the decades since the first legislation on endangered species, scientists have come to realize that individual species cannot be saved without saving their habitats. The focus is therefore shifting to habitat preservation.

THOUGHT QUESTIONS

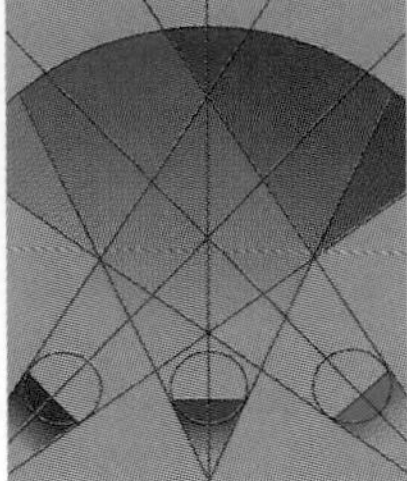

1. Experts still disagree on whether the great extinctions of the past occurred suddenly or gradually. If a new article claimed new evidence for either of these hypotheses, how would you go about evaluating this evidence?

2. The presence or absence of dinosaur fossils is used to determine whether a certain bed of rock is Cretaceous or Tertiary. How would this practice influence research on the question of whether dinosaur extinction came about gradually (at different times in different places) or suddenly (simultaneously everywhere)?

3. In 1995, the U.S. Department of the Interior downgraded the legal status of bald eagles (*Haliaeetus leucocephalus*) from 'endangered' to 'threatened,' and in 1999 they were removed from the endangered species list altogether. Why is the U.S. Government involved in this matter? Bald eagles also live in Canada and part of northern Mexico. Does the U.S. Government have any authority to declare the bald eagle 'endangered' or 'threatened' in those places?

SOME ENTIRE HABITATS ARE THREATENED

Among the most serious threats that any species faces is the destruction of its habitat. Of all the endangered species in the world, an estimated 73% are endangered because of the destruction of their habitats. Habitat destruction threatens many species at once, often thousands at a time. Norman Myers, an ecologist, has identified 18 areas of the world as 'hot spots' (Figure 15.8) where threatened habitats put many thousands of species at the risk of extinction simultaneously.

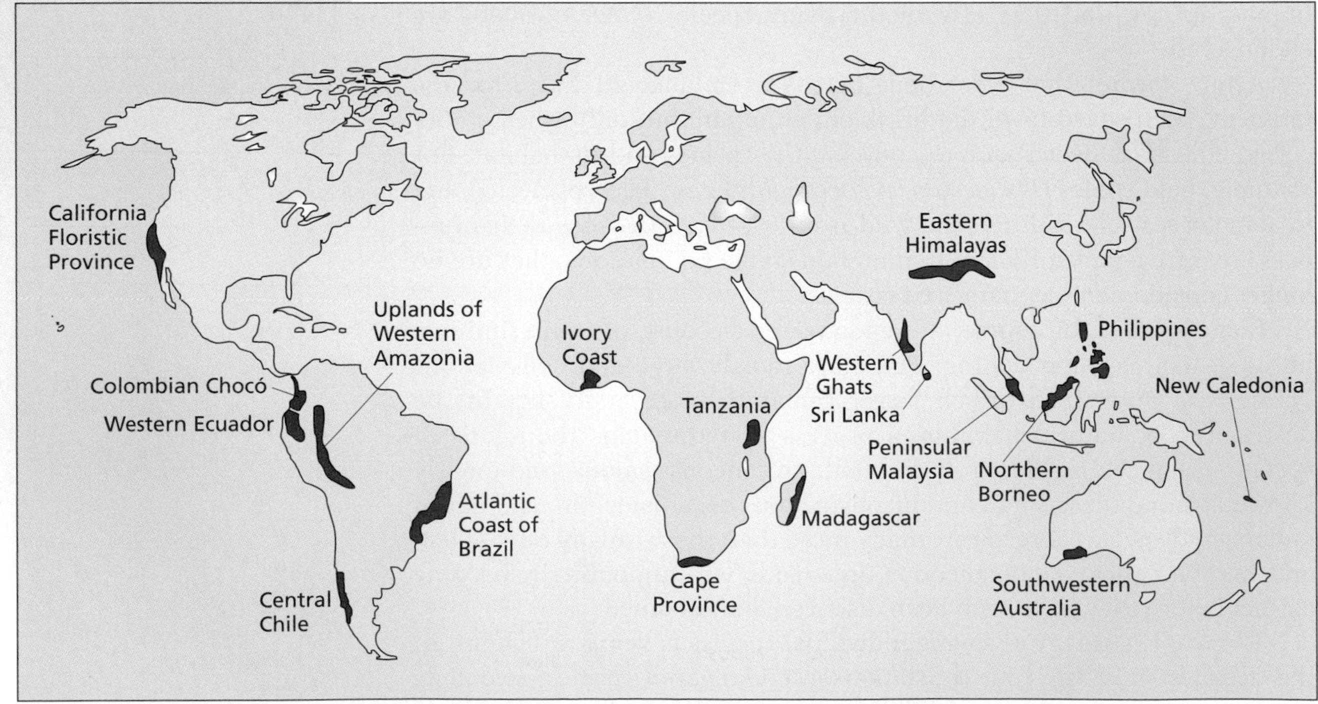

FIGURE 15.8

'Hot spots': habitats containing many species found nowhere else and threatened with extinction from human activity.

What impact does habitat destruction have on human lives? How should humans value habitat? In this section we examine two specific types of habitat destruction: the destruction of tropical rainforests and desertification.

Tropical rainforest destruction

As we have seen, the tropics contain many more species than do temperate and polar regions. Habitat destruction in the tropics therefore affects many more species than it does at higher latitudes. Tropical rainforests are particularly vulnerable because they contain so many different habitats, and they are being destroyed at an accelerating rate. Ecologist E.O. Wilson estimates that one Peruvian farmer clearing some rainforest land to grow food for his family will cut down more kinds of trees than are native to all of Europe.

Habitats do not exist independently but are associated into ecosystems. Groups of ecosystems that are similar in type are called **biomes**. Each biome contains ecosystems in different parts of the world that are similar in climate and contain similar habitats and species assortments. Notice that in Figure 15.9 most biomes are restricted to certain bands of

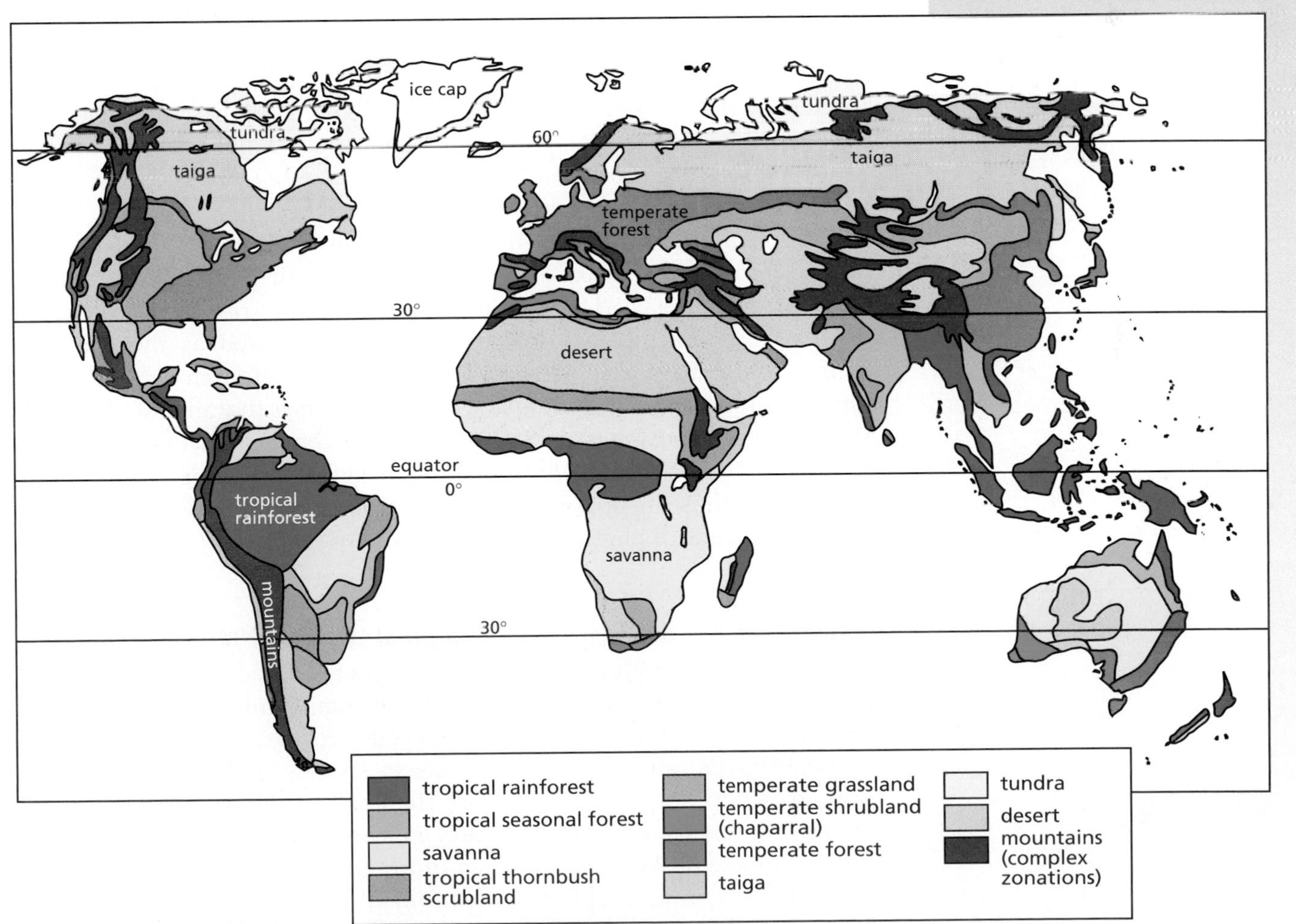

FIGURE 15.9

The world's biomes.

latitude, sometimes in both the Northern and Southern Hemispheres. Individual species differ from place to place, but a particular biome has certain proportions of ecological types, such as large herbivores and tall grasses on tropical grassland (savanna), or large stands of pine and other evergreen conifers on taiga. Temperate shrubland (also called chaparral or Mediterranean) has mild, rainy winters and supports many shrubs and small trees, such as olives. Other examples of biomes are desert, rainforest, and tundra (a cold, treeless land with low-growing vegetation only).

The tropical rainforest biome. Tropical rainforests are a biome that encompasses habitats of low latitude (from about 15° S to 25° N), continually warm temperatures, and year-round high precipitation. Average temperatures are usually about 25 °C, typically fluctuating only from about 22 to 27 °C with seasonal extremes no lower than 20 °C. Precipitation is high throughout the year, with annual totals of 1800 mm (about 71 inches) or more. Although there is seasonal variation, no month averages less than 60 mm (2.36 inches) of rainfall. Under these conditions, humidity stays moderate to high at all times.

The most conspicuous rainforest vegetation consists of tall trees, 30 to 60 m (about 100 to 200 feet) in height, up to the height of a 17-story building. As Figure 15.10 shows, their leafy tops make a **continuous canopy** through which tree-living animals can roam widely without ever descending to the ground. The tallest of the trees may protrude above this canopy level and are called **emergents**. The taller trees are often buttressed at their bases with a variety of woody supports that give their base a fluted rather than cylindrical shape (see Figure 15.10A and C). Much of the rainforest receives little direct sunlight below the canopy, and most of the plants here are shade tolerant.

Tropical rainforests are vital to the biosphere because almost half of all the photosynthesis that plants perform on our planet takes place in tropical rainforests. This photosynthesis is the principal safeguard against a global increase in carbon dioxide (as well as the principal source of atmospheric oxygen). Unfortunately, human activities are already producing carbon dioxide at a rate so fast that global photosynthesis rates cannot keep up. Scientists believe that the increasing levels of atmospheric carbon dioxide are leading to global warming. Tropical rainforests are thus an important global resource whose continued health benefits the entire planet, not just the countries in which the rainforests are found.

Rainforest biodiversity and ecological diversity. Rainforests have a high diversity of both plant and animal species. In addition to the many hundreds of species of trees and other plants, rainforests are home to several hundred kinds of birds, numerous reptiles, amphibians, fishes, and mammals, and thousands of insect species (Figure 15.11). Some biologists have estimated that more than half of the world's species live in the rainforests and nowhere else. For example, Costa Rica, a nation about the size of West Virginia, has more species of birds than the whole of the United States and Canada combined!

FIGURE 15.10

Rainforests.

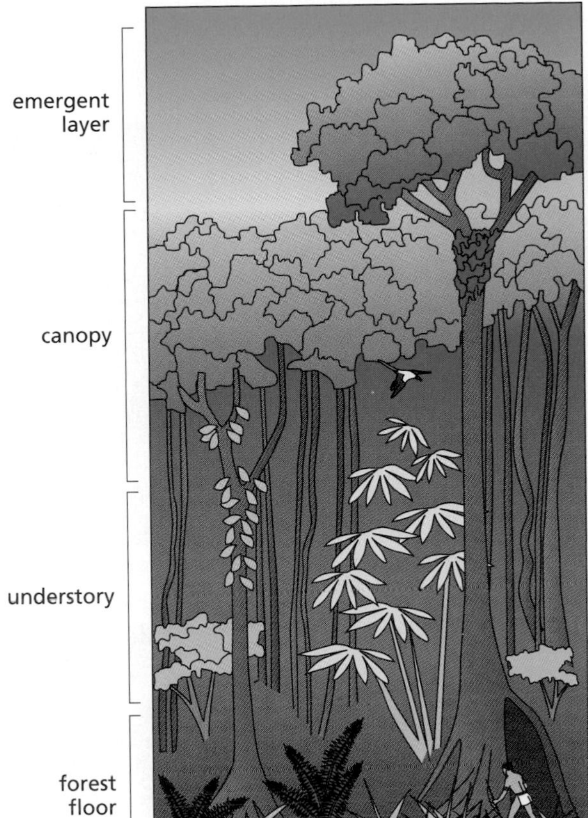

(A) Profile view of a rainforest, showing vertical stratification.

(B) The interior of a Costa Rican rainforest. Notice the marked differences in light levels from the canopy above to the understory below.

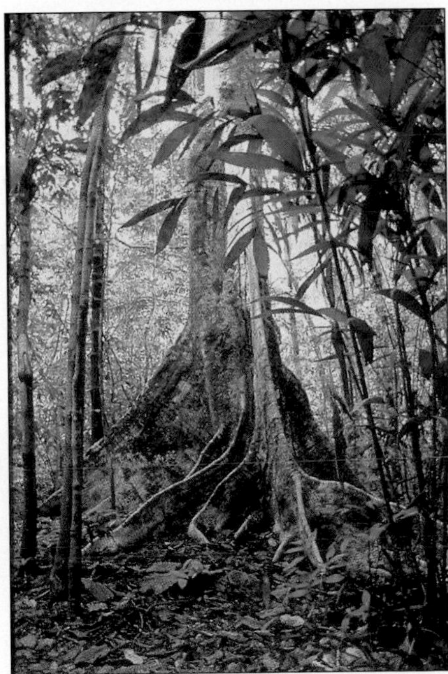

(C) Trees in tropical rainforests often have buttress supports at the base of their trunks.

FIGURE 15.11

The biodiversity of tropical rainforests.

Guzmania nicaraguensis, showing bright yellow flowers surrounded by modified leaves called bracts, whose bright red color attracts the hummingbirds that pollinate this species. Bright red colors are common in bird-pollinated plants.

Spathophyllum, showing leaves damaged by the feeding activities of herbivores.

Piper, a tree whose inflorescences reach upward.

A tarantula from Costa Rica.

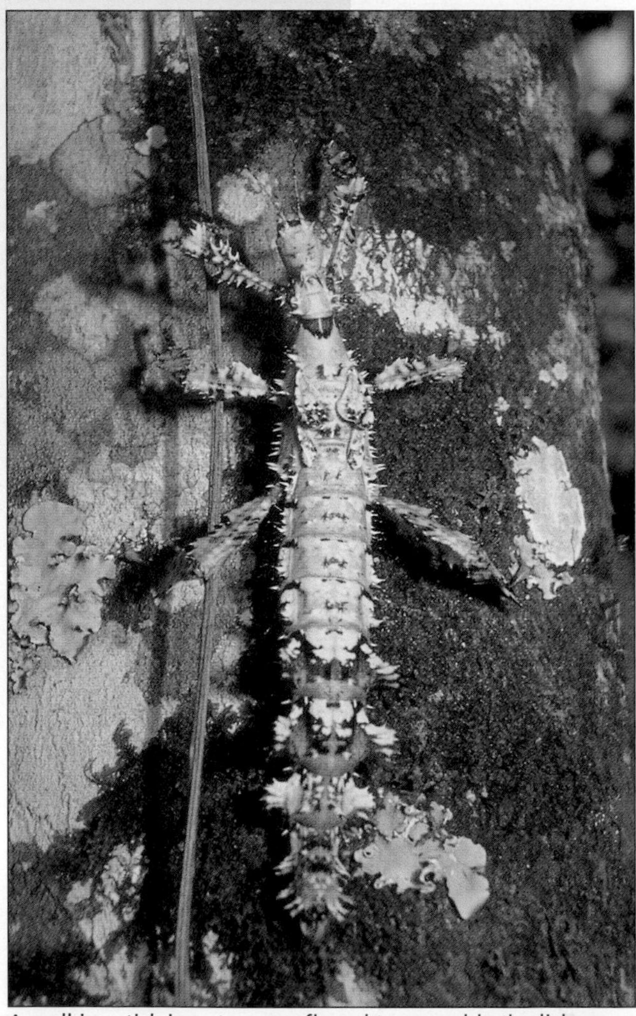

A walking stick insect, camouflaged to resemble the lichens that grow on many tree trunks.

A butterfly feeding on a colorful flower.

Bronze beetle.

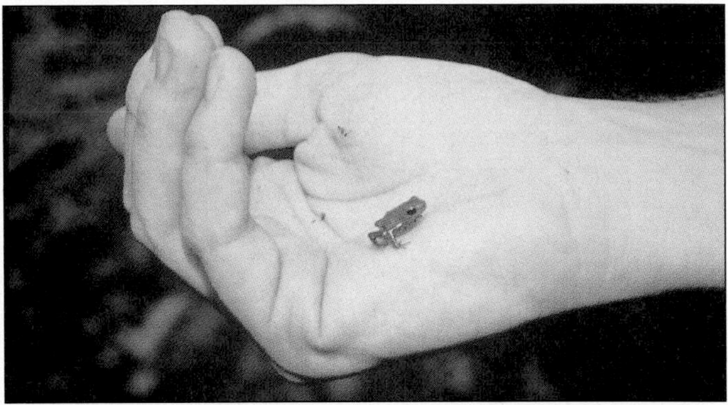

Strawberry frog, one of many poison arrow frogs (*Dendrobates*) whose skin secretes distasteful chemicals that deter predators. Many of these skin secretions are used as dart poisons or arrow poisons by Native Americans because they contain curare, a poison that can paralyze muscles by inhibiting acetylcholinesterase.

Emerald boa, one of many tropical snakes.

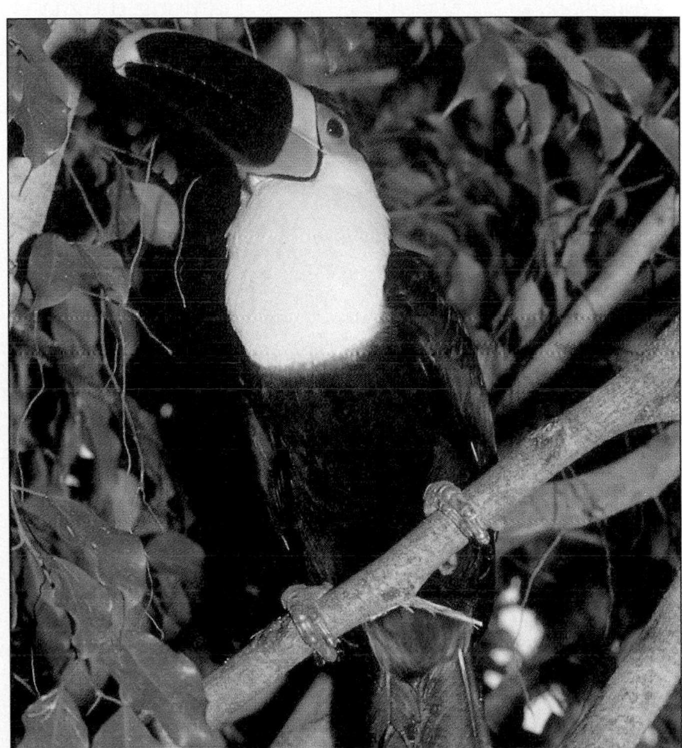

White-throated toucan, a large bird of the parrot family.

A tree sloth and her young, who spend most of each day hanging as you see here.

Jaguar.

The variety of rainforest trees makes for a variety of habitats for **epiphytes**—plants that use other plants instead of the ground for support, such as orchids perched high in the trees (Figure 15.12). There are also many different habitats for other plant and animal species that have adapted to life on the forest floor, in the canopy, in the understory, or in the well-lit clearings created by river banks, landslides, or fallen trees. The great diversity of small-scale habitats within a rainforest contributes to the maintenance of biodiversity among species. To collect samples for the study of this small-scale diversity, tropical biologists may spray the trees above them with a biodegradable insecticide (such as Malathion) and use cone-shaped traps to catch the insects and other small animals that fall.

Complex interactions among species. One result of this great diversity in tropical rainforests is the large number of ecological interactions among the many species present, including predation, competition, and **mutualism**. Most tropical plants have elaborate mechanisms to ensure seed dispersal by animals. For example, shrubs of the genus *Piper* (see Figure 15.11) have fruits that are eaten by bats, which disperse the seeds

FIGURE 15.12

Life on rainforest tree trunks.

Orchids and other epiphytes growing on a tree trunk in Central America.

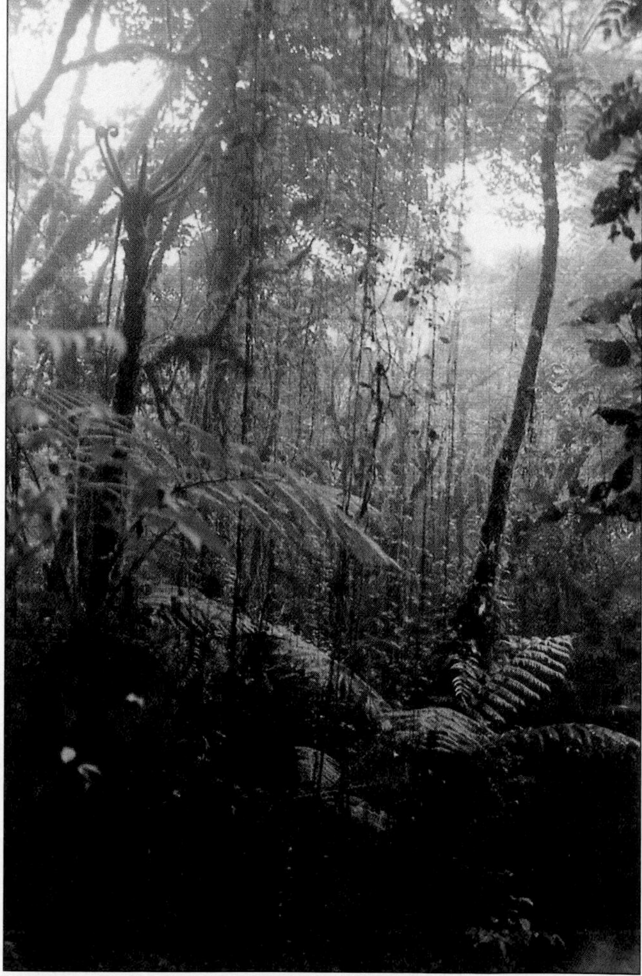

Lianas (woody vines) growing in Costa Rica along with the aerial roots of epiphytic plants growing high above.

in their feces. More elaborate are the many mutualisms between ants (kingdom Animalia, phylum Arthropoda, class Insecta, order Hymenoptera) and the plants that they inhabit. Many plants provide their resident ant populations with food and with places to live. In the case of the bull's horn acacia (*Acacia cornigera*, kingdom Plantae, phylum Anthophyta, class Dicotyledonae), ants of the genus *Pseudomyrmex* live in the base of swollen thorns, and the plants provide them with sugary food from nectaries and protein-rich globules at the tips of the leaflets (Figure 15.13). In return, the ants vigorously defend the plants, swarming to bite any herbivore that would feed on the plant more destructively. Acacia trees of this species whose resident ant populations have been experimentally removed are soon eaten by goats, deer, or other mammalian herbivores.

As an example of an even more complex interaction, consider the figs of the genus *Ficus* (kingdom Plantae, phylum Anthophyta, class Dicotyledonae), an extremely successful group of tropical shrubs and trees that often grow to great heights. The edible part of a fig is an aggregate fruit called a receptacle; this fruit is unusual in that several dozen flowers are contained within it (Figure 15.14). Many of these flowers are home to the tiny fig wasps (phylum Arthropoda, class Insecta, order Hymenoptera, family Agaontidae, genus *Blastophaga*) that pollinate the plants.

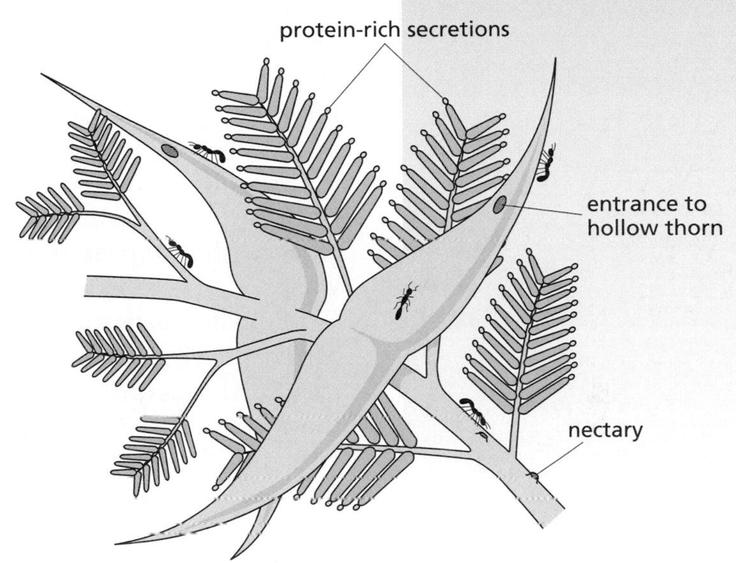

FIGURE 15.13

Mutualism between the bull's horn acacia (*Acacia cornigera*) and ants of the genus *Pseudomyrmex*.

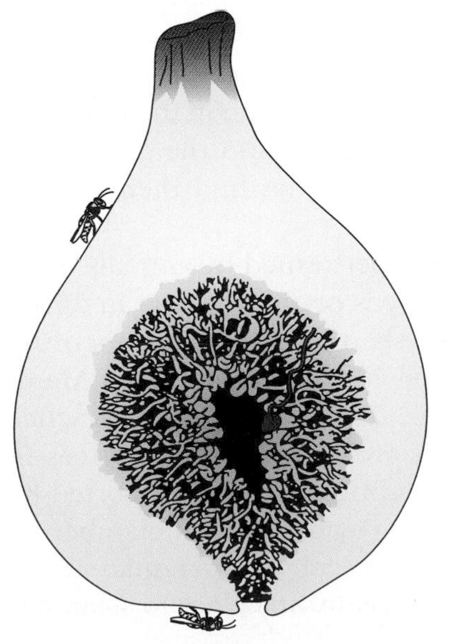

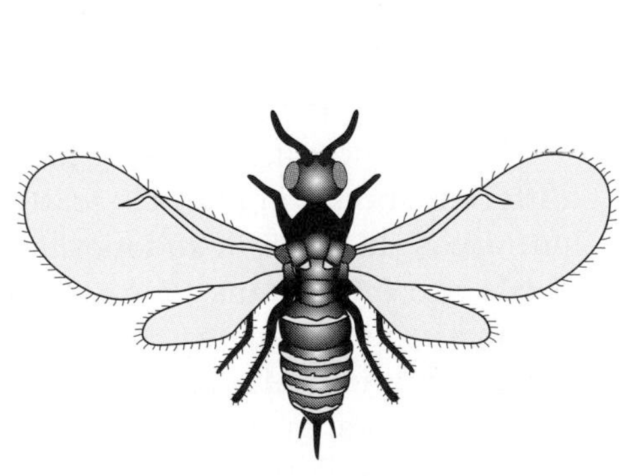

FIGURE 15.14

Cutaway view of a fig (*Ficus*), showing a female fig wasp (*Blastophaga*) ready to enter. An enlarged view of the female wasp is shown; the male has a similar body but is wingless. One of the many flowers inside the fig is shown in red.

Figs produced in the cooler months bear winter receptacles containing mostly sterile flowers but a few fertile male flowers. A female wasp lays her fertilized eggs in the winter receptacle before she dies. The wasps develop within the sterile flowers inside the fig throughout the winter months and in the spring emerge from their pupae (cocoons). The male wasps emerge first, wingless and nearly blind. They move around inside the fig looking for female wasps, which are still in their pupae. A male then chews his way into a female pupa and inseminates the female before she emerges; he then dies without ever leaving the fig. The female wasp emerges later; as she leaves the fig, she picks up pollen from the male flowers located near the exit.

The newly emerged female wasp flies around in search of fresh spring-season figs, which have a second type of receptacle containing both fertile female flowers and sterile flowers. The female wasp enters a fig and roams around inside as she lays her eggs by the hundreds and meanwhile pollinates the female flowers with pollen she has picked up from the male flowers in the winter receptacle. The tiny new wasps that emerge then repeat the process and produce a second generation of wasps.

Female wasps emerging late in the year lay their eggs in a third type of receptacle that contains only sterile flowers. The wasps emerging as this third generation lay their eggs in the winter receptacles, completing the yearly cycle. Three types of receptacles are thus home to three generations of fig wasps each year, with male flowers appearing only in the first (winter) type of receptacle and female flowers (and thus seeds) only in the second. All three types of receptacles contain sterile flowers, which alone support the development of new fig wasps. The wasps develop only within these sterile flowers, which they also use as food. The figs are pollinated only by the wasps, each species of figs generally supporting its own species of wasps.

This story of complex interactions does not end there, for the seeds of the figs will not grow if they fall beneath the tree that bore them. The established trees have such an overwhelming competitive advantage that the offspring have little or no chance of competing successfully for moisture and nutrients if they fall and germinate near their parents. The seeds that succeed are therefore the ones that have dispersed. The seeds of some fig species germinate and grow first as epiphytes upon the branches of other trees and later grow roots reaching down into the ground; to germinate properly, the seeds of these species must find their way to above-ground perches.

The service of dispersing the fig seeds is performed by animals, with different species of animals scattering the seeds of different figs in different areas. In parts of Indonesia and Malaysia, the best dispersers of fig seeds are orangutans, *Pongo pygmaeus* (phylum Chordata, class Mammalia, order Primates, family Pongidae). These large apes practice **quadrumanual clambering**, a form of locomotion through the trees in which the orangutan's weight hangs from a branch, often supported by the feet as well as the hands (Figure 15.15). Quadrumanual clambering requires a lot of energy, especially for a large animal. An orangutan, if it is to avoid starving, must therefore eat enough in any one place to sustain it on its high-energy journey to the next place, which may be miles—and days—away. Wild orangutans are nearly always hungry and are continually wandering in search of energy-rich foods, the most preferred of which are usually figs.

The wide-roaming habits of orangutans are ideal for the figs, for the orangutans consume hundreds of figs, containing tens of thousands of seeds, then wander for miles through the forest. When an orangutan defecates amidst the branches, it leaves behind hundreds of fig seeds, together with a supply of moist, nutrient-rich fertilizer that helps the seeds to sprout and establish themselves and eventually replenish the supply of fig trees in a forest.

The orangutans must cope with the fact that the fig trees flower and bear fruit at different times of the year. Because there are many species of figs, each bearing fruits in a different seasonal pattern, a resourceful orangutan can usually manage to find some trees bearing fruit in almost every month of the year. Among orangutans, there is thus a selective advantage in having a good spatial memory—a mental map of the forest covering many square miles. The orangutan with the best chance of survival is the one whose intelligence allows it to remember where to find the most fig trees, when each was last visited, how far along its figs were at the time, and when the time would be optimal to visit each particular tree again.

The interrelatedness of the lives of figs, wasps, and orangutans (and goats, birds, monkeys, humans, and other seed-dispersing species in the places where the orangutans do not live) shows how complex life may be among the species of the rainforest. Destroy a few fig trees, and many orangutans may starve. The removal of a few orangutans may decrease the ability of fig trees to disperse their seeds, which would also diminish their ability to provide homes for fig wasps and food for the insect-eating species that feed on the wasps. The destruction of a portion of the rainforest thus has consequences far beyond the portion actually cleared, for it diminishes the health of the whole ecosystem for miles around.

FIGURE 15.15

The orangutan, *Pongo pygmaeus*, showing quadrumanual clambering.

Slowness of ecological succession. When a large tree falls in the rainforest, it creates a small clearing. Human activity may create larger clearings, some of which are later abandoned. Certain pioneer species of shade-intolerant plants are adapted to take advantage of such clearings and create a new ground cover. These species take advantage of the open sunlight while they can, only to be replaced at a later stage by the more shade-tolerant species. Only after many years can any of these grow back to the height of the original tree that had fallen. In this way, entire communities are replaced by a **succession** of other communities that take over, one after another. Although a few pioneer species will colonize an area in a few months or years, tropical botanists have found that it may take centuries for the rainforest to again reach its former canopy height and species density. Because the small-scale conditions may be different from those in previous successions, tropical rainforests, even if they regrow, may not become the same community as the forest that was there before. Small clearings in the rainforest are much more likely to grow back than large ones. Many large clearings may not grow back at all.

Deforestation. Around the world, rainforests are being destroyed at an alarming rate. One expert on tropical rainforests estimates that they are disappearing at the rate of 150,000 km^2/year (410 km^2/day), an area equivalent to the size of Manhattan island every 3.5 hours), and that at least 40% of the world's rainforests have already been destroyed!

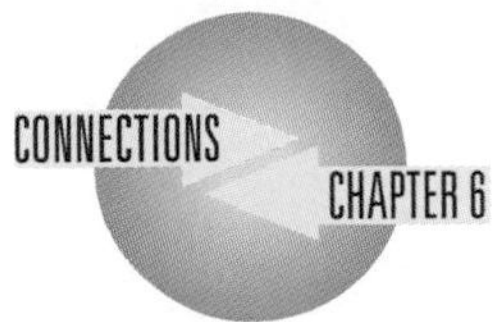

Tropical deforestation has many causes. Among the causes that have been identified are the growth of human populations (see Chapter 6), the spread of agriculture, the desire of farmers to earn a better living, the attitude that humans are entitled to dominion over nature, and the quest for corporate profits from timber, minerals, or agricultural activities. In the early twentieth century, the United Fruit Company (Chiquita Banana) cleared large rainforest areas in Central America and replaced them with banana plantations, while the Goodyear Rubber Company cleared large parts of west African rainforests to establish rubber plantations.

Among the most destructive uses of the rainforest are the exploitation of nonrenewable resources on a one-time basis. Some rainforests are cleared because of the mineral wealth that lies beneath them. In other cases, it is the trees themselves that are being harvested for timber (Figure 15.16). Trees are renewable given a sufficient length of time, but regrowth of tropical rainforests is usually very slow. Many logging operations destroy the trees much faster than they can grow back, so that the forest cannot long sustain continued logging on such a scale.

The largest amount of rainforest destruction is carried out for agricultural reasons. The land cleared from the rainforests is put to agricultural use as grazing land for cattle or as farmland for crops, either for local human consumption or for commerce. In many places, new fields are cleared by cutting down and burning forest (see Figure 15.16), a practice called **slash-and-burn agriculture**. In many cases, fields created in this way are fertile for only a decade or two, so new fields are then cleared to replace them.

It may seem odd that rainforests that appear so lush will not grow back quickly and will not support agriculture for long. This is because the soil beneath many rainforests is very poor in nutrients, for several reasons. One reason is that tropical ecosystems recycle most nutrients in the forest litter before they reach the soil. Another reason is that the continuous and heavy rainfall washes soluble and partially soluble minerals from the upper layers of the soil, a process called leaching, which leaves few mineral nutrients behind. What remains behind after extensive leaching is in many cases a dark or reddish, very hard, mineral-poor soil called **laterite**. Most attempts to grow crops on lateritic soils have been very disappointing because of their low nutrient content and frequent mudslides.

In rainforest regions with lateritic soils, agricultural use of the land quickly exhausts the low nutrient content of the soil. The amount of rainfall is often too high for most crops, and problems with drainage and erosion also arise in many places where the soil holds water poorly. In Madagascar, Haiti, and many other places, rainforests cleared for agricultural use have given way to widespread erosion in which thousands of tons of red, lateritic soil wash annually into the sea. Madagascar also has one of the world's fastest growing populations; its 1999 population of 14 million is expected to double in 22 years, putting even more demands on precious tropical habitat.

Rainforest destruction is often permanent. Because rainforest vegetation contributes (through leaf transpiration) to the rain clouds that maintain the rainy climate, any large-scale clearing of the rainforest is bound to alter the delicately balanced water cycle. Large areas cleared

FIGURE 15.16

Tropical deforestation.

Trees cleared to make way for agriculture.

Timber ready for transport. These operations destroy many hectares of forest at a time; the results include a loss of topsoil and extensive erosion, often leaving the land unsuited for agriculture or human habitation.

Slash-and-burn agriculture quickly depletes the soil of its nutrients, requiring the fields to be abandoned and new fields to be cleared by burning more forest.

These slopes in Costa Rica were once forested, but the harvesting of trees has increased the frequency of landslides and the rate of erosion.

from rainforests tend to suffer greatly reduced precipitation, and the land soon becomes unsuitable for crops or even for cattle grazing.

Deforestation is a problem at all latitudes, not just in the tropics. Norway, Russia, Canada, and the United States all have vast forests that consume carbon dioxide and contribute oxygen and water to the atmosphere. These forests are also being harvested for their timber and cleared for agricultural and other uses. Northern forest timber resources can be renewable if properly managed, especially if only some of the trees are cut in any one location. In contrast, removal of all the trees (clear-cutting) from large tracts of land is damaging to forest ecosystems and to the global atmosphere that all forests support. Even if the forest is replanted, it is often with a single species of tree. This is a form of monoculture (see Chapter 14, p. 610) and, while it may maintain a supply of that tree species, it does not restore biodiversity or habitat variety. Species displaced from their habitat may not come back because the new habitat is very different from the old habitat. Some species, such as the spotted owl mentioned earlier, are particularly sensitive to these changes and can thus be used as 'sentinel species' that allow us to monitor the health of whole ecosystems, as explained later. Of course, species driven to extinction are lost forever.

Desertification

Land that supports the richness of life of a rainforest can be transformed into a desert capable of supporting very little life. Destruction of rainforest can begin this process, which is called **desertification**. To understand it, we need to learn more about zones of climate and vegetation in tropical regions. In this section we first use Africa as an example and take a close look at desertification there. We then extend our view worldwide and consider whether humans can reverse desertification.

FIGURE 15.17

Global patterns of prevailing winds. Notice that winds blow away in both directions from latitudes 30° S and 30° N, carrying moisture away from these latitudes and creating desert regions.

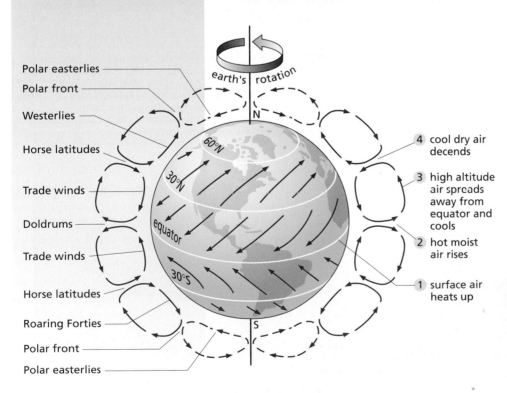

Climatic zones of Africa. The global atmospheric patterns shown in Figure 15.17 conspire to rob the regions around 30° N and 30° S latitude of all its moisture on a continuing basis: prevailing high air pressure creates winds that evaporate moisture from the land and transport it away from these regions. The northern half of Africa lies in the belt at 30° N latitude and is occupied by the world's largest desert, the Sahara.

To the south of the Sahara, much of Africa is characterized by a series of parallel zones differing in moisture and thus differing in vegetation (Figure 15.18): there is tropical rainforest along the south coast; then, heading north, patches of forests interrupted by more open land; then a more open woodland with scattered trees and shrubs only, giving way to a tropical grassland (savanna) further inland, then a type of dry pastureland, the Sahel; and finally the desert. Each band has more rainfall than the one to its north, and less than the band to its south. The overall pattern has local exceptions where the land is mountainous, but in general it prevails across most of Africa north of the Equator. Each band supports a distinctive kind of vegetation and a distinctive human culture. Each of these bands is also a biome, similar to other ecosystems in other parts of the world (see Figure 15.9).

These zones of vegetation and rainfall are not static, but are slowly changing. The Sahara is very slowly advancing southward by the process of desertification, and the other vegetation zones are moving southward with it. Desertification has also taken place in other directions where the Sahara reaches westward to the shores of the Atlantic and eastward to the shores of the Red Sea and beyond into the Arabian Peninsula. Archaeological excavations confirm that these lands were all much wetter and the vegetation was lush only a few thousand years ago, as the bones (and cave paintings) of hippopotami and crocodiles attest.

How does desertification take place? At least within the Sahel, an important factor promoting desertification is the overgrazing of pasture lands by flocks of domestic animals—goats, cattle, camels, sheep, and other species—that removes the land's vegetative cover. Without the many plant roots that held the soil and its moisture, the precious topsoil blows away. The land, which can no longer support plant life, becomes a desert.

Another important process takes place farther south, where rainforests and tropical woodlands are cleared for agricultural use, often by the slash-and-burn agriculture described earlier. After a few decades of agricultural use, the land is abandoned and new land is cleared. There are several unfortunate consequences of this method of agriculture: the abandoned land is never totally reclaimed by forest ecosystems; the agricultural land has

FIGURE 15.18

African vegetation zones.

The Sahel, with a dust storm in the distance.

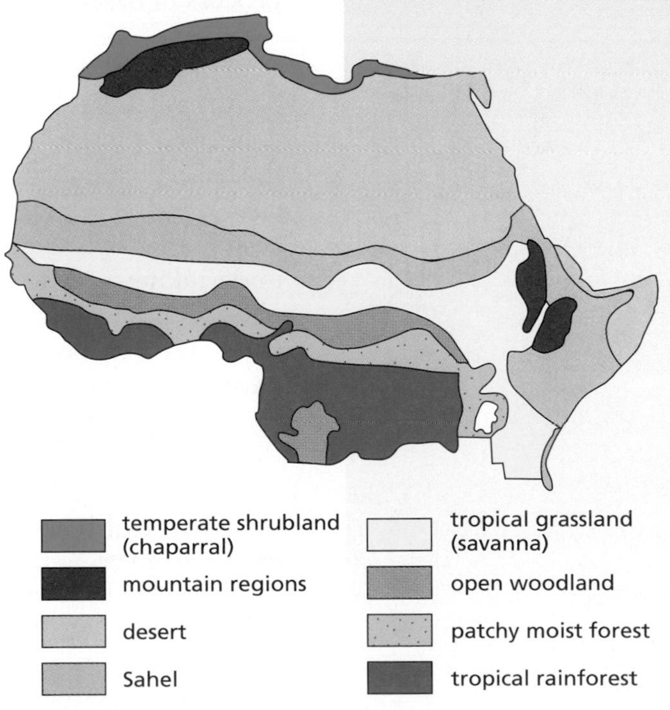

Distribution of vegetation zones in Africa, north of the Equator.

much less ability to retain moisture than the forests that it replaces; and the dry pasture of the Sahel replaces much of the abandoned fields.

Desertification around the world. The problem of desertification is not limited to Africa, although the advance of the Sahara claims more new land each year than all the other deserts of the world combined. The Mojave Desert in Southern California and the Great Indian Desert (along the India–Pakistan border) are two other deserts that are advancing on adjacent agricultural land. The situation in India may be broadly similar to that in much of Africa. The situation in the western United States is somewhat different because desertification is only in its early stages in most places and because most of the problems seem to be associated with the use of underground water reserves for irrigation and for domestic use in cities like Los Angeles. Farmers and ranchers throughout the western United States use aquifers (underground water deposits) for irrigation. In many cases, these aquifers are either shrinking or becoming saltier as ocean waters (e.g., from the Gulf of California) encroach farther inland. If water use exceeds the natural capacity of aquifers to refill, desertification will result.

The immediate effects of desertification are the loss of cropland and rangeland, an effect that is felt keenly but locally. In a few cases, there is also increasing conflict over water rights—for example, between California and Arizona over the use of the Colorado River and between Turkey and Syria over the use of the Euphrates River. Much more serious, however, are the long-term effects of desertification. With reduced vegetation cover, the ground retains less moisture. This means that the air above can become drier, and rain clouds are far less likely to form. The absence of rain clouds results in reduced rainfall, which in turn accelerates the process of desertification.

Prospects for reversing desertification. Can desertification be arrested or turned back? Yes, but only very slowly, very expensively, and with a concerted effort over many years. Israel has had great success in 'making the desert bloom,' turning desert and scrubland (like the Sahel) into agricultural land. One key to this process is irrigation, using water from rivers or lakes or desalinated sea water. Irrigation is always an expensive undertaking—particularly so in a dry climate—and natural water supplies set limits to what can be sustainably farmed with the help of irrigation. Many of the world's desert regions are in poor nations that do not have the financial resources to repeat Israel's successful experiment in reclaiming desert lands for agricultural use.

Valuing habitat

Although we have examined only two of the world's many biomes, some of the conclusions that we have drawn pertain to other types of ecosystems as well. In particular, habitat destruction threatens many ecosystems, whether they be coastal wetlands, pine forests, or the African Sahel. Why does this habitat destruction continue?

Humans often make decisions by first assigning a value, consciously or unconsciously, to such things as happiness, land, money, and even life itself. Decisions are then made by choosing the alternative that has maximum value. Under this system, bad decisions often result from attributing too much or too little value to something, and conflicts may arise if different people assign very different values to the same thing. For example, different people attribute different values to rainforest habitats and to the need to sustain the habitability of our planet.

Ways of assessing value. Philosophers often distinguish between **intrinsic value**, the value that something has as an end in itself, and **instrumental value**, the value that something has as a means to some other end. Dollar bills, for example, have no intrinsic value; they are valued only because of what we can buy with them, and they would be useless in a society that did not accept them in trade. We will soon examine the instrumental value of various habitats as places where valuable resources can be obtained. Before we do so, let us also point out that many people also value other living species, and entire ecosystems, as having a high intrinsic value. The habitat that sustains living ecosystems is likewise valued intrinsically by many people. Also, in the view of many people, no species has a right to destroy another species or to deprive it of its habitat or its means for continued existence.

Value does not only include the value of something to our species alone. Another type of value may be **biological value**, the interdependence of species, genetic diversity, and the dynamic stability of the biological community that results from this underlying diversity.

Habitat destruction and ethics. Habitat destruction can take many forms, including the clear-cutting of forests and the draining of swamps. In some cases, the destruction takes place to permit the building of housing tracts or shopping centers. In other cases, land is cleared for agricultural use. In still others, extractive industries such as mining or logging simply exploit the land on a one-time basis for its mineral wealth or its standing crop of trees. Many of the social, political, and economic forces that impinge on tropical rainforests often threaten the destruction of other habitats as well.

When we pause to consider what the many cases of habitat destruction have in common, we soon realize that the same ethical issues recur in case after case. How important are natural communities? How important are their habitats? Is it more important to leave nature undisturbed or to feed an expanding human population? Is it more important to preserve natural habitat or to satisfy people's demands for agricultural land, timber, or housing? To what extent do the answers to the previous questions depend on the quality of the soil, the economics of the country in question, or other factors? To what extent do the answers depend on how much we value other species in addition to our own? Do other species have value apart from their relationships to humans?

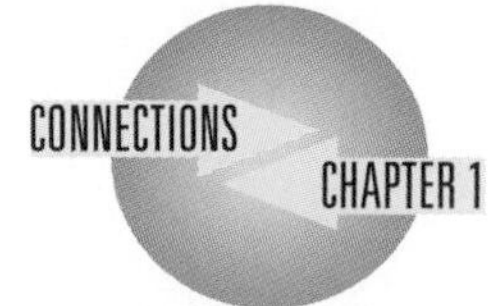

All of these are basically ethical questions (see Chapter 1), or parts of a larger, all-embracing ethical question: is it better to preserve a particular ecosystem in its 'natural' state, or is it better to convert the area into agricultural or similar use? Viewed one nation at a time, the forces that push toward one alternative or the other weigh heavily against many natural ecosystems. The pressure of human population, the need for land and food, the need for income, and the need for economic development are all obvious to the people living near rainforests and other important habitats. Measured against all these forces is the value of undisturbed wilderness, a value that is not always obvious or locally appreciated. Of course, things are never that simple: many Brazilians want to preserve their rainforest habitats, even if this means that economic development cannot proceed quite as fast as others would like.

When we consider the worldwide ecosystem of the Earth as a whole, however, the balance seems to shift in the other direction, in favor of preserving the natural environment. The advance of the Sahara, or the destruction of rainforests, threatens the planet with consequences far greater than the continuation of poverty and underdevelopment in any one country. The case for Brazil can easily be argued in these terms: the preservation of the Amazon rainforest is best for the planet as a whole, and the economic best interests of Brazil would be viewed as secondary if the good of the planet were given priority. Perhaps this makes sense to North American environmentalists and philosophers, but it is certain to be a very unpopular attitude in Brazil! It is the Brazilians who are largely in control of their rainforest, and they are likely to resent any suggestion that they sacrifice the well-being of their nation's economy for the 'greater good' of a global environment that the wealthier nations of the north have already started to destroy. A similar argument can be directed against the industrial nations of North America and Europe: a reduction of resource consumption by these nations would reduce pollution, reduce the trend toward global warming, and benefit the planet as a whole. Large tracts of land in Australia, Argentina, and the United States are devoted to cattle ranching, an activity that produces far less food per unit area than if that same land were used for growing crops. More of the world's hungry could be fed if lands now used for cattle ranching were instead used to raise wheat or corn, but ranchers (whose interests are often supported by their governments) are not likely to give up their way of life for that reason alone. They argue that much of the land now used for ranching is so used precisely because it is unsuitable for growing crops economically.

It is easy, on utilitarian principles (see Chapter 1), to argue that the good of the planet should take precedence over the economic well-being of any single nation or occupational class. However, on just about any principle of fairness, it is just as easy for Brazilians to argue that they should not bear the entire burden for a sacrifice that benefits the whole world. If the world benefits from the Brazilian rainforest, then the world should somehow pay to maintain it in its natural state. If rainforests offer such good protection against global warming, then all nations should contribute to rainforest conservation, perhaps in proportion to the amount of carbon dioxide that they generate. Currently, the United States produces the largest amount of carbon dioxide in relation to its population of any nation on Earth. Also, Brazilians can point to the logging operations that destroy forests at an alarming rate in the United States and other northern countries.

Habitat destruction versus sustainable use. It does not take long to realize that many forms of habitat destruction are driven by very short-sighted goals. As an example, the harvesting of slow-growing trees brings only short-term gains, and only to a small number of people (those in a single industry, sometimes only a single company), but the damage that it causes both to the local economy and to the biosphere as a whole may be irreversible. The same can be said of cutting down the rainforest for the planting of those crops that grow poorly on lateritic soils.

Easter Island in the southeast Pacific Ocean shows us a particularly gruesome lesson in the consequences of habitat destruction. When humans first arrived, the island supported a rainforest that was home to many edible plant species as well as numerous species of birds and other animals. The Easter Islanders, a Polynesian people, prospered for several hundred years, building the large stone statues for which the islands are now famous. But instead of conserving the rainforest and living off its rich resources, the Easter Islanders cut much of it down, until too little was left to sustain the edible bird and plant species, which slowly disappeared along with the forest itself. Like other rainforests, the one on Easter Island had created its own rain clouds, producing the conditions for high rainfall, but when it was destroyed, less rain fell on agricultural crops. With vanishing timber supplies, the Easter Islanders could scarcely find enough wood to build the boats needed to sustain their fishing activities. As all food supplies dwindled, the Easter Islanders began to starve and the last survivors resorted to cannibalism before they were rescued by the arrival of Europeans. Nor is this case unique. The Maori, a Polynesian people of New Zealand, hunted many of the native species of their islands to extinction or nearly so, and were beginning to show signs of starvation and decline when Europeans arrived. Other islands in the Pacific were subject to similar exploitation, although nowhere else did the process go as far as it did on Easter Island.

What we need instead of destructive uses are **sustainable uses** of forests, uses that allow people to derive profit from maintaining the rainforest instead of from destroying it. Most nations that contain rainforests are nonindustrialized, and many of them are also poor. A sustainable economic use of the rainforest would be an economic incentive to maintain the forest rather than to destroy it. It is therefore in the long-term best interests of all nations and all people to help tropical nations to develop such sustainable uses.

One example of sustainable use is the gathering of small amounts of high-income rainforest products such as pharmaceutical plants. For example, the anticancer drugs vinblastine and vincristine, derived from a rainforest plant (the rose periwinkle of Madagascar), account for sales of $180 million a year. In 1991, the pharmaceutical firm of Merck and Company entered into a million dollar agreement with Costa Rica's National Institute of Biodiversity. Scientists working for the Institute were to identify as many rainforest plants as they could (the total number is estimated to be 12,000), extract samples from them, and send the more promising ones to Merck for tests of their medicinal value. Merck's $1 million investment should be compared with their annual sales close to $15 billion in 1994. In 1990, Merck sold $735 million worth of just one drug, Mevacor (lovastatin), a cholesterol-lowering drug derived from a soil fungus. If even one new drug discovered by scientists under this agreement brings in a small fraction of this amount, Merck will recoup its original investment many times over.

Certain kinds of rainforest agriculture can be sustainable, but others are not, and much remains to be learned from experimentation and frequent reevaluation. Most promising are the attemps at mixed uses of rainforest habitats, allowing tall trees, shorter trees and shrubs, and smaller plants to persist side by side. Coffee, vanilla, cocoa, cashews, bananas, and certain spices are potential candidates for such experimental attempts, and tropical botanists can help to identify others. Many uses of rainforest plants are traditional, but they could benefit economically from improvements in harvesting, transport, and marketing. Agricultural scientists and business interests could help to develop new markets, which would bring much-needed income and provide local people with an economic incentive to maintain the forest ecosystem. Tropical plants that could easily be marketed more widely include amaranth (a nutritious and drought-resistant grain), fruits such as durians and mangosteens, and the winged bean (*Psophocarpus tetragonolobus*) of New Guinea (kingdom Plantae, phylum Anthophyta, class Dicotyledonae). This last species is a fast-growing plant that produces spinachlike leaves, young seed pods that resemble green beans, and mature seeds similar to soybeans, all without the use of fertilizers. Hundreds of other fruits are grown and eaten in the tropics but are only rarely exported.

An important goal of such efforts would be to identify which plants might profitably be grown or harvested in a given region without harm to the environment. New ways must be found to exploit the rainforests without destroying them—to develop rainforest ecosystems into sustainable resources for both local and worldwide benefit.

Sustainable use and habitat preservation make sense economically in nearly all climates, not just in rainforests. The city of New York was planning to spend between $6 billion and $8 billion to build a new water filtration and treatment plant, until officials discovered that they could accomplish the same goals by spending only $1.5 billion to help preserve the natural watersheds of the Catskills and the Delaware Basin, two sources of naturally filtered water supplies. In Hawaii, the Maui Pineapple Company runs the Pu'u Kukui Watershed Preserve, in which they protect native plants that maintain both soil and groundwater that sustains nearby pineapple plantations. "Without it," says a company official, "rain would run off into the ocean."

Ecotourism. If Brazilians and other tropical nations are to preserve the rainforest instead of destroying it, they must have economic incentives to do so. One type of economic incentive is the small but growing market for ecologically based tourism, also called green tourism or ecotourism. Ideally, this type of tourism seeks to make as little impact on the natural environment as possible. Ecotourism is now Costa Rica's second largest source of foreign income, behind coffee and ahead of bananas. Ecotourism is also a major source of revenue in Kenya. By comparison, the rainforests of Brazil afford a largely untapped tourist resource. Of course, some land would have to be set aside for airports, roads, and hotels. Beyond this initial investment, however, ecotourism would provide economic incentives for leaving the rest of the rainforest untouched. Ecotourism not only provides a country with income, it also gives that country an economic incentive to preserve its own natural heritage for the benefit of all.

As we search for ways to stop the destruction of ecosystems, we must realize that no solution will work if the rich and poor nations continue at odds with one another. If battles continue over short-range economic interests, the planetary ecosystem will undoubtedly be the loser.

THOUGHT QUESTIONS

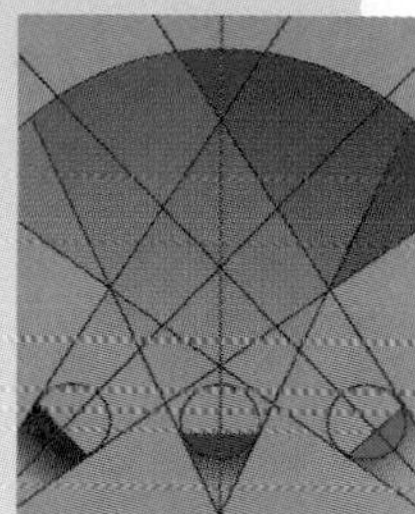

1. Is a biodegradable insecticide completely harmless? If it were used over a wide area of a forest, might it cause the extinction of an insect species confined to that area? What effect would such a loss have on other species? Do you think the use of such insecticides in research carries a certain amount of risk? If so, how can this risk be minimized?
2. Do you think undisturbed habitats have intrinsic value, or only instrumental value? In other words, is habitat valuable as an end in itself, or only because of the uses to which it might be put? Think of other things that you intrinsically value, such as close family members. Does habitat have the same kind of intrinsic value? In what ways are the values similar? In what ways are they different?
3. Are there things of intrinsic value that you personally would be willing to 'do without' if it meant preserving more habitat for other species? Will it be possible to preserve habitat for other species without humans changing their use of habitat?
4. How much habitat destruction do you think takes place at the hands of wealthy people, and how much at the hands of poor people? (The answer may differ from country to country.) What would it take to secure the cooperation of both rich and poor people in an effort to halt desertification or rainforest destruction? Could you easily appeal to both rich and poor together, or would it be easier to appeal to the two groups separately?
5. The ecologist E.O. Wilson has estimated that one Peruvian farmer clearing land to grow food cuts down more species of trees than are native to all of Europe. Do you think there may have been greater numbers of trees in Europe before human populations grew to their present density? More species of trees?
6. Are habitats being destroyed near where you live or go to college? What are they? What factors contribute to the destruction?
7. Does ecotourism sometimes do harm? How? Can the harm be minimized? Can it be eliminated entirely? Think of some examples of ecotourism and other forms of rural tourism. On the whole, do you think that the benefits of ecotourism outweigh the harm?

POLLUTION THREATENS MUCH OF LIFE ON EARTH

You have probably heard of toxic dumps and of the *Exxon Valdez* oil spill. Toxic dumps and oil spills are two of the many kinds of pollution that threaten our environment. The original meaning of the verb *pollute* was to contaminate or make dirty. Today, **pollution** may be defined as anything that is present in the wrong quantities or concentrations, in the wrong place, or at the wrong time. Although there is room for people to disagree about acceptable quantities, usually there is general agreement that pollution exists when it affects human health or kills other organisms. Oil spilled from tankers or drilling operations can kill thousands of aquatic birds, mammals, fish, and other organisms. Lead, cadmium, and other heavy metals in drinking water, food, or house paints can cause brain damage and other neurological defects. Toxic dumps can poison people and raise cancer rates. In some of the worst cases, numerous unidentified chemicals were dumped together into the same toxic waste sites, forming a 'witch's brew' that underwent further and often unpredictable chemical reactions to produce additional hazards. In many countries, it is now illegal to dispose of many chemicals except by government-approved methods.

When we flush the toilet or send our garbage to a landfill, we are contributing to the accumulation of solid and liquid wastes. When we drive our cars, we are contributing to the pollution of the air that we all must breathe. All of us contribute to pollution in many different ways. Even our breathing releases carbon dioxide into the atmosphere.

Does this mean that every act mentioned in the previous paragraph is an immoral act? Certainly not. In order to live, we must breathe, eat, urinate, and defecate. When we eat, we throw away inedible parts (skin, bones, pits, rinds, shells), packaging materials, and unfinished remains. To go on living, we must continue to pollute in certain ways. So why is there such a fuss?

Detecting, measuring, and preventing pollution

Pollution in most cases is a problem of quantities, and sometimes also of location and rates of accumulation. Clearly, you don't want your garbage to accumulate in your living room. Suppose, for the moment, that garbage disposal services were not available, and you had to dispose of your household garbage yourself, as many people in rural places still do. You could perhaps bury it in the back yard, or just toss it away. If you only tossed away bones, rinds, and other biodegradable materials (things that can be broken down by bacteria, fungi, and other decomposer organisms), then you might be able to dispose of your own garbage in this way. Up to a point, that is. Your back yard might have enough decomposer organisms to break down and recycle your own personal wastes, or perhaps your family's wastes. Of course, this depends in part on the size of your family, and also on the size of your back yard. Clearly, your back yard would not be able to handle the garbage produced by an entire town or city.

Pollution: a problem of quantities. Just about every known pollutant is harmless in some sufficiently small quantity. Pollutants become bothersome or toxic as quantities increase. In most cases, measuring pollution means measuring quantities, for pollution is a matter of quantities.

How do we know whether a habitat is polluted? If land, air, or water is polluted, how can we tell to what extent? If pollution cleanup is attempted in a particular place, how do we measure the success of the cleanup effort? Most of the specific answers depend upon the measuring of particular chemical substances. First, a particular chemical pollutant or breakdown product must be identified as an indicator of the pollution in question. Second, the concentration of this particular chemical must be measured repeatedly at various places and times. The field of **toxicology** deals with the damage done to human (or animal) health by various quantities of poisons, including environmental pollutants.

Sentinel species. More general indicators of pollution are also needed, especially pollution from hazards that have not yet been clearly identified. A number of environmentally sensitive species have been suggested as possible general indicators of pollution. These **sentinel species** serve the same role as the canaries that coal miners often took with them into the mines. Because the canaries were extremely sensitive to methane and other dangerous gases present in coal mines, the health of the canary reassured the miners, and the sickness or death of a canary was always viewed by the miners as a danger signal. Frogs and other amphibians are sentinel species that can warn us of the deterioration of freshwater habitats, just as the spotted owl can be considered a sentinel species whose numbers are indicative of the health of old growth Pacific forests. Dolphins have occasionally been suggested as sentinel species for marine habitats because environmental pollution can stress the immune systems of these marine mammals (see Chapter 12) and raises their rate of infectious diseases. The grounding of marine mammals on beaches may also be a stress-related phenomenon that reflects marine pollution.

Pollution prevention. Pollution awareness is a crucial component in the prevention of pollution. Unless we realize how we are polluting, it is unlikely that we will take any corrective action to pollute less. We all should inform ourselves about the disposal of our garbage and our industrial waste, about the emissions from our cars and from nearby (and distant) smokestacks, and the cleanliness of our beaches, playgrounds, drinking supplies, and foods.

Because pollution is a consequence of the ways in which we live and work, there are certain things that we can each do to reduce the amount of pollution that we cause. Most of these measures also have further benefits, such as saving money or contributing to human health. For example, car-pooling saves money, while bicycling to work saves even more money and contributes to health and fitness as well. Recycling saves money, too, and we can all recycle such items as paper, bottles, and cans. Making an aluminum soda can out of recycled materials requires only a small fraction of the electricity needed to make the same can from aluminum ore. Many industries are now responding to market forces by using recycled materials in their products and by advertising their use of biodegradable or other 'Earth-friendly' materials.

Air pollution

Air pollution affects the air that we breathe, both indoors and outdoors. Automobile exhausts and industrial plants are major sources of outdoor air pollution. Large forest fires can also pollute the atmosphere sporadically in the regions where they occur. Many outdoor air pollutants are oxides of carbon (carbon monoxide and carbon dioxide), oxides of nitrogen, and oxides of sulfur released as combustion products. Ozone is released to the air by many processes, but automobile exhausts release more ozone than any other source. Certain other gases, such as chlorine, benzene, and hydrogen sulfide, are also released by some industrial processes. When the concentrations of these pollutants get high, many people begin to suffer from respiratory ailments. Ozone close to ground level is particularly harmful to lungs. (In contrast, ozone in the stratosphere protects us from dangerously high levels of cancer-causing ultraviolet radiation.) In addition to the gases that we have mentioned, outdoor air pollution can also include certain bacterial pathogens, mold spores, and allergens such as pollen.

Air pollutants can become trapped in the ventilation systems of buildings and cause indoor air pollution. Indoor air pollution can include many of the same components as outdoor air pollution (including pollen), plus additional bacteria, plus pollutants such as asbestos, benzene, or formaldehyde formed by the decomposition of materials used in building construction. Buildings with poorly designed ventilation systems can impair the health of the people working in them, a phenomenon sometimes called 'sick building syndrome.' This may be one of the causes of a 70% increase in the incidence of allergies in the United States over the past two decades. Studies in England show that people working in older buildings with open windows as a source of ventilation suffer fewer respiratory illnesses than people working in newer buildings with recirculated air. Second-hand cigarette smoke is another form of indoor air pollution. A large-scale English study showed that cancer rates were much higher among nonsmoking married women whose husbands smoked than among a matched group of women married to nonsmokers, presumably because the women in the first group were breathing the polluted indoor air containing the second-hand smoke (see Chapter 9). Campaigns to banish smoking from restaurants and other public places are motivated by such health considerations.

Acid rain

One of the most widespread and best understood pollution problems is that of acid rain. (We should really speak of 'acid deposition,' because much of the problem comes from acid snow, acid fog, and dry deposition of acid dust or condensation.) The acidity of a substance is measured on a standard pH scale (see Figure 8.3, p. 319), where the lower the value, the higher the acidity (the more hydrogen ions). Rainwater of pH 5 or below is considered acid rain, and values as low as 2.7 have been measured on occasion. Recall that the pH scale is a logarithmic scale, so that a pH of 3.0 is 100 times as acidic as a pH of 5. Because most enzymes work optimally only within a very narrow pH range, and because the structure of biological molecules may change at different pH values, a small change in pH can have an enormous effect on living organisms (Figure 15.19).

Chemical tests of acid rain show that the source of most of the acidity is either sulfuric acid (H_2SO_4) or nitric acid (HNO_3). Acid rain across the eastern United States and the Scandinavian countries is primarily from sulfuric acid, while the acid content of acid rain in western United States locations such as Denver, Colorado, is more than 50% nitric acid, derived primarily from the nitric oxides produced by automobile exhausts.

Sulfuric acid pollution begins when materials containing sulfur are burned in air, forming sulfur dioxide (SO_2). The sulfur dioxide combines with additional oxygen to make sulfur trioxide, which then combines with water to make sulfuric acid. Sulfur is a common impurity in many

FIGURE 15.19

Some of the effects of acid rain.

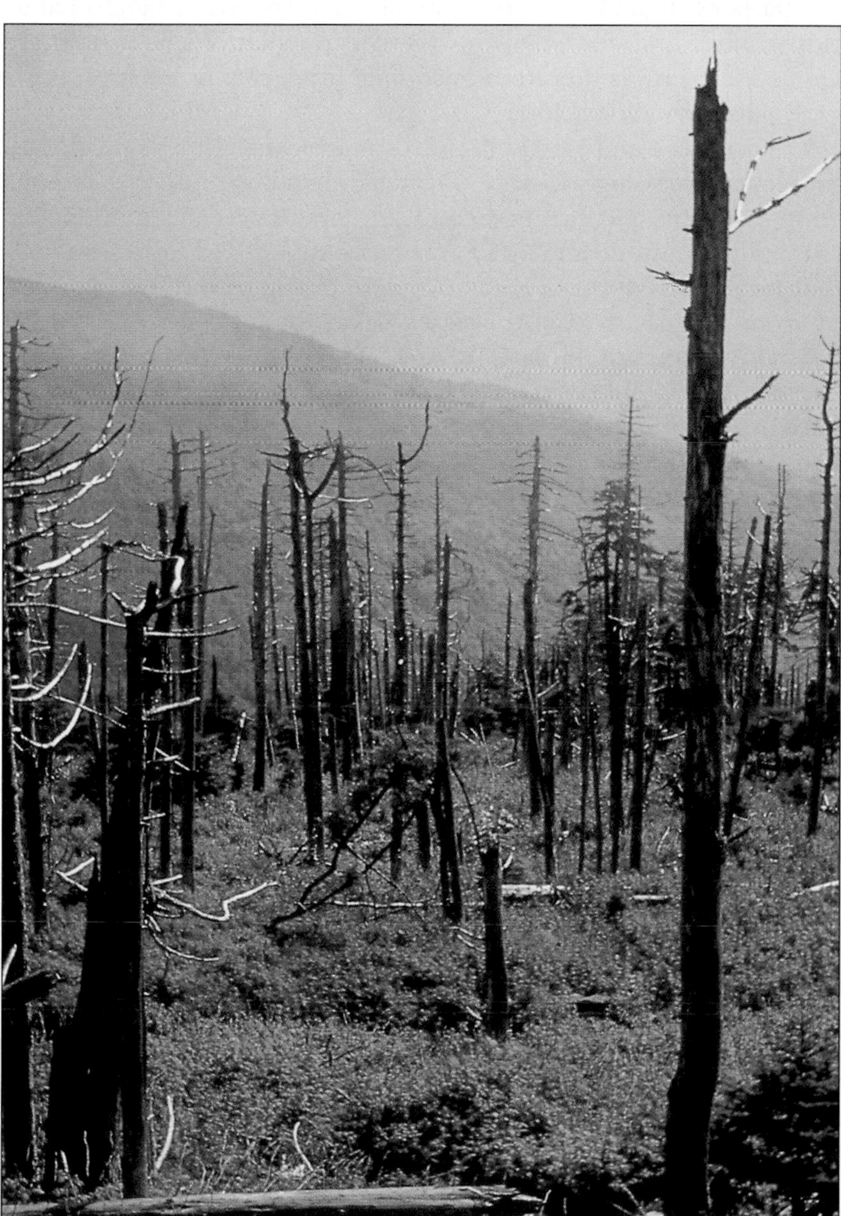

trees killed by acid rain in North Carolina

trout taken from a lake in Ontario, Canada, in 1979, at pH 5.4

trout taken from the same lake in 1982, at pH 5.1

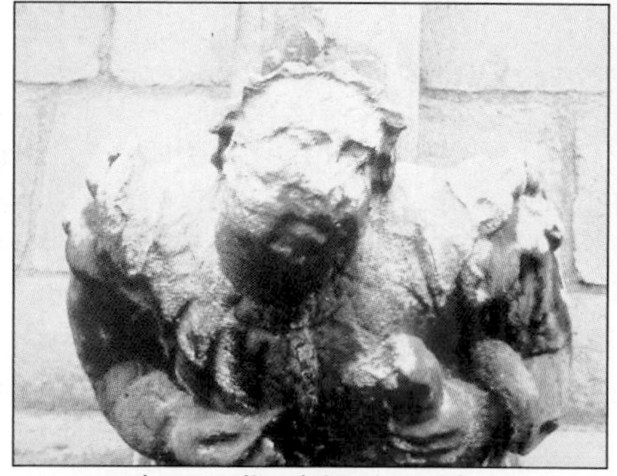

stone sculpture dissolving in acid rain

coal deposits, and is also present in many of the ores from which lead, zinc, nickel, and certain other minerals are commonly obtained. In many parts of the United States, Canada, England, and Germany, the mining of metals from sulfur ores generates sulfur dioxide that ends up as acid rain in such downwind locations as the northeastern United States and Sweden. The burning of high-sulfur coal in electrical power plants is an even larger source of sulfur dioxide pollution that eventually falls as acid rain. This type of acid rain is a political as well as an environmental problem because the governments of New York and of Sweden are relatively powerless to control pollution that originates in Illinois, Indiana, or Germany. The United States and Canada have accused one another of being major sources of cross-border acid rain pollution.

Although it is not much in the news, acid rain continues to be a problem wherever there are large numbers of factories or automobiles. In many parts of Asia and Latin America, countries burn more coal as they industrialize and more gasoline as their populations become more prosperous. Acid rain is therefore becoming more of a problem in these countries, most acutely in China.

Acid rain erodes and slowly dissolves marble and limestone statues and buildings (see Figure 15.19), including those of the famous Parthenon in Athens, Greece. In localities where the rock formations are predominantly limestone and other carbonate rocks, acid rain is neutralized as it runs through these rocks or as it percolates through the soils derived from the weathering of these rocks. However, granitic rocks, such as those that predominate in New England, northern New York State, Scandinavia, and parts of Ontario, have little or no capacity to neutralize acid rain. As a result, acid deposition in those areas accumulates in ponds and lakes to levels that kill fish (see Figure 15.19). The Adirondacks of New York State contain hundreds of lakes and ponds that once teemed with fish, but whose fish populations have completely died out because of the acidity of the water.

In addition to air pollution, there are many other types of pollution, including groundwater contamination, oil spills on oceans and lakes, and fertilizer buildup in soils. Some of these can be remediated biologically, which is the subject of the next section.

THOUGHT QUESTIONS

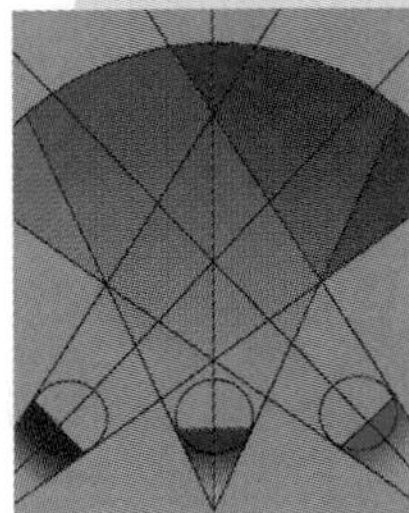

1. How does your town or city dispose of trash? Is there a recycling program? What gets recycled? What is not currently recycled in your community, but could be?

2. Is it possible to live in the industrial world without polluting?

POLLUTED HABITATS CAN BE RESTORED

Some types of pollution can be reduced, and habitats restored, with the help of living organisms. **Bioremediation** is an approach in which the decomposition activities of living organisms are put to work in cleaning up contaminated soil and water. **Biodegradation** refers to the natural decomposition processes that go on without human intervention; bioremediation, in contrast, implies the manipulation of biodegradative processes by humans. Both have been used in the cleanup of oil and chemical spills and in wastewater treatment.

Bioremediation of oil spills

Bioremediation is often applied to problems of oil contamination on land, but in this section we discuss the challenge of cleaning up aquatic oil spills. As you have seen several times in this book, we often need to examine some basic chemistry before we can understand the biological dimensions of an issue. We therefore start with the chemistry of oil and water; then we see how bioremediation can be used to clean up oil spills.

Oil and water. When oil is spilled at sea, little oil dissolves in the water. Oil is a mixture of many different chemical compounds, the exact mixture depending on where the oil came from and whether or not it has been refined. Most of the compounds are **nonpolar**, and so are not very soluble in water, which is very **polar** (see Chapter 8, p. 320). Because crude oil is not very soluble in water and because it is less dense than water, it floats on top, gradually spreading across the surface to form a 'slick.' Although slicks may be only a few molecules thick, they interfere with organisms that need to absorb oxygen at the water surface. Evaporation removes the smaller, lighter-weight components of the slick; the components left behind may then be more dense than water and sink, eventually blocking bottom-dwelling organisms' access to oxygen.

Evaporation, sinking, or dispersal of oil may make it disappear from view, but the oil molecules are unchanged. The strategy of bioremediation is to change the molecules into something harmless. Changing the molecules entails chemical reactions called degradation, at which decomposer organisms are very effective.

Oil-degrading microorganisms. As we saw in Chapter 14 (p. 585), certain organisms derive their energy by breaking down complex molecules. These are the decomposers, mostly bacteria and fungi, which keep both energy and matter cycling through the biosphere.

Oil-degrading bacteria and fungi are found in all types of aquatic habitats, both freshwater and marine, and include representatives of over 70 different genera. No one species can degrade all the molecular compounds in oil. Even when bacteria capable of producing oil-degrading enzymes are present where oil has spilled, the rate and extent of biodegradation depends on many environmental factors, including temperature, the amount of oxygen, and the availability of other nutrients.

Because many of these factors are unpredictable, the success of biodegradation at any given site is also unpredictable.

Probably the most critical factor for biodegradation is the availability of nutrients such as nitrogen, phosphorus, and iron. These elements are necessary for microbial synthesis of proteins and nucleic acids, and as enzyme cofactors, much as they are in other living organisms (see Chapter 8). This microbial requirement is harnessed in bioremediation, as we discuss next.

Bioremediation by nutrient enrichment. Bioremediation attempts to enhance biodegradation. Three basic strategies are employed: (1) enrichment of rate-limiting nutrients, (2) introducing bacteria to a spill site, and (3) introducing genetically engineered bacteria to a site.

In concept, nutrient enrichment is much like the fertilization of soil (see Chapter 14, p. 604). It assumes that oil-degrading microorganisms adapted to the local conditions are already present. Supplying nutrients supports the faster reproduction of degradative bacteria. The hope is that the bacterial population will grow quickly and to a great enough density to overcome the large quantity of oil. After the *Exxon Valdez* spill in 1989, the concept was extensively tested by the Environmental Protection Agency (EPA), in conjunction with Exxon and the state of Alaska. In these tests, the rate of biodegradation on a 110-mile stretch of nutrient-enriched beach was accelerated twofold to fourfold for a period of at least 30 days. A second application of nutrients after 3 to 5 weeks accelerated the rate even more.

Introduction of bacteria. The bacterial species indigenous to a spill area may not have the right enzymes to degrade the compounds in the type of oil in that spill. Each decomposer species is able to make specific enzymes that can degrade some specific shapes of molecules but not others. Even if the right species are present, their numbers may not be sufficient to degrade the oil at an appreciable rate. Hence, it may be advantageous to introduce into a spill site a mixed bacterial population containing many species able to digest more of the oil components than would any single species. Some people advocate using genetic engineering to produce bacteria that can make a greater range of enzymes. The first patented bacterial species was one that had been genetically engineered to degrade oil. This approach may also have potential for remediation of biohazardous wastes other than oil.

Although the concept of introducing bacteria into oil spills seems plausible, we have not yet perfected the process. Two species of commercially available bacteria were introduced on beaches polluted by the *Exxon Valdez* spill, but there was no significant enhancement of biodegradation, compared with that on untreated beaches. It is possible that introduced bacteria might possibly increase to numbers great enough to upset the ecological balance of other organisms, but it is more likely that the introduced population would die off as soon as the oil was used up. Initial experience suggests that a greater problem than overgrowth may be getting an introduced population to survive long enough to degrade all the oil.

Bioremediation of wastewater

Wastewater is all water that has been used, for whatever purpose. All wastewater treatment depends on the biological activity of microorganisms.

Sewage is water that contains human fecal wastes; if not treated, sewage is a major route for the spread of infectious diseases. Ingestion of fecally contaminated water spreads bacterial diseases such as shigella, typhoid fever, and cholera, and viral diseases such as hepatitis A. Swimming or wading in contaminated water puts people at risk of such parasitic diseases as schistosomiasis. Fecally contaminated water is concentrated by filter-feeding animals, so that mollusks collected from polluted waters may be highly infectious to people who eat them.

Gray water is water that has been used for bathing, washing clothes, or other uses in which there is no contact with human wastes. In virtually all homes and other buildings, gray water joins sewage in common drains; in some cities, rainwater from storm drains and industrial effluents also joins sewage drains. Thus, chemical pollutants and fecal pollutants are mixed together and all water must be treated as wastewater, although the particulate, or solid, content of the wastewater may be very low (about 0.03%).

Leach beds. Streams and soil often contain both minerals that trap some pollutants and bacteria capable of biodegrading waste materials. For very low human population densities of times past, natural filtering and biodegradation by streams and soil was adequate sewage treatment. Even today, in many areas where soil is suitable and population density is not too great, the wastes from homes can be collected and treated by septic tanks and leach beds (Figure 15.20). The treated water goes back into the ground, not back into the home. This type of system can only be effective when the wastewater input does not exceed the biodegradation capacity of the soil bacteria. Putting things down the drain that kill the soil bacteria make the system nonfunctional.

FIGURE 15.20

Septic tanks and leach beds.

grease and large solids are trapped, settling in septic tank
water enters perforated pipes, which allow it to slowly enter soil

Wastewater lagoons. A slightly larger-scale sewage system can separate fecal material from wastewater in a shallow lagoon in which the wastewater can be contained for about 30 days. This is sufficient time for solids to settle and for sunlight, air, and microorganisms to kill the bacteria and viruses from human and animal sources, including many microorganisms that cause human disease. Wastewater lagoons are actually complex ecosystems. Algae grow in a wastewater lagoon, using carbon dioxide to conduct photosynthesis and producing oxygen in the process. The oxygen produced by the algae keeps the lagoon aerated, allowing the growth of aerobic bacteria that digest organic matter and kill fecal bacteria.

Simple sewage systems work well if they are not overloaded. They can be overloaded by excess rainwater, an increase in users from a growing population, or an increase in wastewater production per person.

Three-stage wastewater treatment. Where simple systems prove inadequate, centralized water treatment plants are needed. Municipal water-treatment plants are further examples of bacterial biodegradation; they treat water in two or three stages.

Primary wastewater treatment is essentially the same as the settling process in a wastewater lagoon. The part that does not settle quickly, which is called the effluent, goes on to secondary treatment. Here, in large aeration tanks, air or oxygen is bubbled through the effluent, allowing aerobic bacteria to remove up to 90% of the organic wastes. The sludge that settles undergoes further digestion by anaerobic bacteria (bacteria that do not require oxygen). Some of these bacteria degrade the organic matter to organic acids and carbon dioxide. Methane-producing anaerobic bacteria digest the organic acids and release methane gas (CH_4), which is used as fuel for running the treatment plant. The sludge that remains is dried and incinerated or put in landfills. It contains valuable nutrients, about one-fifth the amount in an equal volume of commercial lawn fertilizer, and soil conditioners, much the same as compost, and is sometimes used as such.

All municipalities in the United States must have at least primary and secondary wastewater treatment. Where secondary treatment is the last step, the secondary treatment effluent is disinfected with chlorine and discharged into some body of water or sprayed onto fields designated for that purpose.

In some cities, the secondary effluent goes on to tertiary treatment, where nitrogen can be removed by denitrifying bacteria (see Chapter 14, pp. 591–592). These bacteria convert nitrogen compounds to gaseous ammonia, which is evaporated into the air. The remaining organic material is removed by nonbiological means such as fine filtration or the electrical precipitation of dissolved ions. Water is then chlorinated before being discharged. Tertiary treatment, although very costly, results in water that is once again fit for drinking.

Treatment by marshlands. Some towns are experimenting with using marshlands to treat wastewater. Because many natural wetlands have been destroyed by pollution or development, towns have created new wetlands. Wastewater goes through primary treatment and then is pumped into these marshes. Marsh plants remove nitrogen and phosphorus, break down sewage, and even filter out toxic chemicals. Preliminary treatment may be done in greenhouses. In tanks containing such plants as water hyacinths and cattails, algae and microorganisms decompose the organic materials, which are then used as nutrients by the plants. The water then passes to other tanks containing snails and zooplankton that consume the algae and microorganisms; the zooplankton are then eaten by fish, and the water is further purified by marshlands. Such systems work well but will only work if the quantities of wastewater do not exceed the capacity of the treatment ecosystem. The demonstrated success of such small-scale treatment facilities emphasizes that there are many different, ecologically sound ways to treat wastewater.

Treatment of drinking water

Although wastewater that has passed through tertiary treatment is theoretically fit to drink, municipal drinking water usually comes from other sources and is treated in separate treatment plants. Incoming water, which usually comes from natural streams and lakes, is stored in reservoirs to allow the particulates to settle. Particles of dirt, particularly of clay, as well as bacteria and viruses that are too fine to settle, are removed by flocculation with aluminum potassium sulfate (alum), a process that was known to reduce the incidence of cholera long before it was known that cholera is caused by bacteria. Water is then passed over beds of sand or diatomaceous earth that adsorb microorganisms on their surfaces (Figure 15.21). Flocculation and filtration remove several pathogens (such as the protozoan called *Giardia lamblia*) that are not killed by chlorination.

Drinking water is usually treated with chlorine (Cl_2) to kill disease-causing microorganisms. Chlorine is very reactive and toxic to other organisms, including humans, if its concentrations are too high. The risk of these toxic effects is currently considered acceptable because the benefits of disinfection are so much greater. However, chlorine levels in the water must be carefully monitored to be sure that they are high enough to effectively disinfect the water, yet not so high as to be toxic. Some water-treatment facilities have begun to use ozone as a disinfectant, producing the ozone electrically on site. This is highly effective, although very expensive.

Many treatment facilities also add fluoride (Fl^-) to drinking water (see Chapter 8, p. 347). The fluoride ion can substitute for phosphate in the hydroxyapatite crystal that makes up tooth enamel. The resultant enamel is less soluble than ordinary enamel in the acidic pH produced by dental cavity-causing bacteria. Fluoride has the greatest effect in children as their adult teeth are forming through their early years, but it has decay-preventing benefits in adults as well, because enamel is constantly remodeled even in adults.

FIGURE 15.21

Diatoms: single-celled, ocean-dwelling organisms (kingdom Plantae, division Chrysophyta) with elaborate silicon-impregnated cell walls. Diatoms accumulate in huge numbers and can be mined as diatomaceous earth from land that was formerly on the ocean floor. The enormous surface area of the many separate microscopic pieces gives diatomaceous earth a great capacity for adsorption; hence its use in water purification in water treatment plants and in swimming pools.

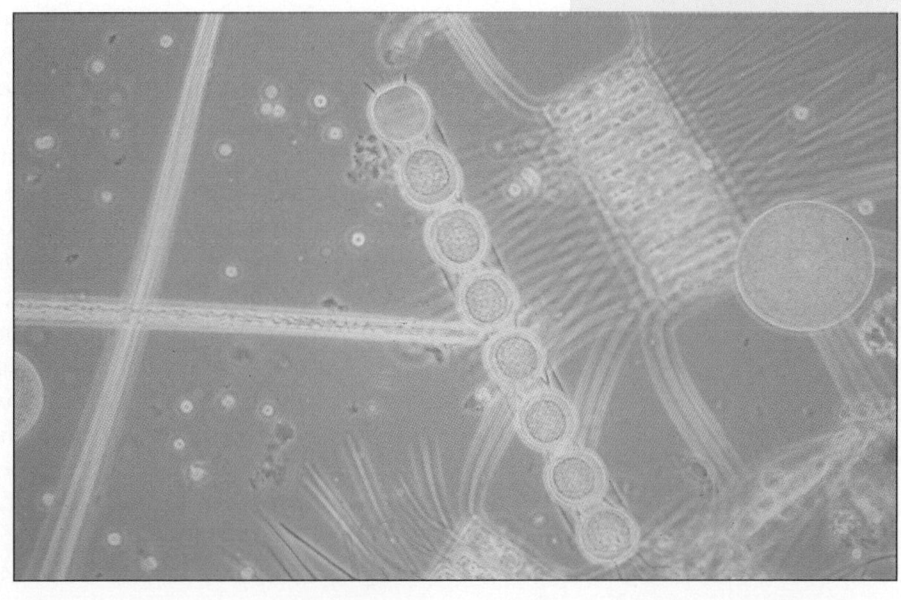

Costs and benefits

We have said that ethical decisions about pollution are decisions about quantities because the problem of pollution is, in large part, a problem of quantities. Expanding on this, we can say that the determination of both costs and benefits associated with pollution depends upon the measurement of quantities. Whether costs are measured in dollars, in lives lost, or in reduced human health, what counts is that they are measured. The same is true of the benefits of preventing or cleaning up pollution, which can be measured in dollars, in lives saved, or in increased health or enjoyment.

One general problem is that costs are often easier to identify and measure than benefits, and are subject to much less uncertainty. The costs of curtailing or modifying a manufacturing process can easily be measured in terms of costs of new equipment, costs to operate the equipment, and either increased wages (if additional employees are needed) or reduced employment (if an activity is curtailed). These costs are fairly certain, and are easily measured. The benefits to the ecosystem are less certain and are harder to measure: if a particular set of changes is implemented, will the ecosystem recover? What levels will the pollutants reach at some distance from the source? How many lives will be saved or how much disease prevented as a result? What dollar value should be placed on the prevention of disease, on the bird or fish population, or on the health of the environment?

Translating 'quality' of health or enjoyment into something that can be quantified is a relatively new and important specialty within economics. One way of measuring these intangibles (certainly not the only way, but definitely one of the easiest ways) is by measuring the average dollar amount that people are willing to pay to obtain them. To take one small example, the value of a better neighborhood can be measured by the additional amount that people are willing to pay to live in it, compared with some other neighborhood. The value of a clean environment can be measured by the amounts of money that people are willing to sacrifice in order to live in it, work in it, or visit it.

Attempts to measure environmental quality can sometimes be misunderstood. When we try to measure environmental quality, we need to consider the value of the entire ecosystem, not just the indicator species used as a measuring stick to monitor quality. Salmon have a certain value as a commercial food species, but they have a far greater importance as a general indicator of pollution or of the health of a river ecosystem. If salmon populations are used to measure pollution or pollution abatement in freshwater ecosystems, it is the value of the entire ecosystem that should be counted as a benefit, not just the commercial value of the salmon fishery. Likewise, bald eagles and spotted owls (neither of which have commercial value) can be used to indicate the general health of ecosystems, and it is the health of these entire ecosystems that should be counted as a benefit, not just the value that we place on the eagles or owls. The death of the canary in the mine means much more to the miners than the price of a replacement bird.

THOUGHT QUESTIONS

1. How can one persuade legislators in one jurisdiction to spend money on anti-pollution measures if most of the damage occurs in other jurisdictions? For example, if acid rain in Norway and Sweden originates mostly in Germany, how can the people in Norway and Sweden influence legislators to change German laws or industrial practices?

2. Materials that cannot easily be recycled are generally taken to places specifically set aside as dumps. If such facilities are needed, where should they be located? Because dumps are generally unsightly, unclean (sometimes toxic), and subject to noisy traffic, people generally don't want to live near them. The motivating desire to have these facilities somewhere else is often symbolized by the phrase 'not in my back yard' (NIMBY). One result is that dumps are often located wherever people have the least political clout, sometimes giving rise to charges of 'ecological racism.' What further social problems arise from the widespread application of the NIMBY principle? Is there any socially responsible way to locate an undesirable facility? What is the best way to reduce the need for such facilities in the first place?

3. In what ways could temperature affect the efficiency of biodegradation of an oil spill?

4. When oil from the tanker *Aragon* washed ashore on beaches in Spain, nutrients were sprayed on the beaches as part of an attempt at bioremediation. The results were not as good as in the case of the *Exxon Valdez* oil spill. What factors might explain why nutrient enrichment might work for one oil spill and not for another?

5. In what ways is wastewater treatment efficient? In what ways is it inefficient? Would it increase efficiency to separate gray water from wastewater? Could this easily be done?

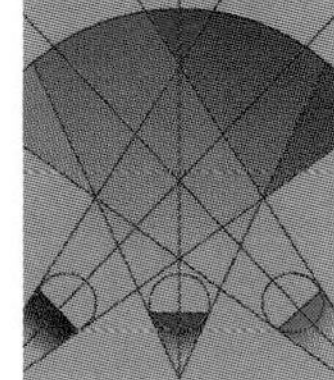

HUMAN ACTIVITY IS CONTRIBUTING TO A RAPID DECLINE in biodiversity comparable to the extinction rates of the mass extinctions in the geological past. The high rate of extinction of species is made even higher by the destruction of entire habitats, such as in rainforests. The quest for corporate profits motivates some of this destruction, but so does the need for food and living space for human populations. The reduction of biodiversity makes ecosystems less stable and makes our planet less habitable for humans and for many other species. Pollution adds to these threats, and the continued buildup of carbon dioxide and the consequent global warming is altering the entire biosphere. The causes of pollution are many: industrial activity certainly contributes, but so do our own everyday activities such as driving our cars and heating our homes. Fortunately, many forms of ecological damage can be remediated once the damage has stopped. For example, the world's forests, especially the rainforests, contain plants capable of reducing global carbon dioxide levels, given enough time. To give plants a chance to halt the process of global warming, we must give these natural ecosystems a chance to work. Can we learn to live in ways that pollute less? Can we value the continued existence of other species, not just humans? Can we learn that the continued existence of humans depends on our living in natural balance with other species? The survival of our planet and its ecosystems depends on the choices that we make. To many people, 'long-range planning' means thinking only one year into the future, but the choices that we make today often have consequences that last for decades, centuries, or even longer. Technology may soon link humans everywhere into a single global community, but we should also remember that ecosystems link our species to the rest of life.

CHAPTER SUMMARY

- **Biodiversity** is measured by the number and variety of species, of which 70% are either insects or vascular plants.
- Each species occupies a **niche** within a **community**. A community and its physical environment interact as an **ecosystem**.
- Speciation increases biodiversity, while **extinction** decreases it.
- **Endangered species** are those threatened with extinction, often because their populations are too small and genetically too homogeneous to adapt to change.
- **Habitats** support the interdependent lives of biological communities of species.
- Destruction of a habitat threatens the survival of all the species that live in it. Overuse, rather than **sustainable use**, is endangering many habitats worldwide.
- Biodiversity is greatest in tropical rainforests and in coral reefs, largely because of the great amount of solar energy and climate stability near the Equator. The expansion of human populations and agricultural lands threatens many rainforests. Because a majority of the world's photosynthesis and oxygen production occurs in rainforests, their destruction is a worldwide threat to the atmosphere and to the entire global ecosystem or **biosphere**.
- **Pollution** is a matter of quantities. Ethical decisions about pollution are also decisions about quantities because neither costs nor benefits can be measured without measuring all the quantities involved. Local sources of pollution can have widespread effects on the biosphere. Sustainable practices can reduce or prevent pollution.
- Living organisms can restore polluted areas by **biodegradation** and **bioremediation**, but prevention of pollution is far less costly than restoration afterwards.

KEY TERMS TO KNOW

biodegradation (p. 677)
biodiversity (p. 636)
biome (p. 653)
bioremediation (p. 677)
biosphere (p. 638)
community (p. 638)
ecosystem (p. 638)
endangered species (p. 650)
extinction (p. 642)
habitat (p. 636)
lineage (p. 642)
niche (p. 638)
pollution (p. 672)
sustainable use (p. 669)

CONNECTIONS TO OTHER CHAPTERS

Chapter 1: Pollution and habitat destruction violate several ethical injunctions including the principle to do no harm to others.

Chapter 3: Genetic diversity is an important component of biodiversity.

Chapter 4: Species arise through evolution.

Chapter 5: Gene frequencies change in populations.

Chapter 6: Carrying capacity is one factor that regulates population size.

Chapter 12: Pollution stresses the immune systems of many species.

Chapter 14: Plants are energy producers in every ecosystem and thus support many other species.

PRACTICE QUESTIONS

1. Which has greater biodiversity, a habitat in which there are 1500 resident species, most of which are procaryotes, or a habitat in which there are 1500 species, most of which are plants?
2. Rainforest is being destroyed at the rate of 410 km^2/day. How many square miles per day is that? How many acres? How big is your campus, town or city?
3. Acidity is measured by pH: the lower the pH the more hydrogen (H^+) ions in relation to the number of hydroxide (OH^-) ions. The pH scale is a logarithmic scale so that a change of 1 on the pH scale means a tenfold change in the H^+ concentration. At a pH of 6, for example, there are ten times more H^+ ions than at a neutral pH of 7. How many times more H^+ ions are there in acid rain at pH 5 than in water at neutral pH? How many times more at pH 5 than in ordinary water at pH 6?
4. Which is more acidic, acid rain or stomach acid (see Figure 8.3)? How much more acidic?
5. Which is more acidic, pH 4 acid rain composed mostly of sulfuric acid, or pH 4 acid rain composed mostly of nitric acid?
6. What percentage of animal species are birds? How many times more species of insects are there than species of birds?
7. What biome is your college located in? Where on the globe are comparable biomes located?
8. How do wind patterns contribute to deserts being located at 30° N and S latitudes and rainforests being located along the Equator?
9. What percentage of the world's bird species live in Costa Rica? What percentage live in Hawaii?
10. If 500,000 genera were living in the Permian period, and if 85% of them became extinct in the last 5 million years of that period, then how many genera became extinct, on average, during each million-year interval? If there were 5 species per genus, on average, then how many species per million years became extinct? How does this rate compare with the estimated 100,000 species extinctions for the decade of the 1980s?
11. What are the similarities between bioremediation and biodegradation? What are the differences?

Glossary

Only 'key terms' (represented in the text by **boldface type**) are listed here. A much more extensive glossary is available on-line at http://www.garlandscience.com/biologytoday

Abortion: Expulsion or removal of a fetus from the womb prematurely.

Action potential (spike): A large reversal of polarization in a nerve cell membrane, resulting in a nerve impulse.

Active transport: Use of energy to transport a substance, often from an area where it is in lower concentration to an area where it is in higher concentration. Active transport is performed by membrane proteins called transporters.

Acute effects: Short-term effects, such as those of a drug that disappear once the drug has been cleared from the body.

Adaptation: (1) Any trait that increases fitness or increases the ability of a population to persist in a particular environment. (2) A physiological change in response to a stimulus that prepares the body to better withstand or react more vigorously to similar stimuli.

Addiction: A strong psychological and physiological dependence.

Additive effect: A physiological response produced by two drugs given together that is the same as the sum of the effects of each drug given separately.

Age pyramid: A diagram that represents the age distribution of a population by a stack of rectangles, each proportional in size to the percentage of individuals in a particular age group.

Agonist: A drug that stimulates a particular receptor or that has a stated effect.

AIDS (Acquired ImmunoDeficiency Syndrome): Impairment of most parts of the immune system resulting from infection with human immunodeficiency virus, accompanied by opportunistic infections or rare cancers and leading, in most cases, to death.

Allele: One of the alternative DNA sequences of a gene.

Allele frequency: The frequency of an allele in a population, or the fraction of gametes that carry a particular allele.

Allen's rule: In any warm-blooded species, populations living in warmer climates tend to have longer and thinner protruding parts (legs, ears, tails, etc.), while the same parts tend to be shorter and thicker in colder climates.

Altruism: Any act that increases another individual's fitness but lowers or endangers one's own fitness.

Anemia: A disease symptom in which the blood has a decreased ability to carry oxygen.

Anisogamy: A condition in which the two types of gametes (eggs and sperm) differ in size and other characteristics.

Antagonist: A drug that inhibits another or that inhibits a particular receptor.

Antagonistic interaction: A combined effect in which two drugs together produce less of a physiological response than either drug given separately.

Antibodies: Proteins, secreted by lymphocytes during an immune response, that bind specifically to the type of molecule that induced their secretion, thus helping to protect the body from disease.

Antigen: Any molecule or part of a cell that is detected by the immune system.

Antioxidant: A substance that prevents oxidation of a molecule by an oxidizing agent.

Artificial selection: Consistent differences in the contributions of different genotypes to future generations, brought about by intentional human activity.

Asexual reproduction: Reproduction (i.e., increase in the number of individuals) without the recombination of genes.

Atherosclerosis: Deposits of fat and cellular debris, which may become calcified, on the interior walls of arteries.

ATP: Adenosine triphosphate, a molecule that stores energy in its chemical bonds.

Autonomic nervous system: Part of the peripheral nervous system that regulates 'involuntary' physiological processes of the body; consists of the parasympathetic and the sympathetic divisions.

Autotroph: An organism capable of making its own energy-rich organic compounds from inorganic compounds.

Axon: An extension of a nerve cell that carries an impulse away from the nerve cell body.

B lymphocyte (B cell): A type of lymphocyte that makes antibodies.

Balanced polymorphism: A situation in which different alleles of a gene persist in a population because of the superior fitness of the heterozygous condition.

Basal metabolic rate: The rate at which the body uses energy when awake but lying completely at rest.

Bergmann's rule: In any warm-blooded species, populations living in colder climates tend to have larger body sizes, compared with smaller body sizes in warmer climates.

Biodegradation: Breaking down a chemical by biological action.

Biodiversity: The number and variety of biological species, their alleles, and their communities.

Biological determinism: The belief that one's physical characteristics and behavior are unalterably determined by one's genetic makeup.

Biology: The scientific study of living systems.

Biomagnification: The increasing concentration of pollutants as one proceeds up the food energy pyramid from one trophic level to the next.

Biome: A group of similar ecosystems in various locations around the world.

Bioremediation: Human manipulation of naturally occurring biodegradation processes, as in the cleanup of environmental pollution.

Biosphere: The ecosystem that includes the whole Earth and its atmosphere.

Birth control: Any measure intended to prevent unwanted births or to reduce the birth rate.

Birth rate (*B*): The number of births in a given period divided by the number of individuals in a population at the beginning of that period.

Cancer: A group of diseases characterized by DNA mutations in growth control genes and in which some cells divide without regard to the growth control signals of other cells.

Carbohydrates: Polar molecules used by organisms as energy sources and consisting of carbon, hydrogen, and oxygen, with hydrogen and oxygen atoms in a 2:1 ratio.

Carcinogen: A physical, chemical, or viral agent that induces cancer; its action is called carcinogenesis.

Carrying capacity (*K*): The maximum population size that can persist in a given environment.

Cell: The smallest unit of living systems that shows the characteristics of life; it can be either free-living or part of a multicelled organism.

Cell cycle: The process by which a cell divides into two cells.

Central nervous system: The brain and spinal cord.

Chemical digestion: The use of enzymes and chemical reactions to break down food into molecules absorbable by cells.

Chlorophyll: A green pigment molecule that traps light in the light reactions of photosynthesis.

Chloroplasts: Photosynthetic organelles containing chlorophyll.

Cholesterol: A lipid with a multiringed structure, found in the cell membranes of most animal cells.

Chromosomes: Elongated structures that contain DNA; in eucaryotic cells, the chromosomes are located in the nucleus and contain protein as well as DNA.

Chronic effects: Lasting or life-long effects.

Circadian rhythm: A biological change whose pattern repeats approximately every 24 hours.

Classical conditioning: A form of learning in which one stimulus (the conditioned stimulus) that repeatedly precedes or accompanies another (the unconditioned stimulus) becomes capable of evoking the response originally elicited only by the unconditioned stimulus.

Classification: An arrangement of larger groups of species that are subdivided into smaller groups on the basis of some organizing principle or theory.

Cline: A gradual geographic variation of a trait within a species.

Clone: The genetically identical cells or organisms derived from a single cell or individual by cell division or asexual reproduction.

Codon: A coding unit of three successive nucleotides in a messenger RNA molecule that together determine an amino acid.

Community: A group of species that interact in such a way that a change in the population of one species has consequences for the other species in the community.

Complete protein: A protein that contains all of the amino acids considered essential for human nutrition.

Concentration gradient: A situation in which the concentration of a substance is different in different locations or on opposite sides of a membrane.

Contraceptive: Any method that prevents conception (fertilization).

Control group: In an experiment, a group used for comparison. For example, if animals are experimentally exposed to a drug, then a control group might consist of similar animals not exposed to the drug but treated the same in every other way.

Cytokines: Chemicals that carry information from one cell to another but have no nutritional value or enzymatic activity of their own.

Cytoplasm: The portion of a cell outside the nucleus but within the plasma membrane.

Cytotoxic (CD8) T cells: Lymphocyte cells of the immune system that react specifically to a nonself molecule, becoming activated to kill cells bearing that molecule.

Data: Information gathered so as to permit the testing of hypotheses.

Death control: Any measure that reduces the death rate.

Death rate (*D*): The number of deaths in a given period divided by the number of individuals in the population at the beginning of that period.

Declarative learning: Conscious remembrance of persons, places, things, and concepts, requiring the actions of the hippocampus and the temporal regions of the brain.

Deduction: Logically valid reasoning of the 'if…then' form.

Demographic momentum: A temporary population increase that can be predicted in a population that has more prereproductive members and fewer postreproductive members than a population with a stable age distribution would have.

Demographic transition: An orderly series of changes in population structure in which the death rate decreases before a similar decrease occurs in the birth rate, resulting in a population increase during the transition period.

Demography: The mathematical study of populations.

Dendrites: Nerve cell processes that receive signals and respond by conducting impulses toward the nerve cell body.

Deontological: A type of ethics in which the rightness or wrongness of an act is judged without reference to its consequences.

Deoxyribonucleic acid (DNA): A nucleic acid containing deoxyribose sugar and usually occurring as two complementary strands arranged in a double helix.

Dependence: Inability to perform normal physiological functions without a particular drug.

Determined: A state of development in which the future identity of a cell's progeny is predictable.

Differentiation: The process of becoming different; a restriction on the set of future possibilities for a cell's progeny.

Diffusion: A process in which molecules move randomly from an area of high concentration to an area of low concentration until they are equally distributed.

Diploid: Possessing chromosomes and genes in pairs, as in all somatic cells.

DNA: See *Deoxyribonucleic acid*.

Dominant: A trait that is expressed in the phenotype of heterozygotes; an allele that expresses its phenotype even when only one copy of the allele is present.

Dose: The amount of a drug given at one time.

Doubling time: The time required for the number of individual units in a population to double.

Drug: Any chemical substance that alters the function of a living organism other than by supplying energy or needed nutrients.

Drug abuse (substance abuse): Excessive use of a drug, or use causing harm to the individual or society.

Economic impact level (EIL): The smallest population level of a pest species that reduces crop yields by an unacceptable amount.

Ecosystem: A biological community interacting with its physical environment.

Egg: The female gamete, nonmotile and larger than the sperm because it contains more cytoplasm.

Electrical potential: A form of stored (potential) energy consisting of a separation of electrical charges.

Endangered species: A species threatened with extinction.

Endocrine glands: Glands that secrete their products, called hormones, into the bloodstream rather than into a duct.

Endocytosis: Bringing a particle into a cell by surrounding it with cell membrane.

Endosymbiosis: A theory that explains the origin of eucaryotic cells from large procaryotic cells that engulfed and maintained smaller procaryotic cells inside the larger cells.

Enzyme: A chemical substance (nearly always a protein) that speeds up a chemical reaction without getting used up in the reaction; a biological catalyst.

Epidemiology: The study of the frequency and patterns of disease in populations.

Ethics: The study of moral rules and moral codes.

Eucaryotic: Cells with the following properties: they contain various organelles bounded by membranes, including nuclei surrounded by a nuclear envelope; their chromosomes are usually multiple and contain protein as well as nucleic acids; they have a cytoskeleton that is composed of structural and/or contractile fibers of protein.

Eugenics: An attempt to change allele frequencies through selection or changes in fitness. Raising the fitness of desired genotypes is called 'positive' eugenics; lowering the fitness of undesired genotypes is called 'negative' eugenics.

Eusocial: A form of social organization characterized by overlapping generations (parents coexist with offspring), strictly delimited subgroups (castes), and cooperative care of eggs and young larvae.

Evolution: The process of permanent change in living systems, especially in genes or in the phenotypes that result from them.

Excretion: The production of waste products, especially by the kidneys, and their subsequent removal from the body.

Experiment: An artificially contrived situation in which hypotheses are tested by comparison with some known condition called the control condition.

Experimental sciences: Sciences that rely primarily on hypothesis testing by means of experiments.

Exponential growth: A form of geometric growth without any limit, according to the equation

$$dN/dT = r\,N.$$

Extinction: Termination of a lineage without any descendents.

Facilitated diffusion: The use of membrane proteins to hasten diffusion.

Fairness: The principle that all individuals in similar circumstances should receive similar treatment.

False negative: A negative test result in a sample that actually has the condition being tested for; indicates a lack of sensitivity of the test.

False positive: A positive test result in a sample that does not actually have the condition being tested for; indicates a lack of specificity of the test.

Falsifiable: Capable of being proved false by experience.

Feedback system: Any process in which a later step modifies or regulates an earlier step in the process.

Females: Individuals who produce large gametes (eggs).

Fertilizer: Any substance artificially furnished to promote the growth of crops.

Fetal alcohol syndrome: Permanent brain damage and mental retardation, accompanied by abnormal facial features, caused by fetal exposure to alcohol while in the uterus.

Fitness: The ability of a particular individual or genotype to contribute genes to future generations, as measured by the relative number of viable offspring of that genotype in the next generation.

Fossils: The remains or other evidence of life forms of past geological ages.

Founder effect: A type of genetic drift in which the allele frequencies of a population result from the restricted variation present in a small number of founders of that population.

Gametes: Reproductive cells (eggs or sperm), containing one copy of each chromosome.

Gene: A portion of DNA that determines a single protein or polypeptide. In earlier use, a hereditary particle.

Gene expression: Transcription and translation of a gene to its protein product.

Gene therapy: Introduction of genetically engineered material into an individual for the purpose of curing a disease or a genetic defect.

Genetic drift: Changes in allele frequencies in populations of small to moderate size as the result of random processes.

Genetic engineering: Direct and purposeful alteration of a genotype.

Genetics: The study of heredity, including genes and hereditary traits.

Genome: The total genetic makeup of an individual, including its entire DNA sequence.

Genotype: The hereditary makeup of an organism as revealed by studying its offspring.

Gloger's rule: In any warm-blooded species, populations living in warm, moist climates tend to be darkly colored or black; populations living in warm, arid climates tend to have red, yellow, brown, or tan colors; and populations living in cold, moist climates tend to be pale or white in color.

Glycogen: A carbohydrate consisting of many glucose units linked together, used as a storage molecule in animals and certain microorganisms.

Group selection: Selection that operates by differences in fitness between social groups.

Growth rate: The population increase during a specified time interval (usually a year) divided by the population size at the beginning of that time interval.

Habitat: The place and environmental conditions in which an organism lives.

Habituation: A form of learning in which an organism learns not to react to a stimulus that is repeated without consequence.

Half-life: For a drug, the time that it takes for the level of the drug in the body to be reduced by half.

Haplodiploidy: A form of sex determination characteristic of the insect order Hymenoptera, in which males have one copy of each chromosome (haploidy) while females have a pair of each type of chromosome (diploidy).

Haploid: Containing only unpaired chromosomes, as in gametes or procaryotic organisms.

Hardy–Weinberg equilibrium: A genetic equilibrium formed in large, randomly mating populations in which selection, migration, and mutation do not occur or are balanced.

Health: The ability of an organism to maintain homeostasis or to return to homeostasis after disease or injury.

Helper (CD4) T cells: Lymphocyte cells of the immune system that react specifically to a nonself molecule by secreting interleukin-2, a cytokine needed for the full activity of either B cells or CD8 T cells.

Hereditarian: A position attributing biological differences to heredity rather than to environment.

Heterotroph: An organism not capable of manufacturing its own energy-rich organic compounds and therefore dependent on eating other organisms or their parts to obtain those compounds.

Heterozygous: Possessing two different alleles of the same gene in a genotype.

High-risk behaviors: Behaviors or actions that increase the probability of undesirable outcomes, such as the transmission of a disease.

High-risk groups: A subpopulation of people who share some behavioral, geographic, nutritional, or other characteristic and who have a higher frequency of a particular disease than the general population.

HIV: See *Human immunodeficiency virus*.

Homeostasis: The ability of a complex system (such as a living organism) to maintain conditions within narrow limits. Also, the resulting state of dynamic equilibrium, in which changes in one direction are offset by other changes that bring the system back to its original state.

Homology: Shared similarity of structure resulting from common ancestry.

Homozygous: Possessing two like alleles of the same gene in a genotype.

Hormone: A chemical messenger, transported through the blood, that affects the activity of the cells in a target tissue.

Human Immunodeficiency Virus (HIV): The virus that causes AIDS by infecting and inactivating cells of the immune system that bear a molecule called CD4.

Hypothesis: A suggested explanation that can be tested.

Immune system: A diffuse organ system consisting of the lymphatic circulation, bone marrow, skin, white blood cells, and lymphoid tissues, and functioning to discriminate nonself from self and to maintain health by eliminating nonself and by inducing tissue healing.

Immunization (vaccination): Artificial exposure to an antigen that evokes a protective immune response against a potential disease-causing antigen similar in structure to the antigen in the vaccine.

Immunodeficiency: A decreased activity of some part of the immune system as the result of genetic, infectious, or environmental factors.

Immunosuppression: Decreasing the strength of future immune functions in any manner that is not antigen-specific.

Inclusive fitness: The total fitness of all individuals sharing one's genotype, including fractional amounts of the fitness of individuals sharing fractions of one's genotype.

Independent assortment, law of: Genes carried on different chromosomes segregate independently of one another; the separation of alleles for one trait has no influence on the separation of alleles for traits carried on other chromosomes. Also called 'Mendel's second law.'

Induction: Reasoning from specific instances to general principles, which can sometimes be unreliable, as in 'these five animals have hearts, so all animals must have hearts.'

Inflammation: A physiological response to cellular injury that includes capillary dilation, redness, heat, and immunological activity that stimulates healing and repair.

Informed consent: A voluntary agreement to submit to certain risks by a person who knows and understands those risks.

Innate immunity: Nonspecific immunity that exists before any exposure to environmental antigens, including the activities of macrophage cells, neutrophils, and complement.

Instinct: Complex behavior that is innate and need not be learned.

Integrated pest management (IPM): An approach to the management of pest populations that emphasizes biological controls and frequent monitoring of pest populations.

Interbreeding: The mating of unrelated individuals or the exchange of genetic information between populations.

Isogamy: A condition in which gametes are all similar in size.

***K*-selection:** Natural selection that characterizes populations living at or near the carrying capacity (K) of their environments by favoring adaptations for parental care and efficient exploitation of resources.

Karyotype: The chromosomal makeup of an individual.

Kilocalories (kcal): The amount of energy required to raise the temperature of 1000 grams of water by 1 degree Celsius; equal to 1000 calories.

Kin selection: Selection that favors characteristics that decrease individual fitness but are nevertheless favored because they increase inclusive fitness.

Koch's postulates: A set of test results that must be obtained to demonstrate that a particular species of microorganism is the cause of a particular infectious disease.

Krebs cycle: A series of biochemical reactions that break apart pyruvate and use the chemical bond energy to make some ATP and NADH from ADP and NAD.

Learning: The modification of behavior or of memory on the basis of experience.

Limiting amino acid: An amino acid present in small amounts that, when used up, prevents the further synthesis of proteins requiring that amino acid.

Limiting nutrient: Any nutrient whose amounts constrain the growth of an organism or population; supplying greater amounts of this nutrient therefore allows a population of organisms to increase or grow more vigorously.

Lineage: A succession of species in an ancestor-to-descendent sequence.

Linear dominance hierarchy: A social organization in which one individual is dominant to all others, a second individual to all others except the first, and so on; also called a 'pecking order' among birds.

Linkage: An exception to the law of independent assortment in which genes carried on the same pair of chromosomes tend to assort together, with the parental combinations of genes predominating.

Lipids: Nonpolar molecules formed primarily of carbon and hydrogen, occurring in cell membranes and also used as energy sources.

Locus: The location of a gene on a chromosome.

Logistic growth: Growth that begins exponentially but then levels off to a stable population size (K), according to the equation

$$dN/dT = rN\,(K{-}N)/K.$$

Lymphatic circulation: An open circulatory system in vertebrate animals that gathers intracellular fluid and returns it to the blood along with cells of the immune system.

Males: Individuals who produce small gametes (sperm).

Mechanical digestion: Breaking food into smaller particles by physical means such as chewing and churning, exposing new surfaces for chemical digestion.

Meiosis: A form of cell division in which the chromosome number is reduced from the diploid to the haploid number. Compare *mitosis*.

Memory: The ability to recall past learning.

Messenger RNA (mRNA): A strand of RNA that leaves the nucleus after transcription and passes into the cytoplasm, where it functions in protein synthesis.

Mimicry: A situation in which one species of organisms derives benefit from its deceptive resemblance to another species.

Minerals: Inorganic (non-carbon-containing) atoms or molecules needed to regulate chemical reactions in the body.

Mitosis: The usual form of cell division, in which the number of chromosomes does not change. Compare *meiosis*.

Model: A mathematical, pictorial, or physical representation of how something is presumed to work.

Monoculture: Growth of only one species in a particular place, as in a field planted with a single crop.

Monogamy: A mating system in which each adult forms a mating pair with only one member of the opposite sex.

Morals: Rules governing human conduct.

mRNA: See *Messenger RNA*.

Mutagen: An agent that causes mutation in DNA.

Mutation: A heritable change in a DNA sequence or gene.

Mutualism: A type of symbiosis in which each of two species benefits the other.

Myelin sheath: A lipid-rich covering that surrounds and insulates many neurons.

Natural selection: A naturally occurring process by which different genotypes consistently differ in fitness, i.e., in the number of copies of themselves that they pass to future generations.

Naturalistic sciences: Sciences in which hypotheses are tested by the observation of naturally occurring events under conditions in which nature is manipulated as little as possible.

Negative reinforcement: The removal of an unpleasant stimulus, which may result in learning whatever behavior preceded the removal.

Nerve: A bundle of axons outside the central nervous system.

Neuroendocrine system: The nervous system and the endocrine system considered as an interactive whole.

Neuron: A nerve cell.

Neurotransmitter: Any chemical that transmits a nerve impulse from one cell to another.

Niche: The way of life of a species, or its role in a community.

Nitrogen cycle: A cyclical series of chemical reactions occurring naturally in which nitrogen compounds are built up, broken down, and changed from one form into another with the help of living organisms.

Nonpolar: Having a molecular structure in which electric charges are evenly distributed (or nearly so) across chemical bonds; nonpolar substances are not stable in water because water is a polar solvent.

Normal science: Science that proceeds step-by-step within a paradigm.

Nucleic acids: DNA and RNA; compounds containing phosphate groups, five-carbon sugars, and nitrogen-containing bases.

Nucleus: (1) The central part of an animal cell, plant cell, or other eucaryotic cell, containing the chromosomes. (2) Also, a clump of nerve cell bodies in the central nervous system.

Obesity: A condition in which ideal body weight is exceeded by at least 20%.

Organ: A group of tissues working together structurally and functionally.

Osmosis: Diffusion of water molecules across a semipermeable membrane in response to a concentration gradient of some other molecule or ion.

Ovulation: The release of an egg from the ovary.

Oxidation: Removal of electrons from an atom or a molecule.

Paradigm: A coherent set of theories, beliefs, values, and vocabulary terms used to organize scientific research.

Parasympathetic nervous system: A division of the autonomic nervous system that brings about the relaxation response and secretes acetylcholine as its final neurotransmitter.

Parental investment: The energy or resources that a parent invests in the production of offspring and the raising of offspring.

Passive immunity: Antigen-specific immunity acquired by one organism and then transferred to another organism in the form of antibodies or specific immune cells.

PCR: See *Polymerase chain reaction*.

Pedigree: A chart showing inheritance of genetic traits within a family.

Peripheral nervous system: The nervous system except for the brain and spinal cord.

Phenotype: The visible or biochemical characteristics or traits of an organism.

Pheromones: Chemical signals by which organisms communicate with other members of their species.

Photosynthesis: A process by which plants and certain other organisms use energy captured from sunlight to build energy-rich organic compounds, especially carbohydrates.

Placebo: Something that lacks the ingredient thought to be effective, usually given as an experimental control for comparison with an active drug given to other subjects.

Placebo effect: Physiological response to a placebo that does not result from the chemistry of the placebo but that often produces the response expected by the subject.

Polar: Having a molecular structure in which most bonds have electrons shared unevenly, producing one part of the bond that has more negative charge than another part; water is polar, so other polar molecules do not spontaneously separate out from water.

Pollution: Contamination of an environment by substances present in undesirable quantities or locations.

Polyandry: An uncommon mating system in which each mating unit consists of one female and many males.

Polygyny: A mating system in which each mating unit consists of one male and many females.

Polymerase chain reaction (PCR): An artificial replication process in which many copies are made of specific DNA regions.

Polymorphism: The persistence of several alleles in a population at levels too high to be explained by mutation alone.

Population: A group of organisms capable of interbreeding among themselves and often sharing common descent as well; a group of individuals within a species living at a particular time and place.

Population ecology: The study of populations and the forces that control them.

Population genetics: The study of genes and allele frequencies in populations.

Positive reinforcement: A pleasant or pleasurable stimulus that results in learning.

Potentiality: The range of possible futures for a cell's progeny.

Procaryotic: Cells containing no cytoskeleton and no internal membrane-bounded organelles, but having a simple nuclear region that is never surrounded by a nuclear envelope and a single chromosome (usually circular) containing nucleic acid only and no protein.

Procedural learning: Learning how to do things, a process that does not require the hippocampus and is not necessarily conscious.

Promiscuity: A mating system in which no permanent mating units are formed and in which each adult of either sex mates with many individuals of the opposite sex.

Proteins: Molecules built of amino acids linked together in straight chains, which then fold themselves to produce complex shapes, functioning most often as enzymes or as structural materials in or around cells or their membranes.

Psychoactive drug: Any chemical substance that alters consciousness, mood, or perception.

Psychoneuroimmunology: A theory that postulates that the mind and the body are a single entity interconnected through interactions of the nervous, endocrine, and immune systems.

***r*-selection:** Natural selection that characterizes populations living far below the carrying capacity of their environments and that favors high rates of reproduction (high *r*) and maximum dispersal ability.

Race: A geographic subdivision of a species distinguished from other subdivisions by the frequencies of a number of genes; a genetically distinct group of populations possessing less genetic variability than the species as a whole. This concept is called the *population genetics race concept* and is distinguished from other, older race concepts by defining race as a characteristic that can apply only to populations, not to individuals. Important older meanings include the following.

Socially constructed race concept: A definition of an oppressed group and the individuals in that group by their oppressors, using whatever cultural or biological distinctions the oppressors find convenient.

Morphological (typological) race concept: A definition of each race by its physical characteristics, based on the assumption that each characteristic is unvarying and reflects an ideal type or *form* shared by all members of the group.

Receptor: A protein or other molecule that binds with a specific drug or other chemical substance and responds to the binding by initiating some cellular activity.

Recessive: A trait that is not expressed in heterozygotes; an allele that expresses its phenotype only when no dominant allele of the same gene is present.

Relaxation response: A voluntary, self-induced stimulation of the parasympathetic nervous system in which the stress response is ended, blood pressure and breathing are reduced, the threshold of excitation of nerve cells becomes higher, and digestive activity is stimulated.

Replication: A process in which DNA is used as a template to make more DNA.

Reproductive isolating mechanism: Any biological mechanism that hinders the interbreeding of populations belonging to different species.

Reproductive isolation: The existence of biological barriers to interbreeding.

Reproductive strategy: A pattern of behavior and physiology related to reproduction.

Resting potential: The difference in electric charge maintained by a nerve cell membrane in the absence of a nerve impulse.

Restriction enzyme: An enzyme that cuts nucleic acids by attacking certain DNA sequences only.

Restriction-Fragment Length Polymorphism (RFLP): Variation among individuals in the lengths of selected DNA restriction fragments.

Reverse transcription: Transcription of complementary DNA from a template of RNA.

Ribonucleic acid (RNA): A nucleic acid containing nucleotides with ribose sugar and usually existing in single-stranded form.

Rights: Any privilege to which individuals automatically have a just claim or to which they are entitled out of respect for their dignity and autonomy as individuals.

Risk: The probability of occurrence of a specified event or outcome.

Risk behaviors: Behaviors classified according to the likelihood of disease transmission. See *High-risk behaviors*.

RNA: See *Ribonucleic acid*.

Rough-and-tumble play: Play in which large muscle groups are used in pushing, pulling, climbing, and mock fighting with other individuals.

Saturated fats: Lipids with no double bonds between their carbon atoms.

Science: An endeavor in which falsifiable hypotheses are systematically tested.

Scientific revolution: The establishment of a new scientific paradigm, including the replacement of earlier paradigms.

Second messenger: Molecules within the cytoplasm of a cell that carry information from membrane receptors to other locations in the cell.

Segregation, law of: When a heterozygous individual produces gametes, the different alleles separate so that some gametes receive one allele and some receive the other, but no gamete receives both.

Sensitivity: The smallest amount of some substance that can be detected by a clinical or other test.

Sensitization: A form of learning in which an intense and often aversive stimulus increases subsequent responses to other stimuli.

Sex chromosome: One of the chromosomes that differ between the sexes, usually distinguished as X and Y.

Sex-linked: Carried on the X chromosome.

Sexual reproduction: Reproduction in which recombination of alleles occurs.

Sexual selection: A process by which different genotypes leave unequal numbers of progeny to future generations on the basis of their success in attracting a mate and in reproducing.

Side effect: A drug effect other than the one for which the drug was intended.

Social behavior: Any behavior that influences the behavior of other individuals of the same species.

Social organization: A set of behaviors that define a social group and the role of individuals within that group.

Sociobiology: The biological study of social groups and social behavior and their evolution.

Sodium–potassium pump: A group of membrane proteins that can actively transport sodium ions from the inside to the outside of a cell, such as a nerve cell, while actively transporting potassium ions in the opposite direction.

Speciation: The process by which a new species comes into being, especially by a single species splitting into two new species.

Species: Reproductively isolated groups of interbreeding natural populations.

Specific (acquired) immunity: An acquired antigen-specific ability to react to a previously encountered antigen.

Specificity: The degree to which a test detects only the molecule that it is meant to detect and does not detect other molecules.

Sperm: The male gamete, smaller and more motile than the egg in most species.

Stem cell: An undifferentiated cell in an adult organism that retains the ability to divide and differentiate but is still under the regulatory influence of other cells.

Stress; stress response: A physiological response or state of heightened activity brought about by the sympathetic nervous system and maintained for a longer time by the immune and endocrine systems.

Susceptibility: (1) The likelihood that a person who is exposed to a microorganism will become infected with that organism. (2) The probability that a person will get a particular disease.

Sustainable use: Use that can be maintained without time limits because only renewable resources are used, at rates that do not exceed their rates of renewal.

Sympathetic nervous system: A division of the autonomic nervous system that brings about the fight-or-flight response and that typically secretes norepinephrine as its final neurotransmitter.

Synapse: A meeting of cells in which a nerve cell stimulates another cell by secreting a neurotransmitter; the postsynaptic cell must have a receptor to which the neurotransmitter binds.

Synergistic interaction: A combination of two causes that lead to an effect greater than the sum of the effects that would have been produced by the two causes independently; for example, a combined effect in which two drugs together produce a greater physiological response than the sum of the effects of each drug given separately.

T lymphocyte: A type of white blood cell responsible for one type of specific immunity. See *Cytotoxic (CD8) T cells* and *Helper (CD4) T cells*.

Taxon: A species or any other collective group of organisms.

Taxonomy: The study of how taxa are recognized and how classifications are made.

Theory: A coherent set of well-tested hypotheses that guide scientific research.

Threshold: (1) The minimum level of a drug below which no physiological response can be detected. (2) The minimum level of a stimulus that is capable of producing an action potential.

Tissue: A group of similar cells and their extracellular products that are built together (structurally integrated) and that function together (functionally integrated).

Tolerance: (1) A condition in which a greater amount of a drug is required to produce the same physiological effect that a smaller amount produced originally. (2) In immunology, acquired unreactivity to a specific antigen after repeated contact with that antigen.

Transcription: A process in which DNA is used as a template to guide the synthesis of RNA.

Transformation: (1) The multistage process that a cell undergoes in changing from a normal cell to an unregulated, less-differentiated, immortal cell lacking contact inhibition and anchorage dependence. (2) In bacteria, a hereditary change caused by the incorporation of DNA fragments from outside the cell.

Transgenic: Containing genes from another species.

Translation: A process in which amino acids are assembled into a polypeptide chain (part or all of a protein) in a sequence determined by codons in a messenger RNA molecule.

Transmission: The transfer of microorganisms from one individual to another; does not imply any particular route by which the transfer may occur.

Tumor: A solid mass of transformed cells that may also contain induced normal cells such as blood vessels.

Unsaturated fats: Lipids with one or more double bonds between their carbon atoms.

Utilitarian: A system of ethics in which the rightness or wrongness of an act is judged according to its consequences.

Vascular plants: Plants containing tissues that efficiently conduct fluids from one part of the plant to another.

Verifiable: Capable of being proved true by experience.

Vestigial structures: Organs reduced in size and nonfunctional, but often showing resemblance to functional organs in related species.

Virulence: The ability of a microorganism to cause a disease.

Virus: A particle of nucleic acid (RNA or DNA) enclosed in a protein coat that cannot replicate itself but can cause a cell to replicate it.

Vitamins: Carbon-containing molecules needed in small amounts to facilitate certain chemical reactions in the body.

Withdrawal: Physiological changes or unpleasant symptoms associated with the cessation of drug taking.

Zygote: The cell that results when a sperm fuses with an egg, doubling the number of chromosomes.

Credits

CHAPTER 1

Figure 1.1. ROSE IS ROSE © United Feature Syndicate. Reprinted by Permission.

Figure 1.4. Charles Darwin courtesy of The Burndy Library, Dibner Institute for the History of Science and Technology, Cambridge, MA; Gregor Mendel courtesy of The Moravian Museum, Brno, Czech Republic; Ernest Everett Just courtesy of the Marine Biological Laboratory, Woods Hole Oceanographic Institution Library, Woods Hole, MA; Barbara McClintock courtesy of Marjorie Bhavnani; Luis and Walter Alvarez courtesy of the Lawrence Berkeley National Laboratory, Berkely, CA; Sarah B. Hrdy courtesy of Anula Jayasuriya.

Figure 1.6. Courtesy of the National Archives and Records Administration.

Figure 1.7. Photograph by Norbert Schäfer/The Stock Market Photo Agency Inc.

CHAPTER 2

Chapter Opener. Courtesy of Everard Williams Jr./Sharpe and Associates.

Figures 2.1 and 2.2. Portions modified from Postlethwait, Hopson, and Veres, *Biology! Bringing Science to Life* (McGraw-Hill, 1991), p. 137, Figure 8.3.

Figure 2.4. Modified from Sinnott, Dunn, and Dobshansky, *Principles of Genetics*, 5th ed. (McGraw-Hill, 1958), p. 72, Figure 6.1.

Figure 2.5. Courtesy of Dan Friend.

Figure 2.7. Photographs courtesy of Conly Reider.

Figure 2.10. Modified from Sinnott, Dunn, and Dobshansky, *Principles of Genetics*, 5th ed. (McGraw-Hill, 1958), p. 163, Figure 13.2.

Figures 2.13 and 2.14. Modified from Pelczar, Chan, and Krieg, *Microbiology Concepts and Applications* (McGraw-Hill, 1993), pp. 44–45, Figures 1.22, 1.23, 1.24.

Figure 2.16. Modified from Postlethwait and Hopson, *The Nature of Life*, 3rd ed. (McGraw-Hill, 1992), p. 245, Figure 10.7B.

Figure 2.18. Modified from Postlethwait and Hopson, *The Nature of Life*, 3rd ed. (McGraw-Hill, 1992), p. 246, Figure 10.8.

Figure 2.19. Right-hand figure modified from Postlethwait and Hopson, *The Nature of Life*, 3rd ed. (McGraw-Hill, 1992), p. 249, Figure 10.11B.

Figure 2.20. Photograph by E. Kiselva and D. Fawcett/Visuals Unlimited.

CHAPTER 3

Chapter Opener. Ronnie Kaufman/The Stock Market Photo Agency Inc.

Figures 3.1, 3.4 and 3.6 (right). Photographs courtesy of Laurent J. Beauregard.

Figure 3.6 (left). Courtesy of the March of Dimes National Foundation.

Figure 3.7C. Modified from Sinnott, Dunn, and Dobshansky, *Principles of Genetics*, 5th ed. (McGraw-Hill, 1958), p. 60, Figure 5.1.

Figure 3.13. In part from Alberts et al., *Essential Cell Biology* (Garland Publishing, 1998), p. 316, Figure 10-2.

Figure 3.15. Modified from Postlethwait and Hopson, *The Nature of Life*, 3rd ed. (McGraw-Hill, 1995), p. 287, Figure 11.18.

Figure 3.16. Modified from Alberts et al., *Essential Cell Biology* (Garland Publishing, 1998), p. 333, Figure 10-22.

CHAPTER 4

Chapter Opener. Lester Lefkowitz/The Stock Market Photo Agency Inc.

Figure 4.3. Photographs courtesy of Gary Feldman.

Figure 4.4. Modified from Postlethwait and Hopson, *The Nature of Life*, 3rd ed. (McGraw-Hill, 1995), p. 860, Figure 39.11.

Figure 4.5. Photographs from the estate of E.B. Ford, courtesy of J.S. Haywood.

Figure 4.6. From Postlethwait and Hopson, *The Nature of Life*, 3rd ed. (McGraw-Hill, 1995), p. 371, Figure 15.7.

Figures 4.7 and 4.10. Modified from Moore, Lalicker, and Fisher, *Invertebrate Fossils* (McGraw-Hill, 1957), pp. 336, 338, 334–345, 364, 394, Figures 9-2, 9-3, 9-4, 9-5, 9-23, 9-47.

Figure 4.11. Modified from Colbert and Morales, *Evolution of the Vertebrates*, 4th ed. (Wiley-Liss, 1991), p. 186, Figure 14-2, as redrawn from G. Heilmann, *Origins of Birds* (D. Appleton and Co., 1927).

Figures 4.12 and 4.13. Modified in part from Hopson and Wessells, *Essentials of Biology* (McGraw-Hill, 1990), pp. 736, 734, Figures 40-3, 40-1.

Figure 4.14. From Postlethwait and Hopson, *The Nature of Life*, 3rd ed. (McGraw-Hill, 1995), p. 404, Figure 16.18.

Figure 4.18. *Palmaria* and *Ascophyllum* courtesy of R. Thomas; *Volvox* by Richard Gross/Biological Photography.

Figure 4.19. *Poyltrichum* and *Nephrolepis* by Richard Gross/Biological Photography; *Cypripedium* and *Equisetum* courtesy of R. Thomas; *Rosa* courtesy of Roberts Botany Slide Collection, Bowdoin College, Brunswick, ME.

Figure 4.20. Photographs courtesy of R. Thomas.

Figure 4.22. Jellyfish and hydra by Richard Gross/Biological Photography; copepod courtesy of Wilhelm Hagen; brittle stars and crinoid courtesy of Dieter Piepenburg; other photographs by Frederick D. Atwood.

Figure 4.23. Angle fish courtesy of Will Ambrose; Penguins courtesy of Cecilie H. von Quillfeldt; other photographs by Richard Gross/Biological Photography.

Figure 4.24. Loris by David Agee/AnthroPhoto; squirrel

monkey by Irven DeVore/AnthroPhoto; other photographs by Peggy Bauer/Wildstock.

Figure 4.25. *Australopithecus* (front view) by K. Cannon-Bonventre/Anthrophoto; *Australopithecus* (side view) by Steve Ward/AnthroPhoto.

Figure 4.26. *Homo erectus* by D.Cooper/AnthroPhoto; *Homo sapiens* by B. Vandermeersch/AnthroPhoto.

Box 4.1. Portions modified from Postlethwait and Hopson, *The Nature of Life*, 3rd ed. (McGraw-Hill, 1995), pp. 71, 75, Table 3.1, Figure 3.12, and from Krieg, *Microbiology Concepts and Applications* (McGraw-Hill, 1993), p. 58, Figure 2.2.

Box 4.2. Modified from Hopson and Wessells, *Essentials of Biology* (McGraw-Hill, 1990), p. 337, Figure 19-10.

CHAPTER 5

Chapter Opener. Rob Lewine/The Stock Market Photo Agency Inc.

Figure 5.3. Modified from Buettner-Janusch, *Origins of Man* (John-Wiley, 1966), pp. 499, 500, 501.

Figure 5.4. Portions modified from Hopson and Wessells, *Essentials of Biology* (McGraw-Hill, 1990), p. 191, Figure 11-6.

Figure 5.5. Modified from Carola, Harley, and Noback, *Human Anatomy and Physiology*, 2nd ed. (McGraw-Hill, 1992), p. 571, Figure 18.11.

Figure 5.7. Modified from Stein and Rowe, *Physical Anthropology*, 5th ed. (McGraw-Hill, 1993), p. 190, Figure 8.9.

Figure 5.9. Photographs courtesy of Patricia N. Farnsworth.

Figure 5.11. Modified from Stein and Rowe, *Physical Anthropology*, 5th ed. (McGraw-Hill, 1993), pp. 126–127, Figure 6.6.

Figure 5.12. Modified from Stein and Rowe, *Physical Anthropology*, 5th ed. (McGraw-Hill, 1993), p. 177, Figure 8.2.

Figure 5.13. Courtesy of Glenbow Archives, Calgary, Canada (ND-7-714).

CHAPTER 6

Chapter Opener. Lester Lefkowitz/The Stock Market Photo Agency Inc.

Figure 6.1. Photograph by Jeff Greenberg/Visuals Unlimited. Graph modified in part from Postlethwait and Hopson, *The Nature of Life*, 3rd ed. (McGraw-Hill, 1995), p. 843, Figure 38.15A.

Figure 6.8. Photograph courtesy of David Epel.

Figure 6.9. Modified from Postlethwait and Hopson, *The Nature of Life*, 3rd ed. (McGraw-Hill, 1995), p. 335, Figure 14.3A.

Figure 6.10. Modified from Postlethwait and Hopson, *The Nature of Life*, 3rd ed. (McGraw-Hill, 1995), p. 338, Figures 14.5A and B.

Figure 6.11. Modified from Postlethwait and Hopson, *The Nature of Life*, 3rd ed. (McGraw-Hill. 1995), p. 340, Figure 14.6.

Figure 6.12. Modified from Audeskirk and Audeskirk, *Biology: Life on Earth*, 4th ed. (Prentice Hall, 1996), p. 778, Figure 35-18B.

Figure 6.14. Photograph courtesy of Dennis Graflin.

Figure 6.15B. Photograph by Frederick D. Atwood.

CHAPTER 7

Chapter Opener. Courtesy of David Baker.

Figure 7.1. From Postlethwait, Hopson and Veres, *Biology! Bringing Science to Life* (McGraw-Hill, 1991), p. 614, drawings 1, 2, 3.

Figures 7.2 and 7.3. By Peggy Bauer/Wildstock.

Figure 7.4. Pelicans by Richard Gross/Biological Photography; minnows by John D. Cunningham/Visuals Unlimited; wildebeest by Frederick D. Atwood; gannets courtesy of David Baker.

Figure 7.5. Modified from Hopson and Wessells, *Essentials of Biology* (McGraw-Hill, 1990), p. 840, Figure 45-9.

Figure 7.6. From Postlethwait and Hopson, *The Nature of Life*, 3rd ed. (McGraw-Hill, 1995), p. 929, Figure 42.20; photograph courtesy of Paul W. Sherman.

Figure 7.8. Honeybees by David S. Addison/Visuals Unlimited; ants by Frederick D. Atwood.

Figure 7.9. Canada geese by Irene Vandermolen/Visuals Unlimited; fur seals by Leonard Lee Rue III/Visuals Unlimited; baboons by James R. McCullagh/Visuals Unlimited.

Figure 7.10. Courtesy of Harlow Primate Laboratory, University of Wisconsin.

Figure 7.11. Modified from Hinde, *Biological Basis of Human Social Behaviour* (McGraw-Hill, 1974), pp. 294–295), Figures 18.1, 18.2.

Figure 7.12. Hamadryas threat by Joseph Popp/AnthroPhoto; baboons grooming by Irven DeVore/ AnthroPhoto; rhesus grooming by Jane Teas/ AnthroPhoto.

CHAPTER 8

Chapter Opener. Mark Cooper/The Stock Market Photo Agency Inc.

Figure 8.1. From Alberts et al., *Essential Cell Biology* (Garland Publishing, 1998), p. 169, Figure 5-27; photograph courtesy of Richard J. Feldmann.

Figure 8.2. Portions modified from Vander, Sherman, and Luciano, *Human Physiology*, 6th ed. (McGraw-Hill, 1994), p. 562, Figure 8.1, and Carola, Harley, and Noback, *Human Anatomy and Physiology*, 2nd ed. (McGraw-Hill, 1992), p. 778, Figure 8.1.

Figure 8.3. Modified from Starr and Taggart, *Biology: The Unity and Diversity of Life*, 8th ed. (Wadsworth Publishing, 1998), p. 32, Figure 2.19.

Figure 8.6. Portions modified from Van Wynsberghe, Noback, and Carola, *Human Anatomy and Physiology*, 3rd ed. (McGraw-Hill, 1995), p. 63, Figure 3.4.

Figure 8.7. Modified from Postlethwait and Hopson, *The Nature of Life*, 3rd ed. (McGraw-Hill, 1995) p. 588, Figure 25.10A.

Figure 8.8. Modified from Hopson and Wessells, *Essentials of Biology* (McGraw-Hill, 1990), p. 519, Figure 29-5.

Figure 8.9. Portions modified from Postlethwait, Hopson, and Veres, *Biology! Bringing Science to Life* (McGraw-Hill, 1991), p. 37, Figure 2.20.

Figure 8.10. Portions modified from Postlethwait, Hopson, and Veres, *Biology! Bringing Science to Life* (McGraw-Hill, 1991), p. 38, Figure 2.21.

Figure 8.12. Portions modified from Carola, Harley, and Noback, *Human Anatomy and Physiology*, 2nd ed. (McGraw-Hill, 1992), p. 58, Figure 3.2, and from Postlethwait and Hopson, *The Nature of Life*, 3rd ed. (McGraw-Hill, 1995), p. 76, Figure 3.13.

Figures 8.14 and 8.15. In part modified from Postlethwait, Hopson, and Veres, *Biology! Bringing Science to Life* (McGraw-Hill, 1991), p. 100, Figure 5.16.

CHAPTER 9

Figure 9.2. Modified from Pelczar, Chan, and Krieg, *Microbiology Concepts and Applications* (McGraw-Hill, 1993), p. 274, Figure 10.1; photograph of *Dictyostelium* fruiting body courtesy of John Bonner; photograph of *Dictyostelium* slug courtesy of David Francis.

Figure 9.4. Modified from Postlethwait and Hopson, *The Nature of Life*, 3rd ed. (McGraw-Hill, 1995), p. 179, Figure 7.13A.

Figure 9.5. Photographs courtesy of G. Steve Martin and J. Guyden.

Figure 9.8. Modified from Postlethwait and Hopson, *The Nature of Life*, 2nd ed. (McGraw-Hill, 1992), p. 220, Figure 10.14A.

Figure 9.9. Part A modified from Hopson and Wessells, *Essentials of Biology* (McGraw-Hill, 1990), p. 280, Figure 16-12b. Part B modified from Hopson and Wessells, *Essentials of Biology* (McGraw-Hill, 1990), p. 289, Figure 17-2.

Figure 9.10. Modified from Gurdon, *Gene Expression During Cell Differentiation* (Oxford University Press, 1973).

Figure 9.11. From Alberts et al, *Molecular Biology of the Cell*, 3rd ed. (Garland Publishing, 1994), p. 1086, Figure 21-58; photographs courtesy of H.G. Frohnhöfer, R. Lehmann, and C. Nüsslein-Volhard, *J. Embryol. Exp. Morphol.* 97[Suppl]: 169–179, 1986, by permission of the Company of Biologists Ltd.

Figure 9.12. Modified from Hopson and Wessells, *Essentials of Biology* (McGraw-Hill, 1990), p. 291, Figure 17.4; photograph courtesy of Jonathan Slack.

Figure 9.13. Photograph courtesy of J.P. Revel and S. Brown, Caltech.

Figure 9.14. Modified from the *New Eng. J. Med.* 332: 986, Figure 1, 1995.

Figure 9.15. A–D from Alberts et al, *Molecular Biology of the Cell*, 3rd ed. (Garland Publishing, 1994), p. 1262, Figure 24-10; photographs E and F courtesy of Andrew Hanby.

Figure 9.17. Modified from Alberts et al, *Essential Cell Biology* (Garland Publishing, 1998), p. 201, Figure 6-25.

Figure 9.18. Death rate data from the National Center for Health Statistics. Cigarette consumption data from the Centers for Disease Control and Prevention.

Figure 9.19. Data from *Sci. Am.* September 1996: 90–91.

Figure 9.20. Modified from R. Doll and A.B. Hill, *BMJ* 1: 1399–1410, 1964.

Figure 9.21. Data from W.J. Blot et al, *Cancer Res.* 48: 3282–3287, 1988.

Table 9.1. Modified from Hopson and Wessells, *Essentials of Biology* (McGraw-Hill, 1990), p. 292, Table 17-2.

CHAPTER 10

Chapter Opener. Howard Sochurek/The Stock Market Photo Agency Inc.

Figure 10.2B and C. Modified from Vander, Sherman, and Luciano, *Human Physiology*, 6th ed. (McGraw-Hill, 1994), p. 182, Figure 8-3.

Figure 10.3A. Modified from Hopson and Wessells, *Essentials of Biology* (McGraw-Hill, 1990), p. 616, Figure 34-3a.

Figure 10.5. Modified from Postlethwait and Hopson, *The Nature of Life*, 3rd ed. (McGraw-Hill, 1995), p. 699, Figure 31.5.

Figure 10.9B. Modified from Campbell et al, *Biology: Concepts & Connections*, 2nd ed. (Benjamin/Cummings, 1997), p. 581, Figure B.

Figure 10.10. Part A modified from Vander, Sherman, and Luciano, *Human Physiology*, 6th ed. (McGraw-Hill, 1994), p. 260, Figure 9-36. Parts B and C modified from Vander et al, *Human Physiology*, 6th ed. (McGraw-Hill, 1994), p. 262, Figure 9-40. Part D photographs courtesy of C.G. Wright, from Rowland et al, (eds.), *Hearing Loss* (Thieme Medical Publishing, 1997), p. 208, Figures 7.8a and c.

Figure 10.12. Modified from Tullar, *The Human Species* (McGraw-Hill, 1977), p. 18, Figure 1-8.

Figure 10.13. Modified from Noback and Demarest, *The Human Nervous System*, 3rd ed. (McGraw-Hill, 1981), p. 6, Figure 1-6.

Figure 10.18. Part A courtesy of Adrian J. Williams, The Lane Fox Unit, St. Thomas' Hospital. Part C modified from Vander, Sherman, and Luciano, *Human Physiology*, 6th ed. (McGraw-Hill, 1994), p. 372, Figure 13-6.

CHAPTER 11

Chapter Opener. Matthias Kulka/The Stock Market Photo Agency Inc.

Figure 11.1. Portions modified from Vander, Sherman, and Luciano, *Human Physiology*, 6th ed. (McGraw-Hill, 1994), pp. 475, 477, Figures 15-1, 15-3B.

Figure 11.2. Modified from Carlson, *Patten's Foundations of Embryology*, 5th ed. (McGraw-Hill, 1988), pp. 279, 281, Figures 7-21C, 7-22.

Figure 11.3. Top-left and top-right figures modified from Hopson and Wessells, *Essentials of Biology* (McGraw-Hill, 1990), p. 599, Figure 33-5a and b; bottom figure modified from Postlethwait and Hopson, *The Nature of Life*, 3rd ed. (McGraw-Hill, 1995), p. 666, Figure 29.8B.

Table 11.3. From Segal, *Drugs and Behaviour* (Gardner Press, 1988), pp. 258–259, 369, Tables 11-2, 15-1.

CHAPTER 12

Chapter Opener. Tom and Dee Ann McCarthy/The Stock Market Photo Agency Inc.

Figure 12.1. Modified from Van Wynsberghe, Noback, and Carola, *Human Anatomy and Physiology*, 3rd ed. (McGraw-Hill, 1995), p. 711, Figure 22.1A.

Figure 12.3. Modified from Van Wynsberghe, Noback, and Carola, *Human Anatomy and Physiology*, 3rd ed. (McGraw-Hill, 1995), p. 740, Figure 23.10.

Figure 12.4. Portion redrawn from Van Wynsberghe, Noback, and Carola, *Human Anatomy and Physiology*, 3rd ed. (McGraw-Hill, 1995), pp. 66–67, Figure 3.5.

Figure 12.8. Modified from Hopson and Wessells, *Essentials of Biology* (McGraw-Hill, 1990), p. 548, Figure 30-14.

Figure 12.9. Modified from Hopson and Wessells, *Essentials of Biology* (McGraw-Hill, 1990), p. 626, Figure 34-14, and Hole, *Human Anatomy and Physiology*, 6th ed. (WCB/McGraw-Hill, 1993), p. 404, Figure 11.39.

Figure 12.10. Modified from Carola, Harley, and Noback, *Human Anatomy and Physiology*, 2nd ed. (McGraw-Hill, 1992), p. 534, Figure 17.13.

Figure 12.12. Courtesy of David and Suzanne Felten.

CHAPTER 13

Chapter Opener. Courtesy of the CDC National Prevention Information Network.

Figure 13.2. Courtesy of the The Names Project Foundation (photograph by Mark Theissen).

Figure 13.3. Modified from Van Wynsberghe, Noback, and Carola, *Human Anatomy and Physiology*, 3rd ed. (McGraw-Hill, 1995), p. 752, unnumbered Figure.

Figure 13.4. Modified from Postlethwait and Hopson, *The Nature of Life*, 3rd ed. (McGraw-Hill, 1995), p. 614, Figure 26.14.

Figure 13.5. Photographs courtesy of Cynthia Goldsmith, Erskine Palmer, and Paul Feorino, Centers for Disease Control and Prevention (CDC).

Figure 13.6. Courtesy of Alyne Harrison, Erskine Palmer, and Paul Feorino, Centers for Disease Control and Prevention (CDC).

Figure 13.8. Modified from Postlethwait and Hopson, *The Nature of Life*, 3rd ed. (McGraw-Hill, 1995), p. 232, Box 9.1, Figure 1.

Figure 13.9. Data from the Centers for Disease Control and Prevention (CDC).

Figure 13.11. Data from Stine, *AIDS update 1999* (Prentice Hall, 1999), p. 238, Figure 8-15.

Figure 13.12. Data from Stine, *AIDS Update 1999* (Prentice Hall, 1999), p. 309, Figure 10-5 and WHO.

CHAPTER 14

Chapter Opener. Courtesy of Nigel Cattlin/Holt Studios.

Figure 14.2. Modified from Hopson and Wessells, *Essentials of Biology* (McGraw-Hill, 1990), p. 139, Figure 8-6.

Figure 14.3. Modified from Hopson and Wessells, *Essentials of Biology* (McGraw-Hill, 1990), p. 138, Figure 8-4.

Figure 14.4. Modified from Hopson and Wessells, *Essentials of Biology* (McGraw-Hill, 1990), p. 143, Figure 8-9.

Figure 14.5. Modified from Starr and Taggart, *Biology: The Unity and Diversity of Life*, 8th ed. (Wadsworth Publishing, 1998), p. 115.

Figure 14.6. Modified from Chrispeels and Sadava, *Plants, Genes and Agriculture* (Jones and Bartlett, 1994), p. 214, Figure 7.16.

Figure 14.7. By C.P. Vance/Visuals Unlimited.

Figure 14.8. Top left photograph by Richard Gross/Biological Photography; drawings modified from Slack, *Carnivorous Plants* (MIT Press, 1980), p. 126; bottom left photograph courtesy of Claude Nuridsany/Science Photo Library; bottom right photograph courtesy of Marie Perenou/Science Photo Library.

Figure 14.9. Modified from Hopson and Wessells, *Essentials of Biology* (McGraw-Hill, 1990), p. 485, Figure 27-1.

Figure 14.12. Modified from Hopson and Wessells, *Essentials of Biology* (McGraw-Hill, 1990), p. 486, Figure 27-2.

Figure 14.13. Modified from Hopson and Wessells, *Essentials of Biology* (McGraw-Hill, 1990), p. 489, Figure 27-5.

Figure 14.14. Modified from Hopson and Wessells, *Essentials of Biology* (McGraw-Hill, 1990), p. 488, Figure 27-4.

Figure 14.15. Left photograph by Richard Gross/Biological Photography; right photograph courtesy of Robert Thomas.

Figure 14.16. Courtesy of David W. Schindler.

Figure 14.18. By Richard Gross/Biological Photography.

Figure 14.19. Left photograph courtesy of the Nebraska Wheat Board; right photograph courtesy of David Handley.

Figure 14.22. A, B, and C courtesy of David Handley; D by Richard Gross/Biological Photography.

Figure 14.23. Data from *Agronomy Journal*, 44: 61, 1952, by permission of the American Association of Agronomy.

Figure 14.24. Modified from Chrispeels and Sadava, *Plants, Genes and Agriculture* (Jones and Bartlett, 1994), p. 404, Figure 15.2.

CHAPTER 15

Chapter Opener. Carl Ganter/The Stock Market Photo Agency Inc.

Figure 15.1. Modified from Wilson, *The Diversity of Life* (The Belknap Press of Harvard University Press), p. 134.

Figure 15.2. Photograph by Hal Beral/Visuals Unlimited.

Figure 15.3. Modified from Moore, Lalicker, and Fisher, *Invertebrate Fossils* (McGraw-Hill, 1957), portions of Figures 3-4, 4-17, 4-29, 5-9, 6-20, 6-24, 6-36, 13-7, 13-20, 18-20, 18-29, 23-1.

Figure 15.4. *Ginkgo* and *Limulus* modified from Palmer and Fowler, *Fieldbook of Natural History*, 2nd ed. (McGraw-Hill, 1974), pp. 112, 433; *Lingula* modified from Hyman, *The Invertebrates*, vol. 5 (McGraw-Hill, 1959), p. 519, Figure 183C; *Latimeria* modified from Weichert, *Anatomy of the Chordates* (McGraw-Hill, 1970), p. 27, Figure 2.18.

Figure 15.5. Ammonoids modified from Moore, Lalicker, and Fisher, *Invertebrate Fossils* (McGraw-Hill, 1957), Figures 9-40, 9-41; photograph courtesy of Don Prothero; painting by Rudolph Zallinger, courtesy of the Peabody Museum of Natural History, Yale University, New Haven, Connecticut.

Figure 15.7. Modified from Wilson, *The Diversity of Life* (The Belknap Press of Harvard University Press), p. 252.

Figure 15.8. Modified from Wilson, *The Diversity of Life* (The Belknap Press of Harvard University Press, pp. 262–263.

Figure 15.9. From Hopson and Wessells, *Essentials of Biology* (McGraw-Hill, 1990), p. 756, Figure 41.7.

Figure 15.10. Part A modified from Postlethwait and Hopson, *The Nature of Life*, 3rd ed. (McGraw-Hill, 1995), p. 896, Figure 41.7A. Part B courtesy of Jane Mackarell. Part C by Richard Thom/Visuals Unlimited.

Figure 15.11. *Spathophyllum*, *Piper*, *Guzmania*, and strawberry frog courtesy of Sharon Kinsman; tarantula courtesy of Jane Mackarell; stick insect, butterfly, bronze beetle, emerald boa, white-throated toucan, and tree sloths by Frederick D. Atwood; jaguar by Peggy Bauer/Wildstock.

Figure 15.12. Orchids by Max and Bea Hunn/Visuals Unlimited; lianas courtesy of Jane Mackarell.

Figure 15.13. Redrawn and modified from multiple sources.

Figure 15.15. By Kjell B. Sandved/Visuals Unlimited.

Figure 15.16. Left-hand photographs by G. Prance/Visuals Unlimited; right-hand photographs courtesy of Sharon Kinsman.

Figure 15.17. Modified from Hopson and Wessells, *Essentials of Biology* (McGraw-Hill, 1990), p. 755, Figure 41-6.

Figure 15.18. Photograph by Leonard Lee Rue, III/Visuals Unlimited.

Figure 15.19. Trees killed by acid rain by Richard Gross/Biological Photography; pH-affected trout courtesy of David Schindler, from D. Schindler et al, Science, 228: 1395–1401 (1985), © 1985 by the AAAS; stone sculpture by John. D. Cunningham/Visuals Unlimited.

Figure 15.21. Photograph courtesy of Elin Haugen.

Index

Note: page numbers in *italics* refer to figures, tables or boxed material

A

ablation technique in brain 428
ABO blood groups 196, *198*, 199, 202–203
abortion 104, 105, 259–260, 261–263
  ethical considerations 263
  induction 260
abstinence contraceptive methods 256–257
abstraction 439
acacia, bull's horn (*Acacia cornigera*) 659
accessory pigments 586–587
acetyl CoA *339*, 340
acetylcholine 418, 419, 432, 433–434, 447, 467
  Alzheimer's disease 439
acid bogs 593, *594*, 595
acid rain 674–676
acquired characteristics 126
acquired immunity 501–503
actin 433
action potentials 416, *417*, 447
active transport 322, *323*, 354
adaptation 126, 183, 228
addiction
  psychoactive drugs 472–477
  psychological dependence 473
addictive potential 456
Addison's disease 525
adenoids 498–499, 501
adenosine deaminase (ADA) gene therapy 107, *108*, 109
adenosine triphosphate *see* ATP
Ader, Robert 528
adopted children 80–81, 253
adoption studies 279–280
  alcoholism 307
adrenaline *see* epinephrine
adrenocorticotrophic hormone (ACTH) 516, *517*
  receptor 522
adult testing 102
Africa
  AIDS prevalence 575
  climatic zones 664–666
  desertification 664–666
  malnutrition 351–352
  social environment 531–532
African Americans 221
  G6PD deficiency 345
African ancestry 215
age
  cancer risk 387–388
  pyramid 242–243
aggregates 361–362
aggression, human 306
aggressive threats 280
agonists 463
agricultural development 240
  rainforest clearance 662, *663*
agriculture 582
  efficiency 619
  rainforest 670
Agrobacterium plasmid 627
AIDS 257, 389, 397, 505, 536, 537, 576
  antiviral drugs 560–561
  biofeedback 530
  causality criteria 544–545
  cause 538
  dementia 554
  drug therapy 560–561
  epidemiology 539
  HIV connection discovery 539–542
  HIV progression 552–555
  immune system 537
  lifestyle hypothesis 540
  marijuana use 480
  needle sharing 457
  opportunist infections 554, 561
  prevalence 575
  quilt 540
  risk avoidance 562–575
  vaccine 558–560
  viral hypothesis 540–542
AIDS-related complex (ARC) 541, 554–555
air pollution 674
alarm 516–517
  calls 287, *288*
albinism 79, *80*, 85, 114
  discontinuous variation 190
alcohol 395, 476
  abuse 472
  blood levels 470, *471*
  blood–brain barrier 457, 470
  cancer risk 393–394
  CNS depressant 469–470, *471*
  dependence 473
  half-life 462
  harm 487
  HIV risk 569
  immune system suppression 511
  metabolic breakdown 458–459
  pregnancy 484
  sleep effects 443
  synergism with tobacco 479
  user health 478–479
alcohol dehydrogenase 459
alcoholism 306–307, 344, 478–479
alfalfa 605, 617
alfalfa weevil 617
algae 163, *164*
  wastewater lagoons 680
alkaptonuria 79, 85, 87
allele frequencies 190, 228
  human populations 209–210
alleles 42, 43, 68
  defective 102–103
  dominant 189
  exchange 147
  recessive 189
  single recessive 87
Allen M.C. 263
Allen's rule 222, 228
allergies 510–511, 674
allomothering 301, *302*, 303
alloparenting 288, 301, 303
Altmann, Jeanne 303
altruism 274, 285–288, 310
  kin selection 286
aluminium potassium sulfate 681
Alvarez, Luis and Walter *18*, 647
Alvarez hypothesis 647–648
alveoli 454, *455*
Alzheimer's disease 84, 410, 439–440
Amazon rainforest 668

Ames test *394–395*
amino acids 152, 589
  dietary 336–337
  essential 337, 338
  limiting 337
  point mutations 64, *66*
  sequence 62
amino groups 589
gamma-aminobutyric acid *see* GABA
ammonia 151, 152, 589
ammonification 592
ammonium 593
ammonoids 140, 645, *646*
amniocentesis 72, 100–101
Amoeba *162*, 163
amphetamines 443, 446, 468, 476
Amphibia 174, 175
amphioxus 173
ampicillin 628
amylase 318
amyloid deposits 440
anabolic steroids 483
anal intercourse 563
anaphase *47*, 48
anchorage dependence 366
anemia 211–212
  hemolytic 345
  microcytic 344
  see also sickle-cell anemia
angiogenesis inhibitors 400
angiogenic growth factors 383
angiosperms *165*, 597
animal cell *156*, *157*
Animal Liberation Front (ALF) 28–29
animal proteins 337, 338, 351
animal rights movement 28–29
Animalia *160–161*, 164–167, *168–169*, 170–175
animals
  experimentation 27–28, 33
  rearing conditions 279
  short-lived 280
  use in science 27–29
anisogamy 294
Annelida *169*, 172
anorexia nervosa 316, 349–350
Antabuse 459
antagonists 463
anthrax 542
anti-abortion extremists 263
antibiotic resistance
  genes *622*, 623
  genetic engineering 628
antibodies 501, 532
  antigen removal 503–505
  cancer treatment 400, 401
  maternal 506–507
anticodon 62, 63, *64*
antigen–antibody complex 503
antigens 499, 532
  destruction 501
  nonself 501
  removal mechanisms 503–505
anti-immigrationists 194
anti-inflammatory agent 524
anti-inflammatory feedback inhibition 524–525
anti-nausea drugs 403
antioxidants 345, 404
antisense gene 625
antitoxins 505
antiviral drugs 560–561
ants 659
  colony 291
anus 171, 172, 324
apes *177*, 178
  large 660–661
aphids 293
apoptosis 551
aquifers 666
arable land loss 266
Archaebacteria 155, 158–159
Archaeopteryx 141
Archosauria 175
aromatic amines 391–392
Arracenia psittacina 594
arsenic 391
arteries 325
Arthropoda 168, 172
artificial insemination 252, 564
artificial selection 129, 632
  plant alteration 617–619
asbestos 391, 392, 395, 674
Ascomycota 164
asexual reproduction 292, 293, 310
Ashkenazi Jews 103, 104
Asia
  HIV incidence 575
  population 345
aspirin 583
assimilation 592
asteroid impact hypothesis 647–648
atherosclerosis 331–333, 335, 336
atmosphere of Earth 151–152
  evolution 153
  stability 641
atomic structure hypothesis 6
ATP 60, 322, *323*, 340, *341*, 354, 433
  photosynthesis 588
atria of heart 326
attention deficit hyperactive disorder 468
auras 432
Australopithecus 178–179, 183
autoimmune disease 510
autoimmunity 512
automobile accidents 470
autonomic nervous system 513, *514*, 515–516, 532
  process regulation 528–529
  triggering by emotional stimuli 516
autonomic processes, conscious control 519
autotrophs 585, 631
Avery, Oswald 53
Aves *174*, 175
axons 411, 447
  giant 413
  nerve impulse measurement 414, *415*
Azotobacter 592
AZT *see* zidovudine

B

B lymphocytes 380, 499, 500–501, 532, *539*
  antibody-producing 524
  receptors 522
BABI (Blastomere Analysis Before Implantation) 252–253
baboons *296*
  harems 303
  social organization 301, *302*, 303
baby boom 243
Bacillus anthracis 542
Bacillus thuringiensis 616, 624
backcrossing 50, 51
bacteria
  cloning 97
  genetics 11, *12*
  gut 324
  ice-minus 625
  nitrogen-fixing 592–593
  oil spill area 678
  oil-degrading 677–678
  photosynthesis 154, 587
  symbiotic in gut 324
  transformation 53, *54*
  virulence 52–53, *54*
bacterial plasmids 621
balance 425, *426*, 427
band patterns, DNA markers 93–94
barbiturates 443, 464
  pregnancy 485
  tolerance 477
barrier contraceptive methods 257
basal metabolic rate 328
Basidiomycota 164, *166*
Bateson, William 50
bats 658–659
bean, winged 670
beans 218, 219, 605
  protein 338
bear *174*
bedbugs *565*
beetles, carrion-feeding 294
behavior
  drug-seeking 474
  genetic strains 279
  high-risk 477
  human 305–308

innate 276–277, 278
instinctive 280–281, 291
learned 275–277, 310
mating 280, 281
nesting 282
parenting 301, *302*, 303
primate 298–301
risk 563
stereotyped 281
territorial 282
see also social behavior
behavioral isolation 147–148
Bentham, Jeremy 24
benzene 391, 392, 674
benzo(α)pyrene 391
Bergmann's rule 222, 228
beri-beri 344
Bernard, Claude 363, 531
Besant, Annie 264
beta carotene 345, 393, 404
Bible 22, 143
bilateral symmetry 167, 170
bilayer membranes 334
bile 319, 320
synthesis 341
binge eating 350
biodegradation 684
biodiversity 636, 684
distribution 638
ecological processes 637–642
energy 638–639
evolutionary processes 637–642
extinction effects 642–651
human interdependence 639–641
biofeedback 529–530
bioinformatics 98
biological control 615
biological determinism 118
biological rhythms 440–444
biological value 667
biology 32
biomagnification 612–613, 614, 632
biomass 639
biomes 653–654
tropical rainforest 654
bioremediation 684
nutrient enrichment 678
oil pollution 677–678
wastewater 679–681
biosphere 638, 684
biotin 324
birds *174*, 175
egg-shell thinning 613
fossil record 140–141
male mating rituals 281
birth, drug persistence 485
birth control 254–261, 269
cultural opposition 260–261
ethics 260–263
pills 257–258
post-fertilization methods 259–260
see also oral contraceptives
birth rate 234, 235, 242, 246, 269
Bishop, J. Michael 379
Biston betularia 131–132
bladder cancer 391–392
blastocyst 251
Blastophaga 659–660
blastula 164, 167, 369, *370*
blood 325–326
handling guidelines 571–572
blood cell differentiation 499–500
blood donors 571–572
blood groups 202–205
ABO 196, *198*, 199, 202–203
alleles 196, *198*, 199–200, 202–203
disease correlation 220
Duffy system 205
Dunkers 208
geographic subgroups 205
geographic variation in frequency 205, 220
MN system 205
Rh system 203–204
blood pressure 346
blood transfusion
HIV infection 564
HIV transmission risk 571–572
passive immunity 507
blood vessels 325, 326
blood–brain barrier 431, 457
alcohol 470
blue-green bacteria 153, 154
Blumenbach, Johann 195–196
body cavity evolution 171
body size 222
body–mind interactions 494, 495–496
bogs, acid 593, *594*, 595
Bohr, Niels 6
bollworm, pink 617
bone density 84–85, 347
bone marrow 499, 500
bone mass loss 350
bone meal 605
bottleneck effect 207, 208
botulism toxin 483
Bowman's capsule 459, *460*
Bradlaugh, Charles 264
brain 410, 411
activity storing 435–437
amyloid deposits 440
anatomy 428–431
dopamine pathways 419–421
electrical activity during sleep 441–443, 445
fetal tissue implants 420
message input 422–425, *426*, 427–428
message processing 428–432
message routing 422–425, *426*, 427–434
mind 465
pleasure centers 477
primate 176
reward centers 473–477
studying 428
tracts 413
brainstem 474–476
branching descent hypothesis 125, 136–137
Brassica napus 624
Brazil, Amazon rainforest 668
BRCA1 and *BRCA2* genes 386–387
predisposition mutations 402
breast cancer 386–387, 395, 404
age 387–388
predisposition mutations 402
risk 403
survival 529
breast feeding, HIV infection 562, 563
breast milk 506
breast self-examination 401
bronchi 454, *455*
bronze beetle *657*
broom, Scotch 482
Brownmiller, Susan 308
Bruch, Hilde 349
Bryophyta 164, *165*
bulimia 350
Burkitt's lymphoma 380, 389
butterfly *656*

C

cadmium 391, 392
caffeine 443, 469, 476, 479
pregnancy 483–484
psychological dependence 473
calciferol *343*, 346
calcium *343*, 347
caloric intake 328
Calvaria 651
Camellia sinensis 469
Canada geese *296*
canaries 673
canary 682
cancer 360
age 385
causes 384–393
childhood 385
depression 526
detection 401–403
diet 393–394, 404, 406
economic factors 397–398
environmental factors 385–386, 406
epidemiology 384, 385
genetic predisposition tests 402–403
homeostasis disruption 363
hormones 395
imaging techniques 519
increasing age 387–388
inherited predisposition 386–387
internal resistance 395–397

lifestyle factors 385–386
management 403
new treatments 400–401
predisposition 401–403
prevention 404
progression to 382–383
resistance variation 395
risk factors 384–393, 406
social factors 397–398
somatic cells 381
survival rate 401–402, 403, 529
treatment 399–400, 406
viruses 388–389
cancer cells 376
properties 377–379
cancer-forming cells *366*
Candida 554
Cann, Rebecca 13
cannabinoids 480
canola plant 624, 626, 629
canopy, continuous 654, *655*
Cape Verde Islands 128
capillaries 325, 454, *455*
carbohydrates 329–331, 354
absorption 331
chemical structure 329, *330*
deficiency 351
dietary need 331
digestion 330
carbon dioxide 153, 154, 588
fixation 588
pollution 674
stomata 601
tropical rainforest photosynthesis 654
U.S. production 668
carbon monoxide 674
carcinogens 389–392, 406, 467
chemical 391–392
marijuana 480
physical 390–391, 392
screening *394–395*
synergistic effect 392
tobacco 479
carcinoma 385
cardiovascular disease 327
cholesterol reduction 341
dietary fats 333
smoking 391
carnivorous plants 593, *594*, 595
carotenes 586
see also beta carotenes
carrier proteins 322, *323*
carriers 106
carrying capacity 239–240, 269
limits 267–268
Carson, Rachel 611
Cartesian dualism 496
case cluster 539
castration 254
catalysis 317
categorical imperative 22
Catholic church 260–262
cattle 294
ranching 668
cause and effect 542–545
Cavalli-Sforza, Luigi 205, 209–210
CD4 T cells 537, 576
count 553
HIV infection 548
CD8 T cells 537
cecum 324
cell cycle 364–365, 367, 406
cell division 45–46, 362–363
anchorage dependence 366
cellular differentiation 369–375
contact inhibition 365
continuous 380
eucaryotic 364–365
human lifetime 381
limits 375–376
regulation 364–376
signaling *367*
tissue formation 369–375
uncontrolled 377–383
cell membranes 333–335
cell phones 391
cells 361, 406
cancer-forming *366*
cooperation 363
differentiation 362–363, 406
drug actions 462–463
fate determination 371
fusion 552
immortal 376
immortalization 381
integration 363
life span 375
migration 375
potentiality 369–371
structure 45–46
see also receptors, cellular
cellular adhesion 375
cellular differentiation 369–375
Celsus 508
census 234
central nervous system 411, 447
depressants 469–470, *471*
stimulants 468
Cephalochordata 173
Cephalopoda 136–137, 172
fossils 139–141
giant axons 413
cerebellum 431
cerebral cortex 176, 431, 439
cerebral hemispheres 430
cerebrospinal fluid 431
cerebrum 430
cervical cancer 389
cervical caps 257
cervical smear test 401
Chargaff, Erwin 57
Chargaff's rules 57, 59
Chase, Martha 55
chatinine 482
cheating 286
checking proteins *382*, 386
chemical signal reception 427–428
chemokine receptors 555
chemotaxis 508
chemotherapy 406
cancer treatment 399–400
chemotrophs 585
child abuse 308
children
cancer 385
undernutrition 352–353
chimpanzee *177*
brain *429*
China
Harvest Moon phenomenon 530
population control 264
Chinese astrology 530
chlamydia 573–574
Chlamydomonas 294
chloride *343*, 346
chlorine 674, 681
chlorophyll 153, 586, 587
chloroplasts 155, 587
chlorpromazine 446
cholesterol 333–336
reduction 341
cholinesterase 419, 433–434
inhibitors 439
chondrodystrophy 181
Chordata 150, 173, *174*
chorionic villus sampling *100*, 101
Christianity 496
chromium 391
chromosomal aberrations 66
chromosomal theory of inheritance 46
confirmation 51–52
chromosomes 46, 68
autosomal 73
fragments 66
homologous pairs 46, 49–50, 73
markers 81
maternal 372
meiosis 49–51
pairs 68
variation 76–78
cigarette smoke 467
pollution 674
cilia 163
circadian rhythms 440–441, 447
circulation 326, 495–496
mammals at birth 142–143
circulatory system 325–326
cities, migration to 266
classical conditioning 436–437, 447, 528, *529*
classification 133, 150
social construction 158

Claviceps purpurea 468–469
climate, body size relationship 222
clinal variation 199
clines 196, 199, 228
  skin color 224
clone 292, 501, 506
cloning 112, 119, 371
  bacteria 97
  plants 623
Clostridium 592
Cnidaria 166, *168*, *169*, 362
coal burning 676
cobalt 348
coca leaves 456
cocaine 455–456, 476
  half-life 461
  HIV infection 569
  psychological dependence 473
cochlea 425, *426*, 427
cocoa plant 469
code 16
codominance 203
codons 16, 62, 64, *65*, 68
coelom *167*, 171
coenzymes 342, 344
Coffea arabica 469
coffee plant 469
Cohen, Nicholas 528
coitus interruptus 256–257
cola tree (*Cola acuminata*) 469
cold-capping 403
Coley, William 512
colon cancer 84, 324, 386, 404
Colorado River (USA) 666
combination therapy 561
communicability of HIV infection 564–567
community 638, 684
compartmentalization 361–362
competition 241
  males 294
  primates 305
complement 503, *504*, 507
composting 608
computer technology development 97
concentration gradient 322, *323*
concordance rate 81
conditioned learning
  drug addiction 476
  immune system 528, *529*
conditioned withdrawal syndrome 473
condoms 257, 563, 574
cones 425
consciousness 428
  psychoactive drugs 465
consort pairs 303
constipation 466
consumers 585
consumption patterns 266–267
contact inhibition 365
continuous variation 189–190
  human populations 192
contraceptive measures 254–258
control groups 11, 33, 519
controlled substances 485, *486*
Controlled Substances Act 485
convergence 135
Copernicus 17
copper deficiency 511–512
corn
  Africa 352
  artificial selection 618–619
  gene linkage 50
  protein 338
  wild relatives 640
corn earworm *615*
corpus luteum 249
corticosteroids 524
corticotropin-releasing hormone 516
cortisol 517
Costa Rica 670
cost–benefit analyses 23
cotton pests 616–617
counseling, genetic 100–106
crack cocaine 569
creation science 144
creationism 142–145
  early twentieth-century 143–144
  modern 144–145
Creighton, Harriet 51
Cretaceous period 645
  extinctions 646–647
cri du chat syndrome 78
Crick, Francis 57, *58*, 59
cricket, Mormon 294
Cro-Magnons 180
crop yields 604–621, *622*, 623–629
  chemical pest control 610–614
  fertilizers 604–607
  hydroponics 609–610
  irrigation 608
crops 582
  major 582
  rotation 611
crossing-over 49
  frequency 81
cross-pollination 40
Cruciferae 404
crude oil 677
cultural innovation 181
Curie, Marie 390
Cushing's syndrome 525
cyanobacteria 153, 155
  nitrogen-fixing 592
  photosynthesis 587
cyclic AMP 469
cyclins 367
  nuclear 369
cyclophosphamide 528
cystic fibrosis 84, 87, 102, 103, 181
  frequency 192
  tuberculosis protection 220
cytisine 482
cytokines 366, 499, 507, 508, 532
  functions 523–524
  growth factors 369
  hypothalamus 518
  immune 523
  lymphocyte interactions *539*
  organizer 373–375
  receptors 522
  shared 522
cytoplasm 45
cytoskeleton 155
cytotoxic T lymphocytes 503, *504*, 510, 524, 537, *539*, 576
  HIV infection 551

D

Dale, Henry H. 416
Daly, Martin 308
dark reactions 588–590
Darwin, Charles *18*, 125
  arguments against creationism 142–143
  Cape Verde Islands 128
  development of ideas 126–128
  Galapagos Islands 128
  natural selection ideas 128–133
Darwinian evolution 15
Darwinian paradigm 125–141
Darwinism, descent with modification 133–137
data 5
DDT 611, 612
  biomagnification 613
  bird egg-shell thinning 613
deafness *116*
death control 241–242, 530
death rate 234, 235, 241–242, 246, 269
decomposers 585
deduction 6, 32
deer, red 297
dementia, AIDS 554
Demerol 481
democracy 25
demographic momentum 244
demographic transition 241–246, 269
  stages 241–242
demography 233–246
dendrites 411
denitrification 592
dental care 564
  HIV infection 570
Denver (USA) 675
deontological ethics 22, 23, 33
  genetic engineering 626–627
deoxyribonucleic acid *see* DNA
dependence
  addictive drugs 472–473
  psychological 473
depolarization *415*, 416, *417*, 432, 447

depressed mental state 353
depression 410, 445–446, 465
  cancer 526
  immune function 526
  manic 84
Descartes 17
descent with modification 133–137
desertification 351–352, 664–666
designer drugs 481
determination 369–371
deuterostomes 172–173
developed world, population growth 239
developing world
  genetic engineering impact 629
  population growth 239
diabetes mellitus
  insulin-dependent 80, 325–326, 510
  Native Americans 223
  reproductive rate 181
  thrifty genes 223
diatoms 681
Dictyostelium 164, 361
didanosine 561
diet
  cancer risk 393–394, 404, 406
  physiological variation 222
  poor 351–352
  requirements for health 328–349
diethylstilbestrol (DES) 395
differentiation 362, 406
  tissue 371–372
diffusion 322, *323*
  passive 354
digestion
  chemical processes 317, 354
  mechanical processes 317, 354
  Nematoda 170–171
digestive enzymes 320–322
digestive system 317–322, *323*, 324
dinosaurs 645
Dionaea muscipula 602–603
dioxin 395
diploidy 46, 50, 68
disability 570
disaccharides 330
discontinuous variation 189, 190
disease
  control of process 530
  exposure 567
  genetic heritage 494
  inherited 79–88
  rational study 495
  severity prediction 102
  stress 525–526
  susceptibility 79, 88
distal tubules 461
disulfiram 459
dizygous twins 81
DNA 15–16, 55, 68
  amplification 101
  building blocks 152
  chemical composition 56–57
  complementary bases 59
  confirmation as genetic material 55, *56*
  damage repair 526
  defect correcting 109
  double helix 59
  genetic transformation 53, 55
  mismatch repair genes 402
  mismatch repairs 382
  molecular markers 83
  mutations 382–383, 392, *394*
  nitrogenous bases 56, 57
  nongene 90
  probes 83, 90
  repair mechanism defect 386
  replication 59–61, 381
  sequence mutations 63–64, *65*, 66
  sequencing 90, *91*
  synthesis 364
  testing 94
  three-dimensional structure 57, 59
  transcription 62, *63*
  viral 548
  X-ray diffraction *58*
  see also mitochondrial DNA
DNA markers 85, 92
  band patterns 93–94
  identification of individuals 92–94
DNA polymerase 60
DNA sequences 89–90, 97
  human populations 209–210
DNase 53
Dobzhansky, Theodosius 115, 124
dodo 651
Dolly (sheep) 371
dolphins 673
dominance hierarchy 297
  linear 285
dominant traits 41, 189
dopamine pathways 446, 476
  brain 419–421
double blind studies 519
doubling number 375
doubling time 237
Down's syndrome 78, 101
dreams 444–445
Drew, Charles 203
drinking water treatment 681
drip irrigation 608
driving, alcohol 470
Drosera 594
Drosophila
  cluster of species *in statu nascendi* 149
  egg-polarity genes 372
drug abuse 456–457, 472–477, 490
  education 487
  health impairment 478–485, *486*, 487
  injected 563
  public health problem 485–487
drug addiction
  conditioned learning 476
  hypotheses 473–474
  rehabilitation 476
Drug Enforcement Agency (DEA) 485
drug therapy for AIDS 560–561
drug–drug interactions 464
drugs 452, 489
  action location 462–463
  activity 453
  additive effects 463, 489
  additives 481–482
  antagonistic 489
  combinations in pregnancy 484–485
  contaminants 481–482
  dependence 472–473, 489–490
  distribution in body 457
  doses 453, 489
  effective dose 453, 455–457
  elimination 457–462
  embryonic effects 483–485
  excretion 489
  fetal development 483–485
  half-lives 461–462, 464, 489
  injected 456
  interactions 463–464
  margin of safety 453
  metabolism 458–459
  persistence after birth 485
  plant-derived 582–583
  potentiation 463
  receptors 462–463, 489
  routes of entry 454, 455–457
  side effects 463–464, 489
  synergistic interactions 463, 489
  threshold response 463–464
  tolerance 477, 489–490
  withdrawal 472–473, 490
drug-seeking behavior 474
Duchenne muscular dystrophy 83, 87
Duesberg, Peter 543
Duffy blood group system 205
Dunkers 208
durians 670
dwarfism 96, 181
dye manufacture 391–392
dynamic equilibrium 497
dystrophin 87

E

eagle, bald 651, 682
ear
  balance 425, *426*, 427
  sound reception 425, *426*, 427
earthworm *169*, 172
Easter Island (Polynesia) 669
eating disorders 349–350
echinacea 483

Echinodermata *168*, 173
Eco R1 95, 96
ecological change, mass extinctions 646
ecological isolation 147
ecological succession in tropical rainforest 661–662
economic factors in cancer 397–398
economic impact level of integrated pest management 616
ecosystems 638, 684
- diversity 641
- stability preservation 640–641

ecotourism 670–671
ecstasy 468
ectoderm 166, *167*, 170, 373
education
- HIV infection 569, 572
- women 264–266

effector molecules 368
effector organ 411
egg 294
egg-polarity genes 372
eggs
- cholesterol content 335
- shell thinning 613

Ehrenreich, Barbara 308
Ehrlich, Paul 267
Eijkman, Christiaan 344
Einstein, Albert 6
Elavil 445
elderly people, stress 525
electric fields, low-frequency 390–391
electrical potentials 413–414, 416
- muscle cell 432

electroencephalograms (EEGs) 428, 441–442
electrolytes *343*, 346
electromagnetic radiation 391
electron transport chain 340, *341*
electron transport proteins 588
electrophoresis 82
elephant bird 649
ELISA test 555, *556–557*, 558
embryo 251
emergents 654
emigration 235–236, 242
emotional states 438–439
- autonomic nervous system triggering 516
- health 494
- immune system 496

emphysema 391, 479
employer discrimination 106
emulsification 320
endangered species 684
endocrine glands 513, 532
endocytosis 322, *323*, 336, 354
endoderm 166, *167*, 170
endorphins 466, 518, 523
endosymbiosis 155, 158, 183
energy
- biodiversity 638–639
- cellular 339–340, 354
- flow 585, *586*
- harvesters 585
- muscle contraction 433
- need 328
- pigments 586–587
- plant source 582
- producers 585
- production 155
- storage 589

energy–stability–area theory 639
enkephalins 466, 518, 523
- receptors 522

environmental determinism 126, 127, 128
environmental factors
- cancer 385–386, 406
- traits 114–115, 117

environmental manipulation 115
environmental pollutants 511
environmental quality 682
enzymes 53, 317, 354
Eohippus 642
Ephedra sinica 468
epidemic 550
epidemiology 333, 576
- AIDS 539
- cancer 384, 385
- HIV infection 569

epilepsy 431–432
epinephrine 467, 516, *517*
- stress response 516, 517

epiphytes 658
Epstein–Barr virus 380–381, 389
erysipelas 512
erythrocytes 325
Erythroxylon coca 455–456
Escherichia coli 11
- viral infection 55, *56*

esophageal cancer 401
esophagus 319
Essay on Population 130
Essay on the Principle of Population 238
estradiol 464
estrogen 74, 249, 395
- contraception 257–258

estrous period 303
ethanol consumption 85
ethical arguments 21–22
ethical decision-making 25–26
ethical questions 27–31
ethical systems 22
ethics 4, 20–26, 33
- abortion 263
- birth control 260–261
- deontological 22, 23, 33
- descriptive 20
- eugenics 111
- gene therapy 110
- genetic engineering 626
- genetic screening 105
- habitat destruction 667–668
- Human Genome Project 92
- normative 20
- surrogacy 253
- utilitarian 23–25, 33

ethnic groups, genetic screening 105
ethyl alcohol *see* alcohol
Eubacteria 159, *160–161*
eucaryote diversity 162–167, *168–169*, 170–175
eucaryotic cells 155, *156*, 183
eugenics 119
- biological objections 113–114
- negative 112–113, 114
- positive 111

euphenics 115
euphoria 466
Euphrates River (Turkey–Syria) 666
eupsychics 115
eusociality 288–291
Eustachian tube 425, *426*
euthenics 115
eutrophication 606–607
Eve hypothesis 13
evolution 125, 183
- complex structures 135–136
- ongoing 180–181
- teaching in U.S. 144, 145

evolutionary paradigm
- modern synthesis 146
- species 146–150

excretion 459
excretory system, drug elimination 458, 459–461
exhaustion 518
exoskeleton 172
experimental design, prospective 526
experimental science 10–12, 32–33
- ethical questions 27–31

experiments 10–12
exponential growth of human population 236–237
exposure to disease 567
expressed sequence tags (ESTs) 92
extinction(s) 642–651, 684
- asteroid impact hypothesis 647–648
- Cretaceous 646–647
- human role 648–650
- hunting 649
- mass 645–648
- models 644
- niche 651
- patterns 643–650
- predictors 650
- Quaternary 648
- random 643
- species in danger 651

species threatened 650–651
true 642, 644
types 642
extracellular matrix 366
extreme environments 158–159
Exxon Valdez 678
eye, light reception 424–425

F

facilitated diffusion 322, *323*
failure to thrive 353
fairness 25
falsifiable statements 5
family 150
fat consumption 7–8, 316
absorption disorders 343
cancer incidence 393
dietary 404
vegetarian diet 338
fatty acids 333–335
favism 218–219
feathers 175
feedback inhibition 505–506
anti-inflammatory 524–525
feedback mechanism 250, 447
feedback systems 420–421
feline leukemia vaccine 400–401
feline leukemia virus 388
Felten, David 522, 523
female athlete triad 350
females 294–295, 310
reproductive organs 248–249
ferns *165*
fertilization 40
human 250–251
nuclear fusion 46
fertilizers 604–607, 631
chemical 605–606
eutrophication 606–607
inorganic 604–606
organic 604–605, 606
runoff 606–607
soil buildup 676
fetal alcohol syndrome 484, 490
fetus
drug effects 490
nicotine 484
fever 508
fiber, dietary 340–341, 354
cancer incidence 393
Ficus 659–660
fig wasps 659–660, 661
fight or flight 515, 516
figs 659–660
seed dispersal 660–661
filter-feeding 173
filtration 681
finches of Galapagos Islands 129, 180
fireflies 281
fish 173, 174–175
lateral line system 424
schooling 283–284
fish meal 605
fish oils 226, 333
Fisher R.A. 52
fitness 130, 183, 285–286
inclusive 286, 287, *288*
male 294
pathogens 549
reproductive strategies 292–297
flagellum 163, 166
flatworms 167, *169*, 170
Flavr-Savr™ 625
Fleming, Alexander 6
flocculation 681
flowering plants *165*
specialization 597, *598*, 599
flowers 597
flukes 605
fluoride *343*, 347, 681
follicle-stimulating hormone (FSH) 247, 249, 250
food
pyramid 612, *613*
safety with genetic engineering 628
salty 393
supplements 483
Food and Drug Administration 482
foot racing 223
Ford E.B. 131
forebrain 429–431
forelimb homology 133–134
forest destruction 668
forest fires 674
formaldehyde 391, 674
fossil fuels 582
fossil record 137–141
fossils 183
chemical 153
correlation by 138–139
living 644–645
reptiles 175
founder effect 207, 228
fragile X syndrome 88
fragrances 583, *584*
frameshift mutations 64–65, *66*
Franklin, Benjamin 237
Franklin, Rosalind *58*
free radicals 345
Freud, Sigmund 298
frog *174*, 175, 673
strawberry *657*
fruit flies
evolution 180
gene linkage 50
fruits 599
Fulani 351
fundamentalists 143–144
fungal infection, plant resistance 626
Fungi *160–161*, 164, *165*
fungicides 611
fur seals 296–297

G

GABA 420–421, 431, 447
Galapagos Islands 128
finches 129, 180
Galileo 17, 19, 496
gall bladder 320
Gallo, Robert 388, 541
Galton, Francis 113
game theory 288
gametes 42, 43, 68
changing genotypes 109
haploidy 46
gamma globulins 204, 507
Gandhi, Mahatma 264
ganglia 413, 515
gannets *284*
Garrod, Archibald 79, 85
gastrula 166, *167*, 369, *370*
gender bias 31
gene expression 61, 406
regulation 367–369
variation 371
gene pool 200
alteration 111–114
gene therapy 106, 110, 118
general adaptation syndrome 516
generalization 439
generator potential 424
genes 38, 41–42, 68
hereditary disease association 85–88
jumping 51
linkage 46, 49, 50–51
location 46
locus 61
mapping 81
nucleotide sequences 89–90
recombination frequency 52
sex-linked 75, 118
silencing 625
susceptibility to disease 88
thrifty 223
transcription 61–63
translation 61–63
genetic conditions, prenatal detection 100–102
genetic counseling 100–106
genetic defects 100
genetic determinism 277–279
genetic drift 206–209, 228, 650
human populations 207, 208–209
genetic engineering 94–97, 118, 619–621, *622*, 623–629
antibiotic resistance 628
benefits 629
developing world impact 629

ethics 626
food safety 628
gene insertion by bacterial plasmids 621
gene insertion by plant viruses 620–621
gene insertion methods 620–621, *622*, 623
gene silencing 625
herbicide-resistant plants 624–625
ice-minus bacteria 625
opposition 626–629
pest-resistance 624
plant cloning 623
plant products 626
plants with altered nutritional content 623–624
restriction enzymes 96–97
screening with antibiotic resistance genes *622*, 623
sustainable agriculture 629
genetic factors, traits 114–115, 117
genetic heritage 494
genetic information misuse 106
genetic screening 105
genetic strains 279
genetic testing 100–106
ethics 104–106
information use 104
informed consent 102, 103
genetic transformation, DNA 53, 55
genetic variation in Protista 162–163
genetically engineered proteins 110
genetics 38
bacteria 11, *12*
see also molecular genetics
Geneva Convention 29
genital warts 389
genome 89, 118
genotype 41–44, 68
alteration of individual 106
humans 181
genus 150
geographic isolation 148
geographic proximity 133
geological time scale *138*
germ cell recombination therapy 110
giant cell, multinucleated 552
Giardia lamblia 681
gill slits 173
gills 172
Gingko biloba 483, 644
Gish, Duane 145
glaciation 648
glaucoma 479
glial cells 522
global warming 654
Gloger's rule 224–225, 228
glucose 588, 589
glutamate 436
reception enhancement 439
glycogen 325
glycolysis *339*, 340
gonorrhea 573–574
Goodall, Jane 303, 305
gossypol 258
gossypure 617
G6PD deficiency 218–219, 345
GRAS (Generally Regarded as Safe) list 482
Great Indian Desert (India–Pakistan) 666
green manure 593, 605, 607
green revolution 619, 629
greenhouses 609–610
grief therapy 403
Griffith, Frederick 52–53, *54*
grooming 301, *302*
ground squirrels 287, *288*
groundwater contamination 676
group selection 286
growth factor receptors
altered 379, *380*
antibody blocking 401
growth factors 366, 369, 508
growth hormone 96
GTP (guanosine triphosphate) 60
guard cells 601
Gurdon J.B. 371
gut 318
bacteria 324
Guzmania nicaraguensis 656
gynecomastia 480

H

habitat 636, 684
preservation 651
restoration of polluted 677–682
threatened 652–654, 655–657, 658–671
valuing 666–671
habitat destruction 646, 652–653, 667–668
indigenous people 669
sustainable use 669–670
tropical rainforest 653–654, *655–657*, 658–664
habituation 435–436, 447
Hae III 95, 96
Haitians 568
Haliaeetus leucocephalus 651
hallucinations
LSD 469
visual 432
hallucinogens 476
haloperidol 446
Hamilton, William D. 287
haplodiploidy 289–290
haploidy 46, 50, 68
Hardy–Weinberg principle/equilibrium 206–207, 228
harems 297
baboons 303
Harlow, Harry 298–300
harm reduction 485–486
Harvest Moon phenomenon 530
Harvey, William 495–496
Hasidic Jews 104
Hayflick L. 375
Hayflick limit 375, 376, 377
Hb^{S} allele 215, 216, *217*, 219
head 170
healing 507–508
health 316, 495
dietary requirements 328–349
ecosystem diversity 641
emotional states 494
immune system 497–508, *509*, *510–512*
mental state 494, 496
poverty 352–353
stress 518
wars 352
health care access 572
Healy, Bernadette 31
heart 325, 326–327
contractions 327
see also cardiovascular disease
heart attacks 327
rate 7–8
heart valve, sticky 142–143
height
continuous trait 189–190
human variation 192
helper T lymphocytes 505, 524, 537, *539*
HIV budding 548, *549*, *551*
HIV infection 548, 551–552
heme 214
hemoglobin 325
abnormal 214
destruction 212
gene 214–215
hemoglobin S 214, 216
hemolytic anemia 218
hemophilia 540–541
hepatitis 457
hepatitis B 389
hepatitis C 389
heptachlor 617
herbal medicines 482–483, 583
herbicides, resistant transgenic plants 624–625
Herceptin 401
hereditarianism 194, 196, *197*
hereditary conditions 100
HER2/neu 401
Herodotus 17
heroin 477

use 486–487
Herrick, Charles 212
Hershey, Alfred 55
heterotrophs 585, 631
heterozygosity 42, 68, 189
heterozygotes 106
high-density lipoproteins 335–336
high-risk behaviors 577
groups 567–568
HIV infection 567
hindbrain 429, 431
hippocampus 430, 438, 439
Hippocrates 495
histamine *509*, 510
Histoplasmodium 554
historical controversies, DNA testing 94
historical science 13
HIV-2 555
HIV infection 505, 536, 538
AIDS causation 544–545
AIDS connection discovery 539–542
biofeedback 530
blood handling guidelines 571–572
blood transfusion 564, 571–572
breast feeding 563
cell fusion 552
cocaine use 569
communicability 564–567
course *553*
drug therapy 560–561
education 569
educational campaigns 572
epidemic 550
epidemiology 569
false positive/negative test results 555–557, 558
health care access 572
health maintenance 555
helper T lymphocytes 551–552
high-risk 567
injected drugs 563
Koch's postulates 543–544
mosquitoes 564, *565*
notification 570–571
opportunistic infections 554
organ transplantation 564
pandemic 573, 575, 577
patterns 568
probability 565–566
progression to AIDS 552–555
public health 569–572
public policy 569–572
risk 563
sexual partner notification 571
stages 552–554
susceptibility 567
testing 570–571, 576
tests 555–558
transmission 562–575, 577
variations in progression 554–555
viral load 566
worldwide incidence 574–575
worldwide patterns 573–575
HIV-positivity 555
Hodgkin's lymphoma 390, 400
homeostasis 363, 406, 497, 532
homeostatic compensation 477
hominids 178
Homo 150, 179–180, 183
Homo erectus 179–180
Homo sapiens 179, 180, 181, 191
homogentisic acid 85, *86*
homologous pairs of chromosomes 46, 49–50
homology 133–134, 183
homosexuality 307–308
HIV infection rate 568
Kaposi's sarcoma incidence 539
homozygosity 42, 68, 102, 189
honeybees *291*
hormonal contraception 257–258
hormones 320
cancer 395
infertility treatment 252
pituitary gland 523
horsetails *165*
Hrdy, Sarah B. *18*, 288, 304, 305
human behaviors 305–308
human experimentation 29–31, 33
guidelines 31
subjects 30–31
human factors engineering 201
human genetics 72, 73–74
human genome 118
mapping 92
Human Genome Project
ethics 21–22, 92
human genome sequencing 90–91
legal issues 92
human immunodeficiency virus (HIV) 541, 576
budding 548, *549*, *551*
evolution 550
immune responses 559
life cycle 546, *547*, 548
regulatory genes 548
size 545–546
structure 546
see also HIV infection
human papilloma virus 389
human population 228
age structure 242–244
census 234
continuous variation 192
control movements 263–264
demographic transition 241–246
explosion 232
exponential growth 236–237, 269
genetic drift 207, 208–209
growth control 254–268
isolated 206–209
logistic growth 239–241
positive checks 238, 239, 241
pressure changes 238–239
preventive checks 238
reconstructing history 209–210
size prediction 233–246
world estimates 244–246
zero growth 242
human T-cell leukemia virus (HTLV) 388, 541
human variation 188
continuous 189–190
discontinuous 189, 190
physical traits 191–192
physiology/physique 221–223
population 191–196, *197–198*, 199–200
population genetics 202–210
study 200–201
humanness, biological definitions 262–263
humans 173, 175, 176–181, 294
biodiversity interdependence 639–641
evolution 181
extinction role 648–650
genotypes 181
hunting 646
migration 200
species 150
Hume, David 237
humors 495, 496
humus 607–608
hunting
extinctions 649
Huntington's disease 84, 87–88, 419, 420–421
dopamine 420
genetic testing 104–105
Huxley, Thomas Henry 140
Hydra *169*
hydrocortisone 524
hydrogen 151, 152
hydrogen bond 602
hydrogen sulfide 154, 674
hydroponics 609–610
Hymenoptera 288
eusociality 289–291
hyperpolarization 416, *417*
hypertension 346
hyphae 164, *165*
hypothalamus 441, 515
cytokines 518
ecstasy 468
stress response 516, *517*
THC receptors 467
hypotheses 5–8, 124
deduction 6
falsifiable 32
falsified 5–6, 11–12
general 5–6

living systems 6–8, 9–10
specific 5–6
testing in experimental science 10–12
testing in living systems 6–8
testing in naturalistic science 13
Hyracotherium 642
hysterectomy 255

I

ice-minus bacteria 625
ileum 322
imaging 519, 529
immigration 235–236, 242
immortal cells 377
immune cells
development 499–501
neuroendocrine effects 523–524
immune deficiency, hereditary 107
immune organs
nerve endings 522–523
immune potentiation 512
immune response
activation 506
antigen-specific 511
harmful 508, 510–511
turning off 505–506
immune responses
HIV 559
immune system 496, 532
AIDS 537
boosting 400–401
conditioned learning 528, *529*
emotional states 496
health maintenance 497–508, *509*, *510–512*
lymphatic circulation 497–499
mind 496
neuroendocrine system interaction 521–530
plasticity 511–512
stress 524–526
voluntary control 528–530
immune system suppression 397
immunity
acquiring specific 501–503
innate 507
passive 506–507
specific 532
immunization 503, 532
immunodeficiency 537, 576
immunoglobulin E (IgE) 510
immunohistochemistry 522–523
immunological memory 501–503
immunological tolerance 511
immunosuppression 511–512, 532
corticosteroids 524
Kaposi's sarcoma incidence 539
implantation 251
prevention 259
in vitro fertilization 252–253
inactivation, indirect 552
inborn error of metabolism 85, *86*, 100
inbreeding depression 650
inclusive fitness 310
independent assortment 43–44, 68
India, population control 264
indicator species 682
induction 6, 32
industrial chemicals 391–392
industrial melanism 131–132
infanticide 260, 308
primates 305
infants
attachment to mother 298–299
care-givers 353
infective dose 566
infertility 251
inflammation 507–508, *509*, 532
informed consent 33, 102, 119
genetic testing 102, 103
voluntary 30–31
inheritance 38
chromosomal basis 45–46, *47*, 48–52
chromosomal theory 46
molecular basis 52–53, *54*, 55–66
single traits 41–43
traits 43–44
inherited behavior 275–277
inherited diseases 79–88
innate behavior 276–277, 278
innate immunity 507, 532
insecticides
neuromuscular junction blocking 433–434
insects
eusocial 288
pollination 599
instinctive behavior 291
instincts 280–282
insulin 96, 325
genetically engineered 97, *98*
injection 456
insulin gene splicing 106
insurance policies 105, 106
insurer discrimination 106
integrated pest management 614–617, 632
economic impact level 616
predator species introduction 616–617
intelligence 113
heritability 196, *197*
interbreeding 147
interferon 507
interleukin-1 (IL-1) 508, 517
functions 523, 524
receptor 522
interleukin-2 (IL-2) 505, 517
functions 523, 524
receptor 522
stress 524
International Convention on Human Rights 29
International Olympic Committee 74, 75
interphase *47*, 48
interstitial fluid 497–498
intrauterine device (IUD) 259
intravenous injection 456–457
introgression 626
Inuit 222, 225–226
diet 333
inversion 66
invertebrates 166, *168*
iodine 348
ionizing radiation 390, *392*, 406
IQ 114
tests 113
iris diaphragm 424–425
iron *343*, 346–347
deficiency 352–353
irrigation 608
isogamy 294
isolation 206–209
experiments 440
human populations 208–209
Isoptera 288

J

jaguar *657*
Jalisco corn 640
Japanese people
atomic bomb survivors 390
diet 333
stomach cancer 393
jay, Florida scrub 287, *288*
Jay, Phyllis 303
Jefferson, Thomas 94
jellyfish 166, *168*, *169*
Jensen, Arthur 196
jet lag 441
Jewish populations 530
see also Ashkenazi Jews; Hasidic Jews; Sephardic Jews
Just, Ernest Everett *18*

K

Kant, Immanuel 22
Kaposi's sarcoma 389, 539
karyotype 73
Kenya 670
Kettlewell H.D.B. 131
kidneys 459–461
kilocalories 328, 331, 354
kin recognition in plants 297
kin selection 286–288
altruism 286
kingdom 150, 158, *160*

kissing 564
Klebsiella 592
Klinefelter's syndrome 76, 78, 101
Knowlton, Charles 264
Koch, Robert 542
Koch's postulates 542–544, 576
  HIV infection 543–544
  limitations 542–543
Koran 22
Krebs cycle 155, 339, 340, *341*, 344, 354, 589
K-selection 293, 294, 310
  plants 297
Kuhn, Thomas 14, 125
kwashiorkor 351, 352–353

L

labor shortages 193
laboratory animals 28, 33
lactase 321
lactose intolerance 322
laetrile 401
Lake Erie (USA) 650
Lake Victoria (Africa) 650
Lamarck, Jean-Baptiste 126
landfills 266–267, 608
Landsteiner, Karl 202
Lappé, Frances Moore 267
large intestine 324
lateral line system 424
laterite 663
Latimeria 644
lauric acid 624
law of independent assortment 44
law of segregation 42
law of unintended consequences 605, 611–612
L-DOPA 420, 446
leach beds 679
learned behavior 275–277, 310
learned helplessness 403
learning 435–437
  declarative 435, 437, 447
  procedural 435–437, 447
leaves 597
Lederberg, Joshua and Esther 11
legal issues of Human Genome Project 92
legal status of women 265
legislation, personhood 262
LeGros Clark W.E. 178
legs, arthropod 172
legumes 607
  green manure 605
Leguminosae 593
lemur 177
lens 425
leucocytes 325
leukemia 385, 390, 400
LeVay, Simon 307
lianas *658*
life, biological definitions 261–262
life cycles, sexual 49–50
life events 526, 527
life expectancy 244, *245*
life on Earth
  evidence 152–154
  origins 151–155, *156–157*, 158
lifestyle factors for cancer 385–386
light
  reactions 587–588
  reception 424–425
lignin 601
limes 345
limestone 676
Limulus 644
lineage 642
linear hierarchies 285
Lingula 644
linkage 46, 68
  maps 82
  studies 81–82
Linnaeus 195
lipases 321
lipids 354
  dietary 331–336
  transport 335, 336
literacy rate, female 264
liver 319, 326
liver cancer 389, 392
liverworts 597
living systems
  characteristics 9–10
  hypotheses 6–8, 9–10
  theories 9–10
locus 68
Loewi, Otto 418
logging 668
  see also timber
logistic growth 239–241, 269
loop of Henle 460
Lorenz, Konrad 306
loris 177
low-density lipoproteins 335–336
Lucy skull 178
lung cancer 390, 392, 404, 479
luteinizing hormone (LH) 249
lymph 498
lymph nodes 498, 501
lymphatic circulation 532
  immune system 497–499
lymphocytes 498
  cytokine interactions *539*
  population selection 500–501
  receptors 522
  self molecules 501–503
lymphoma 385
lysergic acid diethylamide (LSD) 468–469, 481–482
lysine 619
lysozyme 318, 621

M

MacLeod, Colin 53
macronutrients 329
  conversion to cellular energy 339–340
macrophages 498, 508, *509*
  HIV infection 548, 553–554
magnesium deficiency 511–512
magnetic fields 390–391
mah huang plant 468
maize *see* corn
malaria 211–219
  Epstein–Barr virus 389
  G6PD deficiency 218–219
  mortality 211
  Plasmodium life cycle 211–212, *213*
  population genetics of resistance 219
males 294–295, 310
  competition 294
  contraceptive pill 258
  fitness 294
  reproductive organs 247–248
malignancy *378*, 383
malignant melanoma 225, 390
malnutrition 328–329, 349–354
  Africa 351–352
  micronutrients 353
  protein deficiency 350–351
  see also anorexia nervosa
Malthus, Thomas Robert 130, 237–239
Mammalia *174*, 175
mammals
  South American 127
  taxonomy 150
mammograms 401
Mangold, Hilde 373
mangosteens 670
manioc 582
Mann, Jonathan 576
mantle 172
Maori people 669
marasmus 351, 352–353
marijuana 403, 467, 476
  medicinal use 479, 480
  pregnancy 485
  psychological dependence 473
  user health 479–480
marshland, wastewater treatment 680–681
Masai 351
mast cells 499, 510
mastectomy, prophylactic 403
Mastigophora 163
mate choice 294
maternal effect genes 372
mating
  behaviors 280, 281
  rituals 281
  signals 281
  systems 296–297

mayflies 280
McCarty, Maclyn 53
McClintock, Barbara *18*, 51
mechanical isolation 148
medicines
  genetically engineered 620–621
  see also drugs
meditation 519
medulla oblongata 431
medusa 166, *169*
meiosis 46, 48–50, 68, 248–249
  Protista 162
melanin 79, 225
melanism, industrial 131–132
melatonin 440, 441
membrane polarization 414
membrane potential 346
membrane proteins 334
membrane receptors 375
membrane transport systems 377
memory 437–439, 447
  consolidation 438, 439
  declarative 438
  formation 439
  long-term 437–439
  short-term 437
memory cells 399, 501
Mendel, Gregor *18*, 39–45
Mendel's laws 79, 87
menstrual cycle 249–250
  anorexia 350
  marijuana use 467
mental attitude to cancer 403
mental 'defectives' 113
mental illness 445–446, 465
  neurotransmitters 465
mental imaging 519, 529
mental state 494, 496
meperidine 481
Merck and Company 669–670
mesoderm 170
messenger RNA 61, 62, 63, *65*, 68, 367
  translation 368
  viral 548
metaphase *47*, 48
metastases 383
methane 151, 152, 158, 680
methionine 626
methylenedioxymethamphetamine (MDMA) 468
methylphenidate 468
micronutrients 329, 354
  deficiency 511–512
  malnutrition 353
microorganisms
  oil-degrading 677–678
  symbiotic 289
microsatellites 81
microwaves 391
midbrain 429, 431
mifepristone 259, 260
Mill, John Stuart 24
Miller, Stanley 151–152
millet 352
mimicry 130–131
Mimosa pudica 603
mind 410
  body interactions 494, 495–496
  brain 465
  immune system 496
  psychoactive drugs 465–470, *471*
mind equals brain theory 465
mineral wealth 662
minerals *343*, 346–348, 354
  trace 347–348
minnows *284*
misinformation 26
mismatch repairs 382
mitochondria 354
mitochondrial DNA 13, 155
mitosis 46, *47*, 48, 68, 364–365
  Protista 162
MN blood group system 205
moa, giant 649
models 32, 521
  extinction 644
  population 233–234
  theoretical 8–9
Mojave Desert (USA) 666
molds 164, *166*
molecular biology 89
  computer science 98
molecular farming 620
molecular genetics 15–16
Mollusca *168*
  extinctions 645
mollusks 136–137, 171–172
molybdenum 348
monarch butterfly 131
monkey *177*, 178
  langur 304
  oddity problems 439
  rhesus 298–300
monoamine oxidase (MAO) 419, 445
  inhibition 468
monoculture 610, 632
  pests 610–611
monogamy 296, 310
monosaccharides 329, *330*
monosodium glutamate 436
monozygous twins 81
Monsanto Corporation 629
Montagnier, Luc 541, 544
Montagu, M.F. Ashley 200
Moore G.E. 24
moral code 21, 22
moral conflict, resolving 21–22
morals 33
morning-after pill 259
Morowitz, Harold 262–263
morphine 466
  dependence 472–473
morphological species concept 191
Morris, Henry 145
mosaicism 75
mosquitoes and HIV transmission 564, *565*
mosses *165*, 597
mother-in-training behavior 301, *302*, 303
motor neurons 411, 432
mouth 171, 318
MPPP 481
MPTP 481
Muller H.J. 111, 113
multicellular organisms 361–363
multiple sclerosis 412–413, 510
muscle
  contraction 432–434
  opposing sets 433, *434*
  relaxants 443
  relaxation 433–434
muscular dystrophy 83, 84, 181
mushrooms 164, *166*
musk oxen 283
mutagens 392, 394–395
mutations 63–64, *65*, 66, 68, 392, *394*
  accumulation 381–382
  cell immortalization 381
  inheritance 381
  transformation to cancer cell 381
mutualism 324, 593, 631, 658–659
Mycobacterium tuberculosis 542
  HIV infection 554
Mycota *160–161*, 164, *165*
myelin sheath 412
myosin 433
Myxomycota 164

N

NADH *339*, 340, *341*
NADPH 587, 588, *589*
naloxone 466
naltrexone 466
Namibia 351
narcotics 466
National Institutes of Health 31
Native Americans 203, 205, 209, 221
  diabetes 223
  dietary protein 338
natural enemies 615, 616
natural killer cells 507
natural predators 616
natural remedies 482–483
natural selection 125, 128–133, 137, 183
  agents 132–133
  creationist arguments 142–143
  defined 130
  population variation 221–226
  skin color 223–226
Natural Theology movement 126
naturalistic sciences 32–33
  hypothesis testing 13
nautilus 140

Nazis 113, 194, 199
Neanderthals 180
needle sharing 457
needle-stick injuries 564, 566
negative reinforcement 473–474
Nematoda 170–171
  resistant transgenic plants 624
Neolithic people 180–181
Neopilina 644
nephrons 459
nerve endings, immune organs 522–523
nerve impulses 411, 413–414, *415*, 416
  measurement 414, *415*, 416
  threshold 416
nerves 413
nervous system 410, 411–413
  somatic 513
  see also central nervous system; peripheral nervous system
nesting behavior 282
neural tube formation 373, 374
neuroendocrine system 513, *514*, 515–520
  immune system interactions 521–530
neuroglia 413
neuromuscular junction 432, *433*
  blocking by insecticides 433–434
neurons 411–413, 447
  depolarization *415*, 416
  electrical excitation 416, *417*
  parasympathetic nervous system 515
  sympathetic nervous system 515
neurotransmitters 416, 418–419, 447
  Alzheimer's disease 439
  mental illness 445–446, 465
  pesticide actions 613–614
  receptor binding 462–463
  receptors 522
  reuptake 419
  sleep control 443
  types 418–419
neutrophils 499, 508
  mental imaging technique 529
New Zealand 669
newborn testing 102
Newton, Isaac 17, 19
niacin *343*, 344
niche 684
  extinction 651
nickel 348, 391
nicotine 467, 476, 479
  fetal effects 484
  psychological dependence 473
nicotinic receptors 467
nighthawks 285
nightmares 445
nitrates 592, 593, 605–606
  limiting nutrient 589
nitric acid 675
nitric oxide 436
nitrification 592
nitrites 592, 593
nitrogen
  atmospheric 591–592
  cycle 159, 591–592, 631
  fixation 592–593
  plant requirements 589–593, *594*, 595
nitrogen-poor soils 593, *594*, 595
nitrosamines 391
NMDA receptor 436
no races concept 200
nodes of Ranvier 412
noncoding regions 90–91
nondisjunction 76–78
non-governmental organizations (NGOs) 569
nonpolar bonds 320
nonrenewable resources 266–267
nonself 497
nontarget species 611–612
nonvascular plant tissues 596–597
norepinephrine (noradrenaline) 419, 467, 515, 516, *517*
  agonists 468
  receptors 522
  stress response 516, 517
normal science 14–17, *18*, 19, 33
nose 427–428
Nostoc 592
notification of HIV infection 570–571
notochord 173, 373, 374
nuclear fusion 46
nuclear membranes 49
nucleic acids 53, 55, 68
  biosynthesis 592
  translation *65*
nucleotides 56–57
  sequences 83, 89–90
nucleus 45–46
  CNS 413
Nuer people 4
nutrient cycles 591
nutrients
  enrichment of polluted areas 678
  limiting 589, 631
nutrition 316

O

oat bran 341
obesity 84–85, 328
  immune system suppression 511
objectivity 10–11
oceans, origins 151
oddity problems 439
odor cues 287–288
odors 427–428
Ohio River (USA) 650
oil spills 676
  bioremediation 677–678
Old, Lloyd 512
omega-3 fatty acids 333
On Aggression 306
oncogenes 379–381
Oparin, Aleksandr 151
operant conditioning 476
opiates 466–467, 476
  endogenous 466–467
  receptors 466–467
opium 466, 482, 483
opportunistic infections 576
  AIDS 554
oral contraceptives 257–258, 395–396, 464
oral sex 562, 564
orangutan 660–661
orchids *658*
organ transplantation 564
organelles 155
organisms, kingdoms 158, *160*
organizer region 373, 374
Origin of the Species 125
osmosis 599–601, 631
osteoporosis 350
ovarian cancer 395–396
ovarian follicle 249
ovariectomy 395
ovaries 247
  surgical removal 255
overgrazing 665
overnutrition 511
over-the-counter drugs 481
ovulation 248–249
owl, spotted 664, 673, 682
oxidation 345, 354
oxygen 153, 154
  photosynthesis 153
  stomata 601
  water molecule 602
ozone 674, 681

P

p53 oncogene 402
pain perception 423–424
pair-bond, permanent 296
Paleozoic extinctions *643*
Paley, William 126, 142–143
palindrome 96
Palmer, Craig T. 308
pancreas 320–321
pandemic, HIV infection 573, 575, 577
Pap smear 401
Papaver somniferum 466
paper 582
papilloma virus 389
paradigm shift 15
paradigms 14, 15, 33
  molecular genetics 15–16

parasites 604–605
parasympathetic nervous system 513, *514*, 515, 532
  relaxation response 518
Paregoric 466
parental care 176–177
parental investment 294, 310
parenting behavior, learning 301, 303
Parkinsonism 410, 419–420
passive immunity 506–507
Passover 530
Patau's syndrome 78
pathogens 543
  drinking water treatment 681
  fitness 549
Pavlov, Ivan 437
Pavlovian conditioning 436–437, 528, *529*
pea plants traits 39–45
pecking order 285
pedigrees 79–80, 118
pelicans *284*
pellagra 342
Penfield, Wilder 439
penicillin discovery 6
People for the Ethical Treatment of Animals (PETA) 28
peppered moth 131–132
pepsin 319
peptidases 321
peptides 319
perception 465–466
perch, Nile 650
peripheral nervous system 411, 447
  neurotransmitters 418–419
personality profiles 526
personhood, legal definitions 262
Pert, Candace 522
pesticides 611–612
  accumulation 613
  biomagnification 612–613, 614
  bird egg-shell thinning 613
  insect neurotransmitter blocking 613–614
  negative consequences 612
  poisoning 614
pest-resistant plants 624
pests
  chemical control 610–614
  eradication 614–615
pets 519
Peyer's patches 500
pharmaceutical plants 669
pharmacology 453
pharynx 173
phencyclidine (PCP) 485
phenobarbital 395
phenocopy 115
phenotypes 41–44, 68, 114
  modification 115
phenotypic frequency 190
phenylalanine 85, 86–87
phenylketonuria (PKU) 85–87, 102
  misuse of genetic information 106
pheromones 289, 291, *615*
  integrated pest management 617
phloem 596
phosphates 606–607
phospholipids 333–335
phosphorus fertilizers 605
photodynamic therapy 401
photosynthesis 153, 583–590, 631
  bacterial 154
  dark reactions 588–590
  evolution 153–154
  light reactions 587–588
  nonvascular plants 597
  pigments 586–587
  stomata 601
  tropical rainforest 654
  vascular plants 597
  water requirement 599
phototrophs 585
phyla 150
phyletic transformation 642
phylogeny 139–141
Piaget, Jean 17
pickled foods 393
pigeon, passenger 649
pigments, energy 586–587
pineal body 441
pioneer species 661
Piper *656*, 658–659
Pisum sativum 39
pitcher plant *594*, 595
pituitary gland 518
  hormones 523
placebo effect 519–520
placenta 457, *458*
plant breeding, selective 618–619
plant cell *156*, *157*
  osmosis 599–601
plant lice 293
plant proteins 337–339, 351
Plantae *160–161*, 163–164, *165*
plants 582
  alteration through artificial selection 617–619
  cell walls 597
  cloning 623
  gene silencing 625
  gene transfer 626
  genetic engineering 619–621, *622*, 623–629
  genetically engineered products 626
  growth 589
  pest-resistant 624
  products of use to humans 582–590
  rapid movement 602–603
  tissue specialization 596–597, *598*, 599, 631
  water transport 599–603
plasma 325
plasma membrane 322, *323*
plasmids 97, 154–155, 621
plasmodesmata 603
Plasmodium
  life cycle 211–212, *213*
plastids 155, 158
platelets 325
Platyhelminthes 167, *169*, 170, 605
play 299–300
  rough and tumble 300–301
pleasure centers 477
pluralism 25
pluripotent cells 370–371, 378–379
Pneumocystis carinii pneumonia 539, 554, 561
pneumonia 52–53, *54*
point mutations 64, *66*
poisons 582
polar bonds 320
policy
  decisions 25
  issues 26
pollination 40, 597, 599
  bird *656*
  fig wasps 660, 661
pollution 672–676, 684
  air 674
  benefits of cleaning up/preventing 682
  costs 682
  detection 672–673
  fertilizers 606–607
  habitat restoration 677–682
  industrial melanism 131–132
  measuring 672–673
  pesticides 612
  population growth 266
  prevention 672–673
polyandry 297, 310
polychlorinated biphenyls (PCBs) 391
polygynous mating units 297
polygyny 296, 310
polyhydroxybutyrate 626
polymerase chain reaction (PCR) 101, 119
polymorphism 219, 228
  balanced 219, 228
polyps 166
polysaccharides 330
polyunsaturated fats 332
polyurethane elastomer 256
Pongo pygmaeus 660–661
Popovic, Michael 541
poppy, opium 466
population 146–147, 183, 188, 228
  biological 191
  change in size 235
  crash 240
  density 235, *236*

doubling time 237
explosion 232
growth rate 234–236
human variation 191–196, *197–198*, 199–200
impact control 266–268
models 233–234
variation with natural selection 221–226
zero growth 242
population control 254
movements 263–264
programs 265
population ecology 233, 269
population genetics 190, 196, *198*, 199–200, 228
malaria resistance 219
sickle-cell anemia 215–216, *217*
positive checks 238, 239, 241
positive reinforcement 473, 474, *475*
positive reward centers 431
positron emission tomography (PET) 428
postmating mechanisms 147
postranslational processing 368
potassium *343*, 346
active transport *413*, 414
potatoes, transgenic 623–624
potentiation 463
poverty, health 352–353
power lines 390
pre-Darwinian thought 126
predation defense 283
predators 130–131, 283
integrated pest management 616
pregnancy
alcohol 484
barbiturates 485
caffeine 483–484
drug circulation 457, *458*
drug combinations 484–485
drugs 490
genetic disease detection 105
marijuana 485
nicotine 484
Rh factor 204
surrogate 253
premating mechanisms 147
premature babies, survival 263
prenatal detection of genetic conditions 100–102
prenatal screening 105–106
preventive checks 238
prey species 130–131
primates 176, *177*
adult behavior development 299–300
competition 305
infant attachment to mother 298–299
infanticide 305
play 299–301
reproductive cycles 303
reproductive fitness maximization 304
reproductive strategies 303–305
social behavior 298–301
social rank 305
sociobiology 298–301, *302*, 303–308
prior-mutation hypothesis 11–12
procaryotic cells 154–155, *156*, *157*, 183, 587
procaryotic organisms 158–159, *160–161*
producers 585
progesterone 74, 249–250, 395
contraception 257–258
Prohibition 487
promiscuity 297, 310
promoters 62, 367
prophase *47*, 48
prostate cancer 392, 402
prostate gland 248
prostate-specific antigen (PSA) test 402
prostitutes *see* sex workers
protease inhibitors 561
proteases 321
protein(s) 336–339, 354
biosynthesis 592
complete 337–338
deficiency 350–351, 351
folding 62, 336
incomplete 337–338
requirements 336, 337
starvation 352
structure 336, *337*
synthesis 16
undernutrition 511
vegetarian diets 338–339
Protista *160–161*, 162–163
protists 155
proton gradient 340, *341*
proto-oncogenes 379–381
translocation 380
uncorrected mutations 382
protostome phyla 171–173
proximal tubules 460
Prozac 445
pseudocoel *167*, 171
pseudoextinction 642, 644
Pseudomonas syringae 625
Pseudomyrmex 659
pseudopods 163
Psilophyton 644
Psophocarpus tetragonolobus 670
psyche 496
psychoactive drugs 465–470, *471*, 489–490
addiction 472–477
long-term use 476–477
psychological dependence 473
psychoneuroimmunology 495, 521–522, 531–532
psychosis, amphetamines 468
puberty 247
public health 238–239
drug abuse 485–487
HIV infection 569–572
public policy, HIV infection 569–572
Punnett square 43
pure lines 40
pyridoxine 342, *343*
pyruvate 340
Pythagoras 219

Q

quadrumanual clambering 660–661
quantitative-trait locus analysis 84–85
Quaternary extinctions 648

R

race(s) 188, 228
cancer rates 397
concept 192–196, *197*, 198–200
cultural characteristics 193–194
cultural concept 194
IQ 196, *197*
morphological concept 194–196
Nazis 199
physiological differences 221–222
population genetics 196, *198*, 199–200
pure 195–196
typological concept 194–196
racial classification 196
racism 113, 193, 194, 196
radiation 390–391
cancer treatment 399–400
radon 390, *392*
rain clouds 663–664
desertification 666
rain water, pH 674–675
rape 308
rapid eye movement (REM) sleep 442–443, 448
electrical activity in brain 445
Raup, David 646
reading frame 16, 64
reasoning 6
rebound effect 515
receptors 322, *323*
cellular 366, 462–463
recessive traits 41, 103, 189
recombinant DNA technology 94
recombinant DNA therapy 110
recombinant human insulin 97
recombination therapy, germ cell 110
recommended daily allowances (RDAs) of vitamins 342
rectal cancer 404

rectum 324
red blood cells
  destruction 212
  sickle-cell anemia 213
red–green colorblindness 75
refusal skills 569
regulatory proteins 372
reinforcement 446
relaxation response 518–519, 532
religions, sacred texts 22
replica-plating experiment 11, *12*
replication 16
repressor molecules 367–368
reproductive anatomy 247–251
reproductive cycles, primates 303
reproductive fitness 304
reproductive isolation 147, 183
  mechanism 147–148
reproductive maturation 247
reproductive organs
  female 248–249
  male 247–248
reproductive physiology 247–251
reproductive strategies 310
  fitness 292–297
  primates 303–305
reproductive technologies 251–253
reptiles
  extinctions 645
  fossil record 140–141
  Mesozoic *646*
Reptilia *174*, 175
resource use 294
respiratory system 454, *455*
respiratory tract disease 479
rest and ruminate 515–516
resting potential *413*, 414, 416, *417*, 447
restriction enzymes 90, 95, 118
  genetic engineering 96–97
restriction fragments 95
restriction-fragment length polymorphisms (RFLPS) 81, 82–83, *84*, 85, 90, 92, 118
reticular activating system 443–444
reticular formation 431, 469, 470
retina 425
retinoblastoma 386
retinol *343*, 345
retroviruses 541–542, *547*, 548
reverse transcriptase inhibitors 560–561
reverse transcription 548, 550, 576
revolutionary science 14–17, *18*, 19
Rh blood group system 203–204
Rhizobium 592, 593
rhythm method 256
riboflavin *343*
ribonucleic acid *see* RNA
ribosomal RNA 62, 63
ribosomes 63
rice
  improvement 619
  polished 344
rickets 225
Rifkin, Jeremy 626
rights concept 22–23
risk 118, 563
  absolute 403
  behaviors 563
  defective alleles 102–103
  hereditary 100
  relative 403
  statistical 80
Ritalin 468
RNA 15–16, 55, 68
  building blocks 152
  structure *61*
  synthesis 61
  translation to protein 62
RNA polymerase 62, 367–368
  inhibition 368
rods 425
Rollin, Bernard 29
root nodules, mutualistic 592–593
roots 596–597
rotenone 582
Rotifera *169*
Roundup™ 629
Rous, Peyton 388
Rous sarcoma virus 388
royal jelly 291
Royal Society 17
r-selection 293–294, 297, 310
RU-486 259
rubber manufacture 391–392
rubber plantations 662
runoff of fertilizers 606–607
rye 468

S

saccharin 395, 528
saddleback caterpillar *615*
safety of gene therapy 110
Sahara desert 351, 665
Sahel 665
saliva 318, 562, 564
salmon 682
salty foods 393
sampling error 8
Sanger, Margaret 264
Santorio 496
Sarcodina 163
sarcoma 385
saturated fats 332
saturated fatty acids 335
saw palmetto 483
schizophrenia 446
science 4, 32
  definition 8
  policy issues 26
Scientific Committee on Problems of the Environment (SCOPE) 627
scientific community 17, *18*, 19
scientific method 496
scientific revolutions 15, 33
Scopes, John H. 144
sea-squirts 173, *174*
second law of thermodynamics 145
second messengers 367, 469
  altered 379, *380*
secondary sexual characteristics 247
seeds 597, *598*, 599
  dispersal *598*, 599
segmentation 171–172
segregation 42
seizures 431
selection, species level 285–286
selective breeding 618–619
selective estrogen receptor modulators (SERMS) 400
selenium deficiency 511–512
self, maintenance 497
self molecules 501–503
selfishness 286
self-pollination 40, 43
Selye, Hans 516
semicircular canals 425, *426*, 427
seminal fluid 248
sensation 465
sensitive plant 603
sensitization 436, 447
sensory neurons 411
sensory perception 422
sensory reception 422
sentinel species 664, 673
Sephardic Jews 345
Sepkoski, John 646
sequence tags 81
serotonin 445–446, 468
  agonists 469
setae 172
severe combined immune deficiency syndrome (SCIDS) 107
sex chromosomes 73–74, 118
sex determination 74–75
sex workers 555, 568
sexes, differences 294–295
sex-linkage 75, 118
sex-linked traits 75
sexual intercourse 563
sexual life cycles 49–50
sexual orientation 307–308
sexual partners, HIV notification 571
sexual recombination in Protista 162
sexual reproduction 292–293, 310
  adaptive advantage 293
sexual selection 133, 183
sexual status 303
sexually transmitted diseases 257, 536
  incidence 573–574

Shanklin, Eugenia 194
sheep, bighorn 297
shell, mollusk 172
shrubland, temperate 654
sick building syndrome 674
sickle-cell allele
  frequency maps 215–216, *217*
  testing 103
sickle-cell anemia 212–216
  frequency 192
  genetics 214–215
  population genetics 215–216, 217
  symptoms 215
Sideroxylon 651
Silent Spring 611
silicon 348
single trait inheritance 41–43
single-nucleotide polymorphisms (SNPs) 81, 92
skeletal muscles 433, *434*
skin, sensory nerve endings 423–424
skin cancer 192, 225
skin color
  cline 224
  natural selection 223–226
  sunlight 224–225
slash-and-burn agriculture 662, *663*, 665
slavery 193
sleep 441–444, 448
  cycles 442–443
  drug alteration of pattern 443
slime molds 164, 361–362
sloth, tree *657*
Slusher, John 145
small intestine 319–322, *323*
  digestive enzymes 320–322
  fat processing 319–320
  nutrient absorption 322, *323*
smallpox vaccination 559
smell sense 427–428
smoke particles 456, 479
smoking 391
  cancer risk 404
  cocaine 456
  passive 479
  synergism with alcohol consumption 393–394
  synergism with oral contraceptives 395
  see also tobacco
snake venom 483
Snyder, Solomon 522
social behavior 274, 310
  development of adult 300–301
  learned components 299–300
  primates 298–301
social factors in cancer 397–398
social groups
  biological advantages 283, *284*
  size 285
  turnover and replacement 303
social organization 284, 310
  adaptive 283–291
  primates 301, *302*, 303
  simple 284–285
social policy 25
social rank, primates 305
social status 294
  harem-forming species 297
Society for the Prevention of Cruelty to Animals (SPCA) 28
sociobiology 274, 310
  paradigm 277–279
  primates 298–301, *302*, 303–308
  research methods 279–280
sodium *343*, 346
sodium ions
  active transport *413*, 414
  channels 416, *417*
sodium–potassium pumps 414, 416, *417*, 447
soil
  conservation 607–608
  erosion prevention 607
  formation *607*
  improvement 607–608
  laterite 663
  nonrenewable resource 608
soldier bug 616
Somalia 351
somatic cells 46, 49–50
  cancer 381
somatic nervous system 513
somatostatin 96
somites 171–172
sound reception 425, *426*, 427
soy protein 337
soybeans 626
Spathophyllum 656
specialization 362–363
speciation 146, 183
  geographic theory 149
species 183
  biological value of preserving 639–640
  danger of extinction 651
  definition 147
  endangered 684
  evolutionary paradigm 146–150
  morphological 191
  number 637–638
  origination 148–149
  patterns of distribution 128
  pioneer 661
  sentinel 673
  threatened with extinction 650–651
species level selection 285–286
Spemann, Hans 373
sperm 248, 250–251, 294
  production 247–248
sperm duct blockage 256
spermicidal agents 257
Sphenodon 644
spices 583, 584
spinal cord 411
spleen 500
sponges 166, *169*, 362, *643*
Sporozoa 163
sporozoites 212, *213*
sry gene 74, 75
SRY protein 74, 91
St John's wort 483
starch, chemical digestion 318
starvation 328–329
  protein 352
Statement on Race and Racism (UN) 199
statistics 7–8
stem cells 406, 499–500
  transformation 378
stems 596
stereotyped behavior 281
sterilization 254–256
  mental 'defectives' 113
Stern, Curt 51
steroid hormones 516–517
  abuse 483
  glucocorticoid 523–524
  receptors 522
  stress response 516, 518
stick insect *656*
sticky ends *95*, 96
stomach 319
  cancer 393
stomata 601
stratigraphic sequence *139*
stratigraphy 137–139
streptococcal toxin 512
streptomycin 11
stress 397, 516
  disease incidence 525–526
  elderly people 525
  health 518
  immune system 524–526
  pets 519
  T lymphocytes 525
stress response 516–518, 532
  exhaustion 518
  individual variation 526–527
  resistance 517
stressors 516, 526–527
stroke 327
stroma 587
strychnine 481, 483
subspecies 194
substance abuse 472–477, 490
  see also alcoholism; drug abuse
substance P 466
substantia nigra 419, 421
succession 662
sudden infant death syndrome (SIDS) 404, 479
sugar, tooth decay *329*

sulfur 611
sulfur dioxide 675–676
sulfuric acid 675
sundew *594*, 595
sunlight 585, *586*
  skin color selection 224–225
support groups 403, 529
suprachiasmatic nucleus 441
surgery in cancer treatment 399–400
surrogacy 253
susceptibility 577
  HIV infection 567
sustainable agriculture 629
sustainable energy use 267
sustainable practices 608
sustainable use 684
  habitat destruction 669–670
Sutton, Walter 45, 50
symbiosis 289
  gut bacteria 324
symmetry
  bilateral 167, 170
  radial 167
sympathetic nervous system 513, *514*, 515, 532
synapse 411, 447
  neurotransmitter removal *418*, 419
synergism
  alcohol and tobacco 393–394
  carcinogens 392
  drugs 463
syphilis 573–574

T

T lymphocytes 499, 500–501, 532
  antigen removal 503
  stress 525
tamblaocoque tree 651
tamoxifen 395
tandem repeats 83
tapeworms 294
tarantula *656*
taste 427
tattooing 564
taxon 149–150
taxonomy 149–150, 183
Tay–Sachs disease 72, 102, 181
  discontinuous variation 190
  genetic testing 104, 105
  testing 103
tea plant 469
tears 562, 564
tela choroidea 431
telomerase 376
telomeres 375–376
telophase *47*, 48
temperature perception 423–424
termites, eusociality 288–289
territorial behavior 282
territorial defense 286
terrorists 28–29
testes 247
  removal 254
testicular cancer 400
testis-determining factor (TDF) 74
testosterone 74, 75, 247–248
  marijuana use 467
Δ9-tetrahydrocannabinol (THC) 467, 480
Thailand, HIV incidence 575
thalassemia
  allele carrying 103
  malaria protection 217–218
Thea sinensis 469
Theobroma cacao 469
theories 8–10, 32, 124
  living systems 9–10
thiamine *343*, 344
Thornhill, Randy 308
thrifty genes, diabetes mellitus 223
thylakoids 587, *588*
thymus 500
ticks *565*
timber 662, *663*, 664
tin 348
tissue induction, organizer cytokines 373–375
tissues 362, 406
  differentiation 371–372
tobacco 395, 479
  plants 620–621
  smoke 391, 394–395, 404, 406
  smokeless 391
  synergism with alcohol 479
  see also smoking
tobacco budworm 616
tobacco mosaic virus 620
tocopherol *343*, 345
Tofranil 445
tolerance, immunological 511
tomatoes, gene silencing 625
tongue 427
tonsils 498–499, 501
tooth decay *329*
totipotent cells 369, 371
toucan *657*
touch reception 423, 424
toxicology 673
toxins *504*, 505
Toxoplasmodium 554
trace minerals 347–348
trachea 454, *455*
tracts 413
traits
  environmental factors 114–115, 117
  genetic causes 79–85
  genetic factors 114–115, 117
  heterozygous carriers 103
  population frequency *102*, 103
  recessive 103
  specific genes 83–84
transcription 16, 61–63, 68
  regulation 367–368
transfer RNA 62–63, *64*, *65*
transformation 53, *54*, 377, 392, *394*, 406
transformed cells 377–378
  continued division 382
  solid tumor formation 383
transgenic plants 632
  altered nutritional content 623–624
  herbicide-resistant 624–625
  pest-resistant 624
translation 16, 68
translocation 74, 380
transpiration 601, 602
transpiration-pull theory 601–602
transporter proteins 322, *323*
transposable elements 51
Trefil, James 262–263
triglycerides 331, 335
trinucleotide repeat diseases 88
trisomies 78
trophic pyramid 612
tropical rainforest
  agriculture 670
  biodiversity 654, *655–657*
  biome 654
  clearance 665–666
  clearings 662
  deforestation 662–664
  destruction 653–654, *655–657*, 658–664
  high-income products 669
  species interactions 658–661
  tree buttresses *655*
  vertical stratification *655*
tropical regions 224
trust, basic 353
tsetse flies 351
tubal ligation 255, 256
tuberculosis 220, 479, 542
tumor initiators 394–395, 404
tumor necrosis factor (TNF) 512
tumor promoters 394–395, 404
tumor suppressor genes 379, 381
  repressor activity 381
  uncorrected mutations 382
tumor-associated antigens 400–401
tumors 377–378, 406
  benign 383
  malignant *378*, 383
  spread 383
tunicates 173
turgor pressure *600*, 601, 603
  regulation 603–604
Turner's syndrome 76, 78, 101
twin studies 80–81, 280
  alcoholism 307
  homosexuality 307–308
tympanic membrane 425, *426*
tyrosine 85, 86

U

ultraviolet light 225, 321
ultraviolet radiation 390, 406
unicellular organisms 361–362
Union of Concerned Scientists 627, 628, 629
unsaturated fats 332
urbanization 266
urine 461
  excretion 459
Urochordata 173, *174*
uterus
  cervical cancer 389
  surgical removal 255
utilitarian ethics 23–25, 33
  genetic engineering 626, 627
utilitarianism 23–25

V

vaccination 503
  smallpox 559
vaccines
  AIDS 558–560
  antitoxins 505
  cancer 400–401
vacuum aspiration 259–260
vaginal diaphragms 257
vaginal intercourse 563
  HIV infection 568–569
vaginal sponges 257
vagus nerve 418
valerian 482
value 667
vanadium 348
Varmus, Harold 379
vas deferens 248
vascular plants 631
  tissues 596–597
vasectomy 254–255, 256
vectors 107
vegans 338–339
vegetable consumption 393, 404
vegetarian diets 338–339
veins 325, 326
ventral tegmental area 431, 474, 475–476
ventricles of heart 326
Venus flytrap 595, 602–603
verifiable statements 5
Vertebrata 173–175
vestigial structures 135
viceroy butterfly 131
villi 322
vinyl chloride 391, 392
viral infection, plant resistance 626
virulence 52–53, *54*, 576
  evolution 549–550
virus vectors 107
viruses 545–546, *547*, 548
  killing 501
  life cycle 545–546
  new gene insertion 620–621
  nucleic acid 546
  replication 546
vision, binocular 176
visual structure evolution 135–136
vitamin A 342, *343*, 345, 393, 404
  deficiency 511–512
vitamin B 348
vitamin B_1 *343*, 344
vitamin B_6 342, *343*
  deficiency 344
vitamin B_{12} 348
  deficiency 345
vitamin C *343*, 344–345, 347, 393, 404
  deficiency 511–512
  megadoses 345
vitamin D 225, 228, 321, 342, *343*, 346
  calcium metabolism 347
  nutritional source 225–226
vitamin E *343*, 345, 348, 393, 404
vitamin K 324, *343*, 346
vitamins 342–346, 354
  antioxidant 345
  deficiencies 342–344
  deficiency diseases 344
  fat soluble 342, *343*
  overdoses 342–344
  water soluble 342, *343*
voluntary movements 432
Volvox 163, *164*

W

warbler, Bachman's 650
wars 352
waste disposal 266–267
wastewater
  bioremediation 679–681
  lagoons 679–680
  marshland treatment 680–681
  three-stage treatment 680
water
  cycle 663–664
  drinking 681
  evaporation 602
  pesticide contamination 612
  splitting in photosynthesis 154
  transport in plants 599–603
  underground reserves 666
  vapor 151, 152
Watson, James 57, *58*, 59, 90
weaning, delayed 258
weaver birds *284*
weeds 294
  genetically altered strains 627–628
weevils 617
Western blot test 555, *556–557*, *558*
Western science, beginning 17
wheat, improvement 619
white acid 481
white blood cells 499, 501, 508
wildebeest 283, 284
Wilkins, Maurice H.F. 57, *58*
willow bark 583
Wilson, Edward O. 277, 653
Wilson, Margo 308
Wilson Dam (USA) 650
winds, prevailing *664*
withdrawal, addictive drugs 472–473
withdrawal syndrome, conditioned 473
women
  education 264–266
  legal status 265
wood 582
  digestion by termites 289
World Health Organization (WHO) AIDS criteria 574–575
world population estimates 244–246
Wynne-Edwards V.C. 286

X

X chromosomes 73–74
  genes 75
xanthophyll 586
xeroderma pigmentosa 386
X-rays 390
  diagnostic breast 401
XX males 74–75
XXX chromosomal abnormality 77
XY females 74–75
xylem 596

Y

Y chromosomes 73–74
yeasts 164
Yoga 518–519
Yucatan peninsula (Mexico) 648

Z

Zea diploperennis 640
zidovudine 560–561, 563
zinc 348
  deficiency 511–512
Zygomycota *166*
zygotes 49, 251
  diploid 162

ENVIRONMENTAL SYSTEMS

TITLES OF RELATED INTEREST

Aeolian geomorphology
W. G. Nickling (ed.)

Catastrophic flooding
L. Mayer & D. Nash (eds)

The changing climate
M. J. Ford

The climatic scene
M. J. Tooley & G. M. Sheail (eds)

Elements of dynamic oceanography
D. Tolmazin

Environmental change and tropical geomorphology
I. Douglas & T. Spencer (eds)

Environmental chemistry
P. O'Neill

Environmental magnetism
F. Oldfield & R. Thompson

The face of the Earth
G. H. Dury

Fundamentals of physical geography
D. Briggs & P. Smithson

Geomorphological field manual
R. Dackombe & V. Gardiner

Geomorphological techniques
A. S. Goudie (ed.)

Geomorphology: pure and applied
M. G. Hart

Hillslope processes
A. D. Abrahams (ed.)

The history of geomorphology
K. J. Tinkler (ed.)

Image interpretation in geology
S. Drury

Inferential statistics for geographers
G. B. Norcliffe

Introduction to theoretical geomorphology
C. E. Thorn

Karst geomorphology and hydrology
D. C. Ford & P. W. Williams

Models in geomorphology
M. Woldenberg (ed.)

Pedology
P. Duchaufour

Planetary landscapes
R. Greeley

A practical approach to sedimentology
R. C. Lindholm

Quaternary environments
J. T. Andrews (ed.)

Quaternary paleoclimatology
R. S. Bradley

Rock glaciers
J. Giardino *et al.* (eds)

Rocks and landforms
A. J. Gerrard

Tectonic geomorphology
M. Morisawa & J. T. Hack (eds)

ENVIRONMENTAL SYSTEMS

An Introductory Text

I. D. White
Portsmouth Polytechnic

D. N. Mottershead
Edge Hill College of Higher Education

S. J. Harrison
University of Stirling

London
UNWIN HYMAN
Boston Sydney Wellington

Published by the Academic Division of
Unwin Hyman Inc.
15-17 Broadwick Street, London W1V 1FP, UK

Unwin Hyman Inc.
955 Massachusetts Avenue, Cambridge, MA 012139, USA

Allen & Unwin (Australia) Ltd,
8 Napier Street, North Sydney, NSW 2060, Australia

Allen & Unwin (New Zealand) Ltd in association with the
Port Nicholson Press Ltd,
Compusales Building, 75 Ghuznee Street, Wellington 1, New Zealand

First published in 1984
Fifth impression 1990

British Library Cataloguing in Publication Data

White, I. D.
Environmental systems
I. Physical geography
I. Title II. Mottershead, D. N.
III. Harrison, S. J.
910′.02 GB54.5
ISBN 0-04-551064-4
ISBN 0-04-551065-2 Pbk

Printed in Great Britain by Butler & Tanner, Frome, Somerset

Preface

The late 1960s were a time of rapid, profound and stimulating development in geography as a whole and in physical geography and the environmental sciences in particular. Hand in hand with the beginnings of the mathematical and methodological revolutions, which accelerated change in both disciplines, went a quest for a more meaningful conceptual framework. The systems approach quickly emerged as the most successful and appealing candidate. It was not a new approach, for from its origin in physics it had already been assimilated by most of the natural sciences and many of the social sciences; indeed, geographers were already toying with its possibilities. However, in the latter half of the decade its time had come, in both philosophical and technological terms. In the environmental sciences it suddenly blossomed and produced both a fervently held and generally applicable philosophy, which abandoned its ivory tower to become public property with the rise of environmentalism, and a methodological blueprint which was harnessed to increasingly sophisticated instrumentation, to the rigour of mathematics and to the power of the computer.

It was at this time that the three of us converged on Portsmouth and found ourselves writing a foundation course in physical geography for the first honours degree in geography to be validated by the British Council for National Academic Awards. Given our backgrounds, our inclinations and the climate of the moment, it was almost preordained that the course should adopt an integrated systems approach. The long road which was to culminate in this book had begun.

It is a book that is aimed primarily at the foundation year student studying geography or environmental science as was the course from which it developed. Nevertheless, it is hoped that it will appeal to a wider readership and that its use will extend into the second year of degree courses and, in time, into secondary schools. The rationale behind the adoption of the systems approach is further expounded in Chapter 1, which also explains the organisation of the book and how best to use it. Essentially, we have attempted to present models of environmental systems at a variety of scales from the planetary to the local landscape and at levels of discrimination appropriate to those scales.

However, one of the difficulties of adopting a systems approach is that the definition of the boundary of any system is necessarily arbitrary. This difficulty is akin to that which always bedevilled traditional regional geography – namely the delimitation of a region. As the subject of this book is so wide it is not surprising that there are a large number of ways in which system boundaries could have been drawn. The divisions adopted are, we believe, valid and useful, though of course they are not the only definitions possible and, as with any attempt at classification, they produce some difficulties (some would say anomalies). In the first place, the systems defined at various scales appear to reinforce the traditional systematic subdivisions of 'physical geography'. This is in fact no bad thing, for it will provide many readers with a familiar framework on which to build. However, the familiarity is often illusory. For example, the treatment of weathering in Chapter 11 is far removed from the usual approach in geography texts. Life on our planet, considered in Chapters 6 and 7 under the headings of 'The biosphere' and 'The ecosphere', will expose readers to a view which normally they will not associate with physical geography. Although the soil is considered as a separate system in Chapter 20, much of what would be included traditionally in this branch of physical geography appears in other chapters as the component elements or processes of other systems. In a similar way, the oceans have been relegated to a subordinate role, since our prime intention is to convey a 'picture' that promotes an understanding of Man's terrestrial environment.

In order to convey this understanding we have attempted to write not simply a conceptual or methodological text, nor to apply a thin but fashionable conceptual veneer to a standard treatment, but to re-assess and re-interpret the subject matter of our disciplines in some detail while retaining an overall unity of approach. This has been attained not only by preserving the integrity of the systems point of view, but also by adopting a consistent level of scientific treatment.

To facilitate this strategy it has been necessary to include, where appropriate, back-up material from the physical and life sciences. Nevertheless we have remained conscious of the range of experience that our potential readership will possess, and have strived, therefore, to reach a compromise between a superficial treatment and the production of a learned but inscrutable treatise. We hope that the outcome will be used and enjoyed at several levels and we will be satisfied if we have managed to convey some of the insight, the excitement and the pleasure that we have derived from this way of looking at our environment.

I. D. White, D. N. Mottershead and S. J. Harrison

Contents

Preface *page* vii
List of tables xiii
List of boxes xv

Part A A systems framework: to have a picture 1

I SYSTEMS, MAN AND ENVIRONMENT

1 Why a systems approach? 6

1.1 Man, environment and geography 6
1.2 Systems 8
1.3 Environmental systems as energy systems 9
1.4 Systems and models 12
1.5 About the book and how to use it 14
Further reading 16

2 Matter, force and energy 17

2.1 The nature of matter 17
2.2 Fundamental forces 24
2.3 Work and energy 28
Further reading 34

Part B A systems model: a partial view 35

II THE PLANET EARTH

3 Energy relationships 40

3.1 The closed system model 40
3.2 The atmospheric system 46
3.3 Earth surface systems 50
3.4 Energy balance of the Earth–atmosphere system 55
3.5 The Earth's interior systems *page* 58
3.6 Energy conversion 63
Further reading 65

III GLOBAL SYSTEMS

4 The atmosphere 68

4.1 Structure of the atmospheric system 68
4.2 The transfer of energy and mass in the atmospheric system 76
4.3 The primary circulation system: approximations to a successful model 83
4.4 Water in the Earth–atmosphere system 90
Further reading 106

5 The lithosphere 107

5.1 The crustal system 108
5.2 System structure 108
5.3 The crustal system: transfer of matter and energy 116
5.4 The implications of the geochemical model 126
Further reading 127

6 The biosphere 128

6.1 The biosphere and ecosphere 128
6.2 The nature of life 129
6.3 Functional organisation and activity of the cell 136
6.4 The cell, the organism and the community 145
Further reading 145

7 The ecosphere 146

7.1 The organism, the population and the community 146
7.2 A functional model of the ecosphere 153

7.3 The ecosphere and the evolution of life *page* 164
Further reading 166

Part C Open system model refined: environmental systems 167

IV ATMOSPHERIC SYSTEMS

8 The atmosphere and the Earth's surface 172

8.1 Introduction 172
8.2 The physical properties of a surface 172
8.3 Surface roughness 177
8.4 Topography 178
8.5 Surface–air interaction 179
Further reading 182

9 Secondary and tertiary circulation systems 183

9.1 Secondary circulation systems 183
9.2 Tertiary circulation systems 196
9.3 Linkages between atmospheric systems 202
Further reading 203

V DENUDATION SYSTEMS

10 The catchment basin system 206

10.1 The structure of the catchment basin system: functional organisation 206
10.2 System structure: spatial organisation 208
10.3 System structure: material components 216
10.4 System function: energy flow 219
10.5 System function: material flows 221
10.6 Special cases of the model 222
Further reading 224

11 The weathering system *page* 225

11.1 A process–response model of the weathering system 226
11.2 The final state of the system: the regolith 242
11.3 The generalisation of the model: other rock types 245
11.4 Exogenous variables and the control of weathering 248
11.5 Conclusion: further perspectives on weathering 248
Further reading 249

12 The slope system 250

12.1 The initial state of the system 250
12.2 The operation of the slope system: the transfer of water 251
12.3 The operation of the slope system: the transfer of minerals 255
12.4 The balance of slope process and slope form 262
12.5 Complexity of hillslope process–form relationship 266
Further reading 266

13 The fluvial system 267

13.1 Energy and mass transfer in channel systems 267
13.2 Channel dynamics: flow of water 271
13.3 Channel dynamics: erosion and transportation processes 274
13.4 Depositional processes 277
13.5 Channel form 277
Further reading 280

14 The glacial system 281

14.1 Systems function 281
14.2 The transfer of mass: glacier flow 285
14.3 Glacier morphology 286
14.4 Erosion processes 290
14.5 Transfer of materials: transportation by glaciers 292
14.6 Deposition processes 294
Further reading 295

15 Spatial variations in denudation systems *page* 296

15.1 Spatial variations in denudation output 296
15.2 Individual factors controlling denudation 299
15.3 Spatial variations in denudation system form 302

VI ECOLOGICAL SYSTEMS

16 The ecosystem 306

16.1 The ecosystem concept 306
16.2 The structural organisation of the ecosystem 307
16.3 Functional activity of the ecosystem: the transfer of energy and matter 318
Further reading 321

17 The primary production system 322

17.1 Functional organisation and activity of the green plant 322
17.2 Ecosystem primary production 331
17.3 Regulation and limits to photosynthesis and primary production 333
17.4 Geographical variation and comparison of ecosystem primary production 346

18 The grazing–predation system 348

18.1 The heterotroph 348
18.2 Modelling ecosystem animal production 348
18.3 Energy flow and population regulation 353

19 The detrital system 355

19.1 Decomposition, weathering and the soil 355
19.2 The input to the detrital system: the supply of organic matter *page* 355
19.3 Decomposition: a process–response model 361
19.4 Decompositión: a trophic model and the pathway of energy flow 364
19.5 Decomposition: a pedological model 366
Further reading 367

20 The soil system 368

20.1 Defining the soil system 368
20.2 The control and regulation of pedogenesis 376
20.3 Formal processes of pedogenesis 379
Further reading 382

Part D Systems and change 383

VII THE NATURAL WORLD

21 Change in environmental systems 386

21.1 Equilibrium concepts and natural systems 386
21.2 Thermodynamics, equilibrium and change 388
21.3 Manifestation of change 390

22 Change in physical systems 393

22.1 Change in climatic systems 393
22.2 Feedback mechanisms 400
22.3 The form of change 401
22.4 Zonal and dynamic change in the denudation system 402

23 Change in living systems 412

23.1 Inherent change in the ecosystem 412
23.2 Cliseral change 426
23.3 Evolutionary change 432

VIII MAN'S IMPACT

24 Man's modification of environmental systems *page* 436

24.1 Deforestation 439
24.2 Agriculture 444
24.3 Urbanisation and industrialisation 453
24.4 Control of environmental systems by Man: some examples 461

Conclusion *page* 471

25 Systems retrospect and prospect 473

Bibliography 476
Acknowledgements 483
Index 487

List of tables

2.1 The number of electrons in electron shells *page* 18
2.2 Electron configuration of the first 50 elements 18
2.3 The Bohr periodic table 19
3.1 Energy sources and energy stores 40
3.2 The absorption of radiation by the principal gases of the atmosphere 47
3.3 Infrared emissivities 53
3.4 Mean latitudinal values of the components of the energy balance equation for the Earth's surface 59
3.5 The major radioactive elements 60
3.6 Spatial variation in heat flow 61
3.7 Relative magnitudes of solar and terrestrial energy 62
4.1 Constituents of dry air 68
4.2 Sources of materials in the atmosphere 70
4.3 Characteristic dimensions of atmospheric motion systems 83
4.4 Characteristics of major global winds 84
4.5 Physical constants of pure water 92
4.6 Cloud types and their main characteristics 101
4.7 Precipitation types and their characteristics 102
4.8 Water balance of the Earth's surface 106
5.1 Structure and composition of the Earth 107
5.2 Elemental composition in percentages of the Earth's crust 108
5.3 Minerals exposed on the land surfaces of the Earth 112
5.4 Rock types exposed on the land surfaces of the Earth 112
5.5 Type of symmetry about mid-ocean ridges 114
5.6 The exploitation of mineral resources 126
6.1 Percentage atomic composition of the biosphere, lithosphere, hydrosphere and atmosphere 129
7.1 Comparison of the various production estimates for the Earth 157
7.2 The evolution of the atmosphere and the fossil record 165
8.1 Albedo of various surfaces 173
8.2 Thermal properties of selected materials 175
8.3 Radiant energy balance above and within a forest 176
8.4 Roughness length of natural surfaces 177
9.1 Middle-latitude depression and tropical hurricane: a comparison of scale 192
9.2 Areas of the world most subject to hurricane activity 193
10.1 River Wallington: network characteristics 210
10.2 Average porosities for different materials *page* 216
10.3 Strength values for different rock types 219
11.1 A classification of weathering processes 226
11.2 Levels of supersaturation of various salts in water 232
11.3 Volume increase during hydration of selected salts 232
11.4 (a) Characteristics of granite at different grades of weathering; (b) Material properties of granite at different weathering grades 244
11.5 Geochemical types of rock weathering related to climatic conditions 248
11.6 Some characteristics of the principal hydrological zones 249
12.1 Ratio of creep : wash in different environments 263
12.2 The nine-unit land surface model: slope components and associated processes 265
13.1 Estimated velocities along different flow routes in the catchment 267
14.1 Classification of ice masses based on their relationship with relief 286
15.1 Estimates of denudation losses from the continents 296
15.2 Denudation losses for the year 1975 from adjacent catchments in a humid temperate region, North York Moors, England 298
15.3 Denudation losses from small sample areas 298
15.4 Estimates of erosion rates for limestone areas under different climatic conditions 299
15.5 Lithology and sediment yield 300
15.6 Morphometric properties of drainage basins on Dartmoor and adjacent areas 302
16.1 Taxonomic classification 307
16.2 Cover/abundance scales 315
17.1 The characteristics of plants with high and low production capacities 335
17.2 The relationship between certain structural characters of plants and water loss by transpiration 339
17.3 Mineral nutrition 340–1
17.4 Distribution of all major nutrient elements in store in three forest types 344
17.5 Net primary production of the Earth 346
19.1 Variations in annual litter production under different vegetation types 356
19.2 Sequence of utilisation of substrate and waves of decomposers in the breakdown of plant litter 363

20.1 The oxygen and carbon dioxide composition as percentage by volume of the gas phase of well aerated soils, compared with that of dry air *page* 371
22.1 Representative rates of relative land and sea level changes 405
22.2 Zones of intensity of glacial erosion 408
23.1 Types of interspecific population interaction 418
24.1 Nutrient budgets for catchments at Hubbard Brook 441
24.2 Life forms found in chalk grassland, acid grassland and neutral grassland 449
24.3 The effect of an urban area on local climates 454
24.4 Contamination of the urban atmosphere compared to that over rural areas *page* 454
24.5 Radiative properties of typical urban materials and areas 455
24.6 Major flood events over the first six months of 1978 462
24.7 Adjustments to the flood hazard 463
24.8 Methods of frost protection 467
24.9 Minimum soil temperature at a depth of 2.5 cm with and without black plastic mulch at Thorsby, Alabama 468

List of boxes

2.1 Electron configuration and the periodic table *page* 19
2.2 Ions 20
2.3 The states of matter 22–3
2.4 Heat and temperature 30
2.5 Chemical reactions 30–1
2.6 Free energy and entropy change in chemical reactions 32
3.1 Electromagnetic radiation 41
3.2 Solar energy 41
3.3 Radiation laws 42
3.4 Lambert's cosine law 44
3.5 Absorptivity and emissivity 53
3.6 Conduction and convection 56
3.7 Energy balances 57
3.8 Formation and structure of the Earth 59
4.1 The hydrostatic equation 72
4.2 Vertical pressure changes in the atmosphere: I 73
4.3 Vertical pressure changes in the atmosphere: II 74
4.4 Coriolis deflection 79
4.5 The geostrophic wind 80
4.6 Angular momentum 87
4.7 The dishpan experiment 89
4.8 Heat and momentum transfer in wave motion 91
4.9 The water molecule 91
4.10 Water vapour in the atmosphere 92
4.11 Adiabatic temperature changes 99
4.12 Methods of measuring evaporation and potential evapotranspiration based upon the surface water balance equation 105
5.1 Silicate mineral structures 109
5.2 Major rock types *page* 110
5.3 The geological timescale 112
5.4 Isostasy 119
6.1 Carbon chemistry 131
6.2 Organic macromolecules 133
6.3 Information 134
6.4 ATP–ADP energy transfer system 137
6.5 Enzymes 139
6.6 DNA replication and transcription 141
6.7 Chlorophyll 144
7.1 Nitrogen fixation 161
8.1 Heat parameters 174
9.1 Changes associated with the passage of warm and cold fronts 187
10.1 Drainage basin morphometry 211
10.2 Shear strength 217–18
10.3 Comparison of catchment water balances 223
11.1 Hydrolysis of silicate minerals 236
11.2 Acidity of natural waters 238
11.3 Clay minerals 240
12.1 Water potential 251
12.2 Soil water potential 252
12.3 Fundamental forces acting on the hillslope 258
13.1 River channel variables 271
13.2 Hydraulic radius 272
13.3 Flow in open channels 273
14.1 Glacier ice 284
16.1 The ecosystem concept 307
16.2 The species concept 310
17.1 Carbon dioxide compensation point 329
17.2 Plant water potential 338
18.1 Warm-blooded and cold-blooded animals 352
19.1 Soil organisms 358–9
23.1 Phytochrome P_r and P_{fr} 412

Part A

A systems framework: to have a picture

The available information would suggest that the appearance of Neanderthal man was dominated by an ugly and repulsive strangeness enhanced by his low forehead, protruding brows, ape neck, and accentuated by his low stature and extreme hairiness.

William Golding chose to take such unpromising beings as the central characters of his book, *The inheritors*. Their dialogue, however, is restricted by the intellectual capacity of the extended family to which they belong. They think in pictures; pictures sometimes linked, sometimes isolated . . .

> 'At last Mal . . . began to straighten himself by bearing down on the thorn bush and making his hands walk over each other up the stick. He looked at the water then at each of the people in turn, and they waited.
> "I have a picture."
> He freed a hand and put it flat on his head as if confining the images which flickered there . . . His eyes deep in their hollows turned to the people imploring them to share a picture with him . . .
> Fa put her hand flat on top of her head.
> "I have a picture."
> She scrambled out of the overhang and pointed back towards the forest and the sea.
> "I am by the sea and I have a picture. This is a picture of a picture. I am . . ." She screwed up her eyes and scowled ". . . thinking."'

Though the similarities between us and Neanderthal man are slight, thinking is still, as for Mal and Fa, the seeing of pictures, the ability to construct and manipulate mental images. In part these are pictures assembled from empirical facts, observed with our two eyes – seeing is believing. However, the kind of pictures with which William Golding's primitive heroes struggle, and the kind of pictures *we* wish to share with *you*, are seen with another eye . . . the mind's eye. To see with the mind's eye is a much more complex process, for here our pictures may be painted not only with observed visual fact but with words, symbols, ideas, logic and inference, and even with feeling. They may be pictures of pictures, perhaps a collage of different images, some based on our own observation and understanding, some the pictures others have seen and shared with us.

In this book we are going to turn the mind's eye on the world about us, on the natural world at first – Man's natural environment. Images of this environment and of the landscapes through which it is perceived abound in geography; indeed, to many they *are* geography. However, it is one particular image that we have in our mind's eye – an image of our planet and the landscapes we inhabit at its surface, as ordered and functioning systems. This image is not new, but it has become increasingly significant as a way of looking at the world, and it pervades not only geography but much of the rest of science. It is by no means the only valid image of our environment, but it has a strength which has enabled it to 'have illuminated thought, clarified objectives, and cut through the theoretical and technical undergrowth in the third quarter of the twentieth century in a most striking manner' (Bennett & Chorley 1978).

So, like Mal, we have a picture and through the images conveyed by this book we hope to share that picture with you, so that you may say with Fa, 'I have a picture . . . it is a picture of a picture'.

I SYSTEMS, MAN AND ENVIRONMENT

1 Why a systems approach?

1.1 Man, environment and geography

Man is an animal. We are similar in anatomy and physiology to our mammalian cousins and, like all other animals, we must eat food and breathe air to survive. Nevertheless, we *are* different, but this difference lies in accomplishment and behaviour. We possess not only a genetic inheritance which determines our animal nature, but also a cultural inheritance of knowledge and custom transmitted by language and symbols. As with other animals, however, we occupy a habitat which exists in both space and time and forms our environment. Although this is a physical, chemical and biological environment, it is also a cultural environment with social, political, economic and technological dimensions.

This twofold distinction is reflected in a dichotomy in the study of Man–environment relationships, with the natural sciences concerned with the natural environment, and the social sciences and applied sciences concerned with the cultural environment. Geography – the study of Man's interaction with his environment – reflects this dichotomy. In reality these two components of Man's environment are inseparable; the natural environment cannot be fully understood in isolation from Man and his interactions with it. But before Man's cultural relationships with his natural environment can be considered in anything but a superficial way, we *must* understand that environment and, for this reason, this book is concerned mainly with the natural environment – its physical, chemical and biological components, the relationships between these components and those between them and Man.

How should we begin this formidable task? We could start by taking the environment to pieces to see what it is composed of and how it works, much as we might dismantle a piece of machinery. This process can be called the *analysis* of the environment. It is one of the traditional approaches of science, whereby different sciences (or different branches of a science) separate different types of pieces and study them in isolation. Thus, oceanography is concerned with the sea, meteorology with the air, hydrology with water, geology with rocks, and biology with organisms; physics and chemistry take in all of these things but in terms of smaller, more fundamental pieces.

The problem with this approach becomes apparent when trying to put the environment back together again: when turning from analysis to *synthesis*. The selected pieces have been looked at in different ways, with different degrees of rigour, and different terminologies have been developed, so we may no longer recognise the pieces as being part of the original whole. The process of reassembling the pieces becomes more difficult because we can no longer remember how they fitted together, or we failed to find out before we dismantled the environment. Also, some of the pieces may have been overlooked in the process of analysis, because none of the specialist sciences regarded them as their concern. The real problem, therefore, is the lack of any true integration. What are the links between a kingfisher, a developing thunder cloud, a breaking wave and a landslide? Much may be known about each of these pieces, but how do they fit together into the environmental jigsaw puzzle? You might ask whether it really matters. Is it necessary to synthesise, to reassemble the pieces, as long as our knowledge of them is pursued in sufficient depth? These questions can perhaps best be answered visually. Figure 1.1a portrays a section of the environment – a landscape in the Scottish Highlands. This is the scale at which Man's environment is normally observed and most of the components which make up the familiar concept of the natural environment can be seen. There are mountains and hills, valleys, streams, a lake (in fact, a sea loch), plants and animals, and the sky.

The view in Figure 1.1b is from the top of one of

Figure 1.1 Different perspectives on our environment. (a) A sea loch on the west coast of Scotland. (b) The same landscape type seen from the summit of an adjacent mountain, (c) from a satellite and finally (d) the planet Earth.

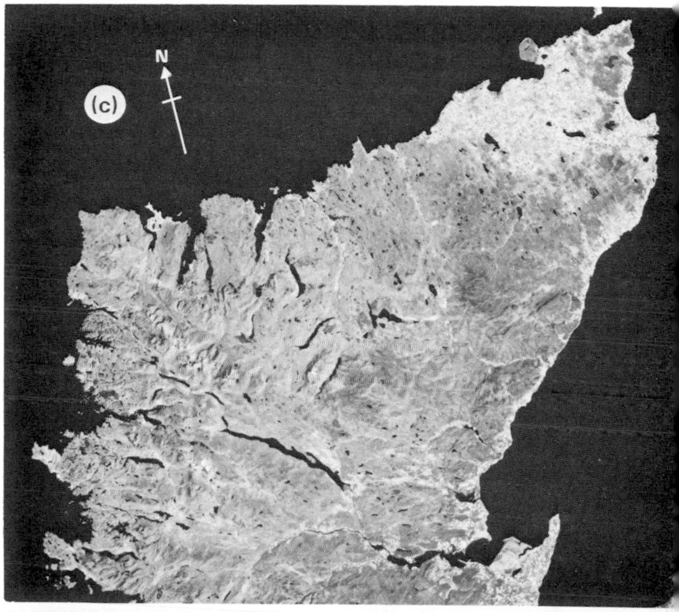

(b)

(c)

N

(d)

the mountains. The perspective has not really altered very much. Most of the components are still visible, though some are now less clear. The distinction between plants and animals is perhaps more blurred at this distance, but some relationships become more obvious. The spatial relationships between the mountain streams, the river, the loch and the sea can be seen more clearly.

The perspective has altered enormously in Figure 1.1c. The view here is the one obtained at an altitude of 435 km. Nevertheless, the Highland environment is still visible, but discernible at an entirely different scale. All the components are there; the cows are still grazing by the loch shore. However, they are no longer perceptible as separate entities. It is only the major topographic features that can be distinguished: mountain ridges, valleys, the major drainage network and the sea. Living things are present, but they cannot be discerned clearly.

Finally, in Figure 1.1d the perspective, at an altitude of 1110 km, has changed so much that only the planet Earth is discernible, though the land, the sea and the atmosphere are still recognisable. All perception of the individual components of the Highland landscape visible in Figure 1.1a has been lost. Here is visual confirmation that the various components of this original environment, and indeed of the environment as a whole, cannot really be viewed in isolation. There is an integrated unity about the natural environment, which is evident in Figure 1.1d as the planet Earth – what Kenneth Boulding (1966) has called 'Spaceship Earth'. It is obvious then that to understand the Earth, its environments and the components of these environments, as well as how they function, we must study the interrelationships of all of these components – a synthesis is required. However, in the late 20th century there is a less philosophical, more pressing and practical reason for such a synthesis.

Our finite natural environment represents not only living space and resources for Man and his animal needs but, because Man is a cultural animal, it must also provide for our developing cultural needs. As our cultural complexity, social organisation and technological capacity have increased through time, we have made progressively greater demands on the environment. Also, as cultural development has made us progressively less dependent on our immediate environment, so the nature of our perception of it, and of our interactions with it, have changed. We have exploited, modified and damaged or destroyed it to an extent not equalled by any other animal. In the 20th century our demands have begun to outstrip the capacity of the environment, and it is our ability to manage intelligently, protect and conserve the environment that has become paramount. It is also the recognition of this need that has precipitated the increased awareness of, and support for, conservation policies at all levels. However, before we can manage our environment we must understand it. It is this urgent need more than any philosophical consideration that has given added impetus to the integrated study of Man's natural environment. More than this, it has emphasised the fact that the separation of the natural from the cultural is not only unreal, but also dangerous. Our interaction with the natural environment can only be understood with reference to our perception of it and behavioural response to it, both of which are conditioned by our complex cultural environment.

1.2 Systems

There are, then, at least two reasons for needing to know more about the physical, chemical and biological environment. As the trustees and potential managers of that environment, and certainly as users of its resources, we need to know how it works. A fragmented and piecemeal approach is at best unsatisfactory and does not impart such understanding, but to create a truly comprehensive and integrated picture – our 'picture of a picture' – is a colossal task. What is needed is some sort of model or framework – a plan, a map, a wiring diagram – to show how to dismantle the environment for analysis and, perhaps more important, to show how to reassemble it so that the results of analysis can be incorporated in an integrated synthesis. Such a framework is provided by the concept of the **system**.

The word 'system' may be very familiar. You, the readers, are probably involved in some part of an education system. The room in which you are sitting probably has a heating system, perhaps even an air conditioning system. Depending on the time of day, you might be about to close this book and go off to satisfy your digestive system. This might involve travelling home via a public transport system, or, if you have a car or motorcycle you will start it by using the ignition system. What does the word 'system' mean? Why is it used to refer to such dissimilar things? In the first instance, all of these things consist of collections of other things – sets of objects. They may be schools, colleges, universities, classrooms, lecture theatres; or mouth, oesophagus, stomach, colon and rectum; or boiler, pipes, radiators and thermostats; or rails, roads, trains, buses, stations and termini. In all cases, from education to transport systems, we are referring to sets of objects. However, the word 'system' is more than just a collective noun, for it also tells us that the objects are organised in some way, that there are connections and links between the units. So the roads or rails pass through the stations which are arranged along them between the termini, pipes connect the boiler to the radiators, and so on.

The term transport system, however, does not tell us whether we travel in a bus or train, let alone whether it is painted red or green. The term heating system does not specify whether it is gas- or oil-fired, nor in fact that it has pipes and radiators; it could be ducted hot air. In other words, the term system is non-specific and the concept can be applied, in these examples, to all such arrangements of units which have a transporting or heating function. The term implies *generality*.

Here another feature of the use of the word 'system' becomes apparent. A system actually functions in some way as a whole. A transport system is not merely a static arrangement of units; it transports people and goods. A heating system circulates hot water or air and warms the room. A common feature of the working of systems is the transfer of some 'material' between the units of the system. This material may be obvious, like the water in the pipes and radiators or the food in the digestive system, or it may be more abstract. In an education system the material may be thought of as a generation of children passing upwards from class to class, or from school to college. Alternatively, one might equally identify the material not as the children but as the transfer of ideas, the flow of knowledge. A political system is concerned not only with the institutions of that system and the transfer of such 'real' materials as people or votes or tax revenues, but also with the transfer of more intangible materials such as directives, decisions, ideas and ideologies.

Such movements of materials require some motivation. In some systems this motivation is again obvious, such as the heat applied to the central heating boiler, or the electricity supply to a hi-fi system. In others it is less obvious: the impetus of the forces of supply and demand on people and goods, for example, or the desire for the acquisition of knowledge and understanding in an education system. In this last example one might be less idealistic and think of the motivation as being provided by the need to obtain qualifications in order to gain employment. But the point remains the same: in all systems there is a driving force that makes them work.

The common characteristics of systems are summarised below.

(a) All systems have some structure or organisation.
(b) They are all to some extent generalisations, abstractions or idealisations of the real world.
(c) They all function in some way.
(d) There are, therefore, functional as well as structural relationships between the units.
(e) Function implies the flow and transfer of some material.
(f) Function requires the presence of some driving force, or source of energy.
(g) All systems show some degree of integration.

1.3 Environmental systems as energy systems

There is nothing new in the concept of the system used in the broad way encountered so far. Indeed, the antiquity and widespread use of the concept reflect the ability of the human mind to perceive things as a whole, and there is no reason to assume that this ability is the preserve of late 20th-century Man. Although all systems have certain characteristics in common – of which organisation, generalisation and integration are perhaps the most important – this book will be concerned with systems of a particular kind. These are **thermodynamic systems**, and their original description and formalisation took place in physics.

Literally, thermodynamics means 'the study of heat as it does work', but this narrow definition is rather misleading. Perhaps thermodynamics could more aptly be termed **energetics**, as it is now concerned not only with heat but also with all other forms of energy. So thermodynamic systems might more profitably be called **energy systems**. Such an energy system is merely a *defined system of matter, the energy content of that system of matter, and the exchange of energy between that system and its surroundings.* This defined system of matter may be a leaf, a rock-forming mineral, a tree, a length of river channel, a slope segment or a parcel of air. It is merely that part of the physical universe whose properties are under investigation. Although such definitions appear vague, they emphasise two important facts: the first is the importance attached to defining the system; the second is that the definition of the system is somewhat arbitrary. Thus, an energy system is confined to a definite place in space by the **boundary** of the system, whether this is natural and real like a cell wall or watershed, or arbitrary, though still real, like the walls of a test-tube or vessel in a laboratory, or arbitrary and intangible like the boundary of a cloud. The boundary of a system separates it from the rest of the universe, which is known as the **surroundings**.

Within its defined boundary the system has three kinds of properties. The **elements** of the system are the kinds of substances composing the system. They may be atoms or molecules, or larger bodies of matter – sand grains, raindrops, grass plants, rabbits – but each is a unit which exists in both space and time. Each element has a set of **attributes** or **states**. These elements and their attributes may be perceived by the senses, or made perceptible by measurement or experiment. In the case of such measurable attributes as number, size, pressure, volume, temperature, colour or age, a numerical **value** can be assigned by direct or indirect comparison with a standard. Between two or more elements, or two or more states or attributes, there are **relationships** which serve to define the states of aggregation of the elements, or the **organisation of the system**.

The **state of the system** is defined when each of its properties (**variables**), i.e. elements, attributes and relationships, has a definite value. These definitions apply to all energy systems, but several distinct types of system can be distinguished on the basis of the behaviour of the system boundary (Fig. 1.2).

In an **isolated system** there is no interaction with the surroundings across the boundary. Such systems are encountered only in the laboratory, but they are important in the development of thermodynamic concepts.

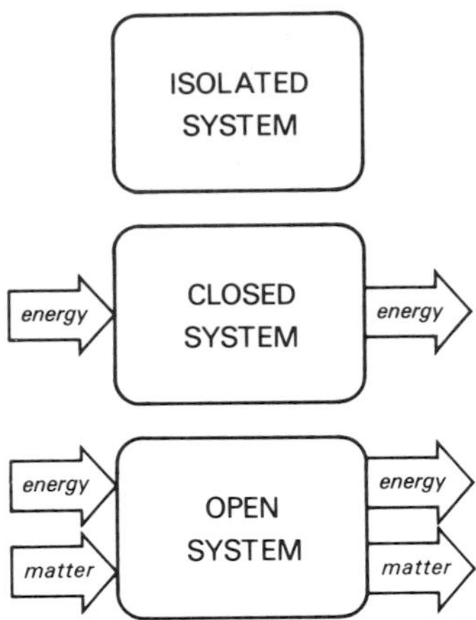

Figure 1.2 Isolated, closed, and open energy systems.

Closed systems are closed with respect to matter, but energy may be transferred between the system and its surroundings. On the Earth closed systems are rare. It is important, however, to be able to analyse complex environmental systems in terms of simpler component systems and it is often useful to be able to treat these as closed systems. They are also important because much of thermodynamic theory was developed in relation to closed systems.

Open systems are those in which both matter and energy can cross the boundary of the system and be exchanged with the surroundings. In such systems the transfer of matter itself represents the transfer of energy, as matter possesses energy by virtue of its organisation (potential chemical energy, for example). All environmental systems are open systems and are characterised by the maintenance of structure in the face of continued throughputs of both matter and energy.

It is not only static systems, in which only the attributes and structural relationships of the elements are defined, that are of interest in the natural environment. Of far greater significance are dynamic functioning systems. Changes in the system are, therefore, important. These can best be thought of as *changes in the state of the system*. Theoretically, the change in the state is completely defined when the *initial* and *final states* of the system are specified. In practice, however, the *pathway* of change of state is often of interest, and to specify such a pathway it is necessary to know not only the initial and final states but also the sequence of *intermediate states* in order. The way in which a change in state is effected is called a **process**.

The maintenance of structural organisation in the face of throughputs of matter and energy is a critical characteristic of almost all environmental open systems. A drainage system maintains the organisation of the stream and river channel network and of the contributing slopes in spite of the continuous throughput of water. Living organisms, including Man, are inconceivable without the maintenance (within very narrow limits) of the extremely complex structural and functional organisation of their bodies, though a regular throughput of food materials and energy exists. In other words, such systems must maintain a more or less stable state defined in terms of their elements, attributes and relationships through time. These are **equilibrium states** and they will be considered in more detail in Chapter 21, but the characteristic equilibrium of an open system is termed a **steady state** and to maintain it the system must possess the capacity for **self-regulation**. In terms of the definition of process above, this means that the net effect of the operation of processes must be to return the system to the initial state.

Self-regulation in environmental systems is effected by **negative-feedback** mechanisms or, as they are called in the life sciences, **homeostatic** mechanisms. These are able to damp down change and as such are control mechanisms. They operate in much the same way as a thermostat in a heating system, in that they are able to sense or interpret the output or effect of some operation, just as the thermostat senses temperature change. They then feed back this information to influence the operation of the process concerned, in the same way as the thermostat switches on or shuts down the source of heat. The number, type and degree of sophistication of negative-feedback mechanisms in environmental systems vary considerably. Many physical systems can accommodate quite wide fluctuations about a mean equilibrium state without jeopardising their function. The detailed properties of a length of river channel may vary widely about the most effective state but it will still transfer water and debris. However, self-regulation in living systems has to be far more precise and sophisticated. The temperature of your body, for example, is regulated by a host of interrelated feedback mechanisms, but if it is allowed to fluctuate widely you feel decidedly ill and if control is not established you die. On death your body ceases to function as a living system and the steady state of the system degenerates.

Nevertheless, natural systems do exhibit change. Indeed, one of their most important properties is the tendency to evolve, or develop, such ordered change in systems state through time. This is the case with the extension of an ice sheet, the development of a cyclone or the growth of an organism. These are not random events but instances of directional change and they are also regulated by feedback mechanisms. Here they do not stabilise the system, but have a cumulative effect and they reinforce particular directions of change. They are termed **positive-feedback** mechanisms. In practice the regulation of natural systems states involves the linking of several feedback mechanisms, some positive and some negative, in complex **feedback loops** (Fig. 1.3).

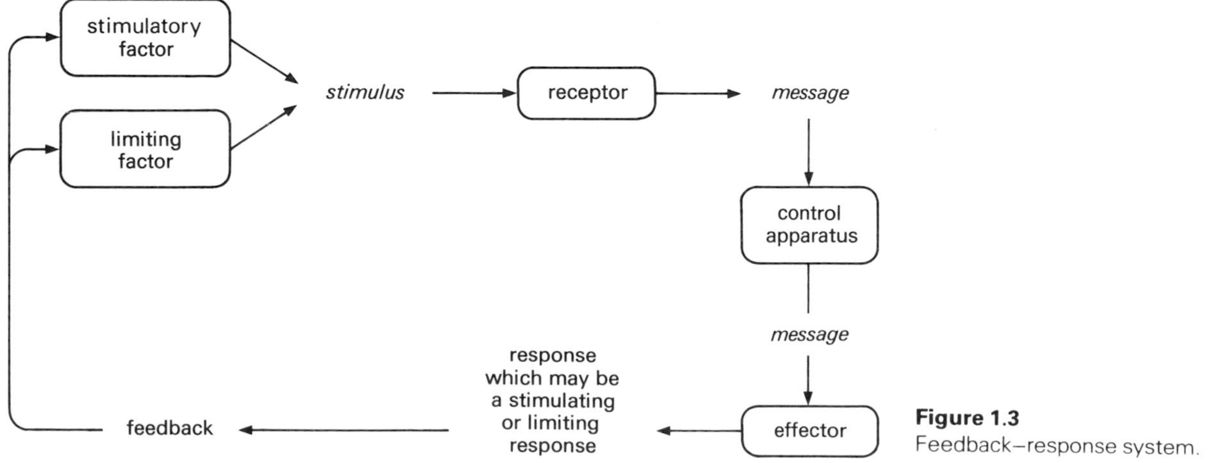

Figure 1.3
Feedback–response system.

Where negative feedback predominates, the overall effect is the maintenance of a steady state. Where positive feedback is dominant and leads to an increase in the order and complexity of the system state, then we speak of the growth or development of the system. Where positive feedback has the cumulative effect of progressively destroying the organisation of the system, it leads to a retrogressive and often irreversible change in state.

Associated with positive feedback is the existence of **thresholds**. These are state variables which, when they assume certain values, are capable of initiating often sudden and sometimes dramatic changes of state. For example, it is possible to define threshold slope angles which, if exceeded, trigger rapid mass-movement processes that result in the complete reorganisation of the state of the slope system. In a similar way, in a soil undergoing decalcification by leaching, the moment the calcium content reaches zero the system will be transformed in state and the chemical environment for soil processes will change rapidly. Such thresholds are really only extreme cases, because in all feedback processes certain state variables play similar key roles in controlling the operation of processes and, in this more general case, we refer to them as **regulators**.

1.4 Systems and models

Much of the discussion of energy systems has been concerned with abstract theoretical systems and with fundamental definitions. The rest of this book will be concerned with systems in the real world. The use of the word 'real' is important, for although these systems may be submitted to precise thermodynamic definition and their functional similarity stressed, they are none the less real. The precipitation input to the denudation system is still the rain you can feel on your face. The throughput of water in the channel system is still the river in which you may swim. The energy store in the biomass of the forest is still the trees you might have climbed as a child. The problem is that the real world is extremely complex and, although the recognition that it consists of thermodynamic or energy systems helps to structure our approach to this complexity, of itself it does nothing to simplify it. The specification of the system, its elements, states, relationships and processes, is in the last analysis the specification of the real world in all its complexity. The dilemma is that of being caught between perceiving the system as a whole and the near impossibility of seeing its complete complexity, a dilemma well expressed by Arnold Schultz (1969) when he wrote '. . . the sheer bulk of information is so overwhelming we can make little use of it. At the other end all the states (of the system) are fused into one grand platitudinous expression and all you can say for it is "There it is".' Clearly there is a need to simplify and here the answer to the dilemma lies in the concept of the **model** and the techniques of **modelling**.

1.4.1 Models

The word 'model' is used in everyday speech with three distinct meanings: 'a replica', 'an ideal' and 'to display'. The concept of the model as adopted here combines aspects of all three meanings. In order to simplify environmental systems, models or replicas of them can be constructed. To be useful, these models must idealise the system and they must display or make clear its structure and how it works.

1.4.2 Systems as models

The concept of the system is itself a model. When referring to central heating systems, or transport systems, or thermodynamic systems, what is being presented is an idealised view that is generally applicable to all real situations having the same general character. The London Underground and the nineteenth century canal network of Britain, for example, are clearly different and unique entities. To regard both merely as transport systems is to strip away all that is special (but irrelevant) and to idealise their structures and relationships in a generally applicable model – the transport system. In the same way, objects as dissimilar as a bunsen-burner flame and a plant or animal cell can both be modelled as open thermodynamic systems with inputs and outputs of matter and energy. When one begins to regard the natural environment in terms of systems, the process of generalisation and idealisation has started.

1.4.3 Models of systems

Wishing to simplify the complexity of the real world involves making models of the system itself. Let us for the moment return to upland Britain as portrayed in Figure 1.1a. Everything visible consists of atoms and their combination in molecules and compounds. These in turn are present as rock-forming minerals, as the water in the loch, as the organic compounds of plants and animals, as the water vapour in the clouds, and so on. Nevertheless, the molecules and compounds cannot be discerned at this scale, but the physical objects of the rocks outcropping in the distance, the grass and the cows, the water, the soil, the sky and clouds. Even these components become lost in the broader perception of the land, water, vegetation and atmosphere as the perspective changed, until in the final picture from space all of this matter was perceived simply as the planet Earth. What was happening, apart from changing the perspective, was that the parts of the system were being resolved at different levels until finally there was no resolution of them at all.

The atoms, molecules, cows and rocks, however, are all elements and their states in this system. The third component of a system is the relationship between the elements. So a model of a system needs to incorporate relationships. A topographic map is a model of the terrain it represents. It distinguishes the elements of that terrain by means of symbols for height, rivers, forests, towns, roads and its power of resolution of these elements varies with scale from, for example, a 1:25000 sheet map to a school atlas map. However, a map goes further, for it models not just the elements but also their geographical relationships – the distance and direction between any two elements. All modelling can in fact be viewed as a mapping process. The relationships in the system and hence in the model can be of several types: spatial distance, causation, conjunction, succession of events. They can be expressed in words such as 'cow eats grass', or 'river takes soil' through statistical statements of the probability of such events, and through quantitative measurements of, for example, transfers of matter and energy between the elements.

As with the topographic map, some models are static, representing the structure of the system rather than the processes going on within it, and between it and its surroundings. However, to understand the functioning of systems we need dynamic models that are capable of identifying processes and modelling their effect on the system. This is particularly the case if the model is to be used to predict the behaviour of the system, for example, a drainage basin model used to predict flood hazard.

In the sections that follow, models of environmental systems are what can be called **homomorphic models** (*homo* is Greek for similar) in that they are imperfect representations of reality. Rarely if at all are the models **isomorphic** (*iso* is Greek for *same* or *equal*) in the sense that for every element, state, relationship and process there is a corresponding component in the model. The level of resolution of the models will vary with scale, the system being discriminated at some level appropriate to the scale. At the fine end some realism is retained, and at the coarse end there is a gain in generality. However, no matter what level of discrimination is adopted, because the model cannot be truly isomorphic it will be lumping together elements and relationships in single compartments of the model and a wealth of information inside each compartment is conveniently ignored. This is what Egler (1964) called the 'meat grinder' approach and it introduces another term for these homomorphic models, viz. **compartment models**.

Each compartment of these models is treated as a **black box**, which can be defined as any unit whose function may be evaluated without specifying the contents. At low resolution the models of environmental systems contain a small number of relatively large black box compartments; indeed the whole system may be treated as a black box. As the level of discrimination increases, so these compartments are progressively split into subcompartments which are in turn treated as black boxes. At intermediate levels of discrimination the model of the whole system has become a partial view of the system, its structure, relationships and processes, and is called a **grey box** model. Finally, as realism increases towards a truly isomorphic model, the grey box becomes a **white box** with most of the elements, states, relationships and processes of the system identified and incorporated in the model (Fig. 1.4). Even here some black box compartments will remain, for no model is a complete representation of reality.

This approach to modelling has a number of advantages. It provides a hierarchy of models of different levels of discrimination and complexity, which are appropriate to different scales of analysis of the system. It also allows the knowledge and results of specialist work on any part of the system to be co-ordinated and coupled into the model at the appropriate level in the hierarchy.

These compartment models may be static or dynamic and they may vary from box and arrow (flow chart) models to quantitative mathematical models at one end, to scaled hardware models at the other. None is necessarily better than another. However, the most important criterion in assessing the validity of a model is its ability to predict the behaviour of the system. In this context it is only dynamic models that have predictive power. Of these it is only some hardware and mathematical models that represent quantitative relationships and predict the behaviour of the system in quantitative terms. Static and non-predictive dynamic models, including those used in this book, are nevertheless very valuable aids to communication and understanding.

1.5 About the book and how to use it

The layout of this book reflects the views of the environment that we obtained in Figure 1.1 and the hierarchy of models alluded to above, each with a different level of discrimination and complexity. In Part B we begin with a picture of the

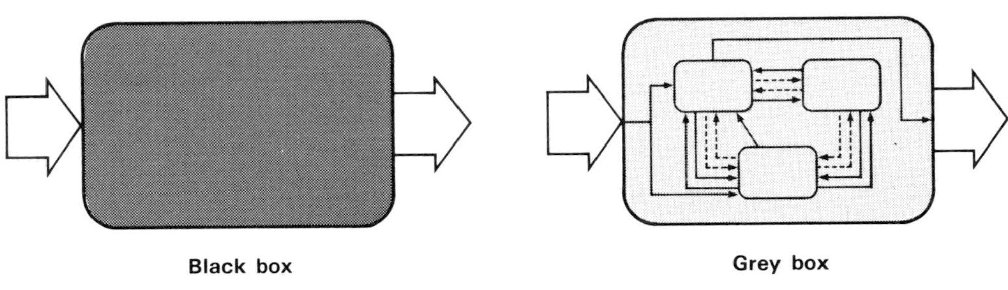

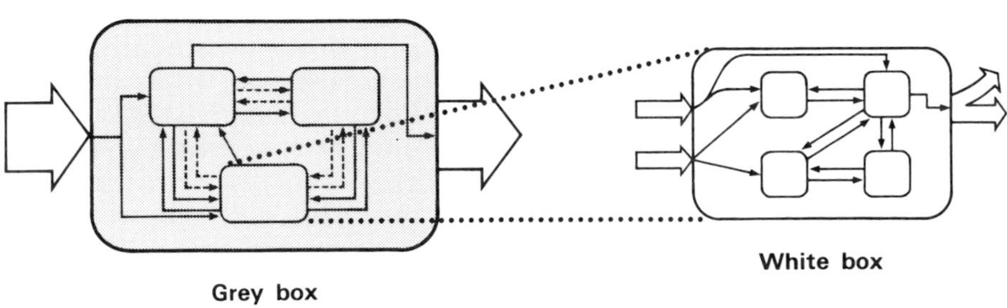

Figure 1.4 Levels of resolution in the modelling of systems.

whole planet (cf. Fig. 1.4) and regard it at first as a closed system. By adopting a black box approach, Section II concentrates on the transfer of energy across the boundary of the system. Before long, however, it becomes evident that we need to 'open up' the black box and examine, at least in a broad way, the energy exchange between subsystems of the model in order to relate energy input and output by means of an initial appreciation of throughput in the system. In Section III the closed system model is abandoned and a partial view of the major subsystems of the Earth is presented where each is modelled as an open system involving transfers of mass as well as energy. For the most part, the models have a high degree of generalisation and the perspective of Section III is a global one.

With Part C not only does the focus of the book change but the spatial scale of the systems narrows down to that of Figure 1.1a & b; in other words, it is the scale of the landscape that predominates. In particular, the catchment basin and the ecosystem are functional models applicable at a scale which we all experience and they are concerned with interlinked geographical units having 'both spatial magnitude and location' (Chorley & Kennedy 1971). The resolution of the models and the level of detail involved increase considerably over that employed in Section II and represent the white box, approach or near-isomorphic model. Although the chapters of Sections IV, V and VI are concerned with dynamic systems, their time-bound development is not explicitly stressed. Instead the equilibrium relationships and the steady state characteristics of these environmental open systems are considered under the prevailing sets of external conditions of existence. Nevertheless, the models presented retain a considerable element of generality in spite of some inclusion of specific spatial or geographical variation.

The developmental or evolutionary tendency of these systems is considered in Part D. Change in the system states is viewed as both an internal readjustment to reach some new equilibrium and as a response to changes in input, both of which are natural characteristics of environmental systems. In Section VIII we also consider the interaction of Man with the environment, in terms of disturbance by inadvertent human intervention and of deliberate and purposeful regulation of natural systems so that some perceived benefit accrues. Here we meet the wider application of a systems approach, recognising that the environmental systems discussed in earlier sections are only part of larger *geographical systems* which also have a socio-economic dimension.

Treating the world about us as a large number of thermodynamic or energy systems, capable of definition and susceptible to modelling at a variety of scales, implies that we shall be concerned with systems of matter, their energy content and the transfer of energy and mass, both within the system and between it and other systems. To do this at anything but a superficial level requires an understanding of the nature of matter and the relationships between the fundamental units of matter and the macroscopic objects such as rocks, trees and rivers that we perceive as the elements of environmental systems. We need to understand what energy is and to be aware of the laws that govern its transfer. We need some fundamental scientific awareness of the nature of things. Some readers will already possess this awareness and platform of knowledge on which to build. For those who do not, or for whom the platform is a bit shaky, Chapter 2 is included to cover matter, force and energy. It is *not* intended to be a condensed basic course in science. It is not comprehensive and the level of treatment varies from topic to topic in relation to its significance in the context of this book. Therefore the reader may treat Chapter 2 in two ways: to be read in sequence or regarded as a reference chapter.

Throughout the book another device is used to convey information, elaborating specific points and providing background material or scientific definition and derivation. This involves boxes which separate such information from the main flow of the narrative but make it available at the point in the text where it is most pertinent. The strength of this approach is that it should allow readers to follow the main argument without being confused by tangential detail. It should also allow readers to choose the level of treatment they require, for in many cases the boxed material can be omitted on first reading if desired.

Earlier in this chapter it was stressed that there is an integrated unity about the natural environment and indeed it is the fervent hope of the authors that, taken as a whole, the book conveys that unity. Nevertheless, the organisation of the text into systems models of varying complexity, applicable at a variety of scales to the major subsystems of our environment, allows a number of possible

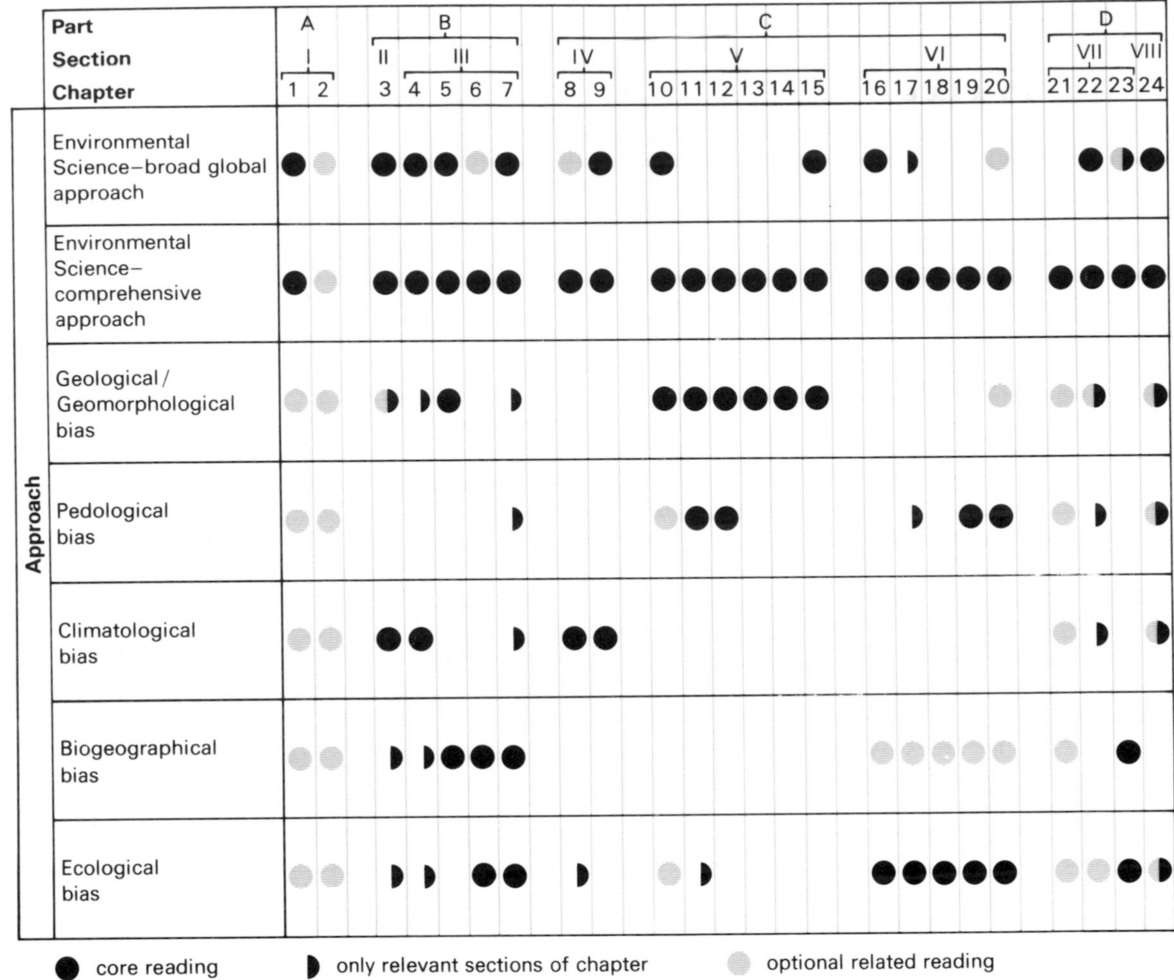

Figure 1.5 Possible pathways through the book.

routes to be taken. This permits readers, particularly lecturers or students intending to use the book as a course text, to devise the chapter sequence which best suits their requirements in terms of subject matter and level. Two broad treatments are possible. First, the holistic approach can be preserved but the level varied. Secondly, particular biases in systematic treatment can be accommodated by following specific groups of systems which accord with, for example, a geomorphological or ecological bias through to an appropriate level. Some suggestions as to possible sequences are given in Figure 1.5, but it is hoped that the majority of users will wish to preserve a breadth of view that will enable them to see 'a picture of a picture' that in its scope and insight is truly comprehensive.

Further reading

Beishon, J. and G. Peters 1972. *Systems behaviour*. London: Harper & Row.

Checkland, P. B. 1971. A systems map of the universe. *J. Systems Engng* **2**, 2.

Churchman, C. W. 1968. *The systems approach*. New York: Delacorte Press.

Emery, F. E. 1969. *Systems thinking*. London: Penguin.

2 Matter, force and energy

2.1 The nature of matter

In Chapter 1 an energy system was formally defined as a specified system of matter, the energy content of that system and the exchange of energy between the system and its surroundings. In this chapter we shall explore in more detail the nature of both the matter and energy involved in such systems.

The matter in environmental energy systems can be recognised as the 'real' physical objects of that environment: rocks, soil, water, plants, animals and the gases of the atmosphere. Here, however, matter will be discussed at a fundamental level, for it is only by so doing that the structural unity of these apparently diverse entities can be appreciated.

2.1.1 *The structure of the atom*

Atomic theory is the key to the current view of the nature of matter. According to this view, matter is composed of very small particles (10^{-7}–10^{-10} m in diameter) – **atoms** – which have a complex internal structure. They consist of a positively charged central **nucleus** around which negatively charged **electrons** orbit (Fig. 2.1). The nucleus has a diameter approximately 10^{-5} that of the whole atom and in this small volume nearly all of the mass of the atom is concentrated. The nucleus is composed of two different types of particle of almost identical mass: the **protons** are positively charged and the **neutrons** carry no charge, but they are each normally present in approximately equal numbers.

The nucleus of an atom contains a definite number of protons – its **atomic number** – and this defines the charge on the nucleus and ultimately the chemical nature of the atom. The protons and neutrons jointly constitute the nuclear mass, so that the number of protons plus the number of neutrons gives the **mass number** of an atom. A **chemical element**, which is the *basic unit of matter*, is defined as a substance in which all of the atoms have the *same atomic number* and which cannot be decomposed by chemical reaction into substances of simpler composition. Therefore, although all atoms of an element have the same number of protons and hence the same atomic number and nuclear charge, their masses may not be equal. This is because the number of neutrons present in the nuclei of the atoms of an element may vary.

Groups of atoms which differ in their mass numbers are called **isotopes**. Hydrogen, for

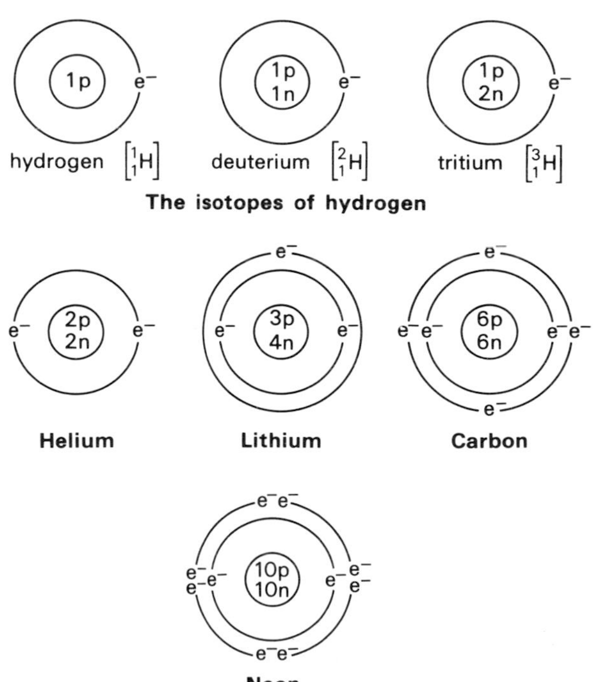

Figure 2.1 The structures of the hydrogen, helium, lithium, carbon and neon atoms.

example, exists as three isotopes with 0, 1 or 2 neutrons in the nucleus respectively, viz:

$$^{1}_{1}H,\ ^{2}_{1}H,\ ^{3}_{1}H.$$

Here the symbol for the element is expanded by a superscript which shows the mass number (protons + neutrons) and a subscript which indicates the atomic number (protons). Most of the chemical elements consist of mixtures of isotopes. Oxygen has three isotopes $^{16}_{8}O$, $^{17}_{8}O$ and $^{18}_{8}O$, as does carbon $^{12}_{6}C$, $^{13}_{6}C$, $^{14}_{6}C$. The separation of isotopes is difficult, for their chemical properties are virtually identical as they have the same nuclear charge and the same number of electrons per neutral atom and, therefore, they react in the same way. However, the nuclei of some isotopes may be unstable and liable to disintegrate spontaneously.

The mass of an electron is only 5.45×10^{-4} that of a proton, but it has a charge which is equal, though opposite in sign, to that of a proton. In the electrically neutral atom, therefore, the nucleus attracts a number of negatively charged electrons just equal to the positive nuclear charge, which is in turn determined by the number of protons (atomic number). In the helium atom, for example,

$$2 \text{ protons} + 2 \text{ electrons} = \text{no charge}$$

$$(2+) \quad + \quad (2-) \quad = 0$$

The electrons of an atom occupy certain orbits, or **electron shells**, and there is a definite limit to the number of electrons each shell can contain (Table 2.1). For example, lithium has its third electron placed in a second shell that can contain a maximum of eight electrons (see Fig. 2.1). All electron shells beyond the first can be subdivided. The second shell has *s* and *p* subshells, the third has *s*, *p* and *d* subshells, and the fourth, fifth and sixth have *s*, *p*, *d* and *f* subshells. The distribution of the electrons of an atom between these shells and their subshells is known as the **electron configuration** of the atom (Table 2.2 and Box 2.1). It is the outermost shell – sometimes called the **valence shell** –

Table 2.1 The number of electrons in electron shells.

shell	1	2	3	4	5	6
electrons (max.)	2	8	8	18	18	32

Table 2.2 Electron configuration of the first 50 elements.

	Shell:	1	2		3			4				5		
	Subshell:	1s	2s	2p	3s	3p	3d	4s	4p	4d	4f	5s	5p	5d
Element	Atomic no.													
H	1	1												
He	2	2												
Li	3	2	1											
Be	4	2	2											
B	5	2	2	1										
C	6	2	2	2										
N	7	2	2	3										
O	8	2	2	4										
F	9	2	2	5										
Ne	10	2	2	6										
Na	11	neon core of 10 electrons			1									
Mg	12				2									
Al	13				2	1								
Si	14				2	2								
P	15				2	3								
S	16				2	4								
Cl	17				2	5								
A	18				2	6								
K	19	argon core of 18 electrons						1						
Ca	20							2						
Sc	21						1	2						
Ti	22						2	2						
V	23						3	2						
Cr	24						5	1						
Mn	25						5	2						
Fe	26						6	2						
Co	27						7	2						
Ni	28						8	2						
Cu	29						10	1						
Zn	30						10	2						
Ga	31						10	2	1					
Ge	32						10	2	2					
As	33						10	2	3					
Se	34						10	2	4					
Br	35						10	2	5					
Kr	36						10	2	6					
Rb	37	krypton core of 36 electrons										1		
Sr	38											2		
Y	39									1		2		
Zr	40									2		2		
Cb	41									4		1		
Mo	42									5		1		
Tc	43									6		1		
Ru	44									7		1		
Rh	45									8		1		
Pd	46									10				
Ag	47									10		1		
Cd	48									10		2		
In	49									10		2	1	
Sn	50									10		2	2	

Box 2.1

ELECTRON CONFIGURATION AND THE PERIODIC TABLE

In 1869, the Russian chemist Mendeleev devised the **periodic table** in which he arranged the elements by atomic weight (relative atomic mass) in such a way that the similar properties of certain groups of elements became apparent. In modern versions of the periodic table (*Bohr periodic table*, see Table 2.3) the elements are in order of their atomic numbers. The vertical columns – **groups** – contain elements which, though similar, vary in their properties down the columns. For example, *Group IV* elements become more metallic in character. The horizontal rows – **periods** – contain series of elements, the chemical properties of which change in discrete steps. Some elements occupy intermediate positions between *Groups II* and *III*. The most important of these in the natural environment belong to the series known as the **transition elements**.

It is now known that this arrangement reflects the electron configuration of the atoms of the elements. In *Groups I, II, III, IV, V, VI, VII,* and *O*, as the atomic number (number of protons) increases along the periods so the electron shells are progressively filled, from the inner shell outwards. For example, in the second period, carbon, nitrogen, and oxygen appear in sequence. All have 2 electrons filling the first shell ($1s^2$) but carbon has 4 ($2s^2\ 2p^2$), nitrogen 5 ($2s^2\ 2p^3$) and oxygen 6 ($2s^2\ 2p^4$) electrons in the second shell. The last element in each period (i.e. *Group O* elements) has no vacancies in its outer shell or subshell. For example, neon in the second period has a full complement of 8 electrons ($2s^2\ 2p^6$) in the second shell.

The elements of all of the groups discussed above are termed **typical elements** and have in common the fact that the *s* and *p* subshells are filled progressively. The remaining elements, the *transition elements*, and those of the *lanthanide series* (the rare earths) and the *actinide series* (including the transuranium elements) differ in that they have incomplete inner shells. They form series where electrons are added to the *d* (transition elements) and *f* (lanthanide and actinide series) subshells of the penultimate shell (see Table 2.2).

Table 2.3 The Bohr periodic table.

Group	I	II		rare earths (lanthanides) / actinides	transition elements									III	IV	V	VI	VII	0
1st period				1 H															2 He
2nd period	3 Li	4 Be												5 B	6 C	7 N	8 O	9 F	10 Ne
3rd period	11 Na	12 Mg												13 Al	14 Si	15 P	16 S	17 Cl	18 Ar
4th period	19 K	20 Ca	21 Sc		22 Ti	23 V	24 Cr	25 Mn	26 Fe	27 Co	28 Ni	29 Cu	30 Zn	31 Ga	32 Ge	33 As	34 Se	35 Br	36 Kr
5th period	37 Rb	38 Sr	39 Y		40 Zr	41 Nb	42 Mo	43 Tc	44 Ru	45 Rh	46 Pd	47 Ag	48 Cd	49 In	50 Sn	51 Sb	52 Te	53 I	54 Xe
6th period	55 Cs	56 Ba	57 La	58 *Ce* 59 *Pr* 60 *Nd* 61 *Pm* 62 *Sm* 63 *Eu* 64 *Gd* 65 *Tb* 66 *Dy* 67 *Ho* 68 *Er* 69 *Tm* 70 *Yb* 71 *Lu*	72 Hf	73 Ta	74 W	75 Re	76 Os	77 Ir	78 Pt	79 Au	80 Hg	81 Tl	82 Pb	83 Bi	84 Po	85 At	86 Rn
7th period	87 Fr	88 Ra	89 Ac	90 *Th* 91 *Pa* 92 *U* 93 *Np* 94 *Pu* 95 *Am* 96 *Cm* 97 *Bk* 98 *Cf* 99 *Es* 100 *Fm* 101 *Md* 102 *No* 103 *Lw*	104	105													

actinides; transition elements

that determines the chemical behaviour of an element (with the exception of some elements where an inner subshell is also involved – see Box 2.1) and the number of electrons in the valence shell is important in relation to the type of chemical reaction with which an element may be concerned.

Some elements have their valence shells fully occupied. These are all inert gases (the **noble gases**) and the first five – helium, neon, argon, krypton and xenon – all occur in the atmosphere and make up about 1% (by volume) of air. Their inertness would suggest that they have a very *stable* electron configuration. Elements that do not possess this stable, inert gas configuration may be expected to show a tendency to attain it. One way in which this can be accomplished is for an element to gain or lose electrons so that its valence shell is full and corresponds to that of one of the inert gases. However, electrons added to or removed from a neutral atom give to it a net negative or positive charge respectively. These charged particles are known as **ions** – negatively charged **anions** and positively charged **cations** (Box 2.2).

2.1.2 Molecules and compounds

About 92 elements have been recognised in nature, most of which occur in combination with one another. Even elements such as oxygen and carbon, which occur naturally uncombined with other elements, do not do so as individual atoms. A particle of this type, made up of two or more atoms of either the same or different elements, is called a **molecule**. Atoms are held together to form molecules by the sharing of electrons between adjacent atoms in **covalent** and **metallic** bonding, and by the **transfer** of electrons in **ionic** bonding. Molecules showing charge separation (polar molecules) may form **hydrogen bonds** because of the resultant electrostatic attraction.

Box 2.2

IONS

The elements of *Group I* of the periodic table (see Table 2.3) have the inert gas structure plus an extra electron. For example, sodium ($1s^2\ 2s^2\ 2p^6\ 3s$) has the same configuration as neon ($1s^2\ 2s^2\ 2p^6$) plus one electron in the *s* subshell of the third shell. If it were to *lose* this extra electron it would attain the inert gas configuration. When this happens in a chemical reaction a *positively* charged **ion** – **cation** – is formed.

$$Na - e^- \rightarrow Na^+$$

In a similar way, a *Group II* element, such as magnesium ($1s^2\ 2s^2\ 2p^6\ 3s^2$), has to *lose* two electrons from the valence shell to attain the configuration of neon.

$$Mg - 2e^- \rightarrow Mg^{2+}$$

Group VII elements, on the other hand, are one electron short of inert gas configurations and so must *gain* an electron. For chlorine ($1s^2\ 2s^2\ 2p^6\ 3s^2\ 3p^5$) to reach the configuration of argon ($1s^2\ 2s^2\ 2p^6\ 3s^2\ 3p^6$) it requires the addition of an electron which forms a *negatively* charged ion – **anion**.

$$Cl + e^- \rightarrow Cl^-$$

In *Group VI*, oxygen ($1s^2\ 2s^2\ 2p^4$) requires two electrons to reach the electron configuration of neon ($1s^2\ 2s^2\ 2p^6$).

$$O + 2e^- \rightarrow O^{2-}$$

The charge carried by an ion, which reflects the number of electrons gained or lost by an element in forming it, is known as the **electrovalency** of the element. In the above examples, therefore, sodium has an electrovalency of +1, while in oxygen the electrovalency is −2. However, some elements form ions that may only approximate to the inert gas configuration. This is especially common among the ions of the transition elements and the rare earths, although it is also true of some elements of *Groups V* and *VI*. These elements may, therefore, form more than one type of ion with different electrovalencies (or oxidation states) that represent different approximations to the appropriate inert gas configuration. For example, iron, the fourth most abundant element in the lithosphere, occurs in two electrovalency states – *iron II*, or *ferrous* (Fe^{2+}), and *iron III*, or *ferric* (Fe^{3+}).

As with the formation of ions, these bonding mechanisms can be viewed as strategies by which the constituent atoms jointly attain stable, inert gas electron configurations in the compound. In ionic (**electrovalent**) compounds the bonding mechanism arises from *electron transfer*, effectively forming oppositely charged ions that are held together by electrostatic attraction. However, in the compound so formed each ion has the electron configuration of an inert gas. For example, in sodium chloride (common salt) (see Boxes 2.1 & 2 for explanation of notation).

Na	+	Cl	$\rightarrow$	Na^+Cl^-	
$1s\,2s^2\,2p^6\,3s$		$1s\,2s^2\,2p^6\,3s^2\,3p^5$		$1s\,2s^2\,2p^6$ neon configuration	$1s\,2s^2\,2p^6\,3s^2\,3p^6$ argon configuration

However, the transfer of electrons in this way cannot be involved in the formations of many compounds. Gaseous oxygen exists in the Earth's atmosphere as molecules (O_2) each containing two atoms of oxygen that cannot reach the stable configuration of an inert gas by electron transfer. However, they can attain such a configuration by *sharing electrons* in pairs, each atom contributing one electron to the pair and each shared pair constituting a covalent bond.

$$\ddot{\underset{..}{.O.}} + \ddot{\underset{..}{.O.}} \rightarrow \ddot{\underset{..}{O}} :: \ddot{\underset{..}{O}}$$

$$O + O \rightarrow O_2$$

(. represents an electron in the valence shell)

The pairs of electrons between the two atoms are counted towards the electron configuration of both atoms, both having the electron structure of the inert gas neon. Furthermore, the position of the electron pairs is localised and it fixes the positions of the atoms quite rigidly.

When two atoms of hydrogen and one of oxygen combine to form water (H_2O), they do so by strong electron pair or covalent bond. The resultant water molecule is stable and, as a whole, electrically neutral. The electrons, however, have an asymmetrical distribution within the molecule which behaves as a **molecular dipole**, or **polar molecule**, due to the **charge separation**. This partial ionic character of the O–H bond in water lends the hydrogen atom some positive character, permitting electrons from another atom to approach closely to the proton, even though the proton is already bonded. This allows a second, weaker, link to be formed – the hydrogen bond (Fig. 2.2).

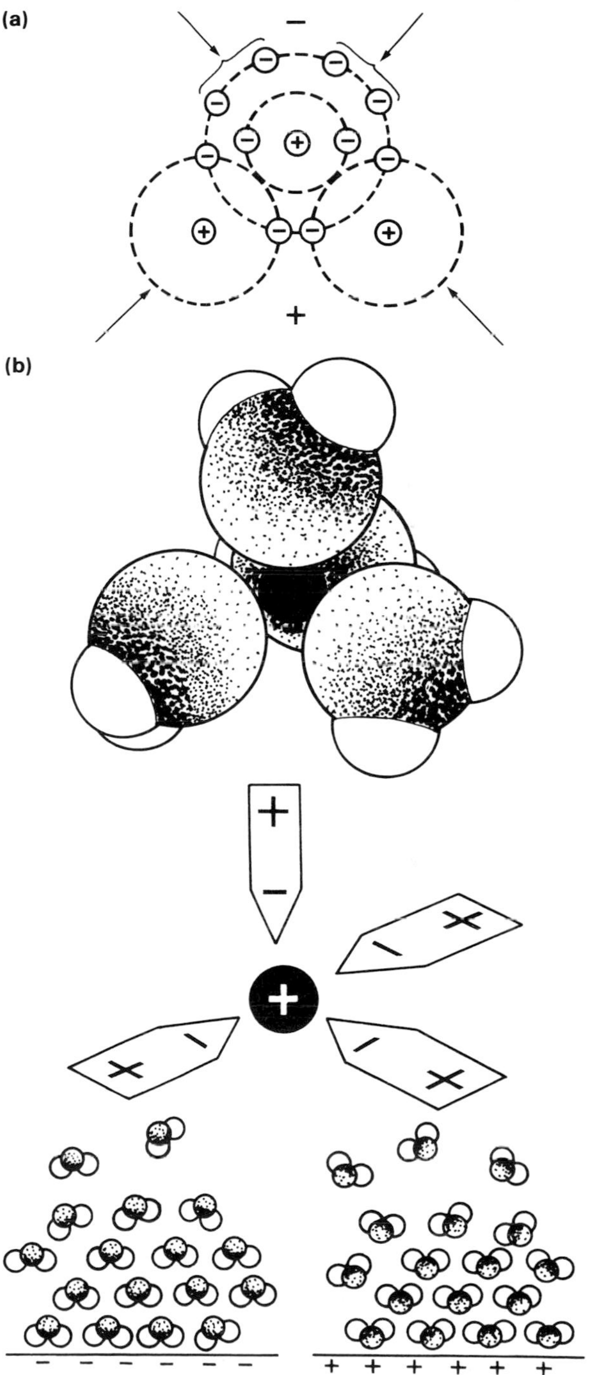

Figure 2.2 The dipolar water molecule: (a) charge separation; (b) the formation of hydration shells.

Hydrogen bonds are found between the atoms of only a few elements, the most common being those in which hydrogen connects two atoms from the group consisting of fluorine, oxygen, nitrogen and, less commonly, chlorine. Hydrogen bonds are always associated with charge separation, but as well as being intermolecular bonds they can also occur between the atoms within a molecule, as in many important organic compounds such as proteins. In fact, hydrogen bonds play a crucial role in biological systems (see Ch. 6).

Because of the presence of two unshared pairs of electrons and two protons, the water molecule can form up to four hydrogen bonds with other molecules. These bonds are arranged in the form of a tetrahedron, as in the regular crystal structure of ice (see Box 14.1). In liquid water this arrangement becomes more irregular and the hydrogen bonds are reduced as the water temperature rises. Hydrogen bonds not only form between water molecules but also between them and other charged particles (ions) or surfaces (soil colloids, for example – see Ch. 20). The polar water molecules orient themselves around such charged particles and surfaces in layers, the thickness of which depends on the intensity of charge. These layers are called **hydration shells** (Fig. 2.2).

In covalent bonding the positions of the electron

Box 2.3

THE STATES OF MATTER

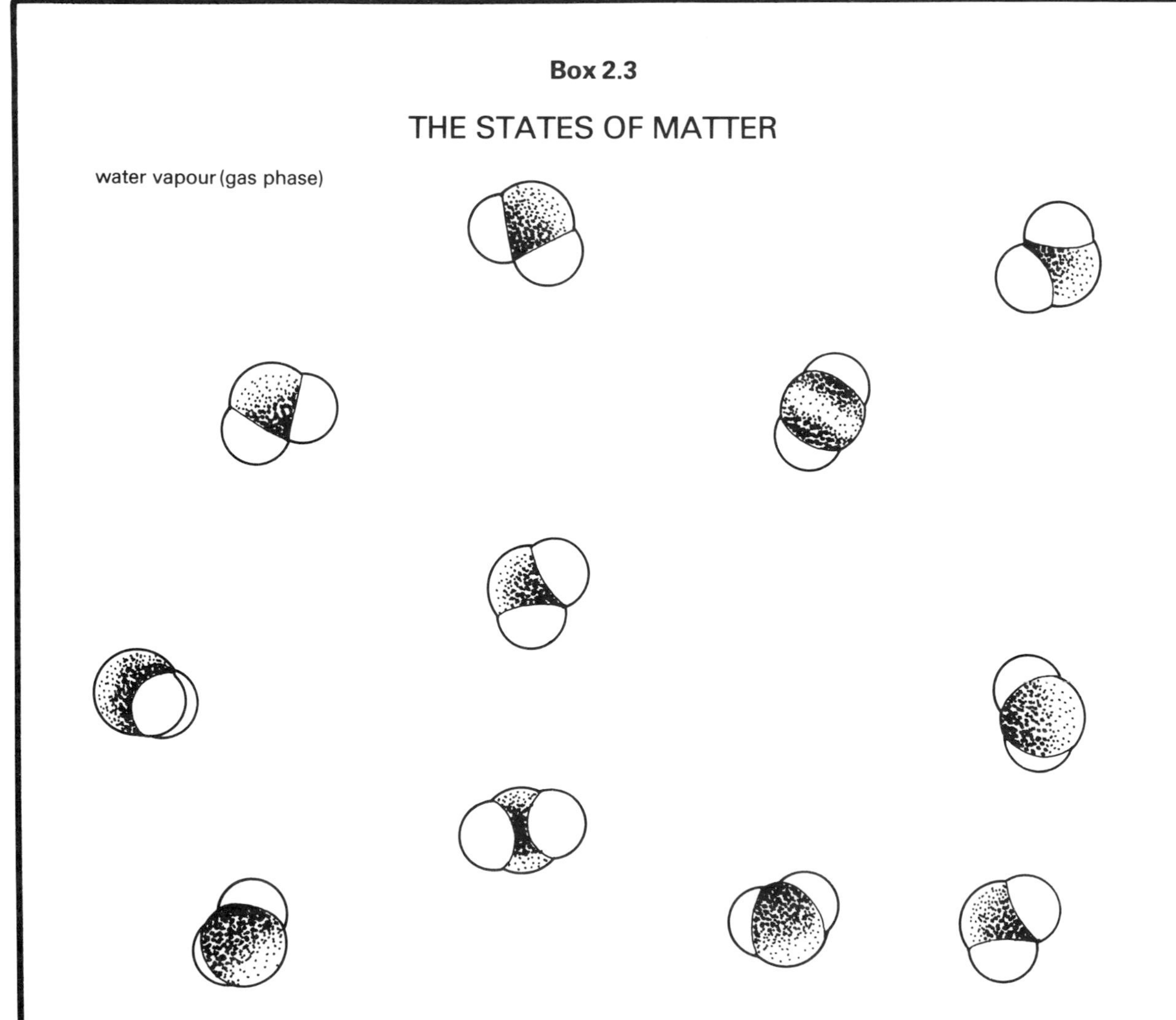

pairs are localised, but in metallic bonding each atom contributes one or more electrons to a 'sea' of electrons which tends to move relatively freely throughout the aggregate of atoms. This 'sea' of negatively charged electrons holds the atoms together, although their relative positions are not as rigidly defined as in covalent bonding. It also accounts for many of the properties of metals, such as their malleability and their behaviour as conductors of an electric current.

Matter exists in three states (phases) – gas, liquid (together called fluid states) and solid – which are distinguished by the relative motion of the molecules or other units of matter (Box 2.3). In fact, most naturally occurring gases do consist of molecules, but some may be ionised: that is, the molecules may have gained or lost electrons from their component atoms and carry a net charge. Many liquids and all solutions contain ions, while in solids the units may be individual atoms, ions, molecules or arbitrary groups of atoms.

In the gaseous and liquid states, the covalent bonding of the molecules is the only bonding mechanism of significance, but in solids all three types are responsible for the orderly arrangement of atoms. The configuration of crystalline arrangement is known as the lattice structure of the crystal. The topic of crystal structure will be

In the gaseous state the molecules are in constant motion and are continually in collision with each other. This movement is sufficient to overcome the main forces of intermolecular attraction which would lead to cohesion between the molecules. A gas can expand to fill the volume available to it and, conversely, it is susceptible to compression because the volume of the molecules is normally small compared with the total volume of the gas.

liquid water (liquid phase)

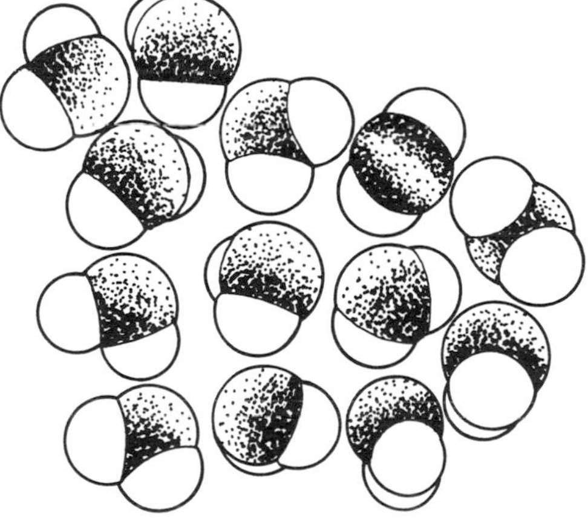

In a liquid the molecules are in motion, but not sufficiently for them to get away from each other. They remain attracted to each other, moving past and round each other in a fluid manner. Liquids differ from gases in that their volumes normally remain constant. Their shape, however, is determined by the shape of the 'container', while they characteristically retain a free upper surface.

ice (solid phase)

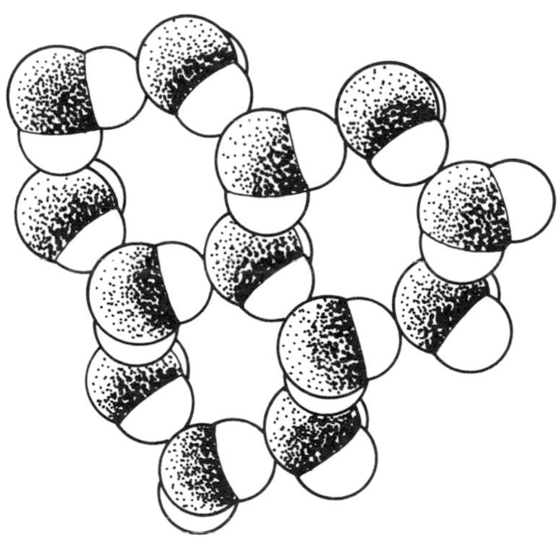

The molecules in a solid, although capable of vibration, are no longer moving as in a liquid, but take up fixed positions in relation to each other. Solids maintain a particular external shape, resisting change in shape and volume.

considered again, in relation to rock-forming minerals, in Chapter 5.

The view of the nature of matter which emerges from this discussion implies two important points. It is immaterial whether an atom, a quartz crystal or a tree is under discussion; the nature of a substance depends first on the kinds of particles of which it is composed and secondly on their arrangement or organisation. All matter can be considered as a more or less orderly arrangement of particles at various scales from the sub-atomic to the supra-molecular. The nature of these particles and the kind of organisation they possess determine the properties and behaviour of that unit of matter.

2.2 Fundamental forces

The fact that matter exists as orderly arrangements of particles at a variety of scales suggests that something is responsible for holding the particles together and maintaining such organisation. There must be some force of attraction between the particles. There are four, apparently distinct, categories of natural forces. Two of these classes are relatively familiar and they account for all forces normally experienced. When particles of matter carry opposite electrical charges they exert an attraction on each other. Such a force of *attraction* is known as an **electrostatic force** and it can be shown to be proportional to the charges carried by the particles and inversely proportional to the square of the distance separating their centres:

$$F_e \propto Q_1 Q_2 L^{-2}$$

$$\text{or} \quad F_e = E Q_1 Q_2 L^{-2}$$

where F_e = electrostatic force, Q_1,Q_2 = magnitude of the charges on the two particles, L = length (distance), and E = proportionality constant.

However, when the charges carried by particles are both of the same type or sign, the electrostatic force is one of *repulsion*, not attraction. This attractive or repulsive force is called **electrostatic** because the charges on the particles can be considered at rest. However, similar attractive and repulsive forces exist between current-carrying conductors and, in this case, because the electric charge is in motion relative to an observer, the forces are termed **electrodynamic**. Both sets of forces are collectively referred to as **electromagnetic forces**. As we have seen, the forces holding an atom together (those binding atoms into molecules, and molecules into liquids and solids) and the contact forces affecting macroscopic bodies are all electromagnetic.

The second familiar force is **gravitational attraction**. Every particle of matter exerts this force on every other particle and it is directly proportional to the masses of the particles and inversely proportional to the square of the distances separating them:

$$F_g \propto M_1 M_2 L^{-2}$$

$$\text{or} \quad F_g = G M_1 M_2 L^{-2}$$

where F_g = force (gravitational), M_1,M_2 = masses of the two bodies, L = length (i.e. distance) separating them, and G = proportionality constant. This force acts as if the masses of the two bodies were concentrated at their centres, and of course is most familiar as the gravitational attraction of the Earth, which we call **gravity**.

The remaining two categories of force are experienced only at the subatomic level. It will be recalled that the protons in the nucleus of the atom are positively charged. We would therefore expect that they would be driven apart by electrostatic repulsion. It is feasible that they are held firmly together in the nucleus by the gravitational attraction of their masses. In fact, it can be shown that the electrostatic repulsive force is very much greater (10^{36} times as great approximately) than the gravitational attraction. Clearly some new force appears to be involved. This is the **nuclear** (or **strong**) **force** and it is some hundred times stronger than the electrostatic repulsion, although its effect is limited to a short range. The fourth category, the **weak force**, need not concern us in the context of this book except in so far as it is responsible for radioactive decay.

2.2.1 *Defining force*

So far we have considered force as a quantity acting on particles of matter, but we have not defined it. Before doing so, however, it is important to note that force is a **vector** quantity; that is, a force has not only *magnitude* but also *direction*. If we return to gravitational force, considered above, and imagine a particle of matter released in a situation where it would have no significant gravitational interaction with any other body, it would fall towards the centre of the Earth. The force therefore operates in a specific direction. The distance the particle falls in a unit of time defines its **velocity**:

$$\text{velocity} = \text{distance per unit time}$$

$$v = LT^{-1} = \text{m s}^{-1}$$

where v = velocity, L = length (distance) and T = time. The product of its mass and its velocity defines the **momentum** of the particle:

$$\text{momentum} = \text{mass} \times \text{velocity}$$

$$p = MLT^{-1} = \text{kg m s}^{-1}$$

where p = momentum, M = mass, L = length (distance) and T = time. Because the force of gravitational attraction is inversely proportional to the square of the distance separating two bodies – in this case the particle of matter and the Earth – the smaller the distance becomes, the greater is the force of attraction between them, and the velocity of fall progressively increases. This rate of change of velocity resulting from the application of a force is called the **acceleration** of the particle, in this earthly example the *acceleration due to gravity* (9.81 m s^{-2}):

$$\text{acceleration} = \text{velocity change per unit time}$$

$$a = LT^{-1}T^{-1}$$

$$= LT^{-2} = \text{m s}^{-2}.$$

The momentum of the particle, however, is a function of its velocity and, as the velocity changes, so too will the momentum. As the rate of change of velocity is defined by the acceleration of the particle, so the rate of change of its momentum will be the product of the mass of the particle and its acceleration. This rate of change of momentum is used to define the magnitude of the force, in this case the gravitational force:

$$\text{rate of change of momentum} = \text{mass} \times \text{acceleration}$$
$$= \text{force}$$

so,

$$\text{force} = \text{mass} \times \text{acceleration}$$

$$F_g = MLT^{-2} = \text{kg m s}^{-2} = \text{N (Newton)}.$$

On Earth this is called the **force of gravity**, and the force of the Earth's attraction for an object is called the **weight** of the object. On the Moon, however, the same object would have a different weight because the acceleration would now be the acceleration due to the gravitational attraction of the Moon, not the Earth. However, the *mass* of the object would remain constant.

2.2.2 *Force and motion*

From this initial consideration of force it should be apparent that the principal effect of a force is to accelerate bodies of matter. If such a body is at rest, it will remain at rest unless acted on by a force. If the body is already in motion, it will continue to move in a straight line at constant velocity unless it is acted on by a force. These assumptions together represent a statement of **Newton's First Law of Motion**. In 1687, Isaac Newton proposed three basic laws defining the relationships between forces and the motions that forces produce. These laws describe our observations of motion and they also predict such motions. They became the basis of **Newtonian mechanics**, a system of mechanics applicable at velocities that are considerably less than the velocity of light.

In the natural environment all objects are being acted on by forces all the time, the same fundamental forces as are discussed in Section 2.2. If these forces are not balanced, the motion of the body is changed in a particular way. The effect of a given force on the motion depends on the mass of that body, for this determines the body's resistance to a change in its motion. If you think of applying your muscle power to two boulders of different mass, this would appear to be self evident. Furthermore, the effect of the force you apply is to accelerate the boulders from rest and, as long as their mass does not change, for each boulder the acceleration you achieve is proportional to the force you apply and in the same direction as the force. This is the recognition of **Newton's Second Law**, which relates the force acting on the mass of a body to the change in velocity, or the acceleration of the body. Stated formally the second law reads: the acceleration of a body is directly proportional to the **resultant force** acting on it and inversely proportional to the mass of the body:

$$a \propto F/m \qquad \text{(see definition of force, Sec. 2.2.1)}$$

The term resultant force is used because under most real situations several component forces will be acting on a body, each of different magnitude and acting in a different direction or sense. It is therefore the net effect of these components – the resultant force – that appears to be responsible for the acceleration of the body.

Newton's Third Law is derived from the everyday experience that, when a force is applied (as for example pushing against a boulder) the boulder appears to be acting back with an equal force. Your subjective impression in this situation is that action is met by reaction. Furthermore, action appears to equal reaction, for if you push harder the reaction from the boulder will be greater. This assumption is Newton's Third Law and it is stated formally as: when one body exerts a force on another, the second body exerts an equal and opposite force on the first – *action equals reaction*.

One of the consequences of Newton's Third Law of motion concerns the effect of action–reaction forces on particles of matter involved in collisions. This is a common occurrence during environmental processes when either one or both of two bodies are in motion. During any small time interval in the collision, the faster body (x) will be acting on the slower body (y) with a force which tends to accelerate it. Over the same short time interval the slower (or stationary) body (y) will be reacting with a force that tends to slow down the faster-moving body (x), so the effect is to increase the velocity of y and decrease the velocity of x. As force is the product of the mass of the body and its acceleration (change in velocity), we can define the force of x acting on y ($_xF_y$) as

$$_xF_y = m_y\, a_y$$

and the force of y acting on x as

$$_yF_x = m_x\, a_x.$$

But Newton's Third Law tells us that these forces are equal so,

$$m_y\, a_y = m_x\, a_x.$$

However, the product ma is the rate of change of momentum, so because the forces, though equal, are opposite in sense (i.e. y has speeded up and x slowed down) the momentum gained by y equals that lost by x. Therefore, the total momentum of x and y has been conserved through the whole collision sequence. This is the **Law of Conservation of Momentum** – in this particular case of linear momentum (Fig. 2.3).

We have been concerned so far mainly with forces acting in what a physicist would term a static frame of reference. However, the Earth and its atmosphere rotate and therefore it is often necessary to be able to account for the behaviour of objects located in a rotating frame. For example, to an observer an athlete throwing the hammer clearly has to pull, or exert an inward force, on the wire as he rotates in the throwing circle to keep the hammer head moving in a circular path. This is the **centripetal force** and is defined as

$$\text{centripetal force} = mv^2/r$$

where r is the radius of the circle. However, the athlete himself, who would be in a rotating frame of reference, experiences the mass of the hammer head trying to accelerate outwards in a radial direction as if acted on by some invisible force. This force is the **centrifugal force** and so the athlete applies a force through the hammer wire just sufficient for the radial movement of the head to cease, and from his point of view there is no

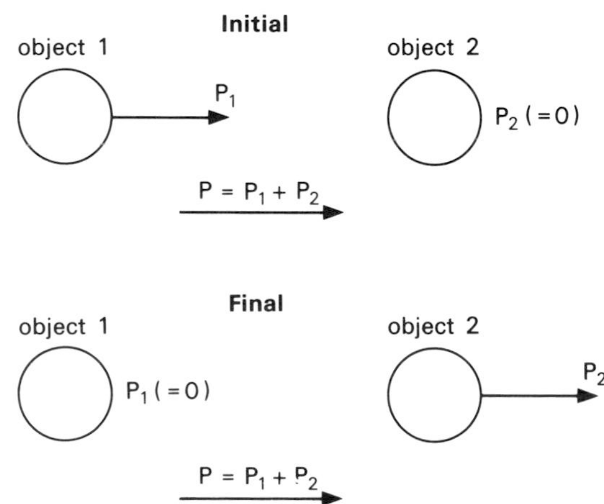

Figure 2.3 The conservation of linear momentum.

apparent acceleration of the hammer. In other words, the centripetal force is equal but opposite to the centrifugal force which must also equal mv^2/r. These forces are important in understanding the motion of the Earth's atmosphere and of the circulation cells within it (see Chs 4 & 9). In these situations it is necessary to redefine velocity and momentum in terms of the period of rotation, and the radius of the circle as the angular velocity and angular momentum (see Box 4.6).

2.2.3 Applied force: stress and resistance

The concept of **stress** is fundamental to the understanding of the many mechanical processes operating within the natural environment, particularly in denudation systems. Stress in this context is largely concerned with mechanical forces. The basic principles involved underlie studies of fluid, soil and rock mechanics, which embrace the displacement of particles or masses of soil and rock, often associated with moving water.

Stress is defined as force applied per unit area:

$$\text{Stress} = \frac{\text{force}}{\text{area}} = \frac{\text{kg m}^2\,\text{s}^{-2}}{\text{m}^2} = \frac{\text{N}}{\text{m}^2} = \text{N m}^{-2}.$$

Thus, a block of rock with a weight (mass × gravitational acceleration) of 1000 kg and a basal area of 2 m^2, resting on a horizontal surface exerts a **normal** (or vertical) **stress** (or pressure) of 500 kg m^{-2} or 500 N m^{-2} on the surface beneath Providing that the underlying surface does not deform or yield, then a stress or **resistance** is set up within it, of equal magnitude to the applied stress. In this case the surface has a resistance, or **strength**, at least equal to the applied stress (Newton's Third Law of Motion).

Normal stress and **pressure** are the same thing and therefore the pressure exerted by the atmosphere on the Earth's surface – **atmospheric pressure** – can be defined as the downward force of the atmosphere per unit area of surface. This force is equal to the product of the mass of the atmosphere above the area of surface and the acceleration due to gravity (g). For a fluid, however, the mass term can be replaced by the product of the density (ρ) and the height, or depth, (h) of the fluid, so that pressure is defined as:

$$P = \rho hg.$$

This is known as the **hydrostatic equation** and it can be modified to apply to the atmosphere. These relationships will be explored further in Chapter 4, particularly in Boxes 4.1 and 4.2.

Different types of stress can be identified in terms of the relationship between the direction of applied force and the plane or body on which it is acting (Fig. 2.4). Where the applied stress is normal to the surface, as in the example cited above of the block of rock on a horizontal plane, it is termed a **compressive stress**. On a plane oblique to the direction of applied stress, as on a slope, the stress becomes a **shear stress**. In this case, the normal stress is attenuated by the obliquity of the plane concerned and there is a tendency for translational movement (shearing) to take place along this plane. A stress exerted within a mass and tending to prise it apart, is a **tensile stress**.

Examples of all of these types of applied force exist in denudation systems. Tensile stresses, for example, are involved in many processes of rock weathering. Most important, however, are the shear stresses, since these exist whenever gravitational forces operate on gradients (in slope systems, fluvial systems and glacial systems) to provide a gravitational component of shear stress applied to the sloping plane.

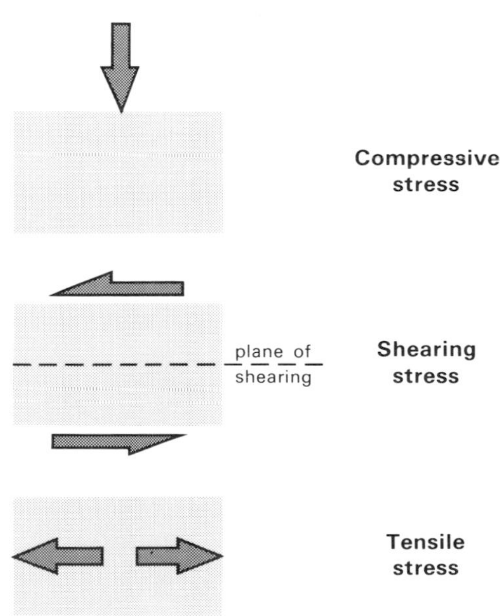

Figure 2.4 Applied force.

The effect of applied stress is to cause a displacement or deformation of the mass to which it is applied. The deformation, or **strain**, is measured as a linear or volumetric proportion of the dimension of the initial body. Different types of behaviour of solid materials under applied stress can be identified (Fig. 2.5). In a rigid solid, insignificant strain takes place at low to moderate stress levels until a critical stress value is reached at which **failure** takes place in the form of **brittle fracture**, and strain increases very rapidly. An elastic material, on the other hand, will show considerable strain before failure takes place. If the applied stress is less than that required to cause failure, the body will rebound to its original form once the stress is removed. A plastic material will show little strain until a critical threshold stress level is reached, and beyond that will deform in direct proportion to the stress applied. These three modes of behaviour represent idealised and arbitrary states, and most materials exhibit a combination of them. Indeed, a particular material may behave in different ways under different conditions, e.g. according to the rate at which stress is applied, or under different moisture conditions.

Most materials in the natural environment exhibit mainly rigid or plastic behaviour. For example, an indurated rock will tend to possess the properties of a rigid solid, and a moist clay will tend to behave in a plastic manner.

The concept of **strength** is intimately related to that of failure. The strength of a material can be defined as the resistance of that material to failure, whether by brittle fracture or plastic deformation.

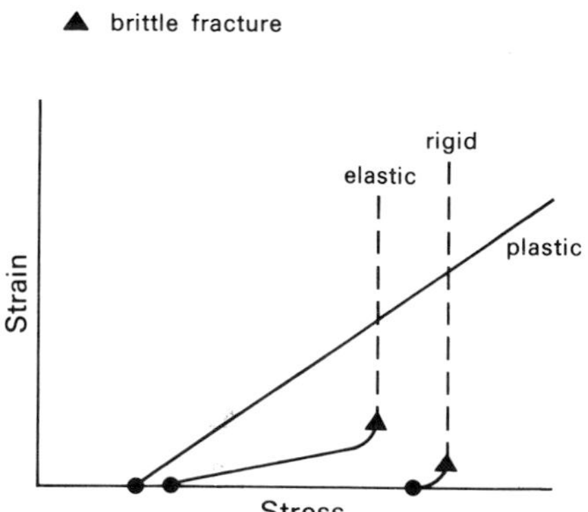

Figure 2.5 Stress–strain relationships.

Strength is therefore equal to the force applied at failure and is measured in terms of applied force per unit area (N m^{-2}).

The behaviour of Earth surface materials under different stress conditions is fundamental to the operation of denudation systems. Properties of these materials relevant to their behaviour under applied stresses, and the nature of the applied stresses, are further considered in Chapters 10 to 15, where the applications of these basic concepts are taken up.

2.3 Work and energy

Every particle of matter is acted on by forces all the time. Some of these forces may be external to it, such as gravitational and some electromagnetic forces, but others exist inside the particle and result in the aggregation of its constituent parts. When these internal forces are balanced, the body of matter is stable. Equally, when the external forces acting on a body exactly balance each other, the motion of the body will not change. If at rest it will remain at rest and if moving it will continue to move in a straight line with the same speed, unless acted on by a force (see Newton's Laws of Motion, p. 25).

If a force is applied to a body of matter, a certain amount of work will be done against the resistance to the force. The effect of the force is to move its point of application through a distance, but the work done varies not only with the resistance overcome but with the distance through which it is overcome. In other words, the work performed on the body is expressed through the distance moved. So the work done by a force (or against the resistance of a force) is the product of the force and the distance:

$$\text{work} = \text{force} \times \text{distance}$$

But, as a force has been defined as mass times acceleration:

$$\text{work} = \text{mass} \times \text{acceleration} \times \text{distance}$$

$$W = MLT^{-2}L$$

$$= ML^2T^{-2} = \text{kg m}^2\ \text{s}^{-2} = \text{Nm (newton metre)}.$$

To this point the existence of these forces has been assumed and they have been regarded as freely being available to do work. It is at this stage, however, that the concept of energy must be introduced.

Energy is often defined as the capacity to do work. It is perhaps better thought of as that 'quantity' which diminishes when work is done. This immediately implies that the unit of energy should be the same as that of work:

$$\begin{aligned} \text{energy} &= \text{force} \times \text{distance} \\ &= ML^2T^{-2} \\ &= \text{kg m}^2\,\text{s}^{-2} = \text{Nm} = \text{J (joule)}. \end{aligned}$$

The newton metre and the joule are therefore equivalent units of energy. The newton metre is applied to energy in its mechanical form, while the joule is used for heat and potential energy.

All particles or bodies of matter, therefore, possess this 'quantity' or capacity. Just as the existence of forces is related to the relative positions and relative motions of such particles and bodies, so energy can be understood in the same terms.

2.3.1 Potential energy and kinetic energy

Any body of matter at rest will possess an amount of potential energy by virtue of both the configuration of its constituent particles and its position relative to other bodies of matter. Such potential energy can be viewed simply as stored energy (or fuel) which is potentially available to be transferred or to do work. An important example of potential energy is the potential gravitational energy possessed by a body relative to the gravitational attraction between its mass and that of another body, such as the Earth.

$$\begin{aligned} PE_g &= \text{mass} \times \text{acceleration due to gravity} \\ &\quad (9.81\ \text{m s}^{-1}) \times \text{height} \\ &= ML^2T^{-2} \\ &= \text{kg m}^2\,\text{s}^{-2} \qquad \text{Nm} = \text{J}. \end{aligned}$$

Bodies of matter also possess potential chemical energy. It is, in fact, electromagnetic potential energy and it reflects the charges carried by the constituent particles of matter and their relative positions. This is then the binding energy which produces the internal forces that hold the nucleus and electrons together in the atom and the constituent atoms together in molecules. In the same way potential nuclear energy is responsible for the nuclear forces that bind the protons and neutrons together in the atomic nucleus.

This concept of stored energy or potential energy has been applied to bodies of matter at rest. What notion of energy can be applied to bodies in motion? Suppose a force of F newtons acts on a body initially at rest and having a mass m kilograms. If v is the velocity acquired by the mass in t seconds under the action of the forces, F is defined as mvt^{-1}. In displacing the body through d metres, the force F does, by definition, Fd joules of work. But d is $\frac{1}{2}vt$, the average velocity ($\frac{1}{2}v$) multiplied by the time (t). The product Fd is, therefore, $\frac{1}{2}mv^2$ ($MLT^{-2}\,\frac{1}{2}LT^{-1}\,T = \frac{1}{2}ML^2\,T^{-2}$). This is the work done by the force in accelerating the body from rest, and it is now the energy resident in the moving mass. This energy of motion is called the kinetic energy (KE) and, at low velocities:

$$KE = \tfrac{1}{2}mv^2 = \tfrac{1}{2}ML^2\,T^{-2} - \text{kg m}^2\,\text{s}^{-2}\ \text{(joules)}.$$

All familiar forms of energy – light, mechanical, radiant, heat or thermal (Box 2.4), nuclear, chemical and electrical energy (actually some of these forms are identical) – can be regarded either as forms of potential or kinetic energy, or as some combination of the two.

2.3.2 Energy transfer, entropy and the laws of thermodynamics

Energy is associated with the organisation of matter, the relative positions and relative motions of particles of matter. When a change in the state of a system of matter occurs through the operation of a process, some re-organisation of the particles of matter occurs, as, for example, during a chemical reaction, or the precipitation of a raindrop. Such changes of state involve, therefore, a redistribution of energy between the particles of matter that compose the system or between the system and its surroundings. This redistribution of energy involves both transfers and transformations of energy.

In a chemical reaction, for example, the potential chemical energy of the reactants (bonding energy plus thermal energy) is transformed to kinetic energy and transferred to the molecules in

Box 2.4

HEAT AND TEMPERATURE

The heat energy content of a body is due directly to the velocity of vibration of the molecules of which it is formed. Temperature is a measure of the amount of this heat energy which a body contains.

Faster vibrating molecules transmit some of their kinetic energy of motion to adjacent slower-moving molecules, a process which is referred to as conduction. In this way, heat energy is transmitted through a body.

In order to measure the temperature of a body, heat energy is exchanged with an indicator, usually another substance which undergoes a known mode of (mechanical) deformation in response to such exchanges. Most substances experience expansion or contraction in response to an input or output of heat energy, according to their coefficient of thermal expansion, which is the deformation produced per unit volume (or length) of the material per unit change in temperature.

If we were to insert a thin-walled glass bulb full of a fluid such as mercury into a medium such as water, air or soil, kinetic energy will be exchanged between the molecules of the medium and those of the glass and the mercury. If the mercury takes up energy from the medium, it will respond by expanding its volume. This will continue until there exists a dynamic equilibrium between the kinetic energy of the molecules of the substances on either side of the thin wall of the glass bulb.

Should there be a decrease in the kinetic energy of the molecules of the medium, then energy will be transferred away from the mercury, resulting in a contraction in its volume. This change in volume is, therefore, an expression of the change in heat energy content of the medium or a change in its temperature. This is the basic principle behind the measurement of changes in temperature using thermometers.

Box 2.5

CHEMICAL REACTIONS

Chemical reactions can be studied from various points of view. Attention can be directed to the bulk changes of the chemical substances involved. This approach is termed the **stoichiometry** of the reaction and expresses the *active masses* of the reactants and products in terms of gram molecular weights (**moles**). As the **law of conservation of mass** requires, the mass of reactants and of products must balance as there should be no loss of mass in a chemical reaction. For example,

$$2Na^+OH^- + (H^+)_2SO_4^{2-} \rightarrow 2H_2O(1) + (Na^+)_2SO_4^{2-}$$

80 g	98 g	36 g	142 g
sodium hydroxide	sulphuric acid	water (liquid)	sodium sulphate

Alternatively, attention can be centred on the mechanism by which the reaction works, and by which the chemicals come together and how the products are formed. Here the recognition of intermediate products which may be formed momentarily becomes important. So, for example, the stoichiometric equation for the reaction between hydrogen and chlorine molecules, $H_2 + Cl_2 \rightarrow 2HCl$, becomes broken down into stages and indicates that the reaction in this case is a chain mechanism where the stages can be repeated many times, viz.

$$Cl_2 \rightarrow Cl + Cl,$$

$$Cl + H_2 \rightarrow HCl + H,$$

$$H + Cl_2 \rightarrow HCl + Cl.$$

However, chemical reactions occur when the chemical bonds of the reacting substances are broken and reformed to produce the reaction products. The breaking and making of bonds involves the transfer of energy. The likelihood of a reaction occurring and the speed with which it occurs will depend on the energetics of the reaction. This is a third approach, and one that is particularly important in understanding the energy transfers in natural systems. Reactions involving covalently bonded molecules concern collisions between them. In practice this means two body collisions, for the probability of more than two molecules colliding instantly is extremely small. The occurrence and speed of a reaction will depend not only on the fact that molecular collisions occur but also on two other factors. First, their **energy of collision** must be sufficient for specific bonds to be broken, and secondly, they must collide in such a way (the **collision geometry**) that these specific bonds will in fact be broken. Together collision energy and collision geometry affect the **activation energy** of the reaction – the minimum energy necessary to break bonds and form products. We can graph the energy changes involved in a reaction in the following way:

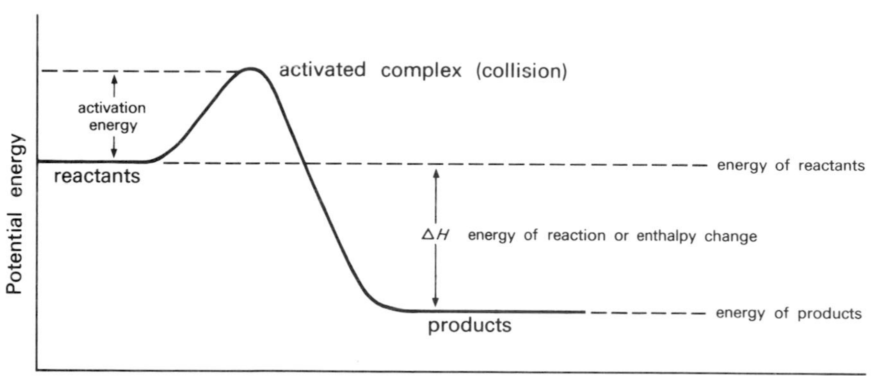

Here, as the reactants approach each other their potential energy increases to a maximum at the point of collision when an **activated complex** of the reactants is formed. The height of the activation energy here shows us the ease or difficulty of a chemical reaction and if the energy and geometry of collision are favourable the activated complex will break up to form the products. If it does not, it will reform the original reactants. In this diagram the energy of the products is lower than that of the reactants. This energy difference is called the **energy** of **reaction** or the **enthalpy** change (ΔH) and, because here energy has been lost during the reaction, ΔH is negative and the reaction is **exothermic**. If the energy level of the products had been higher than that of the reactants ΔH would be positive and the reaction would be termed **endothermic**. All chemical reactions can be classified as to whether they absorb energy (usually heat) from, or evolve energy (again, usually heat) to, the surrounding medium. Outward signs of changes in heat energy in exothermic and endothermic reactions are not the only factor to consider in reaction energetics. The concepts of **free energy** (F, or G) and **entropy** (S) must also be introduced (see Box 2.6).

Certain substances – **catalysts** – are able to increase the speed of reactions by influencing the collision geometry of the molecules involved, hence increasing the number of favourable collisions and lowering the activation energy. Catalytic reactions are especially important in living systems where a host of highly specific proteins – **enzymes** – are the catalysts involved in biochemical reactions (see Ch. 6). (Largely after Ashby *et al.* 1971, see Further reading.)

motion during the reaction (Box 2.5). Some of this kinetic energy is transformed to heat during the collision between these molecules and, when the reaction stops, some again becomes potential chemical energy – of the reaction products this time. The kinetic energy of the falling raindrop will be transferred as mechanical energy to the soil on impact and will perform work by dislodging soil particles. Some will, however, be transformed to heat through friction. Also, as the droplet falls, some of the kinetic energy will be transformed to heat due to the frictional drag of the air.

What these examples suggest is that there is no change in the absolute quantity of energy involved. Energy may be transformed in kind, and transferred, both within the system and between the system and surroundings, but when both system and surroundings are considered it is found that energy has been neither created nor destroyed. This is the **First Law of Thermodynamics**, and is known as the **Law of Conservation of Energy**. There is no exception known: it is a generalisation from experience, not derivable from any other principle, and it states that the total energy of the universe remains constant.

The **Second Law of Thermodynamics** is concerned with the direction of naturally occurring, or real, processes. In combination with the First Law it allows not only the prediction of this direction but also the equilibrium state that will result. To choose a familiar example, if the system consists of a petrol tank and a motor mounted on wheels, the Second Law allows us to predict that the natural sequence of events is: consumption of petrol, the production of carbon dioxide and water and the forward motion of the whole device. From the Second Law, the maximum possible efficiency of the conversion of the chemical energy of the petrol into mechanical energy can be calculated. The Second Law also predicts that one cannot manufacture petrol by feeding carbon dioxide and water into the exhaust and pushing the contraption along the road, not even if it is pushed along backwards!

Natural processes are irreversible. They proceed with an increase in the disorder (randomness) of matter and energy. Order or structure in the arrangement of matter disintegrates and concentrations disperse. Energy is degraded from high

Box 2.6

FREE ENERGY AND ENTROPY CHANGE IN CHEMICAL REACTIONS

The enthalpy change ΔH (Box 2.5) does not tell us the amount of work that can be obtained from a chemical reaction. This is given by the **free energy change** of the reaction, ΔF, or ΔG (after Willard Gibbs), where the free energy is the 'useful energy' capable of doing work on the system. Free energy change is related to **enthalpy** change and **entropy** change by the following equation,

$$\Delta G = \Delta H - T\Delta S$$

where T = absolute temperature (degrees Kelvin) (°K = °C + 273), and ΔS = the change in the entropy (randomness) of the system.

Therefore, the change in the free energy of the system is equal to the change in the heat energy of the system minus the amount of energy used to change the order or randomness of the system. As an expression of the **Second Law of Thermodynamics** it tells us the direction that the reaction will follow spontaneously, i.e. it will proceed in the direction that results in a more disordered system. For the entropy of the system to increase in this way there must be a decline in the free energy, and ΔG must be negative. Such a spontaneous reaction is known as an **exergonic** reaction (cf. exothermic, see Box 2.5). In this case, the free energies of formation of the products must be less than those of the reactants in the following equation,

$$\Delta G = \underset{\text{(products)}}{\Sigma\Delta G_f} - \underset{\text{(reactants)}}{\Sigma\Delta G_f}.$$

Where the free energy change of the reaction, ΔG, is positive the reaction will not proceed spontaneously and will require an input of energy to start it, i.e. it is **endergonic**. Note, however, that in indicating whether or not the reaction will proceed spontaneously these thermodynamic considerations do not take into account the activation energy (see Box 2.5).

levels to low. No spontaneous transformation of energy during the operation of a natural process is 100% efficient. Some is dissipated as heat energy (the random motion of matter) and is unavailable to do work, as we have seen in earlier examples. The energy available to be used is continually being diminished through the operation of natural processes. Energy that becomes unavailable to do work is related to the increase in the **entropy** (S) of the system (Box 2.6). Entropy is a measure of the disorder of the system, but it can never be absolutely quantified. However, entropy always increases as a result of the operation of any natural process, so $\Delta S \geqslant 0$. Therefore, change in entropy (ΔS) can be quantified, though not in absolute terms. The Second Law tells us that all physical and chemical processes proceed towards maximum entropy. At this point there is thermodynamic equilibrium.

2.3.3 Energetics of systems

It is the central premise of this book that energy and its transformations may be viewed as the best way both to systematise and to synthesise the facts and theories of physical geography and environmental science. Therefore, thermodynamic or energy systems become the most fundamental way of analysing all environmental processes, just as they have long been accepted and used as the basic approach to the analysis of physical, chemical and, more recently, biological processes.

These thermodynamic principles may appear to be abstract and formidable, but the approach and working philosophy of thermodynamics is really quite simple. Furthermore, it is necessary to master only a few such principles in order to examine in a broad way the nature of environmental energy systems and the energy transformations associated with them. The most fundamental of these principles are the First and Second Laws of Thermodynamics, which we have encountered already.

It follows from the First Law that when an energy system changes from an initial state to a final state, it may either receive energy from its surroundings or give it up to the surroundings. The difference in energy content between the initial and final states must be balanced by an equal but inverse change in the energy content of the surroundings. In this case the system is deemed to be in equilibrium when the initial and final states are specified; that is, the process of change has not yet either started nor ceased entirely. Classical equilibrium thermodynamics is not concerned with the time taken to bring the system from the initial to the final state, nor with the rate of change. Equally it is not concerned with the pathways or processes by which physical or chemical change occurs, but merely with the energy difference between the initial and final equilibrium states of the system. As an analogy we can say that if a man travels from London to Edinburgh his change in location is completely specified by stating his initial latitude and longitude and his final latitude and longitude. It does not matter how long the journey took, nor what route he followed.

This is the method of approach of thermodynamics to the analysis of physical or chemical changes, but the determination of the total energy content of a system in either the initial or final state is in practice formidable. Even in the laboratory it is only possible with very simple systems of gases, so in complex environmental energy systems it is virtually impossible. However, it is the *changes* in the energy content which are our primary concern and such changes are more easily visualised and measured. If we know the type and magnitude of the energy exchanged with the world outside the system (its surroundings) as it proceeds from an initial to a final state then we can undertake a thermodynamic analysis of the process. This in practice becomes the method of thermodynamics.

Our reasoning so far has been based on the First Law alone, but the Second Law tells us that no spontaneous physical or chemical change in a system is completely efficient. Such processes have a direction. They lead towards system states where the elements and properties of the system are distributed at random. The energy content of the system is progressively degraded to a random state (which was the way we defined entropy) and is unavailable to do work. So the Second Law tells us that all physical and chemical processes proceed in such a way that the entropy of the system is maximised, and at this point we have equilibrium.

However, these principles were developed to apply to closed systems which do not exchange matter with their surroundings. We are concerned with open systems, which do exchange matter with their surroundings. Such systems do not attain a thermodynamic equilibrium of maximum entropy, but are maintained in a dynamic steady state by a throughput of matter and energy. Nevertheless, they are irreversible systems and, although they may appear to violate the Second Law in that their entropy may remain constant or even decrease as they become more ordered, actually it does not. The entropy of the system or any part of the universe may decrease as long as there is a corresponding and simultaneous increase in the entropy of some other part – total entropy increases.

$$\Delta S_{\text{total}} = \Delta S_{\text{system}} + \Delta S_{\text{surroundings}}.$$

An open system, by definition, maintains a steady state by the transfer of matter and energy by real or irreversible processes with the surroundings. Therefore, time and rate become critical variables in the energetics of entropy production, which as a consequence is itself time-dependent. The most probable state of an open system is one in which the production of entropy per unit of energy flow through the system is at a minimum. Such a state is most likely to persist.

Further reading

Ashby, J. F., D. I. Edwards, P. J. Lumb and J. L. Tring 1971. *Principles of biological chemistry*. Oxford: Blackwell Scientific.

Carson, M. A. 1971. *The mechanics of erosion*. London: Pion.

Davidson, D. A. 1978. *Science for physical geographers*. London: Edward Arnold.

Duncan, G. 1975. *Physics for biologists*. Oxford: Blackwell Scientific.

Gymer, R. G. 1973. *Chemistry: an ecological approach*. New York: Harper & Row.

Stamper, J. G. and N. A. Stamper 1971. *Chemistry for biologists*. London: George Allen & Unwin.

White, E. H. 1964. *Chemical background for the biological sciences*. Englewood Cliffs, NJ: Prentice-Hall.

Part B

A systems model: a partial view

As we drew back from the landscape seen in Figure 1.1a and the cattle by the loch shore, the hillslopes and mountain summits receded and our eyes remained fixed on the surface of the planet. As the perspective on our Earthly environment changed we recognised the value of modelling its component systems at different spatial scales such that the properties of each could be discriminated at a level appropriate to a particular scale. In this way we could arrive at a nested hierarchy of systems models applicable to the entire planet at one end of the scale continuum and to the sand grains on the loch shore at the other. Before we embark on the task of constructing such an hierarchy, let us pause and instead of looking inwards at the Earth and its surface environment which are the home of Man, shift our gaze toward the immensity of space. As we do so it becomes clear that our hierarchy of systems of matter does not stop at the planetary level.

The Earth and its solitary Moon are part of the solar system. The Earth is one of nine planets moving round the central star – the Sun. The solar system is an orderly, harmonious system of matter, but it too is only one of countless star systems with their satellite planets which compose our galaxy. This galaxy, however, is itself but one of an estimated hundred thousand million (10^{11}) galaxies within the maximum radius of the observable universe. The planet Earth viewed in this context becomes an insignificant speck in a system of matter and energy, the scale of which is almost beyond comprehension. Within our solar system, however, the Earth is far from insignificant. It occupies a unique position amongst the planets and its distance from the Sun, almost midway through the zone where water can exist as a liquid, its rate of rotation, its receipt of solar radiation, its surface temperature and the presence of an atmosphere, all combine to create conditions favourable to the development of life, while the existence of life has itself transformed the surface of the planet. Around some distant star in some far off galaxy another planet may be orbiting under a similar chance combination of conditions. However, as to whether life has evolved elsewhere, we have no way of knowing.

As we turn away from the rest of the universe and fix our gaze once more on our Earth and begin the task of building systems models of the planet, we would do well to remember this larger perspective. In the first place it sets not only the spatial but also the temporal scales of our models in a context. Secondly, many of the properties of the Earth are, as we have seen, related to its place in the solar system, and to its interactions with other bodies in that system, while other properties are inherited from the time of its origin and that of the solar system. The regular rotation of the Earth about its axis and its motion relative to the Sun imposes rhythmical changes in exposure to light and dark, day and night, summer and winter. Both the Sun and the Moon impose tidal rhythms on the atmosphere, on the oceans, and even on the exposed crust as the Earth rotates. The spin of the Earth distorts patterns of circulation in the gaseous atmosphere and the waters of the oceans, while in combination with the liquid metallic core it sets up a strong magnetic field. In attempting to model the planet as an energy or thermodynamic system in Chapter 3, however, it is the continuous output of energy from the Sun which is the single most important characteristic of the solar system, and it represents the major energy source for our planet. Many of the attributes of the Earth's surface environments and of the living systems which inhabit them reflect the magnitude of this energy input and mirror both the solar and lunar rhythms they experience.

In Chapter 3 it will become apparent that to understand the energy relationship of the Earth even at the global scale it is necessary to recognise the existence of a number of major subsystems and to abandon the black box model of the planet with which the chapter commences. Only by identifying the major stores of matter and energy within these subsystems and by considering the pathways and magnitude of transfers between them can we begin to understand the planetary energy balance. We shall therefore arrive at a model of the Earth which gives us a broad, but partial view of the structure of the system and of its functional relationships, particularly in terms of the energy cascade through the system.

In Section III this model will be refined even further. The atmosphere, Earth surface, and Earth's interior – the subsystems of Chapter 3 – will be redefined and each will be treated as a system in its own right. Chapter 4 will consider the atmosphere and Chapter 5 the Earth's interior, but the Earth-surface subsystem of Chapter 3 will be divided into the physiochemical systems operating at the surface of the planet, the denudation system and the living systems which inhabit this surface. Since the denudation system has strong functional links with, and is partly the surface expression of, the geophysical and geochemical activities of the crustal system, they are both treated under the heading of The lithosphere in Chapter 5. As far as we are aware, the existence of life is probably unique to the planet Earth, at least in our solar system, and therefore because of this, and because of the far reaching effect that life has had on the history of our planet, the living systems of the biosphere are considered separately in Chapter 6. The evolution and survival of life on the Earth are inconceivable without interactions with the non-living environment and, for this reason, Chapter 7 is entitled The ecosphere, for it is concerned not only with the organisms of the biosphere but with these interactions with the environment.

The focus in Section III is still at the global scale and remains a partial view of the systems concerned. It is, however, a partial view not of the entire planet, as in Chapter 3, but of each of the major subsystems. The models developed, therefore, are largely at an intermediate level of sophistication, but they nevertheless increase our understanding of the Earth as the level of resolution has improved considerably over that employed in Chapter 3. There are, however, variations in the scale at which these models will be developed. Chapters 4 and 5 are mainly concerned with broad global models of storage and transfer of matter and energy in the physicochemical systems of the atmosphere and lithosphere. Even so, the scale at which particular processes, such as the change of state of water in the atmosphere, or the crystallisation of igneous rock in the crust, are considered may be at the molecular scale, or at that of the individual crystal. In Chapter 6 the decision is taken to model the organisation and basic functional activity of living systems, first of all at the molecular and cellular scales. The temporal scale will also, of course, vary from the almost instantaneous operation of the biochemistry of the living cell to the millions of years over which movements of crustal plates must be considered. Nevertheless, in all of these cases the understanding which emerges is generalised and the perspective of the models as a whole remains a global one.

II THE PLANET EARTH

3 Energy relationships

3.1 The closed system model

One of the advantages of building systems models is that they simplify the complexity of the real world. In this chapter we shall model the Earth as a closed system; that is, a system of matter whose only interaction with its surroundings is the transfer of energy across its boundary (see Ch. 1). Such a model of the planet is obviously simplified. Meteorites, for example, can penetrate the Earth's atmosphere from space, illustrating that matter also crosses the boundary of this system. So do Man's space vehicles, projected into space perhaps to be lost forever (as with the Pioneer and Voyager vehicles to other planets) or perhaps to return as debris after re-entry, as with the American Skylab in 1979. Nevertheless, at the planetary scale the closed system model is a useful one for it focusses attention on the energy input and output across the boundary of the system, on the sources of that energy, and on the net energy balance of the whole system. Critical to this approach, however, is the definition of the system. For our present purposes the boundary of the planet Earth will be taken as the outer surface of the atmosphere and our model of a closed system Earth will be treated initially as a black box (Fig. 3.1).

Table 3.1 Energy sources and energy stores (after Campbell 1977, Sellers 1965).

	Energy × 10^{20}J
total daily receipt of solar energy by the Earth	149
energy released by 1976 Chinese earthquake	5006
combustive energy stored in Earth's coal reserves	1952
combustive energy stored in Earth's oil reserves	179
combustive energy stored in Earth's natural gas reserves	134
latent heat absorbed by global snow/ice melt in Spring	15
North Sea oil reserves (presently known)	3
annual USA consumption of energy (1970)	0.75
annual UK consumption of energy (1972)	0.09
heat flux from Earth's interior	0.027
total radiation from the Moon	0.006
total energy content of annual grain crop in Britain	0.006
energy released in Krakatoa eruption 1883	1.49×10^{-3}
dissipation of mechanical energy of meteorites	0.89×10^{-5}
total radiation from stars	0.60×10^{-5}

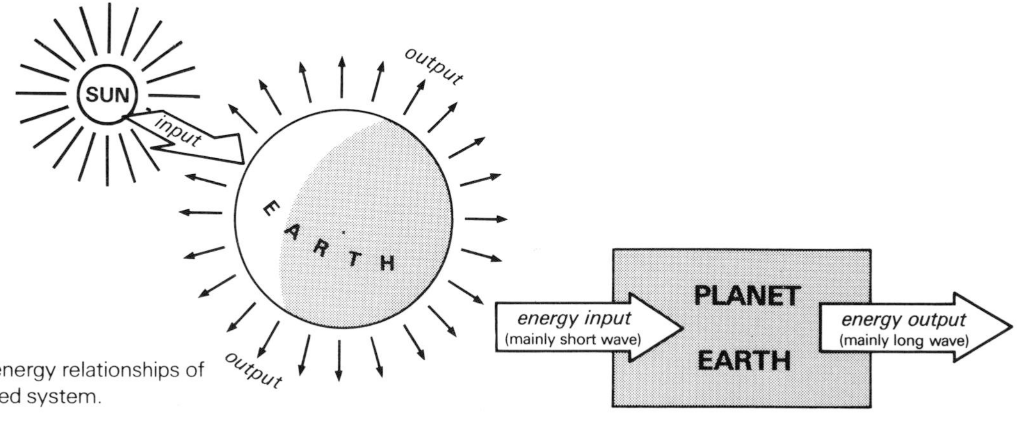

Figure 3.1 The energy relationships of the Earth as a closed system.

3.1.1 Energy input: solar radiation

Energy crosses the boundary of the system in several forms, but the most significant input is the radiant energy received from the Sun (Table 3.1). The Earth also receives electromagnetic energy from other bodies in space and it experiences the gravitational energy associated with their masses. Table 3.1 makes clear that these other inputs, although not unimportant, are nevertheless of a different order of magnitude. So the main input of energy to the black-box closed system model of the planet is in the form of electromagnetic radiation from the Sun (Box 3.1). Its source is the enormous energy locked up in the nuclei of hydrogen atoms, part of which is released by nuclear fusion in the immensely high temperatures of the Sun (Box 3.2). The uncontrolled release of energy by the same process has given mankind the awesome

Box 3.1

ELECTROMAGNETIC RADIATION

There are two ways of visualising electromagnetic radiation. The first is due to Max Planck (a German physicist) who in 1900 suggested that radiation is emitted as a stream of pulses of energy, referred to as quanta – a view supported by the work of Einstein in 1905. The fundamental equation relating the energy content of a single indivisible quantum (E) is given by:

$$E = h\nu$$

where h is Planck's constant (6.60×10^{-34} J s^{-1}) and ν (nu) is the frequency of radiation quanta.

The adoption of this quantum approach allows the calculation of direct measures of the energy of radiation, which is essential when considering energy conversion within the system.

The alternative is to view radiation as a sine wave form. Here, a critical attribute of electromagnetic radiation is its wavelength and the range of wavelength values is represented in the electromagnetic spectrum from the short waves of gamma rays to long radio waves. The wave form of electromagnetic radiation is perhaps more commonly adopted in the environmental sciences and particularly in meteorology, as it lends itself more readily to treatments of transmission, reflection and refraction of radiation.

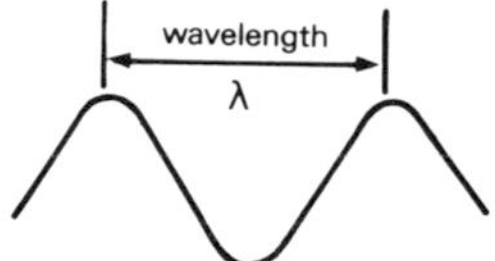

Box 3.2

SOLAR ENERGY

The radiant solar energy received by the Earth arises from nuclear energy. In the immensely high temperatures of the Sun (surface temp of 6000°K) a part of the enormous energy locked up in the nucleus of the hydrogen atom is released by nuclear fusion. In this process four hydrogen nuclei (four protons) are fused to form a helium nucleus (two protons + two neutrons):

$$4\,{}^{1}_{1}H \rightarrow {}^{4}_{2}He + {}^{0}_{1}e + h\nu_0.$$

The mass of the helium nucleus is about 0.7% *less* than the sum of the four hydrogen nuclei. This lost mass is converted to a quantum of energy (see Box 2.1) in the form of gamma radiation. This is represented above by the term $h\nu$, in which h is Planck's constant and ν (nu) is the frequency of gamma radiation. After a complex series of reactions, the gamma radiation is emitted again in the form of photons or quanta of light energy.

capacity for self-destruction which is inherent in the hydrogen (or fusion) bomb. Paradoxically, it is also this same process of nuclear fusion which may one day be harnessed for the peaceful generation of power without the attendant dangers of nuclear waste associated with present nuclear fission technologies.

If we assume the Sun to have an estimated surface temperature of 6000°K and to be an ideal, or full, radiator, then we can derive some of the characteristics of the radiation energy emitted by the Sun from three temperature-based radiation laws (Box 3.3). The first of these laws tells us that the amount, or intensity, of solar radiation is relatively high, for it is a power function of the Sun's surface temperature. The second shows indirectly a relationship between the wavelength distribution of solar radiation and the Sun's surface temperature. This is because wave propagation is a function of molecular oscillation and this increases with temperature, and a hot body such as the Sun has a high frequency of wave propagation and hence a shorter wavelength emission than does a cold body. Indeed, the peak of the wavelength distribution – the wavelength of maximum emission – is short, centering round the visible part of the electromagnetic spectrum, for the third law shows that it is inversely proportional to the surface temperature of the Sun. The complete solar radiation spectrum is seen in Figure 3.2 to consist of ultraviolet (short wavelength, 0.2–0.4 μm) which constitutes on average 7% of the total emission; the visible wavebands (0.4–0.7 μm) which constitute 50%; and infrared radiation (long wavelength 0.7–4.0 μm) which constitutes 43%. Because of the dominance of short wavelengths, solar radiation is typically referred to as shortwave radiation.

Only 0.002% of the total radiation emitted by the Sun forms the input to the system and is received by the Earth. The average amount received per unit area (over a plane at right angles to the solar beam) per unit time at the outermost boundary of the atmosphere is referred to as the **solar**

Box 3.3

RADIATION LAWS

Stefan–Boltzmann's Law

The intensity of radiation emitted by a body (I) is proportional to the fourth power of its absolute temperature (T):

$$I = \sigma T^4$$

where σ = Stefan-Boltzmann constant = 5.57 × 10^{-8} Wm^{-2} K^{-4} and K = degrees absolute or degrees Kelvin = °C + 273.16.

Wavelength–frequency relationship

The radiation emitted by a body has a wavelength distribution which is related to its surface temperature, as wave propagation is a function of molecular oscillation which increases with temperature rise. The equation linking wavelength and frequency is:

$$\lambda = c/\nu$$

where λ = wavelength (cm), c = velocity of light (cm s^{-1}) and ν = frequency (cycles s^{-1}).

As frequency (ν) increases with rise in temperature, so wavelength (λ) decreases.

Wien's Displacement Law

The wavelength of maximum emission (λ_m) is inversely proportional to the absolute temperature (T) of the radiating body:

$$\lambda_m = w/T$$

where w = Wien's constant = 2.897 μm K, μm = 10^{-6} m.

Solids, liquids and gases are all capable of emitting radiation over a range of wavelengths, but these two laws relate to physical ideals known as full radiators, or black bodies. These bodies emit radiation at the maximum possible intensity for every wavelength and absorb all incident radiation. For convenience, it is usually assumed that both the Sun and the Earth radiate as black bodies, although at certain wavelengths they are imperfect radiators, or grey bodies.

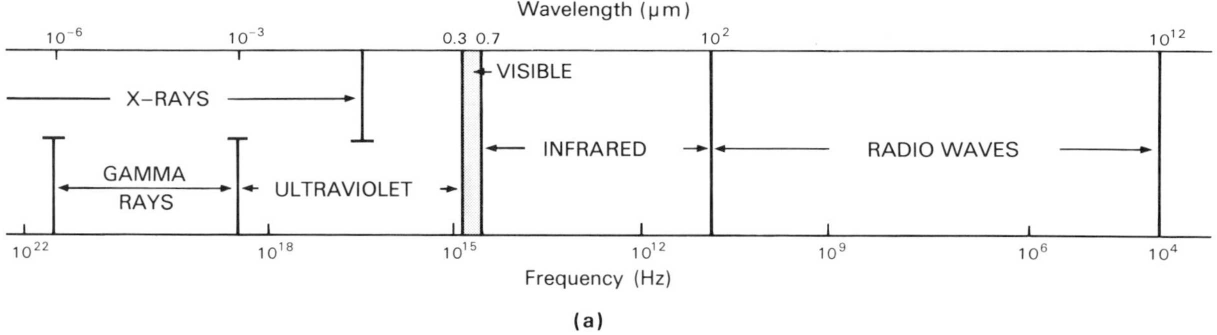

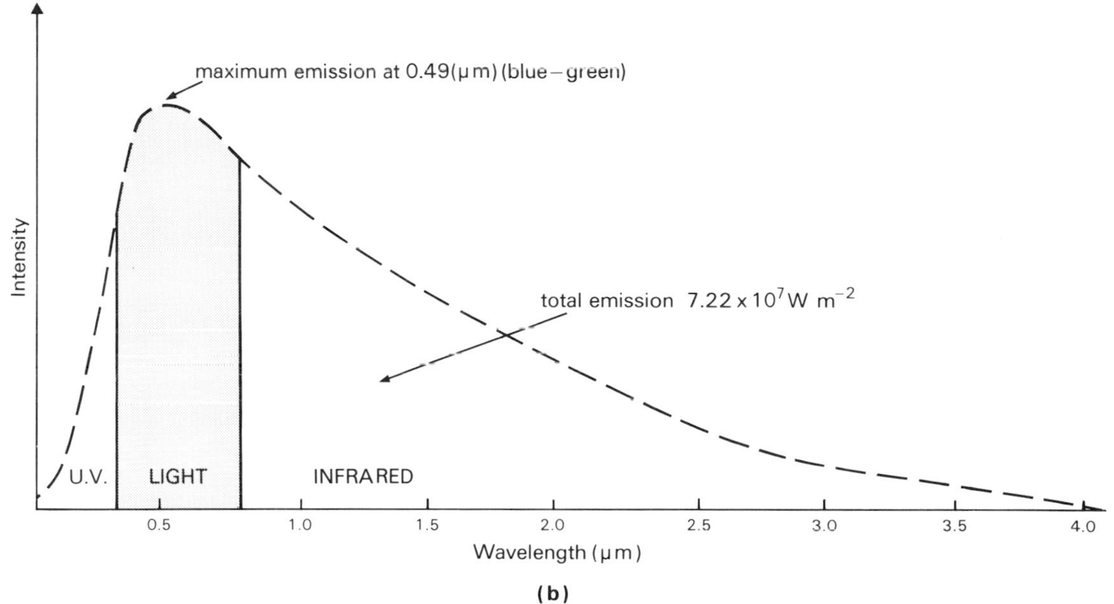

Figure 3.2 (a) Electromagnetic radiation spectrum. (b) Solar radiation spectrum.

constant and is usually given the value 1360 Wm^{-2}. However, the notion of the solar constant belies the fact that the effective input to the system varies considerably over space and time. In the first instance, there is a small degree of variation in the amount and nature of radiation from the Sun's surface. The range of variation is relatively small, fluctuating within 1–2% of the solar constant. Unfortunately it falls within the range of error in the radiation-measuring techniques employed at the outer boundary of the atmosphere. It has frequently been suggested that a potential cause of changes in the value of the solar constant is the occurrence of darker areas on the Sun's surface, referred to as sunspots. Although there is a certain degree of correlation between an observed 11-year cycle of sunspot frequency and the value of the solar constant, the relationship remains somewhat tenuous.

The effective input to the planet will also be conditioned by the movement of the Earth itself and its motion relative to the Sun. The distance between the two varies, as the Earth pursues its elliptical path round the Sun, and is greatest in early July, at 152×10^6 km (aphelion), and least in early January, at 147×10^6 km (perihelion). Because of this, the Earth receives slightly more solar radiation in January than in July. The boundary of our model of the Earth as a closed system, however, has been taken to be the outer surface of the atmosphere. As the solar beam can be assumed to consist of parallel rays, the intensity of radiation received on such a horizontal surface is directly related to the angle of incidence of the rays upon it (Box 3.4). This angle of incidence is determined by the angle of latitude, the tilt of the Earth's axis with respect to its orbital plane, and the rotation of the Earth about its own axis.

Using these relationships, it is possible to calculate the solar radiation received at any point and at any time for a planet Earth which is pursuing a known path round the Sun and whose bounding surface is assumed to be completely uniform (Fig. 3.3). Such calculations show that seasonal variation in radiation receipt on a horizontal surface is greatest in polar latitudes and smallest in tropical latitudes (Fig. 3.3). This implies that, although there is the expected latitudinal variation in energy received, the exact nature of this variation will differ according to the time of year as there is no common seasonal cycle. So, returning to our black box model of the Earth as a closed system (see Fig. 3.1), it is evident that the input of shortwave solar radiation across the boundary of the system varies in a complex manner, both spatially and through time.

3.1.2 Energy output and the planetary radiation balance

The output of energy from the system is also in the form of electromagnetic radiation, for the planet and its atmosphere not only absorb radiation but act themselves as radiators emitting radiation. However, applying the basic radiation laws (see Box 3.3) to the Earth, whose surface temperatures and those of its atmosphere are generally below 300°K, reveals that this radiation is of low intensity and its entire spectrum lies in the infrared waveband. Consequently, the emission of radiant energy across the boundary of the system is characteristically longwave radiation, in contrast to the shortwave input of solar radiation. However, the radiation laws (see Box 3.3) apply to ideal radiators or perfect **'black bodies'** which totally absorb all incident radiation. The system in our model, the planet Earth, does not behave as a black body but as an imperfect absorber and emitter of radiation. Such a **'grey body'**, placed in a stream of radiant energy, will absorb some of it, often differentially in various parts of the spectrum, and transmit and/or reflect the rest. Longwave radiation, therefore, forms only part of the total radiation output of the system, for a proportion of the initial input of shortwave radiation is lost by reflection and is not absorbed.

With the development of satellites it has been possible to measure the radiation output near to the outer boundary of the Earth's atmosphere quite accurately (Fig. 3.4). Not only is this radiation loss a combination of short- and longwave radiation, but it varies in a broadly latitudinal

Box 3.4

LAMBERT'S COSINE LAW

The radiant intensity emitted in any direction from a unit radiating surface varies as the cosine of the angle between the normal to the surface and the direction of the radiation beam.

A corollary relates solar radiation received upon a surface (I) to the angle between the radiation beam and the normal to that surface by:

$$I = I_0 \cos \alpha$$

where I_0 represents the intensity of the solar beam.

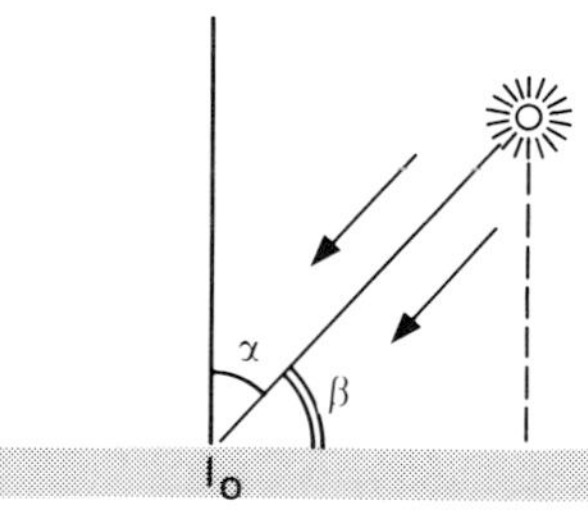

The angular elevation of the Sun above the horizon (β), referred to as the solar elevation, can be introduced:

$$I = I_0 \sin \beta.$$

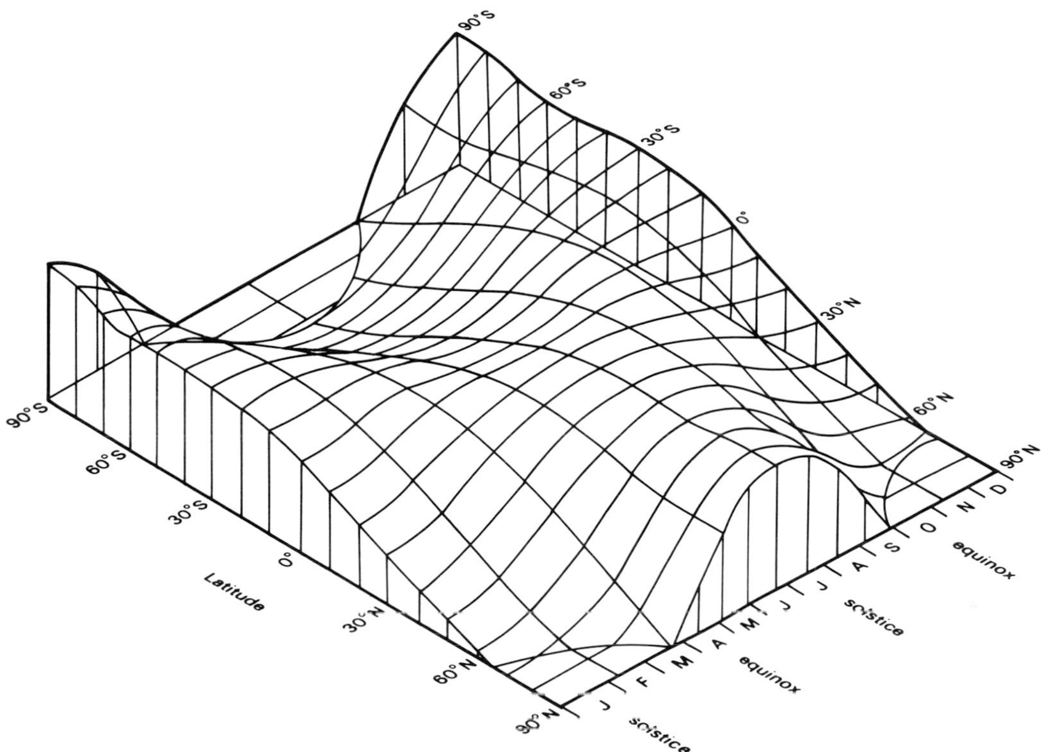

Figure 3.3 Variations in solar radiation received at the Earth's surface in the absence of an atmosphere.

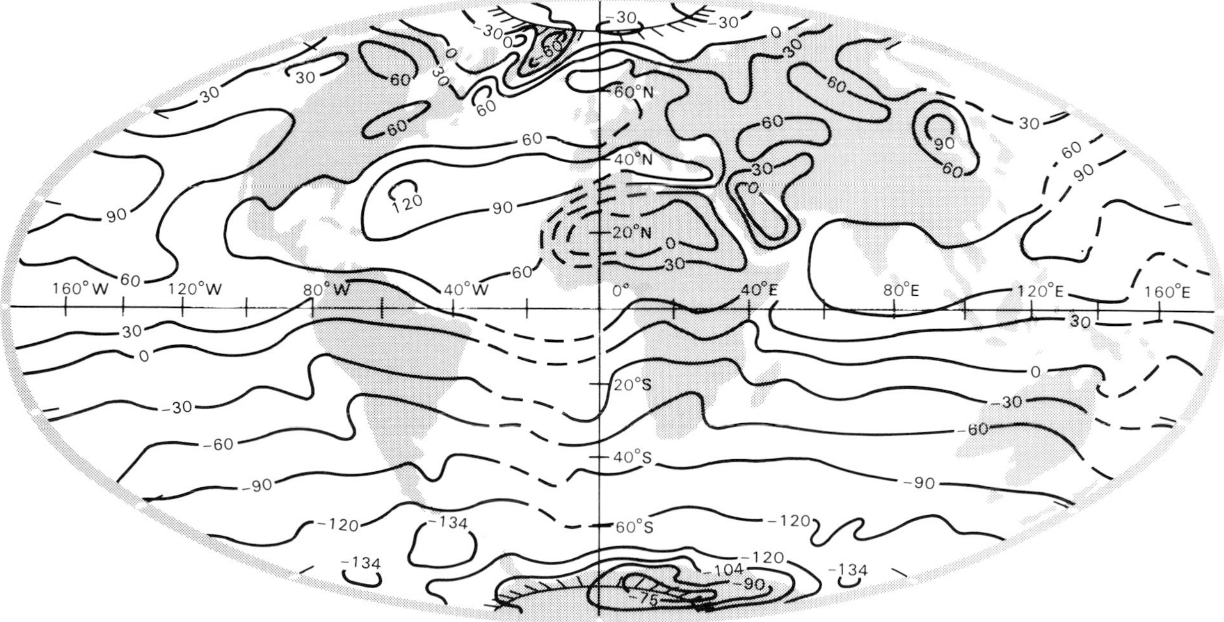

Figure 3.4 Net radiation (W m^{-2}) balance at the top of the atmosphere (from Nimbus II evidence on 1–15 June 1966, after Barrett 1974).

fashion (though this breaks up into a cellular pattern) with a tendency towards outward flux at high latitudes and a net input at low latitudes. However, for the system as a whole – that is, the entire planet and its atmospheric envelope – there is a long-term balance between incoming and outgoing radiation. If this were not the case the total energy content of the system would increase or decrease through time and the Earth would be experiencing a progressive warming or cooling. We should, therefore, be able to represent this balance as:

net radiation = incoming solar radiation (mainly shortwave) minus outgoing radiation (mainly longwave) = 0.

This would be true if we were dealing with finite quantities of radiant energy exchanged over protracted periods of time. In terms of the operation of the system, however, it is more pertinent to consider the rate of flow of radiant energy, and here we need to consider not only inputs and outputs but also the throughput of energy. This is the energy which, at any point in time, is being transferred or stored within the Earth or its atmosphere. This is because within any closed system the balance of energy exchange is not only a function of relative surface temperatures and absorbance and emittance characteristics of all parts of the system, but it also depends on modes of energy transfer other than radiation. These are conduction and convection as well as energy-dependent changes of state, such as the evaporation of water, involving latent heat exchange. So, although a long-term energy balance may exist for the whole system, there will still be energy imbalances, transfers and changes in energy storage within the system.

Therefore, to understand more fully the energy balance of the Earth we need to abandon the black box approach and improve the fidelity of our model in order to follow the way energy is transferred or cascaded through the major subsystems of the model, each of which can be considered to be an energy store. Figure 3.5, therefore, contains separate compartments for the atmosphere, for Earth surface systems and for the Earth's interior. Each of these compartments can be considered as an open system in its own right with transfers of matter as well as energy across its boundaries. In this chapter, however, we shall continue to concentrate on energy exchanges and confine our model to the planetary scale.

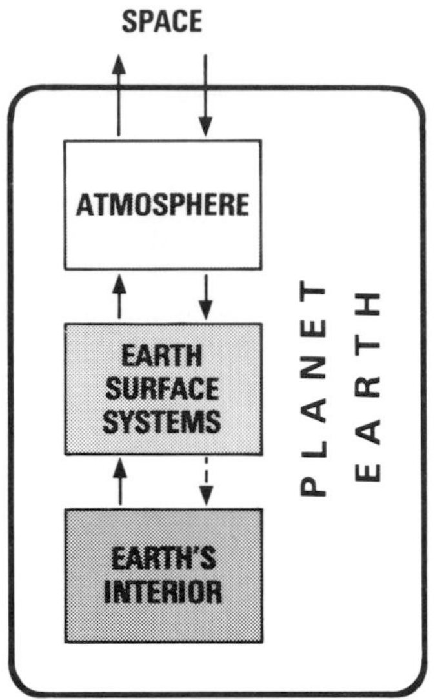

Figure 3.5 Subsystems of the model.

3.2 The atmospheric system

3.2.1 Exchange and throughput of solar radiation

The input of shortwave solar radiation at the outer surface of the atmosphere has already been considered. During its passage through space, solar radiation loses little energy, but on entering the atmosphere the solar beam encounters molecules of gases, liquids and solids, all of which are able partly to absorb or reflect it.

The gases of the atmosphere and suspended liquid and solid matter (see Ch. 4) absorb selectively from the solar beam. Ozone (O_3) and water vapour (H_2O), for example, are major absorbers of radiation but affect different parts of the solar spectrum (Table 3.2). Ozone, which occurs in the upper atmosphere, absorbs a large proportion of ultraviolet radiation in the waveband 0.23–0.32 μm. This absorption of lethal ultraviolet radiation has had profound effects on the development of life on Earth (see Ch. 6).

Water vapour, which is at its greatest concentration near to the Earth's surface, absorbs in the

Table 3.2 The absorption of radiation by the principal gases of the atmosphere.

Gas	Wavelengths of greatest absorption
nitrogen (N_2)	no absorption
oxygen (O_2)	0.69 μm and 0.76 μm (visible – red)
carbon dioxide (CO_2)	12 μm to 18 μm (infrared)
ozone (O_3)	0.23 μm to 0.32 μm (ultraviolet)
water vapour (H_2O)	5 μm to 8 μm
	11 μm to 80 μm (infrared)
also	
liquid water (clouds)	3 μm, 6 μm, 12 μm, to 18 μm (infrared)

infrared sector of the solar spectrum in a series of wavebands of high absorption. Between these are bands of low absorption which are referred to as **radiation windows**.

In addition to ozone and water vapour, both oxygen and carbon dioxide absorb radiation. Carbon dioxide, however, has little effect on the solar beam as it absorbs mainly in the infrared range between wavelengths 12 μm and 18 μm. Further absorption also takes place due to the presence of suspended liquid and solid particles. Ice crystals and water droplets suspended in clouds also absorb in the infrared range at wavelengths 3, 6 and 12 μm.

The total effect of all these agents of absorption can be seen in Figure 3.6. High values, approaching 1 on the curve, indicate a high proportion of absorption. There is considerable loss by absorption in the short- and longwave parts of the solar spectrum, with relatively little absorption of visible radiation, while selective absorption by water vapour, for example, produces a discontinuous solar spectrum at the Earth's surface.

In the atmosphere, gases and suspended matter can both disperse incident solar radiation. A unidirectional solar beam is partially transformed into a multidirectional scatter of radiation, some of which ultimately passes back out of the atmosphere into space. Scattering by gases is related to the wavelength (λ) of the radiation by the expression:

$$\text{degree of scattering} = \frac{\text{a constant}}{\lambda^4}.$$

In the visible part of the solar spectrum, blue light is scattered to a greater extent than other wavelengths, resulting in the predominantly blue colour of the sky.

Scattering by materials suspended in the atmosphere is more correctly termed **diffuse reflection**. Multiple reflection of that part of the solar beam incident upon particle surfaces inevitably results in a scattering effect. The amount of scattering that takes place depends on the size of the particles, particle density in the air, and the distance radiation travels through the atmospheric layer containing the particles. An example of scattering loss may be found in the observed reductions in solar radiation reaching the Earth's surface due to volcanic dust in the atmosphere. Dyer and Hicks (1965) suggested a reduction as high as 25% due to the presence of such materials up to 25 km above the Earth's surface.

The intervention of a band of cloud across the solar beam effectively reduces the amount of radiation reaching the Earth's surface. The upper

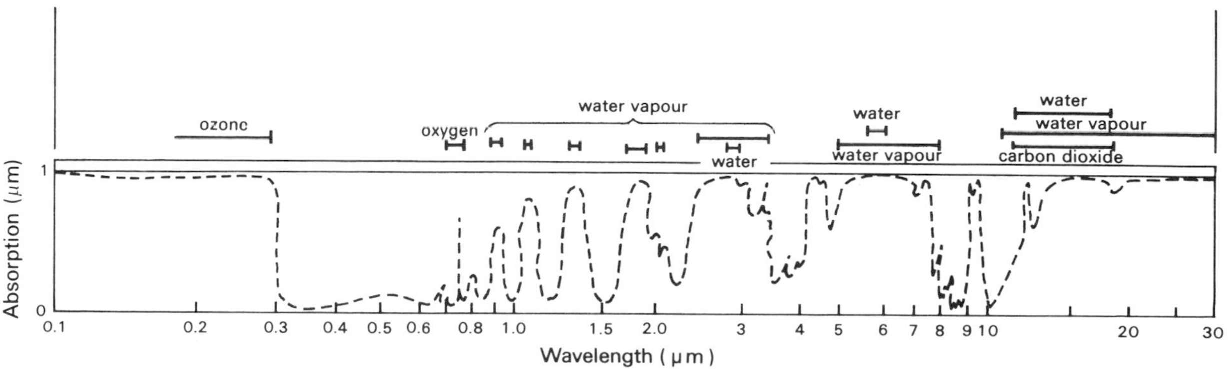

Figure 3.6 Absorption spectrum of the atmosphere. Total absorption = 1, no absorption = 0 (after Fleagle & Businger 1963).

surfaces of clouds are good reflectors of radiation, the amount reflected being dependent upon cloud type, cover and thickness. A light cover of cirrus cloud is a relatively ineffective reflector, whereas dense stratiform and cumuliform cloud may reflect more than 50% of solar radiation incident upon them. Reflection from heavy storm clouds may be as high as 90%. Some of the interrelationships between cloud cover and solar radiation are examined further in Figure 3.7.

The atmosphere directly absorbs a small proportion, some 17 units on average (K_A), of the solar radiation input (100 units), mainly in the short and long wavelengths (see Fig. 3.6). The proportion is small because its major constituents are not efficient absorbers in the wavelengths

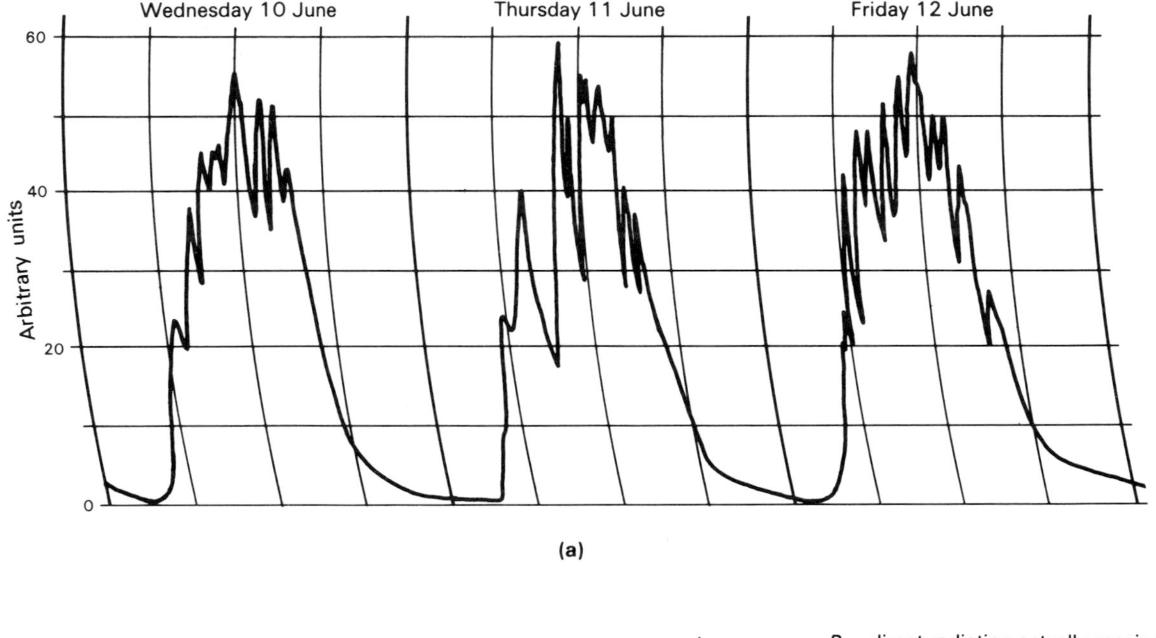

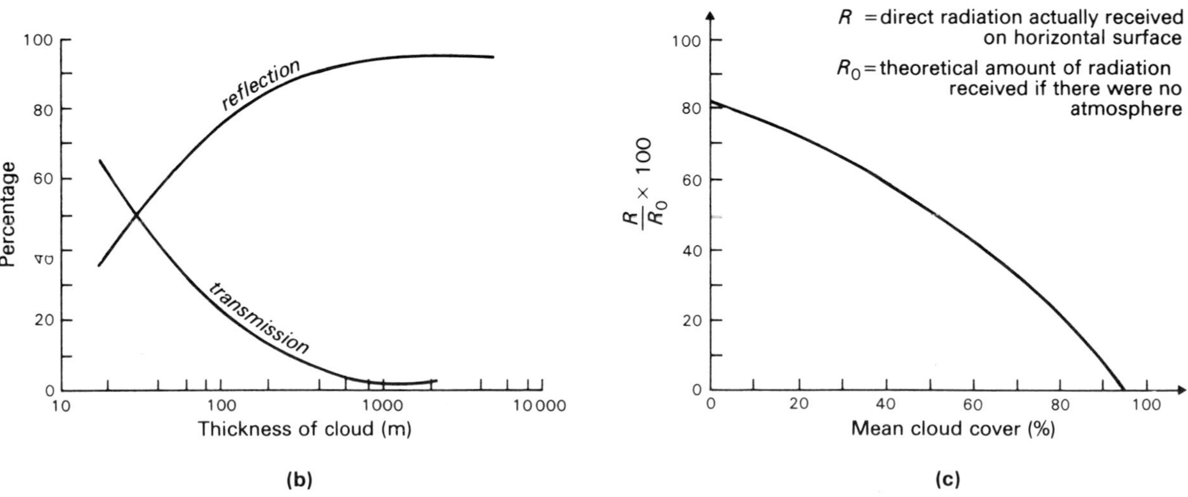

Figure 3.7 The effects of cloud cover: (a) on variations in solar radiation received on a horizontal surface (Portsmouth 1976); (b) on reflectance and transmission in clouds of different thickness (Hewson & Longley 1944); and (c) on radiation receipt (Black *et al.* 1954).

covered by the solar spectrum. However, this absorption of solar radiation contributes to an increase in the internal energy store of the atmosphere. Of the remaining solar radiation, 29 units are lost as output to space by reflection. This is made up of 6 units lost by scattering ($K\uparrow_{Aa}$) and 23 units lost through cloud reflection ($K\uparrow_{Ac}$). A further 54 units of radiant energy are output to Earth surface systems and can be divided into 36 units ($K\downarrow_{D}$) of direct radiation and 18 units ($K\downarrow_{d}$) of diffuse radiation, again as a result of scattering. This throughput of solar radiation forms part of what is known as the solar energy cascade (Fig. 3.8). Solar radiation which is dominantly shortwave is not the only radiation input to the atmospheric system, for it also receives longwave terrestrial radiation from Earth surface systems.

3.2.2 Exchange and throughput of longwave radiation

Gases such as carbon dioxide and water vapour, together with suspended water droplets and ice crystals in clouds, are more effective absorbers in the infrared part of the electromagnetic spectrum (see Table 3.2), particularly for wavelengths greater than 8 μm. As the spectrum of longwave terrestrial radiation from the Earth's surface extends over wavelengths 3 μm to 30 μm, with a maximum emission at 10 μm, a significant proportion will therefore be absorbed in the atmosphere.

In fact only about 7% of terrestrial radiation passes directly into space, the remainder being absorbed. That which is transmitted through the atmosphere does so within a narrow range of wavelengths where little absorption takes place – these wavelengths are referred to as radiation windows. An extremely important window occurs in the absorption spectrum of water vapour between wavelengths 8.5 μm and 11 μm, within which falls the waveband of maximum emission in the terrestrial radiation spectrum. The atmosphere therefore allows most solar radiation to pass through it, but it inhibits the passage of terrestrial radiation. This is commonly referred to as the **greenhouse effect**. As we shall see in Chapter 24, Man can alter this effect considerably by, for example, adding to the amount of carbon dioxide in the atmosphere.

The absorbed radiation is re-emitted, as the atmosphere acts as a radiator. Because of relatively low temperatures, generally less than

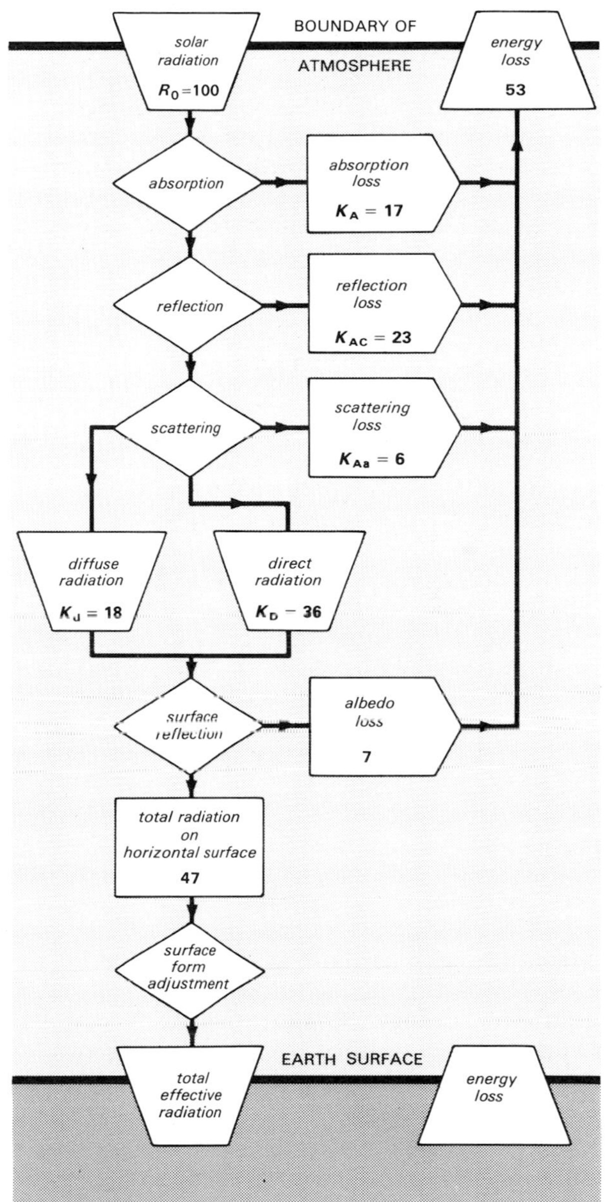

Figure 3.8 The solar energy cascade.

300 K, atmospheric radiation is longwave and of low intensity. Some of the emission passes into space, but approximately 60% returns to the Earth's surface systems as counter radiation. As most of the carbon dioxide, water vapour and clouds are in the lowest 10 km of the atmosphere, it is at this level that greatest absorption and emission

take place. Nearly all counter radiation emanates from the lowest 4 km (Fig. 3.9).

3.2.3 The radiation balance of the atmosphere

The atmosphere receives shortwave radiation from the Sun and by reflection from the Earth's surface. Of this it absorbs only a small proportion (K_A) (see Fig. 3.8). It also receives longwave terrestrial radiation, a large proportion of which it absorbs ($L\uparrow_{EA}$), the remainder passing directly to space through radiation windows. The atmosphere radiates at long wavelengths to space ($L\uparrow_A$) and the ground surface ($L\downarrow$). The majority of the latter takes part in the complex radiant energy exchanges between surface and atmosphere. These input and outputs of energy are represented in Figure 3.10. The equation for the balance of radiation (Q^*_A) may be written:

$$Q^*_A = K_A + L\uparrow_{EA} - L\uparrow_A - L\downarrow.$$

For the whole atmosphere, assuming a solar constant of 100 units, 17 units are absorbed from the solar beam and 91 from terrestrial radiation, 57 units are radiated into space and 78 units back to the ground surface. The atmosphere thus receives 108 units while losing 135, which represents a net radiation balance of −27 units. In reality this energy loss does not take place because other heat transfer processes are at work maintaining an energy balance, as we shall see later.

The net radiation balance of the atmosphere varies little with latitude, as is indicated in Figure 3.11. This is largely due to the localised effects of radiation exchange with the ground surface, which is primarily controlled by variation in carbon dioxide, atmospheric water vapour content and cloud cover. The radiation balance of the atmosphere does not appear to explain the marked latitudinal distribution of balance for the whole Earth–atmosphere system.

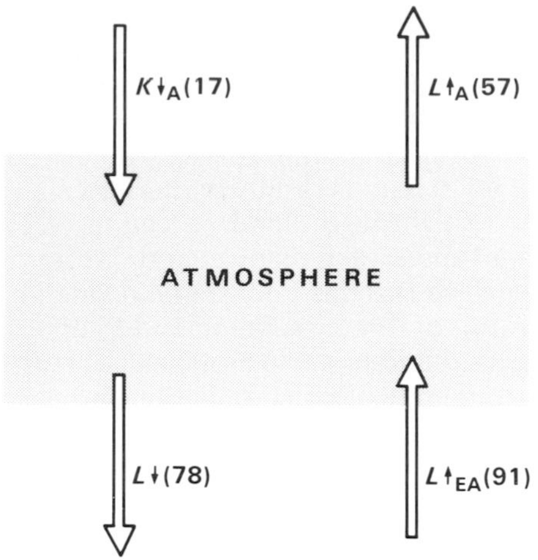

Figure 3.10 Balance of radiation in the atmosphere.

3.3 Earth surface systems

3.3.1 Radiation exchange and throughput

The input of shortwave solar radiation transferred from the atmosphere to Earth surface systems consists of both direct and diffuse radiation ($K\downarrow_D + K\downarrow_d$) where:

$$(K\downarrow_D + K\downarrow_d) = K\downarrow = K\downarrow_0 - K_A - K\uparrow_{Aa} - K\uparrow_{Ac}.$$

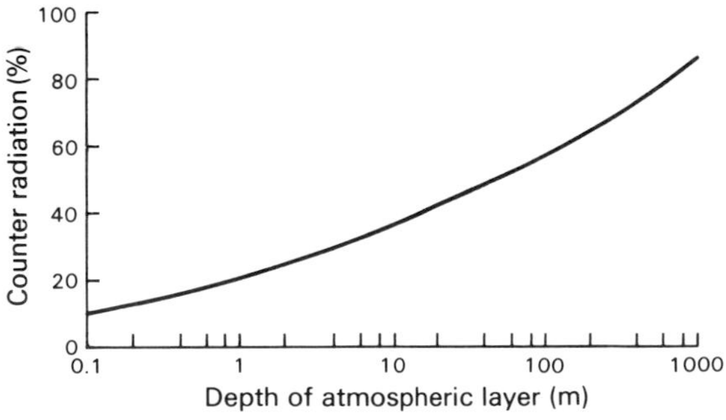

Figure 3.9 The origin of counter radiation arriving at the Earth's surface.

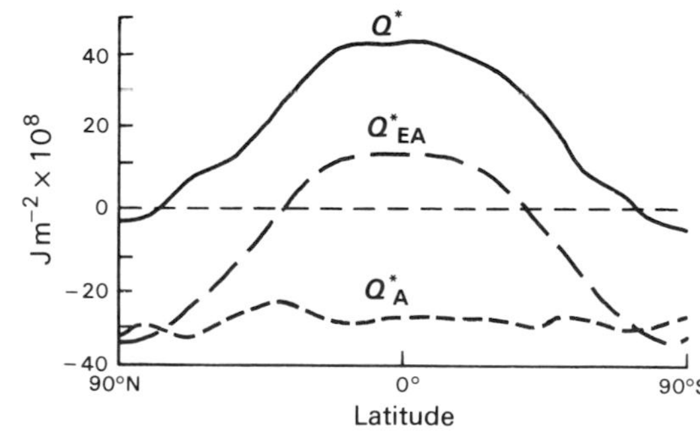

Figure 3.11 Average latitudinal distribution of the radiation balance of the Earth–atmosphere system (Q^*_{EA}), of the Earth's surface (Q^*), and of the atmosphere (Q^*_A) (after Sellers 1965).

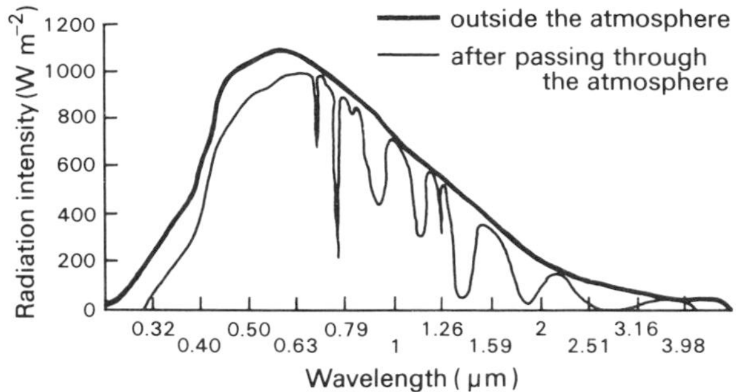

Figure 3.12 The solar spectrum before and after passing through the atmosphere (after Lamb 1972).

The spectrum of $K\downarrow$ differs significantly from that of $K\downarrow_0$ (Fig. 3.12). Not only is there an obvious general reduction in radiant intensity, but in certain wavebands this has been severely reduced, mainly through absorption in the atmosphere. That part of the solar beam which experiences scattering, but which subsequently retains a generally earthward direction of motion, reaches the Earth's surface as diffuse radiation. The proportion of diffuse radiation varies considerably, but for the whole Earth surface it constitutes some 33% of total radiation reaching horizontal surfaces. At the Earth's surface the direct solar beam produces zones of full illumination and full shade. Illumination of the shaded zones is, therefore, dependent on diffuse radiation, and its significance can be appreciated by considering the important role it plays in light penetration into the photosynthesising canopy of vegetation (see Ch. 17).

The distribution of average annual solar radiation falling on horizontal surfaces shows that this does not actually fall into a convenient latitudinal pattern but tends to develop cellular patterns (Fig. 3.13). The role of cloudiness and atmospheric humidity in the development of this pattern can readily be illustrated by examining the distribution of solar radiation over Africa. Relatively low values over the cloudy and humid Zaire Basin, in equatorial latitudes, contrast with the much higher values around the Tropics over the Sahara and Kalahari deserts. In the shorter term, the lifting of

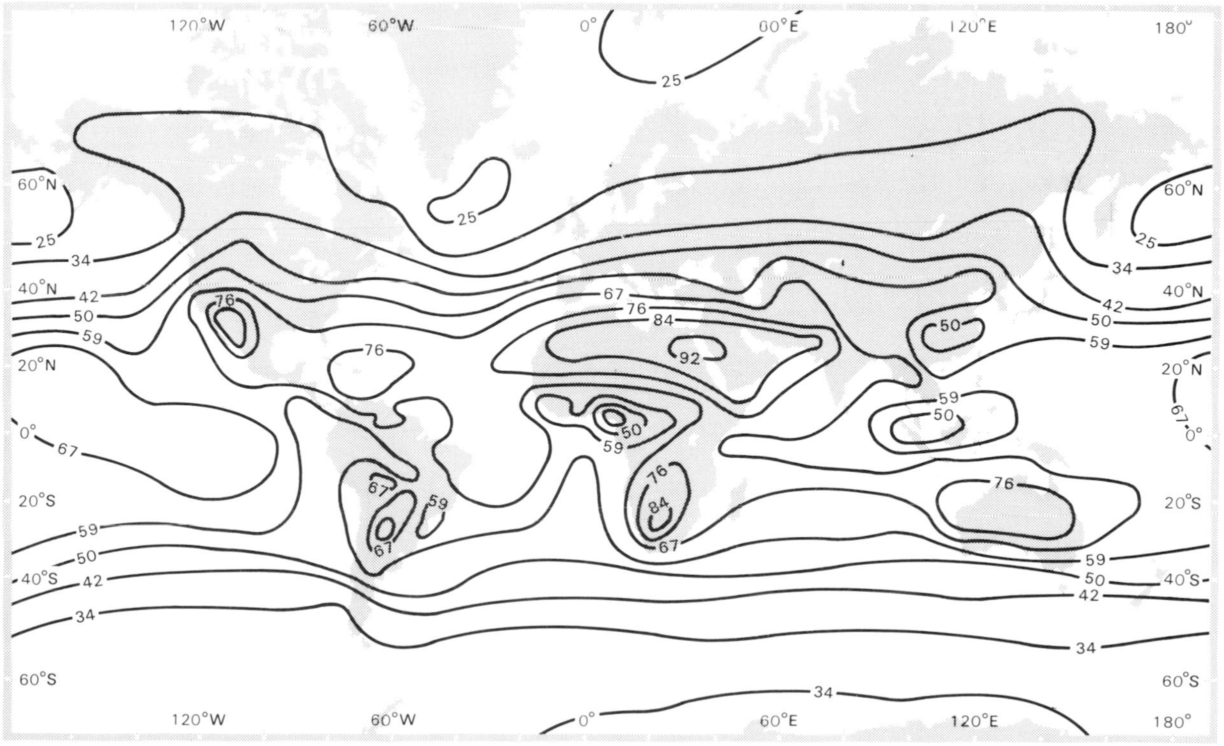

Figure 3.13 The average annual solar radiation on a horizontal surface at the ground, in MJ m^{-2} × 10^2.

dust into the atmosphere by desert dust storms, or by soil erosion arising from poor land management, greatly influences the transmission of solar energy through the atmosphere. Increasing densities of dust in the atmosphere may be instrumental in producing significant long-term changes in radiation, such as that observed in India (Bryson & Baerreis 1967).

The Earth's surface returns some of the solar radiation incident upon it by reflection back through the atmosphere. In terms of the energy cascade, an average of 13% is reflected, which represents 7 units of radiation lost from the surface. Effective solar radiation input on a horizontal surface is, therefore, reduced to 47 units. If r represents the proportion of radiation reflected, then available net shortwave solar radiation at the Earth's surface (K^*) may be written:

$$K^* = K\downarrow - K\downarrow.r \; (\text{or } K\downarrow - K\uparrow).$$

The amount of longwave counter radiation reaching the Earth's surface depends ultimately upon the absorptive properties of the lower atmosphere. Thus areas of the Earth's surface above which the atmosphere is usually humid and cloudy, such as equatorial regions, receive considerable amounts of counter radiation. Over the whole of the Earth's surface, counter radiation constitutes, on average, about 62% of total radiant energy received. The net balance of longwave radiation from the ground surface may be expressed in simple terms in the equation:

$$L^* = L\uparrow - L\downarrow$$

where $L\uparrow$ represents the longwave terrestrial radiation emitted from the ground surface and $L\downarrow$ that returned by counter radiation. This absorption of terrestrial radiation by the atmosphere is of particular significance in that it creates an insulating blanket (the greenhouse effect) above the Earth's surface, effectively inhibiting heat loss. Without such insulation, the temperature of the surface would be some 40°C lower than it is at present.

Earth surface systems also receive energy from the Earth's interior as geothermal heat flow to the surface and a proportion of this is output to the atmosphere as longwave radiation. From the point of view of the atmosphere, it is quantitatively so insignificant as to be lost in the total flux of longwave terrestrial radiation.

The average temperature of the Earth's surface ranges from about 225°K in high latitudes to 300°K in lower latitudes. If we assume an average surface temperature of 288°K and also that the Earth radiates as a black body, then the characteristics of Earth-surface or terrestrial radiation may be determined from the basic radiation laws in Box 3.3. Emission from the Earth's surface is 383 Wm^{-2} and the wavelength of maximum emission is 10 μm. Radiation of this wavelength falls within the infrared sector of the electromagnetic spectrum. Indeed, the entire spectrum of terrestrial radiation lies within the infrared and is consequently referred to as longwave radiation. This is true for the full range of Earth-surface temperatures (Fig. 3.14).

The changes in surface temperature between polar and tropical regions result in variation in the emission of longwave radiation from the Earth's surface. Average rates of emission are less than 150 Wm^{-2} in extremely high latitudes, while values in excess of 400 Wm^{-2} are typical of low latitudes.

The Earth's surface is, however, an imperfect emitter and absorber of radiation. **Emissivity** and

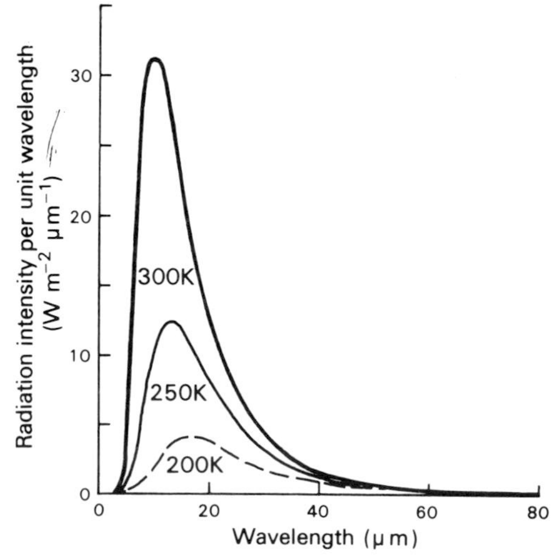

Figure 3.14 The spectra of black body radiation at temperatures of 200 K, 250 K, and 300 K (after Neiburger *et al.* 1971).

Box 3.5

ABSORPTIVITY AND EMISSIVITY

Absorptivity is the fraction of incident radiation absorbed.
Emissivity is the ratio of actual radiation emitted to a theoretical maximum.

Kirchhoff's Law states absorptivity is always equal to emissivity. For a black body:

absorptivity = emissivity = 1.

absorptivity (Box 3.5), instead of having the black body value of 1, have grey body values of less than 1. For example, the ocean surfaces have an emissivity of between 0.92 and 0.96, while for most land surfaces this may fall to below 0.90 (Table 3.3).

Table 3.3 Infrared emissivities.

ice		0.96
water		0.92–0.96
moist ground		0.95–0.98
desert		0.90–0.91
pine forest		0.90
leaves	$0.8\ m^{-1}$	0.05–0.53
	$10.0\ m^{-1}$	0.97–0.98

3.3.2 *The radiation balance of the Earth's surface*

The Earth's surface receives shortwave radiation from the Sun (some of which is lost through reflection) and longwave counter radiation from the atmosphere, while itself emitting longwave radiation. These flows are represented in Figure 3.15. The equation for the net radiation balance (Q^*) may be derived from the short- and longwave radiation balances:

$$Q^* = K^* + L^*$$

For the whole of the Earth, the surface receives 125 units of radiation, while losing only 98. The balance of +27 units implies that the surface is gaining in energy. In reality, this does not take place, so, by inference, there must be a mechanism for dissipating this excess energy. The equality of atmospheric deficit and Earth-surface surplus of radiant energy suggests that the mechanism is one which transfers energy from the surface into the atmosphere above it.

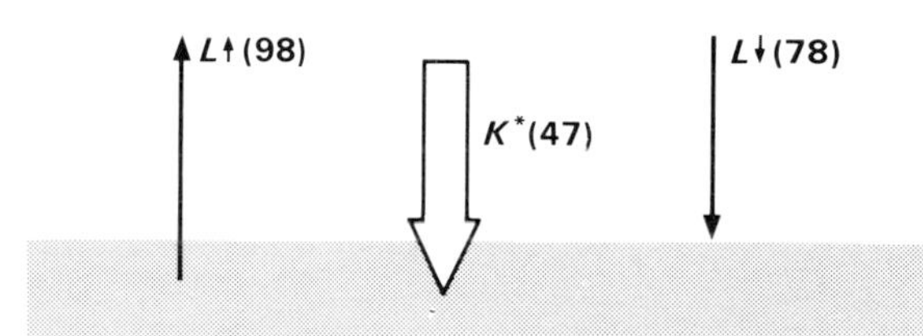

Figure 3.15 Balance of radiation at the Earth's surface.

Unlike that for the atmosphere, the radiation balance of the Earth's surface varies considerably, as indicated in Figure 3.11. There is a general trend towards a positive balance or surplus in low latitudes and a negative balance or deficit in high latitudes. Factors such as cloudiness and atmospheric humidity, which affect the transmission of both solar and terrestrial radiation, and the radiative properties of the surface, create the spatial variation in radiation balance over the Earth's surface. Figure 3.16 is largely based upon estimates by Budyko (1958), and it shows that this spatial variation is not entirely latitudinal. Furthermore, it shows that the net radiation balance of ocean surfaces is generally greater than that of land surfaces at the same latitude. The low reflection and high absorption of solar radiation at ocean surfaces largely accounts for this disparity.

During the year, net radiation changes in value. Solar radiation reaches its maximum intensity at the summer solstice, while net loss of radiation from the surface reaches its maximum during August (Fig. 3.17). Between **y** and **x**, surface losses exceed gains, so the net radiation balance is negative – producing cooling – while between **x** and **y** it is positive – producing warming. In a broadly similar manner, changes in the net radiation balance also occur on a diurnal time scale.

We can see from these hypothetical cases that the radiation balance of the Earth's surface is an important factor in determining the distribution of surface temperature. However, as we have already seen, in the long term there is a net radiation imbalance in the atmosphere and upon the Earth's surface. The long-term net radiation balance of the combined Earth–atmosphere system, however, was shown earlier in this chapter to equal zero. Indeed, if we were able to measure accurately the radiation arriving at the top of the atmosphere, from the Sun and from the solids and fluids of the Earth and its atmosphere, we should expect to receive the following.

Incoming (from space)	*Outgoing (from Earth and atmosphere)*
$K{\downarrow}_0$ solar radiation	$K{\uparrow}_{Ac}$ radiation reflected from clouds
	$K{\uparrow}$ radiation reflected from the Earth's surface
	$K{\uparrow}_{Aa}$ radiation scattered by the atmosphere
	$L{\uparrow}$ terrestrial radiation
	$L{\uparrow}_A$ atmospheric radiation

Figure 3.16 The radiation balance of the Earth's surface in MJ m^{-2} a^{-1} (Budyko 1958).

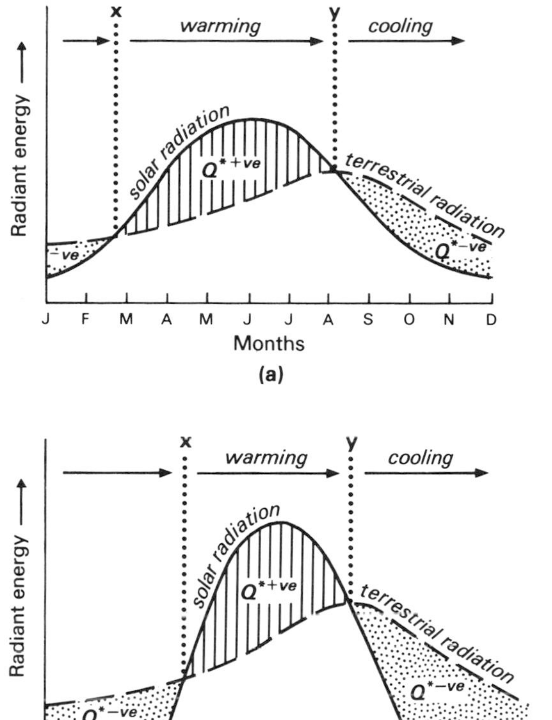

Figure 3.17 Hypothetical (a) annual and (b) diurnal variation in incoming solar radiation and outgoing terrestrial radiation at the Earth's surface.

Expressing this in the form of an equation:

$$Q^* = K{\downarrow}_0 - (K{\uparrow}_{Ac} + K{\uparrow} + K{\uparrow}_{Aa} + L{\uparrow} + L{\uparrow}_A).$$

net radiation at outer boundary of the atmosphere

Over the whole of the atmosphere boundary, the long-term average value of this balance is zero, as indicated in Figure 3.18. Therefore, the existence of a long-term net radiation imbalance in the atmosphere and at the Earth's surface implies that energy is being transferred by processes other than those of electromagnetic radiation, in order to maintain the overall net balance. We must therefore consider alternative forms of energy exchange in the form of an energy balance.

3.4 Energy balance of the Earth–atmosphere system

If we consider the simplest of situations – a completely dry, uniform solid surface – solar and atmospheric radiation is absorbed into it, some of which is then reradiated. The heat energy which passes into the surface (Q_G) does so by **conduction**, while it is also transferred into the atmosphere by conduction and **convection** (Box 3.6). The atmosphere is an extremely poor conductor of heat but, being a fluid, may be set into convectional motion. Heat energy is conducted into a thin air layer in contact with the surface and then redistributed by free or forced convection. Thus some of the available heat energy at the surface is utilised in heating the atmosphere in this manner (Q_H). If we have Q^* units in our net radiation balance, we can express its expenditure as a simple equation:

$$Q^* = Q_H + Q_G.$$

The signs of Q_H and Q_G may be either positive or negative, in that heat may pass also from atmosphere to ground surface or from subsurface to surface.

For surfaces that contain water, there is the

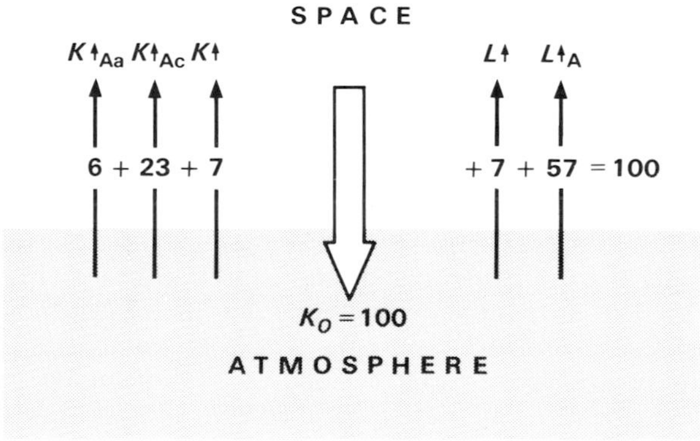

Figure 3.18 Balance of radiation at the outer boundary of the atmosphere.

Box 3.6

CONDUCTION AND CONVECTION

Thermal conduction is the process by which heat energy is transferred from molecule to molecule and which involves no movement of mass.

The rate of conduction of heat through still air is low and, in terms of large-scale heat transfers within the free atmosphere, it can be regarded as being of negligible significance.

Convection is the transfer of heat within a fluid medium involving the transfer of mass. This is of considerable significance in the fluid medium of the atmosphere. It may take the form of free or forced convection. In free convection, the movement of air is in the form of density currents as it is heated from below. Heated air rises and is replaced by cooler air.

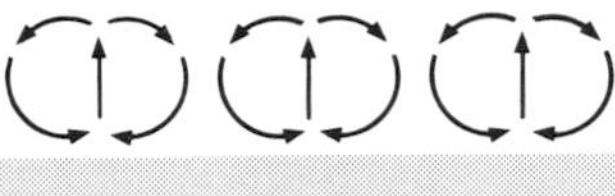

In forced convection air movement is in the form of mechanical turbulence in the air flowing across a surface.

added complication of heat gain or loss as phase changes take place (see Ch. 4). For example, in evaporation, when water changes from liquid to gas, heat energy is consumed (Q_E). So the simple equation becomes:

$$Q^* = Q_H + Q_G + Q_E.$$

The distribution of heat energy along these pathways varies according to the type of surface. On dry desert surfaces, for example, little energy will normally be consumed in evaporation (Fig. 3.19), while most energy is used to heat the atmosphere. In direct contrast, in a moist environment a large proportion of energy is consumed in evaporation at the expense of heating the atmosphere. The basic energy balance equation may be modified to apply to most of the major surface types occurring over the Earth (Box 3.7). However, it must be borne in mind that these are simplifications of what is often a considerably more complex energy balance.

For land surfaces we can ignore, for the sake of simplicity, lateral heat transfer beneath the surface. Not only are rock and soil poor conductors of heat, but temperature gradients within them are also relatively modest. Under a fluid surface, however, a lateral transport of heat energy does take place. The fluid moves in response to free convection or may be driven by the movement of the atmosphere across its surface. The basic energy balance equation for the oceans may be modified to include a lateral movement of heat energy (Q_F), thus:

$$Q^* = Q_H + Q_G + Q_E + Q_F.$$

Such transfers occur on a large scale in the oceans

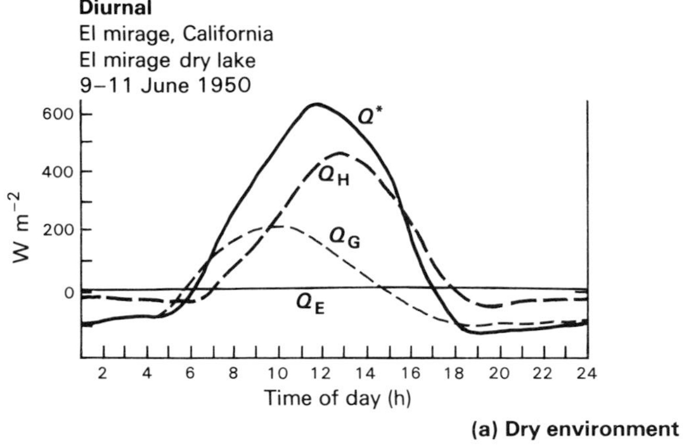

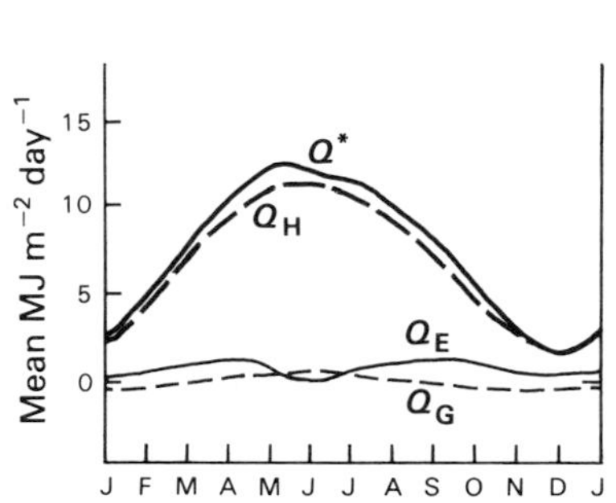

(a) Dry environment

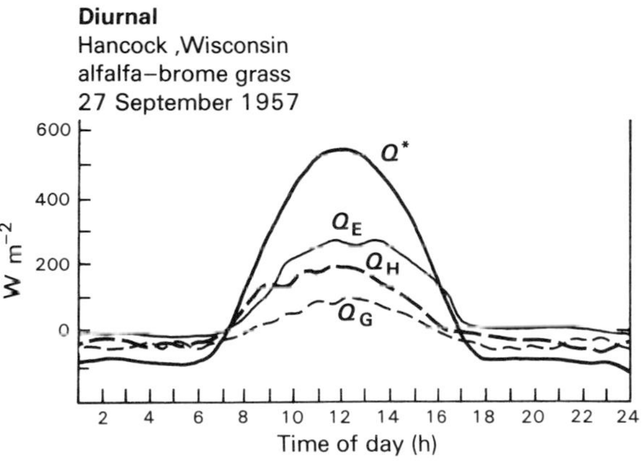

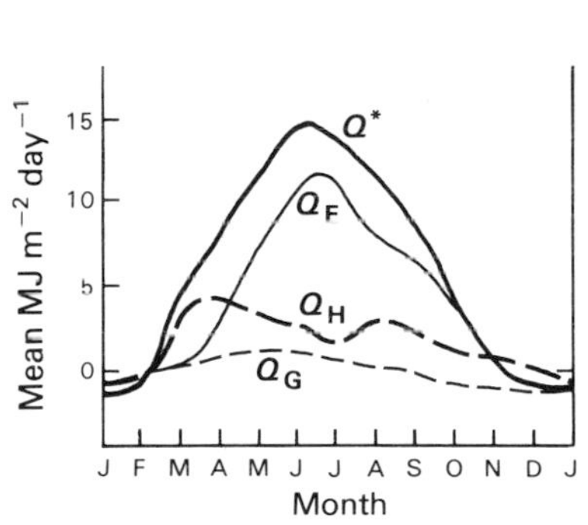

(b) Moist environment

Figure 3.19 Diurnal and annual variation of heat-energy balances for (a) dry and (b) moist surfaces. Q^* = radiation balance, Q_H = heat transferred into air, Q_E = heat utilised in evapotranspiration, Q_G = heat transferred into subsurface.

Box 3.7

ENERGY BALANCES

Vegetated surfaces

The energy balance equation has to include heat consumed in moisture loss from plant surfaces; transpiration (Q_T) (see Ch. 4) and energy consumed in photosynthesis (Q_P):

$$Q_Y^* = Q_E + Q_T + Q_H + Q_G + Q_P + C.$$

C is a complex term as it includes the absorption of heat into the plant itself.

Ice surfaces

The equation has to take into account the melting of ice, which requires heat energy (Q_M). Freezing actually releases heat energy ($-Q_M$):

$$Q_{ice}^* = Q_E + Q_G + Q_H \pm Q_M.$$

in the form of ocean currents. Figure 3.20 illustrates the complex nature of the heat energy balance in ocean waters. Very large amounts of heat energy are supplied beneath the surface to the extent of exceeding the contribution made by net radiant energy.

A summary of global variation in the elements of the energy balance appears in Table 3.4. The significance of subsurface energy flow in the oceans can be seen. Latitudes greater than 30° experience a net gain of energy in this way, while those in lower latitudes are losing energy. The oceans therefore play an important role in the redistribution of heat energy in the Earth–atmosphere system.

So far we have not looked at energy flow in the other fluid – the atmosphere. This will be considered in more detail in Chapter 4, but at this stage it is worth noting that the atmosphere also redistributes heat. We have seen that over the global surface there is a net surplus of radiant energy in low latitudes and a deficit in high latitudes. This imbalance is redressed by heat transfer in a moving atmosphere, through a complex circulatory system.

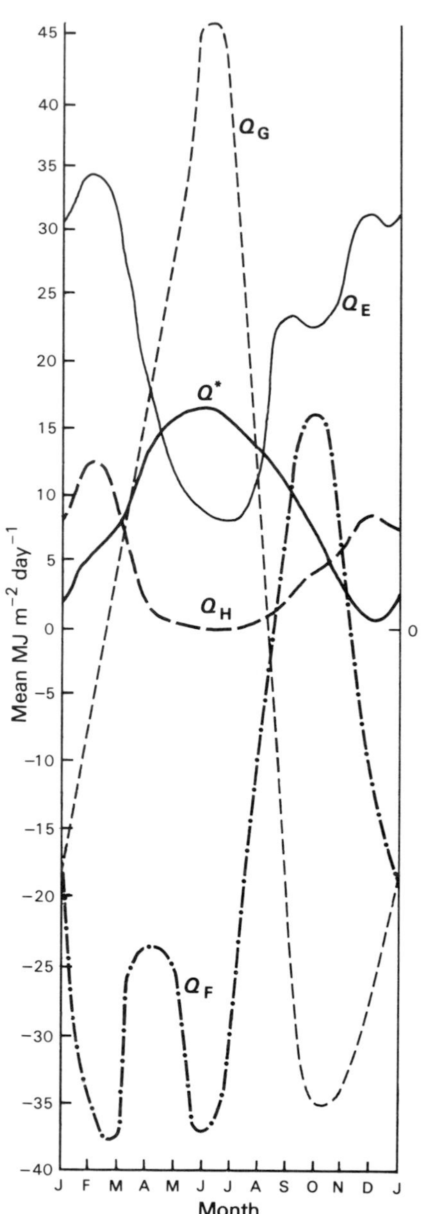

Figure 3.20 Average annual variation in surface heat energy. Q_F = horizontal heat transfer, Q^* = radiation balance, Q_G = heat transferred into the subsurface, Q_H = heat transferred into lower air layers, Q_E = heat utilised in evaporation.

3.5 The Earth's interior systems

At this point we have considered only two of the three subsystems depicted in Figure 3.5, the atmosphere and Earth-surface systems. The third subsystem is the Earth's interior, and here the principal processes of energy transfer are conduction and convection. As we have seen, the energy exchanges of surface and atmospheric systems are dominated by the massive flux of solar energy. However, this has little direct effect on the Earth's interior for it does not penetrate significantly beneath the surface. Diurnal temperature variations seldom exceed 1°C at a depth of 1 m, while seasonal temperature changes penetrate to 30 m at the most, and therefore processes beneath the immediate surface layer are controlled by energy transfers which reflect changes in storage of the Earth's internal energy.

3.5.1 Sources of the Earth's internal energy

Although current theories (Box 3.8) hold that the original Earth-forming materials were fairly cool,

Table 3.4 Mean latitudinal values of the components of the energy balance equation for the Earth's surface, in MJ m^{-2} (after data from Sellers 1965). N.B. Data have been converted to SI form and rounded to the nearest whole number. In some cases, therefore, the energy balance equation may not balance.

Latitude zone	Oceans				Land			Globe			
	Q	Q_E	Q_H	Q_F	Q^*	Q_E	Q_H	Q^*	Q_E	Q_H	Q_F
80–90°N								−38	13	42	8
70–80								4	38	−4	−29
60–70	97	139	67	−109	84	59	25	88	84	42	−38
50–60	122	164	67	−109	126	80	46	126	118	59	−50
40–50	214	223	59	−67	189	101	88	202	160	71	−29
30–40	349	361	55	−67	252	97	155	307	248	101	−42
20–30	475	441	38	−4	290	84	206	403	307	101	−4
10–20	500	416	25	59	298	122	176	445	340	67	38
0–10	483	336	17	130	302	202	101	441	302	46	92
0–90°N								302	231	67	4
0–10	483	353	17	113	302	210	92	441	319	42	80
10–20	475	437	21	17	307	172	134	437	378	46	13
20–30	424	420	29	−25	294	118	176	395	349	67	−21
30–40	344	336	34	−25	260	118	143	336	311	46	−21
40–50	239	231	38	−29	172	88	84	235	223	42	−29
50–60	118	130	42	−55	130	84	46	118	130	46	−59
60–70								55	42	46	−34
70–80								−8	13	−17	−4
80–90								−46	0	−46	0
0–90°S								302	260	46	−4
globe	344	310	34	0	206	105	101	302	247	55	0

Box 3.8

FORMATION AND STRUCTURE OF THE EARTH

Current theories suggest that the Earth formed by accretion of cosmic dust and particles, in a relatively cool state – perhaps 600–1000°C. This process of accretion implies that originally the Earth was homogeneous throughout, as fundamentally similar material was continually added. In contrast, the present structure of the Earth is a series of concentric shells consisting of different mineral types forming crust, mantle and core.

Crust, mantle and core are differentiated by characteristic densities and mineral compositions. Seismic investigations have revealed marked density differences at the crust/mantle boundary (the Mohorovičić discontinuity) and the mantle/core boundary (the Gutenberg discontinuity). From their density and inferred temperatures, reasonable speculations can be made about the mineral composition of the Earth's interior. The composition of crust, of course, can be observed directly at the land surface and by drilling beneath the land surface and ocean floor.

Crust: consisting largely of silicate- and aluminium-rich minerals (sial). It is divided into two forms. The continental crust, fragmented and comprising the land masses, consists of granitic rock with a veneer of sedimentary rocks and possesses a mean density of 2.8. Oceanic crust, flooring the ocean basins and also underlying the continents, is composed of basalt, slightly denser than the continental crust.

Mantle: composed largely of silicate and mafic (magnesium and iron-rich) minerals. Its density ranges from 3.3 to 5.5 and it exhibits high temperatures and pressures. Slow viscous movements take place within the mantle.

Core: probably composed of solid and liquid metals, iron and nickel, at very high temperatures and density.

Recent studies have suggested that temperatures in the core may exceed 5000°C. During the Earth's existence, processes leading to chemical differentiation have been at work, as minerals have separated out and become concentrated in various layers. In addition, the conversion of other energy forms to heat has taken place, causing the Earth's interior temperatures to rise. It is to these processes that we must look for possible sources of the Earth's present terrestrial energy.

various processes associated with formation of the Earth can be inferred to have contributed to heat generation. Small bodies arriving at the Earth's surface would possess kinetic energy, and this would be dissipated as heat energy around the point of impact. Although much of this heat would be radiated back to space, a proportion would be transferred to the Earth's interior. As the Earth grew, due to the continuing addition of such material, its interior would become progressively more compacted and dense, and internal temperatures would rise as a result of the increased pressure. If 0.1% of the kinetic energy became trapped, it can be calculated theoretically that the internal temperature would be raised by 30°C, and compression could account for an increase of up to 900°C at the centre. These processes can account for only a fraction of the Earth's internal heat. It is likely, however, that they have contributed in a small way to the Earth's present internal temperatures. Since heat transfer from the interior to the surface is a very slow process, it is possible that a small proportion of the Earth's present heat output may have derived from these events very early in its history.

As the denser elements, originally scattered throughout the homogeneous Earth, concentrated out and settled towards the Earth's centre, large amounts of gravitational (potential) energy must have been released. This would have been in the form of heat, some of which would have contributed to raising the temperature of the core (Box 3.8), but much of which would be available to raise the Earth's temperature. Calculations, based on the volume of the core and the density of minerals within it, show that the amount of heat released would be sufficient to raise the temperature of the Earth by 1500°C. Clearly this is likely to have been an important heat source. It is not yet known when core formation took place, whether early or late in the Earth's history; therefore it is not possible to say how much of the heat from this source has already been lost by radiation, or how much remains.

When a radioactive isotope decays, there is a loss of nuclear energy which is dissipated as heat in the vicinity of the decaying isotope. There are four major radioactive isotopes which produce heat in significant quantities at the present time. These are shown in Table 3.5. The **half-life** is the length of time required for half of the original parent isotope to decay. Only isotopes with a half-life of 10^9–10^{10} years are important in the present context. Short-lived isotopes are very low in abundance and longer-lived isotopes produce too little heat per unit of time to be significant. The quantity of heat released by these isotopes can be determined experimentally in the laboratory, and their abundance in the field can be measured in terms of their concentration in different types of rock. It is possible, therefore, to calculate their potential as heat producers. Such calculations reveal that radioactive decay in this way could easily account for the current observed terrestrial heat flow. In addition it seems likely that other, short-lived, isotopes formerly existed within the Earth. These too would have decayed to produce heat and may have contributed to present Earth temperatures, even though the parent isotopes, having decayed, no longer exist.

Although there are other possible contributors, these are the probable major sources of the Earth's internal heat energy. As yet, we are uncertain of their exact contributions to terrestrial energy, for many uncertainties and assumptions exist about conditions and processes operating during the Earth's early history. Additionally, of course, our knowledge concerning the Earth's interior is largely inferential.

3.5.2 Energy transfer from the Earth's interior to the surface

Much of the Earth's internal energy is in the form of heat and is transferred to the surface by conduction, but also in part by convection (see Box 3.6). It seems likely that slow upward movements occur with at least part of the mantle convecting heat to the base of the crust, through which it is transferred mainly by conduction. Studies of heat flow did not begin to take place until 1950, and our current knowledge of the subject is still far from complete. Nevertheless, by indicating the method by which it can be measured and the results obtained, we can go some way towards an

Table 3.5 The major radioactive elements.

		Half-life (years)	Heat production ($J\ gm^{-1}\ yr^{-1}$)
uranium	^{238}U	4.5×10^9	2.97
	^{235}U	0.71×10^9	18.00
thorium	^{232}Th	13.9×10^9	0.84
potassium	^{32}K	1.3×10^9	0.88

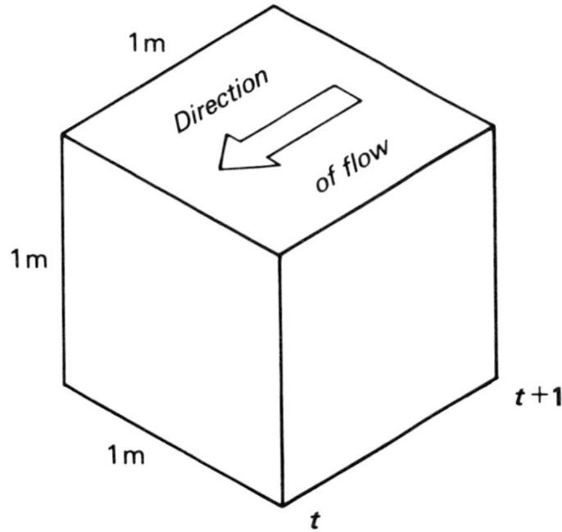

$$Q = \frac{K.A[(t+1)-t]}{L}$$

$$\therefore K = \frac{Q.L}{A[(t+1)-t]}$$

where Q = rate of heat flow ($J\ s^{-1}$)
A = cross section area
L = length of gradient
K = thermal conductivity of material
t = temperature (°C)

Figure 3.21 Thermal conductivity.

understanding of the magnitude and distribution of terrestrial energy inputs to surface systems.

Heat flow per unit area of the Earth's surface at a particular point is defined as follows:

$$q = Kv$$

where q = heat flow, $10^{-6}\ J\ m^{-2}\ s^{-1}$, K = thermal conductivity, $10^{-6}\ J\ m^{-1}\ s^{-1}\ °C^{-1}$ and v = vertical temperature gradient, $°C\ m^{-1}$.

This requires the assessment of the vertical temperature gradient within the crust, obtained by measuring the difference in temperature between two points of differing depth within a borehole in the crust. The average temperature gradient in the upper part of the crust is around $3°C\ km^{-1}$, although significant variations do occur. Also required is the thermal conductivity of the crustal material; that is, the quantity of heat (in joules) flowing across an area of 1 square metre in 1 second in a material where the temperature gradient is 1°C per metre (Fig. 3.21). Since the crust consists of solid materials, this transfer of heat takes place by conduction. Thus heat flows along the geothermal gradient (Fig. 3.22) from the warm interior out towards the cool surface. Heat flow is measured in heat flow units (HFU) where

$$1\ \text{HFU} = 4.19 \times 10^{-6}\ J\ cm^{-2}\ s^{-1}.$$

The mean world heat transfer to the surface is 1.5 HFU, an amount sufficient to melt approximately 5 mm thickness of ice per year, and representing around 1×10^{-3} of the amount of incoming solar radiation. This mean value masks considerable spatial variations in heat flow, as shown in Table 3.6.

The values of heat flow for crust of continental and oceanic provinces are of the same order of magnitude. Yet the composition of the crust varies considerably between the continental and oceanic realms. Continental crust contains a relative abundance of radiogenic materials which are sufficient to account for heat flow observed there. Oceanic crust contains only sparse concentrations of such minerals and the magnitude of heat flow in such regions must be accounted for by the presence of hot mantle material not far beneath the crust. This in turn leads to a consideration of processes within the mantle and, as will be demonstrated in Chapter 5, large-scale slow circulatory movements occur there bringing hot mantle

Table 3.6 Spatial variation in heat flow, shown as mean values for different orogenic regions (after Sass 1971). (1 HFU = $4.19 \times 10^{-6}\ J\ cm^{-2}\ s^{-1}$.)

	HFU
ocean basins	1.27
mid-ocean ridges	1.91
Precambrian shields	0.98
Palaeozoic orogenic belts	1.44
Tertiary orogenic belts	1.77

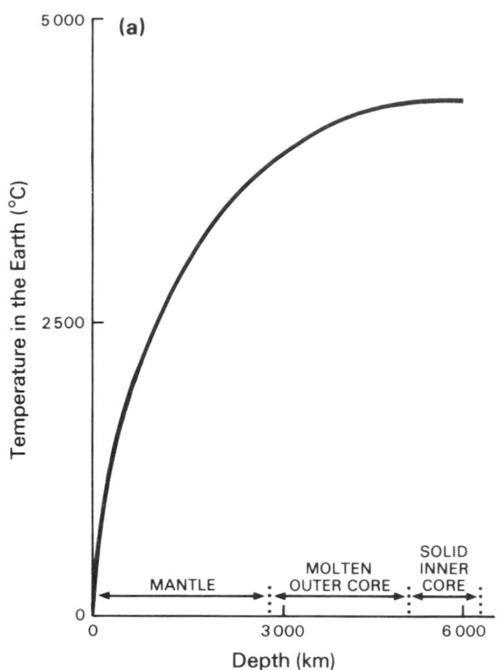

Table 3.7 Relative magnitudes of solar and terrestrial energy.

	Joules/year
solar energy received	5.6×10^{24}
geothermal energy loss by radiation	1.3×10^{21}
volcanic outflows	5.3×10^{18}
earthquakes	1.0×10^{18}

Figure 3.22 (a) Geothermal gradient. (b) A volcanic cone: this represents an accumulation of lava and ash ejected at the Earth's surface; steam and gases are still issuing from the vent, indicating contemporary transfer of heat from within the crust (Mt. Ngauruhoe, New Zealand).

(b)

material up from greater depths. In this way heat transfer by convection takes place through the mantle towards sections of the oceanic crust, through which it is then conducted.

Small quantities of terrestrial heat also arrive at the surface directly by convection. This results from volcanic and geothermal processes, as a result of which hot materials (lava, steam) from the Earth's interior are ejected directly at the surface. The magnitude of such losses is shown in Table 3.7.

In the wake of the energy crises of the 1970s renewed interest has been shown by the developed nations in the potential of geothermal energy (particularly geothermal waters and steam) (Fig. 3.23). So far, the primary use is for the generation of electricity. World electrical generation from geothermal energy in 1971 was approximately 800 MW or about 0.08% of the total world electrical capacity from all modes of generation. Geothermal resources have other uses, although to date they have been minor. They include local space heating and horticulture. Much of Reykjavik (Iceland), Rotorua (New Zealand), Boise (Idaho, USA), Klamath Falls (Oregon, USA) and several towns in Hungary and the Soviet Union are heated by geothermal water (Muffler & White 1975).

Figure 3.24 represents, in a simplified manner, the cascade of energy within the Earth, a flux of energy which represents a long-term change in the internal potential energy store of the system and a long-term net negative energy balance. The implications of this transfer of energy to the crust and surface systems, where further energy conversions take place, will be considered briefly in the following section and in more detail in Chapters 5 and 10.

Figure 3.23 Tapping of subterranean geothermal heat for energy use (Wairakei, New Zealand).

3.6 Energy conversion

The Earth–atmosphere system uses heat energy which it receives from the Sun and, to a lesser extent, from within the Earth. This heat energy provides an input to both Earth-surface and atmospheric systems, thereby increasing their own internal energy. In the case of the atmosphere, this may be seen as increasing its potential energy. The conversion of this potential energy into kinetic energy takes place as the internal energy imbalances of the atmosphere are redressed. In Chapters 4 and 8 we shall see that this conversion is essential to the maintenance of circulation systems and to power the hydrological cascade transferring water from the atmosphere to the Earth's surface and back again. A fraction of the solar radiation input is converted directly to chemical energy by the photochemical reaction of photosynthesis in green plants. This fraction is very small, however, reaching perhaps 2–5% in efficient communities but only a fraction of 1% on a global basis. Nevertheless, the consequences of photosynthesis for the organisms of the biosphere (Ch. 6) are enormous, and the influence of the surface energy balance on water loss and leaf temperature of plants partially determines the overall productivity of the biosphere.

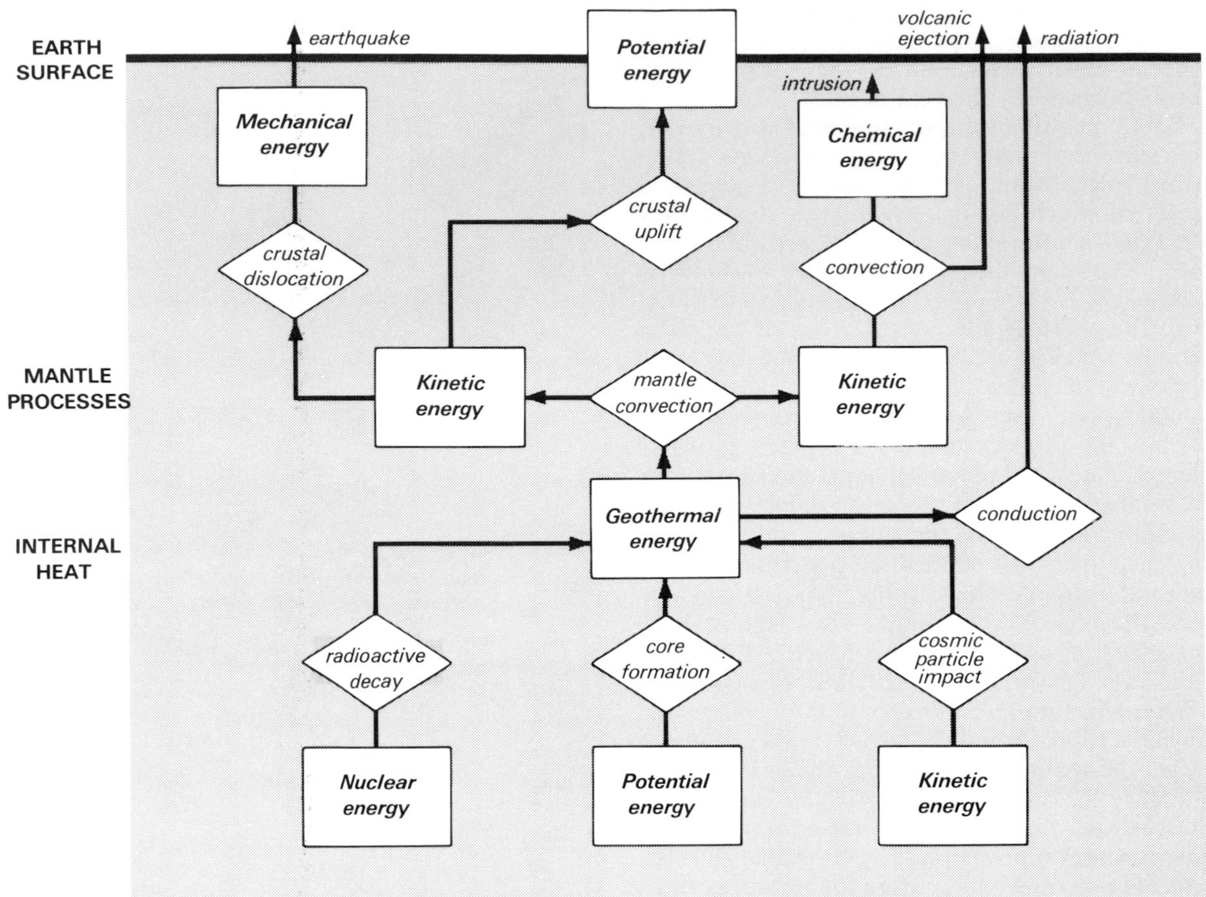

Figure 3.24 Geothermal energy cascade.

Although operating over an infinitely longer time scale, the mobility of the Earth's mantle provides yet another input of potential energy to Earth-surface systems. These slow large-scale mantle currents, representing a conversion from heat to kinetic energy, are responsible for both vertical and lateral displacement of the overlying crust and the conversion of kinetic to mechanical energy which may be released in earthquakes (Fig. 3.24). Vertical displacements of the crust, however, represent a conversion of convective kinetic energy to the potential gravitational energy associated with relief. This potential energy is again converted to kinetic energy which, in combination with the kinetic energy of water moving through the hydrological cascade, is transformed to mechanical energy and heat as surface materials are transferred to lower elevations by the processes of denudation (see Chs 5 & 10).

In accordance with the Second Law of Thermodynamics, none of the energy transformations referred to above is 100% efficient. At each conversion some energy is dissipated as heat, ultimately as longwave radiation output to space. The continued operation of the subsystems of our model (see Fig. 3.5) therefore depends on a continued flux of energy from the Sun and from the Earth's interior.

The net loss of geothermal energy to space represents an irreversible depletion of the Earth's energy store which is itself a legacy of the events that caused the formation of the Earth. Ultimately this store of energy within the planet must become exhausted. The present state of the crustal system, therefore, and the processes operating within it, cannot be regarded as permanent. It is but one phase in the geological evolution of the Earth. In the long term, the depletion of the Earth's internal energy will result in a running down of the system – a cooling down as the heat energy declines, a reduction in the mobility of the mantle and ultimately solidification. The timescale over which this can be expected to occur, however, need not be a source of anxiety for the Earth's present inhabitants!

The same is also true of the external source of energy – the Sun. It too can be regarded as a finite store of energy which is becoming progressively depleted through time. At its present rate of energy conversion, however, the mass of the Sun (which is being transformed to energy by nuclear fusion) will diminish by only one millionth part of its mass in 15 million years. We are probably safe, therefore, in making the assumption, as we have done in this chapter with our closed system model, that the flux of energy across the boundary of the system and within it maintains a long-term average steady state.

Further reading

Good general accounts of the Earth's external energy relationships are given in the following.

Eliassen, A. and K. Pedersen 1977. *Meteorology: an introductory course*, Vol. 1, *Physical process and motion.* Oslo: Scandinavian University Books.

Gates, D. M. 1962. *Energy exchange in the biosphere.* New York: Harper & Row.

Oke, T. R. 1978. *Boundary layer climatology.* London: Methuen.

The internal energy of the Earth is covered by the following.

Clark, S. P. 1971. *Structure of the Earth*, Ch. 8. Englewood Cliffs, NJ: Prentice-Hall.

Gaskell, T. F. 1967. *The Earth's mantle*, Ch. 9. London: Academic Press.

Smith, P. 1973. *Topics in geophysics*, Ch. 3. Milton Keynes: Open University Press.

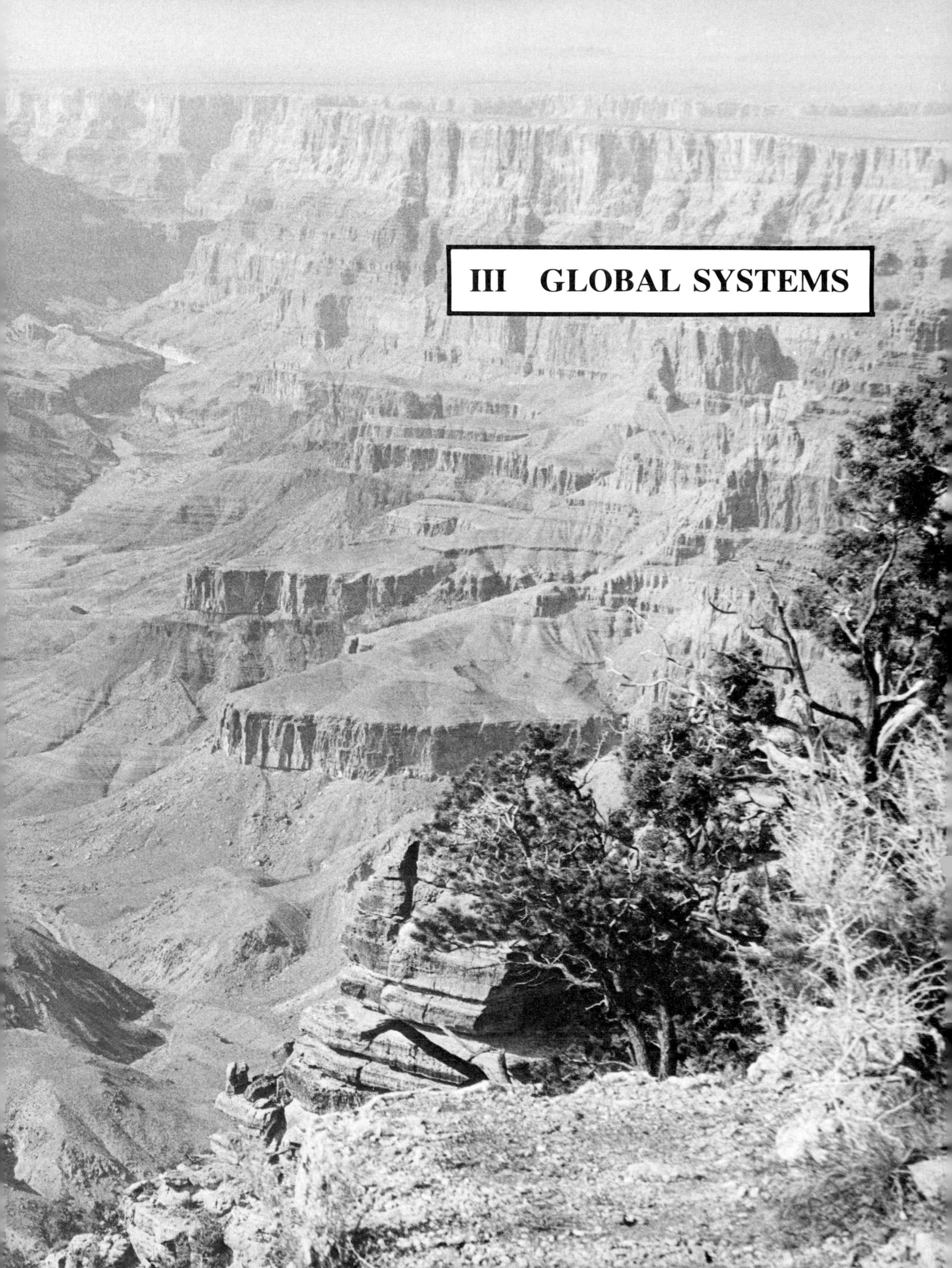

III GLOBAL SYSTEMS

4 The atmosphere

The Earth's atmosphere, a complex fluid system of gases and suspended particles, did not have its origins in the beginnings of the planet. The atmosphere of today has been derived from the Earth itself by chemical and biochemical reactions (see Ch. 7). Although this fluid system forms a gaseous envelope around the Earth, its boundaries are not easily defined. They can be arbitrarily delimited as the Earth/atmosphere interface and atmosphere/space interface, but it must be recognised that these are oversimplifications. For example, there is no outer edge to the atmosphere, but a zone of transition where the Earth and solar atmospheres merge. Similarly, at the Earth/atmosphere interface the atmosphere penetrates into the voids and pore spaces between the particles of soil and regolith and is continuous with the so-called soil atmosphere (see Ch. 20).

4.1 Structure of the atmospheric system

Just five gases – nitrogen, oxygen, argon, carbon dioxide and water vapour – together make up 99.9% of the total volume of the atmosphere. Together with suspended particles such as water droplets, dust and soot, and the minor gases, these represent the elements of the system. Omitting water vapour (which fluctuates considerably in quantity) and the suspended particles, the gases are present in dry air in the proportions indicated in Table 4.1. As most of these constituent elements of the system are in a gaseous phase, they are present for the most part as separate molecules, atoms or even ions, and are in a state of constant motion. The organisation of these units into more complex compounds and into larger aggregates of matter to form observable structures, such as in the lithosphere and biosphere, therefore, is largely lacking in the atmosphere. Structure exists, but it has to be seen partly in terms of the differential distribution of the constituents within the system and partly in terms of the variations in the measurable attributes of the system, such as temperature and pressure.

The proportions of the various gases change very slowly over time, but they do show distinct vertical distributions through the atmosphere. In the lowest 10 km the proportions change little due to the turbulent mixing of the atmosphere up to this height. Above this, layering is more in evidence with gases and suspended particles organised into bands. Before examining this vertical structure we shall consider briefly the distribution of the more important gases in the atmosphere.

Nitrogen. Nitrogen occupies over three-quarters of the atmosphere and occurs throughout a deep layer extending over 100 km from the Earth's surface. The greatest concentrations of molecular nitrogen (N_2) occur in the lowest 50 km, while atomic nitrogen (N) is the more prevalent between 50 and 100 km. Atomic nitrogen is subjected to shortwave cosmic radiation in the upper atmosphere which produces an unstable isotope of carbon (^{14}C) referred to as radiocarbon. This combines with oxygen in the lower atmosphere to produce radiocarbon dioxide which is assimilated by living systems at the Earth's surface. There is normally one molecule of radiocarbon dioxide to 10^{12} molecules of normal (^{12}C) carbon dioxide. The unstable ^{14}C isotope gradually reverts back to nitrogen which passes back into the atmosphere.

Table 4.1 Constituents of dry air.

	Mol. weight	Percentage of volume	Percentage of mass
nitrogen	28.01	78.09	75.51
oxygen	32.00	20.95	23.15
argon	39.94	0.93	1.23
carbon dioxide	44.01	0.03	0.05

This radiocarbon decay has provided man with a means of dating organic deposits at the Earth's surface. Although it occupies such a large volume, nitrogen has little effect on the global radiation balance and, although essential for living systems, it can not be assimilated directly by either plants or animals.

Oxygen. Oxygen occurs throughout the lowest 120 km of the atmosphere and occupies a little more than one-fifth of its total volume. Below 60 km it exists mainly as molecular oxygen (O_2), while above this the dissociated atomic oxygen (O) is the more prevalent. The latter is brought about by the effects of cosmic radiation on the oxygen molecule at high elevations in the atmosphere. Oxygen, which exists in its gaseous state in the atmosphere, represents only a part of the total stored in the Earth–atmosphere system. Animals and plants store oxygen as a component of organic molecules during their lives, while in the rocks of the lithosphere it is bound into chemical compounds such as oxides and carbonates.

Ozone is present in only very small quantities in the atmosphere but its impact on the Earth–atmosphere system is considerable, particularly in the absorption of shortwave radiation. It is produced as a result of the dissociation of molecular oxygen into atomic oxygen by radiation of less than 0.24 μm wavelength. The dissociated atoms recombine with molecular oxygen to produce ozone.

$$O_2 + O \rightarrow O_3.$$

Maximum production of ozone occurs between 30 and 40 km above the Earth's surface, but maximum concentrations occur some 10 km lower than this. At extremely high elevations (greater than 60 km) there are insufficient oxygen molecules to maintain a high rate of production of ozone, while below 10 km there is insufficient shortwave radiation as this has already been absorbed at high elevations. Ozone is also unstable and readily breaks down into molecular oxygen when combined with further atomic oxygen.

$$O_3 + O \rightarrow O_2 + O_2.$$

The level of maximum concentration thus represents a balance between the processes of creation and destruction of ozone and its mixing into the atmosphere.

Carbon dioxide. Carbon dioxide forms a very small proportion of the atmosphere, but it is essential to life on Earth. Because of its close association with the biosphere it is most prevalent in the lowest 50 km of the atmosphere, and particularly the lowest 2 km. It is produced by respiration and by the oxidation of carbon compounds, including combustion. Although such combustion sources as forest fires are natural, the burning by Man of carbon compounds which would otherwise have been effectively immobilised in the lithosphere as fossil fuels has had measurable effects on atmospheric carbon dioxide concentrations. Man adds approximately 1.4×10^{13} kg of carbon dioxide to the atmosphere every year in this way, with the result that its volumetric proportion of the atmosphere has increased from 0.029% to 0.033% during the first half of the 20th century.

Water vapour. Water vapour is the gaseous phase of water and is derived from the diffusion of water molecules from their liquid state at the Earth's surface. Because of this, it is concentrated in the lowest 10 km of the atmosphere, with approximately 90% of it occurring below 6 km. It is one of the most variable constituents of the atmosphere, contributing between 0.5% and 4.0% of the volume of moist air. Figure 4.1 shows a typical distribution of water vapour through the lowest 8 km of the atmosphere.

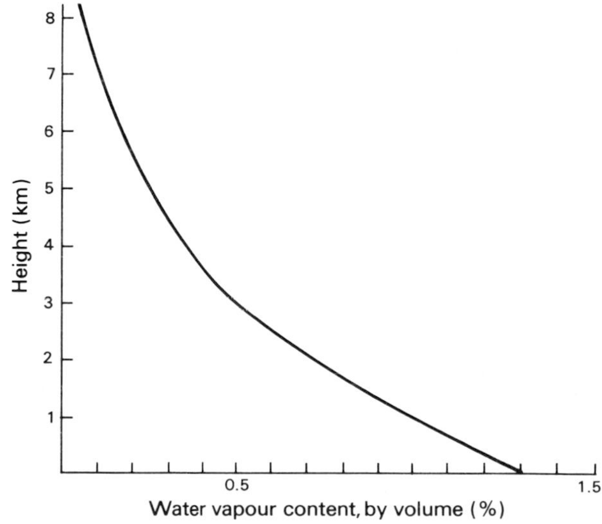

Figure 4.1 Decrease in water vapour content of the atmosphere with height.

Non-gaseous particles. In addition to the major gases, the atmosphere contains materials of non-gaseous nature which are held in suspension or maintained aloft by turbulent mixing. The most important of these are the water droplets and ice crystals that occur in clouds. These are important in that they affect both shortwave (reflection) and longwave (absorption) radiation. The amount of cloud in the atmosphere is highly variable, but there are areas such as equatorial and middle latitudes which experience greater cloud cover than elsewhere (Fig. 4.2).

About 90% of the remaining particles are natural in origin, such as dust from volcanic eruptions, smoke from forest fires, sea-spray and pollen (Table 4.2). These are increasingly augmented by the products of human activity, such as dust from soil erosion in areas of poor land management and smoke from the incomplete combustion of fossil fuels.

All particles are carried into the atmosphere by turbulent mixing and may remain there for considerable periods of time, although the heavier particles (>100 μm radius) tend to fall back to the ground through the effects of gravity. About 80% of particles rise no further than 1 km from the surface, but small quantities of fine particles with radii of the order of 10 μm or less may rise to heights of between 10 and 15 km where they may stay for several years.

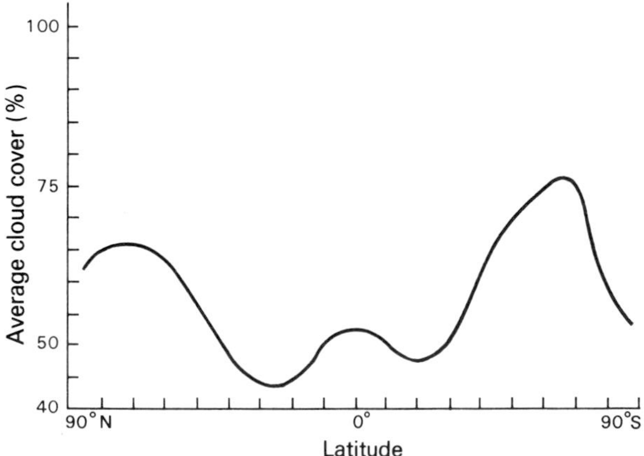

Figure 4.2 Latitudinal variation in average percentage cloud cover (after data from Sellers 1965).

Table 4.2 Sources of materials in the atmosphere (from Smith 1975, after Varney and McCormac 1971).

Type	Source
particles	volcanoes combustion wind action industry sea spray forest fires
hydrocarbons	internal combustion engine bacteria plants
sulphur compounds (SO_2, H_2S, H_2SO_4)	bacteria burning fossil fuels volcanoes sea spray
nitrogen compounds	bacteria combustion

4.1.1 Systems structure: a model of the vertical structure of the atmosphere

It has already been stressed that in the fluid system of the atmosphere structure has to be perceived in terms of the distribution of the properties of the system, particularly those attributes that can be treated as measurable variables. Certainly the preceding discussion has suggested that the atmosphere is not homogeneous and the distribution of the constituent elements indicates that it has a layered structure. This vertical component of the structure of the atmosphere can be related primarily to the distribution of radiant energy absorption and can be described in terms of the variable of temperature.

Below 60 km there are two main zones of absorption: at the Earth's surface and in the ozone layer. The absorbed energy is redistributed by re-radiation, conduction and convection. There are, therefore, two temperature maxima: at the Earth's surface and at an elevation of *c.* 50 km. Above each of these maxima there is mainly convectional mixing. Temperatures in these mixing layers decrease with height above the heat source. The lower of these two zones (Fig. 4.3) is referred to as the **troposphere**, and the upper the **mesosphere**.

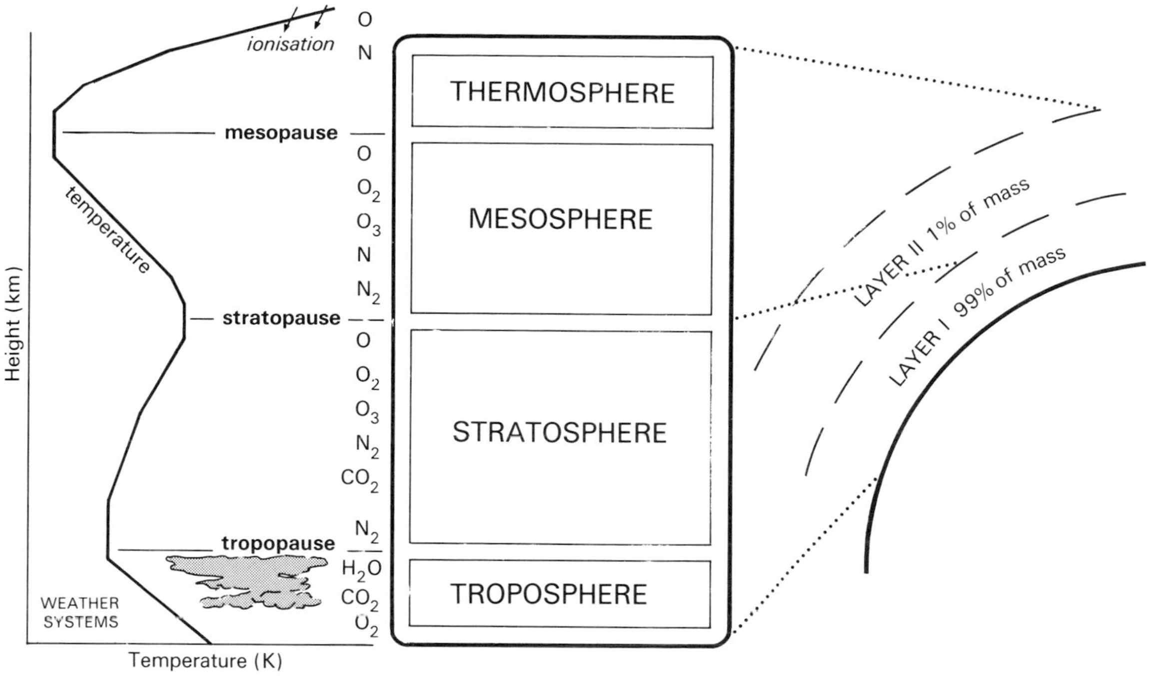

Figure 4.3 Models of the vertical structure of the atmosphere.

These are separated by a layer of little mixing in which the atmosphere tends towards a layered structure referred to as the **stratosphere**. Between the troposphere and the stratosphere is the **tropopause** which marks the approximate upper limit of mixing in the lower atmosphere. The average height of this is usually given as 11 km, but this varies over the Earth. In tropical latitudes its average height is 16 km and in polar latitudes it is only 10 km. There is one further zone of heating, above the mesosphere and more than 90 km from the Earth's surface, where shortwave ultraviolet radiation is absorbed by any oxygen molecules present at this height. This is referred to as the **thermosphere**. Within this layer, ionisation occurs which produces charged ions and free electrons. Beyond the thermosphere, at a height of approximately 700 km, lies the **exosphere** where the atmosphere has an extremely low density. At this level there are increasing numbers of ionised particles which are concentrated into bands referred to as the **Van Allen belts**.

This model of vertical structure can be simplified to provide a model of the atmosphere as two concentric shells, the boundaries of which are defined by the **stratopause**, at approximately 50 km above the Earth's surface, and a hypothetical outer limit of the atmosphere, at approximately 80000 km. Below the stratopause, in the stratosphere and troposphere, there is 99% of the total mass of the atmosphere and it is at this level that atmospheric circulatory systems operate.

Beyond the stratopause a layer of nearly 80000 km thick contains only 1% of total atmospheric mass and experiences ionisation by high energy, short wavelength solar radiation. Its contribution to atmospheric and Earth-surface systems is not fully understood and will not be discussed further in this book.

4.1.2 Atmospheric pressure: the refinement of the model of vertical structure

Atmospheric pressure is defined as the force exerted by the atmosphere per unit area of surface. If a fluid is at rest, because of the random motion of its molecules it exerts uniform pressure in all directions. The pressure such a fluid would exert under the influence of gravity is given by the

hydrostatic equation (Box 4.1.). This defines pressure as the product of the density (ρ) of the fluid, its depth (or height above the surface, h) and the acceleration due to gravity (g):

$$P = \rho h g. \qquad (4.1)$$

The unit of pressure is the Newton per square metre ($N\ m^{-2}$) or the pascal (P). In the case of atmospheric pressure the unit still commonly used is the millibar, which is equal to $100\ N\ m^{-2}$.

If we assume that the density (ρ) of the atmosphere does not vary with height above the Earth's surface, we can infer that pressure decreases uniformly through the atmosphere (Box 4.2). However, observation has shown that atmospheric

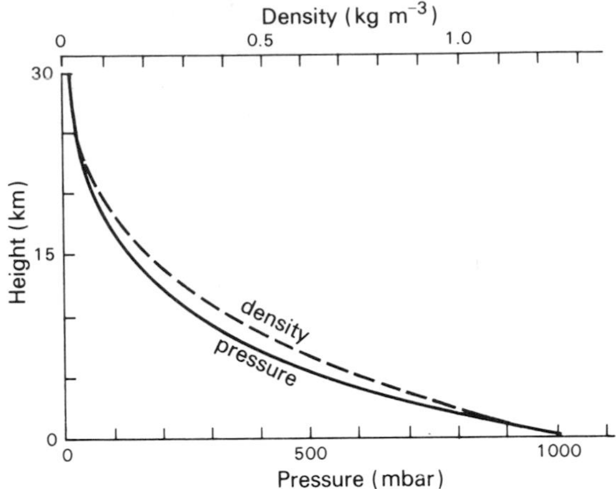

Figure 4.4 Variation of pressure and density with height in a standard atmosphere.

Box 4.1

THE HYDROSTATIC EQUATION

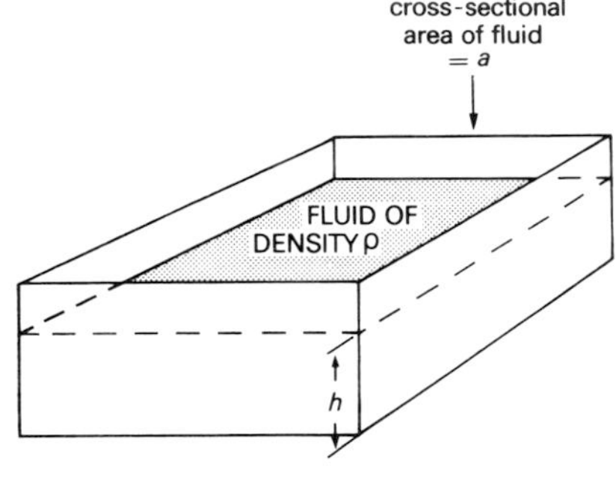

The pressure exerted by a fluid at rest upon the surface beneath it is given by:

$$P = \frac{\text{weight of fluid}}{\text{cross-sectional area}}$$

$$= mg/a \qquad (A)$$

where m = mass of fluid, a = cross-sectional area and g = acceleration due to gravity.

Mass can be expressed in terms of the density (ρ) and the volume (v) of the fluid as:

$$m = \rho v. \qquad (B)$$

The volume occupied by the fluid is given by the product of its cross-sectional area and its depth (h) thus:

$$v = ah. \qquad (C)$$

Substituting (C) into (B) gives:

$$m = \rho a h. \qquad (D)$$

Substituting (D) into (A) gives:

$$P = \rho a h g/a$$

$$P = \rho h g.$$

This is referred to as the **hydrostatic equation**.

pressure does not decrease uniformly with elevation (Fig. 4.4). In the lowest 5 km decrease in pressure is nearly uniform at a rate of approximately 100 mb km^{-1}. Above this it decreases progressively less rapidly, being about 270 mb at 10 km, 125 mb at 15 km and 56 mb at 20 km. The curve approximates to an exponential change in pressure with height. The reasons for this are that the density of the atmosphere is not constant with height. The atmosphere is readily compressed by the overlying air such that its density is greatest at the Earth's surface and then decreases rapidly (Fig. 4.4). The relationship between pressure (P), density (ρ) and temperature (T) for dry air is given in the gas equation:

$$P = R\rho T \qquad (4.2)$$

where R is the gas constant and is equal to 287 J kg^{-1} K^{-1}.

Using this equation we can modify the hydrostatic equation to provide a simple illustration of the effects of decreasing density on atmospheric

Box 4.2

VERTICAL PRESSURE CHANGES IN THE ATMOSPHERE: I

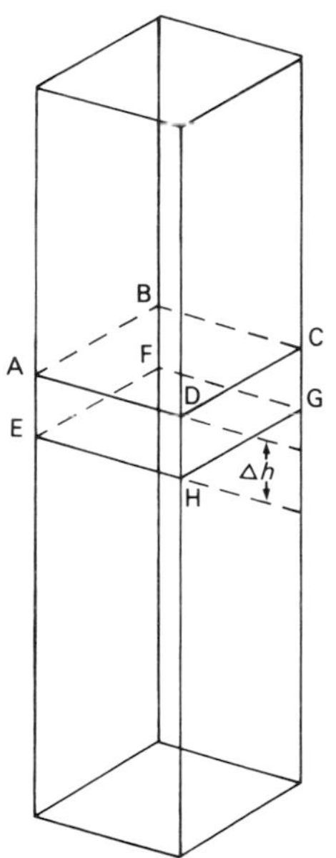

Consider a column of air of unit cross-sectional area and of density ρ.

Consider the air pressure (P_1) on surface EFGH, some height above the base of the column. This will be equal to the weight of the column of air above it (which we can call W).

If we now consider the air pressure on surface ABCD (P_2) this will be equal to the weight of the column of air above EFGH minus the weight of thickness Δh of the column.

The weight of this small section of the column is equal to $\rho g \Delta h$ (see Box 4.1). Thus:

$$P_2 = W - \rho g \Delta h = P_1 - \rho g \Delta h.$$

Therefore

$$P_2 - P_1 = -\rho g \Delta h$$

which may be rewritten as:

$$\Delta P = -\rho g \Delta h$$

where ΔP represents a small incremental change in pressure.

Dividing throughout by Δh:

$$\Delta P/\Delta h = -\rho g$$

which, in the form of a differential equation, is

$$dp/dh = -\rho g.$$

If ρ remains constant, the rate of change of pressure upwards from the base of the column is uniform and decreases as height increases.

pressure (Box 4.3). The relationship between pressure and height may be exponential rather than linear, which approaches the observed relationship.

4.1.3 Systems structure: the horizontal component
The model of the vertical structure of the atmosphere developed so far has implied the existence of concentric shells around the Earth; shells which differ in their relative composition and in the distribution of their properties, particularly in the measurable attributes of temperature, density, volume and pressure. The fact that the vertical model varies both in different localities and with time has already been hinted at. The property which best serves to illustrate the horizontal dimensions of this variation in atmospheric properties is again atmospheric pressure.

The horizontal distribution of atmospheric pressure at the Earth's surface is represented in the form of isobars (lines of equal pressure) of sea-level pressure. Pressures recorded over land surfaces are corrected to sea level using Equation 4.13 in Box 4.3. On maps of surface pressure there are areas of relatively higher or relatively lower pressure which are referred to as high or low pressures, although there is no strictly quantitative definition of either and they are in no sense absolute terms. The location of areas of high or low pressure varies over time and space, but for convenience two types of pressure are usually identified. The first are the semi-permanent areas, which appear on maps showing seasonal average pressure, whose locations are moderately predictable. The second are the ephemeral and mobile pressure cells which appear only on daily pressure charts. The former provide a key to the large-scale circulation of the atmosphere, while the latter (see Ch. 9) are more readily associated with smaller-scale movements of air.

If the Earth's surface were completely uniform, it would be possible to identify, in both hemispheres, a simple zonal pressure pattern of equatorial low (0° to 15°), subtropical high (15° to 40°), middle latitude low (40° to 65°) and polar high (65° to 90°). However, as the Earth's surface is an assemblage of diverse land surfaces and large expanses of ocean, atmospheric pressure does not follow a convenient latitudinal zonation.

In the distribution of sea-level pressure in July (Fig. 4.5a) values are generally low throughout the

Box 4.3

VERTICAL PRESSURE CHANGES IN THE ATMOSPHERE: II

From Box 4.2 we have:

$$dP = -\,dH\rho g. \qquad \text{(A)}$$

The basic gas equation for an ideal gas is:

$$P = R\rho T. \qquad \text{(B)}$$

From (B):

$$\rho = P/RT.$$

Substituting this in (A):

$$dP = -\,dH\frac{P}{RT}g$$

$$\frac{dP}{P} = \frac{-g}{RT}\,dH. \qquad \text{(C)}$$

If we assume that for relatively small changes in height H, changes in T are relatively insignificant. That is, T is not considered to be a function of H.

Then, by integrating equation (C) to solve for P:

$$\log P = \frac{g}{RT}H + C_1 \qquad \text{(D)}$$

where C_1 = constant.

Therefore, there is a logarithmic relationship between P and H. To determine C_1 we require a boundary condition. If $H = 0$ (i.e. sea level) when $P = P_0$ (which we shall call sea level pressure) substituting this in equation (D)

$$\log P_0 = 0 + C_1$$
$$C_1 = \log P_0.$$

Thus equation (D) becomes:

$$\log P = \log P_0 - \frac{g}{RT}H. \qquad \text{(4.13)}$$

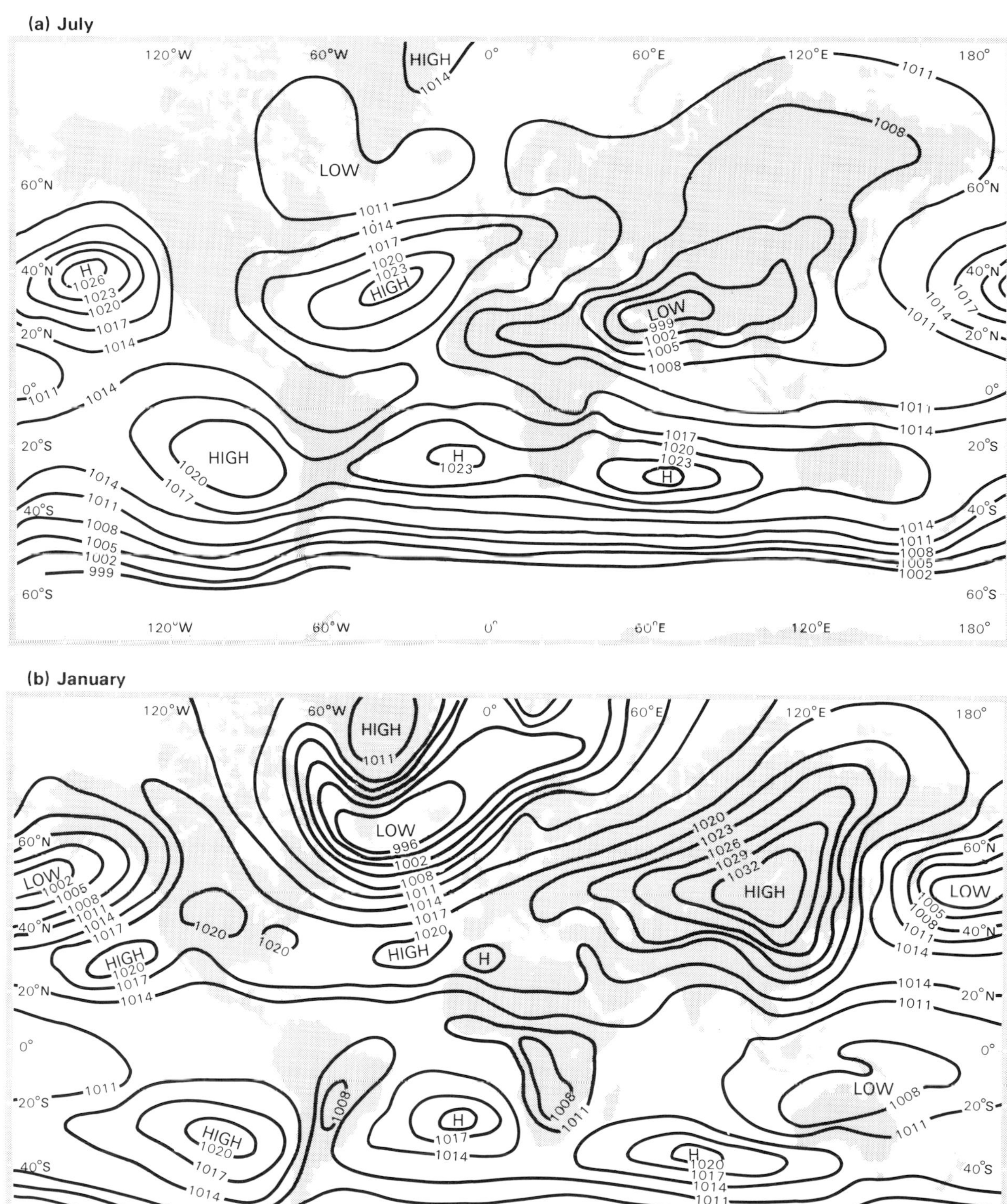

Figure 4.5 Mean sea-level atmospheric pressure (millibars).

equatorial regions. To the north and south of this are the subtropical high pressure areas which, in the southern hemisphere, form a nearly continuous zonal belt. However, in the northern hemisphere, because of the presence of large areas of land surface, high pressure exists over the Pacific and Atlantic oceans while over much of Asia and North America there are continental low pressure areas. In the middle latitudes pressure is, on average, low. However, the use of averages conceals the great temporal variation of pressure that occurs in these disturbed latitudes. In polar latitudes pressure is not consistently high, but averages tend to indicate slightly higher pressure than in middle latitudes.

In January the major pressure areas move southwards, towards the southern hemisphere (Fig. 4.5b). The major changes from July are in the northern hemisphere, the pressure in the southern hemisphere remaining in a zonal pattern. The northern subtropical high pressure is close to exhibiting a zonal distribution, but in the middle latitudes continental high pressure over North America and Asia contrasts with the oceanic low pressures over the Pacific and the Atlantic.

The distribution of sea-level pressure is, therefore, not necessarily one of zonality, although in the southern hemisphere (81% of which is ocean) this is closely approached. In the northern hemisphere, however, zonality is disrupted by the complex arrangement of land and sea, only 61% being ocean surface. As continental pressure centres are relatively shallow, usually less than 2 km, it is possible to filter them out of the global pressure patterns by considering pressure distribution at some height above the Earth's surface. Such distributions are not represented in terms of isobaric maps but as contoured isobaric surfaces. Contours are drawn based on the heights at which certain values of atmospheric pressure are reached. The data from these are readily available from soundings taken through the atmosphere. Pressure surfaces may, for example, be drawn for 1000 mb (surface), 700 mb (about 3 km), 500 mb (5–6 km) and 300 mb (9–10 km).

Employing such a device, it is possible to re-examine the complex pressure pattern of the northern hemisphere using either 700 mb or 500 mb surfaces, as these should lie above the continental pressure areas which extend only to the 850 mb level. The 500 mb surface in Figure 4.6 contrasts with the distribution of sea-level pressure and indicates a meridional pressure gradient southwards from the North Pole, shown by increasing contour values. This pattern, the circumpolar vortex, has a considerable bearing on the circulation of the atmosphere (Sec. 4.3).

4.2 The transfer of energy and mass in the atmospheric system

In Chapter 3 we have already seen that there is an uneven distribution of net radiation over the Earth's surface (Sec. 3.3.2), some of which was seen to be redistributed by heat energy transfer in the oceans. However, there still remains a large surplus of heat energy in tropical latitudes which is redistributed polewards by the atmosphere (Fig. 4.7). This energy is transferred as sensible heat in the poleward movement of warm air, as the kinetic energy of motion and in the form of the latent heat of vaporisation in water vapour (Sec. 4.4). The function of mass transfer, or atmospheric circulation, is to carry out this transfer of energy. However, before attempting to produce a functional model of the way in which this is achieved we will begin by considering the forces which act on the atmosphere in the horizontal plane, and the pattern of air flow over the Earth's surface which results.

4.2.1 Forces acting on the atmosphere in the horizontal plane

In accordance with Newton's First Law of Motion (see Ch. 2) the atmosphere will remain in a state of rest or uniform motion unless a horizontal force is applied to it. The nature of the motion that ensues when such forces are applied is related very closely to their magnitude and their direction of operation. The atmosphere, in reality, is rarely in a state of rest, but moves over the Earth's surface in response to the application of a number of forces, principal of which are **pressure gradient force**, **Coriolis force**, and **frictional force**.

Pressure gradient force. A difference in atmospheric pressure between two points on the same horizontal plane implies that a greater force is being exerted on one than on the other. In order to compensate for this imbalance, air is accelerated from the higher to the lower pressure, in accordance with Newton's Second Law. However, the

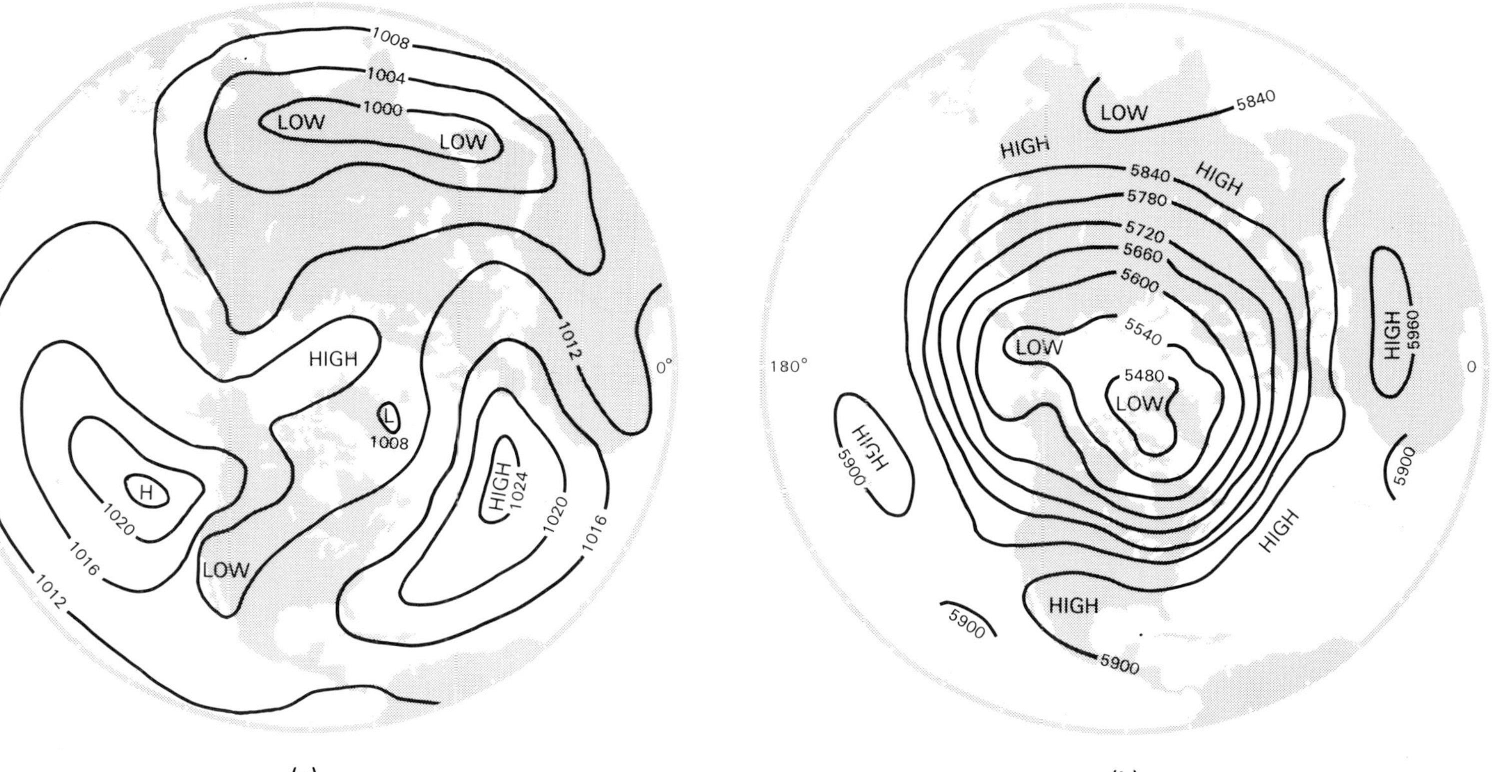

Figure 4.6 Distribution of atmospheric pressure over the Northern Hemisphere in July (after Byers 1974). (a) Average sea-level atmospheric pressure; (b) Average height (m) of the 500 mb isobaric surface (after Neiburger *et al.* 1971).

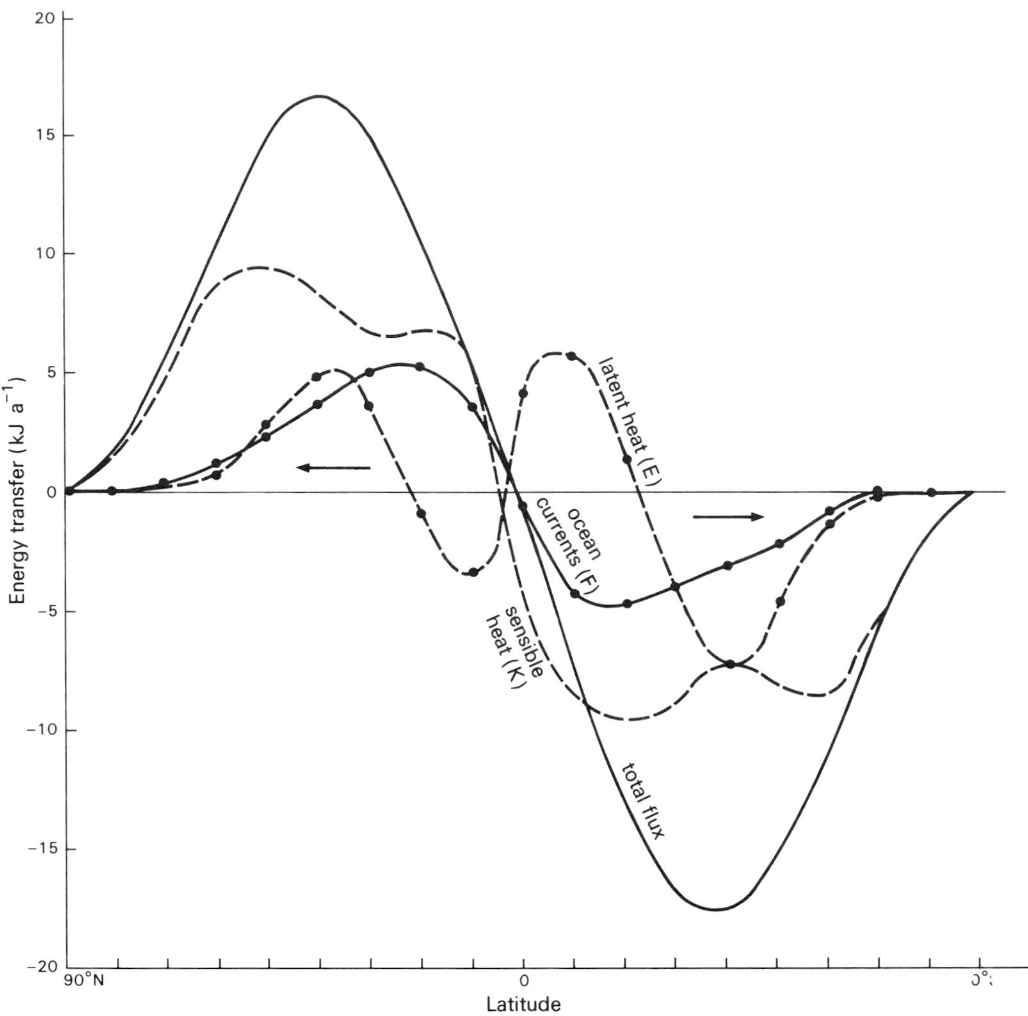

Figure 4.7 Poleward flow of energy in the oceans (F) and the atmosphere (E & K) (after Sellers 1965).

same law states that this acceleration is directly proportional to the magnitude of the force. Magnitude, in this case, is expressed in terms of the difference in pressure between the two points, or more specifically, in terms of the pressure gradient between them.

Thus, the pressure gradient force operates from high to low pressure at right-angles to the isobars and is given by the equation:

$$F_P = \frac{1}{\rho}\frac{dP}{dx} \tag{4.3}$$

where F_P = pressure gradient force; ρ = density of the atmosphere; dP/dx = rate of change of pressure (P) over distance (x) in the direction high to low pressure. In simplest terms this is indicated by the closeness of isobars on the pressure map.

If this were the only force in operation, air would flow directly from high to low pressure, and equalisation of pressure differences would be extremely rapid. The fact that neither takes place indicates the existence of other forces operating upon the air, modifying the effects of the pressure gradient force.

Coriolis force. The rotation of the Earth about its own axis introduces another force affecting the movement of air over its surface. This is referred to as the Coriolis force, named after a 19th-century French mathematician. For objects moving across the Earth's surface, it operates at right-angles to the direction of motion. It introduces, therefore, a sideward deflection, which in the northern hemisphere is to the right and in the southern hemisphere is to the left, irrespective of the initial direction of motion (Fig. 4.8). The acceleration produced by this Coriolis force is, however, only apparent and it arises because air motion must be related to a moving surface and not a stationary one (Box 4.4).

If we were able to view the North and South Poles from the upper atmosphere, the Earth would appear as a disc, rotating anticlockwise around the former and clockwise around the latter. Applying our observations from the disc, Coriolis deflection will be to the right around the North Pole and to the left around the South Pole. However, these polar discs are not flat but are hemispherical surfaces. The value of Coriolis acceleration on these

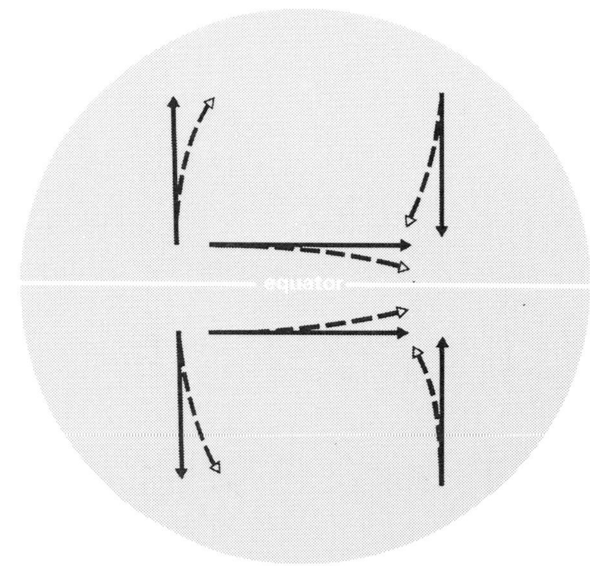

Figure 4.8 Coriolis deflection of air flow over the Earth's surface.

Box 4.4

CORIOLIS DEFLECTION

There are two large rotating discs, one rotating anticlockwise about A, the other clockwise about B. If an object is projected towards X from either A or B, it will not reach its destination. By the time it reaches the outer rim of the disc, X will have moved to a new position X′ because of the rotation of the disc. The object's path relative to a fixed point on the surface of the disc will be a curve, as shown by the dashed line in the figure. The deflection appears to be the result of a sideward acceleration of the object. In the case of A, the apparent deflection is to the right, and of disc B, to the left.

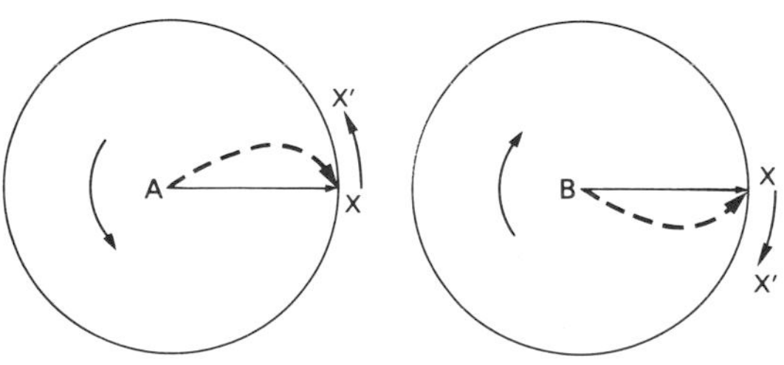

The value of the apparent acceleration experienced by the object is given by:

$$\text{acceleration } (a) = 2v\omega \qquad (4.14)$$

where v = velocity at which the object is moving and ω = rate of spin (angular velocity) of the disc.

hemispheres decreases outwards from the Poles, and Equation 4.14 (Box 4.4) is rewritten as:

$$a = 2v\omega \sin \phi \qquad (4.4)$$

where ϕ is angle of latitude. For the Earth $\omega = 7.29 \times 10^{-5}$ radians s^{-1}. As we normally refer to force operating upon unit mass:

$$\text{Coriolis force } (F_C) = 2v\omega \sin \phi. \qquad (4.5)$$

At the Poles, $\phi = 90°$, so $F_C = 2v\omega$, its maximum value. At the Equator, $\phi = 0°$, so $F_C = 0$, its minimum value.

There are now two forces operating in the atmosphere, F_P and F_C. If these are equal, the net result of their action is to produce airflow which is parallel to the isobars and referred to as the **geostrophic wind** (Box 4.5, Fig. 4.9a). Should the isobars be curved, airflow remains parallel to them, held in a curved path by a centripetal force acting towards the pressure centre, and is known as the **gradient wind**. This will be discussed further in Chapter 9.

If $F_C = F_P$, then, from Equations 4.3 and 4.5:

$$2\omega v_g \sin \phi = \frac{1}{\rho}\frac{dP}{dx}$$

where v_g = geostrophic wind velocity;

$$v_g = \frac{1}{\rho}\frac{dP}{dx}\frac{1}{2\omega \sin \phi}. \qquad (4.6)$$

Frictional force. The Earth exerts a retarding effect on air motion across it which is due to the transfer of energy into the surface. Such frictional resistance reduces wind velocity in the lowest 1000 m of the atmosphere. The greatest reduction occurs in the air layers in immediate contact with the surface where wind speeds will tend towards zero. Away

Box 4.5

THE GEOSTROPHIC WIND

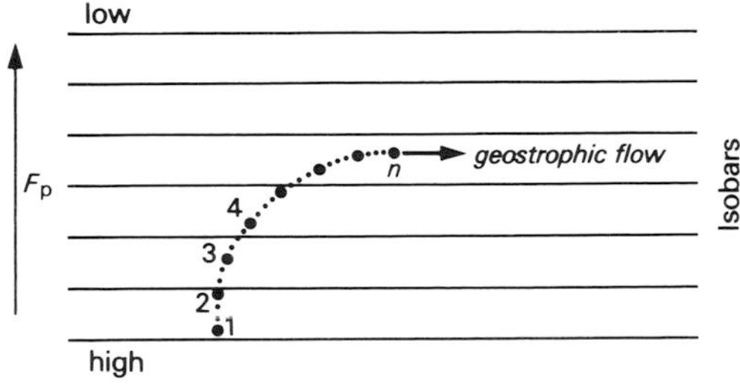

Consider a parcel of air at rest at point 1 which lies on a pressure gradient in the northern hemisphere. As velocity is zero, there is no Coriolis force but it is acted upon by a pressure gradient force. Therefore, it is accelerated according to Newton's First Law of Motion. After travelling the infinitely small distance to point 2, it has acquired velocity and there is, therefore, a Coriolis force acting to the right of its direction of motion. The air parcel is thereby deflected from its path normal to the isobars. By point 3, it has been accelerated further by the two forces applied to it and it has an even greater velocity. Coriolis deflection is greater so the parcel experiences further deviation from its original path. By point 4, the deviation is greater again, and so on to a point *n*. Coriolis force thus increases as long as the parcel experiences a resultant accelerating force. By point *n*, however, the pressure gradient and Coriolis forces are acting in opposite directions to each other and, if they are assumed to be equal, the resultant force on the parcel of air is zero. Therefore, according to Newton's First Law, the parcel will continue in uniform motion in a straight line which, in this case, is parallel to the isobars. This is the **geostrophic wind**.

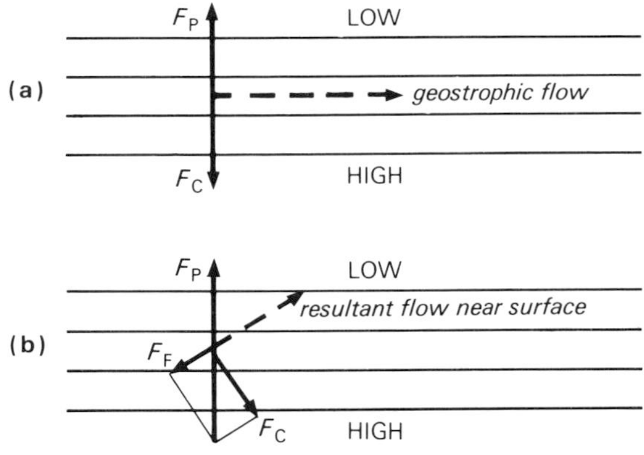

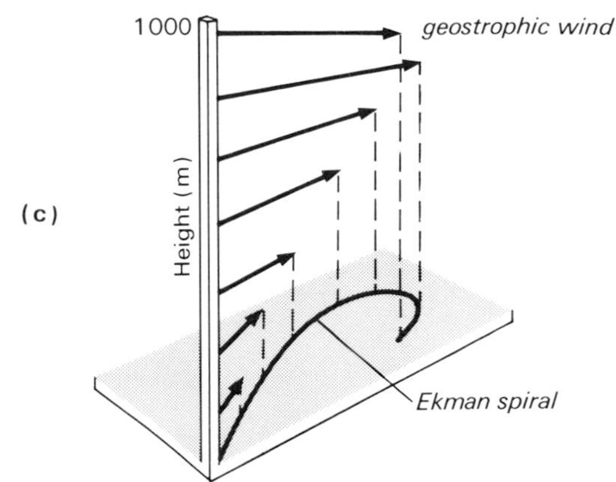

Figure 4.9 Forces acting on air: (a) at 1000 m and (b) 10 m above the Earth's surface (F_P = pressure gradient force, F_C = Coriolis force, F_f = friction force). (c) The effects of friction on horizontal air flow in the lower atmosphere.

from the surface, frictional effects decrease and there is an approximately exponential increase in wind speed with height (see Ch. 8).

As there is a reduction in wind speed, this must also (from Eqn 4.5) reduce the Coriolis force to a value less than that of the pressure gradient force. The resultant airflow will, therefore, no longer be geostrophic and will be directed towards the lower pressure (Fig. 4.9b). As velocity increases away from the surface, Coriolis force also increases and thus winds will move progressively toward geostrophic velocity and direction. This is represented in Figure 4.9c and is referred to as the Ekman spiral.

The frictional drag exerted by the surface is affected by its aerodynamic roughness and is not constant. Over oceans, for example, frictional drag produces a 10° to 20° change in wind direction near to the surface and a 40% reduction in velocity, in relation to the geostrophic wind. The greater frictional drag over land surfaces produces changes in direction between 25° and 35° and a velocity reduction of about 60%.

The effects of these three forces is to produce patterns of airflow related to areas of high or low pressure (Fig. 4.10). Thus air converges on centres of low pressure and diverges from centres of high pressure. However, it must be remembered that the atmosphere is a complex fluid moving across a surface of highly variable physical characteristics.

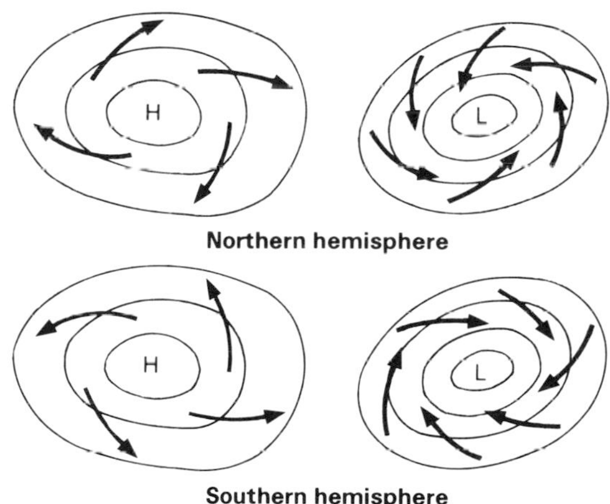

Figure 4.10 Air flow associated with areas of high and low pressure.

The forces operating within it do not, therefore, produce well ordered patterns of airflow but rather a diversity of airflows which, on first sight, appear to possess no obvious spatial organisation. However, it is possible to identify movements which are spatially organised rather than random. These are divided for convenience into **primary**, **secondary** and **tertiary circulation systems** according to the space and time scales over which they operate

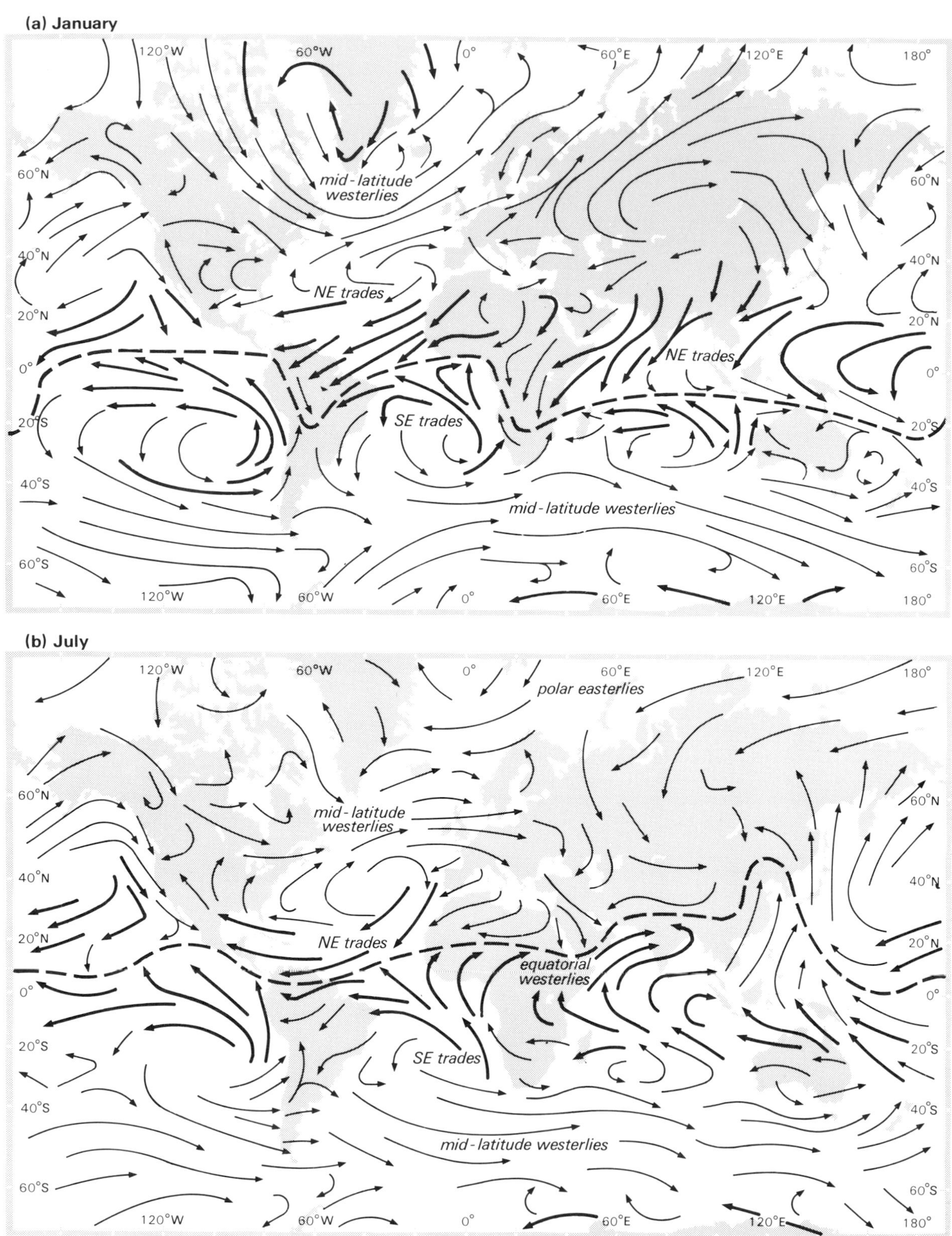

Figure 4.11 Mean surface winds over the Earth, 1900–1950 (after Lamb 1972).

(Table 4.3). The primary system is the movement of the whole atmosphere and the poleward transport of energy from tropical latitudes. Secondary systems operate within this primary system and are associated with the ephemeral pressure cells referred to on page 74, while tertiary systems operate on the smallest spatial scale and may be referred to as being local circulations. Both secondary and tertiary systems will be considered in detail in Chapter 9. We shall first consider more fully the primary circulation system of the atmosphere.

4.2.2 Airflow over the Earth's surface

A simplified view of the distribution of airflow over the surface of the Earth would identify four main wind zones in each hemisphere for an ideal global surface. Around the Equator lies an area of slack winds which is associated with a broad area of low pressure, the **intertropical convergence zone** (ITCZ). Converging on this are the **trade winds** from north-east and south-east, renowned for their consistent speed and direction. To the north and south of these, beyond latitude 40°, lie the **westerlies**, associated with weather disturbances. In the polar regions are the **polar easterlies**, highly variable winds but generally northeasterly or southeasterly (northern and southern hemispheres) in direction.

The movement of air across the Earth's surface is by no means this simple, as is evident from Figure 4.11. However, it is still possible to identify a number of global wind zones (Table 4.4). This latitudinal zonation of surface airflow is based on variation in easterly and westerly components (Fig. 4.12). Any model of the primary circulation system of the atmosphere must be capable of accommodating this observed pattern of surface air movement as well as achieving the Equator to Pole redistribution of energy.

Table 4.3 Characteristic dimensions of atmospheric motion systems (from Barry 1970, Smagorinsky 1979). (Note figures are only accurate to an order of magnitude.)

	Spatial scale (km) Vertical	Horizontal	Time scale (seconds)
PRIMARY			
jet streams tropospheric long waves global surface winds	10^1	5×10^3	7×10^6 to 10^7
SECONDARY			
extratropical cyclones tropical cyclones anticyclones	10^1	2×10^3 to 5×10^2	3×10^5
TERTIARY			
squall lines thunderstorms sea breezes mountain and valley winds	$10^0 - 10^1$	$10^0 - 10^2$	$10^2 - 10^4$

4.3 The primary circulation system: approximations to a successful model

The fact that the atmosphere operates as a large heat engine transferring heat from a source (the Tropics) to a sink (Poles) should, therefore, make

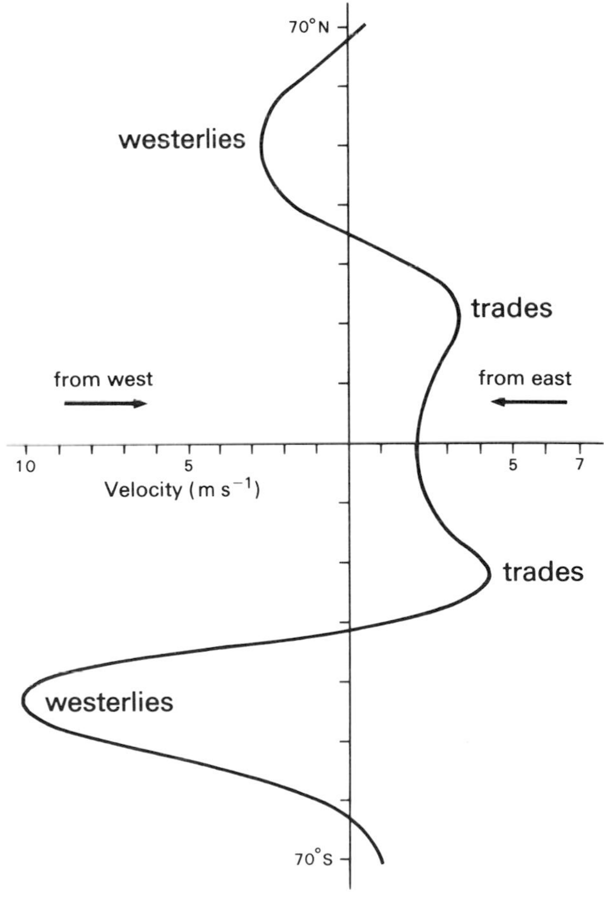

Figure 4.12 The major zonal winds of the world (after Riehl 1965).

Table 4.4 Characteristics of major global winds.

	Latitudinal zone	Associated sea level pressure conditions	Surfaces over which most distinctly developed	Average direction near to surface	Average velocity near to surface	Seasonal variation	Constancy of velocity and direction
doldrums	equatorial – may be displaced 20° from the Equator, particularly in the northern hemisphere	generally low pressure with slack pressure gradients	over ocean surfaces as a discontinuous zone	highly variable	less than 3 m s^{-1}	contiguity of the doldrum zone varies, most extensively developed in March and April; disjointed in August	highly variable
equatorial westerlies	equatorial, extending to 28°, particularly over the Indian subcontinent	low, particularly monsoonal (summer) pressure conditions	primarily oceanic, but of major importance over land areas of West Africa and India	SW in northern hemisphere; NW in southern hemisphere	less than 6 m s^{-1}	best developed in the summer hemisphere	locally very consistent in speed and direction
trade winds	40° to Equator, reaching greatest velocity 5 – 20°	subsiding air associated with subtropical anticyclone	core areas located in the eastern parts of major oceans but also blowing over subtropical land surfaces	NE in northern hemisphere; SE in southern hemisphere	5 to 8 m s^{-1}	core areas most extensive in winter hemisphere, but greatest velocities in summer	remarkably consistent; core areas over 70% of recorded wind direction from east; over 50% in most other trade wind areas
middle latitude westerlies	40° to 65°, greatest velocities 40° to 50°	variable, but generally low, steep meridional pressure gradients	over oceans; disrupted by large land masses in northern hemisphere	SW to W in northern hemisphere; W to NW in southern hemisphere	to 10 m s^{-1}, highest velocities in southern hemisphere	strongest in summer when meridional pressure gradient at its steepest	consistent over southern hemisphere oceans where 75% of winds are between S and SW and N and NW; in northern hemisphere less than 50% over ocean areas down to less than 25% over continental interiors
polar easterlies	poleward of 65°	variable, occur in the zone between polar high pressure mid-latitude low pressure; well developed when strong, anticyclonic conditions prevail over polar region	subpolar oceans and continental margins	variable, generally from the easterly quadrant – depending on synoptic conditions	variable, little information		variable; dependent very much on synoptic conditions in either middle latitude or over polar ice caps; modified by local katabatic flow in Antarctica

it possible to represent its circulation as a simple source–sink–source movement of air, as suggested by Hadley in 1735 (Fig. 4.13). In this single-cell model, warm air rises in the Tropics, travels northwards in the upper atmosphere and returns southwards as cooled air across the Earth's surface. These flows are adjusted to the Earth's rotating motion. As the trade winds blow against the direction of rotation of the Earth, they exert a frictional drag which must be counteracted if rotation is not to be decelerated. The westerly winds of the middle latitudes provide such a compensating drag in the direction of rotation.

This simple model may be explained by referring to Figure 4.14. In (a) there is a uniform decrease in pressure above A, B and C which lie on a uniform horizontal plane. If the surface is heated at B and cooled at A and C, the air column above the former expands, while above the latter it contracts. The isobaric surfaces become distended (as in (b)) producing a pressure gradient at the upper levels. The air begins to move in the direction of this pressure gradient, creating divergent airflow at B^1 and convergent airflow at A^1 and C^1(c). The divergence at B^1 induces convergence at B while convergence at A^1 and C^1 induces divergence at A and C(d). We have, therefore, a convectional circulation of air initiated by

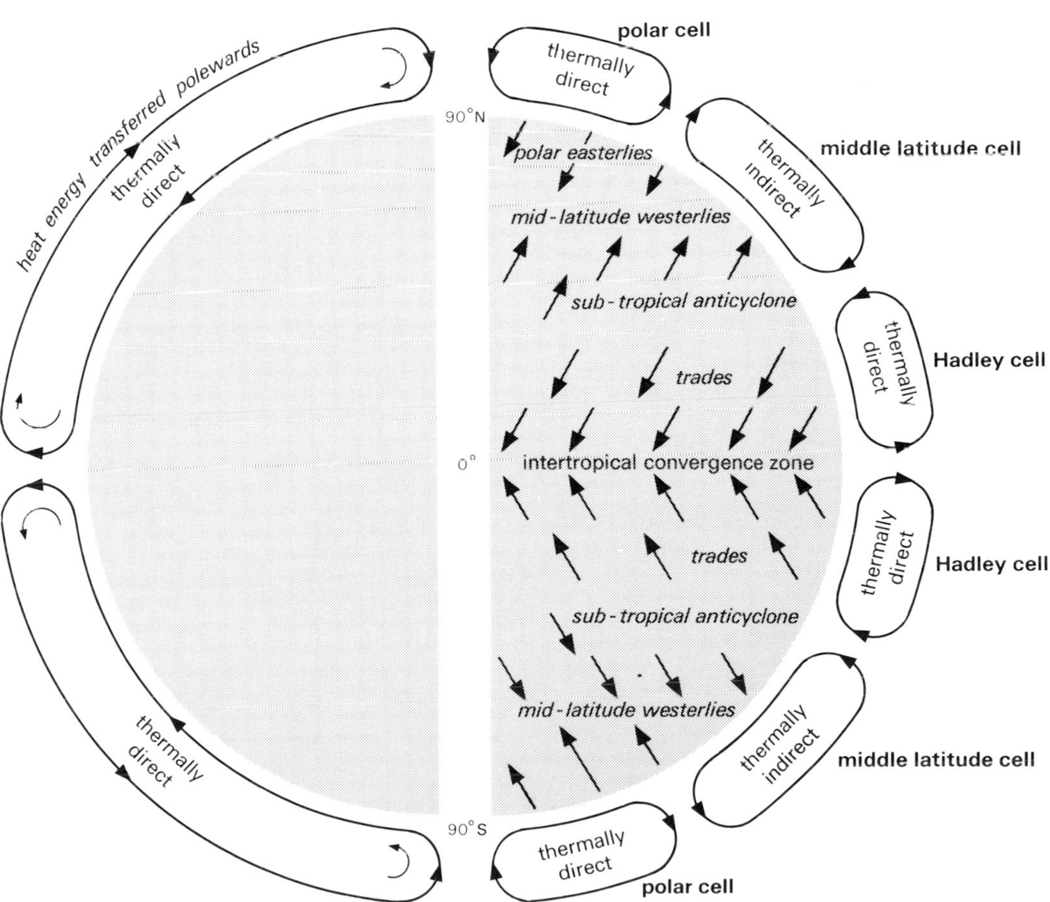

Figure 4.13 Single-cell and three-cell models of the primary circulation of the atmosphere.

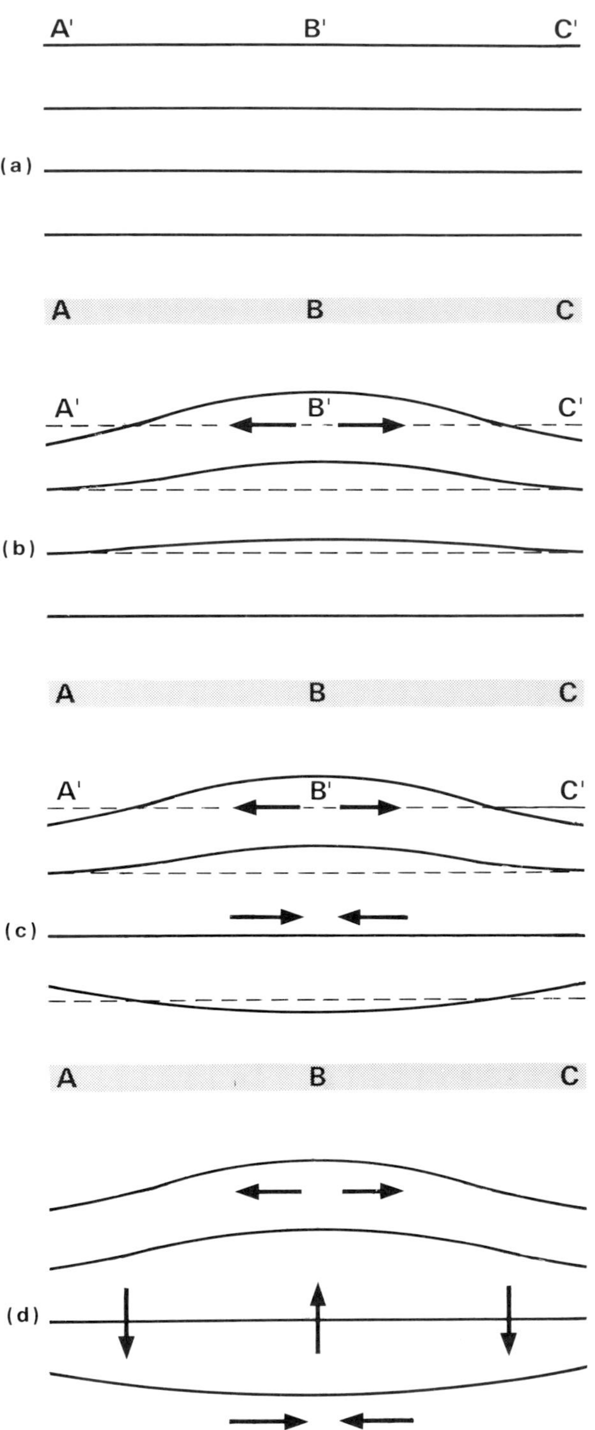

Figure 4.14 The development of convectional cells.

surface temperature differences between points.

This theory can be applied to the atmosphere above the curved surface of the Earth. Heating at the Equator and cooling at the Poles should produce two convectional cells, as indicated in Figure 4.13. This single-cell circulation is accelerated by a heat energy imbalance and decelerated by friction.

The heating of air from below, in causing the vertical expansion of the overlying air column, is effectively increasing its potential energy which is then converted into kinetic energy as motion begins. The simple primary circulation system is represented in Figure 4.15. A state of dynamic equilibrium is reached in this circulation such that the motion of the air is just adequate to maintain a mean heat balance in the atmosphere. Should the meridional temperature gradient change, negative feedback operates to restore equilibrium. Thus an increased temperature gradient speeds up the circulation of the air and hence the redistribution of heat energy, restoring dynamic equilibrium. A decrease in temperature gradient slows down the circulation and thereby limits the redistribution of heat energy, again restoring equilibrium.

Unfortunately this single-cell model does not explain fully the surface distribution of winds. The main flaw lies in its consideration of the effects of the Earth's rotation. As air moves polewards the radius of its rotation about the Earth's axis decreases. In order to conserve angular momentum the air would have to experience a progressive increase in eastward velocity relative to the Earth's surface (Box 4.6). If there were only one cell,

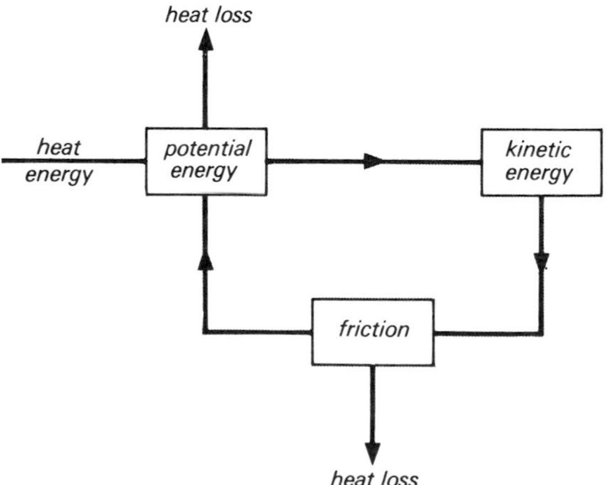

Figure 4.15 System model of single-cell circulation of the atmosphere.

poleward-moving air would be flowing at such extreme velocity by latitude 30° that it would become turbulent, the meridional flow breaking down into eddies. The **Hadley cell** must, therefore, be limited to low latitudes, and alternative mechanisms must be found to achieve the necessary poleward transport of angular momentum.

An alternative to the single-cell model view of the circulation of the atmosphere is the three-cell model where, in place of a single convectional cell, there are three interlocking cells between the Tropics and the Poles (see Fig. 4.13). The first of these is the low-latitude Hadley cell operating between equatorial regions and 30° north and south, which is a convectional or thermally direct cell. Warm air rises in the intertropical convergence zone, travels polewards and subsides in the subtropical anticyclones, whence it returns westwards to low latitudes. Winds on the surface would be the easterly trades and, in the upper atmosphere, the westerly antitrades. The second cell is a **middle-latitude cell** operating between latitudes 30° and 60°. Air diverges from the subtropics and flows polewards and eastwards across the surface, eventually converging with cold polar air at the polar front where it rises to the upper troposphere. From here it returns equatorwards aloft. As warm air is effectively sinking at 30° and cold air rising at 60°, this cell is thermally indirect. The third cell is the **polar cell** which is thermally direct, relatively warm air rising at latitude 60° and cold air sinking over the Poles. The upper airflow is westerly, while the surface flow is easterly.

The operation of this three-cell model appears to explain the global distribution of surface airflow in that the Hadley cells provide the two trade

Box 4.6

ANGULAR MOMENTUM

Momentum is a property which a body possesses by virtue of its motion and it is given by:

momentum = mass × velocity (see Ch. 2).

The law of conservation of momentum states that momentum can be neither created nor destroyed.

Angular momentum is a property which a body possesses by virtue of its motion around a central point.

If there is no relative motion between the atmosphere and the Earth, they are both in solid rotation. In this case, the angular momentum of unit mass of the atmosphere near to the Earth's surface relative to this axis is given by:

$$\text{angular momentum} = \Omega r^2 \cos^2 \phi$$

where Ω is angular velocity of the Earth's rotation, r is the radius of the Earth, and ϕ is the angle of latitude. There is, however, atmospheric motion relative to the Earth. If air is moving zonally with velocity u, then it possesses angular momentum relative to the Earth given by:

$$\text{angular momentum} = ru \cos \phi.$$

Therefore, if angular momentum is to be conserved by air moving zonally, there must be an increase in velocity if there is a poleward component of motion.

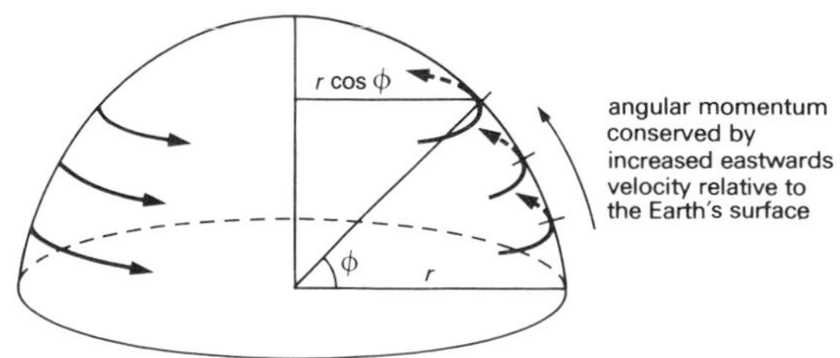

winds and the intertropical convergence zone, the middle-latitude cell the westerlies and the polar front, and the polar cell the polar easterlies. Surface winds accord with the observed distribution of sea-level atmospheric pressure and the operation of pressure gradient, Coriolis and frictional forces.

In terms of the operation of the primary circulation system, energy transfer is not in a simple single circuit (Fig. 4.15) but is a more complex series of transfers. The two thermally direct cells may be represented as simple energy circuits, but the middle-latitude cell is thermally indirect and must be driven by the other two cells. Thus the main input here is of kinetic energy transferred from Hadley and polar cells.

During the 20th century, as more information has been gained about the winds in the upper troposphere, the limitations of the three-cell model have become apparent. Observed airflow in the upper troposphere does not match that of the model, particularly in the low and middle latitudes. Above the surface westerlies in middle latitudes, winds also blow from a westerly direction and not easterly as the three-cell model suggests. Above the trade winds the antitrades do show the required reversal of direction but are often extremely weak, clearly not matching their surface counterparts.

In order for the three-cell model to work, it requires that the middle-latitude cell be driven by the polar and Hadley cells. Not only is the polar cell too weak to fulfil this function, but meridional pressure gradients are at their steepest in both the lower and upper troposphere in middle latitudes. The middle-latitude cell is thus, in contradiction of the three-cell model, the strongest element of the atmospheric circulation and is fundamental to the poleward redistribution of energy.

Examination of the isobaric surfaces in the upper troposphere has revealed a pressure distribution in the form of a circumpolar vortex (Sec. 4.1.3), in which pressure decreases away from the Poles. Pressure reaches a maximum at latitude 20° whence there is a small decrease into equatorial latitudes. Geostrophic winds blow as circumpolar westerlies between the Poles and latitude 20°, and as an easterly flow over equatorial latitudes.

Within the circumpolar westerlies there are two well defined bands of extremely strong winds referred to as the **jet streams**. One of these, the subtropical jet stream, has a circumpolar path between latitudes 20° and 35°. The other, the polar front jet stream, has a path between latitudes 35° and 65° (Fig. 4.16). The subtropical jet stream blows between 12 km and 15 km above the Earth's surface and, within its narrow core, average wind speed is in excess of 65 m s^{-1}. During winter it is well defined, but in summer it is considerably weakened and may become discontinuous.

As air moves polewards in the Hadley cells, it acquires an increased eastward velocity of the

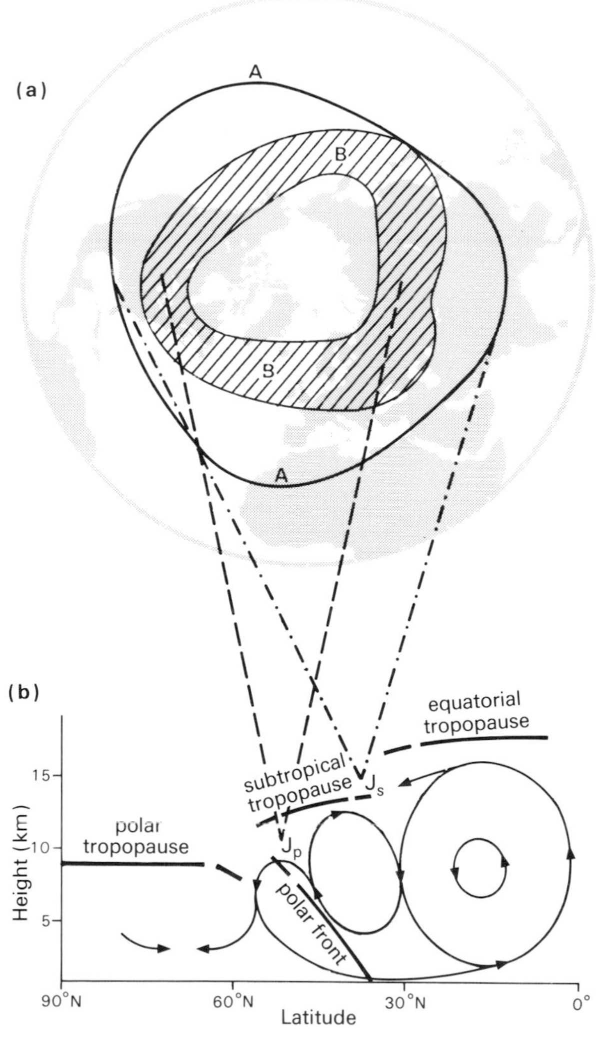

Figure 4.16 The jet streams of the northern hemisphere. (a) Mean winter position of the axis of the subtropical jet stream, J_s (A) and the area of activity of the polar front jet stream, J_p (B) (after Riehl 1965). (b) Mean meridional movement of air between Equator and Pole (after Palmen 1951).

order of 60 m s^{-1} by latitude 30° in conserving angular momentum, relative to the Earth's surface. As already outlined (Box 4.6), the poleward transfer of angular momentum by increasing velocity of airflow in this way must be limited to lower latitudes, the high-velocity subtropical jet stream marking this limit. As the subtropical jet stream lies above high surface pressure, the air within it subsides towards the surface where its energy is either dissipated by friction or transferred polewards or equatorwards. The polar front jet stream is located where meridional temperature gradients are steep in higher latitudes, generating a strong upper westerly flow. Mean wind speeds in the jet stream core, which lies between 10 and 12 km above the ground surface, are 25 m s^{-1}. It tends to be discontinuous and it follows a meandering, highly variable path.

Thus a more realistic representation of the atmospheric circulation in its meridional plane is that given by Palmen (1951) and shown in Figure 4.16. In the northern hemisphere the major feature of the circulation is the Hadley cell, while north of this the thermally indirect cell of the middle latitudes is now much smaller and is confined to the middle and upper troposphere. However, this circulation in itself does not explain entirely how poleward transport of angular momentum and heat energy is achieved beyond latitude 30°. Beyond this, the atmosphere develops eddying motions. It is these which effect the poleward transport of heat and momentum. By using a laboratory model, Hide (1969) has provided an indication of what form this eddying motion may take (Box 4.7). For slow rotation, the movement of the fluid in the dishpan followed a single con-

Box 4.7

THE DISHPAN EXPERIMENT

Two concentric cylinders enclose a fluid (usually water or glycerol) in which there are small polystyrene granules. The outer cylinder is warmed and the inner cooled, thereby establishing a temperature gradient across the fluid which is analogous to that which exists in the atmosphere between tropical and polar latitudes in one hemisphere.

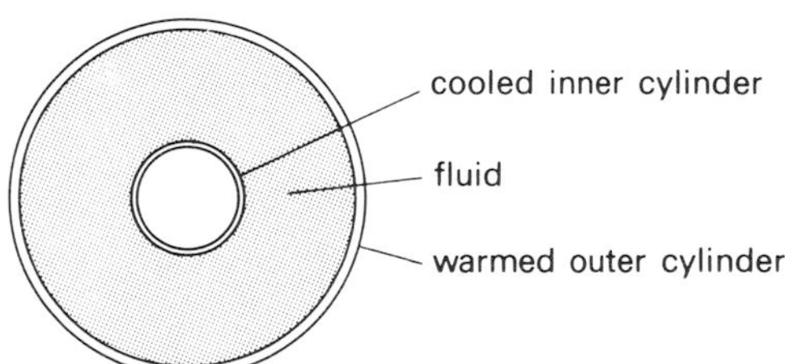

The movement of the fluid is recorded by photographing the paths taken by the polystyrene granules as the temperature gradient is kept constant and the cylinders rotated at various speeds.

At low rates of rotation, in the region of 0.5 rad s^{-1}, the movement of the fluid in the chamber is symmetrical about its central axis. In contrast, at a rotation of 4 rad s^{-1} there is a well developed wave motion in the fluid.

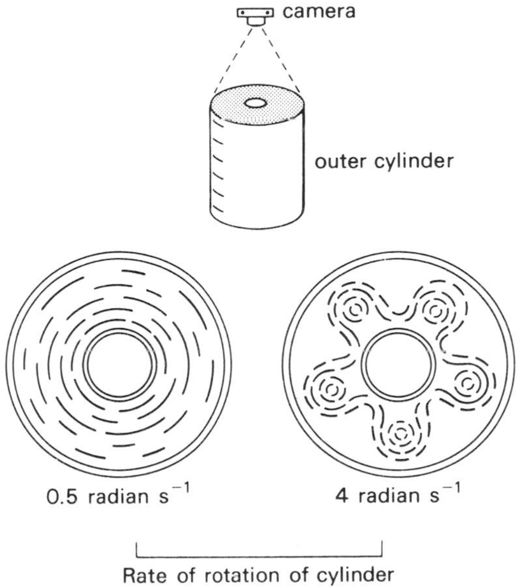

vectional cell, simulating a Hadley cell operating between Equator and Poles. However, as the rotational velocity was increased, thereby changing the Coriolis parameter, the convectional circulation was unable to maintain the flow of heat and momentum from the warm rim to the cool centre of the dish. In place of the convectional cell, a wave motion was initiated which effectively carried out these transfers.

The isobaric surfaces for the upper troposphere reveal a number of identifiable waves in the upper westerlies, referred to as **Rossby waves**. The number of these waves and the degree of their latitudinal oscillation vary considerably over time. They can develop and decay over relatively short periods of time – as little as a month or two (Fig. 4.17). However, it is possible to identify average positions of the waves over extended periods of time, and this wave pattern is essential to the atmospheric circulation. Although net meridional airflow is zero in these waves, they fulfil the functions of transferring heat energy and angular momentum away from the subtropics (Box 4.8).

The primary circulation system of the atmosphere is thus considerably more complex than is implied by the single- and three-cell models. However, whatever the model, it can never fully represent what is, in reality, an extremely efficient yet mechanically complex heat engine. The atmosphere manages to maintain a global heat-energy balance by effectively redistributing the surplus from lower latitudes. That it does so above a surface whose physical characteristics are highly variable over both time and space adds further to the intricacy of its mechanism.

We have so far considered only the movements of dry air. The presence of water vapour in the air involves additional exchanges of heat energy, in that its extraction from the air releases heat and its addition consumes heat. Thus, in transporting water vapour the atmosphere is also transporting a potential source of heat energy.

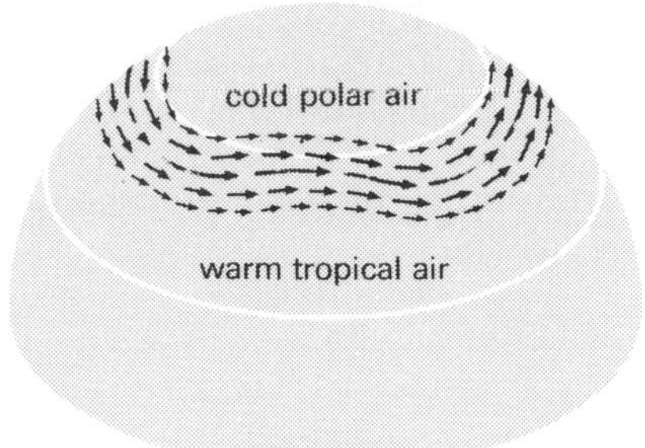

(a) Jet stream begins to undulate

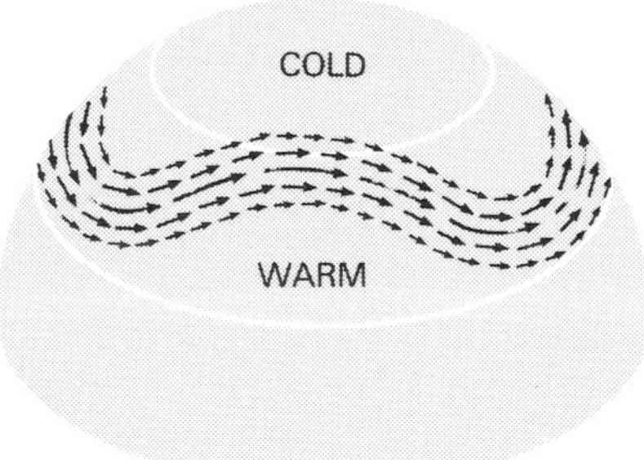

(b) Rossby waves begin to form

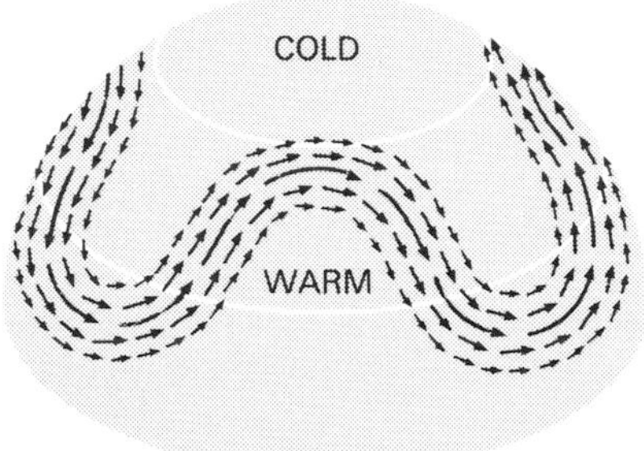

(c) Waves strongly developed

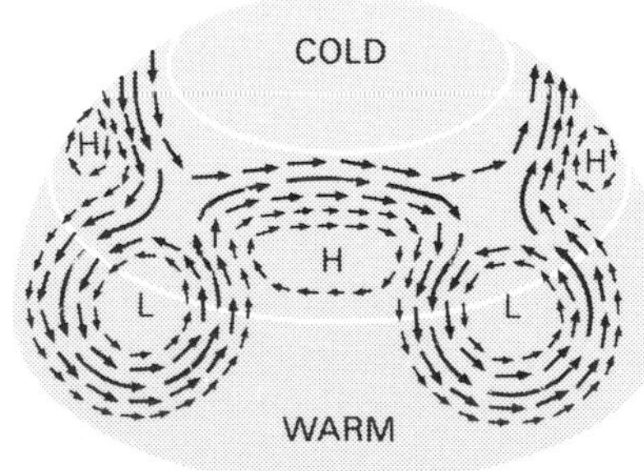

(d) Cells of cold and warm air bodies formed

Figure 4.17 Waves in the upper westerlies (Strahler & Strahler 1973).

4.4 Water in the Earth–atmosphere system

4.4.1 The nature of water

The water molecule is stable and it forms only weak bonds with neighbouring molecules of other substances (Box 4.9). Water exists in three phases – gas (water vapour), liquid and solid (ice) – and it

Box 4.8

HEAT AND MOMENTUM TRANSFER IN WAVE MOTION

Consider air moving through the wave below, in the northern hemisphere.

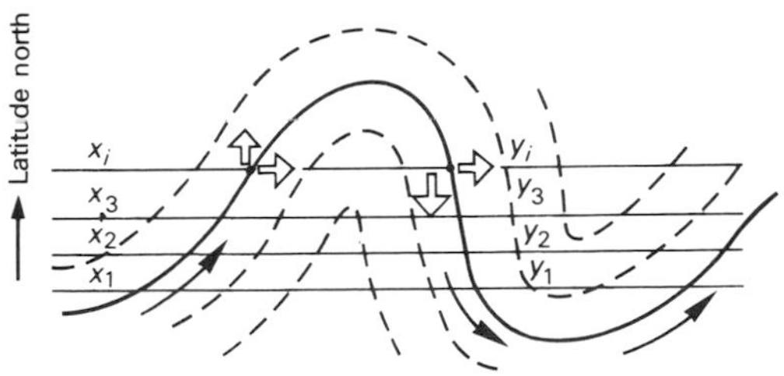

(a) If air temperatures at latitude points x_1, x_2, x_3 are greater than at y_1, y_2, y_3, there must be a net loss of heat in the poleward movement of the air.
(b) If we consider the meridional and zonal components of air motion in the wave at points x_1 and y_1, then along the rising limb of the wave air has both northward and eastward components. Along the falling limb, the air has southward and eastward components. We usually find that the northward component along the rising limb and the southward component along the falling limb are roughly equal, meaning that there is no net meridional flow of air, i.e. there is no loss of momentum in this plane. However, if we were to compare the eastward, or zonal, components we would find a considerable reduction in velocity and hence in momentum. In travelling through the wave the air will, therefore, have lost eastward momentum, effectively transferring this in a poleward direction.

Box 4.9

THE WATER MOLECULE (after Sutcliffe 1968)

Hydrogen and oxygen atoms combine to form water, the electrons being shared between them as in the diagram. There is electrostatic attraction between molecules because of the asymmetry of the distribution of electrons which leaves one side of each molecule with a positive charge. The water molecule can form four such hydrogen bonds, which are relatively weak. In a solid state, as ice, the tetrahedral arrangement of this bonding produces a tetrahedral crystalline structure. In a fluid state, increases in temperature weaken the hydrogen bonding.

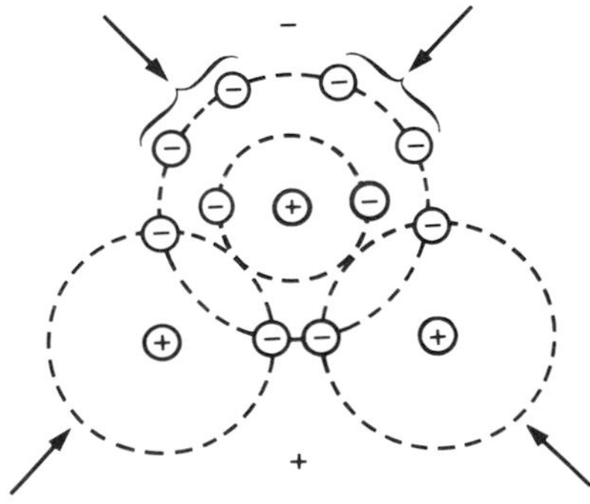

is present in the Earth–atmosphere system in all three phases. Most water vapour is found within the atmosphere; the oceans store most of the water in its liquid phase. At the Poles and in high alpine regions there are stores of water in the form of ice.

Water possesses physical properties that are unique when it is compared with substances of similar molecular mass. It has, for example, the highest specific heat of any known substance, which means that temperature changes take place very slowly within it. It also has a high viscosity and a higher surface tension than most common liquids. Its boiling point at standard atmospheric pressure (1013.25 mb) is 100°C and its melting point 0°C (Table 4.5). In solid and liquid phases, the physical properties of water vary with its temperature. Of these, the variation in density differs markedly from the behaviour of most liquids, which usually reach their maximum density at their freezing point. A maximum density at 4°C means that freezing to great depths in lake and ocean waters will be suppressed by water cooled below 4°C which floats to the surface.

In the gaseous phase, water vapour constitutes less than 4% of the total atmospheric volume (Sec. 4.1). The amount present may be expressed in terms of absolute humidity, specific humidity or mixing ratio (Box 4.10). Water vapour exerts a partial pressure in the atmosphere which is referred to as its vapour pressure (e), which normally varies between 5 mb and 30 mb. When the atmosphere lies above a liquid water surface, water molecules are constantly being exchanged between them. If the atmosphere is dry, the rate of uptake of molecules is greater than the rate of return to the surface. When a point of equilibrium is reached where the number of molecules leaving the surface is equal to the number arriving, the vapour pressure of the air has reached saturation vapour pressure with respect to water. Subsequent additions of water molecules to the air are balanced by deposition onto the surface.

Table 4.5 Physical constants of pure water (Sutcliffe 1968).

specific heat (15°C)	4.18	$Jg^{-1}deg^{-1}$
latent heat of melting	334.4	Jg^{-1}
latent heat of vaporisation (15°C)	2462	Jg^{-1}
surface tension	7340	$mNm^{-2}cm^{-1}$
tensile strength	1418.5	$kNm^{-2}cm^{-2}$
melting point (1013 mb)	0°C	
boiling point (1013 mb)	100°C	

Box 4.10

WATER VAPOUR IN THE ATMOSPHERE

Absolute humidity (χ)
Mass of water vapour in a given volume of air at a given temperature:

$$\chi = m_v/v$$

where m_v = mass of vapour and v = volume of air in which it is contained.

Specific humidity (q)
The proportion by mass of water vapour in moist air:

$$q = \frac{m_v}{(m_v + m_a)}$$

where m_a = mass of moist air.

Saturation deficit
Saturation deficit at temperature T_a is given by:

$$S = (e_s - e_a)$$

where e_a is vapour pressure, and e_s is saturation vapour pressure at temperature T_a.

Relative humidity

$$RH = \left[\frac{e_a}{e_s} \times 100\right]\%.$$

The value of vapour pressure at which saturation occurs is dependent on air temperature, as indicated in the **saturation vapour pressure curve** (Fig. 4.18). If we have a parcel of dry air at point X with temperature T_a and vapour pressure e_a, we can derive measures of its humidity. If its temperature remains constant and more water molecules are added, saturation is reached at Y where saturation vapour pressure is e_s. Saturation deficit and relative humidity for air at temperature T_a may be derived as indicated in Box 4.10. If vapour pressure remains constant but temperature is reduced, saturation is eventually reached at temperature T_d, which is referred to as the dew-point. The difference between T_a and T_d may be used also as an indication of the humidity of the air.

Should the surface underlying the air be ice, then the saturation vapour pressure is slightly less than that over a water surface at the same temperature, as shown in Figure 4.18. If air lies over both water and ice simultaneously, and is cooled from temperature T_b to T_i, then, with respect to the water surface, the air is unsaturated and will accept more water molecules from it. However, with respect to the ice surface, it is at its dew point and it must deposit any further water molecules that it receives. Thus there is a simultaneous withdrawal of water molecules from the liquid surface and deposition on the ice surface. This process is of particular importance in the development of precipitation from clouds (Sec. 4.4.5).

The processes which bring about such changes in

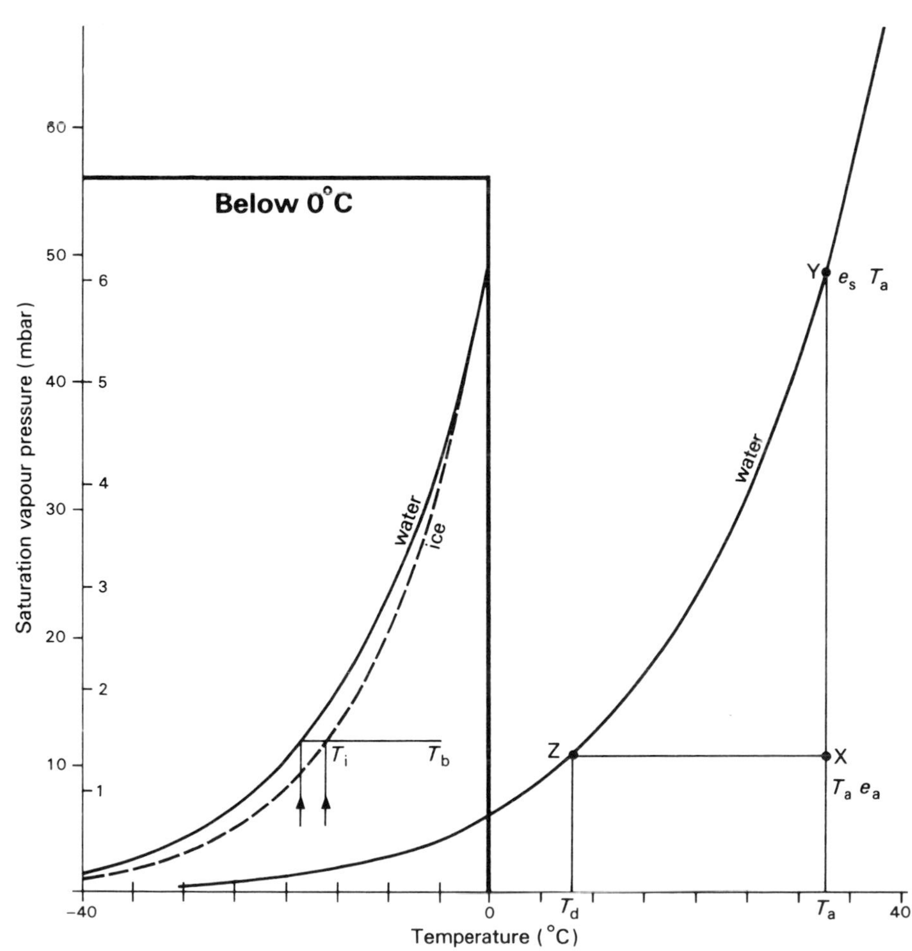

Figure 4.18 Saturation vapour pressure curves with respect to water and ice surfaces.

phase are indicated in Figure 4.19a. The operation of these processes and the transport of water in vapour or liquid form within the Earth–atmosphere system is commonly represented in simplified form as a hydrological cycle (Fig. 4.19b).

4.4.2 The movement of water in the Earth–atmosphere system

As no water is exchanged between the atmosphere and space, we can regard its movement taking place within the Earth–atmosphere system as a closed system. The atmosphere alone must, however, be viewed as an open system, because the movement of water across the Earth/atmosphere interface represents a transfer of both matter and energy across the boundary of the system. The movement of water across this boundary and within the system is initiated and maintained by a flow of energy through the system (Fig. 4.19b). Within this hydrological system there are a number of locations at which water is stored. The major stores are the atmosphere, the land, the oceans and the polar ice caps. The distribution of water between these is indicated in Figure 4.20. The smallest store is the atmosphere, which contains 0.001% of total water in the Earth–atmosphere system, while the greatest is the ocean store, which has 97.6%. Transfers between stores are carried out by the processes of evaporation, condensation, precipitation, runoff and freezing and melting. The greatest exchanges are those between ocean and atmosphere: 86% of evaporation takes place over oceans while 78% of precipitation occurs over them (Baumgartner & Reichel 1975).

4.4.3 Evaporation

Evaporation is the process by which liquid water is changed into its gaseous phase (water vapour). If heat energy is supplied to a water surface, it has the effect of weakening the bonding between water molecules and increasing their kinetic energy. The faster-moving molecules thus have a greater capacity to break away from the water surface and enter the air above. This transfer is partly offset by water molecules returning to the surface, and the net loss represents the rate of evaporation from the water surface.

The change in phase from liquid to gas requires an input of heat energy which is referred to as the latent heat of vaporisation. At 0°C, 2501 J are required to evaporate 1 g of water, while at 40°C this is reduced to 2406 J. If heat energy ceases to be supplied from an external source, the energy for evaporation will be drawn from the remaining water, thus having the effect of reducing its temperature and suppressing further evaporation.

There are a number of factors which affect the rate of evaporation, the most important of which are the heat energy supply, the humidity of the air, the characteristics of the airflow across the surface and the nature of the evaporating surface.

The rate at which water molecules diffuse into the lower layers of the atmosphere is directly governed by the heat energy made available to them. As a component of the heat balance equation (see Ch. 3), evaporation depends not upon solar energy input but on net radiation. For open water and wet soils, most of this net radiation is consumed in evaporation. The spatial and temporal variation of net radiation are thus imposed upon evaporation, producing, for example, summer and early-afternoon maxima.

The balance between water molecules leaving the surface and returning to it depends on the number of them in the overlying air. If the air is relatively dry, their numbers will be small in relation to those leaving the surface and thus the rate of evaporation will be high. If, however, the

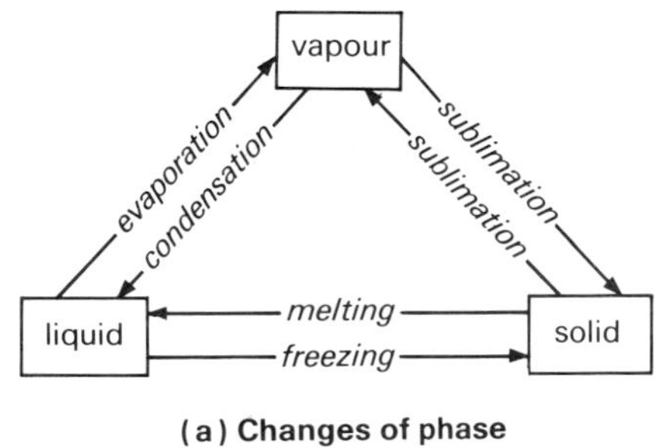

(a) Changes of phase

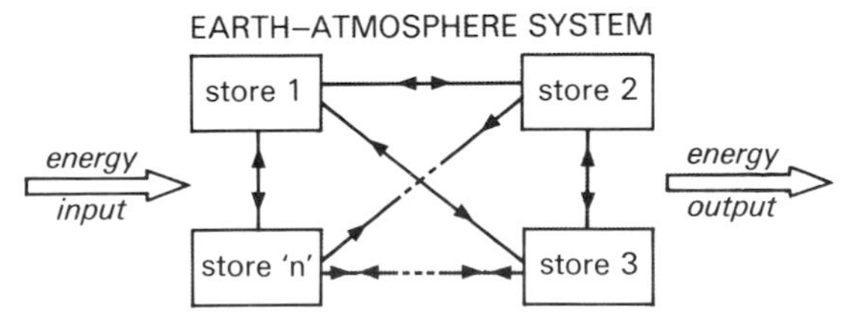

(b) The movement of water in a closed system

Figure 4.19 The movement of water in the Earth–atmosphere system.

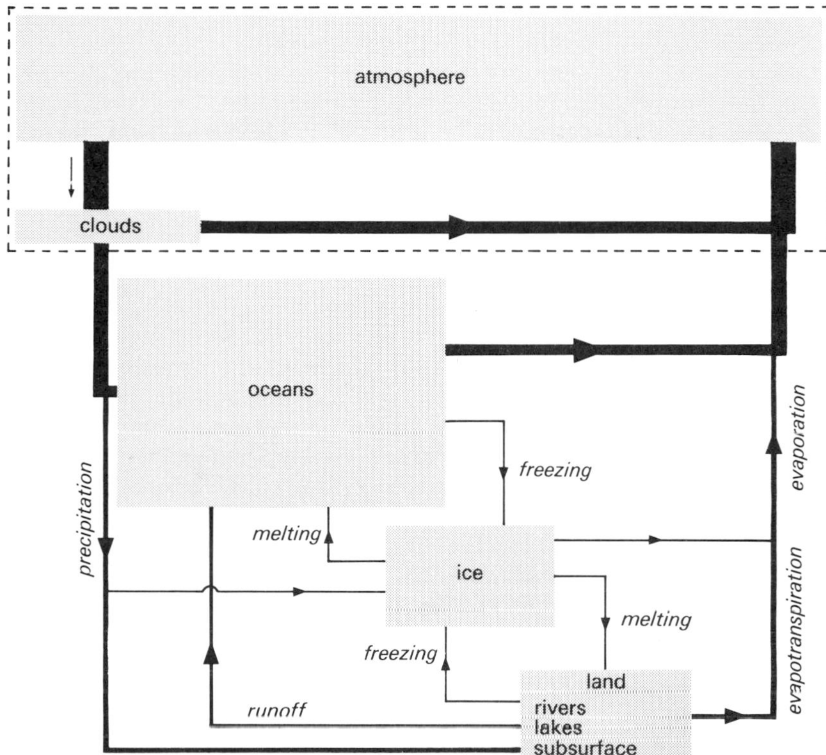

Figure 4.20 The hydrological system.

air approaches saturation, it has a higher population of water molecules from which those returning to the surface will be only slightly fewer than those leaving, and consequently the rate of evaporation will be low. The rate of evaporation is related to the difference between saturation vapour pressure (e_s) at surface temperature and the vapour pressure of the air above (e_d):

$$\text{evaporation } (E) = \text{constant} \times (e_s - e_d). \quad (4.7)$$

Molecules will continue to leave the surface while e_d is less than e_s. However, the effect of this is to increase e_d until it eventually reaches e_s. At this point the rate of evaporation will be zero. So, unless there is a mechanism for mixing the air and thereby redistributing water molecules, evaporation will always exhibit decrease with time.

The mixing of the lower atmosphere by vertical and horizontal motion effectively results in the replacement of air before saturation is reached. Thus evaporation is maintained at a higher rate than if there were no such replacement, as e_d no longer increases towards e_s. Increases in horizontal wind speed cause increases in the rate of evaporation, but only up to a point where further increases have little effect. This maximum is determined by the heat energy available for evaporation and by the humidity of the air. Equation 4.7 may be rewritten to take into account mean wind speed ($\bar{u}$):

$$E = Bf(\bar{u})\,(e_s - e_d) \quad (4.8)$$

where B is a constant, and $f(\bar{u})$ expresses a function of wind speed.

Where a water surface is exposed to the atmosphere, evaporation may take place, whether this surface be of an ocean or a glass of water. However, the rate of evaporation is determined by the rate at which molecules lost to the atmosphere can be replaced from the subsurface store. In the case of an open water surface there is effectively an unlimited supply of water freely available for evaporation. However, in a soil there is usually a limited supply available for evaporation, thereby suppressing rates of loss. Should the soil be saturated, for example after a long period of rain, then evaporation may be considerably higher than from an equal area of open water, because of the water film held on soil particles near to the surface. The quality of the water also affects the rate of evaporation in that it is reduced by the presence of impurities such as dissolved salts. The greatest reduction is from the salt water of, for example, the oceans where there is, as a rough approximation, a 1% reduction in evaporation for a 1% increase in salinity.

If there is vegetation growing upon the surface, this provides an extra pathway for water molecules transferring from the ground surface into the atmosphere. Water vapour is diffused through pores (stomata) on leaf surfaces and is transferred into the atmosphere (see Fig. 17.3). This is associated with a suction pressure inside the plant causing it to withdraw water from the soil through its root system. This transfer of water into the atmosphere is referred to as **transpiration** and it accounts for a large proportion of moisture losses from vegetated surfaces (see Ch. 17).

The rate of transpiration loss is governed by two groups of factors: those already discussed – the extrinsic factors which affect rates of evaporation from water surfaces – and the factors intrinsic to the plant. For example, transpiration loss from most plants will be large when the atmosphere is relatively dry. A good example of this is the effect of the dry Mistral wind of the Rhône Valley in France, which causes harmful increases in transpiration loss from crops. In most plants, leaf stomata open under daylight conditions and close at night, thereby introducing a clear diurnal variation in transpiration losses. However, should the plant be unable to withdraw adequate water through its roots, to match transpiration loss, internal stresses may be set up. In response to this there may be a partial or complete closing of the stomata in order to limit transpiration.

Over vegetated surfaces with a dense plant cover, water losses to the atmosphere are largely accounted for by transpiration. In a dense forest, for example, over 60% of water loss will be achieved through transpiration. If evaporation of precipitation intercepted by the trees is included, over 80% of water transferred to the atmosphere may be due entirely to the presence of a vegetation cover. In semi-arid areas where there is virtually no surface water, transpiration will account for the entire surface-to-atmosphere water transfer.

Most surfaces are neither absolutely bare nor completely vegetated but possess elements of both. Such is the complexity of these surfaces that separation of evaporation and transpiration is impracticable and the two are combined in the one term, **evapotranspiration**. As we have already seen, a shortage of soil moisture limits the rate at which evaporation and transpiration can take place. Under such 'limiting' conditions, actual evapotranspiration is occurring. If, on the other hand, no such limiting condition prevails, then evapotranspiration can take place at its maximum rate within the constraints placed upon it by the availability of heat energy, atmospheric humidity and wind speed. Under these 'non-limiting' conditions, we have potential evapotranspiration. This concept is used extensively in irrigation studies, for it represents the worst possible situation with regard to water loss from the ground surface.

Over the Earth's surface, actual evapotranspiration is greater over oceans than over land surfaces with maxima occurring not at the Equator but over the tropical oceans between latitudes 10° and 40° (Fig. 4.21). The meridional cross section of average rates of evapotranspiration illustrates the operation of the various factors that control it. Taking account only of heat balance we should expect a clear low-latitude maximum and high-latitude minimum. However, this simple distribution is modified in the zones of persistent winds, for example the subtropical trade winds. These winds are warm and dry and they clearly increase evapotranspiration, while in humid equatorial regions low wind speeds and high atmospheric humidity tend to limit evapotranspiration which is, to some extent, offset by the high rates of transpiration from equatorial forests.

4.4.4 Condensation

When the number of water molecules returning to a surface exceeds those leaving, there is a net deposition of water from the air, which is referred to as **condensation**. The implication of this statement is that a necessary condition for condensation is saturation, since a condition of non-saturation produces a net loss of water molecules from the underlying water surface, as we have already seen. Therefore in order for air at point X in Figure 4.18 to produce condensation, it must first achieve a state of saturation. For simplicity we can consider the two alternatives represented by paths XY and XZ. By physically adding more water vapour to the air, while keeping its temperature constant,

Figure 4.21 Global distribution of annual evaporation (E) and evapotranspiration (r); values in millimetres (after Barry 1970).

saturation is eventually achieved at Y, at which point any further addition will initiate condensation. By cooling the air to its dew-point, saturation is achieved at Z and condensation is initiated by further cooling. In the atmosphere, the latter is responsible for the majority of condensation forms.

If heat energy is withdrawn from the water vapour molecules, the subsequent decrease in their kinetic energy and strengthening of intermolecular bonding prevent them remaining in the gaseous state. They return to liquid water and in so doing release the latent heat of vaporisation they gained during evaporation (Sec. 4.4.3).

If moist air is in contact with a water surface, the molecules of the cooling air are readily absorbed into it. In the free atmosphere, however, there are, apparently, no such surfaces over which condensation takes place. If pure air, devoid of any suspended foreign matter, is cooled to its dew-point, no condensation occurs and it may continue to take up water molecules to a state of supersaturation. Under laboratory conditions it has been possible to reach relative humidities of over 400% before condensation into water droplets takes place.

In reality, the Earth's atmosphere contains a number of impurities resulting from natural events or from Man's activities (see Table 4.2). These provide surfaces upon which condensation may take place. The relative humidity at which this occurs is largely dependent upon the nature and number of the particles, which are referred to as condensation nuclei. Condensation onto these results in the rapid growth of water droplets. The radii of nuclei range in size from less than 10^{-3} μm to more than 10 μm. These two extremes in the range of sizes are the least effective in initiating condensation, the former producing unstable droplets readily evaporated back into the atmosphere and the latter falling quickly to the ground under the influence of gravity. The most effective particles are in the range 10^{-1} μm to 1.0 μm, some of which may also be **hygroscopic** (attracting water molecules to them) and they include substances such as sodium chloride (sea salt) and ammonium sulphate.

The presence of hygroscopic nuclei may initiate condensation in free air well before it reaches its saturation vapour pressure, at relative humidities as low as 80%. Other less hygroscopic or non-hygroscopic materials, such as terrestrial dust, are less effective, but when present in large numbers they encourage condensation in slightly super-saturated atmospheres.

Condensation takes place as a result of the cooling of the atmosphere in the presence of a receiving surface, whether this be a terrestrial surface or a suspended condensation nucleus. The condensation forms produced may be classified according to the nature of this cooling process – the most important of which are **contact cooling**, **radiation cooling**, **advection cooling** and **dynamic cooling**.

Contact cooling. When terrestrial surfaces lose heat rapidly by radiation, this produces a lowering of their temperatures. Heat is then conducted from the air to the surface. As air is an extremely poor conductor of heat, this cooling does not penetrate far above the surface, particularly if there is little or no air movement. Should the temperature of the surface fall below the dew-point of this thin layer of cooled air, direct condensation of water droplets onto it occurs in the form of dew. If the dew-point temperature is less than 0°C, deposition occurs in the form of ice crystals of hoar frost. If, however, there is a small amount of turbulence in the air layers near to the ground, there will be limited mixing and a cooling to greater heights. In this case, reduction of air temperature below its dew point produces a radiation, or ground, fog which will persist until it is thoroughly mixed with the drier air that overlies it, or until temperatures rise above dew point through solar heating.

The downslope drainage of air cooled by contact may lead to its accumulation in lower-lying areas, especially where there is standing water. Such movement may create deep fogs, commonly referred to as valley fogs.

Radiation cooling. The atmosphere also experiences direct loss of heat by radiation. However, because of the slow rate of cooling that arises from this, it is rarely the sole cause of condensation. It can, however, enhance the cooling of the air already in contact with cold ground surfaces.

Advection cooling. Cooling may result from a horizontal mixing of air which is referred to as **advection**. Condensation forms can be produced when two streams of air mix together if both have vapour pressures approaching saturation and if there is a relatively large temperature difference between them. If, for example, warm moist air is transported over a cool moist surface, it mixes with the shallow layer of cooled air associated with the surface. When this has taken place, the mixture may have a saturation vapour pressure, determined by its temperature, which is less than its actual vapour pressure and it is therefore supersaturated. This excess is condensed in the form of **advection fog**. These fogs may be associated with relatively turbulent airflow and may therefore be deep. The cold surface may be either sea or land, although most advection fogs are associated with sea surfaces. For example, warm moist air passing over a cool ocean current will produce advection fog. An excellent example of this occurs off the coast of Newfoundland where warm air associated with the Gulf Stream mixes with the cold air over the Labrador Current. Warm tropical air approaching the British Isles from the south crosses cool ocean waters, particularly in spring, and it may also produce thick advection fog.

One further form of advection fog occurs where cold air flows over warm waters. Water evaporating from the surface condenses in the cold air above and produces a fog which resembles smoke or steam rising. This usually occurs where there are large temperature differences between surface and air, such as in the polar latitudes where cold air flowing from the ice caps over warmer seas produces **sea smoke**. In terms of the basic condensation process, it is produced by an addition of water vapour to the air rather than a cooling process.

Dynamic cooling. If air is forced to rise, it is subjected to **adiabatic cooling** (Box 4.11). The form of the condensation resulting from this depends upon the magnitude and rapidity of the enforced rise and the stability of the air. A gradual rising of relatively stable air will produce less spectacular condensation forms than a violent uplift of unstable air. Uplift may be the result of the passage of air over mountain ranges, of localised convection

Box 4.11

ADIABATIC TEMPERATURE CHANGES

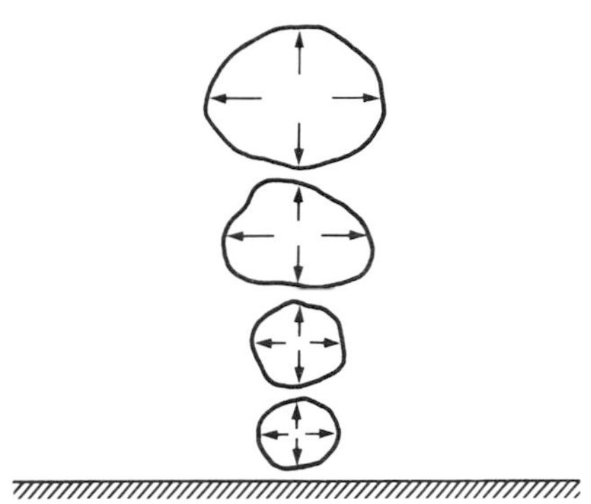

Consider a parcel of air at the ground surface which is forced to rise. As it does so, it expands in response to decreasing pressure. In so doing, the parcel does work and therefore expends energy. If the parcel is a closed system and there is no heat exchanged between the parcel and its surroundings, then the energy for work is drawn from inside it, causing its temperature to decrease. This temperature decrease is referred to as **adiabatic**.

In the atmosphere, when the air is unsaturated, the rate of temperature decrease is referred to as the dry adiabatic lapse rate and has a value of 9.8°C km^{-1}. If the air reaches saturation point, further cooling results in condensation which releases latent heat and so offsets the rate of temperature decrease. Thus an average value of the saturated adiabatic lapse rate is 6.5°C km^{-1}, but its value varies because of the relationship between saturation vapour pressure and temperature.

Stability and instability

Consider two separate vertical temperature profiles, *xy* and ST, referred to as environmental lapse rates. Consider also two points on these profiles, A and B. If a parcel of air at A is forced to rise it will cool at the dry adiabatic lapse rate. At A′ it will be warmer than its surroundings and will, therefore, continue to rise. If the same parcel is forced downwards, it warms at the dry adiabatic lapse rate. At A′′ it will be cooler than its surroundings and will continue to fall. Thus, parcel A shows a disinclination to return to its former position once given an initial upward or downward displacement, and it is, therefore, unstable.

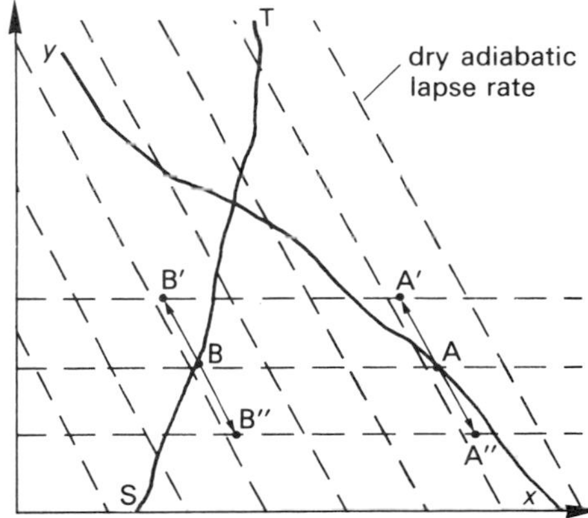

In contrast, if the air at B is forced to rise, by B′ it is cooler than its surroundings and it will fall back to B. If forced downwards, by B′′ it is warmer than its surroundings and will rise back to B. The parcel B thus shows an inclination to return to its former position given an initial displacement and it is, therefore, stable.

In the atmosphere, stability and equilibrium are not synonymous. The atmosphere may be in a state of either stable or unstable dynamic equilibrium, its stability determining the disruption caused by an external impetus. An unstable atmosphere may continue in a state of disequilibrium once displaced, while a stable atmosphere may readily return to equilibrium.

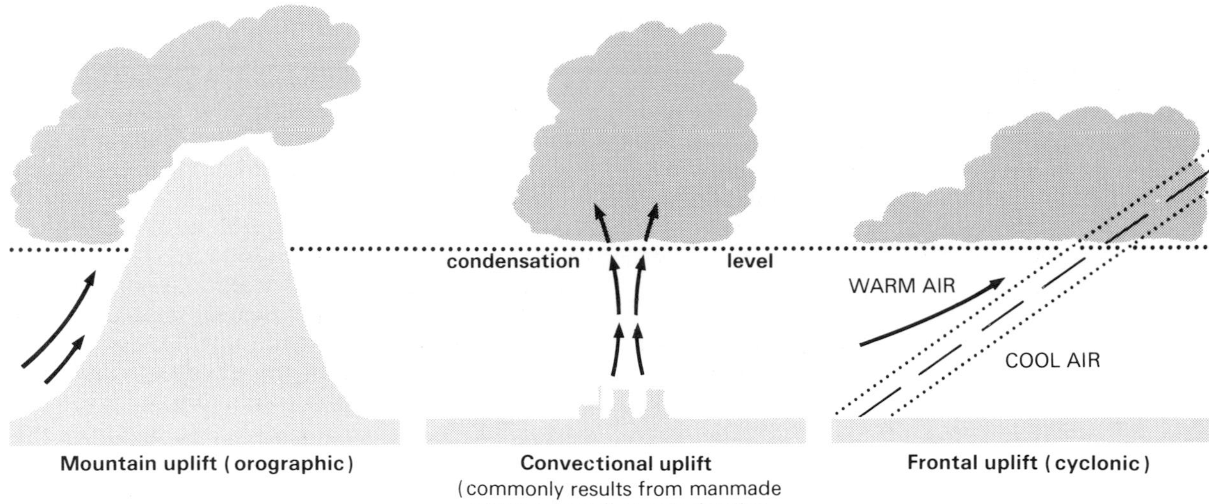

Figure 4.22 A simplified representation of dynamic cooling.

or of frontal contact between masses of air of differing temperatures at fronts (Fig. 4.22). In all cases the air rises and cools at the dry adiabatic lapse rate until it reaches saturation. Above this the air continues to cool at the saturated adiabatic lapse rate, while condensation takes place in the form of clouds. The shape and depth of these clouds vary from shallow layered forms (stratus) characteristic of stability, to clouds with great vertical extent (cumulus) that are associated more with unstable than stable air. The types of cloud are listed and described in Table 4.6.

4.4.5 Precipitation

Water that falls from clouds to the ground in either solid or liquid form is referred to as **precipitation**. Although this is derived from condensed water vapour, it is not true to say that condensation automatically leads to precipitation.

There are, initially, two forces operating on a cloud droplet, these being a gravitational attraction towards the Earth and the frictional force between the droplet and the air through which it is moving. These are represented by *G* and *F* in Figure 4.23a. When these two forces balance each other, in accordance with Newton's First Law of Motion, the droplet falls towards Earth at a

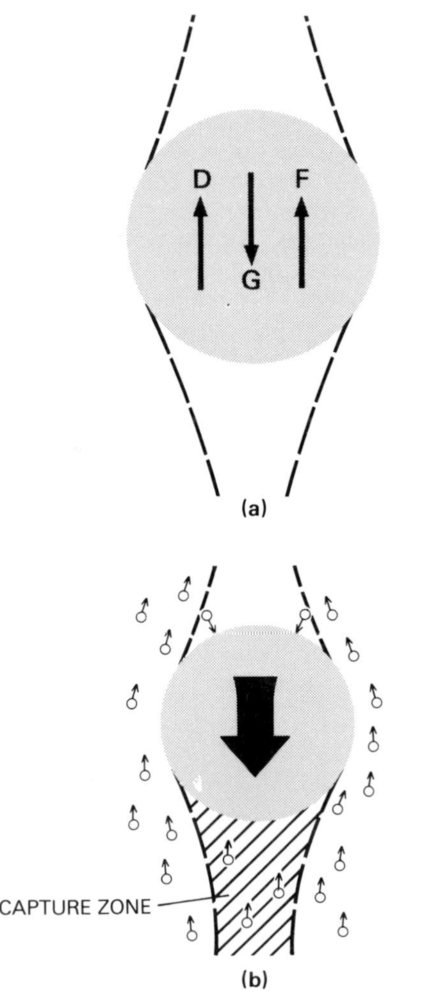

Figure 4.23 (a) Forces acting on a cloud droplet (see text) and (b) collision and coalescence between a large falling drop and smaller cloud droplets.

constant velocity – its **terminal velocity** – which is directly related to droplet size. The radius of cloud droplets ranges from 1 μm to 20 μm which, in still air, would have terminal velocities between 0.0001 and 0.05 m s^{-1}. However, the air is not still, for there are updraughts in clouds, which may reach speeds of 9 m s^{-1}, which represent a further force (D) acting upon the cloud droplets.

The terminal velocities of cloud droplets are so small in relation to those of the updraughts that they are held within the cloud mass and are not precipitated. Precipitation is not usually initiated until drops have reached a radius of approximately 1000 μm, although fine drops of radius as low as 200 μm may be precipitated in particularly calm conditions. The average cloud droplet (of radius 10 μm) must, therefore, increase its volume a millionfold to reach the average precipitation drop of radius 1000 μm. There have been a number of explanations offered as to why this growth should take place, of which two are discussed here.

Ice crystal process. If ice crystals and supercooled water occur together in a cloud at temperatures between −10 and −25°C, water vapour is transferred directly from the water surface and is deposited as ice on the crystal surface (Sec. 4.4.1). These crystals may be the result of the freezing of supercooled water droplets at upper levels in cumulus clouds or they may have fallen from the higher cirrus clouds which are composed almost entirely of ice crystals. Ice crystal growth takes place which increases their earthward velocity. The growing crystal may grow further as it falls and

Table 4.6 Cloud types and their main characteristics. The height and temperature values given refer to the British Isles (from the *International cloud atlas*, WMO 1956).

Cloud genera	Abbreviation	Characteristics	Height of base (km)	Temperatures at base (°C)
cirrus	Ci	white filaments – fibrous appearance or silky sheen, or both	5 – 13	−20 to −60
cirrocumulus	Cc	thin white patch or sheet cloud without shading; composed of very small elements	5 – 13	−20 to −60
cirrostratus	Cs	transparent whitish cloud veil of fibrous or smooth appearance	5 – 13	−20 to −60
altocumulus	Ac	white and/or grey, patch, sheet or layer of cloud; composed of elements	2 – 7	+10 to −30
altostratus	As	greyish cloud sheet; layered appearance – totally or partly covering the sky	2 – 7	+10 to −30
nimbostratus	Ns	grey layer cloud associated with continual falling rain or snow; thick enough to blot out the sun	1 – 3	+10 to −15
stratocumulus	Sc	grey or whitish, patch, sheet or layer; formed of rounded masses	0.5 – 2	+15 to −5
stratus	St	grey cloud layer with uniform base which may give drizzle	0 – 0.5	+20 to −5
cumulus	Cu	detached cloud, dense with sharp outline; vertically developed in rising towers resembling a cauliflower	0.5 – 2	+15 to −5
cumulonimbus	Cb	heavy and dense cloud with considerable vertical extent; precipitation falling	0.5 – 2	+15 to −5

collides or coalesces with other crystals. If the crystal melts, it continues its fall as a water droplet and will grow as a result of coalescence with much smaller water droplets in its path (Fig. 4.23b). On reaching a certain critical size, the droplet becomes unstable and divides into many medium-size droplets which will collide and coalesce in turn, thereby providing an efficient precipitation process operative over short intervals of time.

Collision and coalescence. Although the ice crystal process explains precipitation from clouds that have temperatures below −10°C, it fails to explain why precipitation should occur from clouds whose temperatures may be as high as 5°C, such as those in tropical latitudes. The movement of air within the clouds produces water droplets of different sizes through chance collisions and coalescence. As terminal velocities are directly related to droplet size there will be relative motion between the droplets, in which case the process of coalescence illustrated in Figure 4.23 will take place and cause the growth of some droplets at the expense of smaller ones. Such growth may eventually produce droplets large enough to fall to the ground.

The main forms of precipitation are listed in Table 4.7. A broader classification can also be developed from the nature of the condensation process from which precipitation is derived. Thus we have **orographic** (mountain), **convectional** and **cyclonic** precipitation.

Over the Earth's surface there is considerable variation in precipitation which may be attributed to a number of factors operating over a range of spatial scales. On a global scale, the two areas associated with convergent airflow at the surface, the intertropical convergence and the middle latitudes, are both areas of relatively higher precipitation, arising from the consequent uplift of air and its adiabatic cooling. In contrast, the sub-tropics and the polar regions have surface airflow of a divergent type and hence have much lower precipitation due to the subsidence of air. This elementary zonation is apparent in the distribution of mean annual precipitation especially over ocean areas (Fig. 4.24).

The mean atmospheric circulation is, however, only one of a number of controls on the global distribution of precipitation. We must also consider, for example, the thermal structure of the atmosphere – whether it is stable or unstable. Precipitation also depends upon the number and type of condensation nuclei in the air and the presence of conditions conducive to the growth of cloud droplets.

The land masses, particularly those in the northern hemisphere, exert a marked control on the distribution of precipitation. As most atmospheric water vapour is derived from the oceans, the continental interiors may experience much drier atmospheric conditions than do locations near to coasts. This is reflected in the interior of

Table 4.7 Precipitation types and their characteristics.

Type	Characteristics	Cloud type from which derived	Measurement
rain	waterdrop radius greater than 250 μm	Ns, As, Sc, Ac	rain gauge
drizzle	fine waterdrop radius less than 250 μm	St, Sc	rain gauge
snow	loose aggregates of ice crystals	Ns, As, Sc, Cb	snow gauge; snow run; photogrammetric
sleet	partly melted snowflakes or rain and snow falling together	as above	rain gauge
hail	pieces of ice of radius 2500 to 25 000 μm; concentric shells of ice	Cb	rain gauge; hail gauge
direct precipitation	directly on to surfaces under low-lying cloud – referred to as horizontal interception	St	interception gauges (experimental)

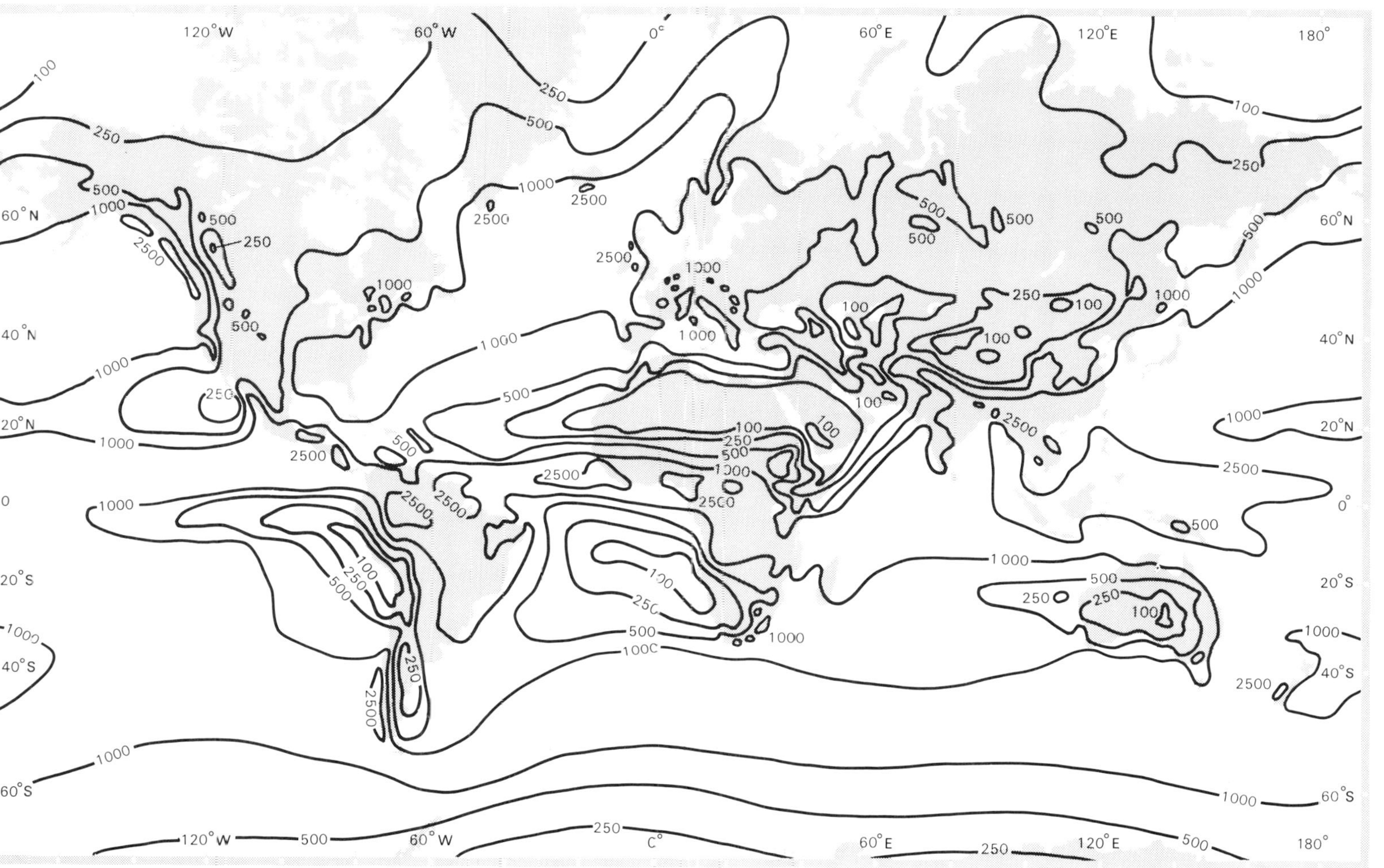

Figure 4.24 Global distribution of average annual precipitation totals, in millimetres (after Lamb 1972).

Eurasia, for example, which derives much of its moisture from the Atlantic ocean, conveyed inland by the middle-latitude westerlies. As this air travels eastwards over the continent it loses more water to the surface than it gains through evapotranspiration. There is thus an eastward decrease in precipitation, especially in winter when frozen surfaces give up little water to the atmosphere and the divergent airflow of the Siberian high pressure restricts the penetration of moist oceanic air.

One further modification to the global distribution of precipitation results from airflow over major mountain ranges. The intensification of both condensation and precipitation processes as air is forced to rise over, for example, the Rockies or the Andes, produces higher precipitation totals.

4.4.6 *The surface water balance*

Of the precipitation that falls upon the Earth's surface, some is returned to the atmosphere by evaporation and transpiration while the remainder is stored on or within the surface and is available for use in surface systems. We can express this distribution in terms of the **surface water balance equation**:

$$\text{(precipitation) } P = E + T + \Delta S + \Delta G + R \quad (4.9)$$

where E is evaporation; T is transpiration; ΔS is the change in soil water storage – water is held between the particles of the soil by retentive forces – some of the incident liquid precipitation will infiltrate into available spaces in the soil; ΔG is the change in groundwater storage – some water will percolate from the soil into the deeper groundwater reservoir; R represents overland flow across the surface, initially in rills but ultimately in streams and rivers. The structure of this balance varies over both time and space according to the prevailing moisture conditions in both the atmosphere and the surface. However, on a global scale, we can for simplicity consider the two basic surface forms of open water and land.

The water balance equation (4.9) may be rewritten for an open water surface, such as a large pan, in the simple form:

$$E = P + \Delta V. \quad (4.10)$$

We may assume that there is no runoff from a water surface ($R = 0$) and that there is no transpiration loss ($T = 0$). ΔV represents a change in the volume of water contained within the reservoir and is the equivalent of storage changes under land surfaces ($\Delta V = \Delta S + \Delta G$). We can use this equation to calculate open water evaporation by recording changes in the level of water in an open pan exposed to the atmosphere (Box 4.12).

In the case of the oceans we can ignore the R term but there is an extra inflow of water in the form of runoff from the surrounding land masses by way of rivers and of flow from one ocean area to another. By considering all the oceans as one surface we can effectively ignore interocean flow of water, and changes in storage over time are negligible. The surface water balance equation, therefore, becomes:

$$P = E - \Delta F \quad (4.11)$$

where ΔF is the flow of water from land to ocean. The average precipitation depth over the oceans amounts to 1066 mm or 385.0×10^{12} m^3 of water while evaporation amounts to 1176 mm or 424.7×10^{12} m^3 (Baumgartner & Reichel 1975). Over the ocean surfaces there is a net loss of 39.7×10^{12} m^3 of water returned through ΔF.

The water balance equation for a land surface must remain as written, as all components are present. However, for practical purposes, evaporation and transpiration are considered together as evapotranspiration. The equation is therefore:

$$P = E_t + \Delta S + \Delta G + R. \quad (4.12)$$

Indirect measurements may be made of both potential and actual evapotranspiration using this equation. If we assume that our surface is perfectly horizontal, then there is no gravitational acceleration of water drops along the surface, only into it. Thus R is zero. If soil moisture levels are maintained such that evapotranspiration takes place at its potential maximum rate, then, effectively $\Delta S = 0$. In this case the evapotranspirometer (Box 4.12) may be used to determine potential evapotranspiration. If irrigation of the soil tanks is not carried out, and a method of monitoring soil moisture changes (ΔS) is incorporated, actual evapotranspiration may be determined. Over the global land surface, the average depth of precipitation amounts to 746 mm, or a total volume of 111.1×10^{12} m^3, while evapotranspiration amounts to 480 mm, or 71.4×10^{12} m^3, which means that there is a surplus of 39.7×10^{12} m^3 at the surface. As storage factors ΔS and

Box 4.12

METHODS OF MEASURING EVAPORATION (a) AND POTENTIAL EVAPOTRANSPIRATION (b) BASED UPON THE SURFACE WATER BALANCE EQUATION

(a) Standard evaporation pan

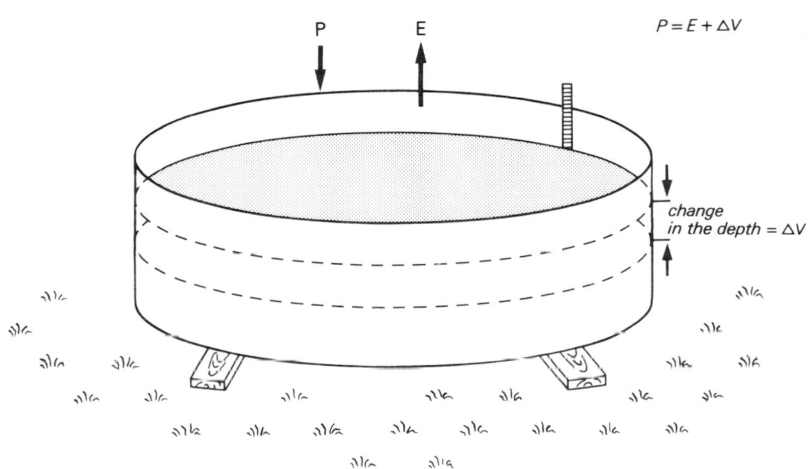

(b) Evapotranspirometer

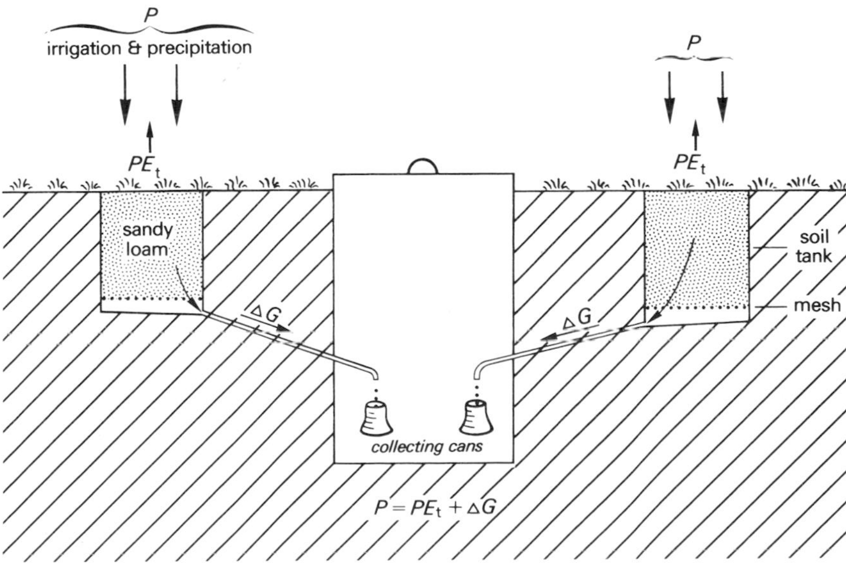

ΔG may be regarded as negligible in the long-term balance, this surplus is discharged in the form of runoff.

Within these broad global balances there are considerable differences in the structure of water balances, between oceans and between continents (Table 4.8). Of particular note is the high proportion of precipitation consumed in evapotranspiration over Africa (84%) and Australia (94%) which are predominantly tropical continents, compared to the 17% over Antarctica.

4.4.7 The atmospheric water balance

Our discussion of water balances would be incomplete without a brief reconsideration of the atmosphere, which is receiving water from the surface beneath it and returning it through precipitation. The horizontal motion of the atmosphere transports water vapour so that the surfaces from which it is derived are not necessarily those to which it returns as precipitation. This is best illustrated by reference to the data in Table 4.8. Over the oceans, the atmosphere receives 424×10^{12} m^3 of water, yet it releases only 385×10^{12} m^3, while over the land it receives only 71.4×10^{12} m^3 and releases 111.1×10^{12} m^3. The implication is that an exchange of water exists between the two, part of which will be through the flow of water vapour from over the oceans to over the land, through movement of the atmosphere.

This transport of water vapour is of significance in the distribution of heat energy in the atmospheric circulation. Water vapour transported in the atmosphere and subsequently condensed has, in releasing its latent heat of vaporisation, acted as a vehicle for heat-energy transfer. The meridional transfer of latent heat (see Fig. 4.7) is equatorwards between latitudes 30°N and 30°S where the trade winds blow across the ocean surfaces. In the middle latitudes, between 30° and 65°, there is a net poleward transport. This form of heat transfer is clearly greater in the southern hemisphere where there is a larger proportion of ocean surfaces acting as sources of water vapour.

Table 4.8 Water balance of the Earth's surface (Baumgartner & Reichel 1975).

	Water volume (10^3 km^3)			Proportion (%) of precipitation	
	Precipitation	Evaporation	Runoff	Evaporation	Runoff
Europe	6.6	3.8	2.8	57	43
Asia	30.7	18.5	12.2	60	40
Africa	20.7	17.3	3.4	84	16
Australia (without islands)	3.4	3.2	0.2	94	6
North America	15.6	9.7	5.9	62	38
South America	28.0	16.9	11.1	60	40
Antarctica	2.4	0.4	2.0	17	83
Arctic Ocean	0.8	0.4	0.4	55	45
Atlantic Ocean	74.6	111.1	−36.5	149	−49
Indian Ocean	81.0	100.5	−19.5	124	−24
Pacific Ocean	228.5	212.6	15.9	93	7
all land	111.1	71.4	39.7	64	36
all ocean	385.0	424.7	−39.7	110	−10

Further reading

Bannister, P. 1976. Water relations of plants. In *Introduction to physiological plant pathology*, Ch. 6. Oxford: Blackwell.

Baumgartner, A. and E. Reichel 1975. *The world water balance*. Amsterdam: Elsevier.

Dobson, G. M. B. 1963. *Exploring the atmosphere*. Oxford: Clarendon Press.

Eliassen, A. and K. Pedersen 1977. *Meteorology: an introductory course*, vol. 1. *Physical process and motion*. Oslo: Scandinavian University Books.

Lockwood, J. G. 1979. *Causes of climate*. London: Edward Arnold.

Mason, B. J. 1975. *Clouds, rain and rainmaking*. Cambridge: Cambridge University Press.

Miller, D. H. 1977. *Water at the Earth's surface: an introduction to ecosystem hydrodynamics*. New York: Academic Press.

Ministry of Agriculture, Fisheries and Food 1967. *Potential transpiration*. Tech. Bull. no. 16. London: HMSO.

Ward, R. C. 1975. *Principles of hydrology*, 2nd edn. Maidenhead: McGraw-Hill.

5 The lithosphere

In Chapter 3 we considered the gross structure of the Earth, albeit briefly, in order to understand the sources of the planet's internal energy. This structure of concentric shells, differing in chemical and mineralogical composition and in physical properties, evolved from the original uniform aggregation of particles by the process of differentiation under the influence of gravity. The least dense materials became segregated in the thin outer shell of the crust, while the denser iron and nickel settled towards the centre of the Earth to form the core. Table 5.1 illustrates the composition of the Earth and the depth of the layers within it. On the basis of mineralogy, a distinction is made between crust, mantle and core, each separated by a major boundary and each further subdivided. The crust, it should be noted, is very thin in comparison with the other layers, and constitutes only 1.55% of the Earth's total volume. It is this thin layer, and primarily its surface characteristics, with which this book is largely concerned. The bulk of the Earth's mass is composed of mantle, of a density between that of crust and core.

Within the mantle, at a depth of *ca* 50 km, geologists recognise a discontinuity separating the rigid crust and upper mantle from the lower mantle, which displays the dual properties of a viscous fluid and an elastic solid. This chapter is largely concerned with the upper zone (or **lithosphere**) but consideration will also be given to the plastic zone of the mantle which extends down to *ca* 250 km and is termed the **asthenosphere**. Below this is the **mesosphere**. It is important to note that these major subdivisions are based on the

Table 5.1 Structure and composition of the Earth.

Layer	Depth to boundary (km)	Percentage volume	Composition	Major subdivisions
crust (continental)	av. 33	1.55	mainly granitic	lithosphere
(oceanic)	10–11		basaltic	
	Mohorovičić discontinuity			
upper mantle	*ca* 50		peridotite	
viscous layer	*ca* 250		partly fused peridotite	asthenosphere
lower mantle	2900	82.25	high density peridotite	
	Gutenberg discontinuity			
outer core	5000		iron–nickel, liquid	mesosphere
		16.20		
inner core	6371 (centre)		iron–nickel, solid	

mode of behaviour of the materials involved and are quite distinct from the mineralogical subdivision of crust, mantle and core (Fig. 5.1).

5.1 The crustal system

The lithosphere can be regarded as a system whose concentric boundaries are the outer surface of the solid Earth and the discontinuity within the mantle. The outer boundary forms a complex interface with the atmosphere and hydrosphere and is also the environment in which life has evolved. The inner boundary is adjacent to rock which is near its melting point and which is capable of motion relative to the lithosphere above. For the remainder of the book we shall refer to the system just defined as the **crustal system**. It is an open system and hence there are exchanges of matter as well as energy across both its outer and inner boundaries. At its outer boundary the crustal system is responsible for the structure and distribution of the continents and ocean basins and for the major relief units within them. As a prelude, therefore, to a closer examination of the systems operating at the Earth's surface, we shall consider the transfer of matter and energy within the crustal system and across its boundaries. In particular we shall consider on a broad scale the functional relationships between it and the denudation systems at the surface of the planet.

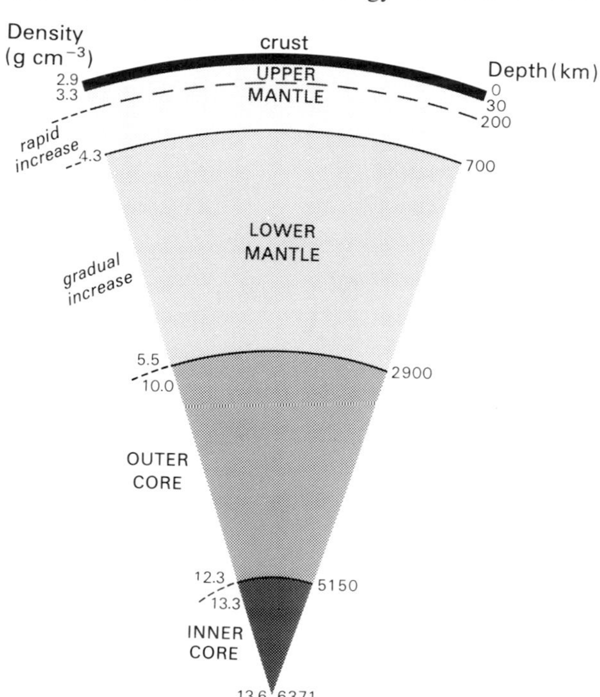

Figure 5.1 A schematic section through the crust, mantle and core.

5.2 System structure

5.2.1 *Chemistry and mineralogy of the lithosphere*

The elements of the crustal system are the chemical elements from the atoms of which the minerals and rocks of the lithosphere are composed. Table 5.2 shows the elemental composition of the crust for the eight most abundant elements as a percentage by weight, atomic composition and volume. Oxygen is by far the most abundant. For a volatile element such as oxygen to be held in the solid crust, in quantities which greatly exceed its presence as free molecules in the atmosphere, it must be firmly bonded to other constituents of the lithosphere. Indeed, the atoms of oxygen and those of the other elements are combined, largely by covalent and ionic bonds (see Ch. 2), into rock-forming minerals. These in turn combine in characteristic mineral assemblages to form the major rock types. The second most abundant element in the lithosphere is silicon, so it is not surprising that most of the oxygen is in combination with it to form silicate minerals. The basic unit of these minerals is the silica tetrahedron (Box 5.1).

The subsequent combination of these units in silicate minerals represents a compromise between the geometrical constraints imposed by the tetrahedral structure and those associated with the net

Table 5.2 Elemental composition in percentages of the Earth's crust (after Mason 1952).

Element	Weight (%)	Volume (%)	Atomic composition (%)
oxygen	46.6	93.77	62.5
silicon	27.7	0.86	21.22
aluminium	8.13	0.47	6.47
iron	5.00	0.43	1.92
magnesium	2.09	0.29	1.84
calcium	3.63	1.03	1.94
sodium	2.83	1.32	2.64
potassium	2.59	1.82	1.42
others*	1.43	0.01	0.05

* In this category is included a wide variety of elements of very low abundance. These are termed **trace elements**, and are often of great significance in the biosphere.

Box 5.1

SILICATE MINERAL STRUCTURES

Most of the common rock-forming minerals of the Earth's crust are silicates, formed by the combination of silicon and oxygen with one or more of the abundant metals.

The fundamental unit of silicate minerals is the silica tetrahedron, in which a central silicon atom is linked to four oxygen atoms. The small silicon ion (Si^{4+}) and four large oxygen ions (O^{2-}) pack together to form a pyramid (SiO^4). This forms a strongly bonded structure and the bonds are usually considered to be covalent.

The silica tetrahedron is a complex ion in which the association of four oxygen ions, each with two units of negative charge, and one silicon ion with four positive units, leaves the resultant tetrahedral ion with a net negative charge of four units. To form an electrically neutral unit, either the tetrahedral ion must be bonded with additional positive ions (for instance magnesium, Mg^{2+}, or iron, Fe^{2+} in the case of the mineral olivine, $(Mg, Fe)_2SiO_4$), or it must share the oxygen ions at the corners with adjacent tetrahedra.

Adjacent silica tetrahedra can be linked in different ways by the sharing of two or more oxygen atoms (Fig. 5.2). Various degrees of oxygen sharing occur, the result of which is the grouping of the tetrahedra into chains (single or double), sheets or three-dimensional lattices. In combination with different mineral elements these varied structures give rise to a wide variety of rock-forming minerals.

Inosilicates

In the **inosilicate** group of minerals, the silica tetrahedra are linked into chains. In the **pyroxenes**, single chains are formed by the sharing of two oxygens between adjacent tetrahedra. The chains are of indefinite length with an Si:O ratio of 1:3. This leaves a double negative charge for each tetrahedron which is satisfied by a single divalent ion such as Mg^{2+}, Al^{2+} or Fe^{2+}. In the **amphibole** group of minerals the lattices of the tetrahedra are linked into double chains, again of indefinite length. This gives an Si:O ratio of 4:11 and the electrical balance is formed by the addition of metallic cations, together with hydroxyl (OH^-) groups.

Phyllosilicates

In the **phyllosilicate** group, the tetrahedra are linked to form sheets of indefinite lateral extent, with an Si:O ratio of 2:5. This commonly results in a hexagonal pattern of atomic structure in the plane of the sheet. Cations and water are accommodated between the sheets. This group contains a large number of commonly occurring minerals – the clay minerals – and mica and possesses a very strong cleavage parallel to the plane of the sheets.

Tectosilicates

The greatest degree of oxygen sharing is attained in the **tectosilicates**, in which every oxygen is shared between adjacent tetrahedral groups to form a three-dimensional lattice. The Si:O ratio is thus 1:2 and at its simplest this forms the mineral quartz (SiO_2). Structurally similar is the abundant feldspar group of minerals, in which a proportion of the silicons are replaced by metallic cations. Thus, in the alkali feldspars one silicon in four is substituted by aluminium. This causes a deficiency of one positive unit of charge per four tetrahedra. The electrical balance is achieved by a single K^+ ion in the case of the mineral **orthoclase** ($KAlSi_3O_8$) and Na^+ in the case of **albite** ($NaAlSi_3O_8$) which are accommodated within the lattice structure. Tectosilicate minerals tend to be hard and do not cleave easily.

Box 5.2
MAJOR ROCK TYPES

The rocks of the Earth's crust can be conveniently grouped into three major classes, according to their mode of origin.

Igneous rocks are formed by the cooling and crystallisation of molten rock (magma) or mineral fluids derived from the mantle. They are formed either as intrusive rocks, where the magma is injected into existing crustal material, or as extrusive rocks, which are formed by ejection at the Earth's surface, as for example volcanic material. Intrusive rocks form at depth and they crystallise slowly, forming large crystals and a coarse-grained rock. The extrusive rocks, cooling rapidly at the Earth's surface, are generally fine textured.

With few exceptions, igneous rocks are composed of interlocking crystals of primary rock-forming minerals. The mineral composition may vary considerably depending on the particular minerals present and a distinction is frequently made between acidic and basic rocks.

The mode of formation of igneous rocks means that there are few voids within the rock, and this low porosity is often associated with a high mechanical strength.

Sedimentary rocks are formed by the denudation of existing rocks, broken down by weathering processes and transported in the form of detrital grains. The sediments are transported across land masses and the bulk of them is carried to the oceans where they accumulate on the continental margins. More restricted and isolated sediment traps exist on the continental surfaces, for example lakes and river valleys.

The detrital grains of which sedimentary rocks are formed are frequently composed of resistant minerals, such ás quartz. Other sedimentary rocks are formed of the accumulation of organic materials, for example chalk and limestone, which consist of the calcareous shells and skeletons of marine creatures.

Sediments are transformed into sedimentary rocks by the process of **lithification**, the compression and compaction of the sediments associated with the extrusion of water. The detrital grains may subsequently become cemented by deposition of salts from percolating waters to form an indurated rock.

Sedimentary rocks are typically stratified, variations in composition and texture resulting from changes in depositional conditions.

Metamorphic rocks are formed by alteration of previously existing rocks, themselves of igneous, sedimentary or even metamorphic origin. Conditions of high temperature and/or high pressure may develop in the crust and these cause mechanical deformation or, more commonly, chemical recombination of the elements in the rock-forming minerals. Metamorphic rocks exist as many types, depending on the original minerals present and the type of metamorphism (whether thermal or dynamic) or a combination of both. A great variety of both mineralogy and texture is possible within the metamorphic group. Typical examples are slates and schists, representing lower and higher grades of metamorphism respectively.

charge on the unit. There are two such compromises. Either the oxygens of adjacent units are shared – hence reducing the negative charge deficit – or a neutral mineral is formed by using the positive charges of other metallic cations to balance the negative charge on the terahedra. In practice, both occur together in almost all silicate minerals. The degrees to which tetrahedra are linked in a mineral structure form a sequence from unlinked tetrahedra, through single chains, double chains and sheets, to continuous three-dimensional networks (Box 5.1 & Fig. 5.2). The more the tetrahedra are linked in a mineral structure, the greater is the reduction in the negative charge imbalance and the smaller is the number of additional cations necessary to produce a neutral mineral (Fig. 5.2).

The manner in which these mineral structures are formed and their relation to the processes operating in the crustal system will be discussed in this chapter. For the moment, suffice it to say that they occur in particular combinations and amounts in response to the conditions under which they formed and it is these mineral assemblages that we recognise as rock types. Of the rocks composing the lithosphere, over 95% by volume are of igneous origin (Box 5.2) and consist of primary

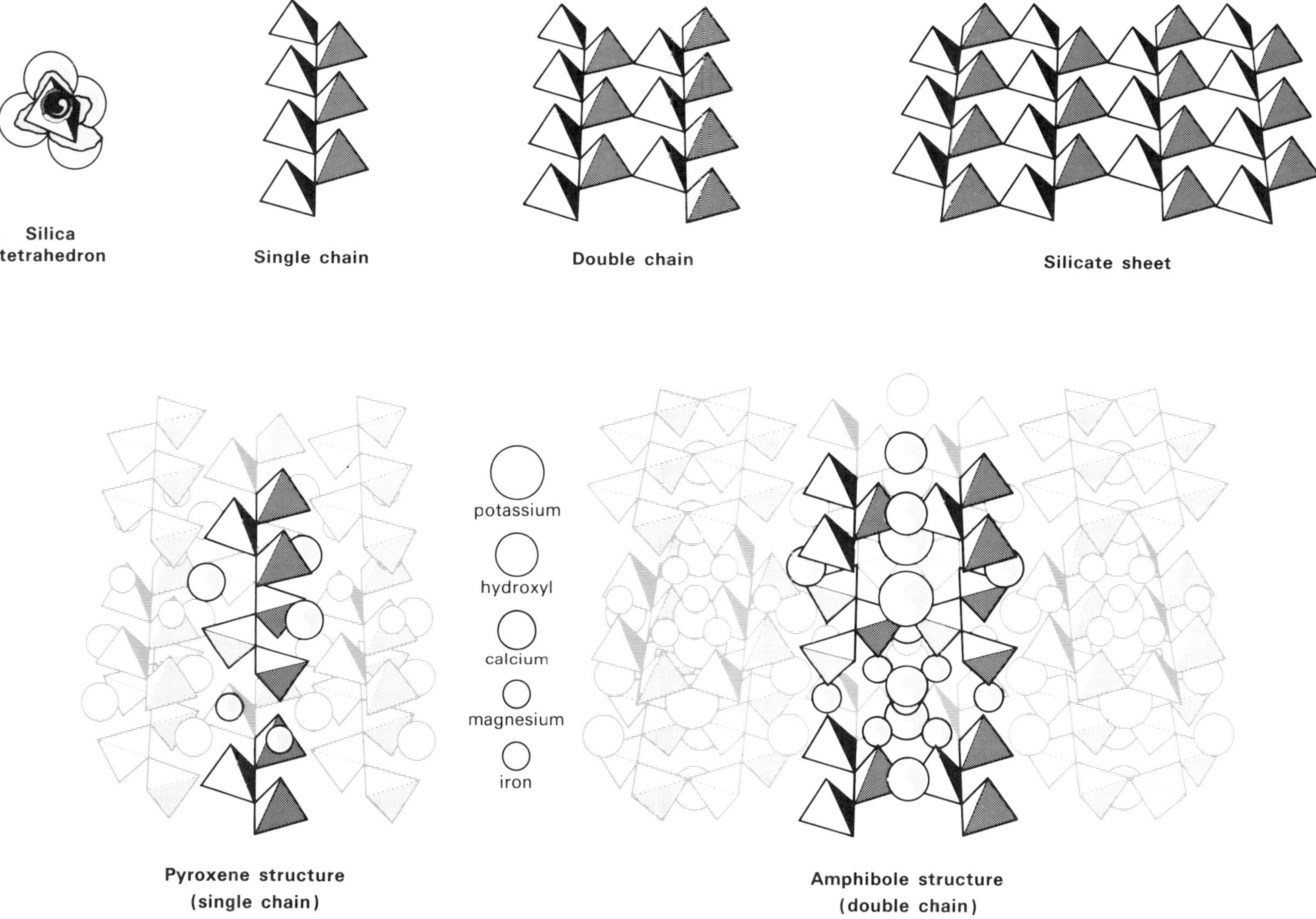

Figure 5.2 The structure of silicate minerals, showing the silicate framework built up from silica tetrahedra and the inclusion of non-framework ions (see text).

silicate minerals (Table 5.3). Metamorphic and sedimentary rocks which make up the remaining 5%, however, can also be thought of as having been derived from these same primary silicate minerals of igneous rock. Igneous rocks are dominant in terms of volume, but the sedimentary rocks (shale, sandstone and limestone) are exposed over 70% of the land surface area of the crust where they occur as a thin veneer, with igneous rocks occupying only 18% of the area (Table 5.4). In addition to their chemical and mineralogical properties, igneous, metamorphic and sedimentary rocks all possess important physical and mechanical attributes which also govern their response to processes operating both within the crustal system and at its interface with the atmosphere. However, a consideration of such properties will be deferred until later in this chapter. Partly on the basis of rock type and mineralogy, the lithosphere is divided into three gross structural units: the **continental crust**, the **oceanic crust** and the **upper mantle**.

5.2.2 *Gross structure of the system: major components*

Continental crust. The continental crust is composed of granitic rocks rich in silicon and aluminium and with a mean density of 2.8 (Box 5.2). These rocks form a discontinuous outer shell to the planet and they underlie the continents. Each continent has a core of ancient Precambrian crystalline rocks of metamorphic or igneous origin,

Table 5.3 Minerals exposed on the land surfaces of the Earth.

Mineral	Percentage of exposed area
feldspars	30
quartz	28
clay minerals and mica	18
calcite	9
iron oxides	4
others	11

Table 5.4 Rock types exposed on the land surfaces of the Earth.

Rock type	Percentage of exposed area
shale	52
sandstone	15
granite	15
limestone	7
basalt	3
others	8

Box 5.3

THE GEOLOGICAL TIMESCALE

Era	*Period*	*Age of base* (10^6 a)
Cenozoic	Quaternary	2
	Tertiary	65
Mesozoic	Cretaceous	135
	Jurassic	200
	Triassic	240
Palaeozoic	Permian	280
	Carboniferous	370
	Devonian	415
	Silurian	445
	Ordovician	515
	Cambrian	600
Precambrian	Proterozoic	2500
	Archaean	3900

Sedimentary rocks occupy the bulk of the exposed rocks in the geological succession from the Cambrian period onwards. The Archaean basement rocks are generally metamorphic rocks of ancient and complex history. The oldest known crustal rocks, from Greenland, have an age of 3.9×10^9 a. The age of the Earth is estimated at 4.5×10^9 a. The majority of the world's present major zones of high relief date from orogenic constructional episodes during the Tertiary period, and surface erosion and deposition during the Quaternary.

which are structurally very stable and are termed **cratons**. This long-term stability is a property which has recently prompted their exploration as possible sites for the disposal of nuclear waste. The rocks of these core (or shield) areas are the oldest exposed at the surface, often with ages in excess of 2000 Ma (Box 5.3). In North America this ancient core is represented by the Laurentian Shield, in Europe by the Fenno-Scandian Shield, and much of central Africa consists of three major shields. Sometimes these crystalline shields have a veneer of derived sedimentary rocks. This mantle of younger sediments is, however, largely undeformed since they rest on a stable crustal base and are hence protected from lateral pressure. This is not true of sediments deposited at the margins of cratons.

The cratons are rimmed and separated by more mobile portions of crust. These zones (up to 300 km across) are known as **orogens**. They are more readily deformed by crustal pressure and are the sites of seismic activity and crustal folding – termed **orogeny**. Here sedimentary rocks are deformed into fold mountains and are metamorphosed and intruded by igneous rocks. These areas of complex geological structure generally form strongly linear patterns in their surface expression and, although old fold mountains may have been reduced to low elevations, the zones of more recent activity stand out as areas of high relief (>4000 m relative relief). Indeed, the surface of the continents which, remember, is part of the outer boundary of the crustal system, is highly varied in topography. The mean land elevation of 870 m above sea level (asl) (Fig. 5.3) is strongly influenced by these orogenic zones, for 70% of the continental surface lies below 1000 m asl. The highly skewed distribution of land surface elevation in Figure 5.3 highlights the existence of steep gradients in certain continental areas.

Oceanic crust. The oceanic crust differs petrologically from the continental crust above it. It is basaltic in composition, consisting of more basic minerals, and has a mean density of 3.0. It is structurally simple and nowhere are oceanic basalts older than the Mesozoic (225–65 Ma) (see Box 5.3). This relative youth of the oceanic crust is emphasised by the fact that no sediments older than Jurassic rest upon it. It forms the outer boundary of the system only on the floor of the

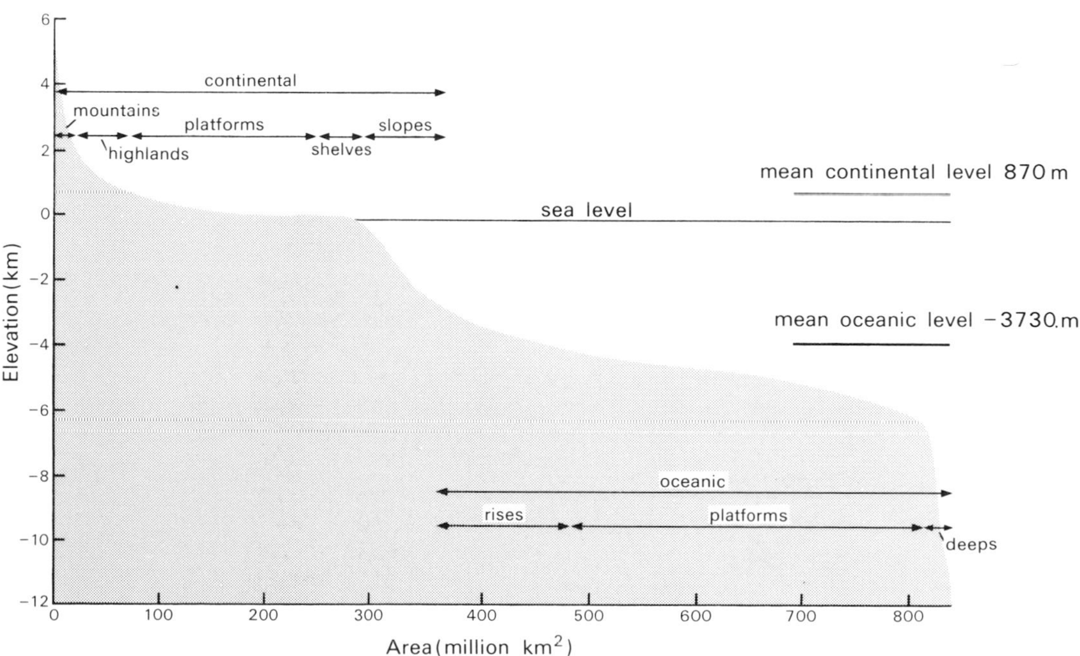

Figure 5.3 Hypsographic curve of the Earth.

ocean basins, which by implication are considerably younger features than the continents.

The topography of the floor of the ocean basins is generally much more subdued than that of the continents and does not have the local variations of relief in relation to area which the continents possess. Submarine elevations fall off quite sharply around the margins of the continents, but 85% (by area) of the ocean floor lies between 3 km and 6 km below sea level. There are, however, two features of ocean floor topography that are quite distinctive – the mid-ocean ridges and the deep submarine trenches. The mid-ocean ridges form an interconnecting worldwide chain over 60000 km in length found in all major ocean basins, although not always in a central position (Fig. 5.4). Between 500 and 1000 km in breadth and reaching heights of up to 3000 m, with a central trough or rift along the centre line, they are less pronounced features than most of the orogenic zones of the continents. One characteristic feature of these ridges is that they

Table 5.5 Type of symmetry about mid-ocean ridges.

age of oceanic crust rocks	become progressively older away from ridges, particularly well exemplified by accessible rocks of oceanic islands
age of ocean floor sediments	become progressively older away from ridges; repeats and confirms pattern of age symmetry displayed by crustal material
magnetic anomalies	variations in Earth's magnetic field arranged in bands parallel to each other and to ridge axis
magnetic polarity reversals	patterns of normal and reversed polarity, as preserved in the palaeomagnetism of oceanic crustal rocks; show bilateral symmetry about ridges
continental margins	topographic and stratigraphic congruence of some opposed continental margins

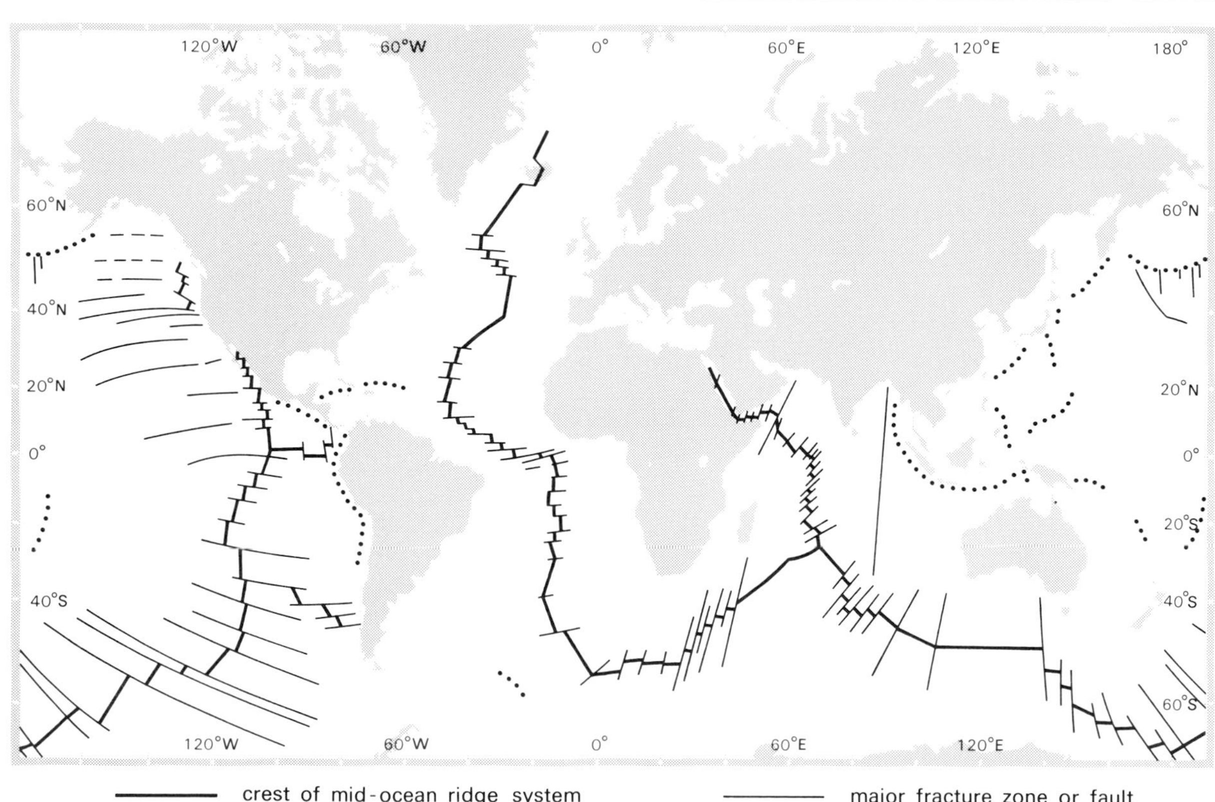

Figure 5.4 Structural and topographic elements in the ocean basins (from Continental drift by J. Tuzo Wilson, copyright © 1963 by Scientific American, Inc., all rights reserved).

are associated with anomalously high geothermal heat flows. Also strongly linear features, but of more restricted area (1% of the ocean floor) are the deep ocean trenches. Located close to the margins of the ocean basins, particularly around the Pacific ocean, they attain depths of 10 000 m to 15 000 m. Here the trenches often contrast strongly with the emergent tips of submarine mountains with which they are associated and which protrude above sea level as island arcs. That ocean trenches are zones of crustal instability is indicated by their coincidence with plots of deep earthquake epicentres. Perhaps the most remarkable feature of the ocean basins is not the mid-oceanic ridges, or the deep trenches, but the bilateral symmetry which they show in relation to these features and particularly to the ridges (Table 5.5). This is best exemplified by the Atlantic ocean basin (Fig. 5.5).

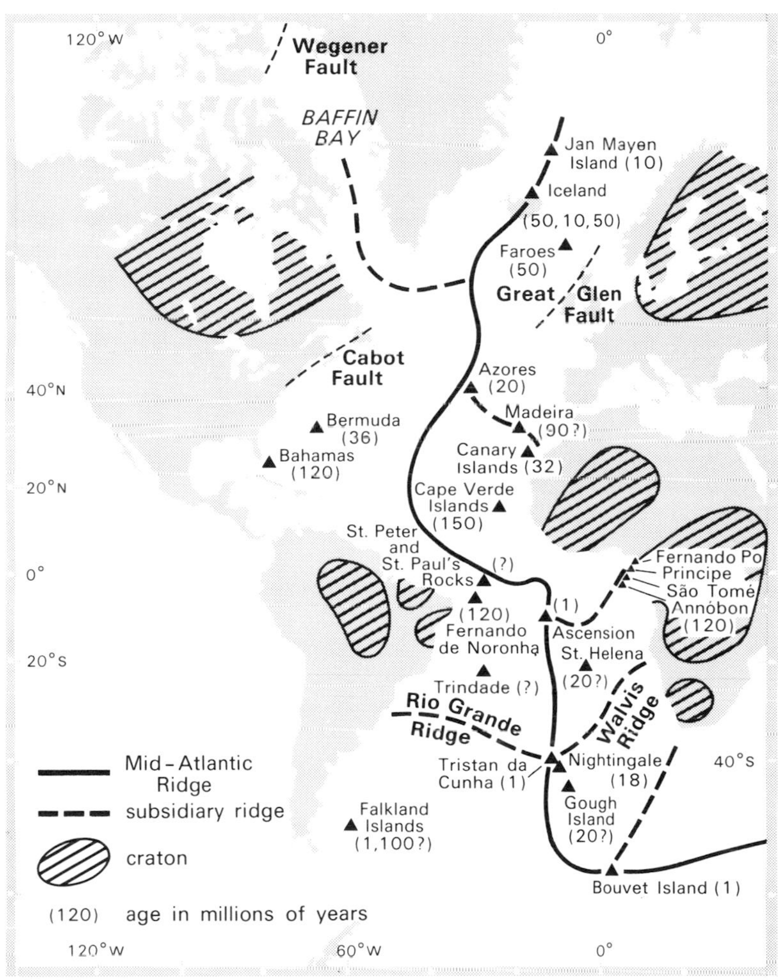

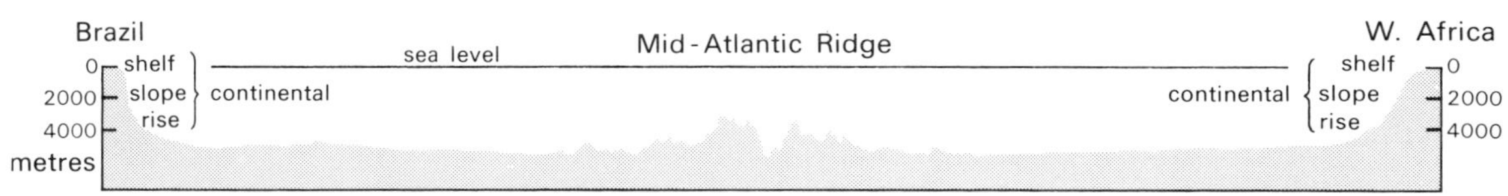

Figure 5.5 Symmetry of the Atlantic ocean basin and age of Atlantic islands (from *Continental drift* by J. Tuzo Wilson, copyright © 1963 by Scientific American, Inc., all rights reserved).

Now that we have discussed the structure of the crustal system at scales appropriate to both the individual rock-forming mineral and the continental land mass, we are in a position to look at the working or function of the system.

5.3 The crustal system: transfer of matter and energy

5.3.1 A geophysical model

The gross structural units described in the preceding section are so organised as to combine to form a large-scale functioning system which is responsible for the formation and destruction of crustal material and for the fragmentation and distribution of the continents. The processes responsible for the transfer of crustal material on such a large scale operate but slowly on the human timescale. Consequently, they require sophisticated scientific observation for their detection and it is only as a result of recent suboceanic research that they have begun to be appreciated. Even now, processes deep within and beneath the crust (especially the continental crust) are in part the subject of speculation, but logic suggests that they must take place in order to account for the directly observed events, while the ancient rocks themselves provide evidence for deep-seated processes.

The key to the operation of the crustal system is the recognition of the process of sea-floor spreading. New basaltic material from the mantle is intruded into the crust along the central rift of a mid-ocean ridge. There it solidifies to form new crustal rock. As this new material is added to the crust, existing crust is pushed aside laterally in both directions in order to accommodate it. Thus, the high heat flow of the ocean ridge is explained by the proximity or presence of hot mantle material beneath the ridge. The pattern of progressively older crustal material (and sedimentary material overlying it) away from the ridge axis is explained by the lateral translation of crustal material in that direction. The mirror image pattern of both magnetic anomalies and magnetic reversals is due to the same process, as the rocks record magnetic conditions at the time of their formation and continue to display them as they are transferred laterally away from the ridge axis in opposite directions.

This continuous formation of new crustal material at the mid-ocean ridges, and the transfer away from them by sea-floor spreading, implies that all the ocean floor is in motion. It also explains the relative youth of the oceanic crust. It is possible to ascertain the speed of sea-floor spreading by dating rocks at varying distances from the ridge axes. Figure 5.6 indicates that in the case of the North Atlantic the average spreading rate is *ca* 2 cm a^{-1}. The Pacific is more active, with a spreading rate commonly of 4.5 cm a^{-1}. Overall, the creation of oceanic crust falls within the range 1–8 cm a^{-1}.

Since the Earth is not expanding, the creation of new crustal material must be balanced by the destruction of crustal material elsewhere at the same rate. This occurs beneath the deep ocean trenches, where ocean crust slides down along a shear zone (the **Benioff zone**) inclined at *ca* 45° beneath the adjacent section of crust. The shearing of the descending section of crustal plate creates friction and sets up earthquake epicentres, often as deep as 700 km. The seismic instability of such zones is reflected in volcanic and geothermal activity at the surface above.

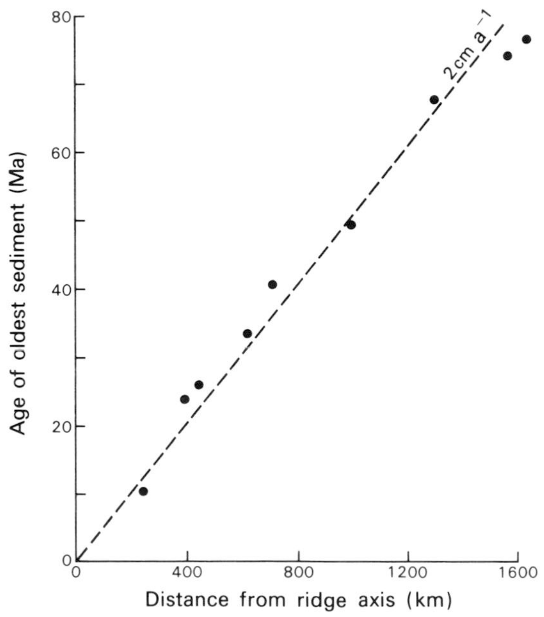

Figure 5.6 Age of oldest sediment against distance from ridge axis compared with a spreading rate of 2 cm a^{-1} (after Maxwell *et al.* 1970).

Thus, distributed across the surface of the Earth are zones where new crustal material is rising to the surface and other zones where it is descending towards the mantle where it is consumed. The lateral movement of crustal material across the surface by sea-floor spreading is presumed to be balanced by a counter movement in the opposite direction at depth within the mantle. Therefore, on the basis of these observable movements at the surface, it seems probable that there is a convective movement of material within the mantle which has widespread and large-scale repercussions on the crustal surface. A series of convective cells, driven by terrestrial energy, appears to exist within the asthenosphere, which over a long period of time behaves in a plastic or semi-fluid manner (Fig. 5.7). The pattern of convective cells is irregular and there is no reason to assume that it has been constant throughout time. It seems that old rises and sinks can die out while new ones become active. One can draw an analogy with a pan of gently simmering, viscous fluid – for example custard – in which the pattern of convective cells and rises and sinks changes through time. For instance, a new rise appears to be developing beneath East Africa, causing arching up of the continental crust, which has fractured to form the East African rift system. This feature is analogous to a mid-ocean ridge in the oceanic crust and appears to herald the break-up of the African continent.

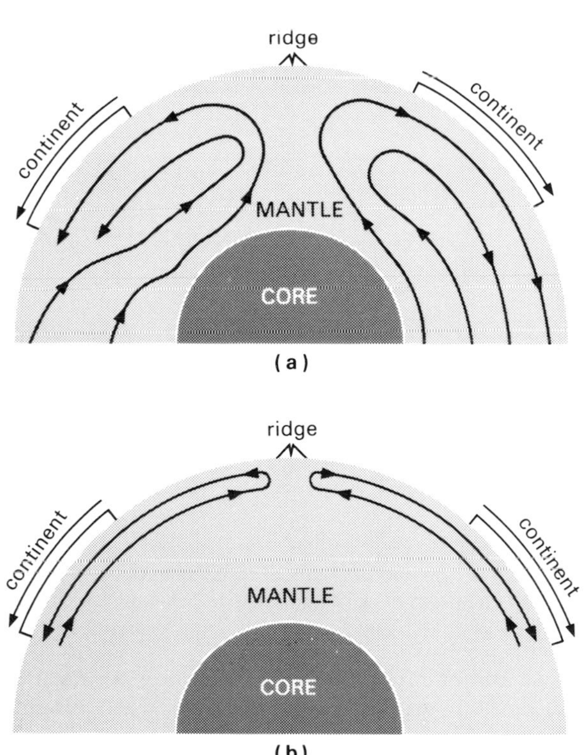

Figure 5.7 Convective cells: (a) in the mantle and (b) confined to the asthenosphere.

Through the operation of the crustal system, then, there is a circulation of oceanic crustal material, from the mantle to the surface and back again. Lateral movements of oceanic crustal material carry sections of continental crust across the surface of the globe.

The surface of the Earth can be conveniently thought of as a series of plates, of which there are six major ones and several minor ones (Fig. 5.8). Most of the major plates carry a continent, though this is not an essential feature. The plate margins are either rises where crustal material is generated (**constructive margins**), sinks where crustal material is consumed (**destructive margins**), or **transform faults** where two plates slide laterally past one another. Events occurring at plate margins have a very significant effect on the form of the Earth's surface topography. Constructive margins beneath the oceans create the mid-ocean ridges, described in detail earlier. A constructive margin developing beneath a continent causes the rifting and eventual fragmentation of that continent. At destructive margins, where there is a convergent movement of surface crust, significant major relief forms are developed. Three types of destructive margins can be identified. Where two oceanic plates converge, one may slide down beneath the other giving rise to deep-seated volcanic activity. Such eruptions may cause volcanic accumulations which break the ocean surface to form island arcs, common around the Pacific plate, as for example the Aleutian Islands. At a continental/oceanic plate boundary, a greater amount of sediment may be available as a result of denudation of the adjacent continent. Thus, a major mountain chain may develop as the sediment is piled up against the advancing continental plate, as in the examples of the Andes and Cordillera of South and North America respectively. The addition of a major mass of rock material to the continental crust may cause a depression of the

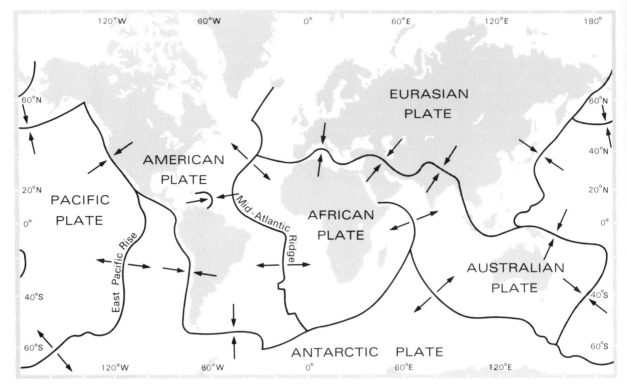

Figure 5.8 Crustal plates.

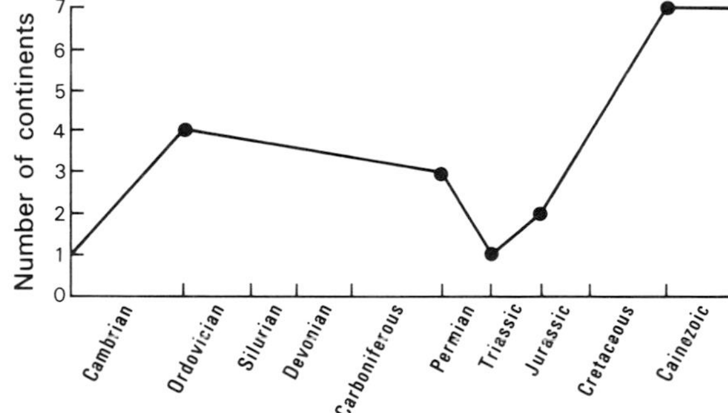

Figure 5.9 Variation in number of continents through geological time (from *Plate tectonics* by J. F. Dewey, copyright © 1972 by Scientific American, Inc., all rights reserved).

latter by isostasy into the lower crust beneath (Box 5.4).

The third type of destructive margin – continent to continent – also results in major mountain chains. The movement of India northwards into continental Asia resulted in the massive uplift of former marine sediments to form the Himalayas.

The operation of the crustal system, therefore, has enormous implications for the distribution of continents and relief across the Earth's surface. Lateral transfer of continents has moved them to their present positions and indeed is continuing to move them yet further. Continents may become fragmented and rejoined as the pattern of convective cells and the position of rises and sinks change through time. There have been times during the geological record when all the continental crust coalesced to form one or two supercontinents. At other times, such as the present, continental crust is fragmented and the continents become scattered (Fig. 5.9, and also see Fig. 7.4). In addition to these major lateral movements, the operation of the crustal system also effects vertical movements of crustal material. In addition, such transfers of mass involve changes in the potential gravitational energy of these materials. For example, orogenic activity transfers crustal rocks and sediments to

Box 5.4

ISOSTASY

The fragments of continental crust float upon the denser oceanic crust at a level determined by the density difference between the two. This is known as isostatic equilibrium.

A simple analogy would be a cube of ice of density 0.9, floating nine-tenths submerged in water of density 1.0. Its isostatic equilibrium would be upset if light downward pressure were exerted, in effect increasing the load by pushing it down. When the pressure is removed, the ice cube returns to its equilibrium level.

So it is with continents. Any increase in mass, by mountain building, local accumulation of sediment or development of an ice sheet, imposes an increased load and the continent subsides. Processes of denudation, or the decay of an ice sheet, will decrease the load and cause isostatic rebound, as the lightened crust regains a new equilibrium level. Whereas, in the example of the ice cube, rebound is almost instantaneous, with crustal material it takes place much more slowly. In several regions from which glaciers have decayed within the past 10 000 years, contemporary isostatic rebound is measured at a rate of several millimetres per year. Likewise, the continuing process of erosion of the continental surface is counterbalanced by slow and continuous isostatic rebound.

Isostatic adjustment also implies some rearrangement of material at depth beneath the continental crust. This may be accounted for by a lateral movement of crustal material at depth (a), counterbalancing the erosion and deposition of material at the surface. Alternatively, it may be accommodated by a phase change in the rock-forming minerals in the mantle. It is known that under increasing pressure olivine, which forms some 70% of the upper mantle, becomes altered to a more dense form without change in chemical composition. Thus, an increase or decrease in pressure at depth could be accommodated by a phase change and a vertical movement of a phase boundary (b).

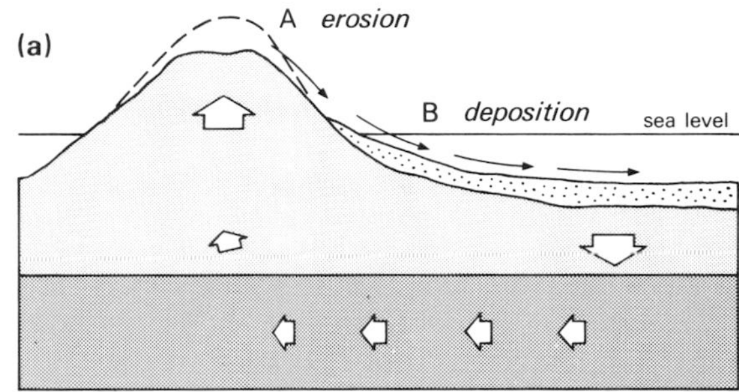

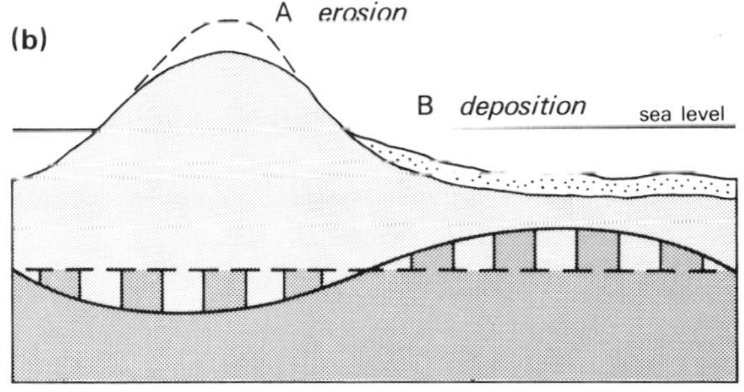

higher elevations, thereby increasing their potential energy. These vertical movements, however, are complicated by isostatic reactions, for the fact that the continents float on the denser lower crust permits them to rise or fall according to whether mass is added or removed from them. These isostatic movements also involve major transfers of gravitational potential energy (Box 5.4).

5.3.2 The denudation system: the exchange of matter and energy at the lithosphere/atmosphere interface

Interaction between the crustal system of the lithosphere and the atmosphere and biosphere takes place where continental crust is exposed above sea level. At the land/air interface crustal material becomes exposed to inputs of solar radiant energy, precipitation and atmospheric gases. These inputs are often modified by, or operate through, the effects of the living systems of the biosphere (see Chs 6 & 7). Under the influence of these inputs, crustal rocks are broken down by weathering processes and are transferred downslope by erosion processes. The effect is the denudation of the continental landscapes. These interactions and processes occurring at the Earth's surface operate within, and define, the denudation system. It is an open system, therefore, which receives inputs of matter and energy from the crustal system and from the atmosphere and biosphere, and its output can be viewed as the products of rock breakdown and erosion delivered as sediments or solutes to the oceans. In this chapter we shall concentrate on the external relationships of the denudation system, while its internal operation will be discussed in detail in Chapters 10 to 15.

The elevation of crustal material above sea level (the base level of land surface erosion) imparts potential energy to rock at the surface, related to its height above sea level. Repeated detailed geodetic surveys of the land surface may reveal significant changes in elevation over quite a short timescale. Figure 5.10 shows contemporary vertical crustal movement on a continental scale in the USA. Both positive and negative movements are shown to be taking place at rates of up to more than 10 mm a^{-1}. Figure 5.11 shows crustal updoming on a regional scale up to 9 mm a^{-1}, as deduced from tide gauges in the Baltic. In this case the uplift is isostatic rebound following the decay of Quaternary ice sheets. Detailed surveys on a more local scale have shown rates of vertical displacement in California of up to 26 mm a^{-1}, in a tectonically highly unstable zone adjacent to the San Andreas fault system.

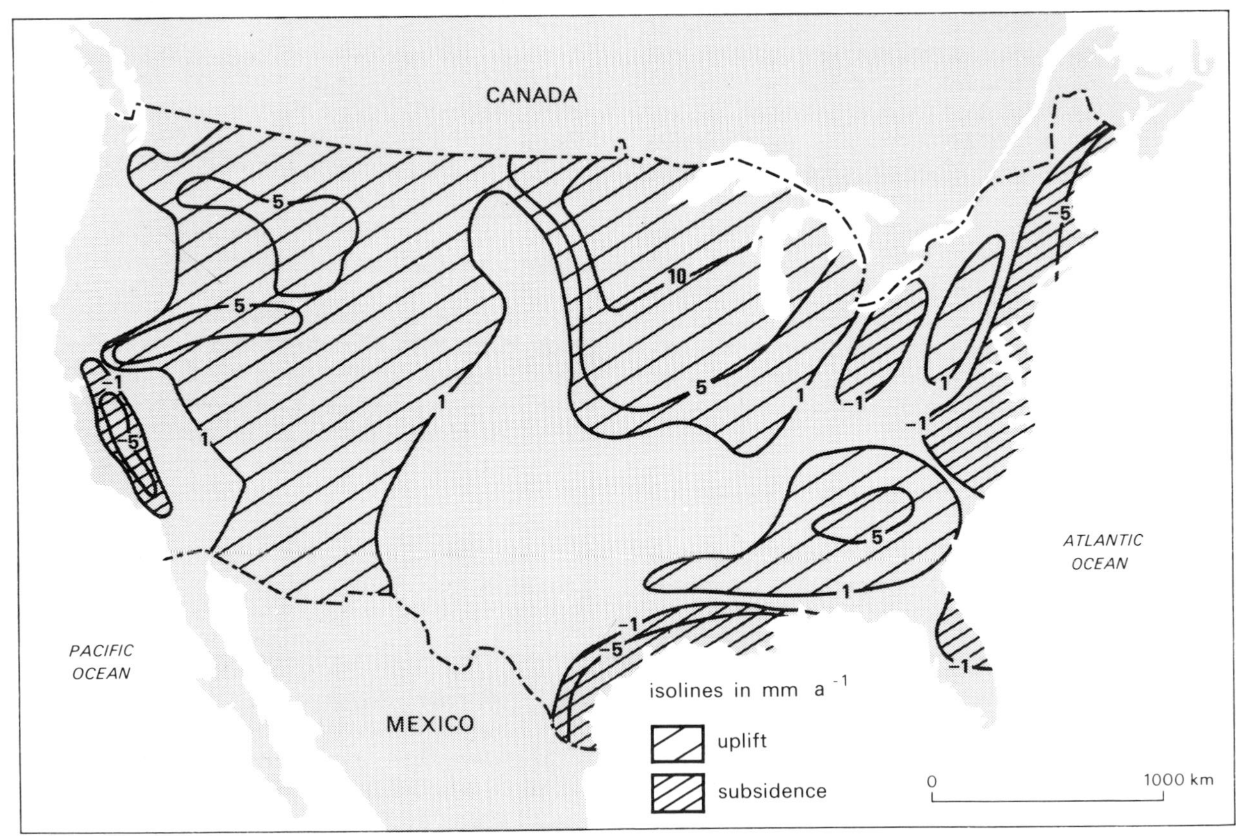

Figure 5.10 Contemporary crustal movement in the United States (adapted from *Fundamentals of geomorphology*, by R. J. Rice, Longman 1977).

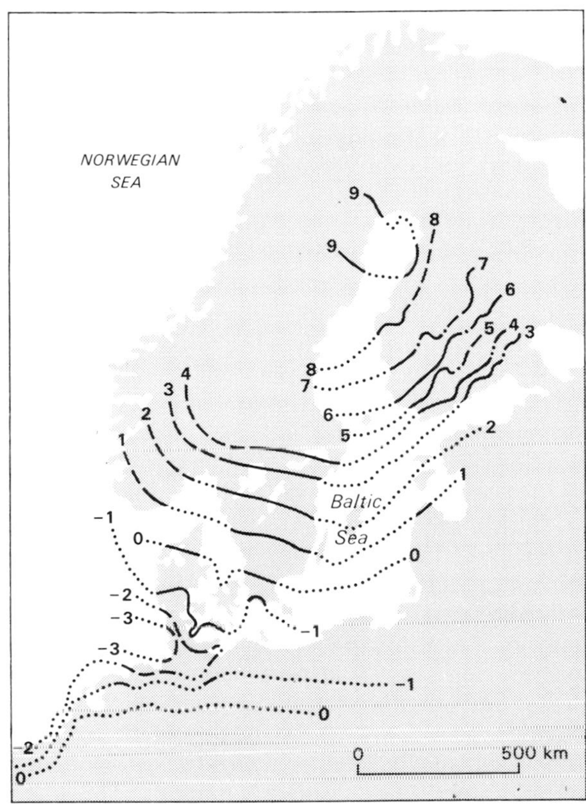

Figure 5.11 Contemporary uplift and downwarping in Scandinavia. The isobases represent rate of change (+ or −) in mm a^{-1}, determined from tide gauge records. Dashed isobases are less certain and the dotted isobases are based on interpolation. (After Fromm 1953.)

In this way, crustal uplift can impart potential mechanical energy to the land mass in calculable amounts. Assuming a mean rock density of 2.8, the increase in potential energy of a 1 m^3 mass of rock elevated through 1 mm is as follows:

$$\Delta E_P = mg\Delta h$$

where m (mass) = 2.8×10^3 kg; g (gravitation acceleration) = 9.81 m s^{-2}; Δh = elevation = 0.001 m.

Therefore

$$\Delta E_P = 2.8 \times 10^3 \times 9.81 \times 1 \times 10^{-3} \text{ kg m}^2 \text{ s}^{-2} \text{ (J)}$$
$$= 27.47 \text{ J}.$$

Such uplift, however, does not simply affect a cube, but a column of rock 1 m^2 in area through the crust from the land surface down to base level. Total increase in potential energy per square metre will, therefore, depend on the total thickness of rock being uplifted. Within a given region, the total potential energy of uplift must be related to the area of crust being uplifted. Clearly, when the length of the column and the total area are taken into account, the increase in potential mechanical energy associated with a crustal uplift of as little as 1 mm a^{-1} is very considerable.

In addition to potential gravitational energy imparted to rock masses by uplift, the rocks themselves and their constituent minerals represent further inputs of potential energy to the denudation system. The internal mechanical forces associated with the packing of the individual mineral grains within the rock represent a potential strain energy inherited from the confining pressure experienced during magmatic crystallisation, metamorphism or lithification. This energy is released in the denudation system with the expansion, fracture and disaggregation of the rock under conditions of low pressure and temperature at the Earth's surface. Furthermore, the organisation of the constituent atoms of the rock-forming minerals into complex structural configurations (see Box 5.1) represents a further input to the denudation system, this time of potential chemical energy. The most significant component of this chemical energy is the bond energy associated with the interatomic and intermolecular bonds in the lattice structure of rock-forming minerals. This energy is inherited from that involved in the chemical reactions under which the minerals originally formed in the crust. In the denudation system this potential chemical energy is released, at least in part, during the chemical breakdown of mineral structures as they undergo weathering reactions at the surface of the Earth.

Rock materials, however, represent an exchange of matter between the crustal system and the denudation system, and the potential energy content of these rock materials is only one aspect of their properties. Their elemental chemical composition, chemical structure and physical and mechanical properties are also important in influencing their behaviour in the denudation system. Like the forms of potential energy these rocks represent, many of their properties are also a legacy of the processes they experienced in the crustal system.

The frequency and distribution of fractures and fissures in the rock mass – referred to as its pattern of jointing – may have resulted from contraction during the cooling of magmatic igneous rocks; in sedimentary rocks, joints may have developed along bedding planes, or as a result of compaction and lithification, normal to the bedding. Superimposed on all these joint patterns, crustal movements may create a variety of further fractures as a result of tension, compression, wrenching, or torsional forces. The physical properties of a rock will also reflect its origin and antecedent history. There is, for example, a clear difference between igneous rocks, whose formation involves the growth of crystals until they are often in intimate contact, and sedimentary rocks, where loose detrital grains accumulate by accretion. Unconsolidated sediments will incorporate a much higher proportion of voids, for example, than those which have undergone lithification.

These then are the principal inputs of matter and energy to the denudation system from the crustal system. The prime source of energy input from the atmospheric system is solar energy. The solar energy cascade has been considered in Chapter 3 (see Fig. 3.8). Of the total radiant energy arriving at the Earth's surface, some is employed in sustaining the biosphere, while the remainder is used in heating the ground surface and causing evaporation of soil and surface moisture. Figure 3.16 shows Budyko's (1958) estimate of solar energy received at the Earth's surface. Clearly, higher values occur in the Tropics than in lower latitudes. The highest values, however, are found in the desert regions where the filtering effect of cloud and vegetation is minimised. Of the energy actually incident at the ground surface, between 15% (in deserts) and 50% (in humid regions) is required to evaporate surface moisture. The remainder raises the ground temperature.

The major material input from the atmosphere is water. The mean annual world precipitation is estimated at 857 mm. This generates a total volume of runoff from the continents of 37×10^3 km^3 with a total mass of 37×10^{15} kg. If, for the sake of simplicity, we assume that this mass falls from the mean continental altitude of 870 m, the potential energy associated with precipitation is over 35×10^{18} kg m, or 34×10^{19} J. As this massive volume of water drains off across the continental surfaces, much is lost directly to the atmosphere by evaporation, but of the remainder the potential energy is converted to kinetic energy. Thus, only a portion of the potential energy is ultimately employed in the mechanical transfer of debris. It is shown in Chapter 15 that a reasonable average rate of erosion of the Earth's continental surface is 0.05 mm a^{-1}. This involves the removal of some 20×10^{12} kg of rock debris per year. Assuming that it descends from the mean continental elevation, the potential energy thus released from the land masses is *ca* 18×10^{15} J, less than one-thousandth of the potential energy available from precipitation.

Within the framework of these massive energy transfers, the interaction between lithosphere, biosphere and atmosphere results in the processes of continental denudation. The presence of heat energy and moisture causes rock breakdown by weathering into transportable debris. Running water and gravitational energy combine to transfer clastic material downslope and downvalley towards the oceans, where it is deposited as sediment, mainly along the continental margins. This interaction, the denudation system, can therefore be seen as a conflict between the constructive forces of the lithosphere and the destructive forces of the atmosphere.

5.3.3 A geochemical model: the cycling of rock-forming materials

Implicit in earlier sections of this chapter is the concept that rock-forming materials are circulated both within the crustal system and between it and the denudation system. Indeed, massive amounts of energy are employed in the operation of the crustal and denudation processes that accomplish this circulation.

In Figure 5.12 the primary source of rock-forming materials can be seen to be the upper mantle from which molten magma rises to form primary igneous rocks. These basaltic magmas from the mantle are complex mixtures of many elements, but, because they are dominated by silicon and oxygen, they are often also referred to as silicate magmas. At certain temperatures and pressures in the crust these magmas start to crystallise. Crystallisation proceeds along reaction series

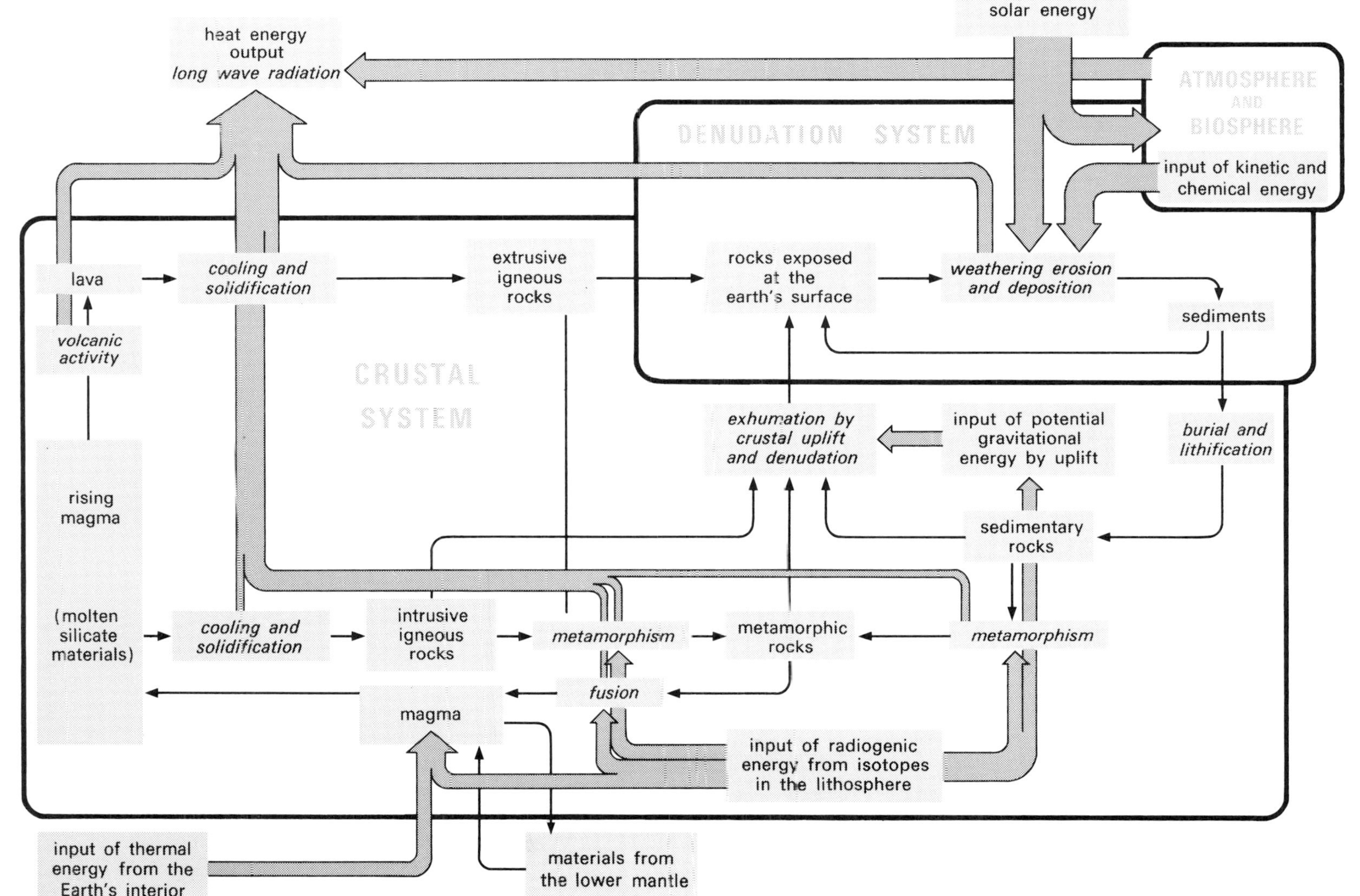

Figure 5.12 A geochemical model of the lithosphere. Stippled pipes represent energy flow and arrowed lines the pathways of material circulation.

(Bowen 1928; Fig. 5.13) with the earlier-formed minerals either changing gradually as the magma cools (continuous series), or being dissolved and reconstituted as different minerals at lower temperatures (discontinuous series). Alternatively, the earlier-formed and denser minerals may sink through the magma, or may be left behind as it migrates upwards (fractionation) and give rise to more basic rock types such as peridotite and gabbro. The composition of the molten fraction is now different, being relatively richer in such elements as silicon, aluminium and potassium, so that the later-formed minerals and the rocks they produce will be less dense and more acidic, such as granite. Finally, residual watery silicate fluids remain and cool and solidify in fissures and cavities.

Where magmatic injection, differentiation and solidification occur at depth in the crust, intrusive igneous rocks result, but magma may be ejected at the surface and cool rapidly in contact with air or water and form extrusive igneous rocks (see Fig. 5.14). The mineral composition of these extrusive rocks will reflect the composition of the extruded magma and this will depend on the degree of magmatic fractionation that has occurred before extrusion.

These primary igneous rocks are relatively heterogeneous in their elemental composition, though magmatic fractionation may concentrate certain elements. Intrusive igneous rocks are exposed to denudation at the surface, either after the erosion of overlying rocks or by a combination of uplift and exhumation. Extrusive rocks are, by definition, exposed to denudation processes almost immediately. Under the influence of energy and material inputs from the atmosphere and biosphere, both types of igneous rock undergo weathering reactions at the surface. Interestingly enough, Goldich (1938) maintained that the susceptibility of the primary minerals of these rocks to such reactions is in the reverse order to that of the Bowen series (Fig. 5.13 & Ch. 11).

The products of weathering are the more resistant of the original minerals, secondary minerals derived from them, and soluble and volatile products. Particular elements may be concentrated preferentially in residual weathering products, whether of primary or secondary origin, and accumulate in the weathered mantle or **regolith**. Erosion and transport, however, ultimately deliver rock debris and soluble products to the oceans, along whose margins they may be deposited or precipitated to form sediments. Other minor sediment traps (environments where sediment may collect) exist at a variety of scales on the way. Soluble products may also be reprecipitated before reaching the oceans. Meander cores, river flood plains, lake basins, glacier forelands and desert basins are all terrestrial environments where

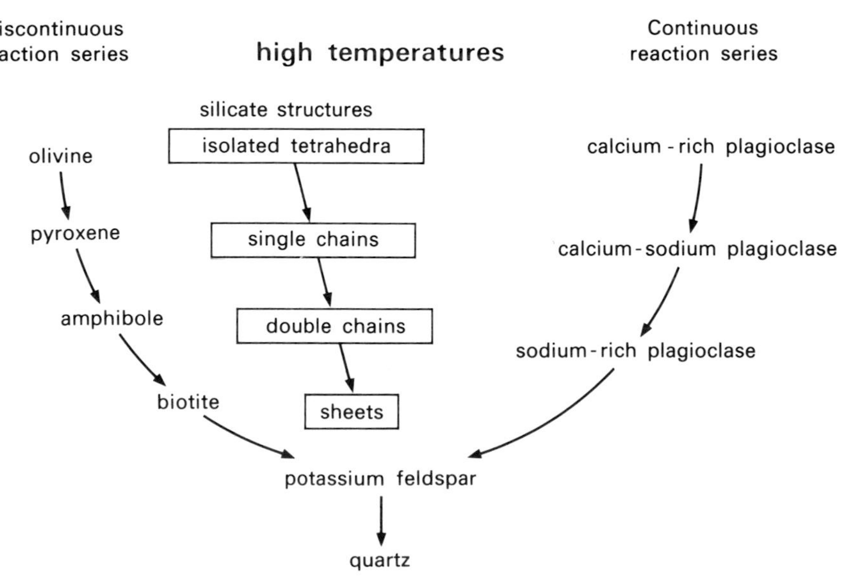

Figure 5.13 The Bowen reaction series.

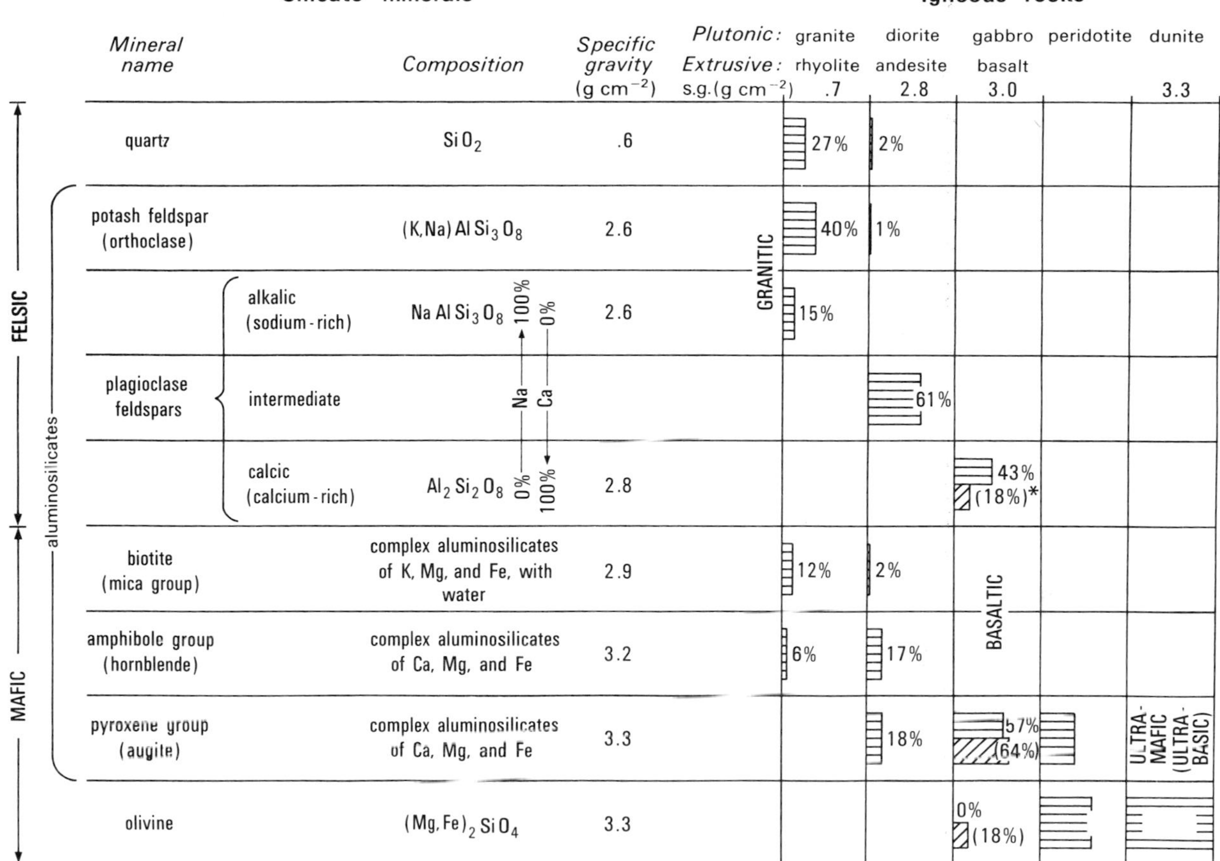

Figure 5.14 The mineralogy of igneous rocks (adapted from *Planet Earth* by A. N. Strahler, copyright © 1972 by Arthur N. Strahler).

significant volumes of deposition may occur. Eventually most sediments become lithified due to compaction by overburden and by internal redistribution of minerals by percolating solutions to produce sedimentary rocks. Transport processes, the environment of sedimentation and redistribution during lithification, all tend to act selectively and, therefore, enhance the likelihood that the elements of the original igneous rock will become further segregated into distinct facies of sedimentary rock.

Sediments and sedimentary rocks may in their turn be exposed to denudation by exhumation and uplift to complete one possible geochemical cycle. Alternatively, sedimentary and igneous rocks may undergo metamorphism. Here rocks experience crushing, recrystallisation, and/or recombination under conditions of high temperature or pressure (or both) within the crust and are converted to metamorphic rocks. Indeed, existing metamorphic rocks may themselves be converted into new types as a result of the same processes. Again metamorphism often has the effect of segregating the elements present in the original rock, sometimes to form relatively rare minerals.

Metamorphic rocks, of course, like their sedimentary and igneous counterparts, may also be exposed at the surface and be subject to denudation and hence complete a second type of cycle. A third type of geochemical cycle is completed when rocks of sedimentary, igneous and metamorphic origin are carried down the Benioff zone towards the mantle. The fractionation of their constituent minerals is lost as they undergo a phase change, fuse or melt and are absorbed back into the largely homogeneous magma in the upper mantle, perhaps to be intruded into the crust again at a later date.

5.4 The implications of the geochemical model

It is evident from the geochemical model developed above, and from Figure 5.12, that the elements of the crust are involved in several alternative cycles between a number of stores in the lithosphere. It is important, however, to realise that the scales of these cycles, both spatial and temporal, may vary considerably. The cycle within the denudation system from weathered rock to unconsolidated sediment and once more to weathering at the surface may be relatively rapid on a geological timescale. Over the northern hemisphere, for example, many superficial deposits of Quaternary age – glacial tills, fluvioglacial, solifluction and interglacial terrace deposits – are all undergoing contemporary weathering and erosion. The cycle, therefore, is being completed in between 10^3 and 2×10^6 years. The same point is true for some of the unconsolidated sediments of arid and semi-arid areas and for large areas of recent alluvium. At least with some of these sediments, such as solifluction or glacial lake deposits, the spatial scale of this pathway of circulation may be very localised. In contrast, the route via lithification and consolidated sedimentary rock usually takes considerably longer ($>10^7$ a) and operates on a larger spatial scale. Finally, the subduction

Table 5.6 The exploitation of mineral resources depends on their occurrence as concentrations which render them economically viable. Such concentrations reflect both crustal and denudational processes that promote the segregation (fractionation) of materials involved in geochemical cycles.

Process		Examples
MINERAL DEPOSITS ORIGINATING IN THE CRUSTAL SYSTEM		
magmatic segregation	(a) settling of early formed minerals to base of magma	magnetite, chromite, platinum (South Africa)
	(b) settling of immiscible sulphide or oxide melts, sometimes with injection along fissures	copper–nickel (Norway and Canada); magnetite (Sweden)
contact metasomatic	wall rocks of intrusion replaced by minerals derived from magma	magnetite and copper (Utah and Arizona)
deposit from watery silicate fluids	filling fissures in wall or outer part of intrusion (pegmatites)	mica (New Mexico)
deposited from hot watery fluids	filling fissures in, and replacing, wall and outer part of intrusion (hydrothermal)	lead–copper–zinc (Cornwall and northern Pennines)
MINERAL DEPOSITS ORIGINATING IN THE DENUDATION SYSTEM		
sedimentary	(a) evaporation of saline waters	salt and potash (Northumberland)
	(b) precipitation from solution	iron (Northamptonshire)
	(c) deposition and sorting of detrital grains	placer gold (Australia, California and Alaska); titanium (India and Australia); diamond (Namibia)
residual	weathering leaving concentrations of insoluble elements in residual material	aluminium (bauxite) (USA, France, Jamaica and Guyana)
secondary enrichment	precipitation at depth from groundwater in a mineral deposit	copper (Arizona and Miami)

and fusion of rock-forming materials into the mantle, and the upwelling of magma to form igneous rocks, is a circulation route which operates over a very long timescale and at the spatial scale of the crustal plate and convection cell.

The conclusion is that all of the routes taken as elements circulate within the crustal system are relatively slow and they involve considerable residence times in the various compartments or stores in the model. However, two alternative routes remain by which elements circulate and they are much more rapid in their operation. Both are coupled to the cycles in the denudation and crustal systems which we have been considering. The first of these alternative routes is through the atmosphere and hydrological cascade to the denudation system, and the second is via the living systems of the biosphere as the mineral nutrients essential to life. In the latter case, the model becomes one of biogeochemical cycles. The atmospheric route has already been encountered in Chapter 4 and the biospheric pathway will be discussed in Chapters 6 and 7.

There is a further implication of the geochemical model which has profound significance, both for the organisms of the biosphere and for Man and his developed technology. The elements of the crust and the minerals and rocks they form are critical resources, both as the nutrient elements on which life depends and as the raw materials of human technologies. Magmatic differentiation, metamorphism and denudation have all been shown to act to fractionate and concentrate rock-forming materials. From the point of view of the organisms of the biosphere, this means that essential mineral nutrients are not uniformly available across the Earth's surface. The normal situation is rather that the rocks, from which nutrients are released by weathering, display either deficiencies or excesses of some essential elements. It is also the concentrations of certain elements in particular minerals and rock types which form economically viable ore deposits, and their discovery and exploitation depends on an understanding of the processes operating in the crustal system (Table 5.6). Although the existence of some ore bodies is predictable, many of the large concentrations of mineral resources represent freak geological events. This fact, coupled with the immense timescale involved in many crustal processes, underlines why mineral resources are considered to be non-renewable.

Further reading

The substance of Chapter 5 largely concerns the field of geology, which is covered by many and varied student texts. Many geological terms are used in this chapter, particularly the names of rock types, and the reader is referred to the appropriate texts for further explanation.

A basic source on mineral structures and mineralogy is:

Read, H. H. 1970. *Rutley's elements of mineralogy*, 26th edn. London: George Allen & Unwin.

The nature of rocks themselves is covered by:

Nockolds, S. R., R. W. O'B. Knox and G. A. Chinner 1978 *Petrology for students*. Cambridge: Cambridge University Press.

and

Brownlow, A. H. 1978. *Geochemistry*. Englewood Cliffs, NJ: Prentice-Hall.

Various aspects of the Earth's internal dynamics and structure are covered in

Gass, I. G., P. J. Smith and R. C. L. Wilson (eds) 1972. *Understanding the Earth*. Horsham, Sussex: Artemis.

While

Wilson, J. T. 1976. *Continents adrift and continents aground*. San Francisco: W. H. Freeman.

is largely concerned with crustal processes.

The cycling of crustal minerals is well treated in

Garrels, R. M. and F. T. Mackenzie 1971. *Evolution of sedimentary rocks*, Chs 4, 5 & 10. New York: Norton.

6 The biosphere

6.1 The biosphere and ecosphere

Between the lithosphere and the atmosphere lies a transition zone that both contains and is modified by an enigmatic arrangement of matter which we know as life. Life in impressive profusion, diversity and ingenious complexity extends through this zone from the limits of root penetration to the tips of the tallest aerial shoots – and beyond. Similarly, in the hydrosphere living forms are to be found from the air/water interface to the bottom of deep ocean trenches. Not only is this transition zone characterised by life but also by the dead and decaying remains of what was once alive. These remains may accumulate on the land surface, be incorporated into the uppermost part of the lithosphere, or in aquatic environments 'rain down' to the bottom to decay or to augment the accumulation of sediment.

The existence of a global veneer of life is the most profound feature of the Earth's surface, its significance far outweighing its small mass in the effect it has on the nature of the lithosphere, hydrosphere and atmosphere. The term **biosphere** is used to describe *either* this veneer of life, *or* these organisms together with the surface environments in which they live and with which they interact (Hutchinson 1970). In this chapter, however, we shall restrict 'biosphere' to the first of these meanings and employ the term **ecosphere** to express the second (Cole 1958). A generalised model of the relationship between the biosphere and ecosphere, used in this sense, is represented diagrammatically in Figure 6.1. Some compartments of the model include the physical systems already discussed – the upper part of the lithosphere, a major part of the atmosphere and most of the hydrosphere – in so far as they have functional links involving transfers of energy and matter with the living material of the 'biosphere'. At the core of this model, however, are these living systems. It is necessary, therefore, to consider at the outset the nature of life.

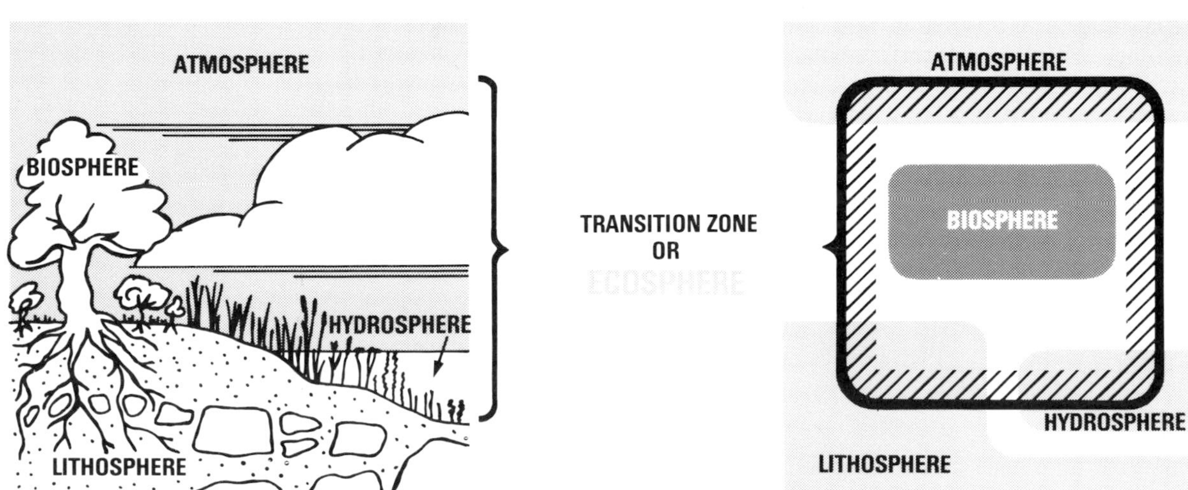

Figure 6.1 Defining the biosphere and the ecosphere.

6.2 The nature of life

In this chapter we shall isolate the compartment labelled 'biosphere' from the rest of our model of the ecosphere and treat it as one large living system. We shall not define the boundaries of this system too closely, for that would involve a precise definition of what life is and what it is not. Admittedly most of us would claim intuitively to be able to decide if an object is alive or not, but that is not quite the same thing, and this is not the place to embark on a philosophical discussion as to the nature of life. So we shall make the *a-priori* assumption that a distinction exists between life and non-life and that our 'biospheric system' contains all living material on the planet. What then are the elements composing this system, the biosphere? In what ways are these systems elements aggregated and organised? What are their properties and what are their links with each other and with the other systems that form the compartments of the ecosphere model?

These are awesome questions when one reflects that estimates of the number of different kinds of organisms constituting the biosphere range between two and four million and that the notion of 'life' encompasses a vast spectrum of phenomena associated with all of these organisms. Fortunately, there are certain characteristics to be found in the simplest of them which also prove essential for the most complex. The simplest organisms consist of a single cell and even the most complex consist of comparatively few cell types. The cell has therefore come to be regarded as the simplest independent structure that possesses all of the necessary properties of life. In attempting to answer these questions, therefore, it is appropriate to model the biosphere in terms of the structural and functional organisation of the living cell. To do this involves a consideration of the molecular basis of cellular activity but, fortunately, cells contain relatively few types of molecule and, although these include the most complex molecular structures known, many are universal in their occurrence in the biosphere. The generality of a biosphere model at the cellular level should, therefore, be assured.

6.2.1 The elemental chemistry of life

Cells and their molecular components consist of atoms. So at one level, the systems elements of the cell and hence the biosphere are the chemical elements from the atoms of which the cell is constructed. In Table 6.1, lists of such elemental constituents are presented with, for comparison, corresponding data for the hydrosphere and atmosphere.

Life is made up of familiar inorganic, chemical elements. At this level there is nothing inherently unique about the cell or about life. These elements are those abundant in the non-living part of the ecosphere. The elements of the air, rocks and waters of the Earth are those of living cells and organisms. Furthermore, with certain exceptions, they are present overall in much the same proportions. Elements common on the Earth's surface are abundant in living things, while those rare at the surface are rare in organisms. These observations are not merely coincidence, but they provide a first insight into the relationships of living systems with their environment and offer a first clue as to the nature and origin not only of life but also of some of the chemical characteristics of that environment (see 7.3).

Although estimates of the relative proportions of the elements in the biosphere vary (and these proportions also vary from organism to organism) four elements – hydrogen, carbon, nitrogen and oxygen – are universally present in living systems. They are amongst the lightest elements and they have atomic numbers (see Ch. 2, Table 2.3) of 1, 6, 7 and 8 respectively. Together they account for more than 99% of the atoms of the biosphere. Of those elements with atomic numbers between 9 and 20, sodium (11), magnesium (12), phosphorus

Table 6.1 Percentage atomic composition of the biosphere, lithosphere, hydrosphere and atmosphere, for the first ten elements.

Biosphere		Lithosphere		Hydrosphere		Atmosphere	
H	49.8	O	62.5	H	65.4	N	78.3
O	24.9	Si	21.22	O	33.0	O	21.0
C	24.9	Al	6.47	Cl	0.33	Ar	0.93
N	0.27	H	2.92	Na	0.28	C	0.03
Ca	0.073	Na	2.64	Mg	0.03	Ne	0.002
K	0.046	Ca	1.94	S	0.02		
Si	0.033	Fe	1.92	Ca	0.006		
Mg	0.031	Mg	1.84	K	0.006		
P	0.030	K	1.42	C	0.002		
S	0.017	Ti	0.27	B	0.0002		

(15), sulphur (16), chlorine (17), potassium (19) and calcium (20) are also universally present, but each contributes less than 1% (and most less than 0.1%) of the atoms of the biosphere. Fluorine (9) and silicon (14) are, however, better considered with the remaining eleven elements now known to be present in either plant or animal cells. Eight of these eleven have atomic numbers between 23 and 34 – vanadium (23), chromium (24), manganese (25), iron (26), cobalt (27), copper (29), zinc (30) and selenium (34) – while the remaining three – molybdenum, tin and iodine – have atomic numbers of 42, 50 and 53 respectively. All of these heavier elements, plus fluorine and silicon, are commonly but not universally present in living cells. They are found, however, in such small quantities (usually less than 0.001%) that they are called trace elements. The remaining elements of the atomic series, mainly the heavy elements, are highly toxic to living systems, as too are the trace elements when present in excess. The overwhelming impression is that living systems are almost exclusively composed of the lighter and more reactive elements with small but essential traces of heavier elements.

In spite of the broad similarity in elemental composition between the biosphere and its inanimate surroundings at the surface of the planet, there remain significant points of contrast. Just as the similarities in composition may suggest clues as to the nature and origin of life, so too do the differences. Silicon, aluminium and iron are, after oxygen, the most abundant elements in the lithosphere, yet they are either absent or are present only in very small quantities in the biosphere. Carbon is scarce in the lithosphere, atmosphere and hydrosphere, yet it is the third most abundant element in the biosphere. In an even more striking way phosphorus, the sixth most abundant element in living systems, is rare in the inorganic environment of these systems (see Table 6.1).

Certainly one implication of the elemental chemistry of the biosphere is that life has, in a sense, been derived from the elements available at the Earth's surface. Nevertheless, the figures in Table 6.1, together with the contrasts in composition alluded to above, indicate that the chemistry of life is not just a reflection of that of the environment. Living systems have concentrated some elements, such as carbon, in proportions far beyond any obvious source at the planet's surface, but have rejected other naturally occurring elements, such as silicon, even when they have been readily available (Box 6.1). Now, not only does this selection raise questions concerning the reasons why certain elements have been incorporated in living systems at the expense of others, but also prompts the inquiry 'What effect has this differential selection had on the elemental composition of the atmosphere, hydrosphere and lithosphere?' (see 7.3).

There are limits, however, to the usefulness of these elemental inventories in providing answers to such questions and to what they can tell us about the structure of the biosphere and the nature of life. We need to know how the atoms of the elements are organised, what molecules and compounds they form and what structures these units build to produce the living cell.

6.2.2 *The molecular basis of life*

As we have seen, there is nothing unique in the chemical elements, the atoms of which form the building blocks of living systems. Although their relative abundance and selection in the biosphere may be significant, they all occur in the environment. Equally, when we turn to the molecular chemistry of life we find that living systems can be described in terms of their constituent chemicals and are subject to the same physical and chemical laws that govern non-living systems.

The constituent chemical elements of the biosphere can be visualised as being combined in living systems into both relatively simple inorganic molecules (often ionised in solution) and organic carbon compounds. Hydrogen and oxygen, the most abundant elements of the biosphere, are present in roughly the same proportions as in the water molecule. It is not surprising, therefore, that the living cell is between 60% and 90% water. By weight, vertebrates are 66% and mammals on average 85% water. Even wood is 60% water by weight, while apparently dry seeds are 10% water. Indeed, all living matter is dispersed in water: it is the essential medium of all life on Earth. Its unique properties have already been dealt with, but water also controls the effectiveness of many of the other compounds essential to life through their response to it: whether or not they are soluble, whether they ionise in solution and exist as charged particles, or whether they affect the physical properties of water. In spite of its fundamental importance in the cell and the biosphere at the molecular level, water is still a simple inorganic molecule.

When we come to carbon, the third most abundant element, the situation is very different. The vast majority of naturally occurring compounds containing carbon are truly organic (see Box 6.1) and they attain a level of complexity and sophistication in the arrangement and organisation of their molecules not experienced in non-living systems. However, this complexity is achieved by carbon in combination with only some half dozen elements – mainly hydrogen, oxygen and nitrogen. The most important of these organic carbon compounds in the living cell are the fatty acids, simple sugars (monosaccharides), mononucleotides,

Box 6.1

CARBON CHEMISTRY

There are more compounds of carbon than of all of the other 102 elements, and, unlike inorganic compounds, many organic carbon compounds number the atoms in their molecules by the hundreds, or even thousands. The reason for this behaviour lies in three properties of the carbon atom.

(1) The high covalency of carbon (four) permits the attachment of a large number of groups to carbon in a large number of different combinations.
(2) The carbon–carbon bond has great strength and permits the formation of chains of carbon atoms of unlimited length.
(3) The formation of multiple bonds by carbon further increases the number of organic compounds.

A few of the other elements, for example silicon, possess one or two of these properties, but none has all three. These properties allow carbon to form stable chains containing from two or three to scores of atoms. Practically any number of simple or branched chains may be derived by repetition, while the possibilities are increased by other factors including the introduction of double or even triple bonds between adjacent carbon atoms at certain positions in the structure. Such compounds are referred to as open-chain compounds. The carbon bonding can also form closed chains of carbon atoms forming rings – the cyclic compounds. Where the ring is composed entirely of carbon atoms (usually five or six), the compound is said to be monocyclic, but where besides carbon atoms the rings contain other elements (chiefly nitrogen, oxygen or sulphur) they are heterocyclic. Both chain and ring compounds often have attached to the rather inert carbon core of the molecule other groups of atoms such as OH, Cl, NO_2 which are known as functional groups. It is here that some of the other elements listed in Table 5.1 become important, for it is often the functional groups which determine the properties of the compound.

Many of these carbon compounds can exist as **isomers**; that is, distinct substances sharing the same molecular formula, but differing, often subtly, in the spatial arrangement of their constituent atoms, especially the functional groups. This may produce asymmetric molecular structures. Many natural organic compounds are asymmetric isomers, but for each kind of compound, nature usually specialises in producing a limited number, often only one, of the many isomers possible. Here we can see not only the further scope for diversity endowed by isomerism but the selective precision with which it is used in the living cell.

amino acids and heterocyclic bases (Fig. 6.2). These in turn form the precursors of very complex molecules indeed: the lipids (fats), the polysaccharides (complex sugars, starch, cellulose), nucleic acids (RNA, DNA) and proteins (Box 6.2). These large molecules are all polymers and their structures are illustrated in Figure 6.2. In the process of polymerisation the linear or cyclic units of the precursor molecules (monomers) are linked so that carbon chains may be joined to carbon chains, to rings or even rings to rings. These large molecules are known to chemists as high polymers. The artificial synthesis of such compounds is the basis of the synthetic fibre and plastics industries, but in nature there seems to be an almost infinite variety of compounds formed in this way.

The search for structure at the molecular level in the cell is, however, not finished yet. There are in fact three levels of structure. The first we have just dealt with (primary). This level involves the

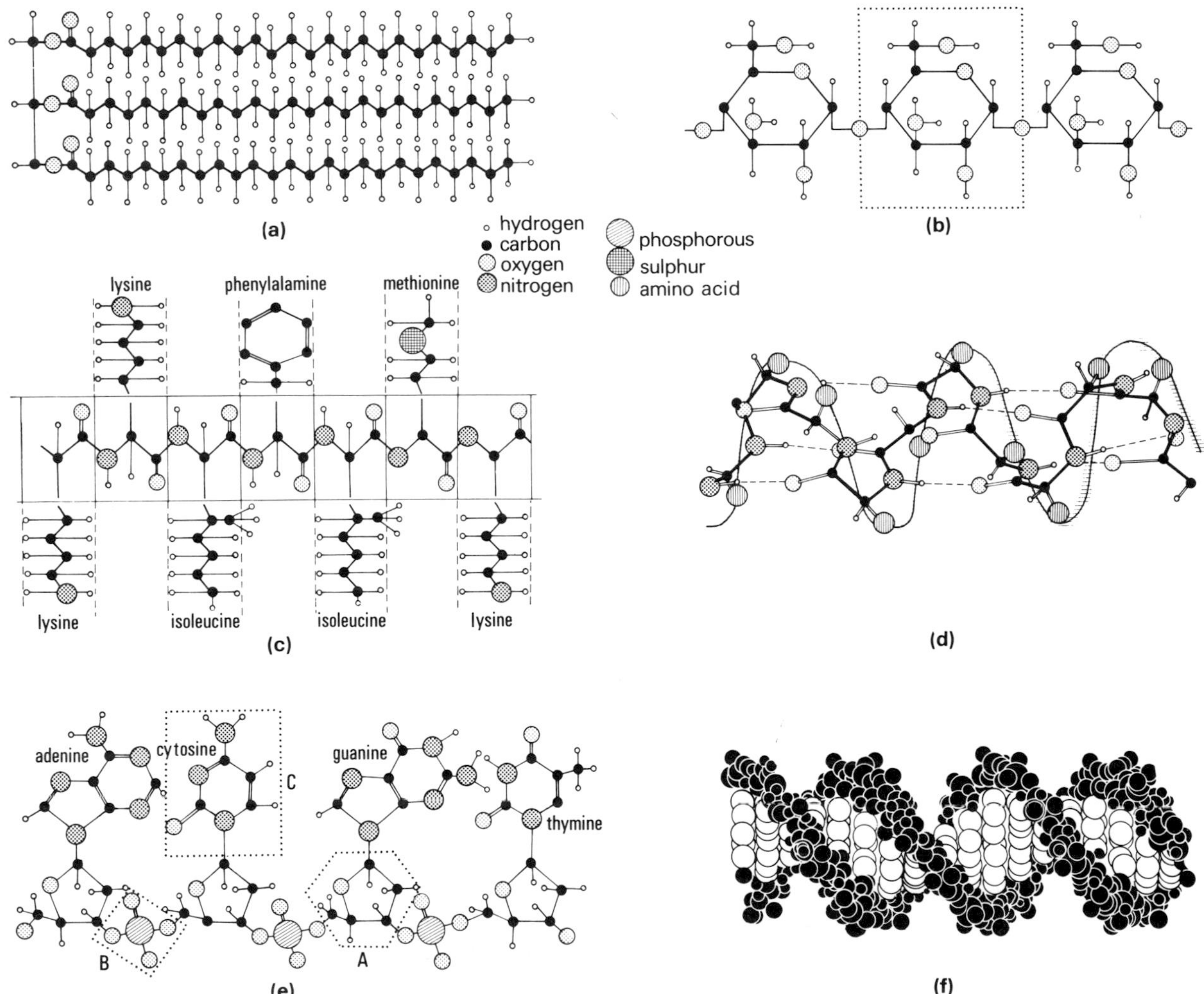

Figure 6.2 Organic macromolecules. (a) Polymerised hydrocarbon chains forming a fat – a triglyceride. (b) Glucose monomers (one is boxed) linked to form a polysaccharide chain. (c) Amino acids linked to form the polypeptide chain of a protein (sequence of seven amino acids from near the *N*-terminal end of cytochrome c from human tissue). (d) Alpha-helix configuration of protein polypeptide chain. (e) One strand (part) of the nucleic acid DNA showing the sugar(A)–phosphate(B) units in ester linkage and the four heterocyclic bases: adenine, cytosine (C), guanine and thymine. (f) Double helix configuration of part of DNA molecule.

arrangement of the monomers (sugars, amino acids or nucleotides) to form the polymer chain. The secondary structure is the way in which the chain itself is coiled or folded (in the proteins, for example, this often involves bonds between the sulphur atoms of some of the constituent amino acids). The tertiary level describes the way in which several polymers may come together, again with a definite three-dimensional arrangement. Because of the carbon bond angle of 110°, these arrangements are often helical or spiral in configuration. Giant molecules formed in this way are known as **macromolecules** and they have high molecular weights (haemoglobin, the protein in your red blood cells, has a molecular weight of 68 000).

Amongst these macromolecules of the cell there is an important distinction between the proteins and nucleic acids and the simpler polymers. This is not because the types of chemical bond linking the building blocks together are particularly complicated, but because these precursor molecules are more diverse and because they occur in the structure of the molecules in an exact and specific order, or sequence. These sequences mean that such macromolecules carry an enormous amount of information, a property of profound significance in the functional activity of the cell. It can be said that

Box 6.2

ORGANIC MACROMOLECULES

The organic compounds found in the living cell mainly fall into four great classes: fats, carbohydrates, proteins and nucleic acids. Structurally the fats are simplest. They are constructed from the hydrocarbon chains of fatty acids (usually three) joined separately to the main part of a molecule of glycerol. They are present principally as energy reserves, but one group of related compounds (the phospholipids) in which one fatty acid is replaced by a phosphoric acid group (H_3PO_4) which in turn is linked to another compound, has a critical role in the formation of biological membranes. Simple sugars consist of pentagonal or hexagonal rings made up of either four or five carbon atoms and one oxygen atom with attached hydroxyl (OH) and CH_2OH side groups. Two such units may form a sugar molecule or they may be linked in chains to form complex sugars or polysaccharides, especially cellulose and starch in plants and glycogen in animals. Starch, which may be up to 500 or so units long, and glycogen (1000+) serve mainly as energy stores, but cellulose with its much longer polymers (8000 units) is the main structural material in plants and combines in threads and meshes of great complexity.

Even greater complexity is encountered in the nucleic acids which form large structures built of at least four types of units, known as nucleotides, each of which consists of a sugar phosphate group (an ester) and a base. These are present in varying proportions and a great variety of sequences. The most important of these molecules are DNA and RNA and the significance of these information-bearing macromolecules is considered further in Box 6.4.

However, it is with the protein molecules that variety and specificity are seen to be most highly developed. These are the largest and most complex molecules known and they are made up of about 25 different amino acids linked to a carbon backbone. They form chains hundreds to thousands of units long, in different proportions, in all kinds of sequences and with a huge variety of folding and branching.

if the word 'life' has meaning, it is at the level of these information macromolecules that it begins to take effect (Box 6.3).

6.2.3 Structure at the cellular level

The decision to model the biosphere at the cellular level was taken earlier in this chapter because the organisms of the biosphere all share a common type of structure. Each is an aggregate of small units – cells – in some cases a single cell. Of course, these cells differ considerably in size, shape and many other characteristics, even within the same organism. This specialisation of cells, however, merely represents the extreme development of a particular property or function which in principle all cells possess or have possessed at some time. In other words, at some stage in their development all cells have many features in common. Therefore, we are justified in considering what might be called the generalised cell as the model of the basic unit of living systems. Indeed, 'the cell' can be considered to be a living system in its own right and, for the time being, that is the way we shall regard it.

A model of such a system is shown in Figure 6.3, but this diagram also serves to make clear the existence of a level of organisation between the molecular and the cellular. There are a number of subcellular structures known as organelles, but here we can make a more basic subdivision of the cell. The first of these subdivisions is the cytoplasm, which is a viscous fluid, largely water, but with inorganic ions, simple organic molecules and macromolecules dispersed through it and forming a colloidal suspension. It has an enigmatic physical condition best described by the physical state known as 'liquid crystal' as it displays properties of both fluid and crystalline states. This cytoplasmic matrix contains the subcellular organelles, one category of which – the cell membrane – encloses it and penetrates deeply and intimately through it. All of these organelles have definite structures built up of combinations of complex structural macromolecules and are best thought of as biochemical compartments within the cell. The second subdivision – the nucleus – is also really an organelle and it too is a biochemical compartment. Its importance, however, lies in the fact that it is the prime site in the cell of the information macromolecules of the nucleic acids.

We shall leave the structure of the cell at this point and return to the model of the ecosphere with which this chapter started (see Fig. 6.1). This model can now be expanded, for we have described the structure of the living system of the biosphere in an hierarchical manner starting with the elemental composition and ending with the living cell (Fig. 6.4). It is possible to continue to integrate the elements of the system into still-higher levels of the hierarchy, but for the moment we have chosen to stop the generalisation of the model at the cellular level. This approach is sometimes called reductionism and it has demonstrated

Box 6.3

INFORMATION

The information content of protein and DNA molecules can be appreciated by employing the techniques of information theory. These techniques, which are related both to thermodynamics and probability theory, have been developed in relation to computers and communication networks. In information theory the basic unit of information is the binary digit, or bit. The information content of one page of the *Encyclopaedia Britannica* expressed in these terms is of the order of 10^6 bits. A single volume of 1000 pages would therefore contain 10^9 bits of information. A modern desk-top computer would be capable of storing between 7000 and 300 000 bits of information. A DNA molecule with a molecular weight of 10^6 and about 4000 nucleotide units contains 8000 bits, while a protein molecule with a unit molecular weight of 10^5 contains 8000 bits of information in the amino acid sequence. A single bacteria cell only 2 μm in diameter and 6×10^{-13} g in weight contains in total approximately 10^{12} bits, or 1000 times more information than one volume of the *Encyclopaedia Britannica*.

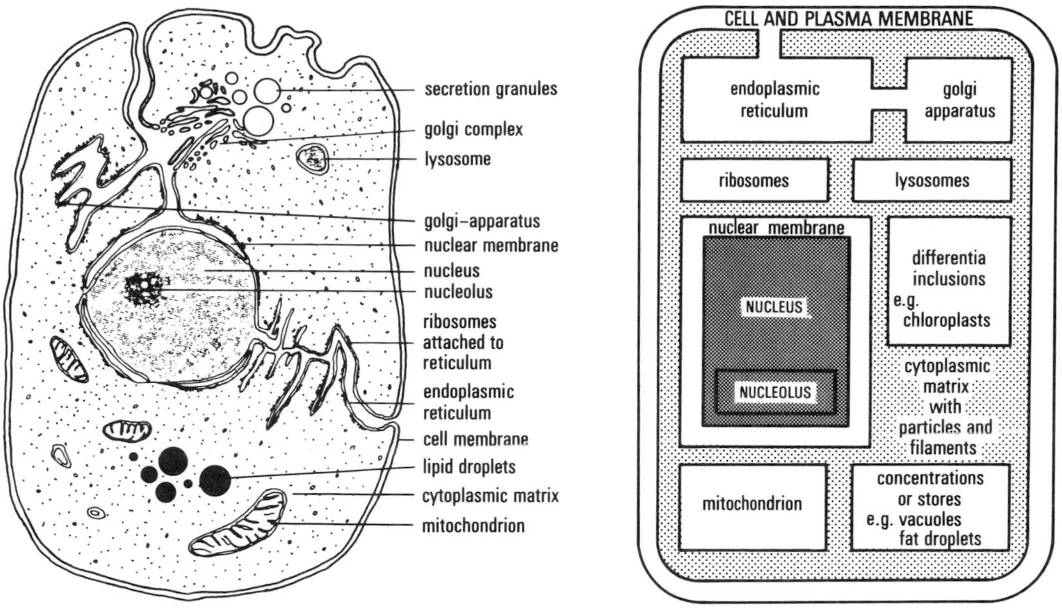

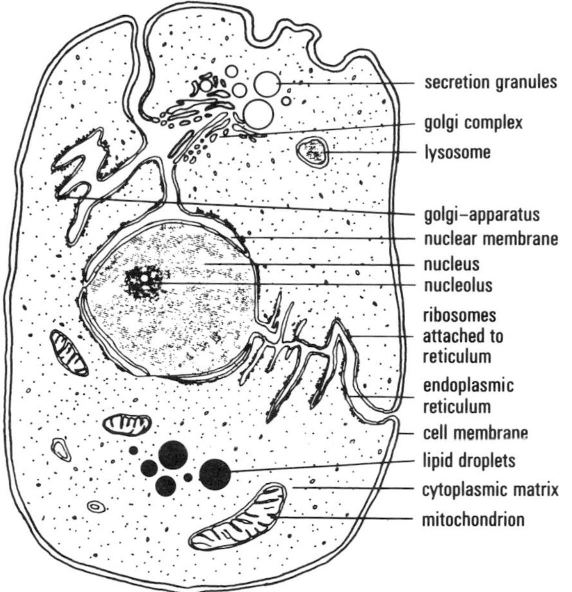

Figure 6.3 Models of the structure of the generalised cell.

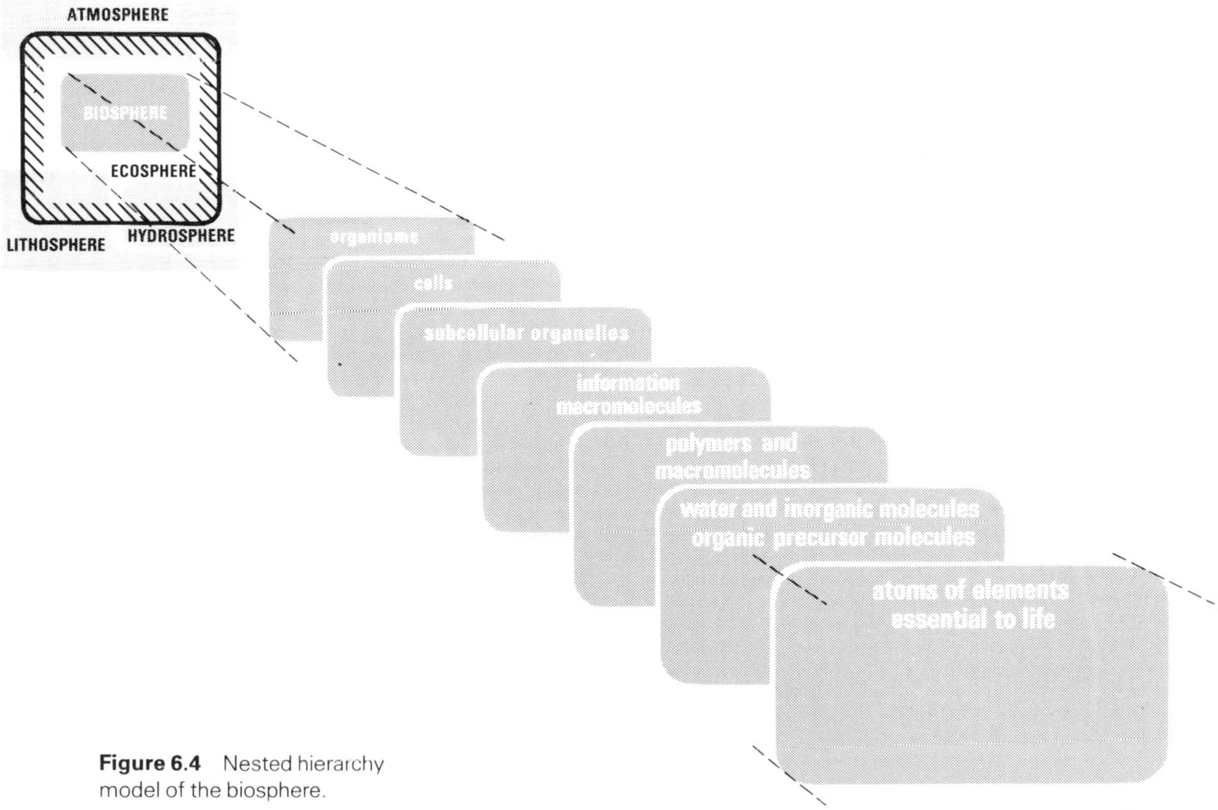

Figure 6.4 Nested hierarchy model of the biosphere.

that life can be described in terms of collections of molecules. It has also shown us that, in the structure of the living cell, a level of organisation is reached which is unmatched in the non-living world. If the *a-priori* assumption that there is a distinction between life and non-life is true, it is not one of kind but one of level. It is in the level of complexity and precision in the internal organisation of the cell and the organism that the distinction must lie. But this is not the complete story, for the model of living systems developed so far is a static one. The precise and definite structures made up of a multitude of complex chemicals have equally precise functions. The nature of life cannot be fully appreciated until we have considered the functional organisation and activity of the cell.

6.3 Functional organisation and activity of the cell

Function, of course, implies the performance of work and hence the utilisation of energy. This has led many authors to draw an analogy between the living cell and a manufacturing plant. Although perhaps overworked, this analogy remains useful as there are indeed parallels in the functional organisation and activities of the cell and those of a factory. Like a factory, the cell needs raw materials from which to manufacture its products. It requires an energy supply to power the production line and to transport both materials and products as well as to dispose of waste. However, in a factory the whole process works only because it has been designed to do so. By analogy, the cell must also possess such design specifications and blueprints.

Three kinds of work take place in this factory. Chemical work is performed both for cell maintenance and during active growth, when the cell makes and assembles all of the complex components required for repair or to make a new cell from comparatively simple substances. These processes are collectively called **biosynthesis**. Secondly, there is the work of transport where materials are moved within the cell or across its bounding membrane. Such work often involves changes in the relative concentration of materials and may take place against gradients of concentration or electrical potential. Finally, there is mechanical work, most obvious in the muscle cells of animals but performed in all cells and associated with contractile filaments. The performance of all these kinds of work involves energetically uphill, or **endergonic**, processes (see Ch. 2), and therefore they all require an investment of energy.

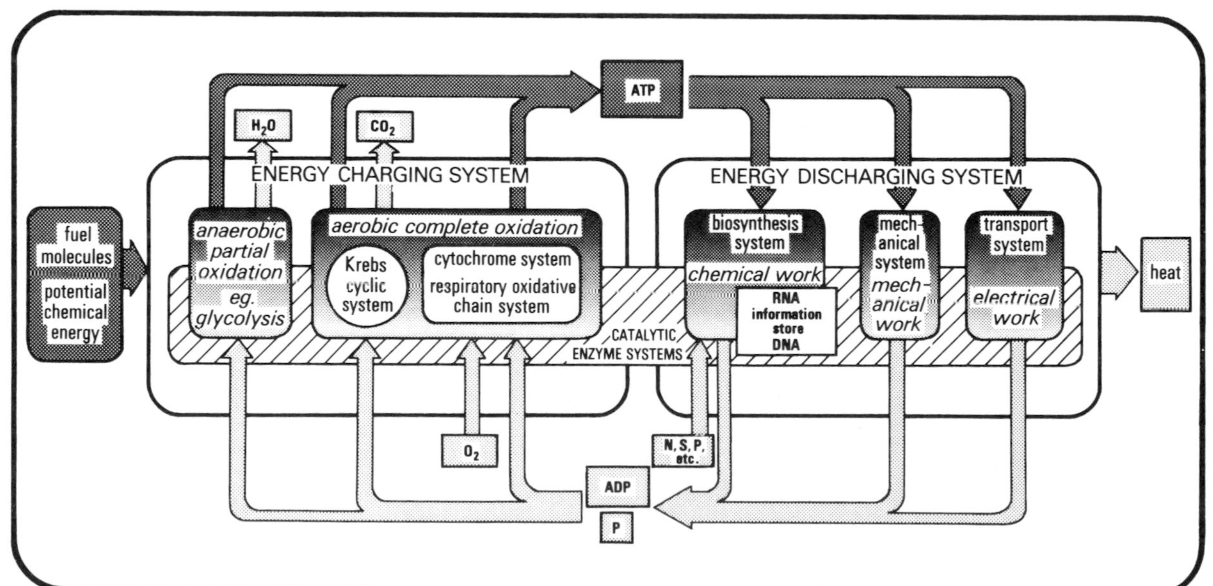

Figure 6.5 The ADP–ATP energy transfer system of the cell (developed from an idea in Lehninger 1965).

6.3.1 The transfer of matter and energy in the cell
In our cellular factory the necessary energy is stored in what can be thought of as molecular batteries, which can be moved around the cell to the places where the energy is needed. Here they are coupled to the mechanism concerned in the processes of cellular activity and the energy they carry is discharged. These spent batteries are then recharged using the chemical energy of fuel molecules. These are relatively complex molecules, particularly carbohydrates (starch, glycogen) and the fats or lipids, reserves of which are often stored in specialised regions of the cell.

This charge–discharge cycle is illustrated in Figure 6.5 and, in the cell, some of the energy of the fuel molecule released during its breakdown or oxidation is conserved when the compound adenosine diphosphate (ADP) is converted to adenosine triphosphate (ATP) in coupled reactions. It is these compounds that act as the mainline energy store and transfer system of the cell, in other words as our cellular batteries. ADP represents the exhausted, discharged state and ATP the fully charged state of the battery (Box 6.4).

The charging of the batteries, or the generation

Box 6.4

ATP–ADP ENERGY TRANSFER SYSTEM

The phosphate compounds of adenosine are all mononucleotides. The structure of these molecules is given in the adjacent diagram (see Fig. 6.2 for key to symbols). The heterocyclic aromatic ring of adenine is linked to the five-carbon sugar D-ribose, to which are attached one, two or three phosphate groups by ester linkage at the 5′ position. It is adenosine triphosphate (ATP) which forms the main energy transfer system of the cell. Adenosine triphosphate is unique in that it is intermediate in position in the energy level scale of cellular phosphate compounds and, therefore, acts as a go-between for phosphate transfer from high energy to low energy forms. The energy level of ATP, however, is not just due to the bond energy of the terminal phosphate group. Because at pH 7 each of the three phosphate groups is completely ionised, four negative charges exist in close proximity and, therefore, repel each other. The transfer of the terminal phosphate (hydrolysis) relieves some of this electrostatic stress. The like charges are thus separated as ADP^{---} and HPO_4^{--} which have little tendency to approach each other and undergo stabilisation. The energy level, in terms of electron configuration, of these products once separated is considerably lower than when part of ATP. It is this difference in energy content between the initial reactants and the final products which is measured as the free energy of hydrolysis of ATP. At standard conditions of pH 7 and 25°C and with one molal concentrations of reactant and products, the free energy change:

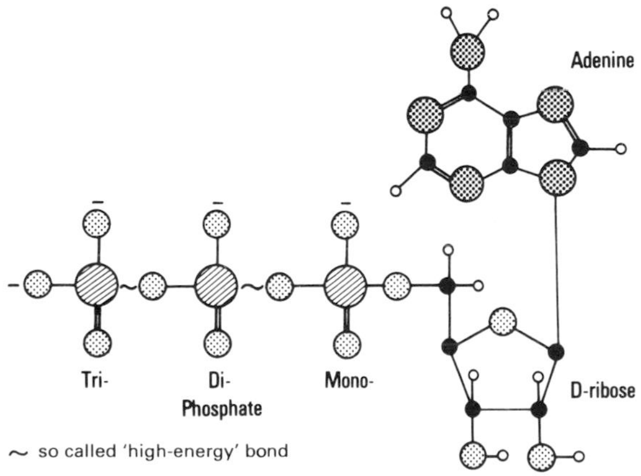

$$\Delta G = -2.929 \times 10^4 \text{ J/mole.}$$

In the cell there is reason to believe it is as high as -12×10^3 cal/mole (-5.016×10^4 J/mole).

of ATP from ADP and inorganic phosphate, is not a simple process. It consists of two distinct sets of operations both involving oxidation–reduction or electron transfer reactions (Fig. 6.6). In aerobic cells all three major fuel molecules – carbohydrates, fats and amino acids – are ultimately oxidised during what is known as the Krebs cycle, but each must first undergo certain enzymatic reactions (Box 6.5) which reduce the carbon core of the molecules to two-carbon pieces. In Figure 6.6 these preliminary reactions are shown for carbohydrate only, in terms of the glycolysis of glucose which takes place in the cytoplasm without the direct involvement of oxygen and by means of eleven enzymatic steps (Box 6.5).

The two-carbon compounds which result from these preliminary reactions are combined in an enzymatically activated form with a four-carbon compound as they enter the Krebs cycle to form a six-carbon compound. Through the cycle, this is progressively broken down to regenerate the initial four-carbon compound – so completing the cycle. The two carbon atoms and the oxygens released appear as two molecules of carbon dioxide and the hydrogen atoms combine with two electron-carrier molecules, NAD and FAD, converting them to their reduced form. The Krebs cycle, therefore, shows the fate of the carbon core of the fuel molecule, but is not directly involved in the mechanism of energy conservation as ATP. This mechanism operates when the reduced electron carriers donate the electrons they derived from the hydrogen atoms to a series of respiratory enzymes. These are electron-transferring enzymes called cytochromes and they have active groups called hemes which resemble the red blood cell pigment, haemoglobin. Like haemoglobin, hemes contain iron which can pass between the oxidised Fe^{+++} (iron III) and reduced Fe^{++} (iron II) states. Each cytochrome can therefore accept an electron when in the oxidised state, so becoming reduced, and then can pass this electron on to the oxidised

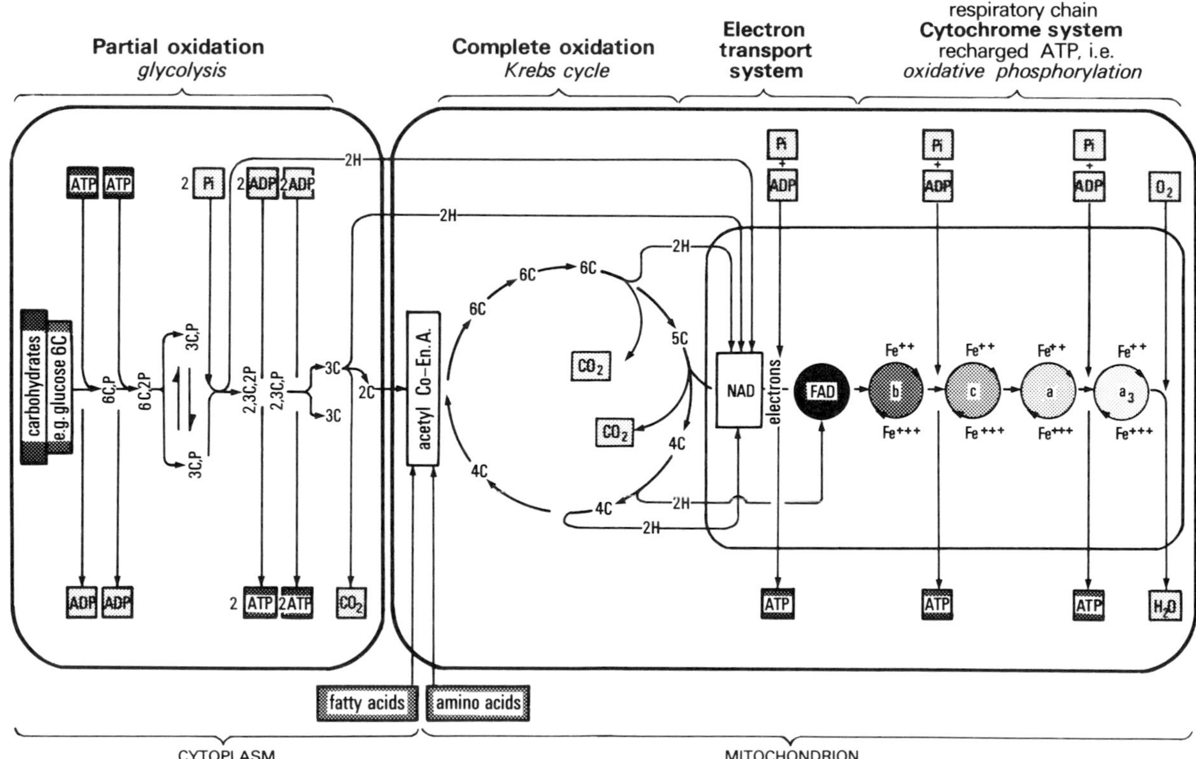

Figure 6.6 The energy transfer pathways in respiration, showing the fate of the carbon core of the fuel molecules and the conservation of energy in the ATP molecules. (The symbol Pi represents inorganic phosphorus.)

Box 6.5

ENZYMES

The catalysts that facilitate metabolic reactions in the living cell are called enzymes. They are true catalysts in that they lower the activation energy of the reaction they catalyse and hence have the capacity to accelerate the reaction. They do not influence the point of equilibrium of the reaction, nor are they used up during catalysis. All enzymes are protein molecules or their derivatives and they have molecular weights varying from 10 000 to 500 000; many contain metal atoms such as iron molybdenum and manganese in combination with the protein. In some it is combined with larger organic groups such as vitamins. Enzyme molecules can be regarded as having two parts: the active group or site which combines transiently with the substrate (substance on which the enzyme acts) and the larger proteinaceous part of the molecule known as the activating group. The catalytic property resides in the active group and is associated with the particular shape of the molecule which allows a 'lock and key' fit with the substrate molecule. Different enzymes have different shapes and are therefore highly specific to particular substrate molecules and hence catalyse only specific reactions. The seat of each kind of reaction and of its enzyme catalyst is usually highly localised in the cell, often associated with a particular organelle, so that thousands of reactions can go on in one cell at the same time, each controlled by specific enzymes. When a number of enzymes act in sequence they form enzyme systems.

One possible classification of enzymes is based on the kind of reaction catalysed, viz.:

'splitters'	breaking of bonds, hydrolysis;
'transferases'	transfer of a group from one molecule to another;
'adders'	add one molecule to another;
'oxidation–reduction'	transfer of electrons between molecules.

form of the next cytochrome. In this way electrons can travel the length of the cytochrome chain, finally being given up to molecular oxygen.

This respiratory chain is the final common pathway by which electrons from the different fuels of the cell flow to the final electron acceptor of aerobic cells – oxygen – which is itself reduced to water. The electrons entering the chain have a relatively high oxidation–reduction potential and free energy. As they flow down this potential gradient, the decline in free energy that occurs in step-like fashion is harnessed to the production of ATP from ADP and inorganic phosphate. Each electron flowing along the chain generates three molecules of ATP, which conserve 40–60% of the energy released. The total output from this oxidative respiratory system for the complete oxidation of one molecule of glucose is 38 molecules of ATP. The batteries have been recharged! Apart from the fuel molecule, the process requires one atom of oxygen for each pair of electrons which pass along the respiratory chain. At the same time, one molecule of water is produced by this chain, while two molecules of carbon dioxide are released for each revolution of the Krebs cycle. Finally, as the conservation of the energy released is not 100% efficient, some is lost as heat, in accordance with the Second Law of Thermodynamics. This is known as respiratory or catabolic heat loss and it will be discussed further in Chapter 7. The Krebs cycle and respiratory chain are located in the mitochondria, which have therefore been called the power houses of the cell.

The discharge of the batteries in cellular work is illustrated by the chemical work of biosynthesis in

Figure 6.7. Here, ATP is the energy input to the biosynthetic system, one of three illustrated in Figure 6.5. The phosphate group transfer potential energy of these molecules is either donated directly, or through intermediate phosphate compounds, to the synthesis of component macromolecules from both inorganic and organic precursor molecules at precise locations in the cell, often particular organelles. Both the process of synthesis and the rate of formation are controlled and regulated by an army of specific enzymes. Only in this way can thousands of individual chemical reactions go on in the cell at the same time, at room temperature and in a medium of liquid water. These enzymes, themselves proteins, are the mechanism by which biological information is expressed in the cell. In terms of our factory analogy, they are the shop-floor directives, standing orders and production targets by means of which the objectives of the plant manager and production manager are realised. The information contained in master plans and production schedules is represented in the cell by the information coded in the molecules of desoxyribose nucleic acid (DNA) in the chromosomes of the nucleus. These are the control systems which direct all of the functional activities of the cell.

The DNA molecule is a double helix polymer, each strand of which (Box 6.6) consists of sugars (desoxyribose), phosphate groups and four kinds of heterocyclic base, all linked by strong covalent bonds. The two strands are held together more loosely by hydrogen bonds (see Ch. 2) between pairs of bases. Each base, however, can only legitimately pair with one other and it is this sequence that constitutes the information stored in the DNA molecule. This information is passed on by transcription (Box 6.6), to another nucleic acid (RNA) and hence to the ribosomes – the seats of protein synthesis. Here the information transcribed to RNA controls the sequence of amino acids being assembled and through them the kind and shape of the protein synthesised.

Enzyme proteins then receive their instructions, when they are synthesised, from the coded information carried by RNA, information derived (with the aid of ATP energy), from the base sequence of DNA in the information store of the

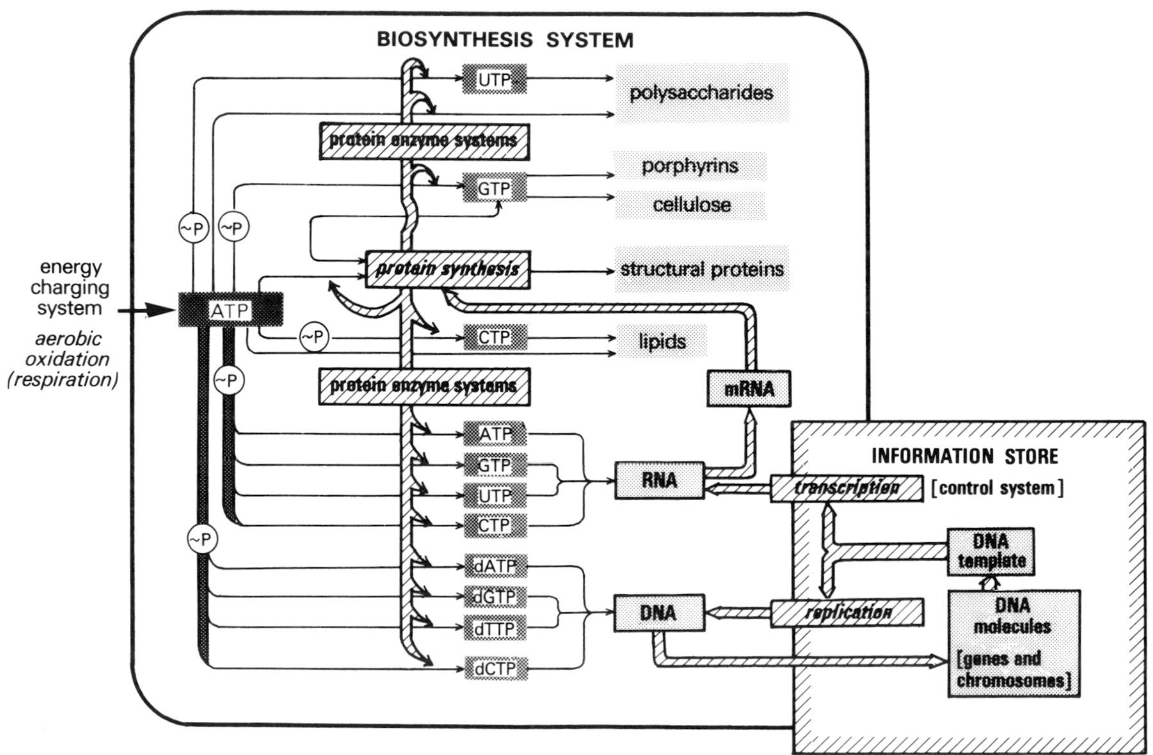

Figure 6.7 The chemical work of biosynthesis. mRNA = messenger RNA; P = phosphate transfer energy; UTP, GTP etc. = intermediate phosphate compounds. (Modified after Lehninger 1965.)

Box 6.6

DNA REPLICATION AND TRANSCRIPTION

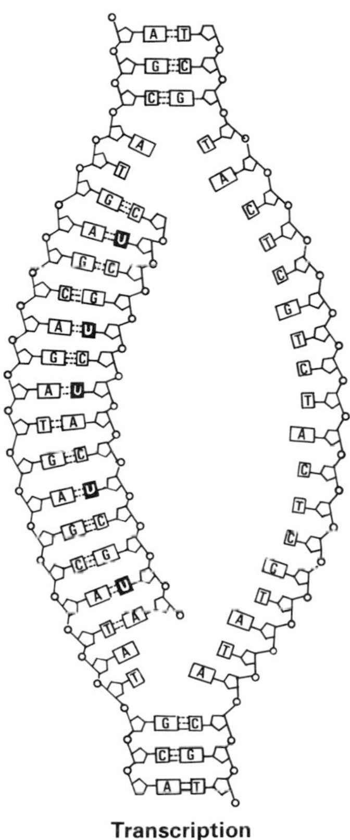

Transcription

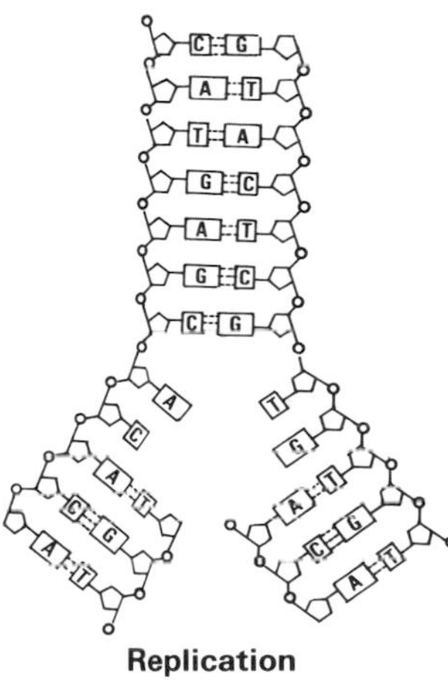

Replication

A Adenine
T Thymine
C Cytosine
G Guanine
U Uracil

The two covalently bonded strands of the DNA molecule are held together by weaker hydrogen bonds between the pairs of bases. Because of the strict base pairing (see text) the sequence of bases on one strand of the double helix predetermines the sequence on the other. It is this base sequence which holds the information stored in the DNA molecule, each group of three bases being known as a codon. If the two strands of one DNA molecule are separated, each can serve as a template from which to produce a replica. This **replication** capacity satisfies the need to duplicate information so that it may be passed on to daughter cells during cell division. The other activity of DNA – the control of cellular activity – is, however, accomplished by **transcription**. Here a strand of ribosenucleic acid (RNA) is able to pair and coil round a single strand of DNA to form a double helix, known as a DNA–RNA duplex. In order to pair, the base sequence of the DNA is copied by the RNA, but not exactly as in duplication. It is instead a transcription into a variant of the base code language, for in RNA the base thymine is replaced by uracil. Once synthesised, the RNA uncoils and leaves the DNA, becoming a messenger taking the information beyond the nucleus to the ribosomes. Here this messenger RNA (mRNA) becomes the template for the assembly of amino acids in protein synthesis.

nucleus. Although the factory analogy has proved useful, it sometimes seems inadequate in the face of such a complex, precise and sophisticated array of mechanisms.

6.3.2 *The steady state of the living cell*

This complex synthesis of cell parts would be more acceptable if it took place once only – if the appropriate molecules were assembled and put together to produce the appropriate structure, and that was that. Resynthesis or replacement would then be necessary only after the structure in question had worn out. Experiments with radioactive tracers, however, have proved that the cell is constantly being rebuilt over and over again. In this continuous state of flux the molecule which survives for more than a few days is exceptional. Clearly, one major function of the cell is to recreate itself constantly, but its ability to synthesise its component parts is not only harnessed to this dynamic turnover at the molecular level. Biosynthesis can be cumulative, giving cells the capacity of growth and thereby increasing their structural complexity. Cells, however, do not grow indefinitely – they divide. The resultant cells then grow, perhaps developing specialised structures and functions and/or dividing further. It is true to say that all organisms, indeed the entire biosphere, have resulted from endless cycles of regulated cell growth and division.

If, for a moment, we look at these functional activities of the cell in thermodynamic terms, then we are dealing with systems that possess relatively little *entropy* (see Ch. 2) compared with the universe around them: they are highly ordered systems. In fact, when cells are dividing and growing rapidly they actually decrease their entropy. Furthermore, our knowledge of evolution implies that, in general, species have evolved from lower primitive to higher, more complex forms. Over time, the biosphere as a whole has progressively decreased its internal entropy as it has increased in diversity. The laws of thermodynamics, however, would suggest just the opposite, for the Second Law leads us to expect the entropy of the universe to increase as no spontaneous energy transformation is 100% efficient.

There are two answers to this apparent paradox of life. First, one of the secrets of life is the way that the cell is able to build complex structures and to carry out equally complex functional activities by diverting part of the natural downhill thermodynamic trend of energy transfer to uphill energy-demanding processes. It does this, as we have seen, by means of a vast array of coupled chemical reactions where each stage of each reaction is catalysed by specific enzymes and energy is conserved and redirected in a highly efficient manner. This does not mean, however, that the cell escapes the implications of the Second Law.

Here we must turn to the second secret of life. Although the entropy of the cell (the system) decreases, it does so at the expense of the surroundings, which gain in entropy. The result is that the total entropy of the system and surroundings increases in accordance with the Second Law, even though the system itself, the cell (or biosphere), is becoming more ordered and hence decreasing its internal entropy. This gain in entropy by the surroundings occurs, of course, because cells are taking low entropy fuel or food molecules representing high-grade potential chemical energy from their surroundings (their environment). They return to it, however, simple inorganic molecules (CO_2, H_2O) and low-grade heat energy which have higher entropy.

Cells and all living systems are then open systems. They maintain their internal structures in the face of continual throughputs of matter and energy. This internal structure is, however, constantly being broken down and resynthesised and is, therefore, maintained in a steady state equilibrium. Only by the continuous self-adjustment of the steady state, facilitated by enzyme systems, can the cell, and indeed all living systems, keep the production of entropy or the tendency towards disorder at a minimum. The steady state is the orderly state of an open system.

The model of the cell used so far is adequate for most cells and for a large proportion of the organisms of the biosphere composed of such cells, which live by breaking down molecules that they or other cells or organisms have manufactured using energy from earlier generations of fuel molecules. But there is a catch in the use of this model, for it implies that the biosphere is in this sense devouring itself: it is living on capital. Such a closed system model is untenable on thermodynamic grounds. The dissipation as heat of some proportion of the available energy during each transfer in respiration, biosynthesis or any other kind of cellular work would inevitably mean a running down and eventual exhaustion of the fuel, in a way analogous to Man's experience with coal and oil reserves. If the cell and the organism are truly an open system maintaining a steady state, then there must be an external energy source somewhere which represents a continuous and renewable energy input to the system.

There is another snag in the use of the model of what is known as the heterotrophic cell, which relies on preformed organic molecules for chemical energy. Such cells also depend on these food molecules for materials which they cannot obtain as simple inorganic substances from their environment and cannot synthesise themselves. For example, heterotrophic cells cannot synthesise all of the 20 essential amino acids of protein. The cells of the human body, for example, can only manufacture 12. Also, for each heterotrophic organism there are certain essential organic substances required in small quantities which the organism cannot obtain from the environment and either cannot synthesise at all or cannot do so in sufficient quantities to meet its requirements. These are known as vitamins and they contribute to enzyme systems. Again, somewhere there must be cells which can produce these compounds and make them available to heterotrophic cells. The conclusion is, therefore, that our model of the generalised cell is incomplete and either needs refinement or we need a model of a different kind of cell.

6.3.3 The autotrophic cell model

Autotrophic cells are those which can produce organic fuel molecules (usually carbohydrates) from inorganic molecules using some external source of energy. There are two such sources utilised by living systems. First, there are those cells and organisms that can use the bond energy of inorganic compounds to reduce carbon dioxide to form organic carbon compounds. These are known as **chemotrophs** and the process as **chemosynthesis**. Secondly, there are cells and organisms that are capable of using light energy for the process of **photosynthesis**.

At the cellular level, photosynthesis can be viewed as a reversal of the pathway by which carbohydrates are decomposed to water and carbon dioxide during respiration (Fig. 6.8). Just as the outcome of respiration is the conversion of ADP to ATP (oxidative phosphorylation, see Fig. 6.6) which is then available for cellular work, so in photosynthesis ADP is converted to ATP which in the autotrophic cell donates energy to the building of carbohydrates and other organic molecules starting from carbon dioxide and water. The critical difference is that the energy that converts ADP to ATP in photosynthesis is entirely derived from absorbed light energy and the process is, therefore, known as **photophosphorylation**.

The trapping and utilisation of this light energy

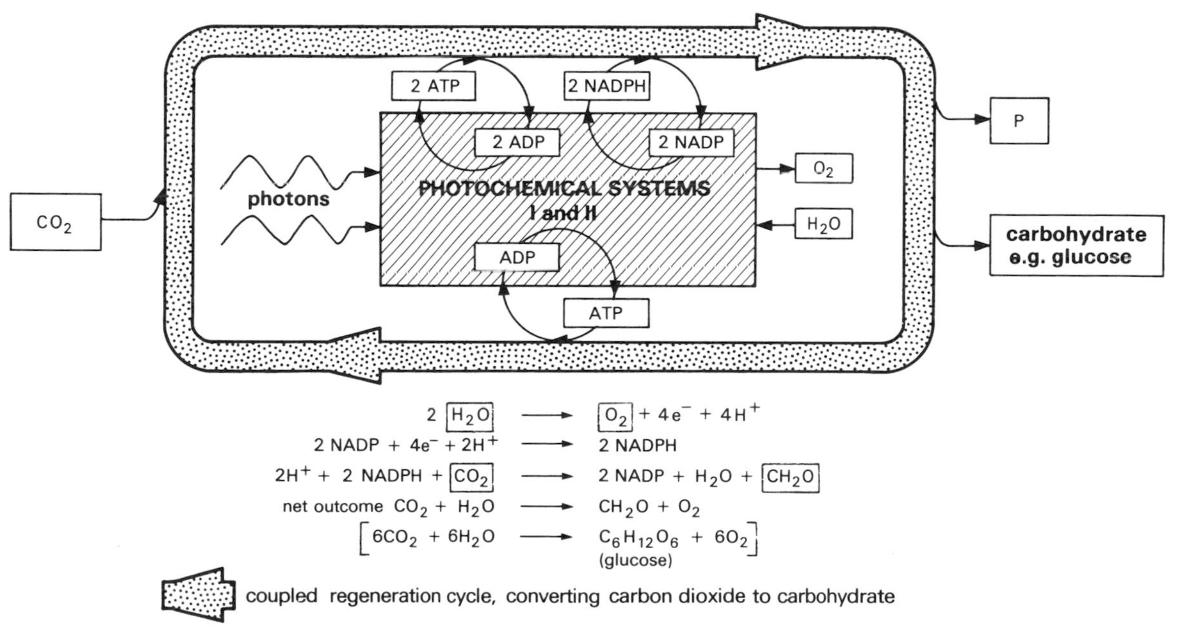

Figure 6.8 Schematic representation of the process of photosynthesis.

Box 6.7

CHLOROPHYLL

There are several kinds of chlorophyll, but in the higher green plants chlorophyll *a* and chlorophyll *b* are the most important. They are similar in molecular structure, both consisting of four pyrrole units arranged to form a porphyn ring with a magnesium atom at its centre and with a long hydrocarbon side chain. The only difference is the replacement of the *methyl* group of chlorophyll *a* by a *formyl* group in chlorophyll *b*. The pair of bonds holding the magnesium atom in the centre of the ring alternate between the four available nitrogens, a phenomenon known as **resonance**.

There are other photosynthetic pigments – the carotenoids in all photosynthetic plants and the phycobilins in blue-green algae and red algae. Chlorophylls absorb mainly red and blue–violet wavelengths, reflecting green light. Carotenoids absorb strongly in the blue–violet range and are therefore yellow–orange, red or brown pigments. Often masked by chlorophyll, they are responsible for the autumn colours of leaves when chlorophyll is lost, as well as for the colours of many non-photosynthesising tissues such as flowers and fruits and, of course, the orange-red of the carrot. Some phycobilins absorb strongly in the green wavelengths, some in the orange and red, and hence appear red and blue respectively and are responsible for the colour of red and blue-green algae.

These light-absorbing pigments are distributed in two photochemical, or pigment, systems – PSI and PSII – in both of which chlorophyll *a* is the primary pigment in what are called the trapping centres of the system. In PSI, only chlorophyll *a* pigments are utilised, but in PSII chlorophyll *b* and other accessory pigments gather absorbed light energy. In both PSI and PSII this energy is passed to modified molecules of chlorophyll *a* which are termed photoreactive or trapping centres. Each system has a variant of chlorophyll *a* with a different absorption peak (PSI 700 nm and PSII 680 nm).

(a)

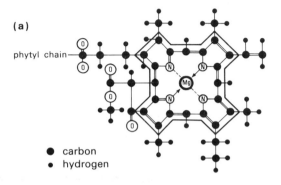

phytyl chain

O

Mg

N

● carbon

• hydrogen

(b)

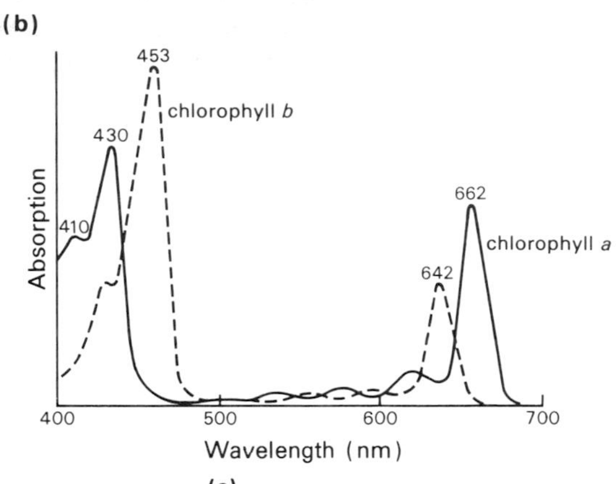

(c)

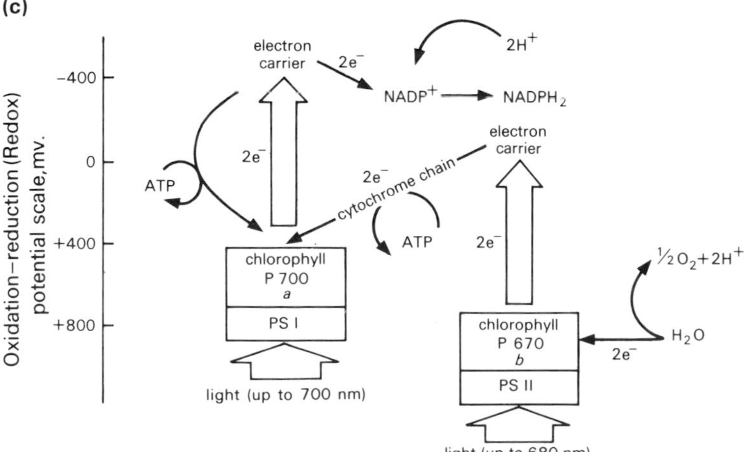

takes place in most autotrophic cells in the specialised organelles – the chloroplasts – which contain the pigment chlorophyll (Box 6.7). Because chlorophyll absorbs the red and blue wavelengths of the visible part of the spectrum, it appears green. The action of the absorbed light is to raise the energy level of the electrons in chlorophyll and their energy is in turn redirected through phosphate compounds as chemical energy to further chemical reactions. When a photon of light representing a quantum of energy falls on the chlorophyll molecule, it excites one of the electrons sufficiently for it to be donated from the chlorophyll to a series of electron acceptors, or enzyme catalysts. The electron passes along this chain which, like the respiratory chain (see Fig. 6.6), contains cytochromes, generating ATP before being returned to the chlorophyll molecule. This is the so-called light phase of the photosynthetic process and, although many autotrophic cells are aerobic, it does not require the presence of oxygen. The synthesis of organic carbon compounds – mainly, but not exclusively, carbohydrates from carbon dioxide and water (or, in some bacteria, hydrogen sulphide) – is termed the dark phase and involves the biochemical reduction of these inorganic molecules using the energy of intermediate phosphorous compounds which in turn have derived their energy from the ATP molecules produced by the light phase. This biochemical reduction liberates molecular oxygen from the water molecules (or sulphur from hydrogen sulphide).

6.4 The cell, the organism and the community

In this chapter the functional organisation and activity of the living systems of the biosphere have been discussed in terms of models constructed at the cellular level. Such models may have seemed at first sight to be far removed from the global perspectives of physical systems presented in the other chapters in this section. The universality of these cellular models is really quite remarkable, in spite of the astounding richness and diversity of life forms on this planet. It is precisely this generality that makes an understanding of the fundamental organisation and functioning of the cell a necessary prerequisite for a consideration of living systems at the level of the organism, the population and the community.

Further reading

There are many introductory biology and biochemistry texts which cover the subject of this chapter, only a small selection of which is given here.

Berrill, M. J. 1967. *Biology in action*. London: Heinemann.
Gerking, S. D. 1969. *Biological systems*. Philadelphia: Saunders.
Novikoff, A. B. and E. Holtzman 1970. *Cells and organelles*. New York: Holt, Rinehart & Winston.
Rose, S. 1970. *The chemistry of life*. London: Penguin.
Watson, J. D. 1970. *The double helix*. London: Penguin.

7 The ecosphere

7.1 The organism, the population and the community

It is at the level of the individual organism that we encounter *discrete* living systems in direct contact with their non-living environment. In single-celled creatures such as the protozoa or the unicellular algae, the individual may differ little from the generalised model of the cell presented in Chapter 6. Most organisms, however, are multicellular. In such cases, cells specialised both structurally and functionally form equally specialised tissues, organs and organ systems that together constitute and function as discrete organisms – a crocus, or a giant redwood, a housefly, or a horse – each of which exchanges matter and energy with its inorganic environment – to which it is functionally linked. For example, Gates (1962) illustrates part of this pattern of exchange when he discusses the heat transfer with which an animal living on the surface of the Earth is involved (Fig. 7.1). The animal is receiving direct and diffuse sunlight from the sky. It receives reflected light from the ground, from vegetation and from everything around it, including the sky. It receives thermal or infrared radiation from the atmosphere, clouds, ground and surrounding vegetation. But the animal is a hot object re-radiating to the sky, to space and to the ground and thereby losing heat by radiation in all directions. There will also be convective heat loss which may be considerable, especially under

Figure 7.1 The energy budget of an animal at the Earth's surface (Gates 1962).

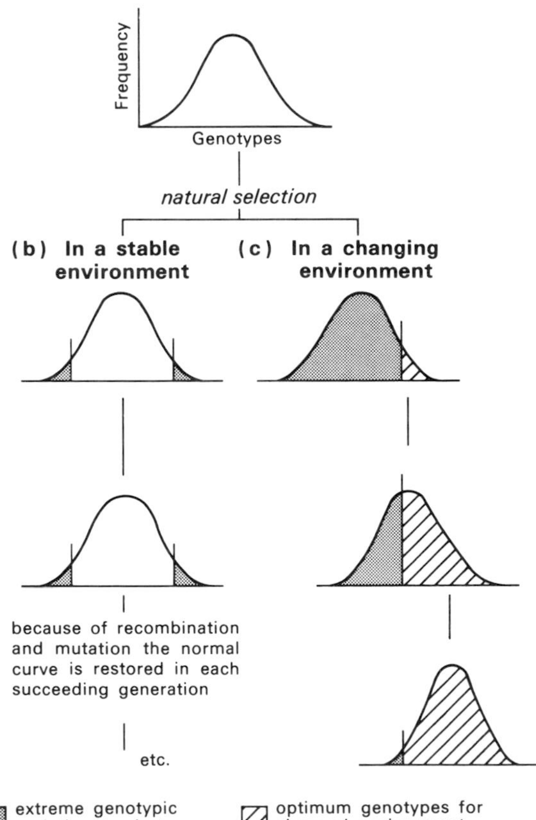

Figure 7.2 Selection of genotypes in a stable and changing environment (Heywood 1967).

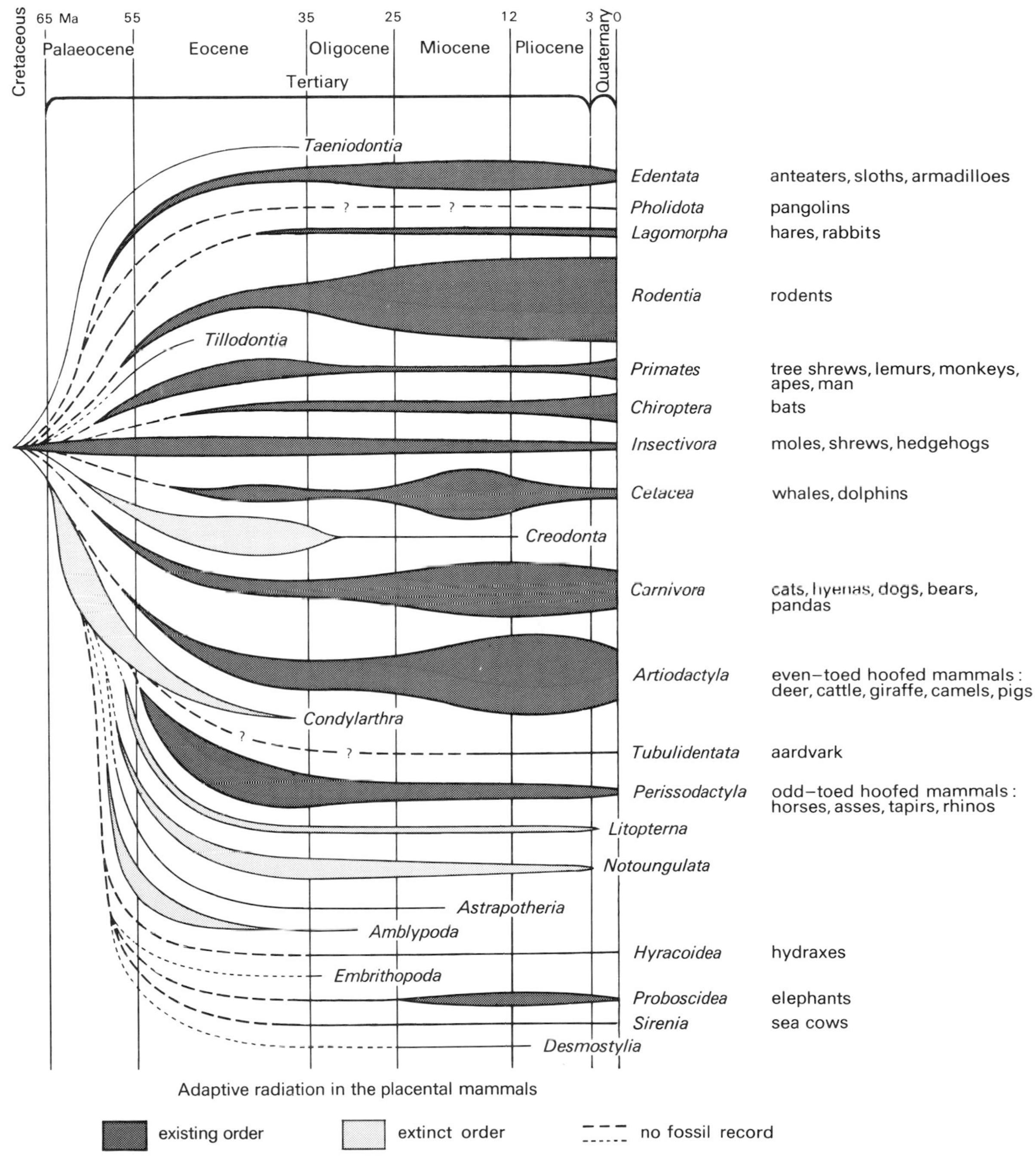

Figure 7.3 Adaptive radiation of evolutionary change: the example of the placental mammals (from *Continental drift and evolution* by B. Kurtén, copyright © 1969 by Scientific American, Inc., all rights reserved).

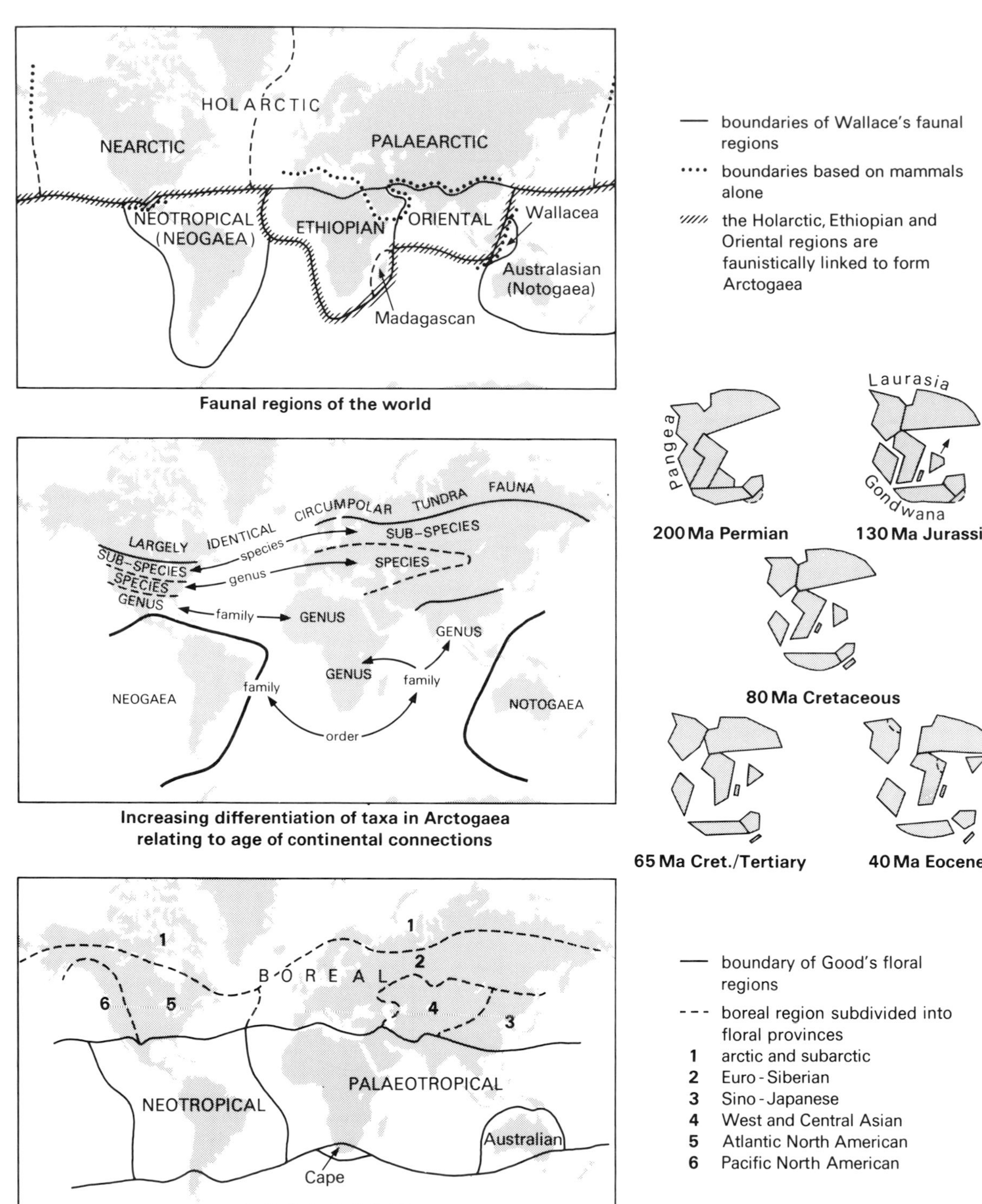

Figure 7.4 Types of plant and animal distribution pattern (compiled from various sources).

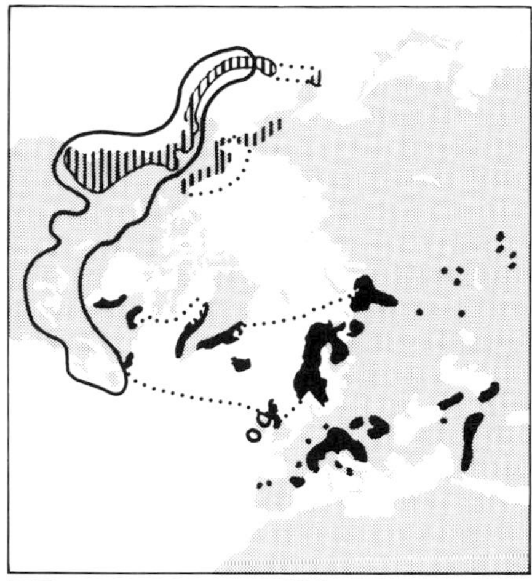

Luzula piperi and *L. wahlenbergii* (woodrushes) – transBeringian distribution

Potentilla crantzii (alpine cinquefoil) – amphi–Atlantic disjunction

Spiranthes romanzoffiana (drooping ladies' tresses) – amphi–Atlantic disjunction

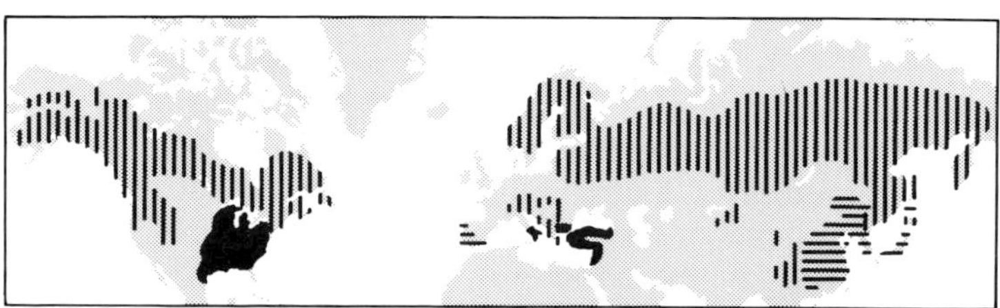

circumboreal – *Picoides tridactylus* (three–toed woodpecker)

Europe – E. Asiatic disjunction – *Cyanopica cyanus* (azure–winged magpie)

E. N. American – Asia Minor disjunction – *Platanus* (plane) *Platanus occidentalis* endemic to N. E. America, *Pl. orientalis* to S. E. Europe and Asia Minor

Symphonia – American – African tropical discontinuity at generic level

Ancystrocladus – Africa – Asia tropical discontinuity at generic level

Buddleia – pantropical discontinuous distribution

Palmae – pantropical distribution at family level

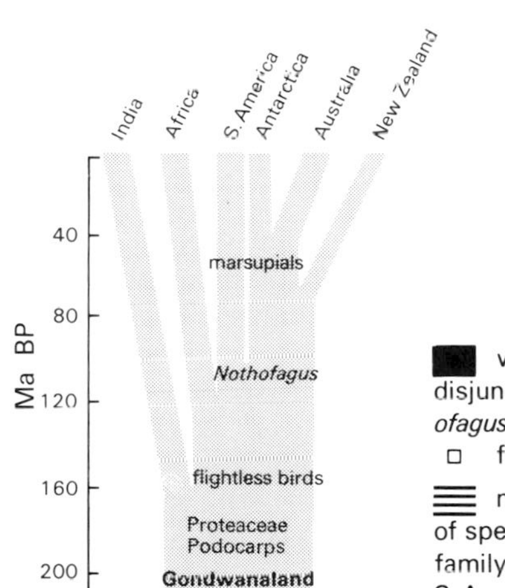

Time of arrival of certain taxa in relation to the fragmentation of Gondwanaland

wide S. hemisphere disjunction of living *Nothofagus* (southern beeches)

□ fossil locations

main concentration of species of the ancient family Proteaceae omitting S. American species

▲ present and △ fossil distribution of the genus *Dachrydium* of the southern conifers, the Podocarpaceae

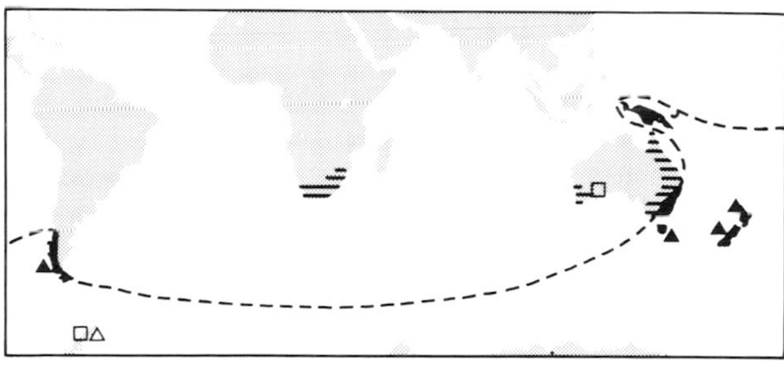

N. E. America – E. Asia disjunction of the genus *Liriodendron* *L. tulipifera* endemic to N. E. America *L. chinense* endemic to China

Sequoia sempeviriens and *Sequoiadendron gigantea* (redwoods) both endemic to W. N. America

○ fossil redwood locations

■ *Metasequoia glyptostroboides* (dawn redwood) – endemic to Yangtse Valley

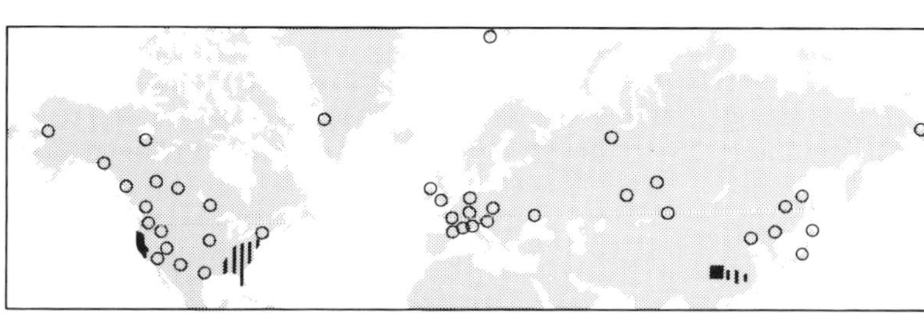

forced convection in a wind and, if the animal perspires, it will experience evaporative heat loss (see Ch. 4). Clearly, it is with the organism that we begin to appreciate more fully the interaction of living systems with the global physical systems of the ecosphere model (see Fig. 6.1).

7.1.1 Species and populations

The Earth is inhabited by an enormous number of different kinds of organism and each is in some sense unique. These are known as **species** and they are populations of organisms that show an overall resemblance, are distinct and reproductively isolated from any other group, and have a degree of constancy through time. At a conservative estimate there are about 400 000 species of plant living today. Among the animals, more than a million species are animals without backbones (invertebrates) and 850 000 of these are insects. The mammals are represented by about 4000 species, the birds by approximately 8000 species and the fishes by more than 20 000 species, and there are about 6000 species of reptiles and amphibians. All of these species are the result of a long-continued process of evolution and speciation. Each of these contemporary species of organism, therefore, has arisen by an interplay between changes that have occurred to the genetic information (genotype) that it carries encoded in the DNA molecules in the nuclei of its cells (see Box 6.6) and the selective pressures exerted by its physicochemical and biotic environment.

Because the genetic information controls, through transcription, the metabolic activity, growth and development of each individual, changes in genotype give rise to parallel changes in the morphology or anatomy of the organism, or to changes in the way that it functions – physiology – or to changes in its behaviour, or to some combination of all three. The result of these changes is the occurrence of hereditable variation in the population of that species – variation that arises at random but that survives in the population only if it has no fatal or deleterious effect on the organisms concerned. It may, however, endow some individuals with a selective advantage in their interactions with the inorganic environment and with both other members of the same species and members of other species. This process – the selection of adaptive variation in the population – is known as the environmental sieving of genotypes. The adaptive significance of such variation may be apparent immediately in the generation in which it appears and in relation to the existing environment. It may be, however, that the variation, though reappearing from one generation to the next, only becomes of adaptive significance when the environmental conditions change or the species experiences a new environment as it spreads by dispersal and migration (Fig. 7.2). This is Darwin's process of natural selection and it ultimately leads to speciation, either abruptly or gradually, as new species arise from old and successful variants either replace ancestral populations or become reproductively segregated and geographically isolated from them as they spread out colonising new environments. This process has been repeated over millions of years and has led, through the **adaptive radiation** of evolutionary change, to the great diversity of organisms that exist today and to those that, though now extinct, existed in the past (Fig. 7.3).

Each of these species exists in a whole complex of environmental interactions. It exists because it can survive and compete successfully under the particular conditions of existence to which it is adapted or of which it is tolerant, and the presence of a species population in any location will reflect the often long and complex history of evolution, dispersal and migration of that species. It is this history that, at least partially, explains the present distribution of plant and animal species (Fig. 7.4). Some are young species with restricted distributions. Some are very ancient species whose area has become drastically reduced to a single relict distribution. However, both are described as narrowly **endemic**. Other species may be present in almost all of the world's land masses and hence are known as **cosmopolitan**, while others may have markedly discontinuous (**disjunct**) distributions. Some of these distribution patterns reflect not only the evolution and migration of the species but also changes in the distribution of land and sea, changes in climate, the appearance of mountain ranges and the extension and retreat of continental ice sheets. In other words, distribution reflects the history of the environment as well as that of the species and in many ways it embodies the response of the species to environmental change through time (see Ch. 23).

The pattern of floral and faunal regions of the world can be seen in Figure 7.4 to reflect in part the changes in the relative positions of the continents through geological time. The greater floral and faunal affinities in the northern hemisphere are associated with the fact that dispersal and migration routes have existed between the northern land masses until relatively recently. In contrast, evolution and speciation has proceeded in isolation for longer on the land masses of the southern hemisphere resulting in distinctive floral and faunal regions with affinities often only apparent at higher taxonomic levels. Nevertheless, problematic disjunct distributions in the southern hemisphere can also be explained with reference to the timing of the fragmentation of an earlier super-continent. Thus the marsupials which reached Australia via Antarctica from South America are absent from New Zealand, Africa and India, all of which had separated before their arrival. Some northern hemisphere disjunctions such as the amphi-atlantic distribution may also reflect the fragmentation of a once more continuous distribution by continental drift, but many are associated with the effects of late Tertiary and Quaternary climatic change on plant and animal distributions which had been circumboreal in the early and mid Tertiary (e.g. the redwoods). In such cases recent divergent evolution of these isolated descendants of common ancestors has resulted in species with endemic distributions today (e.g. the plane tree, and the tulip tree, *Liriodendron*).

7.1.2 Communities, ecosystems and the ecosphere

Wherever they occur, species of plants and animals rarely do so as pure populations. Normally populations of different species grow and live together as members of plant and animal communities. The variety and diversity of such communities is almost as great as that of their component species and, furthermore, the community concept can be applied at a range of scales from that of a small pond to the thousands of square kilometres of the Amazonian rainforest. Like the organism, the community as a whole is adapted to and reflects the conditions imposed by the external environment in which it exists and with which it interacts. This adaptation is partly the sum total of the individual adaptive strategies displayed by the species which make it up, but it is also adaptation at a higher level than the organism. This community-wide adaptation to environment involves the structural and functional organisation of the entire community.

If we now extend the model of the biosphere developed in Figure 6.4 to incorporate the organism, the population and the community, it becomes clear that they are merely higher levels of integration in the model (Fig. 7.5). Just as organelles function as integrated units in a model at the cellular level, or as cells are functional units in the organisation of the multicellular organism, so individual organisms and species populations are functional units in a model at the level of the community. The biosphere is, in its turn, composed of all of the different community types that

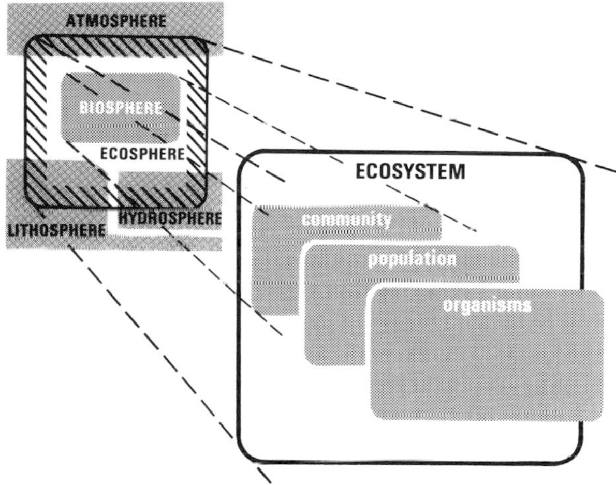

Figure 7.5 Nested hierarchy model of the ecosphere.

occur at the Earth's surface. But from the level of the organism upwards these models involve direct interactions with the non-living environment. At the global level this is recognised by the ecosphere model. At the community level these environmental or ecological interactions are incorporated by modelling the community as an ecological system, or **ecosystem**. By grouping ecosystems which are broadly similar in their structure (particularly of the vegetation) and which occur in similar environments, we can define the major broad subdivisions of the ecosphere. These are known as **biomes** and biome-types, and in the northern hemisphere we can talk of a broad-leaved

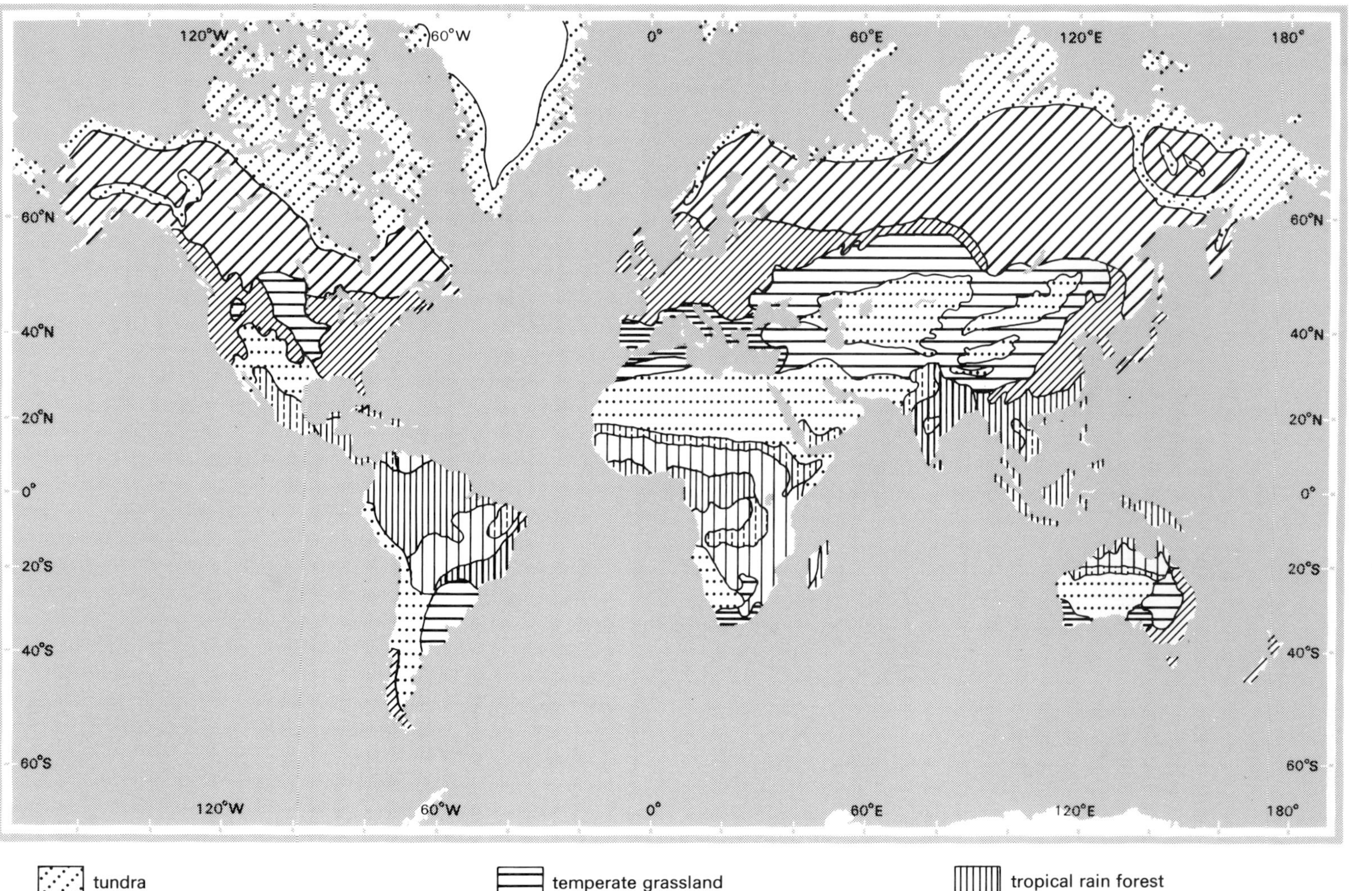

Figure 7.6 Major terrestrial biomes of the world.

deciduous or summer forest biome or a tundra biome (Fig. 7.6).

7.2 A functional model of the ecosphere

In spite of the vast diversity that we see displayed in the real world at the levels of the organism and of the community, it is still possible to present a generalised model of the functional activity of the ecosphere. This is possible because such activity, as in the case of the cell discussed in Chapter 6, remains the exchange of matter and energy between the system and its surroundings and the pathways and processes of energy transfer within the living system, whether that system is an organism, a complex community of organisms, or the entire biosphere. The distinction between the autotrophic and heterotrophic functions of the cell will also, of course, hold true for the organism. Plants for the most part are autotrophic and animals are heterotrophic systems. Not all of the cells of plants, however, are autotrophic. The light energy fixed during photosynthesis and transformed to the chemical energy of the synthesised carbohydrate molecules may be transferred by respiration to ATP and hence to biosynthesis and other activities in the autotrophic cells of these plants. Alternatively, the products of photosynthesis may be translocated within the plant and the energy they represent may be utilised by non-photosynthesising heterotrophic cells elsewhere in the plant. Heterotrophic organisms – animals – will obtain the products of both photosynthesis and subsequent plant biosynthesis and hence the chemical energy represented by these compounds when they consume parts of the plant as food, either directly or indirectly.

The functional organisation that emerges at the organism and community level is, therefore, hierarchical. It reflects the energy flux from radiant light energy via photosynthesis in the presence of chlorophyll in green plants to chemical energy subsequently ingested and assimilated by herbivorous animals whose food the plants represent, and then to carnivorous animals, again as chemical energy when they devour their prey. Because communities, and indeed the whole of the biosphere, are composed of both autotrophic and heterotrophic organisms, the functional organisation of the ecosphere must also reflect this energy flux.

7.2.1 The trophic model

This hierarchical organisation is the food chain. Its formulation as a series of thermodynamically valid steps in the energy flow through the biosphere was first developed in a now classic paper by Lindeman in 1942. In this model, organisms are lumped together into compartments depending on their place in the food chain or on their energy source. These compartments are called **trophic levels** (Fig. 7.7). In accordance with the Second Law of Thermodynamics, as the original radiant energy input is passed from one trophic level to the next some of the energy is dissipated at each step as unavailable heat energy. This is the expression of the catabolic or respiratory heat loss associated with cellular activity, but manifested here at the level of the organism.

Figure 7.7, however, also emphasises the fact that the biosphere contains not only living but also dead organic matter. Furthermore, there is a compartment in the model which we have not considered when taking the simple food-chain viewpoint. Assigned to this compartment are organisms whose energy source is not the chemical energy stored in the tissues of living plants and animals, but that which remains after their death or is excreted or shed by them while still alive. These are the **decomposers** that feed on dead and decaying organic matter. The most important are the bacteria and fungi, both of which belong to groups that on evolutionary grounds are best regarded as being distinct from the plant and animal kingdoms. It will become apparent in Chapter 19 that this decomposer trophic level is an oversimplification, for some decomposer organisms are in a specialised category of herbivore – the detritivores – while others are carnivorous. Nevertheless, the inclusion of the decomposers emphasises the fact that the energy flow through the biosphere can take one of two routes: a fairly direct pathway through the so-called grazing food chain and a more indirect pathway, often involving a timelag, through the detrital food chain. Whichever pathway is followed, however, the ultimate output is unavailable heat energy.

In common with most hierarchies, the character of the individual is lost in this trophic model of the biosphere. All organisms, from algae to forest trees, are lumped together in a trophic level, in this case as autotrophic. Each organism is significant only in so far as it is a contribution to the total mass of organic matter (**biomass**) and total energy con-

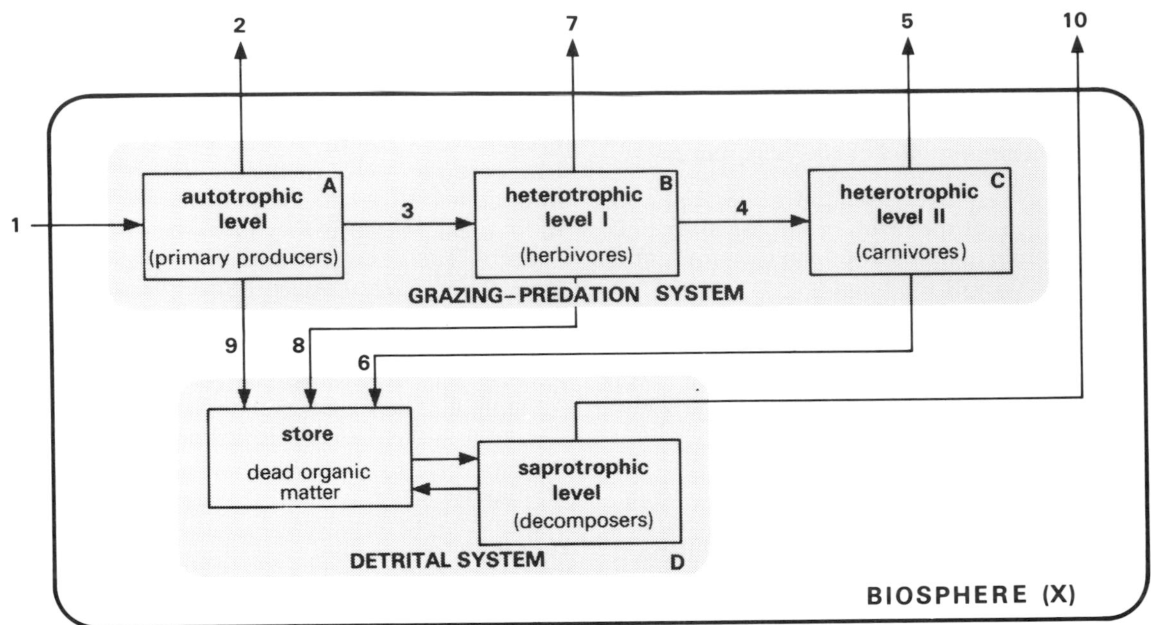

Figure 7.7 Compartment model of the biosphere (X), which has been divided into compartments A, B, C, D (trophic levels). The energy transfers into and out of these compartments are represented by arrows numbered from 1 to 10. The energy budgets of the individual compartments or trophic levels are:

$$1 = 2 + 3 + 9 \pm \Delta A$$
$$3 = 4 + 7 + 8 \pm \Delta B$$
$$4 = 6 + 5 \pm \Delta C$$
$$9 + 8 + 6 = 10 \pm \Delta D$$

and for the entire biosphere:

$$1 = 2 + 5 + 7 + 10 \pm \Delta X$$

where ΔA, etc., are the changes in the energy content of the various compartments (derived from Chapman 1976).

tent of that trophic level per unit area of the Earth's surface at a moment in time. This is because, from the functional point of view, each level in this trophic model represents a temporary store of potential chemical energy.

Ignoring the diurnal cycle for the moment, we can regard the input of radiant energy as a continuous process. The same is true for the metabolic activities of both plants and animals. The energy transfers involved in photosynthesis – the ingestion and assimilation of food, maintenance, activity, growth and reproduction – are taking place all the time, so there will be a continuous loss of energy as heat. Over a period of time (such as a year) a balance will exist for each trophic level between the gains (or inputs) of energy and the losses (or outputs) of energy. This balance is the net accumulation or deposit of energy over that period of time (production, P) and it is manifest as a change in the biomass of the trophic level in question (ΔB). For the autotrophic level this is referred to as the net primary production (P_n) but for heterotrophic levels, although termed secondary production, it is more strictly called conversion. Figure 7.8 shows in more detail the flux of energy through both autotrophic and heterotrophic levels and the quantities that contribute to production.

Production expressed as a rate of change in the energy content of a trophic level is termed **productivity** and is expressed as mass or its energy equivalent per unit area of the Earth's surface per unit time (kg m^{-2} a^{-1} or kJ m^{-2} a^{-1}). The biomass of a trophic level then, whether expressed in units of mass or energy, is a measure of its 'bulk' in terms of the amount of organic matter present at that point in time. It is of course this 'bulk' that is often impressive to an observer. Who would not be

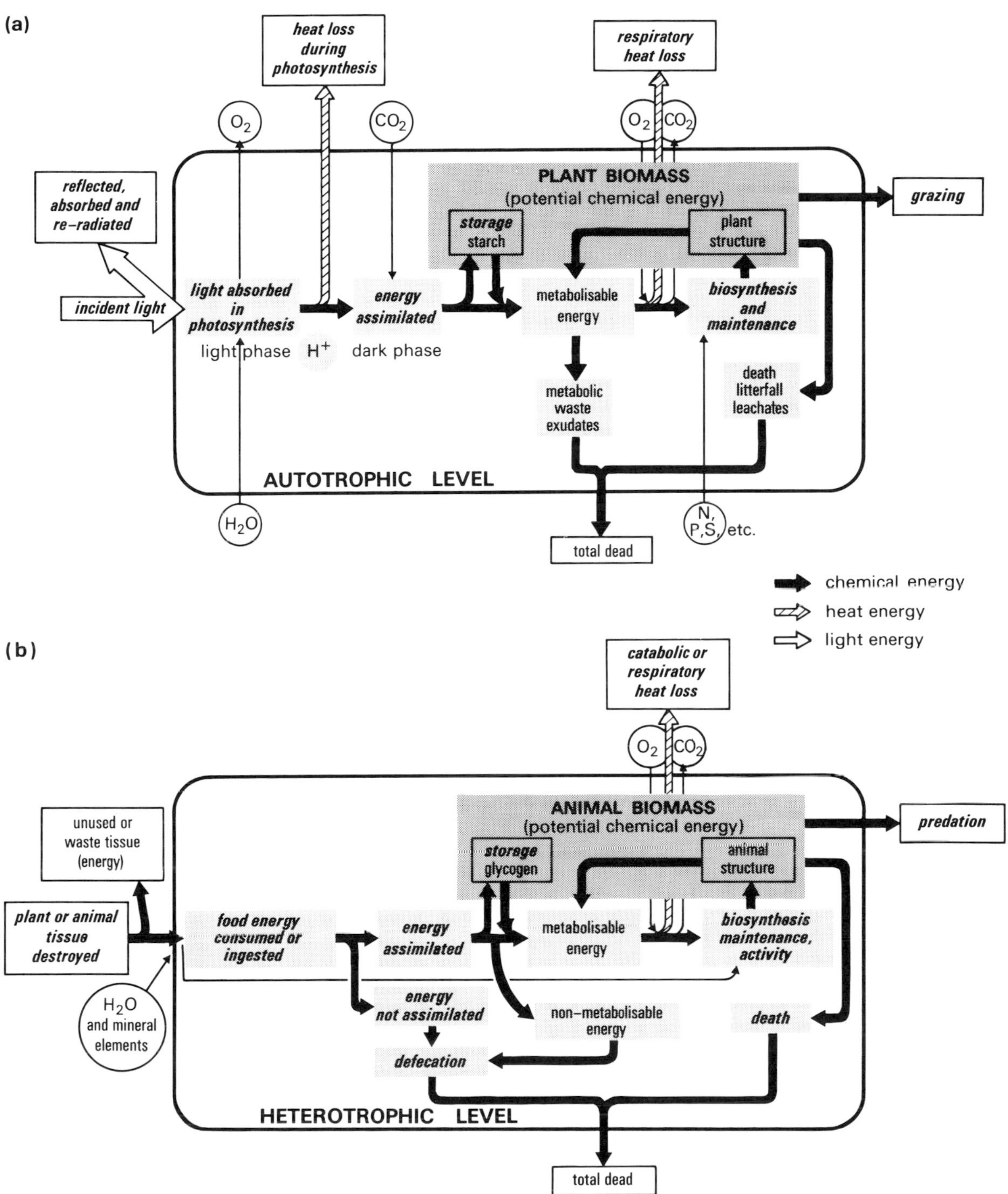

Figure 7.8 Energy flow through the (a) autotrophic and (b) heterotrophic levels.

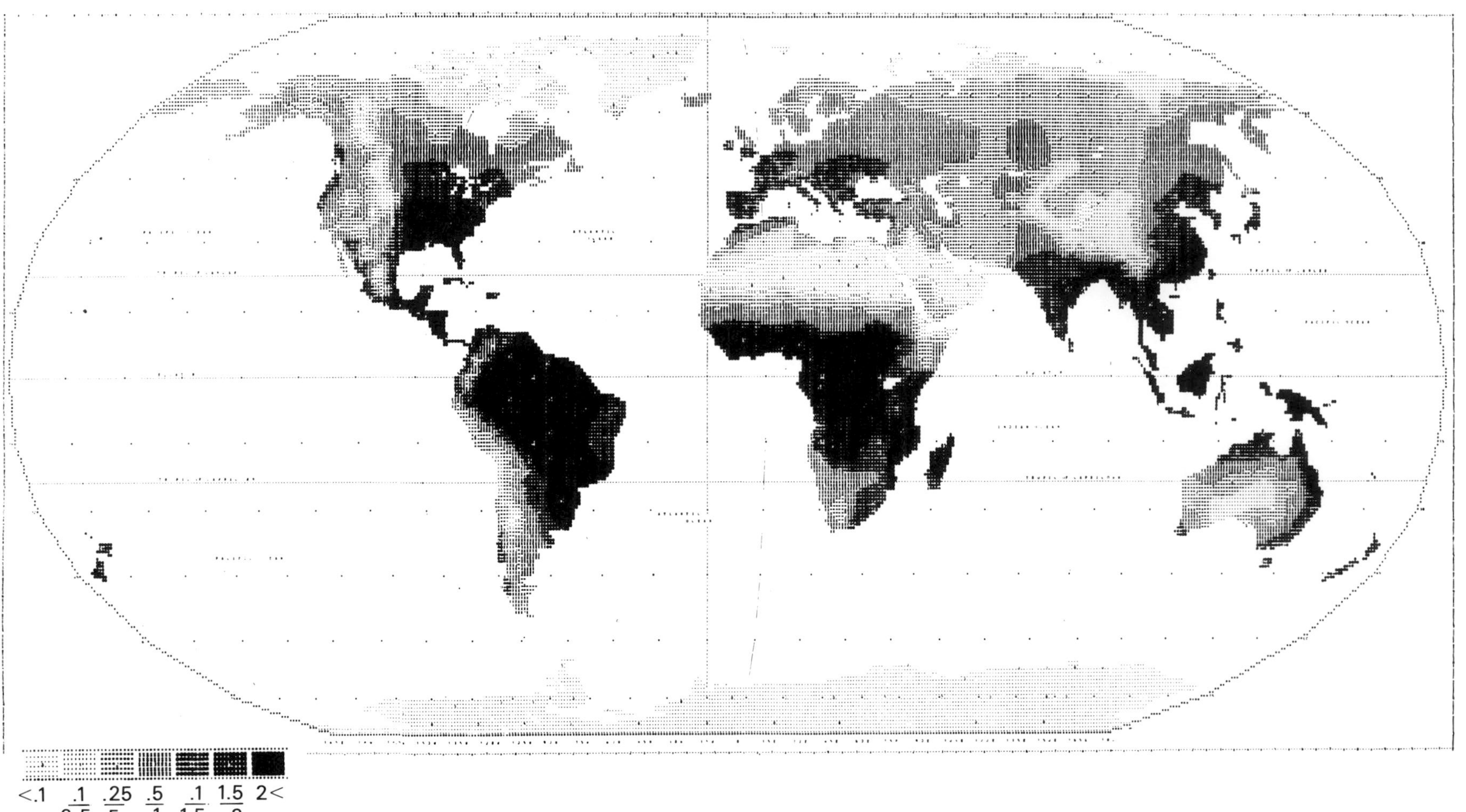

Figure 7.9 Computer-generated productivity map of the land areas of the world (after Lieth 1975).

impressed by the bulk of organic matter represented by the plants of an undisturbed forest? However, the net productivity of a trophic level is a measure of its efficiency, of the net rate of accumulation of this biomass. Of course, this rate may bear little or no relation to the biomass actually present at one moment in time (Macfadyen 1964; see Ch. 17).

As the higher trophic levels in our model (see Fig. 7.7) all depend on the autotrophs for their energy supply, the net primary productivity and its integral net primary production of the biosphere are important parameters. Estimates of the total net primary production of the land and ocean areas of the Earth are given in Table 7.1 (see also Fig. 7.9). If we take Lieth's (1971 and 1973) estimates of 100.2×10^{12} kg a^{-1} for the continents and 55×10^{12} kg a^{-1} for the oceans, these figures represent 178.24×10^{23} J a^{-1} fixed in net primary production on land and 109.2×10^{23} J a^{-1} in the oceans. The total net primary production of the globe is therefore 287.44×10^{23} J a^{-1}. This quantity of energy, fixed annually, is the equivalent of the total generating capacity of approximately 300 million 1000 mW power stations and compares with 2.2×10^{18} J released by one of the largest series of volcanic eruptions this century at Kamchatka in the Soviet Union in 1955–6.

Photosynthesis not only requires light, for it is also dependent on the availability of water and it is sensitive to temperature. As all three factors are far from constant over the Earth (see Ch. 4), it is not surprising that the global figures quoted above conceal a good deal of variation (Fig. 7.10). The

Table 7.1 Comparison of the various production estimates for the Earth (after Box 1975).

Area (10^6 km²)	NPP est. (10^{12} kg yr^{-1})	Method
LAND		
140	96 (38.4×10^{12} kg C)	planimetering Lieth's (1964) productivity map and checking it against annual, global CO_2 fluctuation
149	109.0	sum of estimates by means for major vegetation types
149	100.2	sum of estimates by means for major vegetation types
149	116.8	sum of estimates by means for major vegetation types
140.2	104.9	evaluation of Innsbruck productivity map (Lieth 1972)
140.2	124.5	evaluation of Miami model (Lieth 1972)
140.3	118.7	evaluation of Montreal model (Lieth & Box 1973)
149	121.7	sum of estimates by means for major vegetation types
OCEAN		
332	46–51 (23×10^{12} kg C)	sum of estimates by means for major zones
361	55.0	evaluation of major zones
361	43.8	evaluation of Oceans Productivity Map (Lieth & Box 1972)
361	55.0	sum of estimates of major zones

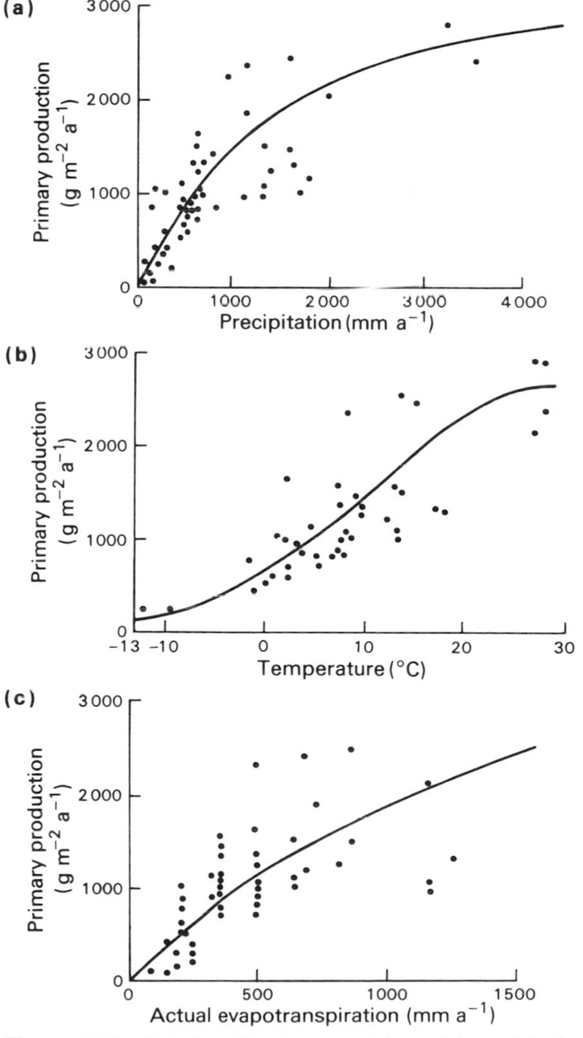

Figure 7.10 Relationship of productivity to (a) precipitation, (b) temperature and (c) evapotranspiration.

nature of this will be discussed further in Chapter 17, but it is worth noting here that even when such variation is taken into account considerable dangers remain in the use of global estimates of production. These have been lucidly reviewed by Newbould (1971), who counsels caution in what he calls the 'numbers game'. Even so, the estimates used in Table 7.1 are probably sufficiently valid at the level of generalisation of the ecosphere model used in this chapter.

However, one further point needs to be made about these figures. The production figures in Table 7.1 are expressed as weight of 'dry matter'. This is the weight after drying in an oven at 105°C until no further loss of weight occurs. In addition to organic compounds dry matter contains some inorganic material. Ignoring this we have the organic weight, which for most plants is between 75% and 95% dry weight, and of this only 45–48% is carbon. The remainder is mainly oxygen and hydrogen, but as we saw at the beginning of the chapter, it also contains small but significant quantities of other elements. Here, then, we will shift our perspective on the stored chemical energy, which the biomass of the trophic levels of the biosphere represents, to the composition of those same chemical compounds. By doing so we can recognise that the net production transferred from one trophic level to the next is also a transfer of materials. When a herbivore grazes a plant, it gains not only energy but also the elements it needs to build new cells and for growth, elements that were first fixed in organic compounds by green plants using the energy of sunlight. But where did these elements come from and where do they go to after the death of the plants or animals?

7.2.2 Transfer of matter in the ecosphere

Figure 7.11 shows the major pathways taken by the elements necessary for life as they are exchanged between the living systems of the biosphere and their non-living environment (see also Fig. 5.13). Although at first sight a complex diagram, it is still basically the simple compartment model of Figure 6.1, for the extent of the interaction of the biosphere with the atmosphere, hydrosphere and lithosphere delimited by these pathways helps to define the boundary of the ecosphere.

If the three kinds of pathway identified in Figure 7.11 are traced through the diagram, it will be found that they pass one into the other. Therefore, at least some of the elements travelling these routes are involved in enormous global cycles through the lithosphere, atmosphere and hydrosphere. Many of these pathways and the cycles they form have already been considered implicitly in Chapters 4 and 5. However, those chemical elements necessary to the functioning of living systems can be seen to take a detour, to follow an alternative pathway through the biosphere. Here too all three kinds of pathway converge to follow a common route as organic molecules through living systems.

The continuity of the movement of elements along these pathways is, however, broken by processes, the operation of which involves transfers of energy, as was stressed in Chapter 2. Again, many of these processes have been examined in Chapters 4 and 5 and are therefore already familiar. Some of the processes carry a positive and some a negative sign. These signs refer to the potential energy level of the materials involved in the process. If, after completion of the process, the material is at a lower energy level, then the sign is negative. If, on the other hand, the operation of the process increases the energy level, the sign is positive. The energy level in question refers in some cases to the potential chemical energy of the molecules of which the element is a part, as for example the decline in potential chemical energy from primary rock-forming minerals to weathering products during the process of weathering. In other cases there may be no change in chemical energy, but merely a change in potential energy by virtue of a change in relative position – as when rocks are uplifted.

It is by means of these processes that the one-way flow of energy, required by the laws of thermodynamics, is coupled into and powers the closed circulation of matter within and between the atmospheric, hydrospheric, lithospheric and biospheric systems. Here the distinction made in Figure 7.11 between the positive, energy-demanding (endergonic) processes and the negative, energy-yielding (exergonic) processes is critical. Without the uphill processes the cycles would not exist; the pathways would not close to turn full circle. From the point of view of the ecosphere, however, these processes have a further significance in that they function as rate-limiting mechanisms. This means that the rate at which they take place determines the speed with which materials flow along the pathways. For example, the rate at which weathering processes

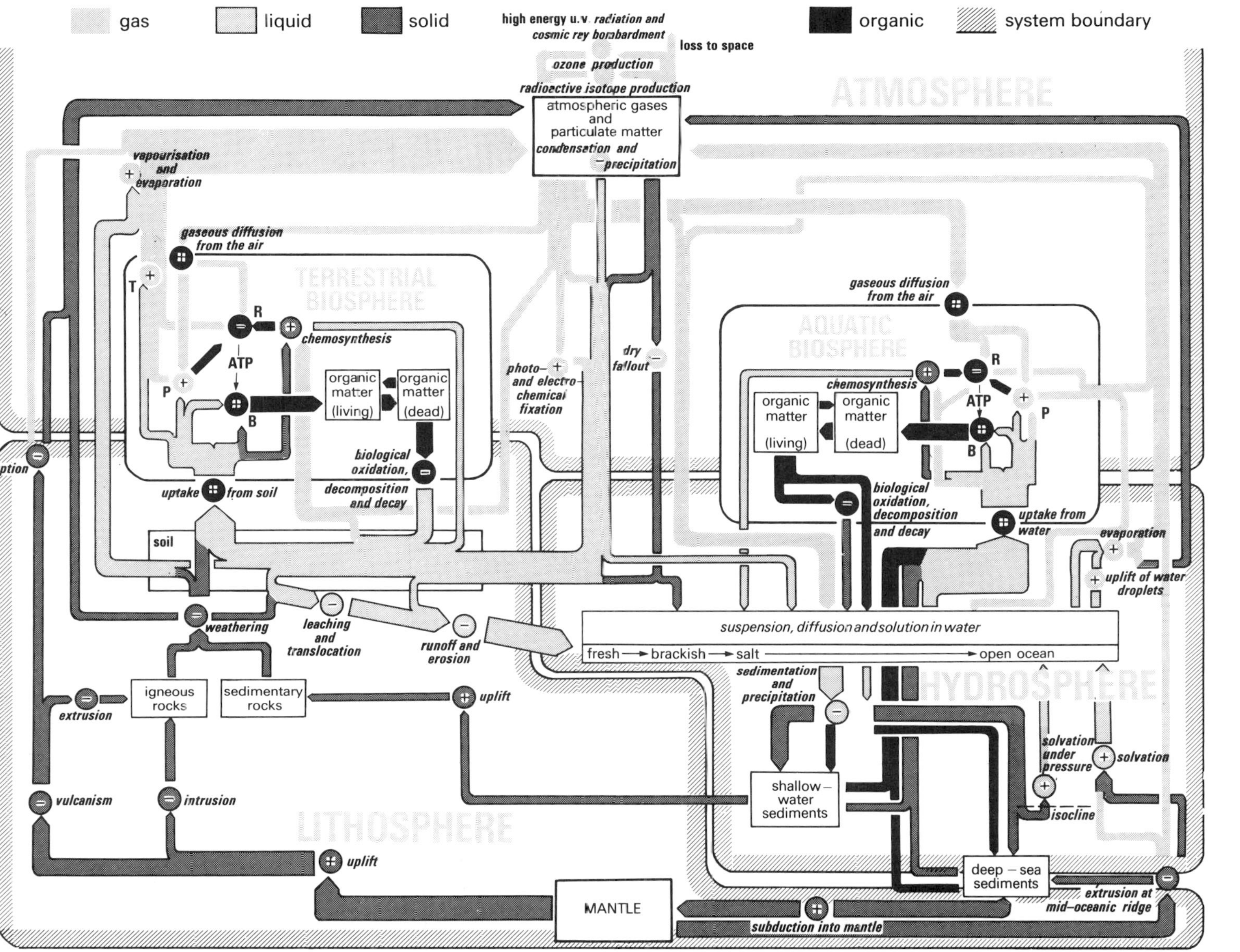

Figure 7.11 The transfer of matter in the ecosphere.

occur (see Ch. 11) determines the rate at which elements are made available for uptake by plant roots. Now some of these processes, evaporation for example, take place rapidly, while others, such as the subduction of ocean sediments into the mantle, take a long time. The speed of transfer between the different compartments in Figure 7.11 will, therefore, vary considerably. Again, from the ecosphere's point of view, some of these stores may act as sinks for some elements and effectively immobilise them. This is the case if the residence time of an element in a compartment is long and its cycling is slow. Here, however, our viewpoint is relative. This statement would be true at one scale for a critical element such as phosphorus, immobilised in undecomposed soil organic matter for a matter of years, and at another for those elements bound up in deep-sea sediments perhaps for millions of years.

For matter moving through these pathways to be raised to a higher energy level requires an input of energy, of course. In the atmosphere this input is radiant solar energy, as for example in the vaporisation of water during evaporation (see Ch. 4). In the lithosphere the energy source is both terrestrial energy in the crustal system and solar energy in the denudation system (see Ch. 5). An input of energy is also required for the transfer of elements at a relatively low energy level (oxidised) in the environment to a high energy level (reduced) in the cells of the living systems of the biosphere.

7.2.3 The energetics of matter transfer through the ecosphere

As we have seen, the organisms of the biosphere are structurally, at least, mainly carbon, hydrogen and oxygen (see Ch. 6). The dominant process that transfers these elements from the environment to living systems is, of course, the photoreduction of carbon dioxide to carbohydrate with the liberation of molecular oxygen from water during photosynthesis. From autotrophic organisms carbon, hydrogen and oxygen are passed to heterotrophs along the food chain. However, light is not the only source of energy necessary for photosynthesis. The carbon dioxide enters the leaves in response to diffusion gradients which require the expenditure of metabolic energy to establish and maintain while water molecules are lifted from the roots by the transpiration stream which is maintained by radiant energy.

The other elements essential to life enter the high energy route through living systems largely as ions in solution in water. They may be passively transported by the transpiration stream, but to enter plant cells, whether in response to diffusion gradients or by 'active ion pumps', involves the expenditure of energy ultimately derived from respiratory oxidation. These essential elements are only then available for the synthesis of organic compounds using ATP energy during biosynthesis. Like carbon, hydrogen and oxygen these elements are passed on to heterotrophs as the constituent atoms of the organic molecules of plant tissue consumed as food by animals.

Living plants and animals return to their environment some of the elements they have drawn from it. Carbon dioxide is released to the atmosphere during respiration, and oxygen during photosynthesis. Some bacteria, using nitrates or sulphates as a substitute for oxygen in respiration, reduce them to nitrogen gas (dinitrogen) and hydrogen sulphide respectively, both of which are released to the atmosphere. Yet other inorganic and organic compounds are excreted by living organisms as waste products. On death, however, the life-sustaining input of energy and matter stops and the spontaneous, if sometimes slow, breakdown of the complex molecules of the organism and its constituent cells begins. These processes of decay and decomposition are oxidation reactions and they proceed with a net loss of free energy and release the constituent elements of the organic molecules back into the environment as simple inorganic compounds. However, much of this dead organic matter forms the food supply of the decomposers. The elements it contains may recirculate through this compartment of the ecosphere many times before being released. Even then the more resistant organic residues may accumulate as, for example, soil humus or peat in terrestrial environments or as organic sediments in aquatic environments, so immobilising the elements they contain.

For all of the elements essential to living systems on land, including hydrogen and oxygen combined as water, the soil is the immediate store on which the biosphere draws and to which it returns these elements. The only exceptions are photosynthetic carbon dioxide and respiratory oxygen which are both obtained from and returned to the lower atmosphere. In aquatic environments the biosphere draws all elements from those

Box 7.1
NITROGEN FIXATION

Due in part to the very stable triple bond ($N \equiv N$) that links the two atoms of nitrogen in the dinitrogen molecule (N_2) of atmospheric nitrogen gas, it is normally very unreactive. However, biological nitrogen fixation succeeds in making the molecule reactive at normal temperatures and pressures. As with other biochemical reactions, this is only possible because the reaction is catalysed by an enzyme – in this case the enzyme **nitrogenase**. Although the mechanisms are as yet not fully understood, nitrogenase is known to be a complex of two proteins, both of which have attached iron and sulphur atoms. The larger of the two proteins also has two atoms of molybdenum per molecule. In the nitrogen fixation process an activated complex of the two proteins is formed with the nitrogen molecule bound to the molybdenum of the larger protein and ATP (see Box 6.4), as a monomagnesium salt is bound to the smaller protein. Electron transfer takes place between the iron atoms of the two proteins (a redox reaction) assisted by the energy from the ATP–Mg which in turn is converted to ADP and Mg^{++}. The result of these electron transfers is the splitting of the molecular bond in dinitrogen and the release of the reduced product of the enzyme reaction as ammonia (NH_3).

It is now known that a very large number of organisms contain nitrogenase and are capable of nitrogen fixation, but they are all primitive groups – bacteria and blue-green algae. They exist both as free-living organisms and in symbiotic relationships with higher plants (in the case of the legumes and *Alnus*-type associations of bacteria) and some lesser plants (as with the lichens where the blue-green algae *Nostoc* and *Calothix* grow within a fungal matrix). None of the 4×10^{12} kg of nitrogen gas in the atmosphere, or of the 2×10^{14} kg bound in rocks, is available to plants until it is fixed by these organisms – with the exception of the application of industrially produced nitrogen fertilizer – but even under agriculture the contribution of nitrogen-fixing micro-organisms is enormous.

Estimated N-input into plant crops in 1971–2 ($\times 10^9$ kg N) (Postgate 1978) is as follows:

Source of N	*UK*	*USA*	*Australia*	*India*
fixation in legumes	0.4	8.6	12.6	0.9
fixation by free-living micro-organisms	<0.04	1.4	1.0	0.7
take up from N fertilizer	0.6	4.9	0.1	1.2

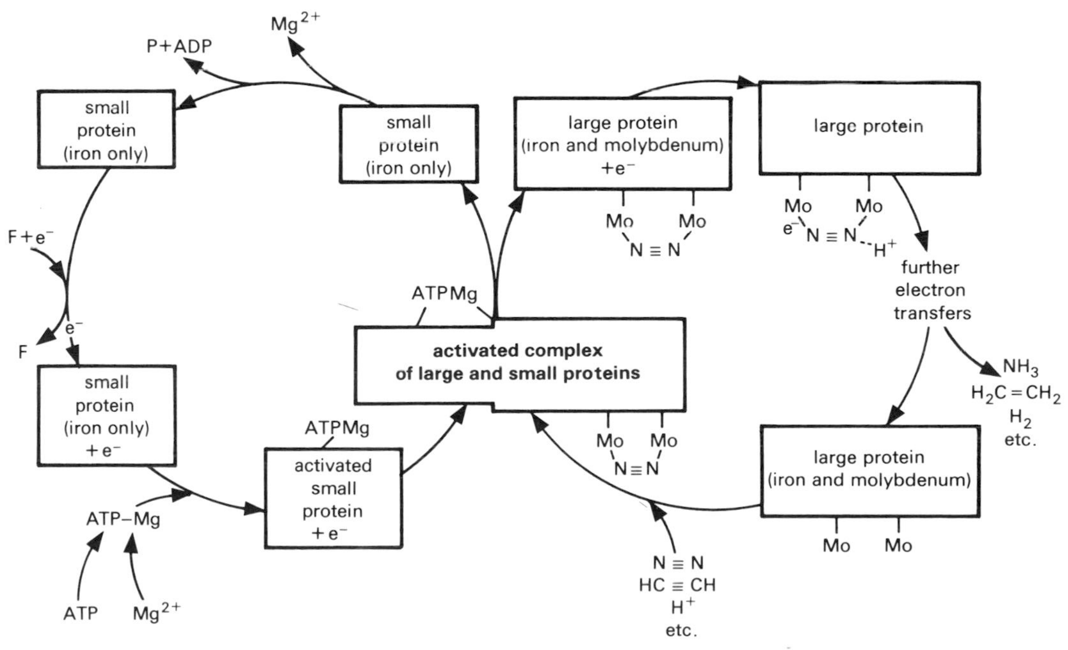

dissolved, suspended or diffused in water, even carbon dioxide for photosynthesis and oxygen for respiration, and returns them to the water.

If the rate of uptake from the soil or water is balanced by the rate of release of elements by decomposition, then the circulation rate is limited by the rate of energy flow from photosynthesis through autotrophic to heterotrophic organisms, i.e. the biosphere energy flux. On the other hand, this flow of energy, and hence the productivity of the biosphere, may itself be rate-limited if the rate of return of essential elements through decomposition is inadequate. For many elements this is indeed the case, for not only may they be immobilised in undecomposed organic matter but even when released they may be converted to a largely unavailable inorganic form. For example, in well aerated and alkaline calcareous soils the availability of copper, manganese and iron decreases as they are readily oxidised or precipitated in such environments and are not easily absorbed by plant roots.

These pathways and processes of transfer of elements between the living systems of the biosphere and their immediate environment can be thought of as the ecosphere loop of the circulation of materials depicted in Figure 7.11. This diagram, however, also shows us that the ecosphere loop is linked to the pathways of transfer of matter through the atmosphere, hydrosphere and lithosphere. Elements are lost from the soil, mainly via the denudation system (see Chs 5 & 10), ultimately to sediments, and in aquatic environments by precipitation and deposition, again ultimately to sediments. There are also elements lost from the ecosphere loop in a volatile or gaseous state. Carbon dioxide released during respiration by autotrophic, heterotrophic and decomposer organisms is returned to the atmosphere. The same is true for the nitrogen gas (dinitrogen) and hydrogen sulphide released by nitrate- and sulphate-reducing bacteria respectively, as well as for the oxygen evolved during photosynthesis.

The pathways by which the losses of these volatile elements, and of those which follow the denudation system, are made good differ fundamentally not only in the routes taken but also in timescale involved. For these oxidised gaseous elements to be returned to the biosphere requires the investment of energy to raise them to a higher energy level or to a chemically reduced form. In the case of nitrogen, ultraviolet radiation or lightning may provide sufficient energy, if only for a fraction of a second, for nitrogen to combine with oxygen or with the hydrogen of water. The oxides and hydride of nitrogen so formed dissolve in rainwater to reach the soil or the oceans as nitrate and ammonium ions available for absorption by plants. The main route, however, is the biological fixation of gaseous nitrogen by both free-living and symbiotic bacteria and algae (Box 7.1), which ultimately make it available to higher plants as nitrate (Fig. 7.12).

Oxygen is coupled to respiratory oxidation as an electron acceptor and is reduced to water. In the case of sulphur, it enters the cells of plants as an inorganic sulphate ion in a similar way to nitrate, but it would not be recycled through the atmosphere without the bacterial reduction to sulphides under anaerobic conditions. When not precipitated as iron sulphide, sulphur escapes to the atmosphere as H_2S which is then reoxidised to sulphate. Finally, carbon as CO_2 is photochemically reduced to carbohydrate during photosynthesis, which is where we started. The cycling of these gaseous elements between the atmosphere and the ecosphere depends, therefore, on four reduction reactions. All require an investment of energy which, in green plants and autotrophic bacteria and algae, is derived from an external source, but in the case of the heterotrophic bacteria involved in nitrogen and sulphur reduction must be supplied from food energy.

All of the remaining elements essential for life are lost from the ecosphere by leaching and they enter the denudation system, eventually reaching the ocean sink. Their recycling only becomes possible over a long geological timescale, following sedimentation, lithification, uplift, erosion and weathering. The only way that this route can be short circuited is for mineral dust from the land surface, or droplets of sea water as spray, to enter the atmosphere where they are mixed and circulated. The elements contained eventually fall out, or are washed out by precipitation, thereby completing the cycle. However, they only enter the biosphere when taken up by plants or ingested by animals and incorporated in organic compounds using energy originally fixed during photosynthesis (Fig. 7.12).

Carbon cycle

mixing & convection
CO2 in upper atmosphere
wind
CO2 in oceans
CARBON DIOXIDE in lower atmosphere
respiration
PLANT carbon fixation in photosynthesis (& by chemosynthetic organisms)
ANIMALS grazing - predation pathway
AIR
SOIL
volcanic eruption
DEAD ORGANIC RESIDUES organic carbon
weathering
combustion
UNAVAILABLE C peat, lignite, coal, carbonate deposits, chalk, limestone
immobilisation
microbial decomposition
crustal carbon

Oxygen cycle

ionisation
OZONE SCREEN $O_2 + O \rightarrow O_3$
high energy input - ultra violet radiation
H_2O
ATMOSPHERIC OXYGEN gaseous O_2
CO_2
CO_2
photosynthesis
respiration
respiration
volcanic eruption
PLANTS
ANIMALS
AIR
SOIL
oxidation
reduction
SILICATES etc in crust
UNAVAILABLE OXYGEN in sediments, as oxides carbonates etc
immobilisation in weathering & decay reactions

Nitrogen cycle

DINITROGEN in lower atmosphere
electro- & photochemical N fixation
biological N fixation
bacterial denitrification
N compounds in precipitation
PLANT amino acid & protein synthesis
ANIMALS grazing - predation pathway
AIR
SOIL
symbionts bacteria fungi
free-living bacteria & algae
DEAD ORGANIC RESIDUES organic N
volcanic eruption
NITRATES
leaching
microbial decomposition
AMMONIA NH_3 AMMONIUM NH_4^+
nitrification (nitrate bacteria)
NITRITES
nitrification (nitrite bacteria)
leaching
N in the oceans
N in sediments and in the crust

Phosphorus cycle

atmospheric phosphorus particles
recycled by airborne dust and sea spray
PLANT amino acid & protein synthesis
ANIMALS grazing - predation pathway
AIR
SOIL
DEAD ORGANIC RESIDUES organic & inorganic P
precipitation
fungal mycorrhiza
direct root uptake
solid inorganic phosphorus
AVAILABLE P orthophosphate in solution
microbial decomposition (mineralisation)
exchangeable P on soil colloids
immobilisation by micro-organisms
leaching
weathering
inorganic P in the crust & sediments
PHOSPHORUS in the oceans

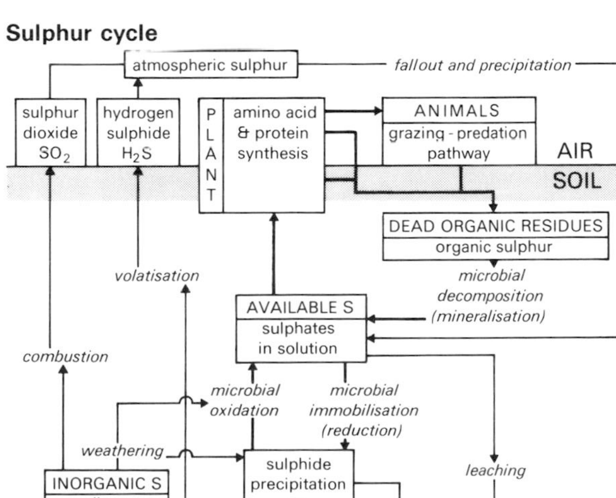

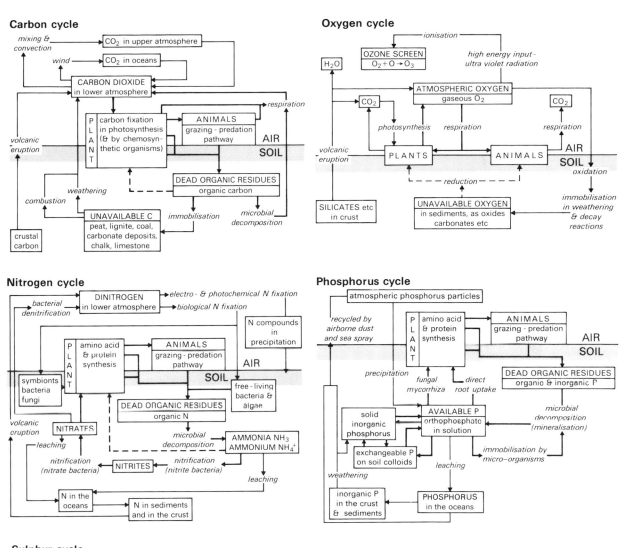

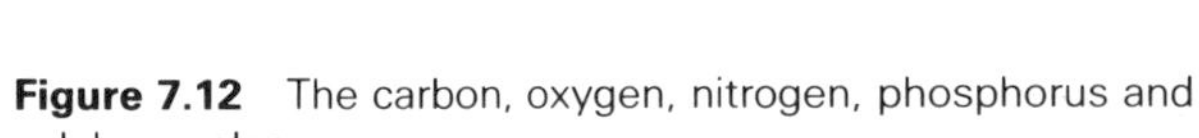
Figure 7.12 The carbon, oxygen, nitrogen, phosphorus and sulphur cycles.

7.3 The ecosphere and the evolution of life

We are now in a position to address again those questions raised in Chapter 6 about the effects that the evolution of life has had on the development of our planet. There is ample evidence to suggest that life evolved in an environment very different from that which exists today. The atmosphere had a secondary origin from the gases evolved from the Earth's interior by volcanism and it lacked free oxygen. Furthermore, as the ozone shield did not exist, destructive ultraviolet radiation was received at the Earth's surface. In this atmosphere, and particularly in the sterile oxygen-free seas, molecules would have been continually circulated and mixed, and collisions between them may well have facilitated the production of the precursors of 'organic molecules'. That such molecules could form without life as we know it has been demonstrated by Miller (1953) in what is now a classic experiment. This involved circulating methane (CH_4), ammonia (NH_3) and water vapour – all of which were present in the primitive atmosphere – over an electric spark for a week. The spark was to simulate the energy input of electrical discharges in the atmosphere. At the end of the period, glycine and alanine (the two simplest amino acids) and other 'organic' compounds had been synthesised.

At some stage the spontaneous but ordered arrangement of these precursor molecules reached a complexity which could be considered to be the first living organism. From that moment on, there began an interaction between life and its inorganic environment which was to change that environment completely. This interaction was to have three major steps.

The first organisms were undoubtedly heterotrophs with no option but to live on the 'organic' molecules from which they themselves had evolved. As we have seen when considering the heterotrophic cell model, such a strategy implies living on capital, while in the absence of oxygen the only process by which the energy of the food molecules could be released was fermentation. Fermentation is a partial oxidation, essentially similar to the first stage of the respiration process outlined in 6.3.1. Compared with total oxidation, however, fermentation is extremely inefficient (Fig. 7.13), while the products are waste which must be disposed of.

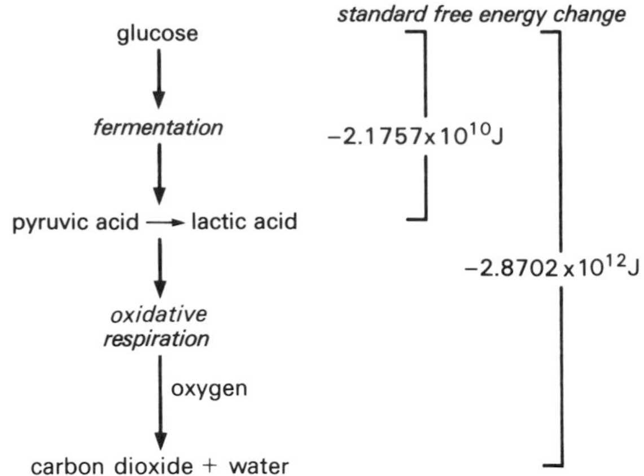

Figure 7.13 The energy yield from fermentation and from oxidative respiration (after Lehninger 1965).

Fortunately, these primitive forms of life developed the capacity of manufacturing their own organic food molecules before they exhausted the capital on which they had been living, i.e. pre-existing 'organic' molecules in the primaeval seas. To do this they utilised carbon dioxide (one of the waste products of fermentation), which had been accumulating in the atmosphere and oceans. Initially they derived the energy necessary for the synthesis of organic carbon compounds from the bond energy of inorganic compounds, but *ca* 3×10^9 years ago the first organisms appeared which were capable of utilising light energy from the Sun. Photosynthesis had arrived.

At first, however, the impact of photosynthesis was limited. These early autotrophs were not equipped to deal with free oxygen and they may have acquired electrons for photosynthesis from substances other than water. Even when water was split during photosynthesis to release molecular oxygen, it has been suggested that the early photosynthetic organisms used inorganic compounds as oxygen acceptors. The potential destructive effects of free oxygen were therefore avoided by these anaerobic organisms. Eventually, protective enzymes were evolved and life became not only autotrophic but for the first time aerobic as well, and oxygen began to enter the atmosphere in gradually increasing quantities (Table 7.2).

The presence of oxygen had two important effects. First of all the early photosynthesisers were limited by fermentation in the energy yield they could win from the fuel molecules they produced. The presence of free oxygen allowed the

Table 7.2 The evolution of the atmosphere and the fossil record.

Period	Age in years	Solar radiation reaching Earth	Terrestrial atmosphere H_2 NH_3 CH_4 N_2 H_2O CO_2 O_2	Metabolic systems	Structural fossils
birth of the Earth	5.0×10^9				
	4.5×10^9			abiogenic production of organic compounds; origin of life?	
early Precambrian	3.5×10^9	visible light; longwave ultraviolet; shortwave ultraviolet		anaerobic heterotrophs; anaerobic heterotrophs using solar energy to produce ATP	Prokaryotic organisms; Bacteria
	3.3×10^9			chemical remains of bacterial-type systems	Blue-green algae – coccoid
	3.2×10^9			water-dissociating organisms	
	3.1×10^9		H_2	anaerobic photoautotrophs using solar energy to produce ATP	
	2.8×10^9			microbial photoautotrophic aerophiles producing more ATP	Stromatolites; Blue-green algae – filamentous
	2.7×10^9				
	2.3×10^9				
	2.2×10^9				
middle Precambrian	2.0×10^9				
	1.9×10^9			O_2 gradually becoming more tolerable constituent of atmosphere	Eukaryotic organisms
	1.6×10^9				
	1.3×10^9			aerobic respiration established	
late Precambrian	1.1×10^9		NH_3 CH_4		Metazoans
	1.0×10^9				
	5.8×10^8		1% PAL	organisms with modern type of metabolism and sexual reproduction	Invertebrates; Higher algae – marine evolution
Cambrian					
	4.0×10^8		10% PAL		Vascular plants
Palaeozoic and Devonian younger	3.4×10^8		20% PAL	O_2 production by photosynthesis at maximum	Terrestrial animals
Carboniferous			N_2 H_2O Co_2 O_2		

- - - -, possible occurrence; — — —, probable occurrence; ———, established occurrence; PAL, present atmospheric level.

development of complete oxidative respiration which, by releasing a far greater amount of energy from the fuel molecules, enabled the biosphere to invest the surplus in an accelerated process of evolution, diversification and accumulation of biomass. Here the second effect of free oxygen becomes important, for ultraviolet radiation caused some of the molecular oxygen to dissociate and the subsequent recombination of the highly reactive atomic oxygen formed ozone. So the developing ozone screen high in the atmosphere (Chs 3 & 4) absorbed the lethal ultraviolet waves of solar radiation and this allowed life to emerge from the protective sediments and waters of the oceans and to colonise the land surface.

The changes in the chemistry of the atmosphere and oceans which accompanied these milestones in evolution had profound effects on the lithosphere. The presence of free oxygen in the surface waters of the Earth changed the nature of sedimentary rocks by altering the sedimentation environment and affecting the solubility of many inorganic compounds; iron-rich sediments, for example, first appeared 1.8×10^9 years ago. The immobilisation of carbon in organic compounds and in insoluble carbonates by the photosynthetic withdrawal of carbon from atmospheric carbon dioxide and from soluble bicarbonates not only reduced the carbon dioxide content of the atmosphere but produced most of the limestones of the stratigraphic column. Today the carbon dioxide concentration of the atmosphere (if we ignore the burning of fossil fuels) is in equilibrium with the photosynthetic system of the biosphere. In organic compounds the immobilisation of elements other than carbon has also affected the processes of erosion and denudation, and the sediment loads delivered to the oceans are very different from those which would arrive in the absence of terrestrial organisms. Finally, weathering and soil formation are both processes whose character has been determined by the biosphere.

The organisms of the biosphere, therefore, inhabit an environment which they themselves have largely produced. Why this should have come about can be understood if we return to the simple systems model of Figure 6.1b. The appearance and evolution of life can be regarded as an apparent reversal of the trend to maximum entropy of the universe. As we have seen, however, for the central compartment in our model – the biosphere – to have reduced its internal entropy throughout evolution there must have been a corresponding gain in entropy by the surroundings. In our model the surroundings can be thought of as that part of the atmosphere, hydrosphere and lithosphere included in the ecosphere with which the organisms of the biosphere exchange matter and energy. Therefore, the changes discussed above can all be seen as an expression of a progressive gain in entropy by the surroundings of the biosphere.

Further reading

Many general biogeography texts give an introductory coverage of the topics treated in this chapter, for example:

Collinson, A. S. 1977. *Introduction to world vegetation*. London: George Allen & Unwin.

Watt, D. 1971. *Principles of biogeography. An introduction to the fundamental mechanisms of ecosystems*. Maidenhead: McGraw-Hill.

At a high level:

Odum, E. P. 1971. *Fundamentals of ecology*, 3rd edn. Philadelphia: Saunders.

Part C

Open system model refined: environmental systems

A number of important conceptual ideas have emerged from the pursual of a systems approach to the natural environment in the first three sections of this book. First of all we have realised that all natural systems are open systems, but more than this we have seen that they exist as components of cascades through which matter and energy flow. The output of one system forms the input to the next. The solar radiant energy which is output from the atmosphere to the Earth's surface is the energy input to the autotrophic cells of the biosphere. Part of the biomass of producer organisms is the energy and nutrient input to the consumer organisms of the trophic model of the biosphere. The precipitation that falls from the atmosphere is the input of water to the denudation system.

Within any component system of such cascades, however, part of the flow of matter and energy through the system is transferred temporarily to stores. Energy and chemical elements flowing through the biosphere are diverted temporarily to the biomass store of the organisms of the different trophic levels. The flow of geothermal heat energy and geochemical elements through the crustal system is diverted to the store of potential and chemical energy and rock-forming materials in the uplifted relief of the Earth's surface. Some energy or matter will entirely bypass these stores, but this, together with the transfer of matter and energy into and out of storage and between stores, represents the throughput of the component systems of the cascade.

We have also seen that the destiny of energy and matter as they cascade through these systems is fundamentally different. As each energy transfer is completed, part of the original input of energy is dissipated, in accordance with the Second Law of Thermodynamics, as unavailable heat energy ultimately radiated to the energy sink of space. Energy flow is, therefore, a one-way process. When we consider the transfer of matter, we have found that ultimately it turns full circle and the cascades of matter in our global models have become cycles. The completion of these cycles, however, only occurs because they are interlocked with and driven by the cascade of energy.

In Part C we are going to increase the resolution and fidelity of our models of natural systems considerably and alter both our spatial and temporal perspectives. From broad global vistas, with one eye on geological time, we shall move to the scale of the local landscape which is both familiar and accessible. We shall wade in streams, walk through woodlands and dig through leaf litter to the soil beneath. We shall don our raincoats as a frontal depression approaches and we shall feel a sea breeze on our faces. For the most part, we shall be concerned with much shorter spans of time – with the day, the year and perhaps the century. Nevertheless, we shall still be concerned with open systems functioning as parts of energy and mass cascades. Although it is possible to consider each type of cascade separately as the solar energy cascade, the hydrological cascade and the debris cascade, for example, we shall define the component systems of these cascades and examine the ways in which the throughputs of matter and energy are interlocked as they cascade through and between these systems. As our models of these systems increase in their level of resolution we shall identify nested hierarchies of systems operating at different spatial and temporal scales. For example, the denudation system is modelled in terms of the catchment or drainage basin, but within it we shall recognise weathering, slope and channel subsystems, and within them further subsystems such as the soil aggregate breakdown/raindrop impact system on a slope or the chemical reaction system in rock weathering. Primary productivity of the ecosystem will be considered at the scale of the individual leaf, the plant and the entire vegetation canopy, while in the atmosphere we shall recognise secondary and tertiary circulation systems within the general pattern of global circulation.

IV ATMOSPHERIC SYSTEMS

8 The atmosphere and the Earth's surface

8.1 Introduction

The lowest 10 km of the atmosphere – the **troposphere** – are characterised by turbulent mixing of constituents and exchange of gases, liquids and solids, as well as heat energy, between the air and the Earth's surface. Water molecules, for example, are transferred from surface to air and air to surface. Similarly, materials such as dust and soot are derived from the surface, to which they eventually return. Heat energy transferred from the surface into the air (see Ch. 3) is convectionally mixed through the troposphere. A net transfer back to the surface tends to engender stability in the lower troposphere which inhibits convectional mixing.

The continuous exchange of mass and energy between the Earth's surface and the atmosphere (and within the atmosphere) represents the functioning of circulatory systems. In these, the movement of mass is initiated and sustained by inputs of energy. While these inputs are ultimately derived from the solar energy cascade, it is possible to be a little more specific as to their origins within the Earth–atmosphere system.

An uptake of heat energy from the Earth's surface in the form of sensible heat or latent heat (see Ch. 4) increases the potential energy of the atmosphere which, in a circulation system, is converted into the kinetic energy of motion. The motion of large volumes of the lower atmosphere is a result of differences in potential energy arising from simple contrasts in sensible heat content, expressed in terms of horizontal temperature gradients, or latent heat content, or of a combination of the two.

In examining the general circulation of the atmosphere (see Ch. 4), we made broad generalisations concerning latitudinal variation in potential energy. If, however, we are to consider circulation systems which operate over smaller temporal and spatial scales, we must first look a little more closely at the characteristics of the Earth's surface which affect the exchanges of mass and energy between it and the atmosphere. The most important of these are its physical properties, its roughness and topography.

8.2 The physical properties of a surface

Shortwave radiation incident upon a surface may be absorbed, reflected or transmitted. The proportion reflected depends upon the shortwave reflectivity (the **albedo**) of the surface and upon the wavelength and angle of incidence of solar radiation.

The albedo values in Table 8.1 are for wavelengths in the solar spectrum and are generalised for a range of solar elevations. Snow and peat are at opposite ends of the range of albedo values for natural surfaces. Freshly fallen snow may reflect 95% of incident shortwave radiation, while for the moorland peat this is only 10%. Weller and Holmgren (1974) have provided a useful illustration of this contrast over tundra surfaces in Alaska. Because of its dark colour, the tundra surface has a relatively low albedo (between 15 and 20%), but after snowfall this rises rapidly to over 80% (Fig. 8.1).

For other surfaces, less drastic but no less significant fluctuations in albedo may occur, some of which are related to moisture content. Many

Table 8.1 Albedo of various surfaces (wavelengths less than 4.0 μm) from Monteith 1973, Sellers 1965, Lockwood 1974, Oke 1978).

Surface	albedo
SNOW	
fresh snow	0.80–0.95
old compacted/dirty snow	0.42–0.70
ICE	
glacier ice	0.20–0.40
WATER	
calm, clear sea water	
solar elevation 60°	0.03
30°	0.06
10°	0.29
SOILS	
dry, wind-blown sand	0.35–0.45
wet, wind-blown sand	0.20–0.30
silty loam (dry)	0.15–0.60
silty loam (wet)	0.07–0.28
peat	0.05–0.15
PLANTS	
short grass (0.02 m)	0.26
long grass (1.0 m)	0.16
heather	0.10
deciduous forest (in leaf)	0.20
deciduous forest (bare)	0.15
pine forest	0.14
field crops	0.15–0.30
sugar beet (spring)	0.17
sugar beet (early summer)	0.14
sugar beet (mid-summer)	0.26
MAN-MADE	
asphalt	0.05–0.20
concrete	0.10–0.35
brick	0.20–0.40

soils tend to have lower albedo values when moist, and over plant surfaces there may be changes in albedo relating to the annual growth cycle.

In the case of water, both **reflection** and **refraction** take place at its surface, with radiation penetrating into the water body. The amount of shortwave radiation reflected depends largely upon the angle at which it strikes the surface. When solar elevation is low, reflection is of the order of 40%, while for high solar elevation, reflection is relatively small at 3% (Table 8.1). In the latter case, much of the radiation is transmitted

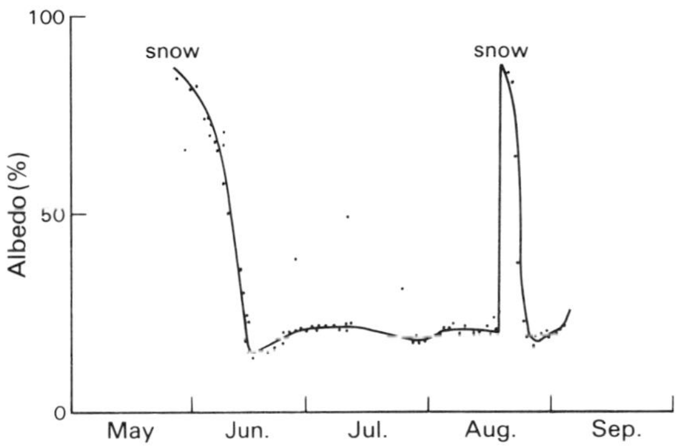

Figure 8.1 The albedo of the tundra surface (Weller & Holmgren 1974).

into the water and is progressively absorbed by it. In clear waters, by a depth of only 5 m, 70% of radiation has been absorbed, mainly in the infrared wavelengths. In turbid waters 70% has been absorbed by a depth of less than 1 m.

For all surfaces the absorbed solar radiation is redistributed by radiation, conduction and convection and the relative importance of these processes varies according to the physical character of the surface. In the absorption and emission of longwave radiation, for example, we have already seen that natural surfaces act as grey rather than black bodies (see Ch. 3).

The rate at which heat is transferred downwards into the subsurface is directly related to the nature and efficiency of the distribution mechanisms. In solids, heat is redistributed by conduction, and the rate at which this takes place depends upon

thermal conductivity (Box 8.1). Most naturally occurring materials, such as soil and rock, have low conductivities and, in relation to commonly used metals (Table 8.2), can be regarded as insulators rather than conductors. The penetration of heat is, therefore, to relatively shallow depths. For example, on a clear summer's day, a typical diurnal range of temperature at a depth of 0.01 m in a sandy soil may be of the order of 33 C°, while at a depth of 0.03 m it is only 17 C° and at 0.3 m it is less than 1 C°. The greatest rates of heating and cooling are experienced in the upper layers of the soil.

The rate at which temperature changes take place in a material depends upon its **thermal capacity** (Box 8.1). For example, it requires less than 1.2×10^3 J of heat energy to raise the temperature of 1 m^3 of still air by 1 C°, while to achieve a similar increase in soil temperature, it requires 2.5×10^6 J. Conversely, when air is cooled by 1 C° it releases less heat than soil cooled by the same amount. Thermal capacity affects the amount of heat energy transferred into and out of the subsurface store.

A vegetation canopy presents to the atmosphere not a simple solid/fluid interface but an ill defined transition zone between free air and air contained in pockets amongst the foliage of the plants. Both heat and moisture are transferred across this zone by the movement of air. The effect a vegetation cover has on soil surface temperatures beneath it and air temperatures within it is determined

Box 8.1

HEAT PARAMETERS

Thermal capacity is the amount of heat required to raise the temperature of unit volume of a substance by one degree. It is given by:

$$s = \rho \text{ (density)} \times c \text{ (specific heat)}.$$

Units are joules per cubic metre per degree Celsius (J m^{-3} °C^{-1}).

Thermal conductivity determines the rate at which heat flows through a substance and is defined as the rate of flow of heat through a unit area of plate of unit thickness when the temperature difference between the faces is unity.

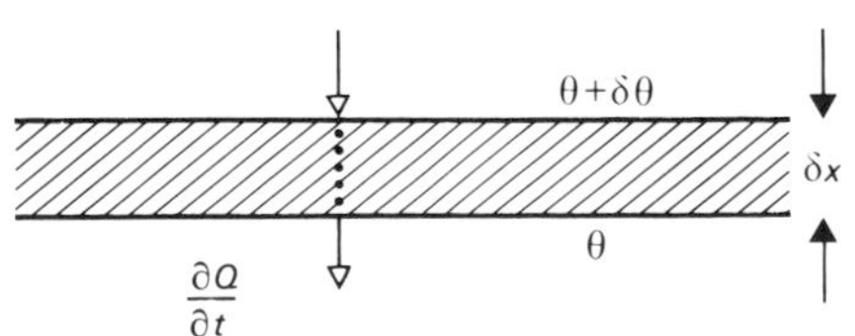

Under steady-state conditions, ∂Q units of heat flow through a plate of thickness ∂x, whose faces differ in temperature by $\partial\theta$, in time ∂t. The rate of flow of heat ($\partial Q/\partial t$) is related to the temperature gradient ($\partial\theta/\partial x$) and thermal conductivity (k) by

$$\frac{\partial Q}{\partial t} = -k\frac{\partial \theta}{\partial x}.$$

Units of k are Watts per metre per degree Celsius (W m^{-1} °C^{-1}).

Thermal diffusivity: a major problem in soils is that steady-state conditions are rarely achieved. An alternative parameter, thermal diffusivity (α), is used which is given by:

$$\alpha = k/s.$$

Units are metres squared per second (m^2 s^{-1}). Thermal diffusivity, for a homogeneous medium, defines the rate at which temperature changes ($\partial\theta/\partial t$) take place:

$$\frac{\partial \theta}{\partial t} = \frac{\partial^2 \theta}{\partial x^2}\alpha.$$

Table 8.2 Thermal properties of selected materials.

	Thermal conductivity $W\ m^{-1}\ °C^{-1}$	Thermal capacity $J\ m^{-3}\ °C^{-1} \times 10^6$
still clear water	0.57	4.18
pure ice	2.24	1.93
still air	0.025	0.0012
fresh snow	0.08	0.21
moist sand	2.20	2.96
dry sand	0.30	1.08
moist peat	0.50	4.02
dry peat	0.06	0.58
iron	87.9	3.47
granite	4.61	2.18

largely by the type of plant and the number of individual plants growing within a certain area. Should the plant cover density be relatively sparse, the soil surface is not shielded from direct radiant energy exchange with the free atmosphere. If, however, dense foliage does shield the surface, intracanopy radiant energy exchanges are largely responsible for canopy and subcanopy temperature variation.

In dense woodland as little as 10% or less of solar radiation may reach the underlying soil surface, mainly as diffuse radiation and in selected wavebands transmitted through leaves. Most incoming solar radiation is absorbed in the canopy (Fig. 8.2). The absorption of this radiation by the leaves and twigs of the woodland canopy raises their own surface temperature. The longwave radiation emitted from these surfaces is absorbed in the air above the trees and by the air pockets within the foliage. The whole of the canopy space then acts as an emitter of longwave radiation, some of which is transmitted to the trunk space and the soil. Heating of this lower zone within the vegetation is, therefore, not by direct absorption of solar radiation.

Figure 8.3 illustrates the effect of these radiant energy exchanges upon net radiation at various

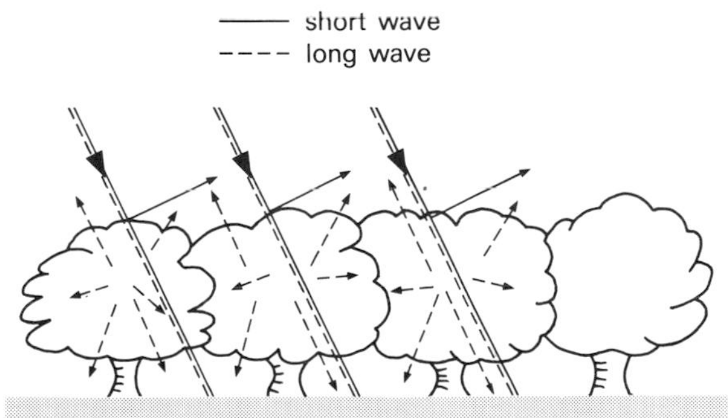

Figure 8.2 Model of radiant energy exchanges in a deciduous woodland.

(a)

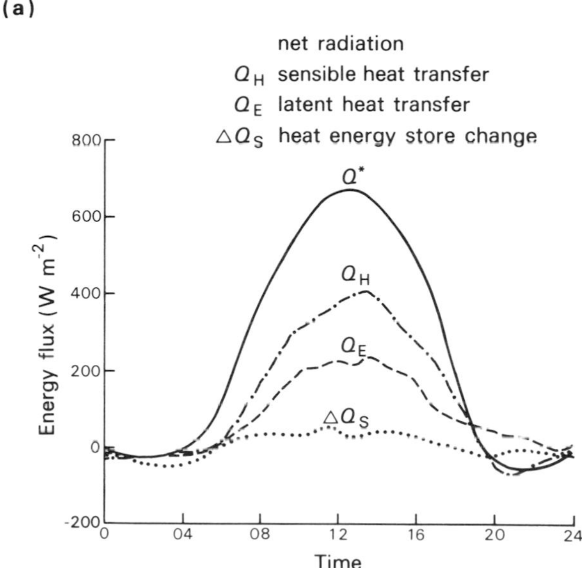

(b)

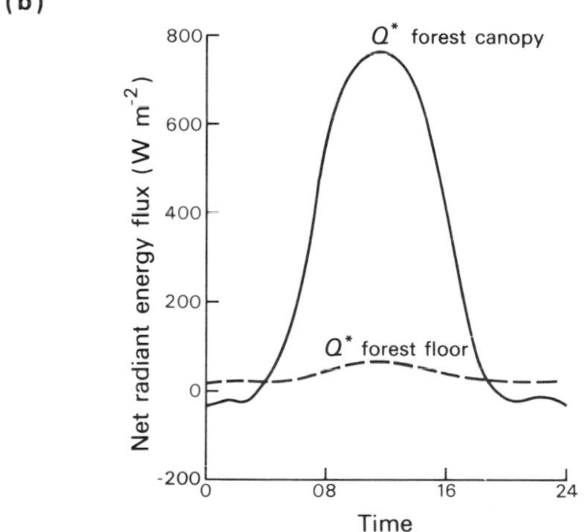

Figure 8.3 Net radiation and heat-energy balances in a forest canopy. (a) Diurnal energy balance of a Scots and Corsican Pine forest at Thetford, England (after Oke 1978) and (b) diurnal variation in net radiation in a spruce forest (after Lee 1978).

levels within a woodland. Above the canopy there is a large diurnal variation with Q^* reaching a relatively large positive value around mid-day and falling below zero for much of the hours of darkness. At the woodland floor, however, Q^* varies little over the 24-hour period and has a large long-wave component (Table 8.3). Of the net radiation available at the top of the canopy, most is transferred as either latent heat or sensible heat into the air above. Only a small amount is exchanged with the canopy heat store (Fig. 8.3).

A marked contrast in thermal properties exists between soil and water surfaces and between the land and the oceans (see Table 8.2). The difference in thermal conductivity is small, but in thermal capacity it is relatively large, that of the oceans being approximately twice that of the land. Cooling by 1 C° releases 4.18×10^6 J of heat energy from water, compared to 2.5×10^6 J from an average soil.

Because it is a fluid, water also redistributes heat by convection. As heating is from above, most of this heat redistribution is by forced convection by mechanical mixing in the upper layers of the water body. At temperatures below 4°C where water reaches its maximum density, cooler yet less dense water rises to the surface in free convectional mixing. By these mechanisms, heat is transferred away from the surface and redistributed through the water body. The significance of this is illustrated by the annual temperature regimes beneath ocean and ground surfaces (Fig. 8.4). While annual variation in terrestrial temperature is negligible at a depth of only 10 m, in oceans there is little difference between temperatures near to the surface and those at a depth of 25 m.

Table 8.3 Radiant energy balance above and within a forest.

	Above		Forest floor	
	Daily total (MJ m^{-2})	Relative to net radiation flux	Daily total (MJ m^{-2})	Relative to net radiation flux
K↓	18.2	3.19	1.0	3.33
K↑	1.8	0.32	0.1	0.33
K*	16.4	2.87	0.9	3.00
L↓	24.6	4.32	29.5	98.33
L↑	35.3	6.19	30.1	100.33
L*	−10.7	−1.87	−0.6	−2.00
Q*	5.7	1.0	0.3	1.0

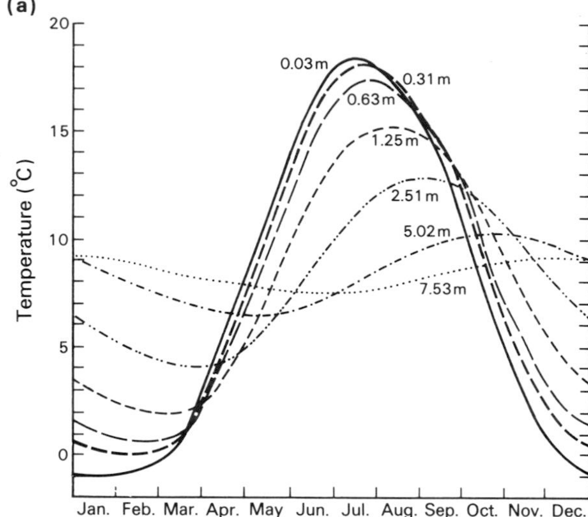

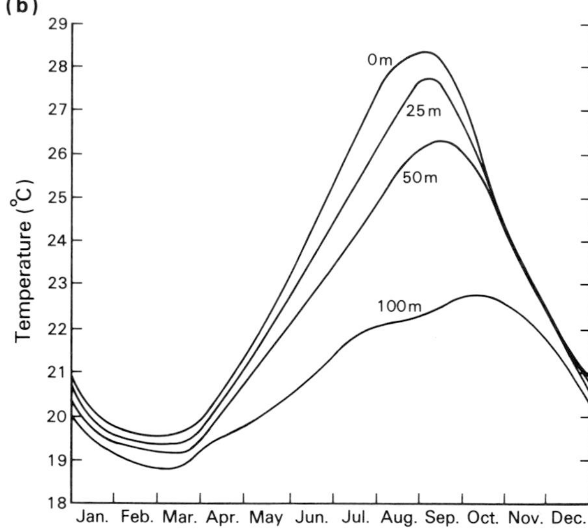

Figure 8.4 (a) Annual variation in Earth temperature at Königsberg (after Schmidt & Leyst, from Geiger 1965). (b) Annual variation in sea temperature off the south coast of Japan (after Sverdrup 1945, from Harvey 1976).

8.3 Surface roughness

The relationship between mean wind speed ($\bar{u}$) and height (z) above a surface approximates to the logarithmic form:

$$\bar{u} = \text{constant} \times \log(z/z_0). \qquad (8.1)$$

In this equation z_0 is the roughness of the surface, referred to as its roughness length or roughness parameter. The values of z_0 for various surfaces are given in Table 8.4. A particularly significant contrast in roughness is that between a water surface (z_0 is approximately 0.5×10^{-5} m) and the land surface (z_0 is approximately 0.1×10^{-2} m), the latter exerting considerably greater frictional drag on air flowing across it (Fig. 8.5). In the case of the sea surface, the friction layer – defined by the height at which gradient wind velocities are attained – is 270 m deep over the water surface. Over the much rougher rural land surface, this increases to 400 m, or 520 m over urban areas.

Because of the penetration of airflow into a vegetation canopy, wind velocities do not tend towards a zero value at the canopy surface. The form of the velocity profile depends on the morphology of the canopy surface and, in particular, its permeability to airflow. For example, in the case of a stand of conifers (Fig. 8.6) the foliage of the canopy presents resistance to airflow, while in the more open trunk space it is possible for airflow to penetrate at this level.

Air flowing across a vegetated surface causes some disruption of the canopy. Sway of trees, for example, while absorbing momentum from the air, also allows turbulent downdraughts to penetrate through the canopy. Grass surfaces, on the other hand, often become aerodynamically less resistant at moderate wind velocities as the sward bends before the wind.

Table 8.4 Roughness length of natural surfaces.

Surface	Roughness length, z_0 (m)
still water	0.1×10^{-5}
ice, mudflats	0.1×10^{-4}
fresh snow	0.1×10^{-2}
sand	0.3×10^{-4}
soils	$0.1 \times 10^{-3} - 0.1 \times 10^{-2}$
short grass (less than 0.01 m)	0.1×10^{-2}
tall grass (up to 0.1 m)	0.2×10^{-1}
forest	4.0

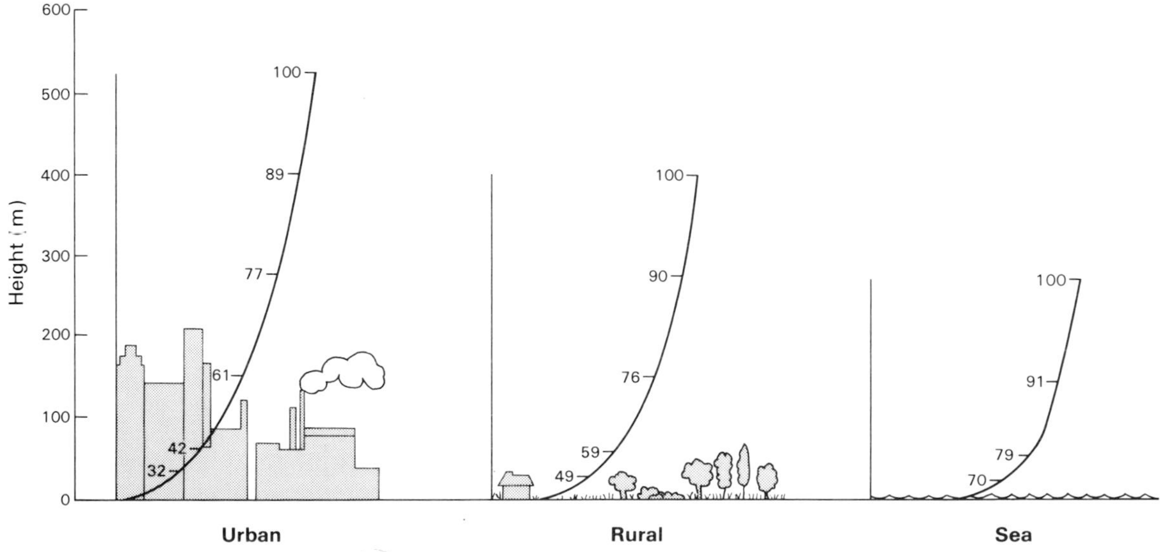

Figure 8.5 Typical wind velocity profiles over three contrasting surfaces. Numbers refer to mean horizontal wind speed expressed as a percentage of the gradient wind speed. (After Leniham & Fletcher 1978.)

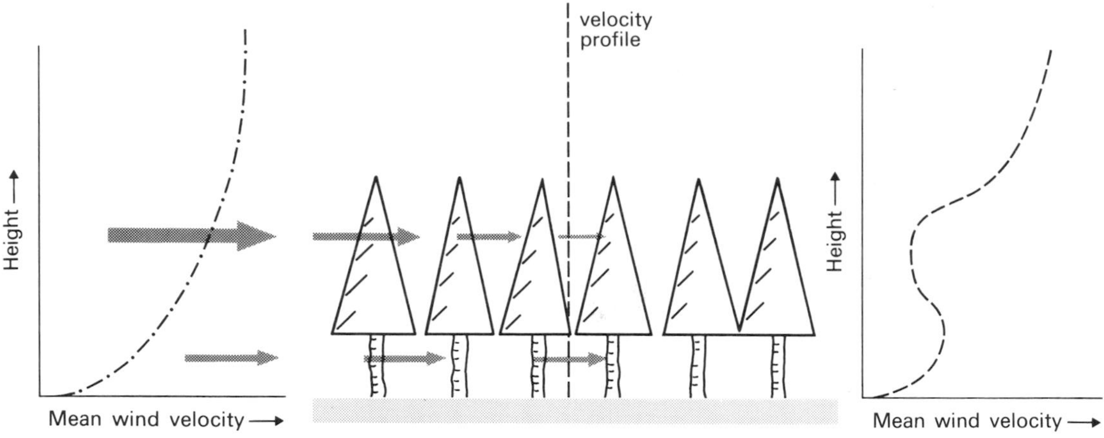

Figure 8.6 The effect of a stand of conifers on the profile of mean wind velocity.

8.4 Topography

The slope angle, aspect and elevation of surfaces affect the amount of direct solar radiation which is incident upon them. The effect of slope may be related to Lambert's Cosine Law (Box 3.4) by which, for direct solar radiation, the angle between the solar beam and the sloping surface determines its intensity. Thus, in Figure 8.7 the intensity of radiation incident upon slope A is greater than that on slope B. The orientation of such a slope, referred to as its aspect, also affects the intensity of solar radiation upon it. For example, in the northern hemisphere, a slope of southerly aspect will receive solar radiation when solar elevation is greatest. Conversely, a slope of northerly aspect will receive a low intensity, if indeed it receives any at all.

The simple example in Figure 8.7 illustrates how slope and aspect combine to determine the intensity of direct solar radiation falling upon sloping surfaces. Over subdued relief, aspect and slope combine to produce greater solar heating of slopes X than of slopes Y. If slope angles are increased, X still receives direct solar radiation and is referred to as the **adret** slope. Y, however, receives no direct solar radiation and is referred to as the **ubac** slope. An illustration of this shading effect may be seen in many of the north-facing mountain corries of Scotland and Wales. Only small amounts of direct solar radiation are received on the steep back wall of corries and this results in delayed snow melt which, in cooler climatic

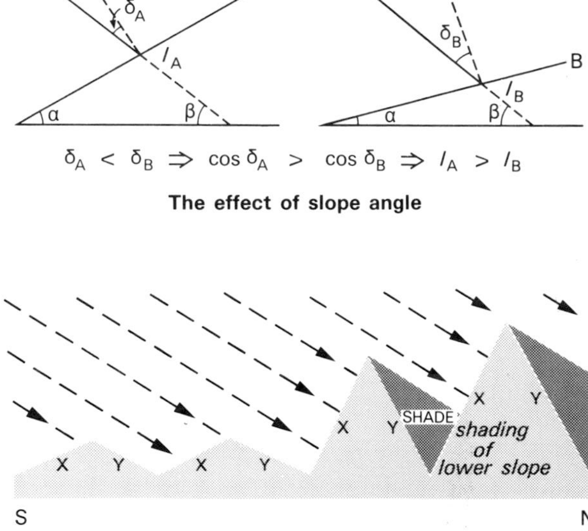

Figure 8.7 The effect of topography on the intensity of direct solar radiation arriving at a surface.

epochs, has been critical in the development of these features. This topographic variation in solar heating of surfaces produces complex patterns of soil and air temperatures and rates of evaporation. Jackson (1967) has illustrated how monthly potential evapotranspiration varies according to aspect, slope, surface albedo and season (Fig. 8.8). A reduction in available heat energy on cooler south-facing slopes in the southern hemisphere inhibits both evaporation and transpiration. This is particularly noticeable at times of relatively lower solar elevation in June and September.

The relationship between surface elevation and direct solar radiation is complex and it depends largely upon atmospheric transparency. Under cloudless skies the effect of increases in surface elevation is to increase direct solar radiation received, because of shortened path through the atmosphere. High alpine areas may receive as much as 90% of potential solar radiation, whereas areas at sea level in similar latitudes may receive only 54%. However, under atmospheric conditions conducive to orographic cloud formation, the relationship changes. The increasing cloud amounts characteristic of the maritime uplands of Britain produce a decrease in solar radiation received as elevation increases. Hughes and Munro (1968) have shown that between elevations 34 m and 335 m above sealevel in mid-Wales there is a 23% decrease in incoming solar radiation.

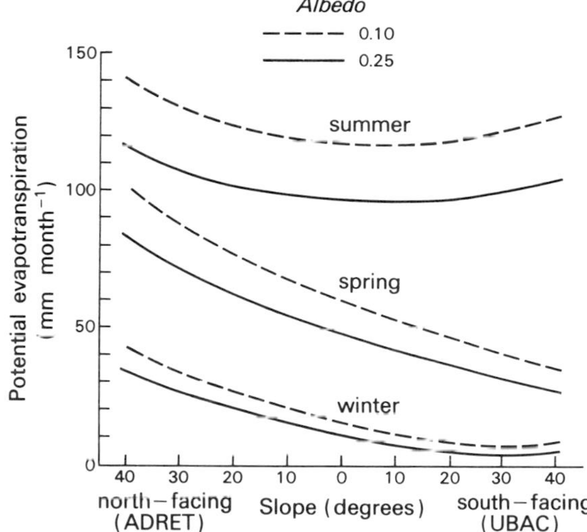

Figure 8.8 Potential evapotranspiration of north- and south-facing slopes of two different albedo values; New Zealand (from Jackson 1967).

8.5 Surface–air interaction

The operation of the factors discussed above creates energy imbalances over much smaller distances and time intervals than those already discussed in Chapter 4. Along coasts, for example, there are two fundamentally different materials lying next to each other which gives rise to localised differences in surface heat energy balance. However, there is a complex relationship between the distribution of heat energy in the lower atmosphere and these contrasts in surface energy balance which depends upon the nature of both horizontal and vertical air movements.

Over relatively short distances of the order of 1 km, air blowing from over a smooth, warm and dry surface to a moist vegetated surface undergoes adjustment to these new conditions (Fig. 8.9). This

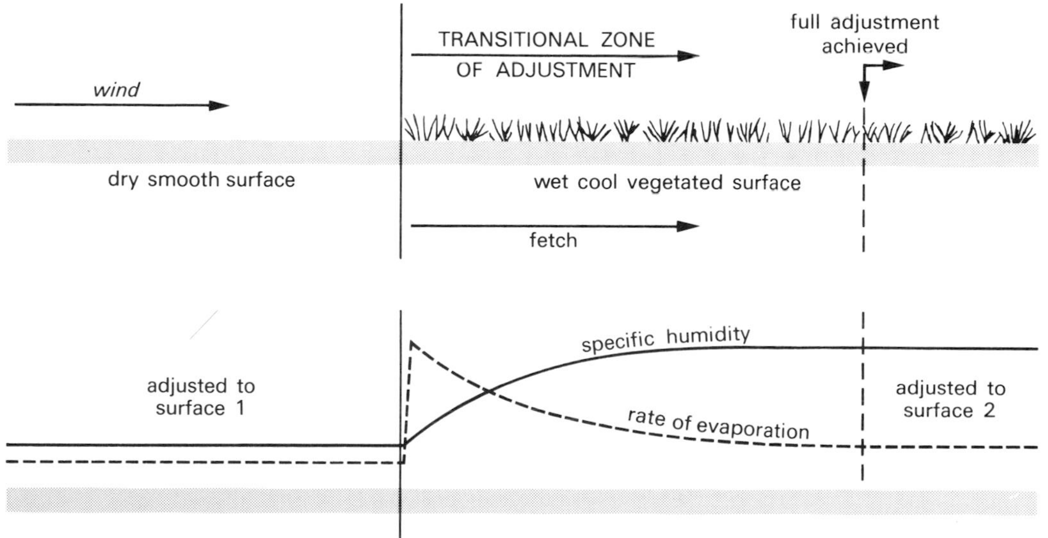

Figure 8.9 Adjustment of air after crossing a boundary between two surfaces (after Oke 1978).

adjustment is not instantaneous but occurs gradually with increasing distance from the point at which the surface character changed; it is referred to as the **fetch** of airflow.

Along coasts, air blowing inland from the sea possesses distinctly maritime properties and has a mean wind profile adjusted to the low roughness parameter of the water surface. As it moves inland there is progressive readjustment to the new surface. This gradual change is evident in the decreasingly oceanic character of climate away from coasts. For example, on a relatively large scale, under a middle-latitude wind system there is a clear west to east change in air temperature regimes (Fig. 8.10). Equability of temperature variation, typical of sea surfaces, is evident on the coast of Ireland while extreme temperature variation, typical of land surfaces, is characteristic of central and eastern Europe.

Over the surface of the Earth there are extensive masses of air which have acquired temperature and moisture characteristics closely related to the surface beneath them. These air masses are associated with extensive source regions. There are two principal groups of air masses associated with high latitudes (polar and arctic air) and with low latitudes (tropical air). These may be further subdivided into maritime or continental, according to whether their source areas are over oceans or land masses. In the middle latitudes, movement of these air masses from their source areas generates the characteristic variability of weather. In addition to the fundamental contrasts in source area characteristics, there are the modifications to which the air masses are subjected in moving over the Earth's surface. Both direction of movement and seasonal timing are important in this respect. This can be illustrated best by considering examples of the principal air masses which affect the British Isles (Fig. 8.11).

Polar maritime air, which is the most frequently occurring air mass, has its source area over the North Atlantic ocean, off the coast of Greenland. In travelling southeastwards towards Britain it

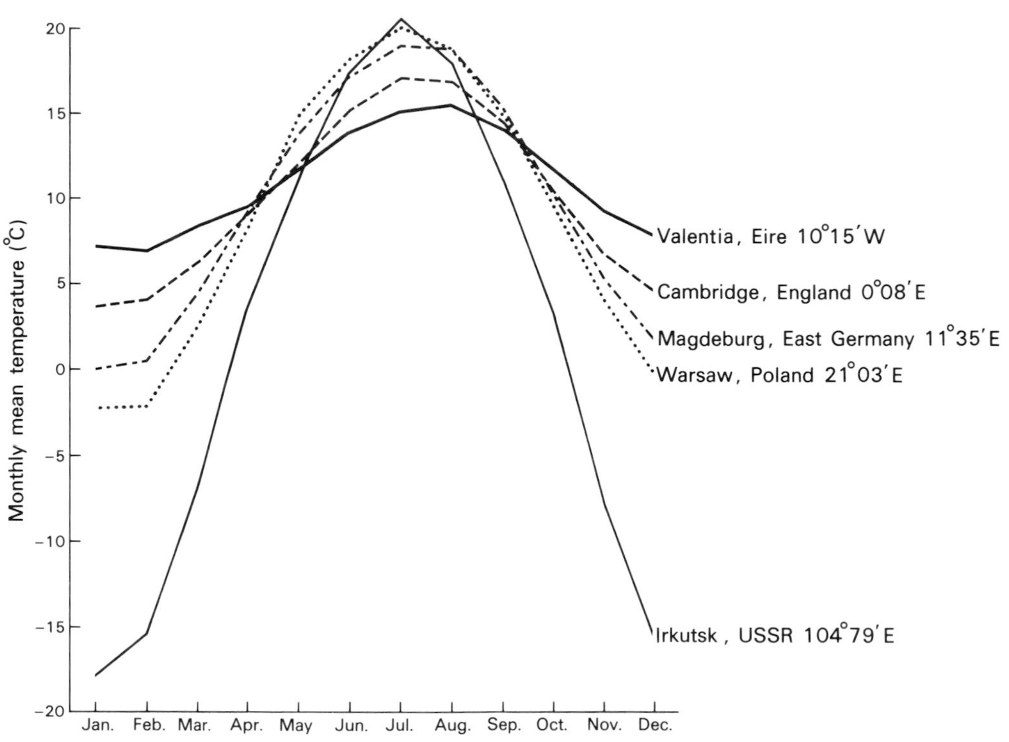

Figure 8.10 Annual temperature regimes for five stations along latitude 52°N with mean monthly temperatures corrected to sea level (Meteorological Office 1972, and Lydolph 1977).

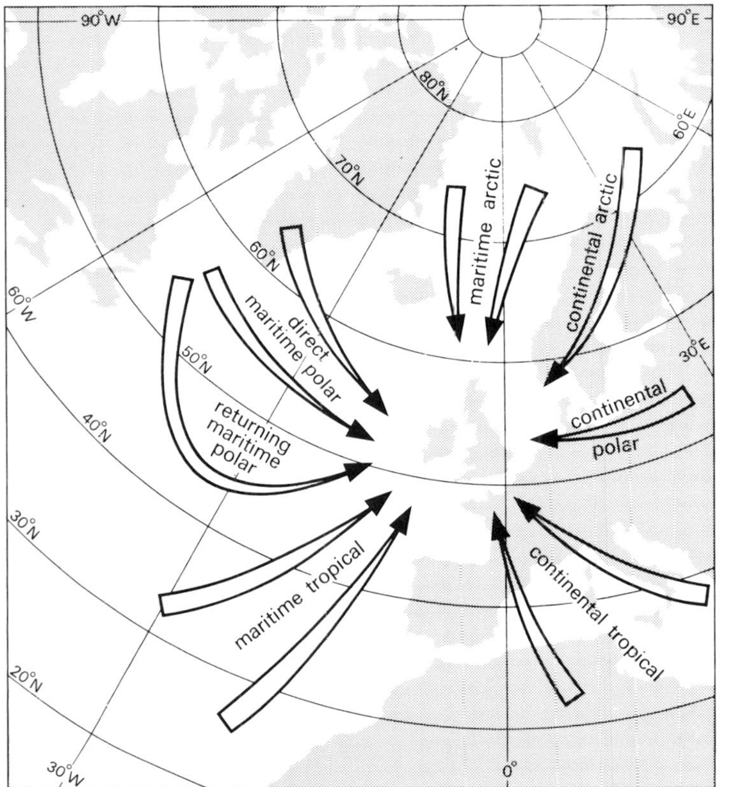

Figure 8.11 The principal air masses affecting the British Isles (from Barry & Chorley 1971).

	Source	Source characteristics	Weather associated
TROPICAL			
maritime	sub-tropical North Atlantic, Azores	warm and moist in lower layers; relatively drier above	mild, grey, cloudy, often with fog or drizzle
continental	North Africa	hot and dry	rare, very mild weather in winter and cloudy; thunder in summer
POLAR			
maritime	North Atlantic	cold, <20°C humid, >60%	unstable cool weather, clear periods, heavy showers
continental	Eastern Europe and Siberia	very cold, relatively dry	extremely cold in winter, clear at source but bringing cloud and snow to east; relatively warmer in summer
ARCTIC			
maritime	Arctic ocean and Spitzbergen	very cold, humid	very low temperatures, <0°C; unstable – frequent heavy showers, usually of snow
continental	North USSR	very cold, relatively dry	extremely cold, snow prolonged at times

passes over relatively warmer ocean surfaces and is gradually heated from below. This causes it to become unstable so that when it reaches Britain it is frequently associated with deep cumulus cloud and a showery type of precipitation. If, however, the air initially travels southwards from its source and subsequently approaches Britain from the south-west, it has been slightly recooled from below in returning northwards and is, therefore, more stable. This returning polar air is usually associated with higher air temperatures than is the direct polar maritime and it provides more continuous precipitation.

Tropical continental air is the least frequent air mass to affect Britain and has its source region over North Africa, whence it travels northwestwards. It is initially hot, dry and unstable but on travelling northwards it may be cooled from below and become more stable. In summer, this cooling may be negligible as the air passes over the warm land surfaces of western Europe and, during this season, it may be sufficiently unstable to give thunderstorms in Britain. In winter, however, a much greater degree of cooling from below makes the lower air very stable and, upon reaching Britain, tropical continental air is associated with stratus cloud and generally dull weather.

The movement of air masses towards Britain and in all other latitudes not only affects the heat and moisture balances of the atmosphere and surface but is also associated with the development of atmospheric circulations within the primary circulation system. For example, where polar air meets tropical air at the polar front (see Ch. 9), the steep temperature gradient between them gives rise to the development of cyclonic storms. Where tropical maritime air moves over warm seas around the western flanks of the subtropical high pressure areas, storms develop in the form of the tropical cyclone.

Spatial variation in the characteristics of the Earth's surface generates energy imbalances over a range of space and time scales, from the larger-scale development and modification of air masses to the smaller-scale differences between land and sea. The air motion resulting from these imbalances may be considered in terms of secondary and tertiary circulation systems respectively.

Further reading

Bannister, P. 1976. *Introduction to physiological plant ecology*. Oxford: Blackwell Scientific.

Geiger, R. 1965. *The climate near the ground*. Cambridge, Mass.: Harvard University Press.

Lockwood, J. G. 1979. *The causes of climate*. London: Edward Arnold.

Monteith, J. L. 1973. *Principles of environmental physics*. London: Edward Arnold.

Oke, T. R. 1978. *Boundary layer climates*. London: Methuen.

Rosenberg, N. J. 1974. *Microclimate: the biological environment*. New York: John Wiley.

9 Secondary and tertiary circulation systems

9.1 Secondary circulation systems

9.1.1 The pressure cell

Over the surface of the Earth there are cells of high and low atmospheric pressure which are both ephemeral and mobile, being apparent on the daily weather map but frequently undetectable in the distribution of seasonal means of atmospheric pressure. These are identified as areas of closed isobars in the form of low-pressure cells (cyclones) and high-pressure cells, or anticyclones. Both these are seen on a typical weather map for the eastern Atlantic ocean and western Europe (Fig. 9.1).

Air converging upon a low-pressure cell rises and diverges aloft, while air diverging from a high-pressure cell is subsident with convergence aloft. In the former, air is cooled as it rises and generates cloud, in contrast to the warming and drying of subsiding air in an anticyclone. This contrast, together with the steeper pressure gradients into the low-pressure cells, means that in terms of weather events, cyclones are the more active features.

9.1.2 The extratropical cyclone

In the middle latitudes there are mobile cyclones (depressions) which move eastwards under the influence of a westerly component of atmospheric circulation and which possess an organised pattern of air movement. These features were identified by Abercromby in 1883 (Fig. 9.2), who noted that they contained areas of greater cloudiness and precipitation, and well defined zones where both air temperature and wind direction changed rapidly over short distances. Work by Norwegian meteorologists in the early 20th century identified these zones, or discontinuities, as the leading edge (warm front) and trailing edge (cold front) of a sector of warm air. These fronts are the principal cloud and precipitation zones of the depression (Fig. 9.2). A satellite photograph of a well developed cyclone in the North Atlantic (Fig. 9.3) clearly shows its associated spiralling bands of cloud and the denser cloud along the frontal zones.

At both fronts air is being lifted aloft over cooler air and is thereby subjected to adiabatic cooling. At the warm frontal zone warmer air overrides the cooler air in advance of the depression. The weather conditions associated with its passage across the surface are mainly of increasing amounts of cloud, precipitation and wind speed in advance of its arrival (Box 9.1). In its wake, cloudiness may persist but precipitation becomes intermittent and wind speeds decrease.

The cold frontal zone is located where advancing polar air undercuts the air of the warm sector. The enforced uplift of air is rapid, resulting in deep cumulus cloud and intense precipitation. After its passage, precipitation may continue in the form of heavy showers. The depresson is therefore asymmetric, the distribution of heat and moisture and the nature of airflow depending upon position relative to its centre, as suggested by Abercromby. Speed of eastward movement of the whole system is variable, each depression having its own individual character, but an average speed would be in the region of 11.5 m s^{-1}.

The Norwegian model of cyclone development was based upon airmass interaction along the Polar Front, which most commonly separates polar mari-

(a)

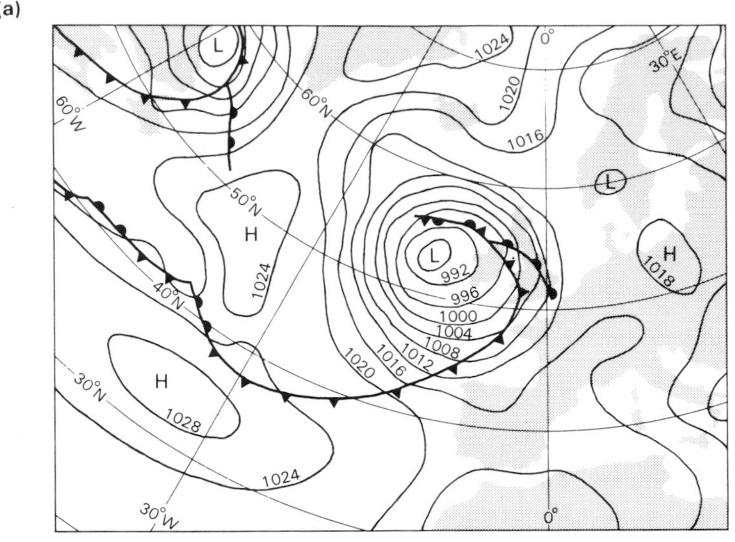

(b)

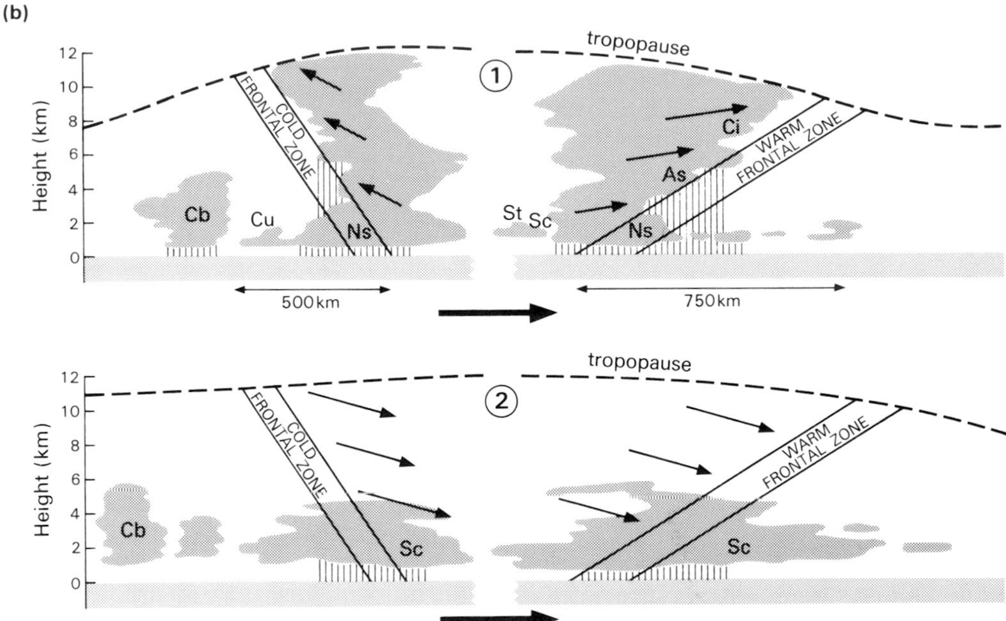

Figure 9.1 A typical weather map: (a) for the North Atlantic Ocean and Western Europe showing an extratropical cyclone approaching the British Isles (weather log 12 GMT, 24 August 1977) (reproduced by permission of the Meteorological Office); (b) cross sections through the extratropical cyclone where warm air is (1) rising and (2) falling relative to the frontal zone. (See Table 4.6 for cloud abbreviations.)

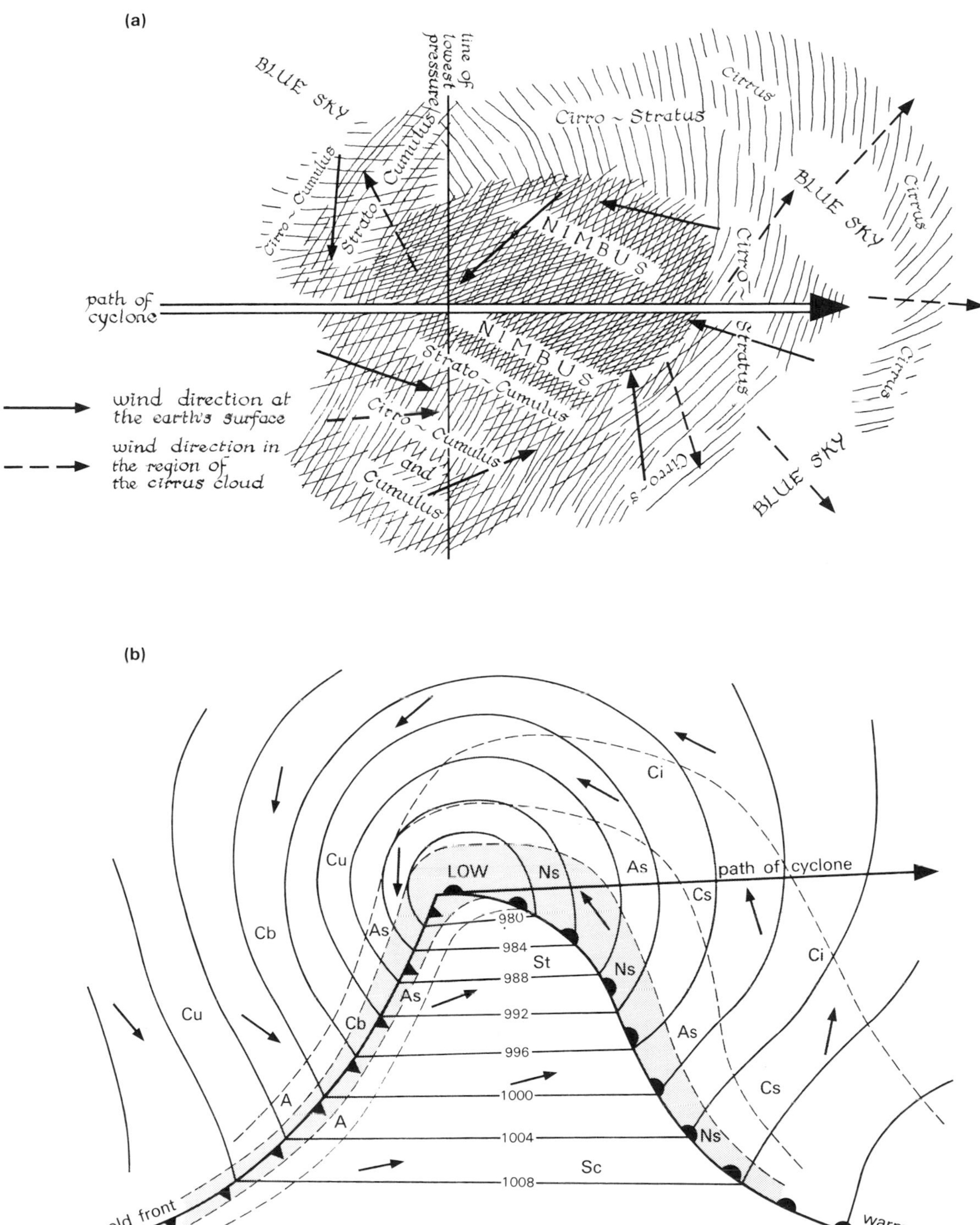

Figure 9.2 Representations of cloud distribution in an extratropical cyclone by (a) Abercromby (1883) and (b) Norwegian meteorologists (1914 onwards).

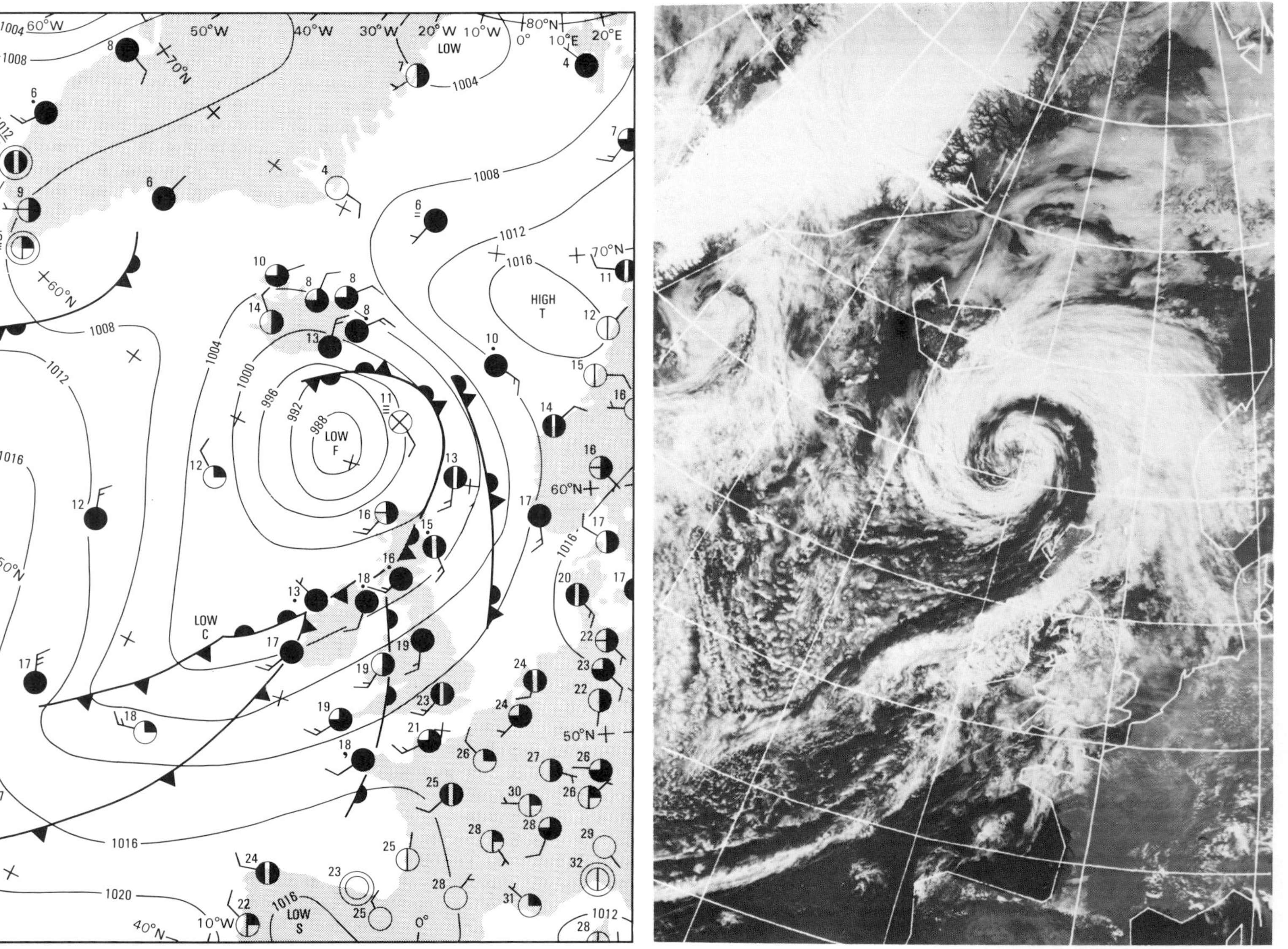

Figure 9.3 Satellite photograph and daily weather map for the North Atlantic ocean and Western Europe, 6 August 1979 (reproduced by permission, University of Dundee and the Meteorological Office).

Box 9.1

CHANGES ASSOCIATED WITH THE PASSAGE OF WARM AND COLD FRONTS (from Meteorology Office, Aviation Meteorology)

Element	*In advance*	*At the passage*	*In the rear*
warm front			
pressure	steady fall	fall arrested	little change or slow fall
wind	backing and increasing	veer and decrease	steady direction
temperature	steady, or slow rise	rise	little change
dew-point	rise in precipitation	rise	steady
relative humidity	rise in precipitation	may rise further if not already saturated	little change; may be saturated
cloud	Ci, Cs, As, Ns in succession; St fra, Cu fra below As and Ns	low Ns and St fra	St or Sc may persist; perhaps some Ci
weather	continuous rain (or snow)	precipitation almost or completely stops	dry, or intermittent slight precipitation
visibility	good, except in rain (or snow)	poor, often mist or fog	usually moderate or poor; mist or fog may persist
cold front			
pressure	fall	sudden rise	rise continues more slowly
wind	backing and increasing, becoming squally	sudden veer, perhaps squall	backing a little after squall, then fairly steady or veering further in later squalls
temperature	steady, but fall in prefrontal rain	sudden fall	little change, variable in showers
dew-point	little change	sudden fall	little change
relative humidity	may rise in prefrontal precipitation	remains high in precipitation	rapid fall as rain (or snow) ceases; variable in showers
cloud	St or Sc, Ac, As then Cb	Cb with St fra, Cu fra or very low Ns	lifting rapidly, followed for a short period by As, Ac and later further Cu or Cb
weather	usually some rain perhaps thunder	heavy rain (or snow) perhaps thunder and hail	heavy rain (or snow) usually for short period, but sometimes more persistent; then fine but followed by further showers
visibility	moderate or poor, perhaps fog	temporary deterioration followed by rapid improvement	very good

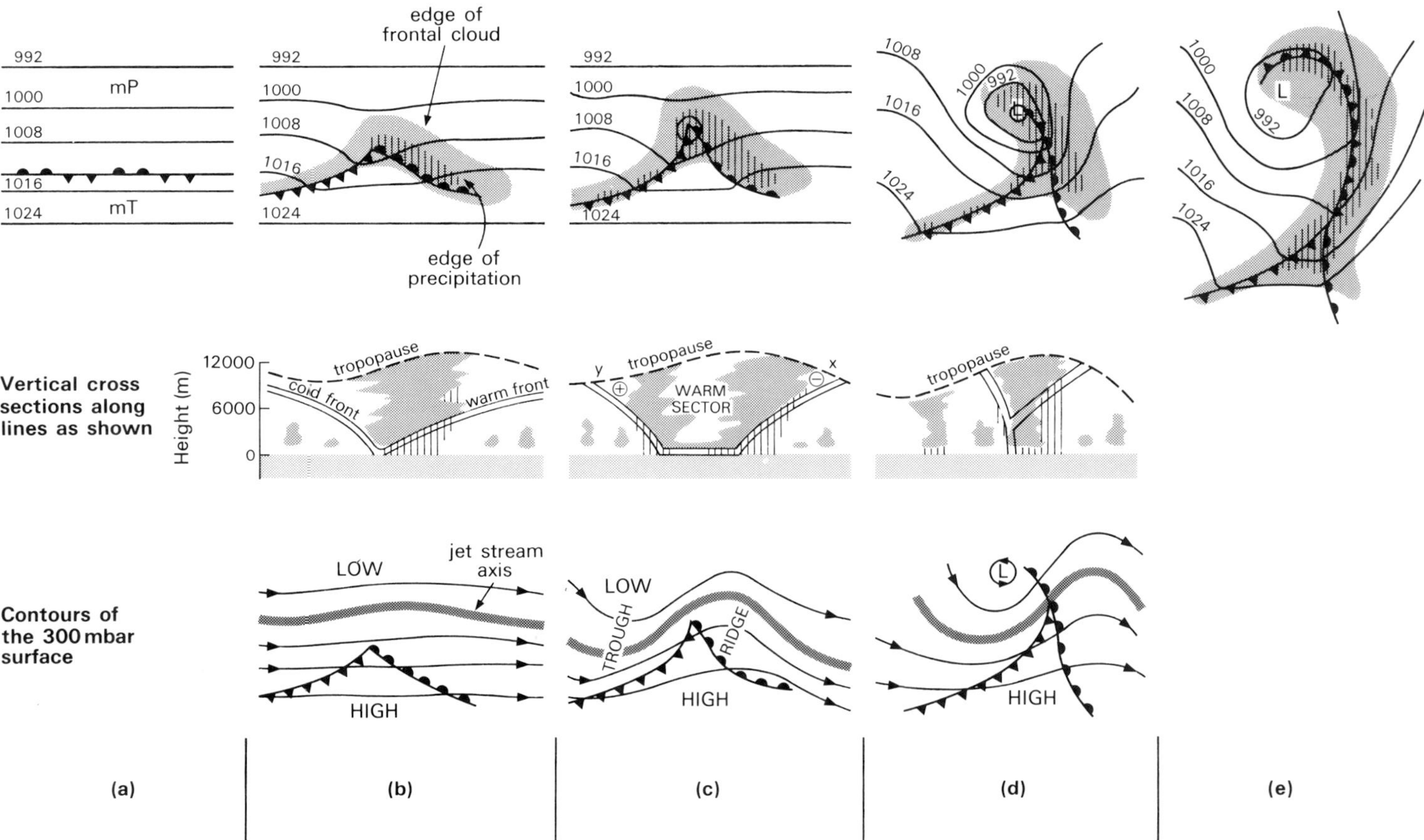

Figure 9.4 Stages in the development of an extratropical cyclone (isobars at 8 mb intervals; from Pedgley 1962).

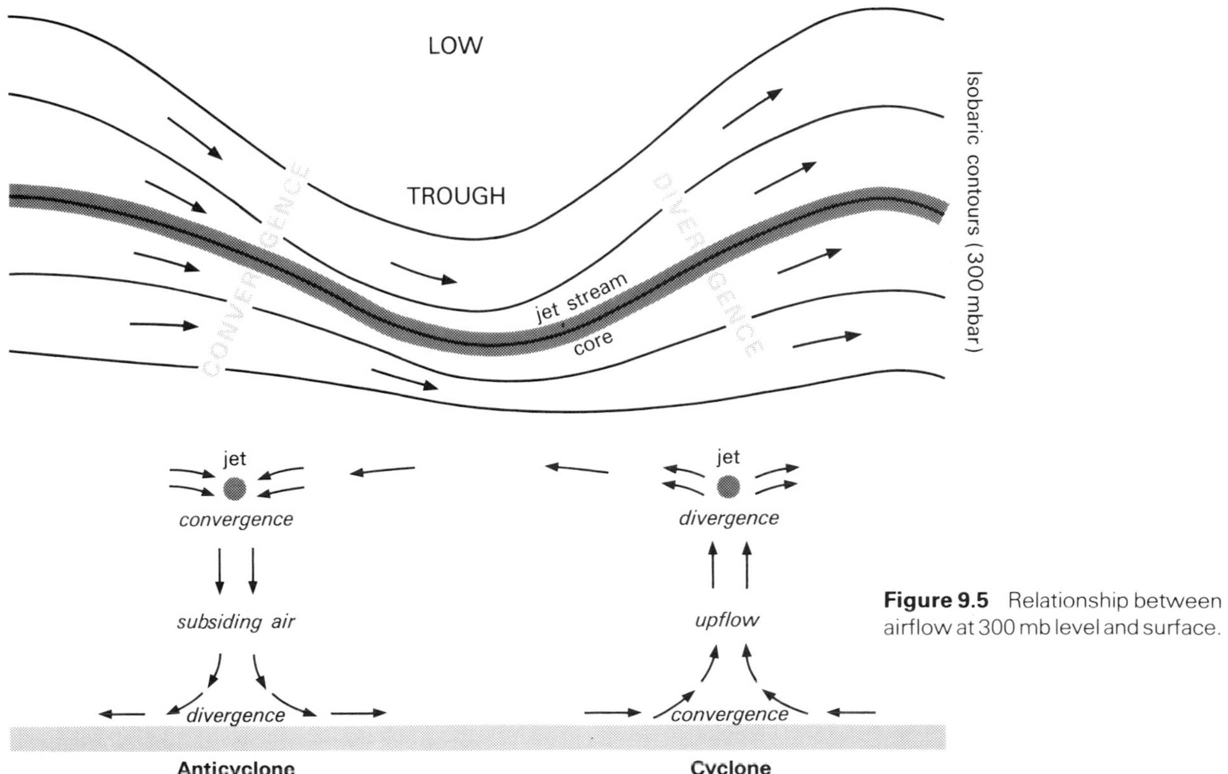

Figure 9.5 Relationship between airflow at 300 mb level and surface.

time and tropical maritime airmasses (Fig. 9.4a). Along this zone of contact a small wave, or perturbation, may develop as warm air begins to move polewards into the cooler air (Fig. 9.4b). A distinct warm sector develops between warm and cold fronts. As the former moves forwards at a slower speed than the latter, the polar air is, in effect, executing a pincer movement on the air of the warm sector which is, therefore, being laterally constricted. Thus, there is a net convergence of air in the lower troposphere and an uplift of warmer air, which together produce decreasing atmospheric pressure at the apex of the wave. The depression develops as this low pressure deepens (Fig. 9.4c).

The meeting at the surface of cold and warm fronts marks the maximum development of the depression when pressure at its centre has reached a minimum (see Fig. 9.4d). After this point, decay, or **occlusion**, begins to occur as the warm sector is lifted clear of the surface. With continued lifting its remaining heat energy is dissipated and the depression weakens, although precipitation may continue to fall from the occluded front (Fig. 9.4e). Eventually the perturbation of the polar front has been removed and only a weak cyclonic cell of cool air in the lower troposphere remains. In so doing, the atmosphere has released the considerable amount of heat energy contained within the sector of warm air.

This polar front model matches closely the observed growth and decay of a typical middle-latitude depression. However, inspection of weather charts for the north Atlantic does not always reveal a clear developmental pattern of depressions and associated fronts. Waves on the polar front may develop and decay rapidly in the western Atlantic or may be still intensifying when they reach Britain. The development of waves along the trailing cold front of an Atlantic depression over the Celtic sea gives rise to secondary depressions. These often intensify and produce deteriorating weather conditions as they move eastwards.

The increasing number of meteorological observations made in the middle and upper troposphere, particularly during the past 20 years, have revealed a close relationship between the upper westerly flow and the formation of extra-tropical cyclones. The contours of the 300-mb surface above the cyclone in Figure 9.5 indicate a wave pattern in the upper westerly airflow. Indeed, if we were to look at the circumpolar vortex in the northern hemisphere (Fig. 4.17), we would find that as it moves northwards through the waves, cyclone development (**cyclogenesis**) commonly occurs. On the western limb of a trough in one such wave (Fig. 9.5) airflow is convergent, indicated by a closer proximity of the isobaric contours. In order to compensate for the net inflow of air at this level (some 9000 m above sea level), underlying subsiding flow towards the Earth's surface develops. The resulting inflow of air into the lower troposphere gives rise to divergent anticyclonic flow across the surface. On the eastern limb of the trough, airflow is divergent, indicated by a spreading out of the isobaric contours. Divergence of airflow at this level results in an upward replacement flow from the lower troposphere. This upflow of air from the Earth's surface is replaced by a convergent cyclonic inflow across the surface.

Thus, to the west of wave troughs convergence in the upper troposphere results in anticyclone development (**anticyclogenesis**) in the lower troposphere, while to the east divergence results in cyclogenesis (Fig. 9.4). The latter affects very markedly the development of extratropical cyclones along the polar front over both Atlantic and Pacific oceans, particularly when there are well developed troughs in the upper westerlies.

As a circulation system, the middle-latitude depression results from a combination of strong horizontal temperature gradients across the polar front and the external impetus provided by the upper westerly airflow. Its operation relies on the continued conversion of potential energy into kinetic energy (Fig. 9.6). Potential energy is

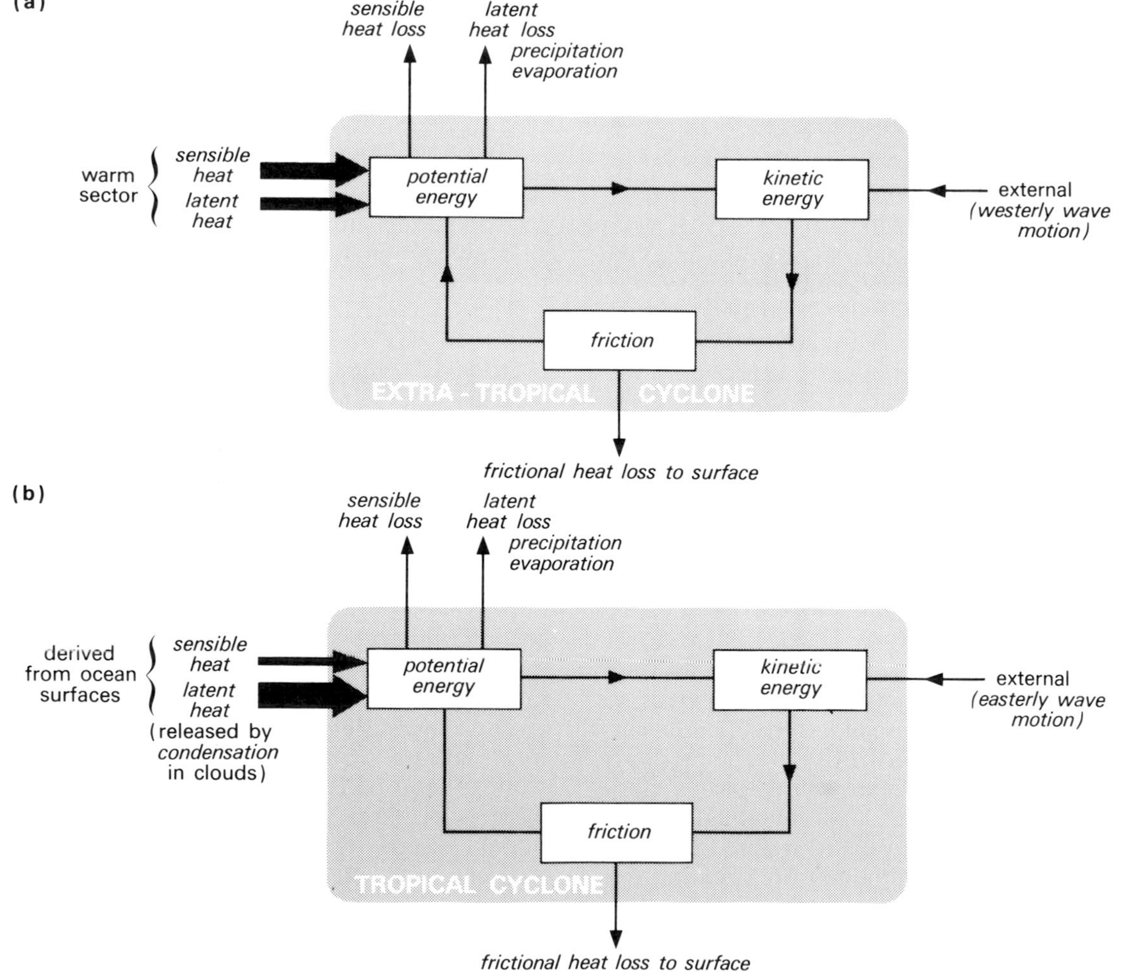

Figure 9.6 Cyclones as energy systems: (a) extratropical cyclone, (b) tropical cyclone.

derived from the temperature contrast between the warm and cold airmasses, and its conversion into kinetic energy is initiated by the warm air being lifted aloft. Heat energy is also gained from the surface in the form of latent heat, this being released by condensation of water vapour in the warm sector. Although some of this energy will be returned into the atmosphere by evaporation, a net loss of water by precipitation will ensure a net gain of energy through condensation.

The kinetic energy of air motion is dissipated through the effects of friction both within the air itself as turbulent friction or at the Earth's surface. Kinetic energy is transferred directly to the surface (for example, in ocean waves) or is converted into heat energy. In the case of the latter it is returned to the potential energy store or is lost from the circulation system.

The vigour of the depression measured in terms of the velocity of both horizontal and vertical air motion ultimately depends on the balance between the rate at which energy is converted from potential into kinetic, and the rate of frictional energy losses. In the depression's occluded state, the supply of potential energy from the elevated warm sector has been exhausted and kinetic energy is rapidly depleted. The depression thus weakens and pressure begins to rise at its centre.

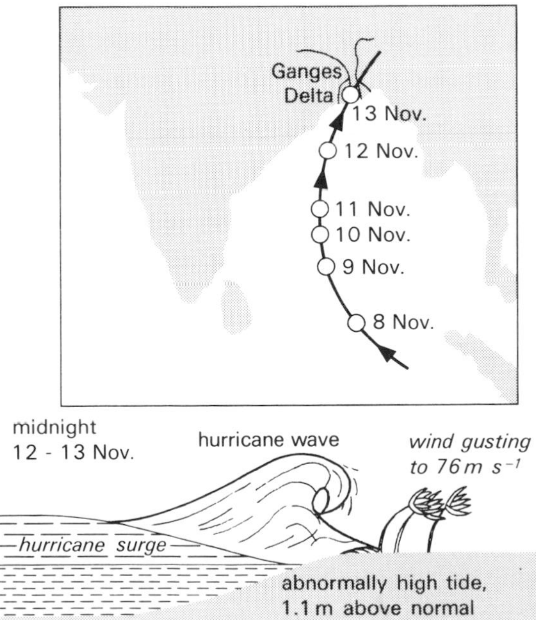

Figure 9.7 Bay of Bengal typhoon, November 1970.

9.1.3 The tropical cyclone

In tropical latitudes, cyclones occasionally take the form of the rotating storm known as the hurricane or typhoon. These are variously described as severe, violent, destructive or devastating. Wind speeds frequently exceed 50 m s^{-1}, causing considerable damage, and in advance of the hurricane there is a storm surge at sea which frequently causes heavy flooding along low coasts. Such a storm surge was responsible for over a quarter of a million deaths in the Ganges Delta when a Bay of Bengal typhoon struck on 13 November 1970 (Fig. 9.7). High tide was 1.13 m above normal and a tidal wave 7.5 m high combined with this to cause widespread flooding in advance of the typhoon which developed wind speeds in excess of 65 m s^{-1} when it eventually arrived.

The hurricane appears on the weather map as a series of concentric and closely spaced isobars (Fig. 9.8) at the centre of which sea level atmos-

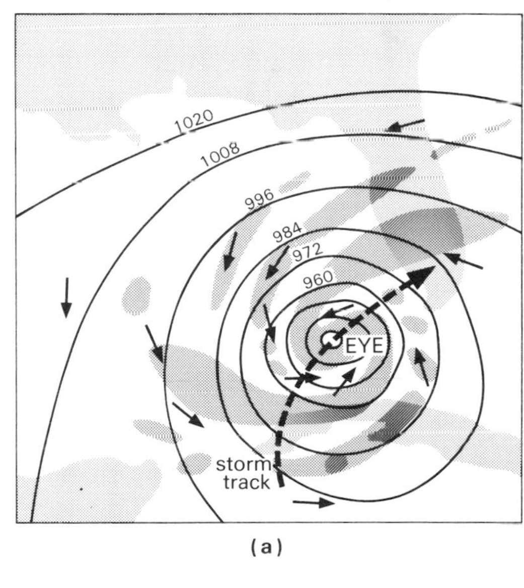

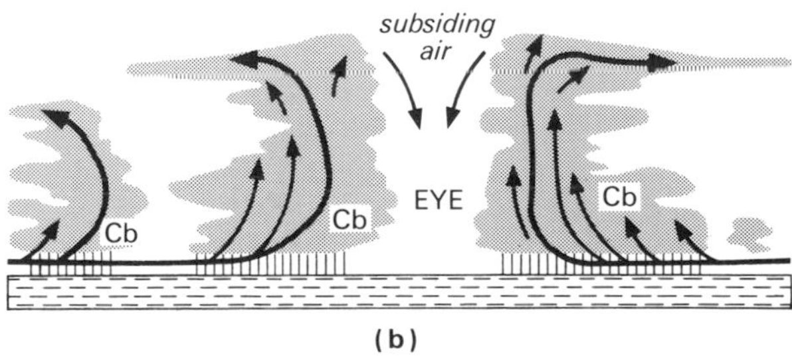

Figure 9.8 (a) Tropical cyclone approaching Florida; (b) vertical section.

pheric pressure usually falls to 950 mb and, in exceptional cases, to 900 mb. The whole system has a diameter of 200–1000 km and is at its most active over a period of between 3 and 14 days.

Within the eye, air is subsiding, maintaining relatively high temperatures and low relative humidity through adiabatic warming (Fig. 9.8). The skies here are relatively clear, there being only a thin veil of high-altitude cirrus cloud. Around this there is a towering wall of cumulonimbus cloud rising as high as 15000 m, from which falls most of the cyclone's precipitation. Precipitation intensities greater than 50 mm h^{-1} have been recorded in this zone of the cyclone. The zonation of clouds within the tropical cyclone is apparent in satellite photographs of hurricanes. Both the central eye and the surrounding ring of cumulus are particularly obvious.

The tropical hurricane may be compared to the middle-latitude depression in that it is an area of low pressure which gives rise to airflow spiralling into its centre and within which precipitation occurs. However, in most other respects it is strikingly different (Table 9.1). In addition to having only half the overall diameter and developing twice the wind speed of the middle-latitude depression, the hurricane possesses no discontinuities. The weather pattern is roughly symmetrically organised around its central eye.

The main hurricane areas are at latitudes greater than 5° over the tropical oceans, along the western flanks of the subtropical high-pressure areas, as indicated in Table 9.2. Most occur in the northern hemisphere, only the southwestern Indian ocean and the offshore waters of northern Australia being greatly affected in the southern hemisphere. Hurricanes usually occur during the late summer and autumn when sea temperatures are high and they usually develop from less violent tropical disturbances which form in tropical maritime air at low latitudes. Many of these small tropical disturbances from which hurricanes develop are formed within westward-moving waves in the tropical easterly airflow. To the east of a wave trough there is divergence in the middle or upper tropospheric flow which encourages convergence at sea level and the development of deep cumulus cloud. Many such tropical lows form, some of which develop into tropical storms of moderate intensity with wind speeds in excess of 17 m s^{-1}, but only a few develop further to form hurricanes.

From a vertical section (Fig. 9.8) we can see that within the hurricane there is a strong convergence and uplift of air near the sea surface, the latter developing to 15000 m. A vital feature which differentiates the hurricane from the tropical storm, is well developed divergence in the upper troposphere, beneath which there is a warm core of subsiding air. Air over the ocean surface takes up both heat and moisture which are then carried aloft in the strong vertical currents. Condensation and precipitation at upper levels cause the release of massive amounts of latent heat which maintains a supply of warm air for the core. An adequate uptake of heat and moisture is, therefore, essential and it is supplied by warm ocean surfaces. A critical threshold ocean temperature of 27°C is commonly accepted and the world distribution of hurricanes is closely related to this isotherm.

As we have already seen in Chapter 4, the Coriolis force has a value of zero at the Equator but it increases rapidly as the sine of latitude. The absence of a strong Coriolis force inhibits the development of a cyclonic vortex and it is not until latitude 5° that it becomes sufficiently large for hurricane development. In migrating northwards in the northern hemisphere, air moving within the hurricane experiences an increasing Coriolis force and may, therefore, be expected to intensify. However, as the ocean surface gradually becomes cooler, the supply of sensible heat and moisture decreases and consequently there is less energy being released in the upper troposphere and the hurricane cannot be sustained. Should this hurricane move over land surfaces, the supply of sensible heat may be maintained, but the supply of moisture, and hence latent heat, is restricted. The total energy input is therefore inadequate to sustain the hurricane. There is also a considerable

Table 9.1 Middle-latitude depression and tropical hurricane: a comparison of scale (partly from Eliassen & Pedersen 1977).

System	Horizontal scale (diameter) (km)	Vertical scale (km)	Wind speed (mean) (m s^{-1})	Average lifetime (days)
tropical hurricane	200–1000	15	30	7
middle-latitude depression	1000–3000	10	10–20	7

Table 9.2 Areas of the world most subject to hurricane activity (after Trewartha 1961, Barry & Chorley 1976).

Area		Hurricane season	Mean annual frequency
I	south and south west Atlantic ocean		
	Cape Verde Islands	August and September	
	east and north of West Indies	June to October	
	north Caribbean	May to November	4.6 (1901–63)
	southwestern Caribbean	June and October	
	Gulf of Mexico	June to October	
II	North Pacific ocean (west coast of Mexico)	June to November	2.2 (1910–40)
III	southwestern North Pacific ocean (including China Sea, Japan)	May to December	19.4 (1924–53)
IV	north Indian ocean		
	Bay of Bengal	April to December	4.7 (1890–1950)
	Arabian Sea	April to June	
		September to December	0.7 (1881–1937)
V	south Indian ocean (eastward from Madagascar)	November to April	4.7 (1848–1935)
VI	south Pacific ocean (eastward from Australia)	December to April	4.0 (1940–56)

increase in energy loss due to greater frictional drag over the aerodynamically rougher land surface. Thus, the hurricane's energy supply is reduced in leaving tropical latitudes, with the result that there is a gradual reduction in energy exchange within it, bringing a decrease in wind speed and precipitation intensity. Hurricanes which have followed the warm Gulf Stream of the western Atlantic eventually develop middle-latitude characteristics as they meet cold polar maritime air. In many cases these become deep depressions which bring stormy weather to Britain.

As a circulation system, the tropical hurricane is maintained by large inputs of heat energy derived from condensation of water vapour. The uptake of sensible and latent heat from the oceans provides potential energy which is converted into kinetic energy. As is the case with the middle-latitude depression, the operation of the system relies upon this conversion (see Fig. 9.5). The brake on the system is similarly provided by friction within the atmosphere and at the surface.

While the system is located over warm ocean areas there is a surplus of potential energy uptake over frictional loss and this maintains an intensifying circulation. Reduction of potential energy and increase in frictional energy losses slow down the circulation. Once energy uptake is exceeded by energy loss, the hurricane begins to decay unless a fresh input of potential energy becomes available.

Although wind speeds and destructive capacity are greater in a hurricane than in a middle-latitude depression, its total kinetic energy is considerably less. The kinetic energy of a developing hurricane is of the order of 10^{16} J, increasing to 10^{18} J at maturity. In comparison, a relatively intense middle-latitude depression develops 10^{19} J. Related to these energy exchanges in the Earth–atmosphere system, the energy of a devastating earthquake, at point 8 or more on the Richter scale, is 10^{18} J and an average earthquake, at point 6, 2×10^{13} J. The problem in developing such a comparison, however, lies in the concentration of energy. Clearly, the hurricane at any one time is a more locally concentrated phenomenon than the more diffuse middle-latitude depression.

9.1.4 The anticyclone

No discussion of circulation systems would be complete without some consideration of the anticyclone, which is the less energetic counterpart of the cyclone. It is associated with relatively shallow pressure gradients and consequently much lighter winds. Air in the high-pressure centre is subsiding and hence it experiences increasing atmospheric pressure as it moves towards the Earth's surface, which results in adiabatic warming (see Box 4.11). As the air temperature increases, it is less likely that the air will become saturated unless extra water vapour is added, and any suspended water droplets are likely to be evaporated. We may, therefore, expect anticyclones to bring cloudless weather with little likelihood of cloud or precipitation. Indeed, we immediately associate high atmospheric pressure during the middle-latitude summer with clear skies and hot weather. The winter anticyclone with its attendant clear skies brings rapid nocturnal radiative cooling of ground surfaces and a high risk of frost.

There are, however, occasions when anticyclones bring long periods of dull cloudy weather, often referred to as anticyclonic gloom. The occurrence of shallow stratus and stratocumulus clouds is closely related to the marked inversion of temperature in the lower atmosphere which is a distinctive feature of the anticyclone. This layer, in which air temperatures increase rather than decrease with height above the Earth's surface, may be in immediate contact with the surface or may be elevated at some distance from it. The cold winter anticyclone of Asia is an example of the former, the warm Azores anticyclone of the latter.

Subsiding air begins to slow down as it approaches the Earth's surface, and air in the lower 1000 m or so may not experience subsidence. The subsiding air diverges above a layer of relatively cooler air in contact with the surface, in which an input of extra moisture may encourage condensation to take place. Radiation fogs are a

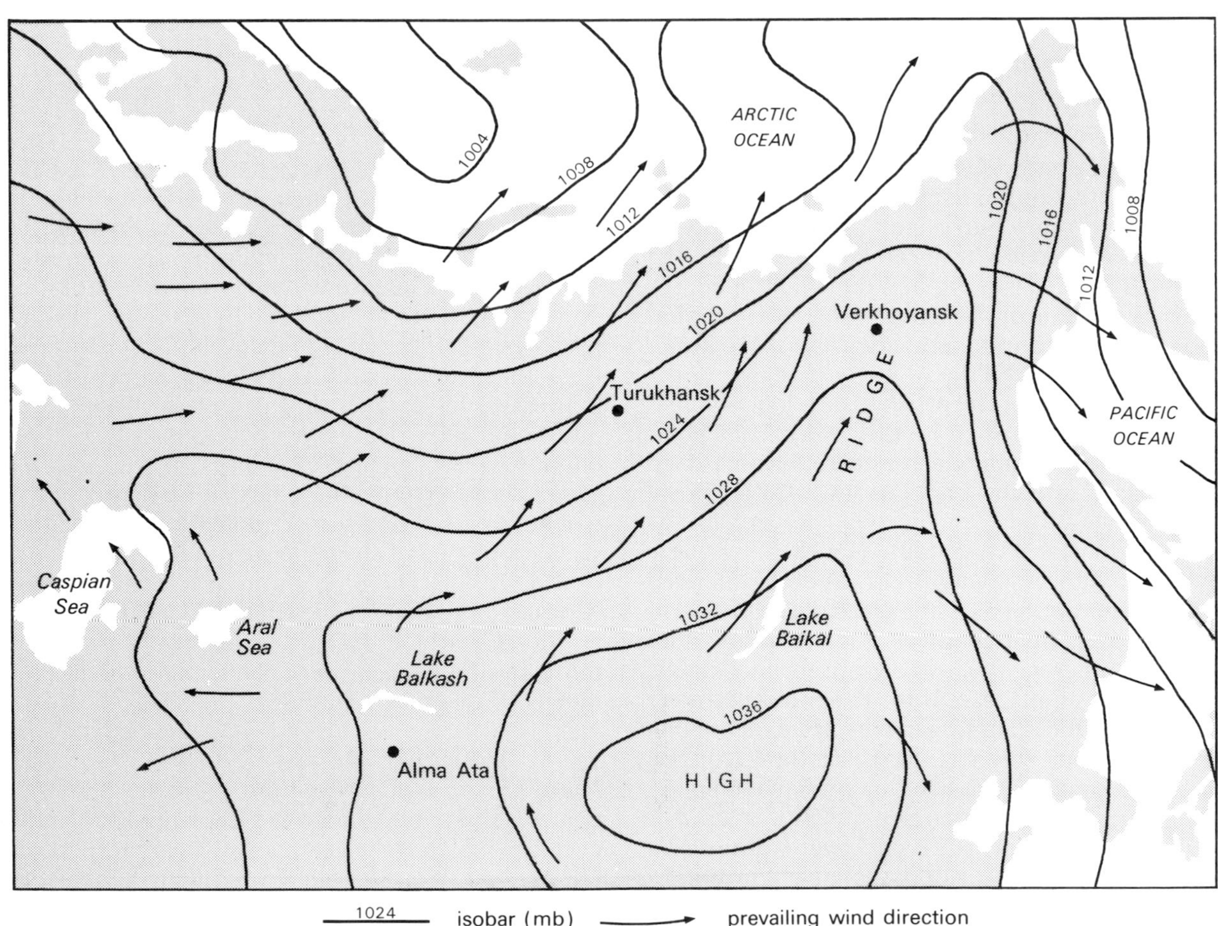

Figure 9.9 Mean sea-level atmospheric pressure and prevailing wind direction over Asia in January (after Borisov 1965).

result of cooling of the ground surface (usually under anticyclonic conditions) and subsequently of the air above it, which reduces the temperature of moist air to its dew-point. The resulting fog thickens as the temperature inversion in the lower atmosphere deepens, and its depth of development is limited to the vertical extent of the inversion layer. In other situations, turbulent mixing of the lower atmosphere may take place beneath an inversion layer not in contact with the surface. Where turbulent uplift of air is localised, as in the case of thermal convection, adiabatic cooling may be sufficient to initiate condensation. This results in isolated cumuliform cloud capping the thermal upcurrents which are limited in their vertical development by the inversion layer. A moist lower atmosphere which has experienced more extensive vertical mixing may produce more widespread stratus or stratocumulus cloud, which may occasionally produce low-intensity precipitation.

In the Earth–atmosphere system there are a number of readily identifiable types of anticyclone which, while sharing the common factor of subsiding air, have fundamentally different origins. Some are mobile features, such as those of the middle latitudes, while others move little, either from day to day or month to month, such as the subtropical anticyclones over the Atlantic and Pacific oceans. Some have a deep vertical development extending through the troposphere; others extend vertically for only 1 or 2 km.

During winter intense radiational cooling over the interior of high-latitude continents remote from oceanic sources of heat energy results in the cooling of a shallow layer of the lower atmosphere. The cold air has a relatively high density which results in an increase in the pressure which the atmosphere exerts on the surface beneath it. The resulting anticyclone is a shallow feature extending vertically for less than 2 km.

Although winter anticyclones are a feature of synoptic charts for both North America and Asia, the one which develops over the latter is the more persistent. The 0°C isotherm of mean January temperature encloses most of Soviet Asia, with mean temperatures in many easterly locations falling below −40°C. The Asian anticyclone has its centre in the vicinity of Mongolia and it extends northeastwards in a ridge across eastern Siberia (Fig. 9.9). There is a well developed temperature inversion layer in the lower atmosphere which extends upwards to the 850 mb level (Fig. 9.10). Surface winds indicate a divergence of airflow from the high-pressure centre, and to the west westerly winds are clearly deflected northwards into the Arctic ocean. However, at the 800 mb level there is little trace of any deflection from a circumpolar westerly airflow. While high pressure dominates the synoptic charts, temperatures remain low and precipitation is infrequent, although low stratus cloud does bring long periods of poor visibility. Occasional incursions of cyclonic activity bring most of the winter precipitation.

In direct contrast to these cold anticyclones, those which occur in subtropical latitudes exhibit deep development through the troposphere and are detectable on pressure surfaces from sea level up to the tropopause. They also vary in position by as little as 5° of latitude during the year. As areas of subsidence and surface divergence, they are associated with some of the Earth's most arid areas, such as the Sahara Desert. A characteristic of the lower atmosphere in these warm anticyclones is a marked inversion of temperature. Along the eastern flank of the anticyclone this inversion is relatively near to the surface, but away from this area it is an elevated feature and it lies over a turbulently mixed air layer. The Cape Verde Islands lie to the south of the main axis of subtropical high pressure in the north Atlantic and

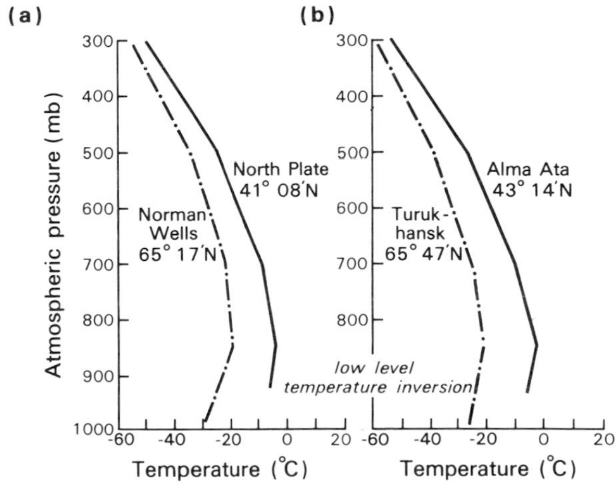

Figure 9.10 Vertical profiles of temperature in the winter anticyclones of (a) North America and (b) Asia (data after Crowe 1971).

the temperature inversion here lies between the 925 mb and 825 mb levels (Fig. 9.11).

9.2 Tertiary circulation systems

9.2.1 Scale

Within the secondary circulation systems are yet smaller-scale circulations of air which operate over relatively short distances, usually less than 160 km, and over short periods of time. They develop limited kinetic energy, of the order of 10^{13} J.

Such 'local' or tertiary circulations may be thermally direct, that is, convectional cells which arise from differential heating of the Earth's surface. In these cases, energy transfers may be likened to those of the single-cell model of the primary atmospheric circulation systems (Ch. 4). They may also operate as modifications of established larger-scale airflows which are referred to here as 'regional' winds. The topography of the ground surface greatly modifies the characteristics of the air flowing across it. In this case, kinetic energy is derived directly from the regional wind. In the succeeding sections we shall consider examples of both types of tertiary circulation.

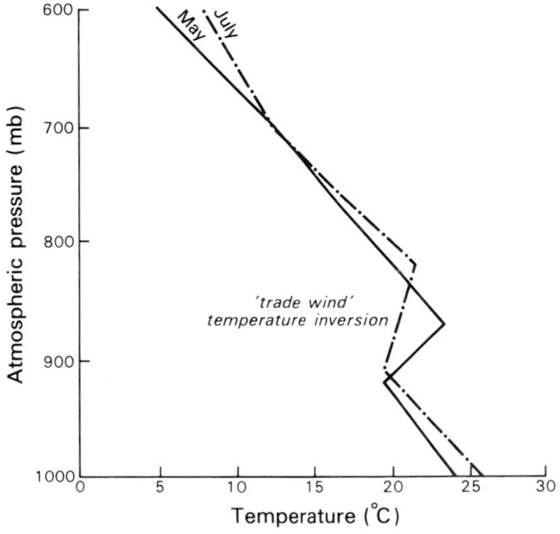

Figure 9.11 Vertical temperature profiles in the subtropical anticyclones of the North Atlantic ocean; Sal, Cape Verde Island (16° 44′ N) (data after Crowe 1971).

9.2.2 The sea breeze

Characteristic of many coastal sites is a diurnal change in wind speed and direction which results from local circulations of air. Analysis of hourly mean wind speeds over a period from 1 June 1975 to 19 August 1975 at Portsmouth, England, has shown that there was a marked diurnal change from a minimum 2.9 m s^{-1} in the early morning to a maximum of 5.4 m s^{-1} in the early afternoon. Wind roses for these two times show a clear difference in dominant wind direction (Fig. 9.12). In the morning there is a high frequency of northerly winds which are offshore along the Hampshire coast. In the afternoon these have been replaced by dominant southerly, or onshore, winds. There is, therefore, evidence for the presence of weak offshore winds in the early morning and strong onshore winds in the mid-afternoon.

Peters (1938) identified this sea breeze in south Hampshire, detecting its onset by noting rapid changes in wind direction associated with decrease in air temperature and increases in relative humidity. On this evidence he identified a sea breeze season extending from March to September. A more extensive study by Simpson (1964) revealed a considerable inland penetration of sea breezes, of the order of 40 km. Using a rise in dew-point temperature as the indicator of sea breeze arrival at a point, he determined that the breeze is established during the late morning in the immediate coastal zone, but then develops landwards. The incursion of sea air over the land surface may do so in the form of a weak cold front, at which cumulus cloud develops.

Broadly similar sea breeze characteristics have been found for many coastal areas in both temperate and tropical latitudes. A typical sea breeze develops during the late morning at the coast, then

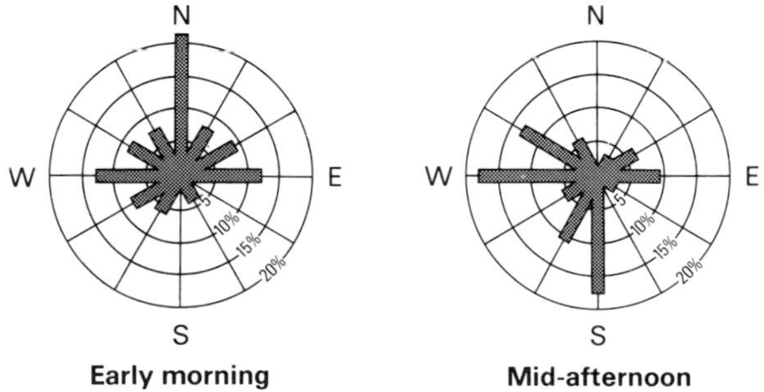

Figure 9.12 Wind roses for Portsmouth (Lion Terrace) in the summer of 1975.

gradually extends its influence inland, when geostrophic wind velocities are relatively low. Wind speed increases rapidly during the late morning and into the early afternoon when it reaches its maximum, which in Britain would be of the order of 4 m s^{-1}. The onshore flow of air extends upwards from the surface to a height of 750 m. In addition to this, there is a weaker offshore return flow aloft which reaches its maximum velocity some 2000 m above the surface.

A simplified view of the sea breeze as a localised circulation of air is as a thermally direct cell. During the daylight hours land surface temperatures increase more rapidly than those of the sea surface, due to a difference in heat distribution mechanisms. This differential heating creates a single thermally direct cell. Air rises above the warmer land surface and is replaced at lower levels by an inflow of air from the sea. To compensate for this, there is a return circulation aloft and a sinking of air over the sea (Fig. 9.13). On this basis, the greatest differences in surface temperature should give rise to the strongest circulation.

The pressure gradient force in the simple model operates directly from sea to land and thus wind direction may be mistakenly assumed to be normal to the coast. All moving bodies are, however, subjected to a Coriolis force which deflects their motion away from the direction of the pressure gradient force. A sea breeze blowing northwards across a coast in the northern hemisphere will, for example, develop an eastward component of motion. Defant (1951) has illustrated the combined effect of Coriolis force and friction upon sea breezes along the Massachusetts coast. Using both theoretical calculation and actual observation, he has shown that the sea breeze is deflected as much as 45° away from normal to the coastline.

Sea breezes are most frequently developed under warm coastal conditions during the middle-latitude summer and in tropical latitudes, which tends to support a simple interpretation based

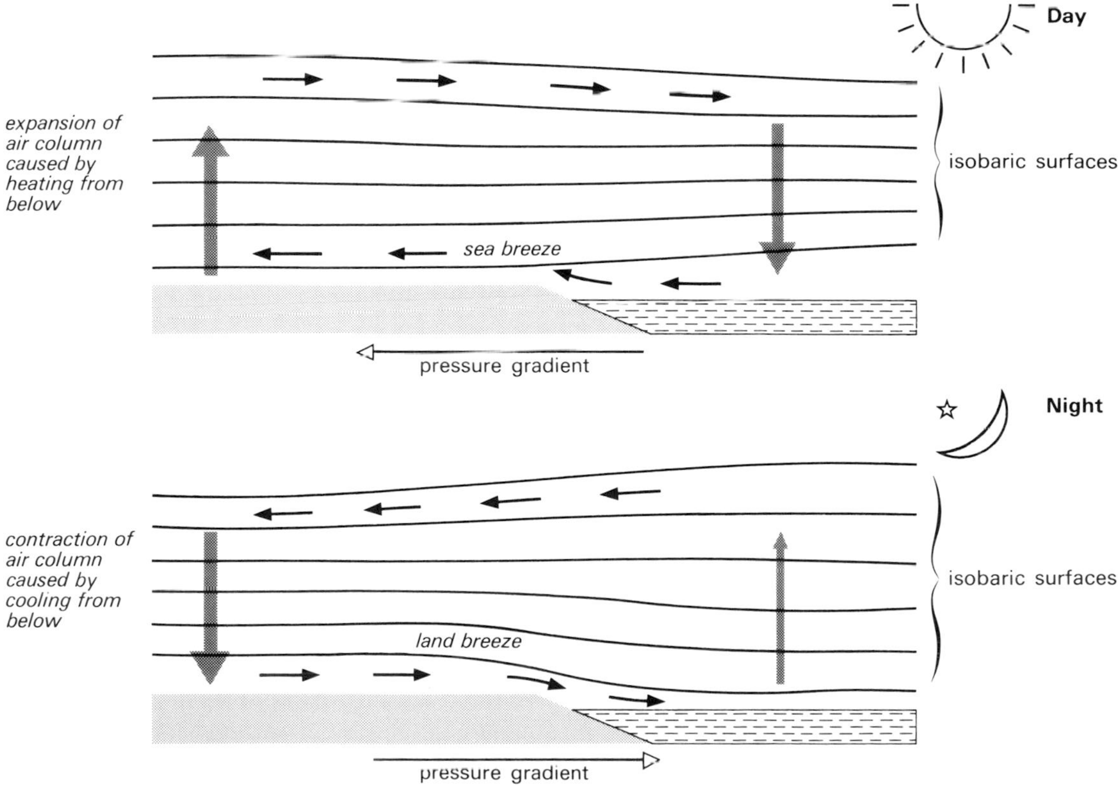

Figure 9.13 Simple model of the circulation of air over coasts.

upon temperature differences. However, observation has shown that the breeze may not develop despite there being relatively large temperature differences between land and sea, or may develop when there are much smaller differences. Watts (1955) determined that along the Sussex coast, at Thorney Island, the timing of the onset of the sea breeze was not related to the magnitude of temperature differences alone but also to the stability of the atmosphere. Under unstable atmospheric conditions the onset was rapid, whereas under stable conditions only gradual changes in wind direction were observed. The inference is that sea breezes are most readily developed where there are atmospheric conditions conducive to the rapid uplift of air over the land surface.

As the air is heated over the land surface under 'suitable' atmospheric conditions, there is rapid vertical expansion of the overlying air and, at a height of about 2 km, pressure begins to rise. This results in air flowing seawards at this level (Fig. 9.13). This divergence aloft causes pressure at the land surface to fall as air rises to take its place, while the arrival of air above the sea surface causes a convergence aloft. Air subsides beneath this convergence causing an increase in atmospheric pressure at sea level. The development of a surface pressure gradient gives rise to the landward air flow of the sea breeze.

In terms of energy flow within a circulation system, the sea breeze derives potential energy from the greater release of heat into the lower atmosphere over land surfaces than over the sea surfaces. This is converted into kinetic energy which is partly dissipated by friction. As sea breezes may develop when there is a regional wind blowing, this may add a direct input of kinetic energy.

A land breeze which blows in the opposite direction to the sea breeze is often similarly represented as a thermally direct cell. Cooler land temperatures at night cause the air to subside and diverge at the surface, while over the sea surface there is convergence and uplift (Fig. 9.13). The land breeze begins to blow offshore before midnight and develops to reach a maximum velocity immediately after sunrise. The low velocity of the land breeze, usually less than 2 m s^{-1}, may be a result of the low magnitude of vertical uplift in this circulation system and the slow progress of subsidence over the land. In many cases, however, there is no obvious return circulation aloft, this being weak and readily obscured by the regional wind. Alternatively, the land breeze may be interpreted as a gravitational downslope movement of cold dense air, related as much to inland topography as to temperature differences between land and sea surfaces. This type of airflow will be considered in more detail in the following section.

9.2.3 Slope winds

Over sloping terrain, local winds are generated as an upslope (**anabatic**) flow or a downslope (**katabatic**) flow of air. If we consider a uniform slope receiving direct solar radiation (Fig. 9.14), its surface is heated which in turn heats the overlying air. The result of this is the vertical expansion of the air and the creation of an upslope pressure gradient. The resulting upslope flow is compensated by a downward sinking of air over the foot of the slope and a flow away from the slope summit, completing a circulation. Over a slope receiving solar radiation directly after sunrise, the unstable anabatic wind begins to blow within an hour of initial slope heating and continues to intensify until maximum slope temperatures are reached in the early afternoon.

Nocturnal cooling of the same slope causes a vertical contraction of the air above it (Fig. 9.14) thereby developing a downslope pressure gradient along which there is a katabatic flow of air. This is compensated by a net inflow of air at the top of the slope and a rising of air above the slope foot, thereby completing a circulation. The stable katabatic wind begins to blow about one hour after sunset and it gradually intensifies as cooling proceeds, to reach a maximum velocity around dawn.

Both upslope and downslope flows are best developed under clear skies, commonly under anticyclonic conditions. Over extensive slopes anabatic winds may reach maximum speeds of about 4 m s^{-1} compared to the 2 m s^{-1} of the katabatic wind. The former develop to depths of 200 m or more while the latter are more closely confined to a layer up to 150 m thick. Within these moving layers of air, maximum velocity is usually attained a little way from the surface, due largely to the retarding effects of friction (Fig. 9.14).

The development of these slope winds is closely related to the radiation balance of the slope surface which, as can be inferred from Chapter 8 (p. 178) is extremely complex. Aspect, slope angle

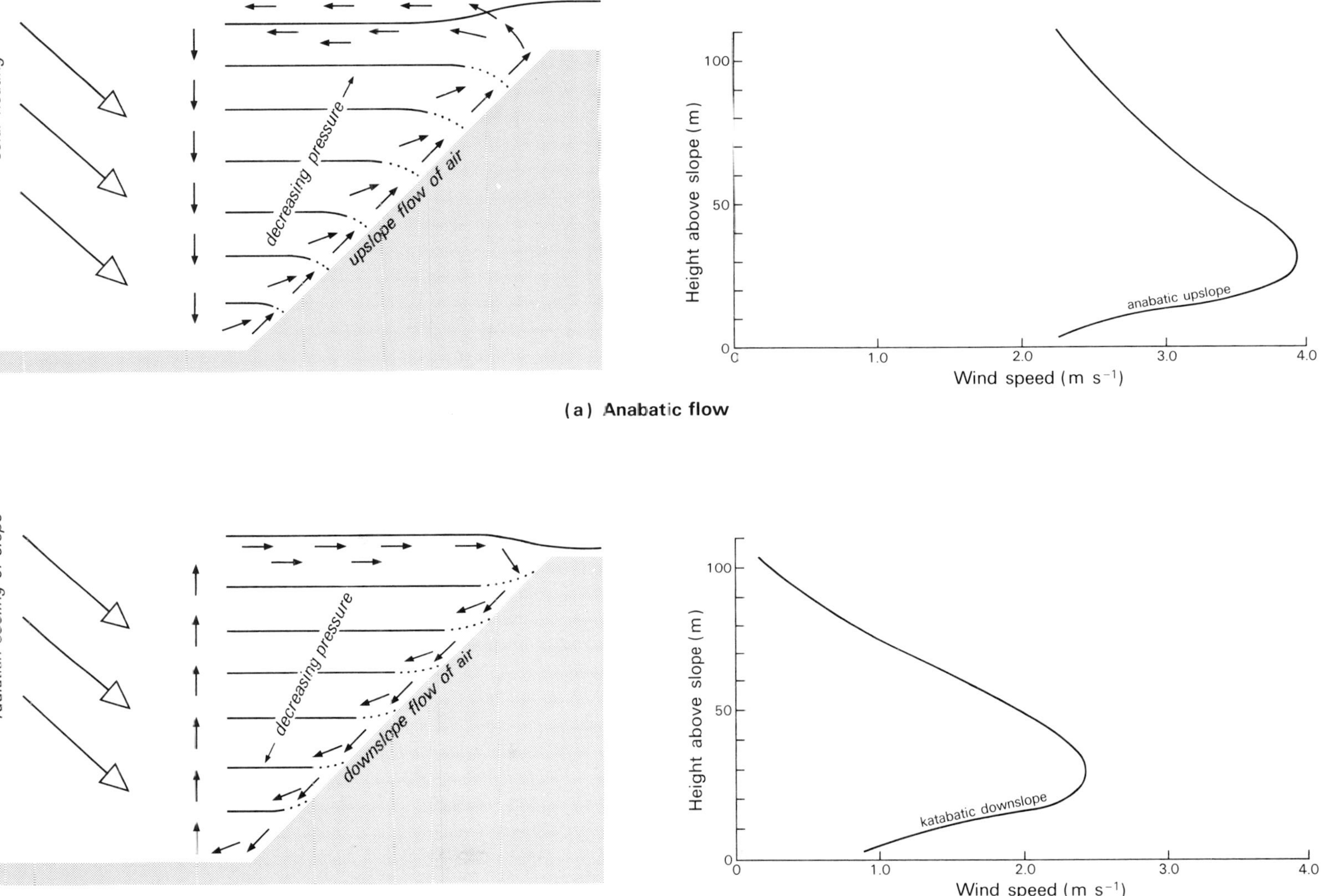

Figure 9.14 Development of anabatic and katabatic airflow over a uniform slope.

and degree of shading by nearby topographic features will affect the character of airflow over slopes.

As a circulation system, the anabatic wind is thermally direct (Fig. 9.15). Potential energy is derived from the heating of the slopes and the raising of the temperature of air in contact with them. There is, as a result, a temperature difference between the air layer in contact with the slope and air in the free atmosphere at the same elevation above the slope foot. In the case of katabatic winds, the potential energy is derived from a reversal of this temperature difference as the slopes are cooled. This potential energy is converted into kinetic energy which is partly dissipated by friction, particularly that between slope surface and the air.

If we consider two slopes forming a symmetrical valley, each will develop a thermal slope wind (Fig. 9.16). Assuming that the slopes are equally heated, the daytime movement of air, particularly under clear skies and relatively calm weather conditions, will be up slope (Fig. 9.16a–d). This creates divergence in the centre of the valley where air subsides to take its place (Fig. 9.16e,f). If we also assume that the slopes cool at similar rates, then the resulting downslope flow of air should create convergence in the valley floor (Fig. 9.16g,h). However, as the air is usually cold and dense, uplift above the valley floor is restricted. The cold air may remain in the valley, accumulating there as a cold pool in which temperatures may fall below 0°C.

If the radiation balances of the two slopes are dissimilar, the two-cell circulation becomes asymmetric. For example, a valley which trends north to south will experience roughly similar solar heating and cooling on its slopes and will develop airflow similar to that represented in Figure 9.16. However, valleys that trend east–west have a warm and a cold slope. In this case, the cold slope develops a weak circulation during the day and valley circulation is dominated by that above the warm slope.

The slope of the valley floor itself, from mountain to plain, superimposes another circulation upon that within valleys. The temperature contrasts between the air over mountain slopes and over the distant plains generate a larger-scale circulation of air which uses the valleys as natural channels of flow. These are referred to as mountain-valley winds. The flow of air within these valleys will be a complex combination of mountain to plain (and vice versa) and valley-side flows, which have been represented in simplified form by Defant (1951) (Fig. 9.16).

9.2.4 *The Föhn wind*

Air which is forced to flow over hills or a mountain range does not necessarily generate localised circulations, but may be modified to produce distinctive changes in the characteristics of flow. One such modification is the Föhn wind which blows as a warm, dry and blustery wind down lee slopes. The name Föhn was originally applied in the Austrian Tyrol. There is also a variety of local names including the Chinook of the Rockies and the Zonda of the Andes.

The onset of the wind is usually accompanied by extremely rapid increases in temperature.

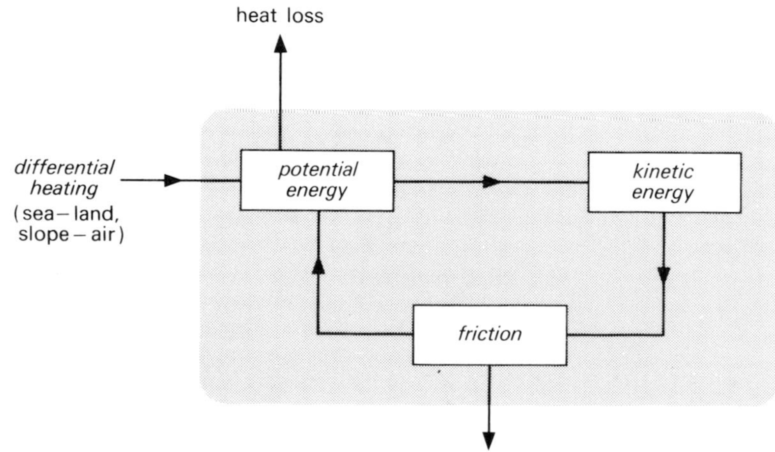

Figure 9.15 Tertiary circulation of a thermally direct type as a simple energy system.

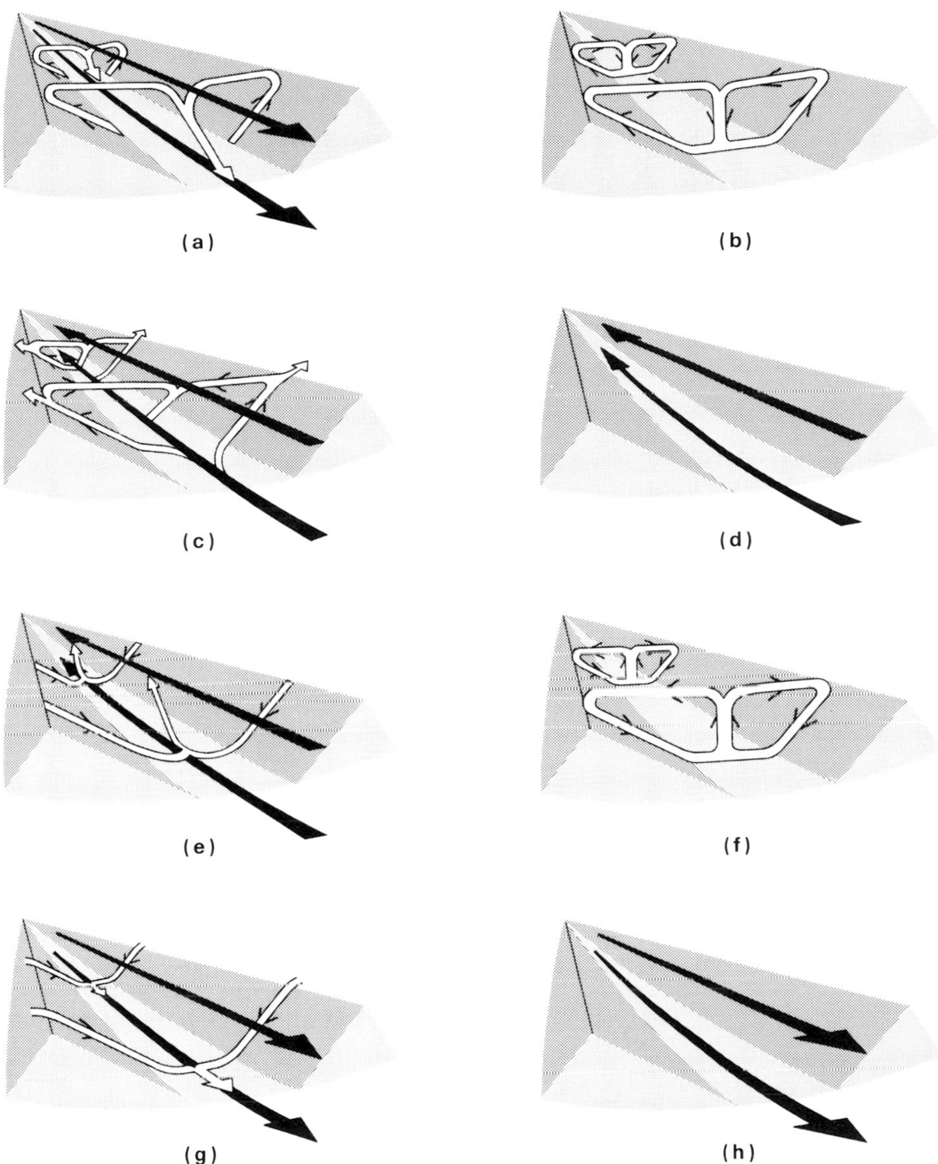

Figure 9.16 Schematic illustration of the normal diurnal variation of air currents in a valley (after Defant 1951). (a) Sunrise: onset of upslope winds (white arrows), continuation of mountain wind (black arrows), valley cold, plains warm. (b) Morning (*c.* 0900h): strong slope winds, transition from mountain to valley wind, valley temperature same as the plains. (c) Noon and early afternoon: diminishing slope winds, fully developed valley wind, valley warmer than plains. (d) Late afternoon: slope winds have ceased, valley wind continues, valley continues warmer than plains. (e) Evening: onset of downslope winds, diminishing valley wind, valley only slightly warmer than plains. (f) Early night: well developed downslope winds, transition from valley to mountain wind, valley and plains at the same temperature. (g) Middle of night: downslope winds continue, mountain wind fully developed, valley colder than plains. (h) Late night to morning: downslope winds have ceased, mountain wind fills valley, valley colder than plains.

Changes of 21 C° in four minutes have been recorded in Canada at the onset of the Chinook. On a smaller scale, Lockwood (1962) has investigated the occurrence of Föhn winds in Britain which produce unseasonal high temperatures. Weather records for Kinloss in Scotland contain evidence of a southerly Föhn wind, most probably emanating from the Cairngorms, its arrival being marked by a distinct decrease in relative humidity and an increase in air temperature (Fig. 9.17).

One explanation of the Föhn effect is outlined in Figure 9.18a. Conditionally unstable air at temperature T_1 is forced to rise over the mountain ridge and it cools at the dry adiabatic lapse rate (AB). After cooling to condensation level, further uplift causes cooling to continue at the slower saturated adiabatic lapse rate (BC). This produces orographic cloud which presents a cumulus wall to the lee side – known as the Föhn wall. If precipitation falls from these clouds, the air experiences a net loss of water and a net gain of sensible heat through the release of latent heat of condensation. As water has been lost there is little consumed in re-evaporation once the air begins to travel down the lee side of the ridge and to warm adiabatically. In its descent, the air is, therefore, warmed at the dry adiabatic lapse rate (CD) to temperature T_2, which is greater than T_1. The unmodified airflow contains both potential (internal heat and gravitation) and kinetic energy. As the air is forced to rise over the mountain barrier, kinetic energy is converted into potential. As the air flows down the lee slope, this is converted back into kinetic. The outflow of energy in the modified flow will be equal to that of the unmodified flow, with the exception of energy lost through friction.

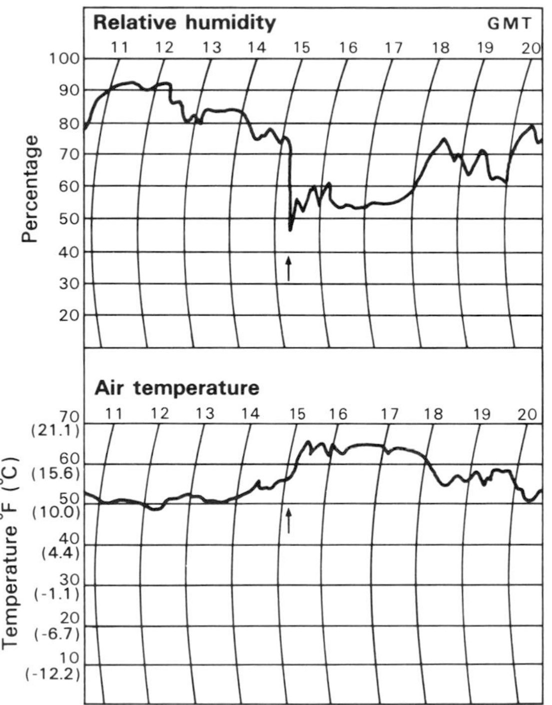

Figure 9.17 Extracts from humidity and air temperature traces (Kinloss, Morayshire, Scotland, 12th March 1957) showing the onset of a Föhn wind (from Lockwood 1962).

As the temperature increase is directly attributed to a net gain from latent heat release, this model of the Föhn requires precipitation to fall from the orographic cloud. In this respect, theory does not match observation, as Föhn winds often occur without orographic loss of moisture. Therefore, more recent theories have tended towards the alternative view that the Föhn wind is generated by the forced descent of upper air rather than the ascent and descent of lower air (Fig. 9.18b). This relatively dry air may be forced to descend and warm adiabatically with the formation of a lee wave downwind of the mountain barrier. This alternative thus involves no release of latent heat. Descending air converts potential energy to kinetic and, as warming is adiabatic, no internal heat is either lost or gained by the air.

9.3 Linkages between atmospheric systems

The movement of air within the Earth–atmosphere system, represented here as secondary and tertiary circulation systems, arises most frequently from spatial and temporal variation in heat energy and moisture supply from the underlying water or ground surface. The range of scales of such variation results in both air movements of short duration operating over small distances, such as the sea breeze, and movements of longer lifespan such as the extratropical cyclone which influences weather patterns over extensive ocean and land surfaces. However, irrespective of scale, there is a basic similarity in all these systems in that they all involve the conversion of potential into kinetic energy and its dissipation by both internal and external friction. Essential to their continued operation is the maintenance of inputs of adequate potential energy. However, we have also seen that, while inputs of potential energy are freιG

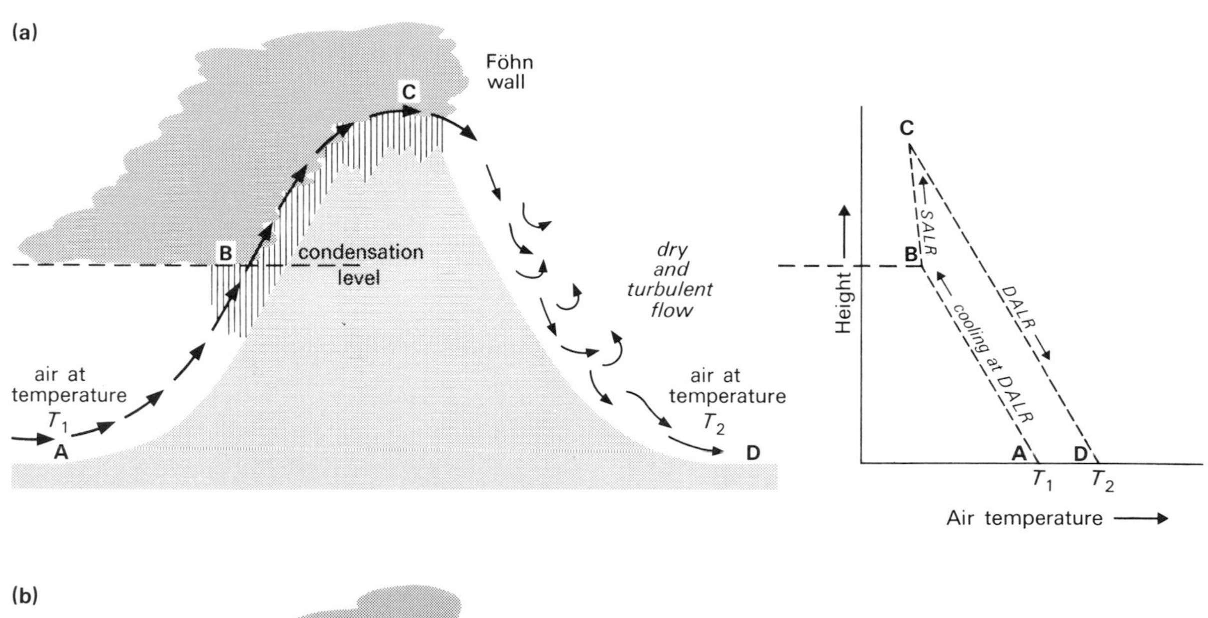

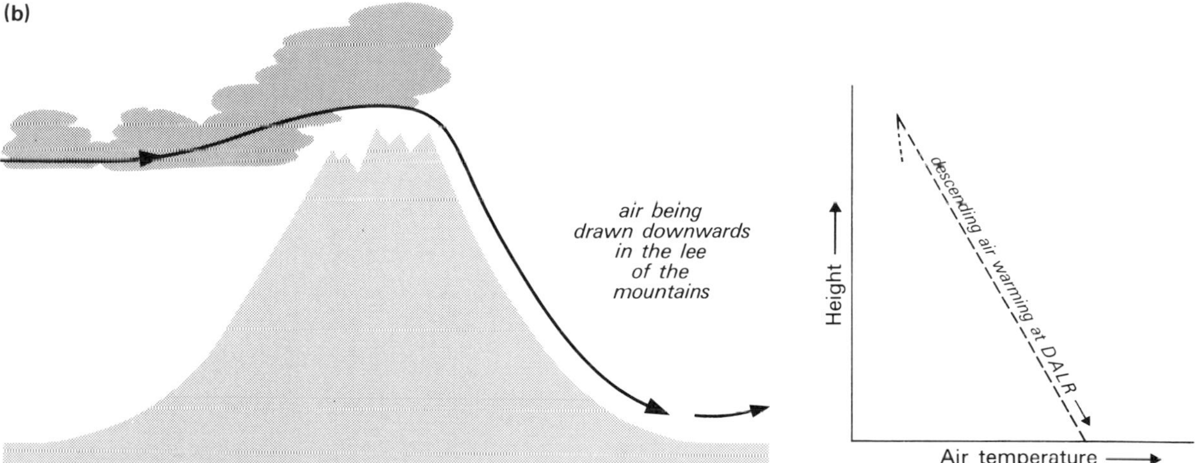

Figure 9.18 Alternative models of the Föhn wind.

quently a result of interaction between the atmosphere and the underlying surface, the operation of circulation systems is also conditioned by the larger scales of atmospheric motion within which they develop. For example, the development of the extratropical cyclone is very closely related to the waves of planetary scale in the upper westerlies of middle latitudes. Similarly, the smaller-scale sea breeze circulation requires certain conditions of atmospheric stability and relatively low gradient wind velocity in order to develop. We have, therefore, a series of mutually dependent – rather than self-contained – atmospheric systems.

Further reading

Atkinson, B. W. (ed.) 1981. *Dynamical meteorology*. London: Methuen.

Barry, R. G. and R. J. Chorley 1976. *Atmosphere, weather and climate*, 3rd edn. London: Methuen.

Chandler, T. J. and S. Gregory (eds) 1976. *The climate of the British Isles*. London: Longman.

McIntosh D. H. and A. S. Thom 1972. *Essentials of meteorology*. London: Wykeham Publications.

Riehl, H. 1979. *Climate and weather in the tropics*. New York: Academic Press.

V DENUDATION SYSTEMS

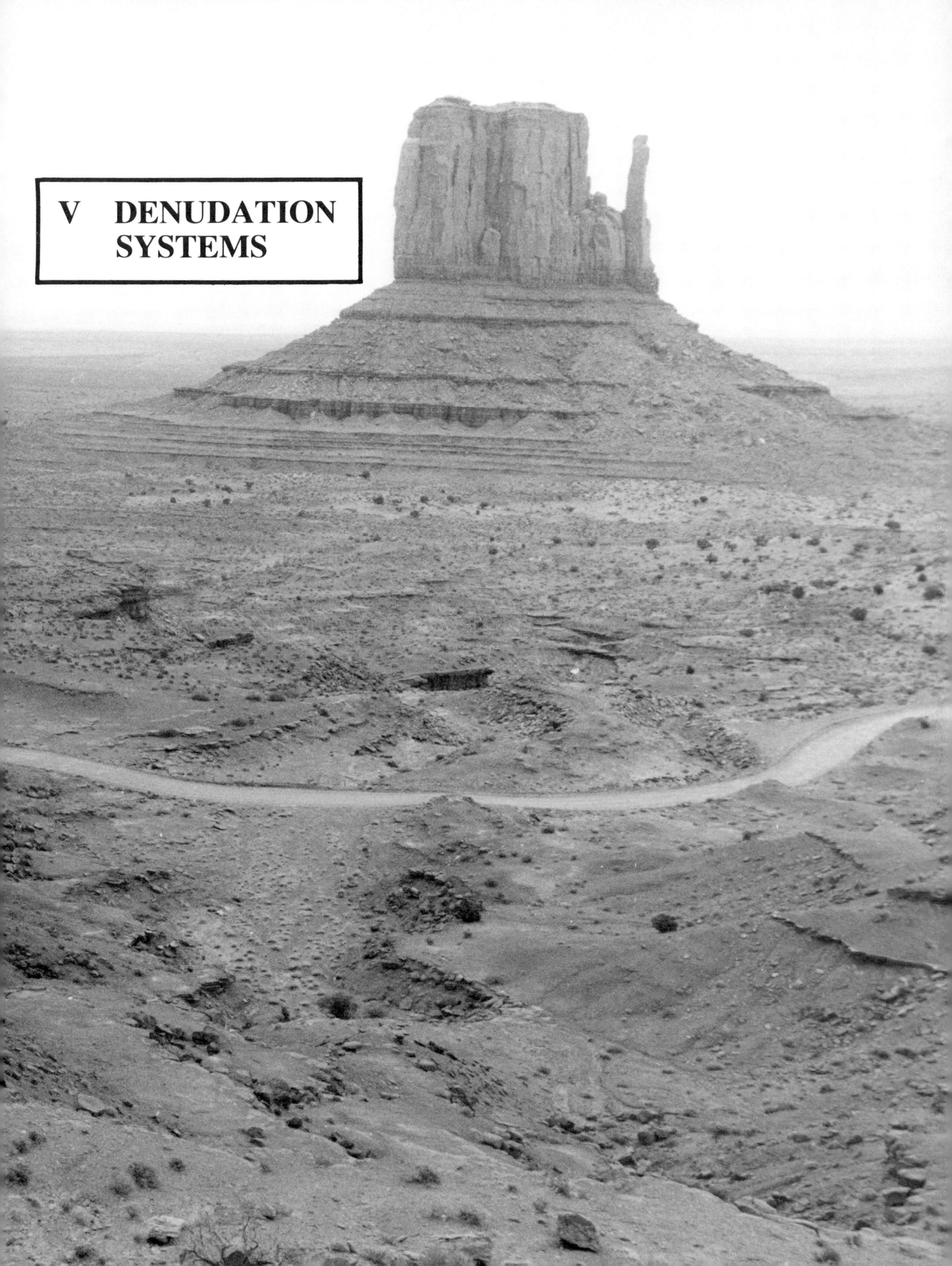

10 The catchment basin system

Figure 10.1 shows a scene typical of upland landscapes in a humid temperate region. In it we can discern a number of features, some of which we can conceptualise as systems and some we can regard as elements (or components) of these systems. The overcast sky represents the atmospheric system containing water vapour which we have considered in Chapters 8 and 9. However, we can see that the form of the land surface is composed of a number of elements. In the foreground a river of significant dimensions sweeps by and an accumulation of fine sediment is visible on the inside of the meander bend, in addition to some large boulders scattered in the channel. The river is flanked by slopes of very low gradient, where again large boulders can be seen on the ground contained within the meander bend. Rising above the valley floor are slopes of moderate gradient, whose complex three-dimensional form can be described in terms of both vertical profile and horizontal curvature. Small outcrops of solid bedrock are visible towards the top of the cliff at the left-hand side of the scene, but otherwise the slopes are mantled by weathered bedrock and soil, not visible here because of the sward of vegetation. The relationships between these elements – the arrangement of the slopes composing the land surface, the river, the solid bedrock and the sediments – represent the denudation system. The land surface is mantled by an almost continuous cover of vegetation, for the most part grassland and heath but with significant areas of woodland. This of course represents the ecosystem.

In this chapter we shall be concerned with the relationships at the interface between these major systems – the atmospheric system, the denudation system, and the ecosystem. At this interface vast quantities of water are held in, and transferred between, various stores (see Ch. 4). Here the rocks of the Earth's crust and their mineral constituents are broken down physically and chemically and are transferred to lower elevations. All of these processes, involving both water and debris, take place in the denudation system. Indeed, it is the operation of the denudation system on relief created by the crustal system that determines the morphology of landscape and provides the functional environment that sustains life. However, at the scale at which we perceive landscape – the scale represented by Figure 10.1 – it is more appropriate to model the operation of the denudation system as the **catchment basin system**. Here, therefore, we shall consider in a broad way the organisation of the catchment basin system, the elements of which it is composed, and their functional relationships. In Chapters 11 to 14 individual subsystems of the model are isolated and considered in detail. However, the inputs of matter and energy to the catchment basin system show significant spatial variations and it is these that in turn are largely responsible for spatial variations in the state of the denudation system and hence in the form of the world's landscapes. In Chapter 15, therefore, we shall examine such spatial variation.

10.1 The structure of the catchment basin system: functional organisation

The basic functional unit through which the denudation system operates is the drainage or catchment basin. For the purposes of this chapter, the catchment basin can be regarded as a system in

Figure 10.1 A typical fluvial landscape in an upland temperate region: West Dart River, Devon, England.

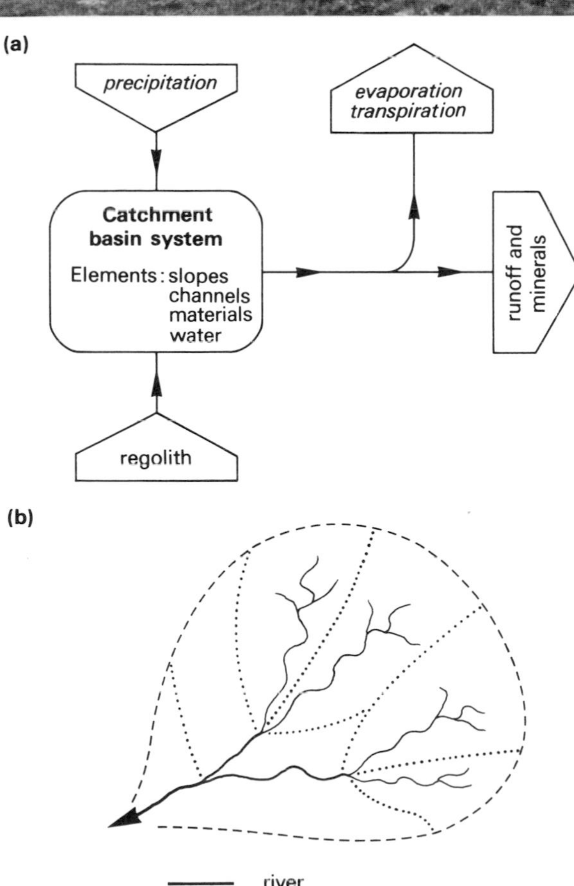

its own right. As such, it has a well defined boundary and its component elements show clear relationships both structurally in terms of its morphology and functionally by virtue of the flow of matter and energy through the system. The inputs and outputs across the system boundary can also be clearly distinguished (Fig. 10.2a).

The boundary of the catchment basin system can in part be regarded as the ground surface and surface of water bodies within the catchment. Its areal limits are set by a major watershed, while minor watersheds define subcatchments within it (Fig. 10.2b). Its lower boundary within the lithosphere is more difficult to define. For the moment this can be thought of as a surface defining both the lower limit of the weathering system and of the active transfer of water within the system.

Figure 10.2 (a) Simplified schematic model of the catchment basin system; (b) delimitation of catchment watersheds.

The major elements in the catchment system are channels, slopes, bedrock and regolith, and water. These are related in an organisation of slopes and converging channels which facilitates the basic function of the system: to evacuate runoff and debris from the catchment basin. If the simple model of the catchment basin system in Figure 10.2a is expanded and refined, these elements can be considered as composing several functional units or subsystems. Each is related to the others and all are an integral part of the function of the basin as a whole.

In Figure 10.3a the slope system, channel system and weathering system have been separated and the relationships between them indicated in a more sophisticated model. The slope system is functionally linked to the weathering system by the transfer of materials both directly as weathering products from the regolith (Fig. 10.3b) and indirectly through the soil and the ecosystem (see Ch. 19) in the form of litter and decaying organic matter. Precipitation forms a link between the slope system and the atmospheric system, as also do the dissolved materials in that precipitation and the dry fallout from the atmosphere. In addition, some elements from the atmosphere are fixed in organic compounds by the organisms of the ecosystem, and these form inputs to the slope system when the organisms die and decay. Water output occurs as evaporation and transpiration back to the atmosphere, as surface runoff into channels and as infiltration entering the soil and regolith as either soil water or ground water. Surface runoff will remove material as solid sediment particles and also in solution.

The outputs from the slope system form the major link with the channel system. Overland flow of water down hillslopes during storms feeds directly into the river channel, and the bulk of the percolating soil water and ground water finds its way into the channel, sustaining channel flow between rainstorms. The third water input to the channel (volumetrically the least important) is direct precipitation onto the river channel surface. Output of water from the channel is partly by direct evaporation from the water surface, although this is normally limited except under high temperatures and very broad channel sections, as for example where the river passes through a lake. The bulk of the water output from the channel is via runoff which contains sediments and solutes. This channel runoff normally flows to the ocean, where it can be regarded as input to the ocean store, which contains 97% of the Earth's water.

The weathering, slope and channel systems, therefore, form a cascading system organised so that the outputs of one are the inputs of another, and together they form the catchment basin system. This model of the denudation system can be applied even to deserts and to landscapes experiencing active glaciation, as well as to fluvial landscapes. These can be regarded as special cases which differ only in the inputs to the system and in the pathways and rates of matter and energy flow through the system. The subsystems of this model are treated in more detail in Chapters 11 to 15.

10.2 System structure: spatial organisation

So far, the structure of the denudation system has been treated in terms of its functional organisation. Its structure, however, can equally be described in terms of its spatial organisation. Although it often parallels functional organisation, this is a particularly useful approach for it enables the structure of the system to be interpreted in terms of the surface morphology or form of the catchment basin.

If we turn again to the simple model of the catchment basin system (see Fig. 10.2), both slopes and channels can be regarded not as subsystems but as elements in the system. These elements can be described in terms of their attributes and, in particular, their spatial relationships. The quantitative treatment of these attributes is known as **morphometric analysis**. Measures have been devised which analyse the linear aspects of the catchment, its areal properties and its relief characteristics. Each of these will be examined in turn, but initially we shall consider the basic morphological element of catchment basins – slopes.

According to the simple classification scheme of Young (1972), based on surface gradient and slope plan form, any catchment basin consists of no more than five fundamental types of slope. Young classifies these into two groups: flats (or nearly so)

(a)

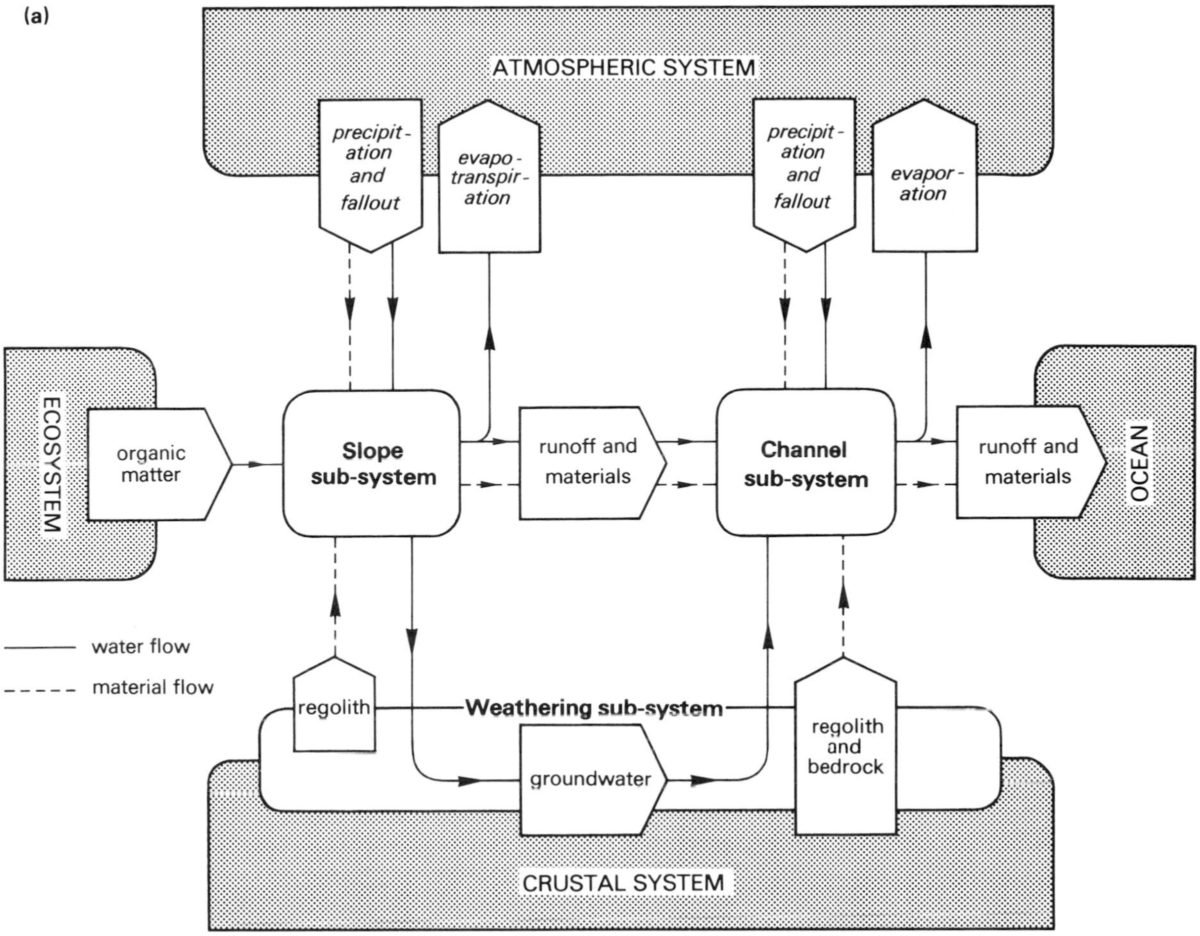

ATMOSPHERIC SYSTEM

precipitation and fallout

evapotranspiration

precipitation and fallout

evaporation

ECOSYSTEM

organic matter

Slope sub-system

runoff and materials

Channel sub-system

runoff and materials

OCEAN

water flow

material flow

regolith

Weathering sub-system

regolith and bedrock

groundwater

CRUSTAL SYSTEM

Figure 10.3 (a) A schematic model of the functional organisation of the catchment basin system. The intimate functional relationship between the slope and channel systems is clearly to be seen in the accompanying photograph (b) of the Colorado River, Arizona, USA, where rock debris is fed directly into the river channel from the steep rock walls and talus slopes.

and valley slopes (Fig. 10.4). Flats are found in two locations: as interfluve remnants they can represent undissected portions of an original surface into which the catchment basin is incised, and can contribute to catchment processes by groundwater flow; flats also occur on the valley floor, where they normally represent the flood plain of the river. In terms of the mass movement of materials by slope processes, flats are insignificant, since the effects of gravitational energy are nullified by such low gradients.

Valley slopes are more important in terms of erosional processes, for on them the movement of material by gravitational processes is facilitated. Slope gradient is important in this respect, since gravitational force is proportional to the sine of slope angle (see Ch. 12). Also important is the form of the slope in plan. Three variations are possible: valley-side slopes, valley-head slopes and spur-end slopes. Valley-side slopes are straight in plan. They can be considered in profile as a simple linear form since lateral movement across them can be neglected. Valley-head slopes are concave in plan and they converge towards the base. The opposite is the case on spur-end slopes, convex in plan, where divergence occurs down slope. The plan of the slope in respect of divergence or convergence towards the base has important implications for the balance of materials and processes on the slope and it leads to differing basal conditions.

This arrangement of morphological components holds true for all landscapes on Earth. Even plainlands are rarely entirely flat, but consist of gently shelving slopes separated by divides. According to Young, only 7% of the Great Plains of the USA is truly flat land, and of the Mato Grosso plateau of Brazil only 5% is truly flat. At the other end of the scale are mountain landscapes which consist entirely of steep valley-side slopes separated by sharp crested divides. In such landscapes, plateau remnants have been entirely removed and flood plains have not developed. The five basic morphological components, therefore, represent the possible variety of slope types, though they may not all be present within an individual catchment basin system.

Table 10.1 River Wallington: network characteristics.

Stream order	No. of streams	Bifurcation ratio	Total stream length (km)	Mean stream length (km)	Ratio of stream lengths
1	69	3.6	26.3	0.38	2.8
2	19	3.8	20.5	1.08	1.4
3	5	2.5	9.8	1.96	1.7
4	2	2.0	6.5	3.25	2.0
5	1		6.5	6.50	
		$\bar{R}_b = 2.97$	$L = 69.6$		$R_L = 1.97$

The spatial organisation of the system – that is, the form of the landscape – can be described and measured in various ways. Any landscape possesses linear, areal and relief properties, and we shall examine each of these in turn.

10.2.1 Linear properties

The linear properties of the system primarily concern the distribution in plan of channel and valley networks. The network consists of a series of channels converging towards a single outlet. The organisation of the network can be analysed by allocating each channel segment a rank according to its position in the network (Box 10.1).

Data for the network of the River Wallington in Hampshire (England) are set out in Table 10.1,

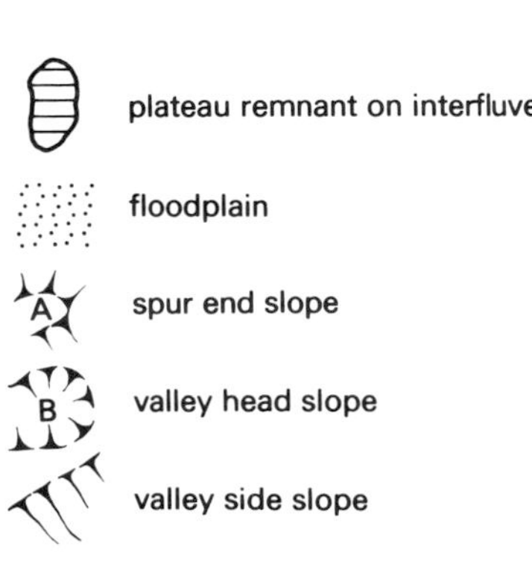

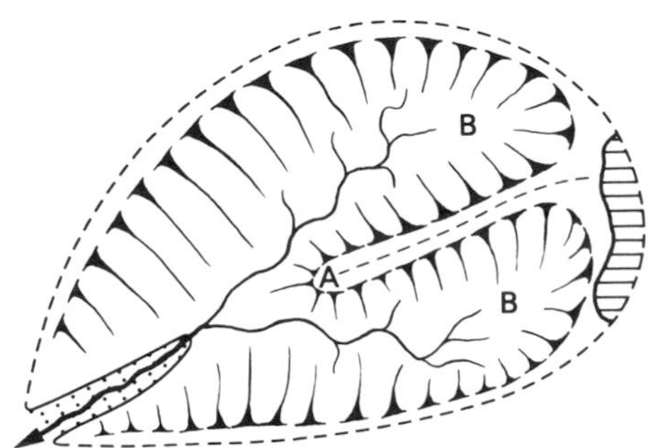

Figure 10.4 Morphological components of the catchment basin (after Young 1972).

Box 10.1

DRAINAGE BASIN MORPHOMETRY

Various methods of network description have been put forward, but the one most widely used is that of A. N. Strahler (1952). In this method, headwater streams which receive no tributaries are designated as first order. Where two first-order streams are confluent they form a second-order stream. The confluence of two second-order streams forms a third-order stream and so on, so that a stream of given order can receive streams of any lower order without its own order being raised. Stream order is raised only when two streams of the same order converge. Finally, the basin order is defined by the highest-order stream within it. Applying this form of analysis to the Wallington system it is shown to be a fifth-order basin. Within it are subcatchments of successively lower orders forming a nested hierarchy.

Analysis of the network can be carried out by tabulating the number of streams in each order. The number of streams decreases progressively through the higher orders in a geometric manner. The rate of change of stream number with stream order is described by the bifurcation ratio (R_b). This is calculated by dividing the number of streams in a given order by the number in the next (higher) order. The mean value of the ratios so calculated is the mean bifurcation ratio ($\bar{R}_b$) and it is characteristic of the network. The bifurcation ratio normally lies in the range 2–5. There is thus a simple geometric relationship between stream number and stream order, of the form:

$$N_u = \bar{R}_b^{(s-u)}$$

where N_u = number of streams of order u, $\bar{R}_b$ = mean bifurcation ratio and s = basin order. For the River Wallington, this equation becomes:

$$N_u = 2.97^{(5-u)}.$$

These data are shown graphically in Figure 10.6.

The values of mean stream length also show a progressive change with stream order. In the same way as the bifurcation ratio describes the rate of change between orders, so the ratio of mean stream lengths can be calculated and a characteristic value derived for the basin. The relationship between mean stream length and stream order shows an increase in the direction of higher orders. It is described by the following equation:

$$\bar{L}_u = \bar{L}_1 \,.\, R_L^{(u-1)}$$

where $\bar{L}_u$ = mean length of streams of order u, $\bar{L}_1$ = mean length of first-order stream and R_L = ratio of mean stream lengths. For the River Wallington, this relationship becomes:

$$\bar{L}_u = 0.38 \times 1.97^{(u-1)} \quad \text{(see Fig 10.6).}$$

Thus, the simple geometric relationship between stream order and stream number is repeated with stream length. Similar geometric relationships exist between stream order and basin area (increasing with stream order), and stream order and stream gradient (decreasing with stream order).

The major advantage of the Strahler method is that it is easy to apply and it has also gained widespread acceptance. Within any given network, streams and basins of each successive order can be found up to the largest value present and this facilitates morphometric comparisons of populations of basin systems. There are, however, some limitations to this method which prompted Shreve (1966) to put forward an alternative method of network analysis.

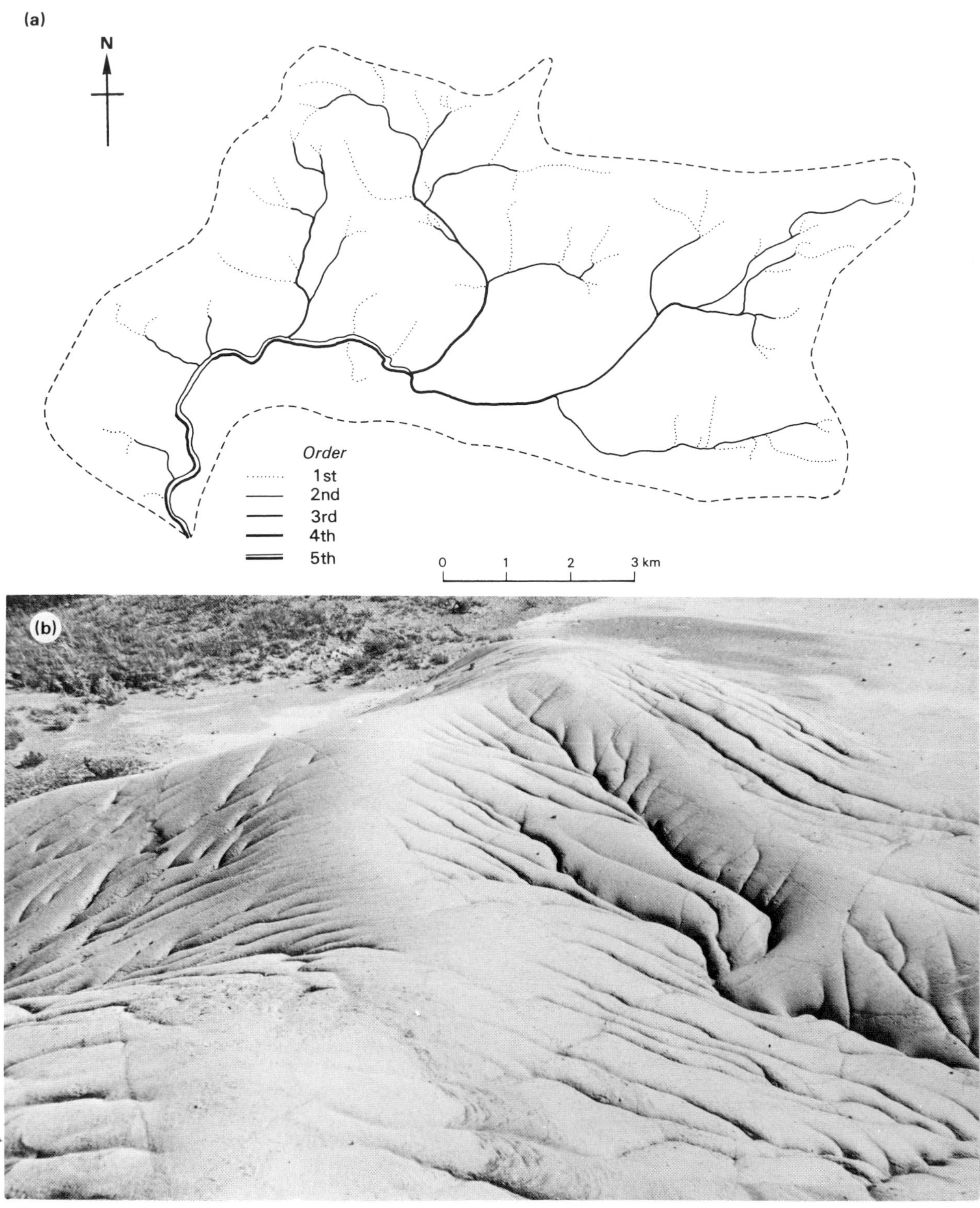

Figure 10.5 (a) The drainage net of the River Wallington (Hampshire, England) showing stream orders according to the Strahler system. (b) Small rills can be seen forming an integrated drainage network (Alberta, Canada).

and the network is shown in Figure 10.5. Analysis of the network shows that it is a fifth-order basin. It is shown to have an internal geometry with consistent relationships between stream order, stream number and stream length, and as such it is characteristic of stream networks in general (Fig. 10.6).

This kind of analysis can be applied also to valley networks, which are generally more extensive than the networks of perennial streams. This is particularly the case in regions of highly permeable rocks, seasonal climates, or where a former climatic regime created a more humid environment with greater surface runoff. Accordingly, it is sometimes more valuable to analyse the valley network, as indicated by contour lines. In this way, for example, the linear patterns of chalk and limestone terrains may be assessed.

10.2.2 Areal properties

Descriptive measures of the areal attributes and relationships of catchment basins are numerous, and only a representative selection of the more significant measures will be discussed here. Gardiner (1974) offers a comprehensive review.

One of the most fundamental catchment characteristics is that of drainage density. This is defined as the total length of drainage channel divided by drainage area, thus:

$$Dd = \frac{L}{A}$$

where Dd = drainage density, L = sum of total stream lengths and A = catchment basin area; and is expressed as a ratio in kilometres per square kilometre. Using the data for the Wallington system presented above, drainage density is shown to be 1.22 km km^{-2}. Comparative values for other areas are discussed in Chapter 15.

The value derived for drainage density will depend closely on the origin of the data used. Often based on cartographic data, it is common to use the mapped stream network for morphometric analysis. Clearly, the correspondence between this network and functioning channels in the field will depend on both the scale and the cartographic conventions of the map employed. For instance, the Mississippi system on an atlas map at a scale of 1:12 500 000 may be shown as a fourth-order basin, whereas Leopold, Wolman and Miller (1964) estimate that its network of perennial channels forms a 12th-order system. By omitting headwater tributaries, stream order is lowered throughout the system and drainage density is correspondingly reduced. Within the British Isles, Gardiner (1974) has shown that the 1:25 000 Second Series maps of the Ordnance Survey offers the most reliable cartographic source of morphometric data.

The most accurate method of determining drainage density is field mapping of actually functioning channels. This underlines the fact that drainage density and basin order are not constant values for a given system, but they vary through time. The amount of water within a catchment basin, recorded in groundwater levels and extent of the drainage net, will vary according to variations in precipitation input. These variations will occur throughout individual storms and also from

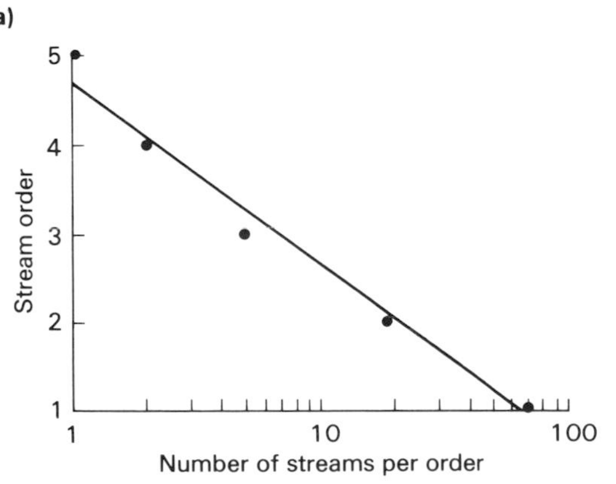

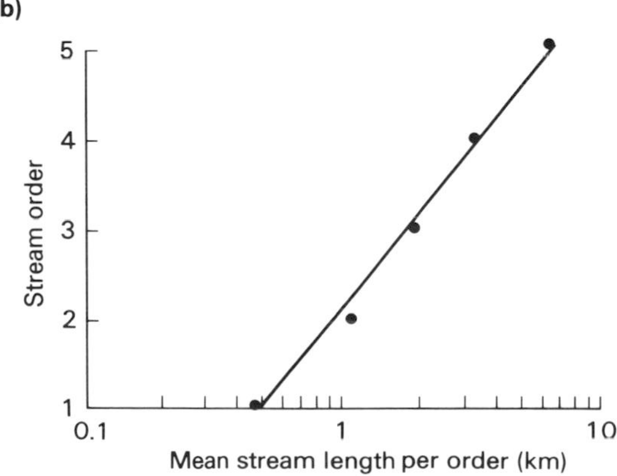

Figure 10.6 Morphometric relationships of the River Wallington (Hampshire, England).

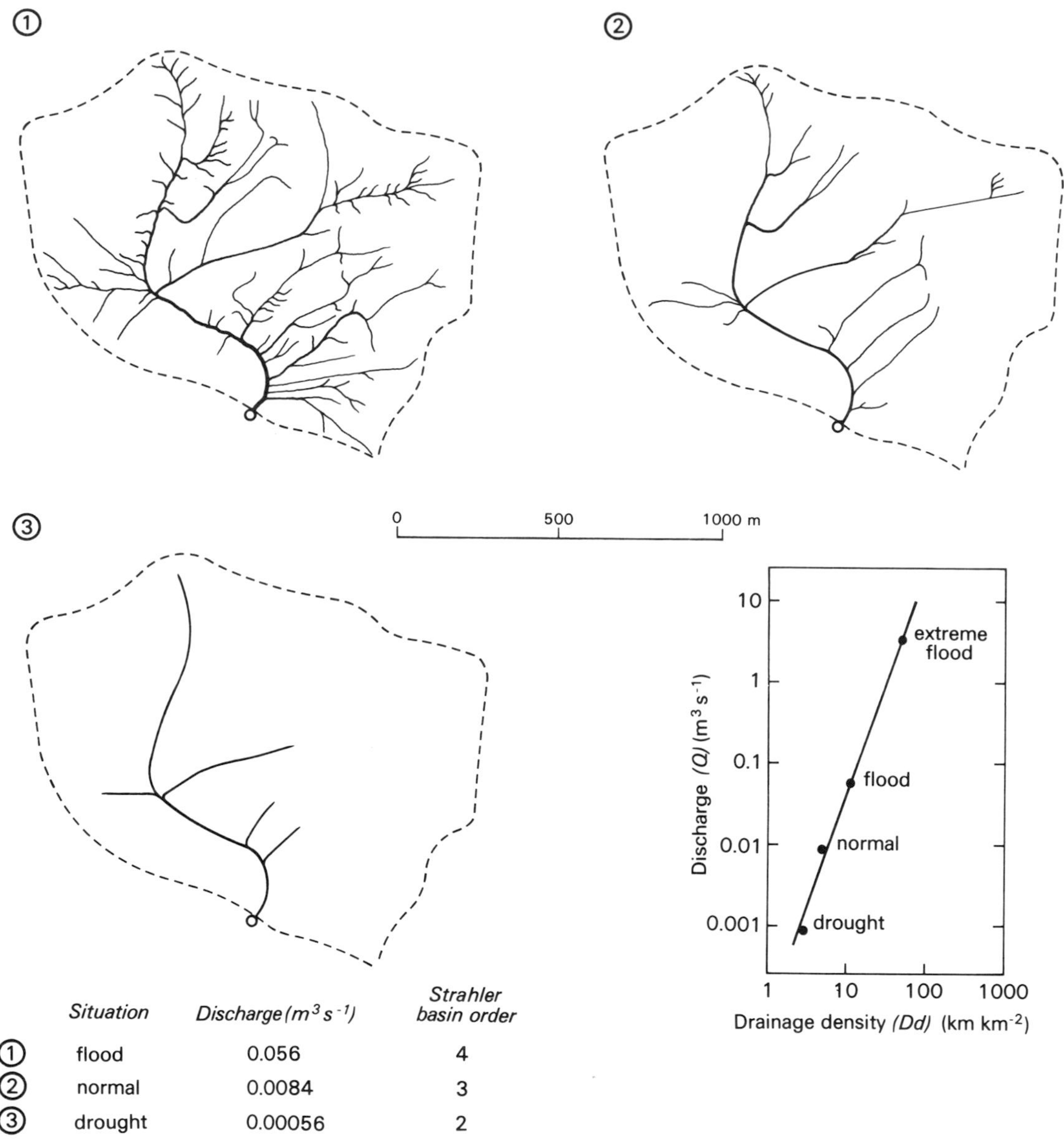

	Situation	*Discharge ($m^3 s^{-1}$)*	*Strahler basin order*
①	flood	0.056	4
②	normal	0.0084	3
③	drought	0.00056	2

Figure 10.7 Seasonal variation of drainage density (Swildon stream, Somerset, England; after Hanwell & Newson 1970).

season to season. Drainage density and basin order are therefore dynamic variables and are related to basin discharge. Hanwell and Newson (1970) show the variations in drainage density in a small catchment (1.18 km^2) at different levels of flow between extreme flood and drought (Fig. 10.7). There is a 20-fold variation in drainage density. As a result of this variation, the basin ranges from second order at low flows to fourth order at high flows.

Simple descriptive measures of basin shape are manifold, but at the same time are beset by problems of definition. Gardiner (1974) recommends the **elongation ratio**, defined as:

$$E = A/L$$

where E = elongation ratio, A = basin area and L = basin length. This yields values ranging from zero for a circular basin to values approaching 1 for more elongated basins. Basin shape is important in influencing the travel time of runoff from various parts of the basin to its outlet.

Basin relief can be expressed by a variety of indices. A simple measure is the **relief ratio**:

$$R_h = H/L$$

where R_h = relief ratio, H = vertical difference between highest and lowest points in the basin and L = maximum basin length. As such, it is a generalised measure of gradient and therefore of potential energy of the basin. It is thus a fundamental influence on denudation processes in the drainage basin, both on slopes and in river channels.

10.2.3 Relief properties

More detailed information concerning the distribution of altitude within the drainage basin can be gained from the hypsometric curve in which altitude is plotted against area. This is derived by measuring the area between successive contours throughout the altitudinal range of the catchment. The data can be plotted in dimensionless form where percentage height is plotted cumulatively against percentage area (Fig. 10.8a). The area beneath the curve, expressed as a proportion of the total volume, represents that volume of land remaining, assuming that the catchment basin has been eroded from a rectangular block of relief equal to the highest point on the watershed. This technique has been widely used as an index of the proportion of the basin removed by erosion.

A more flexible way of handling the same data is to plot relief against percentage area as a frequency distribution of altitude (Fig. 10.8b). From such data, mean relief, standard deviation and skewness can be calculated, giving a better statistical description of the distribution of relief within the catchment system. A negatively skewed distribution, for instance, would indicate a plateau with narrow incised valleys, and positive skewness would result from a broad lowland area with isolated residual hills.

To be able to assign numerical values to the spatial attributes and relationships of the system and its elements allows us not only to specify the state of the system's surface morphology but also to make meaningful comparisons between different systems on the basis of morphometric parameters.

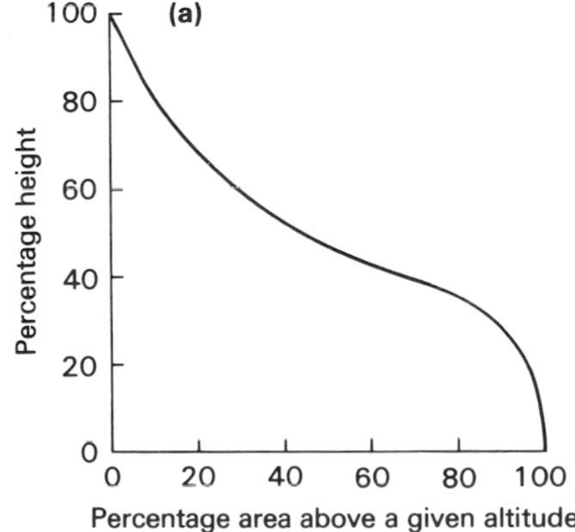

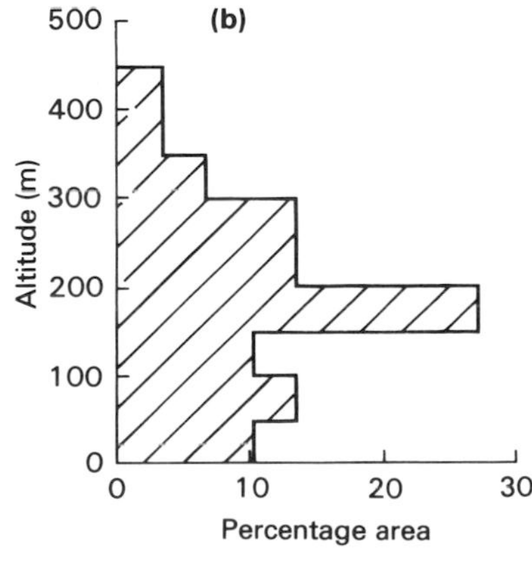

Figure 10.8 (a) Hypsometric integral; (b) altitude frequency distribution.

As a result of the operation of the catchment system, its geometrically distinctive spatial organisation will change with time and, as it changes, so the morphology of landscape will evolve. The state of the system's spatial organisation at any moment in time will, therefore, reflect the inputs to the system, and the relative roles and rates of the processes operating within it. As the latter are to a large extent conditioned by the former, the spatial organisation of the system will be determined mainly by the major environmental inputs – precipitation, lithology, relief, vegetation cover and time.

10.3 System structure: material components

The mineral components of the denudation systems are the rocks and materials derived from them by weathering processes. Their mineralogical characteristics have been outlined in Chapter 5. We are here concerned with their physical and mechanical characteristics, for it is these that govern their behaviour in relation to the applied forces of denudation processes.

The physical nature of the mineral components can vary widely, from solid rock to loose particulate material. Solid (indurated) rock behaves mainly in a rigid manner, whereas soil, or soil–rock mixtures, tend to display plastic or elastic behaviour. There is, then, a fundamental distinction based on the degree of induration, but we can nevertheless derive descriptive parameters that can be used to characterise the wide variety of states of mineral material in the system.

10.3.1 Basic properties

All rocks and soils can be considered to be composed of minerals and voids. Thus:

$$V = V_s + V_v$$

where V = total volume of the mass, V_s = volume of solids and V_v = volume of voids. The voids may be occupied either by air or by water. Thus:

$$V_v = V_w + V_a$$

where V_w = volume of water and V_a = volume of air. Thus we can define the material as comprising three elements: solids, air and water. The relative proportions of these elements permit the derivation of basic descriptive parameters.

(a) Voids ratio (e) is defined as the ratio of voids to solids:

$$e = V_v/V_s.$$

(b) Porosity (n) is defined as the ratio of voids to total volume:

$$n = V_v/V.$$

There is wide variation in porosity, particularly between consolidated and unconsolidated materials (Table 10.2).

(c) Moisture content (m) is defined as the ratio of weight of water to weight of solids, expressed as a percentage:

$$m = \frac{W_w}{W_s} \times 100$$

where W_w = weight of water and W_s = weight of solids.

These basic descriptive properties are used in relation to both indurated and unconsolidated materials. Voids ratio and porosity are fundamental in two ways: they determine the amount of water which a given material is capable of absorbing and they are an indirect measure of the extent to which adjacent grains or crystals are in contact, which contributes in part to the strength of the

Table 10.2 Average porosities for different materials (after Leopold, Wolman & Miller 1964 from various sources).

	Porosity (%)
CONSOLIDATED ROCKS	
granite	1
basalt	1
limestone	10
sandstone	18
shale	18
UNCONSOLIDATED MATERIALS	
gravel	25
sand	35
silt	40
clay	45

material. Furthermore, moisture content closely affects the way in which materials, especially unconsolidated ones, behave under stress.

10.3.2 Strength properties

It is essential to be able to determine strength properties if the way materials in the field behave under stress is to be understood. Engineering practice has developed a number of standardised procedures which permit strength to be determined in relation to the various types of applied stress (see Ch. 2). Since the most common stresses in the natural environment are shear stresses, we shall confine discussion to these alone, while bearing in mind that it is possible also to test for compressive and tensile strength (Duncan 1969). Strength properties are applicable to both consolidated and unconsolidated materials.

Values of shear strength can be determined under controlled conditions in the laboratory (Box 10.2). These also permit the components of strength to be isolated, in addition to other factors which influence strength properties.

Rigid materials (indurated rock) will yield under

Box 10.2

SHEAR STRENGTH

The most basic method of testing shear strength is the simple shearbox, consisting of two separate halves, the lower one of which is fixed, while the upper half is free to slide over it under an applied force. The applied force (stress) is measured through a proving ring, and strain is measured as the distance travelled during shear.

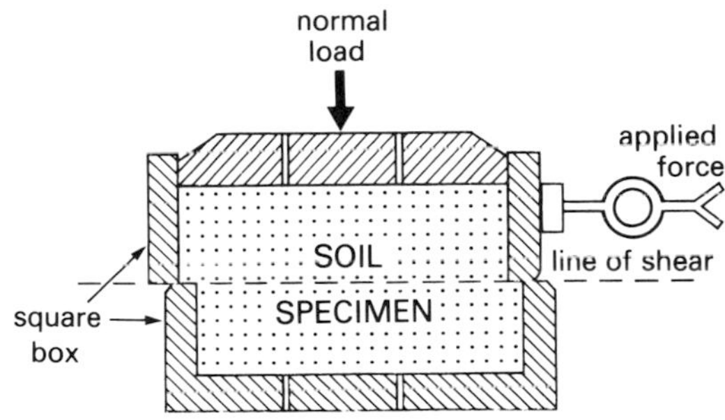

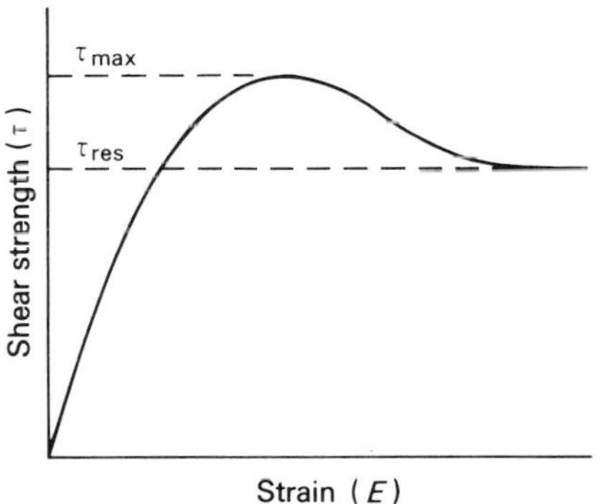

The behaviour of material under shear can be described by plotting stress against strain. For an unconsolidated material of mixed calibre the following can be expected to occur.

At low stress, little strain takes place until the point of failure. This defines the **maximum shear strength** (τ_{max}). Thereafter, the shearing force required to cause further strain diminishes to a steady level, defining the **residual shear strength** (τ_{res}).

The effect of consolidation of the material on its shear strength can be examined by varying the load on the sample, normal to the shear plane and repeating the test. The relationship between τ_{max} and normal load permits the identification of the two major components of shear strength – cohesion and friction – since the effect of increasing the load is to increase interparticle contact, thereby increasing friction. Shear strength at zero load is assumed to be due to cohesion alone.

For most mixed materials, the shear strength : normal load relationship is shown in Figure a over. Strength has a finite value at zero load and increases linearly as load increases. This rate of increase in strength with load is due to the

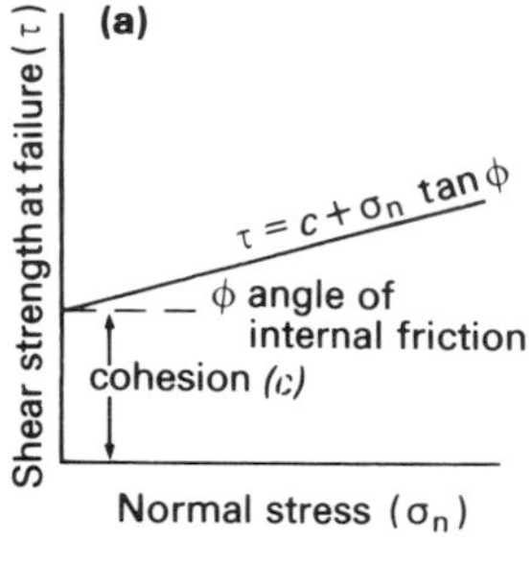

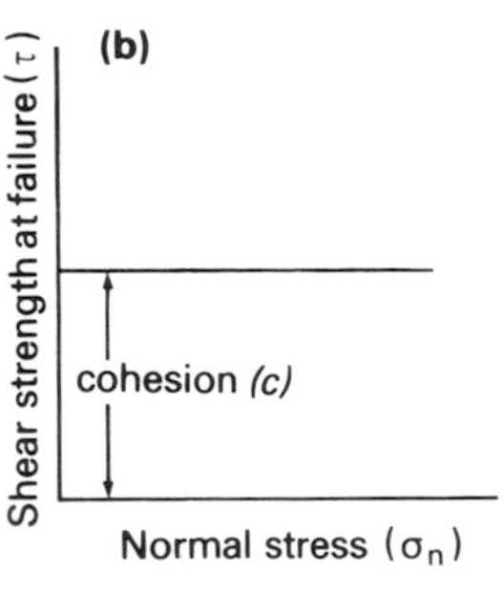

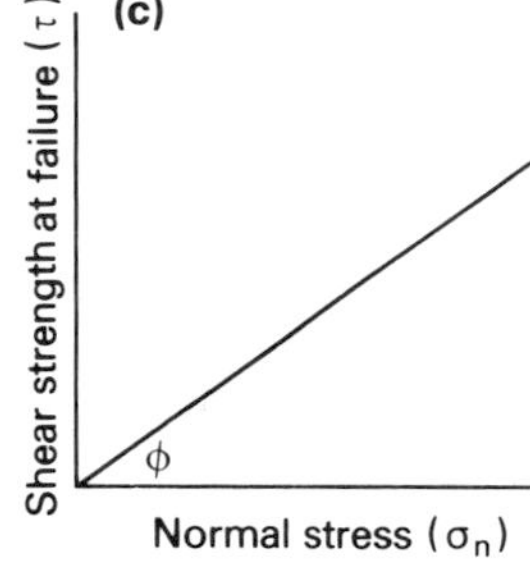

increase in friction. This can be expressed by Coulomb's equation (1776), as follows:

$$\tau = c + \sigma_n \tan \phi$$

where τ = shear strength (N m^{-2}), c = cohesion (N m^{-2}), σ_n = load normal to shear plane (N m^{-2}) and ϕ = angle of shearing resistance.

The two components of shear strength are, therefore, cohesion (c) and friction ($\sigma_n \tan \phi$). A continuum exists between frictionless materials, such as some clays, where loading does not increase the strength ($\phi = 0$) and all the strength is due to cohesion, and cohesionless materials, such as loose sand ($c = 0$), in which strength is assumed to be due entirely to friction.

A further refinement of the shear strength equation can be made by taking into account the effect of pore water. This exerts a force on adjacent particles (pore water pressure) which operates against the normal force, thereby altering the friction component of strength. This is expressed in the equation

$$\tau = c + (\sigma_n - u) \tan \phi$$

where u = pore water pressure (positive or negative).

Thus, with positive pore water pressure (greater moisture content) shear strength is reduced.

stress by brittle fracture and, after fracture has taken place, the shear strength of the body of material can be considered to be insignificant. However, unconsolidated materials exhibit shear strength after failure has taken place. In these materials, therefore, we can recognise two strength values; maximum shear strength (τ_{max}), and residual shear strength (τ_{res}). The latter value is of importance in considering unconsolidated material on slopes, across which shearing has already taken place. In this case stability is related to residual (not maximum) shear strength.

The effect of normal stress is important. Box 10.2 shows that strength increases directly with normal stress. Thus, material at depth beneath the surface will be further consolidated by the pressure of overburden exerting a normal stress, thereby increasing the shear strength at depth. The effect of moisture is to decrease shear strength, as shown by Terzaghi's equation (Box 10.2). Thus moist materials will have a lower shear strength than dry ones.

Representative values of strength for a variety of indurated rocks are presented in Table 10.3. These data should be treated with some degree of caution since they are drawn from a variety of sources and are probably derived from a variety of test conditions, which may influence the values obtained. They do, however, demonstrate the range and order of magnitude of rock strength, and show the variation between different types of rock. They show also the different degrees of resistance possessed by rock in relation to the type of stress. All of the materials illustrated are most resistant to compressive stresses, moderately resistant to shear stresses and least resistant to

Table 10.3 Strength values for different rock types (in MN m^{-2}) (after Billings 1954, from various sources).

	Compressive		Shear		Tensile	
	Mean	Range	Median	Range	Median	Range
granite	145	36–372	22	15–29	4	3–5
sandstone	73	11–247	10	5–15	2	1–3
limestone	94	6–353	15	10–20	4.5	3–6
marble	100	30–257	20	15–25	6	3–9
serpentine	121	62–121	25	18–33	8.5	6–11

tensile stresses, with an order of magnitude difference between the values of each.

It should be borne in mind that these values relate to small samples of intact rock and are not representative of the rock mass as a whole. The overall strength of a mass of rock in the field is dependent on the presence, abundance and orientation of planes of fracture or discontinuities within it and it will be lower than the value obtained for a small sample.

Strength values for unconsolidated materials are, of course, much lower. Data quoted by Carson and Kirkby (1972) for the shear strength of residual soils and taluvial material range from 0–85 kN m^{-2}, several orders of magnitude less than the values of solid rock. Under zero normal load (Box 10.2) many unconsolidated materials may have no measurable shear strength, i.e. they possess no cohesion. The difference in strength between solid rock and soil is therefore shown to be very great indeed.

10.3.3 Consistency properties

Properties of consistency are a measure of the physical reaction of unconsolidated materials to moisture content. Known as **Atterberg limits**, arbitrarily defined tests indicate the moisture content at which the material changes from the solid to the plastic state (**plastic limit**: *PL*) and from the plastic to the liquid state (**liquid limit**: *LL*). The plastic limit is defined as the minimum moisture content at which the soil can be rolled to a thread 3 mm in diameter without breaking. The liquid limit is the minimum moisture content at which the soil flows under its own weight, as defined by a standard test procedure.

The **plasticity index** (*PI*) is defined as the range of moisture contents at which the soil behaves in a plastic manner, and is defined as:

$$PI = LL - PL$$

(all terms expressed as % moisture content).

Atterberg limits vary widely between different soils, depending on the percentage of clay and the type of clay minerals present. Whalley (1976) quotes the following values as typical of London Clay:

PL	30–45%
LL	70–105%
PI	35–65%.

Values are lower with a lower clay content and, for cohesionless materials, *PI* becomes zero.

These material properties have considerable influence on the behaviour of rock and soil under the applied forces of denudation. They are essentially descriptive parameters, yet at the same time they are essential to an understanding of the resistance of materials to denudation and are valuable tools in the analysis of denudation processes.

10.4 System function: energy flow

When considering the functional organisation of the denudation system in the context of its component catchment basins, the links between the subsystems of the model were presented mainly as pathways of material transfer (see Fig. 10.3). These links are also pathways of energy flow, while the processes associated with the operation of the subsystems involve the transformation and transfer of energy. The two fundamental sources for all Earth surface systems are terrestrial energy and solar energy.

Input of the geothermal component of terrestrial energy to denudation systems is so small that its effect on most aspects of systems operation can be regarded as negligible, with the significant exception of the glacial system (see Ch. 14). The expression of terrestrial energy as inputs of potential energy, however, is of great significance in the operation of the denudation systems. First, the primary minerals of the rocks forming the crust can all be considered as possessing potential chemical energy by virtue of their ordered structure and of the atomic and molecular bonds that maintain it. In the weathering system this energy is released as weathering reactions proceed and complex compounds are broken down to produce structurally simpler weathering products. Secondly, and of great significance, is the potential energy of relief allied to gravitational force. Where gradients exist, as on hillslopes and in river channels, the motion of both solid and fluid materials results in the conversion of the potential energy to kinetic energy. It is clear that velocity is the major component of kinetic energy (see Ch. 2). Thus the more rapid movements of materials (e.g. flowing water in channels and rapid mass movements) possess higher levels of kinetic energy. The more slowly moving materials (e.g. soil creep) also involve kinetic energy, although of much less magnitude.

The input of both direct and diffuse solar radiation manifests itself in a number of ways in denudation systems. In the weathering system, absorbed solar radiation raises the temperature of both soil and regolith and, in most cases, increases the rate of chemical reactions. It is responsible for the evaporation of water from land and water surfaces and in part for the transpiration of plants. The latent heat of evaporation means that 2450 J are lost to the denudation system per gramme of water evaporated at 20°C. After condensation in the atmosphere, the gravitational potential energy of each water droplet becomes the kinetic energy of the falling raindrop, defined by its mass and velocity. Accordingly, each rainstorm produces its own specific kinetic energy input, related to its intensity and duration. This kinetic energy is converted to mechanical energy on impact and it performs work on the inorganic and organic particles of the land surface – the work of erosion. Since work is defined as the product of force and distance, the application of this erosional energy results in the movement of materials. These materials, therefore, possess kinetic energy as they move down slope through the slope system, as does the water flowing on the surface and below it. Some of this kinetic energy becomes the input to the next system in the cascade as runoff and debris are delivered to the channel system. Some of the incident solar energy, however, is absorbed by the ecosystem, particularly during photosynthesis by plants, and is subsequently stored as the potential chemical energy of organic compounds. This energy subsequently becomes available in the soil to the weathering system, while the vegetation

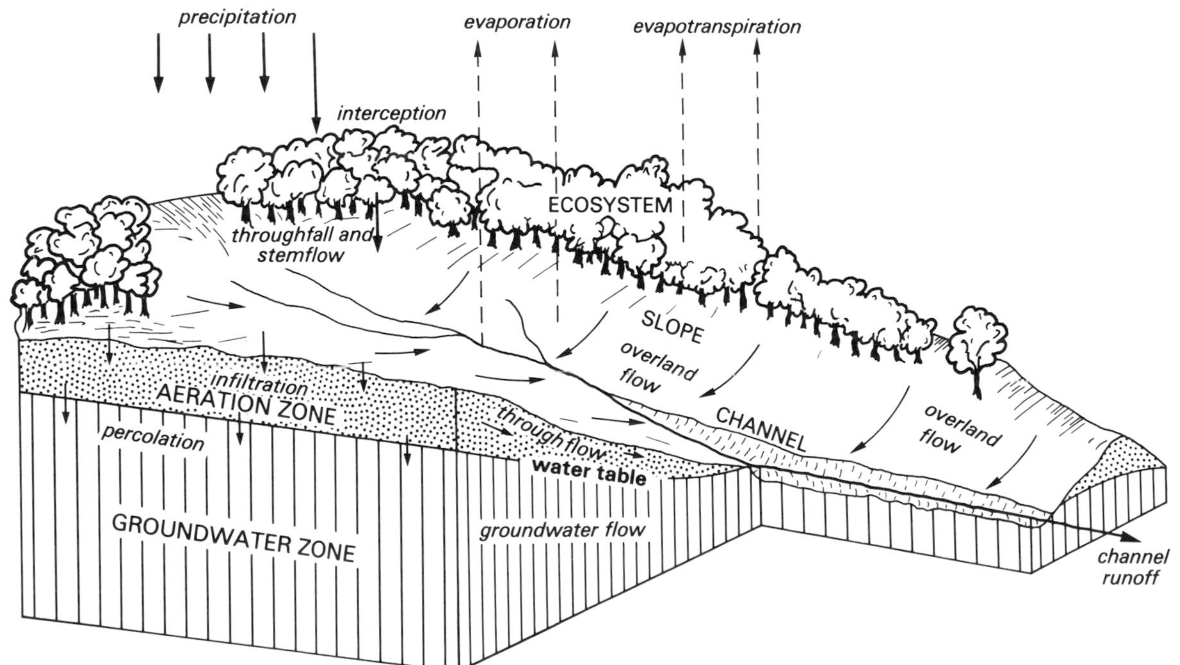

Figure 10.9 Block diagram of the catchment basin system.

itself has profound effects on slope erosional processes and on the catchment water balance.

Outputs of energy from the denudation system occur mainly in the form of heat energy. Direct heat loss from the ground surface by radiation occurs particularly at night. As we have seen, evaporation involves considerable energy loss as latent heat, while much of the kinetic energy involved in erosional processes becomes frictional heat loss to the atmosphere as the material in motion moves relative to static material. Flows of energy through denudation systems, however, although understood in principle, have as yet undergone little quantitative study.

10.5 System function: material flows

The flow of materials through denudation systems involves both water and mineral elements. The flow of water is fundamental to catchment basin processes, since water acts either directly as the transporting medium or plays an important role in most other modes of movement of mineral materials.

The dynamics of the flow of water through the catchment basin are shown in simplified form in Figures 10.9 and 10.10. The input of precipitation is distributed by a series of transfers through a number of stores to outputs of channel runoff and evaporation. Precipitation falls directly upon vegetation where present on the ground surface and on the water surface of channels and lakes. Vegetation controls the interception store and transfers water via evaporation back to the atmosphere and via stemflow and throughfall to the ground surface beneath. From the surface store, water can be transferred to the channel storage by overland flow, evaporated directly to the atmosphere, or transferred by infiltration to the zone of aeration beneath the ground surface. This zone of aeration may consist of soil, regolith and bedrock, and it represents that part of the subsurface, above the permanent water table, which is intermittently wet; in other words, the pore spaces are occupied sometimes by air and sometimes by water. Evaporation and transpiration via vegetation can transfer water from the zone of aeration back to the atmosphere. Lateral flow within this zone (throughflow) can transfer water to the river channel, and deep percolation can carry water down to the groundwater zone. From here, deep transfer takes place by groundwater flow into river channels.

In terms of one element (water), the change in state of the denudation system per unit time can be specified if the transfers between the stores can be given numerical values and related to inputs and outputs. This can be done by reference to the water balance equation (see Ch. 4):

$$P = R + (E + T) + \Delta S + \Delta G$$

where P = precipitation, R = runoff, E = evaporation, T = transpiration, ΔS = change in soil water and ΔG = change in ground water.

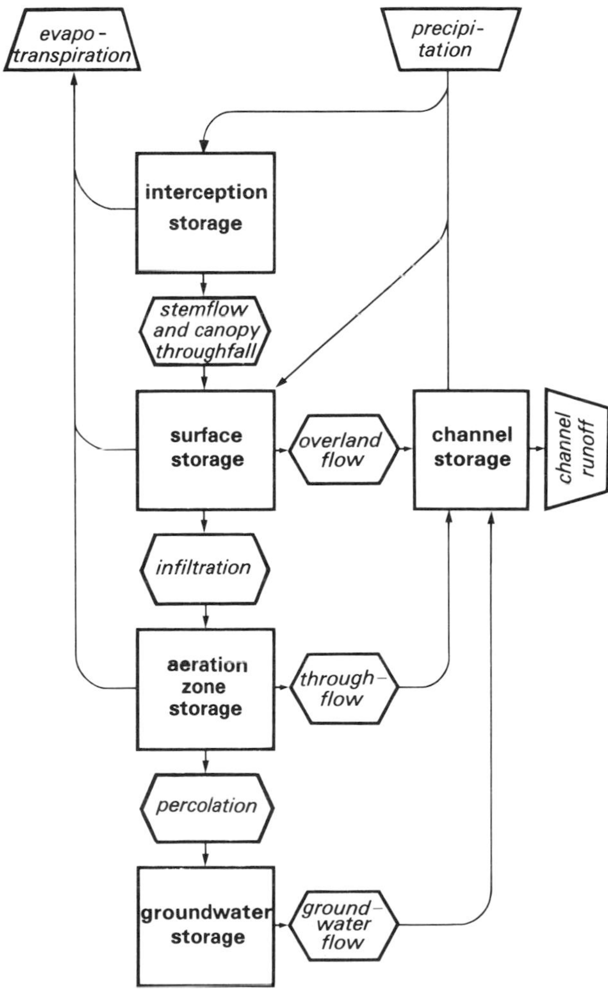

Figure 10.10 Hydrological processes in the catchment basin system.

By combining the elements in brackets above, the equation can be simplified to the following form:

precipitation = runoff + evaporation + infiltration.

This water balance, or budget, approach can be used at all scales from the global to the local in order to illustrate spatial differences in the operation of the denudation system (Box 10.3).

The movements of mineral elements through denudation systems follow, for the most part, the same pathways as those of water, which itself acts as the main transporting agent. Until they pass into solution as charged ions, or into suspension as fine particles, the speed of transfer is much slower than with the flow of water. In other words, the residence time of mineral elements in the stores of the systems is much longer. In the weathering system, the products of weathering accumulate as regolith until surface erosion brings them within reach of mass movement processes which move them down slope. Here fluvial erosion may ultimately transfer them from the slope to the channel subsystem to become part of the sediment load of the river. Other weathering products pass into solution and follow the flow of water, either as overland flow and throughflow, again down slope to the channel system, or as percolation to the groundwater storage and hence to the channel baseflow. Some mineral elements are held temporarily in the soil in the exchange complex (see Ch. 20), while both these exchangeable elements and those free in the soil water may be taken up by plant roots and temporarily immobilised in the biomass of the ecosystem (see Ch. 17). These elements may subsequently be returned to the slope system in the stemflow and throughfall components of precipitation, where they join materials already dissolved in rainfall. Alternatively, they may remain in organic compounds until death and litter fall followed by decomposition release them to the slope system once more. The major inputs of materials, therefore, are from the atmosphere, dissolved in precipitation or as dry fallout; from the crust through the weathering system as the weathering front progresses; or directly to the channel subsystem where the erosion of bedrock occurs. The main output of mineral elements from the catchment basin system is the mineral load in channel runoff, although some elements which can exist in gaseous form may go direct to the atmosphere in a way analogous to water vapour during evaporation.

10.6 Special cases of the model

The basic functional model of the catchment basin system developed so far can be modified to cater for situations where the system operates under more extreme sets of conditions which radically affect the magnitude of the inputs to the system and the distribution of matter and energy through its components.

In glacial landscapes, the precipitation input is largely in the form of snow, and runoff is mainly in the form of slowly moving streams of ice, supplemented by meltwater streams in marginal environments. Thus the glacier takes the place of the river channel and the rate of throughput is much slower than in a normal fluvial basin. Mineral material can enter glacier systems either from above, where slopes overlook the glacier surface, or from below as the glacier erodes material from its margins.

Desert landscapes are characterised by very high values of evaporation loss, often reaching 100%, which result in a zero net surface runoff. Consequently, deserts often contain a series of enclosed catchment basins with centripetal drainage systems. When runoff occurs, it is shortlived and streams ultimately peter out as infiltration and evaporation take their toll. Often, therefore, there is no net outflow of surface runoff.

Limestone (and chalk) landscapes are characterised by high rates of infiltration. Streams on limestone disappear underground in swallet holes, while percolation is more diffuse in chalk. Thus the amount of surface runoff is strictly limited as catchment basins in such areas drain underground, and infiltration and ground water play an important role.

All of these variations can be accommodated by the basic functional model. It is merely that climatic and lithological variations involve differences in the relative significance of different processes and the relative importance of different pathways of energy and matter transfer through the denudation system. These variations are no more, therefore, than particular cases of a general denudation system model.

Box 10.3

COMPARISON OF CATCHMENT WATER BALANCES

Stoddart (1969) quotes values in mm depth of water for the processes of precipitation, runoff, evapotranspiration and infiltration for the total land surface areas of the world, as follows:

$$P = R + E_T + I$$

$$730 = 171 + 478 + 81$$

where E_T = total evaporation loss and I = infiltration. It should be remembered that these values are extremely generalised and are based on estimates from many regions of the Earth for which few accurate data yet exist. If we assume that the water infiltrating rejoins river channels and reappears as surface runoff, then a total of 252 mm (171 mm + 81 mm), out of the original precipitation of 730 mm, ultimately flows from the Earth's land areas. This represents a total proportion of only 34% of the precipitation value and puts into perspective the relative importance of the paths of water flow shown in Figure 10.3.

These general figures for water balance processes on the world scale obscure the considerable differences that exist between continents, shown below (from Budyko 1958):

	Precipitation (mm)	*Evaporation* (mm)	*Runoff* (mm)
Europe	600	360	240
S. America	1350	860	490
Australia	470	410	60
Africa	670	510	160

Clearly, these figures also must be regarded as first-order estimates, but they do reveal important differences between the continents. The temperate mid-latitude continent of Europe exhibits moderate values throughout. South America has a high precipitation value resulting in high runoff. The arid continent of Australia, on the other hand, shows a relatively low rainfall and a high annual evaporation loss due to high temperatures, resulting in a low rate of runoff. The values for Africa show a moderate precipitation and a high value for evaporation, since the continent straddles the Tropics and contains large arid regions.

The same model of water balance can be applied at the scale of the individual catchment. Values in millimetres for a small sample of catchment basins in selected areas of Britain are tabulated below:

	Precipitation	*Evaporation*	*Runoff*	*Runoff* (%)
E. Riding[1]	645	458	187	28.9
S. Hampshire[2]	926	439	487	52.6
E. Devon[3]	1033	551	467	45.2

[1]Pegg & Ward (1971); [2]Mottershead & Spraggs (1976); [3]Gregory & Walling (1973).

Hanwell and Newson (1973) present a map of selected catchments throughout Britain showing variations in mean annual runoff as a percentage of mean annual precipitation. Values range from as low as 10–20% in catchments in Suffolk to 80–90% in the west Highlands of Scotland. Clearly then, great variations exist between different individual catchments in the amount of water flowing along the various pathways of the denudation system. These differences can be related both to the nature of the precipitation input and to the catchment basin characteristics.

Further reading

A wide range of general geomorphological texts is available to cover the range of material presented in this chapter and Chapters 11 to 15.

A useful elementary introduction to geomorphological systems is
Bloom, A. L. 1969. *The surface of the Earth.* Englewood Cliffs, NJ: Prentice-Hall.

A good general text is
Rice, R. J. 1977. *Fundamentals of geomorphology.* London: Longman.

A modern account of geomorphological processes is
Derbyshire, E., K. J. Gregory and J. R. Hails 1979. *Geomorphological processes.* Folkestone: Dawson.

A succinct and useful introduction to the same subject, stressing the mechanics and dynamics of geomorphological processes is
Statham, I. 1977. *Earth surface sediment transport.* Oxford: Oxford University Press.

A more advanced text along the same theme is
Embleton, C. and J. B,. Thornes (eds) 1979. *Process in geomorphology.* London: Edward Arnold.

Useful chapters are also to be found in
Cooke, R. U. and J. C. Doornkamp 1974. *Geomorphology in environmental management.* Oxford: Oxford University Press.

11 The weathering system

Rocks formed within the lithosphere under conditions of high temperature, high pressure, or both, are relatively stable physically and chemically until exposed at the Earth's surface (see Ch. 5). Here they encounter entirely different conditions. For example, pressure is reduced, temperature is subject to considerable fluctuation, and both oxygen and water are abundant. Rocks then become altered by weathering processes into forms which are stable under these new conditions. So, weathering can be defined as the response of materials, which were at equilibrium in the lithosphere, to new conditions at or near the contact with the atmosphere and biosphere.

This response is seen on the left of Figure 11.1. On the right, however, you can see that a similar process is affecting organic compounds which are also being exposed to a parallel change in conditions when, on death or excretion, they cease to be components of living systems. These two directions of alteration and adjustment merge in the soil system where the processes of weathering and decay, together with the properties of the weathering products and organic residues, interact in soil formation (**pedogenesis**). In a sense, then, the soil can be viewed as a new equilibrium state established by these adjustments in both inorganic and organic materials in response to changes in their conditions of existence.

It should be clear from Figure 11.1 that it is difficult to separate weathering, decay and soil-forming processes and to delimit the systems within which they operate. The three clearly overlap and there are intersections between the sets of variables which characterise the systems and interactions in the operation of the processes. In this chapter we consider the weathering system set, as far as possible in isolation. The pedogenesis system set will be treated in part in Chapter 12 and again in Chapter 20. We are concerned here with the overall process of adjustment that the rocks of the lithosphere undergo in order to attain a new equilibrium with the conditions they experience at and near the Earth's surface.

The consequence of the operation of the weathering system is, therefore, the production of new materials, which differ both chemically and mechanically from the original parent rock. Substances are produced which are less massive, less indurated and possess much lower mechanical strength. As such, they become available for transportation by denudation processes, for their

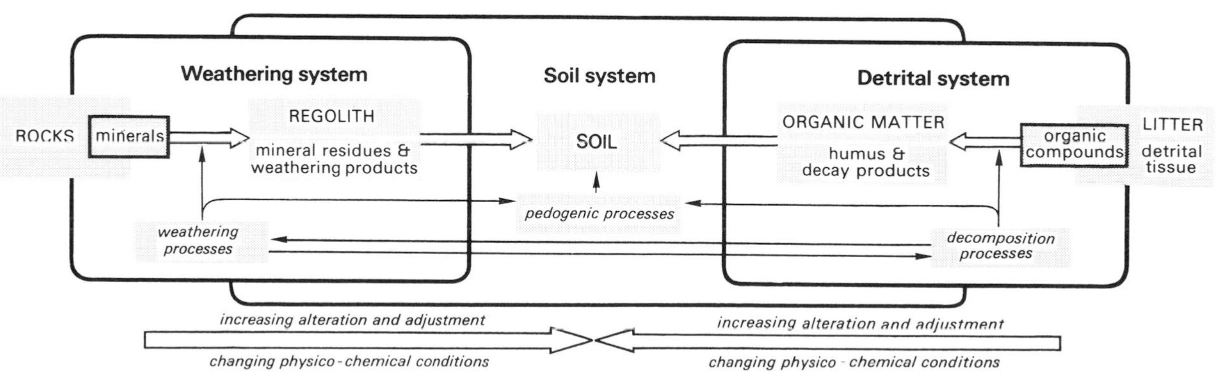

Figure 11.1 Model of the weathering system and its relations with the soil and detrital systems.

resistance falls to a level at which the energy available for erosion is sufficient to transport them. This material is the solid weathering residue. In addition, soluble substances are produced which are readily removed by circulating water. These minerals are, therefore, rendered highly mobile and made available to both the ecosystem and denudation system. As such, they are removed readily, thus being released from storage in the lithosphere and transferred in geochemical or biogeochemical cycles.

11.1 A process–response model of the weathering system

Geographers have traditionally divided weathering into mechanical (physical) processes and chemical processes, with some acknowledgement of the role of the biotic agents in both of these categories. When the fundamentals are considered, however, it becomes apparent that the chemical processes are in fact governed by the basic laws of physics operating at the molecular and submolecular scales. The distinction between chemistry and physics is, therefore, essentially one of scale. Accordingly, we adopt here an original approach to the study of weathering, based on the scale at which breakdown occurs. The model of the weathering system which we shall develop is a process–response model (Fig. 11.2). The response of the system will be viewed as a progression of states, defined in terms of their mineralogical, petrological and mechanical properties, as they replace each other through time. The processes are both those directly responsible for the adjustment and reaction to forces brought to bear on the rock when exposed at or near the surface, and the processes that activate or enable rock breakdown to occur. In this model it is contended that there are two, and only two, **primary mechanisms** of rock breakdown. These are **brittle fracture** and **crystal lattice breakdown**, the former operating at the scale of the rock mass, clast or crystal, and the latter at the molecular and submolecular scales. These primary mechanisms are triggered by environmental inputs of various kinds and promoted by the operation of the second group of processes alluded to above; these we shall term the **activating processes**, which in turn are promoted by a range of **activating agencies** (Table 11.1).

Table 11.1 A classification of weathering processes.

Primary mechanisms	Activating processes	Activating agencies
brittle fracture	strain release (dilatation)	unloading
	applied mechanical stress	thermoclasty haloclasty gelifraction biomechanical forces
crystal lattice breakdown	modification of electrostatic forces, by hydration hydrolysis and redox reactions	exposure of mineral surfaces supply of water supply of hydrogen ions (pH) oxidation/reduction status of weathering environment (Eh) agencies removing weathering products (e.g. by leaching, cheluviation)

The nature of the environmental inputs will vary from one place to another across the Earth. Some activating processes require hot conditions, others cold. Some are restricted to humid environments, others to arid ones. There are surface processes and subsurface processes. Therefore, the combination of different activating processes in operation at a particular location will in turn reflect the conditions under which the system functions – the **weathering environment**.

Approximately one-third of all minerals making up about 90% of the Earth's crust belong to the silicate group (see Box 5.1). For the purposes of this model, therefore, we shall consider a massive indurated rock composed of primary silicate minerals as the initial state of the system. The first compartment in the model (the initial state), therefore, is the unaltered rock, retaining original structures, composed of primary minerals and with characteristic mechanical properties of permeability, porosity and strength. From the point of view of the model, it is regarded as a system organised hierarchically from the atoms of the constituent elements into the orderly arrangement of these atoms and the molecules they form in crystal lattice structures, to the individual mineral particles they produce. These mineral particles are

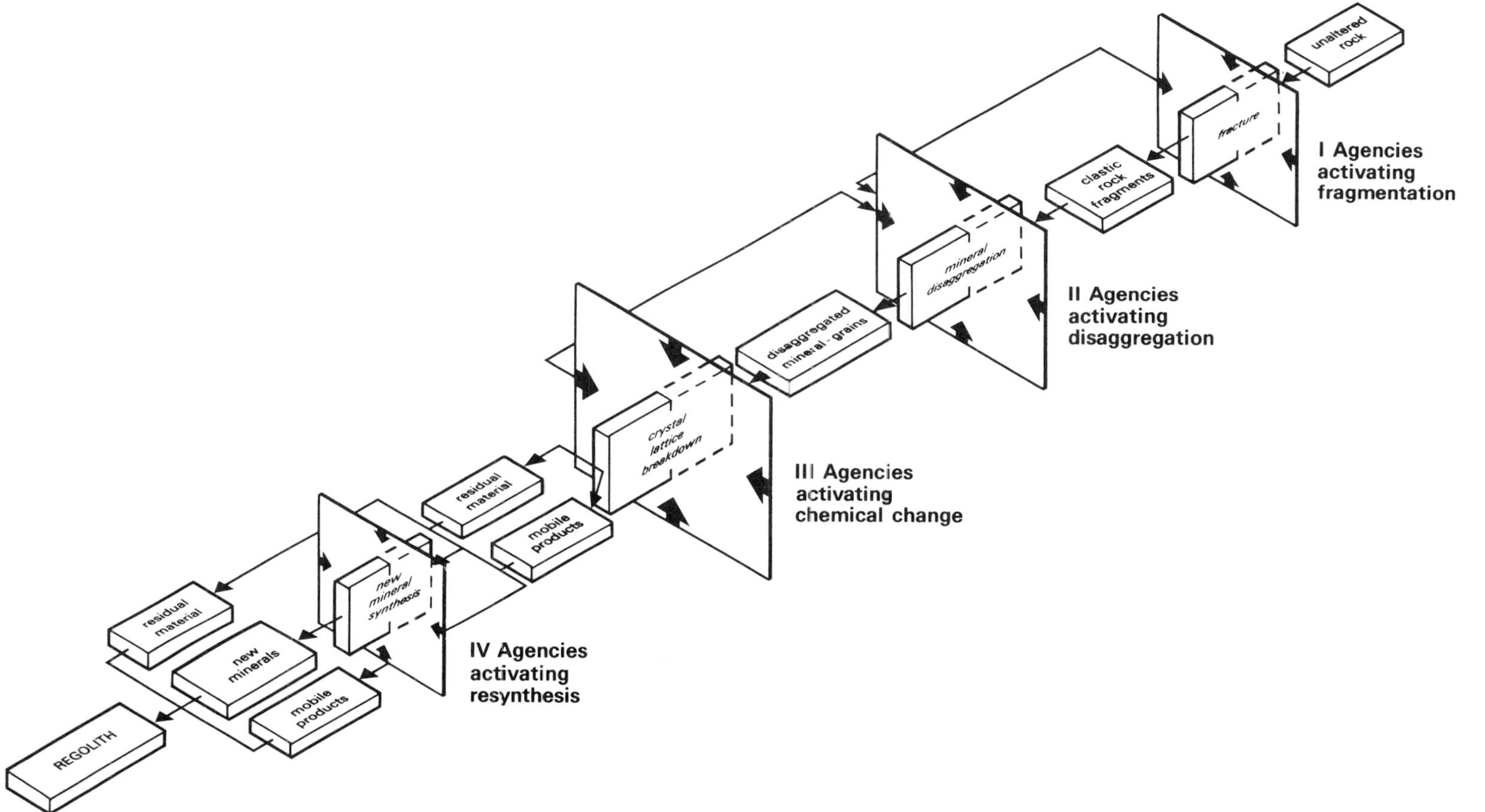

Figure 11.2 A process–response model of the weathering system. Note that the primary mechanisms in disaggregation can be either fracture or chemical breakdown.

organised in turn in a three-dimensional mosaic to form massive rock (see Ch. 5). This structure is maintained by internal mechanical forces due to the packing of the grains and by the chemical bonding forces within the crystal lattices and between the crystals. These forces are the internal or potential chemical energy of the system.

Under the influence of the primary mechanisms of weathering, this initial state is replaced in time by clastic fragments of rock, still composed of primary minerals but with different mechanical properties and with a greater surface area. The next state is one where the minerals of these fragments are altered chemically and the weathered residues of the primary minerals now exist in an environment containing the mobile products of weathering, usually in solution. Also new minerals, not directly derived from primary minerals, may have appeared. The physicochemical properties of the mineral residues and of these new minerals, together with the mechanical characteristics of the regolith they form, may differ radically from those of the original rock and primary rock-forming minerals. Particle size will be much reduced, exposed surface-area:volume ratios will have increased enormously, density and mass are reduced and strength parameters will have changed. The smallest (colloidal) particles will exhibit various degrees of mobility and will have marked ion exchange properties.

11.1.1 The primary mechanisms of weathering: brittle fracture

Brittle fracture occurs when the rock experiences tensile and shear stresses. Such stresses will place under tension the bonds between mineral grains and those within crystal lattices. This in turn stretches or deforms these bonds, placing them under strain or giving to them a certain **strain energy**. This strain energy is dissipated as the bonds break and a crack or fissure forms. Although usually regarded as a macroscopic process operating on massive rock, the actual mechanism of fracture is located at the apex of the crack and it operates at the molecular level where the strain is concentrated.

The development of most cracks and fissures in rocks probably takes place along grain boundaries, while the site of initiation and the direction of propagation will be determined by existing points, lines and planes of relatively weak bonding within the granular mosaic of the rock. Under a maintained stress such cracks can spread very rapidly. Fracture, however, also takes place within the crystal lattices of rock-forming minerals. Such crystals are rarely perfect and strain can be concentrated at defective points in the lattice, or along displaced layers of the lattice known as dislocations. Once initiated, such crystal fracture will be guided preferentially by cleavage planes. Although fracture can overcome the physical and chemical forces holding the lattice together, it does not alter its nature but merely produces smaller fragments of crystal and appears, at its lower limit, to be restricted to the formation of silt size particles. Beyond this threshold of rock dislocation we move into the realm of the second primary mechanism of weathering, the fundamental breakdown of the crystal lattice itself. Lattice breakdown, in contrast, involves basic changes in both the composition of the lattice and in its structural configuration, thereby producing chemical change in the mineral concerned.

11.1.2 Activating agencies of brittle fracture

The immediate processes causing fracture are either the release of internal strain or the application of external stress to the rock mass (Fig. 11.3). The expression of both, however, is controlled by the mechanical properties of the rock in question and of its constituent minerals. Therefore, these mechanical properties – particularly strength parameters – act as regulators which, by assuming threshold values, determine the incidence of fracture under any given stress. As the rock is subject to progressive fracture and fragmentation, so the values of these regulatory parameters change and a negative-feedback loop is established which reduces the effectiveness of fracture until a size threshold is passed, beyond which weathering can proceed only by chemical change.

Internal strain is present in rock masses because of the compressive stress experienced during their formation under conditions of high pressure in the crust, or as a result of the normal compressive stress of overlying rocks. The enacting process of **strain release** is the erosion and removal of overburden leading to unloading and the reduction of normal stress and consequent expansion of the rock. Rocks rarely expand uniformly and the usual result is the production of joint planes (**dilatation joints**), which are subparallel to the topographic surface which has experienced unloading. These joints are often spaced 1–3 m apart and may affect

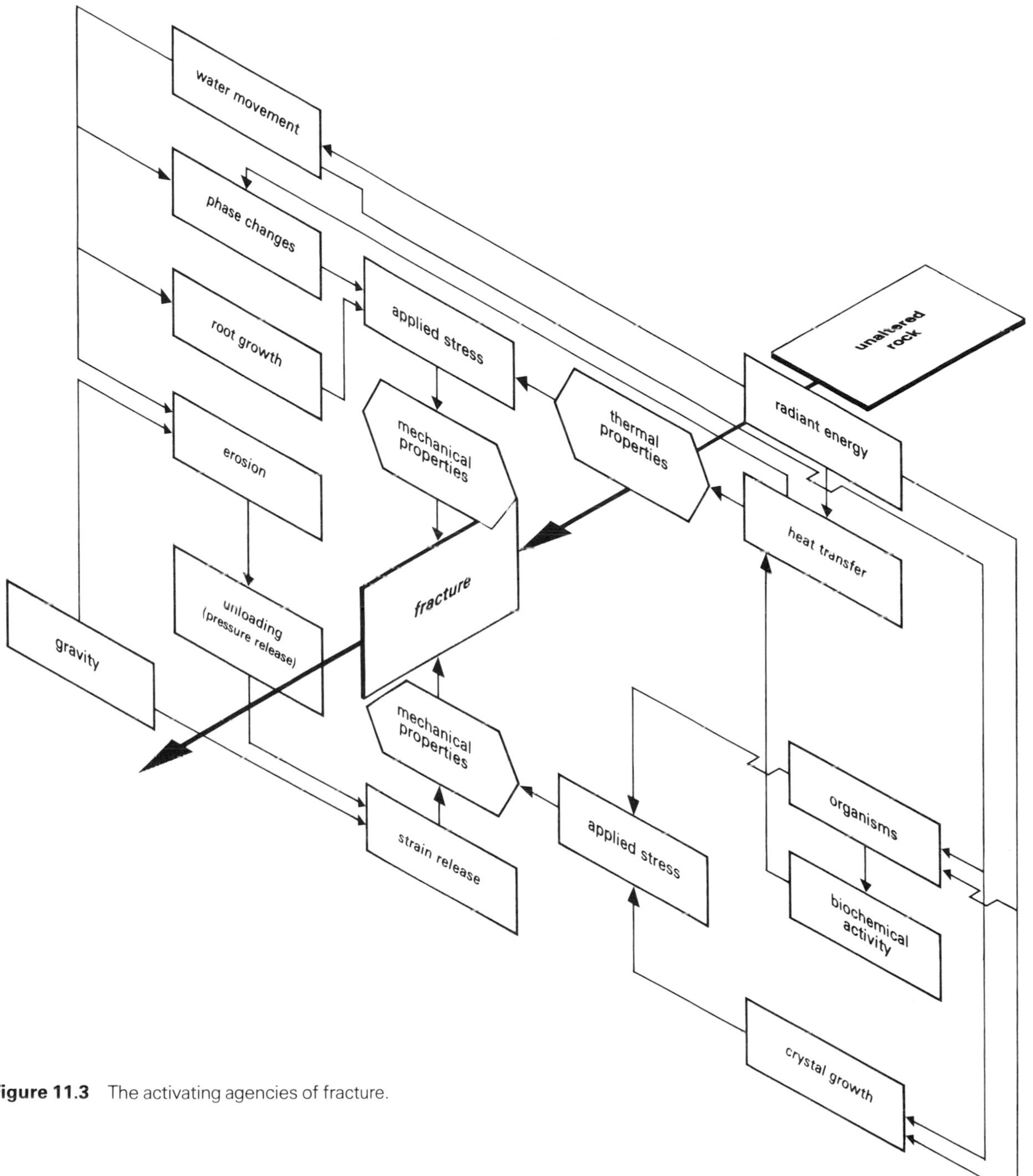

Figure 11.3 The activating agencies of fracture.

the rock to a depth of 20 m or more and they can result in large sheets or slabs becoming detached from the rock mass. Massive igneous rocks (Fig. 11.4) display dilatation joints most clearly, particularly when forming rock domes, or **inselbergs** (residual hills) from which the surrounding rock has been denuded. Such joints are also well displayed in glacial troughs and corries, where again bedrock has responded in this way to the removal of overburden by denudation. Indeed, Whalley (1976) suggests that mountain landscapes may owe far more to strain-release fracture than is usually realised.

The enaction of rock fracture by applied mechanical stress can be divided into two groups of processes. First, there are stresses set up in the

Figure 11.4 Dilatation jointing. (a) Sheet joints nearly parallel to the land surface above (after Chapman & Rioux 1958). (b) Large-scale sheet structures and exfoliation processes in massive Colorado Plateau sandstones; arrows show inferred directions of dilatation; X is an exfoliation dome, Y an exfoliation cave (after Bradley 1963). (c) Dilatation sheets in massive igneous rocks (Yosemite, California, USA).

rock as a direct or indirect result of heat transfer; and secondly, stresses that result from processes of growth and expansion within pores and fissures of the rock.

Because of the poor thermal conductivity of rock (see Ch. 3), the transfer of heat (derived from solar energy) by conduction from the surface is slight, and steep temperature gradients can exist through only a limited thickness of rock. The surface layer responds to heat by expanding, while adjacent mineral grains within it, having different coefficients of expansion, expand (or strain) at different rates and cause increased mechanical stress. These stresses can eventually produce thermal disintegration (**thermoclasty**), a process which was formerly thought to be a major cause of breakdown on exposed bedrock surfaces in desert regions where high air temperatures are attained. However, laboratory experiments have failed to reproduce thermal disintegration without the presence of moisture. This has led to the conclusion that thermal disintegration is limited in occurrence and is normally limited to situations where moisture has previously weakened intergranular and intercrystalline bonds by chemical change, notably by hydration and some hydrolysis along grain and crystal boundaries, thereby lowering the tensile strength of the rock and emphasising the feedback relationship between mechanical and chemical processes.

A more indirect effect of heat transfer (in this case the latent heat of fusion), however, is associated with a phase change affecting the water occupying joints, fissures and pores in the rock. Water freezing to form ice crystals undergoes a volume expansion by 9% and can develop pressures in a confined space theoretically in excess of 200 MN m^{-2}. In reality, it is doubtful whether such values are reached, since water within rocks will not be in a totally confined space. Nevertheless, when compared to the tensile strengths of rocks, normally in the range 1–10 MN m^{-2} (see Table 10.3), it can be seen that there is a sufficient margin for ice crystal growth to be very effective. Rock breakdown by this process (**gelifraction**) will tend to be most effective in conditions with frequent alternations of temperature about 0°C. It is clearly more effective in montane periglacial environments, which may experience frost more than 300 times per year, than in the arctic latitudes where freezing and thawing may be limited to short periods in the spring and autumn. Short-term fluctuations about the freezing point penetrate not more than a few centimetres beneath the ground surface. Effective frost weathering is, therefore, a surface phenomenon. Long-term freezing cycles may also be expected to cause rock fracturing, but they are events of limited frequency and hence may be restricted in their overall effect. Freezing within pore spaces tends to pulverise rock into individual grains and small angular fragments. Chalk, for example, responds in this way. The effect of ice formation in the joints of well consolidated rocks is to prise away joint-bounded blocks.

The growth of ice crystals, of course, can also be thought of as belonging to the group of processes that involve the growth and expansion of material in pores and fissures. The other mechanisms concerned in this second group of processes are the growth of salt crystals and the growth and activity of living organisms (**biomechanical forces**), both of which exert tensile stresses.

Salt crystal growth, normally thought of as a chemical process, can exert mechanical forces within rocks (Winkler & Wilhelm 1970) – a conversion of chemical to mechanical energy. The effectiveness of salt weathering processes – known as **haloclasty** – in the breakdown of building stone has long been appreciated in the literature of building technology, although its significance in geomorphology had been underestimated in the main until the comprehensive review by Evans (1969).

Salts may be emplaced within pore spaces in rocks from saline solutions, especially ground water and sea water. Saline ground water is drawn to the surface by evaporation and capillary action in desert and semi-arid environments, while inundation of rocks by sea water spray is an everyday occurrence in coastal environments. The evaporation of these saline solutions results in the precipitation of crystalline minerals within the pore spaces of surface rocks and often forms visible efflorescences of salts.

Salt weathering is caused commonly by three types of process. First, as salt crystals grow by crystallisation out of saline solutions they exert a **force of crystallisation**. The forces developed are related to the density and molar volume of the salt, the degree of supersaturation of the saline solution, and the temperature at which the process takes place, according to the model of Correns (1949). The force of crystallisation at 20°C is calculated for several salts and is shown in Table 11.2 in relation to the supersaturation of the saline solution. Extremely modest levels of supersaturation are required to generate a force of 1 MN m^{-2}, and a force of 10 MN m^{-2} is attained by only moderate levels of supersaturation. Halite (NaCl) is shown to be the most effective of the salts tabulated. These levels of tensile stress are sufficient to cause disaggregation of most rocks. Secondly, anhydrous salts (for example, anhydrite, $CaSO_4$), already emplaced within the rock, may absorb water and become hydrated (to form gypsum $CaSO_4.2H_2O$). In order to accommodate the water they modify their lattice structure and expand (Table 11.3). The growth of crystals due to hydration generates **hydration pressure**. This is particularly common in desert environments where low night temperatures cause high humidity and the formation of dew on exposed rock surfaces. Winkler and Wilhelm calculated hydration pressures for several salts and show that values of 10 MN m^{-2} are exceeded easily in many hydration reactions. Thirdly, differential thermal expansion of entrapped salts may cause rock breakdown by thermoclasty. Cooke and Smalley (1968) point out that many of the common salts have higher coefficients of expansion than most rocks. Sodium chloride, for example, expands volumetrically three times as rapidly as granite with increasing temperature. In rocks exposed to a high diurnal temperature range this may well lead to damaging levels of tensile stress.

Rock weathering by salts is a topic that lends itself well to experimental investigation in the laboratory. Figure 11.5a illustrates an attempt to isolate experimentally the elements of the weathering environment responsible for the rapid breakdown of a schist on an exposed coast. The experiment sought to identify whether it was the presence of water alone, or the salts dissolved in

Table 11.2 Levels of supersaturation (in %) of various salts in water required to generate crystallisation pressures of 1 MN m^{-2} and 10 MN m^{-2} at 20°C.

	1 MN m^{-2}	10 MN m^{-2}
NaCl	1.16	12.18
$MgCl_2$	1.69	18.34
$MgSO_4$	1.87	20.42
$CaSO_4$	1.91	20.78
Na_2SO_4	2.19	24.30
$CaSO_4.2H_2O$	2.28	25.33

Table 11.3 Volume increase during hydration of selected salts (after Goudie 1977).

Salt	Hydrate	% volume increase
Na_2CO_3	$Na_2CO_3.10H_2O$	374.7
Na_2SO_4	$Na_2SO_4.10H_2O$	315.4
$CaCl_2$	$CaCl_2.2H_2O$	241.1
$MgSO_4$	$MgSO_4.7H_2O$	223.2
$MgCl_2$	$MgCl_2.6H_2O$	216.3
$CaSO_4$	$CaSO_4.2H_2O$	42.3

sea water, or alternations of wetting and drying which was responsible for rapid weathering. Accordingly, rock samples were immersed in either deionised water or salt water for one hour daily and allowed to stand in a 100% humid atmosphere or to air-dry for the remainder of the time. Over a period of 180 days a clear distinction was revealed by the samples immersed in sea water and air-dried, as weight loss took place as a result of weathering. The treatment to which this particular set of samples was subjected permitted the activity of salt crystallisation and it was concluded that this was the dominant weathering process. As a result, flakes of rock became detached and mechanical disaggregation took place. Since weathering took place from one surface only, a rate of weathering of 0.78 mm a^{-1} could be calculated – a clear demonstration of the effectiveness of salt weathering (Mottershead 1982).

Finally, the mechanical effect of plant growth is also to exert tensile stress on rock and to promote fracture. The most significant effect of these mechanisms is the growth of roots as they penetrate joints in bedrock. Research in the field of agricultural technology has demonstrated the power of root growth. Taylor and Burnett (1964) demonstrate that penetration of roots into soil is sometimes possible with a soil strength of 2–2.5 MN m^{-2} and always possible when soil strength is less than 1.9 MN m^{-2}. Thus, root growth is possible into materials at the lower end of the scale of the tensile strength of rocks. Taylor and Ratcliff (1964) show that advancing roots and rootlets are capable of exerting a continuous radial pressure of 1.5 MN m^{-2}, probably sufficient to prise apart already existing fractures.

The below-ground biomass of most plant communities is at least equal to, and in some cases more than, that of the visible parts of the plants, so

(a)

Air temperature (°C)

maximum	31.0	33.0	36.0	27.0	24.5	20.5	20.5	18.5	15.0
minimum	17.0	15.0	16.0	13.5	12.0	5.5	4.5	4.5	5.0

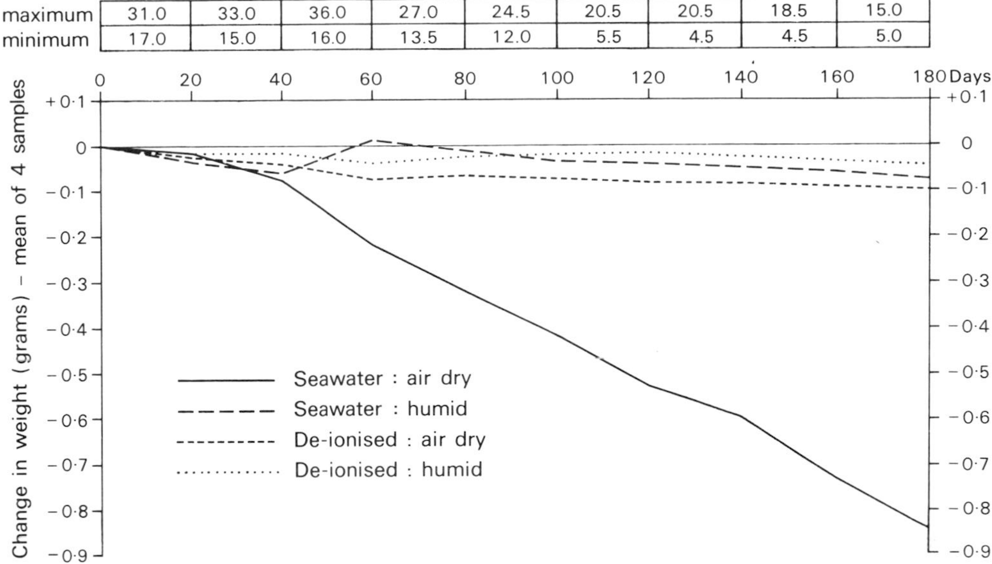

Figure 11.5 (a) The weight loss shown by four samples of schist when subjected to different weathering treatments (see text for explanation). (b) Crystallisation of salts in brickwork as evidenced by salt efflorescences: the surface of the bricks has been disrupted by flaking while the mortar is largely unaffected and stands proud to produce an unusual form of differential weathering (Hurst Castle, Hampshire, England).

it is clear that plants are capable of expending massive amounts of mechanical energy on the bedrock below (Fig. 11.6).

11.1.3 The primary mechanisms of weathering: lattice breakdown

In order to comprehend the breakdown of the crystal lattices of rock-forming minerals we must consider the primary mechanism at the scale of the component atoms. Lattice breakdown is accomplished by chemical reaction. Such reactions occur when the chemical bonds of the reacting substances are broken and reformed to produce the reaction products. The breaking and making of bonds involves the transfer of energy. Reactions which proceed with the aid of a net input of energy are **endothermic** while those that proceed with a net loss of energy are **exothermic**. Most weathering reactions that occur spontaneously are exothermic, although an energy input is still required in many cases to overcome the activation barrier of the reaction (see Box 2.5). Lattice breakdown, therefore, involves the breaking of bonds in the lattice and the formation of new bonds between both the original constituents of the crystal and other reactants, but with a net loss of free energy. Because of the nature of chemical bonds (see Ch. 2), weathering reactions must involve the rearrangement of electrostatic forces within the lattice. This in turn will involve changes in the disposition of the protons and electrons, the fundamental particles responsible for these forces. Before we can see how this is accomplished, we must give more consideration to the nature of the principal reactants.

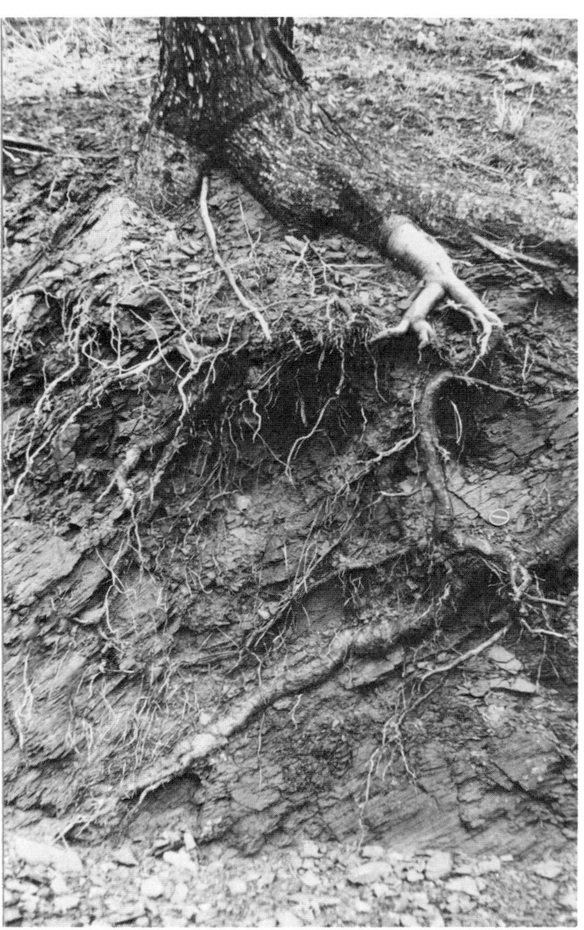

Figure 11.6 Tree roots penetrating fractured and partly weathered rock. Clearly, as the roots grow and expand radially, further tensile stress is exerted on the surrounding rock. (Mallorca)

It will be recalled that in the structure of silicate minerals (see Box 5.1) silica tetrahedra form the lattice framework which increases in three-dimensional complexity from ring silicates such as olivine to the framework silicates such as the feldspars and quartz (note that in some minerals there is some substitution of silicon by aluminium in the tetrahedra). Occupying spaces of various kinds within these structures are various non-framework ions – of which the most important are iron, aluminium, magnesium, potassium, sodium, calcium and hydroxyl ions. These crystal lattices are held together by both ionic and covalent bonds. The silicon and oxygen in the tetrahedral framework of silicate minerals are bound by covalent bonds, many of the non-framework ions by ionic bonds, while elements such as iron and aluminium form intermediate bonds, showing in different proportions tendencies to be ionic or covalent. The strength of the ionic bonds in silicate mineral lattices, however, varies considerably. It is best expressed by the relationship between the charge on the ion – its *valency* – and the number of ions surrounding it, and with which the valency charge is shared – its **co-ordination number**. Although in theory the configuration of the tetrahedra and/or the inclusion of non-framework ions produces an electrically neutral structure, this is rarely the case in real crystals. Certain sites in the lattice remain vacant or contain impurities, while the faces and edges of crystals, which will rarely be perfect, consist of incomplete units. These facts mean that the lattice as a whole will carry a charge, usually a net negative charge.

It was stressed at the beginning of this chapter that the almost ubiquitous presence of water is one of the properties of the environment within which the weathering system operates. In the present context, water possesses two important attributes. First, because of the charge separation (polarisation) of water molecules, they behave as electrical dipoles (polar molecules) (see Ch. 2). Hydrogen bonds formed between these molecules give some crystallinity even to liquid water, though this decreases as temperature rises. Hydrogen bonds also form between water molecules and other charged particles or surfaces, producing layers of orientated water molecules. This is the process of **hydration**, and the thickness of the layers depends on the density of charge carried by the particles or surfaces, which in turn are said to be hydrated. Secondly, in pure water, the molecules dissociate to a small extent into $H^+_{(aq)}$ and $OH^-_{(aq)}$ ions. (aq) stands for aqueous and tells us that these ions are hydrated – they have water molecules attracted to them. However, the hydrogen ion is a single proton and is usually regarded as being attached to one water molecule forming the hydroxonium ion (H_3O^+) or a hydrated proton ($H^+ + H_2O$):

$$H_2O_{(1)} \rightleftharpoons H^+_{(aq)} + OH^-_{(aq)}$$

$$H_2O_{(1)} + H_2O_{(1)} \rightleftharpoons H_3O^+_{(aq)} + OH^-_{(aq)}.$$

Bearing in mind these properties of silicate minerals and water, the presence of a charged mineral surface in contact with water will result in the formation of a layer of orientated water molecules surrounding it. A similar layer of preferentially oriented water molecules will be formed on internal surfaces within the lattice wherever water can penetrate, as for example along cleavage planes. The first result of this process of hydration is that, as a polar solvent such as water begins to surround the lattice, there is a weakening of the interionic (electrostatic) attraction within the lattice. This effect, due to the relative permittivity of the water, aids the detachment of non-framework ions from external and internal surfaces of the lattice. Their removal is effected as ion–water molecule bonds are established with a consequent release of energy. This release of energy facilitates the detachment of ions from the lattice and they diffuse as hydrated ions into the free water away from the crystal surface. This process where water is the solvent is also known as hydration, but is more generally termed **solvation** and the ions are said to be solvated.

The effect of the presence of layers of oriented water molecules does not stop at simple hydration and limited solvation of non-framework ions exposed at mineral surfaces, but leads to the initiation of **hydrolysis**. In this process, hydrogen ions from a variety of sources, including the dissociation of water molecules, are able to penetrate the hydrated lattice as hydroxonium ions. Here the hydrogen protons compete with and tend to replace non-framework ions in the crystal structure. These displaced ions then pass into solution as hydrated cations. When this happens and the lattice begins to become saturated with hydrogen, it is no longer stable and the mineral may disintegrate, although

(a)

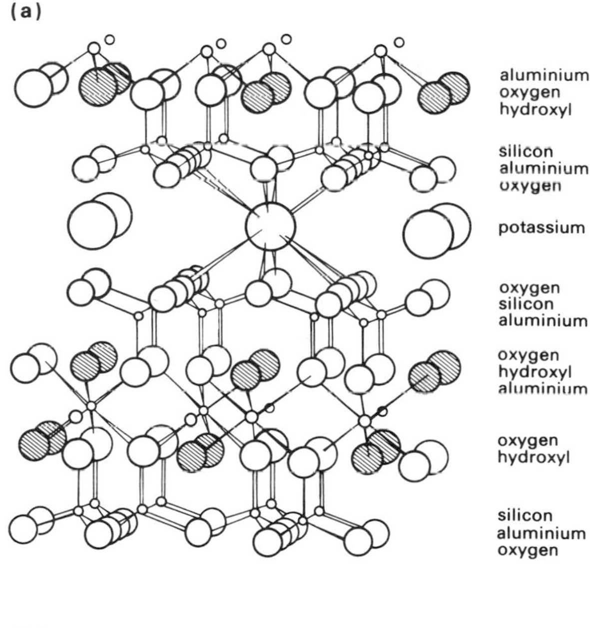

(b)

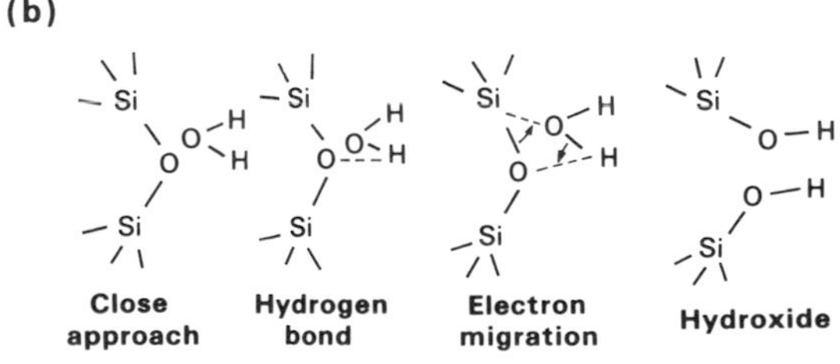

Figure 11.7 (a) Lattice structure of muscovite; note the potassium ion with a co-ordination number of twelve. (b) Silicate hydration (after Curtis 1976).

initially it may retain its silicate structure (Box 11.1).

In both hydration and hydrolysis the electrostatic forces of the lattice are disturbed and rearranged by the hydrogen ion which, remember, is a single proton. In the only other processes of lattice breakdown the electrostatic forces of the lattice are disrupted by the transfer of the electron, the other elemental particle.

Substances that lose electrons are oxidised, while those gaining electrons are said to be reduced. Processes involving the transfer of electrons are known as oxidation–reduction (or **redox**) reactions. The oxidation state of an element describes the charge that an atom of that element appears to have in a compound and this reflects its electron configuration. Now, many elements can exist in more than one oxidation state. For example, iron has oxidation states of +2 and +3 and, in the compounds it forms in many rock-forming minerals such as olivine (where iron is a non-framework ion), it exists in the lower oxidation state, as ferrous iron Fe^{++} (iron II). The transfer of an electron from iron II to an electron acceptor such as oxygen converts (or oxidizes) iron II to iron III (ferric iron), distorting the lattice and leading to the disintegration of the crystal. Here oxidation is an enacting process of lattice breakdown in an iron-bearing silicate mineral and, in other non-silicate minerals such as pyrites (iron sulphide), oxidation is even more effective as an enacting process of weathering. Redox reactions, however, are also

Box 11.1

HYDROLYSIS OF SILICATE MINERALS

Because of their size there is almost a one to one relationship between the water molecules and the oxygens of the tetrahedral units of the lattice at the crystal/water interface. The single unsatisfied negative charge of each oxygen will, therefore, be in proximity to the double positive charge of each orientated water molecule. This produces an accumulation of excess positive charge at this interface. Now some of these water molecules will be the 'hydration' water of hydrogen ions, that is they will be hydroxonium ions (H_3O^+) and the extra proton will add further to this excess positive charge. In order to correct this charge imbalance, these hydrogen ions (protons) and those from both additional dissociation of water molecules and from other sources tend to penetrate into the lattice passing from one water molecule to another, each of which temporarily behaves as a hydroxonium ion.

Within the lattice, hydrogen ions (protons) compete with non-framework ions (such as the potassium ion seen in the muscovite lattice in Figure 11.7a) to neutralise the charge on the silicate framework. Each valency charge on the non-framework ions, however, is often shared with a relatively large number of co-ordinating ions (e.g. 12 in the case of potassium in Fig. 11.7a) so that the bonding is weak. Because of its small size, the hydrogen ion can co-ordinate with only one or two of these neighbouring oxygens and hence its single charge is shared to a much smaller extent and the bonds formed are, therefore, stronger. Not surprisingly then, in competition hydrogen ions will tend to replace non-framework ions. These diffuse as hydrated cations through the lattice to the free water beyond, ultimately leaving the silicate framework at least partially saturated with hydrogen. However, the mineral is now unstable, for small units of the tetrahedral framework of the lattice individually have their negative charge imbalance neutralised by hydrogen and they are no longer ionically bonded to other units of the lattice. In addition some of the oxygens of the silica tetrahedra may carry unsatisfied excess negative charges, thereby setting up forces of electrostatic repulsion as like charges repel. Therefore, the overall effect is the disintegration of the lattice structure of the original mineral, though the covalently bonded silicate units may initially remain intact as detached but hydrated chains, sheets, or individual tetrahedra. However, these too may be disrupted by the breaking of the O–Si–O bond (the fundamental structural unit of the silicate framework) as the hydrogen bonding of hydration water changes through electron migration ultimately to produce silicic acid which passes into solution (Fig. 11.7b).

important as secondary reactions associated with lattice breakdown. This is because the solubility of such elements as iron varies with the oxidation of their ions. In its ferrous (iron II) state iron is soluble and mobile, but when oxidised to the ferric (iron III) state it is precipitated, and is insoluble. Consequently, ferrous ions released by hydrolysis from silicate structures may be oxidised immediately to the iron III state and hence rendered immobile. The nature and proportions of the oxidising and reducing substances in the weathering system or in the soil determine the redox potential (Eh), which is a measure of the tendency of the system to receive or supply electrons.

11.1.4 Activating agencies of lattice breakdown
We have already seen that fracture-activating processes (Fig. 11.3) function as regulatory mechanisms which, together with the physico-chemical properties of the rock and its minerals, control the type and rate of process active in weathering. The same is true of lattice breakdown (Fig. 11.8). It begins as a surface reaction at the

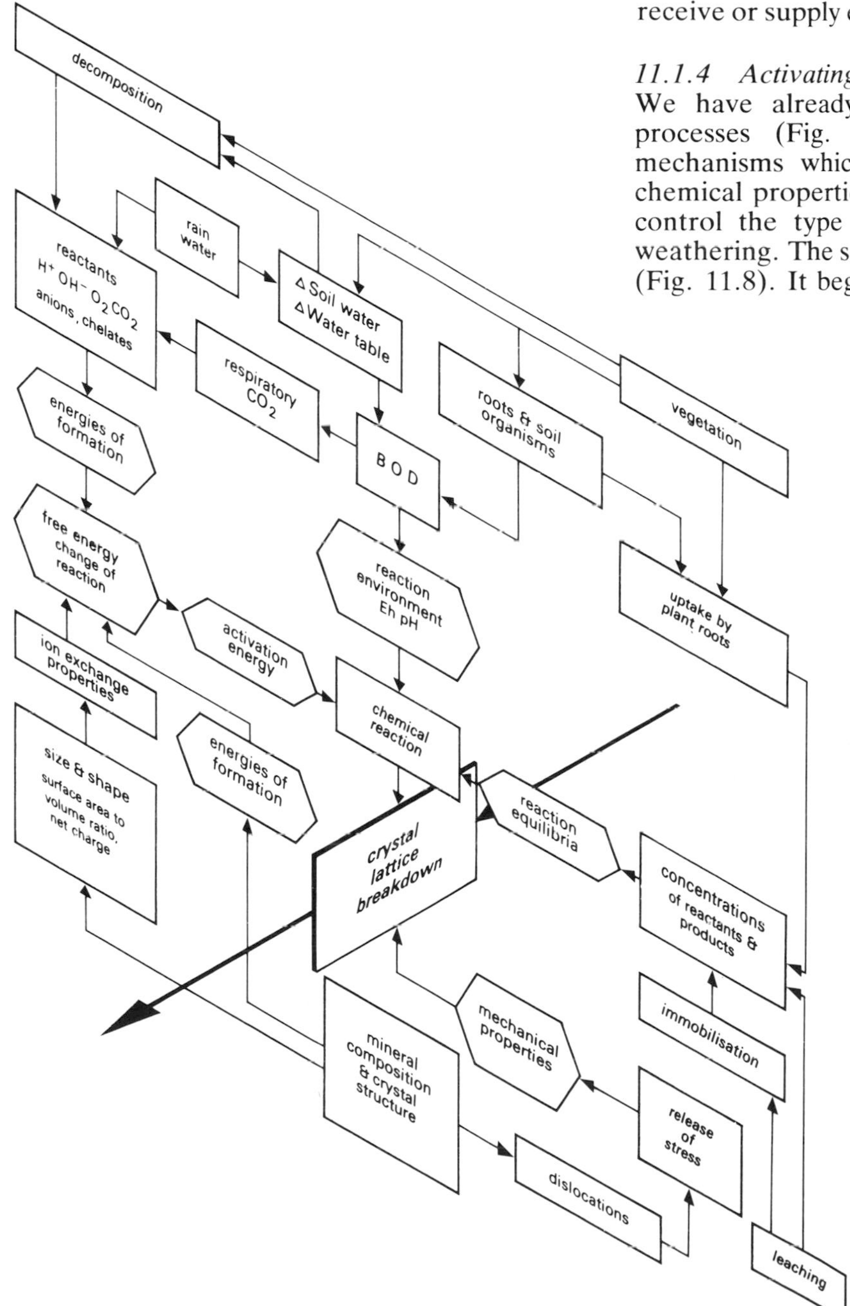

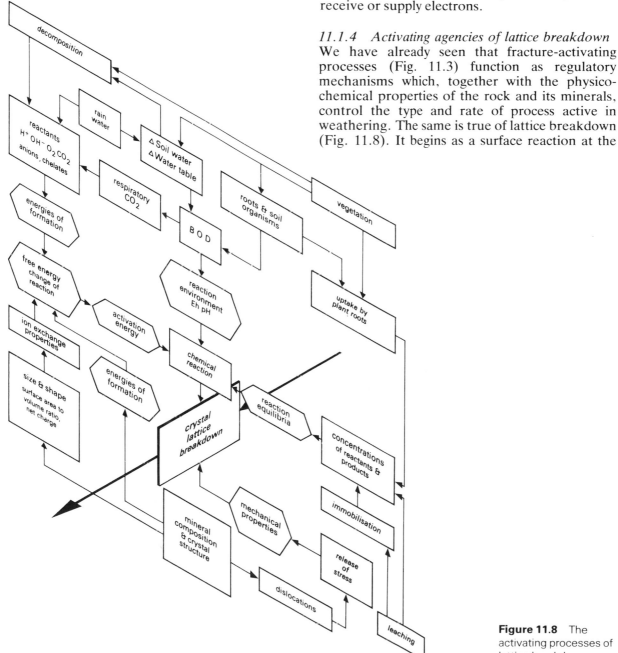

Figure 11.8 The activating processes of lattice breakdown.

mineral/water interface. Therefore, any process that increases the rate at which mineral surfaces are exposed will be an activating agency of lattice breakdown. At the macro-scale, the amount of exposed mineral surface will depend on the mechanical properties of jointing and porosity, while the rate at which surfaces are produced will reflect the effectiveness of fracture and mechanical disintegration, which increases the rock's exposed surface area. At the micro-scale, the ratios of surface-area:volume for particular minerals, the existence of cleavage planes and the voids within the lattice become important and reflect the properties of the crystal itself. Processes which remove the products of lattice breakdown, such as leaching and nutrient uptake by plant roots, will also be important activating agencies, because the removal of products from a weathered mineral surface will expose fresh surfaces of the lattice to chemical reaction. Removal of soluble products, particularly the hydrated non-framework ions released by weathering, depends on the maintenance of concentration (or diffusion) gradients (actually gradients of electrical potential) between the mineral and the free water. In this context it is essential that the voids and pores in soil and regolith are regularly flushed by percolating water.

The removal of reaction products, particularly in solution by leaching, is important for another reason. In any chemical process with a given active mass of reactants, there is an inherent negative feedback at work which brings the reaction to a point where no further chemical change takes place. This point of chemical equilibrium is defined by the thermodynamic equilibrium constant of the reaction and it is conditioned by the energy transfers involved. To prevent this negative feedback and to stop the reaction reaching equilibrium, there must be a continuous replenishment of reactants as well as a continual removal of reaction products.

We have already considered how the supply of fresh mineral surfaces is maintained, but what of the other reactants? In *hydration* and *hydrolysis* the active agents of lattice breakdown are the water molecule and the hydrogen ion or proton dissociated in water. Processes that ensure the movement of water through the weathering zone will, therefore, activate and promote lattice breakdown. Natural waters (rain, soil and ground water), however, are not pure; they are aqueous solutions containing molecules and ions of other elements. Furthermore, natural waters are normally acidic (if only slightly), that is, their hydrogen ion concentration is higher than pure water (Box 11.2).

Box 11.2

ACIDITY OF NATURAL WATERS

The acidity of rain water and stream waters is associated with the presence of hydrogen ions derived from the dissolution of carbon dioxide, sulphur dioxide and the dissociation of organic acids. The presence of aluminium ions also has the effect of increasing the hydrogen ion concentration. In soils, which can be thought of as a solid phase dispersed through a liquid phase – i.e. a soil/water suspension – hydrogen and aluminium ions adsorbed on the solid phase (the exchange complex) and the existence of an equilibrium solution of hydrogen ions in the soil water are responsible for the **reserve** and **active acidity** of the soil respectively. Therefore, acidity can be measured in terms of the hydrogen ion concentration (strictly activity) expressed as pH.

pH is defined as the negative logarithm of hydrogen ion activity (or effective concentration in g ions l^{-1})

$$pH = -\log_{10}[H^+].$$

In pure water the hydrogen ion and hydroxyl ion activities are equal and have a value of 10^{-7} g ions l^{-1}. Therefore, the pH of pure water is 7. As the hydrogen ion activity increases, the pH value falls. In most natural soils the pH value lies in the range 4–8. However, the chemical definition of pH which refers to simple aqueous solutions is complicated by the presence of charged solid particles in soil water suspensions which give to the soil the capacity to resist changes in its active acidity by ion exchange – the so-called **buffer capacity** of the soil.

But just as the dissociated hydrogen ions in pure water are balanced by hydroxyl ions, so in natural waters the excess of hydrogen ions must be balanced by other anions. The source of these excess hydrogen ions, and their balancing anions, is the dissociation of acids in water. Processes that deliver these acids to percolating water are, therefore, highly significant activating agencies in weathering which, by increasing hydrogen ion concentrations, promote the breakdown of lattice structures by *hydrolysis*.

By far the most important process in this category is the dissolution of carbon dioxide to give carbonic acid, which dissociates as the hydrogen ion (H^+) and the bicarbonate ion (HCO_3^-):

$$H_2O + CO_2 \rightleftharpoons H_2CO_3 \rightleftharpoons H^+ + HCO_3^-.$$

Some atmospheric carbon dioxide will dissolve directly in rain water, but the major source of carbonic acid is carbon dioxide photosynthetically withdrawn from the atmosphere and released again by root respiration and by the respiration of detritivores and decomposers during the decomposition of organic matter (see Ch. 19) to become dissolved in percolating waters. In addition, organic compounds released into solution (such as simple carbohydrates) undergo further oxidation and produce carbon dioxide. Other organic compounds – the organic acids – dissociate in water to yield hydrogen ions directly. The decomposition of organic matter is, therefore, an activating agency of the most profound significance, and, as we shall see, its regulatory role does not stop at the provision of hydrogen ions for hydrolysis.

The other fundamental enacting process of weathering concerns redox reactions, which are regulated by the presence of oxidising and reducing agents and by the redox environment (Eh). The principal variable here is the degree of aeration of the weathering zone, so that processes and conditions which maintain the availability of free oxygen will activate oxidation reactions, while conversely, those producing anaerobic conditions will promote reduction reactions. The permeability and drainage relationships of the rock and regolith relative to the water table and the quantity and characteristics of precipitation input will be important regulators of aeration, while the diffusion coefficients of oxygen in air and water will determine the degree of oxygenation of the reaction environment. In addition, the biological oxygen demand (total requirement of respiratory oxygen) will also affect oxygen availability and, even in aerated soils and regoliths, it can produce localised reducing conditions. Conversely, the radial loss of oxygen from roots can initiate localised oxidation, as is evident in the mottling of waterlogged soils.

The redox environment is important not only in regulating active weathering reactions but also in controlling the mobility and hence the removal of some weathering products (Fig. 11.9). Such elements as iron and manganese are markedly more mobile in their lower reduced valency state – Fe^{2+} (Fe II, ferrous ion) and Mn^{2+} (Mn II, manganous ion) – and pass into solution under reducing conditions. Iron, in its Fe^{3+} (Fe III, ferric) valency state, together with aluminium, silicon and titanium, precipitates to form insoluble compounds, mainly complexed oxides and hydroxides, in the normal pH and Eh environments of weathering. In the absence of reducing conditions or extremes of pH, the removal of these elements appears to depend on the supply of organic complexing agents released by the decomposition of organic matter. These are known as chelating

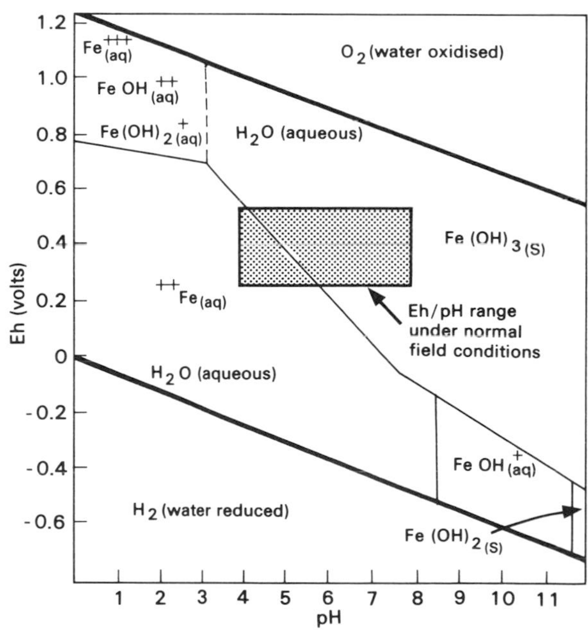

Figure 11.9 Stability areas of the Fe II–Fe III aqueous system. Note that under normal field conditions insoluble ferric compounds would occur.

agents and they consist of organic molecules, particularly of phenolic compounds capable of forming at least two co-ordinate covalent bonds with a central metallic ion to form a complex chelate. The charge on the complex may be positive, negative or zero, but it carries into solution such metallic ions as iron and aluminium which would not otherwise be soluble. **Chelation** is often cited as an active weathering process, but its significance lies not in its contribution to lattice breakdown but in its ability to remove elements already partially released by hydrolysis, which would otherwise accumulate as insoluble products. As such, it is a leaching, erosion, or pedogenic process and, in the weathering system, it functions as an activating agency and regulator.

11.1.5 Resynthesis and the formation of new minerals

The immediate result of lattice breakdown is the appearance of stable but highly mobile cations such as calcium, magnesium, potassium and ferrous iron in aqueous solution. Some silicon (as the monomolecular silicic acid, H_4SiO_4) and a limited amount of aluminium will also exist in solution, though most will be present as colloidal particles. The larger of these particles will partially retain the silicate structure of the original lattice, usually as chains of tetrahedra. The chemical environment within the disintegrating lattice will be alkaline because of the high concentration of basic cations, but as they and the silicon and aluminium in solution diffuse into the soil water,

Box 11.3

CLAY MINERALS

There are many different kinds of clay mineral, but all form very small, flat lamellar crystals with a very large surface-area:volume ratio and resemble the micas such as muscovite and biotite. They are hydrated alumino-silicate minerals with a lattice structure composed of two basic units: the silica tetrahedron and the aluminium octahedron. In the first the silicon atom is equidistant from, and at the centre of, four oxygen atoms in tetrahedral co-ordination, and in the second, aluminium atoms are equidistant from, and at the centre of, six oxygen atoms or hydroxy ions in octahedral co-ordination.

These basic units join by covalent bonding to form tetrahedral and octahedral sheets respectively. In the lattice structure of clay minerals these sheets are combined in various ways to form layers. In minerals such as kaolinite the two sheets are hydrogen bonded in the ratio 1:1, tetrahedral sheet: octahedral sheet. This structure does not swell on wetting as hydration water molecules cannot penetrate the lattice layers. In 2:1 lattice clay minerals such as vermiculite and montmorillonite a single octahedral sheet lies between two tetrahedral sheets. The result is that the mineral swells on wetting as water of hydration penetrates between successive 2:1 layers, carrying with it magnesium and calcium respectively. In the hydrous mica clay minerals such as illite the 2:1 layers are held together by potassium ions with no hydration water in the inter-layer space. With loss of this structural potassium, illite weathers to montmorillonite.

One of the most important properties of clay minerals is that they carry a large surface charge. This is a net negative charge and much of it is a **permanent charge** originating within the lattice structure by **isomorphic substitution**. Here atoms or ions of similar size, but lower valency, replace some of the silicon and aluminium in the tetrahedral and octahedral sheets without straining their structure, leaving unsatisfied negative valency charges associated with the oxygens and hydroxyls. In addition broken bonds occur, particularly at the imperfect edges of the crystals, but these charged sites (which may be either positive or negative) are pH dependent. The surface charge of clay minerals means that they can attract and adsorb on their large external and, in the case of expanding 2:1 lattice clays, internal surfaces a swarm of cations. This phenomenon of **cation adsorption** accounts for their importance in **ion exchange** processes in soils and regoliths.

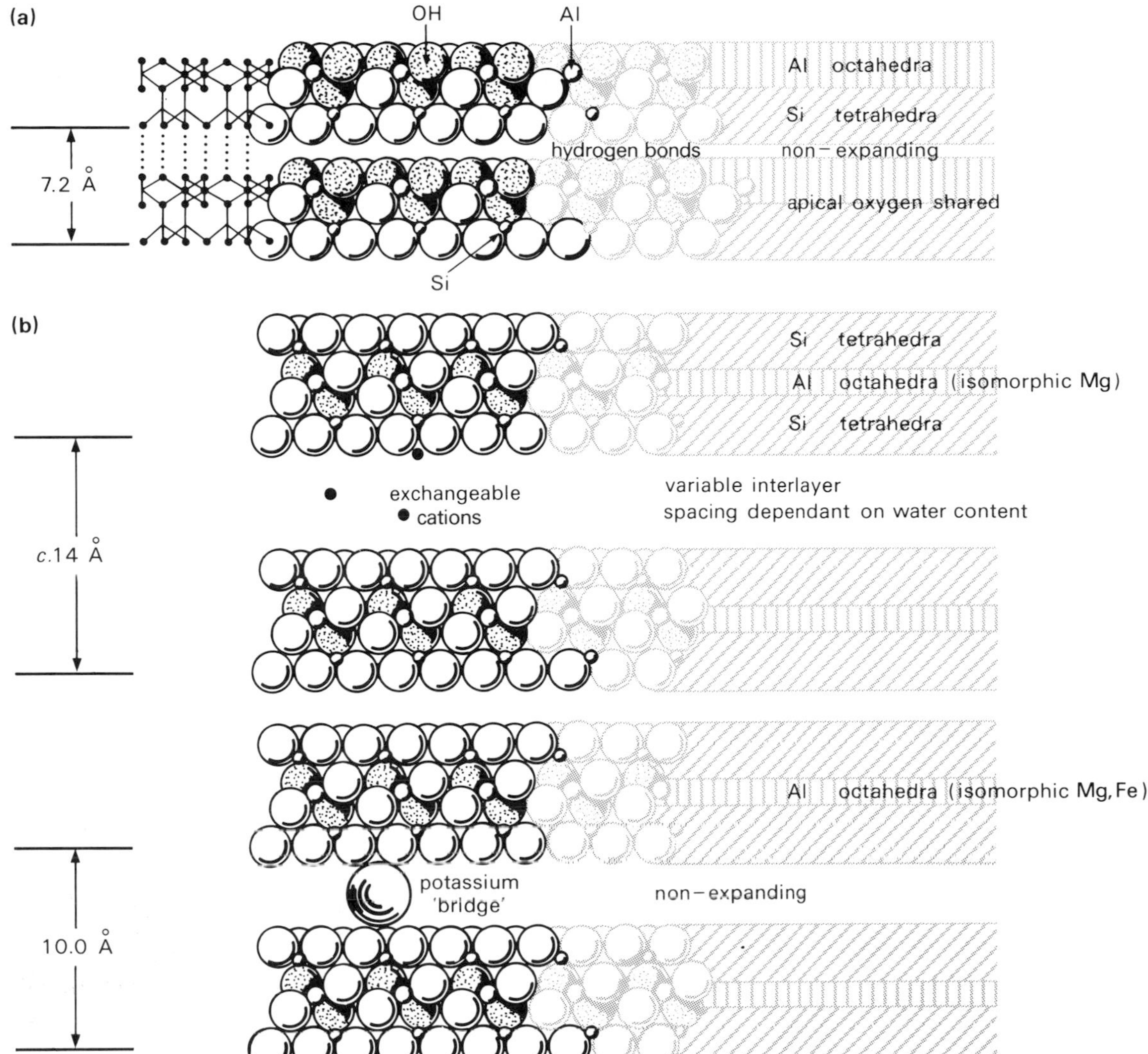

Figure 11.10 Crystal structure of (a) 1 : 1 and (b) 2 : 1 lattice clay minerals.

the environment becomes more acid. It is at this point that the first category of new minerals appears.

As we have seen already, iron and aluminium, and to a lesser extent silicon, are soluble only under reducing conditions and/or extremes of acidity and alkalinity. In the oxidising and mildly acid environment of aerated weathering zones they are immobile. Iron and aluminium, therefore, are precipitated as complexed oxides and hydroxides (e.g. goethite FeO.OH, and gibbsite $Al(OH)_3$) which also usually contain some silicon as an impurity. These form amorphous or microcrystalline particles, normally but not always of colloidal dimensions (<2 μm). They tend to carry a net positive charge in acid environments and may be adsorbed to form coatings on the fragments of the silicate framework of the original mineral.

The second group of new minerals begins to form as a consequence of the same processes. Monomolecular silicon in solution becomes polymerised (see Ch. 7) under the more acid environment and precipitates together with much of the aluminium to give, as we have just seen, mixed aluminium hydroxide and silicic acid particles. At the same time silicon and aluminium are present together in the residual silicate framework fragments. These two entities then form the precursors of a group of new or secondary minerals (crystalline hydrated alumino-silicate minerals) known as clay minerals (Box 11.3). They are sheet silicates (phyllosilicates) with a structure similar to the micas (Fig. 11.10). The mechanisms by which aluminium and silicon in weathering residues become re-organised to form the silica tetrahedral and aluminium octahedral layers of the developing clay minerals is complex and not fully understood. In some cases, however, as in regoliths where the unaltered rock contained a large proportion of sheet silicates such as biotite and muscovite, relatively little modification of the lattice is sufficient to convert these micas to secondary clay minerals. The type of clay mineral produced, however, depends on the ions available in the regolith to be incorporated in the lattice as non-framework ions. As Figure 11.11 indicates, this depends on the intensity of the two processes which control the rate at which these ions (and the precursors of the alumino-silicate sheets) become available and the rate at which they are removed, i.e. lattice breakdown and leaching, respectively. Indeed, the combination of these processes can lead in time to the conversion of one type of clay mineral to another, e.g. illite to montmorillonite to kaolinite.

The properties of clay minerals differ from those of primary minerals in several important respects. The particles they form are smaller, giving to them colloidal properties; their composition is more heterogeneous, even within the same crystal, and at least some have expanding lattices (Box 11.3) giving them enormous surface areas in contact with soil water. Clay particles carry a net negative charge (surface charge), partly originating in unsatisfied charges within the silicate and aluminium sheets (permanent charge) and partly due to the dissociation of hydrogen ions and to other broken bonds (pH-dependent charge). This surface charge is compensated for by a layer of oriented water molecules and adsorbed ions, largely cations (volume charge) (Fig. 11.12). Cations adsorbed on clay minerals in this manner are protected against removal by leaching and the process of ion adsorption is an important regulator of both leaching loss to the solute load of drainage water and of nutrient availability to plants. Both are controlled by the complex process of cation exchange.

In arid environments a third type of new mineral is formed, in addition to clay minerals, for here leaching does not carry away excess ions. Instead, they are precipitated in the form of salts such as carbonates and sulphates of calcium and sodium as soil water evaporates. The mechanical effects of such salt crystal growth have already been discussed.

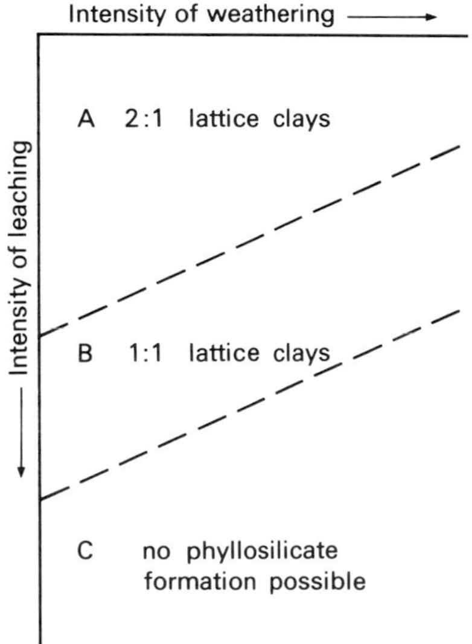

Figure 11.11 The type of secondary mineral formed in relation to the weathering : leaching ratio.

11.2 The final state of the system: the regolith

As a response to the primary mechanisms of weathering, the final state of the system will be characterised by the following.

(a) A group of minerals which have proved resistant to dissolution, acid hydrolysis and redox reactions. These will exist as grains released by the disaggregation of the rock, but will be altered little chemically and structurally. In 1938, Goldich suggested that the persistence of primary silicate minerals under weathering was the reverse of their order of formation as predicted by the Bowen series (see Ch. 5). This order of persistence is in broad agreement with the energy change undergone during weathering, so that minerals with a high loss of free energy on breakdown are less stable than those with a small negative free-energy change. Free-energy change correlates with the bond energy and the energy of formation of primary minerals, thereby providing the thermodynamic link with Goldich's empirical weathering sequence (this does not hold, however, for minerals where oxidation is the prime weathering reaction). On both thermodynamic and empirical grounds, quartz (crystalline silica) appears to be the most resistant mineral under most weathering regimes and it tends to dominate this residual category (Fig. 11.13).

(b) Resynthesised or secondary minerals, of which the hydrous oxides and hydroxides of iron and aluminium, the clay minerals and, in some environments, accumulations of soluble salts are the most important. At any particular moment, however, most regoliths will contain also materials which represent transient or intermediate states of the system (Fig. 11.2).

Therefore, in addition to the two categories above there may also be:

(c) some minerals present as individual grains or aggregates of grains which are still actively undergoing chemical breakdown;

(d) rock fragments (clasts) of various sizes and shapes which have resulted from mechanical fracture but which have not yet undergone significant chemical change.

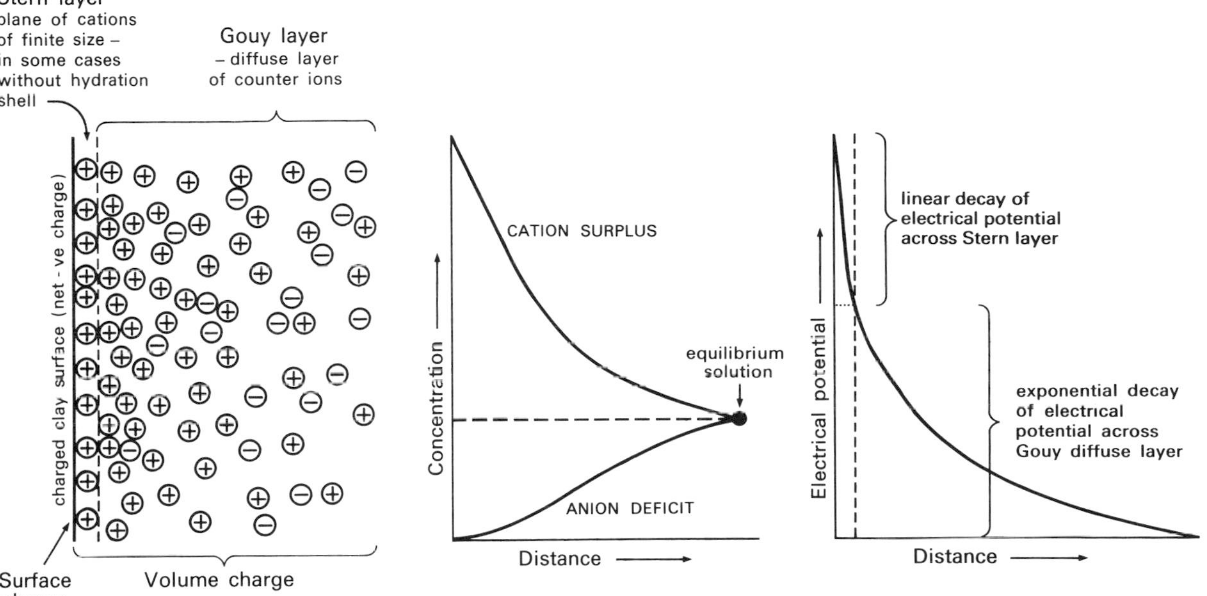

Figure 11.12 The ionic distribution and electrical potential gradients at a negatively charged planar clay surface.

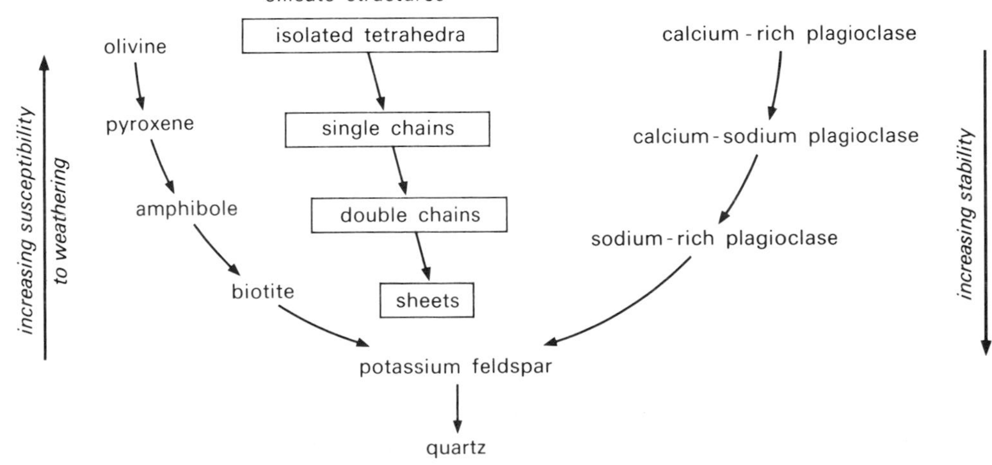

Figure 11.13 The Goldich weathering series.

These four categories differ considerably in particle size. The quartz residue, for example, largely exists as particles in the range 0.02–2.0 mm (sand-size fraction); the secondary oxides and clays are all smaller than 0.002 mm and most are colloidal; mineral grains actively undergoing weathering, though variable in size, largely fall in the range 0.02–0.002 mm (the silt-size class); and finally, the clastic fragments can vary from 2.00 mm to sizeable stones and small boulders. It is the relative proportions of these size fractions that determine the texture of the regolith. The more advanced weathering is and the closer it approaches the final state, the more it will be dominated by the quartz and clays and the more bimodal its texture will become. In other words, the nature of the regolith as it approaches the final state of our model is governed by elapsed time.

The changes wrought by the operation of the weathering system include a reduction in induration, or a change from the massive to the clastic state, in addition to the chemical changes. These inevitably cause changes in the mechanical properties as the initial rock changes its state. Dearman, Baynes & Irfan (1978) describe the changes which affect a granite on weathering. Of the three main constituent minerals, the quartz remains essentially unchanged. The biotite becomes bleached as iron is removed in solution to form chlorite and other clay minerals. Feldspars become altered to kaolinite and associated clay minerals. The rock passes through a series of physical states, from the original sound rock, through stages of discoloration indicating the removal of solubles, to decomposed rock. In the latter state, the original structure is still visible in the form of individual crystals and macrostructures and as joints and dykes. Yet the substance of the material has altered and what remains are unaltered quartz grains set in a matrix of clay minerals which have replaced, yet retain the outlines of, the original parent minerals. As breakdown takes place and mobile products are removed, so porosity increases and mechanical strength decreases. These various grades of weathering are set out in Table 11.4. The relationships between mechanical strength, weathering grade and porosity are illustrated in Figure 11.14

We can, however, substitute space for time and represent the system as a vertical series of horizontal zones forming a profile from the surface to unweathered bedrock at depth. This, of course, is the traditional approach to the description of the weathered mantle or regolith and in the upper zones it becomes the classic model of the soil in terms of the soil profile and soil horizons (Fig. 11.14 and Ch. 20).

In this model it is assumed that the zones nearer the surface, in closer proximity to inputs from the atmosphere and biosphere, will be in a more advanced state of adjustment. Conversely, zones at depth will show less alteration and resemble most closely the original bedrock. Intermediate zones will, therefore, represent various states of

Table 11.4

(a) Characteristics of granite at different grades of weathering (after Fookes *et al.* 1971).

Grade	Description	Physical characteristics
VI	soil	completely weathered, original structure and fabric lost
V	completely weathered	completely weathered, friable, original structure and fabric intact
IV	highly weathered	rock discoloured, discontinuities extensive <50% original rock intact
III	moderately weathered	rock discoloured, discontinuities open, >50% original rock intact
II	slightly weathered	discoloration by bleaching or staining, discontinuities opening
I	fresh rock	no sign of weathering

(b) Material properties of granite at different weathering grades (after Dearman *et al.* 1976).

Grade	Material	Uniaxial compressive strength (MNm^2 $N\ m^{-2}$)	Effective porosity (%)
I	fresh granite	246	0.11
IIii	partly discoloured	219	0.57
IIiv	completely discoloured	165	1.52
V	intact soil	3.5	9.98

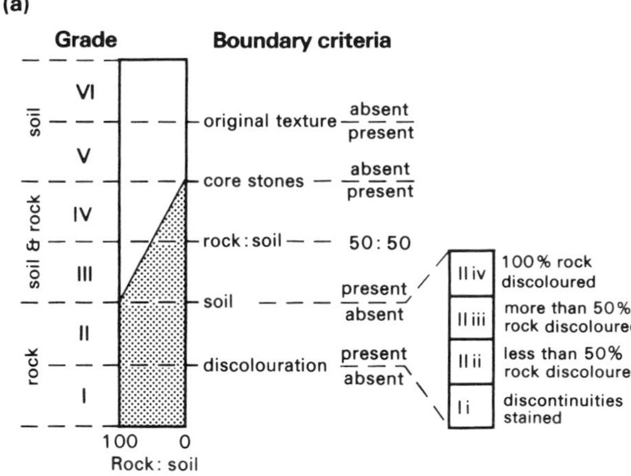

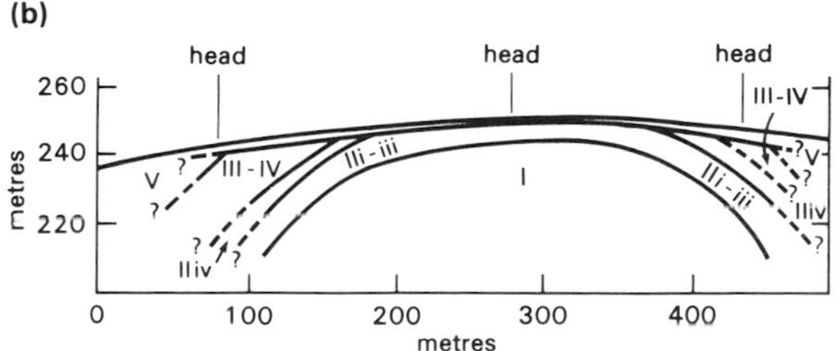

Figure 11.14 (a) Sequence of weathering grades of granite – an engineering classification based on field properties. (b) Cross section through a quarry face showing distribution of weathering grades. (After Dearman, Baynes & Irfan 1976.)

transition. Following this gradient in the states of the regolith are parallel gradients in physico-chemical conditions and properties (including particle size) and in the combination and relative significance of weathering processes. The model is not entirely static, for with time the depth of the most altered zone increases in relation to the rest of the regolith as a weathering front advances down through the profile.

As we have seen, the upper part of this vertical model is the soil profile and it is really impossible to separate weathering from pedogenesis, except in the few environments that are totally devoid of life. Under these pedogenic influences, therefore, the regolith will be further modified by the incorporation of organic matter and the activity of soil organisms, by aggregation and the formation of organomineral complexes and by translocation and horizon development (see Ch. 20).

11.3 The generalisation of the model: other rock types

So far, the model of the weathering system has been restricted to igneous rocks. It can be generalised to cope successfully with both sedimentary and metamorphic rocks, by the inclusion of what are essentially feedback loops. One of these loops, which operates within the weathering system itself, has already been considered implicitly. This is the pathway whereby the products of weathering are resynthesised to produce new minerals such as clays, which can then experience renewed or continued weathering *in situ* to produce a series of possible final states of the mineral, depending on the intensity of weathering and/or the length of time over which it is effective. This concept can be extended to incorporate cyclic pathways outside the weathering system itself.

As we have seen in Chapter 10, the products of weathering are ultimately destined to be removed, and deposited by the processes of denudation. Mobile products in solution can be transported rapidly and continuously and their denudation is separated in time from that of the residual products forming the regolith, the rate of removal of which is conditioned by the slower processes of mass movement on slopes. Eventually, after deposition and sedimentation, perhaps followed by lithification, these weathering products, now components of sediments or sedimentary rocks such as mudstones, shales and sandstones, may be exposed to another cycle of weathering after uplift and exhumation. Such a cycle can be modified to include a diversion through living organisms and the genesis of biogenic sediments incorporating siliceous or carbonate skeletal remains. It can also be extended to include metamorphism of not only sediments but also the igneous rocks. In fact, such an approach is the geochemical model developed in Chapter 5 when considering the crustal system and its relationship to the denudation system.

The generalisation of the model to other rock types in this way must necessarily take into account the differing mineralogical composition and mechanical structure of those rocks. There is a very wide range of variability in the properties of both sedimentary and metamorphic groups. Factors such as porosity, permeability, the degree of fracturing and induration all influence the access

(a)

Figure 11.15 Lithology and weathering; different rock types react in different ways to weathering processes. (a) Massive limestone – joints have opened up as a result of solution of the calcium carbonate composing the rock (Malham, Yorkshire, England). (b) Weakly cemented sandstone – weathering of the cement produces disintegration of the rock; less resistant lenses in the faced blocks have also been etched out (Taunton, Somerset, England). (c) Regolith formed from massive dolerite – weathering and formation of new minerals have taken place along former joints, leaving massive corestones of unaltered rock set in a matrix of weathered material; the exposed face is several metres high (North Queensferry, Fife, Scotland).

and passage of weathering agents (Fig. 11.15).

Sedimentary rocks generally include a lower proportion of primary rock-forming minerals than the igneous rocks from which they were originally derived. Therefore, the most abundant minerals – quartz and white mica – in sedimentary rocks are also the most resistant. On the other hand, less chemically stable minerals such as feldspar are less abundant. Arenaceous sedimentary rocks are commonly indurated with a cement consisting of a metallic compound (e.g. Fe_3O_4, $CaCO_3$) which is more susceptible to weathering than the resistant detrital grains. The weathering of such rocks commonly involves the chemical breakdown of the cementing material by, for example, redox reactions or solvation, and the release of the constituent detrital grains. In the case of argillaceous sedimentary rocks – mudstones and shales – the resistant detrital grains are usually set in a matrix of finely divided material of secondary origin (resynthesised minerals) with a high proportion of base cations. Such material is of course potentially highly vulnerable to weathering reactions, yet the rate of weathering may be strictly limited by the restricted access of water in materials of such low permeability.

Sedimentary rocks consisting almost entirely of soluble minerals constitute a special case. Limestones are the most important of this group; less commonly exposed is rock salt. Here weathering proceeds by the direct solvation of the bulk of the rock, leaving only a small residue of insoluble impurities.

Metamorphic rocks possess such great variety that it is difficult to make valid generalisations. Suffice it to say that many contain minerals that are structurally similar to the primary minerals of igneous rocks as well as some relatively rare minerals which are potentially unstable under most weathering regimes.

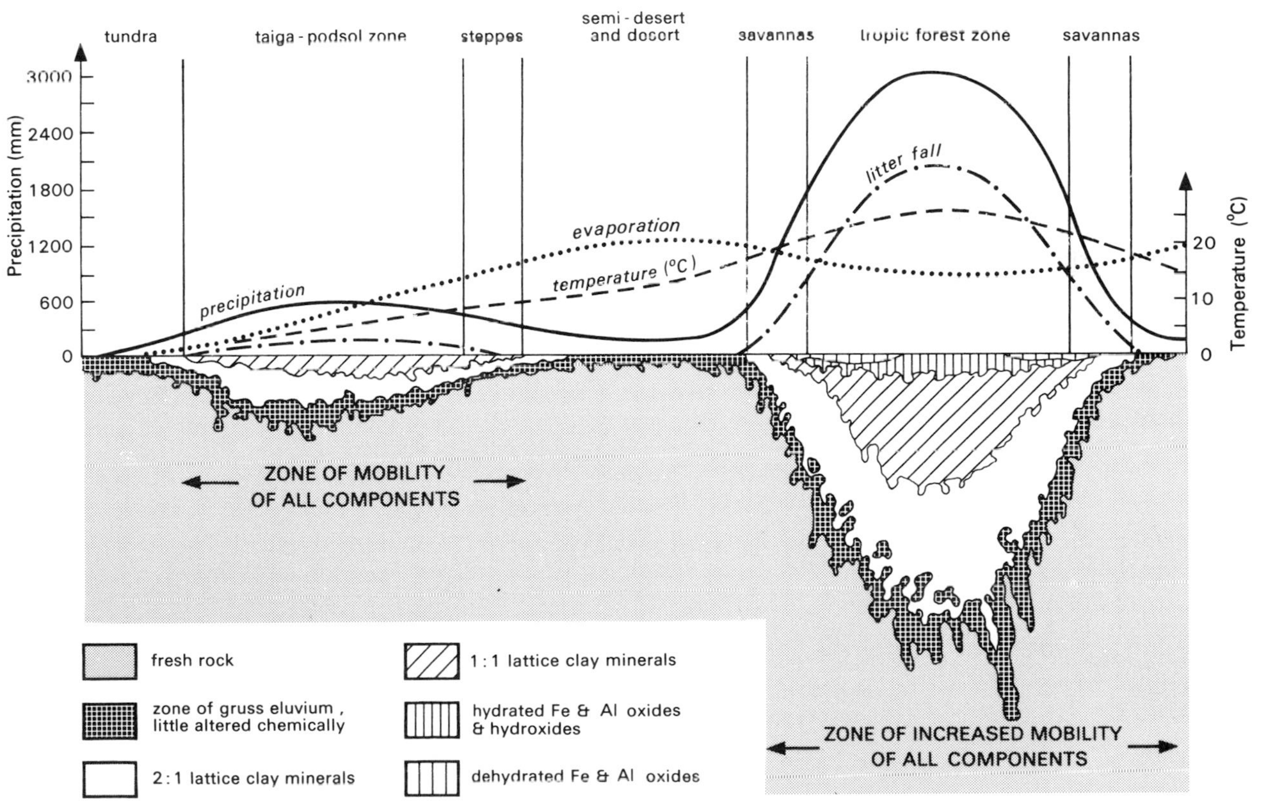

Figure 11.16 Generalised weathering profiles in different climatic zones (after Strakhov 1967).

11.4 Exogenous variables and the control of weathering

The process–response model (see Fig. 11.2) of weathering which we have been considering, applies not only to all rock types but also in all environments. It will have become evident, however, that the activation and regulation of weathering processes is ultimately conditioned by the exogenous (input) variables to the model. It is these variables that define the conditions of existence under which the system operates. In the weathering system, as in the denudation system as a whole, these variables are climate, lithology, vegetation and time, and the control they exert at any site will be modified by its position in a landscape, particularly its position in relation to slope. That the model can accommodate the lithological variable has already been demonstrated, and the effects of climatic and biotic controls at a global scale, are summarised in Figure 11.16 and Table 11.5. The effects of both vegetation and climate have an important manifestation through the control they exert on the hydrological relationships of the weathering environment: that is to say the disposition of the locus of weathering in relation to the movement of water. Table 11.6 shows a classification of hydrological zones in relation to conditions that affect weathering.

Weathering of rock in the sub-aerial zone occurs only where the rock is exposed at the surface, and for any given rock its rate will depend on the nature and quantity of the reactants available. Weathering is probably most intense in the percolation zone where renewal of reactants and removal of weathering products is continuously facilitated. This zone is also subject to considerable fluctuations in temperature. In the subaqueous zone weathering is probably limited. Since in this zone the voids are permanently waterfilled conditions are anaerobic and reducing. Furthermore, the ground water will be highly charged with solutes derived both *in situ* and particularly from above. Under these conditions reaction rates will be slow and the slow rate of removal of weathering products by groundwater seepage will limit the rate of weathering.

11.5 Conclusion: further perspectives on weathering

Weathering has traditionally been separated from erosion, but such separation is misleading and has

Table 11.5 Geochemical types of rock weathering related to climatic conditions (after Lukashev 1970).

Type of residual weathering product	Geochemical nature of the process	Weathering environment and solute transfer conditions
skeletal, clastic	formation of mixture of debris; slight removal of solutes	low temperature, slight chemical and biological breakdown of rocks
siallitic-argillaceous (iron-pan type)	SiO_2 and Al_2O_2 hydrate, mixtures formed with accumulation of SiO_2 in podzol horizons and removal of Al_2O_3 and Fe_2O_3 to underlying horizons, leaching of such elements as Cl, Na, Ca, Mg and K	moderate humidity and temperature; active organic and humic acids; downward migration of solutes
siallitic-carbonatic (calcrete-type)	silica, iron and aluminium hydrates formed, together with accumulation mainly of calcium carbonate, but also Mg, K and Na carbonates	Mediterranean and related semi-arid, seasonal climates; organic and humic activity; both upward and downward migration of solutes
siallitic-chloride-sulphate type (gypsum-type)	formation of hydrated weathering products (siallites); high mobility of SiO_2, accumulation of chloride and sodium, calcium and magnesium sulphates	warm, arid conditions; upward migration of solutes dominant; greatly reduced organic activity
siallitic-ferritic and allitic (ferricrete-bauxite type)	accumulation of iron and aluminium with general loss of silica and more soluble elements	hot, wet climates; widespread leaching and migration of solutes

Table 11.6 Some characteristics of the principal hydrological zones (after Keller 1957).

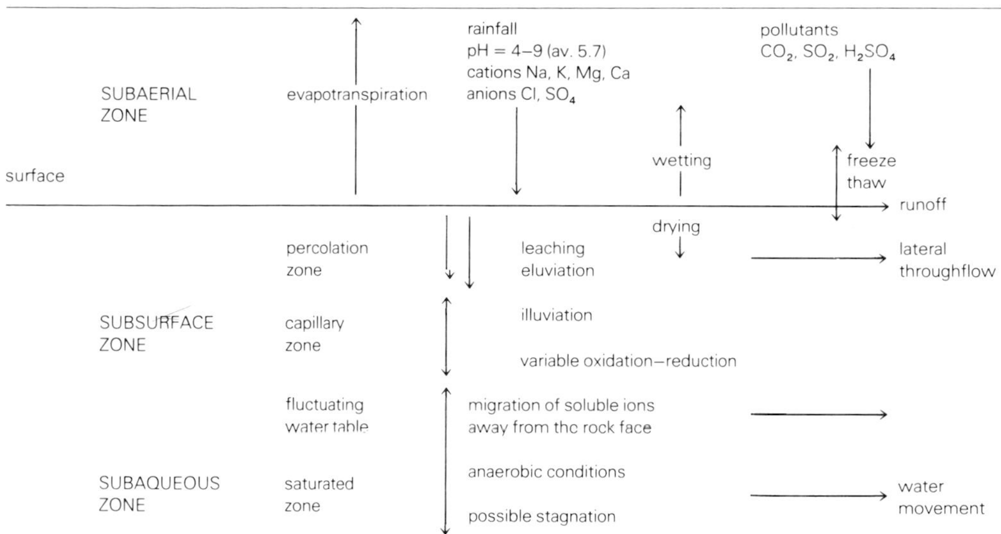

caused some confusion. Weathering is not restricted to material *in situ*; it also affects debris being transported actively. Furthermore, as we have seen, the operation of weathering processes is partially regulated by the removal of weathering products in order to prevent the establishment of equilibrium conditions. Here erosion and transport are seen to be integral processes in the weathering system, and this particularly applies to leaching. Because of its association with soil and the two-dimensional model of the soil profile, it has been the vertical component of movement that has been stressed in relation to the leaching process. However, leaching and erosion by solution are really synonymous. The significance of throughflow in the vadose zone to the solute load of streams is now appreciated and this is, therefore, directly related to weathering on the contributing slopes of their catchments.

The debris cascade is only one perspective on the weathering system. The ecologist's view is different. In his eyes weathering, pedogenesis and ecosystem function are intimately linked and, as we have seen, many of the rate-limiting regulators and thresholds that control the operation of the weathering system are at least indirectly biological. The same is true for the control of the throughput of weathering products on slopes. Indeed, the ecosystem operates to conserve and regulate not only the release of nutrient ions but also their circulation and their loss to the denudation system. The functional link between weathering, soil, the ecosystem and denudation clearly falls within the domain of the slope system and the processes operating within it.

Further reading

The original approach adopted in this chapter makes the recommendation of a comprehensive source difficult, although many useful sources exist on different aspects of weathering.

Ollier, C. 1969. *Weathering*. Edinburgh: Oliver & Boyd, is useful in providing a geomorphological context.

Basic texts on processes of chemical weathering are

Keller, W. D. 1957. *The principles of chemical weathering*. Columbia, Missouri: Lucas.

Loughnan, F. C. 1969. *Chemical weathering of the silicate minerals*. New York: Elsevier.

A thermodynamic approach is provided by

Curtis, C. D. 1976. Chemistry of rock weathering: fundamental reactions and controls. In *Geomorphology and climate*, E. Derbyshire (ed), Chichester: Wiley.

Other useful sources are

Garrels, R. M. and F. T. Mackenzie 1971. *Evolution of sedimentary rocks*, Ch. 6. New York: Norton.

Mottershead, D. N. 1982. Coastal spray weathering of bedrock in the supratidal zone at East Prawle, South Devon. *Field Studies*, 5, pp. 663-84.

Paton, T. R. 1978. *The formation of soil material*, Chs 2–4. London: George Allen & Unwin.

Trudgill, S. T. 1977. *Soil and vegetation systems*, Ch. 3. Oxford: Oxford University Press.

Winkler, E. M. 1975. *Stone: properties, durability in Man's environment*, 2nd edn. Berlin: Springer-Verlag.

12 The slope system

From the functional point of view the slope system is a cascade, in the literal as well as the conceptual sense. Its inputs are from atmospheric, weathering and biological systems and its outputs principally to the channel system, but also to the atmosphere and to the vegetation growing on the slope (Fig. 12.1). Its functional role is one of throughput: the evacuation of rock and debris prepared by weathering and of water carrying elements in solution and suspension. In many ways the visible form of a slope is a balance between the rates of input of these materials and the rates of throughput and output as regulated by the interplay of denudation forces and the resistance of the slope materials.

The disposition of slopes and the spatial relationships of individual slope facets are fundamental units of all landforms and hence create the character of the landscapes they form. Important relationships exist between the state of the system, in terms of slope form, and the processes operating in it. There is a complex feedback between the two. On the one hand, slope form (in terms of angle, profile and depth and disposition of regolith) is a consequence of the past operation of processes, yet at the same time it strongly influences the operation of contemporary processes.

12.1 The initial state of the system

The form of a slope at any moment in time is a function of initial form, geological structure and subsequent modification by denudation processes.

Slopes may be initiated in a number of ways. First, uplift of a relief mass may take place as a result of the activity of the crustal system, by means of folding or faulting. Orogenic activity of this nature has been measured at a variety of locations, commonly yielding values of several millimetres per year. Secondly, linear erosion by rivers or glaciers can incise into a landscape to produce valley-side slopes. At Grand Canyon, the

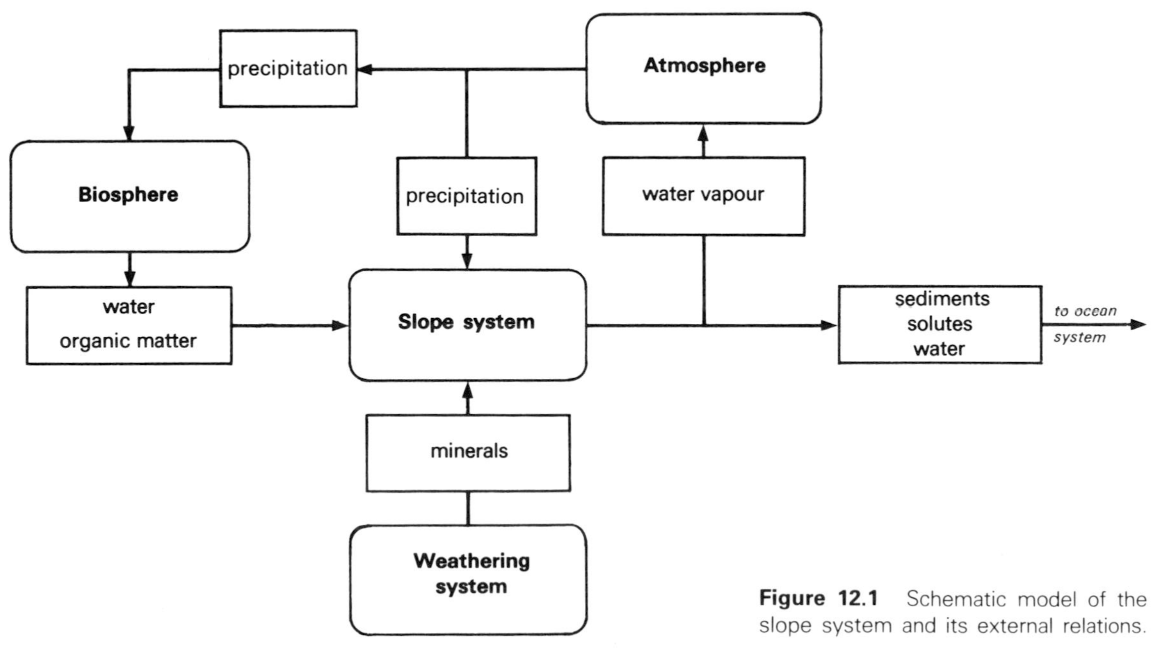

Figure 12.1 Schematic model of the slope system and its external relations.

Colorado River has incised itself at a mean rate of 0.25 mm a^{-1} over approximately 8 Ma. Thirdly, marine action can create slopes either by erosion, whereby the margins of a landmass are trimmed back by cliff formation, or by a fall in sea level which results in the exposure of the slopes of the former sea floor. These various modes of slope initiation are means by which relief is created faster than it is being destroyed. Furthermore, the initiation of slopes is not an instantaneous process but a gradual one, and therefore doubt is cast on models of slopes that assume an instantaneous origin. The number of ways in which slopes can be initiated reveals that a considerable variation may be expected in initial slope form, ranging from the vertical cut of a rapidly incised river to a gently shelving former sea floor.

12.2 The operation of the slope system: the transfer of water

Fundamental to the understanding of slope systems is the recognition of the modes of transfer of water down slope both at and beneath its surface. This will be considered in the present section, followed by a study of the transfer of minerals in Section 12.3. The retention and movement of water in the slope system, both across the surface and within the soil and regolith, is best understood by reference to the forces acting on the water at any point in the system. The ability of the water to do work under the constraints of these forces is an expression of the potential energy of the water in the system (specific free energy) or the **water potential**, ψ (Box 12.1). By treating the water in the slope system in terms of water potentials, we are able to specify the magnitude and direction of the flow of water, for it will tend to move from regions of high potential to low potential, just as heat is transferred from regions of high to low temperatures.

Soil water potential (Box 12.2) is a composite term whose components are related to the chemical and mechanical forces concerned. Movement of water within the system will depend on the difference in potential between any two points, which in turn is a function of soil moisture conditions on the

Box 12.1

WATER POTENTIAL

Water potential (ψ) is a term which expresses the difference between the chemical potential of water at any point in the system (μ_ω) and that of pure free water under standard conditions of temperature and pressure or elevation ($\mu_\omega{}^0$). This difference ($\mu_\omega - \mu_\omega{}^0$) is an indication of the ability of water in the system to do work, as compared with that of pure free water. Chemical potential, however, is not easily measured in absolute terms but water potential can readily be determined because:

$$\psi = \mu_\omega - \mu_\omega{}^0 = RT \ln (e/e^0)$$

where R = universal gas constant (J mole^{-1} degree^{-1}), T = absolute temperature (°K), e = vapour pressure of water in the system at temperature T, e^0 = vapour pressure of pure water at the same temperature and elevation as water in the system, and ln = the natural logarithm, $\log_e$. The units of $RT \ln (e/e^0)$ are J mole^{-1}. Pure free water is arbitrarily defined as having zero water potential under standard conditions. When the ratio e/e^0 is less than one, $\ln e/e^0$ is negative and water potential is a negative quantity. Water potential is increased by increase in pressure or temperature, but decreased by the presence of solutes, hydrostatic or capillary forces and by electrostatic attraction to charged surfaces. Water potential can be expressed in pressure units by dividing by the partial molar volume of water ($\bar{V}_\omega$):

$$\psi = \frac{\mu_\omega - \mu_\omega{}^0}{\bar{V}_\omega}$$

$$\psi = \frac{RT \ln (e/e^0)}{\bar{V}_\omega}$$

A conversion table for various expressions of water potential assuming 1 cm^3 of water weighs 1 g (after Bannister 1976) follows:

Atmospheres (STP)	*Bar*	*N m^{-2}*	*J g^{-1}*	*m of water*
1	1.013	1.013×10^5	0.1013	10.33
0.987	1	10^5	0.1	10.17
9.87×10^{-6}	10^{-5}	1	10^{-6}	1.017×10^{-4}
9.70×10^{-2}	9.833×10^{-2}	9.833	9.833×10^{-3}	1

Box 12.2

SOIL WATER POTENTIAL

Soil water potential (ψ_s) is a composite quantity to which different types of force, both mechanical and chemical, contribute. It is defined as follows:

$$\psi_s = \psi_m + \psi_\pi + \psi_p + \psi_g$$

where ψ_m = matric potential, ψ_π = osmotic potential, ψ_p = pressure potential, ψ_g = gravitational potential.

The first of these is the surface tension force in the water menisci of the pore spaces between soil particles. There are also forces of adsorption between water molecules and charged particles and surfaces, particularly in the colloidal fraction of the soil. Because both of these forces are associated with the solid phase of the soil (the soil matrix), both inorganic and organic, they are combined to form one component – the matric (or matrix) potential, ψ_m. Which force predominates, however, depends on soil textural properties and soil water content, but usually adsorption forces increase as soil water content decreases. In dry soils, therefore, adsorption forces alone largely determine the matric potential.

The second component contributing to the water potential in soils is the concentration of solutes in the soil solution and this determines the osmotic potential (ψ_π). In moist soils, when the matric potential approaches zero as saturation is approached, the osmotic potential may be the predominant component of ψ_s.

A third component of total soil water potential is a pressure potential (ψ_p) which is directly proportional to the excess hydrostatic pressure exerted over atmospheric pressure by the column of soil water above a point. Because of its relation to height of water column, it can be of some importance at depth, but only below a water table in saturated soils. In contrast to the other components mentioned so far, it is positive (see Box 12.1) and it acts to increase the total soil water potential. A related component is the gravity potential (ψ_g) which is relatively insignificant in dry soil, but of some importance in saturated soils on slopes and below the water table. It is proportional to the density of the water, the height of a point above its base level and to the acceleration due to gravity (see Box 1.4).

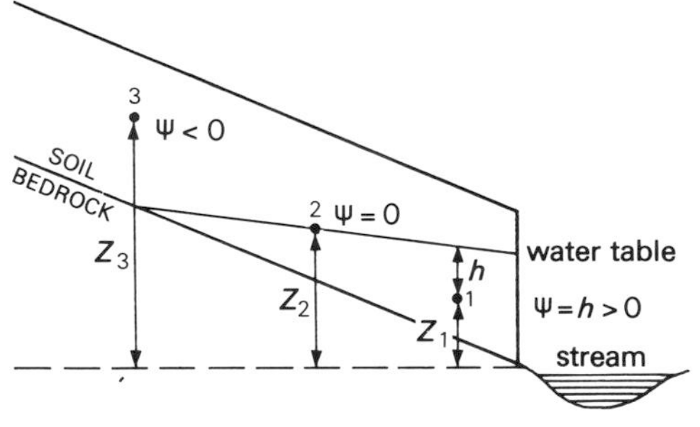

datum Z = 0 at base of slope

In slope hydrology it can be said that differences in ψ_m and ψ_π determine the direction and rate of water movement in unsaturated conditions, whereas ψ_p and ψ_g become the dominant controls under saturated conditions. Where the interest is in saturated flow alone, the matrix and osmotic potential components are often ignored, and the sum of pressure and gravitational potentials is referred to as the total hydraulic potential (ϕ). The calculation of ϕ for various points on the slope relative to elevation and water table is shown in the adjacent diagram (after Atkinson 1978).

	Hydraulic potential	=	*Gravitational potential*	+	*Pressure potential*
point 1	ϕ_1	=	gz_1	+	h
point 2	ϕ_2	=	gz_2	+	0
point 3	ϕ_3	=	gz_3	+	ψ

slope. At any given time these will be determined partly by the duration and intensity of rainfall input, or the time elapsed since the last input, and partly by the characteristics of soil and regolith, which act as regulators of water transfer.

The initial movement of water into the soil is by **infiltration**, though some precipitation may be temporarily held in microtopographical irregularities at the ground surface as **depression storage**. Infiltration is measured in mm h^{-1} and the **infiltration capacity** is the maximum possible rate of infiltration for a particular soil in a specified condition. During infiltration three moisture zones can be distinguished within the soil:

(a) the thin saturated zone at the surface whose water content decreases rapidly with depth to pass into
(b) the transmission zone where water content diminishes more gradually, and
(c) the wetting zone and wetting front, where it decreases rapidly once more.

Water movement into the soil through these transition zones takes place in response to a gradient of matric potential (ψ_m) with high potentials at the surface and low potentials at the wetting front. Although soil properties control the nature of the matric potentials concerned, normally the surface tension or capillarity component of matric potential is most significant in infiltration. The capillary properties of the soil (**hydraulic conductivity**) depend partly on such physical properties as texture and on the presence and type of vegetation because of the effects of roots and organic matter on hydraulic conductivity. The most important regulator of infiltration in a particular soil, however, is its initial or antecedent moisture content, for water movement is proportional to total soil potential differences and these will be less in a wet soil than in a dry soil receiving precipitation. So infiltration capacity may be as high as 50 mm h^{-1} initially for a dry sandy soil, declining to a few millimetres per hour for a wet soil or clay.

Movement of water in soils is a continuous process dependent upon differences in soil water potential (ψ_s). Within the intergranular pore spaces and voids of the soil this movement has both a vertical and lateral component and it is characterised by a diffuse or matrix flow. It may take place under saturated or unsaturated conditions. Lateral transfer of water down slope by this pathway is referred to as **matrix throughflow** (Fig. 12.2). In

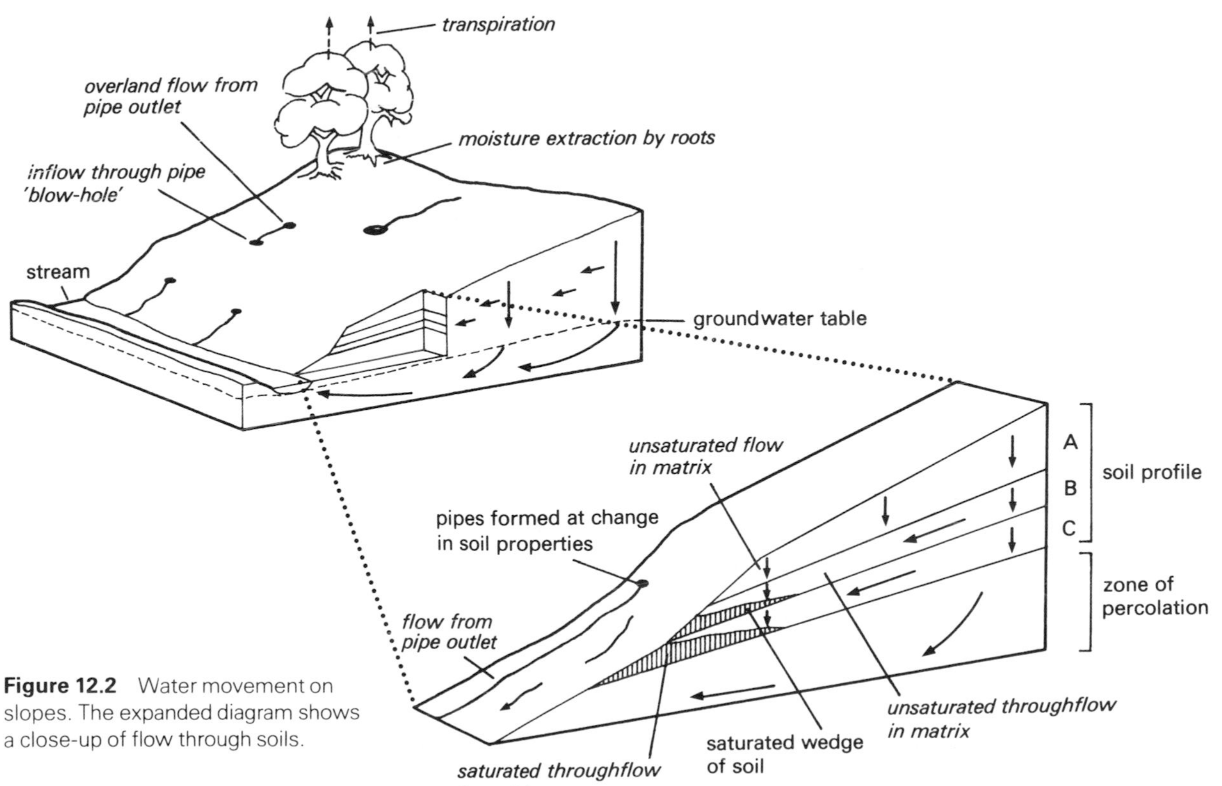

Figure 12.2 Water movement on slopes. The expanded diagram shows a close-up of flow through soils.

unsaturated throughflow, differences in matric potentials contribute most to the potential gradient on the slope and control the direction and rate of diffusion, perhaps with a minor contribution from osmotic or solute potential differences (ψ_π). When the matrix is saturated with water, however, the gravity potential differences and the hydrostatic pressure potential become the determinants of hydraulic potential gradients (Box 12.2). Saturated throughflow can take place within the soil where saturated conditions develop as a result of differences in the permeability of soil horizons. These conditions are most typical at the foot of the slope where a wedge of saturated soil develops and extends up slope as water accumulates in the soil (Fig. 12.2). Whether saturated or unsaturated, throughflow will be conditioned and regulated by the characteristics of the soil as a matrix for water movement, by the forces of moisture retention and by the antecedent moisture conditions.

Under certain soil conditions the diffuse movement of water through the soil matrix may be supplemented by a more concentrated throughflow pathway (Fig. 12.2). Large voids exist in many soils and may be enlarged further by soil fauna and the growth and decay of roots. These may concentrate throughflow movement into networks of soil pipes (Fig. 12.3) through which water is transmitted by turbulent flow. Discharge in completely filled pipes varies in response to pressure and gravity potentials and in partially filled pipes (which behave as channels) in response to the slope of the water surface. Pipeflow velocity is usually much more rapid than matrix flow. Weyman (1975) quotes estimates of pipeflow velocity of 50–500 m h^{-1} and matrix flow 0.005–0.3 m h^{-1} (see Table 13.1). The remaining water which has infiltrated passes down to the water table and is stored there as ground water, ultimately returning to the channel by deep percolation.

After a period of prolonged rainfall, all the pore spaces may become filled with water, thus saturating the soil. At this point the water table has effectively risen to the surface and the infiltration capacity is reduced to zero. The soil is incapable of absorbing any further water, and subsequent rainfall runs off directly across the surface of the slope as **saturated overland flow**. This situation is likely to come about towards the base of a slope where both local infiltration and throughflow received from higher up the slope contribute to soil moisture.

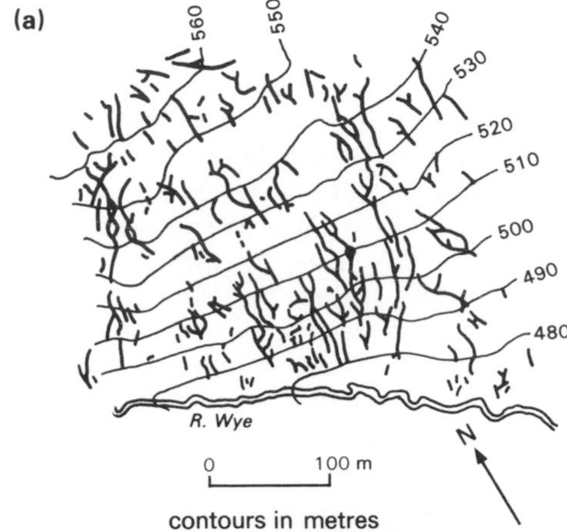

Figure 12.3 (a) Pipe network, Plynlimon, Wales (after Atkinson 1978). (b) Soil pipe developed at the interface between organic soil and the underlying mineral soil, exposed in a shallow cutting. Immediately after a heavy storm a strong discharge is issuing from the pipe. A trickle of overland flow is falling from the vegetated surface of the organic soil. (Sutherland, Scotland.)

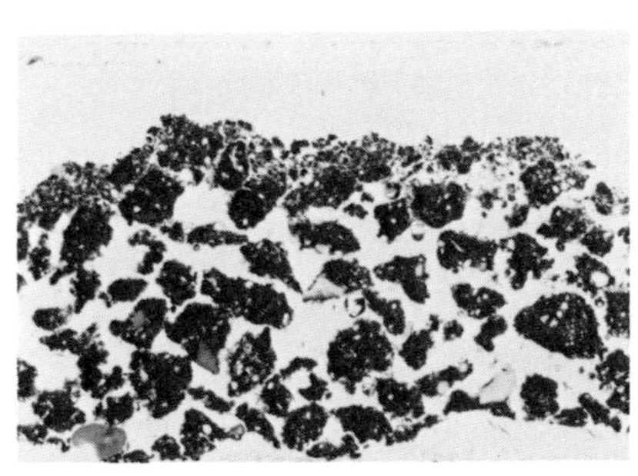

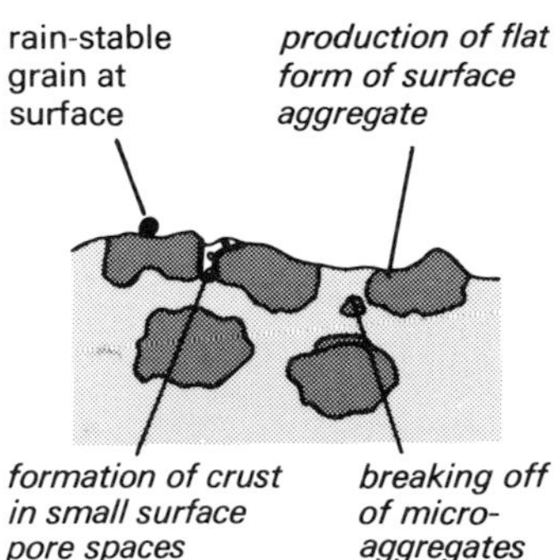

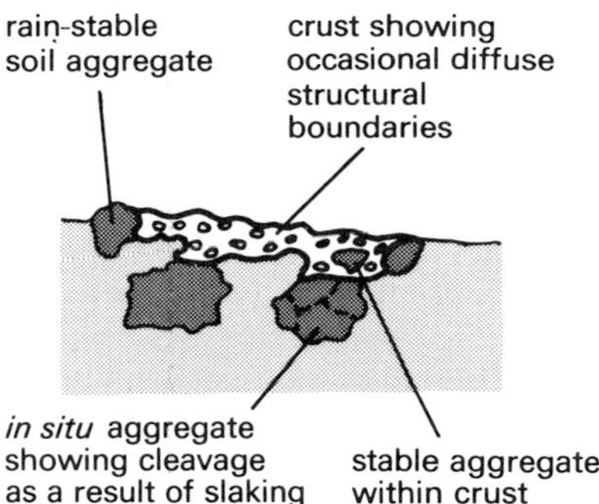

Figure 12.4 Crusting at the soil surface produced by raindrop impact (after Farres 1978).

Under certain circumstances the rate of precipitation may exceed the infiltration capacity of the soil, even when the latter is not saturated. In this case the excess of precipitation runs off down slope as **overland flow**. This appears to be a common process in semi-arid regions, where precipitation intensities are high and infiltration capacity of the sparsely vegetated soils is low. It is further encouraged by the development of a crust (Fig. 12.4) on the soil as the surface layer becomes compacted and the pores blocked as a result of the redistribution of soil particles following raindrop impact (Farres 1978). This kind of overland flow seems absent in temperate environments with only modest precipitation rates and well structured soils, except under certain conditions of cultivation (see Ch. 24).

12.3 The operation of the slope system: the transfer of minerals

Transfer of minerals down slope takes place in a wide variety of ways. Carson & Kirkby (1972) distinguish between **mass movement**, **particulate movement** and **movement in solution**. This forms a convenient classification for discussion, but it must be recognised that the distinctions are not always clear cut. The difference between particle movement and mass movement is that in the latter the debris moves as a coherent mass and, in the former, particles may move as individual bodies constantly changing their positions in relation to their neighbours.

12.3.1 Mass movement

Three basic mechanisms of mass movement can be identified – slide, flow and heave (Fig. 12.5). In the

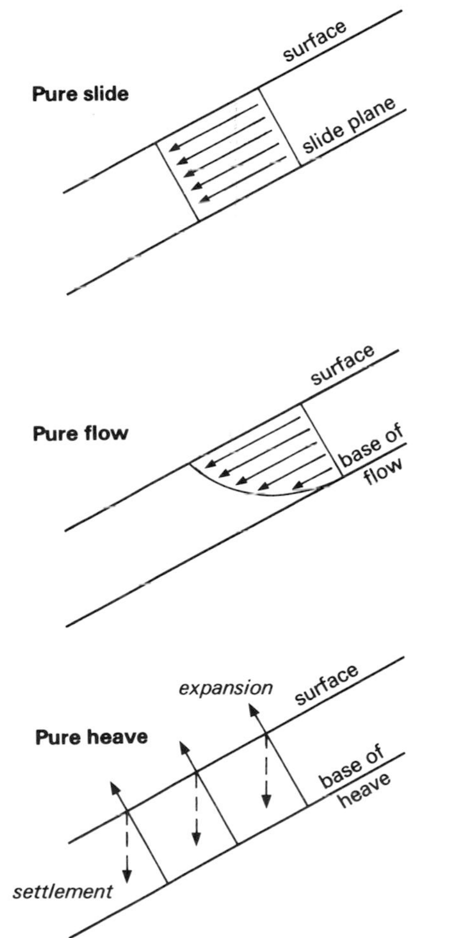

Figure 12.5 Basic mechanisms of mass movement on slopes.

case of a slide, the mass moves as a coherent unit with minimal internal dislocation along a discrete plane of failure (the slide plane). A flow, in contrast, suffers internal deformation as it takes place. Velocity decreases downwards through the flow towards its bed, in much the same way as in a river channel (see Ch. 13). Heave processes are characterised by expansion movement normal to the surface and subsequent contraction causing alternate elevation and lowering of the ground surface.

In practice, however, most processes of mass movement involve a combination of these three basic mechanisms. Accordingly they can be used as a basis for classification (Fig. 12.6). Different types of mass movement are located on the triangular diagram according to the relative proportions of flow, slide and heave involved. An additional feature of this classification is that it shows a gradation from slow to rapid movements in the direction from heaves to slides and flows. It shows also a gradation of increasing moisture along the slide–flow axis. Within this framework we shall now consider the major forms of mass movement.

Soil creep. Soil creep is caused by the disturbance and subsequent settlement of soil particles in the regolith cover of a slope (Fig. 12.7). The net result is the slow continuous downslope movement of the debris sheet en masse. The fundamental mechanism of creep is heaving. This raises particles in a direction normal to the slope and they subsequently tend to settle back in a vertical direction to a position down slope from their original location. The cumulative effect of such repeated disturbance is a ratchet-like downslope movement, affecting the whole of the surface layer of the debris mantle.

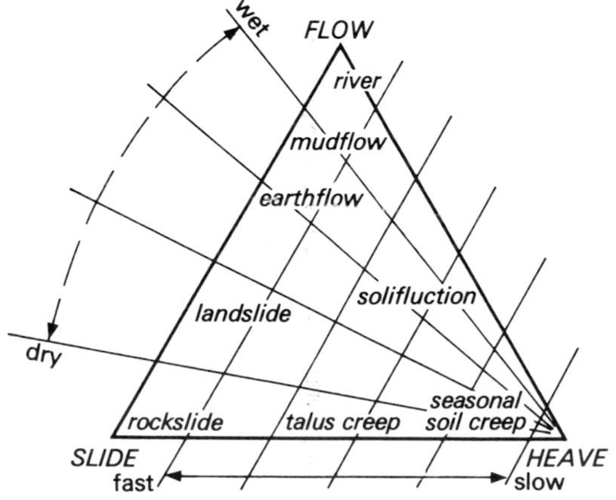

Figure 12.6 Classification of mass movement processes (after Carson & Kirkby 1972).

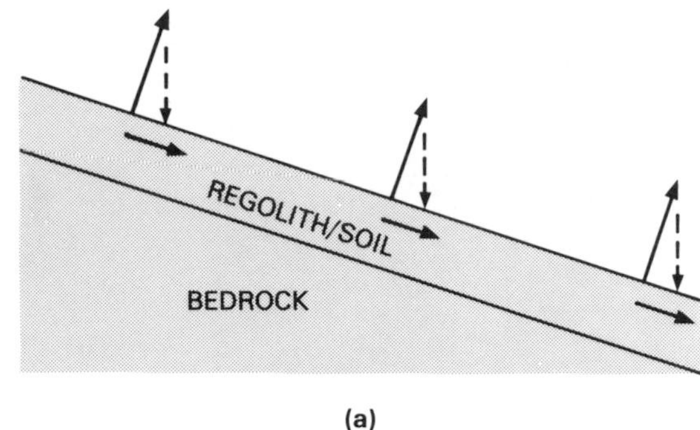

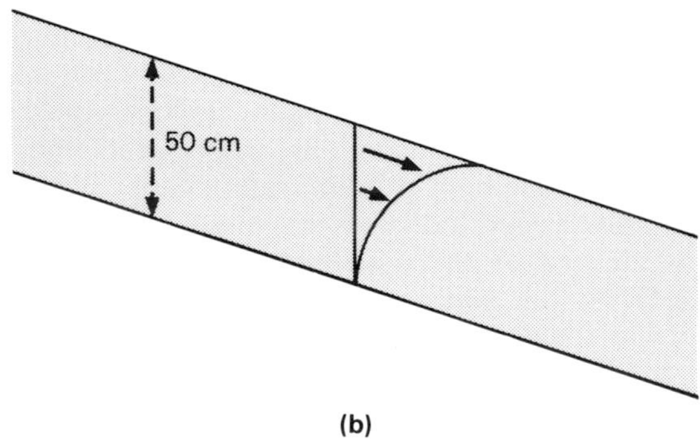

Figure 12.7 Soil creep: (a) fundamentals of movement, (b) velocity distribution.

The motive forces initiating the heave are many and varied. Alternate wetting and drying of soils, especially those with a significant clay component, can cause repeated expansion and contraction. Freezing and thawing of moist and wet soils occurs seasonally, or intermittently in temperate environments, and causes similar effects. The growth of plant roots and the activities of burrowing animals apply much mechanical energy to the soil layer and contribute considerably to the disturbance of soil particles. It has been estimated, for example, that in humid temperate environments the entire surface layer to a depth of 10 cm is entirely ingested and passed through the gut of earthworms in a period ranging from 18 to 64 years. Paton (1978, see Ch. 8) gives a good review of the role of soil fauna in subsurface soil processes.

Measurement of soil creep over a twelve-year period on a 25° slope in humid temperate conditions shows a mean downslope movement by creep of 0.4 mm a^{-1} throughout the top 20 cm of soil (Young 1978). Rates of movement in the range 0.2–3.0 mm a^{-1} have been reported from a variety of sites in humid temperate environments.

Frost creep and gelifluction. In periglacial environments freeze–thaw processes are much more frequent and soil moisture values are generally higher. Accordingly, heaving of the ground surface may attain values of 5–20 cm. During thaw, the water content of the upper thawed layer of soil, unable to drain through the still-frozen ground below, is close to the liquid limit and it tends to flow (**gelifluction**). The combined effects of frost creep and gelifluction under such conditions (they are difficult to separate) have been shown by many studies to attain velocities of several centimetres per year, and they generally affect soils down to a depth of 50 cm.

The consequence of these creep processes is to produce a gradual downslope movement of the surface layers of the debris mantle, often with little visible effect. Where a section is cut into a slope, the dragging-over of steeply dipping strata is sometimes evident. Surface effects include the accumulation of soil up slope of retaining walls or tree trunks, whereas a hollow often develops down slope of a tree trunk. Where the surface of a creep layer is mantled by a turf cover, the latter is sometimes ruptured to form terracettes. Under the more rapid movement typical of gelifluction, terrace and lobe features develop, penned back by a barrier of turf or stones.

Rockslides and landslips. Sliding involves the sudden and rapid downslope movement along a discrete plane of an intact mass of rock on unconsolidated material. In the case of rockslides, this plane may be a major joint or a bedding plane between two adjacent strata. In homogeneous materials such as clays, an arcuate failure plane is liable to develop, leading to a rotational slide. Whenever the shear force comes to exceed the resistance to shear along any potential failure plane, sliding will be initiated.

Slides can be thought of as a consequence of inherent and initiating factors. Inherent factors that predispose slopes to instability and failure include the steepness and height of the slope, which impose high shear stresses towards its base (Fig. 12.8). They also include properties of the materials composing the slope itself, such as the low shear strength of the material involved or pre-existing planes of weakness. Initiating factors are trigger mechanisms which push the slope beyond

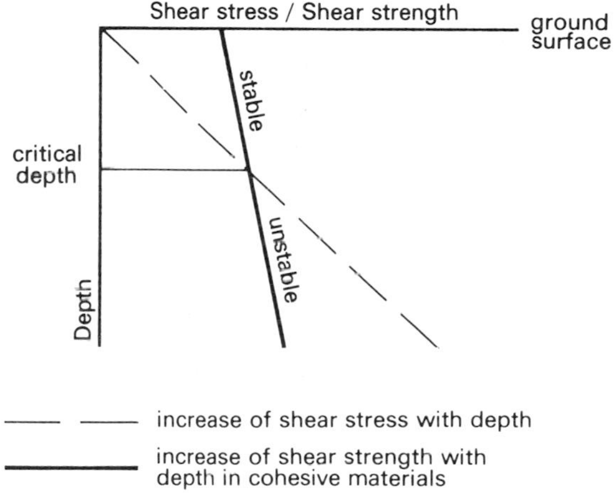

Figure 12.8 Shear stress and strength in relation to depth within a slope.

the threshold of stability (Box 12.3). These include the excessive loading of the slope after prolonged rainfall which increases the overburden pressure and hence shear stress, at the same time reducing the shear strength by raising the pore water pressure. Long-term weathering at the surface of the slope can cause a reduction in shear strength and consequent failure. The removal of supporting material by erosion at the base of a slope is a further cause of instability. This is exemplified by the active undercutting of slopes by fluvial or marine action.

Box 12.3

FUNDAMENTAL FORCES ACTING ON THE HILLSLOPE

Denudation processes involving the transport of solid material down hillslopes can be resolved in terms of fundamental mechanical forces. The movement of material depends upon the nature and magnitude of these forces. There are forces that tend to promote movement and those that resist movement. Whether or not movement takes place depends upon the balance between these two sets. These fundamental mechanical principles apply both to individual particles resting at the surface and to the movement of masses of material comprising the hillslope.

The basic gravitational force (g), operating vertically downwards, can be resolved into a downslope force (or shear stress), $g \sin \alpha$, and a force normal to the slope, $g \cos \alpha$, where α is the slope gradient (see (a)).

An individual particle

In the case of an individual particle of mass M, resting on a slope of gradient α, the downslope shear stress is given by $Mg \sin \alpha$. This shear stress acts as a driving force and may be augmented by an external agent, for example raindrops, flowing water or wind, and produce a tendency to downslope movement (see (b)).

Resistance to movement is the normal stress (i.e. normal to the slope surface) which causes a tendency for the particle to subside into the slope, or at least retain its position. The resistance to movement will be complemented by the frictional resistance between the particle and the surface on which it rests. If the particle is partially embedded in the slope, or is angular in form and interlocked with adjacent particles, it will clearly possess a much greater frictional resistance.

A rock or soil mass

Mass movement takes a wide variety of forms, depending on the induration of the material concerned and the presence of discontinuities within it. However, for the purpose of explaining the basic principles we shall consider the simplest case of an inclined failure plane parallel to the slope surface, and ignore side and edge effects (see (c)). This is a simple shallow planar slide. Assuming a slope of unit width, gradient α and horizontal length l, the vertical force will be distributed over an area $l/\cos \alpha$. Vertical stress is therefore

$$\frac{\gamma z}{l/\cos \alpha} = \gamma z \cos \alpha$$

where γ = unit weight of soil mass and z = depth of mass.

This can be resolved into downslope shear stress,

$$S = \gamma z \cos \alpha \sin \alpha$$

and normal stress (τ) = $\gamma z \cos^2 \alpha$. The **driving force** is the shear stress. The **resisting force** is the shear strength, given by the Coulomb equation (Box 10.2) distributed over the area of the failure plane. Substituting for τ in the Coulomb equation gives:

$$S = \tau + (\gamma z \cos^2 \alpha - u) \tan \phi.$$

The ratio between resisting force and driving force is defined by the **factor of safety** (F). Thus

$$F = \frac{\tau + (\gamma z \cos^2 \alpha - u) \tan \phi}{\gamma z \cos \alpha \sin \alpha} \quad \frac{\text{(resisting force)}}{\text{(driving force)}}$$

where u = pore water pressure.

When $F > 1$ the slope is stable, but whenever conditions lead to $F < 1$ failure will occur.

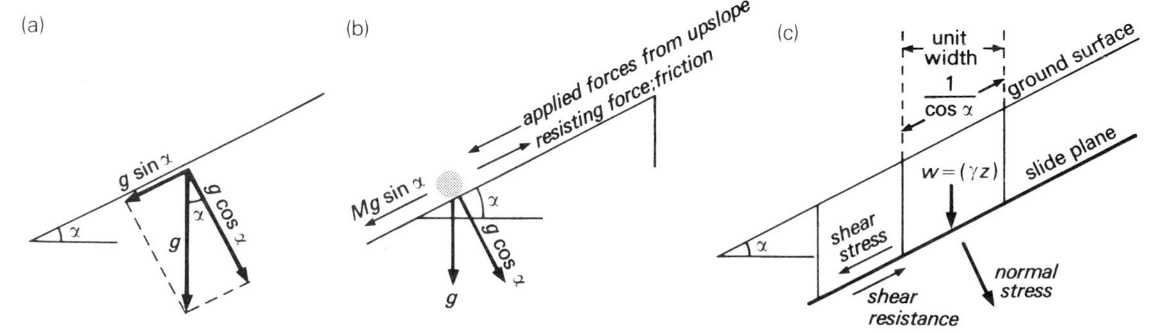

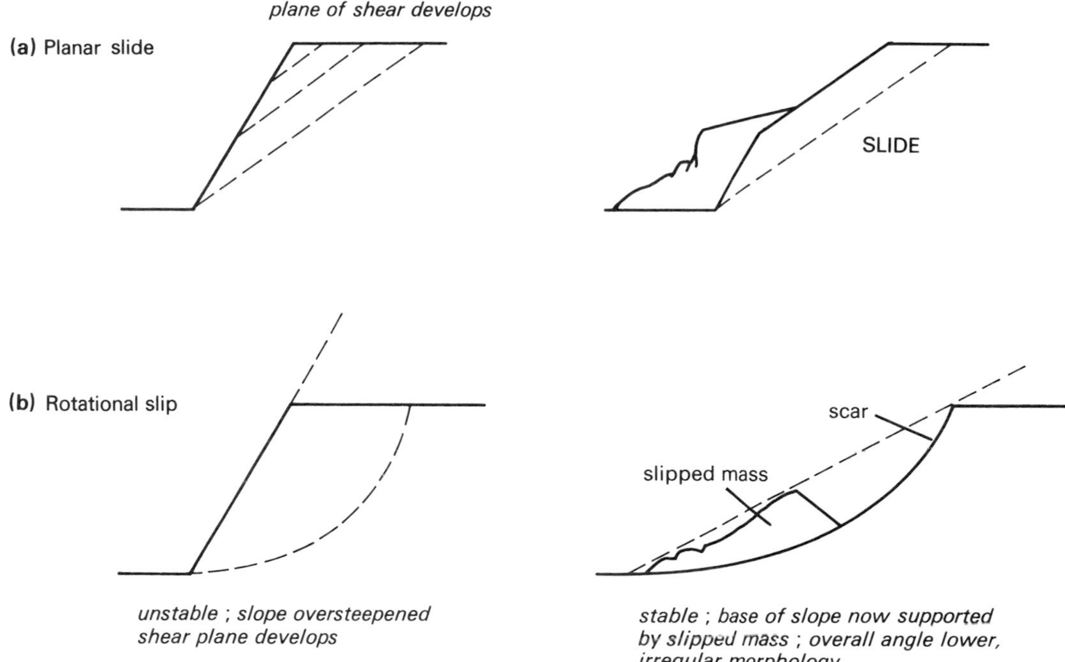

Figure 12.9 (a) Planar slide development on a slope containing discontinuities; (b) rotational slip development in homogeneous material.

There are many varieties of slide and slip (Fig. 12.9). Deep planar slides, often more than 10 m in depth, are usually related to inherent structural weakness within the slope. Shallow planar slides are usually caused by loss of strength in the surface layers of the slope caused by weathering. Deep rotational slips, along an arcuate failure plane, may be simple or multiple in occurrence and they occur in homogeneous materials on slopes upon which the factor of safety has been exceeded.

The morphological effect of rockslides and landslips is to produce an upslope scar, from which the slipped material has been detached, and an accumulation of transported material on the lower slope to form a very irregular topography.

Earthflows and mudflows. Earthflows and mudflows occur when non-indurated materials, usually with a high clay component, attain a high water content. Consequently, the pore water pressure increases, thereby decreasing cohesion and causing loss of strength. Flows are therefore closely related to rainfall events. The material involved has a high mobility, and velocities often reach several metres per second.

Morphologically there are three components to an earthflow or mudflow (Fig. 12.10). The source area, from which the flow originates, is usually a hollow in the hillslope where moisture accumulates. A scar is formed in the hillslope by the removal of material, which moves down slope along a narrow flow track. At the base of the slope the material then fans out to form a broad lobate toe, where the flow comes to rest.

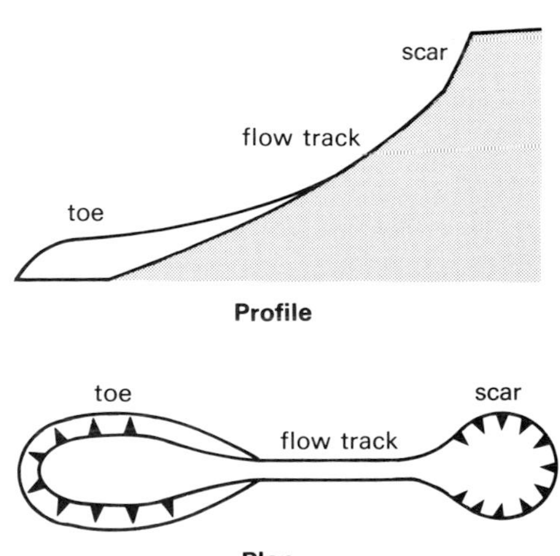

Figure 12.10 Simplified diagram of an earth flow.

Although flows, slides and creep have been discussed as separate processes, in reality individual mass movements in the field usually incorporate a combination of these mechanisms. In a particular instance of slope failure one mode may be dominant, but not exclusively so. Landslides, for example, rarely retain the intact form of the original mass and some deformation usually occurs. In the case of flows, the shear may be limited to the lower portion of the material in transit, which may carry a raft of intact material at the surface of the flow.

12.3.2 Particulate movement

Movement of individual particles through slope systems can be caused by gravitational forces alone, or by the applied forces of falling and running water. In the present context, gravitational forces alone are normally an effective agent only on vertical or near-vertical faces and they cause **rockfall** or **debris fall**. The applied force of kinetic energy of raindrops falling on a slope produces **rainsplash erosion**, and overland flow produces **surface wash**.

Fall processes. Large joint-bounded slabs of rock, or individual small particles on a steep rock face, may become detached by weathering processes. Once detached, they fall freely under the influence of gravity to the slope beneath, accumulating there as a **talus slope**. Larger particles, by virtue of their greater kinetic energy, tend to travel further on landing. All particles tend to roll after falling onto the accumulation slope until they encounter particles of similar size, where they tend to lodge. There is thus a sorting on the talus slope, with finer particles at the top and coarser material distributed down slope. Such slopes are commonly linear in profile, sometimes with a basal concavity. They normally possess gradients of 30–35°, the angle of repose of coarse angular loosely packed debris.

Rainsplash. As a raindrop falls on the ground surface, a proportion of its kinetic energy is transferred to loose sediment particles lying on the surface. These particles are consequently disturbed and, assuming vertical rainfall, they tend to be thrown out radially in all directions equally from the point of impact. Where this happens on a slope, the gradient has the effect of lengthening the travel of a particle thrown in the downslope direction, as compared to a particle thrown up slope with a similar trajectory (Fig. 12.11). Thus the effect of a rainstorm operating on a slope is to produce a net downslope transfer of surface sediment as a result of the impact of myriads of individual raindrops. The kinetic energy of a raindrop is related to its terminal velocity and size. Raindrops vary in diameter from 0.2–5.0 mm according to rainfall type and they possess terminal velocities ranging from 1.5–9.0 m s^{-1}. Thus the applied kinetic energy for a storm of known size and intensity can be calculated and its effects measured in the field. Sand- and silt-size particles are readily removed, whereas clays, which possess cohesion, are more resistant to this process.

(a)

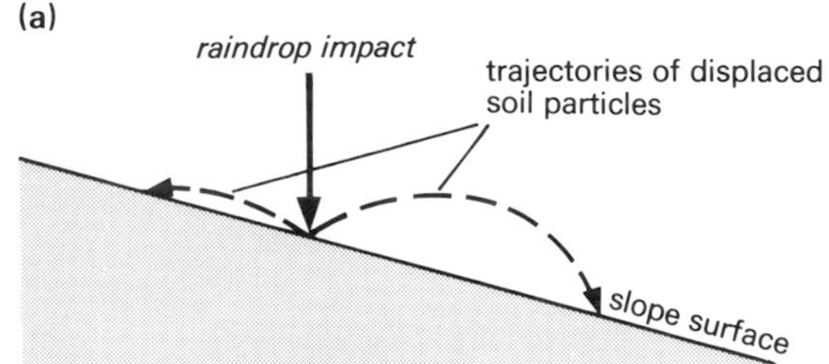

(b)

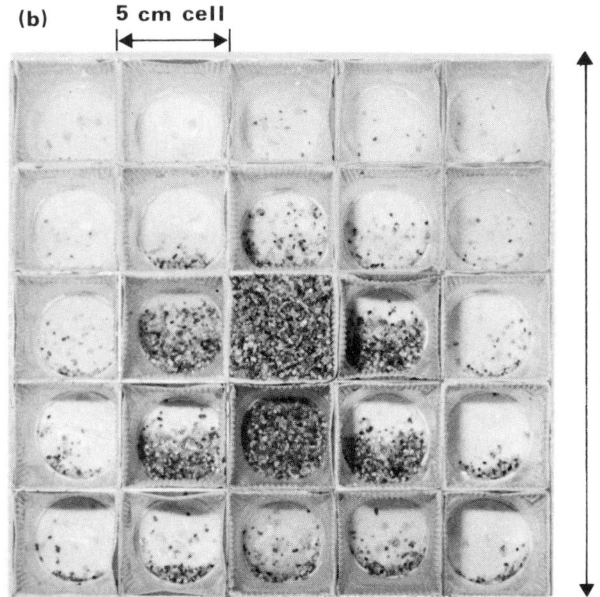

Figure 12.11 Rainsplash erosion. (a) Schematic diagram of the displacement of particles by rainsplash. (b) The result of a laboratory simulation of rainsplash: particles have been displaced outwards from the central cup, preferentially in the downslope direction.

For a given energy input, the magnitude of rainsplash erosion is directly proportional to the sine of slope angle, but it is clearly effective only where loose sediment is exposed at the surface and there is no vegetation cover to absorb the raindrop impact.

Surface wash. Overland flow can be generated both as excess of rainfall over infiltration and as a result of soil saturation. Water flowing down slope applies a shear stress to particles in its path (Box 12.3) proportional to its velocity, which is governed by the Manning equation (see Box 13.3):

$$V = \frac{1.009}{n} R^{2/3} s^{1/2}.$$

In the case of **sheetwash**, flow down a slope plane of great width and minimal depth, the hydraulic radius (see Box 13.2), is effectively equal to the depth of flow, and velocity is closely determined by surface gradient and the roughness (microtopography) of the slope. Since the catchment area increases in the downslope direction, so distance and depth of flow increase. Thus the flow accelerates to a threshold velocity at which entrainment of soil particles begins. Erosion by sheetwash is, therefore, negligible at the top of a slope, increasing in the downslope direction once the critical shear stress has been attained for the soil particles present.

Surface irregularities and vegetation cause the flowing water to be concentrated into anastomosing threads, which then incise themselves into confined channels to form **rills**. The flow of water is now more efficient due to the increased hydraulic radius, and higher velocities are attained. Rillwash, therefore, is a much more effective erosional process than sheetwash. The laws of hydraulic geometry cause a tendency for rills to develop concave long profiles, just as in river channels (see Ch. 13) and this in turn creates a tendency towards concavity in the lower part of the hillslope profile. Rills are ephemeral features forming in response to individual storms and being obliterated subsequently by collapse, creep, frost processes and agricultural practices between storms.

The effectiveness of overland flow as an agent of downslope transfer of particles in suspension is visibly demonstrated by the sheets or ribbons of dirty sediment-laden water traversing slopes when this process occurs. An informative account of soil erosion by surface wash is contained in Morgan (1979).

12.3.3 Movement in solution

Rainwater passing through the slope system initially has a high potential (see Box 12.1) and possesses only low concentrations of ions in solution. As a consequence, it is far from being saturated in respect of ionic minerals. In flowing through the slope system via various routes – overland flow, matrix throughflow, pipeflow and groundwater flow – it comes into contact with the minerals of the slope system.

As yet, very few data have been published on the acquisition of solutes by water moving through slopes, as compared to solution in river channels (see Ch. 13). On theoretical grounds one would expect that the longer the water is in contact with slope minerals, the greater is the opportunity for equilibrium concentrations to be attained. Accordingly, it may be expected that the rapid velocities associated with overland flow and pipeflow allow insufficient scope for the acquisition of substantial concentrations of minerals in solution. In the case of matrix throughflow and groundwater flow, water is in contact with minerals for a much longer period and significant removal of ions in solution may be expected. Certainly, solute concentrations in ground water, as measured both in wells and in emergent springs, indicate significant solution loss at depth within slopes. Concentration of solutes in chalk springs, for example, may be in excess of 300 ppm.

Although few data exist concerning the process of solution in terms of solute concentrations on a slope, a study by Young (1978) has indicated the possible magnitude of its effect. The relative effectiveness of solutional removal on a slope in a humid temperate environment has been estimated in comparison with surface creep. Markers set within the soil showed movement not only in the downslope direction but also subsidence into the slope. This was interpreted as indicating solution loss from the surface layer, causing inward settlement of the remaining undissolved material. This particular study indicates that solution subsidence is a more effective process than downslope creep in the surface soil layers. As undersaturated water of high chemical potential falls on the slope, it acquires minerals in solution and removes them by throughflow, causing surface subsidence. The effect is much less marked below 20 cm depth, since presumably the water is then saturated and less chemically aggressive. Although it is premature to put too much reliance on one specific case study, it is clear that, in humid environments at least, solution loss from surface layers of a slope may be very effective. Clearly this is one topic of study that merits much further investigation.

It is apparent from this review of the major modes of downslope transfer of minerals that our present understanding of such processes is very uneven. Our knowledge of the mechanical processes of mass movement and individual particle transport, underpinned by field and laboratory studies of rock and soil mechanics, is relatively advanced. Chemical denudation of slope systems, however, has been relatively neglected, in contrast to chemical denudation in fluvial systems (see Ch. 13) and the catchment basin as a whole (see Chs 10 & 16).

12.4 The balance of slope process and slope form

The operation of slope processes is closely governed by the input and output of material. The input from the weathering system is either shed directly on to the slope from exposed bedrock or accumulated at the base of the regolith as the weathering front penetrates more deeply to produce a thickening of the weathered mantle. Material undergoing transport can be regarded as throughput (the migratory layer) and may continue to undergo weathering and breakdown during transport. If the rate of transport and rate of weathering are in balance, then the debris mantle on the slope will be maintained through time (Fig. 12.12).

Output of mineral material takes place at the foot of the slope by processes of basal removal

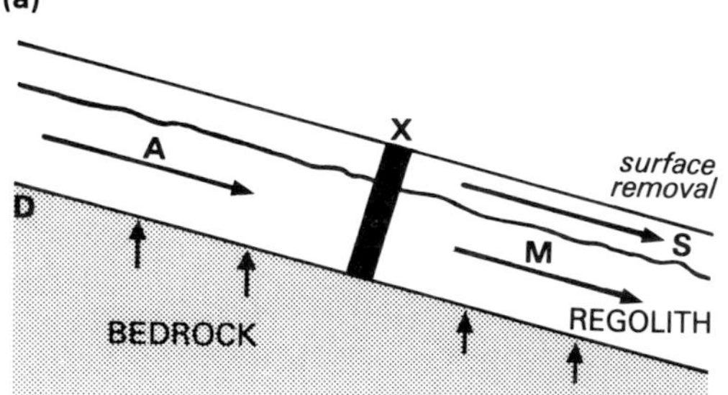

Figure 12.12 (a) The budget of regolith on a slope: S = surface removal of material; M = removal of material by mass movement; A = material input from upslope; D = input of regolith from weathering. Budget in column X is therefore given by D + A = S + M. (b) Resistant outcrop of massive rock (Navajo Sandstone) forming near-vertical slope facet. This is an example of a weathering-limited slope facet, from which debris is released by rockfall to the talus slope beneath. (Monument Valley, Utah, USA.)

where the operative agent is usually the stream channel. Varying degrees of basal activity can be recognised. At the most continuously active end of the scale is the situation where a perennial stream channel undercuts the base of a slope. This facilitates continuous removal of slope material by causing collapse of the channel banks and by receiving mineral-laden water flowing directly off the slope. Several cases exist where basal removal is intermittent. In chalk landscapes, channel flow may be seasonal (in winter only) as is also the case with tundra landscapes, where channel flow takes place only in summer. Under arid conditions, streamflow is intermittent and often less than annual. In the same category are slopes bordering floodplains. These suffer direct basal removal only when the outside of a channel meander undercuts them (Fig. 12.13). As the meander belt shifts downstream, an individual slope profile will be abandoned by the channel until the next meander sweeps by. Basal removal is at a minimum on slopes in dry valleys. In this situation, removal takes place by slow downvalley creep along the valley floor. The magnitude and frequency of basal removal can therefore vary widely under different conditions.

We have seen that there are many different kinds of processes responsible for downslope transfer of debris. They can be classified broadly into two groups: the slow continuous processes such as creep and wash, and the rapid intermittent processes of falls, sliding and flowage. The former can be considered as erosional events of low magnitude and high frequency, whereas the latter occur with low frequency and high magnitude.

It is pertinent to consider which of the groups of processes has the greater overall effect in the long term. Data on the rate of operation of these processes are gradually accumulating and are as yet insufficiently complete for large-scale generalisations to be made. One clue may be gleaned from slope profile morphology. The rapid mass movement processes tend to produce dissected slopes with an irregular profile. The continuously operating processes tend to produce slope profiles of a smooth nature, with little irregularity of form. Taken overall, more slopes conform to the latter form, which suggests that creep and wash processes are more dominant on the world scale.

Data determined by field experiment permit comparisons to be made between the relative effectiveness of creep and wash processes in different environments. The ratio of creep to wash in several environments is shown in Table 12.1.

There are clearly enormous variations between these environments, depending on inputs of erosional agents. The input of precipitation is particularly important in determining the nature of erosional process: the total influences the amount of vegetation cover and total available water, and

Table 12.1 Ratio of creep : wash in different environments.

	Creep	:	Wash
humid temperate (UK)	10–20	:	1
savanna	1	:	5
warm temperate (Australia)	1	:	7
semi-arid (New Mexico)	1	:	98

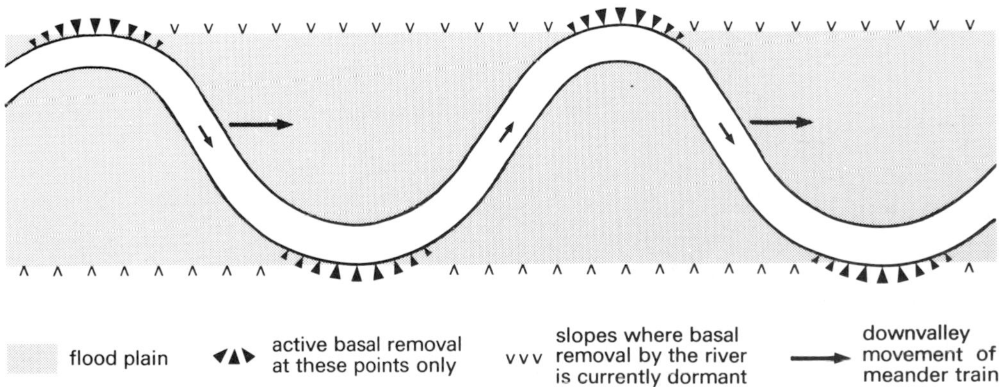

Figure 12.13 Basal removal from slopes affected by meander migration.

rainfall intensity determines the erosional energy of the individual storm. The humid temperate environment, with abundant vegetation and well distributed rainfall of low intensity, favours creep. In semi-arid environments, where opposite conditions prevail, wash processes dominate.

The analysis of slope form has two aspects. The first is the plan form of a slope, whether it be concave, linear or convex. It is an important influence on the concentration or dispersion of materials moving down it (see Ch. 10). Secondly, and easier to handle, is the linear aspect of the slope long profile. Reducing the slope to two dimensions, by assuming it to be linear in plan, simplifies the study of both process and consequent form.

Several approaches have been employed in modelling slope form. At the most basic level there is the descriptive model, as exemplified by a simple two-dimensional hillslope long profile, which may be characterised in descriptive terms, e.g. convexo-concave or rectilinear.

At a greater level of sophistication is what we might call the interpretive model, as exemplified by the nine-unit land surface model of Dalrymple *et al.* (1968) (Fig. 12.14). This scheme identifies nine possible components of which a hillslope may be composed and relates them to the processes that may be expected to dominate on each facet (Table 12.2). As such, it is capable of widespread application under varying conditions of lithology, climate and process.

Individual slopes of greater or lesser complexity can easily be encompassed by the model. For instance, if units 2 and 4 are omitted (Fig. 12.15a), the result is a simple convexo-concave slope profile. Alternatively, a more complex profile in which a series of resistant beds outcrop to form scarps would be represented by repetitions of unit 4 at intervals (Fig. 12.15b), as, for example, in the slopes of Grand Canyon, Arizona.

In theory, therefore, any hillslope profile can be interpreted in terms of this model as a particular combination of the units identified. It should be noted, finally, that units 8 and 9 (channel wall and channel bed) properly belong to the fluvial system in the framework adopted in this text.

On any hillslope there is a relationship between

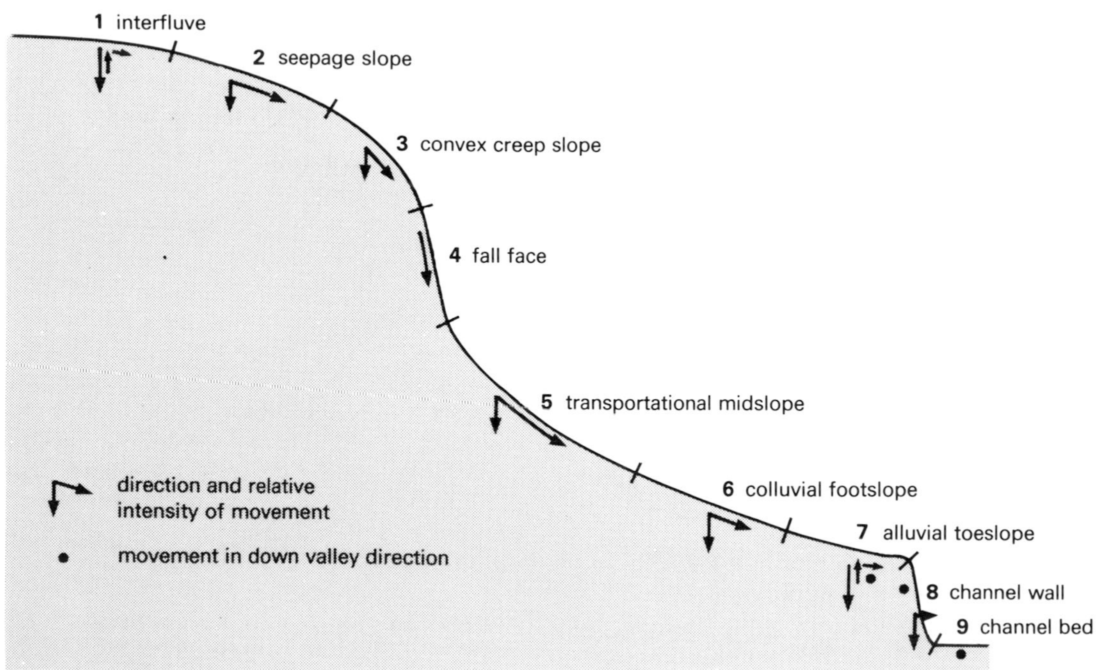

Figure 12.14 The nine-unit land surface model (after Dalrymple *et al.* 1968).

Table 12.2. The nine-unit land surface model; slope components and associated processes.

Slope facet	Dominant processes
(1) interfluve	pedogenic processes, vertical subsurface water movement
(2) seepage slope	mechanical and chemical eluviation by lateral subsurface water movement
(3) convex creep slope	soil creep
(4) fall face	rockfall, rockslide, chemical and physical weathering
(5) transportational midslope	transportation by mass movement, creep flow and wash
(6) colluvial footslope	deposition of material by mass movement, creep flow and wash; limited transportation by creep, wash and subsurface water
(7) alluvial toeslope	alluvial deposition, subsurface water processes
(8) channel wall	corrasion, slumping, fall
(9) channel bed	transportation of material down valley by surface water processes, chemical and mechanical

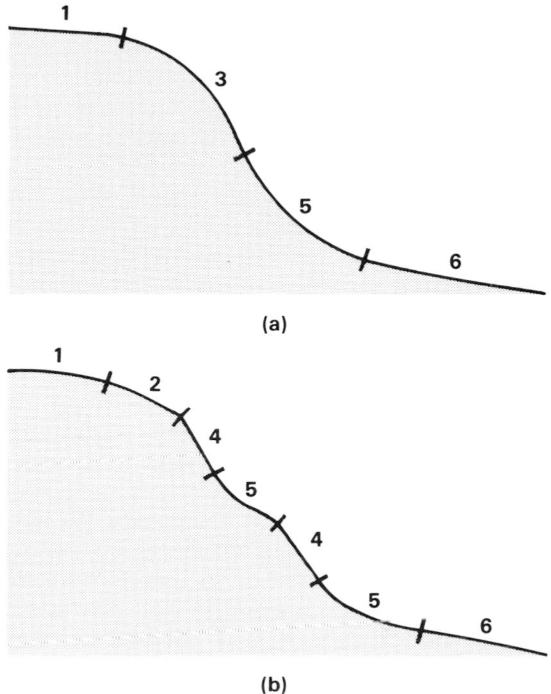

Figure 12.15 Alternative slope forms derived by combination of units from the nine-unit model: (a) simple slope, (b) complex slope.

the weathering rate and the transportation of debris. Clearly this relationship will depend on the properties of the bedrock and the energy available for both weathering and transportation processes. On the one hand, weathering may be relatively slow in relation to transportation processes, such that the latter are able to remove all weathered material as rapidly as it is produced. Thus the factor limiting the rate of development is weathering and this is known as the **weathering-limited** case. Alternatively, it may be that transportation processes are insufficiently vigorous to remove all the weathered material, and the weathered mantle consequently becomes progressively deeper. As the solid bedrock is buried more deeply, so weathering is reduced. The bedrock is insulated from the effects of surface weathering processes, and chemical weathering agents are used up within the weathered mantle. In this case it is the slow transport of debris that inhibits hillslope development, and this is the **transport-limited** case.

Weathering-limited slopes. Weathering-limited slopes tend to develop in regions where climatic inputs favour transportation processes and inhibit weathering processes, or where resistant bedrock exists. It is considered that the weathering-limited condition leads to straight slope profiles. A vertical cliff is the extreme case of a weathering-limited slope, where weathered rock is completely and immediately removed by rock fall, thereby preventing the accumulation of regolith and maintaining the vertical slope form. A similar condition exists on many upland areas formed by resistant rock where abundant precipitation ensures rapid debris removal by creep and other mass movements and even overland flow. In this case a thin debris mantle overlies the long straight midslope profile. Soluble rocks such as limestone and chalk, which yield very little clastic debris since weathered material is carried away almost entirely in solution, also support weathering-limited slopes.

Transport-limited slopes. Transport-limited slopes develop in areas of weak rocks that weather readily and produce lowland topography with limited relief. This condition also tends to suppress transportation processes, which will be inhibited by low hillslope gradients.

Let us assume an initial straight slope with limited transport capability. The rate of soil creep will tend to be constant throughout because of the constant gradient. Since an increasing volume of regolith to be removed exists down slope, the limited transport will lead to a thickening of the regolith in that direction. Consequently, weathering is inhibited on the lower slope. The more rapid weathering continuing on the upper section will produce debris available for transport and lead to the development of an upslope convexity (Fig. 12.16). On the convex slope the rate of transport by creep will be proportional to gradient and there will be an increasing transporting capacity in the downslope direction corresponding with the increased amount of debris available. Accordingly, a state of balance is reached in which gradient and creep velocity are closely related to the rate of weathering over the whole slope. A convex segment thus becomes self-perpetuating.

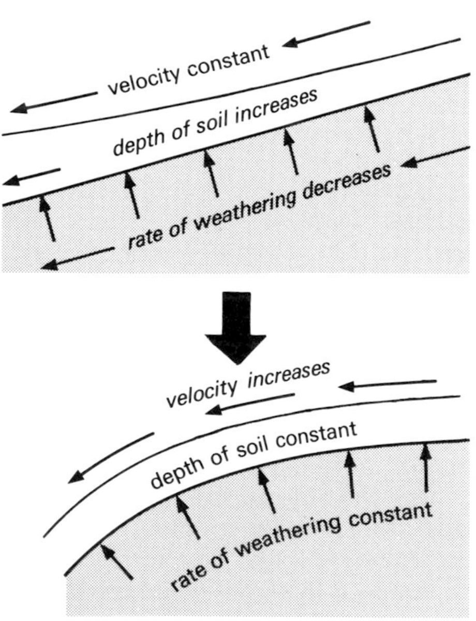

Figure 12.16 The development of slope convexity as a result of transport-limited conditions.

12.5 Complexity of hillslope process–form relationships

Hillslope form and process depend on a number of variable inputs. Variables which have their origin outside the slope are those of temperature, precipitation and vegetation. Initial slope form, geological structure and the nature of the weathering products are the internal variables.

In the present state of knowledge, all hillslope models are at best partial, in that they usually describe only a limited range of processes, or they rest on assumptions of the initial slope form. In reality we rarely have the opportunity to observe the initial slope. A second problem is that on any slope a number of different processes will be operating in combination during a particular period of time. The effective role played by each constituent process will vary depending on the inputs and it is very difficult to assess. Models based on the operation of one process alone are, therefore, limited in application. To some extent the transport–weathering model avoids this problem by treating transport and weathering as grey boxes and generalising the results of the individual processes within them. Although this may simplify the model, it renders it more difficult to verify by field measurement.

A further source of complexity in hillslope process–form relationships is the fact that the form itself may change through time. This in turn affects process, particularly the gradient-dependent processes. Thus there is feedback from form to process.

Further reading

A good general account is
Young, A. 1972. *Slopes.* Edinburgh: Oliver & Boyd.

A more advanced and mathematical treatment is provided by
Carson, M. A. and M. J. Kirkby 1972. *Hillslope form and process.* Cambridge: Cambridge University Press.

Statham, I. 1977. *Earth surface sediment transport.* Oxford: Oxford University Press.
gives a technical explanation of many hillslope processes, and
Kirkby, M. J. (ed.) 1978. *Hillslope hydrology.* Chichester: Wiley.
offers advanced discussion of hydrological aspects of hillslope processes.

Brunsden, D. (ed.) 1971. *Slopes form and processes.* Inst. Br. Geogs Sp. Publ. No. 3.
includes several interesting case studies.

13 The fluvial system

Rivers exhibit a wide variety of forms. They may range from small turbulent mountain streams flowing down irregular beds mantled with coarse boulders and plunging over more resistant outcrops of rock as waterfalls, to the broad lowland river flowing in a tranquil manner in a channel incised into alluvium and often meandering in form. Nevertheless, all streams are governed by the laws of hydraulics, which apply to stream channels at all scales from fingertip tributaries to major trunk streams. As it is possible to enter and observe channels (particularly small ones) and, because their boundaries are easily defined and channel processes operate over short timescales (facilitating monitoring and measurement), our knowledge of river channels is in many respects more advanced than is the case with other types of system.

Channel flow takes place when the major stores within the catchment (canopy interception, surface depression, soil moisture and ground water) have been satisfied, or when a critical regulator such as infiltration capacity has been exceeded. Water may, therefore, arrive at the channel by a variety of routes, along which it may have flowed at widely differing rates (Table 13.1). A proportion of water in the channel is derived by direct precipitation on the channel surface, while water falling on slopes may run off directly as overland flow, or more slowly as soil throughflow (see Ch. 12) or by seepage from ground water. We can therefore regard channel flow as the sum of the outputs of these various stores.

Since these major inputs to the channel operate at different rates and come into play at different times, there are important temporal variations in the magnitude of channel flow, and these are recorded graphically in the **hydrograph**, in which discharge is plotted against time. It is convenient to separate the hydrograph into two components: **direct runoff** (water which finds its way rapidly into the channel) and **indirect runoff** (water which has followed a slower route).

13.1 Energy and mass transfer in channel systems

As in the operation of all systems, energy is the fundamental motive force. The stream channel processes are closely related to the magnitude and expenditure of this energy. At each point along the stream channel, the total effective energy will be the sum of the potential gravitational and kinetic components. Total energy can be related to an individual particle of water or, more usefully in geomorphological terms, to the total mass of water flowing in the channel at any one point. In this respect the mass of water is fundamental and, expressed as **discharge**, it is the major independent controlling variable in the operation of stream channel processes. Velocity is a major component of kinetic energy and is closely related to discharge, though not in a simple way.

The expenditure of energy in the channel takes place in two main ways. More than 95% is used in

Table 13.1 Estimated velocities along different flow routes in the catchment (after Weyman 1975, from various sources).

	Flow routes	Velocities ($m\ h^{-1}$)
	channel flow	300–10 000
surface	overland flow	50–500
soil flow	pipeflow	50–500
	matrix throughflow	0.005–0.3
groundwater flow	limestone (jointed)	10–500
	sandstone	0.001–10
	shale	10^{-8}–1

overcoming the frictional drag of the channel margin on the flowing water. The actual amount will vary according to channel size and shape and to the roughness of the bed and banks. Energy used in this way is converted to heat and lost to the surroundings by radiation and conduction, though it is barely measurable. The remainder is converted to mechanical energy and is used in transporting mineral sediments, which form part of the load of the river. Therefore, the proportion of the energy of the river actually used in erosion and transportation is only small.

The inputs and outputs of water to the river channel system are shown in Figure 13.1. There are four major contributors to the discharge of water in the channel which is measured in terms of volume per unit time – usually cubic metres per second (cumecs).

During rainfall, a proportion of the rain falls directly on the surface of the river channel. Normally this contributes only a small volume of water to discharge, but it will be larger when the channel has a large effective surface area, as when it incorporates a lake. Water falling on slopes may reach the channel by overland flow, or more slowly by way of throughflow. That proportion of water which percolates to ground water may return to the channel by seepage from the groundwater store. This is known as an **influent channel** and it is typical of humid environments.

The outputs of water from the channel system are, firstly, evaporation directly from the channel surface. This clearly becomes more important when the channel has a large surface area exposed to the atmosphere. Secondly, under certain conditions, when the water table is below the level of the channel floor, water may be lost from the channel to ground water. This condition, defined as an **effluent channel**, is sometimes found in arid environments and in areas of permeable rocks such as limestone. The major form of output from the channel system, however, is normally runoff, the process by which water is transferred along the channel system into the ocean store.

The magnitude of runoff varies considerably in both space and time. Within a catchment basin, for example, the discharge through any channel cross-section is closely related to the catchment area drained through that section (Fig. 13.2). More important in the operation of channel dynamics is the temporal variation in discharge. This results from the fact that the four major inputs of water to the channel system are of different magnitudes and they come into play at different times. The inputs consequent upon precipitation (direct runoff) are of high magnitude and are closely related to the periodic distribution of rainfall events. Inflow from ground water (indirect runoff) is normally of much lower magnitude, but is distributed continuously through time.

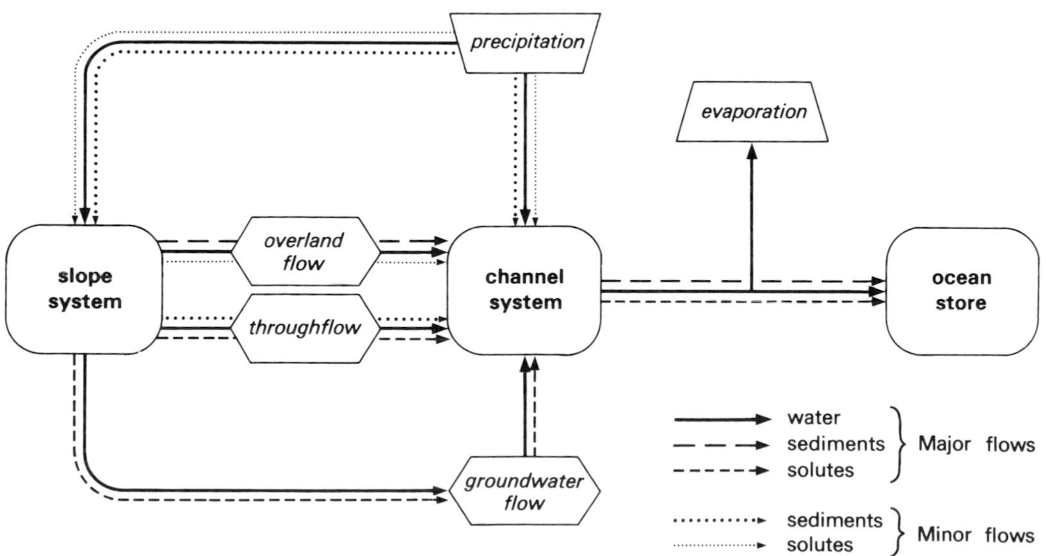

Figure 13.1 Pathways of water and mineral flow through the catchment basin system. (This is a more detailed model than that in Figure 10.2.)

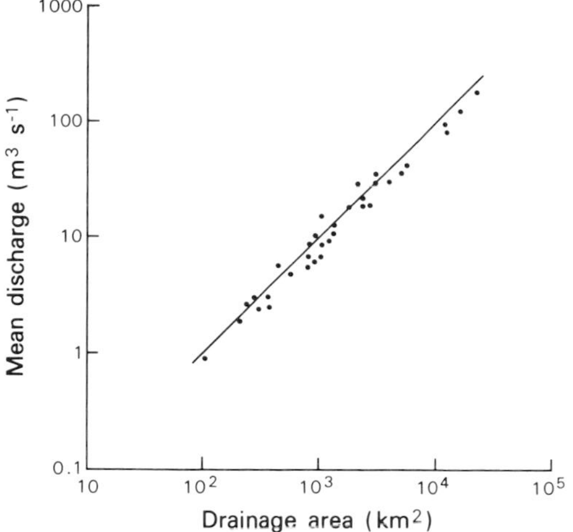

Figure 13.2 Relationship between discharge and drainage area for gauging stations on the Potomac River, USA (after Hack 1957).

Figure 13.3 shows a storm hydrograph in relation to the rainfall input. Following rainfall, discharge rises sharply to a peak, then falls gradually. The length of time between maximum rainfall intensity and peak discharge is the **lag time**. The section of the hydrograph showing increasing discharge is the **rising limb**, and the **recession curve** describes the decrease in discharge. The area under the peak measures the volume of direct runoff, as opposed to indirect runoff, upon which the former is superimposed. Thus the processes associated with a rainfall event lead to storm runoff, while groundwater seepage contributes to baseflow. The magnitude and shape of the hydrograph will vary according to many factors associated with the nature of the rainfall and the characteristics of the catchment.

Figure 13.3 plots an idealised hydrograph of an individual storm flow. In reality, of course, a hydrograph consists of a series of flood peaks distributed through time and superimposed upon base flow. Figure 13.4 shows two such hydrographs for a one-year period. Two adjacent rivers in southern England, of similar catchment area, are contrasted in their responses to the same rainfall inputs. The River Wallington, draining a catchment based mainly on clay, shows a series of sharp flood peaks superimposed upon a baseflow which is higher in winter than in summer, reflecting seasonal differences in groundwater storage level. Because of the impermeable base of the catchment, runoff response to storm rainfall is immediate and highly peaked. The River Meon, with a chalk catchment which precludes direct runoff due to high infiltration, exhibits a much more subdued regime and is fed mainly by groundwater flow.

The flows of water described in the previous paragraphs are, of course, also associated with flows of materials (in both solid and dissolved forms) and are shown in Figure 13.7 together with the state of the materials.

Rain water is not pure water, for it contains a range of ions in solution as well as particulate matter washed out of the atmosphere. Overland flow is capable of transporting both sediment and solutes to the river channel, though the former is dominant. Elements are therefore washed into the channel from adjacent slopes by this process. Throughflow contributes elements in solution derived from regolith and soil. Sometimes deposits of elements transported by throughflow can be seen in the channel banks. A reddish brown stain, for instance, may mark the redeposition of formerly dissolved iron where the throughflow emerges. Similarly, groundwater seepage contributes solutes to the channel.

Mineral sediment is contributed in considerable measure by erosion of the banks of the channel, which collapse into the stream when undermined. Fresh erosion scars are often seen on the outside of bends in the channel.

Once in the channel, solutes tend to be transported continuously by the stream. Sediments, on

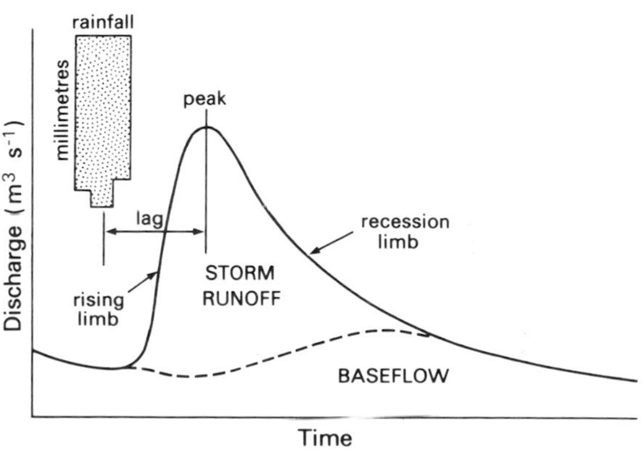

Figure 13.3 Characteristics of the flood hydrograph.

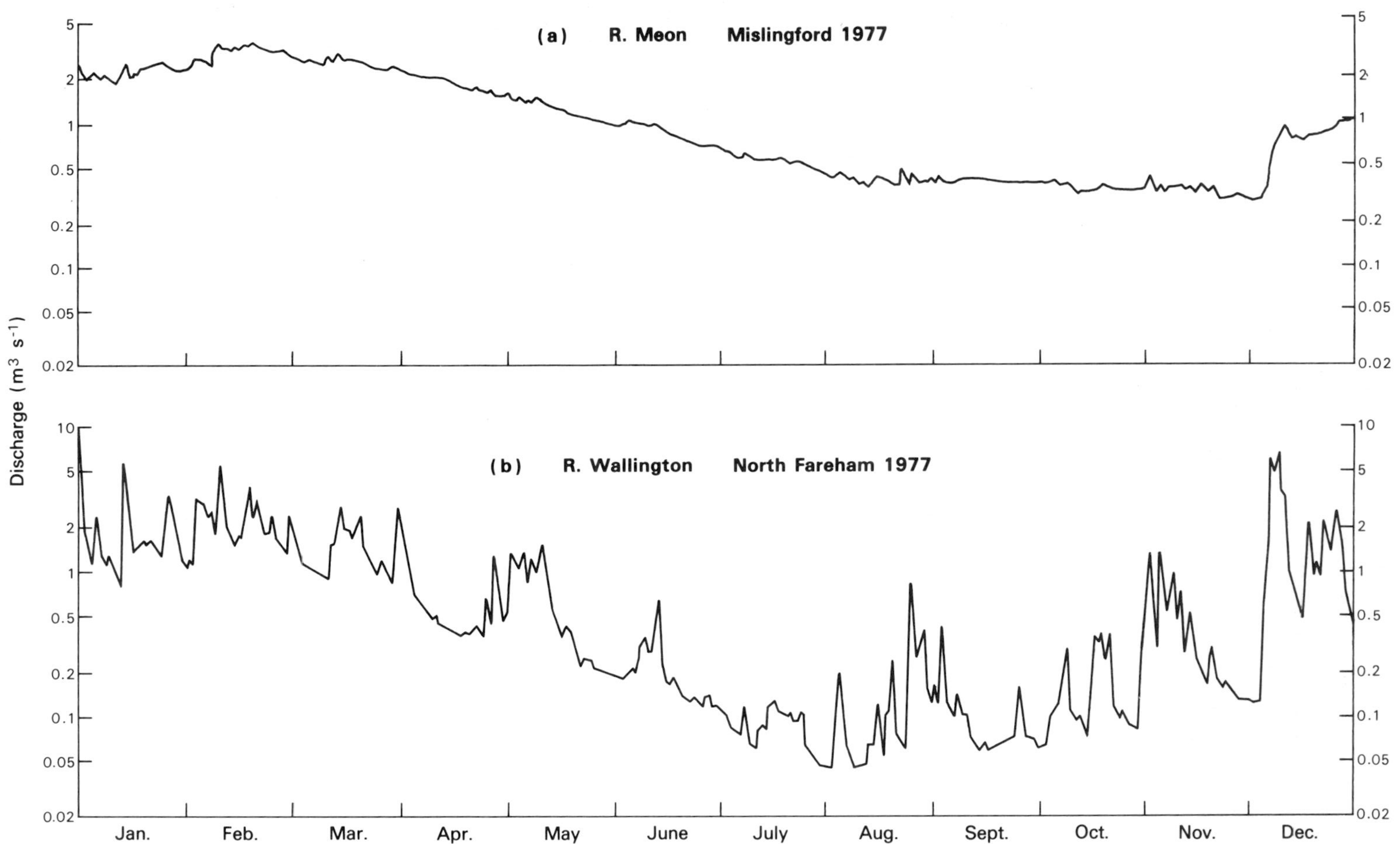

Figure 13.4 Comparison of the annual hydrograph for two adjacent and contrasting catchments for the same year: (a) the Meon, fed mainly by baseflow from the Chalk, shows mainly seasonal var ations; (b) the Wallington, draining a largely clay-based catchment, is much more flashy and responds strongly to individual storms. (Hampshire, England)

the other hand, are largely intermittent in their movement. High concentrations may be moved in times of flood, afterwards being deposited on the channel bed until the next flood. Accordingly, there is a two-way exchange of sediment between the channel bed and the stream. The relationships between erosion, transportation and discharge will be examined in more detail below.

13.2 Channel dynamics: flow of water

In this section we examine channel processes and form in relation to the moving body of water. The channel can be defined by a system of variables (Boxes 13.1 & 2). Water is a Newtonian fluid which is unable to resist shearing stresses and it will deform at a rate directly proportional to the applied stress. Another way of expressing this is to say that water will flow under its own weight within the solid boundaries that contain it.

The flow of water can be expressed as a balance between impelling forces and resisting forces (Box 13.3). In this way basic hydraulic relationships can be developed which express the velocity of flow in relation to channel form characteristics. Expressions such as the Manning equation (Box 13.3), however, deal only with the mean velocity and they neglect important variations which occur through the channel cross-section.

There is resistance to flow not only at the channel boundary but also within the fluid mass itself. The adjacent layers of water are able to slip past each other with differing velocities. Accordingly, velocity in a vertical profile increases away from the bed where the shearing resistance is greatest, and the highest velocities are to be found at the

Box 13.1

RIVER CHANNEL VARIABLES

The geometry of a river channel can be defined in terms of a number of variables.

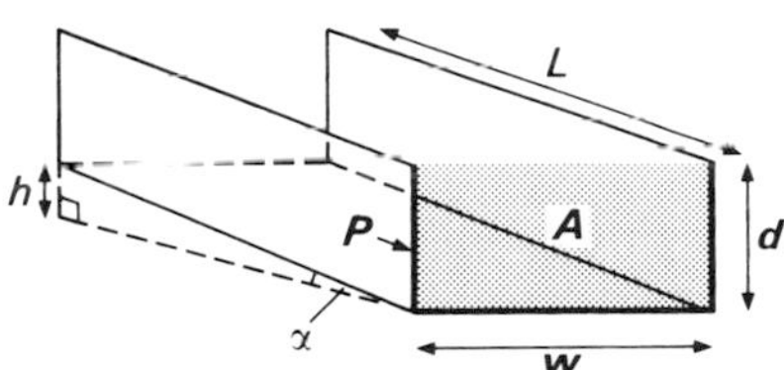

(a) *Channel width* (w): measured at the water surface.

(b) *Channel depth* (d): except in the case of the rectangular channel cross section illustrated, this varies across the section and is normally expressed as mean depth ($\bar{d}$).

(c) *Cross-section area* (A): calculated as the product of width and mean depth; thus $A = \bar{d}w$.

(d) *Wetted perimeter* (P): the length of channel margin in the cross section in contact with flowing water; thus $P = 2d + w$.

(e) *Channel gradient* (S): expressed as change in elevation per unit length; thus

$$S = \frac{h}{h \cos \alpha}, \text{ or } \tan \alpha.$$

(Note that for small gradients S approximates to $\sin \alpha$.)

(f) *Velocity of flow* (V): distance travelled per unit of time (m s^{-1}).

(g) *Discharge* (Q): volume of water passing through a cross section per unit time:

$$Q = w\bar{d}v \text{ (m}^3\text{ s}^{-1}\text{)}.$$

(h) *Channel length* (L): slope length of channel (see Box 13.3).

Box 13.2

HYDRAULIC RADIUS

Hydraulic radius is a measure of the proportion of water in the channel cross section in contact with the channel margins (the wetted perimeter). As such, it is a measure of frictional retardation and therefore of channel efficiency.

By taking the case of the simplest channel cross section, rectangular in form, and substituting different values for width and depth, significant differences in channel efficiency can be demonstrated.

$$\text{Hydraulic radius } (R) = \frac{A}{2d + w}.$$

Case 1

With $d = 2$, $w = 4$; $A = 8$ and $R = 1.0$

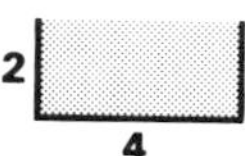

Case 2

With $d = 1$, $w = 8$; $A = 8$ and $R = 0.8$

Thus a wide, shallow channel is less efficient than a more compact one.

Case 3

With $d = 4$, $w = 8$; $A = 32$ and $R = 2.0$

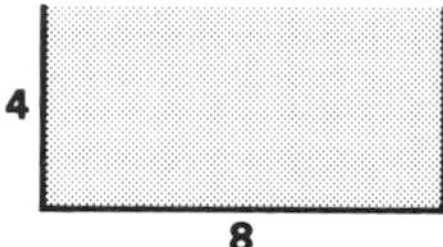

Thus a large channel, of the same proportions but of greater magnitude than Case 1, is more efficient.

A corollary of these relationships is that river channels become more efficient in times of flood, when depth increases to fill the channel and render it greater in area and more compact in form. The channel is most efficient at the bankfull stage, for above this level it spreads out across the floodplain, much increasing its wetted perimeter and thereby decreasing the hydraulic radius.

It follows also that since cross-section area increases downstream with increasing discharge, rivers will tend to become more efficient in the downstream direction.

water surface. This is known as **laminar flow** (Fig. 13.5a). When the bed is rough and velocity is high, this simple pattern of parallel streamlines breaks down and is replaced by **turbulent flow** in which jets of water travel obliquely to the general direction of flow, and eddies develop causing greater mixing of the water and a more even distribution of shear across the channel (Fig. 13.5b).

The relationship in the channel between water and cross-section form is described by **hydraulic geometry**, and the response of channel-form variables to changes in discharge can be assessed by field investigation. Discharge (the controlling variable) varies through time at any section in

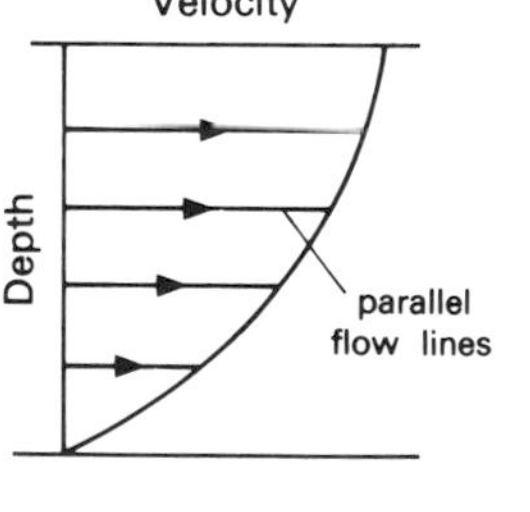

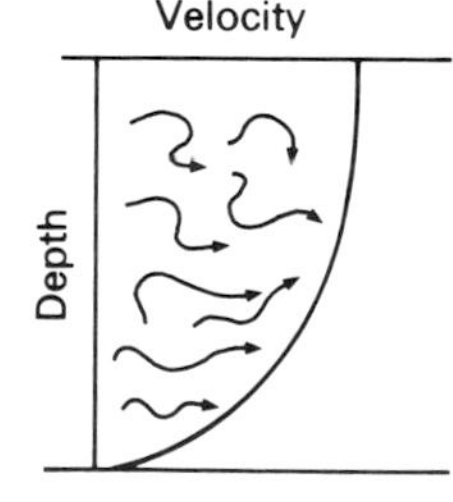

Figure 13.5 Contrast in flow patterns between (a) laminar and (b) turbulent flow.

Box 13.3

FLOW IN OPEN CHANNELS

The flow of a fluid in an open channel can be expressed as a balance between impelling forces and resisting forces.

In a section of channel (see Box 13.1) the tractive force operating in a body of water in the direction of flow is the downslope component of its weight:

$$T = \rho g\, L d w \sin \alpha$$

where ρ = density of the mass, g = gravitational acceleration, L = slope length of channel, and other variables are as defined in Box 13.1.

The resisting force (F_R) is the stress per unit area (τ) multiplied by the area of the boundary over which it is applied. Thus, for a channel of rectangular cross section:

$$R = \tau\,(2d + w)\,L.$$

If there is no acceleration along the unit reach then:

$$F_T = F_R.$$

Thus:

$$\rho g\, L d w \sin \alpha = \tau\,(2d + w)\,L.$$

Simplifying, since $A = \bar{d}w$, and $\sin \alpha = s$ for small angles:

$$\rho g A s = \tau\,(2d + w).$$

Therefore:

$$\tau = \frac{\rho g s A}{(2d + w)}.$$

Since

$$\frac{A}{(2d + w)} = \text{hydraulic radius } (R)$$

then

$$\tau = \rho g R s.$$

In hydraulics, the resistance to flow is related to the square of velocity, thus:

$$\text{resistance} = \tau / v^2$$

or

$$\tau = k v^2$$

where k is a constant.

By substitution:

$$\rho g R s = k v^2.$$

Thus, by setting

$$\sqrt{\frac{\rho g}{k}} = C,$$

$$V = C\sqrt{Rs}.$$

This is known as the **Chezy equation** and expresses mean velocity as a function of hydraulic radius and channel gradient.

A similar, though more refined, version (the **Manning equation**) is more frequently employed in studies of channel flow:

$$V = \frac{1.009}{n} \quad R^{2/3}\, s^{1/2}$$

where n is a measure of channel roughness. It is in part a function of the magnitude of the wetted perimeter and in part an expression of the nature of the channel boundary in terms of material calibre and the amount of vegetation. The rougher the nature of the channel margins, the greater will be the frictional resistance to flow. In natural stream channels, Manning's n varies from 0.03 in straight clean channels, to 0.10 for densely vegetated channels.

the channel and varies also in the downstream direction.

Let us consider the effect of increasing discharge at one point in the channel. The first visible effect is an increase in channel depth as the water level rises and this is usually accompanied by an increase in width, since most channels are trapezoidal in cross-section. These responses determine an increase in channel cross-section, resulting in greater hydraulic radius and channel efficiency. The gradient of the channel remains essentially constant. The more efficient the channel the less the energy of the stream is spent in overcoming the frictional resistance of the channel margins, which permits higher velocity. Relationships between discharge, and width, depth and velocity as determined in the field, are shown in Figure 13.6.

The increased mass and velocity of the water result in a greater capacity for work. For these reasons, most work is carried out at times of high discharge, during floods. As discharge falls after a storm flood, so the channel variables respond in reverse sequence. Width, depth and cross-section area diminish as does hydraulic radius, and velocity decreases. Thus the channel can be considered as an open system, in a state of dynamic equilibrium, responding instantaneously to changes in input conditions and fluctuating about a steady-state condition.

Similar relationships exist between discharge and other channel variables in the downstream direction (see 13.5).

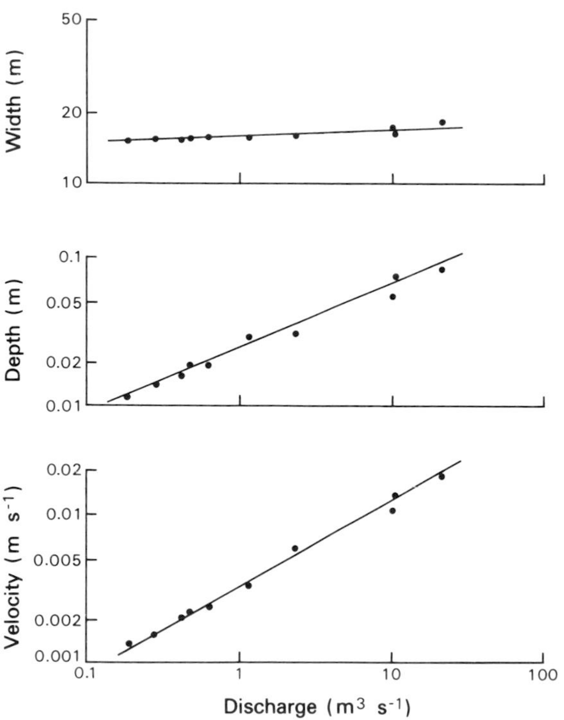

Figure 13.6 Hydraulic geometry of a stream channel: at-a-station changes in velocity, depth and width with varying discharge (after Wolman 1955).

13.3 Channel dynamics: erosion and transportation processes

The energy available in the channel after friction has been overcome is used in eroding the channel margin and transporting mineral debris.

Where the stream flows directly over bedrock it may erode that bedrock in various ways. The hydraulic force applied by the water may detach rock fragments, particularly if the rock is already fractured. Where a pebble lies in a hollow in the channel floor it may be swirled around mechanically, boring into the channel bed to form a pothole. The transport of a mass of coarse sediment along the channel will have an abrasive effect known as **corrasion**. Clearly, all three processes depend on mechanical energy, which is more effective with the higher velocities associated with flood discharges. When the channel margins are formed of the less resistant unconsolidated materials such as alluvium, channel erosion by mechanical processes will be enhanced. In addition, the banks may become undercut at the waterline and collapse into the channel. The effect of animals trampling the river banks may hasten this process.

Debris within the channel may be set in motion as a result of the application of force by the flowing water and thus become part of the transported load of the stream. Material can be carried in three ways. **Bed load** is material, usually of coarse calibre, rolled along the floor of the channel; **suspended load** is finer particles carried along in suspension within the water mass; and the **solution load** consists of minerals dissolved in the water.

As velocity and flow increase, so the shear stress exerted by the water on the upstream side of a sediment particle on the channel floor is increased.

The force required to move a given particle (the critical tractive force) depends on the size and shape of the particle and the gradient of the bed. Once this threshold value is attained the particle is set in motion and rolls along the channel bed in the downstream direction. Clearly, tractive force will increase with velocity, and the bed load accordingly increases at higher levels of discharge. Troake and Walling (1973) have shown that for a Devon (England) stream bedload transport is zero until a critical threshold of discharge is attained, above which it increases rapidly (Fig. 13.7c).

Finer particles (sand, silt and clay) when disturbed, may be lifted above the channel bed by the upward currents of water involved in turbulent flow and carried downstream before being redeposited. The net effect of myriad fine particles being picked up, swirled along and falling to the bed is that a mass of sediment is entrained, particularly in the lower part of the water body, as the suspended load. Provided that suitable fine-grained material is available, there will always be some material in suspension. Since, however, the amount of energy available increases with discharge, the amount of suspended load increases exponentially during flood (Fig. 13.8). Concentrations of suspended sediment load typically vary over at least two orders of magnitude, in the range 10–1000 milligrams per litre (mg l^{-1}), though higher values are by no means uncommon during peak discharges.

The dissolved load of a river depends largely on the solubility of the rock-forming minerals in its catchment and the weathering regime under which it occurs. In a large drainage basin with varied geological outcrops, considerable spatial variation in solute load may exist, as in the example of the

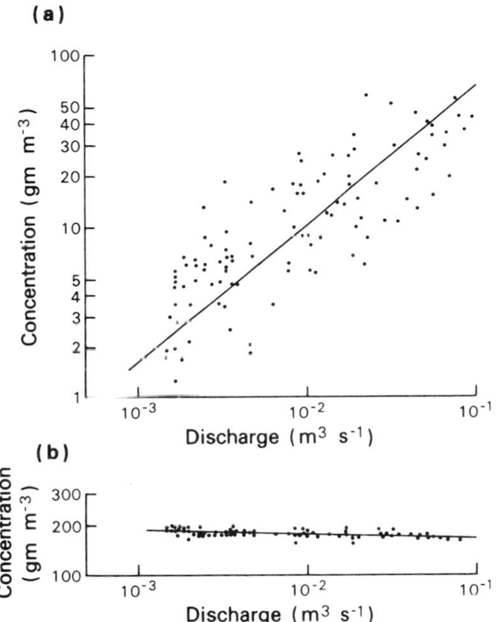

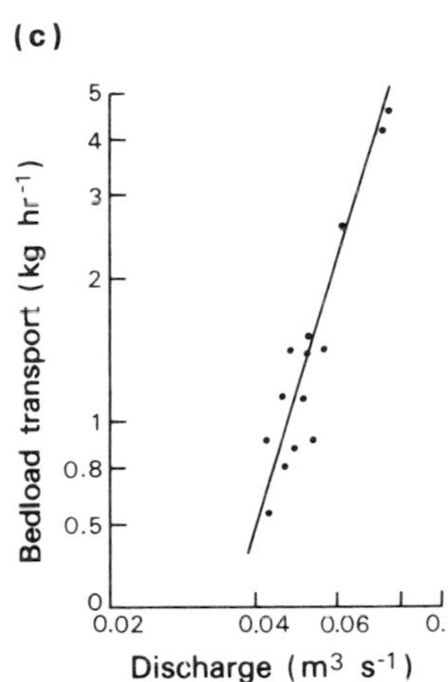

Figure 13.7 Relationships between the various components of load and discharge for a small Devon (England) stream (after Troake & Walling 1973). (a) sediment (b) solutes (c) bedload

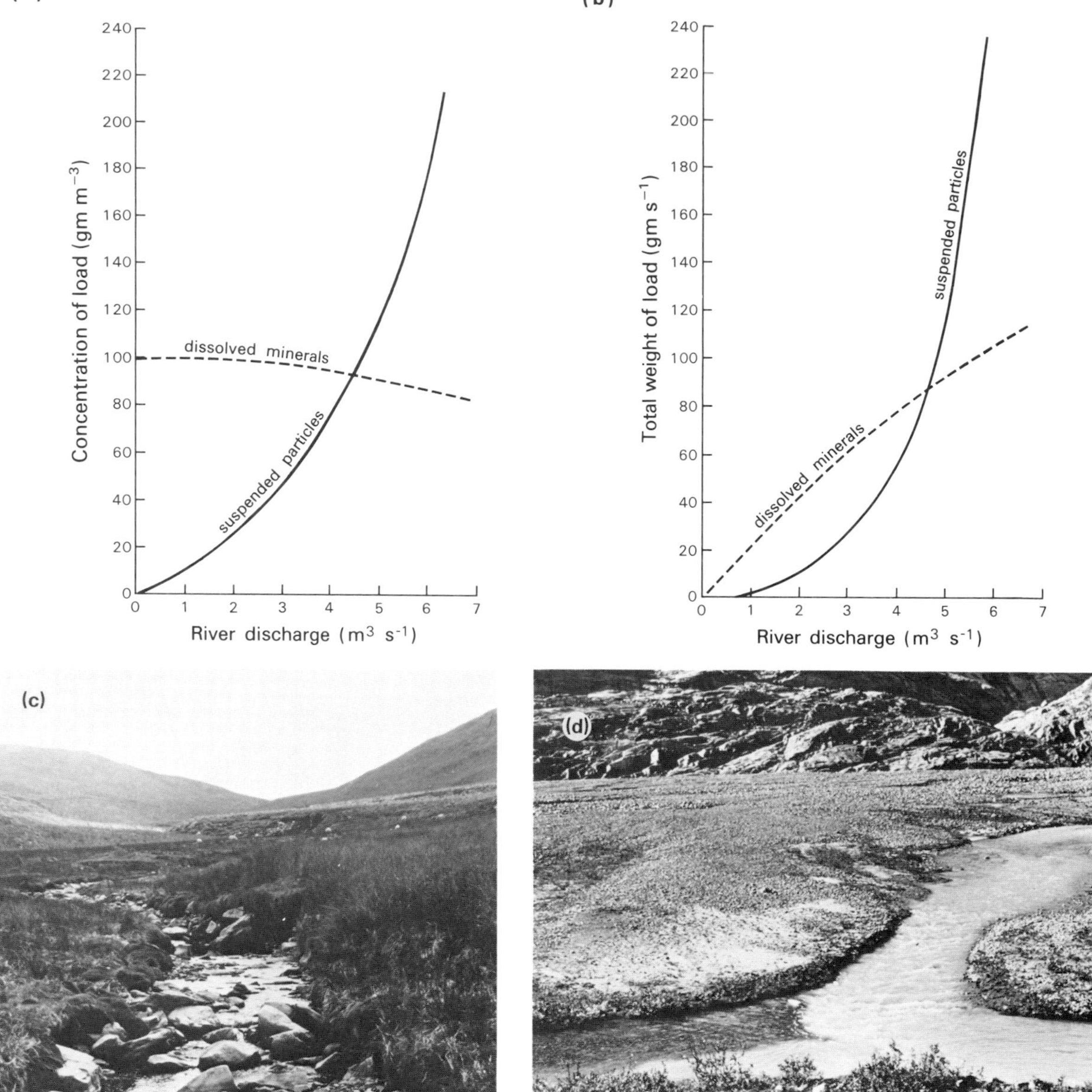

Figure 13.8 (a & b) Comparison between concentration and total load of a stream with varying discharge. (c) A small upland stream with abundant coarse sediment resting in the channel; the coarsest sediment is probably transported only at very infrequent intervals during extreme floods (Upper Rheidol Valley, Mid-Wales). (d) Confluence between sediment-laden glacial outwash stream on the right and a sediment-free stream flowing from the left; the glacial meltwater with an abundant supply of fine sediment is very cloudy (Tunsbergdalen, Norway).

River Exe, Devon (England) (Fig. 13.9) draining areas underlain by highly soluble rocks such as limestone, which tends to have a high solute load (300–400 ppm). Elements in solution reach the river channel mainly by groundwater seepage with additional contributions by soil throughflow and overland flow.

The concentration of solutes is greatest at low flows, when the water in the channel is derived entirely from groundwater seepage. In this situation percolating water spends a greater length of time in intimate contact with the rock-forming minerals and thereby attains higher solute concentrations. During periods of higher discharge when the waters of throughflow and overland flow are contributing to channel flow, the water in the channel is diluted and the overall solute concentration decreases (see Fig. 13.7). At times of flood, however, this dilution effect is more than offset by the increased discharge, such that the total quantity of dissolved materal transported per unit time increases.

The relative importance of these three modes of transport varies from river to river, depending on the nature of the available load. In any one river channel it will vary through time, with the solute load being more important at low flows, while suspended sediment is transported in greater abundance at times of flood, and bedload moves only once a threshold level of discharge has been attained.

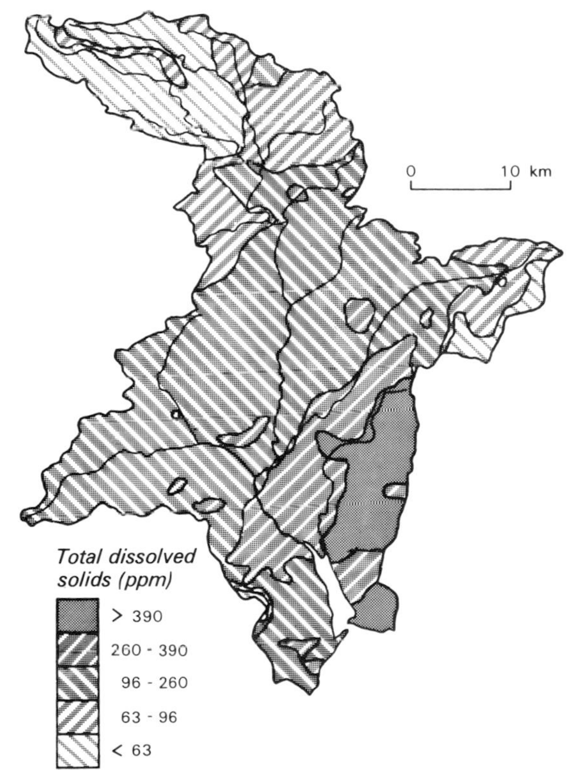

Figure 13.9 Spatial variation in concentration of solutes within the catchment of the River Exe, Devon, England (after Walling & Webb 1975).

13.4 Depositional processes

It has been shown that the bulk of the solid materials in a channel are set in motion at times of flood discharge. As discharge declines following a flood peak, so turbulence and tractive force decrease and the solid particles are redeposited, coarse material first, while finer particles may continue in transport at lower flow velocities. Thus while solutes, once in the channel, tend to be flushed right out of the system, the solid load is conveyed down stream at time of flood and is subsequently redeposited.

Within the channel, abundant deposits of solid particulate material await the next flood. These may be a simple uniform spread of gravel across the channel, or be arranged in more distinctive form. Point bars, for example, represent aggradation on the insides of meander bends, and both longitudinal and transverse bars represent temporary stores of sediment in transit.

13.5 Channel form

The form of the river channel represents the means by which the discharge is accommodated. It can be regarded as the response by the channel to its inputs, which can be considered in terms of adjustments to the long profile, the cross section and to the form of the channel in plan. The three are closely related in many ways which can be explained by the interaction of hydraulic variables.

It has been recognised for some time that the long profiles of streams tend to be concave in profile, as gradient gradually decreases downstream. Major changes may take place in the downstream direction in drainage area and lithology. Drainage area and therefore discharge progressively increase, and in its lower reaches the river channel tends to be cut into fine-grained alluvial deposits, in contrast to the bedrock channel of the upper reaches. Accordingly, as channel cross-section area increases down stream in order to accommodate the increasing discharge, the channel becomes more efficient, i.e. it has a greater hydraulic radius. This tendency is reinforced by the change to a smooth channel margin in finer-grained alluvial sediments, and the stream flows along the lower gradient channel with increased efficiency. In this way the tendency to decreasing gradient downstream is a means by which channel form accommodates the increasing discharge.

Only rarely, and for a variety of reasons, is the long profile of a stream smoothly concave. First, the downstream increase of discharge is not gradual but incremental. Increase in discharge takes place mainly step-wise at tributary junctions. Secondly, increases in discharge at tributary junctions may not be accompanied by a proportional increase in sediment load. Thus the whole sediment–discharge relationship may be altered in the main channel, resulting in local adjustments of gradient by erosion or aggradation. Thirdly, variations in the resistance of bedrock to channel erosion may cause local variations in the rate of gradient adjustment. Thus an outcrop of a particularly resistant bed of rock may create a waterfall. All these circumstances are quite normal occurrences whose effects are superimposed upon the overall trend of decreasing gradient down stream.

It is clear that discharge increases in a downstream direction according to the increase in drainage area. The channel adjusts to the change in the controlling variable by adjusting its own form. Thus it becomes wider and deeper, and so possesses a greater cross-section area. In so doing, it develops a greater hydraulic radius and becomes more efficient. The downstream changes in channel variables are exemplified in Figure 13.10 and are easy to demonstrate by simple field measurements. Not only does the size of the channel change down stream but so do its proportions. It is clear from Leopold and Maddock's (1953) data that width increases more rapidly than depth. Thus, as the width: depth ratio increases down stream, so the channel becomes proportionately wider, accommodating the increasing discharge and energy and dissipating a greater proportion of energy by friction at the channel margins as hydraulic radius decreases.

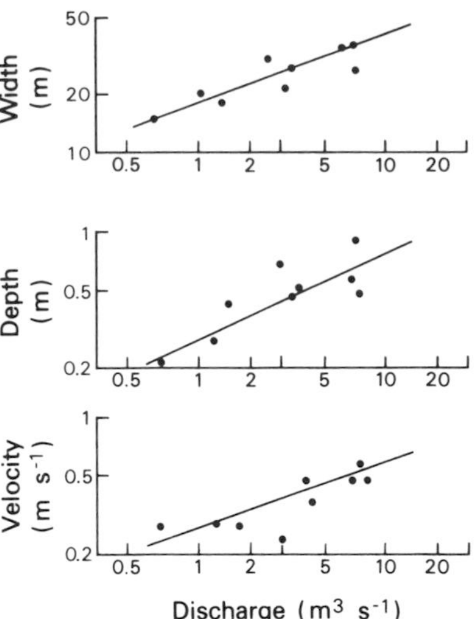

Figure 13.10 Hydraulic geometry of a stream channel: downstream changes in width, depth and velocity with varying discharge (after Wolman 1955).

The third major attribute of channel form is its plan. This represents yet another way in which the river adjusts to inputs of discharge and sediment. Four plan forms are normally recognised. These are straight, sinuous, meandering and braided channels (Fig. 13.11).

Straight natural channels are rare. Even within straight channel sections the flow of water is sinuous or helical. Even in controlled laboratory experiments straight channels cut in uniform sediment last only for a short period before non-straight flow patterns develop. It would appear, therefore, that straight flow is not a normal mode of behaviour for running water.

More common is the sinuous stream pattern, in which the trend of the river channel fluctuates irregularly about its mean. Figure 13.11a shows examples of sinuous stream patterns, in which sinuosity is measured as total channel length divided by valley length.

A particularly characteristic channel plan form is the meandering habit. Meanders are well defined features possessing a high degree of geometric regularity clearly related to discharge. Figure 13.12a shows the significant geometric

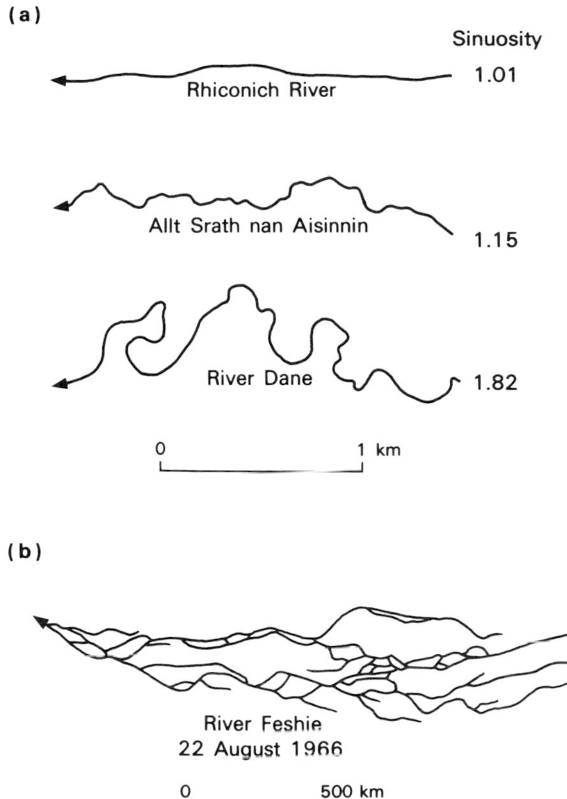

Figure 13.11 Channel patterns: (a) single channels of varying sinuosity; (b) a braided channel.

features of a meandering channel. Data from meandering channels of a wide range of sizes show that meander wavelength and radius of curvature are closely and directly related to channel width and thus to discharge. Amplitude is correlated only poorly with channel width, suggesting that local topographic or structural factors may be more important in influencing this characteristic. A study by Langbein and Leopold (1966) suggests that the form of the meander curve can be described by a sine function. In such curves the rate of change of curvature is minimised, i.e. there is no concentration of curvature. This can be interpreted as leading to the even distribution of stress and therefore energy loss along the channel length. Thus a meandering channel can be interpreted as being adjusted to promote evenly distributed energy loss, given the discharge and load. The effect of meanders is to lengthen the stream channel through increased sinuosity, thereby decreasing its gradient.

Although the precise reason why meanders form is not well understood, their development can be reproduced in miniature form in the laboratory (Fig. 13.12b). Starting with a straight channel in non-cohesive sediments, a sequence of deeps (pools) and shallows (riffles) develops in the channel. This is an unstable form. The pools and riffles are spaced such that a channel length of 10–14 times channel width contains two pools and two riffles. These cause the flow within the channel

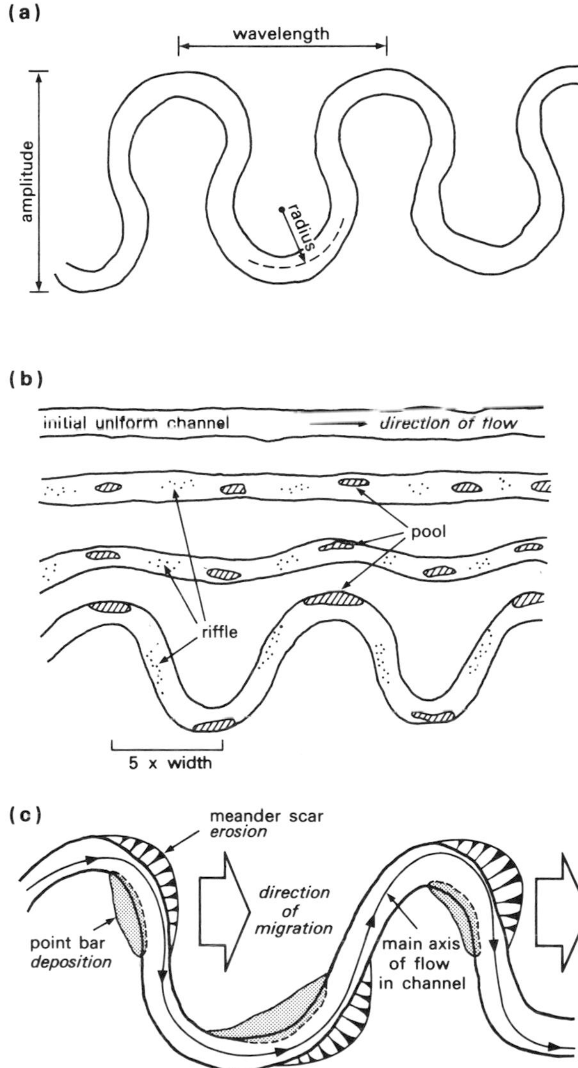

Figure 13.12 Meandering channels: (a) geometric definitions; (b) development of meanders from an initial straight channel; (c) geomorphological features of a meandering channel.

to become sinuous, resulting in differential erosion of the channel margins. The channel form itself becomes sinuous and ultimately meandering.

The axis of flow in a meandering channel is located at the outsides of the meander bends. It crosses over at the point of inflection. Thus the meanders develop by eroding the outside of their channel just down stream of the meander apex and by depositing sediment on the inside of the meander in the form of a point bar. Figure 13.12c shows the lateral migration of a meandering channel. In this way the meanders migrate down valley, ultimately having occupied successive portions across the whole valley floor.

Braided streams (Fig. 13.11b) are associated with the deposition of bedload to form bars within the confines of the channel. Thus the thread of water within the channel bifurcates to flow around the bar. Complex braided (anastomosing) streams have an abundance of bars and a multiplicity of interconnecting channels. They appear to be related to highly variable flows and easily eroded channel margins providing an abundance of bedload. Thus, prolific transport of bedload takes place at high discharge, but it is followed by deposition as discharge subsides. In this way the bars are deposited and the dwindling water supply flows around and between them.

Channel plan form, as expressed by meandering and braiding, is clearly a complex and multivariate phenomenon. It must be considered in relation to the hydraulic variables which describe channel processes. Several empirical relationships have been put forward (Fig. 13.13) which relate channel plan form to gradient but neglect the effect of bedload calibre. Thus, for a given discharge the less efficient braided channel requires a higher

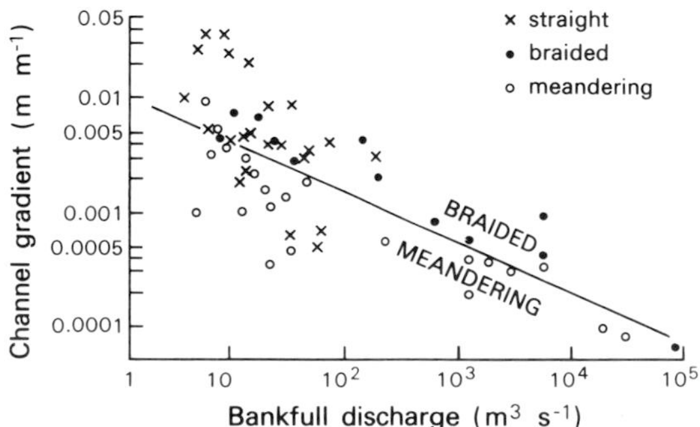

Figure 13.13 Channel pattern relationships: bankfull discharge plotted against channel gradient for selected channels of differing type (after Leopold & Wolman 1957).

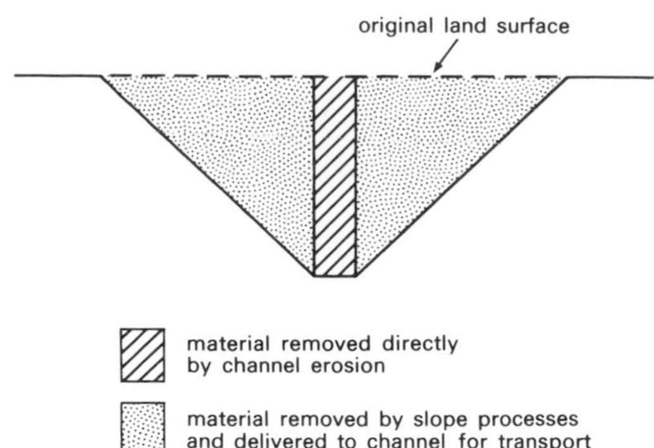

Figure 13.14 The role of river channel processes in relation to landscape denudation.

gradient to do its work. Conversely, for a given gradient a decrease in discharge will convert a braided stream into a meandering one.

The ultimate cause of meandering and braiding has not been established but it is reasonably clear that channel plan form represents a further mode of adjustment to prevailing conditions of discharge and load. In this way it cannot be dissociated from other more easily described aspects of channel form such as gradient, cross-section and bedload calibre.

Fluvial erosion resulting from river channel processes removes only a thin section of mass directly from the landscape (Fig. 13.14). It is the operation of the tributary slope processes in opening out the valley away from the channel margin that gives fluvial landscapes their distinctive form. The fluvial system is the dominant mode of removal of mineral material from the world's land surfaces. The magnitude and frequency of fluvial events vary widely in different environments, as do fluvial forms. However, the principles underlying their operation are fundamentally the same.

Further reading

Useful introductions to stream channel processes are

Morisawa, M. 1968. *Streams – their dynamics and morphology*. New York: McGraw-Hill.

For a comprehensive treatment of fluvial processes in the context of the drainage basin, the reader is referred to

Gregory, K. J. and D. E. Walling 1973. *Drainage basin form and process*. London: Edward Arnold.

In addition, the general geomorphological texts referred to at the end of Chapter 10 all contain sections relating to fluvial processes.

14 The glacial system

Glacial systems develop when winter precipitation is in the form of snow and where the total annual received radiation is insufficient to melt the total annual snowfall. Thus snow and ice are permitted to accumulate in large bodies at the Earth's surface. Glaciers are therefore found in cold and humid environments, in polar regions and at high altitudes in temperate latitudes.

The major material input to the glacial system is precipitation in the form of snow (Fig. 14.1), which in the glacial system becomes transformed into glacier ice. The fragile hexagonal crystals that form snowflakes become compressed into compact granular crystals. Output of mass from the glacier may be in the form of melt water, some of which may evaporate directly to the atmosphere or, when the glacier terminates in a water body, large fragments of ice may break off as icebergs (the process of **calving**).

The glacial system can be considered to act as a store of water at the Earth's surface, since glaciers and ice caps contain some 75% of the water in the Earth's land systems. In its general structure, the glacial system closely resembles other denudation systems, with the major exceptions that (a) water passing through it is in the form of ice, and (b) vegetation has negligible influence.

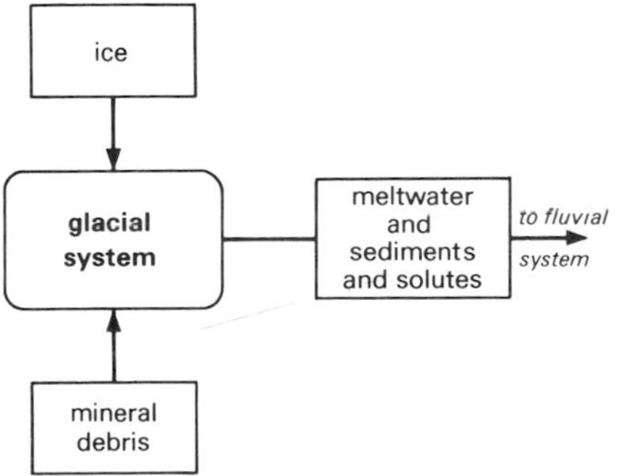

Figure 14.1 Simplified schematic model of the glacial system.

Rock debris is supplied to glacial systems by erosion from the glacier bed, or by rockfalls from steep slopes overlooking the glacier surface. This is transported within the system and output by deposition may be effected directly by glacier ice or indirectly by the action of melt water.

14.1 Systems function

The overall activity of a glacier is described by its budget – the balance between input (accumulation) and output (ablation). The budget reflects the state of a glacier and determines whether it is in equilibrium or increasing or decreasing in size.

In the case of a northern hemisphere glacier, the budget year (Fig. 14.2a) begins at the first snowfall, the start of the year's accumulation. Accumulation takes place throughout the winter, tailing off towards summer. Ablation is concentrated largely in the summer period, although limited ablation may occur sporadically throughout the winter, if temperatures happen to rise above zero. The vertical axis expresses glacier mass in terms of volume (water equivalent). If the area beneath both accumulation and ablation curves is equal, then the budget is balanced and the glacier is in equilibrium. If accumulation is in excess, then the glacier has a positive budget and is growing. If the budget is negative, then the glacier is shrinking.

It is important to note that the budget refers to the throughput of the glacier, not its volume. It is possible to distinguish high-budget glaciers characterised by high input and output, such as are found in some temperate mountains with high precipitation. These are active systems, often small in size, which transfer large volumes of ice rapidly, in con-

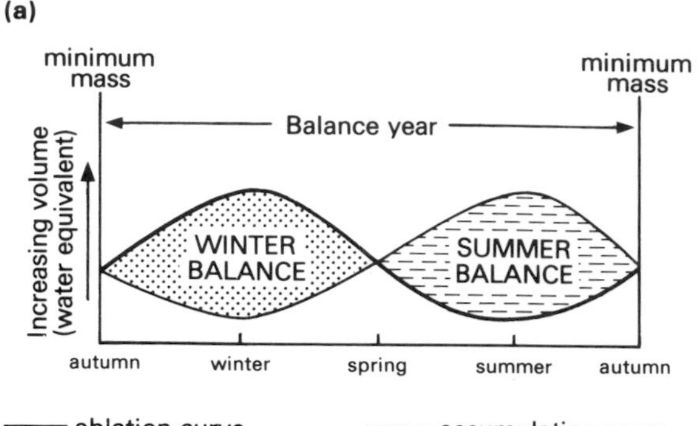

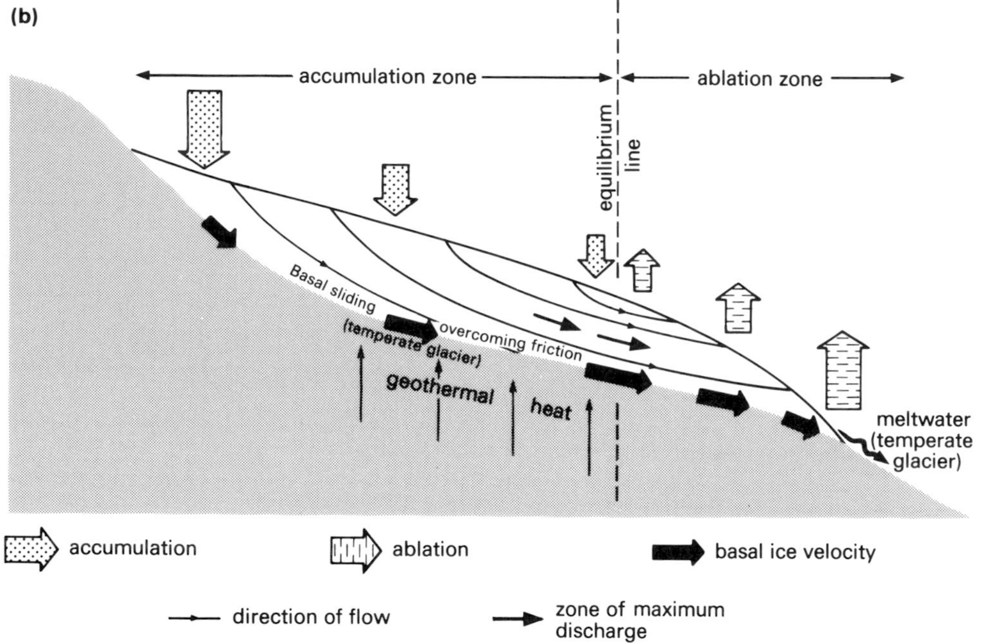

Figure 14.2 (a) The balance year of a glacier: the positive winter balance and the negative summer balance are combined to produce the annual balance. (b) Inputs and outputs of mass in a glacier system (after Sugden & John 1976).

trast to the low-budget systems of polar latitudes where both precipitation and temperatures are lower.

Just as accumulation and ablation are distributed unevenly through time, so also are they distributed unevenly in space over the glacier surface. Accumulation is greater at higher altitudes, where ablation processes are more limited due to lower temperatures. Ablation dominates over accumulation in the lower part of the glacier (Fig. 14.2b). Thus the long profile of a glacier may be divided into an upper accumulation zone and a lower ablation zone. The point at which ablation and accumulation are balanced is the equilibrium line. The maximum discharge of ice takes place through this point in the glacier channel, since down glacier from here ablation continuously reduces the discharge. In order to preserve a stable equilibrium form, the glacier transfers excess accumulation from the upper part by flowage, and it is only this transport of mass which sustains the glacier body within the ablation zone.

In the accumulation zone there is a downward component of motion as accumulation buries the pre-existing ice. The opposite occurs in the ablation zone as underlying ice is revealed by ablation, resulting in an upward component of motion. This is best visualised in the case of a small cirque glacier (Fig. 14.3) where both accumulation and ablation increments are shown as being wedge

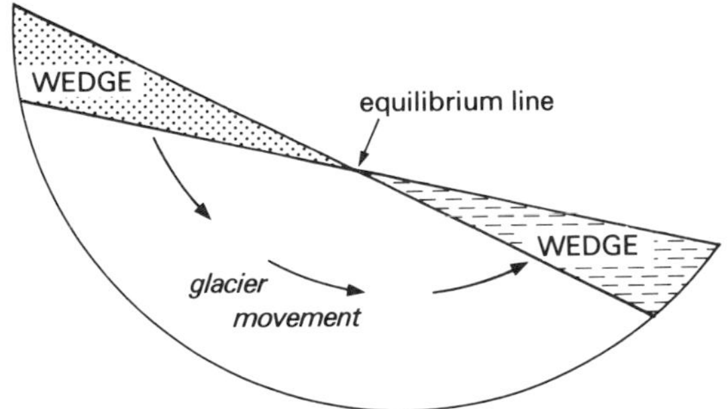

Figure 14.3 Idealised diagram of a small glacier showing wedge-shaped accumulation and ablation zones. Flow of ice is necessary to maintain an equilibrium surface profile. (After Sugden & John 1976.)

shaped in long profile. Rotational movement takes place in order to redistribute the mass more evenly. In the case of longer glaciers, this model of simple rotational movement becomes attenuated and modified by changes in bed gradient, often caused by variation in lithology or the form of the underlying topography. Nevertheless, the terminal zones of most land-based glaciers where ablation is dominant show an upward component of motion.

The behaviour of a glacier is closely related to its temperature. Heat within a glacier is derived from three sources: solar radiation, terrestrial radiation (geothermal heat) and heat derived from the friction created by internal movement and basal sliding. On the basis of temperatures generated by this heat energy, we can distinguish between temperate ice, which is at pressure melting point throughout, and cold ice which is always at temperatures below the pressure melting point.

In cold polar environments surface temperatures of glacier ice are very low, since they are in equilibrium with the local air temperature. Because of continuously low air temperatures, surface melting is negligible and the ice body remains cold and the bedrock beneath is commonly below 0°C. The temperature profile through a cold-based glacier is shown in Figure 14.4b. Since temperature decreases from the base towards the surface, terrestrial heat, flowing from warm to cold, is conducted through the glacier body and radiated to the atmosphere. Accordingly, terrestrial energy is transmitted through the glacier which remains frozen to its bed. In the case of a temperate glacier, temperature decreases downwards within the glacier due to increasing pressure of overlying ice (Fig. 14.4a). (As pressure increases, melting-point temperature decreases.) Thus there is a positive temperature gradient between the glacier's bed and its surface, and geothermal heat cannot be conducted upwards towards the atmosphere. The geothermal heat energy is therefore dissipated by the melting of basal ice. In regions of average levels of geothermal heat flow, the amount of basal melting thus caused is approximately 5 mm a^{-1}.

Therefore, the distinction between cold- and warm-based glaciers is one of thermal condition, and a particular glacier need not fall entirely into one category or the other. Conditions may vary from season to season with polar characteristics in winter and temperate ones in summer. Many glaciers are part temperate and part polar in type, conditions at any one point depending on local climate and the pressure of overlying ice.

As a broad generalisation, warm-based glaciers tend to occur in temperate environments or where

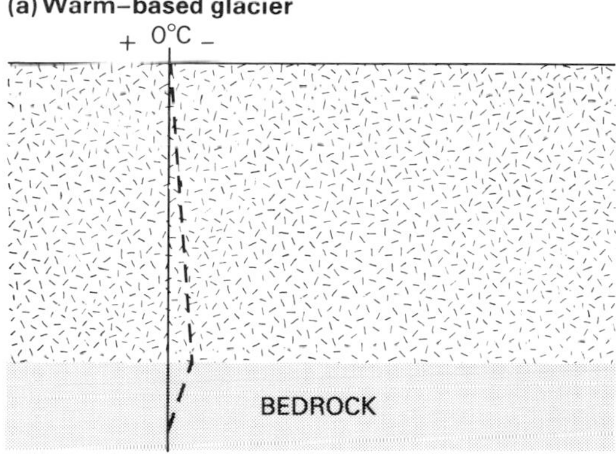

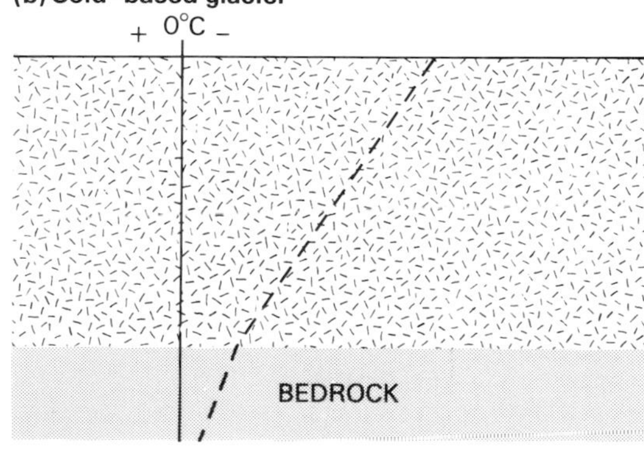

Figure 14.4 Temperature profiles through (a) a warm-based glacier, and (b) a cold-based glacier.

Box 14.1

GLACIER ICE

Glacier ice is a polycrystalline substance derived from the compaction of snow crystals. In contrast to water in the liquid form (Box 4.9), all possible hydrogen bonds are operative. The negative end of each molecule (the oxygen atom) is attracted to the positive end (a hydrogen atom) of an adjacent molecule. This constitutes a molecular bond – a form of bonding less strong than other types of bond. We may expect ice, therefore, to possess lower strength than other more strongly bonded materials, such as many rock-forming minerals.

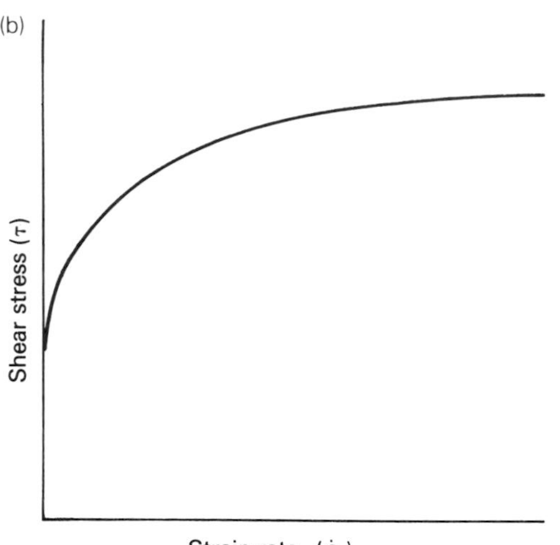

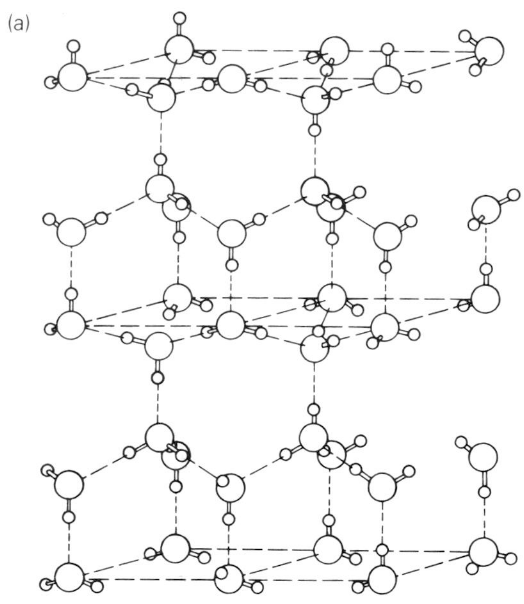

In ice each oxygen atom is enclosed tetrahedrally by four other oxygen atoms to form a regular crystal structure (Fig. a). The open nature of the ordered lattice structure requires a greater volume than the disordered liquid structure; hence water expands on freezing to form ice, which has a density of 0.9 g cm^{-3}. An ice crystal has a strong basal cleavage, determined by the layered lattice structure. As such it can be considered to resemble a pack of cards, with thin layers of card separated by cleavage planes.

The physical properties of ice can be determined by experiment under controlled conditions. The behaviour of ice under stress is illustrated in the accompanying diagram b. It deforms slowly under low applied stresses but more rapidly under higher stresses. Therefore, its behaviour differs from both a Newtonian fluid and a brittle solid. The behaviour of ice is generally described as viscoplastic. The relationship between applied stress and strain for single crystals of ice has been experimentally determined as:

$$\epsilon = k\tau^n$$

where ϵ = strain rate, τ = effective shear stress, k and n are constants.

This expression is known as Glen's Flow Law (1954). The value of k is directly related to temperature and the value of n is generally shown to be around 3.

Ice thus deforms more readily at higher temperatures and under higher levels of stress – relationships which have important consequences in terms of glacier motion.

Deformation of ice can take place at very low stress levels, but it is strictly limited. It deforms readily at a stress level of 1 bar (100 kNm^{-2}), yielding a strain rate of 0.1 a^{-1} at temperatures close to 0°C. This severely limits the magnitude of shear stresses that the glacier is capable of transmitting through to its bed, for stresses above this level tend to be accommodated by shearing within the ice itself. Calculated values of shear stress at the base of sliding glaciers generally yield values in the range of 0.5–1.5 bars.

the depth of ice is great. The cold-based condition is favoured by polar environments and thin ice. Most temperate glaciers are of the warm-based type, as also are some areas of particularly thick polar ice sheets. The geomorphological significance of glacier type lies in the effect this has on movement and erosion processes.

14.2 The transfer of mass: glacier flow

The movement of glaciers involves two main components – internal shearing and basal sliding. Internal deformation is caused by shear stresses built up within the ice by the weight of overlying ice (overburden pressure) as a result of which the ice becomes displaced either by creep, as individual crystals deform along their cleavage planes, or by fracture. Movement takes place in the direction of the ice surface gradient (the pressure gradient) and is described by Glen's Flow Law:

$$\epsilon = k\tau^n \qquad \text{(Box 14.1)}$$

Since

$$\tau = \rho g h \sin \alpha$$

where ρ = density of the ice, g = gravitational acceleration, h = glacier depth, α = slope of glacier surface, τ = shear stress and ϵ = strain rate; then

$$\epsilon = k\,(\rho g h \sin \alpha)^n.$$

Internal shearing, or **intragranular creep**, is favoured when temperatures are relatively warm, as in temperate glaciers, and in polar glaciers when overburden pressure is particularly high beneath a thick mass of ice.

Internal deformation can also take place along discrete shear planes, rather like thrust faults, within the glacier body. These planes are associated with an increase or decrease in velocity down stream, often related to a change in bed gradient. Thus, where the bed steepens the glacier is under tension and the ice cross-section thins as the glacier accelerates. Conversely, where the gradient slackens the ice is compressed and thickens as it rides up thrust planes which dip up glacier. These phenomena have been termed extending and compressive flow (Fig. 14.5).

Basal sliding (the sliding of the glacier over its bed) takes place when the shear stress at the sole of the glacier is greater than the frictional resistance across the ice/rock interface. Clearly, resistance is much greater in the case of a polar glacier, frozen to its bed, and it is commonly assumed that movement in polar glaciers is accounted for entirely by internal deformation close to the bed where the ice is least cold and shear stresses are highest. In the case of a temperate glacier with a film of melt water at the ice/rock interface, the frictional resistance is much lower and basal sliding may take place freely.

Therefore, the total surface velocity of a glacier comprises the two components of basal sliding and internal shearing. The distribution of velocity in vertical profile of a temperate glacier is shown in

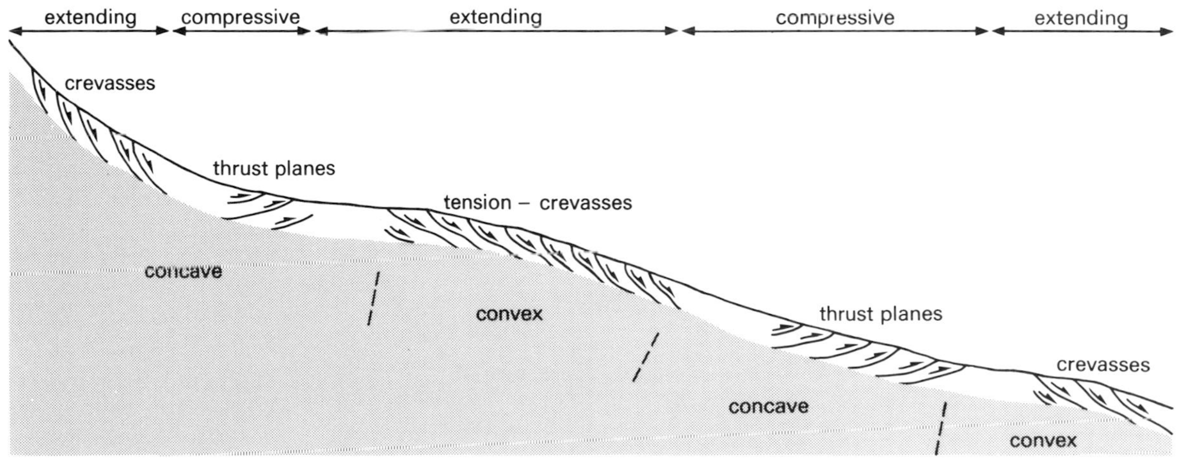

Figure 14.5 Extending and compressive flow down a glacier bed of varying gradient.

Figure 14.6b. In the case of cold glaciers, movement may be restricted to zone of shearing at depth within the ice, causing the ice above to be carried along. In the absence of basal sliding, surface velocity is ascribed to internal deformation alone and in large polar ice caps, where both temperatures and ice surface gradients may be very low, surface velocities are accordingly very slow.

The pattern of glacier movement at the surface can be readily observed by recording the movement of markers in the surface (Fig. 14.6a). The parabolic plan profile shows the effect of both side slip and internal shear. The latter increases toward the centre of the glacier where the frictional effect of the rough margin is least and there is a greater depth of ice, thus permitting the greatest velocity in the central part.

Variation in velocity may also be expected along the long profile of a glacier. Since the motive force at any one point is dependent on the mass of ice, assuming the surface gradient to be constant, velocity will tend to be greatest where the greatest thickness of ice is present. This will tend to be at the point of maximum discharge – the equilibrium zone. Locally, channel characteristics will influence velocity, with increasing bed gradient or decreasing cross-section causing an increase in velocity, just as in a river channel. In contrast to rivers, however, glacier velocity tends to decrease in the lower reaches as ablation reduces the glacier mass or the ice spreads out over a lowland, thus increasing its cross-sectional area.

14.3 Glacier morphology

Morphologically, ice masses can be classified according to their relationship with relief (Table 14.1). Those dominating relief are the ice sheets and domes. The distinction between the two is one of scale. Large masses of ice of continental scale which completely submerge the bedrock topography are commonly termed **ice sheets**, while

Table 14.1 Classification of ice masses based on their relationship with relief.

Unconstrained by relief	Constrained by relief
ice sheet	icefields
ice cap	valley glaciers
	cirque glaciers
	other small glaciers

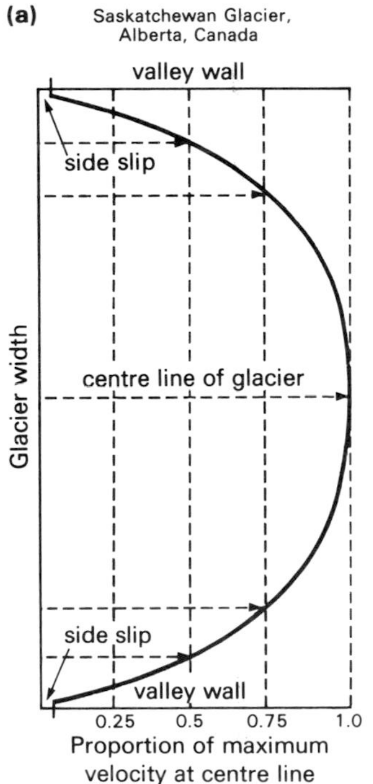

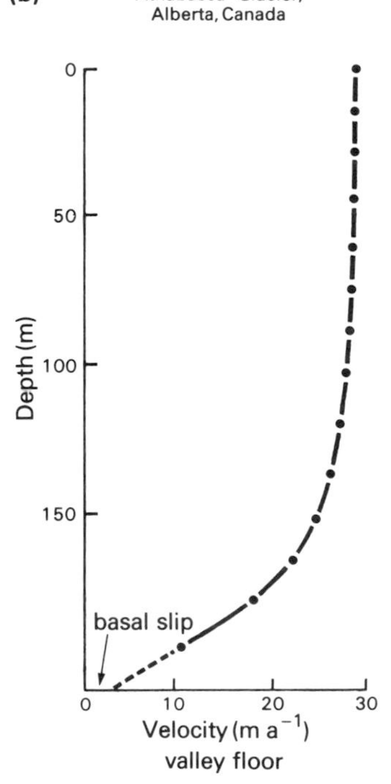

Figure 14.6 Glacier velocity profiles: (a) surface velocity across a valley glacier; (b) vertical velocity profile of a warm-based glacier.

masses of a more regional scale sitting on areas of high relief are **ice caps**. On a more local scale are the glacier types constrained by relief, whose outlines are largely determined by the topography of the land surface. Thirdly, there are the **ice shelves**, which float on water bodies, usually the sea. Since these do not interact directly with the land surface, we shall not consider them further. We shall discuss the land-based glaciers in ascending order of scale.

Cirque glaciers develop in hollows in regions of accentuated relief. Snow will tend to collect in sites sheltered from solar radiation and the prevailing wind. Accordingly, wind and avalanching will concentrate snow into sheltered hollows and the snow will lie longest in sites sheltered from solar radiation. Cirque glaciers are generally small, and wide in relation to their length, with a steep surface gradient. The relative proximity of the zone of

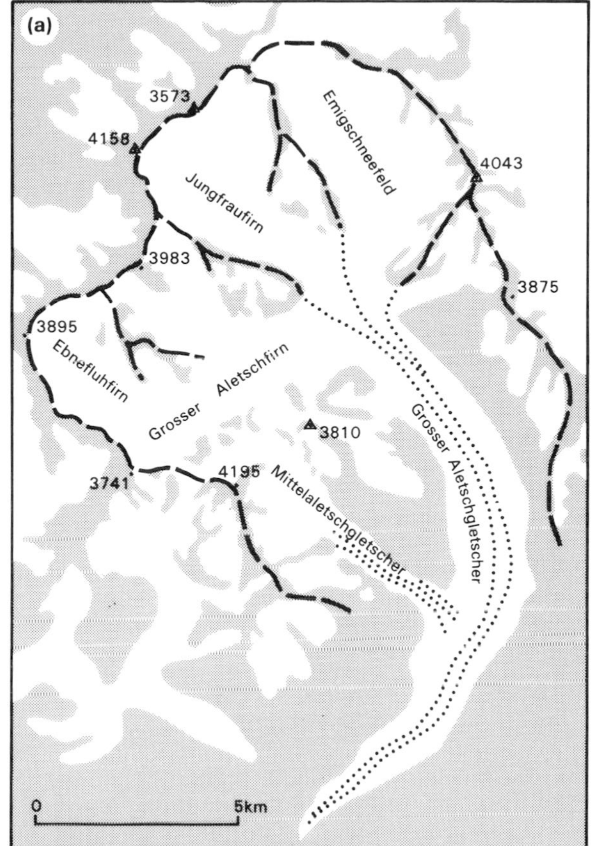

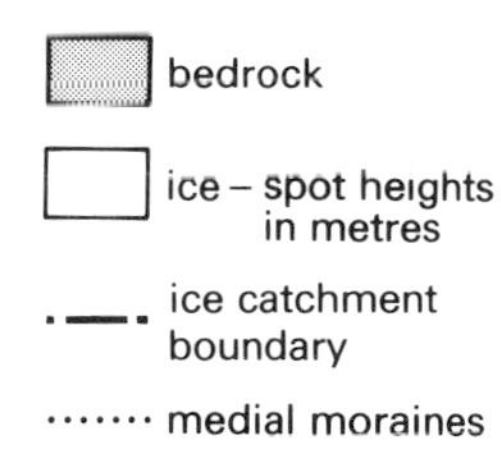

Figure 14.7 Trough glacier receiving tributaries from corries. Most of the foreground is covered by snow, which, down valley, has ablated away to reveal glacier ice beneath. The firn line, delimiting the zones of ablation and accumulation, lies across the lower middle of the frame. Downglacier, convergent tributary glaciers form medial moraines. (Aletschgletscher, Switzerland.)

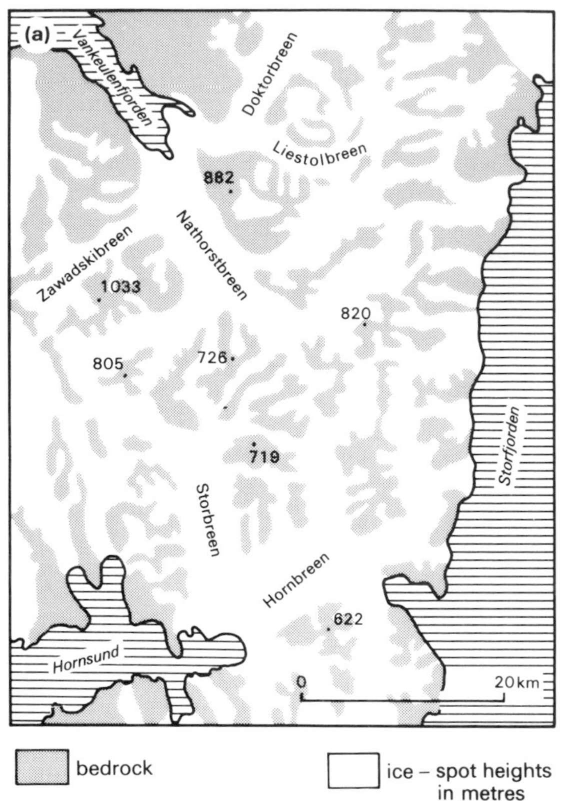

accumulation, with its downward motion, to the ablation zone, with its upward flow movement, gives the incipient cirque glacier its characteristic rotational movement (see Fig. 14.3).

Trough glaciers form where one or more cirque glaciers have extended to become confluent. The well developed trough glacier constrained within a valley may attain a length of many kilometres and receive as tributaries both cirque glaciers and other trough glaciers. In plan it may develop the same kind of network form as a river system. The Aletsch Glacier in Switzerland is just one example in the Alps of a well developed trough glacier system (Fig. 14.7).

A more intense form of glaciation is the coalescent condition. Here the ice partially submerges the underlying landscape by overtopping the topographic divides. This can take place in two

Figure 14.8 Diffluent glaciers. (a) Interconnecting ice streams formed by diffluent glacier flow dissect the landscape (Spitzbergen). (b) Most of the landscape is submerged beneath glacier ice and only isolated peaks of bedrock protrude above the glacial surface (Icefield Ranges, Alaska, USA).

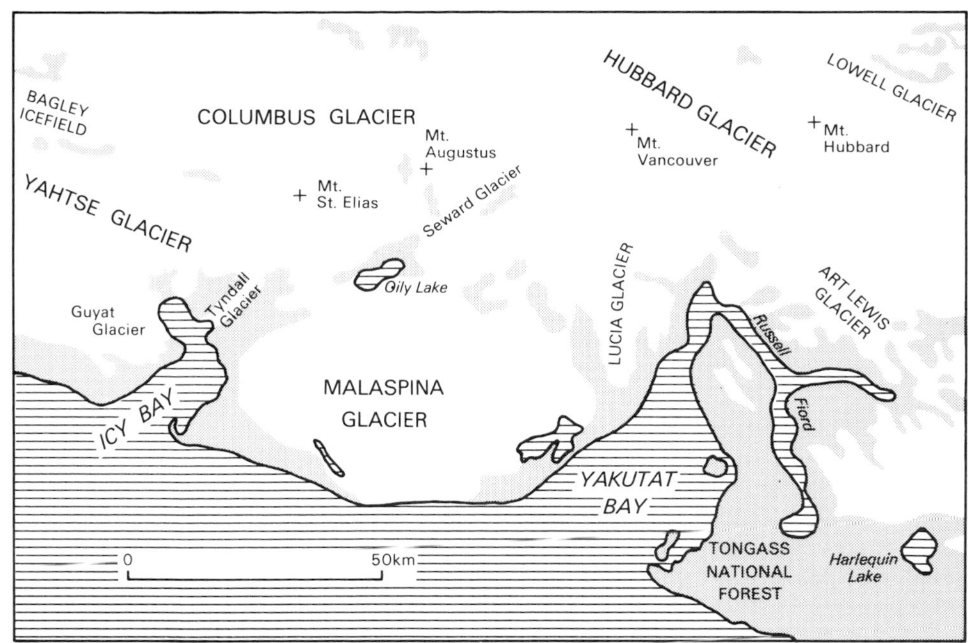

Figure 14.9 Piedmont glacier, spilling out from the upland and spreading out across the coastal plain (Malaspina Glacier, Alaska).

ways. If a trough glacier attains such a thickness that the bedrock topography is barely sufficient to contain it, then the ice will begin to spill over the divides at low points to coalesce with adjacent glaciers. This is known as **diffluent** flow (Fig. 14.8).

A second form of coalescence occurs when valley glaciers emerge from an upland on to a plain, where they spread laterally, fanning out until they coalesce. Here again the local relief becomes totally submerged beneath glacier ice (Fig. 14.9).

An ice mass of regional dimensions in an upland area which generated the precipitation to nourish it, is termed an ice cap. This consists of a central ice dome with marginal outlet glaciers which drain to the surrounding lower ground. Ice caps range up to a diameter of *ca* 250 km and completely submerge bedrock topography (except around their margins where isolated rock masses may protrude as **nunataks** and define the outlet troughs). The ice dome tends to be symmetrical in plan and convex in profile and it normally receives its greatest precipitation in the central summit area. Ice flow is radial outwards from the centre. Vatnajökull in southeastern Iceland is a representative example (Fig. 14.10a).

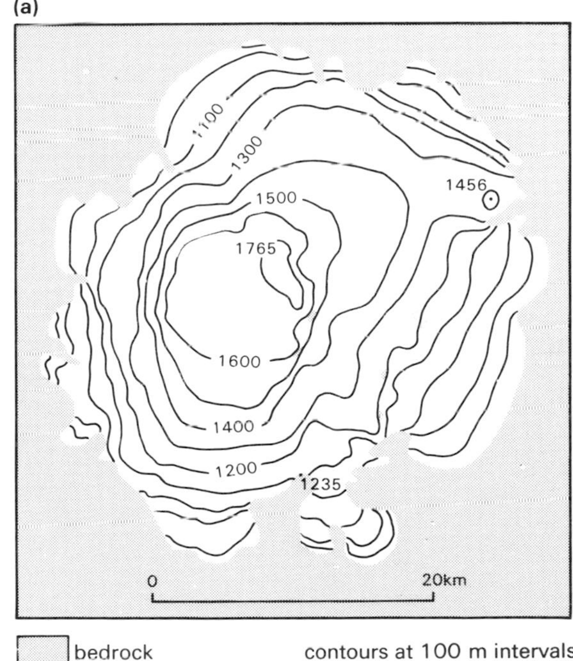

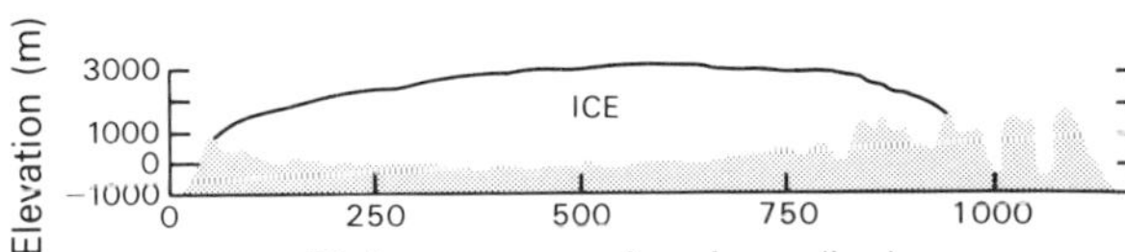

Figure 14.10 (a) Small ice cap, Hofsjökull, Iceland. (b) Off-central ice dome, Greenland. Note the discordance between the glacier surface and the bedrock topography beneath.

An even more extensive ice cover is termed an ice sheet, which may attain continental dimensions. Ice bodies of this magnitude become independent of the topographic surface beneath, such that the highest area of the symmetrical ice sheet may not be coincident with the highest subjacent relief. Figure 14.10b illustrates this using the Greenland ice sheet. Since flow is radial outwards from the central and thickest part of the ice sheet, ice streams pass right through the underlying major topographic divide dissecting it deeply – a feature known as **transfluent** flow.

The various stages of glacier development have been defined in purely descriptive terms in relation to the extent of ice cover of the landscape. It seems possible, however, that there may be a generic link between them in that all ice bodies may have small beginnings – the perennial snowpatch – and subsequently develop by thickening and extending to their ultimate form. In so doing they will tend to assume, in so far as bedrock topography permits, the various forms described above from cirque glacier, through the trough glacier stage, coalescing and perhaps developing as far as the ice cap.

14.4 Erosion processes

It is at the contact of the glacier and its bed that erosion takes place. Glacial erosion has long been considered as consisting of abrasion and plucking. The latter, however, is now recognised (Sugden & John 1976) to embrace a variety of component processes. Accordingly, we shall consider erosion under the headings of abrasion, rock fracturing and melt water, and deal with debris entrainment separately.

14.4.1 Abrasion

Abrasion takes place when the base (or sole) of the glacier, often containing debris of various sizes, slides over the bedrock surface. In so doing it acts like a giant sander, grinding and scratching the rock surface, producing fine debris (rock flour). Clearly, this process is confined to warm-based glaciers. Abrasion produces a variety of small-scale features such as polished, striated and fluted rock surfaces which bear a testimony to its operation.

The presence of coarse clasts in glacial deposits points to the fact that glacial erosion is also capable

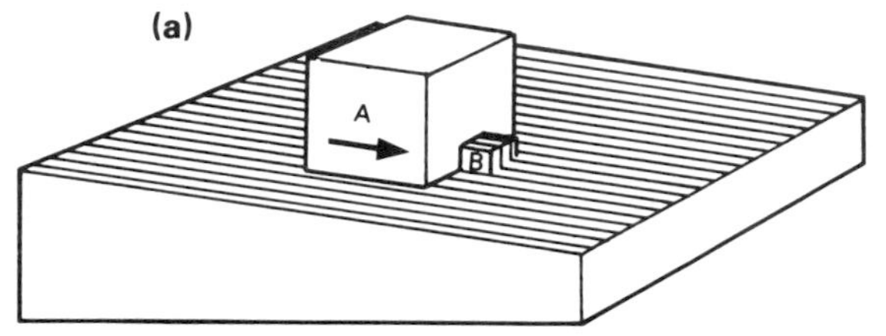

Figure 14.11 (a) Shearing of bedrock by basal debris: shearing stress transmitted through block A in the direction arrowed is sufficient to shear bedrock protuberance B (after Embleton & Thornes 1979). (b) The sole of a glacier: coarse clasts and fine sediment are contained within the ice; this debris acts as an abrasive on the bedrock surface beneath; as ablation proceeds the sediment is released and deposited (Tunsbergdalen, Norway).

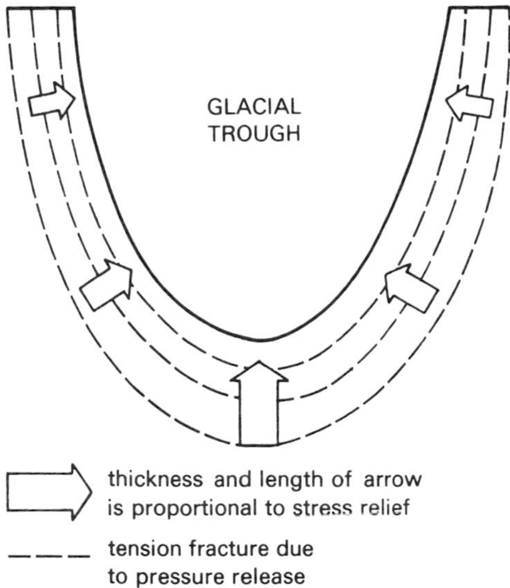

Figure 14.12 Dilatation joints formed as a result of glacier trough erosion (after Sugden & John 1976).

of removing coarse debris. The origin of this material has, however, provoked much controversy. It appears that a variety of mechanisms are available by which detachment of coarse fragments of rock may be effected and these will be considered together as rock fracturing.

14.4.2 Rock fracturing

First, weathering plays a role in providing a pre-existing regolith, prior to glacial action. This may take the form of a temperate weathering profile, which contains coarse fractured blocks in its lower part, or a periglacial profile which may be expected to contain mechanically fractured rock throughout. The amount of material removed by glacial erosion, however, often far exceeds the possible depth of such a regolith, and clearly the direct action of glacier ice is responsible for further breakdown of solid rock.

Rock fracture by shearing or crushing beneath a glacier is another probable mechanism. Since ice deforms readily at stresses greater than 1 bar (100 kN m^{-2}), the direct shear or compressive stress that ice can exert on bedrock is limited by the low shear strength of the ice. At stresses greater than this the ice itself will deform, while the bedrock remains intact. In other words, rock is stronger than ice. If, however, such a stress is transmitted through a layer of rock debris at the glacier bed, then much greater local stresses may be exerted on the bedrock floor. A stress applied across the surface of a large block may be sufficient for that block effectively to concentrate the energy to a level sufficient to exceed the shearing strength of solid rock and thereby shear off a smaller rock protuberance on the glacier bed. McCall (1960) has calculated that the downstream stress on a 1 m^3 block of granite will permit that block to shear off a block of bedrock with an upstream face of 160 cm^2 in area in contact with it (Fig. 14.11a). This process multiplied across the total bed area of the glacier, therefore, has considerable potential for erosion.

A third process of rock fracture likely to contribute to glacial erosion is the development of dilatation joints (Fig. 14.12). Waters (1954) noted the frequency of sheet joints and slabs of rock parallel to surfaces of glacial erosion in cirques and glacial troughs. It seems that once glacial erosion has removed a significant depth of overburden, which is replaced by a less dense body of ice, the underlying rock will respond to stress relief by expanding and fracturing. In this way, fractured rock becomes readily available for transport at the glacier bed. This process can also operate in subglacial cavities on the downstream side of upstanding bedrock masses. Where the sole of the glacier becomes detached from the bed (Fig. 14.13a), stress relief encourages the rock to expand and fracture (Fig. 14.13b).

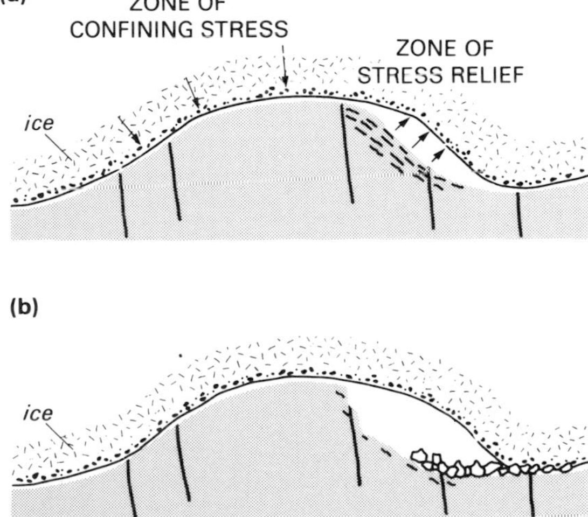

Figure 14.13 Stress relief causing failure in bedrock beneath a subglacial cavity (after Derbyshire *et al.* 1979).

Fourthly, there is the possible effect of freeze and thaw causing subglacial weathering of bedrock. Measurements of air temperatures in subglacial cavities have, however, suggested that freeze–thaw cycles are of very limited occurrence. If the basal ice is close to pressure melting point, it is possible that pressure variations may cause thawing and freezing. As ice flows over an upstanding mass of bedrock, basal pressure is increased, leading to pressure melting. The water produced flows around the protrusion and refreezes on the lee side where pressure is less. This process is known as **regelation** and, where the rock is already jointed by stress relief, it may be effective in detaching small rock fragments. Figure 14.14 compares debris incorporation by regelation in warm- and cold-based glaciers.

14.4.3 Meltwater erosion

Towards the snout of temperate glaciers, melt water may be present in considerable quantities subglacially. Flowing in channels, often under considerable hydrostatic pressure, water is capable of erosion by scouring and corrasion, particularly if loaded with mineral debris or ice fragments, leading to the formation of a variety of bedrock channels.

Once debris has been made available by abrasion and fracturing processes, it can then be entrained by and incorporated into the glacier. The downward pressure of ice on the particle may cause pressure melting of the ice in contact with the particle, which becomes absorbed into the glacier's sole. Particles of sand size and finer are more likely to become entrained by regelation. Pressure melting of the basal ice provides a film of water that will engulf fine particles. As pressure is decreased the water freezes back on to the glacier sole incorporating the debris particles. A temperate glacier flowing over an uneven bed, where melting and refreezing are continuous processes, may possess a layer of basal regelation ice containing fine debris many centimetres deep.

Finally, the effects of melt water in removing debris must be mentioned. Rock flour will be transported wherever melt water is present but, where discharge becomes concentrated into channels, high velocities may develop and coarse sand and even cobbles may be removed.

We have considered glacial erosion processes in a systematic manner on the small scale, looking at the stress–resistance relationships of the individual rock particles. It should be emphasised, however, that the processes described above may operate around the entire perimeter of the glacier/bedrock contact and along a considerable portion of the glacier's long profile. The individual processes, multiplied across the area of the glacier bed, can add up to considerable amounts of erosion over a period of time.

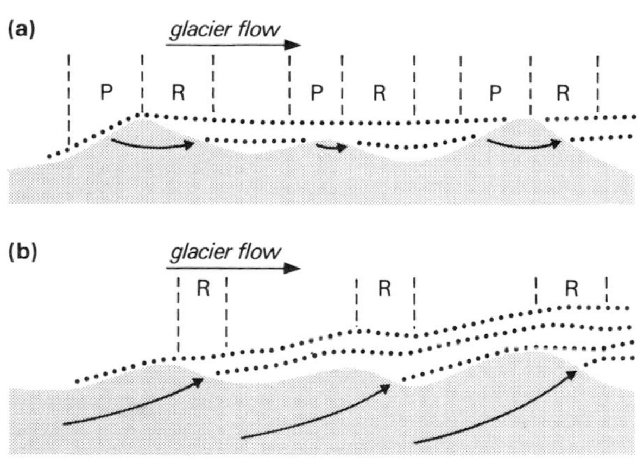

Figure 14.14 Incorporation of debris into the base of a glacier by regelation. (a) In a warm-based glacier the regelation layer remains thin, repeatedly melting and reforming as the ice moves across an uneven bed. (b) In a cold-based glacier the successive freezing on of regelation layers builds up a thick layer of regelation ice. (After Boulton 1972.)

14.5 Transfer of materials: transportation by glaciers

We can assess the ability of a glacier to perform erosional work in terms of power – that is, energy per unit of time (J s^{-1}). Power per unit area (W_T) can be defined as the product of bed shear stress (τ, which $=\rho gh \sin \alpha$) and mean velocity ($\bar{U}$) thus:

$$W_T = \tau\bar{U} \qquad \text{(Andrews 1972)}$$

where W_T is in J s^{-1} m^{-2}, τ is in Nm^{-2} = J m^{-1}, since 1 Nm = 1 J (see Ch. 2) and $\bar{U}$ is in m s^{-1}.

The major controlling variables of glacier power at a point are, therefore, mass, gradient and basal sliding velocity.

Figure 14.15 Glacial transport. (a & b) Debris-rich layers of ice-carrying sediment to the surface of the terminal zone of a glacier (Tunsbergdalen, Norway). (c) Cones of fine debris emerging from the glacier surface and a large isolated boulder. The latter stands on a plinth of ice which it has protected from ablation (Breidamerkurjökull, Iceland). (d) Englacial bands of fine sediment are exposed at the ice cliff forming the terminus of the glacier (Kviarjökull, Iceland).

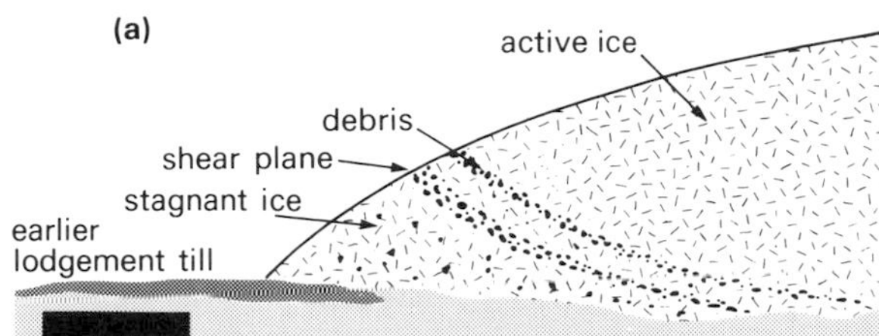

There is, therefore, a considerable distinction to be drawn betwen cold- and warm-based glaciers in terms of the amount of erosional work of which they are capable. With cold-based glaciers basal sliding is inhibited and erosional processes severely limited, restricted probably to the shearing of masses of unconsolidated rock frozen to the base of the glacier. Warm-based glaciers, on the other hand, with a film of melt water at their base, slide readily over the bedrock beneath. This permits them to apply considerable shearing forces to the bed, thereby encouraging erosion.

Once entrained, debris is carried along within the glacier towards the snout. The distribution of the load throughout the glacier's mass is important in influencing the mode of deposition at the end of its journey.

Debris (usually coarse rock fragments), fed directly onto the glacier surface by rockfall from bedrock slopes above, may simply be carried along on the glacier surface. Some debris is always present in the base and margins of the glacier, entrained in the regelation layer. A considerable portion of the total load may become embodied deep within the glacier. There are two principal ways in which this comes about. First, when two glaciers form a confluence, the lateral moraine present along their adjacent margins becomes absorbed in the confluent flow down stream as medial moraine. Clearly, where a trunk glacier is served by many tributaries, a considerable amount of moraine may be incorporated in this way. Secondly, shear planes developed under compressive flow, which often takes place towards the terminal zone of a glacier, may carry debris-rich basal ice upwards. In this way a succession of debris-rich bands of ice may outcrop at the surface of the glacier near its snout (Fig. 14.15).

At the snout, then, debris may be distributed throughout the glacier's cross-section. Concentrations may occur in the sole and at the surface, while vertical concentration of medial moraine and horizontal concentrations of thrust plane debris may occur within the body of the glacier.

14.6 Deposition processes

Deposition takes place at the glacier margin. It embraces a variety of processes which involve first the release of debris as the ice ablates and then distribution or reworking of that debris by falling, flowage, glacial overriding or running water. The mode of deposition will be strongly influenced by the location in which the debris is released from the ice.

14.6.1 Deposition by ice

Beneath an active sliding glacier, pressure melting of basal ice leads to the release of debris particles carried in the sole of the glacier. These particles become plastered onto the bedrock floor as lodgement till. Where the glacier is stagnant, subglacial melting takes place caused by geothermal heat, and mineral debris accumulates by accretion on the glacier bed. On the glacier surface, melt-out of debris takes place as a result of heating by solar radiation. In the case of a steep glacier front, debris so released may fall or roll directly from the glacier margin. Where the terminal slope is more gentle, the debris accumulates as a carpet on the glacier surface. It may then become saturated by water released by the melting of ice and flow down the glacier surface to the foreland areas as a flow till. All materials that accumulate at the glacier margin may be subject to reworking by glacial pushing or overriding as the margin fluctuates in position (Fig. 14.16).

It is clear, therefore, that glacial deposition involves a variety of complex processes, and debris may be reworked several times before it finally comes to rest. It follows, therefore, that the characteristics of tills may vary considerably. However, tills have a number of common sediment characteristics:

(a) they are generally unstratified;
(b) they are generally poorly sorted, and contain clasts of all sizes set in a matrix of fine material;
(c) they are composed of a variety of minerals and rock types, many fresh and unweathered;
(d) the clasts are variable in shape, sometimes sub-angular or, when further transported, with smoothed and rounded facets due to abrasion;
(e) they may have a preferred orientation of the elongated particles, in which a high proportion of particles lie with their long axes in a restricted azimuthal range.

14.6.2 Meltwater deposition

In contrast to the unstratified till deposits are the sediments laid down by melt water. Mineral sedi-

Figure 14.16 Glacier terminus. Contorted dirt bands are visible at the glacier surface. The snout zone is completely covered by debris released by ablation. Melt water released by ablation is stored in proglacial lakes drained by meltwater streams which carry fine sediment to the sea. (Breidamerkurjökull, Iceland.)

ments may be picked up by meltwater streams flowing in a **supraglacial** (on the glacier surface), **englacial** (within the glacier) or **subglacial** (beneath the glacier) course and deposited subglacially or, more widely, in a proglacial location. Meltwater deposition follows the same general principles as other fluvial deposition. Its distinctive characteristics are caused by (a) the widely fluctuating nature of meltwater discharge, which depends on daily variation in temperature and consequent ablation; and (b) the physical constraints on the meltwater channel, which may be constricted within a subglacial tunnel or may braid widely across the glacier foreland.

Meltwater deposits, then, are sorted and stratified. Yet they may vary considerably in calibre over a short distance, due to the variations in discharge and velocity. Where laid down originally in contact with the glacier, they may be affected by subsequent faulting and slumping structures as the supporting ice melts. Alternatively, they too, like till, may be overridden and reworked by the glacier itself.

14.6.3 Conclusion

In its fundamental effect (the transportation of water and erosion of mineral sediment from the land surface) the glacial system resembles the fluvial system. Yet the forms adopted by glacier ice flowing across the land surface, and its erosive power, lead to a variety of distinctive landforms of both erosional and depositional origin. Where glacial occupance of a landscape has been long continued, distinctive and large-scale landforms may be formed – the often spectacular landscapes of glacial erosion. The more subtle lowland landscapes of glacial deposition are probably even more extensive, yet equally distinctive.

For more detailed discussion of glacial modifications of landscapes the reader is referred to the sources listed below.

Further reading

For a modern and thorough treatment of glaciers and their effects, the reader is referred to

Sugden, D. E. and B. S. John 1976. *Glaciers and landscape: a geomorphological approach*. London: Edward Arnold.

A useful introductory handbook on glaciers, mainly descriptive, is

Sharp, R. P. 1960. *Glaciers*. Portland: Oregon State System of Higher education, Condon Lectures.

An interesting collection of key historical studies of glaciers is brought together in

Embleton, C. (ed.) 1972. *Glaciers and glacial erosion*. London: Macmillan.

15 Spatial variation in denudation systems

As we have seen in Chapters 5 and 10, the major energy inputs into denudation systems are precipitation and the potential energy of relief. These forces of denudation operate on the materials exposed at the surface of the lithosphere. Since precipitation, relief and surface materials vary widely across the Earth's land surface, one may expect consequent spatial variations in denudation systems. This spatial variation is expressed both in the functioning of denudation systems and also the land surface form which they produce. We have examined the various components of the denudation system in terms of their internal operation and have seen how they respond to inputs. This chapter, therefore, attempts a comparison of denudation systems operating under different environmental conditions with different inputs.

The output from a catchment basin can be established by a monitoring programme involving measurement of runoff, together with sediment and solute concentrations. From these data the total annual volumetric loss of mineral material can be calculated. Dividing the volumetric loss by the area of the catchment from which it is derived yields a mean erosion rate expressed in millimetres per year. Clearly, this is generalised measure of surface lowering and it does not represent uniform lowering throughout the source area, within which local variations in lithology, gradient and land use can suffer wide variations in erosion. It does, however, serve as a useful standard for comparison of spatial variations in the denudation system.

Routine measurements of denudation have been made for several decades in certain regions, notably the USA. In less developed areas of the world, such as the Tropics and high-latitude lands, data are rather sparse. Therefore, the total world picture of denudation rates remains somewhat speculative, and considerable differences of opinion exist at the present time. Nevertheless, our understanding is such that certain significant relationships have emerged which can be related to world and regional variations in denudation system output.

As indicated in Chapter 10, the major controlling variables in the operation of the catchment basin system are climate, relief, lithology and land use (or vegetation type and cover). We shall examine the effects of these variables first by comparing denudation systems at different spatial scales and then by attempting to isolate the effect of individual variables as demonstrated by specific case studies. Finally, having considered variations in denudation system output, we shall make observations on denudation system form.

15.1 Spatial variations in denudation output

15.1.1 The global scale

Estimates of the amount of mineral material currently being denuded from the continents are presented in Table 15.1. Data of this kind

Table 15.1 Estimates of denudation losses from the continents (adapted after Garrels & Mackenzie 1971, data from various sources).

	Solution loss ($t\ km^{-2}\ a^{-1}$)	Sediment loss ($t\ km^{-2}\ a^{-1}$)	Total loss ($t\ km^{-2}\ a^{-1}$)	Solute: sediment ratio	Surface lowering ($mm\ a^{-1}$)
N. America	33	86	119	0.4	0.044
S. America	28	56	84	0.5	0.031
Asia	32	310	342	0.1	0.127
Africa	24	17	41	1.4	0.015
Europe	42	27	69	1.8	0.025
Australia	2	27	29	0.1	0.011

necessarily present a very generalised picture, since a great variety of climatic, topographic and geological conditions may exist within the bounds of a single continent. The data only include losses in the form of solutes and suspended sediment and they omit bedload, which is very difficult to measure in major rivers. As such, they represent necessarily an underestimate of total denudation.

Significant differences between the continents are, however, apparent. Total denudation, expressed as mean surface lowering, differs by an order of magnitude between Australia, a mainly arid and low-relief continent, and Asia, which has a significant proportion of very high relief and a number of major rivers which traverse alluvial lowlands with a large available supply of sediment.

The relative proportions of solute and sediment losses vary even more widely. In Africa and Europe, both of which are of modest relief and comprise large areas of humid environments, solution losses exceed those of sediment. In other words, chemical denudation is more significant than mechanical denudation. In all the other continents mechanical denudation is dominant, accounting for as much as 90% of the denudation losses in Asia and Australia.

At this scale of generalisation, then, the major controls on denudation appear to be climate and relief. Indeed, Garrels and Mackenzie (1971) suggest that while solution losses do not vary widely between continents, sediment losses appear to be positively related by a power function to mean continental elevation.

Various attempts have been made to map denudation rates at the global scale, based on relationships between climate, relief and denudation in sample areas. Figure 15.1 illustrates J. Corbel's (1964) estimate of global denudation. This is based on studies of total denudation rates in different temperature zones, with varying humidity and relief. Denudation is shown to vary inversely with temperature, but directly with humidity in each temperature zone and with relief. This results in his calculated erosion rates being at their highest in areas of high relief and in mid-latitude humid environments.

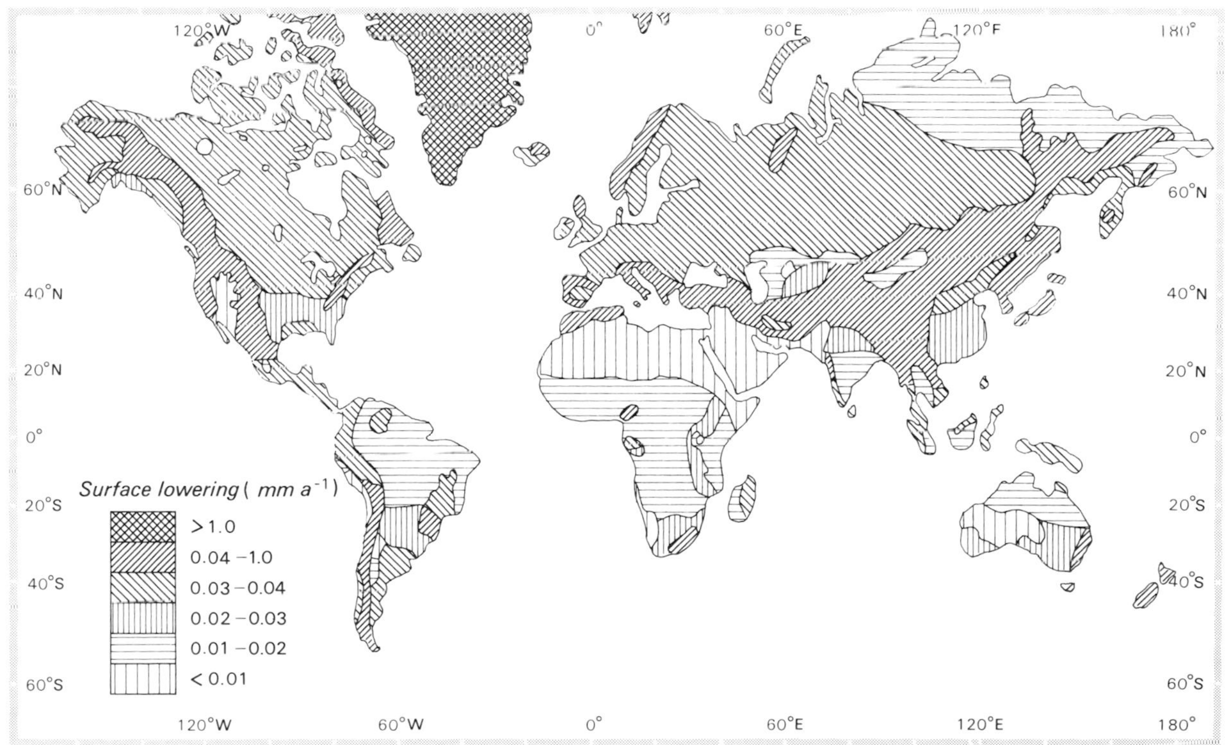

Figure 15.1 Current rates of world denudation. Based on the estimates of Corbel 1964, and adapted from *Fundamentals of geomorphology* by R. J. Rice, Longman 1977.

Table 15.2 Denudation losses for the year 1975 from adjacent catchments in a humid temperate region, North York Moors, England (after Arnett 1979).

Catchment basin	Area (km^2)	Suspended load (%)	Dissolved load (%)	Total yield ($t\ km^{-2}$)	Surface lowering ($mm\ a^{-1}$)
1	11.1	1.5	98.5	115.0	0.043
2	24.2	7.3	92.7	61.3	0.023
3	85.0	5.1	94.9	52.4	0.019
4	11.6	5.1	94.9	37.0	0.014
5	46.2	9.0	91.0	46.5	0.017
6	37.2	5.7	94.3	36.7	0.013
7	22.0	2.0	98.0	40.7	0.015
8	130.7	7.0	93.0	57.1	0.021
9	18.8	3.2	96.8	77.9	0.029
10	13.6	2.7	97.3	62.8	0.023
11	9.7	7.1	92.9	30.0	0.011
12	15.1	16.8	83.2	75.5	0.028
13	19.7	30.8	69.2	104.0	0.038
14	299.4	12.4	87.6	55.6	0.020
15	155.8	3.1	96.9	71.5	0.026

15.1.2 The small catchment basin scale

A study by R. R. Arnett (1979) of a series of contiguous basins in a humid temperate environment permits comparisons to be made at a more local scale. Fifteen adjacent catchment basins were monitored and their outputs assessed over a one-year period (Table 15.2). At this scale the climatic input can be regarded as being essentially uniform, and differences in denudation outputs may be considered largely to be a function of catchment basin form, lithology and land use. Several significant conclusions may be drawn from these data. There is, even within a small region, a fourfold variation in the rate of denudation. By far the greater losses are in the form of dissolved minerals, in line with the data for the humid continent of Europe as a whole. There is, furthermore, a twentyfold variation in the proportion of sediment loss between catchments.

The major variables controlling denudation losses at this scale were found to be land use, lithology and drainage density.

15.1.3 The local scale

At this scale, which embraces both actively eroding gully systems and experimental plots, controlling variables are different again and rates of denudation may be much higher. Erosion from localised plots can take place at rates two to three orders of magnitude higher than from catchment basins as a whole (Table 15.3). In all cases the process of denudation is either sheetwash over an unvegetated soil surface or active gully development. All show rates greater than 10 mm a^{-1}. This emphasises the protective role of vegetation as a control on surface sediment removal. It also points to the conclusion that sediment sources may be highly localised within catchment basins, and that these very high local denudation rates are offset by much lower rates over wide areas.

It is clear from this brief survey of denudation rates at different scales, that many of the functions

Table 15.3 Denudation losses from small sample areas.

	Range ($mm\ a^{-1}$)	Mean ($mm\ a^{-1}$)	Source
mid-Wales*	0–75	15	Slaymaker (1972)
Penninest†	—	14.8	Harvey (1974)
South Wales*	2–24	10.6	Bridges & Harding (1971)
North York Moors			
(vegetated) site	—	+1.35 (gain)‡	Imeson (1974)
(unvegetated)	—	38.10§	

* Sample plots. † Area of gully erosion. ‡ Mean of 7 values. § Mean of 3 values.

controlling denudation are themselves closely interrelated. Relief, for instance, is positively correlated with precipitation in causing orographic rainfall. Zones of exceptionally high relief are commonly young fold mountains, often comprised of relatively erodible sedimentary rocks of Cenozoic age, thus effecting a correlation between relief and lithology. At the global scale there is a correlation between precipitation and vegetation type. These correlations between the major controlling factors in denudation mean that it is often difficult to isolate the effect of individual factors.

15.2 Individual factors controlling denudation

The effects of individual factors on denudation losses, however, can be deduced by comparing similar regions or catchments which differ in their inputs in just one factor. We shall investigate first the effects of the dominant external factor of climate, and then examine the effects of more local factors such as lithology and relief.

15.2.1 Climate

Working on a sample of drainage basins from the USA, Langbein and Schumm (1958) have established a relationship between precipitation and denudation (Fig. 15.2). Referring to sediment

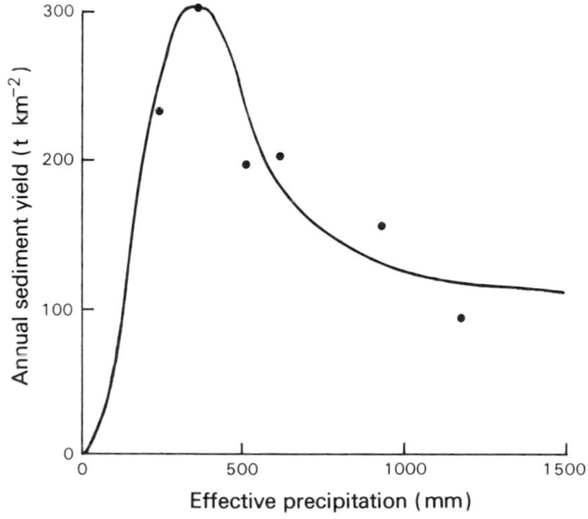

Figure 15.2 Sediment yield in relation to effective precipitation (after Langbein & Schumm 1958).

removal alone, and considering a region with mean annual temperature of 10°C, the maximum denudation takes place with an annual effective precipitation (i.e. runoff) of 300 mm. Other authors have shown a similar peak in sediment yield in semi-arid environments, although differing somewhat in absolute value. Thus, Douglas (1967) demonstrates a maximum denudation with 50 mm mean annual runoff in eastern Australia, while Dunne (1979) shows a peak denudation at *ca* 100 mm annual runoff for catchment basins in Kenya. In truly arid conditions erosion is less because of the lower erosional energy and less frequent runoff, whereas in moister environments vegetation cover increases and inhibits surface erosion of sediments. The critical combination of increasing erosion potential and increasing surface resistance causes denudation to peak under semi-arid conditions.

The influence of climate on denudation can be examined in another way, by comparing the denudation of a particular rock type in different environments. Limestone, a rock which is denuded almost exclusively by solution, is one such rock which has been studied in different environments. Table 15.4 is compiled from rates of solutional lowering which have been obtained from different climatic environments. The values of regional denudation are modest in relation to some already quoted. There is, however, an apparent increase in denudation rate from the tropical to the arctic–alpine environment, suggesting that limestone denudation is in general inversely proportional to mean annual temperature. The distinction, however, is far from clear cut for there is considerable overlap between climatic environments and these data mask the effects of other variables such as vegetation, precipitation and type of limestone.

Denudation rates under glacial conditions may be significantly higher than in fluvial conditions. In

Table 15.4 Estimates of erosion rates for limestone areas under different climatic conditions (simplified after Smith & Atkinson 1976).

	Mean (mm a^{-1})	Standard deviation	Number in sample
tropical	0.017	0.0125	18
temperate	0.021	0.0158	87
arctic–alpine	0.023	0.0141	24

1939, Thorarinsson estimated that the loss of sediment in the melt water from the Hoffelsjökull ice cap in Iceland represented a bedrock lowering rate of 2.8 mm a^{-1}, a rate some five times higher than on an adjacent non-glaciated catchment. This study, however, is based on waterborne sediment alone from a limited number of samples. A more detailed study, by Corbel in 1964 of the St Sorlin glacier in the French Alps, embraced material issuing from the glacier both in the meltwater stream and as morainic deposition. He recorded a bedrock lowering rate of 2.2 mm a^{-1}, again many times higher than adjacent unglaciated terrain. It is not possible to apportion precisely the roles of abrasion, joint block removal or melt water as erosional processes in this value, which represents the total effect of erosion under glaciated conditions. Nevertheless, these results prompt the conclusion that, under active temperate glaciers at least, denudation rates may be an order of magnitude higher than in fluvially eroded catchments in similar terrain.

15.2.2 Lithology

The effects of lithology on both runoff and denudation yield of sediment can be shown by comparing catchments developed on different lithologies within one region. Table 15.5 shows such a comparison from the Cheyenne River basin (S. Dakota, USA), derived from Hadley and Schumm (1961).

Five different rock formations are examined, with widely varying surface permeability as indicated by the mean infiltration rate. Infiltration rate is seen to be inversely related to drainage density. The higher density, reflecting higher surface runoff, is closely related to sediment yield. Clearly then, rock types that encourage greater surface runoff denude more rapidly.

15.2.3 Relief

The role of relief as a control on denudation rate is of paramount importance. High relief is associated with steeper slopes and therefore more active erosion, particularly the mechanical removal of material in sediment form. Figure 15.3 shows the relationship between sediment loss and relief (as expressed by the relief ratio) for small drainage basins in the western USA, and Figure 15.4 shows a similar relationship from a study in Papua New Guinea (Ruxton & McDougall 1967). Young (1972) has collated experimental data on surface lowering from a variety of environments (Fig. 15.5), which strongly suggest a significant distinction between areas of steep relief (mountain areas

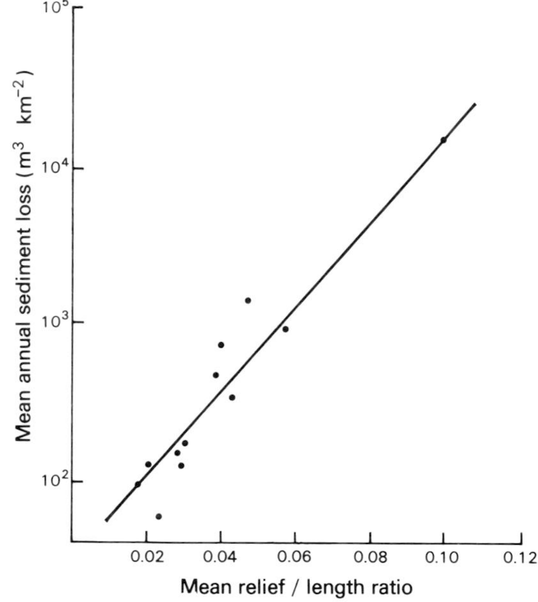

Figure 15.3 Sediment loss as a function of basin relief (after Hadley & Schumm 1961).

Table 15.5 Lithology and sediment yield (from Hadley & Schumm 1961).

Lithological unit	Rock type	Mean infiltration ($cm\ hr^{-1}$)	Sediment yield ($mm\ a^{-1}$)	Drainage density ($km\ km^{-2}$)
Wasatch Formation	incoherent sand	23.0	0.088	8.6
Lance Formation	sandy loam	12.5	0.337	11.4
Fort Union Formation	sandy clay loam	3.2	0.876	18.2
Pierre Shale	sandy clay loam	2.5	0.943	25.8
White River Group	silt and clay loams	0.4	1.213	413

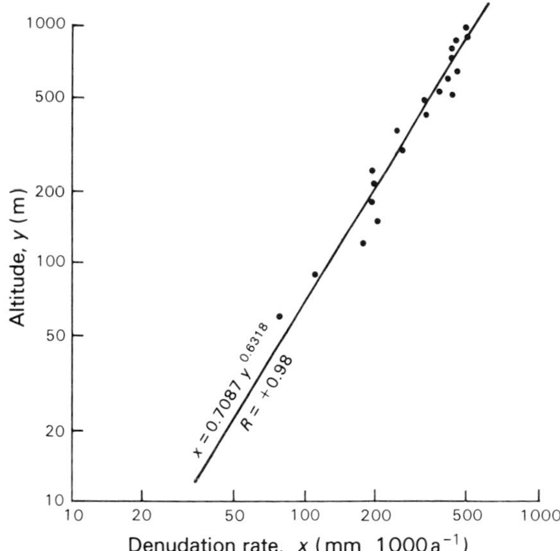

Figure 15.4 Denudation rate as a function of altitude, eastern Papua New Guinea (after Ruxton & McDougall 1967).

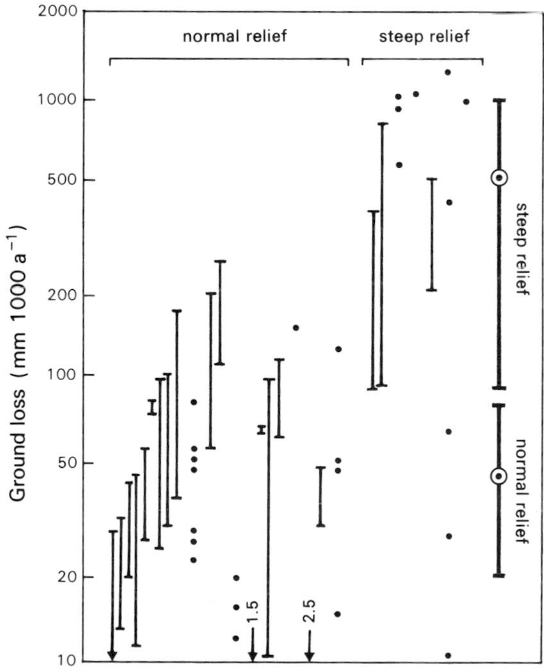

Figure 15.5 Rates of lowering of the ground surface according to type of relief (after Young 1969).

and individual steep slopes) and normal relief (plains, moderately dissected areas and gentle to moderate slopes). There appears to be an order of magnitude difference between the two, with normal relief showing a median denudation rate of 0.046 mm a^{-1} and steep relief 0.5 mm a^{-1}, although there is a degree of overlap in the range of values in the two categories.

These results have two important implications in spatial and temporal terms. First, there is a tendency for small drainage basins to exhibit higher gradients, since they usually form the headwater sections of larger catchments. Thus higher denudation rates tend to be associated with smaller catchments, rather than larger catchments which incorporate significant areas of lowland. Secondly, on the temporal scale, it is implied that a recently uplifted landmass will be denuded more rapidly and that the rate of denudation will slow down as the relief is lowered. The form of the relationship indicates that, in the absence of further crustal uplift, the rate of erosion will decline exponentially through time.

It is clear that significant empirical relationships exist between catchment parameters (climatic, morphological and lithological) and rates of denudation. Spatial variations in these parameters are reflected in denudation rates. However, the relationships between gross climatic parameters and rates of denudation are still not clearly understood. The relationship between mean annual rainfall, for example, and denudation rate is far from clear. In theory, for a given rainfall input, the amount off runoff available for erosion is related to mean temperatures, since these will govern evaporative losses. Some authors (e.g. Fournier 1960) have argued that the seasonal concentration of rainfall is more important than the annual total, since a dry season will permit drying and fragmentation of the soil, and the concentrated heavy storms of the wet season will bring more kinetic energy to bear on the land surface than more evenly distributed rainfall. The search for a realistic explanation of denudation patterns on a worldwide scale will depend on a more thorough understanding of such relationships, which can come about only as the result of a more representative spread of monitored catchment basins.

A final point to understand is that many of the denudation rates so far obtained cannot be considered to be the results of 'natural' processes. Many of the catchments monitored have been affected by Man's alteration of the ground surface, by stripping forests and imposing agricultural practices, thereby exposing bare soil surfaces and modifying the hydrological balance. A more detailed examination of the impact of Man on the denudation system follows in Chapter 24.

15.3 Spatial variation in denudation system form

Since Chapter 10, we have largely concentrated on processes in denudation systems. We shall now examine the effects of the operation of these processes in producing the form of the landscape. Landscape form can be regarded as a function of climatic and lithological inputs, and over a period of time it may reasonably be expected that a landscape will be produced which is adjusted to these major controls, operating through the individual and collective processes of denudation.

The status of landscape morphology depends very much on the scale at which it is considered. In the terms outlined above, at a broad spatial and temporal scale it is clearly a dependent entity. At smaller spatial and temporal scales, however, at which the operation of specific denudation processes are considered, the landscape form may become an independent variable. At this scale, for example, drainage density may become an important control on hydrograph form, and relief and gradient are major controls on slope processes. In this section, however, we shall concern ourselves with spatial variations in landscape patterns, dependent on the major climatic and geological inputs.

The form of a landscape can be assessed by applying morphometric techniques (see Ch. 10). In this way we can analyse and compare landscapes formed under different environmental conditions and observe the consequences of a variation in the major controls. Two examples follow, which demonstrate the effects of lithology and climate upon fluvially eroded terrains.

The influence of lithology alone can be demonstrated by comparing drainage basins developed upon different rocks within the same climatic region. Brunsden (1968) offers morphometric comparison between three sets of drainage basins on and around Dartmoor, England. Each set is developed on a different lithological base, the three types being granite, Devonian slates and grits, and Culm Measures which include significant quantities of shales. The drainage basins are standardised by considering fourth-order basins only, and values are tabulated for the characteristics of basin area, total stream length, mean bifurcation ratio and drainage density (Table 15.6). Three of these basins are illustrated in Figure 15.6. Although the size of the sample is small, there are clear differences between the basins developed on the various rock types in respect of two of the four descriptive parameters – basin area and drainage density. Total stream length and mean bifurcation ratio also show some differences between groups, although there is some overlap in these two parameters. The granite terrain is characterised by large drainage basins with a lower drainage density, while the impermeable shales of the Culm Measures produce small basins with a high drainage density. It can be concluded, therefore, that the morphometric characteristics of these basins are closely influenced by lithology.

The influence of climate on landscape form can be illustrated by comparing landscapes developed under different climatic inputs. The study by Gregory (1976) of variations in one morphometric characteristic (drainage density) in Britain in relation to precipitation is useful in this respect (Fig. 15.7). Drainage density is assessed for 15 sample areas and plotted against mean annual

Table 15.6 Morphometric properties of drainage basins on Dartmoor and adjacent areas (after Brunsden 1968).

	Basin area (km^2)	Total stream length (km)	Mean bifurcation ratio	Drainage density ($km\ km^{-2}$)
GRANITE				
Swincombe	18.8	38.4	3.9	2.0
Cherry Brook	13.2	29.3	3.5	2.2
Walla Brook	13.7	22.4	3.6	1.6
DEVONIAN				
Gatcombe	9.3	27.4	3.8	2.9
Badda Brook	8.0	24.3	4.2	3.0
Dittisham Creek	12.4	35.6	3.9	2.8
CULM				
Yeo River	5.4	20.3	3.6	3.7
Shippen Brook	4.1	20.0	3.7	4.8
Dorna Brook	7.5	24.3	3.4	3.2

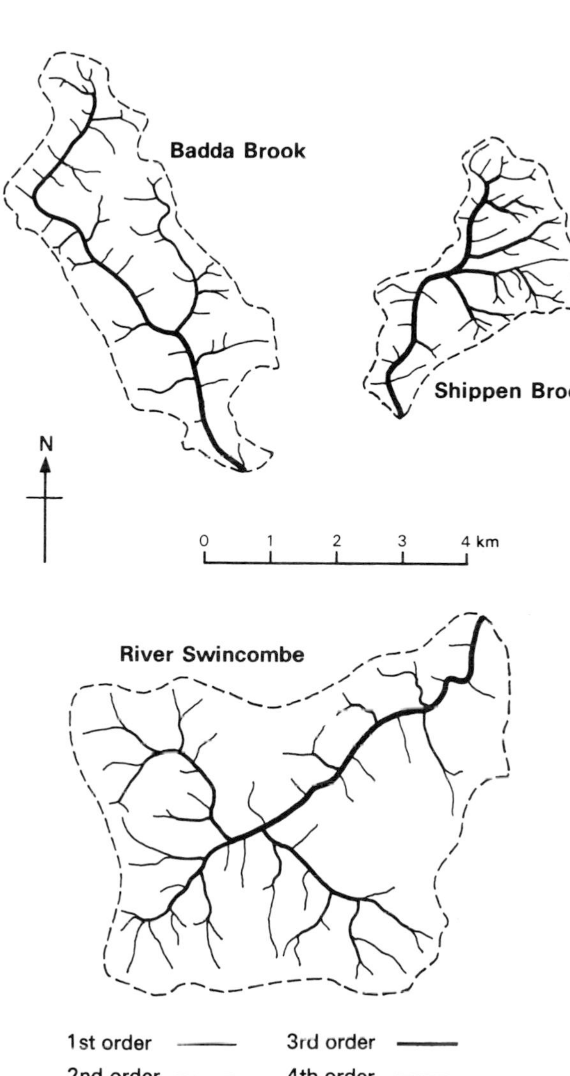

Figure 15.6 Drainage basin morphometry on different lithologies: (a) Badda Brook (Devonian), (b) Shippen Brook (Culm), (c) River Swincombe (granite) (after Brunsden 1968).

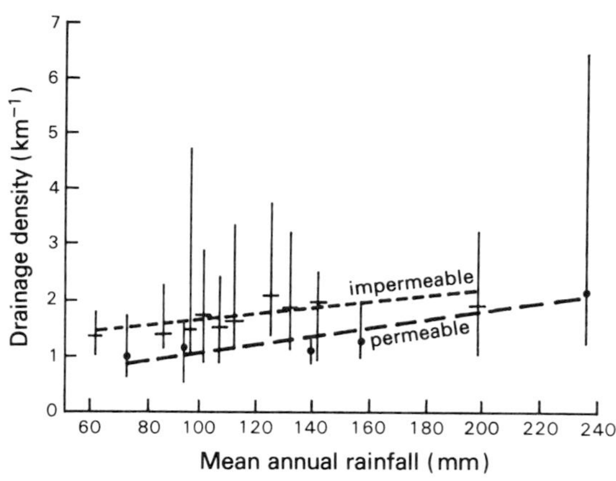

Figure 15.7 Drainage density in relation to mean annual precipitation in Britain (after Gregory 1976).

rainfall. Although there is considerable range demonstrated in some of the sample areas, there is nevertheless a clear tendency for drainage density to increase in direct proportion to rainfall. Furthermore, there is a distinction between permeable and impermeable rocks, with higher drainage densities occurring on the latter.

These two illustrations suffice to show that the characteristics of landscape form can be related to the major environmental inputs. There is clearly scope for further study in this field, particularly in respect of the relationships between landscape and climate. Indeed, on a worldwide scale Gregory (1976) demonstrates relationships between drainage density and annual precipitation, and also a more refined measure of rainfall input – that of rainfall intensity. Such studies, and closely controlled morphometric studies of landscapes developed on similar rock types but under different climatic regimes, point the way to a fuller understanding of the relationships between landscape, lithology and climate. In particular, there is a paucity of morphometric information on the morphology of glacially eroded terrain, although individual landforms, such as cirque basins, have been studied in some detail.

VI ECOLOGICAL SYSTEMS

16 The ecosystem

16.1 The ecosystem concept

The deciduous woodland depicted in Figure 16.1 is an oak/beech wood in southern England. It is, of course, a segment of the global ecosphere considered in Chapter 7, although it is doubtful whether this is the first reaction it evokes. This is simply because here we are encountering the ecosphere at a more familiar and accessible scale. We have taken a step down from the abstraction of a global model to something real and comprehensible. For a moment, let us imagine that we can walk into this woodland.

Initially we are aware of a reduction in light, though our eyes soon compensate and look down as dead wood and dry leaf litter crack under our feet. A disturbed fallen branch reveals woodlice beneath it and on its underside the lacework of fungal threads stands out white against the rotting wood. Here and there, beech seedlings rise through the mantle of fallen leaves, while, near the boles of the trees, green cushions of moss contrast with the browns of the litter. Elsewhere, in shade, wood ruff, dog's mercury, wood anemone and Solomans seal compete for space.

From the massive roots of beech and oak reaching down like flared fingers into the earth, the eyes follow the towering trunks as they reach up in a colonnade to support the vaulting tracery of the branches. Here the leaves lie in a precise mosaic against the Sun, a mosaic more complete above the beech with few gaps and little overlap, but more open above the oak. Epiphytic mosses and lichens cover these trunks and branches, while a dead bough is resplendent with the colourful, if grotesque, fruiting bodies of fungi.

Across a glade, a deer stops browsing and, raising an antlered head to read the air, silently merges into the sunflecked birch scrub at the margin of the clearing. Startled by the resonant drumming of a woodpecker somewhere close at hand, we become aware, in the relative silence that follows, of the songs and calls of other birds, high in the canopy. The subtle sounds of myriads of insects impinge on our consciousness, and looking closely we see further evidence of their existence. Leaves are trimmed by wintermoth, dunbar, or green tortrix larvae while some have neat discs removed by leaf-cutting bees. Oak *apples* formed by gall wasps hang amongst the foliage, abandoned by their former tenants.

As we emerge from this woodland, let us take stock and attempt to impose some order on these images. First, the woodland consists partly of a large number of different organisms, both plants and animals, each being represented by a population of individuals. Each population has at any moment in time a particular spatial distribution so that the total three-dimensional space is partitioned between the organisms present. Although these organisms form a complex woodland community, our image of it is of more than a collection of organisms growing and living together. The still, sun-shafted air, cool within the trunkspace, the soil which tethers the tall trunks, the smooth stream with trees on either side, the still pool and clean gravel and the rotting remains of once freshly fallen leaves are all just as important to our image of the woodland as the living plants and animals. In other words, it is the community and its immediate environment which together make up our perception of the woodland, for both are inextricably linked, not only in our mind's eye but also functionally. In Chapter 7 this unity was recognised at the global scale, where the concept of the ecosphere was used to encompass the living systems of the biosphere and those parts of the atmosphere, lithosphere and hydrosphere with which they exchange matter and energy. At the scale of the plant and animal community, the term *ecosystem* (ecological system) is used to embody the same concept and what we have been describing is, therefore, a woodland ecosystem (Box 16.1).

16.2 The structural organisation of the ecosystem

16.2.1 The elements of the ecosystem and their attributes

From this initial discussion of a woodland ecosystem, it is clear that the structure of the ecosystem has two components: the living component (the organisms themselves) and the non-living (environmental) component. The structural organisation of the environment is dealt with in other chapters of this book. In this chapter, therefore, we shall be concerned largely with the way in which the living organisms of the system are disposed. In the context of an ecosystem model these organisms are the elements of the system. How, then, do we describe or categorise these elements; what are their attributes? As living systems, all organisms in an ecosystem can be regarded initially as having two sets of attributes. First, they possess a set of genetic attributes enshrined in the DNA molecules in the nuclei of their cells (see Ch. 6). This is their **genotype**. It controls their development and activity and it is manifest as the second set of attributes: morphological, physiological and behavioural attributes – their **phenotype**. There are three ways in which we can choose to use these attributes to categorise the organisms of an ecosystem: taxonomically, structurally and functionally. As we shall see, the choice is largely conditioned by the type of relationship between the elements (organisms) of the system (ecosystem) that we wish to stress.

Both genotype and phenotype can be used to define the types of organism present in an ecosystem by identifying them as **species**, **subspecies** or **varieties**, and placing them in a natural hierarchical taxonomic classification (Table 16.1). The

Table 16.1 Taxonomic classification. An example of the Linnaean classification scheme applied to an animal – the red deer. Carolus Linnaeus was a Swedish physician (1707–78) and naturalist. He was Professor of Medicine at the University of Uppsala, but in 1758 he published the 10th edition of his *Systema naturae* which, for the first time, embodied a uniform system of classification for plants and animals alike and from which all other classifications have stemmed.

kingdom	Animalia	animals
subkingdom	Metazoa	multicelled animals
phylum	Chordata	animals with notochords
subphylum	Vertebrata	animals with backbones
class	Mammalia	mammals
subclass	Theria	non-egg laying mammals
infraclass	Entheria	placental mammals
order	Artiodactyla	even-toed hoofed mammals
family	Cervidae	deer – 40 species
genus	*Cervus*	
species	*Cervus elaphus*	red deer

Box 16.1

THE ECOSYSTEM CONCEPT

The concept of the ecosystem as an ecological unit composed of living and non-living components interacting to produce a stable system is not new. It has a long history in the biological and ecological literature, sometimes explicitly stated under a different terminology (e.g. biocoenosis), sometimes merely implied. In 1935 Sir Arthur Tansley, an important figure in the development of ecology in the British Isles, proposed that the term should be used to describe 'not only the organism complex but also the whole complex of physical factors forming what we call the environment' (Tansley 1935). However, the main theoretical development of the concept, and the implementation of research associated with that development, occurred in the period since 1940, the major impetus taking place in the 1950s. This followed Raymond L. Lindeman's formulation of the trophic–dynamic view of the ecosystem, which came to provide a conceptual framework and stimulation for research in ecology (Lindeman 1942).

Figure 16.1 Images of a temperate deciduous forest ecosystem.

underlying assumption is that such a classification reflects the evolutionary relationships between the organisms of the biosphere and it should, therefore, be based on genetic criteria. Nevertheless, inconsistency and ambiguity can become apparent (Box 16.2), particularly when phenotypic characters are used as a basis for classification. Although the 'species list' is the common starting point for the categorisation of the organisms present in an ecosystem, it is not the only approach, nor necessarily the most useful.

The phenotypic attributes of organisms, particularly their morphology and behaviour, can be used to order the elements of the system. For example, plants are often grouped into classes based on life form or growth habit, an approach intuitively adopted and recorded in the vernacular by the use of such terms as trees, shrubs, bushes, herbs and mosses. It is also an approach which lends itself readily to the analysis of the structural relationships of plants (particularly in space) because the life form of a plant is the vegetative form of the plant body. Plants that show the same general vegetative features belong to the same life-form class irrespective of their genetic and taxonomic relationships. The life form, however, is also considered to be an hereditary adjustment to environment and it can therefore reflect functional relationships (Fig. 16.2).

The use of the attributes of organisms to define structural entities is more usual and more developed in plant ecology. In animal ecology the alternative to taxonomy most often adopted is the use of functional entities, particularly those based on the trophic relationships of animals (herbivore, carnivore, omnivore, detritivore and so on). Some functional categories have also been employed for plants, though with less precision. Apart from the trophic counterparts, such as producer (or autotroph) and saprophyte, examples of functional entities from the plant kingdom usually concern adaptation to environment – as in the case of hydrophyte, xerophyte, halophyte, epiphyte and succulent – and they tend to overlap with life-form categories (Fig. 16.2).

16.2.2 Structural relationships in the ecosystem: structure in space

Once we know what organisms are present in an ecosystem and we have defined them in relation to their attributes, we need to be able to describe the

Box 16.2

THE SPECIES CONCEPT

The concept of the species as a unit in the classification of plants and animals was introduced in Chapter 7. Its use is far from simple and the term is used to describe some quite different entities. Ideally, a species is defined as a group of organisms that interbreed to produce fertile offspring. Members of different species do not normally interbreed and, if they do, the progeny are sterile. Not all species have been tested in the context of such a definition. Furthermore, it is inapplicable to species that habitually self-fertilise or reproduce only asexually. Species are perhaps best designated in accordance with the criteria used to define them:

Taxonomic species meets all criteria and satisfies the rules of international nomenclature.

Biospecies fulfils the breeding requirements of the definition and is therefore restricted to sexual reproduction and cross-fertilisation.

Ecospecies is a group of ecotypes (variants or races of a species adapted to different environments which can still cross even though some of their adaptations may be hereditable) within a species which can cross with one another to produce fertile offspring.

Coenospecies are species which belong to a group which can intercross to form hybrids which are sometimes fertile.

Morphospecies are named only on morphological evidence – often these are modified in the light of new evidence.

Agamospecies are species with asexual reproduction only, therefore treated as morphospecies.

Palaeospecies are extinct organisms, therefore the only evidence is derived from fossils, again subject to modification.

Phanerophytes (trees and shrubs)

perenating buds or shoot apices borne on aerial shoots

evergreen, with or without bud scales

deciduous

nanophanerophytes	<2 m high
micro	2 – 8 m
meso	8 – 30 m
mega	>30 m

Myrica gale nanophanerophyte

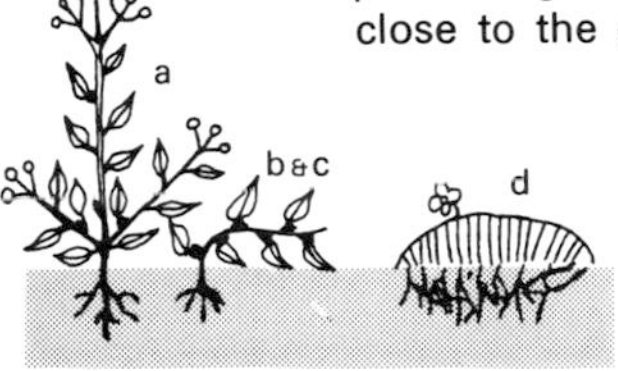

Chamaephytes

perenating buds or shoot apices close to the ground:

(a) on lower portion of erect stems which die back during unfavourable season

(b) passive prostrate stems

(c) persistently prostrate stems

(d) cushion plants – reduced and compact vegetative growth

Silene aucaulis cushion chamaephyte

Hemicryptophytes

perenating bud at ground level

some possess stolons or runners

some are rosette or partial rosette plants

also includes grasses and sedges

Saxifraga sp. rosette hemicryptophyte

Cryptophytes

perenating bud below ground or below water

geophytes with bulbs, tubers or rhizomes

marsh plants (helophytes) with perenating bud below water, but shoot above

hydrophytes – water plants with perenating bud below water leaves submerged or floating

Menyanthes trifoliata helophyte

Therophytes

annual plants – life history complete in the favourable season, 'seed is perenating bud'

Succulents and Epiphytes

usually function as phanerophytes because perenating buds are often borne aloft
frequently used as special categories in life form analysis

Figure 16.2 Life-form classification systems. Note that *Myrica gale*, growing here as a nanophanerophyte, can attain heights greater than 2 m under more optimum conditions.

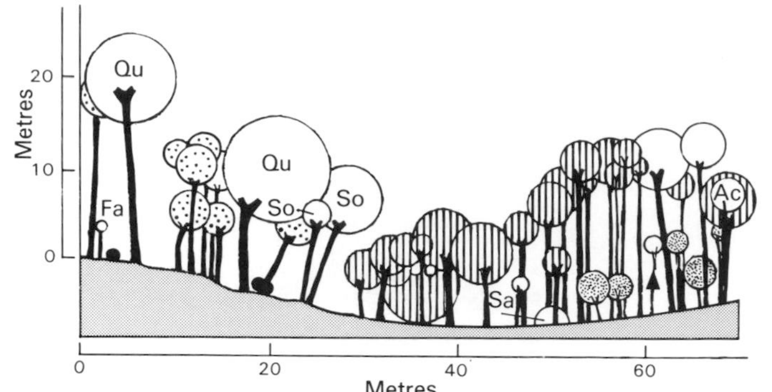

Deciduous woodland on calcareous soil
(canopy representation schematised)

Sa	*Salix* sp.	willow
Fa	*Fagus sylvatica*	beech
Qu	*Quercus* sp.	oak
So	*Sorbus alba*	whitebeam
Ac	*Acer campestre*	field maple
☐	*Fraxinus excelsior*	ash
(stippled)	*Corylus avellana*	hazel
(vertical lines)	*Alnus glutinosa*	alder
(dotted)	*Carpinus betulus*	hornbeam
■	*Ilex aquifolium*	holly
▲	*Euonymus europaeus*	spindle

Tropical rainforest with two canopy strata plus emergent crowns

Two strata in tropical montane forest

Tropical tree savanna

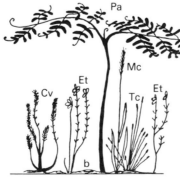

Au	*Arctostaphylos uva-ursi*	bearberry
Vm	*Vaccinium myrtillus*	bilberry
a	pleurocarpus (feather) mosses	
b	acrocarpus (mat) mosses	
Cv	*Calluna vulgaris*	heather
Cl	*Cladonia*	lichen
Ec	*Erica cinerea*	bell heather
Et	*Erica tetralix*	cross-leaved bell heather
Mc	*Molinia caerulea*	purple moor grass
Tc	*Trichophorum cespitosum*	deer sedge
Df	*Deschampsia flexuosa*	wavy hair grass
Pa	*Pteridium aquilinum*	bracken

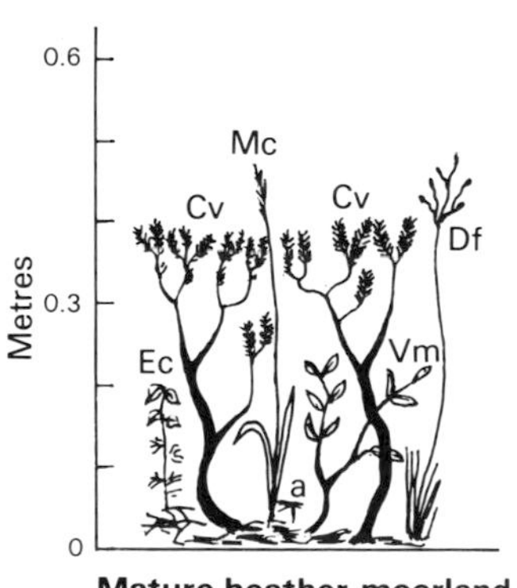

Mature heather moorland

Pa
Mc
Et
Cv
Tc
Et
b

Damp heathland

Montane dwarf shrub heath

Figure 16.3 Vegetation stratification.

relationships between these elements and between their attributes. Structurally, the most obvious organisational relationships are spatial. Here the plants and the vegetation they form are paramount. In most (and certainly in terrestrial) ecosystems, it is the sedentary plant kingdom that forms the spatial framework of ecosystem structure. Plant biomass represents not only the standing crop of the primary producers but also living space, forming (as in our deciduous woodland, Fig. 16.1) the habitat of most consuming organisms. The way in which the vegetation partitions the three-dimensional space it occupies is traditionally approached by considering separately the vertical structure (stratification) and the horizontal structure (areal distribution) of the organisms concerned.

The deciduous woodland described above, with its herb layer and tree canopy, and here and there a discontinuous shrub layer, has a vertical structure consisting of two or three strata. In more complex forest types, such as those of the humid tropical lowlands, the stratification can become much more complex. Nevertheless, as Robert Louis Stevenson's tin soldier could attest lying 'in the forests of the grass', this same concept can be applied to vegetation of smaller stature; to grassland, dwarf shrub heath and aquatic communities (Fig. 16.3). However, stratification not only describes structural relationships but has functional implications as well, because what we are considering is the stratification of the photosynthesising tissue of the primary producers. It can be seen, therefore, as a solution optimising the utilisation of available light energy. Notice that the term 'optimising' is used, not 'maximising', because a price must be paid, at least in terrestrial ecosystems, for the creation of such an elaborate energy filter as, for example, a forest canopy. This price is exacted in the form of the energy diverted to the maintenance of non-photosynthesising support tissue, represented in our woodland ecosystem by the massive biomass of wood. Also associated with a plant's place in the stratification is a host of essentially functional adaptive features related to the microclimatic gradients that the stratification itself creates (see Ch. 8). Indeed, gradients of environmental parameters are not restricted to the lower layers of the atmosphere, and the below-ground biomass of the ecosystem also shows a distinct stratification. The stratification of organisms inhabiting litter and humus and of the roots of higher plants all reflect gradients in physicochemical conditions and in the values of water and nutrient storage parameters in the soil (see Chs 11 and 20).

The aim of the description of the horizontal structure of vegetation is the location of each individual of all species in a two-dimensional framework. Such exact mapping is attempted only for small areas or along short transects in structurally simple communities and usually to monitor short-term change. The more usual approach is to describe statistically the horizontal **pattern** of species populations or of the vegetation, particularly in terms of its departure from a theoretical random distribution that would occur in the absence of intrinsic or extrinsic causal factors (Fig. 16.4). The statistical detection of pattern, therefore, implies that there is an underlying cause. Some pattern occurs in response to variations in environmental conditions, such as soil properties or surface microrelief; some is sociological in the sense that it is caused by interactions, often but not always competitive, between different species, or even between individuals of the same species. Yet other instances of pattern reflect the growth morphology of the species – especially in vegetatively reproducing plants – or its seed dispersal capacity. The investigation of pattern is further complicated by the fact that individual species, or the vegetation as a whole, may display pattern at several different spatial scales and at each scale the causes may be different. At a small scale we may be dealing with morphological pattern, at an intermediate scale with some sociological interaction, and at a larger scale with an imposed environmental pattern.

A third element in the description of spatial structure remains and that is the degree of presence of each species, or a measure of its actual contribution to the community (Fig. 16.5). The most objective and quantitative measure of this contribution is the number of individuals of each species per unit area – the density. However, there are practical difficulties associated with density counts, and subjective estimates of abundance are often employed as an alternative. To accommodate differences in size and growth form between different plant species, measurements or estimates of a species cover can be made. Cover is an area equivalent to the vertical projection of the above-ground parts of the plant onto the ground surface and is expressed as a percentage of the total area; it is related to the share that any species has of the total incoming light energy. Because the vertical

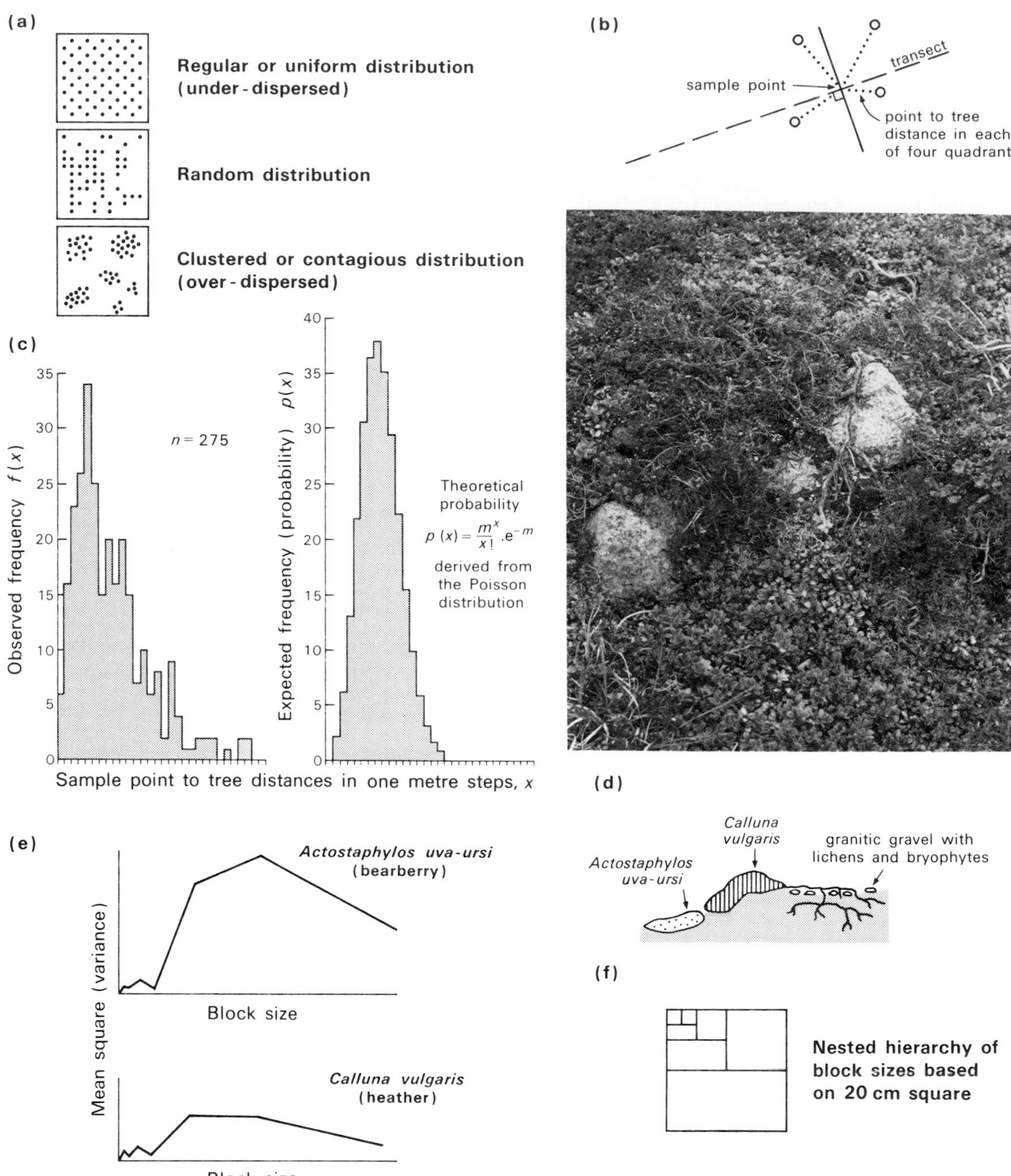

Figure 16.4 The detection and analysis of pattern. (a) Types of spatial pattern. (b, c) The use of the point-centred quarter method in a pine wood to test the pattern of tree distribution against that expected if the trees were distributed at random. Note that the sample frequency distribution in (c) is overdispersed relative to the expected (Poisson) distribution, indicating that the trees show some degree of clustering. (d, e & f) The application of pattern analysis to a *Calluna*/*Arctostaphylos* montane dwarf shrub heath to detect the occurrence of pattern at different spatial scales (block sizes).

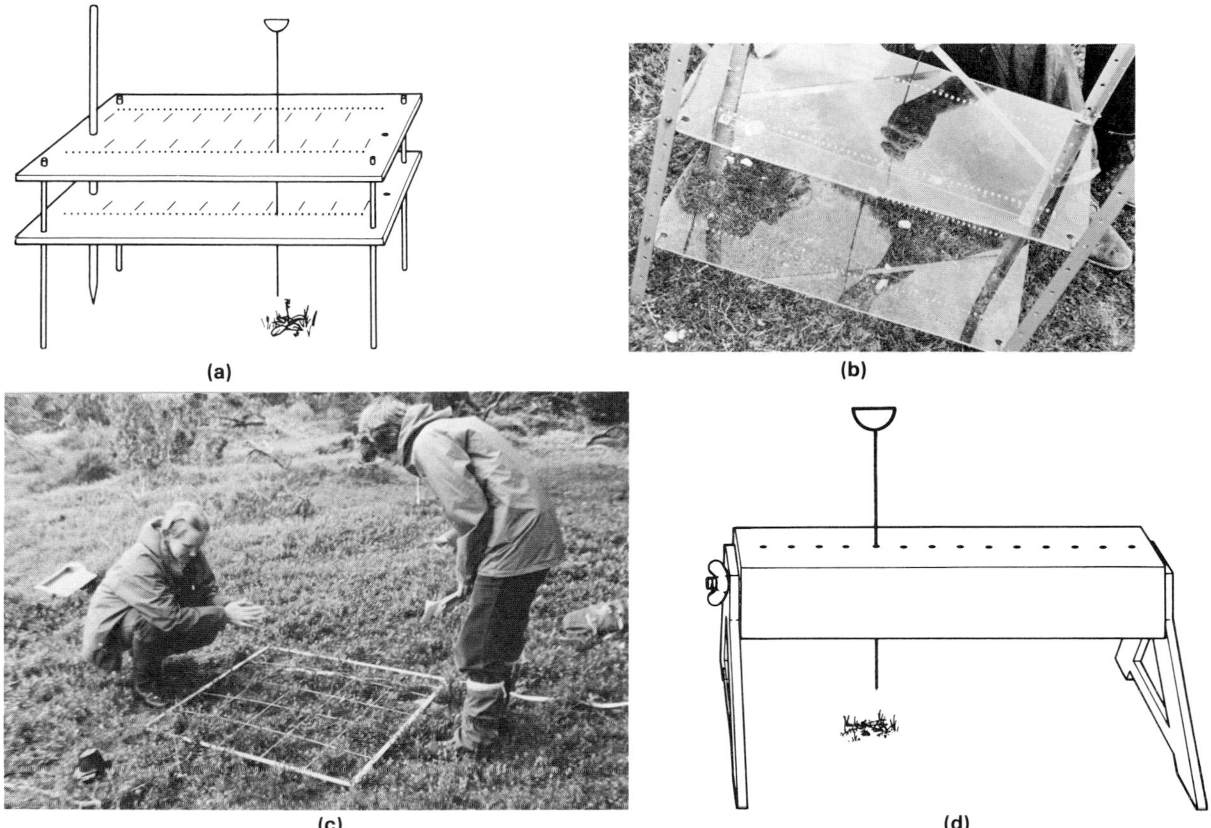

Figure 16.5 The measurement of cover. (a, b) Kershaw's pinframe for transects. (c) A 1 m² quadrat for cover estimation. (d) A simple pinframe.

separation implicit in stratified communities allows overlapping areas, the total cover of all species will normally exceed 100%. Quantitative measurements of cover are difficult to obtain in all but simple communities, but estimates of cover can be combined with those of abundance to give synthetic scales (Table 16.2). Closely related to cover is the use of above-ground biomass measurements to determine the contribution of each species on the assumption that those with the largest biomass have acquired the greatest share of space, energy and other resources.

Animal communities also exhibit a spatial structure, vertically and horizontally, but it tends to reflect that of the vegetation itself, or the disposition of gradients of habitat conditions which the vegetation creates. The bird populations in our deciduous woodland, for example, occupy distinct strata in the woods. This vertical segregation partly reflects their feeding habits and food availability, but also the occurrence of suitable cover, nesting sites and roosting perches. The wren, for example,

Table 16.2 Cover/abundance scales. When the Domin scale is examined with respect to either cover or frequency data alone, it is found to be non-linear. Domin values have therefore been transformed using the equation $y = 0.0428\,x$ to provide a scale with the required linear characteristics (Bannister 1966). The Braun–Blanquet scale has the same deficiency and Moore (1962) has provided a suitable transform.

	Braun–Blanquet	Domin	Transformed Domin (Bannister 1966)
cover about 100%	5	10	8.4
cover about 75%		9	7.4
cover 50–75%	4	8	5.9
cover 33–50%		7	4.6
cover 25–33%	3	6	3.9
abundant; cover about 20%		5	3.0
abundant; cover about 5%	2	4	2.6
scattered; cover small		3	0.9
very scattered; cover small	1	2	0.4
scarce; cover small		1	0.2
isolated; cover small	+	+	0.04

is restricted in nesting sites to thick understorey scrub, while the great and blue tits and perhaps the starling may occupy holes bored and previously occupied by woodpeckers. Thc strength of the branches of the trees of the main canopy supports the large nests of carrion crow and beneath these the uneven extended canopy of oak provides nesting sites for wood pigeon, doves and some thrushes. MacArthur (1958) describes similar stratification in different species of insectivorous warblers in evergreen forest in New England which feed exclusively within a narrow vertical zone (Fig. 16.6). It is in tropical forests, however, that the stratification of animal communities is best exemplified where, for example in Guyana, 61% of the forest mammals are arboreal and each species is restricted to a particular level in the canopy. Aquatic ecosystems also show stratification.

However, the spatial organisation of animals is generally much more fluid than that of plants. Their location and pattern of distribution may change throughout the day or the year, depending on the particular activity in which they are engaged (Fig. 16.7a). Hunting and browsing, nesting and mating, or simply resting, are all activities which may produce different patterns of distribution in the same animal species. Even explicitly spatial phenomena, such as the territorial behaviour of many animals and birds (Fig. 16.7b) may only be established for part of the breeding cycle and may change radically from year to year in response to complex interactions involving, for example, food availability or quality, population numbers, hormonal control of behavioural traits such as aggression, and variations in habitat diversity (see for example the red grouse: Jenkins, Watson & Miller 1967, and Moss 1969).

It would be wrong, however, to think of the vegetation as being static. Some plants may be annuals, some perennial, but all have a life cycle from germination and establishment to death. Each exhibits temporal variation in growth, flowering, fruiting and other functional activities (its **phenological characteristics**) and, as it does so, its place changes in the spatial structure of the community (Fig. 16.8). Also, as each generation succeeds its predecessors, so the spatial pattern of each species population changes with time. So too do the absolute numbers of each species, whether plant or animal, for populations increase and decline through time, perhaps with random fluctuations, perhaps with an identifiable periodicity, perhaps with a recognisable long-term trend. We shall consider all of these aspects of the ecosystem's behaviour with time in more detail later (Ch. 23), but in the present context they can be seen to have an expression in terms of changes in

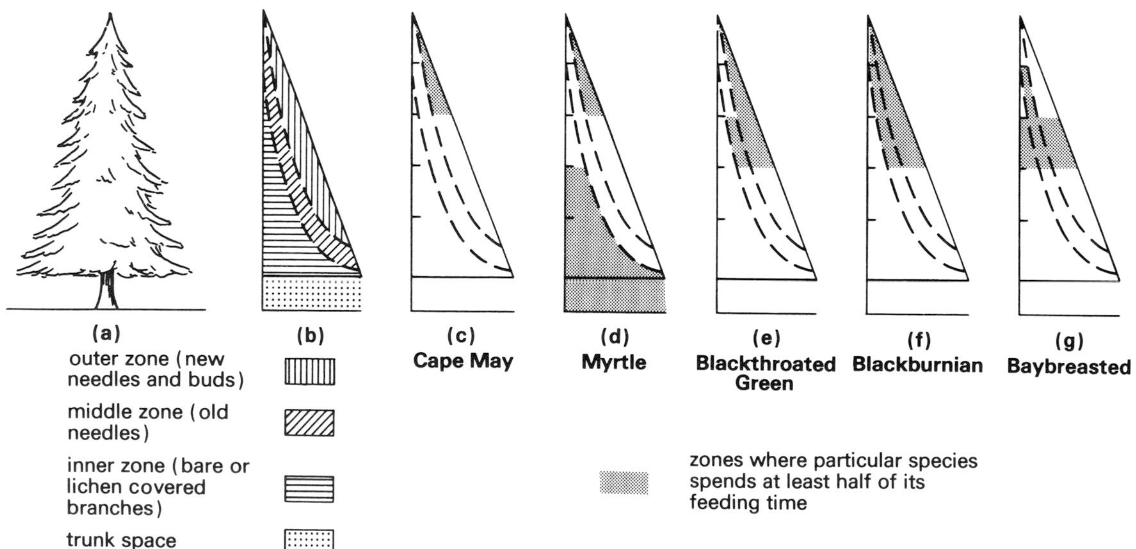

Figure 16.6 Niche separation among species of warbler (after MacArthur 1958). (a & b) Canopy structure; (c–g) distribution.

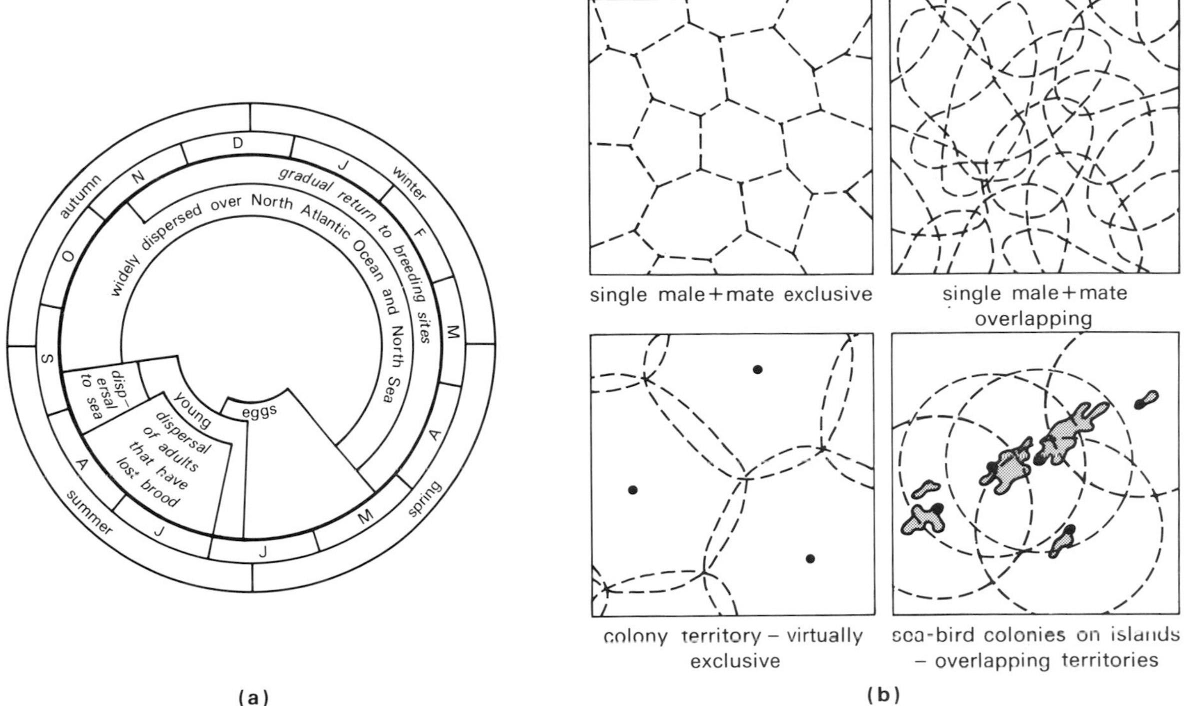

Figure 16.7 (a) The life cycle of the fulmar (after Fisher 1954). (b) Territorial behaviour in animals and birds (after Wynne-Edwards).

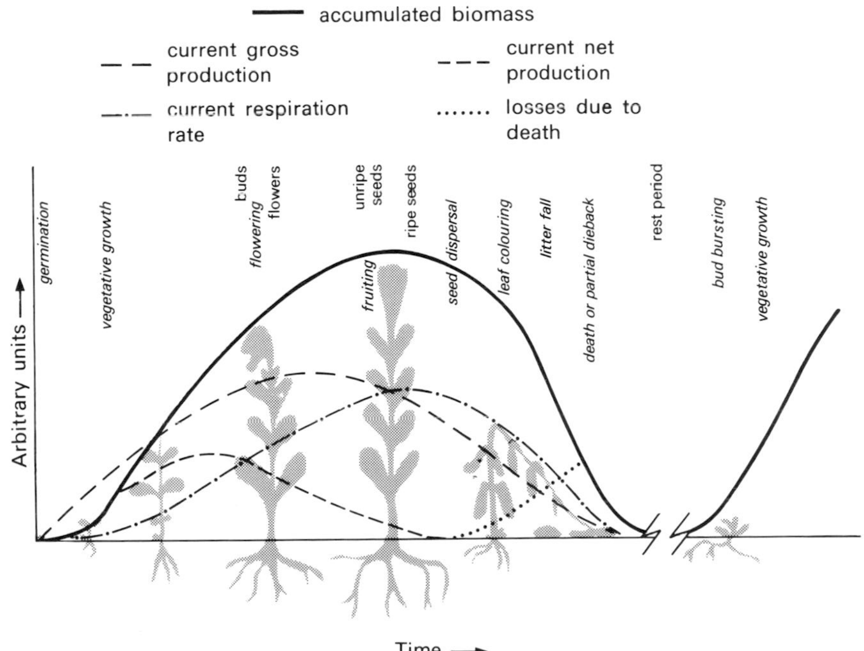

Figure 16.8 Life cycle and plant phenology (after Lieth 1970).

spatial organisation. The continuity of an ecosystem's structure, like the solid permanence of the woodland in Figure 16.1, is as much a figment of our imagination as is the myth of the everlasting hills. Not only are we looking at a structural mosaic in space but also in time.

16.2.3 Structural organisation of the ecosystem: functional relationships

The organisation of the ecosystem is fundamentally functional. As well as spatial relationships between the organisms of an ecosystem there are also functional relationships, for they not only partition the available space but also the energy and resources that sustain life. These functional relationships facilitate the transfer of energy and matter, both between the living components of the system and between them and their non-living environment. The key to this functional structure is the trophic organisation of producer, consumer and decomposer levels, introduced in Chapter 7. However, the trophic structure of real ecosystems is often extremely complex, as in the case of the deciduous forest considered at the beginning of this chapter (Fig. 16.9a). Nevertheless, these patterns of functional organisation provide the framework that makes it possible both to understand the role and significance of each organism in the ecosystem's function and to begin to interpret and monitor the pathways of energy and matter transfer. The trophic model is fundamental to the analysis of function and is applicable in one form or another to all ecosystems. Of equal generality is the fact that the distribution of biomass between trophic levels conforms to the model first introduced in terms of numbers by Charles Elton in 1927, whereby there is a decline in standing biomass from the producer level (t_1) to the top carnivore level (t_n) as the energy available to support it diminishes (Fig. 16.9b). By combining this description of the ecosystem in terms of the distribution of biomass (pyramid of biomass) with the functional relationships of trophic organisation, we arrive at the **trophic–dynamic models** inspired by Lindeman (1942) and developed by Howard and Eugene Odum (e.g. H. T. Odum 1957, 1960; E. P. Odum 1964, 1971) and others (Fig. 16.9c).

The place of each organism in the food economy or trophic organisation of the ecosystem is referred to as its **trophic niche**, a term also introduced by Elton (1927). The concept of the **ecological niche** as the functional role occupied by an organism has been extended and is now used in a broader sense to encompass not just an organism's trophic relationships but also its position in the community in terms of all kinds of relationships with other organisms. Looked at in another way, the ecological niche of an organism is a description of the extent to which it is functionally specialised and the number of distinct niches in an ecosystem is a measure of its functional complexity. However, it remains an elusive concept (and perhaps rightly so) in terms of strict definition and is best captured by Hutchinson's notion of niche space (1965). This he describes as an abstract multi-dimensional space defined by axes that represent a large number of biological and environmental variables affecting species interactions. Each organism can be conceived, therefore, as occupying part of this space to which it is adapted. The extent to which these portions of the total niche space are mutually exclusive is a measure of the degree of **niche segregation** displayed by the community.

The concepts of ecological niche and species diversity are closely linked, for as the number of species in the community increases, so the niche space is divided between more and more organisms. Consequently, if the ecological niche of each species is not to overlap with that of another, it must become smaller and more narrowly defined, leading to greater niche segregation. These are not simple relationships however, for increases in species diversity can themselves generate new and unexploited niche space. Such complications will be considered more fully in Chapter 23.

16.3 Functional activity of the ecosystem: the transfer of energy and matter

The fundamental functional activities of living systems have already been identified, whether at the level of the cell, the organism or indeed the plant and animal community, as the assimilation and utilisation of energy (food) and respiration (i.e. eating and breathing), and all living systems must perform these activities to live, grow and reproduce. This one-way flow of energy powers, and is coupled to, the closed circulation of the elements essential to life within the ecosystem

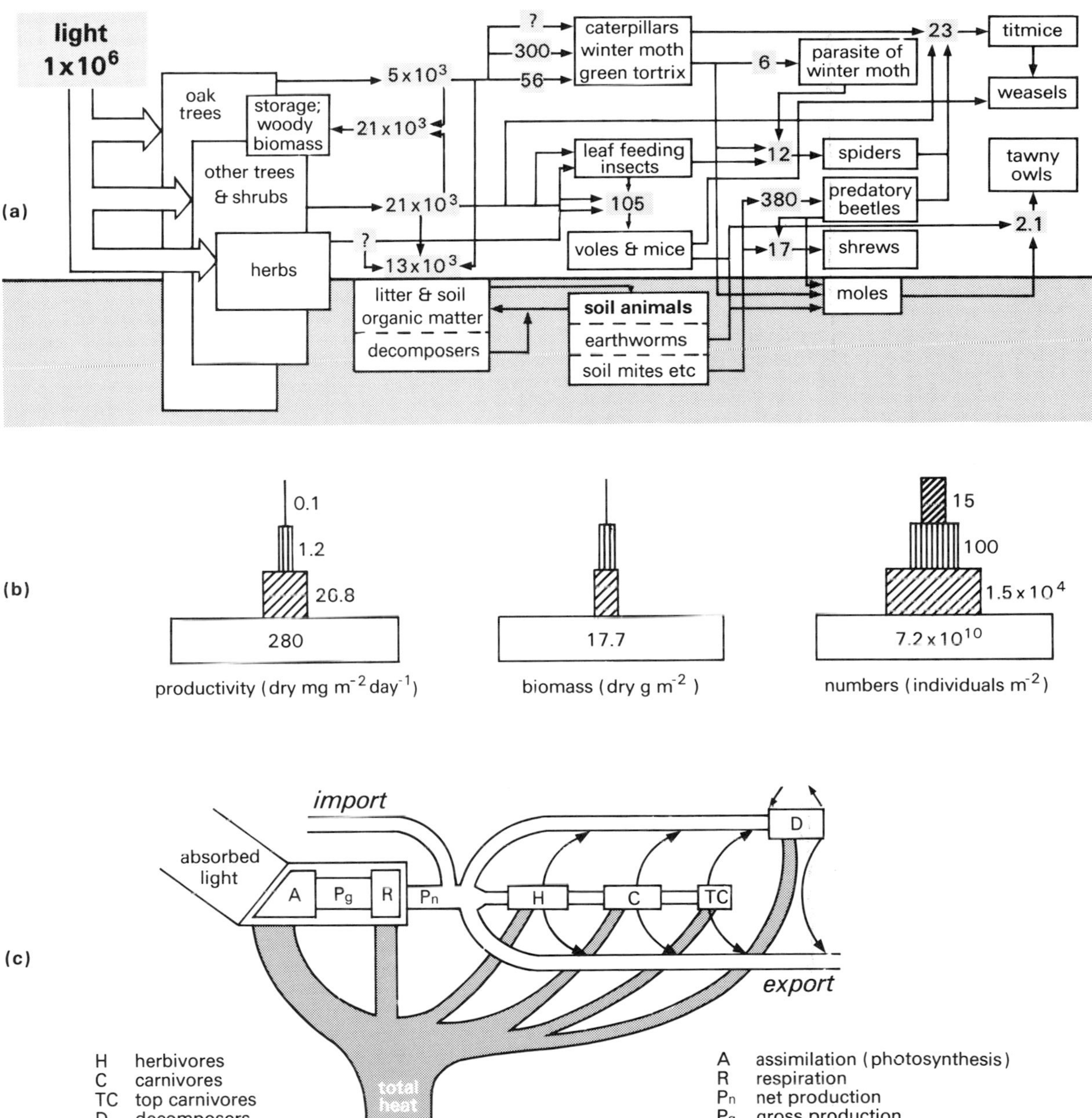

Figure 16.9 (a) Trophic model of an oak woodland (after Varley 1970). (b) Pyramids of productivity, biomass and number (after Whittaker 1975). (c) Energy flow model for Silver Springs, Florida (after Odum 1957).

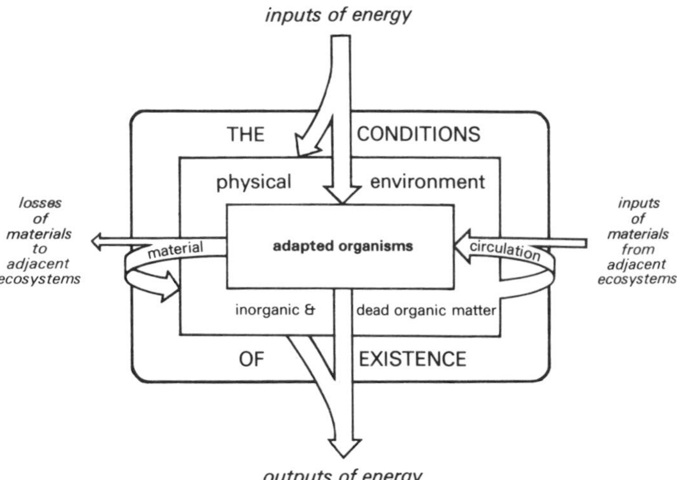

Figure 16.10 Generalised ecosystem model.

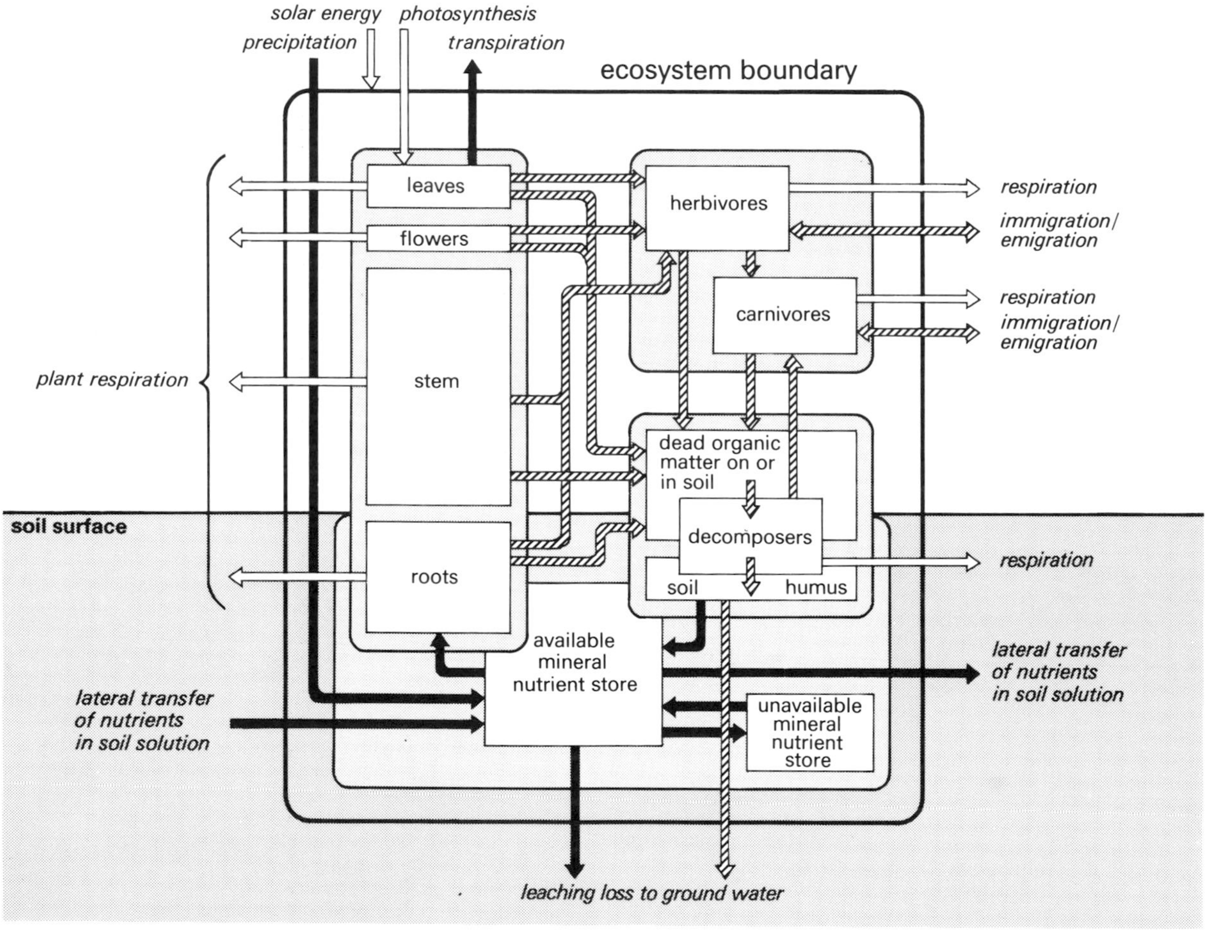

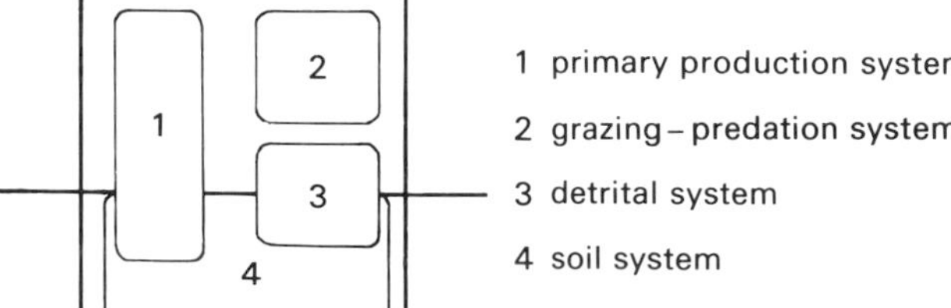

1 primary production system

2 grazing – predation system

3 detrital system

4 soil system

Figure 16.11 Refined ecosystem model showing the primary production, grazing–predation, detrital and soil subsystems.

(Fig. 16.10) and, because it is an open system, between it and adjacent systems. The trophic model of functional organisation is, therefore, an appropriate starting place, but it needs to be extended to incorporate those parts of the inorganic environment that are also involved in the functional activities of energy transfer and nutrient circulation.

In Figure 16.11 the model has been divided into a number of subsystems through which energy and nutrients cascade. The first of these, the *primary production system*, contains as its principal compartment the living plant, rooted in the soil and with its aerial shoots held aloft in the lower atmosphere. The plant is itself an open system, diverting part of the solar energy flux to the manufacture of organic compounds from carbon dioxide, water and mineral nutrients. In terrestrial ecosystems this is possible only because of the transport of water and essential elements from the soil via the transporting tissues of the plant to the sites of photosynthesis in the leaves. To model this subsystem, therefore, we must also incorporate the soil adjacent to the roots and the air in contact with the leaf surfaces, as well as the plant that connects them, for together they form a soil–plant–air continuum. This continuum is the key to our understanding of the way in which water and nutrient transfers take place between the plant and its environment.

The second subsystem maps the transfer of energy and materials (organic and inorganic) through the *grazing, predation pathway*. The understanding of the mechanisms involved in these transfers, and of the strategies adopted by herbivores and carnivores, are of considerable importance to Man, partly because we ourselves fall into one of these two compartments and derive a direct harvest from both producers and consumers, but also because today we often seek to manage and conserve natural populations of animals.

The third subsystem, the *detrital system*, sees the final release of energy originally trapped by the primary producers and provides the potential for the circulation of nutrient elements to be completed by root uptake, as organic compounds undergo the complex processes of decomposition.

Decomposition in terrestrial ecosystems takes place in the soil. The final subsystem, the *soil system*, is therefore the functional connection that completes the ecosystem model depicted in Figure 16.11. However, it is more than a link in the circulation of nutrients, for it couples the functional activities of the ecosystem to the cascades of water and debris in the denudation system and particularly to the processes operating in the weathering and slope systems. The soil system, therefore, must be the focus of attention in any attempt to integrate ecosystem models with those of the denudation system, and this remains as true for managed agricultural systems as it is for natural ecosystems.

Because the flow of energy is linked with a cyclic transfer of matter between these subsystems, it does not really matter where we break into the model, for ultimately, if we follow it through, we shall have considered all four subsystems and be able to appreciate the function of the ecosystem as a whole. Nevertheless, it is logical to start with the primary production system and with the initial input of energy, which will be considered in detail in Chapter 17, while the remaining subsystems of Figure 16.11 will be considered in turn in Chapters 18, 19 and 20.

Further reading

The structural aspects of ecosystem description, particularly in space, are well treated in:

Goldsmith, F. B. and C. M. Harrison 1976. Description and classification of vegetation. In *Methods in plant ecology*, S. B. Chapman (ed.). Oxford: Blackwell Scientific.

Kershaw, K. A. 1973. *Quantitative and dynamic plant ecology*, 2nd edn. London: Edward Arnold.

Shimwell, D. W. 1971. *The description and classification of vegetation*. London: Sidgwick & Jackson.

17 The primary production system

17.1 Functional organisation and activity of the green plant

Before we can begin to examine the transfer of matter and energy through the plant we need to consider more fully its functional organisation. The terrestrial plant can be divided into a root system, a photosynthesising system consisting normally (but not always) of a leaf canopy, and, linking them, a support and transporting system (Fig. 17.1).

Leaves are a system of cells organised to expose the chloroplasts to light in the most efficient manner, to ensure an adequate supply of water, nutrients and carbon dioxide and finally to remove the surplus products of photosynthesis. A leaf resembles a sandwich whose filling consists of large (often columnar) cells occupied by streaming cytoplasm packed with chloroplasts. These are **mesophyll** cells, closely arranged at the top of the filling but with extensive air space between them lower down. Running through them are bundles of conducting tissue – the veins – so spaced that all parts of the mesophyll are near them. The top and bottom of the sandwich are layers one cell thick, transparent and waterproof, called the **epidermis**. The outer surface of the epidermis has a wax coating (**cuticle**). The continuity of the epidermis is broken in the leaves of all plants from ferns to broad-leaved flowering plants, by slit-like pores (**stomata**) between pairs of guard cells which behave as valves. Stomata are usually restricted to the lower surface of the leaf, but in some plants may be sparsely or even equally distributed on the upper surface.

The conducting tissue of the leaf veins is continuous with that of the stem. Stems, whether they are those of forest trees or small herbaceous plants, are essentially **fibrovascular** systems, with a dual function of support and transport. The cylindrical stem is bounded by an epidermis which in trees is replaced by dead cork cells forming the bark. The vascular tissue is of two types, the **xylem** tubes and the **phloem**. The xylem occupies the inner part of the vascular bundles and consists of the lignified (cellulose and lignin in the combination we call wood) walls of dead elongated cells or vessels which connect longitudinally by large perforations and laterally by pores and diaphragms called pits. The lignified xylem elements have a supporting function, especially in the more primitive plants such as the gymnosperms, but in more advanced xylem systems some become specialised to this function alone as wood fibres.

The phloem, in contrast, consists of living thin-walled cells which are elongated also but connected by strands of cytoplasm passing through the small perforations in the sieve plates separating them. Phloem cells or sieve tubes are unusual in that they lack a nucleus, and the cytoplasm is constantly streaming and circulating. The sieve tubes are always associated with nucleated companion cells and very long thick-walled cells, known as phloem fibres, which add support. The remainder of the stem tissue, which we will regard simply as the matrix, has two main functions: storage (**parenchyma**) and support (**collenchyma**). This matrix forms the **cortex** and the fibrovascular tissue the **stele** enclosed in an **endodermis**, though this may not be discernible.

The structure of roots, the absorption system, is essentially similar to that of the stem: an outer epidermis, a cortex and a stele containing the vascular tissue. At the root extremity, however, these distinctions are less clear. The root tip is the area of rapid cell division and growth and it is usually protected by a **root cap**. Behind the tip are regions of cell elongation and differentiation as cells develop specialised functions appropriate to their position in the root cylinder. Further behind

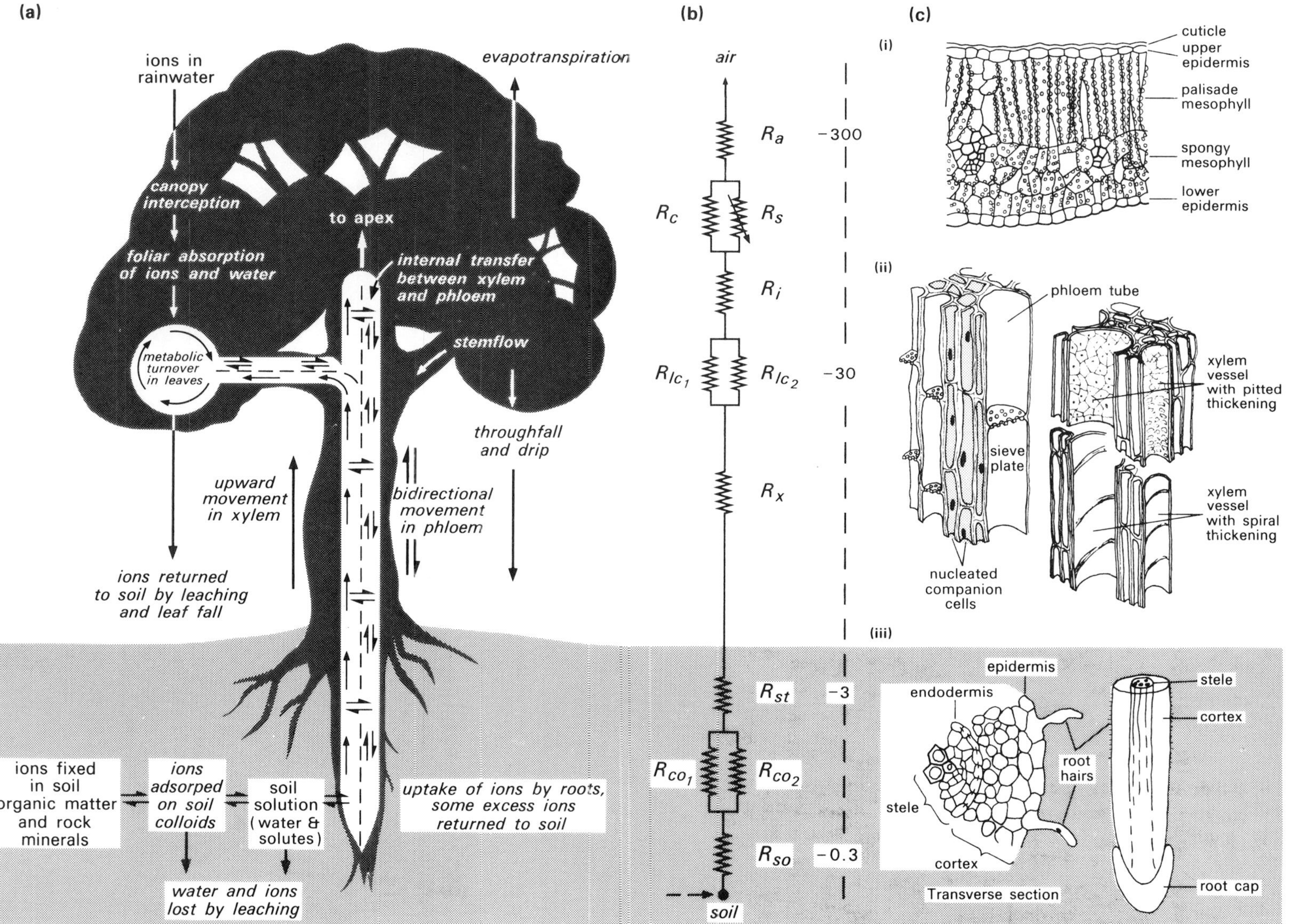

Figure 17.1 (a) Transfer pathways in the plant. (b) An electrical analogue model of resistances to water movement in the plant, where R_{so} is soil resistance, R_{co_1}, R_{co_2} are root cortex free and non-free space resistances, R_{st} is root stele resistance, R_x is xylem resistance, R_{lc} is leaf cellular free and non-free space resistances, R_i is leaf intercellular space resistance, R_c is leaf cutiole resistance, R_s is stomatal resistance, and R_a is atmospheric resistance. Note that only the stomata provide a variable resistance to water movement. (c) Cell structure of the (i) photosynthetic (leaf), (ii) transport (xylem and phloem) and (iii) absorption (root) systems of the plant.

Agave sp.

Zeltova sp.

Betula nana

Gynkgo biloba

Juniperus communis

Victoria amazonica

Platanus orientalis

Pinus

Figure 17.2 Variations in plant growth morphology, illustrated here by variations in leaf shape.

the tip, where the root tissues have become distinguished, is an important (but not the only) absorptive zone of the root where the epidermal cells develop slender outgrowths known as **root hairs**.

There are many variations, of course, on the expression of this theme of root, stem, leaf amongst terrestrial plants and there is often some implied adaptive significance to these variations thrown up by evolution in different environments (Fig. 17.2). Leaf size, shape and detailed morphology show enormous variety. The large obvate-lanceolate leaves of tropical rainforest trees have a thick waxy cuticle and elongate drip tips; the slender leaves of the grasses and sedges are sometimes fine or in-rolled with protected stomata, sometimes wide and pubescent; the needle leaves of the conifers have sunken stomata; while the leaves of heather shoots are tiny and scale-like. Some plants even lack true leaves or have leaves that have taken on an entirely different function, such as the spines of barberry and perhaps stem succulents like the Cactaceae and Euphorbiaceae. In such cases as these the photosynthetic function may be taken over by other organs, such as the leaf petiole (stalk) or 'phyllode' in some Australian species of *Acacia*. Stem systems show variation too. At one extreme are the reduced hemicryptophyte stems of the grasses from which the leaf sheaths arise; at the other the massive trunks of forest trees reaching heights of 100 m or more. In between, every conceivable intermediate seems to exist, such as the prostrate stems of woody chamaephytes like the dwarf willow (*Salix herbacea*) or the pliable but thickened stems of tropical lianas, and to be made more confusing by the variety of growth patterns and branching habits. Roots, however, show fewer examples of detailed refinements in their cellular structure, perhaps because conditions are more uniform below ground, although the velamen of the aerial roots of tropical epiphytic orchids forming a sponge-like tissue absorbing water from the air is one example of root specialisation. Their cortex is photosynthetic, so the root appears white when the velamen is full of air but green when full of water. Nevertheless, although individual roots are less variable, the complete root systems do show variety from dense adventitious systems to deep tap roots, not to mention peculiarities such as stilt roots and breathing roots (**pneumatophores**) so prominent in descriptions of tropical rain forests.

In spite of this fascinating but perplexing diversity, for our present purposes we shall consider the root–stem–leaf system in terms of a generalised model of transfer pathways through the plant. These pathways can be separated into a free space pathway through intercellular voids, through the lumina of xylem vessels and through the microporous structure of cell wall material, and a non-free space pathway involving the crossing of biological membranes and transfer through the cytoplasm of living cells (see Fig. 17.3).

17.1.1 Throughput of water and mineral nutrients
Water held in the soil by surface tension (capillary water) and by adsorption on soil colloids, will enter the root and move across the cortex to the conducting elements of the xylem as long as a water potential gradient exists across the root. Such movement occurs in response to negative hydrostatic pressure potentials developed in the xylem by the transpiration (see Figs 17.1 & 3) of water from the leaves and transmitted to the roots, and to an osmotic potential difference caused by higher solute concentrations in the xylem sap than in the external soil solution.

Nutrient ions enter the root passively with the mass flow of water but, because the concentrations of free ions in the soil water are generally low, other uptake mechanisms must exist. Two hypotheses are normally invoked. Both are processes of ion exchange (see Ch. 11) and both probably occur in many soils. The first hypothesis involves close contact between the root hairs and the soil colloids, so close in fact that the oscillation volume of ions adsorbed on the root surface (particularly hydrogen ions) overlaps with that of ions adsorbed on soil colloids. Under such conditions, ion exchange occurs without the ion appearing free in the soil solution and, not surprisingly, the process is referred to as contact exchange (Fig. 17.4a). The second hypothesis is really hydrolysis of the soil colloids and it occurs when hydrogen ions (H^+) produced by the dissociation of carbonic acid in the soil water exchange with metallic cations adsorbed on the colloid surfaces. These released ions then diffuse to the root surface. The carbonic acid involved in this process – called the 'carbonic acid exchange hypothesis' – is derived from the dissolution of respiratory carbon dioxide (Fig. 17.4b).

The transfer of water takes place across the root by two alternative pathways. The first is a passive

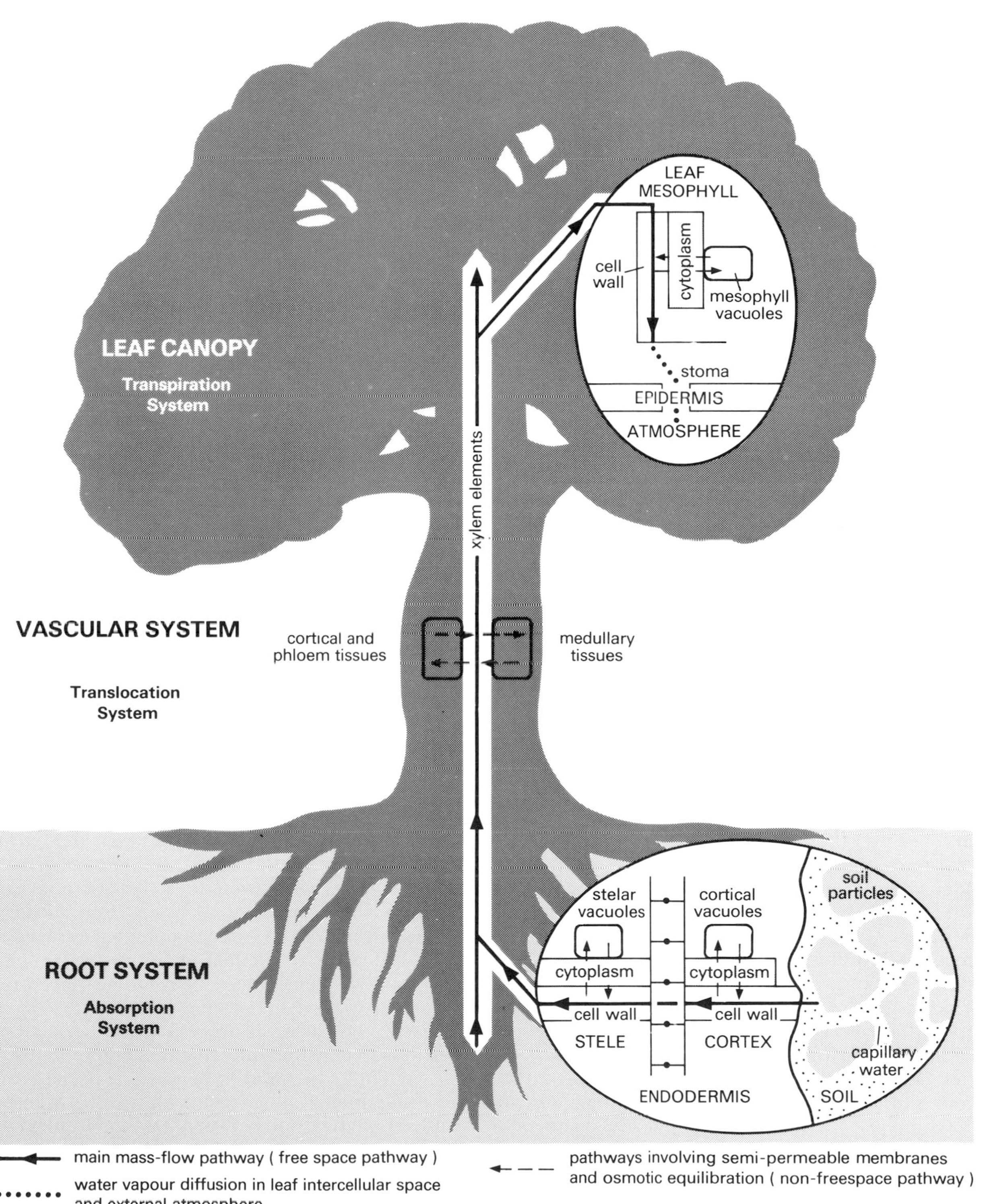

Figure 17.3 Mass-flow and diffusion pathways of water movement in the plant (after Weatherly 1969).

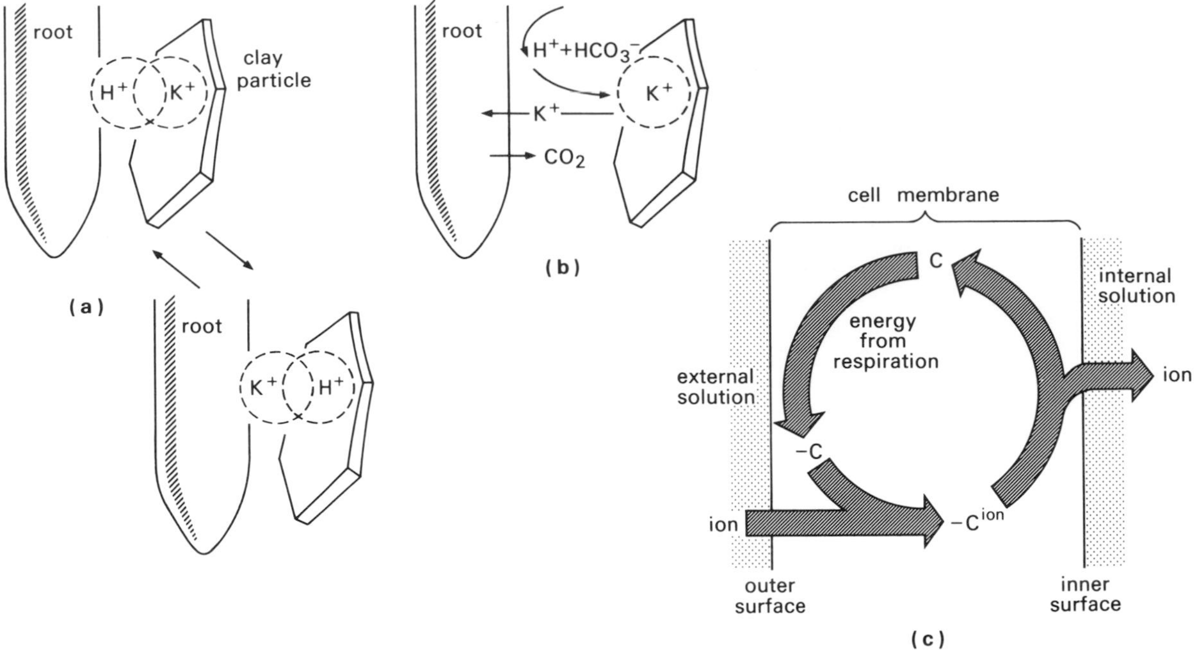

Figure 17.4 (a) Contact hypothesis of ion exchange between root and soil particles. (b) Carbon dioxide hypothesis of ion exchange. (c) Carrier hypothesis of ion movement across cell membranes.

mass flow pathway through the freely permeable intercellular spaces and the microporous structure of the cell walls, i.e. the **root free space** largely in response to the pressure potential gradient (Fig. 17.3). Secondly, water enters the cells and moves through the cytoplasm and vacuoles and across the cell-wall membranes (**non-free space**) by diffusion. Although the entry of some nutrient ions may be a mass flow phenomenon in the free space most are adsorbed on the cell walls of the cortex or cross the cell membranes to join the non-free space pathway. The transfer of water and nutrient ions through the cortical cells is a passive response to diffusion gradients, particularly those across differentially permeable cell membranes. These are electrochemical gradients associated with the charges on the ions, the fixed charges on the membrane and the relative concentrations of ions, and the mechanism of transfer is **electro-osmosis**. There is a great deal of evidence to suggest that ions can also move against such diffusion gradients by means of active ion pumps. These are now known to involve enzyme carrier molecules in the cell membrane which harness metabolic ATP energy (see Ch. 6) to accomplish the work of transporting ions against the resistance of electrochemical potentials (see Fig. 17.4c).

The free space and non-free space pathways through the cortex join in order to cross the endodermis, which represents a barrier to free space transfer for the partial thickening (**suberization**) of endodermal cell walls renders them impermeable. Diffusion, however, can still take place from cortical cells to the stele by the non-free space pathway from which both water and nutrient ions are released into the xylem elements.

As water evaporates from the leaf mesophyll cells during transpiration, their solute concentration increases and lowers their osmotic potential (ψ_π) causing movement of water from adjacent cells. In this way an osmotic potential gradient is established across the leaf and draws water from the leaf xylem elements. The resulting pull on the water in the xylem, placing it under tension or negative pressure, draws water up the stem in response to the pressure potential difference (Fig. 17.5). This is believed to be possible because the cohesive forces and adhesive forces, due to intermolecular attraction between water molecules and between them and the sides of the xylem capillaries, allow the maintenance of unbroken water columns from root to leaf. Most nutrient ions move passively with this transpiration stream in the same form as they were absorbed from the soil.

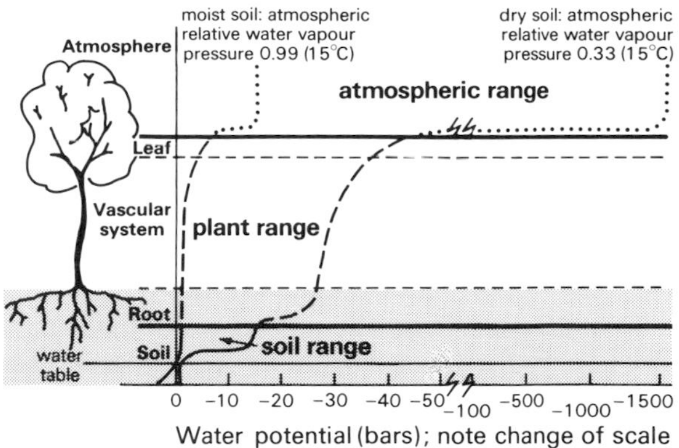

Figure 17.5 Water potential gradient in the soil–plant–air continuum (after Etherington 1975).

Others, notably nitrogen, phosphorus and sulphur, are translocated as organic derivatives which must have formed in the root, while such elements as iron, which tend to precipitate in high-valency forms (see Ch. 11), may be aided in their mobility by forming complexes with organic chelating agents. Some ions do not make the complete trip to the leaves, for during upward translocation they may pass selectively across cell membranes to enter other cells by either passive diffusion or active transfer.

17.1.2 Photosynthesis and plant primary production

The concentration of chloroplasts in the mesophyll of leaves is the site of photosynthesis. These cells are supplied with water and mineral nutrients via the transpiration stream in the xylem of stems and leaves and by alternative pathways across the mesophyll itself similar to those described for the root cortex. In the case of the leaf, movement of water in the free space around the leaves seems to be the principal pathway. Only a very small percentage of the water absorbed by roots is used for the vital functions of the plant in which water is involved and of that a fraction of 1% is used directly in photosynthesis. The remainder is evaporated from the micropores of the mesophyll and cuticular cell walls and escapes to the atmosphere as water vapour by diffusion across the intercellular air space to the stomata and thence by either molecular diffusion in calm conditions, or turbulent diffusion over the leaf surface in moving air.

Carbon dioxide, the remaining raw material for photosynthesis, is apparently freely available in the atmosphere surrounding the leaf, but in practice CO_2 concentrations can become rate limiting for photosynthesis. This is because it is the concentrations at the chloroplasts that are important, and to reach them CO_2 has to diffuse through the stomata and the mesophyll air spaces, across the cell walls and through the cytoplasm. So CO_2 supply to the chloroplasts is determined by the rate of these diffusion processes and this depends on the maintenance of CO_2 concentration gradients between the air outside and the chloroplasts (Box 17.1).

Finally, the photosynthetic process requires the

Box 17.1

CARBON DIOXIDE COMPENSATION POINT

The carbon dioxide compensation point is the carbon dioxide concentration at which the rate of uptake during photosynthesis is exactly balanced by the rate of carbon dioxide output during respiration. The carbon dioxide compensation point for most plants is high – *ca.* 50 ppm CO_2 owing to the occurrence of photorespiration which releases carbon dioxide in the light and is wasteful of energy during the C-3 Calvin type of carbon dioxide fixation, hence reducing the efficiency of photosynthesis. Some plants, such as certain algae and C-4 Hatch–Slack plants in which photorespiration is negligible, have low carbon dioxide compensation points – *ca.* 5 ppm CO_2.

energy input of solar radiation. Figure 17.6 illustrates the complex energy balance of a single leaf, at night and during the day (see Ch. 3). Of the incident visible light, normally less than 20% is reflected (higher in the presence of waxy or light-coloured layers), but reflectivity rises sharply for long-wave radiation with 40–60% of infrared radiation being reflected. Little transmission of energy takes place through the leaf, although again it is higher for infrared (*ca.* 30%) than visible light (<10%), and of course it varies with leaf thickness. The remaining radiation input to the leaf is absorbed (50% of total radiation) but differentially, with *ca.* 80% of visible wavelengths absorbed as opposed to 10% of infrared. High absorption by such thin structures as leaves is due to multiple reflection by cell/water interfaces in the mesophyll. Only *ca.* 2% of absorbed energy (in the photosynthetically active wavelengths) is used in photosynthesis, the remaining energy, which may be as high as *ca.* 80% of total incoming energy (*ca.* 26×10^6 J m^{-2} day^{-1} in mid-summer), is available to raise the temperature of the leaf.

To avoid cell death by high temperatures, much of this heat load must be dissipated by the leaf. Cooling of the leaf takes place by convective cooling particularly forced convection but also free convection, when there is little or no air movement. Evaporative cooling also occurs as long as the stomata are open and transpiration is not limited by soil or plant moisture deficits. These cooling mechanisms have already been discussed in some detail in Chapter 4. Finally, the leaves will emit long-wave radiation, particularly at night (see Ch. 3).

We have seen how the reactants and the energy necessary for photosynthesis to proceed are brought together in the leaf canopy. However, a further consideration remains, for high rates of photosynthesis can occur only if the products (organic assimilates) of photosynthesis are removed from their source in the leaves. These are the fuel molecules on which the metabolism of the rest of the plant and its growth depends and, of course, they are translocated through the plant by the bidirectional diffusion in the phloem sieve tubes (see Fig. 17.1). Nevertheless, they must not be allowed to accumulate at their 'source' and, when produced in greater quantities, the plant's immediate metabolic needs would require that they be removed to a 'sink' in some storage tissue.

The process of photosynthesis itself has been

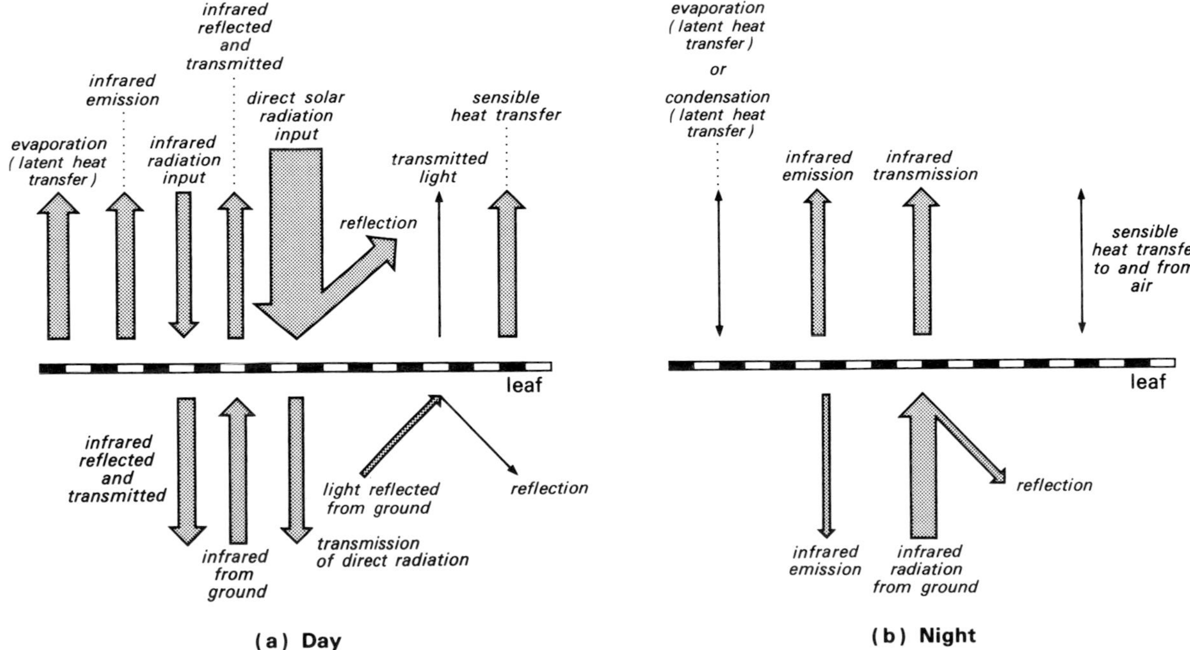

Figure 17.6 The energy budget of a leaf, by (a) day and (b) night (after Meidner & Sheriff 1976).

treated at the biochemical level when we considered the autotrophic cell model in Chapter 6. Because it is essentially a process of carbon fixation or assimilation, the rate of photosynthesis is usually expressed as the rate of CO_2 uptake by the photosynthesising system (g CO_2 m^{-2} s^{-1}) irrespective of whether we are considering a single leaf, a plant or an entire canopy. The net photosynthetic rate is the difference between gross photosynthetic rate and respiration. The net photosynthesis (i.e. the net photosynthetic rate integrated over time) which accumulates in the plant (or stand of vegetation) with the addition of nutrient elements in protein synthesis represents the net production of the plant or stand. Similarly, the accumulation of gross photosynthesis is gross production (see Ch. 7).

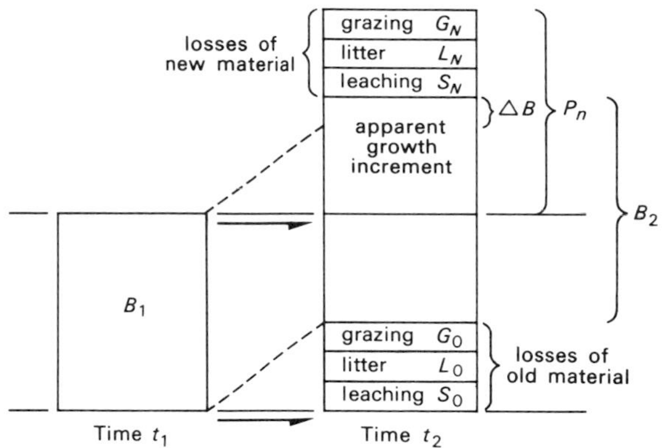

Figure 17.7 The components of biomass change with time and their relationship to production (after Chapman 1976).

17.2 Ecosystem primary production

The net primary production over a period of time of an ecosystem such as the woodland seen in Figure 16.1 is, therefore, partly represented by the change in the biomass during that period ($t_1 - t_2$) (Fig. 17.7)

$$P_n = \Delta B$$

Such a view is, of course, an oversimplification. If we refer to the woodland ecosystem, we can represent the biomass of all the primary producers at the beginning of the time period (i.e. at t_1) as B_1. At the end of the period t_2, some of this initial biomass will have died and been shed as litter (L_O); some will have been consumed by the herbivore population (whether by defoliating insects or the browsing of deer is unimportant) and can be represented as a grazing loss (G_O). Finally, a proportion of this initial biomass will have been leached from the canopy (S_O) (see Ch. 19). During the same period of time ($t_1 - t_2$) an increment of new biomass will have been added. This of course is the net production (P_n), but it too will have suffered grazing (G_N) and some new tissue will have died, perhaps as a result of late frosts or wind damage to young shoots, and been shed as litter (L_N). In addition, some of the constituents of the new biomass will have been leached (S_N). So at the end of the time $t_1 - t_2$, not only will the initial biomass have been reduced but the new biomass remaining will be less than that created by net production. The change in biomass will fall short of net production by the same amount as the sum of the old and new biomass lost to litter, grazing and leaching. A more accurate view of net primary production, in terms of changes in accumulated biomass is therefore:

$$\begin{aligned} P_n &= \Delta B + (L_O + L_N) + (G_O + G_N) + (S_O + S_n) \\ &= \Delta B + L_{total} + G_{total} + S_{total}. \end{aligned}$$

(see Fig. 17.7)

Bearing this relationship in mind, we might make some observations concerning our woodland ecosystem. The thick carpet of leaf litter, together with the inconspicuous nature of the herbivore element and the existence of a lush green canopy with no obvious evidence of heavy grazing pressure, suggests that $L + S$ in the above equation is probably more important than G. Is this the case in all ecosystems? Furthermore we might observe, particularly if we were regular visitors to the woodland, that the biomass of the woodland, especially of the trees, represents a massive accumulation of organic matter, but apart from the predictable seasonal changes in appearance it would seem not to change greatly through time. Why should this be so? This might further prompt the observation that nevertheless, at some time there has been an investment of net production in accumulating the biomass that we see now. When was this and why is biomass apparently no longer accumulating?

These are profound observations and the answers to the questions they pose provide some of the most important theoretical statements con-

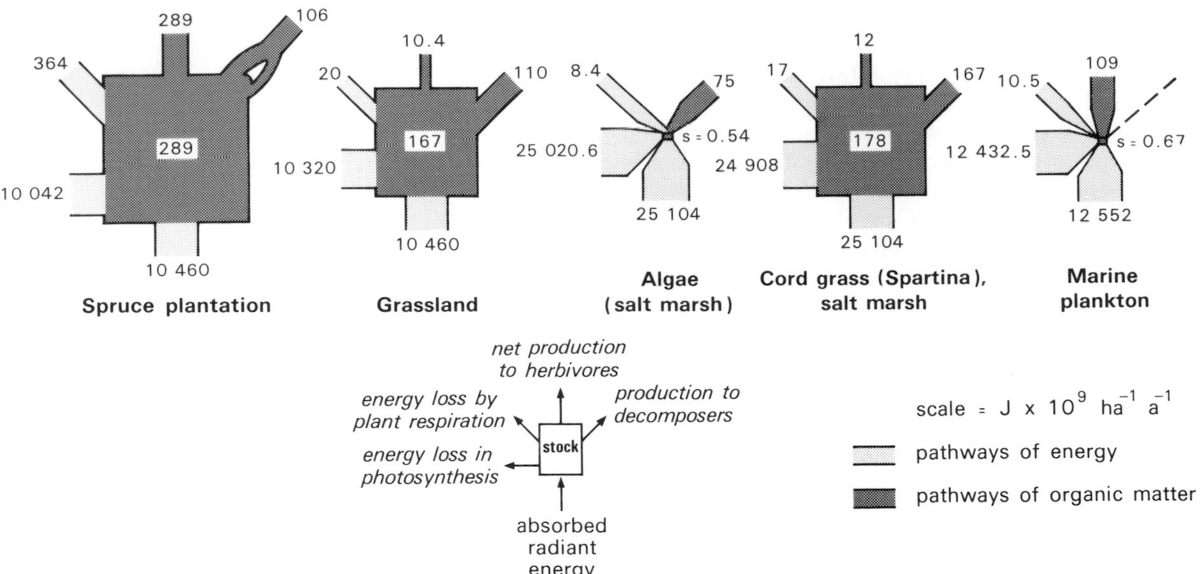

Figure 17.8 Primary production in different ecosystems (note that production to herbivores in the case of the spruce plantation is misleadingly high for it includes a timber crop extracted by man) (after Macfadyen 1964).

cerning production. Figure 17.8 shows the energy flow through the primary producer trophic level of several ecosystems. Here we can see the answer to the first question. There appears to be a distinction between those ecosystems with a high proportion of plant-derived dead organic matter and a low herbivore element and those with little plant-derived dead organic matter and a high herbivore element. Under natural conditions most terrestrial ecosystems, including our woodland, fall into the former category and most of the animal biomass is to be found in the detrital system. Because they are not being destroyed by grazing, the plants (particularly the perennial ones) accumulate a proportion of their net production over time in support tissue and structures such as the wood of the forest trees, a necessary feature of adaptation to terrestrial environments. This accumulation is capital and it must be paid for (see Ch. 16). Much of this tissue in trees is dead wood, but it also contains the living cells of the transport system and **cambial** cells that permit further growth. These non-photosynthesising cells are a respiratory drain on the total assimilated energy or gross production. The greater the biomass the greater the maintenance cost that the plant sustains in terms of respiratory heat loss. In our woodland the law of diminishing returns is operating because, for every increment of photosynthesising surface, an even greater maintenance respiration is required to support it. Initially, the increase in leaf-area index (ratio of leaf surface to ground area) will more than compensate, but, as total biomass increases, the respiratory load increases faster than the gross production, and net production falls off (Fig. 17.9). Here then is part of the answer to the remaining questions, namely that maintenance costs increase in a mature ecosystem, leaving a smaller proportion of assimilated energy as net production. In addition, there is evidence that the photosynthetic efficiency (CO_2 fixed per unit of energy input, g CO_2 J^{-1}) of old plants declines so that both net and gross production are reduced. Finally, we can say that if the woodland is to maintain a steady state (i.e. $\Delta B/\Delta t = 0$), then the net production must be just sufficient to offset total losses by grazing and by leaching and litterfall, i.e. $P_n = G + L + S$,

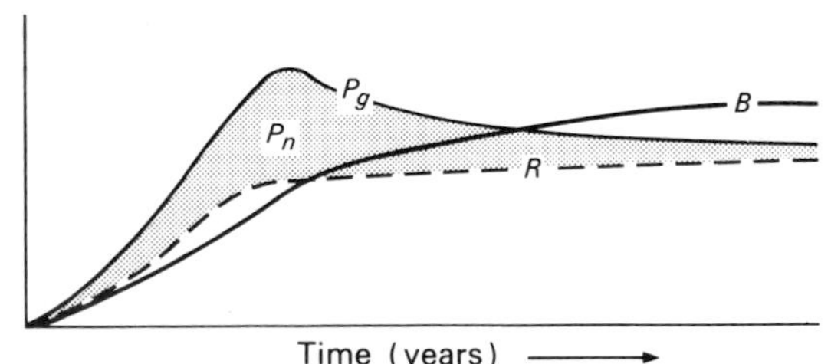

Figure 17.9 Change in ecosystem gross primary production (P_g), plant respiration (R), net primary production (P_n) and plant biomass (B) with time (from Odum 1971, after Kira & Shidei 1967).

so $\Delta B = P_n - (G + L + S) = 0.$

In terrestrial ecosystems then, a large proportion of the energy flow through the primary producer level is trapped or stored in non-productive tissues, destined to be released after a considerable time lag by decomposition. Up to a point, the same is true of the nutrient elements immobilised in the organic compounds of the woodland biomass, although nutrient availability is ensured by other strategies, as we shall see. In the aquatic ecosystems represented in Figure 17.8 by the marine phytoplankton, all the living tissue is productive and hence there is no respiratory sink. They are small organisms with high surface-area:volume ratios, a high metabolic rate and high productivity. They may be grazed many times over or, to put it another way, several generations may be grazed in a year so that annual production figures are cumulative. Both the energy input and energy output to herbivores are high and because life cycles are short there is little or no accumulation of biomass, even if the algal cells escape grazing, as in a sewage pond, death and decomposition are also rapid. In contrast, in the woodland most of the plant material remains free of herbivore attack over the same period. The grassland represents an intermediate situation, though closer to the woodland, for the ratio of net production to biomass (P_n:B) is low (0.06), compared with the high figure for the phytoplankton (162.7). This P_n:B ratio is a useful index of the amount of non-productive tissue in the plants.

17.3 Regulation and limits to photosynthesis and primary production

Primary production is, of course, directly linked to photosynthesis, so the mechanisms regulating and the factors limiting the photosynthetic rate will ultimately affect the primary productivity of the ecosystem. These can be divided into endogenous – **genotypic** and **phenotypic** – variables which affect the photosynthetic rate per unit of photosynthetic tissue, and exogenous – **environmental** – variables which regulate the size of the photosynthesising system by their effect on the plant's metabolism and growth. The first category can be divided further into morphological, physiological and biochemical factors.

17.3.1 Genotypic and phenotypic variables

Morphological variables. Morphological variables are manifest at two scales. At the scale of the individual leaf, certain variables become important such as leaf thickness and volume, number and distribution of chloroplasts, morphology of the mesophyll, and number and position of the stomata. The variation in leaf size, form and structure mentioned earlier in this chapter will influence photosynthetic rate and may represent the evolutionary selection of leaf characteristics of adaptive significance. In many species distinct **sun** and **shade** leaves can be distinguished. The former are exposed to full sunlight, often being smaller but thicker than the latter which show reciprocal adaptations to low light intensities. At the scale of the whole plant the number, arrangement and length of life of the leaf canopy are important variables. Leaf-area index (the ratio of leaf area to ground area) interacts with leaf position and angle of leaf insertion (Fig. 17.10) to govern light penetration into the canopy. High values of leaf-area index inevitably mean the existence of shelf shading. Diffuse radiation is reduced exponentially from the surface of the canopy towards the ground, but the effect on photosynthesis is complicated by changes in the spectrum of radiation inside the canopy as the photosynthetically active wavelengths (red and blue) are selectively absorbed and by the fact that the shade leaves may show physiological and biochemical adaptations to low light intensities. Steep angles of insertion of leaves, as in grasses, will favour more effective penetration of light than will leaves held horizontally, but it must be remembered that leaves of many species can be orientated in relation to light (**phyllotaxy**). The duration of maximum leaf-area index is also an important factor which may compensate for other variables.

Physiological variables. Physiological controls are many and complex, but amongst the most important are the plant's resistances to diffusion and flow of water and to the diffusion of CO_2 in the leaf (see Figs 17.1 & 4). One of the most important resistances is that of the stomata and here physiological factors become significant, such as the existence of endogenous rhythms of stomatal opening and closing. Although the concentration of CO_2 and light and temperature are environmental variables, the fact that they have a physiological expression justifies considering them under

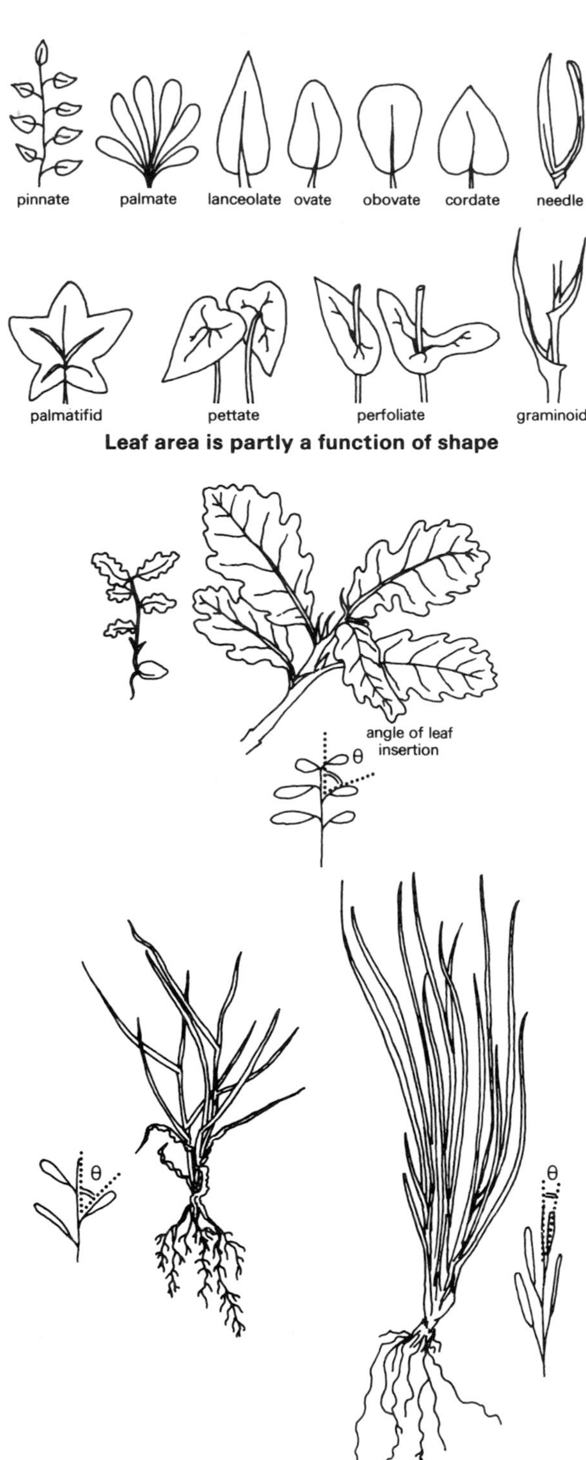

Figure 17.10 Leaf area and angle of leaf insertion, which contribute to the leaf-area index.

this heading. Photosynthesis ceases under conditions of too much or too little heat and it is optimal under a particular temperature range for each species. The high and low temperature extremes are known as **temperature compensation points**. Similarly, there is a low light intensity, the **light compensation point**, below which there is a net loss of CO_2 by respiration which is not compensated for by carbon fixation in photosynthesis. Indeed, leaves low in vegetation canopies may be close to or below their light compensation points and represent a respiratory energy sink. One way in which plants overcome this problem is to shed leaves and branches from the lower part of the canopy as they grow. The upper extreme of light intensity above which photosynthesis ceases is the **light saturation point**. Light and temperature are not independent, however, for at high light intensities the optimum temperature for photosynthesis seems to be higher and the temperature range is increased (Pisek *et al.* 1969). Atmospheric carbon dioxide concentrations are not normally limiting, except perhaps in full sunlight at the top of leaf canopies, but often the diffusion gradient from the air to the chloroplasts is limiting. In this context the **carbon dioxide compensation point** (see Box 17.1), which is a measure of the plant's ability to re-use CO_2 released during respiration and to take up CO_2 from the air, is an important physiological property.

Biochemical variables. The principal biochemical controls on photosynthesis, and hence on productivity, concern the carbon fixation pathways present in the plants of the ecosystem. There are two such pathways, the C-4 Hatch–Slack pathway and the C-3 Calvin (or normal) pathway. Hatch–Slack plants are characterised by low CO_2 compensation points, a lack of photorespiration and, usually, high light saturation points, all of which are responsible for a photosynthetic rate two or three times greater than in Calvin plants. These high-capacity producers are all plants of the Tropics and Subtropics and they include several important crop species such as maize (*Zea mais*) and sugar cane (*Saccharum officinarum*). In contrast, the presence of photorespiration causes high CO_2 compensation points (see Box 17.1) and reduces the efficiency of carbon fixation in Calvin plants by liberating CO_2 into the leaf intercellular space at the same time as photosynthesis is removing it. The combination of morphological,

Table 17.1 The characteristics of plants with high and low production capacities (modified from Black 1971).

	High-capacity producers	Low-capacity producers
GENERAL TYPE OF PLANT	herbaceous and mostly grasses or sedges	herbs, shrubs or trees from all plant families
MORPHOLOGY		
(1) leaf characters	bundle sheath cells around vascular bundles packed with chloroplasts	no chloroplasts in the bundle sheath cells
PHYSIOLOGY		
(2) rate of photosynthesis	40–80 mg CO_2 dm^{-1} h^{-1} in full sunlight; no light saturation	10–35 mg CO_2 dm^{-1} h^{-1} in full sunlight; light saturation at 10–25% full sunlight
(3) response to temperature	growth and photosynthesis optimal at 30–45°C	growth and photosynthesis optimal at 10–25°C
(4) CO_2 compensation point	0–10 p.p.m. CO_2	30–70 p.p.m. CO_2
(5) sugar transport out of leaves	rapid and efficient: 60–80% in 2–4 h (at high temperatures)	slower and less efficient: 20–60% in 2–4 h
(6) water requirements (g water needed to produce 1 g dry matter)	260–350	400–900
BIOCHEMISTRY		
(7) carbon fixation	Hatch–Slack (or C-4) and Calvin cycle (or C-3) pathways	Calvin cycle (C-3) pathway only
(8) photorespiration	not detected	present

physiological and biochemical factors which characterise high and low capacity producers respectively is shown in Table 17.1.

17.3.2 *Environmental variables*

All of the endogenous variables that influence the plant's performance as a primary producer are characters that have appeared during the course of evolution and have been selected, as far as we can judge, because they have adaptive significance. The existence of these characters not only sets limits to the plant's photosynthesis and production but, by their relation to and interaction with exogenous or environmental variables, they control and regulate the plant's response to its environment, as expressed in its growth and productivity.

Light. The prime environmental variable is the radiation flux density or net radiation receipt per unit area of surface, and it is under the control of climate. The input in the photosynthetically active wavelengths (0.4–0.7 μm) at any point on the Earth's surface is the maximum energy potentially available for photosynthesis and production. The variation in energy receipt over the Earth's surface has already been considered at length in Chapter 3 and, although there is a correlation between net radiation and productivity, it is complicated by other relationships, particularly with temperature and water availability (see Fig. 7.9). However, light has effects other than direct energy input, for adapted plants are able to respond to variations in light intensity both spatially and temporally. For example, plants adapted to shade have light saturation points and intensities only a small fraction of full sunlight, while *sun plants* have higher light saturation points. In the case of some tropical plants with the C-4 Hatch–Slack carbon fixation pathway, no light saturation appears to exist and they can photosynthesise at very high light intensities. The horizontal and vertical disposition of plants in an ecosystem can be viewed, at least in part, as a response of species adapted in these ways to the three-dimensional distribution of light intensities within the canopy microclimate.

Light availability and intensity vary with time as well as with space. There are, of course, diurnal variations in radiation received (see Fig. 17.6) and these are mirrored in diurnal patterns of photosynthesis, respiration and production. Such diurnal rhythms may be very important and predominant in tropical ecosystems which lack any marked seasonality. In seasonal environments the growth response of many plants, such as the

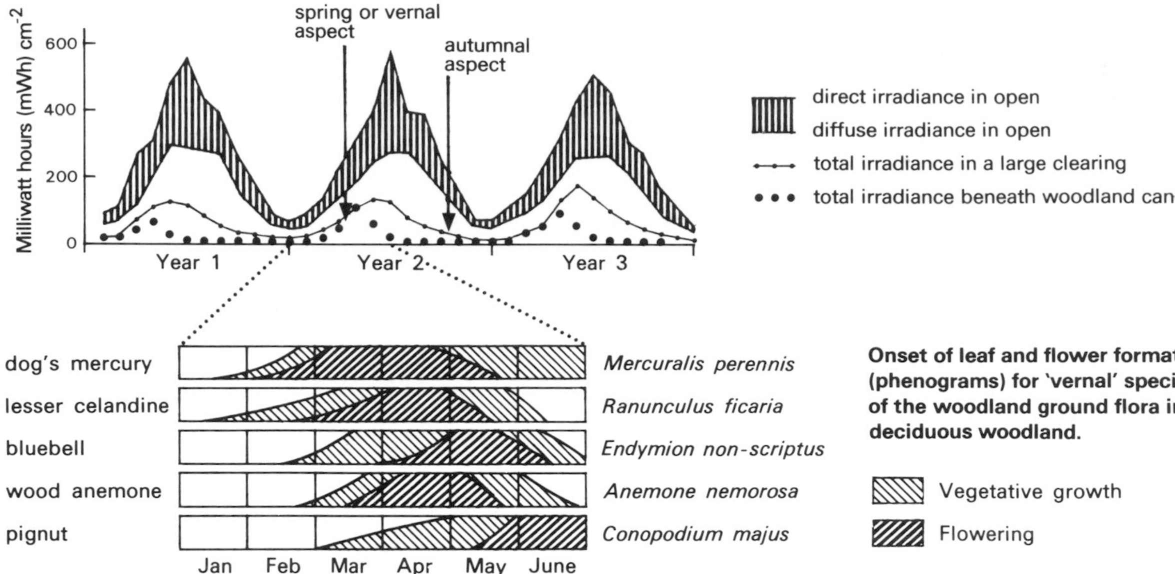

Figure 17.11 Plant phenology and woodland light climate (Anderson 1964).

ground flora herbs in our woodland ecosystem is regulated by the seasonal periodicity of radiation input which defines the ecosystem light climate (Fig. 17.11). The life cycles of different species (phenology) are adapted to different phases of the light climate and they attain optimal leaf-area indices and high daily rates of production at different times. Most natural ecosystems in seasonal climates will, therefore, show a succession of small peaks of production attributable to different species. This is in strong contrast to agricultural crops which rise from zero leaf-area index to a high value and a high single peak of daily production – at which point they may be harvested. In the natural system, however, the production is not all present at one time (which means it is difficult to harvest) but there is a more consistent, prolonged but lower daily rate of production.

The regulatory role of light is to elicit photoperiodic responses from adapted species which result from **circadian rhythms** of light and dark, or variations in day length (see Box 23.1). Their significance lies not only in their control of ecosystem productivity over the annual cycle but also in the regulation they exert on reproductive processes and hence on the dynamics of species populations. In plants this is usually expressed in the time of flowering, and the existence of **short-day** and **long-day** plants is well documented and of economic significance in plant breeding, horticulture and agriculture. There are also similar responses in animal populations, but these will receive some attention in the next section.

Temperature. So far we have considered radiation input in terms of the visible waveband, but the limiting and regulating effects extend to the exchange of longer wavelengths and involve the plant's heat balance (Fig. 17.12). Air, leaf and soil temperatures are, therefore, important environmental variables. High and low temperature extremes are lethal and result in cell and plant death. At the physiological level the mechanism in both cases involves damage (**denaturing**) to enzyme and structural proteins, particularly those associated with membranes, as a result of mechanical stress. At high temperatures this is a consequence of increased kinetic energy at the molecular level, but at low temperatures it is due to dehydration as water is withdrawn from cells by extracellular freezing. The effect seems to be the loss of cellular organisation as biological membranes break down and become permeable. Between these extremes, temperature affects all metabolic processes and hence growth and production, because for every 10°C rise in temperature the rate of chemical reaction roughly doubles (the **Arrhenius relationship**). Biochemical reactions conform to this relationship but, because they are catalysed by enzymes, biochemical reaction rates progressively

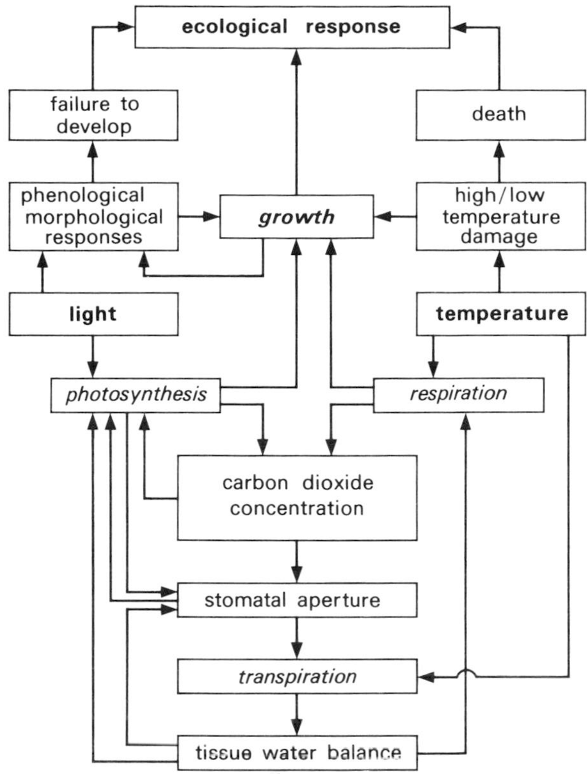

Figure 17.12 Plant response to light and temperature (after Bannister 1976).

decline above 40°C, simply because many of the enzymes are damaged and become inactive above this temperature.

The net production of an ecosystem is, however, a function of both photosynthesis and respiration, and because these processes respond in slightly different ways to temperature its differential effect can greatly influence net production. Respiration rate is affected by temperature in the way outlined above, rising up to 35–40°C, increasing initially at higher temperatures, but then falling rapidly. As was pointed out when discussing temperature compensation points, photosynthesis responds rather differently. This is because in reality it is two reactions: light and dark (see Ch. 6). The light reaction rate is little influenced by temperature (i.e. it has a low temperature coefficient), whereas the dark phase responds to temperature in a way analagous to respiration. Therefore, under low light intensities when light is limiting, the dark reaction rate will not respond appreciably to temperature for it is rate-limited by the supply of reactants from the light reaction. Photosynthesis will increase only with rise in temperature when light is not limiting. The implications are twofold. At the macro-scale the photosynthetic rate will reflect this interaction of light and temperature as expressed in the macroclimate, but perhaps more important, within a vegetation canopy sun and shade leaves will experience different microclimates, and hence combinations of light and temperature, and display different patterns of photosynthesis. Secondly, under conditions of high temperature but low light intensities, the respiration rate can exceed photosynthesis and the leaves will be below their compensation points. These interactions of light and temperature within vegetation canopies are not constant of course; they will change throughout the day and from season to season, and affect the overall net production of the plant community (Fig. 17.13). As with light, temperature is not just a limiting variable, but it acts to regulate metabolic activity, growth and production by the response of plants through inherited adaptive characters to temperature fluctuations. However, perhaps one of

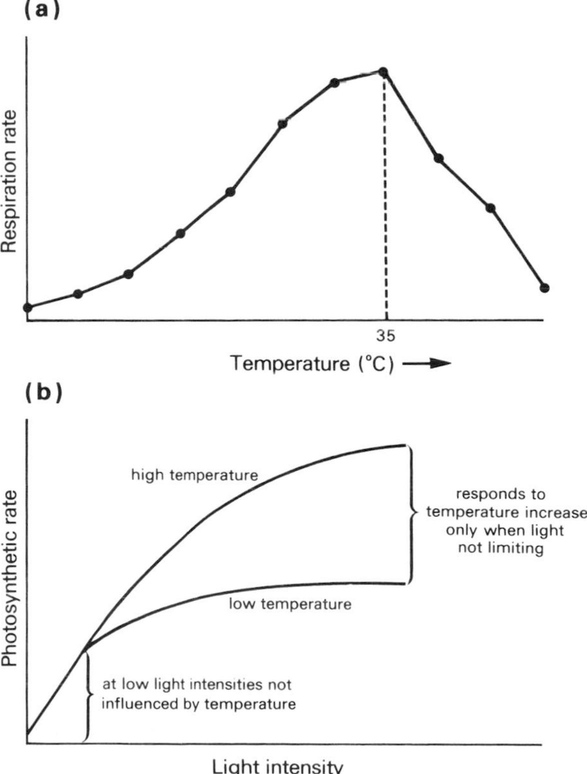

Figure 17.13 The effects of temperature on (a) respiration and (b) photosynthesis.

Box 17.2 (see also Boxes 12.1 and 12.2)

PLANT WATER POTENTIAL

Water in the plant is held by forces of retention and the potential free energy of that water can be expressed in terms of water potential. Water potential in living cells results from a balance between two forces, one due to the presence of solutes in the cell sap and one due to the multi-directional hydrostatic pressure within the cell, the turgor pressure.

$$\underset{C}{\psi_{cell}} = \underset{\pi}{\psi_{osmotic}} + \underset{\substack{\text{(turgor)}\\ p}}{\psi_{pressure}}$$

In living cells turgor pressure is positive, i.e. it enhances cell water potential operating in the direction opposite to osmotic pressure. Plant water potential (ψ_{plant}) is therefore a mean value of the cell water potential in all tissues of the plant, plus the matric potential of the plant matrix, i.e. the intercellular spaces and microporous structure of the cell walls as well as perhaps some xylem vessels (Meidner & Sheriff 1976). ψ_{plant} also includes the contribution of negative hydrostatic pressure in the non-living matrix (the free space), especially in the xylem.

the most important aspects of radiation (long and short wavelength) is its indirect effect on the plant and on production through its regulation of the surface water balance (see Ch. 4).

Water. The development of a plant water deficit has far-reaching physiological implications affecting many of the plant's metabolic activities and retarding growth and production (Box 17.2). Quite small water deficits are now known to reduce photosynthesis, even though a very small proportion of the absorbed water is directly used in photosynthesis. The reasons are complex, but they include the effect on CO_2 diffusion of stomatal closure under water stress, the reduced permeability of membranes and hence a slower rate of removal of photosynthetic products, some adverse effects on the photosynthesis process *per se*, and the reduction of leaf-area index because of loss of cell **turgor** and reduced tissue expansion. Respiration is also reduced by water deficit, but not at the same rate as the reduction of photosynthesis. In some circumstances respiration rate may even increase. In either case, however, a situation can arise where weight loss occurs as respiration rate exceeds that of photosynthesis.

The plant water balance is maintained by the transpiration stream in response to the water potential gradient from soil to free atmosphere (see Fig. 17.4). The rate of uptake from the soil will therefore depend on the rate of water loss by transpiration and the availability of soil water. As we saw in Chapter 4, the transpiration rate will depend on the same environmental factors that determine evaporation from any surface, namely the heat energy supply, the vapour pressure (or pressure potential) difference between the surface and the air, and air movement across that surface, in this case the leaf. Transpiration, therefore, increases with increases in vapour pressure difference, which itself increases with temperature. Again, in Chapter 4, it was pointed out that evaporation rate would also depend on the properties of the evaporating surface. It is in this connection that the characteristics of plant structure affect transpiration rates, and some of the most important of these characters are listed in Table 17.2. The regulation of transpiration is governed mainly by the stomatal resistance (see Fig. 17.1) and its relationship to the resistance of the air above the leaf. In fact, in some circumstances it is the air resistance (R_a) which is predominant (Fig. 17.14). Xerophytes lose as much water in still air as mesophytes, for under these conditions transpiration is controlled by R_a. In a wind, however, the anatomical features of xerophyte leaves (such as hairiness, thick cuticle and sunken stomata) retain a layer of moist air in proximity to the stomata and R_s becomes the regulating resistance (Fig. 17.1).

Mineral nutrition. Of course, water can be considered to be a nutrient because it is essential to plant growth, though it is usually treated separately. Mineral nutrition is, however, a further

Table 17.2 The relationship between certain structural characters of plants and water loss by transpiration.

leaf area	Plants with large foliage area transpire more rapidly; plants with small foliage area transpire less rapidly. Shedding of leaf area and reduction of leaf area/plant are characteristics of xerophytes.
stomata	The number, distribution and size of stomata affect transpiration rate. Stomatal transpiration is proportional to number and size of stomata and to the linear perimeter of the pores not to surface area.
leaf intercellular space	Transpiration is higher in leaves with large surface areas of mesophyll cells exposed, i.e. open spongy mesophyll promotes high rates. Contrasting mesophyll structures may occur on the same plant in sun and shade leaves.
cuticle	Nature of cuticle important, particularly at night when stomata are closed. In some shade plants up to 30% of total water loss may occur via cuticle. In xerophytes this may be reduced to zero.
leaf hairs, scales, leaf rolling and furrows	All these features help retain a layer of moist air in contact with leaf and reduce transpiration. Some have additional effects, such as increasing reflection or radiation.
root : shoot ratio	Plants of dry situations, particularly desert plants, have high root area to shoot area ratios to ensure adequate water absorption.

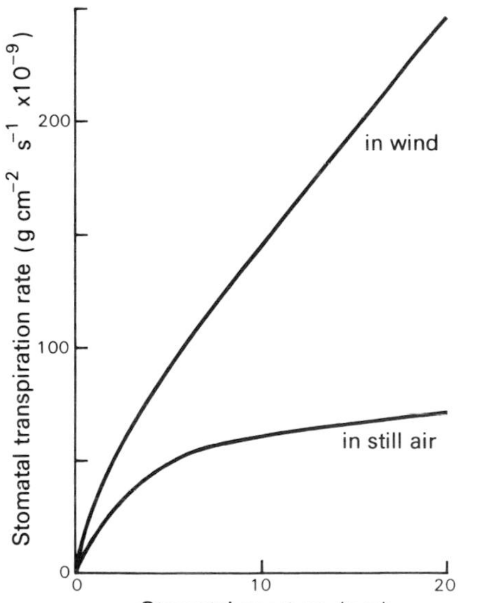

Figure 17.14 Stomatal water loss and air movement (after Sutcliffe 1968).

exogenous variable, the effect of which may be to limit production (Fig. 17.15). The principal nutrients are listed in Table 17.3 together with their functions and sources, and the broad outline of nutrient circulation as global biogeochemical cycles was discussed in Chapter 7. These nutrients are essential requisites for plant growth, either as components of major structural compounds or as critical constituents of enzymes and other compounds active in biochemical reactions. Photosynthesis would be impossible without the magnesium in chlorophyll. Similarly, there would be no nitrogen fixation without molybdenum. If we discount carbon, hydrogen and oxygen, then, with the partial exception of mineral nitrogen and to some extent sulphur, the availability of mineral nutrients to primary producers depends on the microbial decomposition of organic matter and the detrital system (see Ch. 19), with rock weathering and precipitation input representing secondary sources. If these sources are inadequate, nutrient deficiency will occur, and growth and productivity will be limited even when other factors are favourable. Most natural soils show some nutrient deficiency and respond to artificial applications of

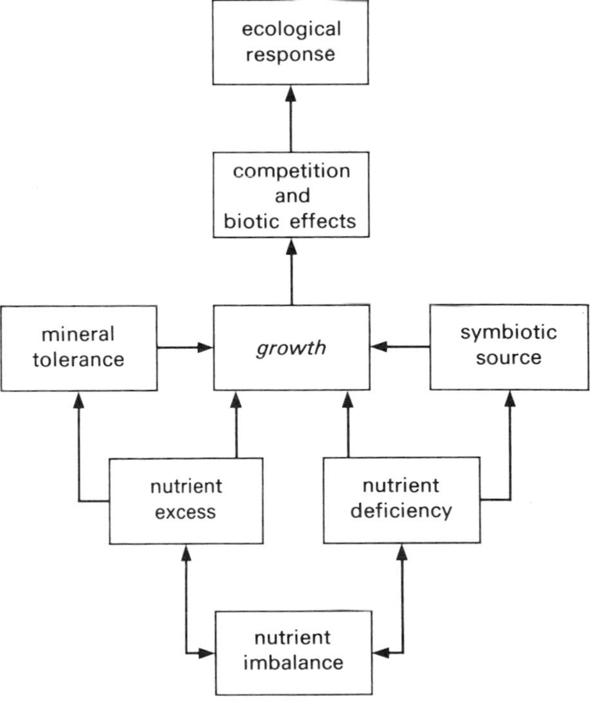

Figure 17.15 Mineral nutrition and ecological response (after Bannister 1976).

Table 17.3 Mineral nutrition (from Collinson 1977).

Nutrient	Physiological function	Sources	Environmental effects
oxygen	respiratory metabolism	photosynthesis of green plants	Important determinant of plant distribution in relation to soil aeration. Where soil is badly aerated or permanently water-logged anaerobic conditions prevent root growth in plants without special adaptations and toxic materials, e.g. H_2S may be generated.
carbon dioxide	source of carbon in photosynthesis	decay, respiration and the oceans release some 10^{12} tons/year	Some slight natural variations of atmospheric concentration but with little effect on plant growth and distribution. Important ecological effects on soil acidity. Total concentration of CO_2 (320 p.p.m.) sets ultimate limit on photosynthesis. Plants can increase photosynthesis up to 3 × normal with increasing concentration.
nitrogen	essential element of proteins. Can only be absorbed in fixed form (NH_4, NO_2, NO_3)	drawn from the atmosphere by a host of microbes, and lightning	In most well aerated soils of intermediate acidity fixed N usually freely available. Deficiencies associated with cold, wet soils, very porous soils and tropical soils where vegetation cover is cleared. Where destruction of organic matter is slow or dead material highly lignified, acid 'mor' peat may accumulate as humification may be inhibited.
sulphur	essential for protein synthesis and vitamin synthesis	sulphates in well aerated soils, pyrites and gypsum in arid lands, H_2S and reduced sulphur in airless soils	Cycled rapidly by micro-organisms similarly to nitrogen. 'Downhill' losses replaced by weathering, airborne dust, salt spray and volcanic gases. In arid regions strong concentrations of SO_4^- ions exist which select for tolerance. Pollutant sources – some 146 000 000 tons annually of SO_2 – increasingly added to biosphere.
phosphorus	ncorporated into many organic molecules, essential for metabolic energy use	Fe, Al and Ca phosphates; free anions in solution (H_2PO_4 in acid, HPO_4 in alkaline conditions)	Great differences in demand between species and hoarded tenaciously in most ecosystems. Cycled on a world scale with downhill losses replaced similarly to sulphur above. Oceanic reservoir returns deepwater reserve along cold currents via plankton, fish and guano of fish-eating birds.
calcium	essential to metabolism but not incorporated into fabric molecules of living matter	feldspars, augite, hornblende, limestone, and sulphates and phosphates in arid lands	Strong selective effects in all habitats – lakes, marshes, grasslands, forests, rock outcrops. Important determinant of prime physicochemical characteristics of soil. Antagonistic to toxic effects of K, Mg and Na. Retention of ions by colloids closely related to climate, especially rainfall.

potassium	essential to many metabolic reactions, especially protein-building and transphosphorylation	feldspars, micas, clay minerals	Deficiency has marked effects on carbon assimilation thus lowering production and biomass. Certain crops – beet, cotton, vine, legumes – are very sensitive.
magnesium	vital constituent of chlorophyll	biotite, olivine, hornblende, augite, dolomite, and clays of the montmorillonite group	Excess produces serpentine barrens, e.g. in California, Spain, New Jersey, southern Urals, Japan, New Zealand. Natural climax is replaced on these by impoverished, often scrubby vegetation commonly with specialised ecotypes, e.g. *Quercus durata* in California.
iron	oxidation and reducing reactions in respiration	iron silicates, iron sulphates, free ions chelated with organic molecules	Calcareous or alkaline soils may be deficient as iron may be precipitated as insoluble hydroxides. May also be deficient where copper or manganese is present to excess. Vines and fruit crops may be easily affected by iron deficiency.
manganese	minute amounts needed for certain enzymatic reactions	ferromagnesian minerals; absorption dependent on other metallic cations	Deficiencies noted in mid-latitudes especially. Tropical soils, especially feralites, may have excess manganese which has toxic effects.
zinc	enzymatic metabolism	zinc-bearing vein minerals	Often leached out of the soil profile in acid soils. May be insoluble in alkaline soils. Certain species, e.g. *Viola calaminaria* of the Harz Mountains in Germany, are endemic to zinc-rich soils.
copper	essential for respiratory metabolism	copper-bearing vein minerals	Deficiency frequent in alkaline soils. Any excess has strong selective effects, e.g. in Katanga, the 'copper flower', *Haumaniastrum robertii*, has 50 × the normal copper content in its leaves; also *Becium homblei* cannot germinate without 50 p.p.m. of copper at least in the soil. The latter is a reliable prospecting index for mineral veins.
boron	necessary for successful cell division during growth	soluble borates are the only assimilable form	May be leached out in acid soils. Some crop plants – beet, potato, cauliflower – show considerable sensitivity to any deficiency.
molybdenum	essential for nitrogen fixation and assimilation	vein minerals	Deficiency in acid soils frequent and also in certain tropical soils on ancient land surfaces.

nutrients with increased production. However, the situation is far from simple, for it is often the case that additions of nutrients lead to a change in species composition, a fact which would suggest that the original species were genetically adapted to grow in nutrient-deficient soils. Sometimes such an effect is indirect, acting through competition. Such a case is reported by Jeffrey and Piggott (1973) where the survival of *Kobresia simpliciuscula*, a rare sedge in the relict flora of Upper Teesdale (England), is due not to a positive tolerance of soil phosphorus deficiency but to the fact that it restricts competition.

The reasons for nutrient deficiency are very complex, but soil type, soil texture, ion exchange capacity, weathering and decomposition rates are all important controlling variables. As in the case cited above, phosphorus is perhaps the most widespread soil mineral nutrient deficiency and it is in short supply in almost all soils. Nitrogen too is often limiting, particularly where decomposition is slow, as for example in cool, acid or waterlogged soil environments where C:N ratios are high (Fig. 17.16). Many plants are adapted to such environments by the presence of nitrogen-fixing bacterial or fungal symbionts associated with their roots, or perhaps more spectacularly by insectivorous habits which enhance their nitrogen supply. In cases such as these, and more particularly that of phosphorus where an element is limited in its availability in the soil and the rate of replenishment by weathering and precipitation is low, the amount of the element in the living and dead biomass of an ecosystem becomes critical. Furthermore, the circulation of the element between them becomes a negative-feedback loop to which the productivity of the vegetation must become adjusted by the conservation of that element. The circulation of nutrients such as these, where the bulk of the ecosystem's stock is held in the organisms themselves and the fraction in the soil is subject to rapid turnover between release from decaying litter and uptake again by roots, can be said to be tight.

In order to conserve such tight circulation elements against loss there is a redistribution and re-use of mobile ions of these elements within the plant prior to any senescence. Ions of elements such as nitrogen, phosphorus and potassium move readily from older tissues to more metabolically active sites. Phosphorus is particularly mobile and an atom of phosphorus may be incorporated and released continuously and make several complete circuits of the plant in one day (Biddulph 1959). As the leaves of deciduous perennial plants senesce and die, considerable quantities of nutrients (of which N, P, K, S, Cl and perhaps Fe and Mg are most important) are withdrawn from them to minimise losses. Calcium, silicon, boron and manganese are translocated to the leaves before abscission, and litter fall occurs enabling the plant to shed excess nutrients not used in metabolic processes. Key nutrients may also be preferentially accumulated in storage organs such as rhizomes to be available for rapid vegetative growth in perennials, while these same elements are concentrated in seeds in both annual and perennial plants to be remobilised and transported into the growing embryo during germination. In this way even the

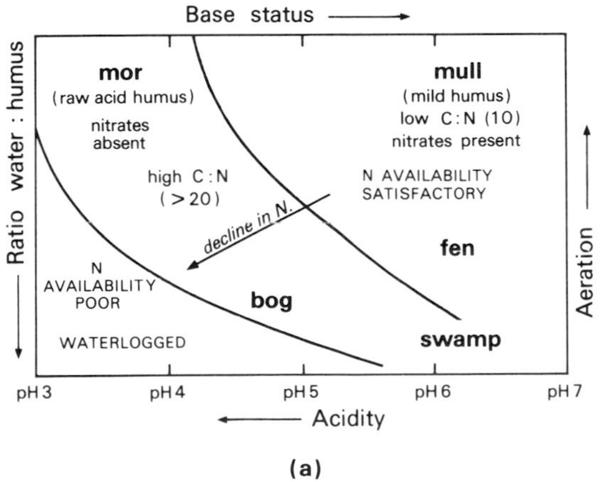

(a)

(b)

Figure 17.16 (a) Carbon : nitrogen ratios in relation to soil water and aeration status. (b) The sundew (*Drosera*), an insectivorous plant.

death of an individual plant need not represent a complete loss of the nutrient elements it contained, for some will have been passed on to the next generation.

Other elements which are present in the environment, and which are released by weathering or supplied by precipitation in appreciable quantities, can be said to be in fairly loose circulation. In such cases the bulk of the ecosystem's store of the element is usually in the soil, not in the plant biomass, and hence availability is not so critically affected by turnover rates and decomposition. The ecosystem can afford to be more careless in their use and it does not actively conserve against loss to the hydrological cascade through the denudation system. Analysis of the solute loads of streams, such as the figures quoted for the Hubbard Brook catchment in Chapter 24 (see Fig. 24.7 and Table

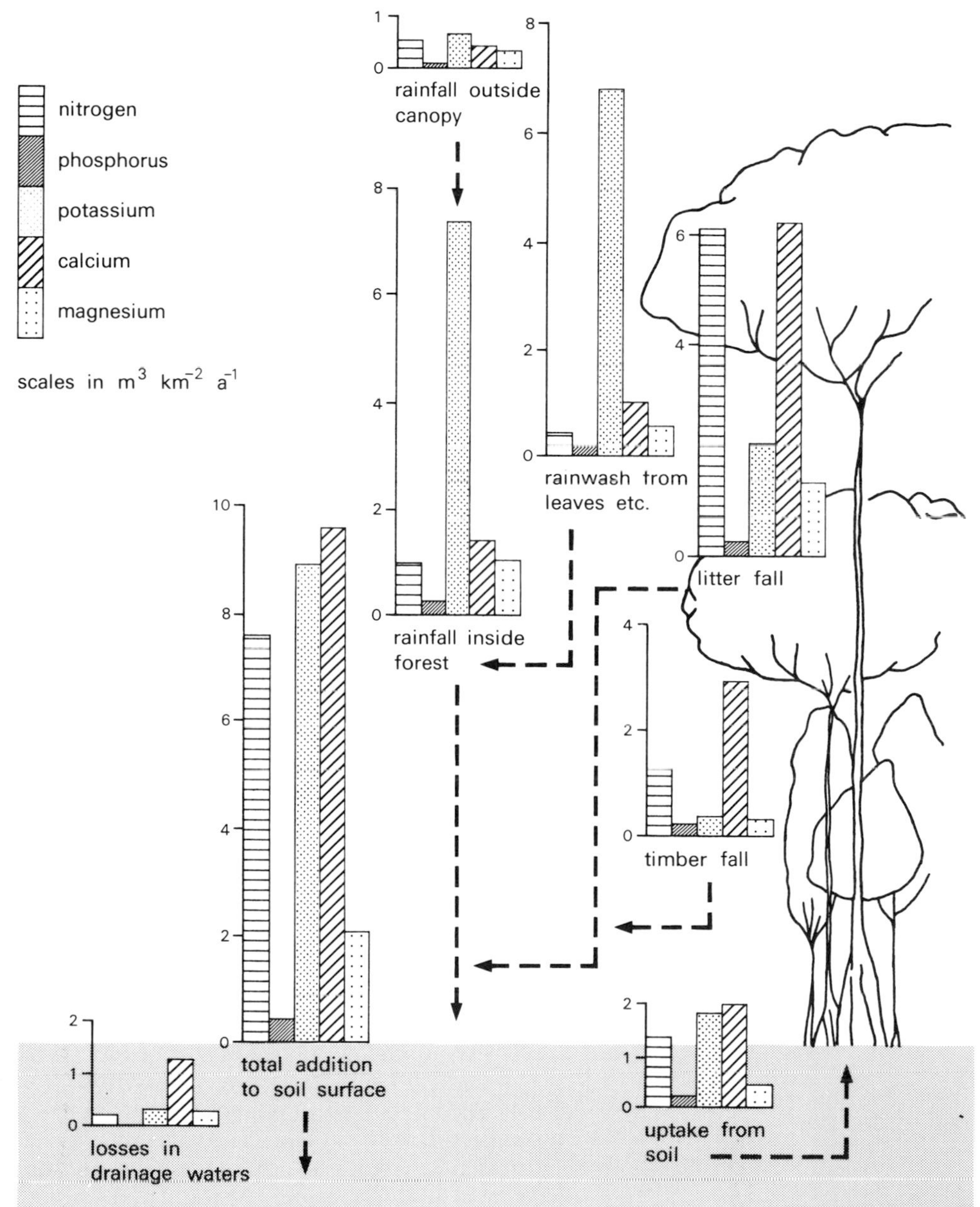

Figure 17.17
Nutrient circulation for tropical rainforest.

24.1) often shows a net loss of such elements (e.g. Na, K, Ca) when precipitation input and stream discharge output are compared. If the ecosystem is actively growing and accumulating biomass or just maintaining a steady-state biomass, this difference must be made good by weathering.

These concepts of tight and loose nutrient circulations are important, for not only may the circulation type be different for different elements in the same ecosystem, but ecosystems in contrasting environments may be characterised by distinct types of nutrient circulation as a whole. For example, tropical forest ecosystems have tight circulations for all nutrients when compared with temperate forests (Fig. 17.17 and Table 17.4).

It is not only the macronutrients that may be in short supply. Deficiencies of trace elements also occur and limit growth and production, but perhaps the most spectacular effects of deviations from normal trace concentrations by these elements are when they are present in abnormally high concentrations. Most of these trace elements are metals with high atomic numbers and they give rise to heavy metal toxicity, of which many examples are documented. Some are entirely natural, like the sparse unproductive vegetation of copper-rich soils of Zambia and the nickel–chromium toxicity and calcium–magnesium imbalance of serpentine soils (see Ch. 5) with their unique and largely unproductive flora. Other sites are the result of environmental pollution by toxic mining and smelting wastes. In all cases, however, some species or ecotypes have evolved tolerance mechanisms to specific metal toxicity which appear to involve the immobilisation of the metallic ion by complexing at cell walls (by chelating agents), thereby keeping it away from metabolically active sites where it would block enzyme activity. Such adapted species have been used often as indicator species in prospecting and, more recently, have received renewed attention as indicators, this time of sources of watersupply pollution by metals such as lead, as awareness of potential health hazard has increased.

Table 17.4 Distribution of all major nutrient elements in store in three forest types, (a) by weight and (b) as a proportion of total biomass.

Nutrients (kg ha^{-1})	Pine forest a	Pine forest b	Beech forest a	Beech forest b	Tropical rainforest a	Tropical rainforest b
forest biomass	1112	1.0	4196	1.0	11 081	1.0
annual litter	40	0.036	352	0.084	1540	0.138
soil and partly decomposed litter	649	0.58	1000	0.28	178	0.0016

There are, however, other examples of soil:nutrient imbalance which are more common. The most familiar is the acid–alkaline continuum in soil reaction, or soil chemical environment. At both extremes nutrient imbalances occur to which particular groups of plants have evolved tolerance, the so-called **calcifuge** and **calcicole** species respectively (Fig. 17.18). This topic has been the subject of a considerable volume of research and the mechanisms involved are not simply tolerance of high acidity at one end and of high calcium carbonate levels at the other. Acid soils are generally leached and nutrient deficient but, more important, at low pH aluminium is more mobile so that adaptation to acid environments involves tolerance of aluminium toxicity as well as of low nutrient status. Basic soils with high pH and well drained oxidising environments, on the other hand, lead to iron deficiency as it is present in an unavailable Fe III state, in addition to the excess of calcium.

The variation in mineral nutrient levels, therefore, not only sets limits to plant growth and primary production but also regulates the distribution, composition and productivity of the community through the ecological response of adapted or tolerant species. Such response, as with water availability, will depend on the **phenotypic plasticity** of the species concerned, or the extent to which they can accommodate variations in nutrition without genetic change. It will also depend on the genetic flexibility of the species and the speed with which new races or ecotypes are produced and selected for their tolerance of particular nutrient regimes.

17.3.3 Primary production and plant competition

So far in this chapter we have been considering the structural organisation and functional activity of the primary production system of the ecosystem. We have seen the way in which these features reflect the interplay between the endogenous variables which characterise the system itself (i.e. the inherited genetic and phenotypic attributes of the plants themselves) and the exogenous, or environmental, variables which define the conditions under which the primary production system

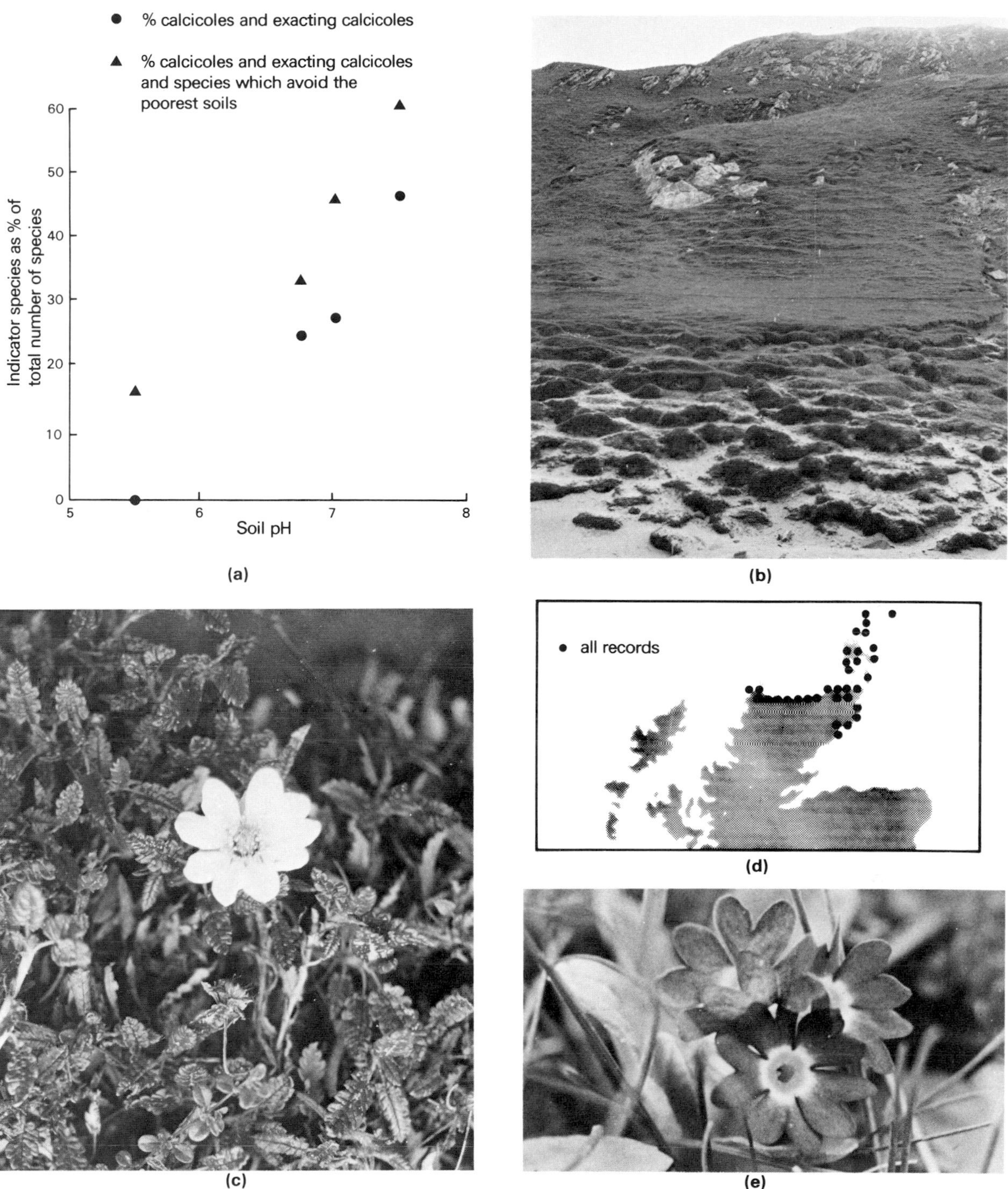

Figure 17.18 Soil base status and the distribution of calcicolous species. (a) The increase in the representation of calcicoles as soil pH increases at sites showing a progressive increase in the effect of wind-blown shell sand on soil calcium. (b) Wind-blown shell sand banked against the headland and colonised by *Dryas octapetala*, forming distinct hummocks. (c) *Dryas octapetala*, the mountain avens, a calcicole. (d) The distribution of an exacting and rare calcicole, *Primula scotica* (e), correlating with coastal sites enriched with shell sand. (All at Bettyhill, Sutherland, Scotland.)

operates. We have seen how this interplay controls and regulates the primary productivity of the system with the result that any ecosystem operating under a given set of conditions will arrive at an optimum state for a sustained production under those constraints. The optimum state is achieved, therefore, by competition between the available species for light, space, water and nutrients over the duration of their life cycles. Such competition is for the primary producer niche in the ecosystem and it leads to niche differentiation and segregation, not only in space but also in time so that a diverse, but integrated, efficient and stable community emerges.

17.4 Geographical variation and comparison of ecosystem primary production

In Chapter 7, global figures for the net primary production of the land and oceans were given together with a warning that many such figures are mere estimates, some almost guesses. Nevertheless, we shall consider here to what extent generalisations can be drawn from the pattern of primary production in different ecosystems at the surface of the planet. From the figures shown in Table 17.5 the annual production of dry matter can be divided into four main groups. At the high end of the range (2.0 kg m^{-2} a^{-1}) are some tropical forests, some developmental or successional communities (see Ch. 23) and some swamp and marsh communities (annual values for *Spartina* and *Phragmites* reach 150–180 kJ m^{-2} a^{-1}) and intensive all-year-round tropical agriculture with such crops as sugar cane, which holds the record for the highest annual production (Java), and maize, which holds the daily rate record (in subtropical Israel). However, the highest transient productivities in natural ecosystems occur in shallow marine waters over coral reefs, estuaries and perhaps some shallow freshwater springs. These ecosystems with dry matter production 1–2 kg m^{-2} a^{-1} include the remaining tropical forests, most temperate forests, some grasslands, wetland habitats and highly productive agriculture, mainly the energy-subsidised (fuel, fertilisers, etc.) agriculture of the temperate zone. Below this there is a middle range of the order 0.25–1.0 kg m^{-2} a^{-1} which contains a variety of communities, grasslands, shrublands, some woodlands and the continental shelf and, interestingly enough, most cereal crops. Finally, there is a group of habitats and plant communities with low net annual production figures up to 0.25 kg m^{-2} a^{-1}. These are the extreme habitats of aridity, and high or low temperatures, desert, semi-desert and tundra on land, and also much of the open ocean.

Table 17.5 Net primary production of the Earth (modified after Whittaker & Woodwell 1971 and Whittaker & Likens 1975).

	Average P_n (kg m^{-2} a^{-1})	Average P_n (10 J m^{-2} a^{-1})	Area (10^6 km^2)	World P_n (10^{19} J a^{-1})	Average B (kg dry wt m^{-2})	Ratio $P:B$
tropical rainforest	2.2	3780	17	75.6	45	84
temperate deciduous forest	1.2	2475	7	44.6	30	92
boreal coniferous forest	0.8	1512	12	18.1	20	76
woodland and shrubs	0.7	1134	8.5	7.9	6	189
tropical savanna	0.9	1323	15	19.8	4	331
temperate grassland	0.6	945	9	8.5	1.6	630
high-latitude and alpine tundra	0.14	265	8	2.1	0.6	442
desert and semi-desert shrub	0.09	132	18	2.4	0.7	189
rock, sand and ice desert	0.003	6	24	0.14	0.002	300
agriculture	0.65	1229	14	17.2	1.0	1229
marsh and swamp	2.0	3780	2	7.6	15	315
Total or average (land)	0.77	1380	149	205.6	12.3	110
open ocean	0.125	242	332	80.3	0.003	80 667
continental shelf	0.36	662	27	17.9	0.01	6620
estuaries and littoral	2.5	3780	2	7.6	1.0	3780
Total or average (ocean)	0.15	293	361	105.8	0.009	32 556

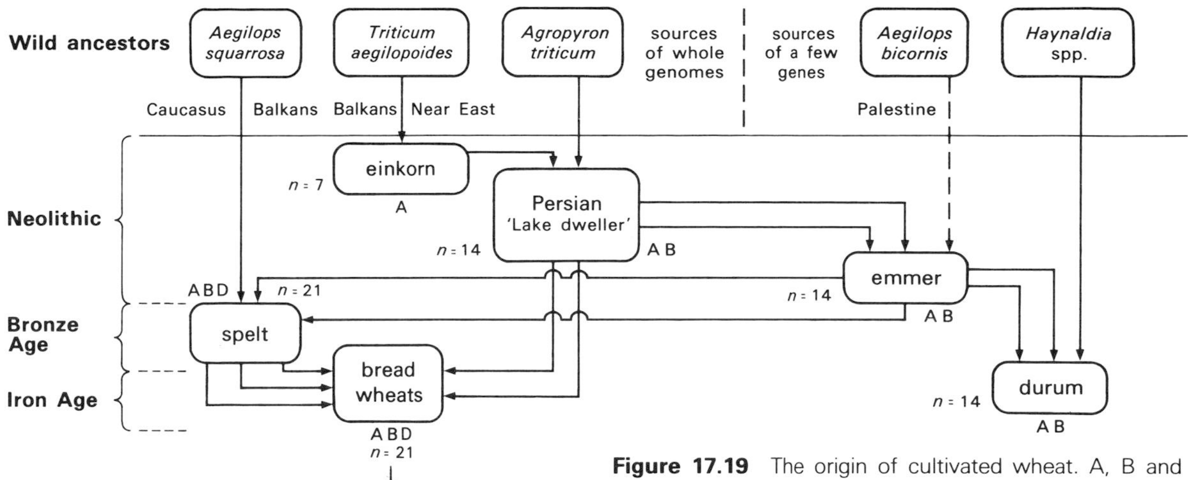

Figure 17.19 The origin of cultivated wheat. A, B and D represent the ancestral genomes *Triticum aegilopoides, Agropyron triticum* and *Aegilops squarrosa* respectively.

These groups reflect fairly clearly the integrated effects of the limiting and regulating factors that we have already discussed as controlling photosynthesis and productivity. Light energy is probably not the major variable here, for the variations in net photosynthetic rate or carbon fixation per unit of photosynthesising surface do not vary enough to account for variation in production figures. Total photosynthesising surface, however, does vary and it is reflected in the leaf-area index. Both this and the variation in productivity reflect gradients of temperature, moisture and nutrient availability. The high production categories above are all associated with non-limiting conditions of temperature, water and nutrients, and all carry large leaf areas; and, in the case of sugar cane, maize and reedswamp, the efficiency of light penetration is enhanced by steep angles of leaf insertion. In conifers and evergreen forests the high leaf-area indices, and particularly the leaf duration, probably compensate for less favourable climatic and environmental conditions. The steep angle of leaf insertion may also partly account for the very high production of some grass and grass-like ecosystems.

These are structural considerations and their effect is emphasised if we look at the production to biomass ratio (P_n:B). Immediately the picture changes with aquatic ecosystems having the highest ratios, then agriculture, then grassland, wetland, desert and tundra communities, savanna and finally forests (with the lowest ratios). As we have seen earlier, this ratio reflects the extent to which the ecosystem carries a large surplus of non-photosynthesising and respiring tissue, which explains the low ratios for forests. Xeromorphic communities with their sclerophytic (thickened) or succulent habits also suffer from this handicap and are therefore inherently inefficient.

The interpretation of production figures will, in the last analysis, depend on the point of view adopted. For example, if we are concerned with the production per unit area of the Earth's surface (P_n m^{-2}), tropical forests, reedswamps and marshes head the list. If, however, it is production per unit biomass that interests us, then it is the oceans, or rather their productive margins, that come out on top, for reasons already discussed. Alternatively, we may be interested in production of the total ecosystem area relative to the total area of the Earth's surface. Tropical rainforest again scores here, for although the oceans occupy two-thirds of the surface they contribute only one-third of global primary production because the large area of deep ocean has low values due to nutrient limitations. Indeed, terrestrial forests contribute something of the order of 45% of total world primary production, a fact of some significance when we reflect on the extent of past deforestation and its currently accelerating rate (see Ch. 24).

We shall continue to explore some of the implications of production data in Chapter 18, but first we must consider the secondary production of the animals of the grazing and predation chain. For the moment, suffice it to observe that from Man's standpoint, moisture and nutrients appear to be the principal limits to natural primary production, other things being equal; that herbaceous communities are more productive than forests under non-limiting conditions; and that even with massive technological investment and energy subsidies, agriculture is only slightly more productive than natural communities, and much agriculture is actually less productive. However, in this context it must be said that crop plants are more useful and more easily harvested, facts which reflect centuries of cultural selection of genotypes and programmes of crop breeding (Fig. 17.19).

18 The grazing–predation system

18.1 The heterotroph

Part of the net primary production of an ecosystem represents the energy and material input to the heterotrophic or consuming organisms of that system – the animal community. As we saw in Chapter 6 when discussing the heterotrophic cell model and heterotrophic level, animals rely either directly (herbivore) or indirectly (carnivore) for all their carbohydrates on those originally produced by plants. Much of the plant carbohydrate material is, however, unavailable to animals in the sense that although they may eat it they cannot use it. The cellulose of plant cell walls cannot be digested by most animals except by those (such as cattle) which possess **cellulase**-secreting bacteria living symbiotically in their digestive tracts. The lignin of thickened woody plant tissue is virtually completely indigestible, though here again a few animals have symbiotic gut fungi which can degrade lignin. In both cases the animal only digests the simple products of degradation.

All animals can use their digested carbohydrate to manufacture other related compounds, for structural tissue, for energy-yielding reactions or for storage, usually as fats. They can also convert their carbohydrates to organic acids which can react with ammonia to form amino acids (the building blocks of protein) with the elimination of oxygen. However, the extent to which animals produce amino acids in this way is negligible when compared with plants, mainly because they have very small quantities of the enzyme systems necessary to do so; but given a few amino acids to start with they can manufacture many others. Some exceptions remain – those amino acids which require the synthesis of hydrocarbon rings – and these they must acquire in their food, from plants. Some amino acids, therefore, are essential components of the animal diet, as too are vitamins (see Ch. 6), of which about 16 are needed by animals.

Therefore, animals are totally dependent on plants. They cannot manufacture carbohydrates or proteins (except to a very limited extent) from any materials except other carbohydrates and proteins, and such substances as vitamins they may not be able to manufacture at all. So the transfer of food from producer to consumer is not only a transfer of an energy source – a respiratory substrate – but also the transfer of essential nutrient elements, as components of essential compounds. Although animals can and do obtain mineral nutrients directly from the inorganic environment, as for example in drinking water or the 'salt licks' put out by the farmer for domestic herbivores, it is usually the nutrient content of their plant food which is critical. Moss (1969) has shown conclusively the effects of variation in the nutrient status of heather shoots in the diet of the hen red grouse (*Lagopus lagopus scoticus*) and ptarmigan (*L. mutus*) on the properties of the egg and chicks. Fluctuations in lemming (*Lemmus* sp.) populations in Alaska have been partly explained in terms of nutrient deficiencies in the vegetation and hence in the milk of the females during lactation, which affects the survival rate of the offspring (Shultz 1964). The production of the red deer (*Cervus elaphus*) in the Scottish Highlands is higher where grazing over basic rock types than over acid rocks.

18.2 Modelling ecosystem animal production

The framework for the discussion of the functional activities of the ecosystem in this section has been the model of trophic structure. In Chapter 17 it was relatively easy to consider the organisation of the individual plant and of the primary producer trophic level from the point of view of energy flow and throughput of nutrients. As we shall also appreciate with trophic models of the detrital system in Chapter 19, the modelling of consuming systems is much more difficult. In the first place the consumer, or grazing–predation system, is concerned with more than one trophic level. This would be unimportant if each organism could be classified unequivocally as a herbivore or carnivore. Even when this is possible it is difficult to define the diet, for very few species are restricted to a single food type (**stenophagus**), as for example the Koala which feeds exclusively on *Eucalyptus* leaves. In such species, which are limited to one food source, the distribution of that food presents a geographical barrier limiting dispersal and speciation. The juniper moth, whose larvae are restricted (in the field) to the juniper as a food plant, have become discontinuous in their distribution in the British Isles (S.E. England, Lake District, N. Pennines and Scotland) as the distribution of the shrub has contracted (Huxley 1942). Most animals, however, have varied diets even when food is abundant and they can show considerable variations in food taken at different times or from place to place. Such behaviour is referred to as **eurphagous** or **polyphagous**. Indeed, many species will regularly, occasionally, or under the pressure of limited food, behave as **omnivores** or **diversivores**. In other words, it is often extremely difficult to assign organisms to particular trophic levels and food-web relationships and hence the pathways of energy and nutrient transfer become both diverse and complex. Many of the small animals in the woodland ecosystem described in Chapter 16 (see Fig. 16.1) such as mice and voles, will, though mainly herbivores, also eat insects, particularly in summer. The blackbirds will feed on worms and insects in the summer because they are plentiful, but in winter they eat fruit, berries and seeds. Even the red grouse mentioned above, with a diet consisting almost exclusively of young heather shoots, will take appreciable quantities of insects during the summer. Now, not only may the food taken vary from season to season as in these examples, but it may also change through the life cycle of an organism, as in the common frog which for a brief interval functions as a herbivore while still a tadpole, before leaving the pond to exist on a carnivorous diet of slugs, worms and insects. Unfortunately, the confusion of trophic levels does not stop with the herbivore and carnivore levels, for as you may have noticed in some of these examples the detritivore level (represented here by earthworms) was being tapped. Other examples of similar confusion arise with animals such as hyenas and vultures which take dead flesh as food. Are they carnivores or detritivores? Indeed, perhaps there should be a separate level – the scavengers – for many carnivores will behave in this fashion, especially under food stress, even the so-called king of beasts, the lion.

In spite of these observations, however, most animals show to a greater or lesser degree some food preference under normal conditions and this is, of course, an important factor in the definition of their ecological niche, particularly the trophic niche (Fig. 18.1). Nevertheless, it remains true that although the concept of trophic levels is helpful, any study of energy flow through the grazing–predation system will require due attention to be paid to food sources, preferences, feeding habits and life histories. However, there is one further difficulty in modelling consumer production and that is the mobility of many animals. As with primary production, the biomass and production of animal communities is expressed as mass or energy per unit area of the Earth's surface (and in the case of production, per unit time). Unfortunately, unlike plants, animals will not necessarily wait to be measured: they will enter and leave the area of study. To a small extent such movements do affect measurements of primary production. This is especially true in aquatic ecosystems where input and output in running water may be important, but it also occurs in terrestrial ecosystems as, for example, when leaf litter is carried out of a woodland by the wind. In studies of animal pro-

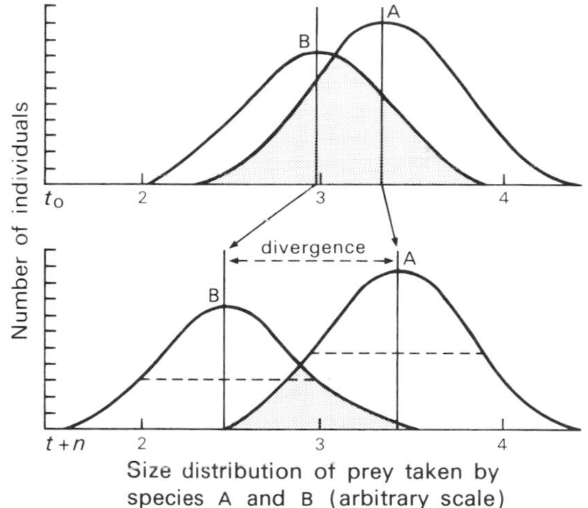

Figure 18.1 Prey size and niche differentiation. When two species overlap in mean prey size, competition favours those individuals taking prey of a size where there is no overlap, so that mean prey sizes for the two species diverge, over the time interval $t_0 - (t + n)$, as indicated by the arrows. (After Whittaker 1975.)

duction these factors may be very significant indeed but, as in the case of migratory species, their effect may be difficult to assess. For example, in the arctic tundra the relationship between the primary production of a relatively small area of vegetation, say 1 km^2, and the production of a herd of caribou or reindeer, may be very misleading (Fig. 18.2). In reality these migratory herbivores will range over perhaps hundreds of square kilometres and be dependent on the production of not only a vast area of tundra vegetation but in winter they may also crop the forest–tundra zone.

Because of these difficulties we shall not attempt to model the trophic structure of real grazing–predation chains too closely, but shall concentrate on some of the more important theoretical relationships of such models. For similar reasons it is not easy to present a model of the transfer pathways of energy and nutrients through the individual animal, as was the case with plants. The great diversity of animal trophic niches, food type and availability means that a large variety of morphophysiological mechanisms have evolved in the animal kingdom affecting feeding mechanisms, structure and shape of the digestive system, and the nature of the digestive process including the types of enzymes involved. In very general terms, however, it might be possible to draw functional parallels between the digestive and respiratory systems of animals and the xylem and stomatal pathways in plants, and the blood vascular systems of animals and the phloem pathway in plants, at least as far as the movement of materials is concerned.

18.2.1 Secondary production: the individual

To begin with, we shall consider the conversion of primary production to secondary, or animal production in relation to the individual animal, because there are additional problems involved if we attempt to discuss the production of the total animal community of an ecosystem. The term conversion is used in preference for consumers as we are not really dealing with production. This distinction is perhaps clarified by an economic analogy: winning a ton of coal is primary production in the economic sense and is comparable, therefore, to photosynthesis (indeed photosynthesis *is* economic primary production in the case of crop plants); the output of coal is said to be consumed by the economy, but this consumption will involve waste, for some coal will be left at the pit unexploited or in tip heaps or lost as coal dust; the use of the coal to smelt iron ore, to generate electricity and to manufacture a motor car is strictly conversion, some of it useful conversion of one material to another, some of it waste such as the heat lost to cooling water (unless used for domestic heating) or the ash from the power station and slag from the blast furnace.

While feeding, the animal will destroy (D) a certain amount of the biomass of the previous trophic level, but it will not necessarily consume (C) this amount. Herbivores often damage or

Figure 18.2 Reindeer, relatively conspicuous herbivores in an open landscape.

destroy more vegetation than they actually eat. Sometimes this may be directly related to feeding, as when the squirrel scatters shells and frass (W) when eating buds and seeds, or in other cases it may be indirect, such as trampling by large herbivores. After the kill, many carnivores do not consume their prey in its entirety. Part of the carcass may be left to scavengers, and even these may leave the skin, hair, teeth and parts of the skeleton (W). We can summarise these activities as

$$D = C + W \quad \text{so} \quad C = D - W.$$

Not all of the energy in the food eaten or consumed by the animal will be absorbed or assimilated. Some food will pass through the digestive system without any chemical change, while other food substances, though experiencing chemical breakdown in the gut, are nevertheless not absorbed into the animal's cells. In both cases these materials and the chemical energy they represent are discarded as faeces (F). The analysis of faecal pellets can be the only evidence as to the food of consumers and it is particularly useful in reconstructing the diet of polyphagous carnivores from the fragments of fur, bone, feathers and the chitin exoskeletons of insects they may contain.

The absorbed food is used for growth, for the repair, maintenance and turnover of cell constituents, and also for the breakdown of fuel molecules and cell components in respiration, thereby releasing the energy necessary for vital functions which in animals, of course, include their movement (see Ch. 6). Compounds absorbed in food in excess of the animal's immediate metabolic needs are stored in the body, as glycogen in the case of carbohydrates. However, excess protein broken down into its constituent amino acids is not stored in this way. The amino acids are de-aminated to form fatty acids which are stored as body fat, and relatively simple nitrogenous compounds which are excreted as urine (U). The compounds in urine, therefore, are not available as sources of energy to the consumer and are usually added to the energy loss in faeces. The remaining food absorbed by the animal is assimilated energy (A):

$$A = C - F - U \quad \text{and} \quad C = A + F + U.$$

But, as we have just seen, some of the assimilated food is oxidised and lost during respiration (R) (catabolic heat loss), while the remainder adds to the biomass of the animal and is production (P). So

$$A = P + R$$

where A (assimilation) is the equivalent of gross primary production by plants and P (production) is the equivalent of net primary production. Therefore

$$\underset{\text{consumption}}{C} = \underset{\text{production}}{P} + \underset{\text{respiration}}{R} + \underset{\text{excretion}}{U} + \underset{\text{egestion}}{F}$$

and

$$P = C - R - U - F.$$

In animals, however, neither R nor P is a simple parameter. In warm-blooded animals R is a complex parameter involving heat used in the maintenance of body temperature (**thermoregulation**) (Box 18.1, Fig. 18.3). In these animals, R varies

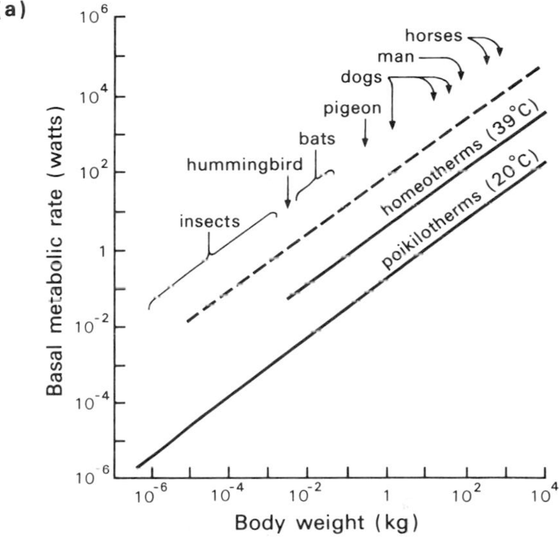

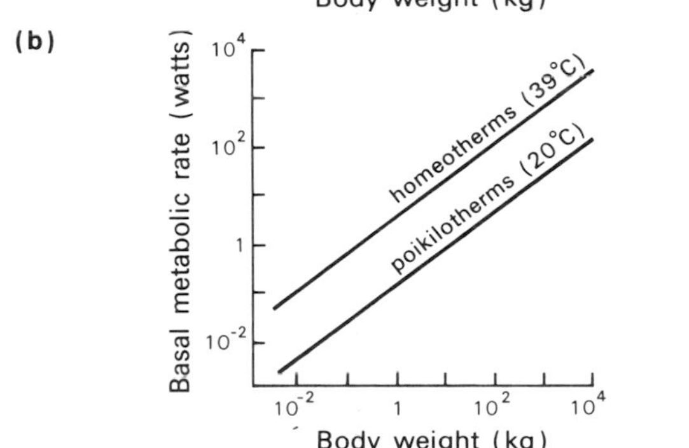

Figure 18.3 (a) Relation between basal metabolic rate of homeotherms, maximum metabolic rate for sustained work by homeotherms (pecked line) and basal metabolic rate for poikilotherms at 20°C. (b) Relation between basal metabolic rates for homeotherms and poikilotherms. (Both after Hemmingsen 1960.)

Box 18.1

WARM-BLOODED AND COLD-BLOODED ANIMALS

The nature of the respiratory heat loss is different in warm- and cold-blooded animals. Warm-blooded animals (**endotherms** or **homeotherms**) which maintain a body temperature (35°–42°C) that is usually higher than ambient environmental temperatures, need to produce heat in amounts that will partly depend on the temperature difference between their body and that of the environment and partly on their insulation and **thermoregulation** mechanisms. Heat generated during the performance of vital metabolic activities will include a proportion used to maintain body temperature, even when the animal is at rest. The production of heat by resting animals is known as their **basal metabolic rate** (BMR). Heat is also produced, however, by what is known as the **specific dynamic action** of food (SDA) and is due to the metabolic processes which follow the absorption of digested food. The SDA is not restricted to warm-blooded animals for it also occurs in cold-blooded creatures (**ectotherms** or **poikilotherms**) but represents wasted energy as it is unavailable for growth or movement and is therefore either neglected or included in a general term for respiration. In warm-blooded animals, however, the heat generated by SDA of the food assimilated is useful heat and can be used to maintain body temperature and therefore reduces the demand for heat from basal metabolism (BMR) or from regulatory movements such as shivering. Ultimately, heat from both sources is lost from the animal, but the rate of loss will be controlled by insulation and other adaptations for thermoregulation. So, in warm-blooded animals the respiratory heat loss (R) can be split into two parts, R_1 (the SDA of assimilated food) + R_2 (the energy loss during vital activities, including movement).

proportionally with the surface-area:volume ratio of the body and with age, tending to be higher in the young than in adults. R is also influenced by activity and behaviour. For example, the congregation of animals will reduce energy consumption as will the ability to maintain normal life activity while increasing the economy of energy expenditure (by lowered metabolic rate), as in the native species of the tundra such as the lemming. Many tundra natives, however, have an enlarged heart relative to related species and are hence endowed with the ability to increase their motor activity (muscular activity, movement) when needed and yet cope with the associated increase in oxygen demand.

Production is also a composite term, for some of the assimilated energy will contribute to growth of the individual while some will be diverted to the development of sexual functions and products (eggs or sperms) which are then lost to the individual but represent the initial energy input into the next generation. So

$$P = \underset{\text{growth}}{P_g} + \underset{\text{reproduction}}{P_r}.$$

The laying of eggs or the growth of young in the uterus, and even after birth during lactation, represent huge investments of energy in P_r for the female. For example, many birds lay the equivalent of their own body weight of eggs over short periods of time. Reproduction, therefore, is very closely correlated with food availability and quality, and this goes some way to explain the territorial and social behaviour of many species during the breeding season. It also brings us to the relationship between animals or their life cycles and production.

In young animals assimilation will exceed respiration ($A>R$) leaving a positive value of P to be channelled into growth of the individual (P_g). At sexual maturity and during breeding, A may still be greater than R but, as we have just seen, the excess production will be accounted for not by P_g (the growth of the individual) but by P_r (reproduction). The rearing of young, particularly in birds, will also involve the expenditure of energy by the parents in food gathering, energy ultimately derived from their own food consumption. So maturity and breeding tend to lead to a reduction or even cessation of growth and, in some species, the exploitation of food reserves laid down in body fat earlier. Although growth may cease with maturity (mammals and birds cease skeletal growth) the number of breeding events and number of offspring will affect the calorific value of the body and it may fluctuate with the storage or dissipation of fat and other food reserves. The value of P, therefore, may be either positive or negative at any particular time in the adult.

18.2.2 Secondary production: the animal community

When we come to apply the relationships considered above for individual animals to complete animal communities, the task of estimating or measuring secondary production or conversion becomes even more difficult. In the first place, there are additional variables to consider. The first of these is the need to measure population size and this is far from easy, except in the case of large and conspicuous animals. Although it may be relatively easy to count polar bear or caribou from the air in an open tundra landscape, such situations are comparatively rare and in theory at least we need such census material on all species in the community. Number alone, however, is not sufficient, for the production of any population will also depend on its age structure, each age class displaying a different productivity relationship. Furthermore, it will be necessary to know the rate at which each population is being added to by both birth and immigration, and diminished by both death and emigration. To measure all of these parameters for populations of all animal species in the community is a daunting task, rarely undertaken. In practice, effort is concentrated on a single or small number of species which are quantitatively most important in functional terms, or we rely on theoretical relationships to estimate the values of variables such as birth rate, using constants based on limited empirical data. Finally, even when these additional population parameters have been measured or estimated, they are used to multiply up or extrapolate production figures obtained from a small number of individuals under controlled conditions and there is no guarantee that such figures will always be valid under the variety of conditions experienced by populations in the field.

It is not surprising, therefore, for all of these reasons, that our knowledge of secondary production on a global scale is still very far from complete and in most ecosystems it is very sketchy indeed. Because of this we shall make no attempt to summarise world secondary production as we have done for primary production. We will consider instead what is known of the efficiency with which energy fixed by photosynthesis is utilised by grazing–predation systems as it is passed to successively higher trophic levels.

18.2.3 Secondary production: the efficiency of energy transfer

We already know that the ratio of gross primary production to incident light is very low indeed, being of the order of 1–4%. The ratio of herbivore production to green plant production is often less than 10% as far as the available figures from natural ecosystems are concerned. The reasons would appear to be twofold. First, the efficiency with which plant tissue is assimilated by grazing animals is reduced by the biochemical differences between plants and animals. In other words, the herbivore's food differs significantly from the tissues that the animal needs to build from it. We have already seen that the consequence of these differences is the indigestibility of such compounds as cellulose and lignin and therefore relatively low net assimilation efficiencies (20–50%). Secondly, consumption by herbivores acts directly to reduce the size of the photosynthetic system, thereby decreasing the primary production rate. The 10% figure may, therefore, represent, as it were, an agreed compromise between a sustained food yield to the herbivore and irreparable damage to the vegetation and seriously reduced primary productivity.

Higher trophic levels fare rather better and the ratio of predator production to prey production is usually higher than 10%. Carnivore net assimilation rates are of the order of 50–90%, which in part reflects the fact that their food is more digestible and closer biochemically to their own tissues. The percentage of energy lost in respiration, however, tends to increase along the food chain from plant to carnivore.

18.3 Energy flow and population regulation

In Chapter 17 and in this chapter we have been tracing the way the original light energy is transferred and partitioned between the trophic levels of the ecosystem, thereby sustaining the populations of plants and animals which make up the living community. These populations are not static, as we have noted on several occasions: they respond to variations in their energy supply or to variations in other conditions such as light and temperature. The organisms of the detrital pathway, which we shall consider in detail in Chapter 19, increase their population size in response to the supply of energy as dead organic detritus. Therefore, they tend to consume their food or energy source almost as fast as it is supplied to them, although the picture may be complicated by seasonal time lags. The result is that under most circumstances undecomposed litter does not accumulate excessively at the soil surface. The only exceptions occur where some other environmental factor such as waterlogging restricts their numbers and rate of increase, so restricting decomposition and promoting the accumulation of peat. Herbivores, however, cannot expand their population size in direct response to the green plant tissue available to them as food. If they did so, the surface of the planet would be denuded of its 'verdant pastures' and 'majestic forests'! In natural ecosystems, for some reason, herbivores settle for about 10% or less of the primary production, and expand their populations only to a size at which this level of cropping is sustained. In other words, some factor other than food or energy supply would appear to be regulating their populations.

There is a considerable amount of evidence, however, that predator populations do increase in response to increased availability of energy as production by prey species. In some cases maximum predator numbers may be limited by direct competition for this food or for living space, but in the higher animals the situation is complicated by social interactions and conventions. Nevertheless, it is broadly true to say that in terrestrial ecosystems decomposer and carnivore numbers are directly limited or regulated by food resources. The same is true, of course, for the primary producers, but in the case of plants we have to read food resources to mean solar energy, water and mineral nutrients. Herbivore numbers, however, are not directly linked to the food resources available to them as vegetation. In contrast, their numbers appear to be mainly regulated by predation. Now, the significance of these relationships can perhaps be realised when it is appreciated that a community where herbivores are held down to population levels which are not damaging to the vegetation (itself resource limited) is most likely to persist or remain stable through time. Conversely, a community where space is continually being made available by excessive grazing pressure from high herbivore populations is most likely to run the risk of replacement and suffer invasion and change in plant composition, i.e. is potentially unstable. The relationship between pathways of energy and matter transfer, the mechanisms of population regulation and the maintenance of a stable community structure are complex subjects and we shall return to them in Chapter 23.

19 The detrital system

19.1 Decomposition, weathering and the soil

After death or excretion, organic compounds cease to be components of living systems and they undergo the processes of alteration and adjustment collectively known as decomposition. In terrestrial ecosystems these processes take place largely on or within the soil. Here too an analogous sequence of processes affect the rocks of the lithosphere and their constituent minerals (see Ch. 11). Both these processes of weathering and those of decomposition proceed towards the establishment of new equilibrium states for materials of inorganic and organic origin respectively. These states are weathered residues and secondary minerals on the one hand, and humus on the other. However, the mobile materials released by weathering and by decomposition are of equal importance for they are the nutrients upon which plant growth depends. In Chapter 7 it became evident that both rock weathering and decomposition, therefore, are critical and potentially rate-limiting steps in the circulation of materials in the ecosphere.

The detrital system of the soil can be modelled in the same way as was done with the weathering system in Chapter 11, by distinguishing between the processes and conditions that promote or activate decomposition and the primary mechanisms of decay. Similarly, these primary mechanisms can be viewed as forming a sequence from the initial breakdown of fresh organic detritus to the production of relatively stable decomposition products. Before considering such a model, however, we shall begin by looking at the nature of the input; that is, at the supply of organic matter.

19.2 The input to the detrital system: the supply of organic matter

The sources of organic matter to the decomposition system of the soil are shown schematically in Figure 19.1. The soil surface in terrestrial ecosystems receives a supply of litter, the major proportion of which is plant debris of varying sizes from leaves, inflorescences, twigs, bark and bud scales to the so-called macrolitter of branches and fallen tree trunks. It also includes the carcasses of animals as well as their metabolites and excreta,

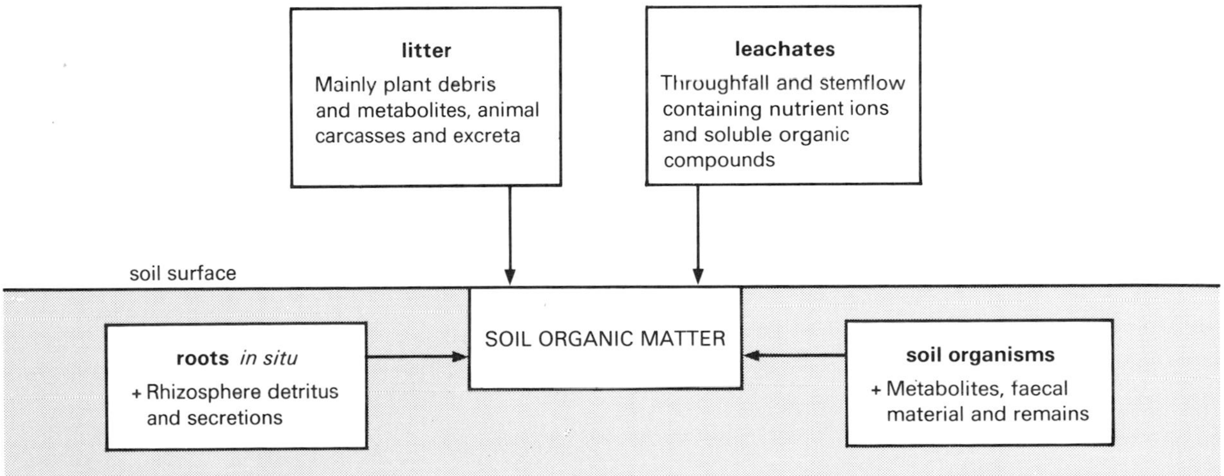

Figure 19.1 The sources of soil organic matter.

though this 'animal litter' is normally quantitatively less significant. The exceptions are those ecosystems with a very large herbivore element, such as managed grazing land where the detrital input from dung can be of great importance. Apart from reproductive structures such as seeds and spores, all of this material is potentially available for decomposition. The litter present at any particular time can be expressed as the biomass of litter in kilograms per square metre or in terms of its energy equivalent in joules per square metre. The amount of litter shed by an ecosystem in a period of time such as a year is the litter production in kilograms per square metre per unit time. Of course, this is not the total increment of dead material, for some will remain above the soil surface (for example the dead limbs of trees) and this is known as the **standing dead biomass**.

The quantity and type of litter will vary from one ecosystem to another (Table 19.1) and, although generally proportional to the biomass of the vegetation, it also responds to variations in environment. Within any one ecosystem the litter supply will vary, not only spatially with variations in canopy density and composition but also temporally both in type and quantity in accordance with the life cycle of the community (Fig. 19.2).

In addition to the litter components, the soil surface also receives organic compounds dissolved or suspended in water that has been intercepted by the vegetation canopy. This rain is, of course, not pure water: it will already contain inorganic material. That portion of the rainfall which reaches the ground directly through gaps in the canopy and as drip from leaves and stems is known

Table 19.1 Variations in annual litter production under different vegetation types (Mason 1970, after various sources).

Trees	Location	Litter production kg/m²/a
Norway spruce	Norway	150
oak	England	300
temperate oakwood	Netherlands	354
	—	440
beech	England	580
beechwood		
evergreen oak	S. France	380–700
tropical forests	Ghana	1055
	Thailand	2330
blanket bog	England	3 tonnes ha^{-1} a^{-1}

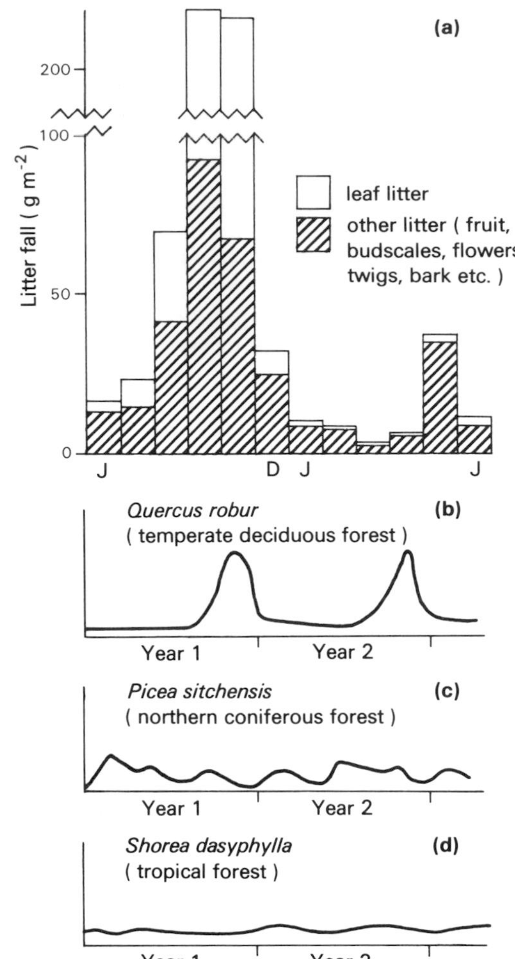

Figure 19.2 Temporal variations in litter supply (a) variations in both quantity and type of litter received by the soil surface over an annual cycle under beech woodland (after Mason 1977). (b), (c), and (d) the annual pattern of litter supply under temperate deciduous, northern coniferous and tropical forest, respectively. Note the autumnal bias in the temperate environment, the regularity of litter supply and the lack of seasonality in the Tropics.

as **throughfall**, while the portion that converges to run down the stems is termed **stemflow**. Both components are involved in a complex exchange system between the canopy and rainfall. Although the canopy may gain water and mineral nutrients through foliar absorption, there is often a net loss of nutrients leached from the leaves as well as substantial losses of soluble carbohydrates. The rain reaching the soil, therefore, is considerably enriched (Fig. 19.3) and it differs from

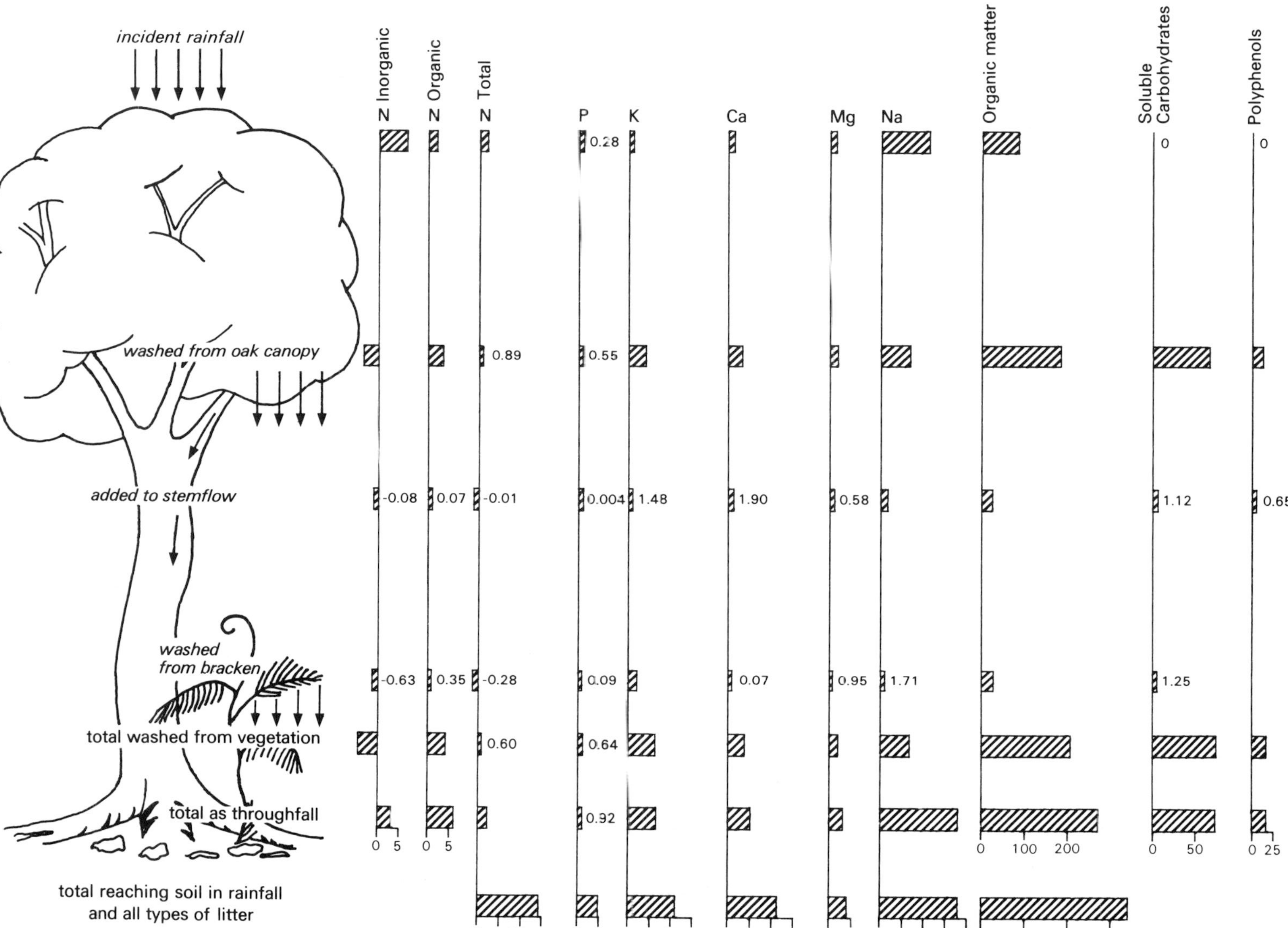

Figure 19.3 Nutrients washed from vegetation and reaching the soil surface in rain water and litter in a sessile oakwood (*Quercus petraea*) (kg ha^{-1} a^{-1}) (after Carlisle *et al.*, 1967).

pre-interception rain water in relative composition.

Beneath the soil surface the death of roots represents a supply of organic matter to the soil detrital system *in situ*. In theory, it should be possible to measure both the biomass and production of these dead roots, but in practice it proves extremely difficult. Nevertheless, it seems probable that large amounts of root tissue become available for decomposition. Even before death, however, live roots are a source of organic compounds in the form of root exudates, root-cap cells, moribund root hairs and epidermal and cortical cells. In some soils, perhaps half the dry weight of lateral roots might be shed in this way.

Although all of this material represents the primary substrate for the production of humus, it is also the food supply or energy source that sustains the final contributor to the organic matter in soils: the organisms of the detrital food chain (Box 19.1,

Box 19.1

SOIL ORGANISMS

There are several bases for the classification of soil organisms, ranging from their habitat preferences and activity to their trophic relationships, but perhaps the most often used schema is based on size.

micro	20–200 μm
meso	200 μm–1 cm
macro	>1 cm

The soil microfauna is made up mainly of **protozoa** – uni- or non-cellular animals including many motile forms with cilia or flagella. There are also many **amoeboid** forms, sometimes with chitin or silica sheaths, and some of these can encyst and survive desiccation for long periods. A third group of protozoa includes the **rotifers**, which have a more differentiated cell structure and which characteristically construct mucous nests from soil particles and faeces. Protozoa are very numerous in soil, with figures of tens of thousands per gram of soil being quite common.

The mesofauna contains the smaller types of worms such as the **nematodes** (eelworms) and **turbellarians**, both unsegmented worms, 1 mm long and 2 μm in diameter which inhabit soil water. They are very numerous in certain soils (several million per hectare) and feed on decaying organic matter and fungi, but many are parasitic and infect plant roots. Slightly larger are the **enchytraeids** or potworms. These are small, segmented worms common in acid and organic soils. Figures of 200 000 m^{-2} have been recorded from heathland and coniferous woodland soils. They feed on algae, fungi, bacteria and organic matter in various states of decay. They are particularly significant in acid soils where they replace earthworms.

Two important groups of small arthropods occur in the mesofauna. These are the **Acari** or mites, common in acid litter and organic matter where they can constitute up to 80% of the soil fauna. Their diet varies greatly, but though some consume litter, many feed on fungal hyphae and spores. Similar food preferences are shown by the **Collembola** or springtails, also arthropods, though some are carnivorous. The **myriapods** include two important groups of soil mesofauna, the vegetarian millipedes (**Diplopoda**) and the active predatory centipedes (**Chilopoda**). Smaller forms of several groups of animals such as ants, beetles, molluscs and particularly the larval stages of many insects also fall into the category of mesofauna. The main role of the group as a whole is to function as activating mechanisms in the fragmentation of litter by consuming plant detritus and its attached bacteria and fungi, reducing it to colloidal dimensions, providing a substrate for humification and moving it deeper into the soil.

On size grounds alone some of the larger members of groups considered above, such as the beetles, are placed in the soil macrofauna. However, the most important members of the macrofauna are the earthworms. These large segmented or **annelid** worms belonging to the Lumbricidae are extremely significant both in the decomposition of litter and in the mixing of mineral and organic components of the soil. There are 25 British species of which only ten are common and, although some species can tolerate mild acidity, earthworms are rare in soils with a pH value below 4.5 and under anaerobic conditions. They ingest both litter and soil so that the comminuted organic fragments and the soil minerals are intimately mixed in their gut, promoting the formation of organomineral complexes. In addition, by secreting calcium from a calciferous gland, the pH of these complexes is raised and they are egested as water-stable crumbs forming worm casts.

The only other members of macrofauna that will be mentioned here are the **Isopoda** or woodlice. These are particularly common in dry and acid litter and soil organic horizons where they replace earthworms as litter destroyers.

The soil microflora is dominated by **bacteria** and **fungi**, though their inclusion in the flora is more convention than anything else, for perhaps neither group should be described as plants. The bacteria are 1–5 μm across and they occur in various shapes, while some are active forms with cila or tufts of flagella. Many secrete polysaccharide gums – perhaps to form a protective sheath which binds and holds small clay and iron oxide particles. These gums are of great significance in soil aggregation. It is their activity that determines their grouping into aerobes and anaerobes, autotrophic, chemosynthetic and heterotrophic groups, while many play important roles in the cycling of elements such as nitrogen and sulphur – some developing symbiotic roles with the roots of higher plants. However, in decomposition their importance lies in their role as efficient decomposers and in the promotion of humification and the resynthesis of humic polymers.

The fungi, which can be very numerous, have two important effects. One is mechanical, as fungal hyphae and mycelia develop and push through decaying litter aiding fragmentation. The other consists of complex chemical effects, as they secrete enzymes which digest organic matter. In acid soils they are often more important and more effective than bacteria, but even on the better soils they may be as important as the bacteria. Some fungi are symbiotic, forming **mycorrhizas** with the roots of species such as the Scots pine. Closely related to both the fungi and the bacteria are the **Actinomycetes** with their fine, branching filaments similar to, but finer than, fungal mycelia. They are aerobic organisms and can survive in dry soils. Most are saprophytic and some, such as **Streptomycetes**, produce antibiotic substances.

The final group of soil microflora is the **algae**. Blue-green algae are most numerous in neutral soils, green algae in more acid environments. Both occur as single-celled micro-organisms, as colonies or as filaments. Because they contain chlorophyll and photosynthesise, most occur near or at the surface and of course some are nitrogen fixers of great significance.

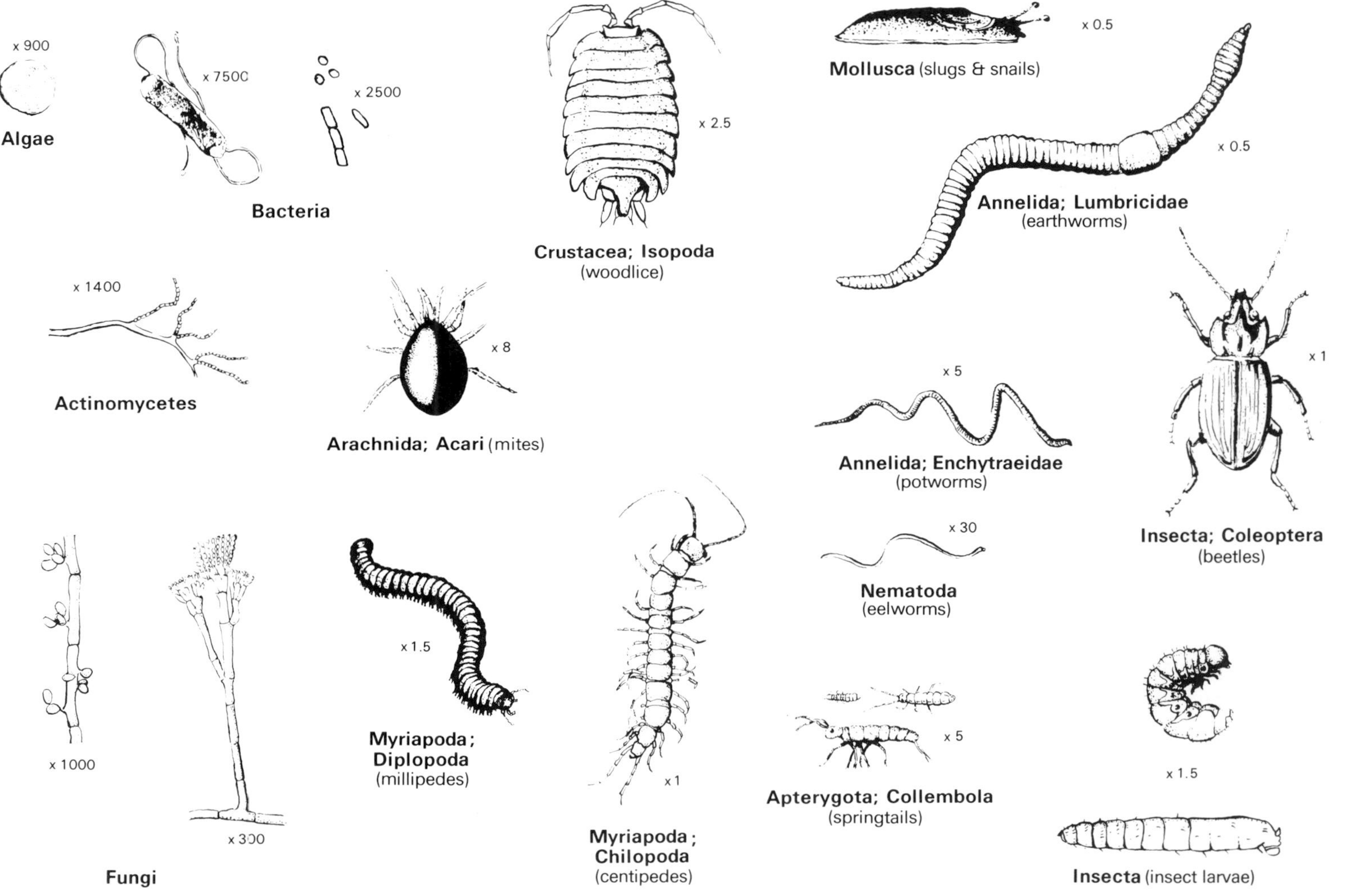

Figure 19.4 The major groups of soil organisms (mostly detritivores), and decomposers and some of the carnivorous groups dependent on them.

Fig. 19.4). On their death these soil organisms represent, like the roots of plants, a direct *in-situ* supply of organic detritus. The same is true of the metabolites and faeces they release into the soil during life. Their overwhelming importance to the functioning of the system, however, is due more to their role as the principal activating mechanism of the decomposition process. Without soil organisms the breakdown of organic detritus would be very much slower and far less complete. The importance of this is clearly evident in environments unfavourable to large and diverse populations of soil organisms, such as the arctic tundra, where largely undecomposed litter accumulates on the soil surface.

19.3 Decomposition: a process–response model

Although the model depicted in Figure 19.5 is similar to that of the weathering system (Fig. 11.2), and in both the active processes are those of fragmentation and the breakdown of chemical structure, the decomposition system is much more complex. The reasons are twofold. In the first instance, the input of dead organic tissue – i.e. the initial state of the system (see Fig. 6.2) – is chemically and structurally more complicated than the equivalent rock and mineral input to the weathering system. Secondly, the role of soil organisms makes decomposition more difficult to model for, in addition to the inorganic processes of weathering, decay involves extremely complex biochemical reactions, many of which are still little understood. Furthermore, and even more so than was the case with the weathering system, the sequential arrangement of these processes is merely an artificial device to aid the clarity of explanation. In the real world they may be proceeding simultaneously; the interactions of populations of the various soil organisms involved and the complementary nature of their roles in decomposition can be very intricate.

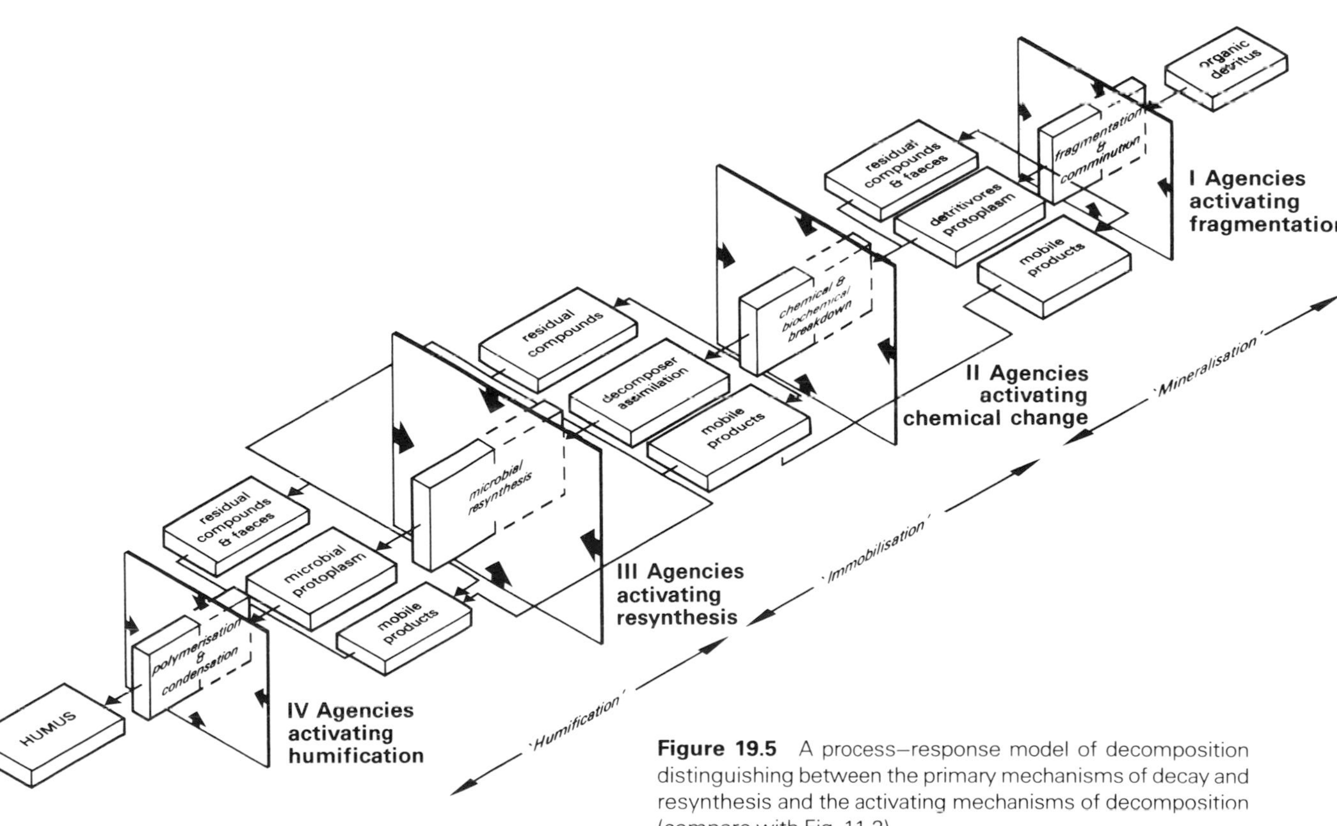

Figure 19.5 A process–response model of decomposition distinguishing between the primary mechanisms of decay and resynthesis and the activating mechanisms of decomposition (compare with Fig. 11.2).

19.3.1 Fragmentation: the mechanical comminution of litter

The fragmentation of litter is activated in part by purely physical processes such as raindrop impact, expansion and contraction and freeze/thaw. Indeed, some of these processes may have started before the litter reached the soil surface. With the onset of senescence, for example, the moisture content of leaves decreases and they become dry, brittle and susceptible to abrasion as a result of movement by wind. Once on the soil surface, trampling can further break up large fragments of litter. By far the most important process activating fragmentation, however, is the maceration and comminution of litter by soil detritivores. These are soil organisms which consume particulate organic detritus and include members of several groups of invertebrate animals. Perhaps the most important are the segmented worms (Annelida), including earthworms and enchytraeid worms, millipedes (Myriapoda), insect larvae (Insecta), mites (Acari), springtails (Collembola) and woodlice (Isopoda) (see Box 19.1). Of course, this digestion by detritivores also involves biochemical change, for the organism digests and assimilates some constituents of the detrital tissue. These are usually only the simpler constituents, for the enzyme systems of these soil invertebrates are apparently incapable of attacking more complex structural molecules such as cellulose, lignin and phenolic complexes. As a consequence, the assimilation efficiencies (food assimilated as a percentage of that consumed) of these organisms are low: 1–3% in earthworms, 6–15% in millipedes, 10%+ in mites, and 15–30% in woodlice. The net effect, therefore, is the physical comminution of the detritus until much of its original character is lost, it is reduced to colloidal dimensions and its total surface area is vastly increased. In this form it is returned to the soil incorporated in faecal pellets, where it may be intimately mixed with inorganic soil particles. These pellets have a better moisture/aeration status than the original litter and a higher nitrogen content as a result of the digestive release of nitrogen (mainly as ammonia) from organic compounds. Such faecal material forms an ideal substrate for microbial activity.

19.3.2 Structural breakdown and microbial decomposition

The initial processes of chemical change are the straightforward inorganic reactions of oxidation, hydration, hydrolysis and solvation, as the dead organic material is exposed to the atmosphere, to light and to water. Such reactions and the release of readily volatisable and soluble products (*ca* 70% of the K and Na in litter is leached into the soil during the first two to three months after deposition) are aided by the initial mechanical break up of the litter. The majority of the litter and detritus, however, consists of complex structural molecules, and the unaided progression of these inorganic reactions would be slow and many of the original compounds would persist. This does not happen, largely because of microbial decomposition which may precede, run parallel with or follow, the initial weathering, leaching and fragmentation by detritivores. It is these decomposer micro-organisms (mainly bacteria and fungi) which permit the detrital substrate to be reduced to a small but highly resistant residue, or to be completely decomposed.

The primary mechanism of microbial decomposition is the enzymatic cleavage of the structural macromolecules of organic detritus. These molecules are long polymers with high molecular weights and are large relative to the size of microbial cells. To utilise the molecules or their components as food, the micro-organisms secrete extracellular degradative enzymes, either on the surface of their cell walls or 'free' into the soil. Here they may be adsorbed onto the organic substrate, or on the surfaces of clay minerals.

These extracellular enzymes are highly specific, not just to the substrate on which they act but to particular linkages. They are often sensitive to the presence of particular functional groups on the molecule, and some can even distinguish between different isomers of the same substance. Many of the enzymes are inductive – that is, they are produced only in the presence of particular substrates. The products of these enzymatically catalysed reactions are the constituent monomers which made up the original polymer, or groups of such units. These are now available to be transported across the microbial cell wall, usually by other carrier enzymes, again highly specific, before they can form a respiratory substrate for the organism. Such products of enzymatic reactions are also available to other soil organisms which do not possess the enzymes necessary to release them and, of course, they are also available to be leached by percolating water (see Ch. 11).

The complexity of these processes has certain

important implications. First, any one organism is unlikely to produce the many different enzymes necessary to degrade the large variety of molecules present in the natural substrate. Complete breakdown, therefore, will depend on a succession of micro-organisms, each contributing particular enzymes and catalysing particular reactions. Secondly, the susceptibility of an organic polymer to degradation is inversely proportional to its heterogeneity. In other words, the more diverse its composition and the more diverse the bonds and linkages involved, then the less likely it is to be successfully degraded by microbial enzymes.

Changes in the biochemistry of decomposing leaf litter suggest that components disappear in the following order: soluble sugars, hemicellulose, cellulose and finally lignin (Burges 1958) (Fig. 19.6). This supports the view that a succession of micro-organisms colonises the substrate in waves, each wave capable of decomposing a particular component. So sugar fungi and other organisms utilising simple carbohydrates are replaced in time by decomposers of cellulose or comparable polysaccharides and finally by decomposers of lignin. This successional concept is not entirely simple, for it is related not only to the physical and nutritional status of the substrate but also to differential growth rates of the species concerned. Nevertheless, such schemes are useful (Table 19.2) and the soil animals can be incorporated in them, for not only does their comminution of detritus pave the way for microbial colonisation but some microbial decomposition of, for example, leaf cuticles may be necessary for detritivores to gain access to ingestible tissue. Furthermore, some detritivores possess bacterial gut floras which secrete cellulase, the degradative enzyme of cellulose, and therefore cannot be clearly separated from other microbial activity (see Ch. 18).

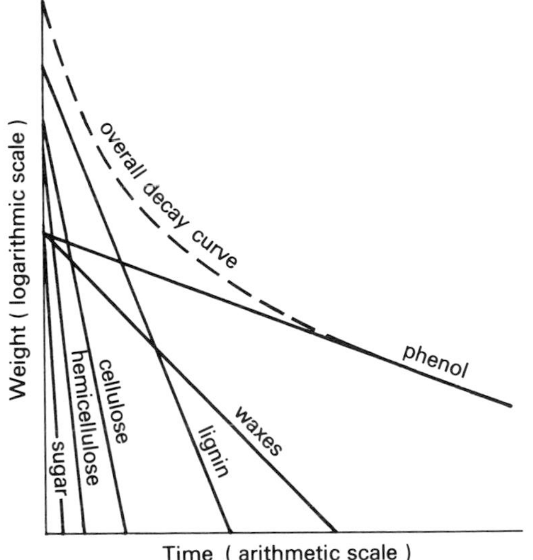

Figure 19.6 Decay curves for the principal constituents of plant litter (after Minderman 1968).

Table 19.2 Sequence of utilisation of substrate and waves of decomposers in the breakdown of plant litter (after Garrett 1981 and Frankland 1966).

Implied succession	Decomposer colonists	Substrate utilisation
1st stage (0–1 year)	parasites and primary colonisers including fungi and procaryotes*	sugars utilised and other simple carbohydrates (incorporated into fungal mycelia)
2nd stage (1–2 to 3 years)	cellulolytic soil saprophytes including soil fungi, protistans, soil invertebrates and procaryotes*	cellulose decomposition (excess simple carbohydrates produced are used by non-cellulolytic fungi)
3rd stage (2 to 3–5 to 6 years)	lignin-decomposing saprophytes, fungi and procaryotes*	lignins, tannins and other complex compounds, later chitin of fungal cell walls utilised

*Bacteria and/or actinomycetes.

19.3.3 *Resynthesis: the final state of the system, humification and the overall effects of decomposition*

The overall effects of decomposition can be summarised as the disappearance of litter and detritus and the associated release of CO_2, H_2O and mineral nutrients (**mineralisation**). These phenomena are accompanied by the appearance of decomposer (particularly microbial) protoplasm and the **immobilisation** of some of these nutrients. Also associated with these events is the appearance of a residuum of organic compounds largely resistant to further breakdown. This is the process of **humification** and the residual material is **humus**. Humus is a mixture of complex compounds which are variable in composition and amorphous, lacking crystallinity. These properties have ensured that, in spite of continuing research efforts extending over a hundred years, the precise nature of soil humus is still obscure. Indeed, it has been suggested that no two molecules of humus may be exactly alike (Fledgmann & George 1975).

We can, however, arrive at some broad conclusions regarding the origin, composition and properties of humus. It appears to consist of insoluble heterogeneous polymers which have been produced within a microbial cell, either when alive or from its protoplasm on its death. These polymers form particles of colloidal dimensions (<2 μm) which are relatively stable and comparatively resistant to acid hydrolysis and to further microbial decomposition because of their heterogeneity relative to microbial enzymes. Humic polymers are believed now to result from reactions between polyphenolic compounds of both plant and/or microbial (especially fungal) origin with amino acids, peptides and protein, also from plant and/or microbial sources. Humus may also contain some of the more resistant polysaccharides from litter, but much of this carbohydrate material in humus is derived from microbial cell walls (e.g. chitin from fungal cell walls) and from bacterial extracellular gums.

Colloidal humus particles have a vast surface-area:volume ratio and considerable ion exchange properties. Like clay minerals, they carry a net negative charge, but because of the dissociation of hydrogen ions from the OH (hydroxyls) of the carboxylic and phenolic active groups under different conditions of soil acidity this charge is pH dependent (Fig. 19.7). These humic particles and clay minerals together make up the major part of the colloidal size fraction of the soil, although under most circumstances they exist not as separate particles but as intimately associated organomineral complexes. Here the humic colloid is adsorbed on the surface of the clay mineral, and perhaps within its lattice, by various forces (such as hydrogen bonds, co-ordination to exchangeable cations or van der Waal's forces) which overcome the inherent electrostatic repulsion of the net negative charge carried by each particle. The importance of these organomineral complexes to the retention of water and mineral nutrients in the soil cannot be overstated. Together with immobilisation in microbial tissue, ion adsorption by colloids constitutes the most important regulator of leaching loss to the solute throughput of the denudation system and of nutrient availability to higher plants.

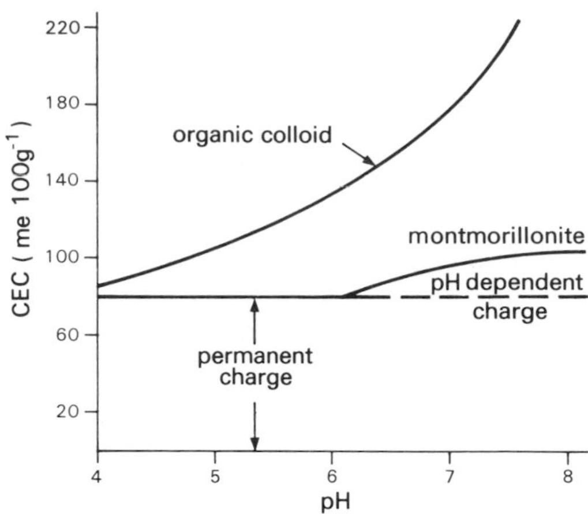

Figure 19.7 Cation exchange capacity (CEC) plotted against pH for an organic colloid and the 2:1 lattice clay mineral, montmorillonite. The CEC that reflects the number of negatively charged adsorption sites on the organic colloid is pH-dependent throughout the soil pH range. (Note CEC units, me = milliequivalent – a quantity chemically equivalent to one milligram of hydrogen) (after Buckman & Brady 1974).

19.4 Decomposition: a trophic model and the pathway of energy flow

The detritivores and decomposers considered above as activating mechanisms in the process–response model of decomposition obtain the energy for their metabolism from dead organic matter. They either ingest this dead material whole or absorb food molecules and nutrients previously made available by the decomposer activity of micro-organisms, a process that may take place outside or inside the digestive system. There are, however, other soil organisms that prey on living detritivores and decomposers, and those that feed on their dead remains. These constitute secondary grazing–predation and secondary decomposer systems respectively.

Many groups of soil animals, for example the mites, graze either selectively or indiscriminately on bacterial cells and fungal hyphae. Some of the larger soil invertebrates, such as the centipedes, are voracious carnivores preying on mites, springtails and nematodes. These trophic relationships are further complicated by antagonistic and parasitic relationships, particularly those involving fungi. Although it is possible to construct trophic dynamic models of these detritivore–decomposer–carnivore food webs (Fig. 19.8), they prove to be

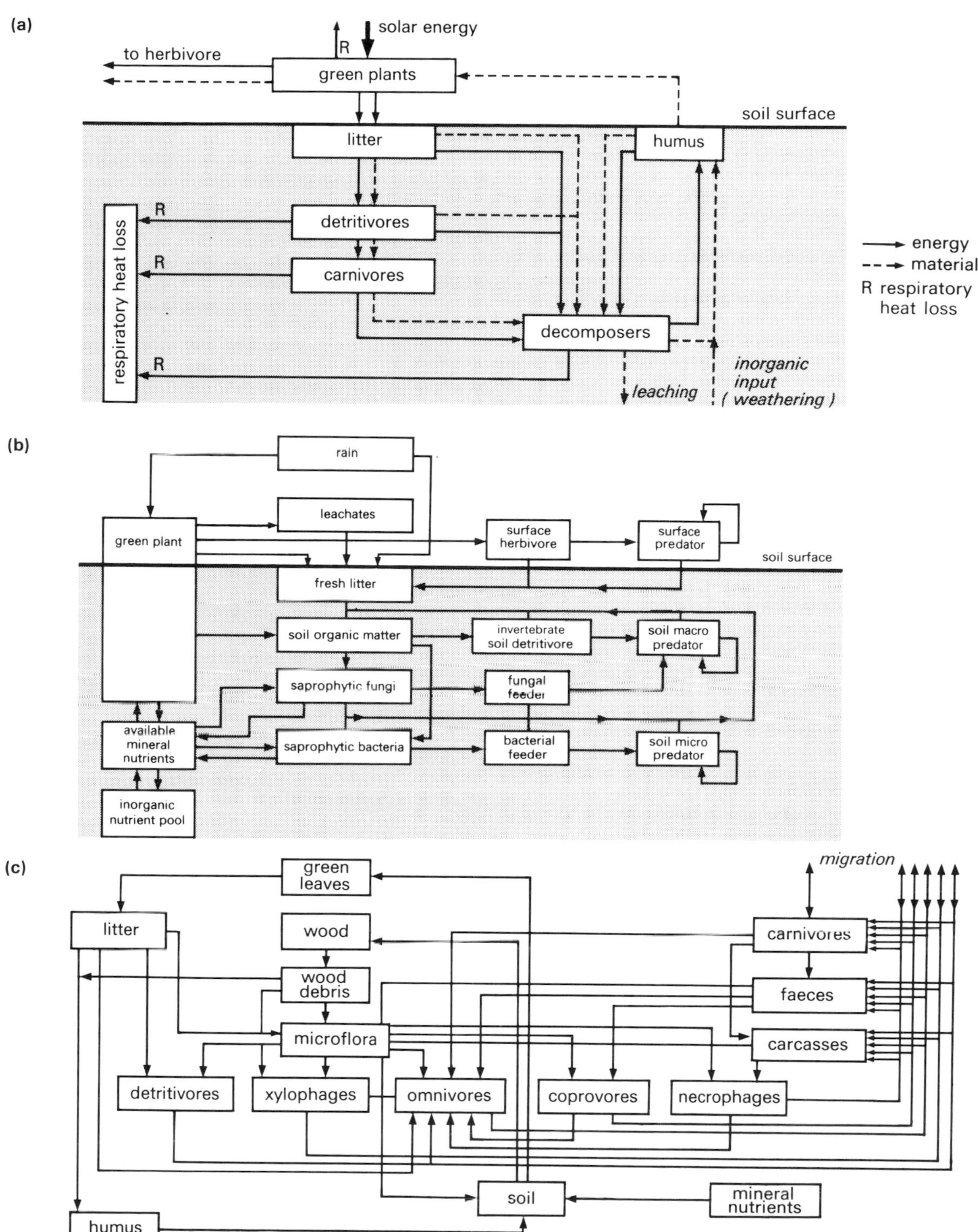

Figure 19.8 Three attempts to depict the trophic relationship of soil organisms and the pathways of matter and energy transfer in the detrital system, (a) after Wallwork 1970 (b) after Fortescue & Martin 1970 and (c) after Edwards *et al.* 1970.

extremely complex to unravel and as yet we know very little in detail of the way energy is partitioned and transferred between the different trophic levels. Nevertheless, some important conclusions can be made.

From the point of view of the process–response model of decomposition developed above, decay is a process of mineralisation or release of the chemical elements present in the detritus and their conversion to a more oxidised state, ultimately as simple inorganic compounds such as carbon dioxide. The detritivores and decomposers, however, and the carnivores dependent on them, assimilate organic compounds originally derived from detritus and take in mineral nutrients both as constituents of these compounds and from soil water. Using respiratory energy they synthesise from these components the constituents of their own protoplasm. So, from the point of view of the trophic model of decomposition, decay is a process of fixation or immobilisation of these same chemical elements in the protoplasm of the organisms of the detrital food web. (Indeed some of these organisms can use their respiratory energy to fix dinitrogen gas from the soil atmosphere.) All of these elements are, therefore, effectively immobilised, albeit temporarily, and are unavailable either to the roots of higher plants or to be lost by leaching. They only become available with the demise of the decomposers concerned and the digestion or breakdown of their protoplasm. Here, then, is a further way in which the decomposers are seen to act as a regulator on the rate of turnover of nutrient elements.

In thermodynamic terms, the organisms of the detrital food web decrease their internal entropy as they develop, grow and reproduce, sustained by the flow of chemical energy and materials from the litter. In accordance with the Second Law of Thermodynamics, however, the decomposing litter increases in entropy as it is degraded to simpler constituents and energy is dissipated as heat – think of the high temperatures in the centre of a compost heap – during the exothermic reactions of decay and by the respiration of the detrital population.

19.5 Decomposition: a pedological model

In Chapter 11 the weathering system was modelled as a profile through the weathered mantle or regolith. Similarly, decomposition can be seen in the same terms with different soil layers or horizons corresponding to different stages in decomposition. Surface horizons in closer proximity to the input of litter will be dominated by relatively fresh organic remains. Conversely, lower horizons will show progressively more advanced alteration, more humified material and more mixing of humus with mineral soil and regolith. Following this gradient in the state of the detritus are parallel gradients in physicochemical conditions, in populations of soil organisms and in their habitat conditions. It must be noted, however, that in certain circumstances such distinctions may be obliterated by the mixing activities of the larger soil animals.

The extent to which this horizonation is

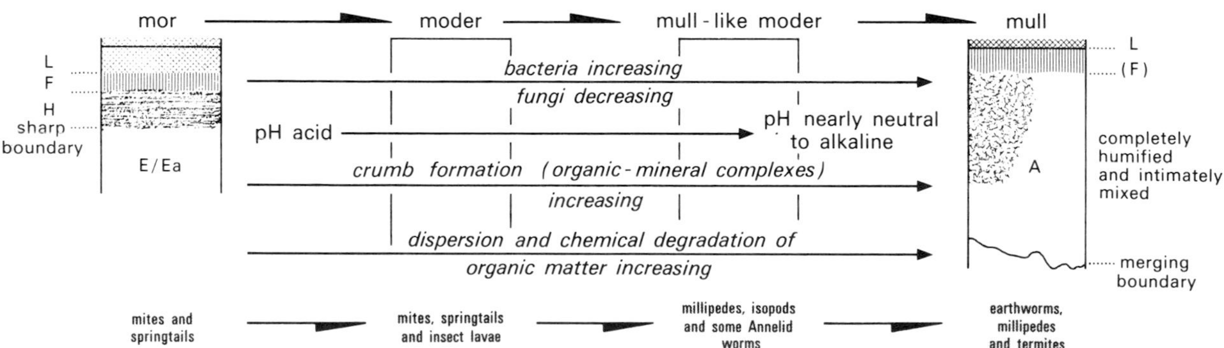

Figure 19.9 The pedogenic model of the detrital system and its expression in the soil profile as a series from mull to mor organic horizons shown in relation to the controlling gradients of soil conditions and changes in the dominant groups of soil organisms (partly after Wallwork 1970).

developed will depend on the nature of the initial input of litter, on the properties of regolith and mineral soil and, above all, on the climatic environment of soil formation. These factors will regulate the number, type and activity of the soil population, which in turn will regulate the rate of decomposition and nutrient turnover. Together all of these factors will influence the state of the soil profile (Fig. 19.9, see Ch. 20).

An understanding of the detrital system is important to mankind in several ways, apart from its obvious contribution to soil fertility. Even here, however, modern farming has opted to reject so-called 'organic farming' in favour of massive applications of fertiliser. Loss of organic matter affects soil structure detrimentally and reduces its capacity to retain these fertilisers. The result – a vicious circle. Society also uses the decomposition process directly in the treatment of sewage and increasingly in the composting of domestic refuse. The effluents released from sewage works, however, are still rich in nutrients, particularly nitrates and phosphates. These, together with fertilisers leached from agricultural soils, have been responsible for many enrichment or eutrophication problems in rivers and lakes, with their associated algal blooms. Conversely, the overloading of natural detrital systems with untreated waste allows the oxygen demand of decomposer organisms to depress oxygen levels to a point where vegetation and fish die of asphyxiation.

Further reading

Accessible introductory texts are

Jackson, R. M. and F. Shaw 1966. *Life in the soil.* London: Edward Arnold.

Mason, C. F. 1977. *Decomposition.* London: Edward Arnold.

More advanced treatments will be found in

Kononova, M. M. 1966. *Soil organic matter*, 2nd edn. Oxford: Pergamon.

Burges, A. 1958. *Micro-organisms in the soil.* London: Hutchinson.

while particular groups of soil organisms are dealt with in

Griffin, D. M. 1972. *Ecology of soil fungi.* London: Chapman & Hall.

Wallwork, J. A. 1970. *The ecology of soil animals.* Maidenhead: McGraw-Hill.

20 The soil system

20.1 Defining the soil system

We have encountered the soil, so far, in several different contexts. For example, its relationship with the regolith was discussed under the heading of the weathering system in Chapter 11. The decomposition of organic matter was seen to take place largely on or within the soil, which was also seen to form the habitat of the organisms of the detrital system in Chapter 19. We also recognised, in Chapter 12, that the soil is intimately associated with the movement of water and debris in the operation of denudation processes on slopes, but at the same time it was seen to function as a reservoir of water and mineral nutrients exploited by plant roots in Chapter 17. Clearly, the soil has been a part – either explicit or implicit – of many environmental systems that we have considered and has been involved in the cascades of matter and energy with which they are concerned (Fig. 20.1). But can we define the soil as a component system of such cascades in its own right?

20.1.1 Modelling the soil: its three-dimensional organisation

From our experience of the soil, we know that it is a three-phase system: solid, liquid and gaseous. The solid phase is in part made up of inorganic material ranging from clastic fragments of largely unaltered rock, through disaggregated mineral grains actively undergoing weathering, to the secondary minerals derived from weathering products. As we saw in Chapter 11, this inorganic

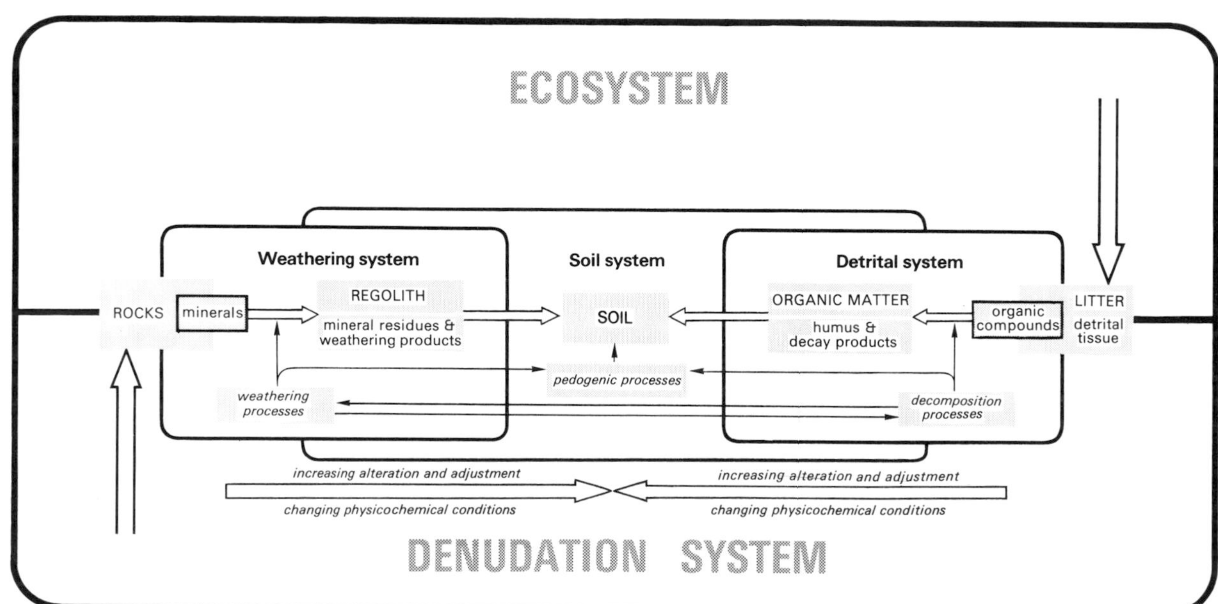

Figure 20.1 The relationships between the soil system and other systems in the environment.

fraction occurs in a range of size classes and the relative proportions of these size categories define the texture of the soil (Fig. 20.2). The second component of the solid phase consists of organic matter in various states of decay, the humic substances resynthesised from the products of decay (and perhaps also the populations of soil organisms directly or indirectly involved in decomposition). The solid phase of the soil, however, does not usually consist of discrete particles, whether organic or inorganic. More usually these particles are intimately associated to form **soil aggregates** or **peds**. They result from the forces of attraction between soil particles and involve the formation of hydrogen and ionic bonds, and may be further promoted by the wetting and drying of the soil, by the pressures exerted by developing roots and by the presence of polysaccharide gums secreted by

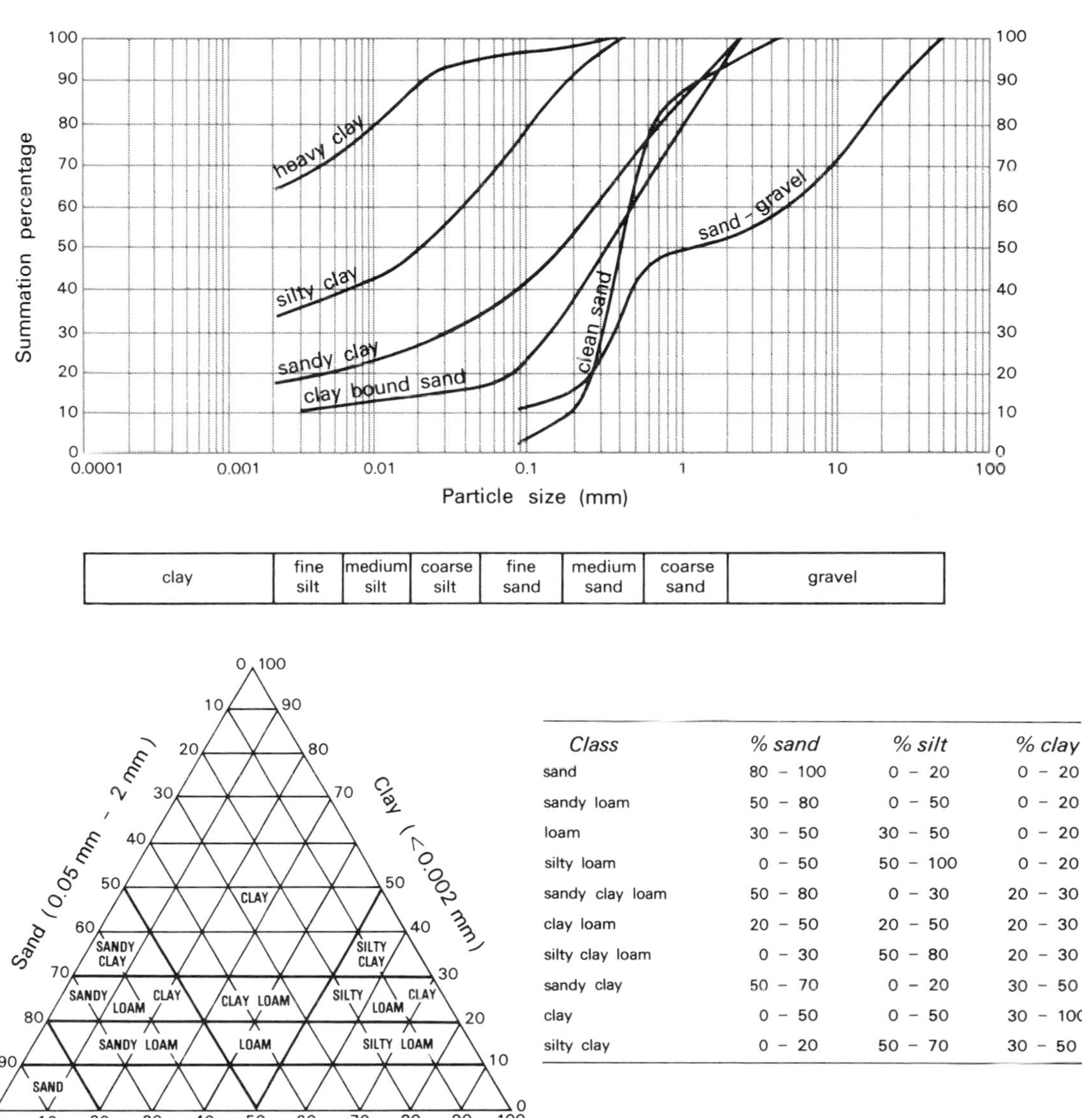

Class	*% sand*	*% silt*	*% clay*
sand	80 - 100	0 - 20	0 - 20
sandy loam	50 - 80	0 - 50	0 - 20
loam	30 - 50	30 - 50	0 - 20
silty loam	0 - 50	50 - 100	0 - 20
sandy clay loam	50 - 80	0 - 30	20 - 30
clay loam	20 - 50	20 - 50	20 - 30
silty clay loam	0 - 30	50 - 80	20 - 30
sandy clay	50 - 70	0 - 20	30 - 50
clay	0 - 50	0 - 50	30 - 100
silty clay	0 - 20	50 - 70	30 - 50

Figure 20.2 Soil texture.

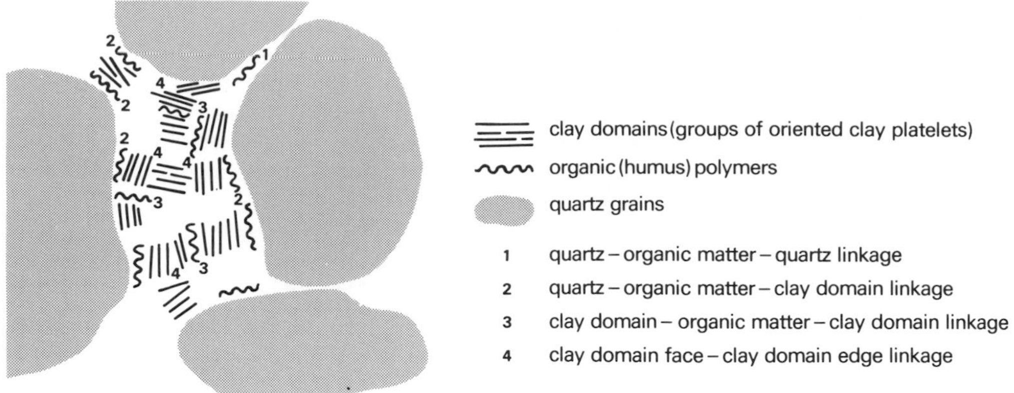

Figure 20.3 The Emerson model of soil aggregation.

Depth (m)
Texture
Structure
0
0.2
0.4
0.6
0.8
silty loam
silt loam
silt loam
silt loam
granular
granular and weak platy
platy and angular blocky
prismatic and angular blocky
(a)
(b)

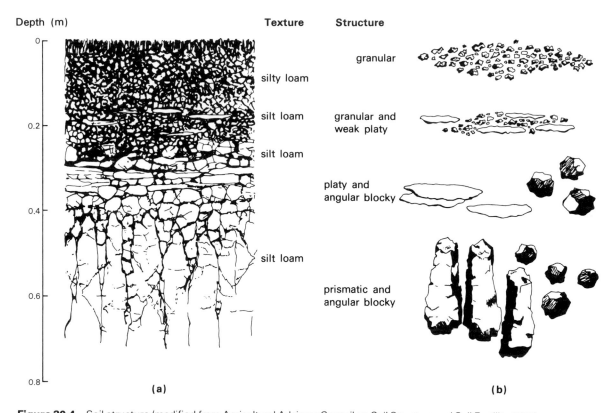

Figure 20.4 Soil structure (modified from Agricultural Advisory Council on Soil Structure and Soil Fertility 1971).

the soil microflora (Fig. 20.3). Between these soil aggregates and within them there are voids of various kinds. A network of micropores (2–20 μm in diameter) lies between individual particles or between microaggregates within the peds, while in clay soils these micropores may be as narrow as 10 nm. Between major soil aggregates, macropores (7200 μm in diameter) occur as well as larger voids and cracks (Fig. 20.4). Sometimes interconnecting tubular passages, probably biotic in origin, may also occur and their significance has already been alluded to in Chapter 12, where they were referred to as soil pipes. Together, the kind, size and distribution of soil aggregates and soil voids and pores determine the structure or fabric of the soil.

The liquid and gas phases are represented by the soil water and air which occupy the pores and voids. We have already seen that water entering the soil by infiltration is subject to certain forces which determine the soil water potential (ψ_s) (Box 12.2). Some components of soil water potential such as the matrix (ψ_m) and osmotic potentials (ψ_π) are due to retention forces and are responsible for holding water in the partially air-filled pores above the water table. Immediately following precipitation and below the water table, the pores may be saturated, being entirely water filled. Soil water, however, is not pure water. As we know, it contains a whole variety of substances in solution, particularly various cations and anions which are potentially available as plant nutrients. These solutes are derived from a number of sources. Some are already present in rain water, some originate in the vegetation canopy, while others have passed into solution in the soil as a result of weathering, decomposition and ion-exchange processes. Indeed, soil water is often referred to as the soil solution. However, some of the soil constituents themselves, although not in true solution, are so small (<2 μm) that they may occur as a colloidal suspension. From the point of view of soil chemistry, it is common to regard the whole of the solid and liquid phases as behaving as a soil–water suspension. Most prominent amongst the colloidal fraction are the hydrated aluminium silicate clay minerals, the humic polymers and the hydrated amorphous oxides and hydroxides of iron, manganese, aluminium and titanium (see Ch. 11, Box 11.2 & Ch. 19).

When the pores are not full of water they are partly or wholly air filled, but the composition of this soil atmosphere may differ significantly from the free atmosphere above. Furthermore, these differences may increase with depth and fluctuate with time. The main difference is that oxygen concentrations are somewhat lower than in the atmosphere and carbon dioxide concentrations significantly higher (Table 20.1). Both fluctuate, however, and depend on the respiratory demand for oxygen and the rate of carbon dioxide evolution by roots and soil organisms. The relationship is complicated by the very different diffusion coefficients of the two gases in air and water and by their solubilities in water.

The soil is a three-dimensional natural body, occurring at the surface of the Earth, reaching vertically to about the lower limit of root penetration and extending laterally as a component of the landscape (Fig. 20.5). However, solid, liquid and gas phases do not form a random mosaic in these three dimensions but are more or less organised to impart a definite vertical and lateral structure to the system. The properties of the soil solids, and of the voids and their associated soil water and soil air, vary both vertically and laterally. Layers are differentiated vertically within the soil body and they differ in the relative proportions of the soil-forming materials and in their characteristics, so that each layer presents a different set of physical, chemical and biological attributes. These sets of attributes reflect the processes operating, or which have operated, in the particular soil layer concerned, but they also form the conditions, or environment, under which contemporary processes take place. This vertical organisation of the soil is referred to as the **soil profile** and the layers as **soil**

Table 20.1 The oxygen and carbon dioxide composition as percentage by volume of the gas phase of well aerated soils, compared with that of dry air (modified from Russell 1973).

	Oxygen (%)	Carbon dioxide (%)
dry air	20.95	0.03
arable soil		
fallow	20.7	0.1
unmanured	20.4	0.2
manured	20.3	0.4
manured sandy soil		
cropped with potatoes	20.3	0.6
pasture	18–20	0.5–1.5

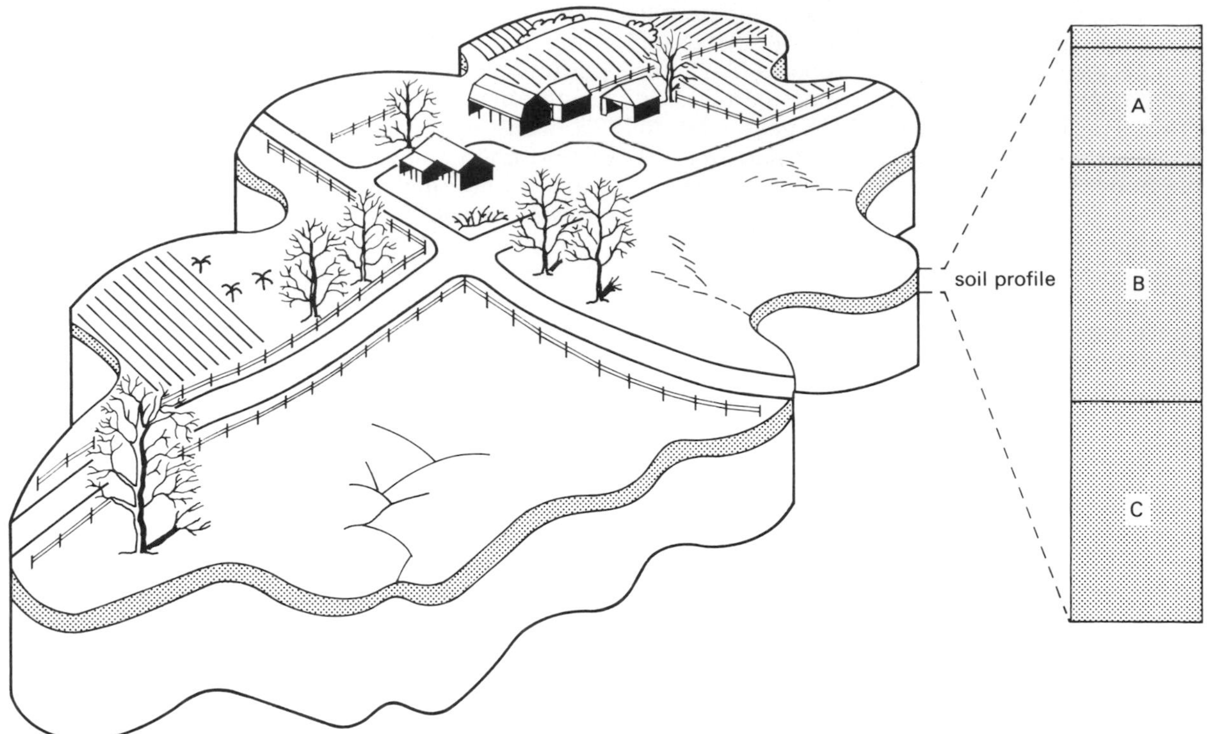

Figure 20.5 Soil as a component of landscape (adapted from R. W. Simonson, *Soil Sci. Soc. Am. Proc.* **23**, 152–6 by permission of the publishers).

horizons (Fig. 20.6). Laterally, too, there is some semblance of organisation for the pattern of horizonation and the characteristics of the horizons vary laterally in a largely predictable manner in response to the soil's position in a landscape, particularly its position relative to slope (see Fig. 20.12).

20.1.2 A process model of the soil

The three-dimensional organisation of the soil system, and particularly the presence of horizons, is the result of a complex suite of processes. Some of these processes have been treated explicitly in other chapters, but their role within the soil system means that they can all be considered as pedogenic processes and contribute to pedogenesis or soil formation. In detail the pattern, type and degree of expression of horizonation displayed by the soils of the world are extremely varied. Nevertheless, the presence of horizons in all genetically developed soils (i.e. where sufficient time has elapsed to allow soil development) suggests that certain processes are common to the development of all soils and that each soil is not, therefore, the product of a set of processes unique to it. Indeed, Simonson (1959) states that a soil forms 'as a result of an aggregate of many physical, chemical and biological processes, all *potential* contributors to the development of *every* soil'. Before we consider the implications of this statement, we need to consider what these pedogenic processes are in the context of a systems model of the soil (Fig. 20.7).

In terms of an open-system model, the characteristics of a soil body – its elements, their attributes and their relationships – represent the state of the system. This will depend on the inputs of matter and energy to the soil system – the **input processes**. For example, these include the addition of organic matter from the vegetation above, the addition of rain water by infiltration, the input of regolith by the weathering of bedrock and the increment of materials supplied from up slope by mass move-

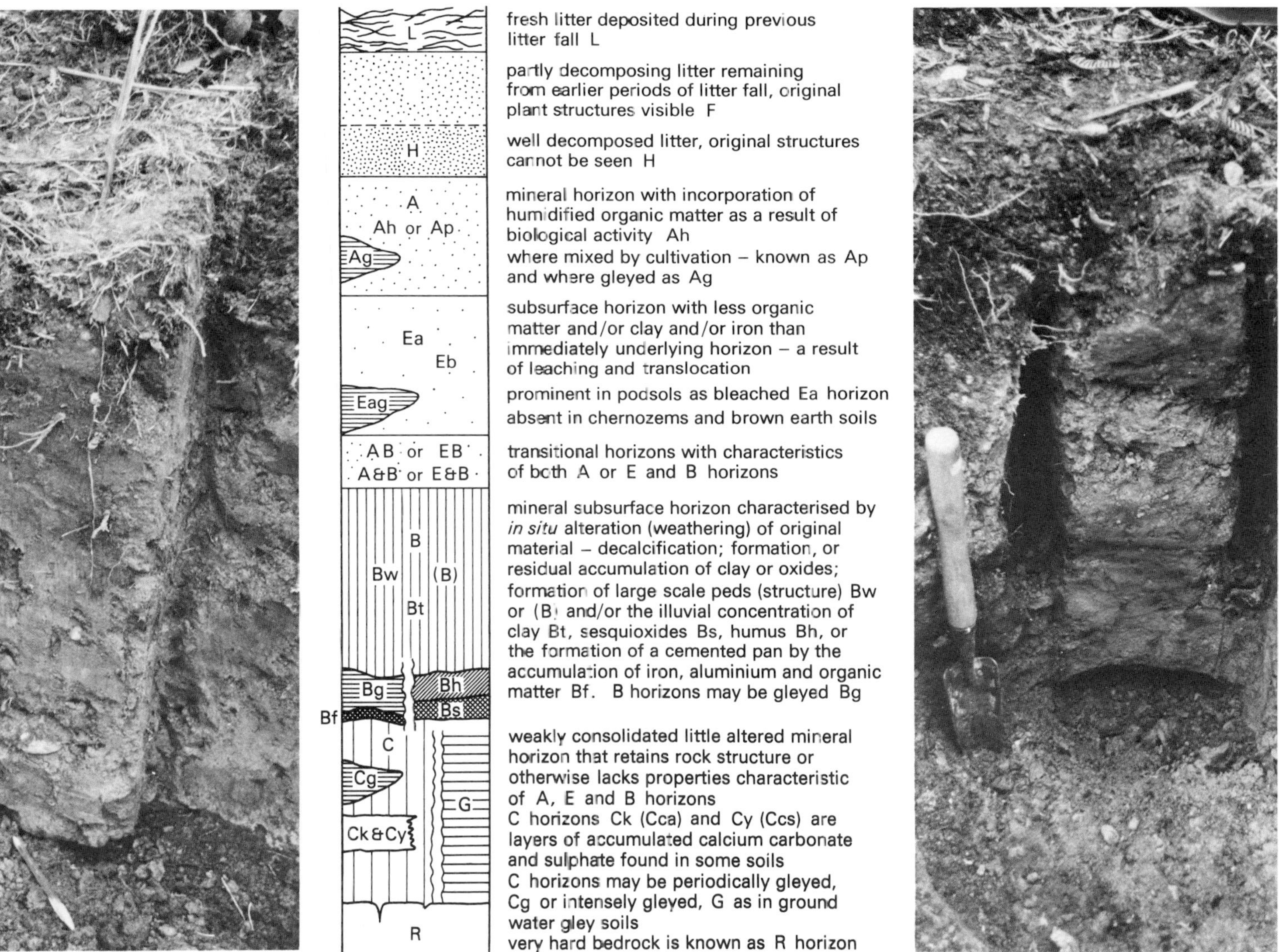
L
H
A
Ah or Ap
Ag
Ea
Eb
Eag
AB or EB
A&B or E&B
B
Bw
(B)
Bt
Bg
Bh
Bs
Bf
C
Cg
G
Ck&Cy
R
fresh litter deposited during previous litter fall L
partly decomposing litter remaining from earlier periods of litter fall, original plant structures visible F
well decomposed litter, original structures cannot be seen H
mineral horizon with incorporation of humidified organic matter as a result of biological activity Ah
where mixed by cultivation – known as Ap and where gleyed as Ag
subsurface horizon with less organic matter and/or clay and/or iron than immediately underlying horizon – a result of leaching and translocation
prominent in podsols as bleached Ea horizon absent in chernozems and brown earth soils
transitional horizons with characteristics of both A or E and B horizons
mineral subsurface horizon characterised by *in situ* alteration (weathering) of original material – decalcification; formation, or residual accumulation of clay or oxides; formation of large scale peds (structure) Bw or (B) and/or the illuvial concentration of clay Bt, sesquioxides Bs, humus Bh, or the formation of a cemented pan by the accumulation of iron, aluminium and organic matter Bf. B horizons may be gleyed Bg
weakly consolidated little altered mineral horizon that retains rock structure or otherwise lacks properties characteristic of A, E and B horizons
C horizons Ck (Cca) and Cy (Ccs) are layers of accumulated calcium carbonate and sulphate found in some soils
C horizons may be periodically gleyed, Cg or intensely gleyed, G as in ground water gley soils
very hard bedrock is known as R horizon

Figure 20.6 The soil profile and soil horizons. Note that the recognition and description of horizons is not as straightforward as the schematic diagram would suggest for, as shown in the photographs, the boundaries often merge. Note also that all of the horizons shown in the diagram would never occur together in the real world.

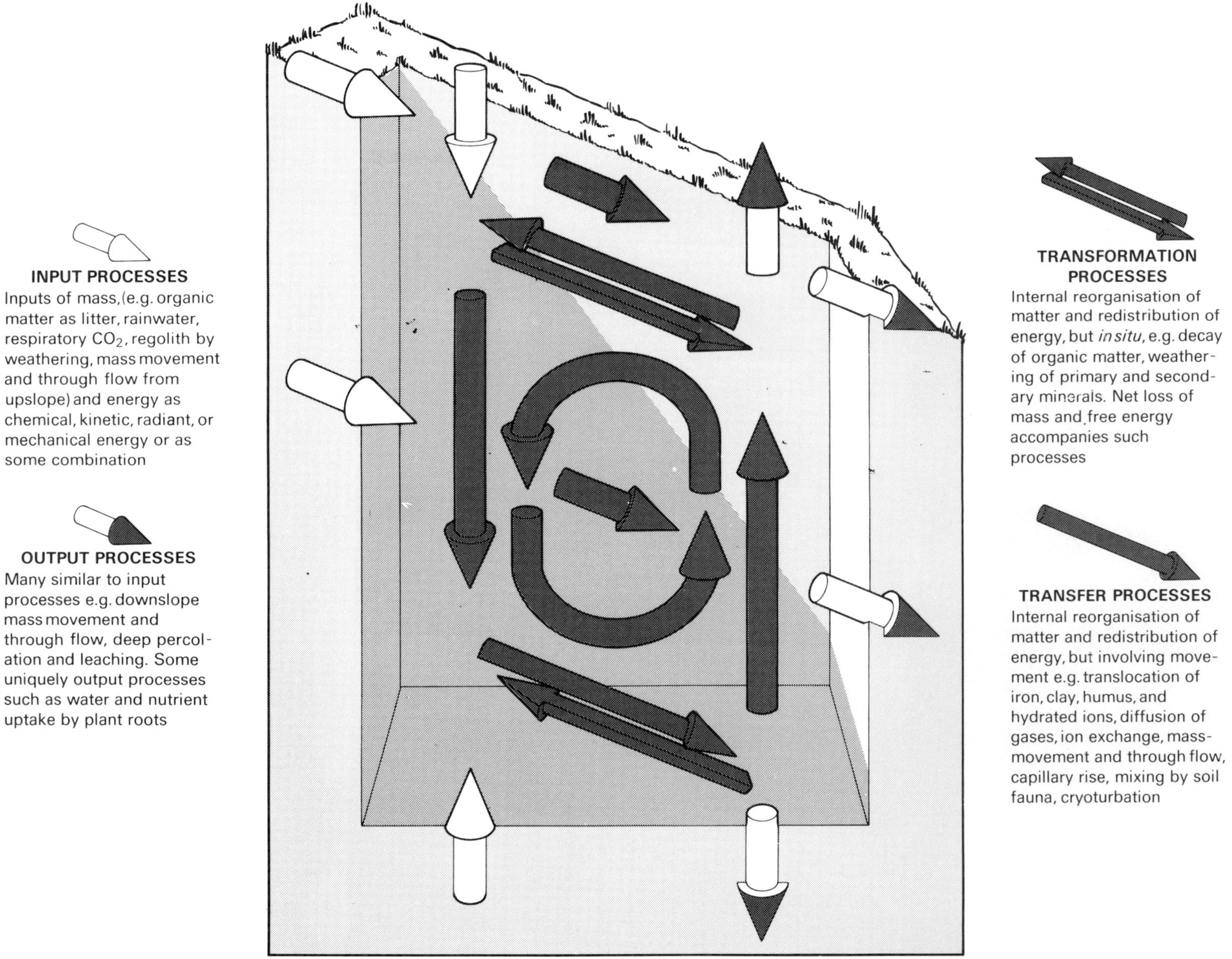

INPUT PROCESSES

Inputs of mass,(e.g. organic matter as litter, rainwater, respiratory CO_2, regolith by weathering, mass movement and through flow from upslope) and energy as chemical, kinetic, radiant, or mechanical energy or as some combination

OUTPUT PROCESSES

Many similar to input processes e.g. downslope mass movement and through flow, deep percolation and leaching. Some uniquely output processes such as water and nutrient uptake by plant roots

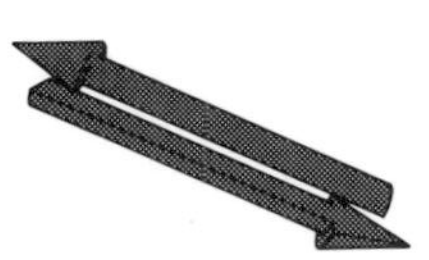

TRANSFORMATION PROCESSES

Internal reorganisation of matter and redistribution of energy, but *in situ*, e.g. decay of organic matter, weathering of primary and secondary minerals. Net loss of mass and free energy accompanies such processes

TRANSFER PROCESSES

Internal reorganisation of matter and redistribution of energy, but involving movement e.g. translocation of iron, clay, humus, and hydrated ions, diffusion of gases, ion exchange, mass-movement and through flow, capillary rise, mixing by soil fauna, cryoturbation

Figure 20.7 Soil processes

ment and throughflow. All of these materials represent inputs of chemical energy, while materials in motion, such as percolating water, will represent kinetic energy inputs capable of doing work on the system. Solar radiation will be absorbed by the soil and heat energy transferred, particularly by conduction. Even the height of the soil body endowing each soil particle with gravitational potential energy can be thought of as an energy input inherited from the geomorphological event that was responsible for the elevation of the land surface.

The state of the system (the soil) will also depend on processes operating within it either to maintain or to change its state. These processes are of two types. The **transformation processes** involve the reorganisation of matter and redistribution of energy, but largely *in situ* and within particular horizons. The **transfer processes** are associated with the pathways of throughput within the system and involve both vertical and lateral transfer between different stores of matter and energy, usually between different horizons. Perhaps the best examples of *in-situ* reorganisation are the transformations of organic matter during decomposition and of primary minerals during weathering, and secondary mineral and humus formation within the soil. Both involve the structural reorganisation of matter but they do so with a net loss of mass and dissipation of free energy, as heat, for as spontaneous processes they proceed in accordance with the Second Law of Thermodynamics. The lost mass is accounted for by the products of weathering and decomposition which are removed and become involved in transfer processes.

Indeed, one major class of transfer processes within the soil system is concerned with the translocation of the products of pedogenic processes. For example, the movement of hydrated ions in solution through the soil, their exchange between the soil solution, the surfaces of colloids and plant roots, the diffusion of respiratory carbon dioxide, the mechanical translocation of clay minerals through soil pores (Fig. 20.8), or the translocation of iron as hydrated hydroxides in colloidal suspension or as complexed ions by cheluviation – are all transfer processes involving the mobile products of weathering and decomposition. Gravity is the major control of such transfers and the usual result is movement to lower horizons, or laterally down slope, as with the movement of water by matrix throughflow. Nevertheless, other forces may outweigh the attraction due to gravity and, under certain circumstances, an upward movement can occur. Such a situation is involved in the capillary rise of water in soil micropores which, under suitable conditions, may result in the enrichment of horizons near the surface with precipitated soluble salts derived from depth.

All of these transfer processes are relatively specific, but there are others which have a more general effect on the state of the system. While many transfer processes are selective and result in the differentiation of specific horizons, others operate in a largely non-selective manner and militate against horizonation, they may even destroy what horizons exist. Mixing by the soil fauna, particularly the macrofauna (especially earthworms in temperate soils and termites in the Tropics), brings organic matter, particularly humified material, into intimate contact with the mineral soil and eliminates any clear distinction between organic and mineral horizons. In appropriate environments, **cryoturbation** (frost heave) will have a similar effect in that it will disrupt the development of horizons. However, the latter group of processes may give rise to distinct patterns within the soil. These are a result of the differential responses of each size fraction to freeze/thaw (frost sorting) and, although genetic features, they are

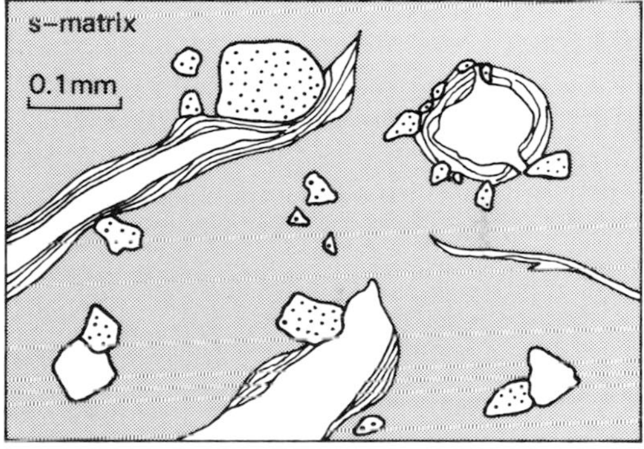

Figure 20.8 Clay skins (cutans) formed by the mechanical translocation of clay minerals.

not true soil horizons (Fig. 20.9). Finally, any mass movement process, particularly if rapid, will tend to blur if not obliterate the differentiation of vertical horizons as the whole soil matrix is subject to lateral movement down slope.

The last group of processes which will determine the nature of a soil or the state of the system are the **output processes**, by which matter and energy are transferred across the boundary of the three-dimensional soil system. As transfer processes, many are similar to those responsible for input or for translocation within the system; indeed many are the same. The downslope mass movement and throughflow processes that deliver material to a soil body from up slope are also the processes that transfer that material through the soil and across its boundary in a downslope direction. The leaching and translocation processes that transfer elements between soil horizons are also responsible for loss to drainage water at depth or by throughflow down slope. However, some processes, such as nutrient and water uptake by plant roots, are uniquely output pathways and processes.

If we now return to Simonson's contention that all of these groups of processes are potential contributors to the development of any soil, it becomes

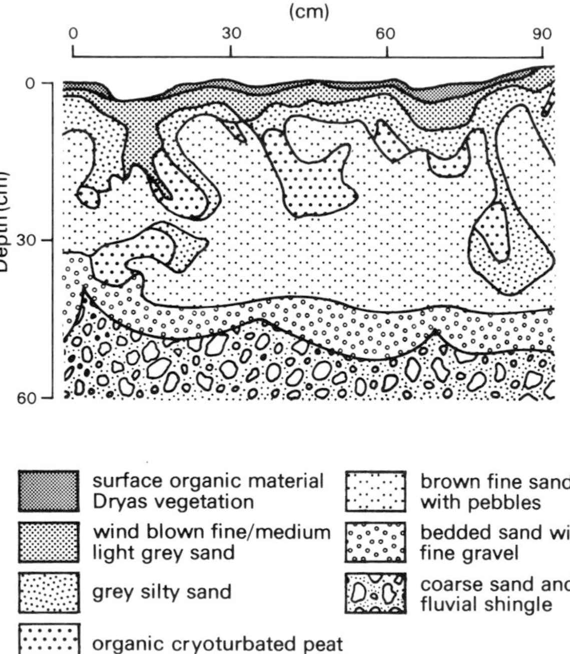

Figure 20.9 Cryoturbation effects, upper Kellett River basin, Banks Island, Northwest Territories, Canada.

clear that a soil developed under tropical rainforest in Africa, a prairie soil in Saskatchewan, and a soil under an old permanent pasture in the English Midlands, are not in any true sense unique. Though the differences between them are real, they merely reflect differences in the magnitude, in the rate of operation, and in the relative combination of the processes outlined above. As any of these properties can change over relatively short distances, it is not surprising that real soils show such bewildering variation and have largely defied attempts at successful classification. The state of each of these soils, indeed of any soil, will represent to some extent the balance or the steady state attained by the system in response to the particular set of processes it experiences. However, the questions arise: what controls the way specific processes are combined to form a set of input, transformation, transfer and output processes; what controls the magnitude and rate of these four suites of processes; and what therefore controls the nature of the steady state soil?

20.2 The control and regulation of pedogenesis

From the beginnings of modern soil science and pedology, soils have been described as a function of five factors of soil formation (Jenny 1941). These are an inorganic factor, an organic factor, a climatic factor, a relief or topographic factor and a time factor, and they can be integrated with the open-system model of the soil by regarding them as sets of exogenous variables which define the operating conditions of the system. It is this whole complex of environmental conditions that determines the characteristics of any suite of processes and of the equilibrium state attained by the soil.

The nature of the parent material (the inorganic factor) controls the properties of the inorganic input to the soil and largely determines the initial textural state of the soil and, by so doing, regulates many of the properties and processes operating in a soil. The mineralogy of the parent material controls in part the initial chemical state of the soil – whether free $CaCO_3$ is present, for example – and this influences many soil properties and processes, such as weathering, the dominant ions in the exchange complex and the presence of earthworms, to quote only a few examples. The chemistry of the parent material also represents part of the initial potential chemical energy of the system.

Climate controls through the soil radiation balance the input of radiant energy and, through the surface water balance, the input of water. Through the magnitude of these inputs, climate also largely determines the weathering regime for they function as activating processes in weathering. Through its integrated effect on plant growth, climate also indirectly regulates the supply of organic matter from the vegetation, and in part the types and activity of soil organisms. Climate not only controls the input of water to the soil but also partly regulates its movement in the soil, again through the surface water balance. Discounting runoff, change in soil moisture (ΔS) and change in ground water storage (ΔG) will depend on the precipitation:evapotranspiration ratio (P:E_T). Where $P>E_T$ there will be a net downward movement of water in the soil profile promoting a whole complex of transfer processes. Where $P<E_T$ water infiltrating into the soil will soon be evaporated and soluble salts will be precipitated at a percolation or evaporation front within the profile.

The organic factor acting through the vegetation regulates the amount and composition of organic matter supplied to the soil in the form of dead litter and hence part of the input of matter and chemical energy to the system. Within the system the number and type of soil organisms, themselves partly determined by the quantity and kind of litter input, largely control the pathways by which this matter and energy is transformed and transferred during decomposition. In addition, through its interception of precipitation and its absorption, transmission, reflection and redistribution of radiant energy, the microclimate of the vegetation canopy regulates the effects of the climatic factor. Finally, the root systems of plants apply mechanical stress which may help to control the development

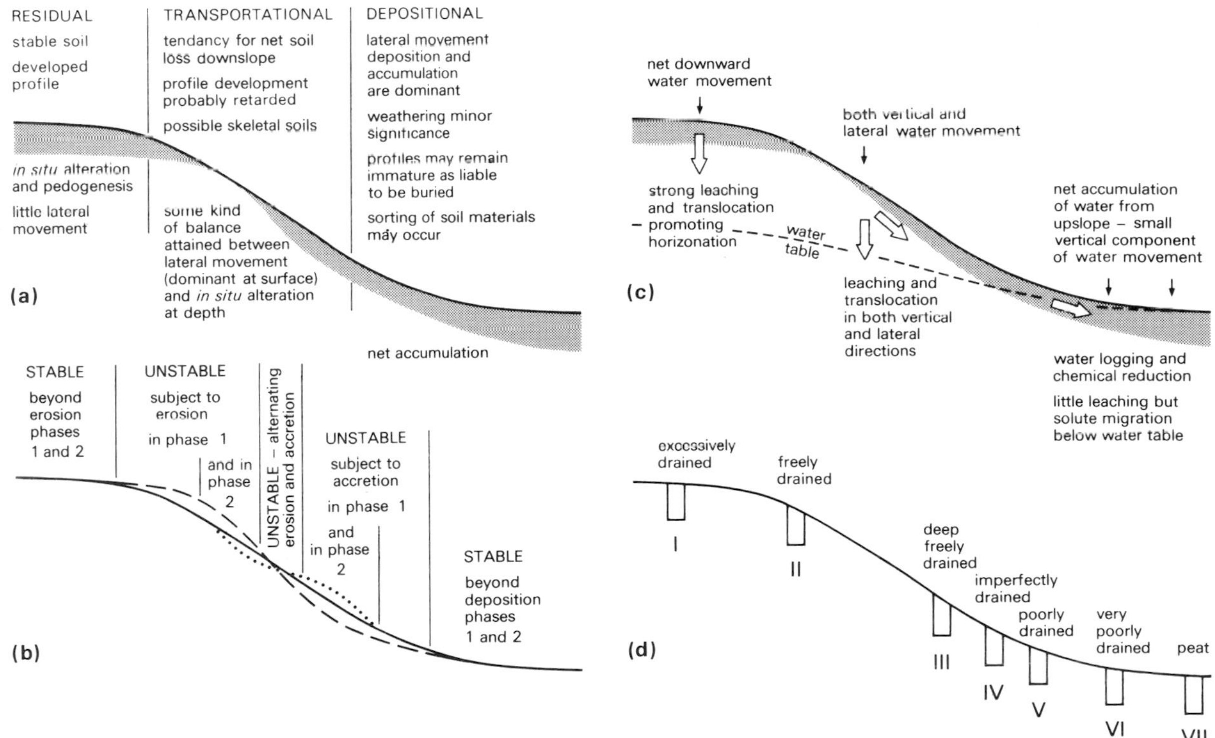

Figure 20.10 Soil–slope relationships (partly after Gentworth & Dion 1949 and Butler 1959). (a) Stressing the erosional and depositional relationships of soil profiles developing at different positions on a slope, but in (b) introducing two phases of erosion and deposition. (c) The drainage relationships of soil profiles developing at different positions on a slope. (d) Such a hydrological sequence of profiles used as a basis of soil classification.

of soil aggregates as well as promoting weathering by fracture. Roots also affect the distribution of voids and soil aeration and, certainly in the volume of soil in close contact with them (the **rhizosphere**), can control the soil chemical environment.

The development and maintenance of a steady state (or equilibrium) soil depend on the long-term stability of the land surface on which it occurs. In other words, it will depend on the balance between on the one hand the pedogenic processes promoting the existence of an ordered three-dimensional soil system and the processes of erosion tending to remove it and on the other those of deposition tending to bury and obliterate it. This balance is controlled, therefore, by the overall effect of the geomorphological factor and, as stated earlier, it is closely related to the soil's position in a landscape – notably its relationship to slope and to the processes operating on slopes. Slope will also control the hydrological characteristics of the site, the drainage relationships and the position of the water table. These factors will in turn influence the operation of pedogenic processes such as leaching and waterlogging. Indeed, the control exercised by slope has long been recognised in the literature in the catena concept of Milne (1947) and the toposequences and hydrological sequences of workers such as Fitzpatrick (1971) and Glentworth & Dion (1949) (Fig. 20.10). These models are closely related to those of slope form such as the nine-facet model of Dalrymple, Blong and Conacher (1968), and of slope process (see Ch. 12).

Finally, the time factor can be built into this factorial approach, for soils are dynamic bodies and the state of the system will also depend on their stage of development or on how far any soil has travelled along the path to equilibrium. However, the dependence of many properties of the system on elapsed time is not straightforward, neither is their equilibrium state a simple function of time (Fig. 20.11). Some properties adjust slowly and, although their initial rate of change may be rapid, it gradually becomes asymptotic and they become virtually time-independent over relatively short time intervals. Such properties only change very slowly, reaching a steady state after perhaps 10^3 or 10^4 years. The development of the textural properties of a soil and soil fabric (Kubiena 1938, Brewer 1964) would fall into this category. In contrast, other properties adjust very rapidly in their response to the external conditions of existence. The pH of the soil solution is such a property which, because of the buffer capacity of the soil solid phase, can be adjusted rapidly to re-establish an equilibrium after a disturbance such as a rainfall event. The occurrence of reversible redox reactions allows rapid adjustments to fluctuations in the redox environment of the soil associated particularly with waterlogging (Fig. 20.12). The soil population of decomposers can also respond

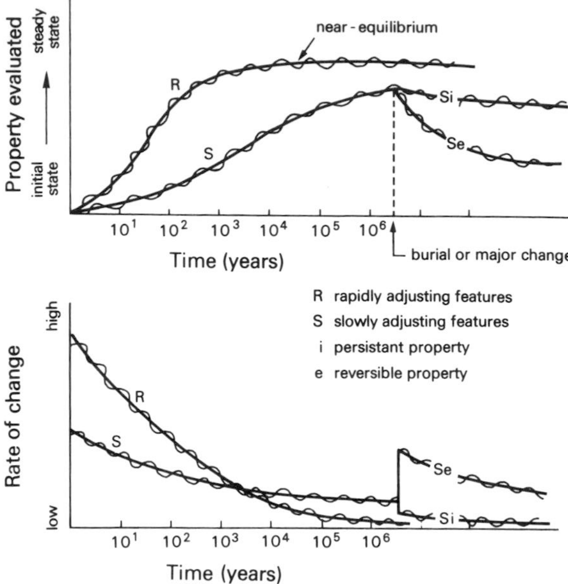

Figure 20.11 Time and the equilibria of soil-forming processes (after Yaalon 1971).

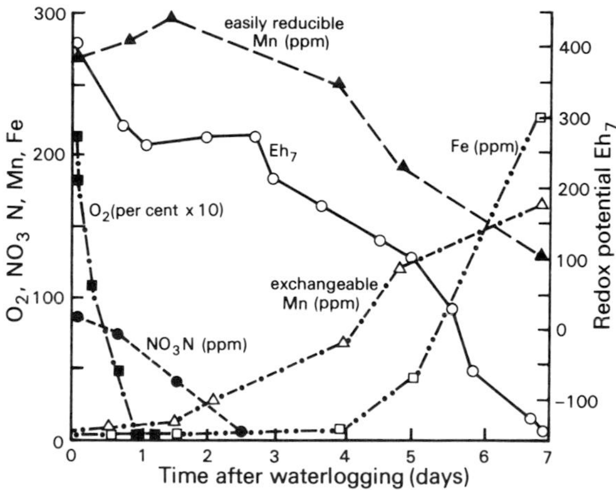

Figure 20.12 Changes in oxygen, nitrate, manganese, iron, and redox potential of a silt clay after waterlogging (Armstrong 1975, after Patrick & Turner 1968).

rapidly to changes in the amount or type of litter supplied to the soil by increasing their number to approach the limits of their food supply. These rapidly adjusting features, and the processes involved, are seen to be maintaining a steady state even over a short interval of time. However, some processes are irreversible and self-terminating, and are dominated by positive feedback. The weathering of primary silicate minerals to clay minerals is this kind of process, for when the reserve of unaltered primary minerals is exhausted the process stops and the clay minerals will not spontaneously reconstitute to form feldspars, for example. The development of an iron or clay pan is in a similar way irreversible and it terminates when all of the iron or clay has been translocated from higher horizons. The equilibrium these properties attain is thermodynamic. Often these irreversible processes convey the soil system as a whole through a threshold which initiates a new system state. For example, the production of an impermeable pan at depth restricts percolation of water, produces a perched water table within the profile, alters the chemical environment, restricts the effectiveness of leaching and translocation processes to that part of the profile above this water table and initiates reducing conditions below it. The new state is very different from the freely draining soil which encouraged the leaching and translocation that led to pan development in the first place. The response of the system with time, therefore, will reflect a complex combination of the behaviour of all three kinds of process with time (see Ch. 21).

20.3 Formal processes of pedogenesis

In the real world, pedogenic processes are not combined at random. They occur consistently in particular combinations determined and controlled by re-occurring sets of conditions. Each of these commonly recurrent combinations is known collectively as a **formal process** of soil formation. Since the ultimate factor regulating pedogenesis is often climate, particular formal processes are broadly correlated with particular climates. We shall not consider these formal processes in great detail here, but they are treated in several of the texts recommended as further reading at the end of the chapter. In the brief treatment that follows, the formal processes have been grouped to provide models of pedogenesis in temperate, tropical and arid environments.

20.3.1 A model of temperate pedogenesis

In a humid temperate environment, on parent materials with a moderate to rich reserve of weatherable minerals and supporting a woodland or grassland community supplying a litter rich in base cations and maintaining an efficient nutrient circulation, a soil develops in which many of the soil constituents have become **stabilised** (Fig. 20.13a). In such **brown earth** soils the dominant process is leaching, specifically **decalcification** for calcium tends to be the most important cation saturating the exchange complex. This tendency is normally very slight for it is offset by the *in-situ* weathering of the parent material (*B*w horizon) and the release of calcium and other cations from the decomposition of litter. Such soils support a large and diverse soil population with bacteria as the principal microbial decomposers. They are characterised by efficient humification, deep and intimate mixing of humus and mineral soil by the soil fauna and the formation of a deep mull horizon (see Fig. 19.9), with a well developed crumb structure promoted by the flocculation effects of divalent cations, such as calcium on the clays, and by the presence of bacterial polysaccharide gums.

Where the parent material has poorer reserves of weatherable minerals, the movement of water through the profile is greater, or the rate of cation replacement from litter is inadequate to offset leaching, the exchange complex will become at least partially dominated by hydrogen ions. Under these conditions aggregate stability will decrease, clay will deflocculate and in the dispersed state will be subject to mechanical translocation through the profile to accumulate as a finer textured Bt horizon at depth. Such a suite of processes is the formal process of **lessivage** (Fig. 20.13b) and the soils are known as **sols lessivée**. A sure indication of the existence of this process is the occurrence of orientated clay skins (**cutans**, see Fig. 20.8) on the sides of soil pores and on the faces of the peds.

On siliceous parent materials with few weatherable minerals, which are freely drained and support a vegetation which supplies a base-deficient litter and promotes the formation of raw acid mor organic horizons, the formal process of **podsolisation** occurs (Fig. 20.13c). Here the base cations are leached rapidly, particularly under a cool temperate climate. These cations are not replaced either

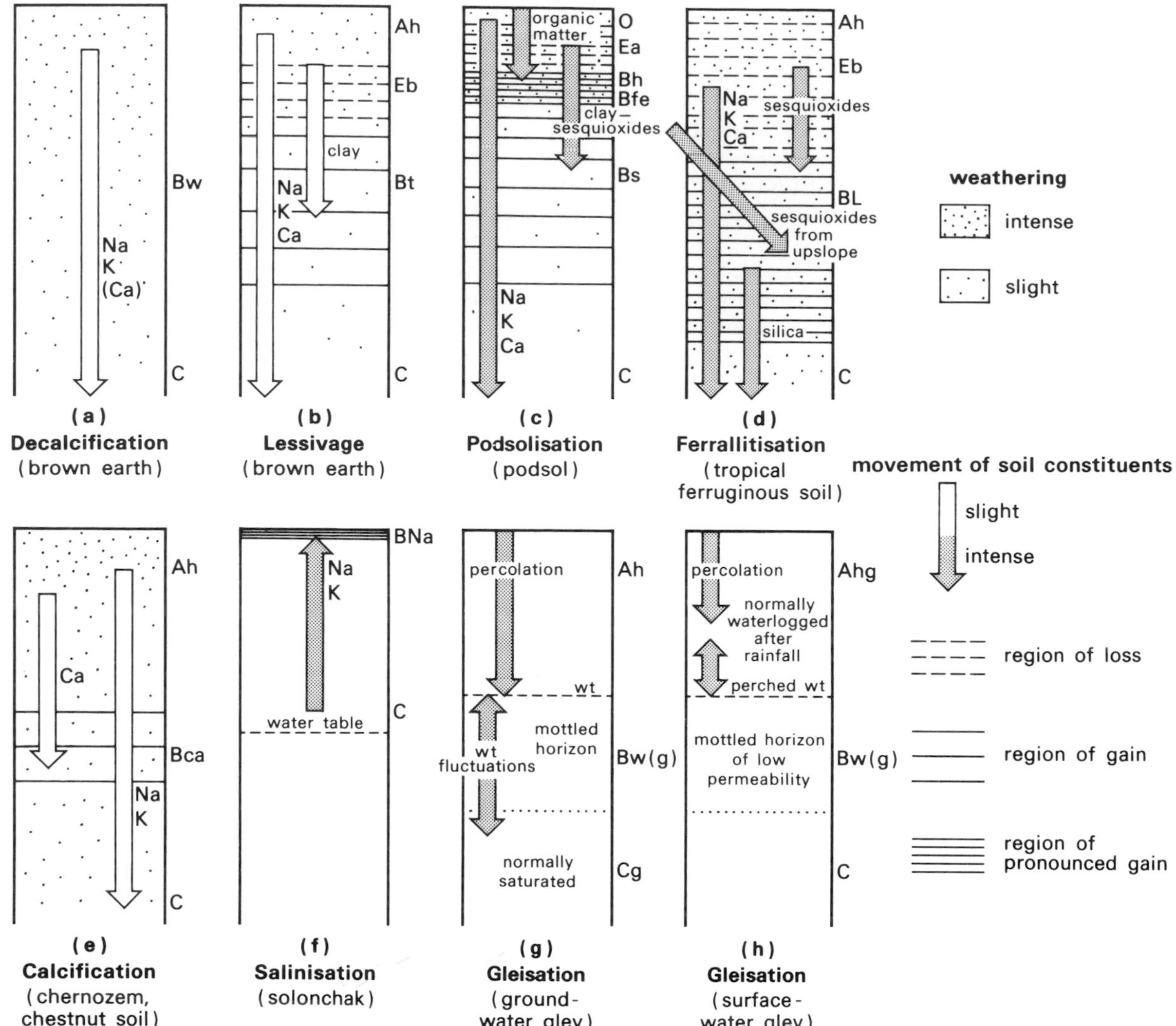

Figure 20.13 The formal processes of pedogenesis (after Knapp 1979).

by *in-situ* weathering or from decomposing litter. The exchange complex is dominated by hydrogen and clay minerals – never present in great quantities in such parent materials – become unstable and are broken down. In addition the soil population, dominated by arthropods and fungi, is less efficient than that inhabiting brown earth soils, with the result that the decomposition and humification of the acid litter remains incomplete. Many organic compounds, particularly the polyphenols, are not incorporated in humic polymers and are leached from the organic horizons and become available as chelating agents. In the mineral soil they complex iron and aluminium oxides which pass into solution as metal ion chelates and are translocated down the profile, leaving a bleached EA horizon of uncoated quartz particles. At depth these metal ion chelates become unstable both as a result of microbial activity and because the soil chemical environment becomes less acid in proximity to the unweathered parent material, and the iron and aluminium are precipitated in their insoluble higher valency state. The iron III oxides may form diffuse horizons (Bs) of iron accumulation or thin discrete horizons (Bfe) with an indurated surface creating iron pans. Above these iron accumulation horizons, translocated humic compounds are trapped to form Bh horizons

Some workers have placed brown earths, sols lessivée and **podsols** in a developmental sequence. However, in most situations the controlling variable is not time but the initial characteristics of the parent material and the type of vegetation, which act as the primary regulators operating within the overall control of a humid temperate climate.

20.3.2 A model of tropical pedogenesis

The dominant formal process in the humid Tropics is **ferrallitisation** (Fig. 20.13d). Where leaching is not excessive, and intense *in-situ* weathering under a hot and humid climate leads to the breakdown of clay minerals, soil profiles are characterised by the accumulation of 1:1 kaolinite clay and hydrated oxides of iron and aluminium. The type of iron compound is different from that of temperate environments and it imparts a rich red colour to these **ferruginous** soils, a process known as **rubification**. With greater leaching intensity the stability of ferruginous soils is disrupted and lessivage of clay minerals in particular and some hydrated oxides occurs leading to a textural B horizon and the formation of **ferrisol** soils.

However, where both leaching and weathering are intense the lattice breakdown of clay minerals proceeds further and deeply weathered soil profiles develop. These soils have a bimodal texture dominated by residual quartz of sand size and hydrated oxides of iron and aluminium of clay size together with some kaolinite. Such **ferrallitic** soils become highly acidic (pH 4) and under these conditions amorphous silica is removed from the profile, thereby enhancing the relative accumulation of the iron and aluminium oxides (BL horizon). However, in many such soils the amount of iron present cannot be explained by *in-situ* weathering of the parent material, or by translocation from the upper horizons (Eb), and their occurrence on footslopes points to the reprecipitation of iron gathered by laterally moving soil water. Indeed, ferrallitic soils seem to have developed best in areas of fluctuating water table under alternately wet and dry tropical seasonal climates. These conditions allow iron III oxides to pass into solution as reduced iron II ions at times of high water table and to migrate laterally down slope to reprecipitate in an oxidising environment when the water table falls. With desiccation (perhaps associated with climatic change) or with erosion the BL horizon may undergo irreversible dehydration leading to the formation of indurated **laterites**.

20.3.3 A model of arid pedogenesis

In semi-arid and arid environments where effective precipitation is insufficient, leaching of solutes from the profile is incomplete. In midcontinental grassland environments of semi-aridity, precipitation is usually sufficient for sodium and potassium to be removed from the profile. However, calcium is precipitated as calcium carbonate (**calcification**, Fig. 20.13e) forming a BCa horizon as carbon dioxide concentrations fall below the rooting zone, or as percolating water begins to evaporate. Under more severe aridity, intense evaporation means that only the most mobile ions – sodium and potassium – pass into solution, but even these are soon precipitated at a percolation front. In extreme cases where texture permits and the water table depth is not great, saline ground water may be drawn up in response to strong surface evaporation leading to the precipitation of sodium and potassium salts (**salinisation**, Fig. 20.13f). Indeed, where saline ground water is too close to the surface, **solonisation** occurs, producing saline soils of very high pH where high concentrations of exchangeable sodium have the effect of deflocculating and dispersing the clay minerals, allowing their translocation and the loss of soil aggregation.

20.3.4 Gleisation

Certain formal processes are controlled less by climate than by other dominant sets of conditions. **Gleisation** is such a process where the drainage relationships of the soil and the type and position of the water table are the paramount controls (Figs. 20.13g & h). Because appropriate conditions can occur under a variety of climatic regimes, gleying can occur in conjunction with several other formal processes. Gleying occurs in soils which experience either periodic or permanent waterlogging. It is associated with the presence of high groundwater tables (**groundwater gleys**) lying within the profile, or with the impedance of soil drainage due to the existence of horizons of low permeability and perhaps the formation of perched water tables (**surface water gleys**). In these circumstances anaerobic conditions are established rapidly as available oxygen is depleted by the respiratory demand of aerobic soil organisms and plant roots.

These are replaced by anaerobic organisms and as the redox potential falls and chemically reducing conditions are established many redox couples – both inorganic and organic – are converted to their reduced forms. The most obvious sign of these changes in soil chemical environment is the conversion of reddish-brown iron III or ferric compounds to greyish iron II or ferrous compounds. These pass into solution as mobile ferrous ions and migrate both within soil aggregates forming coatings on the ped faces and out of the profile with flowing ground water. Where oxygen penetrates the gley horizons, secondary oxidation takes place producing characteristic reddish mottling. This happens particularly where the water table fluctuates, perhaps seasonally, or where radial oxygen loss from live roots occurs.

Further reading

Useful introductory texts are

Bridges, E. M. 1970. *World soils.* Cambridge: Cambridge University Press.

Courtney, F. M. and S. T. Trudgill 1976. *The soil: an introduction to soil study in Britain.* London: Edward Arnold.

Knapp, B. J. 1979. *Soil processes.* London: George Allen & Unwin.

and, at a more advanced level

White, R. E. 1979. *Introduction to the principles and practice of soil science.* Oxford: Blackwell Scientific.

Buringh, P. 1970. *Introduction to the study of soils in tropical and subtropical regions.* Wageningen: Centre for Agricultural Publishing and Documentation.

Russell, E. W. 1973. *Soil conditions and plant growth.* Harlow: Longman.

Part D

Systems and change

In modelling the organisation and operation of Earth and environmental systems we have chosen to ignore the explicit expression of change and have not attempted to isolate those mechanisms and processes leading to changes in systems states except in so far as they are implicit in the normal operation of these systems. In Chapter 1 it was stressed that one of the important properties of any model, including a systems model, is its generality. The models we use to represent environmental systems should, therefore, be applicable to any broadly similar situation. The drainage basin, the ecosystem, the pressure cell – indeed all of the models we have used – are generalisations that are valid at different points in space and time to systems of matter which share the same general form, and which function in the same way. Such generality is possible because our models are abstractions from the real world. In the real world the systems with which we have been concerned vary enormously in the details of their structural and functional organisation, and system processes vary in their magnitude, relative significance and rates of operation. There are two directions that such variation can take, one in space and one in time. To a certain extent we have examined variation in space in all of the models used so far. Geographical or areal variations have been stressed repeatedly, in the disposition of systems elements and their attributes, and in the magnitude of inputs, or processes. We have considered spatial variations in energy and water balances, and in the operation of the denudation system, compared geographical variations in ecosystem primary productivities, and considered the way in which the soil system responds to spatial variations in its operating conditions. However, in each case, we have treated each system as if it were in some broadly stable equilibrium state, and although our attention has been directed to the dynamics of systems operation, apart from a few exceptions, we have not considered explicitly the dimension of time. In Part D, therefore, the behaviour of environmental systems with time will be explored at some length.

VII THE NATURAL WORLD

21 Change in environmental systems

21.1 Equilibrium concepts and natural systems

The equilibrium state of thermodynamic open systems (and that is the way that we have modelled environmental systems throughout this book) is a steady state. However, the notion of equilibrium is more complex than this tacit assumption would suggest and furthermore it is intimately associated with ideas of change, or indeed lack of change, through time. What then have we meant by the concept of a steady state equilibrium as we have used it in earlier chapters? Central to the concept is the maintenance of an average condition of the system, the trajectory of which remains unchanged in time (Fig. 21.1a). This average condition can be defined in terms of both the disposition of the morphological components of the system and the flows of mass and energy. At any instant, the actual condition of the system will only approximate to the average state and over a period of time will be seen to fluctuate about, but may never actually accord with, this average state. One immediate implication of this view is that the recognition that a system is maintaining a steady state is dependent on the time scale over which it is considered. Indeed, the importance of both time scale and spatial scale in the interpretation of

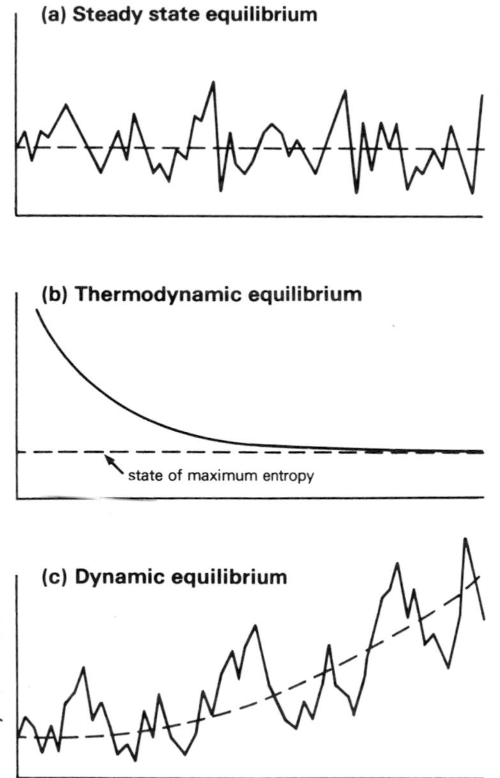

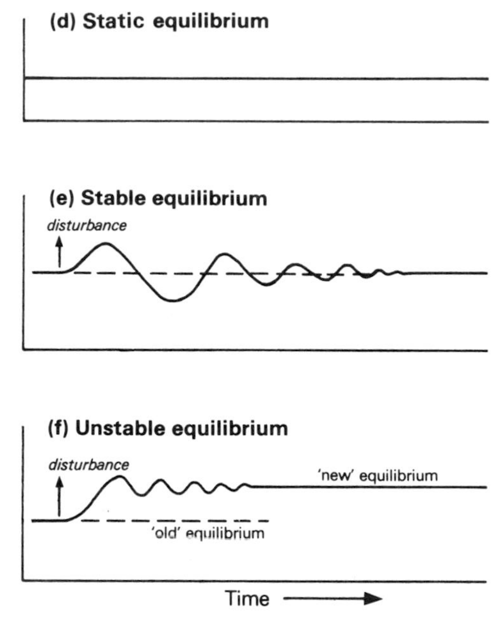

Figure 21.1 Types of equilibrium.

equilibrium conditions in environmental systems is a subject to which we shall return shortly.

Also central to the concept of steady state equilibrium is the significance of self regulation and the ability of the system, consequent on the operation of negative feedback loops and the action of regulators, to damp the amplitude of the cyclic departures from an average state. Therefore, a dynamic open system is able to maintain an orderly state within controlled limits determined by the precision and efficacy of its capacity for self regulation. Furthermore, this orderly state is maintained by, and in the face of, often massive throughputs of mass and energy.

There are, however, other conceptions of equilibrium which prove useful in understanding the relationships between equilibrium and change in environmental systems. For example, embodied in the classic concepts of isolated system thermodynamics is the tendency towards maximum entropy which we have already encountered in Chapter 2. When a system attains such a distribution of mass and energy it is said to be in thermodynamic equilibrium (Fig. 21.1b). Implicit in this concept is irreversible change marked by a decline in free energy capable of doing work on the system, and a corresponding increase in the entropy of the system. Also emphasising progressive changes of state, but this time explicitly, is the concept of dynamic equilibrium. Here, as in a steady state equilibrium, controlled fluctuations occur about average systems states. However, in this case they are unrepeated average states through time (Fig. 21.1c). The existence of dynamic equilibrium is not always easy to recognise because, often, directional change in the average condition of the system is obscured by the greater magnitude and rate of change of the fluctuations about it. Over short time periods, therefore, the system appears to maintain a steady state.

Three further definitions of equilibrium remain that are best applied to relatively simple mechanical and chemical systems and they are defined by the system's response to limited external forces. The first is static equilibrium where force and reaction are balanced and no resultant force exists, and the properties of the system to which the concept is applied remain unchanged or static through time. The second and third definitions are stable and unstable equilibrium respectively. Here the system displays tendencies either to return to (stable) or to be displaced further from (unstable) the initial equilibrium state in response to an externally applied disturbance (Figs 21.1 d, e & f). The notion of balance as applied in static equilibrium is familiar in a number of situations, but one example of its application that we have encountered is the balance between reactants and products when a weathering reaction reaches equilibrium. It can be applied also to systems where a balance pertains between inputs and outputs, as in the slope system. Stable and unstable equilibria are familiar both in relation to systems of gases when considering atmospheric stability and in relation to simple mechanical systems concerned with the erosion and transport of rock materials. Related to the last two of these definitions is the concept of meta-stability. This is the tendency of a system to move from one equilibrium state to another when an external force causes it to cross a threshold beyond which the probability of recovery becomes remote. These concepts of stable, unstable and meta-stable equilibria have been used to some effect by Trudgill (1977) to model the response of soil–vegetation systems to disturbance (Fig. 21.2).

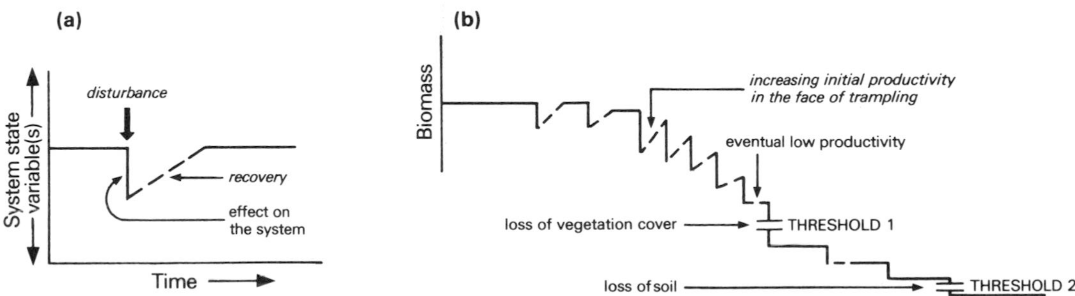

Figure 21.2 (a) General disturbance and recovery model; (b) the effect of trampling on vegetation biomass (both after Trudgill 1977).

21.2 Thermodynamics, equilibrium and change

In the above discussion it will have become obvious that almost all views of equilibrium demand an appreciation of the manifestations and mechanisms of change in the state of the system through time. However, before we turn our attention to change it will be useful to remember that environmental systems are energy systems and are subject to the laws of thermodynamics. In accordance with the Second Law (Ch. 2), the operation of natural irreversible processes leads to a decrease in the free energy and an increase in the entropy of an isolated system. It was the Austrian physicist, Ludwig Boltzman, who recognised, in 1896, that the entropy of a system described in this way is a measure of the way in which the total energy of a system is distributed amongst its constituent elements (atoms or molecules). More precisely, he demonstrated that the entropy of a system (S) is dependent on the number of statistically independent ways of distributing the elements amongst a number of energy levels (or quantum states, P).

$$S = k \log_e P$$

where k = Boltzman constant = 1.38054×10^{-23} J K^{-1}. At thermodynamic equilibrium the entropy of the system is at a maximum and the distribution of elements amongst the energy levels conforms to the most probable state, that is at random.

This statistical or probablistic view of entropy has been extended beyond Boltzman's original formulation and linked with Shannon's work on communication and information theory (Jaynes 1957) so that the entropy (S) of a system can be expressed as the sum of the probabilities of possible states or properties of the system, $p_1, p_2, p_3 \ldots p_N$:

$$S = -k \sum_{i=1}^{N} p_i \log_e p_i$$

$$\text{where} \sum_{i=1}^{N} p_i = 1.$$

The entropy of the system is at a maximum when all of the states or properties have an equal probability ($p = 1/N$) or are distributed at random in time and space. Conversely, entropy is at a minimum when only one state is possible and all others have a probability of zero, i.e. the most organised state. Therefore, entropy is seen to embrace concepts of order and disorder and, as we saw in Chapter 6, the information content of the system. (Shannon showed that the total information content of a system could be given by log N, and that if the base 2 is used instead of 10, or e, then the smallest possible unit of information is the 'bit' (see Box 6.7).) The Second Law, therefore, gives us a direction to natural processes: from high to low energy levels, from organised (ordered) to random (disordered) configurations, and from high to low information content.

However, these concepts were developed originally to apply to isolated systems. Environmental systems are open systems. In such systems the input of mass and energy – precipitation and kinetic and potential energy in channel systems, food molecules and chemical energy in animals, water vapour and thermal energy in atmospheric systems – increases or maintains the free energy, the organisation and information content of the system, and in so doing reduces its internal entropy. How then can these facts be equated with the view of equilibrium, in thermodynamic terms, as a state of maximum entropy and, in probability terms, as the most probable state?

First, the creation and maintenance of order and information in an open system is dependent on the transformation of free energy by the system. Nevertheless, this consumption of free energy in doing work on the system by irreversible processes will lead inevitably to the production of entropy. So secondly, individual processes will proceed in the direction of increasing entropy given the constraints imposed by, for example, the energy environment under which the system operates. However, under open system conditions a true thermodynamic equilibrium cannot be reached. Therefore, the key to understanding the equilibrium state of an open system is to grasp the system's need to minimise the rate of entropy increase per unit of preserved structure, where

rate of increase in the entropy of the system = rate of internal generation of entropy by the system − rate of outflow of entropy to the surroundings.

In theory, the equilibrium state of an open system should be one in which the rate of increase in entropy is zero. That is, the rate of internal generation of entropy by the processes maintain-

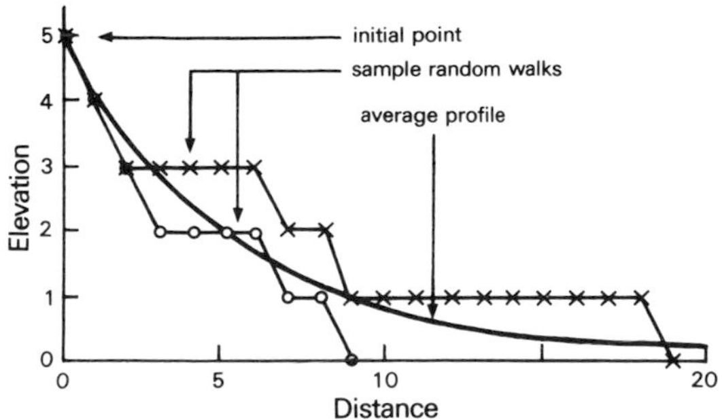

Figure 21.3 Sample random walks generating an average stream long profile (after Leopold & Langbein 1962).

ing an ordered or low entropy state is balanced by the outflow of entropy to the surroundings. The most probable state to satisfy these equilibrium conditions is one where the rate of internal entropy generation is minimised. Such a state can be equated with a configuration of the system which requires the minimum work and minimum expenditure of free energy to maintain its structure.

This concept of minimum energy expenditure and minimum internal production of entropy has been applied to generate the most probable long profile of a stream. Here adjustments in the hydraulic geometry of the channel along its length are seen to establish the most efficient form of long profile for the prevailing conditions of energy and mass flow such that the expenditure of energy on work is minimised. In such a state there is a uniform rate of internal entropy generation per unit stream length and this rate per unit discharge rate equals the rate of outflow of entropy as heat. This form, therefore, represents the most efficient and probable long profile and accords well with natural stream profiles. The fact that such a profile can be simulated by purely random processes, apart from the constraint introduced by the need for the stream to flow down hill, confirms that it is the outcome of the minimum work tendency of the system (Leopold & Langbein 1962, and 1964; Fig. 21.3).

This example, and similar work such as Yang's attempt (1970, 1971) to use the same work-minimising reasoning partially to explain meander form, concern relatively simple physical systems. However, perhaps the most profound insight we gain from these entropy considerations is the beginnings of an explanation of the capacity of living systems to maintain complex information-rich structures. If we return to the cellular level of

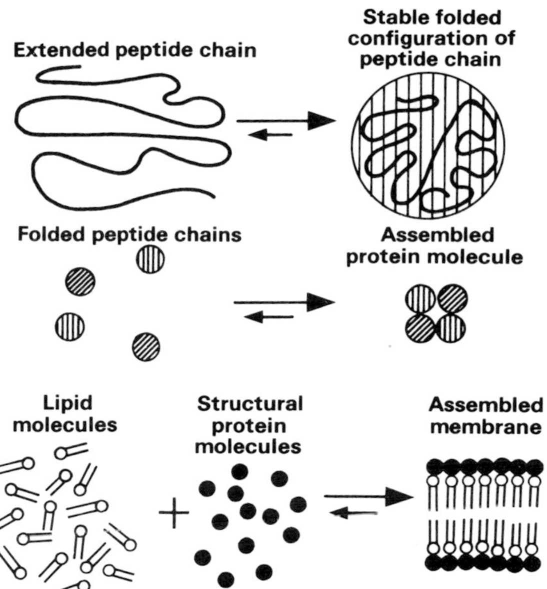

Figure 21.4 The natural tendency of supra-molecular organisation (modified from Lehninger 1965).

Chapter 6 we can re-interpret the complexity of organic macromolecules and the cellular structures they build in the light of an understanding of irreversible open system thermodynamics.

We can see now that the apparently highly improbable complexity of the living cell is the automatic outcome of the chance occurrence of the information-rich programming system represented by the DNA molecule. We have seen already that DNA (Box 6.6) programmes the amino acid sequence of protein polypeptide chains, and that protein enzyme systems programme the synthesis of other molecules. The enzyme proteins depend on their three-dimensional structure for their function, and this structure can be seen as the result of the tendency displayed by all systems to seek that state which possesses the least free energy, i.e. an entropy maximising state. In this case the amino acid side chains tend to arrange themselves in relation to their neighbours in such a way as to minimise the energy content of the structure. The result is that the whole polypeptide chain bends and coils so as to arrive at the most stable arrangement with the lowest energy content under the constraining conditions of pH, temperature and ionic composition that pertain in the cell. The DNA, therefore, has not only coded the amino acid sequence, but by so doing has automatically determined the three-dimensional geometry of the chain and ultimately of the protein molecule (Fig. 21.4). The same reasoning can be applied to supra-molecular structures such as enzyme systems, or biological membranes (Fig. 21.4). Again, because of the large negative free-energy change associated with certain arrangements, it is these configurations that are the most stable, most probable and therefore the automatic outcome of cellular reactions. Even more importantly, they are the configurations that require the least expenditure of energy to maintain under open system conditions and are, therefore, most likely to persist. There is no reason why these thermodynamic considerations may not be valid, in a broad way, at the level of the individual organism or, as we shall see, at the level of the entire ecosystem.

However, the flow of mass and energy through an open system, although maintaining its organisation, also has a disruptive effect on the existence of a constant equilibrium state. This is partly because the inputs to environmental open systems are never uniformly distributed in either time or space. Furthermore, the complexity of the pathways of throughput, the residence times of mass and energy in the various stores and the lags experienced in the response to processes all mean that at any instant the rate of entropy increase is unlikely to be zero. Over a period of time the system will fluctuate about this statistical average or most probable state. In other words, on thermodynamic grounds the equilibrium state of an open system is a steady state of minimum work, minimum internal generation of entropy, maximum preserved order per unit of energy flow and with an average increase in the entropy of the system equal to zero.

21.3 Manifestation of change

The above discussion of the equilibrium state attained by environmental open systems allows us to recognise and describe the types of change that we should look for in the following chapters. Chorley and Kennedy (1971) recognise four classes of change: first, changes in the energy content and energy distribution within the system; secondly, those due to changing inputs or input/output relationships of both mass and energy; thirdly, those associated with shifts in the internal organisation or integration of the system itself; and fourthly, those associated with the development of energy and mass stores which introduce time lags in the operation of systems processes, usually acting as buffers. However, it is important to realise that the operation of all environmental processes involves change in the system; indeed, in Chapter 1 we defined a process as merely the method of operation by means of which a change in state is effected. Furthermore in Chapter 2 and again in this chapter we have recognised that all processes occurring in open systems are time dependent, for open systems maintain a steady state by the transfer of mass and energy by real or irreversible processes with the surroundings. Time and rate are critical variables in the energetics of such systems, while entropy production itself is time dependent (see Ch. 2).

However, as we have seen in Section 21.2, the maintenance of a steady state implies that the net effect of the operation of these irreversible processes is to return the system to the most probable state, or at least keep it fluctuating about such a state. As we know, this is accomplished by self regulation under the control of negative feedback mechanisms. Therefore the first kind of change we should look for is fluctuation about the steady state and indeed much of Parts B & C has done just that. These fluctuations may be associated with all four of the causes of change classified by Chorley and Kennedy. Both the atmospheric energy balance and the atmospheric circulation discussed in Chapters 3 and 4 reflect changes in the energy content and energy distribution of the atmosphere and the feedback mechanisms that redress imbalances. The hydraulic geometry of a stream channel changes as it adjusts in response to fluctuating water and sediment discharge characteristics, themselves a reflection of changing input/output relationships. The steady state condition may in many cases be that which is adjusted to bankfull discharge, but in others the mean annual flood, or even baseflow may be the condition to which the channel returns. Variations in hydrograph characteristics reflecting differences in the contributing areas of a drainage basin under different rainfall conditions is an example of fluctuation reflecting changes in the organisation of internal functional linkages within the basin. The immobilisation of nutrient elements in litter and soil organic matter represents the development of a storage compartment with a lagged output which influences the rate of nutrient turnover.

These causes of variation in the state of the system may themselves possess recognisable periodicities to which the system can become adjusted and for which it develops **memory**. Variations in input, in patterns of energy distribution, in functional linkages or in changes in storage, may all reflect predictable diurnal or seasonal fluctuations. Sometimes the periodicity may be longer and simply be termed cyclic, while in other cases variation may be apparently random. The magnitude of the response of the system, and the length of the **relaxation time** (time over which adjustments take place) will vary from system to system but will reflect the memory and efficiency of negative feedback control that the system has developed.

However, not all systems which we chose to define in the natural environment will have attained a steady state equilibrium. We must expect to encounter systems which are immature, which represent merely transient states in a sequence of states that form part of a developmental sequence. Here we must be prepared to recognise directional change – specifically evolutionary change – as the system seeks to attain a steady state under the prevailing conditions. Such change will be dominated by the cumulative effect of positive feedback propelling the system towards its most probable state. The growth of an immature organism, the attainment of a 'graded' stream long profile, the development of a mature ecosystem on an initially virgin surface, and the increasing integration of a developing drainage network – are all changes which can be viewed in these terms. Nevertheless, they are all the inevitable result of inherent properties of the systems concerned and represent the diversion of surplus energy throughput towards the development of more integrated and efficient functional organisations and may be manifest as any combination of the four classes of change recognised by Chorley and Kennedy. As such changes in systems states approach the most probable steady state condition, we would expect negative feedback self-regulatory mechanisms to become dominant, and directional change to be replaced by fluctuation about a mean condition.

Finally, we must expect that the capacity of any defined environmental system exhibiting a steady state to resist or accommodate changes in the energy inputs to it, or changes in the energy environment of its surroundings to be limited and dependent on the efficiency and resilience of its self-regulatory mechanisms. When this capacity is exceeded we must anticipate that the system would again show directional change with a sequence of transitory states replacing each other through time

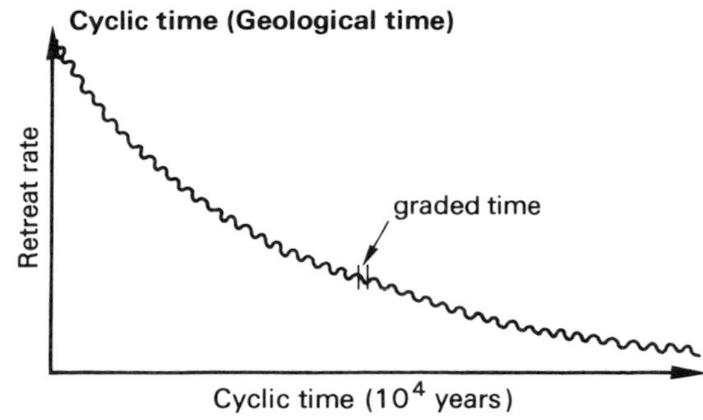

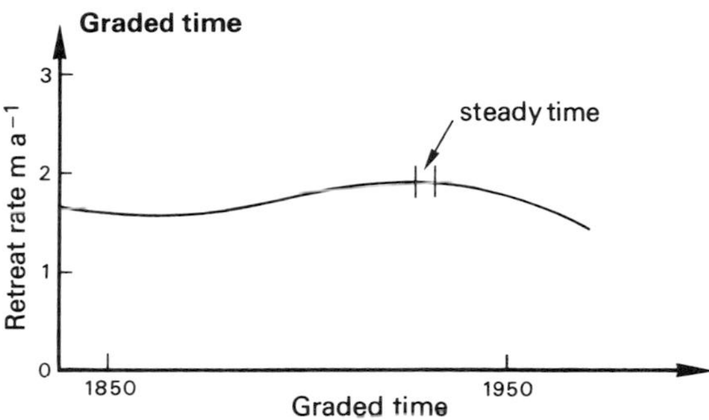

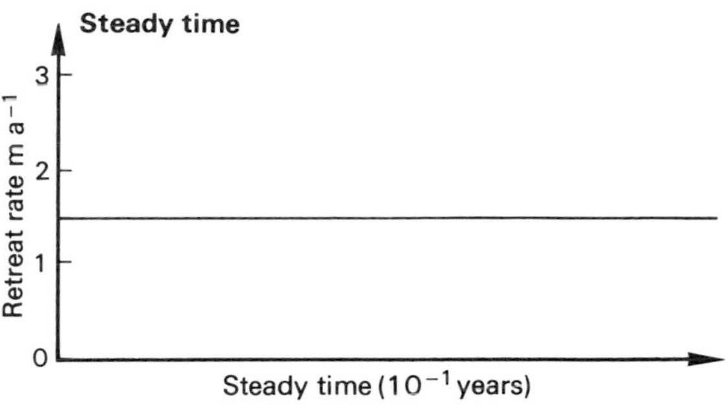

Figure 21.5 Time, space and causality in geomorphic systems (after Schumm & Lichty 1965).

as the system moves towards a new equilibrium determined by the new level of inputs or new energy environment it experiences. Again the dominant control of change will be positive feedback, but this time externally induced. Adjustments in the energy and mass balance of the atmosphere resulting from changes in planetary inputs, or changes in denudation and ecosystem function reflecting climatic change, are examples of such externally induced change (to the system as defined).

All of these manifestations of change in systems states are examined in detail in the chapters that follow, but their recognition depends on a clear definition of the system being considered and of the timescale concerned. This timescale must be appropriate to the scale at which the system is defined and, in the case of environmental systems particularly, its spatial scale. This important point was first explicitly recognised in geomorphology in the now classic paper by Schumm and Lichty (1965) (Fig. 21.5).

22 Change in physical systems

22.1 Change in climatic systems

Climate is defined in terms of statistical averages of, for example, air temperature, hours of bright sunshine and precipitation. Within any climate so defined there are periodic changes in these and other parameters over diurnal and annual time scales, which are an integral part of the operation of the Earth–atmosphere system and do not involve it in long-term structural changes. However, it is possible to identify longer-term changes in climate occurring over periods of the order of hundreds of years. Some of these changes represent considerable modifications to the mode of operation of Earth-surface and atmospheric systems.

The Earth's surface reveals evidence of periods of extreme cold when large areas were under thick sheets of ice. Areas which are deserts today contain landforms that are clearly fluvial in origin, suggesting periods of wetter climatic conditions. Measurements of climatic parameters (which have been available only since the early 18th century) reveal, for example, general increases in global air temperatures over the period from the late 19th to the middle of the 20th century and a reversal of this trend since 1950. More recently, the greater frequency of failure of monsoon rains to reach north-west India, the extension of desert conditions along the south side of the Sahara, and the increasing difficulty of keeping polar sea routes open, all indicate that major changes in the Earth–atmosphere system are currently taking place.

These changes can be viewed alternatively as a result of the changing operation of factors external to the Earth–atmosphere system, or as modifications of the disposition of energy and matter within it. For example, changes in the temperature of the lower troposphere may be due to a changing radiant energy exchange between the Earth–atmosphere system and its surroundings (space). Alternatively, it may be that structural modification within the system has resulted in more or less energy being stored as sensible heat within the atmosphere.

If we view the Earth–atmosphere system as a black box (Ch. 3), then our first alternative concerns us with a discussion of relationships between input and output and we may make general statements regarding the throughput. Changes within the system concern its response to energy inputs, in which case we must consider the pathways along which energy and matter are transferred, the nature of the stores and the rate at which energy and matter are transferred between them. Critical in this respect are the regulators which are the control points within the system.

22.1.1 External changes

If the Earth and its atmosphere were full radiators (Ch. 3), we should expect that any change in the amount of radiation received from the Sun would be matched by increased absorption and emission. A net radiative balance or dynamic equilibrium would thus be maintained. This assumes that the Earth–atmosphere system is capable of establishing such a radiative equilibrium with its surroundings. However, as radiant energy passes through a complex energy cascade, such changes give rise to a temporary state of disorder within the system which results in changes in surface temperatures.

If we were to deal only with radiative exchanges, then a change in the solar constant (ΔR_0) would produce a change in Earth-surface temperature (ΔT_s) given by:

$$\Delta T_s = \frac{1}{4}\left\{\frac{R_0(1-r)}{4\sigma}\right\}^{1/4}\left\{\frac{\Delta R_0}{R_0}\right\}$$

where r = planetary albedo and σ = Stefan's constant = 5.57×10^{-8} W m^{-2} K^{-4}. If the planetary albedo is assumed to be 0.3, then there would be increases in temperature of 0.06°C for every 1% increase in the solar constant.

The information available on solar radiation received by the Earth is limited in quality and quantity for mainly technical reasons. Only in the past two decades has it been possible to monitor incoming solar radiation above the filtering effects of the atmosphere, by using satellites. The alternative is to use high alpine observatories which are still subject to problems of atmospheric transparency. Surface observations of direct solar radiation made at mountain observatories between 30° and 60°N (Fig. 22.1a) show variation within 10% of an overall mean value, while mean daily hours of bright sunshine in Southampton, at only 20 m above sea level, show a haphazard scatter about the mean probably due to the much greater influence of atmospheric transparency (Fig. 22.1b).

In the absence of an atmosphere, long-term fluctuations in solar radiation arriving at the Earth's surface may be due to changes either in emission from the Sun or in the characteristics of the Earth's orbit. The gaseous structure of the Sun undergoes continuous modification through time which produces changes in total emission. Variations of this nature are relatively small in relation to the solar constant. However, the occurrence of dark areas on the solar surface does produce discernible changes in the radiation the Earth receives. These sunspots are the centres of localised disturbances in the centre of which the Sun's

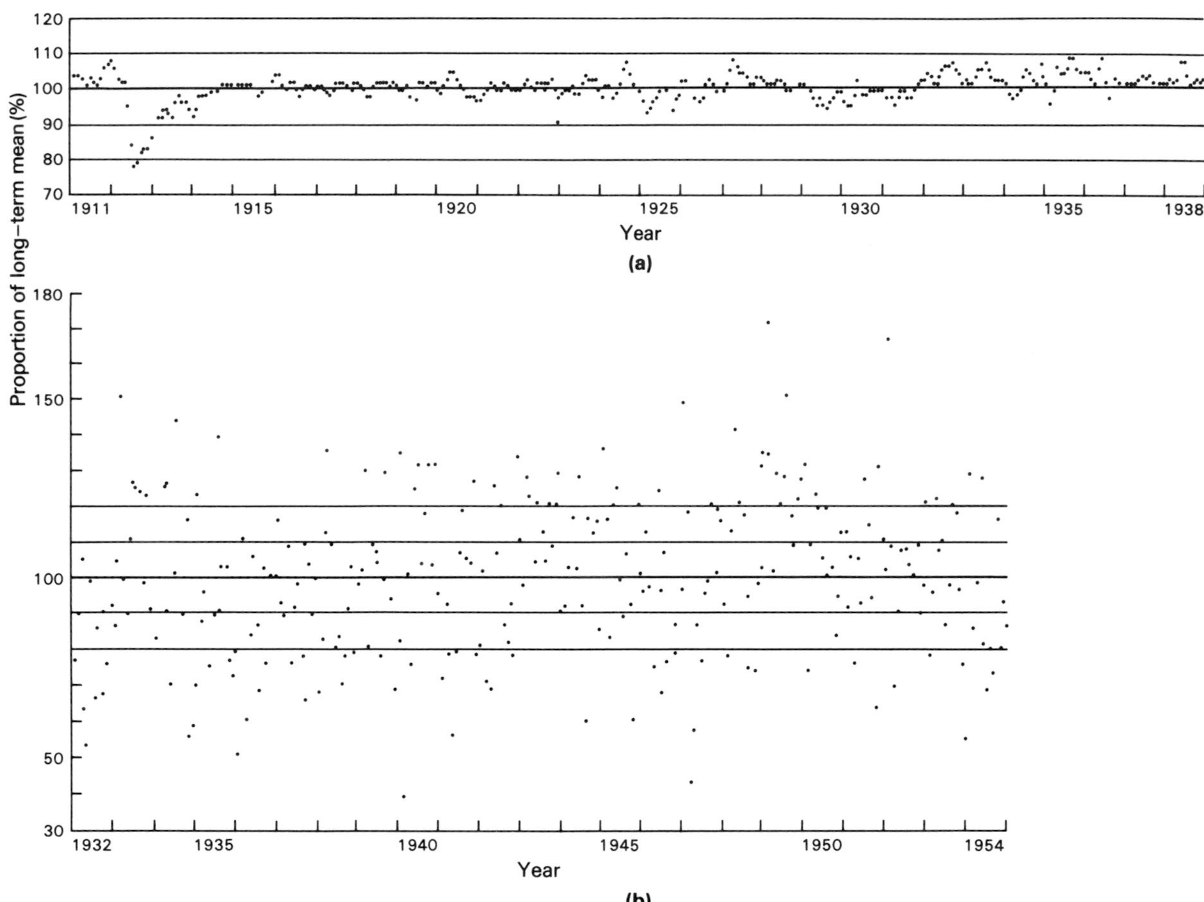

Figure 22.1 (a) Mean monthly values of the strength of direct solar radiation from observations at mountain observatories between 30°N and 60°N in America, Europe, Africa, and India from 1883 to 1938 as a percentage of the long-term average (after Lamb 1972). (b) Mean monthly hours of bright sunshine at Southampton (East Park) expressed as percentage of the long-term (1932–54) mean.

surface temperature is reduced by as much as 2000 K. Around them are brighter zones, referred to as faculae, from which there is intensified radiation. Associated with the occurrence of these disturbances are increases in the emission of both ultraviolet and infra-red radiation, and of solar particles, which directly affect the upper atmosphere. The increase in high-energy radiation increases the probability of dissociation of oxygen molecules in the upper atmosphere, which has the effect of increasing the rate of production of the unstable ozone (O_3). The presence of more ozone and the absorption of the extra ultraviolet radiation produces a slight increase in temperatures in the stratosphere. The number of sunspots on the solar surface exhibits a well marked periodicity, the time between successive maxima being about eleven years. Several attempts have been made to relate these sunspot cycles to surface weather patterns, but these have been largely inconclusive or, at best, tentative.

There are three characteristics of the Earth's orbit which vary over periods of time measured in tens of thousands of years. These affect mainly the seasonal and geographical distribution of solar radiation over the Earth's surface and not the total energy received. The tilt of the Earth's axis relative to its plane of orbit changes over a period of approximately 40000 years. The probable range of variation lies between 21°48′ and 24°24′. It is currently 23°27′ and is decreasing from a maximum value of some 10000 years ago. The significance of this variation lies in the latitudes of tropical and polar circles. An increase in the angle of tilt reduces the latitude of the polar circles and raises the latitude of the tropical circles (Fig. 22.2). With regard to the former, this implies an increase in the area of the Earth's surface which experiences day-long polar nights and days. The effect of changes in tilt of the Earth's axis is, therefore, to alter the seasonal variation in solar radiation over its surface. The shape of the ellipse which describes the Earth's orbit around the Sun changes over periods of approximately 100000 years. The more elliptical the orbit becomes, the more marked are the differences between perihelion and aphelion (Ch. 3). If the Earth described a circular orbit, this difference would be zero. At its most elliptical there is a ± 15% variation either side of an annual mean amount of solar radiation received, while current variation is ± 3.5%. The Earth's orbit itself rotates around the Sun, which causes a change in the timing of perihelion and aphelion, relative to the Earth's seasons. Perihelion currently occurs on 3 January, during the southern hemisphere summer. It takes approximately 21000 years for the cycle to be completed, which means that perihelion in the Earth's orbit occurs one day later every 58 years.

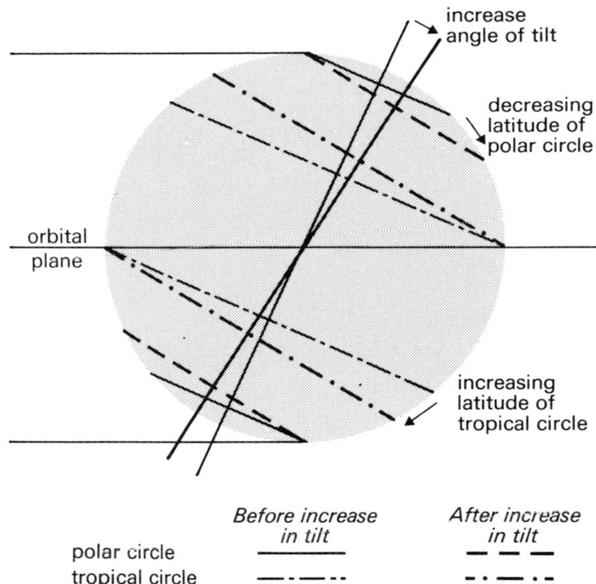

Figure 22.2 The effect of changing the tilt of the Earth's axis on the polar and tropical circles.

Milankovitch (1930) has estimated the net effect of these three changes in the Earth's orbit (Fig. 22.3) which shows that they have resulted in long-term fluctuations in solar radiation received at the Earth's surface. These variations were linked tentatively with the periods of expansion of the polar ice caps during the ice ages. In excluding the transfers of heat by conduction and convection, and by the complex radiational exchanges that occur at or near the Earth's surface, Milankovitch's estimates of temperature changes resulting from changes in radiation input tend to overestimate fluctuations that would take place. However, more recently, the analysis of the isotope ^{18}O in cores taken from ocean core sediments (Shackleton & Opdyke 1973) has indicated some degree of coincidence with the chronology of climatic changes suggested by the Milankovitch model.

Equivalent modern latitude
60°
70°
80°
500
450
400
350
300
250
200
150
100
50
0
1000
950
900
850
800
750
700
650
600
550
500
Years before AD 1950 (x10³)

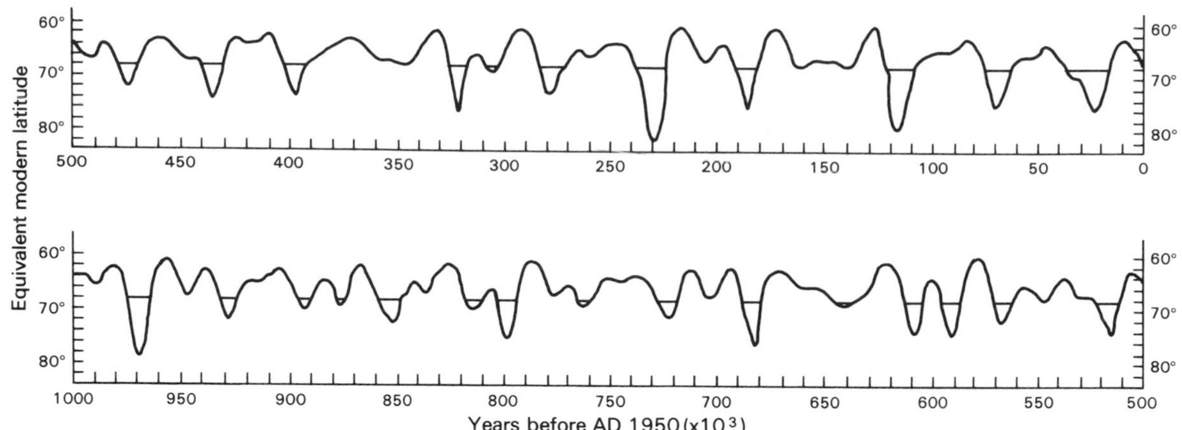

Figure 22.3 Amounts of solar radiation available in the summer half of the year at latitude 65°N over the past million years, expressed as equivalent to the radiation available at various latitudes at the present day (after Milankovitch 1930).

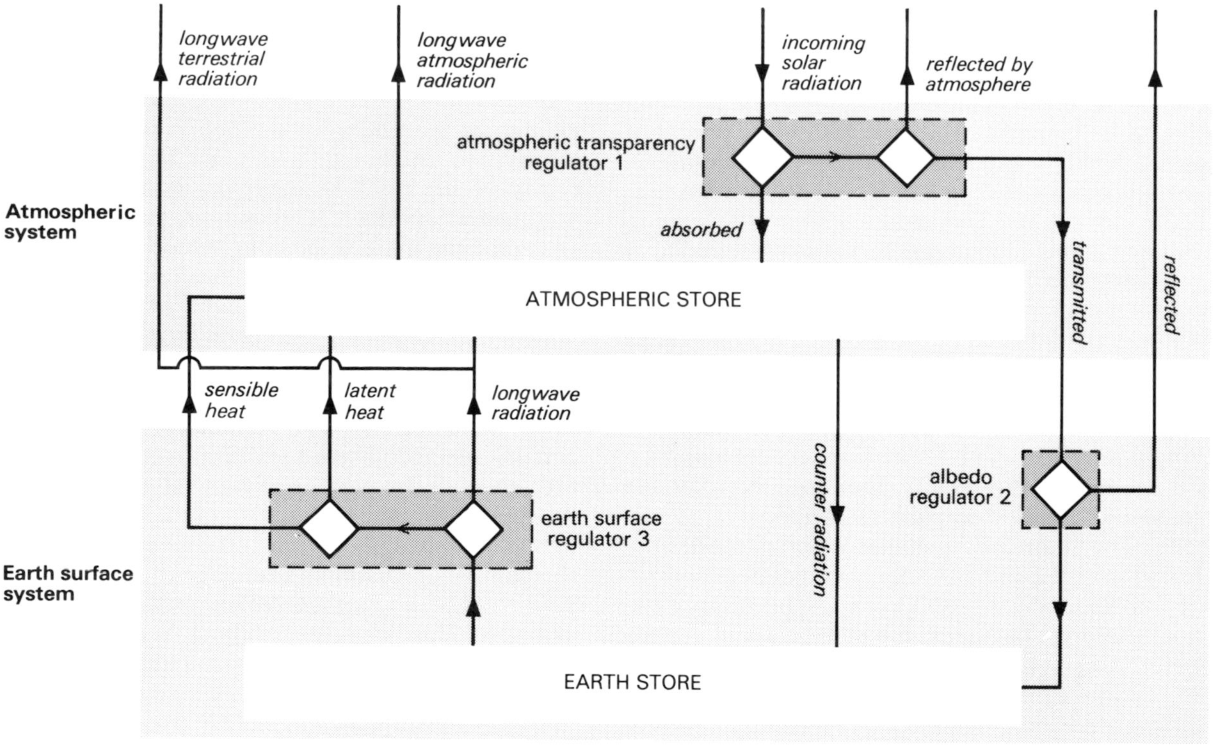

longwave terrestrial radiation
longwave atmospheric radiation
incoming solar radiation
reflected by atmosphere
atmospheric transparency regulator 1
Atmospheric system
absorbed
transmitted
reflected
ATMOSPHERIC STORE
sensible heat
latent heat
longwave radiation
counter radiation
albedo regulator 2
earth surface regulator 3
Earth surface system
EARTH STORE

Figure 22.4 The function of regulators in the solar energy cascade.

22.1.2 Internal changes

The response of the Earth–atmosphere system to inputs of solar energy is determined by the operation of regulators. The most important of these are the transparency of the atmosphere to solar radiation, the albedo of the surface and the physical properties of the subsurface (Fig. 22.4). Climatic changes may be a result of modifications to one or all of these.

The operations of regulators, while directly affecting energy distribution through the Earth–atmosphere system, are themselves inextricably linked to the disposition of matter in the stores within the system. For example, an increase in water storage in the form of ice increases the albedo of the Earth's surface. The materials that have the greatest bearing upon regulators are solids and water, together with carbon dioxide.

Solid materials. Small particles in the atmosphere are derived from three major sources: volcanic eruptions, surface dust and the by-products of combustion (Ch. 4). The force of ejection and the strong thermal convection associated with a volcanic eruption send large quantities of dust and debris into the atmosphere (Fig. 22.5). Most of this falls back to the surface, but some rises to 25 km into the stratosphere. Although initially a localised input, the dust is distributed around the world by the upper winds. This forms a dust veil which directly affects the transparency of the atmosphere by reflecting incident shortwave solar radiation. Because of the scattering of radiation by the dust, the ratio of diffuse to direct solar radiation is increased. Following the Bali eruption in 1963, observations of solar radiation made in southern Australia revealed a distinct decrease in direct and increase in diffuse components indicating scattering by the dust veil (Fig. 22.6). The resulting reduction in solar heating of the surface results in a decrease in its temperature which tends to persist

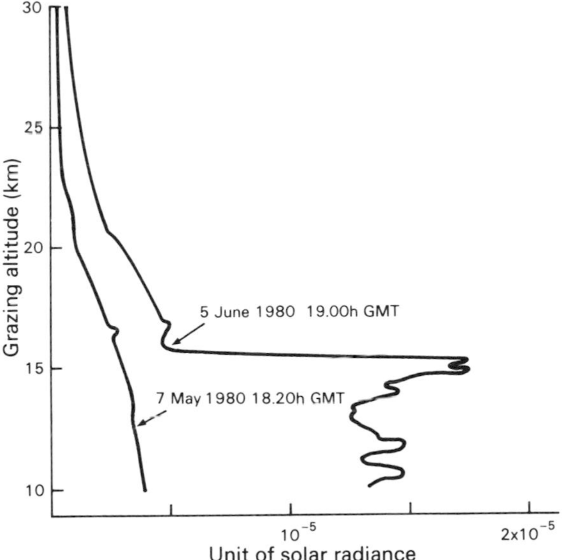

Figure 22.5 Earth limb radiance in the azimuth of the Sun and in units of solar radiance against altitude of the grazing line of sight. Note the large and sharply cut radiance increase between 15.6 and 15.2 km which is due to the stratospheric loading of volcanic materials over Europe following the eruption of Mt St Helens in North America (after Akerman *et al.* 1980).

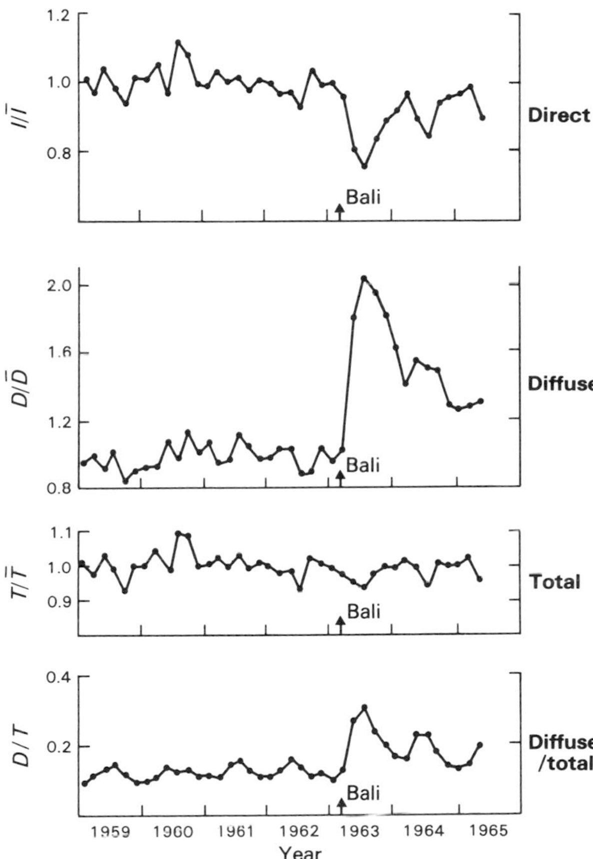

Figure 22.6 Direct (*I*), diffuse (*D*) and total (*T*) solar radiation, and the contribution of the diffuse (sky) radiation to the total at Aspendale, Melbourne (38°S, 145°E) after the Bali eruption in 1963; fractions of the 1959–62 averages (after Dyer & Hicks 1965).

until the dust veil has cleared. Many of the cool summers in western Europe during the past 200 years have followed major periods of volcanic activity elsewhere in the world. However, it is questionable whether volcanic activity would cause a sufficiently persistent reduction in temperatures to initiate a major growth of ice sheets. In order to produce cooling over a long period, the veil would have to be replenished continuously by fresh eruptions, as even the finest dust remains in the stratosphere for only a few years. On the basis of the evidence of the past 200 years, it seems improbable that major eruptions would have occurred with sufficiently high frequency to maintain a persistent dust veil.

Dust may be lifted from the land surface also by the action of wind and may be carried aloft by turbulent mixing of the atmosphere. Much of this is derived from the deserts but some may be a product of soil erosion. Poor land management may lead to rapid erosion such as that which occurred in the 1930s in the American 'dust-bowl'. Bryson and Baerreis (1967) have suggested that dust from dry land surfaces in north-west India directly enhances subsidence in the lower atmosphere, effectively limiting precipitation from the advancing monsoon. As most terrestrial dust is confined to the troposphere, it is likely to be washed out by precipitation. While it may produce localised and short-term modifications to atmospheric transparency, its long-term effects upon climate may be negligible.

Water. Water is stored within the Earth–atmosphere system in oceans, ice deposits and the atmosphere and on the land surfaces (Ch. 4). Changes may occur in the amount stored in any one of these and also in the rate at which water is transferred between them. The amount of water in the form of ice has fluctuated during the Earth's history, reaching several distinct maxima during Ice Ages when polar ice caps and valley glaciers expanded to cover large areas (see Sec. 22.4). These periods were certainly accompanied by a general lowering of surface and air temperatures and possibly may have experienced greater amounts of precipitation in the form of ice crystals. Climatic data for the latter part of the period 1650–1850, during which there were advances of valley glaciers, suggest that these were accompanied by reduced temperatures and increased precipitation.

The melting of large quantities of ice requires a supply of latent heat at the rate of 2.63 J kg^{-1}. During the last ice advances, some 10000 years ago, there was approximately 72×10^6 km^3 of ice over the Earth's surface, compared with the 33×10^6 km^3 of today. Melting of the 39×10^6 km^3 would have consumed approximately 10^{17} J which, in the absence of an external source of energy, would have to be found from elsewhere in the Earth–atmosphere system.

The increase in albedo, as ice covers the ground and ocean surfaces, reduces the amount of solar energy they absorb. The cold surface reduces the temperature of the air in contact with it which, under conditions of weak atmospheric circulation, encourages stability in the lower atmosphere and the formation of a shallow anticyclone. The divergent nature of airflow prevents the advection of heat over the ice surface and also spreads cold air beyond the ice limits. The advent of a more vigorous atmospheric circulation breaks down the weak anticyclone and initiates ice melt. Under these conditions, air temperatures around the ice-limits undergo very marked increases. During the first half of the 20th century, a decrease in the extent of north polar ice produced amelioration of winter mean temperatures in Spitzbergen of the order of 6 C° over less than 20 years.

Changes in atmospheric moisture may arise from a general increase or decrease in surface and air temperatures which enhance or restrict not only evaporation but also the moisture-holding capacity of the air. The atmosphere represents only a very small store of water (Ch. 4), in the form of water vapour, but changes in its contents have far-reaching effects upon surface climates. Increases in atmospheric moisture content are associated with greater amounts of cloud and possibly also of precipitation. As the major absorber of terrestrial radiation, atmospheric moisture also has a considerable bearing upon the **greenhouse effect**.

Atmospheric moisture content is only one of the factors affecting rates of precipitation. The intensity of dynamic cooling of the atmosphere, particularly by cyclonic mechanisms, and the presence of conditions conducive to the growth of cloud droplets are important considerations.

The evidence for long-term changes in precipitation is inconclusive. In Britain, for example, the annual rainfall totals at Southampton and Dumfries (Fig. 22.7a) do not indicate any long-

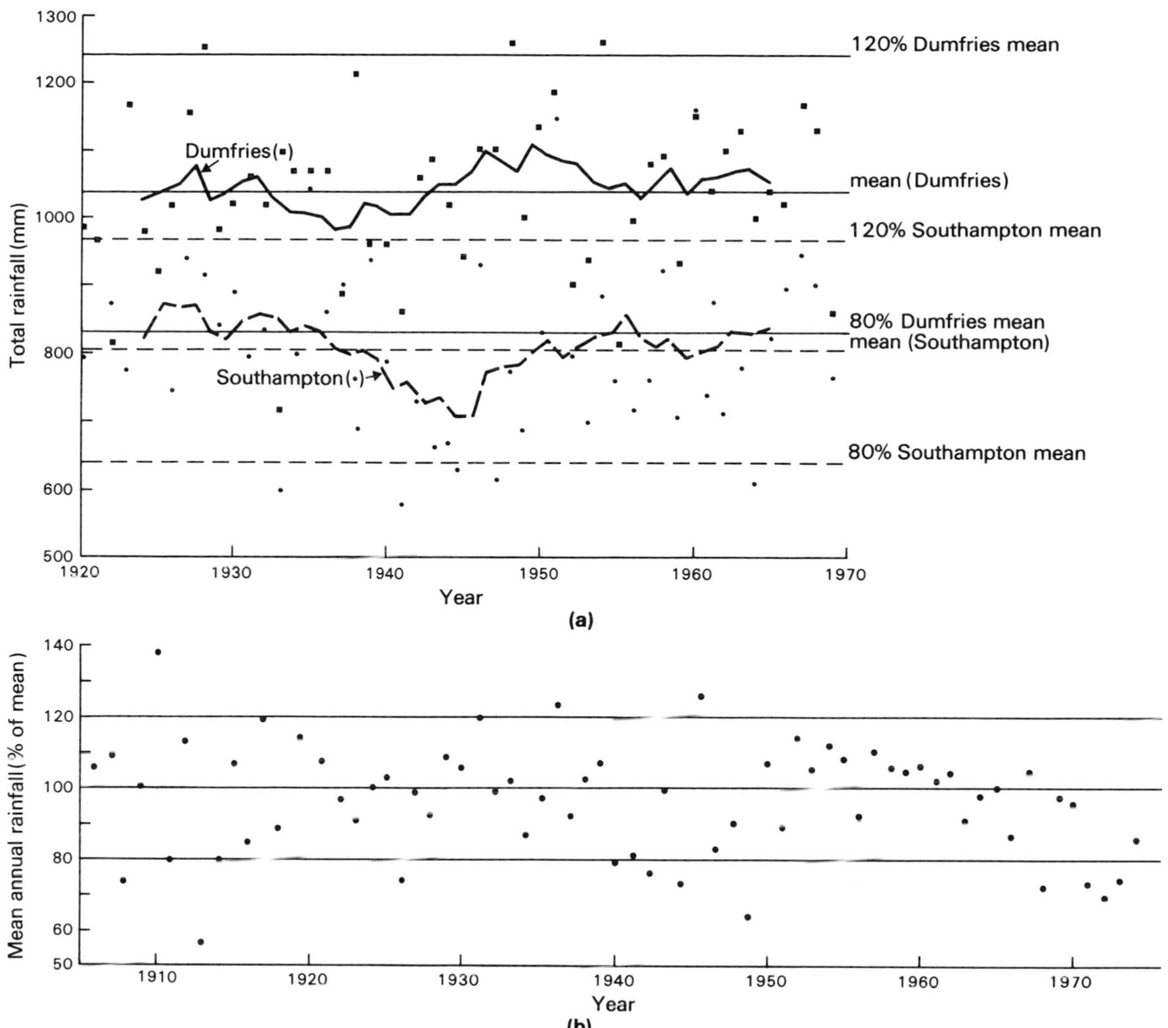

Figure 22.7 (a) Ten-year running means of annual rainfall at Dumfries and Southampton. (b) Mean annual rainfall for five stations in Western Africa expressed as a percentage of the 1931–60 mean (after Bunting *et al.* 1976).

term trends over the period 1920–69 although there appears to be a *decrease* between 1925 and 1945 at the former. Analysis of annual rainfall over the West African Sahel (Bunting *et al.* 1976), where recent droughts have caused considerable human distress, also shows no clear pattern of change even when subjected to rigorous mathematical criteria (Fig. 22.7b).

The hydrological system clearly undergoes changes through time which affect the disposition of energy through the Earth–atmosphere system. The form of these changes is, however, complex and is a result of interrelated internal modifications to the content of stores and the exchange of water between them. The operation of the system has been related to the solar energy input using mathematical models such as that developed by Wetherald and Manabe (1975) who have derived a clear and positive relationship between the solar constant and precipitation rates.

Carbon dioxide. Carbon dioxide (CO_2) represents a relatively minor store of carbon within the Earth–atmosphere system (Ch. 4), but nevertheless it has a significant effect upon the heat-energy balance of the lower atmosphere. The amount in the atmosphere at any point in time or space is largely determined by the dynamic balance between production by living organisms, removal, mainly by photosynthesis, in plants, and uptake into stores, particularly the oceans. This balance varies periodically over both diurnal and annual timescales.

Monthly mean concentrations measured at 3352 m above sea level at Mauna Loa Observatory in Hawaii reveal this annual cycle of change (Fig. 22.8). They also indicate that there has been a longer-term increase in atmospheric carbon dioxide of nearly 4% over the period 1958–74. This trend is a continuation of a general increase which has been taking place since the middle of the 19th century when concentrations were approximately 280 p.p.m. Therefore, in a little over a century there has been a 16% increase in atmospheric carbon dioxide. Most of this increase has been due to the combustion of coal and oil which has, to some extent, been offset by absorption into the ocean store.

Carbon dioxide absorbs infra-red radiation at wavelengths within the spectrum of terrestrial radiation. Therefore increases in its concentration in the lower atmosphere should result in increases in temperatures at this level. Coincidentally, air temperatures in the northern hemisphere increased by approximately 0–6 C° between the middle of the 19th century until around 1940. This can be identified readily from the temperature records for stations in Europe (Fig. 22.9).

The inference is that there is a cause-and-effect relationship between changes in carbon dioxide and in air temperature. The evidence, however,

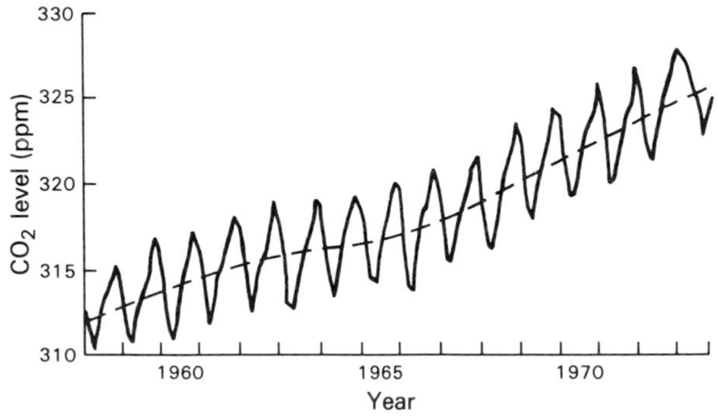

Figure 22.8 Carbon dioxide levels at Mauna Loa observatory, Hawaii.

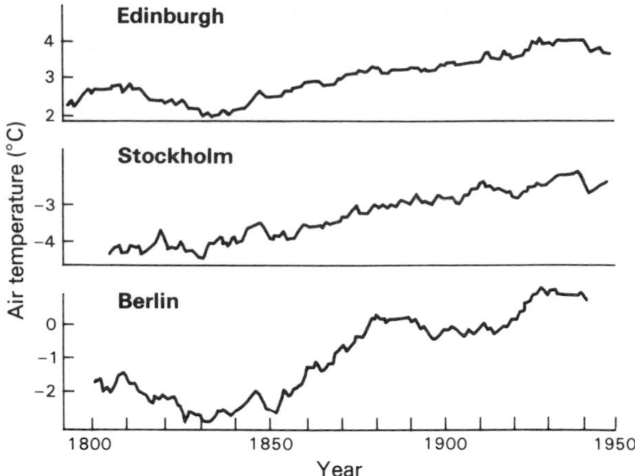

Figure 22.9 Thirty-year running means of air temperature at three western European stations (from Barry & Chorley 1968).

does not fully support this hypothesis, for three main reasons:

(a) The greatest increases in temperature have been recorded in high latitudes where carbon dioxide concentrations are lowest.
(b) Although carbon dioxide absorbs radiation in the infrared wavelengths, most absorption in the atmosphere in these wavelengths is effected by water vapour. Relatively large increases in carbon dioxide may produce negligible increases in absorption.
(c) Temperatures since 1950 have declined while carbon dioxide concentration has continued to increase.

These facts suggest that carbon dioxide concentrations may not be the prime cause of changes in air temperatures. On the other hand, it may be that compensating mechanisms operate in the Earth–atmosphere system to nullify the effects of any increases.

22.2 Feedback mechanisms

External and internal changes to the Earth–atmosphere system are linked inextricably and it is difficult to define singular cause and effect relationships. Mitchell (1976) has referred to a resonance between external and internal forces of change. Each observed effect is the end-product of a complex chain of causes and effects. Some of these chains are assembled in such a way as to produce a compensating mechanism for any initial changes imposed on the Earth–atmosphere system. This negative feedback provides for long-term systems stability. In others, an initial impetus may produce increasing degrees of change within the system as positive feedback occurs.

If we consider a change in the solar constant, or energy input into the system, it could be argued that the resulting increase in surface heating would increase rates of evaporation. The consequently larger amounts of moisture in the atmosphere would result in increasing cloud cover and hence a decrease in solar radiation reaching the surface (Fig. 22.10a). The Earth–atmosphere system thus compensates for change in a negative feedback loop which may be joined at any point.

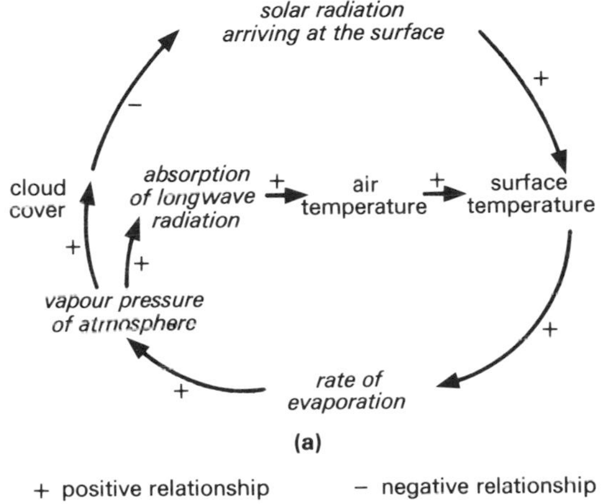

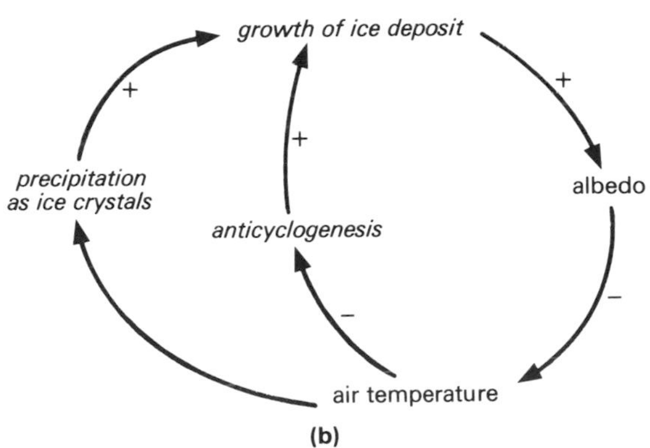

Figure 22.10 (a) Feedback in the Earth–atmosphere system; (b) positive feedback in the growth of an ice sheet.

If there were an increase in carbon dioxide, this would cause an increase in the absorption of terrestrial heat energy in the lower atmosphere which would inhibit the cooling of the Earth's surface. The resulting increase in surface temperature could encourage evaporation, increase the water vapour content of the atmosphere and thereby reduce the amount of radiation reaching the surface. Thus lower heat loss is balanced by lower heat input. Alternatively, we can see the effect of changes in solar radiation as initiating positive feedback. For example, the increases in surface temperature would greatly enhance the greenhouse effect. This would cause further increases in surface temperature through inhibiting heat loss (Fig. 22.10a).

The covering of a surface by ice increases albedo, thereby reducing solar heating. The air above is cooled and this generates surface divergence of airflow which restricts advectional inflow of heat. Therefore, the ice-sheet would extend its area, not by further inputs but by restricting ablation, thereby exhibiting positive feedback (Fig. 22.10b). An alternative view is that cooling of the air increases the likelihood of precipitation in the form of ice crystals, thereby adding to the ice deposit.

These feedback loops are clearly oversimplifications of the very complex interrelationships that exist between meteorological variables within the Earth–atmosphere system. They do, however, illustrate that the system is, to a large extent, self regulating in that it is capable of modifying its mode of operation to accommodate changes in inputs or its own regulators.

22.3 The form of change

Changes in climate are expressed normally in terms of trends. Over short periods, measured in decades, monotonic trends may be identified, such as the increase in air temperatures which took place during the early part of the 20th century. However, most analyses of long-term climatic changes have sought to identify oscillations in the data. Such oscillatory changes take place about a mean long-term value. Most climatic data may not exhibit a single periodicity similar to that of the sunspot cycle, but may be compounded of several interlocking periodic changes through time. For example, Milankovitch's calculations of solar radiation represent a superimposition of three periodicities. An analysis of ice cores from the Greenland ice sheets has identified two possible periodicities in global temperature of 80 and 180 years, over the past 800 years.

In the absence of recognisable periodicities there is the possibility that climatic change is irregular or possibly random as suggested by Curry (1962). In the established presence of regular changes in the orbital characteristics of the Earth and of the gaseous content of its atmosphere it seems unlikely that climatic changes could be random. However, it may prove difficult to assign causes to all observed variation in, for example, air temperature and precipitation.

22.4 Zonal and dynamic change in the denudation system

The normal operation of the denudation system itself will bring about change. Long-continued denudation processes will modify a land mass, reducing its elevation and decreasing the potential energy available for erosion. Thus a feedback loop operates between denudation and morphology, which continually and gradually modifies the conditions under which denudation operates. These long-term changes in potential energy input to the denudation system form the subject of the first part of this section.

In Chapters 10 to 15 we considered the operation of various aspects of the denudation system, with the underlying assumption that systems inputs remained constant through time. In fact, most inputs to Earth systems vary continuously, on different timescales. For instance, solar radiation input varies hourly, diurnally and seasonally. Precipitation, of course, varies day by day and week to week. In this section, however, we shall concern ourselves with changes which take place at a much larger scale, over hundreds or thousands of years. As such, they affect the state of the systems with which they are concerned and cause major alterations in systems operation. These can be regarded as externally imposed, since the means of change is external to the denudation system itself.

Neglecting for the moment the influence of Man in changing systems operation (since this is discussed in Ch. 24) we are concerned here with changes in atmospheric inputs and changes in potential energy. Inputs from the atmosphere – that is, solar energy and precipitation – may vary in magnitude and distribution with climatic change, as we have just seen. This is one kind of change in input. During the Quaternary, climatic zones have undergone considerable latitudinal changes. This means that a specific point on the Earth's surface may have experienced considerable changes in climatic inputs. These we shall term **zonal changes**.

A second type of change in input is the **dynamic change** associated with potential energy and related to the base level of erosion. Any change in the position of sea level relative to the land will result in either a positive or negative change in the potential energy of erosion, leading to increased or decreased denudation respectively.

The change in system inputs leads to the production of new morphological forms. Under the changed regime a new suite of landforms may begin to develop. Thus, in a recently changed denudation system newly formed features may coexist with landforms created during the previous regime. These inherited features are known as **relict** landforms and they will persist until destroyed by denudation under the new conditions. The immense changes in denudation systems which have occurred within the Quaternary, many of them during the past 500000 years, mean that there are many inherited landscape forms in existence at the present time.

22.4.1 Long-term denudation effects

As we have seen in Chapter 10, the effect of the operation of the denudation system is to strip rock debris from the surface of the continents and deliver it to the oceans. Monitoring of the denudation processes involved has provided information on the rate at which this takes place, and is shown to be considerably slower than the rate of uplift by orogenic processes. Accordingly, it would appear that land masses are affected by periods of relatively rapid uplift, creating relief, followed by long-continued periods of denudation, which gradually reduce it.

A major effect of the denudation process–response system is that denudation itself affects the relief upon which it is operating. This further modifies the nature and rate of operation of denudation processes. Thus there exists a positive

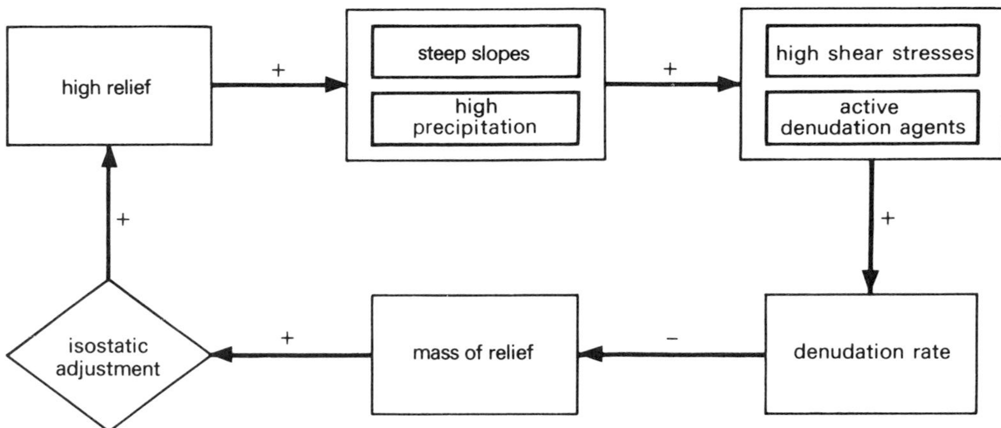

Figure 22.11 Feedback between denudation processes and relief.

feedback loop between denudation processes and relief (Fig. 22.11). As shown in Figure 22.12, denudation rate is closely related to relief (Ahnert 1970). Relief in turn is related to gradient, which controls many of the denudation processes, especially the physically based ones. Assuming constant basin area, the lower the absolute relief in the basin, then the lower its mean gradient will be both in channels and on slopes. Thus potential energy also is reduced, which in turn leads to a decrease in the rate of operation of denudation processes. This positive feedback loop, then, leads in the direction of progressive change, towards lower relief.

The decline in denudation through time is suggested by the empirical evidence in Figure 22.12. Using data derived from a sample of mid-latitude humid drainage basins, the rate of denudation is shown to be directly and linearly related to relief. This can be expressed simply as:

$$d = 0.0001535\,h$$

where h = mean relief in metres and d = mean denudation rate in mm a^{-1}.

Such a relationship assumes that erosion is distributed evenly throughout the catchment and is based on the premise that the higher the relief, the greater the tendency for steep gradients and corresponding erosional stresses. Since the rate of denudation is functionally related to relief, and the latter is reduced through time, it is possible to calculate on the basis of the above relationship that after 11 Ma a land mass would be reduced to 10%

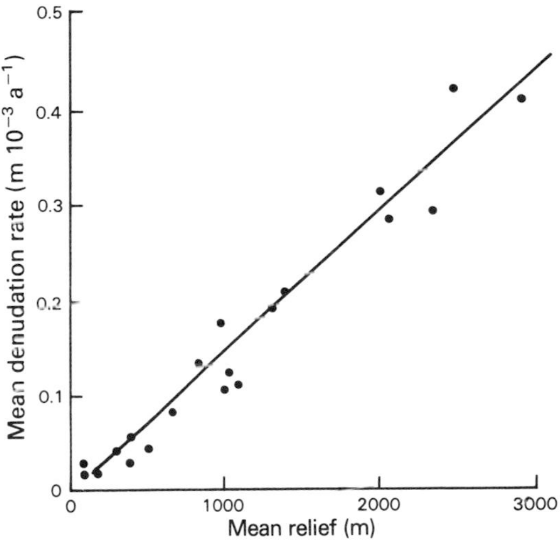

Figure 22.12 Denudation rate as a function of relief (after Ahnert 1970).

of its original relief, and after 22 Ma it would be reduced to 1%, as the rate of denudation declines with both time and relief.

Denudation rates, however, refer only to material stripped from the continental surfaces. In considering destruction of relief we must also take into account isostatic rebound, as the gradually thinning continent becomes more buoyant on the underlying mantle. Isostatic rebound can be calculated thus:

$$h = Br/A$$

where h = isostatic compensation, B = specific

gravity of surface rocks removed, A = specific gravity of material replacing at depth and r = thickness of surface layer removed. Assuming A = 3.4 and B = 2.6, then $h = 0.76\,r$.

Assuming then that isostatic rebound in response to erosional losses is a continuous and widespread process, then three-quarters of the relief removed within a given period of time is thus replaced. Accordingly, the real length of time taken for the destruction of the relief of a land mass will be longer, and Ahnert calculates that, with isostatic compensation, 18.5 Ma will be required for reduction to 10% of initial relief and 37 Ma to 1% (Fig. 22.13).

All this presupposes that external conditions remain essentially constant throughout the operation of the denudation process over periods of millions of years. Yet, as we shall see now, considerable variations in systems input can take place over much shorter timescales.

22.4.2 *Dynamic change*

Changes in the relative levels of sea and land masses can be brought about by changes in either factor. Land level changes (isostatic) are outlined in Chapter 5. The chief causes are orogenic uplift of fold mountain belts, uplift or depression of fault blocks as a result of rifting, epeirogenic deformation of broad continental areas, and downwarping of continental margins adjacent to basins of sedimentation. Locally, the development of an ice sheet can increase the load on a continent and depress its elevation, which returns to its original level subsequent to the glacial episode. While in the long term this can be viewed as a temporary effect, there are several areas which are at the present time recovering from glacial loading and being uplifted at a rate of several millimetres per year. These processes operate independently of sea level and they affect land masses unequally across their area (see Fig. 5.10 for the contemporary deformation of the USA). Accordingly, the changes in potential energy are distributed unevenly, and within one land mass some areas may be increasing and others decreasing in potential energy.

Sea level changes (eustatic) operate on a worldwide scale, affecting all land masses equally and simultaneously. There are three main causes:

(a) There is the probability, though it is difficult to demonstrate with certainty, of change in the form and volume of the ocean basins. The mobility of the oceanic crust renders it inherently unlikely that the ocean basins can retain a constant form. Assuming a constant mass of water in the ocean basins, any change in basin volume would result in a rise or fall of water level relative to the continents.

(b) Sediment eroded from the continents accumulates in the ocean basins. Over a long period of time this tendency to fill the ocean basin will displace the water upwards relative to the continents, leading to marine transgression.

(c) The transfer of large masses of water during glaciation, from storage in ocean basins, to storage in ice-sheets, has resulted in great changes in the volume of water in the ocean basins.

The relative effects of (a) and (b) above are difficult to determine and isolate, although estimates of their magnitude can be made from tide gauge records in areas of continental crust of contrasting stability. Stable coastlines will record only eustatic changes, while in unstable areas local

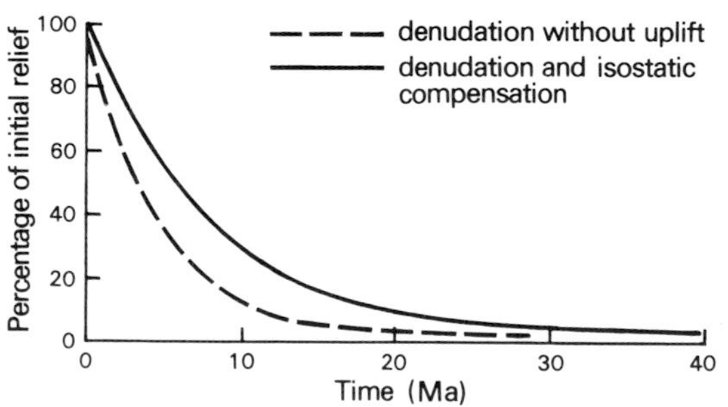

Figure 22.13 Time needed for relief reduction (after Ahnert 1970).

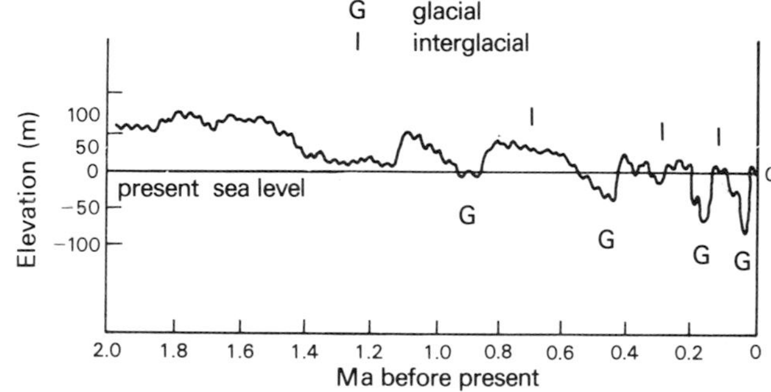

Figure 22.14 General model of sea-level change during the Quaternary (after Fairbridge 1971).

effects will be superimposed on the worldwide process.

Glacially controlled changes in sea level may attain a range of at least 100 m between phases of interglacial high sea level, when the water is in the oceans, and phases of glacial low sea level, when much more of the water is in glacial storage on the land masses.

A general model of sea level change during the Quaternary is shown in Figure 22.14 (Fairbridge 1961). It shows the level falling consistently from an early Quaternary high of +230 m towards the present level, and superimposed upon this general trend the glacio-eustatic fluctuations of the latter part of the past 500 000 years. Clearly this is a highly generalised picture, and doubtless minor oscillations occurred on a smaller timescale. This can be illustrated with reference to the last major (postglacial) rise in sea level, in Britain termed the **Flandrian transgression** (Fig. 22.15). More detailed evidence is available for this event and shows a sea level rising with marked oscillations towards the present level at 6000 BP (before present).

On any one land mass the overall change of elevation in relation to sea level will be the result of the combined effect of eustatic and isostatic changes, superimposed upon each other. The relative rates of change generated by these different mechanisms varies considerably (Table 22.1). The most rapid changes appear to be associated with glacial eustatic sea levels. In a crustally stable region, such as southern Britain, the last major event of this kind, with the most recent effects, was the Flandrian trangression, which terminated as recently as 6000 years ago, although the current rate of eustatic rise is by no means negligible.

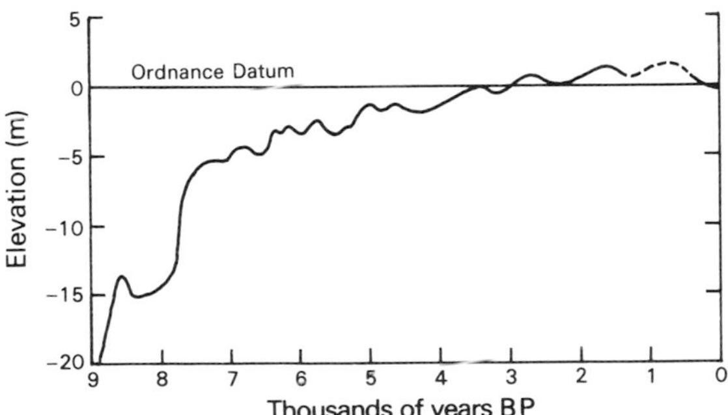

Figure 22.15 Flandrian sea-level change in north-west Britain (after Tooley 1974).

Table 22.1 Representative rates of relative land and sea level changes (after Carson & Kirkby 1972, from various sources).

	mm a^{-1}
glacial eustatic	up to 25
current eustatic rise	1.2
orogenic uplift	
California	3.9–12.6
Japan	0.8–7.5
Persian Gulf	3.0–9.9
epeirogenic uplift	0.1–3.6
isostatic	
Fennoscandia	10.8
Southern Ontario	4.8

The effect of a change in base level is felt most immediately in the coastal zone. Thus the coastline will tend to advance or retreat, and when sea level is stabilised again coastal erosion and deposition processes will make their mark at the new stable level.

The effects of increased potential energy caused by uplift are transmitted through the denudation system. Initially, river channel gradients down stream are locally steepened, leading to increased velocity and erosion and the incision of the river channel. Figure 22.16 shows how this can occur at

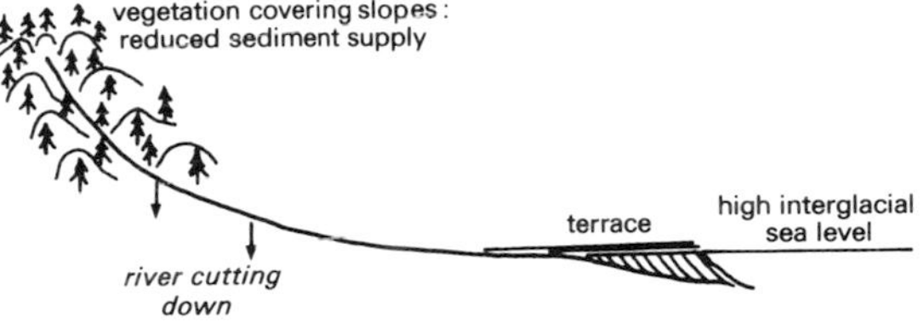

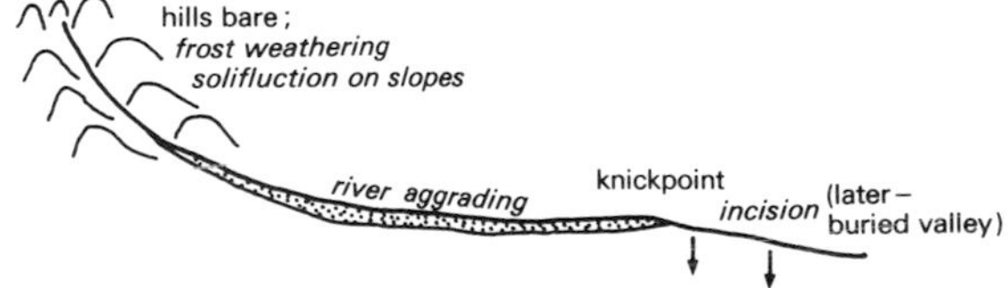

Figure 22.16 The incision of river channels in both inland and coastal locations and under different climatic controls (after Clayton 1977).

both coastal and inland locations. Incision of the channels has the effect of both lengthening and steepening valleyside slopes, thus imposing greater erosional stresses on them. Accordingly, there will be greater volume of eroded material from the slopes delivered to the river channel and the rate of operation of the denudation system, i.e. the rate of erosion, increases. Down stream, a large river will tend to become incised into its former floodplain, which is left upstanding as a dissected river terrace, while the river forms a new floodplain at a lower level. Repeated incision in this way may result in a flight of terraces, each one bearing the deposits of a former floodplain. Figure 22.17 shows the terrace sequence of the Thames formed in this way.

A rise in sea level relative to land will tend to affect denudation systems more locally. First, the seaward positions of river valleys may become drowned, particularly if such valleys are deeply incised. Thus all the major rivers of southern Britain have deep buried channels, formed in response to low glacial sea levels. Such buried valleys become sediment traps, and as the river's velocity is reduced, as it flows into the lower part of its valley being drowned by the rising sea level, sediment accumulates in the submerging valley.

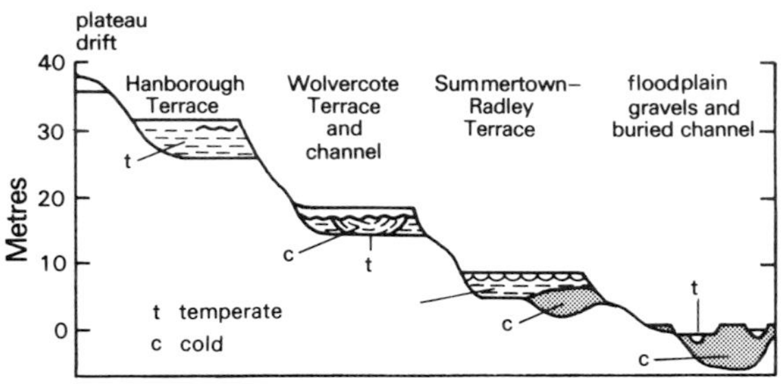

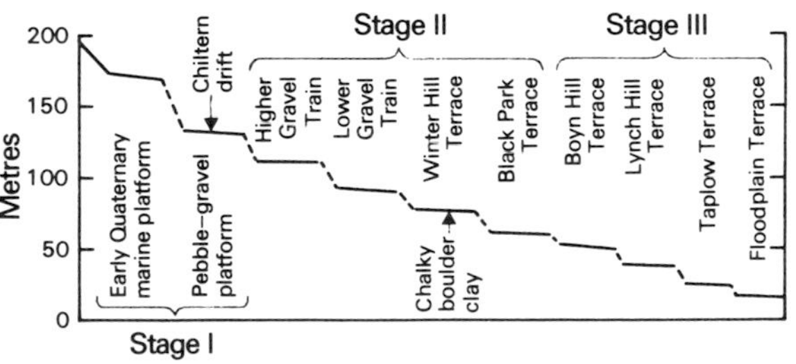

Figure 22.17 Terrace sequences of the River Thames; above in the upper Thames (after Sandford 1954) and below in the middle and lower Thames (after Wooldridge 1960).

However, such effects are not transmitted up stream in the same way as incision and the upstream operation of the denudation system may continue as before.

22.4.3 Zonal change

Although the overall world pattern is still far from being completely understood, it is clear that considerable zonal changes have taken place since the middle of the Quaternary era. On the geomorphological timescale these changes are sufficiently recent to have formed features in the landscape that are clearly visible today. Currently temperate environments may have experienced phases of glacial or periglacial conditions during this time. In subtropical latitudes there were fluctuations between arid and moist (pluvial) phases, although they were not necessarily synchronous throughout the Tropics, nor in step with the changes in temperate latitudes. The Quaternary fluctuations in the world circulation pattern produced effects that would vary from place to place, and which remain to be worked out. It would appear, however, that the zonal changes were most pronounced in mid-latitude and marginal environments, while the core of the equatorial zone remained largely unaffected by climatic change. This general equatorward shift of climatic belts means that a particular point on the Earth's surface would experience a change in climatic inputs. Depending on the existing regime and the nature of the changed conditions, a wide variety of responses took place in denudation systems.

A trend toward greater humidity in a cold environment may lead to the development of glaciation, if winter snow accumulation exceeds summer ablation. In humid and arid environments greater river runoff will tend to occur, with increased frequency of floods. Increased humidity would tend to allow the development of more luxuriant vegetation and accelerate soil and weathering processes. In arid and semi-arid environments the increased precipitation and runoff would lead to an increase in the volume of inland lakes. Trends towards aridity result in a decrease in vegetation cover and an increased susceptibility of surface material to transport. Under these more arid conditions the transporting agent is likely to be wind, and this results in the loess accumulations of the tundra zone and sand-dune activity in the low-latitude arid zones.

Under the changed environmental conditions a variety of new features may develop. Altered denudation processes may produce new landforms, and new agents and environments of deposition may produce new kinds of sedimentary deposit. Changed weathering processes may modify weathering mantles and soils. Features may remain in the landscape which are the product of previous conditions. These are **relict landforms** or, in the case of fossil soils, **palaeosols**. Together they constitute evidence from which the history of climatic change can be reconstructed. They will tend to persist in the landscape until they become obliterated under the new conditions. Fortunately, such major changes in environmental inputs occurred throughout the Quaternary, and so little time has elapsed since that abundant relict forms currently exist in many different environments.

It is marginal areas, at the boundary between climatic zones or systems, that are most sensitive to change, and in such environments it is possible to monitor zonal change even on a short timescale. Glacier fluctuations and changes in desert margins represent two environments where change is sufficiently rapid to do this.

Within the timescales available to modern methods of scientific measurement of environmental variables, changes can be demonstrated in the arid zone boundary in West Africa south of the Sahara. Rainfall records over the period indicate a progressive decrease in precipitation. In such a marginal environment a desiccation of this magnitude brings about a change from semi-arid condition to desert. It is estimated that the desert boundary has shifted southwards and extended the desert in this region by up to 100 km during the period 1968–76. Within this marginal zone, therefore, there has been a change from semi-arid to arid environmental conditions.

Another type of system in which zonal change can be identified readily over a contemporary and historic timescale is the glacial system. This is simplified by the fact that the system boundary is identified easily, i.e. the glacier margin. Figure 22.18 shows the changes in the margin of the valley glacier Tunsbergdalsbreen, in southern Norway (Mottershead & Collin 1976). These have been identified on the basis of historical and botanical evidence, notably the use of lichens as indicators of ground surface age (Mottershead & White 1972, 1973). The evidence indicates that the glacier has receded some 2.4 km since the mid-18th century, at an ever increasing rate.

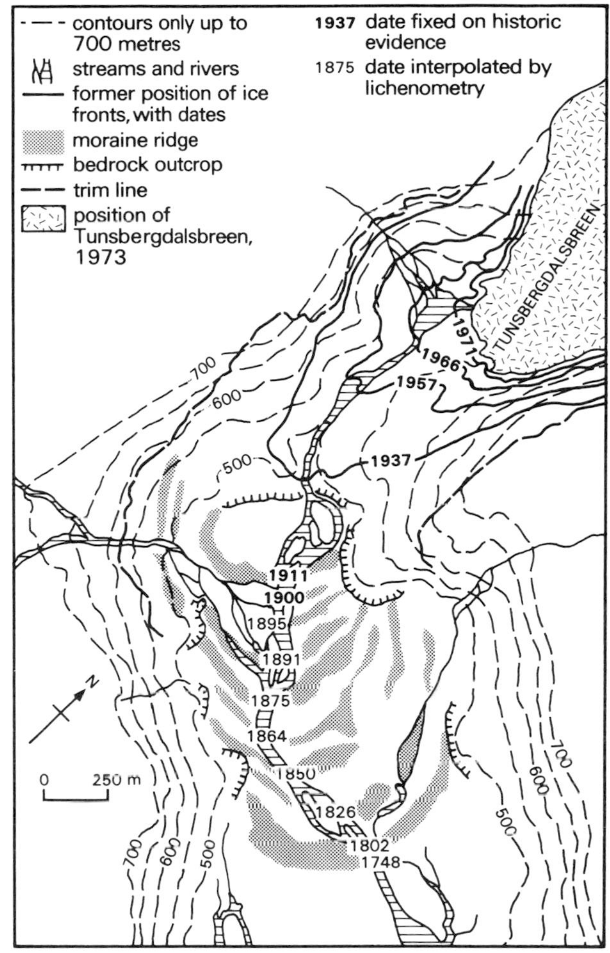

Figure 22.18 Tunsbergdalen, southern Norway, showing recessional moraines and lichenometric dates (after Mottershead & White 1973).

The foregoing are just two examples of environmental change currently taking place in the vicinity of system boundaries. During Quaternary time, and earlier, changes of much greater magnitude have taken place, causing zonal shifts of hundreds of kilometres in some cases. The operation of denudation systems no longer active in these regions formed suites of landforms which are out of phase with contemporary conditions – fossil or relict forms. We shall examine two cases which show the contrasting effect of environmental change. In Britain, where glacial and periglacial systems were formerly active, contemporary temperate conditions of denudation have permitted the survival of relict glacial and periglacial forms. In arid regions of Africa, landforms created under pluvial conditions persist in the landscape, preserved by the limited effects of arid zone denudation.

Glacial features in the temperate zone. Figure 22.19 shows a reconstruction of conditions in the British Isles during the late Devensian glacial maximum – *ca* 18000 BP (Boulton *et al.* 1977). Glacial conditions are seen to cover much of Britain north of a line from the Severn to the Trent. Periglacial conditions existed to the south of the ice margin. In these two zones relict glacial and periglacial landscapes can be detected readily today, in evidence of former conditions.

The late Devensian ice-sheet measured 1000 km from north to south, and 600 km west to east, where it became confluent with the Scandinavian ice cap. It accumulated to a maximum depth of over 1800 m in the Scottish Highlands with subsidiary highs extending across north-west Ireland and southwards from the southern upland towards Wales. This reconstruction of the ice sheet can be matched with patterns of glacial erosion (Fig. 22.20), using a scale devised by Clayton (1974) and summarised in Table 22.2.

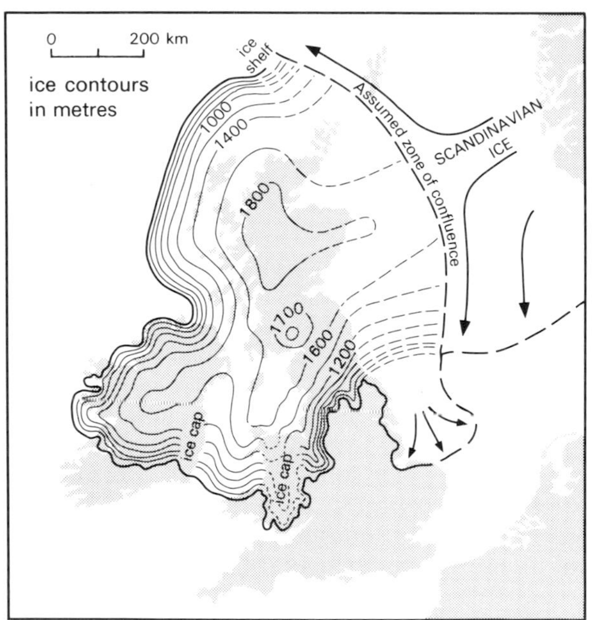

Figure 22.19 The Devensian ice sheet in the British Isles (after Boulton *et al.* 1977).

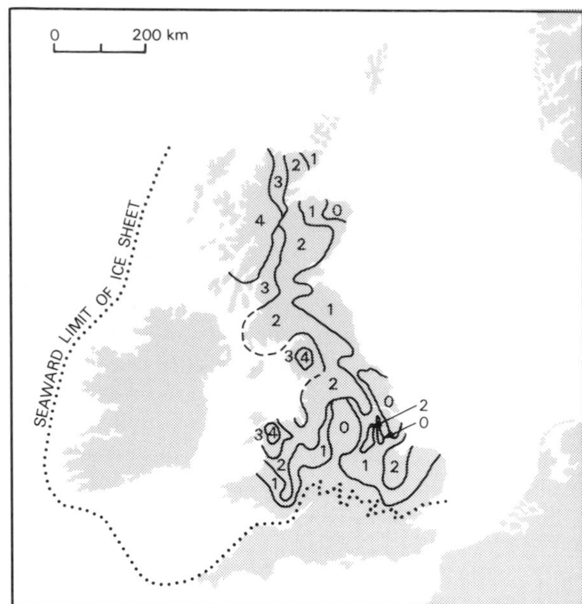

Figure 22.20 Zones of glacial erosion in the British Isles, using the scale devised by Clayton (1974) and summarised in Table 22.2.

Table 22.2 Zones of intensity of glacial erosion (after Clayton 1974).

0	no erosion
1	ice erosion confined to detailed or subordinate modification
2	extensive excavation along main lines of flow, glacial troughs common
3	comprehensive modification of preglacial forms, scoured landscapes common in lowlands, interconnecting systems, troughs extensive in uplands
4	ice-moulded streamlined forms dominant in the landscape

There is a general gradient from landscapes most intensely affected by glacial erosion close to the centre of the ice sheet, with the least affected landscapes towards the margins where the effective pressure of ice on the bed is least. It should be remembered, however, that this ice-sheet model applies only to the maximum glacial extent. Glacial conditions will have lasted longer in the core regions of the highlands, both preceding and postdating the glacial maximum. Furthermore, during the early and late stages of glaciation the ice mass was of a different form – i.e. as cirques and valley glaciers often in discrete bodies – in these

Figure 22.21 Landscape of glacial ice sheet erosion. The lowland shows evidence of widespread glacial scouring with isolated knolls of rock swept clean of regolith, and enclosed rock basins containing lakes. Elongated remnants of high land were streamlined by the ice which flowed from right to left across this landscape. (Sutherland, Scotland.)

regions. These may explain the intensely glaciated landscapes in Snowdonia and the Lake District.

In regions of intense glacial erosion, spectacular assemblages of glacial erosional landforms remain. Since such regions in Britain are invariably highlands developed on resistant bedrock, the glacial landforms have survived the 10000 years since deglaciation in almost pristine condition (Fig. 22.21).

Periglacial features in the temperate zone. Beyond the limits of glaciation in southern England (the most recent (Devensian) and slightly more extensive earlier glaciations) periglacial conditions existed, probably at several different periods. Evidence of permafrost is present in the form of fossil tundra polygons and ice wedges. More widespread, however, and with a more significant morphological impact on the landscape, are the effects of accelerated mass movement. These are particularly prevalent on the chalklands and in south-west England. This can be illustrated with reference to the landscape of the South Devon coast to the west of Start Point (Mottershead 1971).

Here periglacial conditions brought about a considerable metamorphosis in the landscape. Mass movement caused the transfer of regolith down bedrock slopes of 20–30° to form sediment accumulations at the base. These periglacial sediments are typically unsorted, with angular frost-shattered clasts, and are very stony. Consisting of locally derived debris, the deposits are in part bedded, suggesting that the material was transported in sheets or layers, as in solifluction terraces.

The accumulation of slope-foot sediments form gently sloping terraces, concave in profile and thinning towards the sea. Where sediments have accumulated at the base of two opposed slopes confined in a valley, a wedge-shaped fill of material has accumulated on the valley floor, as in some of the small valleys running down towards the sea. The removal of regolith from the upper parts of the slopes has resulted in the exposure of salients of unweathered bedrock to form tors. Their craggy outlines, defined by the fracture patterns in the schists of which they are composed, dominate the crestlines. Periglacial frost shattering of exposed rock outcrops resulted in their partial modification and the production of angular rubble which strews the slopes beneath to form blockfields. This assemblage of tor, blockfield and solifluction deposit represents a periglacial landscape in microcosm (Fig. 22.22). The degree of modification of the landscape by periglacial conditions can be estimated from the depth of accumulated sediments, in places up to 25 m. Thus the elevation of the pre-periglacial valley floors would have been lower by such an amount, while the crests may well have carried a mantle of regolith, increasing their elevation slightly. The degree of modification of relative relief under periglacial conditions, then, may locally be in excess of 25 m.

That the periglacial forms are not in equilibrium with contemporary conditions is shown by the fact that they are currently being destroyed by erosion. Marine erosion is trimming back cliffs cut in the coastal solifluction deposits, while contemporary streams are incising themselves into the valley fills to form valley-side terraces.

The legacy of periglacial conditions in this small area is therefore considerable, and indeed well preserved. However, this example is by no means untypical. Similar landscapes are preserved in many locations in south-west England, while the chalklands of southern England have their own distinctive periglacial features (French 1973).

Pluvial features in the arid zone. Geomorphological evidence of former conditions both more and less arid than the present is to be found in the Sahel region to the south of the Sahara (Fig. 22.23).

Dune systems indicate the extent of truly arid conditions. Under such conditions a perennial cover of closed vegetation cannot exist and loose surface sediment is distributed readily by wind to create dune forms. The present limit of active dunes is shown in Figure 22.23, while extending over 500 km to the south is a zone of old vegetation-covered dunes. In this zone clearly arid conditions existed, permitting the extension of the desert. A subsequent change to more humid conditions permitted the development of vegetation which covered the old dunes, stabilising them and creating a fossil dune landscape.

Evidence of former pluvial conditions is pre-

Figure 22.22 Relict periglacial landscape. The vegetated cliffs represent the accumulation of periglacial solifluction deposits derived from the slopes above, which are crowned with salients of bedrock (tors). A dissected raised shore platform is indicative of a former high sea level, predating the periglacial conditions. (Devon, England.)

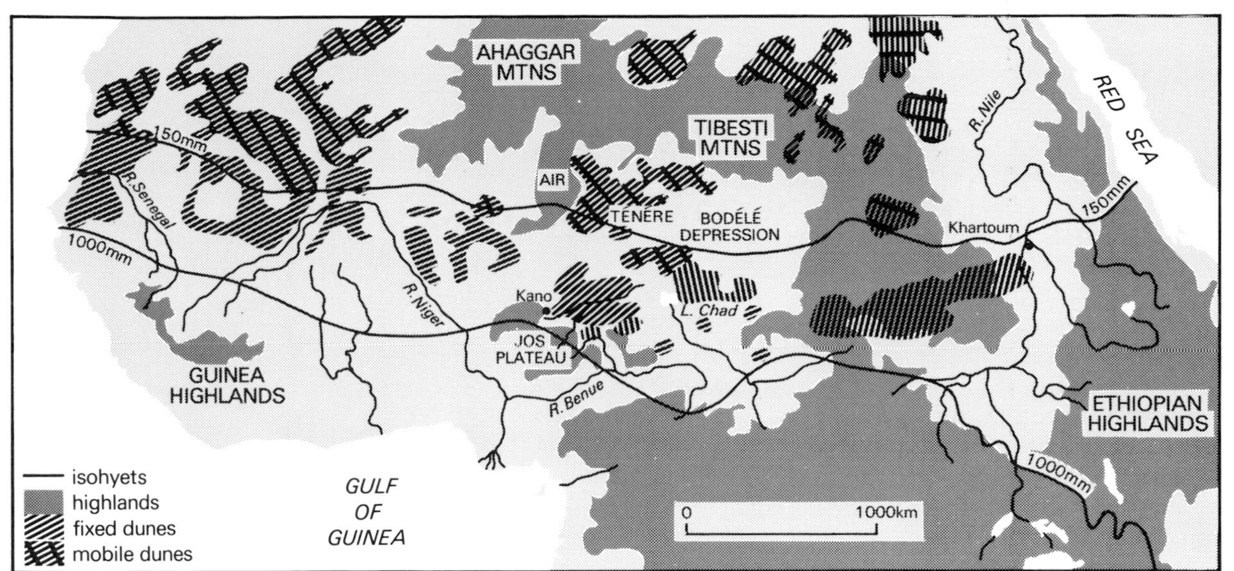

Figure 22.23 The distribution of mobile and fixed dunes in relation to present rainfall in the Sahel and southern Sahara (adapted from Grove & Warren 1968, *Geog. J.* **134**, 194–208, by permission of A. T. Grove).

served around Lake Chad, a basin of inland drainage (Fig. 22.24). This is a system in which inputs are derived mainly from surface runoff from adjacent highland regions augmented by direct precipitation, and are balanced by outputs in the form of direct evaporation from the water surface. The lake, therefore, acts as a store, sensitive to climatic changes which affect both precipitation and evaporation. The morphology of the lake basin is such that small changes in the depth of water present results in large variations in surface area. Its present mean depth ranges between 3 m and 7 m, and its area varies from 10 000 km² to 25 000 km². The shallowness of the basin means that its area is very sensitive to changes in lake volume.

Past pluvial conditions are indicated by elevated strandlines, sand bars and deltaic features. A strandline at 50 m above the present lake would have been associated with a former lake with an area of 400 000 km². Such a lake would have suffered evaporative losses of about 16 times the present rate and would be balanced by surface runoff of the rivers into it. Clearly the maintenance of the lake store at this high level requires conditions which are considerably more humid than those prevailing at the present, and Grove (1967) suggests that such conditions existed as recently as 10 000 or even 5000 years ago.

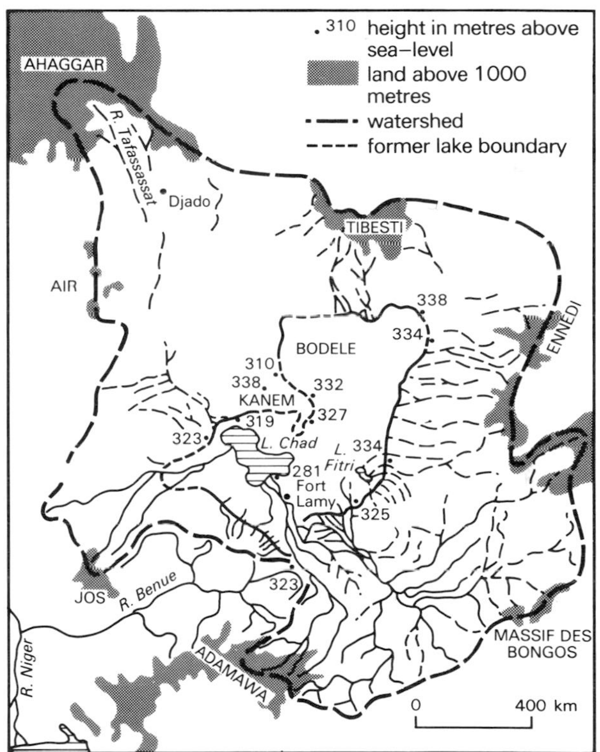

Figure 22.24 The present extent of Lake Chad compared with the old shore lines of the former 'Mega-Chad' at about 320 metres (after Grove 1967).

It should be stressed that these isolated case studies at different scales and from disparate regions are illustrative of a widespread phenomenon. Most of the world's landscapes bear at least some inherited features, indicating different zonal conditions in the past. The range of possible inherited forms is quite considerable, from fossil soils through sediments, which may be of sufficient volume to have morphological expression, to the spectacular landforms of glacial erosion.

Change in living systems 23

23.1 Inherent change in the ecosystem

Biological (living) systems are characterised by a capacity for progressive directional change – the cumulative manifestation of positive feedback. The functional activities of cells are concerned not only with the maintenance of a steady state but also with the investment of surplus matter and energy in cell division and growth. As we saw in Chapter 6, the accumulation of mass, the construction of complex ordered structures, such as organic macromolecules, and their intricate and precise arrangement to form the living cell – are all the expression of the inherent capacity of living systems to decrease their *internal entropy* and to increase in complexity and order as the result of controlled growth. In the following sections we shall examine these processes at the level of the population, community and ecosystem, but first we shall consider the individual organism. Many of the features of the growth model developed for the organism can still be recognised when applied to higher levels.

23.1.1 Growth and development of the organism
In multicellular organisms organised cell growth and division take place rapidly at first and lead to the appearance of specialised regions of differentiated cells. For example, in flowering plants terminal or axillary buds and root tips are growing masses of cells, sources for the production of initially unspecialised cells from which multiple structures develop successively. As long as these shoot and root **meristems** continue this function, new stem, new leaves and new roots will be added to the growing plant. Such growth, however, is controlled. In this example, a combination of external regulating factors, such as temperature and photoperiod, and internal control mechanisms which include the effect of the **auxins** (plant growth hormones) and particularly the **phytochrome P_r and P_{fr} system**, will lead at some time to the production of a flower and the cessation of growth in that system (Box 23.1).

In animals the developing embryo gradually produces specialised regions of differentiated cells which assume different functions and it progres-

Box 23.1

PHYTOCHROME P_r and P_{fr}

Phytochrome is a highly reactive protein molecule capable of changing its shape to exist in two distinct forms (isomers) each with a characteristic absorption spectrum. A blue form P_r responds to red light (wavelength 650–660 nm) and changes to a blue–green form P_{fr}. The reverse change from P_{fr} to P_r occurs on illumination with far–red light (wavelength 725–730 nm) or spontaneously but slowly in the dark. Phytochrome is now known to be the fundamental mechanism that switches on and off many biological responses to light in higher plants. These include many rhythmic responses associated with the alternation of light and dark in the 24-hour cycle and with variations in daylength over the annual period. Phytochrome plays a part in the phasing of many diurnal rhythms such as stomatal movements, photosynthesis, respiration, ion uptake and cell division. Photoperiodic responses involving phytochrome systems are at least partially responsible for seasonal changes such as bud dormancy, internode elongation and stem growth, and of course the initiation of flowering (Kendrick & Frankland 1976).

sively develops towards an identifiable individual organism. Growth continues until sexual maturity is reached (Ch. 18), but in many animals (for example most vertebrates) at this point there is no further increase in size. In the adult animal, the organism as a living system is simply maintaining a steady state. However, as it passes into a post-mature phase this steady state takes on a downward trend as production per unit biomass falls and ageing becomes apparent (Fig. 23.1). The timescale of individual growth masks the activity going on within the organism at the molecular and cellular levels. Tissue growth occurs on a longer timescale than cell growth so that the tissues of an organism outlive individual cells. (Note that there are important exceptions such as the cells of the central nervous system and the human retina, which are not replaced through the life of the individual.) For example, in Man the 4.5×10^6 red blood cells present in every cubic millimetre of blood have a life of only 120 days or so but are being replaced continuously by the 8×10^6 to 10×10^6 that are produced in the bone marrow every hour (see Ch. 6), while as many are removed by the spleen and liver. As in the case of plants, growth in animals is under the control of hormone systems that regulate the direction and rate of growth and co-ordinate growth in different tissues. In mammals the anterior lobe of the pituitary gland secretes both general growth hormones or acts indirectly by stimulating other endocrine systems.

From the very start (even before birth in mammals) the potential for ageing is present and is associated always with growth and development in the organism. If growth is viewed as an accumulation of excess cell multiplication over cell death, then in maturity this ratio must level off to a steady state relationship maintained by a continual turnover at the cellular level. As ageing becomes apparent this turnover slows down: repair and replacement slow down; skin wounds, for example, heal less readily, tissues and organs lose weight, cell death rate increases and blood volume decreases. This inherent ageing process is evident in the decline (by 50%) in the physical capacity of Man during his life. The human heart beats 140 times per minute at birth but the rate declines to 70 at the age of 25 years. At 20 years the blood takes up an average 4 litres of O_2 min^{-1} but by 75 years this is reduced to 1.5 litres of O_2 min^{-1}. The brain decreases in weight by 10% between the ages of 30 years and 90 years. The inescapable conclusion of the cumulative effect of ageing is ultimately a breakdown of bodily function that leads either directly or indirectly to the death of the individual.

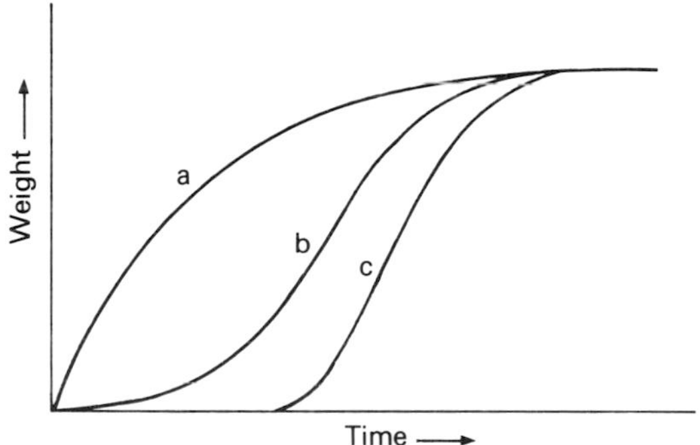

Figure 23.1 Three different theoretical curves, based on different mathematical assumptions, that have been used to describe biological growth. Curve b is the logistic growth curve that will be encountered later in this chapter, a is the monomolecular curve assuming constant rates of addition and subtraction, and c is the more complex Gompertz curve.

This view of the life and death of an individual organism can be simplified into three stages, each characterised by different patterns of control exerted by both external and internal feedback mechanisms. First, the growth stage where the balance of control is dominated by positive feedback reinforcing the upward trend of growth. Secondly, a stage where negative feedback or conservative mechanisms gain the upper hand, slowing the rate of growth and establishing a controlled steady state – the mature organism. Ultimately, conservative control breaks down and once more a stage dominated by positive feedback takes over as ageing progresses. This time it is a downward accelerating trend and finally it causes the system to cross a threshold and take on an entirely new state: death. Although this model is true for the individual organism, the steady state of its adult stage can be envisaged as being perpetuated in its offspring. So although the individual organism does not escape the consequences of the Second Law of Thermodynamics (i.e. death) the capacity for self-replication and reproduction allows the steady state to be maintained, but at the level of the entire species population.

23.1.2 Population dynamics

One of the most important properties of any species population is its size – the number of individuals in the population. At any moment these individuals will be in different stages of their life cycles, so that the population as a whole will have an age structure composed of several overlapping generations of individuals. (Sometimes, however, particularly amongst insects, the life span of the species is the same as its period of reproductive maturity and the population is composed of generations that replace each other and scarcely overlap in time.) For each age group the rate of addition by birth (in mammals) and subtraction by death will be age specific. Birth rate and death rate will change with age. In Man, for example, death rate is highest in the first year and in old age and birth rate is highest at about 20 years, but lower before and after, while death rate is lowest at approximately 11 or 12 years. If these age-specific birth and death rates remain constant, or relatively so, then the population will exhibit a stationary age structure, and if overall the total birth rate and total death rate are the same, then the population size remains constant. Such a situation implies that each individual is replaced only once in its life time. Figure 23.2 illustrates this theoretical state for an idealised bird population. In reality the input and output never quite balance and the population size fluctuates within narrow limits, maintaining a steady state.

From the time of Darwin it has been recognised that most species possess an impressive ability to increase in number. Evolution has provided most species populations with reproductive strategies endowing them with the capacity to produce many more offspring than would be necessary to replace the population. Chapman (1928) called this multiplicative tendency inherent in the reproductive process the **biotic potential** of the species. This kind of growth can be described by the following equation:

$$\mathrm{d}N/\mathrm{d}t \propto N$$

where the change in the number of individuals with time $\mathrm{d}N/\mathrm{d}t$ (i.e. the increase in the population) is proportional to the number of individuals already there (N). This is a classic self-reinforcing positive feedback loop. We can remove the proportionality sign and replace it with an equals sign if we insert a proportionality constant in the equation to give

$$\mathrm{d}N/\mathrm{d}t = rN$$

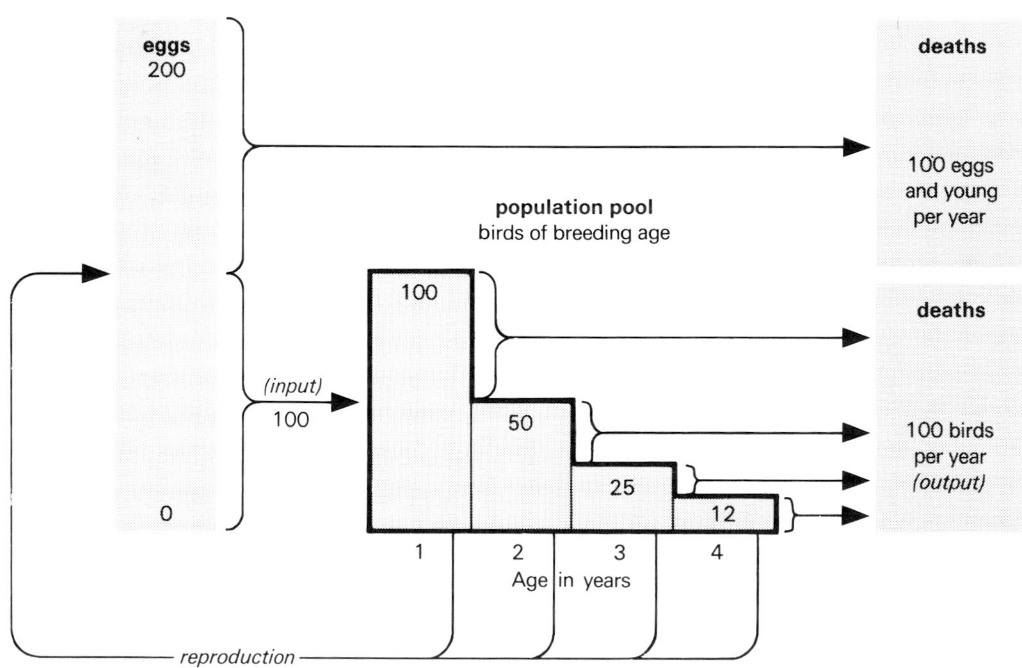

Figure 23.2 Steady state for a theoretical bird population (after Whittaker 1975).

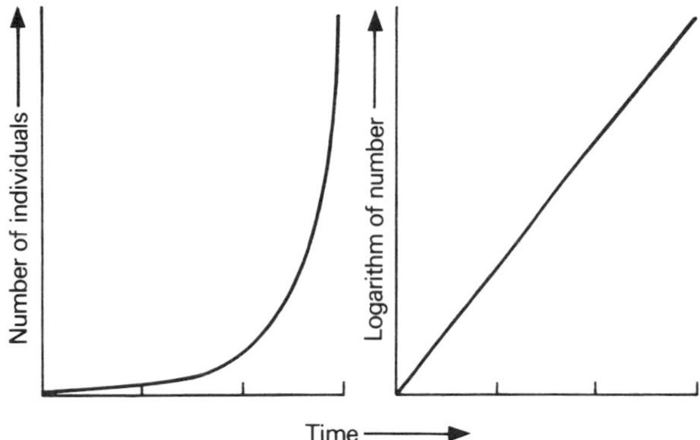

Figure 23.3 Curves showing exponential population growth on both arithmetic and logarithmic scales.

where the rate of population growth is given (r) which is known as the **intrinsic** or **instantaneous rate of increase**. The results of such growth are graphed in Figure 23.3 on both an arithmetic and a semi-logarithmic scale. In the former this **exponential growth** is seen as a characteristic U-shaped curve as population increase accelerates upwards, but in the latter it appears as a straight line (notice the similarity to graphs of stream order against number of streams in Ch. 10).

There are many amusing theoretical calculations in the literature on population dynamics which demonstrate the consequences of applying this concept of exponential growth to populations of various kinds, but of course it rarely occurs in the real world, and where it does it is short lived. Exponential growth is valid only in the absence of crowding, competition or resource depletion. The importance of exponential growth is that it is a model of the capacity for increase possessed by most species. 'Real' growth curves represent the result of a conflict between this capacity – Chapman's biotic potential – and what he called the **environmental resistance**. This term embraces the limits set by resources such as light, space, nutrients, food and by intraspecific (usually called crowding in this context) and interspecific competition which any species encounters in the real world. In combination these limits help to define the maximum **carrying capacity** of the environment for any particular species population. So the population growth equation becomes

$$dN/dt = rN\left(\frac{K-N}{K}\right)$$

where K is the upper limit of population growth or carrying capacity. When N is small, $(K - N)/K$ approaches 1 and the equation reduces to the equation for exponential growth. As N increases and approaches the value of K, $(K - N)/K$ becomes a smaller and smaller fraction and hence the rate of increase slows down. When graphed, this model gives an S-shaped (sigmoidal) curve (Fig. 23.4).

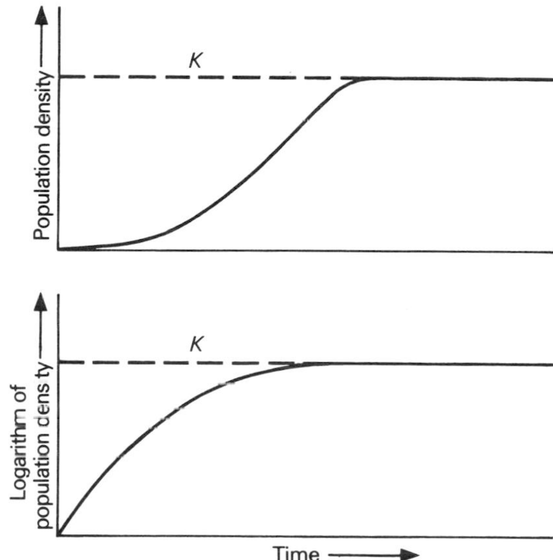

Figure 23.4 Curves showing sigmoidal (logistic) population growth on both arithmetic and logarithmic scales.

When the growth curves of natural populations are compared with this theoretical model, although some accord reasonably well with it, many do not. This is because the feedback mechanisms that operate on real populations do not act as instantaneous switching mechanisms, but as components of complex feedback loops. This introduces delay and lag effects such as those that occur when a population, having reached its carrying capacity K, continues to increase until the cumulative effect of resource depletion (for example) produces a decline in population numbers. Overshooting in this way is very common in the real world (Fig. 23.5) and the subsequent decline sometimes continues to low population densities before recovery begins and another overshoot–recovery cycle starts. In other cases, after the initial overshoot the population numbers drop to and oscillate about K with decreasing amplitude until stabilised. Furthermore, there is some evidence, particularly amongst the higher animals with well developed social organisation, that population densities are stabilised at a level somewhat below the maximum carrying capacity. Such a situation could be interpreted as a tendency to maximise the quality of life (i.e. share of space, resources, etc.) per individual, rather than to maximise the quantity of individuals per unit area.

In species that display relatively unstable growth curves with irregular or periodic fluctuations, the

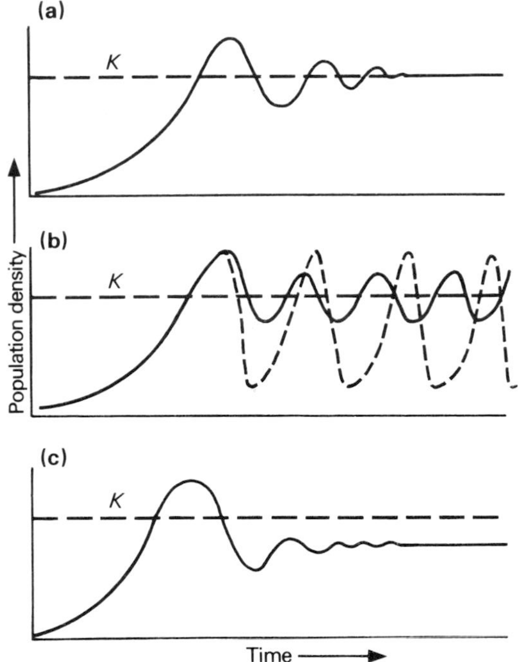

Figure 23.5 Models of population growth in relation to the carrying capacity (K). (a) Initial overshoot followed by damped oscillations of declining amplitude finally stabilising at an upper limit determined by the value of K. (b) Initial overshoot followed by regular cyclic fluctuations; in some cases (pecked line) the recovery limb of each cycle may approximate to exponential growth. (c) Initial overshoot followed by damped oscillations of decreasing amplitude but stabilising at an upper limit below the theoretical value of K, i.e. selection for quality of existence.

capacity for reproductive success, for wide dispersal and for rapid population growth are all of adaptive significance in ensuring their survival. Consequently, such species are grouped on the basis of 'r' selection. On the other hand, species that exhibit stable populations at or near their K values, and which coexist with other species in stable ecosystems where they occupy specialised defined niches to which evolution has made them well suited, are said to be 'K' selected species.

If population fluctuations were random, and increases and decreases occurring in response to change in some environmental factor were entirely independent of density, then sooner or later numbers would fluctuate down to a level where reproduction would fail or the last remaining individuals would be eliminated, leading to extinction. Whittaker (1975) expresses this conclusion succinctly when he maintains that 'in principle, a population that randomly walks in time, without some density-dependent limitation, must walk randomly to extinction. . . . Influences limiting fluctuations are necessary to the long-term survival of populations.' So again, as we saw in Chapter 21, random tendencies are constrained and feedback interactions maintain an organised state, this time a population density regulated within limits. Density-dependent mechanisms must regulate the upper and lower limits of populations by

(a) increasing the mortality of individuals proportionally or decreasing the birth rate per individual as the population grows, and by
(b) decreasing the mortality of individuals proportionally or increasing the birth rate per individual as the population declines.

All natural populations show some degree of regulation or control and the principal interactions involved are shown in Figure 23.6. Competitive or crowding interactions between individuals in a population often result in density-dependent regulation as resource limits are approached. However, this regulation is evident rarely as starvation and death, and even undernourishment and stunted growth, which may occur in plants, are not observed usually in natural populations of animals. In the case of animals, Wynne-Edwards

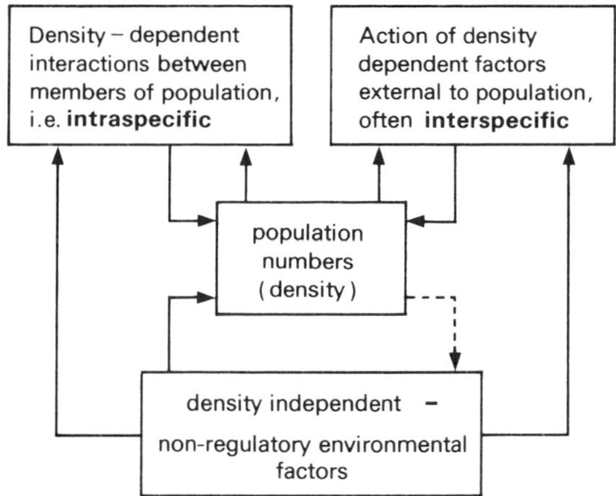

Figure 23.6 Density-dependent, density-independent and environmental interactions in the regulation of population size (after Solomon 1969).

(1965) cites two strategies of regulation. Both are mediated through physiological/behavioural changes and are under the control of social feedback mechanisms.

These strategies are, first, the limitation on the number of individuals allowed to breed and, secondly, influences on the number of young each pair are permitted to produce. They are implemented by social behaviour such as the establishment of territories (particularly amongst birds) which impose a ceiling density for a particular habitat and thereby exclude a non-breeding surplus population that may perish, but in any event does not replace itself by successful mating. Other mechanisms include the adoption of limited acceptable or traditional breeding sites (as with colony nesting birds or seal breeding grounds) where again a non-breeding surplus is excluded. The development of social hierarchies inhibits the onset of sexual maturity in young non-dominant adults in extended family groups, while harem systems will exclude unsuccessful males from mating. Pressure from crowding may increase socially induced mortality through, for example, a decline in the maternal care given to the eggs, or to the young, and in rarer cases may be manifest as cannibalism. However, all of these mechanisms are more complex than they appear at first sight and they usually involve regulation at the biochemical level as hormonal systems control physiological and behavioural responses.

Interspecific competition, on the other hand, appears initially to have a destabilising effect, at least on one of the competing populations. If we first modify the equation for sigmoidal growth to incorporate competitive interactions between two species, this effect will become apparent (Lotka–Volterra competition equations, Hutchinson 1965):

$$\frac{dN_1}{dt} = r_1 N_1 \frac{(K_1 - N_1 - aN_2)}{K_1}$$

$$\frac{dN_2}{dt} = r_2 N_2 \frac{(K_2 - N_2 - bN_1)}{K_2}.$$

Here, N_1 and N_2 are the numbers of species 1 and 2 respectively, r_1 and r_2 their relative growth rates, and K_1 and K_2 the carrying capacity or saturation densities. Expressing through aN_2 and bN_1, the effects of population change of one species on the population of the other are the competition coefficients a and b. The critical parameters in these equations are these competition coefficients and the carrying capacity densities, and the model predicts that all relationships between the values of these parameters, but one, will lead to the extinction of one or other of the competitors. This prediction has become known as the **competitive exclusion principle**. It implies that if two species are direct competitors utilising the same resource at the same time, at the same location – i.e. occupying the same niche – then as their populations reach equilibrium in a stable community one of the species will become extinct. Therefore, competition tends to bring about the ecological separation of closely related or otherwise similar species. That is, it promotes niche differentiation to avoid extinction.

There is, however, one condition where the two species can coexist. In 1962 Slobodkin showed that if the values of the coefficients were small in relation to the ratios of the carrying capacity densities $a<K_1/K_2$ and $b<K_2/K_1$ – both species could survive. In effect each species is inhibiting its own population growth by intraspecific density-dependent regulation more than that of the other species by interspecific competition. Therefore, each population is kept at a level below that which would culminate in an exclusive struggle for survival. These relationships have also been modelled more explicitly in terms of feedback by Margalef

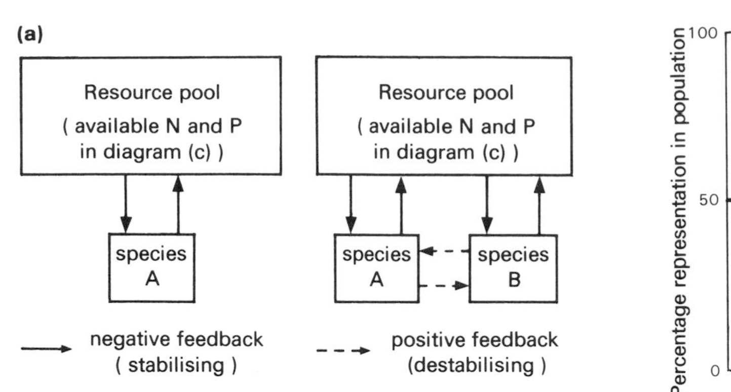

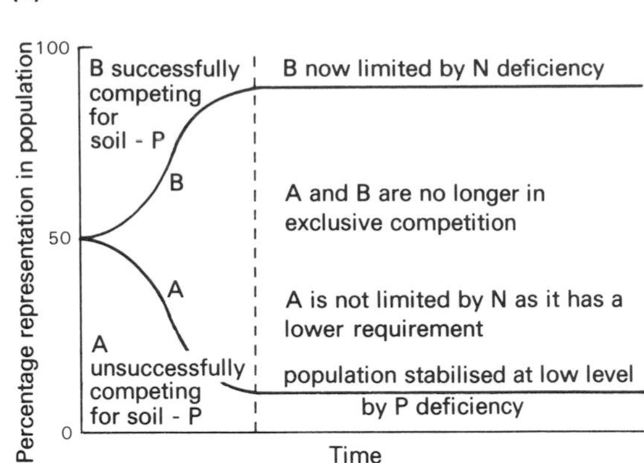

Figure 23.7 (a) The establishment of positive and negative feedback relationships between competing species, and between them and a mutual resource (after Margalef 1968). (b) The interpretation of Margalef's model in terms of the coexistence of two plant species in relation to limitation by two nutrient elements. Species B is a better competitor for soil phosphorus than species A but has a higher nitrogen requirement and so cannot entirely exclude species A by competitive stress (i.e. the development of intraspecific density dependence in B before interspecific competition eliminates A) (after Etherington 1978).

Table 23.1 Types of interspecific population interaction (after Odum 1971).

Type of interaction	General nature of interaction
neutralism	neither population affects the other
competition: direct interference type	direct inhibition of each species by the other
competition: resource use type	indirect inhibition when common resource is in short supply
parasitism	the parasite generally smaller than the host
predation	the predator generally larger than the prey
commensalism	the commensal population benefits while the host is not affected
protoco-operation	interaction favourable to both but not obligatory
mutualism	interaction favourable to both and obligatory

(1968; Fig. 23.7). Here, intraspecific density-dependent regulation is shown by negative feedback loops established between each population and a mutual resource, and interspecific competition by a positive feedback loop between the two species populations. This positive feedback is therefore potentially a self-reinforcing tendency towards the exclusion of one species unless the negative feedback succeeds in regulating the densities of each species in the way outlined above.

There are many other kinds of interspecific interaction apart from competition and all can involve a degree of density-dependent regulation, but equally under many conditions they may not (Table 23.1). For example, in the case of predation, stability is possible if the proportional loss to predation increases as the prey population grows, but if a lag occurs in the predators' response to feedback from prey numbers, then such a predator–prey system is likely to be inherently unstable. However, to consider a single predator–prey system in isolation is, in most circumstances, unreal. The prey, if a herbivore, may be limited directly by food resources or indirectly by breeding sites or territory size; it may even be the prey of more than one predator. Therefore, the two-part predator–prey system is really one aspect of a larger network of interactions and it may indeed exhibit stability as a composite result of all interactions – some of which are density dependent. In fact, many density-dependent mechanisms involve combinations of different kinds of effects. For example, the kind of intraspecific competition that may result in the exclusion of surplus individuals (e.g. territorial behaviour) initially results in emigration and then dispersal of the surplus, but its ultimate effect may be increased mortality of these individuals by predation.

Rarely is a single mechanism responsible for the regulation of natural populations and, although the notion of the steady state population is useful, it is commonly not the case. It is true that many species populations are maintained at or near their upper limits by self-regulatory mechanisms, but many others show wide fluctuations in numbers. However, that they survive is due again to some density-dependent factor which operates at the lower limits to prevent complete extinction.

23.1.3 Succession and climax

The complex, apparently stable and persistent ecosystems described in Chapter 16, and on a global scale in Chapter 7, do not 'arise as it were ready made'. Such mature communities with the highly developed interdependence of their constituent species and their complex network of interaction with the environment are the result of inherent processes of change – directional change akin to the growth and development of the organism. This sequence of changes has been known traditionally as **succession** and it is not caused by any change in the external environment. It consists of a number of successive developmental stages – **seral stages** – which together form an identifiable sequence – a **sere** – from the initial colonisation of a vacant habitat to the fully developed ecosystem.

During the initial stages of colonisation the environment will be the dominant influence on the success or failure of the plants and animals of the pioneer phase. To a greater or lesser extent this environment will be 'extreme' in that these pioneers will have to cope with a new mineral regolith, bare rock, not a *true* soil, with the full range of temperature fluctuations, with the full impact of precipitation and perhaps also desiccation. They are usually species with wide and effective dispersal mechanisms. Amongst the plants, most pioneers of terrestrial environments have light wind-borne seeds or spores which are produced in great numbers, as for example *Epilobium* in Figure 23.8 which produces *ca.* 80000 viable seeds in one year. Once established, however, many show impressive rates of population increase often associated with vegetative or asexual reproduction, all characteristics of '*r*' selection species. Although such vegetative clones are at an evolutionary disadvantage in that they consist of genetically identical individuals, they possess two immediate advantages when colonising new environments (Davis & Heywood 1963). First, they can spread rapidly over an unsaturated habitat where competition is lacking and, secondly, they avoid the risks of elimination that accompany fertilisation, seed dispersal, germination and establishment in sexually reproducing plants. Some species that occur during the pioneer stage, however, are ephemerals (usually annuals) which avoid the worst extremes of the environment by passing the unfavourable season as dormant seeds. Such is the case with many of the driftline plants which are the initial colonists of the sand dune succession (Fig. 23.9). Adaptation amongst pioneer species, therefore, is primarily adaptation to survive in an unfavourable environment rather than to confer an adaptive advantage in competition with other species.

Although pioneers characteristically display wide tolerance limits for external environmental parameters, they are not normally tolerant of competition. Most are 'open habitat' species that do not tolerate shade, while equally many require good root aeration with low levels of soil carbon dioxide and little root competition (for example, marram grass (*Ammophila arenaria*) in Fig. 23.9). Indeed *intraspecific* competition is more important during this phase than *interspecific* competition while the habitat remains unsaturated.

The structural simplicity of these early communities means that the extent to which they can control fluctuations in the environment by any sort of negative feedback is severely limited. Consequently, it is often the case that the magnitude of such fluctuations exceeds the survival capacity of the pioneer community, which is destroyed, and colonisation begins again. Often the destruction is not total and, for example, roots and humus and perhaps propagules survive in the substrate to initiate growth once more. In time the instability of the environment is modified by the developing community and its susceptibility to disruption by extreme events is reduced. Nevertheless, the superimposition of such cyclic processes on the general successional trend remains a feature of the earlier part of succession, although the amplitude of the oscillations decreases with time (Fig. 23.10).

As the pioneer species produce a more closed vegetation cover they alter the nature of the surface. They provide some shelter and conserve moisture for the germination and establishment of other plant species and create microhabitats for

(a)
(b)
(c)
100
50
0
Mean total cover (%)
0
100
200
260
Surface age (years)
(d)
(e)

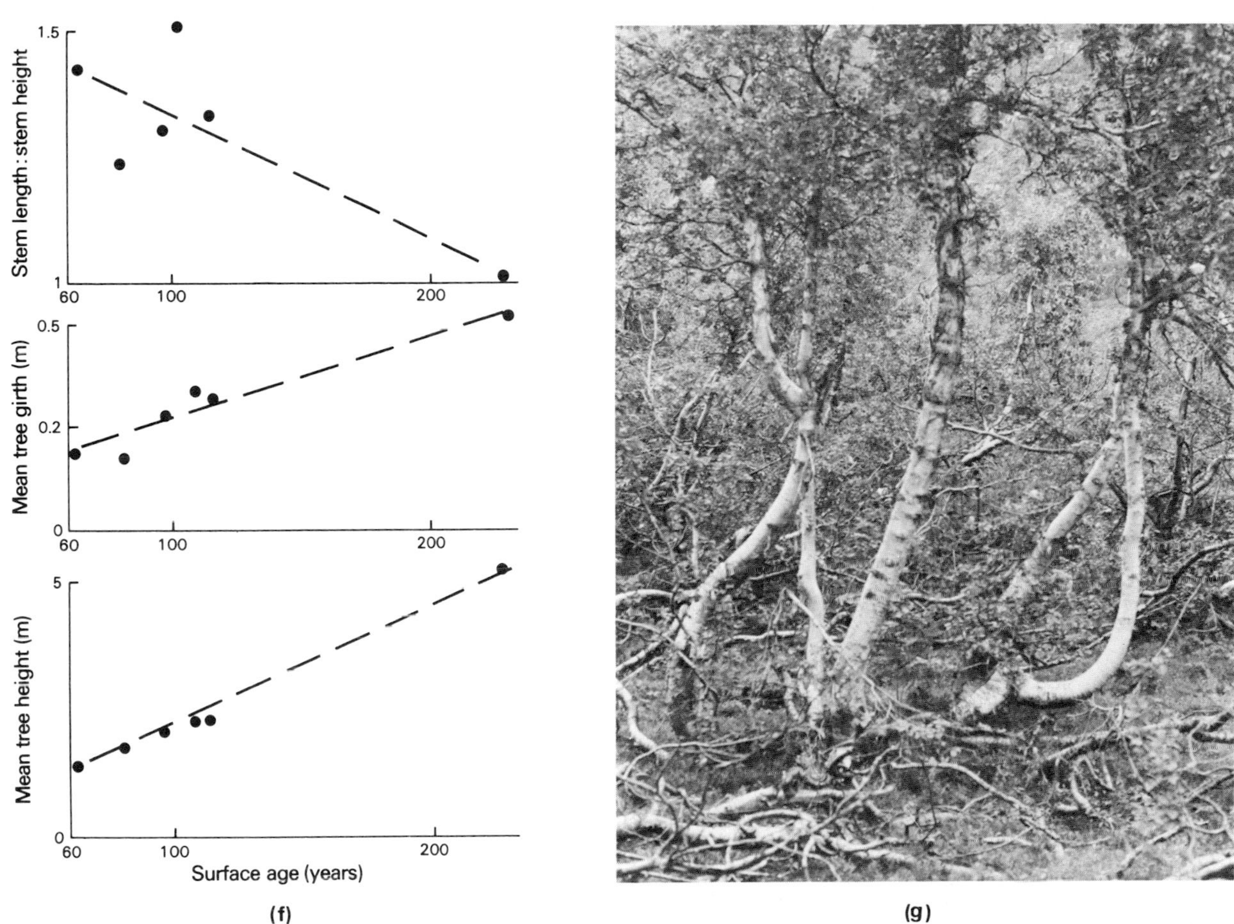

Figure 23.8 Plant succession in front of the Tunsbergdalsbreen (glacier), Norway. (a) The glacial foreland and the moraine sequence (see Fig. 22.18). (b) The initial surface available for colonisation. (c) Three pioneer species: *Epilobium anagallidifolium, Oxyria digyna,* and *Saxifraga groenlandica*. (d) Increase in percentage cover of the vegetation with time. (e) Dwarf shrubs dominate and trees begin to invade as the vegetation closes. Tree growth is at first tortuous and prostrate, but the ratio of stem length to height approaches 1 as a more erect growth habit is assumed, as here on the 1743 moraine, (f) and (g). (White 1973.)

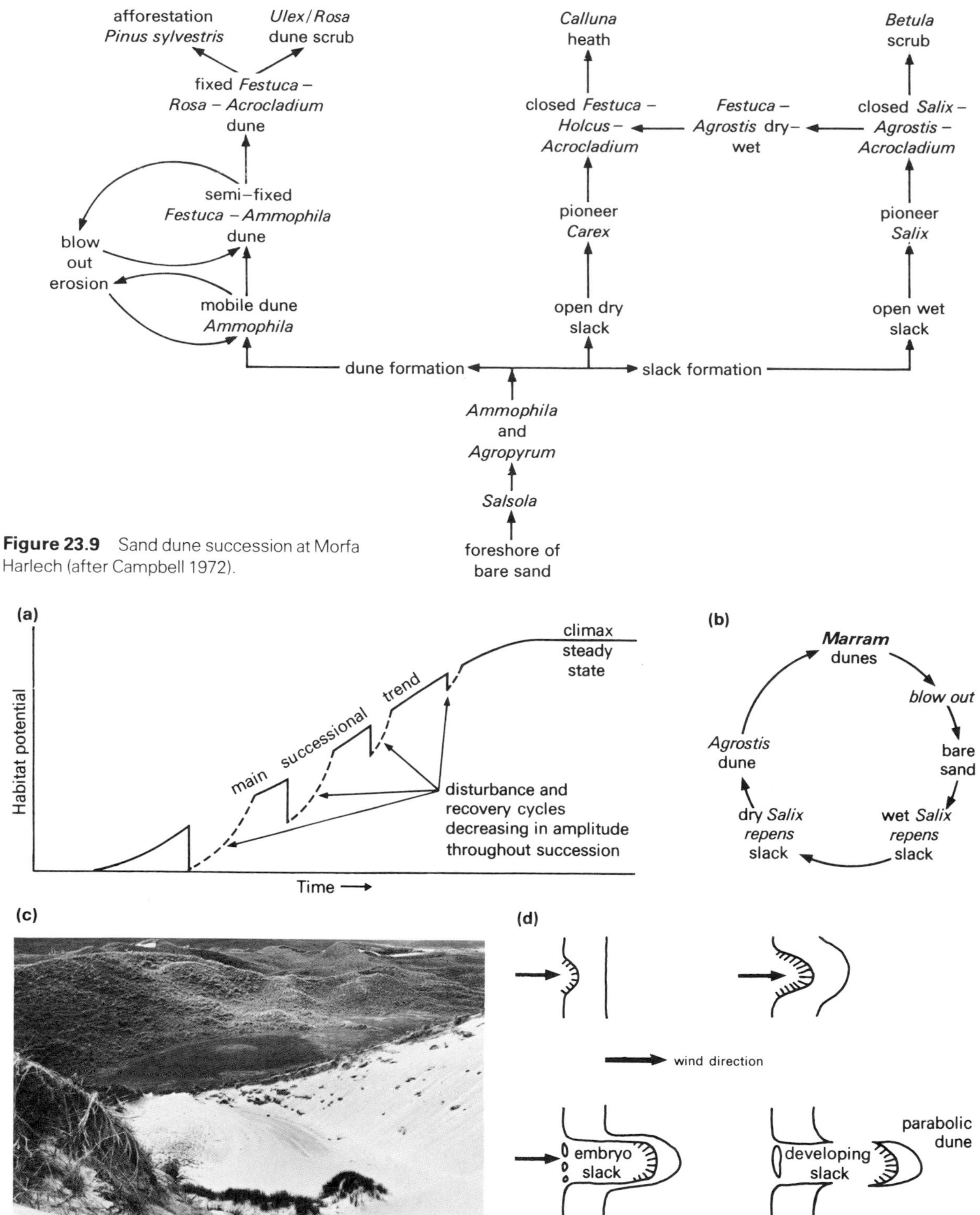

Figure 23.9 Sand dune succession at Morfa Harlech (after Campbell 1972).

Figure 23.10 (a) Cyclic changes superimposed on directional succession; characteristically the amplitude of the cyclic departures declines as the community approaches the climax state. (b) Sand dune erosion–recolonisation cycle. (c) Dune erosion at Balnakiel, Sutherland. (d) Formation and development of a parabolic dune (after Ranwell 1972).

animals, particularly insects, arthropods and small herbivores. Below ground their roots and the decay of the litter they produce begin to create a true soil (Fig. 23.11). They have begun to modify their environment, but in such a way as to allow other species to enter the community. This is a positive feedback process reinforcing change, and it leads to an increase in species diversity, slowly at first and then more rapidly. Although the newcomers compete more successfully in the modified environment than the species of the pioneer stage, some pioneer species, particularly those with the wider tolerance ranges, will persist for some time though their population numbers decline, and often show more marked fluctuations.

As species diversity increases rapidly during the

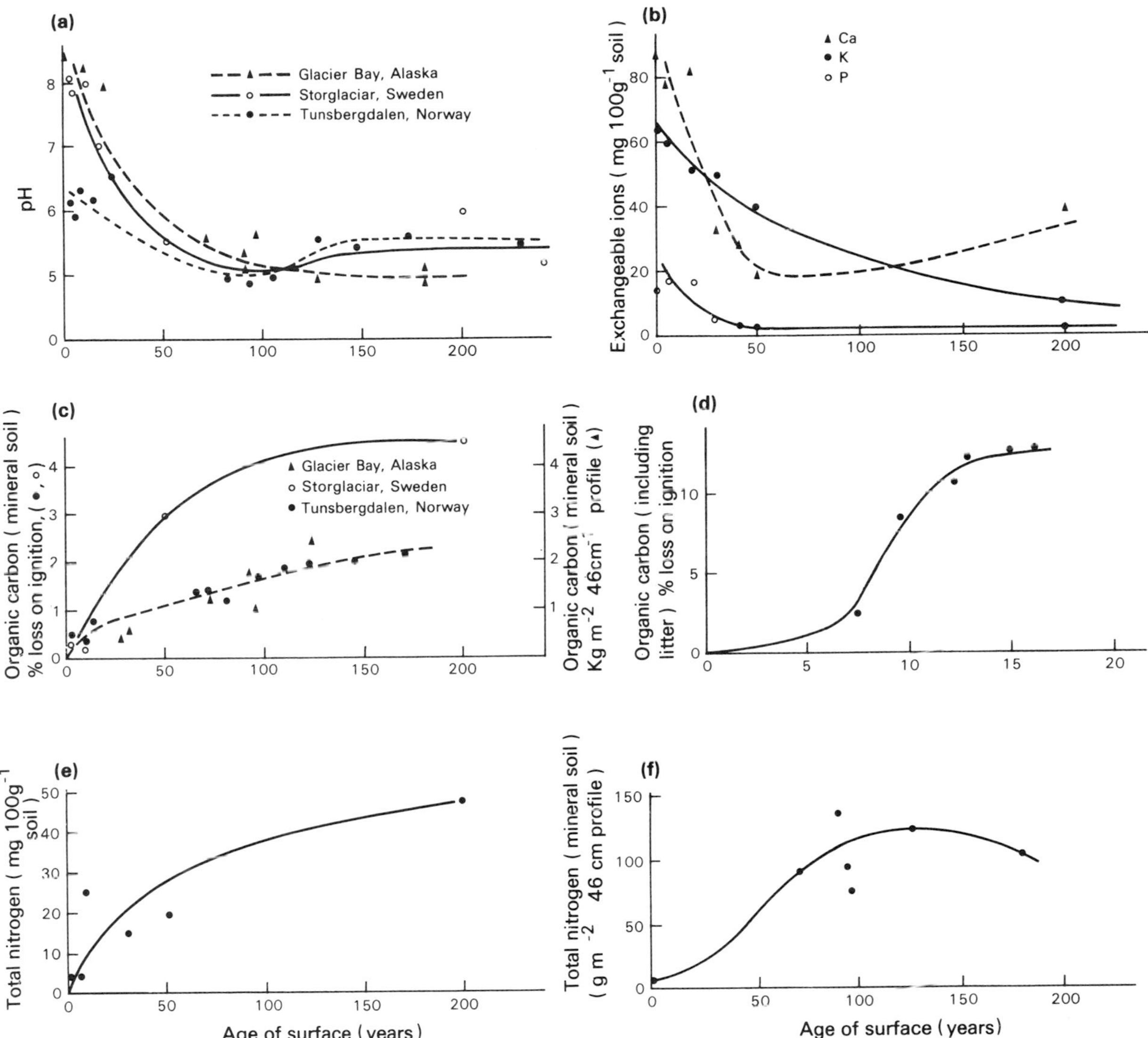

Figure 23.11 Changes in soil parameters through succession. (a & b) Decline in pH and exchangeable ions in soils from glacial foreland successions reflecting progressive leaching of originally fresh morainic material (after Crocker & Major 1955, Stork 1963, and White 1973). Increases in soil organic matter in (c) primary and (d) secondary succession (after Crocker & Major 1955, Stork 1963, White 1973, and Maris 1980), and in total soil nitrogen, (e & f) (after Crocker & Major 1955 and Olson 1958).

middle phase of succession, the tolerance range shown by each species entering the community becomes narrower. This is because, for a species to enter the community and survive in competition with those already there, it must be able to occupy an unexploited functional niche, or to partition an existing niche by being more specialised in part of that role than the current occupant. So this middle phase of succession is one of intense *interspecific* competition and of increasing diversification and segregation of ecological niche, and it affects plants and animals alike. It is marked by the development of negative feedback loops, both between the species of the community and between them and the environment, so that the level of integration and degree of self-regulation increase.

Ultimately a more stabilised community emerges, particularly amongst the plants, when the structural complexity of the vegetation reaches its maximum. This is associated with the development of the most complex stratification of the plant community possible under the prevailing environment. However, with shrubs and trees providing a range of new arboreal habitats, a renewed phase of diversification amongst the animals ensues as they compete for new niche space in the developing canopy. Furthermore, the existence of wood and dead timber, and increases in the amount as well as changes in the type of leaf litter, all provide additional niches. These are exploited by an increasing number of specialised micro-organisms, saprophytes and saprovores. This late increase in the potential for invasion, niche segregation and competition gives a final spurt to species diversity that thereafter tends to decline slightly as some of the species of earlier stages are finally eliminated. Although the diversity remains high at the end of succession because of intense competition, the population density of each species is relatively low. These successful species, which contribute to the establishment of a relatively stable community, tend to be genetically adapted to be tolerant of competition and specialisation, which in conjunction with resources define the carrying capacity for them; in other words, they show '*K*' selection.

Throughout succession the kind and intensity of the interactions between the community and its non-living environment change. The soil develops under changing conditions as the degree of control and regulation exerted by the community increases. Organic matter content increases through succession (Fig. 23.11). The development of organic horizons and the complexity of the soil population and decomposition process parallel changes in the plant and animal communities. The balance of weathering processes changes in response to increases in carbon dioxide from soil respiration, and to increased availability of organic acids and complexing or chelating agents. Soil water and drainage regimes adjust to the increased demand from transpiration. The chemistry of the soil reaches a balance that reflects the nutrient demand of the vegetation and the return of nutrients by litter fall and decomposition (a classic case of a negative feedback loop). Surface and subsurface erosional processes are affected also, particularly by the increased stabilisation of soil

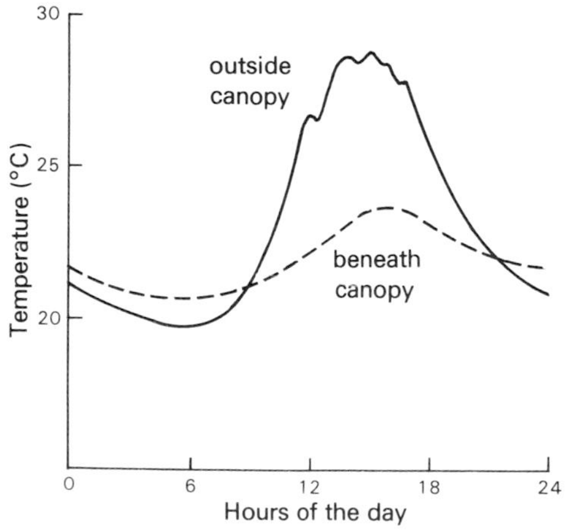

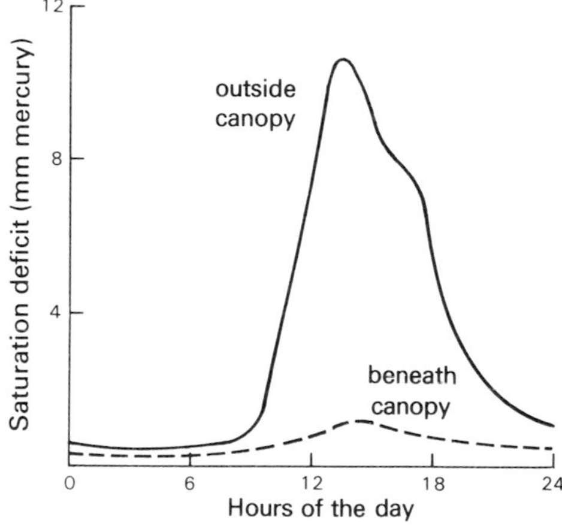

Figure 23.12 The effect of a climax forest canopy on microclimatic parameters (after Hopkins 1962).

and regolith effected by the developing root system and organic horizons, and the reduction in the soluble loss to streams as elements are immobilised or conserved in the community (again negative feedback). Also the effects of climate are modified progressively by the developing community until the canopy microclimate is controlled by the characteristics of the community itself. Such an effect can be seen in Figure 23.12 where diurnal and seasonal temperature and humidity ranges are clearly damped down.

The end-point of succession is a mature, self-regulating ecosystem which maintains a steady-state equilibrium through time. It is characterised by a complex, highly integrated community structure with high species diversity, but relatively low individual population densities not subject to serious fluctuation. The alternative pathways by which energy and matter are transferred through the system are also complex and diverse and not easily disrupted. Furthermore, this community has largely created its own environment and is buffered against reasonable environmental fluctuations. This steady-state ecosystem is a **climax** ecosystem. It is often stated that the relative stability of such a climax ecosystem is due to its complexity (Elton 1958, Hutchinson 1959). However, May (1971a, 1971b, 1972), by the use of mathematical models based on Volterra's equations but varying the coefficients of interaction randomly, was able to show that complexity, in terms of either number of species or number of interactions, tends to produce instability in the system as a whole. The implication is that if complex natural ecosystems are as stable as many observations would suggest, then the interactions occurring in them are highly non-random (Maynard-Smith 1974).

So far the process of succession has been discussed in terms of changes in the structural and functional organisation of the ecosystem and the climax has been defined in the same terms. However, our consideration of succession would not be complete without examining changes in the dynamics of the system through succession, and particularly changes in the pattern of energy flow.

In the early stages of succession, production exceeds respiration ($P > R$) and the excess (net production) is channelled into growth and the accumulation of biomass through time (Fig. 23.13). Therefore a large proportion of the energy input to the system is being diverted into store, and

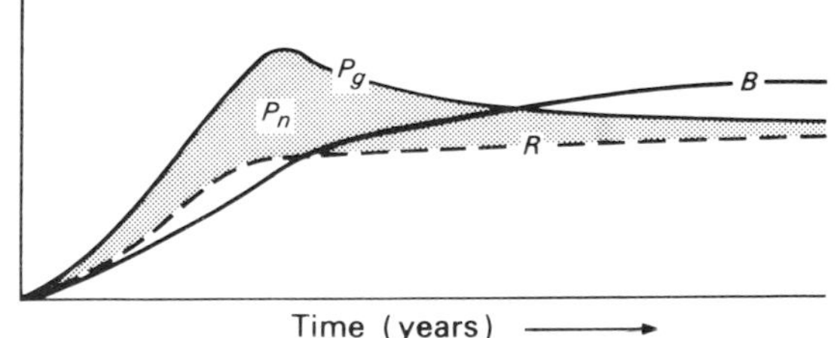

Figure 23.13 Changes in gross (P_g) and net (P_n) production, respiration (R) and biomass (B) through succession.

the output as respiratory heat loss is relatively small. The processes by which species interactions improve the habitat potential and promote increased species diversity has been recognised already as positive feedback. Its effect is to increase both the capacity and complexity of the energy storage compartments (as the total biomass of all species and trophic levels) as well as the complexity of energy transfer pathways. Some of this increasing energy store will be represented by increments of photosynthesising tissue and will feed back to increase the energy input. However, the rate of this increase will decline through succession as the vegetation reaches its maximum potential. Additional biomass or stored energy will continue to accumulate, but much of it will consist of non-photosynthetic plant tissue (particularly wood) and the biomass of heterotrophic organisms. A maintenance must be paid for this additional biomass as respiration, which therefore increases through succession. However, as long as $P > R$, biomass will continue to accumulate, but this ratio progressively decreases through succession and approaches a value of 1, i.e. $P = R$. By this time, competition, the stabilisation of niche structure and of population densities, as negative feedback processes, have established, and continue to maintain, a community steady state with little or no change in store (biomass) through time, but in equilibrium with the available energy flow.

The implications of this view of succession are very far reaching. In the first place the amount of biomass supported per unit of energy flow $B:E$ (where $E = P + R$), increases to a maximum in climax ecosystems. Secondly, the ratio of total respiration (or maintenance cost) to biomass (or structure) – $R:B$ – decreases and is at a minimum in climax ecosystems. Both of these observations would suggest that the climax ecosystem is energetically efficient. Margalef (1968) couches these same observations in the language of information theory and entropy. He views succession as a process of progressive accumulation of information, which we know means order or organisation (see Box 6.6). In these terms, R is representing the tendency of the system to produce entropy, because the respiratory heat loss is unusable energy unavailable to do work on the system, and B represents the ability of the system to preserve an ordered structure because the biomass represents the complex structural and functional organisation of the living systems of the ecosystem. The second ratio therefore tells us that the production of entropy per unit of preserved and transmitted information is at a minimum. The ratio $B{:}E$ tells us that it takes less energy to maintain a complex information-rich system than is necessary for the relatively simple systems characteristic of the early stages of succession.

23.2 Cliseral change

In the life of an individual organism, its growth and development is set against an environment subject to fluctuation and change. External environmental events vary in magnitude and frequency of occurrence, but most are predictable changes and they conform to an established diurnal and seasonal pattern. Indeed, such patterns would have been 'learned' long ago and accommodated in the evolution of adaptation and tolerance by the species that experienced them. This accommodation of predictable rhythmic patterns of change in the external environment is equally true at the level of the community and ecosystem (Fig. 23.14). For this reason they have been largely ignored in the preceding sections and we have assumed that all the manifestations of the inherent processes of change in populations, communities and ecosystems have occurred against a background of a 'constant' external environment through time.

However, we know that such an assumption is an over-simplification. Regulation and control mechanisms in physicochemical systems are generally less sophisticated, less precise and less able to maintain a stable condition than their counterparts in living systems. Although the state of climatic, hydrological and erosional systems can be defined by statistical averages, or by statements of statistical probability, they are nevertheless subject to considerable variation about any steady-state condition. Events of high magnitude but low frequency, such as large-scale flooding, periods of pronounced drought, particularly cold springs, or the sudden occurrence of slope failure, are all examples of the extremes of such variation. However, the effects they have on the ecosystem vary widely, as changes in the inputs to it or to its conditions of existence.

In a climax seasonal forest ecosystem, for example, fluctuations in the extent of the summer water deficit or in the air temperature during the growing season may be recorded merely as fluctuations in the radial growth increment as reflected in the annual ring widths of the tree species (Fig. 23.15). Alternatively, such fluctuations may

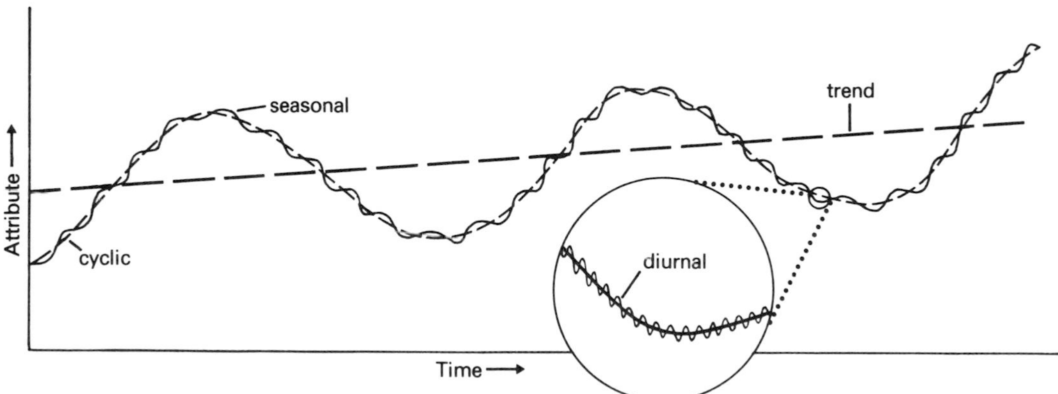

Figure 23.14 Predictable rhythmic change in environmental parameters accommodated by the community adaptation of ecosystems, but often, as here, masking an underlying trend.

Figure 23.15 (a) Taking an increment core from which (b) a plot of annual growth increments against time can be constructed. Unusually narrow or wide rings represent environmental variation as reflected in the pattern of tree growth.

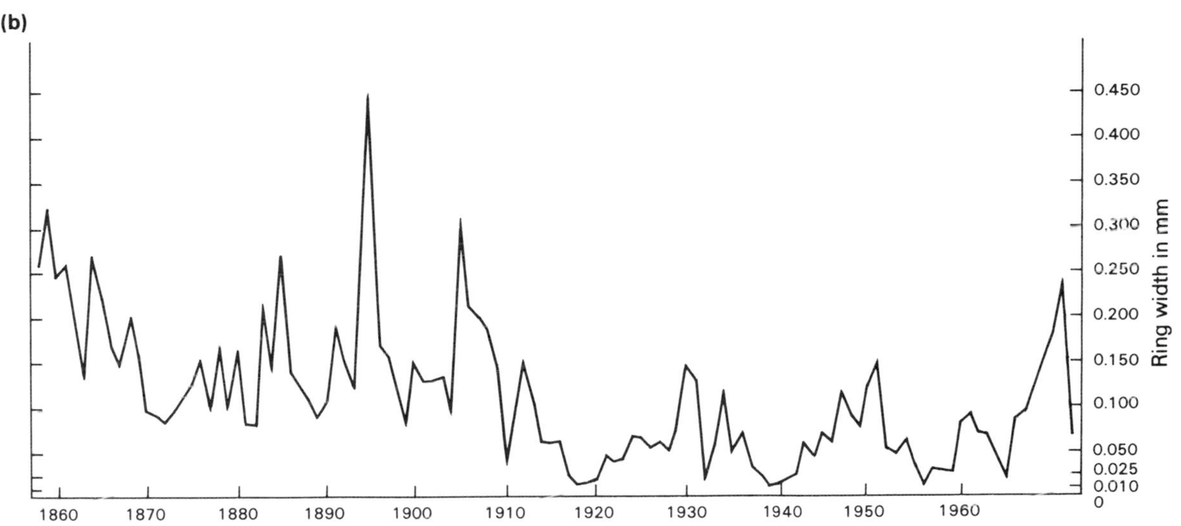

modify the community directly or indirectly by the elimination of one or more species. In extreme cases fluctuations in external environmental conditions can lead to a total, although usually localised, destruction of the ecosystem, as for example with the occurrence of a natural forest fire or a period of rapid mass movement (Fig. 23.16). However, when such changes in the conditions of existence of the ecosystem are fluctuations – albeit sometimes violent – about some steady state, their effect tends to be short lived. The resilience of living systems soon re-establishes the *status quo*.

As we have seen already in Chapter 22, events of the kind considered above may be more than fluctuations about some average steady state. They may instead mask some long-term trend of climatic and environmental change (see Fig. 23.14). Temporary setbacks to the growth and vitality of individual organisms, or minor modifications to species composition, give way to wholesale change in the ecosystem as the underlying trend of

Figure 23.16 Localised destruction of vegetation and burying of soil following rapid mass movement; solifluction lobe on Ben Arkle, Sutherland (after White & Mottershead 1972).

environmental change asserts itself. The effects of such trends of climatic and environmental change are well illustrated by the Quaternary period, to which we have already alluded several times in Chapter 22.

During the Quaternary, any point on the land surface of northern Europe and northern North America will have passed through a sequence of vegetation change, perhaps from glacial conditions devoid of vegetation, through tundra vegetation under a periglacial environment, then through a pre-temperate stage of birch woodland, through an interglacial deciduous forest, to the boreal coniferous forest of a post-temperate stage, back to open tundra and perhaps again to ice. This broad sequence can be regarded as externally induced change in response to broad categories of climatic change. We may envisage, therefore, the climatic zones of the northern hemisphere moving south, then north and then south again with the climax ecosystems or biomes – tundra, boreal coniferous forest and summer deciduous forest – following in sequence (Figs. 23.17 & 18). This concept of zonal ecosystem migration in response to changed climatic inputs is known as **cliseral change** (climatic sere), but its attractive simplicity is complicated by several factors.

It is true that climatic shifts result in shifts of vegetation and animal populations under stress from climatic change, but, in the case of plants, migration is dependent on the production of propagules, their quantity and their dispersal capacity. Some species are prolific seed producers

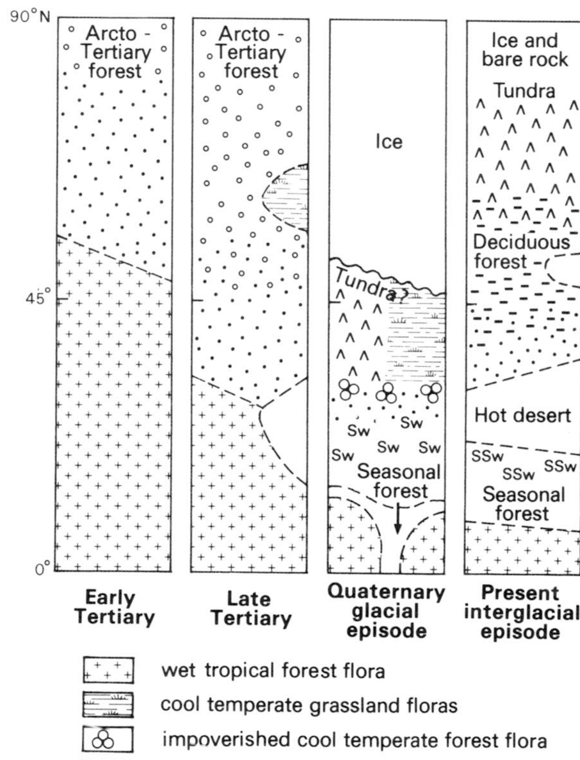

Figure 23.17 Zonal or cliseral change in the northern hemisphere during the Tertiary and Quaternary (partly after Collinson 1978).

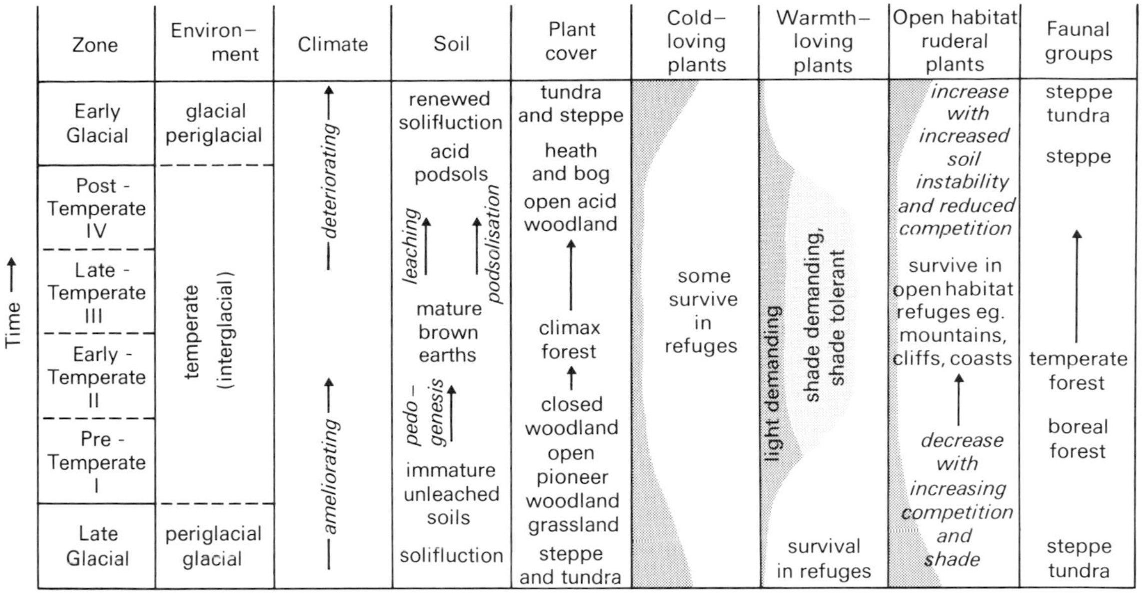

Zone	Environment	Climate	Soil	Plant cover	Cold-loving plants	Warmth-loving plants	Open habitat ruderal plants	Faunal groups
Early Glacial	glacial periglacial	deteriorating ↑	renewed solifluction	tundra and steppe			*increase with increased soil instability and reduced competition*	steppe tundra
Post-Temperate IV	temperate (interglacial)		acid podsols	heath and bog; open acid woodland ↑				steppe
Late-Temperate III			*leaching* ↑ *podsolisation* ↑; mature brown earths	↑	some survive in refuges	light demanding; shade demanding, shade tolerant	survive in open habitat refuges eg. mountains, cliffs, coasts	↑
Early-Temperate II			*pedogenesis* ↑	climax forest ↑			↑	temperate forest
Pre-Temperate I		ameliorating ↑	immature unleached soils	closed woodland; open pioneer woodland			*decrease with increasing competition and shade*	boreal forest
Late Glacial	periglacial glacial		solifluction	grassland; steppe and tundra		survival in refuges		steppe tundra

Time ↑

Figure 23.18 Floral, faunal and environmental changes associated with Quaternary climatic oscillations (modified from Sparks & West 1972).

and these seeds may be adapted beautifully to wide dispersal. Other species produce fewer seeds and may not set viable seed every year, particularly near the limits of their geographical range, while the seeds of some species may be heavy and not easily dispersed over large distances. As a result, even within the same climax ecosystem, species will possess different rates of potential migration. However, the possibility of migration under climatic stress depends, for both plants and animals, on the disposition of barriers in relation to potential migration routes (Fig. 23.19).

The absence of the spruce from the native British flora in the Postglacial (Flandrian) and the impoverished Irish flora relative to the rest of the British Isles are ascribed to the barriers presented by the drowning of the English Channel–southern North Sea and Irish Sea respectively by the eustatic rise in sea level after the last glaciation (Devensian). The differential distribution of barriers can be invoked also, for example, to explain the differences in the floristic diversity of the temperate deciduous forests of North America, Europe and eastern Asia. The key here is the differences in the alignment of the potential topographic barriers in relation to the direction of

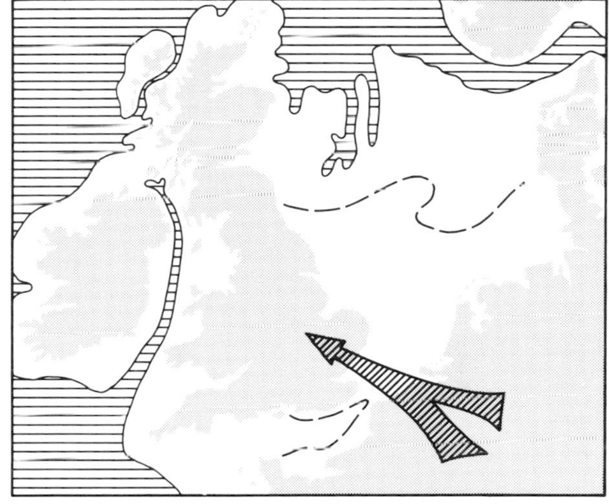

Figure 23.19 The English Channel, Irish Sea, and southern North Sea as barriers to floral migration during Quaternary interglacials. In the Flandrian at 9000 years BP Ireland would already have been cut off while a land connection still existed in the southern North Sea. By 7500 years BP something approaching the present coastline would have been established.

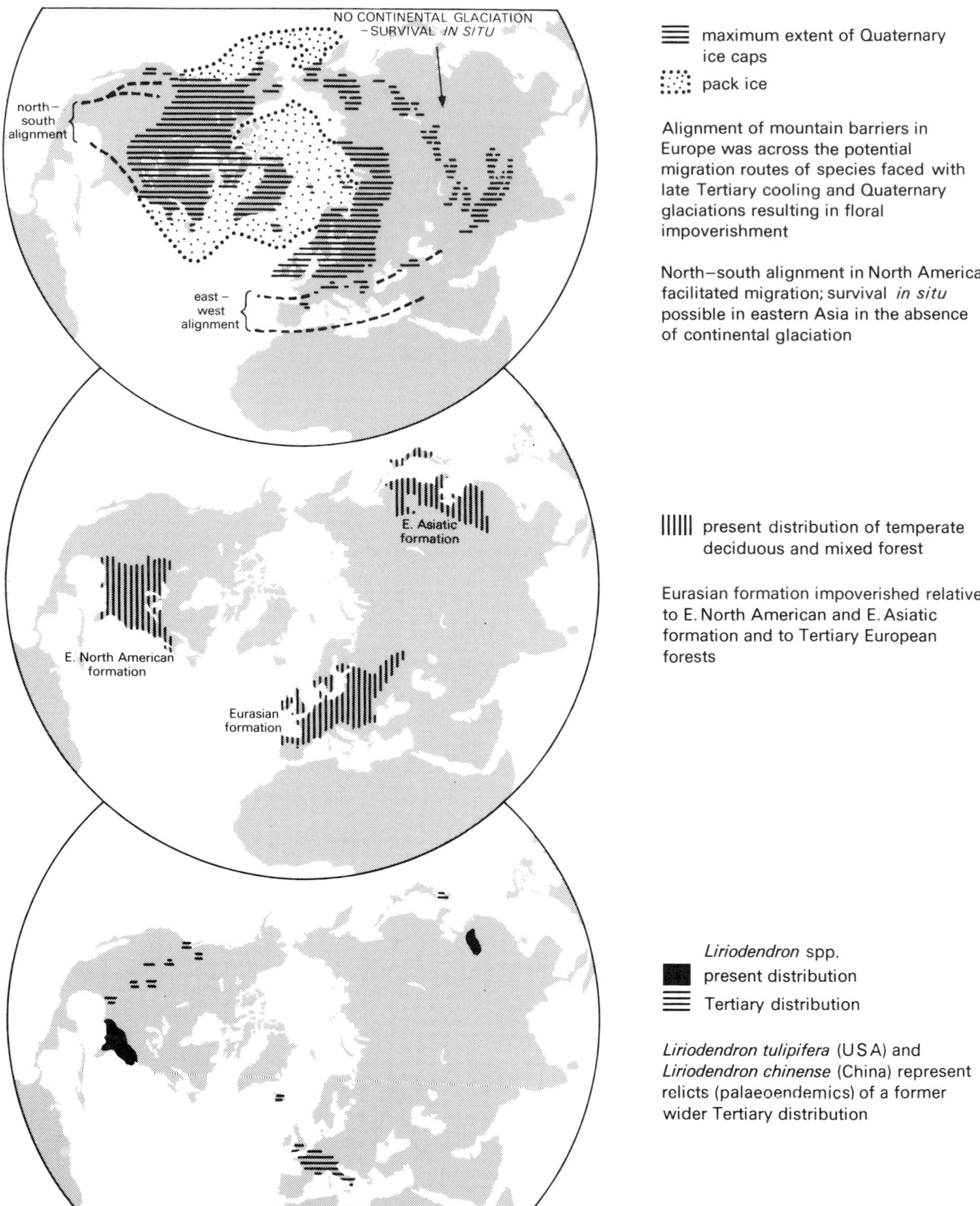

Figure 23.20 Barriers to the migration of mid-latitude forest floras during the late Tertiary and Quaternary in the northern hemisphere. The differential disposition of the barriers in part explains the differential patterns of elimination in the three continents.

migration (Fig. 23.20). Even when migration routes are available, they rarely offer the same opportunities to all species and in practice often act as differential filters.

For plants it is not only dispersal of the seed but successful germination, establishment and propagation that are also necessary before migration can be said to have taken place. So the rate of establishment of vegetation becomes another important consideration. In the Flandrian, Iversen (1964) has shown that forest communities actually became established a long time after climatic conditions had become favourable for forest growth. This is partly because of the time lag before propagules become available (through dispersal and migration), but mainly because the development of a climax forest ecosystem in a postglacial environment requires not only a favourable climate but also favourable soil conditions. Yet these soil conditions can be developed only under a succession of vegetation types. In such cases it may be difficult to disentangle the effects of autogenic succession from those of cliseral change. In a similar way, animal migrants are only vagrants until they become successfully established breeding populations and this will depend on the existence of the correct habitat conditions and hence on vegetation development.

Since extremes, and not average conditions, are critical to vegetation, broad climatic variations are often of less importance than specific features (such as degree of slope, aspect exposure, or proximity to the sea or fresh water) that may modify regional climate. Therefore, during an overall climatic change local conditions could either hasten or prevent a related vegetation change depending on whether they cause a local- or microclimate to cross a threshold of vegetation change.

In addition, closed climax ecosystems, particularly forest ecosystems, offer little scope for invaders as the niche space is fully occupied. They therefore possess an enormous inertia (Smith 1961). For example, a climatic change may prevent regeneration but not kill the existing population so that the effect of the change is delayed for the life span of those individuals, which in the case of some tree genera could be more than a hundred years. In certain circumstances such an inertia may have an effect longer than the life of the individuals concerned. The effect of a climatic change is not uniform and in some regions local ecological conditions may compensate for it, as we have seen. So although under some stress, relict communities may survive long after the initial climatic change. In other regions the threshold for survival may have been crossed and that community (or species) may have disappeared (see Fig. 23.21).

There is one final complication to the simple view of cliseral change: the plants and animals involved in these migrations are not static entities. They are subject to change in their genotype, largely by mutation and gene recombination. Before genetic change has culminated in the differentiation of a new species, however, small changes in genotype may be sufficient to alter the tolerance of the population and thereby affect its subsequent behaviour under environmental stress. For example, it has been suggested that the tolerance of the hazel (*Corylus avellana*) changed during the Quaternary, for it re-enters Britain progressively earlier with each successive inter-

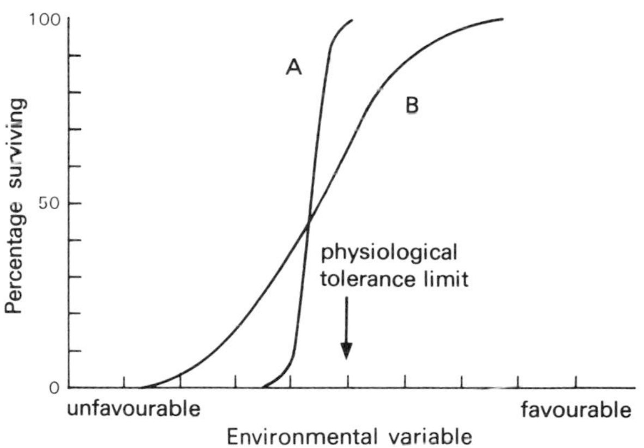

Figure 23.21 Unbuffered and buffered population response to an unfavourable environment. A represents an experimental genetically homogeneous population for which an environmental variable (such as temperature) becomes increasingly unfavourable past a physiological tolerance limit or threshold T. Past this limit the population declines rapidly to extinction. B represents a buffered gradual population decline in which some individuals are more vulnerable and others less so, because of genetic differences among them or differences in the microenvironments they occupy, or both. Genetic or microenvironmental heterogeneity can permit the population to survive environmental fluctuations that would make the population extinct in a homogeneous environment. (After Whittaker 1975.)

glacial (see Fig. 23.22). Deacon (1974) suggests that this is because it was able to survive the glacial stages in refuges progressively closer to the ice limits as its cold tolerance evolved throughout the 2 Ma of the Quaternary. This last example introduces the final manifestation of change in the ecosystem, namely evolutionary change.

23.3 Evolutionary change

As we saw at the end of Chapter 7, the immediate precursor of life on Earth was the appearance in the 'primordial soup' of molecules or molecular concentrations which were autocatalytic in some manner. This self-reproducing tendency is not an uncommon chemical property. However, the appearance of a molecule that would produce chance variations among its 'offspring' and pass such variations on to the next 'generation' must have been a very rare event even in the complex chemical mix of the early seas. Nevertheless, once

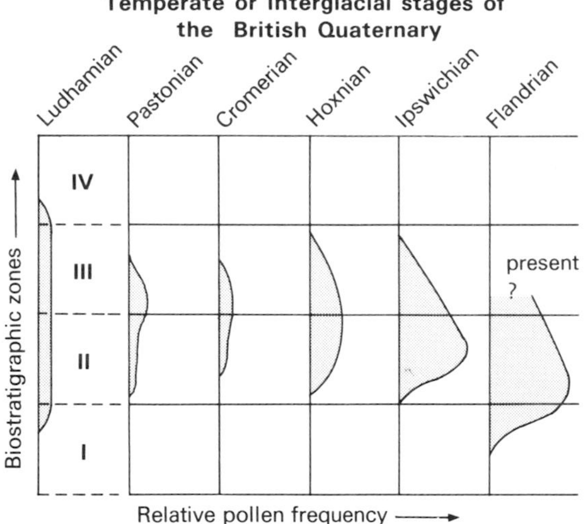

Figure 23.22 The changing behaviour of the hazel (*Corylus avellana*) in the temperate stages of the Quaternary. The hazel, an understorey shrub of mixed oak forest, arrives in the British Isles progressively earlier, finally arriving in the Flandrian well before the tree dominants of the forest of which it is normally a part. There is a massive expansion in its representation in the pollen diagrams as it spreads successfully on fresh unleached soils and flowers freely in the absence of competition. The explanation of this behaviour appears to reflect genetic change in the species itself; see text. (After Walker & West 1970, and Deacon 1970.)

such a system had arisen by chance, natural selection could operate on this variation; its consequence – adaptation – would appear and the Earth would harbour life for the first time. Accepting this view of the origin of life emphasises that from the very beginning three features have characterised living systems:

(a) hereditable variation arising at random;
(b) natural selection;
(c) the appearance of adaptation.

Today, the sources of variation in the genetic information are the segregation of chromosomes derived from different parents, the recombination of chromosome segments by crossing over, the spontaneous multiplication of chromosome sets and the occurrence of mutations. It is important to realise that, in principle, all of these mechanisms appear to operate at random. During the formation of the male and female gametes, chromosomes that originated from the two parents and were pooled at fertilisation are separated into two groups. This separation or segregation is at random so that for n pairs of chromosomes there are 2^n combinations of differently endowed gametes possible. However, before segregation the chromosomes come together in pairs, one from each of the parents, and exchange segments of the DNA-encoded information that they carry, so giving rise to new combinations of genetic information. There is effectively no limit to the number of possible recombinations that may be produced in this way. Polyploidy (the production of multiple chromosome sets) is a complex group of processes involving some form of failure of cell division, apparently at random, during the production of the gametes. It is a very important and abrupt source of speciation in plants. These three sources of variation all involve the re-arrangement of existing genetic information in the offspring. Mutations, on the other hand, are random alterations of this information itself. More specifically, they are changes in the base sequence of the DNA molecule.

In practice, this capacity for random genetic variation is constrained and regulated by natural selection that determines which genotypes survive and by so doing determines the rate and direction of evolution. We have seen already in Chapter 7 that in a stable environment it is only the best-suited genotypes that are preserved. If such an environment remains unchanged for a long period and the ecosystem is a stable climax with no available niche space, then stabilising selection dominates and inhibits evolutionary change in the population, even though mutation and genetic change will still occur amongst its individual members. Under most conditions these variants would be eliminated. However, Federov (1966) argues that, where the environment is genial and there is a degree of reproductive isolation experienced by individuals of a population in a diverse community (as in tropical rainforest), then mutations might persist and accumulate at random in the population. He calls this phenomenon **genetic drift** and uses it to explain the variety of physiognomic features of tropical rainforest. Many workers, notably Ashton (1969), would not agree with this view. Nevertheless, Simpson (1953) maintained that evolutionary stabilisation is characteristic of organisms that occupy continuously available, generalised or homogeneous environments such as the major forests and oceans where the range of variation in the environment does not greatly affect the capacity of the available genotypes to exploit it. Of course, in theory, the reverse should also be true. That is, in such generalised environments a greater range of genotypes would be expected to survive as selective pressures are also more general, and hence some genetic drift might be expected. However, many of these hypotheses are rather speculative because of the nature of the evidence and of the timescales involved. Even so it is safe to say that, in general, stable environments tend to promote evolutionary stabilisation of populations, whether or not one accepts the possibility of random drift round this stabilised genotype.

Again in Chapter 7, we noted that, in contrast, environmental variability, whether in time or space, allows natural selection to promote differentiation of populations in response to these environmental changes. Environmental change in time or variation in space will lead, by the selection from random recombination of existing genotypes, to adaptive shifts in the population to occupy either new or formally unoccupied niche space. If a stable environment becomes diversified or migration to a new and diverse environment occurs, diversifying selection leads to adaptive radiation. If the trajectory of environmental change is maintained for a long enough period of time (millions of years), then successive adaptive shifts can produce what Simpson (1953) called directional selection. Mutations contribute to these changes largely indirectly by their interaction with other genes and by producing new genetic combinations which may enhance the adaptive value of the whole genotype. In plants, polyploidy has been a very effective and rapid contributor to adaptive shifts under selective pressure from environmental change.

In a lucid review of evolution Stebbins (1974) arranges adaptive shifts in what he terms a spectrum, from those with a simple genetic base such as shifts in the frequency of particular genes which promote survival in an altered habitat, citing the case of industrial melanism in moths. The spectrum is followed through single phenotypic characters controlled by several genes to the differentiation of subspecies and ecotypes with more complex alterations affecting many characters. Ultimately, if such shifts affect reproductive isolating mechanisms, restricting gene flow, then new evolutionary lines emerge at the species level and above. Therefore, there is a unity in the evolutionary process which differs not in kind but in magnitude of effect at different taxonomic levels.

VIII MAN'S IMPACT

24 Man's modification of environmental systems

The interactions of Man with the environment are enormously complex. At one level *Homo sapiens* is just one kind of organism, a primate distinguished by an upright posture, sparse body hair and a highly developed brain. As such we may regard him as fitting naturally into the structural and functional organisation of the ecosystems in which he occurs. Indeed, such an approach is perfectly valid and the treatment of relatively simple human societies in this way has helped to improve our understanding of Man–environment relationships. Figure 24.1 shows the trophic relationships of Mesolithic Man in the forests of the British Isles during the climatic optimum of the Flandrian; similar relationships involving *Homo sapiens* at the end of the Devensian glaciation are depicted in Figure 24.2. This application of systems modelling to complete human ecosystems is not restricted to primitive societies and is equally valid with advanced technological societies, often shedding new light on the energy relationships and transfers that pertain in these complex systems (Fig. 24.3).

However, in the introduction to this book it was stressed that Man is not merely an omnivorous animal occupying a trophic niche in the food web of natural ecosystems: he is conditioned also by a complex and heterogeneous cultural inheritance. We stated there that Man's interaction with his natural environment can be understood only by considering his perception of it and his behavioural responses to it, both of which are conditioned by his cultural environment. The implication is that to do justice to Man–environment interactions we have to be able to mesh models of social and economic systems with our models of natural systems so that the linkages between the activities of these systems become clear. We must be able to accommodate flows of investment, technology, decision making, generated energy and other elements of what have been called human information and control systems in a comprehensive Man–environment system model (Chorley 1973). Perhaps this is one of the most important academic challenges that exist today. There have been attempts at such complete models, for example the world models of Forrester (1971) and Meadows and the 'limits to growth' team (Meadows *et al.*

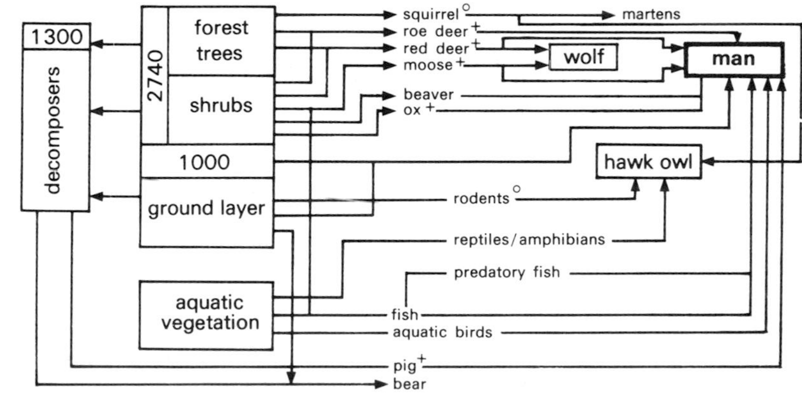

+ total *c.* 6 kg/ha
○ total *c.* 5 kg/ha

Figure 24.1 The trophic relationships of Mesolithic man in the early Flandrian forests of Britain (after Simmons 1973).

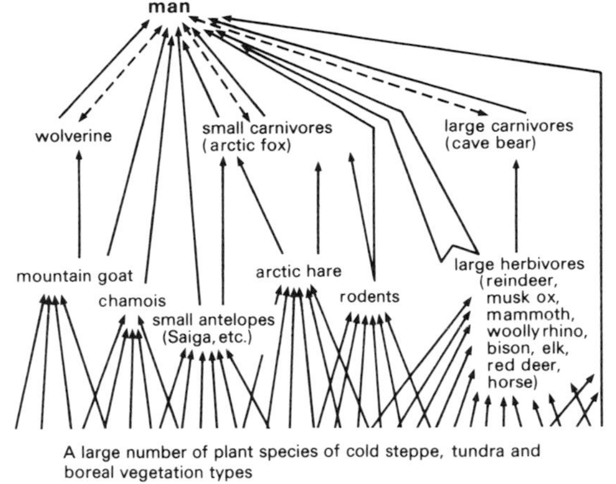

Figure 24.2 The trophic relationships of Man at the end of the Devensian glaciation, inferred from accumulations of animal bones (after Cox *et al.* 1976).

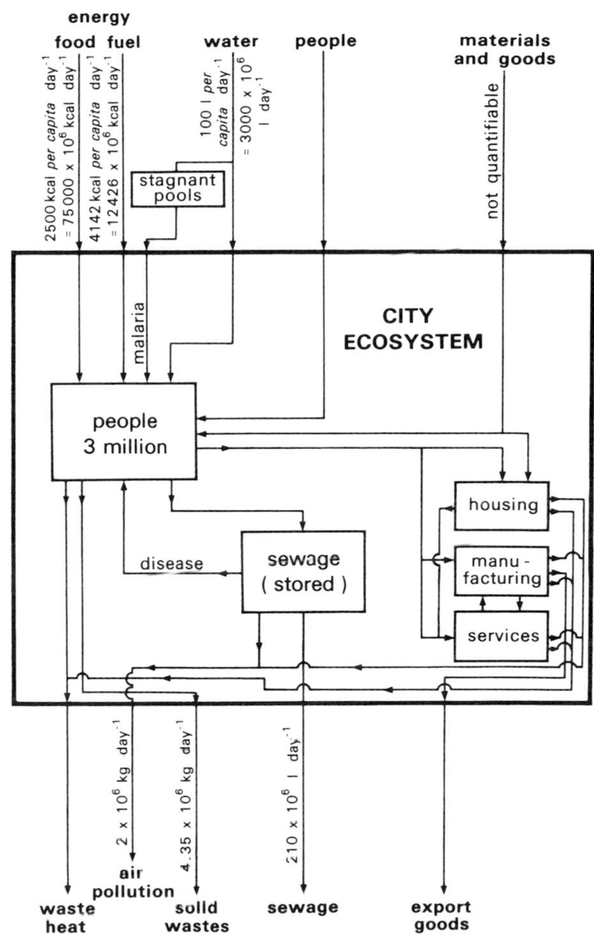

Figure 24.3 A simplified model of the metropolitan area of greater Calcutta viewed as an ecosystem (after Learmonth 1977).

1971), but this kind of comprehensive approach is beyond the scope of this chapter, as is the detailed consideration of information-control systems (Fig. 24.4). What these observations make clear, however, is that many of the models we have used to explain and understand natural environmental systems do not, as they stand, accommodate the cultural activities of Man. Such activities must be considered therefore as exogenous variables, as inputs to our models of environmental systems, and this approach will predominate in our treatment of Man and his interactions with his environment.

The effect of Man's activities as inputs to environmental systems is to redistribute matter and energy between the stores of the system and to alter both the magnitude of, and the pathways by which, energy and matter are transferred. Such interference with transfer pathways applies not only to those within a particular system, but also between systems. Because natural systems are open, and are organised (no matter at what scale we may view them) as energy and mass cascades, the consequences of human intervention in the operation of a system will almost inevitably have ramifications beyond the boundaries of that system.

Even the planting of human feet on the ground causes localised compression of the soil and a consequent reduction in the amount of rainfall that will infiltrate it. For example, multiplied manyfold, the trampling of footpaths creates localised high surface runoff and hence a more consequential modification of the hydrological system. Indeed, Man's activities can change the nature of the land surface in ways that are quite fundamental to the operation of natural systems by, for example, the removal or modification of the accumulated living and dead biomass of the natural vegetation and the disruption of the ecosystem's functional organisation. As a result, natural regulators to the flow of water through the system are fundamentally modified, such as vegetation and the ability of the land surface to store water. However, the extent to which the hydrological properties of the land surface are altered will reflect a sequence of changes in land use, initially by deforestation, then

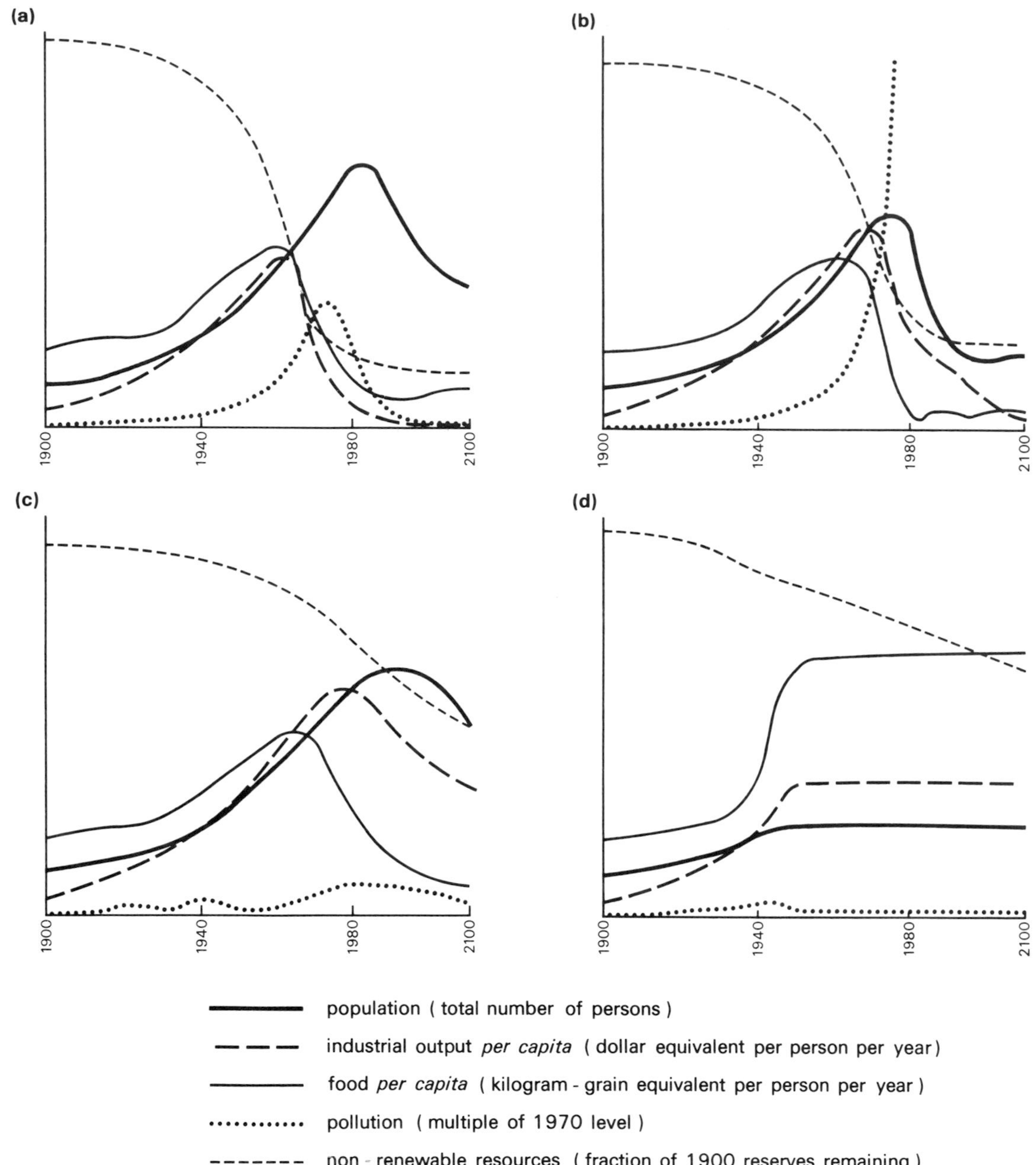

Figure 24.4 The Forrester–Meadows world model. (a) A standard run of the computer simulation model based on 1970 values. (b) Pollution-induced collapse predicted even when known natural resource reserves are doubled. (c) Collapse due to population growth even though resources are set as unlimited and pollution controls are assumed. (d) Stabilised model producing an equilibrium state sustainable into the future. However, assumptions include birth rate set equal to death rate, capital investment equal to capital depreciation, and that a range of technological policies are implemented, including resource recycling, pollution control, and restoration of eroded and infertile soils. (After Meadows 1971.)

by the creation of arable land with its exposed soil surfaces, and in more recent times the spread of urban and industrial areas with paved surfaces.

These modifications lead to changes in the water balance of catchment areas, usually increasing the proportion of surface runoff. In addition the distribution of runoff through time is changed also, normally producing concentration of runoff into higher peak flows. These hydrological changes in turn permit greater removal of mineral material from the land surface in the form of increased rates of erosion. In order to accommodate these increased flows of water and minerals, river channel adjustments take place. Such changes in process are normally progressive, as the negative feedback processes prevalent in natural systems are replaced by positive feedback mechanisms.

We have referred already in our discussion in the preceding chapters to some of the modifications Man may make to atmospheric systems. Man's ability to effect major long-term changes is limited by the ability of the Earth–atmosphere system to compensate for the relatively small temporal and spatial scales of change which arise from his activities. Most of these are localised and are usually short lived. The nuclear bomb represents an event of the highest order of magnitude which Man is capable of generating. Although the localised effect is total devastation, modifications to the whole Earth–atmosphere system of a single explosion are small. Most of the changes that Man makes are not, in fact, a result of such high-magnitude events but of continuous and cumulative modifications to the operation of regulators within the Earth–atmosphere system.

The effects of Man's activities, some of which we have just considered, are brought about either as an inadvertent by-product or through his conscious efforts to manipulate his environment, though the latter are often accompanied by the former. In both cases, as we have seen already, the mechanisms by means of which change is effected are the system regulators – the parameters that dominate the feedback relationships controlling the operation of the system. Nevertheless, the impact of human intervention in the dynamics of natural systems, whether purposeful or unwitting, is so widespread that to present a comprehensive picture of its extent would require us to retrace our steps through this book, identifying at almost every turn the effect that Man has had, is having, or could have in the future. In the sections that follow, therefore, we can provide only a glimpse of the profound and pervasive influence of this strange, erect primate with the highly developed brain.

We shall begin by looking at some of the accidental changes in the operation of environmental systems that have resulted from and accompanied different levels of modification and alteration that Man has exerted on the environment. For convenience we shall consider such levels of interaction under the headings of deforestation, agriculture and urbanisation, although each of these terms will be interpreted in a rather broad way.

24.1 Deforestation

The potential natural climax of most of the land areas of the globe that today support high human populations is some type of forest ecosystem, or at least an ecosystem in which tree cover forms a significant component. Therefore, it is justifiable to begin by considering the effects that Man's long history of modification, exploitation and clearance of these forests has had both on the ecosystem itself and on its functional relationships with the denudation system.

The biomass of the living vegetation of the ecosystem and the dead biomass store represented by the litter and organic matter of the soil are critical regulators of catchment hydrology, both directly and indirectly through their control of interception, surface and soil moisture storage, infiltration and evapotranspiration. Therefore, under natural conditions mature forest ecosystems regulate the two-way transfer of energy and water between the soil and the atmosphere. They regulate weathering and slope processes and conserve elements, both in the biomass of the forest itself and in its massive closed nutrient circulation, against leaching and erosion loss. Only in recent decades have controlled scientific experiments permitted an assessment of the magnitude of change in system operation following deforestation. In this way, flows of water and nutrients through a forested catchment can be measured during a calibration period, and then the forest is cut and the ensuing effect on runoff monitored. Alternatively, a pair of similar catchments may be monitored simultaneously. One of them can be deforested and the ensuing effects can be compared against its still-forested neighbour.

The primary effect of deforestation is to decrease the loss of water by evapotranspiration as the biomass is reduced. The fundamental change in water balance leads to an increase in surface runoff. This effect is illustrated dramatically by results from the Coweeta catchment in the Appalachian Mountains (Fig. 24.5), operated by the US Forest Service. With an annual precipitation of approximately 2000 mm, clear felling of the hardwood forest produced an increased surface runoff equivalent to 373 mm depth of precipitation. As the forest grew again, the excess water yield declined exponentially with time over a period of 20 years, until a second clear felling operation produced a similar increase in surface runoff.

A similar effect was observed at the Hubbard's Brook catchment in New Hampshire, USA (Bormann & Likens 1970). An overall increase in surface runoff of 40% was recorded after deforestation, with the increased stream discharge particularly marked in summer when evapotranspiration would have been at its highest. The summer period showed a fourfold increase in surface runoff, which would be achieved in large measure by increased flood peaks and increased velocity of streamflow.

Removal of solutes was also monitored in the Hubbard's Brook experiment. Figure 24.6 shows the increase in concentrations of calcium and potassium in stream water consequent upon felling. On average, concentrations of dissolved elements increased fourfold to produce a net output 14.6 times greater than when forested. The increased availability of unprotected surface soil and the greater energy of the channel runoff led in addition to a fourfold increase in sediment output from the basin. Table 24.1 shows the budget for several elements for both the forested and felled catchments at Hubbard Brook and emphasises the control the forest exerts over nutrient loss to drainage under undisturbed conditions. In the forested catchment, there is a net input of nitrogen from total nitrogen fixation in the ecosystem, and a conservation of the element in the nitrogen cycle. In the cleared catchment, the nitrogen cycle is broken and organic nitrogen and ammonia are oxidised rapidly to nitrate-nitrogen and lost to the stream. The increased concentration of the nitrate anion affects the ionic balance of soil- and stream water and increases their ability to transport metallic cations.

Such clear felling of an entire catchment illustrates the magnitude of the effect of a forest cover on denudation and ecosystem processes in terms of nutrient cycling, surface runoff and rate of erosion. Studies such as these have prompted a revaluation of the long-term benefits of conservation and intelligent management of natural ecosystems by Man.

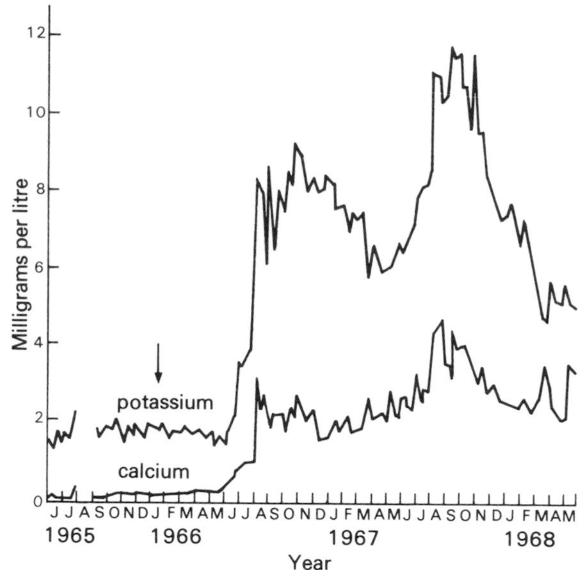

Figure 24.6 Increased output of potassium and calcium in stream water from watershed no. 2 following deforestation (arrowed), Hubbard Brook, New Hampshire, USA (modified from *The nutrient cycles of an ecosystem* by F. H. Bormann and G. E. Likens, copyright © 1970 by Scientific American, Inc., all rights reserved).

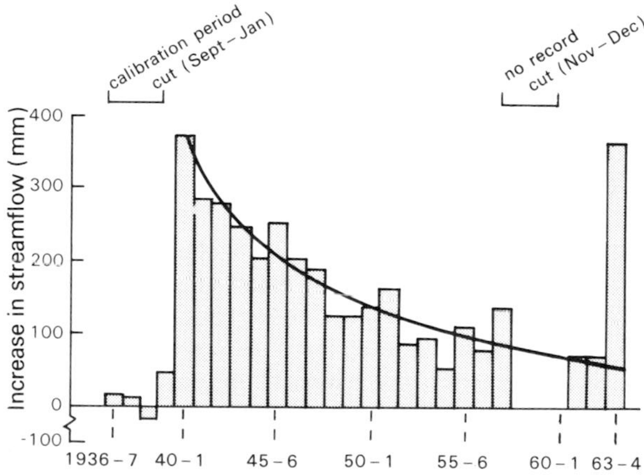

Figure 24.5 Increase in water yield after clear felling in the Coweeta catchment (after Hibbert 1967).

Table 24.1 Nutrient budgets for catchments at Hubbard Brook (after Likens & Bormann 1972).

	Precipitation input (kg $ha^{-1}a^{-1}$)	Streamflow – precipitation net output (kg $ha^{-1}a^{-1}$) Forested catchment	Cutover catchment
calcium	2.6	9.1	77.9
sodium	1.5	5.3	15.4
magnesium	0.7	2.1	15.6
potassium	1.1	0.6	30.4
NH_4 nitrogen	2.1	−1.8	1.6
NO_3 nitrogen	3.7	−1.7	114.0

We are learning to look on and conserve ecosystems not from the point of view of productivity and economic return alone, but in terms of the control that their highly complex regulatory mechanisms exert over environmental processes. The original undisturbed natural forests of the world were cleared gradually and partially and often allowed to regenerate (Fig. 24.7). Forest ecosystems display a remarkable resilience in the face of human intervention as long as the degree of exploitation is limited and does not exceed the threshold beyond which recovery is impossible. Their species diversity, the complexity of their functional organisation and the concentration of resources and information in the living forest all mean that they are to an extent buffered against change; they possess the capacity for self-regulation and regeneration. Therefore, they can coexist with the limited demands of shifting cultivation, limited grazing pressure, limited timber exploitation and settlement. Indeed, much of the floral and faunal diversity and aesthetic wealth of woodlands as perceived in the temperate zone of the northern hemisphere – the small grassy clearing, the rather open canopy, the variety of woodland birds and the flush of spring flowers – is a reflection of centuries of low-level exploitation and structural modification (Streeter 1974, Rackham 1971) (Fig. 24.7). Wholesale commercial timber exploitation and clear felling of virgin forest is an entirely different matter.

As we have seen in Chapter 21, there are considerable theoretical and empirical arguments indicating that complex ecosystems are dynamically fragile. They may well possess great stability in the relatively narrow range of environments within which they have evolved, but faced with disturbances wrought by Man they are inherently unstable. Such a view applies particularly to primary tropical rainforest which has proved much less resistant to intervention by Man than have simpler and more robust temperate ecosystems. The reasons again lie with regulatory mechanisms. May (1975) has argued that evolution has selected organisms which show genetic morphological and physiological characters that regulate population densities at or near equilibrium values, i.e., $N \simeq K$ (where N is the population number and K is the carrying capacity). Animals and birds produce fewer progeny than their temperate counterparts (Southwood 1974) but occupy more precisely defined niches and are more successful competitors in the sense that they therefore avoid direct competition. Plants are selected also for competitive ability in the face of extreme interspecific competition for space and resources in a genial environment. In the face of habitat disturbance, such organisms are poorly equipped to respond: they are not characterised by opportunism, by wide tolerance limits, or by a potential for rapid population growth. They are adapted instead to survive and compete in a niche precisely defined by the structural and functional complexity of the forest itself; destroy that and you destroy the organism's capacity to survive. The seeds of many species, for example, have little or no dormancy period and are adjusted to the moist cool shade of the forest and the stable microclimate of the forest soil (23–26°C). Soil heating on clearance therefore results in seedling death and an inability to recolonise, once sufficiently large areas have been cleared. These characteristics have led Gomez Pompa (1972) to refer to tropical rainforest as a non-renewable resource, and one which is in fact disappearing as a result of timber operations in South-East Asia and of the opening of Amazonia to commercial ranching following the road building programme.

The plight of the rainforests of the humid Tropics is perhaps an extreme and urgent contemporary case, but to exploit any natural or semi-natural forest ecosystem, particularly for timber, is to decrease its complexity. Unwanted trees are eliminated; access is improved by reducing understorey growth; other herbivores competing with Man are culled, or excluded by fencing; uniformity is increased by planting policies; and animal diversity declines further as the range of available habitats is reduced. Managing and cropping such

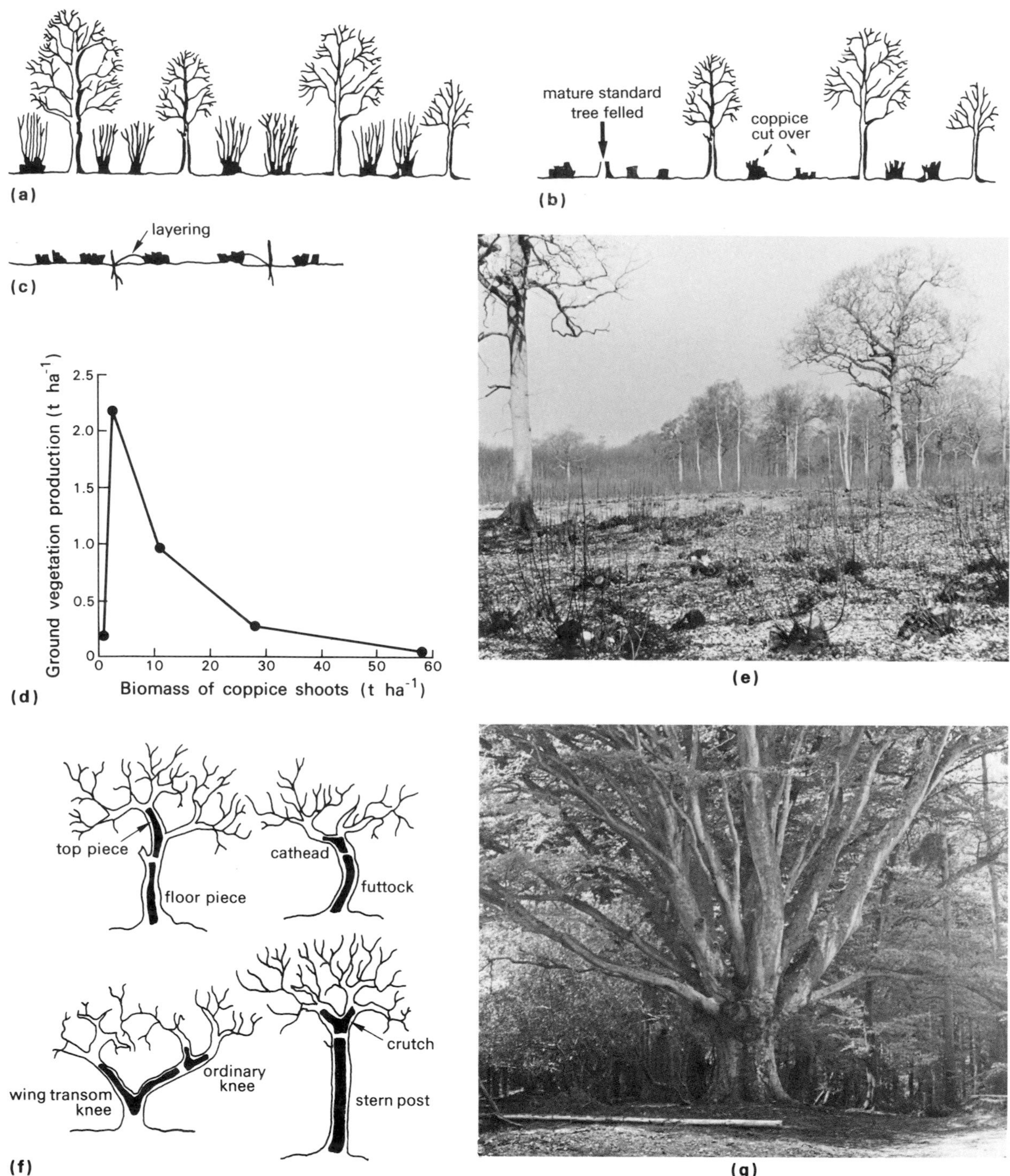

Figure 24.7 Coppice with standards system of woodland management, (a) before and (b) after cropping – note that one standard tree has been felled. (c) Layering of coppice to produce new stools (all after Ovington 1965). (d) Relationship between annual production in the herb layer and biomass of coppice shoots at different stages in the coppice cycle (after Ford & Newbould 1977). (e) Coppice with standards, Stansted Forest, Hampshire, England. (f) Training of oak growth to produce timber suitable for wooden ships and for timber frame housing (after Albion 1926). (g) Old pollarded – cut back to the top of the bole-beech, New Forest, England.

(a)

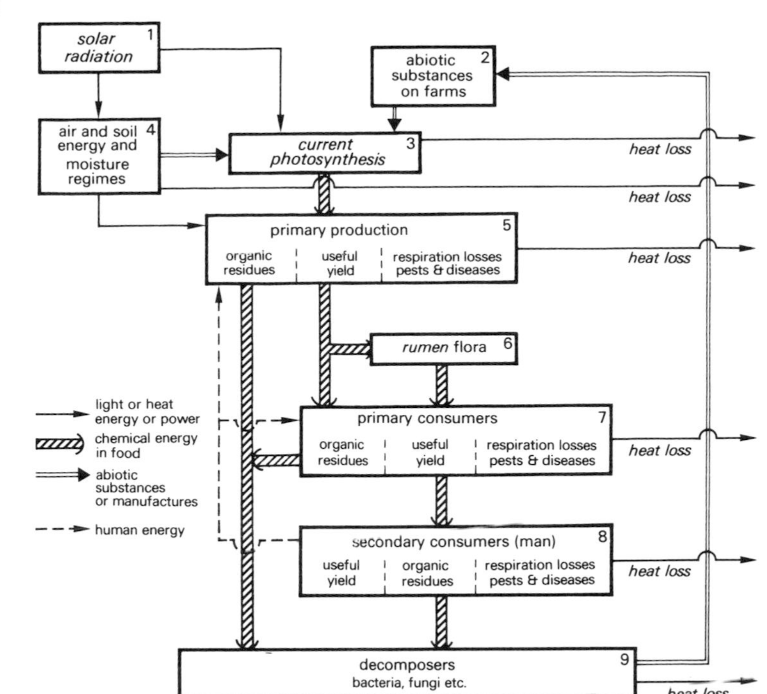

(b)

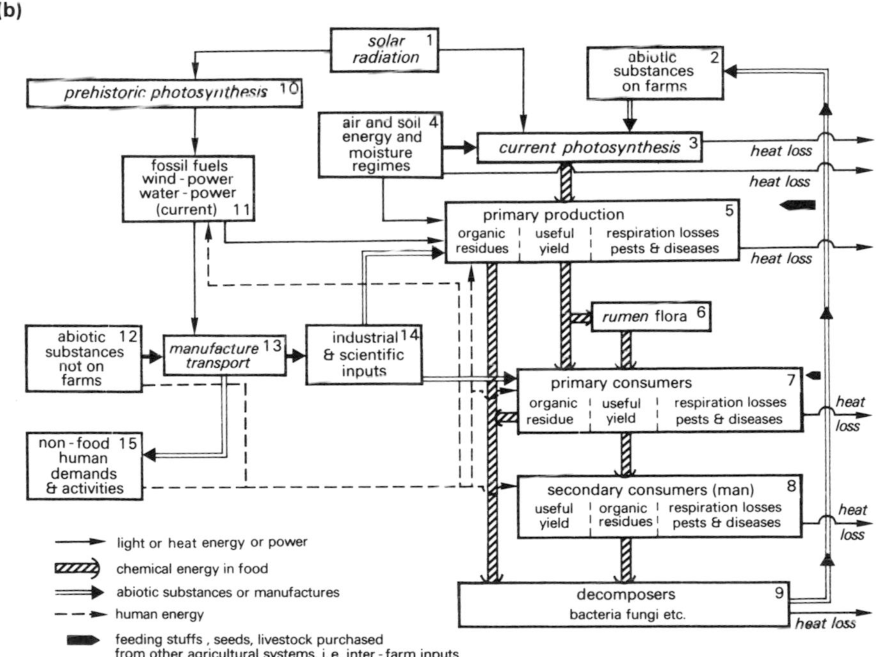

Figure 24.8 Flows of mass and energy in (a) a simple farming system in Uganda cropping domesticated herbivores and (b) a complex farming system in the UK, linked to the urban–industrial economy (after Duckham & Masefield 1970).

forests for a sustained timber yield has been likened to the management of permanent pasture for grazing by domesticated herbivores. Afforestation, on the other hand, is the equivalent of arable agriculture in the sense that seedlings raised in nurseries from selected seeds are planted and maintained with the aid of fertilisers, pesticides and a considerable manpower and management input. Like field crops, and for the same economic reasons, these new forests are often monocultures, although mixed planting is employed sometimes with, for example, a slow-growing hardwood species such as beech (*Fagus sylvatica*) planted with a faster-maturing conifer such as the larch (*Larix decidua*).

However, there are problems associated with all intensively managed forests, especially with new commercial afforestation, which indicate their long-term ecological instability. First, the uniformity of dense stands of a single tree species makes them particularly susceptible to explosive increases in pest populations and to outbreaks of disease. Combating such problems in the absence of natural regulation mechanisms is a costly operation often involving the aerial spraying of pesticides, sometimes with unforeseen side effects. Secondly, and again because of their uniformity, commercial plantations and managed semi-natural forests are more prone to fire hazard than are most natural forest ecosystems. Finally, the sustained cropping of timber represents a break in the natural nutrient circulation that, together with the accelerating effects of timber extraction operations on runoff and erosion, is a progressive drain on the fund of available plant nutrients. This is especially true of softwood forests in the northern hemisphere with short cropping cycles of 30–40 years, established on soils of low inherent fertility, and ultimately yields may decline unless the nutrient status is maintained by chemical fertilisation.

24.2 Agriculture

All agricultural activity can be seen in the widest sense as a deliberate attempt on the part of Man to modify and manipulate the trophic relationships that pertain in nature (Fig. 24.8). In almost all cases the effect is to simplify the complexity of

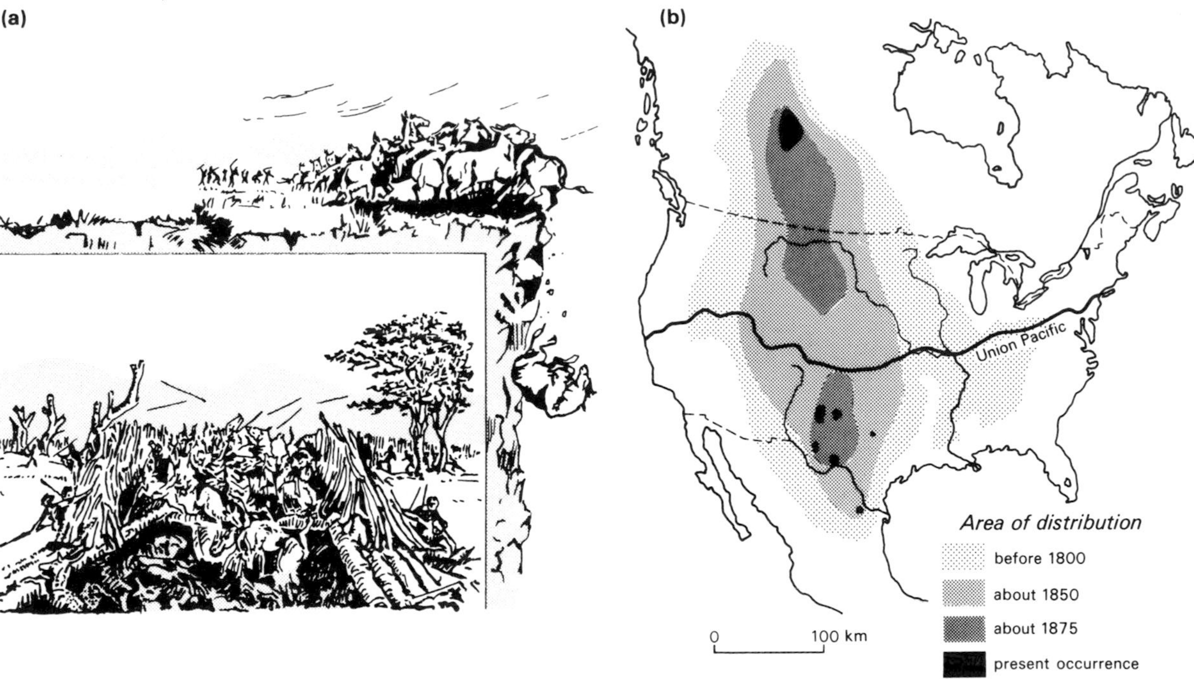

Figure 24.9 (a) Co-operative hunting by driving and pitfall trapping and by driving over a cliff. (b) The dramatic contraction in the distribution of the bison in North America following the opening of the Union Pacific railway and the extension of commercial hunting (from Illes 1974, after Ziswiler 1965).

natural ecosystems in order to increase the direct harvest to Man. By so doing, the ecological balance of the natural community is modified or destroyed as the complex but stable network of interactions is broken within that community and between its organisms and their abiotic environment.

As a hunter–gatherer, Man was (and in some parts of the world still is) an integrated part of the dynamic balance of the steady-state climax ecosystem. However, with increasing social organisation and co-operative capacity, and with the development of tools and weapons, the magnitude of Man's effect on natural herbivores and carnivores increased as he upset the regulation of population densities inherent in many natural grazing–predation systems. The extinction of many large Quaternary mammals in North America, Africa and, to a lesser extent, Eurasia has been attributed to the co-operative hunting techniques of Palaeolithic Man (Fig. 24.9). Although there is some debate as to whether environmental change could account adequately for these extinctions, it remains true (as we shall see) that the extinction of some and the reduction in the numbers of many

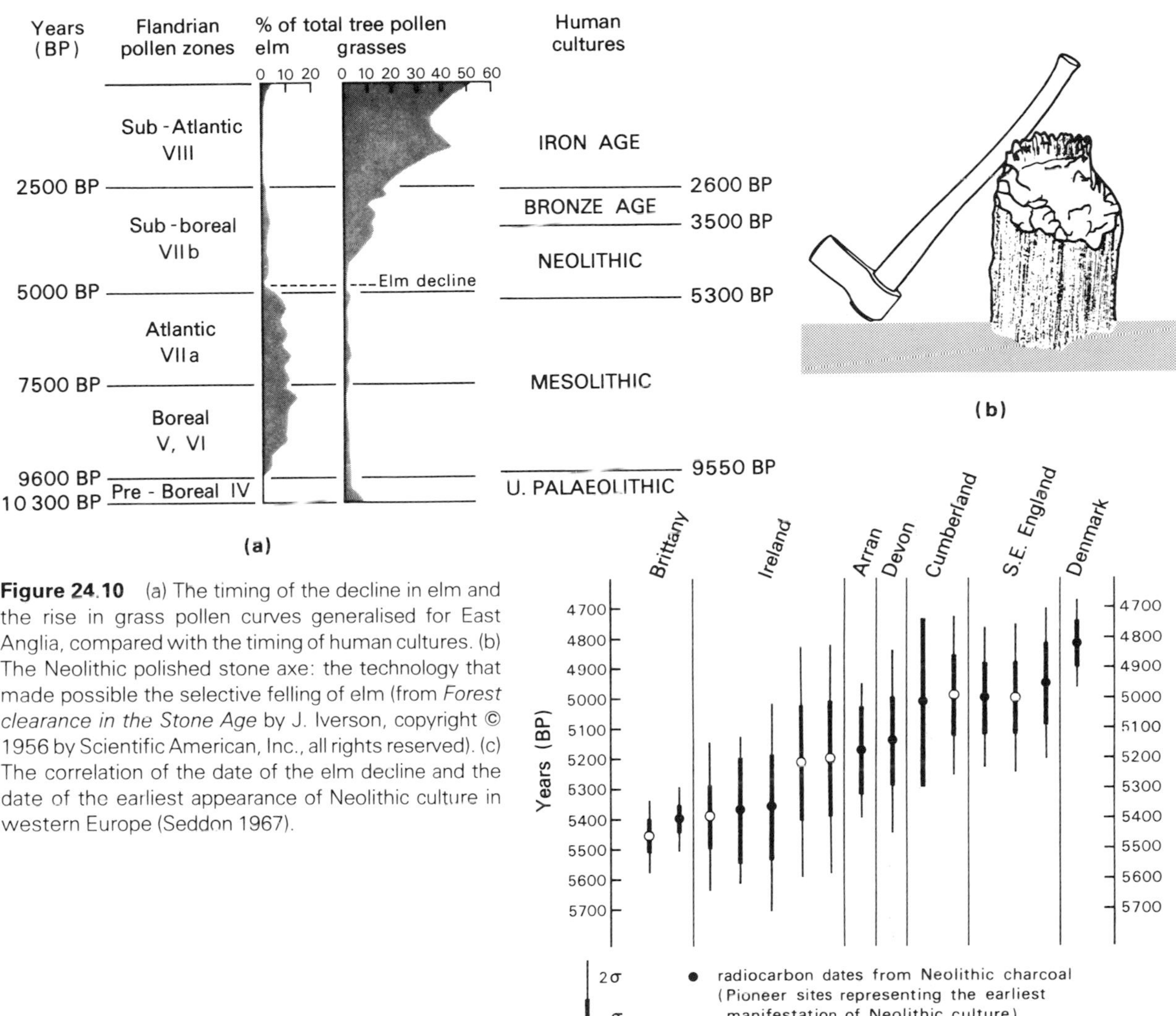

Figure 24.10 (a) The timing of the decline in elm and the rise in grass pollen curves generalised for East Anglia, compared with the timing of human cultures. (b) The Neolithic polished stone axe: the technology that made possible the selective felling of elm (from *Forest clearance in the Stone Age* by J. Iverson, copyright © 1956 by Scientific American, Inc., all rights reserved). (c) The correlation of the date of the elm decline and the date of the earliest appearance of Neolithic culture in western Europe (Seddon 1967).

animal species are due directly to hunting by Man.

The relationships between Man and his fellow heterotrophs changes fundamentally with the domestication of grazing herbivores and the development of pastoral economies. The stocking densities of domesticated herbivores are usually higher – often very much higher – than those of natural herbivores. The consequences of such densities are to reduce the standing biomass and to alter the structure and composition of the primary production system. As we have already seen vegetation change may initially be promoted directly by human action. The selective lopping of elm branches for fodder by Neolithic Man is thought partially to explain the decline in elm pollen in the Flandrian (Fig. 24.10). However, the clearance of forest, often using fire, is the ultimate expression of such modification motivated by a desire to extend pasture, and at least some parts of the major grassland biomes of savanna and prairie have probably been created in this way. Indeed, as we have just seen, it is still continuing on a large scale in the case of the Amazonian cattle ranches. Nevertheless, the large domesticated herbivore population itself brings about more subtle changes in the structure and composition of the sward.

Free range or extensive grazing of unimproved or rough pasture allows the grazing preference of the herbivore to be expressed. For example, the extensive studies of Hunter (1962 a & b) and others have shown that in the neutral and acid grasslands of upland Britain sheep have recognisable preferences which change seasonally and influence the way in which they interact with a largely edaphically controlled vegetation mosaic (Fig. 24.11). Where socially determined feeding ranges cover a variety of sward types, selective grazing pressure, particularly if combined with high stocking densities, can lead to an extension of the less palatable herbage species such as mat-grass (*Nardus stricta*) and purple moor grass (*Molinia caerulea*). In addition, every grazing animal has a distinctive way of obtaining herbage, and the sheep can graze very close indeed. Combined with selective grazing, such intense defoliation results in lower yield, a reduction in the recovery capacity of the sward, and ultimately in overgrazing. The existence of an unsaturated niche then allows

(a)

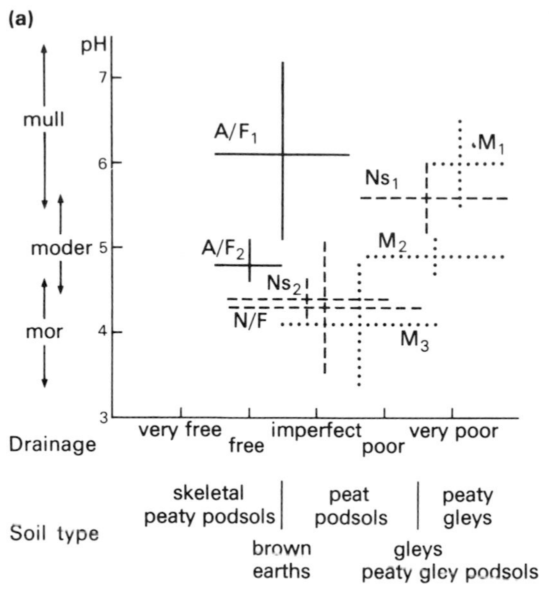

(b)

A/F_1 species rich *Agrostis / Festuca*
A/F_2 species poor *Agrostis / Festuca*
Ns_1 species rich *Nardus stricta* (sub - alpine)
Ns_2 species poor *Nardus stricta* (sub - alpine)
N/F *Nardus – Festuca* (with *Deschampsia*)
M_1 *Molinia – Agrostis*
M_2 *Molinia – Festuca / Agrostis*
M_3 *Molinia – Festuca / Deschampsia*

Figure 24.11 (a) Habitat ranges in terms of pH, humus type, soil type, and drainage class for neutral and acid grassland communities in the British uplands (after Burnett 1964). (b) Extensive sheep grazing in the Scottish Highlands. Sheep stocking rates and grazing pressure, acting in conjunction with the edaphic gradients depicted in (a), largely determine the sward types present.

The twayblade, *Listera ovata* (L.) R. Br.

The presence and size of these ant hills at Noar Hill, Hampshire indicates that the pasture has remained undisturbed by the plough for a considerable time.

The sites of former chalk pits at Noar Hill.

The milkwort, *Polygala vulgaris* L.

The germander speedwell, *Veronica chamaedrys* L.

The pyramidal orchid, *Anacamptis pyramidalis*.

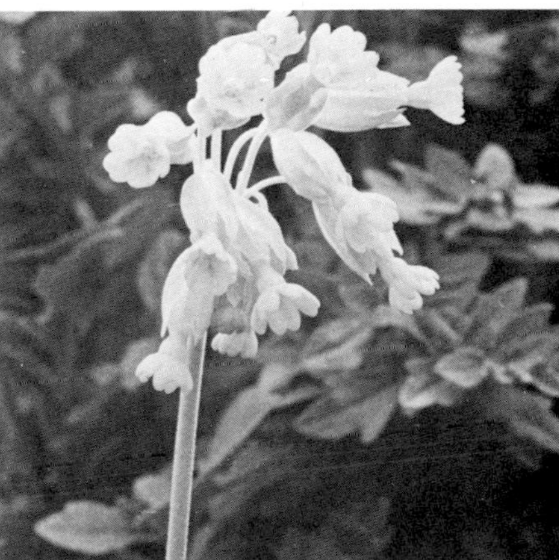

The cowslip, *Primula veris* L.

Figure 24.12 The grassland sward of the English chalklands, with its rich complement of flowering herbs and short springy turf, is a classic example of a community whose character has been largely produced and maintained by grazing.

opportunities for invasion by coarser and less palatable species. Virtually all of the dwarf shrub moor and grassland communities so characteristic of the British upland landscape have originated as by-products of their exploitation for grazing after forest clearance and from the use of fire as a management tool. They represent a balance between edaphic and climatic factors on the one hand and the differential effect of the particular kind and intensity of grazing both now and in the past on the other.

The complete plant assemblage, as well as many of the characteristics of the soils of the short-cropped downland turf of south-east and south-central England, are also accidental products of centuries of grazing, particularly by sheep and rabbits (Fig. 24.12). The latter were introduced by the Normans as a semi-domesticated herbivore, but subsequently they spread naturally. The reduction in grazing intensity and the temporary, virtual annihilation of rabbit populations by myxomatosis in 1954 have made this status clear as areas of downland turf have reverted through tall grass communities to scrub and woodland (Thomas 1963). Comparable examples of the effect of grazing can be seen in many other parts of the world. The destruction of the sclerophyllous evergreen oak forests of the Mediterranean by goats and its replacement by maquis and garrique communities, if not by virtual desert, is a well documented example (Fig. 24.13). In New Zealand the natural tussock grasses of South Island were invaded and replaced by introduced species of grass under grazing pressure from sheep and introduced wild herbivores. Thorsteinsson (1971) recounts the deforestation of Iceland by sheep and the impoverishment of the vegetation and erosion of up to 40% of the soil cover that resulted from it.

The effect of grazing on largely unmanaged pastures is to convert the ecosystem to something very similar to the early stages of succession. Apart from the reduction in above-ground producer biomass, many of the features of the plants themselves, such as the position of the perenating bud (Table 24.2), the presence of protective spines, and the prevalence of annuals or plants with vegetative means of reproduction, equip them to withstand grazing. The last-named adaptation is associated often with below-ground food storage organs such as rhizomes endowing the plant with the capacity for rapid growth and recovery. Indeed, the below-ground production of pasture grassland is often several times higher than the above-ground production which is available to herbivores – for example, there is twice as much annual production below ground in the Missouri prairies (Kucera *et al.* 1967). Most, if not all, of the adaptations that prove to be advantageous in the face of grazing are also those that would be important in the relatively extreme environments experienced by pioneer plant species in the early stages of natural succession.

There are parallel changes in soil conditions accompanying grazing. Exposure under a thin grass cover may lead to a greatly changed soil

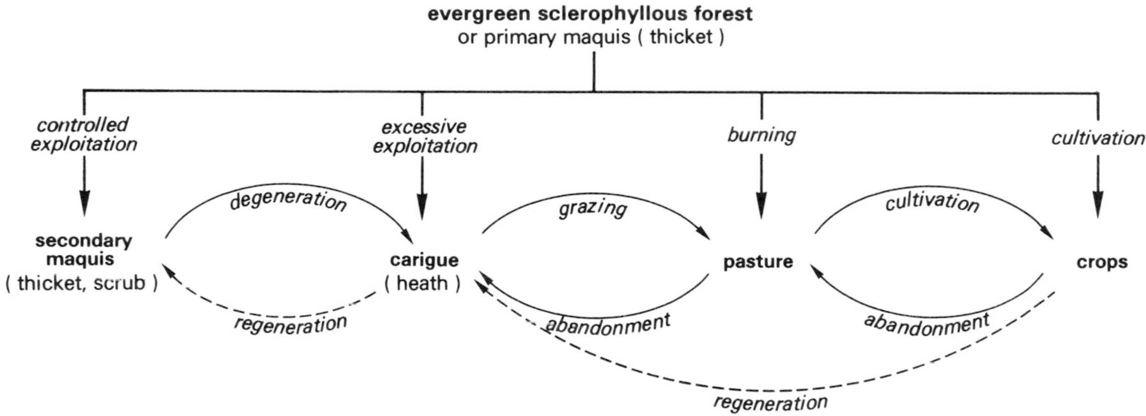

Figure 24.13 Stages in the degeneration and regeneration of the plant communities of the Mediterranean. The long history of modification by man from at least Classical times to the present has replaced almost all of the natural evergreen sclerophyllous forests by secondary communities with associated soil erosion and environmental deterioration. (After Polunin & Huxley 1967.)

Table 24.2 Life forms found in chalk grassland, acid grassland and neutral grassland.

Life form	Position of overwintering bud	Chalk grass-land (%)	Acid grass-land (%)	Neutral grass-land (%)
chamaephytes	soil surface to 25 cm	7.6	20.8	7.2
hemicryptophytes	at soil surface	67.6	54.6	70.3
geophytes	below soil surface	16.5	5.6	13.5
theraphytes	as seeds	8.2	18.8	8.6

microclimate resulting in greater soil aridity. Alternatively, under a wetter climate reduced interception and transpiration can lead to localised waterlogging. This tendency will be promoted further where burning is practised, for the induration of the soil surface that results will increase the runoff component by changing the infiltration capacity regulator. The area occupied by gley and peaty gley soils in depressions and along drainage lines in upland Britain has probably been extended in this way. Even in lowland Britain under a drier climate the same process can be invoked to explain partially the valley mires of the New Forest (Fig. 24.14). Perhaps the most important soil modification, however, is the drain of grazing on its inherent fertility. The removal of most of the above-ground primary production, and hence the nutrient elements that it contains, as a crop first to domesticated herbivores and eventually to Man, represents a loss to the detrital and soil systems. Although the deposition of dung and urine can to some extent offset this loss, it can lead to alterations in the carbon : nitrogen ratio and may lead to the breakdown of effective nutrient cycles. Furthermore, the dense adventitious root system of the pasture plants is shallower than that of the woodland or forest it may have replaced and it cannot draw on the nutrient reserves of the subsoil. Again the result is an alteration in the pattern of nutrient cycling in the soil.

Animals are affected also by changes in the habitat associated with extensive grazing. For example, with the creation of bare areas, perhaps by localised overgrazing or by excessive trampling, the habitat becomes unsaturated, and available niches appear which can be exploited by burrowing animals such as the rabbit in Britain or the gopher in the American Midwest. Carnivores also are affected partly by changes in the habitat, but by hunting and trapping as well, as Man tries to control the loss of his domestic herbivores to predators (and, in the case of Britain particularly, of his wild but partially managed game species, e.g. pheasant, grouse, deer). The wolf and the golden eagle in Britain, the lion and leopards in Africa, the jaguar, puma and coyote in North America, and a marsupial 'wolf' in Tasmania are all examples of predators that have been exterminated or reduced drastically in number (except perhaps in national parks). Natural herbivores such as many species of deer, the bison and some of the larger species of kangaroo have also been culled to reduce competition with domestic herbivores or to reduce damage to arable crops.

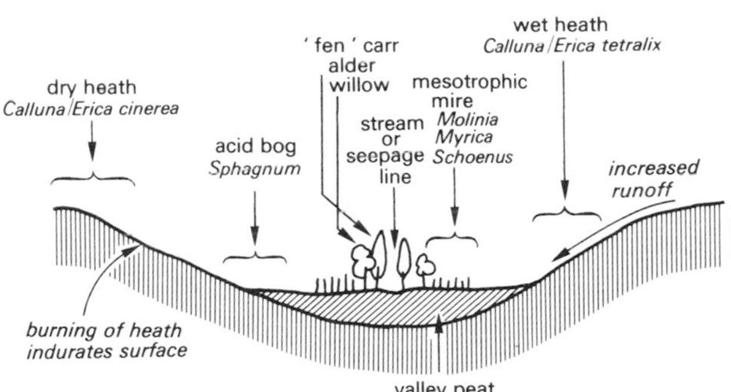

Figure 24.14 A New Forest valley 'bog'. Strictly this community is a soligenous mire maintained by impeded downvalley seepage of water. Its nutrition is minerotrophic (soil water) and of moderate base status. The removal of forest from the interfluves and the maintenance of heath communities under a grazing and burning regime has led to intense podsolisation, induration of the soil surface, reduced infiltration and accelerated runoff. As a result marginal acid bog communities have developed, characterised by ombrotrophic (rain water) nutrition, low base status and low pH.

Ironically it is now realised that, because they are adapted to utilise a larger proportion of the herbage, many of these natural herbivores have a higher productivity than domesticated herbivores, especially where the latter have been introduced.

Apart from the alterations to natural ecosystems that result from grazing, it is important to realise that the kind of pastoral grazing system which we have been discussing is also inherently inefficient. The animals are too selective in their intake, too much energy is dissipated in their metabolism and locomotion, and lost as faeces, and as a result the conversion of plant production to animal protein is too wasteful. However, the need for some animal protein in Man's diet has led to the development of more efficient 'pastoral' systems which are best considered under the heading of intensive agricultural systems.

If extensive grazing management leads to the modification of natural ecosystems, then intensive grazing, including store-fed and factory livestock systems, and arable agriculture lead to the establishment of almost entirely artificial systems. In fact the modification of the landscape to suit agricultural practice can involve not only the formation of artificial ecosystems of great simplicity, but also profound alteration of the land surface itself. Even in the early days of agriculture, quite prodigious feats of civil engineering were carried out using human energy aided by simple tools. Many primitive agricultural civilisations adopted hillside terracing for cultivation or irrigation of steep hillslopes. The conversion of even a slope of modest gradient (15°) to terraces 2 m in width is illustrated in Figure 24.15. For each terrace is 0.134 m^3 per metre width of slope. For a completely terraced hillside this represents a transfer of soil of 67 000 m^3 km^{-2}. The widespread extent of such terracing in hilly country in such regions as the Himalayas, Andes and Japan is an indication of the total volume of material transferred, volumes which are even greater on steeper slopes. The extensive ridge and furrow agricultural systems characteristic of the English Midlands involve modifications of similar magnitude. A wavelength crest to crest of 9 m and an average amplitude of 0.5 m requires a transfer of earth of 62 500 m^3 km^{-2}. In north Buckinghamshire alone some 35 km^2 of ridge and furrow forms exist, representing the uplift of well over 3.5 million tons of earth by an average of half a metre.

The modifications to the operation of environmental systems that result from bringing land into arable cultivation are many, but a major characteristic of arable land is that the biomass present in cereals, root crops or grass is considerably less than in natural ecosystems. Accordingly, there will be a lower rate of evapotranspiration and greater runoff. Often this is facilitated by the creation of artificial drainage ditches, which in effect extend the natural drainage network. Thus drainage density, which in a humid temperate environment such as Britain is normally in the range 1.5 to 3.5 km km^{-2}, may be raised to between 5 and 10 km km^{-2}. Drainage may be effected further by tile drains beneath the soil, which are spaced at 1–3 m depending on the soil texture. These drains may raise the drainage density further at the times when water is flowing in them to between 100 and 350 km km^{-2}. The effect of this increasingly efficient evacuation of water from the soil surface is to transfer it more rapidly to the main river channels and thus to increase peak flows.

The conversion of forest to arable land also destroys the forest litter and lays bare the surface of the soil. This is particularly the case when the land is fallow or before the crop has matured fully. Even then, with the exception of pasture, most crops still leave a proportion of soil directly exposed, and rainsplash erosion, rill action and overland flow can detach soil particles readily and carry them towards the river channel, increasing sediment concentration and rates of erosion. Thus Evans and Morgan (1974) observed soil erosion under immature crops in an area of Cambridgeshire of 3.3 t ha^{-1} representing a surface lowering of 0.25 mm in a few days. Douglas (1967) quotes an increase in sediment yield in Java from 900 m^3 km^{-2} yr^{-1} in 1911 up to 1900 m^{-3} km^{-2} yr^{-1} in 1934 as a result of increasing deforestation and an extension of cultivation. Arable agriculture, therefore, can lead to increased total runoff, increased peak flows and increased sediment yield.

The loss of organic matter from the agricultural soil will promote this soil loss by erosion further as the structure of the soil is lost. This loss of structure results from the reduction of aggregating forces within soil crumbs as organic matter is lost. No longer water stable, soil aggregates disintegrate on raindrop impact and the now-discrete soil particles accumulate in the surface pores forming a soil crust, reducing the infiltration capacity and initiating overland flow (see Fig. 12.4c). The breakdown of organomineral complexes and loss of structure

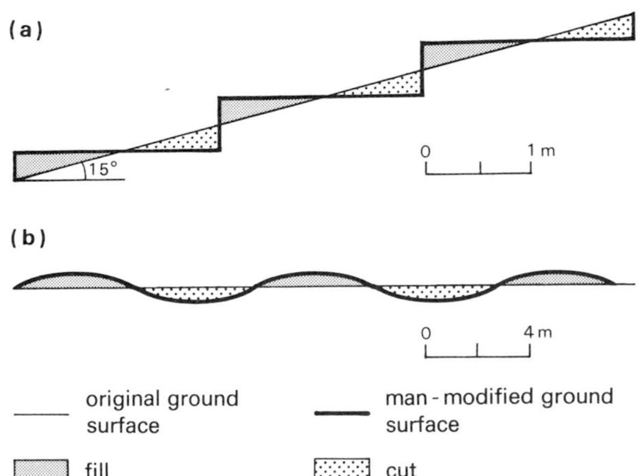

Figure 24.15 Man's modification of the ground surface by (a) hillslope terracing and (b) ridge & furrow. (c) An entire landscape artificially modified by terracing (Nepal).

(c)

reduce the nutrient and water retention capacity of the soil and, together with the progressive drain on soil nutrient reserves that cropping represents, lead inevitably to reduced fertility. The addition of manure, chemical fertilisers and perhaps, in periods of summer water deficit, irrigation water then becomes a necessity for sustained crop yields. However, low soil-storage capacity means that leaching losses of fertiliser and liquid manure can be high and, when they reach drainage channels, this can result in pollution and eutrophication problems in the aquatic ecosystems of streams and lakes. Soil drainage and soil aeration also can be affected adversely by agricultural practice, particularly where heavy machinery is used consistently and produces soil compaction. Repeated ploughing to the same depth has a similar effect with the development of a plough pan at the depth of the furrow, leading to impeded drainage (Fig. 24.16).

From the ecological point of view, the single most important effect of cultivation is the destruction of habitats. This reduction in habitat and species diversity in the rural landscape increases alongside the trend for greater mechanisation, reduced labour costs and greater intensification of production. It is well illustrated by the removal of hedges in parts of rural England (Fig. 24.17). Between 1950 and 1970, 4000 to 7000 miles of hedge were lost in Britain, representing between 3000 and 5000 acres of habitat. Ecologically, the hedge and its associated timber simulate the woodland edge or clearing habitat both in structure and species composition. Hedgerow removal, therefore, can mean a reduction in the species diversity of an average rural 10-km grid square in Britain of about 30%, affecting not only the plants but the insects, birds, and small mammals dependent on them (Hooper 1970).

Again in Britain, the use of mechanical excavators has accelerated the trend towards the drainage of potential agricultural land, as well as improving

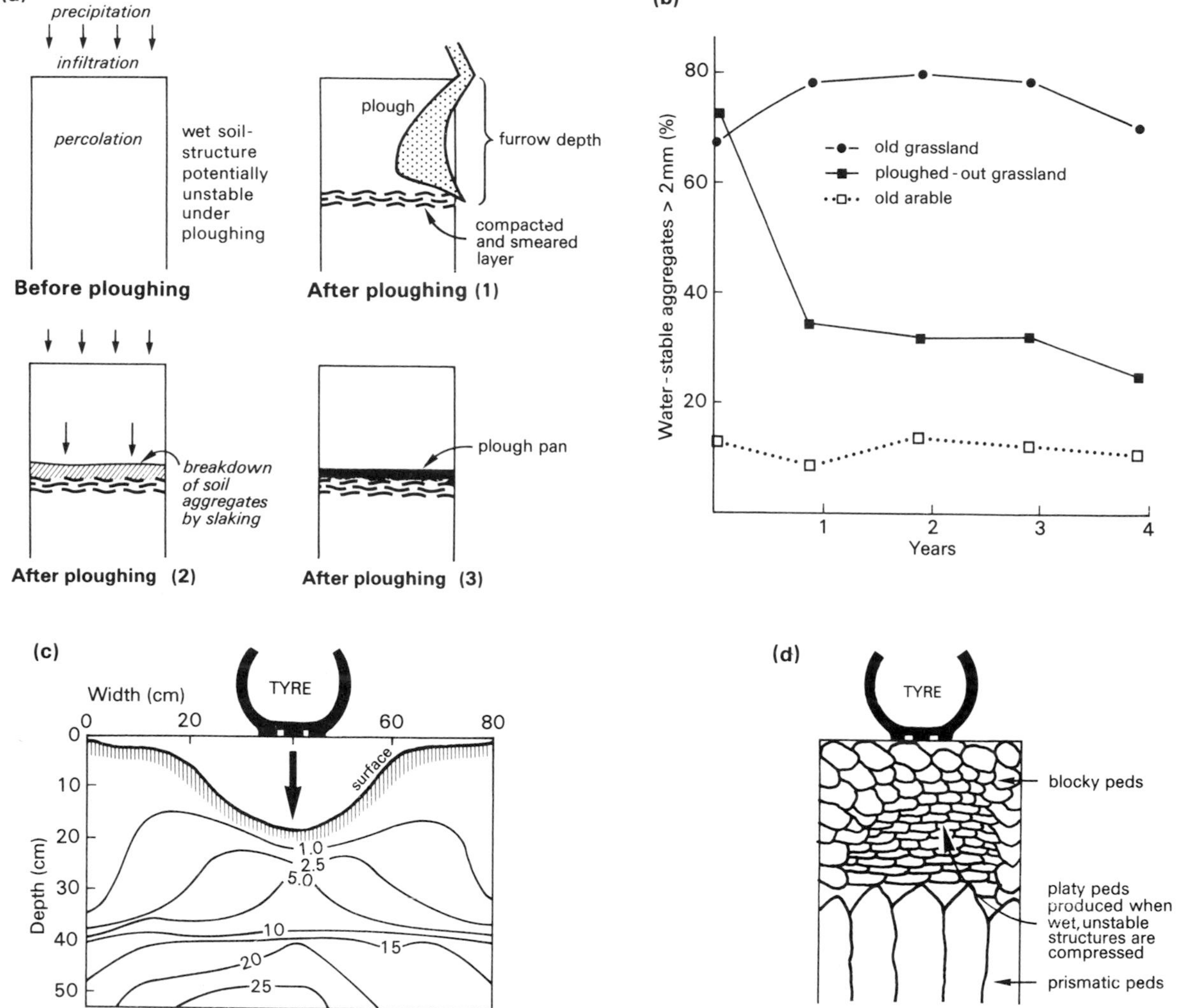

Figure 24.16 (a) The formation of a plough pan. (b) Changes in the degree of soil aggregation under different land uses (after Low 1972). (c) Distribution of cone resistance (bar) after passage of a test tyre; contours show compression of the soil under tyre track and sideways displacement of surface soil (after Soane 1973). (d) Alteration of soil structure by the passage of a tractor wheel.

Figure 24.17 The reduction in the estimated mileage, and change in the pattern of hedges, in part of Huntingdonshire 1364–1965 (after Moore *et al.* 1967).

the drainage of that already under cultivation. The introduction of herbicides, providing a cost-effective method of controlling ditch bank vegetation and aquatic weeds, has helped to reduce labour costs. Even so, mechanised farming with increased field sizes, requiring the removal of hedges and the filling in of ditches, has removed drainage water to mole and tile drain systems below the surface. Finally, the misuse of herbicides and pesticides, some of which persist in the habitat, has further contributed to an impoverishment of the wild flora and fauna of agricultural landscapes.

24.3 Urbanisation and industrialisation

The agglomeration of industry and housing in large urban complexes creates local modifications to the Earth's surface and the atmosphere above it which may extend beyond the immediate urban area. The main characteristics of these urban climates are outlined in Table 24.3. The reason for these effects lies in fundamental modification to systems regulators. The amount of suspended particles in the urban atmosphere is much higher than in the surrounding suburban and rural areas (Table 24.4). In the solar energy cascade this directly affects atmospheric transparency, reducing the amount of direct radiation reaching the ground surface. Much is lost by reflection, scattering and absorption in the urban atmosphere.

Table 24.3 The effect of an urban area on local climates (after Landsberg 1960).

Element	Comparison with rural environs
RADIATION	
total on horizontal surface	15–20% less
ultraviolet, winter	30% less
ultraviolet, summer	5% less
CLOUDINESS	
clouds	5–10% more
fog, winter	100% more
fog, summer	30% more
PRECIPITATION	
amounts	5–10% more
days with 5 mm	10% more
TEMPERATURE	
annual mean	0.5–0.8°C more
winter minima	1.0–1.5°C more
RELATIVE HUMIDITY	
annual mean	6% less
winter	2% less
summer	8% less
WIND SPEED	
annual mean	20–30% less
extreme gusts	10–20% less
calms	5–20% more

Table 24.4 Contamination of the urban atmosphere compared to that over rural areas.

Contaminants	
dust particles	10 times more
sulphur dioxide	5 times more
carbon dioxide	10 times more
carbon monoxide	25 times more

Another regulator of processes in the atmosphere is the number of hygroscopic nuclei in the air affecting directly the amounts of cloud and possibly also of precipitation. The most notable form of condensation in urban areas has been the urban fog for which London was well known until the 1956 Clean Air Act gradually brought a reduction in air pollution concentrations. During the London smog of December 1952, the atmosphere contained approximately 276 gm km^{-3} of smoke and sulphur dioxide and 124200 gm km^{-3} of condensed water (Fig. 24.18). Figure 24.19 shows the smoke concentrations in central London in 1958. Los Angeles (California, USA) still has a major urban fog problem, but this is due largely to the photochemical effects of solar radiation upon the gases emitted from vehicle exhausts and not to the action of hygroscopic nuclei.

The effects of increased numbers of condensation nuclei in Rochdale (Lancashire, England) were noted by Ashworth (1929), who suggested that lower average rainfall on Sundays was a direct consequence of cleaner air while local mills were closed. More recently, Atkinson (1975) and others have suggested that free and forced convection over the aerodynamically rough and relatively

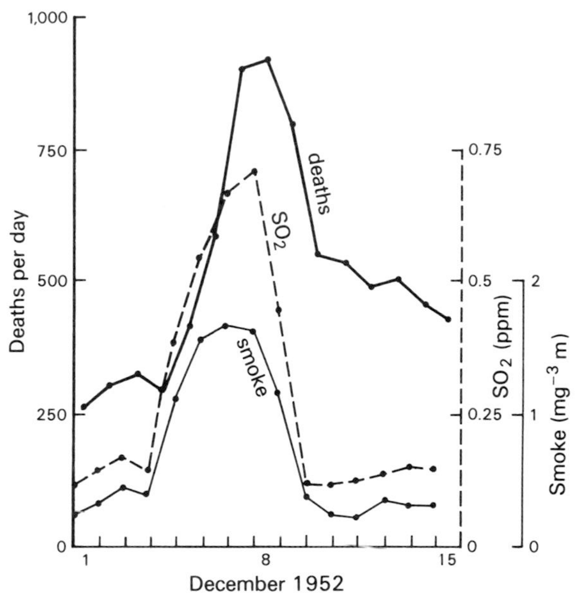

Figure 24.18 Deaths and pollution levels for smoke and sulphur dioxide during the London fog of December 1952 (after the Royal College of Physicians 1970).

Figure 24.19 Average distribution of smoke concentrations in London, April 1957 – March 1958 in mg 100 m^{-3} (after Chandler 1965).

Table 24.5 Radiative properties of typical urban materials and areas (from Oke 1978, after Threlkeld 1962, Sellers 1965, van Straaten 1967, Oke 1974).

Surface	α Albedo	ε Emissivity	Surface	α Albedo	ε Emissivity
roads			windows		
asphalt	0.05–0.20	0.95	clear glass		
			zenith angle less than 40°	0.08	0.87–0.94
walls			zenith angle 40–80°	0.09–0.52	0.87–0.92
concrete	0.10–0.35	0.71–0.90			
brick	0.20–0.40	0.90–0.92			
stone	0.20–0.35	0.85–0.95	paints		
wood		0.90	white, whitewash	0.50–0.90	0.85–0.95
			red, brown, green	0.20–0.35	0.85–0.95
roofs			black	0.02–0.15	0.90–0.98
tar and gravel	0.08–0.18	0.92			
tile	0.10–0.35	0.90			
slate	0.10	0.90	urban areas*		
thatch	0.15–0.20		range	0.10–0.27	0.85–0.95
corrugated iron	0.10–0.16	0.13–0.28	average	0.15	?

* Based on mid-latitude cities in snow-free conditions.

warmer urban surface may be instrumental in producing higher urban rainfall totals.

The surfaces of urban areas have physical properties that contrast sharply with their rural surroundings. Albedo may be slightly higher due to the large amounts of highly reflective concrete and glass used in construction (Table 24.5). More important, however, is the fact that the urban surface has a highly variable morphology and that amongst the buildings there are multiple reflections from their vertical faces and shading of the streets below, which results in complex radiation balances.

The relatively high thermal capacity of the fabric of older urban structures and the large amounts of heat lost from heating systems operating within buildings combine to create an urban '**heat island**' effect. In most urban areas, air temperatures recorded near their centres of activity, where building density and height are usually greatest, are frequently higher than in the surrounding suburban areas. This temperature anomaly is well developed during the late evening when solar heat stored in the urban fabric during the day is being released and heat is still being supplied to the urban atmosphere from heating systems. Indeed, as Table 24.5 shows, the latter may constitute a major proportion of heat inputs into the urban environment. Chandler (1965) has shown that, under calm anticyclonic weather conditions, London develops a well defined heat island (Fig. 24.20) in the centre of which air temperatures at street level are more than 6 C° higher than in surrounding areas of lower building density and stature.

Therefore, in constructing large urban agglomerations, Man has modified the character of both the Earth's surface and the atmosphere above it. The distinctive local climates that are generated are a product of consequential changes to regulation within the Earth–atmosphere system. The alteration of the Earth's surface mentioned above represents the direct reshaping of landform by urbanisation, if not the creation of entirely artificial landforms. Constructions associated with prehistoric settlement sites can involve substantial accumulation of material. Even using primitive techniques, man-made landforms of the magnitude of the pyramids of Egypt and Silbury Hill (Wiltshire) were constructed. The latter, some 10 km west of Marlborough, is 40 m high and covers a basal area of 2.1 ha. It is estimated that 350 000 m^3 of material were raised during its construction. Even the more modest artefacts such as hill forts required the transfer of substantial volumes of earth in the creation of their rampart and ditch systems. It is estimated that the surface of central London has been raised by the accumulation of construction materials and waste by an average of 3.5 m. If we assume that London is 2000 years old, then made ground has accumulated at an average rate of 1750 $m^3\,km^{-2}\,a^{-1}$.

In more recent times, the development of extensive transport systems and the expansion of

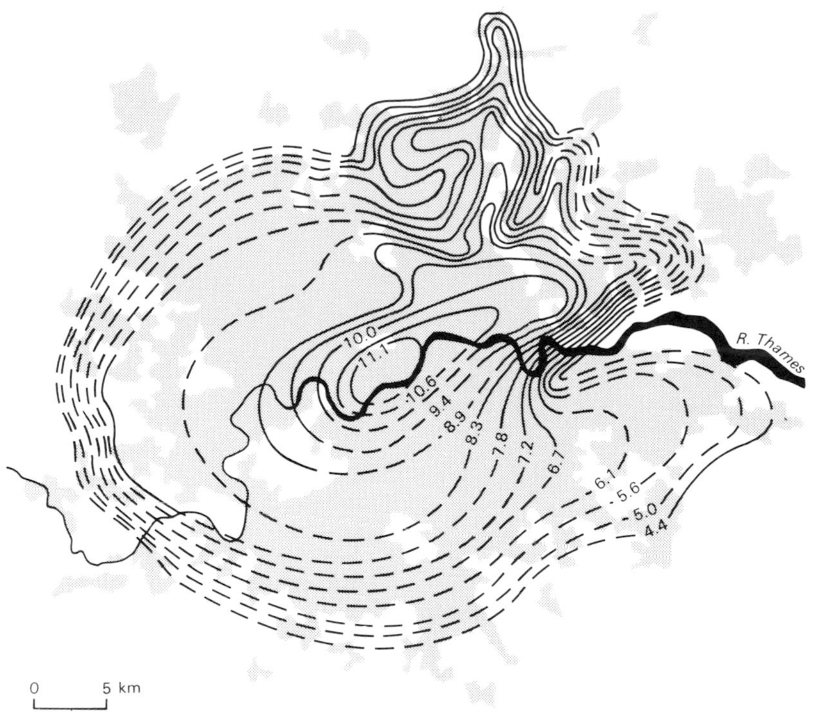

Figure 24.20 Distribution of minimum air temperature (°C) in London, 14 May 1959 (after Chandler 1965).

Figure 24.21 A manmade landform: spoil from the extraction of china clay. Note the extensive modification by sub-aerial denudation resulting in gullying (Dartmoor, England).

urban areas associated with advancing technology have led to large-scale exercises in land levelling. In this way deep cuttings are made through areas of positive relief and the material removed is used to form embankments across low-lying areas. Railway and motorway construction, requiring the maintenance of modest gradients, are examples of this kind of modification. The Panama Canal can be considered as one of the more spectacular modern examples of such engineering.

Restricted urban sites can be extended by building land out into the sea. Material can be acquired for this purpose either by quarrying from higher ground or by dredging from river or nearshore channels. Some 11% of the land area of the county borough of Belfast was man-made and Hong Kong Airport is constructed entirely on reclaimed land. Spectacular extension of land surface area can be achieved also by building dykes in order to extend the coastline seawards, as in the classic case of the Dutch Polders.

Extractive industries also modify land surface form. Opencast working of building stone, coal, sand & gravel, and raw materials such as lime and iron ore produces pits which scar the landscape. Many of these activities are associated with adjacent accumulations of spoil, the unwanted by-product. In prehistoric times the extensive cave system of Grimes Graves was created by flint mining. In Mediaeval times the extraction of chalk for agricultural purposes in East Anglia created 30 000 pits and ponds and the extraction of peat led to the formation of the Norfolk Broads. The magnitude of contemporary extractive processes in Britain is illustrated by the facts that some 8 km^2 of land is excavated annually for sand and gravel, 4 km^2 for chalk and limestone and 1.8 km^2 for clay, to provide for the needs of the brick industry. In addition to the pits directly produced by the extraction process, the accumulation of adjacent spoil heaps can produce landforms of considerable magnitude, as illustrated by the china clay mining landscape on the southwestern fringe of Dartmoor and around St Austell in Cornwall (Fig. 24.21). Underground extraction of salt has led to surface subsidence and the formation of water-filled hollows – the 'flashes' of Cheshire. A salutary reminder of the magnitude of extractive processes is the quarrying of limestone in the Mendip Hills of Somerset. Some 10×10^6 t a^{-1} are currently

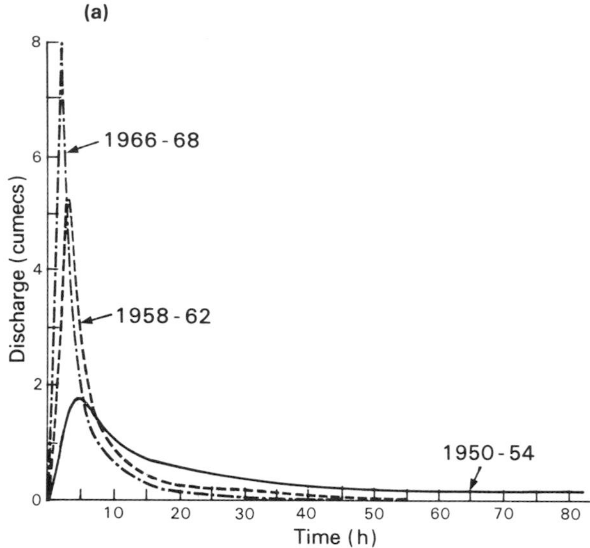

Figure 24.22 (a) Mean unit hydrographs for three stages in the urbanisation of the Canon's Brook (after Hollis 1974). (b) River channel artificially straightened and confined in training walls to prevent lateral channel erosion and diminish overbank flooding (Afon Wyre, Llanrhystud, Dyfed, Wales).

being removed, representing a loss of 800 m^3 km^{-2} a^{-1} over the limestone outcrop as a whole. Losses of rock by erosional processes are in the range of 50–100 m^3 km^{-2} a^{-1}, indicating that the extractive industry is eroding the landscape at a rate eight to sixteen times greater than natural erosion (Smith & Newson 1974).

These are mostly highly localised examples of human activity in directly creating landforms and they are of limited area and volume in relation to the land surface as a whole. The greatest contrasts in artificially created relief are usually the juxtaposition of a pit and a spoil heap. Of more lasting significance, and affecting a much broader surface area, is the effect of Man on the dynamics of surface processes by altering the nature of the land surface. In this way hydrological and denudation processes may be changed permanently and at the scale of the drainage basin.

The process of urban development usually involves the encroachment of the built area onto former agricultural land. Initially, there is a period of construction, during which there is much mechanical disturbance of the ground surface as drains, roads and foundations are prepared. Ultimately, the hydrological surface of the urban area becomes increasingly impervious, consisting of a high proportion of paved surfaces – roads, footpaths and roofs – which drain via a system of gutters and storm sewers into pre-existing natural drainage channels. Such paved surfaces (with the exception of leaking roofs) have an infiltration capacity of virtually zero and therefore zero storage. A high proportion of precipitation leaves as direct runoff via the storm drainage system, and storage and evaporation are limited to the soil and vegetation surfaces of parks and gardens.

Under urban conditions, therefore, the water balance will be modified again, leading to increased surface runoff and more rapid response of runoff to storm inputs. These effects have been confirmed by Gregory (1974) from observation of a small catchment (0.26 km^2) on the urban fringe of Exeter. In four years (1968–72) urbanisation extended to cover 12.2% of the catchment area. It was found that total runoff increased between two- and three-fold and that peak discharge increased in the same proportion, while the lag time of floods decrease from an average 70–80 minutes to 35 minutes. Similar results were recorded by Hollis (1974) from the growth of Harlow New Town in Essex to cover 21.4% of the catchment of Canon's Brook. Over an 18-year period the mean hydrograph progressively increased in discharge, while time to rise decreased. The catchment thus became increasingly flashy in character (Fig. 24.22).

A further effect of urbanisation may be the addition of imported water. Water pumped from groundwater storage, or transferred to the urban area from an adjacent catchment, may be discharged as effluent into urban rivers. This in effect increases the natural catchment area or leads to a further increase in total discharge in the urban river channel.

Important effects on water quality may also accompany urban development. During the construction phase the disturbance of the soil surface by machinery may contribute sediment to the river channel, and values of sediment concentration in excess of 3000 ppm have been recorded. Once construction has ceased and paved surfaces are extensive, the scope for sediment erosion is much reduced and sediment concentration in streams may be lower than in the initial natural forested state. Water quality may be affected also by industrial processes. Effluent from industry may be discharged into river channels increasing the concentration of dissolved solids.

The effects of this sequence of land use change have been summarised by Wolman (1967) who studied its effect on sediment yield (Fig. 24.23) in the eastern USA. Initially under forest, yields are estimated at 250 t km^{-2} a^{-1}. As forest is replaced by arable farming, this rises to 2000 t km^{-2} a^{-1}, declining as the urban fringe approaches and farmland is degraded to grazing and forest. The massive disturbance associated with construction may produce erosion rates of 250 000 t km^{-2} a^{-1} over small areas, falling to an estimated 125 t km^{-2} a^{-1} when urbanisation is complete.

Thus, profound changes in the water balance can be brought about as a result of Man's activities changing the land use and consequently the hydrological characteristics of catchment areas. These lead frequently to increased channel runoff, increased peak discharges and increased sediment and solute loads. In the case of urban development, sediment loads, in contrast, are normally reduced. Since the form of river channels is adjusted naturally to the output of runoff and sediment, it follows that river channel changes can be expected to accompany changes in land use and flow régime.

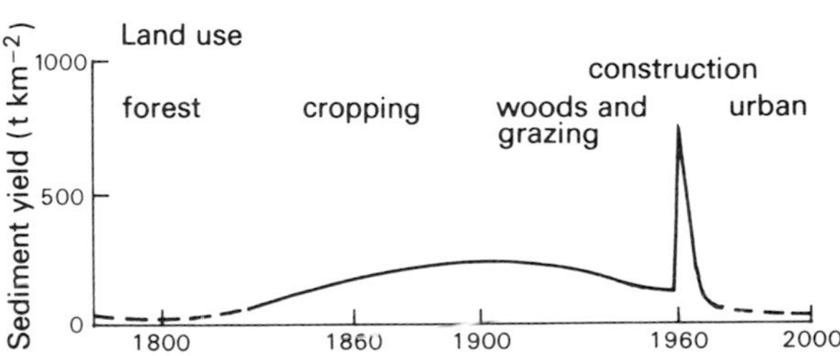

Figure 24.23 Land-use change and sediment yield in the Piedmont region of Maryland (after Wolman 1967).

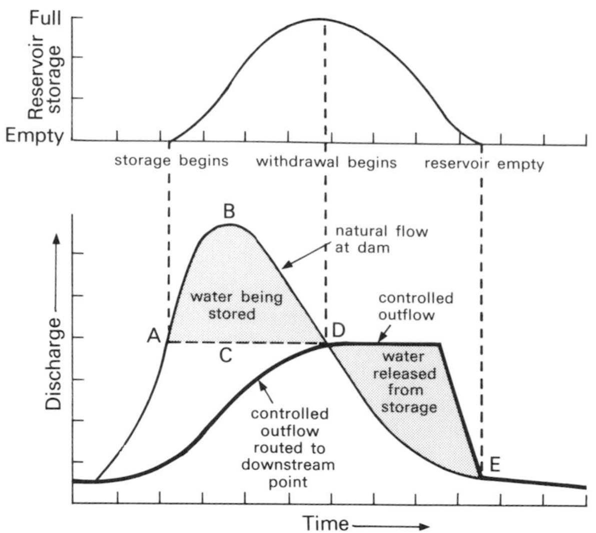

Figure 24.24 The use of a reservoir as a regulator in flood control (after Linsley *et al.* 1949).

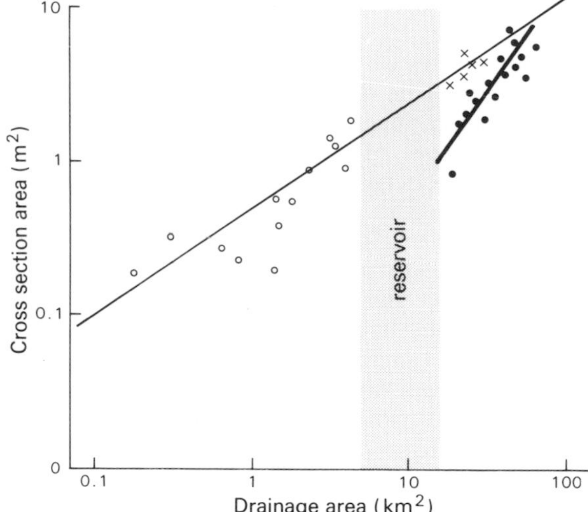

Figure 24.25 Relationship between channel capacity and drainage area above and below a reservoir, River Tone, Somerset, England (after Gregory & Park 1974).

Since channel form is regarded as a function of the hydrological characteristic of the catchment area, it is possible to establish a relationship between channel capacity and drainage area. Thus channel capacity is shown to be related closely to drainage area, increasing in direct proportion in the downstream direction as discharge increases. The precise nature of this relationship will, of course, vary from river to river, but the same general tendency holds true. It is then possible to use the relationship established for a particular river in order to predict expected channel capacity further down stream. If actual channel capacity is measured down stream, below a point encompassing a major change in land use within the catchment area, then the effect of the land use change can be assessed by comparing the predicted channel capacity against that measured. In this way Gregory and Park (1976a) show that stream channel capacity below the Angram reservoirs on the River Nidd in north-west Yorkshire has been reduced significantly (Fig. 24.25). On the basis of the channel capacity–drainage area relationship established above the reservoir, former bankfull capacity measured below the reservoir is shown to fit in well. The current channel capacity below the reservoir is reduced to 50–80% of its former value and this reduction persists down stream until that part of the catchment area controlled by the reservoir comprises an insignificant proportion of the total drainage area.

A similar approach can be adopted to demonstrate natural channel adjustments below an urban area. In the same region as the previous example, the River Swale shows an increase in channel capacity down stream of the urbanised area of Catterick. Within this urban channel capacities average 1.66 times that predicted from rural channels up stream. Down stream from the urbanised area channel capacity averages 2.62 times that expected.

On a smaller scale, Gregory and Park (1976b) demonstrate the dramatic effect of diverting runoff from a paved road surface via a storm sewer into a small natural channel in Devon. While not increasing the drainage area of 0.55 km^2, the paved road locally decreases infiltration and storage capacity and contributes high peakflow discharges of sediment-free water into the existing natural channel. This has led to rapid channel enlargement over a period of 29 years from an estimated original average channel cross section of 0.39 m^2 to a value of 2.07 m^2 over a length of 500 m of channel (see Fig. 24.30).

24.4 Control of environmental systems by Man: some examples

Most human activities and the consequent decision-making processes are sensitive to environment. Choice of home, place of work and recreation, in addition to the choice and success of agricultural enterprise, are affected to varying degrees by the environment. Kates (1970) has represented this relationship between Man and his environment as an interaction between a 'human-use system' and a 'natural-events system' (Fig. 24.26). An event in the natural-events system which directly disrupts Man's activities is commonly referred to as 'a natural hazard'.

The definition of a hazard involves the identification of critical magnitudes and frequencies of events. The snow and low temperatures which disrupted transport throughout Western Europe during the winters 1962–3 and 1978–9 clearly constituted a hazard in these areas. However, if we examine the climatic data for northern Canada or Siberia, we would expect low air temperatures and attendant snowfall during most winters, but considerably less disruption of normal patterns of human activity. In considering the interaction between Kates' human-use and natural-events systems therefore, we must consider expectation of events in the latter and human adjustment to them. In the preceding example the expectation of severe cold in Western Europe is relatively low, yet its impact in terms of disruption of human activity is great.

If we consider the alternatives in Figure 24.26, it

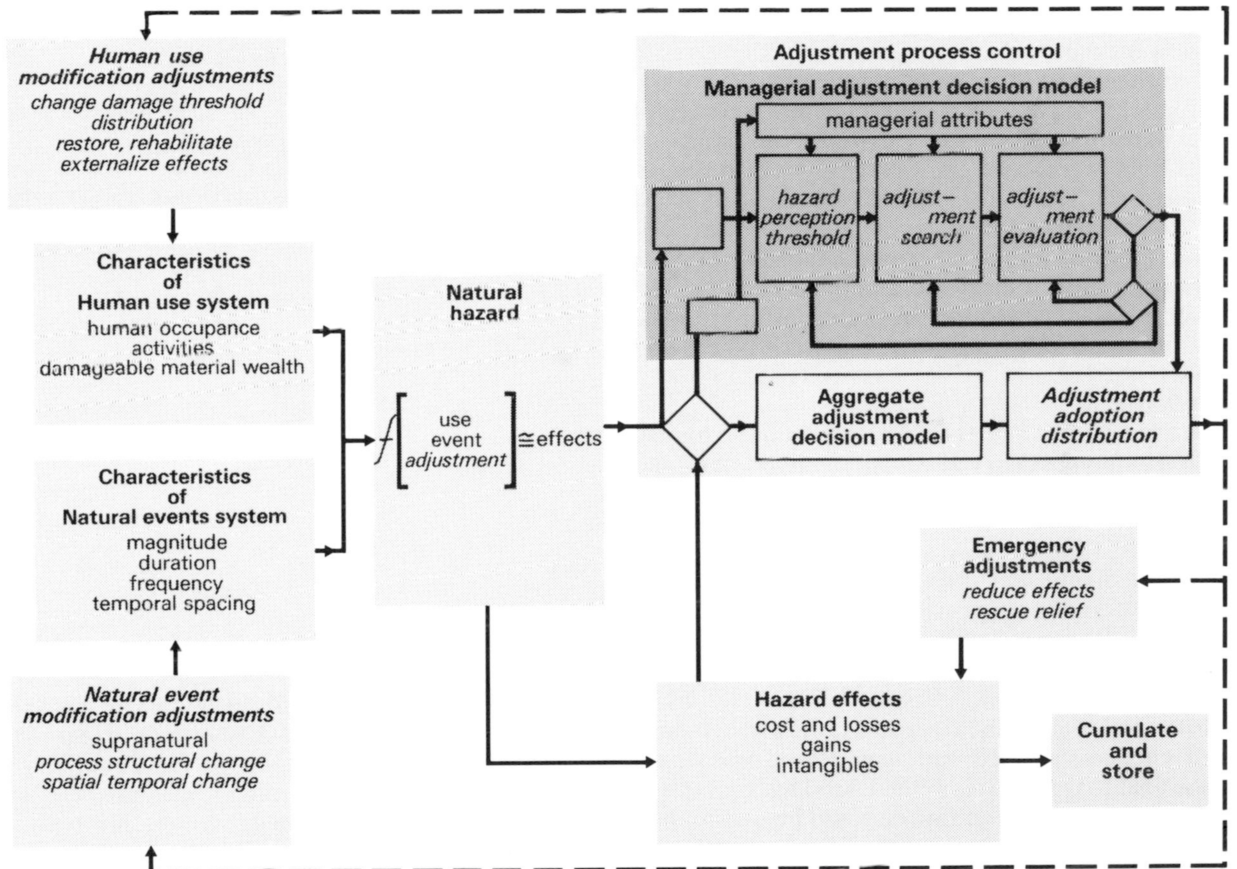

Figure 24.26 Natural hazards represented as an interaction between a human-use system and a natural-events system (after Kates 1970).

appears that there is a range of human adjustments to events such as snowfall, floods, drought or gale. In the short term, the effects of the event may be tolerated as a short-term inconvenience, perhaps due to the infrequency of their occurrence or their minimal long-term impact upon human activity. However, should there be a fund of experience of the event, then human adjustment is likely to take a relatively well organised form. The impact of the event is evaluated and an adjustment policy is formulated. In the case of, for example, urban snowfall this may incorporate a range of adjustments from greater investment in snow-moving machinery to long-term modifications to the design of buildings and modes of transport. The outcome of decision making inevitably results in the adoption of one of two alternatives – either to modify the 'human-use system' to reduce the impact of the hazardous event, or to modify the 'natural-events system'.

If we consider agriculture, the recurrence of, for example, damaging frosts and droughts may lead to changes in the types of crops grown. To some extent this may be avoided by the development of less sensitive hybrid crop varieties, an approach which is being adopted in many Third World countries. The alternative is to seek to exert a degree of control over Earth-surface and atmospheric systems to create a more favourable environment for crop and animal husbandry. The basic principles adopted in such control are illustrated in the following examples.

Flooding

For many reasons the concentration of human activity along river valleys and on floodplains has inevitably increased their susceptibility to flood damage. During the first six months of 1978, for example, floods caused widespread loss of human life and livelihood throughout the world (Table 24.6). A number of possible human adjustments to this flood hazard are listed in Table 24.7, one of which is to 'modify the flood'.

The essence of flood control is to reduce the peak discharge of the river to a level where it does not exceed the maximum capacity of its channel (see Ch. 13). In the simplified flood hydrograph in Figure 24.27 the river banks would be overtopped between X and Y. A more desirable form shows a delayed rise, a protracted recession and, most important of all, a reduced peak discharge. If we examine the hydrographs for the River Derwent in Derbyshire (Fig. 24.28), for example, we can see that the presence of the Ladybower Reservoir clearly introduces a damping of the flood peak. The construction of reservoirs, which increase the storage capacity of a drainage system, thus results in a degree of control over river discharge. In the case of the River Severn, adjustments to frequent flooding have included both the development of a flood-warning scheme and the employment

Table 24.6 Major flood events over the first six months of 1978.

January	9–20	Severe floods in southern Iran; 142 villages affected, roads washed away; 20 000 families homeless; ten dead.
	13–16	Rain and flash floods in Brazil; 1400 homeless; 26 dead.
February	10–13	Heavy rain, up to 300 mm in Uruguay causes floods.
	21–26	Widespread flood in south-west England; roads and homes flooded; one death.
	24–25	Severe rainstorm causes serious flooding to 2000 km² in Western Australia; livestock losses; roads cut.
March	1–4	Floods in Tijuana City, Mexico, leave 15 dead; 20 000 homeless.
	15	Worst flood for 21 years in Brazil; 22 000 homeless.
	15–24	Northern Argentina floods leave 11 000 homeless; extensive crop damage; eight dead.
	20	Floods, the worst for 40 years, in Omaha and Nebraska, USA; 2000 homeless; one dead.
	21–31	Heavy rains and floods along the River Zambesi in Mozambique; 250 000 homeless; 45 dead; 56 000 hectares of agricultural land destroyed; damage estimated $70 million.
	24	Floods up to 2 m deep in Indiana, USA; damage estimated $10 million.
April	2	Worst floods for 23 years in Paris.
May	3	225 mm rain falls in 5 hours in New Orleans, USA; flood 1.5 m deep in the city; two dead.
	22–26	Heavy rain and serious floods in south-west Germany; roads blocked; damage estimated at 100 million D-marks.
	26–27	125 mm rain in West Texas; flash flood up to 4 m deep; homes, cars, campsites washed out; three dead.
June	26	Heavy rain in Japan; 259 mm rain in 24 hours; eight dead.

Table 24.7 Adjustments to the flood hazard (after Beyer 1974, adapted from Sewell 1964 and Sheaffer *et al.* 1970).

Modify the flood	Modify the damage susceptibility	Modify the loss burden	Do nothing
flood protection (channel phase) dykes flood walls channel improvement reservoirs river diversions watershed treatment (land phase) modification of cropping practices terracing gully control bank stabilisation forest fire control revegetation weather modification	land-use regulation and changes statutes zoning ordinances building codes urban renewal subdivision regulations government purchase of lands and property subsidised relocation floodproofing permanent closure of low-level windows and other openings waterproofing interiors mounting store counters on wheels installation of removable covers closing of sewer valves covering machinery with plastic structural change use impervious material for basements and walls seepage control sewer adjustment anchoring machinery underpinning buildings land elevation and fill	flood insurance tax write-offs disaster relief volunteer private activities government aid emergency measures removal of persons and property flood fighting rescheduling of operations	bear the loss

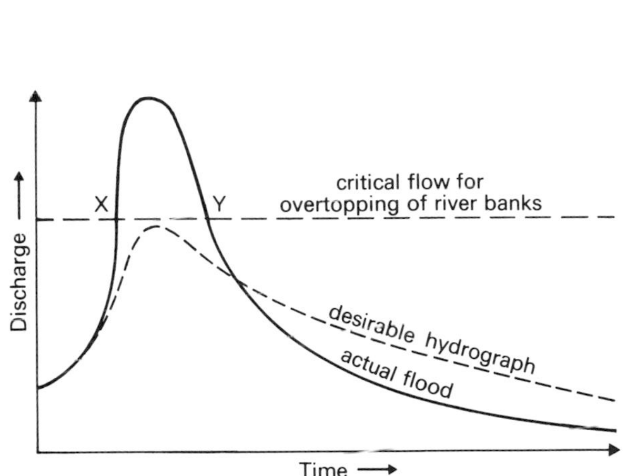

Figure 24.27 Hypothetical flood hydrograph and a desirable modified flood hydrograph.

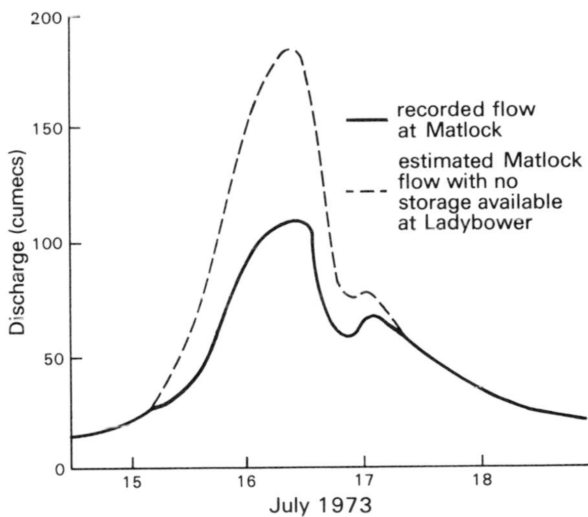

Figure 24.28 The effect of storage in the Ladybower reservoir on the flood hydrograph of the River Derwent at Matlock (after Richards & Wood 1977) (see Fig. 24.24).

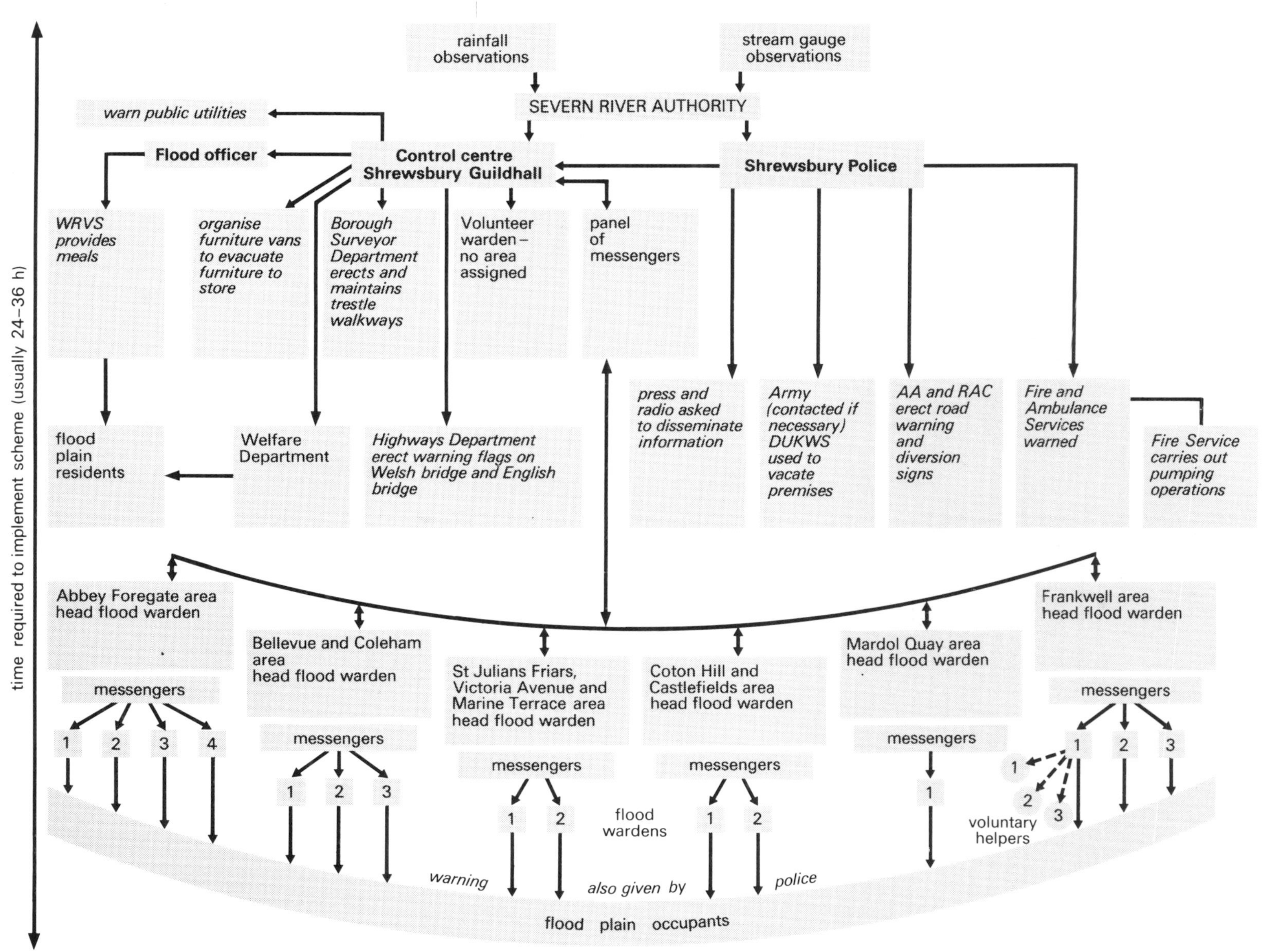

time required to implement scheme (usually 24–36 h)

rainfall observations

stream gauge observations

SEVERN RIVER AUTHORITY

warn public utilities

Flood officer

Control centre Shrewsbury Guildhall

Shrewsbury Police

WRVS provides meals

organise furniture vans to evacuate furniture to store

Borough Surveyor Department erects and maintains trestle walkways

Volunteer warden – no area assigned

panel of messengers

press and radio asked to disseminate information

Army (contacted if necessary) DUKWS used to vacate premises

AA and RAC erect road warning and diversion signs

Fire and Ambulance Services warned

Fire Service carries out pumping operations

flood plain residents

Welfare Department

Highways Department erect warning flags on Welsh bridge and English bridge

Abbey Foregate area head flood warden

messengers

1 2 3 4

Bellevue and Coleham area head flood warden

messengers

1 2 3

St Julians Friars, Victoria Avenue and Marine Terrace area head flood warden

messengers

1 2

Coton Hill and Castlefields area head flood warden

messengers

1 2

Mardol Quay area head flood warden

messengers

1

Frankwell area head flood warden

messengers

1 2 3

1 2 3

voluntary helpers

flood wardens

warning also given by police

flood plain occupants

Figure 24.29 The Shrewsbury (Shropshire, England) flood warning scheme (after Harding 1974).

of flood-control measures. In the flood-warning scheme, information on exceptional falls of rain and critical river levels is fed into a well organised network for its dissemination (Fig. 24.29). A fund of experience in Shrewsbury apparently has led to a well organised form of human response.

The effects of such reservoir and dam construction are to modify the pattern of channel discharge so that total discharge and peak discharge may be either increased or decreased by deliberate regulation. Deliberate acts of manipulation inevitably generate side effects. The increase in open water surface area in reservoirs results in greater loss by evaporation and hence discharge down stream from the reservoir is reduced in the long term. Furthermore, the control of discharge will precipitate changes in channel form as a response to Man-induced changes in the surface characteristics of the basin. We have seen already the response where natural channels adjust themselves freely to Man-induced changes in the surface of the contributing catchment area. However, artificial channel modification can take place also as a deliberate policy, as part of urban drainage or flood relief. In this way new channels may be cut or existing channels may be artificially enlarged and constrained by the construction of fixed artificial channels. Thus both channel cross section and planform are modified deliberately in order to permit high discharges and consequently reduce overbank flows and flooding. This may be accomplished in a number of ways. The roughness of the cross section may be decreased by clearing the banks of vegetation and other obstacles and by lining the channel with a hydraulically smooth surface such as concrete. The channel capacity may be increased by widening and deepening or by the construction of marginal embankments. Alternatively, the gradient can be increased, shortening the channel by straightening bends and cutting out meanders. Many river channels in and down stream of urban areas and in low-lying agricultural land may be modified in these ways (Fig. 24.30, see page 460).

Drought

Water shortages have serious implications for agriculture and, as in the case of floods, there are a range of human adjustments to it. Man engages in a number of control activities, principal amongst which are the construction of reservoirs to manipulate the seasonal discharge of drainage systems, which we have just considered, and the enhancement of precipitation through the seeding of clouds.

The initiation of precipitation from clouds was explored first between 1940 and 1950, and most attempts were based upon the Bergeron–Findersen ice-crystal process (see Ch. 4). In the USA in 1946, Schaefer discovered that, at temperatures less than −39°C, there was spontaneous freezing of supercooled water in the free atmosphere. If objects at temperatures less than −39°C were introduced to air in the laboratory, ice crystals were formed. In clouds these could initiate the process of ice crystal growth and precipitation. Altostratus clouds at −20°C were seeded at an elevation of 5000 m using dry ice (solid carbon dioxide), which resulted in

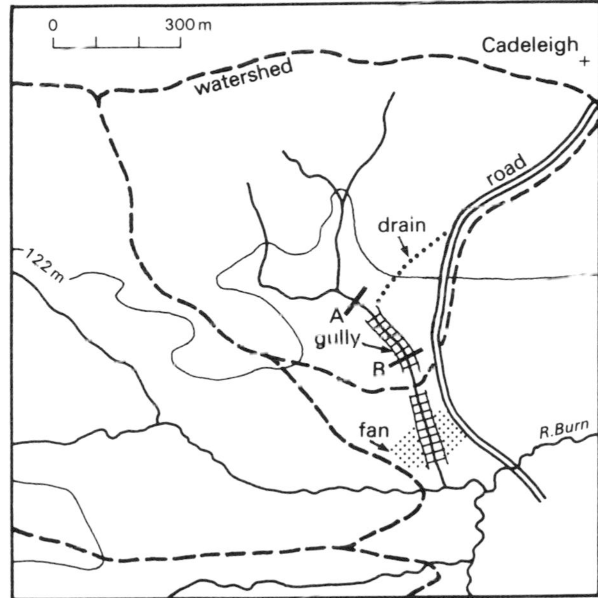

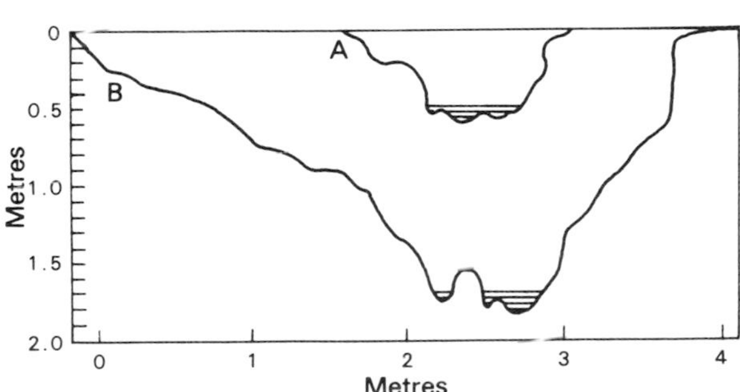

Figure 24.30 Sections in a gully in the valley of the River Burn, Devon, England (after Gregory & Park 1976b).

the precipitation of snow which re-evaporated before reaching the ground. Vonnegut, also in the USA, experimentally introduced silver iodide smoke to supercooled clouds and observed that this caused snowflakes to form. The silver iodide provided freezing nuclei upon which ice crystals formed within the cloud. Successful experiments were conducted in Australia in the 1950s in the seeding of selected cumulus clouds using silver iodide ejected from aircraft. Of clouds with temperatures below −5°C in their upper levels, 72% produced precipitation within 20–25 minutes of seeding and 21% evaporated. Similar seeding of randomly selected clouds was less successful. Recent research in the USSR into the seeding of cumulus clouds has suggested increases of up to 30% in rainfall over the Ukraine (Battan 1977).

The production of precipitation from cumulus clouds is regulated largely by cloud temperature, the number and rate of production of ice crystals and the number of water droplets. In seeding clouds, Man directly influences the second of these, but so far his impact has been relatively minor. The resulting amounts of precipitation are small and localised, due mainly to the physical limitations on the rates at which silver iodide may be added and to the constraints on the other regulators. The latter suggests that successful seeding, based upon the ice-crystal process, is limited at present to deep cumulus clouds. Five seeding trials carried out on Hurricane Debbie during August 1969 produced an apparent decrease in maximum wind speed of as much as 35% (Gentry 1970; Fig. 24.31). This represents a significant reduction in the kinetic energy of the storm. Although results as yet are inconclusive, any measure of control of such an energetic system represents a considerable achievement. However, the additional possibility of steering these storms by seeding has revealed a more sinister aspect of weather control – that of weather warfare. Such has been the degree of concern that international agreements have been reached on the limitation of this aspect of environmental warfare.

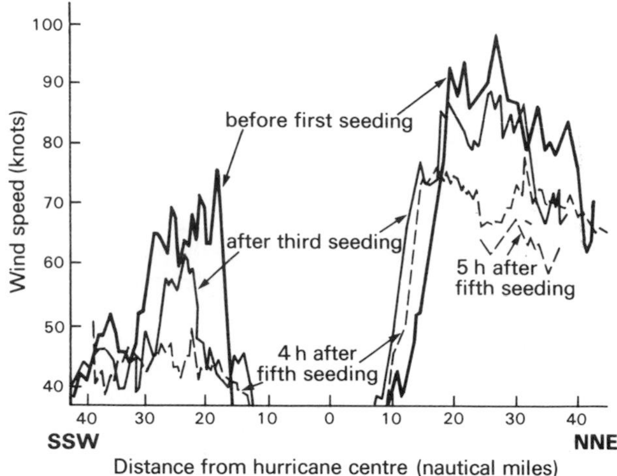

Figure 24.31 Windspeed changes with time at 36 000 m in hurricane 'Debbie' on 18th August 1969 (after Gentry 1970).

Frosts

The control of the effect of frost upon crops provides us with one of the best illustrations of ameliorative modifications to microclimate that Man can make. An air and ground surface temperature which falls below 0°C constitutes a frost hazard to crops. Severe frosts cause crop damage and consequent financial loss. There are two main types: advection and radiation frosts. The former results from an inflow of severely cold air; the latter is a result of intense nocturnal cooling of the ground surface under clear skies (see Ch. 4). Heat is transferred from the atmosphere to the cooling surface, resulting in a lowering of air temperatures. As the air is cooled from below, a temperature inversion develops, the air is stable and there is little turbulent mixing. In the absence of an inflow of heat, ground and air temperatures continue to decrease until mixing takes place by either a freshening wind or solar heating. Lowest temperatures are usually reached in the early morning.

Winter frosts are often a combination of both advection and radiation frost. For example, cold continental air (cP and cA) moving westwards across Europe during December 1962 brought extremely severe frosts. Radiation cooling of the ground surface under clear skies intensified the already low temperatures of the air above it.

Hogg (1970) has examined the frequency of frosts in Somerset, England (Fig. 24.32) and has shown that the probability of frost occurrence decreases rapidly through spring until there is statistically an extremely low risk of frost in late May and June. However, the damage caused by these late frosts is not a function of frequency but of timing. Taylor (1970), for example, has described an extremely late frost in 1957 which decimated the early potato crop in south-west Lancashire as late as mid-June.

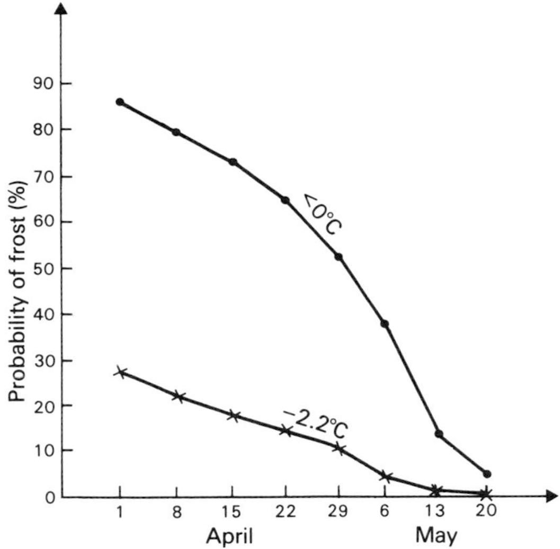

Figure 24.32 Variation in frost risk. Number of days in a hundred with frost between the stated dates and the end of the frost season.

Human response to this frost hazard follows a range of options, for example changing the agricultural system (i.e. the 'human-use system') or practising some form of frost control (the 'natural-events system'). As most late frosts (which cause the most damage) are due largely to radiation cooling, attempts at frost control are directed most particularly at modification of surface radiation balance and vertical heat exchange in the atmosphere. A number of frost-control measures are commonly practised, some of which are listed in Table 24.8. The use of cloches, greenhouses and artificial fogs modify atmospheric transparency to longwave terrestrial radiation. Glass and clear plastic or polythene permit solar radiation to pass through and warm the plant and ground surfaces but inhibit both radiational and convectional heat loss (Fig. 24.33a). Artificial fogs usually take the form of dense smoke, often from burning oil or tyres (Fig. 24.33b). Apart from being a severe pollution hazard, a uniformly dense smoke cover will produce, at best, a 50% reduction in terrestrial heat loss. As an alternative, if dense fogs comprising water droplets of diameter 15 to 17 μm could be produced, these could cause a 99% reduction in heat loss (de Boer 1965). For small

Table 24.8 Methods of frost protection.

Modify	Modify
LONGWAVE RADIATION BALANCES	THERMAL CHARACTERISTICS OF SUBSURFACE
radiation covers	improve thermal diffusivity by
cloches	compression (rolling)
Dutch lights	soil mixing (e.g. sand with peat)
greenhouses	subsurface heating
artificial fogs	
smoke	VERTICAL HEAT EXCHANGE IN LOWER ATMOSPHERE
water-based clouds	heating
	orchard burners
THERMAL CHARACTERISTICS OF SURFACE	stoves
mulching	mixing
straw	mechanical methods
newspaper	
plastic sheeting	
foam	
flooding	
spraying with water	
surface albedo changes	

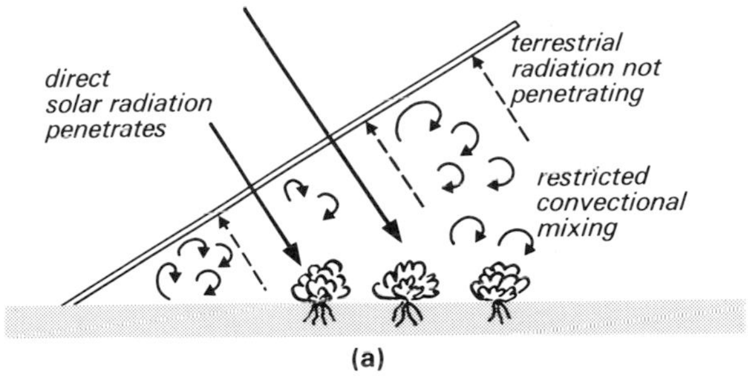

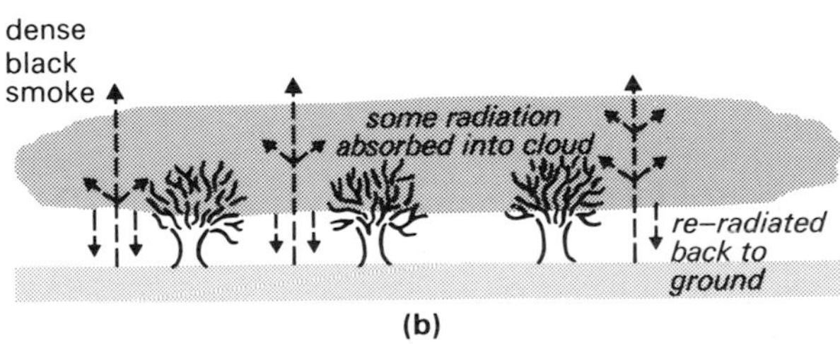

Figure 24.33 Frost control by the modification of atmospheric transparency to terrestrial radiation: (a) use of cloche; (b) use of dense smoke.

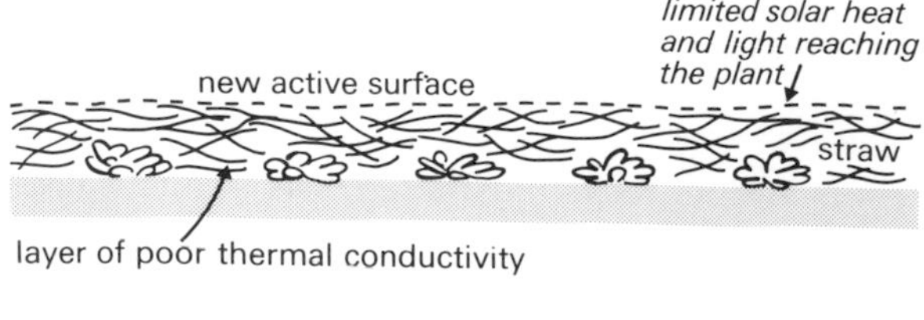

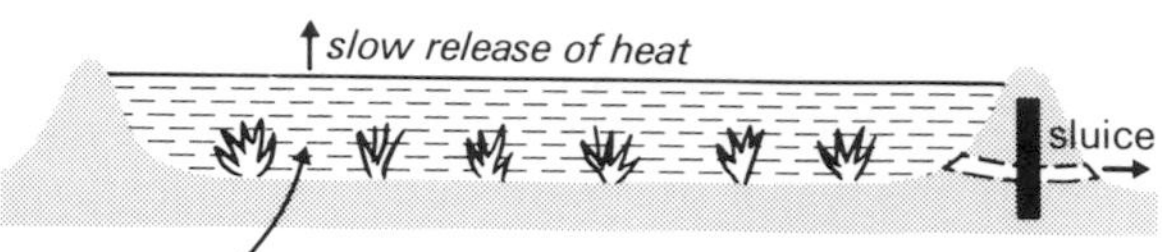

Figure 24.34 Surface temperature control methods.

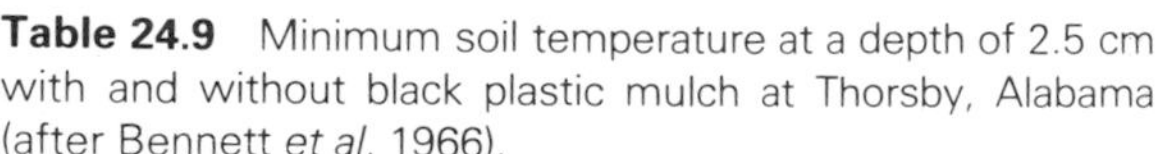

Table 24.9 Minimum soil temperature at a depth of 2.5 cm with and without black plastic mulch at Thorsby, Alabama (after Bennett *et al.* 1966).

Bare soil	Plastic covered	Difference due to plastic	Cloud cover
7.8	12.8	−5.0	clear, partly cloudy
16.1	18.3	−2.2	mostly cloudy
16.7	18.9	−2.2	cloudy, rain
18.3	20.6	−2.3	partly cloudy
17.8	21.7	−3.9	clear, partly cloudy
20.6	23.9	−3.3	clear
21.7	25.0	−3.3	clear
22.8	26.7	−3.9	mostly cloudy
19.4	23.3	−3.9	clear
22.2	24.4	−2.2	clear, rain
23.3	25.0	−1.7	partly cloudy, rain

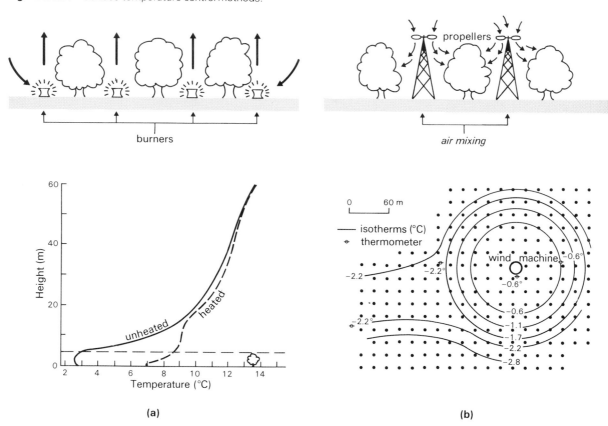

(a)

(b)

Figure 24.35 Mixing of the lower atmosphere as a frost protection: (a) the effect of fuel burners on the temperature profile in an orchard (after Kepner 1951); (b) the effect of a wind machine on minimum temperatures in an orchard during a radiation frost (after Bates 1972).

plants, the use of mulches is a common form of frost protection. The physical principle of the method is to cover the soil and plant surface with a layer of material of poor thermal conductivity – in other words, an insulator. In terms of the radiant energy cascade, Man is acting directly on the character of the surface, which is one of the main regulators. Air is one naturally occurring material which is readily available. The placing of layers of, for example, straw or newspaper on the surface traps pockets of air around the plant which, in effect, insulate it from severe heat losses. The flooding of crops such as cranberries in the marshes of the eastern USA also interposes an insulating water layer between plant and the cold atmosphere (Fig. 24.34 & Table 24.9). The mixing of air above the cold surface either by free or forced convection is used in combating air frosts in orchards. Air being cooled from below will not exhibit free convectional mixing naturally but, should moderately large sources of heat be introduced, this takes place readily. Orchard heaters (commonly peat stoves) generate free convectional movement of warmed air and hence reduce the risk of fog (Fig. 24.35a). The warmer air above and the cold air near the surface may be mixed mechanically and the risk of frost reduced without the use of extraneous sources of heat. One method of achieving this is by using propellers mounted on towers, as illustrated in Figure 24.35b. The net result is a breakdown of the cold inversion layer.

Conclusion

25 Systems retrospect and prospect

If we were now to look again on the view of the Earth from space portrayed in Figure 1.1d, the image in our mind's eye would contain more than an evocative impression of the 'blue planet'. Our picture of a picture is enhanced and our awareness heightened by insight and understanding. It is now not only aesthetically appealing but also intellectually stimulating. Instantaneously we are aware of water molecules breaking free from the ocean's surface, cloud droplets growing, colliding and falling as raindrops. We see the enormous and inevitable evacuation from the continents of this water and the erosive load it entrains. We see the colossal and continuous gaseous exchange between the atmosphere and biosphere, as well as the energy flux that sustains it. Not only do we perceive the planet as it is, but also as it was. Like some vast time-lapse sequence we can see the extension of ice sheets, the contraction of deserts and the continents riding on their plates – colliding or cleaving – progressing like giant Noah's Arks carrying their cargoes of creatures through millions of years of evolution.

Our mind's eye is equally active at a different scale. As well as the delight that the tranquillity of the highland glen of Figure 1.1a evokes, we recognise the valley for what it is – a relict of glacial overdeepening now drowned by eustatic rise in sea level. At the same time it interprets the tilted platform above the loch side as evidence of differential isostatic uplift. The saturated, anaerobic peat is seen as a store of organic matter and mineral nutrients imprisoned in a slowly decomposing blanket across the hillside. Simultaneously we see the turbulent pipeflow draining the peat, the tenacious trickle of water emerging along stream sides and the seemingly oily smear of the sulphide-stained surface of bog pools in terms of slope hydrology. The cow quietly cropping the grass by the loch shore is seen as part of the energy flow from producer to consumer. This same acid and neutral grassland and the dwarf shrub moor are perceived not only as an integral component in the scale and appeal of the landscape, but as communities established and maintained by the anthropogenic imposition of grazing and burning regimes that modify and control their ecology. Indeed, the treelessness of the landscape can roll back before our mind's eye and reveal the oceanic birch and oak woods which once held sway along the lochside, or the magnificent pines which, on the higher slopes, were once the Atlantic bulwark of the Caledonian Forest. Our understanding can embrace both the complex functional relationships of these natural forest ecosystems and the labyrinth of interactions with the cultures which at first modified, then exploited and finally destroyed them. Today we can recognise the conflicting pressures, assess the points of view and make informed value judgements as to the social and economic effects, as well as the environmental impact of an extension of commercial afforestation into the valley.

In short, our mental picture is a montage of images at once comprehensive in the panorama it commands and incisive in the perspective it produces. Just as the fragments of pigment in a painting, resonating individually but responding and relating to each other, are held together by the weft and warp of the supporting canvas, so our images of environment are underpinned and supported by a framework of systems and a fabric of scientific law and principle. Breadth of understanding becomes possible without superficiality, and detailed knowledge without the isolation of specialisation.

As a way of looking at our world and as a framework for thought the systems approach is richly rewarding, but it is also undeniably an attitude of mind – some would say a philosophy. Nevertheless, as a *modus operandi* it has its shortcomings. Perhaps, therefore, in conclusion we should take a brief retrospective view of its application to the natural environment before looking ahead at the prospects for its continued application and development.

In this book we have adopted a largely pragmatic approach to the application of systems thinking, opting to present models at a variety of convenient and useful scales that for the most part have been purely descriptive and explanatory. We have concentrated on the structural and functional organisation of environmental systems and stressed the functional relationships within and between these systems. The flow of energy and the transfer of matter have been the unifying themes in an essentially thermodynamic perspective. To this end the underlying scientific laws and principles that condition the properties and behaviour of natural phenomena, and govern the operation of natural processes, have been emphasised repeatedly. The focus has remained the explanation of the real world by modelling it as linked, ordered systems.

However, the formal development of the systems approach requires that it be seen as more than a framework for thought and for the conveyance of explanation. Indeed, the broader applications of systems thinking across the sciences and social sciences and the claims embodied in General Systems Theory lead some to maintain that it represents a philosophy of science. That we have hardly mentioned this dimension is perhaps excusable, for it is at this level that the most vociferous criticism has been generated. A second omission is more significant, for ultimately the systems approach must involve reducing the description of the system, the analysis of its organisation and the prediction of its behaviour to the language and rigour of mathematics; for the most part, and in spite of a largely scientific treatment, we have avoided this. Nevertheless, the full power of the qualitative models of environmental systems that we have discussed can be realised only when they are translated into quantitative mathematical models.

This process began early in some branches of the environmental sciences such as meteorology (with its roots in physics) and in soil mechanics and hydrology with their parent disciplines of civil and hydraulic engineering respectively. In other areas such as biology it came later, and most recently it has begun to transform physical geography. Mathematical models represent the simplification and abstraction of the underlying rationale of the real world in its purest form. Nevertheless, like the empirical and intuitive models, the inductive hypotheses and conceptual theories on which they build, they are not the complete answer to our future understanding of the environment. They model environmental systems in terms of sets of equations at the appropriate level of complexity. Some are equations of state, perhaps specifying dependent endogenous or state variables as vectors of state and relating these to exogenous independent variables. Processes bringing about change in state variables, or factors influencing such changes, may be modelled as transfer functions, or where not affected by the components of the system as vectors of input or forcing functions. Relationships may be expressed through the symbolism of matrix algebra and sets of difference equations, or where appropriate both linear and higher-order differential equations may be used to describe mathematically the way the system changes through time. These express the rates of change of a number of state variables as functions of each other and of other characteristics of the system. When combined differential models are written in matrix form they allow many of the properties of the model to be determined by the methods of applied mathematics. However, because of the complexity of real environmental systems these models begin to approach the limits of analytical techniques.

Most such models are **deterministic** in that they allow the prediction of the outcome of some operation performed on the system. However, others are **stochastic** in the sense that the model incorporates some element of randomness or uncertainty and is able only, therefore, to predict the probability of the outcome of an operation. These models recognise that certain variables are truly random, and that in the case of others their variability is so complex that our only option is to treat their behaviour as if it occurred at random. As with deterministic models, so the sophistication of stochastic models can vary. At the simple level, an otherwise deterministic model may be improved by one or more inputs being generated at random. On another level, the 'mathematical techniques and assumptions are incorporated in the model' in order to 'provide the uncertainty from within' (Brunsden & Thornes 1977). However, the entropy modelling developed from statistical mechanics and considered in Chapter 21 perhaps represents the most abstract of these stochastic approaches.

As with the conceptual models that we have used, mathematical models vary in their realism, in their resolution of systems attributes and in their completeness or the number of processes and interactions they incorporate. They also vary in their generality and breadth of application. There is, however, no best fit, maximum or optimum model, although, unlike conceptual models, quantitative mathematical models can be evaluated by their precision and their numerical predictive power. To increase this power there is always a tendency for models to be expanded: that is, the independent variables of simple models may be made more realistic by recognising their dependence on processes going on in the system and by introducing new functions. Alternatively, descriptive state variables may be refined by subdivision, or the number of variables influencing each process may be increased by the inclusion of more detail. Mechanisms which have been treated as 'black box' compartments may be opened up, again in the interests of greater realism and greater predictive precision.

In spite of the progress that has been made in mathematical systems analysis and modelling, if it is not to become a sterile exercise it must retain a continuous and constant dialogue with reality. To validate any such model requires real data, data that can be provided only by the study of the real world. That the age of the computer and of simulation has dawned over the environmental sciences does not mean that the age of exploration and discovery has set forever. Field measurement and monitoring, the description of spatial variation, laboratory experimentation and hardware modelling are just as important as they ever were. The constants (or 'parameters', as they are known) in the elegant equations of our models must be estimated with as much precision as possible to improve prediction. Although values may be arrived at by trial and error or by regression, measurement and experiment are the only real answer. Equally, the expansion of models must be based on a better understanding of what goes on in the real world.

This understanding of the real world is still far from complete and certainly far from uniform. Our detailed knowledge of the accessible and heavily populated developed world with its post-Renaissance tradition of scientific enquiry is far greater than that of the less accessible and less developed parts of the globe. This state of knowledge has two important consequences. First, many of the mathematical models that are part of the accepted wisdom of physical geography and environmental science are based on empirical observation and the construction of inductive hypotheses that occurred particularly in Europe and North America. Secondly, the validation and testing of these models has taken place largely in these same environments. There is no *a priori* guarantee that their generality can be extended to other environments except in the most abstract way. Some of the assumptions of these models might well be questioned in, for example, the humid Tropics. Indeed by their very nature, abstract mathematical models tend to obscure the rich geographical variety of real-world systems. One of the most rewarding challenges is to modify, extend and refine the model as our knowledge of the environment increases. An example of the sterility which can creep in if a fundamental model fails to develop towards greater realism is afforded by Hubbell's criticism (1971) of the use of Lindemann's trophic–dynamic model in productivity studies. He maintains that 'the prevalent treatment of organisms as passive agents has hindered further developments in the field of ecological bioenergetics by producing few significant questions as to what living systems are really doing with energy'. By failing to do so, 'the capacity of living organisms to regulate, within the bounds established by the laws of thermodynamics, the rates at which they accumulate and dissipate energy' is overlooked. This criticism is akin to Thornes' plea (1978) that the tendency of many geomorphic models to concentrate on the steady or equilibrium states of systems denies the importance of understanding transient states in both time and space.

Perhaps the most important development of systems modelling, however, is what Bennett and Chorley (1978) call **systems interfacing**. The interfacing of 'natural' physico-ecological and 'human' socio-economic systems involves an understanding of the links and interactions which explain the way human activities mesh with environmental systems. It is necessary to be able to model these interfaces quantitatively because we need to project them into the future and to predict the probable outcome of alternative strategies available to civilisation. The study of such interfacing, whether it be in terms of intervention or harmonious symbiosis, is the only way to generate the understanding necessary to formulate sound policies to direct and secure the future for both mankind and for the environment. The decisions to develop and implement such policies will not rest with environmental scientists, but with politicians and governments. If they can say with William Golding's Neanderthal heroine 'I have a picture . . . it is a picture of a picture . . . I am . . . thinking', then perhaps there is hope for the future.

Bibliography

Ackerman, H., C. Lippens and M. Lechevallier 1980. Volcanic material from Mount St. Helen's in the stratosphere over Europe. *Nature* **287**, 614–16.

Agricultural Advisory Council on Soil Structure and Soil Fertility 1971. *Modern farming and the soil.* London: HMSO.

Ahnert, F. 1970. Functional relationships between denudation, relief and uplift in large mid-latitude drainage basins. *Am. J. Sci.* **268**, 243–63.

Albion, R. G. 1926. *Forests and sea power.* Cambridge, Mass.: Harvard University Press.

Anderson, M. C. 1964. Studies on the woodland light climate II seasonal variations in light climate. *J. Ecol.* **52**.

Andrews, J. T. 1972. Glacier power, mass balance, velocities and erosional potential. *Z. Geomorph.* N.F. **13**, 1–17.

Armstrong, W. 1975. Waterlogged soils. In *Environment and plant ecology*, J. R. Etherington and W. Armstrong (eds). Chichester: Wiley.

Arnett, R. R. 1979. The use of differing scales to identify factors controlling denudation rates. In *Geographical approaches to fluvial processes*, A. F. Pitty (ed.), 127–47. Norwich: Geo Books.

Ashton, P. S. 1969. Speciation among tropical forest trees; some deductions in the light of recent evidence. *Biol J. Linn.* **1**, 155–96.

Ashworth, J. R. 1929. The influence of smoke and hot gases from factory chimneys on rainfall. *Q.J. Meteorol. Soc.* **55**, 341–50.

Atkinson, B. W. 1975. *The mechanical effect of an urban area on convective precipitation.* Occasional paper 3. Department of Geography, Queen Mary College, University of London.

Atkinson, T. C. 1978. Techniques for measuring subsurface flow on hillslopes. In *Hillslope hydrology*, M. J. Kirkby (ed.), 73–120. Chichester: Wiley.

Bange, G. G. J. 1953. *Acta Bot. Neerl.* **2**, 225.

Bannister, P. 1966. The use of subjective estimates of cover-abundance as a basis for ordination. *J. Ecol.* **54**, 665–74.

Bannister, P. 1976. Physiological ecology and plant nutrition. In *Methods in plant ecology*, S. B. Chapman (ed.). Oxford: Blackwell Scientific.

Barry, R. G. 1970. A framework for climatological research with particular reference to scale concepts. *Trans Inst. Br. Geogs* **49**, 61–70.

Barry, R. G. and R. J. Chorley 1971. *Atmosphere, weather and climate.* London: Methuen.

Bates, E. M. 1972. Temperature inversion and freeze protection by wind machine. *Agric. Meteorol.* **9**, 335–46.

Battan, L. 1977. Weather modification in the Soviet Union. *Bull. Am. Meteorol. Soc.* **58**, 4–19.

Baumgartner, A. and E. Reichel 1975. *The world water balance.* Amsterdam: Elsevier.

Bennett, D. L., D. A. Ashley and B. D. Doss 1966. Cotton responses to black plastic mulch and irrigation. *Agron. J.* **58**, 57–60.

Bennett, R. J. and R. J. Chorley 1978. *Environmental systems philosophy, analysis and control.* London: Methuen.

Beyer, J. L. 1974. Global summary of human response to natural hazards; floods. In *Natural hazards*, G. F. White (ed.). New York: Oxford University Press.

Biddulph, O. 1959. Translocation of inorganic solutes. In *Plant physiology – a treatise, Vol II*, F. C. Steward (ed.). New York: Academic Press.

Billings, M. P. 1954. *Structural geology*, 2nd edn. Englewood Cliffs, NJ: Prentice-Hall.

Black, C. C. 1971. Ecological implications of dividing plants into groups with distinct photosynthetic production capacities. *Adv. Ecol Res.* **7**, 87–114.

Black, J. N., C. W. Bonyphon and J. A. Prescott 1954. Solar radiation and the duration of sunshine. *Q.J. Meteorol. Soc.* **80**, 231–5.

Bloom, A. L. 1969. *The surface of the Earth.* Englewood Cliffs, NJ: Prentice-Hall.

de Boer, H. J. 1965. Comparisons of water vapour water and water drops of various sizes as means of preventing night frost. *Agric. Meteorol.* **2**, 247–58.

Borisov, A. A. 1945. *Climates of the USSR* (transl. R. A. Ledward). Edinburgh: Oliver & Boyd.

Bormann, F. H. and G. E. Likens 1970. The nutrient cycles of an ecosystem. *Scient. Am.* 92–101.

Boulding, K. E. 1966. The economics of the coming spaceship Earth. In *Environmental quality in a growing economy: resources for the future*, 3–14. Baltimore, Md: Johns Hopkins University Press.

Boulton, G. S. 1972. Rôle of thermal regime in glacial sedimentation. In *Polar geomorphology*, R. J. Price and D. E. Sugden (eds), 1–19. Inst. Br. Geogs Sp. Publ. 4.

Boulton, G. S., A. S. Jones, K. M. Clayton and M. J. Kenning 1977. A British ice sheet model and patterns of glacial erosion and deposition in Britain. In *British Quaternary studies: recent advances*, F. W. Shotton (ed.), 231–46. Oxford: Oxford University Press.

Bowen, N. L. 1928. *The evolution of the igneous rocks.* Princeton, NJ: Princeton University Press.

Box, E. 1975. Quantitative evaluation of global primary productivity models generated by computers. In *Primary productivity of the biosphere*, H. Lieth and R. H. Whitaker (eds), 266–83. New York: Springer-Verlag.

Bradley, W. C. 1963. Large scale exfoliation in massive sandstones of the Colorado plateau. *Geol Soc. Am. Bull.* **74**, 519–28.

Brewer, R. 1964. *Fabric and mineral analysis of soils.* New York: Wiley.

Bridges, E. M. and D. M. Harding 1971. Microerosion processes and factors affecting slope development in the Lower Swansea Valley. In *Slopes: form and process*, D. Brunsden (ed.), 65–80. Inst. Br. Geogs Sp. Publ. 3.

Brown, E. H. 1970. Man shapes the earth. *Geog. J.* **136**(1), 74–85.

Brunsden, D. 1968. *Dartmoor.* Sheffield: Geographical Association.

Brunsden, D. 1979. Weathering. In *Process in geomorphology*, C. Embleton and J. B. Thornes (eds), 73–129. London: Edward Arnold.

Bryson, R. A. and D. A. Baerreis 1967. Possibilities of major climatic modification and their implications; north-west India: a case study. *Bull. Am. Meteorol. Soc.* **48**, 136–42.

Budyko, M. I. 1958. *The heat balance of the Earth's surface* (transl. N. A. Strepanova). Washington, DC: US Dept. of Commerce.

Bunting, A. H., M. D. Dennett, J. Elston and J. R. Milford 1976. Rainfall trends in West African Sahel. *Q.J. Meteorol. Soc.* **102**, 59–64.
Burges, A. 1958. *Micro-organisms in the soil.* London: Hutchinson.
Burnett, J. H. (ed.) 1964. *The vegetation of Scotland.* Edinburgh: Oliver & Boyd.
Butler, B. E. 1959. *Periodic phenomena in landscapes as a basis for soil studies.* Soil Publ. CSIRO, Australia.
Byer, H. R. 1974. *General meteorology*, 4th edn. New York: McGraw-Hill.

Campbell, D. A. 1972. *Morfa Harlech: the dune system and its vegetation.* B.A. Hons CNAA Geography dissertation, Portsmouth Polytechnic.
Campbell, I. M. 1977. *Energy and the atmosphere: a physical chemical approach.* Chichester: Wiley.
Carlisle, A., A. H. F. Brown and E. J. White 1967. The nutrient content of tree stemflow and ground flora litter and leachates in a sessile oak (*Quercus petraea*) woodland. *J. Ecol.* **55**, 615–27.
Carson, M. A. and M. J. Kirkby 1972. *Hillslope form and process.* Cambridge: Cambridge University Press.
Chandler, T. J. 1965. *The climate of London.* London: Hutchinson.
Chapman, C. A. and R. L. Rioux 1958. Statistical study of topography, sheeting, and jointing in granite, Acadia National Park, Maine. *Am. J. Sci.* **256**, 111–27.
Chapman, R. N. 1928. The quantitative analysis of environmental factors. *Ecology* **9**, 111–22.
Chapman, S. B. 1976. *Methods in plant ecology*, 229–95. Oxford: Blackwell Scientific.
Chorley, R. J. 1973. Geography as human ecology. In *Directions in geography*, R. J. Chorley (ed.), 155–59. London: Methuen.
Chorley, R. J. and B. Kennedy 1971. *Physical geography: a systems approach.* Hemel Hempstead, England: Prentice-Hall.
Clayton, K. M. 1974. Zones of glacial erosion. In *Progress in geomorphology*, E. H. Brown and R. S. Waters (eds), 163–76. Inst. Br. Geogs Sp. Publ. 7.
Clayton, K. M. 1977. River terraces. In *British Quaternary studies, recent advances*, F. W. Shotton (ed.), 153–67. Oxford: Oxford University Press.
Cole, L. C. 1958. The ecosphere. *Scient. Am.* **198**(4), 83–92.
Cooke, R. U. and J. C. Doornkamp 1974. *Geomorphology in environmental management.* Oxford: Oxford University Press.
Cooke, R. U. and I. J. Smalley 1968. Salt weathering in deserts. *Nature* **220**(1), 226–7.
Corbel, J. 1964. L'érosion terrestre, étude quantative (Méthodes – techniques – résultats). *Ann. Geog.* **73**, 385–412.
Correns, C. W. 1949. *Growth and dissolution of crystals under linear pressure.* Disc. Faraday Soc. **5**, 271–97.
Cox, C. B., I. N. Healey and P. D. Moore 1973. *Biogeography: an ecological and evolutionary approach.* Oxford: Blackwell Scientific.
Crisp, P. J. 1964. *Grazing in terrestrial and marine environments.* Oxford: Blackwell Scientific.
Crocker, R. L. and B. A. Dickinson 1957. Soil development on the recessional moraines of the Herbert and Mendenhall glaciers, S.E. Alaska. *J. Ecol.* **45**, 169–85.
Crocker, R. L. and J. Major 1955. Soil development in relation to vegetation and surface age at Glacier Bay, Alaska. *J. Ecol.* **43**, 427–48.
Crompton, E. 1960. *The significance of the weathering leaching ratio in the differentiation of major soil groups.* Trans 7th Int. Cong. Soil Sci. **4**, 406–12.
Crowe, P. R. 1971. *Concepts in climatology.* London: Longman.
Curry, L. 1962. Climatic change as a random series. *Ann. Assoc. Am. Geogs* **52**, 21–31.
Curtis, C. D. 1976. Chemistry of rock weathering: fundamental reactions and controls. In *Geomorphology and climate*, E. Derbyshire (ed.). Chichester: Wiley.

Daily Telegraph 1976. USA accused of fouling Cuba's weather. 28 June.
Dalrymple, J. B., R. L. Blong and A. J. Conacher 1968. A hypothetical nine unit land surface model. *Z. Geomorph.* **12**, 60–76.
Davis, P. H. and V. H. Heywood 1963. *Principles of angiosperm taxonomy.* Edinburgh: Oliver & Boyd.
Deacon, J. 1974. The location of refugia of *Corylus avellana* L. during the Weichselian glaciation. *New Phytol.* **73**, 1055–63.
Dearman, W. R., F. J. Baynes and T. Y. Irfan 1976. Practical aspects of periglacial effects on weathered granite. *Proc. Ussher Soc.* **3**(3), 373–81.
Dearman, W. R., F. J. Barnes and T. Y. Irfan 1978. Engineering grading of weathered granite. *Q.J. Engng Geol.* **12**, 345–74.
Defant, F. 1951. Local winds. In *Compendium of meteorology*, T. F. Malone (ed.). Boston, Mass.: Am. Meteorol. Soc.
Derbyshire, E., K. J. Gregory and J. R. Hails 1979. *Geomorphological processes.* London: Butterworth.
Dewey, J. F. 1972. Plate tectonics. *Scient. Am.* **226** (May), 56–66.
Douglas, I. 1967. Man, vegetation, and the sediment yields of rivers. *Nature* **215**, 925–8.
Douglas, I. 1967. Natural and manmade erosion in the humid tropics of Australia, Malaysia and Singapore. In *Symposium on river morphology*, 17–29. Inst. Ass. Sci. Hydrol.
Duckham, A. N. and G. B. Masefield 1970. *Farming systems of the world.* London: Chatto & Windus.
Duncan, N. 1969. *Engineering geology and rock mechanics.* London: Leonard Hill.
Dunne, T. 1979. Sediment yield and land use in tropical catchments. *J. Hydrol.* **42**, 281–300.
Dyer, A. J. and B. B. Hicks 1965. Stratospheric transport of volcanic dust inferred from solar radiation measurements. *Nature* **208**, 131–3.

Edwards, C. A., D. E. Reichle and D. A. Crossley 1970. The role of soil invertebrates in turnover of organic matter and nutrients. In *Analysis of temperate forest ecosystems*, D. E. Reichle (ed.), Ch. 12. London: Chapman & Hall.
Egler, F. E. 1964. Pesticides in our ecosystem. *Am. Sci.* **52**, 110–36.
Eliassen, A. and K. Pedersen 1977. *Meteorology; an introductory course.* Vol. 1: *Physical processes and motion.* Oslo: Universitetsforlaget.
Elton, C. 1927. *Animal ecology.* London: Sidgwick & Jackson. (Paperback edn Methuen 1966.)
Elton, C. S. 1958. *The ecology of invasion by animals and plants.* London: Methuen.

Embleton, C. (ed.) 1972. *Glaciers and glacial erosion*. London: Macmillan.

Embleton, C. and J. B. Thornes (eds) 1979. *Process in geomorphology*. London: Edward Arnold.

Etherington, J. R. 1975. *Environment and plant ecology*. Chichester: Wiley.

Etherington, J. R. 1978. *Plant physiological ecology*. London: Edward Arnold.

Evans, I. S. 1970. Salt crystallisation and rock weathering: a review. *Rev. Géomorph. Dyn.* **19**, 153–77.

Evans, R. and R. P. C. Morgan 1974. Water erosion of arable land. *Area* **6**(3), 221–5.

Fairbridge, R. H. 1961. Eustatic changes of sea-level. In *Physics and chemistry of the Earth*, Vol. 4, L. H. Ahrens *et al.* (eds), 99–185. Oxford: Pergamon.

Farres, P. J. 1978. The rôle of time and aggregate size in the crusting process. *Earth Surf. Proc.* **3**, 243–54.

Federov, An. A. 1966. The structure of tropical rainforest, and speciation in the humid Tropics. *J. Ecol.* **54**, 1–11.

Fisher, J. 1954. *Bird recognition: 1. Sea birds and waders*. London: Penguin.

Fitzpatrick, E. A. 1971. *Pedology*. Edinburgh: Oliver & Boyd.

Flegmann, A. F. and R. A. T. George 1975. *Soils and other growth media*. London: Macmillan.

Fookes, P. G., W. R. Dearman and J. A. Franklin 1971. Some engineering aspects of rock weathering with field examples from Dartmoor and elsewhere. *Q.J. Engng Geol.* **4**, 139–85.

Ford, E. D. and P. J. Newbould 1977. The biomass and production of ground vegetation and its relation to tree cover through a deciduous woodland cycle. *J. Ecol.* **65**, 201–12.

Forrester, J. W. 1971. *World dynamics*. Cambridge, Mass.: Wright Allen.

Fortescue, J. A. C. and G. G. Martin 1970. Micronutrients: forest ecology and systems analysis. In *Analysis of temperate forest ecosystems*, D. E. Reichle (ed.). London: Chapman & Hall.

Fournier, F. 1960. *Climat et érosion: la relation entre l'érosion du sol par l'eau et les précipitations atmosphériques*. Paris: Presses Univ. de France.

Frankland, J. C. 1966. Succession of fungi on decaying petioles of *Pteridium aquilinum*. *J. Ecol.* **54**, 41–63.

French, H. M. 1973. Cryopediments on the chalk of southern England. *Bull. Periglac.* **22**, 149–56.

Gardiner, V. 1974. *Drainage basin morphometry*. Tech. Bull. No. 14. British Geomorph. Res. Group.

Garrels, R. H. and F. T. Mackenzie 1971. *Evolution of sedimentary rocks*. New York: Norton.

Garrett, S. D. 1981. *Soil fungi and soil fertility*, 2nd edn. Oxford: Pergamon.

Gates, D. M. 1962. *Energy exchange in the biosphere*. New York: Harper & Row.

Geiger, R. 1965. *The climate near the ground*. Cambridge, Mass.: Harvard University Press.

Gentry, R. C. 1970. Hurricane Debbie modification experiments. *Science* **168**, 473–5.

Glentworth, R. and H. G. Dion 1949. The association or hydrologic sequence in certain soils of the pedsolic zone of N.E. Scotland. *J. Soil Sci.* **1**, 35–49.

Goldich, S. S. 1938. A study in rock weathering. *J. Geol.* **46**, 17–58.

Golding, W. 1961. *The inheritors*. London: Faber & Faber.

Gomez Pompa, A., C. Vàzquez-Yanes, S. Guevara 1972. The tropical rainforest: a non-renewable resource. *Science* **177**, 762–5.

Goudie, A. S. 1977a. Sodium sulphate weathering and the disintegration of Mohenjo-daro, Pakistan. *Earth Surf. Proc.* **2**, 75–86.

Goudie, A. S. 1977b. *Environmental change*. Oxford: Oxford University Press.

Gregory, K. J. 1974. Streamflow and building activity. In *Fluvial processes in instrumented watersheds*, K. J. Gregory and D. E. Walling (eds). Inst. Br. Geogs Sp. Publ. 6.

Gregory, K. J. 1976. Drainage networks and climate. In *Geomorphology and climate*, E. Derbyshire (ed.), 289–315. Chichester: Wiley.

Gregory, K. J. and C. C. Park 1974. Adjustments of river channel capacity down stream from a reservoir. *Water Resources Res.* **10**, 870–3.

Gregory, K. J. and C. C. Park 1976a. Stream channel morphology in N.W. Yorkshire. *Rev. Géomorph. Dyn.* **25**(2), 63–72.

Gregory, K. J. and C. C. Park 1976b. The development of a Devon gully and man. *Geography* **61**, 77–82.

Gregory, K. J. and D. E. Walling 1973. *Drainage basin form and process*. London: Edward Arnold.

Grove, A. T. 1967. The last 20000 years in the tropics. In *Tropical geomorphology*, A. M. Harvey (ed.). BGRG Spec. Publ. 5.

Grove, A. T. and A. Warren 1968. Quaternary landforms and climate on the south side of the Sahara. *Geog. J.* **134**(2), 194–208.

Guardian, The (London) 1977. Convention bans weather war. 19 May.

Hack, J. T. 1957. *Studies of longitudinal stream profiles in Virginia and Maryland*. USGS Prof. Paper 294B.

Hadley, R. F. and S. A. Schumm 1961. Sediment sources and drainage basin characteristics in upper Cheyenne River basin. In *USGS Water Supply Paper* 1531-B, 137–96.

Hanwell, J. D. and M. D. Newson 1970. *The storms and floods of July 1968 on Mendip*. Wessex Cave Club Occ. Publ. Ser. 1. 2.

Hanwell, J. D. and M. D. Newson 1973. *Techniques in physical geography*. London: Macmillan.

Harding, D. M. and D. J. Parker 1974. Flood hazard at Shrewsbury. In *Natural hazards*, G. F. White (ed.). New York: Oxford University Press.

Harvey, A. M. 1974. Gully erosion and sediment yield in the Howgill Fells, Westmorland. In *Fluvial processes in instrumented watersheds*, K. J. Gregory and D. E. Walling (eds), 45–58. Inst. Br. Geogs Sp. Publ. 6.

Harvey, J. G. 1976. *Atmosphere and ocean; our fluid environments*. Sussex: Artemis Press.

Hemmingsen, A. M. 1960. *Energy metabolism as related to body size and respiratory surfaces*. Rep. Steno Meml Hosp., Copenhagen 9, part 2, 7.

Hewson, E. W. and R. W. Longley 1944. *Meteorology, theoretical and applied*. New York: Wiley.

Heywood, V. H. 1967. *Plant taxonomy*. London: Edward Arnold.

Hide, R. 1969. Some laboratory experiments on free thermal convection in a rotating fluid subject to a horizontal temperature gradient and their relation to the theory of global atmospheric circulation. In *The global circulation of the atmosphere*, G. A. Corby (ed.). London: R. Meteorol. Soc.

Hogg, W. H. 1970. Basic frost irrigation and degree-day data for planning purposes. In *Weather economics*, J. A. Taylor (ed.). Oxford: Pergamon.

Hollis, G. 1974. The effect of urbanisation on floods in the Canon's Brook, Harlow, Essex. In *Fluvial processes in instrumented watersheds*, K. J. Gregory and D. E. Walling (eds), 123–39. Inst. Br. Geogs Sp. Publ. 6.

Hooper, M. 1970. Dating hedges. *Area* **4,** 63–5.

Hopkins, B. 1965. *Forest and savanna.* London: Heinemann.

Hubbell, S. P. 1971. Of sowbugs and systems: the ecological energetics of a terrestrial isopod. In *Systems analysis and simulation ecology*. New York: Academic Press.

Hughes, R. and J. M. M. Munro 1968. Climatic and soil factors in the hills of Wales in their relation to the breeding of special herbage varieties. In *Hill land productivity.* Occ. Symp. no. 4, British Grassland Soc.

Hunter, R. F. 1962a. Hill sheep and their pasture. A study in sheep grazing in S.E. Scotland. *J. Ecol.* **50**, 651–80.

Hunter, R. F. 1962b. Home range behaviour in hill sheep. In *Grazing*, D. J. Crisp (ed.), Proc. 3rd Symp. on grazing, Bangor. Br. Ecol Soc. Oxford: Blackwell Scientific.

Hutchinson, G. E. 1959. Homage to Santa Rosalia, or why are there so many kinds of animals? *Am. Nat.* **93**, 145–59.

Hutchinson, G. E. (ed.) 1965. *The ecological theatre and the evolutionary play.* New Haven, Conn.: Yale University Press.

Hutchinson, G. E. 1970. The biosphere. *Scient. Am.* **223**(3), 45–53.

Illies, V. 1974. *Introduction to zoogeography* (transl. W. D. Williams). London: Macmillan.

Imeson, A. C. 1974. The origin of sediment in a moorland catchment with particular reference to the rôle of vegetation. In *Fluvial processes in instrumented watersheds*, K. J. Gregory and D. E. Walling (eds). Inst. Br. Geogs Spec. Publ. 6.

Iverson, J. 1956. Forest clearance in the Stone Age. *Scient. Am.* **194**, 36–41.

Iverson, J. 1964. Retrogressive vegetational succession in the post-glacial. *J. Ecol.* **52** (suppl.), 59–70.

Jackson, R. J. 1967. The effect of slope aspects and albedo on PET from hillslopes and catchments. *N.Z. J. Hydrol.* **6**, 60–9.

Jaynes, E. T. 1957. Information theory and statistical mechanics. *Phys. Rev.* **104**.

Jeffery, D. W. and C. D. Pigott 1973. The response of grassland on sugar limestone to applications of phosphorus and nitrogen. *J. Ecol.* **61**, 85–92.

Jenkins, D., A. Watson and G. R. Miller 1967. Population fluctuations in the red grouse (*Lagopus lagopus scoticus*). *J. Anim. Ecol.* **36**, 97–122.

Jenny, H. 1941. *Factors of soil formation.* New York: McGraw-Hill.

Kates, R. W. 1970. *Natural hazard in human ecological perspective: hypothesis and models.* Nat. Hazards Res. Working Paper, no. 14.

Keller, W. D. 1957. *The principles of chemical weathering.* Columbia, Miss.: Lucas.

Kendrick, R. E. and B. Frankland 1976. *Phytochrome and plant growth.* London: Edward Arnold.

Kepner, R. A. 1951. *Effectiveness of orchard heaters.* Bulletin 723, California Agric. Exp. Stat.

Kira, T. and T. Shidei 1967. Primary production and turnover of organic matter in different forest ecosystems of the western Pacific. *Jap. J. Ecol.* **17**, 70–87.

Knapp, B. J. 1979. *Soil processes.* London: George Allen & Unwin.

Kramer, P. J. 1969. *Plant and soil water relationships.* New York: McGraw-Hill.

Kubiena, W. L. 1938. *Micropedology.* Ames, Iowa: Collegiate Press.

Kucera, C. L., R. C. Dahlmann and M. R. Krelling 1967. Total net productivity and turnover on an energy basis for tall grass prairie. *Ecol.* **48**, 536–41.

Kurtén, B. 1969. Continental drift and evolution. *Scient. Am.* **220**(3), 54–64.

Lamb, H. H. 1972. *Climate present, past and future.* Vol. 1: *Fundamentals and climate now.* London: Methuen.

Landsberg, H. E. 1960. *Physical climatology*, 2nd edn. Dubois Penn.: Gray Printing Co.

Langbein, W. B. and L. B. Leopold 1964. Quasi equilibrium states in channel morphology. *Am. J. Sci.* **262**, 782–94.

Langbein, W. B. and L. B. Leopold 1966. *River meanders – the theory of minimum variance.* USGS Prof. Paper 422-H.

Langbein, W. B. and S. A. Schumm 1958. Yield of sediment in relation to mean annual precipitation. *Trans Am. Geophys. Union* **39**, 1076–84.

Learmonth, A. 1977. *Man–environment relationships as complex ecosystems.* Milton Keynes: Open University Press.

Lee, R. 1978. *Forest microclimatology.* New York: Columbia University Press.

Lehninger, A. L. 1965. *Bioenergetics. The molecular basis of biological energy transformations.* New York: Benjamin.

Lenihan, J. and W. W. Fletcher 1978. *The built environment.* Vol. 8: *Of environment and man.* Glasgow: Blackie.

Leopold, L. B. and W. B. Langbein 1962. *The concept of entropy in landscape evolution*, 20. USGS Prof. Paper 500-A.

Leopold, L. B. and T. Maddock 1953. *The hydraulic geometry of stream channels and some physiographic implications.* USGS Prof. Paper 252.

Leopold, L. B. and M. G. Wolman 1957. *River channel patterns: braided, meandering and straight.* USGS Prof. Paper 282-B.

Leopold, L. B., M. G. Wolman and J. P. Miller 1964. *Fluvial processes in geomorphology.* San Francisco: W. H. Freeman.

Lieth, H. 1964. Versuch einer kartographischen Darstellung der produktivität der Pflanzendecke auf der Erde. In *Geographisches Taschenbuch 1964/65*, 72–80. Wiesbaden: Steiner.

Lieth, H. 1970. Phenology in productivity studies. In *Analysis of temperate forest ecosystems*, D. E. Reichle (ed.), Ch. 4. London: Chapman & Hall.

Lieth, H. 1971. The net primary productivity of the Earth with special emphasis on land areas. In *Perspectives on primary productivity of the Earth*, R. Whittaker (ed.). Symp. AIBS 2nd Natl Congr., Miami, Florida, October.

Lieth, H. 1972. Über die primärproduktion der Pflanzdecke der Erde. Symp. Deut. Bot. Gesell. Innsbruck, Austria, September 1971. *Z. Angew. Bot.* **46**, 1–37.

Lieth, H. 1975. Modelling the primary productivity of the world. In *Primary productivity of the biosphere*, H. Lieth and R. H. Whitaker (eds), 237–63. New York: Springer-Verlag.

Lieth, H. and E. Box 1972. Evapotranspiration and primary productivity; C. W. Thornthwaite memorial model. In

Papers on selected topics in climatology, J. R. Mather (ed.), 37–46. Elmer, NJ: C. W. Thornthwaite Associates.

Likens, G. E. and F. H. Bormann 1972. Nutrient cycling in ecosystems. In *Ecosystem structure and function*, J. H. Wrens (ed), 25–67. Oregon State Univ. Ann. Biol. Colloquia 31.

Lindeman, R. L. 1942. The trophic dynamic aspects of ecology. *Ecology*, **23**, 399–418.

Linsley, R. K., M. A. Kohler and J. L. H. Paulhus 1949. *Applied hydrology*. New York: McGraw-Hill.

Lockwood, J. G. 1962. The occurrence of Föhn winds in the British Isles. *Meteorol. Mag.* **91**, 57–65.

Lockwood, J. G. 1974. *World climatology: an environmental approach*. London: Edward Arnold.

Loughnan, F. C. 1969. *Chemical weathering of the silicate minerals*. New York: Elsevier.

Low, A. J. 1972. The effect of cultivation on the structure and other physical characteristics of grassland and arable soils (1945–70). *J. Soil Sci.* **23**, 363–80.

Lukashev, K. I. 1970. *Lithology and geochemistry of the weathering crust*. Jerusalem: Israel Program for Scientific Translation.

Lydolph, P. E. 1977. *Climates of the Soviet Union: world survey of climatology*, vol. 7. Amsterdam: Elsevier.

MacArthur, R. H. 1958. Population ecology of some warblers of northeastern coniferous forests. *Ecology* **39**, 599–619.

Macfadyen, A. 1964. Energy flow in ecosystems and its exploitation by grazing. In *Grazing in terrestrial and marine environments*, D. J. Crisp (ed.), 3–20. Oxford: Blackwell Scientific.

Margalef, R. 1968. *Perspectives in ecological theory*. Chicago: University of Chicago Press.

Maris, S. L. 1980. *A study of the initial stages of succession on six disused railway tracks in Warwickshire and Leicestershire*. B.A. Hons CNAA Geography dissertation, Portsmouth Polytechnic.

Mason, B. 1952. *Principles of geochemistry*. New York: Wiley.

Mason, C. F. 1977. *Decomposition*. London: Edward Arnold.

Maxwell, A. E. *et al.* 1970. Deep sea drilling in the South Atlantic. *Science* **168**, 1047–59.

May, R. M. 1971a. Stability in model ecosystems. *Proc. Ecol. Soc. Aust.* **6**, 18–56.

May, R. M. 1971b. Stability in multispecies community models. *Bull. Math. Biophys.* **12**, 59–79.

May, R. M. 1975. Will a complex system be stable? *Nature* **238**, 413–14.

Maynard-Smith, J. 1974. *Models in ecology*. Cambridge: Cambridge University Press.

McCall, J. G. 1960. The flow characteristics of a cirque glacier and their effect on cirque formation. In *Investigations on Norwegian cirque glaciers*, J. G. McCall and W. V. Lewis (eds), 39–62. R. Geog. Soc. Res. Series IV.

Meadows, D. H., D. L. Meadows, J. Randers and W. W. Behrens 1971. *The limits to growth. Report of the Club of Rome*. London: Earth Island.

Meidner, H. and D. W. Sheriff 1976. *Water and plants*. London: Blackie.

Menard, H. W. and Smith 1966. Hypsometry of ocean basin provinces. *J. Geophys. Res.* **7**, 4305–25.

Meteorological Office 1972. *Tables of temperature, humidity, precipitation and sunshine for the world, Part III. Europe and the Azores*. London: HMSO.

Milankovitch, M. 1930. Mathematische klimalehne und astronomische theorie der Klimaschwaukungen. In *Handbuch der Klimatologie 1*. W. Köppen and R. Geiger (eds). Berlin: Borntraeger.

Miller, S. L. 1953. A production of amino acids under possible primitive Earth conditions. *Science* **117**, 528.

Milne, G. 1947. A soil reconnaissance journey through parts of Tanganyika territory, December 1935–February 1936. *J. Ecol.* **35**, 192–265.

Minderman, G. 1968. Addition, decomposition, and accumulation of organic matter in forests. *J. Ecol.* **56**, 355–62.

Mitchell, J. M. 1976. An overview of climatic variability and its causal mechanisms. *Quatern. Res.* **6**, 481–94.

Monteith, J. L. 1973. *Principles of environmental physics*. London: Edward Arnold.

Moore, J. J. 1962. The Braun–Blanquet system: a reassessment. *J. Ecol.* **50**, 761–9.

Moore, N. W., M. D. Hooper and B. N. K. Davis 1967. Hedges: 1. Introduction and reconnaissance studies. *J. Appl. Ecol.* **4**, 201–20.

Morgan, R. P. C. 1979. *Soil erosion*. London: Longman.

Morisawa, M. 1968. *Streams – their dynamics and morphology*. New York: McGraw-Hill.

Moss, R. 1969. A comparison of red grouse stocks with the production and nutritive value of heather. *J. Anim. Ecol.* **38**, 103–22.

Mottershead, D. N. 1971. Coastal head deposits between Start Point and Hope Cove, Devon. *Field Studies* **3**, 433–53.

Mottershead, D. N. and R. L. Collin 1976. A study of Flandrian glacier fluctuations in Tunsbergdalen, southern Norway. *Norsk Geol. Tids.* **56**, 417–36.

Mottershead, D. N. and G. E. Spraggs 1976. An introduction to the hydrology of the Portsmouth region. In *Portsmouth Geographical Essays II*, D. N. Mottershead and R. C. Riley (eds), 76–93. Portsmouth Polytechnic, Department of Geography.

Mottershead, D. N. and I. D. White 1972. The lichonometric dating of glacier succession: Tunsbergdalen, southern Norway. *Geog. Ann.* **54**(A), 47–52.

Mottershead, D. N. and I. D. White. Lichen growth in Tunsbergdalen – a confirmation. *Geog. Ann.* **54**(A), 3–4.

Muffler, L. J. P. and D. E. White 1975. Geothermal energy. In *Perspectives on energy: issues, ideas and environmental dilemmas*, L. C. Ruedisili and M. W. Firebaugh (eds), 352–8. New York: Oxford University Press.

Neiburger, M., J. G. Edinger and W. D. Bonner 1971. *Understanding our atmospheric environment*. San Francisco: W. H. Freeman.

Newbould, P. J. 1971. Comparative production of ecosystems. In *Potential crop production*, P. F. Waring and J. P. Cooper (eds), 228–38. London: Heinemann.

Odum, E. P. 1964. The new ecology. *Biol Sci.* **14**, 14–16.

Odum, E. P. 1971. *Fundamentals of ecology*, 3rd edn. Philadelphia: Saunders.

Odum, H. T. 1957. Trophic structure and productivity of Silver Springs, Florida. *Ecol. Monog.* **27**, 55–112.

Odum, H. T. 1960. Ecological potential and analogue circuits for the ecosystem. *Am. J. Sci.* **48**, 1–8.

Oke, T. R. 1974. *Review of urban climatology 1968–73*. World Met. Organisation, Tech. Note 134. Geneva: WMO.

Oke, T. R. 1978. *Boundary layer climates*. London: Methuen.

Ollier, C. 1969. *Weathering*. Edinburgh: Oliver & Boyd.

Ovington, J. T. 1966. *Woodlands*. London: English Universities Press.

Palmen, E. 1951. The rôle of atmospheric disturbances in the general circulation. *Q. J. R. Meteorol Soc.* **77**, 337.

Paton, T. R. 1978. *The formation of soil material*. London: George Allen & Unwin.

Pauling, L. 1970. *General chemistry*, 3rd edn. San Francisco: W. H. Freeman.

Pedgley, D. E. 1962. *A course in elementary meteorology*. London: HMSO.

Pegg, R. K. and R. C. Ward 1971. What happens to rain? *Weather* **26**, 88–97.

Peters, S. P. 1938. *Sea breezes at Worthy Down, Winchester*. Met. Office Prof. Notes, No. 86. London: HMSO.

Pisek, A., W. Larcher, W. Moser and I. Pack 1969. Kardinale temperaturbereiche und genztemperaturen des lebens der blätter verschiedener spermatophyten III temperaturabhängigkeit und optimaler temperaturbereich der nettophotosynthese. *Flora*, Jena **158**, 608–30.

Polunin, O. and A. Huxley 1967. *Flowers of the Mediterranean*. London: Chatto & Windus.

Postgate, J. R. (ed.) 1971. *The chemistry and biochemistry of nitrogen fixation*. New York: Plenum.

Postgate, J. R. 1978. *Nitrogen fixation*. London: Edward Arnold.

Proctor, J. 1971. The plant ecology of serpentine. *J. Ecol.* **59**, 375–410.

Rackham, O. 1971. Historical studies and woodland conservation. In *The scientific management of animal and plant communities for conservation*, E. Duffey and A. S. Watt (eds), 563–80. Oxford: Blackwell Scientific.

Ranwell, D. 1972. *Ecology of salt marshes and sand dunes*. London: Chapman & Hall.

Rice, R. J. 1977. *Fundamentals of geomorphology*. London: Longman.

Richards, K. S. and T. R. Wood 1977. Urbanisation, water redistribution and their effect on channel processes. In *River channel changes*, K. J. Gregory (ed.), 369–88. Chichester: Wiley.

Riehl, H. 1965. *Introduction to the atmosphere*, 3rd edn. New York: McGraw-Hill.

Ruxton, B. P. and I. McDougall 1967. Denudation rates in north-east Papua from K–Ar dating of lavas. *Am. J. Sci.* **265**, 545–61.

Sandford, K. S. 1954. *The Oxford region*. Oxford: Oxford University Press.

Sass, J. H. 1971. The Earth's heat and internal temperatures. In *Understanding the Earth*, I. G. Gass, P. J. Smith and R. C. L. Wilson (eds), 81–7. Sussex: Artemis Press.

Schultz, A. M. 1964. The nutrient-recovery hypothesis for arctic microtine cycles. In *Grazing in terrestrial and marine environments*, D. J. Crisp (ed.), 57–68. Oxford: Blackwell Scientific.

Schultz, A. M. 1969. The study of an ecosystem: the arctic tundra. In *The ecosystem concept in natural resource management*, G. Van Dyne (ed.), 77–93. New York: Academic Press.

Schumm, S. A. and R. W. Lichty 1965. Time space and causality in geomorphology. *Am. J. Sci.* **263**, 110–19.

Seddon, B. 1967. Prehistoric climate and agriculture: a review of recent palaeoecological investigations. In *Weather and agriculture*, J. A. Taylor (ed.). Oxford: Pergamon.

Sellers, W. D. 1965. *Physical climatology*. Chicago: University of Chicago Press.

Shackleton, N. J. and N. D. Opdyke 1973. Oxygen isotope and palaeomagnetic stratigraphy of equatorial pacific core V.28–238: oxygen isotope temperature and ice volumes on a 10^5 and 10^6 year scale. *Quatern. Res.* **3**, 39–55.

Sharp, R. P. 1960. *Glaciers*. Gondon Lectures, Oregon State System of Higher Education, Oregon.

Shreve, R. L. 1966. Statistical laws of stream numbers. *J. Geol.* **74**, 17–37.

Simmons, I. G. 1975. The ecological setting of Mesolithic man in the Highland zone. In *The effect of man on the landscape: the Highland zone*, J. G. Evans, S. Limbrey and H. Cleere (eds). Res. Rep. No. 11. Council for British Archaeology.

Simonson, R. W. 1959. Outline of a generalized theory of soil genesis. *Proc. Soil Sci. Soc. Am.* **23**, 152–6.

Simpson, G. G. 1953. *The major features of evolution*. New York: Columbia University Press.

Simpson, J. 1964. Sea breeze fronts in Hampshire. *Weather* **19**, 208–20.

Skipworth, J. P. 1974. Continental drift and the New Zealand biota. *NZ J. Geog.* **57**, 1–13.

Slaymaker, H. O. 1972. Patterns of present sub-aerial erosion and landforms in mid-Wales. *Trans Inst. Br. Geogs* **55**, 47–68.

Smagorinsky, J. 1979. Topics in dynamical meteorology: 10, a perspective of dynamical meteorology. *Weather* **34**, 126–35.

Smith, A. G. 1961. The Atlantic–sub-boreal transition. *Proc. Linn. Soc.* **172**, 38–49.

Smith, D. I. and T. C. Atkinson 1976. Process, landform and climate in limestone regions. In *Geomorphology and climate*, E. Derbyshire (ed.), 367–409. Chichester: Wiley.

Smith, D. I. and M. D. Newson 1974. The dynamics of solutional and mechanical erosion in limestone catchments on the Mendip Hills, Somerset. In *Fluvial processes in instrumented watersheds*, K. J. Gregory and D. E. Walling (eds), 155–67. Inst. Br. Geogs Sp. Publ. 6.

Smith, K. 1975. *Principles of applied climatology*. London: McGraw-Hill.

Soane, B. D. 1973. Techniques for measuring changes in the packing state and cone resistance of soil, after the passage of wheels and tracks. *J. Soil Sci.* **24**, 311–23.

Solomon, M. E. 1969. *Population dynamics*. London: Edward Arnold.

Southwood, T. R. E., R. M. May, M. P. Hassell, and G. R. Conway 1974. Ecological strategies and population parameters. *Am. Nat.* **108**, 791.

Sparks, B. W. and R. G. West 1972. *The ice age in Britain*. London: Methuen.

Statham, I. 1977. *Earth surface sediment transport*. Oxford: Oxford University Press.

Stebbins, G. L. 1974. Adaptive shifts and evolutionary novelty. In *Studies in the philosophy of biology*, 285–306.

Stoddart, D. R. 1969. World erosion and sedimentation. In *Water, earth and man*, R. J. Chorley (ed.), 43–64. London: Methuen.

Stork, A. 1963. Plant immigration in front of retreating glaciers, with examples from the Kebnekajse area, Northern Sweden. *Geog. Ann.* **XLV** (1), 1–22.

Strahler, A. N. 1952. Hypsometric analysis of erosional topography. *Geol Soc. Am. Bull.* **63**, 923–38.

Strahler, A. N. 1972. *Planet Earth: its physical systems through geological time*. New York: Harper & Row.

Strahler, A. N. and A. H. Strahler 1973. *Environmental geoscience*. California: Hamilton.

Strakhov, N. M. 1967. *Principles of lithogenesis*. Vol. 1; *Consultants bureau*. New York, London: Oliver & Boyd.

Streeter, D. T. 1974. Ecological aspects of oak woodland conservation. In *The British oak*, M. G. Morris and F. H. Perring (eds). Bot. Soc. Brit. Isles Conf. Report 14. Faringdon: Classey.

Sunday Times, The (London) 1970. The catastrophe everyone saw coming. 22 November.

Sugden, D. E. and B. S. John 1976. *Glaciers and landscape. A geomorphological approach*. London: Edward Arnold.

Sutcliffe, J. 1968. *Plants and water*. London: Edward Arnold.

Sutcliffe, R. C. 1948. *Meteorology for aviators*. Met. Office 432. London: HMSO.

Tansley, Sir A. G. 1935. The use and abuse of vegetational concepts and terms. *Ecology* **16**, 284–307.

Taylor, A. M. and E. Burnett 1964. Influence of soil strength on the root growth habits of plants. *Soil Sci.* **98**, 178–80.

Taylor, A. M. and L. F. Ratcliff 1969. Root growth pressure of cotton peas and peanuts. *Agron. J.* **61**, 389–402.

Taylor, J. A. 1970. The cost of British weather. In *Weather economics*, J. A. Taylor (ed.). Oxford: Pergamon.

Thomas, A. J. 1978. Worldwide weather disasters. *J. Meteorol.* (UK), **3**.

Thomas, A. S. 1963. Further changes in the vegetation since the advent of myxomatosis. *J. Ecol.* **51**, 151–86.

Thorarinsson, S. 1939. Observations on the drainage and rates of denudation in the Hoffelsjökull district. *Geog. Ann.* **21**, 19–215.

Thornes, J. B. 1978. The character and problems of theory in contemporary geomorphology. In *Geomorphology, present problems and future prospects*, C. Embleton, D. Brunsden, and D. K. C. Jones (eds). Oxford: Oxford University Press.

Thornes, J. B. and D. Brunsden 1977. *Geomorphology and time*. London: Methuen.

Thorsteinsson, I., G. Olafsson and G. M. Van Dyne 1971. Range resources of Iceland. *J. Range Mgmt* **24**, 86–93.

Tooley, M. J. 1974. Sea-level changes during the last 9000 years in north-west England. *Geog. J.* **140**, 18–42.

Trewartha, G. T. 1961. *The earth's problem climates*. Madison: University of Wisconsin Press.

Troake, R. P. and D. E. Walling 1973. The natural history of Slapton Ley Nature Reserve. VII The hydrology of the Slapton Wood stream; a preliminary report. *Field Studies* **3**, 719–40.

Trudgill, S. 1977. *Soil and vegetation systems*. Oxford: Oxford University Press.

Valentine, J. W. and E. M. Moores 1970. Plate tectonics regulation of faunal diversity and sea level: a model. *Nature* **228**, 657–9.

Walker, D. and R. G. West (eds) 1970. *Studies in the vegetational history of the British Isles*. Cambridge: Cambridge University Press.

Walling, D. E. and B. W. Webb 1975. Spatial variation of river water quality: a survey of the River Exe. *Trans Inst. Br. Geogs* **65**, 155–71.

Wallwork, J. A. 1970. *The ecology of soil animals*. Maidenhead, England: McGraw-Hill.

Waters, R. S. 1954. Pseudobedding in the Dartmoor granite. *Trans R. Geol Soc.* **18**, 456–62.

Watts, A. J. 1955. Sea breeze at Thorney Island. *Meteorol. Mag.* **84**, 42–8.

Weatherly, P. E. 1969. Ion movement within the plant and its integration with other physiological processes. In *Ecological aspects of the mineral nutrition of plants*, I. H. Rorison (ed.), 323–40. Oxford: Blackwell Scientific.

Weller, G. and B. Holmgren 1974. The microclimates of the arctic tundra. *J. Appl. Meteorol.* **13**, 854–62.

Wetherald, R. T. and S. Manabe 1975. The effects of changing the solar constant on the climate of a general circulation model. *J. Atmos. Sci.* **32**, 2044–9.

Weyman, D. R. 1975. *Run-off processes and streamflow modelling*. Oxford: Oxford University Press.

Whalley, W. B. 1976. *Properties of materials and geomorphological explanation*. Oxford: Oxford University Press.

White, I. D. and D. N. Mottershead 1972. Past and present vegetation in relation to solifluction on Ben Arkle, Sutherland. *Trans Bot. Soc. Edin.* **41**, 475–89.

White, I. D. 1973. *Plant succession and soil development on the recessional moraines*. Tunsbergdalen Res. Expedition Report No. 2. Dept. of Geography, Portsmouth Polytechnic.

White, I. D. 1973. *Preliminary report on the results of dendrochronological studies*. Tunsbergdalen Res. Expedition Report No. 2. Dept. of Geography, Portsmouth Polytechnic.

White, R. E. 1979. *Introduction to the principles and practice of soil science*. Oxford: Blackwell Scientific.

Whittaker, R. H. 1975. *Communities and ecosystems*. New York: Macmillan.

Whittaker, R. H. and G. E. Likens 1975. The biosphere and man.

Whittaker, R. H. and G. M. Woodwell 1971. Measurement of net primary production of forests. In *Productivity of forest ecosystems*, P. Davigneaud (ed.), 159–75. Paris: Unesco.

Wilson, J. T. 1963. Continental drift. *Scient. Am.* **208**(April), 86–102.

Wilson, K. 1960. The time factor in the development of dune soils at South Haven Peninsula, Dorset. *J. Ecol.* **48**, 341–59.

Winkler, E. M. and E. J. Wilhelm 1970. Salt burst by hydration pressures in architectural stone in urban atmosphere. *Geol Soc. Am. Bull.* **81**, 567–72.

Winstanley, D. 1978. The drought that won't go away. *New Scient.* **164**, 57.

Wolman, M. G. 1955. *The natural channel of Brandywine Creek, Pennsylvania*. USGS Prof. Paper 271.

Wolman, M. G. 1967. A cycle of erosion and sedimentation in urban river channels. *Geog. Ann.* **49**(A), 385–95.

Wooldridge, S. W. 1960. The Pleistocene succession in the London Basin. *Proc. Geol Assoc.* **71**(2), 113–29.

World Meteorological Organisation 1956. *International cloud atlas*. Geneva: WMO.

Wynne-Edwards, V. C. 1965. Self-regulating systems in populations of animals. *Science* **147**, 1543–8.

Yaalon, D. H. 1971. *Palaeopedology: origin, nature and dating of palaeosols*. Jerusalem: Israel University Press.

Yang, C. T. 1970. On river meanders. *J. Hydrol.* **13**, 231–53.

Yang, C. T. 1971. Potential energy and stream morphology. *Water Resources Res.* **7**(2), 311–22.

Young, A. 1969. Present rate of land erosion. *Nature* **224**, 851–2.

Young, A. 1972. *Slopes*. London: Longman.

Young, A. 1978. A twelve-year record of soil movement on a slope. *Z. Geomorph.* N.F. Suppl. **29**, 104–10.

Acknowledgements

The following individuals and organisations are thanked for permission to reproduce text illustrations and tables (numbers in parentheses refer to text figures unless otherwise stated):

Figures 3.11, 4.2, 4.7 and Tables 3.1, 3.4 and 8.1 reproduced from *Physical climatology* (W. D. Sellers) by permission of The University of Chicago Press, © 1965 University of Chicago; Figures 3.12, 4.11, 4.24 and 22.1a reproduced from *Climate, present, past and future* (H. H. Lamb) by permission of Methuen and C. W. Allen; Figure 3.16 reproduced by kind permission of the US Department of Commerce; Table 3.6 reproduced from *Understanding the Earth* (I. G. Gass *et al.*, eds) by permission of the Open University, © 1971 The Open University Press.

McGraw-Hill (4.6, 19.8a, 19.9, 24.24); Figures 4.12, 4.16a reproduced from *Introduction to the atmosphere*, 3rd edn (H. Riehl) by permission of McGraw-Hill, © 1965 McGraw-Hill; Figure 4.16b reproduced from E. Palmen, *Q. J. R. Meteorol. Soc.* **77**, 337–54 by permission of the Royal Meteorological Society; Figures 4.17 and 5.14 reproduced from Figures 4.19 (p. 87) and 7.3 (p. 149) of *Planet Earth* (A. N. Strahler) by permission of Harper & Row and author, © 1972 Arthur N. Strahler; Figure 4.21 and Table 4.3 reproduced from Tables I and II of R. G. Barry, *Trans Inst. Br. Geogs* **49**, 61–70 by permission of the Institute of British Geographers; D. Reidel Publishing Company (Table 4.2); Table 4.3 taken from J. Smagorinsky, *Weather* **34**, 126–35 by permission of the Royal Meteorological Society and Methuen; Table 4.5, Box 4.9 and Figure 17.14 reproduced from *Plants and water*, Studies in biology 14 (J. Sutcliffe) by permission of Edward Arnold; World Meteorological Organisation (Table 4.6); Elsevier Scientific (Table 4.8, 24.36); Figure 5.3 reproduced from Menard & Smith, *J. Geophys. Res.* **71**, 4305–25, © 1966 by the American Geophysical Union; Figures 5.4 and 5.9 reproduced from J. F. Dewey, *Scient. Am.* 1972 by permission of W. H. Freeman; Figure 5.5 reproduced from J. Tuzo Wilson, *Scient. Am.* 1963 by permission of W. H. Freeman; Figure 5.6 reproduced from A. E. Maxwell, *Science* **168**, 1047–59 by kind permission of the publisher and author, © 1970 by the American Association for the Advancement of Science; Figure 5.8 reprinted by permission of the publisher and author from *Nature* **228**, 657–9, © 1970 Macmillan Journals; Figures 5.10 and 15.1 reproduced from *Fundamentals of geomorphology* (R. J. Rice) by permission of Longman; Figure 5.11 reproduced from *Pleistocene geology and biology*, 2nd edn (R. G. West) by permission of Longman; Table 5.2 reproduced from *Principles of geochemistry* (B. Mason) by permission of John Wiley & Sons, © 1952 John Wiley & Sons; Tables 5.3, 5.4 and Figure 10.2 reproduced from *Fluvial processes in geomorphology* (L. B. Leopold *et al.*) by permission of W. H. Freeman; Benjamin Cummings (6.5, 6.7, 7.13, 21.4); Open University Press (Box 6.6); Figure 7.1 reproduced by permission of the publisher from Figure 5 (p. 13) of *Energy exchange in the biosphere* (D. M. Gates), Harper & Row Biological Monographs, ed. A. H. Brown, © 1962 Harper & Row; Figure 7.2 reproduced from *Plant taxonomy*, Studies in biology 5 (V. H. Heywood), by permission of Edward Arnold; Figure 7.2 reproduced from B. Kurten, *Scient. Am.* 1969 by permission of W. H. Freeman; Figure 7.9 and Table 7.1 reproduced from *Primary productivity of the biosphere*, Ecological studies 14 (H. Lieth and R. H. Whittaker, eds), by permission of Springer-Verlag, H. Lieth and E. Box; Box 7.1 reproduced from *Nitrogen fixation*, Studies in biology 92 (J. Postgate), by permission of Edward Arnold.

Figure 8.1 reproduced from G. Weller & B. Holmgren, *J. Appl. Meteorol.* 1974 by permission of the American Meteorological Society; Figures 8.3a, 8.9 and Tables 8.1 and 24.5 reproduced from *Boundary layer climates* (T. R. Oke) by permission of Methuen, Figure 8.3a also by permission of J. B. Stewart; R. Lee (8.3b); VGB-Kraftwerkstechnik GMBH (8.4a); Artemis Press (8.4b); Figure 8.5 reproduced by kind permission of The Blackie Publishing Group; Figure 8.10 reproduced from *Tables of temperature, humidity, precipitation and sunshine for the world: Part III, Europe and the Azores* (Meteorological Office) by permission of Her Majesty's Stationery Office, and from *Climates of the Soviet Union: World survey of climatology* vol. 7 (P. E. Lydolph) by permission of Elsevier Scientific; Table 8.1 reproduced from *Principles of environmental physics* (J. L. Monteith) by permission of Edward Arnold; Figure 9.3 by permission of the University of Dundee and the Meteorological Office; Figure 9.4 reproduced from *A course in elementary meteorology* (D. E. Pedgley) by permission of the Controller of Her Majesty's Stationery Office; Oliver & Boyd and the Copyright Agency of the USSR (9.9); Figures 9.10 and 9.11 reproduced from *Concepts in climatology* (P. R. Crowe) by permission of Longman; Figure 9.16 reproduced from *Compendium of meteorology* (T. F. Malone, ed.) by permission of the American Meteorological Society; Figure 9.17 reproduced from J. G. Lockwood, *Meteorol. Mag.* **91**, 57–65 by permission of the Controller of Her Majesty's Stationery Office; Universitets forlaget Oslo (Table 9.1); Table 9.2 and Figure 22.9 taken from *Atmosphere, weather and climate* (R. G. Barry and R. J. Chorley) by permission of Methuen.

Longman (10.4); J. D. Hanwell (10.7); Table 10.3 taken from *Structural geology* (M. P. Billings) by permission of Prentice Hall © 1954 Prentice Hall, original data compiled from *International critical tables* **2**, 47–9 by permission of McGraw-Hill and from *Geol. Soc. Am. Special Paper 36* p. 111 (Birch *et al.*) by permission of the Geological Society of America; Box 10.3 reproduced from R. K. Pegg and R. C. Ward, *Weather* **26**, 88–97 by permission of the Royal Meteorological Society; Figure 11.4a reproduced from C. A. Chapman and R. L. Rioux, *Am. J. Sci.* **256**, 111–27 by permission of the publisher and authors; W. C. Bradley (11.4b); Ussher Society (11.14 and Table 11.4b); Figure 11.16 reproduced from *Principles of lithogenesis* vol. 1 (N. M. Strakhov) by permission of Oliver & Boyd; Table 11.3 reproduced from A. S. Goudie, *Earth surface processes* **2**, 75–86 by permission of John Wiley & Sons, © 1977 John Wiley & Sons Ltd; Table 11.4a reproduced by permission of the Geological Society of London and P. J. Fookes; Tables 11.5 and 15.4 reproduced from *Geomorphology and climate* (E. Derbyshire, ed.) by permission of John Wiley & Sons, © 1976 John Wiley & Sons Ltd; Table 11.6 and Figure 14.11 reproduced from *Processes in geomorphology* (C. Embleton and J. Thornes, eds) by permission of Edward Arnold; Figure 12.3a and Box 12.1 reproduced from *Hillslope hydrology* (M. J. Kirkby, ed.) by permission of John Wiley & Sons, © 1978 John Wiley & Sons Ltd; Figures 12.6 and 22.1 reproduced from *Hillslope form and process* (M. A. Carson and M. J. Kirkby) by permission of Cambridge University Press; Gebrüder Borntraeger (12.14, 22.3); United States Geological Survey (13.2, 13.6, 13.10, 13.13, 15.2, 15.3, 21.3, Table 15.5); Field Studies Council (13.7); Figure 13.9 reproduced from D. E. Walling and B. W. Webb, *Trans Inst. Br. Geogs* **65**, 155–71, Figure 5 by permission of the Institute of British Geographers; Oxford University Press (21.2, 22.17, 24.16b, 24.29, Table 13.1); Figures 14.2, 14.3 and 14.12 reproduced from *Glaciers and landscape* (D. E. Sugden and B. S. John) by permission of Edward Arnold; Butterworths, E. Derbyshire, K. V. Gregory and J. R. Hails (14.13); Figure 14.14b reproduced from *Polar geomorphology* (R. U. Price and D. E. Sugden (eds)) Inst. Br. Geogs Special Publication 4, 1–19 by permission of the Institute of British Geographers.

Figure 15.4 reproduced with kind permission from B. P. Ruxton and I. McDougall, *Am. J. Sci.* **265**, 545–61; Figure 15.5 reprinted by permission of the publisher and author from *Nature* **224**, 851–2, © 1969 Macmillan Journals Ltd; Geographical Association and D. Brunsden (15.6, Table 15.6); Table 15.1 reproduced from *Evolution of sedimentary rocks* (R. H. Garrels and F. T. Mackenzie) by permission of W. W. Norton; Table 15.2 reproduced from *Geographical approaches to fluvial processes* (A. Pitty, ed.) by permission of Geo Abstracts and editor; Figure 16.6 reproduced from R. H. MacArthur, *Ecology* **39**, 599–619; Executors of the Estate of J. Fisher (16.7a); V. C. Wynne-Edwards (16.7b); Figures 16.8 and 19.8c reproduced from *Analysis of temperate forest ecosystems*, Ecological studies 1 (D. E. Reichle, ed.), by permission of Springer-Verlag and authors; Figure 16.9a reproduced from *Animal populations in relation to their food resources* (A. Watson, ed.) by permission of Blackwell Scientific Publications; Figures 16.9b, 18.1, 23.2 and 23.21 reproduced from Figure 5.7 p. 215, Figure 3.6 p. 79, Figure 2.3 p. 12 and Figure 2.12 p. 44 respectively of *Communities and ecosystems*, 2nd edn (R. H. Whittaker) by permission of Macmillan Publishing, © 1975 by R. H. Whittaker; Figure 16.9c reproduced from H. T. Odum, *Ecol. Monogr.* **27**, 55–112 by permission of the Ecological Society of America and author, © 1957 The Ecological Society of America; Figures 16.11, 17.7, 17.12 and 17.15 reproduced from

Methods in plant ecology (S. B. Chapman, ed.) by permission of Blackwell Scientific Publications; Figure 17.3 reproduced from *Ecological aspects of the mineral nutrition of plants* (I. H. Rorison, ed.) by permission of Blackwell Scientific Publications; Figures 17.5 and 20.12 reproduced from *Environment and plant ecology* (J. R. Etherington) by permission of John Wiley & Sons; Figures 17.6a & b reproduced from H. Meidner and D. W. Sherriff, *Water and plants* Blackie, Glasgow, 1976 by permission of the publisher; Figure 17.8 reproduced from *Grazing in terrestrial and marine environments* (D. J. Crisp, ed.) by permission of Blackwell Scientific Publications; Figure 17.11 reproduced from *J. Ecol.* 1964 by permission of Blackwell Scientific Publications; Pergamon Press (Table 17.1); Figure 19.2a and Table 19.1 reproduced from *Decomposition* (C. F. Mason) by permission of Edward Arnold; Figure 19.6 reproduced from *J. Ecol.* 1968 by permission of Blackwell Scientific Publications; Figure 19.7 reproduced from Figure 2.7 p. 39 and Figure 4.9 p. 97 of *The nature and properties of soils* (H. O. Buckman and N. C. Brady) by permission of Macmillan Publishing, © 1960, 1969 by Macmillan Publishing; Table 19.2 reproduced from *Soil fungi and soil fertility*, 2nd edn (S. D. Garrett) by permission of Pergamon Press, and from *J. Ecol.* 1966 by permission of Blackwell Scientific Publications; Figure 20.5 adapted from *Soil Sci. Soc. Am. Proc.* **23**, 152–6 (R. W. Simonson) by permission of the publisher; International Society of Soil Science (20.11); Table 20.1 reproduced from *Soil conditions and plant growth* (Russell) by permission of Longman.

Figure 21.5 reproduced from S. A. Schumm and R. W. Lichty, *Am. J. Sci.* **263**, 110–19, by permission of the American Journal of Science and the authors; Figure 22.5 reprinted by permission of publisher and author from *Nature* **287**, 614–16, © 1980 Macmillan Journals; Figure 22.6 reprinted by permission of publisher and author from *Nature* **208**, 131–3,© 1965 Macmillan Journals; Figure 22.7b reproduced from A. H. Bunting *et al. Q. J. R. Met. Soc.* **102**, 59–64 by permission of the Royal Meteorological Society; Figures 22.12 and 22.13 reproduced from F. Ahnert, *Am. J. Sci.* **268**, 243–63 by permission of the American Journal of Science and author; Figure 22.14 reproduced from *Physics and chemistry of the Earth*, Vol. 4 (L. H. Ahrens, ed.) by permission of Pergamon Press; The Geographical Journal and M. J. Tooley (22.15); Figures 22.16 and 22.19 reproduced from *British Quaternary Studies* (F. W. Shotton, ed.) by permission of Oxford University Press, © 1977 Oxford University Press; The Geologists Association (22.17); Figure 22.20 and Table 22.2 reproduced from *Progress in geomorphology* (E. H. Brown and R. S. Waters, eds) by permission of the Institute of British Geographers; A. T. Grove (22.23, 22.24); Figure 23.6 reproduced from *Population dynamics*, Studies in biology 18 (M. E. Solomon), by permission of Edward Arnold; Figures 23.7a reproduced from *Perspectives in ecological theory* (R. Margalef) by permission of The University of Chicago Press, © 1968 University of Chicago; Figure 23.7b&c reproduced from *Plant physiological ecology*, Studies in biology 98 (J. R. Etherington), by permission of Edward Arnold; Portsmouth Polytechnic (23.9); Chapman and Hall (23.10); Figure 23.11a–f adapted from *J. Ecol.* 1955 by permission of Blackwell Scientific Publications, a–d also adapted from *Geografiska Annales* **XLV**(1), 1–22 (A. Stork) by permission of the publisher, c & d reproduced by permission of Portsmouth Polytechnic; Figure 23.12 reproduced from *J. Ecol.* 1939 by permission of Blackwell Scientific Publications; Figures 23.18 and 23.19 reproduced from *The Ice Age in Britain* (B. W. Sparks and R. G. West) by permission of Methuen; Figure 23.22 taken from *Studies in vegetative history of the British Isles* (D. Walker and R. G. West, eds) by permission of Cambridge University Press.

Council for British Archaeology and I. G. Simmons (24.1); Figure 24.2 reproduced from *Biogeography, an ecological and evolutionary approach* (C. B. Cox *et al.*) by permission of Blackwell Scientific Publications; Figure 24.3 reproduced from Unit 8 *D204 Fundamentals of human geography: Man–environment relationships as complex ecosystems* by permission of The Open University, © 1977 The Open University Press; Figure 24.4 adapted from *The limits to growth: A report from the Club of Rome's project on the predicament of mankind* (D. H. Meadows, D. L. Meadows, J. Randers, W. W. Behrens III), a Potomac Associates book published by Universe Books, NY, 1972, Graphics by Potomac Associates; Figure 24.5 reproduced from *Proc. Int. Symp. on Forest Hydrology* (Sopper and Lull, eds) by permission of Pergamon Press; Figure 24.6 reproduced from F. H. Bormann and G. E. Likens, *Scient. Am.* 1970 by permission of W. H.

Freeman; Figure 24.7d reproduced from *J. Ecol.* 1977 by permission of Blackwell Scientific Publications; Figure 24.8 adapted from *Farming systems of the world* (A. N. Duckham and G. B. Masefield) by permission of Chatto & Windus; Westermann Verlag (24.9); Figure 24.10b reproduced from J. Iversen, *Scient. Am.* 1956 by permission of W. H. Freeman; Figure 24.10c reproduced from *Weather and agriculture* (J. A. Taylor, ed.) by permission of Pergamon Press; J. King and I. A. Nicholson (24.11a); Figure 24.13 reproduced from *Flowers of the Mediterranean* (O. Polunin and A. Huxley) by permission of Chatto & Windus; Figure 24.16c reproduced from *J. Soil Sci.* 1973 by permission of Blackwell Scientific Publications; Figure 24.17 reproduced from *J. Appl. Ecol.* 1967 by permission of Blackwell Scientific Publications; Figures 24.19 and 24.20 reproduced from *The climate of London* (T. J. Chandler) by permission of Hutchinson; Figure 24.22a reproduced from *Fluvial processes in instrumental watersheds*, Inst. Br. Geogs Special Publication 6, 123–39 (K. J. Gregory and D. E. Walling, eds) by permission of the Institute of British Geographers; the Editor *Geografiska Annales* (24.23); Figure 24.25 reproduced from K. J. Gregory and C. C. Park, *Water Resources Res.* **10**, 870–3, 1974, © by the American Geophysical Union; Figure 24.26 reproduced from R. W. Kates, *Econ. Geog.* **47**, 438–51 by permission of the Editor, *Economic Geography*; Figure 24.28 reproduced from *River channel changes* (K. J. Gregory, ed.) by permission of John Wiley & Sons, © 1977 John Wiley & Sons Ltd; Figure 24.30 reproduced from K. J. Gregory and C. C. Park, *Geography* **61**, 77–82 by permission of the Geographical Association; Figure 24.31 reproduced by permission of the publisher and author R. C. Gentry, *Science* **168**, 473–5, © 1970 by the American Association for the Advancement of Science; Oregon State University (Table 24.1); Gray Printing Company (Table 24.3); Table 24.7 reproduced with permission from *Natural hazards* (G. F. White, ed.), copyright by Oxford University Press.

Index

Page numbers in **bold** type refer to boxed material.

ablation 282–3, 286–8, 295, 401, 406
abrasion (glacial) 290, 294
absolute humidity 92
absorption spectrum 47, 49
absorptivity 53
abundance 313
acceleration 25
accretion 59
activation complex **30–1**
activation energy **30–1**
active mass **30–1**
adaptive radiation 150
adaptive variation 150
adenosine diphosphate (ADP) 137–40, **137**
adenosine triphosphate (ATP) 137–40, **137**
adhesive force 328
adiabatic temperature change 98–100, 183, 192–4, 202
advection fog 98
advectional mixing 98
aeration, zone of 221
aerodynamic roughness 81, 176
afforestation 444
aggregates, soil 369
agricultural effects on ecosystems 444
agricultural systems
  intensive 450
  ridge and furrow 450–1
  terraced hillside 450–1
air masses 180–2, 189
albedo 172–3, 178, 394–8, 401, 455
albite 109
alpine areas, radiation receipt 179
aluminium 108–9, 112, 124, 126, 234–5, 238–43, 248
amphibole 109
anabatic airflow 198–9
analysis 6
angular momentum 27, 87, 89
angular velocity 27
animal communities
  secondary production 353
  spatial structure 315–16
animals
  cold blooded **352**
  warm blooded **352**
anions 20, 371
annual temperature regime 180
anticyclogenesis 190
anticyclone 83, 183, 193–6, 398
aphelion 43, 395
applied force 27
argon 68
Arrhenius relationship 336
artificial ecosystems 450
aspect 178–9
assimilation 351
  efficiencies 362
asthenosphere 107, 117
atmosphere 158, 164–6
  circulation in 76, 102, 106
  density of 72
  pollution of 454
  pressure of 27, 72–5, 194
  soil 371
  stability of 99, 100, 102, 182, 198, 202, 398
  system 46, 68, 203, 393, 439
  vertical structure 71, 100
  water balance in 106
atom 17
  electron 17
  electron shells 18–20
  neutrons 17
  nucleus 17
  protons 17
atomic number 17
atomic structure 17
Atterberg limits 219
attributes of systems 10
autotrophic cell 143, 153
auxins 412

bankfull capacity 460
basal metabolic rate (BMR) **352**
basal sliding 283–6
baseflow 269
bedload 274–7, 280, 297
Benioff Zone 116, 125–6
bifurcation ratio 210–11, 302
biochemical controls on photosynthesis 334
biomass 153–4, 331–3, 349–50, 439, 446, 448, 450
  standing dead 356
biome types 151
biomechanical forces 226, 231
biosphere 128–45, 151
biosynthesis 136, 153
biotic potential 414
bit 388
black-body radiation 42, 44, 52
black-box
  approach 393
  models 13
blockfield 410
bond, molecular 284
bonding
  covalent 20–3, 108, 234, 240
  hydrogen 20–1, 235, 284, 364, 369
  ionic 20, 108, 234, 369
  metallic 20, 23
boundary of systems 10
Bowen Series 124, 242
boxes 15
brittle fracture 28, 218, 226, 228
brown earth 379
buffer capacity 238
buried channels 406

C-3 Calvin cycle **329**, 334
C-4 Hatch–Slack pathway **329**, 334
calcicoles 344
calcification 381
calcifuge 344
calcium 108, 234, 240, 242, 248
Calvin C-3 pathway **329**, 334
calving (glacier) 281
carbon compounds
  cyclic **131**, **133**
  heterocyclic **131**, **133**
  monocyclic **131**, **133**
  open chain **131**, **133**
carbon dioxide 47, 68–9, 238–9, 397–401
  compensation point **329**, 334
  diffusion 329
carbon : nitrogen ratio 449
carbonic acid exchange hypothesis 326
carotenoids **144**
carrying capacity 415, 441
catabolic (respiratory) heat loss 139, 153
catalysts **30–1**, 362
catchment nutrient budgets 441
catena, soil 378
cation
  adsorption 240
  exchange 242
  metallic 109
cations 20, 236, 240–1, 242
cell
  autotrophic 143, 153
  functional organisation 136
  heterotrophic 143, 153, 348
  living 129, 134–6, 145
  membrane 134
  nucleus 133, 140
  potential **338**
cellular
  activity 136–7
  organelles 134
  structure 133–4, 136
cellulase 348, 363
centrifugal force 26
centripetal force 26
change in systems 10, 12, 383–92
channel
  capacity 460, 462–5
    rural 460
    urban 460

effluent 268
form 460
influent 268
charge separation 21
chelating agents 329, 380
cheluviation 226, 240, 375
chemistry
carbon **131**
of life 129
chemical element 17
chemical elements of biosphere 129
chemical energy 29
chemical reactions **30–1**
chemosynthesis 143
chemotrophs 143
Chezy equation 273
Chinook 200
see also Föhn
chlorophyll (*a* and *b*) **144**, 145
chromosomes 140
circadian rhythms 336
circulation
system 63, 83, 172, 190, 193–4, 198–200, 203
tight and loose 343–4
circumpolar vortex 76, 88, 189
cirque 282, 286–91, 303
clastic fragments 368
clay
minerals 109, 112, 240–4, 371
skins 375, 379
Clean Air Act 454
clear felling 440–1
climatic change 401–2, 406–7
climatic optimum 436
climax ecosystem 419–26, 445
cliseral change 426–32
closed systems 10, 34, 40, 142
cloud
cover 70
form (type) 100, 182, 187
seeding 465–6
coalescence 100, 102
coefficient of thermal expansion **30**
cohesive force 328
cold-blooded animals **352**
cold front 183, 187, 196
collision 100, 102
energy **30–1**
geometry **30–1**
colloids, soil 326, 364, 371
communication theory 388
community 145–6, 151, 315–16, 353
compartment models 13
compensation point
carbon dioxide **329**, 334
light 334
temperature 334
competition 346, 416–17, 419–26
competitive exclusion principle 417
complexing agents 329, 380
compounds
chemical 20
electrovalent **20**, 21
inorganic 130–1
organic **131–3**, 131–3
compressive stress 27
condensation 94, 96–100, 193–5, 202
nuclei 98, 102, 454
conduction **30**
conductivity, thermal 231
conservation 440
of mass **30–1**
construction industry 456
constructive margin 117
consumption 351–2
contact cooling 98
contact exchange hypothesis 326
continental crust 59, 61
continental drift 151
convectional cells 86, 196
convectional precipitation 102
convective cooling 330
convergent airflow 85, 183, 189–90, 192, 198–200
conversion 350
cooling
convective 330
evaporative 330
coordination number 234
core (Earth) 107–8
Coriolis deflection 79, 80
Coriolis force 76–81, 192, 197
corrasion 274, 292
corrie (cirque) 178, 408
cortex 322
cosmic radiation 68–9
cosmopolitan distribution 150
Coulomb's equation 218
counter-radiation 49, 50, 52
covalent bonding 20–3
cover 313–15
craton 113
crust
Earth 107, 113, 116, 222
oceanic 107, 112–14, 117
continental 107, 112, 118–20
crustal system 108–19, 121–2, 127, 245, 250
cryoturbation 375
crystal lattice breakdown 226
cultural environments 6–8
cutans 375, 379
cycle
biogeochemical 127, 226
geochemical 125–6, 226
cycling, elemental 158–60
cyclogenesis 190
cyclone 183, 189, 191, 193–4
cyclonic precipitation 102
cytochromes 138–9, 145

decalcification 379
decomposers 153, 239, 362–6
decomposition 355
microbial 362–4
pedological model 366–7
process–response model 361
deforestation 439–40, 448, 450
degeneration of plant communities 448
deglaciation 409
delta 410–11
density 313
density–dependent regulation 416
denudation
rate 403
system 402, 405–8, 439
deoxyribose nucleic acid (DNA) 132, 140, **141**, 150, 389
depression storage 253
deserts
boundary changes 407
energy balance 56, 58
destructive margin 117
detrital food chain 153
detrital system 310, 355–67
organic matter input 355
detritivores 153, 239, 362, 364
dew 98
point 92–3, 97–8, 187, 195–6
diffluent flow 289–90
diffusion, carbon dioxide 329
dilatation 226, 228–30, 291
discharge
basin 213–15
glacier 282–3, 286–7
river channel 268–9, 271–80, 292, 295
discontinuous distribution 150
dishpan experiment 89–90
disjunct distribution 150
distribution
cosmopolitan 150
discontinuous 150
disjunct 150
diurnal patterns 335
divergent airflow 83, 183, 190, 192, 194–5, 198, 200, 398, 401
diversivores 349
doldrums 84
downwarping 404
drainage
area 460
basin 403
density 213–15, 300, 302–3, 450
drought hazard 465
dry adiabatic lapse rate 98–9, 202
dry fallout 208, 222
dune systems 410–11
dynamic
change 402, 404
cooling 98, 100, 398
equilibrium 86, 99, 393

Earth
–atmosphere system 69, 86, 193–5, 202–3, 393, 397–9, 401–2, 439, 456
core 59–60
crust 59–60
interior systems 58
internal energy 58–65
mantle 59, 64–5, 404

orbit 395
structure 59
surface systems 50–1, 393
temperature 59–60
earthflow 259
earthquakes 40, 64, 193
earthworms 362
ecological balance 445
ecological niche 318
ecosphere 128, 146–66
transfer of matter in 158–62
ecosystem 151–3, 206, 208, 220, 226, 249, 306–21, **307**
functional relationships 318, 321
spatial structure 310–13
ecosystems
comparison of primary production 346–7
transfer of energy in 154, 318–21
efficiency of 353
ecotypes 344
ectotherms **352**
efficiencies, assimilation 362
efficiency of energy transfer in ecosystems 353–4
egestion 351
Ekman spiral 81
electrical energy 29
electrodynamic force 24
electromagnetic force 24
electromagnetic spectrum 41–3, 49
electron
atomic 17
configuration 18–20, **19**, **20**
sharing 21
shells, atomic 18–20
transfer 20–1, **20**
electro-osmosis 328
electrostatic force 24
electrovalent compounds **20**, 21
element, chemical 17
elemental cycling 158–62
elements of systems 10
elm decline 446
elongation ratio 215
eluviation 249, 265
emissivity 52–3
Enchytraeid worms 362
endemic 150
endemism 150
endergonic reactions **32**, 136, 158
endothermic reaction **30–1**, 234
endotherms **352**
energy 17, 28–9, 136
balance
deserts 56
Earth–atmosphere system 55
Earth's surface 53, 59, 63, 179
equation 55, 57
forest 175
global variation 58
ice surface 57
oceans 56–8
vegetated surface 57
bond 242
chemical 29, 63, 121, 220, 228, 231
conservation 63
electrical 29
electromagnetic 41
flow
atmosphere 78
oceans 58
free **30–1**, **32**
gravitational 41, 60, 119, 122, 210
heat 29, 122, 221
kinetic 29, 86, 122, 191, 193, 196, 198, 200, 202, 220–1, 260, 267, 301, 466
mechanical 220, 231, 234, 256, 268, 274
nuclear 29
of collision **30–1**
of reaction **30–1**
potential 29, 86, 118–22, 172, 190–1, 193, 198, 200, 202, 215, 219–20, 251, 296, 402, 404–5
radiant 29, 122
solar 40, 62, 119–22, 219–20
strain 121, 228
systems 9–10, 33
terrestrial 117, 219, 283
transfer 29
efficiency of 353
in ecosystems 154, 318–21
englacial meltwater 295
enthalpy (change) **30–1**, **32**
entropy 29, **30–1**, **32**, 33–4, 142–3, 166, 366, 387–8
environment
cultural 6, 8
natural 6–8
environmental controls on photosynthesis 335–44
environmental resistance 415
environmental system 9, 33
enzymatic cleavage of organic molecules 362–3
enzymes **30–1**, 138, **139**, 140, 161, 362–3, 389
epeirogenic deformation 404
equatorial low 76
equatorial westerlies 84
equilibrium
metastable 387
stable 387
states of systems 10–11, 33, 386–7
static 387
thermodynamic 33, 387
unstable 387
eurphagous 349
eutrophication 367
evaporation 56, 94–6, 208, 220–3, 231, 401
pan 105
evaporative cooling 330
evapotranspiration
actual 96, 104–6, 439–40, 450
potential 96, 104, 178–9
evapotranspirometer 104, 105
evolution 142, 151, 164
evolutionary change 391, 432–3
exchange
carbonic acid hypothesis 326
complex 222
contact hypothesis 326
excretion 351
exergonic reactions **32**, 137, 158
exosphere 71
exothermic reaction **30–1**, 234
exponential growth 415
exponential wind profile 81
extra-tropical cyclone 83, 183, 185, 189, 202
extractive industry 457–8

fabric (soil) 371
factor of safety 25
failure 28
farming
organic 367
system
complex 443
simple 443
fault blocks 404
feedback 401
loops 12, 245
negative 11, 379
positive 11–12, 379
ferisols 381
ferrallitisation 381
ferruginous soils 381
fetch 180
Flandrian transgression 405
flood
control 460, 462–5
events 462
hazard 462–3
hydrograph 461–2
flow
laminar 272
turbulent 272, 275
flowchart models 14
fluid system 68, 70
Föhn wind 200–2
fold mountains 404
foliar absorption 356
food
chain 153
detrital 153
preference 349
force 17, 24–5, 28
adhesive 328
applied 27
centrifugal 26
centripetal 26
cohesive 328
driving 258
electrodynamic 24
electromagnetic 24
electrostatic 24
impelling 271, 273
nuclear, strong 24

of crystallisation 231
of gravitational attraction 24
of gravity 24–5
resisting 258, 271, 273
resultant 26
tractive 273, 275–7
weak 24
forced convection 56, 454, 469
forces
fundamental 24
Van der Waals 364
forest
clearance 439–40
ecosystem 439–41
forested catchment 440
Forrester–Meadows world model 438
fossil dune landscape 410
fracture, brittle 28
fragmentation 362
free convection 56, 454, 469
free-energy **30–1**, **32**, 388
free space 328
freezing nuclei 466
friction layer 177
frictional force 76, 80–1, 191–3, 197–200, 202
frost
advection 466
control 467
creep 257
hazard 466–7
radiation 466–7
functional relationships of ecosystem 318, 321

garrique 448
gelifluction 257
gelifraction 226, 231
genetic drift 433
genetic flexibility 344
genetic information 150
genotype 150, 307
genotypic limits to photosynthesis 333
geographical systems 15
geostrophic wind 80–1, 88, 197
geothermal energy 62–4
cascade 64
geothermal gradient 61–2
geothermal heat 52
flow 40, 60–1
glacial erosion 408–9
zones of intensity 408
glacial landscape 408
glaciation 405–6
glei (gley) soils 381
gleisation 381–2
gley (glei) soils 381
global systems 67
global winds 83–4
glycolysis 138
gradient wind 80
grassland
prairie 446
savanna 446
gravity 24–5
grazing–predation system 321, 348–54, 445
grazing pressure 441, 446
greenhouse effect 49, 52, 398, 401
grey-body radiation 42, 44
grey box 266
models 13
gross production 330
groundwater 104, 208–10, 221–2, 231, 238, 248, 254, 261–2, 267–9, 277, 459
glei (gley soils 381
growth 412–13
Gutenberg discontinuity 59

habitat 452–3
Hadley
cell 87–9
single-cell model 85, 196
haloclasty 226, 231
hardware models 14
heat energy 29, **30**
heterotroph 348
heterotrophic cell 143, 153, 348
hoar frost 98
homeostatic mechanism 11
homeotherms **352**
homomorphic models 13
horizonation 372–82
horizons, soil 366, 371–82
human
adjustments to hazards 461–3
ecosystems 436, 440
humification 363–4
humus 363–4, 369
hurricane season 193
hydration 226, 231–2, 235–6, 238
shells 21
pressure 232
hydraulic conductivity 253
hydraulic geometry 261, 272, 274, 278
hydraulic radius 261, 272–4, 277–8
hydrogen 226, 236–9, 242
bonds 21–2, 235, 284, 364, 369
ion potential (pH) 386
hydrograph 267–9, 302
hydrological
cascade 64
cycle 94
sequence, soil 378
stores 94, 398, 459
system 95, 399, 437
hydrolysis 226, 231, 235–6, 238–9, 242
hydrostatic equation 72–3
hypsometric curve 215

ice
ages 398
cap 286–90
crystal process 101, 465–6
sheet 286–7, 290
sheets 398, 404, 408–9
shelf 287
surface
energy balance 57
saturation vapour pressure 93
wedges 409
igneous rocks 110, 112–13, 124–5, 245
immobilisation 363, 366
impelling force 271, 273
inert (noble) gases 20
infiltration 208, 221–3, 253–4, 261, 269, 300, 439, 460
capacity 253–5, 267, 449, 451
information theory 133–4, **134**, 388
inosilicate 109
insect larvae (Insecta) 362
inselberg 230
interception 221, 267, 439, 449
intertropical convergence zone 83, 88, 102
intragranular creep 285
intrinsic rate of increase (of population number) *see* population dynamics
ion exchange 228, 240, 371
ionic balance
soil water 440
stream water 440
ionic bonding 20, 108, 234, 369
ionisation 71
ions **20**, 20, 326, 369–71
iron 107–8, 126, 234, 236–7, 239–41, 243–4, 248
island arc 115, 117
isobaric surface 76, 85, 88, 90, 190
isolated systems 10
isomerism **131**, 362
isomers **131**, 362
isomorphic models 13
isomorphic substitution 240
isostasy 118–19, 404–5
isotopes 17–18

jet streams 83, 88
see also polar front jet stream, subtropical jet stream

'K' selection 416, 424
katabatic airflow 198–200
kinetic energy 29
Kirchhoff's law 53
Krebs cycle 138

lag time 269
Lambert's cosine law 44, 178
land
breeze 198
reclamation 457
landscape 6
latent heat
of fusion (melting) 40, 398
of vaporisation 94, 97, 106
transfer 46, 106, 172, 176, 190, 192–3, 202
laterites 381
law
of conservation of mass **30–1**
of conservation of momentum 26

laws
  Newton's of motion 25–6, 28
  of thermodynamics 29–33, **32**, 142, 153, 388
leaching 238, 240, 242, 248–9, 364, 366, 376
leaf
  angle of insertion 333
  area index (LAI) 333, 336
  duration 333
  energy balance 330
  structure 322, 326
lessivage (*sols lessivée*) 379
levels, trophic 153–4, 157, **307**, 318, 348–9, 364–6
lichenometric dates 407
lichens 407
life
  chemistry of 129
  -form 310
  forms 449
  molecular basis of 130
  nature of 129
light
  compensation point 334
  photosynthesis and 335–6
  saturation point 334
liquid limit 219, 257
lithification 110, 122, 125–7, 245
lithosphere 107–10, 122, 130, 158, 207, 225–6, 291
litter 355–61
living cell 128–9, 133–6, 145
living systems 129–30, 142
  change in 412
loess 406
loose circulation 343–4

macro-fauna **358–9**
macromolecules 133
magma 110, 122–4, 126
magnesium 108, 240
man 127, 302
  –environment relationship 436–7
manmade landform 457
Manning equation 261, 271, 273
mantle 107–8, 119, 122, 126
maquis 448
marine transgression 404
mass 25
  active **30–1**
  flow 328
  law of conservation of **30–1**
  movement 222, 245, 255, 263, 265, 372–6, 409
  number 17
mathematical models 14
matrix (matric) potential (ψ_m) 252–4, 371
matrix throughflow 253–4, 261, 267
matter 17, 24
  states (phases) of **22–3**, 23, 368
  transfer in ecosphere 158–64
mean relief 403
meander 279–80
meltwater, glacial 281, 285, 290, 292–5, 299
membrane, cell 134
memory 391
meridional temperature gradient 86, 89
meristems 412
meso(meio)-fauna **358–9**
Mesolithic man 436
mesosphere 70–1, 107
metallic bonding 20, 23
metamorphic rocks 110, 112, 125, 245, 247
micro-fauna 358–9
micro-flora 358–9
microbial decomposition 362–4
mid-latitude cell 87–8
mid-latitude depression (extra-tropical cyclone) 183–93
mid-latitude low 76
mid-latitude westerlies 83–5
mid-ocean ridge 114–17
millipedes (Myriapods) 362
mineral nutrition and photosynthesis 338–9
mineralisation 363
minerals, clay 371
mistral wind 96
mites (Acari) 362, 364
mixing ratio 92
model
  soil 368
  trophic 153–4, **307**, 318, 364–6
  trophic–dynamic **307**, 318, 364
models
  black box 13
  compartment 13
  deterministic 474
  flowchart 14
  grey box 13
  hardware 14
  homomorphic 13
  isomorphic 13
  mathematical 14, 474
  modelling 12–13
  stochastic 474
  white box 13
Mohorovičić discontinuity 59
molecules 20–1
  organic 131, **131**, **133**
molecular basis of life 130
molecular dipole 20–1
momentum 25–7, 87
  angular 27
  law of conservation of 26
morphometric analysis 208, 211, 213
motion 25–6, 28
mottling 382
mountain–valley wind 83, 200
movement
  in solution 255, 261
  mass 255, 258, 265
  particulate 255, 260
myxomatosis 448
natural environment 6–8
natural hazard 461
negative feedback 11, 86, 378, 387, 391, 401, 425, 439
Neolithic man 446
net primary production 154–8, 331–3
net radiation 94
  diurnal change 54, 176
  seasonal change 54
neutrons, atomic 17
Newton's laws of motion 25–6, 28
  first 76, 80, 100
niche
  ecological 318
  segregation 318
  space 318
  trophic 318
nine-unit land surface model 264–5
nitrogen
  atmospheric 47, 68, 70
  cycle 440
  fixation **161**, 440
nitrogenase **161**
noble (inert) gases 20
non-free space 328
nuclear energy 29
nuclear fusion 41
nucleus
  atomic 17
  cell 132, 139–40
number, atomic 17
nunatak 289
nutrient
  cycling 158–64, 326, 339–44, 440, 449
  deficiency 342, 344
  excess 339–42, 344
nutrients, transfer in plant 326

^{18}O isotope 395
occlusion 189
ocean
  basins 404
  core sediments 395
  currents 58
oceanic crust 59, 61, 404
oceans 164
  energy balance 56
omnivores 391
open systems 10, 33, 94, 142–3, 386–7
organelles, cellular 134, 145
organic farming 367
organic matter, input to detrital system 355
organic molecules **131**, 132, **133**
organic nitrogen 440
organism 145, 146, 151
organisms, soil **358–9**, 361, 369
orogen 113
orogenic uplift 404
orogeny 113
orographic precipitation 102, 202
orthoclase 109
oscillation volume of ions 326

osmotic (water) potential (ψ_π) 328, **338**, 371
overland flow 208, 221, 254, 261, 267–9, 277
oxidation 239, 249
oxidation–reduction 138–9, 378, 382
oxidative phosphorylation 143
oxygen 108–9, 122, 225, 236, 239–40, 284
  atmospheric 47, 68–70, 395
ozone 47, 69, 164, 395

palaeosols 407
pastoral systems 450
pattern of vegetation 313
peak flows 450
pedogenesis 372, 376–7, 379–82
  formal processes of 225–6, 245, 249, 379–82
pedological model of decomposition 366–7
peds 369
periglacial landscape 408–10
periglacial process 409–10
perihelion 43, 395
periodic table **19**
permafrost 409
permanent charge 240
permeability 245–7, 254
pH, soil 378–9
phases (states) of matter **22–3**, 23, 368–9
phenology 316, 336, 356
phenotype 307
phenotypic controls on photosynthesis 334
phosphorylation, oxidative 143
photochemical fog 454
photophosphorylation 143
photosynthesis 63, 143, 153, 220, 329–31
  biochemical controls on 334
  Calvin C-3 pathway **329**, 334
  environmental controls on 334
  Hatch–Slack C-4 pathway **329**, 334
  light and 335–6
  limits to 333–44
  mineral nutrition and 338–9
  phenotypic controls on 334
  physiological controls on 333–4
  temperature and 336–7
  water and 338
phycobilins **144**
phyllosilicate 109, 241
phylotaxy 333
physiological controls on photosynthesis 333–4
phytochromes **412**, 412
pipeflow 254, 261, 267
pipes, soil 371
Planck's constant 41
plant water potential 325
plastic deformation 28
plastic limit 219
plasticity index 219
plough pan 452
pneumatophores 326
podsol 380
podsolisation 379–80
poikilotherms **352**
polar
  cell 87–8
  easterlies 83–4
  front 182–9
    jet stream 88–9
    model 189
  high 74
  ice caps 395, 398
  molecules 20–1
polyphagous 349
polyploidy 432
population 146–51, 413–19
  dynamics 413–19
  number 413–19, 441
  regulation 353–4, 414–19
    and energy flow 353–4
pore water 218, 259
  pressure 218, 259
pores, soil 371
porosity 216, 238, 244–5
positive feedback 11, 379, 391, 401–3, 439
potassium 108, 124, 234–6, 240
potential
  energy 29, 86, 119–22, 172, 190–3, 198–200, 202, 215, 219–20, 251, 296, 402–5
  hydrogen ion (pH) 378
  osmotic (ψ_π) 328, **338**, 371
  redox 378, 382
  water 251–3
  cell **338**
  matrix (ψ_m) 252–4, 371
    plant **338**
    pressure 338
    soil (ψ_s) 252, 369–71
precipitation 94, 100–6, 120, 122, 182–3, 192, 195, 208, 213, 216, 221–3, 244, 255, 263, 266–7, 281, 289, 296, 299, 303, 398, 402
  types 102
pressure 27
  atmospheric 27
  cell 183
  gradient force 76–7, 81, 197–8
  potential 338
  vapour 338
primary circulation system 81–5, 88, 90, 182, 196
primary production system 321–47, 446
probable state 388
probablistic view 388
process 11, 33, 387
  –response model 226, 248, 361
  –response system 402–3
production 154
  gross 331
  net primary 154, 331–3
  secondary 154, 350–3
productivity 154
  limits to 333–44
  net 157, 331–3
profile, soil 371
proteins 132, 140
protons, atomic 17
pyroxene 109

quantum theory 41
quartz 109, 112, 242–5
Quaternary 402, 404–8, 428–32

'r' selection 413, 419
radiant energy 29
radiation
  absorption 47, 175–6
  balance
    atmosphere 50–1
    Earth–atmosphere system 50, 54
    Earth's surface 50–4, 198, 466
    equation 50, 53–5
    forest 175–6
    ocean surfaces 54
    planetary 44–6
  cooling 98
  diffuse 49, 51, 175, 397, 586
  direct 49, 50, 178, 397, 586
  electromagnetic 41, 42, 44
  fog (ground fog) 194–5
  laws 42, 44
  on a horizontal surface 44, 51–2
  scattering 47, 397
  solar 220, 281, 287
  wavelength 41–2
  windows 47, 49
radioactive decay 60
radioactive half-life 60
radioactive isotopes 60
radiocarbon 68
raindrop 258, 260
rainfall intensity 264, 269, 303
rainsplash erosion 260–1, 450
rainstorm 260
rate-limiting processes 158
reaction energy **30–1**
recession curve 269
recessional moraines 407
redox potential 237, 378, 382
redox reactions 226, 236, 239, 242, 247
reduction 239, 249
reflection
  clouds 48
  diffuse 47
  Earth's surface 52
regelation 292–4
regeneration of plant community 448
regolith 124, 208, 221–2, 228, 238–40, 242–6, 251, 262, 266, 269, 291, 409–10
regulators 12, 228, 239, 249, 253, 267, 387, 393–6, 401, 437–9, 454
regulatory mechanisms 441
relative humidity 92, 93, 187
relaxation time 391

relict landforms 402, 407
relief ratio 215, 301
replication (DNA) **141**
reproduction 352
residence time 160, 390
resistance 27, 28
respiration 138–9, 351–2
respiratory (catabolic) heat 139, 153
resultant force 26
resynthesis 363–4
retention capacity
  nutrient 451
  water 451
rhizosphere 378
rhythms, circadian 336
rifting 404
rigid solid 28
rill 212, 261
  action 450
rising limb 269
river
  channel 405–6, 439, 450–1, 458–60
    gradient 405
  terraces 406, 410
rockfall 260, 262, 265, 281, 294
rockslide 257
root
  cap 322, 359
  hairs 326, 359
  –stem–leaf system 326
  structure 322–6
roots, dead 361
Rossby waves 90
roughness length 177
rubification 381
runoff 94, 104, 406, 437, 440, 449–50, 459
  channel 221–2
  direct 267, 269
  indirect 267, 269
  surface 223, 226, 300

salinisation 381
sand-bars 411
sand-dunes 406
saturated adiabatic lapse rate 99–100, 202
saturated overland flow 254
saturation
  deficit 92–3
  point, light 334
  vapour pressure 93, 95, 98
scavengers 351, 391
sclerophyllous evergreen oak forests 448
sea
  -breeze 83, 196–8, 203
  -level
    change (eustatic change) 404–6
    pressure 74–6
  smoke 98
  temperature 176
second law of thermodynamics 64
secondary circulation system 81, 83, 182–3, 196, 203
secondary depression 189
secondary minerals 368
secondary production 154, 350–3
  of animal communities 353
sediment yield 450, 459
sedimentary basins 404
sedimentary rocks 110, 112–13, 125, 245–7
selective grazing 446
self regulation 11, 142, 387
segmented worms (Annelida) 362
sensible heat transfer 172, 176, 192–3, 202
shade leaves 333, 337
shear stress 27
sheetwash 261, 298
shifting cultivation 441
silica 108
silicate minerals 108–12, 226, 234–6, 242
silicon 108–9, 112, 122, 124, 234, 239–41
slope winds 198–200
smoke concentration 455
snowfall 281
sodium 108, 234, 242
soil
  aggregates 371, 450, 452
  catena 378
  colloids 326, 364, 371
  creep 220, 256
  crust 450
  erosion 70, 398, 448, 450
  fabric 371
  formation 371, 376–7, 379–82
  horizons 366, 371–82
  hydrological sequence 378
  model 368
  moisture 267
    storage 439
  organisms 245, **358–9**, 361, 369
  pH 378
  pipes 371
  pores 371
  potential (ψ_s) 252, 371
  processes 372–7
  profile 371
  system 321, 368–82
  temperature 178, 468
  texture 369
  three-phase system 368
  toposequence 378
  transfer processes 375
  transformation processes 375
  water 104–6
soils
  glei (gley) 376, 381–2, 449
  podsols 381
sol lessivée (lessivage) 379
solar
  constant 42–3, 393–4, 399, 401
  elevation 44, 172, 178–9
  energy cascade 49, 52, 172, 393, 396, 405
  faculae 395
  radiation 40–4, 46, 47, 51, 63, 94, 175, 178, 198, 394–8, 401–2, 467
  spectrum 42, 49, 51, 172
solifluction
  deposits 410
  terrace 409
solonisation 381
solution, dissolved, load 274–5, 298
solvation 235, 247
spatial structure, animal communities 315–16
speciation 151, 349
species 150, 307, **310**
  diversity 441, 452
specific dynamic action (SDA) **352**
specific humidity 92
springtails (Collembola) 362
squall lines 83
standing dead biomass 356
states (phases) of matter **22–3**, 23, 368
states of systems 10
steady state 11, 387
  of living cell 142–3
  of soil 378
Stefan–Boltzmann's law 42
stele 322
stem structure 322–6
stemflow 221, 356–9
stenophagous 349
stocking densities 446
stoichiometry **30–1**
stomata 96, 322
strain 28
  energy 228
  release 226, 228–30, 291–2
strandlines 411
stratification of vegetation 313
stratopause 71
stratosphere 71, 395–8
strength 27–8
  compressive 217–19, 244
  shear 217–19, 257–9, 291
    maximum 217–18
    residual 218
  tensile 217, 219, 231–2
stress 27
  compressive 27, 228, 291
  shear 27, 217–19, 228, 257–8, 261, 271, 274, 284–6, 290, 292
  tensile 27, 228, 231–2, 234
strong nuclear force 24
structure
  atomic 17
  of ecosystems in space 310–13
subglacial meltwater 295
subspecies 307
sub-tropical high pressure 76, 182, 195
sub-tropical jet stream 88
succession 419–25
sun leaves 333, 337
sunspots 43, 394–5
supraglacial meltwater 294–5
surface elevation, effects of 179
surface wash 260–1

surface water
  balance 104–6, 459
  glei (gley) soils 381
surroundings of systems 10, 33
suspended load 274–5, 298
symbiosis **161**, 348
synthesis 6
system
  atmospheric 122, 206–8, 462
  cascading 208
  catchment basin 206–8, 210, 221–2, 270, 296
  channel 208, 220, 250
  crustal 107–18, 121–2, 126–7, 245, 250
  denudation 108, 120, 122, 125–6, 206–8, 219, 221–2, 226, 245, 249, 296
  detrital 310, 355–67
  fluvial 262, 264, 267, 295
  glacial 281
  grazing–predation 321, 348–54
  primary production 322–47
  root–stem–leaf 326
  slope 208, 220, 222, 249–50, 262
  soil 225, 321, 368–82
  weathering 207–8, 220–2, 225, 233, 235, 237, 245, 248, 262
systems 8, 12
  attributes 10
  boundary 10
  changes in 10, 12, 383
  closed 10, 33, 142
  digestive 8–9
  Earth surface 462
  elements 10
  energy 9–10, 33
  environmental 9, 33
  equilibrium states of 11, 33
  geographical 15
  heating 9
  ignition 9
  interfacing 475
  isolated 10, 388
  living 128–30, 142
  open 10, 34, 108, 120, 142–3, 386–7
  political 9
  processes 11, 33
  relationships 10
  states 10
  surroundings 10, 33
  thermodynamic 9, 33
  transport 9

talus slope 209, 260, 262
taxonomy 307
tectosilicate 109
temperature **30**
  compensation point 334
  inversion 193–5, 469
  photosynthesis and 336–7
tensile stress 27
terminal velocity 101–2
terrestrial energy 59–61
terrestrial radiation 49–50, 52–3, 398, 467
  spectrum 52, 400
tertiary circulation system 81, 83, 182, 195–6, 200, 202
tetrahedral crystalline structure 91
texture, soil 368
thermal
  capacity 174–6, 456
  conduction 46, 56, 58, 61, 98, 174
  conductivity 61, 174–6, 469
  convection 46, 56, 58, 60–1, 176–7, 195
  diffusivity 174
  energy 29, **30**
  expansion, coefficient of **30**
thermally direct cell 88, 196–8, 200
thermally indirect cell 88
thermoclasty 226, 231–2
thermodynamic equilibrium 32–4, 387
thermodynamic systems 9, 33
thermodynamics, laws of 29–33, **32**, 142, 153, 388
thermoregulation 351, **352**
thermosphere 71
three-cell circulation model 86–8, 388
thresholds 12, 379, 388
throughfall 221–2, 356
throughfall, soil 222, 249, 254, 261, 267–9, 277, 375–6
thunderstorms 83
tight circulation 342
till 294
tilt of Earth's axis 395
toposequence, soil 378
tor 409–10
trace elements 108
trade winds 83–5, 96
transcription, DNA **141**, 150
transfer
  of energy 29
    in ecosystems 154, 318–21, 353
  of nutrients in plants 326
  processes, soil 375
transfluent flow 290
transform fault 117
transformation processes, soil 375
transient states 391
transition elements **19**
translocation 376
transpiration 96, 208, 220, 329, 338, 449
  loss 96
transport-limited 266
trophic-dynamic model **307**, 318, 366
trophic levels 153–7, **307**, 318, 348–9, 364–6
trophic model 153–4, **307**, 316–18, 364–6
  of decomposition 364
trophic niche 318, 436
tropical cyclone 83, 182, 191–3, 466
tropical easterly waves 192
tropical rain forest 441–4
tropopause 71, 195
troposphere 70, 172, 189–90, 192, 195, 393, 398
tundra 173, 406
  polygons 409
turgor **338**, 338

ubac slope 178
unit hydrograph 458
urban
  atmosphere 454
  climates 453–4
  fog 454–5
  heat island 456
  snowfall 462
urbanisation of catchments 459–60

valence shell 18
valency 234, 236
valley
  bog 449
  fog 98
  glaciers 398, 407–8
Van Allen belts 71
Van der Waals forces 364
vapour pressure 92, 95, 98, 338
varieties of species 307
vector 24
vegetation
  canopy 175
  horizontal structure 313
  pattern 313
  stratification of 313
velocity 25
  angular 27
vitamins 143, 348
voids ratio 216
volcanic activity 40, 62–3, 397–8

warm blooded animals **352**
warm front 183, 187
water
  balance 106, 221, 223, 440
  molecule 21–2, 91
  phases of 92
  photosynthesis and 338
  physical properties of 92
  potential 251–2
    cell **338**
    matrix (ψ_m) 252–3, 371
    osmotic (ψ_π) 328, **338**, 371
    plant **338**
    pressure 338
    soil (ψ_s) 252, 369–71
  throughput in plant 326
  vapour
    atmospheric 47, 69–70, 77, 92, 106, 398, 400–1
  yield 440
wave theory 88
waves in upper westerlies 83, 90–1, 189
weak force 24
weather warfare 466
weathering 119, 121, 126–7, 158, 216, 262, 265, 355, 366, 368

environment 226, 248
limited 265–6
weight 25
white box models 13
Wien's displacement law 42
wind
rose 196
velocity profile 178
woodland management system 442
woodlice (Isopoda) 362
work 28, 136

zonal changes 402, 406–7
zonal pressure pattern 74